OF THE ELEMENTS

(1971 values for atomic weights)

VIII	I*b*	II*b*	III*a*	IV*a*					O

Representative elements (nonmetals)

Noble gases

									2 He 4.0026
			5 B 10.81	6 C 12.011	7 N 14.0067	8 O 15.9994	9 F 18.9984	10 Ne 20.179
			13 Al 26.9815	14 Si 28.086	15 P 30.9738	16 S 32.06	17 Cl 35.453	18 Ar 39.948

Transition Metals

27 Co 58.9332	28 Ni 58.71	29 Cu 63.546	30 Zn 65.38	31 Ga 69.72	32 Ge 72.59	33 As 74.9216	34 Se 78.96	35 Br 79.904	36 Kr 83.80
45 Rh 102.905	46 Pd 106.4	47 Ag 107.868	48 Cd 112.40	49 In 114.82	50 Sn 118.69	51 Sb 121.75	52 Te 127.60	53 I 126.9045	54 Xe 131.30
77 Ir 192.22	78 Pt 195.09	79 Au 196.967	80 Hg 200.59	81 T1 204.37	82 Pb 207.2	83 Bi 208.9804	84 Po (210)	85 At (210)	86 Rn (222)

Representative elements (metals)

Inner-transition metals

62 Sm 150.4	63 Eu 151.96	64 Gd 157.25	65 Tb 158.925	66 Dy 162.50	67 Ho 164.930	68 Er 167.26	69 Tm 168.934	70 Yb 173.04	71 Lu 174.97
94 Pu (242)	95 Am (243)	96 Cm (247)	97 Bk (247)	98 Cf (249)	99 Es (254)	100 Fm (253)	101 Md (256)	102 No (254)	103 Lr (257)

LANGE'S
HANDBOOK OF
CHEMISTRY

LANGE'S
HANDBOOK OF
CHEMISTRY

Editor: JOHN A. DEAN

Professor of Chemistry
University of Tennessee (Knoxville)

Formerly Compiled and Edited by
NORBERT ADOLPH LANGE, Ph.D.

TWELFTH EDITION

McGRAW-HILL BOOK COMPANY
New York   St. Louis   San Francisco   Auckland   Bogotá
Düsseldorf   Johannesburg   London   Madrid   Mexico
Montreal   New Delhi   Panama   Paris   São Paulo
Singapore   Sydney   Tokyo   Toronto

Library of Congress Cataloging in Publication Data

Lange, Norbert Adolph, 1892–1970
  Lange's Handbook of chemistry.

  1. Chemistry—Handbooks, manuals, etc.
I. Dean, John Aurie   II. Title.
III. Title: Handbook of chemistry.
QD65.L36   1978   540′.2′02   78-15335
ISBN 0-07-016191-7

1234567890   KPKP   7865432109

*The editors for this book were Harold B. Crawford and Ruth L. Weine
and the production supervisor was Teresa F. Leaden.
It was set in Bodoni by York Graphic Services, Inc.
Printed and bound by The Kingsport Press.*

*To those workers in science who through
their labors determined the values recorded herein,
this compilation is dedicated. Their devotion
to the search for the constants of nature
and the dissemination of this knowledge are the
foundations upon which rest the achievements
of applied science.*

CONTENTS

*For the detailed contents of any section, consult
the title page of that section. See also the alphabetical index
in the back of this handbook.*

PREFACE TO THE FIRST EDITION

This book is the result of a number of years' experience in the compiling and editing of data useful to chemists. In it an effort has been made to select material to meet the needs of chemists who cannot command the unlimited time available to the research specialist, or who lack the facilities of a large technical library which so often is not conveniently located at many manufacturing centers. If the information contained herein serves this purpose, the compiler will feel that he has accomplished a worthy task. Even the worker with the facilities of a comprehensive library may find this volume of value as a time-saver because of the many tables of numerical data which have been especially computed for this purpose.

Every effort has been made to select the most reliable information and to record it with accuracy. Many years of occupation with this type of work bring a realization of the opportunities for the occurrence of errors, and while every endeavor has been made to prevent them, yet it would be remarkable if the attempts towards this end had always been successful. In this connection it is desired to express appreciation to those who in the past have called attention to errors, and it will be appreciated if this be done again with the present compilation for the publishers have given their assurance that no expense will be spared in making the necessary changes in subsequent printings.

It has been aimed to produce a compilation complete within the limits set by the economy of available space. One difficulty always at hand to the compiler of such a book is that he must decide what data are to be excluded in order to keep the volume from becoming unwieldy because of its size. He can hardly be expected to have an expert's knowledge of all branches of the science nor the intuition necessary to decide in all cases which particular value to record, especially when many differing values are given in the literature for the same constant. If the expert in a particular field will judge the usefulness of this book by the data which it supplies to him from fields other than his specialty and not by the lack of highly specialized information in which only he and his co-workers are interested (and with which he is familiar and for which he would never have occasion to consult this compilation), then an estimate of its value to him will be apparent. However, if such specialists will call attention to missing data with which they are familiar and which they believe others less specialized will also need, then works of this type can be improved in succeeding editions.

Many of the gaps in this volume are caused by the lack of such information in the literature. It is hoped that to one of the most important classes of workers in chemistry, namely the teachers, the book will be of value not only as an aid in answering the most varied questions with which they are confronted by interested students, but also as an inspiration through what it suggests by the gaps and inconsistencies, challenging as they do the incentive to engage in the creative and experimental work necessary to supply the missing information.

While the principal value of the book is for the professional chemist or student of chemistry, it should also be of value to many people not especially educated as chemists. Workers in the natural sciences—physicists, mineralogists, biologists, pharmacists, engineers, patent attorneys, and librarians—are often called upon to solve problems dealing with the properties of chemical products or materials of construction. For such needs this compilation supplies helpful information and will serve not only as an economical substitute for the costly accumulation of a large library of monographs on specialized subjects, but also as a means of conserving the time required to search for information so widely scattered throughout the literature. For this reason especial care has been taken in compiling a comprehensive index and in furnishing cross references with many of the tables.

It is hoped that this book will be of the same usefulness to the worker in science as is the dictionary to the worker in literature, and that its resting place will be on the desk rather than on the bookshelf.

N. A. Lange

Cleveland, Ohio
May 2, 1934

PREFACE TO
THE ELEVENTH EDITION

The new editor has assumed the task of data compilation from the late Dr. Lange, the man who initiated the *Handbook of Chemistry* almost four decades ago. It seems only fitting that his name should be embodied within the new title in recognition for his efforts on the ten preceding editions of this handbook.

Perhaps it would be simplest to begin by stating the ways in which this new edition has not been changed. It remains the one-volume source of factual information for chemists, both professionals and students—the first place in which to "look it up" on the spot. The aim is to provide sufficient data to satisfy all one's general needs.

The changes, however, are both numerous and significant. First of all, there is a basic change in organization. The handbook is now divided into sections—mathematics, general information and conversion tables, atomic and molecular structure, inorganic chemistry, analytical chemistry, electrochemistry, organic chemistry, spectroscopy, thermodynamic properties, physical properties, miscellaneous—and within these sections related groups of factual data are presented. This arrangement, plus the new sectional tables of contents, which provide a complete listing of items within each section, backed up by a thorough and extensively cross-indexed subject index, makes it possible to find the information quickly.

The following subject matter is offered for the first time:

Emission and absorption lines for arc, spark, and flame and atomic absorption—with sensitivities and/or detection limits
Formation constants of metal complexes with organic and inorganic ligands
Mass absorption coefficients of X-ray emission lines commonly used in X-ray absorption work, with coefficients for all elements
Statistical tables
Atomic electron affinities
Electronegativities of the elements

Spatial orientation of common hybrid bonds
Hammett and Taft substituent constants
Selectivity coefficients for ion-exchange resins
Cross contamination and separation factors in separation methods

Also new are self-instructional sections developed for certain areas: measurement of pH, use of statistics, separation methods, and X-ray methods.

Expanded coverage is provided in such important areas as:

Solubility products
Proton transfer reactions (acid dissociation constants) with 1200 entries for organic compounds and 150 for inorganic compounds
Electrode potentials of elements and their compounds listed by element
Bond energies and radii of atoms and ions
Reference electrodes
Reference pH buffers for water, deuterium oxide, and aqueous-organic systems
Approved symbols and abbreviations

Updating has increased the usefulness of such valuable tabulations as:

Physical properties of 4000 inorganic compounds
Nomenclature of inorganic compounds
Heats and free energies of formation, entropies, and heat capacities—incorporating the latest recommended values of the National Bureau of Standards
X-ray emission spectra and X-ray K and L absorption edges, given both as wavelengths and as energies in keV
Critical properties
Limiting equivalent ionic conductances in aqueous solution
Table of nuclides, now 100 pages extending through element 105
Ionization potentials of the elements

Finally, the mathematical section has been expanded from its rather restricted size in recent editions so as to include mathematical information commonly needed by an upper-division or graduate student, or professional, without recourse to other reference sources.

It is hoped that users of this and previous editions will continue to offer friendly criticism and suggestions and call attention to errors.

John A. Dean

Knoxville, Tennessee

PREFACE TO
THE TWELFTH EDITION

In this edition, the second under the aegis of the present editor, new and updated material has become available for incorporation from the numerous compilations of standard reference data that have appeared in the interim since the eleventh edition went to press. In particular, the section on thermodynamic properties has been thoroughly revised to reflect the latest recommended values for heats of formation and Gibbs energies of formation, entropies, heat capacities at five different temperatures, and latent heats of melting, vaporization, and sublimation for approximately 2400 inorganic compounds and 1500 organic compounds.

The following subject matter is offered for the first time:

Structure-correlation tables for proton magnetic resonance and infrared spectroscopy

Ionization potentials of molecular and radical species

Potentials of reference electrodes for water–organic solvent mixtures

Wavenumber/wavelength conversion table

New explanatory sections have been added for thermodynamic relations, column chromatography, Hammett and Taft substituent constants, conductance relations, physical chemistry equations for gases, spectroscopic relationships, computation of pH values of solutions, and relationships involving refractive indices, dipole moments, dielectric constants, viscosity, and surface tension.

Expanded coverage is provided in the areas of:

Conversion factors—1800 entries, incorporating SI units

Atomic electron affinities

pH ranges of buffer solutions for control purposes

Temperature dependence of selected values of pK_a and pK_{sp} in water

Equivalent conductivities of 180 electrolytes in aqueous solutions at 11 concentrations

Concentrations of commonly used acids and bases

Physical properties of 450 organic solvents, a section useful for HPLC applications which lists refractive indices, dielectric constants, dipole moments, surface tensions, and viscosities

Physical properties of 225 inorganic substances listing dielectric constant, dipole moment, surface tension, and viscosity

Cryoscopic constants

Ebullioscopic constants

Drying agents

Solutions for maintaining constant humidity

Vapor pressures of approximately 370 inorganic and 770 organic compounds from data to be fitted into an Antoine or similar type equation

Solubilities of 400 inorganic compounds and metal salts of organic acids in water at 9 fixed temperatures

Updating has increased the usefulness of such valuable tabulations as:

Fundamental physical constants

Physical and chemical symbols and terminology

Hammett and Taft substituent constants

Bond energies

Nuclear properties of the elements

Ionization potentials of the elements

Fluorescent indicators

Oxidation-reduction indicators

Proton-transfer reactions of inorganic materials in water

Potentials of the elements and their compounds

International practical temperature scale, adopted in 1968—the IPTS-68, and the revised thermocouple reference data based on this scale

Finally, the mathematical section has been restructured to exclude data now easily obtained with the ubiquitous hand calculator. Reintroduced are commonly used differential equations and integrals.

It is hoped that users of this edition will continue to offer friendly criticism and suggestions and call attention to errors. These communications should be directed to the editor at his home address.

John A. Dean

1112 W. Nokomis Circle
Knoxville, TN 37919

ACKNOWLEDGMENT

Grateful acknowledgment is hereby made of an indebtedness to those who have contributed to former editions and whose compilations continue in use in this edition. In particular, acknowledgment is made of the contribution of Dr. Joseph R. Peterson, who prepared the expanded Table of Nuclides; also that of Mr. Theodore C. Rains who supplied many of the atomic absorption sensitivities.

LANGE'S
HANDBOOK OF
CHEMISTRY

Section 1

MATHEMATICS

MATHEMATICAL TABLES

LOGARITHMS

PROPERTIES AND USES

Definition of Logarithm. The *logarithm* x of the number N to the base b is the exponent of the power to which b must be raised to give N. That is,

$$\log_b N = x \text{ or } b^x = N.$$

The number N is positive and b may be any positive number except 1.

Properties of Logarithms.

(**a**) *The logarithm of a product is equal to the sum of the logarithms of the factors; thus,*

$$\log_b M \cdot N = \log_b M + \log_b N.$$

(**b**) *The logarithm of a quotient is equal to the logarithm of the numerator minus the logarithm of the denominator; thus,*

$$\log_b \frac{M}{N} = \log_b M - \log_b N.$$

(**c**) *The logarithm of a power of a number is equal to the logarithm of the base multiplied by the exponent of the power; thus,*

$$\log_b M^p = p \cdot \log_b M.$$

(**d**) *The logarithm of a root of a number is equal to the logarithm of the number divided by the index of the root; thus*

$$\log_b \sqrt[q]{M} = \frac{1}{q} \log_b M.$$

Other properties of logarithms:

$\log_b b = 1.$ $\log_b \sqrt[q]{M^p} = \frac{p}{q} \cdot \log_b M.$

$\log_b 1 = 0.$ $\log_b N = \log_a N \cdot \log_b a = \dfrac{\log_a N}{\log_a b}.$

$\log_b (b^N) = N.$ $b^{\log_b N} = N.$

Systems of Logarithms. There are two common systems of logarithms in use: (1) the *natural* (Napierian or hyperbolic) system which uses the base $e = 2.71828 \ldots$; (2) the *common* (Briggsian) system which uses the base 10.

We shall use the abbreviation $\log N \equiv \log_{10} N$ in this section.

Unless otherwise stated, tables of logarithms are always tables of common logarithms.

Characteristic of a Common Logarithm of a Number. Every real positive number has a real common logarithm such that if $a < b$, $\log a < \log b$. Neither zero nor any negative number has a real logarithm.

A common logarithm, in general, consists of an integer, which is called the *characteristic*, and a decimal (usually endless), which is called the *mantissa*. The characteristic of any number may be determined from the following rules.

Rule I. *The characteristic of any number greater than 1 is one less than the number of digits before the decimal point.*

Rule II.* *The characteristic of a number less than 1 is found by subtracting from 9 the number of ciphers between the decimal point and the first significant digit, and writing* -10 *after the result.*

Thus the characteristic of $\log 936$ is 2; the characteristic of $\log 9.36$ is 0; of $\log 0.936$ is $9 - 10$; of $\log 0.00936$ is $7 - 10$.

Mantissa of a Common Logarithm of a Number. An important consequence of the use of base 10 is that the mantissa of a number is independent of the position of the decimal point. Thus 93,600, 93.600, 0.000936, all have the same mantissa. Hence in Tables of Common Logarithms only mantissas are given. A five-place table gives the values of the mantissa correct to five places of decimals.

To Find the Logarithm of a Given Number N. By means of Rules I and II determine the characteristic. Then use the table to find the mantissa.

To find a mantissa when the given number (neglecting decimal point) consists of four, or less, digits (exclusive of zeros at the beginning or end), look in the column marked N for the first three significant digits and pick the column headed by the fourth digit—the mantissa is the number appearing at the intersection of this row and column. Thus to find the logarithm of 64030, first note (by Rule I) that the characteristic

---

* Some writers use a dash over the characteristic to indicate a negative value: for example,

$$\log .004657 = 7.66811 - 10 = \bar{3}.66811.$$

is 4. Next, in the table, find 640 in column marked N and opposite it in column 3 is the desired mantissa, .80638. Hence $\log 64030 = 4.80638$. Likewise, $\log 0.0064030 = 7.80638 - 10$; $\log 0.64030 = 9.80638 - 10$.

Interpolation. The mantissa of a number of more than four significant figures can be found approximately by assuming that the mantissa varies directly as the number in the small interval not tabulated. Thus if N has five digits (significant), and f is the fifth digit of N, the mantissa of N is

$$m = m_1 + \frac{f}{10}(m_2 - m_1),$$

where m_1 is the mantissa corresponding to the first four digits of N, m_2 is the next larger mantissa in the table. $(m_2 - m_1)$ is called a *tabular difference*. The proportional part of the difference $m_2 - m_1$ is called the *correction*. These proportional parts are printed without zeros at the right-hand side of each page as an aid to mental multiplications.

For example, find $\log 64034$. Here $f = 4$. From the table we see $m_1 = .80638$, $m_2 = .80645$, whence $m = .80638 + (4/10)(.00007)$, $\log 64034 = 4 + m = 4.80641$.

To Find the Number N When Its Logarithm Is Known. (The number N whose logarithm is k is called the *antilogarithm* of k.)

Case 1. If the mantissa m is found exactly in the table, join the figure at the top of the column containing m to the right of the figures in the column marked N and in the same row as m, and place the decimal point according to the characteristic of the logarithm.

Case 2. If the mantissa m is not found exactly in the table, interpolate as follows: find the next smaller mantissa m_1 to m; the first four significant digits of N correspond to the mantissa m_1, and the fifth digit f equals the nearest whole number to

$$f = 10\left(\frac{m - m_1}{m_2 - m_1}\right),$$

where m_2 is the next larger mantissa to m appearing in the table. Then locate the decimal point according to the characteristic.

The decimal point may be located by means of the following rules: *Rule III. If the characteristic of the logarithm is positive (then the mantissa is not followed by -10), begin at the left, count digits one more than the characteristic, and place the decimal point to the right of the last digit counted.*

Rule IV. *If the characteristic is negative (then the mantissa will be preceded by an integer n and followed by* -10), *prefix* $(9 - n)$ *zeros and place the decimal point to the left of these zeros.*

Illustrations of the Use of Logarithms.

Example 1.   Given $\log x = 2.91089$, find x. The mantissa .91089 appears in the table. Join the figure 5 which appears at the top of the column to the right of the number 814 in the column N, giving the number 8145. By Rule III, the decimal point is placed to the right of 4, thus giving $x = 814.5$.

Example 2.   Given $\log x = 2.34917$, find x. The mantissa $m = .34917$ does not appear in the table. The next smaller and next larger mantissas are m_1 and m_2,

$$m_1 = .34908, \quad m = .34917, \quad m_2 = .34928.$$

The first four digits of N, corresponding to m_1, are 2234, and the fifth digit is the nearest whole number (5) to

$$10 \left(\frac{m - m_1}{m_2 - m_1} \right) = 10 \left(\frac{.00009}{.00020} \right) = 4.5.$$

By Rule III, we locate decimal point, thus giving $x = 223.45$.

Example 3.   Find $x = (396.21)\,(.004657)\,(21.21)$.
$$\log 396.21 \;=\; 2.59792$$
$$\log .004657 = \;7.66811 - 10$$
$$\log 21.210 \;\;= \;\underline{1.32654} \qquad \text{(add)}$$
$$\log x \qquad = 11.59257 - 10, \quad x = 39.135.$$

Example 4.   Find $x = \dfrac{396.21^*}{24.300}$.
$$\log 396.21 = 2.59792$$
$$\log 24.3 \;\;\; = \underline{1.38561} \;\text{(subtract)}$$
$$\log x \qquad = 1.21231, \quad x = 16.305.$$

Example 5.   Find $x = (3.5273)^4$.
$$\log 3.5273 = 0.54745$$
$$\underline{\qquad\quad 4} \;\text{(multiply)}$$
$$\log x \qquad = 2.18980, \quad x = 154.81.$$

---

* Some writers use *cologarithms*. The cologarithm of a number N is the negative of the logarithm of N: i.e., $\operatorname{colog} N = 10.00000 - \log N - 10$. Adding the cologarithm is equivalent to subtracting the logarithm. Thus in our example, $\operatorname{colog} 24.3 = (10.00000 - 1.38561) - 10 = 8.61439 - 10$. $\log 396.21 - \log 24.3 = \log 396.21 + \operatorname{colog} 24.3 = 2.59792 + 8.61439 - 10 = 11.21231 - 10 = \operatorname{antilog} 16.305$.

Example 6.   Given $\log x = -2.23653$, to find x. To convert this logarithm to one with a positive mantissa, add algebraically -2.23653 to $10.00000 - 10$. Thus

$$10.00000 - 10$$
$$\underline{-2.23653}$$

$\log x \quad = \quad 7.76347 - 10$, hence $x = 0.0058006$.

Example 7.   Find $x = \sqrt[3]{.04657}$.

$\log x \quad = \frac{1}{3} \log (.04657)$

$\quad\quad\quad = \frac{1}{3}(8.66811 - 10) = \frac{1}{3}(-1.33189)$

$\log x \quad = -0.44396 = 9.55604 - 10, \quad x = 0.35978$.

or

$\log x \quad = \frac{1}{3}(8.66811 - 10) = \frac{1}{3}(28.66811 - 30)$

$\log x \quad = 9.55604 - 10, \quad x = 0.35978$.

Example 8.   Find $x = \dfrac{1}{21.210}$.

$\log x = \log 1 - \log 21.210$

$\log 1 \quad\quad = 10.00000 - 10$

$\log 21.210 = \underline{\quad 1.32654 \quad\quad}$ (subtract)

$\log x \quad\quad = \;\; 8.67346 - 10, \quad x = 0.047148$.

Helpful Hints:

1. When connecting numbers to logarithms, use as many decimal places in the mantissa as there are significant digits in the number.

2. When finding the antilogarithm, keep as many significant digits as there are decimal places in the mantissa.

Examples: $\log 10.35 = 1.0149$; antilog $0.065 = 1.16$.

Table 1-1
COMMON LOGARITHMS OF NUMBERS
100–150

N.	0	1	2	3	4	5	6	7	8	9
100	00 000	043	087	130	173	217	260	303	346	389
101	432	475	518	561	604	647	689	732	775	817
102	860	903	945	988	*030	*072	*115	*157	*199	*242
103	01 284	326	368	410	452	494	536	578	620	662
104	703	745	787	828	870	912	953	995	*036	*078
105	02 119	160	202	243	284	325	366	407	449	490
106	531	572	612	653	694	735	776	816	857	898
107	938	979	*019	*060	*100	*141	*181	*222	*262	*302
108	03 342	383	423	463	503	543	583	623	663	703
109	743	782	822	862	902	941	981	*021	*060	*100
110	04 139	179	218	258	297	336	376	415	454	493
111	532	571	610	650	689	727	766	805	844	883
112	922	961	999	*038	*077	*115	*154	*192	*231	*269
113	05 308	346	385	423	461	500	538	576	614	652
114	690	729	767	805	843	881	918	956	994	*032
115	06 070	108	145	183	221	258	296	333	371	408
116	446	483	521	558	595	633	670	707	744	781
117	819	856	893	930	967	*004	*041	*078	*115	*151
118	07 188	225	262	298	335	372	408	445	482	518
119	555	591	628	664	700	737	773	809	846	882
120	918	954	990	*027	*063	*099	*135	*171	*207	*243
121	08 279	314	350	386	422	458	493	529	565	600
122	636	672	707	743	778	814	849	884	920	955
123	991	*026	*061	*096	*132	*167	*202	*237	*272	*307
124	09 342	377	412	447	482	517	552	587	621	656
125	691	726	760	795	830	864	899	934	968	*003
126	10 037	072	106	140	175	209	243	278	312	346
127	380	415	449	483	517	551	585	619	653	687
128	721	755	789	823	857	890	924	958	992	*025
129	11 059	093	126	160	193	227	261	294	327	361
130	394	428	461	494	528	561	594	628	661	694
131	727	760	793	826	860	893	926	959	992	*024
132	12 057	090	123	156	189	222	254	287	320	352
133	385	418	450	483	516	548	581	613	646	678
134	710	743	775	808	840	872	905	937	969	*001
135	13 033	066	098	130	162	194	226	258	290	322
136	354	386	418	450	481	513	545	577	609	640
137	672	704	735	767	799	830	862	893	925	956
138	988	*019	*051	*082	*114	*145	*176	*208	*239	*270
139	14 301	333	364	395	426	457	489	520	551	582
140	613	644	675	706	737	768	799	829	860	891
141	922	953	983	*014	*045	*076	*106	*137	*168	*198
142	15 229	259	290	320	351	381	412	442	473	503
143	534	564	594	625	655	685	715	746	776	806
144	836	866	897	927	957	987	*017	*047	*077	*107
145	16 137	167	197	227	256	286	316	346	376	406
146	435	465	495	524	554	584	613	643	673	702
147	732	761	791	820	850	879	909	938	967	997
148	17 026	056	085	114	143	173	202	231	260	289
149	319	348	377	406	435	464	493	522	551	580
150	609	638	667	696	725	754	782	811	840	869
N.	0	1	2	3	4	5	6	7	8	9

Proportional parts

	44	43	42
1	4.4	4.3	4.2
2	8.8	8.6	8.4
3	13.2	12.9	12.6
4	17.6	17.2	16.8
5	22.0	21.5	21.0
6	26.4	25.8	25.2
7	30.8	30.1	29.4
8	35.2	34.4	33.6
9	39.6	38.7	37.8

	41	40	39
1	4.1	4.0	3.9
2	8.2	8.0	7.8
3	12.3	12.0	11.7
4	16.4	16 0	15.6
5	20.5	20.0	19.5
6	24.6	24.0	23.4
7	28.7	28.0	27.3
8	32.8	32.0	31.2
9	36.9	36.0	35.1

	38	37	36
1	3.8	3.7	3.6
2	7.6	7.4	7.2
3	11.4	11.1	10.8
4	15.2	14.8	14.4
5	19.0	18.5	18.0
6	22.8	22.2	21.6
7	26.6	25.9	25.2
8	30.4	29.6	28.8
9	34.2	33.3	32.4

	35	34	33
1	3.5	3.4	3.3
2	7.0	6.8	6.6
3	10.5	10.2	9.9
4	14.0	13.6	13.2
5	17.5	17.0	16.5
6	21.0	20.4	19.8
7	24.5	23.8	23.1
8	28.0	27.2	26.4
9	31.5	30.6	29.7

	32	31	30
1	3.2	3.1	3.0
2	6.4	6.2	6.0
3	9.6	9.3	9.0
4	12.8	12.4	12.0
5	16.0	15.5	15.0
6	19.2	18.6	18.0
7	22.4	21.7	21.0
8	25.6	24.8	24.0
9	28.8	27.9	27.0

Proportional parts

.00 000–.17 869

Table 1-1 (*Continued*)
COMMON LOGARITHMS OF NUMBERS
150–200

N.	0	1	2	3	4	5	6	7	8	9
150	17 609	638	667	696	725	754	782	811	840	869
151	898	926	955	984	*013	*041	*070	*099	*127	*156
152	18 184	213	241	270	298	327	355	384	412	441
153	469	498	526	554	583	611	639	667	696	724
154	752	780	808	837	865	893	921	949	977	*005
155	19 033	061	089	117	145	173	201	229	257	285
156	312	340	368	396	424	451	479	507	535	562
157	590	618	645	673	700	728	756	783	811	838
158	866	893	921	948	976	*003	*030	*058	*085	*112
159	20 140	167	194	222	249	276	303	330	358	385
160	412	439	466	493	520	548	575	602	629	656
161	683	710	737	763	790	817	844	871	898	925
162	952	978	*005	*032	*059	*085	*112	*139	*165	*192
163	21 219	245	272	299	325	352	378	405	431	458
164	484	511	537	564	590	617	643	669	696	722
165	748	775	801	827	854	880	906	932	958	985
166	22 011	037	063	089	115	141	167	194	220	246
167	272	298	324	350	376	401	427	453	479	505
168	531	557	583	608	634	660	686	712	737	763
169	789	814	840	866	891	917	943	968	994	*019
170	23 045	070	096	121	147	172	198	223	249	274
171	300	325	350	376	401	426	452	477	502	528
172	553	578	603	629	654	679	704	729	754	779
173	805	830	855	880	905	930	955	980	*005	*030
174	24 055	080	105	130	155	180	204	229	254	279
175	304	329	353	378	403	428	452	477	502	527
176	551	576	601	625	650	674	699	724	748	773
177	797	822	846	871	895	920	944	969	993	*018
178	25 042	066	091	115	139	164	188	212	237	261
179	285	310	334	358	382	406	431	455	479	503
180	527	551	575	600	624	648	672	696	720	744
181	768	792	816	840	864	888	912	935	959	983
182	26 007	031	055	079	102	126	150	174	198	221
183	245	269	293	316	340	364	387	411	435	458
184	482	505	529	553	576	600	623	647	670	694
185	717	741	764	788	811	834	858	881	905	928
186	951	975	998	*021	*045	*068	*091	*114	*138	*161
187	27 184	207	231	254	277	300	323	346	370	393
188	416	439	462	485	508	531	554	577	600	623
189	646	669	692	715	738	761	784	807	830	852
190	875	898	921	944	967	989	*012	*035	*058	*081
191	28 103	126	149	171	194	217	240	262	285	307
192	330	353	375	398	421	443	466	488	511	533
193	556	578	601	623	646	668	691	713	735	758
194	780	803	825	847	870	892	914	937	959	981
195	29 003	026	048	070	092	115	137	159	181	203
196	226	248	270	292	314	336	358	380	403	425
197	447	469	491	513	535	557	579	601	623	645
198	667	688	710	732	754	776	798	820	842	863
199	885	907	929	951	973	994	*016	*038	*060	*081
200	30 103	125	146	168	190	211	233	255	276	298
N.	0	1	2	3	4	5	6	7	8	9

Proportional parts

	29	28		27	26		25		24	23		22	21
1	2.9	2.8	1	2.7	2.6	1	2.5	1	2.4	2.3	1	2.2	2.1
2	5.8	5.6	2	5.4	5.2	2	5.0	2	4.8	4.6	2	4.4	4.2
3	8.7	8.4	3	8.1	7.8	3	7.5	3	7.2	6.9	3	6.6	6.3
4	11.6	11.2	4	10.8	10.4	4	10.0	4	9.6	9.2	4	8.8	8.4
5	14.5	14.0	5	13.5	13.0	5	12.5	5	12.0	11.5	5	11.0	10.5
6	17.4	16.8	6	16.2	15.6	6	15.0	6	14.4	13.8	6	13.2	12.6
7	20.3	19.6	7	18.9	18.2	7	17.5	7	16.8	16.1	7	15.4	14.7
8	23.2	22.4	8	21.6	20.8	8	20.0	8	19.2	18.4	8	17.6	16.8
9	26.1	25.2	9	24.3	23.4	9	22.5	9	21.6	20.7	9	19.8	18.9

.17 609–.30 298

Table 1-1 (*Continued*)
COMMON LOGARITHMS OF NUMBERS
200–250

N.	0	1	2	3	4	5	6	7	8	9
200	30 103	125	146	168	190	211	233	255	276	298
201	320	341	363	384	406	428	449	471	492	514
202	535	557	578	600	621	643	664	685	707	728
203	750	771	792	814	835	856	878	899	920	942
204	963	984	*006	*027	*048	*069	*091	*112	*133	*154
205	31 175	197	218	239	260	281	302	323	345	366
206	387	408	429	450	471	492	513	534	555	576
207	597	618	639	660	681	702	723	744	765	785
208	806	827	848	869	890	911	931	952	973	994
209	32 015	035	056	077	098	118	139	160	181	201
210	222	243	263	284	305	325	346	366	387	408
211	428	449	469	490	510	531	552	572	593	613
212	634	654	675	695	715	736	756	777	797	818
213	838	858	879	899	919	940	960	980	*001	*021
214	33 041	062	082	102	122	143	163	183	203	224
215	244	264	284	304	325	345	365	385	405	425
216	445	465	486	506	526	546	566	586	606	626
217	646	666	686	706	726	746	766	786	806	826
218	846	866	885	905	925	945	965	985	*005	*025
219	34 044	064	084	104	124	143	163	183	203	223
220	242	262	282	301	321	341	361	380	400	420
221	439	459	479	498	518	537	557	577	596	616
222	635	655	674	694	713	733	753	772	792	811
223	830	850	869	889	908	928	947	967	986	*005
224	35 025	044	064	083	102	122	141	160	180	199
225	218	238	257	276	295	315	334	353	372	392
226	411	430	449	468	488	507	526	545	564	583
227	603	622	641	660	679	698	717	736	755	774
228	793	813	832	851	870	889	908	927	946	965
229	984	*003	*021	*040	*059	*078	*097	*116	*135	*154
230	36 173	192	211	229	248	267	286	305	324	342
231	361	380	399	418	436	455	474	493	511	530
232	549	568	586	605	624	642	661	680	698	717
233	736	754	773	791	810	829	847	866	884	903
234	922	940	959	977	996	*014	*033	*051	*070	*088
235	37 107	125	144	162	181	199	218	236	254	273
236	291	310	328	346	365	383	401	420	438	457
237	475	493	511	530	548	566	585	603	621	639
238	658	676	694	712	731	749	767	785	803	822
239	840	858	876	894	912	931	949	967	985	*003
240	38 021	039	057	075	093	112	130	148	166	184
241	202	220	238	256	274	292	310	328	346	364
242	382	399	417	435	453	471	489	507	525	543
243	561	578	596	614	632	650	668	686	703	721
244	739	757	775	792	810	828	846	863	881	899
245	917	934	952	970	987	*005	*023	*041	*058	*076
246	39 094	111	129	146	164	182	199	217	235	252
247	270	287	305	322	340	358	375	393	410	428
248	445	463	480	498	515	533	550	568	585	602
249	620	637	655	672	690	707	724	742	759	777
250	794	811	829	846	863	881	898	915	933	950
N.	0	1	2	3	4	5	6	7	8	9

Proportional parts

	22	21
1	2.2	2.1
2	4.4	4.2
3	6.6	6.3
4	8.8	8.4
5	11.0	10.5
6	13.2	12.6
7	15.4	14.7
8	17.6	16.8
9	19.8	18.9

	20
1	2.0
2	4.0
3	6.0
4	8.0
5	10.0
6	12.0
7	14.0
8	16.0
9	18.0

	19
1	1.9
2	3.8
3	5.7
4	7.6
5	9.5
6	11.4
7	13.3
8	15.2
9	17.1

	18
1	1.8
2	3.6
3	5.4
4	7.2
5	9.0
6	10.8
7	12.6
8	14.4
9	16.2

	17
1	1.7
2	3.4
3	5.1
4	6.8
5	8.5
6	10.2
7	11.9
8	13.6
9	15.3

.30 103–.39 950

Table 1-1 (*Continued*)
COMMON LOGARITHMS OF NUMBERS
250–300

N.	0	1	2	3	4	5	6	7	8	9
250	39 794	811	829	846	863	881	898	915	933	950
251	967	985	*002	*019	*037	*054	*071	*088	*106	*123
252	40 140	157	175	192	209	226	243	261	278	295
253	312	329	346	364	381	398	415	432	449	466
254	483	500	518	535	552	569	586	603	620	637
255	654	671	688	705	722	739	756	773	790	807
256	824	841	858	875	892	909	926	943	960	976
257	993	*010	*027	*044	*061	*078	*095	*111	*128	*145
258	41 162	179	196	212	229	246	263	280	296	313
259	330	347	363	380	397	414	430	447	464	481
260	497	514	531	547	564	581	597	614	631	647
261	664	681	697	714	731	747	764	780	797	814
262	830	847	863	880	896	913	929	946	963	979
263	996	*012	*029	*045	*062	*078	*095	*111	*127	*144
264	42 160	177	193	210	226	243	259	275	292	308
265	325	341	357	374	390	406	423	439	455	472
266	488	504	521	537	553	570	586	602	619	635
267	651	667	684	700	716	732	749	765	781	797
268	813	830	846	862	878	894	911	927	943	959
269	975	991	*008	*024	*040	*056	*072	*088	*104	*120
270	43 136	152	169	185	201	217	233	249	265	281
271	297	313	329	345	361	377	393	409	425	441
272	457	473	489	505	521	537	553	569	584	600
273	616	632	648	664	680	696	712	727	743	759
274	775	791	807	823	838	854	870	886	902	917
275	933	949	965	981	996	*012	*028	*044	*059	*075
276	44 091	107	122	138	154	170	185	201	217	232
277	248	264	279	295	311	326	342	358	373	389
278	404	420	436	451	467	483	498	514	529	545
279	560	576	592	607	623	638	654	669	685	700
280	716	731	747	762	778	793	809	824	840	855
281	871	886	902	917	932	948	963	979	994	*010
282	45 025	040	056	071	086	102	117	133	148	163
283	179	194	209	225	240	255	271	286	301	317
284	332	347	362	378	393	408	423	439	454	469
285	484	500	515	530	545	561	576	591	606	621
286	637	652	667	682	697	712	728	743	758	773
287	788	803	818	834	849	864	879	894	909	924
288	939	954	969	984	*000	*015	*030	*045	*060	*075
289	46 090	105	120	135	150	165	180	195	210	225
290	240	255	270	285	300	315	330	345	359	374
291	389	404	419	434	449	464	479	494	509	523
292	538	553	568	583	598	613	627	642	657	672
293	687	702	716	731	746	761	776	790	805	820
294	835	850	864	879	894	909	923	938	953	967
295	982	997	*012	*026	*041	*056	*070	*085	*100	*114
296	47 129	144	159	173	188	202	217	232	246	261
297	276	290	305	319	334	349	363	378	392	407
298	422	436	451	465	480	494	509	524	538	553
299	567	582	596	611	625	640	654	669	683	698
300	712	727	741	756	770	784	799	813	828	842
N.	0	1	2	3	4	5	6	7	8	9

Proportional parts

18		17		16		15		14	
1	1.8	1	1.7	1	1.6	1	1.5	1	1.4
2	3.6	2	3.4	2	3.2	2	3.0	2	2.8
3	5.4	3	5.1	3	4.8	3	4.5	3	4.2
4	7.2	4	6.8	4	6.4	4	6.0	4	5.6
5	9.0	5	8.5	5	8.0	5	7.5	5	7.0
6	10.8	6	10.2	6	9.6	6	9.0	6	8.4
7	12.6	7	11.9	7	11.2	7	10.5	7	9.8
8	14.4	8	13.6	8	12.8	8	12.0	8	11.2
9	16.2	9	15.3	9	14.4	9	13.5	9	12.6

$\log e = 0.43429$

Proportional parts

.39 794–.47 842

Table 1-1 (*Continued*)
COMMON LOGARITHMS OF NUMBERS
300–350

N.	0	1	2	3	4	5	6	7	8	9
300	47 712	727	741	756	770	784	799	813	828	842
301	857	871	885	900	914	929	943	958	972	986
302	48 001	015	029	044	058	073	087	101	116	130
303	144	159	173	187	202	216	230	244	259	273
304	287	302	316	330	344	359	373	387	401	416
305	430	444	458	473	487	501	515	530	544	558
306	572	586	601	615	629	643	657	671	686	700
307	714	728	742	756	770	785	799	813	827	841
308	855	869	883	897	911	926	940	954	968	982
309	996	*010	*024	*038	*052	*066	*080	*094	*108	*122
310	49 136	150	164	178	192	206	220	234	248	262
311	276	290	304	318	332	346	360	374	388	402
312	415	429	443	457	471	485	499	513	527	541
313	554	568	582	596	610	624	638	651	665	679
314	693	707	721	734	748	762	776	790	803	817
315	831	845	859	872	886	900	914	927	941	955
316	969	982	996	*010	*024	*037	*051	*065	*079	*092
317	50 106	120	133	147	161	174	188	202	215	229
318	243	256	270	284	297	311	325	338	352	365
319	379	393	406	420	433	447	461	474	488	501
320	515	529	542	556	569	583	596	610	623	637
321	651	664	678	691	705	718	732	745	759	772
322	786	799	813	826	840	853	866	880	893	907
323	920	934	947	961	974	987	*001	*014	*028	*041
324	51 055	068	081	095	108	121	135	148	162	175
325	188	202	215	228	242	255	268	282	295	308
326	322	335	348	362	375	388	402	415	428	441
327	455	468	481	495	508	521	534	548	561	574
328	587	601	614	627	640	654	667	680	693	706
329	720	733	746	759	772	786	799	812	825	838
330	851	865	878	891	904	917	930	943	957	970
331	983	996	*009	*022	*035	*048	*061	*075	*088	*101
332	52 114	127	140	153	166	179	192	205	218	231
333	244	257	270	284	297	310	323	336	349	362
334	375	388	401	414	427	440	453	466	479	492
335	504	517	530	543	556	569	582	595	608	621
336	634	647	660	673	686	699	711	724	737	750
337	763	776	789	802	815	827	840	853	866	879
338	892	905	917	930	943	956	969	982	994	*007
339	53 020	033	046	058	071	084	097	110	122	135
340	148	161	173	186	199	212	224	237	250	263
341	275	288	301	314	326	339	352	364	377	390
342	403	415	428	441	453	466	479	491	504	517
343	529	542	555	567	580	593	605	618	631	643
344	656	668	681	694	706	719	732	744	757	769
345	782	794	807	820	832	845	857	870	882	895
346	908	920	933	945	958	970	983	995	*008	*020
347	54 033	045	058	070	083	095	108	120	133	145
348	158	170	183	195	208	220	233	245	258	270
349	283	295	307	320	332	345	357	370	382	394
350	407	419	432	444	456	469	481	494	506	518
N.	0	1	2	3	4	5	6	7	8	9

Proportional parts

	15		14		13		12
1	1.5	1	1.4	1	1.3	1	1.2
2	3.0	2	2.8	2	2.6	2	2.4
3	4.5	3	4.2	3	3.9	3	3.6
4	6.0	4	5.6	4	5.2	4	4.8
5	7.5	5	7.0	5	6.5	5	6.0
6	9.0	6	8.4	6	7.8	6	7.2
7	10.5	7	9.8	7	9.1	7	8.4
8	12.0	8	11.2	8	10.4	8	9.6
9	13.5	9	12.6	9	11.7	9	10.8

$\log \pi = 0.49715$

.47 712–.54 518

Table 1-1 (*Continued*)
COMMON LOGARITHMS OF NUMBERS
350–400

N.	0	1	2	3	4	5	6	7	8	9	Proportional parts
350	54 407	419	432	444	456	469	481	494	506	518	
351	531	543	555	568	580	593	605	617	630	642	
352	654	667	679	691	704	716	728	741	753	765	
353	777	790	802	814	827	839	851	864	876	888	
354	900	913	925	937	949	962	974	986	998	*011	**13**
											1 1.3
355	55 023	035	047	060	072	084	096	108	121	133	2 2.6
356	145	157	169	182	194	206	218	230	242	255	3 3.9
357	267	279	291	303	315	328	340	352	364	376	4 5.2
358	388	400	413	425	437	449	461	473	485	497	5 6.5
359	509	522	534	546	558	570	582	594	606	618	6 7.8
											7 9.1
360	630	642	654	666	678	691	703	715	727	739	8 10.4
361	751	763	775	787	799	811	823	835	847	859	9 11.7
362	871	883	895	907	919	931	943	955	967	979	
363	991	*003	*015	*027	*038	*050	*062	*074	*086	*098	
364	56 110	122	134	146	158	170	182	194	205	217	
365	229	241	253	265	277	289	301	312	324	336	**12**
366	348	360	372	384	396	407	419	431	443	455	1 1.2
367	467	478	490	502	514	526	538	549	561	573	2 2.4
368	585	597	608	620	632	644	656	667	679	691	3 3.6
369	703	714	726	738	750	761	773	785	797	808	4 4.8
											5 6.0
370	820	832	844	855	867	879	891	902	914	926	6 7.2
371	937	949	961	972	984	996	*008	*019	*031	*043	7 8.4
372	57 054	066	078	089	101	113	124	136	148	159	8 9.6
373	171	183	194	206	217	229	241	252	264	276	9 10.8
374	287	299	310	322	334	345	357	368	380	392	
375	403	415	426	438	449	461	473	484	496	507	
376	519	530	542	553	565	576	588	600	611	623	
377	634	646	657	669	680	692	703	715	726	738	**11**
378	749	761	772	784	795	807	818	830	841	852	1 1.1
379	864	875	887	898	910	921	933	944	955	967	2 2.2
											3 3.3
380	978	990	*001	*013	*024	*035	*047	*058	*070	*081	4 4.4
381	58 092	104	115	127	138	149	161	172	184	195	5 5.5
382	206	218	229	240	252	263	274	286	297	309	6 6.6
383	320	331	343	354	365	377	388	399	410	422	7 7.7
384	433	444	456	467	478	490	501	512	524	535	8 8.8
											9 9.9
385	546	557	569	580	591	602	614	625	636	647	
386	659	670	681	692	704	715	726	737	749	760	
387	771	782	794	805	816	827	838	850	861	872	
388	883	894	906	917	928	939	950	961	973	984	**10**
389	995	*006	*017	*028	*040	*051	*062	*073	*084	*095	1 1.0
											2 2.0
390	59 106	118	129	140	151	162	173	184	195	207	3 3.0
391	218	229	240	251	262	273	284	295	306	318	4 4.0
392	329	340	351	362	373	384	395	406	417	428	5 5.0
393	439	450	461	472	483	494	506	517	528	539	6 6.0
394	550	561	572	583	594	605	616	627	638	649	7 7.0
											8 8.0
395	660	671	682	693	704	715	726	737	748	759	9 9.0
396	770	780	791	802	813	824	835	846	857	868	
397	879	890	901	912	923	934	945	956	966	977	
398	988	999	*010	*021	*032	*043	*054	*065	*076	*086	
399	60 097	108	119	130	141	152	163	173	184	195	
400	206	217	228	239	249	260	271	282	293	304	
N.	0	1	2	3	4	5	6	7	8	9	Proportional parts

.54 407–.60 304

Table 1-1 (*Continued*)
COMMON LOGARITHMS OF NUMBERS
400–450

N.	0	1	2	3	4	5	6	7	8	9
400	60 206	217	228	239	249	260	271	282	293	304
401	314	325	336	347	358	369	379	390	401	412
402	423	433	444	455	466	477	487	498	509	520
403	531	541	552	563	574	584	595	606	617	627
404	638	649	660	670	681	692	703	713	724	735
405	746	756	767	778	788	799	810	821	831	842
406	853	863	874	885	895	906	917	927	938	949
407	959	970	981	991	*002	*013	*023	*034	*045	*055
408	61 066	077	087	098	109	119	130	140	151	162
409	172	183	194	204	215	225	236	247	257	268
410	278	289	300	310	321	331	342	352	363	374
411	384	395	405	416	426	437	448	458	469	479
412	490	500	511	521	532	542	553	563	574	584
413	595	606	616	627	637	648	658	669	679	690
414	700	711	721	731	742	752	763	773	784	794
415	805	815	826	836	847	857	868	878	888	899
416	909	920	930	941	951	962	972	982	993	*003
417	62 014	024	034	045	055	066	076	086	097	107
418	118	128	138	149	159	170	180	190	201	211
419	221	232	242	252	263	273	284	294	304	315
420	325	335	346	356	366	377	387	397	408	418
421	428	439	449	459	469	480	490	500	511	521
422	531	542	552	562	572	583	593	603	613	624
423	634	644	655	665	675	685	696	706	716	726
424	737	747	757	767	778	788	798	808	818	829
425	839	849	859	870	880	890	900	910	921	931
426	941	951	961	972	982	992	*002	*012	*022	*033
427	63 043	053	063	073	083	094	104	114	124	134
428	144	155	165	175	185	195	205	215	225	236
429	246	256	266	276	286	296	306	317	327	337
430	347	357	367	377	387	397	407	417	428	438
431	448	458	468	478	488	498	508	518	528	538
432	548	558	568	579	589	599	609	619	629	639
433	649	659	669	679	689	699	709	719	729	739
434	749	759	769	779	789	799	809	819	829	839
435	849	859	869	879	889	899	909	919	929	939
436	949	959	969	979	988	998	*008	*018	*028	*038
437	64 048	058	068	078	088	098	108	118	128	137
438	147	157	167	177	187	197	207	217	227	237
439	246	256	266	276	286	296	306	316	326	335
440	345	355	365	375	385	395	404	414	424	434
441	444	454	464	473	483	493	503	513	523	532
442	542	552	562	572	582	591	601	611	621	631
443	640	650	660	670	680	689	699	709	719	729
444	738	748	758	768	777	787	797	807	816	826
445	836	846	856	865	875	885	895	904	914	924
446	933	943	953	963	972	982	992	*002	*011	*021
447	65 031	040	050	060	070	079	089	099	108	118
448	128	137	147	157	167	176	186	196	205	215
449	225	234	244	254	263	273	283	292	302	312
450	321	331	341	350	360	369	379	389	398	408
N.	0	1	2	3	4	5	6	7	8	9

Proportional parts

	11		10		9
1	1.1	1	1.0	1	0.9
2	2.2	2	2.0	2	1.8
3	3.3	3	3.0	3	2.7
4	4.4	4	4.0	4	3.6
5	5.5	5	5.0	5	4.5
'6	6.6	6	6.0	6	5.4
7	7.7	7	7.0	7	6.3
8	8.8	8	8.0	8	7.2
9	9.9	9	9.0	9	8.1

.60 206–.65 408

Table 1-1 (*Continued*)
COMMON LOGARITHMS OF NUMBERS
450–500

N.	0	1	2	3	4	5	6	7	8	9	Proportional parts
450	65 321	331	341	350	360	369	379	389	398	408	
451	418	427	437	447	456	466	475	485	495	504	
452	514	523	533	543	552	562	571	581	591	600	
453	610	619	629	639	648	658	667	677	686	696	
454	706	715	725	734	744	753	763	772	782	792	
455	801	811	820	830	839	849	858	868	877	887	
456	896	906	916	925	935	944	954	963	973	982	
457	992	*001	*011	*020	*030	*039	*049	*058	*068	*077	**10**
458	66 087	096	106	115	124	134	143	153	162	172	1 1.0
459	181	191	200	210	219	229	238	247	257	266	2 2.0 / 3 3.0
460	276	285	295	304	314	323	332	342	351	361	4 4.0
461	370	380	389	398	408	417	427	436	445	455	5 5.0
462	464	474	483	492	502	511	521	530	539	549	6 6.0
463	558	567	577	586	596	605	614	624	633	642	7 7.0
464	652	661	671	680	689	699	708	717	727	736	8 8.0 / 9 9.0
465	745	755	764	773	783	792	801	811	820	829	
466	839	848	857	867	876	885	894	904	913	922	
467	932	941	950	960	969	978	987	997	*006	*015	
468	67 025	034	043	052	062	071	080	089	099	108	
469	117	127	136	145	154	164	173	182	191	201	
470	210	219	228	237	247	256	265	274	284	293	
471	302	311	321	330	339	348	357	367	376	385	**9**
472	394	403	413	422	431	440	449	459	468	477	1 0.9
473	486	495	504	514	523	532	541	550	560	569	2 1.8
474	578	587	596	605	614	624	633	642	651	660	3 2.7 / 4 3.6
475	669	679	688	697	706	715	724	733	742	752	5 4.5
476	761	770	779	788	797	806	815	825	834	843	6 5.4
477	852	861	870	879	888	897	906	916	925	934	7 6.3
478	943	952	961	970	979	988	997	*006	*015	*024	8 7.2
479	68 034	043	052	061	070	079	088	097	106	115	9 8.1
480	124	133	142	151	160	169	178	187	196	205	
481	215	224	233	242	251	260	269	278	287	296	
482	305	314	323	332	341	350	359	368	377	386	
483	395	404	413	422	431	440	449	458	467	476	
484	485	494	502	511	520	529	538	547	556	565	
485	574	583	592	601	610	619	628	637	646	655	**8**
486	664	673	681	690	699	708	717	726	735	744	1 0.8
487	753	762	771	780	789	797	806	815	824	833	2 1.6
488	842	851	860	869	878	886	895	904	913	922	3 2.4
489	931	940	949	958	966	975	984	993	*002	*011	4 3.2 / 5 4.0
490	69 020	028	037	046	055	064	073	082	090	099	6 4.8
491	108	117	126	135	144	152	161	170	179	188	7 5.6
492	197	205	214	223	232	241	249	258	267	276	8 6.4
493	285	294	302	311	320	329	338	346	355	364	9 7.2
494	373	381	390	399	408	417	425	434	443	452	
495	461	469	478	487	496	504	513	522	531	539	
496	548	557	566	574	583	592	601	609	618	627	
497	636	644	653	662	671	679	688	697	705	714	
498	723	732	740	749	758	767	775	784	793	801	
499	810	819	827	836	845	854	862	871	880	888	
500	897	906	914	923	932	940	949	958	966	975	
N.	0	1	2	3	4	5	6	7	8	9	Proportional parts

.65 321–.69 975

Table 1-1 (*Continued*)
COMMON LOGARITHMS OF NUMBERS
500–550

N.	0	1	2	3	4	5	6	7	8	9
500	69 897	906	914	923	932	940	949	958	,966	975
501	984	992	*001	*010	*018	*027	*036	*044	*053	*062
502	70 070	079	088	096	105	114	122	131	140	148
503	157	165	174	183	191	200	209	217	226	234
504	243	252	260	269	278	286	295	303	312	321
505	329	338	346	355	364	372	381	389	398	406
506	415	424	432	441	449	458	467	475	484	492
507	501	509	518	526	535	544	552	561	569	578
508	586	595	603	612	621	629	638	646	655	663
509	672	680	689	697	706	714	723	731	740	749
510	757	766	774	783	791	800	808	817	825	834
511	842	851	859	868	876	885	893	902	910	919
512	927	935	944	952	961	969	978	986	995	*003
513	71 012	020	029	037	046	054	063	071	079	088
514	096	105	113	122	130	139	147	155	164	172
515	181	189	198	206	214	223	231	240	248	257
516	265	273	282	290	299	307	315	324	332	341
517	349	357	366	374	383	391	399	408	416	425
518	433	441	450	458	466	475	483	492	500	508
519	517	525	533	542	550	559	567	575	584	592
520	600	609	617	625	634	642	650	659	667	675
521	684	692	700	709	717	725	734	742	750	759
522	767	775	784	792	800	809	817	825	834	842
523	850	858	867	875	883	892	900	908	917	925
524	933	941	950	958	966	975	983	991	999	*008
525	72 016	024	032	041	049	057	066	074	082	090
526	099	107	115	123	132	140	148	156	165	173
527	181	189	198	206	214	222	230	239	247	255
528	263	272	280	288	296	304	313	321	329	337
529	346	354	362	370	378	387	395	403	411	419
530	428	436	444	452	460	469	477	485	493	501
531	509	518	526	534	542	550	558	567	575	583
532	591	599	607	616	624	632	640	648	656	665
533	673	681	689	697	705	713	722	730	738	746
534	754	762	770	779	787	795	803	811	819	827
535	835	843	852	860	868	876	884	892	900	908
536	916	925	933	941	949	957	965	973	981	989
537	997	*006	*014	*022	*030	*038	*046	*054	*062	*070
538	73 078	086	094	102	111	119	127	135	143	151
539	159	167	175	183	191	199	207	215	223	231
540	239	247	255	263	272	280	288	296	304	312
541	320	328	336	344	352	360	368	376	384	392
542	400	408	416	424	432	440	448	456	464	472
543	480	488	496	504	512	520	528	536	544	552
544	560	568	576	584	592	600	608	616	624	632
545	640	648	656	664	672	679	687	695	703	711
546	719	727	735	743	751	759	767	775	783	791
547	799	807	815	823	830	838	846	854	862	870
548	878	886	894	902	910	918	926	933	941	949
549	957	965	973	981	989	997	*005	*013	*020	*028
550	74 036	044	052	060	068	076	084	092	099	107
N.	0	1	2	3	4	5	6	7	8	9

Proportional parts

	9		8		7
1	0.9	1	0.8	1	0.7
2	1.8	2	1.6	2	1.4
3	2.7	3	2.4	3	2.1
4	3.6	4	3.2	4	2.8
5	4.5	5	4.0	5	3.5
6	5.4	6	4.8	6	4.2
7	6.3	7	5.6	7	4.9
8	7.2	8	6.4	8	5.6
9	8.1	9	7.2	9	6.3

.69 897–.74 107

Table 1-1 (*Continued*)
COMMON LOGARITHMS OF NUMBERS
550–600

N.	0	1	2	3	4	5	6	7	8	9	Proportional parts
550	74 036	044	052	060	068	076	084	092	099	107	
551	115	123	131	139	147	155	162	170	178	186	
552	194	202	210	218	225	233	241	249	257	265	
553	273	280	288	296	304	312	320	327	335	343	
554	351	359	367	374	382	390	398	406	414	421	
555	429	437	445	453	461	468	476	484	492	500	
556	507	515	523	531	539	547	554	562	570	578	
557	586	593	601	609	617	624	632	640	648	656	
558	663	671	679	687	695	702	710	718	726	733	
559	741	749	757	764	772	780	788	796	803	811	
560	819	827	834	842	850	858	865	873	881	889	
561	896	904	912	920	927	935	943	950	958	966	
562	974	981	989	997	*005	*012	*020	*028	*035	*043	
563	75 051	059	066	074	082	089	097	105	113	120	
564	128	136	143	151	159	166	174	182	189	197	
565	205	213	220	228	236	243	251	259	266	274	
566	282	289	297	305	312	320	328	335	343	351	
567	358	366	374	381	389	397	404	412	420	427	
568	435	442	450	458	465	473	481	488	496	504	
569	511	519	526	534	542	549	557	565	572	580	
570	587	595	603	610	618	626	633	641	648	656	
571	664	671	679	686	694	702	709	717	724	732	
572	740	747	755	762	770	778	785	793	800	808	
573	815	823	831	838	846	853	861	868	876	884	
574	891	899	906	914	921	929	937	944	952	959	
575	967	974	982	989	997	*005	*012	*020	*027	*035	
576	76 042	050	057	065	072	080	087	095	103	110	
577	118	125	133	140	148	155	163	170	178	185	
578	193	200	208	215	223	230	238	245	253	260	
579	268	275	283	290	298	305	313	320	328	335	
580	343	350	358	365	373	380	388	395	403	410	
581	418	425	433	440	448	455	462	470	477	485	
582	492	500	507	515	522	530	537	545	552	559	
583	567	574	582	589	597	604	612	619	626	634	
584	641	649	656	664	671	678	686	693	701	708	
585	716	723	730	738	745	753	760	768	775	782	
586	790	797	805	812	819	827	834	842	849	856	
587	864	871	879	886	893	901	908	916	923	930	
588	938	945	953	960	967	975	982	989	997	*004	
589	77 012	019	026	034	041	048	056	063	070	078	
590	085	093	100	107	115	122	129	137	144	151	
591	159	166	173	181	188	195	203	210	217	225	
592	232	240	247	254	262	269	276	283	291	298	
593	305	313	320	327	335	342	349	357	364	371	
594	379	386	393	401	408	415	422	430	437	444	
595	452	459	466	474	481	488	495	503	510	517	
596	525	532	539	546	554	561	568	576	583	590	
597	597	605	612	619	627	634	641	648	656	663	
598	670	677	685	692	699	706	714	721	728	735	
599	743	750	757	764	772	779	786	793	801	808	
600	815	822	830	837	844	851	859	866	873	880	
N.	0	1	2	3	4	5	6	7	8	9	Proportional parts

Proportional parts:

	8
1	0.8
2	1.6
3	2.4
4	3.2
5	4.0
6	4.8
7	5.6
8	6.4
9	7.2

	7
1	0.7
2	1.4
3	2.1
4	2.8
5	3.5
6	4.2
7	4.9
8	5.6
9	6.3

.74 036–.77 880

Table 1-1 (*Continued*)
COMMON LOGARITHMS OF NUMBERS
600–650

N.	0	1	2	3	4	5	6	7	8	9	Proportional parts
600	77 815	822	830	837	844	851	859	866	873	880	
601	887	895	902	909	916	924	931	938	945	952	
602	960	967	974	981	988	996	*003	*010	*017	*025	
603	78 032	039	046	053	061	068	075	082	089	097	
604	104	111	118	125	132	140	147	154	161	168	
605	176	183	190	197	204	211	219	226	233	240	
606	247	254	262	269	276	283	290	297	305	312	
607	319	326	333	340	347	355	362	369	376	383	**8**
608	390	398	405	412	419	426	433	440	447	455	1 0.8
609	462	469	476	483	490	497	504	512	519	526	2 1.6
											3 2.4
610	533	540	547	554	561	569	576	583	590	597	4 3.2
611	604	611	618	625	633	640	647	654	661	668	5 4.0
612	675	682	689	696	704	711	718	725	732	739	6 4.8
613	746	753	760	767	774	781	789	796	803	810	7 5.6
614	817	824	831	838	845	852	859	866	873	880	8 6.4
											9 7.2
615	888	895	902	909	916	923	930	937	944	951	
616	958	965	972	979	986	993	*000	*007	*014	*021	
617	79 029	036	043	050	057	064	071	078	085	092	
618	099	106	113	120	127	134	141	148	155	162	
619	169	176	183	190	197	204	211	218	225	232	
620	239	246	253	260	267	274	281	288	295	302	
621	309	316	323	330	337	344	351	358	365	372	**7**
622	379	386	393	400	407	414	421	428	435	442	1 0.7
623	449	456	463	470	477	484	491	498	505	511	2 1.4
624	518	525	532	539	546	553	560	567	574	581	3 2.1
											4 2.8
625	588	595	602	609	616	623	630	637	644	650	5 3.5
626	657	664	671	678	685	692	699	706	713	720	6 4.2
627	727	734	741	748	754	761	768	775	782	789	7 4.9
628	796	803	810	817	824	831	837	844	851	858	8 5.6
629	865	872	879	886	893	900	906	913	920	927	9 6.3
630	934	941	948	955	962	969	975	982	989	996	
631	80 003	010	017	024	030	037	044	051	058	065	
632	072	079	085	092	099	106	113	120	127	134	
633	140	147	154	161	168	175	182	188	195	202	
634	209	216	223	229	236	243	250	257	264	271	
635	277	284	291	298	305	312	318	325	332	339	**6**
636	346	353	359	366	373	380	387	393	400	407	1 0.6
637	414	421	428	434	441	448	455	462	468	475	2 1.2
638	482	489	496	502	509	516	523	530	536	543	3 1.8
639	550	557	564	570	577	584	591	598	604	611	4 2.4
											5 3.0
640	618	625	632	638	645	652	659	665	672	679	6 3.6
641	686	693	699	706	713	720	726	733	740	747	7 4.2
642	754	760	767	774	781	787	794	801	808	814	8 4.8
643	821	828	835	841	848	855	862	868	875	882	9 5.4
644	889	895	902	909	916	922	929	936	943	949	
645	956	963	969	976	983	990	996	*003	*010	*017	
646	81 023	030	037	043	050	057	064	070	077	084	
647	090	097	104	111	117	124	131	137	144	151	
648	158	164	171	178	184	191	198	204	211	218	
649	224	231	238	245	251	258	265	271	278	285	
650	291	298	305	311	318	325	331	338	345	351	
N.	0	1	2	3	4	5	6	7	8	9	Proportional parts

.77 815–.81 351

Table 1-1 (*Continued*)
COMMON LOGARITHMS OF NUMBERS
650-700

N.	0	1	2	3	4	5	6	7	8	9	Proportional parts
650	81 291	298	305	311	318	325	331	338	345	351	
651	358	365	371	378	385	391	398	405	411	418	
652	425	431	438	445	451	458	465	471	478	485	
653	491	498	505	511	518	525	531	538	544	551	
654	558	564	571	578	584	591	598	604	611	617	
655	624	631	637	644	651	657	664	671	677	684	
656	690	697	704	710	717	723	730	737	743	750	
657	757	763	770	776	783	790	796	803	809	816	
658	823	829	836	842	849	856	862	869	875	882	
659	889	895	902	908	915	921	928	935	941	948	
660	954	961	968	974	981	987	994	*000	*007	*014	
661	82 020	027	033	040	046	053	060	066	073	079	
662	086	092	099	105	112	119	125	132	138	145	
663	151	158	164	171	178	184	191	197	204	210	
664	217	223	230	236	243	249	256	263	269	276	
665	282	289	295	302	308	315	321	328	334	341	
666	347	354	360	367	373	380	387	393	400	406	
667	413	419	426	432	439	445	452	458	465	471	
668	478	484	491	497	504	510	517	523	530	536	
669	543	549	556	562	569	575	582	588	595	601	
670	607	614	620	627	633	640	646	653	659	666	
671	672	679	685	692	698	705	711	718	724	730	
672	737	743	750	756	763	769	776	782	789	795	
673	802	808	814	821	827	834	840	847	853	860	
674	866	872	879	885	892	898	905	911	918	924	
675	930	937	943	950	956	963	969	975	982	988	
676	995	*001	*008	*014	*020	*027	*033	*040	*046	*052	
677	83 059	065	072	078	085	091	097	104	110	117	
678	123	129	136	142	149	155	161	168	174	181	
679	187	193	200	206	213	219	225	232	238	245	
680	251	257	264	270	276	283	289	296	302	308	
681	315	321	327	334	340	347	353	359	366	372	
682	378	385	391	398	404	410	417	423	429	436	
683	442	448	455	461	467	474	480	487	493	499	
684	506	512	518	525	531	537	544	550	556	563	
685	569	575	582	588	594	601	607	613	620	626	
686	632	639	645	651	658	664	670	677	683	689	
687	696	702	708	715	721	727	734	740	746	753	
688	759	765	771	778	784	790	797	803	809	816	
689	822	828	835	841	847	853	860	866	872	879	
690	885	891	897	904	910	916	923	929	935	942	
691	948	954	960	967	973	979	985	992	998	*004	
692	84 011	017	023	029	036	042	048	055	061	067	
693	073	080	086	092	098	105	111	117	123	130	
694	136	142	148	155	161	167	173	180	186	192	
695	198	205	211	217	223	230	236	242	248	255	
696	261	267	273	280	286	292	298	305	311	317	
697	323	330	336	342	348	354	361	367	373	379	
698	386	392	398	404	410	417	423	429	435	442	
699	448	454	460	466	473	479	485	491	497	504	
700	510	516	522	528	535	541	547	553	559	566	
N.	0	1	2	3	4	5	6	7	8	9	Proportional parts

Proportional parts:

	7
1	0.7
2	1.4
3	2.1
4	2.8
5	3.5
6	4.2
7	4.9
8	5.6
9	6 3

	6
1	0.6
2	1.2
3	1.8
4	2.4
5	3.0
6	3.6
7	4.2
8	4.8
9	5.4

.81 291–.84 566

Table 1-1 (*Continued*)
COMMON LOGARITHMS OF NUMBERS
700–.750

N.	0	1	2	3	4	5	6	7	8	9
700	84 510	516	522	528	535	541	547	553	559	566
701	572	578	584	590	597	603	609	615	621	628
702	634	640	646	652	658	665	671	677	683	689
703	696	702	708	714	720	726	733	739	745	751
704	757	763	770	776	782	788	794	800	807	813
705	819	825	831	837	844	850	856	862	868	874
706	880	887	893	899	905	911	917	924	930	936
707	942	948	954	960	967	973	979	985	991	997
708	85 003	009	016	022	028	034	040	046	052	058
709	065	071	077	083	089	095	101	107	114	120
710	126	132	138	144	150	156	163	169	175	181
711	187	193	199	205	211	217	224	230	236	242
712	248	254	260	266	272	278	285	291	297	303
713	309	315	321	327	333	339	345	352	358	364
714	370	376	382	388	394	400	406	412	418	425
715	431	437	443	449	455	461	467	473	479	485
716	491	497	503	509	516	522	528	534	540	546
717	552	558	564	570	576	582	588	594	600	606
718	612	618	625	631	637	643	649	655	661	667
719	673	679	685	691	697	703	709	715	721	727
720	733	739	745	751	757	763	769	775	781	788
721	794	800	806	812	818	824	830	836	842	848
722	854	860	866	872	878	884	890	896	902	908
723	914	920	926	932	938	944	950	956	962	968
724	974	980	986	992	998	*004	*010	*016	*022	*028
725	86 034	040	046	052	058	064	070	076	082	088
726	094	100	106	112	118	124	130	136	141	147
727	153	159	165	171	177	183	189	195	201	207
728	213	219	225	231	237	243	249	255	261	267
729	273	279	285	291	297	303	308	314	320	326
730	332	338	344	350	356	362	368	374	380	386
731	392	398	404	410	415	421	427	433	439	445
732	451	457	463	469	475	481	487	493	499	504
733	510	516	522	528	534	540	546	552	558	564
734	570	576	581	587	593	599	605	611	617	623
735	629	635	641	646	652	658	664	670	676	682
736	688	694	700	705	711	717	723	729	735	741
737	747	753	759	764	770	776	782	788	794	800
738	806	812	817	823	829	835	841	847	853	859
739	864	870	876	882	888	894	900	906	911	917
740	923	929	935	941	947	953	958	964	970	976
741	982	988	994	999	*005	*011	*017	*023	*029	*035
742	87 040	046	052	058	064	070	075	081	087	093
743	099	105	111	116	122	128	134	140	146	151
744	157	163	169	175	181	186	192	198	204	210
745	216	221	227	233	239	245	251	256	262	268
746	274	280	286	291	297	303	309	315	320	326
747	332	338	344	349	355	361	367	373	379	384
748	390	396	402	408	413	419	425	431	437	442
749	448	454	460	466	471	477	483	489	495	500
750	506	512	518	523	529	535	541	547	552	558
N.	0	1	2	3	4	5	6	7	8	9

Proportional parts

7	
1	0.7
2	1.4
3	2.1
4	2.8
5	3.5
6	4.2
7	4.9
8	5.6
9	6.3

6	
1	0.6
2	1.2
3	1.8
4	2.4
5	3.0
6	3.6
7	4.2
8	4.8
9	5.4

5	
1	0.5
2	1.0
3	1.5
4	2.0
5	2.5
6	3.0
7	3.5
8	4.0
9	4.5

.84 510–.87 558

Table 1-1 (*Continued*)
COMMON LOGARITHMS OF NUMBERS
750–800

N.	0	1	2	3	4	5	6	7	8	9	Proportional parts
750	87 506	512	518	523	529	535	541	547	552	558	
751	564	570	576	581	587	593	599	604	610	616	
752	622	628	633	639	645	651	656	662	668	674	
753	679	685	691	697	703	708	714	720	726	731	
754	737	743	749	754	760	766	772	777	783	789	
755	795	800	806	812	818	823	829	835	841	846	
756	852	858	864	869	875	881	887	892	898	904	
757	910	915	921	927	933	938	944	950	955	961	
758	967	973	978	984	990	996	*001	*007	*013	*018	
759	88 024	030	036	041	047	053	058	064	070	076	
											6
760	081	087	093	098	104	110	116	121	127	133	1 0.6
761	138	144	150	156	161	167	173	178	184	190	2 1.2
762	195	201	207	213	218	224	230	235	241	247	3 1.8
763	252	258	264	270	275	281	287	292	298	304	4 2.4
764	309	315	321	326	332	338	343	349	355	360	5 3.0
											6 3.6
765	366	372	377	383	389	395	400	406	412	417	7 4.2
766	423	429	434	440	446	451	457	463	468	474	8 4.8
767	480	485	491	497	502	508	513	519	525	530	9 5.4
768	536	542	547	553	559	564	570	576	581	587	
769	593	598	604	610	615	621	627	632	638	643	
770	649	655	660	666	672	677	683	689	694	700	
771	705	711	717	722	728	734	739	745	750	756	
772	762	767	773	779	784	790	795	801	807	812	
773	818	824	829	835	840	846	852	857	863	868	
774	874	880	885	891	897	902	908	913	919	925	
775	930	936	941	947	953	958	964	969	975	981	
776	986	992	997	*003	*009	*014	*020	*025	*031	*037	
777	89 042	048	053	059	064	070	076	081	087	092	
778	098	104	109	115	120	126	131	137	143	148	
779	154	159	165	170	176	182	187	193	198	204	
											5
780	209	215	221	226	232	237	243	248	254	260	1 0.5
781	265	271	276	282	287	293	298	304	310	315	2 1.0
782	321	326	332	337	343	348	354	360	365	371	3 1.5
783	376	382	387	393	398	404	409	415	421	426	4 2.0
784	432	437	443	448	454	459	465	470	476	481	5 2.5
											6 3.0
785	487	492	498	504	509	515	520	526	531	537	7 3.5
786	542	548	553	559	564	570	575	581	586	592	8 4.0
787	597	603	609	614	620	625	631	636	642	647	9 4.5
788	653	658	664	669	675	680	686	691	697	702	
789	708	713	719	724	730	735	741	746	752	757	
790	763	768	774	779	785	790	796	801	807	812	
791	818	823	829	834	840	845	851	856	862	867	
792	873	878	883	889	894	900	905	911	916	922	
793	927	933	938	944	949	955	960	966	971	977	
794	982	988	993	998	*004	*009	*015	*020	*026	*031	
795	90 037	042	048	053	059	064	069	075	080	086	
796	091	097	102	108	113	119	124	129	135	140	
797	146	151	157	162	168	173	179	184	189	195	
798	200	206	211	217	222	227	233	238	244	249	
799	255	260	266	271	276	282	287	293	298	304	
800	309	314	320	325	331	336	342	347	352	358	
N.	0	1	2	3	4	5	6	7	8	9	Proportional parts

.87 506–.90 358

Table 1-1 (*Continued*)
COMMON LOGARITHMS OF NUMBERS
800–850

N.	0	1	2	3	4	5	6	7	8	9	Proportional parts
800	90 309	314	320	325	331	336	342	347	352	358	
801	363	369	374	380	385	390	396	401	407	412	
802	417	423	428	434	439	445	450	455	461	466	
803	472	477	482	488	493	499	504	509	515	520	
804	526	531	536	542	547	553	558	563	569	574	
805	580	585	590	596	601	607	612	617	623	628	
806	634	639	644	650	655	660	666	671	677	682	
807	687	693	698	703	709	714	720	725	730	736	
808	741	747	752	757	763	768	773	779	784	789	
809	795	800	806	811	816	822	827	832	838	843	
810	849	854	859	865	870	875	881	886	891	897	
811	902	907	913	918	924	929	934	940	945	950	
812	956	961	966	972	977	982	988	993	998	*004	
813	91 009	014	020	025	030	036	041	046	052	057	
814	062	068	073	078	084	089	094	100	105	110	
815	116	121	126	132	137	142	148	153	158	164	
816	169	174	180	185	190	196	201	206	212	217	
817	222	228	233	238	243	249	254	259	265	270	
818	275	281	286	291	297	302	307	312	318	323	
819	328	334	339	344	350	355	360	365	371	376	
820	381	387	392	397	403	408	413	418	424	429	
821	434	440	445	450	455	461	466	471	477	482	
822	487	492	498	503	508	514	519	524	529	535	
823	540	545	551	556	561	566	572	577	582	587	
824	593	598	603	609	614	619	624	630	635	640	
825	645	651	656	661	666	672	677	682	687	693	
826	698	703	709	714	719	724	730	735	740	745	
827	751	756	761	766	772	777	782	787	793	798	
828	803	808	814	819	824	829	834	840	845	850	
829	855	861	866	871	876	882	887	892	897	903	
830	908	913	918	924	929	934	939	944	950	955	
831	960	965	971	976	981	986	991	997	*002	*007	
832	92 012	018	023	028	033	038	044	049	054	059	
833	065	070	075	080	085	091	096	101	106	111	
834	117	122	127	132	137	143	148	153	158	163	
835	169	174	179	184	189	195	200	205	210	215	
836	221	226	231	236	241	247	252	257	262	267	
837	273	278	283	288	293	298	304	309	314	319	
838	324	330	335	340	345	350	355	361	366	371	
839	376	381	387	392	397	402	407	412	418	423	
840	428	433	438	443	449	454	459	464	469	474	
841	480	485	490	495	500	505	511	516	521	526	
842	531	536	542	547	552	557	562	567	572	578	
843	583	588	593	598	603	609	614	619	624	629	
844	634	639	645	650	655	660	665	670	675	681	
845	686	691	696	701	706	711	716	722	727	732	
846	737	742	747	752	758	763	768	773	778	783	
847	788	793	799	804	809	814	819	824	829	834	
848	840	845	850	855	860	865	870	875	881	886	
849	891	896	901	906	911	916	921	927	932	937	
850	942	947	952	957	962	967	973	978	983	988	
N.	0	1	2	3	4	5	6	7	8	9	Proportional parts

Proportional parts:

	6
1	0.6
2	1.2
3	1.8
4	2.4
5	3.0
6	3.6
7	4.2
8	4.8
9	5.4

	5
1	0.5
2	1.0
3	1.5
4	2.0
5	2.5
6	3.0
7	3.5
8	4.0
9	4.5

.90 309–.92 988

Table 1-1 (*Continued*)
COMMON LOGARITHMS OF NUMBERS
850–900

N.	0	1	2	3	4	5	6	7	8	9	Proportional parts
850	92 942	947	952	957	962	967	973	978	983	988	
851	993	998	*003	*008	*013	*018	*024	*029	*034	*039	
852	93 044	049	054	059	064	069	075	080	085	090	
853	095	100	105	110	115	120	125	131	136	141	
854	146	151	156	161	166	171	176	181	186	192	
855	197	202	207	212	217	222	227	232	237	242	
856	247	252	258	263	268	273	278	283	288	293	
857	298	303	308	313	318	323	328	334	339	344	
858	349	354	359	364	369	374	379	384	389	394	
859	399	404	409	414	420	425	430	435	440	445	
860	450	455	460	465	470	475	480	485	490	495	
861	500	505	510	515	520	526	531	536	541	546	
862	551	556	561	566	571	576	581	586	591	596	
863	601	606	611	616	621	626	631	636	641	646	
864	651	656	661	666	671	676	682	687	692	697	
865	702	707	712	717	722	727	732	737	742	747	
866	752	757	762	767	772	777	782	787	792	797	
867	802	807	812	817	822	827	832	837	842	847	
868	852	857	862	867	872	877	882	887	892	897	
869	902	907	912	917	922	927	932	937	942	947	
870	952	957	962	967	972	977	982	987	992	997	
871	94 002	007	012	017	022	027	032	037	042	047	
872	052	057	062	067	072	077	082	086	091	096	
873	101	106	111	116	121	126	131	136	141	146	
874	151	156	161	166	171	176	181	186	191	196	
875	201	206	211	216	221	226	231	236	240	245	
876	250	255	260	265	270	275	280	285	290	295	
877	300	305	310	315	320	325	330	335	340	345	
878	349	354	359	364	369	374	379	384	389	394	
879	399	404	409	414	419	424	429	433	438	443	
880	448	453	458	463	468	473	478	483	488	493	
881	498	503	507	512	517	522	527	532	537	542	
882	547	552	557	562	567	571	576	581	586	591	
883	596	601	606	611	616	621	626	630	635	640	
884	645	650	655	660	665	670	675	680	685	689	
885	694	699	704	709	714	719	724	729	734	738	
886	743	748	753	758	763	768	773	778	783	787	
887	792	797	802	807	812	817	822	827	832	836	
888	841	846	851	856	861	866	871	876	880	885	
889	890	895	900	905	910	915	919	924	929	934	
890	939	944	949	954	959	963	968	973	978	983	
891	988	993	998	*002	*007	*012	*017	*022	*027	*032	
892	95 036	041	046	051	056	061	066	071	075	080	
893	085	090	095	100	105	109	114	119	124	129	
894	134	139	143	148	153	158	163	168	173	177	
895	182	187	192	197	202	207	211	216	221	226	
896	231	236	240	245	250	255	260	265	270	274	
897	279	284	289	294	299	303	308	313	318	323	
898	328	332	337	342	347	352	357	361	366	371	
899	376	381	386	390	395	400	405	410	415	419	
900	424	429	434	439	444	448	453	458	463	468	
N.	0	1	2	3	4	5	6	7	8	9	Proportional parts

Proportional parts:

6
1	0.6
2	1.2
3	1.8
4	2.4
5	3.0
6	3.6
7	4.2
8	4.8
9	5.4

5
1	0.5
2	1.0
3	1.5
4	2.0
5	2.5
6	3.0
7	3.5
8	4.0
9	4.5

4
1	0.4
2	0.8
3	1.2
4	1.6
5	2.0
6	2.4
7	2.8
8	3.2
9	3.6

.92 942–.95 468

Table 1-1 (*Continued*)
COMMON LOGARITHMS OF NUMBERS
900–950

N.	0	1	2	3	4	5	6	7	8	9	Proportional parts
900	95 424	429	434	439	444	448	453	458	463	468	
901	472	477	482	487	492	497	501	506	511	516	
902	521	525	530	535	540	545	550	554	559	564	
903	569	574	578	583	588	593	598	602	607	612	
904	617	622	626	631	636	641	646	650	655	660	
905	665	670	674	679	684	689	694	698	703	708	
906	713	718	722	727	732	737	742	746	751	756	
907	761	766	770	775	780	785	789	794	799	804	
908	809	813	818	823	828	832	837	842	847	852	
909	856	861	866	871	875	880	885	890	895	899	
910	904	909	914	918	923	928	933	938	942	947	
911	952	957	961	966	971	976	980	985	990	995	
912	999	*004	*009	*014	*019	*023	*028	*033	*038	*042	
913	96 047	052	057	061	066	071	076	080	085	090	
914	095	099	104	109	114	118	123	128	133	137	
915	142	147	152	156	161	166	171	175	180	185	
916	190	194	199	204	209	213	218	223	227	232	
917	237	242	246	251	256	261	265	270	275	280	
918	284	289	294	298	303	308	313	317	322	327	
919	332	336	341	346	350	355	360	365	369	374	
920	379	384	388	393	398	402	407	412	417	421	
921	426	431	435	440	445	450	454	459	464	468	
922	473	478	483	487	492	497	501	506	511	515	
923	520	525	530	534	539	544	548	553	558	562	
924	567	572	577	581	586	591	595	600	605	609	
925	614	619	624	628	633	638	642	647	652	656	
926	661	666	670	675	680	685	689	694	699	703	
927	708	713	717	722	727	731	736	741	745	750	
928	755	759	764	769	774	778	783	788	792	797	
929	802	806	811	816	820	825	830	834	839	844	
930	848	853	858	862	867	872	876	881	886	890	
931	895	900	904	909	914	918	923	928	932	937	
932	942	946	951	956	960	965	970	974	979	984	
933	988	993	997	*002	*007	*011	*016	*021	*025	*030	
934	97 035	039	044	049	053	058	063	067	072	077	
935	081	086	090	095	100	104	109	114	118	123	
936	128	132	137	142	146	151	155	160	165	169	
937	174	179	183	188	192	197	202	206	211	216	
938	220	225	230	234	239	243	248	253	257	262	
939	267	271	276	280	285	290	294	299	304	308	
940	313	317	322	327	331	336	340	345	350	354	
941	359	364	368	373	377	382	387	391	396	400	
942	405	410	414	419	424	428	433	437	442	447	
943	451	456	460	465	470	474	479	483	488	493	
944	497	502	506	511	516	520	525	529	534	539	
945	543	548	552	557	562	566	571	575	580	585	
946	589	594	598	603	607	612	617	621	626	630	
947	635	640	644	649	653	658	663	667	672	676	
948	681	685	690	695	699	704	708	713	717	722	
949	727	731	736	740	745	749	754	759	763	768	
950	772	777	782	786	791	795	800	804	809	813	
N.	0	1	2	3	4	5	6	7	8	9	Proportional parts

Proportional parts:

	5
1	0.5
2	1.0
3	1.5
4	2.0
5	2.5
6	3.0
7	3.5
8	4.0
9	4.5

	4
1	0.4
2	0.8
3	1.2
4	1.6
5	2.0
6	2.4
7	2.8
8	3.2
9	3.6

.95 424–.97 813

Table 1-1 (*Continued*)
COMMON LOGARITHMS OF NUMBERS
950–1000

N.	0	1	2	3	4	5	6	7	8	9	Proportional parts
950	97 772	777	782	786	791	795	800	804	809	813	
951	818	823	827	832	836	841	845	850	855	859	
952	864	868	873	877	882	886	891	896	900	905	
953	909	914	918	923	928	932	937	941	946	950	
954	955	959	964	968	973	978	982	987	991	996	
955	98 000	005	009	014	019	023	028	032	037	041	
956	046	050	055	059	064	068	073	078	082	087	
957	091	096	100	105	109	114	118	123	127	132	
958	137	141	146	150	155	159	164	168	173	177	
959	182	186	191	195	200	204	209	214	218	223	
960	227	232	236	241	245	250	254	259	263	268	**5**
961	272	277	281	286	290	295	299	304	308	313	1　0.5
962	318	322	327	331	336	340	345	349	354	358	2　1.0
963	363	367	372	376	381	385	390	394	399	403	3　1.5
964	408	412	417	421	426	430	435	439	444	448	4　2.0
965	453	457	462	466	471	475	480	484	489	493	5　2.5
966	498	502	507	511	516	520	525	529	534	538	6　3.0
967	543	547	552	556	561	565	570	574	579	583	7　3.5
968	588	592	597	601	605	610	614	619	623	628	8　4.0
969	632	637	641	646	650	655	659	664	668	673	9　4.5
970	677	682	686	691	695	700	704	709	713	717	
971	722	726	731	735	740	744	749	753	758	762	
972	767	771	776	780	784	789	793	798	802	807	
973	811	816	820	825	829	834	838	843	847	851	
974	856	860	865	869	874	878	883	887	892	896	
975	900	905	909	914	918	923	927	932	936	941	
976	945	949	954	958	963	967	972	976	981	985	
977	989	994	998	*003	*007	*012	*016	*021	*025	*029	
978	99 034	038	043	047	052	056	061	065	069	074	
979	078	083	087	092	096	100	105	109	114	118	
980	123	127	131	136	140	145	149	154	158	162	**4**
981	167	171	176	180	185	189	193	198	202	207	1　0.4
982	211	216	220	224	229	233	238	242	247	251	2　0.8
983	255	260	264	269	273	277	282	286	291	295	3　1.2
984	300	304	308	313	317	322	326	330	335	339	4　1.6
985	344	348	352	357	361	366	370	374	379	383	5　2.0
986	388	392	396	401	405	410	414	419	423	427	6　2.4
987	432	436	441	445	449	454	458	463	467	471	7　2.8
988	476	480	484	489	493	498	502	506	511	515	8　3.2
989	520	524	528	533	537	542	546	550	555	559	9　3.6
990	564	568	572	577	581	585	590	594	599	603	
991	607	612	616	621	625	629	634	638	642	647	
992	651	656	660	664	669	673	677	682	686	691	
993	695	699	704	708	712	717	721	726	730	734	
994	739	743	747	752	756	760	765	769	774	778	
995	782	787	791	795	800	804	808	813	817	822	
996	826	830	835	839	843	848	852	856	861	865	
997	870	874	878	883	887	891	896	900	904	909	
998	913	917	922	926	930	935	939	944	948	952	
999	957	961	965	970	974	978	983	987	991	996	
1000	00 000	004	009	013	017	022	026	030	035	039	
N.	0	1	2	3	4	5	6	7	8	9	Proportional parts

.97 772–.99 996

Table 1-2
TRIGONOMETRIC FUNCTIONS
At Intervals of 10′

From Baumeister and Marks, "Standard Handbook for Mechanical Engineers," 7th ed., McGraw-Hill Book Company, New York (1967); by permission.

Annex—10 in columns marked *.

De-grees	Ra-dians	Sines Nat.	Sines Log*	Cosines Nat.	Cosines Log*	Tangents Nat.	Tangents Log*	Cotangents Nat.	Cotangents Log		
0° 00′	0.0000	.0000	∞	1.0000	0.0000	.0000	∞	∞	∞	1.5708	90° 00′
10	0.0029	.0029	7.4637	1.0000	.0000	.0029	7.4637	343.77	2.5363	1.5679	50
20	0.0058	.0058	.7648	1.0000	.0000	.0058	.7648	171.89	.2352	1.5650	40
30	0.0087	.0087	.9408	1.0000	.0000	.0087	.9409	114.59	.0591	1.5621	30
40	0.0116	.0116	8.0658	.9999	.0000	.0116	8.0658	85.940	1.9342	1.5592	20
50	0.0145	.0145	.1627	.9999	.0000	.0145	.1627	68.750	.8373	1.5563	10
1° 00′	0.0175	.0175	8.2419	.9998	9.9999	.0175	8.2419	57.290	1.7581	1.5533	89° 00′
10	0.0204	.0204	.3088	.9998	.9999	.0204	.3089	49.104	.6911	1.5504	50
20	0.0233	.0233	.3668	.9997	.9999	.0233	.3669	42.964	.6331	1.5475	40
30	0.0262	.0262	.4179	.9997	.9999	.0262	.4181	38.188	.5819	1.5446	30
40	0.0291	.0291	.4637	.9996	.9998	.0291	.4638	34.368	.5362	1.5417	20
50	0.0320	.0320	.5050	.9995	.9998	.0320	.5053	31.242	.4947	1.5388	10
2° 00′	0.0349	.0349	8.5428	.9994	9.9997	.0349	8.5431	28.636	1.4569	1.5359	88° 00′
10	0.0378	.0378	.5776	.9993	.9997	.0378	.5779	26.432	.4221	1.5330	50
20	0.0407	.0407	.6097	.9992	.9996	.0407	.6101	24.542	.3899	1.5301	40
30	0.0436	.0436	.6397	.9990	.9996	.0437	.6401	22.904	.3599	1.5272	30
40	0.0465	.0465	.6677	.9989	.9995	.0466	.6682	21.470	.3318	1.5243	20
50	0.0495	.0494	.6940	.9988	.9995	.0495	.6945	20.206	.3055	1.5213	10
3° 00′	0.0524	.0523	8.7188	.9986	9.9994	.0524	8.7194	19.081	1.2806	1.5184	87° 00′
10	0.0553	.0552	.7423	.9985	.9993	.0553	.7429	18.075	.2571	1.5155	50
20	0.0582	.0581	.7645	.9983	.9993	.0582	.7652	17.169	.2348	1.5126	40
30	0.0611	.0610	.7857	.9981	.9992	.0612	.7865	16.350	.2135	1.5097	30
40	0.0640	.0640	.8059	.9980	.9991	.0641	.8067	15.605	.1933	1.5068	20
50	0.0669	.0669	.8251	.9978	.9990	.0670	.8261	14.924	.1739	1.5039	10
4° 00′	0.0698	.0698	8.8436	.9976	9.9989	.0699	8.8446	14.301	1.1554	1.5010	86° 00′
10	0.0727	.0727	.8613	.9974	.9989	.0729	.8624	13.727	.1376	1.4981	50
20	0.0756	.0756	.8783	.9971	.9988	.0758	.8795	13.197	.1205	1.4952	40
30	0.0785	.0785	.8946	.9969	.9987	.0787	.8960	12.706	.1040	1.4923	30
40	0.0814	.0814	.9104	.9967	.9986	.0816	.9118	12.251	.0882	1.4893	20
50	0.0844	.0843	.9256	.9964	.9985	.0846	.9272	11.826	.0728	1.4864	10
5° 00′	0.0873	.0872	8.9403	.9962	9.9983	.0875	8.9420	11.430	1.0580	1.4835	85° 00′
10	0.0902	.0901	.9545	.9959	.9982	.0904	.9563	11.059	.0437	1.4806	50
20	0.0931	.0929	.9682	.9957	.9981	.0934	.9701	10.712	.0299	1.4777	40
30	0.0960	.0958	.9816	.9954	.9980	.0963	.9836	10.385	.0164	1.4748	30
40	0.0989	.0987	.9945	.9951	.9979	.0992	.9966	10.078	.0034	1.4719	20
50	0.1018	.1016	9.0070	.9948	.9977	.1022	9.0093	9.7882	0.9907	1.4690	10
6° 00′	0.1047	.1045	9.0192	.9945	9.9976	.1051	9.0216	9.5144	0.9784	1.4661	84° 00′
10	0.1076	.1074	.0311	.9942	.9975	.1080	.0336	9.2553	.9664	1.4632	50
20	0.1105	.1103	.0426	.9939	.9973	.1110	.0453	9.0098	.9547	1.4603	40
30	0.1134	.1132	.0539	.9936	.9972	.1139	.0567	8.7769	.9433	1.4574	30
40	0.1164	.1161	.0648	.9932	.9971	.1169	.0678	8.5555	.9322	1.4544	20
50	0.1193	.1190	.0755	.9929	.9969	.1198	.0786	8.3450	.9214	1.4515	10
7° 00′	0.1222	.1219	9.0859	.9925	9.9968	.1228	9.0891	8.1443	0.9109	1.4486	83° 00′
10	0.1251	.1248	.0961	.9922	.9966	.1257	.0995	7.9530	.9005	1.4457	50
20	0.1280	.1276	.1060	.9918	.9964	.1287	.1096	7.7704	.8904	1.4428	40
30	0.1309	.1305	.1157	.9914	.9963	.1317	.1194	7.5958	.8806	1.4399	30
40	0.1338	.1334	.1252	.9911	.9961	.1346	.1291	7.4287	.8709	1.4370	20
50	0.1367	.1363	.1345	.9907	.9959	.1376	.1385	7.2687	.8615	1.4341	10
8° 00′	0.1396	.1392	9.1436	.9903	9.9958	.1405	9.1478	7.1154	0.8522	1.4312	82° 00′
10	0.1425	.1421	.1525	.9899	.9956	.1435	.1569	6.9682	.8431	1.4283	50
20	0.1454	.1449	.1612	.9894	.9954	.1465	.1658	6.8269	.8342	1.4254	40
30	0.1484	.1478	.1697	.9890	.9952	.1495	.1745	6.6912	.8255	1.4224	30
40	0.1513	.1507	.1781	.9886	.9950	.1524	.1831	6.5606	.8169	1.4195	20
50	0.1542	.1536	.1863	.9881	.9948	.1554	.1915	6.4348	.8085	1.4166	10
9° 00′	0.1571	.1564	9.1943	.9877	9.9946	.1584	9.1997	6.3138	0.8003	1.4137	81° 00′
		Nat.	Log*	Nat.	Log*	Nat.	Log*	Nat.	Log		
		Cosines		Sines		Cotangents		Tangents		Ra-dians	De-grees

Table 1-2 (*Continued*)
TRIGONOMETRIC FUNCTIONS
Annex—10 in columns marked *.

De-grees	Ra-dians	Sines		Cosines		Tangents		Cotangents			
		Nat.	Log *	Nat.	Log *	Nat.	Log *	Nat.	Log		
9° 00′	0.1571	.1564	9.1943	.9877	9.9946	.1584	9.1997	6.3138	0.8003	1.4137	81° 00′
10	0.1600	.1593	.2022	.9872	.9944	.1614	.2078	6.1970	.7922	1.4108	50
20	0.1629	.1622	.2100	.9868	.9942	.1644	.2158	6.0844	.7842	1.4079	40
30	0.1658	.1650	.2176	.9863	.9940	.1673	.2236	5.9758	.7764	1.4050	30
40	0.1687	.1679	.2251	.9858	.9938	.1703	.2313	5.8708	.7687	1.4021	20
50	0.1716	.1708	.2324	.9853	.9936	.1733	.2389	5.7694	.7611	1.3992	10
10° 00′	0.1745	.1736	9.2397	.9848	9.9934	.1763	9.2463	5.6713	0.7537	1.3963	80° 00′
10	0.1774	.1765	.2468	.9843	.9931	.1793	.2536	5.5764	.7464	1.3934	50
20	0.1804	.1794	.2538	.9838	.9929	.1823	.2609	5.4845	.7391	1.3904	40
30	0.1833	.1822	.2606	.9833	.9927	.1853	.2680	5.3955	.7320	1.3875	30
40	0.1862	.1851	.2674	.9827	.9924	.1883	.2750	5.3093	.7250	1.3846	20
50	0.1891	.1880	.2740	.9822	.9922	.1914	.2819	5.2257	.7181	1.3817	10
11° 00′	0.1920	.1908	9.2806	.9816	9.9919	.1944	9.2887	5.1446	0.7113	1.3788	79° 00′
10	0.1949	.1937	.2870	.9811	.9917	.1974	.2953	5.0658	.7047	1.3759	50
20	0.1978	.1965	.2934	.9805	.9914	.2004	.3020	4.9894	.6980	1.3730	40
30	0.2007	.1994	.2997	.9799	.9912	.2035	.3085	4.9152	.6915	1.3701	30
40	0.2036	.2022	.3058	.9793	.9909	.2065	.3149	4.8430	.6851	1.3672	20
50	0.2065	.2051	.3119	.9787	.9907	.2095	.3212	4.7729	.6788	1.3643	10
12° 00′	0.2094	.2079	9.3179	.9781	9.9904	.2126	9.3275	4.7046	0.6725	1.3614	78° 00′
10	0.2123	.2108	.3238	.9775	.9901	.2156	.3336	4.6382	.6664	1.3584	50
20	0.2153	.2136	.3296	.9769	.9899	.2186	.3397	4.5736	.6603	1.3555	40
30	0.2182	.2164	.3353	.9763	.9896	.2217	.3458	4.5107	.6542	1.3526	30
40	0.2211	.2193	.3410	.9757	.9893	.2247	.3517	4.4494	.6483	1.3497	20
50	0.2240	.2221	.3466	.9750	.9890	.2278	.3576	4.3897	.6424	1.3468	10
13° 00′	0.2269	.2250	9.3521	.9744	9.9887	.2309	9.3634	4.3315	0.6366	1.3439	77° 00′
10	0.2298	.2278	.3575	.9737	.9884	.2339	.3691	4.2747	.6309	1.3410	50
20	0.2327	.2306	.3629	.9730	.9881	.2370	.3748	4.2193	.6252	1.3381	40
30	0.2356	.2334	.3682	.9724	.9878	.2401	.3804	4.1653	.6196	1.3352	30
40	0.2385	.2363	.3734	.9717	.9875	.2432	.3859	4.1126	.6141	1.3323	20
50	0.2414	.2391	.3786	.9710	.9872	.2462	.3914	4.0611	.6086	1.3294	10
14° 00′	0.2443	.2419	9.3837	.9703	9.9869	.2493	9.3968	4.0108	0.6032	1.3265	76° 00′
10	0.2473	.2447	.3887	.9696	.9866	.2524	.4021	3.9617	.5979	1.3235	50
20	0.2502	.2476	.3937	.9689	.9863	.2555	.4074	3.9136	.5926	1.3206	40
30	0.2531	.2504	.3986	.9681	.9859	.2586	.4127	3.8667	.5873	1.3177	30
40	0.2560	.2532	.4035	.9674	.9856	.2617	.4178	3.8208	.5822	1.3148	20
50	0.2589	.2560	.4083	.9667	.9853	.2648	.4230	3.7760	.5770	1.3119	10
15° 00′	0.2618	.2588	9.4130	.9659	9.9849	.2679	9.4281	3.7321	0.5719	1.3090	75° 00′
10	0.2647	.2616	.4177	.9652	.9846	.2711	.4331	3.6891	.5669	1.3061	50
20	0.2676	.2644	.4223	.9644	.9843	.2742	.4381	3.6470	.5619	1.3032	40
30	0.2705	.2672	.4269	.9636	.9839	.2773	.4430	3.6059	.5570	1.3003	30
40	0.2734	.2700	.4314	.9628	.9836	.2805	.4479	3.5656	.5521	1.2974	20
50	0.2763	.2728	.4359	.9621	.9832	.2836	.4527	3.5261	.5473	1.2945	10
16° 00′	0.2793	.2756	9.4403	.9613	9.9828	.2867	9.4575	3.4874	0.5425	1.2915	74° 00′
10	0.2822	.2784	.4447	.9605	.9825	.2899	.4622	3.4495	.5378	1.2886	50
20	0.2851	.2812	.4491	.9596	.9821	.2931	.4669	3.4124	.5331	1.2857	40
30	0.2880	.2840	.4533	.9588	.9817	.2962	.4716	3.3759	.5284	1.2828	30
40	0.2909	.2868	.4576	.9580	.9814	.2994	.4762	3.3402	.5238	1.2799	20
50	0.2938	.2896	.4618	.9572	.9810	.3026	.4808	3.3052	.5192	1.2770	10
17° 00′	0.2967	.2924	9.4659	.9563	9.9806	.3057	9.4853	3.2709	0.5147	1.2741	73° 00′
10	0.2996	.2952	.4700	.9555	.9802	.3089	.4898	3.2371	.5102	1.2712	50
20	0.3025	.2979	.4741	.9546	.9798	.3121	.4943	3.2041	.5057	1.2683	40
30	0.3054	.3007	.4781	.9537	.9794	.3153	.4987	3.1716	.5013	1.2654	30
40	0.3083	.3035	.4821	.9528	.9790	.3185	.5031	3.1397	.4969	1.2625	20
50	0.3113	.3062	.4861	.9520	.9786	.3217	.5075	3.1084	.4925	1.2595	10
18° 00′	0.3142	.3090	9.4900	.9511	9.9782	.3249	9.5118	3.0777	0.4882	1.2566	72° 00′
		Nat.	Log *	Nat.	Log *	Nat.	Log *	Nat.	Log		
		Cosines		Sines		Cotangents		Tangents		Ra-dians	De-grees

Table 1-2 (*Continued*)
TRIGONOMETRIC FUNCTIONS
Annex—10 in columns marked *

De-grees	Ra-dians	Sines		Cosines		Tangents		Cotangents			
		Nat.	Log *	Nat.	Log *	Nat.	Log *	Nat.	Log		
18° 00'	0.3142	.3090	9.4900	.9511	9.9782	.3249	9.5118	3.0777	0.4882	1.2566	72° 00'
10	0.3171	.3118	.4939	.9502	.9778	.3281	.5161	3.0475	.4839	1.2537	50
20	0.3200	.3145	.4977	.9492	.9774	.3314	.5203	3.0178	.4797	1.2508	40
30	0.3229	.3173	.5015	.9483	.9770	.3346	.5245	2.9887	.4755	1.2479	30
40	0.3258	.3201	.5052	.9474	.9765	.3378	.5287	2.9600	.4713	1.2450	20
50	0.3287	.3228	.5090	.9465	.9761	.3411	.5329	2.9319	.4671	1.2421	10
19° 00'	0.3316	.3256	9.5126	.9455	9.9757	.3443	9.5370	2.9042	0.4630	1.2392	71° 00'
10	0.3345	.3283	.5163	.9446	.9752	.3476	.5411	2.8770	.4589	1.2363	50
20	0.3374	.3311	.5199	.9436	.9748	.3508	.5451	2.8502	.4549	1.2334	40
30	0.3403	.3338	.5235	.9426	.9743	.3541	.5491	2.8239	.4509	1.2305	30
40	0.3432	.3365	.5270	.9417	.9739	.3574	.5531	2.7980	.4469	1.2275	20
50	0.3462	.3393	.5306	.9407	.9734	.3607	.5571	2.7725	.4429	1.2246	10
20° 00'	0.3491	.3420	9.5341	.9397	9.9730	.3640	9.5611	2.7475	0.4389	1.2217	70° 00'
10	0.3520	.3448	.5375	.9387	.9725	.3673	.5650	2.7228	.4350	1.2188	50
20	0.3549	.3475	.5409	.9377	.9721	.3706	.5689	2.6985	.4311	1.2159	40
30	0.3578	.3502	.5443	.9367	.9716	.3739	.5727	2.6746	.4273	1.2130	30
40	0.3607	.3529	.5477	.9356	.9711	.3772	.5766	2.6511	.4234	1.2101	20
50	0.3636	.3557	.5510	.9346	.9706	.3805	.5804	2.6279	.4196	1.2072	10
21° 00'	0.3665	.3584	9.5543	.9336	9.9702	.3839	9.5842	2.6051	0.4158	1.2043	69° 00'
10	0.3694	.3611	.5576	.9325	.9697	.3872	.5879	2.5826	.4121	1.2014	50
20	0.3723	.3638	.5609	.9315	.9692	.3906	.5917	2.5605	.4083	1.1985	40
30	0.3752	.3665	.5641	.9304	.9687	.3939	.5954	2.5386	.4046	1.1956	30
40	0.3782	.3692	.5673	.9293	.9682	.3973	.5991	2.5172	.4009	1.1926	20
50	0.3811	.3719	.5704	.9283	.9677	.4006	.6028	2.4960	.3972	1.1897	10
22° 00'	0.3840	.3746	9.5736	.9272	9.9672	.4040	9.6064	2.4751	0.3936	1.1868	68° 00'
10	0.3869	.3773	.5767	.9261	.9667	.4074	.6100	2.4545	.3900	1.1839	50
20	0.3898	.3800	.5798	.9250	.9661	.4108	.6136	2.4342	.3864	1.1810	40
30	0.3927	.3827	.5828	.9239	.9656	.4142	.6172	2.4142	.3828	1.1781	30
40	0.3956	.3854	.5859	.9228	.9651	.4176	.6208	2.3945	.3792	1.1752	20
50	0.3985	.3881	.5889	.9216	.9646	.4210	.6243	2.3750	.3757	1.1723	10
23° 00'	0.4014	.3907	9.5919	.9205	9.9640	.4245	9.6279	2.3559	0.3721	1.1694	67° 00'
10	0.4043	.3934	.5948	.9194	.9635	.4279	.6314	2.3369	.3686	1.1665	50
20	0.4072	.3961	.5978	.9182	.9629	.4314	.6348	2.3183	.3652	1.1636	40
30	0.4102	.3987	.6007	.9171	.9624	.4348	.6383	2.2998	.3617	1.1606	30
40	0.4131	.4014	.6036	.9159	.9618	.4383	.6417	2.2817	.3583	1.1577	20
50	0.4160	.4041	.6065	.9147	.9613	.4417	.6452	2.2637	.3548	1.1548	10
24° 00'	0.4189	.4067	9.6093	.9135	9.9607	.4452	9.6486	2.2460	0.3514	1.1519	66° 00'
10	0.4218	.4094	.6121	.9124	.9602	.4487	.6520	2.2286	.3480	1.1490	50
20	0.4247	.4120	.6149	.9112	.9596	.4522	.6553	2.2113	.3447	1.1461	40
30	0.4276	.4147	.6177	.9100	.9590	.4557	.6587	2.1943	.3413	1.1432	30
40	0.4305	.4173	.6205	.9088	.9584	.4592	.6620	2.1775	.3380	1.1403	20
50	0.4334	.4200	.6232	.9075	.9579	.4628	.6654	2.1609	.3346	1.1374	10
25° 00'	0.4363	.4226	9.6259	.9063	9.9573	.4663	9.6687	2.1445	0.3313	1.1345	65° 00'
10	0.4392	.4253	.6286	.9051	.9567	.4699	.6720	2.1283	.3280	1.1316	50
20	0.4422	.4279	.6313	.9038	.9561	.4734	.6752	2.1123	.3248	1.1286	40
30	0.4451	.4305	.6340	.9026	.9555	.4770	.6785	2.0965	.3215	1.1257	30
40	0.4480	.4331	.6366	.9013	.9549	.4806	.6817	2.0809	.3183	1.1228	20
50	0.4509	.4358	.6392	.9001	.9543	.4841	.6850	2.0655	.3150	1.1199	10
26° 00'	0.4538	.4384	9.6418	.8988	9.9537	.4877	9.6882	2.0503	0.3118	1.1170	64° 00'
10	0.4567	.4410	.6444	.8975	.9530	.4913	.6914	2.0353	.3086	1.1141	50
20	0.4596	.4436	.6470	.8962	.9524	.4950	.6946	2.0204	.3054	1.1112	40
30	0.4625	.4462	.6495	.8949	.9518	.4986	.6977	2.0057	.3023	1.1083	30
40	0.4654	.4488	.6521	.8936	.9512	.5022	.7009	1.9912	.2991	1.1054	20
50	0.4683	.4514	.6546	.8923	.9505	.5059	.7040	1.9768	.2960	1.1025	10
27° 00'	0.4712	.4540	9.6570	.8910	9.9499	.5095	9.7072	1.9626	0.2928	1.0996	63° 00'
		Nat.	Log *	Nat.	Log *	Nat.	Log *	Nat.	Log		
		Cosines		Sines		Cotangents		Tangents		Ra-dians	De-grees

Table 1-2 (*Continued*)
TRIGONOMETRIC FUNCTIONS
Annex—10 in columns marked *.

De-grees	Ra-dians	Sines Nat.	Sines Log *	Cosines Nat.	Cosines Log *	Tangents Nat.	Tangents Log *	Cotangents Nat.	Cotangents Log		De-grees
27° 00′	0.4712	.4540	9.6570	.8910	9.9499	.5095	9.7072	1.9626	0.2928	1.0996	63° 00′
10	0.4741	.4566	.6595	.8897	.9492	.5132	.7103	1.9486	.2897	1.0966	50
20	0.4771	.4592	.6620	.8884	.9486	.5169	.7134	1.9347	.2866	1.0937	40
30	0.4800	.4617	.6644	.8870	.9479	.5206	.7165	1.9210	.2835	1.0908	30
40	0.4829	.4643	.6668	.8857	.9473	.5243	.7196	1.9074	.2804	1.0879	20
50	0.4858	.4669	.6692	.8843	.9466	.5280	.7226	1.8940	.2774	1.0850	10
28° 00′	0.4887	.4695	9.6716	.8829	9.9459	.5317	9.7257	1.8807	0.2743	1.0821	62° 00′
10	0.4916	.4720	.6740	.8816	.9453	.5354	.7287	1.8676	.2713	1.0792	50
20	0.4945	.4746	.6763	.8802	.9446	.5392	.7317	1.8546	.2683	1.0763	40
30	0.4974	.4772	.6787	.8788	.9439	.5430	.7348	1.8418	.2652	1.0734	30
40	0.5003	.4797	.6810	.8774	.9432	.5467	.7378	1.8291	.2622	1.0705	20
50	0.5032	.4823	.6833	.8760	.9425	.5505	.7408	1.8165	.2592	1.0676	10
29° 00′	0.5061	.4848	9.6856	.8746	9.9418	.5543	9.7438	1.8040	0.2562	1.0647	61° 00′
10	0.5091	.4874	.6878	.8732	.9411	.5581	.7467	1.7917	.2533	1.0617	50
20	0.5120	.4899	.6901	.8718	.9404	.5619	.7497	1.7796	.2503	1.0588	40
30	0.5149	.4924	.6923	.8704	.9397	.5658	.7526	1.7675	.2474	1.0559	30
40	0.5178	.4950	.6946	.8689	.9390	.5696	.7556	1.7556	.2444	1.0530	20
50	0.5207	.4975	.6968	.8675	.9383	.5735	.7585	1.7437	.2415	1.0501	10
30° 00′	0.5236	.5000	9.6990	.8660	9.9375	.5774	9.7614	1.7321	0.2386	1.0472	60° 00′
10	0.5265	.5025	.7012	.8646	.9368	.5812	.7644	1.7205	.2356	1.0443	50
20	0.5294	.5050	.7033	.8631	.9361	.5851	.7673	1.7090	.2327	1.0414	40
30	0.5323	.5075	.7055	.8616	.9353	.5890	.7701	1.6977	.2299	1.0385	30
40	0.5352	.5100	.7076	.8601	.9346	.5930	.7730	1.6864	.2270	1.0356	20
50	0.5381	.5125	.7097	.8587	.9338	.5969	.7759	1.6753	.2241	1.0327	10
31° 00′	0.5411	.5150	9.7118	.8572	9.9331	.6009	9.7788	1.6643	0.2212	1.0297	59° 00′
10	0.5440	.5175	.7139	.8557	.9323	.6048	.7816	1.6534	.2184	1.0268	50
20	0.5469	.5200	.7160	.8542	.9315	.6088	.7845	1.6426	.2155	1.0239	40
30	0.5498	.5225	.7181	.8526	.9308	.6128	.7873	1.6319	.2127	1.0210	30
40	0.5527	.5250	.7201	.8511	.9300	.6168	.7902	1.6212	.2098	1.0181	20
50	0.5556	.5275	.7222	.8496	.9292	.6208	.7930	1.6107	.2070	1.0152	10
32° 00′	0.5585	.5299	9.7242	.8480	9.9284	.6249	9.7958	1.6003	0.2042	1.0123	58° 00′
10	0.5614	.5324	.7262	.8465	.9276	.6289	.7986	1.5900	.2014	1.0094	50
20	0.5643	.5348	.7282	.8450	.9268	.6330	.8014	1.5798	.1986	1.0065	40
30	0.5672	.5373	.7302	.8434	.9260	.6371	.8042	1.5697	.1958	1.0036	30
40	0.5701	.5398	.7322	.8418	.9252	.6412	.8070	1.5597	.1930	1.0007	20
50	0.5730	.5422	.7342	.8403	.9244	.6453	.8097	1.5497	.1903	0.9977	10
33° 00′	0.5760	.5446	9.7361	.8387	9.9236	.6494	9.8125	1.5399	0.1875	0.9948	57° 00′
10	0.5789	.5471	.7380	.8371	.9228	.6536	.8153	1.5301	.1847	0.9919	50
20	0.5818	.5495	.7400	.8355	.9219	.6577	.8180	1.5204	.1820	0.9890	40
30	0.5847	.5519	.7419	.8339	.9211	.6619	.8208	1.5108	.1792	0.9861	30
40	0.5876	.5544	.7438	.8323	.9203	.6661	.8235	1.5013	.1765	0.9832	20
50	0.5905	.5568	.7457	.8307	.9194	.6703	.8263	1.4919	.1737	0.9803	10
34° 00′	0.5934	.5592	9.7476	.8290	9.9186	.6745	9.8290	1.4826	0.1710	0.9774	56° 00′
10	0.5963	.5616	.7494	.8274	.9177	.6787	.8317	1.4733	.1683	0.9745	50
20	0.5992	.5640	.7513	.8258	.9169	.6830	.8344	1.4641	.1656	0.9716	40
30	0.6021	.5664	.7531	.8241	.9160	.6873	.8371	1.4550	.1629	0.9687	30
40	0.6050	.5688	.7550	.8225	.9151	.6916	.8398	1.4460	.1602	0.9657	20
50	0.6080	.5712	.7568	.8208	.9142	.6959	.8425	1.4370	.1575	0.9628	10
35° 00′	0.6109	.5736	9.7586	.8192	9.9134	.7002	9.8452	1.4281	0.1548	0.9599	55° 00′
10	0.6138	.5760	.7604	.8175	.9125	.7046	.8479	1.4193	.1521	0.9570	50
20	0.6167	.5783	.7622	.8158	.9116	.7089	.8506	1.4106	.1494	0.9541	40
30	0.6196	.5807	.7640	.8141	.9107	.7133	.8533	1.4019	.1467	0.9512	30
40	0.6225	.5831	.7657	.8124	.9098	.7177	.8559	1.3934	.1441	0.9483	20
50	0.6254	.5854	.7675	.8107	.9089	.7221	.8586	1.3848	.1414	0.9454	10
36° 00′	0.6283	.5878	9.7692	.8090	9.9080	.7265	9.8613	1.3764	0.1387	0.9425	54° 00′
		Nat.	Log *	Nat.	Log *	Nat.	Log *	Nat.	Log		
		Cosines		Sines		Cotangents		Tangents		Ra-dians	De-grees

Table 1-2 (*Continued*)
TRIGONOMETRIC FUNCTIONS
Annex—10 in columns marked *.

Degrees	Radians	Sines		Cosines		Tangents		Cotangents			
		Nat.	Log *	Nat.	Log *	Nat.	Log *	Nat.	Log		
36° 00′	0.6283	.5878	9.7692	.8090	9.9080	.7265	9.8613	1.3764	0.1387	0.9425	54° 00′
10	0.6312	.5901	.7710	.8073	.9070	.7310	.8639	1.3680	.1361	0.9396	50
20	0.6341	.5925	.7727	.8056	.9061	.7355	.8666	1.3597	.1334	0.9367	40
30	0.6370	.5948	.7744	.8039	.9052	.7400	.8692	1.3514	.1308	0.9338	30
40	0.6400	.5972	.7761	.8021	.9042	.7445	.8718	1.3432	.1282	0.9308	20
50	0.6429	.5995	.7778	.8004	.9033	.7490	.8745	1.3351	.1255	0.9279	10
37° 00′	0.6458	.6018	9.7795	.7986	9.9023	.7536	9.8771	1.3270	0.1229	0.9250	53° 00′
10	0.6487	.6041	.7811	.7969	.9014	.7581	.8797	1.3190	.1203	0.9221	50
20	0.6516	.6065	.7828	.7951	.9004	.7627	.8824	1.3111	.1176	0.9192	40
30	0.6545	.6088	.7844	.7934	.8995	.7673	.8850	1.3032	.1150	0.9163	30
40	0.6574	.6111	.7861	.7916	.8985	.7720	.8876	1.2954	.1124	0.9134	20
50	0.6603	.6134	.7877	.7898	.8975	.7766	.8902	1.2876	.1098	0.9105	10
38° 00′	0.6632	.6157	9.7893	.7880	9.8965	.7813	9.8928	1.2799	0.1072	0.9076	52° 00′
10	0.6661	.6180	.7910	.7862	.8955	.7860	.8954	1.2723	.1046	0.9047	50
20	0.6690	.6202	.7926	.7844	.8945	.7907	.8980	1.2647	.1020	0.9018	40
30	0.6720	.6225	.7941	.7826	.8935	.7954	.9006	1.2572	.0994	0.8988	30
40	0.6749	.6248	.7957	.7808	.8925	.8002	.9032	1.2497	.0968	0.8959	20
50	0.6778	.6271	.7973	.7790	.8915	.8050	.9058	1.2423	.0942	0.8930	10
39° 00′	0.6807	.6293	9.7989	.7771	9.8905	.8098	9.9084	1.2349	0.0916	0.8901	51° 00′
10	0.6836	.6316	.8004	.7753	.8895	.8146	.9110	1.2276	.0890	0.8872	50
20	0.6865	.6338	.8020	.7735	.8884	.8195	.9135	1.2203	.0865	0.8843	40
30	0.6894	.6361	.8035	.7716	.8874	.8243	.9161	1.2131	.0839	0.8814	30
40	0.6923	.6383	.8050	.7698	.8864	.8292	.9187	1.2059	.0813	0.8785	20
50	0.6952	.6406	.8066	.7679	.8853	.8342	.9212	1.1988	.0788	0.8756	10
40° 00′	0.6981	.6428	9.8081	.7660	9.8843	.8391	9.9238	1.1918	0.0762	0.8727	50° 00′
10	0.7010	.6450	.8096	.7642	.8832	.8441	.9264	1.1847	.0736	0.8698	50
20	0.7039	.6472	.8111	.7623	.8821	.8491	.9289	1.1778	.0711	0.8668	40
30	0.7069	.6494	.8125	.7604	.8810	.8541	.9315	1.1708	.0685	0.8639	30
40	0.7098	.6517	.8140	.7585	.8800	.8591	.9341	1.1640	.0659	0.8610	20
50	0.7127	.6539	.8155	.7566	.8789	.8642	.9366	1.1571	.0634	0.8581	10
41° 00′	0.7156	.6561	9.8169	.7547	9.8778	.8693	9.9392	1.1504	0.0608	0.8552	49° 00′
10	0.7185	.6583	.8184	.7528	.8767	.8744	.9417	1.1436	.0583	0.8523	50
20	0.7214	.6604	.8198	.7509	.8756	.8796	.9443	1.1369	.0557	0.8494	40
30	0.7243	.6626	.8213	.7490	.8745	.8847	.9468	1.1303	.0532	0.8465	30
40	0.7272	.6648	.8227	.7470	.8733	.8899	.9494	1.1237	.0506	0.8436	20
50	0.7301	.6670	.8241	.7451	.8722	.8952	.9519	1.1171	.0481	0.8407	10
42° 00′	0.7330	.6691	9.8255	.7431	9.8711	.9004	9.9544	1.1106	0.0456	0.8378	48° 00′
10	0.7359	.6713	.8269	.7412	.8699	.9057	.9570	1.1041	.0430	0.8348	50
20	0.7389	.6734	.8283	.7392	.8688	.9110	.9595	1.0977	.0405	0.8319	40
30	0.7418	.6756	.8297	.7373	.8676	.9163	.9621	1.0913	.0379	0.8290	30
40	0.7447	.6777	.8311	.7353	.8665	.9217	.9646	1.0850	.0354	0.8261	20
50	0.7476	.6799	.8324	.7333	.8653	.9271	.9671	1.0786	.0329	0.8232	10
43° 00′	0.7505	.6820	9.8338	.7314	9.8641	.9325	9.9697	1.0724	0.0303	0.8203	47° 00′
10	0.7534	.6841	.8351	.7294	.8629	.9380	.9722	1.0661	.0278	0.8174	50
20	0.7563	.6862	.8365	.7274	.8618	.9435	.9747	1.0599	.0253	0.8145	40
30	0.7592	.6884	.8378	.7254	.8606	.9490	.9772	1.0538	.0228	0.8116	30
40	0.7621	.6905	.8391	.7234	.8594	.9545	.9798	1.0477	.0202	0.8087	20
50	0.7650	.6926	.8405	.7214	.8582	.9601	.9823	1.0416	.0177	0.8058	10
44° 00′	0.7679	.6947	9.8418	.7193	9.8569	.9657	9.9848	1.0355	0.0152	0.8029	46° 00′
10	0.7709	.6967	.8431	.7173	.8557	.9713	.9874	1.0295	.0126	0.7999	50
20	0.7738	.6988	.8444	.7153	.8545	.9770	.9899	1.0235	.0101	0.7970	40
30	0.7767	.7009	.8457	.7133	.8532	.9827	.9924	1.0176	.0076	0.7941	30
40	0.7796	.7030	.8469	.7112	.8520	.9884	.9949	1.0117	.0051	0.7912	20
50	0.7825	.7050	.8482	.7092	.8507	.9942	.9975	1.0058	.0025	0.7883	10
45° 00′	0.7854	.7071	9.8495	.7071	9.8495	1.0000	0.0000	1.0000	0.0000	0.7854	45° 00′
		Nat.	Log *	Nat.	Log *	Nat.	Log *	Nat.	Log		
		Cosines		Sines		Cotangents		Tangents		Radians	Degrees

Table 1-3
DERIVATIVES AND DIFFERENTIATION

From Baumeister and Marks, "Standard Handbook for Mechanical Engineers," 7th ed., McGraw-Hill Book Company, New York (1967); by permission.

To find the derivative of a given function at a given point: (1) If the function is given only by a curve, measure graphically the slope of the tangent at the point in question; (2) if the function is given by a mathematical expression, use the following rules for differentiation. These rules give, directly, the differential, dy, in terms of dx; to find the derivative, dy/dx, divide through by dx.

Rules for Differentiation. (Here u, v, w, ... represent any functions of a variable x, or may themselves be independent variables. a is a constant which does not change in values in the same discussion; $e = 2.71828$.)

1. $d(a + u) = du$. 2. $d(au) = a \, du$.

3. $d(u + v + w + \cdots) = du + dv + dw + \cdots$.

4. $d(uv) = u \, dv + v \, du$.

5. $d(uvw \ldots) = (uvw \ldots)\left(\dfrac{du}{u} + \dfrac{dv}{v} + \dfrac{dw}{w} + \cdots\right)$.

6. $d\dfrac{u}{v} = \dfrac{v \, du - u \, dv}{v^2}$. 7. $d(u^m) = mu^{m-1} \, du$.

Thus, $d(u^2) = 2u \, du$; $d(u^3) = 3u^2 \, du$; etc.

8. $d\sqrt{u} = \dfrac{du}{2\sqrt{u}}$. 9. $d\left(\dfrac{1}{u}\right) = -\dfrac{du}{u^2}$.

10. $d(e^u) = e^u \, du$. 11. $d(a^u) = (\ln a)a^u \, du$.

12. $d\ln u = \dfrac{du}{u}$. 13. $d\log_{10} u = (\log_{10} e)\dfrac{du}{u} = (0.4343\ldots)\dfrac{du}{u}$.

14. $d\sin u = \cos u \, du$. 15. $d\csc u = -\cot u \csc u \, du$.

16. $d\cos u = -\sin u \, du$. 17. $d\sec u = \tan u \sec u \, du$.

18. $d\tan u = \sec^2 u \, du$. 19. $d\cot u = -\csc^2 u \, du$.

20. $d\sin^{-1} u = \dfrac{du}{\sqrt{1 - u^2}}$. 21. $d\csc^{-1} u = -\dfrac{du}{u\sqrt{u^2 - 1}}$.

22. $d\cos^{-1} u = -\dfrac{du}{\sqrt{1 - u^2}}$. 23. $d\sec^{-1} u = \dfrac{du}{u\sqrt{u^2 - 1}}$.

24. $d\tan^{-1} u = \dfrac{du}{1 + u^2}$. 25. $d\cot^{-1} u = -\dfrac{du}{1 + u^2}$.

26. $d\ln\sin u = \cot u \, du$. 27. $d\ln\tan u = \dfrac{2 \, du}{\sin 2u}$.

28. $d\ln\cos u = -\tan u \, du$. 29. $d\ln\cot u = -\dfrac{2 \, du}{\sin 2u}$.

30. $d\sinh u = \cosh u \, du$. 31. $d\operatorname{csch} u = -\operatorname{csch} u \coth u \, du$.

32. $d\cosh u = \sinh u \, du$. 33. $d\operatorname{sech} u = -\operatorname{sech} u \tanh u \, du$.

34. $d\tanh u = \operatorname{sech}^2 u \, du$. 35. $d\coth u = -\operatorname{csch}^2 u \, du$.

36. $d\sinh^{-1} u = \dfrac{du}{\sqrt{u^2 + 1}}$. 37. $d\operatorname{csch}^{-1} u = -\dfrac{du}{u\sqrt{u^2 + 1}}$.

38. $d\cosh^{-1} u = \dfrac{du}{\sqrt{u^2 - 1}}$. 39. $d\operatorname{sech}^{-1} u = -\dfrac{du}{u\sqrt{1 - u^2}}$.

40. $d\tanh^{-1} u = \dfrac{du}{1 - u^2}$. 41. $d\coth^{-1} u = \dfrac{du}{1 - u^2}$.

42. $d(u^v) = (u^{v-1})(u \ln u \, dv + v \, du)$.

Table 1-4
INTEGRALS

From Baumeister and Marks, "Standard Handbook for Mechanical Engineers," 7th ed., McGraw-Hill Book Company, New York (1967); by permission.

An integral of $f(x)\,dx$ is any function whose differential is $f(x)\,dx$, and is denoted by $\int f(x)\,dx$. All the integrals of $f(x)\,dx$ are included in the expression $\int f(x)\,dx + C$, where $\int f(x)\,dx$ is any particular integral, and C is an arbitrary constant. The process of finding (when possible) an integral of a given function consists in recognizing by inspection a function which, when differentiated, will produce the given function; or in transforming the given function into a form in which such recognition is easy. The most common integrable forms are collected in the following brief table; for a more extended list, see Peirce, "Table of Integrals," Ginn, or Dwight, "Table of Integrals and other Mathematical Data," Macmillan.

GENERAL FORMULAS

1. $\int a\,du = a\int du = au + C.$ 2. $\int (u + v)\,dx = \int u\,dx + \int v\,dx.$

3. $\int u\,dv = uv - \int v\,du.$ 4. $\int f(x)\,dx = \int f[F(y)]F'(y)\,dy,\ x = F(y).$

5. $\int dy \int f(x, y)\,dx = \int dx \int f(x, y)\,dy.$

FUNDAMENTAL INTEGRALS

6. $\int x^n\,dx = \dfrac{x^{n+1}}{n + 1} + C$, when $n \neq -1.$

7. $\int \dfrac{dx}{x} = \ln x + C = \ln cx.$ 8. $\int e^x\,dx = e^x + C.$

9. $\int \sin x\,dx = -\cos x + C.$ 10. $\int \cos x\,dx = \sin x + C.$

11. $\int \dfrac{dx}{\sin^2 x} = -\cot x + C.$ 12. $\int \dfrac{dx}{\cos^2 x} = \tan x + C.$

13. $\int \dfrac{dx}{\sqrt{1 - x^2}} = \sin^{-1} x + C = -\cos^{-1} x + c.$

14. $\int \dfrac{dx}{1 + x^2} = \tan^{-1} x + C = -\cot^{-1} x + c.$

RATIONAL FUNCTIONS

15. $\int (a + bx)^n\,dx = \dfrac{(a + bx)^{n+1}}{(n + 1)b} + C.$

16. $\int \dfrac{dx}{a + bx} = \dfrac{1}{b}\ln(a + bx) + C = \dfrac{1}{b}\ln c(a + bx).$

17. $\int \dfrac{1}{x^2}\,dx = -\dfrac{1}{x} + C.$ 18. $\int \dfrac{dx}{(a + bx)^2} = -\dfrac{1}{b(a + bx)} + C.$

19. $\int \dfrac{dx}{1 - x^2} = \tfrac{1}{2}\ln\dfrac{1 + x}{1 - x} + C = \tanh^{-1} x + C$, when $x < 1.$

20. $\int \dfrac{dx}{x^2 - 1} = \tfrac{1}{2}\ln\dfrac{x - 1}{x + 1} + C = -\coth^{-1} x + C$, when $x > 1.$

Table 1-4 (*Continued*)
INTEGRALS

21. $\int \dfrac{dx}{a + bx^2} = \dfrac{1}{\sqrt{ab}} \tan^{-1}\left(\sqrt{\dfrac{b}{a}}\, x\right) + C$

22. $\int \dfrac{dx}{a - bx^2} = \dfrac{1}{2\sqrt{ab}} \ln \dfrac{\sqrt{ab} + bx}{\sqrt{ab} - bx} + C$ $\quad [a > 0,\, b > 0]$.

$\qquad = \dfrac{1}{\sqrt{ab}} \tanh^{-1}\left(\sqrt{\dfrac{b}{a}}\, x\right) + C$

23. $\int \dfrac{dx}{a + 2bx + cx^2} = \dfrac{1}{\sqrt{ac - b^2}} \tan^{-1}\dfrac{b + cx}{\sqrt{ac - b^2}} + C \Big\}$ $[ac - b^2 > 0]$.

$\qquad = \dfrac{1}{2\sqrt{b^2 - ac}} \ln \dfrac{\sqrt{b^2 - ac} - b - cx}{\sqrt{b^2 - ac} + b + cx} + C$

$\qquad = -\dfrac{1}{\sqrt{b^2 - ac}} \tanh^{-1}\dfrac{b + cx}{\sqrt{b^2 - ac}} + C$ $\quad [b^2 - ac > 0]$.

24. $\int \dfrac{dx}{a + 2bx + cx^2} = -\dfrac{1}{b + cx} + C$, when $b^2 = ac$.

25. $\int \dfrac{(m + nx)\, dx}{a + 2bx + cx^2} = \dfrac{n}{2c} \ln (a + 2bx + cx^2) + \dfrac{mc - nb}{c} \int \dfrac{dx}{a + 2bx + cx^2}$.

26. In $\int \dfrac{f(x)\, dx}{a + 2bx + cx^2}$, if $f(x)$ is a polynomial of higher than the first degree, divide by the denominator before integrating.

27. $\int \dfrac{dx}{(a + 2bx + cx^2)^p} = \dfrac{1}{2(ac - b^2)(p - 1)} \times \dfrac{b + cx}{(a + 2bx + cx^2)^{p-1}}$

$\qquad + \dfrac{(2p - 3)c}{2(ac - b^2)(p - 1)} \int \dfrac{dx}{(a + 2bx + cx^2)^{p-1}}$.

28. $\int \dfrac{(m + nx)\, dx}{(a + 2bx + cx^2)^p} = -\dfrac{n}{2c(p - 1)} \times \dfrac{1}{(a + 2bx + cx^2)^{p-1}}$

$\qquad + \dfrac{mc - nb}{c} \int \dfrac{dx}{(a + 2bx + cx^2)^p}$.

29. $\int x^{m-1}(a + bx)^n\, dx = \dfrac{x^{m-1}(a + bx)^{n+1}}{(m + n)b} - \dfrac{(m - 1)a}{(m + n)b} \int x^{m-2}(a + bx)^n\, dx$

$\qquad = \dfrac{x^m(a + bx)^n}{m + n} + \dfrac{na}{m + n} \int x^{m-1}(a + bx)^{n-1}\, dx$.

IRRATIONAL FUNCTIONS

30. $\int \sqrt{a + bx}\, dx = \dfrac{2}{3b} \sqrt{(a + bx)^3} + C$.

31. $\int \dfrac{dx}{\sqrt{a + bx}} = \dfrac{2}{b} \sqrt{a + bx} + C$.

32. $\int \dfrac{(m + nx)\, dx}{\sqrt{a + bx}} = \dfrac{2}{3b^2} (3mb - 2an + nbx) \sqrt{a + bx} + C$.

33. $\int \dfrac{dx}{(m + nx) \sqrt{a + bx}}$; substitute $y = \sqrt{a + bx}$, and use 21 and 22.

Table 1-4 (*Continued*)
INTEGRALS

34. $\int \dfrac{f(x, \sqrt[n]{a + bx})}{F(x, \sqrt[n]{a + bx})}\, dx$; substitute $\sqrt[n]{a + bx} = y$.

35. $\int \dfrac{dx}{\sqrt{a^2 - x^2}} = \sin^{-1}\dfrac{x}{a} + C = -\cos^{-1}\dfrac{x}{a} + c.$

36. $\int \dfrac{dx}{\sqrt{a^2 + x^2}} = \ln\,(x + \sqrt{a^2 + x^2}) + C = \sinh^{-1}\dfrac{x}{a} + c.$

37. $\int \dfrac{dx}{\sqrt{x^2 - a^2}} = \ln\,(x + \sqrt{x^2 - a^2}) + C = \cosh^{-1}\dfrac{x}{a} + c.$

38. $\int \dfrac{dx}{\sqrt{a + 2bx + cx^2}} = \dfrac{1}{\sqrt{c}}\ln\,(b + cx + \sqrt{c}\sqrt{a + 2bx + cx^2}) + C$, when $c > 0$.

$$= \dfrac{1}{\sqrt{c}}\sinh^{-1}\dfrac{b + cx}{\sqrt{ac - b^2}} + C,\ \text{when } ac - b^2 > 0.$$

$$= \dfrac{1}{\sqrt{c}}\cosh^{-1}\dfrac{b + cx}{\sqrt{b^2 - ac}} + C,\ \text{when } b^2 - ac > 0.$$

$$= \dfrac{-1}{\sqrt{-c}}\sin^{-1}\dfrac{b + cx}{\sqrt{b^2 - ac}} + C,\ \text{when } c < 0.$$

39. $\int \dfrac{(m + nx)\, dx}{\sqrt{a + 2bx + cx^2}} = \dfrac{n}{c}\sqrt{a + 2bx + cx^2} + \dfrac{mc - nb}{c}\int \dfrac{dx}{\sqrt{a + 2bx + cx^2}}.$

40. $\int \dfrac{x^m\, dx}{\sqrt{a + 2bx + cx^2}} = \dfrac{x^{m-1}X}{mc} - \dfrac{(m - 1)a}{mc}\int \dfrac{x^{m-2}\, dx}{X}$

$$-\dfrac{(2m - 1)b}{mc}\int \dfrac{x^{m-1}\, dx}{X},\ \text{when } X = \sqrt{a + 2bx + cx^2}.$$

41. $\int \sqrt{a^2 + x^2}\, dx = \dfrac{x}{2}\sqrt{a^2 + x^2} + \dfrac{a^2}{2}\ln\,(x + \sqrt{a^2 + x^2}) + C$

$$= \dfrac{x}{2}\sqrt{a^2 + x^2} + \dfrac{a^2}{2}\sinh^{-1}\dfrac{x}{a} + c.$$

42. $\int \sqrt{a^2 - x^2}\, dx = \dfrac{x}{2}\sqrt{a^2 - x^2} + \dfrac{a^2}{2}\sin^{-1}\dfrac{x}{a} + C.$

43. $\int \sqrt{x^2 - a^2}\, dx = \dfrac{x}{2}\sqrt{x^2 - a^2} - \dfrac{a^2}{2}\ln\,(x + \sqrt{x^2 - a^2}) + C$

$$= \dfrac{x}{2}\sqrt{x^2 - a^2} - \dfrac{a^2}{2}\cosh^{-1}\dfrac{x}{a} + c.$$

44. $\int \sqrt{a + 2bx + cx^2}\, dx = \dfrac{b + cx}{2c}\sqrt{a + 2bx + cx^2}$

$$+ \dfrac{ac - b^2}{2c}\int \dfrac{dx}{\sqrt{a + 2bx + cx^2}} + C.$$

TRANSCENDENTAL FUNCTIONS

45. $\int a^x\, dx = \dfrac{a^x}{\ln a} + C.$

46. $\int x^n e^{ax}\, dx = \dfrac{x^n e^{ax}}{a}\left[1 - \dfrac{n}{ax} + \dfrac{n(n - 1)}{a^2 x^2} - \cdots \pm \dfrac{n!}{a^n x^n}\right] + C.$

Table 1-4 (*Continued*)
INTEGRALS

47. $\int \ln x \, dx = x \ln x - x + C.$

48. $\int \dfrac{\ln x}{x^2} \, dx = -\dfrac{\ln x}{x} - \dfrac{1}{x} + C.$

49. $\int \dfrac{(\ln x)^n}{x} \, dx = \dfrac{1}{n+1} (\ln x)^{n+1} + C.$

50. $\int \sin^2 x \, dx = -\tfrac{1}{4} \sin 2x + \tfrac{1}{2} x + C = -\tfrac{1}{2} \sin x \cos x + \tfrac{1}{2} x + C.$

51. $\int \cos^2 x \, dx = \tfrac{1}{4} \sin 2x + \tfrac{1}{2} x + C = \tfrac{1}{2} \sin x \cos x + \tfrac{1}{2} x + C.$

52. $\int \sin mx \, dx = -\dfrac{\cos mx}{m} + C.$ 53. $\int \cos mx \, dx = \dfrac{\sin mx}{m} + C.$

54. $\int \sin mx \cos nx \, dx = -\dfrac{\cos (m+n)x}{2(m+n)} - \dfrac{\cos (m-n)x}{2(m-n)} + C.$

55. $\int \sin mx \sin nx \, dx = \dfrac{\sin (m-n)x}{2(m-n)} - \dfrac{\sin (m+n)x}{2(m+n)} + C.$

56. $\int \cos mx \cos nx \, dx = \dfrac{\sin (m-n)x}{2(m-n)} + \dfrac{\sin (m+n)x}{2(m+n)} + C.$

57. $\int \tan x \, dx = -\ln \cos x + C.$ 58. $\int \cot x \, dx = \ln \sin x + C.$

59. $\int \dfrac{dx}{\sin x} = \ln \tan \dfrac{x}{2} + C.$ 60. $\int \dfrac{dx}{\cos x} = \ln \tan \left(\dfrac{\pi}{4} + \dfrac{x}{2} \right) + C.$

61. $\int \dfrac{dx}{1 + \cos x} = \tan \dfrac{x}{2} + C.$ 62. $\int \dfrac{dx}{1 - \cos x} = -\cot \dfrac{x}{2} + C.$

63. $\int \sin x \cos x \, dx = \tfrac{1}{2} \sin^2 x + C.$ 64. $\int \dfrac{dx}{\sin x \cos x} = \ln \tan x + C.$

65.* $\int \sin^n x \, dx = -\dfrac{\cos x \sin^{n-1} x}{n} + \dfrac{n-1}{n} \int \sin^{n-2} x \, dx.$

66.* $\int \cos^n x \, dx = \dfrac{\sin x \cos^{n-1} x}{n} + \dfrac{n-1}{n} \int \cos^{n-2} x \, dx.$

67. $\int \tan^n x \, dx = \dfrac{\tan^{n-1} x}{n-1} - \int \tan^{n-2} x \, dx.$

68. $\int \cot^n x \, dx = -\dfrac{\cot^{n-1} x}{n-1} - \int \cot^{n-2} x \, dx.$

69. $\int \dfrac{dx}{\sin^n x} = -\dfrac{\cos x}{(n-1) \sin^{n-1} x} + \dfrac{n-2}{n-1} \int \dfrac{dx}{\sin^{n-2} x}.$

70. $\int \dfrac{dx}{\cos^n x} = \dfrac{\sin x}{(n-1) \cos^{n-1} x} + \dfrac{n-2}{n-1} \int \dfrac{dx}{\cos^{n-2} x}.$

71.* $\int \sin^p x \cos^q x \, dx = \dfrac{\sin^{p+1} x \cos^{q-1} x}{p+q} + \dfrac{q-1}{p+q} \int \sin^p x \cos^{q-2} x \, dx$

$\quad = -\dfrac{\sin^{p-1} x \cos^{q+1} x}{p+q} + \dfrac{p-1}{p+q} \int \sin^{p-2} x \cos^q x \, dx.$

72.* $\int \sin^{-p} x \cos^q x \, dx = -\dfrac{\sin^{-p+1} x \cos^{q+1} x}{p-1} + \dfrac{p-q-2}{p-1} \int \sin^{-p+2} x \cos^q x \, dx.$

*If n, p, or q is an odd number, substitute $\cos x = z$ or $\sin x = z$.

Table 1-4 (*Continued*)
INTEGRALS

73.* $\int \sin^p x \cos^{-q} x \, dx = \dfrac{\sin^{p+1} x \cos^{-q+1} x}{q-1} + \dfrac{q-p-2}{q-1} \int \sin^p x \cos^{-q+2} x \, dx.$

74. $\int \dfrac{dx}{a + b \cos x} = \dfrac{2}{\sqrt{a^2 - b^2}} \tan^{-1}\left(\sqrt{\dfrac{a-b}{a+b}} \tan \tfrac{1}{2} x \right) + C,$ when $a^2 > b^2$

$$= \dfrac{1}{\sqrt{b^2 - a^2}} \ln \dfrac{b + a \cos x + \sin x \sqrt{b^2 - a^2}}{a + b \cos x} + C$$

$$= \dfrac{2}{\sqrt{b^2 - a^2}} \tanh^{-1}\left(\sqrt{\dfrac{b-a}{b+a}} \tan \tfrac{1}{2} x \right) + C$$

$[a^2 < b^2].$

75. $\int \dfrac{\cos x \, dx}{a + b \cos x} = \dfrac{x}{b} - \dfrac{a}{b} \int \dfrac{dx}{a + b \cos x} + C.$

76. $\int \dfrac{\sin x \, dx}{a + b \cos x} = -\dfrac{1}{b} \ln (a + b \cos x) + C.$

77. $\int \dfrac{A + B \cos x + C \sin x}{a + b \cos x + c \sin x} \, dx = A \int \dfrac{dy}{a + p \cos y}$

$+ (B \cos u + C \sin u) \int \dfrac{\cos y \, dy}{a + p \cos y} - (B \sin u - C \cos u) \int \dfrac{\sin y \, dy}{a + p \cos y},$

where $b = p \cos u,\ c = p \sin u$ and $x - u = y.$

78. $\int e^{ax} \sin bx \, dx = \dfrac{a \sin bx - b \cos bx}{a^2 + b^2} e^{ax} + C.$

79. $\int e^{ax} \cos bx \, dx = \dfrac{a \cos bx + b \sin bx}{a^2 + b^2} e^{ax} + C.$

80. $\int \sin^{-1} x \, dx = x \sin^{-1} x + \sqrt{1 - x^2} + C.$

81. $\int \cos^{-1} x \, dx = x \cos^{-1} x - \sqrt{1 - x^2} + C.$

82. $\int \tan^{-1} x \, dx = x \tan^{-1} x - \tfrac{1}{2} \ln (1 + x^2) + C.$

83. $\int \cot^{-1} x \, dx = x \cot^{-1} x + \tfrac{1}{2} \ln (1 + x^2) + C.$

84. $\int \sinh x \, dx = \cosh x + C.$ 85. $\int \tanh x \, dx = \ln \cosh x + C.$

86. $\int \cosh x \, dx = \sinh x + C.$ 87. $\int \coth x \, dx = \ln \sinh x + C.$

88. $\int \operatorname{sech} x \, dx = 2 \tan^{-1}(e^x) + C.$ 89. $\int \operatorname{csch} x \, dx = \ln \tanh\left(\dfrac{x}{2}\right) + C.$

90. $\int \sinh^2 x \, dx = \tfrac{1}{2} \sinh x \cosh x - \tfrac{1}{2} x + C.$

91. $\int \cosh^2 x \, dx = \tfrac{1}{2} \sinh x \cosh x + \tfrac{1}{2} x + C.$

92. $\int \operatorname{sech}^2 x \, dx = \tanh x + C.$ 93. $\int \operatorname{csch}^2 x \, dx = -\coth x + C.$

*If p or q is an odd number, substitute $\cos x = z$ or $\sin x = z.$

SURFACE AREAS AND VOLUMES*

Let a, b, c, d, and s denote lengths, A denote areas, and V denote volumes.

Triangle.

$A = bh/2$, where b denotes the base and h the altitude.

Rectangle.

$A = ab$, where a and b denote the lengths of the sides.

Parallelogram (opposite sides parallel).

$A = ah = ab \sin \theta$, where a and b denote the sides, h the altitude, and θ the angle between the sides.

Trapezoid (four sides, two parallel).

$A = \frac{1}{2}h(a + b)$, where a and b are the sides and h the altitude.

Regular Polygon of n Sides. (Fig. 1-1).

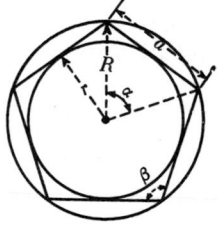

Fig. 1-1

$A = \dfrac{1}{4} n\, a^2 \operatorname{ctn} \dfrac{180°}{n}$,   where a is length of side.

$R = \dfrac{a}{2} \csc \dfrac{180°}{n}$,   where R is radius of circumscribed circle.

$r = \dfrac{a}{2} \operatorname{ctn} \dfrac{180°}{n}$,   where r is radius of inscribed circle.

$\alpha = \dfrac{360°}{n} = \dfrac{2\pi}{n}$,   radians.

$\beta = \left(\dfrac{n-2}{n}\right) \cdot 180° = \left(\dfrac{n-2}{n}\right)\pi$ radians where α and β are the angles indicated in Fig. 1-1.

$a = 2\, r \tan \dfrac{\alpha}{2} = 2R \sin \dfrac{\alpha}{2}$.

Circle. (Fig. 1-2).

Let C = circumference, S = length of arc subtended by θ,
R = radius, l = chord subtended by arc S,
D = diameter, h = rise,
A = area, θ = central angle in radians.

* Adapted by permission from Burington, "Handbook of Mathematical Tables and Formulas," 3d ed., McGraw-Hill Book Company, New York (1959).

$$C = 2\pi R = \pi D, \ \pi = 3.14159\cdots.$$

$$S = R\theta = \tfrac{1}{2}D\theta = D\cos^{-1}\frac{d}{R}.$$

$$l = 2\sqrt{R^2 - d^2} = 2R\sin\frac{\theta}{2} = 2d\tan\frac{\theta}{2}.$$

$$d = \tfrac{1}{2}\sqrt{4R^2 - l^2} = R\cos\frac{\theta}{2} = \tfrac{1}{2}l\,\text{ctn}\,\frac{\theta}{2}.$$

$$h = R - d.$$

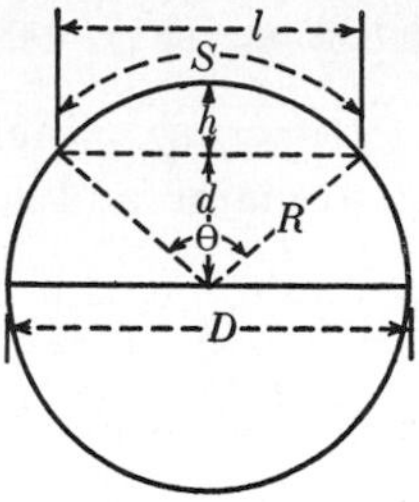

Fig. 1-2

$$\theta = \frac{S}{R} = \frac{2S}{D} = 2\cos^{-1}\frac{d}{R} = 2\tan^{-1}\frac{l}{2d} = 2\sin^{-1}\frac{l}{D}.$$

$$A\ (\text{circle}) = \pi R^2 = \tfrac{1}{4}\pi D^2.$$

$$A\ (\text{sector}) = \tfrac{1}{2}Rs = \tfrac{1}{2}R^2\theta.$$

$$A\ (\text{segment}) = A\ (\text{sector}) - A\ (\text{triangle}) = \tfrac{1}{2}R^2\,(\theta - \sin\theta)$$

$$= R^2\cos^{-1}\frac{(R - h)}{R} - (R - h)\sqrt{2Rh - h^2}.$$

Perimeter of n-sided regular polygon inscribed in a circle

$$= 2n\,R\sin\frac{\pi}{n}.$$

Area of inscribed polygon $= \tfrac{1}{2}nR^2\sin\dfrac{2\pi}{n}.$

Perimeter of n-sided regular polygon circumscribed about a circle

$$= 2nR\tan\frac{\pi}{n}.$$

Area of circumscribed polygon $= nR^2\tan\dfrac{\pi}{n}.$

Radius of circle inscribed in a triangle of sides a, b, and c is

$$r = \sqrt{\frac{(s - a)(s - b)(s - c)}{s}}, \quad s = \tfrac{1}{2}(a + b + c).$$

Radius of circle circumscribed about a triangle is

$$R = \frac{abc}{4\sqrt{s(s - a)(s - b)(s - c)}}.$$

Ellipse. (Fig. 1-3).

$A = \pi ab$, where a and b are lengths of semi-major and semi-minor axes respectively.

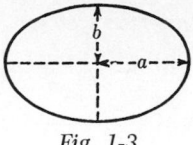

Fig. 1-3

Parabola. (Fig. 1-4).

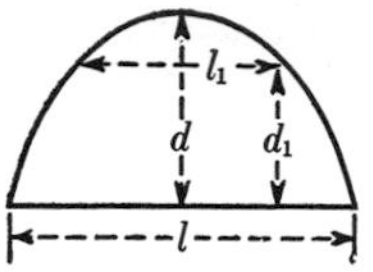

$A = 2ld/3$.

Height of $d_1 = \dfrac{d}{l^2}(l^2 - l_1^2)$.

Width of $l_1 = l\sqrt{\dfrac{d - d_1}{d}}$.

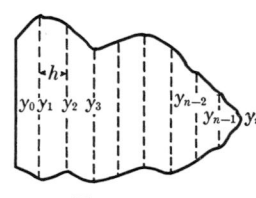

Fig. 1-4

Length of arc $= l\left[1 + \dfrac{2}{3}\left(\dfrac{2d}{l}\right)^2 - \dfrac{2}{5}\left(\dfrac{2d}{l}\right)^4 + \cdots\right]$.

Area by Approximation. (Fig. 1-5). If $y_0, y_1, y_2, \cdots, y_n$ be the lengths of a series of equally spaced parallel chords, and if h be their distance apart, the area enclosed by boundary is given approximately by any one of the following formulae:

Fig. 1-5

$A_T = h[\tfrac{1}{2}(y_0 + y_n) + y_1 + y_2 + \cdots + y_{n-1}]$. (Trapezoidal Rule.)

$A_D = h[0.4(y_0 + y_n) + 1.1(y_1 + y_{n-1}) + y_2 + y_3 + \cdots + y_{n-2}]$.
(Durand's Rule.)

$A_S = \tfrac{1}{3}h[(y_0 + y_n) + 4(y_1 + y_3 + \cdots + y_{n-1})$
$\quad + 2(y_2 + y_4 + \cdots + y_{n-2})]$.   (*n* even, Simpson's Rule.)

In general, A_S gives the most accurate approximation.
The greater the value of n, the greater the accuracy of approximation.

Cube.

$V = a^3$; $d = a\sqrt{2}$; total surface area $= 6a^2$, where a is length of side and d is length of diagonal.

Rectangular Parallelopiped.

$V = abc$; $d = \sqrt{a^2 + b^2 + c^2}$; total surface area $= 2(ab + bc + ca)$, where a, b, and c are the lengths of the sides and d is length of diagonal.

Prism or Cylinder. $V = $ (area of base) $\cdot$ (altitude).

Lateral area $= $ (perimeter of right section) $\cdot$ (lateral edge).

Pyramid or Cone. $V = \frac{1}{3}$ (area of base) · (altitude).

Lateral area of regular pyramid
$$= \frac{1}{2} \text{ (perimeter of base)} \cdot \text{(slant height)}.$$

Frustum of Pyramid or Cone.

$V = \frac{1}{3}(A_1 + A_2 + \sqrt{A_1 \cdot A_2})\, h$, where h is the altitude and A_1 and A_2 are the areas of the bases.

Lateral area of a regular figure
$$= \frac{1}{2} \text{ (sum of perimeters of base)} \cdot \text{(slant height)}.$$

Prismoid. $V = \dfrac{h}{6}(A_1 + A_2 + 4A_3)$, where $h = $ altitude, A_1 and A_2 are the areas of the bases, and A_3 is the area of the midsection parallel to bases.

Area of Surface and Volume of Regular Polyhedra of edge l.

Name	Type of Surface	Area of Surface	Volume
Tetrahedron	4 equilateral triangles	$1.73205l^2$	$0.11785l^3$
Hexahedron (cube)	6 squares	$6.00000l^2$	$1.00000l^3$
Octahedron	8 equilateral triangles	$3.46410l^2$	$0.47140l^3$
Dodecahedron	12 pentagons	$20.64578l^2$	$7.66312l^3$
Icosahedron	20 equilateral triangles	$8.66025l^2$	$2.18170l^3$

Sphere. (Fig. 1-6)

A (sphere) $= 4\pi R^2 = \pi D^2$.

A (zone) $= 2\pi Rh_1 = \pi Dh_1$.

V (sphere) $= \frac{4}{3}\pi R^3 = \frac{1}{6}\pi D^3$.

V (spherical sector) $= \frac{2}{3}\pi R^2 h_1 = \frac{1}{3}\pi D^2 h_1$.

V (spherical segment
of one base) $= \frac{1}{6}\pi h_3 (3r_3^2 + h_3^2)$.

V (spherical segment
of two bases) $= \frac{1}{6}\pi h_2 (3r_3^2 + 3r_2^2 + h_2^2)$.

A (lune) $= 2R^2\theta$, where θ is angle in radians of lune.

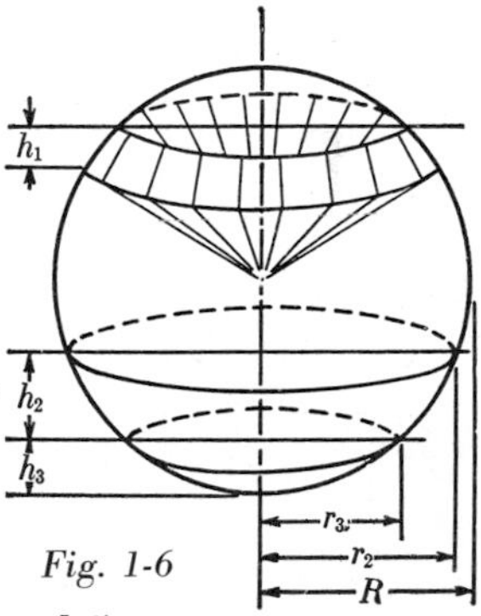

Fig. 1-6

Ellipsoid.

$V = \frac{4}{3}\pi abc$, where a, b, and c are the lengths of the semi-axes.

Torus. (Fig. 1-7). $V = 2\pi^2 Rr^2$.

Area of surface $= S = 4\pi^2 Rr$.

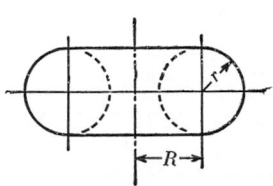

Fig. 1-7

TRIGONOMETRIC FUNCTIONS OF AN ANGLE α

Let α be any angle whose initial side lies on the positive x-axis and whose vertex is at the origin, and (x, y) be any point on the terminal side of the angle. (x is positive if measured along OX to the right, from the y-axis; and negative, if measured along OX' to the left from the y-axis. Likewise, y is positive if measured parallel to OY, and negative if measured parallel to OY'.) Let r be the positive distance from the origin to the point. The trigonometric functions of an angle are defined as follows:

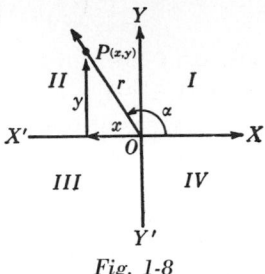

Fig. 1-8

sine α	=	sin α	=	$\dfrac{y}{r}$
cosine α	=	cos α	=	$\dfrac{x}{r}$
tangent α	=	tan α	=	$\dfrac{y}{x}$
cotangent α	=	ctn α = cot α	=	$\dfrac{x}{y}$
secant α	=	sec α	=	$\dfrac{r}{x}$
cosecant α	=	csc α	=	$\dfrac{r}{y}$
exsecant α	=	exsec α	=	$\sec α - 1$
versine α	=	vers α	=	$1 - \cos α$
coversine α	=	covers α	=	$1 - \sin α$
haversine α	=	hav α	=	$\frac{1}{2}$ vers α

Signs of the Functions.

Quadrant	sin	cos	tan	ctn	sec	csc
I..........	+	+	+	+	+	+
II	+	−	−	−	−	+
III........	−	−	+	+	−	−
IV........	−	+	−	−	+	−

Relations between the Functions of a Single Angle*

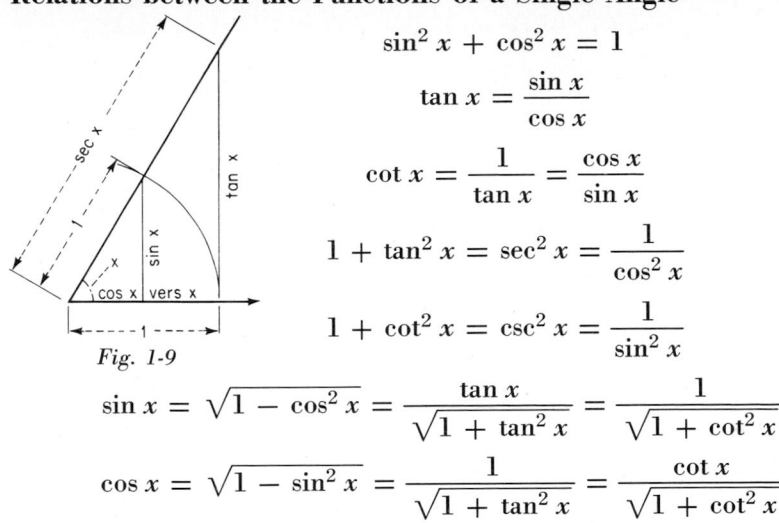

$$\sin^2 x + \cos^2 x = 1$$

$$\tan x = \frac{\sin x}{\cos x}$$

$$\cot x = \frac{1}{\tan x} = \frac{\cos x}{\sin x}$$

$$1 + \tan^2 x = \sec^2 x = \frac{1}{\cos^2 x}$$

$$1 + \cot^2 x = \csc^2 x = \frac{1}{\sin^2 x}$$

Fig. 1-9

$$\sin x = \sqrt{1 - \cos^2 x} = \frac{\tan x}{\sqrt{1 + \tan^2 x}} = \frac{1}{\sqrt{1 + \cot^2 x}}$$

$$\cos x = \sqrt{1 - \sin^2 x} = \frac{1}{\sqrt{1 + \tan^2 x}} = \frac{\cot x}{\sqrt{1 + \cot^2 x}}$$

Functions of Negative Angles. $\sin(-x) = -\sin x$; $\cos(-x) = \cos x$; $\tan(-x) = -\tan x$.

Functions of the Sum and Difference of Two Angles

$\sin(x + y) = \sin x \cos y + \cos x \sin y$.
$\cos(x + y) = \cos x \cos y - \sin x \sin y$.
$\tan(x + y) = (\tan x + \tan y)/(1 - \tan x \tan y)$.
$\cot(x + y) = (\cot x \cot y - 1)/(\cot x + \cot y)$.
$\sin(x - y) = \sin x \cos y - \cos x \sin y$.
$\cos(x - y) = \cos x \cos y + \sin x \sin y$.
$\tan(x - y) = (\tan x - \tan y)/(1 + \tan x \tan y)$.
$\cot(x - y) = (\cot x \cot y + 1)/(\cot y - \cot x)$.
$\sin x + \sin y = 2 \sin \frac{1}{2}(x + y) \cos \frac{1}{2}(x - y)$.
$\sin x - \sin y = 2 \cos \frac{1}{2}(x + y) \sin \frac{1}{2}(x - y)$.
$\cos x + \cos y = 2 \cos \frac{1}{2}(x + y) \cos \frac{1}{2}(x - y)$.
$\cos x - \cos y = -2 \sin \frac{1}{2}(x + y) \sin \frac{1}{2}(x - y)$.

$$\tan x + \tan y = \frac{\sin(x + y)}{\cos x \cos y}; \quad \cot x + \cot y = \frac{\sin(x + y)}{\sin x \sin y}.$$

$$\tan x - \tan y = \frac{\sin(x - y)}{\cos x \cos y}; \quad \cot x - \cot y = \frac{\sin(y - x)}{\sin x \sin y}.$$

$\sin^2 x - \sin^2 y = \cos^2 y - \cos^2 x = \sin(x + y)\sin(x - y)$.
$\cos^2 x - \sin^2 y = \cos^2 y - \sin^2 x = \cos(x + y)\cos(x - y)$.

* From Baumeister and Marks, "Standard Handbook for Mechanical Engineers," 7th ed., McGraw-Hill Book Company, New York (1967); by permission.

$\sin (45° + x) = \cos (45° - x); \tan (45° + x) = \cot (45° - x).$
$\sin (45° - x) = \cos (45° + x); \tan (45° - x) = \cot (45° + x).$

In the following transformations, a and b are supposed to be positive, $c = \sqrt{a^2 + b^2}$, $A =$ the positive acute angle for which $\tan A = a/b$, and $B =$ the positive acute angle for which $\tan B = b/a$:

$a \cos x + b \sin x = c \sin (A + x) = c \cos (B - x).$
$a \cos x - b \sin x = c \sin (A - x) = c \cos (B + x).$

EXPANSION IN SERIES

From Baumeister and Marks, "Standard Handbook for Mechanical Engineers," 7th ed., McGraw-Hill Book Company, New York (1967); by permission.

The range of values of x for which each of the series is convergent is stated at the right of the series.

Exponential and Logarithmic Series

$$e^x = 1 + \frac{x}{1!} + \frac{x^2}{2!} + \frac{x^3}{3!} + \frac{x^4}{4!} + \cdots \qquad (-\infty < x < +\infty)$$

$$a^x = e^{mx} = 1 + \frac{m}{1!}x + \frac{m^2}{2!}x^2 + \frac{m^3}{3!}x^3 + \cdots$$

$$[a > 0, -\infty < x < +\infty]$$

where $m = \ln a = (2.3026)(\log_{10} a)$.

$$\ln (1 + x) = x - \frac{x^2}{2} + \frac{x^3}{3} - \frac{x^4}{4} + \frac{x^5}{5} \cdots \qquad [-1 < x < +1]$$

$$\ln (1 - x) = -x - \frac{x^2}{2} - \frac{x^3}{3} - \frac{x^4}{4} - \frac{x^5}{5} - \cdots \qquad [-1 < x < +1]$$

$$\ln \left(\frac{1 + x}{1 - x}\right) = 2\left(x + \frac{x^3}{3} + \frac{x^5}{5} + \frac{x^7}{7} + \cdots\right) \qquad [-1 < x < +1]$$

$$\ln \left(\frac{x + 1}{x - 1}\right) = 2\left(\frac{1}{x} + \frac{1}{3x^3} + \frac{1}{5x^5} + \frac{1}{7x^7} + \cdots\right)$$

$$[x < -1 \text{ or } +1 < x]$$

$$\ln x = 2\left[\frac{x - 1}{x + 1} + \frac{1}{3}\left(\frac{x - 1}{x + 1}\right)^3 + \frac{1}{5}\left(\frac{x - 1}{x + 1}\right)^5 + \cdots\right] \quad [0 < x < \infty]$$

$$\ln (a + x) = \ln a + 2\left[\frac{x}{2a + x} + \frac{1}{3}\left(\frac{x}{2a + x}\right)^3 + \frac{1}{5}\left(\frac{x}{2a + x}\right)^5 + \cdots\right]$$

$$[0 < a < +\infty, -a < x < +\infty]$$

Series for the Trigonometric Functions. In the following formulas, *all angles must be expressed in radians.* If D = the number of degrees in the angle, and x = its radian measure, then $x = 0.017453D$.

$$\sin x = x - \frac{x^3}{3!} + \frac{x^5}{5!} - \frac{x^7}{7!} + \cdots \qquad\qquad [-\infty < x < +\infty]$$

$$\cos x = 1 - \frac{x^2}{2!} + \frac{x^4}{4!} - \frac{x^6}{6!} + \frac{x^8}{8!} - \cdots \qquad\qquad [-\infty < x < +\infty]$$

$$\tan x = x + \frac{x^3}{3} + \frac{2x^5}{15} + \frac{17x^7}{315} + \frac{62x^9}{2835} + \cdots$$

$$[-\pi/2 < x < +\pi/2]$$

$$\cot x = \frac{1}{x} - \frac{x}{3} - \frac{x^3}{45} - \frac{2x^5}{945} - \frac{x^7}{4725} - \cdots \qquad [-\pi < x < +\pi]$$

$$\sin^{-1} y = y + \frac{y^3}{6} + \frac{3y^5}{40} + \frac{5y^7}{112} + \cdots \qquad\qquad [-1 \leqq y \leqq +1]$$

$$\tan^{-1} y = y - \frac{y^3}{3} + \frac{y^5}{5} - \frac{y^7}{7} + \cdots \qquad\qquad [-1 \leqq y \leqq +1]$$

$$\cos^{-1} y = \tfrac{1}{2}\pi - \sin^{-1} y; \quad \cot^{-1} y = \tfrac{1}{2}\pi - \tan^{-1} y.$$

Reversing a Series. If $y = x + bx^2 + cx^3 + dx^4 + ex^5 + \cdots$, then $x = y - by^2 + (2b^2 - c)y^3 - (5b^3 - 5bc + d)y^4 + (14b^4 - 21b^2c + 6bd + 3c^2 - e)y^5 + \cdots$, provided the latter series is convergent.

Fourier's Series. Let $f(x)$ be a function which is finite in the interval from $x = -c$ to $x = +c$ and whose graph has finite arc length in that interval.* Then, for any value of x between $-c$ and c,

$$f(x) = \tfrac{1}{2}a_0 + a_1 \cos \frac{\pi x}{c} + a_2 \cos \frac{2\pi x}{c} + a_3 \cos \frac{3\pi x}{c} + \cdots$$

$$+ b_1 \sin \frac{\pi x}{c} + b_2 \sin \frac{2\pi x}{c} + b_3 \sin \frac{3\pi x}{c} + \cdots$$

where the constant coefficients are determined as follows:

$$a_n = \frac{1}{c} \int_{-c}^{c} f(t) \cos \frac{n\pi t}{c}\, dt; \qquad b_n = \frac{1}{c} \int_{-c}^{c} f(t) \sin \frac{n\pi t}{c}\, dt.$$

In case the curve $y = f(x)$ is symmetrical with respect to the origin, the a's are all zero, and the series is a sine series. In case the curve is

---

* If $x = x_0$ is a point of discontinuity, $f(x_0)$ is to be defined as $\tfrac{1}{2}[f_1(x_0) + f_2(x_0)]$, where $f_1(x_0)$ is the limit of $f(x)$ when x approaches x_0 from below, and $f_2(x_0)$ is the limit of $f(x)$ when x approaches x_0 from above.

symmetrical with respect to the y-axis, the b's are all zero, and a cosine series results. (In this case, the series will be valid not only for values of x between $-c$ and c, but also for $x = -c$ and $x = c$.) A Fourier series can always be integrated term by term; but the result of differentiating term by term may not be a convergent series.

Table 1-5
SOME CONSTANTS

Constant	Number	Log$_{10}$ of Number
Pi (π)	3.14159 26535 89793 23846	0.49714 98726 94133 85435
Napierian Base (e)	2.71828 18284 59045 23536	0.43429 448
$M = \log_{10}e$	0.43429 44819 03251 82765	9.63778 43113 00536 78912 $-$ 10
$1 \div M = \log_e 10$	2.30258 50929 94045 68402	0.36221 569
$180 \div \pi =$ degrees in 1 radian	57.2957 795	1.75812 263
$\pi \div 180 =$ radians in $1°$	0.01745 329	8.24187 737 $-$ 10
$\pi \div 10800 =$ radians in $1'$	0.00029 08882	6.46372 612 $-$ 10
$\pi \div 648000 =$ radians in $1''$	0.00000 48481 36811 095	4.68557 487 $-$ 10

STATISTICAL TABLES

STATISTICS

Raw data are collected observations that have not been organized numerically. An array is an arrangement of raw numerical data in ascending or descending order of magnitude. An *average* is a value that is typical or representative of a set of data. Several averages can be defined, the most common being the arithmetic mean (or briefly the mean), the median, the mode, and the geometric mean.

The *mean* of a set of N numbers, $x_1, x_2, x_3, \ldots, x_N$ is denoted by $\bar{x}$ and is defined as

$$\bar{x} = \frac{x_1 + x_2 + x_3 + \cdots + x_N}{N}$$

It is an estimate of the unknown true value μ of an infinite population.

The *median* of a set of numbers arranged in order of magnitude is the middle value or the arithmetic mean of the two middle values. The median allows inclusion of all data in a set without undue influence from outlying values; it is preferable to the mean for small sets of data.

The *mode* of a set of numbers is that value which occurs with the greatest frequency (the most common value). The mode may not exist, and even if it does exist it may not be unique. The empirical relation that exists between the mean, the mode, and the median for unimodal frequency curves which are moderately asymmetrical is

$$\text{Mean} - \text{mode} = 3(\text{mean} - \text{median})$$

The *geometric mean* of a set of N numbers is the Nth root of the product of the numbers:

$$\sqrt[N]{x_1 x_2 x_3 \ldots x_N}$$

Distribution of Measurements

When a large number of measurements are made, the individual measurements are not all identical and equal to the accepted value μ, which is the mean of an infinite population or universe of data, but are scattered about μ, owing to random error. If the magnitude of any single measurement is the abscissa and the relative frequencies (i.e., the probability) of occurrence of different-sized measurements are the ordinate, the smooth curve drawn through the points (Fig. 1-10) is the *normal distribution curve* (also called the *Gaussian distribution curve*, the *error curve*, or the *probability curve*). The term "error curve" arises when one considers the distribution of errors $(x - \mu)$ about the true value.

The breadth or spread of the curve indicates the precision of the measurements, and is determined by and related to the standard deviation, a relationship that is expressed in the equation for the normal curve,

$$Y = \frac{1}{\sigma\sqrt{2\pi}}\, e^{-1/2(x-\mu)^2/\sigma^2}$$

where σ is the standard deviation of the infinite population. The population mean μ expresses the magnitude of the quantity being measured; the standard deviation σ expresses the scatter. When $(x - \mu)/\sigma$ is replaced by the standardized variable z, then

$$Y = \frac{1}{\sqrt{2\pi}}\, e^{-1/2z^2}$$

The standardized variable measures the deviation from the population mean in units of standard deviation. Y is 0.399 for the most probable value, μ.

Table 1-6 lists the height of an ordinate (column Y) as a distance z from the mean, and the area under the normal curve (column A) at a distance z from the mean, expressed as fractions of the total area, 1.000.

Example 1. The true value of a quantity is 30.00, and σ for the

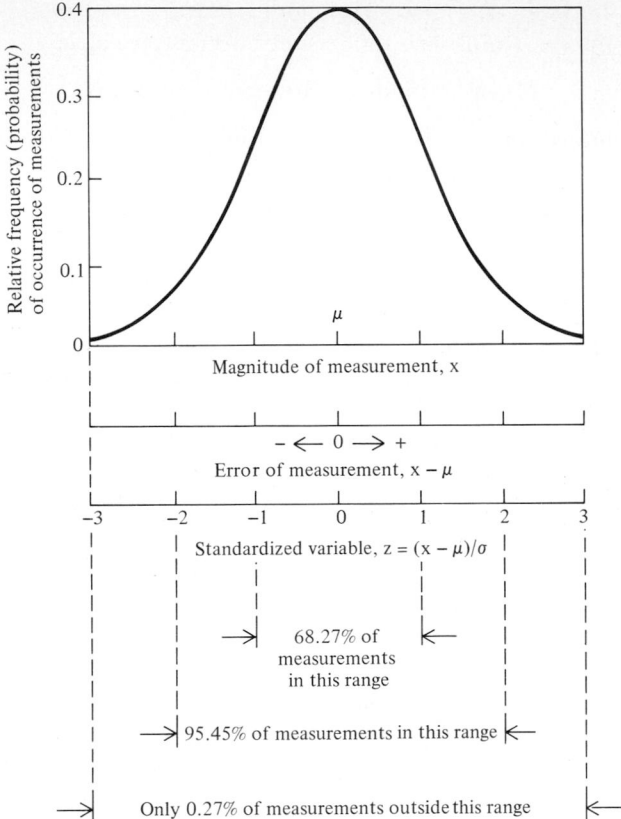

Fig. 1-10. The normal distribution curve.

method of measurement is 0.30. What is the probability that a single measurement will have a deviation from the mean greater than 0.45; that is, what percentage of results will fall outside the range 30.00 ± 0.45?

$$z = \frac{x - \mu}{\sigma} = \frac{0.45}{0.30} = 1.5$$

From Table 1-6 the area under the normal curve from -1.5σ to $+1.5\sigma$ is 0.866, meaning that 86.6 per cent of the measurements will fall within the range 30.00 ± 0.45 and 13.4 per cent will lie outside this range. Half of these measurements, 6.8 per cent, will be less than 29.55; and a similar percentage will exceed 30.45. In actuality the uncertainty in z is about 1 in 15; therefore, the value of z could lie between 1.4 and 1.6; the corresponding areas under the curve could lie between 84 and 89 per cent.

Example 2. In the foregoing example, what expected number of samples will be 29.40 or less from a total of 400 samples?

$$z = \frac{29.40 - 30.00}{0.30} = 2.0$$

The area under the negative portion of the normal curve corresponds to $0.500 - 0.477 = 0.023$. Only $0.023(400) = 9.2$ or nine samples will be 29.40 or less.

Example 3. If the mean value of 500 determinations is 151 and $\sigma = 15$, how many results lie between 120 and 155 (actually any value between 119.5 and 155.5)?

$$z = \frac{119.5 - 151}{15} = -2.10 \qquad \text{Area: } 0.482$$

$$z = \frac{155.5 - 151}{15} = 0.30 \qquad \qquad \underline{0.118}$$

$$\text{Total area: } 0.600$$

$500(0.600) = 300$ results

Measures of Dispersion

Several ways may be used to characterize the spread or dispersion in the original data. The *range* is the difference between the largest value and the smallest value in a set of observations. However, almost always the most efficient quantity for characterizing variability is the *standard deviation* (also called the *root mean square*).

The standard deviation is the square root of the average squared difference between the individual observations and the population mean:

$$\sigma = \sqrt{\frac{\sum\limits_{i=1}^{N} (x_i - \mu)^2}{N}}$$

The standard deviation may be estimated by calculating s drawn from a sample set as follows:

$$s = \sqrt{\frac{\sum\limits_{i=1}^{N} (x_i - \overline{x})^2}{N - 1}} \qquad \text{or}$$

$$s = \sqrt{\frac{x_1^2 + x_2^2 + \cdots - [(x_1 + x_2 + \cdots)^2]/N}{N - 1}}$$

where $x_i - \bar{x}$ represents the deviation of each number in the array from the arithmetic mean. Since two pieces of information, namely s and $\bar{x}$, have been extracted from the data, we are left with $N - 1$ independent data available for measurement of precision. The divisor is termed the *degrees of freedom*. If a relatively large sample of data corresponding to $N > 30$ is available, its mean can be taken as a measure of μ, and $s \simeq \sigma$.

So basic is the notion of a statistical *estimate* of a physical parameter that statisticians use Greek letters for the *parameters* and English letters for the estimates.

For many purposes, one uses the *variance*, which for the sample is s^2 and for the entire population is σ^2. The variance s^2 of a finite sample is an unbiased estimate of σ^2, whereas the standard deviation s is not an unbiased estimate of σ.

When a series of observations can be logically arranged into k subgroups, the variance is calculated by summing the squares of the deviations for each subgroup, and then adding all the k sums and dividing by $N - k$ because one degree of freedom is lost in each subgroup. For two groups of observations consisting of N_A and N_B members, of standard deviations s_A and s_B, respectively, the variance is given by

$$s^2 = \frac{(N_A - 1)s_A^2 + (N_B - 1)s_B^2}{N_A + N_B - 2}$$

Another measure of dispersion is the *coefficient of variation*, which is merely the standard deviation expressed as a percentage of the arithmetic mean, viz.: $\dfrac{100s}{\bar{x}}$. It is useful mainly to show whether the relative or the absolute spread of values is constant as the values are changed.

Theoretical Distributions and Tests of Significance

If the data contained only random (or chance) errors, the cumulative estimates $\bar{x}$ and s would gradually approach the limits μ and σ. The distribution of results would be normally distributed with mean μ and standard deviation σ. Were the true mean of the infinite population known, it would be expected that the averaged means for each group of data would also have some symmetrical type of distribution centered around μ. However, it would be expected that the dispersion or spread of this dispersion about the mean would depend on the sample size. The standard deviation of the distribution of means equals $\sigma/\sqrt{N}$. A distribution of this type is called the "Student's distribution"; and the corresponding test of significance, a measure of error between μ and $\bar{x}$, the t test. The Student t takes into account both the possible variation of the value of $\bar{x}$ from μ on the basis of the expected variance

$\sigma^2/\sqrt{N}$ and the reliability of using s in place of σ. The distribution of the statistic is

$$\pm t = \frac{\overline{x} - \mu}{s/\sqrt{N}} \qquad \text{or} \qquad \mu = \overline{x} \pm \frac{ts}{\sqrt{N}}$$

This distribution is symmetrical about zero, and its dispersion is a function of the degrees of freedom $N - 1$. Its limits are called *confidence limits*. The percentage probability that μ lies within this interval is called the *confidence level*. The *level of significance* or *error probability* (100 − confidence level or 100 − α) is the per cent probability that μ will lie outside the confidence interval, and represents the chances of being incorrect in stating that μ lies within the confidence interval. Values of t are in Table 1-7 for any desired degrees of freedom and various confidence levels.

Statistical methods are frequently used to give a yes or no answer to a particular question concerning the significance of data. The answer is qualified by a confidence level indicating the degree of certainty of the answer. A common procedure is to set up a *null hypothesis*, which states that there is no significant difference between two sets of data or that a variable exerts no significant effect. Generally confidence levels of 95 and 99 per cent are chosen to express the probability that the answer is correct. These are also denoted as the 0.05 and 0.01 level of significance, respectively. When the hypothesis can be rejected at the 0.05 level of significance, but not at the 0.01 level, we can say that the sample results are probably significant. If, however, the hypothesis is also rejected at the 0.01 level, the results become highly significant. For a small number of samples we replace z, obtained from Table 1-6, by t from Table 1-7, and we replace σ by $[\sqrt{N/(N - 1)}]s$.

Example 4. In the past a method gave $\mu = 0.050$ per cent. A recent set of 10 results gave $\overline{x} = 0.053$ per cent and $s = 0.003$ per cent. Is everything satisfactory at a level of significance of 0.05? Of 0.01? We wish to decide between the hypotheses

H_0: $\mu = 0.050\%$ and the method is working properly; and
H_1: $\mu \neq 0.050\%$ and the method is not working properly.

A *two-tailed test* is required; that is, both tails on the distribution curve are involved:

$$t = \frac{0.053 - 0.050}{0.003}\sqrt{10 - 1} = -3.00$$

Enter Table 1-7 for 9 degrees of freedom under the column headed $t_{.975}$ for the 0.05 level of significance, and the column $t_{.995}$ for the 0.01 level of significance. At the 0.05 level, accept H_0 if t lies inside the interval $-t_{.975}$ to $t_{.975}$, that is, within -2.26 and 2.26; reject otherwise.

Since $t = -3.00$, we reject H_0. At the 0.01 level of significance, the corresponding interval is -3.25 to 3.25, which t lies within, indicating acceptance of H_0. Because we can reject H_0 at the 0.05 level but not at the 0.01 level of significance, we can say that the sample results are probably significant and that the method is not working properly.

Example 5. Six samples from a bulk chemical shipment averaged 77.50 per cent active ingredient with $s = 1.45$ per cent. The manufacturer claimed 80.00 per cent. Can his claim be supported? A one-tailed test is required.

$$t = \frac{77.50 - 80.00}{1.45} \sqrt{6 - 1} = -3.86$$

Since $t_{.95} = -2.01$, and $t_{.99} = -3.36$, the hypothesis is rejected at both the 0.05 and the 0.01 levels of significance. It is extremely unlikely that the claim is justified.

It is possible to compare the means of two relatively small sets of observations when the variances within the sets can be regarded as the same, as indicated by the F test. Also the t test can be applied to differences between pairs of observations. Perhaps only a single pair can be performed at one time, or possibly one wishes to compare two methods using samples of differing analytical content. It is still necessary that the two methods possess the same inherent standard deviation. An average difference $\bar{d}$ is calculated, and individual deviations from $\bar{d}$ are used to evaluate the variance of the differences.

Example 6. From the following data do the two methods actually give concordant results?

Sample	Method A	Method B	Difference
1	33.27	33.04	$d_1 = 0.23$
2	51.34	50.96	$d_2 = 0.38$
3	23.91	23.77	$d_3 = 0.14$
4	47.04	46.79	$d_4 = 0.25$
			$\bar{d} = 0.25$

$$s_d = \sqrt{\frac{\Sigma(d - \bar{d})^2}{N - 1}} = 0.099$$

$$t = \frac{0.25}{0.099} \sqrt{4 - 1} = 4.30$$

From Table 1-7, $t_{.975} = 3.18$ (at 95 per cent probability) and $t_{.995} = 5.84$ (at 99 per cent probability). The difference between the two methods is probably significant.

One can consider the distribution involving estimates of the true variance. With s_1^2 determined from a group of N_1 observations and s_2^2

from a second group of N_2 observations, the distribution of the ratio of the sample variances is given by the F test:

$$F = \frac{s_1^2}{s_2^2}$$

The larger variance is placed in the numerator. For example, the F test allows judgment regarding the existence of a significant difference in the precision between two sets of data or between two analysts. The hypothesis assumed is that both variances are indeed alike and a measure of the same σ.

Example 7. Suppose Analyst A made five observations and obtained a standard deviation of 0.06 whereas Analyst B with six observations obtained $s_B = 0.03$. The experimental variance ratio is

$$F = \frac{(0.06)^2}{(0.03)^2} = 4.00$$

From Table 1-9, with four degrees of freedom for A and five degrees of freedom for B, the value of F would exceed 5.19 five per cent of the time. Therefore, the null hypothesis is valid, and comparable skills are exhibited by the two analysts. As applied, the F test was one-tailed. The F test may also be applied as a two-tailed test in which the alternative to the null hypothesis is $\sigma_1^2 \neq \sigma_2^2$. This doubles the probability that the null hypothesis is invalid and has the effect of changing the confidence level, in the above example, from 95 to 90 per cent.

If improvement in precision is claimed for a set of measurements, the variance for the set against which comparison is being made should be placed in the numerator, regardless of magnitude. An experimental F smaller than unity indicates that the claim for improved precision cannot be supported.

As we have seen, for each group of samples a standard deviation can be calculated. These estimates of σ possess a distribution called the chi-square (χ^2) distribution:

$$\chi^2 = \frac{s^2}{\sigma^2/df}$$

The upper and lower confidence limits for the standard deviation are obtained by dividing $(N - 1)s^2$ by two numbers taken from Table 1-8.

Example 8. The variance obtained for 10 samples is $(0.65)^2$. How reliable is s^2 as an estimate of σ^2? σ^2 is known to be $(0.75)^2$.

$$\frac{s^2(N-1)}{\chi^2_{.975}} < \sigma^2 < \frac{s^2(N-1)}{\chi^2_{.025}}$$

$$\frac{(0.65)^2(10-1)}{19.02} < \sigma^2 < \frac{(0.65)^2(10-1)}{2.70}$$

$$0.20 < \sigma^2 < 1.43$$

Thus, only one time in 40 will $9s^2/\sigma^2$ be less than 2.70 by chance alone. Similarly, only one time in 40 will $9s^2/\sigma^2$ be greater than 19.02. Consequently, it is not unlikely that s^2 is a reliable estimate of σ^2. Stated differently:

Upper limit: $\quad \sigma^2 = \dfrac{9s^2}{2.7} = 3.3s^2$

Lower limit: $\quad \sigma^2 = \dfrac{9s^2}{19.02} = 0.48s^2$

Ten measurements give an estimate of σ^2 that may be as much as 3.3 times or only about one-half the true variance.

Table 1-6
ORDINATES AND AREAS BETWEEN ABSCISSA VALUES −z and +z OF THE NORMAL DISTRIBUTION CURVE

From Perry, Chilton, and Kirkpatrick, "Chemical Engineers' Handbook," 4th ed., McGraw-Hill Book Company, New York (1963); by permission.

z	X	Y	A	1 − A	z	X	Y	A	1 − A
0	μ	0.399	0.0000	1.0000	±1.50	μ ± 1.50σ	0.1295	0.8664	0.1336
±.05	μ ± 0.05σ	.398	.0399	0.9601	±1.55	μ ± 1.55σ	.1200	.8789	.1211
±.10	μ ± .10σ	.397	.0797	.9203	±1.60	μ ± 1.60σ	.1109	.8904	.1096
±.15	μ ± .15σ	.394	.1192	.8808	±1.65	μ ± 1.65σ	.1023	.9011	.0989
±.20	μ ± .20σ	.391	.1585	.8415	±1.70	μ ± 1.70σ	.0940	.9109	.0891
±.25	μ ± .25σ	.387	.1974	.8026	±1.75	μ ± 1.75σ	.0863	.9199	.0801
±.30	μ ± .30σ	.381	.2358	.7642	±1.80	μ ± 1.80σ	.0790	.9281	.0719
±.35	μ ± .35σ	.375	.2737	.7263	±1.85	μ ± 1.85σ	.0721	.9357	.0643
±.40	μ ± .40σ	.368	.3108	.6892	±1.90	μ ± 1.90σ	.0656	.9426	.0574
±.45	μ ± .45σ	.361	.3473	.6527	±1.95	μ ± 1.95σ	.0596	.9488	.0512
±.50	μ ± .50σ	.352	.3829	.6171	±2.00	μ ± 2.00σ	.0540	.9545	.0455
±.55	μ ± .55σ	.343	.4177	.5823	±2.05	μ ± 2.05σ	.0488	.9596	.0404
±.60	μ ± .60σ	.333	.4515	.5485	±2.10	μ ± 2.10σ	.0440	.9643	.0357
±.65	μ ± .65σ	.323	.4843	.5157	±2.15	μ ± 2.15σ	.0396	.9684	.0316
±.70	μ ± .70σ	.312	.5161	.4839	±2.20	μ ± 2.20σ	.0355	.9722	.0278
±.75	μ ± .75σ	.301	.5467	.4533	±2.25	μ ± 2.25σ	.0317	.9756	.0244
±.80	μ ± .80σ	.290	.5763	.4237	±2.30	μ ± 2.30σ	.0283	.9786	.0214
±.85	μ ± .85σ	.278	.6047	.3953	±2.35	μ ± 2.35σ	.0252	.9812	.0188
±.90	μ ± .90σ	.266	.6319	.3681	±2.40	μ ± 2.40σ	.0224	.9836	.0164
±.95	μ ± .95σ	.254	.6579	.3421	±2.45	μ ± 2.45σ	.0198	.9857	.0143
±1.00	μ ± 1.00σ	.242	.6827	.3173	±2.50	μ ± 2.50σ	.0175	.9876	.0124
±1.05	μ ± 1.05σ	.230	.7063	.2937	±2.55	μ ± 2.55σ	.0154	.9892	.0108
±1.10	μ ± 1.10σ	.218	.7287	.2713	±2.60	μ ± 2.60σ	.0136	.9907	.0093
±1.15	μ ± 1.15σ	.206	.7499	.2501	±2.65	μ ± 2.65σ	.0119	.9920	.0080
±1.20	μ ± 1.20σ	.194	.7699	.2301	±2.70	μ ± 2.70σ	.0104	.9931	.0069
±1.25	μ ± 1.25σ	.183	.7887	.2113	±2.75	μ ± 2.75σ	.0091	.9940	.0060
±1.30	μ ± 1.30σ	.171	.8064	.1936	±2.80	μ ± 2.80σ	.0079	.9949	.0051
±1.35	μ ± 1.35σ	.160	.8230	.1770	±2.85	μ ± 2.85σ	.0069	.9956	.0044
±1.40	μ ± 1.40σ	.150	.8385	.1615	±2.90	μ ± 2.90σ	.0060	.9963	.0037
±1.45	μ ± 1.45σ	.139	.8529	.1471	±2.95	μ ± 2.95σ	.0051	.9968	.0032
±1.50	μ ± 1.50σ	.130	.8664	.1336	±3.00	μ ± 3.00σ	.0044	.9973	.0027
					±4.00	μ ± 4.00σ	.0001	.99994	.00006
					±5.00	μ ± 5.00σ	.000001	.999994	.000006
±0.000	μ	0.3989	0.0000	1.0000	±1.036	μ ± 1.036σ	0.2331	0.7000	0.3000
±.126	μ ± 0.126σ	.3958	.1000	.9000	±1.282	μ ± 1.282σ	.1755	.8000	.2000
±.253	μ ± .253σ	.3863	.2000	.8000	±1.645	μ ± 1.645σ	.1031	.9000	.1000
±.385	μ ± .385σ	.3704	.3000	.7000	±1.960	μ ± 1.960σ	.0584	.9500	.0500
±.524	μ ± .524σ	.3477	.4000	.6000	±2.576	μ ± 2.576σ	.0145	.9900	.0100
±.674	μ ± .674σ	.3178	.5000	.5000	±3.291	μ ± 3.291σ	.0018	.9990	.0010
±.842	μ ± .842σ	.2890	.6000	.4000	±3.891	μ ± 3.891σ	.0002	.9999	.0001

Table 1-7
VALUES OF t

From Perry, Chilton, and Kirkpatrick, "Chemical Engineers' Handbook," 4th ed., McGraw-Hill Book Company, New York (1963); by permission.

df	$t_{.60}$	$t_{.70}$	$t_{.80}$	$t_{.90}$	$t_{.95}$	$t_{.975}$	$t_{.99}$	$t_{.995}$
1	0.325	0.727	1.376	3.078	6.314	12.706	31.821	63.657
2	.289	.617	1.061	1.886	2.920	4.303	6.965	9.925
3	.277	584	0.978	1.638	2.353	3.182	4.541	5.841
4	.271	.569	.941	1.533	2.132	2.776	3.747	4.604
5	.267	.559	.920	1.476	2.015	2.571	3.365	4.032
6	.265	.553	.906	1.440	1.943	2.447	3.143	3.707
7	.263	.549	.896	1.415	1.895	2.365	2.998	3.499
8	.262	.546	.889	1.397	1.860	2.306	2.896	3.355
9	.261	.543	.883	1.383	1.833	2.262	2.821	3.250
10	.260	.542	.879	1.372	1.812	2.228	2.764	3.169
11	.260	.540	.876	1.363	1.796	2.201	2.718	3.106
12	.259	.539	.873	1.356	1.782	2.179	2.681	3.055
13	.259	.538	.870	1.350	1.771	2.160	2.650	3.012
14	.258	.537	.868	1.345	1.761	2.145	2.624	2.977
15	.258	.536	.866	1.341	1.753	2.131	2.602	2.947
16	.258	.535	.865	1.337	1.746	2.120	2.583	2.921
17	.257	.534	.863	1.333	1.740	2.110	2.567	2.898
18	.257	.534	.862	1.330	1.734	2.101	2.552	2.878
19	.257	.533	.861	1.328	1.729	2.093	2.539	2.861
20	.257	.533	.860	1.325	1.725	2.086	2.528	2.845
21	.257	.532	.859	1.323	1.721	2.080	2.518	2.831
22	.256	.532	.858	1.321	1.717	2.074	2.508	2.819
23	.256	.532	.858	1.319	1.714	2.069	2.500	2.807
24	.256	.531	.857	1.318	1.711	2.064	2.492	2.797
25	.256	.531	.856	1.316	1.708	2.060	2.485	2.787
26	.256	.531	.856	1.315	1.706	2.056	2.479	2.779
27	.256	.531	.855	1.314	1.703	2.052	2.473	2.771
28	.256	.530	.855	1.313	1.701	2.048	2.467	2.763
29	.256	.530	.854	1.311	1.699	2.045	2.462	2.756
30	.256	.530	.854	1.310	1.697	2.042	2.457	2.750
40	.255	.529	.851	1.303	1.684	2.021	2.423	2.704
60	.254	.527	.848	1.296	1.671	2.000	2.390	2.660
120	.254	.526	.845	1.289	1.658	1.980	2.358	2.617
∞	.253	.524	.842	1.282	1.645	1.960	2.326	2.576
df	$-t_{.40}$	$-t_{.30}$	$-t_{.20}$	$-t_{.10}$	$-t_{.05}$	$-t_{.025}$	$-t_{.01}$	$-t_{.005}$

When the table is read from the foot, the tabled values are to be prefixed with a negative sign. Interpolation should be performed using the reciprocals of the degrees of freedom.

Table 1-8
PERCENTILES OF THE χ^2 DISTRIBUTION

From Perry, Chilton, and Kirkpatrick, "Chemical Engineers' Handbook," 4th ed., McGraw-Hill Book Company, New York (1963); by permission.

df	0.5	1	2.5	5	10	90	95	97.5	99	99.5
					Per cent					
1	0.000039	0.00016	0.00098	0.0039	0.0158	2.71	3.84	5.02	6.63	7.88
2	.0100	.0201	.0506	.1026	.2107	4.61	5.99	7.38	9.21	10.60
3	.0717	.115	.216	.352	.584	6.25	7.81	9.35	11.34	12.84
4	.207	.297	.484	.711	1.064	7.78	9.49	11.14	13.28	14.86
5	.412	.554	.831	1.15	1.61	9.24	11.07	12.83	15.09	16.75
6	.676	.872	1.24	1.64	2.20	10.64	12.59	14.45	16.81	18.55
7	.989	1.24	1.69	2.17	2.83	12.02	14.07	16.01	18.48	20.28
8	1.34	1.65	2.18	2.73	3.49	13.36	15.51	17.53	20.09	21.96
9	1.73	2.09	2.70	3.33	4.17	14.68	16.92	19.02	21.67	23.59
10	2.16	2.56	3.25	3.94	4.87	15.99	18.31	20.48	23.21	25.19
11	2.60	3.05	3.82	4.57	5.58	17.28	19.68	21.92	24.73	26.76
12	3.07	3.57	4.40	5.23	6.30	18.55	21.03	23.34	26.22	28.30
13	3.57	4.11	5.01	5.89	7.04	19.81	22.36	24.74	27.69	29.82
14	4.07	4.66	5.63	6.57	7.79	21.06	23.68	26.12	29.14	31.32
15	4.60	5.23	6.26	7.26	8.55	22.31	25.00	27.49	30.58	32.80
16	5.14	5.81	6.91	7.96	9.31	23.54	26.30	28.85	32.00	34.27
18	6.26	7.01	8.23	9.39	10.86	25.99	28.87	31.53	34.81	37.16
20	7.43	8.26	9.59	10.85	12.44	28.41	31.41	34.17	37.57	40.00
24	9.89	10.86	12.40	13.85	15.66	33.20	36.42	39.36	42.98	45.56
30	13.79	14.95	16.79	18.49	20.60	40.26	43.77	46.98	50.89	53.67
40	20.71	22.16	24.43	26.51	29.05	51.81	55.76	59.34	63.69	66.77
60	35.53	37.48	40.48	43.19	46.46	74.40	79.08	83.30	88.38	91.95
120	83.85	86.92	91.58	95.70	100.62	140.23	146.57	152.21	158.95	163.64

For large values of degrees of freedom the approximate formula

$$\chi_\alpha^2 = n\left(1 - \frac{2}{9n} + z_\alpha\sqrt{\frac{2}{9n}}\right)^3$$

where z_α is the normal deviate and n is the number of degrees of freedom, may be used. For example, $\chi_{.99}^2 = 60[1 - 0.00370 + 2.326(0.06086)]^3 = 60(1.1379)^3 = 88.4$ for the 99th percentile for 60 degrees of freedom.

Table 1-9
F DISTRIBUTION

From Perry, Chilton, and Kirkpatrick, "Chemical Engineers' Handbook," 4th ed., McGraw-Hill Book Company, New York (1963); by permission.

Upper 5% Points ($F_{.95}$)

Denom.	\\	1	2	3	4	5	6	7	8	9	10	12	15	20	24	30	40	60	120	∞
							Degrees of freedom for numerator													
1		161	200	216	225	230	234	237	239	241	242	244	246	248	249	250	251	252	253	254
2		18.5	19.0	19.2	19.2	19.3	19.3	19.4	19.4	19.4	19.4	19.4	19.4	19.4	19.5	19.5	19.5	19.5	19.5	19.5
3		10.1	9.55	9.28	9.12	9.01	8.94	8.89	8.85	8.81	8.79	8.74	8.70	8.66	8.64	8.62	8.59	8.57	8.55	8.53
4		7.71	6.94	6.59	6.39	6.26	6.16	6.09	6.04	6.00	5.96	5.91	5.86	5.80	5.77	5.75	5.72	5.69	5.66	5.63
5		6.61	5.79	5.41	5.19	5.05	4.95	4.88	4.82	4.77	4.74	4.68	4.62	4.56	4.53	4.50	4.46	4.43	4.40	4.37
6		5.99	5.14	4.76	4.53	4.39	4.28	4.21	4.15	4.10	4.06	4.00	3.94	3.87	3.84	3.81	3.77	3.74	3.70	3.67
7		5.59	4.74	4.35	4.12	3.97	3.87	3.79	3.73	3.68	3.64	3.57	3.51	3.44	3.41	3.38	3.34	3.30	3.27	3.23
8		5.32	4.46	4.07	3.84	3.69	3.58	3.50	3.44	3.39	3.35	3.28	3.22	3.15	3.12	3.08	3.04	3.01	2.97	2.93
9		5.12	4.26	3.86	3.63	3.48	3.37	3.29	3.23	3.18	3.14	3.07	3.01	2.94	2.90	2.86	2.83	2.79	2.75	2.71
10		4.96	4.10	3.71	3.48	3.33	3.22	3.14	3.07	3.02	2.98	2.91	2.85	2.77	2.74	2.70	2.66	2.62	2.58	2.54
11		4.84	3.98	3.59	3.36	3.20	3.09	3.01	2.95	2.90	2.85	2.79	2.72	2.65	2.61	2.57	2.53	2.49	2.45	2.40
12		4.75	3.89	3.49	3.26	3.11	3.00	2.91	2.85	2.80	2.75	2.69	2.62	2.54	2.51	2.47	2.43	2.38	2.34	2.30
13		4.67	3.81	3.41	3.18	3.03	2.92	2.83	2.77	2.71	2.67	2.60	2.53	2.46	2.42	2.38	2.34	2.30	2.25	2.21
14		4.60	3.74	3.34	3.11	2.96	2.85	2.76	2.70	2.65	2.60	2.53	2.46	2.39	2.35	2.31	2.27	2.22	2.18	2.13
15		4.54	3.68	3.29	3.06	2.90	2.79	2.71	2.64	2.59	2.54	2.48	2.40	2.33	2.29	2.25	2.20	2.16	2.11	2.07
16		4.49	3.63	3.24	3.01	2.85	2.74	2.66	2.59	2.54	2.49	2.42	2.35	2.28	2.24	2.19	2.15	2.11	2.06	2.01
17		4.45	3.59	3.20	2.96	2.81	2.70	2.61	2.55	2.49	2.45	2.38	2.31	2.23	2.19	2.15	2.10	2.06	2.01	1.96
18		4.41	3.55	3.16	2.93	2.77	2.66	2.58	2.51	2.46	2.41	2.34	2.27	2.19	2.15	2.11	2.06	2.02	1.97	1.92
19		4.38	3.52	3.13	2.90	2.74	2.63	2.54	2.48	2.42	2.38	2.31	2.23	2.16	2.11	2.07	2.03	1.98	1.93	1.88
20		4.35	3.49	3.10	2.87	2.71	2.60	2.51	2.45	2.39	2.35	2.28	2.20	2.12	2.08	2.04	1.99	1.95	1.90	1.84
21		4.32	3.47	3.07	2.84	2.68	2.57	2.49	2.42	2.37	2.32	2.25	2.18	2.10	2.05	2.01	1.96	1.92	1.87	1.81
22		4.30	3.44	3.05	2.82	2.66	2.55	2.46	2.40	2.34	2.30	2.23	2.15	2.07	2.03	1.98	1.94	1.89	1.84	1.78
23		4.28	3.42	3.03	2.80	2.64	2.53	2.44	2.37	2.32	2.27	2.20	2.13	2.05	2.01	1.96	1.91	1.86	1.81	1.76
24		4.26	3.40	3.01	2.78	2.62	2.51	2.42	2.36	2.30	2.25	2.18	2.11	2.03	1.98	1.94	1.89	1.84	1.79	1.73
25		4.24	3.39	2.99	2.76	2.60	2.49	2.40	2.34	2.28	2.24	2.16	2.09	2.01	1.96	1.92	1.87	1.82	1.77	1.71
30		4.17	3.32	2.92	2.69	2.53	2.42	2.33	2.27	2.21	2.16	2.09	2.01	1.93	1.89	1.84	1.79	1.74	1.68	1.62
40		4.08	3.23	2.84	2.61	2.45	2.34	2.25	2.18	2.12	2.08	2.00	1.92	1.84	1.79	1.74	1.69	1.64	1.58	1.51
60		4.00	3.15	2.76	2.53	2.37	2.25	2.17	2.10	2.04	1.99	1.92	1.84	1.75	1.70	1.65	1.59	1.53	1.47	1.39
120		3.92	3.07	2.68	2.45	2.29	2.18	2.09	2.02	1.96	1.91	1.83	1.75	1.66	1.61	1.55	1.50	1.43	1.35	1.25
∞		3.84	3.00	2.60	2.37	2.21	2.10	2.01	1.94	1.88	1.83	1.75	1.67	1.57	1.52	1.46	1.39	1.32	1.22	1.00

Degrees of freedom for denominator

Table 1-9 (Continued)
F DISTRIBUTION
Upper 1% Points ($F_{.99}$)

Degrees of freedom for denominator	\	1	2	3	4	5	6	7	8	9	10	12	15	20	24	30	40	60	120	∞
											Degrees of freedom for numerator									
	1	4052	5000	5403	5625	5764	5859	5928	5982	6023	6056	6106	6157	6209	6235	6261	6287	6313	6339	6366
	2	98.5	99.0	99.2	99.2	99.3	99.3	99.4	99.4	99.4	99.4	99.4	99.4	99.4	99.5	99.5	99.5	99.5	99.5	99.5
	3	34.1	30.8	29.5	28.7	28.2	27.9	27.7	27.5	27.3	27.2	27.1	26.9	26.7	26.6	26.5	26.4	26.3	26.2	26.1
	4	21.2	18.0	16.7	16.0	15.5	15.2	15.0	14.8	14.7	14.5	14.4	14.2	14.0	13.9	13.8	13.7	13.7	13.6	13.5
	5	16.3	13.3	12.1	11.4	11.0	10.7	10.5	10.3	10.2	10.1	9.89	9.72	9.55	9.47	9.38	9.29	9.20	9.11	9.02
	6	13.7	10.9	9.78	9.15	8.75	8.47	8.26	8.10	7.98	7.87	7.72	7.56	7.40	7.31	7.23	7.14	7.06	6.97	6.88
	7	12.2	9.55	8.45	7.85	7.46	7.19	6.99	6.84	6.72	6.62	6.47	6.31	6.16	6.07	5.99	5.91	5.82	5.74	5.65
	8	11.3	8.65	7.59	7.01	6.63	6.37	6.18	6.03	5.91	5.81	5.67	5.52	5.36	5.28	5.20	5.12	5.03	4.95	4.86
	9	10.6	8.02	6.99	6.42	6.06	5.80	5.61	5.47	5.35	5.26	5.11	4.96	4.81	4.73	4.65	4.57	4.48	4.40	4.31
	10	10.0	7.56	6.55	5.99	5.64	5.39	5.20	5.06	4.94	4.85	4.71	4.56	4.41	4.33	4.25	4.17	4.08	4.00	3.91
	11	9.65	7.21	6.22	5.67	5.32	5.07	4.89	4.74	4.63	4.54	4.40	4.25	4.10	4.02	3.94	3.86	3.78	3.69	3.60
	12	9.33	6.93	5.95	5.41	5.06	4.82	4.64	4.50	4.39	4.30	4.16	4.01	3.86	3.78	3.70	3.62	3.54	3.45	3.36
	13	9.07	6.70	5.74	5.21	4.86	4.62	4.44	4.30	4.19	4.10	3.96	3.82	3.66	3.59	3.51	3.43	3.34	3.25	3.17
	14	8.86	6.51	5.56	5.04	4.70	4.46	4.28	4.14	4.03	3.94	3.80	3.66	3.51	3.43	3.35	3.27	3.18	3.09	3.00
	15	8.68	6.36	5.42	4.89	4.56	4.32	4.14	4.00	3.89	3.80	3.67	3.52	3.37	3.29	3.21	3.13	3.05	2.96	2.87
	16	8.53	6.23	5.29	4.77	4.44	4.20	4.03	3.89	3.78	3.69	3.55	3.41	3.26	3.18	3.10	3.02	2.93	2.84	2.75
	17	8.40	6.11	5.18	4.67	4.34	4.10	3.93	3.79	3.68	3.59	3.46	3.31	3.16	3.08	3.00	2.92	2.83	2.75	2.65
	18	8.29	6.01	5.09	4.58	4.25	4.01	3.84	3.71	3.60	3.51	3.37	3.23	3.08	3.00	2.92	2.84	2.75	2.66	2.57
	19	8.19	5.93	5.01	4.50	4.17	3.94	3.77	3.63	3.52	3.43	3.30	3.15	3.00	2.92	2.84	2.76	2.67	2.58	2.49
	20	8.10	5.85	4.94	4.43	4.10	3.87	3.70	3.56	3.46	3.37	3.23	3.09	2.94	2.86	2.78	2.69	2.61	2.52	2.42
	21	8.02	5.78	4.87	4.37	4.04	3.81	3.64	3.51	3.40	3.31	3.17	3.03	2.88	2.80	2.72	2.64	2.55	2.46	2.36
	22	7.95	5.72	4.82	4.31	3.99	3.76	3.59	3.45	3.35	3.26	3.12	2.98	2.83	2.75	2.67	2.58	2.50	2.40	2.31
	23	7.88	5.66	4.76	4.26	3.94	3.71	3.54	3.41	3.30	3.21	3.07	2.93	2.78	2.70	2.62	2.54	2.45	2.35	2.26
	24	7.82	5.61	4.72	4.22	3.90	3.67	3.50	3.36	3.26	3.17	3.03	2.89	2.74	2.66	2.58	2.49	2.40	2.31	2.21
	25	7.77	5.57	4.68	4.18	3.86	3.63	3.46	3.32	3.22	3.13	2.99	2.85	2.70	2.62	2.53	2.45	2.36	2.27	2.17
	30	7.56	5.39	4.51	4.02	3.70	3.47	3.30	3.17	3.07	2.98	2.84	2.70	2.55	2.47	2.39	2.30	2.21	2.11	2.01
	40	7.31	5.18	4.31	3.83	3.51	3.29	3.12	2.99	2.89	2.80	2.66	2.52	2.37	2.29	2.20	2.11	2.02	1.92	1.80
	60	7.08	4.98	4.13	3.65	3.34	3.12	2.95	2.82	2.72	2.63	2.50	2.35	2.20	2.12	2.03	1.94	1.84	1.73	1.60
	120	6.85	4.79	3.95	3.48	3.17	2.96	2.79	2.66	2.56	2.47	2.34	2.19	2.03	1.95	1.86	1.76	1.66	1.53	1.38
	∞	6.63	4.61	3.78	3.32	3.02	2.80	2.64	2.51	2.41	2.32	2.18	2.04	1.88	1.79	1.70	1.59	1.47	1.32	1.00

Interpolation should be performed using reciprocals of the degrees of freedom.

Section 2

GENERAL INFORMATION
AND CONVERSION TABLES

Table 2-1
FUNDAMENTAL PHYSICAL CONSTANTS

E. R. Cohen and B. N. Taylor, *J. Phys. Chem. Reference Data*, **2** (No. 4), 663 (1973)

A. Defined Values:

---

1. SI Base Units

Meter (metre)*	m = 1 650 763.73 wavelengths in vacuum of the orange-red line of the spectrum of krypton-86
Kilogram	kg = mass of a cylinder of platinum-iridium alloy kept at Paris
Second	s = duration of 9 192 631 770 cycles of the radiation associated with a specified transition of the cesium atom
Ampere	A = magnitude of the current that, when flowing through each of two long parallel wires separated by one meter in free space, results in a force between the two wires 2×10^{-7} newton for each meter of length
Kelvin (degree Kelvin)	K = defined in the thermodynamic scale by assigning 273.16°K to the triple point of water (freezing point, 273.15°K = 0°C)
Candela	cd = luminous intensity of 1/600 000 of a square meter of a radiating cavity at the temperature of freezing platinum (2042°K)
Mole	mol = amount of substance which contains as many specified entities (molecules, atoms, ions, electrons, photons, etc.) as there are atoms of carbon-12 in exactly 0.012 kg of that nuclide

2. Supplementary SI Units

Radian	rad = the plane angle between two radii of a circle which cut off on the circumference an arc equal in length to the radius
Steradian	sr = the solid angle which, having its vertex in the center of a sphere, cuts off an area of the surface of the sphere equal to that of a square with sides of length equal to the radius of the sphere

---

* In the United States the preferred spelling is "meter."

Table 2-1 (*Continued*)
FUNDAMENTAL PHYSICAL CONSTANTS
B. Derived SI Units

Quantity, Symbol	Name of SI Unit	Symbol & Definition
Capacitance (electric), C	farad	$F = C \cdot V^{-1}$
Charge (electric), quantity of electricity, Q	coulomb	$C = A \cdot s$
Conductance (electric), $G(= 1/R)$	siemens	$S = \Omega^{-1}$
Energy, work, quantity of heat, H	joule	$J = kg \cdot m^2 \cdot s^{-2}$
Force	newton	$N = kg \cdot m \cdot s^{-2}$
Frequency	hertz	$Hz = s^{-1}$
Illuminance, illumination	lux	$lx = lm \cdot m^{-2}$
Inductance, L	henry	$H = \Omega \cdot s$
Luminous flux	lumen	$lm = cd \cdot sr$
Magnetic flux	weber	$Wb = V \cdot s$
Magnetic flux density	tesla	$T = Wb \cdot m^{-2}$
Potential difference, E	volt	$V = kg \cdot m^2 \cdot s^{-3} \cdot A^{-1} = J \cdot A^{-1} \cdot s^{-1}$
Power, radiant flux	watt	$W = kg \cdot m^2 \cdot s^{-3} = J \cdot s^{-1}$
Pressure, stress	pascal	$Pa = N \cdot m^{-2} = kg \cdot m^{-1} \cdot s^{-2}$
Resistance (electric), R	ohm	$\Omega = V \cdot A^{-1} = kg \cdot m^2 \cdot s^{-3} \cdot A^{-2}$

C. Recommended Consistent Values of Constants

The digits in parentheses following a numerical value represent the standard deviation of that value in terms of the final listed digits.

Anomalous electron moment correction	$(\mu_e/\mu_0) - 1 = 1.159\ 615(15) \times 10^{-3}$
Atomic mass unit	$u = (10^{-3}\ kg \cdot mol^{-1})/N_A = 1.660\ 566(9) \times 10^{-27}\ kg$
Avogadro constant	$N_A = 6.022\ 045(31) \times 10^{23}\ mol^{-1}$
Bohr magneton	$\mu_B = e\hbar/2m_ec = 9.274\ 078(36) \times 10^{-24}\ J \cdot T^{-1}$
Bohr radius	$a_0 = \alpha/4\pi R_\infty = 0.529\ 177\ 06(44) \times 10^{-10}\ m$
Boltzmann constant	$k = R/N_A = 1.380\ 662(44) \times 10^{-23}\ J \cdot K^{-1}$
Charge-to-mass ratio for electron	$e/m_e = 1.758\ 805(5) \times 10^{11}\ C \cdot kg^{-1}$
Compton wavelength of electron	$\lambda_c = \alpha^2/2R_\infty = 2.426\ 309(4) \times 10^{-12}\ m$
	$\lambdabar_c = \lambda_c/2\pi = \alpha a_0 = 3.861\ 591(6) \times 10^{-13}\ m$
Compton wavelength of neutron	$\lambda_{c,n} = h/m_nc = 1.319\ 591(2) \times 10^{-15}\ m$
Compton wavelength of proton	$\lambda_{c,p} = h/m_pc = 1.321\ 410(2) \times 10^{-15}\ m$
Diamagnetic shielding factor, spherical H_2O molecule	$1 + \sigma(H_2O) = 1.000\ 025\ 64(7)$
Electron g-factor	$g_e/2 = \mu_e/\mu_B = 1.001\ 159\ 657(4)$
Electron magnetic moment	$\mu_e = 9.284\ 832(36) \times 10^{-24}\ J \cdot T^{-1}$
Electron radius (classical)	$\alpha\lambdabar_c = \mu_0e^2/4\pi m_e = r_e = 2.817\ 938(7) \times 10^{-15}\ m$
Electron rest mass	$m_e = 0.910\ 953(5) \times 10^{-30}\ kg$
	$= 5.485\ 803(2) \times 10^{-4}\ u$
Elementary charge	$e = 1.602\ 189(5) \times 10^{-19}\ C$
Faraday constant	$N_Ae = F = 9.648\ 456(27) \times 10^4\ C \cdot mol^{-1}$
Fine structure constant ($\mu_0ce^2/2h$)	$\alpha = 0.007\ 297\ 351(6)$
	$1/\alpha = 1.370\ 360(1)$
First radiation constant	$2\pi hc^2 = c_1 = 3.741\ 83(2) \times 10^{-16}\ W \cdot m^2$
Gas constant (molar)	$R = P_0V_m/T_0 = 8.314\ 41(26)\ J \cdot mol^{-1} \cdot K^{-1}$
	$= 82.0568(26)\ cm^3 \cdot atm \cdot mol^{-1} \cdot K^{-1}$
	$= 1.987\ 19(6)\ cal \cdot mol^{-1} \cdot K^{-1}$
Gravitational constant	$G = 6.672(4) \times 10^{-11}\ N \cdot m^2 \cdot kg^{-2}$
Gyromagnetic ratio of proton	$\gamma_p = 2.675\ 199(8) \times 10^8\ s^{-1} \cdot T^{-1}$
(uncorrected for diamagnetism of H_2O)	$\gamma'_p = 2.675\ 130(8) \times 10^8\ s^{-1} \cdot T^{-1}$
Josephson frequency-voltage ratio	$2e/h = 4.835\ 939(13) \times 10^{14}\ Hz \cdot V^{-1}$

Table 2-1 (*Continued*)
FUNDAMENTAL PHYSICAL CONSTANTS

Magnetic flux quantum	$\Phi_0 = h/2e = 2.067\ 851(5) \times 10^{-15}$ Wb
Molar standard volume, ideal gas	$V_m = RT_0/P_0 = 0.022\ 413\ 8(7)$ m$^3 \cdot$ mol^{-1}
Muon g-factor	$e\hbar/2m_\mu c = g_\mu/2 = 1.001\ 166\ 16(31)$
Muon magnetic moment	$\mu_\mu = 4.490\ 474(18) \times 10^{-26}$ J $\cdot$ T^{-1}
Muon rest mass	$m_\mu = 1.883\ 566(11) \times 10^{-28}$ kg
Neutron rest mass	$m_n = 1.674\ 954(9) \times 10^{-27}$ kg
Normal volume, perfect gas	$V_0 = 2.241\ 36(30) \times 10^4$ cm$^3 \cdot$ mol^{-1}
Nuclear magneton	$\mu_N = e\hbar/2m_p c = 5.050\ 824(20) \times 10^{-27}$ J $\cdot$ T^{-1}
Permeability of vacuum	$\mu_0 = 4\pi \times 10^{-7}$ H $\cdot$ m^{-1}
Permittivity of vacuum	$\varepsilon_0 = (\mu_0 c^2)^{-1} = 8.854\ 187\ 82(7) \times 10^{-12}$ F $\cdot$ m^{-1}
Planck constant	$h = 6.626\ 176(36) \times 10^{-34}$ J $\cdot$ s
	$\hbar = h/2\pi = 1.054\ 589(6) \times 10^{-34}$ J $\cdot$ s
Proton magnetic moment:	$\mu_p = 1.410\ 617(5) \times 10^{-26}$ J $\cdot$ T^{-1}
in Bohr magnetons	$\mu_p/\mu_B = 1.521\ 032\ 209(16) \times 10^{-3}$
in nuclear magnetons	$\mu_p/\mu_N = 2.792\ 845\ 6(11)$
Proton rest mass	$m_p = 1.672\ 649(9) \times 10^{-27}$ kg
Quantum-charge ratio	$h/e = 4.135\ 701(11) \times 10^{-15}$ J $\cdot$ Hz$^{-1} \cdot$ C^{-1}
Quantum of circulation	$h/m_e = 7.273\ 89(1) \times 10^{-4}$ J $\cdot$ s $\cdot$ kg^{-1}
Ratio, electron to proton magnetic moments	$\mu_e/\mu_p = 6.582\ 106\ 88(7) \times 10^2$
Ratio, kxu (Siegbahn) to Angstrom	$= 1.000\ 020\ 5(56)$
Ratio, muon moment to proton moment	$\mu_\mu/\mu_p = 3.183\ 340(7)$
Rydberg constant	$R_\infty = 1.097\ 373\ 18(8) \times 10^7$ m^{-1}
Second radiation constant	$c_2 = hc/k = 1.438\ 786(45) \times 10^{-2}$ m $\cdot$ K
Speed of light in vacuum	$c = 2.997\ 924\ 58(12) \times 10^8$ m $\cdot$ s^{-1}
Stefan-Boltzmann constant	$\sigma = (\pi^2/60)k^4/\hbar^3 c^2 = 5.670\ 3(7) \times 10^{-8}$ W $\cdot$ m$^{-2} \cdot$ K^{-4}
Thomson cross section	$\sigma_e = 8\pi r_e^2/3 = 6.652\ 448(33) \times 10^{-28}$ m^2
Voltage-wavelength product	$V\lambda = 1.239\ 852(3) \times 10^{-6}$ eV $\cdot$ m
Wien displacement constant	$b = 0.289\ 78(4)$ cm $\cdot$ K
Zeeman splitting constant	$\mu_B/hc = 4.668\ 58(4) \times 10^{-5}$ cm$^{-1} \cdot$ G^{-1}
Energy equivalents:	
1 u	$= 931.501\ 6(26)$ MeV
1 proton mass (m_p)	$= 938.279\ 6(27)$ MeV
1 neutron mass (m_n)	$= 939.573\ 1(27)$ MeV
1 muon mass (m_μ)	$= 105.659\ 48(35)$ MeV
1 electron mass (m_e)	$= 0.511\ 003\ 4(14)$ MeV
1 electronvolt	1 eV$/k = 1.160\ 450(36) \times 10^4$ K
	1 eV$/hc = 8.065\ 479(21) \times 10^3$ cm^{-1}
	1 eV$/h = 2.417\ 970(6) \times 10^{14}$ Hz

Table 2-2
PHYSICAL AND CHEMICAL SYMBOLS AND TERMINOLOGY

Symbols separated by commas represent equivalent recommendations. Symbols preceded by three dots are alternatives to be used only when there is some reason for not using a symbol before the three dots. Symbols for physical and chemical quantities should be printed in *italic* type (and therefore must be underlined in typewritten material). Subscripts and superscripts which are themselves symbols for physical quantities should be italicized; all others should be in Roman type. References: Manual of Physico-chemical Symbols and Terminology (International Union of Pure and Applied Chemistry, M. L. McGlashan, *Pure Appl. Chem.*, **21**, 1 (1970); Recommendations on Nomenclature and Presentation of Data in Gas Chromatography (IUPAC Division of Analytical Chemistry), *Pure Appl. Chem.*, **8**, 553 (1964).

A. Atomic and Molecular Quantities; Chemical Reactions

Quantity	Symbol	Quantity	Symbol
Affinity of a reaction $(-\Sigma \nu_B \mu_B)$	A	Molality of substance B	m_B
Atomic mass number	A	Molar concentration of substance B	c_B, $[B]$, $c(B)$
Avogadro number (constant)	N_A	Molarity	M
Binding energy of an electron in an atom	B_e	Mole fraction of substance B	x_B, y_B
Bohr (or orbital) magneton $(= eh/4\pi mc)$	μ_B	Molecular concentration	C
Boltzmann constant	k	Molecular mass	m
Characteristic temperature	Θ	Neutron flux	ϕ
Charge-to-mass ratio for electron	e/m_e	Neutron rest mass	m_n
		Nuclear cross section	σ
Collision diameter of a molecule	d, σ	Nuclear magneton	$\mu_N \ldots \mu N$
Collision number (collisions per unit volume and unit time)	Z	Number of molecules	N
		Normality	N
		Number of moles	n
Concentration of solute substance B	c_B	Osmotic pressure	Π
		Planck constant	h
Decay constant	λ	Proton moment	μ_p
Degree of reaction (e.g., degree of dissociation)	α	Proton rest mass	m_p
Diameter of molecule	$\sigma \ldots D$	Quantum-charge ratio	h/e
Diffusion coefficient	D	Quantum number:	
Dirac $\hbar$ $[= h/2\pi]$	$\hbar$	Azimuthal or orbital	l
Disintegration constant $[N = N_0 e^{-\lambda t}]$	λ	Azimuthal or orbital, total	L
Electron radius	r_e	Inner	j
Electron rest mass	m_e	Inner, total	J
Equilibrium constant	K	Magnetic	m
Equivalent weight	M/Z	Magnetic, total	M
Equilibrium quotient (or equilibrium product of molalities)	Q	Principal	n
		Rotational	R
		Spin	s
		Spin, nuclear	I
Extent of reaction $(dn_B = \nu_B d\xi)$	ξ	Spin, total	S
		Rate constant	k
Fine structure constant	α	Rate constant corresponding rate Z	z
Force constant	k	Rate of reaction:	$d\xi/dt$
Gyromagnetic ratio	g, γ	Reduced mass	
Loschmidt's number (2.68715 × 10^{19} molecules per cm³)	n_0	Rotation, specific $(= \theta/lc)$	α
		Rydberg constant for infinite mass	R_∞
Mass fraction of substance B	w_B	Rydberg constant for hydrogen	R_H
Mean free path	l	Solubility product	K_{sp}

Table 2-2 (*Continued*)
PHYSICAL AND CHEMICAL SYMBOLS AND TERMINOLOGY

Quantity	Symbol	Quantity	Symbol
Statistical weight	g	Transition probability of induced or stimulated emission ($m \rightarrow n$ energy level)	B_{nm}
Surface concentration, surface excess	Γ		
Symmetry number	σ, s		
Transition probability of spontaneous emission ($m \rightarrow n$ energy level)	A_{nm}	Volume fraction of substance B	ϕ_B
		Valence	z
Transition probability of absorption ($n \rightarrow m$ energy level)	B_{mn}	Work function	ϕ

B. Chromatography

Quantity	Symbol	Quantity	Symbol
Adjusted retention time	t'_R	Partition coefficient	K, K_d
Adjusted retention volume	V'_R	Peak resolution	R_{ij}
Average linear gas velocity in column	$\bar{u}$	Plate height	H
Capacity, volume	Q_v	Relative retention	r_{is}
Capacity, weight	Q_w	Relative retention ratio	α
Capacity or partition ratio	k	Retardation value	R
Column inlet pressure	p_i	Retardation factor based on relative distances traveled	R_f
Column length	L		
Column outlet pressure	p_o	Retention distance on chromatogram	d_R
Column pressure gradient correction factor	j	Retention time	t_R, t_{max}
		Retention volume	V_R, V_{max}
Column temperature	T_c	Retention index (isothermal gas chromatography)	I_x
Concentration at peak maximum	C_{max}		
Corrected retention time	t_R°	Retention index (linear programmed-temperature gas chromatography)	I_{PT}
Corrected retention volume	V_R°		
Density of liquid phase	ρ_L	Selectivity coefficient (equivalents of M replacing hydrogen ion)	$E_H^{M/n}$
Distribution ratio (extraction)	D		
Diffusion coefficient in liquid film	D_f	Specific retention volume	V_g
Diffusion coefficient in mobile phase	D_M	Temperature, retention	T_R
		Theoretical plates (plate number)	n, N
Diffusion coefficient within resin bead	D_r	Volume, bed	V_b
		Volume distribution ratio	D_v
Diffusion coefficient in stationary phase	D_s	Volume, inner	V_i
		Volume of liquid phase in column	V_L
Elution volume (exclusion chromatography)	V_e	Volume of mobile phase (carrier gas) in column	$V_M, (V_G)$
Gas flow rate from column	F_c	Volume, outer or interstitial	V_o
Gas holdup time	t_M	Volume, resin or gel	V_r
Gas holdup volume	V_M	Volume, supporting solid in column	V_s
Gas/liquid volume ratio	β	Zone width at baseline	W
Height equivalent to a theoretical plate	H	Zone width at C_{max}/e or $0.368C_{max}$	W_e
		Zone width at peak height/2	$W_{1/2}$
Net retention volume	V_N		

Table 2-2 (*Continued*)
PHYSICAL AND CHEMICAL SYMBOLS AND TERMINOLOGY

C. Electrical Units

Quantity	Symbol & Definition	SI Unit
Admittance (reciprocal impedance)	$Y = \sqrt{G^2 + B^2}$	ohm^{-1}, Ω^{-1}
Attenuation		decibel, db
Capacitance	C	farad, F
Charge, electric	$Q = It$	coulomb, C
Charge density, surface	σ_s	
Charge density, volume	σ_v	
Conductance	$G = 1/R = \kappa A/l$	siemens, S or Ω^{-1}
Conductivity (formerly specific conductance)	κ	
Current	$I = Q/t$	ampere, A
Current density	$j = Il^{-2}$	ampere per square meter, A$\cdot$m^{-2}
Dielectric constant	ε	
Dielectric polarization	$P = D - \epsilon_0 E$	debye
Dipole moment	p, μ (permanent); p_e (electric); p_i (induced)	
Displacement	D	
Electric charge	e	
Electric field strength	$E = V/l$	volt per meter, V$\cdot$m^{-1}
Electric flux density	$D = Q/l^2$	coulomb per meter squared, C$\cdot$m^{-2}
Electric potential	V, ϕ	volt, V
Electric tension*	$U = IR$	volt, V
Electrode potential	$\mathcal{E} \dots \mathcal{U}$	volt, V
Electrokinetic potential	ζ	
Electromotive force (emf)	E	volt, V
Energy	$W = elt$	joule, J
Frequency	$f = t^{-1}$	hertz, Hz
Impedance	$Z = \sqrt{R^2 + (X_L + X_C)^2}$	ohm, Ω
Inductance, mutual	$L_{12}, M = \sqrt{L_1 L_2}$	henry, H
Inductance, self	L	henry, H
Moment, electric quadrupole	Q	
Period	$t = f^{-1}$	second, s
Power, active	$P = EI \cos \theta$	watt, W
Power, apparent	$P = EI$	watt, W
Power, reactive	$jQ = EI \sin \theta$	watt, W
Power factor	$pf = P/EI = \cos \theta$	
Quantity of electricity	$Q = It$	coulomb, C
Reactance, capacitive	$X_C = (2\pi fC)^{-1}$	ohm, Ω
Reactance, inductive	$X_L = 2\pi fL$	ohm, Ω
Resistance	$R = \rho l/A$	ohm, Ω
Resistivity, volume (formerly specific resistance)	$\rho = RA/l$	ohm meter, $\Omega \cdot$m
Susceptance	$B = X/Z^2$	ohm^{-1}, Ω^{-1}

* American usage is often E.

Table 2-2 (*Continued*)
PHYSICAL AND CHEMICAL SYMBOLS AND TERMINOLOGY

D. Electrochemistry

Quantity	Symbol	Quantity	Symbol
Anodic current	i_a	Half-neutralization potential	**HNP**
Cathodic current	i_c	Half-wave potential	$E_{1/2}$
Charge number of an ion, plus or minus	$z_+ z_-$	Faraday's constant (the faraday)	F, **F**
Degree of electrolytic dissociation	α	Ionic conductance	λ
Diffusion coefficient	D	Ionic strength	$I \ldots \mu$
Diffusion current	i_d	Limiting current	i_l
Effective ionic radius	$\mathring{a}$	Mobility	$\mu_+ \mu_-$
Electric potential, inner	ϕ	Overpotential (or overvoltage)	η
Electric potential, outer	ψ	Residual current	i_r
Electrolytic conductivity or specific conductance ($\kappa = l/\rho = l/RA$)	κ	Standard potential of half-reaction	E°
		Thickness of diffusion layer	δ
Equivalent conductance of electrolyte or ion	Λ	Transport number [transference number $= \mu_\pm/(\mu_+ + \mu_-)$]	$t_+ t_-$

E. Optical Units

Quantity	Symbol & Definition	SI Unit
Absorbance	$A = -\log T$	
Absorption coefficient	$\kappa = -(\ln T)/bc$	
Absorption factor (fraction of incident radiant power which is absorbed)	$\alpha = \Phi_a/\Phi_0$	
Absorptivity (specific absorbance or decadic absorption coefficient)	$a = A/bc$	$\text{liter} \cdot \text{g}^{-1} \cdot \text{cm}^{-1}$
Angle, plane	$\alpha, \beta, \gamma, \delta, \theta, \phi$	rad
Angle, solid	ω, Ω	sr
Angle of diffraction	θ	
Angle of incidence	ϕ	
Angle of optical rotation	$\alpha \ldots \theta$	
Angle of reflection	r	
Area	S, A	m^2
Breadth (width)	b	m
Emissivity (radiant)	$J = \Phi/V\omega$	$\text{W} \cdot \text{sr}^{-1} \cdot \text{m}^{-3}$
Energy density (radiant)	$u = Q/V$	$\text{J} \cdot \text{m}^{-3}$
Exposure (radiant)	$H = \int_0^t E\, dt$	$\text{J} \cdot \text{m}^{-2} = \text{W} \cdot \text{s} \cdot \text{m}^{-2}$
Focal length	f	m
Irradiance	$E = d\Phi/dS$	$\text{W} \cdot \text{m}^{-2}$
Luminance	L, B	$\text{W} \cdot \text{sr}^{-1} \cdot \text{m}^{-2}$
Molar absorptivity	$\varepsilon = A/bc$	$\text{dm}^3 \cdot \text{mol}^{-1} \cdot \text{cm}^{-1} = \text{liter} \cdot \text{mol}^{-1} \cdot \text{cm}^{-1}$
Optical conductance	$G = \Phi/B\tau$	$\text{sr} \cdot \text{m}^2$
Planck's constant	h	$\text{J} \cdot \text{s}$
Planck's constant divided by 2π	$\hbar$	$\text{J} \cdot \text{s}$
Radial frequency	$\omega = 2\pi f$	
Radiance	$B = \Phi(S\,\omega \cos \varepsilon)^{-1}$	$\text{W} \cdot \text{sr}^{-1} \cdot \text{m}^{-2}$

Table 2-2 (*Continued*)
PHYSICAL AND CHEMICAL SYMBOLS AND TERMINOLOGY

E. Optical Units

Quantity	Symbol & Definition	SI Unit
Radiant energy	$Q = \int_0^t \Phi \, dt$	$J = W \cdot s$
Radiant intensity	$I = d\Phi/d\omega$	watt per steradian, $W \cdot sr^{-1}$
Radiant power (or flux)	$\Phi = \partial Q/\partial t$	watt, W
Reciprocal linear dispersion	$d\lambda/dx$	
Reflection factor or luminous flux (fraction of incident radiant power which is absorbed)	$\rho = \Phi_r/\Phi_0$	
Refraction index	n	
Refractivity	r	
Relaxation time	τ	s
Resolution, practical	$R = \lambda/\partial\lambda$	
Resolving power	$R_0 = \lambda/\partial_0\lambda$	
Space coordinates	x, y, z	
Spectral bandwidth	$\Delta\lambda$	m
Speed of light in vacuum	c, c_0	$m \cdot s^{-1}$
Transmission factor (fraction of incident radiant power which is transmitted)	$\tau = \Phi_t/\Phi_0$	
Transmittance	$T = I/I_0 = P/P_0$	
Velocity	v, u	
Velocity (average)	$\overline{v}, \overline{u}$	
Wavelength	λ	
Wavenumber	ν or $\sigma = \lambda^{-1}$.	

See also *Appl. Spectrosc.*, **28,** 398–410 (1974).

F. Magnetic Units

Quantity	Symbol & Definition	SI Unit
Field strength, magnetic	$H, \int H \, \partial l = nI$	$A \cdot m^{-1}$
Flux, magnetic	$\Phi = \mathscr{F}/\mathscr{R}$	Wb
Flux density (induction)	$B = \Phi/l^2$	$Wb \cdot m^{-2} = T$
Magnetization	$M = (B/\mu) - H$	
Magnetomotive force	$\mathscr{F} = \phi\mathscr{R} = 0.4\pi nI$	A
Permeability	$\mu = B/H$	$H \cdot m^{-1}$
Permeability of vacuum	μ_0	
Permittivity	$\epsilon = D/E$	$F \cdot m^{-1}$
Permittivity of vacuum	ϵ_0	
Reluctance (magnetic resistance)	$\mathscr{R} = l/A\mu$	$A \cdot Wb^{-1}$
Susceptibility, specific	$\chi = k/\rho$	
Susceptibility, volume	$k = M/H$	

n = number of turns
See also Section C., Electrical Units, for meaning of symbols.

Table 2-2 (*Continued*)
PHYSICAL AND CHEMICAL SYMBOLS AND TERMINOLOGY

G. Mechanical and Related Quantities

Quantity	Symbol	Quantity	Symbol
Angle of contact	θ	Pressure	p, P
Compressibility $\left[-\frac{1}{V}\left(\frac{\partial V}{\partial P}\right)_T\right]$	κ	Shear modulus of elasticity	G
		Shear stress	τ
Compression modulus, bulk modulus	$1/\kappa,\ \mathbf{K}$	Surface tension	$\gamma \ldots \sigma$
Force	F	Traction	σ
Force due to gravity (weight)	$G \ldots W$	Viscosity, dynamic	η
Fluidity $(= 1/\eta)$	ϕ	Viscosity, kinematic $(= \eta/\rho)$	ν
Friction coefficient	f	Volume expansivity, coefficient of	β
Modulus (Young's) of elasticity	E, Y	volume expansion	
Moment of force	M	Weight	$G, (W)$
Momentum	f		

H. Space, Time, Mass, and Related Quantities

Quantity	Symbol	Quantity	Symbol
Acceleration	a	Path, length of arc	s
Acceleration of free fall (gravity)	g	Period (any characteristic interval) $[= 1/f$ or $1/\nu]$	T, τ
Angular frequency	ω	Plane angle	$\alpha, \beta, \gamma, \theta, \phi, \psi$
Angular velocity $(= 2\pi f);\ d\phi/dt$	ω	Radius	r
Area	A, S	Rectangular coordinates	x, y, z
Density	ρ	Relative density	d
Diameter	d	Solid angle	ω
Frequency $(= 1/T)$	ν, f	Specific volume	ν
Height	h	Time	t
Length	l	Time constant	τ
Mass	m	Velocity	v, u, w, c
Mass, reduced $[1/\mu = (1/m_1 + 1/m_2)]$	μ	Volume	V
		Wavelength	λ
Moment of inertia	I	Wavenumber	$\sigma, \bar{\nu}$

I. Thermodynamic and Related Quantities

Quantity	Symbol	Quantity	Symbol
Activity, absolute	λ	Gibbs (free) energy $[= H - TS]$	G, Gf
Activity, relative	a	Heat	q, Q
Activity coefficient, mole fraction basis	f_B	Heat capacity at constant pressure	C_p
Activity coefficient, molality basis	f_B	Heat capacity at constant volume	C_v
Activity coefficient, concentration basis	y_B	Helmholtz energy (Gibbs ψ) $[= U - TS]$	A
Amplitude of wave associated with moving particle	Ψ	Joule-Thompson (Kelvin) coefficient $[= (\partial T/\partial P)_H]$	μ
Boltzmann constant	k	Osmotic coefficient	ϕ
Chemical potential	μ	Partition function	Q
Energy, internal (Gibbs ε)	$U \ldots E$	Planck function: $-G/T$	Y
Energy, kinetic	E_k, T, K	Ratio of specific heats $(= C_p/C_v)$	γ, κ
Energy, potential	E_p, V, Φ	Temperature, Celsius	t, θ
Enthalpy, heat content (Gibbs χ) $[H = U + PV]$	H, Hf	Temperature, absolute or thermodynamic	T
Entropy (Gibbs η)	S	Thermal conductivity	λ, K
Fugacity	f	Work	w, W
Gas constant (molar) $(= PV/nT)$	$R, \mathbf{R}$		

Table 2-3
CONVERSION FACTORS

Relations which are exact are indicated by an asterisk (*).
I.T. = International Steam Table.

To Convert	Into	Multiply by
Abampere	ampere*	10
	statampere	$2.997\ 96 \times 10^{10}$
Abampere-turn	ampere-turn	10
	gilbert	12.566
Abcoulomb	ampere-hour	0.002 777 8
	coulomb*	10
	statcoulomb	$2.997\ 96 \times 10^{10}$
Abcoulomb per kilogram	statcoulomb per dyne	30 577
Abcoulomb per pound	statcoulomb per dyne	67 411
Abfarad	farad*	1×10^{9}
	statfarad	$8.987\ 76 \times 10^{20}$
Abhenry	henry*	1×10^{-9}
	stathenry	$1.112\ 63 \times 10^{-21}$
Abmho	mho*	1×10^{9}
	mho (int)(1948)	$1.000\ 495 \times 10^{9}$
Abmho per cubic centimeter	mho per mil foot	166.2
Abohm	ohm*	1×10^{-9}
	statohm	$1.112\ 63 \times 10^{-21}$
Abvolt	statvolt	$3.335\ 60 \times 10^{-11}$
	volt*	1×10^{-9}
Abvolt per centimeter	volt per centimeter*	1×10^{-8}
	volt per inch	$2.540\ 005 \times 10^{-8}$
Abvolt per °C	microvolt per °F	0.005 555 6
Abvolt per °F	microvolt per °C	0.018
Abvolt per inch	volt per centimeter	$3.937\ 0 \times 10^{-9}$
Acre (British)	acre (U.S.)	0.999 994
	rood (British)	4
	square meter	4 046.849
	square yard (British)	4 840
Acre (U.S.)	acre (British)	1.000 006
	are (square dekameter)	40.468 73
	hectare (square hectometer)	0.404 687 3
	square chain (Gunter's)	10
	square foot	43 560
	square link (Gunter's)	1×10^{5}
	square meter	4 046.856 422 4
	square mile	0.001 562 5
	square rod	160
	square yard	4 840
Acre-foot	cubic foot	43 560
	cubic meter	1 233.49
	cubic yard	1 613.3
	gallon (U.S.)	325 851
Acre-foot per day	cubic foot per second	0.504 17
	gallon (U.S.) per minute	226.3
Acre-inch	cubic foot	3 630
Aeon	year	1×10^{9}
Ampere	abampere	0.1
	ampere (int)(1948)	1.000 165
	faraday per second (chemical)	$1.036\ 316 \times 10^{-5}$
	faraday per second (based on carbon-12)	$1.036\ 435 \times 10^{-5}$
	statampere	$2.997\ 96 \times 10^{9}$

Table 2-3 (*Continued*)
CONVERSION FACTORS

To Convert	Into	Multiply by
Ampere per square centimeter	ampere per square inch	6.452
	ampere per square meter*	1×10^4
Ampere per square inch	ampere per square centimeter	0.155 0
	ampere per square meter	1 550.0
Ampere per square meter	ampere per square centimeter*	1×10^{-4}
	ampere per square inch	6.452×10^{-4}
Ampere-hour	coulomb	3 600.0
	faraday	0.037 31
Ampere-turn	abampere-turn	0.1
	gilbert	1.256 6
Ampere-turn per centimeter	ampere-turn per inch	2.540
	ampere-turn per meter	100.0
	gilbert per centimeter	1.256 6
	oersted	1.256 6
Ampere-turn per inch	ampere-turn per centimeter	0.393 7
	ampere-turn per meter	39.37
	gilbert per centimeter	0.495 0
Ampere-turn per meter	ampere-turn per centimeter*	0.01
	ampere-turn per inch	0.025 4
	gilbert per centimeter	0.012 566
Angstrom (Angstrom unit)	inch	393.7×10^{-8}
	meter*	1×10^{-10}
	micron*	1×10^{-4}
	nanometer (millimicron)*	0.1
	kxu (Siegbahn, 1000x units)	0.997 984
Apostilb	meter-lambert	1
Are	acre (British)	0.024 710 58
	acre (U.S.)	0.024 710 44
	hectare (square hectometer)	0.01
	rood (British)	0.098 842
	square foot	1 076.4
	square meter	100.0
	square mile	$3.861 0 \times 10^{-5}$
	square yard	119.60
Assay ton	gram	29.166 7
Astronomical unit	kilometer	1.495×10^8
	light-year	$1.580 4 \times 10^{-5}$
	mile	9.290×10^7
	parsec	4.848×10^{-6}
Atmosphere (normal = 760 torr)	bar	1.013 250
	centimeter of mercury (0°C)	76.0
	dyne per square centimeter	$1.013 250 \times 10^6$
	foot of water (4°C)	33.90
	foot of water (39.2°F)	33.899
	gram per square centimeter	1 033.227
	inch of mercury (0°C)	29.921 26
	kilogram-force per square meter	$1.033 227 \times 10^4$
	millimeter of mercury (0°C)*	760
	newton per square meter	$1.013 250 \times 10^5$
Atmosphere	pascal	$1.013 25 \times 10^5$
	pound-force per square foot	2 116.22
	pound-force per square inch	14.695 95
	ton (short) per square foot	1.058 1
	ton (short) per square inch	0.007 348 2
Atmosphere (technical = 1 kgf/cm²)	pascal	$9.806 650 \times 10^4$

Table 2-3 (*Continued*)
CONVERSION FACTORS

To Convert	Into	Multiply by
Avogram	gram	$1.660\ 36 \times 10^{-24}$
Bag (British)	bushel (British)	3
	cubic meter	0.109 107
Bag, cement	pound of cement	94
Bar	atmosphere	0.986 92
	dyne per square centimeter*	1×10^6
	kilogram per square meter	10 197.1
	millimeter of mercury (0°C)	750.062
	pascal*	1×10^5
	pound-force per square foot	2 089
	pound-force per square inch	14.504
Barleycorn (British)	centimeter	0.846 67
	inch	0.333
Barn	square centimeter*	1×10^{-24}
	square meter*	1×10^{-28}
Barrel (British, dry)	cubic meter	0.163 66
	gallon (British)	36
Barrel, cement	pound of cement	376
Barrel, oil	gallon oil	42.0
Barrel (U.S., dry)	bushel (U.S.)	3.281
	cubic inch	7056
	cubic meter	0.115 62
	quart (U.S., dry)	105
Barrel (U.S., liquid)	cubic meter	0.115 62
	gallon (U.S.)	31.5
Barye	bar*	1×10^{-6}
	dyne per square centimeter	1
	pascal*	0.1
Board foot	cubic centimeter	2 359.8
	cubic foot	$\frac{1}{12}$
	cubic inch	144
Boiler horsepower	Btu (mean) per hour	33 479
	kilowatt	9.803
	pound of water evaporated from (at 212°F)	34.507
Bolt (U.S., cloth)	linear foot	120
	meter	36.576
Bougie decimale	candle (int)	1
British thermal unit, Btu (I.T.)	calorie	252.164 4
	calorie (I.T.)	251.995 8
	centigrade heat unit (chu)	0.555 56
	cubic foot-atmosphere	0.376 6
	cubic foot $\times$ (pound per square inch)	5.403 953
	foot-pound	777.98
	foot-poundal	25 030.7
	horsepower-hour	$3.929\ 2 \times 10^{-4}$
	joule*	1 055.056
	kilogram-meter	107.565
	kilowatt-hour	$2.930\ 711 \times 10^{-4}$
	liter-atmosphere	10.412 59
	pound of carbon to CO_2	6.88×10^{-5}
	pound of water evaporated (at 212°F)	0.001 036
	watt-hour	0.293 071
	watt-second	1 054.8

Table 2-3 (*Continued*)
CONVERSION FACTORS

To Convert	Into	Multiply by
Btu per cubic foot	calorie, kilogram per cubic meter	8.899
Btu per hour	boiler horsepower	$2.986\ 4 \times 10^{-5}$
	Btu per minute	0.016 667
	foot-pound per second	0.216 2
	gram-calorie per second	0.070 0
	horsepower	0.023 575
	horsepower-hour	3.929×10^{-4}
	watt	0.293 071
Btu per (hour × square foot × °F)	calorie per (second × square centimeter × °C)	0.000 135 62
Btu per (hour × square foot × °F/in)	Btu per (hour × square foot × °F/ft)	$\frac{1}{12}$
	calorie, gram (15°C) per (second × square centimeter × °C/cm)	$3.444\ 8 \times 10^{-4}$
	chu per (hour × square foot × °C/in)	1
	joule per (second × square centimeter × °C/cm)	1.442×10^{-3}
	kiloergs per (second × square centimeter × °C/cm)	14.42
	watt per (square centimeter × °C/cm)	1.442×10^{-3}
Btu per minute	foot-pound per second	12.96
	horsepower	0.023 575
	kilowatt	0.017 57
	watt	17.57
Btu per pound	calorie per gram	0.555 927 3
	calorie (int) per gram	0.555 555. . .
	joule per gram	2.326
	kilowatt-hour per pound	$2.390\ 711 \times 10^{-4}$
Btu per (pound × °F)	calorie, gram per (gram × °C)*	1
	joule per (gram × °C)*	4.186
Btu per second	horsepower	1.414 5
	watt	1 054.8
Btu per square foot per minute	horsepower per square foot	0.023 575
	kilowatt per square foot	0.175 80
	watt per square inch	0.122 08
Bucket (British, dry)	cubic centimeter	1.8184×10^4
	gallon (British)	4
Bushel (British)	bag (British)	$\frac{1}{3}$
	bushel (U.S.)	1.032 05
	cubic centimeter	36 369
	cubic foot	1.284 3
	cubic inch	2 219.3
	gallon (British)	8
	liter	36.367 704 8
	quarter (British, capacity)	0.125
Bushel (U.S.)	barrel (U.S., dry)	0.304 785
	bushel (British)	0.968 946
	cubic centimeter	35 239.070 166 88
	cubic foot	1.244 4
	cubic inch	2 150.42
	cubic meter	0.035 239

Table 2-3 (*Continued*)
CONVERSION FACTORS

To Convert	Into	Multiply by
Bushel (U.S.) (*Cont.*)	gallon (U.S.)	9.309 2
	liter	35.239
	peck (U.S.)	4
	pint (U.S., dry)	64
	quart (U.S., dry)	32
Butt (British)	cubic foot	20.228 5
	cubic meter	0.572 81
	gallon (British)	126
Cable length (U.S. or British)	foot	720
	meter	219.46
	mile (U.S., nautical)	0.118 417
	mile (U.S., statute)	0.136 364
Caliber	meter*	0.000 254
Calorie (I.T.)	Btu (I.T.)	$3.968\ 321 \times 10^{-3}$
	calorie	1.000 669 22
	joule	4.186 8
	kilowatt-hour	$1.163\ 000 \times 10^{-6}$
	liter-atmosphere	0.041 320 50
Calorie (thermochemical)	Btu (I.T.)	$3.965\ 667 \times 10^{-3}$
	calorie (I.T.)	0.999 331 2
	cubic foot-atmosphere	0.001 459
	cubic foot × (pound per square inch)	0.021 430 28
	foot-pound	3.087 4
	foot-poundal	99.334
	horsepower-hour	$1.559\ 3 \times 10^{-6}$
	joule*	4.184
	kilogram-meter	0.426 85
	liter-atmosphere	0.041 320 50
	watt-hour	0.001 162 2...
Calorie (15°C/°C)	Btu (60°F/°F)	0.002 204 6
	joule per °C	4.185
Calorie per gram	Btu per pound	1.8
	joule per gram	4.186
Calorie, kilogram per minute	foot-pound per second	51.457
	horsepower	0.093 558
	watt	69.769
Calorie, kilogram per second	kilowatt	4.186
Candela (also candle)	candle (pentane)	1 (approx)
	carcel unit	0.104
	Hefner unit	1.11
	lumen per steradian	1
Candela per square centimeter (or stilb)	candela per square foot	929
	foot-lambert	2 918.5
	lambert	3.1416
	meter-lambert	3.1416×10^{4}
Candela per square foot	foot-lambert	3.1416
	lambert	0.003 381 6
	meter-lambert	33.804
Candela per square inch	foot-lambert	452.39
	lambert	0.486 95
Candela per square meter (or nit)	candela per square foot	0.0929
Candle power (spherical)	lumen	12.566
Carat (metric)	grain	3.086 48
	gram*	0.2

Table 2-3 (*Continued*)
CONVERSION FACTORS

To Convert	Into	Multiply by
Carcel unit	candela (or candle)	9.61
	Hefner unit	10.66
Celsius (Centigrade) scale (°C)	Fahrenheit scale (°F)	see Conversion of Thermometer Scales, this section
Cental	kilogram	45.359 2
	pound	100
Centare	square meter	1.0
Centesimal minute	grade	0.01
Centigrade heat unit (chu)	Btu	1.8
	calorie, gram (15°C)	453.59
	joule	1 897.8
Centiliter	cubic inch	0.610 3
	dram	2.705
	ounce (U.S., fluid)	0.3382
Centimeter	angstrom unit*	1×10^{-8}
	foot	0.032 808
	inch	0.3937
	kilometer*	1×10^{-5}
	ligne (Paris line)	4.433 0
	micron*	1×10^5
	mil	393.7
	mile (U.S., statute)	$6.213\ 699 \times 10^{-6}$
	nanometer (millimicron)*	1×10^7
	palm (British)	0.131 23
	pica (printers' type)	2.362 22
	point (printers' type)	28.346 4
	pouce (Paris inch)	0.369 413
	yard	0.010 936
Centimeter of mercury (0°C)	atmosphere	0.013 158
	bar	0.013 332
	dyne per square centimeter	$1.333\ 22 \times 10^4$
	foot of water (39.2°F)	0.446 04
	gram per square centimeter	13.595
	kilogram per square meter	135.95
	pascal	1 333.22
	pound per square foot	27.845
	pound per square inch	0.193 37
Centimeter of water (4°C)	dyne per square centimeter	980.639
	pascal	98.063 8
Centimeter per °C	foot per °F	0.018 227
	inch per °F	0.218 722
Centimeter per second	foot per minute	1.968 5
	foot per second	0.032 81
	kilometer per hour	0.036
	knot	0.1943
	meter per minute	0.6
	mile per hour	0.022 369
	mile per minute	$3.728\ 2 \times 10^{-4}$
Centimeter per second squared	foot per second squared	0.032 81
	kilometer per hour per second	0.036
	meter per second squared*	0.01
	mile per hour per second	0.022 369
Centimeter-dyne	centimeter-gram	1.020×10^{-3}
	meter-kilogram	1.020×10^{-8}
	pound-foot	7.376×10^{-8}

Table 2-3 (*Continued*)
CONVERSION FACTORS

To Convert	Into	Multiply by
Centimeter-gram	centimeter-dyne	980.7
	meter-kilogram*	1×10^{-5}
	pound-foot	7.233×10^{-5}
Centipoise	centistoke	(density of liquid)$^{-1}$
	poise*	0.01
Centistoke	centipoise	density of liquid
Chain (engineers' or Ramsden's)	foot	100
	meter	30.48
Chain (surveyors' or Gunter's)	foot	66
	furlong	0.1
	link (Gunter's)	100
	meter	20.117
	mile	0.012 5
	rod	4
	yard	22.00
Chaldron (British)	bushel (British)	32
	cubic meter	1.163 8
Chaldron (U.S.)	bushel (U.S.)	36
	cubic meter	1.268 6
Cheval-vapeur	horsepower	0.986 32
	kilowatt	0.735 499
Cheval-vapeur-heure	joule	$2.647\ 8 \times 10^{6}$
Circular inch	circular mil*	1×10^{6}
	square centimeter	5.067 1
	square inch	0.785 40
	square mil	7.854×10^{5}
Circular mil	square centimeter	$5.067\ 1 \times 10^{-6}$
	square inch	7.854×10^{-7}
	square mil	0.785 40
Circular millimeter	square millimeter	0.785 40
Circumference	degree	360
	grade	400
	radian	6.283 19
Clove (British stones)	kilogram	3.628 7
	pound	8
Coomb (British)	bushel (British)	4
	cubic meter	0.145 48
Cord	cord foot	8
	cubic foot	128
	cubic meter	3.625
Cord foot	cord	0.125
	cubic foot	16
Coulomb	abcoulomb	0.1
	coulomb (int)(1948)	1.000 165
	electronic charge	6.281×10^{18}
	faraday	1.036×10^{-5}
	statcoulomb	$2.997\ 96 \times 10^{9}$
Coulomb (int)(1948)	coulomb	0.999 835
Coulomb per kilogram	statcoulomb per dyne	3 057.7
Coulomb per square centimeter	coulomb per square inch	6.4516
	coulomb per square meter*	1×10^{4}
Coulomb per square inch	coulomb per square centimeter	0.155 0
	coulomb per square meter	1.550×10^{3}
Coulomb per square meter	coulomb per square centimeter*	1×10^{-4}
	coulomb per square inch	6.452×10^{-4}

Table 2-3 (*Continued*)
CONVERSION FACTORS

To Convert	Into	Multiply by
Cubic centimeter	board foot	4.2377×10^{-4}
	bucket (British, dry)	$5.499\ 3 \times 10^{-5}$
	bushel (British)	$2.749\ 6 \times 10^{-5}$
	cubic foot	$3.531\ 5 \times 10^{-5}$
	cubic inch	0.061 023
	cubic meter*	1×10^{-6}
	cubic yard (U.S.)	$1.307\ 9 \times 10^{-6}$
	drachm (British, fluid)	0.281 57
	dram (U.S., fluid)	0.270 53
	gallon (British)	$2.199\ 7 \times 10^{-4}$
	gallon (U.S., liquid)	$2.641\ 7 \times 10^{-4}$
	gill (British)	0.007 039 0
	gill (U.S.)	0.008 453 5
	liter*	0.001
	milliliter*	1
	minim (British)	16.894
	minim (U.S.)	16.231
	ounce (British, fluid)	0.035 196
	ounce (U.S., fluid)	0.033 814
	pint (U.S., dry)	0.001 816 2
	pint (U.S., liquid)	0.002 113 4
	quart (British, liquid)	$8.798\ 8 \times 10^{-4}$
	quart (U.S., dry)	$9.080\ 8 \times 10^{-4}$
	quart (U.S., liquid)	0.001 056 7
Cubic centimeter-atmosphere	joule	0.101 325
	kilowatt-hour	$2.814\ 6 \times 10^{-8}$
Cubic centimeter per gram	cubic foot per pound	0.016 019
Cubic centimeter per second	cubic foot per minute	0.002 118 6
Cubic foot (British)	cubic foot (U.S.)	0.999 991 6
	cubic meter	0.028 316 77
Cubic foot (U.S.)	acre-foot	$2.295\ 68 \times 10^{-5}$
	board foot	12
	bushel (British)	0.778 61
	bushel (U.S.)	0.803 57
	cord	0.007 812 5
	cord foot	0.0625
	cubic centimeter	28 317.016
	cubic foot (British)	1.000 008 4
	cubic inch	1 728
	cubic meter	0.028 317 016
	cubic yard	0.037 037
	gallon (British)	6.228 8
	gallon (U.S.)	7.480 52
	hogshead (U.S.)	0.118 739
	liter	28.316
	pint (U.S., liquid	59.84
	quart (U.S., dry)	25.714
	quart (U.S., liquid)	29.922
Cubic foot-atmosphere	Btu	2.720 3
	calorie, gram	680.74
	foot-pound	2 116.3
	joule	2 869.4
	kilogram-meter	292.59
	kilowatt-hour	7.968×10^{-4}
	liter-atmosphere	28.316

Table 2-3 (*Continued*)
CONVERSION FACTORS

To Convert	Into	Multiply by
Cubic foot (common brick)	pound	120
Cubic foot of water (60°F)	pound	62.37
Cubic foot per minute	cubic centimeter per second	471.95
	gallon per second	0.124 68
	liter per second	0.471 93
Cubic foot per pound	cubic meter per kilogram	0.062 428
Cubic foot per second	acre-foot per day	1.983 4
	acre-inch per hour	0.991 8
	cubic yard per minute	2.222 22
	gallon per minute	448.831
	liter per minute	1 698.96
	million gallons per day	0.646 316 9
Cubic foot × (pound per square inch)	Btu (I.T.)	0.185 049 7
	calorie	46.662 95
	calorie (I.T.)	46.631 74
	joule	195.327 8
	kilowatt-hour	$5.423\ 272 \times 10^{-5}$
	liter-atmosphere	1.926 847
Cubic inch (British)	bushel (British)	$4.505\ 9 \times 10^{-4}$
	cubic centimeter	16.387 025
	cubic foot	$5.787\ 04 \times 10^{-4}$
	gallon (British)	0.003 604 64
	liter	0.016 385 58
	ounce (British, fluid)	0.576 74
	peck (British)	0.001 803 1
	pint (British)	0.028 837
Cubic inch (U.S.)	board foot	0.006 944 44
	bushel (U.S.)	$4.650\ 25 \times 10^{-4}$
	cubic centimeter	16.387 162
	cubic foot (U.S.)	$5.787\ 04 \times 10^{-4}$
	cubic inch (British)	1.000 008 4
	cubic meter	$1.638\ 716\ 2 \times 10^{-5}$
	cubic yard	$2.143\ 347 \times 10^{-5}$
	dram (fluid)	4.432 88
	gallon (U.S.)	0.004 329 0
	liter	0.016 386 73
	ounce (U.S., fluid)	0.554 11
	peck (U.S.)	0.001 860 10
	pint (U.S., dry)	0.029 761 6
	quart (U.S., dry)	0.014 880 8
	quart (U.S., liquid)	0.017 316
Cubic kilometer	cubic mile	0.239 911
Cubic meter	acre-foot	$8.107\ 1 \times 10^{-4}$
	barrel (U.S., liquid)	8.386 5
	bushel (U.S.)	28.377 6
	cord	0.275 9
	cubic centimeter*	1×10^{6}
	cubic foot (British)	35.314 77
	cubic foot (U.S.)	35.314 445
	cubic inch (U.S.)	61 023.38
	cubic yard (U.S.)	1.307 942 8
	gallon (British)	219.969
	gallon (U.S.)	264.173
	liter	999.973
	pint (U.S., liquid)	2 113.4

Table 2-3 (*Continued*)
CONVERSION FACTORS

To Convert	Into	Multiply by
Cubic meter (*Cont.*)	quart (U.S., liquid)	1 056.7
	register ton	0.353 14
Cubic meter per minute	acre-foot per day	1.167 4
Cubic millimeter	cubic centimeter*	0.001
	cubic inch	$6.102\ 3 \times 10^{-5}$
	minim (British)	0.016 894
	minim (U.S.)	0.016 231
Cubic yard (British)	cubic foot (British)	27
	cubic inch (British)	46 656
	cubic meter	0.764 552 85
	cubic yard (U.S.)	0.999 991 6
	gallon (British)	168.18
Cubic yard (U.S.)	acre-foot	$6.198\ 5 \times 10^{-4}$
	bushel (U.S.)	21.696
	cubic centimeter	764 559.45
	cubic foot (U.S.)	27
	cubic inch	46 656
	cubic meter	0.764 555
	cubic yard (British)	1.000 008 4
	gallon (U.S.)	210.97
	liter	764.54
	pint (U.S., liquid)	1 615.8
	quart (U.S., liquid)	807.9
Cubic yard of sand	pound	2 700
Cubic yard per minute	cubic foot per second	0.45
	gallon per second	3.367
	liter per second	12.74
Cubit	centimeter	45.720
	foot	1.5
	inch	18
Curie	disintegration per second*	3.70×10^{10}
Dalton	gram	$1.649\ 8 \times 10^{-24}$
	mass of atom of oxygen	$\frac{1}{16}$
Day (mean solar)	hour	24
	minute	1 440
	second	86 400
Day (sidereal)	second	86 164
Debye unit (dipole moment)	electrostatic unit	1×10^{18}
Decimeter	meter	0.1
Decistere	cubic meter	0.1
Degree (angular = $\frac{1}{90}$ right angle)	circumference	0.002 777 78
	minute	60
	quadrant	$\frac{1}{90}$
	radian	$\frac{\pi}{180} = 0.017\ 453\ 3$
	revolution	0.002 777 78
	second	3600
Degree Celsius (°C)	kelvin*	1
	see Table 2-6 for t°C	
Degree Fahrenheit (°F)	kelvin*	$\frac{5}{9}$
	see Table 2-6 for t°F	
Degree Rankine	kelvin	$\frac{5}{9}$
Degree Réaumur	see Conversion of Thermometer scales, this section	
Degree per foot	radian per centimeter	$5.726\ 1 \times 10^{-4}$
Degree per inch	radian per centimeter	0.006 871 4

Table 2-3 (*Continued*)
CONVERSION FACTORS

To Convert	Into	Multiply by
Degree per second	radian per second	0.017 453 3
	revolution per minute	0.166 667
	revolution per second	0.002 777 8
Dekameter	meter*	10
Drachm (British, fluid)	cubic centimeter	3.551 6
	minim (British)	60
	ounce (British, fluid)	0.125
Dram (apothecaries' or troy)	dram (avoirdupois)	2.194 286
	grain	60
	gram	3.887 935
	ounce (avoirdupois)	0.137 142 9
	ounce (troy)	0.125
	pennyweight	2.5
	pound (avoirdupois)	0.008 571 429
	pound (troy)	0.010 416 67
	scruple	3
Dram (avoirdupois)	dram (troy or apothecaries')	0.455 729 2
	grain	27.343 75
	gram	1.771 845 195
	ounce (avoirdupois)	0.062 5
	ounce (troy)	0.056 966 146
	pennyweight	1.139 323
	pound (avoirdupois)	0.003 906 25
	pound (troy)	0.004 747 179
	scruple	1.367 188
Dram (U.S., fluid)	cubic centimeter	3.696 691 195
	cubic inch	0.225 570
	gallon (U.S.)	$9.765 \ 6 \times 10^{-4}$
	gill (U.S.)	0.031 25
	milliliter	3.696 61
	minim (U.S.)	60
	ounce (fluid)	0.125
	pint (U.S., liquid)	0.007 812 5
	quart (U.S., liquid)	0.003 906 25
Dyne	grain	0.015 736 8
	gram	0.001 019 72
	newton*	1×10^{-5}
	poundal	$7.233 \ 0 \times 10^{-5}$
	pound	$2.248 \ 09 \times 10^{-6}$
Dyne per centimeter	erg per square centimeter*	1
	milligram per inch	2.590 1
	milligram per millimeter	0.101 97
	pascal*	10
Dyne per square centimeter	atmosphere	$9.869 \ 23 \times 10^{-7}$
	bar*	1×10^{-6}
	centimeter of water (4°C)	0.001 019 74
	gram-force per square centimeter	0.001 019 716
	inch of mercury (32°F)	$2.953 \ 0 \times 10^{-5}$
	inch of water (39.2°F)	$4.014 \ 8 \times 10^{-4}$
	kilogram-force per square meter	0.010 197 16
	millimeter of mercury (0°C)	$7.500 \ 617 \times 10^{-4}$
	newton per square meter*	10
	pound per square foot	0.002 088 6
	pound per square inch	$1.450 \ 4 \times 10^{-5}$

Table 2-3 (*Continued*)
CONVERSION FACTORS

To Convert	Into	Multiply by
Dyne-centimeter (torque)	kilogram-meter	$1.019\ 7 \times 10^{8}$
	pound-foot	$7.375\ 6 \times 10^{-8}$
	pound-inch	$8.851\ 1 \times 10^{-7}$
	poundal-foot	$2.373\ 0 \times 10^{-6}$
Electromagnetic cgs unit of magnetic permeability	electrostatic cgs unit of magnetic permeability	$1.112\ 79 \times 10^{-21}$
Electromagnetic cgs unit of mass resistance	ohm-meter-gram	$9.994\ 8 \times 10^{-6}$
Electromagnetic foot-pound-second unit of magnetic permeability	electromagnetic cgs unit of magnetic permeability	$0.001\ 076\ 4$
	electrostatic cgs unit of magnetic permeability	$1.033\ 82 \times 10^{-18}$
Electronic charge	abcoulomb	$1.592\ 1 \times 10^{-20}$
	coulomb	$1.592\ 1 \times 10^{-19}$
	statcoulomb	4.774×10^{-10}
Electronic charge per kilogram	statcoulomb per dyne	4.868×10^{-16}
Electronvolt	centimeter^{-1}	$8\ 065.48$
	erg	$1.602\ 0 \times 10^{-12}$
	joule	$1.602\ 19 \times 10^{-19}$
	kilocalorie per mole	$23.060\ 9$
	kilojoule per mole	$9.648\ 46$
	mass unit	$1.073\ 7 \times 10^{-9}$
	Rydberg unit of energy	$0.073\ 86$
Electrostatic cgs unit of dipole moment	debye unit*	1×10^{-18}
Electrostatic cgs unit of Hall effect	electromagnetic cgs unit of Hall effect	$2.696\ 2 \times 10^{31}$
Electrostatic cgs unit of magnetic flux density	maxwell per square centimeter	$2.998\ 6 \times 10^{10}$
Electrostatic cgs unit of magnetomotive force	gilbert	$3.335\ 60 \times 10^{-11}$
Electrostatic cgs unit of mass resistance	ohm-meter-gram	$8.986\ 9 \times 10^{15}$
Electrostatic foot-pound-second unit of charge	abcoulomb	117.58
	coulomb	$1.195\ 2 \times 10^{-6}$
	statcoulomb	$3\ 583.9$
Electrostatic foot-pound-second unit of magnetic permeability	electrostatic cgs unit of magnetic permeability	929.03
Ell	centimeter	114.30
	inch	45
	yard	1.25
Em, pica (printers' type)	centimeter	$0.423\ 33$
	inch	$\frac{1}{6}$
Erg	Btu	$9.480\ 5 \times 10^{-11}$
	calorie, gram	$2.388\ 9 \times 10^{-8}$
	calorie, kilogram	$2.388\ 9 \times 10^{-11}$
	electronvolt	$6.242\ 2 \times 10^{11}$
	foot-pound	$7.375\ 6 \times 10^{-8}$
	foot-poundal	$2.373\ 0 \times 10^{-6}$
	gram-centimeter	$0.001\ 019\ 72$
	horsepower-hour	$3.725\ 0 \times 10^{-14}$
	joule*	1×10^{-7}
	kilogram-meter	$1.019\ 72 \times 10^{-8}$

Table 2-3 (*Continued*)
CONVERSION FACTORS

To Convert	Into	Multiply by
Erg (*Cont.*)	mass unit	6.702×10^4
	watt-hour	2.778×10^{-11}
Erg per second	Btu per minute	$5.688\ 3 \times 10^{-9}$
	calorie, kilogram per minute	$1.433\ 3 \times 10^{-9}$
	foot-pound per minute	$4.425\ 4 \times 10^{-6}$
	foot-pound per second	$7.375\ 6 \times 10^{-8}$
	horsepower	$1.341\ 0 \times 10^{-10}$
	watt*	1×10^{-7}
Erg per square centimeter	dyne per centimeter*	1
Erg-second	Planck's constant	$1.509\ 6 \times 10^{26}$
Fahrenheit degree to °C	see Table 2-6	
Farad	abfarad*	1×10^{-9}
	farad (int)(1948)	1.000 495
	microfarad*	1×10^{-6}
	statfarad	$8.987\ 7 \times 10^{11}$
Farad (int)(1948)	farad	0.999 505
Faraday (based on carbon-12)	coulomb	96 484.56
(chemical)	coulomb	96 495.7
(physical)	coulomb	96 521.9
Faraday per kilogram	statcoulomb per dyne	$2.950\ 7 \times 10^8$
Fathom	cable length	0.008 333 3
	foot	6
	meter	1.828 8
Fermi	meter*	1×10^{-15}
Firkin (British)	gallon (British)	9
	liter	34.068
Foot (British)	centimeter	30.479 97
	foot (U.S.)	0.999 997 2
	pace (British)	0.4
Foot (U.S.)	centimeter	30.480 060 95
	chain (Gunter's)	0.151 515
	chain (Ramsden's)	0.01
	fathom	$\frac{1}{6}$
	foot (British)	1.000 002 8
	furlong	0.001 515 15
	hand (U.S.)	3
	inch	12
	link (Gunter's)	1.515 15
	link (Ramsden's)	1
	meter	0.304 8
	mile (nautical)	$1.645\ 8 \times 10^{-4}$
	mile (statute)	$1.893\ 939 \times 10^{-4}$
	rod	0.060 606 1
	yard	$\frac{1}{3}$
Foot (U.S., survey)	meter	$\frac{1200}{3937}$
Foot of air (1 atm, 60°F)	pound per square inch	5.30×10^{-4}
Foot of water (39.2°F)	atmosphere	0.029 499
	inch of mercury (32°F)	0.882 65
	kilogram per square meter	304.79
	pascal	2 988.98
	pound per square foot	62.427
	pound per square inch	0.433 52
Foot per °F	centimeter per °C	54.864
Foot per minute	centimeter per second	0.508 001
	kilometer per hour	0.018 288

Table 2-3 (*Continued*)
CONVERSION FACTORS

To Convert	Into	Multiply by
Foot per minute (*Cont.*)	meter per second	0.005 080 01
	mile per hour	0.011 364
Foot per second	centimeter per second*	30.48
	kilometer per hour*	1.097 28
	knot	0.592 1
	meter per minute	18.288
	mile per hour	0.681 82
	mile per minute	0.113 64
Foot per second squared	centimeter per second squared	30.48
	kilometer per hour per second	1.097 3
	mile per hour per second	0.681 82
Foot-candle	lumen per square foot*	1
	lumen per square meter	10.763 910
	lux	10.763 910
	phot	0.001 076 4
Foot-lambert	candela per square centimeter	$3.426\ 3 \times 10^{-4}$
	candela per square foot	0.318 31
	candela per square meter	3.426 3
	lambert	$1.076\ 39 \times 10^{-3}$
	meter-lambert	10.76
Foot-pound	Btu	0.001 285 4
	calorie, gram	0.323 89
	cubic foot-atmosphere	$4.725\ 3 \times 10^{-4}$
	erg	$1.355\ 82 \times 10^{7}$
	foot-poundal	32.174
	gram-centimeter	13 825.5
	horsepower-hour	$5.050\ 5 \times 10^{-7}$
	joule	1.355 82
	kilogram-meter	0.138 255
	kilowatt-hour	$3.766\ 2 \times 10^{-7}$
	liter-atmosphere	0.013 381
Foot-pound per minute	Btu per minute	0.001 285 4
	calorie, kilogram per minute	$3.238\ 9 \times 10^{-4}$
	foot-pound per second	0.016 667
	horsepower	$3.030\ 3 \times 10^{-5}$
	horsepower (metric)	$3.072\ 3 \times 10^{-5}$
	kilowatt	$2.259\ 7 \times 10^{-5}$
Foot-pound per second	Btu per minute	0.077 124
	calorie, kilogram per minute	0.019 433 4
	horsepower	0.001 818 2
	kilowatt	0.001 355 82
Foot-poundal	Btu	$3.995\ 1 \times 10^{-5}$
	calorie, gram	0.010 067
	erg	$4.214\ 01 \times 10^{5}$
	joule	0.042 140 1
	kilogram-meter	0.004 297 2
	liter-atmosphere	$4.158\ 8 \times 10^{-4}$
Foot-pound-force	joule	1.355 82
Force de cheval	see Horsepower (metric)	
Furlong	chain (Gunter's)	10
	chain (Ramsden's)	660
	foot	660
	meter	201.168
	mile	0.125

Table 2-3 (*Continued*)
CONVERSION FACTORS

To Convert	Into	Multiply by
Furlong (*Cont.*)	rod	40
	yard	220
Gal (acceleration)	centimeter per second squared*	1
Gallon (British)	barrel (British,dry)	0.027 778
	bushel (British)	0.125
	cubic centimeter	4 546.087
	cubic foot	0.160 54
	cubic inch	277.418
	gallon (U.S.)	1.200 94
	gill (British)	32
	liter	4.545 963 1
	ounce (British, fluid)	160
	peck (British)	0.5
	pint (British, liquid)	8
	pound (avoirdupois) of water at 62°F	10
	quart (British, liquid)	4
Gallon (imperial)	see Gallon (British)	
Gallon (U.S., dry)	cubic centimeter	4 404.884
Gallon (U.S., liquid)	acre-foot	$3.068\ 89 \times 10^{-6}$
	barrel (U.S., liquid)	0.031 746
	cubic centimeter	3 785.434
	cubic foot	0.133 680 5
	cubic inch	231
	cubic meter	0.003 785 434
	cubic yard	0.004 951 1
	gallon (British)	0.832 68
	gill (U.S.)	32
	liter	3.785 33
	minim	61 440
	ounce (U.S., fluid)	128
	pint (U.S., liquid)	8
	pound (avoirdupois) of water at 60°F	8.337 0
	quart (U.S., liquid)	4
Gallon (U.S.) per minute	acre-foot per day	0.004 419
	cubic foot per hour	8.020 8
	cubic foot per second	0.002 228 0
	liter per second	0.063 088
Gallon (U.S.) per second	cubic yard per minute	0.297 06
Gallon (U.S.) of water per minute	ton of water per day	6.002 4
Gamma (magnetism)	tesla*	1×10^{-9}
(weight per volume)	microgram per milliliter*	1
(weight per weight)	microgram per gram*	1
Gauss	electrostatic cgs unit of magnetic flux density	$3.335\ 85 \times 10^{-4}$
	line per square centimeter	1
	maxwell per square centimeter*	1
	maxwell per square inch	6.451 6
	tesla*	1×10^{-4}
	weber per square centimeter*	1×10^{-8}
	weber per square meter*	1×10^{-4}
Geepound	slug	1
Gilbert	abampere-turn	0.079 577
	ampere-turn	0.795 775
	electrostatic cgs unit of magnetomotive force	$2.997\ 7 \times 10^{10}$

Table 2-3 (*Continued*)
CONVERSION FACTORS

To Convert	Into	Multiply by
Gilbert per centimeter	ampere-turn per centimeter	0.795 775
	ampere-turn per inch	2.021
	oersted*	1
Gill (British)	cubic centimeter	142.07
	gallon (British)	0.031 25
	liter	0.142 06
	ounce (British, fluid)	5
	pint (British, liquid)	0.25
Gill (U.S.)	cubic centimeter	118.294
	cubic inch	7.218 75
	dram (fluid)	32
	gallon (U.S.)	0.031 25
	liter	0.118 292
	minim	1 920
	ounce (U.S., fluid)	4
	pint (U.S., liquid)	4
	quart (U.S., liquid)	0.125
Grade	centesimal minute	100
	circumference	0.002 5
	degree (angular)*	0.9
	minute	54
	radian	$\pi/200 = 0.015\ 708$
	second	3 240
Grain	carat (metric)	0.324 0
	dram (avoirdupois)	0.036 571 43
	dram (troy)	0.016 667
	dyne	63.545 3
	gram*	0.064 798 91
	ounce (avoirdupois)	0.002 285 7
	ounce (troy)	0.002 083 3
	pennyweight	0.041 666 7
	pound (avoirdupois)	$1/7000 = 1.428\ 6 \times 10^{-4}$
	pound (troy)	$1/5760 = 1.736\ 1 \times 10^{-4}$
	scruple	0.05
Grain per gallon (British)	part per million	14.254
Grain per gallon (U.S.)	gram per liter	0.017 118
	part per million	17.118
	pound per million gallons (U.S.)	142.86
Gram	avogram	$6.022\ 8 \times 10^{23}$
	carat (metric)	5
	dram (avoirdupois)	0.564 383
	dram (troy)	0.257 206
	dyne	980.665
	grain	15.4324
	joule per centimeter	9.807×10^{-5}
	milligram*	1 000
	ounce (avoirdupois)	0.035 273 96
	ounce (troy)	0.032 150 7
	pennyweight	0.643 014 9
	pound (avoirdupois)	0.002 204 623
	pound (troy)	0.002 679 23
	poundal	0.070 931 4
	scruple	0.771 618
	ton (metric)*	1×10^{-6}
	ton (short)	$1.102\ 311 \times 10^{-6}$

Table 2-3 (*Continued*)
CONVERSION FACTORS

To Convert	Into	Multiply by
Gram per centimeter-second	centipoise*	100
	poise*	1
	pound per foot-second	0.067 197
Gram per cubic centimeter	kilogram per cubic meter*	1 000
	kilogram per liter*	1
	pound per cubic foot	62.428
	pound per cubic inch	0.036 127
	pound per gallon (British)	10.022 4
	pound per gallon (U.S.)	8.345 4
	pound per mil-foot	3.405×10^{-7}
Gram per cubic meter	grain per cubic foot	0.437 00
Gram per liter	grain per gallon (U.S.)	58.417
	gram per cubic centimeter*	1
	part per million*	1 000
	pound per cubic foot	0.062 427
	pound per 1000 gallon (U.S.)	8.345 4
Gram per square centimeter	atmosphere	$9.678 4 \times 10^{-4}$
	bar	$9.806 65 \times 10^{-4}$
	dyne per square centimeter	980.665
	foot of water at 60°F	0.032 843
	inch of mercury at 32°F	0.028 959
	kilogram per square meter*	10
	millimeter of mercury (0°C)	0.735 56
	pound per square foot	2.048 17
	pound per square inch	0.014 223 4
Gram per ton (long)	milligram per kilogram	0.984 21
Gram per ton (short)	milligram per kilogram	1.102 3
Gram-centimeter	Btu	$9.296 667 \times 10^{-8}$
	calorie, gram	$2.342 7 \times 10^{-5}$
	erg	980.665
	foot-pound	$7.233 0 \times 10^{-5}$
	joule	$9.806 65 \times 10^{-5}$
	kilogram-meter*	1×10^{-5}
	pound per inch	5.600×10^{-3}
Gram-centimeter per second	watt	$9.806 65 \times 10^{-5}$
Gram-force per centimeter	pascal*	98.066 50
Gram-second per square centimeter	poise	980.665
Gram-square centimeter (moment of inertia)	pound-square foot	$2.373 05 \times 10^{-6}$
	pound-square inch	$3.417 2 \times 10^{-4}$
Gravity	centimeter per second squared	980.665
	foot per second squared	32.174
Hand	centimeter	10.160
	inch	4
Hectare	acre (British)	2.471 058
	acre (U.S.)	2.471 044
	are*	100
	square foot	107 638.7
	square meter*	1×10^4
	square rod (U.S.)	395.367
	square yard (U.S.)	11 959.85
Hefner unit	candle (int)	0.9
Hemisphere	sphere	0.5
	spherical right angle	4
	steradian	6.283 2

Table 2-3 (*Continued*)
CONVERSION FACTORS

To Convert	Into	Multiply by
Henry	abhenry*	1×10^9
	henry (int)(1948)	0.999 505
Henry (int)(1948)	henry	1.000 495
Hogshead (British)	cubic foot	10.114
	cubic meter	0.286 40
	gallon (British)	63
	liter	286.40
Hogshead (U.S.)	cubic foot	8.421 84
	cubic meter	0.238 481
	gallon (U.S.)	63
	liter	238.481
Horsepower	Btu per hour	2 545.08
	Btu per minute	42.418
	Btu per second	0.706 97
	calorie, kilogram per minute	10.688
	foot-pound per minute	33 000
	foot-pound per second	550.0
	horsepower (metric)	1.013 9
	joule per second	745.70
	pound of carbon to CO_2 per hour	0.175
	pound of water evaporated per hour from (at 212°F)	2.622
	watt (**g** = 980.665)	745.70
	watt (**g** = 980)	745.2
Horsepower (boiler)	Btu per hour	33.479
	watt	9 809.50
Horsepower (electric)	foot-pound per second	550.22
	watt*	746
Horsepower (metric)(542.5 foot-pounds per second)	foot-pound per minute	32 549
	foot-pound per second	542.48
	horsepower (550 foot-pounds per second)	0.986 32
	kilogram-meter per second	75
Horsepower (water)	watt	746.043
Horsepower-hour	Btu	2 545.08
	calorie, kilogram	641.304
	erg	$2.684\ 5 \times 10^{13}$
	foot-pound	1.98×10^6
	joule*	$2.684\ 5 \times 10^6$
	kilogram-meter	$2.737\ 4 \times 10^5$
	kilowatt-hour	0.745 7
	pound of carbon to CO_2	0.175
	pound of water evaporated from (at 212°F)	2.622
	pound of water raised from 62° to 212°F	17.0
Horsepower-hour (electric)	joule*	$2.685\ 6 \times 10^6$
Hour (mean solar)	day*	$\frac{1}{24}$
	minute*	60
	second*	3 600
	week	0.005 952 4
Hour (sidereal)	second (mean solar)	3 590.170 4
Hundredweight (long)	kilogram*	50.802 345 44

Table 2-3 (*Continued*)
CONVERSION FACTORS

To Convert	Into	Multiply by
Hundredweight (long) (*Cont.*)	pound	112
	ton (long)	0.05
Hundredweight (short)	kilogram	45.359 237
	ounce (avoirdupois)	1600
	pound	100
	ton (long)	0.044 642 9
	ton (metric)	0.045 359 2
	ton (short)	0.05
Inch (British)	barleycorn	3
	centimeter*	2.54
	foot (British)	0.083 333
	inch (U.S.)	0.999 997 2
	span	$\frac{1}{9}$
	yard (British)	$\frac{1}{36}$
Inch (U.S.)	angstrom unit*	2.540×10^8
	centimeter*	2.540
	chain (Gunter's)	0.001 262 63
	chain (Ramsden's)	$8.333 333 \times 10^{-4}$
	foot (U.S.)	$\frac{1}{12}$
	inch (British)	1.000 002 8
	link (Gunter's)	0.126 263
	link (Ramsden's)	0.083 333 3
	mil	1000
	mile	$1.578 28 \times 10^{-5}$
	millimeter*	25.40
	pica (printers' type)	6
	point (printers' type)	72
	rod	0.005 050 51
	yard	$\frac{1}{36}$
Inch of mercury at 32°F	atmosphere	0.033 421 05
	dyne per square centimeter	33 863.88
	foot of water (39.2°F)	1.132 99
	kilogram per square meter	345.315 5
	pascal ($N \cdot m^{-2}$)	3 386.388
	pound per square foot	70.727
	pound per square inch	0.491 154 1
Inch of mercury at 60°F	pascal ($N \cdot m^{-2}$)	3 376.85
Inch of water at 39.2°F (4°C)	atmosphere	0.002 458 3
	dyne per square centimeter	2 490.82
	gram per square centimeter	2.539 9
	inch of mercury at 32°F	0.073 554
	kilogram per square meter	25.399
	ounce per square inch	0.578 02
	pascal ($N \cdot m^{-2}$)	249.082
	pound per square foot	5.202 2
	pound per square inch	0.036 126
Inch of water at 60°F	pascal	248.84
Inch per °F	centimeter per °C	4.572 0
Inch per hour	centimeter per minute	0.042 333
	foot per minute	0.001 388 89
Inch per minute	centimeter per hour	152.400
	foot per hour	5
	mile per hour	9.469×10^{-4}
Inch per second	centimeter per minute	152.400
Joule	Btu (I.T.)	$9.478 172 \times 10^{-4}$
	calorie	0.239 005 7

Table 2-3 (*Continued*)
CONVERSION FACTORS

To Convert	Into	Multiply by
Joule (*Cont.*)	calorie (I.T.)	0.238 845 9
	centigrade heat unit (chu)	$5.269\ 8 \times 10^{-4}$
	cubic foot-atmosphere	$3.485\ 3 \times 10^{-4}$
	cubic foot × (pound per square inch)	$5.121\ 960 \times 10^{-3}$
	erg*	1×10^{7}
	foot-pound	0.737 56
	foot-poundal	23.730
	gram-centimeter	10 197.2
	horsepower-hour	$3.725\ 08 \times 10^{-7}$
	joule (int)(1948)	0.999 835
	kilogram-meter	0.101 972
	kilowatt-hour	$2.777\ 78 \times 10^{-7}$
	liter-atmosphere	0.009 869 233
	watt-hour	$2.777\ 78 \times 10^{-4}$
	watt-second*	1
Joule (int)(1948)	joule	1.000 165
Joule per abcoulomb per °F	joule per coulomb per °C	0.18
Joule per ampere-hour	joule per abcoulomb	0.002 777 8
	joule per statcoulomb	$9.263\ 6 \times 10^{-14}$
Joule per centimeter	dyne*	1×10^{7}
	gram	1.020×10^{4}
	joule per meter*	100
	newton*	100
	pound	22.48
	poundal	723.3
Joule per coulomb	joule per abcoulomb	10
Joule per coulomb per °F	joule per coulomb per °C	1.8
Joule per °C	Btu (60°F) per °F	$5.267\ 9 \times 10^{-4}$
	calorie per °C	0.239 005 7
Joule per electron	joule per abcoulomb	$6.281\ 1 \times 10^{19}$
Joule per electron per °C	joule per coulomb per °C	$6.281\ 1 \times 10^{19}$
Joule per faraday	joule per abcoulomb	$1.036\ 3 \times 10^{-4}$
Joule per faraday per °C	joule per coulomb per °C	$1.036\ 3 \times 10^{-5}$
Joule per gram	Btu per pound	0.429 922 6
	calorie per gram	0.239 005 7
	calorie (int) per gram	0.238 845 9
	kilowatt-hour per pound	$1.259\ 979 \times 10^{-4}$
Joule per gram per °C	Btu per pound per °F	0.238 89
	calorie per gram per °C	0.239 005 7
Joule per statcoulomb per °C	joule per coulomb per °C	$2.998\ 6 \times 10^{9}$
Joule per statcoulomb per °F	joule per coulomb per °C	$5.397\ 5 \times 10^{9}$
Joule-second	quanta	$1.525\ 8 \times 10^{33}$
Karat (1 of gold to 24 of mixture)	milligram per gram	41.667
Kayser	meter^{-1}	100
Kilderkin (British)	cubic meter	0.081 830
	gallon (British)	18
	liter	81.827
Kilogram	dram (avoirdupois)	654.383
	dram (troy)	257.206
	dyne	980 665
	grain	15 432.4
	gram*	1000
	hundredweight (long)	0.019 684 1
	hundredweight (short)	0.022 046 223

Table 2-3 (*Continued*)
CONVERSION FACTORS

To Convert	Into	Multiply by
Kilogram (*Cont.*)	newton (joule per meter)*	9.806 65
	ounce (avoirdupois)	35.273 96
	ounce (troy)	32.150 742
	pennyweight	643.014 9
	pound (avoirdupois)	2.204 623
	pound (troy)	2.679 228 5
	poundal	70.931 4
	scruple	771.618
	ton (long)	$9.842\ 07 \times 10^{-4}$
	ton (metric)*	0.001
	ton (short)	0.001 102 311
Kilogram per cubic meter	gram per cubic centimeter*	0.001
	pound per cubic foot	0.062 428
	pound per cubic inch	$3.612\ 8 \times 10^{-5}$
Kilogram per meter	gram per centimeter*	10
	pound per foot	0.671 97
	pound per inch	0.559 98
	pound per mile	3 548.0
Kilogram-force per square centimeter	atmosphere	0.967 841 1
	bar	0.980 655
	inch of mercury at 0°C	28.959 03
	millimeter of mercury at 0°C	735.559 2
	pascal	$9.806\ 650 \times 10^4$
	pound per square inch	14.223 341 1
Kilogram-force per square meter	atmosphere	$9.678\ 41 \times 10^{-5}$
	dyne per square centimeter	98.066 5
	foot of water at 39.2°F	0.003 280 9
	gram per square centimeter*	0.1
	inch of mercury at 32°F	0.002 895 903
	millimeter of mercury at 0°C	0.073 555 92
	pascal	9.806 650
	pound per square foot	0.204 817
	pound per square inch	0.001 422 34
Kilogram-meter	Btu	0.009 296 67
	calorie, gram	2.342 7
	cubic foot-atmosphere	0.003 417 7
	erg	$9.806\ 65 \times 10^7$
	foot-pound	7.233 0
	foot-poundal	232.71
	gram-centimeter*	1×10^5
	horsepower-hour	3.653×10^{-6}
	joule	9.806 65
	kilowatt-hour	$7.724\ 07 \times 10^{-6}$
	liter-atmosphere	0.096 781
	watt-hour	0.002 724 07
Kilogram-meter (torque)	dyne-centimeter	$9.806\ 65 \times 10^7$
	pound-foot	7.233 0
	pound-inch	86.796
Kilogram-square centimeter (moment of inertia)	pound-square foot	0.002 373 05
	pound-square inch	0.341 719
Kiloline	maxwell*	1000
	weber*	1×10^{-5}
Kilometer	astronomical unit	6.689×10^{-9}

Table 2-3 (*Continued*)
CONVERSION FACTORS

To Convert	Into	Multiply by
Kilometer (*Cont.*)	centimeter*	1×10^5
	chain (Gunter's)	49.709 6
	chain (Ramsden's)	32.808 33
	foot (British)	3280.843
	foot (U.S.)	3280.833
	light-year	$1.056\ 7 \times 10^{-13}$
	mile (British)	0.621 373
	mile (nautical)	0.539 957
	mile (U.S.)	0.621 369 946
	parsec	3.243×10^{-14}
	rod	198.836
	yard (British)	1 093.61
	yard (U.S.)	1 093.61
Kilometer per hour	centimeter per second	27.777 8
	foot per minute	54.681
	foot per second	0.911 34
	knot	0.539 593
	meter per minute	16.667
	meter per second	0.277 778
Kilometer per hour per second	centimeter per second squared	27.777 8
	foot per second squared	0.911 34
	meter per second squared	0.277 778
	mile per hour per second	0.621 4
Kilowatt	boiler horsepower	0.101 92
	Btu per hour	3 413.04
	Btu per minute	56.884
	Btu per second	0.948 068
Kilowatt-hour	Btu	3 413.0
	calorie	$8.600\ 1 \times 10^5$
	foot-pound	$2.655\ 2 \times 10^6$
	horsepower-hour	1.341 0
	joule	3.6×10^6
	kilogram-meter	$3.670\ 9 \times 10^5$
Kilowatt-hour per pound	Btu (int) per pound	3 414.425
	calorie per gram	1 896.903
	calorie (int) per gram	1 895.643
	calorie, kilogram per minute	14.333 4
	calorie, kilogram per second	0.238 89
	foot-pound per hour	$2.655\ 2 \times 10^6$
	foot-pound per minute	44 254
	foot-pound per second	737.56
	horsepower	1.341 0
	horsepower (metric)	1.359 6
	joule (reciprocal)	7 936.641
	pound of carbon to CO_2 per hour	0.227 5
	pound of water evaporated per hour from (at 212°F)	3.517
Kip	newton	4 448.221 615
Kip per square inch	pascal	$6.894\ 757 \times 10^6$
Knot	centimeter per second	51.479
	foot per hour	6 080.20
	foot per second	1.688 94
	kilometer per hour*	1.852
	mile (nautical) per hour*	1
	mile (statute) per hour	1.151 55
	yard per hour	2 027

Table 2-3 (*Continued*)
CONVERSION FACTORS

To Convert	Into	Multiply by
kxu (Siegbahn)	angstrom	1.002 02
Lambert	candela per square centimeter	0.318 310
	candela per square inch	2.054
	candela per square meter*	$(1/_\pi) \times 10^4$
	foot-lambert	929.03
	lumen emitted per square centimeter of a perfectly diffusing surface*	1
	lumen per square foot	929.03
Langley	joule per square meter	4.184×10^4
Last (British)	bushel (British)	80
	cubic meter	2.909 5
League (British, nautical)	kilometer	5.559 552
	mile (nautical)	3
League (int, nautical)	meter	5 556
League (statute)	meter	4 828.032
	mile (statute)	3
Light-year	astronomical unit	63 274
	kilometer	$9.459\ 9 \times 10^{12}$
	mile	$5.878\ 1 \times 10^{12}$
Ligne (Paris line)	centimeter	0.225 583
	pouce (Paris inch)	$1/_{12}$
Line	maxwell*	1
	weber*	1×10^{-8}
Line per square centimeter	gauss*	1
Line per square inch	gauss	0.155 0
	weber per square centimeter	1.550×10^{-9}
Link (Gunter's or surveyors')	centimeter	20.116 8
	chain (Gunter's)	0.01
	foot	0.66
	inch	7.92
	mile	1.25×10^{-4}
	rod	0.04
	yard	0.22
Link (Ramsden's or engineers')	chain (Ramsden's)	0.01
	foot	1
	meter	0.304 8
Liter (before 1964)	cubic centimeter	1 000.028
Liter (after 1964, 1 dm³)	bushel (British)	0.027 496
	bushel (U.S.)	0.028 377
	cubic centimeter*	1 000
	cubic foot	0.035 315
	cubic inch	61.024
	cubic meter*	1×10^{-3}
	cubic yard	0.001 308
	dram (U.S., fluid)	270.52
	gallon (British)	0.219 970
	gallon (U.S.)	0.264 170
	gill (British)	7.039 033
	gill (U.S.)	8.453 44
	milliliter*	1 000
	minim	16 230.6
	ounce (British)	35.196
	ounce (U.S., fluid)	33.813
	peck (British)	0.109 984

Table 2-3 (*Continued*)
CONVERSION FACTORS

To Convert	Into	Multiply by
Liter (after 1964, 1 dm^3) (*Cont.*)	peck (U.S.)	0.113 509
	pint (British, liquid)	1.7598
	pint (U.S., dry)	1.816 141
	pint (U.S., liquid)	2.113 30
	quart (British, liquid)	0.879 88
	quart (U.S., dry)	0.908 071
	quart (U.S., liquid)	1.056 789
Liter per minute	cubic foot per hour	2.118 9
	cubic foot per second	5.886×10^{-4}
	gallon (U.S.) per second	0.004 403
Liter per second	cubic foot per minute	2.118 9
	cubic yard per minute	0.078 48
	gallon (U.S.) per minute	15.851
Liter-atmosphere	Btu (I.T.)	0.096 037 57
	calorie	24.217 26
	calorie (I.T.)	24.201 06
	cubic foot-atmosphere	0.035 316
	cubic foot $\times$ (pound per square inch)	0.518 982 5
	foot-pound	74.735
	foot-poundal	2 404.5
	horsepower-hour	$3.774 5 \times 10^{-5}$
	joule	101.325 0
	kilogram-meter	10.333
	kilowatt-hour	$2.814 583 \times 10^{-5}$
Lumen	candle power (spherical)	0.079 58
	watt (maximum visible radiation)	0.001 47
Lumen per square centimeter	lambert*	1
	phot*	1
Lumen per square centimeter per steradian	lambert	3.1416
Lumen per square foot	foot-candle*	1
	lumen per square meter	10.764
Lumen per square foot per steradian	millilambert	3.381 6
Lumen per square meter	lumen per square foot (or foot-candle)	0.092 902
	meter-candle*	1
	phot*	1×10^{-4}
Lux	foot-candle	0.092 902
	lumen per square meter*	1
	meter-candle*	1
	milliphot	0.1
	nox*	10^3
Mass of oxygen atom	dalton	16
Mass unit	electronvolt	$9.313 8 \times 10^8$
	erg	$1.492 1 \times 10^{-5}$
Maxwell	electrostatic cgs unit of magnetic flux	$3.335 6 \times 10^{-11}$
	gauss-square centimeter*	1
	kiloline	0.001
	weber*	1×10^{-8}
Maxwell per square centimeter	electromagnetic cgs unit of magnetic flux density	1
	electrostatic cgs unit of magnetic flux density	$3.335 9 \times 10^{-11}$
	maxwell per square inch	6.451 6

Table 2-3 (*Continued*)
CONVERSION FACTORS

To Convert	Into	Multiply by
Maxwell per square inch	maxwell per square centimeter	0.155 0
Megohm	ohm*	1×10^6
Mercury at 0°C	gram per cubic centimeter	13.595 1
Meter	angstrom unit*	1×10^{10}
	cable length	0.004 556 6
	chain (Gunter's)	0.049 709 6
	chain (Ramsden's)	0.032 808 3
	fathom	0.546 806
	foot (British)	3.280 843
	foot (U.S.)	3.280 833
	furlong	0.004 970 96
	inch (British)	39.370 113
	inch (U.S.)	39.37
	kilometer*	0.001
	link (Gunter's)	4.970 960
	link (Ramsden's)	3.280 83
	micron (micrometer)*	1×10^{-6}
	mile (nautical)	$5.395\ 93 \times 10^{-4}$
	mile (statute)	$6.213\ 7 \times 10^{-4}$
	nanometer (millimicron)*	1×10^{-9}
	pied (French foot)	3.078 34
	rod (U.S.)	0.198 838
	yard (British)	1.093 614
	yard (U.S.)	1.093 611
Meter per hour	centimeter per minute	1.666 67
	foot per minute	0.054 681
	foot per second	$9.111\ 34 \times 10^{-4}$
Meter per minute	centimeter per second	1.666 67
	foot per second	0.054 681
	kilometer per hour	0.06
	knot	0.032 376
	mile per hour	0.037 282
Meter per second	foot per minute	196.85
	kilometer per hour	3.6
	kilometer per minute	0.6
	knot	1.942 54
	mile per hour	2.236 93
	mile per minute	0.037 282
Meter per second squared	centimeter per second squared*	1×10^2
	foot per second squared	3.281
	kilometer per hour per second	3.6
	mile per hour per second	2.236 9
Meter-candle	lumen per square meter*	1
	lux*	1
	milliphot	0.1
Meter-lambert (or apostilb)	candela per square centimeter	$0.318\ 31 \times 10^{-4}$
	candela per square foot	0.029 572
	candela per square meter	0.318 31
	foot-lambert	0.092 9
	lambert*	1×10^{-4}
Mho (ohm⁻¹)	abmho*	1×10^{-9}
	mho (int)	1.000 52
	statmho	$8.986\ 45 \times 10^{11}$
Microfarad	abfarad*	1×10^{-15}
	farad*	1×10^{-6}
	statfarad	$8.987\ 76 \times 10^5$

2-35

Table 2-3 (*Continued*)
CONVERSION FACTORS

To Convert	Into	Multiply by
Micron (micrometer)	angstrom unit*	1×10^{-4}
	centimeter*	1×10^{-4}
	inch	3.937×10^{-5}
	mil	0.039 37
Microvolt per °C	microvolt per °F	0.555 56
Microvolt per °F	microvolt per °C	1.8
Mil	centimeter	0.002 54
	inch*	0.001
	micron (micrometer)	25.40
Mile (British)	kilometer	1.609 343
Mile (int, nautical)	foot	6 076.098 97
	kilometer	1.852
	league (nautical)	$\frac{1}{3}$
	mile (U.S., statute)	1.150 82
	yard	2 025.366 3
Mile (U.S., statute)	chain (Gunter's)	80
	chain (Ramsden's)	52.8
	foot	5 280
	furlong	8
	inch	63 360
	kilometer	1.609 344
	light-year	1.691×10^{-13}
	link (Gunter's)	8 000
	link (Ramsden's)	5 280
	meter	1 609.344
	mile (British)	1.000 002 8
	mile (nautical)	0.868 39
	parsec	$5.281\ 7 \times 10^{-14}$
	rod	320
	yard	1760
Mile per hour	centimeter per second	44.704 1
	foot per minute	88
	foot per second	1.466 7
	kilometer per hour*	1.609 344
	kilometer per minute	0.026 822
	knot	0.868 39
	meter per minute	26.822
	mile per minute	0.016 666 7
Mile per hour per minute	centimeter per second squared	0.745 07
Mile per hour per second	centimeter per second squared	44.704 1
	foot per second squared	1.466 7
	meter per second squared	0.447 041
Mile per minute	centimeter per second	2 682.2
	foot per second	88
	kilometer per hour	96.561
	knot	52.103 6
Millibar	pascal*	100
Millier	see Ton (metric)	
Milligram	carat (metric)	0.005
	dram (avoirdupois)	$5.643\ 83 \times 10^{-4}$
	dram (troy)	$2.572\ 06 \times 10^{-4}$
	grain	0.015 432
	ounce (avoirdupois)	$3.527\ 396 \times 10^{-5}$
	ounce (troy)	$3.215\ 074 \times 10^{-5}$
	pennyweight	$6.430\ 15 \times 10^{-4}$

Table 2-3 (*Continued*)
CONVERSION FACTORS

To Convert	Into	Multiply by
Milligram (*Cont.*)	pound (avoirdupois)	$2.204\ 62 \times 10^{-6}$
	pound (troy)	$2.679\ 23 \times 10^{-6}$
	scruple	$7.716\ 18 \times 10^{-4}$
Milligram per assay ton	milligram per kilogram	34.286
	ounce (troy) per ton (short)	1
Milligram per centimeter	pound per inch	$5.599\ 7 \times 10^{-6}$
Milligram per inch	dyne per centimeter	0.386 09
Milligram per liter	grain per gallon (U.S.)	0.058 417
	part per million (microgram per milliliter)*	1
	pound per cubic foot	$6.242\ 7 \times 10^{-3}$
Millimeter of mercury at 0°C	atmosphere	0.001 315 789 5
	bar	0.001 333 22
	dyne per square centimeter	1 333.224
	foot of water at 39.2°F	0.044 604
	gram per square centimeter	1.359 509 9
	kilogram per square meter	13.595 099
	pascal ($N \cdot m^{-2}$)	133.322 4
	pound per square foot	2.784 5
	pound per square inch	0.019 336 78
Millimicron	See Nanometer	
Minim (British)	cubic centimeter	0.059 192
Minim (U.S.)	cubic centimeter	0.061 612
	dram (U.S., fluid)	$\frac{1}{60}$
	gallon (U.S.)	$1.627\ 6 \times 10^{-5}$
	gill (U.S.)	$5.208\ 33 \times 10^{-4}$
	milliliter	0.061 61
	ounce (U.S., fluid)	0.002 083 3
	pint (U.S., fluid)	$1.302\ 08 \times 10^{-4}$
Minute (angle)	circumference	$4.629\ 63 \times 10^{-5}$
	degree	$\frac{1}{60}$
	quadrant	$1.851\ 85 \times 10^{-4}$
	radian	$2.908\ 88 \times 10^{-4}$
	second	60
Minute (mean solar)	day	$6.944\ 444 \times 10^{-4}$
	hour	$\frac{1}{60}$
	second*	60
	week	$9.920\ 635 \times 10^{-5}$
Minute (sidereal)	second (mean solar)	59.836 174
Minute per centimeter	radian per centimeter	$2.908\ 88 \times 10^{-4}$
Month (lunar)	day, hour, minute	29 d, 12 h, 44 min
Month (mean calendar)	day	30.420 2
	hour	730.085
	minute	43 805.1
	second (mean solar)	2.628×10^{6}
Nail (British)	centimeter	5.715
	inch	2.25
Nanometer (millimicron)	angstrom unit*	10
	centimeter*	1×10^{-7}
	inch (U.S.)	3.937×10^{-8}
	meter*	1×10^{-9}
Neper	decibel	8.686
Newton	dyne*	1×10^{5}
Newton per square meter	see Pascal	
Nit	candela per square meter	1

Table 2-3 (*Continued*)
CONVERSION FACTORS

To Convert	Into	Multiply by
Noggin (British)	cubic centimeter	142.07
	gallon (British)	$\frac{1}{32}$
Nox	lux*	1×10^{-3}
Oersted	ampere per meter	785.398
	electromagnetic cgs unit of magnetizing force	1
	electrostatic cgs unit of magnetizing force	$2.997\ 74 \times 10^{10}$
Ohm	abohm*	1×10^{9}
	ohm (int)(1948)	0.999 505
	statohm	$1.112\ 63 \times 10^{-12}$
Ohm (int)(1948)	ohm	1.000 495
Ohm per 100 feet	ohm per kilometer	32.808 3
Ohm-meter-gram	ohm-mile-pound	5 710
Ounce (avoirdupois)	dram (avoirdupois)	16
	dram (troy)	7.291 66
	grain	437.5
	gram	28.349 52
	hundredweight (short)	6.25×10^{-4}
	ounce (troy)	0.911 458 3
	pennyweight	18.229 17
	pound (avoirdupois)	$\frac{1}{16} = 0.062\ 50$*
	pound (troy)	0.075 955
	scruple	21.875
	ton (long)	$2.790\ 179 \times 10^{-5}$
	ton (metric)	$2.834\ 95 \times 10^{-5}$
	ton (short)	3.125×10^{-5}
Ounce (troy or apothecaries')	dram (avoirdupois)	17.554 28
	dram (troy)	8
	grain	480
	gram	31.103 481
	ounce (avoirdupois)	1.097 14
	pennyweight	20
	pound (avoirdupois)	0.068 571 43
	pound (troy)	$\frac{1}{12}$
	scruple	24
	ton (short)	$3.428\ 57 \times 10^{-5}$
Ounce (U.S., fluid)	cubic centimeter	29.573 7
	cubic inch	1.804 69
	dram (fluid)	8
	gallon (U.S.)	$\frac{1}{128}$
	gill (U.S.)	0.25
	liter	0.029 573
	minim (U.S.)	480
	pint (U.S., liquid)	$\frac{1}{16}$
	quart (U.S., liquid)	0.031 25
Ounce (avoirdupois) per square inch	dyne per square centimeter	4 309.2
Ounce (avoirdupois) per ton (long)	milligram per kilogram	27.901 8
Ounce (avoirdupois) per ton (short)	milligram per kilogram	31.250
Ounce-force	newton	0.278 014
Pace	centimeter	76.2
	foot	2.5
	inch	30
Palm (British)	centimeter	7.62
	inch	3

Table 2-3 (*Continued*)
CONVERSION FACTORS

To Convert	Into	Multiply by
Parsec	astronomical unit	$2.062\ 65 \times 10^5$
	light year	3.260
	meter	$3.085\ 68 \times 10^{16}$
	mile	$1.916\ 3 \times 10^{13}$
Part per million (ppm)	grain per gallon (British)	0.070 155
	grain per gallon (U.S.)	0.058 417
	microgram per gram*	1
	microgram per milliliter*	1
	milligram per kilogram*	1
	milligram per liter*	1
	pound per million gallons (U.S.)	8.345 2
Pascal	atmosphere (760 torr)	$9.869\ 233 \times 10^{-6}$
	bar*	1×10^{-5}
	barye*	10
	centimeter of mercury (0°C)	$7.500\ 64 \times 10^{-4}$
	centimeter of water (4°C)	0.010 197 44
	dyne per square centimeter*	0.1
	foot of water at 39.2°F	$3.345\ 623 \times 10^{-4}$
	gram-force per centimeter	0.010 197 16
	inch of mercury at 32°F	$2.952\ 998 \times 10^{-4}$
	inch of mercury at 60°F	$2.961\ 340 \times 10^{-4}$
	inch of water at 39.2°F	$4.014\ 742 \times 10^{-3}$
	inch of water at 60°F	$4.018\ 6 \times 10^{-3}$
	kip per square inch	$1.450\ 377 \times 10^{-7}$
	millimeter of mercury (0°C)	$7.500\ 615 \times 10^{-3}$
	poundal per square foot	0.671 969
	pound-force per square foot	0.020 885 43
	pound-force per square inch	$1.450\ 377 \times 10^{-4}$
	torr	$7.500\ 638 \times 10^{-3}$
Pascal-second	poise*	10
Peck (British)	bushel (British)	0.25
	cubic inch	554.84
	gallon (British)	2
	liter	9.091 9
Peck (U.S.)	bushel (U.S.)	0.25
	cubic inch	537.605
	liter	8.809 6
	pint (U.S., dry)	16
	quart (U.S., dry)	8
Pennyweight	dram (avoirdupois)	0.877 714
	dram (troy)	0.4
	grain	24
	gram	1.555 174
	ounce (avoirdupois)	0.054 857
	ounce (troy)	0.05
	pound (avoirdupois)	0.003 428 57
	pound (troy)	0.004 166 67
Perch	meter	5.029 2
	rod	1
Perch (masonry)	cubic foot	24.75
Phot	foot-candle	929.03
	lumen per square centimeter*	1
	lumen per square foot	929.03
	lux (lumen per square meter)*	1×10^4

Table 2-3 (*Continued*)
CONVERSION FACTORS

To Convert	Into	Multiply by
Pica (printers' type)	centimeter	0.421 752
	inch	$\frac{1}{6}$
	point (printers' type)	12
Pied (French foot)	meter	0.324 85
Pint (British, liquid)	cubic centimeter	568.26
	gallon (British)	0.125
	gill (British)	4
	quart (British)	0.5
Pint (U.S., dry)	bushel (U.S.)	0.015 625
	cubic centimeter	550.610 471
	cubic inch	33.600 3
	liter	0.550 610
	peck (U.S.)	0.062 5
	quart (U.S., dry)	0.5
Pint (U.S., liquid)	cubic centimeter	473.176 473
	cubic foot	0.016 710
	cubic inch	28.875
	cubic yard	$6.188\ 91 \times 10^{-4}$
	dram (fluid)	128
	gallon (U.S.)	0.125
	gill	4
	liter	0.473 176
	minim (U.S.)	7 680
	ounce (U.S., fluid)	16
	quart (U.S., liquid)	0.5
Planck's quantum	erg-second	$6.624\ 2 \times 10^{-27}$
Point (printers' type)	centimeter	0.035 278
	inch	0.013 888 9
	pica (printers' type)	$\frac{1}{12}$
Poise	cgs unit of absolute viscosity	1
	gram per centimeter-second*	1
	pascal-second	0.1
	stoke per density in g/cc	1
Poise-cubic centimeter per gram	square centimeter per second	1
Poise-cubic foot per pound	square centimeter per second	62.43
Poise-cubic inch per gram	square centimeter per second	16.387
Pole (British)	meter	5.029 2
Pottle (British)	cubic centimeter	2 273.04
	gallon (British)	0.5
Pouce (Paris inch)	centimeter	2.707 00
	ligne	12
	pied	0.083 333
Pound (avoirdupois)	cubic inch of water at 4°C and 760 mm of mercury	27.687 3
	dram (avoirdupois)	256
	dram (troy)	116.666 7
	dyne	$4.448\ 22 \times 10^5$
	grain	7 000
	gram*	453.592 4
	hundredweight (long)	0.008 922 857
	hundredweight (short)	0.01
	newton (joule per meter)	4.448 221
	ounce (avoirdupois)	16
	ounce (troy)	14.583 3

Table 2-3 (*Continued*)
CONVERSION FACTORS

To Convert	Into	Multiply by
Pound (avoirdupois) (*Cont.*)	pennyweight	291.666 7
	pound (troy)	1.215 278
	poundal	32.174
	scruple	350
	stone (British)	0.071 428 6
	ton (long)	$4.464\ 286 \times 10^{-4}$
	ton (metric)	$4.535\ 924 \times 10^{-4}$
	ton (short)	5×10^{-4}
Pound (troy)	dram (avoirdupois)	210.651 4
	dram (troy)	96
	grain	5 760
	gram	373.241 726
	ounce (avoirdupois)	13.165 714
	ounce (troy)	12
	pennyweight	240
	pound (avoirdupois)	0.822 857
	scruple	288
	ton (long)	$3.673\ 47 \times 10^{-4}$
	ton (metric)	$3.732\ 42 \times 10^{-4}$
	ton (short)	$4.114\ 29 \times 10^{-4}$
Pound of carbon to CO_2	Btu	14 544
	foot-pound	$1.131\ 5 \times 10^7$
	horsepower-hour	5.71
	kilowatt-hour	4.26
	pound of water evaporated from (at 212°F)	15
Pound of water (39.2°F) per minute	cubic foot per minute	0.016 021
	gallon (U.S.) per second	0.001 997 3
Pound per circular mil foot	gram per cubic centimeter	$2.936\ 96 \times 10^6$
Pound per cubic foot	gram per cubic centimeter	0.016 018 5
	pound per cubic inch	5.787×10^{-4}
Pound per cubic inch	gram per cubic centimeter	27.679 742
	pound per cubic foot	1 728
	pound per mil-foot	9.425×10^{-6}
Pound per foot	gram per centimeter	14.881 6
Pound per gallon (U.S.)	gram per cubic centimeter	0.119 826
	ton (short) per cubic yard	0.100 987
Pound per hour	gram per minute	7.559 87
	kilogram per day	10.886
Pound per inch	gram per centimeter	178.579
Pound per minute	kilogram per hour	27.215 6
Pound per second	gram per minute	27 215.6
Poundal	dyne	13 825.5
	gram	14.098 1
	newton (joule per meter)	0.138 254 954
	pound (avoirdupois)	0.031 081
Poundal per square foot	pascal	1.488 164
Poundal-foot (torque)	dyne-centimeter	$4.214\ 0 \times 10^5$
Pound-foot (torque)	dyne-centimeter	$1.355\ 8 \times 10^5$
	newton-meter	1.355 82
Pound-force per square foot	atmosphere	$4.725\ 41 \times 10^{-4}$
	bar	$4.788\ 0 \times 10^{-4}$
	dyne per square centimeter	478.801
	foot of water at 39.2°F	0.016 018

Table 2-3 (*Continued*)
CONVERSION FACTORS

To Convert	Into	Multiply by
Pound-force per square foot (*Cont.*)	gram per square centimeter	0.488 241
	inch of mercury at 32°F	0.014 139 0
	millimeter of mercury at 0°C	0.359 131
	pascal	47.880 3
	pound per square inch	0.006 944 5
Pound-force per square inch (psi)	atmosphere	0.068 045 96
	bar	0.068 947 57
	centimeter of mercury at 0°C	5.171 493
	dyne per square centimeter	68 947 57
	foot of water at 39.2°F	2.306 6
	gram per square centimeter	70.306 96
	inch of mercury at 32°F	2.036 021
	millimeter of mercury at 0°C	51.714 93
	pascal $(N \cdot m^{-2})$	6 894.757
	pound per square foot	144
Pound-inch (torque)	dyne-centimeter	$1.129\ 8 \times 10^6$
Pound-square foot (moment of inertia)	gram-square centimeter	$4.212\ 0 \times 10^5$
	pascal	47.880 3
	pound-square inch	144
Pound-square inch (moment of inertia)	pound-square foot	0.006 944 44
Pound-weight-second per square foot	poise	478.8
Pound-weight-second per square inch	poise	6.895×10^4
Proof (U.S.)	percent alcohol by volume	0.5
Puncheon (British)	cubic meter	0.318 23
	gallon (British)	70
	wine gallon	84
Quadrant	degree	90
	grade	100
	minute	5 400
	radian	1.570 80
Quart (British, liquid)	cubic centimeter	1 136.521
	gallon (British)	0.25
	liter	1.136 50
	pint (British, liquid)	2
Quart (U.S., dry)	bushel (U.S.)	0.031 25
	cubic centimeter	1 101.221
	cubic foot	0.038 889
	cubic inch	67.200 625
	liter	1.101 20
	peck (U.S.)	0.125
	pint (U.S., dry)	2
Quart (U.S., liquid)	cubic centimeter	946.352 95
	cubic foot	0.033 420
	cubic inch	57.75
	dram (fluid)	256
	gallon (U.S.)	0.25
	gill	8
	liter	0.946 353
	ounce (U.S., fluid)	32
	pint (U.S., liquid)	2
	quart (British)	0.832 674

Table 2-3 (*Continued*)
CONVERSION FACTORS

To Convert	Into	Multiply by
Quarter (British, capacity)	bushel (British)	8
	liter	290.94
Quarter (British, mass) (long)	pound	28
Quarter (British, mass) (short)	pound	25
Quarter (U.S., mass) (long)	pound	560
Quarter (U.S., mass) (short)	pound	500
Quartern (British, dry)	cubic centimeter	2 273.1
	gallon (British)	0.5
Quartern (British, liquid)	cubic centimeter	142.07
	gallon (British)	0.031 25
Quintal (metric)	hundredweight (long)	1.968 41
	kilogram	100
	pound (avoirdupois)	220.462
Quintal (long) (U.S. or British)	pound	112
Quire	sheet	24 or 25
Rad (radiation dose absorbed)	joule per kilogram*	1×10^{-2}
Radian	circumference	0.159 16
	degree	57.295 78
	degree, minute, second	57°, 17′, 44.8″
	minute	3 437.75
	quadrant	0.636 62
	revolution	0.159 16
	second	$2.062\ 65 \times 10^{5}$
Radian per centimeter	degree per centimeter	57.295 78
	degree per foot	1 746.4
	degree per inch	145.53
	minute per centimeter	3 437.75
Radian per second squared	revolution per minute squared	572.96
Ream	quire	20
	sheet	480 or 500
Register ton (British)	cubic foot	100
	cubic meter	2.831 7
Revolution	degree	360
	quadrant	4
	radian	6.283 2
Revolution per minute	degree per second	6
	radian per second	0.104 72
Revolution per minute squared	radian per second squared	0.001 745 3
	revolution per second squared	$2.777\ 8 \times 10^{-4}$
Revolution per second squared	revolution per minute squared	3 600
Reyn	centipoise	$6.894\ 76 \times 10^{6}$
	pound-second per square inch	1
Rhe	square meter per (newton × second)	10
Right angle	radian	1.570 796
Rod (British, volume)	cubic foot	1 000
	cubic meter	28.317
Rod (surveyors' measure)	chain (Gunter's)	0.25
	chain (Ramsden's)	0.165
	foot	16.5
	furlong	0.025
	inch	198
	link (Gunter's)	25
	link (Ramsden's)	16.5
	meter	5.029 2

Table 2-3 (*Continued*)
CONVERSION FACTORS

To Convert	Into	Multiply by
Rod (surveyors' measure) (*Cont.*)	mile	0.003 125
	perch	1
	yard	5.5
Roentgen	coulomb per kilogram*	2.58×10^{-4}
Rood (British)	acre	0.25
	are	10.117
	square perch	40
	square yard	1 210
Rope (British)	foot	20
	meter	6.096 0
Rydberg unit of energy	electronvolt	13.54
Sack (British)	bushel (British)	3
	cubic meter	0.109 11
Scruple	dram (avoirdupois)	0.731 428 6
	dram (troy)	$\frac{1}{3}$
	grain	20
	gram	1.295 978 4
	ounce (avoirdupois)	0.045 714 3
	ounce (troy)	0.041 666 7
	pennyweight	0.833 333
	pound (avoirdupois)	0.002 857 14
	pound (troy)	0.003 472 22
Second	degree	$2.777\ 78 \times 10^{-4}$
	minute	$\frac{1}{60}$
	radian	$4.848\ 14 \times 10^{-6}$
Second (mean solar)	second (sidereal)	1.002 737 91
Section	square meter	$2.589\ 988 \times 10^{5}$
	square mile	1
Shake	second	1×10^{-8}
Siegbahn unit (kxu)	angstrom unit	1.002 02
Siemens	ohm (int)	0.940 73
Sign	degree	30
Skein	foot	360
	meter	109.728
Slug	geepound	1
	kilogram	14.593 903
	pound	32.174
Slug per cubic foot	gram per cubic centimeter	0.515 38
Span	fathom	0.125
	foot	0.75
	inch	9
	meter	0.228 6
	yard	0.25
Square centimeter	circular mil	$1.973\ 50 \times 10^{5}$
	circular millimeter	127.32
	square foot (U.S.)	0.001 076 387
	square inch (U.S.)	0.154 999 69
	square meter*	1×10^{-4}
	square mil	$1.549\ 997 \times 10^{5}$
	square rod	$3.953\ 67 \times 10^{-6}$
	square yard	$1.196\ 0 \times 10^{-4}$
Square degree	steradian	$3.046\ 2 \times 10^{-4}$
Square foot (British)	square meter	0.092 902 89
Square foot (U.S.)	acre	$2.295\ 68 \times 10^{-5}$
	are	$9.290\ 341 \times 10^{-4}$

Table 2-3 (*Continued*)
CONVERSION FACTORS

To Convert	Into	Multiply by
Square foot (U.S.) (*Cont.*)	square centimeter	929.034 1
	square chain (Gunter's)	$2.295\ 68 \times 10^{-4}$
	square chain (Ramsden's)*	1×10^{-4}
	square inch	144
	square link (Gunter's)	2.295 68
	square link (Ramsden's)*	1
	square meter	0.092 903 41
	square mile	$3.587\ 01 \times 10^{-8}$
	square rod	0.003 673 09
	square yard	$\frac{1}{9}$
Square inch (U.S.)	circular mil	$1.273\ 240 \times 10^6$
	square centimeter	6.451 6
	square foot	$\frac{1}{144}$
	square mil*	1×10^6
	square yard	$7.716\ 05 \times 10^{-4}$
Square kilometer	acre (British)	247.105 8
	acre (U.S.)	247.104 4
	are*	1×10^4
	hectare*	100
	square foot	$1.076\ 4 \times 10^7$
	square mile (British)	0.386 102 8
	square mile (U.S.)	0.386 100 6
	square yard	$1.196\ 0 \times 10^6$
Square meter	acre (U.S.)	$2.471\ 044 \times 10^{-4}$
	are*	0.01
	hectare*	1×10^{-4}
	square centimeter*	1×10^4
	square chain (Gunter's)	0.002 471 04
	square foot (U.S.)	10.763 87
	square inch	1 550.0
	square kilometer*	1×10^{-6}
	square link (Gunter's)	24.710 4
	square mile	$3.861\ 0 \times 10^{-7}$
	square rod (U.S.)	0.039 536 7
	square yard (U.S.)	1.195 985
Square mil	circular mil	1.273 24
	square centimeter	$6.451\ 62 \times 10^{-6}$
	square inch*	1×10^{-6}
	square millimeter	$6.451\ 62 \times 10^{-4}$
Square mile	acre	640
	square chain (Gunter's)	640 0
	square foot	$2.787\ 84 \times 10^7$
	square kilometer	2.589 998
	square meter	$2.589\ 998 \times 10^6$
	square rod	$1.024\ 00 \times 10^5$
	square yard	$3.097\ 6 \times 10^6$
	township	0.027 777 8
Square millimeter	circular mil	1 973.52
	clrcular millimeter	1.273 24
	square centimeter*	0.01
	square inch	0.001 550 0
	square mil	1 550.0
Square rod	acre	0.006 25
	are	0.252 929 5
	square chain (Gunter's)	0.062 5

Table 2-3 (*Continued*)
CONVERSION FACTORS

To Convert	Into	Multiply by
Square rod (*Cont.*)	square chain (Ramsden's)	0.027 225
	square foot	272.25
	square inch	$3.920\ 4 \times 10^4$
	square link (Gunter's)	625
	square meter	25.292 95
	square mile	$9.765\ 625 \times 10^{-6}$
	square yard	30.25
Square yard (U.S.)	acre (U.S.)	$2.066\ 12 \times 10^{-4}$
	are	0.008 361 27
	square centimeter	8 361.27
	square chain (Gunter's)	0.002 066 12
	square foot	9
	square inch	1 296
	square link (Gunter's)	20.661 2
	square meter	0.836 127
	square mile	$3.228\ 31 \times 10^{-7}$
	square rod	0.033 057 85
Statampere	abampere	$3.335\ 64 \times 10^{-11}$
	ampere	$3.335\ 64 \times 10^{-10}$
Statcoulomb	abcoulomb	$3.335\ 64 \times 10^{-11}$
	coulomb	$3.335\ 64 \times 10^{-10}$
	electronic charge	$2.094\ 7 \times 10^9$
Statcoulomb per kilogram	statcoulomb per dyne	$1.019\ 7 \times 10^{-6}$
Statcoulomb per pound	statcoulomb per dyne	$2.248\ 1 \times 10^{-6}$
Statfarad	abfarad	$1.112\ 65 \times 10^{-21}$
	farad	$1.112\ 650 \times 10^{-12}$
Stathenry	abhenry	$8.987\ 554 \times 10^{20}$
	henry	$8.987\ 554 \times 10^{11}$
Statmho	mho	$1.112\ 65 \times 10^{-12}$
Statohm	abohm	$8.987\ 554 \times 10^{20}$
	ohm	$8.987\ 554 \times 10^{11}$
Statvolt	abvolt	$2.997\ 925 \times 10^{10}$
	volt	299.792 5
Statweber	electromagnetic cgs unit of magnetic flux	$2.997\ 96 \times 10^{10}$
	electrostatic cgs unit of magnetic flux	1
	maxwell	$2.997\ 96 \times 10^{10}$
	weber	299.796
Steradian	hemisphere	0.159 16
	solid angle	0.079 578
	sphere	0.079 578
	spherical right angle	0.636 64
	square degree	3 282.8
Stere	cubic meter*	1
	liter*	1
Stilb	candela per square centimeter	1
	candela per square meter*	1×10^4
Stokes	cgs unit of kinematic viscosity	1
	poise-cubic centimeter per gram	1
	square meter per second*	1×10^{-4}
Stone (British)	kilogram	6.350 3
	pound (avoirdupois)	14

Table 2-3 (*Continued*)
CONVERSION FACTORS

To Convert	Into	Multiply by
Strike (British)	bushel (British)	2
	cubic meter	0.072 737
Tablespoon	cubic centimeter	14.786 76
	teaspoon	3
Teaspoon	cubic centimeter	4.928 92
Tesla	weber per square meter*	1
Toise (French)	foot (Paris) (pied)	6
Ton (assay)	gram	29.167
Ton (long)	dyne	$9.964\ 0 \times 10^8$
	hundredweight (short)	22.400
	kilogram	1 016.046 9
	ounce (avoirdupois)	35 840
	pound (avoirdupois)	2 240
	pound (troy)	2 722.22
	ton (metric)	1.016 047
	ton (short)	1.12
Ton (metric)	gram*	1×10^6
	hundredweight (short)	22.046
	kilogram*	1×10^3
	ounce (avoirdupois)	35 273.96
	pound (avoirdupois)	2 204.623
	pound (troy)	2 679.23
	ton (long)	0.984 207
	ton (short)	1.102 311
Ton (register)	cubic meter	2.831 685
Ton (short)	dyne	$8.896\ 44 \times 10^8$
	hundredweight (short)	20
	kilogram	907.184 7
	ounce (avoirdupois)	3.200×10^4
	ounce (troy)	$2.916\ 7 \times 10^4$
	pound (avoirdupois)*	2×10^3
	pound (troy)	$2.430\ 56 \times 10^3$
	ton (long)	0.892 857
	ton (metric)	0.907 184 7
Ton (long) per square foot	dyne per square centimeter	$1.072\ 51 \times 10^6$
Ton (long) per square inch	dyne per square centimeter	$1.544\ 4 \times 10^8$
Ton (short) of water per 24 hours	cubic foot per hour	1.334 9
	gallon (U.S.) per minute	0.166 43
Ton (short) per square foot	atmosphere	0.945 09
	dyne per square centimeter	$9.576\ 07 \times 10^5$
	kilogram per square meter	9 764.87
	pound per square inch	13.888 9
Ton (short) per square inch	dyne per square centimeter	$1.378\ 95 \times 10^8$
	kilogram per square meter	$1.406\ 135 \times 10^6$
Ton-force	newton	$9.964\ 02 \times 10^3$
Tonne (metric)	kilogram*	1×10^3
Torr (millimeter of mercury, 0°C)	pascal	133.322
Township (U.S.)	acre	23 040
	square kilometer	93.240
	square mile	36
Tun	gallon	252
Unit pole	weber	$1.256\ 637 \times 10^{-7}$
Volt	abvolt*	1×10^8
	statvolt	0.003 333 56
Volt (int) (1948)	volt	1.000 330

Table 2-3 (*Continued*)
CONVERSION FACTORS

To Convert	Into	Multiply by
Volt per inch	volt per centimeter	0.393 701
Volt per mil	kilovolt per centimeter	0.393 701
Volt-electronic charge-second	quanta	$2.429\ 2 \times 10^{14}$
Volt-faraday-second	quanta	$1.472\ 4 \times 10^{38}$
Volt-second	maxwell*	1×10^8
Volt per °C	joule per coulomb per °C	1
Watt	Btu per hour	3.413 04
	Btu per minute	0.056 884
	calorie, kilogram per minute	0.014 333
	erg per second*	1×10^7
	foot-pound per minute	44.254
	foot-pound per second	0.737 56
	horsepower	0.001 341 0
	horsepower (electric)	0.001 340 5
	horsepower (metric)	0.001 359 6
	joule per second*	1
	kilogram-meter per second	0.101 97
	kilowatt*	0.001
Watt (int) (1948)	watt	1.000 165
	kilogram-meter	367.09
	kilowatt-hour*	0.001
Watt of maximum visible radiation	lumen	680
Watt per square inch	Btu per square foot per minute	8.191 3
	foot-pound per square foot per minute	6 372.6
	horsepower per square foot	0.193 10
Watt-hour	Btu	3.413 0
	calorie, gram	860.01
	foot-pound	2 655.3
	horsepower-hour	0.001 341 0
	joule	3600
Weber	electromagnetic cgs unit of magnetic flux*	1×10^8
	electrostatic cgs unit of magnetic flux	0.003 333 59
	line*	1×10^8
	maxwell*	1×10^8
	statweber	0.003 333 59
Weber per square inch	gauss	1.550×10^7
	weber per square centimeter	0.155 0
Weber per square meter	gauss*	1×10^4
	weber per square inch	6.452×10^{-4}
Week	day	7
	hour	168
	minute	$1.008\ 0 \times 10^4$
	second	$6.048\ 00 \times 10^5$
Wey (British, capacity)	bushel (British)	40
Wey (British, mass)	pound	252
Yard (British)	barleycorn (British)	108
	centimeter	91.439 92
	pole (British)	0.181 818
	quarter (British, linear)	4
	yard (U.S.)	0.999 997 2

Table 2-3 (*Continued*)
CONVERSION FACTORS

To Convert	Into	Multiply by
Yard (U.S.)	centimeter	91.44
	chain (Gunter's)	0.045 454 5
	chain (Ramsden's)	0.03
	fathom	0.5
	foot	3
	furlong	$\frac{1}{220}$
	inch	36
	link (Gunter's)	4.545 45
	meter	0.914 4
	mile (statute)	$5.681\ 82 \times 10^{-4}$
	rod	0.181 818
Year (calendar)	day	365
	second (mean solar)	$3.155\ 815\ 0 \times 10^7$
Year (light)	meter	$9.460\ 55 \times 10^{15}$
Year (sidereal)	day (mean solar)	365.256 4
	day (sidereal)	366.256 4
	second (mean solar)	$3.155\ 815\ 0 \times 10^7$
Year (tropical, mean solar)	day (mean solar)	365.242 2
	hour (mean solar)	8 765. 812 8
	second (mean solar)	$3.155\ 692\ 6 \times 10^7$
	year (sidereal)	1.002 737 80

Table 2-4
ABBREVIATIONS AND STANDARD LETTER SYMBOLS

Abampere	abamp	Beta particle	β
Absolute	abs	Body-centered cubic	bcc
Acetyl-	Ac	Boiling point	bp
Alcohol	alc	British thermal unit	Btu
Alkaline	alk	Butyl-	Bu
Alpha particle	α	Calorie	cal
Alternating current	ac	Calorie, international	cal_{IT}
Ampere	A	steam table	
Amplification factor	μ	Calorie, thermochemical	cal_{th}
Anhydrous	anhyd	Candela	cd
Angstrom	Å, A	Centimeter-gram-second (system)	cgs
Approximate (*circa*)	*ca*	Chemically pure	CP
Aqueous	aq	*circa*	*ca*
Atmosphere	atm	Citrate	Cit
Atomic mass unit	amu	Confer (compare)	cf.
Atomic per cent	at.%	Counts per minute	cpm
Atomic weight	at. wt.	Coulomb	C
Average	av	Critical temperature	t_c
Barn	b	Crystalline	cryst
Barrel	bbl	Cubic	cub
Base of natural logarithms	*e*	Curie	Ci
[= 2.718..]		Cycles per second (hertz)	Hz

Table 2-4 (*Continued*)
ABBREVIATIONS AND STANDARD LETTER SYMBOLS

Day	d	Horsepower	hp
Decibel	db	Hour	h, hr
Decompose	dec	Hygroscopic	hygr
Density, critical	d_c	*ibidem* (in the same place)	*ibid.*
Degrees Baumé	°Bé	*id est* (that is)	i.e.
Degrees Celsius	°C	Ignition	ign
Degrees Fahrenheit	°F	Inch	in.
Degrees Kelvin	°K	Indices of a family of	(*hkl*)
Degrees Rankine	°R	crystallographic planes	
Degrees Réaumur	°Ré	Infrared	ir
Determine(d); detect	det(d)	Inorganic	inorg
Determination	detn	Inside diameter	i.d.
Deuteron	d	Insoluble	insol
Diameter	diam	Joule	J
Differential thermal analysis	DTA	Kilocalorie	kcal
Dilute	dil	Kilogram	kg
Dilution	diln	Kilogram-force	kg-f
Direct current	dc	Kilowatt-hour	kWh
Disintegrations per minute	dpm	Limit (math.)	lim
Distil(led)	dist(d)	Liquid	liq, *l*
Dropping mercury electrode	dme	Liter	l
Dyne	dyn	*loco citato* (in the place cited)	*loc cit*
Electromagnetic unit	emu	Logarithm (common)	log
Electromotive force	emf	Logarithm (natural)	ln
Electron	e^-, e	Lumen	lm, lum
Electronvolt	eV	Lux	lx
Electrostatic unit	esu	Maximum	max
Entropy unit	eu, e.u.	Maxwell	Mx
Equivalent weight	equiv wt	Meter-kilogram-second (system)	mks
Especially	esp.	Melting point	mp
et alii (and others)	*et al*	Metallic	met
et cetera	etc.	Metastable	*m*
Ethyl-	Et	Meter	m
Ethylenediamine	en	Methyl-	Me
Ethylenediaminetetraacetic acid	EDTA	Micron (micrometer)	μm
exempli gratia (for example)	e.g.	Miller indices	(*hkl*)
Experiment(al)	expt(l)	Milliequivalent	meq
Exponential	exp	Millimeters of mercury	mm Hg
Face-centered cubic	fcc	Millimicron (nanometer)	nm
Farad	F	Millimole	mmol, mM
Focal length	*f*	Minimum	min
Foot	ft	Minute	min, m
Formal (concentration)	*F*	Mixture	mixt
Freezing point	fp	Molal	*m*
Gallon	gal	Molar	*M*
Gamma radiation	γ	Mole	mol
Gas (physical state)	*g*	Molecular weight	mol wt
Gauss	G	Mole per cent	mol %
Gram	g	Monoclinic	mn
Gram-atom	g-atom	Naperian base	*e*
Gram-calorie	g-cal	Negative	neg
Gram-mole	g-mol	Neutralization equivalent	neut equiv
Gravimetric	grav	Neutrino	ν
Henry	H		
Hexagonal	hex		

Table 2-4 (*Continued*)
ABBREVIATIONS AND STANDARD LETTER SYMBOLS

Newton	N	Saturated calomel electrode	SCE
Normal (concentration)	*N*	Second	s, sec
Nuclear magnetic resonance	nmr	Section	sect
Nuclear magneton	μ_N	Separation	sepn
Oersted	Oe	Slightly	sl
Ohm	Ω	Solid	s, c
opere citato (in the work cited)	*op cit*	Soluble	sol
		Solution	soln
Optical speed	*f*/number	Solvent	solv
Organic	org	Specific gravity	sp gr
Orthorhombic	o-rh	Square	sq
Ounce	oz	Standard	std
Outer diameter	o.d.	Standard hydrogen electrode	SHE
Oxalate	Ox	Standard temperature and pressure	STP
Oxidant	ox		
Page(s)	p. (pp.)	Steradian	sr
Parts per billion, volume	ng/ml	Stoke(s)	St
Parts per billion, weight	ng/g	Substance	subst
Parts per million, volume	μg/ml	Symmetrical	sym
Parts per million, weight	μg/g	Tartrate	Tart
Per cent	%	Temperature	temp
Phenyl-	Ph, ϕ	Temperature at boiling point	T_b
Photon	γ	Temperature, critical	t_c
Physical	phys	Tesla	T
Poise	P	Tetragonal	tetr
Positive	pos	Thermogravimetric analysis	TGA
Positron	β^+	Thin-layer chromatography	TLC
Pound	lb	Tonne	t, ton
Powder	pwd, powd	Torr (mm of mercury)	Torr
Precipitate(d)	ppt(d)	Transconductance	g_m
Pressure, critical	p_c	Transition	tr
Propyl-	Pr	Triclinic	tric
Proton	p	Trigonal	trig
Proton magnetic resonance	pmr	Triton (tritium)	t
Pyridine	py	Ultrahigh frequency	uhf
Qualitative	qual	Ultraviolet	uv
Quantitative	quant	United States Pharmacopoeia	USP
Quantum (energy)	$h\nu$	Vacuum	vac
Quart	qt	Vapor pressure	vp
Radian	rad	Versus	*vs*
Radiofrequency	rf	Volt	V
Recrystallized	recryst	Volt-ampere-reactive	var
Reductant	red	Volume	vol
Reference	ref	Volume per cent	vol %
Revolutions per minute	rpm	Volume per volume	v/v
Rhombic	rh, rhb	Watt	W
Rhombohedral	rh-hed	Weber	Wb
Roentgen	R	Weight	wt
Root-mean-square	rms	Weight per cent	wt %
Saturated	satd	Weight per volume	w/v

Table 2-5
MATHEMATICAL SYMBOLS AND ABBREVIATIONS

$+$	Plus or Positive	sec	Secant
$-$	Minus or Negative	csc	Cosecant
$\pm$	{Plus or minus / Positive or negative	vers	Versed sine
		covers	Coversed sine
$\mp$	{Minus or plus / Negative or positive	exsec	Exsecant
$\times$ or $\cdot$	Multiplied by	$\sin^{-1}a$ or arc sin a	{Anti-sine a / Angle whose sine is a / Inverse sine a
$\div$ or $:$	Divided by		
$=$ or $::$	Equals, as	sinh	Hyperbolic sine
$\neq$, $\;\not\equiv$	Does not equal	cosh	Hyperbolic cosine
		tanh	Hyperbolic tangent
$\simeq$, $\cong$	{Equals approximately / Congruent	$\sinh^{-1}a$ or arc sinh a	{Anti-hyperbolic sine a / Number whose hyper- / bolic sine is a
$>$	Greater than		
$<$	Less than	$P(x,y)$	Rect. coord. of point P
$\geq$, $\geqq$	Greater than or equal to	$P(r,\theta)$	Polar coord. of point P
$\leq$, $\leqq$	Less than or equal to	$f(x)$, $F(x)$ or $\Phi\,(x)$	{Function of x
$\sim$	Similar to		
$\therefore$	Therefore	Δy	Increment of y
$\sqrt{\;}$	Square root	$\doteq$ or $\rightarrow$	Approaches as a limit
$\sqrt[n]{\;}$	nth root	Σ	Summation of
a^n	nth power of a	∞	Infinity
$\log$ or $\log_{10}$	{Common logarithm / Briggsian "	dy	Differential of y
$\ln$ or $\log_e$	{Natural logarithm / Hyperbolic " / Napierian "	$\dfrac{dy}{dx}$ or $f'\,(x)$	Derivative of $y = f(x)$ with respect to x
e or ϵ	Base (2.718) of natural system of logarithms	$\dfrac{d^2y}{dx^2}$ or $f''\,(x)$	Second deriv. of $y = f(x)$ with respect to x
π	Pi (3.1416)	$\dfrac{d^ny}{dx^n}$ or $f^{(n)}\,(x)$	nth deriv. of $y = f(x)$ with respect to x
$\angle$	Angle		
$\perp$	Perpendicular to	$\dfrac{\partial z}{\partial x}$	Partial derivative of z with respect to x
$\parallel$	Parallel to		
a°	a degrees (angle)	$\dfrac{\partial^2 z}{\partial x\,\partial y}$	Second partial deriv. of z with respect to y and x
a'	{a minutes (angle) / a prime		
a''	{a seconds (angle) / a second / a double-prime	$\displaystyle\int$	Integral of
a'''	{a third / a triple-prime	$\displaystyle\int_a^b$	Integral between the limits a and b
a_n	a sub n		
$\sin$	Sine	j or i	{Imaginary quantity / $(\sqrt{-1})$, $i^2 = -1$
$\cos$	Cosine		
$\tan$	Tangent	$x = a + jb$	Symbolic vector notation
$\cot$ or ctn	Cotangent	$n! = 1\cdot2\cdot3\;\cdots\;n$	

Greek Alphabet

A	α	Alpha	N	ν	Nu
B	β	Beta	Ξ	ξ	Xi
Γ	γ	Gamma	O	o	Omicron
Δ	δ	Delta	Π	π	Pi
E	ε	Epsilon	P	ρ	Rho
Z	ζ	Zeta	Σ	σ	Sigma
H	η	Eta	T	τ	Tau
Θ	θ	Theta	Υ	υ	Upsilon
I	ι	Iota	Φ	φ	Phi
K	κ	Kappa	X	χ	Chi
Λ	λ	Lambda	Ψ	ψ	Psi
M	μ	Mu	Ω	ω	Omega

Prefixes for Naming Multiples and Submultiples of Units
For example: 10^{-9} gram is one nanogram or 1 ng.

Factor	Prefix	Symbol	Factor	Prefix	Symbol
10^{12}	tera	T	10^{-2}	centi	c
10^{9}	giga	G	10^{-3}	milli	m
10^{6}	mega	M	10^{-6}	micro	μ
10^{3}	kilo	k	10^{-9}	nano	n
10^{2}	hecto	h	10^{-12}	pico	p
10	deka	da	10^{-15}	femto	f
10^{-1}	deci	d	10^{-18}	atto	a

The prefix "myria" is sometimes used for 10^4 and "lakh" for 10^5.

Numerical Prefixes

½—Hemi	20—Eicosa	41—Hentetraconta
1—Mono	21—Henicosa	42—Dotetraconta
1½—Sesqui	22—Docosa	43—Tritetraconta
2—Di or Bi	23—Tricosa	44—Tetratetraconta
3—Tri	24—Tetracosa	45—Pentatetraconta
4—Tetra	25—Pentacosa	46—Hexatetraconta
5—Penta	26—Hexacosa	47—Heptatetraconta
6—Hexa	27—Heptacosa	48—Octatetraconta
7—Hepta	28—Octacosa	49—Nonatetraconta
8—Octa	29—Nonacosa	50—Pentaconta
9—Nona or Ennea	30—Triaconta	51—Henpentaconta
10—Deca or Deka	31—Hentriaconta	52—Dopentaconta
11—Undeca or Henadeca	32—Dotriaconta	53—Tripentaconta
12—Dodeca	33—Tritriaconta	54—Tetrapentaconta
13—Trideca	34—Tetratriaconta	55—Pentapentaconta
14—Tetradeca	35—Pentatriaconta	56—Hexapentaconta
15—Pentadeca	36—Hexatriaconta	57—Heptapentaconta
16—Hexadeca	37—Heptatriaconta	58—Octapentaconta
17—Heptadeca	38—Octatriaconta	59—Nonapentaconta
18—Octadeca	39—Nonatriaconta	60—Hexaconta
19—Nonadeca	40—Tetraconta	

CONVERSION OF THERMOMETER SCALES*

The following abbreviations are used: F, degrees Fahrenheit; C, degrees Celsius; Ré, degrees Réaumur; K, degrees Kelvin; R, degrees Rankine; Z, degrees on any scale; (fp)"Z", the freezing point of water on the Z scale; (bp)"Z", the boiling point of water on the Z scale. Cf. Dodds, *Chemical and Metallurgical Engineering*, Vol. 38, p. 476 (1931).

$$\frac{F - 32}{180} = \frac{C}{100} = \frac{Ré}{80} = \frac{K - 273}{100} = \frac{R - 492}{180} = \frac{Z - (fp)"Z"}{(bp)"Z" - (fp)"Z"}$$

Examples: (1) To find the Fahrenheit temperature corresponding to $-20°C$:

$$\frac{F - 32}{180} = \frac{C}{100} \text{ or } \frac{F - 32}{180} = \frac{-20}{100}; \text{ i.e., } -20°C = -4°F$$

(2) To find the Réaumur temperature corresponding to $20°F$:

$$\frac{F - 32}{180} = \frac{Ré}{80} \text{ or } \frac{20 - 32}{180} = \frac{Ré}{80}; \text{ i.e., } 20°F = -5.33°Ré$$

(3) To find the correct temperature on a thermometer reading 80°C and which shows a reading of $-0.30°C$ in melting ice and 99.0°C in steam at 760 mm pressure:

$$\frac{C}{100} = \frac{Z - (fp)"Z"}{(bp)"Z" - (fp)"Z"} = \frac{80 - (-0.30)}{99.0 - (0.30)}; \text{ i.e., } C = 80.87° \text{ (corrected)}$$

The reading of any inaccurate thermometer can be corrected and converted in this manner into a reading on any of the other scales; thus, in the example (3) above, to convert into a Fahrenheit scale:

$$\frac{F - 32}{180} = \frac{80 - (-0.30)}{99.0 - (-0.30)}; \text{ i.e., } F = 177.56° \text{ (corrected)}$$

Celsius to Fahrenheit
Formula: $(C \times 9/5) + 32 = F$ Examples: (1) To convert 60°C to F, $(60 \times 9/5) + 32 = 140°F$
(2) To convert $-20°C$ to F, $(-20 \times 9/5) + 32 = -4°F$
(See also special table)

Celsius to Réaumur
Formula: $C \times 4/5 = Ré$ Examples: (1) To convert 40°C to Ré $40 \times 4/5 = 32°Ré$
(2) To convert $-30°C$ to Ré, $-30 \times 4/5 = -24°Ré$

Celsius to Kelvin
Formula: $C + 273.1 = K$ Examples: (1) To convert 20°C to K, $20 + 273.1 = 293.1°K$
(2) To convert $-20°C$ to K, $-20 + 273.1 = 253.1°K$

Fahrenheit to Celsius
Formula: $(F - 32) \times 5/9 = C$ Examples: (1) To convert 140°F to C, $(140 - 32) \times 5/9 = 60°C$
(2) To convert $-76°F$ to C, $(-76 - 32) \times 5/9 = -60°C$
(See also special table)

Fahrenheit to Réaumur
Formula: $(F - 32) \times 4/9 = Ré$ Examples: (1) To convert 41°F to Ré, $(41 - 32) \times 4/9 = 4°Ré$
(2) To convert $-13°F$ to Ré, $(-13 - 32) \times 4/9 = -20°Ré$

Fahrenheit to Rankine
Formula: $F + 459.58 = R$ Examples: (1) To convert 20°F to R, $20 + 459.58 = 479.58°R$
(2) To convert $-20°F$ to R $-20 + 459.58 = 439.58°R$

Réaumur to Celsius
Formula: $Ré \times 5/4 = C$ Examples: (1) To convert 32°Ré to C, $32 \times 5/4 = 40°C$
(2) To convert $-24°Ré$ to C, $-24 \times 5/4 = -30°C$

Réaumur to Fahrenheit
Formula: $(Ré \times 9/4) + 32 = F$ Examples: (1) To convert 8°Ré to F, $(8 \times 9/4) + 32 = 50°F$
(2) To convert $-8°Ré$ to F, $(-8 \times 9/4) + 32 = 14°F$
(3) To convert $-20°Ré$ to F, $(-20 \times 9/4) + 32 = -13°F$

*SI usage for degrees Kelvin is kelvins, abbreviated K.

Table 2-6
TEMPERATURE CONVERSION TABLE

The column of figures in bold and which is headed "Reading in °F. or °C. to be converted" refers to the temperature either in degrees Fahrenheit or Celsius which it is desired to convert into the other scale. If converting from Fahrenheit degrees to Celsius degrees, the equivalent temperature will be found in the column headed "°C."; while if converting from degrees Celsius to degrees Fahrenheit, the equivalent temperature will be found in the column headed "°F." This arrangement is very similar to that of Sauveur and Boylston, copyrighted 1920, and is published with their permission.

°F.	Reading in °F. or °C. to be converted	°C.	°F.	Reading in °F. or °C. to be converted	°C.
....	−458	−272.22	....	−358	−216.67
....	−456	−271.11	....	−356	−215.56
....	−454	−270.00	....	−354	−214.44
....	−452	−268.89	....	−352	−213.33
....	−450	−267.78	....	−350	−212.22
....	−448	−266.67	....	−348	−211.11
....	−446	−265.56	....	−346	−210.00
....	−444	−264.44	....	−344	−208.89
....	−442	−263.33	....	−342	−207.78
....	−440	−262.22	....	−340	−206.67
....	−438	−261.11	....	−338	−205.56
....	−436	−260.00	....	−336	−204.44
....	−434	−258.89	....	−334	−203.33
....	−432	−257.78	....	−332	−202.22
....	−430	−256.67	....	−330	−201.11
....	−428	−255.56	....	−328	−200.00
....	−426	−254.44	....	−326	−198.89
....	−424	−253.33	......	−324	−197.78
....	−422	−252.22	....	−322	−196.67
....	−420	−251.11	....	−320	−195.56
....	−418	−250.00	....	−318	−194.44
....	−416	−248.89	....	−316	−193.33
....	−414	−247.78	....	−314	−192.22
....	−412	−246.67	....	−312	−191.11
....	−410	−245.56	....	−310	−190.00
....	−408	−244.44	....	−308	−188.89
....	−406	−243.33	....	−306	−187.78
....	−404	−242.22	....	−304	−186.67
....	−402	−241.11	....	−302	−185.56
....	−400	−240.00	....	−300	−184.44
....	−398	−238.89	....	−298	−183.33
....	−396	−237.78	....	−296	−182.22
....	−394	−236.67	....	−294	−181.11
....	−392	−235.56	....	−292	−180.00
....	−390	−234.44	....	−290	−178.89
....	−388	−233.33	....	−288	−177.78
....	−386	−232.22	....	−286	−176.67
....	−384	−231.11	....	−284	−175.56
....	−382	−230.00	....	−282	−174.44
....	−380	−228.89	....	−280	−173.33
....	−378	−227.78	....	−278	−172.22
....	−376	−226.67	....	−276	−171.11
....	−374	−225.56	....	−274	−170.00
....	−372	−224.44	−457.6	−272	−168.89
....	−370	−223.33	−454.0	−270	−167.78
....	−368	−222.22	−450.4	−268	−166.67
....	−366	−221.11	−446.8	−266	−165.56
....	−364	−220.00	−443.2	−264	−164.44
....	−362	−218.89	−439.6	−262	−163.33
....	−360	−217.78	−436.0	−260	−162.22

Table 2-6 (*Continued*)
TEMPERATURE CONVERSION TABLE

°F.	Reading in °F. or °C. to be converted	°C.	°F.	Reading in °F. or °C. to be converted	°C.
−432.4	−258	−161.11	−216.4	−138	−94.44
−428.8	−256	−160.00	−212.8	−136	−93.33
−425.2	−254	−158.89	−209.2	−134	−92.22
−421.6	−252	−157.78	−205.6	−132	−91.11
−418.0	−250	−156.67	−202.0	−130	−90.00
−414.4	−248	−155.56	−198.4	−128	−88.89
−410.8	−246	−154.44	−194.8	−126	−87.78
−407.2	−244	−153.33	−191.2	−124	−86.67
−403.6	−242	−152.22	−187.6	−122	−85.56
−400.0	−240	−151.11	−184.0	−120	−84.44
−396.4	−238	−150.00	−180.4	−118	−83.33
−392.8	−236	−148.89	−176.8	−116	−82.22
−389.2	−234	−147.78	−173.2	−114	−81.11
−385.6	−232	−146.67	−169.6	−112	−80.00
−382.0	−230	−145.56	−166.0	−110	−78.89
−378.4	−228	−144.44	−162.4	−108	−77.78
−374.8	−226	−143.33	−158.8	−106	−76.67
−371.2	−224	−142.22	−155.2	−104	−75.56
−367.6	−222	−141.11	−151.6	−102	−74.44
−364.0	−220	−140.00	−148.0	−100	−73.33
−360.4	−218	−138.89	−144.4	−98	−72.22
−356.8	−216	−137.78	−140.8	−96	−71.11
−353.2	−214	−136.67	−137.2	−94	−70.00
−349.6	−212	−135.56	−133.6	−92	−68.89
−346.0	−210	−134.44	−130.0	−90	−67.78
−342.4	−208	−133.33	−126.4	−88	−66.67
−338.8	−206	−132.22	−122.8	−86	−65.56
−335.2	−204	−131.11	−119.2	−84	−64.44
−331.6	−202	−130.00	−115.6	−82	−63.33
−328.0	−200	−128.89	−112.0	−80	−62.22
−324.4	−198	−127.78	−108.4	−78	−61.11
−320.8	−196	−126.67	−104.8	−76	−60.00
−317.2	−194	−125.56	−101.2	−74	−58.89
−313.6	−192	−124.44	−97.6	−72	−57.78
−310.0	−190	−123.33	−94.0	−70	−56.67
−306.4	−188	−122.22	−90.4	−68	−55.56
−302.8	−186	−121.11	−86.8	−66	−54.44
−299.2	−184	−120.00	−83.2	−64	−53.33
−295.6	−182	−118.89	~79.6	−62	−52.22
−292.0	−180	−117.78	−76.0	−60	−51.11
−288.4	−178	−116.67	−72.4	−58	−50.00
−284.8	−176	−115.56	−68.8	−56	−48.89
−281.2	−174	−114.44	−65.2	−54	−47.78
−277.6	−172	−113.33	−61.6	−52	−46.67
−274.0	−170	−112.22	−58.0	−50	−45.56
−270.4	−168	−111.11	−54.4	−48	−44.44
−266.8	−166	−110.00	−50.8	−46	−43.33
−263.2	−164	−108.89	−47.2	−44	−42.22
−259.6	−162	−107.78	−43.6	−42	−41.11
−256.0	−160	−106.67	−40.0	−40	−40.00
−252.4	−158	−105.56	−36.4	−38	−38.89
−248.8	−156	−104.44	−32.8	−36	−37.78
−245.2	−154	−103.33	−29.2	−34	−36.67
−241.6	−152	−102.22	−25.6	−32	−35.56
−238.0	−150	−101.11	−22.0	−30	−34.44
−234.4	−148	−100.00	−18.4	−28	−33.33
−230.8	−146	−98.89	−14.8	−26	−32.22
−227.2	−144	−97.78	−11.2	−24	−31.11
−223.6	−142	−96.67	−7.6	−22	−30.00
−220.0	−140	−95.56	−4.0	−20	−28.89

Table 2-6 (*Continued*)
TEMPERATURE CONVERSION TABLE

°F.	Reading in °F. or °C to be converted	°C.	°F.	Reading in °F. or °C to be converted	°C.
−0.4	−18	−27.78	+116.6	+47	+8.33
+3.2	−16	−26.67	+118.4	+48	+8.89
+6.8	−14	−25.56	+120.2	+49	+9.44
+10.4	−12	−24.44	+122.0	+50	+10.00
+14.0	−10	−23.33	+123.8	+51	+10.56
+17.6	−8	−22.22	+125.6	+52	+11.11
+19.4	−7	−21.67	+127.4	+53	+11.67
+21.2	−6	−21.11	+129.2	+54	+12.22
+23.0	−5	−20.56	+131.0	+55	+12.78
+24.8	−4	−20.00	+132.8	+56	+13.33
+26.6	−3	−19.44	+134.6	+57	+13.89
+28.4	−2	−18.89	+136.4	+58	+14.44
+30.2	−1	−18.33	+138.2	+59	+15.00
+32.0	±0	−17.78	+140.0	+60	+15.56
+33.8	+1	−17.22	+141.8	+61	+16.11
+35.6	+2	−16.67	+143.6	+62	+16.67
+37.4	+3	−16.11	+145.4	+63	+17.22
+39.2	+4	−15.56	+147.2	+64	+17.78
+41.0	+5	−15.00	+149.0	+65	+18.33
+42.8	+6	−14.44	+150.8	+66	+18.89
+44.6	+7	−13.89	+152.6	+67	+19.44
+46.4	+8	−13.33	+154.4	+68	+20.00
+48.2	+9	−12.78	+156.2	+69	+20.56
+50.0	+10	−12.22	+158.0	+70	+21.11
+51.8	+11	−11.67	+159.8	+71	+21.67
+53.6	+12	−11.11	+161.6	+72	+22.22
+55.4	+13	−10.56	+163.4	+73	+22.78
+57.2	+14	−10.00	+165.2	+74	+23.33
+59.0	+15	−9.44	+167.0	+75	+23.89
+60.8	+16	−8.89	+168.8	+76	+24.44
+62.6	+17	−8.33	+170.6	+77	+25.00
+64.4	+18	−7.78	+172.4	+78	+25.56
+66.2	+19	−7.22	+174.2	+79	+26.11
+68.0	+20	−6.67	+176.0	+80	+26.67
+69.8	+21	−6.11	+177.8	+81	+27.22
+71.6	+22	−5.56	+179.6	+82	+27.78
+73.4	+23	−5.00	+181.4	+83	+28.33
+75.2	+24	−4.44	+183.2	+84	+28.89
+77.0	+25	−3.89	+185.0	+85	+29.44
+78.8	+26	−3.33	+186.8	+86	+30.00
+80.6	+27	−2.78	+188.6	+87	+30.56
+82.4	+28	−2.22	+190.4	+88	+31.11
+84.2	+29	−1.67	+192.2	+89	+31.67
+86.0	+30	−1.11	+194.0	+90	+32.22
+87.8	+31	−0.56	+195.8	+91	+32.78
+89.6	+32	±0.00	+197.6	+92	+33.33
+91.4	+33	+0.56	+199.4	+93	+33.89
+93.2	+34	+1.11	+201.2	+94	+34.44
+95.0	+35	+1.67	+203.0	+95	+35.00
+96.8	+36	+2.22	+204.8	+96	+35.56
+98.6	+37	+2.78	+206.6	+97	+36.11
+100.4	+38	+3.33	+208.4	+98	+36.67
+102.2	+39	+3.89	+210.2	+99	+37.22
+104.0	+40	+4.44	+212.0	+100	+37.78
+105.8	+41	+5.00	+213.8	+101	+38.33
+107.6	+42	+5.56	+215.6	+102	+38.89
+109.4	+43	+6.11	+217.4	+103	+39.44
+111.2	+44	+6.67	+219.2	+104	+40.00
+113.0	+45	+7.22	+221.0	+105	+40.56
+114.8	+46	+7.78	+222.8	+106	+41.11

Table 2-6 (*Continued*)
TEMPERATURE CONVERSION TABLE

°F.	Reading in °F. or °C. to be converted	°C.	°F.	Reading in °F. or °C. to be converted	°C.
+224.6	+107	+41.67	+332.6	+167	+75.00
+226.4	+108	+42.22	+334.4	+168	+75.56
+228.2	+109	+42.78	+336.2	+169	+76.11
+230.0	+110	+43.33	+338.0	+170	+76.67
+231.8	+111	+43.89	+339.8	+171	+77.22
+233.6	+112	+44.44	+341.6	+172	+77.78
+235.4	+113	+45.00	+343.4	+173	+78.33
+237.2	+114	+45.56	+345.2	+174	+78.89
+239.0	+115	+46.11	+347.0	+175	+79.44
+240.8	+116	+46.67	+348.8	+176	+80.00
+242.6	+117	+47.22	+350.6	+177	+80.56
+244.4	+118	+47.78	+352.4	+178	+81.11
+246.2	+119	+48.33	+354.2	+179	+81.67
+248.0	+120	+48.89	+356.0	+180	+82.22
+249.8	+121	+49.44	+357.8	+181	+82.78
+251.6	+122	+50.00	+359.6	+182	+83.33
+253.4	+123	+50.56	+361.4	+183	+83.89
+255.2	+124	+51.11	+363.2	+184	+84.44
+257.0	+125	+51.67	+365.0	+185	+85.00
+258.8	+126	+52.22	+366.8	+186	+85.56
+260.6	+127	+52.78	+368.6	+187	+86.11
+262.4	+128	+53.33	+370.4	+188	+86.67
+264.2	+129	+53.89	+372.2	+189	+87.22
+266.0	+130	+54.44	+374.0	+190	+87.78
+267.8	+131	+55.00	+375.8	+191	+88.33
+269.6	+132	+55.56	+377.6	+192	+88.89
+271.4	+133	+56.11	+379.4	+193	+89.44
+273.2	+134	+56.67	+381.2	+194	+90.00
+275.0	+135	+57.22	+383.0	+195	+90.56
+276.8	+136	+57.78	+384.8	+196	+91.11
+278.6	+137	+58.33	+386.6	+197	+91.67
+280.4	+138	+58.89	+388.4	+198	+92.22
+282.2	+139	+59.44	+390.2	+199	+92.78
+284.0	+140	+60.00	+392.0	+200	+93.33
+285.8	+141	+60.56	+393.8	+201	+93.89
+287.6	+142	+61.11	+395.6	+202	+94.44
+289.4	+143	+61.67	+397.4	+203	+95.00
+291.2	+144	+62.22	+399.2	+204	+95.56
+293.0	+145	+62.78	+401.0	+205	+96.11
+294.8	+146	+63.33	+402.8	+206	+96.67
+296.6	+147	+63.89	+404.6	+207	+97.22
+298.4	+148	+64.44	+406.4	+208	+97.78
+300.2	+149	+65.00	+408.2	+209	+98.33
+302.0	+150	+65.56	+410.0	+210	+98.89
+303.8	+151	+66.11	+411.8	+211	+99.44
+305.6	+152	+66.67	+413.6	+212	+100.00
+307.4	+153	+67.22	+415.4	+213	+100.56
+309.2	+154	+67.78	+417.2	+214	+101.11
+311.0	+155	+68.33	+419.0	+215	+101.67
+312.8	+156	+68.89	+420.8	+216	+102.22
+314.6	+157	+69.44	+422.6	+217	+102.78
+316.4	+158	+70.00	+424.4	+218	+103.33
+318.2	+159	+70.56	+426.2	+219	+103.89
+320.0	+160	+71.11	+428.0	+220	+104.44
+321.8	+161	+71.67	+431.6	+222	+105.56
+323.6	+162	+72.22	+435.2	+224	+106.67
+325.4	+163	+72.78	+438.8	+226	+107.78
+327.2	+164	+73.33	+442.4	+228	+108.89
+329.0	+165	+73.89	+446.0	+230	+110.00
+330.8	+166	+74.44	+449.6	+232	+111.11

Table 2-6 (*Continued*)
TEMPERATURE CONVERSION TABLE

°F.	Reading in °F. or °C. to be converted	°C.	°F.	Reading in °F. or °C. to be converted	°C.
+453.2	+234	+112.22	+669.2	+354	+178.89
+456.8	+236	+113.33	+672.8	+356	+180.00
+460.4	+238	+114.44	+676.4	+358	+181.11
+464.0	+240	+115.56	+680.0	+360	+182.22
+467.6	+242	+116.67	+683.6	+362	+183.33
+471.2	+244	+117.78	+687.2	+364	+184.44
+474.8	+246	+118.89	+690.8	+366	+185.56
+478.4	+248	+120.00	+694.4	+368	+186.67
+482.0	+250	+121.11	+698.0	+370	+187.78
+485.6	+252	+122.22	+701.6	+372	+188.89
+489.2	+254	+123.33	+705.2	+374	+190.00
+492.8	+256	+124.44	+708.8	+376	+191.11
+496.4	+258	+125.56	+712.4	+378	+192.22
+500.0	+260	+126.67	+716.0	+380	+193.33
+503.6	+262	+127.78	+719.6	+382	+194.44
+507.2	+264	+128.89	+723.2	+384	+195.56
+510.8	+266	+130.00	+726.8	+386	+196.67
+514.4	+268	+131.11	+730.4	+388	+197.78
+518.0	+270	+132.22	+734.0	+390	+198.89
+521.6	+272	+133.33	+737.6	+392	+200.00
+525.2	+274	+134.44	+741.2	+394	+201.11
+528.8	+276	+135.56	+744.8	+396	+202.22
+532.4	+278	+136.67	+748.4	+398	+203.33
+536.0	+280	+137.78	+752.0	+400	+204.44
+539.6	+282	+138.89	+755.6	+402	+205.56
+543.2	+284	+140.00	+759.2	+404	+206.67
+546.8	+286	+141.11	+762.8	+406	+207.78
+550.4	+288	+142.22	+766.4	+408	+208.89
+554.0	+290	+143.33	+770.0	+410	+210.00
+557.6	+292	+144.44	+773.6	+412	+211.11
+561.2	+294	+145.56	+777.2	+414	+212.22
+564.8	+296	+146.67	+780.8	+416	+213.33
+568.4	+298	+147.78	+784.4	+418	+214.44
+572.0	+300	+148.89	+788.0	+420	+215.56
+575.6	+302	+150.00	+791.6	+422	+216.67
+579.2	+304	+151.11	+795.2	+424	+217.78
+582.8	+306	+152.22	+798.8	+426	+218.89
+586.4	+308	+153.33	+802.4	+428	+220.00
+590.0	+310	+154.44	+806.0	+430	+221.11
+593.6	+312	+155.56	+809.6	+432	+222.22
+597.2	+314	+156.67	+813.2	+434	+223.33
+600.8	+316	+157.78	+816.8	+436	+224.44
+604.4	+318	+158.89	+820.4	+438	+225.56
+608.0	+320	+160.00	+824.0	+440	+226.67
+611.6	+322	+161.11	+827.6	+442	+227.78
+615.2	+324	+162.22	+831.2	+444	+228.89
+618.8	+326	+163.33	+834.8	+446	+230.00
+622.4	+328	+164.44	+838.4	+448	+231.11
+626.0	+330	+165.56	+842.0	+450	+232.22
+629.6	+332	+166.67	+845.6	+452	+233.33
+633.2	+334	+167.78	+849.2	+454	+234.44
+636.8	+336	+168.89	+852.8	+456	+235.56
+640.4	+338	+170.00	+856.4	+458	+236.67
+644.0	+340	+171.11	+860.0	+460	+237.78
+647.6	+342	+172.22	+863.6	+462	+238.89
+651.2	+344	+173.33	+867.2	+464	+240.00
+654.8	+346	+174.44	+870.8	+466	+241.11
+658.4	+348	+175.56	+874.4	+468	+242.22
+662.0	+350	+176.67	+878.0	+470	+243.33
+665.6	+352	+177.78	+881.6	+472	+244.44

Table 2-6 (*Continued*)
TEMPERATURE CONVERSION TABLE

°F.	Reading in °F. or °C. to be converted	°C.	°F.	Reading in °F. or °C. to be converted	°C.
+885.2	+474	+245.56	+1101.2	+594	+312.22
+888.8	+476	+246.67	+1104.8	+596	+313.33
+892.4	+478	+247.78	+1108.4	+598	+314.44
+896.0	+480	+248.89	+1112.0	+600	+315.56
+899.6	+482	+250.00	+1115.6	+602	+316.67
+903.2	+484	+251.11	+1119.2	+604	+317.78
+906.8	+486	+252.22	+1122.8	+606	+318.89
+910.4	+488	+253.33	+1126.4	+608	+320.00
+914.0	+490	+254.44	+1130.0	+610	+321.11
+917.6	+492	+255.56	+1133.6	+612	+322.22
+921.2	+494	+256.67	+1137.2	+614	+323.33
+924.8	+496	+257.78	+1140.8	+616	+324.44
+928.4	+498	+258.89	+1144.4	+618	+325.56
+932.0	+500	+260.00	+1148.0	+620	+326.67
+935.6	+502	+261.11	+1151.6	+622	+327.78
+939.2	+504	+262.22	+1155.2	+624	+328.89
+942.8	+506	+263.33	+1158.8	+626	+330.00
+946.4	+508	+264.44	+1162.4	+628	+331.11
+950.0	+510	+265.56	+1166.0	+630	+332.22
+953.6	+512	+266.67	+1169.6	+632	+333.33
+957.2	+514	+267.78	+1173.2	+634	+334.44
+960.8	+516	+268.89	+1176.8	+636	+335.56
+964.4	+518	+270.00	+1180.4	+638	+336.67
+968.0	+520	+271.11	+1184.0	+640	+337.78
+971.6	+522	+272.22	+1187.6	+642	+338.89
+975.2	+524	+273.33	+1191.2	+644	+340.00
+978.8	+526	+274.44	+1194.8	+646	+341.11
+982.4	+528	+275.56	+1198.4	+648	+342.22
+986.0	+530	+276.67	+1202.0	+650	+343.33
+989.6	+532	+277.78	+1205.6	+652	+344.44
+993.2	+534	+278.89	+1209.2	+654	+345.56
+996.8	+536	+280.00	+1212.8	+656	+346.67
+1000.4	+538	+281.11	+1216.4	+658	+347.78
+1004.0	+540	+282.22	+1220.0	+660	+348.89
+1007.6	+542	+283.33	+1223.6	+662	+350.00
+1011.2	+544	+284.44	+1227.2	+664	+351.11
+1014.8	+546	+285.56	+1230.8	+666	+352.22
+1018.4	+548	+286.67	+1234.4	+668	+353.33
+1022.0	+550	+287.78	+1238.0	+670	+354.44
+1025.6	+552	+288.89	+1241.6	+672	+355.56
+1029.2	+554	+290.00	+1245.2	+674	+356.67
+1032.8	+556	+291.11	+1248.8	+676	+357.78
+1036.4	+558	+292.22	+1252.4	+678	+358.89
+1040.0	+560	+293.33	+1256.0	+680	+360.00
+1043.6	+562	+294.44	+1259.6	+682	+361.11
+1047.2	+564	+295.56	+1263.2	+684	+362.22
+1050.8	+566	+296.67	+1266.8	+686	+363.33
+1054.4	+568	+297.78	+1270.4	+688	+364.44
+1058.0	+570	+298.89	+1274.0	+690	+365.56
+1061.6	+572	+300.00	+1277.6	+692	+366.67
+1065.2	+574	+301.11	+1281.2	+694	+367.78
+1068.8	+576	+302.22	+1284.8	+696	+368.89
+1072.4	+578	+303.33	+1288.4	+698	+370.00
+1076.0	+580	+304.44	+1292.0	+700	+371.11
+1079.6	+582	+305.56	+1295.6	+702	+372.22
+1083.2	+584	+306.67	+1299.2	+704	+373.33
+1086.8	+586	+307.78	+1302.8	+706	+374.44
+1090.4	+588	+308.89	+1306.4	+708	+375.56
+1094.0	+590	+310.00	+1310.0	+710	+376.67
+1097.6	+592	+311.11	+1313.6	+712	+377.78

Table 2-6 (*Continued*)
TEMPERATURE CONVERSION TABLE

°F.	Reading in °F. or °C. to be converted	°C.	°F.	Reading in °F. or °C. to be converted	°C.
+1317.2	+714	+378.89	+1533.2	+834	+445.56
+1320.8	+716	+380.00	+1536.8	+836	+446.67
+1324.4	+718	+381.11	+1540.4	+838	+447.78
+1328.0	+720	+382.22	+1544.0	+840	+448.89
+1331.6	+722	+383.33	+1547.6	+842	+450.00
+1335.2	+724	+384.44	+1551.2	+844	+451.11
+1338.8	+726	+385.56	+1554.8	+846	+452.22
+1342.4	+728	+386.67	+1558.4	+848	+453.33
+1346.0	+730	+387.78	+1562.0	+850	+454.44
+1349.6	+732	+388.89	+1565.6	+852	+455.56
+1353.2	+734	+390.00	+1569.2	+854	+456.67
+1356.8	+736	+391.11	+1572.8	+856	+457.78
+1360.4	+738	+392.22	+1576.4	+858	+458.89
+1364.0	+740	+393.33	+1580.0	+860	+460.00
+1367.6	+742	+394.44	+1583.6	+862	+461.11
+1371.2	+744	+395.56	+1587.2	+864	+462.22
+1374.8	+746	+396.67	+1590.8	+866	+463.33
+1378.4	+748	+397.78	+1594.4	+868	+464.44
+1382.0	+750	+398.89	+1598.0	+870	+465.56
+1385.6	+752	+400.00	+1601.6	+872	+466.67
+1389.2	+754	+401.11	+1605.2	+874	+467.78
+1392.8	+756	+402.22	+1608.8	+876	+468.89
+1396.4	+758	+403.33	+1612.4	+878	+470.00
+1400.0	+760	+404.44	+1616.0	+880	+471.11
+1403.6	+762	+405.56	+1619.6	+882	+472.22
+1407.2	+764	+406.67	+1623.2	+884	+473.33
+1410.8	+766	+407.78	+1626.8	+886	+474.44
+1414.4	+768	+408.89	+1630.4	+888	+475.56
+1418.0	+770	+410.00	+1634.0	+890	+476.67
+1421.6	+772	+411.11	+1637.6	+892	+477.78
+1425.2	+774	+412.22	+1641.2	+894	+478.89
+1428.8	+776	+413.33	+1644.8	+896	+480.00
+1432.4	+778	+414.44	+1648.4	+898	+481.11
+1436.0	+780	+415.56	+1652.0	+900	+482.22
+1439.6	+782	+416.67	+1655.6	+902	+483.33
+1443.2	+784	+417.78	+1659.2	+904	+484.44
+1446.8	+786	+418.89	+1662.8	+906	+485.56
+1450.4	+788	+420.00	+1666.4	+908	+486.67
+1454.0	+790	+421.11	+1670.0	+910	+487.78
+1457.6	+792	+422.22	+1673.6	+912	+488.89
+1461.2	+794	+423.33	+1677.2	+914	+490.00
+1464.8	+796	+424.44	+1680.8	+916	+491.11
+1468.4	+798	+425.56	+1684.4	+918	+492.22
+1472.0	+800	+426.67	+1688.0	+920	+493.33
+1475.6	+802	+427.78	+1691.6	+922	+494.44
+1479.2	+804	+428.89	+1695.2	+924	+495.56
+1482.8	+806	+430.00	+1698.8	+926	+496.67
+1486.4	+808	+431.11	+1702.4	+928	+497.78
+1490.0	+810	+432.22	+1706.0	+930	+498.89
+1493.6	+812	+433.33	+1709.6	+932	+500.00
+1497.2	+814	+434.44	+1713.2	+934	+501.11
+1500.8	+816	+435.56	+1716.8	+936	+502.22
+1504.4	+818	+436.67	+1720.4	+938	+503.33
+1508.0	+820	+437.78	+1724.0	+940	+504.44
+1511.6	+822	+438.89	+1727.6	+942	+505.56
+1515.2	+824	+440.00	+1731.2	+944	+506.67
+1518.8	+826	+441.11	+1734.8	+946	+507.78
+1522.4	+828	+442.22	+1738.4	+948	+508.89
+1526.0	+830	+443.33	+1742.0	+950	+510.00
+1529.6	+832	+444.44	+1745.6	+952	+511.11

Table 2-6 (*Continued*)
TEMPERATURE CONVERSION TABLE

°F.	Reading in °F. or °C. to be converted	°C.	°F.	Reading in °F. or C°. to be converted	°C.
+1749.2	+954	+512.22	+2498.0	+1370	+743.33
+1752.8	+956	+513.33	+2516.0	+1380	+748.89
+1756.4	+958	+514.44	+2534.0	+1390	+754.44
+1760.0	+960	+515.56	+2552.0	+1400	+760.00
+1763.6	+962	+516.67	+2570.0	+1410	+765.56
+1767.2	+964	+517.78	+2588.0	+1420	+771.11
+1770.8	+966	+518.89	+2606.0	+1430	+776.67
+1774.4	+968	+520.00	+2624.0	+1440	+782.22
+1778.0	+970	+521.11	+2642.0	+1450	+787.78
+1781.6	+972	+522.22	+2660.0	+1460	+793.33
+1785.2	+974	+523.33	+2678.0	+1470	+798.89
+1788.8	+976	+524.44	+2696.0	+1480	+804.44
+1792.4	+978	+525.56	+2714.0	+1490	+810.00
+1796.0	+980	+526.67	+2732.0	+1500	+815.56
+1799.6	+982	+527.78	+2750.0	+1510	+821.11
+1803.2	+984	+528.89	+2768.0	+1520	+826.67
+1806.8	+986	+530.00	+2786.0	+1530	+832.22
+1810.4	+988	+531.11	+2804.0	+1540	+837.78
+1814.0	+990	+532.22	+2822.0	+1550	+843.33
+1817.6	+992	+533.33	+2840.0	+1560	+848.89
+1821.2	+994	+534.44	+2858.0	+1570	+854.44
+1824.8	+996	+535.56	+2876.0	+1580	+860.00
+1828.4	+998	+536.67	+2894.0	+1590	+865.56
+1832.0	+1000	+537.78	+2912.0	+1600	+871.11
+1850.0	+1010	+543.33	+2930.0	+1610	+876.67
+1868.0	+1020	+548.89	+2948.0	+1620	+882.22
+1886.0	+1030	+554.44	+2966.0	+1630	+887.78
+1904.0	+1040	+560.00	+2984.0	+1640	+893.33
+1922.0	+1050	+565.56	+3002.0	+1650	+898.89
+1940.0	+1060	+571.11	+3020.0	+1660	+904.44
+1958.0	+1070	+576.67	+3038.0	+1670	+910.00
+1976.0	+1080	+582.22	+3056.0	+1680	+915.56
+1994.0	+1090	+587.78	+3074.0	+1690	+921.11
+2012.0	+1100	+593.33	+3092.0	+1700	+926.67
+2030.0	+1110	+598.89	+3110.0	+1710	+932.22
+2048.0	+1120	+604.44	+3128.0	+1720	+937.78
+2066.0	+1130	+610.00	+3146.0	+1730	+943.33
+2084.0	+1140	+615.56	+3164.0	+1740	+948.89
+2102.0	+1150	+621.11	+3182.0	+1750	+954.44
+2120.0	+1160	+626.67	+3200.0	+1760	+960.00
+2138.0	+1170	+632.22	+3218.0	+1770	+965.56
+2156.0	+1180	+637.78	+3236.0	+1780	+971.11
+2174.0	+1190	+643.33	+3254.0	+1790	+976.67
+2192.0	+1200	+648.89	+3272.0	+1800	+982.22
+2210.0	+1210	+654.44	+3290.0	+1810	+987.78
+2228.0	+1220	+660.00	+3308.0	+1820	+993.33
+2246.0	+1230	+665.56	+3326.0	+1830	+998.89
+2264.0	+1240	+671.11	+3344.0	+1840	+1004.4
+2282.0	+1250	+676.67	+3362.0	+1850	+1010.0
+2300.0	+1260	+682.22	+3380.0	+1860	+1015.6
+2318.0	+1270	+687.78	+3398.0	+1870	+1021.1
+2336.0	+1280	+693.33	+3416.0	+1880	+1026.7
+2354.0	+1290	+698.89	+3434.0	+1890	+1032.2
+2372.0	+1300	+704.44	+3452.0	+1900	+1037.8
+2390.0	+1310	+710.00	+3470.0	+1910	+1043.3
+2408.0	+1320	+715.56	+3488.0	+1920	+1048.9
+2426.0	+1330	+721.11	+3506.0	+1930	+1054.4
+2444.0	+1340	+726.67	+3524.0	+1940	+1060.0
+2462.0	+1350	+732.22	+3542.0	+1950	+1065.6
+2480.0	+1360	+737.78	+3560.0	+1960	+1071.1

Table 2-6 (*Continued*)
TEMPERATURE CONVERSION TABLE

°F.	Reading in °F. or °C. to be converted	°C.	°F.	Reading in °F. or °C. to be converted	°C.
+3578.0	+1970	+1076.7	+4604.0	+2540	+1393.3
+3596.0	+1980	+1082.2	+4622.0	+2550	+1398.9
+3614.0	+1990	+1087.8	+4640.0	+2560	+1404.4
+3632.0	+2000	+1093.3	+4658.0	+2570	+1410.0
+3650.0	+2010	+1098.9	+4676.0	+2580	+1415.6
+3668.0	+2020	+1104.4	+4694.0	+2590	+1421.1
+3686.0	+2030	+1110.0	+4712.0	+2600	+1426.7
+3704.0	+2040	+1115.6	+4730.0	+2610	+1432.2
+3722.0	+2050	+1121.1	+4748.0	+2620	+1437.8
+3740.0	+2060	+1126.7	+4766.0	+2630	+1443.3
+3758.0	+2070	+1132.2	+4784.0	+2640	+1448.9
+3776.0	+2080	+1137.8	+4802.0	+2650	+1454.4
+3794.0	+2090	+1143.3	+4820.0	+2660	+1460.0
+3812.0	+2100	+1148.9	+4838.0	+2670	+1465.6
+3830.0	+2110	+1154.4	+4856.0	+2680	+1471.1
+3848.0	+2120	+1160.0	+4874.0	+2690	+1476.7
+3866.0	+2130	+1165.6	+4892.0	+2700	+1482.2
+3884.0	+2140	+1171.1	+4910.0	+2710	+1487.8
+3902.0	+2150	+1176.7	+4928.0	+2720	+1493.3
+3920.0	+2160	+1182.2	+4946.0	+2730	+1498.9
+3938.0	+2170	+1187.8	+4964.0	+2740	+1504.4
+3956.0	+2180	+1193.3	+4982.0	+2750	+1510.0
+3974.0	+2190	+1198.9	+5000.0	+2760	+1515.6
+3992.0	+2200	+1204.4	+5018.0	+2770	+1521.1
+4010.0	+2210	+1210.0	+5036.0	+2780	+1526.7
+4028.0	+2220	+1215.6	+5054.0	+2790	+1532.2
+4046.0	+2230	+1221.1	+5072.0	+2800	+1537.8
+4064.0	+2240	+1226.7	+5090.0	+2810	+1543.3
+4082.0	+2250	+1232.2	+5108.0	+2820	+1548.9
+4100.0	+2260	+1237.8	+5126.0	+2830	+1554.4
+4118.0	+2270	+1243.3	+5144.0	+2840	+1560.0
+4136.0	+2280	+1248.9	+5162.0	+2850	+1565.6
+4154.0	+2290	+1254.4	+5180.0	+2860	+1571.1
+4172.0	+2300	+1260.0	+5198.0	+2870	+1576.7
+4190.0	+2310	+1265.6	+5216.0	+2880	+1582.2
+4208.0	+2320	+1271.1	+5234.0	+2890	+1587.8
+4226.0	+2330	+1276.7	+5252.0	+2900	+1593.3
+4244.0	+2340	+1282.2	+5270.0	+2910	+1598.9
+4262.0	+2350	+1287.8	+5288.0	+2920	+1604.4
+4280.0	+2360	+1293.3	+5306.0	+2930	+1610.0
+4298.0	+2370	+1298.9	+5324.0	+2940	+1615.6
+4316.0	+2380	+1304.4	+5342.0	+2950	+1621.1
+4334.0	+2390	+1310.0	+5360.0	+2960	+1626.7
+4352.0	+2400	+1315.6	+5378.0	+2970	+1632.2
+4370.0	+2410	+1321.1	+5396.0	+2980	+1637.8
+4388.0	+2420	+1326.7	+5414.0	+2990	+1643.3
+4406.0	+2430	+1332.2	+5432.0	+3000	+1648.9
+4424.0	+2440	+1337.8	+5450.0	+3010	+1654.4
+4442.0	+2450	+1343.3	+5468.0	+3020	+1660.0
+4460.0	+2460	+1348.9	+5486.0	+3030	+1665.6
+4478.0	+2470	+1354.4	+5504.0	+3040	+1671.1
+4496.0	+2480	+1360.0	+5522.0	+3050	+1676.7
+4514.0	+2490	+1365.6	+5540.0	+3060	+1682.2
+4532.0	+2500	+1371.1	+5558.0	+3070	+1687.8
+4550.0	+2510	+1376.7	+5576.0	+3080	+1693.3
+4568.0	+2520	+1382.2	+5594.0	+3090	+1698.9
+4586.0	+2530	+1387.8	+5612.0	+3100	+1704.4

Table 2-7
HYDROMETER CONVERSION TABLE

This table gives the relation between density (c.g.s.) and degrees on the Baumé and Twaddell scales. The Twaddell scale is never used for densities less than unity. See also section on *Hydrometers*.

Density	Degrees Baumé Bu. Stand. Scale	Degrees Baumé A.P.I. Scale	Density	Degrees Baumé Bu. Stand. Scale	Degrees Baumé A.P.I. Scale
0.600	103.33	104.33	0.895	26.42	26.60
0.605	101.40	102.38	0.900	25.56	25.72
0.610	99.51	100.47	0.905	24.70	24.85
0.615	97.64	98.58	0.910	23.85	23.99
0.620	95.81	96.73	0.915	23.01	23.14
0.625	94.00	94.90	0.920	22.17	22.30
0.630	92.22	93.10	0.925	21.35	21.47
0.635	90.47	91.33	0.930	20.54	20.65
0.640	88.75	89.59	0.935	19.73	19.84
0.645	87.05	87.88	0.940	18.94	19.03
0.650	85.38	86.19	0.945	18.15	18.24
0.655	83.74	84.53	0.950	17.37	17.45
0.660	82.12	82.89	0.955	16.60	16.67
0.665	80.52	81.28	0.960	15.83	15.90
0.670	78.95	79.69	0.965	15.08	15.13
0.675	77.41	78.13	0.970	14.33	14.38
0.680	75.88	76.59	0.975	13.59	13.63
0.685	74.38	75.07	0.980	12.86	12.89
0.690	72.90	73.57	0.985	12.13	12.15
0.695	71.43	72.10	0.990	11.41	11.43
0.700	70.00	70.64	0.995	10.70	10.71
0.705	68.57	69.21	1.000	10.00	10.00
0.710	67.18	67.80			
0.715	65.80	66.40			
0.720	64.44	65.03			

DENSITIES GREATER THAN UNITY

Density	Degrees Baumé Bu. Stand. Scale	Degrees Twaddell
1.00	0.00	0
1.01	1.44	2
1.02	2.84	4
1.03	4.22	6
1.04	5.58	8
1.05	6.91	10
1.06	8.21	12
1.07	9.49	14
1.08	10.78	16
1.09	11.97	18
1.10	13.18	20
1.11	14.37	22
1.12	15.54	24
1.13	16.68	26
1.14	17.81	28
1.15	18.91	30
1.16	20.00	32
1.17	21.07	34
1.18	22.12	36
1.19	23.15	38
1.20	24.17	40
1.21	25.16	42
1.22	26.15	44
1.23	27.11	46
1.24	28.06	48
1.25	29.00	50
1.26	29.92	52
1.27	30.83	54
1.28	31.72	56
1.29	32.60	58
1.30	33.46	60
1.31	34.31	62

Left lower continuation:

Density	Degrees Baumé Bu. Stand. Scale	Degrees Baumé A.P.I. Scale
0.725	63.10	63.67
0.730	61.78	62.34
0.735	60.48	61.02
0.740	59.19	59.72
0.745	57.92	58.43
0.750	56.67	57.17
0.755	55.43	55.92
0.760	54.21	54.68
0.765	53.01	53.47
0.770	51.82	52.27
0.775	50.65	51.08
0.780	49.49	49.91
0.785	48.34	48.75
0.790	47.22	47.61
0.795	46.10	46.49
0.800	45.00	45.38
0.805	43.91	44.28
0.810	42.84	43.19
0.815	41.78	42.12
0.820	40.73	41.06
0.825	39.70	40.02
0.830	38.68	38.98
0.835	37.66	37.96
0.840	36.67	36.95
0.845	35.68	35.96
0.850	34.71	34.97
0.855	33.74	34.00
0.860	32.79	33.03
0.865	31.85	32.08
0.870	30.92	31.14
0.875	30.00	30.21
0.880	29.09	29.30
0.885	28.19	28.39
0.890	27.30	27.49

Table 2-7 (*Continued*)
HYDROMETER CONVERSION TABLE

Density	Degrees Baumé Bu. Stand. Scale	Degrees Twaddell	Density	Degrees Baumé Bu. Stand. Scale	Degrees Twaddell
1.32	35.15	64	1.67	58.17	134
1.33	35.98	66	1.68	58.69	136
1.34	36.79	68	1.69	59.20	138
1.35	37.59	70	1.70	59.71	140
1.36	38.38	72	1.71	60.20	142
1.37	39.16	74	1.72	60.70	144
1.38	39.93	76	1.73	61.18	146
1.39	40.68	78	1.74	61.67	148
1.40	41.43	80	1.75	62.14	150
1.41	42.16	82	1.76	62.61	152
1.42	42.89	84	1.77	63.08	154
1.43	43.60	86	1.78	63.54	156
1.44	44.31	88	1.79	63.99	158
1.45	45.00	90	1.80	64.44	160
1.46	45.68	92	1.81	64.89	162
1.47	46.36	94	1.82	65.31	164
1.48	47.03	96	1.83	65.77	166
1.49	47.68	98	1.84	66.20	168
1.50	48.33	100	1.85	66.62	170
1.51	48.97	102	1.86	67.04	172
1.52	49.60	104	1.87	67.46	174
1.53	50.23	106	1.88	67.87	176
1.54	50.84	108	1.89	68.28	178
1.55	51.45	110	1.90	68.68	180
1.56	52.05	112	1.91	69.08	182
1.57	52.64	114	1.92	69.48	184
1.58	53.23	116	1.93	69.87	186
1.59	53.80	118	1.94	70.26	188
1.60	54.38	120	1.95	70.64	190
1.61	54.94	122	1.96	71.02	192
1.62	55.49	124	1.97	71.40	194
1.63	56.04	126	1.98	71.77	196
1.64	56.58	128	1.99	72.14	198
1.65	57.12	130	2.00	72.50	200
1.66	57.65	132			

BAROMETRY AND BAROMETRIC CORRECTIONS

In principle, the mercurial barometer balances a column of pure mercury against the weight of the atmosphere. The height of the column above the level of the mercury in the reservoir can be measured and serves as a direct index of atmospheric pressure. The space above the mercury in a barometer tube should be a Torricellian vacuum, perfect except for the practically negligible vapor pressure of mercury. The perfection of the vacuum is indicated by the sharpness of the click noted when the barometer tube is inclined. A barometer should be in a vertical position, suspended rather than fastened to a wall, and in a good light but not exposed to direct sunlight or too near a source of heat. The standard conditions for barometric measurements are 0°C., and gravity as at 45° latitude and sea level. There are numerous sources of error, but corrections for most of these are readily applied. Some of the corrections are very small, and their application may be questionable in view of the probably larger errors. The degree of consistency to be expected in careful measurements is about 0.13 mm with a 6.4-mm tube, increasing to 0.04 mm with a tube 12.7 mm in diameter.

In reading a barometer of the Fortin type (the usual laboratory instrument for precision measurements), the procedure should be as follows: (1) Observe and record the temperature as indicated by the thermometer attached to the barometer. The temperature correction is very important and may be affected by heat from the observer's body. (2) Set the mercury in the reservoir at zero level, so that the point of the pin above the mercury just touches the surface, making a barely noticeable dimple therein. Tap the tube at the top and verify the zero setting. (3) Bring the vernier down until the view at the light background is cut off at the highest point of the meniscus. Record the reading.

The corrections to be made on the reading are as follow: (1) Temperature, to correct for the difference in thermal expansion of the mercury and the brass (or glass) to which the scale is attached.—This correction converts the reading into the value of 0°C. The brass scale table is applicable to the Fortin barometer. See tables. (2a) Latitude-gravity correction, and (2b) altitude-gravity correction, to compensate for differences in gravity, which would affect the height of the mercury column by variation in mass.—If local gravity is unknown, an approximate correction may be made from the tables. Local values of gravity are often subject to irregularities which lead to errors even when the corrections here provided are made. It is, therefore, advisable to determine the local value of gravity, from which the correction can be effected in the following manner:

$$Bt = Br + \left(\frac{g_1 - g_0}{g_0}\right) \times Br$$

in which Bt and Br are the true and the observed heights of the barometer respectively, g_0 is standard gravity (980 665 cm · sec^{-2}), and g_1 is the local gravity. It may be noted that for most localities, g_1 is smaller than g_0, which makes the correction negative. These corrections compensate the reading to gravity at 45° latitude and sea level. (3) Correction for capillary depression of the level of the meniscus.—This varies with the tube diameter and actual height of the meniscus in a particular case. Some barometers are calibrated to allow for an average value of the latter and approximating the correction. See table. (4) Correction for vapor pressure of mercury.—This correction is usually negligible, being only 0.001 mm at 20°C. and 0.006 mm at 40°C. This correction is added. See table of vapor pressure of mercury.

The corrections above do not apply to aneroid barometers. These instruments should be calibrated at regular intervals by checking them against a corrected mercurial barometer.

For records on weather maps, meteorologists customarily correct barometer readings to sea level, and some barometers may be calibrated accordingly. Such instruments are not suitable for laboratory use where true pressure under standard conditions is required. Scale corrections should be specified in the maker's instructions with the instrument, and are also indicated by the lack of correspondence between a gauge mark usually placed exactly 76.2 cm from the zero point and the 76.2-cm scale graduation.

Table 2-8
BAROMETER TEMPERATURE CORRECTION—METRIC UNITS
A. Glass Scale

The values in the table below are to be subtracted from the observed readings to correct for the difference in the expansion of the mercury and the glass scale at different temperatures.

Temp. °C.	Observed barometer height in millimeters						
	700	730	740	750	760	770	800
	mm.	mm.	mm.	mm.	mm.	mm.	mm.
0	0.00	0.00	0.00	0.00	0.00	0.00	0.00
1	0.12	0.13	0.13	0.13	0.13	0.13	0.14
2	0.24	0.25	0.26	0.26	0.26	0.27	0.27
3	0.36	0.38	0.38	0.39	0.40	0.40	0.42
4	0.49	0.51	0.51	0.52	0.53	0.53	0.55
5	0.61	0.63	0.64	0.65	0.66	0.67	0.69
6	0.73	0.76	0.77	0.78	0.79	0.80	0.83
7	0.85	0.89	0.90	0.91	0.92	0.93	0.97
8	0.97	1.01	1.03	1.04	1.05	1.07	1.11
9	1.09	1.14	1.15	1.17	1.18	1.20	1.25
10	1.21	1.26	1.28	1.30	1.32	1.33	1.39
11	1.33	1.39	1.41	1.43	1.45	1.47	1.52
12	1.45	1.52	1.54	1.56	1.58	1.60	1.66
13	1.58	1.64	1.67	1.69	1.71	1.73	1.80
14	1.70	1.77	1.79	1.82	1.84	1.87	1.94
15	1.82	1.90	1.92	1.95	1.97	2.00	2.08
16	1.94	2.02	2.05	2.08	2.10	2.13	2.21
17	2.06	2.15	2.18	2.21	2.23	2.26	2.35
18	2.18	2.27	2.30	2.33	2.37	2.40	2.49
19	2.30	2.40	2.43	2.46	2.50	2.53	2.63
20	2.42	2.52	2.56	2.59	2.63	2.66	2.77
21	2.54	2.65	2.69	2.72	2.76	2.79	2.90
22	2.66	2.78	2.81	2.85	2.89	2.93	3.04
23	2.78	2.90	2.94	2.98	3.02	3.06	3.18
24	2.90	3.03	3.07	3.11	3.15	3.19	3.32
25	3.02	3.15	3.20	3.24	3.28	3.32	3.45
26	3.14	3.28	3.32	3.37	3.41	3.46	3.59
27	3.26	3.40	3.45	3.50	3.54	3.59	3.73
28	3.38	3.53	3.58	3.63	3.67	3.72	3.87
29	3.50	3.65	3.70	3.75	3.80	3.85	4.00
30	3.62	3.78	3.83	3.88	3.93	3.99	4.14
31	3.74	3.90	3.96	4.01	4.06	4.12	4.28
32	3.86	4.03	4.08	4.14	4.20	4.25	4.42
33	3.98	4.15	4.21	4.27	4.33	4.38	4.55
34	4.10	4.28	4.34	4.40	4.46	4.51	4.69
35	4.22	4.40	4.47	4.53	4.59	4.65	4.83

Table 2-8 (*Continued*)
BAROMETER TEMPERATURE CORRECTION—METRIC UNITS
B. Brass Scale

The values in the table below are to be subtracted from the observed readings to correct for the difference in the expansion of the mercury and the brass scale at different temperatures.

Temp. °C.	Observed barometer height in millimeters						
	640	650	660	670	680	690	700
	mm.	mm.	mm.	mm.	mm.	mm.	mm.
0	0.00	0.00	0.00	0.00	0.00	0.00	0.00
1	0.10	0.11	0.11	0.11	0.11	0.11	0.11
2	0.21	0.21	0.22	0.22	0.22	0.23	0.23
3	0.31	0.32	0.32	0.33	0.33	0.34	0.34
4	0.42	0.42	0.43	0.44	0.44	0.45	0.46
5	0.52	0.53	0.54	0.55	0.55	0.56	0.57
6	0.63	0.64	0.65	0.66	0.66	0.67	0.68
7	0.73	0.74	0.75	0.76	0.78	0.79	0.80
8	0.84	0.85	0.86	0.87	0.89	0.90	0.91
9	0.94	0.95	0.97	0.98	1.00	1.01	1.03
10	1.04	1.06	1.07	1.09	1.11	1.12	1.14
11	1.15	1.16	1.18	1.20	1.22	1.24	1.25
12	1.25	1.27	1.29	1.31	1.33	1.35	1.37
13	1.35	1.38	1.40	1.42	1.44	1.46	1.48
14	1.46	1.48	1.50	1.53	1.55	1.57	1.59
15	1.56	1.59	1.61	1.64	1.66	1.68	1.71
16	1.67	1.69	1.72	1.74	1.77	1.80	1.82
17	1.77	1.80	1.82	1.85	1.88	1.91	1.94
18	1.87	1.90	1.93	1.96	1.99	2.02	2.05
19	1.98	2.01	2.04	2.07	2.10	2.13	2.16
20	2.08	2.11	2.15	2.18	2.21	2.24	2.28
21	2.18	2.22	2.25	2.29	2.32	2.35	2.39
22	2.29	2.32	2.36	2.40	2.43	2.47	2.50
23	2.39	2.43	2.47	2.50	2.54	2.58	2.62
24	2.49	2.53	2.57	2.61	2.65	2.69	2.73
25	2.60	2.64	2.68	2.72	2.76	2.80	2.84
26	2.70	2.74	2.79	2.83	2.87	2.91	2.96
27	2.81	2.85	2.89	2.94	2.98	3.02	3.07
28	2.91	2.95	3.00	3.05	3.09	3.14	3.18
29	3.01	3.06	3.11	3.15	3.20	3.25	3.29
30	3.12	3.16	3.21	3.26	3.31	3.36	3.41
31	3.22	3.27	3.32	3.37	3.42	3.47	3.52
32	3.32	3.37	3.43	3.48	3.53	3.58	3.63
33	3.42	3.48	3.53	3.59	3.64	3.69	3.75
34	3.53	3.58	3.64	3.69	3.75	3.80	3.86
35	3.63	3.69	3.74	3.80	3.86	3.91	3.97

Table 2-8 (*Continued*)
BAROMETER TEMPERATURE CORRECTION—METRIC UNITS
B. Brass Scale

Observed barometer height in millimeters								Temp. °C.
710	720	730	740	750	760	770	780	
mm.	mm.	mm.	mm.	mm.	mm.	mm.	mm.	
0.00	0.00	0.00	0.00	0.00	0.00	0.00	0.00	0
0.12	0.12	0.12	0.12	0.12	0.12	0.13	0.13	1
0.23	0.23	0.24	0.24	0.24	0.25	0.25	0.25	2
0.35	0.35	0.36	0.36	0.37	0.37	0.38	0.38	3
0.46	0.47	0.48	0.48	0.49	0.50	0.50	0.51	4
0.58	0.59	0.59	0.60	0.61	0.62	0.63	0.64	5
0.69	0.70	0.71	0.72	0.73	0.74	0.75	0.76	6
0.81	0.82	0.83	0.84	0.86	0.87	0.88	0.89	7
0.93	0.94	0.95	0.96	0.98	0.99	1.00	1.02	8
1.04	1.06	1.07	1.08	1.10	1.11	1.13	1.14	9
1.16	1.17	1.19	1.21	1.22	1.24	1.25	1.27	10
1.27	1.29	1.31	1.33	1.34	1.36	1.38	1.40	11
1.39	1.41	1.43	1.45	1.47	1.48	1.50	1.52	12
1.50	1.52	1.54	1.57	1.59	1.61	1.63	1.65	13
1.62	1.64	1.66	1.69	1.71	1.73	1.75	1.78	14
1.73	1.76	1.78	1.81	1.83	1.85	1.88	1.90	15
1.85	1.87	1.90	1.93	1.95	1.98	2.00	2.03	16
1.96	1.99	2.02	2.05	2.07	2.10	2.13	2.16	17
2.08	2.11	2.14	2.17	2.20	2.22	2.25	2.28	18
2.19	2.22	2.25	2.29	2.32	2.35	2.38	2.41	19
2.31	2.34	2.37	2.41	2.44	2.47	2.50	2.54	20
2.42	2.46	2.49	2.53	2.56	2.59	2.63	2.66	21
2.54	2.57	2.61	2.65	2.68	2.72	2.75	2.79	22
2.65	2.69	2.73	2.77	2.80	2.84	2.88	2.91	23
2.77	2.81	2.85	2.88	2.92	2.96	3.00	3.04	24
2.88	2.92	2.96	3.00	3.05	3.09	3.13	3.17	25
3.00	3.04	3.08	3.12	3.17	3.21	3.25	3.29	26
3.11	3.16	3.20	3.24	3.29	3.33	3.38	3.42	27
3.23	3.27	3.32	3.36	3.41	3.45	3.50	3.54	28
3.34	3.39	3.44	3.48	3.53	3.58	3.62	3.67	29
3.46	3.50	3.55	3.60	3.65	3.70	3.75	3.80	30
3.57	3.62	3.67	3.72	3.77	3.82	3.87	3.92	31
3.68	3.74	3.79	3.84	3.89	3.94	4.00	4.05	32
3.80	3.85	3.91	3.96	4.01	4.07	4.12	4.17	33
3.91	3.97	4.02	4.08	4.13	4.19	4.24	4.30	34
4.03	4.09	4.14	4.20	4.26	4.31	4.37	4.43	35

C. Correction of a Barometer for Capillarity
(*Smithsonian Tables*)

Diameter of tube, millimeters	Height of meniscus in millimeters							
	0.4	0.6	0.8	1.0	1.2	1.4	1.6	1.8
	Correction to be added in millimeters							
4	0.83	1.22	1.54	1.98	2.37	....	....	....
5	0.47	0.65	0.86	1.19	1.45	1.80	....	....
6	0.27	0.41	0.56	0.78	0.98	1.21	1.43	....
7	0.18	0.28	0.40	0.53	0.67	0.82	0.97	1.13
8	....	0.20	0.29	0.38	0.46	0.56	0.65	0.77
9	....	0.15	0.21	0.28	0.33	0.40	0.46	0.52
10	....	....	0.15	0.20	0.25	0.29	0.33	0.37
11	....	....	0.10	0.14	0.18	0.21	0.24	0.27
12	....	....	0.07	0.10	0.13	0.15	0.18	0.19
13	....	....	0.04	0.07	0.10	0.12	0.13	0.14

Table 2-9
BAROMETRIC LATITUDE-GRAVITY TABLE—METRIC UNITS
(*Smithsonian Tables*)

The values below are to be subtracted from the barometric reading for latitudes from 0 to 45° inclusive, and are to be added from 46 to 90°.

Deg. Lat.	Barometer readings, millimeters					
	680	700	720	740	760	780
	mm.	mm.	mm.	mm.	mm.	mm.
0	1.82	1.87	1.93	1.98	2.04	2.09
5	1.79	1.85	1.90	1.95	2.00	2.06
10	1.71	1.76	1.81	1.86	1.92	1.97
15	1.58	1.63	1.67	1.72	1.77	1.81
20	1.40	1.44	1.49	1.53	1.57	1.61
21	1.36	1.40	1.44	1.48	1.52	1.56
22	1.32	1.36	1.40	1.44	1.48	1.51
23	1.28	1.31	1.35	1.39	1.43	1.46
24	1.23	1.27	1.30	1.34	1.37	1.41
25	1.18	1.22	1.25	1.29	1.32	1.36
26	1.13	1.17	1.20	1.23	1.27	1.30
27	1.08	1.12	1.15	1.18	1.21	1.24
28	1.03	1.06	1.09	1.12	1.15	1.18
29	0.98	1.01	1.04	1.07	1.10	1.12
30	0.93	0.95	0.98	1.01	1.04	1.06
31	0.87	0.90	0.92	0.95	0.98	1.00
32	0.82	0.84	0.86	0.89	0.91	0.94
33	0.76	0.78	0.80	0.83	0.85	0.87
34	0.70	0.72	0.74	0.76	0.79	0.81
35	0.64	0.66	0.68	0.70	0.72	0.74
36	0.58	0.60	0.62	0.64	0.65	0.67
37	0.52	0.54	0.56	0.57	0.59	0.60
38	0.46	0.48	0.49	0.51	0.52	0.53
39	0.40	0.42	0.43	0.44	0.45	0.46
40	0.34	0.35	0.36	0.37	0.38	0.39
41	0.28	0.29	0.30	0.30	0.31	0.32
42	0.22	0.22	0.23	0.24	0.24	0.25
43	0.16	0.16	0.16	0.17	0.17	0.18
44	0.09	0.10	0.10	0.10	0.10	0.11
45	0.03	0.03	0.03	0.03	0.03	0.04
46	0.03	0.03	0.03	0.03	0.04	0.04
47	0.09	0.10	0.10	0.10	0.10	0.11
48	0.16	0.16	0.17	0.17	0.18	0.18
49	0.22	0.23	0.23	0.24	0.25	0.25
50	0.28	0.29	0.30	0.31	0.31	0.32
51	0.34	0.35	0.36	0.37	0.38	0.39
52	0.40	0.42	0.43	0.44	0.45	0.46
53	0.46	0.48	0.49	0.51	0.52	0.53
54	0.52	0.54	0.56	0.57	0.59	0.60
55	0.58	0.60	0.62	0.64	0.65	0.67
56	0.64	0.66	0.68	0.70	0.72	0.74
57	0.70	0.72	0.74	0.76	0.78	0.80
58	0.76	0.78	0.80	0.82	0.85	0.87
59	0.81	0.84	0.86	0.89	0.91	0.93
60	0.87	0.89	0.92	0.94	0.97	1.00
61	0.92	0.95	0.98	1.00	1.03	1.06
62	0.97	1.00	1.02	1.05	1.08	1.11
63	1.03	1.06	1.09	1.12	1.15	1.18
64	1.08	1.11	1.14	1.17	1.20	1.23
65	1.13	1.16	1.19	1.22	1.26	1.29
66	1.17	1.21	1.24	1.28	1.31	1.35
67	1.22	1.25	1.29	1.33	1.36	1.40
68	1.26	1.30	1.34	1.37	1.41	1.45
69	1.31	1.34	1.38	1.42	1.46	1.50
70	1.35	1.39	1.43	1.47	1.51	1.55
72	1.42	1.47	1.51	1.55	1.59	1.63
75	1.53	1.57	1.62	1.66	1.71	1.75
80	1.66	1.71	1.76	1.81	1.86	1.90
85	1.74	1.79	1.84	1.90	1.95	2.00
90	1.77	1.82	1.87	1.93	1.98	2.03

Table 2-9 (*Continued*)
BAROMETRIC CORRECTION FOR GRAVITY—METRIC UNITS

The values in the table below are to be subtracted from the readings taken on a mercurial barometer to correct for the decrease in gravity with increase in altitude.

Height above sea-level meters	Observed barometer height in millimeters								
	400 mm.	450 mm.	500 mm.	550 mm.	600 mm.	650 mm.	700 mm.	750 mm.	800 mm.
100	....	....	....	....	....	....	0.02	0.02	0.02
200	....	....	....	....	....	....	0.04	0.05	0.05
300	....	....	....	....	....	....	0.07	0.07	0.07
400	....	....	....	....	....	....	0.09	0.10	0.10
500	....	....	....	....	....	....	0.11	0.12	0.13
600	....	....	....	....	....	0.12	0.13	0.14	....
700	....	....	....	....	....	0.14	0.15	0.16	....
800	....	....	....	....	....	0.16	0.18	0.19	....
900	....	....	....	....	....	0.18	0.20	0.22	....
1000	....	....	....	0.18	0.19	0.20	0.22	0.24	....
1100	....	....	....	0.19	0.21	0.22	0.24	....	....
1200	....	....	....	0.21	0.23	0.24	0.26	....	....
1300	....	....	....	0.22	0.24	0.26	0.29	....	....
1400	....	....	....	0.24	0.26	0.28	0.31	....	....
1500	....	....	0.24	0.26	0.28	0.30	0.33	....	....
1600	....	....	0.25	0.28	0.30	0.32	....	....	....
1700	....	....	0.27	0.30	0.32	0.34	....	....	....
1800	....	....	0.28	0.31	0.34	0.36	....	....	....
1900	....	....	0.30	0.33	0.36	0.39	....	....	....
2000	....	0.28	0.31	0.34	0.38	0.41	....	....	....
2100	....	0.30	0.33	0.36	0.40	....	....	....	....
2200	....	0.31	0.35	0.38	0.41	....	....	....	....
2300	....	0.32	0.36	0.40	0.43	....	....	....	....
2400	....	0.34	0.38	0.42	0.45	....	....	....	....
2500	0.31	0.35	0.39	0.43	0.47	....	....	....	....
2600	0.33	0.37	0.41	....	....	....	....	....	....
2800	0.35	0.40	0.44	....	....	....	....	....	....
3000	0.38	0.42	0.47	....	....	....	....	....	....
3200	0.40	0.46	....	....	....	....	....	....	....
3400	0.43	0.48	....	....	....	....	....	....	....

Table 2-10
ATMOSPHERIC DATA

J. Chem. Ed., **31**, 115 (1954); *Phys. Rev.*, **88**, 1027 (1952); *Mon. Not. Roy. Astron. Soc.*, **112**, 101 (1952).

Altitude km	Temperature ($\mu = 28.966$) °K	Pressure mm Hg	Density g/cc	* Mean free path, cm
0	288.0	7.6×10^2	1.1×10^{-3}	8.6×10^{-6}
10	230.8	2.1×10^2	4.2×10^{-4}	2.1×10^{-5}
20	212.8	4.2×10^1	9.3×10^{-5}	9.7×10^{-5}
30	231.7	9.5×10^0	1.9×10^{-5}	4.8×10^{-4}
40	262.5	2.4×10^0	4.2×10^{-6}	2.2×10^{-3}
50	270.8	7.6×10^{-1}	1.2×10^{-6}	7.8×10^{-3}
60	252.8	2.2×10^{-1}	3.5×10^{-7}	2.6×10^{-2}
70	218.0	5.5×10^{-2}	9.7×10^{-8}	9.3×10^{-2}
80	205.0	1.1×10^{-2}	2.1×10^{-8}	4.3×10^{-1}
90	217.0	2×10^{-3}	4.1×10^{-9}	2.1×10^0
100	240.0	6×10^{-4}	8.6×10^{-10}	9.5×10^0
110	270.0	2×10^{-4}	2.0×10^{-10}	3.8×10^1
120	330.0	6×10^{-5}	5.6×10^{-11}	1.3×10^2
130	390.0	2×10^{-5}	1.9×10^{-11}	3.7×10^2
140	447.0	7×10^{-6}	7.6×10^{-12}	8.7×10^2
150	503.0	3.7×10^{-6}	3.4×10^{-12}	1.8×10^3
160	560.0	2×10^{-6}	1.6×10^{-12}	3.6×10^3
180	676.9	7×10^{-7}	4.8×10^{-13}	1.0×10^4
200	792.5	3×10^{-7}	1.7×10^{-13}	3.0×10^4
220	906.6	1.4×10^{-7}	7.0×10^{-14}	8.7×10^4

* The average distance a particle travels between collisions with other particles; not the distance between particles.

Table 2-11
REDUCTION OF THE BAROMETER TO SEA LEVEL

A barometer located at an elevation above sea level will show a reading lower than a barometer at sea level by an amount approximately 2.5 millimeters (0.1 inch) for each 30.5 meters (100 feet) of elevation. A closer approximation can be made by reference to the following tables, which take into account (1) the effect of altitude of the station at which the barometer is read, (2) the mean temperature of the air column extending from the station down to sea level. (3) the latitude of the station at which the barometer is read, and (4) the reading of the barometer corrected for its temperature, a correction which is applied only to mercurial barometers since the aneroid barometers are compensated for temperature effects.

Example.—A barometer which has been corrected for its temperature read 650 mm at a station whose altitude is 1350 meters above sea level and at a latitude of 30°. The mean temperature (outdoor temperature) at the station is 20°C.

Table A (metric units) gives for these conditions a temperature-altitude factor of.. 135.2
The Latitude Factor Table gives for 135.2 at 30° lat. a correction of........+0.17
Therefore, the corrected value of the temperature-altitude factor is...... 135.37

Entering Table B (metric units), with a temperature-altitude factor of 135.37 and a barometric reading of 650 mm. (corrected for temperature), the correction is found to be... 109.6
Accordingly the barometric reading reduced to sea level is 650 + 109.6 = 759.6 mm.

Latitude Factor—English or Metric Units

For latitudes 0°-45° add the latitude factor, for 45°-90° subtract the latitude factor, from the values obtained in Table A.

Temp.—Alt. Factor From Table A	Latitude				
	0°	10°	20°	30°	45°
50	0.1	0.1	0.1	0.1	0.0
100	0.3	0.3	0.2	0.1	0.0
150	0.4	0.4	0.3	0.2	0.0
200	0.5	0.5	0.4	0.3	0.0
250	0.7	0.6	0.5	0.3	0.0
300	0.8	0.8	0.6	0.4	0.0
350	0.9	0.9	0.7	0.5	0.0
	90°	80°	70°	60°	45°

Table 2-11 (*Continued*)
REDUCTION OF THE BAROMETER TO SEA LEVEL—METRIC UNITS
A. Values of the Temperature-Altitude Factor for
Use in Table B
(From *Smithsonian Meteorological Tables*, 3d Ed., 1907.)

Altitude in Meters	Mean Temperature of Air Column in Centigrade Degrees										
	−16°	−8°	−4°	0°	6°	10°	14°	18°	20°	22°	26°
10	1.2	1.1	1.1	1.1	1.1	1.0	1.0	1.0	1.0	1.0	1.0
50	5.8	5.6	5.5	5.4	5.3	5.2	5.1	5.0	5.0	5.0	4.9
100	11.5	11.2	11.0	10.8	10.6	10.4	10.3	10.1	10.0	9.9	9.8
150	17.3	16.7	16.5	16.2	15.9	15.6	15.4	15.1	15.0	14.9	14.7
200	23.0	22.3	22.0	21.6	21.1	20.8	20.5	20.2	20.0	19.9	19.6
250	28.8	27.9	27.5	27.0	26.4	26.0	25.6	25.2	25.0	24.9	24.5
300	34.5	33.5	33.0	32.5	31.7	31.2	30.7	30.3	30.1	29.8	29.4
350	40.3	39.0	38.5	37.9	37.0	36.4	35.9	35.3	35.1	34.8	34.3
400	46.0	44.6	43.9	43.3	42.3	41.6	41.0	40.4	40.1	39.8	39.2
450	51.8	51.3	49.4	48.7	47.6	46.8	46.1	45.4	45.1	44.8	44.1
500	57.5	55.8	54.9	54.1	52.9	52.0	51.2	50.5	50.1	49.7	49.0
550	63.3	61.4	60.4	59.5	58.1	57.2	56.4	55.5	55.1	54.7	53.9
600	69.0	66.9	65.9	64.9	63.4	62.4	61.5	60.6	60.1	59.7	58.8
650	74.8	72.5	71.4	70.3	68.7	67.6	66.6	65.6	65.1	64.6	63.7
700	80.6	78.1	76.9	75.7	74.0	72.9	71.7	70.7	70.1	69.6	68.6
750	86.3	83.7	82.4	81.1	79.3	78.1	76.9	75.7	75.1	74.6	73.5
800	92.1	89.2	87.9	86.5	84.6	83.3	82.0	80.8	80.1	79.6	78.4
850	97.8	94.8	93.4	92.0	89.8	88.5	87.1	85.8	85.2	84.5	83.3
900	103.6	100.4	98.9	97.4	95.1	93.7	92.2	90.8	90.2	89.5	88.2
950	109.3	106.0	104.4	102.8	100.4	98.9	97.4	95.9	95.2	94.5	93.1
1000	115.1	111.5	109.8	108.2	105.7	104.1	102.5	100.9	100.2	99.4	98.0
1050	120.8	117.1	115.3	113.6	111.0	109.3	107.6	106.0	105.2	104.4	102.9
1100	126.6	122.7	120.8	119.0	116.3	114.5	112.7	111.0	110.2	109.4	107.8
1150	132.3	128.3	126.3	124.4	121.6	119.7	117.9	116.1	115.2	114.4	112.7
1200	138.1	133.8	131.8	129.8	126.8	124.9	123.0	121.1	120.2	119.3	117.6
1250	143.8	139.4	137.3	135.2	132.1	130.1	128.1	126 2	125.2	124.3	122.5
1300	149.6	145.0	142.8	140.6	137.4	135.3	133.2	131.2	130.2	129.3	127.4
1350	155.3	150.6	148.3	146.0	142.7	140.5	138.4	136.3	135.2	134.2	132.3
1400	161.1	156.2	153.8	151.4	148.0	145.7	143.5	141.3	140.2	139.2	137.2
1450	166.8	161.7	159.3	156.8	153.3	150.9	148.6	146.4	145.3	144.2	142.1
1500	172.6	167.3	164.8	162.3	158.5	156.1	153.7	151.4	150.3	149.1	147.0
1550	178.3	172.9	170.2	167.7	163.8	161.3	158.8	156.4	155.3	154.1	151.8
1600	184.1	178.5	175.7	173.1	169.1	166.5	164.0	161.5	160.3	159.1	156.7
1650	189.8	184.0	181.2	178.5	174.4	171.7	169.1	166.5	165.3	164.1	161.6
1700	195.6	189.6	186.7	183.9	179.7	176.9	174.2	171.6	170.3	169.0	166.5
1750	201.4	195.2	192.2	189.3	185.0	182.1	179.3	176.6	175.3	174.0	171.4
1800	207.1	200.8	197.7	194.7	190.2	187.3	184.5	181.7	180.3	179.0	176.3
1850	212.9	206.3	203.2	200.1	195.5	192.5	189.6	186.7	185.3	183.9	181.2
1900	218.6	211.9	208.7	205.5	200.8	197.7	194.7	191.8	190.3	188.9	186.1
1950	224.4	217.5	214.2	210.9	206.1	202.9	199.8	196.8	195.3	193.9	191.0
2000	230.1	223.0	219.7	216.3	211.4	208.1	204.9	201.9	200.3	198.8	195.9
2050	235.9	228.6	225.1	221.7	216.7	213.3	210.1	206.9	205.3	203.8	200.8
2100	241.6	234.2	230.6	227.1	221.9	218.5	215.2	211.9	210.4	208.8	205.7
2150	247.4	239.8	236.1	232.5	227.2	223.7	220.3	217.0	215.4	213.8	210.6
2200	253.1	245.4	241.6	237.9	232.5	228.9	225.4	222.0	220.4	218.7	215.5
2250	258.9	250.9	247.1	243.4	237.8	234.1	230.6	227.1	225.4	223.7	220.4
2300	264.6	256.5	252.6	248.8	243.1	239.3	235.7	232.1	230.4	228.7	225.3
2350	270.4	262.1	258.1	254.2	248.3	244.5	240.8	237.2	235.4	233.6	230.2
2400	276.1	267.7	263.6	259.6	253.6	249.7	245.9	242.2	240.4	238.6	235.1
2450	281.9	273.2	269.1	265.0	258.9	254.9	251.0	247.3	245.4	243.6	240.0
2500	287.6	278.8	274.5	270.4	264.2	260.1	256.2	252.3	250.4	248.5	244.9
2550	293.4	284.4	280.0	275.8	269.5	265.3	261.3	257.3	255.4	253.5	249.8
2600	299.1	290.0	285.5	281.2	274.8	270.5	266.4	262.4	260.4	258.5	254.7
2650	304.9	295.5	291.0	286.6	280.0	275.7	271.5	267.4	265.4	263.4	259.6
2700	310.6	301.1	296.5	292.0	285.3	280.9	276.6	272.5	270.4	268.4	264.5
2750	316.4	306.7	302.0	297.4	290.6	286.1	281.8	277.5	275.4	273.4	269.4
2800	322.1	312.3	307.5	302.8	295.9	291.3	286.9	282.6	280.4	278.3	274.3
2850	327.9	317.8	313.0	308.2	301.2	296.5	292.0	287.6	285.4	283.3	279.2
2900	333.6	323.4	318.4	313.6	306.4	301.7	297.1	292.6	290.4	288.3	284.1
2950	339.4	329.0	323.9	319.0	311.7	306.9	302.2	297.7	295.5	293.3	289.0
3000	345.1	334.5	329.4	324.4	317.0	312.1	307.4	302.7	300.5	298.2	293.8

Table 2-11 (*Continued*)
REDUCTION OF THE BAROMETER TO SEA LEVEL—METRIC UNITS
B. Values in Millimeters to Be Added
(From *Smithsonian Meteorological Tables*, 3d Ed., 1907.)

Left portion

Temp.—Alt. Factor	Barometer Reading in Millimeters						
	790	770	750	730	710	690	670
1	0.9	0.9	0.9	0.8	0.8	0.8	
5	4.6	4.4	4.3	4.2	4.1	4.0	
10	9.1	8.9	8.7	8.5	8.2	8.0	
15	13.8	13.4	13.1	12.7	12.4	12.0	
20	18.4	17.9	17.5	17.0	16.5	16.1	
25		22.5	21.9	21.3	20.7	20.1	
30		27.1	26.4	25.7	25.0	24.2	
35		31.7	30.8	30.0	29.2	28.4	
40		36.3	35.3	34.4	33.5	32.5	31.6
45			39.9	38.8	37.8	36.7	35.6

Temp.—Alt. Factor	750	730	710	690	670	650	630
50	44.4	43.3	42.1	40.9	39.7		
55	49.0	47.7	46.4	45.1	43.8		
60	53.6	52.2	50.8	49.3	47.9		
65	58.3	56.7	55.2	53.6	52.1		
70		61.3	59.6	57.9	56.2		
75		65.8	64.0	62.2	60.4		
80		70.4	68.5	66.6	64.6	62.7	60.8
85		75.0	73.0	70.9	68.9	66.8	64.8
90			77.5	75.3	73.1	71.0	68.8
95			82.1	79.7	77.4	75.1	72.8

Temp.—Alt. Factor	710	690	670	650	630	610
100	86.6	84.2	81.8	79.3	76.9	
105	91.2	88.7	86.1	83.5	81.0	
110	95.9	93.2	90.5	87.8	85.1	
115	100.5	97.7	94.8	92.0	89.2	
120		102.2	99.3	96.3	93.3	
125		106.8	103.7	100.6	97.5	94.4
130		111.4	108.2	104.9	101.7	98.5
135		116.0	112.7	109.3	105.9	102.6
140		120.7	117.2	113.7	110.2	106.7
145			121.7	118.1	114.5	110.8

Temp.—Alt. Factor	670	650	630	610	590	570
150	126.3	122.5	118.8	115.0		
155	130.9	127.0	123.1	119.2		
160	135.5	131.5	127.4	123.4		
165	140.2	136.0	131.8	127.6		
170		140.5	136.2	131.9	127.5	123.2
175		145.1	140.6	136.2	131.7	127.2
180		149.7	145.1	140.5	135.9	131.3
185		154.3	149.5	144.8	140.0	135.3
190		158.9	154.0	149.2	144.3	139.4
195			158.6	153.5	148.5	143.5

Right portion

Temp.—Alt. Factor	Barometer Reading in Millimeters						
	630	610	590	570	550	530	510
200	163.1	157.9	152.8	147.6			
205	167.7	162.4	157.1	151.7			
210	172.3	166.8	161.4	155.9			
215	176.9	171.3	165.7	160.1	154.5	148.9	
220		175.8	170.1	164.3	158.5	152.8	
225		180.4	174.5	168.5	162.6	156.7	
230		184.9	178.9	172.8	166.7	160.7	
235		189.5	183.3	177.1	170.9	164.7	
240		194.1	187.8	181.4	175.0	168.7	
245			198.8	192.3	185.7	179.2	172.7

Temp.—Alt. Factor	590	570	550	530	510
250	196.8	190.1	183.4	176.8	
255	201.3	194.5	187.7	180.8	
260	205.9	198.9	191.9	185.0	178.0
265	210.5	203.3	196.2	189.1	181.9
270	215.1	207.8	200.5	193.2	185.9
275	219.8	212.3	204.9	197.4	190.0
280		216.8	209.2	201.6	194.0
285		221.4	213.6	205.8	198.1
290		225.9	218.0	210.1	202.1
295		230.5	222.4	214.3	206.3

Temp.—Alt. Factor	570	550	530	510	490
300	235.1	226.9	218.6	210.4	
305	239.8	231.4	223.0	214.6	206.1
310		235.9	227.3	218.7	210.1
315		240.4	231.7	222.9	214.2
320		245.0	236.1	227.2	218.3
325		249.6	240.5	231.4	222.4
330		254.2	244.9	235.7	226.5
335		258.8	249.4	240.0	230.6
340		263.5	253.9	244.4	234.8
345			258.4	248.7	238.9

Table 2-12
VISCOSITY CONVERSION TABLE

Poise = c. g. s. unit of absolute viscosity.   Centipoise = 0.01 poise.
Stoke = c. g. s. unit of kinematic viscosity.   Centistoke = 0.01 stoke.
Centipoises = centistokes × density (at temperature under consideration).
Reyn (1 lb. sec. per sq. in.) = 69 × 10⁵ centipoises.

Centistokes to Saybolt, Redwood, and Engler Units

Cf. *Jour. Inst. Pet. Tech., Vol. 22, p. 21 (1936); Reports of A. S. T. M. Committee D-2, 1936 and 1937.*

The values of Saybolt Universal Viscosity at 100°F and at 210°F are taken directly from the comprehensive *ASTM Viscosity Tables, Special Technical Publication No. 43A* (1953) by permission of the publishers, American Society for Testing Materials, 1916 Race St., Philadelphia 3, Pa.

Centistokes	Saybolt Universal Viscosity at			Redwood Seconds at			Engler Degrees at all Temps.
	100°F.	130°F.	210°F.	70°F.	140°F.	200°F.	
2.0	32.62	32.68	32.85	30.2	31.0	31.2	1.14
3.0	36.03	36.10	36.28	32.7	33.5	33.7	1.22
4.0	39.14	39.22	39.41	35.3	36.0	36.3	1.31
5.0	42.35	42.43	42.65	37.9	38.5	38.9	1.40
6.0	45.56	45.65	45.88	40.5	41.0	41.5	1.48
7.0	48.77	48.86	49.11	43.2	43.7	44.2	1.56
8.0	52.09	52.19	52.45	46.0	46.4	46.9	1.65
9.0	55.50	55.61	55.89	48.9	49.1	49.7	1.75
10.0	58.91	59.02	59.32	51.7	52.0	52.6	1.84
11.0	62.43	62.55	62.86	54.8	55.0	55.6	1.93
12.0	66.04	66.17	66.50	57.9	58.1	58.8	2.02
14.0	73.57	73.71	74.09	64.4	64.6	65.3	2.22
16.0	81.30	81.46	81.87	71.0	71.4	72.2	2.43
18.0	89.44	89.61	90.06	77.9	78.5	79.4	2.64
20.0	97.77	97.96	98.45	85.0	85.8	86.9	2.87
22.0	106.4	106.6	107.1	92.4	93.3	94.5	3.10
24.0	115.0	115.2	115.8	99.9	100.9	102.2	3.34
26.0	123.7	123.9	124.5	107.5	108.6	110.0	3.58
28.0	132.5	132.8	133.4	115.3	116.5	118.0	3.82
30.0	141.3	141.6	142.3	123.1	124.4	126.0	4.07
32.0	150.2	150.5	151.2	131.0	132.3	134.1	4.32
34.0	159.2	159.5	160.3	138.9	140.2	142.2	4.57
36.0	168.2	168.5	169.4	146.9	148.2	150.3	4.83
38.0	177.3	177.6	178.5	155.0	156.2	158.3	5.08
40.0	186.3	186.7	187.6	163.0	164.3	166.7	5.34
42.0	195.3	195.7	196.7	171.0	172.3	175.0	5.59
44.0	204.4	204.8	205.9	179.1	180.4	183.3	5.85
46.0	213.7	214.1	215.2	187.1	188.5	191.7	6.11
48.0	222.9	223.3	224.5	195.2	196.6	200.0	6.37
50.0	232.1	232.5	233.8	203.3	204.7	208.3	6.63
60.0	278.3	278.8	280.2	243.5	245.3	250.0	7.90
70.0	324.4	325.0	326.7	283.9	286.0	291.7	9.21
80.0	370.8	371.5	373.4	323.9	326.6	333.4	10.53
90.0	417.1	417.9	420.0	364.4	367.4	375.0	11.84
100.0*	463.5	464.4	466.7	404.9	408.2	416.7	13.16

* At higher values use the same ratio as above for 100 centistokes; *e.g.,* 102 centistokes = 102 × 4.635 Saybolt seconds at 100 F.
To obtain the Saybolt Universal viscosity equivalent to a kinematic viscosity determined at *t*°F., multiply the equivalent Saybolt Universal viscosity at 100°F. by 1 + (t − 100) 0.000064; *e.g.,* 10 centistokes at 210°F are equivalent to 58.91 × 1.0070, or 59.32 Saybolt Universal Viscosity at 210°F.

Table 2-13
CONVERSION OF WEIGHINGS IN AIR TO WEIGHINGS IN VACUO

If the mass of a substance in air is M_f, its density D_m, the density of weights used in making the weighing D_w, and the density* of air D_a, the true mass of the substance in vacuo, M_{vac} is

$$M_{vac} = M_f + D_a M_f\left(\frac{1}{D_m} - \frac{1}{D_w}\right)$$

For most purposes it is sufficient to assume a density of 8.4 for brass weights, and a density of 0.0012 for air under ordinary conditions. The equation then becomes

$$M_{vac} = M_f + 0.0012 M_f\left(\frac{1}{D_m} - \frac{1}{8.4}\right)$$

This table which follows gives the values of k (buoyancy reduction factor), which is the correction necessary because of the buoyant effect of the air upon the object weighed; the table is computed for air with the density of 0.0012; m is the weight in grams of the object when weighed in air; weight of object reduced to "in vacuo" $= m + \dfrac{km}{1000}$.

Density of object weighed	Buoyancy reduction factor, k			
	Brass weights, density =8.4	Pt or Pt-Ir weights, density =21.5	Al or quartz weights, density =2.7	Gold weights, density =17
0.2	5.89	5.98	5.58	5.97
0.3	3.87	3.96	3.56	3.95
0.4	2.87	2.95	2.55	2.94
0.5	2.26	2.35	1.95	2.34
0.6	1.86	1.95	1.55	1.93
0.7	1.57	1.66	1.26	1.65
0.75	1.46	1.55	1.15	1.53
0.80	1.36	1.45	1.05	1.43
0.82	1.32	1.41	1.01	1.39
0.84	1.29	1.37	0.98	1.36
0.86	1.25	1.34	0.94	1.33
0.88	1.22	1.31	0.91	1.29
0.90	1.19	1.28	0.88	1.26
0.92	1.16	1.25	0.85	1.24
0.94	1.13	1.22	0.82	1.21
0.96	1.11	1.20	0.80	1.18
0.98	1.08	1.17	0.77	1.16
1.00	1.06	1.15	0.75	1.13
1.02	1.03	1.12	0.72	1.11
1.04	1.01	1.10	0.70	1.08
1.06	0.99	1.08	0.68	1.06
1.08	0.97	1.06	0.66	1.04
1.10	0.95	1.04	0.64	1.02
1.12	0.93	1.02	0.62	1.00
1.14	0.91	1.00	0.60	0.98
1.16	0.89	0.98	0.58	0.96
1.18	0.87	0.96	0.56	0.95
1.20	0.86	0.95	0.55	0.93
1.25	0.82	0.91	0.51	0.89
1.30	0.78	0.87	0.47	0.85
1.35	0.75	0.83	0.44	0.82
1.40	0.71	0.80	0.40	0.79
1.50	0.66	0.74	0.35	0.73
1.6	0.61	0.69	0.30	0.68
1.7	0.56	0.65	0.25	0.64
1.8	0.52	0.61	0.21	0.60
1.9	0.49	0.58	0.18	0.56
2.0	0.46	0.54	0.15	0.53
2.2	0.40	0.49	0.09	0.48
2.4	0.36	0.44	0.05	0.43

* See special table: *Specific Gravity of Air.*

Table 2-13 (*Continued*)
CONVERSION OF WEIGHINGS IN AIR TO WEIGHINGS IN VACUO

Density of object weighed	Buoyancy reduction factor, k			
	Brass weights, density = 8.4	Pt or Pt-Ir weights, density = 21.5	Al or quartz weights density = 2.7	Gold weights, density = 17
2.6	0.32	0.41	0.01	0.39
2.8	0.29	0.37	−0.02	0.36
3.0	0.26	0.34	−0.05	0.33
3.5	0.20	0.29	−0.11	0.27
4	0.16	0.24	−0.15	0.23
5	0.10	0.18	−0.21	0.17
6	0.06	0.14	−0.25	0.13
7	0.03	0.12	−0.28	0.10
8	0.01	0.09	−0.30	0.08
9	−0.01	0.08	−0.32	0.06
10	−0.02	0.06	−0.33	0.05
12	−0.04	0.04	−0.35	0.03
14	−0.06	0.03	−0.37	0.02
16	−0.07	0.02	−0.38	0.00
18	−0.08	0.01	−0.39	0.00
20	−0.08	0.00	−0.39	−0.01
22	−0.09	0.00	−0.40	−0.02

TOLERANCES FOR VOLUMETRIC BURETS AND PIPETS

From Meites, *Handbook of Analytical Chemistry*, 1963, McGraw-Hill Book Company; by permission

Table 2-14
PERMISSIBLE DEVIATIONS FROM NOMINAL CAPACITY

Capacity, up to and including, ml	Deviations, ml, permitted for			
	Volumetric flasks, calibrated		Transfer pipets and burets	Measuring pipets
	To contain	To deliver		
1	±0.01	......	....	......
2	......	......	±0.006*	±0.01
3	0.015	......	....	......
5	0.02	......	0.01	0.02
10	0.02	±0.04	0.02	0.03
25	0.03	0.05	....	......
30	......	......	0.03	0.05
50	0.05	0.10	0.05	0.08
100	0.08	0.15	0.08†	0.15
200	0.10	0.20	0.10*	......
300	0.12	0.25	....	......
500	0.15	0.30	....	......
1000	0.30	0.50	....	......
2000	0.50	1.00	....	......
3000	0.75	1.50	....	......
4000	1.00	2.0	....	......
5000	1.2	2.4	....	......

* Applies to pipets only.
† ±0.10 ml. for burets.

Table 2-15
TOLERANCES FOR ANALYTICAL WEIGHTS
By Alan D. Westland with Fred E. Beamish

This table gives the individual and group tolerances established by the National Bureau of Standards (Washington, D.C.) for classes M, S, S-1, and P weights. Individual tolerances are "acceptance tolerances" for new weights. Group tolerances are defined by the National Bureau of Standards as follows: "The corrections of individual weights shall be such that no combination of weights that is intended to be used in a weighing shall differ from the sum of the nominal values by more than the amount listed under the group tolerances."

For class S-1 weights, two-thirds of the weights in a set must be within one-half of the individual tolerances given below. No group tolerances have been specified for class P weights. See *Natl. Bur. Standards Circ.* 547, sec. 1 (1954).

Denomination	Class M		Class S		Class S-1, individual tolerance, mg	Class P, individual tolerance, mg
	Individual tolerance, mg	Group tolerance, mg	Individual tolerance, mg	Group tolerance, mg		
100 g	0.50		0.25	None	1.0	2.0
50 g	0.25	None	0.12	specified	0.60	1.2
30 g	0.15	specified	0.074		0.45	0.90
20 g	0.10		0.074	0.154	0.35	0.70
10 g	0.050		0.074		0.25	0.50
5 g	0.034		0.054		0.18	0.36
3 g	0.034		0.054		0.15	0.14
2 g	0.034	0.065	0.054	0.105	0.13	0.26
1 g	0.034		0.054		0.10	0.20
500 mg	0.0054		0.025		0.080	0.16
300 mg	0.0054	0.0105	0.025	0.055	0.070	0.14
200 mg	0.0054		0.025		0.060	0.12
100 mg	0.0054		0.025		0.050	0.10
50 mg	0.0054		0.014		0.042	0.085
30 mg	0.0054	0.0105	0.014	0.034	0.038	0.076
20 mg	0.0054		0.014		0.035	0.070
10 mg	0.0054		0.014		0.030	0.060
5 mg	0.0054		0.014		0.028	0.055
3 mg	0.0054		0.014		0.026	0.052
2 mg	0.0054	0.0105	0.014	0.034	0.025	0.050
1 mg	0.0054		0.014		0.025	0.050
½ mg	0.0054		0.014		0.025	.....

Table 2-16
FACTORS FOR SIMPLIFIED COMPUTATION OF VOLUME

The volume is determined by weighing the water, having a temperature of $t\,°C$, contained in or delivered by the apparatus at the same temperature. The weight of water, w grams, is obtained with brass weights in air having a density of 1.20 mg/ml.

For apparatus made of soft glass, the volume contained or delivered at 20°C is given by

$$v_{20} = wf_{20} \quad \text{ml}$$

where v_{20} is the volume at 20° and f_{20} is the factor (apparent specific volume) obtained from the table below for the temperature t at which the calibration is performed. The volume at any other temperature t' may then be obtained from

$$v' = v_{20}[1 + 0.00002(t' - 20)] \quad \text{ml}$$

For apparatus made of any other material, the volume contained or delivered at the temperature t is

$$v_t = wf_t \quad \text{ml}$$

where w is again the weight in air obtained with brass weights (in grams), and f_t is the factor given in the fourth column of the table for the temperature t. The volume at any temperature t' may then be obtained from

$$v'_t = v_t[1 + \beta(t' - t)] \quad \text{ml}$$

where β is the cubical coefficient of thermal expansion of the material from which the apparatus is made. Approximate values of β for some frequently encountered materials are given in Table 2-16.

Table 2-16 (*Continued*)
FACTORS FOR SIMPLIFIED COMPUTATION OF VOLUME

t, °C	f_{20}	$\log_{10} f_{20}$	f_t	$\log_{10} f_t$
0	1.0017	0.00074	1.0012	0.00052
1	16	70	11	49
2	15	67	11	47
3	15	65	11	46
4	15	63	11	46
5	1.0015	0.00063	1.0011	0.00046
6	15	63	11	47
7	15	63	11	49
8	15	64	12	51
9	15	66	13	54
10	1.0016	0.00069	1.0013	0.00058
11	17	72	14	62
12	17	75	15	67
13	18	79	17	72
14	19	84	18	78
15	1.0020	0.00089	1.0020	0.00084
16	22	095	21	091
17	23	101	23	098
18	25	108	24	106
19	27	115	26	114
20	1.0028	0.00123	1.0028	0.00123
21	30	130	30	132
22	32	139	33	142
23	34	149	35	152
24	37	159	38	163
25	1.0039	0.00168	1.0040	0.00174
26	41	178	43	185
27	44	189	45	197
28	46	200	48	209
29	49	211	51	222
30	1.0052	0.00224	1.0054	0.00235
31	54	236	57	248
32	57	248	60	262
33	60	261	64	276
34	63	275	67	290
35	1.0067	0.00289	1.0070	0.00305
36	71	307	74	319
37	74	320	77	335
38	77	333	81	350
39	81	350	85	367
40	1.0085	0.00368	1.0089	0.00383

Table 2-17
CUBICAL COEFFICIENTS OF THERMAL EXPANSION

This table lists values of β, the cubical coefficient of thermal expansion, taken from "Essentials of Quantitative Analysis," by Benedetti-Pichler, and from various other sources. The value of β represents the relative increase in volume for a change in temperature of 1°C at temperatures in the vicinity of 25°C, and is equal to 3α, where α is the linear coefficient of thermal expansion. Data are given for the types of glass from which volumetric apparatus is most commonly made, and also for some other materials which have been or may be used in the fabrication of apparatus employed in analytical work.

Material	β
Glasses	
Alkali-resistant, Corning 728	1.90×10^{-5}
Geräteglas, Schott G20	1.47
Kimble KG-33 (borosilicate)	0.96
N-51A ("Resistant")	1.47
R-6 (soft)	2.79
Pyrex, Corning 744	0.96
Vitreous silica	0.15
Vycor, Corning 790	0.24
Metals	
Brass	*ca.* 5.5
Copper	5.0
Gold	4.3
Monel metal	4.0
Platinum	2.7
Silver	5.7
Stainless steel	*ca.* 5.3
Tantalum	*ca.* 2.0
Tungsten	1.3
Plastics and other materials	
Hard rubber	24×10^{-5}
Polyethylene	45–90
Polystyrene	18–24
Porcelain	*ca.* 1.2
Teflon (polytetrafluoroethylene)	16.5

Table 2-18
MOLAR EQUIVALENT OF ONE LITER OF GAS AT VARIOUS TEMPERATURES AND PRESSURES

The values in this table, which give the number of moles in one liter of gas, are based on the properties of an "ideal" gas and were calculated by use of the formula:

$$\text{moles/liter} = \frac{P}{760} \times \frac{273}{T} \times \frac{1}{22.40}$$

where P is the pressure in millimeters of mercury and T is the temperature in kelvins ($= t°C + 273$).

To convert to moles per cubic foot multiply the values in the table by 28.316.

Pressure mm of mercury	Temperature °C					
	10°	12°	14°	16°	18°	20°
655	0.03712	0.03686	0.03660	0.03634	0.03610	0.03585
660	3731	3714	3688	3662	3637	3612
665	3768	3742	3716	3690	3665	3640
670	3796	3770	3744	3718	3692	3667
675	3825	3798	3772	3745	3720	3695
680	0.03853	0.03826	0.03800	0.03773	0.03747	0.03694
685	3881	3854	3827	3801	3775	3749
690	3910	3882	3855	3829	3802	3776
695	3938	3910	3883	3856	3830	3804
700	3967	3939	3911	3884	3858	3831
702	0.03978	0.03950	0.03922	0.03895	0.03869	0.03842
704	3989	3961	3934	3906	3880	3853
706	4000	3972	3945	3917	3891	3864
708	4012	3984	3956	3929	3902	3875
710	4023	3995	3967	3940	3913	3886
712	0.04035	0.04006	0.03978	0.03951	0.03924	0.03897
714	4046	4018	3989	3962	3935	3908
716	4057	4029	4001	3973	3946	3919
718	4068	4040	4012	3984	3957	3930
720	4080	4051	4023	3995	3968	3941
722	0.04091	0.04063	0.04034	0.04006	0.03979	0.03952
724	4103	4074	4045	4017	3990	3963
726	4114	4085	4057	4028	4001	3973
728	4125	4096	4068	4040	4012	3984
730	4136	4108	4079	4051	4023	3995
732	0.04148	0.04119	0.04090	0.04062	0.04034	0.04006
734	4159	4130	4101	4073	4045	4017
736	4171	4141	4112	4084	4056	4028
738	4182	4153	4124	4095	4067	4039
740	4193	4164	4135	4106	4078	4050
742	0.04204	0.04175	0.04146	0.04117	0.04089	0.04061
744	4216	4186	4157	4128	4100	4072
746	4227	4198	4168	4139	4111	4083
748	4239	4209	4179	4151	4122	4094
750	4250	4220	4191	4162	4133	4105
752	0.04261	0.04231	0.04202	0.04173	0.04144	0.04116
754	4273	4243	4213	4184	4155	4127
756	4284	4254	4224	4195	4166	4138
758	4295	4265	4235	4206	4177	4149
760	4307	4276	4247	4217	4188	4160
762	0.04318	0.04287	0.04258	0.04228	0.04199	0.04171
764	4329	4299	4269	4239	4210	4181
766	4341	4310	4280	4250	4221	4192
768	4352	4321	4291	4262	4232	4203
770	4363	4333	4302	4273	4243	4214
772	0.04375	0.04344	0.04314	0.04284	0.04254	0.04225
774	4386	4355	4325	4295	4265	4236
776	4397	4366	4336	4306	4276	4247
778	4409	4378	4347	4317	4287	4258
780	4420	4389	4358	4328	4298	4269

Table 2-18 (*Continued*)
MOLAR EQUIVALENT OF ONE LITER OF GAS

Pressure mm of mercury	Temperature °C					
	22°	24°	26°	28°	30°	32°
655	0.03561	0.03537	0.03515	0.03490	0.03467	0.03444
660	3588	3564	3541	3516	3493	3470
665	3614	3591	3568	3543	3520	3496
670	3642	3618	3595	3569	3546	3523
675	3669	3645	3622	3596	3572	3549
680	0.03697	0.03672	0.03649	0.03623	0.03599	0.03575
685	3724	3699	3676	3649	3625	3602
690	3751	3726	3702	3676	3652	3628
695	3778	3753	3729	3703	3678	3654
700	3805	3780	3756	3729	3705	3680
702	0.03816	0.03790	0.03767	0.03740	0.03715	0.03691
704	3827	3801	3777	3750	3726	3701
706	3838	3812	3788	3761	3736	3712
708	3849	3823	3799	3772	3747	3722
710	3860	3834	3810	3783	3758	3733
712	0.03870	0.03844	0.03820	0.03793	0.03768	0.03744
714	3881	3855	3831	3804	3779	3754
716	3892	3866	3842	3815	3789	3765
718	3902	3877	3853	3825	3800	3775
720	3914	3888	3863	3836	3811	3786
722	0.03925	0.03898	0.03874	0.03847	0.03821	0.03796
724	3936	3909	3885	3857	3832	3807
726	3947	3920	3896	3868	3842	3817
728	3957	3931	3906	3878	3853	3828
730	3968	3941	3917	3889	3863	3838
732	0.03979	0.03952	0.03928	0.03900	0.03874	0.03849
734	3990	3963	3938	3910	3885	3859
736	4001	3974	3949	3921	3895	3870
738	4012	3985	3960	3932	3906	3880
740	4023	3995	3971	3942	3916	3891
742	0.04033	0.04006	0.03981	0.03953	0.03927	0.03901
744	4044	4017	3992	3964	3938	3912
746	4055	4028	4003	3974	3948	3922
748	4066	4039	4014	3985	3959	3933
750	4077	4049	4024	3996	3969	3943
752	0.04088	0.04060	0.04035	0.04006	0.03980	0.03954
754	4099	4071	4046	4017	3991	3964
756	4110	4082	4056	4028	4001	3975
758	4121	4093	4067	4038	4012	3985
760	4131	4103	4078	4049	4022	3996
762	0.04142	4114	4089	4060	4033	4006
764	4153	4125	4099	4070	4043	4017
766	4164	4136	4110	4081	4054	4027
768	4175	4147	4121	4092	4065	4038
770	4186	4158	4132	4102	4075	4048
772	0.04197	0.04168	0.04142	0.04113	0.04086	0.04059
774	4207	4179	4153	4124	4096	4070
776	4218	4190	4164	4134	4107	4080
778	4229	4201	4175	4145	4117	4091
780	4240	4211	4185	4155	4128	4101

Table 2-19
FACTORS FOR REDUCING GAS VOLUMES TO NORMAL (STANDARD) TEMPERATURE AND PRESSURE (760 mm Hg)

Examples: (*a*) 20 ml of dry gas at 22°C and 730 mm = 20 × 0.8888 = 17.78 ml at 0°C and 760 mm. (*b*) 20 cc of a gas over water at 22° and 730 mm = 20 × (factor corrected for aqueous tension; i.e., 730 − 19.8 or 710.2 mm) = 20 ml of dry gas at 22° and 710.2 mm = 20 × 0.86475 = 17.30 ml at 0°C and 760 mm. Weight in milligrams of 1 ml of gas at N. T. P.: acetylene, 1.173; carbon dioxide, 1.9769; hydrogen, 0.0899; nitric oxide (NO), 1.3402; nitrogen, 1.25057; oxygen, 1.42904.

Pressure mm of mercury	Temperature °C							
	10°	11°	12°	13°	14°	15°	16°	17°
670	0.8504	0.8474	0.8445	0.8415	0.8386	0.8357	0.8328	0.8299
672	0.8530	0.8500	0.8470	0.8440	0.8411	0.8382	0.8353	0.8324
674	0.8555	0.8525	0.8495	0.8465	0.8436	0.8407	0.8377	0.8349
676	0.8580	0.8550	0.8520	0.8490	0.8461	0.8431	0.8402	0.8373
678	0.8606	0.8576	0.8545	0.8516	0.8486	0.8456	0.8427	0.8398
680	0.8631	0.8601	0.8571	0.8541	0.8511	0.8481	0.8452	0.8423
682	0.8657	0.8626	0.8596	0.8566	0.8536	0.8506	0.8477	0.8448
684	0.8682	0.8651	0.8621	0.8591	0.8561	0.8531	0.8502	0.8472
686	0.8707	0.8677	0.8646	0.8616	0.8586	0.8556	0.8527	0.8497
688	0.8733	0.8702	0.8672	0.8641	0.8611	0.8581	0.8551	0.8522
690	0.8758	0.8727	0.8697	0.8666	0.8636	0.8606	0.8576	0.8547
692	0.8784	0.8753	0.8722	0.8691	0.8661	0.8631	0.8601	0.8572
694	0.8809	0.8778	0.8747	0.3717	0.8686	0.8656	0.8626	0.8596
696	0.8834	0.8803	0.8772	0.8742	0.8711	0.8681	0.8651	0.8621
698	0.8860	0.8828	0.8798	0.8767	0.8736	0.8706	0.8676	0.8646
700	0.8885	0.8854	0.8823	0.8792	0.8761	0.8731	0.8700	0.8671
702	0.8910	0.8879	0.8848	0.8817	0.8786	0.8756	0.8725	0.8695
704	0.8936	0.8904	0.8873	0.8842	0.8811	0.8781	0.8750	0.8720
706	0.8961	0.8930	0.8898	0.8867	0.8836	0.8806	0.8775	0.8745
708	0.8987	0.8955	0.8924	0.8892	0.8861	0.8831	0.8800	0.8770
710	0.9012	0.8980	0.8949	0.8917	0.8886	0.8856	0.8825	0.8794
712	0.9037	0.9006	0.8974	0.8943	0.8911	0.8880	0.8850	0.8819
714	0.9063	0.9031	0.8999	0.8968	0.8936	0.8905	0.8875	0.8844
716	0.9088	0.9056	0.9024	0.8993	0.8961	0.8930	0.8899	0.8869
718	0.9114	0.9081	0.9050	0.9018	0.8987	0.8955	0.8924	0.8894
720	0.9139	0.9107	0.9075	0.9043	0.9012	0.8980	0.8949	0.8918
722	0.9164	0.9132	0.9100	0.9068	0.9037	0.9005	0.8974	0.8943
724	0.9190	0.9157	0.9125	0.9093	0.9062	0.9030	0.8999	0.8968
726	0.9215	0.9183	0.9151	0.9118	0.9087	0.9055	0.9024	0.8993
728	0.9241	0.9208	0.9176	0.9144	0.9112	0.9080	0.9049	0.9017
730	0.9266	0.9233	0.9201	0.9169	0.9137	0.9105	0.9073	0.9042
732	0.9291	0.9259	0.9226	0.9194	0.9162	0.9130	0.9098	0.9067
734	0.9317	0.9284	0.9251	0.9219	0.9187	0.9155	0.9123	0.9092
736	0.9342	0.9309	0.9277	0.9244	0.9212	0.9180	0.9148	0.9117
738	0.9368	0.9334	0.9302	0.9269	0.9237	0.9205	0.9173	0.9141
740	0.9393	0.9360	0.9327	0.9294	0.9262	0.9230	0.9198	0.9166
742	0.9418	0.9385	0.9352	0.9319	0.9287	0.9255	0.9223	0.9191
744	0.9444	0.9410	0.9377	0.9345	0.9312	0.9280	0.9248	0.9216
746	0.9469	0.9436	0.9403	0.9370	0.9337	0.9305	0.9272	0.9240
748	0.9494	0.9461	0.9428	0.9395	0.9362	0.9329	0.9297	0.9265
750	0.9520	0.9486	0.9453	0.9420	0.9387	0.9354	0.9322	0.9290
752	0.9545	0.9511	0.9478	0.9445	0.9412	0.9379	0.9347	0.9315
754	0.9571	0.9537	0.9504	0.9470	0.9437	0.9404	0.9372	0.9339
756	0.9596	0.9562	0.9529	0.9495	0.9462	0.9429	0.9397	0.9364
758	0.9621	0.9587	0.9554	0.9520	0.9487	0.9454	0.9422	0.9389
760	0.9647	0.9613	0.9579	0.9546	0.9512	0.9479	0.9446	0.9414
762	0.9672	0.9638	0.9604	0.9571	0.9537	0.9504	0.9471	0.9439
764	0.9698	0.9663	0.9630	0.9596	0.9562	0.9529	0.9496	0.9463
766	0.9723	0.9689	0.9655	0.9620	0.9587	0.9554	0.9521	0.9488
768	0.9748	0.9714	0.9680	0.9646	0.9612	0.9579	0.9546	0.9513
770	0.9774	0.9739	0.9705	0.9671	0.9637	0.9604	0.9571	0.9538
772	0.9799	0.9764	0.9730	0.9696	0.9662	0.9629	0.9596	0.9562
774	0.9825	0.9790	0.9756	0.9721	0.9687	0.9654	0.9620	0.9587
776	0.9850	0.9815	0.9781	0.9746	0.9712	0.9679	0.9645	0.9612
778	0.9875	0.9840	0.9806	0.9772	0.9737	0.9704	0.9670	0.9637
780	0.9901	0.9866	0.9831	0.9797	0.9763	0.9729	0.9695	0.9662
782	0.9926	0.9891	0.9856	0.9822	0.9788	0.9754	0.9720	0.9686
784	0.9952	0.9916	0.9882	0.9847	0.9813	0.9778	0.9745	0.9711
786	0.9977	0.9942	0.9907	0.9872	0.9838	0.9803	0.9770	0.9736
788	1.0002	0.9967	0.9932	0.9897	0.9863	0.9828	0.9794	0.9761

Table 2-19 (*Continued*)
FACTORS FOR REDUCING GAS VOLUMES TO NORMAL (STANDARD) TEMPERATURE AND PRESSURE

Pressure mm of mercury	Temperature °C							
	18°	19°	20°	21°	22°	23°	24°	25°
670	0.8270	0.8242	0.8214	0.8186	0.8158	0.8131	0.8103	0.8076
672	0.8295	0.8267	0.8239	0.8211	0.8183	0.8155	0.8128	0.8100
674	0.8320	0.8291	0.8263	0.8235	0.8207	0.8179	0.8152	0.8124
676	0.8345	0.8316	0.8288	0.8259	0.8231	0.8204	0.8176	0.8149
678	0.8369	0.8341	0.8312	0.8284	0.8256	0.8228	0.8200	0.8173
680	0.8394	0.8365	0.8337	0.8308	0.8280	0.8252	0.8224	0.8197
682	0.8419	0.8390	0.8361	0.8333	0.8304	0.8276	0.8249	0.8221
684	0.8443	0.8414	0.8386	0.8357	0.8329	0.8301	0.8273	0.8245
686	0.8468	0.8439	0.8410	0.8382	0.8353	0.8325	0.8297	0.8269
688	0.8493	0.8464	0.8435	0.8406	0.8378	0.8349	0.8321	0.8293
690	0.8517	0.8488	0.8459	0.8430	0.8402	0.8373	0.8345	0.8317
692	0.8542	0.8513	0.8484	0.8455	0.8426	0.8398	0.8369	0.8341
694	0.8567	0.8537	0.8508	0.8479	0.8451	0.8422	0.8394	0.8366
696	0.8591	0.8562	0.8533	0.8504	0.8475	0.8446	0.8418	0.8390
698	0.8616	0.8587	0.8557	0.8528	0.8499	0.8471	0.8442	0.8414
700	0.8641	0.8611	0.8582	0.8553	0.8524	0.8495	0.8466	0.8438
702	0.8665	0.8636	0.8606	0.8577	0.8547	0.8519	0.8490	0.8462
704	0.8690	0.8660	0.8631	0.8602	0.8572	0.8543	0.8515	0.8486
706	0.8715	0.8685	0.8655	0.8626	0.8597	0.8568	0.8539	0.8510
708	0.8740	0.8710	0.8680	0.8650	0.8621	0.8592	0.8563	0.8534
710	0.8764	0.8734	0.8704	0.8675	0.8645	0.8616	0.8587	0.8558
712	0.8789	0.8759	0.8729	0.8699	0.8670	0.8640	0.8611	0.8582
714	0.8814	0.8783	p.8753	0.8724	0.8694	0.8665	0.8636	0.8607
716	0.8838	0.8808	0.8778	0.8748	0.8718	0.8689	0.8660	0.8631
718	0.8863	0.8833	0.8802	0.8773	0.8743	0.8713	0.8684	0.8655
720	0.8888	0.8857	0.8827	0.8797	0.8767	0.8738	0.8708	0.8679
722	0.8912	0.8882	0.8852	0.8821	0.8792	0.8762	0.8732	0.8703
724	0.8937	0.8906	0.8876	0.8846	0.8816	0.8786	0.8757	0.8727
726	0.8962	0.8931	0.8901	0.8870	0.8840	0.8810	0.8781	0.8751
728	0.8986	0.8956	0.8925	0.8895	0.8865	0.8835	0.8805	0.8775
730	0.9011	0.8980	0.8950	0.8919	0.8889	0.8859	0.8829	0.8799
732	0.9036	0.9005	0.8974	0.8944	0.8913	0.8883	0.8853	0.8824
734	0.9060	0.9029	0.8999	0.8968	0.8938	0.8907	0.8877	0.8848
736	0.9085	0.9054	0.9023	0.8992	0.8962	0.8932	0.8902	0.8872
738	0.9110	0.9079	0.9048	0.9017	0.8986	0.8956	0.8926	0.8896
740	0.9135	0.9103	0.9072	0.9041	0.9011	0.8980	0.8950	0.8920
742	0.9159	0.9128	0.9097	0.9066	0.9035	0.9005	0.8974	0.8944
744	0.9184	0.9153	0.9121	0.9090	0.9059	0.9029	0.8998	0.8968
746	0.9209	0.9177	0.9146	0.9115	0.9084	0.9053	0.9023	0.8992
748	0.9233	0.9202	0.9170	0.9139	0.9108	0.9077	0.9047	0.9016
750	0.9258	0.9226	0.9195	0.9164	0.9132	0.9102	0.9071	0.9041
752	0.9283	0.9251	0.9219	0.9188	0.9157	0.9126	0.9095	0.9065
754	0.9307	0.9276	0.9244	0.9212	0.9181	0.9150	0.9119	0.9089
756	0.9332	0.9300	0.9268	0.9237	0.9206	0.9174	0.9144	0.9113
758	0.9357	0.9325	0.9293	0.9261	0.9230	0.9199	0.9168	0.9137
760	0.9381	0.9349	0.9317	0.9286	0.9254	0.9223	0.9192	0.9161
762	0.9406	0.9374	0.9342	0.9310	0.9279	0.9247	0.9216	0.9185
764	0.9431	0.9399	0.9366	0.9335	0.9303	0.9272	0.9240	0.9209
766	0.9456	0.9423	0.9391	0.9359	0.9327	0.9296	0.9265	0.9233
768	0.9480	0.9448	0.9415	0.9383	0.9352	0.9320	0.9289	0.9258
770	0.9505	0.9472	0.9440	0.9408	0.9376	0.9344	0.9313	0.9282
772	0.9530	0.9497	0.9464	0.9432	0.9400	0.9369	0.9337	0.9306
774	0.9554	0.9522	0.9489	0.9457	0.9425	0.9393	0.9361	0.9330
776	0.9579	0.9546	0.9514	0.9481	0.9449	0.9417	0.9385	0.9354
778	0.9604	0.9571	0.9538	0.9506	0.9473	0.9441	0.9410	0.9378
780	0.9628	0.9595	0.9563	0.9530	0.9498	0.9466	0.9434	0.9402
782	0.9653	0.9620	0.9587	0.9555	0.9522	0.9490	0.9458	0.9426
784	0.9678	0.9645	0.9612	0.9579	0.9546	0.9514	0.9482	0.9450
786	0.9702	0.9669	0.9636	0.9603	0.9571	0.9538	0.9506	0.9474
788	0.9727	0.9694	0.9661	0.9628	0.9595	0.9563	0.9531	0.9499

Table 2-19 (*Continued*)
FACTORS FOR REDUCING GAS VOLUMES TO NORMAL (STANDARD) TEMPERATURE AND PRESSURE

Pressure mm of mercury	Temperature °C							
	26°	27°	28°	29°	30°	31°	32°	33°
670	0.8049	0.8022	0.7996	0.7969	0.7943	0.7917	0.7891	0.7865
672	0.8073	0.8046	0.8020	0.7993	0.7967	0.7940	0.7914	0.7889
674	0.8097	0.8070	0.8043	0.8017	0.7990	0.7964	0.7938	0.7912
676	0.8121	0.8094	0.8067	0.8041	0.8014	0.7988	0.7962	0.7936
678	0.8145	0.8118	0.8091	0.8064	0.8038	0.8011	0.7985	0.7959
680	0.8169	0.8142	0.8115	0.8088	0.8061	0.8035	0.8009	0.7982
682	0.8193	0.8166	0.8139	0.8112	0.8085	p.8059	0.8032	0.8006
684	0.8217	0.8190	0.8163	0.8136	0.8109	0.8082	0.8056	0.8029
686	0.8241	0.8214	0.8187	0.8160	0.8133	0.8106	0.8079	0.8053
688	0.8265	0.8238	0.8211	0.8183	0.8156	0.8129	0.8103	0.8076
690	0.8289	0.8262	0.8234	0.8207	0.8180	0.8153	0.8126	0.8100
692	0.8313	0.8286	0.8258	0.8231	0.8204	0.8177	0.8150	0.8123
694	0.8338	0.8310	0.8282	0.8255	0.8227	0.8200	0.8174	0.8147
696	0.8362	0.8334	0.8306	0.8278	0.8251	0.8224	0.8197	0.8170
698	0.8386	0.8358	0.8330	0.8302	0.8275	0.8248	0.8221	0.8194
700	0.8410	0.8382	0.8354	0.8326	0.8299	0.8271	0.8244	0.8217
702	0.8434	0.8406	0.8378	0.8350	0.8322	0.8295	0.8268	0.8241
704	0.8458	0.8429	0.8401	0.8374	0.8346	0.8319	0.8291	0.8264
706	0.8482	0.8453	0.8425	0.8397	0.8370	0.8342	0.8315	0.8288
708	0.8506	0.8477	0.8449	0.8421	0.8393	0.8366	0.8338	0.8311
710	0.8530	0.8501	0.8473	0.8445	0.8417	0.8389	0.8362	0.8335
712	0.8554	0.8525	0.8497	0.8469	0.8441	0.8413	0.8386	0.8358
714	0.8578	0.8549	0.8521	0.8493	0.8465	0.8437	0.8409	0.8382
716	0.8602	0.8573	0.8545	0.8516	0.8488	0.8460	0.8433	0.8405
718	0.8626	0.8597	0.8569	0.8540	0.8512	0.8484	0.8456	0.8429
720	0.8650	0.8621	0.8592	0.8564	0.8536	0.8508	0.8480	0.8452
722	0.8674	0.8645	0.8616	0.8588	0.8559	0.8531	0.8503	0.8475
724	0.8698	0.8669	0.8640	0.8612	0.8583	0.8555	0.8527	0.8499
726	0.8722	0.8693	0.8664	0.8635	0.8607	0.8579	0.8550	0.8522
728	0.8746	0.8717	0.8688	0.8659	0.8631	0.8602	0.8574	0.8546
730	0.8770	0.8741	0.8712	0.8683	0.8654	0.8626	0.8598	0.8569
732	0.8794	0.8765	0.8736	0.8707	0.8678	0.8649	0.8621	0.8593
734	0.8818	0.8789	0.8759	0.8730	0.8702	0.8673	0.8645	0.8616
736	0.8842	0.8813	0.8783	0.8754	0.8725	0.8697	0.8668	0.8640
738	0.8866	0.8837	0.8807	0.8778	0.8749	0.8720	0.8692	0.8663
740	0.8890	0.8861	0.8831	0.8802	0.8773	0.8744	0.8715	0.8687
742	0.8914	0.8884	0.8855	0.8826	0.8796	0.8768	0.8739	0.8710
744	0.8938	0.8908	0.8879	0.8849	0.8820	0.8791	0.8762	0.8734
746	0.8962	0.8932	0.8903	0.8873	0.8844	0.8815	0.8786	0.8757
748	0.8986	0.8956	0.8927	0.8897	0.8868	0.8838	0.8809	0.8781
750	0.9010	0.8980	0.8950	0.8921	0.8891	0.8862	0.8833	0.8804
752	0.9034	0.9004	0.8974	0.8945	0.8915	0.8886	0.8857	0.8828
754	0.9058	0.9028	0.8998	0.8968	0.8939	0.8909	0.8880	0.8851
756	0.9082	0.9052	0.9022	0.8992	0.8962	0.8933	0.8904	0.8875
758	0.9106	0.9076	0.9046	0.9016	0.8986	0.8957	0.8927	0.8898
760	0.9130	0.9100	0.9070	0.9040	0.9010	0.8980	0.8951	0.8922
762	0.9154	0.9124	0.9094	0.9064	0.9034	0.9004	0.8974	0.8945
764	0.9178	0.9148	0.9118	0.9087	0.9057	0.9028	0.8998	0.8969
766	0.9202	0.9172	0.9141	0.9111	0.9081	0.9051	0.9021	0.8992
768	0.9227	0.9196	0.9165	0.9135	0.9105	0.9075	0.9045	0.9015
770	0.9251	0.9220	0.9189	0.9159	0.9128	0.9098	0.9069	0.9039
772	0.9275	0.9244	0.9213	0.9182	0.9152	0.9122	0.9092	0.9062
774	0.9299	0.9268	0.9237	0.9206	0.9176	0.9146	0.9116	0.9086
776	0.9323	0.9292	0.9261	0.9230	0.9200	0.9169	0.9139	0.9109
778	0.9347	0.9316	0.9285	0.9254	0.9223	0.9193	0.9163	0.9133
780	0.9371	0.9340	0.9308	0.9278	0.9247	0.9217	0.9186	0.9156
782	0.9395	0.9363	0.9332	0.9301	0.9271	0.9240	0.9210	0.9180
784	0.9419	0.9387	0.9356	0.9325	0.9294	0.9264	0.9233	0.9203
786	0.9443	0.9411	0.9380	0.9349	0.9318	0.9287	0.9257	0.9227
788	0.9467	0.9435	0.9404	0.9373	0.9342	0.9311	0.9281	0.9250

Table 2-19 (*Continued*)
FACTORS FOR REDUCING GAS VOLUMES TO NORMAL (STANDARD) TEMPERATURE AND PRESSURE

Pressure mm of mercury	Temperature °C.		
	34°	35°	36°
670	0.7839	0.7814	0.7789
672	0.7863	0.7837	0.7812
674	0.7886	0.7861	0.7835
676	0.7910	0.7884	0.7858
678	0.7933	0.7907	0.7882
680	0.7956	0.7931	0.7905
682	0.7980	0.7954	0.7928
684	0.8003	0.7977	0.7951
686	0.8027	0.8001	0.7975
688	0.8050	0.8024	0.7998
690	0.8073	0.8047	0.8021
692	0.8097	0.8071	0.8044
694	0.8120	0.8094	0.8068
696	0.8144	0.8117	0.8091
698	0.8167	0.8141	0.8114
700	0.8190	0.8164	0.8137
702	0.8214	0.8187	0.8161
704	0.8237	0.8211	0.8184
706	0.8261	0.8234	0.8207
708	0.8284	0.8257	0.8230
710	0.8307	0.8281	0.8254
712	0.8331	0.8304	0.8277
714	0.8354	0.8327	0.8300
716	0.8378	0.8350	0.8323
718	0.8401	0.8374	0.8347
720	0.8424	0.8397	0.8370
722	0.8448	0.8420	0.8393
724	0.8471	0.8444	0.8416
726	0.8495	0.8467	0.8440
728	0.8518	0.8490	0.8463
730	0.8541	0.8514	0.8486
732	0.8565	0.8537	0.8509
734	0.8588	0.8560	0.8533
736	0.8612	0.8584	0.8556
738	0.8635	0.8607	0.8579
740	0.8658	0.8630	0.8602
742	0.8682	0.8654	0.8626
744	0.8705	0.8677	0.8649
746	0.8729	0.8700	0.8672
748	0.8752	0.8724	0.8695
750	0.8775	0.8747	0.8719
752	0.8799	0.8770	0.8742
754	0.8822	0.8794	0.8765
756	0.8846	0.8817	0.8788
758	0.8869	0.8840	0.8812
760	0.8892	0.8864	0.8835
762	0.8916	0.8887	0.8858
764	0.8939	0.8910	0.8881
766	0.8963	0.8934	0.8905
768	0.8986	0.8957	0.8928
770	0.9009	0.8980	0.8951
772	0.9033	0.9004	0.8974
774	0.9056	0.9027	0.8998
776	0.9080	0.9050	0.9021
778	0.9103	0.9074	0.9044
780	0.9127	0.9097	0.9067
782	0.9150	0.9120	0.9091
784	0.9173	0.9144	0.9114
786	0.9197	0.9167	0.9137
788	0.9220	0.9190	0.9160

Table 2-20
VALUES OF ABSORBANCE FOR PERCENT ABSORPTION

To convert percent absorption ($\%A$) to absorbance, find the percent absorption to the nearest whole digit in the left-hand column; read across to the column located under the tenth of a percent desired, and read the value of absorbance. The value of absorbance corresponding to 26.8% absorption is thus 0.1355.

%A	.0	.1	.2	.3	.4	.5	.6	.7	.8	.9
0.0	.0000	.0004	.0009	.0013	.0017	.0022	.0026	.0031	.0035	.0039
1.0	.0044	.0048	.0052	.0057	.0061	.0066	.0070	.0074	.0079	.0083
2.0	.0088	.0092	.0097	.0101	.0106	.0110	.0114	.0119	.0123	.0128
3.0	.0132	.0137	.0141	.0146	.0150	.0155	.0159	.0164	.0168	.0173
4.0	.0177	.0182	.0186	.0191	.0195	.0200	.0205	.0209	.0214	.0218
5.0	.0223	.0227	.0232	.0236	.0241	.0246	.0250	.0255	.0259	.0264
6.0	.0269	.0273	.0278	.0283	.0287	.0292	.0297	.0301	.0306	.0311
7.0	.0315	.0320	.0325	.0329	.0334	.0339	.0343	.0348	.0353	.0357
8.0	.0362	.0367	.0372	.0376	.0381	.0386	.0391	.0395	.0400	.0405
9.0	.0410	.0414	.0419	.0424	.0429	.0434	.0438	.0443	.0448	.0453
10.0	.0458	.0462	.0467	.0472	.0477	.0482	.0487	.0491	.0496	.0501
11.0	.0506	.0511	.0516	.0521	.0526	.0531	.0535	.0540	.0545	.0550
12.0	.0555	.0560	.0565	.0570	.0575	.0580	.0585	.0590	.0595	.0600
13.0	.0605	.0610	.0615	.0620	.0625	.0630	.0635	.0640	.0645	.0650
14.0	.0655	.0660	.0665	.0670	.0675	.0680	.0685	.0691	.0696	.0701
15.0	.0706	.0711	.0716	.0721	.0726	.0731	.0737	.0742	.0747	.0752
16.0	.0757	.0762	.0768	.0773	.0778	.0783	.0788	.0794	.0799	.0804
17.0	.0809	.0814	.0820	.0825	.0830	.0835	.0841	.0846	.0851	.0857
18.0	.0862	.0867	.0872	.0878	.0883	.0888	.0894	.0899	.0904	.0910
19.0	.0915	.0921	.0926	.0931	.0937	.0942	.0947	.0953	.0958	.0964
20.0	.0969	.0975	.0980	.0985	.0991	.0996	.1002	.1007	.1013	.1018
21.0	.1024	.1029	.1035	.1040	.1046	.1051	.1057	.1062	.1068	.1073
22.0	.1079	.1085	.1090	.1096	.1101	.1107	.1113	.1118	.1124	.1129
23.0	.1135	.1141	.1146	.1152	.1158	.1163	.1169	.1175	.1180	.1186
24.0	.1192	.1198	.1203	.1209	.1215	.1221	.1226	.1232	.1238	.1244
25.0	.1249	.1255	.1261	.1267	.1273	.1278	.1284	.1290	.1296	.1302
26.0	.1308	.1314	.1319	.1325	.1331	.1337	.1343	.1349	.1355	.1361
27.0	.1367	.1373	.1379	.1385	.1391	.1397	.1403	.1409	.1415	.1421
28.0	.1427	.1433	.1439	.1445	.1451	.1457	.1463	.1469	.1475	.1481
29.0	.1487	.1494	.1500	.1506	.1512	.1518	.1524	.1530	.1537	.1543
30.0	.1549	.1555	.1561	.1568	.1574	.1580	.1586	.1593	.1599	.1605
31.0	.1612	.1618	.1624	.1630	.1637	.1643	.1649	.1656	.1662	.1669
32.0	.1675	.1681	.1688	.1694	.1701	.1707	.1713	.1720	.1726	.1733
33.0	.1739	.1746	.1752	.1759	.1765	.1772	.1778	.1785	.1791	.1798
34.0	.1805	.1811	.1818	.1824	.1831	.1838	.1844	.1851	.1858	.1864
35.0	.1871	.1878	.1884	.1891	.1898	.1904	.1911	.1918	.1925	.1931
36.0	.1938	.1945	.1952	.1959	.1965	.1972	.1979	.1986	.1993	.2000
37.0	.2007	.2013	.2020	.2027	.2034	.2041	.2048	.2055	.2062	.2069
38.0	.2076	.2083	.2090	.2097	.2104	.2111	.2118	.2125	.2132	.2140
39.0	.2147	.2154	.2161	.2168	.2175	.2182	.2190	.2197	.2204	.2211
40.0	.2218	.2226	.2233	.2240	.2248	.2255	.2262	.2269	.2277	.2284
41.0	.2291	.2299	.2306	.2314	.2321	.2328	.2336	.2343	.2351	.2358

Table 2-20 (*Continued*)
VALUES OF ABSORBANCE FOR PERCENT ABSORPTION

% A	.0	.1	.2	.3	.4	.5	.6	.7	.8	.9
42.0	.2366	.2373	.2381	.2388	.2396	.2403	.2411	.2418	.2426	.2434
43.0	.2441	.2449	.2457	.2464	.2472	.2480	.2487	.2495	.2503	.2510
44.0	.2518	.2526	.2534	.2541	.2549	.2557	.2565	.2573	.2581	.2588
45.0	.2596	.2604	.2612	.2620	.2628	.2636	.2644	.2652	.2660	.2668
46.0	.2676	.2684	.2692	.2700	.2708	.2716	.2725	.2733	.2741	.2749
47.0	.2757	.2765	.2774	.2782	.2790	.2798	.2807	.2815	.2823	.2832
48.0	.2840	.2848	.2857	.2865	.2874	.2882	.2890	.2899	.2907	.2916
49.0	.2924	.2933	.2941	.2950	.2958	.2967	.2976	.2984	.2993	.3002
50.0	.3010	.3019	.3028	.3036	.3045	.3054	.3063	.3072	.3080	.3089
51.0	.3098	.3107	.3116	.3125	.3134	.3143	.3152	.3161	.3170	.3179
52.0	.3188	.3197	.3206	.3215	.3224	.3233	.3242	.3251	.3261	.3270
53.0	.3279	.3288	.3298	.3307	.3316	.3325	.3335	.3344	.3354	.3363
54.0	.3372	.3382	.3391	.3401	.3410	.3420	.3429	.3439	.3449	.3458
55.0	.3468	.3478	.3487	.3497	.3507	.3516	.3526	.3536	.3546	.3556
56.0	.3565	.3575	.3585	.3595	.3605	.3615	.3625	.3635	.3645	.3655
57.0	.3665	.3675	.3686	.3696	.3706	.3716	.3726	.3737	.3747	.3757
58.0	.3768	.3778	.3788	.3799	.3809	.3820	.3830	.3840	.3851	.3862
59.0	.3872	.3883	.3893	.3904	.3915	.3925	.3936	.3947	.3958	.3969
60.0	.3979	.3990	.4001	.4012	.4023	.4034	.4045	.4056	.4067	.4078
61.0	.4089	.4101	.4112	.4123	.4134	.4145	.4157	.4168	.4179	.4191
62.0	.4202	.4214	.4225	.4237	.4248	.4260	.4271	.4283	.4295	.4306
63.0	.4318	.4330	.4342	.4353	.4365	.4377	.4389	.4401	.4413	.4425
64.0	.4437	.4449	.4461	.4473	.4485	.4498	.4510	.4522	.4535	.4547
65.0	.4559	.4572	.4584	.4597	.4609	.4622	.4634	.4647	.4660	.4672
66.0	.4685	.4698	.4711	.4724	.4737	.4750	.4763	.4776	.4789	.4802
67.0	.4815	.4828	.4841	.4855	.4868	.4881	.4895	.4908	.4921	.4935
68.0	.4948	.4962	.4976	.4989	.5003	.5017	.5031	.5045	.5058	.5072
69.0	.5086	.5100	.5114	.5129	.5143	.5157	.5171	.5186	.5200	.5214
70.0	.5229	.5243	.5258	.5272	.5287	.5302	.5317	.5331	.5346	.5361
71.0	.5376	.5391	.5406	.5421	.5436	.5452	.5467	.5482	.5498	.5513
72.0	.5528	.5544	.5560	.5575	.5591	.5607	.5622	.5638	.5654	.5670
73.0	.5686	.5702	.5719	.5735	.5751	.5768	.5784	.5800	.5817	.5834
74.0	.5850	.5867	.5884	.5901	.5918	.5935	.5952	.5969	.5986	.6003
75.0	.6021	.6038	.6055	.6073	.6091	.6108	.6126	.6144	.6162	.6180
76.0	.6198	.6216	.6234	.6253	.6271	.6289	.6308	.6326	.6345	.6364
77.0	.6383	.6402	.6421	.6440	.6459	.6478	.6498	.6517	.6536	.6556
78.0	.6576	.6596	.6615	.6635	.6655	.6676	.6696	.6716	.6737	.6757
79.0	.6778	.6799	.6819	.6840	.6861	.6882	.6904	.6925	.6946	.6968
80.0	.6990	.7011	.7033	.7055	.7077	.7100	.7122	.7144	.7167	.7190
81.0	.7212	.7235	.7258	.7282	.7305	.7328	.7352	.7375	.7399	.7423
82.0	.7447	.7471	.7496	.7520	.7545	.7570	.7595	.7620	.7645	.7670
83.0	.7696	.7721	.7747	.7773	.7799	.7825	.7852	.7878	.7905	.7932
84.0	.7959	.7986	.8013	.8041	.8069	.8097	.8125	.8153	.8182	.8210
85.0	.8239	.8268	.8297	.8327	.8356	.8386	.8416	.8447	.8477	.8508
86.0	.8539	.8570	.8601	.8633	.8665	.8697	.8729	.8761	.8794	.8827
87.0	.8861	.8894	.8928	.8962	.8996	.9031	.9066	.9101	.9136	.9172
88.0	.9208	.9245	.9281	.9318	.9355	.9393	.9431	.9469	.9508	.9547
89.0	.9586	.9626	.9666	.9706	.9747	.9788	.9830	.9872	.9914	.9957

Table 2-21
TRANSMITTANCE-ABSORBANCE CONVERSION TABLE

From Meites, *Handbook of Analytical Chemistry*, 1963, McGraw-Hill Book
Company; by permission

This table gives absorbance values to four significant figures corresponding to %
transmittance values which are given to three significant figures. The values of %
transmittance are given in the left-hand column and in the top row. For example, 8.4%
transmittance corresponds to an absorbance of 1.076.

Interpolation is facilitated and accuracy is maximized if the % transmittance is
between 1 and 10, by multiplying its value by 10, finding the absorbance corresponding
to the result, and adding 1. For example, to find the absorbance corresponding to 8.45%
transmittance, note that 84.5% transmittance corresponds to an absorbance of 0.0731,
so that 8.45% transmittance corresponds to an absorbance of 1.0731. For % transmittance values between 0.1 and 1, multiply by 100, find the absorbance corresponding
to the result, and add 2.

Conversely, to find the % transmittance corresponding to an absorbance between 1
and 2, subtract 1 from the absorbance, find the % transmittance corresponding to the
result, and divide by 10. For example, an absorbance of 1.219 can best be converted
to % transmittance by noting that an absorbance of 0.219 would correspond to 60.4%
transmittance; dividing this by 10 gives the desired value, 6.04% transmittance. For
absorbance values between 2 and 3, subtract 2 from the absorbance, find the %
transmittance corresponding to the result, and divide by 100.

% Trans-mittance	0.0	0.1	0.2	0.3	0.4	0.5	0.6	0.7	0.8	0.9
0	......	3.000	2.699	2.523	2.398	2.301	2.222	2.155	2.097	2.046
1	2.000	1.959	1.921	1.886	1.854	1.824	1.796	1.770	1.745	1.721
2	1.699	1.678	1.658	1.638	1.620	1.602	1.585	1.569	1.553	1.538
3	1.523	1.509	1.495	1.481	1.469	1.456	1.444	1.432	1.420	1.409
4	1.398	1.387	1.377	1.367	1.357	1.347	1.337	1.328	1.319	1.310
5	1.301	1.292	1.284	1.276	1.268	1.260	1.252	1.244	1.237	1.229
6	1.222	1.215	1.208	1.201	1.194	1.187	1.180	1.174	1.167	1.161
7	1.155	1.149	1.143	1.137	1.131	1.125	1.119	1.114	1.108	1.102
8	1.097	1.092	1.086	1.081	1.076	1.071	1.066	1.060	1.056	1.051
9	1.046	1.041	1.036	1.032	1.027	1.022	1.018	1.013	1.009	1.004
10	1.000	0.9957	0.9914	0.9872	0.9830	0.9788	0.9747	0.9706	0.9666	0.9626
11	0.9586	0.9547	0.9508	0.9469	0.9431	0.9393	0.9355	0.9318	0.9281	0.9245
12	0.9208	0.9172	0.9136	0.9101	0.9066	0.9031	0.8996	0.8962	0.8928	0.8894
13	0.8861	0.8827	0.8794	0.8761	0.8729	0.8697	0.8665	0.8633	0.8601	0.8570
14	0.8539	0.8508	0.8477	0.8447	0.8416	0.8386	0.8356	0.8327	0.8297	0.8268
15	0.8239	0.8210	0.8182	0.8153	0.8125	0.8097	0.8069	0.8041	0.8013	0.7986
16	0.7959	0.7932	0.7905	0.7878	0.7852	0.7825	0.7799	0.7773	0.7747	0.7721
17	0.7696	0.7670	0.7645	0.7620	0.7595	0.7570	0.7545	0.7520	0.7496	0.7471
18	0.7447	0.7423	0.7399	0.7375	0.7352	0.7328	0.7305	0.7282	0.7258	0.7235
19	0.7212	0.7190	0.7167	0.7144	0.7122	0.7100	0.7077	0.7055	0.7033	0.7011
20	0.6990	0.6968	0.6946	0.6925	0.6904	0.6882	0.6861	0.6840	0.6819	0.6799

Table 2-21 (*Continued*)
TRANSMITTANCE-ABSORBANCE CONVERSION TABLE

% Trans-mittance	0.0	0.1	0.2	0.3	0.4	0.5	0.6	0.7	0.8	0.9
21	0.6778	0.6757	0.6737	0.6716	0.6696	0.6676	0.6655	0.6635	0.6615	0.6596
22	0.6576	0.6556	0.6536	0.6517	0.6498	0.6478	0.6459	0.6440	0.6421	0.6402
23	0.6383	0.6364	0.6345	0.6326	0.6308	0.6289	0.6271	0.6253	0.6234	0.6216
24	0.6198	0.6180	0.6162	0.6144	0.6126	0.6108	0.6091	0.6073	0.6055	0.6038
25	0.6021	0.6003	0.5986	0.5969	0.5952	0.5935	0.5918	0.5901	0.5884	0.5867
26	0.5850	0.5834	0.5817	0.5800	0.5784	0.5766	0.5751	0.5735	0.5719	0.5702
27	0.5686	0.5670	0.5654	0.5638	0.5622	0.5607	0.5591	0.5575	0.5560	0.5544
28	0.5528	0.5513	0.5498	0.5482	0.5467	0.5452	0.5436	0.5421	0.5406	0.5391
29	0.5376	0.5361	0.5346	0.5331	0.5317	0.5302	0.5287	0.5272	0.5258	0.5243
30	0.5229	0.5214	0.5200	0.5186	0.5171	0.5157	0.5143	0.5129	0.5114	0.5100
31	0.5086	0.5072	0.5058	0.5045	0.5031	0.5017	0.5003	0.4989	0.4976	0.4962
32	0.4949	0.4935	0.4921	0.4908	0.4895	0.4881	0.4868	0.4855	0.4841	0.4828
33	0.4815	0.4802	0.4789	0.4776	0.4763	0.4750	0.4737	0.4724	0.4711	0.4698
34	0.4685	0.4672	0.4660	0.4647	0.4634	0.4622	0.4609	0.4597	0.4584	0.4572
35	0.4559	0.4547	0.4535	0.4522	0.4510	0.4498	0.4486	0.4473	0.4461	0.4449
36	0.4437	0.4425	0.4413	0.4401	0.4389	0.4377	0.4365	0.4353	0.4342	0.4330
37	0.4318	0.4306	0.4295	0.4283	0.4271	0.4260	0.4248	0.4237	0.4225	0.4214
38	0.4202	0.4191	0.4179	0.4168	0.4157	0.4145	0.4134	0.4123	0.4112	0.4101
39	0.4089	0.4078	0.4067	0.4056	0.4045	0.4034	0.4023	0.4012	0.4001	0.3989
40	0.3979	0.3969	0.3958	0.3947	0.3936	0.3925	0.3915	0.3904	0.3893	0.3883
41	0.3872	0.3862	0.3851	0.3840	0.3830	0.3820	0.3809	0.3799	0.3788	0.3778
42	0.3768	0.3757	0.3747	0.3737	0.3726	0.3716	0.3706	0.3696	0.3686	0.3675
43	0.3665	0.3655	0.3645	0.3635	0.3625	0.3615	0.3605	0.3595	0.3585	0.3575
44	0.3565	0.3556	0.3546	0.3536	0.3526	0.3516	0.3507	0.3497	0.3487	0.3478
45	0.3468	0.3458	0.3449	0.3439	0.3429	0.3420	0.3410	0.3401	0.3391	0.3382
46	0.3372	0.3363	0.3354	0.3344	0.3335	0.3325	0.3316	0.3307	0.3298	0.3288
47	0.3279	0.3270	0.3261	0.3251	0.3242	0.3233	0.3224	0.3215	0.3206	0.3197
48	0.3188	0.3179	0.3170	0.3161	0.3152	0.3143	0.3134	0.3125	0.3116	0.3107
49	0.3098	0.3089	0.3080	0.3072	0.3063	0.3054	0.3045	0.3036	0.3028	0.3019
50	0.3010	0.3002	0.2993	0.2984	0.2976	0.2967	0.2958	0.2950	0.2941	0.2933
51	0.2924	0.2916	0.2907	0.2899	0.2890	0.2882	0.2874	0.2865	0.2857	0.2848
52	0.2840	0.2832	0.2823	0.2815	0.2807	0.2798	0.2790	0.2782	0.2774	0.2765
53	0.2757	0.2749	0.2741	0.2733	0.2725	0.2716	0.2708	0.2700	0.2692	0.2684
54	0.2676	0.2668	0.2660	0.2652	0.2644	0.2636	0.2628	0.2620	0.2612	0.2604
55	0.2596	0.2588	0.2581	0.2573	0.2565	0.2557	0.2549	0.2541	0.2534	0.2526
56	0.2518	0.2510	0.2503	0.2495	0.2487	0.2480	0.2472	0.2464	0.2457	0.2449
57	0.2441	0.2434	0.2426	0.2418	0.2411	0.2403	0.2396	0.2388	0.2381	0.2373
58	0.2366	0.2358	0.2351	0.2343	0.2336	0.2328	0.2321	0.2314	0.2306	0.2299
59	0.2291	0.2284	0.2277	0.2269	0.2262	0.2255	0.2248	0.2240	0.2233	0.2226
60	0.2218	0.2211	0.2204	0.2197	0.2190	0.2182	0.2175	0.2168	0.2161	0.2154
61	0.2147	0.2140	0.2132	0.2125	0.2118	0.2111	0.2104	0.2097	0.2090	0.2083
62	0.2076	0.2069	0.2062	0.2055	0.2048	0.2041	0.2034	0.2027	0.2020	0.2013
63	0.2007	0.2000	0.1993	0.1986	0.1979	0.1972	0.1965	0.1959	0.1952	0.1945
64	0.1938	0.1931	0.1925	0.1918	0.1911	0.1904	0.1898	0.1891	0.1884	0.1878
65	0.1871	0.1864	0.1858	0.1851	0.1844	0.1838	0.1831	0.1824	0.1818	0.1811

Table 2-21 (*Continued*)
TRANSMITTANCE-ABSORBANCE CONVERSION TABLE

% Trans-mittance	0.0	0.1	0.2	0.3	0.4	0.5	0.6	0.7	0.8	0.9
66	0.1805	0.1798	0.1791	0.1785	0.1778	0.1772	0.1765	0.1759	0.1752	0.1746
67	0.1739	0.1733	0.1726	0.1720	0.1713	0.1707	0.1701	0.1694	0.1688	0.1681
68	0.1675	0.1669	0.1662	0.1656	0.1649	0.1643	0.1637	0.1630	0.1624	0.1618
69	0.1612	0.1605	0.1599	0.1593	0.1586	0.1580	0.1574	0.1568	0.1561	0.1555
70	0.1549	0.1543	0.1537	0.1530	0.1524	0.1518	0.1512	0.1506	0.1500	0.1494
71	0.1487	0.1481	0.1475	0.1469	0.1463	0.1457	0.1451	0.1445	0.1439	0.1433
72	0.1427	0.1421	0.1415	0.1409	0.1403	0.1397	0.1391	0.1385	0.1379	0.1373
73	0.1367	0.1361	0.1355	0.1349	0.1343	0.1337	0.1331	0.1325	0.1319	0.1314
74	0.1308	0.1302	0.1296	0.1290	0.1284	0.1278	0.1273	0.1267	0.1261	0.1255
75	0.1249	0.1244	0.1238	0.1232	0.1226	0.1221	0.1215	0.1209	0.1203	0.1198
76	0.1192	0.1186	0.1180	0.1175	0.1169	0.1163	0.1158	0.1152	0.1146	0.1141
77	0.1135	0.1129	0.1124	0.1118	0.1113	0.1107	0.1101	0.1096	0.1090	0.1085
78	0.1079	0.1073	0.1068	0.1062	0.1057	0.1051	0.1046	0.1040	0.1035	0.1029
79	0.1024	0.1018	0.1013	0.1007	0.1002	0.0996	0.0991	0.0985	0.0980	0.0975
80	0.0969	0.0964	0.0958	0.0953	0.0947	0.0942	0.0937	0.0931	0.0926	0.0921
81	0.0915	0.0910	0.0904	0.0899	0.0894	0.0888	0.0883	0.0878	0.0872	0.0867
82	0.0862	0.0857	0.0851	0.0846	0.0841	0.0835	0.0830	0.0825	0.0820	0.0814
83	0.0809	0.0804	0.0799	0.0794	0.0788	0.0783	0.0778	0.0773	0.0768	0.0762
84	0.0757	0.0752	0.0747	0.0742	0.0737	0.0731	0.0726	0.0721	0.0716	0.0711
85	0.0706	0.0701	0.0696	0.0691	0.0685	0.0680	0.0675	0.0670	0.0665	0.0660
86	0.0655	0.0650	0.0645	0.0640	0.0635	0.0630	0.0625	0.0620	0.0615	0.0610
87	0.0605	0.0600	0.0595	0.0590	0.0585	0.0580	0.0575	0.0570	0.0565	0.0560
88	0.0555	0.0550	0.0545	0.0540	0.0535	0.0531	0.0526	0.0521	0.0516	0.0511
89	0.0506	0.0501	0.0496	0.0491	0.0487	0.0482	0.0477	0.0472	0.0467	0.0462
90	0.0458	0.0453	0.0448	0.0443	0.0438	0.0434	0.0429	0.0424	0.0419	0.0414
91	0.0410	0.0405	0.0400	0.0395	0.0391	0.0386	0.0381	0.0376	0.0372	0.0367
92	0.0362	0.0357	0.0353	0.0348	0.0343	0.0339	0.0334	0.0329	0.0325	0.0320
93	0.0315	0.0311	0.0306	0.0301	0.0297	0.0292	0.0287	0.0283	0.0278	0.0273
94	0.0269	0.0264	0.0259	0.0255	0.0250	0.0246	0.0241	0.0237	0.0232	0.0227
95	0.0223	0.0218	0.0214	0.0209	0.0205	0.0200	0.0195	0.0191	0.0186	0.0182
96	0.0177	0.0173	0.0168	0.0164	0.0159	0.0155	0.0150	0.0146	0.0141	0.0137
97	0.0132	0.0128	0.0123	0.0119	0.0114	0.0110	0.0106	0.0101	0.0097	0.0092
98	0.0088	0.0083	0.0079	0.0074	0.0070	0.0066	0.0061	0.0057	0.0052	0.0048
99	0.0044	0.0039	0.0035	0.0031	0.0026	0.0022	0.0017	0.0013	0.0009	0.0004

Table 2-22
WAVENUMBER/WAVELENGTH CONVERSION TABLE

This table is based on the conversion: wavenumber (in cm^{-1}) $= \dfrac{10\ 000}{\text{wavelength (in } \mu\text{m)}}$.
For example, 15.4 μm is equal to 649 cm^{-1}.

Wavelength (μm)	0	0.1	0.2	0.3	0.4	0.5	0.6	0.7	0.8	0.9 cm^{-1}
1.0	10000	9091	8333	7692	7143	6667	6250	5882	5556	5263
2.0	5000	4762	4545	4348	4167	4000	3846	3704	3571	3448
3.0	3333	3226	3125	3030	2941	2857	2778	2703	2632	2564
4.0	2500	2439	2381	2326	2273	2222	2174	2128	2083	2041
5.0	2000	1961	1923	1887	1852	1818	1786	1754	1724	1695
6.0	1667	1639	1613	1587	1563	1538	1515	1493	1471	1449
7.0	1429	1408	1389	1370	1351	1333	1316	1299	1282	1266
8.0	1250	1235	1220	1205	1190	1176	1163	1149	1136	1124
9.0	1111	1099	1087	1075	1064	1053	1042	1031	1020	1010
10.0	1000	990	980	971	962	952	943	935	926	917
11.0	909	901	893	885	877	870	862	855	847	840
12.0	833	826	820	813	806	800	794	787	781	775
13.0	769	763	758	752	746	741	735	730	725	719
14.0	714	709	704	699	694	690	685	680	676	671
15.0	667	662	658	654	649	645	641	637	633	629
16.0	625	621	617	613	610	606	602	599	595	592
17.0	588	585	581	578	575	571	568	565	562	559
18.0	556	552	549	546	543	541	538	535	532	529
19.0	526	524	521	518	515	513	510	508	505	503
20.0	500	498	495	493	490	488	485	483	481	478
21.0	476	474	472	469	467	465	463	461	459	457
22.0	455	452	450	448	446	444	442	441	439	437
23.0	435	433	431	429	427	426	424	422	420	418
24.0	417	415	413	412	410	408	407	405	403	402
25.0	400	398	397	395	394	392	391	389	388	386
26.0	385	383	382	380	379	377	376	375	373	372
27.0	370	369	368	366	365	364	362	361	360	358
28.0	357	356	355	353	352	351	350	348	347	346
29.0	345	344	342	341	340	339	338	337	336	334
30.0	333	332	331	330	329	328	327	326	325	324
31.0	323	322	321	319	318	317	316	315	314	313
32.0	313	312	311	310	309	308	307	306	305	304
33.0	303	302	301	300	299	299	298	297	296	295
34.0	294	293	292	292	291	290	289	288	287	287
35.0	286	285	284	283	282	282	281	280	279	279
36.0	278	277	276	275	275	274	273	272	272	271
37.0	270	270	269	268	267	267	266	265	265	264
38.0	263	262	262	261	260	260	259	258	258	257
39.0	256	256	255	254	254	253	253	252	251	251
40.0	250									

Section 3
ATOMIC AND MOLECULAR STRUCTURE

Table 3-1
PHYSICAL PROPERTIES OF THE ELEMENTS

Name	Sym-bol	At. No.	Electronic Configuration	Atomic Weight	Density, g·cm⁻³ at 20°C	Melting Point, °C	Boiling Point, °C	Thermal Conductivity, W·cm⁻¹·K⁻¹ at 25°C	Electrical Resistivity, $\mu\Omega\cdot$cm⁻¹ at 0°C
Actinium	Ac	89	$[\text{Rn}]6d\,7s^2$	(227)		1050	(3330)	0.12	
Aluminum	Al	13	$[\text{Ne}]3s^23p$	26.9815	2.6984	660.37	2447	2.37	2.50
Americium	Am	95	$[\text{Rn}]5f^77s^2$	(243)	13.67	994	(2600)	0.1	
Antimony	Sb	51	$[\text{Kr}]4d^{10}5s^25p^3$	121.75	6.684	630.74	1640	0.244	39
Argon	Ar	18	$[\text{Ne}]3s^23p^6$	39.948	0.001 782 4	−189.38	−185.87	0.1772×10^{-3}	26
Arsenic	As	33	$[\text{Ar}]3d^{10}4s^24p^3$	74.9216	5.72 gray 2.026 yel 4.7 black	817 (28 atm)	613 subl	0.502	
Astatine	At	85	$[\text{Xe}]4f^{14}5d^{10}6s^26p^5$	(210)		302	334	0.017	36
Barium	Ba	56	$[\text{Xe}]6s^2$	137.34	3.59	725.1	1849.0	0.184	
Berkelium	Bk	97	$[\text{Rn}]5f^86d7s^2$	(245)	1.86	1285	2970	0.1	2.8
Beryllium	Be	4	$[\text{He}]2s^2$	9.012 18	1.86	1277	2484	2.01	107
Bismuth	Bi	83	$[\text{Xe}]4f^{14}5d^{10}6s^26p^3$	208.9806	9.80	271.44	1579	0.0792	1.8×10^{12}
Boron	B	5	$[\text{He}]2s^22p$	10.81	2.46	2177	3658	0.274	$7.8 \times 10^{18}\,l$
Bromine	Br	35	$[\text{Ar}]3d^{10}4s^24p^5$	79.904	3.1028 l	−7.3	58.76	0.001 22	6.8
Cadmium	Cd	48	$[\text{Kr}]4d^{10}5s^2$	112.40	8.642	321.11	767	0.969	3.2
Calcium	Ca	20	$[\text{Ar}]4s^2$	40.08	1.55	850	1487	2.01	
Californium	Cf	98	$[\text{Rn}]5f^{10}7s^2$	(248)				0.1	1375
Carbon	C	6	$[\text{He}]2s^22p^2$	12.011	2.267 graphite 3.515 diamond	4000 (63 atm)	3930		
Cerium	Ce	58	$[\text{Xe}]4f5d\,6s^2$	140.12	6.77	795	3470	0.113	73
Cesium	Cs	55	$[\text{Xe}]6s$	132.9055	1.8785¹⁵	28.8	678.5	0.359	19
Chlorine	Cl	17	$[\text{Ne}]3s^23p^5$	35.453	0.002 98 g	−101.0	−34.05	0.001 34 l 0.000 89 g (1 atm)	$>10^9\,l$

Element	Symbol	No.	Configuration	Atomic weight	Density	Melting point	Boiling point		
Chromium	Cr	24	$[Ar]3d^54s$	51.996	7.2	1857	2682	0.939	12.7
Cobalt	Co	27	$[Ar]3d^74s^2$	58.9332	8.9	1494	2897.1	1.00	5.6
Copper	Cu	29	$[Ar]3d^{10}4s$	63.546	8.92	1084.5	2582	4.01	1.6
Curium	Cm	96	$[Rn]5f^76d\,7s^2$	(247)	13.51	1500	2600	(0.1)	89
Dysprosium	Dy	66	$[Xe]4f^{10}6s^2$	162.50	8.536	1500	2600	0.107	89
Einsteinium	Es	99	$[Rn]5f^{11}7s^2$	(254)				(0.1)	
Erbium	Er	68	$[Xe]4f^{12}6s^2$	167.26	9.051	1497	2900	0.145	81
Europium	Eu	63	$[Xe]4f^76s^2$	151.96	5.259	826	1440	0.139	89
Fermium	Fm	100	$[Rn]5f^{12}7s^2$	(253)				(0.1)	
Fluorine	F	9	$[He]2s^22p^5$	18.9984	0.001 580 g	-219.62	-188.14	0.277×10^{-3} g	
Francium	Fr	87	$[Rn]7s$	(223)		(27)		(0.15)	
Gadolinium	Gd	64	$[Xe]4f^75d\,6s^2$	157.25	7.895	1306	3000	0.105	126
Gallium	Ga	31	$[Ar]3d^{10}4s^24p$	69.72	5.903	29.78	1980	0.294^{30} l	13.6
Germanium	Ge	32	$[Ar]3d^{10}4s^24p^2$	72.59	5.323	940	2852	0.602	89 000
Gold	Au	79	$[Xe]4f^{14}5d^{10}6s$	196.9665	19.3	1064.43	2808	3.18	2.05
Hafnium	Hf	72	$[Xe]4f^{14}5d^26s^2$	178.49	13.31	2222	4450	0.230	29.6
Helium	He	2	$1s^2$	4.002 60	$0.178\ 47 \times 10^{-4}$ (STP) g	-272.2 (25 atm)	-268.935	1.513×10^{-3} g (1 atm)	
Holmium	Ho	67	$[Xe]4f^{11}6s^2$	164.9303	8.803	1461	2600	0.162	90
Hydrogen	H	1	$1s$	1.0080	0.8987×10^{-4} g	-259.20	-252.87	1.805×10^{-3} g (1 atm)	
Indium	In	49	$[Kr]4d^{10}5s^25p$	114.82	7.28	156.63	2070	0.818	8.0
Iodine	I	53	$[Kr]4d^{10}5s^25p^5$	126.9045	4.660 s	113.6	184.24	4.49^{27}	1.3×10^{15}
Iridium	Ir	77	$[Xe]4f^{14}5d^76s^2$	192.22	22.65	2454	4389	1.47	4.7
Iron	Fe	26	$[Ar]3d^64s^2$	55.847	7.86	1537	2872.3	0.804	8.9
Krypton	Kr	36	$[Ar]3d^{10}4s^24p^6$	83.80	0.0037	-157.2	-153.4	0.0943	
Lanthanum	La	57	$[Xe]5d\,6s^2$	138.9055	6.174	920	3470	0.134	54
Lawrencium	Lr	103	$[Rn]5f^{14}6d\,7s^2$	(257)				(0.1)	
Lead	Pb	82	$[Xe]4f^{14}5d^{10}6s^26p^2$	207.2	11.34	327.50	1751	0.353	19.2
Lithium	Li	3	$1s^22s$	6.941	0.535	180.6	1336	0.848	8.55
Lutetium	Lu	71	$[Xe]4f^{14}5d\,6s^2$	174.97	9.842	1652	3330	0.164	54
Magnesium	Mg	12	$[Ne]3s^2$	24.305	1.74	650	1105	1.56	3.94
Manganese	Mn	25	$[Ar]3d^54s^2$	54.9380	7.30	1244	2120	0.0781	138

Table 3-1 (Continued)
PHYSICAL PROPERTIES OF THE ELEMENTS

Name	Symbol	At. No.	Electronic Configuration	Atomic Weight	Density, g·cm⁻³ at 20°C	Melting Point, °C	Boiling Point, °C	Thermal Conductivity, W·cm⁻¹·K⁻¹ at 25°C	Electrical Resistivity, μΩ·cm⁻¹ at 0°C
Mendelevium	Md	101	$[\text{Rn}]5f^{13}7s^2$	(256)				(0.1)	94 l; 21 s
Mercury	Hg	80	$[\text{Xe}]4f^{14}5d^{10}6s^2$	200.59	13.595	−38.862	356.66	0.0830	5.0
Molybdenum	Mo	42	$[\text{Kr}]4d^55s$	95.94	10.2	2610	4646	1.38	61
Neodymium	Nd	60	$[\text{Xe}]4f^46s^2$	144.24	7.004	1019	3111	0.165	
Neon	Ne	10	$1s^22s^22p^6$	20.179	1.207 _l_ at bp	−248.6	−246.048	0.491×10^{-3} _g_ (1 atm)	
Neptunium	Np	93	$[\text{Rn}]5f^46d\,7s^2$	237.0482	20.45	630	2920	0.063	119[100]
Nickel	Ni	28	$[\text{Ar}]3d^84s^2$	58.70	8.90	1455	2920	0.909	6.2
Niobium	Nb	41	$[\text{Kr}]4d^45s$	92.9064	8.57	2477	4863	0.537	15.2
Nitrogen	N	7	$1s^22s^22p^3$	14.0067	0.001 165 _g_	−209.97	−195.798	0.2583×10^{-3} _g_(1 atm)	
Nobelium	No	102	$[\text{Rn}]5f^{14}7s^2$	(254)				(0.1)	
Osmium	Os	76	$[\text{Xe}]4f^{14}5d^66s^2$	190.2	22.61	~2727	~5500	0.876	8.1
Oxygen	O	8	$1s^22s^22p^4$	15.9994	0.001 331 _g_	−218.787	−182.962	0.2658×10^{-3} _g_(1 atm)	
Palladium	Pd	46	$[\text{Kr}]4d^{10}$	106.4	12.023	1554	2940	0.718	10.0
Phosphorus	P	15	$[\text{Ne}]3s^23p^3$	30.9738	1.828 white 2.34 red 2.699 black	44.2 597 1027	280.3 416 subl 453 subl	0.002 36	10[17]
Platinum	Pt	78	$[\text{Xe}]4f^{14}5d^96s$	195.09	21.45	1772	3824	0.716	9.81
Plutonium	Pu	94	$[\text{Rn}]5f^67s^2$	(242)	19.82	639.6	3235	0.0670	146
Polonium	Po	84	$[\text{Xe}]4f^{14}5d^{10}6s^26p^4$	210	9.20 cub 9.40 rh	254	962	(0.20)	~40
Potassium	K	19	$[\text{Ar}]4s$	39.102	0.87	63.7	765.5	1.025	6.1
Praseodymium	Pr	59	$[\text{Xe}]4f^36s^2$	140.9077	6.782	919	3130	0.125	65

Promethium	Pm	61	$[Xe]4f^5 6s^2$	(147)		1080	(2727)	0.179	50
Protactinium	Pa	91	$[Rn]5f^2 6d\, 7s^2$	231.0359	15.37	(1227)	(4027)	(0.47)	17.7
Radium	Ra	88	$[Rn]7s^2$	226.0254	ca 6	700	1737	0.186^{20}	
Radon	Rn	86	$[Xe]4f^{14}5d^{10}6s^2 6p^6$	(222)	4.4 l at bp	−71	−62	0.0361×10^{-3} g (1 atm)	
Rhenium	Re	75	$[Xe]5f^{14}5d^5 6s^2$	186.207	21.04	3180	5687	0.480	17.2
Rhodium	Rh	45	$[Kr]4d^8 5s$	102.9055	12.41	1963	3727	1.50	4.3
Rubidium	Rb	37	$[Kr]5s$	85.4678	1.53	39.0	694	0.582	11.0
Ruthenium	Ru	44	$[Kr]4d^7 5s$	101.07	12.45	2427	4119	1.17	7.1
Samarium	Sm	62	$[Xe]4f^6 6s^2$	150.4	7.536	1072.1	1803	0.133	91.4
Scandium	Sc	21	$[Ar]3d\, 4s^2$	44.9559	2.992	1539	2730	0.158	50.5
Selenium	Se	34	$[Ar]3d^{10}4s^2 4p^4$	78.96	4.792 hex-rbhed, 4.48 red, mn	221, 170	685	0.005 19 amorp, 0.0452 ∥ c-axis, 0.0131 ⊥ c-axis	1.2
Silicon	Si	14	$[Ne]3s^2 3p^2$	28.086	2.33	1412	2680	1.49	10^5
Silver	Ag	47	$[Kr]4d^{10}5s$	107.868	10.50	961.93	2164	4.29	1.47
Sodium	Na	11	$[Ne]3s$	22.9898	0.97	97.8	881.4	1.42	4.2
Strontium	Sr	38	$[Kr]5s^2$	87.62	2.60	769	1381	0.354	20
Sulfur	S	16	$[Ne]3s^2 3p^4$	32.06	2.08 α, 1.96 β, 1.92 γ	115.21, 115.21, 106.8	444.674	0.002 05 amorp	2×10^{23}
Tantalum	Ta	73	$[Xe]4f^{14}5d^3 6s^2$	180.9479	16.60	2985	5513	0.575	12.3
Technetium	Tc	43	$[Kr]4d^5 5s^2$	98.9062	11.487	2250	4567	0.506	22.6^{100}
Tellurium	Te	52	$[Kr]4d^{10}5s^2 5p^4$	127.60	6.24	449.6	1009	0.0338 ∥ c-axis, 0.0197 ⊥ c-axis	$(5.8\text{-}33) \times 10^3$
Terbium	Tb	65	$[Xe]4f^9 6s^2$	158.9254	8.272	1356	2800	0.111	113
Thallium	Tl	81	$[Xe]4f^{14}5d^{10}6s^2 6p$	204.37	11.80	303.1	1487	0.461	15
Thorium	Th	90	$[Rn]6d^2 7s^2$	232.0381	11.71	1750	4787	0.540	14.7
Thulium	Tm	69	$[Xe]4f^{13}6s^2$	168.9342	9.332	1545	1730	0.169	67
Tin	Sn	50	$[Kr]4d^{10}5s^2 5p^2$	118.69	7.28 white, 4.507 α	231.97	2623	0.668	11.5
Titanium	Ti	22	$[Ar]3d^2 4s^2$	47.90	4.32 β	1660	3318	0.219	39
Tungsten (Wolfram)	W	74	$[Xe]4f^{14}5d^4 6s^2$	183.85	19.35	3407	5663	1.73	4.9

Table 3-1 (*Continued*)
PHYSICAL PROPERTIES OF THE ELEMENTS

Name	Sym-bol	At. No.	Electronic Configuration	Atomic Weight	Density, g · cm^{-3} at 20°C	Melting Point, °C	Boiling Point, °C	Thermal Conductivity, W · cm^{-1} · K^{-1} at 25°C	Electrical Resistivity, $\mu\Omega$ · cm^{-1} at 0°C
Uranium	U	92	[Rn]5$f^3$6d 7s^2	238.029	19.05	1130	3927	0.275	28
Vanadium	V	23	[Ar]3$d^3$4s^2	50.9414	6.1	1917	3421	0.307	18.2
Wolfram (see Tungsten)									
Xenon	Xe	54	[Kr]4d^{10}5$s^2$5p^6	131.30	0.005 897 1^0	−111.8	−108.1	0.0565 × 10^{-3} *g* (1 atm)	
Ytterbium	Yb	70	[Xe]4f^{14}6s^2	173.04	6.977	824	1430	0.349	27.7
Yttrium	Y	39	[Kr]4d 5s^2	88.9059	4.478	1530	3304	0.172	55
Zinc	Zn	30	[Ar]3d^{10}4s^2	65.37	7.1	419.6	911	1.16	5.5
Zirconium	Zr	40	[Kr]4$d^2$5s^2	91.22	6.52	1852	4504	0.226	40

Table 3-2
IONIZATION POTENTIALS OF THE ELEMENTS
In Electronvolts

The minimum amount of energy required to remove the least strongly bound electron from a gaseous atom or ion is called the ionization energy or potential, and is expressed in electronvolts. In Table 3-2 the successive stages of ionization are indicated at the heading of each column: I, denoting first spectra (from neutral atoms)

$$M(\text{gas}) \longrightarrow M^+(\text{gas}) + e^-$$

II, second spectra (from singly ionized atoms), and so on for successive stages of ionization as the outermost electron is removed.

Reference: C. E Moore, *National Standard Reference Data Series* **34**, U.S. Government Printing Office, Washington, D.C., 1970. This reference also lists ionization limits derived from the analysis of optical spectra and the literature references used for each spectrum. W. C. Martin *et al.*, *J. Phys. Chem. Reference Data*, **3**, 771 (1974) for the lanthanoids and actinoids.

At. No.	Element	Spectrum						
		I	II	III	IV	V	VI	VII
1	H	13.598*						
2	He	24.587	54.416					
3	Li	5.392	75.638	122.451				
4	Be	9.322	18.211	153.893	217.713			
5	B	8.298	25.154	37.930	259.368	340.217		
6	C	11.260	24.383	47.887	64.492	392.077	409.981	
7	N	14.534	29.601	47.448	77.472	97.888	552.057	667.029
8	O	13.618	35.116	54.934	77.412	113.896	138.116	739.315
9	F	17.422	34.970	62.707	87.138	114.240	157.161	185.182
10	Ne	21.564	40.962	63.45	97.11	126.21	157.93	207.27
11	Na	5.139	47.286	71.64	98.91	138.39	172.15	208.47
12	Mg	7.646	15.035	80.143	109.24	141.26	186.50	224.94
13	Al	5.986	18.828	28.447	119.99	153.71	190.47	241.43
14	Si	8.151	16.345	33.492	45.141	166.77	205.05	246.52
15	P	10.486	19.725	30.18	51.37	65.023	220.43	263.22
16	S	10.360	23.33	34.83	47.30	72.68	88.049	280.93
17	Cl	12.967	23.81	39.61	53.46	67.8	97.03	114.193
18	Ar	15.759	27.629	40.74	59.81	75.02	91.007	124.319
19	K	4.341	31.625	45.72	60.91	82.66	100.0	117.56
20	Ca	6.113	11.871	50.908	67.10	84.41	108.78	127.7
21	Sc	6.54	12.80	24.76	73.47	91.66	111.1	138.0
22	Ti	6.82	13.58	27.491	43.266	99.22	119.36	140.8
23	V	6.74	14.65	29.310	46.707	65.23	128.12	150.17
24	Cr	6.766	16.50	30.96	49.1	69.3	90.56	161.1
25	Mn	7.435	15.640	33.667	51.2	72.4	95	119.27
26	Fe	7.870	16.18	30.651	54.8	75.0	99	125
27	Co	7.86	17.06	33.50	51.3	79.5	102	129
28	Ni	7.635	18.168	35.17	54.9	75.5	108	133
29	Cu	7.726	20.292	36.83	55.2	79.9	103	139
30	Zn	9.394	17.964	39.722	59.4	82.6	108	134
31	Ga	5.999	20.51	30.71	64			
32	Ge	7.899	15.934	34.22	45.71	93.5		
33	As	9.81	18.633	28.351	50.13	62.63	127.6	
34	Se	9.752	21.19	30.820	42.944	68.3	81.70	155.4

*D = 13.601; T = 13.603.

Table 3-2 (*Continued*)
IONIZATION POTENTIALS OF THE ELEMENTS
In Electronvolts

At. No.	Element	Spectrum						
		I	II	III	IV	V	VI	VII
35	Br	11.814	21.8	36	47.3	59.7	88.6	103.0
36	Kr	13.999	24.359	36.95	52.5	64.7	78.5	111.0
37	Rb	4.177	27.28	40	52.6	71.0	84.4	99.2
38	Sr	5.695	11.030	43.6	57	71.6	90.8	106
39	Y	6.38	12.24	20.52	61.8	77.0	93.0	116
40	Zr	6.84	13.13	22.99	34.34	81.5	(99)	
41	Nb	6.88	14.32	25.04	38.3	50.55	102.6	125
42	Mo	7.099	16.15	27.16	46.4	61.2	68	126.8
43	Tc	7.28	15.26	29.54				
44	Ru	7.37	16.76	28.47				
45	Rh	7.46	18.08	31.06				
46	Pd	8.34	19.43	32.93				
47	Ag	7.576	21.49	34.83				
48	Cd	8.993	16.908	37.48				
49	In	5.786	18.869	28.03	54			
50	Sn	7.344	14.632	30.502	40.734	72.28		
51	Sb	8.641	16.53	25.3	44.2	56	108	
52	Te	9.009	18.6	27.96	37.41	58.75	70.7	137
53	I	10.451	19.131	33				
54	Xe	12.130	21.21	32.1				
55	Cs	3.894	25.1	(35)				
56	Ba	5.212	10.004	(37)				
57	La	5.577	11.06	19.177	49.95			
58	Ce	5.47	10.85	20.198	36.758			
59	Pr	5.42	10.55	21.624	38.98	57.45		
60	Nd	5.49	10.73	22.1	40.4			
61	Pm	5.55	10.90	22.3	41.1			
62	Sm	5.63	11.07	23.4	41.4			
63	Eu	5.666	11.241	24.9	42.6			
64	Gd	6.14	12.09	20.6	44.0			
65	Tb	5.85	11.52	21.9	39.8			
66	Dy	5.928	11.67	22.8	41.5			
67	Ho	6.02	11.80	22.84	42.5			
68	Er	6.10	11.93	22.74	42.7			
69	Tm	6.1844	12.05	23.68	42.7			
70	Yb	6.2539	12.184	25.03	43.7			
71	Lu	5.4259	13.9	20.960	45.19			
72	Hf	6.65	14.9	23.3	33.33			
73	Ta	7.89	(16)					
74	W	7.98	(18)					
75	Re	7.88	13.1	26.0	37.7			
76	Os	8.7	(17)					
77	Ir	9.1	(17)					
78	Pt	9.0	18.563					
79	Au	9.225	20.5					
80	Hg	10.437	18.756	34.2				
81	Tl	6.108	20.428	29.83				

Table 3-2 (*Continued*)
IONIZATION POTENTIALS OF THE ELEMENTS
In Electronvolts

At. No.	Element	Spectrum						
		I	II	III	IV	V	VI	VII
82	Pb	7.416	15.032	31.937	42.32	68.8		
83	Bi	7.289	16.69	25.56	45.3	56.0	88.3	
84	Po	8.42						
85	At							
86	Rn	10.748						
87	Fr							
88	Ra	5.279	10.147					
89	Ac	5.17	12.1					
90	Th	6.08	11.5	20.0	28.8			
91	Pa	5.89						
92	U	6.05	(14.7)					
93	Np	6.19						
94	Pu	6.06						
95	Am	5.993						
96	Cm	6.02						
97	Bk	6.23						
98	Cf	6.30						
99	Es	6.42						
100	Fm	6.50						
101	Md	6.58						
102	No	6.65						
103	Lr							
104								
105								

Table 3-3
IONIZATION POTENTIALS OF MOLECULAR AND RADICAL SPECIES
In Electronvolts

Species	Ionization Potential	Species	Ionization Potential
H_2	15.43	CH_4	13.0
N_2	15.6	C_2H_6	11.7
O_2	12.1	C_2H_4	10.5
F_2	16.5(?)	C_2H_2	11.4
Cl_2	11.5	NH_3	10.15
Br_2	10.6	C_2N_2	13.6
I_2	9.3	HCN	13.9
CH	11.1	H_2O	12.59
OH	12.8(?)	H_2S	10.46
HCl	12.74	CH_3I	9.6
HBr	11.62	CO_2	13.79
HI	10.38	N_2O	12.90
CO	14.0	NO_2	12.0
NO	9.25	SO_2	12.34
CH_2	10.3	CH_3OH	10.85
CH_3	9.9	CH_3NH_2	8.97
C_2H_5	8.7		
t-Butyl	6.9		
CH_3CO	7.9		
HCO	8.8(?)		
NH_2	11.3		
HO_2	11.5		

Table 3-4
ATOMIC ELECTRON AFFINITIES

The electron affinity of an atom (A) is defined as the energy released when an atom and an electron react to form a negative ion in the gas phase at 0 K; viz.,

$$A(g) + e^- = A^-(g)$$

Data are limited to those negative ions which, by virtue of their positive electron affinity, are stable. Uncertainty in the final data figures is given in parentheses.

Reference: H. Hotop and W. C. Lineberger, *J. Phys. Chem. Reference Data*, **4**, 539 (1975).

Element	Electron Affinity, eV	Element	Electron Affinity, eV
Aluminum	0.46(3)	Molybdenum	1.0(2)
Antimony	1.05(5)	Neon	<0
Argon	<0	Nickel	1.15(10)
Arsenic	0.80(5)	Niobium	1.0(3)
Astatine	2.8(2)	Nitrogen	−0.07(8)
Barium	<0	Osmium	1.1(3)
Beryllium	<0	Oxygen	1.462(3)
Bismuth	1.1(2)	Palladium	0.6(3)
Boron	0.28(1)	Phosphorus	0.743(10)
Bromine	3.364(4)	Platinum	2.128
Cadmium	<0	Polonium	1.9(3)
Calcium	<0	Potassium	0.5012(5)
Carbon	1.268(5)	Radon	<0
Cesium	0.4715(5)	Rare earths	≲0.5 estimate
Chlorine	3.615(4)	Rhenium	0.15(10)
Chromium	0.66(5)	Rhodium	1.2(3)
Cobalt	0.7(2)	Rubidium	0.4860(5)
Copper	1.226(10)	Ruthenium	1.1(3)
Fluorine	3.399(3)	Scandium	<0
Francium	(0.456)	Selenium	2.0206(3)
Gallium	0.30(15)	Silicon	1.385(5)
Germanium	1.2(1)	Silver	1.303(7)
Gold	2.3086(7)	Sodium	0.546(5)
Hafnium	>0	Strontium	<0
Helium	<0	Sulfur	2.0772(5)
Hydrogen	0.754 209(3)	Tantalum	0.6(4)
Indium	0.30(15)	Technetium	0.7(3)
Iodine	3.061(4)	Tellurium	1.9708(3)
Iridium	1.6(2)	Thallium	0.3(2)
Iron	0.25(20)	Tin	1.25(10)
Krypton	<0	Titanium	0.2(2)
Lanthanum	0.5(3)	Tungsten	0.6(4)
Lead	1.1(2)	Vanadium	0.5(2)
Lithium	0.620(7)	Xenon	<0
Magnesium	<0	Yttrium	0.0(3)
Manganese	<0	Zinc	≈0
Mercury	<0	Zirconium	0.5(3)

Table 3-5
ELECTRONEGATIVITIES OF THE ELEMENTS

According to Pauling, electronegativity χ is the relative attraction of an atom for the valence electrons in a covalent bond. It is proportional to the effective nuclear charge and inversely proportional to the covalent radius:

$$\chi = \frac{0.31(n + 1 \pm c)}{r} + 0.50$$

where n is the number of valence electrons, c is any formal valence charge on the atom and the sign before it corresponds to the sign of this charge, and r is the covalent radius. Because electronegativity is concerned with atoms in molecules rather than atoms in isolation, it is not possible to define precise electronegativity values. Pauling determined his set of values from bond energy data based on experimentally measured heats of dissociation and formation. Originally the element fluorine, whose atoms have the greatest attraction for electrons, was given an arbitrary electronegativity of 4.0. A revision of Pauling's values based on newer heat data assigns 3.90 to fluorine. Values given in Table 3-5 refer to the common oxidation states of the elements.

The greater the difference in electronegativity, the greater is the ionic character of the bond. A percentage of complete separation, or a "degree of ionic character" of the still covalent bond may be estimated from the approximate rule suggested by Gordy [*Discussions Faraday Soc.*, **19,** 23 (1955)]:

$$\text{Degree of ionic character} = \frac{|\chi_A - \chi_B|}{2}$$

for $|\chi_A - \chi_B| < 2$, and 100 per cent for $|\chi_A - \chi_B| > 2$. When $\Delta\chi = 0$ the bond is completely nonpolar. A more sophisticated expression was proposed by Hannay-Smyth [*J. Am. Chem. Soc.*, **68,** 171 (1946)]:

$$\text{Degree of ionic character} = 0.46|\chi_A - \chi_B| + 0.035(\chi_A - \chi_B)^2$$

The bond stretching force constant k (in units of 10^5 dynes cm^{-1}) can be estimated by the expression for stable molecules exhibiting their normal covalencies:

$$k = 1.67N\left(\frac{\chi_A\chi_B}{d^2}\right)^{3/4} + 0.30$$

where N is the bond order (i.e., the effective number of covalent or ionic bonds acting between the two atoms A and B) and d is the internuclear distance in angstroms.

Electronegativity is also proportional to the work function ϕ, which is the energy necessary to just remove an electron from the metal surface in thermoelectric or photoelectric emission:

$$\chi = 0.44\phi - 0.15$$

Element	χ	Element	χ	Element	χ	Element	χ
H	2.20	Na	0.9	Sc	1.3	Cu(I)	1.9
Li	1.0	Mg	1.2	Ti	1.5	Cu(II)	2.0
Be	1.5	Al	1.5	V	1.6	Zn	1.6
B	2.0	Si	1.90	Cr	1.6	Ga	1.6
C	2.60	P	2.15	Mn	1.5	Ge	1.90
N	3.05	S	2.60	Fe(II)	1.8	As	2.00
O	3.50	Cl	3.15	Fe(III)	1.9	Se	2.45
F	4.00*	K	0.8	Co	1.8	Br	2.85
	3.90†	Ca	1.0	Ni	1.8	Rb	0.8

* Arbitrary reference value.　　† Recent value.

Table 3-5 (*Continued*)
ELECTRONEGATIVITIES OF THE ELEMENTS

Element	χ	Element	χ	Element	χ	Element	χ
Sr	1.0	In	1.7	Ta	1.3	Bi	1.9
Y	1.3	Sn(II)	1.8	W	1.7	Po	2.0
Zr	1.6	Sn(IV)	1.90	Re	1.9	At	2.2
Nb	1.6	Sb	2.05	Os	2.2	Fr	0.65
Mo	1.8	Te	2.30	Ir	2.2	Ra	0.9
Tc	1.9	I	2.65	Pt	2.2	Ac	1.1
Ru	2.2	Cs	0.7	Au	2.4	Th	1.3
Rh	2.2	Ba	0.9	Hg	1.9	Pa	1.5
Pd	2.2	La–Lu	1.1–1.2	Tl	1.8	U	1.7
Ag	1.9	Hf	1.3	Pb	1.8	Np-No	1.3
Cd	1.7						

Mulliken's scale of electronegativities of the elements is based on

$$0.5(I + E)$$

where I is the ionization potential (Table 3-2) and E is the electron affinity (Table 3-4) of an element.

Electronegativities on the Allred-Rochow scale [*J. Inorg. Nucl. Chem.*, **5**, 264, 269 (1958)] are given by

$$0.359 \frac{Z_{eff}}{r^2} + 0.744$$

where Z_{eff} is the effective nuclear charge and r is the atomic radius.

Table 3-6
NUCLEAR PROPERTIES OF THE ELEMENTS

In the following table the magnetic moment μ is in multiples of the nuclear magneton μ_N $(eh/4\pi Mc)$, with diamagnetic correction, the spin I is in multiples of $h/2\pi$, and the electric quadrupole moment Q is in multiples of 10^{-28} m². The sign of μ and Q is uncertain for those nuclides for which no sign is given.

Nuclide	NMR Frequency, MHz at 14 092 G	NMR Frequency, MHz at 23 490 G	NMR Field Values, kG at 60 MHz	NMR Field Values, kG at 90 MHz	Magnetic Moment μ/μ_N	Spin I	Electric Quadrupole Moment Q, 10^{-28} m²
^{1}n	41.102				−1.913 12	$-\frac{1}{2}$	
^{1}H	60.000	100.00	14.092	21.06	+2.792 78	$\frac{1}{2}$	
^{2}H	9.2101	15.352	45.5*		+0.857 42	1	+2.8 × 10^{-3}
^{3}H	45.4131†				+2.9789	$\frac{1}{2}$	
^{3}He	45.7057		18.49	27.75	−2.1276	$-\frac{1}{2}$	
^{6}Li	8.8293		47.88*		+0.822 03	1	−8.0 × 10^{-4}
^{7}Li	23.317	38.867	36.26	54.39	+3.256 36	$\frac{3}{2}$	−0.04
^{9}Be	8.4318	14.055	50.14*		−1.177 45	$-\frac{3}{2}$	0.05
^{10}B	6.4477	10.748			+1.8006	3	+0.085
^{11}B	19.250	32.087	43.92		+2.6885	$\frac{3}{2}$	+0.041
^{13}C	15.086	25.147	56.05		+0.7024	$\frac{1}{2}$	
^{14}N	4.3341	7.2246			+0.403 75	1	+0.01
^{15}N	6.0796	10.134			−0.2831	$-\frac{1}{2}$	
^{17}O	8.134	13.56	51.97*		−1.8937	$-\frac{5}{2}$	−0.026
^{19}F	56.444	94.087	14.98	22.47	+2.6288	$\frac{1}{2}$	
^{21}Ne	4.7365				−0.661 76	$-\frac{3}{2}$	+0.09
^{22}Na	6.248				1.746	3	
^{23}Na	15.870	26.454	53.28		+2.217 40	$\frac{3}{2}$	+0.10
^{24}Na	4.54				1.690	4	
^{25}Mg	3.672				−0.8554	$-\frac{5}{2}$	+0.22
^{27}Al	15.634	26.060	54.08		+3.6413	$\frac{5}{2}$	+0.15
^{29}Si	11.919	19.867	35.47*		−0.555 26	$-\frac{1}{2}$	
^{31}P	24.288	40.485	34.81	52.22	+1.1317	$\frac{1}{2}$	
^{33}S	4.6016	7.6704			+0.6435	$\frac{3}{2}$	−0.055
^{35}S					+1.00 or −1.07	$\frac{3}{2}$	+0.038
^{35}Cl	5.8788	9.7993			+0.821 81	$\frac{3}{2}$	−0.078
^{36}Cl	6.895				+1.2853	2	−0.10
^{37}Cl	4.893	8.156			+0.684 07	$\frac{3}{2}$	−0.079
^{39}K	2.7992				+0.391 43	$\frac{3}{2}$	+0.049
^{40}K	3.481				−1.2981	4	−0.061
^{41}K	1.5365				+0.2149	$\frac{3}{2}$	+0.060
^{43}Ca	4.0379				−1.3172	$\frac{7}{2}$	<0.2
^{45}Sc	14.5759		58.00		+4.7559	$\frac{7}{2}$	−0.22
^{47}Ti	3.3817				−0.788 46	$-\frac{5}{2}$	+0.29
^{49}Ti	3.3826				−1.104 14	$-\frac{7}{2}$	+0.24
^{50}V	5.979				+3.3470	6	0.06
^{51}V	15.77	26.29	53.60		+5.1485	$\frac{7}{2}$	−0.05
^{53}Cr	3.3910				−0.4735	$\frac{3}{2}$	+0.03
^{55}Mn	14.798	24.667	56.85		+3.449	$\frac{5}{2}$	+0.4
^{57}Fe	1.94				+0.090 42	$\frac{1}{2}$	
^{59}Co	14.168	23.617	59.39		+4.616	$\frac{7}{2}$	+0.38
^{61}Ni	5.3617				−0.7498	$\frac{3}{2}$	+0.16
^{63}Cu	15.903	26.508	53.17		+2.2228	$\frac{3}{2}$	−0.211
^{65}Cu	17.036	28.397	49.63		+2.3812	$\frac{3}{2}$	−0.195
^{67}Zn	3.753				+0.875 24	$\frac{5}{2}$	+0.16
^{69}Ga	14.4003		58.71		+2.0145	$\frac{3}{2}$	+0.19

Table 3-6 (*Continued*)
NUCLEAR PROPERTIES OF THE ELEMENTS

Nuclide	NMR Frequency, MHz at 14 092 G	at 23 490 G	NMR Field Values, kG at 60 MHz	at 90 MHz	Magnetic Moment μ/μ_N	Spin I	Electric Quadrupole Moment Q, $10^{-28}\,\mathrm{m}^2$
^{71}Ga	18.2971		46.21		+2.5597	$\frac{3}{2}$	+0.12
^{73}Ge	2.093				−0.879 18	$\frac{9}{2}$	−0.18
^{75}As	10.276	17.129	41.14*		+1.439	$\frac{3}{2}$	+0.29
^{77}Se	11.44	19.07			+0.534	$\frac{1}{2}$	
^{79}Br	15.032	25.057	56.25		+2.1055	$\frac{3}{2}$	+0.37
^{81}Br	16.203	27.009	52.18		+2.2696	$\frac{3}{2}$	+0.31
^{83}Kr	2.3083				−0.9703	$\frac{9}{2}$	+0.26
^{85}Rb	5.793				+1.3524	$\frac{5}{2}$	+0.26
^{87}Rb	19.632	32.724			+2.7500	$\frac{3}{2}$	+0.13
^{87}Sr	2.6001				−1.093	$\frac{9}{2}$	+0.3
^{89}Y	2.9396				−0.137 33	$-\frac{1}{2}$	
^{91}Zr	5.5773				−1.3028	$-\frac{5}{2}$	
^{93}Nb	14.666	24.446	57.65		+6.167	$\frac{9}{2}$	−0.22
^{95}Mo	3.909				−0.9135	$-\frac{5}{2}$	0.12
^{97}Mo	3.991				−0.9327	$\frac{5}{2}$	1.1
^{99}Tc	13.5047				+5.681	$\frac{9}{2}$	+0.3
^{99}Ru	2.67				−0.62	$\frac{5}{2}$	
^{101}Ru	3.0				−0.68	$-\frac{5}{2}$	
^{103}Rh	1.8931				−0.0883	$-\frac{1}{2}$	
^{105}Pd	2.6664				−0.642	$-\frac{5}{2}$	+0.8
^{107}Ag	2.4281				−0.1135	$-\frac{1}{2}$	
^{109}Ag	2.7913				−0.1305	$-\frac{1}{2}$	
^{111}Cd	12.7227		33.23*		−0.594 28	$-\frac{1}{2}$	
^{113}Cd	13.3086		31.77*		−0.6225	$-\frac{1}{2}$	
^{113}In	13.122				+5.5229	$\frac{9}{2}$	+0.82
^{115}In	13.149		32.16*		+5.5348	$\frac{9}{2}$	+0.83
^{115}Sn	19.619				−0.9178	$\frac{1}{2}$	
^{117}Sn	21.375	35.630	39.55	59.33	−1.000	$-\frac{1}{2}$	
^{119}Sn	22.363	37.276	37.81	56.71	−1.0461	$-\frac{1}{2}$	
^{121}Sb	14.358	23.934			+3.3592	$\frac{5}{2}$	−0.28
^{123}Sb	7.777				+2.5466	$\frac{7}{2}$	−0.36
^{123}Te	15.725				−0.736	$\frac{1}{2}$	
^{125}Te	18.958				−0.8872	$-\frac{1}{2}$	
^{127}I	12.004	20.009	35.21*		+2.8091	$\frac{5}{2}$	−0.79
^{129}I	7.989				+2.6174	$\frac{7}{2}$	−0.55
^{129}Xe	16.5974				−0.7768	$-\frac{1}{2}$	
^{131}Xe	4.9184				+0.6908	$\frac{3}{2}$	−0.12
^{133}Cs	7.8699	13.118	53.71*		+2.5779	$\frac{7}{2}$	−0.0030
^{135}Ba	5.9602				+0.8372	$\frac{3}{2}$	+0.18
^{137}Ba	6.6675				+0.9366	$\frac{3}{2}$	+0.28
^{138}La	7.915				+3.704	5	+0.51
^{139}La	8.475				+2.778	$\frac{7}{2}$	+0.22
^{141}Pr	18.3				+4.16	$\frac{5}{2}$	−0.058
^{143}Nd	3.245				−1.063	$-\frac{7}{2}$	−0.48
^{145}Nd	1.993				−0.654	$\frac{7}{2}$	−0.25
^{147}Sm	2.42				−0.813	$\frac{7}{2}$	−0.18
^{149}Sm	1.96				−0.670	$\frac{7}{2}$	+0.052
^{151}Eu	14.59				+3.4631	$\frac{5}{2}$	+1.1
^{153}Eu	6.43				+1.530	$\frac{5}{2}$	+2.8
^{155}Gd	2.0				−0.2584	$\frac{3}{2}$	+1.6
^{157}Gd	2.7				−0.3388	$\frac{3}{2}$	+1.7
^{159}Tb	13.7				2.008	$\frac{3}{2}$	+1.3

Table 3-6 (*Continued*)
NUCLEAR PROPERTIES OF THE ELEMENTS

Nuclide	NMR Frequency, MHz		NMR Field Values, kG		Magnetic Moment μ/μ_N	Spin I	Electric Quadrupole Moment Q, 10^{-28} m^2
	at 14 092 G	at 23 490 G	at 60 MHz	at 90 MHz			
^{161}Dy	1.96				−0.482	$\frac{5}{2}$	+2.4
^{163}Dy	2.73				+0.676	$\frac{5}{2}$	+2.5
^{165}Ho	12.3				+4.12	$\frac{7}{2}$	+2.7
^{167}Er	1.72				−0.564	$\frac{7}{2}$	+2.83
^{169}Tm	4.89				−0.231	$\frac{1}{2}$	
^{171}Yb	10.507				+0.4919	$\frac{1}{2}$	
^{173}Yb	2.889				−0.6776	$\frac{5}{2}$	+3.0
^{175}Lu	6.7908				+2.230	$\frac{7}{2}$	+5.6
^{176}Lu	4.78				3.18	7	+8.0
^{177}Hf	1.8				+0.7902	$\frac{7}{2}$	+4.5
^{179}Hf	1.13				−0.638	$\frac{9}{2}$	+5.1
^{181}Ta	7.17				+2.38	$\frac{7}{2}$	+3
^{183}W	2.4964				+0.1169	$\frac{1}{2}$	
^{185}Re	13.509				+3.172	$\frac{5}{2}$	+2.3
^{187}Re	13.647				+3.204	$\frac{5}{2}$	+2.2
^{187}Os	1.4				+0.0643	$\frac{1}{2}$	
^{189}Os	4.656				+0.6565	$\frac{3}{2}$	+0.8
^{191}Ir	1.146				+0.1454	$\frac{3}{2}$	+1.1
^{193}Ir	1.21				+0.1583	$\frac{3}{2}$	+1.0
^{195}Pt	12.90	21.50	32.78*		+0.6022	$\frac{1}{2}$	
^{197}Au	1.0445				+0.144 86	$\frac{3}{2}$	+0.59
^{199}Hg	10.696	17.829			+0.504 15	$\frac{1}{2}$	
^{201}Hg	3.9597	6.6005			−0.5583	$-\frac{3}{2}$	+0.44
^{203}Tl	34.289	57.156			+1.6115	$\frac{1}{2}$	
^{205}Tl	34.624	57.715			+1.6274	$\frac{1}{2}$	
^{207}Pb	12.553	20.924	33.71*		+0.5783	$\frac{1}{2}$	
^{209}Bi	9.6414	16.071			+4.080	$\frac{9}{2}$	−0.38
^{235}U	1.06				−0.43	$\frac{7}{2}$	4.9

* At 30 MHz   † At 10 000 G.

Table 3-7
TABLE OF NUCLIDES

Introduction

The data presented here were compiled by J. R. Peterson from the *Table of Isotopes*, 6th ed. (C. M. Lederer, J. M. Hollander, and I. Perlman, John Wiley & Sons, New York, 1967), *Chart of the Nuclides*, 3d ed. (W. Seelmann-Eggebert, G. Pfennig, and H. Münzel, Gersbach und Sohn Verlag, Munich, 1968), the "1964 Atomic Mass Table" [J. H. E. Mattauch, W. Thiele, and A. H. Wapstra, *Nucl. Phys.*, **67,** 1 (1965)], and in the cases of the more recently discovered nuclides, both published and unpublished reports. Not all known nuclides were included, those omitted usually having very short half-lives.

Explanation of Column Headings

Nuclide. Each nuclide is identified by its atomic number Z, equal to the number of protons in the nucleus; the corresponding symbol for that element; and the mass number A, equal to the sum of the numbers of protons Z and neutrons N in the

Table 3-7 (*Continued*)
TABLE OF NUCLIDES

nucleus. Thus, $A = Z + N$, or $N = A - Z$. The m following the mass number (e.g., $^{58m}_{27}Co$) indicates an isomer of that nuclide. Isomers differ only in energy, the nuclide denoted by m (metastable) following its mass number being the higher-energy nuclide.

Mass Excess. The mass excess Δ is the difference between the actual mass M and the mass number A, i.e., $\Delta = M - A$. Thus, the actual mass M is obtained by adding the mass number A and the mass excess Δ. The mass excesses are expressed both in micro mass units (μmu; $1\mu mu = 10^{-6}$ mu) and in million electron volts (MeV; 1 MeV = 10^6 eV) to facilitate mass and energy calculations. All values are based on the unified, or carbon-12, mass scale; i.e., the mass of one atom of carbon-12 is taken to be exactly 12 mu: thus $\Delta^{12}C = 0$. For example, the mass of an atom of $^{34}_{16}S$ would be $34 + (-0.032135) = 33.967865$ mu.

Half-life. For the radioactive nuclides this time period corresponds to that during which loss by disintegration of 50% of the nuclide occurs. The units of time are designated by year (y), day (d), hour (h), minute (m), and second (s).

Natural Abundance. The isotopic abundances listed are on an "atom per cent" basis for the stable nuclides present in naturally occurring elements in the earth's crust.

Thermal Neutron Absorption Cross Section. The ease with which a given nuclide can absorb a thermal neutron (energy $\leq 1/40$ eV) and become a different nuclide is indicated by the cross section, given here in units of barns (1 barn = 10^{-24} cm^2). If the mode of reaction is other than (n,γ), it is so indicated, e.g., (n,p) or (n,α), where n = neutron, p = proton, γ = gamma ray, and α = alpha particle (^4_2He). In the case of the heaviest elements, neutron absorption leading to fission, i.e., $(n, fission)$, is designated by (f) following the numerical value of the cross section.

Major Radiations. In this column are listed the principal mode(s) of decay and the energies of the emanating radiations in million electronvolts (MeV). The gamma-ray (γ) intensities, where given, are given to the nearest whole percentage in parentheses following the numerical energy value for that particular γ. In most cases the radiations listed should be sufficient for identification of the particular nuclide.

The following designations are used: negatron (β^-), positron (β^+), conversion electron (e^-), gamma ray (γ), alpha particle (α), spontaneous fission (SF), neutron (n), and proton (p). More detailed information is available in the first source listed above (*Table of Isotopes*).

Nuclide		Mass Excess		Half-life	Natural	Thermal Neutron	Major Radiations,
Z Symbol	Mass Number	μmu	MeV	y = yr, d = day h = hr, m = min s = sec	Abundance, %	Absorption Cross Section σ, barns (mode if not n,γ)	Energies in MeV and γ Intensities, %
$_0n$	1	8,665	8.0714	11.7 m	. . . .	. . . .	β^-, 0.78
$_1H$	1	7,825	7.2890	. . . .	99.985	0.332	. . . .
	2	14,102	13.1359	. . . .	0.015	0.0005	. . . .
	3	16,050	14.9500	12.26 y	. . . .	. . . .	β^-, 0.0186; no γ
$_2He$	3	16,030	14.9313	. . . .	1.3×10^{-4}	5,330(n,p)	. . . .
	4	2,603	2.4248	. . . .	~100	0	. . . .
	6	18,893	17.598	0.797 s	. . . .	. . . .	β^-, 3.508; no γ
	8	37,520	31.7	0.122 s	. . . .	. . . .	β^-, 9.7; γ, 0.98(88)
$_3Li$	6	15,125	14.088	. . . .	7.42	953(n,α)	. . . .

Table 3-7 (*Continued*)
TABLE OF NUCLIDES

Nuclide		Mass Excess		Half-life $y=yr, d=day$ $h=hr, m=min$ $s=sec$	Natural Abundance, %	Thermal Neutron Absorption Cross Section σ, barns (mode if not n,γ)	Major Radiations, Energies in MeV and γ Intensities, %
Z Symbol	Mass Number	μmu	MeV				
	7	16,004	14.907	....	92.58	0.037	....
	8	22,487	20.946	0.841 s	....	....	β^-, 13; α, 1.6
	9	26,802	24.97	0.176 s	....	....	β^-, 13.61; n, 0.76;
$_4$Be	6	19,717	18.37	~0.4 s	....	....	....
	7	16,929	15.769	53.6 d	....	54,000(n,p)	γ, 0.477(10)
	9	12,186	11.351	....	100	0.009	....
	10	13,534	12.607	2.5×10^6 y	....	....	β^-, 0.555; no γ
	11	21,666	20.18	13.6 s	....	....	β^-, 11.5; γ, 2.14(32), 4.67(2), 5.85(2), 6.79(4), 7.99(2)
$_5$B	8	24,609	22.923	0.77 s	....	....	β^+, 14.0; α, 1.6
	10	12,939	12.052	....	19.7	3,837(n,α)	....
	11	9,305	8.6677	....	80.3	0.005	....
$_6$C	9		29.0	0.127 s	....	....	β^+, 3.5; p, 8.2, 1.1; α, 0.05, 1.6
	10	16,810	15.66	19.48 s	....	....	β^+, 1.87; γ, 0.511, 0.717(100), 1.023(2)
	11	11,432	10.648	20.34 m	....	....	β^+, 0.97; γ, 0.511
	12	0	0	....	98.892	0.0034	....
	13	3,354	3.125	....	1.108	0.0009	....
	14	3,242	3.0198	5730 y	....	....	β^-, 0.156; no γ
	15	10,600	9.873	2.5 s	....	....	β^-, 9.82, 4.51; γ, 5.299(68)
	16	14,700	13.69	0.74 s	....	....	....
$_7$N	13	5,738	5.345	9.96 m	....	....	β^+, 1.20; γ, 0.511
	14	3,074	2.8637	....	99.635	1.81(n,p)	....
	15	108	0.100	....	0.365	2.4×10^{-5}	....
	16	6,103	5.685	7.14 s	....	....	β^-, 10.40, 4.27; γ, 2.75(1), 6.13(69), 7.11(5); α, 1.7
	17	8,450	7.87	....	....	....	β^-, 8.68, 7.81, 4.1; γ, 0.87(3), 2.19(1); n, 0.40, 1.21, 1.81
	18		13.1	0.63 s	....	....	β^-, 9.4; γ, 0.82(59), 1.65(59), 1.98(100), 2.47(41)
$_8$O	14	8,597	8.0080	70.91 s	....	....	β^+, 4.12, 1.811; γ, 0.511, 2.312(99)
	15	3,070	2.860	123 s	....	....	β^+, 1.74; γ,0.511
	16	−5,085	−4.7366	....	99.759	0.00018	....
	17	−867	−0.808	....	0.037	0.24(n,α)	....
	18	−840	−0.7824	....	0.204	0.00021	....
	19	3,578	3.333	29.1 s	....	....	β^-, 4.60; γ, 0.197(97), 1.37(59)

Table 3-7 (*Continued*)
TABLE OF NUCLIDES

Nuclide		Mass Excess		Half-life y = yr, d = day h = hr, m = min s = sec	Natural Abundance, %	Thermal Neutron Absorption Cross Section σ, barns (mode if not n,γ)	Major Radiations, Energies in MeV and γ Intensities, %
Z Symbol	Mass Number	μmu	MeV				
	20	4,079	3.80	14 s	....	....	β^-, 2.75; γ, 1.06(100)
$_9$F	17	2,096	1.952	66.6 s	....	....	β^+, 1.74; γ, 0.511
	18	937	0.872	109.7 m	....	....	β^+, 0.635; γ, 0.511
	19	−1,595	−1.486	....	100	0.010	....
	20	−13	−0.012	11.56 s	....	....	β^-, 5.41; γ, 1.63(100)
	21	−49	−0.05	4.35 s	....	....	β^-, 5.4; γ, 0.350, 1.38
	22		4	4.0 s	....	....	β^-, 11; γ, 1.28(100), 2.06(67)
$_{10}$Ne	17		33.9	0.10 s	....	....	p, 4.59
	18	5,711	5.319	1.5 s	....	....	β^+, 3.42; γ, 0.511, 1.04(7)
	19	1,881	1.752	17.4 s	....	....	β^+, 2.22; γ, 0.511
	20	−7,560	−7.042	....	90.92	....	....
	21	−6,151	−5.730	....	0.257	....	....
	22	−8,615	−8.025	....	8.82	0.04	....
	23	−5,527	−5.148	37.6 s	....	....	β^-, 4.38; γ, 0.439(33), 1.64(1)
	24	−6,387	−5.95	3.38 m	....	....	β^-, 1.99; γ, 0.472(100), 0.88(8)
$_{11}$Na	20	8,880	8.3	0.39 s	....	....	β^+, 11.4; γ, 0.511, 1.63; α, 2.14
	21	−2,345	−2.19	23.0 s	....	....	β^+, 2.52; γ, 0.511, 0.350(2)
	22	−5,563	−5.182	2.62 y	....	....	β^+, 1.820, 0.545; γ, 0.511, 1.275(100)
	23	−10,229	−9.528	....	100	0.53	....
	24	−9,038	−8.418	14.96 h	....	....	β^-, 4.17, 1.389; γ, 1.369(100), 2.754(100)
	25	−10,045	−9.36	60 s	....	....	β^-, 3.83; γ, 0.39(14), 0.58(14), 0.98(15), 1.61(6)
	26	−8,260	−7.7	1.04 s	....	....	β^-, 6.7; γ, 1.82(100)
$_{12}$Mg	20		16	0.6 s	....	....	....
	21		10.9	0.121 s	....	....	p, 3.3, 3.8, 4.58, 6.14
	23	−5,875	−5.472	12.1 s	....	....	β^+, 3.03; γ, 0.44(9), 0.511
	24	−14,958	−13.933	....	78.60	0.03	....
	25	−14,161	−13.191	....	10.11	0.3	....
	26	−17,407	−16.214	....	11.29	0.027	....
	27	−15,655	−14.583	9.46 m	....	<0.030	β^-, 1.75; γ, 0.18(1), 0.84(70), 1.013(30)

Table 3-7 (*Continued*)
TABLE OF NUCLIDES

Nuclide		Mass Excess		Half-life y = yr, d = day h = hr, m = min s = sec	Natural Abundance, %	Thermal Neutron Absorption Cross Section σ, barns (mode if not n,γ)	Major Radiations, Energies in MeV and γ Intensities, %
Z Symbol	Mass Number	μmu	MeV				
	28	−16,125	−15.02	21.2 h	....	....	β⁻, 0.46; e⁻, 0.03; γ, 0.031(96), 0.40(30), 0.95(30),1.35(70)
₁₃Al	24	100	−0.1	2.10 s	....	....	β⁺, 8.5; γ, 0.511, 1.368, 2.754, 4.2, 5.3, 7.1; α, 2
	25	−9,588	−8.93	7.24 s	....		β⁺, 3.24; γ, 0.511
	26	−13,109	−12.211	7.4×10⁵ y	....		β⁺, 1.17; γ, 0.511, 1.12(4),1.81(100)
	26 m		−11.982	6.37 s	....	....	β⁺, 3.21; γ, 0.511
	27	−18,461	−17.196	....	100	0.235	....
	28	−18,095	−16.855	2.31 m	....		β⁻, 2.85; γ, 1.780(100)
	29	−19,558	−18.22	6.6 m	....		β⁻, 2.40; γ, 1.28(94), 2.43(6)
	30	−18,410	−17.2	3.3 s	....		β⁻, 5.0; γ, 2.23(61), 3.51(39)
₁₄Si	25		4.0	0.23 s	....		p, 3.34, 4.08, 4.68, 5.39
	26	−7,657	−7.13	2.1 s	....		β⁺, 3.83; γ, 0.511, 0.82(34)
	27	−13,297	−12.386	4.14 s	....		β⁺, 3.85; γ, 0.511
	28	−23,071	−21.490	....	92.18	0.08	....
	29	−23,504	−21.894	....	4.71	0.3	....
	30	−26,237	−24.439	....	3.12	0.11	....
	31	−24,651	−22.96	2.62 h	....		β⁻, 1.48; γ, 1.26
	32	−25,980	−24.08	~650 y	....		β⁻, 0.21; no γ
₁₅P	28	−8,220	−7.7	0.28 s	....		β⁺, 11.0; γ, 0.511, 1.780(75), 2.6, 4.44(10), 4.9, 6.1, 6.7, 7.0, 7.6(5)
	29	−18,192	−16.95	4.45 s	....		β⁺, 3.95; γ, 0.511, 1.28(1), 2.43
	30	−21,683	−20.20	2.50 m	....		β⁺, 3.24; γ, 0.511, 2.23(1)
	31	−26,235	−24.438	....	100	0.19	....
	32	−26,091	−24.303	14.28 d	....		β⁻, 1.710
	33	−28,272	−26.335	24.4 d	....		β⁻, 0.248; no γ
	34	−26,660	−24.8	12.4 s	....		β⁻, 5.1; γ, 2.13(25), 4.0
₁₆S	29		−2.9	0.19 s	....		p, 3.73, 5.40
	30	−15,127	−14.09	1.4 s	....		β⁺, 5.09, 4.42; γ, 0.511, 0.687(80)
	31	−20,389	−18.99	2.72 s	....		β⁺, 4.42; γ, 0.511, 1.27(1)
	32	−27,926	−26.013	....	95.0		....
	33	−28,538	−26.583	....	0.76		....
	34	−32,135	−29.934	....	4.22	0.27	....

Table 3-7 (*Continued*)
TABLE OF NUCLIDES

Nuclide		Mass Excess		Half-life	Natural	Thermal Neutron	Major Radiations,
Z Symbol	Mass Number	μmu	MeV	y = yr, d = day h = hr, m = min s = sec	Abundance, %	Absorption Cross Section σ, barns (mode if not *n*,γ)	Energies in MeV and γ Intensities, %
	35	−30,969	−28.847	87.9 d	....	....	β⁻, 0.167; no γ
	36	−32,910	−30.655	....	0.014	0.14	....
	37	−28,990	−27.0	5.07 m	....	....	β⁻, 4.7, 1.6; γ, 3.09(90)
	38	−28,770	−26.8	2.87 h	....	....	β⁻, 3.0, 1.1; γ, 1.88(95)
₁₇Cl	32	−13,760	−12.8	0.306 s	....	....	β⁺, 9.9; γ, 0.511, 2.24(70), 4.29(7), 4.77(14)
	33	−22,560	−21.01	2.53 s	....	....	β⁺, 4.55; γ, 0.511, 2.9
	34	−26,250	−24.45	1.56 s	....	....	β⁺, 4.46; γ, 0.511
	34 *m*		−24.31	31.99 m	....	....	β⁺, 2.48; e⁻, 0.142; γ, 0.145(45), 0.511, 1.17(12), 2.12(38), 3.30(12)
	35	−31,149	−29.015	....	75.53	44	....
	36	−31,691	−29.520	3.08×10⁵ y	....	100	β⁻, 0.714; γ, 0.511
	37	−34,102	−31.765	....	24.47	0.4	....
	38	−31,995	−29.80	37.29 m	....	....	β⁻, 4.91; γ, 1.60(38), 2.170(47)
	38 *m*		−29.13	0.74 s	....	....	γ, 0.66(100); e⁻, 0.66
	39	−31,992	−29.80	55.5 m	....	....	β⁻, 3.45, 2.18, 1.91; γ, 0.246(44), 1.27(50), 1.52(42)
	40	−29,600	−27.5	1.4 m	....	....	β⁻, 7.5; γ, 1.46, 2.83, 3.10, 5.8
₁₈Ar	33		−9.5	0.18 s	....	....	p, 3.16
	35	−24,746	−23.05	1.83 s	....	....	β⁺, 4.94; γ, 0.511, 1.22(5), 1.76(2)
	36	−32,456	−30.232	....	0.337	6	....
	37	−33,228	−30.951	35.1 d	....	....	Cl X-rays
	38	−37,272	−34.718	....	0.063	0.8	....
	39	−35,683	−33.24	269 y	....	....	β⁻, 0.565; no γ
	40	−37,616	−35.038	....	99.600	0.61	....
	41	−35,500	−33.061	1.83 h	....	0.5	β⁻, 2.49, 1.198; γ, 1.293(99)
	42	−36,952	−34.42	33 y	....	....	....
₁₉K	37	−26,635	−24.79	1.23 s	....	....	β⁺, 5.14; γ, 0.511, 2.79(2)
	38	−30,903	−28.79	7.71 m	....	....	β⁺, 2.68; γ, 0.511, 2.170(100)
	38 *m*		−28.66	0.95 s	....	....	β⁺, 5.0; γ, 0.511
	39	−36,290	−33.803	....	93.22	2.0	....
	40	−36,000	−33.533	1.26×10⁹ y	0.118	70	β⁻, 1.314; β⁺, 0.483; γ, 1.460(11)

Table 3-7 (*Continued*)
TABLE OF NUCLIDES

| Nuclide | | Mass Excess | | Half-life | Natural | Thermal Neutron | Major Radiations, |
Z Symbol	Mass Number	μmu	MeV	y = yr, d = day h = hr, m = min s = sec	Abundance, %	Absorption Cross Section σ, barns (mode if not n,γ)	Energies in MeV and γ Intensities, %
	41	−38,168	−35.552	....	6.77	1.2	....
	42	−37,594	−35.02	12.36 h	....	....	β^-, 3.52; γ, 0.31, 1.524(18)
	43	−39,270	−36.58	22.4 h	....	....	β^-, 1.82, 1.2, 0.83; γ, 0.220(3), 0.373(85), 0.39(18), 0.59(13), 0.619(81), 1.01(2)
	44	−37,960	−36.3	22.0 m	....	....	β^-, 5.2; γ, 1.156(61), 1.74(8), 2.1(37), 2.6(7), 3.7(4)
	45	−39,320	−36.6	16.3 m	....	....	β^-, 4.0, 2.1; γ, 0.175, 0.50, 0.95, 1.23, 1.71, 1.90, 2.10, 2.35, 2.60, 3.1
	47	−38, 910	−36.3	17.5 s	....	....	β^-, 6.1, 4.1; γ, 2.0(84), 2.6(15)
$_{20}$Ca	37		−13.3	0.173 s	....	....	p, 3.10
	38	−23,280	−22	0.66 s	....	....	γ, 0.511. 3.5
	39	−29,309	−27.30	0.87 s	....	....	β^+, 5.49; γ, 0.511
	40	−37,411	−34.848	....	96.97	0.23	....
	41	−37,725	−35.125	8×10^4 y	....	....	K X-rays
	42	−41,375	−38.540	....	0.64	42	....
	43	−41,220	−38.396	....	0.145	....	
	44	−44,510	−41.460	....	2.06	0.7	....
	45	−43,811	−40.809	165 d	....	....	β^-, 0.252
	46	−46,311	−43.14	....	0.0033	0.3	....
	47	−45, 462	−42.35	4.535 d	....	....	β^-, 1.98, 0.67; γ, 0.49(5), 0.815(5), 1.308(74)
	48	−47,469	−44.22	>10^{18} y	0.185	1.1	....
	49	−44,325	−41.29	8.8 m	....	....	β^-, 1.95; γ, 3.10(89), 4.1(10)
	50		−41	9 s	....	....	γ, 0.072, 0.258
$_{21}$Sc	40	−22,430	−20.3	0.179 s	....	....	β^+, 9.1; γ, 0.511, 3.75(100)
	41	−30,753	−28.63	0.60 s	....	....	β^+, 5.47; γ, 0.511
	42	−34,505	−32.109	0.683 s	....	....	β^+, 5.41; γ, 0.511
	42 *m*		−31.58	60.6 s	....	....	β^+, 2.82; γ, 0.438(100), 0.511, 1.22(100), 1.52(100)
	43	−38,835	−36.17	3.92 h	....	....	β^+, 1.20; γ, 0.375(22), 0.511
	44	−40,594	−37.81	3.92 h	....	....	β^+, 1.47; γ, 0.511, 1.159(100)

Table 3-7 (*Continued*)
TABLE OF NUCLIDES

Nuclide		Mass Excess		Half-life y = yr, d = day h = hr, m = min s = sec	Natural Abundance, %	Thermal Neutron Absorption Cross Section σ, barns (mode if not n,γ)	Major Radiations, Energies in MeV and γ Intensities, %
Z Symbol	Mass Number	μmu	MeV				
	44 *m*		−37.54	2.44 d	....	....	γ, 0.271(86), 1.02(1), 1.14(3); e⁻, 0.267
	45	−44,081	−41.061	....	100	13	....
	46	−44,827	−41.756	83.9 d	....	....	β⁻, 1.48, 0.357; γ, 0.889(100), 1.120(100)
	46 *m*		−41.614	19.5 s	....	....	γ, 0.142; e⁻
	47	−47,587	−44.326	3.43 d	....	....	β⁻, 0.600; γ, 0.160(73)
	48	−47,779	−44.51	1.83 d	....	....	β⁻, 0.65; γ, 0.175(6), 0.983(100), 1.040(100), 1.314(100)
	49	−49,974	−46.55	57.5 m	....	....	β⁻, 2.01; γ, 1.76
	50	−48,270	−45.0	1.72 m	....	....	β⁻, 3.6; γ, 0.520(100), 1.12(100), 1.55(100)
	50 *m*		−44.7	0.35 s	....	....	γ, 0.258
₂₂Ti	43	−31,500	−29.3	0.56 s	....	....	β⁺, 5.8; γ, 0.511
	44	−40,428	−37.66	48 y	....	....	γ, 0.068(90), 0.078(98); e⁻, 0.065, 0.073
	45	−41,871	−39.002	3.09 h	....	....	β⁺, 1.04; γ, 0.718, 1.408
	46	−47,368	−44.123	....	7.99	0.6	....
	47	−48,232	−44.927	....	7.32	1.7	....
	48	−52,050	−48.483	....	73.99	8.0	....
	49	−52,130	−48.558	....	5.46	1.9	....
	50	−55,214	−51.431	....	5.25	0.14	....
	51	−53,397	−49.74	5.79 m	....	....	β⁻, 2.14; γ, 0.320(95), 0.605(2), 0.928(5)
₂₃V	46	−39,786	−37.069	0.426 s	....	....	β⁺, 6.03; γ, 0.511
	47	−45,101	−42.01	33 m	....	....	β⁺, 1.89; γ, 0.511, 1.80(1)
	48	−47,741	−44.470	16.0 d	....	....	β⁺, 0.696; γ, 0.511, 0.945(10), 0.983(100), 1.312(97), 2.241(3)
	49	−51,478	−47.950	330 d	....	....	Ti X-rays
	50	−52,836	−49.216	6×10^{15} y	0.25	130	γ, 0.783(30), 1.55(70)
	51	−56,039	−52.199	....	99.75	4.9	....
	52	−55,220	−51.44	3.75 m	....	....	β⁻, 2.47; γ, 1.434(100)

Table 3-7 (*Continued*)
TABLE OF NUCLIDES

Nuclide		Mass Excess		Half-life y=yr, d=day h=hr, m=min s=sec	Natural Abundance, %	Thermal Neutron Absorption Cross Section σ, barns (mode if not n,γ)	Major Radiations, Energies in MeV and γ Intensities, %
Z Symbol	Mass Number	μmu	MeV				
₂₄Cr	53	−56,020	−51.8	2.0 m	....	....	β⁻, 2.50; γ, 1.00(100)
	54	−53,280	−50	55 s	....	....	β⁻, 3.3; γ, 0.84(100), 0.99(100), 2.21(100)
	46			1.1 s	....	....	....
	48	−46,240	−43.1	23 h	....	....	γ, 0.116(98), 0.31(99); e⁻, 0.111, 0.31
	49	−48,729	−45.39	41.9 m	....	....	β⁺, 1.54; e⁻, 0.058, 0.084, 0.148; γ, 0.063(14), 0.091(28), 0.153(13), 0.511
	50	−53,946	−50.249	....	4.31	17	....
	51	−55,232	−51.447	27.8 d	....	....	γ, 0.320(9); e⁻, 0.315
	52	−59,487	−55.411	....	83.76	0.8	....
	53	−59,347	−55.281	....	9.55	18	....
	54	−61,119	−56.931	....	2.38	0.38	....
	55	−59,167	−55.11	3.52 m	....	....	β⁻, 2.59
	56	−59,360	−55.3	5.9 m	....	....	β⁻, 1.5; e⁻, 0.020, 0.077; γ, 0.026, 0.083
₂₅Mn	50	−45,785	−42.648	0.286 s	....	....	β⁺, 6.61; γ, 0.511
	50			2 m	....	....	γ, 0.511, 0.66(25), 0.783(100), 1.11(100), 1.28(25), 1.45(75)
	51	−51,810	−48.26	45.2 m	....	....	β⁺, 2.17; γ, 0.511, 1.56, 2.03
	52	−54,432	−50.70	5.60 d	....	....	β⁺, 0.575; γ, 0.511, 0.744(82), 0.935(84), 1.434(100)
	52 m		−50.32	21.1 m	....	....	β⁺, 1.63; γ, 0.383(2), 0.511, 1.434(100)
	53	−58,705	−54.683	1.9×10⁶ y	....	170	Cr X-rays
	54	−59,638	−55.55	303 d	....	....	γ, 0.835(100); e⁻, 0.829
	55	−61,950	−57.705	....	100	13.3	....
	56	−61,090	−56.904	2.576 h	....	....	β⁻, 2.85; γ, 0.847(99), 1.811(29), 2.110(15)
	57	−61,700	−57.5	1.7 m	....	....	β⁻, 2.55; γ, 0.122,

Table 3-7 (*Continued*)
TABLE OF NUCLIDES

Nuclide Z Symbol	Nuclide Mass Number	Mass Excess μmu	Mass Excess MeV	Half-life y=yr, d=day h=hr, m=min s=sec	Natural Abundance, %	Thermal Neutron Absorption Cross Section σ, barns (mode if not n,γ)	Major Radiations, Energies in MeV and γ Intensities, %
	57			7 d	....	....	0.136, 0.22, 0.353, 0.692 ....
	58	−59,740	−56	1.1 m	....	....	γ, 0.36, 0.41, 0.52, 0.57, 0.82, 1.0, 1.25, 1.4, 1.6, 2.2, 2.8
$_{26}$Fe	52	−51,883	−48.33	8.2 h	....	....	β⁺, 0.80; γ, 0.165(100), 0.511
	53	−54,428	−50.70	8.51 m	....	....	β⁺, 3.0; γ, 0.38(32), 0.511
	54	−60,383	−56.246	....	5.84	2.9	....
	55	−61,701	−57.474	2.60 y	....	....	Mn X-rays
	56	−65,064	−60.605	....	91.68	2.7	
	57	−64,602	−60.176	....	2.17	2.5	....
	58	−66,718	−62.147	....	0.31	1.1	....
	59	−65,122	−60.660	45.6 d	....	....	β⁻, 1.57, 0.475; γ, 0.143(1), 0.192(3), 1.095(56), 1.292(44)
	60	−66,036	−61.51	3×10⁵ y	....	....	....
	61	−63,480	−59	6.0 m	....	....	β⁻, 2.8; γ, 0.13(11), 0.30(48), 1.03(98), 1.20(100)
$_{27}$Co	54	−51,525	−47.99	0.194 s	....	....	β⁺, 7.23; γ, 0.511
	54			1.5 m	....	....	β⁺, 4.3; γ, 0.41(100), 0.511, 1.14(100), 1.41(100)
	55	−57,987	−54.01	18.2 h	....	....	β⁺, 1.50; γ, 0.480(12), 0.511, 0.930(80), 1.41(13)
	56	−60,153	−56.03	77.3 d	....	....	β⁺, 1.49; γ, 0.511, 0.847(100), 1.04(15), 1.24(66), 1.76(15), 2.02(11), 2.60(17), 3.26(13)
	57	−63,704	−59.339	270 d	....	....	γ, 0.014(9), 0.122(87), 0.136(11), 0.692; e⁻, 0.007, 0.013, 0.115, 0.129
	58	−64,239	−59.84	71.3 d	....	2,500	β⁺, 0.474; γ, 0.511, 0.810(99),

Table 3-7 (*Continued*)
TABLE OF NUCLIDES

Nuclide		Mass Excess		Half-life y = yr, d = day h = hr, m = min s = sec	Natural Abundance, %	Thermal Neutron Absorption Cross Section σ, barns (mode if not n.γ)	Major Radiations, Energies in MeV and γ Intensities, %
Z Symbol	Mass Number	μmu	MeV				
	58 *m*		−59.81	9.2 h	....	1.4×10⁵	0.865(1), 1.67(1) e⁻, 0.017, 0.024
	59	−66,811	−62.233	....	100	19	....
	60	−66,187	−61.651	5.263 y	....	6	β⁻, 1.48, 0.314; γ, 1.173(100), 1.332(100)
	60 *m*		−61.593	10.47 m	....	100	β⁻, 1.55; e⁻, 0.051, 0.058; γ, 0.059(2), 1.33
	61	−67,560	−62.93	99.0 m	....	....	β⁻, 1.22; e⁻, 0.059; γ, 0.067(89)
	62	−66,054	−61.53	13.9 m	....	....	β⁻, 2.88; γ, 1.17(180), 1.47(20), 1.74(19), 2.03(7)
	62			1.5 m	....	....	....
	63	−66,470	−61.9	52 s	....	....	β⁻, 3.6; no γ
	63			1.40 h	....	....	....
	64			28 s	....	....	γ, 0.095
	64			7.8 m	....	....	....
	64			2.0 m	....	....	....
₂₈Ni	56	−57,884	−53.92	6.10 d	....	....	γ, 0.163(99), 0.276(31), 0.472(35), 0.748(48), 0.812(85), 1.56(14); e⁻, 0.155
	57	−60,231	−56.10	36.0 h	....	....	β⁺, 0.85; γ, 0.127(14), 0.511, 1.37(86), 1.89(14)
	58	−64,658	−60.23	....	67.76	4.4	....
	59	−65,658	−61.159	8×10⁴ y	....	....	Co X-rays
	60	−69,213	−64.471	....	26.16	2.6	....
	61	−68,944	−64.22	....	1.25	2	....
	62	−71,658	−66.75	....	3.66	15	....
	63	−70,336	−65.52	92 y	....	....	β⁻, 0.067; no γ
	64	−72,042	−67.11	....	1.16	1.5	....
	65	−69,928	−65.14	2.564 h	....	....	β⁻, 2.13; γ, 0.368(5), 1.115(16), 1.481(25)
	66	−70,915	−66.06	54.8 h	....	....	β⁻, 0.20; no γ
	67		−63.2	50 s	....	....	β⁻, 4.1; γ, 0.90(51), 1.26(15)
₂₉Cu	58	−55,459	−51.66	3.20 s	....	....	β⁺, 8.2; γ, 0.511
	58			9.5 m	....	....	....
	59	−60,504	−56.36	81.5 s	....	....	β⁺, 3.7; γ, 0.343(5), 0.463(5), 0.511,

Table 3-7 (*Continued*)
TABLE OF NUCLIDES

Nuclide		Mass Excess		Half-life y = yr, d = day h = hr, m = min s = sec	Natural Abundance, %	Thermal Neutron Absorption Cross Section σ, barns (mode if not n,γ)	Major Radiations, Energies in MeV and γ Intensities, %
Z Symbol	Mass Number	μmu	MeV				
	60	−62,638	−58.35	23.4 m	....	....	0.872(9), 1.305(11), 1.70(1) β⁺, 3.92, 3.00, 2.00; γ, 0.511, 0.85(15), 1.332(80), 1.76(52), 2.13(6), 2.64(5), 3.13(4), 2.52(2), 4.0(1)
	61	−66,543	−61.98	3.32 h	....	....	β⁺, 1.22; e⁻, 0.059; γ, 0.067(4), 0.284(12), 0.38(3), 0.511, 1.19(5)
	62	−67,434	−62.81	9.76 m	....	....	β⁺, 2.91; γ, 0.511, 0.88, 1.17(1)
	63	−70,408	−65.583	....	69.1	4.5	....
	64	−70,241	−65.428	12.80 h	....	....	β⁻, 0.573; β⁺, 0.656; e⁻, 1.33; γ, 0.511, 1.34(1)
	65	−72,214	−67.27	....	30.9	2.3	....
	66	−71,129	−66.26	5.10 m	....	130	β⁻, 2.63; γ, 1.039(9)
	67	−72,241	−67.29	58.5 h	....	....	β⁻. 0.57; e⁻, 0.082, 0.091; γ, 0.092(23), 0.184(40)
	68	−70,230	−65.4	30 s	....	....	β⁻, 3.5; γ, 0.80(17), 1.078(95), 1.24(3), 1.88(5)
₃₀Zn	60			2.1 m	....	....	....
	61	−60,750	−56.6	1.48 m	....	....	β⁺, 4.4; γ, 0.48(11), 0.511, 0.98(3), 1.64(6)
	62	−65,620	−61.12	9.13 h	....	....	β⁺, 0.66; e⁻, 0.033; γ, 0.042(20), 0.51(47), 0.59(22)
	63	−66,794	−62.22	38.4 m	....	....	β⁺, 2.34; γ, 0.511, 0.669(8), 0.962(6), 1.42(1)
	64	−70,855	−66.000	>8×10¹⁵ y	48.89	0.46	....
	65	−70,766	−65.92	245 d	....	....	β⁺, 0.327; e⁻, 1.106; γ, 0.511, 1.115(49)
	66	−73,948	−68.88	....	27.81	....	....
	67	−72,855	−67.86	....	4.11	....	....
	68	−75,143	−69.99	....	18.56	1.0	....
	69	−73,459	−68.43	57 m	....	....	β⁻, 0.90; no γ
	69 *m*		−67.99	13.8 h	....	....	γ, 0.439(95); e⁻, 0.429
	70	−74,666	−69.55	>10¹⁵ y	0.62	0.10	....

Table 3-7 (*Continued*)
TABLE OF NUCLIDES

Nuclide		Mass Excess		Half-life y = yr, d = day h = hr, m = min s = sec	Natural Abundance, %	Thermal Neutron Absorption Cross Section σ, barns (mode if not n,γ)	Major Radiations, Energies in MeV and γ Intensities, %
Z Symbol	Mass Number	μmu	MeV				
	71	−72,490	−67.5	2.4 m	....	....	β⁻, 2.61; γ, 0.120(1), 0.39(1), 0.510(13), 0.92(3), 1.12(1)
	71 *m*		−67.2	3.92 h	....	....	β⁻, 1.46; γ, 0.13(9), 0.385(94), 0.495(75), 0.609(65), 0.76(5), 0.99(8), 1.11(4)
	72	−73,157	−68.14	46.5 h	....	....	β⁻, 0.30; e⁻, 0.005, 0.014; γ, 0.015(8), 0.046, 0.145(90), 0.192(10)
₃₁Ga	63	−60,890	−57	33 s	....	....	....
	64	−63,263	−58.93	2.6 m	....	....	β⁺, 6.05, 2.8; γ, 0.511, 0.80(15), 0.992(43), 1.25(7), 1.38(14), 1.56(7), 1.78(5), 2.18(11), 2.34(9), 3.32(18)
	65	−67,267	−62.66	15.2 m	....	....	β⁺, 2.24, 2.11; e⁻, 0.044, 0.053, 0.105; γ, 0.054(8), 0.061(12), 0.115(55), 0.152(10), 0.206(4), 0.511, 0.75(10), 0.93(3)
	65			8.0 m	....	....	....
	66	−68,393	−63.71	9.45 h	....	....	β⁺, 4.153; γ, 0.511, 0.828(5), 1.039(37), 1.91(3), 2.183(5), 2.748(25), 4.30(5)
	67	−71,784	−66.87	77.9 h	....	....	γ, 0.093(40), 0.184(24), 0.296(22), 0.388(7); e⁻, 0.084, 0.092
	68	−72,008	−67.07	68.3 m	....	....	β⁺, 1.90; γ, 0.511, 0.80, 1.078(4), 1.24, 1.87
	69	−74,426	−69.326	....	60.2	1.9	....
	70	−73,965	−68.90	21.1 m	....	....	β⁻, 1.65; γ, 0.173, 1.040(1)
	71	−75,294	−70.135	....	39.8	5.0	....
	72	−73,628	−68.58	14.12 h	....	....	β⁻, 3.15; γ, 0.601(8),

Table 3-7 (*Continued*)
TABLE OF NUCLIDES

Nuclide Z Sym-bol	Mass Number	Mass Excess μmu	Mass Excess MeV	Half-life y=yr, d=day h=hr, m=min s=sec	Natural Abundance, %	Thermal Neutron Absorption Cross Section σ, barns (mode if not n,γ)	Major Radiations, Energies in MeV and γ Intensities, %
							0.630(27), 0.835(96), 0.894(10), 1.050(7), 1.465(4), 1.60(5), 1.860(5), 2.201(26), 2.50(20)
	73	−74,874	−69.74	4.9 h	....	....	β^-, 1.19; e^-, 0.012, 0.043, 0.053; γ, 0.054(9), 0.295(94), 0.74(6)
	74	−72,810	−67.8	8.0 m	....	....	β^-, 2.5; γ, 0.50(11), 0.60(100), 0.87(9), 1.11(5), 1.20(8), 1.33(5), 1.46(8), 1.76(7), 2.35(45)
	75		−68.5	2.0 m	....	....	β^-, 3.3; γ, 0.58(3)
	76			32 s	....	....	β^-, 6; γ, 0.563, 0.96, 1.12
$_{32}$Ge	65	−60,400	−56	1.5 m	....	....	β^+, 3.7; γ, 0.511, 0.67(3), 1.72(2)
	66	−65,200	−60.7	2.4 h	....	....	β^+, 2.0, 1.3; γ, 0.046(37), 0.068(11), 0.114(22), 0.185(23), 0.245(7), 0.27(19), 0.30(6), 0.34(19), 0.38(48), 0.40(6), 0.47(19), 0.511
	67	−67,060	−62.5	18.7 m	....	....	β^+, 3.1; γ, 0.170(105), 0.511, 0.84(4), 0.92(7), 1.48(5)
	68	−71,470	−67	275 d	....	....	Ga X-rays
	69	−72,037	−67.101	36 h	....	....	β^+, 1.22; γ, 0.511, 0.573(13), 0.872(10), 1.107(28), 1.335(3)
	70	−75,749	−70.558	....	20.55	3.2	....
	71	−75,044	−69.90	11.4 d	....	....	Ga X-rays
	72	−77,918	−72.579	....	27.37	1.0	....
	73	−76,538	−71.293	....	7.67	14	....
	73 *m*		−71.226	0.53 s	....	....	γ, 0.054(9); e^-, 0.012, 0.043,

Table 3-7 (*Continued*)
TABLE OF NUCLIDES

Nuclide		Mass Excess		Half-life	Natural	Thermal Neutron	Major Radiations,
Z Symbol	Mass Number	μmu	MeV	y = yr, d = day h = hr, m = min s = sec	Abundance, %	Absorption Cross Section σ, barns (mode if not n,γ)	Energies in MeV and γ Intensities, %
	74	−78,819	−73.419	....	36.74	0.3	0.053 ....
	75	−77,117	−71.83	82 m	....	....	β^-, 1.19; γ, 0.066, 0.199(1), 0.265(11), 0.427, 0.477, 0.628
	75 *m*		−71.69	48 s	....	....	γ, 0.139(34); e^-, 0.128, 0.138
	76	−78,595	−73.209	$>2 \times 10^{16}$ y	7.67	0.1	....
	77	−76,400	−71.2	11.3 h	....	....	β^-, 2.2; e^-, 0.198, 0.253; γ, 0.21(61), 0.263(45), 0.368(15), 0.417(25), 0.563(18), 0.632(11), 0.73(14), 0.80(6), 0.93(5), 1.09(6)
	77 *m*		−71.0	54 s	....	....	β^-, 2.9; e^-, 0.148, 0.158; γ, 0.159(12), 0.215(21)
	78		−71.8	1.47 h	....	....	β^-, 0.71; γ, 0.277(94)
₃₃As	68			~7 m	....	....	....
	69	−67,850	−63.2	15 m	....	....	β^+, 2.9; γ, 0.23, 0.511
	70	−69,054	−64.32	52 m	....	....	β^+, 2.89, 2.14; γ, 0.511, 0.60(23), 0.67(25), 0.75(23), 0.91(17), 1.040(78), 1.12(23), 1.36(12), 1.42(10), 1.54(7), 1.71(22), 1.80(6), 2.03(19)
	71	−72,887	−67.89	62 h	....	....	β^+, 0.81; e^-, 0.012, 0.022, 0.164; γ, 0.175(90), 0.511
	72	−73,237	−68.22	26 h	....	....	β^+, 3.34, 2.50; e^-, 0.679; γ, 0.511, 0.630(8), 0.835(78)
	73	−76,139	−70.92	80.3 d	....	....	γ, 0.054(9); e^-, 0.012, 0.043, 0.053
	74	−76,067	−70.855	17.9 d	....	....	β^-, 1.36; β^+, 1.54,

Table 3-7 (*Continued*)
TABLE OF NUCLIDES

Nuclide		Mass Excess		Half-life y=yr, d=day h=hr, m=min s=sec	Natural Abundance, %	Thermal Neutron Absorption Cross Section σ, barns (mode if not n,γ)	Major Radiations, Energies in MeV and γ Intensities, %
Z Symbol	Mass Number	μmu	MeV				
							0.95; γ, 0.511, 0.596(61), 0.635(14)
	74 m		−70.572	8.0 s	....	....	γ, 0.283
	75	−78,404	−73.031	....	100	4.5	....
	76	−77,603	−72.29	26.4 h	....	....	β−, 2.97; γ, 0.559(43), 0.657(6), 1.22(5), 1.44(1), 1.789, 2.10(1)
	77	−79,354	−73.92	38.7 h	....	....	β−, 0.68; γ, 0.086, 0.239(3), 0.522(1)
	78	−78,100	−72.8	91 m	....	....	β−, 4.1; γ, 0.614(42), 0.70(15), 0.83(8), 1.31(11)
	78 m			6 m	....	....	....
	79	−79,110	−73.7	9.0 m	....	....	β−, 2.15; γ, 0.36(2), 0.43(2), 0.54(1), 0.73(1), 0.89(1)
	80	−77,030	−71.8	15.3 s	....	....	β−, 6.0; γ, 0.666(42), 0.8(1), 1.22(4), 1.64(4), 1.77(2)
	81		−72.6	33 s	....	....	β−, 3.8; no γ
	85			0.43 s	....	....	....
$_{34}$Se	70			~44 m	....	....	....
	71	−68,160	−63.5	4.5 m	....	....	β+, 3.4; γ, 0.16, 0.511
	72	−72,590	−68	8.4 d	....	....	γ, 0.046(59); e−, 0.034, 0.044
	73	−73,186	−68.17	7.1 h	....	....	β+, 1.30; e−, 0.054, 0.064, 0.347; γ, 0.066(65), 0.359(99), 0.511
	73		−68.2	42 m	....	....	β+, 1.7
	74	−77,524	−72.212	....	0.87	30	....
	75	−77,475	−72.166	120.4 d	....	....	γ, 0.066(1), 0.097(3), 0.121(17), 0.136(57), 0.265(60), 0.280(25), 0.401(12); e−, 0.085, 0.095, 0.109, 0.124, 0.253

Table 3-7 (*Continued*)
TABLE OF NUCLIDES

Nuclide		Mass Excess		Half-life $v = yr.$ $d = day$ $h = hr.$ $m = min$ $s = sec$	Natural Abundance, %	Thermal Neutron Absorption Cross Section σ, barns (mode if not n,γ)	Major Radiations, Energies in MeV and γ Intensities, %
Z Symbol	Mass Number	μmu	MeV				
	76	−80,793	−75.26	....	9.02	63	....
	77	−80,089	−74.60	....	7.58	42	....
	77 *m*		−74.44	17.5 s	....	....	γ, 0.161(50); e^-, 0.148, 0.160
	78	−82,686	−77.021	....	23.52	0.36	....
	79	−81,506	−75.921	≤6.5×10⁴ y	....	....	β^-, 0.16; no γ
	79 *m*		−75.825	3.91 m	....	....	γ, 0.096(9); e^-, 0.083, 0.095
	80	−83,473	−77.753	....	49.82	0.5	....
	81	−82,016	−76.40	18.6 m	....	....	β^-, 1.58; γ, 0.030, 0.28(1), 0.56, 0.83
	81 *m*		−76.29	56.8 m	....	....	γ, 0.103(8); e^-, 0.090, 0.102
	82	−83,293	−77.59	>10¹⁷ y	9.19	0.05	....
	83		−75.4	25 m	....	....	β^-, 1.8; γ, 0.22(44), 0.36(69), 1.88(16), 2.29(9)
	83 *m*		−75.2	70 s	....	....	β^-, 3.8; γ, 0.35(16), 0.65(20), 1.01(100), 2.02(40)
	84			3.3 m	....	....	....
	85			39 s	....	....	....
	87			16 s	....	....	....
₃₅Br	74	−70,220	−65	36 m	....	....	β^+, 4.7; γ, 0.511
	75	−74,553	−69.44	1.7 h	....	....	β^+, 1.70; γ, 0.285, 0.511, 0.62
	76	−75,820	−70.6	16.1 h	....	....	β^+, 3.6; γ, 0.511, 0.559(63), 0.65(19), 0.75(6), 0.85(7), 1.21(13), 1.37(5), 1.47(7), 1.86(11), 2.10(7), 2.39(4), 2.78(5), 2.97(8), 3.57(2)
	77	−78,624	−73.24	57 h	....	....	β^+, 0.34; e^-, 0.229, 0.287, 0.508; γ, 0.24(30), 0.300(6), 0.52(24), 0.58(7), 0.75(2), 0.82(3), 1.00(1)
	77 *m*		−73.13	4.2 m	....	....	γ, 0.108; e^-, 0.094, 0.106
	78	−78,850	−73.45	6.5 m	....	....	β^+, 2.55; γ, 0.511,

Table 3-7 (*Continued*)
TABLE OF NUCLIDES

Nuclide		Mass Excess		Half-life y=yr, d=day h=hr. m=min s=sec	Natural Abundance, %	Thermal Neutron Absorption Cross Section σ. barns (mode if not n.γ)	Major Radiations, Energies in MeV and γ Intensities, %
Z Symbol	Mass Number	μmu	MeV				
79		−81,671	−76.075	. . . .	50.52	8.5	0.614(14) . . . .
	79 m		−75.87	4.8 s	. . . .	. . . .	γ, 0.21
	80	−81,464	−75.882	17.6 m	. . . .	. . . .	β⁻, 2.00; β⁺, 0.87; γ, 0.511, 0.618(7), 0.666(1)
	80 m		−75.796	4.38 h	. . . .	. . . .	γ, 0.037(36); e⁻, 0.024, 0.036, 0.047
	81	−83,708	−77.97	. . . .	49.48	3	. . . .
	82	−83,198	−77.50	35.34 h	. . . .	. . . .	β⁻, 0.444; γ, 0.554(66), 0.619(41), 0.698(27), 0.777(83), 0.828(25), 1.044(29), 1.317(26), 1.475(17)
	82 m		−77.45	6.05 m	. . . .	. . . .	γ, 0.046, 0.777, 1.475
	83	−84,832	−79.02	2.41 h	. . . .	. . . .	β⁻, 0.93; γ, 0.530(1)
	84	−83,450	−77.7	31.8 m	. . . .	. . . .	β⁻, 4.68; γ, 0.81(9), 0.88(51), 1.01(10), 1.21(4), 1.90(18), 2.47(8), 3.93(13)
	84			6.0 m	. . . .	. . . .	β⁻, 1.9; γ, 0.44(68), 0.88(75), 1.46(75), 1.89(16)
	85	−84,470	−78.7	3.00 m	. . . .	. . . .	β⁻, 2.5; no γ
	86	−81,800	−76	54 s	. . . .	. . . .	β⁻, 7.1; γ, 1.29(12), 1.36(39), 1.56(100), 1.97(20), 2.34(20), 2.75(36)
	87		−74.6	55.6 s	. . . .	. . . .	β⁻, 8.0, 2.6; n, 0.3; γ, 1.44(100), 1.85(18), 2.48(18),

Table 3-7 (*Continued*)
TABLE OF NUCLIDES

Nuclide		Mass Excess		Half-life y = yr, d = day h = hr, m = min s = sec	Natural Abundance, %	Thermal Neutron Absorption Cross Section σ, barns (mode if not n,γ)	Major Radiations, Energies in MeV and γ Intensities, %
Z Sym- bol	Mass Number	μmu	MeV				
							2.64(16), 2.98(25), 3.18(16), 3.80(11), 4.19(21), 4.8(17), 5.0(17), 5.2(12)
	88			15.5 s	....	....	γ, 0.76
	89			4.5 s	....	....	n, 0.5
	90			1.6 s	....	....	n
$_{36}$Kr	74	−66,900	−62	20 m	....	....	β⁺, 3.1; γ, 0.511
	75	−69,080	−64	5.5 m	....	....	....
	76	−74,530	−69	14.8 h	....	....	γ, 0.039, 0.104, 0.135, 0.267, 0.316, 0.407, 0.452
	77	−75,520	−70.4	1.19 h	....	....	β⁺, 1.86; e⁻, 0.011, 0.023, 0.094, 0.106, 0.118, 0.136; γ, 0.024, 0.108, 0.131, 0.149, 0.665
	78	−79,597	−74.14	....	0.354	2	....
	79	−79,932	−74.46	34.92 h	....	....	β⁺, 0.60; e⁻, 0.031, 0.043, 0.123, 0.204, 0.248, 0.384; γ, 0.136(1), 0.261(9), 0.398(10), 0.511, 0.606(10), 0.836(2), 1.119(1), 1.336(1)
	79 m		−74.33	55 s	....	....	γ, 0.127; e⁻, 0.113, 0.125
	80	−83,620	−77.89	....	2.27	15	....
	81	−83,390	−77.7	2.1×10⁵ y	....	....	Br X-rays
	81 m		−77.5	13 s	....	....	γ, 0.190(65); e⁻, 0.176, 0.188
	82	−86,518	−80.589	....	11.56	42	....
	83	−85,869	−79.985	....	11.55	180	....
	83 m		−79.943	1.86 h	....	....	γ, 0.009(9); e⁻, 0.007, 0.018, 0.031
	84	−88,497	−82.433	....	56.90	0.10	....
	85	−87,477	−81.48	10.76 y	....	<15	β⁻, 0.67; γ, 0.514

Table 3-7 (*Continued*)
TABLE OF NUCLIDES

Nuclide		Mass Excess		Half-life y = yr, d = day h = hr, m = min s = sec	Natural Abundance, %	Thermal Neutron Absorption Cross Section σ, barns (mode if not n,γ)	Major Radiations, Energies in MeV and γ Intensities, %
Z Symbol	Mass Number	μmu	MeV				
	85 *m*		−81.18	4.4 h	....	....	β⁻, 0.82; e⁻, 0.134, 0.291; γ, 0.150(74), 0.305(13)
	86	−89,384	−83.259	....	17.37	0.06	....
	87	−86,635	−80.70	76 m	....	<600	β⁻, 3.8; γ, 0.403(84), 0.85(16), 2.57(35)
	88	−85,730	−79.9	2.80 h	....	....	β⁻, 2.8; e⁻, 0.013; γ, 0.028, 0.166(7), 0.191(35), 0.36(5), 0.85(23), 1.55(14), 2.19(18), 2.40(35)
	89	−83,400	−78	3.18 m	....	....	β⁻, 4.0; γ, 0.23(85), 0.36(28), 0.43(29), 0.51(42), 0.60(100), 0.74(32), 0.88(65), 1.12(45), 1.29(31), 1.51(88), 1.71(34), 1.93(10), 2.04(16), 2.23(10), 2.42(22), 2.57(10), 2.84(25)
	90	−80,280	−74.8	33 s	....	....	β⁻, 2.80; γ, 0.105(15), 0.120(65), 0.236(16), 0.495(12), 0.536(48), 1.11(48), 1.54(17), 1.79(11), 2.48(4)
	91			9.8 s	....	....	β⁻, 3.6; no γ
	92			3.0 s	....	....	β⁻
	93			2.0 s	....	....	β⁻
	94			1.4 s	....	....	β⁻
	97			~1 s	....	....	β⁻

Table 3-7 (*Continued*)
TABLE OF NUCLIDES

Nuclide		Mass Excess		Half-life y = yr, d = day h = hr, m = min s = sec	Natural Abundance, %	Thermal Neutron Absorption Cross Section σ, barns (mode if not n,γ)	Major Radiations, Energies in MeV and γ Intensities, %
Z Symbol	Mass Number	μmu	MeV				
$_{37}$Rb	79			24 m	....	....	γ, 0.15(73), 0.19(29), 0.511
	80	−78,100	−73	34 s	....	....	β⁺, 4.1; γ, 0.511, 0.618(39)
	81	−80,980	−75.4	4.7 h	....	....	β⁺, 1.03; γ, 0.253, 0.450, 0.511, 1.10
	81 m		−75.3	31 m	....	....	β⁺, 1.4; e⁻, 0.071, 0.083
	82	−82,041	−76.42	1.25 m	....	....	β⁺, 3.15; γ, 0.511, 0.777(9)
	82 m		−76.14	6.3 h	....	....	β⁺, 0.78; γ, 0.511, 0.554(66), 0.619(41), 0.698(27), 0.777(83), 0.828(25), 1.044(29), 1.317(26), 1.475(17)
	83	−85,270	−79	83 d	....	....	γ, 0.53(93), 0.79(1); e⁻, 0.007, 0.52
	84	−85,619	−79.753	33.0 d	....	....	β⁺, 1.66; β⁻, 0.91; γ, 0.511, 0.88(74), 1.01(1), 1.90(1)
	84 m		−79.289	20 m	....	....	γ, 0.216(37), 0.250(65), 0.464(32); e⁻, 0.201, 0.214, 0.449
	85	−88,200	−82.16	....	72.15	0.9	....
	86	−88,807	−82.72	18.66 d	....	....	β⁻, 1.78; γ, 1.078(9)
	86 m		−82.16	1.02 m	....	....	γ, 0.56
	87	−90,814	−84.591	4.8×10¹⁰ y	27.85	0.12	β⁻, 0.274; no γ
	88	−88,730	−82.7	17.8 m	....	1.0	β⁻, 5.3; γ, 0.898(13), 1.863(21), 2.68(2)
	89	−88,350	−82.3	15.4 m	....	....	β⁻, 3.92, 2.9, 1.6; γ, 0.66(17), 1.05(75), 1.26(54), 2.20(14), 2.59(13)
	90	−85,180	−79.3	2.91 m	....	....	β⁻, 6.6; γ, 0.53(4), 0.83(61), 1.03(5),

Table 3-7 (*Continued*)
TABLE OF NUCLIDES

| Nuclide | | Mass Excess | | Half-life | Natural | Thermal Neutron | Major Radiations, |
Z Symbol	Mass Number	μmu	MeV	y = yr, d = day h = hr, m = min s = sec	Abundance, %	Absorption Cross Section σ, barns (mode if not n,γ)	Energies in MeV and γ Intensities, %
							1.11(7), 1.40(5), 1.70(3), 3.07(5), 3.34(15), 3.54(5), 4.13(11), 4.34(18), 4.60(5), 5.2(4)
	91	−83,930	−78	1.2 m	....	....	β^-, 4.6
	91			14 m	....	....	β^-
	92	−80,860	−75	5.3 s	....	....	....
	93			5.6 s	....	....	....
	94			2.9 s	....	....	....
	95			<2.5 s	....	....	....
$_{38}$Sr	80			1.7 h	....	....	γ, 0.58
	81			29 m	....	....	....
	82	−81,610	−76	25.0 d	....	....	Rb X-rays
	83	−82,800	−77	32.4 h	....	....	β^+, 1.15; e^-, 0.025, 0.040; γ, 0.040(24), 0.38(35), 0.511, 0.76(40), 1.16, 1.52
	84	−86,570	−80.638	....	0.56	0.8	....
	85	−87,011	−81.05	64.0 d	....	....	γ, 0.514(100); e^-, 0.499
	85 m		−80.81	70 m	....	....	γ, 0.150(14), 0.231(85); e^-, 0.005, 0.134, 0.215
	86	−90,715	−84.499	....	9.86	1.3	....
	87	−91,108	−84.865	....	7.02	....	....
	87 m		−84.477	2.83 h	....	....	γ, 0.388(80); e^-, 0.372, 0.386
	88	−94,359	−87.89	....	82.56	0.006	....
	89	−92,558	−86.22	52.7 d	....	0.4	β^-, 1.463; γ, 0.91
	89 m			10 d	....	....	....
	90	−92,253	−85.95	27.7 y	....	1	β^-, 0.546; no γ
	91	−89,839	−83.68	9.67 h	....	....	β^-, 2.67; γ, 0.645(15), 0.748(27), 0.93(3), 1.025(30), 1.413(5)
	92	−89,020	−82.9	2.71 h	....	....	β^-, 1.5; 0.55; γ, 0.23(3), 0.44(4), 1.37(90)
	93	−85,290	−79.4	8.3 m	....	....	β^-, 2.9; γ, 0.60, 0.8, 1.2
	94	−84,620	−78.8	1.35 m	....	....	β^-, 2.1; γ, 1.42(100)
	95			0.8 m	....	....	β^-

Table 3-7 (*Continued*)
TABLE OF NUCLIDES

Nuclide		Mass Excess		Half-life y = yr, d = day h = hr, m = min s = sec	Natural Abundance, %	Thermal Neutron Absorption Cross Section σ, barns (mode if not *n*,γ)	Major Radiations, Energies in MeV and γ Intensities, %
Z Sym- bol	Mass Number	μmu	MeV				
₃₉Y	82			12.3 m	....	....	....
	83			7.4 m	....	....	....
	84	−79,810	−74.3	43 m	....	....	$β^+$, 3.5; γ, 0.511, 0.590(15), 0.795(100), 0.982(100), 1.041(50), 1.27(9), 1.47(6)
	85	−83,511	−77.79	5.0 h	....	....	$β^+$, 2.24; e^-, 0.215; γ, 0.231(13), 0.511, 0.77(8), 2.16(9)
	85 m		−77.75	2.68 h	....	....	$β^+$, 1.54; γ, 0.51, 0.92(9)
	86	−85,054	−79.23	14.6 h	....	....	$β^+$, 3.15, 2.34; γ, 0.443(14), 0.511, 0.63(37), 0.704(14), 0.778(21), 0.836(7), 1.026(10), 1.077(82), 1.16(35), 1.857(18), 1.925(24)
	86 m		−79.01	48 m	....	....	γ, 0.208(94); e^-, 0.008
	87	−89,260	−83.2	80 h	....	....	$β^+$; γ, 0.483
	87 m		−82.8	14 h	....	....	$β^+$; e^-, 0.364, 0.379; γ, 0.381(74)
	88	−90,472	−84.27	108.1 d	....	....	$β^+$, 0.76; γ, 0.898(91), 1.836(100)
	89	−94,128	−87.678	....	100	1.3	....
	89 m		−86.77	16.1 s	....	....	γ, 0.91(99); e^-, 0.89
	90	−92,837	−86.50	64.0 h	....	....	$β^-$, 2.27; no γ
	90 m		−85.81	3.1 h	....	....	γ, 0.202(97), 0.482(91), 2.315; e^-, 0.185, 0.465
	91	−92,705	−86.35	58.8 d	....	1.4	$β^-$, 1.545; γ, 1.21
	91 m		−85.80	50.3 m	....	....	γ, 0.551(95); e^-, 0.534
	92	−91,074	−84.83	3.53 h	....	....	$β^-$, 3.63; γ, 0.448(2), 0.560(3), 0.934(14), 1.40(5), 1.83
	93	−90,448	−84.22	10.3 h	....	....	$β^-$, 2.89; γ,

Table 3-7 (*Continued*)
TABLE OF NUCLIDES

Nuclide Z Symbol	Mass Number	Mass Excess μmu	Mass Excess MeV	Half-life (y = yr, d = day, h = hr, m = min, s = sec)	Natural Abundance, %	Thermal Neutron Absorption Cross Section σ, barns (mode if not n,γ)	Major Radiations, Energies in MeV and γ Intensities, %
	94	−88,320	−82.3	20.3 m	....	....	0.267(6), 0.67(1), 0.94(2), 1.42(1), 1.90(2), 2.18 β−, 5.0; γ, 0.56(6), 0.92(43), 1.13(5), 1.65(2), 1.90(2), 2.13(2), 2.57(2), 3.06(1), 3.53(1)
	95	−87,460	−81	10.9 m	....	....	γ, 1.30, 1.80
	96	−84,310	−79	2.3 m	....	....	β−, 3.5; γ, 0.7, 1.0, 1.5
₄₀Zr	86	−83,770	−78	16.5 h	....	....	γ, 0.028(20), 0.243(96), 0.612(5)
	87	−85,510	−79.7	1.6 h	....	....	β+, 2.10; γ, 0.511, 1.2, 2.2
	88	−89,940	−84	85 d	....	....	γ, 0.394(97); e−, 0.377
	89	−91,086	−84.85	78.4 h	....	....	β+, 0.90; e−, 0.89; γ, 0.511, 0.91(99), 1.71(1)
	89 m		−84.26	4.18 m	....	....	β+, 2.40, 0.89; e−, 0.570; γ, 0.588(87), 1.51(6)
	90	−95,300	−88.770	....	51.46	0.1	....
	90 m		−86.45	0.80 s	....	....	γ, 0.133(4), 2.18(14), 2.32(86); e−, 0.115, 0.130
	91	−94,358	−87.893	....	11.23	1	....
	92	−94,969	−88.462	....	17.11	0.2	....
	93	−93,550	−87.11	1.5×10^6 y	....	<4	β−, 0.060; no γ
	94	−93,687	−87.267	....	17.40	0.08	....
	95	−91,965	−85.663	65.5 d	....	....	β−, 0.89, 0.396; γ, 0.724(49), 0.756(49)
	96	−91,714	−85.430	$>3.6 \times 10^{17}$ y	2.80	0.05	....
	97	−89,034	−82.93	17.0 h	....	....	β−, 1.91; γ, 0.747(92)
	98	−88,040	−82	1 m	....	....	....
₄₁Nb	88	−82,210	−77	14 m	....	....	β+, 3.2; γ, 0.076, 0.141, 0.272, 0.399, 0.511, 0.671, 1.058, 1.083
	89	−86,920	−81.0	1.9 h	....	....	β+, 2.9; γ, 0.511, 1.626, 3.577, 3.838
	89 m		−80.2	42 m	....	....	β+, 3.1; e−, 0.570; γ,

Table 3-7 (*Continued*)
TABLE OF NUCLIDES

Nuclide		Mass Excess		Half-life $y = yr, d = day$ $h = hr, m = min$ $s = sec$	Natural Abundance, %	Thermal Neutron Absorption Cross Section σ, barns (mode if not n,γ)	Major Radiations, Energies in MeV and γ Intensities, %
Z Symbol	Mass Number	μmu	MeV				
	90	−88,741	−82.66	14.6 h	....	....	0.511, 0.588(93) β^+, 1.50; e^-, 0.115, 0.123; γ, 0.142(75), 0.511, 1.14(97), 2.18(14), 2.32(82)
	90 *m*		−82.54	24 s	....	....	γ, 0.122(71); e^-, 0.104, 0.120
	91	−93,140	−86.8	long	....	....	....
	91 *m*		−86.6	64 d	....	....	γ, 0.104(1), 1.21(3); e^-, 0.086, 0.102
	92	−92,789	−86.45	>350 y or <1 h	....	....	....
	92 *m*		−86.32	10.16 d	....	....	γ, 0.934(99)
	93	−93,618	−87.204	....	100	1	....
	93 *m*		−87.173	13.6 y	....	....	e^-, 0.011, 0.028
	94	−92,697	−86.35	2.0×10^4 y	....	~15	β^-, 0.49; γ, 0.702(100), 0.871(100)
	94 *m*		−86.31	6.29 m	....	....	γ, 0.871; e^-, 0.023, 0.039
	95	−93,168	−86.784	35.0 d	....	~7	β^-, 0.160; γ, 0.765(100)
	95 *m*		−86.549	90 h	....	....	e^-, 0.216
	96	−91,944	−85.64	23.35 h	....	....	β^-, 0.7; γ, 0.459(28), 0.569(59), 0.778(97), 0.811(14), 0.851(22), 1.092(49), 1.200(21)
	97	−91,904	−85.61	72 m	....	....	β^-, 1.27; γ, 0.665(98)
	97 *m*		−84.86	1.0 m	....	....	γ, 0.747(98); e^-, 0.728
	98	−89,650	−83.5	51 m	....	....	β^-, 3.1; γ, 0.330(9), 0.720(75), 0.787(100), 1.16(30), 1.44(10), 1.52(4), 1.68(10), 1.88(4), 1.93(8)
	98			<2 m	....	....	β^-
	99	−88,950	−83	2.4 m	....	....	β^-, 3.2; γ, 0.100(1), 0.260(1)
	99			10 s	....	....	β^-
	100	−85,980	−80	3.0 m	....	....	γ, 0.140(10), 0.36(55), 0.45(40),

Table 3-7 (*Continued*)
TABLE OF NUCLIDES

Nuclide		Mass Excess		Half-life	Natural	Thermal Neutron	Major Radiations,
Z Sym-bol	Mass Number	μmu	MeV	y = yr, d = day h = hr, m = min s = sec	Abundance, %	Absorption Cross Section σ, barns (mode if not n,γ)	Energies in MeV and γ Intensities, %
	100		−80	11 m	....	....	0.53(100), 0.65, 2.2, 2.3, 2.65, 2.85 β^-, 4.2, 3.5; γ, 0.535(100), 0.62(60), 1.04(10), 1.15(10), 1.47(5)
$_{42}$Mo	88			27 m	....	....	β^+, 2.5; γ, 0.511, 2.69
	89			7 m	....	....	β^+, 4.9; γ, 0.511
	90	−86,060	−80.17	5.67 h	....	....	β^+, 1.2; e^-, 0.104, 0.120, 0.239, 0.255; γ, 0.122(71), 0.257(85), 0.445(9), 0.511, 0.945(10), 1.273(8), 1.389(4), 1.46(4)
	91	−88,350	−82.3	15.49 m	....	....	β^+, 3.44; γ, 0.511
	91 *m*		−81.6	64 s	....	....	β^+, 3.99, 2.78; e^-, 0.638; γ, 0.511, 0.658(54), 1.21(22), 1.53(15)
	92	−93,190	−86.804	>4×10^{18} y	15.86	<0.3	....
	93	−93,170	−86.79	>100 y	....	....	Nb X-rays
	93 *m*		−84.36	6.95 h	....	....	γ, 0.264(58), 0.685(100), 1.479(100); e^-, 0.244, 0.261
	94	−94,910	−88.407	....	9.12	....	....
	95	−94,161	−87.709	....	15.70	14	....
	96	−95,326	−88.794	....	16.50	1	....
	97	−93,979	−87.539	....	9.45	2	....
	98	−94,591	−88.110	....	23.75	0.51	....
	99	−92,280	−85.96	66.7 h	....	....	β^-, 1.23; γ, 0.041(2), 0.181(7), 0.372(1), 0.740(12), 0.780(4)
	100	−92,525	−86.185	≥3×10^{17} y	9.62	0.2	....
	101	−89,647	−83.50	14.6 m	....	....	β^-, 2.23; e^-, 0.170; γ, 0.191(25), 0.51(15), 0.59(21), 0.70(11), 0.89(15),

Table 3-7 (*Continued*)
TABLE OF NUCLIDES

Nuclide		Mass Excess		Half-life y = yr, d = day h = hr, m = min s = sec	Natural Abundance, %	Thermal Neutron Absorption Cross Section σ, barns (mode if not n,γ)	Major Radiations, Energies in MeV and γ Intensities, %
Z Sym- bol	Mass Number	μmu	MeV				
	102	−89,750	−84	11.5 m	....	....	1.02(25), 1.18(11), 1.38(9), 1.56(11), 2.08(16) β⁻, 1.2
	103			62 s	....	....	....
	104			1.1 m	....	....	β⁻, 4.8; γ, 0.070
	105			40 s	....	....	β⁻
₄₃Tc	92	−84,540	−78.8	4.4 m	....	....	β⁺, 4.1; γ, 0.090(20), 0.14(67), 0.24(30), 0.33(90), 0.511, 0.79(95), 1.54(100)
	93	−89,749	−83.60	2.75 h	....	....	β⁺, 0.80; γ, 0.511, 1.35(65), 1.49(33)
	93 m		−83.21	43 m	....	....	γ, 0.390(63), 2.66(18); e⁻, 0.369
	94	−90,337	−84.15	293 m	....	....	β⁺, 0.816; γ, 0.511, 0.702(100), 0.849(100), 0.871(100)
	94 m		−84.04	53 m	....	....	β⁺, 2.47; γ, 0.511, 0.871(91), 1.53(10), 1.87(9), 2.73(5), 3.20(2)
	95	−92,380	−86.05	20.0 h	....	....	γ, 0.768(82), 0.84(11), 1.06(4)
	95 m		−86.01	61 d	....	....	β⁺, 0.68; e⁻, 0.019, 0.036, 0.184; γ, 0.204(70), 0.584(36), 0.78(12), 0.823(9), 0.838(27), 1.042(4)
	96	−92,170	−85.9	4.35 d	....	....	γ, 0.32(5), 0.778(100), 0.81(84) 0.851(100), 1.12(16); e⁻, 0.30, 0.75, 0.79, 0.82
	96 m		−85.8	52 m	....	....	e⁻, 0.013, 0.032
	97	−93,660	−87	2.6×10⁶ y	....	....	Mo X-rays
	97 m		−87	91 d	....	....	e⁻, 0.075, 0.094
	98	−92,890	−86.5	1.5×10⁶ y	....	3	β⁻, 0.30; γ,

Table 3-7 (*Continued*)
TABLE OF NUCLIDES

Nuclide		Mass Excess		Half-life y = yr, d = day h = hr, m = min s = sec	Natural Abundance, %	Thermal Neutron Absorption Cross Section σ, barns (mode if not n,γ)	Major Radiations, Energies in MeV and γ Intensities, %
Z Symbol	Mass Number	μmu	MeV				
							0.66(100), 0.76(100)
	99	−93,751	−87.33	2.12×10⁵ y	....	22	β⁻, 0.292; no γ
	99 m		−87.18	6.049 h	....	....	γ, 0.140(90); e⁻, 0.001, 0.119
	100	−92,160	−85.9	15.8 s	....	....	β⁻, 3.38; γ, 0.540, 0.60, 0.71, 0.81, 0.89, 1.01, 1.31, 1.49, 1.8
	101	−92,674	−86.32	14.0 m	....	....	β⁻, 1.32; γ, 0.13(3), 0.307(91), 0.545(8)
	102	−90,820	−85	5 s	....	....	β⁻, 4.4
	103	−91,170	−84.9	50 s	....	....	β⁻, 2.2; γ, 0.135(17), 0.21(10), 0.35
	104	−88,290	−82.2	18 m	....	....	β⁻, 4.6; γ, 0.36, 0.53, 0.89, 1.15, 1.25, 1.37, 1.6, 1.9, 2.2, 2.7, 3.2, 3.4, 3.7, 4.0, 4.4, 4.7
	105	−88,670	−82.6	7.7 m	....	....	β⁻, 3.4; γ, 0.110
	106			37 s	....	....	....
	107			29 s	....	....	....
₄₄Ru	93			50 s	....	....	β⁺
	94			57 m	....	....	....
	95	−90,199	−84.02	1.65 h	....	....	β⁺, 1.33; γ, 0.340(70), 0.511, 0.625(13), 1.09(21), 1.43(5)
	96	−92,402	−86.07	....	5.46	0.2	....
	97	−92,370	−86	2.88 d	....	....	γ, 0.215(91), 0.324(8); e⁻, 0.194
	98	−94,711	−88.222	....	1.868	<8	....
	99	−94,065	−87.619	....	12.63	11	....
	100	−95,782	−89.219	....	12.53	10	....
	101	−94,423	−87.953	....	17.02	3	....
	102	−95,652	−89.098	....	31.6	1.4	....
	103	−93,694	−87.27	39.5 d	....	....	β⁻, 0.70, 0.21; γ, 0.497(88), 0.610(6)
	104	−94,570	−88.090	....	18.87	0.48	....
	105	−92,321	−86.00	4.44 h	....	0.2	β⁻, 1.87, 1.15; γ, 0.263(6), 0.317(11), 0.40(6), 0.475(20),

Table 3-7 (*Continued*)
TABLE OF NUCLIDES

Nuclide		Mass Excess		Half-life y = yr, d = day h = hr. m = min s = sec	Natural Abundance. %	Thermal Neutron Absorption Cross Section σ. barns (mode if not n.γ)	Major Radiations, Energies in MeV and γ Intensities, %
Z Symbol	Mass Number	μmu	MeV				
₄₅Rh							0.67(16), 0.726(48)
	106	−92,678	−86.33	368 d	....	0.15	β⁻, 0.039; no γ
	107	−89,870	−83.7	4.2 m	....	....	β⁻, 3.2; γ, 0.195(14), 0.37, 0.48, 0.86(7), 0.93(4), 1.03(4), 1.29(4)
	108	−89,900	−84	4.5 m	....	....	β⁻, 1.3; γ, 0.165(28)
	97	−88,620	−83	32 m	....	....	β⁺, 2.47; γ, 0.08, 0.187, 0.255, 0.420, 0.511, 0.86, 1.18, 1.57, 1.70, 1.96, 2.16
	97			1.0 m	....	....	γ, 0.75
	98	−90,200	−84.0	8.7 m	....	....	β⁺, 2.5; γ, 0.65(100)
	99		−85.57	16.1 d	....	....	β⁺, 1.03; γ, 0.090, 0.175, 0.31, 0.354, 0.444, 0.48, 0.511, 0.529
	99	−91,810	−85.52	4.7 h	....	....	β⁺, 0.74; γ, 0.34(70), 0.511, 0.62(20), 0.89, 1.26, 1.41
	100	−91,874	−85.58	20.8 h	....	....	β⁺, 2.62; e⁻, 0.516; γ, 0.444(8), 0.511, 0.540(88), 0.820(25), 1.11(13), 1.35(20), 1.55(23), 1.93(10), 2.37(39)
	101	−93,822	−87.39	3.0 y	....	....	γ, 0.127(88), 0.198(75), 0.325(11); e⁻, 0.105, 0.124, 0.176
	101 *m*		−87.24	4.4 d	....	....	γ, 0.307(83), 0.545(6); e⁻, 0.134, 0.154
	102	−93,158	−86.77	206 d	....	....	β⁻, 1.15; β⁺, 1.29; γ, 0.475(57), 0.511, 0.628(4), 1.103(3), 1.37(1), 1.57
	102			2.9 y	....	....	γ, 0.418(13), 0.475(95), 0.632(54),

Table 3-7 (*Continued*)
TABLE OF NUCLIDES

| Nuclide | | Mass Excess | | Half-life | Natural | Thermal Neutron | Major Radiations, |
Z Symbol	Mass Number	μmu	MeV	y=yr, d=day h=hr, m=min s=sec	Abundance, %	Absorption Cross Section σ, barns (mode if not n,γ)	Energies in MeV and γ Intensities, %
	103	−94,489	−88.014	....	100	144	0.698(41), 0.768(30), 1.05(41), 1.11(22) ....
	103 m		−87.974	57.5 m	....	....	γ, 0.040; e⁻, 0.017, 0.037
	104	−93,341	−86.95	43 s	....	40	β⁻, 2.44; γ, 0.56(2), 1.24
	104 m		−86.82	4.41 m	....	800	γ, 0.051(47), 0.078(3), 0.097(3), 0.56, 0.77; e⁻, 0.028, 0.054, 0.074
	105	−94,329	−87.87	35.88 h	....	15,000	β⁻, 0.568; γ, 0.306(5), 0.319(19)
	105 m		−87.74	45 s	....	....	γ, 0.129; e⁻, 0.106, 0.126
	106	−92,721	−86.37	30 s	....	....	β⁻, 3.54; γ, 0.512(21), 0.622(11), 1.05(2), 1.13(1), 1.55
	106		−86.3	130 m	....	....	β⁻, 1.62, 1.1; γ, 0.220(18), 0.406(18), 0.451(35), 0.512(88), 0.616(29), 0.735(41), 0.82(35), 1.046(25), 1.128(12), 1.223(17), 1.56(18)
	107	−93,247	−86.86	21.7 m	....	....	β⁻, 1.20; γ, 0.305(73), 0.390(11), 0.68(3)
	108	−91,300	−85	16.8 s	....	....	β⁻, 4.5; γ, 0.434(43), 0.51(10), 0.62(22)
	109	−91,360	−85	<1 h	....	....	....
	110	−88,900	−83	5 s	....	....	β⁻, 5.5; γ, 0.374
₄₆Pd	98			17.5 m	....	....	γ, 0.132
	99	−87,730	−81.7	22 m	....	....	β⁺, 2.0; γ, 0.140, 0.275, 0.420, 0.511, 0.67
	100	−91,230	−85	4.0 d	....	....	γ, 0.074(34), 0.084(49),

Table 3-7 (*Continued*)
TABLE OF NUCLIDES

| Nuclide | | Mass Excess | | Half-life | Natural | Thermal Neutron | Major Radiations, |
Z Symbol	Mass Number	μmu	MeV	y = yr, d = day h = hr, m = min s = sec	Abundance, %	Absorption Cross Section σ, barns (mode if not n,γ)	Energies in MeV and γ Intensities, %
	101	−91,930	−85.40	8.4 h	....	....	0.126(16), 0.159(4); e^-, 0.010, 0.019, 0.052, 0.061, 0.071, 0.081 γ, 0.270(8), 0.296(30), 0.511, 0.566(7), 0.590(24), 0.723(5), 0.993(2), 1.20(3), 1.30(3); β^+, 0.78; e^-, 0.021
	102	−94,391	−87.92	....	0.96	4.8	....
	103	−93,893	−87.46	17.0 d	....	....	γ, 0.297, 0.362, 0.498
	104	−95,989	−89.41	....	10.97		....
	105	−94,936	−88.43	....	22.2	....	
	106	−96,521	−89.91	....	27.3	0.29	....
	107	−94,868	−88.368	∼7×10⁶ y	....		β^-, 0.04; no γ
	107 *m*		−88.16	21.3 s	....	....	γ, 0.21; e^-, 0.19, 0.21
	108	−96,109	−89.52	....	26.7	12	....
	109	−94,046	−87.60	13.47 h	....	....	β^-, 1.028; e^-, 0.062, 0.084; γ, 0.088(5), 0.129, 0.31, 0.41, 0.60, 0.64
	109 *m*		−87.41	4.69 m	....	....	γ, 0.188(58); e^-, 0.164, 0.185
	110	−94,836	−88.34	....	11.8	0.2	....
	111	−92,330	−86.0	22 m	....		β^-, 2.2; γ, 0.38(5), 0.60(13), 0.81(1), 1.4(8)
	111 *m*		−85.8	5.5 h	....	....	β^-, 2.0; e^-, 0.148, 0.169; γ, 0.17
	112	−92,614	−86.27	21.0 h	....	....	β^-, 0.28; γ, 0.019(20)
	113			1.4 m	....	....	no γ
	114			2.4 m	....	....	no γ
	115			45 s	....	....	....
₄₇Ag	102	−88,700	−83	15 m	....	....	....
	103	−91,110	−84.9	66 m	....	....	β^+, 1.6; γ, 0.12(26), 0.15(23), 0.24(10), 0.27(34), 0.511, 1.01(10), 1.16(9), 1.28(13)
	103 *m*		−84.7	5.7 s	....	....	γ, 0.138

Table 3-7 (*Continued*)
TABLE OF NUCLIDES

Nuclide		Mass Excess		Half-life	Natural	Thermal Neutron	Major Radiations,
Z Symbol	Mass Number	μmu	MeV	y = yr, d = day h = hr, m = min s = sec	Abundance, %	Absorption Cross Section σ, barns (mode if not n,γ)	Energies in MeV and γ Intensities, %
	104	−91,404	−85.14	66 m	. . . .	. . . .	β⁺, 0.99; e⁻, 0.532, 0.743; γ, 0.511, 0.556(84), 0.764(48), 0.854(30), 1.34(8), 1.53(7), 1.62(8), 1.81(7)
	104 m		−85.12	29.8 m	. . . .	. . . .	β⁺, 2.70; e⁻, 0.532; γ, 0.511, 0.556(100)
	105	−93,540	−87	40 d	. . . .	. . . .	γ, 0.064(10), 0.280(32), 0.344(42), 0.443(10), 0.62–0.68(12), 1.088(2); e⁻, 0.040, 0.060, 0.256, 0.320
	106	−93,339	−86.94	23.96 m	. . . .	. . . .	β⁺, 1.96; γ, 0.511
	106 m		−86.6	8.5 d	. . . .	. . . .	γ, 0.221(9), 0.451(9), 0.512(86), 0.616(23), 0.717(31), 0.748(13), 0.80(41), 1.046(29), 1.128(9), 1.199(9), 1.528(15), 1.58(8), 1.83(3); e⁻, 0.197, 0.382, 0.405, 0.426, 0.487, 0.508, 0.592, 0.693
	107	−94,906	−88.403	. . . .	51.35	35	. . . .
	107 m		−88.310	44.3 s	. . . .	. . . .	γ, 0.094(5); e⁻, 0.068, 0.090
	108	−94,051	−87.61	2.42 m	. . . .	. . . .	β⁻, 1.64; β⁺, 0.90; γ, 0.434(1), 0.511, 0.615, 0.632(2)
	108 m		−87.50	>5 y	. . . .	. . . .	γ, 0.080(5), 0.434(89), 0.614(90), 0.722(90); e⁻, 0.027
	109	−95,244	−88.717	. . . .	48.65	89	. . . .
	109 m		−88.630	39.2 s	. . . .	. . . .	γ, 0.088(5); e⁻, 0.062, 0.084

Table 3-7 (*Continued*)
TABLE OF NUCLIDES

Nuclide		Mass Excess		Half-life y = yr, d = day h = hr, m = min s = sec	Natural Abundance, %	Thermal Neutron Absorption Cross Section σ, barns (mode if not n,γ)	Major Radiations, Energies in MeV and γ Intensities, %
Z Sym-bol	Mass Number	μmu	MeV				
	110	−93,905	−87.47	24.4 s	. . . .	. . . .	β^-, 2.87; γ, 0.658(5)
	110 m		−87.35	255 d	. . . .	80	β^-, 1.5, 0.53, 0.087; e^-, 0.090, 0.113; γ, 0.658(96), 0.68(16), 0.706(19), 0.764(23), 0.818(8), 0.885(71), 0.937(32), 1.384(21), 1.505(11)
	111	−94,684	−88.20	7.5 d	. . . .	. . . .	β^-, 1.05; γ, 0.247(1), 0.342(6)
	111 m		−88.13	74 s	. . . .	. . . .	γ, 0.065
	112	−92,936	−86.57	3.14 h	. . . .	. . . .	β^-, 3.94; γ, 0.617(41), 1.40(5), 1.63(3), 2.11(3), 2.55(2)
	113	−93,444	−87.04	5.3 h	. . . .	γ, ...	β^-, 2.0; γ, 0.12(10), 0.30(100), 0.58(5), 0.67(17), 0.88(4), 0.98(5), 1.18(4)
	113			1.2 m	. . . .	. . . .	β^-, <2.0; γ, 0.14, 0.30, 0.39, 0.56, 0.70
	114	−91,700	−85.4	4.5 s	. . . .	. . . .	β^-, 4.6; γ, 0.57
	115	−91,070	−84.8	20.0 m	. . . .	. . . .	β^-, 3.2; γ, 0.14(12), 0.22(49), 0.28(13), 0.36(11), 0.42(7), 0.47(10), 0.64(4), 1.48(11), 1.66(8), 1.89(10), 2.12(13)
	116	−88,690	−83	2.5 m	. . . .	. . . .	β^-, 5.0; γ, 0.52, 0.70
	117			1.1 m	. . . .	. . . .	. . . .
$_{48}$Cd	103			10 m	. . . .	. . . .	γ, 0.22, 0.511, 0.63, 0.85
	104	−90,120	−84	57 m	. . . .	. . . .	γ, 0.084; e^-, 0.041, 0.058, 0.080
	105	−90,530	−84	55 m	. . . .	. . . .	β^+, 1.69; e^-, 0.282, 0.295, 0.321, 0.408
	106	−93,537	−87.128	. . . .	1.22	1	. . . .
	107	−93,385	−86.99	6.49 h	. . . .	. . . .	β^+, 0.302; γ, 0.511, 0.796, 0.829
	108	−95,813	−89.248	. . . .	0.88	3	. . . .
	109	−95,072	−88.55	453 d	. . . .	. . . .	γ, 0.088; e^-, 0.062,

Table 3-7 (*Continued*)
TABLE OF NUCLIDES

Nuclide		Mass Excess		Half-life y=yr, d=day h=hr, m=min s=sec	Natural Abundance, %	Thermal Neutron Absorption Cross Section σ, barns (mode if not n.γ)	Major Radiations, Energies in MeV and γ Intensities, %
Z Sym-bol	Mass Number	μmu	MeV				
	110	−96,988	−90.342	....	12.39	0.1	0.084 ....
	111	−95,812	−89.246	....	12.75	....	....
	111 m		−88.850	48.6 m	....	....	γ, 0.150(30), 0.247(94); e−, 0.123, 0.146
	112	−97,238	−90.575	....	24.07	0.03	....
	113	−95,592	−89.041	>1.3×10¹⁵ y	12.26	20,000	....
	113 m		−88.77	13.6 y	....	....	β−, 0.58; γ, 0.265
	114	−96,640	−90.018	....	28.86	1.1	....
	115	−94,569	−88.09	53.5 h	....	....	β−, 1.11; γ, 0.230(1), 0.262(2), 0.49(10), 0.53(26)
	115 m		−87.91	43 d	....	....	β−, 1.62; γ, 0.485, 0.935(2), 1.29(1)
	116	−95,238	−88.712	>10¹⁷ y	7.58	1.4	....
	117	−92,761	−86.41	2.4 h	....	....	β−, 2.23; e−, 0.286; γ, 0.089(7), 0.273(31), 0.314(16), 0.345(18), 0.434(13), 0.832(4), 0.880(3), 0.95(4), 1.052(5), 1.303(19), 1.577(17)
	117 m		−86.27	3.4 h	....	....	β−, 0.67; e−, 0.286; γ, 0.273(18), 0.314(8), 0.345(4), 0.434(4), 0.565(6), 0.715(4), 0.880(10), 1.065(9), 1.117(4), 1.24(11), 1.338(8), 1.408(8), 1.433(10), 1.562(6), 1.998(15), 2.319(3)
	118	−93,030	−87	49 m	....	....	β−
	119	−90,260	−84.1	2.7 m, 10 m	....	....	β−, 3.5
	121			12.8 s	....	....	....
₄₉In	106	−86,560	−80.6	5.3 m	....	....	β+, 4.9; γ, 0.511,

Table 3-7 (*Continued*)
TABLE OF NUCLIDES

Nuclide		Mass Excess		Half-life y = yr, d = day h = hr, m = min s = sec	Natural Abundance, %	Thermal Neutron Absorption Cross Section σ, barns (mode if not n,γ)	Major Radiations, Energies in MeV and γ Intensities, %
Z Symbol	Mass Number	μmu	MeV				
	107	−89,640	−83.5	33 m	....	....	0.63, 1.65, 1.85 β+, 2.2; γ, 0.22(46), 0.32, 0.511, 0.73, 0.84, 0.94, 1.05, 1.25
	108		−84.14	57 m	....	....	β+, 1.29; e−, 0.123, 0.147, 0.216, 0.238, 0.260, 0.606, 0.845; γ, 0.150, 0.175, 0.243, 0.511, 0.633, 0.872
	108	−90,280	−84.10	39 m	....	....	β+, 3.50; e−, 0.606; γ, 0.383, 0.511, 0.633, 0.842
	109	−92,904	−86.53	4.3 h	....	....	β+, 0.79; e−, 0.033, 0.056, 0.178, 0.201; γ, 0.205, 0.28, 0.35, 0.65, 0.91
	109 m_1		−85.87	1.3 m	....	....	γ, 0.658; e−, 0.630
	109 m_2		−84.42	0.20 s	....	....	γ, 0.17(12), 0.21(12), 0.40(20), 0.68(100), 1.04(20), 1.43(77)
	110	−92,769	−86.41	66 m	....	....	β+, 2.25; e−, 0.631; γ, 0.511, 0.658(95)
	110			4.9 h	....	....	γ, 0.66(160), 0.91(110); e−, 0.094, 0.558, 0.615, 0.631, 0.653, 0.680, 0.858, 0.910
	111	−94,640	−88.2	2.81 d	....	....	γ, 0.173(89), 0.247(94); e−, 0.146, 0.220, 0.243
	112	−94,456	−87.98	14.4 m	....	....	β−, 0.66; β+, 1.56; γ, 0.511, 0.617(6)
	112 m		−87.83	20.7 m	....	....	γ, 0.156(9); e−, 0.128, 0.152
	113	−95,911	−89.34	....	4.23	8	....
	113 m		−88.95	99.8 m	....	....	γ, 0.393(64); e−, 0.365, 0.389
	114	−95,095	−88.58	72 s	....	....	β−, 1.988; β+, 0.42; γ, 1.299
	114·m		−88.39	50.0 d	....	....	γ, 0.192(17),

Table 3-7 (*Continued*)
TABLE OF NUCLIDES

Nuclide		Mass Excess		Half-life y = yr, d = day h = hr, m = min s = sec	Natural Abundance, %	Thermal Neutron Absorption Cross Section σ, barns (mode if not n,γ)	Major Radiations, Energies in MeV and γ Intensities, %
Z Sym- bol	Mass Number	μmu	MeV				
	115	−96,129	−89.54	6×10¹⁴ y	95.77	154	0.558(4), 0.724(4); e^-, 0.164, 0.188 . . . .
	115 m		−89.21	4.50 h	. . . .	. . . .	β^-, 0.83; e^-, 0.308, 0.331; γ, 0.335(50)
	116	−94,683	−88.20	13.4 s	. . . .	. . . .	β^-, 3.3; γ, 0.434, 0.95, 1.293(1)
	116 m₁		−88.14	54.0 m	. . . .	. . . .	β^-, 1.00; γ, 0.138(3), 0.417(36), 0.819(17), 1.09(53), 1.293(80), 1.508(11), 2.111(20)
	116 m₂		−87.98	2.16 s	. . . .	. . . .	γ, 0.164; e^-, 0.138, 0.160
	117	−95,466	−88.93	45 m	. . . .	. . . .	β^-, 0.74; e^-, 0.132; γ, 0.158(87), 0.565(100)
	117 m		−88.61	1.93 h	. . . .	. . . .	β^-, 1.78; e^-, 0.286; γ, 0.158(14), 0.314(31)
	118	−93,890	−87.5	5.7 s	. . . .	. . . .	β^-, 4.2; γ, 1.230(15)
	118		−87.4	4.35 m	. . . .	. . . .	β^-, 2.0; γ, 0.69(41), 1.05(80), 1.230(97), 2.04(3)
	119	−94,010	−87.6	2.1 m	. . . .	. . . .	β^-, 1.6; γ, 0.82(95)
	119 m		−87.3	17.5 m	. . . .	. . . .	β^-, 2.7; γ, 0.024, 0.30, 0.91
	120	−92,000	−86	3.2 s	. . . .	. . . .	β^-, 5.6; γ, 1.711(15)
	120		−85.8	44 s	. . . .	. . . .	β^-, 3.1; γ, 0.090(12), 0.198(9), 0.71(12), 0.86(34), 0.94(12), 1.02(61), 1.171(100), 1.28(14), 1.47(6), 1.87(7), 2.01(6)
	121	−91,910	−86	30 s	. . . .	. . . .	γ, 0.94
	121		−86	3.1 m	. . . .	. . . .	β^-, 3.7
	122	−89,400	−83	8 s	. . . .	. . . .	β^-, 5; γ, 0.99, 1.14
	123	−89,430	−83	36 s	. . . .	. . . .	β^-, 4.6
	123		−83	10 s	. . . .	. . . .	γ, 1.1
	124	−86,800	−81	~3.6 s	. . . .	. . . .	β^-, 5; γ, 0.99(3),

Table 3-7 (*Continued*)
TABLE OF NUCLIDES

| Nuclide | | Mass Excess | | Half-life | Natural | Thermal Neutron | Major Radiations, |
Z Symbol	Mass Number	μmu	MeV	y = yr, d = day h = hr, m = min s = sec	Abundance, %	Absorption Cross Section σ, barns (mode if not n,γ)	Energies in MeV and γ Intensities, %
$_{50}$Sn	108			9.2 m	....	....	1.13(10), 3.21(3) γ, 0.28, 0.42
	109			18.1 m	....	....	β^+, 1.6; e^-, 0.305, 0.491, 0.86, 1.09; γ, 0.335, 0.521, 0.89, 1.12
	110			4.0 h	....	....	γ, 0.283(95); e^-, 0.255
	111	−91,940	−85.6	35.0 m	....	....	β^+, 1.51; γ, 0.511, 0.75(1), 0.97(1), 1.14(2), 1.54(1), 1.59(1), 1.89(1), 2.11, 2.32
	112	−95,165	−88.64	....	0.95	0.9	....
	113	−94,813	−88.32	115 d	....	....	γ, 0.255(2)
	113 *m*		−88.24	20 m	....	....	γ, 0.079(1); e^-, 0.050, 0.075
	114	−97,227	−90.57	....	0.65	....	....
	115	−96,654	−90.03	....	0.34	....	....
	116	−98,255	−91.523	....	14.24	0.006	....
	117	−97,042	−90.392	....	7.57	....	....
	117 *m*		−90.075	14.0 d	....	....	γ, 0.158(87); e^-, 0.130, 0.155
	118	−98,394	−91.652	....	24.01	0.01	....
	119	−96,687	−90.062	....	8.58	....	....
	119 *m*		−89.973	~250 d	....	....	γ, 0.024(16); e^-, 0.020, 0.026, 0.061
	120	−97,802	−91.100	....	32.97	0.14	....
	121	−95,773	−89.21	27.5 h	....	....	β^-, 0.383
	121 *m*		−89.14	76 y	....	....	β^-, 0.42; γ, 0.037
	122	−96,559	−89.943	....	4.71	0.2	....
	123	−94,262	−87.80	125 d	....	....	β^-, 1.42
	123 *m*		−87.78	39.5 m	....	....	β^-, 1.26; γ, 0.160
	124	−94,728	−88.237	>2×10^{17} y	5.98	0.1	....
	125	−92,254	−85.93	9.4 d	....	....	β^-, 2.34; γ, 0.342, 0.468, 0.811(2), 0.904(1), 1.068(4), 1.17, 1.41, 1.97(1), 2.23
	125 *m*		−85.91	9.5 m	....	....	β^-, 2.04; γ, 0.325(97)
	126	−92,360	−86	~10^5 y	....	....	γ, 0.060, 0.067, 0.092
	127	−89,740	−84	2.05 h	....	....	β^-, 1.45; γ, 0.44, 0.49, 0.82, 1.10, 2.00, 2.32, 2.58, 2.68, 2.82

Table 3-7 (*Continued*)
TABLE OF NUCLIDES

Nuclide		Mass Excess		Half-life y=yr, d=day h=hr, m=min s=sec	Natural Abundance, %	Thermal Neutron Absorption Cross Section σ, barns (mode if not n,γ)	Major Radiations, Energies in MeV and γ Intensities, %
Z Symbol	Mass Number	μmu	MeV				
	127		−83.5	4.1 m	....	....	β⁻, 2.7; γ, 0.49(100)
	128	−89,530	−83.4	59 m	....	....	β⁻, 0.80; γ, 0.044(7), 0.072(19), 0.50(61), 0.57(22)
	129			9 m	....	....	γ, 1.15
	129			1.0 h	....	....	....
	130			2.6 m	....	....	....
	131			3.4 m	....	....	....
	132			2.2 m	....	....	....
₅₁Sb	112			0.9 m	....	....	γ, 0.511, 1.27
	113	−90,014	−83.85	6.4 m	....	....	β⁺, 2.42; γ, 0.32, 0.511, 0.6–0.9, 1.03, 1.2
	114	−90,490	−84.3	3.3 m	....	....	β⁺, 2.7; γ, 0.9, 1.30
	115	−93,401	−87.00	31 m	....	....	β⁺, 1.51; γ, 0.499(100), 0.511, 0.98(5), 1.24(5), 2.22(1)
	116	−93,370	−87.0	16 m	....	....	β⁺, 2.3; γ, 0.511, 0.93(26), 1.293(85), 2.23(14)
	116 *m*		−86.5	60 m	....	....	β⁺, 1.16; e⁻, 0.070, 0.095, 0.111; γ, 0.099(30), 0.140(30), 0.406(36), 0.511, 0.545(68), 0.96(75), 1.06(27), 1.293(100)
	117	−95,088	−88.57	2.8 h	....	....	β⁺, 0.57; γ, 0.158(87), 0.511
	118	−94,426	−87.96	3.5 m	....	....	β⁺, 2.67; γ, 0.511, 0.83, 1.230(3)
	118 *m*		−87.77	5.1 h	....	....	γ, 0.041(29), 0.254(93), 1.049(100), 1.230(100); e⁻, 0.012, 0.036, 0.223
	119	−96,065	−89.48	38.0 h	....	....	γ, 0.024(16); e⁻, 0.020
	120	−94,919	−88.42	15.89 m	....	....	β⁺, 1.70; γ, 0.511, 1.171(1)
	120			5.8 d	....	....	γ, 0.090(81), 0.200(88), 1.03(99),

Table 3-7 (*Continued*)
TABLE OF NUCLIDES

Nuclide		Mass Excess		Half-life y = yr, d = day h = hr, m = min s = sec	Natural Abundance, %	Thermal Neutron Absorption Cross Section σ, barns (mode if not n,γ)	Major Radiations, Energies in MeV and γ Intensities, %
Z Symbol	Mass Number	μmu	MeV				
							1.171(100); e^-, 0.061, 0.096, 0.171, 0.196
	121	−96,184	−89.593	....	57.25	6	....
	122	−94,817	−88.32	2.80 d	....	....	β^-, 1.97; β^+, 0.56; γ, 0.564(66), 0.686(3), 1.14(1), 1.26(1)
	122 m		−88.16	4.2 m	....	....	γ, 0.061(50), 0.075(17); e^-, 0.021, 0.030, 0.045, 0.056, 0.071
	123	−95,787	−89.224	$>1.3 \times 10^{16}$ y	42.75	3.3	....
	124	−94,027	−87.58	60.4 d	....	2000	β^-, 2.31; γ, 0.603(97), 0.644(7), 0.72(14), 0.967(2), 1.048(2), 1.31(3), 1.37(5), 1.45(2), 1.692(50), 2.088(7)
	124 m_1		−87.57	93 s	....	....	β^-, 1.19; e^-, 0.006, 0.009; γ, 0.505(20), 0.603(20), 0.644(20)
	124 m_2		−87.55	21 m	....	....	e^-, 0.021, 0.024; γ
	125	−94,768	−88.28	2.71 y	....	<20	β^-, 0.61; e^-, 0.004, 0.030, 0.144, 0.395; γ, 0.176(6), 0.427(31), 0.463(10), 0.599(24), 0.634(11), 0.66(3)
	126			19.0 m	....	....	β^-, 1.9; γ, 0.41, 0.67
	126	−92,680	−86.3	12.5 d	....	....	β^-, 1.9; γ, 0.41, 0.69
	127	−93,073	−86.70	93 h	....	....	β^-, 1.5; γ, 0.060, 0.25, 0.41, 0.46, 0.68, 0.77, 0.92, 1.10, 1.34
	128	−90,930	−84.7	10.8 m	....	....	β^-, 2.6; γ, 0.320(83), 0.75(200), 1.07(4)
	128			8.6 h	....	....	β^-, 1; γ, 0.314, 0.53, 0.64, 0.75
	129	−90,740	−85	4.3 h	....	....	β^-, 1.87; γ, 0.073, 0.34, 0.460,

Table 3-7 (*Continued*)
TABLE OF NUCLIDES

| Nuclide | | Mass Excess | | Half-life | Natural | Thermal Neutron | Major Radiations, |
Z Symbol	Mass Number	μmu	MeV	y = yr, d = day h = hr, m = min s = sec	Abundance, %	Absorption Cross Section σ, barns (mode if not n,γ)	Energies in MeV and γ Intensities, %
	130	−87,960	−82	33 m	....	....	0.540, 0.81, 0.91, 1.04, 1.24 γ, 0.19, 0.33, 0.82, 0.94
	130		−82	7.1 m	....	....	γ, 0.20, 0.82, 1.03, 1.16
	131			26 m	....	....	γ, 0.64(37), 0.95(48)
	132			2.1 m	....	....	β^-
	133			4.2 m	....	....	β^-
	135			2 s	....	....	....
$_{52}$Te	107			2.2 s	....	....	α, 3.28
	108			5.3 s	....	....	α, 3.08; p, 2.6, 3.4, 3.7
	115		−82.5	6.0 m	....	....	β^+, 2.8; γ, 0.511, 0.72(34), 0.96(6), 1.08(24), 1.28(32), 1.38(32), 1.58(6)
	116	−91,700	−85.4	2.50 h	....	....	γ, 0.094; e^-, 0.063, 0.089
	117	−91,330	−85.1	61 m	....	....	β^+, 1.81; γ, 0.511, 0.72(65), 0.93(6), 1.78(9)
	117			1.9 h	....	....	β^+, 1.7
	118	−94,100	−88	6.00 d	....	....	Sb X-rays
	119	−93,602	−87.19	15.9 h	....	....	β^+, 0.627; γ, 0.645(85), 0.70(11), 1.76(4)
	119 m		−86.9	4.68 d	....	....	γ, 0.153(62), 0.270(25), 0.92–1.14(36), 1.221(67), 2.09(4); e^-, 0.122, 0.133, 0.148, 0.240, 0.266
	120	−95,977	−89.40	....	0.089	2.0	....
	121	−94,801	−88.31	17 d	....	....	γ, 0.508(18), 0.573(80); e^-, 0.007, 0.033, 0.543
	121 m		−88.01	154 d	....	....	γ, 0.212(82), 1.10(3); e^-, 0.007, 0.050, 0.077, 0.180
	122	−96,934	−90.29	....	2.46	2	....
	123	−95,723	−89.16	1.2×10^{13} y	0.87	400	Sb X-rays
	123 m		−88.92	117 d	....	....	γ, 0.159(84); e^-, 0.057, 0.084,

Table 3-7 (*Continued*)
TABLE OF NUCLIDES

Nuclide		Mass Excess		Half-life y = yr, d = day h = hr, m = min s = sec	Natural Abundance, %	Thermal Neutron Absorption Cross Section σ, barns (mode if not n,γ)	Major Radiations, Energies in MeV and γ Intensities, %
Z Sym- bol	Mass Number	μmu	MeV				
	124	−97,158	−90.50	....	4.61	5	0.127 ....
	125	−95,582	−89.03	....	6.99	1.5	....
	125 m		−88.89	58 d	....	....	e⁻, 0.004, 0.030, 0.078, 0.105; γ, 0.035(7), 0.110
	126	−96,678	−90.05	....	18.71	0.9	....
	127	−94,791	−88.30	9.4 h	....	....	β⁻, 0.70; γ, 0.058, 0.21, 0.360, 0.417
	127 m		−88.21	109 d	....	....	γ, 0.059, 0.089, 0.67; e⁻, 0.057, 0.084; β⁻
	128	−95,524	−88.98	....	31.79	0.14	....
	129	−93,425	−87.02	68.7 m	....	....	β⁻, 1.45; e⁻, 0.022, 0.026; γ, 0.027(19), 0.275(2), 0.455(15), 0.81(1), 1.08(2)
	129 m		−86.92	34.1 d	....	....	β⁻, 1.60; e⁻, 0.074, 0.102; γ, 0.69(6)
	130	−93,762	−87.34	8 × 10²⁰ y	34.49	0.2	....
	131	−91,425	−85.16	24.8 m	....	....	β⁻, 2.14; e⁻, 0.116, 0.144; γ, 0.150(68), 0.453(16), 0.493(5), 0.603(4), 0.95(3), 1.00(4), 1.147(6)
	131 m		−84.98	30 h	....	....	β⁻, 2.46, 0.9; e⁻, 0.048, 0.069, 0.149, 0.177; γ, 0.081(2), 0.102(5), 0.200(8), 0.241(8), 0.336(9), 0.78(60), 0.85(31), 1.127(13), 1.206(11), 1.629(3), 1.860(1), 1.965(2)
	132	−91,477	−85.21	77.7 h	....	....	β⁻, 0.22; e⁻, 0.020, 0.048, 0.197; γ, 0.053(17), 0.230(90)
	133			12.5 m	....	....	γ, 0.15, 0.31, 0.41, 0.73, 1.02, 1.33,

Table 3-7 (*Continued*)
TABLE OF NUCLIDES

Nuclide		Mass Excess		Half-life	Natural	Thermal Neutron	Major Radiations,
Z Symbol	Mass Number	μmu	MeV	y = yr, d = day h = hr, m = min s = sec	Abundance, %	Absorption Cross Section σ, barns (mode if not n,γ)	Energies in MeV and γ Intensities, %
	133 *m*			50 m	....	....	1.71, 1.85 β⁻, 2.4; e⁻, 0.303; γ, 0.31(21), 0.432(50), 0.47(22), 0.557(35), 0.63(18), 0.70(24), 0.754(85), 0.91(57), 1.01(10), 1.33, 1.71, 1.85
	134			42 m	....	....	γ, 0.08(13), 0.17(16), 0.204(21), 0.262(19)
	135			<2 m	....	....	β⁻
₅₃I	117			7 m	....	....	β⁺; γ, 0.16, 0.34, 0.522
	117			14.5 m	....	....	....
	118		−81	13.9 m	....	....	γ, 0.511, 0.55, 0.60, 1.15
	119			19.5 m	....	....	γ, 0.26, 0.511, 0.78
	120	−90,180	−83.8	1.35 h	....	....	β⁺, 4.0; γ, 0.511, 0.56, 0.62, 1.52
	121	−92,270	−86.0	2.12 h	....	....	β⁺, 1.2; γ, 0.212(90), 0.27(3), 0.32(6), 0.511
	122	−92,489	−86.15	3.5 m	....	....	β⁺, 3.1; γ, 0.511, 0.564, 0.69, 0.78
	123	−94,270	−88	13.3 h	....	....	γ, 0.159(83); e⁻, 0.127
	124	−93,754	−87.33	4.15 d	....	....	β⁺, 2.14; γ, 0.511, 0.605(67), 0.644(12), 0.73(14), 1.37(3), 1.51(4), 1.69(14), 2.09(2), 2.26(2)
	125	−95,422	−88.88	60.2 d	....	900	γ, 0.035(7); e⁻, 0.004, 0.030
	126	−94,369	−87.90	12.8 d	....	....	β⁻, 1.25; β⁺, 1.13; γ, 0.386(34), 0.667(33)
	127	−95,530	−88.984	....	100	6.4	....
	128	−94,162	−87.71	24.99 m	....	....	β⁻, 2.12; γ, 0.441(14),

Table 3-7 (*Continued*)
TABLE OF NUCLIDES

Nuclide		Mass Excess		Half-life y = yr, d = day h = hr, m = min s = sec	Natural Abundance, %	Thermal Neutron Absorption Cross Section σ, barns (mode if not n,γ)	Major Radiations, Energies in MeV and γ Intensities, %
Z Symbol	Mass Number	μmu	MeV				
	129	−95,013	−88.50	1.7×10^7 y	....	28	0.528(1), 0.743, 0.969 β^-, 0.150; e^-, 0.005, 0.034; γ, 0.040(9)
	130	−93,324	−86.89	12.3 h	....	18	β^-, 1.7, 1.04; γ, 0.419(35), 0.538(99), 0.669(100), 0.743(87), 1.15(12)
	131	−93,873	−87.441	8.05 d	....	~0.7	β^-, 0.806, 0.606; e^-, 0.046, 0.330; γ, 0.080(3), 0.284(5), 0.364(82), 0.637(7), 0.723(2)
	132	−92,019	−85.71	2.26 h	....	....	β^-, 2.12; γ, 0.24(1), 0.52(20), 0.67(144), 0.773(89), 0.955(22), 1.14(6), 1.28(7), 1.40(14), 1.45(1), 1.91(1), 1.99(1)
	133	−92,250	−85.9	20.3 h	....	....	β^-, 1.27; γ, 0.53(90)
	134	−90,150	−84.0	52.0 m	....	....	β^-, 2.43; γ, 0.135(3), 0.41(8), 0.55(8), 0.61(18), 0.85(95), 0.89(65), 1.07(1), 1.15(10), 1.46(4), 1.62(5), 1.79(5)
	135	−89,980	−84	6.68 h	....	....	β^-, 1.4; γ, 0.42(7), 0.86(11), 1.04(9), 1.14(37), 1.28(34), 1.46(12), 1.72(19), 1.80(11)
	136	−85,260	−79.4	83 s	....	....	β^-, 7.0, 5.6; γ, 0.20(12), 0.27(18), 0.39(19), 1.32(95), 2.3(19), 2.63(10), 2.8(8), 3.2(5)
	137			22.0 s	....	....	β^-; n, 0.6 av.
	138			5.9 s	....	....	β^-; n
	139			2.7 s	....	....	β^-; n

Table 3-7 (*Continued*)
TABLE OF NUCLIDES

| Nuclide | | Mass Excess | | Half-life | Natural | Thermal Neutron | Major Radiations, |
Z Symbol	Mass Number	μmu	MeV	y = yr, d = day h = hr, m = min s = sec	Abundance, %	Absorption Cross Section σ, barns (mode if not n,γ)	Energies in MeV and γ Intensities, %
$_{54}$Xe	118			6 m	....	....	γ, 0.05, 0.511
	120			40 m	....	....	γ, 0.055, 0.073, 0.176, 0.76
	121	−88,200	−82.2	39 m	....	....	β^+, 2.8; γ, 0.080, 0.096, 0.132, 0.437, 0.511
	122			20.1 h	....	....	γ, 0.060, 0.090, 0.110, 0.148, 0.180, 0.345; e^-, 0.058, 0.116
	123	−91,270	−85	2.08 h	....	....	β^+, 1.51; e^-, 0.115, 0.144, 0.295; γ, 0.090, 0.110, 0.149, 0.178, 0.329, 0.511, 0.68, 0.90, 1.10
	124	−93,880	−87.5	....	0.096	110	....
	125	−93,380	−87	16.8 h	....	....	γ, 0.055, 0.188, 0.242; e^-, 0.022, 0.050, 0.154, 0.182, 0.209
	125 m			55 s	....	....	γ, 0.075, 0.111
	126	−95,712	−89.15	....	0.090	~2	....
	127	−94,780	−88.54	36.41 d	....	....	γ, 0.058(1), 0.145(4), 0.172(22), 0.203(65), 0.375(20); e^-, 0.024, 0.112, 0.139, 0.170, 0.198
	127 m			75 s	....	....	γ, 0.125, 0.175
	128	−96,460	−89.85	....	1.919	<5	....
	129	−95,216	−88.692	....	26.44	25	....
	129 m		−88.456	8.0 d	....	....	γ, 0.040(9), 0.197(6); e^-, 0.005, 0.034, 0.162, 0.191
	130	−96,491	−89.88	....	4.08	<5	....
	131	−94,915	−88.411	....	21.18	85	....
	131 m		−88.247	11.8 d	....	....	γ, 0.164(2); e^-, 0.129, 0.159
	132	−95,839	−89.272	....	26.89	<5	....
	133	−94,185	−87.73	5.270 d	....	190	β^-, 0.346; e^-, 0.045, 0.075; γ, 0.081(37)
	133 m		−87.50	2.26 d	....	....	γ, 0.233(14); e^-, 0.198, 0.227
	134	−94,603	−88.121	....	10.4	<5	....

Table 3-7 (*Continued*)
TABLE OF NUCLIDES

Nuclide		Mass Excess		Half-life y = yr, d = day h = hr, m = min s = sec	Natural Abundance, %	Thermal Neutron Absorption Cross Section σ, barns (mode if not n,γ)	Major Radiations, Energies in MeV and γ Intensities, %
Z Symbol	Mass Number	μmu	MeV				
	135	−92,980	−86.6	9.14 h	....	2.7×10⁶	β^-, 0.92; e^-, 0.214; γ, 0.250(91), 0.61(3)
	135 m		−86.1	15.6 m	....	....	γ, 0.527(80); e^-, 0.493, 0.522
	136	−92,779	−86.42	....	8.87	0.15	....
	137	−88,900	−82.8	3.9 m	....	....	β^-, 4.1; γ, 0.455(33)
	138	−86,190	−80.9	17.5 m	....	....	β^-, 2.4; γ, 0.16(33), 0.26(100), 0.42(40), 0.51(8), 1.78(66), 2.02(58)
	139	−82,160	−76.5	43 s	....	....	γ, 0.18(41), 0.22(100), 0.30(57), 1.15(23)
	140			16.0 s	....	....	γ, 0.13
	141			1.7 s	....	....	β^-
	142			~1.5 s	....	....	....
	143			1.0 s	....	....	β^-
	144			~1 s	....	....	β^-
₅₅Cs	123			8.0 m	....	....	β^+
	125	−90,090	−84	45 m	....	....	β^+, 2.05; e^-, 0.077, 0.107; γ, 0.112, 0.511
	126	−90,560	−84.4	1.6 m	....	....	β^+, 3.8; γ, 0.386(38), 0.511
	127	−92,520	−86.4	6.2 h	....	....	γ, 0.125(10), 0.406(72), 0.511; e^-, 0.090, 0.119, 0.371; β^+, 1.08
	128	−92,241	−85.92	3.8 m	....	....	β^+, 2.9; e^-, 0.407; γ, 0.441(27), 0.511, 0.528, 0.576, 0.97(1), 1.12(1)
	129	−94,040	−88	32.1 h	....	....	γ, 0.040(2), 0.280(3), 0.320(4), 0.375(48), 0.416(25), 0.550(5); e^-, 0.005, 0.034, 0.057, 0.336, 0.376
	130	−93,280	−86.89	30 m	....	....	β^+, 1.97; β^-, 0.442; γ, 0.511
	131	−94,534	−88.06	9.70 d	....	....	Xe X-rays
	132	−93,607	−87.19	6.59 d	....	....	β^+, 0.40; γ, 0.48(4), 0.668(99), 1.138(1), 1.320(1)
	133	−94,645	−88.16	....	100	28	....

Table 3-7 (*Continued*)
TABLE OF NUCLIDES

Nuclide		Mass Excess		Half-life y = yr, d = day h = hr, m = min s = sec	Natural Abundance, %	Thermal Neutron Absorption Cross Section σ, barns (mode if not n,γ)	Major Radiations, Energies in MeV and γ Intensities, %
Z Symbol	Mass Number	μmu	MeV				
	134	−93,177	−86.79	2.046 y	....	136	β^-, 0.662; γ, 0.57(23), 0.605(98), 0.796(99), 1.038(1), 1.168(2), 1.365(3)
	134 *m*		−86.65	2.895 h	....	....	γ, 0.128(14); e^-, 0.005, 0.009, 0.092, 0.122; β^-, 0.55
	135	−94,230	−87.8	3.0×10⁶ y	.... 8.7		β^-, 0.21; no γ
	135 *m*		−86.2	53 m	....	....	γ, 0.781(100), 0.840(96); e^-, 0.745, 0.775, 0.804
	136	−92,660	−86.6	13.7 d	....	....	β^-, 0.657, 0.341; e^-, 0.116, 0.126, 0.158, 0.302; γ, 0.067(11), 0.086(6), 0.16(36), 0.273(18), 0.340(53), 0.818(100), 1.05(82), 1.25(20)
	137	−93,230	−86.9	30.0 y	....	0.11	β^-, 1.176, 0.514; e^-, 0.624, 0.656; γ, 0.662(85)
	138	−89,200	−83.7	32.2 m	....	....	β^-, 3.40; γ, 0.463(23), 0.55(8), 1.01(25), 1.426(73), 2.21(18), 2.63(9)
	139	−87,100	−81.1	9.5 m	....	....	γ, 0.50, 0.63, 0.80, 1.28, 1.65, 1.90, 2.08
	140	−82,890	−77	66 s	....	....	γ, 0.59, 0.88, 1.14, 1.62, 1.85, 2.06, 2.32, 2.72, 3.15
	141			24 s	....	....	....
	142			2.3 s	....	....	....
₅₆Ba	143			2.0 s	....	....	....
	123			2.0 m	....	....	....
	125			6.5 m	....	....	....
	126			97 m	....	....	γ, 0.23(100), 0.70(33), 0.9
	127	−88,660	−83	10.0 m	....	....	β^+
	128	−91,490	−85	2.43 d	....	....	γ, 0.134, 0.278; e^-, 0.128, 0.242

Table 3-7 (*Continued*)
TABLE OF NUCLIDES

Nuclide		Mass Excess		Half-life y = yr, d = day h = hr, m = min s = sec	Natural Abundance, %	Thermal Neutron Absorption Cross Section σ, barns (mode if not n,γ)	Major Radiations, Energies in MeV and γ Intensities, %
Z Sym- bol	Mass Number	μmu	MeV				
	129	−91,410	−85	2.61 h	....	....	$β^+$, 1.42; e^-, 0.017, 0.048, 0.093, 0.142, 0.171; γ, 0.129(26), 0.182(100), 0.21(65), 0.511, 1.45(42)
	130	−93,755	−87.33	....	0.101	8.8	....
	131	−93,284	−86.89	12.0 d	....	....	γ, 0.124(28), 0.216(19), 0.25(5), 0.373(13), 0.496(48), 0.60(3), 0.924(1), 1.048(1); e^-, 0.019, 0.042, 0.049, 0.088, 0.097, 0.118, 0.180, 0.460
	131 m		−86.71	14.6 m	....	....	γ, 0.107(40); e^-
	132	−94,880	−88.4	....	0.097	7	....
	133	−94,121	−87.67	7.2 y	....	....	γ, 0.080(36), 0.276(7), 0.302(14), 0.356(69), 0.382(8); e^-, 0.045, 0.075, 0.266, 0.319
	133 m		−87.39	38.9 h	....	....	γ, 0.276(17); e^-, 0.006, 0.011, 0.238, 0.270
	134	−95,388	−88.85	....	2.42	<4	....
	135	−94,450	−88.0	....	6.59	5	....
	135 m		−87.7	28.7 h	....	....	γ, 0.268(16); e^-, 0.231, 0.262
	136	−95,700	−89.1	....	7.81	<1	....
	136 m		−87.1	0.32 s	....	....	γ, 0.164(40), 0.818(100), 1.05(100)
	137	−94,500	−88.0	....	11.32	4	....
	137 m		−87.4	2.554 m	....	....	γ, 0.662(89); e^-, 0.624, 0.656
	138	−95,000	−88.5	....	71.66	0.4	
	139	−91,400	−85.1	82.9 m	....	4	$β^-$, 2.3; e^-, 0.126, 0.159; γ, 0.166(23), 1.43
	140	−89,435	−83.31	12.80 d	....	<20	$β^-$, 1.02; e^-, 0.024, 0.029; γ, 0.030(11),

Table 3-7 (*Continued*)
TABLE OF NUCLIDES

Nuclide		Mass Excess		Half-life y=yr, d=day h=hr, m=min s=sec	Natural Abundance, %	Thermal Neutron Absorption Cross Section σ, barns (mode if not n,γ)	Major Radiations, Energies in MeV and γ Intensities, %
Z Symbol	Mass Number	μmu	MeV				
	141	−85,950	−80.1	18 m	....	....	0.163(6), 0.305(6), 0.438(5), 0.537(34) β⁻, 3.0; γ, 0.118(10), 0.193(100), 0.28(50), 0.35(20), 0.46(30), 0.64(20), 0.73(7), 0.86(6), 0.93(3), 1.19(8), 1.29(3), 1.42(4), 1.65(3)
	142	−83,650	−77.9	11 m	....	....	β⁻, 1.7; γ, 0.080(30), 0.26(100), 0.89(40), 0.97(15), 1.08(10), 1.20(35)
₅₇La	143			12 s	....	....	β⁻
	125			<1 m	....	....	....
	126			1.0 m	....	....	γ, 0.256, 0.511
	127			3.5 m	....	....	....
	128			4.2 m	....	....	γ, 0.279, 0.511
	129	−87,110	−81	10.0 m	....	....	....
	130	−87,740	−82	8.7 m	....	....	γ, 0.356, 0.45, 0.511, 0.55, 0.72, 0.81, 0.91, 1.01, 1.19, 1.45, 1.55
	131	−90,110	−83.9	56 m	....	....	β⁺, 1.94; e⁻, 0.078; γ, 0.115(23), 0.169(5), 0.214(8), 0.285(17), 0.364(20), 0.417(20), 0.455(8), 0.511, 0.597(7), 0.878(4)
	132	−89,700	−83.1	4.5 h	....	....	β⁺, 3.8; γ, 0.47, 0.511, 0.56, 0.66, 0.90, 1.03, 1.22, 1.58, 1.92
	133	−91,760	−85.5	4.0 h	....	....	γ, 0.511, 0.8; β⁺, 1.2; e⁻, 0.26
	134	−91,340	−85.1	6.8 m	....	....	β⁺, 2.7; γ, 0.511, 0.605(6)
	135	−93,110	−87.0	19.4 h	....	....	γ, 0.481(2), 0.588,

Table 3-7 (*Continued*)
TABLE OF NUCLIDES

| Nuclide | | Mass Excess | | Half-life | Natural | Thermal Neutron | Major Radiations, |
Z Symbol	Mass Number	μmu	MeV	y = yr, d = day h = hr, m = min s = sec	Abundance, %	Absorption Cross Section σ, barns (mode if not n,γ)	Energies in MeV and γ Intensities, %
	136	−92,620	−86.3	9.5 m	....	....	0.87; e⁻, 0.181, 0.444, 0.475 β⁺, 1.9; γ, 0.511, 0.818(3)
	137	−93,960	−88	6×10⁴ y	....	....	Ba X-rays
	138	−93,090	−86.7	1.12×10¹¹ y	0.089	....	β⁻, 0.21; γ, 0.81(30), 1.426(70)
	139	−93,860	−87.43	....	99.911	8.9	....
	140	−90,562	−84.36	40.22 h	....	....	β⁻, 2.175, 1.69, 1.36; γ, 0.329(20), 0.487(40), 0.815(19), 0.923(10), 1.596(96), 2.53(3)
	141	−89,172	−83.06	3.87 h	....	....	β⁻, 2.43; γ, 1.37(2)
	142	−86,020	−80.1	92.5 m	....	....	β⁻, 4.51; γ, 0.65(48), 0.90(9), 1.01(5), 1.06(4), 1.55(5), 1.74(5), 1.91(9), 2.06(6), 2.41(15), 2.55(11), 2.99(5), 3.31(2), 3.65(2)
	143	−84,130	−78.4	14.0 m	....	....	β⁻, 3.3; γ, 0.62(100), 0.80(44), 1.07(26), 1.17(57), 1.58(28), 1.98(35), 2.56(27)
₅₈Ce	129			~13 m	....	....	γ, 0.080, 0.32, 0.75
	130			30 m	....	....	γ, 0.13
	132	−88,410	−82	4.2 h	....	....	γ, 0.18
	133	−88,750	−83	6.3 h	....	....	β⁺, 1.3; γ, 0.511, 1.8
	134	−91,190	−84.9	72.0 h	....	....	γ
	135	−90,860	−85	17.0 h	....	....	γ, 0.205(17), 0.265(100), 0.300(56), 0.39(10), 0.52(46), 0.59(98), 0.777(22), 0.821(22), 0.865(14), 0.901(10); e⁻, 0.048, 0.078, 0.166, 0.225, 0.25; β⁺, 0.81

Table 3-7 (*Continued*)
TABLE OF NUCLIDES

Nuclide		Mass Excess		Half-life y = yr, d = day h = hr, m = min s = sec	Natural Abundance, %	Thermal Neutron Absorption Cross Section σ, barns (mode if not n,γ)	Major Radiations, Energies in MeV and γ Intensities, %
Z Sym- bol	Mass Number	μmu	MeV				
	136	−92,900	−86.6	>2.9×10¹¹ y	0.193	6.0	....
	137	−92,670	−86	9.0 h	....	....	γ, 0.446(2), 0.481, 0.698, 0.92; e⁻, 0.408
	137 *m*		−87	34.4 h	....	....	γ, 0.168, 0.255(11), 0.762, 0.825(1); e⁻, 0.214, 0.248
	138	−94,170	−87.7	....	0.250	1.0	....
	139	−93,570	−87.16	140 d	....	....	γ, 0.165(80); e⁻, 0.126, 0.159
	139 *m*		−86.41	54 s	....	....	γ, 0.746(93); e⁻, 0.706, 0.740
	140	−94,608	−88.13	....	88.48	0.6	....
	141	−91,781	−85.49	32.5 d	....	30	β⁻, 0.581; e⁻, 0.104, 0.139; γ, 0.145(48)
	142	−90,860	−84.63	>5×10¹⁶ y	11.07	1	....
	143	−87,673	−81.67	33 h	....	6	β⁻, 1.39; e⁻, 0.015, 0.051, 0.252; γ, 0.057(11), 0.293(46), 0.493(2), 0.668(7), 0.725(8), 0.88(1), 1.10(1)
	144	−86,409	−80.49	284 d	....	1.0	β⁻, 0.31; e⁻, 0.038, 0.092; γ, 0.080(2), 0.134(11)
	145	−82,730	−77	3.0 m	....	....	β⁻, 2.0; γ
	146	−81,330	−75.8	14 m	....	....	β⁻, 0.7; γ, 0.110(20), 0.142(42), 0.22(50), 0.27(12), 0.32(100)
	147			65 s	....	....	β⁻
	148			~43 s	....	....	β⁻
₅₉Pr	134			17 m	....	....	γ, 0.22, 0.30, 0.409, 0.511, 0.639, 0.96
	135			22 m	....	....	β⁺, 2.5; γ, 0.080, 0.22, 0.30, 0.511
	136			1.2 h	....	....	β⁺, 2.0; γ, 0.511
	137	−89,640	−84	1.5 h	....	....	β⁺, 1.7; γ, 0.511
	138	−89,540	−82.9	2.10 h	....	....	β⁺, 1.65; e⁻, 0.258, 0.292; γ, 0.298(77), 0.40(9), 0.511, 0.79(100), 1.04(100)

Table 3-7 (*Continued*)
TABLE OF NUCLIDES

Nuclide		Mass Excess		Half-life y = yr, d = day h = hr, m = min s = sec	Natural Abundance, %	Thermal Neutron Absorption Cross Section σ, barns (mode if not n,γ)	Major Radiations, Energies in MeV and γ Intensities, %
Z Symbol	Mass Number	μmu	MeV				
	139	−91,420	−85.0	4.5 h	....	....	β^+, 1.09; γ, 0.511, 1.35(1), 1.61
	140	−90,993	−84.78	3.39 m	....	....	β^+, 2.32; e^-, 1.862; γ, 0.511, 1.596
	141	−92,404	−86.07	>2×10¹⁶ y	100	12	....
	142	−90,022	−83.85	19.2 h	....	20	β^-, 2.16; γ, 1.57(4)
	143	−89,219	−83.11	13.59 d	....	89	β^-, 0.933; no γ
	144	−86,752	−80.81	17.27 m	....	....	β^-, 2.99; γ, 0.695(2), 1.487, 2.186(1)
	145	−85,524	−79.66	5.98 h	....	....	β^-, 1.80; γ, 0.072, 0.68, 0.75, 0.92, 0.98, 1.05, 1.16
	146	−82,410	−76.8	24.0 m	....	....	β^-, 3.7; γ, 0.455(77), 0.74(16), 0.78(15), 0.92(6), 1.37(6), 1.51(27), 1.72(4), 2.23(4), 2.39(3), 2.73(2)
	147	−81,200	−75.5	12.0 m	....	....	β^-, 2.1; γ, 0.078(17), 0.127(9), 0.32(47), 0.56(39), 0.61(10), 0.65(24), 1.26(11)
	148	−78,090	−72.9	2.0 m	....	....	β^-, 4.2; γ, 0.30
	149			2.3 m	....	....	β^-, 2.8; γ, 0.08, 0.155, 0.325, 0.36, 0.745
₆₀Nd	137			55 m	....	....	β^+, 3; e^-, 0.067
	138			22 m	....	....	β^+, 2.4
	139 m	−88,420	−82	5.5 h	....	....	β^+, 3.1; e^-, 0.072, 0.107, 0.189, 0.226; γ, 0.114(80), 0.327(50), 0.511(1400), 0.73(210), 0.82(70), 0.90(25), 0.983(70), 1.03(30), 1.24(20), 1.34(20), 1.48(10), 1.58(8), 2.05(10)

Table 3-7 (*Continued*)
TABLE OF NUCLIDES

Nuclide		Mass Excess		Half-life y = yr, d = day h = hr, m = min s = sec	Natural Abundance, %	Thermal Neutron Absorption Cross Section σ, barns (mode if not n,γ)	Major Radiations, Energies in MeV and γ Intensities, %
Z Symbol	Mass Number	μmu	MeV				
	140	−90,670	−84	3.3 d	....	....	Pr X-rays
	141	−90,472	−84.27	2.42 h	....	....	β⁺, 0.79; γ, 0.145, 0.511, 1.14(2), 1.30(1)
	141 *m*		−83.52	64 s	....	....	γ, 0.755
	142	−92,337	−86.01	....	27.13	17	....
	143	−90,221	−84.04	....	12.20	330	....
	144	−89,961	−83.80	2.4×10¹⁵ y	23.87	5	α, 1.83
	145	−87,462	−81.47	....	8.29	50	
	146	−86,914	−80.96	....	17.18	2	....
	147	−83,926	−78.18	11.06 d	....	....	β⁻, 0.81; e⁻, 0.046, 0.084; γ, 0.091(28), 0.319(3), 0.43(4), 0.533(13)
	148	−83,131	−77.44	....	5.72	4	....
	149	−79,878	−74.41	1.8 h	....	....	β⁻, 1.5; e⁻, 0.051, 0.068, 0.079, 0.090, 0.165, 0.195; γ, 0.114(18), 0.156(4), 0.210(27), 0.27(26), 0.327(5), 0.424(9), 0.541(10), 0.654(9)
	150	−79,085	−73.67	>10¹⁶ y	5.60	1.5	....
	151	−76,230	−71.0	12 m	....	....	β⁻, 2.0; e⁻, 0.072; γ, 0.086(5), 0.118(40), 0.138(6), 0.174(10), 0.256(11), 0.425(5), 0.737(5), 0.797(3), 1.122(2), 1.180(9)
₆₁Pm	141	−86,590	−80.7	22 m	....	....	β¹, 2.6; γ, 0.195(13), 0.511
	142	−87,180	−81.2	40 s	....	....	β⁺, 3.78; γ, 0.511
	143	−89,010	−82.9	0.73 y	....	....	γ, 0.742(47); e⁻, 0.698
	144	−87,490	−82	0.96 y	....	....	γ, 0.474(45), 0.615(99), 0.695(99); e⁻, 0.430, 0.571, 0.651

Table 3-7 (*Continued*)
TABLE OF NUCLIDES

Nuclide		Mass Excess		Half-life y=yr, d=day h=hr, m=min s=sec	Natural Abundance, %	Thermal Neutron Absorption Cross Section σ, barns (mode if not n,γ)	Major Radiations, Energies in MeV and γ Intensities, %
Z Symbol	Mass Number	μmu	MeV				
	145	−87,309	−81.33	17.7 y	....	....	γ, 0.067(1), 0.072(2); e⁻, 0.023, 0.028, 0.061
	146	−85,368	−79.52	4.4 y	....	....	β⁻, 0.78; γ, 0.453(65), 0.75(65)
	147	−84,892	−79.08	2.62 y	....	120	β⁻, 0.224; no γ
	148	−82,579	−76.89	5.4 d	....	~2000	β⁻, 2.48; γ, 0.551(27), 0.914(15), 1.465(23)
	148 m		−76.75	41.8 d	....	30,000	β⁻, 0.69; e⁻, 0.031, 0.053, 0.091, 0.242, 0.503, 0.583; γ, 0.289(13), 0.413(17), 0.551(95), 0.630(87), 0.727(36), 0.916(21), 1.015(20)
	149	−81,670	−76.07	53.1 h	....	....	β⁻, 1.07; γ, 0.286(2), 0.58, 0.85
	150	−79,040	−73.6	2.68 h	....	....	β⁻, 3.05; γ, 0.334(71), 0.406(7), 0.71(8), 0.831(18), 0.88(12), 1.165(23), 1.33(22), 1.75(10), 1.96(3), 2.06(1), 2.53(1)
	151	−78,802	−73.40	27.8 h	....	....	β⁻, 1.19; e⁻, 0.003, 0.018, 0.053, 0.058; γ, 0.07(5), 0.10(7), 0.17(18), 0.24(5), 0.275(6), 0.340(21), 0.45(5), 0.66(3), 0.72(6)
	152	−76,490	−71	6.5 m	....	....	β⁻, 2.2; γ, 0.122, 0.245
	153	−75,970	−70.8	5.5 m	....	....	β⁻, 1.65; γ, 0.12, 0.18
	154			2.5 m	....	....	β⁻, 2.5
₆₂Sm	142			73 m	....	....	γ, 0.15–0.35, 0.511
	143	−85,450	−79.6	9.0 m	....	....	γ, 0.511

Table 3-7 (*Continued*)
TABLE OF NUCLIDES

| Nuclide | | Mass Excess | | Half-life | Natural | Thermal Neutron | Major Radiations, |
Z Symbol	Mass Number	μmu	MeV	y = yr, d = day h = hr, m = min s = sec	Abundance, %	Absorption Cross Section σ, barns (mode if not *n,γ*)	Energies in MeV and γ Intensities, %
	143 *m*		−78.8	64 s	....	....	γ, 0.748
	144	−88,011	−81.98	....	3.16	∼0.7	....
	145	−86,606	−80.67	340 d	....	∼100	γ, 0.061(13), 0.485; e^-, 0.016, 0.054
	146	−87,008	−81.05	7×10^7 y	$< 2 \times 10^{-7}$	....	α, 2.46
	147	−85,133	−79.30	1.05×10^{11} y	15.07	∼90	α, 2.23
	148	−85,209	−79.37	$> 2 \times 10^{14}$ y	11.27	....	....
	149	−82,820	−77.15	$> 1 \times 10^{15}$ y	13.82	41,500	....
	150	−82,724	−77.06		7.47	100	....
	151	−80,081	−74.59	∼87 y	....	15,000	$β^-$, 0.076; e^-, 0.014, 0.020; γ, 0.022(4)
	152	−80,244	−74.75	....	26.63	210	....
	153	−77,898	−72.56	46.8 h	....	....	$β^-$, 0.80; e^-, 0.022, 0.055, 0.062, 0.095, 0.101; γ, 0.070(5), 0.103(28), 0.41–0.64(1)
	154	−77,718	−72.39	....	22.53	5	....
	155	−75,299	−70.14	23.5 m	....	....	$β^-$, 1.53; e^-, 0.056, 0.097, 0.103; γ, 0.104(73), 0.246(4)
	156	−74,431	−69.33	9.4 h	....	....	$β^-$, 0.72; e^-, 0.014, 0.021, 0.030, 0.039; γ, 0.088(30), 0.166(10), 0.204(20), 0.25(5), 0.291(3)
	157			0.5 m	....	....	γ, 0.57
₆₃Eu	143			2.3 m	....	....	$β^+$, 4.0; γ, 0.511
	144		−75.66	10.5 s	....	....	$β^+$, 5.2; γ, 0.511
	145	−83,610	−77.9	5.9 d	....	....	γ, 0.53, 0.656(30), 0.766(10), 0.894(100), 1.66(16), 2.00(8); e^-, 0.063, 0.103, 0.847
	146	−82,862	−77.18	4.59 d	....	....	γ, 0.511, 0.634(77), 0.666(12), 0.71(13), 0.749(100), 0.90(8), 1.058(7), 1.16(6), 1.298(6), 1.408(5), 1.535(8); $β^+$,

Table 3-7 (*Continued*)
TABLE OF NUCLIDES

| Nuclide | | Mass Excess | | Half-life | Natural | Thermal Neutron | Major Radiations, |
Z Symbol	Mass Number	μmu	MeV	y = yr, d = day h = hr, m = min s = sec	Abundance, %	Absorption Cross Section σ, barns (mode if not n,γ)	Energies in MeV and γ Intensities, %
	147	−83,200	−77.5	21.5 d	....	....	2.11, 1.47; e⁻, 0.586, 0.702 γ, 0.122(20), 0.198(24), 0.600(7), 0.680(11), 0.800(6), 0.957(9), 1.079(9), 1.25(1); e⁻, 0.030, 0.075, 0.114, 0.151; α, 2.91
	148	−81,890	−76.26	54 d	....	....	γ, 0.413(18), 0.551(120), 0.62(90), 0.72(18), 0.872(7), 0.917(5), 0.967(5), 1.033(7), 1.16(5), 1.345(8), 1.62(11); e⁻, 0.02–0.04, 0.51, 0.193, 0.366, 0.505, 0.544, 0.584; β⁺, 0.92; α, 2.63
	149	−82,000	−76	106 d	....	....	γ, 0.277(10), 0.328(10); e⁻, 0.015, 0.021, 0.230, 0.281
	150	−80,311	−74.81	12.55 h	....	....	β⁻, 1.01; β⁺, 1.24; γ, 0.334(4), 0.406(3), 0.511, 0.619, 0.713, 0.831(1), 0.921, 1.165, 1.224, 1.630, 1.964
	150			~5 y	....	....	γ, 0.334(96), 0.439(86), 0.584(60), 0.74(21), 1.049(9), 1.248(5), 1.347(4); e⁻, 0.287, 0.327, 0.392
	151	−80,162	−74.67	....	47.77	5,900	....
	152	−78,251	−72.89	12.7 y	....	5,000	β⁻, 1.48; e⁻, 0.075,

Table 3-7 (*Continued*)
TABLE OF NUCLIDES

Nuclide		Mass Excess		Half-life y = yr, d = day h = hr, m = min s = sec	Natural Abundance, %	Thermal Neutron Absorption Cross Section σ, barns (mode if not n,γ)	Major Radiations, Energies in MeV and γ Intensities, %
Z Sym- bol	Mass Number	μmu	MeV				
	152 m_1		−72.84	9.3 h	. . . .	. . . .	0.115, 0.120; β^+, 0.71; γ, 0.122(37), 0.245(8), 0.344(27), 0.779(14), 0.965(15), 1.087(12), 1.113(14), 1.408(22) β^-, 1.88; e^-, 0.075, 0.115, 0.120; β^+, 0.89; γ, 0.122(8), 0.344(3), 0.842(13), 0.963(12), 1.315(1), 1.389(1)
	152 m_2		−72.74	96 m	. . . .	. . . .	γ, 0.090(74); e^-, 0.010, 0.016, 0.032, 0.039
	153	−78,758	−73.36	. . . .	52.23	320	. . . .
	154	−76,947	−71.68	16 y	. . . .	1,400	β^-, 1.85, 0.87; e^-, 0.073, 0.115, 0.122; γ, 0.123(38), 0.248(7), 0.593(6), 0.724(21), 0.759(5), 0.876(12), 1.00(31), 1.278(37)
	155	−77,070	−71.79	1.811 y	. . . .	13,000	β^-, 0.25; e^-, 0.011, 0.017, 0.036, 0.054, 0.078, 0.082; γ, 0.087(32), 0.105(20)
	156	−75,198	−70.05	15.4 d	. . . .	. . . .	β^-, 2.45; e^-, 0.039, 0.081, 0.087; γ, 0.089(8), 0.646(7), 0.723(6), 0.812(9), 1.07(11), 1.15(14), 1.24(16), 1.97(7), 2.098(3), 2.19(5)
	157	−74,610	−69.43	15.1 h	. . . .	. . . .	β^-, 1.3; e^-, 0.004, 0.014, 0.046,

Table 3-7 (*Continued*)
TABLE OF NUCLIDES

Nuclide		Mass Excess		Half-life y = yr, d = day h = hr, m = min s = sec	Natural Abundance, %	Thermal Neutron Absorption Cross Section σ, barns (mode if not n,γ)	Major Radiations, Energies in MeV and γ Intensities, %
Z Symbol	Mass Number	μmu	MeV				
	158	−72,060	−67.1	46 m	....	....	0.056; γ, 0.055(5), 0.064(27), 0.32(5), 0.37(14), 0.413(27), 0.477(5), 0.623(6) β⁻, 2.5; e⁻; γ, 0.080(100), 0.182, 0.52(25), 0.61(8), 0.95(95), 1.11(11), 1.19(16)
	159	−71,160	−66.02	18.1 m	....	....	β⁻, 2.6; γ, 0.07(42), 0.09(18), 0.15(14), 0.22(5), 0.67(21), 0.73(10), 0.8(11), 1.1(11), 1.5(5)
₆₄Gd	160	−69,000	−64	~2.5 m	....	....	β⁻, 3.6; no γ
	145			25 m	....	....	β⁺, 2.4; γ, 0.511, 0.80(9), 1.03(10), 1.75(100)
	146	−81,680	−76	50 d	....	....	γ, 0.078(30), 0.115(100), 0.155(45); e⁻, 0.066, 0.106
	147	−80,830	−75	35 h	....	....	γ, 0.229(150), 0.39(85), 0.64(70), 0.77(60), 0.932(60), 1.10(19); e⁻, 0.181, 0.221, 0.321, 0.348, 0.388
	148	−81,899	−76.29	84 y	....	....	α, 3.18
	149	−80,700	−75.2	9.5 d	....	....	γ, 0.150(48), 0.299(26), 0.347(25), 0.750(11), 0.790(10), 0.94(5); e⁻, 0.101, 0.142, 0.250, 0.298; α, 3.01
	150	−81,395	−75.82	2.1×10⁶ y	....	....	α, 2.73
	151	−79,730	−74	120 d	....	....	γ, 0.0216(3), 0.154(7), 0.175(3), 0.244(7),

Table 3-7 (*Continued*)
TABLE OF NUCLIDES

Nuclide		Mass Excess		Half-life y = yr, d = day h = hr, m = min s = sec	Natural Abundance, %	Thermal Neutron Absorption Cross Section σ, barns (mode if not n,γ)	Major Radiations, Energies in MeV and γ Intensities, %
Z Symbol	Mass Number	μmu	MeV				
							0.308(1); e⁻, 0.014, 0.020, 0.105, 0.127, 0.167; α, 2.60
	152	−80,206	−74.71	1.1×10¹⁴ y	0.20	<180	α, 2.1
	153	−78,497	−73.12	242 d	....	....	γ, 0.070(2), 0.099(55); e⁻, 0.021, 0.049, 0.065, 0.101
	154	−79,071	−73.65	....	2.15	....	....
	155	−77,336	−72.04	....	14.7	58,000	....
	156	−77,825	−72.49	....	20.47	....	....
	157	−75,975	−70.77	....	15.68	2.4×10⁵	....
	158	−75,822	−70.63	....	24.9	3.4	....
	159	−73,632	−68.59	18.0 h	....	....	β⁻, 0.95; e⁻, 0.006, 0.049, 0.056; γ, 0.058(3), 0.363(9)
	160	−72,885	−67.89	....	21.9	0.8	....
	161	−70,280	−65.5	3.6 m	....	....	β⁻, 1.6; e⁻, 0.005, 0.026, 0.049, 0.055, 0.263, 0.309; γ, 0.102(11), 0.284(8), 0.315(25), 0.361(66)
₆₅Tb	147			24 m	....	....	γ, 0.305, 0.511
	148	−75,870	−70.7	70 m	....	....	β⁺, 4.6; γ, 0.511, 0.78, 1.12
	149	−76,650	−71.4	4.10 h	....	....	γ, 0.16, 0.35; e⁻, 0.115, 0.127, 0.157, 0.301, 0.338, 0.587; α, 3.95
	149 *m*			4.3 m	....	....	α, 3.99
	150	−76,252	−71.03	3.1 h	....	....	β⁺, 3.6; γ, 0.511, 0.637(100), 0.93(35)
	151	−76,850	−71.6	18 h	....	....	γ, 0.108(35), 0.18(18), 0.252(35), 0.288(32), 0.40, 0.44, 0.48, 0.60, 0.72, 0.87; e⁻, 0.058, 0.100, 0.130, 0.202, 0.237; α, 3.42
	152	−75,720	−70.5	17.4 h	....	....	β⁺, 2.82; e⁻, 0.221, 0.263, 0.294, 0.336, 0.382,

Table 3-7 (*Continued*)
TABLE OF NUCLIDES

Nuclide		Mass Excess		Half-life	Natural Abundance, %	Thermal Neutron Absorption Cross Section σ, barns (mode if not n,γ)	Major Radiations, Energies in MeV and γ Intensities, %
Z Sym-bol	Mass Number	μmu	MeV	y = yr, d = day h = hr, m = min s = sec			
							0.536, 0.565, 0.607; γ, 0.271(13), 0.344(100), 0.411(6), 0.586(14), 0.779(14), 0.974(10), 1.12(10), 1.31(11), 1.60(7), 1.95(8), 2.40(9), 2.70(6)
	152			4.0 m	....	....	γ, 0.14, 0.23, 0.511
	153	−76,510	−71	55 h	....	....	γ, 0.083(11), 0.11(12), 0.17(9), 0.212(30), 0.250, 0.33, 0.88; e⁻, 0.012, 0.034, 0.037, 0.040, 0.044, 0.052, 0.057, 0.162
	154	−75,420	−70	21.0 h	......	....	γ, 0.123, 0.248, 0.30, 0.347, 0.53, 0.65; e⁻, 0.073, 0.115, 0.122, 0.198
	154		−70	8.5 h	....	....	γ, 0.123, 0.248, 0.53, 0.65; e⁻, 0.073, 0.115, 0.122, 0.198
	155	−76,370	−71	5.6 d	....	....	γ, 0.087(37), 0.105(25), 0.163(8), 0.180(8), 0.262(7), 0.368(4); e⁻, 0.011, 0.034, 0.053, 0.078, 0.110, 0.129, 0.210
	156	−75,250	−70	5.1 d	....	....	γ, 0.089(17), 0.199(40), 0.356(13), 0.535(70), 1.065(12), 1.16(17), 1.22(29), 1.42(15), 1.65(5), 1.85(4); e⁻,

Table 3-7 (*Continued*)
TABLE OF NUCLIDES

Nuclide		Mass Excess		Half-life y=yr, d=day h=hr, m=min s=sec	Natural Abundance, %	Thermal Neutron Absorption Cross Section σ, barns (mode if not n,γ)	Major Radiations, Energies in MeV and γ Intensities, %
Z Sym-bol	Mass Number	μmu	MeV				
	156 *m*		−70	5.5 h	....	....	0.039, 0.081, 0.087, 0.149 γ; e⁻, 0.036, 0.081
	157	−75,910	−70.71	1.5×10² y	....	....	Gd X-rays
	158	−74,536	−69.43	1.2×10³ y	....	....	β⁻, 0.85; e⁻, 0.029, 0.044, 0.072, 0.078, 0.092, 0.132; γ, 0.080(12), 0.182(10), 0.782(10), 0.95(69), 1.110(2), 1.190(2)
	158 *m*		−69.32	10.5 s	....	....	e⁻, 0.060, 0.102; γ, 0.110(1)
	159	−74,649	−69.53	>5×10¹⁶ y	100	46	....
	160	−72,854	−67.85	72.1 d	....	525	β⁻, 1.74, 0.86; e⁻, 0.033, 0.079, 0.085; γ, 0.087(12), 0.197(6), 0.299(30), 0.879(31), 0.966(31), 1.178(15), 1.272(7)
	161	−72,428	−67.47	6.9 d	....	....	β⁻, 0.59, 0.52; e⁻, 0.017, 0.040, 0.048; γ, 0.026(21), 0.049(19), 0.057(5), 0.075(10)
	162	−70,190	−65	7.48 m	....	....	γ, 0.040(17), 0.081(8), 0.140(6), 0.180(26), 0.258(100), 0.81(44), 0.89(54)
	162		−65	2.24 h	....	....	....
	163	−69,440	−64.7	6.5 h	....	....	β⁻, 1.65; γ, 0.025, 0.235, 0.330, 0.510
₆₆Dy	164	−66,720	−62	23 h	....	....	....
	149			10–20 m	....	....	....
	150	−74,410	−69	7.2 m	....	....	γ, 0.39, 0.511; α, 4.23
	151	−73,750	−69	18.0 m	....	....	α, 4.06; γ, 0.145, 0.511

Table 3-7 (*Continued*)
TABLE OF NUCLIDES

Nuclide		Mass Excess		Half-life	Natural	Thermal Neutron	Major Radiations,
Z Symbol	Mass Number	μmu	MeV	y=yr, d=day h=hr, m=min s=sec	Abundance, %	Absorption Cross Section σ, barns (mode if not n,γ)	Energies in MeV and γ Intensities, %
	152	−75,271	−70.11	2.41 h	....	....	γ, 0.257; α, 3.65
	153	−74,260	−69.2	6.4 h	....	....	γ, 0.08; e⁻, 0.029, 0.047, 0.072, 0.091, 0.192, 0.202; α, 3.48
	154	−75,650	−70.5	>10 y	....	....	α, 2.85
	154 m			13 h	....	....	α, 3.37
	155	−74,120	−69	10.2 h	....	....	γ, 0.227(68), 0.52(8), 0.65(5), 0.74(4), 0.91(5), 1.000(6), 1.091(5), 1.16(6), 1.250(4), 1.39(3), 1.45(4), 1.66(2); β⁺, 1.08, 0.85; e⁻, 0.013, 0.038, 0.057, 0.175
	156	−76,070	−70.9	>1×10¹⁸ y	0.0524	~3	....
	157	−74,730	−70	8.1 h	....	....	γ, 0.326(91); e⁻, 0.009, 0.031, 0.052, 0.074, 0.274
	158	−75,551	−70.37	....	0.0902	100	....
	159	−74,241	−69.15	144 d	....	....	γ, 0.058(4), 0.348; e⁻, 0.006, 0.049, 0.056
	160	−74,798	−69.67	....	2.294	....	....
	161	−73,055	−68.05	....	18.88	600	....
	162	−73,197	−68.18	....	25.53	140	....
	163	−71,245	−66.36	....	24.97	130	....
	164	−70,800	−65.95	....	28.18	2,000	....
	165	−68,184	−63.51	139.2 m	....	4,700	β⁻, 1.29; e⁻, 0.039, 0.085; γ, 0.095(4), 0.280(1), 0.361(1), 0.633(1), 0.716(1)
	165 m		−63.40	1.26 m	....	....	β⁻, 1.04, 0.89; e⁻, 0.054, 0.100, 0.106; γ, 0.108(3), 0.152, 0.362(1), 0.514(2)
	166	−67,193	−62.59	81.5 h	....	....	β⁻, 0.48, 0.40; e⁻, 0.019, 0.027, 0.046; γ, 0.082(12), 0.372(1), 0.426(1)
	167			4.4 m	....	....	....
₆₇Ho	150			~20 s	....	....	....
	151			35.6 s	....	....	α, 4.51

Table 3-7 (*Continued*)
TABLE OF NUCLIDES

| Nuclide | | Mass Excess | | Half-life | Natural | Thermal Neutron | Major Radiations, |
| | | | | y = yr, d = day | Abundance, | Absorption Cross | Energies in MeV |
Z Symbol	Mass Number	μmu	MeV	h = hr, m = min s = sec	%	Section σ, barns (mode if not n,γ)	and γ Intensities, %
	151			42 s	....	....	α, 4.60
	152			52.3 s	....	....	α, 4.45
	152	−68,440	−63.8	2.4 m	....	....	α, 4.38
	153	−69,730	−65.0	9 m	....	....	α, 3.92
	154	−69,740	−65	7 m	....	....	γ, 0.335, 0.511
	155			50 m	....	....	β+, 2.1; γ, 0.092, 0.138, 0.511
	156			55 m	....	....	γ, 0.138(100), 0.266(99), 0.367(23), 0.511, 0.685, 0.89, 1.20, 1.41; e−, 0.084, 0.130, 0.213; β+, 2.9, 1.8
	157			14 m	....	....	γ, 0.087, 0.152, 0.190, 0.227, 0.511, 0.71, 0.86, 0.90, 1.20
	158	−71,210	−66.33	11.5 m	....	....	γ, 0.099, 0.218, 0.329, 0.412, 0.52, 0.647, 0.73, 0.86, 0.940, 1.21, 1.47, 1.6, 1.8, 2.05, 2.21, 2.87, 3.1; e−, 0.045, 0.062, 0.091, 0.097, 0.164
	158 m		−66.26	29 m	....	....	γ, 0.099, 0.218, 0.32, 0.356, 0.412, 0.46, 0.52, 0.63, 0.73, 0.85, 0.95, 1.21, 1.47, 1.60, 1.80, 2.06, 2.20, 2.62; e−, 0.029, 0.044, 0.072, 0.078, 0.092, 0.132; β+, 1.32
	159	−72,310	−67	33 m	....	....	γ, 0.057, 0.080, 0.13, 0.18, 0.253, 0.309; e−, 0.048, 0.071, 0.121, 0.198, 0.243, 0.256, 0.300
	159 m		−67	6.9 s	....	....	γ, 0.206; e−, 0.150, 0.197
	160	−71,260	−66.4	25.6 m	....	....	see 160mHo
	160 m		−66.3	5.0 h	....	....	γ, 0.087(14), 0.197(20),

Table 3-7 (*Continued*)
TABLE OF NUCLIDES

Nuclide		Mass Excess		Half-life	Natural	Thermal Neutron	Major Radiations,
Z Symbol	Mass Number	μmu	MeV	y = yr, d = day h = hr, m = min s = sec	Abundance, %	Absorption Cross Section σ, barns (mode if not n,γ)	Energies in MeV and γ Intensities, %
	161	−72,200	−67	2.4 h	....	....	0.539(5), 0.646(20), 0.729(50), 0.880(26), 0.965(37); e^-, 0.033, 0.051, 0.058, 0.079, 0.085, 0.144, 0.188; β^+, 1.9 γ, 0.026(23), 0.075(15), 0.157(1), 0.176(2); e^-, 0.017, 0.024, 0.049, 0.069, 0.076
	161 m		−67	6.1 s	....	....	γ, 0.211(53); e^-, 0.155, 0.202
	162	−70,878	−66.02	15 m	....	....	γ, 0.081(8), 0.511; β^+, 1.10; e^-, 0.27, 0.072, 0.079
	162 m		−65.92	68 m	....	....	γ, 0.081(10), 0.185(26), 0.283(12), 0.940(13), 1.224(24); e^-, 0.027, 0.036, 0.048, 0.072, 0.079, 0.131, 0.177
	163	−71,234	−66.35	>10^3 y	....	....	....
	163 m		−66.05	1.1 s	....	....	γ, 0.305; e^-, 0.249, 0.296
	164	−69,610	−64.84	36.7 m	....	....	β^-, 0.99; e^-, 0.019, 0.034, 0.065, 0.071, 0.083, 0.089; γ, 0.073, 0.091
	165	−69,579	−64.81	>6×10^{16} y	100	64	....
	166	−67,711	−63.07	26.9 h	....	....	β^-, 1.84; e^-, 0.023, 0.072, 0.078; γ, 0.081(5), 1.380(1), 1.582, 1.663
	166 m		−63.06	1.2×10^3 y	....	....	β^-; e^-, 0.023, 0.072, 0.078, 0.127, 0.175; γ, 0.081(12), 0.184(90),

Table 3-7 (*Continued*)
TABLE OF NUCLIDES

Nuclide Z Symbol	Mass Number	Mass Excess μmu	Mass Excess MeV	Half-life y=yr, d=day h=hr, m=min s=sec	Natural Abundance, %	Thermal Neutron Absorption Cross Section σ, barns (mode if not n,γ)	Major Radiations, Energies in MeV and γ Intensities, %
₆₈Er	167	−66,870	−62.3	3.1 h	....	....	0.280(30), 0.412(12), 0.532(12), 0.711(58), 0.810(60), 0.830(11) β⁻, 0.96; e⁻, 0.024, 0.048, 0.073, 0.150, 0.180, 0.199, 0.263; γ
	168	−64,070	−59.7	3.3 m	....	....	β⁻, 2.2; γ, 0.85
	169	−63,140	−58.8	4.8 m	....	....	β⁻, 1.95; γ, 0.15, 0.68, 0.76, 0.84, 0.92
	170	−59,930	−55.8	45 s	....	....	β⁻, 3.1; γ, 0.43
	152			10.7 s	....	....	α, 4.80
	153			36 s	....	....	α, 4.67
	154	−67,240	−63	5 m	....	....	α, 4.15
	157			~25 m	....	....	γ, 0.117, 0.386, 0.511, 1.32, 1.66, 1.82, 2.0
	158			2.3 h	....	....	γ, 0.072, 0.250, 0.315, 0.387, 0.511, 0.875, 0.906, 0.978; e⁻, 0.058, 0.065; β⁺, 0.8
	159			36 m	....	....	γ, 0.206, 0.37, 0.511, 0.62, 0.84, 1.20, 1.40, 1.80, 2.60; e⁻, 0.150, 0.197
	160			29.4 h	....	....	Ho X-rays
	161	−70,050	−65	3.1 h	....	....	γ, 0.211(9), 0.305(3), 0.592(8), 0.826(63), 1.17(8), 1.37(5), 1.66(2); e⁻, 0.059, 0.065, 0.155, 0.202; β⁺, 1.2
	162	−71,260	−66.4	....	0.136	2	....
	163	−69,935	−65.14	75.1 m	....	....	γ, 0.43, 1.10; β⁺, 0.19
	164	−70,713	−65.87	....	1.56	1.7	....
	165	−69,181	−64.44	10.34 h	....	....	Ho X-rays
	166	−69,693	−64.92	....	33.41	12	....
	167	−67,940	−63.29	....	22.94	700	....

Table 3-7 (*Continued*)
TABLE OF NUCLIDES

Nuclide		Mass Excess		Half-life y = yr, d = day h = hr, m = min s = sec	Natural Abundance, %	Thermal Neutron Absorption Cross Section σ, barns (mode if not n,γ)	Major Radiations, Energies in MeV and γ Intensities, %
Z Sym- bol	Mass Number	μmu	MeV				
	167 *m*		−63.08	2.3 s	....	....	γ, 0.208(43); e⁻, 0.150, 0.199
	168	−67,617	−62.98	....	27.07	2	....
	169	−65,390	−60.91	9.6 d	....	....	β⁻, 0.34; e⁻, 0.006; γ, 0.008
	170	−64,440	−60.0	....	14.88	9	....
	171	−61,870	−57.6	7.52 h	....	....	β⁻, 1.49, 1.06; e⁻, 0.004, 0.052, 0.065, 0.102, 0.115; γ, 0.112(25), 0.124(9), 0.296(28), 0.308(63)
	172	−60,670	−56.5	49.5 h	....	....	β⁻, 0.89, 0.37; e⁻, 0.010, 0.020, 0.049, 0.058, 0.348; γ, 0.407(40), 0.610(40)
₆₉Tm	153			1.6 s	....	....	α, 5.10
	154			3.0 s	....	....	α, 5.04
	154			5 s	....	....	α, 4.96
	161	−66,270	−62	32 m	....	....	γ, 0.084, 0.106, 0.112, 0.145, 0.172; e⁻, 0.027, 0.036, 0.050, 0.055, 0.065, 0.075, 0.089, 0.115
	162	−66,010	−61.5	77 m	....	....	γ, 0.102(20), 0.236(10)
	162		−61.5	22 m	....	....	β⁺, 3.82; e⁻, 0.045, 0.093, 0.100
	163	−67,498	−62.87	1.8 h	....	....	γ, 0.104(8), 0.17(1), 0.240(5), 0.29(3), 0.34(3); e⁻, 0.047, 0.095, 0.184; β⁺, 1.1
	164	−66,459	−61.91	2.0 m	....	....	γ, 0.091(4), 0.356, 0.361, 0.391, 0.511, 0.773, 0.862, 0.907, 0.930; β⁺, 2.94; e⁻, 0.034, 0.083, 0.089
	165	−67,460	−62.87	30.1 h	....	....	γ, 0.054, 0.113, 0.243(50), 0.297(35),

Table 3-7 (*Continued*)
TABLE OF NUCLIDES

Nuclide		Mass Excess		Half-life	Natural	Thermal Neutron	Major Radiations,
Z Symbol	Mass Number	μmu	MeV	y = yr, d = day h = hr, m = min s = sec	Abundance, %	Absorption Cross Section σ, barns (mode if not n,γ)	Energies in MeV and γ Intensities, %
	166	−66,490	−61.88	7.7 h	....	....	0.34(10), 0.44(5), 0.70(2), 0.807(15), 1.13(5), 1.30(1); e⁻, 0.038, 0.045, 0.052, 0.056, 0.068, 0.161, 0.185, 0.233, 0.240; β⁺, 0.30 β⁺, 1.94; e⁻, 0.023, 0.072, 0.079, 0.127; γ, 0.081, 0.19, 0.215, 0.46, 0.60, 0.69, 0.78, 1.180, 1.277, 1.378, 1.873, 2.06
	167	−66,970	−62.13	9.6 d	....	....	γ, 0.057(4), 0.208(43), 0.532(2); e⁻, 0.048, 0.150, 0.199
	168	−65,770	−61.27	85 d	....	....	γ, 0.080(11), 0.19(77), 0.448(27), 0.63(14), 0.73(40), 0.82(88), 0.917(4), 1.280(3); e⁻, 0.022, 0.071, 0.077, 0.127, 0.141
	169	−65,755	−61.25	>5×10¹⁶ y	100	125	....
	170	−63,940	−59.6	134 d	....	150	β⁻, 0.97; e⁻, 0.023, 0.075, 0.082; γ, 0.084(3)
	171	−63,470	−59.1	1.92 y	....	....	β⁻, 0.097; e⁻, 0.057, 0.065; γ, 0.067
	172	−61,620	−57.4	63.6 h	....	....	β⁻, 1.88; γ, 0.079(5), 0.181(2), 0.91(1), 1.09(7), 1.39(7), 1.46(7), 1.53(6), 1.61(5)
	173	−60,520	−56.4	8.2 h	....	....	β⁻, 1.3, 0.89; e⁻, 0.008, 0.056, 0.064; γ, 0.066(1), 0.399(89),

Table 3-7 (*Continued*)
TABLE OF NUCLIDES

Nuclide		Mass Excess		Half-life y=yr, d=day h=hr, m=min s=sec	Natural Abundance, %	Thermal Neutron Absorption Cross Section σ, barns (mode if not n,γ)	Major Radiations, Energies in MeV and γ Intensities, %
Z Symbol	Mass Number	μmu	MeV				
	174		−54.6	5.5 m	....	....	0.465(8) β⁻, 2.5; no γ
	174	−58,030	−54.1	5.2 m	....	....	β⁻, 1.2; e⁻, 0.015, 0.067, 0.074; γ, 0.176(67), 0.273(85), 0.366(93), 0.50(15), 0.99(89)
	175	−56,170	−52.3	20 m	....	....	β⁻, 2.0; γ, 0.51
	176	−52,810	−49.2	1.5 m	....	....	β⁻, 4.2; no γ
₇₀Yb	154			0.39 s	....	....	α, 5.33
	155			1.6 s	....	....	α, 5.21
	162			~24 m	....	....	e⁻, 0.032, 0.039
	164			75 m	....	....	Tm X-rays
	165	−64,560	−60	10.5 m	....	....	....
	166	−66,150	−61.6	57.5 h	....	....	γ, 0.082(17); e⁻, 0.023, 0.072
	167	−64,870	−60.17	17.7 m	....	....	γ, 0.113(90), 0.176(15); e⁻, 0.047, 0.055, 0.096
	168	−65,840	−61.3	....	0.140	11,000	....
	169	−64,470	−60	31.8 d	....	....	γ, 0.063(45), 0.110(18), 0.131(11), 0.177(22), 0.198(35), 0.308(10); e⁻, 0.004–0.011, 0.034, 0.050, 0.053, 0.071, 0.100, 0.118, 0.121, 0.139
	169 m		−60	46 s	....	....	e⁻, 0.014, 0.022
	170	−64,980	−60.5	....	3.03	....	....
	171	−63,570	−59.2	....	14.31	....	....
	171 m		−59.1	<<8 d	....	....	γ, 0.019, 0.076; e⁻, 0.010, 0.017, 0.067, 0.074
	172	−63,640	−59.3	....	21.82	....	....
	173	−61,940	−57.7	....	16.13	....	....
	174	−61,260	−57.1	....	31.84	46	....
	175	−58,860	−54.8	101 h	....	....	β⁻, 0.466; γ, 0.114(2), 0.283(4), 0.396(6); e⁻, 0.051, 0.102, 0.112, 0.333
	176	−57,320	−53.4	....	12.73	7	....

Table 3-7 (*Continued*)
TABLE OF NUCLIDES

| Nuclide | | Mass Excess | | Half-life | Natural | Thermal Neutron | Major Radiations, |
Z Symbol	Mass Number	μmu	MeV	y = yr, d = day h = hr, m = min s = sec	Abundance, %	Absorption Cross Section σ, barns (mode if not n,γ)	Energies in MeV and γ Intensities, %
	176 m		−52.4	11.7 s	....	....	γ, 0.19, 0.29, 0.39
	177	−54,590	−50.8	1.9 h	....	....	β⁻, 1.40; γ, 0.122(3), 0.151(16), 1.080(5), 1.241(3); e⁻, 0.059, 0.075, 0.088, 0.110, 0.140
	177 m		−50.5	6.5 s	....	....	γ, 0.104(65), 0.228(13); e⁻, 0.043, 0.094, 0.167, 0.219
	156			0.23 s	....	....	α, 5.54
	156			0.5 s	....	....	α, 5.43
	167	−61,610	−57.1	54 m	....	....	γ, 0.030, 0.18–0.24, 0.278, 0.372, 0.402, 0.511; e⁻, 0.020, 0.028, 0.039, 0.069, 0.076, 0.152, 0.178; β⁺, 1.5
	168	−60,910	−57	7.1 m	....	....	γ, 0.087(7), 0.223, 0.71, 0.90(10), 0.99(13), 1.41, 1.81, 2.1; e⁻; β⁺, 1.2
	169	−62,040	−58	34 h	....	....	γ, 0.063, 0.111, 0.191, 0.577; e⁻, 0.010, 0.014, 0.022, 0.026, 0.050, 0.053, 0.060, 0.066, 0.077; β⁺, 1.2
	169 m		−58	2.7 m	....	....	e⁻, 0.019, 0.027
	170	−61,170	−57.1	2.05 d	....	....	γ, 0.084(13), 0.193, 0.24, 1.01, 1.03, 1.17, 1.27, 1.41, 2.03, 2.32, 2.67, 2.89, 3.09; e⁻, 0.023, 0.075, 0.082; β⁺, 2.4
	170 m		−57.0	0.7 s	....	....	e⁻, 0.036, 0.044
	171	−61,860	−58	8.3 d	....	....	γ, 0.019(20), 0.075(8), 0.668(14), 0.741(68), 0.842(7); e⁻, 0.010, 0.017,

Table 3-7 (*Continued*)
TABLE OF NUCLIDES

| Nuclide | | Mass Excess | | Half-life | Natural | Thermal Neutron | Major Radiations, |
Z Symbol	Mass Number	μmu	MeV	y = yr, d = day h = hr, m = min s = sec	Abundance, %	Absorption Cross Section σ, barns (mode if not n,γ)	Energies in MeV and γ Intensities, %
171 *m*			−58	76 s	....	....	0.057, 0.066, 0.074 γ, 0.071; *e*⁻, 0.061, 0.069
172		−60,740	−57	6.70 d	....	....	γ, 0.079(13), 0.182(26), 0.81(21), 0.90(45), 1.09(60); *e*⁻, 0.017, 0.029, 0.069, 0.077, 0.081, 0.120
172 *m*			−57	3.7 m	....	....	*e*⁻, 0.032, 0.040
173		−61,200	−57.0	1.37 y	....	....	γ, 0.079(14), 0.101(7), 0.17(5), 0.272(18), 0.637(2); *e*⁻, 0.017, 0.039, 0.068, 0.077, 0.090
174		−59,650	−55.6	3.6 y	....	....	γ, 0.076(6), 1.24(9); *e*⁻, 0.015, 0.067, 0.074
174 *m*			−55.4	140 d	....	....	γ, 0.067, 0.176, 0.273, 0.994; *e*⁻, 0.004, 0.034, 0.050, 0.057
175		−59,360	−55.3	>1×10¹⁷ y	97.40	18	....
176		−57,340	−53.4	2.2×10¹⁰ y	2.60	....	β⁻, 0.43; *e*⁻, 0.023, 0.078, 0.086, 0.137; γ, 0.088(15), 0.202(85), 0.306(95)
176 *m*			−53.1	3.69 h	....	....	β⁻, 1.31; *e*⁻, 0.023, 0.078, 0.086; γ, 0.088(10)
177		−56,070	−52.2	6.74 d	....	....	β⁻, 0.497; γ, 0.113(3), 0.208(6); *e*⁻, 0.048, 0.103, 0.111, 0.143
177 *m*			−51.3	155 d	....	....	γ, 0.105(13), 0.113(23), 0.128(17), 0.153(17), 0.174(13), 0.208(62), 0.228(37),

Table 3-7 (*Continued*)
TABLE OF NUCLIDES

Nuclide		Mass Excess		Half-life y = yr, d = day h = hr, m = min s = sec	Natural Abundance, %	Thermal Neutron Absorption Cross Section σ, barns (mode if not n,γ)	Major Radiations, Energies in MeV and γ Intensities, %
Z Sym- bol	Mass Number	μmu	MeV				
	178	−53,700	−50.0	30 m	....	....	0.281(14), 0.319(10), 0.327(18), 0.378(29), 0.414(17), 0.418(21); β^-; e^-, 0–0.47 β^-, 2.25; no γ
	178		−49.6	22.0 m	....	....	β^-, 1.50; e^-, 0.023, 0.028, 0.077, 0.083, 0.091, 0.148, 0.204; γ, 0.089, 0.214, 0.326, 0.427
	178			5 m	....	....	β^-, 2.25; γ, 0.090, 0.22, 0.33, 0.43
	179	−52,530	−48.9	4.6 h	....	....	β^-, 1.35; γ, 0.213
	180	−49,630	−46.2	2.5 m	....	....	β^-, 3.3; no γ
$_{72}$Hf	157			0.12 s	....	....	α, 5.68
	158			3 s	....	....	α, 5.27
	168			22 m	....	....	γ, 0.129, 0.17
	169			1.5 h	....	....	γ, 0.115; β^+, 1.3
	170			12.2 h	....	....	γ, 0.120, 0.165, 0.99, 1.28, 1.65, 2.03, 2.36, 2.52, 2.94; e^-, 0.035, 0.057, 0.102, 0.145
	171			10.7 h	....	....	γ, 0.122, 0.188, 0.29, 0.34, 0.47, 0.66, 0.86, 1.07
	172			5 y	....	....	γ, 0.024(22), 0.082(10), 0.125(21); e^-, 0.014, 0.018, 0.032, 0.040, 0.063
	173			23.6 h	....	....	γ, 0.13(96), 0.162(5), 0.30(52), 0.55(1), 0.898(2), 1.04(1), 1.20; e^-, 0.060, 0.072, 0.076, 0.113, 0.127
	174	−59,640	−55.6	2.0×10^{15} y	0.163	400	....
	175	−58,390	−54.7	70 d	....	....	γ, 0.089(3), 0.343(85), 0.433(1); e^-, 0.026, 0.079,

Table 3-7 (*Continued*)
TABLE OF NUCLIDES

Nuclide		Mass Excess		Half-life	Natural Abundance, %	Thermal Neutron Absorption Cross Section σ, barns (mode if not n,γ)	Major Radiations, Energies in MeV and γ Intensities, %
Z Symbol	Mass Number	μmu	MeV	y = yr, d = day h = hr, m = min s = sec			
							0.280, 0.333
	176	−58,430	−54.4	....	5.21	<30	....
	177	−56,600	−52.7	....	18.56	370	....
	177 *m*		−51.4	1.1 s	....	....	γ, 0.105(17), 0.113(30), 0.128(21), 0.153(22), 0.174(16), 0.208(81), 0.228(48), 0.281(18), 0.327(23), 0.378(37), 0.418(27); *e*⁻, 0–0.47
	178	−56,120	−52.3	....	27.1	50	....
	178 *m*		−51.1	4.3 s	....	....	γ, 0.089(54), 0.093(14), 0.214(75), 0.326(94), 0.427(97); *e*⁻, 0.023, 0.028, 0.077, 0.083, 0.091, 0.148, 0.204
	179	−53,970	−50.3	....	13.75	65	....
	179 *m*		−49.9	18.6 s	....	....	γ, 0.217(94); *e*⁻, 0.096, 0.150
	180	−53,180	−49.5	....	35.22	10	....
	180 *m*		−48.4	5.5 h	....	....	γ, 0.058(48), 0.093(16), 0.215(82), 0.333(93), 0.444(80), 0.501(17); *e*⁻, 0.028, 0.047, 0.055, 0.083, 0.091, 0.150, 0.206, 0.267
	181	−50,895	−47.41	42.5 d	....	~40	β⁻, 0.41; *e*⁻, 0.066, 0.069, 0.122, 0.415; γ, 0.133(48), 0.346(13), 0.482(81)
	182	−49,300	−45.8	9×10⁶ y	....	....	β⁻; γ, 0.271(84)
	183	−46,170	−43.0	65 m	....	....	β⁻, 1.6; γ, 0.46(58), 0.82(100)
₇₃Ta	172			44 m	....	....	γ, 0.092, 0.208,

Table 3-7 (*Continued*)
TABLE OF NUCLIDES

Nuclide		Mass Excess		Half-life	Natural	Thermal Neutron	Major Radiations,
Z Symbol	Mass Number	μmu	MeV	y = yr, d = day h = hr, m = min s = sec	Abundance, %	Absorption Cross Section σ, barns (mode if not n.γ)	Energies in MeV and γ Intensities, %
	173			3.7 h	....	....	0.511 γ, 0.090, 0.170, 0.64, 1.00; e^-, 0.059, 0.069, 0.095, 0.107, 0.161
	174			1.2 h	....	....	γ, 0.091, 0.125, 0.160, 0.205, 0.280, 0.350, 0.511; e^-, 0.026, 0.081, 0.089
	175			10.5 h	....	....	γ, 0.08, 0.13, 0.21, 0.27, 0.35, 0.45, 0.60, 0.83, 1.2, 1.4, 1.7; e^-, 0.016, 0.039, 0.061, 0.070, 0.116, 0.202
	176		−51	8.0 h	....	....	γ, 0.088, 0.202; e^-, 0.023, 0.078, 0.086, 0.137
	177	−55,350	−51.6	56.6 h	....	....	γ, 0.113(6), 0.208(1), 0.425, 0.509, 0.746, 1.058; e^-, 0.048, 0.102, 0.111
	178	−54,070	−50.4	9.35 m	....	....	γ, 0.093(100), 0.511, 1.10(11), 1.18(4), 1.35(46), 1.45(9); $β^+$, 0.89; e^-, 0.028, 0.082
	178			2.1 h	....	....	γ, 0.089(54), 0.093(14), 0.214(75), 0.328(120), 0.427(97); e^-, 0.023, 0.028, 0.077, 0.083, 0.091, 0.148, 0.204
	179	−53,840	−50.2	∼600 d	....	....	Hf X-rays
	180	−52,456	−48.86	>1×10¹² y	0.0123	....	....
	180 *m*		−48.65	8.15 h	....	....	$β^-$, 0.71; e^-, 0.028, 0.083, 0.091; γ, 0.093(4), 0.103(1)
	181	−51,993	−48.43	....	99.9877	21	....
	182	−49,833	−46.35	115.1 d	....	8,000	$β^-$, 1.71, 0.522; e^-, 0.030, 0.044, 0.054, 0.073,

Table 3-7 (*Continued*)
TABLE OF NUCLIDES

Nuclide		Mass Excess		Half-life	Natural	Thermal Neutron	Major Radiations,
Z Symbol	Mass Number	μmu	MeV	y = yr, d = day h = hr, m = min s = sec	Abundance, %	Absorption Cross Section σ, barns (mode if not n,γ)	Energies in MeV and γ Intensities, %
182 *m*			−45.84	16.5 m	....	....	0.089, 0.110; γ, 0.068(42), 0.100(14), 0.152(7), 0.222(8), 1.122(34), 1.189(16), 1.222(27), 1.231(13) γ, 0.147(40), 0.172(40), 0.184(20), 0.319(5), 0.356; e^-, 0.080, 0.105, 0.117, 0.173
183		−48,530	−45.20	5.0 d	....	....	β^-, 0.62; γ, 0.046(5), 0.053(5), 0.099(7), 0.108(11), 0.161(17), 0.246(33), 0.30(11), 0.354(11); e^-, 0.034–0.043, 0.050, 0.073, 0.088, 0.093, 0.177
184		−46,020	−42.9	8.7 h	....	....	β^-, 2.64, 1.76, 1.19; e^-; γ, 0.111(21), 0.16(7), 0.21(7), 0.25(42), 0.30(24), 0.41(71), 0.53(19), 0.79(16), 0.90(49), 0.95(15), 1.16(12)
185		−44,440	−41.3	50 m	....	....	β^-, 1.7; γ, 0.075(5), 0.100(6), 0.175(60), 0.245(5)
186		−41,590	−38.7	10.5 m	....	....	β^-, 2.2; γ, 0.123(18), 0.20(74), 0.30(18), 0.41(15), 0.51(33), 0.61(33),

Table 3-7 (*Continued*)
TABLE OF NUCLIDES

| Nuclide | | Mass Excess | | Half-life | Natural | Thermal Neutron | Major Radiations, |
Z Symbol	Mass Number	μmu	MeV	y = yr, d = day h = hr, m = min s = sec	Abundance, %	Absorption Cross Section σ, barns (mode if not n,γ)	Energies in MeV and γ Intensities, %
$_{74}$W	173			16.5 m	. . . .	. . . .	0.73(48), 0.94(11) e^-
	174			31 m	. . . .	. . . .	. . . .
	175			34 m	. . . .	. . . .	γ, 0.26, 0.80, 1.3, 1.6
	176		−50	2.3 h	. . . .	. . . .	γ, 0.034, 0.100; e^-, 0.017, 0.023, 0.027, 0.033, 0.050, 0.083
	177		−50	135 m	. . . .	. . . .	γ, 0.20, 0.42, 0.62, 0.83, 1.00; e^-, 0.020, 0.028, 0.048, 0.059, 0.068, 0.075, 0.088, 0.119, 0.360
	178		−50	21.5 d	. . . .	. . . .	Ta X-rays
	179		−49	37.5 m	. . . .	. . . .	γ, 0.031(22); e^-, 0.020, 0.029
	179 *m*		−49	5.2 m	. . . .	. . . .	γ, 0.222; e^-, 0.152, 0.211
	180	−53,000	−49.37	>1.1×10^{15} y	0.135	<20	. . . .
	181	−51,789	−48.24	140 d	. . . .	. . . .	γ, 0.006(1), 0.136, 0.152; e^-, 0.004, 0.006
	182	−51,699	−48.16	>2×10^{17} y	26.4	20	. . . .
	183	−49,676	−46.27	>1.1×10^{17} y	14.4	11	. . . .
	183 *m*		−45.96	5.3 s	. . . .	. . . .	γ, 0.046(8), 0.053(11), 0.099(9), 0.102(4), 0.108(19), 0.160(6); e^-, 0.034, 0.040
	184	−48,975	−45.62	. . . .	30.6	2.1	. . . .
	185	−46,481	−43.30	75 d	. . . .	. . . .	$β^-$, 0.429; no γ
	185 *m*		−42.93	1.62 m	. . . .	. . . .	γ, 0.075(8), 0.100(16), 0.13(70), 0.17(100)
	186	−45,560	−42.44	>6×10^{15} y	28.4	40	. . . .
	187	−42,756	−39.83	23.9 h	. . . .	~90	$β^-$, 1.31, 0.63; e^-, 0.063, 0.122; γ, 0.072(11), 0.134(9), 0.479(23), 0.552(5), 0.618(6),

Table 3-7 (*Continued*)
TABLE OF NUCLIDES

Nuclide		Mass Excess		Half-life y = yr, d = day h = hr, m = min s = sec	Natural Abundance, %	Thermal Neutron Absorption Cross Section σ, barns (mode if not n,γ)	Major Radiations, Energies in MeV and γ Intensities, %
Z Sym- bol	Mass Number	μmu	MeV				
	188	−41,184	−38.44	69.4 d	. . . .	. . . .	0.686(27), 0.773(4) β⁻, 0.349; γ, 0.227, 0.290
	189		−35.3	11.5 m	. . . .	. . . .	β⁻, 2.5, 2.0; γ, 0.130(12), 0.178(13), 0.258(100), 0.417(96), 0.55(28), 0.86(20), 0.96(17)
₇₅Re	177		−47	17 m	. . . .	. . . .	. . . .
	178			15 m	. . . .	. . . .	β⁺, 3.1
	179		−46	20 m	. . . .	. . . .	W X-rays
	180			2.4 m	. . . .	. . . .	β⁺, 1.1; γ, 0.11, 0.511, 0.88
	180			20 h	. . . .	. . . .	β⁺, 1.9
	181		−47	18 h	. . . .	. . . .	γ, 0.365; e⁻, 0.008, 0.040, 0.053, 0.296
	182	−48,628	−45.30	12.7 h	. . . .	. . . .	γ, 0.068, 0.100, 1.122, 1.189, 1.23, 2.01, 2.05; β⁺, 1.74; e⁻, 0.015, 0.031, 0.056, 0.089, 0.098
	182			64.0 h	. . . .	. . . .	γ, 0.068, 0.100, 0.15–0.36, 1.08, 1.112, 1.19, 1.22, 1.43; e⁻, 0.015, 0.031, 0.044, 0.061, 0.089, 0.098, 0.100, 0.122, 0.160, 0.187
	183	−48,740	−45	71 d	. . . .	. . . .	γ, 0.046, 0.053, 0.109, 0.209, 0.246, 0.292; e⁻, 0.030, 0.034, 0.040, 0.088, 0.093
	184	−47,220	−44	38 d	. . . .	. . . .	γ, 0.111, 0.78, 0.90; e⁻, 0.042, 0.100
	184 m		−44	169 d	. . . .	. . . .	γ, 0.111, 0.78, 0.90; e⁻, 0.035, 0.042, 0.073, 0.081, 0.100
	185	−46,941	−43.73	. . . .	37.07	110	. . . .

Table 3-7 (*Continued*)
TABLE OF NUCLIDES

Nuclide		Mass Excess		Half-life y=yr, d=day h=hr, m=min s=sec	Natural Abundance, %	Thermal Neutron Absorption Cross Section σ, barns (mode if not n,γ)	Major Radiations, Energies in MeV and γ Intensities, %
Z Symbol	Mass Number	μmu	MeV				
	186	−44,980	−41.9	88.9 h	....	....	β^-, 1.07; e^-, 0.063, 0.125; γ, 0.137(9), 0.632, 0.768
	187	−44,167	−41.14	4.3×10^{10} y	62.93	70	β^-, 0.003
	188	−41,647	−38.79	16.7 h	....	<2	β^-, 2.12; e^-, 0.081, 0.143; γ, 0.155(10), 0.478(1), 0.633(1), 0.829, 0.932
	188 *m*		−38.62	18.7 m	....	....	γ, 0.092(5), 0.106(10); e^-, 0.004, 0.013, 0.021, 0.034, 0.051, 0.061, 0.080, 0.093
	189	−40,630	−37.8	24.3 h	....	....	β^-, 1.00; e^-, 0.023, 0.028, 0.057, 0.074, 0.112, 0.143; γ, 0.150(4), 0.187(3), 0.218(10), 0.245(4)
	189			140 d	....	....	γ, 0.211, 0.57, 0.67
	190	−38,040	−35.4	2.8 m	....	....	β^-, 1.6; γ, 0.191(10), 0.392(10), 0.57(10), 0.83(3)
	190 *m*			2.8 h	....	....	β^-, 1.6; γ, 0.12, 0.19, 0.23, 0.38, 0.56, 0.82
	191		−34.6	9.8 m	....	....	β^-, 1.8
	192			6 s	....	....	β^-, 2.5; γ, 0.20, 0.29, 0.37, 0.48, 0.57
₇₆Os	181		−44	23 m	....	....	e^-, 0.093, 0.101
	181			2.7 h	....	....	γ, 0.23
	182		−44	21.9 h	....	....	γ, 0.180(7), 0.263(1), 0.510(10); e^-, 0.015, 0.025, 0.043, 0.052, 0.108, 0.438
	183		−43	12.0 h	....	....	γ, 0.114(27), 0.168(10), 0.236(5), 0.382(90), 0.48(9), 0.86(5), 1.44(1); e^-, 0.043, 0.102
	183 *m*		−43	9.9 h	....	....	γ, 1.035(6),

Table 3-7 (*Continued*)
TABLE OF NUCLIDES

Nuclide		Mass Excess		Half-life	Natural	Thermal Neutron	Major Radiations,
Z Symbol	Mass Number	μmu	MeV	y = yr, d = day h = hr, m = min s = sec	Abundance, %	Absorption Cross Section σ, barns (mode if not n,γ)	Energies in MeV and γ Intensities, %
							1.105(48); e⁻, 0.055, 0.096, 0.158, 0.168
	184	−47,250	−44.0	··········	0.018	<200	················
	185	−45,887	−42.74	93.6 d	········	············	γ, 0.646(80), 0.875(14); e⁻, 0.059, 0.091, 0.574, 0.634
	186	−46,130	−43.0	··········	1.59	············	················
	187	−44,168	−41.14	··········	1.64	············	················
	188	−43,919	−40.91	··········	13.3	············	················
	189	−41,700	−38.8	··········	16.1	0.008	················
	189 m		−38.8	5.7 h	········	············	e⁻, 0.019, 0.028
	190	−41,370	−38.5	··········	26.4	8.6	················
	190 m		−36.8	9.9 m	········	············	γ, 0.187(70), 0.361(94), 0.502(98), 0.616(99); e⁻, 0.026, 0.036, 0.113, 0.175
	191	−39,030	−36.4	15.0 d	········	············	β⁻, 0.143; e⁻, 0.030, 0.042, 0.053, 0.116, 0.127; γ, 0.129(25)
	191 m		−36.3	13.0 h	········	············	e⁻, 0.062, 0.072
	192	−38,550	−35.9	>10¹⁴ y	41.0	1.6	················
	193	−35,773	−33.32	31.5 h	········	200	β⁻, 1.13; e⁻, 0.060, 0.070; γ, 0.139(3), 0.28(2), 0.322(1), 0.38(2), 0.460(4), 0.558(2)
	194	−34,771	−32.39	6.0 y	········	············	β⁻, 0.053; e⁻; γ, 0.043(10), 0.078
	195	−32,000	−30	6.5 m	········	············	β⁻, 2
₇₇Ir	182		−39	15 m	········	············	γ, 0.133, 0.278, 0.510
	183			0.9 h	········	············	γ, 0.24
	184		−40	3.2 h	········	············	γ, 0.125(100), 0.267(200), 0.392(90), 0.51, 0.83, 0.96, 1.09
	185		−40	14 h	········	············	γ, 0.101, 0.254; e⁻, 0.024, 0.034, 0.047, 0.085, 0.180
	186	−42,010	−39.1	15.8 h	········	············	γ, 0.137(45), 0.297(74), 0.434(35), 0.511,

Table 3-7 (*Continued*)
TABLE OF NUCLIDES

Nuclide		Mass Excess		Half-life $y=yr, d=day$ $h=hr, m=min$ $s=sec$	Natural Abundance, %	Thermal Neutron Absorption Cross Section σ, barns (mode if not n,γ)	Major Radiations, Energies in MeV and γ Intensities, %
Z Sym-bol	Mass Number	μmu	MeV				
	186			1.7 h	....	....	0.64(9), 0.77(8), 1.60–1.75(4); β^+, 1.94; e^-, 0.063, 0.125, 0.135, 0.226 γ, 0.137, 0.295, 0.511, 0.630, 0.77, 0.99; β^+, 2.6; e^-, 0.063, 0.125
	187	−42,440	−40	10.5 h	....	....	γ, 0.18(45), 0.31(14), 0.41(100), 0.50(35), 0.61(45), 0.90(40), 0.98(50); e^-, 0.007, 0.013, 0.053, 0.063, 0.073, 0.104
	188	−40,878	−38.08	41.5 h	....	....	γ, 0.155(34), 0.478(16), 0.633(29), 0.829(7), 1.210(7), 1.717(4), 2.08(16), 2.217(13); β^+, 1.66; e^-, 0.081, 0.143
	189	−41,090	−38	13.3 d	....	....	γ, 0.245(18); e^-, 0.023, 0.046, 0.058, 0.067, 0.171
	190	−39,170	−36.5	11 d	....	....	γ, 0.187(51), 0.37(39), 0.40(39), 0.518(39), 0.56(72), 0.604(47); e^-, 0.113, 0.175
	190 m_1		−36.5	1.2 h	....	....	e^-, 0.015, 0.024
	190 m_2		−36.3	3.2 h	....	....	γ, 0.187(66), 0.361(88), 0.502(92), 0.616(93); e^-, 0.026, 0.036, 0.113, 0.175
	191	−39,360	−36.7	....	38.5	750	....

Table 3-7 (*Continued*)
TABLE OF NUCLIDES

Nuclide Z Symbol	Nuclide Mass Number	Mass Excess μmu	Mass Excess MeV	Half-life y = yr, d = day h = hr, m = min s = sec	Natural Abundance, %	Thermal Neutron Absorption Cross Section σ, barns (mode if not n,γ)	Major Radiations, Energies in MeV and γ Intensities, %
	191 *m*		−36.5	4.9 s	....	....	γ, 0.129(25); e⁻, 0.030, 0.042, 0.053, 0.116, 0.127
	192	−37,300	−34.7	74.2 d	....	700	β⁻, 0.67; e⁻, 0.217, 0.230, 0.239, 0.390; γ, 0.296(29), 0.308(30), 0.317(81), 0.468(49), 0.589(4), 0.604(9), 0.612(6)
	192 *m₁*		−34.7	1.42 m	....	....	γ, 0.058, 0.317, 0.612; e⁻, 0.046, 0.056; β⁻, 1.5
	192 *m₂*		−34.6	>5 y	....	....	e⁻, 0.149, 0.158
	193	−36,988	−34.45	....	61.5	110	....
	193 *m*		−34.37	11.9 d	....	....	e⁻, 0.069, 0.078
	194	−34,875	−32.49	17.4 h	....	....	β⁻, 2.24; γ, 0.328(10), 0.64(1), 0.939, 1.16(1), 1.48(1), 1.7
	194 *m*			47 s	....	....	β⁻, 2.3; γ, 0.13, 0.32, 0.63
	195	−34,110	−31.8	4.2 h	....	....	β⁻, 1.0; γ, 0.10, 0.13, 0.33, 0.37, 0.43, 0.66
	196	−31,750	−29.23	120 m	....	....	β⁻, 0.95; γ, 0.100(33), 0.356(94), 0.39(95), 0.44(95), 0.522(99), 0.65(100)
	197	−30,510	−28.4	7 m	....	....	β⁻, 2.0; γ, 0.50
	198	−27,380	−25.5	50 s	....	....	β⁻, 3.6; γ, 0.78
₇₈Pt	173			short	....	....	α, 6.19
	174			0.7 s	....	....	α, 6.03
	175			2.1 s	....	....	α, 5.95
	176			6.0 s	....	....	α, 5.74
	177			6.6 s	....	....	α, 5.51
	178			21 s	....	....	α, 5.44
	179			33 s	....	....	α, 5.15
	180			50 s	....	....	α, 5.14
	181			51 s	....	....	α, 5.02
	182		−36	3.0 m	....	....	α, 4.84
	183			6.5 m	....	....	α, 4.73

Table 3-7 (*Continued*)
TABLE OF NUCLIDES

| Nuclide | | Mass Excess | | Half-life | Natural | Thermal Neutron | Major Radiations, |
Z Symbol	Mass Number	μmu	MeV	y=yr, d=day h=hr, m=min s=sec	Abundance, %	Absorption Cross Section σ, barns (mode if not n,γ)	Energies in MeV and γ Intensities, %
	197 *m*		−30.02	78 m	. . . .	. . . .	0.063, 0.074, 0.110; γ, 0.077(20), 0.191(6) γ, 0.279(3), 0.346(13); e⁻, 0.040, 0.050, 0.268, 0.332; β⁻, 0.737
	198	−32,105	−29.91	>10¹⁵ y	7.19	4	. . . .
	199	−29,420	−27.40	31 m	. . . .	~15	β⁻, 1.69; γ, 0.075, 0.197(9), 0.245(4), 0.32(8), 0.475(12), 0.540(24), 0.715(3), 0.790(2), 0.960(2)
	199 *m*		−26.98	14.1 s	. . . .	. . . .	γ, 0.393(90); e⁻, 0.018, 0.029, 0.315, 0.381
	200	−28,570	−27	11.5 h	. . . .	. . . .	. . . .
	201	−25,230	−23.5	2.3 m	. . . .	. . . .	β⁻, 2.66; γ, 0.15, 0.23, 1.76
₇₉Au	177			1.4 s	. . . .	. . . .	α, 6.11
	178			2.7 s	. . . .	. . . .	α, 5.91
	179			7.1 s	. . . .	. . . .	α, 5.84
	181			10 s	. . . .	. . . .	α, 5.60, 5.47
	183			44 s	. . . .	. . . .	α, 5.34
	185			4.33 m	. . . .	. . . .	α, 5.07
	186			12 m	. . . .	. . . .	γ, 0.16, 0.22, 0.30, 0.40
	187			8 m	. . . .	. . . .	Pt X-rays
	188			8 m	. . . .	. . . .	γ, 0.25, 0.33, 0.63
	189			30 m	. . . .	. . . .	e⁻, 0.027, 0.036, 0.088, 0.137, 0.154, 0.166, 0.269
	190	−35,290	−33	39 m	. . . .	. . . .	γ, 0.29(100), 0.60(5); e⁻, 0.22, 0.29
	191	−36,450	−34	3.2 h	. . . .	. . . .	γ, 0.14(10), 0.30(60), 0.39(5), 0.48(4), 0.60(10); e⁻, 0.035, 0.046, 0.054, 0.080, 0.089
	192	−35,380	−33.0	4.1 h	. . . .	. . . .	γ, 0.137, 0.158, 0.296, 0.308, 0.317; e⁻, 0.032,

Table 3-7 (*Continued*)
TABLE OF NUCLIDES

Nuclide		Mass Excess		Half-life y = yr, d = day h = hr, m = min s = sec	Natural Abundance, %	Thermal Neutron Absorption Cross Section σ, barns (mode if not n.γ)	Major Radiations, Energies in MeV and γ Intensities, %
Z Sym-bol	Mass Number	μmu	MeV				
	193	−35,760	−33	15.8 h	....	....	0.143, 0.23, 0.30; β⁺, 2.2 γ, 0.114(5), 0.18(11), 0.26(9), 0.378(1), 0.440(3); e⁻, 0.034, 0.095, 0.108, 0.177
	193 *m*		−33	3.9 s	....	....	γ, 0.258(65); e⁻, 0.019, 0.030
	194	−34,582	−32.21	39.5 h	....	....	β⁺, 1.49; e⁻, 0.250, 0.315; γ, 0.294(12), 0.328(68), 1.469(8), 1.596(3), 1.887(4), 2.044(4)
	195	−34,949	−32.55	183 d	....	....	γ, 0.099(10), 0.129(1); e⁻, 0.018, 0.028, 0.085
	195 *m*		−32.23	30.6 s	....	....	γ, 0.261(77); e⁻, 0.044, 0.056, 0.180
	196	−33,445	−31.15	6.18 d	....	....	β⁻, 0.259; e⁻, 0.255, 0.277, 0.343; γ, 0.333(25), 0.356(94), 0.426(6), 1.091
	196 *m*		−30.56	9.7 h	....	....	γ, 0.148(42), 0.188(32), 0.285(5), 0.316(5); e⁻, 0.069, 0.081, 0.094, 0.108, 0.135, 0.160
	197	−33,459	−31.17	....	100	98.8	....
	197 *m*		−30.76	7.2 s	....	....	γ, 0.130(8), 0.279(75); e⁻, 0.050, 0.117, 0.127, 0.198, 0.265
	198	−31,769	−29.59	2.697 d	....	26,000	β⁻, 0.962; e⁻, 0.329, 0.398; γ, 0.412(95), 0.676(1), 1.088
	199	−31,227	−29.09	3.15 d	....	~30	β⁻, 0.46, 0.30; γ, 0.158(37),

Table 3-7 (*Continued*)
TABLE OF NUCLIDES

| Nuclide | | Mass Excess | | Half-life | Natural | Thermal Neutron | Major Radiations, |
Z Symbol	Mass Number	μmu	MeV	y = yr, d = day h = hr, m = min s = sec	Abundance, %	Absorption Cross Section σ, barns (mode if not n,γ)	Energies in MeV and γ Intensities, %
$_{80}$Hg	200	−29,300	−27.3	48.4 m	....	....	0.208(8); e^-, 0.075, 0.125, 0.145 β^-, 2.2; γ, 0.368(24), 1.227(23), 1.593(1)
	201	−28,080	−26.2	26 m	....	....	β^-, 1.5; γ, 0.53
	203	−24,870	−23	55 s	....	....	β^-, 1.9; γ, 0.69
	185			50 s	....	....	....
	186			1.5 m	....	....	γ, 0.125, 0.27, 0.35, 0.44
	187			3 m	....	....	γ, 0.175, 0.255, 0.40
	188			3.7 m	....	....	γ, 0.14
	189			9.6 m	....	....	γ, 0.165, 0.24, 0.32, 0.50
	190		−31	20 m	....	....	γ, 0.14; e^-, 0.015, 0.026, 0.049, 0.062, 0.076
	191			55 m	....	....	γ, 0.26; e^-, 0.170, 0.191, 0.239
	192	−33,840	−32	4.8 h	....	....	γ, 0.114(10), 0.157(20), 0.274(100); e^-, 0.017, 0.028, 0.034, 0.039, 0.077
	193	−33,250	−31	~6 h	....	....	γ, 0.187, 0.574, 0.762, 0.855, 1.04, 1.08; e^-, 0.025, 0.035, 0.108, 0.174
	193 m		−31	10.0 h	....	....	γ, 0.218, 0.258, 0.574; e^-, 0.020, 0.025, 0.029, 0.036, 0.087, 0.178, 0.243
	194	−34,210	−32.2	1.9 y	....	....	Au X-rays
	194 m			0.4 s	....	....	γ, 0.048, 0.134
	195	−33,380	−31	9.5 h	....	....	γ, 0.20, 0.261, 0.59, 0.780, 0.930, 1.110, 1.172; e^-, 0.048, 0.058, 0.099
	195 m		−31	40.0 h	....	....	γ, 0.200(35), 0.261(20), 0.560(20); e^-, 0.0014, 0.013,

Table 3-7 (*Continued*)
TABLE OF NUCLIDES

Nuclide		Mass Excess		Half-life y = yr, d = day h = hr, m = min s = sec	Natural Abundance, %	Thermal Neutron Absorption Cross Section σ, barns (mode if not n,γ)	Major Radiations, Energies in MeV and γ Intensities, %
Z Sym- bol	Mass Number	μmu	MeV				
₈₁Tl	196	−34,180	−31.84	>1×10¹⁴ y	0.146	880	0.022, 0.034, 0.043, 0.048, 0.053, 0.058, 0.109, 0.120, 0.180
	197	−32,640	−30.75	65 h	....	....	γ, 0.077(18), 0.191(2), 0.268; e⁻, 0.064, 0.074
	197 m		−30.45	24 h	....	....	γ, 0.134(42), 0.279(7); e⁻, 0.051, 0.082, 0.120, 0.131, 0.152, 0.162
	198	−33,244	−30.97	....	10.02	0.02	....
	199	−31,721	−29.55	....	16.84	2,000	....
	199 m		−29.01	43 m	....	....	γ, 0.158(53), 0.375(15); e⁻, 0.075, 0.144, 0.285, 0.354
	200	−31,673	−29.50	....	23.13	<50	....
	201	−29,692	−27.66	....	13.22	<50	....
	202	−29,358	−27.35	....	29.80	4	....
	203	−27,120	−25.26	46.9 d	....	....	β⁻, 0.214; e⁻, 0.194, 0.264, 0.275; γ, 0.279(77)
	204	−26,505	−24.69	....	6.85	0.4	....
	205	−23,790	−22.2	5.5 m	....	....	β⁻, 1.7; γ, 0.205
	206	−22,487	−20.95	8.1 m	....	....	β⁻; γ, 0.31
	191			10 m	....	....	γ, 0.511
	192			11 m	....	....	γ, 0.424; e⁻, 0.341
	193			23 m	....	....	γ, 0.158, 0.169, 0.178, 0.187, 0.208, 0.216, 0.238, 0.247, 0.511; e⁻, 0.24
	193 m			2.1 m	....	....	γ, 0.365; e⁻, 0.280
	194	−28,430	−26	33.0 m	....	....	γ, 0.427; e⁻, 0.344
	194 m			32.8 m	....	....	γ, 0.097; e⁻, 0.083
	195	−30,160	−28	1.16 h	....	....	e⁻, 0.022, 0.034; β⁺, 1.8
	195 m		−28	3.5 s	....	....	γ, 0.383(95); e⁻, 0.084, 0.096
	196	−29,240	−27.2	1.84 h	....	....	γ, 0.426; e⁻, 0.343
	196 m		−26.8	1.41 h	....	....	γ, 0.426; e⁻, 0.071, 0.081, 0.107
	197	−30,280	−28.5	2.84 h	....	....	γ, 0.152, 0.426; e⁻, 0.067, 0.137

Table 3-7 (*Continued*)
TABLE OF NUCLIDES

Z Sym-bol	Mass Number	Mass Excess μmu	Mass Excess MeV	Half-life y=yr, d=day h=hr, m=min s=sec	Natural Abundance, %	Thermal Neutron Absorption Cross Section σ, barns (mode if not n,γ)	Major Radiations, Energies in MeV and γ Intensities, %
	197 *m*		−27.9	0.54 s	....	....	γ, 0.222(40), 0.385(90); e⁻, 0.136, 0.207, 0.219, 0.300
	198	−29,530	−27.5	5.3 h	....	....	γ, 0.412(90), 0.65(40), 1.20(21), 1.42(24), 2.01(15), 2.45(5), 2.78(2); β⁺, 2.4; e⁻, 0.111, 0.201, 0.317, 0.329
	198 *m*		−27.0	1.87 h	....	....	γ, 0.283(30), 0.412(45), 0.586(35), 0.635(35); e⁻, 0.033, 0.046, 0.175, 0.197, 0.246
	199	−30,540	−28.5	7.4 h	....	....	γ, 0.158(5), 0.208(12), 0.247(9), 0.455(14); e⁻, 0.035, 0.125, 0.161, 0.193
	200	−29,038	−27.05	26.1 h	....	....	γ, 0.368(88), 0.579(10), 0.829(8), 1.21(35), 1.364(4), 1.410(2), 1.517(4); β⁺, 1.44, 1.07; e⁻, 0.285, 0.354
	201	−29,250	−27.3	74 h	....	....	γ, 0.135(2), 0.167(8); e⁻, 0.016, 0.052, 0.084
	202	−28,050	−26.13	12.0 d	....	....	γ, 0.439(95), 0.522, 0.961; e⁻, 0.356
	203	−27,647	−25.75	....	29.50	11	....
	204	−26,135	−24.34	3.81 y	....	....	β⁻, 0.766
	205	−25,558	−23.81	....	70.50	0.11	....
	206	−23,896	−22.26	4.19 m	....	....	β⁻, 1.52; no γ
	207	−22,550	−21.01	4.79 m	....	....	β⁻, 1.44; γ, 0.897
	207 *m*		−19.67	1.3 s	....	....	γ, 0.35, 1.00
	208	−17,987	−16.76	3.10 m	....	....	β⁻, 1.80; e⁻, 0.187, 0.423, 0.495; γ, 0.511(23),

Table 3-7 (*Continued*)
TABLE OF NUCLIDES

Nuclide		Mass Excess		Half-life y = yr, d = day h = hr, m = min s = sec	Natural Abundance, %	Thermal Neutron Absorption Cross Section σ, barns (mode if not n,γ)	Major Radiations, Energies in MeV and γ Intensities, %
Z Symbol	Mass Number	μmu	MeV				
	209	−14,704	−13.65	2.2 m	....	....	0.583(86), 0.860(12), 2.614(100) β⁻, 1.99; e⁻, 0.03, 0.10; γ, 0.12(50), 0.45(100), 1.56(100)
	210	−9,946	−9.23	1.32 m	....	....	β⁻, 2.3; e⁻, 0.208, 0.28; γ, 0.296(80), 0.795(100), 1.08(19), 1.21(17), 1.31(21), 2.01(7), 2.09(5), 2.36(8), 2.43(9)
₈₂Pb	194			11 m	....	....	γ, 0.204
	195			17 m	....	....	γ, 0.39; e⁻, 0.084, 0.096, 0.30
	196	−26,200	−24	37 m	....	....	γ, 0.192, 0.240, 0.253, 0.367, 0.503; e⁻, 0.155, 0.168
	197	−25,910	−24	....	....	....	γ, 0.386
	197 m		−24	42 m	....	....	γ, 0.085, 0.222, 0.234, 0.386; e⁻, 0.069, 0.136, 0.146, 0.207, 0.219, 0.300
	198	−27,590	−26	2.4 h	....	....	γ, 0.117(3), 0.173(28), 0.259(8), 0.290(16), 0.38(40), 0.575(4), 0.649(2), 0.865(6); e⁻, 0.031, 0.088, 0.159, 0.172, 0.205, 0.270
	199	−27,140	−25	90 m	....	....	γ, 0.353(17), 0.367(80), 0.720(10); e⁻, 0.267; β⁺, 2.8
	199 m		−25	12.2 m	....	....	γ, 0.424(20); e⁻, 0.336, 0.409
	200	−28,030	−26	21.5 h	....	....	γ, 0.109, 0.146, 0.236, 0.26, 0.290, 0.450, 0.605; e⁻, 0.024,

Table 3-7 (*Continued*)
TABLE OF NUCLIDES

Nuclide		Mass Excess		Half-life y=yr, d=day h=hr, m=min s=sec	Natural Abundance, %	Thermal Neutron Absorption Cross Section σ, barns (mode if not n,γ)	Major Radiations, Energies in MeV and γ Intensities, %
Z Symbol	Mass Number	μmu	MeV				
	201	−27,140	−25	9.4 h	....	....	0.06, 0.133, 0.150, 0.172, 0.183 γ, 0.330, 0.361, 0.406, 0.585, 0.766, 0.907, 0.946, 1.30, 1.40; e⁻, 0.244, 0.275, 0.316; β⁺, 0.55
	201 m		−25	61 s	....	....	γ, 0.629(51); e⁻, 0.541, 0.614
	202	−27,997	−26.08	~3×10⁵ y	....	....	Tl X-rays
	202 m		−23.91	3.62 h	....	....	γ, 0.390(7), 0.422(90), 0.460(8), 0.490(10), 0.658(35), 0.787(45), 0.961(90); e⁻, 0.115, 0.126, 0.302, 0.334, 0.699, 0.772
	203	−26,771	−24.94	52.1 h	....	....	γ, 0.279(81), 0.401(5), 0.680(1); e⁻, 0.193, 0.264
	203 m		−24.11	6.1 s	....	....	γ, 0.825(70); e⁻, 0.737, 0.810
	204	−26,956	−25.11	....	1.40	0.7	....
	204 m		−22.92	66.9 m	....	....	γ, 0.375(93), 0.90(189); e⁻, 0.287, 0.360, 0.824, 0.897
	205	−25,520	−23.77	3.0×10⁷ y	....	....	Tl X-rays
	206	−25,532	−23.79	....	25.1	0.03	....
	207	−24,097	−22.45	....	21.7	0.72	....
	207 m		−20.81	0.80 s	....	....	γ, 0.570(98), 1.064(83); e⁻, 0.482, 0.975, 1.048
	208	−23,350	−21.75	....	52.3	0.0005	....
	209	−18,918	−17.63	3.30 h	....	....	β⁻, 0.635; no γ
	210	−15,813	−14.73	20.4 y	....	....	β⁻, 0.061; e⁻, 0.030, 0.043; γ, 0.047(4); α, 3.72
	211	−11,258	−10.46	36.1 m	....	....	β⁻, 1.36; γ, 0.405(3), 0.427(2), 0.702, 0.766(1), 0.832(3)

Table 3-7 (*Continued*)
TABLE OF NUCLIDES

Nuclide		Mass Excess		Half-life	Natural	Thermal Neutron	Major Radiations,
Z Symbol	Mass Number	μmu	MeV	y = yr, d = day h = hr, m = min s = sec	Abundance, %	Absorption Cross Section σ, barns (mode if not n,γ)	Energies in MeV and γ Intensities, %
	212	−8,095	−7.55	10.64 h	....	....	β^-, 0.58; e^-, 0.148, 0.222; γ, 0.239(47), 0.300(3)
	213	−3,710	−3	10.2 m	....	....	....
	214	−234	−0.15	26.8 m	....	....	β^-, 1.03, 0.67; e^-, 0.037, 0.049; γ, 0.053(1), 0.242(4), 0.295(19), 0.352(36)
$_{83}$Bi	197?			8.0 m	....	....	α, 5.81
	≤198			1.7 m	....	....	α, 6.2
	199	−21,560	−20	24.4 m	....	....	α, 5.53
	200	−21,060	−20	35 m	....	....	....
	201	−22,630	−21	1.85 h	....	....	Pb X-rays
	201 *m*			52 m	....	....	α, 5.28
	202	−22,120	−21	95 m	....	....	γ, 0.422, 0.961
	203	−23,350	−21.8	11.8 h	....	....	β^+, 1.35; e^-, 0.045, 0.098, 0.112, 0.176, 0.737; γ, 0.186(6), 0.264(6), 0.381(9), 0.82(78), 1.034(16), 1.52(31), 1.87(35)
	204	−22,190	−21	11.2 h	....	....	γ, 0.21, 0.375, 0.671, 0.91, 0.98, 1.21; e^-, 0.063, 0.075, 0.087, 0.128, 0.133, 0.161, 0.201, 0.287, 0.360, 0.583, 0.811, 0.824, 0.897
	205	−22,618	−21.07	15.31 d	....	....	β^+, 0.98; e^-, 0.011, 0.023; γ, 0.26(3), 0.51(4), 0.57(14), 0.703(28), 0.911(4), 0.988(17), 1.044(8), 1.615(4), 1.766(27), 1.864(6), 1.906(2)
	206	−21,611	−20.18	6.243 d	....	....	γ, 0.184(21), 0.343(26), 0.398(10),

Table 3-7 (*Continued*)
TABLE OF NUCLIDES

| Nuclide | | Mass Excess | | Half-life | Natural | Thermal Neutron | Major Radiations, |
Z Symbol	Mass Number	μmu	MeV	y = yr, d = day h = hr, m = min s = sec	Abundance, %	Absorption Cross Section σ, barns (mode if not n,γ)	Energies in MeV and γ Intensities, %
	207	−21,562	−20.04	30.2 y	....	....	0.497(18), 0.516(46), 0.538(34), 0.803(99), 0.880(72), 0.895(19), 1.019(8), 1.099(13), 1.596(8), 1.720(36); e^-, 0.096, 0.168, 0.255 γ, 0.570(98), 1.063(77), 1.771(9); e^-, 0.482, 0.975, 1.048
	208	−20,269	−18.88	3.68×10^5 y	....	....	γ, 2.614(100)
	209	−19,606	−18.26	$>2 \times 10^{18}$ y	100	0.019	α(?), 3.0
	210	−15,813	−14.79	5.013 d	....	....	β^-, 1.160; α, 4.69, 4.65
	210 m		−14.52	$\sim 2.6 \times 10^6$ y	....	....	α, 4.96, 4.92, 4.57; γ, 0.262(45), 0.30(23), 0.34, 0.61
	211	−12,700	−11.84	2.16 m	....	....	α, 6.62, 6.28; γ, 0.351(14); e^-, 0.265
	212	−8,721	−8.13	60.60 m	....	....	β^-, 2.25; e^-, 0.025, 0.036; α, 6.09, 6.05; γ, 0.040(2), 0.288(1), 0.46(1), 0.727(7), 0.785(1), 1.620(2)
	213	−5,683	−5.24	47 m	....	....	β^-, 1.39; γ, 0.437; α, 5.87
	214	−1,314	−1.19	19.7 m	....	....	β^-, 3.26; γ, 0.609(47), 0.769(5), 0.935(3), 1.120(17), 1.238(6), 1.378(5), 1.40(4), 1.509(2), 1.728(3), 1.764(17), 1.848(2), 2.117(1), 2.204(5),

Table 3-7 (*Continued*)
TABLE OF NUCLIDES

| Nuclide | | Mass Excess | | Half-life | Natural | Thermal Neutron | Major Radiations, |
Z Symbol	Mass Number	μmu	MeV	y = yr, d = day h = hr, m = min s = sec	Abundance, %	Absorption Cross Section σ, barns (mode if not n,γ)	Energies in MeV and γ Intensities, %
	215	1,830	1.7	7 m	....	....	2.445(2); α, 5.51, 5.45 β⁻
₈₄Po	193			short	....	....	α, 7.0
	194			0.5 s	....	....	α, 6.85
	195			3 s	....	....	α, 6.63
	195 m			1.4 s	....	....	α, 6.72
	196			6 s	....	....	α, 6.53
	197			54 s	....	....	α, 6.30
	197 m			25 s	....	....	α, 6.39
	198			1.7 m	....	....	α, 6.16
	199			5.0 m	....	....	α, 5.94
	199 m			4.2 m	....	....	α, 6.05
	200	−17,180	−16	10.5 m	....	....	α, 5.86
	201	−16,980	−16	15.1 m	....	....	α, 5.68
	201 m			8.9 m	....	....	α, 5.78
	202	−18,870	−18	45 m	....	....	α, 5.58
	203	−18,530	−17	42 m	....	....	α, 5.49
	204	−19,540	−18	3.6 h	....	....	α, 5.38
	205	−18,800	−18	1.8 h	....	....	α, 5.25
	206	−19,676	−18.33	8.8 d	....	....	γ, 0.286(35), 0.338(40), 0.51(100), 0.807(60), 1.02(85); e⁻, 0.045, 0.196, 0.248; α, 5.22
	207	−18,442	−17.14	5.7 h	....	....	γ, 0.25(5), 0.35(4), 0.41(13), 0.74(36), 0.95(84), 1.15(6), 1.37(4), 2.06(2); e⁻, 0.159, 0.255, 0.315, 0.652, 0.902; β⁺, 1.14; α, 5.11
	207 m		−15.75	2.8 s	....	....	γ, 0.26(42), 0.31(40), 0.82(100); e⁻, 0.22, 0.24
	208	−18,757	−17.47	2.93 y	....	....	α, 5.11; γ, 0.285, 0.60
	209	−17,574	−16.37	103 y	....	....	α, 4.88; γ, 0.261, 0.91(1); e⁻, 0.173
	210	−17,124	−15.95	138.40 d	....	<0.03	α, 5.305; γ, 0.803
	211	−13,343	−12.43	0.52 s	....	....	α, 7.45; γ, 0.570(1), 0.90(1)
	211 m		−11.00	25 s	....	....	α, 8.88, 7.28; γ, 0.570(92),

Table 3-7 (*Continued*)
TABLE OF NUCLIDES

Nuclide		Mass Excess		Half-life y=yr, d=day h=hr, m=min s=sec	Natural Abundance, %	Thermal Neutron Absorption Cross Section σ, barns (mode if not n,γ)	Major Radiations, Energies in MeV and γ Intensities, %
Z Sym-bol	Mass Number	μmu	MeV				
₈₅At	212 *m*		−7.44	45 s	....	....	1.063(77); e⁻ α, 11.65; γ, 0.57(2), 2.61(3)
	216	1,922	1.78	0.145 s	....	....	α, 6.78
	217	6,060	6	<10 s	....	....	α, 6.55
	218	8,930	8.38	3.05 m	....	....	α, 6.00
	200			0.9 m	....	....	α, 6.47, 6.42
	201			1.5 m	....	....	α, 6.35
	202	−10,200	−10	3.0 m	....	....	α, 6.23, 6.12
	203	−12,290	−11	7.4 m	....	....	α, 6.09
	204	−11,940	−11	9.3 m	....	....	α, 5.95
	205	−13,560	−13	26.2 m	....	....	α, 5.90
	206	−13,210	−12	32.8 m	....	....	α, 5.70; γ, 0.068(10); e⁻, 0.052, 0.064
	207	−14,440	−13.41	1.8 h	....	....	α, 5.76
	208	−13,390	−12	1.6 h	....	....	γ, 0.18(25), 0.25, 0.66(100); α, 5.65
	209	−13,833	−12.89	5.5 h	....	....	γ, 0.195(23), 0.545(62), 0.780(94); e⁻, 0.076, 0.102, 0.178, 0.451, 0.686; α, 5.65
	210	−12,964	−12.12	8.3 h	....	....	γ, 0.245(79), 1.180(100), 1.436(29), 1.483(48), 1.599(14); e⁻, 0.023, 0.031, 0.043, 0.152, 0.229; α, 5.52, 5.44, 5.36
	211	−12,538	−11.64	7.21 h	....	....	α, 5.868; γ, 0.67
	212	−9,276	−8.64	0.30 s	....	....	α, 7.66, 7.60; e⁻, 0.047, 0.059
	212 *m*		−8.42	0.12 s	....	....	α, 7.88, 7.82; e⁻, 0.047, 0.059
	213	−6,930	−6.5	short	....	....	α, 9.2
	218	8,607	8.11	1.5–2.0 s	....	....	α, 6.70, 6.65
	219	11,290	10.5	0.9 m	....	....	α, 6.28
₈₆Rn	202			13 s	....	....	α, 6.64
	203			45 s	....	....	α, 6.50
	203 *m*			28 s	....	....	α, 6.55
	204	−7,700	−7	75 s	....	....	α, 6.42
	205	−7,440	−7	1.8 m	....	....	α, 6.26
	206	−9,420	−9	6.5 m	....	....	α, 6.26
	207	−9,240	−9	11 m	....	....	α, 6.15
	208	−10,210	−10	23 m	....	....	α, 6.15

Table 3-7 (*Continued*)
TABLE OF NUCLIDES

| Nuclide | | Mass Excess | | Half-life | Natural | Thermal Neutron | Major Radiations, |
Z Symbol	Mass Number	μmu	MeV	y=yr, d=day h=hr, m=min s=sec	Abundance, %	Absorption Cross Section σ, barns (mode if not n,γ)	Energies in MeV and γ Intensities, %
	209	−9,580	−9	30 m	....	....	α, 6.04
	210	−10,460	−9.74	2.42 h	....	....	α, 6.04
	211	−9,434	−8.75	15 h	....	....	α, 5.85, 5.78; γ, 0.445(29), 0.680(74), 0.865(18), 0.946(21), 1.13(23), 1.37(38); e^-, 0.053, 0.065, 0.073, 0.153, 0.168, 0.200, 0.237, 0.349, 0.584, 0.665
	212	−9,293	−8.66	25 m	....	....	α, 6.27
	219	9,481	8.85	4.00 s	....	....	α, 6.82, 6.55, 6.42; γ, 0.272(9), 0.401(5); e^-, 0.179, 0.255, 0.308
	220	11,401	10.61	55.3 s	....	<0.2	α, 6.29; γ, 0.55
	221	15,230	14	25 m	....	....	α, 6.0
	222	17,531	16.39	3.8229 d	....	0.7	α, 5.49; γ, 0.510
	223			43 m	....	....	....
	224			1.9 h	....	....	....
$_{87}$Fr	204			2.0 s	....	....	α, 7.03
	205			3.7 s	....	....	α, 6.92
	206	−160	−0	15.8 s	....	....	α, 6.80
	207	−2,270	−2	19 s	....	....	α, 6.78
	208	−2,050	−2	37 s	....	....	α, 6.66
	209	−3,680	−3	55 s	....	....	α, 6.66
	210	−3,430	−3	2.6 m	....	....	α, 6.56
	211	−4,670	−4.3	3.1 m	....	....	α, 6.56
	212	−3,770	−4	19.3 m	....	....	α, 6.42, 6.39, 6.35
	213	−3,816	−3.55	34 s	....	....	α, 6.78
	217	4,750	4.4	short	....	....	α, 8.3
	220	12,337	11.47	27.5 s	....	....	α, 6.68, 6.64
	221	14,183	13.27	4.8 m	....	....	α, 6.34, 6.12; γ, 0.218(14); e^-, 0.122, 0.202
	222	17,630	16.34	14.8 m	....	....	....
	223	19,736	18.40	22 m	....	....	β^-, 1.15; e^-, 0.031, 0.045, 0.062, 0.075; γ, 0.050(40), 0.080(13), 0.234(4)
	224	23,590	22	<2 m	....	....	....
$_{88}$Ra	213	420	−0	2.7 m	....	....	α, 6.91

Table 3-7 (*Continued*)
TABLE OF NUCLIDES

Nuclide		Mass Excess		Half-life	Natural	Thermal Neutron	Major Radiations,
Z Symbol	Mass Number	μmu	MeV	y = yr. d = day h = hr. m = min s = sec	Natural Abundance. %	Absorption Cross Section σ. barns (mode if not $n.\gamma$)	Energies in MeV and γ Intensities, %
	221	13,892	12.96	30 s	....	....	α, 6.76, 6.67, 6.61, 6.59; γ, 0.091(4), 0.151(13), 0.175(2)
	222	15,376	14.32	38 s	....	....	α, 6.56; γ, 0.325(4), 0.473, 0.52, 0.85
	223	18,501	17.26	11.435 d	....	130	α, 5.75, 5.71, 5.61, 5.54; γ, 0.149(10), 0.270(10), 0.33(6); e^-, 0.024, 0.046, 0.056, 0.126, 0.136, 0.171
	224	20,218	18.82	3.64 d	....	12	α, 5.68, 5.45; γ, 0.241(4), 0.29, 0.41, 0.65; e^-, 0.144, 0.225
	225	23,528	22.01	14.8 d	....	....	β^-, 0.36; e^-, 0.021, 0.035; γ, 0.040(33)
	226	25,360	23.69	1,602 y	....	20	α, 4.78, 4.60; γ, 0.186(4), 0.26, 0.42, 0.61; e^-, 0.087, 0.170
	227	29,159	27.18	41.2 m	....	....	β^-, 1.31; e^-, 0.008, 0.023; γ, 0.291(4), 0.498(1)
	228	31,139	28.96	6.7 y	....	~36	β^-, 0.05; e^-, 0.005
	229			short	....	....	....
	230		35	1 h	....	....	β^-, 1.2
$_{89}$Ac	221	15,680	14.6	short	....	....	α, 7.6
	222	17,760	16.55	5.5 s	....	....	α, 7.00
	223	19,144	17.82	2.2 m	....	....	α, 6.66, 6.65, 6.57; γ, 0.082, 0.096
	224	21,690	20.21	2.9 h	....	....	γ, 0.132(28), 0.217(62); e^-, 0.067, 0.080; α, 6.20, 6.14, 6.04
	225	23,153	21.62	10.0 d	....	....	α, 5.83, 5.79, 5.73; γ, 0.099, 0.150, 0.187; e^-, 0.020, 0.032, 0.044, 0.081
	226	26,160	24.31	29 h	....	....	β^-, 1.2; e^-, 0.053, 0.067; γ, 0.158(32), 0.185(9), 0.230(47), 0.253(11); α

Table 3-7 (*Continued*)
TABLE OF NUCLIDES

Nuclide		Mass Excess		Half-life y=yr, d=day h=hr, m=min s=sec	Natural Abundance, %	Thermal Neutron Absorption Cross Section σ, barns (mode if not n,γ)	Major Radiations, Energies in MeV and γ Intensities, %
Z Symbol	Mass Number	μmu	MeV				
	227	27,753	25.87	21.6 y	....	830	β^-, 0.046; e^-, 0.005, 0.010; γ, 0.070, 0.166, 0.190; α, 4.95, 4.86
	228	31,080	28.91	6.13 h	....	....	β^-, 2.11; e^-, 0.040, 0.054, 0.110; γ, 0.34(15), 0.908(25), 0.96(20)
	229	32,800	31	66 m	....	....	....
	230	36,210	34	<1 m	....	....	β^-, 2.2
	231	38,550	35.9	15 m	....	....	β^-, 2.1; γ, 0.185, 0.28, 0.39, 0.71
$_{90}$Th	223		19.5	0.9 s	....	....	α, 7.56
	224		20.00	1.05 s	....	....	α, 7.18, 6.91; γ, 0.177(9), 0.235, 0.297, 0.410(1)
	225		22.30	8.0 m	....	....	α, 6.80, 6.75, 6.50, 6.48, 6.44; γ, 0.246(5), 0.322(27), 0.362(5), 0.45(1), 0.49(1)
	226		23.19	30.9 m	....	....	α, 6.34, 6.22; γ, 0.111(3), 0.131, 0.20, 0.242(1); e^-, 0.094, 0.107
	227		25.82	18.2 d	....	~1,500(f)	α, 6.04, 5.98, 5.76, 5.72; γ, 0.050(8), 0.237(15), 0.31(8); e^-, 0.013, 0.026, 0.044
	228		26.77	1.910 y	....	123; <0.3(f)	α, 5.43, 5.34; γ, 0.084(2), 0.132, 0.167, 0.214; e^-, 0.067, 0.080
	229	31,652	29.61	7,340 y	....	32(f)	α, 5.05, 4.97, 4.90, 4.84, 4.81; γ, 0.137(3), 0.20(10); e^-, 0.006–0.090
	230	33,087	30.87	8.0×10⁴ y	....	23; ≤0.001(f)	α, 4.68, 4.62; γ, 0.068(1), 0.142, 0.184, 0.253; e^-, 0.051, 0.064
	231	36,291	33.83	25.52 h	....	....	β^-, 0.30; e^-, 0.040, 0.054, 0.061; γ,

Table 3-7 (*Continued*)
TABLE OF NUCLIDES

Nuclide		Mass Excess		Half-life $y=yr, d=day$ $h=hr, m=min$ $s=sec$	Natural Abundance, %	Thermal Neutron Absorption Cross Section σ, barns (mode if not n,γ)	Major Radiations, Energies in MeV and γ Intensities, %
Z Symbol	Mass Number	μmu	MeV				
	232	38,124	35.47	1.41×10^{10} y	100	7.4; $<0.0002(f)$	0.026(2), 0.084(10) α, 4.01, 3.95; e^-, 0.042, 0.055
	233	41,469	38.76	22.12 m	....	1,500; 15(f)	β^-, 1.23; e^-, 0.009, 0.024, 0.036, 0.051, 0.067, 0.082; γ, 0.029(2), 0.087(3), 0.171(1), 0.195, 0.453(1), 0.67, 0.895
	234	43,583	40.64	24.10 d	....	1.8; $<0.01(f)$	β^-, 0.191; e^-, 0.012, 0.025, 0.072, 0.088; γ, 0.063(4), 0.093(4)
$_{91}$Pa	224			0.6 s	....	....	α
	225	26,230	25	0.8 s	....	....	....
	226	27,810	25.96	1.8 m	....	....	α, 6.86, 6.82
	227	28,811	26.83	38.3 m	....	....	γ, 0.065(6), 0.110(2); α, 6.47, 6.42, 6.40, 6.36
	228	31,010	28.86	22 h	....	....	γ, 0.14(3), 0.20(9), 0.28(5), 0.33(18), 0.41(13), 0.46(32), 0.95(93), 1.57(7), 1.85(4); e^-, 0.040, 0.054, 0.110; α, 6.11, 6.08, 6.03, 5.80
	229	32,022	29.88	1.5 d	....	....	e^-, 0.023, 0.038; α, 5.67, 5.62, 5.58, 5.54
	230	34,433	32.17	17.7 d	....	1,500(f)	β^-, 0.41; e^-, 0.034, 0.048; γ, 0.45(18), 0.51(8), 0.91(24), 0.954(50); α, 5.26–5.34
	231	35,877	33.44	3.25×10^4 y	....	200; 0.010(f)	α, 5.06, 5.02, 5.01, 4.95, 4.73; γ, 0.027(6), 0.29(6); e^-, 0–0.10, 0.195, 0.323, 0.350
	232	38,612	35.95	1.31 d	....	~760; ~700(f)	β^-, 1.3, 0.32; e^-, 0.028, 0.043, 0.091; γ, 0.107(5), 0.150(12), 0.39(9), 0.46(9),

Table 3-7 (*Continued*)
TABLE OF NUCLIDES

Nuclide		Mass Excess		Half-life y = yr, d = day h = hr, m = min s = sec	Natural Abundance, %	Thermal Neutron Absorption Cross Section σ, barns (mode if not n,γ)	Major Radiations, Energies in MeV and γ Intensities, %
Z Sym- bol	Mass Number	μmu	MeV				
	233	40,132	37.51	27.0 d	....	22; <0.1(f)	0.57(8), 0.87(51), 0.971(40) β⁻, 0.568, 0.257; e⁻, 0.013, 0.023, 0.036, 0.054, 0.065, 0.185, 0.197, 0.291; γ, 0.31(44)
	234	43,298	40.38	6.75 h	....	<5,000(f)	β⁻, 1.3, 1.13, 0.53; e⁻, 0.024, 0.039, 0.080, 0.095, 0.112; γ, 0.100(50), 0.126(26), 0.22(14), 0.36(13), 0.56(15), 0.70(24), 0.90(70), 1.08(12)
	234 m		40.45	1.175 m	....	<500(f)	β⁻, 2.29; γ, 0.765, 1.001(1)
	235	45,420	42.3	23.7 m	....	....	β⁻, 1.4; no γ
	236	49,230	45	12 m	....	....	β⁻, 3.3
	237	51,080	47.7	39 m	....	....	β⁻, 2.3; γ, 0.090(50), 0.145(45), 0.205(55), 0.275(20), 0.330(40), 0.405(30), 0.46(100), 0.55(30), 0.59(25), 0.75(50), 0.80(45), 0.87(100), 0.92(100), 1.04(35), 1.32(10), 1.42(15)
₉₂U	227		29	1.3 m	....	....	α, 6.8
	228	31,387	29.23	9.1 m	....	....	α, 6.69, 6.60; γ, 0.152, 0.187, 0.246
	229	33,481	31.20	58 m	....	....	α, 6.36, 6.33, 6.30
	230	33,937	31.60	20.8 d	....	25(f)	α, 5.89, 5.82; γ, 0.072, 0.156, 0.231; e⁻, 0.054, 0.068
	231	36,270	33.8	4.3 d	....	~400(f)	γ, 0.026, 0.084(7),

Table 3-7 *(Continued)*
TABLE OF NUCLIDES

Nuclide		Mass Excess		Half-life y=yr, d=day h=hr, m=min s=sec	Natural Abundance, %	Thermal Neutron Absorption Cross Section σ, barns (mode if not n,γ)	Major Radiations, Energies in MeV and γ Intensities, %
Z Symbol	Mass Number	μmu	MeV				
	232	37,168	34.60	72 y	....	78; 77(f)	0.218(1); e^-, 0.040, 0.054, 0.063; α, 5.46 α, 5.32, 5.27; γ, 0.058, 0.129, 0.270, 0.328; e^-, 0.040, 0.054
	233	39,522	36.94	1.62×10^5 y	....	49; 524(f)	α, 4.82, 4.78; γ, 0.029(60), 0.042(310), 0.055(68), 0.097(100), 0.119(40), 0.146(35), 0.164(27), 0.22(45), 0.291(23), 0.32(43); e^-, 0.023, 0.038
	234	40,904	38.16	2.47×10^5 y	0.0057	95	α, 4.77, 4.72; γ, 0.053, 0.117, 0.48, 0.58
	235	43,915	40.93	7.1×10^8 y	0.7196	101; 577(f)	α, 4.58, 4.40, 4.37; γ, 0.143(11), 0.185(54), 0.204(5)
	235 m		40.93	26.1 m	....	....	e^-, ≤ 0.0001
	236	45,637	42.46	2.39×10^7 y	....	6	α, 4.49, 4.44; e^-, 0.032, 0.045
	237	48,608	45.41	6.75 d	....	....	β^-, 0.248; e^-, 0.008, 0.011, 0.038, 0.089, 0.186; γ, 0.026(2), 0.060(36), 0.165(2), 0.208(23), 0.267(1), 0.332(1), 0.370
	238	50,770	47.33	4.51×10^9 y	99.276	2.73; <0.0005(f)	α, 4.20, 4.15; e^-, 0.030, 0.043
	239	54,300	50.60	23.54 m	....	22; 14(f)	β^-, 1.29; e^-, 0.011, 0.023, 0.052, 0.069; γ, 0.044(4), 0.075(51)
	240	56,594	52.74	14.1 h	....	....	β^-, 0.36; e^-, 0.022, 0.038
$_{93}$Np	229			4.0 m	....	....	α, 6.89
	230	37,680	35.1	4.6 m	....	....	α, 6.66
	231	38,280	35.7	~50 m	....	....	α, 6.29

Table 3-7 (*Continued*)
TABLE OF NUCLIDES

Nuclide		Mass Excess		Half-life y=yr, d=day h=hr, m=min s=sec	Natural Abundance, %	Thermal Neutron Absorption Cross Section σ, barns (mode if not n,γ)	Major Radiations, Energies in MeV and γ Intensities, %
Z Sym- bol	Mass Number	μmu	MeV				
	232	39,860	37	~13 m	....	....	U X-rays
	233	40,670	38	35 m	....	....	α, 5.54; γ
	234	42,860	40.0	4.40 d	....	~900(f)	γ, 0.109, 0.23, 0.25, 0.45, 0.50, 0.75, 0.95, 1.21, 1.56; e⁻, 0.024, 0.039, 0.696; β⁺, 0.8
	235	44,049	41.05	410 d	....	....	U X-rays; α, 5.02
	236	46,624	43.41	22 h	....	....	β⁻, 0.52; e⁻, 0.025, 0.040; γ, 0.642, 0.688
	236			>5×10³ y	....	2,500(f)	....
	237	48,056	44.89	2.14×10⁶ y	....	170; 0.019(f)	α, 4.79, 4.77; γ, 0.030(14), 0.086(14), 0.145(1); e⁻, 0.009, 0.024, 0.036, 0.051, 0.067, 0.082
	238	50,896	47.47	2.10 d	....	1,600(f)	β⁻, 1.25; e⁻, 0.022, 0.039; γ, 1.01(42)
	239	52,924	49.32	2.346 d	....	35; <1(f)	β⁻, 0.713, 0.437; e⁻, 0.02–0.04, 0.048, 0.088, 0.106, 0.156; γ, 0.106(23), 0.209(4), 0.228(12), 0.278(14)
	240	56,080	52.2	63 m	....	....	β⁻, 0.89; γ, 0.16, 0.25, 0.44, 0.56, 0.60, 0.92, 1.00, 1.16
	240 m		52.3	7.3 m	....	....	β⁻, 2.16; e⁻, 0.022, 0.038; γ, 0.56(21), 0.60(13), 0.92(3), 1.5(3)
	241	58,200	54.3	16 m	....	....	β⁻, 1.4
	241			3.4 h	....	....	....
₉₄Pu	232	41,180	38.4	36 m	....	....	α, 6.59
	233	42,972	40.04	20 m	....	....	α, 6.31
	234	43,315	40.34	9.0 h	....	....	α, 6.20, 6.15; Np X-rays
	235	45,270	42.2	26 m	....	....	Np X-rays; α, 5.86
	236	46,071	42.90	2.85 y	....	170(f)	α, 5.77, 5.72; γ, 0.048, 0.109; e⁻, 0.028, 0.043
	237	48,298	45.12	45.6 d	....	2,500(f)	γ, 0.060(5); e⁻, 0.026, 0.032,

Table 3-7 (*Continued*)
TABLE OF NUCLIDES

Z Symbol	Mass Number	μmu	MeV	Half-life (y = yr, d = day, h = hr, m = min, s = sec)	Natural Abundance, %	Thermal Neutron Absorption Cross Section σ, barns (mode if not n,γ)	Major Radiations, Energies in MeV and γ Intensities, %
							0.038, 0.042, 0.056; α, 5.66, 5.37
	237 m		45.26	0.18 s	········	·············	γ, 0.145(2); e⁻, 0.125, 0.140
	238	49,511	46.18	86.4 y	········	500; 16.8(f)	α, 5.50, 5.46; γ, 0.099, 0.150, 0.77; e⁻, 0.024, 0.039
	239	52,146	48.60	24,390 y	········	274; 741(f)	α, 5.16, 5.11; γ, 0.039, 0.052, 0.129, 0.375, 0.414, 0.65, 0.77; e⁻, 0.008, 0.019, 0.033, 0.047
	240	53,882	50.14	6,580 y	········	286; <0.08(f)	α, 5.17, 5.12; γ, 0.65; e⁻, 0.026, 0.040
	241	56,737	52.98	13.2 y	········	425; 950(f)	β⁻, 0.021; α, 4.90, 4.85; γ, 0.145
	242	58,725	54.74	3.79×10⁵ y	········	19; <0.2(f)	α, 4.90, 4.86
	243	61,972	57.77	4.98 h	········	170	β⁻, 0.58; e⁻, 0.019, 0.036; γ, 0.084(21), 0.381(1)
	244	64,100	59.83	~7.6×10⁷ y	········	1.8	α, 4.58; SF
	245	67,830	63	10.1 h	········	~260	β⁻
	246	70,090	65.3	10.85 d	········	·············	β⁻, 0.33, 0.15; e⁻, 0.020, 0.038, 0.055, 0.156; γ, 0.044(30), 0.180(10), 0.224(25)
₉₅Am	237	49,840	47	~1.3 h	········	·············	α, 6.02
	238	51,940	48	1.9 h	········	·············	γ, 0.36(12), 0.58(29), 0.98(80), 1.35(76)
	239	53,016	49.41	12.1 h	········	·············	γ, 0.209(5), 0.228(18), 0.278(17); e⁻, 0.02–0.04, 0.048, 0.088, 0.106, 0.156; α, 5.78
	240	55,280	51	51.0 h	········	·············	γ, 0.90(23), 1.00(77); e⁻, 0.022, 0.038, 0.079, 0.094
	241	56,714	52.96	433 y	········	700; 3.0(f)	α, 5.49, 5.44; γ, 0.060(36), 0.101,

Table 3-7 (*Continued*)
TABLE OF NUCLIDES

| Nuclide | | Mass Excess | | Half-life | Natural | Thermal Neutron | Major Radiations, |
Z Symbol	Mass Number	μmu	MeV	y = yr. d = day h = hr. m = min s = sec	Abundance, %	Absorption Cross Section σ, barns (mode if not n,γ)	Energies in MeV and γ Intensities, %
	242	59,502	55.48	16.01 h	....	2,900(f)	0.208, 0.335, 0.37, 0.663, 0.722; e$^-$, 0.022, 0.038, 0.054 β^-, 0.67; e$^-$, 0.021, 0.037; Pu X-rays
	242 m		55.52	152 y	....	2,000; 6,000(f)	α, 5.21; e$^-$, 0.028, 0.044; γ, 0.049, 0.087, 0.110, 0.163
	243	61,367	57.18	7.95×10^3 y	....	74; <0.07(f)	α, 5.28, 5.23; γ, 0.044(4), 0.075(50); e$^-$
	244	64,355	59.90	10.1 h	....	2,300(f)	β^-, 0.387; e$^-$, 0.020, 0.037, 0.077, 0.094; γ, 0.099(5), 0.154(19), 0.746(66), 0.900(25)
	244 m		60.02	26 m	....	....	β^-, 1.50; e$^-$, 0.020, 0.037; Cm X-rays
	245	66,340	61.93	2.07 h	....	....	β^-, 0.91; e$^-$, 0.125; γ, 0.253
	246	69,660	64.9	25.0 m	....	....	β^-, 2.10, 1.60; γ, 0.799(29), 1.07(65)
	247	72,090	67	24 m	....	....	β^-; γ, 0.29, 0.23
$_{96}$Cm	238	53,036	49.39	2.5 h	....	....	α, 6.51
	239	54,880	51	2.9 h	....	....	γ, 0.188
	240	55,545	51.72	26.8 d	....	....	α, 6.26, 6.25
	241	57,542	53.73	35 d	....	....	γ, 0.475(95), 0.60; e$^-$, 0.123, 0.350; α, 5.94
	242	58,788	54.82	162.5 d	....	20; <5(f)	α, 6.12, 6.07; γ, 0.044, 0.102, 0.158, 0.58, 0.89; e$^-$, 0.022, 0.039
	243	61,370	57.19	32 y	....	250; 660(f)	α, 6.06, 5.99, 5.79, 5.74; γ, 0.209(4), 0.228(12), 0.278(14); e$^-$, 0.02–0.04, 0.048, 0.088, 0.106, 0.156
	244	62,821	58.47	18.099 y	....	15	α, 5.81, 5.77; γ, 0.043, 0.100, 0.150, 0.262, 0.59, 0.82; e$^-$,

Table 3-7 (*Continued*)
TABLE OF NUCLIDES

Nuclide		Mass Excess		Half-life y = yr, d = day h = hr. m = min s = sec	Natural Abundance. %	Thermal Neutron Absorption Cross Section σ. barns (mode if not n.γ)	Major Radiations, Energies in MeV and γ Intensities, %
Z Symbol	Mass Number	μmu	MeV				
	245	65,371	61.02	8.3×10³ y	....	200; 1,900(f)	0.022, 0.038 α, 5.36, 5.31; γ, 0.13(5), 0.173(14)
	246	67,202	62.64	4.7×10³ y	....	15	α, 5.39, 5.34; γ
	247	70,280	65.56	1.6×10⁷ y	....	180	....
	248	72,220	67.43	3.8×10⁵ y	....	6	α, 5.08, 5.04; γ
	249	75,810	70.8	64 m	....	....	β⁻, 0.9
	250		73	1.7×10⁴ y	....	....	SF
₉₇Bk	243	62,965	58.70	4.6 h	....	....	α, 6.76, 6.72, 6.57, 6.54, 6.21; γ, 0.755, 0.84, 0.946
	244	65,170	61	4.4 h	....	....	α, 6.67, 6.62; γ, 0.145(7), 0.188(16), 0.218(100), 0.334(10), 0.490(14), 0.892(88), 0.922(17), 1.16(11)
	245	66,272	61.84	4.98 d	....	....	α, 6.36, 6.32, 6.15, 6.12, 5.89; γ, 0.253(31), 0.39(3); e⁻, 0.125
	246	68,770	64	1.8 d	....	....	γ, 0.800(40), 1.07(12)
	247	70,260	65.47	1.4×10³ y	....	....	α, 5.68, 5.52; γ, 0.084(40), 0.27(30)
	248	72,960	67.9	16 h	....	....	β⁻, 0.65; γ
	248			>9 y	....	....	....
	249	74,883	69.86	314 d	....	500	β⁻, 0.125; α, 5.42; γ, 0.32
	250	78,270	72.95	193.3 m	....	....	β⁻, 1.76, 0.73; e⁻, 0.019, 0.036; γ, 0.990(47), 1.032(39)
	251	80,810	75	57 m	....	....	β⁻, ~1.0, ~0.5; γ, 0.034, 0.14
₉₈Cf	242			3.4 m	....	....	α, 7.39; e⁻
	243	65,310	61	10.3 m	....	....	e⁻; α, 7.17, 7.06; γ, 0.12
	244	65,969	61.43	19.4 m	....	....	α, 7.21
	245	67,905	63.38	44 m	....	....	α, 7.12
	246	68,766	64.11	35.7 h	....	....	α, 6.76, 6.72; γ
	247	71,070	66	2.5 h	....	....	γ, 0.295(1), 0.417, 0.460; e⁻, 0.164
	248	72,262	67.26	350 d	....	....	α, 6.27, 6.22; γ

Table 3-7 (*Continued*)
TABLE OF NUCLIDES

| Nuclide | | Mass Excess | | Half-life | Natural | Thermal Neutron | Major Radiations, |
Z Symbol	Mass Number	μmu	MeV	y = yr, d = day h = hr, m = min s = sec	Abundance, %	Absorption Cross Section σ, barns (mode if not *n*,γ)	Energies in MeV and γ Intensities, %
	249	74,749	69.74	352 y	....	270; 1,735(*f*)	α, 5.81; γ, 0.333(16), 0.388(72)
	250	76,384	71.19	13.1 y	....	1,500; <350(*f*)	α, 6.03, 5.99; *e*⁻, 0.023, 0.038; γ
	251	79,260	74.15	900 y	....	3,000; 3,000(*f*)	α, 5.85, 5.67; γ, 0.18
	252	81,500	76.05	2.646 y	....	30	α, 6.12, 6.08; *e*⁻, 0.022, 0.038; γ, SF
	253	85,020	79.3	17.6 d	....	....	β⁻, 0.27; α, 5.98
	254		81	60.5 d	....	<2	SF; α, 5.83, 5.79
₉₉Es	245	71,060	66	1.3 m	....	....	α, 7.70
	246	72,430	68	7.3 m	....	....	α, 7.33
	247	73,580	68	5.0 m	....	....	α, 7.33
	248	75,280	70	25 m	....	....	α, 6.88
	249	76,258	71.15	2 h	....	....	α, 6.77
	250	78,610	73	8 h	....	....	γ
	251	79,930	74.5	1.5 d	....	....	α, 6.48
	252	82,810	77.1	~140 d	....	....	α, 6.64, 6.58; γ, 0.074, 0.154, 0.198, 0.228, 0.278, 0.40(1)
	253	84,730	79.03	20.47 d	....	300	α, 6.64; *e*⁻, 0.017, 0.027, 0.035, 0.040; γ, 0.387, 0.429
	254	87,900	82.00	276 d	....	<40	α, 6.44; γ, 0.063(2), 0.27, 0.31, 0.39; *e*⁻, 0.011, 0.018, 0.030, 0.037
	254 *m*		82.10	39.3 h	....	....	β⁻, 1.13, 0.43; *e*⁻, 0.020, 0.038; γ, 0.65(31), 0.69(38)
	255		84	38.3 d	....	....	α, 6.31
₁₀₀Fm	245			4.2 s	....	....	α, 8.15
	246			1.2 s	....	....	α, 8.24; SF
	247			9.2 s	....	....	α, 8.18
	247			35 s	....	....	α, 7.93, 7.87
	248	77,092	72	38 s	....	....	α, 7.87, 7.83; SF
	249	79,140	73.8	~2.5 m	....	....	α, 7.53
	250	79,490	74.10	30 m	....	....	α, 7.44
	251	81,190	76	7 h	....	....	α, 6.90; γ, 0.41
	252	82,562	76.84	22.7 h	....	....	α, 7.06
	253	84,930	80	3 d	....	....	α, 6.96, 6.91
	254	86,839	80.93	3.24 h	....	....	α, 7.20, 7.16; γ; *e*⁻, 0.019, 0.036
	255	89,640	83.82	20.1 h	....	....	α, 7.03; γ, 0.059(1), 0.081(1); *e*⁻,

Table 3-7 (*Continued*)
TABLE OF NUCLIDES

Nuclide		Mass Excess		Half-life y = yr, d = day h = hr, m = min s = sec	Natural Abundance, %	Thermal Neutron Absorption Cross Section σ, barns (mode if not n,γ)	Major Radiations, Energies in MeV and γ Intensities, %
Z Symbol	Mass Number	μmu	MeV				
							0.032, 0.05–0.07
	256		85.44	2.7 h	....	....	SF; α, 6.93
	257		88.6	80 d	....	....	α, 6.53; γ, 0.180(8), 0.242(10); e⁻, 0.037, 0.045, 0.055, 0.106
	258			380 μs	....	....	SF
₁₀₁Md	248			6 s	....	....	α, 8.32
	249			24 s	....	....	α, 8.03
	250			53 s	....	....	α, 7.73
	251			4.0 m	....	....	α, 7.53
	252	86,120	80	~8 m	....	....	....
	254			10 m	....	....	....
	255	90,550	84.4	27 m	....	....	α, 7.34
	256		86.9	77 m	....	....	α, 7.23, 7.17
	257		89	300 m	....	....	α, 7.08
	258			54 d	....	....	α, 6.79, 6.73
₁₀₂No	251			0.8 s	....	....	α, 8.68, 8.60
	252		83	2.4 s	....	....	α, 8.41; SF
	253	91,340	84	105 s	....	....	α, 8.01
	254	91,140	84.8	55 s	....	....	α, 8.10
	255	92,730	87	185 s	....	....	α, 8.11, 7.92, 7.76
	256		87.83	3.2 s	....	....	α, 8.43; SF
	257		90	26 s	....	....	α, 8.32, 8.27, 8.22
	258			<<1 s	....	....	SF
	259			~1 h	....	....	α, 7.50
₁₀₃Lr	253			0.5 s	....	....	α, 8.8
	254			20 s	....	....	....
	255			22 s	....	....	α, 8.37, 8.35
	256			31 s	....	....	α, 8.64, 8.52, 8.48, 8.43, 8.39, 8.32
	257			0.6 s	....	....	α, 8.87, 8.81
	258			4.2 s	....	....	α, 8.68, 8.65, 8.62, 8.59
	259			5.4 s	....	....	α, 8.45
	260			180 s	....	....	α, 8.03
₁₀₄Rf	257			4.5 s	....	....	α, 9.00, 8.95, 8.78, 8.70
	258?			~0.01 s	....	....	SF
	259			3 s	....	....	α, 8.86, 8.77
	260?			0.1 s	....	....	SF
	261			70 s	....	....	α, 8.40–8.25; SF
₁₀₅Ha	260			1.6 s	....	....	α, 9.14, 9.10, 9.06
	261			1.7 s	....	....	α, 8.93
	262			~40 s	....	....	α, 8.66–8.45
	263?			~60 s	....	....	α, 8.8

Rf = rutherfordium; Ha = hahnium; names and symbols not officially accepted by I.U.P.A.C.

Table 3-8
SPATIAL ORIENTATION OF COMMON HYBRID BONDS

On the assumption that the pairs of electrons in the valency shell of a bonded atom in a molecule are arranged in a definite way which depends on the number of electron pairs (coordination number), the geometrical arrangement or shape of molecules may be predicted. A multiple bond is regarded as equivalent to a single bond as far as molecular shape is concerned.

Coordination Number	Orbitals Hybridized	Geometrical Arrangement	Minimum Radius Ratio
2	sp dp	Linear	
	p^2 ds d^2	Bent (angular)	
3	sp^2 ds^2	Trigonal planar	0.155
	p^3 d^2p	Trigonal pyramidal	
4	sp^2d p^2d^2	Square planar	
	sp^3 d^3s	Tetrahedral	0.225
	d^4	Tetragonal pyramidal	
5	sp^3d d^3sp	Trigonal bipyramidal	0.155
6	d^2sp^3	Octahedral	0.414
	d^4sp	Trigonal prism	
7		One atom above the face of an octahedron, which is distorted chiefly by separating the atoms at the corners of this face.	0.592
8	d^4sp^3	Square antiprism (dodecahedral)	0.645
		Cube	0.732
9		Formed by adding atoms beyond each of the vertical faces of a right triangular prism.	0.732
12		Cube-octahedron	1.000

3-117

Table 3-9
CRYSTAL STRUCTURE

Unit cells of the different lattice types in each system are illustrated in Fig. 3-1.

System	Characteristics	Essential Symmetry	Axes in Unit Cell	Angles in Unit Cell
Cubic	Three axes equal and mutually perpendicular	Four threefold axes	$a = b = c$	$\alpha = \beta = \gamma = 90°$
Tetragonal	Two equal axes and one unequal axis mutually perpendicular	One fourfold axis	$a = b \neq c$	$\alpha = \beta = \gamma = 90°$
Orthorhombic (or rhombic)	Three unequal axes mutually perpendicular	Three mutually perpendicular twofold axes, or two planes intersecting in a twofold axis	$a \neq b \neq c$	$\alpha = \beta = \gamma = 90°$
Hexagonal or trigonal	Three equal axes inclined at 120° with a fourth axis unequal and perpendicular to the other three	One sixfold axis or one threefold axis	$a = b \neq c$ $a = b = c$	$\alpha = \beta = 90°;$ $\gamma = 120°$ $\alpha = \beta = \gamma \neq 90°$
Monoclinic	Two axes at an oblique angle with a third perpendicular to the other two	One twofold axis or one plane	$a \neq b \neq c$	$\alpha = \beta = 90°;$ $\gamma \neq 90°$
Triclinic	Three unequal axes intersecting obliquely	No planes or axes of symmetry	$a \neq b \neq c$	$\alpha \neq \beta \neq \gamma \neq 90°$
Rhombohedral	Two equal axes making equal angle with each other			

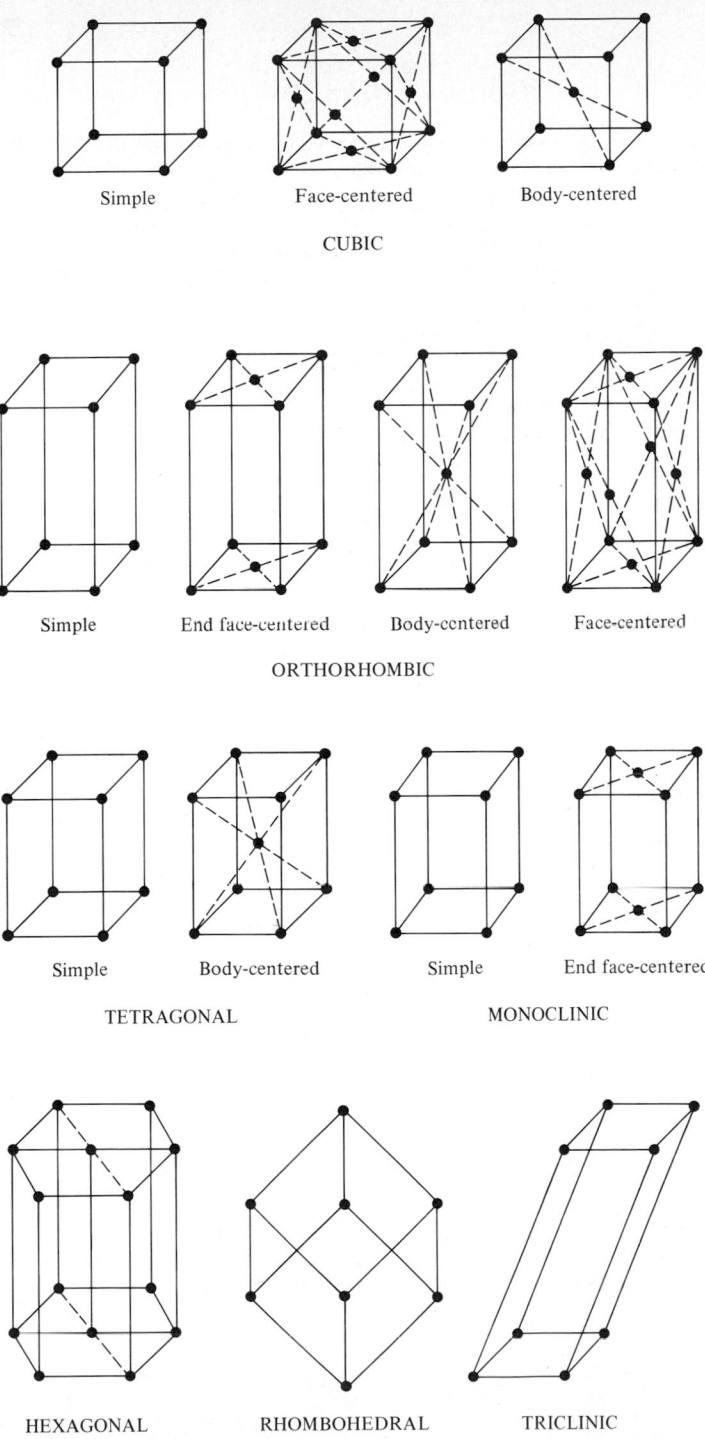

Simple Face-centered Body-centered

CUBIC

Simple End face-centered Body-centered Face-centered

ORTHORHOMBIC

Simple Body-centered Simple End face-centered

TETRAGONAL MONOCLINIC

HEXAGONAL RHOMBOHEDRAL TRICLINIC

Fig. 3-1. Crystal lattice types.

Table 3-10
RADII OF ATOMS AND IONS

The radius of an atom is the closest distance to which it will approach another atom of any size under the circumstances specified.

The *ionic radius* of an atom is the radius it exhibits in an ionic crystal like KCl in which the K^+ and Cl^- ions are packed together with their outermost shells of electrons in contact with each other. Values of the ionic radii are tabulated for a coordination number of 6. For structures with coordination numbers other than 6 the appropriate correction factors given below must be employed:

Coordination Number	Correction Factors	
	Ionic Radius	Metallic Radius
12	1.12	1.00
9	1.05	
8	1.03	0.98
6	1.00	0.96
4	0.94	0.88

The normal *covalent radius* is the effective distance from the center of the nucleus to the outer valence shell of that atom in a typical covalent or coordinate bond, assuming spherical symmetry. For example, it is half the distance between the centers of the two bromine atoms in the Br_2 molecule. The bond length contracts for multiple bonds; the double-bond radius is roughly 86 per cent of the length of a single-bond radius, and the triple bond is 78 per cent.

The *nonbonded* or *van der Waals radius* is the internuclear distance or radius of closest approach of an atom to another with which it forms no bond. For nonmetals the nonbonded radius is close to the anion radius.

One can also speak about the *radius of metal atoms* in a crystalline metal. A correction must be applied for coordination numbers other than 12 (see table above).

The *atomic volume*, a rough relative measure for the space occupied by the atoms in 1 gram atom of the element in question, is given by atomic weight/density.

Element	Crystal Ionic Radius		Covalent Radius, Å	Atomic Radius in Metals, Å	Van der Waals Radius, Å
	Charge	Radius, Å			
Actinium	+3	1.11		1.878	
Aluminum	+3	0.50	1.25	1.431	
Americium	+3	0.99		1.84	
	+4	0.89			
	+5	0.86			
	+6	0.80			
Antimony	−3	2.45	(s) 1.41		2.2
	+1	0.89	(d) 1.31		
	+3	0.9			
	+5	0.62			
Argon	+1 (gas)	1.54	1.74		1.91
Arsenic	−3	2.22	(s) 1.21	1.248	2.0
	+3	0.58	(d) 1.11		
	+5	0.47			

Table 3-10 (*Continued*)
RADII OF ATOMS AND IONS

Element	Crystal Ionic Radius		Covalent Radius, Å	Atomic Radius in Metals, Å	Van der Waals Radius, Å
	Charge	Radius, Å			
Astatine	−1	2.27			
	+5	0.57			
	+7	0.51			
Barium	+2	1.35	1.98	2.173	
Berkelium	+2	1.18			
	+3	0.98			
	+4	0.87			
Beryllium	−1	1.95	0.89	(α) 1.113	
	+2	0.31			
Bismuth	−3	2.13	1.52	1.547	
	+3	0.96			
	+5	0.74			
Boron	+1	0.35	(s) 0.88	0.83	2.08
	+3	0.20	(d) 0.76		
			(t) 0.68		
Bromine	−1	1.96	(s) 1.142		1.95
	+5	0.47	(d) 1.04		
	+7	0.39			
Cadmium	+1	1.14	1.41	1.489	
	+2	0.97			
Calcium	+2	0.99	1.74	(α) 1.973	
				(β) 1.939	
Californium	+2	1.17			
	+3	0.98			
	+4	0.86			
Carbon	−4	2.60	(s) 0.77		1.85
	+4	0.15	(d) 0.67		
			(t) 0.60		
Cerium	+3	1.03	1.646	1.825	
	+4	0.92			
Cesium	+1	1.69	2.35	2.654	2.62
Chlorine	−1	1.81	(s) 0.99		1.81
	+5	0.34	(d) 0.89		
	+7	0.26			
Chromium	+1	0.81	1.17	(α) 1.249	
	+2	0.84		(β) 1.305	
	+3	0.64			
	+4	0.56			
	+6	0.52			
Cobalt	+2	0.74	1.16	1.253	
	+3	0.63			
Copper	+1	0.96	1.17	1.278	
	+2	0.72			
Curium	+2	1.19			
	+3	0.99			
	+4	0.88			
Dysprosium	+3	0.91	1.589	1.773	
Einsteinium	+2	1.16			
	+3	0.98			
	+4	0.85			

Table 3-10 (*Continued*)
RADII OF ATOMS AND IONS

Element	Crystal Ionic Radius		Covalent Radius, Å	Atomic Radius in Metals, Å	Van der Waals Radius, Å
	Charge	Radius, Å			
Erbium	+3	0.88	1.567	1.757	
Europium	+2	1.12	1.850	2.042	
	+3	0.95			
Fermium	+2	1.15			
	+3	0.97			
	+4	0.84			
Fluorine	−1	1.36	(s) 0.64	0.717	1.35
	+7	0.07	(d) 0.54		
Francium	+1	1.76		2.7	
Gadolinium	+3	0.94	1.614	1.802	
Gallium	+3	0.62	1.25	1.221	
Germanium	−4	2.72	(s) 1.22	1.225	
	+2	0.70	(d) 1.12		
	+4	0.53			
Gold	+1	1.37	1.34	1.442	
	+2	1.05			
	+3	0.91			
Hafnium	+4	0.78	1.44	(α) 1.564	
Helium	+1 (gas)	0.93			1.22
Holmium	+3	0.89	1.580	1.766	
Hydrogen	−1	2.08	0.371		1.2
	+1	10^{-5}			
Indium	+1	1.32	1.50	1.626	
	+3	0.81			
Iodine	−1	2.16	(s) 1.333		2.15
	+5	0.62	(d) 1.23		
	+7	0.50			
Iridium	+2	0.89	, 1.26	1.357	
	+3	0.75			
	+4	0.64			
	+6	0.56			
Iron	+2	0.76	1.165	(α) 1.241	
	+3	0.64		(γ) 1.289	
				(δ) 1.27	
Krypton	+1 (gas)	1.69	1.89		1.98
Lanthanum	+3	1.06	1.690	1.877	
Lawrencium	+2	1.12			
	+3	0.94			
	+4	0.83			
Lead	−4	2.15	1.54	1.750	
	+2	1.20			
	+4	0.84			
Lithium	+1	0.68	1.23	1.52	
Lutetium	+3	0.85	1.557	1.734	
Magnesium	+2	0.65	1.36	1.60	
Manganese	+2	0.80	1.17	(α) 1.24	
	+3	0.62		(γ) 1.366	
	+4	0.54		(δ) 1.334	
	+7	0.46			
Mendelevium	+2	1.14			

Table 3-10 (*Continued*)
RADII OF ATOMS AND IONS

Element	Crystal Ionic Radius		Covalent Radius, Å	Atomic Radius in Metals, Å	Van der Waals Radius, Å
	Charge	Radius, Å			
	+3	0.96			
	+4	0.84			
Mercury	+1	1.27	1.44	1.60	
	+2	1.10			
Molybdenum	+4	0.66	1.29	1.362	
	+6	0.62			
Neodymium	+3	1.00	1.642	1.821	
Neon	+1 (gas)	1.12	1.31		1.60
Neptunium	+3	1.01		(α) 1.31	
	+4	0.92		(β) 1.38	
	+5	0.88		(γ) 1.52	
	+6	0.82			
Nickel	+2	0.72	1.15	1.246	
	+3	0.62			
Niobium	+4	0.74	1.34	1.429	
	+5	0.70			
Nitrogen	−3	1.71	(s) 0.70	[N_2 0.549]	1.54
	+1	0.25	(d) 0.60		
	+3	0.13	(t) 0.55		
	+5	0.11			
Nobelium	+2	1.13			
	+3	0.95			
	+4	0.83			
Osmium	+2	0.89	1.26	1.34	
	+3	0.81			
	+4	0.65			
	+6	0.6			
	+8	0.53			
Oxygen	−2	1.40	(s) 0.66	[O_2 0.603]	1.40
	−1	1.76	(d) 0.55		
	+1	0.22	(t) 0.51		
	+6	0.09			
Palladium	+2	0.86	1.28	1.376	
	+4	0.64			
Phosphorus	−3	2.12	(s) 1.10	black crys 1.08	1.9
	+3	0.42	(d) 1.00	yellow crys 0.93	
	+5	0.34	(t) 0.93	amorph red 1.15	
Platinum	+2	0.85	1.29	1.38	
	+4	0.70			
Plutonium	+3	1.00		(γ) 1.51	
	+4	0.90		(δ) 1.64	
	+5	0.87		(ε) 1.58	
	+6	0.81			
Polonium	−2	2.30	1.53	(α) 1.67	
	+4	0.65		(β) 1.68	
	+6	0.56			
Potassium	+1	1.33	2.025	2.272	2.31
Praseodymium	+3	1.01	1.648	1.828	
	+4	0.90			
Promethium	+3	0.98		1.810	

Table 3-10 (*Continued*)
RADII OF ATOMS AND IONS

Element	Crystal Ionic Radius		Covalent Radius, Å	Atomic Radius in Metals, Å	Van der Waals Radius, Å
	Charge	Radius, Å			
Protactinium	+3	1.05		1.606	
	+4	0.96			
	+5	0.90			
Radium	+2	1.40		2.20	
Radon			2.14		
Rhenium	+4	0.72	1.28	1.370	
	+6	0.61			
	+7	0.60			
Rhodium	+2	0.86	1.25	1.345	
	+3	0.75			
	+4	0.67			
Rubidium	+1	1.48	2.16	2.475	2.44
Ruthenium	+3	0.77	1.24	1.325	
	+4	0.63			
	+8	0.54			
Samarium	+2	1.11	1.66	1.802	
	+3	0.96			
Scandium	+3	0.81	1.44	1.606	
Selenium	−2	1.98	(s) 1.17		2.00
	−1	2.32	(d) 1.07		
	+1	0.66			
	+4	0.69			
	+6	0.42			
Silicon	−4	2.71	(s) 1.17		2.0
	−1	3.84	(d) 1.07		
	+1	0.65	(t) 1.00		
	+4	0.41			
Silver	+1	1.26	1.34	1.444	
	+2	0.97			
Sodium	+1	0.95	1.57	1.537	2.31
Strontium	+2	1.13	1.92	(α) 2.151	
				(β) 2.16	
				(γ) 2.10	
Sulfur	−2	1.84	(s) 1.04	[S_2 0.944]	1.85
	−1	2.19	(d) 0.94	[S_8 1.04]	
	+4	0.37	(t) 0.87		
	+6	0.29			
Tantalum	+5	0.7	1.34	1.43	
Technetium	+2	0.95		1.358	
	+4	0.72			
	+7	0.58			
Tellurium	−2	2.21	(s) 1.37	1.432	2.20
	−1	2.50	(d) 1.27		
	+4	0.81			
	+6	0.56			
Terbium	+3	0.92	1.592	1.782	
	+4	0.84			
Thallium	+1	1.44	1.55	(α) 1.704	
	+3	0.95		(β) 1.681	
Thorium	+3	1.08		(α) 1.798	

Table 3-10 (*Continued*)
RADII OF ATOMS AND IONS

Element	Crystal Ionic Radius		Covalent Radius, Å	Atomic Radius in Metals, Å	Van der Waals Radius, Å
	Charge	Radius, Å			
	+4	0.99		(β) 1.78	
Thulium	+3	0.87	1.562	1.746	
	+4	0.94			
Tin	−4	2.94	(s) 1.40	1.405	
	−1	3.70	(d) 1.30		
	+2	1.02			
	+4	0.71			
Titanium	+1	0.96	1.32	(α) 1.448	
	+2	0.90		(β) 1.432	
	+3	0.77			
	+4	0.68			
Tungsten	+4	0.68	1.30	1.370	
	+6	0.65			
Uranium	+3	1.03		(α) 1.385	
	+4	0.93		(β) 1.53	
	+5	0.87			
	+6	0.83			
Vanadium	+2	0.88	1.22	1.321	
	+3	0.74			
	+4	0.60			
	+5	0.59			
Xenon	+1 (gas)	1.90	2.09	2.18	
Ytterbium	+2	1.13	1.699	1.940	
	+3	0.86			
Yttrium	+3	0.93	1.62	1.81	
Zinc	+1	0.88	1.25	1.332	
	+2	0.74			
Zirconium	+1	1.09	1.45	1.60	
	+4	0.79			

(s) single bond; (d) double bond; (t) triple bond.

Table 3-11
BOND ENERGIES

The quantity D_0(A—B) corresponds to the bond dissociation energy at $0°K$, all species considered to be ideal gases, for a bond A—B which is broken through the reaction: AB → A + B, where

$$D_0 = \Delta Hf_0°(A) + \Delta Hf_0°(B) - \Delta Hf_0°(AB)$$

D_0 at $298°K$, or $\Delta Hf_{25°}$, is greater than D_0 at $0°K$ by an amount which lies between RT and $(3/2)RT$, or between 0.6 and 0.9 kcal mol^{-1}. In polyatomic molecules this difference may be somewhat greater. It is important to note that the bond dissociation energy refers to the enthalpy change ΔHf in the dissociation process.

References: T. L. Cottrell, *The Strengths of Chemical Bonds*, 2d ed., Butterworths Scientific Publications, London, 1958; B. deB. Darwent, *National Standard Reference Data Series*, National Bureau of Standards, No. 31, Washington, D.C., 1970; and S. W. Benson, *J. Chem. Educ.*, **42**, 502 (1965).

Bond	$D_0°$	$\Delta Hf_{25°}$	Bond	$D_0°$	$\Delta Hf_{25°}$
Aluminum			BrBa—Br		(143 ± 13)
Al—Al	45 ± 10		Ba—Cl	118 ± 10	
Al—Br	105 ± 2		ClBa—Cl	109 ± 10	
Al—C		61	Ba—F	136 ± 10	
Al—Cl	117 ± 3	118 ± 3	FBa—F	140 ± 10	
ClAl—Cl	95.5 ± 2	96 ± 2	Ba—H	46	
Cl$_2$Al—Cl	89 ± 2	89 ± 2	Ba—O	134 ± 10	
OAl—Cl		123 ± 20	Ba—OH	113 ± 10	
Al—F	157.5 ± 1.5	158.6 ± 1.5	HOBa—OH	110 ± 20	
FAl—F	130 ± 10		Ba—S	95 ± 5	
F$_2$Al—F	130 ± 11		Beryllium		
OAl—F		182 ± 10	Be—Br		89
Al—H	67.0 ± 1.5	68.1 ± 1.5	Be—Cl	92 ± 15	93 ± 15
Al—I	87 ± 1	88 ± 1	ClBe—Cl	128 ± 15	129 ± 15
Al—O	115 ± 2	116 ± 2	Be—F	137 ± 10	138 ± 10
ClAl—O		129 ± 10	FBe—F	165 ± 15	167 ± 15
FAl—O		139	Be—H	53 ± 5	54 ± 5
Al—P	50.8 ± 3		Be—I		69
Al—S	96	97	Be—O	106 ± 5	107 ± 5
Antimony			Bismuth		
Sb—C		47	Bi—Bi	46 ± 1	47 ± 1
Sb—Cl		74	Bi—Br		63 ± 3
Sb—F	92 ± 23		Bi—C		31
Sb—H		61	Bi—Cl		67
Sb—Sb	70.6 ± 1.5	71.6	Bi—F	74 ± 9	
Sb—Te	60.0		Bi—H	58	
Arsenic			Bi—O	85 ± 21	
As—As	91 ± 5	91.5 ± 5	Bi—S		73
As—Br		58	Bi—Se	55	62
As—C		48	Bi—Te	48	54
As—Cl	106	107	Boron		
As—F		111	B—B	70 ± 5	71 ± 5
As—H		70 ± 3	OB—BO	120 ± 20	121 ± 20
As—I		43	B—Br	103.5 ± 5	104 ± 5
As—N	138	139	B—C	106	
As—O	114 ± 2	115 ± 2	B—Cl	127 ± 7	128 ± 7
Astatine			OB—Cl		110 ± 10
At—At	27.7		B—F	181.5 ± 2.5	183 ± 3
Barium			FB—F	125 ± 15	
Ba—Br	(65)	(66)	OB—F		170 ± 10

Table 3-11 (*Continued*)
BOND ENERGIES

Bond	D_0°	ΔHf_{25°	Bond	D_0°	ΔHf_{25°
Boron (*cont.*)			$H_2C{=}CH_2$		163
B—H	78 ± 1		$HC{\equiv}CH$		230
B—N	92 ± 5	93 ± 5	$\dot{C}H_2$—CH_3		96
B—O	187 ± 10	188 ± 10	$\dot{C}H_2CH_2$—CH_3		25.5
ClB—O		171 ± 10	$(\dot{C}H_2)_2C$—CH_3		51
FB—O		175 ± 10	$\dot{C}HCH$—CH_3		32
B—S	118 ± 10	119 ± 10			
B—Si	68		(cyclohexane ring)—CH_3		27.5
Bromine					
Br—Br	45.45 ± 0.01	46.10 ± 0.01	(cyclohexene ring)—CH_3		35
Br—Cl	51.6 ± 0.1	52.3 ± 0.1			
Br—F	67	68	(cyclohexadiene ring)—CH_3		11.5
Br—H	86.6 ± 0.1	87.5 ± 0.1			
Br—O	55.3 ± 0.1	56.2 ± 0.1	$(CH_3)_2C(\dot{C}H_2)$—CH_3		20
Br—Br$^+$		74	$\dot{O}CH_2$—CH_3		12
Br$^+$—H		83	$(CH_3)_2C(\dot{O})$—CH_3		7
Cadmium			$\dot{C}H_2CO$—CH_3		30
Cd—Br	76 ± 10		$\dot{O}C$—CH_3		11
Cd—Cl	(84)		$\dot{O}_2C$—CH_3		−20
Cd—H	15.6		CH_3^+—CH_3		46
Cd—I	(50)		CH_2—CH_3^+		119
Cd—O	30 ± 10		$CH_2{=}CH_2^+$		162
Cd—S	(≤47)		$HC{\equiv}CH^+$		223
Cd—Se	(≤74)		C—Cl	80 ± 10	
Calcium			CCl_3—Cl		70 ± 5
Ca—Cl	102 ± 10		CF_3—Cl		81 ± 3
Ca—F	132 ± 10		CH_3—Cl		81 ± 5
Ca—H	≤39		$(CH_3)_3C$—Cl		78.5
Ca—I	58 ± 23		CH_2Cl—Cl		73.6 ± 2.8
Ca—O	110 ± 20		CH_3CO—Cl		83.5
Ca—OH	100 ± 8		Vinyl—Cl		84
Ca—S	74 ± 5		CH_3^+—Cl		51
Carbon			C—F		116
CH_3—Br	67 ± 2	68 ± 2	CCl_3—F		106 ± 5
CH_2Br—Br		61 ± 3	CF_3—F		125 ± 4
CBr_3—Br	49 ± 3	50 ± 3	CH_3—F		108 ± 5
CCl_3—Br	51 ± 3	52 ± 3	$(CH_3)_3C$—F		105
CF_3—Br		68 ± 3	C_6H_5—F		116
C—C	144 ± 5	145 ± 5	CH_3CO—F		119
CH_3—CH_3		88	C—H	80	81 ± 0.5
$(CH_3)_3C$—CH_3		80	CH—H		106
$(CH_3)_3C$—$C(CH_3)_3$		67.5	CH_2—H	112.3 ± 0.1	113 ± 1
CH_3—C_6H_5		93	CH_3—H	101.6 ± 2.0	102.7 ± 2.0
CH_3—$CH_2C_6H_5$		72	CH_3CH_2—H		98
$(CH_3)_3C$—$C(C_6H_5)_3$		15	$(CH_3)_2CH$—H		94.5
CH_3—allyl		72	$(CH_3)_3C$—H		91
CH_3—vinyl		29	CBr_3—H	88 ± 2	90 ± 2
CH_3—CCH		117	CH_2Br—H	97 ± 5	
CH_2CH—$CHCH_2$		100	CCl_3—H	89 ± 3	90 ± 3
HCC—CCH		150	C_6H_5—H		103
CH_3—CN	119 ± 5	121 ± 5	$(C_6H_5)_3C$—H		75
NC—CN	143 ± 5	144 ± 5	Vinyl—H		103
C_6H_5—C_6H_5		100			
CH_3—CHO		75			
CH_3—CO		11 ± 1			

Table 3-11 (*Continued*)
BOND ENERGIES

Bond	D_0°	ΔHf_{25°	Bond	D_0°	ΔHf_{25°
Carbon (*cont.*)			CH_3CO-OH		108 ± 5
Allyl—H		85	$CH_3CO-OCH_3$		97
HCC—H		~125	CH_3CH_2O-OH		43
(C₆H₅ ring)—H		74	Vinyl—OH		87
NC—H	127 ± 5	129 ± 5	C=O	256.2 ± 0.1	257.3 ± 0.1
OC—H		30.4 ± 2.3	OC=O	125.7 ± 0.1	127.2 ± 0.1
HCO—H		(88)	$CH_2{=}O$		175
CF_3—H	105 ± 3	106 ± 3	SC=O		150
$HOCH_2$—H		93	C≡O		257
HOOC—H		90	CH_3^+—OH		67
$\dot{C}H_2$—H		106	C—S	181 ± 5	182 ± 5
$\dot{C}H$—H		106	CH_3—SH	71 ± 3	73 ± 3
$\dot{C}$—H		81	Cesium		
$\dot{C}H_2CH_2$—H		39	Cs—Br	99.5 ± 3	
$\dot{O}CH_2$—H		22	Cs—Cl	104 ± 5	
$\dot{C}O$—H		19	Cs—Cs	10.4	
$\dot{C}HCH$—H		43	Cs—F	120 ± 10	
$\dot{C}C$—H		~125	Cs—H	42	
$\dot{C}OCH_2$—H		43.5	Cs—I	80 ± 5	
$\dot{C}H_2CO$—H		36	Cs—OH	91 ± 3	
(cyclohexyl radical)—H		40	Chlorine		
(cyclohexenyl radical)—H		47.5	Cl—Cl	57.3 ± 0.1	
(cyclohexadienyl radical)—H		24	Cl—ClO	33.3 ± 1	
C^+—H		85	Cl—F	59.5 ± 0.5	
CH_3^+—H		30	O_3Cl—F		61
$CH_3CH_2^+$—H		29	Cl—H	102.3 ± 0.1	
$CH_2CH_3^+$—H		79	$Cl-NH_2$		(60 ± 6)
CH_3—I	54 ± 3	55.5 ± 3	Cl—O	64 ± 1	
C_2H_5—I		53.5	OCl—O	58 ± 3	
$i\text{-}C_3H_5$—I		53	O_2Cl—O		48 ± 1
t-Bu—I		49.5	Cl—OH		60
CH_3CO—I		52.5	$Cl-Cl^+$ or (Cl_2^+)		94
CH_3CHCH—I		41	Cl^+—H		97
CH_3^+—I		62	Chromium		
C—N	174.5 ± 5	175 ± 5	Cr—Br	78 ± 6	
CH_3-NH_2		79 ± 3	Cr—Cl	87 ± 6	
CH_3-NO_2		59 ± 3	Cr—Cr	44	
$CH_2{=}NH$		154 ± 5	Cr—I	68 ± 6	
HC≡N		224	Cr—O	101 ± 7	
$CH_3^+-NH_2$		100	OCr—O	126 ± 15	
CH_3—OH	88.5 ± 3	90 ± 3	O_2Cr—O	114 ± 20	
C_6H_5—OH		103	Cobalt		
$C_6H_5CH_2$—OH		77	Co—Cl		86
Allyl—OH		109	Co—Co	39.2 ± 4	40
CH_3-OCH_3		80	Copper		
$C_6H_5-OCH_3$		91	Cu—Br	78 ± 6	
OHC—OH		96 ± 3	Cu—Cl	88 ± 4	
			Cu—Cu	45.5 ± 3	
			Cu—F	87 ± 9	
			Cu—H	66 ± 2	
			Cu—I	76	
			Cu—O	61	
			Cu—Sn	41.4 ± 4	
			Dysprosium		
			Dy—O		150

Table 3-11 (*Continued*)
BOND ENERGIES

Bond	D_0°	ΔHf_{25°	Bond	D_0°	ΔHf_{25°
Erbium			In—I	81 ± 4	81.9
Er—O		152	In—In	24 ± 3	24 ± 3
Fluorine			In—O	25	25
F—F	37 ± 1		In—OH	86 ± 5	86 ± 5
F—H	135 ± 1	135.8	In—S		35
F—I		58	In—Sb	35.4 ± 3	38.5
F—O	37 ± 3		Iodine		
OF—F	64 ± 3		I—Br	41.9 ± 0.1	42.5 ± 0.1
O_2F—F	18.4		I—Cl	49.7 ± 0.1	50.5 ± 0.1
F^+—F	>60		I—F	66.4 ± 1	67 ± 1
Gallium			I—H	70.4 ± 0.1	71.3 ± 0.1
Ga—Br	104 ± 4	106 ± 4	I—I	35.60 ± 0.01	36.15
Ga—Cl	114 ± 3		I^+—H		70
Ga—F	144 ± 3		I^+—I		61
Ga—Ga	27 ± 4	28 ± 4	Iron		
Ga—H	65	66	$FeBr_2$—Br		45 ± 5
Ga—I	84 ± 5	85 ± 3	$FeCl_2$—Cl		54 ± 2
Ga—O	59 ± 10		FeI_2—I		~23
Ga—OH	103 ± 5		Fe—O	95 ± 3	
Germanium			Fe—S	≤75	
Ge—Br	60 ± 7		Krypton		
Ge—C	109		Kr—F	12	
Ge—Cl	81 ± 5	82 ± 5	Lanthanum		
Ge—F	115 ± 10		La—La	58 ± 5	
Ge—Ge	65 ± 5	67 ± 5	La—O	187 ± 5	
Ge—H	76	77	La—S	137 ± 6	
H_3Ge—H	87		Lead		
Ge—I		51	Pb—Cl	71 ± 6	
Ge—O	160 ± 5	161 ± 5	Pb—F	74 ± 9	
Ge—S	133 ± 4	134 ± 4	Pb—H		49
Ge—Se	120 ± 6	121 ± 6	Pb—O	90 ± 3	91 ± 3
Ge—Si	71 ± 4	72 ± 4	Pb—S	76 ± 5	77 ± 5
Ge—Te	94 ± 5	95 ± 5	Pb—Se	61.5 ± 2.5	62.4 ± 2.5
Gold			Pb—Te	51.4 ± 2	52.3 ± 2
Au—Au	51.5 ± 1.5	52 ± 1.5	Lithium		
Au—Cl	69 ± 15	70 ± 15	Li—Br	100 ± 5	101 ± 5
Au—Cr	50.5 ± 1.5		Li—Cl	111 ± 3	112 ± 3
Au—Cu	54.5 ± 2.2		Li—F	137 ± 5	138 ± 5
Au—H	68 ± 3	69 ± 3	Li—H	58	
Au—Pd	33.5 ± 5		Li—I	83 ± 3	84 ± 3
Au—Sn	57.5 ± 4		Li—Li	25	
Hafnium			Li—O	79 ± 3	80 ± 3
Hf—O	185 ± 5		Li—OH	101 ± 5	102 ± 5
Holmium			Lutetium		
Ho—O		150	Lu—O		163
Hydrogen			Magnesium		
H—H	103.25 ± 0.01	104.19	Mg—Br	57 ± 23	
H—D	104.07 ± 0.01	105.00	Mg—Cl	62 ± 16	
D—D	105.05 ± 0.01	105.96	Mg—F	110 ± 10	
H^+—H or (H_2^+)		62	MgF—F	135 ± 10	
Indium			Mg—H	46 ± 11	
In—Br	97 ± 5	99 ± 5	Mg—O	90 ± 10	
In—Cl	104 ± 2	105 ± 2	Mg—OH	56 ± 5	
In—F	125 ± 2	126 ± 2	Manganese		
In—H	57 ± 5	59	Mn—Br	70 ± 5	

Table 3-11 (*Continued*)
BOND ENERGIES

Bond	D_0°	ΔHf_{25°	Bond	D_0°	ΔHf_{25°
Manganese			NN—O		40
(*cont.*)			ON—O		73
Mn—Cl	80 ± 10		N—P	~138	
Mn—F	81 ± 23		N—Si	~104	
Mn—I	67 ± 3		HN=CH$_2$		(154 ± 5)
Mn—O	96 ± 3		HN=NH		109 ± 10
Mn—S	71 ± 4		HN=O		115
Mercury			N=CH		224
Hg—Br	16.4 ± 1	17.4 ± 1	N≡N		226
Hg—Cl	23 ± 2	24 ± 2	N—N$^+$ (N$_2^+$)		200
Hg—C	81 ± 2		N—O$^+$		251
Hg—H	8.6		NN—O$^+$		56
Hg—Hg	3.2		N—NO$^+$		56
Hg—S	≤50		N=NO$^+$		155
Molybdenum			ON$^+$—O		56
Mo—I		89	Osmium		
Mo—O	116 ± 15		OsO$_3$—O		72 ± 5
MoO—O	160 ± 20		Oxygen		
MoO$_2$—O	134 ± 20		O—H	101.3 ± 0.5	102.3 ± 0.5
Neodymium			HO—H	118.0 ± 0.2	119.2 ± 0.2
Nd—F		130 ± 3	HOO—H	88.5 ± 2.0	89.5 ± 2
Nd—O	165 ± 10	166 ± 10	O$_2$—H	46 ± 3	47 ± 3
Nickel			O—O (O$_2$)	117.97 ± 0.1	119.11
Ni—Br	85 ± 3		O—OH	63.2 ± 1	
Ni—Cl	83 ± 10		HO—OH	49.5 ± 0.5	51.1 ± 0.5
Ni—H	~60		O—OF	110.7	
Ni—I	69 ± 5		FO—OF		62 ± 20
Ni—Ni	54.5		O—O$^+$ (O$_2^+$)		168
Ni—O	≤99		Ȯ—H		102
Niobium			ȮO—H		47
Nb—O	92 ± 13		ĊH$_2$O—H		31
Nitrogen			ĊOO—H		12
N—Br	68 ± 5		Phosphorus		
ON—Br	27.8 ± 1.5	28.7 ± 1.5	P—Br		63.7
N—Cl		62	P—Cl		78.5
ClN—Cl		67	P—F		117
Cl$_2$N—Cl		91	P—H		79 ± 1
ON—Cl	37 ± 1.5	38 ± 1.5	P—I		44
O$_2$N—Cl	33.0 ± 1	34.0 ± 1	P—O	141.5 ± 1	142.3 ± 1
N—F	71 ± 10	72 ± 10	P—P	115 ± 2	116 ± 2
FN—F	75 ± 5	76 ± 5	P=S	82	
F$_2$N—F	57 ± 2	58 ± 2	Potassium		
ON—F	55.2 ± 1	56.3 ± 1	K—Br	90.5 ± 2	91.5 ± 2
O$_2$N—F	46 ± 5	45 ± 5	K—Cl	101 ± 2	102 ± 2
N—H	85 ± 2	85 ± 2	K—F	117 ± 5	118 ± 5
HN—H	89 ± 2	90 ± 2	K—H	43	
H$_2$N—H	103 ± 2	104 ± 2	K—I	78 ± 3	79 ± 3
H$_3$N$_2$—H		76 ± 2	K—K	11.8	
N—N	225.07 ± 0.01	225.96 ± 0.01	K—OH	81 ± 2	
F$_2$N—NF$_2$	20 ± 1	21 ± 1	Praseodymium		
H$_2$N—NH$_2$		59 ± 3	Pr—O		180
HN—N$_2$		9 ± 1	Rubidium		
ON—N	113.5 ± 1	114.9 ± 1	Rb—Br	92 ± 6	
ON—NO$_2$	8.4 ± 0.2	9.5 ± 0.2	Rb—Cl	106 ± 5	107 ± 5
O$_2$N—NO$_2$	12.7 ± 0.5	13.7 ± 0.5	Rb—F	117 ± 5	118 ± 5

Table 3-11 (*Continued*)
BOND ENERGIES

Bond	D_0°	ΔHf_{25°	Bond	D_0°	ΔHf_{25°
Rubidium			Sr—OH	$\leq 90 \pm 10$	
(*cont.*)			Sr—S	75 ± 10	
Rb—H		119	Sulfur		
Rb—I	79 ± 3	80 ± 3	S—Cl		61
Rb—OH	83 ± 2		F_5S—F	≤ 78	
Ruthenium			O_2S—F	16	
RuO_3—O	104		FO_2S—F	~ 157	
Samarium			S—H	83.5 ± 1.5	84.4 ± 1.5
Sm—O		140	HS—H	90 ± 1	91 ± 1
Scandium			S—N	115	
Sc—F		121 ± 17	S—O	123.6 ± 2	
Sc—O		163	OS—O	130.8 ± 2.0	
Sc—Sc	26 ± 5		O_2S—O	81.9 ± 1	83.2 ± 1
Selenium			S—S	101.5 ± 1.5	102.5 ± 1.5
Se—Cl		58	HS—SH		65 ± 5
Se—F		68	S—Sb	70.6	71.6
Se—O	81 ± 23		S—Te	60	
Se—Se	65		HS^+—H		104
Silicon			HS—H^+		161
Si—Br	69 ± 14		OS—O^+		155
Si—Cl	76 ± 12		Tantalum		
Si—F		135	Ta—O	193 ± 12	
Si—H	74 ± 6		Tellurium		
Si—I		56	Te—F		80
Si—N	~ 104		Te—H		57
Si—O	185 ± 7		Te—O	$(63 - 79)$	
Si—S	147 ± 3		Te—Se	58	
Si—Se	134 ± 6		Te—Te	52 ± 2	
Si—Si		42	Terbium		
Si—Te	122 ± 9		Tb—O		171
Silver			Thallium		
Ag—Ag	38 ± 2		Tl—Br	78 ± 5	79 ± 5
Ag—Au	47.6 ± 2.2		Tl—Cl	87 ± 2	
Ag—Br	69 ± 10		Tl—F	105 ± 5	
Ag—Cl	75 ± 5		Tl—H	46	
Ag—Cu	41 ± 2		Tl—I	67 ± 5	68 ± 5
Ag—H	55 ± 3		Thorium		
Ag—I	68.7		Th—O	196	197
Ag—O	$(32 - 57) \pm 10$		Thulium		
Ag—Sn	31.6 ± 5		Tm—O		138
Sodium			Tin		
Na—Br	87.5 ± 3	88.5 ± 3	BrSn—Br		78
Na—Cl	98 ± 2	98 ± 2	Br_3Sn—Br		65
Na—F	107 ± 6		Sn—Cl		76
Na—H	47		Sn—H		61.0 ± 0.7
Na—I	71 ± 2	72 ± 2	Sn—I		65
Na—K	14.3		Sn—O	130 ± 5	131 ± 5
Na—Na	17.3		Sn—S	111 ± 5	112 ± 5
Na—OH		91 ± 3	Titanium		
Na—Rb	13.1		Ti—Br	78 ± 5	79 ± 5
Strontium			Ti—Cl	87 ± 2	
Sr—Cl	80 ± 20	81 ± 20	Ti—F	105 ± 5	
Sr—F	129 ± 10		Ti—I	67 ± 5	68 ± 5
Sr—H	38		Ti—O	156	157
Sr—O	110 ± 20		Ti—Ti	<58	

Table 3-11 (*Continued*)
BOND ENERGIES

Bond	D_0°	ΔHf_{25°	Bond	D_0°	ΔHf_{25°
Tungsten			Yttrium		
W—F		74	Y—La	47	
W—O	158 ± 10	160 ± 10	Y—Y	37 ± 5	
OW—O	150 ± 20		Zinc		
O_2W—O	142 ± 10		Zn—Cl	(48)	
Uranium			Zn—H	19.6 ± 0.5	
U—O	179 ± 7		Zn—I	41 ± 15	
OU—O	161 ± 14		Zn—O	≤ 92	
O_2U—O	153 ± 21		Zn—S	48 ± 3	
U—S	134 ± 2		Zn—Se	32 ± 3	
Vanadium			Zn—Te	49 ± 6	
V—Cl		92	Zirconium		
V—O	147 ± 5		Zr—Cl		116
OV—O	147 ± 5		Zr—O	180 ± 5	
Xenon					
Xe—F		31 ± 1			

HAMMETT AND TAFT EQUATIONS

Many equilibrium and rate processes can be systematized when the influence of each substituent on the reactivity of substrates is assigned a characteristic constant, σ, and the reaction parameter, ρ, is known. The Hammett equation

$$\log\left(\frac{K}{K^\circ}\right) = \rho\sigma$$

describes the behavior of many meta- and para-substituted aromatic species. In this equation K° is the acid dissociation constant of the reference in aqueous solution at 25°C, and K is the corresponding constant for the substituted acid. Values of σ_{meta} and σ_{para} are listed in Table 3-12.

Taft sigma values, σ^*, perform a similar function with respect to aliphatic and alicyclic systems:

$$\log\left(\frac{K}{K^\circ}\right) = \rho\sigma^*$$

Values of σ^* are also listed in Table 3-12.

The reaction parameter depends upon the reaction series, but not upon the substituents employed. Values of ρ for some aromatic and heterocyclic acids are given in Table 3-13, and for some aliphatic and alicyclic acids in Table 3-14.

Since substituent effects in aliphatic systems and in meta positions in aromatic systems are essentially inductive in character, σ^* and σ_{meta} values are related by the expression: $\sigma_{meta} = 0.217\sigma^* - 0.106$, which enables σ^* values to be estimated from σ_{meta}, and vice versa. Substituent effects fall off with increasing distance from the reaction center; generally a factor of 0.36 corresponds to the interposition of a $-CH_2-$ group to compute σ^* values for $R-CH_2-$ groups, which are not otherwise available.

The Taft equation fails when the substituent enters into resonance with the reaction center to different extents in the initial and final (or transition) states. Because of "through-resonance," different values of σ_{para} are needed for the Hammett equation to handle phenols, anilines, and pyridines (Table 3-15).

To calculate the pK_a value of m-bromobenzoic acid:

$$\log K - \log K^\circ = \rho\sigma = pK_a^\circ - pK_a = (1.00)(0.37)$$

and

$$pK_a = 4.21 - 0.37 = 3.84$$

where pK_a° is 4.21 for benzoic acid, the reaction parameter is 1.00, and σ_{meta} is 0.37.

Similarly, to calculate the pK_a value of bromoacetic acid:

$$\log K - \log K^\circ = \rho\sigma^* = pK_a^\circ - pK_a = (0.67)(2.84)$$

and

$$pK_a = 4.76 - 1.90 = 2.86$$

where pK_a° is 4.76 for acetic acid (RCH_2COOH), the reaction parameter is 0.67, and σ^* is 2.84.

Table 3-12
HAMMETT AND TAFT SUBSTITUENT CONSTANTS

Substituent	Hammett Constants		Taft Constant σ^*
	σ_{meta}	σ_{para}	
—AsO$_3$H$^-$	−0.09	−0.02	0.06
—B(OH)$_2$	0.01	0.45	
—Br	0.39	0.27	2.84
—CH$_2$Br			1.00
—p-C$_6$H$_4$Br		0.08	
—m-C$_6$H$_4$Br		0.09	
—CH$_3$	−0.07	−0.17	0.00
—C$_2$H$_5$	−0.07	−0.15	−0.10
—n-C$_3$H$_7$	−0.05	−0.13	−0.12
—i-C$_3$H$_7$	−0.07	−0.15	−0.19
—n-C$_4$H$_9$	−0.07	−0.16	−0.13
—sec-C$_4$H$_9$		−0.12	−0.19
—i-C$_4$H$_9$	−0.07	−0.12	−0.19
—t-C$_4$H$_9$	−0.10	−0.20	−0.30
—CH$_2$—C(CH$_3$)$_3$		−0.23	−0.12
—C(CH$_3$)$_2$(C$_2$H$_5$)		−0.19	
—n-C$_5$H$_{11}$			−0.25
—n-C$_7$H$_{15}$			−0.37
—cyclohexyl			−0.15
—CH$_2$—cyclohexyl			−0.31
—3,4-[CH$_2$]$_3$-(fused ring)		−0.26	
—3,4-[CH$_2$]$_4$-(fused ring)		−0.48	
—3,4-[CH]$_4$-(fused ring)	0.06	0.04	
—CH=CH$_2$	0.02		0.56
—CH=C(CH$_3$)$_2$			0.19
—CH=CH—CH$_3$ (trans)			0.36
—CH=CH—C$_6$H$_5$			0.31
—CH$_2$—CH=CH$_2$			0.00
—CH$_2$—CH=CH—CH$_3$			0.00
—CH=CH—C$_6$H$_5$	0.14	−0.05	0.41
—C≡CH	0.21	0.23	2.18
—CH$_2$—C≡CH			0.81
—C≡C—C$_6$H$_5$	0.14	0.16	1.35
—C$_6$H$_5$	0.06	0.01	0.60
—p-C$_6$H$_4$CH$_3$		−0.05	
—1-naphthyl (also —2-)			0.75
—CH$_2$—C$_6$H$_5$	−0.05		0.27
—CH$_2$—CH$_2$—C$_6$H$_5$			−0.06
—CH(CH$_3$)—C$_6$H$_5$			0.37
—2-furoyl			0.25
—3-indolyl			−0.06
—2-thienyl			1.31
—2-thienylmethyl			0.31
—CH(C$_6$H$_5$)$_2$			0.41
—CH$_2$—(1-naphthyl)			0.44
—CHO	0.36	0.22	
—CO—CH$_3$	0.38	0.50	1.65
—CO—C$_6$H$_5$	0.34	0.46	2.2
—CO—CF$_3$	0.65		3.7
—CO—NH$_2$	0.28	0.36	1.68
—CO—NH—C$_6$H$_5$			1.56
—COO$^-$	0.01	0.31	−1.06

Table 3-12 (*Continued*)
HAMMETT AND TAFT SUBSTITUENT CONSTANTS

Substituent	Hammett Constants		Taft Constant
	σ_{meta}	σ_{para}	σ^*
—COOH	0.37	0.41	2.08
—CO—OCH$_3$	0.32	0.39	2.00
—CO—OC$_2$H$_5$	0.37	0.45	2.12
—CH$_2$—CO—NH$_2$			0.31
—CH$_2$—CO—NH—C$_6$H$_5$			0.00
—CH$_2$CO—OCH$_3$			1.06
—CH$_2$—CO—OC$_2$H$_5$			0.82
—CH$_2$COO$^-$			−0.06
—CH$_2$—CH$_2$—CO—NH$_2$			0.19
—CH$_2$—CH$_2$—COOH	−0.03	−0.07	
—CH$_2$—CH$_2$—CH$_2$—CO—NH$_2$			0.12
—Cl	0.35	0.30	2.96
—CCl$_3$	0.47		2.65
—CH$_2$Cl	0.12	0.18	1.05
—CHCl$_2$			1.94
—CH$_2$—CCl$_3$			0.75
—CH$_2$—CH$_2$—CCl$_3$			0.25
—CH=CCl$_2$			1.00
—CH$_2$—CH=CCl$_2$			0.19
—p-C$_6$H$_4$Cl		0.08	
—F	0.34	0.06	3.21
—CF$_3$	0.47	0.54	2.61
—CH$_2$—CF$_3$			0.87
—CH$_2$F			1.10
—CHF$_2$			2.05
—CH$_2$—C$_3$F$_7$			0.87
—Ge(CH$_3$)$_3$		0.00	
—Ge(C$_2$H$_5$)$_3$		0.00	
—H			0.49
—I	0.35	0.30	2.46
—CH$_2$I			0.85
—IO$_2$	0.70	0.76	
—N$_2^+$	1.76	1.91	
—N$_3$	0.33	0.08	2.62
—NH$_2$	−0.04	−0.66	0.62
—NH$_3^+$	1.13	1.70	3.76
—CH$_2$—NH$_2$			0.50
—CH$_2$—NH$_3^+$			2.24
—CH$_2$—N(CH$_3$)$_3^+$			1.90
—NH—NH$_2$	−0.02	−0.55	
—CN	0.61	0.66	3.30
—CH$_2$—CN	0.17	0.01	1.30
—NH—CH$_3$	−0.30	−0.84	
—NH—C$_2$H$_5$	−0.24	−0.61	
—NH—n-C$_4$H$_9$	−0.34	−0.51	
—NH(CH$_3$)$_2^+$			4.36
—N(CH$_3$)$_2$	−0.15	−0.83	0.32

Table 3-12 (*Continued*)
HAMMETT AND TAFT SUBSTITUENT CONSTANTS

Substituent	Hammett Constants		Taft Constant
	σ_{meta}	σ_{para}	σ^*
—N(CH$_3$)$_3^+$	0.88	0.82	4.55
—NH$_2$—CH$_3^+$	0.96		3.74
—NH$_2$—C$_2$H$_5^+$	0.96		3.74
—N=N—C$_6$H$_5$		0.64	
—N(CF$_3$)$_2$	0.45	0.53	
—p-C$_6$H$_4$NH$_2$		−0.30	
—NO		0.12	
—NO$_2$	0.71	0.78	4.0
—CH$_2$—CH$_2$—NO$_2$			0.50
—p-C$_6$H$_4$NO$_2$		0.23	
—m-C$_6$H$_4$NO$_2$		0.18	
—NH—OH	−0.04	−0.34	
—NH—CO—CH$_3$	0.21	0.00	1.40
—NH—CO—C$_2$H$_5$			1.56
—NH—CO—C$_6$H$_5$	0.22	0.08	1.68
—NH—CHO	0.25		1.62
—NH—CO—NH$_2$	0.18		1.31
—CH$_2$—CO—NH$_2$			0.31
—CH$_2$—NH—CO—CH$_3$			0.43
—NH—CO—OC$_2$H$_5$	0.33		1.99
—NH—CH$_2$—CO—OC$_2$H$_5$	−0.10	−0.68	
—NH—SO$_2$—C$_6$H$_5$			1.99
—N(COCH$_3$)(COC$_6$H$_5$)			1.37
—N(COCH$_3$)(1-naphthyl)			1.62
—N(COCH$_3$)(2-naphthyl)			1.68
—O$^-$	−0.71	−0.52	
—OH	0.10	−0.37	1.34
—OCH$_3$	0.14	−0.32	1.81
—OC$_2$H$_5$	0.07	−0.24	1.68
—O—n-C$_3$H$_7$	0.00	−0.25	1.68
—O—i-C$_3$H$_7$	0.05	−0.45	1.62
—O—n-C$_4$H$_9$	−0.05	−0.32	1.68
—O—cyclohexyl	0.29		1.81
—O—cyclopentyl			1.62
—O—CH$_2$—cyclohexyl	0.18		1.31
—O—C$_6$H$_5$	0.25	−0.32	2.43
—O—CH$_2$—OC$_6$H$_5$		−0.42	
—3,4-O—CH$_2$—O—		−0.27	
—3,4-O—[CH$_2$]$_2$—O—		−0.12	
—OCF$_3$	0.36	0.32	
—ONO$_2$			3.86
—O—CO—CH$_3$	0.39	0.31	
—O—N=C(CH$_3$)$_2$			1.81
—ONH$_3^+$			2.92
—CH$_2$—OH	0.08	0.08	0.31
—CH$_2$—OCH$_3$			0.44
—CH(OH)—CH$_3$			0.12
—CH(OH)—C$_6$H$_5$			0.50
—p-C$_6$H$_4$OH		−0.24	
—p-C$_6$H$_4$OCH$_3$		−0.09	
—CH$_2$—CH(OH)—CH$_3$			−0.06
—CH$_2$—C(OH)(CH$_3$)$_2$			−0.25

Table 3-12 (*Continued*)
HAMMETT AND TAFT SUBSTITUENT CONSTANTS

Substituent	Hammett Constants		Taft Constant
	σ_{meta}	σ_{para}	σ^*
$-P(CH_3)_2$	0.1	0.05	
$-P(CH_3)_3^+$	0.8	0.9	
$-P(CF_3)_2$	0.6	0.7	
$-PO_3H^-$	0.2	0.26	
$-PO(OC_2H_5)_2$	0.55	0.60	
$-SH$	0.25	0.15	1.68
$-SCH_3$	0.15	0.00	1.56
$-S(CH_3)_2^+$	1.00	0.90	
$-SC_2H_5$	0.23	0.03	1.56
$-S-n-C_3H_7$			1.49
$-S-n-C_4H_9$			1.44
$-S-cyclohexyl$			1.93
$-SC_6H_5$	0.30		1.87
$-S-C(C_6H_5)_3$			0.69
$-S-CH_2-C_6H_5$			1.56
$-S-CH_2-CH_2-C_6H_5$			1.44
$-CH_2-SH$	0.03		0.62
$-CH_2-S-CH_2-C_6H_5$			0.37
$-S-CF_3$	0.35	0.38	
$-SCN$	0.63	0.52	3.43
$-S-CO-CH_3$	0.39	0.44	
$-S-CO-NH_2$	0.34		2.07
$-SO-CH_3$	0.52	0.49	
$-SO-C_6H_5$			3.24
$-SO_2-CH_3$	0.68	0.72	3.68
$-SO_2-C_2H_5$			3.74
$-SO_2-n-C_3H_7$			3.68
$-SO_2-C_6H_5$	0.67		3.55
$-SO_2-CF_3$	0.79	0.93	
$-SO_2-NH_2$	0.46	0.57	
$-CH_2-SO_2-CH_3$			1.32
$-SO_3^-$	0.05	0.09	0.81
$-SO_3H$		0.50	
$-SeCH_3$	0.1	0.0	
$-Se-cyclohexyl$			2.37
$-SeCN$	0.67	0.66	3.61
$-Si(CH_3)_3$	-0.04	-0.07	-0.81
$-Si(C_2H_5)_3$		0.0	
$-Si(CH_3)_2(C_6H_5)$			-0.87
$-Si(CH_3)_2-O-Si(CH_3)_3$			-0.81
$-CH_2-Si(CH_3)_3$	-0.16	-0.21	-0.31
$-CH_2-CH_2-Si(CH_3)_3$			-0.25
$-Sn(CH_3)_3$		0.0	
$-Sn(C_2H_5)_3$		0.0	

Table 3-13
pK_a° AND ρ VALUES FOR HAMMETT EQUATION

Acid	pK_a°	ρ
Arenearsonic acids, pK_1	3.54	1.05
pK_2	8.49	0.87
Areneboronic acids (in aqueous 25% ethanol)	9.70	2.15
Arenephosphonic acids, pK_1	1.84	0.76
pK_2	6.97	0.95
α-Arylaldoximes	10.70	0.86
Benzeneseleninic acids	4.78	1.03
Benzenesulfonamides (20°C)	10.00	1.06
Benzenesulfonanilides (20°C)		
$X—C_6H_4—SO_2—NH—C_6H_5$	8.31	1.16
$C_6H_5—SO_2—NH—C_6H_4—X$	8.31	1.74
Benzoic acids	4.21	1.00
Cinnamic acids	4.45	0.47
Phenols	9.92	2.23
Phenylacetic acids	4.30	0.49
Phenylpropiolic acids (in aqueous 35% dioxane)	3.24	0.81
Phenylpropionic acids	4.45	0.21
Phenyltrifluoromethylcarbinols	11.90	1.01
Pyridine-1-oxides	0.94	2.09
2-Pyridones	11.65	4.28
4-Pyridones	11.12	4.28
Pyrroles	17.00	4.28
5-Substituted pyrrole-2-carboxylic acids	2.82	1.40
Thiobenzoic acids	2.61	1.0
Thiophenols	6.50	2.2
Trifluoroacetophenone hydrates	10.00	1.11
5-Substituted tropolones	6.42	3.10
Protonated Cations of		
Acetophenones	−6.0	2.6
Anilines	4.60	2.90
C-Aryl-N-dibutylamidines (in aqueous 50% ethanol)	11.14	1.41
N,N-Dimethylanilines	5.07	3.46
Isoquinolines	5.32	5.90
1-Naphthylamines	3.85	2.81
2-Naphthylamines	4.29	2.81
Pyridines	5.18	5.90
Quinolines	4.88	5.90

Table 3-14
pK_a° AND ρ VALUES FOR TAFT EQUATION

Acid	pK_a°	ρ
RCOOH	4.66	1.62
RCH$_2$COOH	4.76	0.67
RC≡C—COOH	2.39	1.89
H$_2$C=C(R)—COOH	4.39	0.64
(CH$_3$)$_2$C=C(R)—COOH	4.65	0.47
cis-C$_6$H$_5$—CH=C(R)—COOH	3.77	0.63
trans-C$_6$H$_5$—CH=C(R)—COOH	4.61	0.47
R—CO—CH$_2$—COOH	4.12	0.43
HON=C(R)—COOH	4.84	0.34
RCH$_2$OH	15.9	1.42
RCH(OH)$_2$	14.4	1.42
R$_1$CO—NHR$_2$ (σ^* for R$_1$CO and R$_2$)	22.0	3.1
CH$_3$CO—C(R)=C(OH)CH$_3$	9.25	1.78
CH$_3$CO—CH(R)—CO—OC$_2$H$_5$	12.59	3.44
R—CO—NHOH	9.48	0.98
R$_1$R$_2$C=NOH (R$_1$, R$_2$ not acyl groups)	12.35	1.18
(R)(CH$_3$CO)C=NOH	9.00	0.94
RC(NO$_2$)$_2$H	5.24	3.60
RSH	10.22	3.50
RCH$_2$SH	10.54	1.47
R—CO—SH	3.52	1.62
Protonated Cations of		
RNH$_2$	10.15	3.14
R$_1$R$_2$NH	10.59	3.23
R$_1$R$_2$R$_3$N	9.61	3.30
R$_1$R$_2$PH	3.59	2.61
R$_1$R$_2$R$_3$P	7.85	2.67

Table 3-15
HAMMETT SIGMA CONSTANTS FOR para SUBSTITUENTS IN PHENOLS, ANILINES, AND PYRIDINES

Phenols		Anilines		Pyridines	
Substituent	σ_{para}	Substituent	σ_{para}	Substituent	σ_{para}
—CHO	1.03	—CHO	0.99	—CHO	0.99
—CO—CH$_3$	0.84	—CO—CH$_3$	0.81	—SH	0.65
—CN	0.88	—CN	1.00	—OH	0.35
—OCH$_3$	−0.11	—CF$_3$	0.74	—OCH$_3$	−0.21
—SCH$_3$	0.21	—CO—C$_6$H$_5$	0.83	—SCH$_3$	−0.12
—NO$_2$	1.24	—NO$_2$	1.26	—NO$_2$	1.26
—SO$_3^-$	0.39	—CO—OC$_2$H$_5$	0.72	—CO—OC$_2$H$_5$	0.72
—SO$_2$—CH$_3$	0.92	—SO$_2$—CH$_3$	1.14		
—NH$_2$	−0.15	—CO—OCH$_3$	0.75		
—N(CH$_3$)$_2$	−0.12	—NH$_3^+$	0.56		
—3,4-[CH]$_4$—	0.11	—SO$_2$—NH$_2$	0.80		

Section 4

INORGANIC CHEMISTRY

NOMENCLATURE OF INORGANIC COMPOUNDS

Complete details are available in the report: "Nomenclature of Inorganic Chemistry, Definitive Rules 1970," published in *Pure Appl. Chem.*, **28**, 39 (1971). The following short glossary is not intended to cover all the possible cases contained in the comprehensive 110-page report.

A. WRITING FORMULAS

1. The mass number, atomic number, number of atoms, and ionic charge of an element may be indicated by means of four indices placed around the symbol

mass number SYMBOL ionic charge
atomic number — number of atoms

$^{15}_{7}N^{3-}_{2}$

Ionic charge should be indicated by an Arabic superscript numeral preceding the plus or minus sign: Mg^{2+}, PO_4^{3-}.

2. The electropositive constituent (cation) is placed first in a formula. If the compound contains more than one electropositive or more than one electronegative constituent, the sequence within each class should be in alphabetical order of their symbols. Acids are treated as hydrogen salts; hydrogen is cited last among the cations. The alphabetical order may be different in formulas and names; viz., $NaNH_4HPO_4$, ammonium sodium hydrogenphosphate.

3. For binary compounds between nonmetals, that constituent should be placed first which appears earlier in the sequence: Rn, Xe, Kr, B, Si, C, Sb, As, P, N, H, Te, Se, S, At, I, Br, Cl, O, F.

Examples: $AsCl_3$, SbH_3, H_2Te, BrF_3, OF_2 and N_4S_4.

4. For chain compounds containing three or more elements, the sequence should be in accordance with the order in which the atoms are actually bound in the molecule or ion. *Examples:* SCN^- (thiocyanate), HSCN (thiocyanic acid), HNCO (isocyanic acid), HONC (fulminic acid), HPH_2O_2 (hydrogen phosphinate).

5. A centered period is used to denote water of hydration, other solvates, and addition compounds; viz., $CuSO_4 \cdot 5H_2O$.

6. In the formula of a free radical the unshared electron may be indicated by a point in the middle position (or above the atom where it is localized):

$HO \cdot$ (or $\dot{N}H_3^+$)

7. The structural prefixes *cis, trans, anti-, syn, cyclo, o-, m-*, and *p-* for *ortho-, meta-*, and *para-*; *n-* for normal; *as-* and *s-* for asymmetrical and symmetrical; *d-* and *t-* for deuterium and tritium; *d-* and *l-* for *dextro-* (dextrorotary) and *levo-* (levorotary); *v-* for vicinal, and *dl(meso)* should

be italicized and connected with the chemical formula by a hyphen.

8. Names for gaseous and liquid modifications of elements (allotropes) can be indicated by Greek numerical prefixes: viz., O_3 (trioxygen), P_4 (*tetrahedro*-tetraphosphorus), S_8 (*cyclo*-octasulfur), S_n (*catena*-polysulfur).

9. Isotopically labeled compounds may be described by inserting the italic symbol of the isotope in brackets into the name of the compound; viz., $H^{36}Cl$ is hydrogen chloride[^{36}Cl] or hydrogen chloride-36.

10. Roman D and L are used for configurational relationships.

11. Italicized symbols of the elements are sometimes used to indicate unusual valence; viz., 1,2,3-thia(S^{IV}) diazole for

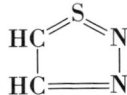

B. NAMING COMPOUNDS

1. Names and symbols for the elements are given in Table 3-1. Wolfram is preferred to tungsten. In forming a complete name of a compound, the name of the electropositive constituent is left unmodified except when it is necessary to indicate the valency (see Stock and Ewens-Bassett systems).

2. The stoichiometric proportions of the constituents in a formula may be denoted by Greek numerical prefixes: mono, di, tri, tetra, penta, hexa, hepta, octa, ennea (or Latin nona), deca, hendeca (or Latin undeca), and dodeca, preceding without hyphen the names of the elements to which they refer. The prefix mono can usually be omitted; occasionally hemi ($\frac{1}{2}$) and sesqui ($\frac{3}{2}$) are used. Beyond 10, prefixes may be replaced by Arabic numerals.

When it is required to indicate the number of entire groups of atoms, the multiplicative numerals bis, tris, tetrakis, and so on, are used, and the whole group to which they refer is placed in parentheses.

In the **Stock System** the oxidation number of an element is indicated by a Roman numeral placed in parentheses immediately following the name of the element. For zero, the cipher 0 is used. When used in conjunction with symbols the Roman numeral may be placed above and to the right.

In the **Ewens-Bassett System** the charge of an ion (rather than the oxidation state), indicated by an Arabic numeral followed by the sign of the charge cited, is placed in parentheses immediately following the name of the ion.

Examples: P_2O_5, diphosphorus pentaoxide or phosphorus(V) oxide or phosphorus(5+) oxide; $Ca[PF_6]_2$, calcium bis(hexafluorophosphate);

$K_4[Fe(CN)_6]$, potassium hexacyanoferrate(II) or potassium hexacyano-ferrate(4−); Hg_2^{2+}, mercury(I) ion or dimercury(2+) ion; $Pb_2^{II}Pb^{IV}O_4$, dilead(II) lead(IV) oxide or trilead tetraoxide.

In indexing it may be convenient to italicize a numerical prefix at the beginning of the name and connect it to the rest of the name with a hyphen.

3. Collective names include: halogens (F, Cl, Br, I, At); chalcogens (O, S, Se, Te, Po); alkali metals (Li to Fr); alkaline-earth metals (Ca, Sr, Ba, Ra); lanthanoids (La to Lu); actinoids (Ac to Lr, or those whose $5f$ shell is being filled); and noble gases (He to Rn). The name rare-earth metals may be used for the elements Sc, Y, and La to Lu inclusive. A transition element is an element whose atom has an incomplete d subshell, or which gives rise to a cation or cations with an incomplete d subshell.

4. The hydrogen isotopes are given special names: ^1H (protium), ^2H or D (deuterium), and ^3H or T (tritium). Other isotopes are designated by mass numbers: ^{10}B (boron-10).

5. Electronegative constituents which are monoatomic or homopoly-atomic are named by stripping the name of the element back to the penultimate consonant (i.e., last syllable omitted) and then adding -ide. A few Latin names are used with affixes: cupr- (copper), aur- (gold), ferr- (iron), plumb- (lead), argent- (silver), and stann- (tin). For binary compounds the name of the element standing later in the sequence in A.3 is modified to end in -ide. Elements other than those in the sequence of A.3 are taken in the reverse order of the following sequence, and the name of the element occurring last is modified to end in -ide.

Examples: KI_3, potassium triiodide; Ni_3As_2, nickel arsenide; sodium plumbide. *Exceptions:* bismuthide, hydride, mercuride, nitride, oxide, phosphide, and zincide.

ELEMENT SEQUENCE

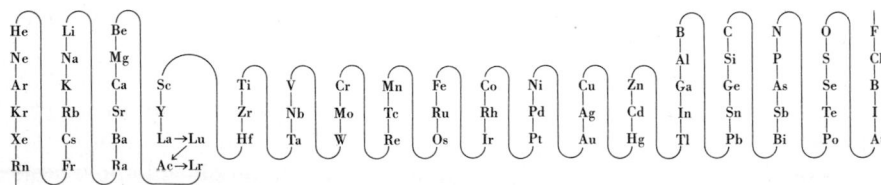

6. A few heteroatomic *anions* have names ending in -ide. These are

—OH, hydroxide	—NH$_2$, amide	—NH—NH$_2$, hydrazide
—CN, cyanide	—NH—, imide	—NHOH, hydroxylamide

Added to these anions are:

—O$_3$, ozonide   —O—O—, peroxide   —N$_3$, azide   —S—S—, disulfide

7. Binary *compounds* of hydrogen with the more electropositive elements are designated hydrides (NaH, sodium hydride). Volatile hydrides, except those of Periodic Group VII and of oxygen and nitrogen, are named by citing the root name of the element (penultimate consonant and Latin affixes, B.5) followed by the suffix -ane. Exceptions are water, ammonia, hydrazine, phosphine, arsine, stibine, and bismuthine.

Examples: B_2H_6, diborane; $B_{10}H_{14}$, decaborane(14); $B_{10}H_{16}$, decaborane(16); P_2H_4, diphosphane; Sn_2H_6, distannane; H_2Se_2, diselane; H_2Te_2, ditellane; H_2S_5, pentasulfane; PbH_4, plumbane.

8. Certain neutral *radicals* have special names ending in -yl:

HO	hydroxyl	CrO_2	chromyl	SO_2	sulfonyl (sulfuryl)
CO	carbonyl	NO	nitrosyl	S_2O_5	disulfuryl
ClO	chlorosyl*	NO_2	nitryl	SeO	seleninyl
ClO_2	chloryl*	PO	phosphoryl	SeO_2	selenonyl
ClO_3	perchloryl*	SO	sulfinyl (thionyl)	UO_2	uranyl
				NpO_2	neptunyl†

* Similarly for the other halogens.
† Similarly for the other actinoid elements.

Radicals analogous to the above containing other chalcogens in place of oxygen are named by adding the prefixes thio-, seleno-, and so on.

C. CATIONS

1. Monoatomic cations are named as the corresponding element; viz., Fe^{2+}, iron(II) ion; Fe^{3+}, iron(III) ion. This principle also applies to polyatomic cations corresponding to radicals with special names ending in -yl; viz., PO^+, phosphoryl cation; NO^+, nitrosyl cation; NO_2^+, nitryl cation; O_2^{2+}, oxygenyl cation.

Use of the Stock system extends the range for radicals: viz., UO_2^{2+}, uranyl(VI) ion; UO_2^+, uranyl(V) ion.

2. Polyatomic cations derived by addition of more protons than required to give a neutral unit to monoatomic anions are named by adding the ending -onium to the root of the name of the anion element; viz., PH_4^+, phosphonium ion; H_2I^+, iodonium ion; H_3O^+, oxonium ion; $CH_3OH_2^+$, methyl oxonium ion. *Exception:* The name ammonium is retained for the ion NH_4^+. Similarly for substituted ammonium ions; viz., NF_4^+, tetrafluoroammonium ion.

Substituted ammonium ions derived from nitrogen bases with names ending in -amine will receive names formed by changing -amine into -ammonium. When known by a name not ending in -amine, the cation name is formed by adding the ending -ium to the name of the base (eliding the final vowel); viz., anilinium, hydrazinium, imidazolium, acetonium, dioxanium.

Exceptions are the names uronium and thiouronium derived from urea and thiourea, respectively.

3. Where more than one ion is derived from one base, the ionic charges are indicated in their names: $N_2H_5^+$, hydrazinium(1+) ion; $N_2H_6^{2+}$, hydrazinium(2+) ion.

D. ANIONS

1. See B.5 and B.6 for naming monoatomic and certain polyatomic anions.

2. Ions such as HSO_4^- are named hydrogensulfate ion with the two words written as one following the usual practice for polyatomic anions.

3. Names for other polyatomic anions consist of the root name of the central atom with the ending -ate and followed by the valence of the central atom expressed by its Stock number. Atoms and groups attached to the central atom are treated as ligands in a complex.

Examples: $[Sb(OH)_6]^-$, hexahydroxoantimonate(V); $[Fe(CN)_6]^{3-}$, hexacyanoferrate(III); $[Fe(CN)_6]^{4-}$, hexacyanoferrate(II); $[PCl_6]^-$, hexachlorophosphate.

Exceptions to the use of the root name of the central atom are: antimonate, bismuthate, carbonate, cobaltate, nickelate (also niccolate), nitrate, phosphate, tungstate (or wolframate), and zincate.

4. Oxygen should be treated in the same manner as other ligands with the number of -oxo groups indicated by a suffix; viz., SO_3^{2-}, trioxosulfate.

The ending -ite, formerly used to denote a lower state of oxidation, may be retained in trivial names in these cases:

AsO_3^{3-}	arsenite	IO^-	hypoiodite	SO_3^{2-}	sulfite
BrO^-	hypobromite	NO_2^-	nitrite	$S_2O_5^{2-}$	disulfite
ClO^-	hypochlorite	$N_2O_2^{2-}$	hyponitrite	$S_2O_4^{2-}$	dithionite
ClO_2^-	chlorite	NOO_2^-	peroxonitrite	$S_2O_2^{2-}$	thiosulfite
				SeO_3^{2-}	selenite

However, compounds known to be double oxides in the solid state are named as such; viz., Cr_2CuO_4 (actually $Cr_2O_3 \cdot CuO$) is chromium(III) copper(II) oxide (and not copper chromite).

5. Isopolyanions are named by indicating with numerical prefixes the number of atoms of the characteristic element. It is not necessary to give the number of oxygen atoms when the charge of the anion or the number of cations is indicated.

Examples: $Ca_3Mo_7O_{24}$, tricalcium 24-oxoheptamolybdate, may be shortened to tricalcium heptamolybdate; the anion, $Mo_7O_{24}^{6-}$, is heptamolybdate(6−); $S_2O_7^{2-}$, disulfate(2−).

When the characteristic element is partially or wholly present in a lower oxidation state than corresponds to its Periodic Group number,

Stock numbers are used; viz., $[O_2HP—O—PO_3H]^{2-}$, dihydrogen-diphosphate(III,V)(2−).

A bridging group should be indicated by adding the Greek letter μ immediately before its name and separating this from the rest of the complex by a hyphen. The atom or atoms of the characteristic element, to which the bridging atom is bonded, is indicated by numbers.

Examples:

$$\left[\begin{array}{c} O \quad\ O \quad\ O \\ | \quad | \quad | \\ O—P—S—P—O—P—O \\ | \quad | \quad | \\ O \quad\ O \quad\ O \end{array} \right]^{5-} , \quad 1,2\text{-}\mu\text{-thiotriphosphate}$$

(5−); $[S_3P—O—PS_2—O—PS_3]^{5-}$, di-$\mu$-oxo-octathiotriphosphate(5−); $[Be_4O(OOCCH_3)_6]$, hexa-μ-acetato-μ_4-oxo-tetraberyllium, which structure consists of a tetrahedron of beryllium ions with an oxygen in the center and one acetate ion bridging the two beryllium ions along each edge.

E. ACIDS

1. Acids giving rise to the -ide anions (B.5) should be named as hydrogen . . . -ide; viz., HCl, hydrogen chloride; HN_3, hydrogen azide.

Names such as hydrobromic acid refer to aqueous solutions, and percentages such as 48% HBr denote the weight/volume of hydrogen bromide in the solution.

2. Acids giving rise to anions bearing names ending in -ate are treated as in E.1; viz., H_4GeO_4, hydrogen germanate; $H_4[Fe(CN)_6]$, hydrogen hexacyanoferrate(II).

3. Acids given in the following table retain their trivial names due to long-established usage. Anions may be formed from these trivial names by changing -ous acid to -ite, and -ic acid to -ate. The prefix hypo- is used to denote a lower oxidation state; the prefix per- designates a higher oxidation state. The prefixes ortho- and meta- distinguish acids of differing water content.

4. When used in conjunction with the trivial names of acids, the prefix peroxo- indicates substitution of —O— by —O—O—.

5. Acids derived from oxoacids by replacement of oxygen by sulfur are called thioacids, and the number of replacements are given by prefixes di-, tri-, and so on. The affixes seleno- and telluro- are used analogously.

6. See the section on Coordination Compounds for acids containing ligands other than oxygen and sulfur (selenium and tellurium).

Trivial Names Retained for Acids
(alphabetically by characteristic element)

H_3AsO_4	arsenic acid	$H_4P_2O_8$	peroxodiphosphoric
H_3AsO_3	arsenious acid	$(HO)_2OP-$	acid
H_3BO_3	orthoboric acid	$PO(OH)_2$	diphosphoric(IV)
	(or boric acid)		acid or hypo-
$(HBO_2)_n$	metaboric acid		phosphoric acid
$HBrO_3$	bromic acid	$(HO)_2P-O-$	diphosphoric(III,V)
$HBrO_2$	bromous acid	$PO(OH)_2$	acid
$HBrO$	hypobromous acid	H_2PHO_3	phosphonic acid
H_2CO_3	carbonic acid	$H_2P_2H_2O_5$	diphosphonic acid
$HOCN$	cyanic acid	HPH_2O_2	phosphinic acid
$HNCO$	isocyanic acid		(formerly hypophos-
$HONC$	fulminic acid		phorous acid)
$HClO_4$	perchloric acid	$HReO_4$	perrhenic acid
$HClO_3$	chloric acid	H_2ReO_4	rhenic acid
$HClO_2$	chlorous acid	H_2SO_4	sulfuric acid
$HClO$	hypochlorous acid	$H_2S_2O_7$	disulfuric acid
H_2CrO_4	chromic acid	H_2SO_5	peroxomonosulfuric
$H_2Cr_2O_7$	dichromic acid		acid
H_5IO_6	orthoperiodic acid	$H_2S_2O_3$	thiosulfuric acid
HIO_4	periodic acid	$H_2S_2O_6$	dithionic acid
HIO_3	iodic acid	H_2SO_3	sulfurous acid
HIO	hypoiodous acid	$H_2S_2O_5$	disulfurous acid
$HMnO_4$	permanganic acid	$H_2S_2O_2$	thiosulfurous acid
H_2MnO_4	manganic acid	$H_2S_2O_4$	dithionous acid
HNO_4	peroxonitric acid	H_2SO_2	sulfoxylic acid
HNO_3	nitric acid	$H_2S_xO_6$	polythionic acid
HNO_2	nitrous acid	$(x = 3,4, \ldots)$	(tri-, tetra-, . . .)
H_2NO_2	nitroxylic acid	$HSb(OH)_6$	hexahydroxoantimonic
$H_2N_2O_2$	hyponitrous acid		acid
$HOONO$	peroxonitrous acid	H_2SeO_4	selenic acid
H_3PO_4	orthophosphoric	H_2SeO_3	selenious acid
	acid or phosphoric	H_4SiO_4	orthosilicic acid
	acid	$(H_2SiO_3)_n$	metasilicic acid
$(HPO_3)_n$	metaphosphoric	$HTcO_4$	pertechnetic acid
	acid	H_2TcO_4	technetic acid
H_3PO_5	peroxomonophos-	H_6TeO_6	orthotelluric acid
	phoric acid		
$H_4P_2O_7$	diphosphoric acid		
	or pyrophosphoric		
	acid		

F. SALTS AND FUNCTIONAL DERIVATIVES OF ACIDS

1. For acid halogenides the name is formed from the corresponding acid radical if this has a special name (B.8); viz., $NOCl$, nitrosyl chloride.

In other cases these compounds are named as halogenide oxides with the ligands listed alphabetically; viz., $BiClO$, bismuth chloride oxide; $VOCl_2$, vanadium(IV) dichloride oxide.

2. Anhydrides of inorganic acids are named as oxides; viz., N_2O_5, dinitrogen pentaoxide.

3. Esters of inorganic acids are named as the salts; viz., $(CH_3)_2SO_4$, dimethyl sulfate. However, if it is desired to specify the constitution of the compound, the nomenclature for coordination compounds should be used.

4. Names for amides are derived from the names of acid radicals (or from the names of acids by replacing acid by amide; viz., $SO_2(NH_2)_2$, sulfonyl diamide (or sulfuric diamide); NH_2SO_3H, sulfamidic acid (or amidosulfuric acid).

5. Salts containing acid hydrogen are named by adding the word hydrogen without a space before the name of the anion; viz., KH_2PO_4, potassium dihydrogenphosphate; $NaHCO_3$, sodium hydrogencarbonate; $NaHPHO_3$, sodium hydrogenphosphonate.

6. Salts containing O^{2-} and HO^- anions are named oxide and hydroxide, respectively. Anions are cited in alphabetical order which may be different in formulas and names.

Examples: $AlCl_2OH$, aluminum dichloride hydroxide; $VO(SO_4)$, vanadium(IV) oxide sulfate.

The multiplicative numerical prefixes bis, tris, etc., are used with certain anions for indicating stoichiometric proportions when di, tri, etc. have been preempted to designate condensed anions; viz., $AlK(SO_4)_2 \cdot 12H_2O$, aluminum potassium bissulfate 12-water.

7. The structure type of crystals may be added in parentheses and in italics after the name; the latter should be in accordance with the structure. When the type-name is also the mineral name of the substance itself, the italics are not used.

Examples: $MgTiO_3$, magnesium titanium trioxide (*ilmenite* type); $FeTiO_3$, iron(II) titanium trioxide (ilmenite).

G. COORDINATION COMPOUNDS

1. To name a coordination compound, the names of the ligands are attached directly in front of the name of the central atom. The ligands are listed in alphabetical order regardless of the number of each with the name of a ligand treated as a unit. Thus "diammine" is listed under "a" and "dimethylamine" under "d." The oxidation number of the central atom is stated last by either the Stock or the Ewens-Bassett system.

2. Whether inorganic or organic, the names for anionic ligands end in -o (eliding the final -e, if present, in the anion name). Enclosing marks are required for inorganic anionic ligands containing numerical prefixes, and for thio, seleno and telluro analogs of oxo anions containing more than one atom.

The following anions do not follow the nomenclature rules:

F^-	fluoro	HO_2^-	hydrogenperoxo
Cl^-	chloro	S^{2-}	thio (only for single sulfur)
Br^-	bromo	S_2^{2-}	disulfido
I^-	iodo	HS^-	mercapto
O^{2-}	oxo	CN^-	cyano
H^-	hydrido (or hydro)	CH_3O	methoxo or methanolato
OH^-	hydroxo	CH_3S	methylthio or methanethiolato
O_2^{2-}	peroxo		

3. Enclosing marks are nested within square brackets as follows:

$$[\,()\,], [\{\,()\,\}], [\{[\,()\,]\}], [\{\{[\,()\,]\}\}].$$

4. Neutral and cationic ligands are used without change in name and are set off with enclosing marks. Water and ammonia as neutral ligands are called "aqua" and "ammine," respectively. The groups NO and CO, when linked directly to a metal atom, are called nitrosyl and carbonyl, respectively.

5. If the coordination entity is negatively charged, the cations paired with the complex anion (with -ate ending) are listed first; if the entity is positively charged, the anions paired with the complex cation are listed immediately afterward.

6. The different points of attachment of a ligand are denoted by adding italicized symbol(s) for the atom or atoms through which attachment occurs at the end of the name of the ligand; viz., glycine-N or glycinato-O,N. If the same element is involved in different possible coordination sites, the position in the chain or ring to which the element is attached is indicated by numerical superscripts; viz., tartrato$(3-)$-O^1,O^2 or tartrato$(4-)$-O^2,O^3 or tartrato$(2-)$-O^1,O^4.

7. Abbreviations for ligand names are used extensively. Except for the general abbreviations for ligand (L) and metal (M), and a few with H, all shall be in lowercase letters and should not involve hyphens. Some common abbreviations are:

acac	acetylacetonato
Hacac	acetylacetone or 2,4-pentanedione
Hbg	biguanide
H_2dmg	dimethylglyoxime
dmg	dimethylglyoximato$(2-)$
Hdmg	dimethylglyoximato$(1-)$
H_4edta	ethylenediaminetetraacetic acid
Hedta or edta	coordinated ions derived from H_4edta
ox	oxalato$(2-)$ from parent H_2ox
bpy	2,2'-bipyridine
dien	diethylenetriamine
en	ethylenediamine
phen	1,10-phenanthroline

pn	propylenediamine
py	pyridine
tren	2,2′,2″-triaminotriethylamine
trien	triethylenetetraamine
ur	urea

Examples: Li[B(NH$_2$)$_4$], lithium tetraamidoborate(1−) or lithium tetraamidoborate(III); [Co(NH$_3$)$_5$Cl]Cl$_2$, pentaamminechlorocobalt(III) chloride or pentaamminechlorocobalt(2+) chloride; K$_3$[Fe(CN)$_5$CO], potassium carbonylpentacyanoferrate(II) or potassium carbonylpenta-cyanoferrate(3−); [Mn{C$_6$H$_4$(O)(COO)}$_2$(H$_2$O)$_4$]$^-$, tetraaquabis[sali-cylato(2−)]manganate(III) ion; [Ni(C$_4$H$_7$N$_2$O$_2$)$_2$] or [Ni(dmg)$_2$], bis(2,3-butanedione dioximato)nickel(II) or bis[dimethylglyoximato-(2−)]nickel(II).

H. ADDITION COMPOUNDS

1. The names of addition compounds are formed by connecting the names of individual compounds by spaced hyphens (or a dash) and indicating the numbers of molecules in the name by Arabic numerals separated by the solidus (diagonal). All molecules are cited in order of increasing number; those having the same number are cited in alphabetic order. However, boron compounds and water are always cited last and in that order.

Examples: 3CdSO$_4$ · 8H$_2$O, cadmium sulfate – water (3/8); Al$_2$(SO$_4$)$_3$ · K$_2$SO$_4$ · 24H$_2$O, aluminum sulfate – potassium sulfate – water (1/1/24); AlCl$_3$ · 4C$_2$H$_5$OH, aluminum chloride – ethanol (1/4); 2CH$_3$OH · BF$_3$, methanol – boron trifluoride (2/1).

Table 4-1
PHYSICAL CONSTANTS OF INORGANIC COMPOUNDS

Names, while following the IUPAC Nomenclature, are often alphabeticized by the central atom to facilitate their location. Solvates are listed under the entry for the anhydrous salt. Acid salts are entered as hydrogen . . .

Formula Weights are based upon the International Atomic Weights of 1973 and are computed to the nearest hundredth. See Table 3-1 for additional significant figures in the atomic weights of the elements.

Refractive Index, unless otherwise specified, is given for the sodium line at 589.6 nm.

Density values are given at room temperature unless otherwise indicated by the superscript figure; thus, 2.487^{15} indicates a density of 2.487 for the substance at 15°C. For gases the values are given as grams per liter (g/L).

Melting Point is recorded in a certain case as 250 d and in some other cases as d 250, the distinction being made in this manner to indicate that the former is a melting point with decomposition at 250°C, while in the latter decomposition only occurs at 250°C and higher temperatures. Where a value such as $-6H_2O$, 150 is given it indicates a loss of six moles of water per formula weight of the compound at a temperature of 150°C.

Boiling Point is given at atmospheric pressure (760 mm of mercury) unless otherwise indicated; thus 82^{15mm} indicates that the boiling point is 82°C when the pressure is 15 mm. Also, subl 550 indicates that the compound sublimes at 550°C.

Solubility is given in parts by weight (of the formula weight) per 100 parts by weight of the solvent (water unless otherwise specified) and at room temperature. Other temperatures are indicated by the small superscript. The symbols of the common mineral acids represent aqueous solutions of these acids.

Synonyms and Mineral Names as well as alternate chemical names for compounds to be found in this table but under different names are listed in Table 4-2.

Abbreviations Used in the Table

a, acid	cr, crystals or crystalline	hygr, hygroscopic
abs, absolute	cub, cubic	i, insoluble
ac a, acetic acid	d, decomposes	ign, ignites
acet, acetone	deliq, deliquescent	leaf, leaflets
al, 95% ethanol	dil, dilute	(Li), lithium line at 671 nm
alk, alkali (aq NaOH or KOH)	disprop, disproportionates	liq, liquid
amorp, amorphous	dk, dark	lt, light
anhyd, anhydrous	effl, effloresces	lust, lustrous
aq, aqueous	eth, ether	MeOH, methyl alcohol
aq reg, aqua regia	EtOH, ethyl alcohol	met, metal, metallic
atm, atmosphere	expl, explodes, explosive	min, mineral
bcc, body-centered cubic	fcc, face-centered cubic	mn, monoclinic
bct, body-centered tetragonal	fum, fuming	nd, needles
blk, black	fus, fusion, fuses	oct, octahedral
bl, blue	g, gas	ol, olive
brn, brown	gel, gelatinous	or, orange
bz, benzene	gelat, gelatin	org, organic
c, solid state	gly, glycerol	o-rh, orthorhombic
ca, approximately	grn, green	oxid, oxidizing
cc, cubic centimeter	h, hot	pl, plates
chl, chloroform	hcp, hexagonal close-packed	pois, poisonous
col, colorless	hex, hexagonal	ppt, precipitate
conc, concentrated	(Hg), mercury line at λ stated	pr, prisms

INORGANIC CHEMISTRY

purp, purple
pwd, powder
py, pyridine
pyr, pyramidal
rh, rhombic
rbhd, rhombohedral
s, soluble
satd, saturated
sens, sensitive
-sh (e.g., yelsh, yellowish)

silv, silver
sl, slightly
slky, silky
sol, solid
soln, solution
solv, solvent(s)
subl, sublimes
sulf, sulfides
tart, tartrate
tetr, tetragonal

tr, transition
tric, triclinic
trig, trigonal
v, very
vac, vacuum
viol, violently
vlt, violet
volat, volatile or volatilizes
wh, white
yel, yellow

Table 4-1 (*Continued*)

PHYSICAL CONSTANTS OF INORGANIC COMPOUNDS

Name	Formula	Formula weight	Color, crystalline form, refractive index	Density	Melting point, °C	Boiling point, °C	Solubility in 100 parts
Actinium	Ac	(227)	silv-wh, fcc	10.07	1050	(3300)	d to Ac(OH)$_3$
bromide	AcBr$_3$	466.7	wh, hex	5.85	subl 800	....	s
bromide oxide	AcBrO	322.9	wh, tetr	7.9	....	....	i; s a
chloride	AcCl$_3$	333.4	wh, hex	4.81	subl 960	....	i; s a
chloride oxide	AcClO	278.3	wh, tetr	9.70	....	....	i
fluoride	AcF$_3$	284.0	wh, hex	7.88	....	....	i
fluoride oxide	AcFO	262.0	wh, cub	8.28	....	....	i; s a
hydroxide	Ac(OH)$_3$	278.0	wh	....	subl 700	....	s
iodide	AcI$_3$	607.7	wh	....	....	....	....
sulfide	Ac$_2$S$_3$	550.2	dark, cub	6.75	....	....	s HCl, H$_2$SO$_4$, alk
Aluminum	Al	26.98	silv, cub	2.70	660.1	2450	s
acetate	Al(C$_2$H$_3$O$_2$)$_3$	204.12	wh pd	....	d	....	s
acetylacetonate	Al(C$_5$H$_7$O$_2$)$_3$	324.31	col, mn	1.27	subl 193	314	i; v s al; s bz eth
ammonium tetrachloride	AlNH$_4$Cl$_4$	186.83	wh cr	....	304	....	s
ammonium bissulfate	AlNH$_4$(SO$_4$)$_2$	237.14	col, hex	2.45^{20}	....	....	7.74^{20}
ammonium bissulfate-12-water	AlNH$_4$(SO$_4$)$_2 \cdot$ 12H$_2$O	453.33	col, cub, 1.459	1.64	93.5	anhyd 250; d > 280	15
arsenate	AlAsO$_4$	165.90	wh, trig. 1.596	3.25	....	....	i; sl s a
benzoate	Al(C$_7$H$_5$O$_2$)$_3$	390.33	wh cr	....	....	....	v sl s
beryllium oxide (1/1) (chrysoberyl)	Al$_2$O$_3 \cdot$ BeO	126.97	col, rh, 1.747 1.748, 1.757	3.76	1873	....	i
beryllium silicon oxide (1/3/6) (beryl)	Al$_2$O$_3 \cdot$ 3BeO $\cdot$ 6SiO$_2$	537.51	col, hex, 1.580, 1.547	2.66	1410	....	i
beryllium silicon oxide (1/2/2) (euclase)	Al$_2$O$_3 \cdot$ 2BeO $\cdot$ 2SiO$_2$	290.17	col, mn, 1.652, 1.655, 1.671	3.1	....	....	i
borate, tetrahydro-	Al[BH$_4$]$_3$	71.53	col liq	....	−64.5	44.5	d viol
boride, di-	AlB$_2$	48.60	redsh-br, hex	3.19	....	....	....
bromate-9-water	Al(BrO$_3$)$_3 \cdot$ 9H$_2$O	572.84	wh cr, hygr	....	62.3	d 100	s

Name	Formula	Mol. wt.	Crystalline form, color, etc.	Density	M.p.	B.p.	Solubility
bromide	$AlBr_3$	266.71	wh, rh, deliq	2.64	97.5	253.3	s
bromide-6-water	$AlBr_3 \cdot 6H_2O$	374.80	wh er, deliq	2.54	93	d 135	s
butoxide, tert-	$Al(C_4H_9O)_3$	246.33	wh er	1.025	subl 180	......	v s org solv
calcium oxide (1/1)	$Al_2O_3 \cdot CaO$	158.04	wh, mn, 1.643	2.981	1605	......	s HCl
calcium oxide (1/3)	$Al_2O_3 \cdot 3CaO$	270.20	wh, cub, 1.710	3.038	1535	......	i; s a
calcium oxide-3-water (1/3/3)	$Al_2O_3 \cdot 3CaO \cdot 3H_2O$	378.29	col, oct, 1.603	2.52^{20}	d 700	......	
calcium iron(III) oxide (1/4/1) (celite)	$Al_2O_3 \cdot 4CaO \cdot Fe_2O_3$	485.97	brn, rh, 1.98, 2.05, 2.08	3.77	1420	......	d a
calcium silicon oxide (1/2/1)	$Al_2O_3 \cdot 2CaO \cdot SiO_2$	274.20	col, tetr, 1.669, 1.658	3.048	1590	......	
calcium silicon oxide (1/1/2) (anorthite)	$Al_2O_3 \cdot CaO \cdot 2SiO_2$	278.21	wh, tric, 1.583	2.765	1551	......	
(tetra) carbide, tri-	Al_4C_3	143.96	yel-grn, hex, 2.70	2.36	2100	d 2200	d to CH_4
cesium bisulfate-12-water	$AlCs(SO_4)_2 \cdot 12H_2O$	568.19	col, cub, 1.4587	1.97	117	......	0.91^{25}
chlorate-6-water	$Al(ClO_3)_3 \cdot 6H_2O$	385.41	col, rhhd, deliq	......	d	......	v s
chloride	$AlCl_3$	133.34	wh, hex, deliq	2.44	$194^{5.2\,atm}$	subl 181	70 (viol); s al, eth
chloride-6-ammonia	$AlCl_3 \cdot 6NH_3$	235.52	col cr, hygr	1.412	d	......	s
chloride-6-water	$AlCl_3 \cdot 6H_2O$	241.43	wh, trig, deliq, 1.56	2.40	d 100	......	83^{20}, 50 al; s eth
chloride oxide	$AlClO$	78.44	wh	......	226	......	
chlorodiethyl-[bis]	$[(C_2H_5)_2AlCl]_2$	241.12	col liq	0.961^{25}	32	100^{17mm}	0.5
chloroethyl-[bis], di-	$[C_2H_5AlCl_2]_2$	253.90	col sol	1.207^{50}	......	100^{30mm}	
citrate	$AlC_6H_5O_7$	216.08	wh	......	......	......	0.5
ethoxide	$Al(C_2H_5O)_3$	162.14	wh cr	1.142	134	200^{8mm}	d; v sl s al, eth
ethyl, tri-	$Al(C_2H_5)_3$	114.17	col liṭ, dimeric	0.832^{25}	......	194	v s
fluoride	AlF_3	83.98	col, tric	2.88	1040	subl 1276	0.67^{20}; i a, alk
fluoride-1-water (fluellite)	$AlF_3 \cdot H_2O$	101.99	col, rã, 1.490	2.17	d	......	sl s
hydroxide	$Al(OH)_3$	78.00	wh, mn	2.42	$-H_2O$, 300	......	i; s a, alk
hydroxide palmitate, di-	$Al(C_{16}H_{31}O_2)(OH)_2$	316.41	wh	1.095	200	......	i; s alk, hydrocarbon
iodide	AlI_3	407.71	brn cr	3.98	191	385	s d; s al, eth
iodide-6-water	$AlI_3 \cdot 6H_2O$	515.79	wh cr, hygr	2.63	d 185	......	v s; s al, CS_2
isobutyl, tri-	$Al(i\text{-}C_4H_9)_3$	198.33	col m:nomeric liq	0.781^{25}	6	86^{8mm}	d; s al, bz, chl
isopropoxide	$Al(i\text{-}C_3H_7O)_3$	204.25	wh cr, hygr	1.0346	118.5	135^{10mm}	i; i a
metaphosphate	$Al(PO_3)_3$	263.90	col, tetr	2.779	......	......	
nitrate-9-water	$Al(NO_3)_3 \cdot 9H_2O$	375.13	col, rh, deliq, 1.54	......	73	d 150	130^{20}, 100 al
nitride	AlN	40.99	wh, hex	3.26	subl 2000	d 2517	d; d a, alk

Table 4-1 (*Continued*)
PHYSICAL CONSTANTS OF INORGANIC COMPOUNDS

Name	Formula	Formula weight	Color, crystalline form, refractive index	Density	Melting point, °C	Boiling point, °C	Solubility in 100 parts
Aluminum							
oxalate-4-water	$Al_2(C_2O_4)_3 \cdot 4H_2O$	390.08	wh pd	....	....	....	i; s a
oxide	Al_2O_3	101.96	col, hex, 1.768, 1.760	3.965	2054	(2980)	i; v sl s a, alk
(α) (corundum)	Al_2O_3	101.96	col, rh, 1.765	3.97	2018	(2980)	i; v sl s a, alk
(γ) (alumina)	Al_2O_3	101.96	wh micr cr, 1.7	3.5–3.9	....	....	i; sl s a, alk
oxide-1-water (boehmite)	$Al_2O_3 \cdot H_2O$	119.98	col, rh, 1.624	3.014	d 360	....	i
oxide-3-water (gibbsite)	$Al_2O_3 \cdot 3H_2O$	156.01	wh, mn, 1.577, 1.577	2.42	....	....	i; s hot a
(bayerite)	$Al_2O_3 \cdot 3H_2O$	156.01	wh cr, 1.583	2.53	....	....	i; s hot a
perchlorate	$Al(ClO_4)_3$	325.34	col, hygr	2.209	....	....	133[20]
perchlorate-6-water	$Al(ClO_4)_3 \cdot 6H_2O$	433.43	col, hygr	2.020	120.8	−6H₂O, 178	177[20]
phenoxide	$Al(C_6H_5O)_3$	306.27	gray-wh	1.23	d 265	....	d; s al, eth, chl
phosphate	$AlPO_4$	121.95	wh, rh, 1.546, 1.556	2.556	1500	....	i; s a, alk
phosphide	AlP	57.96	dk gray or yel cr	2.40	>1000	....	
potassium bisulfate-12-water (kalinite)	$AlK(SO_4)_2 \cdot 12H_2O$	474.39	col, cub, oct	1.757[20]	−9H₂O, 92	−12H₂O,	11.4[20]
propoxide	$Al(C_3H_7O)_3$	204.25	wh cr	1.0578[20]	106	200	d; s al
rubidium bisulfate-12-water	$AlRb(SO_4)_2 \cdot 12H_2O$	520.76	col, cub, 1.457	1.867[0]	99	....	2.56[20]
selenide	Al_2Se_3	290.84	lt brn, hex (*wurtzite* type)	3.437[15]	947	....	d; d a
silicon oxide (1/1) (andalusite, kyanite, sillimanite)	$Al_2O_3 \cdot SiO_2$	162.05	wh, rh, 1.66	3.247	....	....	i; d HF; s fus alk
silicon oxide (3/2) (mullite)	$3Al_2O_3 \cdot 2SiO_2$	426.05	col, rh, 1.638, 1.642, 1.653	3.156	1750	....	i; i a, HF
silicon sodium oxide (1/2/1) (nephelite)	$Al_2O_3 \cdot 2SiO_2 \cdot Na_2O$	284.11	col, hex, 1.537	2.619[21]	1526	....	i; s a
silicon sodium oxide (1/4/1) (jadeite)	$Al_2O_3 \cdot 4SiO_2 \cdot Na_2O$	404.28	col, mn	3.3	1060	....	i; d HCl
silicon sodium oxide (1/6/1) (albite)	$Al_2O_3 \cdot 6SiO_2 \cdot Na_2O$	524.48	col, tric	2.61	1100	....	i; s HCl

Name	Formula	Formula weight	Color, crystalline form, refractive index	Density	Melting point, °C	Boiling point, °C	Solubility in 100 parts
sodium bissulfate-12-water	AlNa(SO$_4$)$_2$ · 12H$_2$O	458.28	col	1.675[20]	61	....	110[15]
stearate	Al(C$_{18}$H$_{35}$O$_2$)$_3$	877.42	wh pd	1.010	117–120	....	i; s al, bz, alk
sulfate	Al$_2$(SO$_4$)$_3$	342.15	wh pd, 1.47	2.71	d 770	....	36.4[20]
sulfate-18-water (alunogenite)	Al$_2$(SO$_4$)$_3$ · 18H$_2$O	666.45	col, mn, 1.474, 1.467	1.69[17]	d 86.5	....	71[20]
sulfide	AlS	59.05	....	....	1197	....	....
(di) sulfide, tri-	Al$_2$S$_3$	150.16	yel, hex	2.02[13]	1097	subl 1500	d; s a
(di) telluride, tri-	Al$_2$Te$_3$	436.76	....	....	897	....	....
thallium bissulfate-12-water	AlTl(SO$_4$)$_2$ · 12H$_2$O	639.66	col, oct, 1.5011	2.325[20]	91	....	9.6[20]
Americium	Am	243	silv, hcp	13.671	994	2607	s dil a
bromide	AmBr$_3$	482.9	wh, rh	6.79	....	subl 850	s
chloride	AmCl$_3$	349.5	pink, hex	5.78	....	subl 850	s
chloride oxide	AmClO	294.4	wh, tetr	8.96	....	....	i; s a
(III) fluoride	AmF$_3$	300.1	pink, hex	9.53	1395	....	i
(IV) fluoride	AmF$_4$	319.0	tan, mn	7.34	....	....	....
hydride, di-	AmH$_2$	245	black, cub	10.7	....	....	....
iodide	AmI$_3$	623.8	yel, hex	6.04	....	subl 900	s
(III) oxide	Am$_2$O$_3$	534.3	brn, cub	....	....	....	s a
(IV) oxide	AmO$_2$	275.1	blk, cub	11.68	....	....	s a
Amidosulfuric acid	NH$_2$SO$_2$H	97.09	col, rh	2.126	205	d	14.63
Ammonia	NH$_3$	17.03	col g, 1.000350 1; col liq, 1.325[16]	0.7188 g/L 20°	−77.75	−33.42	89.9; 13.2 al
-[^{2}H]	N^{2}H$_3$ or ND$_3$	20.05	col g, 1.000347	0.8437 g/L 20°	−74.33	−31.05	....
Ammonium							
acetate	NH$_4$C$_2$H$_3$O$_2$	77.08	wh cr, hygr	1.17[20]	114	d	148[4]; 7.9[15] MeOH
aluminate, tetrachloro-	NH$_4$[AlCl$_4$]	186.84	col	....	304	....	s; s eth
amidosulfate	NH$_3$SO$_3$NH$_4$	114.12	pl, deliq	....	135	d 200	167[10]
antimonate(III), penta-fluoro-	(NH$_4$)$_2$[SbF$_5$]	252.84	col, rh	....	d subl	....	108
arsenate-3-water	(NH$_4$)$_3$AsO$_4$ · 3H$_2$O	247.08	wh, rh	....	d	....	sl s
aurate(III), tetrachloro-	NH$_4$[AuCl$_4$]	356.82	yel, mn or rh	....	....	520	s; sl s al
aurate(I), dicyano-	NH$_4$[Au(CN)$_2$]	267.04	col, cub	....	d 100	....	v s; v s al
aurate(III)-1-water, tetra-cyano-	NH$_4$[Au(CN)$_4$] · H$_2$O	337.09	col	....	d 200	....	v s; v s al

Table 4-1 (*Continued*)
PHYSICAL CONSTANTS OF INORGANIC COMPOUNDS

Name	Formula	Formula weight	Color, crystalline form, refractive index	Density	Melting point, °C	Boiling point, °C	Solubility in 100 parts
Ammonium							
azide	NH_4N_3	60.06	col pl	1.346	160	subl 134 explodes	25.3[20]; i bz, eth
benzoate	$NH_4C_6H_5O_2$	139.16	col, rh	1.260	d 198	subl 160	20[15]
beryllium arsenate-1.5-water	$BeNH_4AsO_4 \cdot 1.5H_2O$	193.02	....	....		d 1200	....
beryllium phosphate-1-water	$BeNH_4PO_4 \cdot H_2O$	140.04	wh, tetr	1.87^{15}	$-H_2O$, 250; $-NH_3$, 400	d 700	0.0012
boranate, tetrafluoro-	$NH_4[BF_4]$	104.84	wh, rh		subl	....	25[16]
borate-4-water, tetra-	$(NH_4)_2B_4O_7$	191.36	col, tetr		....	....	s
bromate	NH_4BrO_3	145.95	col, hex		explodes	....	v s
bromide	NH_4Br	97.95	col, cub, hygr, 1.712	2.429	452 (under pressure)	d 397	76[20]; s acet, al, eth
cadmium trichloride	$CdNH_4Cl_3$	236.79	col, rh	2.93	289	....	33.5[16]
cadmium bissulfate-6-water	$Cd(NH_4)_2(SO_4)_2 \cdot 6H_2O$	448.69	col, mn	2.061^{20}	$-H_2O$, 100	....	s
calcium arsenate-6-water	$CaNH_4AsO_4 \cdot 6H_2O$	305.13	col, mn	1.905^{15}	d 140	....	0.02; s NH_4Cl
calcium phosphate-7-water	$CaNH_4PO_4 \cdot 7H_2O$	279.20	col	1.561^{15}	d	....	i; s a
caprylate	$C_7H_{15}COONH_4$	161.24	col, mn, hygr, dec	....	70–85	....	hyd; s al, ac a
carbamate	$NH_4CO_2NH_2$	78.07	col, rh	....	subl 60	....	v s
carbonate-1-water	$(NH_4)_2CO_3 \cdot H_2O$	114.10	col, cub	....	d 20	....	100[15]
cerate(IV), hexanitrato-	$(NH_4)_2[Ce(NO_3)_6]$	548.23	or, mn	....	....	....	135[20]
cerate(III)-4-water, pentakis-(nitrato)-	$(NH_4)_2[Ce(NO_3)_5] \cdot 4H_2O$	558.28	col, mn	....	74	....	318[20]
cerate(III)-8-water, tetrakis-(sulfato)-	$(NH_4)_2[Ce(SO_4)_4]_2 \cdot 8H_2O$	844.69	mn	2.523	d 100	....	5.5[20]
cerate(IV)-2-water, tetrakis-(sulfato)-	$(NH_4)_4[Ce(SO_4)_4] \cdot 2H_2O$	632.55	....	....	....	....	....
chlorate	NH_4ClO_3	101.49	col, mn	1.80	102 expl	....	28.7[0]

Name	Formula	Formula wt.	Color, crystalline form, refractive index	Density	Melting point, °C	Boiling point, °C	Solubility
chloride	NH_4Cl	53.49	col, cub, 1.642	1.527	520	d 339	37^{20}
chromate(VI)	$(NH_4)_2CrO_4$	152.08	yel, mn	1.91^{12}	d 180		34^{20}
chromium(III) bissulfate-12-water	$CrNH_4(SO_4)_2 \cdot 12H_2O$	478.34	grn or vlt, cub, 1.4842	1.72	94, $-9H_2O$	anhyd 300	7.2^{20}
citrate	$(NH_4)_3C_6H_5O_7$	243.22	wh cr, deliq		d		v s
cobalt(II) bissulfate-6-water	$Co(NH_4)_2(SO_4)_2 \cdot 6H_2O$	395.23	ruby-red, mn, 1.490, 1.495, 1.503	1.902			18^{20}
copper(II) tetrachloride-2-water	$Cu(NH_4)_2Cl_4 \cdot 2H_2O$	277.46	blue, tetr, 1.744, 1.724	1.993	d 110		40.3^{20}
cyanate	NH_4OCN	60.06	wh cr	1.342^{20}	d 60		v s; sl s al
cyanide	NH_4CN	44.06	col, eub	1.02^{100}	d 36	subl 40	v s
dichromate	$(NH_4)_2Cr_2O_7$	252.06	or, mn, flam	2.15	d 170		35.6^{20}, s al
dithiocarbamate	NH_2CSSNH_4	110.19	yel cr, rh	1.451^{20}	99 d		s
dithionate-0.5-water	$(NH_4)_2S_2O_6 \cdot 0.5H_2O$	205.21	col, mn	1.704	d 130		166^{20}
ferrate(II)-3-water, hexacyano-	$(NH_4)_4[Fe(CN)_6] \cdot 3H_2O$	338.15	yel, mn		d		s
ferrate(III), hexacyano-	$(NH_4)_3[Fe(CN)_6]$	266.06	red, rh	1.78^{18}	d	d 160-170	v s
ferrate(III)-3-water, trioxalato-	$(NH_4)_3[Fe(C_2O_4)_3]$	428.09	bright-grn cr, mn		$-3H_2O$, 100		v s
fluoride	NH_4F	37.04	col, hex, deliq	1.315	subl d	d	100^{0}; s al
formate	HCO_2NH_4	63.06	wh, mn, deliq	1.280	116	d 180	143^{20}, s al, eth
gallate(III), hexafluoro-	$(NH_4)_3[GaF_6]$	237.83		wh, oct			
gallium bissulfate-12-water	$GaNH_4(SO_4)_2 \cdot 12H_2O$	496.07	col, cub, 1.468	1.777	d 250		sl s
germanate(IV), hexafluoro-	$(NH_4)_2[GeF_6]$	222.66	col, hex, 1.428, 1.425	2.564			54.6^{20}
hydrogen arsenate, di-	$(NH_4)_2HAsO_4$	176.00	col, mn	1.989	d		s; i al
hydrogen arsenate-3-water	$NH_4H_2AsO_4$	158.98	col, tetr, 1.577, 1.522	2.311^{9}	d		48.7^{20}
hydrogen tetraborate-3-water	$NH_4HB_4O_7 \cdot 3H_2O$	228.33	col cr	2.6	d		10
hydrogen carbonate	NH_4HCO_3	79.06	col, rh or mn, 1.423, 1.536, 1.555	1.58	d 35		21.7^{20}
hydrogen citrate	$(NH_4)_2HC_6H_5O_7$	226.19	wh	1.48			100; sl s al
hydrogen difluoride	NH_4HF_2	57.04	rh or tetr, deliq, 1.390	1.50	125.6	d	v s; sl s al
hydrogen (l)-malate	$NH_4HC_4H_4O_5$	151.08	col, rh	1.5	161		32^{15}
hydrogen oxalate-1-water	$NH_4HC_2O_4 \cdot H_2O$	125.08	col, rh	1.556	d 170		4
hydrogen diphosphate(IV), di-	$(NH_4)_2H_2P_2O_6$	196.04	col cr		170		7

Table 4-1 (Continued)
PHYSICAL CONSTANTS OF INORGANIC COMPOUNDS

Name	Formula	Formula weight	Color, crystalline form, refractive index	Density	Melting point, °C	Boiling point, °C	Solubility in 100 parts
Ammonium							
hydrogen phosphate	$(NH_4)_2HPO_4$	132.05	col, mn, 1.52	1.619	d 100	....	68.9^{20}
hydrogen phosphate, di-	$NH_4H_2PO_4$	115.03	col, tetr, 1.525, 1.479	1.803^{19}	d 150	....	37.4^{20}
hydrogen phosphonate	NH_4HPHO_3	99.03	col, mn	....	123	d 145	83^{20}
hydrogen sulfate	NH_4HSO_4	115.11	col, rh, 1.473	1.78	146.9	d 350	100
hydrogen sulfide	NH_4HS	51.11	wh, rh, 1.74	1.17	d 20	....	128^0
hydrogen sulfite	NH_4HSO_3	99.10	col, mn, 1.561, 1.591	2.03	subl 150	....	267^{10}
hydrogen dl-tartrate	$NH_4HC_4H_4O_6$	167.12		1.636	d 200	....	2.35^{15}
hydroxide	NH_4OH	35.05	dissolved NH_3	....	−77	....	s
iodate	NH_4IO_3	192.94	col, rh or mn	3.42	d 150	....	2.6^{15}
iodide	NH_4I	144.95	col, cub, hygr, 1.7031	2.514	551	....	172^{20}
iodide, tri-	NH_4I_3	398.75	dk brn, rh	3.749	d 175	....	s d
iridate(III)-1-water, hexachloro-	$(NH_4)_3[IrCl_6] \cdot H_2O$	477.05	grn-blk, cub	....	d 350	....	10.5^{20}
iron(III) pentachloride-1-water	$Fe(NH_4)_2Cl_5 \cdot H_2O$	485.93	red, rh, hygr, 1.78	1.99	234	....	v s
iron(III) hexafluoride	$Fe(NH_4)_3F_6$	223.95	lt yel, oct	1.96		....	sl s
iron(III) bisulfate	$FeNH_4(SO_4)_2$	266.01	wh, hex	2.49^{22}	d 420	....	44.15
iron(III) bisulfate-12-water	$FeNH_4(SO_4)_2 \cdot 12H_2O$	482.19	vlt, cub, 1.4854	1.71	39–41	d 230	124
iron(II) bisulfate-6-water	$Fe(NH_4)_2(SO_4)_2 \cdot 6H_2O$	392.14	grn, mn, 1.487, 1.492, 1.499	1.864^{20}	d 100	....	36.4^{20}
lactate	$NH_4C_3H_5O_3$	107.11	lt yelsh liq	1.2^{15}	91–94	....	v s; v s al
magnesium arsenate-6-water	$MgNH_4AsO_4 \cdot 6H_2O$	289.36	col, tetr, 1.608	1.932^{15}	d	....	0.038^{20}
magnesium trichloride-6-water	$MgNH_4Cl_3 \cdot 6H_2O$	256.80	col, rh, deliq	1.456	d 100	....	16.7
magnesium phosphate-6-water (guanite, struvite)	$MgNH_4PO_4 \cdot 6H_2O$	245.41	col, rh, 1.495, 1.496, 1.504	1.711	−6H₂O, 100	d	0.023^0; v s dil a

Name	Formula	Formula weight	Crystalline form, color, refractive index	Density	Melting point	Boiling point	Solubility
magnesium bissulfate-6-water (boussingaultite)	Mg(NH₄)₂(SO₄)₂ · 6H₂O	360.60	col, mn, 1.472, 1.473, 1.479	1.723	120	d 250	25.7[20]
manganese phosphate-1-water	MnNH₄PO₄ · H₂O	185.96	wh cr	....	....	....	0.0031
molybdate(VI)(2−)	(NH₄)₂MoO₄	196.01	col, mn	2.276	....	d	s d; s a
molybdate(VI)(6−)-4-water, hepta-	(NH₄)₆Mo₇O₂₄ · 4H₂O	1235.86	lt yelsh, mn	2.498	−H₂O, 90	d 190	43; s a, al
nickel trichloride-6-water	NH₄NiCl₃ · 6H₂O	291.20	grn, mn, deliq	1.654	150	....	150
nickel bissulfate-6-water	(NH₄)₂Ni(SO₄)₂ · 6H₂O	395.00	dk bl-grn, mn, 1.495, 1.501	1.923	....	....	8.95[20]
nitrate	NH₄NO₃	80.04	col, rh	1.725	169.6	d > 210	192[20]
nitrite	NH₄NO₂	64.04	lt-yelsh cr	1.69	60 expl	subl 30 (vac)	v s
N-nitroso-N-phenylhydroxylamine (cupferron)	C₆H₅N(NO)ONH₄	155.16	creamy-white	....	163–164	....	s; s al
oleate	C₁₇H₃₃COONH₄	299.48	yelsh-brn paste	....	70–72	....	s
oxalate-1-water	(NH₄)₂C₂O₄ · H₂O	142.11	col, rh, 1.439, 1.546, 1.594	1.50	d 70	....	5.1[20]
palladate(II), tetrachloro-	(NH₄)₂[PdCl₄]	284.29	olive grn, tetr	2.17	d	....	s
palladate(IV), hexachloro-	(NH₄)₂[PdCl₆]	355.20	red-brn, cub	2.418	d	....	sl s
palmitate	CH₃(CH₂)₁₄COONH₄	273.45	yel-wh pwd	....	70–73	....	s
perchlorate	NH₄ClO₄	117.50	col, rh, 1.482	1.95	d 240	....	21.7[20]
permanganate	NH₄MnO₄	136.97	purple, rh	2.208[10]	d 110	....	0.8[15]
peroxoborate-0.5-water	NH₄BO₃ · 0.5H₂O	85.86	wh cr	....	d	....	1.55[18]
peroxochromate	(NH₄)₃CrO₈	234.11	red-brn, cub	....	d 40	expl 50	sl s; expl H₂SO₄
peroxodisulfate	(NH₄)₂S₂O₈	228.18	col, mn, 1.498, 1.502, 1.587	1.982	d 120	expl 180	58.2⁰
periodate	NH₄IO₄	208.94	col, tetr	3.056[18]	expl	....	2.7[16]
perrhenate	NH₄ReO₄	268.24	wh, hex	3.97	d	....	12.0[20]
phosphate-3-water	(NH₄)₃PO₄ · 3H₂O	203.13	wh pr	....	....	....	26.1
phosphate, hexafluoro-	NH₄[PF₆]	163.00	col pl	2.180[18]	d	....	75[20], s al, acet
phosphate-molybdenum trioxide (1/12)	(NH₄)₃PO₄ · 12MoO₃	1876.49	yel cr pwd	....	....	....	0.00002[20]
phosphinate	NH₄PH₂O₂	83.03	col, rh	1.634	200	d 240	100; 5 al
picrate	NH₄C₆H₂N₃O₇	246.14	red or yel, rh	1.719	d	expl 423	1.1[20]
plumbate(IV), hexachloro-	(NH₄)₂[PbCl₆]	455.98	yel, cub	2.925	d 120	....	sl s; s a
platinate(IV), hexabromo-	(NH₄)₂[PtBr₆]	710.62	red-brn, cub	4.265	d 145	....	0.6[20]

Table 4-1 (*Continued*)
PHYSICAL CONSTANTS OF INORGANIC COMPOUNDS

Name	Formula	Formula weight	Color, crystalline form, refractive index	Density	Melting point, °C	Boiling point, °C	Solubility in 100 parts
Ammonium							
platinate(II), tetrachloro-	$(NH_4)_2[PtCl_4]$	372.98	red, rh	2.936	d 140	....	s
platinate(IV), hexachloro-	$(NH_4)_2[PtCl_6]$	443.89	yel, cub, 1.8	3.065	d	....	0.5[20]
praseodymium bissulfate-4-water	$NH_4Pr(SO_4)_2 \cdot 4H_2O$	423.13	lt grn cr	2.531[17]	d 170	....	sl s
propionate	$NH_4C_3H_5O_2$	91.11	col, deliq	1.108	45	....	v s
reineckate	$[Cr(NH_3)_2(SCN)_4] \cdot H_2O$	354.44	....	....	d	....	117[7]
selenate(VI)	$(NH_4)_2SeO_4$	179.03	col, mm, 1.561, 1.563, 1.585	2.193[20]	d	....	18.6[20]
silicate, hexafluoro- (crypto-halite)	$(NH_4)_2SiF_6$	178.14	col, oct, 1.3696	2.011	d	....	20
sodium hydrogen phosphate-4-water	$NaNH_4HPO_4 \cdot 4H_2O$	209.07	col, mm, 1.439, 1.442, 1.469	1.574	d 97	....	21.1[0]
sodium hydrogen tartrate-4-water	$NaNH_4HC_4H_4O_6 \cdot H_2O$	261.16	wh, rh	1.590	....	....	
sodium sulfate-2-water	$NaNH_4SO_4 \cdot 2H_2O$	173.12	col	1.63[15]	d 80	....	s
stannate(IV), hexabromo-	$(NH_4)_2[SnBr_6]$	634.22	col, cub	3.50	d	....	v s
stearate	$C_{17}H_{35}COONH_4$	301.50	yel-wh pwd	....	70–75	....	sl s; s al
succinate	$(NH_4)_2C_4H_4O_4$	152.15	col cr	1.37	....	....	s
sulfate (mascagnite)	$(NH_4)_2SO_4$	132.14	col, rh, 1.521, 1.523, 1.533	1.769[20]	d 230	....	75.4[20]
sulfate, fluoro-	NH_4SO_3F	117.10	wh nd	....	d 245	....	s
sulfide	$(NH_4)_2S$	68.14	yel cr, hygr	....	ca −18	d	v s
sulfite-1-water	$(NH_4)_2SO_3 \cdot H_2O$	134.15	col, mm, 1.515	1.41	d 60	....	75[20]
dl-tartrate	$(NH_4)_2C_4H_4O_6$	184.15	col, mm, 1.55	1.601	d	....	58[15]
tellurate(VI)(2−)	$(NH_4)_2TeO_4$	227.67	wh pd	3.024	d	....	s
thioantimonate(V)-4-water	$(NH_4)_3SbS_4 \cdot 4H_2O$	376.18	yel pr	....	d	....	71.2[0]
thiocarbamate	$NH_4CS_2NH_2$	110.20	yel cr	....	d 50	....	v s

Name	Formula	Formula wt.	Color, crystalline form, refractive index	Density	M.P., °C	B.P., °C	Solubility
thiocarbonate, di-	$(NH_4)_2CS_2$	144.28	yel cr, hygr	....	subl	....	v s
thiocyanate	NH_4SCN	76.12	col mn, deliq	1.305	149.6	d 170	128^{0}
thiosulfate	$(NH_4)_2S_2O_3$	148.20	col mn, hygr	1.679	d 100	....	2.15^{15}
uranyl(VI) triscarbonate-2-water	$(NH_4)_4UO_2(CO_3)_3 \cdot 2H_2O$	558.24	yel. mn	2.773	d 100	....	5.8^{20}
uranyl(VI) pentafluoride	$(NH_4)_3UO_2F_5$	419.14	tetr, 1.495	3.186	subl	....	s
vanadate(V)(1−)	NH_4VO_3	116.98	wh cr	2.326	d 130	....	0.48^{20}
vanadium(III) bissulfate-12-water	$NH_4V(SO_4)_2 \cdot 12H_2O$	477.28	red to blue	1.687	49	....	28.5^{20}
zinc bissulfate-6-water	$(NH_4)_2Zn(SO_4)_2 \cdot 6H_2O$	401.66	wh, mn, 1.489, 1.493, 1.499	1.931	d	....	17.1^{20}
zincate, tetrachloro-	$(NH_4)_2[ZnCl_4]$	243.26	wh, rh, hygr	1.879	d 150	subl 341	v s
zirconate, hexafluoro-	$(NH_4)_2[ZrF_6]$	241.29	hex	1.154	....	....	....
Antimonic acid, hexahydroxo-	$HSb(OH)_6$	224.80	wh pd	6.6	d	....	sl s
Antimony	Sb	121.75	silv-wh, rhhd	6.684	630.5	1635	i; s hot conc H_2SO_4
(III) bromide	$SbBr_3$	361.48	col, rh, hygr, 1.74	4.148	95	289	d; s HCl, HBr, CS_2
(III) chloride	$SbCl_3$	228.11	col, rh, deliq	3.06	73.4	220	910^{20}, s al, bz, chl
(V) chloride	$SbCl_5$	299.02	yel liq, trig bipyr	2.336^{20}	2.8	140	d; s HCl, chl
(III) chloride oxide	SbClO	173.20	wh, mn	5.01	d 170	....	i; s HCl, CS_2
(III) fluoride	SbF_3	178.75	col, rh, deliq	4.379^{21}	292	376	444^{20}
(V) fluoride	SbF_5	216.75	col oily liq	2.99^{23}	7.0	150	s; s KF
hydride (stibine)	SbH_3	124.77	inflam gas	4.36^{15}	−91.5	−18.4	20 ml^{0}, s CS_2
(III) iodide	SbI_3	502.46	yel-red, rh or mn	4.766^{22}	171	400	d; s al, KI, HCl
(V) iodide	SbI_5	756.27	dk brn liq	5.2	79	....	d
(III) oxide (senarmontite)	Sb_2O_3	291.50	wh, cub, 2.087	5.67	573	656	v sl s; s KOH, HCl
(III,V) oxide	Sb_2O_4	307.50	wh, rh, 2.18, 2.35, 2.35	5.82	tr 567 to cub	....	....
(V) oxide	Sb_2O_5	323.50	wh, rd, 2.00	3.8	d 930	....	v sl s
(III) dioxide sulfate	$Sb_2O_2SO_4$	371.56	yel-wh pd	4.89	d 380	....	v sl s; sl s KOH
potassium oxide tartrate 0.5-water	$K(SbO)C_4H_4O_6 \cdot 0.5H_2O$	333.93	wh, col cr	2.607	d 100	....	d; s H_2SO_4 8.3^{20}, 6.7 gly
(III) selenide	Sb_2Se_3	480.38	gray, rh	....	615	....	v sl s; s conc HCl
(III) sulfate	$Sb_2(SO_4)_3$	531.68	col slky nd, deliq	3.625^{4}	d	....	hyd; s a

Table 4-1 (Continued)
PHYSICAL CONSTANTS OF INORGANIC COMPOUNDS

Name	Formula	Formula weight	Color, crystalline form, refractive index	Density	Melting point, °C	Boiling point, °C	Solubility in 100 parts
Antimony							
(III) sulfide (stibnite)	Sb_2S_3	339.69	blk, rh (also orange form)	4.64	546	....	0.002^{20}d; s H_2SO_4
(III) telluride	Sb_2Te_3	626.30	gray, rhbd	6.52	618.5	....	i; s HNO_3
Argon	Ar	39.948	col g, 1.000284	1.7824 (g/L)	-189.38	-185.87	3.36 ml^{20}
Arsenic			liq	1.400	....	....	....
			col, fcc	1.7840	....	....	....
	As	74.9216	gray, rh	5.72	$817^{28\,atm}$	subl 612	i; s HNO_3
(III) bromide			yel, cub	2.026^{18}	d 358	....	i
	$AsBr_3$	314.66	col, rh, deliq	3.3972^{25} 3.3282 (liq)	31	221	hyd; s CS_2, HCl
(V) calcium oxide–water (1/2/3) (haidingerite)	$As_2O_5 \cdot 2CaO \cdot 3H_2O$	396.04	col, rh, 1.590, 1.602, 1.638	2.967	....	....	....
(III) chloride	$AsCl_3$	181.28	col liq	2.15^{25}	-16	130	d; s al, eth, HCl
bis(dimethylarsine) oxide	$[(CH_3)_2As]_2O$	225.98	yel liq	....	-25	120	i; s al, bz, HF
(III) fluoride	AsF_3	131.92	col liq	3.01	-5.95	51	s; s al, bz, eth
(V) fluoride	AsF_5	169.91	col gas	7.71 (g/L)	-80.3	-52.6	s al, bz, eth
(III) fluoride oxide	AsFO	144.92	....	....	-68.3	25.6	
(III) hydride (arsine)	AsH_3	77.95	col gas	2.695 (g/L)	-116.9	-62.5	28 ml^{20}; s bz, chl
hydroxydimethylarsine oxide	$(CH_3)_2As(O)OH$	137.99	hygr cr	....	....	....	200; v s al
(III) iodide	AsI_3	455.64	red, hex	4.38^{13}	141.8	403	6^{25}, 5.2 CS_2
(III) oxide dimer (claudetite)	As_4O_6	395.68	col, rh	4.15	278	460	1.82^{20}; s al
(arsenolite)	....	....	col, cub, 1.755	3.865	309	....	1.2^2
(V) oxide	As_2O_5	229.84	wh amorp, deliq	4.32	d 800	....	65.8^{20}; s al

Name	Formula	Mol. wt.	Color, crystalline form, index of refraction	Density	M.P.	B.P.	Solubility
phosphide	AsP	105.90	blk cr	......	subl d	......	i; s HCl
(III) selenide	As$_2$Se$_3$	386.72	brn. rhbd	4.75	377	565	i; s alk
(di-) sulfide, di- (realgar)	As$_2$S$_2$	213.97	brn-red, mn, 2.68	3.254[19]	307	707	i; s alk
(III) sulfide	As$_2$S$_3$	246.04	yel, mn	3.43	312	707	0.0003; s alk, HNO$_3$
(V) sulfide	As$_2$S$_5$	310.16	yel	......	d 500	......	......
(III) telluride	As$_2$Te$_3$	532.64	mn	......	375	165	sl s
tetramethyldiarsine	(CH$_3$)$_2$AsAs(CH$_3$)$_2$	209.96	oily liq, flam	......	−6	......	

Auric and Aurous salts

see under Gold

Barium

Name	Formula	Mol. wt.	Color, crystalline form, index of refraction	Density	M.P.	B.P.	Solubility
	Ba	137.34	silv-met	3.51[20]	725.1	1849.0	d
acetate-1-water	Ba(C$_2$H$_3$O$_2$)$_2$·H$_2$O	273.45	col, tric, 1.500	2.19	d 150	......	76[20]
amide, di-	Ba(NH$_2$)$_2$	169.39	gray-wh cr	......	280	......	d
arsenate	Ba$_3$(AsO$_4$)$_2$	689.86	......	5.10	1600	......	0.055[20]; s a
arsenide, di-	Ba$_3$As$_2$	561.86	brn	4.1[15]	......	......	d
azide	Ba(N$_3$)$_2$	221.38	mn	2.936	d 120	expl	17[17]
boride, hexa-	BaB$_6$	202.21	met.blk, cub	4.36[16]	2270	......	i; s HNO$_3$
bromate-1-water	Ba(BrO$_3$)$_2$·H$_2$O	411.17	col, mn	3.99[18]	d 260	......	0.65[20]; s acet
bromide	BaBr$_2$	297.16	col cr, 1.75	4.781	854	d	45[20]
bromide-2-water	BaBr$_2$·2H$_2$O	333.19	col, mn, 1.713, 1.727, 1.744	3.58	−2H$_2$O, 120	......	49.5[20]
carbide, di-	BaC$_2$	161.36	gray, tetr	3.75	......	......	d
carbonate (witherite)	BaCO$_3$	197.35	wh, hex	4.43	d 1360	......	0.002; s a
			wh, rh, 1.529, 1.676, 1.677	4.43			0.002
chlorate-1-water	Ba(ClO$_3$)$_2$·H$_2$O	322.26	col, mn, 1.562, 1.577	3.18	−H$_2$O, 120	−O$_2$, 250	33.9[20]
chloride	BaCl$_2$	208.25	col, mn, 1.7303, 1.7367, 1.7420	3.856	962	2029	36[20]
chloride-2-water	BaCl$_2$·2H$_2$O	244.28	col, mn, 1.629, 1.642, 1.658	3.097	−2H$_2$O, 113	......	36[20]
chloroanilate	C$_6$BaCl$_2$O$_4$	344.34	......	......	......	......	......
chromate(VI)	BaCrO$_4$	253.33	yel, rh	4.498[15]	......	......	0.001[20]; s a
citrate-7-water	Ba$_3$(C$_6$H$_5$O$_7$)$_2$·7H$_2$O	916.33	wh ɔd	......	−7H$_2$O, 150	......	0.057[25]
cyanide	Ba(CN)$_2$	189.38	wh ɔr pd, d slowly	......	......	......	80[14]
dichromate	BaCr$_2$O$_7$	353.33	red, mn	......	......	......	sl s; s hot conc H$_2$SO$_4$
diphenylaminesulfonate	(C$_6$H$_5$NHC$_6$H$_4$·4·SO$_3$)$_2$Ba	633.93	wh	......	......	......	......

Table 4-1 (*Continued*)
PHYSICAL CONSTANTS OF INORGANIC COMPOUNDS

Name	Formula	Formula weight	Color, crystalline form, refractive index	Density	Melting point, °C	Boiling point, °C	Solubility in 100 parts
Barium							
diphosphate	$Ba_2P_2O_7$	448.62	wh, rh	3.9^{20}	....	....	0.01; s a
dithionate-2-water	$BaS_2O_6 \cdot 2H_2O$	333.50	col, mn, 1.586, 1.595, 1.607	4.536^{14}	d 120	....	18.6^{20}
ferrate(II)-6-water, hexacyano-	$Ba_2[Fe(CN)_6] \cdot 6H_2O$	594.73	yel, mn	2.666	d 80	....	0.17^{15}
fluoride	BaF_2	175.34	col, cub, 1.4741	4.89	1368	2272	0.16^{20}
formate	$Ba(CHO_2)_2$	227.38	col, rh, 1.573, 1.597, 1.636	3.21	d	....	30^{20}
gallate(III)-1-water, hexafluoro-	$Ba_3[GaF_6]_2 \cdot H_2O$	797.46	wh cr	4.06	$-H_2O$, 230	....	i; s HF
gluconate-3-water	$Ba(C_6H_{11}O_7)_3 \cdot 3H_2O$	581.69	col, rh	....	d 120	....	8.7^{25}
hydride	BaH_2	139.36	gray cr	4.21^0	d > 1000	....	d
hydrogen arsenate-1-water	$BaHAsO_4 \cdot H_2O$	295.28	col, mn, 1.635	1.93^{15}	$-H_2O$, 150	....	sl s; s HCl
hydrogen phosphate	$BaHPO_4$	233.32	wh, rh, 1.635, 1.617	4.165^{15}	d 410	....	0.01; s a
hydrogen phosphate, di-	$Ba(H_2PO_4)_2$	331.32	col, tric	2.9^4	....	....	d; s a
hydrogen sulfide-4-water	$Ba(HS)_2 \cdot 4H_2O$	275.56	yel, rh	....	d 50	....	48.8^{20}
hydroxide-8-water	$Ba(OH)_2 \cdot 8H_2O$	315.48	col, mn, 1.471, 1.502, 1.500	2.18^{16}	78	....	3.9^{20}
iodate	$Ba(IO_3)_2$	487.15	col, mn	5.23^{20}	d	....	0.033^{20}, s HCl
iodate-1-water	$Ba(IO_3)_2 \cdot H_2O$	505.16	col, mn	4.657^{18}	$-H_2O$, 200	....	v sl s
iodide	BaI_2	391.15	col cr	5.15	711	2067	204^{20}
iodide-2-water	$BaI_2 \cdot 2H_2O$	427.18	col, rh, deliq	5.15	$-2H_2O$, 540	....	223^{20}
manganate(VI)(2−)	$BaMnO_4$	256.28	gray-grn, hex	4.85	....	....	disprop
molybdate(VI)(2−)	$BaMoO_4$	297.28	wh pd	4.65	1480	....	0.0058^{23}
nitrate (nitrobarite)	$Ba(NO_3)_2$	261.35	col, cub, 1.572	3.24	575	d	9.0^{20}
nitride	Ba_3N_2	440.03	yel-brn	4.783	....	1000^{vac}	d
nitrite	$Ba(NO_2)_2$	229.35	col, hex	3.23	d 220	....	67.5^{20}
nitrite-1-water	$Ba(NO_2)_2 \cdot H_2O$	247.37	lt yelsh, hex	3.173^{20}	d 115	....	73^{20}

oxalate	BaC_2O_4	225.36	col cr	2.658	....	....	0.009
oxide	BaO	153.34	col, cub, 1.98	5.72	2013	3088	3.48^{20}
perchlorate	$Ba(ClO_4)_2$	336.24	col, hex	3.681^{25}	123	d 162	198
perchlorate-3-water	$Ba(ClO_4)_2 \cdot 3H_2O$	390.29	col, hex, 1.533	2.74	d 400	....	336^{20}
permanganate	$Ba(MnO_4)_2$	375.21	brn-vlt cr	3.77	d 200	....	6^{11}
peroxide	BaO_2	169.34	wh-gray pd	4.96	450	$-O_2$, 800	1.5^0
phosphate	$Ba_3(PO_4)_2$	601.96	wh, cub	4.1^{16}	....	....	i; s a
phosphinate-1-water	$Ba(PH_2O_2)_2 \cdot H_2O$	285.33	wh, mn	2.90^{17}	d 100	....	30^{15}
propionate-1-water	$Ba(C_3H_5O_2)_2 \cdot H_2O$	301.50	col, rh, 1.518	....	d 300	....	57^{20}
selenate	$BaSeO_4$	280.30	wh, rh	4.75	d	....	0.008^{25}
selenide	BaSe	216.30	wh, cub, 2.268	5.02	(1830)	....	d
silicate, meta-	$BaSiO_3$	213.42	col, rh, 1.673, 1.674, 1.678	4.399	1605	....	i; s HCl
silicate-6-water, meta-	$BaSiO_3 \cdot 6H_2O$	351.52	col, rh, 1.542, 1.548, 1.548	2.59	....	....	0.17
silicate, ortho-	Ba_2SiO_4	336.77	....	....	1760	....	....
silicate, hexafluoro-	$Ba[SiF_6]$	279.42	col, rh	4.29^{21}	d 300	....	0.025^{25}
stearate	$Ba(C_{18}H_{35}O_2)_2$	704.13	wh pd	....	....	....	0.004^{15}
sulfate (barite)	$BaSO_4$	233.40	wh, rh, 1.637, 1.638, 1.649	4.50^{15}	1580	....	0.0002
sulfide	BaS	169.40	col, cub, 2.155	4.25^{15}	2227	....	7.9 d^{20}
sulfide-1-water, tetra-	$BaS_4 \cdot H_2O$	283.61	red or yel, rh	2.988	d 300	....	4^{15}
sulfite	$BaSO_3$	217.40	col, hex	....	d	....	0.02^{20}
tartrate-1-water	$BaC_4H_4O_6 \cdot H_2O$	303.52	wh cr	2.980^{20}	....	....	0.026^{18}
tellurate(VI)(2−)-3-water	$BaTeO_4 \cdot 3H_2O$	383.07	wh	4.2^{20}	d 200	....	sl s; s HCl
telluride	BaTe	264.94	yel-wh, cub, 2.440	5.13	....	....	d a
thiocarbonate, di-	$BaCS_3$	245.54	yel, hex	....	d	....	1.08^0
thiocyanate-2-water	$Ba(SCN)_2 \cdot 2H_2O$	289.53	col nd, deliq	2.286^{18}	d 160	....	170^{25}
thiosulfate-1-water	$BaS_2O_3 \cdot H_2O$	267.48	wh, rh	3.5^{18}	d 220	....	0.21^{20}
titanate(IV)(2−)	$BaTiO_3$	233.24	tetr & hex, 2.40	6.017^{hex} 5.806^{tetr}	1705 1625	....	....
titanate(IV)(4−)	Ba_2TiO_4	386.58	....	....	1860	....	....
tungstate(VI)(2−)	$BaWO_4$	385.19	col, tetr	5.04	860	....	sl s; d a
vanadate(V)(4−), di-	$Ba_2V_2O_7$	488.56	wh cr	....	986	2970	....
Berkelium	Bk	(245)	α-form: double hcp β-form: fcc	14.78 13.25	....	....	....

Table 4-1 (Continued)
PHYSICAL CONSTANTS OF INORGANIC COMPOUNDS

Name	Formula	Formula weight	Color, crystalline form, refractive index	Density	Melting point, °C	Boiling point, °C	Solubility in 100 parts
Beryllium	Be	9.012	dk gray, hcp	1.86	1277	2484	i; s a, alk
(acetato)-μ_4-oxotetraberyllium, hexakis-	Be$_4$O(OCOCH$_3$)$_6$	406.32	wh ... tetrahedra	1.25	285	330	i; s org solv exc al, eth
boron oxide (1/3)	B$_2$O$_3 \cdot$ 3BeO	144.65	.....	.....	1495	.....	.....
bromide	BeBr$_2$	168.83	wh nd, deliq	3.465^{25}	488	521	s
carbide	Be$_2$C	30.04	yel, hex	1.90^{15}	2127	.....	s a
carbonate-4-water	BeCO$_3 \cdot$ 4H$_2$O	141.08	wh, hex	.....	.....	.....	0.36^0
chloride	BeCl$_2$	79.92	wh, rh, deliq	1.899^{25}	415	532	42
fluoride	BeF$_2$	47.01	wh, deliq	1.986^{25}	552	1175	v s but slow
hydride	BeH$_2$	11.03	wh cr	.....	d 125	.....	d
hydroxide	Be(OH)$_2$	43.03	amorp pwd or cr	1.92	134 d	.....	v sl s; s hot conc a & alk
iodide	BeI$_2$	262.82	col nd, tetr	4.2	480	482	hyd; s al, eth, CS$_2$
nitrate-4-water	Be(NO$_3$)$_2 \cdot$ 4H$_2$O	205.08	wh cr, deliq	1.557	60.5	d 142	166^{20}
nitride	Be$_3$N$_2$	55.05	col, cub	.....	2200	.....	d hot H$_2$O, alk
oxalate-3-water	BeC$_2$O$_4 \cdot$ 3H$_2$O	151.08	col, rh	.....	d 225	.....	63.5^{25}
oxide	BeO	25.01	wh, hex, 1.719, 1.733	3.01	2408 (α) 2447 (β)	3787	s conc H$_2$SO$_4$
perchlorate-4-water	Be(ClO$_4$)$_2 \cdot$ 4H$_2$O	279.97	wh cr	.....	$-$H$_2$O, 100	.....	147^{25}
phosphate-3-water	Be$_3$(PO$_4$)$_2 \cdot$ 3H$_2$O	271.03	wh	.....	.....	.....	s
phosphide	Be$_3$P$_2$	88.96	brn, cub	2.24	.....	.....	.....
selenate-4-water	BeSeO$_4 \cdot$ 4H$_2$O	224.03	col, rh, 1.466, 1.501, 1.503	2.03	$-$4H$_2$O, 213	d 560	49^{25}
selenide	BeSe	87.98	cub	.....	.....	.....	.....
silicon oxide–water (4/2/1) (bertrandite)	4BeO $\cdot$ 2SiO$_2 \cdot$ H$_2$O	238.23	col, rh, 1.591, 1.605, 1.604	2.6	.....	.....	i
silicon oxide (2/1) (phenazite)	2BeO $\cdot$ SiO$_2$	110.11	col, tric, 1.654, 1.670	3.0	1560	.....	i

stearate	Be(C₁₈H₃₅O₂)₂	575.97	wh waxy	....	45	....	i; s CCl₄, eth
sulfate	BeSO₄	105.07	col, tetr	2.443	d 550	....	i
sulfate-4-water	BeSO₄·4H₂O	177.14	col, tetr, 1.472, 1.440	1.713^{11}	−4H₂O, 270	d 580	39.1^{20}
sulfide	BeS	41.08	wh, cub	2.36	....	....	i; s HNO₃
telluride	BeTe	136.62	cub	....	....	....	....
Bismuth	Bi	208.98	silv-wh, rh	9.80	271.5	1579	i; s hot H₂SO₄
arsenate	BiAsO₄	347.90	wh, mn	7.14	....	....	i
bromide, tri-	BiBr₃	448.71	yel, hygr	5.72^{25}	218	460	hyd; s HCl, eth
bromide oxide	BiBrO	304.89	col, cub	8.082^{15}	d	....	i; s a
carbonate dioxide	Bi(CO₃)O₂	509.97	wh pwd	6.86	d	....	i; s a
chloride, tri-	BiCl₃	315.34	wh cr, deliq	4.6	232	447	d; s a, al, eth
chloride oxide	BiClO	260.43	wh pwd	7.72^{15}	red heat	....	i; s a
citrate	BiC₆H₅O₇	398.08	wh cr	3.458	d	....	sl s; s a
fluoride, tri-	BiF₃	265.98	wh, cub, 1.74	5.32^{20}	727	(1027)	i; s a
fluoride, penta-	BiF₅	303.98	col, oct	5.4^{25}	151	230	i; s a
fluoride oxide	BiFO	243.98	wh cr	7.5^{20}	d	....	i; s a
hydride (bismuthine)	BiH₃	212.00	v unstable liq	....	....	16.8	....
hydroxide	Bi(OH)₃	260.00	wh amorp pwd	4.36	−H₂O, 100	....	d; s a
iodide, tri-	BiI₃	589.69	blk, hex	5.8	408	d 500	i; s al, HCl
iodide oxide	BiIO	351.88	red, tetr	7.922	d red heat	....	i; s a
molybdate(VI)(2−)	Bi₂(MoO₄)₃	897.78	lt yel, tetr	6.07	643	....	d; s a, gly, acet
nitrate-5-water	Bi(NO₃)₃·5H₂O	485.07	col, tric	2.83	d 30	....	i; s a
nitrate oxide-1-water	Bi(NO₃)O·H₂O	305.00	col, hex	4.928^{18}	−H₂O, 105	d 260	i; s a
oxide, tri-	Bi₂O₃	495.96	yel, rh	8.76	817	1890	i; s KOH
oxide, penta-	Bi₂O₅	497.96	brn	5.10	d 150	....	s
oxide perchlorate	Bi(ClO₄)O	342.44	wh, rh	....	....	....	i; s HCl
phosphate	BiPO₄	303.95	wh, mn	6.323^{15}	d	d	i; s aq reg
selenide	Bi₂Se₃	654.84	blk, rbhd	7.66^{20}	706	....	i; s HCl
silicon oxide (2/3) (euly-tite)	2Bi₂O₃·3SiO₂	1112.17	yel, cub, 2.05	6.11	....	....	....
sulfate	Bi₂(SO₄)₃	706.14	wh nd, hygr	5.08^{15}	d 405	....	d; s a
sulfide (bismuthinite)	Bi₂S₃	514.15	dk brn, rh	7.39	763	....	i; s HNO₃
tartrate-3-water	Bi₂(C₄H₄O₆)₃·3H₂O	970.27	wh pwd	2.595^{25}	d 105	....	i; s a, alk
tellurate-2-water (montanite)	Bi₂TeO₆·2H₂O	677.59	....	3.79	....	....	i
telluride	Bi₂Te₃	800.76	gray, hex	7.859	588.5	....	i

Table 4-1 (Continued)
PHYSICAL CONSTANTS OF INORGANIC COMPOUNDS

Name	Formula	Formula weight	Color, crystalline form, refractive index	Density	Melting point, °C	Boiling point, °C	Solubility in 100 parts
Boranes							
diborane(6)	B_2H_6	27.67	col g; hex solid	0.578^{-183}	-165.5	-92.5	s NH_4OH, conc H_2SO_4
1-bromodiborane(6)	B_2H_5Br	106.67	col gas	....	-104	ca 10	hyd
tetraborane(10)	B_4H_{10}	53.32	col gas; hex solid	0.652	-120	-18	sl s d; s bz
pentaborane(9)	B_5H_9	63.13	col liq, tetr cr	0.760 (c) 0.637 (liq)	-46.81	62	hyd
pentaborane(11)	B_5H_{11}	65.14	col, mn	0.745	-123	63	....
hexaborane(10)	B_6H_{10}	74.95	col liq; rh cr	0.69^0 (liq)	-65.1	82.2	hyd
nonaborane(15)	B_9H_{15}	112.41	col liq, mm cr	....	2.7	....	sl s; s bz, CS_2, eth
decaborane(14)	$B_{10}H_{14}$	122.22	wh, mn	0.948	98.8	219	....
decaborane(16)	$B_{10}H_{16}$	124.23	wh cr	0.84	196	....	....
icosaborane(26)	$B_{20}H_{26}$	238.41	col, tetr	1.130	....	....	sl d
Borazine	$B_3H_6N_3$	80.53	col liq	0.86^{25}	-58	55	sl d
Boric acids, see under Hydrogen							
Boron							
	B	10.81	brn, rh	2.46	2177	3658	i
arsenate	$BAsO_4$	149.73	wh, tetr, 1.681. 1.690	3.64	subl 700	....	s min a
bromide, tri-	BBr_3	250.57	col fum liq, 1.5312	2.6954^0	-46.0	91.3	d
calcium oxide (1/2)	$B_2O_3 \cdot 2CaO$	181.78	col	....	1310	....	....
calcium oxide (1/3)	$B_2O_3 \cdot 3CaO$	237.86	col	....	1490	....	....
calcium oxide (2/1)	$2B_2O_3 \cdot CaO$	195.32	col	....	990	....	....
carbide	B_4C	55.26	blk, rbhd	2.52	2470	>3500	i; s fus alk
chloride, tri-	BCl_3	117.19	col fum liq, 1.4195	1.373^0	-107	12	d
(di-) chloride, tetra-	$Cl_2B{-}BCl_2$	163.43	col liq	....	-92	65.5 d	....
fluoride, tri-	BF_3	67.81	col gas	0.00299	-128.7	-101	105 ml⁰
fluoride-1-diethyl ether	$BF_3 \cdot O(C_2H_5)_2$	141.94	fum liq	1.125^{25}	-60.4	125.7	hyd
fluoride-1-ammonia	$BF_3 \cdot NH_3$	84.84	....	1.86	163	....	s
fluoride-2-water	$BF_3 \cdot 2H_2O$	103.84	col liq	1.6316^{20}	6	....	i; s eth, dioxane

Name	Formula	Mol wt	Color, crystalline form, etc.	Density	m.p., °C	b.p., °C	Solubility
(di) fluoride, tetra-	B_2F_4	97.62	col, mn	1.92	−56	−34	d; s bz
hydrides, see under Boranes							
iodide, tri-	BI_3	391.25	col pl, hygr	3.35^{50}	50	210	sl s hot a
nitride	BN	24.82	wh, hex	2.25	d 2325	subl 3000	
oxide	B_2O_3	69.62	wh, rh, 1.64, 1.61	2.46	450	2065	(slowly) 3.3
phosphide	BP	41.78	maroon pwd		ign 200		i
selenide	B_2Se_3	258.50	yel-gray pwd	2.52	480		d
(tri) silicide	B_3Si	60.52	blk, rh	2.47			sl s HNO_3
(hexa) silicide	B_6Si	92.95	blk cr				s HNO_3
sulfide, di-	BS_2	74.94		1.55	417		d
(di) sulfide, tri-	B_2S_3	117.82	wh, mn		563		
Bromic acid	$HBrO_3$	128.92	col, known only in soln		d 100		v s
Bromine	Br_2	159.81	red-brn fum liq, 1.654	3.1028	−7.3	58.75	3.58^{20}; v s al, chl, CS_2, eth
azide	BrN_3	121.93	or-red liq, expl		ca −45	expl	i; s eth, KI
chloride	BrCl	115.36	red liq or gas		−54	ca 5	s d; s CS_2, eth
cyanide	BrCN	105.93	col, rh	1.863^{62}	52	61.2	
fluoride	BrF	98.91	dk red		−33	20	d viol; d alk
fluoride, tri-	BrF_3	136.90	straw, 1.4536	2.8030	8.77	125.74	d; s HF
fluoride, penta-	BrF_5	174.90	col fum liq, 1.3529	2.4626	−60.5	41.2	3.86^{mp}
hydrate, octa-	$Br_2 \cdot 8H_2O$	303.95	red, oct	1.49	d 3.84		i; s CCl_4
nitrate	$BrNO_3$	141.91	yel liq		−42		
(di) oxide	Br_2O	175.82	dk brn		−17.5 d	d 0.0	d; s d CCl_4
oxide, di-	BrO_2	111.91	lt yel cr		d −40		i; s CCl_4
sulfate, fluoro-	$Br[SO_3F]$	178.98	dk red liq		59 (vac)		d viol
tris(sulfate, fluoro-)	$Br[SO_3F]_3$	377.11	or, hygr	2.60	−9.0	120.5	d viol
Bromyl fluoride	BrO_2F	130.91	col cr		d −60		
nitrate	BrO_2NO_3	173.91	or			d 20	
Cadmium	Cd	112.40	silv-wh, hex	8.642	320.9	770	i; s a
acetate	$Cd(C_2H_3O_2)_2$	230.50	col	2.341	256	d	v s
acetate-2-water	$Cd(C_2H_3O_2)_2 \cdot 2H_2O$	266.53	col, mn	2.01	−H_2O, 130		v s
amide	$Cd(NH_2)_2$	144.45			d 120		
antimonide	CdSb	234.15		3.05	456		
arsenide	Cd_3As_2	487.04	dk gray, cub	6.21^{15}	721		i; s HNO_3
borate-1-water	$Cd(BO_2)_2 \cdot H_2O$	248.03	wh, rh	3.758	d		125^{17}
bromate-1-water	$Cd(BrO_3)_2 \cdot H_2O$	386.23	col, rh	3.758	d		125

Table 4-1 (*Continued*)

PHYSICAL CONSTANTS OF INORGANIC COMPOUNDS

Name	Formula	Formula weight	Color, crystalline form, refractive index	Density	Melting point, °C	Boiling point, °C	Solubility in 100 parts
Cadmium							
bromide	$CdBr_2$	272.22	lt yel	5.192	570–585	863	99[20]
carbonate	$CdCO_3$	172.41	wh, trig	4.258[4]	d 500	....	i; s a, NH_4OH
chlorate-2-water	$Cd(ClO_3)_2 \cdot 2H_2O$	315.33	col pr, deliq	2.28[18]	80	....	322[20]
chloride	$CdCl_2$	183.32	col, hex	4.047	568	961	120[25]
chloride-2.5-water	$CdCl_2 \cdot 2.5H_2O$	228.36	col, mn, 1.6513	3.327	....	....	113[20]
chromium(III) oxide (1/1)	$CdO \cdot Cr_2O_3$	280.39	dk grn, cub	5.79[17]	....	....	i; i a
cyanide	$Cd(CN)_2$	164.44	col cr	2.226	d 200	....	1.71[18]
diphosphate	$Cd_2P_2O_7$	398.74	wh cr	4.965[15]	....	....	sl s; s a
dithionate-6-water	$CdS_2O_6 \cdot 6H_2O$	380.62	col, tric	2.272[20]	d	....	4.0[20]
fluoride	CdF_2	150.40	wh, cub, 1.56	6.64	1100	1760	17[20]
formate-2-water	$Cd(CHO_2)_2 \cdot 2H_2O$	238.47	col, mn	2.44	d	....	....
hydrogen arsenate-1-water	$CdHAsO_4 \cdot H_2O$	270.34	....	4.164[15]	>120	....	i; s a
hydrogen phosphate-2-water, di-	$Cd(H_2PO_4)_2 \cdot 2H_2O$	342.41	col, tric	2.74[15]	d 100	....	
hydroxide	$Cd(OH)_2$	146.11	wh, trig	4.79[15]	d 300	....	0.00026[20], s a
iodate	$Cd(IO_3)_2$	462.21	wh cr	6.43	d	....	0.097[20]
iodide	CdI_2	366.21	yel-grn pwd	5.670[30]	387	796	84.7[20]
molybdate(VI)(2—)	$CdMoO_4$	272.34	yel pl	5.347	....	....	sl s; s a
nitrate	$Cd(NO_3)_2$	236.41	col	....	350	....	150[20]
nitrate-4-water	$Cd(NO_3)_2 \cdot 4H_2O$	308.47	wh pr nd, hygr	2.455[17]	59.4	132	236[25]
oxalate	CdC_2O_4	200.42	col cr	3.32[18]	d 340	....	0.0033[18], s a
oxide	CdO	128.40	brn, cub, 2.49[LI]	8.15	subl 1497	....	i; s a
permanganate-6-water	$Cd(MnO_4)_2 \cdot 6H_2O$	458.36	....	2.81	d 95	....	v s
phosphate	$Cd_3(PO_4)_2$	527.14	col, amorp	....	1500	....	i; s a
phosphide	Cd_3P_2	399.15	grn, tetr	5.60	700	....	i; s d HCl
platinate(IV)-3-water, hexachloro-	$Cd[PtCl_6] \cdot 3H_2O$	574.25	yel, trig	2.882	d 170	....	s

Name	Formula	Formula weight	Color, crystalline form, index of refraction	Density	Melting point, °C	Boiling point, °C	Solubility
potassium bissulfate-2-water	CdK$_2$(SO$_4$)$_2$ · 2H$_2$O	322.69	col, tric	2.922[16]			42.9[16]
selenate-2-water	CdSeO$_4$ · 2H$_2$O	291.39	col, rh	3.63	−2H$_2$O, 170		73[20]
selenide	CdSe	191.36	grn-brn, hex	5.81[15]	1264		i; d a
silicate, meta-	CdSiO$_3$	188.48	col, rh, 1.739	4.93	1240		v sl s
sulfate	CdSO$_4$	208.46	wh, rh	4.691[20]	1000		76.6[20]
sulfate-1-water	CdSO$_4$ · H$_2$O	226.48	col, mn	3.79[20]			83[25]
sulfate-water (3/8)	3CdSO$_4$ · 8H$_2$O	769.56	col, mn, 1.565	3.09	forms mono-hydrate 80		94.4[25]
sulfide (greenockite)	CdS	144.46	or-yel, hex, 2.506, 2.529	4.82		subl 1380 (in N$_2$)	i; s a
telluride	CdTe	240.00	blk, cub	6.20[15]	1098		i; i a; d HNO$_3$
tungstate(VI)	CdWO$_4$	360.33	yelsh mn cr			1494	i; i dil a; s KCN
Calcium	Ca	40.08	silv-wh, fcc	1.55	850		d; s a
acetate	Ca(C$_2$H$_3$O$_2$)$_2$	158.17	col cr, 1.55, 1.56, 1.57		d		37.4[0]
acetate-2-water	Ca(C$_2$H$_3$O$_2$)$_2$ · 2H$_2$O	194.21	col nd		d 84		34.7[20]
arsenate	Ca$_3$(AsO$_4$)$_2$	398.08	col amorp pwd	3.620			0.013[25]
arsenide	Ca$_3$As$_2$	270.08	red cr	3.031	d		d; d a; s hot HNO$_3$
azide	Ca(N$_3$)$_2$	124.12	col, rh, hygr		expl 144		45[15]
benzoate-3-water	Ca(C$_7$H$_5$O$_2$)$_2$ · 3H$_2$O	336.36	col, rh	1.436	−3H$_2$O, 110		2.72[20]
borate, meta-	Ca(BO$_2$)$_2$	125.70	col, rh, 1.550, 1.660, 1.680		1160		0.31[30]
borate-6-water, meta-	Ca(BO$_2$)$_2$ · 6H$_2$O	233.79	col, tetr, 1.520, 1.502	1.88			0.25[30]
boride, hexa-	CaB$_6$	104.95	blk, cub	2.3[20]	2235		i; s HNO$_3$
bromate-1-water	Ca(BrO$_3$)$_2$ · H$_2$O	313.91	col, mn	3.329	−H$_2$O, 180		230
bromide	CaBr$_2$	199.90	col, rh, deliq	3.353	742	1810	143[20]; s al, acet
bromide-6-water	CaBr$_2$ · 6H$_2$O	307.99	col, hex	2.295	38.2	149	222; s al, acet
carbide, di-	CaC$_2$	64.10	col, tetr, 1.75	2.22	2300		d
carbonate (aragonite)	CaCO$_3$	100.09	col, rh, 1.530, 1.681, 1.685	2.930	d 900		0.0013[20]; s a
(calcite)	CaCO$_3$	100.09	col, rh or hex, 1.6583, 1.4864	2.710[18]	d 900		0.0014; s a
chlorate	Ca(ClO$_3$)$_2$	206.99	wh cr, hygr		340		178; s al, acet
chlorate-2-water	Ca(ClO$_3$)$_2$ · 2H$_2$O	243.01	yelsh, rh or mn	2.711	−2H$_2$O, 100		209[20]
chloride	CaCl$_2$	110.99	col, cub, 1.52	2.15	772	1940	42[20]
chloride-2-water	CaCl$_2$ · 2H$_2$O	147.02	col cr				98[0]

Table 4-1 (Continued)
PHYSICAL CONSTANTS OF INORGANIC COMPOUNDS

Name	Formula	Formula weight	Color, crystalline form, refractive index	Density	Melting point, °C	Boiling point, °C	Solubility in 100 parts
Calcium							
chloride-6-water	$CaCl_2 \cdot 6H_2O$	219.08	col, trig, 1.417, 1.393	1.71	$-6H_2O$, 200	....	74.5[20]
chlorite	$Ca(ClO_2)_2$	174.99	wh, cub	2.71[25]	....	....	d
chromate(VI)-2-water	$CaCrO_4 \cdot 2H_2O$	192.09	yel, mn	....	$-2H_2O$, 200	....	16.6
chromium(III) oxide (1/1)	$CaO \cdot Cr_2O_3$	208.07	olive-grn, cub	4.8[18]	2090	....	i a; s fus K_2CO_3
citrate-4-water	$Ca(C_6H_6O_7) \cdot 4H_2O$	570.51	wh nd	....	$-4H_2O$, 120	....	0.22
cyanamide	$CaCN_2$	80.11	col, hex cr, rhbd	2.29[20]	1340	sub 1200	d evol NH_3
cyanide	$Ca(CN)_2$	92.12	wh pwd	....	d 350	....	d
diphosphate	$Ca_2P_2O_7$	254.10	col, biax, 1.585, 1.604	3.09	1230	....	i; s a
diphosphate-5-water	$Ca_2P_2O_7 \cdot 5H_2O$	344.18	col, mn, 1.539, 1.545, 1.551	2.25	....	....	sl s; s a
dithionate-4-water	$CaS_2O_6 \cdot 4H_2O$	272.27	col, trig, 1.5496	2.176	....	....	33[20]
ferrate(III)-12-water, hexacyano-	$Ca_3[Fe(CN)_6]_2 \cdot 12H_2O$	760.42	red nd, deliq	....	....	....	v s
ferrate(II)-12-water, hexacyano-	$Ca_2[Fe(CN)_6] \cdot 12H_2O$	490.28	yel, tric, 1.570, 1.582, 1.596	1.68	d	....	104[20]
fluoride (fluorite)	CaF_2	78.08	col, cub, 1.434	3.180	1418	2510	0.002[20]; sl s a
fluoride hexakisphosphate, di- (fluoroapatite)	$Ca_{10}F_2(PO_4)_6$	1008.63	....	....	....	....	....
formate	$Ca(CHO_2)_2$	130.12	col, rh, 1.510, 1.514, 1.578	2.015	d	....	16.6[20]
d-gluconate	$[HOCH_2(CHOH)_4COO]_2Ca$	430.38	wh cr pwd	....	....	....	3.72[20]
glycerophosphate	$Ca[C_3H_5(OH)_2]PO_4$	210.16	wh cr pwd, hygr	....	d 170	....	1.66[20]
hydride	CaH_2	42.10	wh, rh	1.902	816 (in H_2)	....	d
hydrogen phosphate-2-water	$CaHPO_4 \cdot 2H_2O$	172.09	wh, tric, 1.5576, 1.5457, 1.5392	2.306[16]	$-2H_2O$, 109	....	0.017[25]; s a
hydrogen phosphate-1-water, di-	$Ca(H_2PO_4)_2 \cdot H_2O$	252.07	col, tric, 1.5292, 1.5176, 1.4392	2.220[15]	$-H_2O$, 109	d 203	1.8[30]

Name	Formula	Mol wt	Crystalline form, color, refractive index	Density	Melting point, °C	Boiling point, °C	Solubility
hydrogen sulfide-6-water	Ca(HS)$_2$ · 6H$_2$O	214.32	col pr		d 15–18		33[20]
hydroxide	Ca(OH)$_2$	74.09	col, hex, 1.574, 1.545	2.24	−H$_2$O, 522		0.17[10]; s a
hydroxide hexakisphosphate, di- (hydroxyapatite)	Ca$_{10}$(OH)$_2$(PO$_4$)$_6$	1004.64					
hypochlorite	Ca(OCl)$_2$	142.99			d 320		d dil a
hyponitrite-4-water (lautarite)	CaN$_2$O$_2$ · 4H$_2$O	172.15	wh cr	1.834	d 540		0.24
iodate (lautarite)	Ca(IO$_3$)$_2$	389.89	col, mn	4.519[15]	d 35		0.24[20]
iodate-6-water	Ca(IO$_3$)$_2$ · 6H$_2$O	497.98	col, rh	3.956	779		67.6[20]; s al, acet
iodide	CaI$_2$	293.89	yelsh-wh, hex	5.08	1250	1755	i; v sl s a
iron(III) oxide (1/1)	CaO · Fe$_2$O$_3$	215.77	dk red, rh, 2.58, 2.43				5.4[15]
lactate-5-water	Ca(C$_3$H$_5$O$_3$)$_2$ · 5H$_2$O	308.30	wh nd, effl		−3H$_2$O, 100		0.004[15]
laurate-1-water	Ca(C$_{12}$H$_{23}$O$_2$)$_2$ · H$_2$O	456.73	wh nd, effl		182		i; s a
lead(IV) oxide (2/1)	2CaO · PbO$_2$	351.35	red-brn cr	5.71	d		0.032[18]; s HCl
magnesium carbonate (dolomite)	CaMg(CO$_3$)$_2$	184.41	col, trig, 1.6817, 1.5026	2.872	d 730		
magnesium silicon oxide (1/1/1) (monticellite)	CaO · MgO · SiO$_2$	156.48					
magnesium silicon oxide (1/1/2) (diopside)	CaO · MgO · 2SiO$_2$	216.56	col, mn, 1.665, 1.672, 1.695	3.275	1390		i; i a
magnesium silicon oxide (2/1/2) (akermanite)	2CaO · MgO · 2SiO$_2$	272.64					
magnesium silicon oxide (3/1/1) (merwinite)	3CaO · MgO · SiO$_2$	328.72	col, mn, 1.708, 1.711, 1.718	3.150			
metaphosphate	Ca(PO$_3$)$_2$	198.02	col, 1.588, 1.595	2.82	975		i; i a
molybdate(VI)(2−) (powellite)	CaMoO$_4$	200.02	col, tetr, 1.967, 1.978	4.4			i; s a
nitrate	Ca(NO$_3$)$_2$	164.09	col, cub, hygr	2.504[18]	561		152[30]
nitrate-4-water	Ca(NO$_3$)$_2$ · 4H$_2$O	236.15	col, mn, 1.465, 1.498, 1.504	1.896 (α), 1.82 (β)	42.7; 39.7	d 132	129[20]
nitride	Ca$_3$N$_2$	148.25	brn, hex	2.63[17]	1195		d; s dil a
nitrite-4-water	Ca(NO$_2$)$_2$ · 4H$_2$O	204.15	col, tetr	1.674[0]	d		84.5[18]
oleate	Ca(C$_{18}$H$_{33}$O$_2$)$_2$	603.01	wh waxy cr		83–84		0.04
oxalate-1-water	CaC$_2$O$_4$ · H$_2$O	146.12	col, cub	2.2	−H$_2$O, 200		0.0006; s a
oxide	CaO	56.08	col, cub, 1.838	3.25	2927	3500	0.13[25]; s a

Table 4-1 (Continued)
PHYSICAL CONSTANTS OF INORGANIC COMPOUNDS

Name	Formula	Formula weight	Color, crystalline form, refractive index	Density	Melting point, °C	Boiling point, °C	Solubility in 100 parts
Calcium							
palmitate	$Ca(C_{16}H_{31}O_2)_2$	550.93	wh fatty pwd	....	d 155	....	0.003
pantothenate(+) (vitamin B-3)	$Ca[O_2CH_2CH_2NHOCH(OH)C(CH_3)_2CH_2OH]_2$	476.55	minute nd, hygr	....	d 195	....	36; s gly
perchlorate	$Ca(ClO_4)_2$	238.98	col cr	2.651[25]	d 270	....	188
permanganate-5-water	$Ca(MnO_4)_2 \cdot 5H_2O$	368.03	purp cr	2.4	d	....	338
peroxide	CaO_2	72.08	wh, tetr, 1.895	2.92	d 275	....	sl s; s a
peroxide-8-water	$CaO_2 \cdot 8H_2O$	216.20	wh, tetr	1.70	$-8H_2O$, 200	expl 275	sl s; s a
phosphate (whitlockite)	$Ca_3(PO_4)_2$	310.18	wh amorp pwd, 1.629, 1.626	3.14	1730	....	0.03[25], s a
phosphate-2-water, fluoro-	$Ca[PO_3F] \cdot 2H_2O$	174.09	mn cr	....	....	....	0.42[27]
phosphide	Ca_3P_2	182.19	gray lumps	2.51	ca 1600	....	d; s a
phosphinate	$Ca(PH_2O_2)_2$	170.06	wh-gray, mn	....	d > 300	....	15.4
salicylate-2-water	$Ca(C_7H_5O_3)_2 \cdot 2H_2O$	350.34	wh, oct	....	$-2H_2O$, 120	d 244	2.85[15], 1.55[17] al
selenate	$CaSeO_4$	183.04	col	2.88	....	d 698	7.4[25]
selenate-2-water	$CaSeO_4 \cdot 2H_2O$	219.07	col, mn	2.68	anhyd 200	d 698	9.2[25]
selenide	$CaSe$	119.04	cub, 2.274	3.57	....	....	....
silicate	Ca_2SiO_4	172.24	col, mn, 1.717, 1.735	3.27	2130	....	i; s a, alk
silicide, di-	$CaSi_2$	96.25	....	2.5	....	....	....
silicon oxide (1/1) (pseudo-wollastonite) (wollastonite)	$CaO \cdot SiO_2$	116.16	col, mn, 1.610, 1.611, 1.664; col, mn, 1.616, 1.629, 1.631	2.905	1540	....	0.0096[17], s HCl
silicon oxide (3/1) (alite)	$3CaO \cdot SiO_2$	228.32	col, mn, 1.718	2.5	1540	....	....
silicon oxide (3/2) (rankinite)	$3CaO \cdot 2SiO_2$	288.41	....	....	....	....	....
silicon titanium oxide (1/1/1) (sphene)	$CaO \cdot SiO_2 \cdot TiO_2$	196.06	....	....	....	....	....

Name	Formula	Formula weight	Color, crystalline form, refractive index	Density	Melting point, °C	Boiling point, °C	Solubility
stearate	$Ca(C_{18}H_{35}O_2)_2$	607.04	wh cr pwd	....	179–180	....	0.004^{15}
succinate-3-water	$CaC_4H_6O_4 \cdot 3H_2O$	212.22	col, 1.460, 1.540, 1.610	....	....	....	1.28^{20}
sulfate (anhydrite)	$CaSO_4$	136.14	col, rh, 1.569, 1.575, 1.613	2.960	1400	....	0.20; s a
(soluble anhydrite)	....	....	col, hex or tric, 1.505, 1.548	2.61	....	....	....
sulfate-hemihydrate (plaster of Paris)	$CaSO_4 \cdot 0.5H_2O$	145.15	wh pwd	....	$-H_2O$, 163	....	0.3^{20}; s a, glyc
sulfate-2-water (gypsum)	$CaSO_4 \cdot 2H_2O$	172.17	col, mn, 1.521, 1.523, 1.530	2.32	$-2H_2O$, 163	....	0.26^{20}; s a, glyc
sulfide (oldhamite)	CaS	72.14	col, cub, 2.137	2.5	2400	....	0.02^{15} d; d a
sulfite-2-water	$CaSO_3 \cdot 2H_2O$	156.17	col, hex	....	$-2H_2O$, 100	....	0.004; s a
d-tartrate-4-water	$CaC_4H_4O_6 \cdot 4H_2O$	260.21	col, rh, 1.525, 1.535, 1.550	....	$-4H_2O$, 200	....	0.034^{20}
dl-tartrate-4-water	$CaC_4H_4O_6 \cdot 4H_2O$	260.21	tric	....	$-4H_2O$, 200	....	0.0045^{25}
meso-tartrate-3-water	$CaC_4H_4O_6 \cdot 3H_2O$	242.20	wh, mn or tric	....	$-3H_2O$, 170	....	0.28^{18}
telluride	$CaTe$	167.68	cub, 2.51, 2.58	4.873	....	....	150, v s al
thiocyanate-3-water	$Ca(SCN)_2 \cdot 3H_2O$	210.29	wh cr, deliq	....	d 160	....	92^{25}; s al
thiosulfate-6-water	$CaS_2O_3 \cdot 6H_2O$	260.30	wh, tric	1.872	d 43–49	....	....
titanate(IV)(2−) (perovskite)	$CaTiO_3$	135.98	col, cub, 2.34	4.10	1970	....	0.0032
titanium oxide (3/2)	$3CaO \cdot 2TiO_2$	328.04	....	....	....	....	....
tungstate(VI)(2−) (scheelite)	$CaWO_4$	287.93	col, tetr, 1.918, 1.934	6.062^{20}	....	....	....
vanadium(V) oxide (1/1)	$CaO \cdot V_2O_5$	237.96	col, mn	....	778	....	....
zirconate(IV)(2−)	$CaZrO_3$	179.30	....	4.78	2340	....	....
Californium	Cf	251	silv-met, fcc & hcp	5.88	900	....	....
chloride	$CfCl_3$	357	grn, hex	....	....	....	....
Carbon							
(diamond)	C	12.011	col, cub, 2.4173	3.5153	3500	3930	i
(graphite)	....	....	blk, hex	2.267	$4000^{63.5\,atm}$	3930	i
(amorphous)	....	....	blk	1.8–2.1	subl 3650	3930	i
bromide, tetra-	CBr_4	331.65	col, mn	3.42	90.1	190	i; s al, eth, chl
chloride, tetra-	CCl_4	153.82	col liq, 1.4601	1.5867	−22.9	76.7	i; s al, eth, chl
fluoride, tetra-	CF_4	88.00	col gas	1.96^{-184}	−183.7	−128.0	sl s
hydride (methane)	CH_4	16.04	col gas	0.415^{-164}	−182.48	−161.49	s bz

Table 4-1 (Continued)
PHYSICAL CONSTANTS OF INORGANIC COMPOUNDS

Name	Formula	Formula weight	Color, crystalline form, refractive index	Density	Melting point, °C	Boiling point, °C	Solubility in 100 parts
Carbon							
iodide, tetra-	CI_4	519.63	dk red, cub	4.34	d 171	....	s al, bz, eth
oxide, mono-	CO	28.01	col gas, 1.000335	1.250 (g/L)(g); 0.793 (liq)	-205.05	-191.49	2.1 ml; s al, bz
oxide, di-	CO_2	44.01	col gas, 1.000449	1.975 (g/L)(g); 1.56^{-79} (liq)	-56.2 subl	-78.44	31 ml^{15}
(tri-) oxide, di-	C_3O_2	68.03	col liq, 1.4538	1.11	-112.19	6.4	d
selenide, di-	CSe_2	169.93	yel liq, 1.845	2.6626^{25}	-43	125.1	s acet, eth, CCl_4
selenide sulfide	$CSSe$	123.04	yel oily liq	1.9874	-85	84.5	s CS_2
sulfide, di-	CS_2	76.14	col liq, 1.6295	1.261^{22}	-111.6	46.26	0.29^{20}; s al, eth
(tri-) sulfide, di-	C_3S_2	100.16	red liq	1.274	-0.5	d 90	....
sulfide telluride	$CSTe$	171.68	yel-red liq	2.9^{-50}	-54	d	s bz, CS_2
Carbonic acid	H_2CO_3	62.03	known in soln only	....	....	....	....
Carbonyl bromide	$COBr_2$	187.83	col liq	1.392	....	60	d; s bz, toluene
chloride (phosgene)	$COCl_2$	98.92	col gas	1.139^{-114}	-127.8	7.6	d
fluoride	COF_2	66.01	col gas	....	-114.0	-83.3	d; s $COCl_2$
selenide	$COSe$	106.97	col gas	1.182 (liq)	-124.4	-21.7	....
sulfide	COS	60.07	col gas	1.073^{0} (g/L)	-138.81	-50.23	54 ml^{20}; s al, CS_2
Cerium							
	Ce	140.12	gray-met, fcc	6.771	795	3470	i; s a
(III) acetate	$Ce(C_2H_3O_2)_3$	317.26	col	....	d 308	....	0.35^{20}
boride, tetra-	CeB_4	183.36	tetr	5.74	....	....	i
boride, hexa-	CeB_6	204.98	bl-viol, cub	4.69	2190	d	....
(III) bromate-9-water	$Ce(BrO_3)_3 \cdot 9H_2O$	685.98	redsh-yel, hex	....	49	....	s
(III) bromide	$CeBr_3$	379.84	col, hex, hygr	5.18	733	1560	v s
carbide, di-	CeC_2	164.14	red, hex	5.23	....	....	d; s a

Name	Formula	Formula weight	Color, crystalline form	Density	Melting point	Boiling point	Solubility
(III) carbonate-5-water	$Ce_2(CO_3)_3 \cdot 5H_2O$	550.37	wh crys	3.92			s a
(III) chloride	$CeCl_3$	246.48	wh, hex, deliq	6.16	817	1730	v s
(III) fluoride	CeF_3	197.12	wh, hex	4.80	1430	2327	i; s H_2SO_4
(IV) fluoride	CeF_4	216.12	wh, mn	5.43	ca 650	d	i
hydride, di-	CeH_2	142.13	fcc	5.5			
hydride, tri-	CeH_3	143.14	dk blue, amorp		ignites		s a
(III) hydroxide	$Ce(OH)_3$	191.14	wh gel				s a
(IV) hydroxide	$Ce(OH)_4$ or $CeO_2 \cdot 2H_2O$	208.15	yel gel				i
(IV) iodate	$Ce(IO_3)_4$	839.73	yel				0.15^{20}
iodide, di-	$[Ce^{3+}(e^-)(I^-)_2]$	393.93	dk bronze		808		
(III) iodide	CeI_3	520.83	yel, rh		766	1400	s
metaphosphate	$Ce(PO_3)_3$	377.04	micro nd	3.272			i a
(III) molybdate(VI)	$Ce_2(MoO_4)_3$	760.05	yel, tetr, 2.019, 2.007	4.83	973		234^{20}
(III) nitrate-3-water	$Ce(NO_3)_3 \cdot 6H_2O$	434.23	col, deliq		−3H₂O, 150	d 200	i; s HCl
(III) oxalate-10-water	$Ce_2(C_2O_4)_3 \cdot 10H_2O$	724.46	wh cr		d		i; s a
(IV) oxide	CeO_2	172.13	yelsh				i; s H_2SO_4
(III) oxide	Ce_2O_3	328.24	gray-grn, hex	6.86	1692		i, s H_2SO_4
(di-) oxide sulfide, di-	Ce_2O_2S	344.30	brn-maroon, hex	6.00	1950		ii; s a
(III) phosphate	$CePO_4$	235.09	red, mn or yel, rh	5.22			35.2^{20}
(III) selenate	$Ce_2(SeO_4)_3$	709.11	rhbd	4.456			
selenide	$CeSe$	219.08	cub	6.53	2177		
(III) selenide	Ce_2Se_3	517.12	bl-blk, cub	5.67			i
silicide, di-	$CeSi_2$	196.29		3.912	d 1000		9.72^{21}
(III) sulfate	$Ce_2(SO_4)_3$	568.42	wh, mn, hygr	3.17	anhyd 250	d > 650	10.2^{45}
(III) sulfate-5-water	$Ce_2(SO_4)_3 \cdot 5H_2O$	658.50	wh, mn	2.831	anhyd 200	d 350	9.8^{20}
(III) sulfate-9-water	$Ce_2(SO_4)_3 \cdot 9H_2O$	730.56	wh, hex	3.91			hyd; s H_2SO_4
(IV) sulfate	$Ce(SO_4)_2$	332.24	yel cr	5.88			i; s dil a
sulfide	CeS	172.18	gold, cub	5.020^{11}	2450		
(III) sulfide, di-	Ce_2S_3	376.43	red-brn, cub		1887		
sulfide, di-	CeS_2	204.25	cub	5.3	1677		
(tri-) sulfide, tetra-	Ce_3S_4	548.60	blk, cub		2050		
telluride	$CeTe$	267.72	cub		1817		
(di-) telluride, tri-	Ce_2Te_3	663.04	yel, tetr	6.77	1089		0.014^{20}
(III) tungstate(VI)(2−)	$Ce_2(WO_4)_3$	1023.78					
Cesium	Cs	132.91	silv-met, hex	1.8785^{15}	28.8	678.5	d; s a
acetate	$Cs(C_2H_3O_2)$	191.95	col, deliq		194		1011^{20}

Table 4-1 (*Continued*)
PHYSICAL CONSTANTS OF INORGANIC COMPOUNDS

Name	Formula	Formula weight	Color, crystalline form, refractive index	Density	Melting point, °C	Boiling point, °C	Solubility in 100 parts
Cesium							
amide	CsNH$_2$	148.93	wh nd	3.44	262		d
aurate(III), tetrachloro-	Cs[AuCl$_4$]	471.68	yel, mn				0.8^{20}
azide	CsN$_3$	174.93	col, nd, deliq		310		307^{15}
borate, tetrafluoro-	Cs[BF$_4$]	219.71	col, rh, 1.350	3.20	550 d		1.63^{20}
borate, tetrahydro-	Cs[BH$_4$]	147.75	col, fcc, 1.498	2.404			
bromate	CsBrO$_3$	260.81	wh, hex, 2.2	4.10^{20}	420 d		3.66^{25}
bromide	CsBr	212.81	col, bcc, 1.6984	4.44	635	1300	107^{18}
carbonate	Cs$_2$CO$_3$	325.82	col cr, deliq		d 610		v s; 11^{20} al
chlorate	CsClO$_3$	216.36	wh cr	3.626^{20}	342		6.2^{20}; s al
chloride	CsCl	168.36	col, cub, 1.6418	3.988	645	1324	187^{20}
chromate	Cs$_2$CrO$_4$	381.80	yel, rh	4.237			71.4^{13}
(hexaaqua)chromium(III) bissulfate-6-water	[Cr(H$_2$O)$_6$]Cs(SO$_4$)$_2$ · 6H$_2$O	593.21	vlt cr	2.064	116		0.94
cyanide	CsCN	158.92	wh cr	2.93			v s
fluoride	CsF	151.90	wh, cub, 1.478	4.115	703	1231	322^{18}
formate	CsCHO$_2$	172.92	col	1.0169^{21}			450^{20}
gallium bissulfate-12-water	CsGa(SO$_4$)$_2$ · 12H$_2$O	610.93	col, cub, 1.4650				1.53^{25}
germanate, hexafluoro-	Cs$_2$[GeF$_6$]	452.39	oct	4.10			sl s; sl s a
hydride	CsH	133.91	wh, cub	3.42	d		d; d a
hydrogen carbonate	CsHCO$_3$	193.92	col, rh		−H$_2$O 175		210^{20}; s al
hydrogen difluoride	CsHF$_2$	171.91	wh nd, deliq		160		v s
hydrogen sulfate	CsHSO$_4$	229.97	col, rh	3.352^{16}	red heat		s
hydrogen tartrate	CsHC$_4$H$_4$O$_6$	281.99	wh, rh	2.59			7.1^{20}
hydroxide	CsOH	149.91	lt yel, deliq	3.675	315	990	386^{15}
iodate	CsIO$_3$	307.81	wh, mn	4.937^{20}	565		2.6^{25}
iodide	CsI	259.81	wh, rh, 1.7876	4.510	621	1280	76.5^{20}
iodide, tri-	CsI$_3$	513.62	blk, rh	4.47	207.5		sl s; s al

Name	Formula	Mol wt	Color, crystalline form, refractive index	Density	Melting point	Boiling point	Solubility
iron(II) bissulfate-6-water	$Cs_2Fe(SO_4)_2 \cdot 6H_2O$	621.87	lt grn, mn, 1.500, 1.504, 1.509	2.791^{20}	ca 70	......	101
iron(III) bissulfate-12-water	$CsFe(SO_4)_2 \cdot 12H_2O$	597.06	lt-viol cr, 1.484	2.061^{20}	ca 90	......	2.80
magnesium bissulfate-6-water	$Cs_2Mg(SO_4)_2 \cdot H_2O$	590.34	col, mn, 1.486, 1.452	2.676^{20}	......	......	53
nitrate	$CsNO_3$	194.91	col, hex or cub, 1.55, 1.56	3.685	414	d 849	23.0^{20}
nitrite	$CsNO_2$	178.91	yel cr	......	......	......	v s
oxalate	$Cs_2C_2O_4$	353.82	......	3.230^{15}	......	......	313
oxide	Cs_2O	281.81	or nd	4.25	490 (in N_2)	......	v s
(di-) oxide, tri-	Cs_2O_3	313.81	choc-brn, cub	4.25	400	......	d; s a
perchlorate	$CsClO_4$	232.36	col, cub, 1.480	3.327^{25}	577 d	......	1.6^{20}
periodate	$CsIO_4$	323.81	wh, rh	4.259^{15}	......	......	2.15^{15}
permanganate	$CsMnO_4$	251.84	purple	3.507	d 320	$-O_2$, 650	0.23^{20}
peroxide	Cs_2O_2	297.81	lt yel nd	4.25	400	......	s
perrhenate	$CsReO_4$	383.11	wh cr	4.76	616	......	0.79^{19}
platinate(IV), hexachloro-	$Cs_2[PtCl_6]$	673.62	yel, cub	4.197	d 570	......	0.0087^{20}
polonium(IV) hexachloride	Cs_2PoCl_6	688.53	cub, 1.86	3.82	......	......	......
rhodium(III) bissulfate-12-water	$CsRh(SO_4)_2 \cdot 12H_2O$	644.12	yel, oct	2.238	111	......	sl s
selenate	Cs_2SeO_4	408.77	col, rh, 1.595, 1.506, 1.596	4.4528^{20}	......	......	244^{12}
silicate, hexafluoro-	$Cs_2[SiF_6]$	407.89	wh, cub	3.372^{17}	......	......	60^{17}
sulfate	Cs_2SO_4	361.87	col, rh, 1.560, 1.564, 1.566	4.243	995	......	179^{20}
sulfide-4-water	$Cs_2S \cdot 4H_2O$	369.94	wh cr, deliq	......	......	......	v s
l-tartrate	$Cs_2C_4H_4O_6$	413.88	col, trig	3.03^{14}	......	......	s
vanadium(III) bissulfate-12-water	$CsV(SO_4)_2 \cdot 12H_2O$	592.15	red. cub, 1.4780	2.033^{20}	82	d 300	1.4^{20}
Chloric acid-7-water	$HClO_3 \cdot 7H_2O$	210.57	known only in soln	1.282^{14}	< −20	d 40	v s
Chlorine	Cl_2	70.91	grn-yel gas, 1.3786	2.98^{20} (g/L) 1.57^{bp}(liq)	−100.99	−34.03	199 ml^{25}, 310 ml^{10}
azide	ClN_3	77.48	col gas, expl	......	......	......	sl s; d alk
fluoride	ClF	54.56	col gas, 1.000494^{25} (Hg 5461)	1.62^{bp}	−155	−90	d viol

Table 4-1 (*Continued*)
PHYSICAL CONSTANTS OF INORGANIC COMPOUNDS

Name	Formula	Formula weight	Color, crystalline form, refractive index	Density	Melting point, °C	Boiling point, °C	Solubility in 100 parts
Chlorine							
fluoride, tri-	ClF_3	92.45	col gas, 1.000633 (Hg 5461)	1.825^{bp}	−76.28	11.74	d viol
fluoride, penta-	ClF_5	130.45	col gas	1.79^{20} (g/L)	−102	−13.9	....
hydrate, octa-	$Cl_2 \cdot 8H_2O$	215.03	lt-yel, rh	1.23	d 9.6	....	i
(di-) oxide	Cl_2O	86.91	yel-brn gas; red-brn liq; expl	3.02^{bp}	−120.6	2.1	$3.5\ g^{20}$
oxide, di-	ClO_2	67.46	red-brn liq	1.642^{0}	−59.6	10.9	$11.6\ g^{10.7}$
oxide, tri- (dimer)	$(ClO_3)_2$	166.90	....	1.92^{20}	3.5	d	d
(di-) oxide, hepta-	Cl_2O_7	182.90	col oily liq, expl	1.805^{25}	−91.5	83.6	d
sulfate, fluoro-	$Cl(SO_3F)$	134.51	....	....	−84	45.1	s; s al, eth
Chloroamine	NH_2Cl	51.48	yel liq	....	−66	....	....
Chlorodifluoroamine	$NClF_2$	87.45	col gas	....	−183/196	−67	....
Chlorofluoroamine, di-	NCl_2F	103.91	col gas	....	−80	−3	....
Chlorosulfonic acid	HSO_3Cl	116.52	col fum liq, 1.437	1.787	−80	158	d to $HCl + H_2SO_4$
Chlorosyl chlorate	$ClOClO_3$	134.91	....	....	−117	−44.4	....
Chlorous acid	$HClO_2$	68.46	col, in soln only	....	....	....	d
Chloryl fluoride	ClO_2F	86.45	gas, fumes in moisture	....	−115	−5.64	d
perchlorate	ClO_2ClO_4	182.90	dk red liq	2.023	3.5	203	d
Chromium	Cr	51.996	steel gray, bcc	7.20	1857	2682	i; s dil HCl
(III) acetate	$Cr(C_2H_3O_2)_3$	229.13	gray-grn pwd	....	....	....	s
(III) acetylacetonate	$(CH_3COCHCOCH_3)_3Cr$	349.33	....	....	216	340	i; s org solv
arsenide	$CrAs$	126.92	gray, hex	6.35^{16}	....	....	i
boride	CrB	62.81	silv cr	6.17	....	....	i; s fus Na_2O_2
(II) bromide	$CrBr_2$	211.81	wh, mn	4.36	842	....	s
(III) bromide	$CrBr_3$	291.72	grn, rhbd	4.25	....	....	i; v s al
hexaaqua(III)-]bromide	$[Cr(H_2O)_6]Br_3$	399.81	bl-gray	5.4^{17}	....	....	v s

Name	Formula	Mol. wt.	Color, crystalline form, refractive index	Density	M.p., °C	B.p., °C	Solubility in g/100 cc
(tri-) carbide, di-	Cr_3C_2	180.01	gray, rh	6.68	1895	3800	i
carbonyl, hexa-	$Cr(CO)_6$	220.06	col, rh	1.77	subl 148.5	210 expl	i; i al, eth
(II) chloride	$CrCl_2$	122.90	wh, rh, hygr	2.878	815	1300	v s
(III) chloride	$CrCl_3$	158.35	vlt, mn	2.76^{15}	877	subl 947	i; i al, acet, eth
[tetraaqua(II) chloride-2-water	$[Cr(H_2O)_4]Cl_2 \cdot 2H_2O$	266.45	vlt, mn	1.76	83		58.5
(III) diphosphate	$Cr_4(P_2O_7)_3$	729.81	lt grn, mn	3.2			
(II) fluoride	CrF_2	89.99	grn, mn	4.11	894		i; s alk
(III) fluoride	CrF_3	108.99	grn, rhbd	3.8	1100	subl 1100–1200	sl s; s hot HCl
(IV) fluoride	CrF_4	127.99	grn	2.89	−28	subl 100 (vac); bp 400	i; s HF
(V) fluoride	CrF_5	147.00	crimson, rh		30	117	s hyd
(VI) fluoride	CrF_6	166.00	lemon-yel		d −100		
fluoride oxide, tetra-	CrF_4O	143.99	dk red, mn		55		hyd
(III) formate	$Cr(HCOO)_3$	187.06	fne grn cr		d > 300		s
(II) hydroxide	$Cr(OH)_2$	86.01	yel-brn		d		d; s a
(III) hydroxide	$Cr(OH)_3$	103.02	grn or bl, gelat		d		i; s a
(II) iodide	CrI_2	305.80	red-brn, mn	5.196	868		
(III) iodide	CrI_3	432.71	blk, hex	4.915	> 600		i a, alk
(di-) iron tetraoxide	Cr_2FeO_4	223.84	brn-blk, cub	4.97^{20}	2180		i; sl s a
(III) nitrate-9-water	$Cr(NO_3)_3 \cdot 9H_2O$	400.15	purple, mn		60	d 100	208^{15}
nitride	CrN	66.00	cub	5.9	d 1282		i; s HCl
(II) oxalate-1-water	$CrC_2O_4 \cdot H_2O$	158.03	yel cr pwd	2.468			sl s; s dil a
(III) oxalate-6-water	$Ca_2(C_2O_4)_3 \cdot 6H_2O$	476.14	red, amorp, hygr		d 120		s, v s al
(II) oxide	CrO	68.00	blk pwd				i
(III) oxide	Cr_2O_3	152.02	grn, hex, 2.551	5.21	2330	3000	i
(IV) oxide	CrO_2	83.99	brn-blk pwd	4.89	−O$_2$, 300		i; s HNO$_3$
(VI) oxide	CrO_3	99.99	red, rh, deliq	2.70	198	d 250 to Cr_2O_3	167^{20}; s al, eth
(III) phosphate-6-water	$CrPO_4 \cdot 6H_2O$	255.06	vlt, tric, 1.568, 1.591, 1.699	2.121^{14}	800 anhyd in 1 hr		
phosphide	CrP	82.97	gray-blk cr	5.7^{15}			i; s HNO$_3$
(tri-) phosphide	Cr_3P	186.97	gray-blk, tetr	6.1	1515		i
potassium bissulfate-12-water	$CrK(SO_4)_2 \cdot 12H_2O$	499.41	vlt-red, cub, 1.4814	1.826	−12H$_2$O, 400		22^{25}

Table 4-1 (Continued)
PHYSICAL CONSTANTS OF INORGANIC COMPOUNDS

Name	Formula	For-mula weight	Color, crystalline form, refractive index	Density	Melting point, °C	Boiling point, °C	Solubility in 100 parts
Chromium							
(tri-) silicide, di-	Cr_3Si_2	212.17	gray, tetr	5.5[0]	....	....	i; s HCl
(II) sulfate-7-water	$CrSO_4 \cdot 7H_2O$	274.17	bl cr	....	....	....	22.9[0]
(III) sulfate	$Cr_2(SO_4)_3$	392.18	vlt pwd	3.012	....	....	i; i a
(III) sulfate-18-water	$Cr_2(SO_4)_3 \cdot 18H_2O$	716.45	bl-vlt, oct, 1.564	1.7	d 100	....	220[20]
(II) sulfide	CrS	84.06	blk, mn	4.85	1567	....	i; v s a
(III) sulfide	Cr_2S_3	200.18	brn-blk, trig	3.77[10]	d 1350	....	i; s HNO_3
Chromium Complexes							
[hexaamminechromium(III)]-trischloride-1-water	$[Cr(NH_3)_6]Cl_3 \cdot H_2O$	278.55	yel cr	1.585	....	....	2.2[17]
[pentaamminechlorochromium(III)]bischloride	$[Cr(NH_3)_5Cl]Cl_2$	243.51	red, oct	1.696	....	....	0.73[20]
Chromyl bromide	CrO_2Br_2	243.80	dk vlt nd, hygr	....	d −70	....	d; s eth
chloride	CrO_2Cl_2	154.90	dk red liq, fum	....	−96.5	117	
fluoride	CrO_2F_2	121.99	red-vlt cr	....	31.6	....	i; s a
Cobalt	Co	58.93	silv-gray, cub	8.9	1494	2897.1	s; 2.1[15] MeOH
acetate-4-water	$Co(C_2H_3O_2)_2 \cdot 4H_2O$	249.08	red-vlt, mn, 1.542	1.705[19]	−4H_2O, 140	....	i; s dil a, NH_4OH
arsenate-8-water	$Co_3(AsO_4)_2 \cdot 8H_2O$	598.75	red-vlt, mn, 1.626, 1.661, 1.669	3.178[15]	−8H_2O, 400	d 1000	i; s HNO_3
arsenide sulfide (cobaltite)	$CoAsS$	156.92	redsh-gray	6.2	d	....	d; s HNO_3
(di-) arsenide	Co_2As	192.79	cr pwd	8.28	950	....	45.5[17]
boride	CoB	69.74	pr	7.25[18]	....	....	112[20]
bromate-6-water	$Co(BrO_3)_2 \cdot 6H_2O$	422.84	red, cub	2.55	....	....	167[20], s al, acet
bromide	$CoBr_2$	218.75	grn, hcp, deliq	4.909[25]	678 (in N_2)	....	0.18[15], s a
bromide-6-water	$CoBr_2 \cdot 6H_2O$	326.84	red-vlt pr, deliq	2.46[25]	−6H_2O, 130	....	i; s CS_2, eth
carbonate (spherocobaltite)	$CoCO_3$	118.94	red, trig 1.855, 1.60	4.13	d	....	i; s al, bz, eth
(di-) carbonyl, octa-	$Co_2(CO)_8$	341.95	or cr	1.73[18]	51	d 52	
carbonylnitroso, tri-	$Co(CO)_3(NO)$	172.97	cherry-red liq	....	−1.05	48.6	

Name	Formula	Formula weight	Color, crystalline form, and refractive index	Density	Melting point, °C	Boiling point, °C	Solubility
chlorate-6-water	$Co(ClO_3)_2 \cdot 6H_2O$	333.93	red, cub, 1.55	1.92	50	d 100	266^{20}
chloride	$CoCl_2$	129.84	blue, hex, hygr	3.356	740	1087	52.9^{20}
chloride-6-water	$CoCl_2 \cdot 6H_2O$	237.93	red, mn	1.924	$-6H_2O$, 110	......	97^{20}
chromate(VI)	$CoCrO_4$	174.93	gray-blk cr	......	d	......	i; s a
citrate-2-water	$Co_3(C_6H_5O_7)_2 \cdot 2H_2O$	591.04	rose-red	1.872	$-2H_2O$, 150	......	0.3^{10}
cyanide	$Co(CN)_2$	110.99	buff	......	d 300	......	0.0042^{18}, s KCN
cyanide-2-water	$Co(CN)_2 \cdot 2H_2O$	147.00	bl-vlt pwd	......	$-2H_2O$, 280	d 300	0.0056^{18}, s KCN
(II) fluoride	CoF_2	96.93	pink, tetr	4.46	1127	1739	1.36^{20}
(II) fluoride-4-water	$CoF_2 \cdot 4H_2O$	168.99	red, rh	2.192	d 200	......	2.6^{20}
(III) fluoride	CoF_3	115.93	brn, rhbd, hygr	3.88	926 ± 200	......	i
formate-2-water	$Co(CHO_2)_2 \cdot 2H_2O$	185.00	red cr	2.129^{22}	$-2H_2O$, 140	d 175	5.03^{20}
(II) hydroxide	$Co(OH)_2$	92.95	rose-red, rh	3.597^{15}	d	......	0.00018^{18}; s a
(III) hydroxide	$Co(OH)_3$	109.96	blk-brn	4.46	$-H_2O$, 100	d	0.00032^{18}; s a
iodate	$Co(IO_3)_2$	408.74	bl-vlt nd	4.931^{25}	d 200	......	1.02^{20}
iodate-6-water	$Co(IO_3)_2 \cdot 6H_2O$	516.83	red, oct	3.689^{21}	d 61	......	1.05^{18}
iodide	CoI_2	312.74	blk, hex, hygr	5.68	505 d	570 (vac)	203
iodide-6-water	$CoI_2 \cdot 6H_2O$	420.83	red-brn, hex, hygr	2.90	$-6H_2O$, 130	......	272
nitrate-6-water	$Co(NO_3)_2 \cdot 6H_2O$	291.04	red, mn, 1.52	1.87	55	d 74	155^{30}
oxalate	CoC_2O_4	146.95	pinkish	......	d 250	......	......
oxalate-2-water	$CoC_2O_4 \cdot 2H_2O$	182.98	pink cr	3.021^{25}	$-2H_2O$, 190	......	0.0026^{18}
(II) oxide	CoO	74.93	grn-brn, fcc	6.45	1805	......	i; s a
(III) oxide	Co_2O_3	165.86	gray-blk, hex	5.18	d 895	......	i; s a
(II,III) oxide	Co_3O_4	240.80	blk, cub	6.07	......	......	i; v sl s a
perchlorate	$Co(ClO_4)_2$	257.83	red nd, 1.510, 1.490	3.327	......	......	104^{18}
perchlorate-6-water	$Co(ClO_4)_2 \cdot 6H_2O$	365.93	red, oct, 1.51	......	d 182	......	148^{18}
phosphate	$Co_3(PO_4)_2$	366.74	redish cr	2.587	......	......	i; s NH_4OH
phosphate-8-water	$Co_3(PO_4)_2 \cdot 8H_2O$	510.87	redwh pwd	2.769	$-8H_2O$, 200	......	sl s; s min a
phosphide	CoP	89.91	gray, rh	6.24	......	......	i
phosphide, tri-	CoP_3	151.85	gray	4.26	......	......	i
(di-) phosphide	Co_2P	148.84	gray, rh	6.4^{15}	1386	......	i; s HNO_3
platinate(IV)-6-water, hexa-chloro-	$Co[PtCl_6] \cdot 6H_2O$	574.83	trig	2.699	d	......	......
potassium bissulfate-6-water	$CoK(SO_4)_2 \cdot 6H_2O$	437.36	red, mn, 1.481, 1.487, 1.500	2.218	anhyd 200	......	20.6^{20}

Table 4-1 (*Continued*)
PHYSICAL CONSTANTS OF INORGANIC COMPOUNDS

Name	Formula	Formula weight	Color, crystalline form, refractive index	Density	Melting point, °C	Boiling point, °C	Solubility in 100 parts
Cobalt							
selenate-6-water	$CoSeO_4 \cdot 6H_2O$	309.98	red, mn, 1.5225	2.25[17]	....	....	205[25]
selenide	$CoSe$	137.89	yel, hex	7.65	....	....	i; s HNO_3
selenide, di-	$CoSe_2$	216.85	cub	....	....	....	i; s HCl
silicate	Co_2SiO_4	209.95	vlt, rh	4.63	1345	....	118[21]
silicate-6-water, hexafluoro-	$Co[SiF_6] \cdot 6H_2O$	309.10	pink, trig, 1.382, 1.387	2.113[19]	....	....	i; s HCl
silicide	$CoSi$	87.03	rh	5.3	1395	....	d a
silicide, di-	$CoSi_2$	115.11	rh	....	1277	....	s slowly at 100°
(di) silicide	Co_2Si	145.95	gray cr	7.28[0]	1327	....	s
sulfate	$CoSO_4$	154.99	red-purp cr, rh	3.71[25]	d 1140	....	
sulfate-1-water	$CoSO_4 \cdot H_2O$	173.01	red cr, 1.603, 1.639, 1.683	3.075[25]	d	....	
sulfate-7-water (bieberite)	$CoSO_4 \cdot 7H_2O$	281.10	pink, mn, 1.477, 1.483, 1.489	1.948	96.8	$-7H_2O$, 420	65.4[20]
sulfide (sycoporite)	CoS	91.00	redsh-silv, hex	5.45[18]	>1100	....	i; sl s a
sulfide, di-	CoS_2	123.06	blk, cub	4.269	777	....	i; s HNO_3
telluride, di-	$CoTe_2$	314.13	cub & hex	....	....	....	....
thiocyanate-3-water	$Co(SCN)_2 \cdot 3H_2O$	229.14	vlt, rh	....	$-3H_2O$, 105	....	7.8[18], s al, eth
titanate(IV)(4−)	Co_2TiO_4	229.76	grn-blk, cub	5.1	1575	....	i; s conc HCl
tungstate(VI)(2−)	$CoWO_4$	306.78	bl-grn, mn	8.42	....	....	i; s hot conc a
Cobalt Complexes							
[hexaamminecobalt(III)] trichloride	$[Co(NH_3)_6]Cl_3$	267.46	red, mn, 1.710	1.804	d 215	....	7.35[20]
[hexaamminecobalt(III)] trisnitrate	$[Co(NH_3)_6](NO_3)_3$	347.13	yel, tetr	....	....	....	1.8[20]
bis[hexaamminecobalt(III)] trissulfate-5-water	$[Co(NH_3)_6]_2(SO_4)_3 \cdot 5H_2O$	700.50	dk yel, mn	1.797 (anhyd)	$-5H_2O$, 150	....	1.4[20]

Copper

Name	Formula	Formula weight	Crystalline form, color, and index of refraction	Specific gravity	Melting point, °C	Boiling point, °C	Solubility
Copper	Cu	63.546	redsh-met, fcc	8.92	1084.5	2575	i; s HNO_3, hot H_2SO_4
acetate-1-water (verdigris)	$Cu(C_2H_3O_2)_2 \cdot H_2O$	199.65	dk-grn pwd, 1.545, 1.550	1.882	115	d 240	8.0; 0.48 MeOH
acetate-metaarsenate(III)(1/3) (Paris green)	$Cu(C_2H_3O_2)_2 \cdot 3Cu(AsO_2)_2$	1013.77	emerald grn pwd		expl		i; s a, NH_4OH
(di-) acetylide, di-	$Cu-C \equiv C-Cu$	151.10	red, amorp, expl			expl 200	
[diammine(II)-] azide	$[Cu(NH_3)_2](N_3)_2$	181.64	dk grn cr, expl	1.91	expl 210		v sl s; s a
[tetraammine(II)-] bisnitrate	$[Cu(NH_3)_4](NO_3)_2$	255.67	dk bl, oct	1.79	d		i; d a
[tetraammine(II)-] sulfate-1-water	$[Cu(NH_3)_4]SO_4 \cdot H_2O$	245.74	dk bl, rh				18^{21}; s
(tri-) antimonide	Cu_3Sb	312.37	gray	8.51	687		
arsenate-4-water	$Cu_3(AsO_4)_2 \cdot 4H_2O$	540.52	blsh-grn		830		i; s a, NH_4OH
(tri-) arsenide (domeykite)	Cu_3As	265.54	hex	8.0			i; s a
(penta-) arsenide, di-	Cu_5As_2	467.54	bl, oct	7.56			i; s a
(I) azide	CuN_3	105.56	col cr, v expl	3.26	d 180		i; s NH_4Cl
borate(1-)	$Cu(BO_2)_2$	149.16	blsh-grn pwd	3.859			s
borate, tetrafluoro-	$Cu[BF_4]_2$	237.16	yel				
(tri-) boride, di-	Cu_3B_2	212.24		8.116			
bromate-6-water	$Cu(BrO_3)_2 \cdot 6H_2O$	427.45	bl-grn, cub	2.583	d 180		v s
(I) bromide	$CuBr$	143.45	wh, cub, 2.116	4.98	488	1318	v sl s; s a
(II) bromide	$CuBr_2$	223.31	blk, mn, deliq	4.71^{20}	498	900	126; s al; acet, py
(I) carbonate	Cu_2CO_3	187.09	yel	4.40	d		i; s a
(II) carbonate-dihydroxide (1/1) (malachite)	$CuCO_3 \cdot Cu(OH)_2$	221.11	dk grn, mn, 1.655, 1.875, 1.909	4.0	d 200		i; s a
(II) carbonate-dihydroxide (2/1) (azurite, chessylite)	$2CuCO_3 \cdot Cu(OH)_2$	344.65	bl, mn, 1.730, 1.758, 1.838	3.88	d 220		i; s NH_4OH
chlorate-6-water	$Cu(ClO_3)_2 \cdot 6H_2O$	338.53	grn, cub, deliq		65	d 100	242^{18}, s al, acet
(I) chloride (nantokite)	$CuCl$	98.99	wh, cub, 1.93	4.14	430	1212	0.024; s HCl
(II) chloride	$CuCl_2$	134.44	brn-yel pwd, hygr	3.386	d > 300	d	73^{20}
(II) chloride-2-water	$CuCl_2 \cdot 2H_2O$	170.47	bl-grn, rh, 1.644, 1.683, 1.732	2.54	$-2H_2O$, 200	d	76.4^{25}
(II) chloride-dihydroxide (1/1)	$CuCl_2 \cdot Cu(OH)_2$	232.00	yel-grn, hex	3.78	$-H_2O$, 250		
(I) chromate(III)(2-)	$Cu_2Cr_2O_4$	295.07	gray-blk, cub	5.24^{20}	d > 900		i; s HNO_3
(II) citrate-water (2/5)	$2Cu_2C_6H_4O_7 \cdot 5H_2O$	720.43	blsh-grn pwd		$-H_2O$, 100		0.17; s a
(I) cyanide	$CuCN$	89.56	wh, mn	2.92	473 (in N_2)	d	0.00026; s HCl, KCN

Table 4-1 (Continued)
PHYSICAL CONSTANTS OF INORGANIC COMPOUNDS

Name	Formula	Formula weight	Color, crystalline form, refractive index	Density	Melting point, °C	Boiling point, °C	Solubility in 100 parts
Copper							
(II) cyanide	$Cu(CN)_2$	115.58	yel-grn pwd	....	d	....	i; s a, alk, KCN
(II) dichromate-2-water	$CuCr_2O_7 \cdot 2H_2O$	315.56	blk cr, deliq	2.283	$-2H_2O$, 100	....	v s
(I) fluoride	CuF	82.54	red cr	....	908	subl 1100	i; s HCl, HF
(II) fluoride	CuF_2	101.54	wh, mn, hygr	4.23	770	1449	0.075; s a
(II) fluoride-2-water	$CuF_2 \cdot 2H_2O$	137.57	bl, mn	2.93	d	....	0.10
formate	$Cu(CHO_2)_2$	153.55	bl, mn	1.831	....	....	12.5
formate-4-water	$Cu(CHO_2)_2 \cdot 4H_2O$	225.61	bl cr	1.81	$-4H_2O$, 130	....	6.2
bis(glycinate)	$(H_2NCH_2COO)_2Cu$	211.66	dk-bl rh nd	....	d 228	....	s
hydride	CuH	64.55	red-brn	6.39	d	....	i; d HCl
hydroxide	$Cu(OH)_2$	97.55	bl-grn pwd	3.368	160	....	i; s a
iodate	$Cu(IO_3)_2$	413.35	grn, mn	4.89^{25}	d	....	0.14^{25}
(i) iodide (marshite)	CuI	190.44	beige, cub, 2.346	5.62^{25}	588	1207	i; s HCl, KI
nitrate-3-water	$Cu(NO_3)_2 \cdot 3H_2O$	241.60	bl cr, deliq	2.32	114.5	d 170	138^0
(tri) nitride	Cu_3N	204.63	dk grn pwd	5.84^{25}	d 300	....	d a
oxalate-hemihydrate	$CuC_2O_4 \cdot 0.5H_2O$	160.57	blsh-wh	....	anhyd 200	d 310	0.002; s NH_4OH
(I) oxide (cuprite)	Cu_2O	143.08	rec, cub, 2.705	6.0^{25}	1236	$-O_2$, 1800	i; s a
(II) oxide (tenorite)	CuO	79.54	blk, mn, 2.63	6.4	d 1122	....	i; s a
perchlorate	$Cu(ClO_4)_2$	262.43	mn, 1.495, 1.505, 1.522	2.225^{23}	....	d > 130	145.7^{30}
perchlorate-6-water	$Cu(ClO_4)_2 \cdot 6H_2O$	370.53	lt bl, tric, 1.505	2.225	82	d 120	206^{30}
phosphate-3-water	$Cu_3(PO_4)_2 \cdot 3H_2O$	434.61	bl, rh	....	d	....	i; s a
phosphide, di-	CuP_2	135.51	gray-blk	4.20	....	....	i
(di-) phosphide	Cu_2P	158.07	gray-met	6.4	....	....	i
(tri) phosphide	Cu_3P	221.59	gray-blk, hex	7.15	1023	....	i; s HNO_3
phthalocyanine	$(C_6H_5C_2H)_4N_4Cu$	576.05	....	....	....	....	....
-potassium chloride-water (1/2/2)	$CuCl_2 \cdot 2KCl \cdot 2H_2O$	319.59	....	....	....	....	....

[dipyridine(II)] chloride	$[Cu(py)_2]Cl_2$	316.67	blsh-grn, mm, 1.60, 1.75	1.76	d 263		s
salicylate-4-water	$Cu(C_7H_5O_3)_2 \cdot 4H_2O$	409.83	bl-grn nd				v s
selenate-5-water	$CuSeO_4 \cdot 5H_2O$	296.57	bl, tric, 1.56	2.559	$-5H_2O$, 150		25.1^{20}
(I) selenide	Cu_2Se	206.04	blk, cub	6.749^{30}	1117		d HCl
(II) selenide	$CuSe$	142.50	grn-blk, hex	5.99	d red heat		i; s hot HNO_3
(tri-) selenide, di-	Cu_3Se_2	348.54	tetr			d 460	i
selenate(IV)(2−)-2-water	$CuSeO_3 \cdot 2H_2O$	226.53		3.31	$-2H_2O$, 260		
silicate-6-water, hexafluoro-	$Cu[SiF_6] \cdot 6H_2O$	313.71	bl, rh, 1.409, 1.408	2.207			124^{20}
(tetra-) silicide	Cu_4Si	282.25	wh-met	7.53	850		i; d HNO_3
(I) sulfate	Cu_2SO_4	223.14	gray pwd, 1.724, 1.733, 1.739				d; s conc a
(II) sulfate	$CuSO_4$	159.61	grn-wh, rh, 1.733	3.603	805 d		14.3^{0}
(II) sulfate-5-water (chalcanthite)	$CuSO_4 \cdot 5H_2O$	249.68	bl, tric, 1.514, 1.537, 1.543	2.284^{16}	$-5H_2O$, 150		32.0^{20}
(I) sulfide (chalcocite)	Cu_2S	159.14	blk, rh	5.6^{20}	1130		i; s HNO_3
(II) sulfide (covellite)	CuS	95.60	blk, hex, 1.45	4.6			i; s HNO_3
(I) sulfite-1-water	$Cu_2SO_3 \cdot H_2O$	225.16	wh, hex	3.83^{15}	d		sl s; s HCl
(II) tartrate-3-water	$CuC_4H_4O_6 \cdot 3H_2O$	211.61	lt bl pwd				0.42^{20}; s a
(I) telluride	Cu_2Te	254.68	bl-blk, hex	7.27	1125		i
(II) telluride	$CuTe$	191.14	rh		340		i
(II) thiocyanate	$CuSCN$	121.62	wh	2.84	1084		0.00044; s NH_4OH
tungstate(VI)(2−)-2-water	$CuWO_4 \cdot 2H_2O$	347.42	lt-grn, oct		d		0.1^{15}
xanthate	$Cu(C_3H_5OS_2)_2$	305.94	yel				i; s NH_4OH
Curium	Cm	247	silv, hcp	13.51	1350		s a
bromide	$CmBr_3$	486.7	wh, rh	6.87	400		s
chloride	$CmCl_3$	353.3	wh, hex	5.81	500		s
fluoride, tri-	CmF_3	304	wh, hex	9.70	1406		i
fluoride, tetra-	CmF_4	323	grnsh-tan, mm	7.49			s
iodide	CmI_3	627.7	wh, hex	6.37			s a
oxide	Cm_2O_3	542	wh				
Disiloxane	$(SiH_3)_2O$	78.22	col gas	3.491 (g/L)	−144	−15.2	v sl s

Table 4-1 (*Continued*)
PHYSICAL CONSTANTS OF INORGANIC COMPOUNDS

Name	Formula	Formula weight	Color, crystalline form, refractive index	Density	Melting point, °C	Boiling point, °C	Solubility in 100 parts
Cyanogen	NC—CN	52.04	col gas	2.335(g/L)	−27.84	−21.15	450 ml[20], 230 ml al
bromide	NC—Br	105.93	....	....	51.4	61.35	....
chloride	NC—Cl	61.48	....	....	−6.90	13.0	....
iodide	NC—I	152.92	....	....	146	subl 140	sl s
Deuterium or hydrogen[²H]	D_2 or $^2H_2$	4.032	col gas	....	−252.89	−248.24	....
bromide	DBr	81.92	col gas, 1.000569	3.39^{20} (g/L)	−87.46	−66.5	....
chloride	DCl	37.47	col gas, 1.000406^{25}	1.49^{25} (g/L)	−114.64	−84.72	....
fluoride	DF	21.02	col gas	....	−83.6	18.65	....
iodide	DI	128.92	col gas	....	−51.87	−35.7	....
oxide (heavy water)	D_2O	20.03	col liq; hex cr, 1.33844	1.1045	3.82	101.43	....
selenide	D_2Se	82.99	col gas	....	−66.92	....	....
sulfide	D_2S	36.09	col gas	....	−86.02	....	d
Diphosphonic acid	$H_2P_2H_2O_5$	145.98	col nd	....	38	d 120	d
Diphosphoric(IV) acid	$(HO)_2OP—PO(OH)_2$	162.01	col, rh, deliq	....	70	d 100	d
Diphosphoric(V) acid	$H_4P_2O_7$	177.98	col nd or liq, hygr	....	61	....	s
Disulfuric acid	$H_2S_2O_7$	178.14	col cr, hygr	1.9^{20}	35	d	d
Disulfuryl dichloride	$S_2O_5Cl_2$	215.03	col liq, 1.937	1.818^{11}	−37.5	152.5	d; d a
chloride fluoride	$Cl—SO_2—O—SO_2—F$	198.57	col liq	1.797^{20}	−65	100.1	s
fluoride, di-	$F—SO_2—O—SO_2—F$	182.12	col liq	1.75^0	−58	51	s
Dysprosium	Dy	162.50	met, hcp	8.536	1500	2600	s a
acetate-4-water	$Dy(C_2H_3O_2)_3 \cdot 4H_2O$	411.64	yel nd	....	d 120	....	s
bromate-9-water	$Dy(BrO_3)_3 \cdot 9H_2O$	708.36	yel nd, hex	....	78	....	v s
bromide	$DyBr_3$	402.23	lt-yel, rhbd	4.78	880	1480	s
chloride	$DyCl_3$	268.86	wh, mn	3.67^0	647	1530	s
chromate-10-water	$Dy_2(CrO_4)_3 \cdot 10H_2O$	853.13	yel cr	....	d 150	....	1.00^{25}
fluoride	DyF_3	219.50	lt-grn, hex	7.465	1154	2230	i
hydroxide	$Dy(OH)_3$	213.52	wh, hex	....	d 310	....	s a

Name	Formula	Mol. wt.	Crystalline form, color	Density	Melting point	Boiling point	Solubility
iodide	DyI$_3$	543.21	dk-grn, hex	....	978	1320	s
nitrate-5-water	Dy(NO$_3$)$_3$·5H$_2$O	438.58	yel cr	....	88.6	....	s
oxalate-10-water	Dy$_2$(C$_2$O$_4$)$_3$·10H$_2$O	769.21	wh pwd	....	....	....	s dil a
oxide	Dy$_2$O$_3$	373.00	wh, bcc	8.15	2340	....	s a
phosphate-5-water	DyPO$_4$·5H$_2$O	347.55	yelsh pwd	....	-5H$_2$O, 200	....	s dil a
sulfate-8-water	Dy$_2$(SO$_4$)$_3$·8H$_2$O	757.31	lt yel	....	-8H$_2$O, 360	....	5.08[20]
sulfide	Dy$_2$S$_3$	421.18	cub	....	1490	....	
telluride	Dy$_2$Te$_3$	707.8	rh	....	1510	2900	s a
Erbium	Er	167.26	gray, hcp	9.051	1497		
acetate-4-water	Er(C$_2$H$_3$O$_2$)$_3$·4H$_2$O	416.48	wh, tric	2.114	....	....	
boride, hexa-	ErB$_6$	232.12	bl, cub	4.61	....	....	
bromide	ErBr$_3$	406.97	vlt-rose, rhhd	4.93	923	1460	s
chloride	ErCl$_3$	273.62	vlt-rose, mn	4.1	774	1500	s
fluoride	ErF$_3$	224.26	pink, hex	7.814	1140	2230	s H$_2$SO$_4$
hydroxide	Er(OH)$_3$	218.28	rose, hex	....	d 315	....	s a
iodide	ErI$_3$	547.97	vlt-red, hex	....	1015	1280	s
nitrate-5-water	Er(NO$_3$)$_3$·5H$_2$O	443.37	red cr	....	d 130	....	i; s a
oxalate-10-water	Er$_2$(C$_2$O$_4$)$_3$·10H$_2$O	778.77	redsh pwd	....	d 575	....	0.0005[25]; sl s a
oxide	Er$_2$O$_3$	382.52	rose red pwd	8.640	....	....	
selenide	Er$_2$Se$_3$	571.40	brn, rh	6.96	....	....	16.0
sulfate-8-water	Er$_2$(SO$_4$)$_3$·8H$_2$O	766.87	rose-red, mn	3.217	-8H$_2$O, 110	d 630	s a
sulfide	Er$_2$S$_3$	430.70	ochre, mn	....	1730	....	s
Europium	Eu	151.96	gray, bcc	5.259	826	1440	s a
(II) bromide	EuBr$_2$	311.78	wh	....	683	1880	s
(III) bromide	EuBr$_3$	391.69	rose, rh	5.40	702	d	s
(II) chloride	EuCl$_2$	222.87	wh, rh	....	731	2027	s
(III) chloride	EuCl$_3$	258.32	yel, hex	4.89	623 d	....	s
(II) fluoride	EuF$_2$	189.96	dk yel, cub	6.495	1416	2427	i
(III) fluoride	EuF$_3$	208.96	wh, hex	6.793	1276	2277	i; s H$_2$SO$_4$
(III) hydroxide	Eu(OH)$_3$	202.96	....	....	....	....	i; s a
(II) iodide	EuI$_2$	405.77	olive-grn, mn	5.50[25]	580	1577	s
(III) iodide	EuI$_3$	532.68	col	....	d	....	s
(III) nitrate-6-water	Eu(NO$_3$)$_3$·6H$_2$O	446.07	lt pink, mn	....	85	....	s
(III) oxide	Eu$_2$O$_3$	351.92	cub	7.27	2050	....	s a
selenide	EuSe	230.92	cub				

Table 4-1 (*Continued*)
PHYSICAL CONSTANTS OF INORGANIC COMPOUNDS

Name	Formula	Formula weight	Color, crystalline form, refractive index	Density	Melting point, °C	Boiling point, °C	Solubility in 100 parts
Europium							
(III) sulfate-8-water	$Eu_2(SO_4)_3 \cdot 8H_2O$	736.23	lt pink	....	$-8H_2O$, 375	d 1600	2.56[20]
sulfide	EuS	184.02	blk, cub	5.75	ign 1527	....	....
telluride	EuTe	279.6	cub	....			d
Fluorine	F_2	38.00	yel-grn gas, 1.000187[25]	1.554[25] (g/L)	-219.70	-188.20	d
fluorosulfate	$F[SO_3F]$	118.06	col gas	1.78[-74] (g/L)	-158.5	-31.3	
nitrate	$F(NO_3)$	81.01	col gas	1.507[bp liq]	-175	-45.9	hyd; s acet
nitride, tri-	FN_3	60.92	grn-yel gas		-154	-82	
perchlorate	$F[ClO_4]$	102.45	col gas, expl	4.85[25] (g/L)	-167.3	-15.9[755mm]	d
Fluoroamine, di-	HNF_2	53.01	unstable gas	....	-116	-23.6	
Fluoroboric acid	$H[BF_4]$	87.81	col liq	1.818	d 130	....	v s
Fluorophosphonic acid	$H_2[PO_3F]$	99.99	col oily liq		-80		v s
Fluorophosphonic acid, di	$H_2[PO_2F_2]$	102.99	col fum liq	1.583	-96.5	115.9	s
Fluorophosphoric acid, hexa-	$H[PF_6]$	145.97	col fum liq	1.65			s
Fluorosilicic acid, hexa-	$H_2[SiF_6]$	144.08	col fum liq, 1.3465 61% soln	1.463	d		s
Fluorosulfonic acid	$H[SO_3F]$	100.07	col liq	1.743[15]	-87.3	165.5	s
Gadolinium	Gd	157.25	met, hcp	7.895	1306	3000	s a
acetate-4-water	$Gd(C_2H_3O_2)_3 \cdot 4H_2O$	406.45	col, tric	1.611			11.6[25]
boride, hexa-	GdB_6	222.11	blue, cub	4.65			
bromide	$GdBr_3$	396.96	col, rhbd	4.57	770	1490	s
chloride	$GdCl_3$	263.61	wh, hex, hygr	4.52[0]	602	1580	s
fluoride	GdF_3	214.25	wh, hex	7.047	1231	2277	i
hydride, di-	GdH_2	159.27	....	7.08			

Name	Formula	Mol. wt.	Color, crystalline form	Density	M.p.	B.p.	Solubility
hydroxide	Gd(OH)$_3$	308.27	wh, hex	....	d 310	....	s a
iodide, di-	[Gd^{3+}(e$^-$)(I$^-$)$_2$]	411.06	bronze	....	831	....	s
iodide, tri-	GdI$_3$	537.96	yel, hex	....	925	1340	v s; s al
nitrate-6-water	Gd(NO$_3$)$_3$ · 6H$_2$O	451.36	col, tric, deliq	2.322	91	....	i
oxalate-10-water	Gd$_2$(C$_2$O$_4$)$_3$ · 10H$_2$O	758.71	col, mn	....	d 110	....	s a
oxide	Gd$_2$O$_3$	362.50	wh, mn, hygr	7.64	2340	....	s a
selenate-8-water	Gd$_2$(SeO$_4$)$_3$ · 8H$_2$O	887.50	pearly, mn	3.309	−8H$_2$O, 130	....	s
selenide	GdSe	236.2	cub	....	1860	....	2.60[20]
sulfate	Gd$_2$(SO$_4$)$_3$	602.68	col	4.139[15]	d 500	....	4.08
sulfate-8-water	Gd$_2$(SO$_4$)$_3$ · 8H$_2$O	746.81	col, mn	3.010[15]	anhyd 400	....	s a
sulfide	GdS	189.3	cub	....	2027	....	s a
(di-) sulfide, tri-	Gd$_2$S$_3$	419.69	yel-brn, cub, hygr	3.8	1885	....	s a
telluride	GdTe	284.85	cub	....	1887	....	....
Gallium	Ga	69.72	gray-blk, rh	5.907 (c); 6.0948 (liq)	29.75	1980	....
antimonide	GaSb	191.47	cub	3.9	712	....	s HCl
arsenide	GaAs	144.64	dk-gray, cub	5.31[25]	1238	....	s HCl
bromide	GaBr$_3$	309.45	wh	3.69	124	280	v s
bromide–ammonia (1/1)	GaBr$_3$ · NH$_3$	326.48	wh	3.112[25]	124	....	d
chloride	GaCl$_3$	176.03	wh nd, hygr	2.47	77.75	201.2	d; s bz, CCl$_4$, CS$_2$
ethyl, tri-	Ga(C$_2$H$_5$)$_3$	146.90	....	1.058[30]	−82.3	142.8	0.002; s HF
fluoride	GaF$_3$	126.72	wh, rhbd	4.47	subl 950	....	d; 4.6 bz
(I) gallate(III), tetrachloro-hydride [digallane(6)]	Ga[GaCl$_4$]	281.25	lt gray, rh, hygr	2.417	170.5	d 200	s coordinating solv
hydride [digallane(6)]	Ga$_2$H$_6$	145.49	vol liq	....	−21.4	139 d	s a, alk
hydroxide	Ga(OH)$_3$	120.74	....	....	d 300	....	....
iodide	GaI$_3$	450.43	col nd	4.15[25]	212	346	v s
methyl, tri-	Ga(CH$_3$)$_3$	114.84	....	1.151[15]	−15.7	55.8	s hot conc H$_2$SO$_4$
nitrate-x-water	Ga(NO$_3$)$_3$ · xH$_2$O	....	wh cr, deliq	....	d 110	....	v s
nitride	GaN	83.73	dk-gray pwd	6.1	subl 800	....	0.4
oxalate-4-water	Ga$_2$(C$_2$O$_4$)$_3$ · 4H$_2$O	475.56	wh cr, hygr	....	−4H$_2$O, 180	d 200	i; s a, alk
(di-) oxide	Ga$_2$O	155.44	blk-brn pwd	4.77[25]	subl 500	....	i; s alk
oxide	Ga$_2$O$_3$	187.44	wh, rh	6.44	1900	....	....
oxide-1-water	Ga$_2$O$_3$ · H$_2$O	205.45	wh, rh, 1.84	5.2	d 400	....	....

Table 4-1 (Continued)
PHYSICAL CONSTANTS OF INORGANIC COMPOUNDS

Name	Formula	Formula weight	Color, crystalline form, refractive index	Density	Melting point, °C	Boiling point, °C	Solubility in 100 parts
Gallium							
phosphide	GaP	100.69	or, cub	....	1465	....	....
potassium bissulfate-12-water	GaK(SO$_4$)$_2$ · 12H$_2$O	517.13	col cr	1.895	....	....	s
selenate-16-water	Ga$_2$(SeO$_4$)$_3$ · 16H$_2$O	856.56	col, mn	....	....	....	18.1[25]
selenide	GaSe	148.68	dk red-brn, hex	5.03[25]	960	d	....
(di-) selenide, tri-	Ga$_2$Se$_3$	376.32	redsh-bl, cub	4.92[25]	1020	....	....
sulfate	Ga$_2$(SO$_4$)$_3$	427.62	wh pwd	....	d 690	....	v s
sulfate-18-water	Ga$_2$(SO$_4$)$_3$ · 18H$_2$O	751.90	col, oct	....	....	....	v s
sulfide	GaS	101.78	yel, hex	3.80[25]	970	d	hyd; s a, alk
(di-) sulfide, tri-	Ga$_2$S$_3$	235.63	yel, cub	3.65[25]	1090	....	....
telluride	GaTe	197.32	blk, mn	5.44[25]	835	....	....
(di-) telluride, tri-	Ga$_2$Te$_3$	522.24	blk, cub	5.57[25]	790	....	....
Germane	GeH$_4$	76.62	col gas	1.523[−142]	−165.9	−88.5	sl s hot HCl
chloro-	GeH$_3$Cl	111.07	col liq	1.75[−25]	−52	28	d
dichloro-	GeH$_2$Cl$_2$	145.51	col liq	1.90[−68]	−68.0	69.5	d
trichloro-	GeHCl$_3$	179.96	col liq	1.93	−71	75.2	hyd; s eth
Germane(6), di-	Ge$_2$H$_6$	151.23	col liq	1.98[−109]	−109	29	i; s CCl$_4$
Germane(8), tri-	Ge$_3$H$_8$	225.83	col liq	2.20[−105.6]	−105.6	110.5	
Germane(10), tetra-	Ge$_4$H$_{10}$	300.44	col liq	....	....	176.9	....
Germane(12), penta-	Ge$_5$H$_{12}$	374.04	col oily liq	....	....	234	i; s hot H$_2$SO$_4$
Germanium	Ge	72.59	gray, cub	5.323	940	2852	s a
(II) bromide	GeBr$_2$	232.41	col nd	....	122	d	hyd; s bz, eth
(IV) bromide	GeBr$_4$	392.23	col, oct, 1.6269	3.132	26.1	186.5	hyd; s bz, eth
(II) chloride	GeCl$_2$	143.50	wh cr	....	d	....	hyd; s dil HCl
(IV) chloride	GeCl$_4$	214.40	col fum liq	1.88	−49.5	83.1	hyd
(di-) chloride, hexa-	Ge$_2$Cl$_6$	358.30	col cr	....	40–42	....	....
(II) fluoride	GeF$_2$	110.59	wh cr, hygr	....	110	d 160	hyd

	Formula	Mol. wt.	Color, crystalline form, etc.	Density	Melting point, °C	Boiling point, °C	Solubility
(IV) fluoride	GeF_4	148.58	col gas, garlic odor	2.46^{-37}	-15^{302mm}	subl -36.5	hyd; s dil HCl
(IV) fluoride-3-water	$GeF_4 \cdot 3H_2O$	202.63	col cr, hygr				s
hydrides, see under Germane							
(II) hydroxide	$Ge(OH)_2$	106.61	yel, easily oxidized		$-H_2O$, 650		i; s a
(II) iodide	GeI_2	326.40	yel, hex	5.37	subl 240		hyd; s conc HI
(IV) iodide	GeI_4	580.21	or, cub, deliq	4.322^{25}	144.0	ca 348	s bz, CS_2, MeOH
(tri-) nitride, di-	Ge_3N_2	245.78	blk		subl 650		i hot a, alk
(tri-) nitride, tetra-	Ge_3N_4	273.80	lt brn pwd	5.25^{25}	d 1000		0.0018
(II) oxide	GeO	88.59	blk nd, 1.607		subl 710		0.43^{20}
(IV) oxide (soluble)	GeO_2	104.59	col, hex, 1.650	4.228^{25}	1115	1200	i
(IV) oxide (insoluble)	GeO_2	104.59	col, tetr	6.239	1086		
(II) selenide	$GeSe$	151.55	brn-blk, rh	5.30	675		sl s alk
(IV) selenide	$GeSe_2$	230.51	yel, hex	4.56^{25}	740		
(IV) sulfate	$Ge(SO_4)_2$	264.71	wh	3.92	d 200		0.24^{20}, s a, alk
(II) sulfide	GeS	104.65	gray-blk, rh	3.54	665		0.45^{20}, s alk
(IV) sulfide	GeS_2	136.72	wh, rh		840		
(II) telluride	$GeTe$	200.19	trig		724		
Gold	Au	196.97	yel-met, fcc	19.3	1063	2808	i; s aq reg, KCN, hot H_2SO_4
antimonide, di-	$AuSb_2$	440.47		7.90	460		i; d a; s KCN
(I) bromide	$AuBr$	276.88	yel-gray pwd		115 d		s; s al, eth, gly
(III) bromide	$AuBr_3$	436.69	brn cr		97.5	d 160	s HCl, HBr
(I) chloride	$AuCl$	232.42	yel, rh	7.4	d 170		68^{20}; s al, eth
(III) chloride	$AuCl_3$	303.33	red, mn	3.9^{20}	subl 180	229	s KCN, NH_4OH
(I) cyanide	$AuCN$	222.98	lt yel pwd	7.12^{20}	d		v s
(III) cyanide-3-water	$Au(CN)_3 \cdot 3H_2O$	329.07	col, hygr		d 50		s KI
(III) fluoride	AuF_3	253.96	or-yel, hex		subl 300	d 500	s KI
(I) iodide	AuI	323.87	grnsh-yel, tetr	8.25	120 d		i; s HCl
(III) iodide	AuI_3	577.68	dk grn				i a
(III) oxide	Au_2O_3	441.93	brn pwd		d 160		s aq reg, alk sulf
(di-) phosphide, tri-	Au_2P_3	486.86	grav pwd	8.12	d		
selenide	$AuSe$	275.93	mn: bl amorp	4.65^{22}	400		
stannide	$AuSn$	315.66			418		
(III) sulfide	Au_2S_3	490.13	brn-blk pwd	8.754	d 197		i; s Na_2S
telluride, di- (krennerite)	$AuTe_2$	452.16	yel. mn	8.2–9.3	464		i

Table 4-1 (Continued)
PHYSICAL CONSTANTS OF INORGANIC COMPOUNDS

Name	Formula	Formula weight	Color, crystalline form, refractive index	Density	Melting point, °C	Boiling point, °C	Solubility in 100 parts
Hafnium	Hf	178.49	met, hcp	13.31	2222	4450	i; s HF
boride, di-	HfB_2	200.11	.....	.....	3335	.....	.....
boronate, tetra-	$Hf[BH_4]_4$	237.85	.....	.....	29.0	118	.....
bromide, tri-	$HfBr_3$	418.20	bl-blk	.....	d 350	.....	hyd
bromide, tetra-	$HfBr_4$	498.13	wh, cub	.....	425	subl 322	v s
carbide	HfC	190.54	wh	12.20	4160	.....	.....
chloride	$HfCl_4$	320.30	.....	.....	432	subl 316	s acet, MeOH
fluoride	HfF_4	254.48	wh, mn, 1.56	.....	.....	.....	.....
hydride, di-	HfH_2	180.51	tetr	11.37	.....	.....	.....
hydride[²H], di-	Hf^2H_2 or HfD_2	182.52	tetr	11.68	.....	.....	.....
iodide	HfI_4	686.11	yel-or, cub	.....	449	subl 400 (vac)	.....
nitride	HfN	192.50	yel-brn, cub	.....	3307	.....	.....
oxide	HfO_2	210.49	wh, cub	9.68^{20}	2910	.....	i
sulfide	HfS_2	242.62	hex	.....	.....	.....	.....
Helium	He	4.003	col gas, 1.000035	0.17847^0 (g/L) 0.1249 (liq)	−272.2 (25 atm)	−268.935	0.861 ml^{20}
Holmium	Ho	164.93	met, hcp	8.803	1461	2600	i; s a
bromide	$HoBr_3$	404.66	lt yel, rhbd	4.86	919	1470	s
chloride	$HoCl_3$	271.29	lt yel, mn	.....	718	1510	s
fluoride	HoF_3	221.93	pink, hex	7.644	1143	2230	i; s H_2SO_4
iodide	HoI_3	545.64	lt yel, hex	.....	994	1300	s
oxalate-10-water	$Ho_2(C_2O_4)_3 \cdot 10H_2O$	774.10	lt tan	.....	d 40	.....	i; s a
oxide	Ho_2O_3	377.86	lt yel, bcc	.....	.....	.....	i; s a
sulfate-8-water	$Ho_2(SO_4)_3 \cdot 8H_2O$	762.23	lt yel cr	.....	.....	.....	8.18^{20}
sulfide	HoS	196.99	cub	.....	.....	.....	.....

Name	Formula	Mol. wt	Crystalline form, color	Density	mp	bp	Solubility
Hydrazine	H_2N-NH_2	32.05	col, hygr, 1.4710	1.0083^{20}	1.54	113.8	v s
hydrate	$H_2N-NH_2 \cdot H_2O$	50.16	col, 1.428	1.038	−51.7	119.4	v s
Hydrazinium(1+) azide	$N_2H_5N_3$	75.07	wh pr, 1.53, 1.76	......	75.4	......	v s
(1+) bromide	N_2H_5Br	112.96	......	......	86.5	......	s
(2+) bromide	$N_2H_6Br_2$	192.87	......	......	195	......	s
(1+) chloride	N_2H_5Cl	68.51	wh nd	......	92.6	d 240	v s
(2+) chloride	$N_2H_6Cl_2$	104.97	col, oct	1.4226^{20}	198	d 200	v s
(1+) chlorate	$N_2H_5ClO_3$	116.50	cr	......	80	......	......
(2+) formate	$N_2H_6(HCO_2)_2$	124.10	......	......	128	......	s
(1+) iodide	N_2H_5I	159.98	col pr	......	127	......	s
(2+) iodide-2-water	$N_2H_6I_2 \cdot 2H_2O$	323.90	cr	......	65	......	......
(1+) nitrate	$N_2H_5NO_3$	95.96	col nd	......	70.71	subl 140	175^{10}
(2+) nitrate	$N_2H_6(NO_3)_2$	158.07	col cr	......	104	......	v s
(1+) oxalate	$(N_2H_5)_2C_2O_4$	154.14	wh nd	......	148	......	200^{35}
(1+) perchlorate	$N_2H_5ClO_4$	132.51	cr, expl	1.939^{15}	137	d 145	d; s al
(1+) sulfate	$(N_2H_5)_2SO_4$	162.18	col cr, hygr	......	85	......	202^{25}
(2+) sulfate	$N_2H_6SO_4$	130.13	col, rh	1.37	254	d	3.415^{20}
(1+) tartrate	$(N_2H_5)_2C_4H_4O_6$	182.13	col cr, $[\alpha]_4^{20} + 22.5°$		183	......	6.0^0
Hydrogen	H_2	2.016	col gas	0.0899 (g/L) 0.07099^{bp} (liq)	−259.19	−252.76	1.9 ml
azide	HN_3	43.03	col liq, v expl	1.126	−80	37	v s
borate(1−)	HBO_2	43.83	wh, cub, 1.619	2.486	236	......	v sl s
borate(3−), ortho-	H_3BO_3	61.83	col, tric, 1.337, 1.461, 1.462	1.435^{15}	171.0	d 300	6.35^{30}
borate(2−), tetra-	$H_2B_4O_7$	157.26	wh pwd	......	......	......	s
borate(1−), tetrafluoro-	$H[BF_4]$	87.81	col liq	3.388^{20}	d 130	......	v s
bromide	HBr	80.92	col gas, 1.0005591^{25}	2.160^{bp} (liq)	−86.81	−66.71	193^{25}
bromide-2-water	$HBr \cdot 2H_2O$	116.95	wh cr, col liq	2.11^{-15}	−11.3	......	s
bromide-3-water	$HBr \cdot 3H_2O$	134.96	col liq, 60% HBr		d −49.6	......	s
bromide-4-water	$HBr \cdot 4H_2O$	152.98	col liq, 53% HBr		−57.9	......	s
bromide-6-water	$HBr \cdot 6H_2O$	189.01	col liq, 43% HBr		d −88.2	......	s

Table 4-1 (Continued)
PHYSICAL CONSTANTS OF INORGANIC COMPOUNDS

Name	Formula	Formula weight	Color, crystalline form, refractive index	Density	Melting point, °C	Boiling point, °C	Solubility in 100 parts
Hydrogen							
bromide	48% HBr + H₂O	......	const. boiling	1.49	−11	126	v s
[2H] bromide	2HBr	81.92	col gas, 1.000569	3.39[20] (g/L)	−87.46	−66.5	v s
chloride	HCl	36.46	col gas, 1.000408[25]	1.526[20] (g/L) 1.187[bp] (liq)	−114.18	−85.00	71.9[20]
chloride-1-water	HCl · H₂O	54.48	col liq	1.48	−15.35	......	v s
chloride-2-water	HCl · 2H₂O	72.49	col liq	1.46[18]	−17.7	d	v s
chloride-3-water	HCl · 3H₂O	90.51	col liq	......	−24.9	d	v s
chloride	20.24% HCl + H₂O	......	constant boiling	1.097	......	110	v s
[2H] chloride	2HCl	37.47	col gas, 1.000406[25]	1.49[25] (g/L)	−114.64	−84.72	v s
chlorite	HClO₂	68.46	known only in soln	......	......	......	......
cyanate	HOCN	43.04	col liq	1.14[20]	......	......	s; s bz, eth
cyanide	HCN	27.06	col liq or gas, liq: 1.26751[10]	0.901 (g/L) 0.699[22] (liq)	−13.24	25.70	v s
deuteride	1H2H or HD	3.02	col gas	......	−256.56	−251.03	......
ferrate(II), hexacyano-	H₄[Fe(CN)₆]	215.99	wh nd	......	d	......	s
ferrate(III), hexacyano-	H₃[Fe(CN)₆]	214.98	grn-brn nd, deliq	......	d	......	s
fluoride	HF	20.01	col gas, 1.90	0.922[0] (g/L) 0.957[bp]	−83.57	19.52	v s
[2H] fluoride	2HF	21.02	col liq, 1.1574[25]	......	−83.6	18.65	s
fluoride	35.35% HF + H₂O	......	constant boiling	......	......	120	v s
hypobromite	HBrO	96.92	known only in soln	......	......	......	s

Name	Formula	Mol. wt.	Color, crystalline form, refractive index	Density	m.p., °C	b.p., °C	Solubility
hypochlorite	$HClO$	52.46	known only in soln				s
hypofluorite	HFO	36.01	known only in soln				s
hyponitrite	$H_2N_2O_2$	62.03	wh		expl		s
iodate	HIO_3	175.91	wh, rh	4.629^{0}	d 110		310^{16}
iodide	HI	127.92	col gas, 1.000853	5.37^{20} (g/L), 2.799^{bp}	−50.79	−35.35	70^{0}
iodide	57% HI + H_2O		lt yel liq, 1.466^{16}, constant boiling	1.70^{15}		127	v s
[^2H] iodide	^2HI	128.92	col gas		−51.87	−35.7	v s
iodide-2-water	$HI \cdot 2H_2O$	163.94	col liq		−43		v s
iodide-3-water	$HI \cdot 3H_2O$	181.96	col liq		−48		v s
iodide-4-water	$HI \cdot 4H_2O$	199.97	col liq		−36.5		
molybdate(VI)(2−)	H_2MoO_4	161.95	sl yelsh, hex	3.112	−H_2O, 70		0.133^{18}, s alk
nitrate	HNO_3	63.02	col liq, 1.3970	1.5027	−41.59	83	v s
nitrate	69% HNO_3 + H_2O		constant boiling	1.41^{20}		120.5	v s
nitrite	HNO_2	47.02	lt bl, known only in soln		d		s
oxide (water)	H_2O	18.02	col, hex, 1.309, 1.313; col liq, 1.333	1.000^{4}	0.00	100.00	
[^2H] oxide (heavy water)	^2H_2O	20.03	col, 1.32844	1.1045	3.82	101.43	s
perchlorate	$HClO_4$	100.46	col oily liq, shock sensitive, 1.380^{25}	1.7756^{20}	−101.2	d	v s
perchlorate-1-water	$HClO_4 \cdot H_2O$	118.47	col nd	1.7677^{20}		50	v s
perchlorate-2-water	$HClO_4 \cdot 2H_2O$	136.49	col, 1.70^{20}, commercial 72% acid	1.67^{20}	−17.8	203	v s
periodate(1−)	HIO_4	191.91	col cr		subl 110	d 138	440^{25}
periodate(5−)	H_5IO_6	227.94	col pr		130	d 140	113
peroxide	H_2O_2	34.02	col liq, 1.414	1.4649^{0}	−0.40	151.2	v s; s al; eth
peroxodisulfate	$H_2S_2O_8$	194.14	col cr, hygr		d 60		v s
phosphate(V)(1−)	HPO_3	79.98	col cr, deliq	2.2-2.5	subl		s
phosphate(V)(3−)	H_3PO_4	98.00	col liq, 1.317; col, 85% commercial	1.88	42.3	d 213	v s
phosphide (phosphine)	PH_3	34.00	col gas	1.529 (g/L), 0.765^{bp}	−133.81	−87.78	26 ml^{17}; s al, eth
selenide	H_2Se	80.98	col gas or liq	2.12^{bp}	−65.73	−42	9.5 ml^{20}, 0.68^{25}
[^2H] selenide	^2H_2Se	82.99	col gas		−66.92		

Table 4-1 (*Continued*)
PHYSICAL CONSTANTS OF INORGANIC COMPOUNDS

Name	Formula	Formula weight	Color, crystalline form, refractive index	Density	Melting point, °C	Boiling point, °C	Solubility in 100 parts
Hydrogen							
sulfide	H_2S	34.08	col gas	1.1906 (g/L)	−85.52	−60.33	0.334[25], 9.5 ml[20]
[²H] sulfide	^2H_2S	36.09	col gas	....	−86.02	....	....
tellurate(IV)(2−)	H_2TeO_3	177.61	wh	....	d to TeO_2	....	0.0007; s a, alk
tellurate(VI)(6−)	H_6TeO_6	229.64	col, cub	2.99	d 120	....	30[18]
telluride	H_2Te	129.63	col gas	6.234 (g/L)	−49	−2	d
tungstate(VI)(2−)	H_2WO_4	249.86	yel pwd, 2.24	5.5	−H_2O, 100	....	i; s alk, HF
tungstate(VI)(2−)-9-water, tetra-	$H_2W_4O_{13} \cdot 9H_2O$	1107.55	col, tetr	3.93	d 50	....	88.6[22]
vanadate(V)(1−)	HVO_3	99.95	yel	....	....	....	i; s a, alk
Hydroxylamine	$HONH_2$	33.03	wh nd or col liq	1.332	33.1	58[22mm]	s; s al, MeOH
Hydroxylammonium							
acetate	$HONH_3C_2H_3O_2$	93.08	col cr	....	87	subl 90	v s
bromide	$HONH_3Br$	113.95	wh, mn	2.35[22]	....	....	v s
chloride	$HONH_3Cl$	69.49	col, mn	1.680[20]	150.5	d	83[17]; 4.4[20] al
iodide	$HONH_3I$	160.94	col nd, hygr	....	83	....	v s
sulfate	$(HONH_3)_2SO_4$	164.14	col, mn	....	d 170	....	68.5[20]
Indium	In	114.82	silv-wh, tetr	7.28	156.3	2070	s a
antimonide	InSb	236.57	cr, cub	5.74[mp]	535	....	i
arsenide	InAs	189.74	met cr	....	943	....	i; i a
bromide	$InBr_3$	354.55	wh nd, deliq	4.75	436	subl 371	571[25]
chloride, di-	$InCl_2$	185.73	wh, rh, deliq	3.655[25]	235	560	d
chloride, tri-	$InCl_3$	221.18	col, hygr	4.0	586	subl 418	212[25]
fluoride	InF_3	171.82	col	4.39[25]	1170	>1200	0.040
fluoride-3-water	$InF_3 \cdot 3H_2O$	225.86	col	....	−3H_2O, 100	....	11.2[25]
hydroxide	$In(OH)_3$	165.84	wh amorp	....	−H_2O, 150	....	s a
iodate	$In(IO_3)_3$	639.53	wh	....	d	....	0.067[20]; s a

Name	Formula	Mol. wt.	Color, form	Density	Melting point	Boiling point	Solubility
iodide	InI_3	495.53	lt yel, hygr	4.69	210	....	d; s a
methyl, tri-	$In(CH_3)_3$	159.93	col	1.568	88.4	135.8	d; s acet, bz
nitrate-3-water	$In(NO_3)_3 \cdot 3H_2O$	354.88	col	....	d 100	....	v s
(di-) oxide	In_2O	245.64	blk	6.99^{25}	subl 565^{vac}	....	s HCl
(III) oxide	In_2O_3	277.64	lt yel, trig	7.179	850	850	s a
perchlorate-8-water	$In(ClO_4)_3 \cdot 8H_2O$	557.29	col, deliq	....	d 200	....	v s
phosphide	InP	145.79	met, brittle	....	1070	....	v sl s a
(di-) selenide	In_2Se	308.60	rh	....	d 540	....	....
selenide	$InSe$	193.78	rbhd, blk cr	5.67^{25}	660	....	s
(di-) selenide, tri-	In_2Se_3	466.52	blk	5.18^{25}	890	....	s
sulfate	$In_2(SO_4)_3$	517.83	lt gray, mn, hygr	3.438	....	d 850	s
sulfide	InS	146.88	wine red, rh	4.9	692	....	s HCl
(di-) sulfide, tri-	In_2S_3	325.83	yel, fcc	6.29^{25}	1050	....	s a
telluride	$InTe$	242.42	dk met, tetr	....	692	....	s HNO$_3$
(di-) telluride	In_2Te	357.24	rh	....	d 462	....	....
(di-) telluride, tri-	In_2Te_3	612.44	blk cr, fcc	5.78	671	....	....
(di-) telluride, penta-	In_2Te_5	867.64	mn	....	d 455	....	....
Iodic acid	HIO_3	175.91	wh, rh	4.629^{0}	d 110 to H_5IO_6	d 195 to I_2O_5	310^{16}
Iodine	I_2	253.82	vlt, rh, 3.34	4.660^{20}	113.60	184.24	0.029^{20}; s al, bz, chl, CS$_2$, CCl$_4$, eth
bromide	IBr	206.81	vlt-bl	4.4157^{0}	42	116 d	s; s al, eth
chloride	ICl	162.36	ruby-red, cub	3.20	27.38	97.8	d; s al, eth
chloride, tri-	ICl_3	233.26	yel nd, deliq	3.202	101 d	....	d; s al, bz, eth
cyanide	ICN	152.92	wh cr	....	....	....	3.86^{20}; s al, eth
fluoride, penta-	IF_5	221.90	col liq	3.252	8.5	102	d
fluoride, hepta-	IF_7	259.89	col gas	2.8^{6} (g/L)	4.5	5.5	....
(tetra-) oxide, nona-	I_4O_9	651.61	pale yel, hygr	....	d 75	....	....
(di-) oxide, penta-	I_2O_5	333.81	wh cr, trimetric	4.799^{25}	d 275	....	187^{13}
sulfate, fluoro-	$I(SO_3F)$	125.97	....	4.2^{10}	51.5	....	d; s H$_2$SO$_4$
Iodosyl iodate	$(IO)IO_3$	158.90	yel	....	....	....	....
Iridium	Ir	192.2	silv-met, fcc	22.65^{20}	2454	4389	s aq reg
(di-) carbonyl, octa-	$Ir_2(CO)_8$	608.48	grn-yel cr	....	subl 160	....	i; s CCl$_4$
carbonyl, tri-x-mer	$Ir(CO)_3I_x$	....	yel, rhbd	....	d 210	....	i CCl$_4$
chloride, tri-	$IrCl_3$	298.56	red, mn	....	d 775	....	i a, alk
fluoride, tri-	IrF_3	249.19	rhbd	....	d 250	....	i

Table 4-1 (*Continued*)
PHYSICAL CONSTANTS OF INORGANIC COMPOUNDS

Name	Formula	Formula weight	Color, crystalline form, refractive index	Density	Melting point, °C	Boiling point, °C	Solubility in 100 parts
Iridium							
fluoride, penta-	IrF_5	287.19	yel, mn	......	104.5	......	hyd
fluoride, hexa-	IrF_6	306.19	yel, cub	6.0	44	53.6	d
iodide, tri-	IrI_3	572.91	grn	......	d	......	sl s
iodide-3-water, tri-	$IrI_3 \cdot 3H_2O$	626.96	yel	......	......	......	s
(III) oxide	Ir_2O_3	432.40	blk pwd	......	d 400	......	i; s a
(IV) oxide	IrO_2	224.20	blk, tetr	3.15	d 1100	......	0.00002[20], s HCl
(IV) oxide-2-water	$IrO_2 \cdot 2H_2O$	260.23	dk bl cr	......	$-2H_2O$, 350	......	i; s HCl
selenide, di-	$IrSe_2$	350.12	dk gray, rh	......	d 600	......	sl s aq reg
sulfide, di-	IrS_2	256.33	brn-blk, rh	8.43	d 300	......	i; s aq reg
(di-) sulfide, tri-	Ir_2S_3	480.59	brn-blk	9.64	d	......	i; s HNO_3, K_2S
telluride, di-	$IrTe_2$	575.00	dk gray, hex	9.5	......	......	i; s hot aq reg
Iron	Fe	55.85	silv-met, bcc	7.86	1537	2872.3	i; s a
(III) acetylacetonate	$Fe(C_5H_7O_2)_3$	353.18	red, rh	1.33	184	......	sl s; s al, bz, chl
(III) arsenate-2-water (scorodite)	$FeAsO_4 \cdot 2H_2O$	230.80	grn, rh, 1.765, 1.774, 1.797	3.18	d	......	i; s HCl
arsenide	$FeAs$	130.77	wh	7.86	1020	......	v sl s
arsenide, di- (arsenoferrite)	$FeAs_2$	205.69	silv-gray, cub	7.4	990	......	i; sl s HNO_3
boride	FeB	66.66	gray cr	7.15[18]	......	......	i; s HNO_3
(II) bromide	$FeBr_2$	215.67	grn-yel, hex	4.636	691	934	117[20]
(III) bromide	$FeBr_3$	295.57	dk red-brn, rh	......	subl	......	s
(III) bromide-6-water	$FeBr_3 \cdot 6H_2O$	403.68	dk grn	......	27	d	v s
carbide, tri-	Fe_3C	179.55	gray, cub	7.694	1227	......	i; s a
(II) carbonate (siderite)	$FeCO_3$	115.85	gray, trig, 1.875 1.633	3.8	d	......	0.0072[18], s a
carbonyl, penta-	$Fe(CO)_5$	195.00	yel liq, viscous	1.49	-21	103	i; s al, bz, eth
(di-) carbonyl, nona-	$Fe_2(CO)_9$	363.79	yel-met, hex	2.09	d 100	......	i; d HNO_3
carbonyl, tetra- trimer	$[Fe(CO)_4]_3$	503.67	dk grn, mn	1.996[18]	d 140	......	i; s org solv
carbonyldinitrosyl, di-	$Fe(CO)_2(NO)_2$	171.88	dk red cr	1.56	18.5	d 50	s org solv

Name	Formula	Formula weight	Color, crystalline form, refractive index	Density	M.P., °C	B.P., °C	Solubility in cold water	Solubility in hot water
(II) chloride (lawrencite)	$FeCl_2$	126.75	grn-yel, hex, 1.567	3.16	677	1024	62.5^{20}	98^{20}
(II) chloride-4-water	$FeCl_2 \cdot 4H_2O$	198.81	bl-grn, mn, deliq	1.93	$-2H_2O$, 105			
(III) chloride (molysite)	$FeCl_3$	162.21	blk-brn, hex	2.898	304	332	74^{0}	
(III) chloride-6-water	$FeCl_3 \cdot 6H_2O$	270.30	brnsh-yel, deliq	1.82	37	280	91.6^{20}	
(III) ethylenediammonium sulfate-4-water	$FeC_2H_4(NH_3)_2(SO_4)_2 \cdot 4H_2O$	382.16	grn c=				s	
(II) ferrate(II), hexacyano-	$Fe_2[Fe(CN)_6]$	323.65	bl-wh, amorp	1.601	d 100		i	
(III) ferrate(II), hexacyano-	$Fe_4[Fe(CN)_6]_3$	859.25	dk bl cr	1.80	d		i; s HCl	
(II) fluoride	FeF_2	93.84	wh, tetr	4.09	1100	1837	sl s; s a	
(II) fluoride-4-water	$FeF_2 \cdot 4H_2O$	165.91	wh, rh	2.095	d		sl s; s a	
(III) fluoride	FeF_3	112.84	grn, rh	3.52	subl 927		0.091^{25}, s a	
(II) fumarate	$C_4H_2O_4Fe$	169.91	redsh-brn pwd	2.435^{25}	>280		0.14^{25}, >0.01 al	
(II) hydroxide	$Fe(OH)_2$	89.86	lt grn, hex	4.28			0.006; s a	
(III) hydroxide oxide (goethite)	$FeO(OH)$	88.85	brn-blk, rh, 2.260, 2.394, 2.400		$-H_2O$, 136		s HCl	
(III) iodate	$Fe(IO_3)_3$	580.55	yel-grn pwd	4.80^{20}	d 130		0.36^{20}	
(III) iodide	FeI_2	309.66	gray, hex, hygr	5.315	587	1093	s	
(II) iodide-4-water	$FeI_2 \cdot 4H_2O$	381.72	gray-blk cr, deliq	2.873	d 90		v s	
(II) nitrate-6-water	$Fe(NO_3)_2 \cdot 6H_2O$	287.95	grn, rh		60.5		134^{10}	
(III) nitrate-9-water	$Fe(NO_3)_3 \cdot 9H_2O$	404.02	lt vlt, mn, deliq	1.684	47		138^{20}	
(di-) nitride	Fe_2N	125.70	gray	6.35	d 200		i; s HCl	
(II) oxalate-2-water	$FeC_2O_4 \cdot 2H_2O$	179.90	lt yel, rh	2.28	d 190		0.044^{18}, s a	
(III) oxalate-5-water	$Fe_2(C_2O_4)_3 \cdot 5H_2O$	465.83	yel		d 100		v s	
(II) oxide (wuestite)	FeO	71.85	blk, cub, 2.32	5.7	1377	3414	i; s a	
(III) oxide (hematite)	Fe_2O_3	159.69	red-brn, trig, 3.01, 2.94 (Li)	5.24	1565		i; s HCl	
(II,III) oxide (magnetite)	Fe_3O_4	231.54	blk, cub, 2.42	5.1	1597		i; s a	
(II) perchlorate-6-water	$Fe(ClO_4)_2 \cdot 6H_2O$	362.84	grn, 1.493, 1.478		d 100		299^{25}	
(II) phosphate-8-water	$Fe_3(PO_4)_2 \cdot 8H_2O$	501.61	wh-ol, mn, 1.579, 1.603, 1.633	2.58			i; s a	
(III) phosphate-2-water	$FePO_4 \cdot 2H_2O$	186.85	pink, mn	2.87	$-2H_2O$, 140		i; s HCl	
phosphide	FeP	86.82	gray, rh	6.07	1370		i; s aq reg	
(di-) phosphide	Fe_2P	142.67	bl-gray	6.77			i	
(di-) phosphide, di-	Fe_2P_2	173.64	gray, rh	4.95			i	
(tri-) phosphide	Fe_3P	198.51	gray, tetr	7.11	1100			
selenide	$FeSe$	134.81	blk-met	6.78	d		i; s a	

Table 4-1 (*Continued*)

PHYSICAL CONSTANTS OF INORGANIC COMPOUNDS

Name	Formula	Formula weight	Color, crystalline form, refractive index	Density	Melting point, °C	Boiling point, °C	Solubility in 100 parts
Iron							
selenide, di-	$FeSe_2$	213.77	rh	....	349	......	......
(II) silicate(2−) (gruenerite)	$FeSiO_3$	131.93	gray-grn, rh, 1.672, 1.697, 1.717	3.5	1140	......	......
(II) silicate(4−) (fayalite)	Fe_2SiO_4	203.78	col, rh	4.34	1220	......	i; d HCl
(II) silicate-6-water, hexa-fluoro-	$Fe[SiF_6] \cdot 6H_2O$	306.01	col, trig, 1.361, 1.385	1.961	......	......	77[25]
silicide	$FeSi$	83.93	yel-gray, oct	6.1	1420	......	i; i a
(II) sulfate-1-water (szomolnokite)	$FeSO_4 \cdot H_2O$	169.96	graysh-wh, mn	2.970	671 d	......	sl s
(II) sulfate-5-water (siderotil)	$FeSO_4 \cdot 5H_2O$	242.02	wh, tric, 1.526, 1.536, 1.542	2.2	−5H$_2$O, 300	......	42[20]
(II) sulfate-7-water	$FeSO_4 \cdot 7H_2O$	278.04	yel, rh, 1.814	1.89	......	......	48[20]
(III) sulfate	$Fe_2(SO_4)_3$	399.88	yel, rh, deliq, 1.552	3.097[18]	d 1178	......	sl s
(III) sulfate-9-water (coquimbite)	$Fe_2(SO_4)_3 \cdot 9H_2O$	562.01	yel, rh, deliq, 1.552, 1.558	2.1	d 175	......	440
sulfide	FeS	87.92	blk-brn, hex	4.82	1195	d	0.0006[18], s a
sulfide, di- (pyrite)	FeS_2	119.98	yel, cub	5.0	1171	d	i; d a
(marcasite)			yel, rh	4.87	1171		i; d a
(di-) sulfide, tri-	Fe_2S_3	207.87	yel-grn	4.246	d	......	i; d a
(II) sulfite-3-water	$FeSO_3 \cdot 3H_2O$	189.96	grnsh-wh cr	....	d 250	......	v sl s
tantalate(V)(1−), bis- (tapiolite)	$Fe(TaO_3)_2$	513.73	lt brn, tetr, 2.27, 2.42 (Li)	7.33	......	......	......
(II) tartrate	$FeC_4H_4O_6$	203.92	......	....	......	......	......
telluride, di-	$FeTe_2$	311.07	rh	....	660	......	v s
(II) thiocyanate-3-water	$Fe(SCN)_2 \cdot 3H_2O$	226.06	grn, rh	....	d	......	v s
(III) thiocyanate	$Fe(SCN)_3$	230.09	red-blk, cub, deliq	....	......	......	v s
(II) thiosulfate-5-water	$FeS_2O_3 \cdot 5H_2O$	258.05	grn cr, deliq	....	......	......	v s

(II) titanate(IV)(2−) (il-menite)	$FeTiO_3$	151.75			1470		i; s a
(II) tungstate(VI)(2−) (fer-berite)	$FeWO_4$	303.69	tetr, 2.40 (Li)	6.64			
(II) vanadate(V)(2−)	$Fe(VO_3)_2$	352.67	graysh-brm pwd				
Krypton	Kr	83.80	col gas, 1.000427 col liq	3.736 (g/L) 2.413^{bp}	−157.2	−153.4	$5.94\ ml^{20}$
fluoride, di-	KrF_2	122.80	col tetr		subl −60		s anhyd HF, SbF_5
Lanthanum	La	138.91	whsh-met, hex	6.174	920	3470	i; s HCl
acetate-1.5-water	$La(C_2H_3O_2)_3 \cdot 1.5H_2O$	343.07	col cr				16.9^{20}
boride, hexa-	LaB_6	203.78	bl-vlt, cub	4.61	2210	d	i
bromate-9-water	$La(BrO_3)_3 \cdot 9H_2O$	684.77	col, hex		37.5	d 100	195^{20}
bromide	$LaBr_3$	378.62	col, hex	5.07	789	1480	s
carbide, di-	LaC_2	162.93	yel	5.35	1800–2100		d; s H_2SO_4
carbonate-8-water	$La_2(CO_3)_3 \cdot 8H_2O$	601.97	wh, rh	2.65			s a
chloride	$LaCl_3$	245.27	wh, hex, deliq	3.818	855	1812	v s
chloride-7-water	$LaCl_3 \cdot 7H_2O$	371.38	wh, tric, hygr		−H_2O, 91		v s
fluoride	LaF_3	195.91	wh, hex	4.49	1493	2327	
hydride, di-	LaH_2	140.92	fcc	5.14			
hydride, tri-	LaH_3	141.93	bl-blk amorp	5.26			
hydroxide	$La(OH)_3$	189.93	wh, hex		−H_2O, 260	>380 to oxide	i; s a
iodate	$La(IO_3)_3$	663.62	col cr	5.63			0.068^{25}
iodide, di-	$[La^{3+}(e^-)(I^-)_2]$	392.71	blk		820		s
iodide, tri-	LaI_3	519.62	gray, rh, hygr	4.77^{16}	766	1405	0.0018^{25}; s HCl
molybdate(VI)(2−)	$La_2(MoO_4)_3$	757.63	wh, tetr		1181		181^{20}
nitrate-6-water	$La(NO_3)_3 \cdot 6H_2O$	433.02	wh, tric, deliq		40	d 126	0.00015^{25}; s a
oxalate-10-water	$La_2(C_2O_4)_3 \cdot 10H_2O$	722.04	wh		d		s a
oxide	La_2O_3	325.82	wh, hcp	6.48	2320	4200	
selenide	LaSe	217.87	cub		ca 1977		
(di-) selenide, tri-	La_2Se_3	514.70	brick-red	6.32	1627		
(tri-) selenide, tetra-	La_3Se_4	732.57	cub		1847		
sulfate	$La_2(SO_4)_3$	566.00	wh, hygr	3.60	d 500		2.33^{20}
sulfate-9-water	$La_2(SO_4)_3 \cdot 9H_2O$	728.14	col, hex, 1.564	2.821	−H_2O, 400		2.92^{20}
sulfide	LaS	170.97	gold, cub	5.75	2327		

Table 4-1 (Continued)
PHYSICAL CONSTANTS OF INORGANIC COMPOUNDS

Name	Formula	Formula weight	Color, crystalline form, refractive index	Density	Melting point, °C	Boiling point, °C	Solubility in 100 parts
Lanthanum							
(di-) sulfide, tri-	La_2S_3	374.01	yel-wh, cub	4.99	2127	....	d; s a
telluride	LaTe	266.51	cub	....	1777	....	....
(di-) telluride, tri-	La_2Te_3	622.37	cub	....	1487	1753	i; s HNO_3
Lead	Pb	207.21	silv-blsh met, fcc	11.34 Ra-Pb: 11.288 U-Pb: 11.296	327.50	....	....
(II) acetate-3-water	$Pb(C_2H_3O_2)_2 \cdot 3H_2O$	379.33	wh, mn	2.55	d 200	....	45.6[15]
(IV) acetate	$Pb(C_2H_3O_2)_4$	443.37	col, mn	2.228[17]	175	....	d; s chl
antimonate(V)(3−) (moni-molite)	$Pb_3(SbO_4)_2$	993.07	or-yel pwd	6.58[20]	....	....	i; v sl s HCl
arsenate(V)(1−)	$Pb(AsO_3)_2$	453.03	col, hex	6.42[15]	....	....	d; s HNO_3
arsenate(V)(3−)	$Pb_3(AsO_4)_2$	899.41	wh cr	7.80	1040	....	i; s HNO_3
arsenate(III)(1−)	$Pb(AsO_2)_2$	421.03	wh pwd	5.85	....	....	i; s HNO_3
azide	$Pb(N_3)_2$	291.23	col pwd	....	expl 350	....	0.023[18], s acetic a
borate(1−)-1-water	$Pb(BO_2)_2 \cdot H_2O$	310.82	wh cr	5.598 (anhyd)	$-H_2O$, 160	....	i; s a
borate, tetrafluoro-	$Pb[BF_4]_2$	380.80	col cr	....	....	....	d
bromate-1-water	$Pb(BrO_3)_2 \cdot H_2O$	481.02	col, mn	5.53	d 180	....	1.38[20]
bromide, di-	$PbBr_2$	367.01	wh, rh	6.67	371	912	0.84[20], s a
bromide, tetra-	$PbBr_4$	526.82	....	....	....	....	....
carbonate (cerussite)	$PbCO_3$	267.20	col, rh, 1.804, 2.076, 2.078	6.6	d 340	....	i; s a, alk
carbonate-hydroxide (2/1) (hydrocerussite)	$2PbCO_3 \cdot Pb(OH)_2$	775.60	wh, hex	6.14	d 400	....	i; s HNO_3
chlorate	$Pb(ClO_3)_2$	374.09	wh, mn, deliq	3.89	d 230	....	140[18]
chlorate-1-water	$Pb(ClO_3)_2 \cdot H_2O$	392.11	wh, mn, deliq	4.037	d 110	....	151[18]

Name	Formula	Mol wt	Crystal form, color, refractive index	Density	m.p., °C	b.p., °C	Solubility	
chloride (cotunite)	PbCl$_2$	278.10	wh, rh, 2.199, 2.217, 2.260	5.85	501	953	0.99^{20}	d; s HCl
(IV) chloride	PbCl$_4$	349.01	yel oily fum liq	3.18^0	−15	expl 105		d; s HCl
chloride fluoride (matlockite)	PbClF	261.66	wh, tetr, 2.145, 2.006	7.05	601	......	0.0325	
chloride-hydroxide (1/1) (laurionite)	PbCl$_2$·Pb(OH)$_2$	519.29	rh	6.24	d 142	......	......	
chloride-oxide (1/2) (mendipite)	PbCl$_2$·2PbO	724.47	yel, rh, 2.24, 2.27, 2.31	7.08	693	......	i; s alk	
chloride-oxide-water (1/1/1) (paralaurionite)	PbCl$_2$·PbO·H$_2$O	519.29	wh, mn, 2.146	6.05^{15}	d 150	......	......	
chloride-oxide-water (2/1/1) (fiedlerite)	2PbCl$_2$·PbO·H$_2$O	797.40	mn, 1.816, 2.102, 2.026	5.88^{15}	d 150	......	0.01^{18}; s HNO$_3$	
chlorite	Pb(ClO$_2$)$_2$	342.09	yel, mn	....	expl 126	......	0.095^{20}; s KOH	
chromate(VI)(2−) (crocoite)	PbCrO$_4$	323.18	yel, mn, 2.31, 2.37 (Li)	6.12^{15}	844	d	i; s a	
cyanide	Pb(CN)$_2$	259.23	yel-wh pwd	....	......	......	sl s; s KCN	
diarsenate(V)(4−)	Pb$_2$As$_2$O$_7$	676.22	col, rh	6.85^{15}	802	......	i; s HCl	
dichromate	PbCr$_2$O$_7$	423.18	red cr	....	......	......	d; s a, alk	
diphosphate(V)(4−)	Pb$_2$P$_2$O$_7$	588.32	wh, rh	5.8^{20}	824	......	i; s a, alk	
dithionate-4-water	PbS$_2$O$_6$·4H$_2$O	439.38	col, trig, 1.635, 1.653	3.22	d	......	115^{20}	
ethyl, tetra-	Pb(C$_2$H$_5$)$_4$	323.45	col liq, 1.5198^{20}	1.653^{20}	......	ca 200	i; s bz, hydrocarbons	
(II) fluoride	PbF$_2$	245.20	col, rh	8.24	830	1303	0.064^{20}	
(IV) fluoride	PbF$_4$	283.21	wh, tetr	6.7	600	......	hyd	
formate	Pb(CHO$_2$)$_2$	297.23	wh, rh, 1.789, 1.852	4.63	d 190	......	1.6^{20}	
hydrogen arsenate (schultenite)	PbHAsO$_4$	347.12	col, mn	5.79	d 280	......	i; s HNO$_3$	
hydrogen arsenate, di-	Pb(H$_2$AsO$_4$)$_2$	489.06	col, tric, 1.74, 1.82	4.46^{15}	d 140	......	i; s HNO$_3$	
hydrogen phosphate	PbHPO$_4$	303.17	wh, rh	5.661^{15}	d	......	i; s HNO$_3$	
hydroxide	Pb(OH)$_2$	241.20	wh, amorp	....	d 145	......	0.0155^{20}; s a, alk	
hydroxide nitrate	Pb(NO$_3$)(OH)	286.20	wh, rh	5.93	d 180	......	19.4^{19}	
iodate	Pb(IO$_3$)$_2$	557.00	wh	6.155^{20}	d 300	......	0.003^{25}	
iodide	PbI$_2$	461.05	yel, hex	6.16	410	832	0.063^{20}; s alk, KI	
methyl, tetra-	Pb(CH$_3$)$_4$	267.35	....	....	......	......	....	
molybdate(VI)(2−) (wulfenite)	PbMoO$_4$	367.16	yelsh-wh, tetr	6.92	1065	......	i; s a, alk	

Table 4-1 (*Continued*)
PHYSICAL CONSTANTS OF INORGANIC COMPOUNDS

Name	Formula	Formula weight	Color, crystalline form, refractive index	Density	Melting point, °C	Boiling point, °C	Solubility in 100 parts
Lead							
nitrate	$Pb(NO_3)_2$	331.23	col, cub, 1.782	4.53^{20}	d 200	......	56^{20}, 1.3 MeOH
oleate	$Pb(C_{18}H_{33}O_2)_2$	770.12	granular wax like	......	......	......	i; s al, bz, eth
oxalate	PbC_2O_4	295.23	wh pwd	5.28	d 300	......	0.00015^{18}; s a, alk
oxide (litharge)	PbO	223.21	yel, tetr	9.53	886	1516	0.0017^{20}; s HNO_3
(massicot)			yel, rh, 2.51, 2.61	8.0	886	1472	0.0023; s HNO_3
(IV) oxide (plattnerite)	PbO_2	239.21	blk, tetr	9.375	d 752	......	i; s HCl
(di-) oxide, tri-	Pb_2O_3	462.42	or-yel pwd, amorp	......	d 370	......	i; d a
(II, IV) oxide (red lead)	Pb_3O_4	685.63	red pwd	9.1	d 530	......	i; s HCl
perchlorate	$Pb(ClO_4)_2$	406.09	wh	3.86^{20}	273	......	440^{25}
perchlorate-3-water	$Pb(ClO_4)_2 \cdot 3H_2O$	460.14	wh, rh	2.6	d 88	......	i; s HNO_3, alk
phosphate	$Pb_3(PO_4)_2$	811.59	wh, hex, 1.970, 1.936	6.9	1014	......	0.88^{15}
picrate-2-water	$Pb(C_6H_2N_3O_7)_2 \cdot 2H_2O$	681.45	yel nd	2.831^{20}	$-H_2O$, 130	expl	i; s HNO_3
selenide (clausthalite)	PbSe	286.15	gray, cub	8.15	1065	......	i; d a
(II) silicate(2−) (alamosite)	$PbSiO_3$	283.27	wh, mn	6.49	764	......	......
(II) silicate(4−)	Pb_2SiO_4	506.49	......	......	743	......	......
(IV) silicate(4−)	$PbSiO_4$	299.29	......	......	1538	......	......
stearate	$Pb(C_{18}H_{35}O_2)_2$	774.15	wh pwd	......	ca 125	......	0.05^{35}; s hot al
sulfate (anglesite)	$PbSO_4$	303.28	wh, mn, 1.877, 1.822, 1.894	6.2	1090	......	0.004
sulfide (galena)	PbS	239.28	blk, cub, 3.921	7.5	1113	......	0.0126^{20}; s a
tartrate	$PbC_4H_4O_6$	355.26	wh cr pwd	2.53^{19}	......	......	0.0025; s HNO_3, alk
thiosulfate	PbS_2O_3	319.32	wh cr	5.18	d	......	0.03
titanate(IV)(2−)	$PbTiO_3$	303.09	yel, rh-pyr	7.52	1170	......	i
telluride (altaite)	PbTe	334.82	wh, cub	8.16^{20}	904	......	0.44^{18}; s HNO_3
thiocyanate	$Pb(SCN)_2$	323.35	wh, mn	3.82	d 190	......	i; s KOH
tungstate(VI)(2−) (stolzite) (raspite)	$PbWO_4$	455.04	col, tetr, 2.269, 2.182, col, mn, 2.27, 2.27, 2.30	8.23	1125	......	0.03; d a

Lithium

Name	Formula	Formula weight	Crystalline form, color, refractive index	Density	mp, °C	bp, °C	Solubility
Lithium	Li	6.94	silv-met, bcc	0.535^{20}	180.6	1340	d to LiOH
acetate-2-water	$LiC_2H_3O_2 \cdot 2H_2O$	102.01	wh, rh		70	d	63^{20}
acetylsalicylate	$LiC_9H_7O_4$	186.08	wh				100, 25 al
aluminate(1−)	$LiAlO_2$	65.92	wh pwd	2.554	1610		
aluminate(3−), hexafluoro-	$Li_3[AlF_6]$	161.79	wh		785		
aluminate(1−), tetrahydro-	$Li[AlH_4]$	37.95	wh cr, mn	0.917	d 137	d 430	d; 30 eth, 13 THF
amide	$LiNH_2$	22.96	col, tetr	1.178^{18}	374		d; i bz, eth
antimonide	Li_3Sb	142.57		3.2^{17}	950		
arsenate	Li_3AsO_4	159.74	wh cr pwd	3.07			
benzoate	$LiC_7H_5O_2$	128.05					33; 7.7 al
beryllate(1−), trifluoro-	$Li[BeF_3]$	72.95			377		
beryllate(2−), tetrafluoro-	$Li_2[BeF_4]$	98.88	wh, rh		459.1		
borate(1−), tetrahydro-	$LiBH_4$	21.79	wh, tric	0.666	268	d 380	d; s eth, THF
borate(1−)	$LiBO_2$	49.75	wh cr pwd		844	1719	2.7^{20}
borate(2−), tetra-	$Li_2B_4O_7$	169.12	wh, rh		917		sl s
bromate	$LiBrO_3$	134.85		3.62			179^{20}
bromide	$LiBr$	86.84	wh, fcc, 1.784	3.464	550	1289	164; s al, eth
(di) carbide, di-	Li_2C_2	37.90	wh cr	1.65^{18}		d	d
carbonate	Li_2CO_3	73.89	col, mn, 1.567	2.11^{0}	720		1.3^{20}; i al; s a
chlorate	$LiClO_3$	90.39	col	2.631^{20}	129	d 270	372^{20}
chloride	$LiCl$	42.40	wh, cub, 1.662	2.068	610	1383	77^{20}; s al, acet
chromate(VI)(2−)-2-water	$Li_2CrO_4 \cdot 2H_2O$	165.90	yel, rh, deliq		$-2H_2O$, 130		142^{18}
citrate-4-water	$Li_3C_6H_5O_7 \cdot 4H_2O$	281.98	wh cr		$-4H_2O$, 105		61.2^{15}
dichromate-2-water	$Li_2Cr_2O_7 \cdot 2H_2O$	265.90	yel-red, hygr	2.34^{30}	$-2H_2O$, 110		151^{30}
fluoride	LiF	25.94	wh, cub, 1.3915	2.640^{20}	846	1717	0.13^{25}; s a
formate-1-water	$LiCHO_2 \cdot H_2O$	69.97	col, rh	1.46	$-H_2O$, 94		39.3^{20}
hydride	LiH	7.95	col, bcc	0.780	688.7	d 950	d; no solv known
hydride[^{2}H]	$Li{}^2H$	8.96	col	0.881	686		
hydrogen oxalate-1-water	$LiHC_2O_4 \cdot H_2O$	113.98	col		d		8^{17}
hydrogen phosphate, di-	LiH_2PO_4	103.93	col	2.461	100		126^{0}
hydrogen sulfate	$LiHSO_4$	104.01	col cr	2.123^{13}	171		
hydroxide	$LiOH$	23.95	wh cr	2.54	471.2	1626	12.4^{20}
hydroxide-1-water	$LiOH \cdot H_2O$	41.96	col, mn	1.51^{20}	d		11^{20}
iodate	$LiIO_3$	181.84	col, deliq	4.502^{32}	450		66; i al
iodide	LiI	133.84	col, cub, 1.955	4.061	467	1178	165^{20}; v s al
iodide-3-water	$LiI \cdot 3H_2O$	187.89	wh, mn, deliq	3.5	73	$-3H_2O$, 300	200; 200 al

Table 4-1 (Continued)
PHYSICAL CONSTANTS OF INORGANIC COMPOUNDS

Name	Formula	Formula weight	Color, crystalline form, refractive index	Density	Melting point, °C	Boiling point, °C	Solubility in 100 parts
Lithium							
nitrate	$LiNO_3$	68.94	col, trig, 1.735	2.38	261	....	70[20]; s al
(tri) nitride	Li_3N	34.82	blk	....	849	....	d
nitrite-1-water	$LiNO_2 \cdot H_2O$	70.96	wh nd	1.615[0]	100	....	97[20]
oxalate	$Li_2C_2O_4$	101.88	wh cr	2.121[17]	d	....	8[20]
oxide	Li_2O	29.88	col, 1.644	2.013[25]	1570	2563	forms LiOH
(di) oxide, di-	Li_2O_2	45.88	wh	....	d 195	....	....
perchlorate	$LiClO_4$	106.40	wh cr, deliq	2.43[25]	236	d 400	56[20]
perchlorate-3-water	$LiClO_4 \cdot 3H_2O$	160.44	col, hex	1.84	95	$-3H_2O$, 130	v s
permanganate	$LiMnO_4$	125.87	purp cr	2.06	d 190	....	71.4[16]
phosphate	Li_3PO_4	115.76	wh, rh	2.53[18]	837	....	0.034[18]
phosphate-12-water	$Li_3PO_4 \cdot 12H_2O$	331.97	wh, trig	1.645	100	....	....
platinate(IV)-6-water, hexachloro-	$Li_2[PtCl_6] \cdot 6H_2O$	529.78	or-red, hex	....	$-6H_2O$, 180	....	....
rubidium tetracyanoplatinate(II)	$LiRb[Pt(CN)_4]$	391.58	grn-yel cr	....	....	....	s
salicylate	$LiC_7H_5O_3$	144.16	col	....	d	....	128
selenate(IV)-1-water	$Li_2SeO_3 \cdot H_2O$	151.92	cr, hygr	....	....	....	sl s
selenate(VI)-1-water	$Li_2SeO_4 \cdot H_2O$	167.92	col, mn	2.565	....	....	v s
selenide	Li_2Se	92.84	cub	2.91	1102	....	57.7
silicate(2−)	Li_2SiO_3	89.97	col, rh, 1.548	2.52[25]	1201	....	s; d a
silicate(4−)	Li_4SiO_4	119.84	col, rh, 1.60	2.28	1256	....	....
silicate-2-water, hexafluoro-(hexa)	$Li_2[SiF_6] \cdot 2H_2O$	191.99	col, mn	2.3[12]	$-2H_2O$, 100	d	73[17]
silicide, di-	Li_6Si_2	97.81	bl cr	1.12	d 500	....	ign in air
sulfate	Li_2SO_4	109.88	col, mn, 1.465	2.22	860	....	34.5[20]
sulfate-1-water	$Li_2SO_4 \cdot H_2O$	127.95	col, mn, 1.477	2.06	$-H_2O$, 130	....	38; i al
sulfide	Li_2S	45.94	wh, cub	1.63	950	....	....
(di) sulfide, di-	Li_2S_2	78.00	....	....	370	....	....

Name	Formula	Formula wt.	Color, crystalline form, refractive index	Density	M.p., °C	B.p., °C	Solubility
sulfite	Li₂SO₃	93.94			455 d		s
tartrate-1-water	Li₂C₄H₄O₆·H₂O	179.97	wh pwd				s
telluride	Li₂Te	141.48	wh cr pwd	3.24			v s
thiocyanate	LiSCN	65.22	cub				
titanate(IV)(2−)	Li₂TiO₃	109.78	wh cr, deliq		1547		
Lutetium	Lu	174.97	met, hcp	9.842	1652	3330	i; s a
bromide	LuBr₃	414.70	col, rhbd	5.17	1025	1410	s
chloride	LuCl₃	281.33	wh, mn	3.98	925	1480	s
fluoride	LuF₃	231.97	wh, hex	8.332	1182	2230	i
iodide	LuI₃	555.68	brn, hex		1050	1210	s
nitrate-5-water	Lu(NO₃)₃·5H₂O	451.06	wh				
oxalate-6-water	Lu₂(C₂O₄)₃·6H₂O	722.09	wh cr		−H₂O, 50		i; s a
oxide	Lu₂O₃	397.94	wh, bcc	9.42			i; s a
sulfate-8-water	Lu₂(SO₄)₃·8H₂O	782.25	col cr				57.9[20]
sulfide	Lu₂S₃	446.12	cub				
Magnesium	Mg	24.31	silv-met, hex	1.74[20]	650	1105	i; s a
acetate	Mg(C₂H₃O₂)₂	142.40	wh	1.42	323 d		53.4[20]
acetate-4-water	Mg(C₂H₃O₂)₂·4H₂O	214.46	wh, mn, 1.491	1.454	80		80.6[20]
acetylsalicylate	Mg(C₉H₇O₄)₂	382.61	wh				s; s al
aluminate(2−), di- (spinel)	MgAl₂O₄	142.27	col, cub, 1.723	3.6	2135		i; v sl s HCl
amide	Mg(NH₂)₂	56.37	wh pwd	1.39[25]	d; ign in air		d viol
antimonide	Mg₃Sb₂	316.44	met, hex	4.088	961		i
arsenate-8-water (hoernesite)	Mg₃(AsO₄)₂·8H₂O	494.90	wh, mn	2.60	800		d; d a
arsenide	Mg₃As₂	222.78	brn-red, cub	3.148	ca 200		5; s al
benzoate-3-water	Mg(C₇H₅O₂)₂·3H₂O	374.67	wh pwd				
bismuthide	Mg₃Bi₂	490.90	met, hex	5.945	823		sl s; s a
borate(1−)-8-water	Mg(BO₂)₂·8H₂O	254.06	col, tetr, 1.565, 1.575	2.30			i; s a
borate(3−)	Mg₃(BO₃)₂	190.55	col, rh, 1.6527, 1.6537, 1.6548	2.99[21]			
borate(4−)-1-water, di- (ascharite)	Mg₂B₂O₄·H₂O	168.26	col, rh, 1.54	2.6			
boride, di-	MgB₂	45.93			1047 d		
bromate-6-water	Mg(BrO₃)₂·6H₂O	388.22	col, cub, 1.5136	2.29	−6H₂O, 200	d	58[18]
bromide	MgBr₂	184.13	wh, hex, deliq	3.72	711	1158	101[20]
bromide-6-water	MgBr₂·6H₂O	292.22	col, hex, deliq	2.00	165 d		160[20]; s al
carbonate (magnesite)	MgCO₃	84.32	wh, trig, 1.717, 1.515	2.958	d 402		0.01; s a

Table 4-1 (*Continued*)
PHYSICAL CONSTANTS OF INORGANIC COMPOUNDS

Name	Formula	Formula weight	Color, crystalline form, refractive index	Density	Melting point, °C	Boiling point, °C	Solubility in 100 parts
Magnesium							
carbonate-3-water (nesquehonite)	$MgCO_3 \cdot 3H_2O$	138.37	col, rh, 1.501	1.852	$-H_2O$, 100	....	0.16; s a
carbonate-5-water (lansfordite)	$MgCO_3 \cdot 5H_2O$	174.40	wh, mn, 1.456, 1.476, 1.502	1.73	d in air	....	0.18^8; s HCl
carbonate–hydroxide–water (1/1/3) (artinite)	$MgCO_3 \cdot Mg(OH)_2 \cdot 3H_2O$	196.69	wh, rh, 1.489, 1.534, 1.557	2.02^{20}	....	....	0.04; s a
carbonate–hydroxide–water (3/1/3) (hydromagnesite)	$3MgCO_3 \cdot Mg(OH)_2 \cdot 3H_2O$	365.35	wh, rh, 1.527, 1.530, 1.540	2.16	d	....	....
chlorate-6-water	$Mg(ClO_3)_2 \cdot 6H_2O$	299.31	wh, rh, deliq	1.80	35	d 120	111^{20}
chloride (chloromagnesite)	$MgCl_2$	95.23	col, hex, 1.675, 1.59	2.41	714	1437	54.6^{20}
chloride-6-water (bischofite)	$MgCl_2 \cdot 6H_2O$	203.31	wh, mn, deliq, 1.495, 1.507, 1.528	2.569	118 d	....	116^{20}
chromate(VI)-7-water	$MgCrO_4 \cdot 7H_2O$	266.41	yel, rh, 1.5500, 1.521, 1.568	1.695	$-3H_2O$, 120	....	137^{20}
chromate(III)(2−), di-	$MgCr_2O_4$	192.30	dk-grn, cub	4.6^{20}	....	....	i; s conc H_2SO_4
citrate-5-water	$MgHC_6H_5O_7 \cdot 5H_2O$	304.50	wh		....	....	20
diphosphate-3-water	$Mg_2P_2O_7 \cdot 3H_2O$	276.62	col, mn, 1.602, 1.604, 1.615	2.56	1383	....	i; s a
ferrate(VI)(2−)	$MgFeO_4$	200.00	blk, oct, 2.35	4.5	1750	....	i; s HCl
fluoride (sellaite)	MgF_2	62.31	col, tetr, 1.378, 1.390, 314	3.148	1263	2226	0.013^{25}, s HNO_3
formate-2-water	$Mg(CHO_2)_2 \cdot 2H_2O$	150.38	col, rh	....	$-2H_2O$, 100	....	18.9^{20}; i al
germanide	Mg_2Ge	121.21	cr	....	1115	....	....
hydride	MgH_2	26.34	wh, tetr	1.45	d 287 (vac)	ign in air	d viol; d al
hydrogen arsenate-7-water	$MgHAsO_4 \cdot 7H_2O$	290.35	wh, mn	1.943^{15}	$-5H_2O$, 100	....	d
hydrogen phosphate-3-water (newberylite)	$MgHPO_4 \cdot 3H_2O$	174.34	wh, rh, 1.514, 1.518, 1.533	2.123^{15}	$-3H_2O$, 205	d 550	sl s; s a

Name	Formula	Formula weight	Color, crystalline form, refractive index	Density	Melting point, °C	Boiling point, °C	Solubility
hydrogen phosphate-7-water	$MgHPO_4 \cdot 7H_2O$	246.40	wh, mn	1.729^{15}	$-4H_2O$, 100	....	0.3; s a
hydroxide (brucite)	$Mg(OH)_2$	58.33	col, hex, 1.559, 1.580	2.36	268 d	....	i; s a
iodate-4-water	$Mg(IO_3)_2 \cdot 4H_2O$	446.18	wh, mn	3.3^{13}	$-4H_2O$, 210	d	10.2^{20}
iodide	MgI_2	278.12	wh, hex, deliq	4.43	d 700	....	140^{20}
iodide-8-water	$MgI_2 \cdot 8H_2O$	422.24	wh pwd, deliq	....	d 41	....	81^{20}
lactate-3-water	$MgC_6H_{10}O_6 \cdot 3H_2O$	256.52	wh cr	....	....	....	4; sl s al
mandelate	$MgC_{16}H_{14}O_6$	326.61	wh pwd	....	....	....	0.004^{100}; i al
molybdate(VI)(2−)	$MgMoO_4$	184.25	wh, rh	2.208	....	....	13.7
nitrate-6-water (nitromagnesite)	$Mg(NO_3)_2 \cdot 6H_2O$	256.41	col, mn, deliq	1.464	95	d 129	120^{20}; v s al
nitride	Mg_3N_2	100.95	grn-yel pwd	2.712	d 800	subl 700 (vac)	d; s a
nitrite-3-water	$Mg(NO_2)_2 \cdot 3H_2O$	170.37	wh pr, hygr	....	d 100	....	s
oleate	$MgC_{18}H_{33}O_2)_2$	293.61	yelsh pwd	....	....	....	s; s al, eth
oxalate-2-water	$MgC_2O_4 \cdot 2H_2O$	148.36	wh pwd	2.45	d 150	....	0.042^{36}; s a; i al
oxide (magnesia, periclase)	MgO	40.32	col, cub, 1.736	3.58	2825	3260	i; s a
perchlorate	$Mg(ClO_4)_2$	223.23	wh, hygr	2.21^{20}	d 251	....	49.6
perchlorate-6-water	$Mg(ClO_4)_2 \cdot 6H_2O$	331.31	wh, rh, deliq, 1.482, 1.458	1.970	185	....	v s
permanganate	$Mg(MnO_4)_2$	262.19	blsh-blk cr, deliq	....	....	....	v s
peroxide	MgO_2	56.31	wh pwd	....	d 100	....	i d; s a d
peroxoborate-7-water	$Mg(BO_3)_2 \cdot 7H_2O$	268.10	wh pwd	....	....	....	sl s d; s a
phosphate	$Mg_3(PO_4)_2$	262.88	col, rh	....	1348	....	i
phosphate-5-water	$Mg_3(PO_4)_2 \cdot 5H_2O$	352.98	col, mn	1.64^{15}	$-5H_2O$, 400	....	0.02; s a
phosphate-8-water (bobierrite)	$Mg_3(PO_4)_2 \cdot 8H_2O$	407.00	wh, mn, 1.510, 1.520, 1.543	2.195^{15}	$-8H_2O$, 400	....	i; s HNO_3
phosphide	Mg_3P_2	134.88	yel-grn, cub	2.055	....	....	d; s a
phosphinate-6-water	$Mg(PH_2O_2)_2 \cdot 6H_2O$	262.38	wh, tetr	1.59^{12}	$-6H_2O$, 180	....	20
platinate(IV)-6-water, hexachloro-	$Mg[PtCl_6] \cdot 6H_2O$	540.21	yel, trig	2.437	$-6H_2O$, 180	d	....
salicylate-4-water	$Mg(C_7H_5O_3)_2 \cdot 4H_2O$	370.54	wh pwd, effl	....	....	....	7.7; s al
selenate(IV)-6-water	$MgSeO_3 \cdot 6H_2O$	259.41	rh	....	....	....	i; s a
selenate(VI)-6-water	$MgSeO_4 \cdot 6H_2O$	275.36	col, mn, 1.468, 1.489, 1.491	1.928	....	....	63^{20}
selenide	$MgSe$	103.28	lt brn, cub, 2.44	4.21	d in air	....	d; d a
silicate(2−) (clinoenstatite, amphibole, enstatite)	$MgSiO_3$	100.40	wh, mn, 1.651, 1.660	3.192^{25}	1557	....	i

Table 4-1 (*Continued*)
PHYSICAL CONSTANTS OF INORGANIC COMPOUNDS

Name	Formula	Formula weight	Color, crystalline form, refractive index	Density	Melting point, °C	Boiling point, °C	Solubility in 100 parts
Magnesium							
silicate(4−) (forsterite)	Mg_2SiO_4	140.71	wh, rh, 1.65, 1.66, 1.67	3.21	1898	....	i; d hot HCl
silicate-6-water, hexafluoro-	$Mg[SiF_6] \cdot 6H_2O$	274.48	wh, trig	1.788	d 120	....	50.8[20]
silicate-2-water, di- (serpentine, chrysotile)	$MgSi_2O_5 \cdot 2H_2O$	196.53	fibrous; "asbestos" "florisil"		....	....	i
silicon oxide-water- (3/4/1) (talc)	$3MgO \cdot 4SiO_2 \cdot H_2O$	379.29	grnsh wh, mn 1.539, 1.589, 1.589	2.7–2.8	....	....	i
sulfite-6-water	$MgSO_3 \cdot 6H_2O$	212.47	wh, rh, 1.511	1.725	−6H₂O, 200	d	66[25]
d-tartrate-5-water	$MgC_4H_4O_6 \cdot 5H_2O$	262.46	wh, rh	1.67	−5H₂O, 200	....	1.61[20]; s a
telluride	$MgTe$	151.91	wh, hex	3.86	....	....	d; d a
thiosulfate-6-water	$MgS_2O_3 \cdot 6H_2O$	244.53	col, rh	1.818	−3H₂O, 170	d	50; i al
titanate(IV)(2−)	$MgTiO_3$	120.21	....		1680	....	....
titanate(IV)(4−)	Mg_2TiO_4	160.52	....		1740	....	....
tungstate(VI)(2−)	$MgWO_4$	272.18	col, mn	5.66	....	....	i; d a
Manganese							
	Mn	54.94	gray-met, bcc	7.47	1244	2120	d; s a
acetate-4-water	$Mn(C_2H_3O_2)_2 \cdot 4H_2O$	245.08	rose, mn	1.589	....	....	38[50]
arsenide	$MnAs$	129.86	blk, hex	6.18	d 400	....	i; s HCl
boride	MnB	65.75	pwd	6.2[15]	....	....	....
bromide	$MnBr_2$	214.76	rose cr	4.39	698	1027	147[20]
bromide-4-water	$MnBr_2 \cdot 4H_2O$	286.82	rose, mn, deliq	....	64 d	....	200; s al
(tri-) carbide	Mn_3C	176.83	blk, tetr	6.89[17]	1520	....	d; s a
carbonate (rhodochrosite)	$MnCO_3$	114.94	rose, rhhd	3.125	d	....	0.0065[25]; s a
(di-) carbonyl, deca-	$Mn_2(CO)_{10}$	389.99	gold, mn	1.75[25]	155 (CO atm)	d 110	i; s org solv
chloride (scacchite)	$MnCl_2$	125.84	pink, hex, deliq	2.977	650	1231	74[20]
chloride-4-water	$MnCl_2 \cdot 4H_2O$	197.91	rose, mn	2.01	−4H₂O, 198	....	143; s al
(III) chloride	$MnCl_3$	161.30	brn-blk cr	....	d −40	....	s abs al
(VII) chloride trioxide	$MnClO_3$	138.29	grn-vlt gas	....	....	−50	expl 20
dichromate(III)(2−)	$MnCr_2O_4$	222.93	gray-blk, cub	4.97[20]	....	....	i; i a

Name	Formula	Formula weight	Color, crystalline form, refractive index	Density	Melting point, °C	Boiling point, °C	Solubility
diphosphate(V)	$Mn_2P_2O_7$	283.82	brn-pink, mn, 1.695, 1.704, 1.710	3.707	1196	......	i; s a
dithionate	MnS_2O_6	215.06	rose, tric	1.757	......	......	s
fluoride	MnF_2	92.93	pink, tetr	3.98	856	1820	1.06[20]; s a; i al
(III) fluoride	MnF_3	111.93	red, mn	3.54	d 600	......	hyd
(IV) fluoride	MnF_4	140.93	blue, hygr	......	d 20	......	
(VII) fluoride trioxide	$MnFO_3$	121.84	dk grn liq	1.953	−38	ca 60	
formate-2-water	$Mn(CHO_2)_2 \cdot 2H_2O$	181.00	rose, rh	......	d	......	sl s; s a
hydrogen phosphate-3-water	$MnHPO_4 \cdot 3H_2O$	204.97	pink pwd, 1.656	3.258[13]	d	......	0.002[18]; s a
hydroxide (pyrochroite)	$Mn(OH)_2$	88.95	lt pink, trig, 1.723, 1.681	......	d	......	
(III) hydroxide oxide (manganite)	$Mn(OH)O$	87.94	brn-blk, rh, 2.24, 2.24, 2.53(Li)	4.2–4.4	d	......	i; s a
(IV) hydroxide dioxide	$Mn(OH)O_2$	104.95	blk-brn, amorp	2.58	638	......	i; s HCl
iodide	MnI_2	308.75	pink, hex, deliq	5.0[1]	d	1017	s
iodide-4-water	$MnI_2 \cdot 4H_2O$	380.81	rose, mn, deliq	......	37.1	......	s d
nitrate-4-water	$Mn(NO_3)_2 \cdot 4H_2O$	251.01	lt rose, mn, deliq	2.129	25	129	195[20]
nitrate-6-water	$Mn(NO_3)_2 \cdot 6H_2O$	287.05	rose, mn, deliq	1.8	d 150	......	v s; v s al
oxalate	MnC_2O_4	142.96	wh cr pwd	2.43[22]	......	......	i; s a
oxalate-2-water	$MnC_2O_4 \cdot 2H_2O$	178.98	pink, oct	......	−2H₂O, 100	......	0.035[20]; s a
(II) oxide (manganosite)	MnO	70.94	grn, cub, 2.16	5.44	1785	......	i; s a
(III) oxide (braunite)	Mn_2O_3	157.87	blk, cub	4.50	d 940	......	i; s a
(IV) oxide (pyrolusite)	MnO_2	86.94	blk, rh	5.026	d 530	......	i; s HCl
(II,IV) oxide (hausmannite)	Mn_3O_4	228.79	brn-blk pwd	4.7	......	......	i; s HCl
(VI) oxide	MnO_3	102.94	redsh, deliq	......	d	......	s
(VII) oxide	Mn_2O_7	221.87	dk red oil, hygr, explosive	2.396[20]	ca −20	ca 25	v s
perchlorate-6-water	$Mn(ClO_4)_2 \cdot 6H_2O$	361.95	pink, hex, 1.492	3.102	d 165	......	s
phosphate-3-water	$Mn_3(PO_4)_2 \cdot 3H_2O$	408.80	rose, rh, 1.651, 1.656, 1.683	......	......	......	i; sl s HNO₃
phosphide	MnP	85.91	dk gray, rh	5.34	1150	......	......
(di-) phosphide	Mn_2P	140.85	gray, hex	6.32	1327	......	i; sl s HNO₃
(tri-) phosphide, di-	Mn_3P_2	226.76	blk pwd	5.12[18]	1095	......	15; i al
phosphinate-1-water	$Mn(PH_2O_2)_2 \cdot H_2O$	202.93	pink cr	......	d ign	......	sl s
phosphonate-1-water	$MnPHO_3 \cdot H_2O$	152.93	redsh	......	−H₂O, 200	......	sl s

Table 4-1 (*Continued*)
PHYSICAL CONSTANTS OF INORGANIC COMPOUNDS

Name	Formula	Formula weight	Color, crystalline form, refractive index	Density	Melting point, °C	Boiling point, °C	Solubility in 100 parts
Manganese							
selenide	MnSe	133.90	gray-blk, cub	5.59^{15}	d	....	i; d a
selenide, di-	MnSe$_2$	212.86	cub	....	....	....	i; i HCl
silicate(1−) (rhodonite, tephroite)	MnSiO$_3$	131.02	redsh-yel, tric, 1.733, 1.740, 1.744	3.48	1270	....	i; i HCl
silicate(4−)	Mn$_2$SiO$_4$	201.96	....	....	1340	....	140^{18}
silicate-6-water, hexafluoro-	Mn[SiF$_6$]·6H$_2$O	305.11	rose, hex, 1.357, 1.374	1.903	d	....	i; s HF
silicide	MnSi	83.02	tetrahed	5.90^{15}	1280	....	i; s HF, alk
silicide, di-	MnSi$_2$	111.11	gray, oct	5.24^{13}	....	....	i; s HCl, alk
(di-) silicide	Mn$_2$Si	137.96	pr	6.20^{15}	1315	....	62.9^{20}
sulfate	MnSO$_4$	151.00	pink	3.25	700	d 850	70^{20}
sulfate-1-water (szmikite)	MnSO$_4$·H$_2$O	169.01	pink, mn, 1.562, 1.595, 1.632	2.95	−H$_2$O, 400	....	....
sulfate-4-water	MnSO$_4$·4H$_2$O	223.06	pink, mn, effl, 1.508, 1.522	2.107	d 27	....	93^{20}
sulfate-7-water	MnSO$_4$·7H$_2$O	277.11	red, mn	2.09	−7H$_2$O, 280	....	115^{20}
(III) sulfate	Mn$_2$(SO$_4$)$_3$	398.06	grn, hex, deliq	3.24	d 160	....	d; s HCl
sulfide (alabandite)	MnS	87.00	grn, cub, 2.70(Li)	3.99	1530	....	0.0006^{18}; s a
sulfide, di-	MnS$_2$	119.07	blk, cub, 2.69(Li)	3.462	....	....	i; d HCl
tantalate(V)(1−)	Mn(TaO$_3$)$_2$	512.83	blk, rh, 2.22, 2.25, 2.29	7.03	....	....	....
telluride	MnTe	182.58	hex	....	1165	....	s
thiocyanate-3-water	Mn(SCN)$_2$·3H$_2$O	225.16	pink, deliq	....	−3H$_2$O, 160	....	....
titanate(IV)(2−) (pyrophanite)	MnTiO$_3$	150.84	yel, trig, 2.481, 2.210	4.54	1360	....	....
titanate(IV)(4−)	Mn$_2$TiO$_4$	221.78	....	....	1450	....	....
Mercury (quicksilver)	Hg	200.59	silv-wh, liq	13.5939^{20}	−38.86	356.60	i; s HNO$_3$
(I) acetate	Hg$_2$(C$_2$H$_3$O$_2$)$_2$	519.27	col pwd	....	d	....	1^{21}; s acet a

Name	Formula	Mol. wt.	Color and crystalline form	Density	Melting point, °C	Boiling point, °C	Solubility (in parts per 100 parts)
(II) acetate	$Hg(C_2H_3O_2)_2$	318.70	wh pwd	3.28	178	.....	25^{10}; 7.5^{15} MeOH
(II) acetylide–water (3/1)	$3HgC_2 \cdot H_2O$	691.85	wh pwd	5.3	expl	.....	i; i al
(II) amide chloride	$HgNH_2Cl$	252.09	wh pwd	5.38	infusible	subl dull red heat	i; s HCl, acet a
(I) azide	$Hg_2(N_3)_2$	485.22	wh cr	.....	expl	.....	0.025
(II) benzoate-1-water	$Hg(C_7H_5O_2)_2 \cdot H_2O$	460.84	wh cr	.....	d	.....	1.1
(I) bromate	$Hg_2(BrO_3)_2$	656.99	wh cr	.....	d 130	.....	sl s HNO_3
(II) bromate-2-water	$Hg(BrO_3)_2 \cdot 2H_2O$	492.44	wh, rh	.....	.....	d	0.08; s HCl
(I) bromide	Hg_2Br_2	561.00	wh-yelsh, tetr	7.307	subl 393	.....	i; i al; s a
(II) bromide	$HgBr_2$	360.44	col, rh	6.05	241	subl > 241	0.56^{20}, 20^{25} al
bromide iodide	$HgBrI$	407.40	yel, rh	.....	229	360	i; s al, eth
(I) carbonate	Hg_2CO_3	461.19	yel-brn or	.....	d 130	d	i; s HNO_3
(I) chlorate	$Hg_2(ClO_3)_2$	568.08	wh, rh	6.409	d 250	.....	sl s
(II) chlorate	$Hg(ClO_3)_2$	367.49	wh nd	4.998	.....	.....	25
(I) chloride (calomel)	Hg_2Cl_2	472.09	wh, tetr, 1.973, 2.656	7.150	subl 382	.....	0.00027; s aq reg
(II) chloride	$HgCl_2$	271.52	col, rh, 1.859	5.44	277	304	6.6^{20}, 33 al; 4 eth
(I) chromate	Hg_2CrO_4	517.17	red pwd	.....	d	.....	v sl s; s HCl
(II) chromate	$HgCrO_4$	316.58	red, rh	.....	d	.....	sl s
(II) cyanide	$Hg(CN)_2$	252.65	col, tetr	3.996	d 320	subl 240	9.3^{20}, 8 al, 25 MeOH
(II) dichromate	$HgCr_2O_7$	416.63	red pwd	.....	.....	.....	i; s HCl
(I) fluoride	Hg_2F_2	439.22	yel, cub	8.73^{15}	570 d	.....	hyd
(II) fluoride	HgF_2	238.61	col, cub	8.95^{15}	645	647	hyd; s HF
(II) formate	$Hg_2(CHO_2)_2$	491.22	col	.....	d	.....	0.4^{17}
(II) fulminate	$Hg(ONC)_2$	284.62	wh, cub	4.42	expl	.....	sl s; s al
(I) iodate	$Hg_2(IO_3)_2$	750.99	yelsh	.....	d 250	.....	0.002^{20}
(II) iodate	$Hg(IO_3)_2$	550.45	wh amorp pwd	7.70	d 175	.....	i; s HCl
(I) iodide	Hg_2I_2	654.99	yel amorp pwd	6.28	290 d	.....	v sl s; s KI
(II) iodide	HgI_2	454.45	red, tetr; tr 127° to yel	6.094	259	d 357	0.006^{25}, 1 al, 1.7 acet
(I) nitrate-2-water	$Hg_2(NO_3)_2 \cdot 2H_2O$	561.22	col, mn, effl	4.79^{4}	70 d	350	0.0048; s eth
(II) nitrate	$Hg(NO_3)_2$	324.63	yelsh-wh, deliq	4.3	79	350	hyd; s HNO_3
(I) nitrite	$Hg_2(NO_2)_2$	493.19	yel	7.33	d 100	.....	v s; s acet
nitride	Hg_3N_2	629.78	brn pwd	.....	expl	d	d; d a
(II) oleate	$Hg(C_{18}H_{33}O_2)_2$	763.52	yelsh-brn	.....	.....	.....	i
(I) oxalate	$Hg_2C_2O_4$	489.20	.....	.....	.....	.....	0.001^{20}

Table 4-1 (*Continued*)
PHYSICAL CONSTANTS OF INORGANIC COMPOUNDS

Name	Formula	Formula weight	Color, crystalline form, refractive index	Density	Melting point, °C	Boiling point, °C	Solubility in 100 parts
Mercury							
(II) oxalate	HgC_2O_4	288.61	....	....	d	....	0.011^{20}, s HCl
(II) oxide (montroydite)	HgO	216.61	yel or red, rh, 2.37, 2.5, 2.65	11.14	d 476	....	0.005^{25}, s a
(II) perchlorate-3-water	$Hg(ClO_4)_2 \cdot 3H_2O$	452.57	....	....	....	....	i; s aq reg
(II) selenide (tiemannite)	$HgSe$	279.55	gray, cub	8.266	770	....	5
(II) succinimide	$Hg(C_4H_4NO_2)_2$	396.77	wh cr	....	....	....	0.06^{25}, s HNO_3
(I) sulfate	Hg_2SO_4	497.29	col, mn	7.56	d	d	d; s a
(II) sulfate	$HgSO_4$	296.68	col, rh	6.47	d	....	i; s aq reg
(II) sulfide (cinnabar, vermillion)	HgS	232.68	red, hex, 2.854, 3.201	8.10	subl 583	subl 583	i; s aq reg
(metacinnabar)	....	....	blk, cub	7.73	825	....	i; s aq reg
(II) telluride	$HgTe$	328.19	cub	....	670	....	....
(II) thiocyanate	$Hg(SCN)_2$	316.78	wh pwd	....	d 165	....	0.063^{25}, s HCl
Molybdenum	Mo	95.94	silv-wh, bcc	10.2	2610	4646	i; s hot H_2SO_4, HNO_3
boride, di-	MoB_2	117.56	rh	7.12	2100	....	....
bromide, di-	$MoBr_2$	255.76	or, amorp	4.88^{18}	....	....	i; s alk
bromide, tri-	$MoBr_3$	335.67	dk-grn nd	....	subl 977	....	i; d alk
bromide, tetra-	$MoBr_4$	415.58	blk nd, deliq	....	d	....	v s
carbide	MoC	107.95	gray, hex	8.20	2692	....	i; sl s HNO_3
(di-) carbide	Mo_2C	203.89	wh, hex	8.9	2430	....	i; sl s HNO_3
carbonyl, hexa-	$Co(CO)_6$	264.02	col, rh	1.96	subl 102	156.4	s bz
chloride, di-	$MoCl_2$	166.85	yel	3.714^{25}	d 530	....	s a, alk
chloride, tri-	$MoCl_3$	202.30	red-brn, mn	3.578^{25}	1027	....	i; s HNO_3
chloride, tetra-	$MoCl_4$	237.75	brn pwd, deliq	....	317	407	d; s conc a
chloride, penta-	$MoCl_5$	273.21	blk, mn, deliq	2.928	194	264	hyd; s conc a
chloride, hexa-	$MoCl_6$	308.66	....	....	subl 254	....	hyd; s HCl
chloride dioxide, di-	$MoCl_2O_2$	198.85	wh, mn	3.31^{17}	175	250	hyd; s HCl

Name	Formula	Formula wt	Color, crystalline form	Density	M.p.	B.p.	Solubility
chloride oxide, tri-	$MoCl_3O$	218.30	grn cr		295	d 215	d; s HCl
chloride oxide, tetra-	$MoCl_4O$	253.75	dk-grn		102	197	hyd; s HCl
fluoride, tri-	MoF_3	152.94	tan, rhbd		d 600		
fluoride, penta;	MoF_5	190.94	yel, mn	3.44	64	214	hyd; s alk
fluoride, hexa-	MoF_6	209.95	wh, bcc, hygr	2.543^{19}	17.61	33.89	v s
fluoride dioxide, di-	MoF_2O_2	165.94	wh, hygr	3.494^{25}	subl 270		s
fluoride oxide, tetra-	MoF_4O	187.93	wh, deliq	3.001^{25}	97.2	186	0.2; sl s HCl
hydroxide	$Mo(OH)_3$	146.96	blk pwd		d		i; v sl s a
iodide, di-	MoI_2	349.75	brn pwd	5.278^{25}	d 100		i; s a
iodide, tetra-	MoI_4	603.56	blk cr				i
(III) oxide	Mo_2O_3	239.90	gray-blk				i
(IV) oxide	MoO_2	127.94	gray, tetr	6.47			0.049^{28}, s alk
(V) oxide	Mo_2O_5	271.88	vlt-blk pwd				i
(VI) oxide	MoO_3	143.95	yelsh-wh, rh	4.696^{26}	801	1155	i
phosphide	MoP	126.93	gray-grn, hex	6.58			i
phosphide, di-	MoP_2	157.89		5.35			i
(tri-) phosphide	Mo_3P	318.79	gray, tetr	9.07			i
silicide, di-	$MoSi_2$	152.11	gray-met, tetr	6.31	2375		i; s HNO_3 + HF
sulfide, di-	MoS_2	160.08	dk-gray, hex	5.06^{15}		subl 450	i; s aq reg
sulfide, sesqui-	Mo_2S_3	288.07	gray, tetr	5.91^{15}	1100	1200	i; d hot HNO_3
(di-)sulfide, penta-	MoS_5	352.18	dk-brn pwd				i; s alk
sulfide, tri-	MoS_3	192.13	blk, amorp		d		sl s; s alk
Molybdic acid	H_2MoO_4	161.96	wh, hex	3.112	$-H_2O$, 70		sl s; s alk, NH_4OH
-1-water	$H_2MoO_4 \cdot H_2O$	179.97	yel, mn	3.124^{15}	$-H_2O$, 70		0.133^{18}, s alk
Molybdic phosphoric acid	$H_7[P(Mo_2O_7)_6] \cdot 28H_2O$	2365.71	yel, oct	2.53	78		hyd
Molybdic silicic acid	$H_8[Si(Mo_2O_7)_6] \cdot 28H_2O$	2363.83	yel, tetr		45		v s
Neodymium	Nd	144.24	silv-wh, hex	7.004	1019	3111	i; s hot H_2O, a
acetate-1-water	$Nd(C_2H_3O_2)_3 \cdot H_2O$	339.39					26.2^{25}
boride, hexa-	NdB_6	209.10	bl, cub	4.68			
bromate-9-water	$Nd(BrO_3)_3 \cdot 9H_2O$	690.10	red, hex		66.7	$-9H_2O$, 150	99^{20}
bromide	$NdBr_3$	383.97	red-vlt, rh	5.35	682	1537	
carbide, di-	NdC_2	168.26	yel leaf, hex	5.15	d		sl s
(II) chloride	$NdCl_2$	215.15	grn, rh		841		d; s a
(III) chloride	$NdCl_3$	250.60	purp, hex, hygr	4.134^{25}	760	1690	98^{20}
fluoride	NdF_3	201.24	lilac, hex		1374	2327	i; s H_2SO_4

Table 4-1 (*Continued*)
PHYSICAL CONSTANTS OF INORGANIC COMPOUNDS

Name	Formula	Formula weight	Color, crystalline form, refractive index	Density	Melting point, °C	Boiling point, °C	Solubility in 100 parts
Neodymium							
hydride, di-	NdH$_2$	146.26	....	5.94	....	....	....
hydride, tri-	NdH$_3$	147.26	indigo-bl, amorp	....	....	>320 to oxide	i; s a
hydroxide	Nd(OH)$_3$	234.26	blsh-pink, hex	....	-H$_2$O, 210		i; s a
(II) iodide	NdI$_2$	398.05	red-vlt	....	562	....	
(III) iodide	NdI$_3$	524.95	lt-grn, rh	....	784	1370	s
molybdate(VI)	Nd$_2$(MoO$_4$)$_3$	768.29	tetr, 2.005	5.14	1176	....	0.0019[28]
nitrate-6-water	Nd(NO$_3$)$_3$ · 6H$_2$O	438.35	tric	....	....	....	188[20]
oxalate-10-water	Nd$_2$(C$_2$O$_4$)$_3$ · 10H$_2$O	732.69	rose cr	....	....	....	i; s a
oxide	Nd$_2$O$_3$	336.48	bl, hex	7.28	1900	....	0.00019[28], s a
(di-) oxide sulfide, di-selenide	Nd$_2$O$_2$S	352.54	blsh-wh, hex	6.22	1990	....	....
selenide	NdSe	223.2	cub		1827		
(di-) selenide, tri-	Nd$_2$Se$_3$	525.36	vlt-blk, cub	6.67	1557		
sulfate-8-water	Nd$_2$(SO$_4$)$_3$ · 8H$_2$O	720.79	red, mn, 1.41, 1.55, 1.56	2.85	d 700		8.87[20]
sulfide	NdS	176.30	gold, cub	6.24	2137		i; s a
(di-) sulfide, tri-	Nd$_2$S$_3$	384.67	yel-grn, cub	5.38	2207		
telluride	NdTe	271.84	cub		1755		
(di-) telluride, tri-	Nd$_2$Te$_3$	671.3	rh		1376		
Neon	Ne	20.18	col gas, 1.000067; col liq	0.8990[0] (g/L) 1.207[bp]	-248.6	-246.1	1.05 ml[20]
Neptunium	Np	237.05	silv, rh	20.45	630	....	s HCl
(III) bromide	NpBr$_3$	476.76	grn, hex	6.62	467	subl 800	s
(IV) bromide	NpBr$_4$	556.67	red-brn, mm	....	464	subl 500	
(III) chloride	NpCl$_3$	343.41	grn, hex	5.58	ca 800	subl 500	s
(IV) chloride	NpCl$_4$	378.86	red-brn, tetr	4.92	538	subl 500	

Name	Formula	Mol. wt.	Color, crystalline form, etc.	Density	M.p., °C	B.p., °C	Solubility
(IV) chloride oxide, di-	$NpCl_2O$	323.96	or, cub		subl 500 d		
(III) fluoride	NpF_3	294.05	purp, hex	8.95	1425		i
(IV) fluoride	NpF_4	313.04	grn, mn	6.80			i
(VI) fluoride	NpF_6	351.04	or-brn, rh	5.00	55.1	76.8	hyd
(VI) fluoride dioxide, di-	NpF_2O_2	307.05	pink, rhbd	6.41			s
hydride, di-	NpH_2	239.06	cub				
hydride, tri-	NpH_3	240.07	hex				
(III) iodide	NpI_3	617.76	brn, rh	6.92	770	subl 800	i; s a
(IV) oxide	NpO_2	269.05	lt-grn, cub	11.11			i; s HNO_3
(tri-) oxide, octa-	Np_3O_8	839.14	brn, cub		d 500		i; s HNO_3
Nickel	Ni	58.71	silv-met, fcc	8.90	1455	2920	
acetate-4-water	$Ni(C_2H_3O_2)_2 \cdot 4H_2O$	248.86	grn pr	1.744	d		16; s al
acetylacetonate	$Ni(C_5H_7O_2)_2$	256.93	emerald-grn, rh	1.455^{17}	229	235	s; s al, bz, chl
[hexaammine-] dichloride	$[Ni(NH_3)_6]Cl_2$	231.80	blsh, cub	1.468			s
antimonide (breithauptite)	$NiSb$	180.46	lt-copper-red, hex	7.54	1158	d 1400	i; s a
arsenate	$Ni_3(AsO_4)_2$	453.97	yelsh-grn pwd	4.982			i; s aq reg
arsenide (niccolite)	$NiAs$	133.63	hex	7.57^{0}	968		
(tri-) arsenide, di-	Ni_3As_2	325.97	met, tetr	7.86^{0}	1000		
benzenesulfonate-6-water	$Ni(C_6H_5SO_3)_2 \cdot 6H_2O$	481.10	grn, mn	1.628			14.3^{18}
boride	NiB	69.52	pr	7.39^{18}	d		d; s HNO_3
bromate-6-water	$Ni(BrO_3)_2 \cdot 6H_2O$	422.62	grn, mn	2.60			28^{20}
bromide	$NiBr_2$	218.53	golden-yel, deliq	5.098	963	subl	131^{20}
bromide-3-water	$NiBr_2 \cdot 3H_2O$	272.57	yelsh-grn nd, deliq		−3H₂O, 200		v s
(tri-) carbide	Ni_3C	188.14	dk-gray pwd	7.957			
carbonate	$NiCO_3$	118.72	lt-grn, rh	2.6	d		0.009^{25}, s a
carbonate–hydroxide (1/2) (zaratite)	$NiCO_3 \cdot 2Ni(OH)_2$	304.17	grn, cub, 1.58				i; s a
carbonyl, tetra-	$Ni(CO)_4$	170.75	col liq, inflam, explodes 60°	1.318^{17}	−19.3	42	0.018^{20}, s al, bz, eth
chlorate-6-water	$Ni(ClO_3)_2 \cdot 6H_2O$	333.70	dk red	2.07	d 80		196^{20}
chloride	$NiCl_2$	129.62	yel, deliq	3.55	1030	subl 970	60.8^{20}
chloride-6-water	$NiCl_2 \cdot 6H_2O$	237.70	grn, mn, 1.57		−4H₂O, 200		111^{20}
cyanide-4-water	$Ni(CN)_2 \cdot 4H_2O$	182.81	lt-grn-pwd			d	0.006^{18}, s KCN, NH₄OH
dimethylglyoxime	$Ni(HC_4H_6N_2O_2)_2$	288.91	scarlet cr pwd		subl 250		i; s a, abs al

Table 4-1 (Continued)
PHYSICAL CONSTANTS OF INORGANIC COMPOUNDS

Name	Formula	For-mula weight	Color, crystalline form, refractive index	Density	Melting point, °C	Boiling point, °C	Solubility in 100 parts
Nickel							
fluoride	$NiF_2 \cdot 2H_2O$	96.71	yelsh-grn, tetr	4.72	1450	1740	2.56[20]; i, al
formate-2-water	$Ni(CHO_2)_2 \cdot 2H_2O$	184.78	grn, mn	2.154[20]	−2H$_2$O, 130	d 180	s; i al
hydroxide-1-water	$Ni(OH)_2 \cdot H_2O$	110.74	apple-grn pwd	....	d 230	....	0.0013[20]
iodate	$Ni(IO_3)_2$	408.52	yel nd	5.02[25]	....	....	1.15[30]
iodate-4-water	$Ni(IO_3)_2 \cdot 4H_2O$	480.59	yel, hex	....	d 100	....	1.09[20]
iodide-6-water	$NiI_2 \cdot 6H_2O$	420.64	grn-blk, hex, deliq	5.834	797	....	143[20]
nitrate-6-water	$Ni(NO_3)_2 \cdot 6H_2O$	290.81	grn, mn, deliq	2.05	56.7	136.7	150[20]
oxalate-2-water	$NiC_2O_4 \cdot 2H_2O$	182.76	lt-grn pwd	....	....	....	0.0003[18]; s a
oxide (bunsenite)	NiO	74.71	grn, cub, 2.182	6.67	1984	....	i; s a
oxide, sesqui-	Ni_2O_3	165.42	gray-blk pwd	....	d 600	....	i; s hot HCl
perchlorate-6-water	$Ni(ClO_4)_2 \cdot 6H_2O$	365.70	grn, hex, 1.518, 1.498	....	140	....	156[20], s al, chl
phosphate-8-water	$Ni_3(PO_4)_2 \cdot 8H_2O$	510.20	lt-grn pwd	....	d	....	i; s a
phosphide, di-	NiP_2	120.66	gray, rh	4.62	....	....	....
phosphide, tri-	NiP_3	151.63	gray, hex	4.19	....	....	....
(di-) phosphide	Ni_2P	148.39	gray, hex	7.2	1112	....	s HNO_3 + HF
(tri-) phosphide	Ni_3P	207.10	gray, tetr	7.7	975	....	....
(tri-) phosphide, di-	Ni_3P_2	238.08	dk-grn	5.99	1185	....	....
(penta-) phosphide, di-	Ni_5P_2	355.50	gray nd	7.5	1185	....	....
phosphinate-6-water	$Ni(PH_2O_2)_2 \cdot 6H_2O$	296.78	grn	1.82[20]	d 100	....	s
selenate-6-water	$NiSeO_4 \cdot 6H_2O$	309.76	grn, tetr, 1.5393	2.314	627	....	s
selenide	$NiSe_2$	216.63	cub	....	....	....	
silicate-6-water, hexafluoro-	$Ni[SiF_6] \cdot 6H_2O$	308.88	grn, trig, 1.391, 1.407	2.134	d	....	i; i a
(di-) silicide	Ni_2Si	145.51	....	7.21[7]	1309	....	29.3[0]
sulfate	$NiSO_4$	154.78	yel, cub	3.68	d 848	....	40.1[20]
sulfate-6-water	$NiSO_4 \cdot 6H_2O$	262.86	α: bl-grn, tetr; β: grn, mn, 1.511, 1.487	2.07	53.3 (tr to β); −6H$_2$O, 280	....	44.1[20]
sulfate-7-water (morenosite)	$NiSO_4 \cdot 7H_2O$	280.88	grn, rh, 1.467, 1.489, 1.492	1.948	99	−7H$_2$O, 103	37.7[20]

Name	Formula	Mol. wt.	Crystalline form, color	Density	Melting point, °C	Boiling point, °C	Solubility
sulfide (millerite)	NiS	90.77	blk, trig & amorp	5.3–5.6	797		i; s HNO_3, KHS
(tri) sulfide, di-	Ni_3S_2	240.26	lt-yel, cub	5.82	790		i; s HNO_3
(tri) sulfide, tetra-	Ni_3S_4	304.39	gray-blk, cub	4.7	d 377		i; s HNO_3
sulfite-6-water	$NiSO_3 \cdot 6H_2O$	246.86	grn, tetr				i; s HCl
telluride, di-	$NiTe_2$	313.91	hex		705		
Niobium	Nb	92.906	steel-gray, bcc	8.57	2477	4863	s fus alk
boride, di-	NbB_2	114.53	hex	6.97	3000		
(III) bromide	$NbBr_3$	332.62	blk, rh		subl 400 (vac)		s HCl
(IV) bromide	$NbBr_4$	412.53	blk, rh				
(V) bromide	$NbBr_5$	492.46	vlt-blk, rb, hygr	4.65	227	272	d; s al
carbide	NbC	104.92	blk, cub	7.6	3500		s HNO_3, HF
(di-) carbide	Nb_2C	197.82			3090		
(III) chloride	$NbCl_3$	199.27	blk cr	3.61^{20}	d 100		i; i org solv
(IV) chloride	$NbCl_4$	234.72	vlt-blk, rh & mn	3.23	subl 455		hyd; s HCl
(V) chloride	$NbCl_5$	270.20	yel, mn, deliq	2.75	204	250	s HCl, CCl_4
(V) chloride oxide, tri-	$NbCl_3O$	215.26	wh, tetr	3.27	subl 335		hyd
(IV) fluoride	NbF_4	168.90	blk, tetr, hygr		d 400		
(V) fluoride	NbF_5	187.91	wh, mn	2.70^{80}	76	225	hyd; hyd al
(di-) fluoride, penta-	Nb_2F_5	280.82	cub	4.91	d 700		
fluoride dioxide	$NbFO_2$	143.90	cub	6.6	201		
hydride	NbH	93.91	gray pwd		infusible		s HF + HNO_3
(III) iodide	NbI_3	473.62	hex		d 513		
(IV) iodide	NbI_4	600.53	gray nd, rh		d 450		
(V) iodide	NbI_5	727.43	bronze, mn, hyd air	5.11	327	347	hyd
nitride	NbN	106.91	blk, cub	7.3	2050		s HF + HNO_3
(di-) nitride	Nb_2N	199.82	blk, cub		2400		
(II) oxide	NbO	108.91	bl-blk	7.30	1937		s a, alk
(III) oxide	Nb_2O_3	233.81	blk		1780		
(IV) oxide	NbO_2	124.91		5.9	1902		i a; sl s alk
(V) oxide	Nb_2O_5	265.82	wh, rh	4.6	1512		i; s HF
potassium heptafluoride oxide-1-water	$K_2NbF_7O \cdot H_2O$	338.12	lustrous leaf, mn				7.7
sulfide	NbS	124.97	hex				
Nitric acid	HNO_3	63.02	col liq, 1.3970	1.5027	−41.59	83	v s
-1-water	$HNO_3 \cdot H_2O$	81.04	col liq		−37.68		v s
-3-water	$HNO_3 \cdot 3H_2O$	117.07	col liq	1.41^{20}	−18.47		v s
	68% HNO_3 + H_2O		constant boiling			120.5	v s

Table 4-1 (*Continued*)
PHYSICAL CONSTANTS OF INORGANIC COMPOUNDS

Name	Formula	Formula weight	Color, crystalline form, refractive index	Density	Melting point, °C	Boiling point, °C	Solubility in 100 parts
Nitrogen	N_2	28.01	col gas, 1.0002779	1.165^{20} (g/L)	-210.00	-195.81	1.52 ml^{20}
			col liq	0.808^{bp}			
[^{15}N]	$^{15}N_2$	30.01	col gas, 1.000298	1.25^{20}(g/L)	-209.952	-195.73	....
bromide, tri-	NBr_3	253.73	red-vlt, explosive	....	expl -67	....	....
chloride, tri-	NCl_3	120.37	yel oily liq, explodes 93°	1.653^{20}	-27	71	i; s bz, CS$_2$, CCl$_4$
fluoride, tri-	NF_3	71.01	col gas, 1.0004492	2.96^{20}(g/L)	-206.78	-129.06	i
			col liq	1.885^{bp}			
iodide, tri-	NI_3	394.72	blk, explosive	....	expl	....	s KSCN, Na$_2$S$_2$O$_3$
iodide-ammonia, tri-	$NI_3 \cdot NH_3$	411.75	blk, rh, shock sensitive to explode	....	d	....	....
(di-) oxide (laughing gas)	N_2O	44.02	col gas, 1.0004732	1.8433^{20} (g/L)	-90.85	-88.47	130 ml^0; s al, oils
oxide	NO	30.01	col gas, 1.0002743	1.2488^{20} (g/L)	-163.64	-151.76	7 ml^0
			deep bl liq	1.2906^{bp}			....
(di-) oxide, tri-	N_2O_3	76.02	red-brn gas; d 3.5°	1.447^2	-111	2	s eth
(di-) oxide, tetra-	$N_2O_4 \rightleftharpoons 2NO_2$	92.02	col; brn as NO$_2$	1.447^{20}	-9.3	21.10 d	s HNO$_3$, H$_2$SO$_4$, chl
(di-) oxide, penta-	N_2O_5	108.01	col, hex	2.05^{15}	30	47.0	s; s chl
oxide, tri-	NO_3	62.01	blsh gas	....	d 20	....	s eth
(tetra-) selenide, tetra-	N_4Se_4	371.87	or-red amorp pwd; also mn, hygr; expl	4.2	expl	30(vac)	i; sl s bz, CS$_2$
(di-) sulfide, tetra-	N_2S_4	156.26	red-brn liq, gray cr	1.71^{20}	23	d	s org solv
(di-) sulfide, penta-	N_2S_5	188.33	red liq	1.901^{20}	10	185	s CS$_2$
(tetra-) sulfide, tetra-	N_4S_4	184.28	or-yel, mn, 2.046	2.24^{18}	180	1.5	s org solv
Nitrosyl amide	NON_3	72.02	....	....	....	....	....
bromide	$NOBr$	109.92	dk brn gas & liq	1.592^{bp}	-55.5	19	hyd
chloride	$NOCl$	65.47	redsh-yel gas	....	-61.5	-5.5	hyd

Name	Formula	Mol. wt.	Crystalline form, color	Density	m.p.	b.p.	Solubility
chloride-tin(IV) chloride (2/1)	$2NOCl \cdot SnCl_4$	391.42	ltyel, oct	2.60	180	...	hyd
fluoride	NOF	49.01	col gas	2.788^{20} (g/L)	−132.5	−59.5	...
fluoroborate, tetra-	$NO[BF_4]$	116.83	col liq	1.326^{bp}	subl 250^{10mm}	...	d
hydrogen sulfate	$NOHSO_4$	127.08	col, rh, hygr	2.185^{25}	73.5	...	d: s H_2SO_4
perchlorate-1-water	$NOClO_4 \cdot H_2O$	147.47	col, rh	...	d 100	...	d
sulfate	$NOSO_4$	127.08	rh, deliq	2.169	d 73.5	...	d: s H_2SO_4
sulfuric anhydride	$(NOSO_3)_2O$	236.14	pr	...	217	360	hyd
Nitrous acid	HNO_2	47.02	lt-bl, known only in soln	...	d	...	
Nitryl amide	NO_2NH_2	62.03	unstable weak acid	1.41^{-16}	d 72	...	d
chloride	NO_2Cl	81.46	col gas	2.81^{100} (g/L)	−145	−13.5	d
fluoride	NO_2F	65.00	col gas	2.7^{20} (g/L)	−166.0	−72.4	d
fluoride oxide	NO_2FO	71.00	col gas	1.507^{-181}	−181	−45.9	...
hydrogen disulfate	$NO_2HS_2O_7$	223.14	...	2.1788	106	...	...
nitrosyl trisulfate	$NO_2NOS_3O_{10}$	332.19	...	...	126	...	...
(di-) trisulfate	$(NO_2)_2S_3O_{10}$	302.20	...	...	128	200	...
Osmium	Os	190.2	blsh-wh met, hcp	22.61^{20}	ca 2727	ca 5500	s molten alk or oxid fluxes
chloride, tri-	$OsCl_3$	296.56	dk-gray, cub	4.38^{20}	subl 350	d > 450	hyd; s HNO_3
chloride, tetra-	$OsCl_4$	332.00	red cr	...	subl 450	slow hyd	
chloride oxide, tetra-	$OsCl_4O$	348.01	col nd, dk-yel gas	...	32	ca 200	s
fluoride, tetra-	OsF_4	266.19	yel pwd	...	230	...	hyd
fluoride, penta-	OsF_5	247.19	b.-gray, mn	...	70	225.7	hyd
fluoride, hexa-	OsF_6	304.20	lt-yel, bcc	...	32.1	45.9	hyd
fluoride trioxide, di-	OsF_2O_3	276.20	o-, hex	...	171	100.5	hyd
fluoride oxide, penta-	OsF_5O	263.19	emerald grn, rh	...	59.8	...	hyd
iodide, tetra-	OsI_4	697.82	vlt-blk, hygr	...	...	...	v s
oxide, di-	OsO_2	222.20	blk, tetr	7.91	...	...	s HCl, HF
oxide, tetra-	OsO_4	254.20	lt-yel, mn	4.91	40.6	130.0	7.24^{25}, 375^{25} CCl4
sulfide, di-	OsS_2	254.33	b.-k, cub	9.47	...	...	sl s aq reg
telluride, di-	$OsTe_2$	445.40	gray-blk, cub	...	ca 600	...	s HNO_3

Table 4-1 (Continued)
PHYSICAL CONSTANTS OF INORGANIC COMPOUNDS

Name	Formula	Formula weight	Color, crystalline form, refractive index	Density	Melting point, °C	Boiling point, °C	Solubility in 100 parts
Oxygen	O_2	32.00	col gas, 1.0002713	1.331^{20} (g/L) 1.118^{bp}(liq)	−218.75	−182.96	36 ml^{25}
difluoride	OF_2	54.00	col gas	2.26^{20}(g/L)	−223.8	−144.8	6.8 ml^0
(di-) difluoride	O_2F_2	70.00	lt-yel gas	1.45^{bp}(liq)	−154	−56 d	.....
Oxygenyl hexafluoroplatinate(V)	$O_2[PtF_6]$	275.09	dk red, cub	.....	219 d	.....	hyd
Ozone	O_3	48.00	bl expl gas bl liq	1.998^{20} (g/L) 1.46^{bp}	−192.5	−110.50	49.4 ml°
Palladium	Pd	106.4	silv-wh met, fcc	12.023	1550	2940	s hot HNO$_3$, H$_2$SO$_4$
acetate	$Pd(C_2H_3O_2)_2$	224.49	or-brn cr	.....	205 d	.....	i; i al; s acet, chl
bromide	$PdBr_2$	266.22	red-brn, mn	5.173^{16}	d	.....	d; s HBr
chloride	$PdCl_2$	177.30	red, rh	4.0^{18}	680	d > 680	s
chloride-2-water	$PdCl_2 \cdot 2H_2O$	213.34	dk-brn cr	.....	d	.....	v s; s al, acet
cyanide	$Pd(CN)_2$	158.44	yel-wh	.....	d	.....	i; s KCN, NH$_4$OH
fluoride	PdF_2	144.40	vlt, tetr	5.80	volatile red heat	.....	i; s HF
iodide	PdI_2	360.21	blk pwd	6.003^{18}	d > 350	.....	i; s KI
nitrate	$Pd(NO_3)_2$	230.42	yel-brn, rh, deliq	.....	d	.....	hyd; s HNO$_3$
oxide	PdO	122.40	blk pwd	8.70^{20}	870 d	.....	i; i a
oxide–hydrate, di-	$PdO_2 \cdot (H_2O)_x$	.....	dull red	.....	d	.....	i; s a, alk
selenide	PdSe	185.36	dk-gray, tetr	.....	>960	.....	i; s aq reg
selenide, di-	$PdSe_2$	264.32	olive-gray, rh	.....	.....	.....	i; s aq reg
silicide	PdSi	134.49	cr	7.31^{15}	.....	.....	i
sulfide	PdS	138.46	brn-blk, tetr	6.6	950	.....	sl s HNO$_3$
sulfide, di-	PdS_2	170.53	dk-brn, rh	4.75	972	.....	i; s aq reg, NH$_4$S
sulfate-2-water	$PdSO_4 \cdot 2H_2O$	238.50	red-brn cr, deliq	.....	.....	.....	v s
telluride	PdTe	234.01	hex	.....	745	.....	i; s HNO$_3$
telluride, di-	$PdTe_2$	361.60	silv cr, hex	.....	.....	.....	

Name	Formula	Formula wt.	Color, crystalline form, etc.	Density	Melting point	Boiling point	Solubility
Perchloric acid	HClO$_4$	100.47	col oily liq, hygr, spontaneous & expl decomposition	1.768[22]	−112	19[11mm]	v s
-1-water	HClO$_4 \cdot$ H$_2$O	118.47	col nd, expl & shock sensitive	1.767[20]	−101.2	d	v s
-2-water	HClO$_4 \cdot$ 2H$_2$O	136.49	col liq; 70–72% commercial acid	1.67[20]	−17.8	203	v s
			commercial 60% acid	1.539[15]	......	......	v s
Perchloryl fluoride	ClO$_3$F	102.46	col gas	0.637	−147.47	−46.67	......
Periodic acid, ortho-	H$_5$IO$_6$	227.96	col, mn, hygr	......	122	d 130–140	113; s al
meta-	HIO$_4$	191.91	col cr	......	subl 110	......	440[25]
Peroxodisulfuric acid	H$_2$S$_2$O$_8$	194.14	col cr, hygr	......	d 60	......	v s
Peroxosulfuric acid (Caro's acid)	H$_2$SO$_5$	114.08	col	......	d 45	......	d; s H$_3$PO$_4$
Phospham	PHN$_2$	60.00	wh, amorp	......	infusible	......	i; s conc H$_2$SO$_4$
Phosphine	PH$_3$	34.00	col gas	1.529 (g/L)	−133.81	−87.78	26 ml[17]; s al, eth
Phosphinic acid (hypophosphorous acid)	HPH$_2$O$_2$	66.00	col oily liq or leaflets	1.493[19]	26.5	d 50	s
Phosphonic acid	H$_2$PHO$_3$	82.00	wh cr, hygr & deliq	1.651[21]	ca 73	d > 180	v s; v s al
Phosphonium bromide	PH$_4$Br	114.93	col cr, tetr	2.464 (g/L)(g)	18.5	ca 20	d
chloride	PH$_4$Cl	70.46	col, cub	......	subl 50	......	d
[tetrabromo-] fluoride	[PBr$_4$]F	369.61	......	......	87	......	......
[tetrachloro-] fluoride	[PCl$_4$]F	191.79	cr	......	177	subl 175	d
iodine	PH$_4$I	161.91	col, tetr, deliq	2.86	18.5	80; subl 61.8	d
Phosphoramidic acid (amidophosphoric acid)	NH$_2$PO(OH)$_2$	97.01	wh, cub	......	d	......	v s
Phosphoric acid, meta-	HPO$_3$	79.98	trans, vitr, hygr	2.2–2.5	volatile at red heat	......	slowly hyd; s al
ortho-	H$_3$PO$_4$	98.00	col unstable solid, rh	1.88	42.35	......	v s
			col liq, 85% comm'l	1.685	anhyd 150	to H$_2$P$_2$O$_7$ ca 200; to HPO$_3$ > 300	
difluoro-	H$_2$PO$_2$F$_2$	102.99	col fum liq	1.583	−96.5	115.9	d
fluoro-	H$_2$PO$_3$F	99.99	col oily liq	1.818	−80	......	200; s al, eth
-tungsten oxide-water (2/24/48) (phosphotungstic acid)	H$_3$PO$_4 \cdot$ 24WO$_3 \cdot$ 48H$_2$O	......	yel-grn, effl water content varies				

Table 4-1 (Continued)
PHYSICAL CONSTANTS OF INORGANIC COMPOUNDS

Name	Formula	Formula weight	Color, crystalline form, refractive index	Density	Melting point, °C	Boiling point, °C	Solubility in 100 parts
Phosphorus, white	P	30.974	wh, waxy, cub cr of P_4 molecules, ignites 30° moist air	1.828	44.2	280.3	i; 0.025 al, 1.0 eth, 2.5 chl, bz, 1.25 CS_2
red	P	30.974	red-vlt pwd, tric, ignites 260° in air	2.34	597	subl 416	i
bromide, tri-	PBr_3	270.73	col fum liq, 1.6903	2.85[15]	−40.5	173.2	d; d al; s acet, CS_2
bromide, penta-	PBr_5	430.56	red cr, rh	3.46[20]	>100 d	....	d; s CCl_4, CS_2
bromide difluoride	$PBrF_2$	148.88	col gas	....	−133.8	−16.1	d
bromide fluoride, di-	PBr_2F	209.79	col liq	2.127[20]	−115.0	78.3	d; d al; s bz, chl
chloride, tri-	PCl_3	137.35	col fum liq, 1.515	1.575[20]	−91	75	d; d al; s bz, chl
chloride, penta-	PCl_5	208.27	lt-yel to wh, fum, tetra-hed, deliq	2.119[20]	subl 100 without melting	166 d	hyd; s CCl_4, CS_2
chloride biscyanate	$PCl(OCN)_2$	150.45	col liq	1.50	−50	134.6	....
chloride difluoride	$PClF_2$	104.42	col gas	....	−164.8	−47.3	....
chloride fluoride, di-	PCl_2F	120.88	col gas	−144.0	13.85	215	....
(di-) chloride trioxide, tetra-	$P_2Cl_4O_3$	251.76	col fum liq	1.813[20]	....	....	....
cyanate, tris-	$P(OCN)_3$	157.02	col	1.439	−2	169.3	d; v s eth
cyanide, tri-	$P(CN)_3$	109.03	wh nd	....	subl 130	....	hyd
fluoride, tri-	PF_3	87.98	col gas	3.907(g/L)	−151.30	−101.38	hyd
fluoride, penta-	PF_5	125.98	col gas, fumes	5.805(g/L)	−93.8	−84.6	hyd
hydride, see Phosphine							
iodide, tri-	PI_3	411.68	dk red, hex, deliq	4.18	61.5	d 200	d; s CS_2
(di-) iodide, tetra-	P_2I_4	569.57	or, tric	....	124.5	330	d; s CS_2
(di-) oxide, tetra-	P_2O_4	125.95	col, rh, deliq	2.539[20]	100	180 vac	d
(tetra-) oxide, hexa-	P_4O_6	219.90	wh, mn, 1.540, or col liq	2.136[20]	24	175(N_2 atm) d > 210	hyd; s bz, CS_2

Name	Formula	Mol wt	Color, crystalline form	Density	m.p.	b.p.	Solubility
(tetra-) oxide, deca-	P_4O_{10}	283.88	wh, hex, 1.47	2.30	340	subl 360	d; s H_2SO_4
(tetra-) oxide tetrasulfide, hexa-	$P_4O_6S_4$	343.76	tetr, deliq	....	102	295	d; s CS_2
(di-) selenide, tri-	P_2Se_3	298.84	dk-red mass	....	d	....	d; s KOH
(tetra-) selenide, tri-	P_4Se_3	360.80	or-red cr, flammable in air	1.31	245	360–400	hyd; s bz, chl, acet
(tetra-) sulfide, tri-	P_4S_3	220.09	yel, rh	2.03^{17}	167	407	i; 100^{17} CS_2
(tetra-) sulfide, hepta-	P_4S_7	348.33	col, mn	2.19^{17}	308	523	i; sl s CS_2
(tetra-) sulfide, deca-	P_4S_{10}	444.54	yel-gray, tric	2.09	288	514	hyd; s alk, CS_2
thiocyanate, tris-	$P(SCN)_3$	205.22	....	1.625^{18}	ca −4	265	d; s al, bz, eth, CS_2
Phosphoryl bromide, tri-	$POBr_3$	286.72	faint or tinge, pl, tetrahed	2.822	56	192 d	hyd slowly; s bz, CS_2, eth
bromide chloride, di-	$POBr_2Cl$	242.27	col cr	2.45^{50}	31	165	d
bromide dichloride	$POBrCl_2$	197.79	col liq, 1.5091	2.1167	13	136	d
bromide fluoride, di-	$POBr_2F$	225.81	col liq	2.036^{20}	−117.2	110.1	....
bromide difluoride	$POBrF_2$	154.89	col liq	2.484^{20}	−84.8	31.9	....
bromide chloride fluoride	$POBrClF$	181.33	col liq	1.87^{20}	....	79	....
chloride, tri-	$POCl_3$	153.35	col liq, 1.4866	1.645^{25}	2	105	d; d al
chloride difluoride	$POClF_2$	120.43	col	1.656^{0}	−96.4	3.1	....
chloride fluoride, di-	$POCl_2F$	136.89	col liq	1.5497^{20}	−80.1	52.90	....
cyanate, tris-	$PO(OCN)_3$	173.02	col liq	1.570^{25}	5.0	193.1	....
fluoride, tri-	POF_3	103.97	col gas	4.69 (g/L)	subl −39.5	−39.1	d
thiocyanate, tris-	$PO(SCN)_3$	221.16	col liq	1.484^{25}	13.8	300.1	....
Platinic(IV) acid-6-water, hexachloro-	$H_2[PtCl_6]\cdot 6H_2O$	517.92	red-brn cr, deliq	2.431	60	....	v s
Platinum	Pt	195.09	silv-met, fcc	21.45^{20}	1770	3824	i; s aq reg, fus alk
arsenide, di- (sperrylite)	$PtAs_2$	344.93	gray, cub	11.8	d 800	....	v sl s a
(II) bromide	$PtBr_2$	354.91	brn	6.65^{25}	d 250	....	i; s HBr, KBr
(IV) bromide	$PtBr_4$	514.73	dk-red	5.69^{25}	d 180	....	0.41^{25}; v s al, eth
[bis(carbonyl)(di-)] tetrabromide	$[Pt_2(CO)_2]Br_4$	765.84	lt red nd, hygr	5.115^{25}	d 180	....	s d; s bz, CCl_4
[carbonyl-] dichloride	$[Pt(CO)]Cl_2$	294.00	yel nd	4.2346^{25}	195	subl 240	d; s HCl
(II) chloride	$PtCl_2$	266.00	olive-grn, hex	6.05	d 581	....	i; s HCl, NH_4OH
(III) chloride	$PtCl_3$	301.45	grn-blk	5.256^{25}	435	....	sl s; s hot HCl
(IV) chloride	$PtCl_4$	336.90	red-brn, cub	4.303^{25}	d 370	....	142^{25}
(II) cyanide	$Pt(CN)_2$	247.13	yel-brn cr	....	....	....	i; s KCN

Table 4-1 (*Continued*)
PHYSICAL CONSTANTS OF INORGANIC COMPOUNDS

Name	Formula	Formula weight	Color, crystalline form. refractive index	Density	Melting point, °C	Boiling point, °C	Solubility in 100 parts
Platinum							
(IV) diphosphate	PtP_2O_7	369.03	grn-yel	4.85	d 600	....	v sl s
(II) fluoride	PtF_2	233.09	yel-grn	....	....	....	i
(IV) fluoride	PtF_4	271.08	yel-brn, mn	....	d red heat	....	slow hyd; s a, alk
(V) fluoride	PtF_5	290.08	dk-red	....	80	disproportion-ates	....
(VI) fluoride	PtF_6	309.08	dk-red, rh	3.826 (liq)	61.3	69.1	....
fluoride oxide, tetra-	PtF_4O	286.08	red	....	260	subl 150	....
(II) iodide	PtI_2	448.90	blk pwd	6.403^{25}	d 500	....	i; s HI
(III) iodide	PtI_3	575.80	blk, cub	7.414^{25}	d 230	....	disproportionates
(IV) iodide	PtI_4	702.71	blk cr & brn amorp	6.064^{25}	d > 25	....	hyd; s al, alk, KI
(II) oxide	PtO	211.09	blk-blk	14.9^{15}	d 550	....	i; s HCl
(IV) oxide	PtO_2	227.09	blk	10.2	450	....	i; i aq reg
(II,IV) oxide	Pt_2O_3	438.18	....	....	d	....	i; i aq reg
(VI) oxide	PtO_3	243.09	red-brn pwd	....	....	....	i; s HCl, H_2SO_4
phosphide, di-	PtP_2	257.04	steel gray, cub	9.25	ca 1500	....	v sl s aq reg
selenide, di-	$PtSe_2$	353.01	gray cr, hex	7.65	....	....	i; s aq reg
(IV) sulfate-4-water	$Pt(SO_4)_2 \cdot 4H_2O$	459.27	yel pl	....	....	....	s
sulfide	PtS	227.15	blk, tetr	10.04	d	....	i
sulfide, di-	PtS_2	259.22	slk-brn, hex	7.66	d 225	....	i; s HCl, HNO_3
telluride	$PtTe$	322.7	rh	....	1200–1300	....	i; sl s Na_2S
telluride, di-	$PtTe_2$	450.29	gray, hex	....	....	....	i; s a
Plutonium	Pu	239.05	silv-wh, mn	19.82	639.6	3230	i; s a
(III) bromide	$PuBr_3$	478.79	bl-grn, rh, hygr	6.69	681	d > 1300	s
(III) bromide oxide	$PuBrO$	334.97	dk-grn, tetr	9.07	....	....	i; s a
carbide	PuC	251.06	blk, fcc	13.6	1650 d	....	i; s hot HNO_3 + NaF
(di) carbide, tri-	Pu_2C_3	514.15	dull blk, bcc	12.70	2050 d	....	....
carbide, di-	PuC_2	263.07	blk, tetr	....	2250 d	....	....
(III) chloride	$PuCl_3$	345.42	emerald grn, hex	5.70	760	1767	i; v s a

Name	Formula	Mol. wt.	Color, crystalline form, refr. index	Density	Melting point	Boiling point	Solubility
(III) chloride oxide	$PuClO$	290.51	bl-grn, tetr	8.81	......	......	i; s a
(IV) diphosphate	PuP_2O_7	412.99	col, cub, 1.68	4.37	d 1000	d 2000	hyd
(III) fluoride	PuF_3	296.06	purp, hex	9.32	1425	......	i
(IV) fluoride	PuF_4	315.05	lt-brn-pink, mn	7.00	1037 d	......	......
(VI) fluoride	PuF_6	353.05	yel-brn, rh	4.86	51.59	62.16	......
(III) fluoride oxide	$PuFO$	274.05	met, fcc	9.76	1637	......	hyd; s HCl
hydride, di	PuH_2	241.08	blk, fcc	10.40	ca 727	d 1000	s
hydride, tri-	PuH_3	242.08	blk, hex	9.61	ca 327	d 180	......
(III) iodide	PuI_3	619.77	bl-grn, rh	6.92	ca 777	......	......
(IV) nitrate-5-water	$Pu(NO_3)_4 \cdot 5H_2O$	578.16	grn, rh, 1.554	......	$-H_2O$, 40	......	s
nitride	PuN	253.07	brn-blk, fcc	14.25	2570 (in H_2)	......	......
(III) oxalate-10-water	$Pu_2(C_2O_4)_3 \cdot 10H_2O$	900.33	grn	......	$-10H_2O$, 225	d 300	0.004
(IV) oxalate-6-water	$Pu(C_2O_4)_2 \cdot 6H_2O$	515.19	yel-grn	......	$-6H_2O$, 135	d 150	0.006
(II) oxide	PuO	255.05	blk, fcc	13.9	1900	......	......
(III) oxide	Pu_2O_3	526.12	blk, bcc	10.2	2085 (in He)	......	......
(IV) oxide	PuO_2	271.05	grn-brn, fcc	11.46	2390 (in He)	d 2800	......
(III) phosphate	$PuPO_4$	334.02	ol, mn, 1.86	7.55	d 1000	d 1000	......
silicide	$PuSi$	267.14	met, rh	10.15	1576 (in N_2)	d 700 (air)	4.2
(III) sulfate-7-water	$Pu_2(SO_4)_3 \cdot 7H_2O$	882.40	vlt cr	......	......	......	s
(IV) sulfate	$Pu(SO_4)_2$	431.18	pink, hygr	10.6	2350	......	......
sulfide	PuS	271.12	yel-brn, cub	9.95	1725	......	......
(di-) sulfide, tri-	Pu_2S_3	574.30	blk, cub	......	d 130	......	......
Plutonyl carbonate	PuO_2CO_3	331.06	red	......	......	......	......
chloride-6-water	$PuO_2Cl_2 \cdot 6H_2O$	350.06	yel-grn	......	$-H_2O$, 130	......	35
fluoride	PuO_2F_2	309.05	wh, rhbd	6.50	expl 180	......	......
nitrate-6-water	$PuO_2(NO_3)_2 \cdot 6H_2O$	513.17	grn, rh, 1.554	......	......	d 230	s
oxalate-3-water	$PuO_2C_2O_4 \cdot 3H_2O$	413.12	red	......	......	......	......
Polonium	Po	210	silv-gray, cub	9.20	254	962	sl s; s a
(II) bromide	$PoBr_2$	369.8	purp-brn	......	$110^{0.03mm}$	subl	i; s HBr
(IV) bromide	$PoBr_4$	529.6	bright red, fcc	......	324 (in Br_2 vapor)	......	sl hyd; s al, HBr
(II) chloride	$PoCl_2$	280.9	dk red, rh	6.50	subl 190 d	......	i; s HNO_3
(IV) chloride	$PoCl_4$	351.8	yel, mn, hygr	......	300 (in Cl_2 vapor)	390	sl hyd

Table 4-1 (Continued)
PHYSICAL CONSTANTS OF INORGANIC COMPOUNDS

Name	Formula	Formula weight	Color, crystalline form, refractive index	Density	Melting point, °C	Boiling point, °C	Solubility in 100 parts
Polonium							
(IV) hydroxide	Po(OH)$_4$	278.0	lt yel, fcc	....	d 350	....	i; s alk
(IV) iodate	Po(IO$_3$)$_4$	909.6	wh	....	subl 200 d (in N$_2$)	....	i; s HCl
(IV) iodide	PoI$_4$	717.6	blk	....		....	i; s acet; sl a alk
(IV) oxide	PoO$_2$	242.0	yel fcc (& tetr)	*ca* 9	subl 885	....	v s dil HCl
(IV) sulfate	Po(SO$_4$)$_2$	402.2	purp	....	d 550		
Potassium	K	39.10	silv-wh, bcc	0.856^{20}	63.7	765.5	d to KOH; s a
acetate	KC$_2$H$_3$O$_2$	98.14	wh pwd	1.57^{25}	292	....	256^{20}, 34 al
aluminate-3-water, di-	K$_2$Al$_2$O$_4 \cdot$ 3H$_2$O	250.21	hard lust cr	....		....	s
aluminate, tetrachloro-	K[AlCl$_4$]	207.91	col cr	....	256	....	d
amide	KNH$_2$	55.12	yel-grn	1.64	338	subl 400	
argentate(I), dicyano-	K[Ag(CN)$_2$]	199.01	wh cr, 1.625	2.36		....	25^{30}
arsenate	K$_3$AsO$_4$	256.23	col cr, deliq	....	1310	....	19
aurate(III)-2-water, tetrabromo-	K[AuBr$_4$] $\cdot$ 2H$_2$O	591.74	vlt cr, rh	4.08		....	19.5^{15}
aurate(III), tetrachloro-	K[AuCl$_4$]	377.88	yel, mn	3.75	d 357	....	61.8^{20}
aurate(I)-2-water, dicyano-	K[Au(CN)$_2$] $\cdot$ 2H$_2$O	324.17	col cr, rh	3.45		....	14.3; sl s al
aurate(III)-1.5-water, tetracyano-	K[Au(CN)$_4$] $\cdot$ 1.5H$_2$O	367.16	col pl	....	d 200	....	s
azide	KN$_3$	81.12	col cr, tetr	2.04	350 (vac)	....	51^{20}
beryllate, hexafluoro-	K$_2$[BeF$_6$]	163.21	col cr, rh	....	d red heat	....	2^{20}
boranate, di-	K$_2$B$_2$H$_6$	105.87	wh cub, 1.493	1.18	subl 400	....	d
borate(1-)	KBO$_2$	81.91	col, hex, 1.526, 1.450	....	947	1401	71^{30}
borate-5-water, tetra-	K$_2$B$_4$O$_7 \cdot$ 5H$_2$O	323.52	col, mn	1.74	815	....	26.7^{30}
borate, tetrafluoro-	K[BF$_4$]	125.91	col, cub & rh-bipyr	2.505	530	....	0.45^{20}
borate, tetrahydro-	K[BH$_4$]	53.95	col cr	1.11	d 497	....	21^{25}, 3.5^{20} MeOH
bromate	KBrO$_3$	167.01	col cr, trig	3.27^{17}	350	d 370	6.9^{20}
bromide	KBr	119.01	col, fcc, 1.5594	2.75	734	1398	65^{20}; 0.4 al

cadmate, tetracyano-	$K_2[Cd(CN)_4]$	294.68	col, oct	1.846	450		33
cadmate-2-water, tetraiodo-	$K_2[CdI_4] \cdot 2H_2O$	698.21	col cr, distorted oct	3.359^{21}		d	137^{15}; 71^{15} al
carbonate	K_2CO_3	138.20	wh hygr pwd, 1.531	2.29	901		111^{20}; i al
carbonate-sesquihydrate	$K_2CO_3 \cdot 1.5H_2O$	165.24	wh cr, mn, 1.380, 1.482, 1.573	2.043			129; i al
chlorate	$KClO_3$	122.55	col cr, mn, 1.5167	2.338^{20}	368	d > mp	7.3^{20}, 2 gly
chloride (sylvite)	KCl	74.56	wh cr, cub, 1.4904	1.988	711	1437	34.2^{20}, 7 gly, 0.4 al
chromate (tarapacaite)	K_2CrO_4	194.20	lemon-yel, rh, 1.7261	2.732^{18}	975		63.7^{20}; i al
citrate-1-water	$K_3C_6H_5O_7 \cdot H_2O$	324.42	wh cr or pwd	1.98	d 230		167^{15}
cobaltate(II), hexacyano-	$K_4[Co(CN)_6]$	371.45	red-brn cr, deliq	2.039			v s
cobaltate(III), hexacyano-	$K_3[Co(CN)_6]$	332.34	yel-wh, mn	1.906	melts dec		s
cobaltate(III)-1.5-water, hexanitrito-	$K_3[Co(NO_2)_6] \cdot 1.5H_2O$	479.30	yel, tetr		d 200		0.089^{17}
cyanate	$KOCN$	81.11	wh cr, tetr	2.048	d 700-900		s; sl s al
cyanide	KCN	65.12	wh, cub, 1.410	1.52^{16}	622	1625	50
dichromate	$K_2Cr_2O_7$	294.19	or-red, tric pinacoidal	2.676^{25}	398	d 500	12.3^{20}
diphosphate(V)-3-water	$K_4P_2O_7 \cdot 3H_2O$	384.40	col cr, deliq	2.33	$-3H_2O$, 300		s
disulfate(IV)	$K_2S_2O_5$	222.32	wh cr, flammable if ground				s
dithionate	$K_2S_2O_6$	238.33	col, trig, 1.4550	2.278	d		6.6^{20}
ethyldithiocarbonate	KC_2H_5OCSS	160.30	lt-yel cr	1.558^{22}	d 200	d 200	v s
ferrate(II)-3-water, hexacyano-	$K_4[Fe(CN)_6] \cdot 3H_2O$	422.41	yel, mn, 1.5772	1.853^{17}	$-3H_2O$, 100		28^{20}
ferrate(III), hexacyano-	$K_3[Fe(CN)_6]$	329.26	ruby-red, mn, 1.569	1.89	d	d	84^{20}(slow)
ferrate(III)-3-water, trisoxalato-	$K_3[Fe(C_2O_4)_3] \cdot 3H_2O$	491.25	grn, mn		$-3H_2O$, 100	d 230	4.7^{0}
ferrate(II)-2-water, nitrosylpentacyano-	$K_2[Fe(CN)_5NO] \cdot 2H_2O$	330.18	garnet red hygr cr				100; s al
fluoride	KF	58.10	col, cub, 1.352	2.481	858	1517	95^{20}
formate	$KCHO_2$	84.10	col, rh, deliq	1.91	167.5	d > mp	337^{20}
germanate(IV) (2-)	K_2GeO_3	198.79	wh cr	3.40^{21}	823	835	s
germanate(IV), hexafluoro-	$K_2[GeF_6]$	264.78	wh, hex		730		0.5^{20}
gluconate	$KC_6H_{11}O_7$	234.24	yelsh-wh cr		d 180		v s; i al, bz, chl
hafnate, hexafluoro-	$K_2[HfF_6]$	370.68	wh, mn				3.1^{20}
hydride	KH	40.11	wh nd, cub, 1.453	1.43	417 d		d

Table 4-1 (Continued)
PHYSICAL CONSTANTS OF INORGANIC COMPOUNDS

Name	Formula	Formula weight	Color, crystalline form, refractive index	Density	Melting point, °C	Boiling point, °C	Solubility in 100 parts
Potassium							
hydrogen arsenate-1-water	$K_2HAsO_4 \cdot H_2O$	236.15	col, trig	....	$-H_2O$, 110	....	s
hydrogen arsenate, di-	KH_2AsO_4	180.02	col, tetr, 1.5674	2.867	288	....	19^6; 63 gly; i al
hydrogen carbonate	$KHCO_3$	100.11	col, mn, 1.482	2.17	d 100–200	....	33.7^{20}
hydrogen difluoride	KHF_2	78.11	col, tetr	2.37	238.7	d 478	39.2^{20}; s al
hydrogen bisiodate	$KH(IO_3)_2$	389.92	col, rh & mn	2.044	....	....	1.33^{15}
hydrogen oxalate	KHC_2O_4	128.11	col, mn, 1.382, 1.553, 1.573	2.044	d	....	2.5
hydrogen bisoxalate-2-water, tri-	$KH_3(C_2O_4)_2 \cdot 2H_2O$	254.20	col to wh, tric, 1.415, 1.536, 1.560	1.836	d	....	1.8^{13}
hydrogen phosphate	K_2HPO_4	174.18	wh, hygr	....	d ($K_4P_2O_7$)	....	150
hydrogen phosphate, di-	KH_2PO_4	136.09	col, tetr, deliq	2.338	400	d(KPO_3)	22.6^{20}
hydrogen phthalate	$KHC_8H_4O_4$	204.22	wh cr, rh	1.636^{25}	d	....	10.2; sl s al
hydrogen sulfate (mercal-lite)	$KHSO_4$	136.17	wh, rh, 1.480	2.24	197	d($K_2S_2O_7$)	48^{20}
hydrogen sulfide	KHS	72.17	col, trig, deliq	1.70	455–510	....	s; s al
hydrogen sulfite	$KHSO_3$	120.17	wh, mn	....	d 190	....	45.5^{15}
hydrogen tartrate	$KHC_4H_4O_6$	188.18	col, rh	1.956	....	....	0.5^{20}
hydrogen tellurate-3-water, tetra-	$K_2H_4TeO_6 \cdot 3H_2O$	359.88	col, rh, deliq	....	d 300	....	25
hydroxide	KOH	56.11	wh, rh, deliq	2.044	406	1320	112^{20}; 33 al
iodate	KIO_3	214.02	col, mn	3.89^{25}	560 d	....	8.1^{20}; i al
iodide	KI	166.02	wh, cub, 1.6670	3.12	681	1345	144^{20}; 4.5 al; 1.2 acet
iodide, tri-	KI_3	419.80	dk bl, mn, deliq	3.498^{15}	45	d 225	v s
iridate(IV), hexachloro-	$K_2[IrCl_6]$	483.12	blk, cub	3.546	d	....	1.3^{20}
manganate(VI)	K_2MnO_4	197.12	dk-grn, rh	....	d 190	....	s; stable in KOH
mercurate(II), tetracyano-	$K_2[Hg(CN)_4]$	382.87	wh cr	....	....	....	s
mercurate(II), tetraiodo-	$K_2[HgI_4]$	786.48	yel cr, deliq	....	....	....	v s

Name	Formula	Mol. wt.	Crystalline form, color, refractive index	Density	Melting point	Boiling point	Solubility
molybdate(VI)	K_2MoO_4	238.14	wh pwd, deliq	2.91^{18}	919	d 1400	160
molybdate(IV)-2-water, octacyano-	$K_4[Mo(CN)_8] \cdot 2H_2O$	496.52	yel, rh	2.337	$-2H_2O$, 110	......	v s
nickelate(II)-1-water, tetracyano-	$K_2[Ni(CN)_4] \cdot H_2O$	259.00	or-yel pwd, mn	1.875^{51}	$-H_2O$, 100	......	s
niobate(V)-1-water, pentafluoro-	$K_2[NbOF_5] \cdot H_2O$	300.12	col pl or leaf, mn	......	......	......	7.7
nitrate (saltpeter)	KNO_3	101.10	col, rh, 1.335, 1.5056, 1.5064	2.109^{16}	334.3	d 400	32^{20}; 0.16 al, s gly
nitrite	KNO_2	85.10	yelsh-wh pr, deliq	1.915	441	d 250	306^{20}
osmate(IV), hexachloro-	$K_2[OsCl_6]$	481.13	dk-red, cub	......	d	$d(K_2CO_3)$	s
oxalate-1-water	$K_2C_2O_4 \cdot H_2O$	184.24	wh, mn, 1.4410, 1.485, 1.550	2.127^4	$-H_2O$, 160	......	36^{20}
oxide	K_2O	94.20	wh, cub	2.32^{20}	d 881^{600mm}	d to KOH	
oxide, di-	KO_2	71.10	yel leaf, cub	2.14	509	d	v s d
palladate(II), tetrachloro-	$K_2[PdCl_4]$	326.42	red-brn, tetr	2.67	d 105	......	s
palladate(IV), hexachloro-	$K_2[PdCl_6]$	397.32	red, cub	2.738	d	......	sl s d; sl s HCl
perchlorate	$KClO_4$	138.56	col, rh, 1.4737	2.5298^{25}	525	d 653	1.68^{20}
periodate	KIO_4	230.01	col, tetr, 1.6205	3.618^{15}	582 d	......	0.42^{20}
permanganate	$KMnO_4$	158.03	dk-purp, rh	2.703	240 d	......	6.34^{20}
peroxide	K_2O_2	110.20	wh amorp, deliq	......	490	......	d
peroxoborate-0.5-water	$KBO_3 \cdot 0.5H_2O$	106.92	wh cr	......	$-O_2$, 100	d 150	2.43^{15}
peroxochromate	K_3CrO_8	297.30	brn-red, cub	......	d 170	......	sl s
peroxodicarbonate-1-water	$K_2C_2O_6 \cdot H_2O$	216.24	wh granules	......	......	......	6.5; d hot H_2O
peroxodisulfate	$K_2S_2O_8$	270.32	col, tric, 1.461, 1.467, 1.566	2.477	d 100	......	5.3^{20}
perrhenate	$KReO_4$	289.30	wh, tetr, 1.643	4.38	555	1370	0.99^{20}
phenolsulfonate-1-water	$KC_6H_4(OH)SO_3 \cdot H_2O$	240.28	wh cr	1.87	......	......	s; s al
phosphate	K_3PO_4	212.28	wh deliq cr, rh	2.564^{17}	1340	......	92.3^{20}
phosphate, meta-	KPO_3	118.07	col, 1.458, 1.487	2.45^{20}	807	......	i
phosphinate	KPH_2O_2	104.08	col deliq cr, hex	......	d&ign	......	167; 11 al
phosphonate	K_2PHO_3	158.18	wh deliq pwd	......	d	......	v s
picrate	$KC_6H_2N_3O_7$	267.20	yelsh or grnsh, rh, 1.527, 1.903, 1.952, explodes on shock	1.852	expl 310	......	0.5^{15}; 1.08 acet, 0.27 MeOH
platinate(IV)-3-water	$K_2PtO_3 \cdot 3H_2O$	375.34	yel, rh	......	d	......	s

Table 4-1 (Continued)
PHYSICAL CONSTANTS OF INORGANIC COMPOUNDS

Name	Formula	Formula weight	Color, crystalline form, refractive index	Density	Melting point, °C	Boiling point, °C	Solubility in 100 parts
Potassium							
platinate(IV), hexabromo-	$K_2[PtBr_6]$	752.75	dk red-brn, cub	4.66	d 400	....	2.0[20]
platinate(IV), hexachloro-	$K_2[PtCl_6]$	486.03	or-yel, cub, 1.83	3.499	d 250	....	0.48[12]
platinate(II)-3-water, tetra-cyano-	$K_2[Pt(CN)_4] \cdot 3H_2O$	431.41	lt yel, rh, deliq	2.455[16]	$-3H_2O$, 100	d 400	sl s
platinate(IV), hexaiodo-	$K_2[PtI_6]$	1034.72	blk, cub	4.96	....	....	s
rhenate(IV), hexachloro-	$K_2[ReCl_6]$	477.12	yel-grn, oct	3.34	....	....	0.8
ruthenate(VI)-1-water	$K_2RuO_4 \cdot H_2O$	261.29	blk, rh	....	$-H_2O$, 200	....	v s
ruthenate(III), pentachloro-nitrosyl-	$K_2[RuCl_5(NO)]$	386.55	dk red, rh	....	d	....	12
salicylate	$KC_7H_5O_3$	176.21	wh pwd	....	....	....	118
selenate(VI)	K_2SeO_4	221.15	col, rh, 1.535, 1.539, 1.545	3.066[20]	....	....	110[20]
selenate(IV)	K_2SeO_3	205.16	wh deliq cr	....	d 875	....	203[20]
selenide	K_2Se	157.15	wh deliq cr, cub	2.29	turns red-blk	....	s
selenocyanate	$KSeCN$	144.08	deliq nd	....	d 100	....	s
silicate(2–)	K_2SiO_3	154.29	col, rh, 1.530	2.417	976	....	s
silicate(2–), tetra-	$K_2Si_4O_9 \cdot H_2O$	352.56	col, rh, 1.530	2.27	d 400	....	s
silicate, hexafluoro- (hiera-tite)	$K_2[SiF_6]$	220.25	col, cub, 1.3991		d	....	sl s
sodium carbonate-6-water	$KNaCO_3 \cdot 6H_2O$	230.19	col, mn	1.633	$-6H_2O$, 100	....	185[15]
sodium hexanitritoco-baltate(III)-1-water	$K_2Na[Co(NO_2)_6] \cdot H_2O$	454.18	yel	1.633	d 135	....	0.07
sodium tartrate-4-water (Rochelle salt)	$KNaC_4H_4O_6 \cdot 4H_2O$	282.23	col pwd cr, rh, 1.492, 1.493, 1.496	1.790	70–80	$-4H_2O$, 130, d 220	54[15]
sorbate	$KC_6H_7O_2$	150.22	col cr	1.363	d 270	....	110[20]
stannate(IV)-3-water	$K_2SnO_3 \cdot 3H_2O$	298.94	col cr, trig	3.197	$-3H_2O$, 140	....	100[20]
stannate(IV), hexachloro-	$K_2[SnCl_6]$	409.63	col, cub, 1.657	2.71	....	....	s
stannate(IV)-water, hexa-fluoro-	$K_2[SnF_6] \cdot H_2O$	328.90	wh, mn	3.053	....	....	6.7[18]

Name	Formula	Formula weight	Color, crystalline form	Density	Melting point °C	Boiling point °C	Solubility
stearate	ca $KC_{18}H_{35}O_2$	......	wh pwd	......	......	......	slowly s
sulfate (arcanite)	K_2SO_4	174.27	col, rh. 1.494	2.662	1067	1670	11^{20}; 1.3 gly; i al
sulfite-2-water	$K_2SO_3 \cdot 2H_2O$	194.30	wh cr, rh		d		106^{20}
(di-) sulfide, tetra–	K_2S_4	206.46	red-brn	4.5-5.2	145	d 850	s
tantalate(V), heptafluoro-	$K_2[TaF_7]$	392.14	ccl, rh	1.98			1.28^{20}
tartrate-hemihydrate	$K_2C_4H_4O_6 \cdot 5H_2O$	235.28	wh cr, mn, 1.526		$-H_2O$, 155		138^{17}
tellurate(IV)	K_2TeO_3	253.80	wh pwd		d 200	d 200	s
telluride	K_2Te	205.80	col hygr cr, cub	2.51			s d
thioantimonate(V)-4.5-water	$K_3SbS_4 \cdot 4.5H_2O$	448.38	col to yel cr		300^{0}		300^{0}
thioarsenate(V)	K_3AsS_4	320.48	wh deliq cr		d		v s
thiocarbonate	K_2CS_3	186.41	yel-red, deliq		d		v s
thiocyanate	KSCN	97.18	col, rh pr, deliq	1.886^{14}	173	d 500	217^{20}, 200 acet, 8 al
thiostannate(IV)–3–water	$K_2SnS_3 \cdot 3H_2O$	347.13	dk-brn oil		$-3H_2O$, 100		s
thiosulfate	$K_2S_2O_3$	190.33	col hygr cr, cub	1.847^{18}	d 400		155^{20}
titanate(IV)-1-water, hexa-fluoro-	$K_2[TiF_6] \cdot H_2O$	258.11	wh, mn		$-H_2O$, 100		1.3^{20}
titanate(IV), oxobisoxalato-diaqua-	$K_2[TiO(C_2O_4)_2(H_2O)_2]$	354.18	co. cr	2.992			v s
tungstate(IV)	K_2WO_4	326.06	col deliq cr, mn	3.12	921		51.5
tungstate(VI)-1-water, tetra-fluoro-	$K_2[WO_2F_4] \cdot H_2O$	388.06	col, mn		$-H_2O$, 600		6^{17}
tungstate(VI)-2-water, octacyano-	$K_4[W(CN)_8] \cdot 2H_2O$	584.43	lt yel-grn pwd cr	1.989	$-2H_2O$, 114		130^{18}
uranyl(VI) acetate-1-water	$K(UO_2)(C_2H_3O_2)_3 \cdot H_2O$	504.28	yel, tetr	3.296^{15}	$-H_2O$, 275		s
uranyl(VI) carbonate	$K_4(UO_2)(CO_3)_3$	606.46	yel, hex		d 300		7.4^{15}
uranyl(VI) nitrate	$K(UO_2)(NO_3)_3$	495.19	grn-yel pwd		90		90
uranyl(VI) sulfate-2-water	$K_2(UO_2)(SO_4)_2 \cdot 2H_2O$	576.39	yel, mn	3.363^{19}	$-2H_2O$, 120		11^{25}
vanadium(III) sulfate-12-water	$KV(SO_4)_2 \cdot 12H_2O$	498.35	vlt, cub	1.783^{20}	$-12H_2O$, 230		v s
zinc sulfate-6-water	$K_2Zn(SO_4)_2 \cdot 6H_2O$	443.81	col cr	3.58			s
zirconate, hexafluoro-	$K_2[ZrF_6]$	283.41	col, mn. 1.465				2.7^{20}
Praseodymium	Pr	140.907	lt yel, hex	6.782	919	3130	i; s a & hot H_2O
acetate-3-water	$Pr(C_2H_3O_2)_3 \cdot 3H_2O$	372.09	grn nd		56.5		32^{25}
bromate-9-water	$Pr(BrO_3)_3 \cdot 9H_2O$	686.77	grn, hex			$-7H_2O$, 170	120^{20}
bromide	$PrBr_3$	380.63	yel-grn, hex	5.26	691	1547	sl s d
carbide, di-	PrC_2	164.93	yel cr	5.10	d		s dil a
carbonate-8-water	$Pr_2(CO_3)_3 \cdot 8H_2O$	605.96	grn silky pl		$-6H_2O$, 100		s a

Table 4-1 (*Continued*)
PHYSICAL CONSTANTS OF INORGANIC COMPOUNDS

Name	Formula	Formula weight	Color, crystalline form, refractive index	Density	Melting point, °C	Boiling point, °C	Solubility in 100 parts
Praseodymium							
chloride	$PrCl_3$	247.27	lt grn nd, hex	4.02	769–782	1710	104[13], s al
fluoride, tri-	PrF_3	197.91	grn, hex	....	1395	2327	i; s H_2SO_4
fluoride, tetra-	PrF_4	215.91	wh, mn	4.94	....	....	....
hydride, tri-	PrH_3	143.92	grn amorp	5.5	....	....	....
hydroxide	$Pr(OH)_3$	191.93	lt-grn, hex	....	$-H_2O$, 220	d 340 to Pr_2O_3	i; s a
iodide	PrI_3	521.62	rh	....	737	1380	s
molybdate(VI)	$Pr_2(MoO_4)_3$	761.63	grn, tetr	4.48	1030	....	0.0015[20]
nitrate-6-water	$Pr(NO_3)_3 \cdot 6H_2O$	339.03	grn nd	....	....	....	164[20]
oxalate-10-water	$Pr_2(C_2O_4)_3 \cdot 10H_2O$	726.03	lt grn er	....	oxidizes to Pr_6O_{11}	....	s a
(III) oxide	Pr_2O_3	329.81	lt grn, hex	7.07	tr 350 to Pr_6O_{11}	....	i; s a
(IV) oxide	PrO_2	172.91	bl-blk, bcc	6.82	....	....	....
selenate	$Pr_2(SeO_4)_3$	710.69	....	4.30[15]	....	....	36[0]
selenide	$PrSe$	219.87	cub	....	2097	....	....
(di-) selenide, tri-	Pr_2Se_3	478.69	carmine red, cub	6.64	....	....	....
sulfate	$Pr_2(SO_4)_3$	570.00	lt grn	3.72	....	....	12.7[20]
sulfide	PrS	172.97	cub	....	2227	....	....
(di-) sulfide, tri-	Pr_2S_3	378.01	brn, cub	5.024	1795	....	i; s a
telluride	$PrTe$	268.5	cub	....	1927	....	....
Promethium-147	Pm	146.915	....	....	1080	....	....
bromide	$PmBr_3$	386.7	pink, rh	5.38	727	1667	s
chloride	$PmCl_3$	153.4	lt bl, hex	....	737	1670	s
fluoride	PmF_3	204	purplish-pink, hex	....	1410	2330	....
iodide	PmI_3	527.7	....	....	800	1370	s
nitrate	$Pm(NO_3)_3$	333	pink	....	....	....	s

Name	Formula	Mol wt	Form/color	Density	mp/°C	bp/°C	Solubility
Protoactinium							
	Pa	231.04	gray-met, bc tetr	15.37	(1227)	(4027)	hyd; s al
(V) bromide	$PaBr_5$	630.56	dk red, mn	......	subl 250	......	i; s HCl
carbide	PaC	243.05	blk, fcc	4.72	>1900	......	......
(IV) chloride	$PaCl_4$	372.85	yel-grn, tetr	......	subl 400	......	hyd; s CH_3CN
(V) chloride	$PaCl_5$	408.31	yel, mn, hygr	3.74	301	420	i
(V) chloride oxide, octa-	Pa_2Cl_8O	761.70	wh, bcc, hygr	......	800	......	......
(IV) fluoride	PaF_4	307.03	brn, mn	6.36	d 160	......	hyd; s CH_3CN
(V) fluoride	PaF_5	325.04	wh, tetr	6.28	d 850(vac)	subl 500 (vac)	hyd; s CH_3CN
(V) fluoride oxide, octa-	Pa_2F_8O	619.08	wh, bcc, hygr	......	subl 500	......	......
(IV) iodide	PaI_4	738.66	dk grn	......	......	......	......
(V) iodide	PaI_5	865.56	blk, rh, hygr	......	......	......	......
(IV) oxide	PaO_2	263.04	blk, cub	......	>1550	......	......
(V) oxide	Pa_2O_5	542.14	wh, cub	......	......	......	i; s HF
Radium							
	Ra	226.03	silv-wh met, bcc	5.5	700.1	1737	d; s a
bromide	$RaBr_2$	385.88	yelsh-wh, mn	5.79	728	subl 900	s
carbonate	$RaCO_3$	286.01	wh, mn	......	......	......	i; s a
chloride	$RaCl_2$	296.96	yelsh-wh, mn	4.91	1000	......	s
iodate	$Ra(IO_3)_2$	575.83	wh	......	......	......	0.018^{0}
nitrate	$Ra(NO_3)_2$	350.03	wh cr	......	......	......	13.9^{20}
sulfate	$RaSO_4$	322.09	wh, rh	......	......	......	i; i a
Radon							
	Rn	(222)	col inert gas	9.73^{0} (g/L); 4.4^{bp} (liq)	−71	−62	23 ml^{20}, s org solv
Rhenium							
	Re	186.21	silv-gray met, hcp	21.02	3180	5678	i; s HNO_3
(III) bromide	$ReBr_3$	425.93	brn-blk	......	subl 500	d	v s
(V) bromide	$ReBr_5$	585.72	dk bl	......	ca 28	......	......
(di-) carbonyl, deca-	$Re_2(CO)_{10}$	652.51	col, cub	......	d 250	......	......
(V) chloride	$ReCl_5$	363.47	brn-blk, mn	4.9	ca 260	ca 330	i; s HCl
(VI) chloride	$ReCl_6$	398.92	dk grn	......	25	......	hyd
(VI) chloride oxide, tetra-	$ReCl_4O$	344.01	brn cr	......	34	225	hyd; s CCl_4
(VII) chloride trioxide	$ReClO_3$	269.65	col liq	3.309^{34}	4.5	128	hyd
(IV) fluoride	ReF_4	262.19	lt bl, tetr	5.38	124.5	795	hyd
(V) fluoride	ReF_5	280.20	yel-grn, rh	......	48	d 140	d
(VI) fluoride	ReF_6	300.19	lt yel, bcc, hygr	3.58 liq	18.5	33.8	s HF
(VII) fluoride	ReF_7	319.19	yel, cub	3.65^{52}	48.3	73.7	hyd
fluoride oxide, tetra-	ReF_4O	278.19	bl, mn	4.032	108	171	d

Table 4-1 (Continued)
PHYSICAL CONSTANTS OF INORGANIC COMPOUNDS

Name	Formula	Formula weight	Color, crystalline form, refractive index	Density	Melting point, °C	Boiling point, °C	Solubility in 100 parts
Rhenium							
fluoride oxide, penta-	ReF_5O	297.19	creamy cr or col liq		34.5	73.0	hyd
fluoride dioxide, tri-	ReF_3O_2	275.19	wh		95	(185)	hyd
fluoride trioxide	$ReFO_3$	253.20	yel		147	ca 164	hyd
(IV) iodide	ReI_4	693.82	amorp, hygr				hyd; s acet
(IV) oxide	ReO_2	218.20	blk, mn	11.4	d 1000		i; s HNO_3
(VI) oxide	ReO_3	234.20	red, cub	6.9–7.4	disprop 400(vac)	750	i; s HNO_3
(VII) oxide	Re_2O_7	484.44	yel, hex, deliq	6.1	300.3; subl 250	360.3	v s; s org solv
phosphide	ReP	217.17	gray, cub	12.0			
phosphide, di-	ReP_2	248.14	gray, cub	8.83			
sulfide, di-	ReS_2	250.33	blk, hex	7.5	d 1000		i; s HNO_3
sulfide, hepta-	Re_2S_7	596.85	blk, tetr	4.866	d 460		i; s HNO_3
telluride, di-	$ReTe_2$	441.4	tric		935		
Rhodium							
	Rh	102.91	silv-wh, fcc	12.41^{20}	1966	3727	s fus $KHSO_4$
(di) carbonyl, octa-	$Rh_2(CO)_8$	429.90	or		76 d		s org solv exc hydrocarbons
tetracarbonyldi-μ-chlorodi-	$Rh(CO)_4Cl_2$	388.75	or-red cr		124–125		
(III) chloride	$RhCl_3$	209.28	red pwd, mn		d 450	subl 850	i; s KOH, KCN
(III) fluoride	RhF_3	159.90	red, rhbd	5.38	subl 600		i a, alk
(V) fluoride	RhF_5	197.90	dk red, mn		95.5		hyd viol
(VI) fluoride	RhF_6	216.90	yel, cub		70	d	v s
(III) nitrate-2-water	$Rh(NO_3)_3 \cdot 2H_2O$	324.93	red, deliq				
(III) oxide	Rh_2O_3	253.81	bl-gray, rhbd	8.20	d 1100		i aq reg, KOH
(IV) oxide	RhO_2	134.90	brn				i a, alk
(di) phosphide	Rh_2P	236.79	steel-gray, cub	9.13	ca 1500		

Physical Constants of Inorganic Compounds (continued)

Name	Formula	Mol. wt.	Crystalline form; color; index of refraction	Density	Melting point, °C	Boiling point, °C	Solubility
(III) sulfate-4-water, di-	Rb₂(SO₄)₃·4H₂O	566.05	red	......	......	......	s
selenide, di-	RhSe₂	260.83	cub	......	d	......	i; i aq reg
(III) sulfide	Rh₂S₃	302.00	blk	6.40	......	694	d to RbOH
Rubidium	Rb	85.47	silv-wh, bcc	1.532²⁰; 1.475 (liq)	39.0	......	......
acetate	RbC₂H₃O₂	144.52	col, hygr	......	246	......	86⁴⁵
amide	RbNH₂	115.49	col	2.58	309	......	......
borate, tetrafluoro-	Rb[BF₄]	172.27	wh, rh, 1.333	2.820²⁰	590	......	0.6¹⁷
bromate	RbBrO₃	213.37	wh, hex, 2.2	3.68²²	430 d	d 500	2.9
bromide	RbBr	165.39	wh, fcc, 1.5528	3.35	680	1352	108²⁰
carbonate	Rb₂CO₃	230.97	wh, deliq	......	837	d 900	v s; 0.74¹⁹ al
chlorate	RbClO₃	168.94	wh, trimetric	3.184²⁰	342	......	5.4²⁰
chloride	RbCl	120.94	wh, cub, 1.4936	2.76	715	1381	91²⁰; 1.1 MeOH
chromate	Rb₂CrO₄	286.93	yel, rh, 1.71	3.518	980	......	73.6²⁰
cyanide	RbCN	111.49	col or pwd	2.32	......	......	s
dichromate	Rb₂Cr₂O₇	386.93	red, tric	3.021	400	d 650	5.8²⁰
fluoride	RbF	104.48	wh, cub, 1.396	3.557	775	1390	300¹⁸
hydride	RbH	86.49	wh, bcc	2.595	d 300	......	d
hydrogen carbonate	RbHCO₃	146.49	wh, rh	......	d 175	......	110²⁰
hydrogen phosphate, di-	RbH₂PO₄	182.47	wh cr	2.892¹⁶	840	......	s
hydrogen sulfate	RbHSO₄	182.54	col, rh, 1.473	2.282	d red heat	......	s
hydrogen tartrate	RbHC₄H₄O₆	234.55	trimetric pr	3.203¹¹	201 d	......	1.18
hydroxide	RbOH	102.49	graysh-wh, deliq	......	300	......	180¹⁵; s al
iodate	RbIO₃	260.37	wh, mn & cub	4.56	d	......	1.96
iodide	RbI	212.37	col, cub	3.55	640	1304	144¹⁸
nitrate	RbNO₃	147.47	col, hygr, hex	3.11	310	......	53²⁰
oxide	Rb₂O	186.94	yel, cub	3.72	477 d	......	s
oxide, di-	RbO₂	117.47	dk or	3.53⁰	412	......	......
(di-) oxide, tri-	Rb₂O₃	218.94	blk, cub	3.04²⁰	489	......	......
perchlorate	RbClO₄	184.92	col, rhbd, 1.4701	3.918¹⁶	324	d 606	1.55²⁰
periodate	RbIO₄	276.37	col, tret	3.235¹⁰	d 295	......	0.65¹³
permanganate	RbMnO₄	204.41	vlt cr	......	......	......	1.06²⁰
peroxide	Rb₂O₂	202.94	yel, cub	3.65⁰	600	......	d
perrhenate	RbReO₄	335.68	wh cr	4.73	598	d 1011	0.47³⁰

Table 4-1 (*Continued*)
PHYSICAL CONSTANTS OF INORGANIC COMPOUNDS

Name	Formula	Formula weight	Color, crystalline form, refractive index	Density	Melting point, °C	Boiling point, °C	Solubility in 100 parts
Rubidium							
platinate(IV), hexachloro-	$Rb_2[PtCl_6]$	578.75	yel, cub	3.94^{18}	d	....	0.028^{20}
selenate	Rb_2SeO_4	313.94	col, rh	3.90	....	....	159^{12}
silicate, hexafluoro-	$Rb_2[SiF_6]$	313.02	wh, cub	3.332	....	....	0.16^{20}
sulfate	Rb_2SO_4	267.03	col, rh, 1.5133	3.613^{20}	1060	....	48^{20}
sulfide	Rb_2S	203.00	red, cub, deliq	2.912	530 d (vac)	....	v s
(di) sulfide, penta-	Rb_2S_5	331.26	red, rh, deliq	2.618^{15}	225	....	d; s al
vanadium(III) sulfate-12-water	$RbV(SO_4)_2 \cdot 12H_2O$	544.72	yel, cub, 1.4689	1.915^{20}	$-12H_2O$, 230	d 300	2.56^{10}
Ruthenium	Ru	101.07	silv-met, hcp	12.45^{20}	2427	4119	s fus alk, oxid fluxes
(III) bromide	$RuBr_3$	340.80	dk-brn pwd, hex	....	....	....	i; s al, bz
carbonyl, penta-	$Ru(CO)_5$	241.12	col liq	....	-22	....	i; s HCl, al
(III) chloride	$RuCl_3$	207.47	dk-brn fluffy, hex	3.11	d > 500	....	s al
(IV) chloride-5-water	$RuCl_4 \cdot 5H_2O$	332.96	red-brn cr, hygr	....	d	....	s
[di-μ-oxotetradecaamminetriruthenium(II)]hexachloride-4-water (ruthenium red)	$[(NH_3)_5 \cdot Ru{-}O{-}Ru(NH_3)_4{-}O{-}Ru(NH_3)_5]Cl_6 \cdot 4H_2O$	858.41	red-brn pwd	....	....	....	
(V) fluoride	RuF_5	196.06	dk-grn, mn	3.82	86.5	227	d
(VI) fluoride	RuF_6	215.06	dk-brn, cub	....	54.0	d	
fluoride oxide, tetra-	RuF_4O	193.06	lt-grn cr	....	115	184	
(III) iodide	RuI_3	481.78	bl, hex	....	d 590	....	i; s fus alk
(IV) oxide	RuO_2	133.07	bl, tetr	6.97	d	....	i; s fus alk
oxide, tetra-	RuO_4	165.07	yel nd, mn, volatile	3.29	25.4	40	2.03^{20}; s CCl_4
sulfide, di-	RuS_2	165.20	blk, cub	6.14	d>1000	....	i; s fus alk
Samarium	Sm	150.35	yelsh, rhbd	7.536	1072	1803	i; s a
acetate-3-water	$Sm(C_2H_3O_2)_3 \cdot 3H_2O$	381.53	....	1.94	....	....	15^{25}
bromate-9-water	$Sm(BrO_3)_3 \cdot 9H_2O$	696.21	yel, hex	....	75	$-9H_2O$, 150	81.5^{20}
(II) bromide	$SmBr_2$	310.17	brn	5.1	700	1880	s d

	Formula	Mol. wt.	Crystalline form, color, refractive index	Density	Melting point, °C	Boiling point, °C	Solubility
(III) bromide	$SmBr_3$	390.06	yel, rh	5.40	640	d 900(vac)	d; s a
carbide, di-	SmC_2	174.37	yel, hex	5.86		2030	s d
(II) chloride	$SmCl_2$	221.26	dk-brn, rh	3.687^{22}	848	d	93.4^{20}
(III) chloride	$SmCl_3$	256.71	lt-yel, hex, hygr	4.465	686		0.053^{20}
(III) chromate-8-water	$Sm_2(CrO_4)_3 \cdot 8H_2O$	792.80	yel				i; s H_2SO_4
(II) fluoride	SmF_2	188.35	yel, cub	6.643	1417	2427	i; s a
(III) fluoride	SmF_3	207.35	wh, hex	6.52	1306	2427	
hydride, di-	SmH_2	152.37					hyd
(III) hydroxide	$Sm(OH)_3$	201.37	lt-yel gel ppt, hex		$-H_2O$, 220	d 325 to Sm_2O_3	s
(II) iodide	SmI_2	404.16	dk-grn, mn	5.36	520		s
(III) iodide	SmI_3	531.06	or-yel, hex		850 d		v s
(III) molybdate(VI)	$Sm_2(MoO_4)_3$	780.51	vlt, rh-oct			1583	i; s a
(III) nitrate-6-water	$Sm(NO_3)_3 \cdot 6H_2O$	444.46	lt-yel, tric	2.375	78		s a
oxalate-10-water	$Sm_2(C_2O_4)_3 \cdot 10H_2O$	744.91	wh cr				i
(III) oxide	Sm_2O_3	348.70	yel-tan, mn	8.347	2300		
(di) oxide sulfate, di-	$Sm_2O_2SO_4$	428.76	yel pwd		d 1100		
(di) oxide sulfide, di-	Sm_2O_2S	364.76	lt-tan, hex	6.90	1980		
sulfate-8-water	$Sm_2(SO_4)_3 \cdot 8H_2O$	733.01	lt-yel, mn, 1.543, 1.552, 1.563	2.930^{18}	$-8H_2O$, 450		2.7^{20}
(III) sulfide	Sm_2S_3	396.89	yel-pink, cub	5.729	1780		d
Scandium	Sc	44.956	silv-met, fcc(β)	2.99	1538	2730	s a
bromide	$ScBr_3$	284.68	wh, rh	3.93	948		s
carbonate-12-water	$Sc_2(CO_3)_3 \cdot 12H_2O$	486.13	bulky wh ppt				s alk, hot Na_2CO_3
chloride	$ScCl_3$	151.32	wh, rh, deliq	2.39	960	967	v s
fluoride	ScF_3	101.96	wh, rh	2.5	1515		i; s alk fluorides
hydroxide	$Sc(OH)_3$	95.98	cub & hemihed		$-H_2O$, 550		0.006; s alk
iodide	ScI_3	425.67	wh, hex		920		s a
nitrate-4-water	$Sc(NO_3)_3 \cdot 4H_2O$	303.03	col pr, deliq		d 120		v s
nitride	ScN	58.96	bcc		2650		
oxalate-6-water	$Sc_2(C_2O_4)_3 \cdot 6H_2O$	462.07	wh		d 635 to oxide		
oxide	Sc_2O_3	137.91	wh pwd, bcc	3.864	>2400		s a
sulfate-5-water	$Sc_2(SO_4)_3 \cdot 5H_2O$	468.17	col cr	2.519	$-5H_2O$, 250	d 550	54.6^{25}
sulfide	Sc_2S_3	186.09	yel, rh				i; s a

Table 4-1 (*Continued*)

PHYSICAL CONSTANTS OF INORGANIC COMPOUNDS

Name	Formula	Formula weight	Color, crystalline form, refractive index	Density	Melting point, °C	Boiling point, °C	Solubility in 100 parts
Selenic acid	H_2SeO_4	144.98	wh, hex pr, deliq	2.9508	58	260	567^{20}(viol)
-1-water	$H_2SeO_4 \cdot H_2O$	162.99	wh nd	2.627^{15} / 2.356 liq	26	205	v s
Selenious acid	H_2SeO_3	128.98	col, hex, deliq	3.004^{15}	d 70	....	167^{20}; v s al
Selenium	Se	78.96	gray-met, hex-rh / dk-red, mn(α)	4.81^{20} / 4.48^{20}	221 / 170	685	s CS_2, KOH, KCN
(di-) bromide, di-	Se_2Br_2	317.75	dk red oil liq, hygr, 1.96(Li)	3.604^{15}	....	225–230	d; s chl, CS_2
(IV) bromide	$SeBr_4$	398.62	red to yel cr pwd	4.029	d 70–80	subl 115	d; s HBr, chl, CS_2
(di-) chloride, di-	Se_2Cl_2	228.83	dk-red oily liq. 1.5993	2.789^{20}	−85	127^{733mm}d	s bz, chl, CS_2
(IV) chloride	$SeCl_4$	220.77	col-lt yel cr	2.6	....	subl 196	d
chloride oxide, di-	$SeCl_2O$	165.87	col liq. 1.6516	2.42^{22}	ca 5	180	d; s bz, chl
(IV) fluoride	SeF_4	154.95	col liq	2.732^{20}	−9.5	107.8	hyd viol; s chl, eth
(VI) fluoride	SeF_6	192.96	col gas	2.108^{-10}	−50.8	subl −63.8	i
fluoride oxide, di-	SeF_2O	132.97	fum liq	2.801^{20}	15.0	126	hyd
fluoride dioxide, di-	SeF_2O_2	148.96		3.05^{-73}	−99.5	−8.4	d
fluoride hypofluorite, penta-	$SeF_5(OF)$	208.95	col gas	....	−54	−29	d
(IV) oxide	SeO_2	110.96	wh nd, tetr with molecules joined in chains	3.954^{15}	340	subl 315	38^{14}; 10^{12} MeOH
(VI) oxide	SeO_3	126.96	wh, tetr, hygr	3.6	118	d 180	s
(di-) sulfide, hexa-	Se_2S_6	350.28	lt-or nd	2.44	121.5	....	i; 1.2 bz, s CS_2
(tetra-) sulfide, tetra-	Se_4S_4	444.08	red tabular cr	3.20	113	....	i; 0.04 bz, s CS_2
Silane	SiH_4	32.09	col gas	0.68^{mp}liq	−184.7	−111.9	d slowly
bromo-	SiH_3Br	111.02	col gas, expl air	1.72^{-80}	−94	1.9	
bromo-, di-	SiH_2Br_2	189.92	col liq, flammable	2.17^{0}	−70.1	66	d
bromo-, tri-	$SiHBr_3$	268.82	col liq, flammable	2.7^{17}	−73	109	d

Name	Formula	Formula wt	Form, color	Density	mp	bp	Solubility
bromotrichloro-	$SiBrCl_3$	214.35	col liq	1.826	−62	80.3	d
bromodichloro, di-	$SiBr_2Cl_2$	258.81	col liq	2.172	−45.5	104	d
bromochloro, tri-	$SiBr_3Cl$	303.27	col liq	2.497	−21	127	d
chloro	SiH_3Cl	66.56	col gas	3.033 (g/L)	−118.1	−30.4	....
chloro, di-	SiH_2Cl_2	101.01	col gas	4.60 (g/L)	−122	8.3	d; s bz, chl, CS_2
chloro, tri-	$SiHCl_3$	135.45	col liq	1.34	−126.5	33	d
chlorotrifluoro-	$SiClF_3$	120.53	col gas	5.455 (g/L)	−138	−70.0	d
chlorodifluoro, di-	$SiCl_2F_2$	136.99	col gas	6.278 (g/L)	−144	−31.7	d
chloroiodo, tri-	$SiCl_3I$	261.35	col liq	....	>−60	113.5	d; s toluene
fluoro, tri	$SiHF_3$	86.09	col liq	3.86 (g/L)	−131.4	ca −95	....
iodo-	SiH_3I	158.01	col liq	2.035^{15}	−57.0	45.5	d
iodo, tri-	$SiHI_3$	409.81	red liq	3.314	8	220	d; s bz, CS_2
Silicic acid, meta-	H_2SiO_3	78.10	wh amorp pwd	2.33^{25}	d	....	i; s HF, hot alk
Silicon	Si	28.086	gray-met, cub; also dk-brn amorp form	2.33^{25}	1415	2680	s HF + HNO_3, molten alk oxides
acetate, tetra-	$Si(C_2H_3O_2)_4$	264.27	ccl cr, hygr	....	110	148^{6mm}	d viol; s bz, acet
bromide, tetra-	$SiBr_4$	347.72	col fum liq	2.772	5.4	154	hyd
(di-) bromide, hexa-	Si_2Br_6	535.62	wh, rh	....	95	240	d; s CS_2
(tri-) bromide, octa-	Si_3Br_8	723.6	ccl cr	....	43.3	....	....
carbide	SiC	40.07	grn to bl-blk, hex	3.217	subl 2700	d 2972	s fused alk
chloride, tetra-	$SiCl_4$	169.89	col fum liq	1.48^{20}	−70	57.6	hyd; s bz, eth, CCl_4
(di-) chloride, hexa-	Si_2Cl_6	268.89	col liq	1.562^{15}	−2.5	147	hyd
cyanate, tetra-	$Si(OCN)_4$	196.15	col liq or solid	1.414^{20}	34.5	247.2	d
fluoride, tetra-	SiF_4	104.06	col gas	4.69 (g/L)	-90.3^{1318mm}	subl −95.5	hyd; s HF
(di-) fluoride, hexa-	Si_2F_6	170.16	col gas	7.76 (g/L)	-18^{780mm}	−19	....
hydrides, see under Silane							
iodide, tetra-	SiI_4	535.70	wh, cub	4.198	120.5	288	d; 2.2^{27} CS_2
(di-) iodide, hexa-	Si_2I_6	817.60	wh, hex	....	250	subl 150 (vac)	d; 19 CS_2
isocyanate, tetra-	$Si(NCO)_4$	196.15	col cr	1.434	26	185.6	hyd; s bz, CCl_4, CS_2
isothiocyanate, tetra-	$Si(NCS)_4$	260.40	wh pwd	....	143.8	314.2	d
(tri-) nitride, tetra-	Si_3N_4	140.28	lt-gray amorp pwd	3.44	d 1878	....	i; s HF
oxide	SiO	44.09	brn-blk scales, cub	2.18	>1702	1880	s HF + HNO_3
oxide, di- (high cristobalite)	SiO_2	60.08	col, cub, 1.487	2.32	1723	2230	i; s HF
(lechatelierite)		....	col, 1.459	2.19	....	....	i; s HF
(opal)		....	col amorp, 1.43	2.17–2.20	>1600	....	i; s HF

Table 4-1 (Continued)
PHYSICAL CONSTANTS OF INORGANIC COMPOUNDS

Name	Formula	Formula weight	Color, crystalline form, refractive index	Density	Melting point, °C	Boiling point, °C	Solubility in 100 parts
Silicon oxide, di- (quartz)	....	....	col, hex, 1.544, 1.553	2.635–2.660	1423	2230	i; s HF
(tridymite)	....	....	col, rh, 1.469, 1.470, 1.471	2.26	1703	2230	i; s HF
oxide-tungsten trioxide–water (1/12/26)(silico-tungstic acid)	$SiO_2 \cdot 12WO_3 \cdot 26H_2O$	3310.66	yelsh-wh-cr, deliq				v s; v s al
phosphide	SiP	59.06	lt-yel pwd	2.4			hyd
sulfide	SiS	60.15	yel nd	1.853^{15}	1090		d alk
sulfide, di-	SiS_2	92.21	wh nd, rh	2.02	1090	1130	s d; s d al, bz
(di) telluride, tri-	Si_2Te_3	438.97	hex		892		i; s HNO_3
Silver	Ag	107.868	wh-met, fcc	10.49^{15}	960.15	2164	1.04^{20}
acetate	$AgC_2H_3O_2$	166.92	wh cr pwd	3.259^{15}	d		i; s a
acetylide	Ag_2C_2	239.76	wh		explodes		
antimonide	Ag_3Sb	445.35	....		559		
antimony ditelluride	$AgSbTe_2$	484.82	cub	7.12	573		
arsenate(V)	Ag_3AsO_4	462.53	dk-red, cub	6.657	d		0.00085^{20}; s NH_4OH
arsenate(III)	Ag_3AsO_3	446.53	yel pwd		d 150		0.0011^{20}; s NH_4OH
azide	AgN_3	149.89	wh pr, rh, expl		252	297	i; s KCN, HNO_3
bromate	$AgBrO_3$	235.80	wh pwd, tetr, 1.920, 1.874	5.206^{20}	d		0.16^{20}
bromide (bromyrite)	AgBr	187.80	lt-yel, 2.253	6.473	430	1560	0.000014; s KCN
carbonate	Ag_2CO_3	275.77	lt-yel pwd	6.077	d 220		0.003^{20}
chlorate	$AgClO_3$	191.34	wh, tetr	4.430^{20}	231	d 270	15.3^{20}
chloride (cerargyrite)	AgCl	143.34	wh, cub, 2.071	5.56	455	1564	0.00019; s NH_4OH
chlorite	$AgClO_2$	175.34	yel cr		105 expl		0.45
chromate(VI)	Ag_2CrO_4	331.77	dk red-brn, mn	5.625^{25}			0.002^{20}

Name	Formula	Formula weight	Crystalline form; color; index of refraction	Density	Melting point, °C	Boiling point, °C	Solubility
cyanate	AgOCN	149.89	col	4.00	d	....	0.007[20]
cyanide	AgCN	133.90	wh, hex	3.95	d 320	....	i; s KCN
dichromate	Ag₂Cr₂O₇	413.73	red, tric	4.770	d	....	0.008[20]
diphosphate	Ag₄P₂O₇	605.42	wh	5.306[18]	585	....	i; s a
fluoride	AgF	126.88	yel, cub, hygr	5.852[16]	435	1150	172[20]
(II) fluoride	AgF₂	145.87	wh, rh, hygr	4.57	690	d 700	hyd viol
(di-) fluoride	Ag₂F	234.76	bronze, hex	8.57	d >100	....	d
fulminate	Ag₂(CNO)₂	299.77	nd	......	explodes	....	0.075[13]; s NH₄OH
hyponitrite	Ag₂N₂O₂	275.78	lt-yel cr	5.75[30]	d 110	....	0.0003; d a
iodate	AgIO₃	282.80	wh cr pwd, rh	5.525[20]	>200	d	0.004[20]
iodide (iodyrite)(α)	AgI	234.80	lt-yel pwd, hex, 2.21, 2.22	5.683[30]	558	1505	i; s KCN
mercurate(II), tetraiodo-	Ag₂[HgI₄]	924.05	yel, tetr	6.02	d 158	....	i; s KI, KCN
nitrate	AgNO₃	169.89	col, rh, 1.729, 1.744, 1.788	4.352[19]	210	d 440	216[20]
nitrite	AgNO₂	153.89	pale yel cr, rh	4.453	d 140	....	0.41[25]
oxalate	Ag₂C₂O₄	303.78	wh cr pwd	5.029[4]	expl 140	....	0.004[20]
(II) oxide	Ag₂O	231.76	brn-blk, cub	7.22[25]	d 200	....	0.002[25]
oxide	AgO	123.88	grey-blk, cub	7.483[25]	d 100	....	i; s alk
perchlorate	AgClO₄	207.34	wh cr, deliq	2.806[25]	d 486	....	525[20]; s org solv
periodate	AgIO₄	298.79	yel-or cr, tetr	1.884[22]	d 180	....	d
permanganate	AgMnO₄	226.81	dk-vlt cr, mn	4.49	d	....	0.9; d al
perrhenate	AgReO₄	358.07	wh, tetr & rh	7.05	430	....	0.32[20]
phosphate(1—), meta-	AgPO₃	186.84	wt amorp	6.37	ca 482	....	i
phosphate, ortho-	Ag₃PO₄	418.62	yel pwd, cub	6.370	849	....	0.006
selenate(VI)	Ag₂SeO₄	358.72	wh, rh	5.72	....	....	0.12[20]
selenate(IV)	Ag₂SeO₃	342.72	wh nd	5.93	530	d >530	sl s; s HNO₃
selenide	Ag₂Se	294.72	gray pl, rh & tetr	8.216[15]	897	d	i
sulfate	Ag₂SO₄	311.83	wh, rh, 1.7583, 1.7748, 1.7852	5.45[30]	660	d 1085	0.80[20]
sulfide (acanthite)	Ag₂S	247.83	blk, mn	7.317	838	d	i
(argentite)		....	grey-blk, rh	7.234[20]	838	d	i

Table 4-1 (*Continued*)
PHYSICAL CONSTANTS OF INORGANIC COMPOUNDS

Name	Formula	Formula weight	Color, crystalline form, refractive index	Density	Melting point, °C	Boiling point, °C	Solubility in 100 parts
Silver							
sulfite	Ag_2SO_3	295.83	wh cr	....	d 100	....	v sl s; s a
d-tartrate	$Ag_2C_4H_4O_6$	363.81	wh scales	3.423[15]	d	....	0.2[18]
telluride (hessite)	Ag_2Te	343.34	gray, rh	8.5	957	....	i; s KCN, NH_4OH
thioantimonate(III), tri- (pyrargyrite)	Ag_3SbS_3	541.55	red, trig, 3.084, 2.881(Li)	5.76	486	....	i; s HNO_3
thioarsenate(III), tri- (proustite)	Ag_3AsS_3	494.72	red, trig, 3.088, 2.792	5.49	490	....	i; s HNO_3
thiocyanate	$AgSCN$	165.95	col cr	....	d	....	i; s NH_4OH
thiosulfate	$Ag_2S_2O_3$	327.87	wh cr	....	d	....	sl s
tungstate(VI)	Ag_2WO_4	463.59	lt-yel cr	....	....	....	0.05[15], s HNO_3
Sodium							
	Na	22.99	silv-wh met, bcc	0.968[20]	97.82	881.4	d to NaOH
acetate	$NaC_2H_3O_2$	82.04	wh pwd, mn, 1.464	1.528	324	....	46.5[20]
acetate-3-water	$NaC_2H_3O_2 \cdot 3H_2O$	136.08	col pr, mn, effl	1.45	58	$-3H_2O$, 120	76.2[20], 5.1 al
alizarinesulfonate	$NaC_{14}H_7O_7S$	342.25	or-yel pwd	....	....	....	s; s al
aluminate(1−)	$NaAlO_2$	81.97	wh amorp pwd, hygr	....	>1650	....	v s; i al
aluminate, tetrachloro-	$NaAlCl_4$	191.80	wh	....	151	....	
aluminate, hexachloro-	Na_3AlCl_6	308.70	wh	....	507	....	
aluminate, hexafluoro- (cryolite)	Na_3AlF_6	209.94	wh	....	1012	....	
amide	$NaNH_2$	39.02	wh	....	210	subl 400	d viol
antimonate(V) (leuconine)	$NaSbO_3$	192.74	wh pwd	....	d	....	i; s Na_2S soln
antimonate(III)-1-water	$NaSbO_2 \cdot 3H_2O$	230.78	col, rh	2.864	....	....	d
antimonide	Na_3Sb	190.72	bl, flammable	....	856	....	d
arsenate(V)(1−)	$NaAsO_3$	145.91	wh, rh, 1.479, 1.502, 1.527	2.301	615	....	v s
arsenate(V)(3−)-12-water	$Na_3AsO_4 \cdot 12H_2O$	423.93	col, trig or hex pr, 1.457, 1.466	1.78	86.3	....	26.7[17]

Name	Formula	Formula weight	Crystalline form, properties, refractive index	Density	Melting point, °C	Boiling point, °C	Solubility
arsenate(III)(1−)	$NaAsO_2$	129.91	gray-wh pwd	1.87	......	......	v s
p-aminophenylarsenate	$NH_2C_6H_4AsO(OH)ONa$	311.08	wh cr pwd	......	......	......	16; 1 al
ascorbate	$C_6H_7O_6Na$	198.12	wh cr, $[\alpha]_D^{20}$ +104.4	......	d 218	......	62^{25}
aurate(III)-2-water, tetra-chloro-	$Na[AuCl_4]\cdot 2H_2O$	397.80	or-yel, rh-bipyr, 1.6	......	d >100	......	166^{20}
azide	NaN_3	65.01	col, hex	1.846^{20}	d	d	41^{20}, 0.3 al
barbital	$NaC_8H_{11}N_2O_3$	206.18	wh pwd	......	......	......	20^{25}
benzenesulfonate	$C_6H_5SO_3Na$	180.16	wh cr	......	......	......	35.8^{30}
benzoate	$C_6H_5O_2Na$	144.11	col cr pwd	......	......	......	62.8^{25}, 1.3 al
bismuthate(V)(1−)	$NaBiO_3$	280.00	yel-brn pwd, hygr	......	d	......	i; d a
bismuthide	Na_3Bi	277.95	bl-vlt	......	766	......	d
borate, tetrahydro-	$NaBH_4$	37.84	wh cr, cub, hygr	1.074	497 d	......	55^{25}, 4 al
borate(1−)	$NaBO_2$	65.82	col hex pr	......	966	1476	26^{20}
borate, tetra-	$Na_2B_4O_7$	201.27	col pwd pl	2.367	742.5	......	2.6^{20}
borate-10-water, tetra- (borax)	$Na_2B_4O_7\cdot 10H_2O$	381.37	wh, mn, 1.469	1.73	75	−10H_2O, 320	6.3; 100 gly
borate, tetrafluoro-	$Na[BF_4]$	109.82	wh, rh	2.47^{20}	384	d	108^{27}
bromate	$NaBrO_3$	150.91	col, cub	3.339^{17}	380 d	......	36^{20}
bromide	$NaBr$	102.91	wh pwd, cub, 1.641	3.205^{18}	747	1447	90^{20}, 6 al, 16 MeOH
carbide	Na_2C_2	70.00	brn pwd	1.575^{15}	d	d	d
carbonate	Na_2CO_3	106.00	wh pwd, 1.535	2.533	850.0	......	21.5^{20}, s gly
carbonate-1-water	$Na_2CO_3\cdot H_2O$	124.00	wh, rh, 1.51	2.25	−H_2O, 100	......	33; 14 gly
carbonate-10-water	$Na_2CO_3\cdot 10H_2O$	286.14	wh, mn, 1.425	1.46	34	......	50; s gly
carbonate-hydrogen carbonate-water (1/1/2)(trona)	$Na_2CO_3\cdot NaHCO_3\cdot 2H_2O$	190.00	wh nd, mn	2.112	......	......	13^0
chlorate	$NaClO_3$	106.45	wh, cub & trig, 1.515	2.489	248	d 350	96^{20}, 0.77 al; 25 gly
chloride (halite)	$NaCl$	58.45	col, cub, 1.5443	2.164^{20}	801	1465	35.9^{20}, 10 gly
chlorite	$NaClO_2$	90.45	wh cr	......	d 180–200	......	34^{17}
chromate(VI)	Na_2CrO_4	161.97	yel, rh	2.723	792	......	84^{20}
citrate-2-water	$C_6H_5O_7Na_3\cdot 2H_2O$	294.10	wh cr	......	−2H_2O, 150	......	77^{25}
cobaltate(III), hexanitrito-	$Na_3[Co(NO_2)_6]$	403.98	yel-brn pwd	......	......	......	v s
cyanate	$NaOCN$	65.01	col nd	1.893^{20}	550	......	s d; 0.22^0 al
cyanide	$NaCN$	49.02	wh cub, 1.452	......	562	1530	58.7^{20}
dichromate-2-water	$Na_2Cr_2O_7\cdot 2H_2O$	298.00	red, mn sphenoidal, 1.6994	2.348^{25}	−2H_2O, 84.6; mp: 356	d 400	208^{20}
diethyldithiocarbamate	$(C_2H_5)_2NCS_2Na$	225.31	wh pwd	......	94–96	......	s; s al

Table 4-1 (*Continued*)
PHYSICAL CONSTANTS OF INORGANIC COMPOUNDS

Name	Formula	Formula weight	Color, crystalline form, refractive index	Density	Melting point, °C	Boiling point, °C	Solubility in 100 parts
Sodium							
dimethylarsonate-3-water (cacodylate)	$(CH_3)_2AsO_2Na \cdot 3H_2O$	214.03	wh	....	60	$-3H_2O$, 120	200; 40 al
dimolybdate(VI)	$Na_2Mo_2O_7$	349.86	wh nd	....	612	....	sl s
diphosphate(V)	$Na_4P_2O_7$	265.90	wh	2.45	988	....	2.26^0
diphosphate(V)-10-water	$Na_4P_2O_7 \cdot 10H_2O$	446.06	wh, mm, 1.4525	1.82	d	....	5.4^0
diphosphate(IV)-10-water	$NaOOP\text{-}POONa \cdot 10H_2O$	430.06	wh, mm	1.832	....	....	3.3^{20}
disulfate(IV)	$Na_2S_2O_5$	190.10	wh pwd	....	$-SO_2$, d	d 267	v s
dithionate-2-water	$Na_2S_2O_6 \cdot 2H_2O$	242.13	col, rh, 1.4593	2.189	$-2H_2O$, 110		6.05^{20}
dithionate(III) (hydrosulfite)	$Na_2S_2O_4$	174.13	wh cr pwd, oxid	....	d		22^{20}
dodecylsulfate	$CH_3(CH_2)_{10}CH_2OSO_3Na$	288.38	creamy flakes	....	....		10
ethoxide	C_2H_5ONa	68.06	wh, hygr	....	....		d; s abs al
ethylenebis(iminodiacetate) (edta)	$(OOCCH_2)_2NC_2H_4N\text{-}(CH_2COO)_2Na_4$	380.20	....	....	....		103
ferrate(III)-1-water, hexacyano-	$Na_3[Fe(CN)_6] \cdot H_2O$	298.93	ruby-red, deliq	....	....		18.9^0
ferrate(II)-10-water, hexacyano-	$Na_4[Fe(CN)_6] \cdot 10H_2O$	484.07	lt-yel, mm	1.458	$-10H_2O$, 82	d 435	18.8^{20}
ferrate(III)-2-water, pentacyanonitrosyl- (nitroprusside)	$Na_2[Fe(CN)_5NO] \cdot 2H_2O$	297.65	ruby red, rh	1.72	....		40^{16}
ferrate(III)-5.5-water, tris(oxalato)-	$Na_3[Fe(C_2O_4)_3] \cdot 5.5\ H_2O$	487.96	grn, mm	1.973^{18}	$-4H_2O$, 100	anhyd, 200	32.5^0
fluoride (villiaumite)	NaF	41.99	wh, cub, 1.3258	2.78	996	1787	4^{20}
formate	$NaHCO_2$	68.02	wh, mm, deliq	1.919	253	d>253	81^{20}; s gly
gluconate	$C_6H_{11}O_7Na$	218.13	col cr	....	d>130	....	59^{25}
glycerophosphate	$Na_2C_3H_5(OH)_2PO_4$	216.03	wh scale-like cr	....	d 425	....	60
hydride	NaH	24.00	silv nd	1.396			d viol; d al viol

Name	Formula	Formula weight	Crystalline form, color, index of refraction	Density	Melting point	Boiling point	Solubility
hydrogen arsenate(V)-1-water, di-	$NaH_2AsO_4 \cdot H_2O$	181.94	col rh & mn, 1.507, 1.583, 1.553	2.53	$-H_2O$, 130	d 200	s
hydrogen arsenate(V)-7-water	$Na_2HAsO_4 \cdot 7H_2O$	312.01	col pwd. effl	1.88	$-7H_2O$, 130	d 150	61^{15}; s gly
hydrogen carbonate	$NaHCO_3$	84.01	wh pwd, mn, 1.500	2.20	$-CO_2$, 270	......	9.6^{20}; i al
hydrogen diphosphate(V), di-	$Na_2H_2P_2O_7$	221.97	wh pwd, mn, 1.510	1.862	d 220	......	4.5^{0}
hydrogen difluoride	$NaHF_2$	62.01	col cr	......	d	......	3.7^{20}
hydrogen oxalate-1-water	$NaHC_2O_4 \cdot H_2O$	130.03	wh tric	......	......	......	1.7^{15}
hydrogen diphosphate-(IV)-6-water, di-	$Na_2H_2P_2O_6 \cdot 6H_2O$	314.03	col, mn	1.849	$-6H_2O$, 100; mp: 250(anhyd)	......	2.2^{20}
hydrogen phosphate(V)-1-water, di-	$NaH_2PO_4 \cdot H_2O$	137.99	wh, rh, 1.4852	2.040	$-H_2O$, 100	d 200	71^{0}
hydrogen phosphate(V)-7-water	$Na_2HPO_4 \cdot 7H_2O$	268.07	col, mn, 1.4424	1.679	d	......	185^{40}
hydrogen phosphate(V)-12-water	$Na_2HPO_4 \cdot 12H_2O$	358.14	col, mn, 1.4361	1.52	34.6	$-12H_2O$, 180	33
hydrogen phosphonate-2.5-water	$NaHPHO_3 \cdot 2.5H_2O$	149.01	col pr, mn	......	42	$-H_2O$, 100	78^{10}
hydrogen sulfate	$NaHSO_4$	120.07	col, tric, hygr	2.435	315	d	28.5^{25}; d al
hydrogen sulfate-1-water	$NaHSO_4 \cdot H_2O$	138.07	col, mn, deliq	......	58.5	d	100
hydrogen sulfite	$NaHSO_3$	104.06	col, mn, 1.526	1.48	d	......	29; 1.4 al
hydrogen sulfide-2-water	$NaHS \cdot 2H_2O$	92.09	col nd, deliq	......	55	d	s; s al, eth
hydroxide	$NaOH$	40.01	wh, deliq	2.130^{25}	322	1557	108^{20}, 14 abs al, 24 MeOH, s gly
hydroxybenzenesulfonate-2-water	$C_6H_4(OH)SO_3Na \cdot 2H_2O$	232.18	wh cr	......	......	......	24
hypochlorite	$NaClO$	74.44	arhyd v expl, yel	......	......	......	53^{20}
hyponitrite	$Na_2N_2O_2$	105.99	ccl	2.466^{30}	300 d	......	s
iodate	$NaIO_3$	197.90	wh cr, rh	4.227^{20}	d	......	8.1^{20}
iodide	NaI	149.92	wh, cub, 1.7745	3.667^{0}	660	1304	178^{20}
iridate(IV)-6-water, hexachloro-	$Na_2[IrCl_6] \cdot 6H_2O$	588.99	red, tric	......	d	......	39^{20}
lactate	$CH_3CHOHCOONa$	112.07	col liq	......	d	......	v s; s al
manganate(VI)-10-water	$Na_2MnO_4 \cdot 10H_2O$	345.07	grn pwd	......	17	......	s

Table 4-1 (*Continued*)
PHYSICAL CONSTANTS OF INORGANIC COMPOUNDS

Name	Formula	Formula weight	Color, crystalline form, refractive index	Density	Melting point, °C	Boiling point, °C	Solubility in 100 parts
Sodium							
mercaptoacetate (thioglycolate)	$HSCH_2COONa$	114.11	wh cr, hygr	....	....	....	s; sl s al
methoxide	$NaOCH_3$	54.03	wh pwd	....	....	mp: 687	d; s al
molybdate(VI)-2-water	$Na_2MoO_4 \cdot 2H_2O$	241.95	wh pwd	3.28	$-2H_2O$, 100	....	65[20]
nitrate	$NaNO_3$	85.01	trans cr, trig, 1.5874	2.257	308	d 380	88[20]
nitride	Na_3N	82.98	gray	....	d	....	d
nitrite	$NaNO_2$	69.00	pale yel, rh, hygr	2.168[0]	271	d>320	81[20]
oleate	$C_{17}H_{33}COONa$	304.45	wh pwd	....	....	....	10
oxalate	$Na_2C_2O_4$	134.01	wh cr	2.27	....	....	3.4[20]
oxide	Na_2O	61.98	wh, deliq	2.27	1132	d 1950	d to NaOH
oxide, di-	NaO_2	54.99	....	....	552	....	....
perchlorate	$NaClO_4$	122.44	wh cr, rh, 1.4617	2.499	468	....	201[20]
perchlorate-1-water	$NaClO_4 \cdot H_2O$	140.46	wh hex cr	2.02	d 130	....	66[0]
periodate	$NaIO_4$	213.91	wh, tetr	3.865[16]	d 300	....	10.3[20]
periodate-3-water	$NaIO_4 \cdot 3H_2O$	267.96	wh effl cr, trig	3.219[18]	d 175	....	12.5[20]
permanganate-3-water	$NaMnO_4 \cdot 3H_2O$	195.94	purp, hygr	2.46	d 170	....	v s
peroxide	Na_2O_2	77.99	yelsh-wh pwd	2.805	675	d	v s d
peroxoborate-4-water	$NaBO_3 \cdot 4H_2O$	153.88	wh pwd	....	d 60	....	2.5
peroxochromate(VI)	Na_3CrO_8	248.96	or pl	....	d 115	....	400[0]
peroxodisulfate(VI)	$Na_2S_2O_8$	238.13	wh cr pwd	....	d	....	55
perrhenate	$NaReO_4$	273.19	col pl	5.24	300	....	33[20]
phenoxide	C_6H_5ONa	116.10	wh to redsh granules	....	....	....	v s; s al
phosphate(V)	Na_3PO_4	163.94	wh	2.53[18]	1340	....	12.1[20]
phosphate-12-water	$Na_3PO_4 \cdot 12H_2O$	380.12	wh, trig, 1.4458	1.62	73.4	$-11H_2O$, 100	28.3[15]
phosphate(1-)	$(NaPO_3)_x$	x(101.96)	clear hygr glass	1.478	628	....	21(slowly)
phosphinate-1-water (hypophosphite)	$NaPH_2O_2 \cdot H_2O$	105.99	wh, mn	....	$-H_2O$, 200	d	100[23]; s gly

Name	Formula	Formula weight	Crystalline form, color, index of refraction	Density	Melting point	Boiling point	Solubility
platinate(IV)-6-water, hexabromo-	$Na_2[PtBr_6] \cdot 6H_2O$	828.62	dk-red, tric	3.323	d	....	v s
platinate(IV)-6-water, hexachloro-	$Na_2[PtCl_6] \cdot 6H_2O$	561.88	yel, tric, hygr	2.50	$-6H_2O$, 110	....	v s; s al
salicylate	$C_7H_5O_3Na$	160.11	wh cr	3.098	....	....	95^{20}, 11 al, 25 gly
selenate(VI)	Na_2SeO_4	188.94	wh, rh	....	....	....	27^{20}
selenate(VI)-10-water	$Na_2SeO_4 \cdot 10H_2O$	369.09	wh cr, mn	1.58	....	....	77^{30}
selenide, di-	Na_2Se	124.95	de.iq cr, cub	2.625^{10}	>875	....	d
selenide, di-	Na_2Se_2	203.90	....	....	495	....	
silicate(2—)	Na_2SiO_3	122.08	col. cr, rh glass: 1.520	2.614	1089	....	s; hyd hot water
silicate(2—)-5-water	$Na_2SiO_3 \cdot 5H_2O$	212.14	wh, tric, 1.456	1.749	72.2	$-5H_2O$, 100	v s
silicate(4—)	Na_2SiO_4	184.04	col. hex, 1.530	....	1018	....	s
silicate, hexafluoro-	$Na_2[SiF_6]$	188.05	wh gran pwd, hex, 1.312	2.679	red heat	....	0.44^0
stannate(IV)-3-water	$Na_2SnO_3 \cdot 3H_2O$	266.71	wh hex tablets	....	d 140	....	50^0
stearate	$C_{17}H_{35}COONa$	306.47	wh soap	....	d	....	sl s
sulfate (mirabilite, thenardite)	Na_2SO_4	142.06	col. rh-bipyr, 1.477	2.664	884	....	19.5^{20}
sulfate-10-water (Glauber's salt)	$Na_2SO_4 \cdot 10H_2O$	322.19	ccl. mn, 1.396	1.464	32.4	$-10H_2O$, 100	36^{15}
sulfide	Na_2S	78.05	wh-pink, cub, hygr, v expl upon shock or rapid heating	1.856^{14}	950	....	15.7^{20}
sulfide-9-water	$Na_2S \cdot 9H_2O$	240.18	col deliq cr, tetr	1.427^{16}	ca 50	d	125^{25}
sulfide, di-	Na_2S_2	110.11	tetr	....	480	....	
sulfide, tetra-	Na_2S_4	174.24	yel cub	....	275	....	s
sulfite	Na_2SO_3	126.06	wh hex pr, 1.565	2.633^{15}	d	....	26^{20}
tartrate-2-water	$Na_2C_4H_4O_6 \cdot 2H_2O$	230.08	wh cr, rh	1.818	$-2H_2O$, 120	....	29^6
tellurate(VI)(2—)	Na_2TeO_4	237.58	wh pwd	....	....	....	0.008
telluride	Na_2Te	173.58	cub	....	953	....	
telluride, tri-	$NaTe_3$	405.79	....	436	....	....	
tetraphenylborate	$(C_6H_5)_4BNa$	342.24	wh cr	....	....	....	s; s acet
thioantimonate(V)-9-water	$Na_3SbS_3 \cdot 9H_2O$	481.11	l.t-yel, cub	1.839	....	....	28^{20}
thioarsenate(V)-8-water	$Na_3AsS_4 \cdot 8H_2O$	416.29	mn, 1.6802	....	d	....	s

Table 4-1 (Continued)
PHYSICAL CONSTANTS OF INORGANIC COMPOUNDS

Name	Formula	Formula weight	Color, crystalline form, refractive index	Density	Melting point, °C	Boiling point, °C	Solubility in 100 parts
Sodium							
thiocyanate	NaSCN	81.07	col, rh, deliq, 1.625	....	287	....	134[20]
thiosulfate-5-water	Na$_2$S$_2$O$_3 \cdot$ 5H$_2$O	248.18	wh pr, mn, effl, 1.5079	1.685	$-$5H$_2$O, 100	....	70[20](d slowly)
l-thyroxine	C$_{15}$H$_{10}$I$_4$NO$_4$Na	798.85	creamy pwd, tric	2.381[20]			0.015; s a, alk
tungstate(VI)	Na$_2$WO$_4$	293.83	wh, rh	4.179	695.6	....	73[20]
tungstate(VI)-2-water	Na$_2$WO$_4 \cdot$ 2H$_2$O	329.86	col effl cr, rh	3.245	$-$2H$_2$O, 100	1381	88[0]
Strontium							
acetate	Sr(C$_2$H$_3$O$_2$)$_2$	205.71	wh cr	2.099	d		d to Sr(OH)$_2$
boride, hexa-	SrB$_6$	152.49	blk, cub	3.39[15]	2235		i; s HNO$_3$
bromate-1-water	Sr(BrO$_3$)$_2 \cdot$ H$_2$O	361.45	yelsh-wh, mn, hygr	3.773	$-$H$_2$O, 120		30.9[18]
bromide	SrBr$_2$	247.43	wh, hex, 1.575	4.216	657	2146	100[20]
bromide-6-water	SrBr$_2 \cdot$ 6H$_2$O	355.53	col, hex, hygr	2.386	$-$6H$_2$O, 180		146[20], 63 al
carbide, di-	SrC$_2$	111.64	blk, tetr	3.2	>1700		d; d a
carbonate (strontianite)	SrCO$_3$	147.64	col, rh, 1.516, 1.664, 1.666	3.70	$-$CO$_2$, 1172		0.001[25]; s a
chlorate	Sr(ClO$_3$)$_2$	254.54	col, rh, 1.516, 1.605, 1.626	3.152[20]	d 120		175[18]
chloride	SrCl$_2$	158.52	col, cub, 1.650	3.052	874	2058	52.9[20]
chloride-6-water	SrCl$_2 \cdot$ 6H$_2$O	266.62	col, trig, 1.536, 1.487	1.96	$-$6H$_2$O, 150		88.8[20]
chromate(VI)	SrCrO$_4$	203.64	yel pwd, mn	3.895[15]			0.09[20]; s HCl
cyanide-4-water	Sr(CN)$_4 \cdot$ 4H$_2$O	211.72	wh, rh, deliq	....	d		v s
fluoride	SrF$_2$	125.63	wh pwd, cub, 1.442	4.24	1477	2486	0.011[20]; s hot HCl
formate	Sr(HCO$_2$)$_2$	177.66	col, rh, 1.559, 1.547, 1.598	2.693	71.9		12.7[20]
formate-2-water	Sr(HCO$_2$)$_2 \cdot$ 2H$_2$O	213.69	col, rh, 1.484, 1.521, 1.538	2.25	d		15.3[20]
hydride	SrH$_2$	89.65	rh, hygr	3.269	d>1000		d
hydrogen arsenate-1-water	SrHAsO$_4 \cdot$ H$_2$O	245.56	wh nd, rh	3.606[15]	$-$H$_2$O, 125		0.28[16]; s a
hydrogen phosphate	SrHPO$_4$	183.60	col, rh	3.544[15]			i; s a

Name	Formula	Formula weight	Crystalline form, color, refractive index	Density	Melting point, °C	Boiling point, °C	Solubility
hydroxide	Sr(OH)$_2$	121.64	wh, deliq	3.625	375(in H$_2$)	$-$H$_2$O, 710	1.77^{20}
hydroxide-8-water	Sr(OH)$_2\cdot$8H$_2$O	265.76	col, tetr, deliq, 1.499, 1.476	1.90	$-$8H$_2$O, 100		2
iodate	Sr(IO$_3$)$_2$	437.43	wh, tric	5.045^{15}			0.03^{15}
iodide	SrI$_2$	341.42	col pl	4.549	538	1908	178^{20}
iodide-6-water	SrI$_2\cdot$6H$_2$O	449.52	yelsh-wh, hex	2.672	120		234^{20}
lactate-3-water	Sr(CH$_3$CHOHCOO)$_2\cdot$3H$_2$O	319.81	wh pwd		$-$3H$_2$O, 150		33
molybdate(VI)	SrMoO$_4$	247.56	col, tetr, 1.91	4.54	d	d 1100	0.01^{17}; s a
nitrate	Sr(NO$_3$)$_2$	211.65	wh pwd, cub	2.986	570		69.5^{20}
nitrate-4-water	Sr(NO$_3$)$_2\cdot$4H$_2$O	283.69	col, mn	2.2	$-$4H$_2$O, 100		93^{20}
nitride	Sr$_3$N$_2$	290.87			1030		d; s HCl
nitrite-1-water	Sr(NO$_2$)$_2\cdot$H$_2$O	197.65	col, hex, 1.588	2.408^{0}	$-$H$_2$O, 100	d 240	71.5^{20}
oxalate-1-water	SrC$_2$O$_4\cdot$H$_2$O	193.64	wh cr pwd		$-$H$_2$O, 150		0.006^{20}; s HCl
oxide	SrO	103.63	gray-wh, cub, 1.810	4.7	2665		0.69^{20}
perchlorate	Sr(ClO$_4$)$_2$	286.52	col cr, hygr	3.00^{25}			v s
permanganate-3-water	Sr(MnO$_4$)$_2\cdot$3H$_2$O	379.54	vlt, cub	2.75	d 175		2.6^{0}
peroxide	SrO$_2$	119.63	wh pwd	4.56	d 215		0.018^{20}, d hot H$_2$O
phosphate	Sr$_3$(PO$_4$)$_2$	452.85	wh pwd				i
phosphide	Sr$_3$P$_2$	324.81	brn cr	2.68			hyd
selenate(VI)	SrSeO$_4$	230.59	col, rh	4.25	1580		i; s hot HCl
selenide	SrSe	166.58	wh, cub, 2.220	4.38	>1750		i; s HCl
silicate(2−)	SrSiO$_3$	163.70	col, mn, 1.599, 1.637	3.65			i
silicate(4−)	Sr$_2$SiO$_4$	267.32	col, mn, 1.728, 1.732, 1.758	3.84			
silicate-2-water, hexafluoro-	Sr[SiF$_6$]$\cdot$2H$_2$O	265.73	col, mn	2.99^{18}	d		3.2^{18}
sulfate (celestite)	SrSO$_4$	183.70	wh cr, rh, 1.622, 1.624, 1.631	3.96	1600		0.013^{20}, sl s a
sulfide	SrS	119.70	lt gray, cub, 2.107	3.70^{15}	2002		sl s; d a
tartrate-4-water	SrC$_4$H$_4$O$_6\cdot$4H$_2$O	307.75	wh, mn	1.966			0.11^{0}; s HCl
telluride	SrTe	215.22	wh, cub, 2.408	4.83			v s
thiocyanate-3-water	Sr(SCN)$_2\cdot$3H$_2$O	257.83	wh, deliq		$-$3H$_2$O, 100		
thiosulfate-5-water	SrS$_2$O$_3\cdot$5H$_2$O	289.82	col, mn	2.17^{17}	$-$4H$_2$O, 100	d 160	2.5^{13}
titanate(IV)(2−)	SrTiO$_3$	183.52			1910		
tungstate(VI)	SrWO$_4$	335.47	col, tetr	6.187	d		0.14^{15}
zirconate(IV)(2−)	SrZrO$_3$	226.84			2600		

Table 4-1 (*Continued*)
PHYSICAL CONSTANTS OF INORGANIC COMPOUNDS

Name	Formula	Formula weight	Color, crystalline form, refractive index	Density	Melting point, °C	Boiling point, °C	Solubility in 100 parts
Sulfamic acid	NH_2SO_3H	97.09	col, rh	2.126	d 200	....	14.7
Sulfuric diamide	$SO_2(NH_2)_2$	96.11	col pl, rh	1.807	91.5	d 250	s
Sulfinyl bromide	$SOBr_2$	207.88	red-yel liq	2.67	−49.5	139.7	d
chloride	$SOCl_2$	118.98	col liq	1.656^{15}	−104.5	75.8	hyd
chloride fluoride	$SOClF$	102.51	col liq	1.576^{0}	−139	12.2	....
fluoride	SOF_2	86.06	col gas	3.0^{bp}	−110	−43.8	d; s bz, chl, eth
Sulfonyl diamide	$SO_2(NH_2)_2$	96.11	col, rh	1.807	91.5	d 250	s
bromide fluoride	SO_2BrF	162.98	col liq	2.17^{0}	−86.5	40	d
chloride	SO_2Cl_2	134.98	col liq, 1.4437	1.6674^{20}	−46	69.3	d; s bz
chloride fluoride	SO_2ClF	118.51	col gas	1.623^{0}	−124.7	7.1	d
fluoride	SO_2F_2	102.07	col gas	3.72 (g/L)	−135.8	−55.38	4 ml; 24 ml al; 136 ml CCl_4; 210 ml toluene
Sulfur, cycloocta- (α)	S_8	256.51	amber, rh, 1.940	2.08^{20}	115.21	444.60	i; 23^{0} CS_2; s al, bz
(β)	S_8	256.51	lt yel, mn, 2.017	1.96^{20}	115.21	444.60	i; s org solv
(γ)	S	32.07	yel, mn, 2.216	1.92	106.8	444.60	i
(di-) bromide, di-	S_2Br_2	223.95	dk red liq, 1.73	2.63^{20}	−46	117	d
bromide pentafluoride	$SBrF_5$	206.97	gas	....	−79	3.1	hyd
(di-) chloride, di-	S_2Cl_2	135.03	or fum liq, 1.670	1.6885^{15}	−80	138.1	hyd
chloride, di-	SCl_2	102.97	red liq	1.622^{15}	−78	59	hyd
chloride, tetra-	SCl_4	173.88	lt yel liq	....	−30	d −15	hyd
chloride pentafluoride	$SClF_5$	162.51	gas	....	−64	−21	hyd
(di-) fluoride, di-	FSSF	102.13	col liq	1.5^{-100}	−133	15	d
fluoride, di-	SF_2	70.06	col liq	....	−35	....	d viol; s bz
fluoride, tetra-	SF_4	108.07	col gas	1.919^{-73}	−121	−38	sl s; s al, KOH
fluoride, hexa-	SF_6	146.07	col gas	1.88^{mp}	−50.8	subl 63.8	d fus KOH
(di-) fluoride, deca-	S_2F_{10}	254.11	col liq	2.08^{0}	−52.7	17	d

Name	Formula	Formula weight	Color, crystalline form, refractive index, etc.	Density	Melting point, °C	Boiling point, °C	Solubility
oxide, di-	SO_2	64.07	col gas, 1.000613	2.716^{20} (g/L) 1.46 liq^{bp}	-75.47	-10.01	3937 ml^{20}, 25 ml al, 32 ml MeOH
oxide, tri- (α or I) (β or II)	SO_3	80.07	asbestos-like nd; nd, polymer mol in helical chains		62.3; 32.6	......	......
(γ or III)	.....	.....	col gas; liq 1.410; solid is S_3O_9 rings	1.9225^{20}	16.86	43.4	slowly v s
(di-) oxide, tri-	S_2O_3	112.13	bl liq, hygr	1.663^{20}	d 20		hyd
Sulfuric acid	H_2SO_4	98.08	col liq, 1.4297	1.8318^{20}	10.38	335.5	v s
	$H_2SO_4 \cdot H_2O$	116.09	col liq, 1.438; mn cr	1.788	8.48	290	v s
	$H_2SO_4 \cdot 2H_2O$	134.11	col liq, 1.405	1.650^{0}	-39.47	167	v s
[²H]	$^2H_2SO_4$ or D_2SO_4	100.09	col liq, 1.454	1.8620^{20}	14.35		v s
bromo-	$HOSO_2Br$	240.90	col liq, hygr		-6 to -8	d	hyd
chloro-	$HOSO_2Cl$	116.52	col liq	1.753^{20}	-80	152	d viol
fluoro-	$HOSO_2F$	100.07	col fum liq	1.726^{25}	-88.98	162.6	d viol
Sulfurous acid	H_2SO_3	82.08	in soln only	1.03	......		s
Tantalum	Ta	180.95	gray-blk, bcc	16.69	2985	5513	i; s HF, fus alk
boride, di-	TaB_2	202.57		11.15	3100	......	......
(IV) bromide	$TaBr_4$	500.57	brn-blk cr, rh		392		
(V) bromide	$TaBr_5$	580.49	bronze, rh, hygr	4.989	240	322	hyd; s abs al, eth
carbide	TaC	192.96	blk, cub	13.9	4000	d 5308	sl s HF
(di-) carbide	Ta_2C	373.91			3500		
(III) chloride	$TaCl_3$	287.31	blk cr	5.03^{20}	d 340		i; s hot a
(IV) chloride	$TaCl_4$	322.76	blk, rh & mn	4.35	d		hyd
(V) chloride	$TaCl_5$	358.24	wh, hygr	3.68	216.5	240	hyd; s inert org solv
(V) fluoride	TaF_5	275.95	wh, mn	4.74^{20}	95–97	229	s
fluoride dioxide	$TaFO_2$	231.96	cub		d 850		
(V) iodide	TaI_5	815.47	blk-brn pwd, rh	5.80^{26}	367	397	hyd; s eth
nitride	TaN	194.95	bronze or blk, hex	16.30	3090		sl s aq reg
(IV) oxide	TaO_2	212.95	dk-gray pwd		oxid		i a
(V) oxide	Ta_2O_5	441.90	wh cr, rh	8.2	1785		i; s HF
sulfide, di-	TaS_2	245.08	blk pwd, hex		>1300		i
Technetium	Tc	98.91	hep	11.487	2250 ± 50	4567	i; s HNO_3
(IV) chloride	$TcCl_4$	240.72	red, rh		subl 300		s HCl
(V) fluoride	TcF_5	193.91	yel, rh		50	d 60	hyd

Table 4-1 (Continued)
PHYSICAL CONSTANTS OF INORGANIC COMPOUNDS

Name	Formula	Formula weight	Color, crystalline form, refractive index	Density	Melting point, °C	Boiling point, °C	Solubility in 100 parts
Technetium							
(VI) fluoride	TcF_6	212.91	yel, cub	....	33.4	55.3	s HCl
fluoride oxide, tetra-	TcF_4O	190.91	bl, mn	....	134	....	....
fluoride trioxide	$TcFO_3$	165.91	yel cr or liq	....	18.3	ca 100	hyd
(IV) oxide	TcO_2	130.91	blk	6.9	subl 1000	....	i; s a, alk
(VII) oxide	Tc_2O_7	309.81	lt yel, hygr	....	119.5	310.6	s
(VII) sulfide	Tc_2S_7	422.23	dk brn	....	subl 100	d	i; s H_2O_2 + NH_4OH
Telluric(VI) acid	H_6TeO_6	229.66	wh dimorphic;	3.068 (mn) 3.16 (cub)	$-2H_2O$, 120	d 320 to TeO_3	30^{18}
Telluric(IV) acid	H_2TeO_3	177.63	wh cr pwd	3.0	d to TeO_2	....	0.0007; s a, alk
Tellurium	Te	127.60	silv, hex-rh	6.24^{20}	450	1009	s HNO_3, KOH
bromide, di-	$TeBr_2$	287.42	choc-brn cr	....	d 210	....	sl s a; s eth
bromide, tetra-	$TeBr_4$	447.27	or, mn	4.31^{15}	363	420	sl s; s a, KOH
bromide dichloride, di-	$TeBr_2Cl_2$	357.73	yel pwd, hygr	....	292	415	hyd
chloride, di-	$TeCl_2$	198.51	blk cr; purp gas	....	175	328	d; s a
chloride	$TeCl_4$	269.44	lt yel; maroon liq	3.01	227.9	388	hyd; s HCl, al, bz
fluoride, tetra-	TeF_4	203.60	col nd-like cr	....	129.6	d > 194	d
fluoride, hexa-	TeF_6	241.61	col gas	3.025^{-24} (g/L)	-37.8	-38.9	hyd
(di-) fluoride, deca-	Te_2F_{10}	317.58	col liq	2.852^{20}	-33.7	53	....
(di-) iodide, di-	Te_2I_2	509.00	lt gray glass	5.49	....	....	sl s; s alk, HI
iodide, tetra-	TeI_4	635.29	gray-blk cr	5.145	280 (vac)	....	s HCl, HF
(IV) oxide (tellurite)	TeO_2	159.60	wh, tetr	5.76	732.6	subl 790	s KOH but disprop
(V) oxide	Te_2O_5	335.20	....	4.14	d 450	....	i a; s conc alk
(VI) oxide	TeO_3	175.60	yel amorp pwd	5.075	d 400	....	....
sulfide, di-	TeS_2	191.73	gray cr red (fresh) or blk	6.21	....	....	....

Name	Formula	Mol. wt.	Crystalline form; color; refr. index	Density	M.p., °C	B.p., °C	Solubility
Terbium	Tb	158.93	silv-gray, hcp	8.272	1356	2800	i; s a
bromide	$TbBr_3$	398.65	wh, rh	4.67	828	1490	s
chloride	$TbCl_3$	265.28	wh, rh	4.35^{0}	588	1550	v s
fluoride	TbF_3	215.92	wh, hex	7.236	1172	2280	i; s H_2SO_4
iodide	TbI_3	539.64	wh, hex		957	1330	s
nitrate-6-water	$Tb(NO_3)_3 \cdot 6H_2O$	453.03	col, mn		89.3		s
oxide	Tb_2O_3	365.85	wh, bcc				i; s a
sulfate-8-water	$Tb_2(SO_4)_3 \cdot 8H_2O$	750.16	wh cr		$-8H_2O$, 360		3.56^{20}
sulfide	TbS	190.98	cub		1967		
Thallium	Tl	204.37	bl-wh met, hcp	11.85	303.5	1487	i; s HNO_3
(I) acetate	$TlC_2H_3O_2$	263.43	silky wh nd, deliq	3.765^{137}	131		v s; s al
(I) bromate	$TlBrO_3$	332.28	col nd			825	0.35^{20}
(I) bromide	$TlBr$	284.31	yel-wh, cub	7.54	460	816	0.05^{20}, s al
(I) carbonate	Tl_2CO_3	468.75	wh cr, mn	7.11	272		5.3^{20}
(I) chlorate	$TlClO_3$	287.82	nd	4.58^{20}			3.92^{20}
(I) chloride	$TlCl$	239.85	wh cr pwd, 2.247	7.004^{30}	429		0.33^{20}
(III) chloride	$TlCl_3$	310.73	wh, hex, hygr		33	$-Cl_2$, 40	70
(I) chromate(VI)	Tl_2CrO_4	524.73	yel	5.48^{20}			0.03^{60}, sl s a
(I) cyanate	$TlOCN$	246.39	col nd	6.523			s
(I) cyanide	$TlCN$	230.39		6.786	d		16.8^{28}
(I) diphosphate	$Tl_4P_2O_7$	991.42	pr, mn				40
(I) dithionate	$Tl_2S_2O_6$	568.86	wh, mn	5.57^{20}	d		41.8
(I) divanadate(V)	$Tl_4V_2O_7$	1031.36	lt yel	8.21	454		0.2^{14}
(I) ethoxide	$TlOC_2H_5$	249.43	liq, 1.6714	3.493^{20}	−3	d 130	sl s al, s eth
(I) fluoride	TlF	223.39	col, rh	8.23^{4}	322	700	78^{15}
(III) fluoride	TlF_3	261.39	olive-grn, rh	8.36^{25}	d 550		hyd
(I) formate	$TlHCO_2$	249.39	col nd, hygr	4.967^{103}	101		500^{10}
hydrogen oxalate-2-water, tri-	$TlH_3(C_2O_4)_2 \cdot 2H_2O$	419.46	wh, tric, 1.510, 1.632, 1.654	2.992^{17}	d 100		76.9
(I) hydroxide	$TlOH$	221.38	lt yel nd		d 139		35^{20}
(I) iodate	$TlIO_3$	379.27	wh nd	7.29			0.058^{20}
(I) iodide	TlI	331.31	yel, rh; red, cub	7.098^{15}	440	845	0.006^{20}
(I) nitrate	$TlNO_3$	266.40	col, rh	5.556	206	430	9.55^{20}
(III) nitrate-3-water	$Tl(NO_3)_3 \cdot 3H_2O$	444.43	col, rh, deliq				s

Table 4-1 (*Continued*)
PHYSICAL CONSTANTS OF INORGANIC COMPOUNDS

Name	Formula	Formula weight	Color, crystalline form, refractive index	Density	Melting point, °C	Boiling point, °C	Solubility in 100 parts
Thallium							
(I) nitrite	$TlNO_2$	250.38	yel cr	....	182	....	40.3[20]
(I) oxalate	$Tl_2(C_2O_4)$	496.76	pr, mn	6.31	300	....	1.58[20]
(I) oxide	Tl_2O	424.78	blk, deliq	9.52[16]	300	1080	v s d; s a
(III) oxide	Tl_2O_3	456.78	brn pwd	9.65 (amorp) 10.19[22] (hex)	717	$-O_2$, 875	i; s a
(I) perchlorate	$TlClO_4$	303.82	col, rh	4.959[20]	505	d	13.1[20]
(I) phosphate	Tl_3PO_4	708.08	col nd	6.89[10]	....	....	0.5[15]
(I) platinate(IV), hexa-chloro-	$Tl_2[PtCl_6]$	816.55	l-or cr	5.76[17]	....	....	0.0064[15]
(I) selenate(VI)	Tl_2SeO_4	551.74	rh, 1.949, 1.959, 1.964	6.875	>400	....	2.8[20]
(di-) selenide	Tl_2Se	487.74	dk gray leaf	9.05[25]	390	....	i; i a
selenide	$TlSe$	283.33	tetr	....	330	....	....
(I) sulfate	Tl_2SO_4	504.85	col, rh, 1.860, 1.867, 1.885	6.77	632	d	4.87[20]
(III) sulfate-7-water	$Tl_2(SO_4)_3 \cdot 7H_2O$	823.03	col leaf	....	$-6H_2O$, 220	....	d; s H_2SO_4
sulfide	TlS	236.34	tetr	....	d 247	....	....
(di-) sulfide	Tl_2S	440.85	bl-blk, hex-rbhd	8.39	457	oxid	0.02[20], s a
(tetra-) sulfide, tri-	Tl_4S_3	913.67	....	....	d 314	....	....
(I) *dl*-tartrate	$Tl_2C_4H_4O_6$	556.81	mn	4.659	d 165	....	13.3[15]
(I) thiocyanate	$TlSCN$	262.45	col, tetr	4.956	....	....	0.32
(I) vanadate(V) (1−)	$TlVO_3$	303.41	gray cr	6.09[17]	424	....	0.87[11]
Thiocarbonyl chloride	$CSCl_2$	114.98	yel-red liq, 1.5442	1.509[15]	ca −2	73.5	d; s eth
Thiocyanogen	$(SCN)_2$	116.16	liq or yel sol	....	....	....	d; s al, CS_2, eth
Thionyl-, see under Sulfi-nyl-							
Thiophosphoryl diamide	$PS(NH_2)_2$	111.11	yel-wh amorp	1.7[13]	d 200	....	sl s
bromide, tri-	$PSBr_3$	302.78	yel fin cr, cub	2.85[17]	38.2	206 d	s s eth, CS_2

Name	Formula	Formula weight	Crystalline form; color	Density	M.p., °C	B.p., °C	Solubility
chloride, tri-	$PSCl_3$	169.41	col gas, fum liq	1.635^{25}	-35	125	hyd; s bz, chl, CS_2
cyanate, tris-	$PS(OCN)_3$	237.22	col liq	1.538	8.8	215	.....
fluoride, tri-	PSF_3	120.03	col gas	.....	-148.8	-52.2	hyd
Thiosulfinyl fluoride	$S{=}SF_2$	102.13	col gas	.....	-165	-10.6	s a
Thorium	Th	232.04	gray, fcc	11.71	1755	4787	s a
boride, tetra-	ThB_4	275.28	tetr pr	7.5^{15}	2195	.....	s HNO_3
boride, hexa-	ThB_6	296.92	vlt-blk, cub	6.4^{15}	.....	.....	.....
borate, tetrahydro-	$Th(BH_4)_4$	291.40	tetr	2.50	203	.....	.....
bromide	$ThBr_4$	551.67	wh, rh, deliq	5.55	725	subl 610	v s
carbide, di-	ThC_2	256.06	yel, tetr	8.96^{18}	2660	.....	v sl s conc a
chloride	$ThCl_4$	373.88	wh, tetr, deliq	4.60	770	922	v s; s al
chloride oxide, di-	$ThCl_2O$	318.94	wh, rh	.....	d 550	.....	i
diphosphate(V)	ThP_2O_7	411.98	wh	5.71	.....	.....	i; s a
fluoride	ThF_4	308.09	wh, mn	.....	1110	.....	0.25^{55}
fluoride-4-water	$ThF_4 \cdot 4H_2O$	380.09	cr	.....	$-2H_2O$, 140	.....	i; s H_2SO_4
hydride, di-	ThH_2	234.05	bctetr	9.20	d > 300	.....	hyd
iodate, di-	$Th(IO_3)_4$	931.65	wh	.....	d 550	>675, ThO_2	hyd
iodide, di-	ThI_2	485.85	blk or golden, hex	.....	.....	.....	.....
iodide	ThI_4	739.69	yel-wh, mn	6.00	566	837	191^{20}
nitrate	$Th(NO_3)_4$	480.06	col pl, deliq	.....	d > 216	>630 ThO_2	.....
nitride	ThN	246.04	.....	.....	2630	.....	.....
oxalate	$Th(C_2O_4)_2$	408.08	wh cr	4.63^{16}	d 350	.....	0.0017; sl s a
oxide	ThO_2	264.05	wh, cub	9.86	3220	4400	i; s hot H_2SO_4
oxide sulfide	$ThOS$	280.10	yel cr	6.44^{0}	d	.....	i; s aq reg
phosphate(V) (1—)	$Th(PO_3)_4$	547.93	col rh pr	4.08^{16}	.....	.....	.....
phosphide	ThP	263.01	gray-blk, cub	8.81	.....	.....	.....
(tri-) phosphide, tetra-	Th_3P_4	820.01	gray-blk, cub	8.44	.....	.....	.....
platinate(II)-16-water, tetra-cyano-	$Th[Pt(CN)_4]_2 \cdot 16H_2O$	1118.6	yel cr	.....	.....	.....	sl s
selenate-9-water	$Th(SeO_4)_2 \cdot 9H_2O$	680.09	col, mn	3.026	$-8H_2O$, 200	d 1500	0.65^{0}
selenide	$ThSe$	311.0	cub	.....	1877	.....	.....
silicate(4—)	$ThSiO_4$	324.12	col, tetr, 1.80	6.82^{16}	.....	.....	v sl s; i a
silicide, di-	$ThSi_2$	288.21	blk, tetr	7.96^{16}	.....	.....	i; s hot HCl
sulfate	$Th(SO_4)_2$	424.15	wh cr, hygr	4.225^{17}	.....	.....	2.1^{30}
sulfate-9-water	$Th(SO_4)_2 \cdot 9H_2O$	586.31	wh, mn	2.77	$-9H_2O$, 400	.....	1.57^{25}
sulfide	ThS	264.10	cub	.....	<2200	.....	.....

Table 4-1 (Continued)
PHYSICAL CONSTANTS OF INORGANIC COMPOUNDS

Name	Formula	Formula weight	Color, crystalline form, refractive index	Density	Melting point, °C	Boiling point, °C	Solubility in 100 parts
Thorium							
sulfide, di-	ThS$_2$	296.17	brn-blk, rh	7.30	1905	......	i; s hot aq reg
(di-) sulfide, tri-	Th$_2$S$_3$	560.26	rh	......	1847	......	......
telluride	ThTe	359.64	cub	......	......	......	......
telluride, di-	ThTe$_2$	487.24	hex	......	......	......	......
Thulium							
	Tm	168.93	silv-wh, hcp	9.332	1545	1730	i; s a
bromide	TmBr$_3$	408.66	wh, rbhd	5.02	952	1440	s
chloride	TmCl$_3$	275.29	lt-yel, mn	......	824	1490	s
fluoride	TmF$_3$	225.93	wh, hex	7.971	1158	2230	i; s H$_2$SO$_4$
iodide	TmI$_3$	549.65	yel, hex	......	1021	1260	s
nitrate-4-water	Tm(NO$_3$)$_3 \cdot 4$H$_2$O	499.08	grn, deliq	......	......	......	s
oxalate-6-water	Tm$_2$(C$_2$O$_4$)$_3 \cdot 6$H$_2$O	710.02	wh-grn	......	......	......	i; s alk, oxalate slowly s a
oxide	Tm$_2$O$_3$	385.87	lt-grn, bcc	8.6	......	......	i; s HCl, H$_2$SO$_4$
Tin	Sn	118.69	silv-wh, tetr gray, cub	7.28 5.75	231.89 stable −161 to 13.2	2623	......
(II) acetate	Sn(C$_2$H$_3$O$_2$)$_2$	236.79	wh cr, rh	2.31	182	240	d; s HCl
(II) bromide	SnBr$_2$	278.53	lt yel, rh	5.12	215	622	85°; s al, eth
(IV) bromide	SnBr$_4$	438.36	wh cr, rh-pyr	3.35^{33}	30	207	hyd; s acet
(II) chloride	SnCl$_2$	189.61	wh, rh	3.95	247	652	84°; s al, eth
(II) chloride-2-water	SnCl$_2 \cdot 2$H$_2$O	225.63	wh, mn	2.710^{16}	37.7	d	hyd; s al, eth
(IV) chloride	SnCl$_4$	260.53	col liq	2.226	−34	115	s; s eth
(IV) chromate(VI)	Sn(CrO$_4$)$_2$	350.72	brn-yel cr pwd	4.009^{16}	d	......	s
(II) diphosphate(V)	Sn$_2$P$_2$O$_7$	411.32	wh amorp pwd	4.57^{25}	......	......	s; s conc a
(II) fluoride	SnF$_2$	156.70	wh lamellar pl, mn	4.780^{19}	213	......	30
(IV) fluoride	SnF$_4$	194.70	wh cr, tetr, hygr	......	subl 705		hyd
hydride	SnH$_4$	122.72	col gas	......	−150	−52	s conc alk, H$_2$SO$_4$
(II) iodide	SnI$_2$	372.54	red nd, mn	5.285	320	717 d	0.98^{20}; s HCl
(IV) iodide	SnI$_4$	626.38	or-red, cub, 2.106	4.473^0	144.5	364.5	hyd; s al, bz, chl

(II) nitrate-20-water	$Sn(NO_3)_2 \cdot 20H_2O$	603.07	col leaf	....	-20	....	hyd
(IV) nitrate	$Sn(NO_3)_4$	366.71	silky nd	....	d 50	....	hyd
(II) oxalate	SnC_2O_4	206.72	wh pwd	3.56	....	....	i; i HCl
(II) oxide	SnO	134.70	brn-blk, cub	6.446^0	oxid, 300	....	i; s a, alk
(IV) oxide (cassiterite)	SnO_2	150.70	wh, tetr, 1.997, 2.093	6.95	1630	subl 1900	i
(II) phosphate	$Sn_3(PO_4)_2$	546.01	wh amorp	3.823^{17}	....	....	i; d a
phosphide, tri-	SnP	149.66	silv-wh	6.56	d	....	i; s HCl
phosphide, tri-	SnP_3	211.61	silv-wh	4.10^0	530	....	i; s HCl
(tetra) phosphide, tri-	Sn_4P_3	567.68	silv-wh	5.181	550	....	d HCl
(II) selenide	$SnSe$	197.66	steel-gray pr, rh	6.179^0	880	....	i; s aq reg
(IV) selenide	$SnSe_2$	276.62	red-brn cr, hex	5.133	675	....	i; s alk, aq reg
(II) sulfate	$SnSO_4$	214.77	wh, rh	....	$-SO_2$, 378	....	18.9^{20}
(IV) sulfate-2-water	$Sn(SO_4)_2 \cdot 2H_2O$	346.84	wh, hex, deliq	....	....	....	v s
(II) sulfide	SnS	150.77	gray-blk, rh	5.08	881	1210	i; s conc HCl
(IV) sulfide	SnS_2	182.83	golden yel, hex	4.5	765	d	i; d aq reg
(II) telluride	$SnTe$	246.29	gray, cub	6.445	806	....	i
(II) zirconate(IV), hexafluoro-	$Sn[ZrF_6]$	323.92	wh cr	4.21	....	....	s
Titanium	Ti	47.90	dk-gray, hcp	4.507	1660	3318	s hot a, s HF
boride	TiB	58.71	....	....	d 2227	d 3977	....
boride, di-	TiB_2	69.52	hex	4.50	2920	....	d
(II) bromide	$TiBr_2$	207.72	blk, hex	4.31	subl 935.7	....	....
(III) bromide	$TiBr_3$	287.63	gray-blk, rhbd	4.24	subl 794.1	....	....
(III) bromide-6-water	$TiBr_3 \cdot 6H_2O$	395.72	red-vlt, deliq	....	115	d 400	s
(IV) bromide	$TiBr_4$	367.56	amber-yel, mn, hygr	3.25(c) 2.59^{40}(liq)	38.3	231.0	hyd; 287 al
carbide	TiC	59.91	met-gray, cub	4.938	3017	....	i; s HNO_3 + HF
(II) chloride	$TiCl_2$	118.81	blk cr, hex; ign when heated in air	3.13	1035	subl 1026	d
(III) chloride	$TiCl_3$	154.27	vlt, hex, deliq, pyrophoric	2.71	subl 831 (vac)	d > 500	s; s al
(IV) chloride	$TiCl_4$	189.73	col liq, 1.612	1.726	-24.10	136.4	s(cold); s al
chloride oxide	$TiClO$	99.35	bronze, rh	3.14	ca 700 d	....	i; s HNO_3 + HF
chloride oxide, di-	$TiCl_2O$	134.81	lt yel, cub, hygr	2.44	d 180	....	i
(III) fluoride	TiF_3	104.90	bl cr, rhbd	3.00	subl 1037	....	ii; i al
(IV) fluoride	TiF_4	123.90	wh pwd, v hygr	2.798	subl 285.5	....	s, slow hyd

Table 4-1 (Continued)
PHYSICAL CONSTANTS OF INORGANIC COMPOUNDS

Name	Formula	Formula weight	Color, crystalline form, refractive index	Density	Melting point, °C	Boiling point, °C	Solubility in 100 parts
Titanium							
fluoride oxide, di-	TiF_2O	101.90	yel, cub	3.09	......	......	......
hydride, di-	TiH_2	49.92	gray-blk pwd, fcc	3.752	d 400	......	......
[²H]hydride, di-	Ti^2H_2	51.94	fcc	3.940	......	......	......
(II) iodide	TiI_2	301.70	bl-brn, hex	4.99	1085.1	......	i; s HF
(III) iodide	TiI_3	428.61	dk-vlt nd, hex	4.95	subl 727	......	s
(IV) iodide	TiI_4	555.50	red-yel, cub	4.40	155	379.5	s dry nonpolar solv
(IV) isopropoxide	$Ti[OCH(CH_3)_2]_4$	284.26	fum liq	0.9711^{20}	ca 20	220	......
nitride	TiN	61.91	yel-brn pwd, bcc	5.21	2947(H₂)	d 1800(vac)	......
(II) oxide	TiO	63.90	bronze, cub & mn	4.88	1750	3660	i; s H_2SO_4
(III) oxide	Ti_2O_3	143.80	vlt-blk, trig	4.49	1839	......	s H_2SO_4
(IV) oxide (rutile)	TiO_2	79.90	wh, tetr	4.23	1857	......	s HF
(anatase)	......		wh pwd, tetr	3.90	1676	......	
(brookite)	......		rh	4.13	1770	......	
(tri-) oxide, penta-	Ti_3O_5	223.70	......	......	1770	......	d
oxide sulfate	$TiOSO_4$	159.97	wh pwd	......	......	......	i a
phosphide	$TiP_{0.95}$		gray, hex	3.95	......	......	s HCl
(III) sulfate	$Ti_2(SO_4)_3$	384.00	grn pwd	......	......	......	s conc H_2SO_4
sulfide	TiS	79.96	bronze, hex	4.05	1927	......	s hot H_2SO_4
sulfide, di-	TiS_2	112.03	dk-bronze, hex	3.27	d	......	s conc H_2SO_4
(di-) sulfide, tri-	Ti_2S_3	191.99	gray-blk, hex	3.584	......	613^{mm}	
Trisulfuryl dichloride	$ClSO_2-O-SO_2-O-SO_2Cl$	295.09	col liq	1.90^{20}	18.7	120	
difluoride	$FSO_2-O-SO_2-O-SO_2F$	262.18	col liq	1.83	−67.2	......	
Tungsten (Wolfram)	W	183.85	gray-blk, bcc	19.35	3407	5663	s HNO_3 + HF
boride, di-	WB_2	205.47	silv, oct	10.77	(3195)	......	s aq reg
(V) bromide	WBr_5	583.40	blk, grn iridescence	......	286	360.4	hyd; s chl, eth
(VI) bromide	WBr_6	663.30	bl-blk lustrous nd	6.9	279	subl 327	hyd; s CS_2, eth
bromide oxide, tetra-	WBr_4O	519.49	blk nd, bctetr	......	322.4	331	hyd

Name	Formula	Formula wt	Color, crystalline form	Density	Melting point, °C	Boiling point, °C	Solubility
bromide dioxide, di- carbide	WBr_2O_2	375.67	purp-brn	......	d 200–300	......	hyd
carbide	WC	195.86	blk, hex	15.63^{15}	2785	......	s aq reg
(di-) carbide	W_2C	379.71	blk, hex	17.15	2795	......	s aq reg
carbonyl, hexa-	$W(CO)_6$	351.91	col, rh	2.65	......	175	s fum HNO_3
(II) chloride	WCl_2	254.76	gray amorp	5.436	d 589	......	d
(IV) chloride	WCl_4	325.66	blk, deliq	4.624	d 498	......	d
(V) chloride	WCl_5	361.12	blk, deliq	3.875	253	288.3	hyd
(VI) chloride	WCl_6	396.57	bl-blk	2.721^{282} (liq)	281.5	340.5	hyd; s CS_2, CCl_4
chloride oxide, tetra-	WCl_4O	346.66	or-red, bct	3.22^{209} (liq)	211	220	hyd
chloride dioxide, di-	WCl_2O_2	286.76	wh, rh		subl 260	d 369	hyd; s HCl
(VI) fluoride	WF_6	297.86	wh cr, rh; lt-yel liq; col gas	3.441^{15} (liq); 0.0129	2.05	17.2	hyd
fluoride oxide, tetra-	WF_4O	275.85	col	......	106	187	d
fluoride dioxide, di-	WF_2O_2	253.85	wh	......	d 200–300	......	......
nitride, di-	WN_2	211.86	brn, cub	......	>400 vac	......	s a, KOH
(IV) oxide	WO_2	215.85	brn, cub	12.11	1550(N_2)	d 1724	i
(V) oxide	W_2O_5	447.70	bl-vlt, tric	......	subl 800	......	i; s hot alk
(VI) oxide	WO_3	231.86	yel, rh	7.16	1472	1837	s aq reg
phosphide	WP	214.82	gray pr	8.5	......	......	s aq reg
phosphide, di-	WP_2	245.80	blk cr	5.8	d	......	s HNO_3 + HF
silicide, di-	WSi_2	240.02	bl-gray, tetr	9.4	>900	......	s HNO_3 + HF
sulfide, di-	WS_2	247.98	dk-gray, hex & rbhd	7.5^{10}	d 1250	......	i; s alk, HF, NH_4OH
Tungstic acid(2−)	H_2WO_4	249.88	yel pwd, 2.24	5.5	−H_2O, 100	1473	88.6^{22}
(tetra-)(2−)-9-water	$H_2W_4O_{13} \cdot 9H_2O$	1107.55	col, tetr	3.93	d 50	......	s a
Uranium	U	238.03	silv-wh, rh	19.05	1130	3927	i
boride, di-	UB_2	259.65	hex	12.70	2365	......	s
(III) bromide	UBr_3	477.76	brn-red nd, hex	6.53	730	777	v s
(IV) bromide	UBr_4	557.67	brn, mn, deliq	5.55	519	......	d; s a
carbide	UC	250.04	met	......	2525	......	s
carbide, di-	UC_2	262.05	dk purp nd, hex	11.28^{16}	2350	4370	v s
(III) chloride	UCl_3	344.44	dk grn, tetr, hygr	5.51	837	1657	d; s CS_2; s chl
(IV) chloride	UCl_4	379.90	red-brn, mn	4.725^{25}	590	789	s d
(V) chloride	UCl_5	415.30	grn-blk, hex	......	327	527	......
(VI) chloride	UCl_6	450.75	grn	3.56	179	392	......
chloride oxide, di-	UCl_2O	324.94		......	d 550	......	......

Table 4-1 (*Continued*)
PHYSICAL CONSTANTS OF INORGANIC COMPOUNDS

Name	Formula	Formula weight	Color, crystalline form, refractive index	Density	Melting point, °C	Boiling point, °C	Solubility in 100 parts
Uranium							
chloride oxide, tri-	UCl_3O	360.40	red-brn	....	d 700	....	s al
(III) fluoride	UF_3	295.03	blk, hex	8.95	1427	....	s sl d
(IV) fluoride	UF_4	314.07	grn, mn	6.70	1036	1457	i; s conc a, alk
(V) fluoride	UF_5	333.02	wh, mn, deliq	....	348	....	hyd; s chl
(VI) fluoride	UF_6	352.07		5.09^{20}	64.8	subl 56.5	i
(di-) fluoride, nona-	U_2F_9	646.06		7.06	d	....	....
hydride, tri-	UH_3	241.05	blk, bcc	10.95	....	....	i
(III) iodide	UI_3	618.74	brn-blk, cub	6.76	766	....	s
(IV) iodide	UI_4	745.65	blk, rh	5.6^{15}	506(I_2 atm)	757	i a
nitride	UN	252.04	blk	14.31	2480	....	....
(di-) nitride, tri-	U_2N_3	518.08	brn pwd	11.2	d	....	s HNO_3
(IV) oxide (pitchblende)	UO_2	270.07	brn-blk, cub	10.97	2827	....	s HNO_3
(tri-) oxide, octa-	U_3O_8	842.09	olive-grn-blk	8.30	d 1300 to UO_2	....	....
(VI) oxide	UO_3	286.07	yel-brn pwd	7.29	d	....	i; s HNO_3, HCl
peroxide-2-water	$UO_4 \cdot 2H_2O$	338.06	lt yel, hygr	....	d 90–195 to U_2O_7	d > 200 to UO_3	i; d HCl
phosphide	UP	269.00		10.2	....	....	....
(tri-) phosphide, tetra-	U_3P_4	837.99	gray-blk, cub	10.0	....	....	....
silicide, di-	USi_2	294.20	gray-blk, cub	....	1750	....	....
(IV) sulfate-4-water	$U(SO_4)_2 \cdot 4H_2O$	502.21	grn, rh	....	−$4H_2O$, 300	....	10^{20}
sulfide	US	270.09	blk, cub	10.87	3462	....	i a
(di-) sulfide, tri-	U_2S_3	572.25	gray-blk, cub	....	2027	....	s HNO_3
sulfide, di-	US_2	302.16	gray-blk, rh	7.96	1680	....	s conc HCl
Uranyl(VI) acetate-2-water	$UO_2(C_2H_3O_2)_2 \cdot 2H_2O$	422.13	yel cr pwd, rh	2.893^{15}	−$2H_2O$, 110	d 275	7.69^{15}
chloride	UO_2Cl_2	340.98	yel cr, rh, hygr	5.43	577 d	....	320^{18}
fluoride	UO_2F_2	308.03	lt yel, rbhd, hygr	6.37	d 300	....	v s
formate-1-water	$UO_2(HCO_2)_2 \cdot H_2O$	378.08	yel, oct	3.695	−H_2O, 110	....	7.2^{15}

hydrogen phosphate-4-water	$UO_2HPO_4 \cdot 4H_2O$	438.07	yel pl, tetr	....	d 250	....	i; s HNO_3
iodate	$UO_2(IO_3)_2$	619.83	yel, rh	5.22^{18}	d(in air)	....	0.105
iodide	UO_2I_2	523.84	red, deliq	2.807^{13}	60.2	d 100	i; s al, bz, eth
nitrate-6-water	$UO_2(NO_3)_2 \cdot 6H_2O$	502.13	yel, rh, 1.4967	....	$-3H_2O$, 110	155^{20}	
oxalate-3-water	$UO_2C_2O_4 \cdot 3H_2O$	412.09	yel cr	....	90	d 110	0.58^{20}
perchlorate-6-water	$UO_2(ClO_4)_2 \cdot 6H_2O$	577.02	yel, rh, deliq	....	d 100	....	s
sulfate-3-water	$UO_2SO_4 \cdot 3H_2O$	420.14	lemon-yel cr	3.28^{16}	d 40	....	21; 4 al
sulfide	UO_2S	302.09	brn-blk, tetr	....	....	....	sl s; s a
Vanadic(V) acid(1−)	HVO_3	99.95	yel	....	....	....	i; s a, alk
Vanadium	V	50.94	lt-gray, bcc, 3.028	6.11^{19}	1917	3421	s HNO_3, H_2SO_4
(di-) arsenide	V_2As	176.80	silv-gray	6.39	1345(vac)	....	....
boride, di-	VB_2	72.56	hex	5.10	2400	....	....
(II) bromide	VBr_2	210.76	or-brn, hex, hygr	4.58	subl 800	....	v s
(III) bromide	VBr_3	290.67	dk-gray, hex, deliq	4.84	d 400	....	s
(III) bromide oxide	$VBrO$	146.85	vlt	4.00	d 480	....	sl s; s acet
(IV) bromide oxide, di-	VBr_2O	226.76	brn pwd, deliq	....	d 180	....	s
(V) bromide oxide, tri-	VBr_3O	306.67	dk-red liq	2.932	−59	170	s HNO_3
carbide	VC	62.95	blk, bcc	5.77	2830	3900	....
(II) chloride	VCl_2	121.85	lt-grn, hex, hygr	3.09	subl 1027	....	s d; s al, eth
(III) chloride	VCl_3	157.30	red-vlt, rhbd	2.87	d 425	....	s
(IV) chloride	VCl_4	192.75	dk-brn-redsh liq	1.82	−25.5	152	hyd; s nonpolar solv
(III) chloride oxide	$VClO$	102.39	brn, rh	3.45	620 d(vac)	....	i; s HNO_3
(IV) chloride oxide, di-	VCl_2O	137.86	grn tabular cr	2.88^{13}	disprop 384	....	slowly hyd; s abs al
(V) chloride oxide, tri-	VCl_3O	173.30	yel liq	1.84	−79	127	hyd; s CCl_4, hydrocarbons
(V) chloride dioxide	$VClO_2$	118.39	or, hygr	2.29	d 150	....	i; s THF
(III) fluoride	VF_3	107.95	yel-grn pwd, rbhd	3.363	ca 1406	subl 800	i; i org solv
(IV) fluoride	VF_4	126.95	lime-grn, hex, hygr	3.15	subl 100–120(vac)	....	s; s acet
(V) fluoride	VF_5	145.95	wh, rh	2.502^{20}(liq)	19.5	47.9	hyd; s acet, chl
(IV) fluoride oxide, di-	VF_2O	104.94	yel	3.396^{18}	d	....	i; sl s acet
(V) fluoride oxide, tri-	VF_3O	123.94	lt-yel, hygr	2.459	subl 110	(480)	s
(II) iodide	VI_2	304.74	vlt-rose, hex	5.47	subl 750–800	d 1000	s
(III) iodide	VI_3	431.64	brn-blk, hygr	4.2	d > 280	....	v s
nitride	VN	64.95	gray-vlt, bcc	6.04	d 2346	....	s hot HNO_3
(II) oxide	VO	66.94	gray, bcc	5.55	1790	....	s HCl

Table 4-1 (Continued)
PHYSICAL CONSTANTS OF INORGANIC COMPOUNDS

Name	Formula	Formula weight	Color, crystalline form, refractive index	Density	Melting point, °C	Boiling point, °C	Solubility in 100 parts
Vanadium							
(III) oxide	V_2O_3	149.90	gray-blk pwd	4.87	2067	....	i; s HNO_3, HF
(IV) oxide	VO_2	82.94	dk-bl	4.34	1360	....	s a, alk
(IV) oxide, dimeric	V_2O_4	165.88	indigo-bl cr	....	1545	....	
(V) oxide	V_2O_5	181.90	rust-brn, rh	3.35	670	1690	0.80; s a, alk
(IV) oxide sulfate	$VOSO_4$	163.00	bl cr pwd	....	....	....	v s
(II) sulfate-7-water	$VSO_4 \cdot 7H_2O$	273.11	vlt, mn	....	d(in air)	....	
(III) sulfate	$V_2(SO_4)_3$	390.10	lemon-yel	....	d 410(vac)	....	slowly s
(II) sulfide	VS	83.01	brn-blk, hex	4.51	d	....	s HNO_3
(III) sulfide	V_2S_3	198.10	grn-blk pwd	4.72	d 600	....	s hot a, alk sulfide
sulfide, tetra- (patronite)	VS_4	179.20	blk pwd	2.80	d 500	....	i; s alk
Water, see Hydrogen oxide							
Xenon	Xe	131.30	col gas, 1.000702	5.8971^0 (g/L) 3.057^{bp} (liq)	-111.8	-108.10	10.81 ml^{20}
fluoride, di-	XeF_2	169.30	col, bctetr	3.13^{25}	129.0	subl 114	2.5^0
fluoride, tetra-	XeF_4	207.30	col, mn	3.03^{25}	117.1	subl 116	hyd; s F_3CCOOH
fluoride, hexa-	XeF_6	245.30	col, mn	3.411^{25} (c) 3.173^{55} (liq)	49.5	75.6	hyd
fluoride oxide, tetra-	XeF_4O	223.30	col liq	3.168^0	-41 to -28	....	d
fluoride dioxide, di-	XeF_2O_2	201.30	col cr	....	30.8	....	hyd
oxide, tri-	XeO_3	179.30	col, rh, explosive	4.55	d 40	....	s (xenic acid)
oxide, tetra-	XeO_4	195.30	col gas, expl	....	d -40	....	
Ytterbium	Yb	173.04	silv, fcc	6.977	824	1430	s a
acetate-4-water	$Yb(C_2H_3O_2)_3 \cdot 4H_2O$	422.24	col, hex	2.09	-4H_2O, 100	....	v s
(II) bromide	$YbBr_2$	332.86	grn	5.91	613	1830	s
(III) bromide	$YbBr_3$	412.77	col, rbhd	5.10	940	d	s
(II) chloride	$YbCl_2$	243.95	grn, rh	5.08	702	1930	s
(III) chloride	$YbCl_3$	279.40	wh, mn	....	865	....	
(III) chloride-6-water	$YbCl_3 \cdot 6H_2O$	387.49	grn, rh, deliq	2.575	-6H_2O, 180	....	v s

Name	Formula	Mol. wt.	Crystalline form, color, index of refraction	Density	Melting point	Boiling point	Solubility
(II) fluoride	YbF₂	211.04	cub		1407	2380	i
(III) fluoride	YbF₃	230.04	wh, hex	8.168	1157	2230	i; s H₂SO₄
(III) hydroxide	Yb(OH)₃	224.06	col gel ppt		−H₂O, 190	d 290 to Yb₂O₃	s a
(II) iodide	YbI₂	426.85	blk, hex	5.40	772	1330	s dil a
(III) iodide	YbI₃	553.75	yel, hex		700	d	s
(III) nitrate-4-water	Yb(NO₃)₃·4H₂O	431.12	col hygr pr				s
(III) oxalate-10-water	Yb₂(C₂O₄)₃·10H₂O	790.29	col cr	2.644			s dil a
(III) oxide	Yb₂O₃	394.08	wh, bcc	9.18			s d
(III) selenide	Yb₂Se₃	582.96	vlt-blk, rh	7.33			30[20]
(III) sulfate	Yb₂(SO₄)₃	634.26	col	3.793	d 900		34.8[20]
(III) sulfate-8-water	Yb₂(SO₄)₃·8H₂O	778.39	col pr	3.286			s hot H₂O
Yttrium	Y	88.91	gray, hcp	4.472	1530	3304	9.03
acetate-4-water	Y(C₂H₃O₂)₃·4H₂O	338.10	col, tric		74	−6H₂O, 100	168
bromate-9-water	Y(BrO₃)₃·9H₂O	634.76	hex pr, deliq		913	1470	75[20]
bromide	YBr₃	328.63	col, rh, deliq				d
carbide, di-	YC₂	112.93	yel	4.13			d
carbonate-2-water	Y₂(CO₃)₃·2H₂O	393.86	redsh-wh pwd		−2H₂O, 130		79[20]
chloride	YCl₃	195.26	wh, mn	2.67	721	1510	122[20]
chloride-6-water	YCl₃·6H₂O	303.36	redsh-wh, rh	2.18	162	d	d conc a
fluoride	YF₃	145.90	wh, rh	5.069	1152	2230	s a
hydroxide	Y(OH)₃	139.93	wh gelat, hex		−H₂O, 856		v s
iodide	YI₃	469.62	wh, hex, deliq		965	1307	171[20]
nitrate-6-water	Y(NO₃)₃·6H₂O	383.01	redsh-wh, tric	2.68	−3H₂O, 100		s HNO₃
nitride	YN	102.91	cub	5.60	2670		0.0018[30]; s a
oxalate-9-water	Y₂(C₂O₄)₃·9H₂O	604.01	wh, mn		−9H₂O, 410		s a
oxide	Y₂O₃	225.81	wh, bcc, 1.92	5.03	2420	4300	s a
(di-) oxide sulfide, di-	Y₂O₂S	241.87	gray, hex	4.86	2120		s a
selenide	Y₂Se₃	414.69	gray-blk, rh	5.13			
sulfate	Y₂(SO₄)₃	465.99	wh pwd, hygr	2.52	d > 1000		7.3[20]
sulfate-8-water	Y₂(SO₄)₃·8H₂O	610.12	yelsh-wh, mn, 1.543	2.558	−8H₂O, 400		9.6[20]
sulfide	YS	120.97	red, cub	4.51	2040		s a
(di-) sulfide, tri-	Y₂S₃	273.99	lemon, mn	3.91	1600		s a
Zinc	Zn	65.37	blsh-wh met, hex	7.14[25]	419.6	911	i; s a, alk
acetate	Zn(C₂H₃O₂)₂	183.46	col, mn	1.84	d 200		30[20]
acetate-2-water	Zn(C₂H₃O₂)₂·2H₂O	219.49	col, mn	1.735	237		41.6[20]; 3.3 al

Table 4-1 (*Continued*)
PHYSICAL CONSTANTS OF INORGANIC COMPOUNDS

Name	Formula	Formula weight	Color, crystalline form, refractive index	Density	Melting point, °C	Boiling point, °C	Solubility in 100 parts
aluminum oxide (1/1) (gahnite)	$Al_2O_3 \cdot ZnO$	183.33	grn, cub, 1.78	4.58	....	....	i; sl s alk
Zinc							
amide, di-	$Zn(NH_2)_2$	97.42	wh amorp pwd	2.13	d 200 vac	....	d; i al, eth
antimonide	Zn_3Sb_2	439.61	silv-wh pr, rh	6.33	570	....	d
arsenate(V)-8-water (koettigite)	$Zn_3(AsO_4)_2 \cdot 8H_2O$	618.08	col, mn, 1.662, 1.683, 1.717	3.309^{15}	$-H_2O$, 100	....	i; s HNO_3
arsenate(III)	$Zn(AsO_2)_2$	279.20	wh pwd	....	....	....	s a
arsenate(V)–hydroxide (1/1) (adamite)	$Zn_3(AsO_4)_2 \cdot Zn(OH)_2$	573.34	col, rh	4.475^{15}	d 250	....	
arsenide	Zn_3As_2	345.95	gray-met, tetr	5.528	1015	....	i; d a
bromate-6-water	$Zn(BrO_3)_2 \cdot 6H_2O$	429.28	wh, cub, 1.5452	2.566	100	$-6H_2O$, 200	v s
bromide	$ZnBr_2$	225.21	col, rh, 1.5452	4.22	402	650	446^{20}, 200 al, s eth
carbonate (zincspar, smithsonite)	$ZnCO_3$	125.38	col, trig, 1.818, 1.618	4.398	$-CO_2$, 300	....	0.02^{25}, s a, alk
chlorate-4-water	$Zn(ClO_3)_2 \cdot 4H_2O$	304.33	yelsh-wh, cub	2.15	d 60	....	262^{20}
chloride	$ZnCl_2$	136.29	wh, hex, deliq, 1.681, 1.713	2.907^{25}	318	732	395^{20}, 77 al, 50 gly
[-diammine] chloride	$[Zn(NH_3)_2]Cl_2$	170.34	col, rh, 1.625, 1.590	2.10	210.8	d 271	d
chromate(VI)	$ZnCrO_4$	181.36	lemon-yel pr	3.40	....	....	i; s a
cyanide	$Zn(CN)_2$	117.42	wh pwd, rh	1.852	d 800	....	0.058^{18}, s KCN, alk
dichromate(VI)-3-water	$ZnCr_2O_7 \cdot 3H_2O$	335.40	or-yel cr, hygr	....	....	....	s
diphosphate	$Zn_2P_2O_7$	304.72	wh cr pwd	3.75^{23}	....	....	i; s a
ferrate(II), hexacyano-	$Zn_2[Fe(CN)_6]$	342.68	wh pwd	1.85	....	....	i; s alk
fluoride	ZnF_2	103.38	col, mn	5.00^{25}	872	1500	1.6^{20}
fluoride-4-water	$ZnF_2 \cdot 4H_2O$	175.43	col, rbhd	2.255	$-4H_2O$, 100	d 3000 to ZnO	1.6^{18}
formate	$Zn(HCO_2)_2$	155.41	col cr	2.368	d	....	5.2^{18}

Name	Formula	Formula weight	Color, crystalline form, index of refraction	Density	Melting point, °C	Boiling point, °C	Solubility
formate-2-water	$Zn(HCO_2)_2 \cdot 2H_2O$	191.44	wh, mn, 1.513, 1.526, 1.566	2.207^{20}	$-2H_2O$, 140	d	6.4^{20}
hydroxide	$Zn(OH)_2$	99.38	col, rh	3.053	d 125	....	v sl s; s a, alk
iodate	$Zn(IO_3)_2$	415.22	wh cr pwd	5.063^{25}	d	....	0.87^{20}; s alk, HNO_3
iodate-2-water	$Zn(IO_3)_2 \cdot 2H_2O$	451.21	wh cr pwd	4.223	$-2H_2O$, 200	....	0.88; s HNO_3, NH_4OH
iodide	ZnI_2	319.22	col, hex, hygr	4.7364^{25}	446	730	432^{20}; 50 gly
iron oxide (1/1)	$Fe_2O_3 \cdot ZnO$	241.06	blk, oct	5.33^{20}	1590	....	i; s conc HCl
nitrate-6-water	$Zn(NO_3)_2 \cdot 6H_2O$	297.47	col, tetr	2.065^{14}	36.4	$-6H_2O$, 105–131	146^0
nitride	Zn_3N_2	224.12	gray	6.22	....	....	d; s HCl
oxalate-2-water	$ZnC_2O_4 \cdot 2H_2O$	189.42	wh pwd	3.28	d 100	....	0.002; s a, alk
oxide (zincite)	ZnO	81.37	wh, hex, 2.004, 2.020	5.67	1970	....	i; s a, alk
perchlorate-6-water	$Zn(ClO_4)_2 \cdot 6H_2O$	372.36	wh, hex, deliq, 1.508, 1.480	2.252	106	d 200	s; s al
permanganate-6-water	$Zn(MnO_4)_2 \cdot 6H_2O$	411.33	vlt-blk, deliq	2.47	$-6H_2O$, 100	....	33.3
peroxide	ZnO_2	97.38	yelsh-wh pwd	3.00	d > 150	....	i, d slowly
p-phenolsulfonate-8-water	$Zn[C_6H_4(OH)SO_3]_2 \cdot 8H_2O$	555.83	cr pwd, effl	....	$-8H_2O$, 120	....	63; 56 al
phosphate(V)	$Zn_3(PO_4)_2$	386.05	col, rh	3.998^{15}	900	....	i; s a, NH_4OH
phosphate-4-water (hopeite)	$Zn_3(PO_4)_2 \cdot 4H_2O$	458.11	col, rh, 1.572, 1.591, 1.59	3.04	....	....	i; s a
phosphide	Zn_3P_2	258.09	dk-gray, tetr	4.55	> 420	subl 1100 (in H_2)	d; d viol HCl; s bz, CS_2
propionate	$Zn(C_3H_5O_2)_2$	211.52	col pl tab	2.591^{20}	....	....	32; 2.8 al
selenate(VI)-5-water	$ZnSeO_4 \cdot 5H_2O$	298.40	wh, tric	....	d > 50	....	s
selenide	$ZnSe$	144.34	yel, cub, 2.89	5.42^{15}	1526	....	i; d HNO_3
silicate(2−)	$ZnSiO_3$	141.45	col, rh	3.42	1429	....	i; i a
silicate(4−) (willemite)	Zn_2SiO_4	222.85	col, trig, 1.694, 1.723	4.103	1512	....	i; s a
silicon oxide-water (2/1/1) (hemimorphite)	$2ZnO \cdot SiO_2 \cdot H_2O$	240.84	col, rh & trig, 1.614, 1.617, 1.636	3.45	d 100	....	i
silicate-6-water, hexafluoro-	$Zn[SiF_6] \cdot 6H_2O$	315.54	col hex pr, 1.3824, 1.3956	2.104	ca 120	....	v s
stearate	$Zn(C_{18}H_{35}O_2)_2$	632.33	col pwd	....	....	....	i; s bz; i al, eth
sulfate (zinkosite)	$ZnSO_4$	161.44	col, rh, 1.658, 1.669, 1.670	3.54	1200	d > 500	53.8^{20}
sulfate-7-water (goslarite)	$ZnSO_4 \cdot 7H_2O$	287.54	col, rh, effl, 1.457, 1.480, 1.484	1.957	$-7H_2O$, 280	....	96^{20}; 40 gly; i al

Table 4-1 (*Continued*)
PHYSICAL CONSTANTS OF INORGANIC COMPOUNDS

Name	Formula	Formula weight	Color, crystalline form, refractive index	Density	Melting point, °C	Boiling point, °C	Solubility in 100 parts
Zinc							
sulfide (α) (wurtzite)	ZnS	97.43	graysh-wh, hex, 2.356, 2.378	4.087	1722	....	i; s a
(β) (sphalerite)	......	......	col, cub, 2.368	4.102	....	....	i; s a
sulfite-2-water	$ZnSO_3 \cdot 2H_2O$	181.46	wh cr pwd	......	$-2H_2O$, 100	d 200	0.16; s a
telluride	ZnTe	192.99	brn-red pwd & ruby red cr, cub, 3.56	6.34^{15}	1297	....	d slowly a
thiocyanate	$Zn(SCN)_2$	181.53	wh deliq pwd	......	....	....	0.14^{18}
Zirconium	Zr	91.22	graysh-wh, hcp, & bl-blk amorp pwd	6.52^{30}	1852	4504	s aq reg
borate, tetrahydro-	$Zr[BH_4]_4$	150.58	......	......	28.7	123	....
boride, di-	ZrB_2	112.84	......	6.085	3050	d 4193	....
(II) bromide	$ZrBr_2$	251.04	blk pwd	......	627	1282	d
(III) bromide	$ZrBr_3$	330.95	blk-blk, hex	......	subl 827	....	d
(IV) bromide	$ZrBr_4$	410.86	wh, cub, deliq	......	450	subl 357	sl s conc H_2SO_4
carbide	ZrC	103.23	gray, cub	6.73	3532	5100	d
(II) chloride	$ZrCl_2$	162.13	blk pwd	3.6^{18}	727	1292	d
(III) chloride	$ZrCl_3$	197.58	bl-blk, hex & tetr	3.00^{15}	subl 773	....	d
(IV) chloride	$ZrCl_4$	233.05	wh cr, mn, hygr	2.803^{15}	437	subl 334	hyd; s al, eth
chloride oxide-8-water	$ZrCl_2O \cdot 8H_2O$	322.25	wh nd, tetr, effl	1.91	$-8H_2O$, 210	....	s
(II) fluoride	ZrF_2	129.22	blk, rh	......	902	2256	....
(III) fluoride	ZrF_3	148.52	....	......	1202	1227 d	....
(IV) fluoride	ZrF_4	167.22	wh, mn, 1.59	4.16^{16}	932	subl 906	1.32^{20}
hydride, di-	ZrH_2	93.24	gray-blk, bct	5.61	....	....	i
[²H] hydride, di-	Zr^2H_2	94.25	tetr	5.75	....	....	....
hydroxide	$Zr(OH)_4$	159.25	wh amorp pwd	3.25	$-2H_2O$, 500	....	s a
(II) iodide	ZrI_2	345.04	......	......	427	1027	....
(III) iodide	ZrI_3	471.95	blk-blk, hex	......	727	subl 697	....

(IV) iodide	ZrI$_4$	598.86	or-red to yel pwd	....	499(under pressure)	subl 431	s d; s eth
(IV) nitrate-5-water	Zr(NO$_3$)$_4$ · 5H$_2$O	429.32	col cr, hygr	....	....	....	v s
nitride	ZrN	105.23	yel-brn cr	7.3	2952	....	slowly s a
(IV) oxide (baddeleyite, zirconia)	ZrO$_2$	123.22	col, mn	5.85	2677	4275	s hot H$_2$SO$_4$, HF slowly
oxide sulfide	ZrOS	139.28	yel pwd	4.87	ign in air	....	i
phosphide, di-	ZrP$_2$	153.17	gray	4.77	....	....	s hot conc H$_2$SO$_4$
selenate(IV)	Zr(SeO$_3$)$_2$	345.14	wh cr	4.3	d 400	....	s slowly H$_2$SO$_4$
silicate, di-	ZrSiO$_4$	183.31	col cr, tetr-bipyr	4.56	d 1538	....	v inert
silicide, di-	ZrSi$_2$	147.39	gray, rh	4.88	....	....	s HF
sulfate	Zr(SO$_4$)$_2$	283.34	wh, hygr	3.22[16]	d 410	....	s
sulfate-4-water	Zr(SO$_4$)$_2$ · 4H$_2$O	355.41	wh, rh	3.22[16]	anhyd 380	....	52.5[18]
sulfide, di-	ZrS$_2$	155.35	gray, hex	3.87	1550	....	i

Table 4-2
SYNONYMS AND MINERAL NAMES

Acanthite, see Silver sulfide

Adamite, see Zinc arsenate(V)-hydroxide (1/1)

Agate, see Silicon dioxide-x-water

Akermanite, see Calcium magnesium silicon oxide (2/1/2)

Alabandite, see Manganese sulfide

Alamosite, see Lead(II) silicate(2−)

Albite, see Aluminum silicon sodium oxide (1/6/1)

Alite, see Calcium silicon oxide (3/1)

Altaite, see Lead telluride

Alumina, see Aluminum oxide

Alundum, see Aluminum oxide

Alunogenite, see Aluminum sulfate-18-water

Amphibole, see Magnesium silicate(2−)

Anatase, see Titanium(IV) oxide

Andalusite, see Aluminum silicon oxide (1/1)

Anglesite, see Lead sulfate

Anhydrite, see Calcium sulfate

Anhydrone, see Magnesium perchlorate

Anorthite, see Aluminum calcium silicon oxide (1/1/2)

Aragonite, see Calcium carbonate

Arcanite, see Potassium sulfate

Argentite, see Silver sulfide

Argol, see Potassium hydrogen tartrate

Arkansite, see Titanium(IV) oxide

Arsenoferrite, see Iron diarsenide

Arsenolite, see Arsenic(III) oxide dimer

Arsine, see Arsenic hydride

Artinite, see Magnesium carbonate-hydroxide-water (1/1/3)

Ascharite, see Magnesium diborate (4−)-1-water

Auric and aurous, see under Gold

Azoimide, see Hydrogen azide

Azurite, see Copper(II) carbonate-dihydroxide (2/1)

Baddeleyite, see Zirconium(IV) oxide

Baking soda, see Sodium hydrogen carbonate

Barite (barytes), see Barium sulfate

Bauxite, see Aluminum oxide-2-water

Bayerite, see Aluminum oxide-3-water

Bertrandite, see Beryllium silicon oxide-water (4/2/1)

Beryl, see Aluminum beryllium silicon oxide (1/3/6)

Bieberite, see Cobalt sulfate-7-water

Bischofite, see Magnesium chloride-6-water

Bismuthine, see Bismuth hydride

Bismuthinite, see Bismuth sulfide

Bleaching powder, see Calcium hypochlorite-3-water

Bleaching solution, see Sodium hypochlorite

Blue copperas, see Copper(II) sulfate-5-water

Bobierrite, see Magnesium phosphate-8-water

Boehmite, see Aluminum oxide-1-water

Boracic acid, see Boric acid

Borax, see Sodium tetraborate-10-water

Boussingaultite, see Ammonium magnesium bissulfate-6-water

Braunite, see Manganese(III) oxide

Breithauptite, see Nickel antimonide

Brimstone, see Sulfur

Bromellite, see Beryllium oxide

Bromyrite, see Silver bromide

Brookite, see Titanium(IV) oxide

Brucite, see Magnesium hydroxide

Brushite, see Calcium hydrogen phosphate

Bunsenite, see Nickel oxide

Cacodylate, see Sodium dimethylarsonate-3-water

Caesium, see under Cesium

Calamine, see Zinc carbonate

Calcia, see Calcium oxide

Calcite, see Calcium carbonate

Calomel, see Mercury(I) chloride

Caro's acid, see Peroxosulfuric acid

Cassiopeium, see Lutetium

Cassiterite, see Tin(IV) oxide

Caustic potash, see Potassium hydroxide

Caustic soda, see Sodium hydroxide

Celestite, see Strontium sulfate

Celite, see Aluminum calcium iron(III) oxide (1/4/1)

Cementite, see Iron carbide

Cerargyrite, see Silver chloride

Cerussite, see Lead carbonate

Chalcanthite, see Copper sulfate-5-water

Chalcocite, see Copper(I) sulfide

Chalk, see Calcium carbonate

Chessylite, see Copper(II) carbonate-dihydroxide (2/1)

Chile nitre, see Sodium nitrate

Chile saltpeter, see Sodium nitrate

Chloromagnesite, see Magnesium chloride

Chrysoberyl, see Aluminum beryllium oxide

Cinnabar, see Mercury(II) sulfide

Claudetite, see Arsenic(III) oxide dimer

Clausthalite, see Lead selenide

Clinoenstatite, see Magnesium silicate(2−)

Cobaltite, see Cobalt arsenide sulfide

Columbium, see under Niobium

Cooperite, see Platinum(II) sulfide

Coquimbite, see Iron(III) sulfate-9-water

Corrosive sublimate, see Mercury(II) chloride

Table 4-2 (*Continued*)
SYNONYMS AND MINERAL NAMES

Corundum, see Aluminum oxide
Cotunite, see Lead chloride
Covellite, see Copper(II) sulfide
Cream of tartar, see Potassium hydrogen tartrate
Cristobalite, see Silicon dioxide
Crocoite, see Lead chromate(VI)(2−)
Cryolite, see Sodium hexafluoroaluminate
Cryptohalite, see Ammonium silicate, hexafluoro-
Cupric and cuprous, see under Copper
Cuprite, see Copper(I) oxide

Dakin's solution, see Sodium hypochlorite
Daubreelite, see Chromium(II) sulfide
Dehydrite, see Magnesium perchlorate
Dental gas, see Nitrogen(I) oxide
Diamond, see Carbon
Diaspore, see Aluminum oxide-1-water
Digallane, see Gallium hydride
Diopside, see Calcium magnesium silicon oxide (1/1/2)
Diuretic salt, see Potassium acetate
Dolomite, see Calcium magnesium carbonate (1/1)
Domeykite, see *tri*Copper arsenide
Dry Ice, see Carbon dioxide (solid)

Enstatite, see Magnesium silicate(2−)
Epsom salts, see Magnesium sulfate-7-water
Epsomite, see Magnesium sulfate-7-water
Eriochalcite, see Copper(II) chloride
Erythrite, see Cobalt arsenate-8-water
Euclase, see Aluminum beryllium silicon oxide (1/2/2)
Eulytite, see Bismuth silicon oxide (2/3)

Fayalite, see Iron(II) silicate(4−)
Ferberite, see Iron(II) tungstate(VI)(2−)
Ferric and ferrous, see under Iron
Fiedlerite, see Lead chloride oxide-water (2/1/1)
Fluellite, see Aluminum fluoride-1-water
Fluorine oxide, see Oxygen difluoride
Fluoristan, see Tin(II) fluoride
Fluorite, see Calcium fluoride
Fluoroapatite, see Calcium difluoride hexakisphosphate
Fluorspar, see Calcium fluoride
Forsterite, see Magnesium silicate (4−)
Freezing salt, see Sodium chloride
Fulminating mercury, see Mercury fulminate

Gahnite, see Zinc aluminum oxide (1/1)

Galena, see Lead sulfide
Gibbsite, see Aluminum oxide-3-water
Glauber's salt, see Sodium sulfate-10-water
Goethite, see Iron(III) hydroxide oxide
Goslarite, see Zinc sulfate-7-water
Graham's salt, see Sodium phosphate(1−)
Graphite, see Carbon
Greenockite, see Cadmium sulfide
Gruenerite, see Iron(II) silicate(2−)
Guanajuatite, see Bismuth selenide
Guanite, see Ammonium magnesium phosphate-6-water
Gypsum, see Calcium sulfate-2-water

Haidingerite, see Arsenic(V) calcium oxide-water (1/2/3)
Halite, see Sodium chloride
Hauerite, see Manganese disulfide
Hausmannite, see Manganese(II,IV) oxide
Heavy hydrogen, see Deuterium or Hydrogen[²H]
Heavy water, see Deuterium oxide or Hydrogen[²H] oxide
Heazlewoodite, see Nickel (tri-) disulfide
Hematite, see Iron(III) oxide
Hemimorphite, see Zinc silicon oxide-water (2/1/1)
Hermannite, see Manganese silicate
Hessite, see Silver telluride
Hieratite, see Potassium hexafluorosilicate
Hoernesite, see Magnesium arsenate-8-water
Hopeite, see Zinc phosphate-4-water
Hydroazoic acid, see Hydrogen azide
Hydrocerussite, see Lead carbonate-hydroxide (2/1)
Hydromagnesite, see Magnesium carbonate-hydroxide-water (3/1/3)
Hydrophilite, see Calcium chloride
Hydrosulfite, see Sodium dithionate(III)
Hydroxyapatite, see Calcium dihydroxide hexakisphosphate
Hypo (photographic), see Sodium thiosulfate-5-water
Hypophosphite, see under Phosphinate

Ice, see Hydrogen oxide (solid)
Iceland spar, see Calcium carbonate
Ilmenite, see Iron(II) titanate(IV)(2−)
Iodyrite, see Silver iodide

Jadeite, see Aluminum silicon sodium oxide (1/4/1)
Jeweler's borax, see Sodium tetraborate-10-water

Table 4-2 (*Continued*)
SYNONYMS AND MINERAL NAMES

Jeweler's rouge, see Iron(III) oxide

Kalinite, see Aluminum potassium sulfate
Kernite, see Sodium tetraborate
Kieserite, see Magnesium sulfate-1-water
Koettigite, see Zinc arsenate(V)-8-water
Krennerite, see Gold ditelluride
Kyanite, see Aluminum silicon oxide (1/1)

Lansfordite, see Magnesium carbonate-5-water
Laughing gas, see Nitrogen (di-) oxide
Laurionite, see Lead chloride hydroxide (1/1)
Lautarite, see Calcium iodate
Lawrencite, see Iron(II) chloride
Lechatelierite, see Silicon dioxide
Leuconine, see Sodium antimonate(V)
Lime, see Calcium oxide
Linneite, see Cobalt (tri-) tetrasulfide
Litharge, see Lead(II) oxide
Lithium aluminum hydride, see Lithium tetra-
    hydroaluminate
Lodestone, see Iron(II,III) oxide
Lunar caustic, see Silver nitrate
Lye, see Sodium hydroxide

Magnesia, see Magnesium oxide
Magnesite, see Magnesium carbonate
Magnetite, see Iron(II,III) oxide
Malachite, see Copper carbonate dihydroxide
Manganite, see Manganese(III) hydroxide
    oxide
Manganosite, see Manganese(II) oxide
Marcasite, see Iron disulfide
Marshite, see Copper(I) iodide
Mascagnite, see Ammonium sulfate
Massicotite, see Lead oxide
Matlockite, see Lead chloride fluoride
Mendipite, see Lead chloride-oxide (1/2)
Mercuric and mercurous, see under Mercury
Merwinite, see Calcium magnesium silicon
    oxide (3/1/1)
Metacinnabar, see Mercury(II) sulfide
Metastannic acid, see Tin(IV) oxide hydrate
Microcline, see Aluminum potassium silicate
Microcosmic salt, see Ammonium sodium hy-
    drogen phosphate-4-water
Millerite, see Nickel sulfide
Mirabilite, see Sodium sulfate
Mohr's salt, see Ammonium iron(II) sulfate-6-
    water
Moissanite, see Silicon carbide
Molybdenite, see Molybdenum disulfide
Molybdite, see Molybdenum(VI) oxide
Molysite, see Iron(III) chloride

Monazite, see Cerium(III) phosphate
Monimolite, see Lead antimonate(V)(3−)
Montanite, see Bismuth tellurate-2-water
Monticellite, see Calcium magnesium silicon
    oxide (1/1/1)
Montroydite, see Mercury(II) oxide
Morenosite, see Nickel sulfate-7-water
Mosaic gold, see Tin disulfide
Mullite, see Aluminum silicon oxide (3/2)
Muriatic acid, see Hydrogen chloride, aqueous
    solutions

Nantokite, see Copper(I) chloride
Natron, see Sodium carbonate
Naumannite, see Silver selenide
Nephelite, see Aluminum silicon sodium oxide
    (1/2/1)
Nesquehonite, see Magnesium carbonate-3-
    water
Neutral verdigris, see Copper(II) acetate
Newberyite, see Magnesium hydrogen phos-
    phate-3-water
Niccolite, see Nickel arsenide
Niter (nitre), see Potassium nitrate
Nitric oxide, see Nitrogen oxide
Nitrobarite, see Barium nitrate
Nitromagnesite, see Magnesium nitrate-6-water
Nitroprusside, see Sodium pentacyanonitrosyl-
    ferrate(III)-2-water

Oldhamite, see Calcium sulfide
Opal, see Silicon dioxide
Orpiment, see Arsenic trisulfide
Oxygen powder, see Sodium peroxide

Paralaurionite, see Lead chloride-oxide-water
    (1/1/1)
Paris green, see Copper acetate-arsenate(III)
    (1/3)
Patronite, see Vanadium tetrasulfide
Pawellite, see Calcium molybdate(VI)(2−)
Pearl ash, see Potassium carbonate
Perborax, see Sodium peroxoborate
Periclase, see Magnesium oxide
Perovskite, see Calcium titanate(IV)(2−)
Phenazite, see Beryllium silicon oxide (2/1)
Phosgene, see Carbonyl chloride
Phosphine, see Hydrogen phosphide
Phosphotungstic acid, see Phosphoric acid-
    tungsten oxide-water (2/24/48)
Pickling acid, see Sulfuric acid
Picromerite, see Magnesium potassium sulfate
Pitchblende, see Uranium(IV) oxide
Plaster of Paris, see Calcium sulfate hemi-
    hydrate

Table 4-2 (*Continued*)
SYNONYMS AND MINERAL NAMES

Plattnerite, see Lead(IV) oxide
Polianite, see Manganese(IV) oxide
Polishing powder, see Silicon dioxide
Polydymite, see Nickel (tri-) tetrasulfide
Potash, see Potassium carbonate
Potassium acid phthalate, see Potassium hydrogen phthalate
Poydomite, see Nickel (tri-) tetrasulfide
Proustite, see Silver thioarsenate(III)(3−)
Prussic acid, see Hydrogen cyanide
Pseudowollastonite, see Calcium silicon oxide (1/1)
Pyrargyrite, see Silver thioantimonate(III)(3−)
Pyrite, see Iron disulfide
Pyrochroite, see Manganese(II) hydroxide
Pyrolusite, see Manganese(IV) oxide
Pyrophanite, see Manganese titanate(IV)(2−)
Pyrosulfuric acid, see Disulfuric acid

Quartz, see Silicon dioxide
Quicksilver, see Mercury

Rankinite, see Calcium silicon oxide (3/2)
Raspite, see Lead tungstate(VI)(2−)
Realgar, see Arsenic (di-) disulfide
Red lead, see Lead(II,IV) oxide
Rhodochrosite, see Manganese carbonate
Rhodonite, see Manganese silicate(1−)
Richardite, see Copper telluride
Rochelle salt, see Potassium sodium tartrate-4-water
Rock crystal, see Silicon dioxide
Roesslerite, see Magnesium hydrogen arsenate-7-water
Rutile, see Titanium(IV) oxide

Sal soda, see Sodium carbonate-10-water
Saltpeter, see Potassium nitrate
Scacchite, see Manganese chloride
Scheelite, see Calcium tungstate(VI)(2−)
Schultenite, see Lead hydrogen arsenate
Scorodite, see Iron(III) arsenate-2-water
Sellaite, see Magnesium fluoride
Senarmontite, see Antimony(III) oxide
Serpentine, see Magnesium disilicate-2-water
Siderite, see Iron(II) carbonate
Siderotil, see Iron(II) sulfate-5-water
Silica, see Silicon dioxide
Silicotungstic acid, see Silicon oxide-tungsten trioxide-water (1/12/26)
Sillimanite, see Aluminum silicon oxide (1/1)
Smithsonite, see Zinc carbonate
Soda ash, see Sodium carbonate
Spelter, see Zinc metal
Sperrylite, see Platinum (di-) arsenide

Sphalerite, see Zinc sulfide
Sphene, see Calcium silicon titanium oxide (1/1/1)
Spherocobaltite, see Cobalt(II) carbonate
Spinel, see Magnesium dialuminate(2−)
Stannic and stannous, see under Tin
Stibine, see Antimony hydride
Stibnite, see Antimony(III) sulfide
Stolzite, see Lead tungstate(VI)(2−)
Strengite, see Iron(III) phosphate
Strontianite, see Strontium carbonate
Struvite, see Ammonium magnesium phosphate-6-water
Sugar of lead, see Lead acetate
Sulfamate, see Amidosulfate
Sulfurated lime, see Calcium sulfide
Sulfuretted hydrogen, see Hydrogen sulfide
Sulphur, see Sulfur
Sycoporite, see Cobalt sulfide
Sylvite, see Potassium chloride
Szmikite, see Manganese(II) sulfate-1-water
Szomolnokite, see Iron(II) sulfate-1-water

Talc, see Magnesium silicon oxide-water (3/4/1)
Tapiolite, see Iron(II) tantalate(V)(1−)
Tarapacaite, see Potassium chromate
Tartar emetic, see Animony potassium oxide tartrate-hemihydrate
Tellurite, see Tellurium dioxide
Tenorite, see Copper(II) oxide
Tephroite, see Manganese silicate(1−)
Thenardite, see Sodium sulfate
Thionyl, see Sulfinyl
Thorianite, see Thorium dioxide
Tiemannite, see Mercury(II) selenide
Topaz, see Aluminum hexafluorosilicate
Tridymite, see Silicon dioxide
Troilite, see Iron(II) sulfide
Trona, see Sodium carbonate-hydrogen carbonate-2-water
Tschermigite, see Aluminum ammonium sulfate
Tungstenite, see Tungsten disulfide
Tungstite, see Tungstic acid

Uraninite, see Uranium(IV) oxide

Valentinite, see Antimony(III) oxide
Verdigris, see Copper acetate-1-water
Vermillion, see Mercury(II) sulfide
Villiaumite, see Sodium fluoride
Vitamin B-3, see Calcium pantothenate(+)
Vivianite, see Iron(III) phosphate

Table 4-2 (*Continued*)
SYNONYMS AND MINERAL NAMES

Washing soda, see Sodium carbonate-10-water

White lead, see Lead carbonate dihydroxide

Whitlockite, see Calcium phosphate

Willemite, see Zinc silicate(4−)

Witherite, see Barium carbonate

Wollastonite, see Calcium silicon oxide (1/1)

Wuestite, see Iron(II) oxide

Wulfenite, see Lead molybdate(VI)(2−)

Wurtzite, see Zinc sulfide

Xantheosite, see Nickel arsenate(V)(3−)

Zaratite, see Nickel carbonate dihydroxide

Zincite, see Zinc oxide

Zincosite, see Zinc sulfate

Zincspar, see Zinc carbonate

Zirconia, see Zirconium(IV) oxide

Section 5

ANALYTICAL CHEMISTRY

ACTIVITY COEFFICIENTS

Although it is not possible to measure an individual ionic activity coefficient, f_i, it may be estimated from the following equation of the Debye-Hückel theory:

$$-\log f_i = \frac{Az_i^2 \sqrt{I}}{1 + B\mathring{a}\sqrt{I}}$$

where I is the ionic strength of the medium, and $\mathring{a}$ is the ion-size parameter—the effective ionic radius (Table 5-2). The values of A and B vary with the temperature and dielectric constant of the solvent; values from 0 to 100°C for aqueous medium ($\mathring{a}$ in angstrom units) are listed in Table 5-3. Corresponding values of A and B for unit weight of solvent (when employing molality) can be obtained by multiplying the corresponding values for unit volume (molarity units) by the square root of the density of water at the appropriate temperature.

The ionic strength can be estimated from the summation of the product molarity times ionic charge squared for all the ionic species present in the solution, i.e., $I = 0.5(c_1z_1^2 + c_2z_2^2 + \cdots + c_iz_i^2)$.

Values for the activity coefficients of ions in water at 25°C are given in Table 5-1 in terms of their effective ionic radii.

At moderate ionic strengths a considerable improvement is effected by subtracting a term bI from the Debye-Hückel expression; b is an adjustable parameter which is 0.2 for water at 25°C. Table 5-4 gives the values of the ionic activity coefficients (for z_i from 1 to 6) with $\mathring{a}$ taken to be 4.6Å.

In general, the mean ionic activity coefficient is given by

$$f_\pm = \sqrt[(x+y)]{f_+^x f_-^y}$$

where f_+, f_- are the individual ionic activity coefficients, and x, y are the charge numbers (z_+, z_-) of the respective ions. In binary electrolyte solution,

$$f_\pm = \sqrt{f_+ f_-}$$

In ternary electrolytes, e.g., $BaCl_2$ or K_2SO_4,

$$f_\pm = \sqrt[3]{f_+ f_-^2} \quad \text{or} \quad f_\pm = \sqrt[3]{f_+^2 f_-}$$

In quaternary electrolytes, e.g., $LaCl_3$ or $K_3[Fe(CN)_6]$,

$$f_\pm = \sqrt[4]{f_+ f_-^3} \quad \text{or} \quad f_\pm = \sqrt[4]{f_+^3 f_-}$$

Table 5-1
INDIVIDUAL ACTIVITY COEFFICIENTS
OF IONS IN WATER AT 25°C

Effective Ionic Radii $\mathring{a}$ (in Å)	f_i at Ionic Strength of				
	0.001	0.005	0.01	0.05	0.1
Univalent Ions					
9	0.967	0.933	0.914	0.86	0.83
8	0.966	0.931	0.912	0.85	0.82
7	0.965	0.930	0.909	0.845	0.81
6	0.965	0.929	0.907	0.835	0.80
5	0.964	0.928	0.904	0.83	0.79
4	0.964	0.928	0.902	0.82	0.775
3.5	0.964	0.926	0.900	0.81	0.76
3	0.964	0.925	0.899	0.805	0.755
2.5	0.964	0.924	0.898	0.80	0.75
Divalent Ions					
8	0.872	0.755	0.69	0.52	0.45
7	0.872	0.755	0.685	0.50	0.425
6	0.870	0.749	0.675	0.485	0.405
5	0.868	0.744	0.67	0.465	0.38
4.5	0.868	0.741	0.663	0.45	0.36
4	0.867	0.740	0.660	0.445	0.355
Trivalent Ions					
6	0.731	0.52	0.415	0.195	0.13
5	0.728	0.51	0.405	0.18	0.115
4	0.725	0.505	0.395	0.16	0.095
Tetravalent Ions					
11	0.588	0.35	0.255	0.10	0.065
5	0.57	0.31	0.20	0.048	0.021
Pentavalent Ions					
9	0.43	0.18	0.105	0.020	0.009

Table 5-2

APPROXIMATE EFFECTIVE IONIC RADII IN AQUEOUS SOLUTIONS AT 25°C

Inorganic Ions

$\mathring{a}$ (in Å)	Inorganic Ions
2.5	Rb^+, Cs^+, NH_4^+, Tl^+, Ag^+
3	K^+, Cl^-, Br^-, I^-, CN^-, NO_2^-, NO_3^-
3.5	OH^-, F^-, SCN^-, OCN^-, HS^-, ClO_3^-, ClO_4^-, BrO_3^-, IO_4^-, MnO_4^-
4	Na^+, $CdCl^+$, Hg_2^{2+}, ClO_2^-, IO_3^-, HCO_3^-, $H_2PO_4^-$, HSO_3^-, $H_2AsO_4^-$, SO_4^{2-}, $S_2O_3^{2-}$, $S_2O_8^{2-}$, SeO_4^{2-}, CrO_4^{2-}, HPO_4^{2-}, $S_2O_6^{2-}$, PO_4^{3-}, $Fe(CN)_6^{3-}$, $Cr(NH_3)_6^{3+}$, $Co(NH_3)_5H_2O^{3+}$
4.5	Pb^{2+}, CO_3^{2-}, SO_3^{2-}, MoO_4^{2-}, $Co(NH_3)_5Cl^{2+}$, $Fe(CN)_5NO^{2-}$
5	Sr^{2+}, Ba^{2+}, Ra^{2+}, Cd^{2+}, Hg^{2+}, S^{2-}, $S_2O_4^{2-}$, WO_4^{2-}, $Fe(CN)_6^{4-}$
6	Li^+, Ca^{2+}, Cu^{2+}, Zn^{2+}, Sn^{2+}, Mn^{2+}, Fe^{2+}, Ni^{2+}, Co^{2+}, $Co(en)_3^{3+}$, $Co(S_2O_3)(CN)_5^{4-}$
8	Mg^{2+}, Be^{2+}
9	H^+, Al^{3+}, Fe^{3+}, Cr^{3+}, Sc^{3+}, Y^{3+}, La^{3+}, In^{3+}, Ce^{3+}, Pr^{3+}, Nd^{3+}, Sm^{3+}, $Co(SO_3)_2(CN)_4^{5-}$
11	Th^{4+}, Zr^{4+}, Ce^{4+}, Sn^{4+}

Organic Ions

$\mathring{a}$ (in Å)	Organic Ions
3.5	$HCOO^-$, H_2Cit^-, $CH_3NH_3^+$, $(CH_3)_2NH_2^+$
4	$H_3N^+CH_2COOH$, $(CH_3)_3NH^+$, $C_2H_5NH_3^+$
4.5	CH_3COO^-, $ClCH_2COO^-$, $(CH_3)_4N^+$, $(C_2H_5)_2NH_2^+$, $H_2NCH_2COO^-$, oxalate^{2-}, $HCit^{2-}$
5	Cl_2CHCOO^-, Cl_3COO^-, $(C_2H_5)_3NH^+$, $C_3H_7NH_3^+$, Cit^{3-}, succinate^{2-}, malonate^{2-}, tartrate^{2-}
6	benzoate$^-$, hydroxybenzoate$^-$, chlorobenzoate$^-$, phenylacetate$^-$, vinylacetate$^-$, $(CH_3)_2C{=}CHCOO^-$, $(C_2H_5)_4N^+$, $(C_3H_7)_2NH_2^+$, phthalate^{2-}, glutarate^{2-}, adipate^{2-}
7	trinitrophenolate$^-$, $(C_3H_7)_3NH^+$, methoxybenzoate$^-$, pimelate^{2-}, suberate^{2-}, Congo red anion^{2-}
8	$(C_6H_5)_2CHCOO^-$, $(C_3H_7)_4N^+$

Table 5-3
CONSTANTS OF THE DEBYE-HÜCKEL EQUATION
FROM 0 TO 100°C

$$-\log f_i = \frac{Az_i^2 \sqrt{I}}{1 + B\mathring{a}\sqrt{I}}$$

Temp., °C	Unit Volume of Solvent		Temp., °C	Unit Volume of Solvent	
	A	B		A	B
0	0.4918	0.3248	55	0.5432	0.3358
5	0.4952	0.3256	60	0.5494	0.3371
10	0.4989	0.3264	65	0.5558	0.3384
15	0.5028	0.3273	70	0.5625	0.3397
20	0.5070	0.3282	75	0.5695	0.3411
25	0.5115	0.3291	80	0.5767	0.3426
30	0.5161	0.3301	85	0.5842	0.3440
35	0.5211	0.3312	90	0.5920	0.3456
40	0.5262	0.3323	95	0.6001	0.3471
45	0.5317	0.3334	100	0.6086	0.3488
50	0.5373	0.3346			

The values for unit weight of solvent (molality scale) can be obtained by multiplying the corresponding values for unit volume by the square root of the density of water at the appropriate temperature.

Table 5-4
INDIVIDUAL IONIC ACTIVITY COEFFICIENTS
AT HIGHER IONIC STRENGTHS AT 25°C

The values were calculated from the modified Debye-Hückel equation utilizing the modifications proposed by Robinson and by Guggenheim and Bates:

$$-\frac{\log f_i}{z_i^2} = \frac{0.511I}{1 + 1.5I} - 0.2I$$

where I is the ionic strength and $\mathring{a}$ is assumed to be 4.6 Å.

I	$-\dfrac{\log_{10} f_i}{z_i^2}$	f_i for $z_i =$					
		1	2	3	4	5	6
0.05	0.0756	0.840	0.498	0.209	0.0617	0.0129	0.00190
0.1	0.0896	0.814	0.438	0.156	0.0369	0.00576	0.000595
0.2	0.0968	0.800	0.410	0.138	0.0283	0.00380	0.000328
0.3	0.0936	0.806	0.422	0.144	0.0318	0.00457	0.000427
0.4	0.0858	0.821	0.454	0.169	0.0424	0.00716	0.000815
0.5	0.0753	0.841	0.500	0.210	0.0624	0.0131	0.00190
0.6	0.0631	0.865	0.559	0.270$_5$	0.0978	0.0265	0.00535
0.7	0.0496	0.892	0.633	0.358	0.161	0.0575$_5$	0.0164
0.8	0.0352	0.922	0.723	0.482	0.273	0.132	0.0541
0.9	0.0201	0.955	0.831	0.659	0.477	0.314	0.189
1.0	0.0044	0.900	0.960	0.913	0.850	0.776	0.694

EQUILIBRIUM CONSTANTS

Table 5-5
IONIC PRODUCT CONSTANT OF WATER

This table gives values of pK_w on a molal scale, where K_w is the ionic activity product constant of water. Taking $pK_w = 13.997$ at 25°C and 1 atm, values of pK_w are given at pressures up to 2000 atm.

Temp., °C	pK_w	Temp., °C	pK_w	Pressure, atm	pK_w
0	14.944	60	13.017	1	13.997
5	14.734	65	12.908	250	13.907
10	14.535	70	12.800	500	13.824
15	14.346	75	12.699	750	13.747
20	14.167	80	12.598	1000	13.667
25	13.997	85	12.510	1250	13.585
30	13.833	90	12.422	1500	13.524
35	13.680	95	12.341	1750	13.449
40	13.535	100	12.259	2000	13.394
45	13.396	110	12.126		
50	13.262	120	12.002		
55	13.137	130	11.907		

Table 5-6
SOLUBILITY PRODUCTS

The data refer to various temperatures between 18 and 25°C, and were primarily compiled from values cited by Bjerrum, Schwarzenbach, and Sillen, *Stability Constants of Metal Complexes*, Part II, Chemical Society, London, 1958.

Substance	pK_{sp}	K_{sp}	Substance	pK_{sp}	K_{sp}
Actinium			$BaCrO_4$	9.93	1.2×10^{-10}
$Ac(OH)_3$	15	1×10^{-15}	$Ba_2[Fe(CN)_6] \cdot 6H_2O$	7.5	3.2×10^{-8}
Aluminum			BaF_2	5.98	1.0×10^{-6}
$AlAsO_4$	15.8	1.6×10^{-16}	$BaSiF_6$	6	1×10^{-6}
cupferrate, AlL_3	18.64	2.3×10^{-19}	$Ba(IO_3)_2 \cdot 2H_2O$	8.82	1.5×10^{-9}
$Al(OH)_3$ amorphous	32.9	1.3×10^{-33}	$Ba(OH)_2$	2.3	5×10^{-3}
$AlPO_4$	18.24	6.3×10^{-19}	$Ba(MnO_4)_2$	9.61	2.5×10^{-10}
8-quinolinolate, AlL_3	29.00	1.00×10^{-29}	$BaMoO_4$	7.40	4.0×10^{-8}
Al_2S_3	6.7	2×10^{-7}	$Ba(NbO_3)_2$	16.50	3.2×10^{-17}
Al_2Se_3	24.4	4×10^{-25}	$Ba(NO_3)_2$	2.35	4.5×10^{-3}
Americium			BaC_2O_4	6.79	1.6×10^{-7}
$Am(OH)_3$	19.57	2.7×10^{-20}	$BaC_2O_4 \cdot H_2O$	7.64	2.3×10^{-8}
$Am(OH)_4$	56	1×10^{-56}	$BaHPO_4$	6.5	3.2×10^{-7}
Ammonium			$Ba_3(PO_4)_2$	22.47	3.4×10^{-23}
$NH_4UO_2AsO_4$	23.77	1.7×10^{-24}	$Ba_2P_2O_7$	10.5	3.2×10^{-11}
Arsenic			$BaHPO_3 \cdot 0.5H_2O$	3	1×10^{-3}
$As_2S_3 + 4H_2O \rightarrow$	21.68	2.1×10^{-22}	8-quinolinolate, BaL_2	8.3	5.0×10^{-9}
$2HAsO_2 + 3H_2S$			$Ba(ReO_4)_2$	1.28	5.2×10^{-2}
Barium			$BaSeO_4$	7.46	3.5×10^{-8}
$Ba_3(AsO_4)_2$	50.11	8.0×10^{-51}	$BaSO_4$	9.96	1.1×10^{-10}
$Ba(BrO_3)_2$	5.50	3.2×10^{-6}	$BaSO_3$	6.1	8×10^{-7}
$BaCO_3$	8.29	5.1×10^{-9}	BaS_2O_3	4.79	1.6×10^{-5}
$BaCO_3 + CO_2 +$	4.35	4.5×10^{-5}	Beryllium		
$H_2O \rightarrow Ba^{2+} +$			$BeCO_3 \cdot 4H_2O$	3	1×10^{-3}
$2HCO_3^-$					

5-7

Table 5-6 (*Continued*)
SOLUBILITY PRODUCTS

Substance	pK_{sp}	K_{sp}	Substance	pK_{sp}	K_{sp}
Be(OH)$_2$ amorphous	21.8	1.6×10^{-22}	CaSeO$_3$	5.53	8.0×10^{-6}
Be(OH)$_2$ + OH$^-$ →	2.50	3.2×10^{-3}	CaSiO$_3$	7.60	2.5×10^{-8}
HBeO$_2^-$ + H$_2$O			CaSO$_4$	5.04	9.1×10^{-6}
BeMoO$_4$	1.5	3.2×10^{-2}	CaSO$_3$	7.17	6.8×10^{-8}
Be(NbO$_3$)$_2$	15.92	1.2×10^{-16}	tartrate dihydrate	6.11	7.7×10^{-7}
Bismuth			CaWO$_4$	8.06	8.7×10^{-9}
BiAsO$_4$	9.36	4.4×10^{-10}	Cerium		
cupferrate	27.22	6.0×10^{-28}	CeF$_3$	15.1	8×10^{-16}
Bi(OH)$_3$	30.4	4×10^{-31}	Ce(IO$_3$)$_3$	9.50	3.2×10^{-10}
BiI$_3$	18.09	8.1×10^{-19}	Ce(IO$_3$)$_4$	16.3	5×10^{-17}
BiPO$_4$	22.89	1.3×10^{-23}	Ce(OH)$_3$	19.8	1.6×10^{-20}
Bi$_2$S$_3$	97	1×10^{-97}	Ce(OH)$_4$	47.7	2×10^{-48}
BiOBr	6.52	3.0×10^{-7}	Ce$_2$(C$_2$O$_4$)$_3$ · 9H$_2$O	25.5	3.2×10^{-26}
BiOCl	30.75	1.8×10^{-31}	CePO$_4$	23	1×10^{-23}
BiOOH	9.4	4×10^{-10}	Ce$_2$(SeO$_3$)$_3$	24.43	3.7×10^{-25}
BiO(NO$_2$)	6.31	4.9×10^{-7}	Ce$_2$S$_3$	10.22	6.0×10^{-11}
BiO(NO$_3$)	2.55	2.82×10^{-3}	(III) tartrate	19.0	1×10^{-19}
BiOSCN	6.80	1.6×10^{-7}	Cesium		
Cadmium			CsBrO$_3$	1.7	5×10^{-2}
anthranilate, CdL$_2$	8.27	5.4×10^{-9}	CsClO$_3$	1.4	4×10^{-2}
Cd$_3$(AsO$_4$)$_2$	32.66	2.2×10^{-33}	Cs$_2$[PtCl$_6$]	7.5	3.2×10^{-8}
[Cd(NH$_3$)$_6$](BF$_4$)$_2$	5.7	2×10^{-6}	Cs$_3$[Co(NO$_2$)$_6$]	15.24	5.7×10^{-16}
benzoate · 2H$_2$O	2.7	2×10^{-3}	Cs[BF$_4$]	4.7	5×10^{-5}
Cd(BO$_2$)$_2$	8.64	2.3×10^{-9}	Cs$_2$[PtF$_6$]	5.62	2.4×10^{-6}
CdCO$_3$	11.28	5.2×10^{-12}	Cs$_2$[SiF$_6$]	4.90	1.3×10^{-5}
Cd(CN)$_2$	8.0	1.0×10^{-8}	CsClO$_4$	2.4	4×10^{-3}
Cd$_2$[Fe(CN)$_6$]	16.49	3.2×10^{-17}	CsIO$_4$	2.36	4.3×10^{-3}
Cd(OH)$_2$ fresh	13.6	2.5×10^{-14}	CsMnO$_4$	4.08	8.2×10^{-5}
CdC$_2$O$_4$ · 3H$_2$O	7.04	9.1×10^{-8}	CsReO$_4$	3.40	4.0×10^{-4}
Cd$_3$(PO$_4$)$_2$	32.6	2.5×10^{-33}	Chromium(II)		
quinaldate, CdL$_2$	12.3	5.0×10^{-13}	Cr(OH)$_2$	15.7	2×10^{-16}
CdS	26.1	8.0×10^{-27}	Chromium(III)		
CdWO$_4$	5.7	2×10^{-6}	CrAsO$_4$	20.11	7.7×10^{-21}
Calcium			CrF$_3$	10.18	6.6×10^{-11}
Ca$_3$(AsO$_4$)$_2$	18.17	6.8×10^{-19}	Cr(NH$_3$)$_6$(BF$_4$)$_3$	4.21	6.2×10^{-5}
acetate · 3H$_2$O	2.4	4×10^{-3}	Cr(OH)$_3$	30.2	6.3×10^{-31}
benzoate · 3H$_2$O	2.4	4×10^{-3}	Cr(NH$_3$)$_6$(ReO$_4$)$_3$	11.11	7.7×10^{-12}
CaCO$_3$	8.54	2.8×10^{-9}	CrPO$_4$ · 4H$_2$O green	22.62	2.4×10^{-23}
CaCO$_3$ calcite	8.35	4.5×10^{-9}	violet	17.00	1.0×10^{-17}
CaCO$_3$ aragonite	8.22	6.0×10^{-9}	Cobalt		
CaCrO$_4$	3.15	7.1×10^{-4}	anthranilate, CoL$_2$	9.68	2.1×10^{-10}
CaF$_2$	8.28	5.3×10^{-9}	Co$_3$(AsO$_4$)$_2$	28.12	7.6×10^{-29}
Ca[SiF$_6$]	3.09	8.1×10^{-4}	CoCO$_3$	12.84	1.4×10^{-13}
Ca(OH)$_2$	5.26	5.5×10^{-6}	Co$_2$[Fe(CN)$_6$]	14.74	1.8×10^{-15}
Ca(IO$_3$)$_2$ · 6H$_2$O	6.15	7.1×10^{-7}	Co(NH$_3$)$_6$[BF$_4$]$_2$	5.4	4×10^{-6}
Ca[Mg(CO$_3$)$_2$] dolomite	11	1×10^{-11}	Co(OH)$_2$ fresh	14.8	1.6×10^{-15}
CaMoO$_4$	7.38	4.2×10^{-8}	Co(OH)$_3$	43.8	1.6×10^{-44}
Ca(NbO$_3$)$_2$	17.06	8.7×10^{-18}	Co(IO$_3$)$_2$	4.0	1.0×10^{-4}
CaC$_2$O$_4$ · H$_2$O	8.4	4×10^{-9}	quinaldate, CoL$_2$	10.8	1.6×10^{-11}
CaHPO$_4$	7.0	1×10^{-7}	Co[Hg(SCN)$_4$]	5.82	1.5×10^{-6}
Ca$_3$(PO$_4$)$_2$	28.70	2.0×10^{-29}	α-CoS	20.4	4.0×10^{-21}
8-quinolinolate, CaL$_2$	11.12	7.6×10^{-12}	β-CoS	24.7	2.0×10^{-25}
CaSeO$_4$	3.09	8.1×10^{-4}	8-quinolinolate, CoL$_2$	24.8	1.6×10^{-25}

Table 5-6 (*Continued*)
SOLUBILITY PRODUCTS

Substance	pK_{sp}	K_{sp}	Substance	pK_{sp}	K_{sp}
CoHPO$_4$	6.7	2×10^{-7}	AuI$_3$	46	1×10^{-46}
Co$_3$(PO$_4$)$_2$	34.7	2×10^{-35}	Au$_2$(C$_2$O$_4$)$_3$	10	1×10^{-10}
CoSeO$_3$	6.8	1.6×10^{-7}	Hafnium		
Copper(I)			Hf(OH)$_3$	25.4	4.0×10^{-26}
CuN$_3$	8.31	4.9×10^{-9}	Holmium		
Cu[B(C$_6$H$_5$)$_4$]			Ho(OH)$_3$	22.3	5.0×10^{-23}
tetraphenylborate	8.0	1×10^{-8}	Indium		
CuBr	8.28	5.3×10^{-9}	In$_4$[Fe(CN)$_6$]$_3$	43.72	1.9×10^{-44}
CuCl	5.92	1.2×10^{-6}	In(OH)$_3$	33.2	6.3×10^{-34}
CuCN	19.49	3.2×10^{-20}	quinolinolate, InL$_3$	31.34	4.6×10^{-32}
CuI	11.96	1.1×10^{-12}	In$_2$S$_3$	73.24	5.7×10^{-74}
CuOH	14.0	1×10^{-14}	In$_2$(SeO$_3$)$_3$	32.6	4.0×10^{-33}
Cu$_2$S	47.6	2.5×10^{-48}	Iron(II)		
CuSCN	14.32	4.8×10^{-15}	FeCO$_3$	10.50	3.2×10^{-11}
Copper(II)			Fe(OH)$_2$	15.1	8.0×10^{-16}
anthranilate, CuL$_2$	13.22	6.0×10^{-14}	FeC$_2$O$_4 \cdot$2H$_2$O	6.5	3.2×10^{-7}
Cu$_3$(AsO$_4$)$_2$	35.12	7.6×10^{-36}	FeS	17.2	6.3×10^{-18}
Cu(N$_3$)$_2$	9.2	6.3×10^{-10}	Iron(III)		
CuCO$_3$	9.86	1.4×10^{-10}	FeAsO$_4$	20.24	5.7×10^{-21}
CuCrO$_4$	5.44	3.6×10^{-6}	Fe$_4$[Fe(CN)$_6$]$_3$	40.52	3.3×10^{-41}
Cu$_2$[Fe(CN)$_6$]	15.89	1.3×10^{-16}	Fe(OH)$_3$	37.4	4×10^{-38}
Cu(IO$_3$)$_2$	7.13	7.4×10^{-8}	FePO$_4$	21.89	1.3×10^{-22}
Cu(OH)$_2$	19.66	2.2×10^{-20}	quinaldate, FeL$_3$	16.9	1.3×10^{-17}
CuC$_2$O$_4$	7.64	2.3×10^{-8}	Fe$_2$(SeO$_3$)$_3$	30.7	2.0×10^{-31}
Cu$_3$(PO$_4$)$_2$	36.9	1.3×10^{-37}	Lanthanum		
Cu$_2$P$_2$O$_7$	15.08	8.3×10^{-16}	La(BrO$_3$)$_3 \cdot$9H$_2$O	2.5	3.2×10^{-3}
quinaldate, CuL$_2$	16.8	1.6×10^{-17}	LaF$_3$	16.2	7×10^{-17}
8-quinolinolate, CuL$_2$	29.7	2.0×10^{-30}	La(OH)$_3$	18.7	2.0×10^{-19}
rubeanate	15.12	7.67×10^{-16}	La(IO$_3$)$_3$	11.21	6.1×10^{-12}
CuS	35.2	6.3×10^{-36}	La$_2$(MoO$_4$)$_3$	20.4	4×10^{-21}
CuSeO$_3$	7.68	2.1×10^{-8}	La$_2$(C$_2$O$_4$)$_3 \cdot$9H$_2$O	26.60	2.5×10^{-27}
Dysprosium			LaPO$_4$	22.43	3.7×10^{-23}
Dy$_2$(CrO$_4$)$_3 \cdot$10H$_2$O	8	1×10^{-8}	La$_2$S$_3$	12.70	2.0×10^{-13}
Dy(OH)$_3$	21.85	1.4×10^{-22}	La$_2$(WO$_4$)$_3 \cdot$3H$_2$O	3.90	1.3×10^{-4}
Erbium			Lead		
Er(OH)$_3$	23.39	4.1×10^{-24}	acetate	2.75	1.8×10^{-3}
Europium			anthranilate, PbL$_2$	9.81	1.6×10^{-10}
Eu(OH)$_3$	23.05	8.9×10^{-24}	Pb$_3$(AsO$_4$)$_2$	35.39	4.0×10^{-36}
Gadolinium			Pb(N$_3$)$_2$	8.59	2.5×10^{-9}
Gd(HCO$_3$)$_3$	1.7	2×10^{-2}	Pb(BO$_2$)$_2$	10.78	1.6×10^{-11}
Gd(OH)$_3$	22.74	1.8×10^{-23}	PbBr$_2$	4.41	4.0×10^{-5}
Gallium			Pb(BrO$_3$)$_2$	1.70	2.0×10^{-2}
Ga$_4$[Fe(CN)$_6$]$_3$	33.82	1.5×10^{-34}	PbCO$_3$	13.13	7.4×10^{-14}
Ga(OH)$_3$	35.15	7.0×10^{-36}	PbCl$_2$	4.79	1.6×10^{-5}
8-quinolinolate, GaL$_3$	40.8	1.6×10^{-41}	PbClF	8.62	2.4×10^{-9}
Germanium			PbCrO$_4$	12.55	2.8×10^{-13}
GeO$_2$	57.0	1.0×10^{-57}	Pb(ClO$_2$)$_2$	8.4	4×10^{-9}
Gold(I)			Pb$_2$[Fe(CN)$_6$]	14.46	3.5×10^{-15}
AuCl	12.7	2.0×10^{-13}	PbF$_2$	7.57	2.7×10^{-8}
AuI	22.8	1.6×10^{-23}	PbFI	8.07	8.5×10^{-9}
Gold(III)			Pb(OH)$_2$	14.93	1.2×10^{-15}
AuCl$_3$	24.5	3.2×10^{-25}	PbOHBr	14.70	2.0×10^{-15}
Au(OH)$_3$	45.26	5.5×10^{-46}	PbOHCl	13.7	2×10^{-14}

Table 5-6 (*Continued*)
SOLUBILITY PRODUCTS

Substance	pK_{sp}	K_{sp}	Substance	pK_{sp}	K_{sp}
PbOHNO$_3$	3.55	2.8×10^{-4}	Hg$_2$CO$_3$	16.05	8.9×10^{-17}
PbI$_2$	8.15	7.1×10^{-9}	Hg$_2$(CN)$_2$	39.3	5×10^{-40}
Pb(IO$_3$)$_2$	12.49	3.2×10^{-13}	Hg$_2$Cl$_2$	17.88	1.3×10^{-18}
PbMoO$_4$	13.0	1.0×10^{-13}	Hg$_2$CrO$_4$	8.70	2.0×10^{-9}
Pb(NbO$_3$)$_2$	16.62	2.4×10^{-17}	(Hg$_2$)$_3$[Fe(CN)$_6$]$_2$	20.07	8.5×10^{-21}
PbC$_2$O$_4$	9.32	4.8×10^{-10}	Hg$_2$(OH)$_2$	23.7	2.0×10^{-24}
PbHPO$_4$	9.90	1.3×10^{-10}	Hg$_2$(IO$_3$)$_2$	13.71	2.0×10^{-14}
Pb$_3$(PO$_4$)$_2$	42.10	8.0×10^{-43}	Hg$_2$I$_2$	28.35	4.5×10^{-29}
PbHPO$_3$	6.24	5.8×10^{-7}	Hg$_2$C$_2$O$_4$	12.7	2.0×10^{-13}
quinaldate, PbL$_2$	10.6	2.5×10^{-11}	Hg$_2$HPO$_4$	12.40	4.0×10^{-13}
PbSeO$_4$	6.84	1.4×10^{-7}	quinaldate, Hg$_2$L$_2$	17.9	1.3×10^{-18}
PbSeO$_3$	11.5	3.2×10^{-12}	Hg$_2$SeO$_3$	14.2	8.4×10^{-15}
PbSO$_4$	7.79	1.6×10^{-8}	Hg$_2$SO$_4$	6.13	7.4×10^{-7}
PbS	27.9	8.0×10^{-28}	Hg$_2$SO$_3$	27.0	1.0×10^{-27}
Pb(SCN)$_2$	4.70	2.0×10^{-5}	Hg$_2$S	47.0	1.0×10^{-47}
PbS$_2$O$_3$	6.40	4.0×10^{-7}	Hg$_2$(SCN)$_2$	19.7	2.0×10^{-20}
PbWO$_4$	6.35	4.5×10^{-7}	Hg$_2$WO$_4$	16.96	1.1×10^{-17}
Lead(IV)			Mercury(II)		
Pb(OH)$_4$	65.5	3.2×10^{-66}	Hg(OH)$_2$	25.52	3.0×10^{-26}
Lithium			Hg(IO$_3$)$_2$	12.5	3.2×10^{-13}
Li$_2$CO$_3$	1.60	2.5×10^{-2}	1,10-phenanthroline	24.70	2.0×10^{-25}
LiF	2.42	3.8×10^{-3}	quinaldate, HgL$_2$	16.8	1.6×10^{-17}
Li$_3$PO$_4$	8.5	3.2×10^{-9}	HgSeO$_3$	13.82	1.5×10^{-14}
LiUO$_2$AsO$_4$	18.82	1.5×10^{-19}	HgS red	52.4	4×10^{-53}
Lutetium			HgS black	51.8	1.6×10^{-52}
Lu(OH)$_3$	23.72	1.9×10^{-24}	Neodymium		
Magnesium			Nd(OH)$_3$	21.49	3.2×10^{-22}
MgNH$_4$PO$_4$	12.6	2.5×10^{-13}	Neptunium		
Mg$_3$(AsO$_4$)$_2$	19.68	2.1×10^{-20}	NpO$_2$(OH)$_2$	21.6	2.5×10^{-22}
MgCO$_3$	7.46	3.5×10^{-8}	Nickel		
MgCO$_3 \cdot 3H_2O$	4.67	2.1×10^{-5}	[Ni(NH$_3$)$_6$][ReO$_4$]$_2$	3.29	5.1×10^{-4}
MgF$_2$	8.19	6.5×10^{-9}	anthranilate, NiL$_2$	9.09	8.1×10^{-10}
Mg(OH)$_2$	10.74	1.8×10^{-11}	Ni$_3$(AsO$_4$)$_2$	25.51	3.1×10^{-26}
Mg(IO$_3$)$_2 \cdot 4H_2O$	2.5	3.2×10^{-3}	NiCO$_3$	8.18	6.6×10^{-9}
Mg(NbO$_3$)$_2$	16.64	2.3×10^{-17}	Ni$_2$(CN)$_4$ $\rightarrow$ Ni^{2+} +	8.77	1.7×10^{-9}
Mg$_3$(PO$_4$)$_2$	23–27	10^{-23} to 10^{-27}	Ni(CN)$_4^{2-}$		
8-quinolinolate, MgL$_2$	15.4	4.0×10^{-16}	Ni$_2$[Fe(CN)$_6$]	14.89	1.3×10^{-15}
MgSeO$_3$	4.89	1.3×10^{-5}	[Ni(N$_2$H$_4$)$_3$]SO$_4$	13.15	7.1×10^{-14}
MgSO$_3$	2.5	3.2×10^{-3}	Ni(OH)$_2$ fresh	14.7	2.0×10^{-15}
Manganese			Ni(IO$_3$)$_2$	7.85	1.4×10^{-8}
anthranilate, MnL$_2$	6.75	1.8×10^{-7}	NiC$_2$O$_4$	9.4	4×10^{-10}
Mn$_3$(AsO$_4$)$_2$	28.72	1.9×10^{-29}	Ni$_3$(PO$_4$)$_2$	30.3	5×10^{-31}
MnCO$_3$	10.74	1.8×10^{-11}	Ni$_2$P$_2$O$_7$	12.77	1.7×10^{-13}
Mn$_2$[Fe(CN)$_6$]	12.10	8.0×10^{-13}	8-quinolinolate, NiL$_2$	26.1	8×10^{-27}
Mn(OH)$_2$	12.72	1.9×10^{-13}	quinaldate, NiL$_2$	10.1	8×10^{-11}
MnC$_2$O$_4 \cdot 2H_2O$	14.96	1.1×10^{-15}	NiSeO$_3$	5.0	1.0×10^{-5}
8-quinolinolate, MnL$_2$	21.7	2.0×10^{-22}	α-NiS	18.5	3.2×10^{-19}
MgSeO$_3$	6.9	1.3×10^{-7}	β-NiS	24.0	1.0×10^{-24}
MnS amorphous	9.6	2.5×10^{-10}	γ-NiS	25.7	2.0×10^{-26}
crystalline	12.6	2.5×10^{-13}	Palladium		
Mercury(I)			Pd(OH)$_2$	31.0	1.0×10^{-31}
Hg$_2$(N$_3$)$_2$	9.15	7.1×10^{-10}	Pd(OH)$_4$	70.2	6.3×10^{-71}
Hg$_2$Br$_2$	22.24	5.6×10^{-23}	quinaldate, PdL$_2$	12.9	1.3×10^{-13}

Table 5-6 (*Continued*)
SOLUBILITY PRODUCTS

Substance	pK_{sp}	K_{sp}	Substance	pK_{sp}	K_{sp}
Platinum			$AgBrO_3$	4.28	5.3×10^{-5}
$PtBr_4$	40.5	3.2×10^{-41}	AgBr	12.30	5.0×10^{-13}
$Pt(OH)_2$	35	1×10^{-35}	Ag_2CO_3	11.09	8.1×10^{-12}
Plutonium			$AgClO_2$	3.7	2.0×10^{-4}
PuO_2CO_3	12.77	1.7×10^{-13}	AgCl	9.75	1.8×10^{-10}
PuF_3	15.6	2.5×10^{-16}	Ag_2CrO_4	11.95	1.1×10^{-12}
PuF_4	19.2	6.3×10^{-20}	$Ag_3[Co(NO_2)_6]$	20.07	8.5×10^{-21}
$Pu(OH)_3$	19.7	2.0×10^{-20}	cyanamide, Ag_2CN_2	10.14	7.2×10^{-11}
$Pu(OH)_4$	55	1×10^{-55}	AgOCN	6.64	2.3×10^{-7}
$PuO_2(OH)$	9.3	5×10^{-10}	AgCN	15.92	1.2×10^{-16}
$PuO_2(OH)_2$	24.7	2×10^{-25}	$Ag_2Cr_2O_7$	6.70	2.0×10^{-7}
$Pu(IO_3)_4$	12.3	5×10^{-13}	dicyanimide, $AgN(CN)_2$	8.85	1.4×10^{-9}
$Pu(HPO_4)_2 \cdot xH_2O$	27.7	2×10^{-28}	$Ag_4[Fe(CN)_6]$	40.81	1.6×10^{-41}
Polonium			AgOH	7.71	2.0×10^{-8}
PoS	28.26	5.5×10^{-29}	$Ag_2N_2O_2$	18.89	1.3×10^{-19}
Potassium			$AgIO_3$	7.52	3.0×10^{-8}
$K_2[PdCl_6]$	5.22	6.0×10^{-6}	AgI	16.08	8.3×10^{-17}
$K_2[PtCl_6]$	4.96	1.1×10^{-5}	Ag_2MoO_4	11.55	2.8×10^{-12}
$K_2[PtBr_6]$	4.2	6.3×10^{-5}	$AgNO_2$	3.22	6.0×10^{-4}
$K_2[PtF_6]$	4.54	2.9×10^{-5}	$Ag_2C_2O_4$	10.46	3.4×10^{-11}
K_2SiF_6	6.06	8.7×10^{-7}	Ag_3PO_4	15.84	1.4×10^{-16}
K_2ZrF_6	3.3	5×10^{-4}	quinaldate, AgL	17.9	1.3×10^{-18}
KIO_4	3.08	8.3×10^{-4}	$AgReO_4$	4.10	8.0×10^{-5}
$K_2Na[Co(NO_2)_6] \cdot H_2O$	10.66	2.2×10^{-11}	Ag_2SeO_3	15.00	1.0×10^{-15}
$K[B(C_6H_5)_4]$	7.65	2.2×10^{-8}	Ag_2SeO_4	7.25	5.7×10^{-8}
KUO_2AsO_4	22.60	2.5×10^{-23}	AgSeCN	15.40	4.0×10^{-16}
$K_4[UO_2(CO_3)_3]$	4.2	6.3×10^{-5}	Ag_2SO_4	4.84	1.4×10^{-5}
Praseodymium			Ag_2SO_3	13.82	1.5×10^{-14}
$Pr(OH)_3$	21.17	6.8×10^{-22}	Ag_2S	49.2	6.3×10^{-50}
Promethium			AgSCN	12.00	1.0×10^{-12}
$Pm(OH)_3$	21	1×10^{-21}	$AgVO_3$	6.3	5×10^{-7}
Radium			Ag_2WO_4	11.26	5.5×10^{-12}
$Ra(IO_3)_2$	9.06	8.7×10^{-10}	Sodium		
$RaSO_4$	10.37	4.2×10^{-11}	$Na[Sb(OH)_6]$	7.4	4.0×10^{-8}
Rhodium			Na_3AlF_6	9.39	4.0×10^{-10}
$Rh(OH)_3$	23	1×10^{-23}	$NaK_2[Co(NO_2)_6]$	10.66	2.2×10^{-11}
Rubidium			$Na(NH_4)_2[Co(NO_2)_6]$	11.4	4×10^{-12}
$Rb_3[Co(NO_2)_6]$	14.83	1.5×10^{-15}	$NaUO_2AsO_4$	21.87	1.3×10^{-22}
$Rb_2[PtCl_6]$	7.2	6.3×10^{-8}	Strontium		
$Rb_2[PtF_6]$	6.12	7.7×10^{-7}	$Sr_3(AsO_4)_2$	18.09	8.1×10^{-19}
$Rb_2[SiF_6]$	6.3	5.0×10^{-7}	$SrCO_3$	9.96	1.1×10^{-10}
$RbClO_4$	2.60	2.5×10^{-3}	$SrCrO_4$	4.65	2.2×10^{-5}
$RbIO_4$	3.26	5.5×10^{-4}	SrF_2	8.61	2.5×10^{-9}
Ruthenium			$Sr(IO_3)_2$	6.48	3.3×10^{-7}
$Ru(OH)_3$	36	1×10^{-36}	$SrMoO_4$	6.7	2×10^{-7}
Samarium			$Sr(NbO_3)_2$	17.38	4.2×10^{-18}
$Sm(OH)_3$	22.08	8.3×10^{-23}	$SrC_2O_4 \cdot H_2O$	6.80	1.6×10^{-7}
Scandium			$Sr_3(PO_4)_2$	27.39	4.0×10^{-28}
ScF_3	17.37	4.2×10^{-18}	8-quinolinolate, SrL_2	9.3	5×10^{-10}
$Sc(OH)_3$	30.1	8.0×10^{-31}	$SrSeO_3$	5.74	1.8×10^{-6}
Silver			$SrSeO_4$	3.09	8.1×10^{-4}
AgN_3	8.54	2.8×10^{-9}	$SrSO_3$	7.4	4×10^{-8}
Ag_3AsO_4	22.0	1.0×10^{-22}	$SrSO_4$	6.49	3.2×10^{-7}

Table 5-6 (*Continued*)
SOLUBILITY PRODUCTS

Substance	pK_{sp}	K_{sp}	Substance	pK_{sp}	K_{sp}
SrWO$_4$	9.77	1.7×10^{-10}	Uranium		
Terbium			UO$_2$HAsO$_4$	10.50	3.2×10^{-11}
Tb(OH)$_3$	21.70	2.0×10^{-22}	UO$_2$CO$_3$	11.73	1.8×10^{-12}
Tellurium			(UO$_2$)$_2$[Fe(CN)$_6$]	13.15	7.1×10^{-14}
Te(OH)$_4$	53.52	3.0×10^{-54}	UF$_4 \cdot$2.5H$_2$O	21.24	5.7×10^{-22}
Thallium(I)			UO$_2$(OH)$_2$	21.95	1.1×10^{-22}
TlN$_3$	3.66	2.2×10^{-4}	UO$_2$(IO$_3$)$_2 \cdot$H$_2$O	7.5	3.2×10^{-8}
TlBr	5.47	3.4×10^{-6}	UO$_2$C$_2$O$_4 \cdot$3H$_2$O	3.7	2×10^{-4}
TlBrO$_3$	4.07	8.5×10^{-5}	(UO$_2$)$_3$(PO$_4$)$_2$	46.7	2.0×10^{-47}
Tl$_2$[PtCl$_6$]	11.4	4.0×10^{-12}	UO$_2$HPO$_4$	10.67	2.1×10^{-11}
TlCl	3.76	1.7×10^{-4}	UO$_2$SO$_3$	8.59	2.6×10^{-9}
Tl$_2$CrO$_4$	12.00	1.0×10^{-12}	UO$_2$(SCN)$_2$	3.4	4×10^{-4}
Tl$_4$[Fe(CN)$_6$]$\cdot$2H$_2$O	9.3	5×10^{-10}	Vanadium		
TlIO$_3$	5.51	3.1×10^{-6}	VO(OH)$_2$	22.13	5.9×10^{-23}
TlI	7.19	6.5×10^{-8}	(VO)$_3$PO$_4$	24.1	8×10^{-25}
Tl$_2$C$_2$O$_4$	3.7	2×10^{-4}	Ytterbium		
Tl$_2$SeO$_3$	38.7	2×10^{-39}	Yt(OH)$_3$	23.6	2.5×10^{-24}
Tl$_2$SeO$_4$	4.00	1.0×10^{-4}	Yttrium		
Tl$_2$S	20.3	5.0×10^{-21}	YF$_3$	12.14	6.6×10^{-13}
TlSCN	3.77	1.7×10^{-4}	Y(OH)$_3$	22.1	8.0×10^{-23}
Thallium(III)			Y$_2$(C$_2$O$_4$)$_3$	28.28	5.3×10^{-29}
Tl(OH)$_3$	45.20	6.3×10^{-46}	Zinc		
8-quinolinolate, TlL$_3$	32.4	4.0×10^{-33}	anthranilate, ZnL$_2$	9.23	5.9×10^{-10}
Thorium			Zn$_3$(AsO$_4$)$_2$	27.89	1.3×10^{-28}
ThF$_4 \cdot$4H$_2$O + 2H$^+ \rightarrow$	7.23	5.9×10^{-6}	Zn(BO$_2$)$_2 \cdot$H$_2$O	10.18	6.6×10^{-11}
ThF$_2^{2+}$ + 2HF + 4H$_2$O			ZnCO$_3$	10.84	1.4×10^{-11}
Th(OH)$_4$	44.4	4.0×10^{-45}	Zn$_2$[Fe(CN)$_6$]	15.39	4.0×10^{-16}
Th(C$_2$O$_4$)$_2$	22	1×10^{-22}	Zn(IO$_3$)$_2$	7.7	2.0×10^{-8}
Th$_3$(PO$_4$)$_4$	78.6	2.5×10^{-79}	Zn(OH)$_2$	16.92	1.2×10^{-17}
Th(HPO$_4$)$_2$	20	1×10^{-20}	ZnC$_2$O$_4$	7.56	2.7×10^{-8}
Th(IO$_3$)$_4$	14.6	2.5×10^{-15}	Zn$_3$(PO$_4$)$_2$	32.04	9.0×10^{-33}
Thullium			quinaldate, ZnL$_2$	13.8	1.6×10^{-14}
Tm(OH)$_3$	23.48	3.3×10^{-24}	8-quinolinolate, ZnL$_2$	24.3	5.0×10^{-25}
Tin			ZnSeO$_3$	6.59	2.6×10^{-7}
Sn(OH)$_2$	27.85	1.4×10^{-28}	α-ZnS	23.8	1.6×10^{-24}
Sn(OH)$_4$	56	1×10^{-56}	β-ZnS	21.6	2.5×10^{-22}
SnS	25.0	1.0×10^{-25}	Zn[Hg(SCN)$_4$]	6.66	2.2×10^{-7}
Titanium			Zirconium		
Ti(OH)$_3$	40	1×10^{-40}	ZrO(OH)$_2$	48.2	6.3×10^{-49}
TiO(OH)$_2$	29	1×10^{-29}	Zr$_3$(PO$_4$)$_4$	132	1×10^{-132}

PROTON-TRANSFER REACTIONS

The pK_a values listed in Tables 5-7 and 5-8 are the negative (decadic) logarithms of the acidic dissociation constant, i.e., $-\log_{10}K_a = pK_a$. For the general proton-transfer reaction

$$HB \rightleftharpoons H^+ + B$$

the acidic dissociation constant is formulated as follows:

$$K_a = \frac{[H^+][B]}{[HB]}$$

The most common charge types for the acid (HB) and its conjugate base (B) are as follows:

$CH_3COOH \rightleftharpoons H^+ + CH_3COO^-$ (acetic acid, acetate ion)
$HSO_4^- \rightleftharpoons H^+ + SO_4^{2-}$ (hydrogen sulfate ion, sulfate ion)
$NH_4^+ \rightleftharpoons H^+ + NH_3$ (ammonium ion, ammonia)

Acids that have more than one acidic hydrogen ionize in steps, as shown for phosphoric acid:

$H_3PO_4 \rightleftharpoons H^+ + H_2PO_4^-$ $pK_1 = 2.15$ $K_1 = 7.08 \times 10^{-3}$
$H_2PO_4^- \rightleftharpoons H^+ + HPO_4^{2-}$ $pK_2 = 7.20$ $K_2 = 6.3 \times 10^{-8}$
$HPO_4^{2-} \rightleftharpoons H^+ + PO_4^{3-}$ $pK_3 = 12.38$ $K_3 = 4.4 \times 10^{-13}$

If the basic dissociation constant K_b for the equilibrium such as

$$NH_3 + H_2O \rightleftharpoons NH_4^+ + OH^-$$

is required, pK_b may be calculated from the relationship

$$pK_b = pK_w - pK_a$$

where $K_w = [H^+][OH^-]$ is the ionic product of water and $pK_w = pH + pOH$ (see Table 5-5). Thus, for ammonia

$$pK_b = 14.00 - 9.24 = 4.76$$
or $$K_b = 1.7 \times 10^{-5}$$

If a desired organic acid is not entered in Table 5-8, a useful estimate of its pK_a value can sometimes be obtained by making a comparison with compounds of recognizably similar character for which pK_a values are known: (1) alkyl chains, alicyclic rings, or saturated carbocyclic rings fused to aromatic or heterocyclic rings can be replaced by methyl or ethyl groups; (2) acid-strengthening and base-weakening inductive and mesomeric effects of a nitro group attached to an aromatic ring are very similar to those of a nitrogen atom located at the same position in a heteroaromatic ring (e.g., 3-hydroxypyridine and 3-nitrophenol).

Hammett and Taft substituent constants and, in particular, Tables 3-13 through 3-15 may also prove useful for estimating pK_a values.

Table 5-7
PROTON-TRANSFER REACTIONS OF INORGANIC
MATERIALS IN WATER AT 25°C

Substance	Formula or Remarks	pK_1	pK_2
Aluminum ion (aquo)	Al^{3+} (aquo)	5.01	
Aluminic acid (alumina)	H_3AlO_3	11.2	
Amidophosphoric acid	$H_2NPO(OH)_2$	3.3	8.28
Aminodisulfonic acid	$HN(SO_3H)_2$	pK_3 8.50	
Ammonium ion	NH_4^+	9.24	
Antimonic acid	$HSb(OH)_6$	2.55	
Antimony(III) ion	$SbO^+ + H_2O = HSbO_2 + H^+$	0.87	
Arsenic acid	H_3AsO_4	2.22	6.98
		pK_3 11.50	
Arsenous acid	$HAsO_2$ or $HAs(OH)_4$	9.18	
Barium ion	Ba^{2+} (aquo)	13.36	
Beryllium ion	Be^{2+} (aquo) $= BeOH^+ + H^+$	6.5	
Bismuth ion	Bi^{3+} (aquo) $= BiOH^{2+} + H^+$	1.58	
Boric acid (ortho-)	H_3BO_3	9.24	12.74
		pK_3 13.80	
Boric acid (tetra-)	$H_2B_4O_7$	4	9
Cadmium ion (aquo)	Cd^{2+} (aquo) hydrolysis	9.0	
Calcium ion (aquo)	Ca^{2+} (aquo)	12.8	
Carbonic acid (carbon dioxide)	$CO_2 + H_2O$ (without including dehydration constant)	6.35 3.76	10.33
Cerium(III) ion (aquo)	Ce^{3+} (aquo) hydrolysis	~9	
Cerium (IV) ion (aquo)	Ce^{4+} (aquo) hydrolysis	−0.72	
Chloric acid	$HClO_3$	~−2.7	
Chlorous acid	$HClO_2$	1.94	
Chlorosulfonic acid	$HOSO_2Cl$	−10.43	
Chromic acid	H_2CrO_4	−0.98	6.50
Chromium(III) ion (aquo)	Cr^{3+} (aquo) hydrolysis	4.0	
Cobalt(II) ion (aquo)	Co^{2+} (aquo) hydrolysis	9.85	
Cobalt(III) ion (aquo)	Co^{3+} (aquo) hydrolysis	1.75	
Copper(II) ion (aquo)	Cu^{2+} (aquo) $= CuOH^+ + H^+$	8.0	
Cyanic acid	$HOCN$	3.46	
Deuteroammonium ion	ND_3H^+	9.76	
Deuteroarsenic acid	D_3AsO_4	2.60	
Deuterocarbonic acid	D_2CO_3 $(CO_2 + D_2O)$	6.77	10.96
Deuterohydrazine (18°)	N_2D_4	9.11	
Deuterohydrazoic acid (20°)	DN_3	5.01	
Deuteroiodic acid	DIO_3	1.15	
Deuterophosphoric acid	D_3PO_4	2.35	7.78
Deuterosulfuric acid	D_2SO_4		2.33
Deuterium oxide	D_2O $(pK_1 = 14.96$ molal scale$)$	14.87	
Diamidophosphoric acid	$(H_2N)_2PO_2H$	4.83	
Diimidotriphosphoric acid	$H_5P_3O_8(NH)_2$	~0.5	~2
		pK_3 3.94	pK_4 7.74
		pK_5 9.95	
Dithionic acid	$H_2S_2O_6$	−3.4	−0.2
Dithionous acid	$H_2S_2O_4$	0.35	2.45
Dysprosium(III) ion	Dy^{3+} (aquo)	8.10	
Erbium(III) ion	Er^{3+} (aquo)	7.99	
Europium(III) ion	Eu^{3+} (aquo)	8.3	
Ferricyanic acid	$H_3Fe(CN)_6$	<1	
Ferrocyanic acid	$H_2[Fe(CN)_6]^{2-}$	pK_3 2.57	pK_4 4.35
Fluorophosphoric acid	$FPO(OH)_2$		5.12
Gadolinium(III) ion	Gd^{3+} (aquo)	8.4	
Gallium(III) ion	Ga^{3+} (aquo)	3.4	

Table 5-7 (*Continued*)
PROTON-TRANSFER REACTIONS OF INORGANIC
MATERIALS IN WATER AT 25°C

Substance	Formula or Remarks	pK_1	pK_2
Germanic acid	H_2GeO_3	9.0	12.4
Hexapolyphosphoric acid	$H_8P_6O_{19}$	~2.1	2.19
		pK_3 5.98	pK_4 8.13
Holmium(III) ion	Ho^{3+} (aquo)	8.04	
Hydrazinium(2+) ion	$^+H_3NNH_3^+$	0.27	7.94
Hydrazinosulfuric acid	H_2NNHSO_3H	3.85	
Hydrazoic acid	HN_3	4.72	
Hydrocyanic acid	HCN	9.21	
Hydrogen bromide	HBr	−9	
Hydrogen chloride	HCl	−6.1	
Hydrogen fluoride	HF	3.18	
Hydrogen iodide	HI	−9.5	
Hydrogen peroxide	H_2O_2	11.65	
Hydrogen polysulfide (20°)	H_2S_4	3.8	6.3
Hydrogen selenide	H_2Se	3.89	11.0
Hydrogen sulfide	H_2S	6.97	12.90
Hydrogen telluride (18°)	H_2Te	2.64	11–12
Hydroperoxy radical (23°)	$HO_2\cdot = H^+ + O_2^-$	4.45	
Hydroxylamine-*N,N*-disulfonic acid	$HON(SO_3H)_2$	pK_3 11.85	
Hydroxylamine-*N*-sulfonic acid	$HONHOSO_2H$		~12.5
Hydroxylammonium ion	$HONH_3^+$	5.95	
Hypobromous acid	HBrO	8.62	
Hypochlorous acid	HClO	7.54	
Hypoiodous acid	HIO	10.64	
Hyponitrous acid	HON=NOH	7.05	11.4
Hypophosphoric acid (20°)	$H_4P_2O_6$	2	2.19
		pK_3 6.77	pK_4 9.48
Hypophosphorous acid	HPH_2O_2	1.23	
Hyposulfurous acid	$H_2S_2O_4$	0.35	2.45
Imidodiphosphoric acid	$(HO)_2PONHPO(OH)_2$	~2	2.85
		pK_3 7.08	pK_4 9.72
Indium(III) ion	In^{3+} (aquo) = $InOH^{2+}$ and $In(OH)_2^+$ (successive values for hydrolysis)	4.4	4.4
Iodic acid	HIO_3	0.804	
Iron(II) ion	Fe^{2+} (aquo) hydrolysis	6.74	
Iron(III) ion	Fe^{3+} (aquo) (successive values for hydrolysis)	2.83	4.59
Lanthanum(III) ion	La^{3+} (aquo)	9.03	
Lead(II) ion	Pb^{2+} (aquo) (stepwise hydrolysis)	7.8	9.4
		pK_3 10.8	
Lithium ion	Li^+	13.82	
Lutetium(III) ion	Lu^{3+} (aquo) hydrolysis	7.94	
Magnesium ion	Mg^{2+} (aquo)	11.41	
Manganese(II) ion	Mn^{2+} (aquo) hydrolysis	10.59	
Manganese(III) ion	Mn^{3+} (aquo) hydrolysis	~0	
Manganic acid (35°)	H_2MnO_4		10.15
Mercury(I) ion	Hg_2^{2+} (aquo) hydrolysis	5.0	
Mercury(II) ion	Hg^{2+} (aquo) successive hydrolysis	3.55	2.66
Methyltrioxoarsenic acid	$CH_3AsO(OH)_2$	3.61	8.24
Molybdic acid	H_2MoO_4	~3.6	4.08
Neodymium(III) ion	Nd^{3+} (aquo) hydrolysis	8.5	
Neptunium(IV) ion	Np^{4+} (aquo) hydrolysis	2.30	
Neptunium(V) ion	$NpO_2^+ = NpO_2OH + H^+$	8.9	

Table 5-7 (*Continued*)
PROTON-TRANSFER REACTIONS OF INORGANIC
MATERIALS IN WATER AT 25°C

Substance	Formula or Remarks	pK_1	pK_2
Nickel(II) ion	Ni^{2+} (aquo) hydrolysis	9.86	
Nitramide	O_2NNH_2	6.48	
Nitric acid	$HNO_3 \cdot 3H_2O$	1.44	
Nitrous acid	HNO_2	3.20	
Osmic acid	H_2OsO_5 (mainly OsO_4)	12.0	14.5
α-Oxyhyponitric acid	$H_2N_2O_3$	2.51	9.70
Perboric acid	$H_3BO_3 + H_2O_2 = (H_2BO_3 \cdot H_2O_2)^- + H^+$	7.91	
Perchloric acid	$HClO_4$ (completely dissociated up to 10 M)		
	$HClO_4 \cdot 3H_2O$	−4.8	
	$HClO_4 \cdot 7H_2O$	−2.12	
Perchloryl amide	H_2NClO_2	3.7	8.6
Perchromic acid (22°)	H_2CrO_5	4.30	
Periodic acid (para-)	H_5IO_6	1.55	8.27
Permanganic acid	$HMnO_4$	−2.25	
Peroxydiphosphoric acid	$H_4P_2O_3$	−0.3	0.5
		pK_3 5.18	pK_4 7.67
Peroxymonophosphoric acid	H_3PO_5	1.1	5.5
		pK_3 12.8	
Peroxymonosulfuric acid	H_2SO_5	1.0	9.3
Perrhenic acid	$HReO_4$	−1.25	
Pertechnic acid	$HTcO_4$	0.3	
Perxenic acid	H_4XeO_6	~2	~6
		pK_3 ~10	
Phosphoric acid (ortho-)	H_3PO_4	2.15	7.20
		pK_3 12.38	
Phosphoric acid (pyro-)	$H_4P_2O_7$	1.52	2.36
		pK_3 6.60	pK_4 9.25
Phosphorous acid (20°)	H_3PO_3	1.20	6.70
		pK_3 >14	
Plumbic acid	H_2PbO_2	10.1	
Plutonium(III) ion	Pu^{3+} (aquo) hydrolysis	7.37	
Plutonium(IV) ion	Pu^{4+} (aquo) hydrolysis	1.27	
Plutonium(V) ion	PuO_2^+ hydrolysis	9.7	
Potassium ion	K^+ (aquo)	16.0–16.5	
Praseodymium(III) ion	Pr^{3+} (aquo) to $PrOH^{2+}$	8.55	
Protoactinium(IV) ion	Pa^{4+} (aquo) successive hydrolysis	0.14	0.38
		pK_3 1.25	
Protoactinium(V) ion	$Pa(OH)_3^{2+} + H_2O = Pa(OH)_4^+ + H^+$	1.05	
Rhodium(III) ion	Rh^{3+} (aquo) = $RhOH^{2+} + H^+$	3.43	
Samarium(III) ion	Sm^{3+} (aquo) hydrolysis	8.34	
Scandium(III) ion	Sc^{3+} (aquo) hydrolysis	4.93	
Selenic acid	H_2SeO_4	−3	1.66
Selenous acid	H_2SeO_3	2.64	8.27
Silicic acid	H_2SiO_3	9.77	11.80
Silver(I) ion	Ag^+ (aquo) hydrolysis	≥11.1	
Sodium ion	Na^+ (aquo)	14.77	
Strontium ion	Sr^{2+} (aquo)	13.18	
Sulfamic acid	$HOSO_2NH_2$	0.988	
Sulfuric acid	H_2SO_4	~−3	1.96
	$H_3SO_4^+$	−8.3	
Sulfurous acid	$SO_2 + H_2O$ (includes dehydration constant)	1.89	7.21
Tantalic acid	$HTa(OH)_4$ (structure questionable)	9.6	
Telluric acid	H_6TeO_6	7.61	11.00

Table 5-7 (*Continued*)
PROTON-TRANSFER REACTIONS OF INORGANIC MATERIALS IN WATER AT 25°C

Substance	Formula or Remarks	pK_1	pK_2
Tellurous acid	H_2TeO_3	2.46	7.7
Terbium(III) ion	Tb^{3+} (aquo) hydrolysis	8.16	
Tetrametaphosphoric acid	$H_4P_4O_{12}$	pK_4 2.74	
Tetramolybdic acid	$H_2Mo_4O_{13}$	1.4	1.5
Tetraperoxychromic acid (30°)	H_3CrO_8	7.16	
Tetrapolyphosphoric acid	$H_6P_4O_{13}$	pK_3 1.3	pK_4 2.23
		pK_5 6.63	pK_6 8.34
Thallium(I) ion	Tl^+ (aquo)	13.18	
Thallium(III) ion	Tl^{3+} (aquo) successive hydrolysis	1.14	1.49
Thiocyanic acid	HSCN	−1.85	
Thiosulfuric acid	$H_2S_2O_3$	0.60	1.5–1.7
Thorium(IV) ion	Th^{4+} (aquo) successive hydrolysis	3.89	4.20
Tin(II) ion	Sn^{2+} (aquo) hydrolysis	1.70	
Titanium(III) ion	$Ti^{3+} + H_2O = TiOH^{2+} + H^+$	1.29	
Titanium(IV) ion (18°)	$TiO^{2+} + H_2O = TiO(OH)^+ + H^+$	1.3	
Trimetaphosphoric acid	$H_3P_3O_4$	pK_3 2.0	
Tripolyphosphoric acid	$H_5P_3O_{10}$	~1	~2
		pK_3 2.79	pK_4 6.47
		pK_5 9.24	
Trithiocarbonic acid (20°)	H_2CS_3	2.68	8.18
Tungstic acid (20°)	H_2WO_4	~3.5	~4.6
Uranium(IV) ion	U^{4+} (aquo) $= UOH^{3+} + H^+$	0.68	
Uranyl ion	UO_2^{2+} for hydrolysis	5.82	
	$UO_2(OH)_2$	7.68	
Vanadic acid	H_3VO_4	3.78	7.8
		pK_3 13.0	
Vanadium(III) ion	V^{3+} (aquo) successive hydrolysis	2.92	3.5
Vanadyl ion	$V(OH)_2^{2+} + H_2O = VO_2OH^+ + H^+$	5.36	
Water	H_2O	15.74	
Xenon trioxide	XeO_3 (aquo) $- HXeO_4^- + H^+$	10.8	
Ytterbium(III) ion	Yb^{3+} (aquo) hydrolysis	8.03	
Yttrium(III) ion	Y^{3+} (aquo) hydrolysis	8.34	
Zinc ion	Zn^{2+} (aquo) hydrolysis	8.96	
	H_2ZnO_2	4.95	
Zirconium(IV) ion	Zr^{4+} (aquo) successive hydrolysis	−0.32	0.06
		pK_3 0.35	pK_4 0.64

Table 5-8
PROTON-TRANSFER REACTIONS OF ORGANIC MATERIALS IN WATER AT 25°C

Substance	pK_1	pK_2	pK_3	pK_4
Abietic acid	7.62			
Acetamide (protonated cation)	1.40	12.40		
Acetamidobenzoic acid: ortho-	3.63			
meta-	4.07			
para-	4.28			
Acetamidoglycine (protonated cation) (20°)	7.7			
N-(2-Acetamido)iminodiacetic acid (20°)	6.62			
Acetanilide (protonated cation)	0.4			
Acethydrazidinium ion	3.24			
Acetic acid	4.76			
Acetidinium ion	11.29			
Acetoacetic acid (18°)	3.58			
Acetone dicarboxylic acid	3.10			
Acetoxime	12.42			

Table 5-8 (*Continued*)
PROTON-TRANSFER REACTIONS OF ORGANIC MATERIALS IN WATER AT 25°C

Substance	pK_1	pK_2	pK_3	pK_4
Acetoxybenzoic acid: ortho-	4.56			
meta-	4.00			
para-	4.38			
Acetylacetone	8.2			
N-Acetyl-α-alanine (*dl*)	3.72			
N-Acetyl-β-alanine	4.45			
β-Acetylamino-n-propionic acid	4.45			
Acetyl-α-amino-n-butyric acid	3.72			
Acetylbenzoic acid: ortho-	4.13			
meta-	3.83			
para-	3.70			
2-Acetylcyclohexanone	14.1			
N-Acetylcysteine (30°)	9.52			
Acetylene dicarboxylic acid	1.75	4.40		
N-Acetylglycine	3.67			
N-Acetylguanidinium ion	8.23			
Acetyl-L-histidine	7.08			
Acetylhydroxamic acid (20°)	9.40			
4-Acetyl-β-mercaptoisoleucine (30°)	10.30			
N-Acetyl-2-mercaptoethylamine	9.92			
2-Acetyl-1-naphthol (30°)	13.40			
N-Acetylpenicillamine (30°)	9.90			
Acetylphenol: meta-	9.19			
para-	8.05			
2-Acetylpyridinium ion	3.18			
Acetylsalicylic acid	4.56			
3-Acetylthiophenol (49% aq EtOH)	6.93			
4-Acetylthiophenol (49% aq EtOH)	5.93			
Aconitine (protonated cation)	8.11			
Acridinium ion	5.60			
Acrylic acid	4.26			
Adenine	4.17	9.75		
Adenine-N-oxide	2.69	8.49		
Adenosinediphosphoric acid	3.99	6.35		
Adenosine-3'-phosphoric acid	3.63	5.80		
Adenosine-5'-phosphoric acid	3.81	6.21		
Adenosinetriphosphoric acid	4.05	6.50		
Adipamic acid (monoamide)	4.63			
Adipic acid	4.43	5.41		
α-Alanine (protonated cation)	2.34	9.87		
β-Alanine (protonated cation)	3.55	10.23		
Alanylglycine	8.17			
Alizarin black SN	5.79	12.8	>14	
Alizarin-3-sulfonic acid	5.54	11.01		
Allantoin	8.96			
Allothreonine	2.11	9.10		
Alloxanic acid	6.64			
Allylacetic acid	4.68			
Allylammonium ion	9.69			
β-Allylpropionic acid	4.72			

Table 5-8 (*Continued*)
PROTON-TRANSFER REACTIONS OF ORGANIC MATERIALS IN WATER AT 25°C

Substance.	pK_1	pK_2	pK_3	pK_4
3-Amidotetrazolinium ion	3.95			
α-Aminoacetic acid (*see* Glycine)				
Aminobenzoic acid: ortho-	2.05	4.95		
meta-	3.07	4.74		
para-	2.38	4.89		
4-Aminobenzophenone (protonated cation)	2.17			
Aminobenzosulfonic acid: ortho-	2.48			
meta-	3.73			
para-	3.24			
α-Amino-*i*-butyric acid	2.36	10.21		
α-Amino-*n*-butyric acid	2.29	9.83		
γ-Amino-*n*-butyric acid	4.03	10.56		
ε-Amino-*n*-caproic acid	4.37	10.80		
1-Aminocycloheptanecarboxylic acid	2.59	10.46		
1,1-Aminocyclohexanecarboxylic acid	2.65	10.03		
2-Amino-4,5-dimethylphenol	5.28	10.40		
Aminoethane-β-sulfonic acid (20°)	−0.3	9.08		
2-Aminoethanol-1-phosphoric acid ester		5.84		
2-[2-(2-Aminoethyl)aminoethyl]pyridine H+	3.50	6.59	9.51	
2-Aminoethyldihydrogen phosphate	10.31			
2-Aminoethyldihydrogen sulfate	9.13			
2-Aminoethylphosphonic acid	2.45	7.0	10.8	
2,2'-Aminoethylpyrrolidine	6.56	9.74		
1-Aminoethylsulfonic acid	−0.33	9.06		
2-Amino-2'-hydroxydiethylsulfide	9.27			
4-Aminoisoxazolidine-3-one	7.4			
Aminomalonic acid	3.32	9.83		
Aminomethylphosphonic acid	2.35	5.9	10.8	
2-Amino-2-methylpropan-1,3-diol	8.79			
α-Amino-α-methylpropionic acid	2.48	9.9		
2-Aminomethylpyridinium ion	8.65			
Aminomethylsulfonic acid		5.75		
5-Amino-*n*-pentylsulfonic acid		10.95		
2-Aminophenol (protonated cation)	4.78	9.97		
3-Aminophenol (protonated cation)	4.37	9.82		
4-Aminophenol (protonated cation)	5.29	10.30		
o Aminophenylphosphoric acid	4.10	7.29		
β-Aminopropionic acid	3.55	10.24		
2-Aminopyridine-1-oxide	2.58	16.74		
2-Aminopyridinium ion	6.86			
3-Aminopyridinium ion	5.98			
4-Aminopyridinium ion	9.11			
8-Aminoquinaldinium ion	4.86			
8-Aminoquinolinium ion	4.04			
4-Aminosalicylic acid	1.70	3.90		
5-Aminosalicylic acid	2.74	5.84		
4-Aminothiophenol (40% aq EtOH)	7.95			
o-Aminothiophenol	<2	7.90		
α-Amino-*n*-valeric acid	2.32	9.81		
5-Amino-*n*-valeric acid	4.27	10.77		
n-Amylammonium ion (also *iso*-)	10.60			
Angelic acid	4.30			

Table 5-8 (*Continued*)
PROTON-TRANSFER REACTIONS OF ORGANIC MATERIALS
IN WATER AT 25°C

Substance	pK_1	pK_2	pK_3	pK_4
Anilinediacetic acid (20°)	2.40	4.98		
Aniline-3-sulfonic acid	0.39	3.72		
Aniline-4-sulfonic acid	0.58	3.23		
Anilinium ion	4.60			
Anisic acid: ortho-	4.09			
meta-	4.08			
para-	4.49			
Anisidinium ion: ortho-	4.49			
meta-	4.20			
para-	4.29			
β-Anisylpropionic acid: ortho-	4.80			
meta-	4.65			
para-	4.69			
Anthracenecarboxylic acid −1	3.68			
−2	4.18			
−9	3.65			
Anthranilic acid	2.05	4.95		
Anthraquinonecarboxylic acid −1 (20°)	3.37			
−2 (20°)	3.42			
9,10-Anthraquinone monoxime	9.78			
Apomorphine (protonated cation)	7.00			
Arginine	2.02	9.04	12.48	
Arsenazo (pK_5 10.5; pK_6 12.0)	0.1	1.2	2.7	7.9
Ascorbic acid	4.30	11.82		
Asparagine (protonated cation)	2.02	8.80		
Aspartic acid	2.05	3.86	10.00	
Aureomycine	3.30	7.44	9.27	
Azelaic acid	4.53	5.40		
Aziridinium ion	8.04			
Barbituric acid	4.04			
Benzamide (protonated cation)	−2.16	13–14		
Benzene-1-carboxylic-2-phosphoric acid	1.71	3.78	9.17	
Benzene-1-carboxylic-3-phosphoric acid	1.55	4.03	7.03	
Benzene-1-carboxylic-4-phosphoric acid	1.50	3.95	6.89	
Benzenediazonium ion	11.08			
Benzenehexacarboxylic acid	1.40	2.19	3.31	4.78
(pK_5 5.89; pK_6 6.96)				
Benzenepentacarboxylic acid (pK_5 6.46)	1.80	2.73	3.96	5.25
Benzenesulfinic acid	1.50			
Benzene-1,2,3,5-tetracarboxylic acid	2.38	3.51	4.44	5.81
Benzene-1,2,3,4-tetracarboxylic acid	2.05	3.25	4.73	6.21
Benzene-1,2,4,5-tetracarboxylic acid	1.92	2.87	4.49	5.63
Benzene-1,2,3-tricarboxylic acid	2.80	4.20	5.87	
Benzene-1,2,4-tricarboxylic acid	2.52	3.84	5.20	
Benzene-1,3,5-tricarboxylic acid	2.12	3.89	5.20	
Benzhydroxamic acid (20°)	8.89			
Benzidinium ion	3.63	4.70		
Benzil-α-dioxime	12.0			
Benzilic acid	3.04			
Benzimidazolinium ion (ortho-)	5.30	12.3		
Benzoic acid	4.21			
Benzoic hydrazide	3.03	12.45		

Table 5-8 (*Continued*)
PROTON-TRANSFER REACTIONS OF ORGANIC MATERIALS IN WATER AT 25°C

Substance	pK_1	pK_2	pK_3	pK_4
Benzoquinone monoxime	6.20			
Benzosulfonic acid	0.70			
Benzoylacetone	8.23			
Benzoylammonium ion	9.34			
o-Benzoylbenzoic acid	3.54			
Benzoylglutamic acid	3.49	4.99		
Benzoylpyruvic acid	6.40	12.10		
3-Benzoyl-1,1,1-trifluoroacetone	6.35			
Benztriazolinium ion (20°)	1.6			
Benzylammonium ion	9.35			
Benzylmercaptan	9.43			
Benzylpyrrolidinium ion	9.51			
Benzylsuccinic acid	4.11	5.65		
Betaine (protonated cation) (20°)	12.16			
Biguanide	2.96	11.51		
2,2'-Bipyridylinium ion	4.35			
Bromoacetic acid	2.90			
Bromoanilinium ion: ortho-	2.53			
meta-	3.58			
para-	3.86			
Bromobenzoic acid: ortho-	2.85			
meta-	3.81			
para-	4.00			
α-Bromobutyric acid	2.97			
o-Bromo-(*trans*)-cinnamic acid	4.41			
m-Bromomandelic acid	3.23			
2-Bromo-6-nitrobenzoic acid	1.37			
Bromophenol: ortho-	8.40			
meta-	8.85			
para-	8.25			
Bromophenoxyacetic acid: ortho-	3.12			
meta-	3.09			
para-	3.13			
α-Bromopropionic acid (18°)	2.97			
β-Bromopropionic acid (18°)	4.00			
γ-Bromopropionic acid	4.58			
2-Bromopyridinium ion	0.90			
Bromosuccinic acid	2.55	4.41		
Brucine (protonated cation)	2.3	7.95		
1,4-Butanediammonium ion	9.35	10.82		
2,3-Butanediammonium ion	6.91	10.00		
But-3-ene-1-oic acid	4.68			
i-Butylammonium ion	10.41			
n-Butylammonium ion	10.61			
sec-Butylammonium ion	10.56			
tert-Butylammonium ion	10.45			
2-tert-Butylanilinium ion	3.78			
N-tert-Butylanilinium ion	7.10			
n-Butylarsonic acid	4.23	8.91		
t-Butylbenzoic acid: ortho-	3.57			
meta-	4.18			
para-	4.40			
N-Butylethylenediammonium ion	7.53	10.30		

Table 5-8 (*Continued*)
PROTON-TRANSFER REACTIONS OF ORGANIC MATERIALS IN WATER AT 25°C

Substance	pK_1	pK_2	pK_3	pK_4
t-Butylhydroperoxide	12.80			
p-tert-Butylphenylacetic acid	4.42			
n-Butylphosphinic acid	3.41			
t-Butylphosphinic acid	4.24			
2-*t*-Butylpyridinium ion	5.76			
But-2-yn-1-oic acid	2.66			
But-2-yn-1,4-dioic acid	1.75	4.40		
i-Butyric acid	4.85			
n-Butyric acid	4.83			
Cacodylic acid	6.27			
Caffeine (40°)	10.4			
Calcein: <4, 5.4, 9.0, 10.5, >12				
Calmagite	8.14	12.35		
Camphoric acid	4.57	5.10		
n-Caproic acid (also *iso*-)	4.88			
n-Caprylic acid	4.90			
N-Carbamoylacetic acid	3.64			
N-Carbamoylalanine (*dl*-)	3.89			
N-Carbamoylglycine	3.88			
Carbamylmethylammonium ion	7.93			
β-Carboxymethylaminopropionic acid	3.61	9.46		
Catechol	9.45			
Chloranilic acid	1.09	2.42		
Chloroacetic acid	2.86			
Chloroanilinium ion: ortho-	2.64			
meta-	3.46			
para-	3.99			
Chlorobenzoic acid: ortho-	2.94			
meta-	3.82			
para-	3.99			
2-Chloro-*n*-butyric acid	2.86			
2-Chloro-*iso*-butyric acid	2.98			
3-Chloro-*n*-butyric acid	4.05			
4-Chloro-*n*-butyric acid	4.50			
Chloro-(*trans*)-cinnamic acid: ortho-	4.23			
meta-	4.29			
para-	4.41			
α-Chlorocrotonic acid	3.14			
β-Chlorocrotonic acid	3.84			
4-Chloro-2,6-dinitrophenol	2.97			
α-Chloroisocrotonic acid	2.80			
β-Chloroisocrotonic acid	4.02			
Chloromethylphosphonic acid	1.40	6.30		
2-Chloro-3-nitrobenzoic acid	2.02			
2-Chloro-4-nitrobenzoic acid	1.96			
2-Chloro-5-nitrobenzoic acid	2.17			
2-Chloro-6-nitrobenzoic acid	1.34			
3-Chloro-6-nitrobenzoic acid	1.85			
Chlorophenol: ortho-	8.48			
meta-	9.02			
para-	9.37			

Table 5-8 (*Continued*)
PROTON-TRANSFER REACTIONS OF ORGANIC MATERIALS IN WATER AT 25°C

Substance	pK_1	pK_2	pK_3	pK_4
Chlorophenoxyacetic acid: ortho-	3.05			
meta-	3.07			
para-	3.10			
2-Chlorophenylacetic acid	4.07			
3-Chlorophenylacetic acid	4.14			
4-Chlorophenylacetic acid	4.19			
2-Chlorophenylphosphonic acid	1.63	6.98		
3-Chlorophenylphosphonic acid	1.55	6.65		
4-Chlorophenylphosphonic acid	1.66	6.75		
3-(o-Chlorophenyl)propionic acid	4.58			
3-(m-Chlorophenyl)propionic acid	4.59			
3-(p-Chlorophenyl)propionic acid	4.61			
4-Chlorophthalic acid	1.60			
2-Chloropropionic acid	2.80			
3-Chloropropionic acid	4.00			
2-Chloropyridinium ion	0.49			
3-Chloropyridinium ion	2.84			
Chlorotetracycline	3.30	7.44	9.27	
4-Chlorothlophenol	5.9			
N-Chloro-p-tolylsulfonamide	4.54			
Cholamine chloride (20°)	7.1			
Choline (protonated cation)	8.94			
Chrome Azurol S	2.45	4.86	11.47	
Chrome Dark Blue	7.56	9.3	12.4	
Cinchonidine (protonated cation)	3.92	8.20		
Cinchonine (protonated cation) (15°)	4.04	8.15		
cis-Cinnamic acid	3.88			
trans-Cinnamic acid	4.44			
Citraconic acid	2.29	6.15		
Citric acid	3.13	4.76	6.40	
1-Citrullin	2.43	9.41		
Cocaine (protonated cation)	8.41			
Codeine (protonated cation)	7.95			
Colchicine (protonated cation)	1.65			
Coniine (protonated cation)	11.0			
Creatine (protonated cation) (40°)	3.28			
Creatinine (protonated cation)	3.57			
Cresol: ortho-	10.26			
meta-	10.09			
para-	10.26			
trans-Crotonic acid	4.69			
cis-Crotonic acid (18°)	4.41			
Cuminic acid	4.35			
Cyanamide (protonated cation)	10.27			
N-Cyanoacetamide	4			
Cyanoacethydrazide	2.34	11.17		
Cyanoacetic acid	2.46			
Cyanoanilinium ion: meta-	2.76			
para-	1.74			
Cyanobenzoic acid: ortho-	3.14			
meta-	3.60			
para-	3.55			

Table 5-8 (*Continued*)
PROTON-TRANSFER REACTIONS OF ORGANIC MATERIALS IN WATER AT 25°C

Substance	pK_1	pK_2	pK_3	pK_4
2-Cyano-*i*-butyric acid	2.42			
4-Cyano-*n*-butyric acid	4.44			
2-Cyanoethylammonium ion	7.7			
Cyanomethylammonium ion	5.34			
Cyanophenol: meta-	8.61			
para-	7.95			
Cyanophenoxyacetic acid: ortho-	2.98			
meta-	3.03			
para-	2.93			
2-Cyanopropionic acid	2.37			
3-Cyanopropionic acid	3.99			
3-Cyanopyridinium ion	1.45			
Cyanuric acid	6.78			
trans-Cyclobutanecarboxylic acid	4.78			
cis-Cyclobutane-1,2-dicarboxylic acid	3.90	5.89		
trans-Cyclobutane-1,2-dicarboxylic acid	3.79	5.61		
cis-Cyclobutane-1,3-dicarboxylic acid	4.04	5.31		
trans-Cyclobutane-1,3-dicarboxylic acid	3.81	5.28		
Cyclohexanecarboxylic acid	4.90			
Cyclohexane-1,2-*trans*-diamine-*NNN'N''*-tetraacetic acid	2.4	3.5	6.12	11.58
cis-Cyclohexanediammonium ion	6.43	9.93		
trans-Cyclohexanediammonium ion	6.34	9.74		
Cyclohexane-1,1-dicarboxylic acid	3.45	4.11		
cis-Cyclohexane-1,2-dicarboxylic acid (20°)	4.34	6.77		
trans-Cyclohexane-1,2-dicarboxylic acid (18°)	4.18	5.93		
cis-Cyclohexane-1,3-dicarboxylic acid (16°)	4.10	5.46		
trans-Cyclohexane-1,3-dicarboxylic acid (19°)	4.31	5.73		
trans-Cyclohexane-1,4-dicarboxylic acid (16°)	4.18	5.42		
cis,cis-Cyclohexane-1,3,5-triamine	6.9	8.7	10.4	
Cyclohexanonimine	9.15			
cis-Cyclohex-4-en-1,2-dicarboxylic acid (20°)	3.89	6.79		
trans-Cyclohex-4-en-1,2-dicarboxylic acid (20°)	3.95	5.81		
Cyclohexylacetic acid	4.80			
Cyclohexylammonium ion	10.64			
4-Cyclohexyl-*n*-butyric acid	4.95			
Cyclohexyl-1,1-diacetic acid	3.49	6.96		
cis-Cyclohexyl-1,2-diacetic acid (20°)	4.42	5.45		
trans-Cyclohexyl-1,2-diacetic acid (20°)	4.38	5.42		
3-Cyclohexylpropionic acid	4.91			
Cyclopentanecarboxylic acid	4.99			
Cyclopentane-1,1-dicarboxylic acid	3.23	4.08		
cis-Cyclopentane-1,2-dicarboxylic acid	4.43	6.57		
trans-Cyclopentane-1,2-dicarboxylic acid	3.96	5.85		
cis-Cyclopentane-1,3-dicarboxylic acid	4.26	5.51		
trans-Cyclopentane-1,3-dicarboxylic acid	4.32	5.42		
Cyclopentyl-1,1-diacetic acid	3.80	6.77		
Cyclopropane-1,1-dicarboxylic acid	1.82	5.43		
cis-Cyclopropane-1,2-dicarboxylic acid	3.33	6.47		
trans-Cyclopropane-1,2-dicarboxylic acid	3.65	5.13		
Cumene hydroperoxide	12.60			
Cysteine (protonated cation)	1.96	8.36	10.28(SH)	
Cystine (protonated cation)	1.65	2.26	7.85	9.85

Table 5-8 (*Continued*)
PROTON-TRANSFER REACTIONS OF ORGANIC MATERIALS IN WATER AT 25°C

Substance	pK_1	pK_2	pK_3	pK_4
Dehydroascorbic acid (20°)	3.21	7.92	10.3	
Diacetylacetone	7.42			
1,3-Diamino-2-aminomethylpropane	6.44	8.56	10.38	
3,5-Diaminobenzoic acid	5.30			
α,γ-Diaminobutyric acid	1.85	8.24	10.40	
2,2'-Diaminodiethylsulfide (30°)	8.84	9.64		
3,3'-Diaminodipropylammonium ion (30°)	8.02	9.70	10.70	
Di-(2-aminoethyl)ether	8.62	9.59		
N,N'-Di(2-aminoethyl)ethylenediammonium ion (20°)	3.32	6.67	9.20	9.92
Diaminoglyoxime	2.95			
1,3-Diamino-2-propanol (protonated cation) 20°	7.93	9.69		
Di-*iso*-Amylammonium ion	10.89			
α,α-Dibenzylsuccinic acid (20°)	3.96	6.66		
α,α-Dibromopropionic acid	1.48			
α,β-Dibromopropionic acid	2.17			
α,α'-Dibromosuccinic acid (*meso-*, 20°)	1.47	2.71		
Di-*i*-butylammonium ion	10.59			
Dichloroacetic acid	1.26			
Dichloroacetylacetic acid	2.11			
2,6-Dichlorohydroquinone	7.30	10.0		
Dichloromethylphosphonic acid	1.14	5.61		
2,3-Dichlorophenol	7.44			
2,4-Dichlorophenol	7.85			
2,6-Dichlorophenol	6.78			
3,6-Dichlorophthalic acid	1.46			
Di-2-cyanoethylammonium ion	5.14			
Didodecylammonium ion	10.99			
Diethanolammonium ion	8.88			
3-(Diethoxyphosphinyl)phenol	8.66			
4-(Diethoxyphosphinyl)phenol	8.28			
3-(Diethoxyphosphonyl)benzoic acid	3.65			
4-(Diethoxyphosphonyl)benzoic acid	3.60			
Diethylacetic acid	4.74			
Diethylammonium ion	10.93			
N,N-Diethylanilinium ion	6.56			
5,5-Diethylbarbituric acid	7.98	13.31		
N,N-Diethylbenzylammonium ion	9.48			
Diethylbiguanide (protonated cation) (30°)	2.53	11.68		
N,N'-Diethylethylenediammonium ion	7.70	10.46		
Diethylenetriamine	4.42	9.21	10.02	
Diethylenetriamine-N,N,N',N'',N''-pentaacetic acid (pK_5 10.58)	1.80	2.55	4.33	8.60
β,β-Diethylglutaric acid	3.48	7.12		
N,N-Diethylglycine	2.04	10.47		
Diethylmalonic acid	2.15	7.29		
2,3-Diethylsuccinic acid: (*meso-*)	3.63	6.46		
(*rac-*)	3.54	6.59		
N,N-Diethyl-*o*-toluidinium ion	7.18			
Diglycolic acid	2.96			
Diguanidinium ion	2.95	11.49		

Table 5-8 (*Continued*)
PROTON-TRANSFER REACTIONS OF ORGANIC MATERIALS IN WATER AT 25°C

Substance	pK_1	pK_2	pK_3	pK_4
Dihexylammonium ion	11.0			
Dihydroresorcinol	5.26			
3,4-Dihydroxyalanine (protonated cation)	2.32	8.68	9.87	
2,3-Dihydroxybenzoic acid	2.94	11.76	>13	
2,4-Dihydroxybenzoic acid	3.29	8.98	>13	
2,5-Dihydroxybenzoic acid	2.97	10.50	>12	
2,6-Dihydroxybenzoic acid	1.30			
3,4-Dihydroxybenzoic acid	4.48	8.67	11.74	
3,5-Dihydroxybenzoic acid	4.04			
Dihydroxybenzoquinone	2.71	5.18		
Dihydroxyfumaric acid	1.10			
Dihydroxymaleic acid	1.15			
Dihydroxymalic acid	1.92			
2,4-Dihydroxyoxazolidine (protonated cation)	6.11			
2,4-Dihydroxypteridine (protonated cation)	<1.3	7.92		
Dihydroxytartaric acid	1.95	4.00		
3,5-Diiodotyrosine (protonated cation)	2.12	6.48	7.82	
2,3-Dimercaptopropan-1-ol (BAL)	8.62	10.58		
α,β-Dimercaptosuccinic acid	2.71	3.48	8.89	10.79
2,6-Dimethoxybenzoic acid	3.44			
cis-Dimethylacrylic acid	4.30			
trans-Dimethylacrylic acid	5.00			
β,β-Dimethylacrylic acid	5.12			
Dimethylaminoantipyrine (protonated cation)	4.84			
Dimethylammonium ion	10.77			
2,3-Dimethylanilinium ion	4.70			
2,4-Dimethylanilinium ion	4.89			
2,5-Dimethylanilinium ion	4.53			
2,6-Dimethylanilinium ion	3.95			
3,4-Dimethylanilinium ion	5.17			
N,N-Dimethylanilinium ion	5.21			
1,3-Dimethylbarbituric acid	4.68			
2,3-Dimethylbenzoic acid	3.74			
2,4-Dimethylbenzoic acid	4.18			
2,5-Dimethylbenzoic acid	3.98			
2,6-Dimethylbenzoic acid	3.25			
3,4-Dimethylbenzoic acid	4.41			
3,5-Dimethylbenzoic acid	4.30			
N,N-Dimethylbenzylammonium ion	9.02			
Dimethylbiguanide (protonated cation)	2.77	11.52		
2,2'-Dimethyl-*n*-butyric acid (18°)	5.03			
2,6-Dimethyl-4-cyano-phenol	8.27			
3,5-Dimethyl-4-cyano-phenol	8.21			
N,N-Dimethylethylenediamine-*N,N*-diacetic acid	6.63	9.53		
N,N-Dimethylethylenediamine-*N,N'*-diacetic acid	7.40	10.16		
N,N-Dimethylethylenediamine-*N',N'*-diacetic acid	5.99	9.97		
β,β-Dimethylglutaric acid	3.70	6.34		
N,N-Dimethylglycine (protonated cation)	2.15	9.94		
Dimethylglyoxime	10.60			
2,2-Dimethylhexanedione-3,5	10.01			
5,5-Dimethylhydantoin	9.19			

Table 5-8 (*Continued*)
PROTON-TRANSFER REACTIONS OF ORGANIC MATERIALS IN WATER AT 25°C

Substance	pK_1	pK_2	pK_3	pK_4
2,4-Dimethyl-8-hydroxyquinoline	6.20	10.60		
3,4-Dimethyl-8-hydroxyquinoline	5.89	10.05		
Dimethylimidazolium ion	8.52			
Dimethylmalic acid	3.17	6.06		
Dimethylmalonic acid	3.17	6.06		
α,α-Dimethyloxaloacetic acid	1.77	4.62		
2,3-Dimethylphenol	10.50			
2,4-Dimethylphenol	10.58			
2,5-Dimethylphenol	10.22			
3,4-Dimethylphenol	10.32			
3,5-Dimethylphenol	10.15			
Dimethylphosphoric acid	1.29			
2,3-Dimethylquinoline	4.94			
2,6-Dimethylquinoline	5.46			
2,3-Dimethylsuccinic acid: (*meso-*)	3.77	5.93		
(*rac-*)	3.94	6.20		
N,N-Dimethyltoluidinium ion: ortho-	5.86			
para-	7.24			
Dinicotinic acid	2.80			
2,3-Dinitrobenzoic acid	1.85			
2,4-Dinitrobenzoic acid	1.43			
2,5-Dinitrobenzoic acid	1.62			
2,6-Dinitrobenzoic acid	1.14			
3,4-Dinitrobenzoic acid	2.82			
3,5-Dinitrobenzoic acid	2.82			
Dinitro-o-cresol	4.35			
2,4-Dinitrophenol	4.09			
2,5-Dinitrophenol	5.22			
2,6-Dinitrophenol	3.71			
3,4-Dinitrophenol	5.42			
2,4-Dinitrophenylacetic acid	3.50			
3,5-Dinitrotoluic acid	2.97			
Diphenylacetic acid	3.94			
α,α-Diphenyladipic acid (20°)	4.17	5.40		
β,β'-Diphenyladipic acid	4.22	5.19		
Diphenylammonium ion	0.9			
α,α-Diphenylglutaric acid	3.91	5.38		
1,3-Diphenylguanidinium ion	10.12			
Diphenylketimine	6.82			
α,α-Diphenylsuccinic acid (20°)	3.05	7.3		
α,α'-Diphenylsuccinic acid (*meso-*)	3.48			
Diphenylthiocarbazone	4.82	15		
Dipicrylammonium ion	5.42			
Di-n-propylammonium ion	10.91			
Di-n-propylmalonic acid	2.04	7.51		
Dithiodiacetic acid (18°)	3.07	4.20		
1,4-Dithioerythritol	9.5			
Dithiooxamide	10.89			
1,4-Dithiothreitol	8.9			
Emetine (protonated cation)	7.36	8.23		
ψ-Ephedrinium ion	9.71			

Table 5-8 (*Continued*)
PROTON-TRANSFER REACTIONS OF ORGANIC MATERIALS
IN WATER AT 25°C

Substance	pK_1	pK_2	pK_3	pK_4
Eriochrome black T	6.3	11.55		
Ethane-1,2-dithiol	8.96	10.54		
Ethanolammonium ion	9.50			
Ethoxyacetic acid (18°)	3.65			
Ethoxyanilinium ion: ortho-	4.47			
meta-	4.17			
para-	5.25			
Ethoxybenzoic acid: ortho- (20°)	4.21			
meta- (20°)	4.17			
para- (20°)	4.80			
Ethoxycarbonylacetic acid	3.55			
Ethoxycarbonylethylammonium ion	9.13			
2-Ethoxyethanethiol	9.38			
Ethylacetoacetate	10.68			
Ethylammonium ion	10.63			
N-Ethylanilinium ion	5.11			
Ethylarsonic acid	3.89	8.35		
Ethylbenzoic acid: ortho-	3.79			
para-	4.35			
Ethylbiguanide (protonated cation)	2.09	11.47		
Ethylenebiguanide (protonated cation) (30°)	1.74	2.88	11.34	11.76
Ethylene-*N,N*-diacetic acid	5.58	11.05		
Ethylene-*N,N'*-diacetic acid	6.42	9.46		
Ethylenediamine-*N,N,N',N'*-tetraacetic acid	2.0	2.67	6.27	10.95
Ethylenediammonium ion	6.85	9.93		
trans-Ethyleneoxidedicarboxylic acid	1.93	3.25		
cis-Ethyleneoxidedicarboxylic acid	1.93	3.92		
N-Ethylethylenediammonium ion	7.63	10.56		
β-Ethylglutaric acid	4.28	5.33		
N-Ethylglycine (protonated cation)	2.34	10.23		
Ethylhydroperoxide	11.80			
Ethylmalonic acid	2.95	5.83		
Ethylmercaptan (20°)	10.50			
N-Ethylmercaptoacetamide	8.14			
Ethylmercaptoacetate	7.95			
Ethyl-3-mercaptopropionate	9.48			
N,N-Ethylmethylammonium ion	4.23			
Ethylnitroacetate	5.85			
3-Ethylpentanedione-2,4	11.34			
p-Ethylphenylacetic acid	4.37			
Ethylphosphinic acid	3.29			
Ethylphosphonic acid	2.43	8.05		
Ethyl-*n*-propylmalonic acid	3.14	7.43		
Ethylpyrrolidine	10.43			
S-Ethylthioacetic acid	5.06			
N-Ethyl-*o*-toluidinium ion	4.92			
Eugenol	10.0			
Fluoroacetic acid	2.59			
Fluoroanilinium ion: ortho-	3.20			
meta-	3.59			
para-	4.65			

Table 5-8 (*Continued*)
PROTON-TRANSFER REACTIONS OF ORGANIC MATERIALS
IN WATER AT 25°C

Substance	pK_1	pK_2	pK_3	pK_4
Fluorobenzoic acid: ortho-	3.27			
meta-	3.87			
para-	4.14			
m-Fluoromandelic acid	4.24			
Fluorophenol: ortho-	8.81			
meta-	9.28			
para-	9.81			
Fluorophenoxyacetic acid: ortho-	3.08			
meta-	3.08			
para-	3.13			
p-Fluorophenylacetic acid	4.25			
Fluorophenylalanine: ortho-	2.12	9.01		
meta-	2.10	8.98		
para-	2.18	9.05		
o-Fluorophenylphosphonic acid	1.64	6.80		
2-Fluoropyridinium ion	−0.44			
Folic acid (Pteroylglutamic acid)	8.26			
Formic acid	3.75			
N-Formylglycine	3.43			
Formylphenol: ortho-	8.37			
meta-	9.02			
para-	7.62			
Fumaric acid (*trans-*)	3.02	4.38		
2-Furancarboxylic acid	3.15			
2-Furoic acid	3.16			
Galactose-*l*-phosphoric acid ester	1.00	6.17		
Gallic acid (30°)	4.34	8.85		
Glucoascorbic acid	4.26	11.58		
Gluconic acid	3.86			
Glucose (18°)	12.43			
Glucose-*l*-phosphate	6.50			
Glutaconic acid (*trans*)	3.77	5.08		
Glutamic acid (protonated cation)	2.30	4.28	9.67	
Glutamic acid-α-ethyl half ester	3.85	7.84		
Glutamic acid-γ-ethyl half ester	2.15	9.19		
Glutammonium ion	2.17	9.13		
Glutaramic acid	4.60			
Glutaric acid	4.34	5.42		
Glutarimide	11.43			
Glutathione (30°)	8.56			
Glyceric acid	3.64			
Glycerol	14.15			
Glycero-2-phosphoric acid	1.32	6.65		
Glycine (protonated cation)	2.35	9.78		
Glycine amide hydrochloride	7.99			
Glycine hydroxamic acid	7.10	9.10		
Glycol	14.22			
Glycollic acid	3.82			
Glycylalanine (protonated cation)	3.15	8.17		
Glycylglycine (protonated cation)	3.15	8.25		

Table 5-8 (*Continued*)
PROTON-TRANSFER REACTIONS OF ORGANIC MATERIALS IN WATER AT 25°C

Substance	pK_1	pK_2	pK_3	pK_4
Glycylsarcosine	8.77			
Glycylserine	2.91			
Glyoxalic acid	3.30			
Glyoxaline (protonated cation)	7.03			
Guanine (protonated cation)	3.3	9.2	12.3	
Guanylethyl urea	3.20	11.10		
Guanylurea	1.80	8.20		
Hemimellitic acid	2.80	4.20	5.87	
Heptafluoro-*n*-butyric acid	0.17			
Heptanoic acid	4.89			
Hexahydrobenzoic acid	4.89			
Hexamethylenediammonium ion	9.83	10.93		
i-Hexanoic acid	4.84			
n-Hexanoic acid	4.86			
Hex-2-en-3-oic acid	4.72			
Hex-3-en-4-oic acid	4.58			
Hex-4-en-5-oic acid (*also* Hex-5-en-6-oic acid)	4.74			
Hippuric acid	3.65			
Histamine (protonated cation)	6.14	9.85		
Histidine (protonated cation)	1.82	6.12(Imid)	9.17(NH$_3^+$)	
Hydantoin	9.12			
Hydrastine (protonated cation)	6.23			
Hydrazine-*N,N*-diacetic acid	<0.1	2.8	3.8	
Hydrazine-*N,N'*-diacetic acid	2.40	3.12	7.32	
Hydrazinium ion (20°)	−0.88	8.48		
Hydroquinone (30°)	9.91	12.04		
Hydroxyacetophenone: meta-	9.19			
para-	8.05			
Hydroxybenzoic acid: ortho-	3.00	13.4		
meta-	4.08	9.93		
para-	4.58	9.23		
o-Hydroxybiphenyl	10.01			
2-Hydroxy-5-bromobenzoic acid	2.61			
2-Hydroxybutyric acid (30°)	3.65			
3-Hydroxybutyric acid (30°)	4.41			
4-Hydroxybutyric acid (30°)	4.71			
2-Hydroxy-5-chlorobenzoic acid	2.63			
2-Hydroxy-6-chlorobenzoic acid	2.63			
trans-Hydroxycinnamic acid: ortho-	4.61			
meta-	4.40			
N-(2-Hydroxyethyl)biguanide	2.8	11.53		
N'-(2-Hydroxyethyl)ethylenediamine-*N,N,N'*- triacetic acid	2.39	5.37	9.93	
N-(2-Hydroxyethyl)ethylenediammonium ion	7.21	10.12		
N,N-bis(2-Hydroxyethyl)glycine	8.35			
N-(2-Hydroxyethyl)iminodiacetic acid	2.2	8.73		
β-Hydroxyethylmercaptan	9.43			
N-(2-Hydroxyethyl)piperazine-*N'*-2-ethane sulfonic acid (20°)	7.55			
β-Hydroxyglutamic acid	2.32	4.23	9.56	

Table 5-8 (*Continued*)
PROTON-TRANSFER REACTIONS OF ORGANIC MATERIALS
IN WATER AT 25°C

Substance	pK_1	pK_2	pK_3	pK_4
Hydroxylammonium ion	5.96			
2-Hydroxy-1-methoxybenzylammonium ion	8.89	10.52		
3-Hydroxy-2-methoxybenzylammonium ion	8.94	10.52		
1-Hydroxy-2-methoxybenzylammonium ion	8.70	10.52		
2-Hydroxy-3-methylbenzoic acid	2.99			
2-Hydroxy-4-methylbenzoic acid	3.17			
2-Hydroxy-5-methylbenzoic acid	4.08			
2-Hydroxy-6-methylbenzoic acid	3.32			
N-tris(Hydroxymethyl)methyl-2-aminoethane sulfonic acid (20°)	7.5			
N-tris(Hydroxymethyl)methylglycine (20°)	8.15			
2,2-bis(Hydroxymethyl)2,2',2''-nitrilotriethanol	6.46			
Hydroxymethylphosphonic acid	1.91	7.15		
8-Hydroxy-5-methylquinoline-5-sulfonic acid		4.80	9.30	
2-Hydroxy-3-nitrobenzoic acid	1.87			
2-Hydroxy-4-nitrobenzoic acid	2.23			
2-Hydroxy-5-nitrobenzoic acid	2.12			
2-Hydroxy-6-nitrobenzoic acid	2.24			
Hydroxyproline (protonated cation)	1.92	9.73		
β-Hydroxypropionic acid	3.73			
4-Hydroxypteridine (protonated cation)	<1.3	7.89		
2-Hydroxypyridine (protonated cation) (20°)	0.75	11.62		
3-Hydroxypyridine (protonated cation) (20°)	4.79	8.72		
4-Hydroxypyridine (protonated cation) (20°)	3.20	11.09		
2-Hydroxypyrimidine (protonated cation) (20°)	2.24	9.17		
4-Hydroxypyrimidine (protonated cation) (20°)	1.85	8.59		
8-Hydroxyquinazolinium ion	3.41	8.65		
8-Hydroxyquinoline-5-sulfonic acid	1.3	4.11	8.75	
2-Hydroxyquinolinium ion (20°)	−0.31	11.74		
3-Hydroxyquinolinium ion (20°)	4.30	8.06		
4-Hydroxyquinolinium ion (20°)	2.27	11.25		
5-Hydroxyquinolinium ion (20°)	5.20	8.54		
6-Hydroxyquinolinium ion (20°)	5.17	8.88		
7-Hydroxyquinolinium ion (20°)	5.48	8.85		
8-Hydroxyquinolinium ion (20°)	4.91	9.81		
Hydroxytetracycline	3.27	7.32	9.11	
Hydroxyuracil	8.64			
Hypoxanthine	1.98	8.94		
Imidazolinium ion	7.03			
Iminodiacetic acid	2.54	9.12		
Indanol-4	10.32			
Iodoacetic acid	3.18			
Iodoanilinium ion: ortho-	2.60			
meta-	3.61			
para-	3.78			
Iodobenzoic acid: ortho-	2.86			
meta-	3.85			
para-	3.93			
7-Iodo-8-hydroxyquinoline-5-sulfonic acid	2.51	7.42		

Table 5-8 (*Continued*)
PROTON-TRANSFER REACTIONS OF ORGANIC MATERIALS
IN WATER AT 25°C

Substance	pK_1	pK_2	pK_3	pK_4
m-Iodomandelic acid	3.26			
Iodophenol: ortho-	8.46			
meta-	8.88			
para-	9.20			
Iodophenoxyacetic acid: ortho-	3.17			
meta-	3.13			
para-	3.16			
Iodophenylacetic acid (meta-)	4.16			
2-Iodopropionic acid (18°)	3.11			
3-Iodopropionic acid	4.05			
2-Iodopyridinium ion	1.82			
Isocrotonic acid	4.44			
Isoleucine (protonated cation)	2.32	9.76		
Isophthalic acid	3.62	4.60		
Isothiocyanatoacetic acid	6.62			
Itaconic acid	3.85	5.45		
δ-Ketovaleric acid	4.72			
Kojic acid	7.75			
Lactic acid	3.86			
i-Leucine (protonated cation)	2.32	9.76		
Leucine (protonated cation)	2.33	9.74		
Levulinic acid	4.59			
Lutidine (protonated cation)	2.15			
Lysine (protonated cation)	2.18	8.95(α-NH$_3^+$)	10.53(ε-NH$_3^+$)	
Maleic acid	1.94	6.22		
Malic acid	3.40	5.05		
Malonamic acid	3.64			
Malonic acid	2.86	5.70		
Mandelic acid	3.41			
Mannose (18°)	13.57			
Mellitic acid (pK_5 5.89; pK_6 6.96)	1.40	2.19	3.31	4.78
Mellophanic acid	2.05	3.25	4.73	6.21
Mercaptoacetic acid	3.60	10.55		
2-Mercaptobutyric acid	3.53			
Mercaptodiacetic acid	3.32	4.29		
2-Mercaptoethanesulfonic acid (20°)		9.5 (−SH)		
2-Mercaptoethylammonium ion	8.27	10.53		
2-Mercaptopropionic acid	4.32	10.20		
2-Mercaptoquinoline (20°)	−1.44	10.21		
3-Mercaptoquinoline (20°)	2.33	6.13		
4-Mercaptoquinoline (20°)	0.77	8.83		
Mercaptosuccinic acid	3.30	4.94	10.94	
Mesaconic acid	3.09	4.75		
Mesitylenic acid	4.32			
Metanilic acid	0.39	3.72		
Methionine (protonated cation)	2.28	9.2		
Methoxyacetic acid	3.57			
2-Methoxyanilinium ion	4.49			
3-Methoxyanilinium ion	4.20			
4-Methoxyanilinium ion	5.29			
2-Methoxybenzoic acid	4.09			

Table 5-8 (*Continued*)
PROTON-TRANSFER REACTIONS OF ORGANIC MATERIALS
IN WATER AT 25°C

Substance	pK_1	pK_2	pK_3	pK_4
3-Methoxybenzoic acid	4.08			
4-Methoxybenzoic acid	4.49			
N,N-Methoxybenzylammonium ion	9.68			
2-Methoxycarbonylanilinium ion	2.23			
3-Methoxycarbonylanilinium ion	3.64			
4-Methoxycarbonylanilinium ion	2.38			
Methoxycarbonylmethylammonium ion	7.66			
2-Methoxycinnamic acid (*trans-*)	4.46			
3-Methoxycinnamic acid (*trans-*)	4.38			
4-Methoxycinnamic acid (*trans-*)	4.54			
Methoxyethylammonium ion	9.45			
2-Methoxyphenol	9.98			
3-Methoxyphenol	9.65			
4-Methoxyphenol	10.21			
2-Methylacrylic acid (18°)	4.66			
N-Methylalanine (protonated cation)	2.22	10.19		
2-(*N*-Methylamino)benzoic acid	1.93	5.34		
3-(*N*-Methylamino)benzoic acid		5.10		
4-(*N*-Methylamino)benzoic acid		5.05		
Methylaminodiacetic acid	2.15	10.00		
Methylammonium ion	10.62			
N-Methylanilinium ion	4.85			
Methylarsonic acid (18°)	3.61	8.18		
1-Methylbarbituric acid	4.35			
Methylbiguanidinium ion	3.00	11.44		
2-Methyl-2-butanethiol	11.35			
trans-2-Methyl-*n*-but-1-ene-1-oic acid	5.13			
4-Methylcarboxylphenol	8.47			
trans-2-Methylcinnamic acid	4.50			
trans-3-Methylcinnamic acid	4.44			
trans-4-Methylcinnamic acid	4.56			
1-Methylcyclohexane carboxylic acid	5.13			
cis-2-Methylcyclohexane carboxylic acid	5.03			
trans-2-Methylcyclohexane carboxylic acid	5.73			
cis-3-Methylcyclohexane carboxylic acid	4.88			
trans-3-Methylcyclohexane carboxylic acid	5.02			
cis-4-Methylcyclohexane carboxylic acid	4.88			
trans-4-Methylcyclohexane carboxylic acid	5.03			
3-Methylcyclohexyl-1,1-diacetic acid	3.49	6.10		
4-Methylcyclohexyl-1,1-diacetic acid	3.49	6.10		
Methyldiethylammonium ion	10.43			
2,2'-Methylene-*bis*(4-chlorophenol)	7.6	11.5		
2,2'-Methylene-*bis*(4,6-dichlorophenol)	5.6	10.65		
2,2'-Methylene-*bis*(3,4,6-trichlorophenol)		10.1		
Methylethylacetic acid (18°)	4.81			
Methylethylacrylic acid (*cis-* and *trans-*)	5.15			
N-Methylethylenediammonium ion (20°)	6.86	10.15		
Methylethyl ketoxime	12.45			
β-Methylglutaric acid	4.24	5.41		
N-Methylglycine (protonated cation)	2.35	10.18		
N-Methylguanidinium ion	13.4			
2-Methylhexanedione-3,5	9.43			
3-Methylhistamine (protonated cation)	5.80	9.90		

Table 5-8 (*Continued*)
PROTON-TRANSFER REACTIONS OF ORGANIC MATERIALS IN WATER AT 25°C

Substance	pK_1	pK_2	pK_3	pK_4
2-Methyl-8-hydroxyquinolinium ion	5.53	10.71		
4-Methyl-8-hydroxyquinolinium ion	5.55	10.00		
1-Methylimidazole (protonated cation)	7.06			
4-Methylimidazole (protonated cation)	7.45			
N-Methyliminodiacetic acid	2.15	10.09		
S-Methylisothiourea (20°)	9.83			
O-Methylisourea	9.72			
Methylmalonic acid	3.07	5.76		
Methylmercaptan	10.70			
3-Methylmercaptophenol	9.53			
4-Methylmercaptophenol	9.53			
N-Methylmorpholinium ion	7.13			
2-Methylnaphthoic acid-1	3.11			
N-Methyl-1-naphthylammonium ion	3.70			
2-Methyl-6-nitrobenzoic acid	1.87			
1-Methyl-2-nitroterephthalic acid	3.11			
4-Methyl-2-nitroterephthalic acid	1.82			
3-Methylpentanedione-2,4	10.87			
Methylphosphinic acid	3.08			
Methylphosphonic acid	2.38	7.74		
Methylphthalic acid: ortho-	3.18			
meta- (or *iso*-)	3.89			
N-Methylpiperidine (protonated cation)	10.08			
2-Methyl-2-propanethiol	11.2			
2-Methyl-*n*-propylacetic acid (18°)	4.79			
2-Methylpyridinium ion (20°)	5.97			
3-Methylpyridinium ion (20°)	5.68			
4-Methylpyridinium ion (20°)	6.02			
Methyl-2-pyridyl ketoxime	9.97			
N-Methylpyrrolidinium ion	10.18			
2-Methylquinolinium ion	5.55	10.31		
5-Methylquinolinium ion	4.62			
Methylsuccinic acid	4.13	5.64		
Methylsulfonylacetic acid	2.36			
Methylsulfonylanilinium ion: meta-	2.68			
para-	1.48			
3-Methylsulfonylbenzoic acid	3.52			
4-Methylsulfonylbenzoic acid	3.64			
Methylsulfonylphenol: meta-	9.33			
para-	7.83			
Methylthioacetic acid	3.72			
Methylthioanilinium ion: meta-	4.05			
para-	4.40			
Methylthioglycolic acid (20°)	7.68			
3-(S-Methylthio)phenol (*also* 4-)	9.53			
Methyluracil	9.52			
2-Methyl-*n*-valeric acid (18°)	4.79			
1-Methylxanthine	7.70	12.0		
3-Methylxanthine	8.10	11.3		
7-Methylxanthine	8.33	∼13		
9-Methylxanthine	6.25			
Morphine (protonated cation)	7.87			

Table 5-8 (*Continued*)
PROTON-TRANSFER REACTIONS OF ORGANIC MATERIALS
IN WATER AT 25°C

Substance	pK_1	pK_2	pK_3	pK_4
Morpholinium ion	8.70			
2-(*N*-Morpholino)ethanesulfonic acid (20°)	6.15			
Murexide	0.0	9.20	10.50	
1-Naphthalenesulfonic acid (20°)	0.17			
1-Naphthoic acid	3.70			
2-Naphthoic acid	4.16			
1-Naphthol (20°)	9.30			
2-Naphthol (20°)	9.57			
Naphthoquinone monoxime	8.01			
1-Naphthylacetic acid	4.24			
2-Naphthylacetic acid	4.26			
1-Naphthylammonium ion	3.92			
2-Naphthylammonium ion	4.11			
1-Naphthylarsonic acid	3.66	8.66		
Narceine (protonated cation)	3.3			
Narcotine (protonated cation)	6.18			
Nicotine (protonated cation)	3.15	7.87		
Nicotinic acid	4.82	12.00		
iso-Nicotinic acid	4.84	12.33		
Nicotinic acid amide	3.33			
iso-Nicotinic acid methyl ester	3.26			
Nitrilotriacetic acid (30°)	1.65	2.95	10.23	
Nitroacetic acid (18°)	1.68			
Nitroanilinium ion: ortho-	−0.26			
meta-	2.46			
para-	0.99			
Nitrobenzoic acid: ortho-	2.17			
meta-	3.49			
para-	3.44			
2-Nitro-4-chlorophenol	6.48			
Nitrocinnamic acid (*trans*-): ortho-	4.15			
meta-	4.12			
para-	4.05			
Nitroethane	8.44			
2-Nitrohydroquinone	7.63	10.06		
N-Nitroiminodiacetic acid	2.21	3.33		
Nitromethane	10.21			
Nitrophenol: ortho-	7.23			
meta-	8.40			
para-	7.15			
Nitrophenylacetic acid: ortho-	4.00			
meta-	3.97			
para-	3.85			
m-Nitrophenylphosphonic acid	1.30	6.27		
p-Nitrophenylphosphonic acid	1.24	6.23		
β-(*o*-Nitrophenyl)propionic acid	4.50			
β-(*p*-Nitrophenyl)propionic acid	4.47			
3-Nitrophthalic acid	1.88			
4-Nitrophthalic acid	2.11			
2-Nitropropionic acid	3.79			
N-Nitrosoiminodiacetic acid	2.28	3.38		
Nitroterephthalic acid	1.73			

Table 5-8 (*Continued*)
PROTON-TRANSFER REACTIONS OF ORGANIC MATERIALS
IN WATER AT 25°C

Substance	pK_1	pK_2	pK_3	pK_4
4-Nitro-o-toluic acid	1.86			
Nitrourea	4.15			
Nonanoic acid	4.95			
Norleucine (protonated cation)	2.34	9.83		
Norvaline (protonated cation)	2.29	9.70		
Novocaine (protonated cation)	8.85			
Octanoic acid	4.89			
Octylammonium ion	10.65			
Ornithine (protonated cation)	1.94	8.65	10.76	
Orthanilic acid	2.46			
Oxaloacetic acid	2.56	4.37		
Oxalic acid	1.27	4.27		
Papaverine (protonated cation)	5.90			
Parabanic acid	6.10			
Pelargonic acid	4.95			
Pentenedioic acid	3.77	6.08		
Pent-1-en-oic acid	4.70			
Pent-2-en-oic acid	4.69			
Pent-3-en-oic acid	4.51			
Pent-4-en-oic acid	4.68			
Peracetic acid	8.20			
1,10-Phenanthrolinium ion	4.86			
Phenazinium ion (20°)	1.20			
Phenetidine (protonated cation): ortho-	4.47			
meta-	4.17			
para-	5.34			
Phenol	9.99			
Phenol-3-phosphonic acid	1.78	7.03	10.2	
Phenol-4-phosphonic acid	1.99	7.25	9.9	
p-Phenolsulfonic acid	. . . . . .	9.05		
Phenoxyacetic acid	3.17			
Phenoxybenzoic acid: ortho-	3.53			
meta-	3.95			
para-	4.52			
Phenylacetic acid	4.31			
Phenylalanine (protonated cation)	2.16	9.13		
Phenylanilinium ion: ortho-	3.78			
meta-	4.18			
para-	4.27			
Phenylarsonic acid	3.46	8.48		
Phenylboric acid	8.86			
o-Phenylbenzoic acid	3.46			
α-Phenyl-α-benzylsuccinic acid (20°)	3.69	6.47		
γ-Phenylbutyric acid	4.76			
Phenylenediammonium ion: ortho- (20°)	0.80	4.57		
meta-	2.65	4.88		
para-	3.29	6.08		
β-Phenylethylammonium ion	9.83			
β-Phenylethylboric acid	10.0			
Phenylguanidinium ion	10.77			
Phenylhydrazinium ion	5.20			

Table 5-8 (*Continued*)
PROTON-TRANSFER REACTIONS OF ORGANIC MATERIALS IN WATER AT 25°C

Substance	pK_1	pK_2	pK_3	pK_4
α-Phenyl-α-hydroxypropionic acid	3.53			
β-Phenyl-β-hydroxypropionic acid	4.40			
Phenylmalonic acid	2.58	5.03		
Phenylphenol: ortho-	9.97			
meta-	9.63			
para-	9.55			
Phenylphosphinic acid	2.1			
Phenylphosphoric acid	1.83	7.07		
α-Phenylpropionic acid	4.38			
β-Phenylpropionic acid	4.64			
γ-Phenylpropylammonium ion	10.39			
Phenylselenonic acid	4.79			
Phenylsuccinic acid (20°)	3.78	5.55		
Phloroglucinol	8.45			
Phosphoserine (protonated cation)	2.08	5.65	9.74	
Phthalamic acid	3.79			
Phthalazinium ion (20°)	3.47			
Phthalic acid: ortho-	2.95	5.41		
meta-	3.62	4.60		
para-	3.54	4.46		
Phthalic acid monoamide	3.76			
Phthalimide	9.90			
Physostigmine (protonated cation)	1.76	7.88		
Picolinic acid	3.4	11.0		
Picolinic acid methylester	2.21			
2-Picolinium ion	6.48			
3-Picolinium ion (*and* 4-)	6.00			
Picric acid	-0.22			
Pilocarpine (protonated cation)	1.3	6.85		
Pimelic acid	4.50	5.42		
Piperazine (protonated cation)	5.68	9.82		
Piperazine-2-carboxylic acid	1.5	5.41	9.53	
Piperidinium ion	11.12			
Piperine (protonated cation)	0.0			
Pivalic acid	5.05			
Polyacrylic acid	5.08			
Prehnitic acid	2.38	3.51	4.44	5.81
L-Proline (protonated cation)	1.95	10.64		
1,2-Propanediammonium ion	7.10	9.97		
1,3-Propanediammonium ion (20°)	8.64	10.62		
2-Propanethiol	10.86			
1,2,3-Propanetriammonium ion	3.72	7.95	9.59	
1,2,3-Propanetricarboxylic acid	3.50	4.63	5.95	
Propenylacetic acid	4.51			
Propiolic acid	1.84			
Propionic acid	4.87			
N-n-Propionylglycine (protonated cation)	3.72			
n-Propoxybenzoic acid: ortho- and meta-	4.22			
para-	4.78			
iso-Propylacrylic acid	4.70			
iso-Propylammonium ion	10.53			
N-iso-Propylanilinium ion	5.50			

Table 5-8 (*Continued*)
PROTON-TRANSFER REACTIONS OF ORGANIC MATERIALS
IN WATER AT 25°C

Substance	pK_1	pK_2	pK_3	pK_4
n-Propylarsonic acid	4.21	9.09		
iso-Propylbenzoic acid: ortho-	3.64			
para-	4.36			
β-*n*-Propylglutaric acid	4.31	5.39		
N-*n*-Propylglycine (protonated cation)	2.38	10.19		
n-Propylmalonic acid (also *iso*-)	2.97	5.84		
p-*iso*-Propylphenylacetic acid	4.39			
n-Propylphosphinic acid	3.46			
iso-Propylphosphinic acid	3.56			
n-Propylphosphonic acid	2.49	8.18		
iso-Propylphosphonic acid	2.66	8.44		
Pteroylglutamic acid	8.26			
Purine (protonated cation)	2.52	8.92		
Pyrazinium ion	0.6			
Pyrazolinium ion	2.53			
Pyridazinium ion	2.33			
Pyridine-2-carboxylic acid	1.06	5.37		
Pyridine-3-carboxylic acid	2.07	4.73		
Pyridine-4-carboxylic acid	1.70	4.89		
Pyridine-2,3-dicarboxylic acid	2.36	7.08		
Pyridine-2,4-dicarboxylic acid	2.23	7.02		
Pyridine-2,6-dicarboxylic acid	2.16	6.92		
2-Pyridinium ion	3.28			
3-Pyridinium ion	4.88			
4-Pyridinium ion	6.62			
Pyridinium-*N*-oxide ion	0.79			
Pyrimidinium ion	1.30			
Pyrocathechol (30°)	9.13	12.08		
Pyrocatechol Violet	7.82	9.76	11.73	
Pyrogallol	9.03	12.63		
Pyromellitic acid	1.92	2.87	4.49	5.63
Pyromucic acid	5.15			
Pyroracemic acid	2.49			
Pyroxilidinium ion	11.11			
1-Pyrrolecarboxylic acid	4.45			
2-Pyrrolecarboxylic acid	4.45			
Pyrrolinium ion	−0.27			
Pyruvic acid	2.49			
Quinidine (protonated cation)	4.0	8.54		
Quinine (protonated cation)	4.11	8.0		
Quinol	9.96			
Quinolinium ion	4.88			
Quinoxalinium ion	0.72			
Reductic acid (20°)	4.72			
Resorcinol (30°)	9.15	11.32		
β-Resorcylic acid	3.29	8.98	>13	
Rubeanic acid	10.89			
Saccharin	11.68			
Salicylaldehyde	8.34			

Table 5-8 (*Continued*)
PROTON-TRANSFER REACTIONS OF ORGANIC MATERIALS
IN WATER AT 25°C

Substance	pK_1	pK_2	pK_3	pK_4
Salicylaldoxime	1.37	9.18	12.11	
Salicylamide (30°)	8.36			
Salicylic acid	3.00	13.4		
Sarcosine (protonated cation)	2.25	10.20		
Sarcosine amide (protonated cation)	8.35			
Sarcosine dimethylamide (protonated cation)	8.86			
Sarcosine methylamide (protonated cation)	8.28			
Sarcosylsarcosine (protonated cation)	2.92	9.15		
Sarcosylserine (protonated cation)	3.17	8.63		
Semicarbazide (protonated cation)	3.43			
Serine (protonated cation)	2.20	9.21		
Serine methyl ester (protonated cation)	7.10			
Solanine (protonated cation)	7.34			
Sorbic acid	4.77			
Sorbose (18°)	13.57			
Sparteine (protonated cation)	4.49	11.76		
Strychnine (protonated cation)	2.3	8.0		
Suberic acid	4.52	5.41		
Succinamic acid	4.54			
Succinic acid	4.21	5.64		
Succinimide	9.62			
Sulfamic acid	0.99			
β-(p-Sulfaminophenyl)-alanine	1.99	8.64	10.26	
Sulfamoylbenzoic acid: meta-	3.54			
para-	3.47			
p-Sulfamoylphenylphosphoric acid	1.42	6.38	10.0	
Sulfanilic acid	0.58	3.23		
Sulfoacetic acid	4.07			
Sulfobenzoic acid: meta-	3.78			
para-	3.72			
Sulfophenol: meta-	0.39	9.07		
para-	0.58	8.70		
α-Sulfopropionic acid	1.99			
β-Sulfopropionic acid	3.62			
5-Sulfosalicylic acid	2.49	12.00		
Sylvic acid	7.62			
Tartaric acid: (*meso-*)	3.22	4.81		
(*d-*)	3.04	4.37		
Tartronic acid	2.37	4.74		
Taurine (protonated cation) (20°)	−0.3	9.19		
Terephthalic acid	3.54	4.46		
Terramycine	3.10	7.26	9.11	
Tetracycline	3.30	7.68	9.69	
Tetraethylenepentamine (protonated cation)	2.66	4.70	8.10	9.05; pK_5 9.54
Tetralol-1	10.28			
Tetralol-2	10.48			
Tetramethylenediammonium ion	9.22	10.75		
N,N,N',N'-Tetramethylethylenediammonium ion	6.56	10.13		
N,N'-Tetramethyl-p-phenylenediammonium ion	2.20	6.35		

Table 5-8 (*Continued*)
PROTON-TRANSFER REACTIONS OF ORGANIC MATERIALS
IN WATER AT 25°C

Substance	pK_1	pK_2	pK_3	pK_4
Tetrolic acid	2.66			
2-Thenoic acid	3.53			
3-Thenoic acid	4.10			
Thebaine (protonated cation)	7.95			
Theobromic acid	8 to 10			
Theobromine (protonated cation)	0.68	7.89		
Theophylline	8.80			
Thiazolinium ion	2.53			
Thioacetic acid	3.33			
Thiocyanatoacetic acid	2.58			
Thiodiacetic acid (18°)	3.30	4.50		
Thiodiglycollic acid	3.32	4.29		
Thioglycollic acid	3.60	10.55		
1-Thionylcarboxylic acid	3.53			
2-Thionylcarboxylic acid	4.10			
Thiophenecarboxylic acids (see Thenoic acid)				
Thiophenol	6.50			
Thiourea (protonated cation)	2.03			
Thorin	3.7	8.3	11.8	
Threonine (protonated cation)	2.09	9.10		
Tiglic acid (18°)	5.00			
Tiron	7.66	12.6		
2-Toluenethiol	6.64			
3-Toluenethiol	6.58			
4-Toluenethiol	6.52			
p-Toluenesulfinic acid	1.7			
Toluhydroquinone	10.03	11.62		
Toluic acid: ortho-	3.91			
meta-	4.27			
para-	4.37			
Toluidinium ion: ortho-	4.39			
meta-	4.68			
para-	5.09			
o-Tolylacetic acid (also *p*-) (18°)	4.36			
Tolylphosphonic acid: ortho-	2.10	7.68		
meta-	1.88	7.44		
para-	1.84	7.33		
Triacetylmethane	5.81			
1,2,3-Triaminopropane (protonated cation)	3.80	8.03	9.67	
2,2′,2″-Triaminotriethylamine (protonated cation)	8.64	9.67	10.37	
2,4,6-Tribromobenzoic acid	1.41			
Tri-*i*-butylammonium ion	10.42			
Tricarballylic acid	3.50	4.63	5.95	
Trichloroacetic acid	0.64			
3,3,3-Trichlorolactic acid	2.34			
Trichloromethylphosphonic acid	1.63	4.81		
Trichlorophenol	6.00			
Triethanolammonium ion	7.76			
Triethylammonium ion	10.72			
Triethylenediammonium ion	4.18	8.19		
Triethylenetetrammonium ion (20°)	3.32	6.67	9.20	9.92

Table 5-8 (*Continued*)
PROTON-TRANSFER REACTIONS OF ORGANIC MATERIALS IN WATER AT 25°C

Substance	pK_1	pK_2	pK_3	pK_4
Trifluoroacetic acid	0.25			
γ,γ,γ-Trifluoro-*n*-butyric acid	4.16			
Trifluoromethylanilinium ion: meta-	3.5			
para-	2.6			
β,β,β-Trifluoropropionic acid	3.06			
1,1,1-Trifluoro-3-(2-thenoyl)acetone	5.70			
2,4,6-Trihydroxybenzoic acid	1.68			
3,4,5-Trihydroxybenzoic acid	4.34	8.85		
Trimellitic acid	2.52	3.84	5.20	
Trimesic acid	2.12	3.89	5.20	
Trimethylacetic acid (20°)	5.03			
Trimethylammonium ion	9.80			
2,4,6-Trimethylanilinium ion	4.38			
2,4,6-Trimethylbenzoic acid	3.44			
Trimethylenediammonium ion	10.45			
2,4,6,-Trimethylphenol	10.88			
2,4,6-Trinitrobenzoic acid	0.65			
2,2,2-Trinitroethanol	2.36			
2,4,6-Trinitrophenol	−0.20			
Triphenylacetic acid	3.96			
Tripropylammonium ion	10.70			
Trishydroxymethylamineomethane (protonated cation)	8.08			
Tryptophane (protonated cation)	2.38	9.39		
Tyrosine (protonated cation)	2.20	9.11	10.07(OH)	
Urea (protonated cation)	0.10			
Uric acid	5.40	5.53		
i-Valeric acid	4.78			
n-Valeric acid	4.84			
t-Valeric acid	5.08			
Valine (protonated cation)	2.29	9.74		
Vanillic acid	4.52			
Vanillin	7.40			
o-Vanillin	7.91			
iso-Vanillin	8.90			
Veratrine (protonated cation)	8.85			
Veronal	7.43			
Vinylacetic acid	4.34			
Vinylmethylammonium ion	9.69			
Violuric acid	4.57			
Xanthene (protonated cation) (40°)	0.68			
Xanthine	7.53	11.63		
Xylenol orange (pK_5 10.46; pK_6 12.28)	......	2.58	3.23	6.37
Zincon	4	7.85	15	

Table 5-9

TEMPERATURE DEPENDENCE OF SELECTED EQUILIBRIUM CONSTANTS IN AQUEOUS SOLUTIONS

Substance	Temperature, °C									
	0°	5°	10°	15°	20°	25°	30°	35°	40°	50°
Acetic acid, pK_a	4.78	4.77	4.76	4.76	4.76	4.76	4.76	4.76	4.77	4.79
dl-Alanine, pK_1			2.42		2.35	2.34	2.33	2.33	2.32	2.33
pK_2			10.59		10.01	9.87	9.74	9.62	9.49	9.26
Ammonium ion, pK_a	10.081	9.904	9.731	9.564	9.400	9.245	9.093	8.947	8.805	8.539
Boric acid, pK_a	9.508	9.439	9.380	9.326	9.280	9.237	9.195	9.161	9.128	9.077
Carbon dioxide + water, pK_1	6.58	6.52	6.46	6.42	6.38	6.35	6.33	6.31	6.30	6.29
pK_2	10.63	10.56	10.49	10.43	10.38	10.33	10.29	10.25	10.22	10.17
Citric acid, pK_1	3.22	3.20	3.18	3.16	3.14	3.13	3.12	3.11	3.10	3.10
pK_2	4.84	4.81	4.80	4.78	4.77	4.76	4.76	4.75	4.75	4.76
pK_3	6.39	6.39	6.38	6.38	6.39	6.40	6.41	6.42	6.44	6.48
5,5-Diethylbarbituric acid, pK_a	8.40	8.30	8.22	8.14	8.06	7.98	7.91	7.85	7.79	7.68
Glycerine-2-phosphoric acid, pK_1		1.22	1.25	1.27	1.30	1.33	1.37	1.42	1.46	1.45
pK_2		6.66	6.65	6.65	6.65	6.65	6.66	6.67	6.68	6.71
Glycine, pK_1			2.397	2.380	2.36	2.35	2.34	2.33	2.33	2.32
pK_2			10.193	10.044	9.91	9.78	9.65	9.53	9.41	9.19
Hydrogen cyanide, pK_a			9.63	9.49	9.36	9.21	9.11	8.99	8.88	
Hydrogen peroxide, pK_1	12.23			11.86	11.75	11.65	11.55	11.45		11.21
Hydrogen sulfide, pK_1		7.33	7.24	7.13	7.05	6.97	6.90	6.82	6.79	6.69
pK_2		13.5		13.2		12.90	12.75	12.6		
Hypochlorous acid, pK_a	7.82	7.75	7.69	7.63	7.58	7.54	7.50	7.46		7.05
Lactic acid, pK_a	3.88	3.87	3.87	3.86	3.86	3.86	3.86	3.87	3.87	3.90
Lead sulfate, pK_{sp}	8.01			7.87		7.80		7.73		7.63

	C1	C2	C3	C4	C5	C6	C7	C8	C9	C10
dl-Leucine, pK_1	2.39		2.38			2.33				2.33
pK_2	10.56		10.45			9.74				9.14
Mercury(I) chloride, pK_{sp}			18.65	18.48	18.27	17.88		16.79		
dl-Norleucine, pK_1						2.34				2.33
pK_2						9.83				9.22
Oxalic acid, pK_2	4.21	4.21	4.22	4.23	4.25	4.27	4.29	4.31	4.34	4.40
Phosphoric acid, pK_1	2.056	2.073	2.088	2.107	2.127	2.148	2.171	2.196	2.224	2.277
pK_2	7.313	7.282	7.254	7.231	7.213	7.198	7.189	7.185	7.181	7.183
o-Phthalic acid, pK_1	2.92	2.93	2.93	2.94	2.94	2.95	2.96	2.97	2.98	3.00
pK_2	5.43	5.42	5.41	5.41	5.41	5.41	5.42	5.43	5.44	5.48
Silver bromide, pK_{sp}		13.33		12.83	12.57	12.30		11.83	11.61	11.19
Silver chloride, pK_{sp}		10.595		10.152		9.749	12.07	9.381	9.21	8.88
Silver sulfate, pK_{sp}						4.835				4.62
Succinic acid, pK_1	4.28	4.26	4.24	4.23	4.22	4.21	4.20	4.19	4.19	4.19
pK_2	5.67	5.66	5.65	5.64	5.64	5.64	5.64	5.65	5.65	5.68
Sulfuric acid, pK_2	1.58	1.63				1.96	2.05	2.09	2.17	2.28
Sulfurous acid, pK_1	1.63		1.74	1.80		1.89		1.98		2.12
pK_2						7.21				7.45
d-Tartaric acid, pK_1	3.12	3.10	3.08	3.06	3.04	3.04	3.03	3.02	3.02	3.02
pK_2	4.43	4.41	4.39	4.38	4.37	4.37	4.37	4.37	4.37	4.39
dl-Valine, pK_1	2.32					2.29				2.31
pK_2	10.43					9.74				9.14

ACIDITIES OF VARIOUS COMPOUNDS IN NONAQUEOUS SOLVENTS

The properties of common solvents employed in nonaqueous titrations of acids and bases are given in Table 5-10; these properties include the dielectric constant, the autoprotolysis constant ($-\log K_s$), and the approximate potential span in millivolts for acid-base titrations. Figure 5-1 graphically displays the approximate potential ranges in various solvents; however, the basic end is truncated by the nature of the titrant employed—tetrabutylammonium hydroxide. All values refer to $0.01M$ solutions. These limiting potentials are subject to wide variation with changes in titrant concentration.

Only limited data pertaining to acidity constants in nonaqueous solvents are available. These are summarized in Table 5-11. For computing the approximate half-neutralization potential (**HNP**) values from pK_a values in aqueous solutions (Tables 5-7 and 5-8), a useful equation is

$$\textbf{HNP} = a - b(pK_a)$$

The constants a and b in the equation are given in Table 5-12 for the solvents listed in the first column. The classes of compounds for which the expression is applicable are denoted by the numbers appearing in the last column. The values of **HNP** computed for compounds of classes 1 and 2 are referred to diphenylguanidine ($pK_a = 10.12$; assigned a **HNP** value of 0.0 mV); those for compounds of classes 3 and 4 are referred to benzoic acid ($pK_a = 4.21$; assigned a **HNP** value of 0.0 mV).

The acidity function H_0 is not affected by the dielectric constant and allows a quantitative comparison of acidity in different solvents. H_0 is defined by

$$H_0 = (pK_{HB^+})_{\text{water}} + \log (C_B/C_{HB^+})_{\text{solvent}}$$

where K_{HB^+} refers to the dissociation constant of the protonated cation or cation acid with reference to water as the standard state. Table 5-13 gives H_0 for mineral acids in aqueous solutions; all values are referred to the value of $pK_{HB^+} = 0.99$ for p-nitroaniline. Values for weak bases are given by M. A. Paul and F. A. Long, *Chem. Rev.*, **57**, 1 (1957).

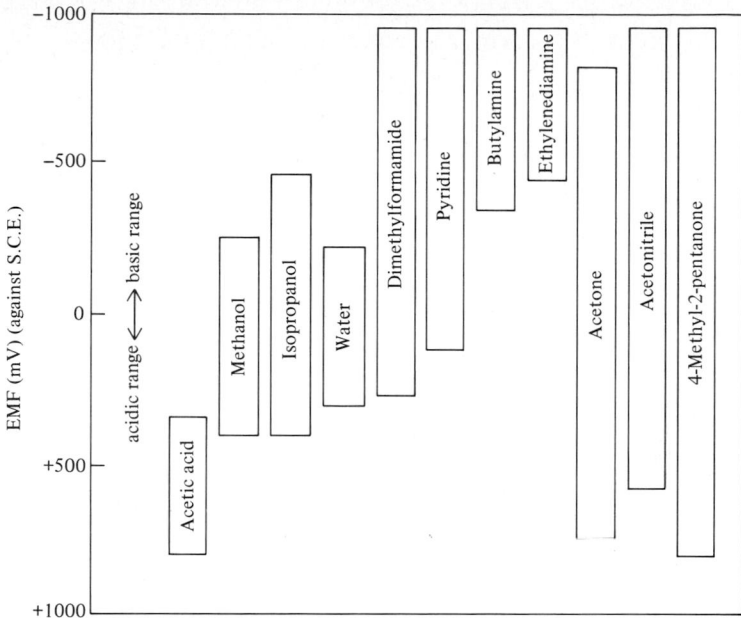

Fig. 5-1.   Approximate potential ranges in nonaqueous solvents.

Table 5-10
PROPERTIES OF COMMON ACID-BASE SOLVENTS

Solvent	Potential Span, mV	$-\log K_s$	Dielectric Constant, 25°C
Acetic acid	400	14.5	6.1(20°)
Acetic anhydride	800	14.5	20.7(20°)
Acetone	1600		20.7
Acetonitrile	1600	26.5	37.5(20°)
Ammonia (at −50°C)		33	22(−33°)
n-Butanol	900		17.1
n-Butylamine	500		4.88(20°)
Chlorobenzene	1500		5.62
N,N-Dimethylformamide	1300	18.0	36.71
Dimethylsulfoxide		17.3	46.6
Ethanol	800	19.1	24.55
Ethanolamine		5.1	37.7
Ethyl acetate	1500		6.02
Ethylenediamine	500	15.3	14.2(20°)
Formic acid	200	6.2	58.5
Methanol	800	16.7	32.7
4-Methyl-2-pentanone (methyl *iso*butyl ketone)	1600	25.0	13.1(20°)
Nitromethane	1000		35.8(30°)
2-Propanol	900		19.92
Pyridine	1000		12.3
Sulfuric acid		3.85	101
Water	800	14.0	78.3

Table 5-11
pK_a VALUES FOR PROTON-TRANSFER REACTIONS
IN NONAQUEOUS SOLVENTS

Acid	Methanol	Ethanol	
Acetic acid	9.52	10.32	11.4[a], 9.75[d]
p-Aminobenzoic acid	10.25		
Ammonium ion	10.7		6.40[b]
Anilinium ion	6.0	5.70	
Benzoic acid		10.72	10.0[a]
Bromocresol purple	11.3	11.5	
Bromocresol green	9.8	10.65	
Bromophenol blue	8.9	9.5	
Bromothymol blue	12.4	13.2	
di-n-Butylammonium ion			10.3[a]
o-Chloroanilinium ion	3.4		
Cyanoacetic acid		7.49	
2,5-Dichloroanilinium ion			9.48[b]
Dimethylaminoazobenzene		5.2	6.32[b]
N,N'-Dimethylanilinium ion		4.37	
Formic acid		9.15	
Hydrobromic acid			5.5[c]
Hydrochloric acid			8.55[b], 8.9[c]
Methyl orange	3.8	3.4	
Methyl red (acid range)	4.1	3.55	
(alkaline range)	9.2	10.45	
Methyl yellow	3.4	3.55	
Neutral red	8.2	8.2	
o-Nitrobenzoic acid	7.6		
m-Nitrobenzoic acid	8.3		
p-Nitrobenzoic acid	8.4		
Perchloric acid			4.87[b]
Phenol	14.0		
Phenol red	12.8	13.4	
Phthalic acid, pK_2	11.65		11.5[d], 6.10[d](pK_1)
Picric acid	3.8	3.8	8.9[c]
Pyridinium ion			6.1[b]
Salicylic acid	8.7	7.9	
Stearic acid	10.0		
Succinic acid, pK_2	11.4		
Sulfuric acid, pK_1			7.24[b,c]
Tartaric acid, pK_2	9.9		
Thymol blue (alkaline range)	14.0	15.2	
(acid range)	4.7	5.35	
Thymolbenzein (acid range)	3.5		
(alkaline range)	13.1		
p-Toluenesulfonic acid			8.44[b]
p-Toluidinium ion		6.24	
Tribenzylammonium ion			5.40[b]
Tropeoline 00	2.2		
Urea (protonated cation)			6.96[b]
Veronal	12.6		

[a] Dimethylsulfoxide. [b] Glacial acetic acid. [c] Acetonitrile. [d] Acetone + 10% water.

Table 5-12
CONSTANTS FOR COMPUTING HNP VALUES
FROM pK_a VALUES

Solvent	Constant a	Constant b	Applicable to Class
Acetic acid*	253	48.5	1,2
Acetic anhydride	520	52.3	1,2
Acetonitrile	805	77.5	1
	665	69.4	2
N,N-Dimethylformamide	601	144	3
	198	37.9	4
Ethyl acetate	664	63.5	1
	730	84.7	2
4-Methyl-2-pentanone	616	145	3
	904	122	4
Nitromethane	784	75.1	1
	635	76.9	2
Pyridine†	646	156	3
	1126	160	4

* All bases with pK_a values above 4.76 are leveled.
† All phenols with pK_a values below 4.1 are leveled.
Class 1. Monofunctional amines.
Class 2. Amides, amines containing oxygen functions, diamines, heterocyclic bases, and ureas.
Class 3. Aliphatic and m- and p-substituted aromatic carboxylic acids.
Class 4. Hydroxyaromatic compounds.

Table 5-13
H_0 FOR MINERAL ACIDS IN AQUEOUS SOLUTIONS AT 25°C

Acid Concentration, moles per liter	Acid						
	$HClO_4$	HBr	H_2SO_4	HCl	H_3PO_4	HNO_3	HF
0.1		+0.98	+0.83	+0.98	+1.45	+0.98	
0.25		+0.44	+0.55	+1.15	+0.55		
0.5	+0.18	+0.20	+0.13	+0.20	+0.97	+0.21	
1.0	−0.22	−0.20	−0.26	−0.20	+0.63	−0.18	+1.20
2.0	−0.78	−0.71	−0.84	−0.69	+0.24	−0.67	+0.91
3.0	−1.23	−1.11	−1.38	−1.05	−0.08	−1.02	+0.60
4.0	−1.72	−1.50	−1.85	−1.40	−0.37	−1.32	+0.40
5.0	−2.23	−1.93	−2.28	−1.76	−0.69	−1.57	+0.28
6.0	−2.84	−2.38	−2.76	−2.12	−1.04	−1.79	+0.15
7.0	−3.61	−2.85	−3.32	−2.50	−1.45	−1.99	+0.02
8.0	−4.33	−3.34	−3.87	−2.86	−1.85		−0.11
9.0	−5.05	−3.89	−4.40	−3.22	−2.22		−0.24
10.0	−5.79	−4.44	−4.89	−3.59	−2.59		−0.36

FORMATION CONSTANTS OF METAL COMPLEXES

Each value listed in Tables 5-14 and 5-15 is the logarithm of the overall formation constant for the cumulative binding of a ligand L to the central metal cation M, viz.:

	Cumulative Formation Constant	Stepwise Stability Constants
$M + L = ML$	K_1	k_1
$M + 2L = ML_2$	K_2	$k_1 k_2$
. . . .		
$M + nL = ML_n$	K_n	$k_1 k_2 \cdots k_n$

As an example, the entries in Table 5-14 for the zinc ammine complexes represent these equilibria:

$$Zn^{2+} + NH_3 = Zn(NH_3)^{2+} \qquad K_1 = \frac{[Zn(NH_3)^{2+}]}{[Zn^{2+}][NH_3]}$$

$$Zn^{2+} + 2NH_3 = Zn(NH_3)_2^{2+} \qquad K_2 = \frac{[Zn(NH_3)_2^{2+}]}{[Zn^{2+}][NH_3]^2}$$

$$Zn^{2+} + 3NH_3 = Zn(NH_3)_3^{2+} \qquad K_3 = \frac{[Zn(NH_3)_3^{2+}]}{[Zn^{2+}][NH_3]^3}$$

$$Zn^{2+} + 4NH_3 = Zn(NH_3)_4^{2+} \qquad K_4 = \frac{[Zn(NH_3)_4^{2+}]}{[Zn^{2+}][NH_3]^4}$$

If the stepwise stability or formation constants of the reactions are desired, for the first step $\log K_1 = \log k_1 = 2.37$. For the second and succeeding steps the equilibria and corresponding constants are as follows:

$Zn(NH_3)^{2+} + NH_3 = Zn(NH_3)_2^{2+} \qquad \log k_2 = \log K_2 - \log K_1 = 2.44$

$Zn(NH_3)_2^{2+} + NH_3 = Zn(NH_3)_3^{2+} \qquad \log k_3 = \log K_3 - \log K_2 = 3.50$

$Zn(NH_3)_3^{2+} + NH_3 = Zn(NH_3)_4^{2+} \qquad \log k_4 = \log K_4 - \log K_3 = 2.15$

The reverse of the association or formation reactions would represent the dissociation or instability constant for the systems, i.e., $-\log K_f = \log K_{instab}$.

The data in the tables generally refer to temperatures of about 20 to 25°C. Most of the values in Table 5-14 refer to zero ionic strength, but those in Table 5-15 often refer to a finite ionic strength.

Table 5-14
CUMULATIVE FORMATION CONSTANTS FOR METAL COMPLEXES WITH INORGANIC LIGANDS

	$\log K_1$	$\log K_2$	$\log K_3$	$\log K_4$	$\log K_5$	$\log K_6$
Ammonia						
Cadmium	2.65	4.75	6.19	7.12	6.80	5.14
Cobalt(II)	2.11	3.74	4.79	5.55	5.73	5.11
Cobalt(III)	6.7	14.0	20.1	25.7	30.8	35.2
Copper(I)	5.93	10.86				
Copper(II)	4.31	7.98	11.02	13.32	12.86	
Iron(II)	1.4	2.2				
Manganese(II)	0.8	1.3				
Mercury(II)	8.8	17.5	18.5	19.28		
Nickel	2.80	5.04	6.77	7.96	8.71	8.74
Platinum(II)						35.3
Silver(I)	3.24	7.05				
Zinc	2.37	4.81	7.31	9.46		
Bromide						
Astatine	2.51 [AtBr]					
Bismuth(III)	4.30	5.55	5.89	7.82		9.70
Bromine	1.24 [Br$_3^-$]					
Cadmium	1.75	2.34	3.32	3.70		
Cerium(III)	0.42					
Copper(I)		5.89				
Copper(II)	0.30					
Gold(I)		12.46				
Indium	1.30	1.88				
Iodine	2.64 [IBr]					
Iron(III)	−0.30	−0.50				
Lead	1.2	1.9		1.1		
Mercury(II)	9.05	17.32	19.74	21.00		
Palladium(II)				13.1		
Platinum(II)				20.5		
Rhodium(III)		14.3	16.3	17.6	18.4	17.2
Scandium	2.08	3.08				
Silver(I)	4.38	7.33	8.00	8.73		
Thallium(I)	0.93					
Thallium(III)	9.7	16.6	21.2	23.9	29.2	31.6
Tin(II)	1.11	1.81	1.46			
Uranium(IV)	0.18					
Yttrium	1.32					
Chloride						
Americium(III)	1.17					
Antimony(III)	2.26	3.49	4.18	4.72		
Bismuth(III)	2.44	4.7	5.0	5.6		
Cadmium	1.95	2.50	2.60	2.80		
Cerium(III)	0.48					
Copper(I)		5.5	5.7			
Copper(II)	0.1	−0.6				
Curium(III)	1.17					
Gold(III)		9.8				
Indium	1.42	2.23	3.23			
Iron(II)	0.36					
Iron(III)	1.48	2.13	1.99	0.01		
Lead	1.62	2.44	1.70	1.60		
Manganese(II)	0.96					

Table 5-14 (*Continued*)
CUMULATIVE FORMATION CONSTANTS FOR METAL COMPLEXES WITH INORGANIC LIGANDS

	$\log K_1$	$\log K_2$	$\log K_3$	$\log K_4$	$\log K_5$	$\log K_6$
Mercury(II)	6.74	13.22	14.07	15.07		
Palladium(II)	6.1	10.7	13.1	15.7		
Platinum(II)		11.5	14.5	16.0		
Plutonium(III)	1.17					
Silver(I)	3.04	5.04		5.30		
Thallium(I)	0.52					
Thallium(III)	8.14	13.60	15.78	18.00		
Thorium	1.38	0.38				
Tin(II)	1.51	2.24	2.03	1.48		
Tin(IV)						4
Uranium(IV)	0.8					
Uranium(VI)	0.22					
Zinc	0.43	0.61	0.53	0.20		
Zirconium	0.9	1.3	1.5	1.2		
Cyanide						
Cadmium	5.48	10.60	15.23	18.78		
Copper(I)		24.0	28.59	30.30		
Gold(I)		38.3				
Iron(II)						35
Iron(III)						42
Mercury(II)				41.4		
Nickel				31.3		
Silver(I)		21.1	21.7	20.6		
Zinc				16.7		
Fluoride						
Aluminum	6.10	11.15	15.00	17.75	19.37	19.84
Beryllium	5.1	8.8	12.6			
Cerium(III)	3.20					
Chromium(III)	4.41	7.81	10.29			
Gadolinium	3.46					
Gallium	5.08					
Indium	3.70	6.25	8.60	9.70		
Iron(III)	5.28	9.30	12.06			
Lanthanum	2.77					
Magnesium	1.30					
Manganese(II)	5.48					
Plutonium(III)	6.77					
Scandium						17.3
Thallium(I)	0.1					
Thallium(III) [TIO$^+$]	6.44					
Thorium	7.65	13.46	17.97			
Titanium(IV) [TiO^{2+}]	5.4	9.8	13.7	18.0		
Uranium(VI)	4.59	7.93	10.47	11.84		
Yttrium	4.81	8.54	12.14			
Zirconium	8.80	16.12	21.94			
Hydroxide						
Aluminum	9.27			33.03		
Antimony(III)		24.3	36.7	38.3		
Arsenic [as AsO$^+$]	14.33	18.73	20.60	21.20		
Beryllium	9.7	14.0	15.2			
Bismuth(III)	12.7	15.8		35.2		
Cadmium	4.17	8.33	9.02	8.62		

Table 5-14 (*Continued*)
CUMULATIVE FORMATION CONSTANTS FOR METAL COMPLEXES WITH INORGANIC LIGANDS

	$\log K_1$	$\log K_2$	$\log K_3$	$\log K_4$	$\log K_5$	$\log K_6$
Cerium(III)	4.6					
Cerium(IV)	13.28	26.46				
Chromium(III)	10.1	17.8		29.9		
Copper(II)	7.0	13.68	17.00	18.5		
Dysprosium	5.2					
Erbium(III)	5.4					
Gadolinium	4.6					
Gallium	11.0	21.7		34.3	38.0	40.3
Indium	9.9	19.8		28.7		
Iodine	9.49	11.24				
Iron(II)	5.56	9.77	9.67	8.58		
Iron(III)	11.87	21.17	29.67			
Lanthanum	3.3					
Lead(II)	7.82	10.85	14.58			61.0
Lutetium	6.6					
Magnesium	2.58					
Manganese(II)	3.90		8.3			
Neodymium	5.5					
Nickel	4.97	8.55	11.33			
Praseodymium	4.30					
Plutonium(III)	7.0					
Plutonium(IV)	12.39					
Plutonium [as PuO_2^{2+}]	8.3	16.6	20.9			
Samarium(III)	4.8					
Scandium	8.9					
Tellurium(IV)			41.6	53.0	64.8	72.0
Thallium(III)	12.86	25.37				
Titanium(III)	12.71					
Uranium(IV)	13.3				41.2	
Uranium(VI) [as UO_2^{2+}]	9.5	22.80		32.4		
Vanadium(III)	11.1	21.6				
Vanadium(IV) [as VO^{2+}]	8.6		[25.8 for $V_2O_4(OH)^-$]			
Vanadium(V) [as VO^{3+}]		25.2		46.2	58.5	
Yttrium	5.0					
Zinc	4.40	11.30	14.14	17.66		
Zirconium	14.3	28.3	41.9	55.3		
Iodide						
Bismuth	3.63			14.95	16.80	18.80
Cadmium	2.10	3.43	4.49	5.41		
Copper(I)		8.85				
Indium	1.00	2.26				
Iodine	2.89	5.79				
Iron(III)	1.88					
Lead	2.00	3.15	3.92	4.47		
Mercury(II)	12.87	23.82	27.60	29.83		
Silver	6.58	11.74	13.68			
Thallium(I)	0.72	0.90	1.08			
Thallium(III)	11.41	20.88	27.60	31.82		
Iodate						
Barium	1.05					
Calcium	0.89					
Magnesium	0.72					

Table 5-14 (*Continued*)
CUMULATIVE FORMATION CONSTANTS FOR METAL COMPLEXES WITH INORGANIC LIGANDS

	$\log K_1$	$\log K_2$	$\log K_3$	$\log K_4$	$\log K_5$	$\log K_6$
Strontium	1.00					
Thorium	2.88	4.79	7.15			
Nitrate						
Barium	0.92					
Beryllium	1.62					
Bismuth(III)	1.26					
Cadmium	0.40					
Calcium	0.28					
Cerium(III)	1.04	2.55				
Curium(III)	0.57					
Hafnium	0.92	2.43	4.32	6.40	8.48	10.29
Iron(III)	1.0					
Lanthanum	0.26	0.69	1.27			
Lead	1.18					
Mercury(II)	0.35					
Neodymium	0.52	1.18				
Neptunium(IV)	0.38					
Plutonium(III)	0.77	1.93	3.09			
Plutonium(IV)	0.54					
Strontium	0.82					
Thallium(I)	0.33					
Thallium(III)	0.92					
Thorium	0.78	1.89	2.89	3.63		
Uranium(IV)	0.20	0.37				
Uranium(VI)	0.34	0.45				
Ytterbium	0.45	1.30	2.42			
Zirconium [as ZrO^{2+}]		1.91		3.54		
Pyrophosphate						
Barium	4.6					
Calcium	4.6					
Cadmium	5.6					
Copper(II)	6.7	9.0				
Lead		5.3				
Magnesium	5.7					
Nickel	5.8	7.4				
Strontium	4.7					
Yttrium		9.7				
Zirconium		6.5				
Sulfate						
Cerium(III)	3.40					
Erbium	3.58					
Gadolinium	3.66					
Holmium	3.58					
Indium	1.78	1.88	2.36			
Iron(III)	2.03	2.98				
Lanthanum	3.64					
Neodymium	3.64					
Nickel	2.4					
Plutonium(IV)	3.66					
Praseodymium	3.62					
Samarium	3.66					
Thorium	3.32	5.50				

Table 5-14 (*Continued*)
CUMULATIVE FORMATION CONSTANTS FOR METAL COMPLEXES WITH INORGANIC LIGANDS

	$\log K_1$	$\log K_2$	$\log K_3$	$\log K_4$	$\log K_5$	$\log K_6$
Uranium(IV)	3.24	5.42				
Uranium(VI)	1.70	2.45	3.30			
Yttrium	3.47					
Ytterbium	3.58					
Zirconium	3.79	6.64	7.77			
Sulfite						
Copper(I)	7.5	8.5	9.2			
Mercury(II)		22.66				
Silver	5.30	7.35				
Thiocyanate						
Bismuth	1.15	2.26	3.41	4.23		
Cadmium	1.39	1.98	2.58	3.6		
Chromium(III)	1.87	2.98				
Cobalt(II)	−0.04	−0.70	0	3.00		
Copper(I)	12.11	5.18				
Gold(I)		23		42		
Indium	2.58	3.00	4.63			
Iron(III)	2.95	3.36				
Mercury(II)		17.47		21.23		
Nickel	1.18	1.64	1.81			
Ruthenium(III)	1.78					
Silver		7.57	9.08	10.08		
Thallium(I)	0.80					
Uranium(IV)	1.49	2.11				
Uranium(VI)	0.76	0.74	1.18			
Vanadium(III)	2.0					
Vanadium(IV)	0.92					
Zinc	1.62					
Thiosulfate						
Cadmium	3.92	6.44				
Copper(I)	10.27	12.22	13.84			
Iron(III)	2.10					
Lead		5.13	6.35			
Mercury(II)		29.44	31.90	33.24		
Silver	8.82	13.46				

Table 5-15
CUMULATIVE FORMATION CONSTANTS FOR METAL COMPLEXES WITH ORGANIC LIGANDS

Temperature is 25°C, and ionic strengths are approaching zero unless indicated otherwise: (*a*) At 20°C, (*b*) at 30°C, (*c*) 0.1*M* uni-univalent salt, (*d*) 1.0*M* uni-univalent salt, (*e*) 2.0*M* uni-univalent salt present.

	$\log K_1$	$\log K_2$	$\log K_3$	$\log K_4$
Acetate				
Ag(I)	0.73	0.64		
Ba(II)	0.41			
Ca(II)	0.6			
Cd(II)	1.5	2.3	2.4	
Ce(III)	1.68	2.69	3.13	3.18
Co(II)	1.5	1.9		
Cr(III)	1.80	4.72		
Cu(II) *a*	2.16	3.20		
Fe(II) *c*	3.2	6.1	8.3	
Fe(III) *a,d*	3.2			
In(III)	3.50	5.95	7.90	9.08
Hg(II)		8.43		
La(III) *a,e*	1.56	2.48	2.98	2.95
Mg(II)	0.8			
Mn(II)	9.84	2.06		
Ni(II)	1.12	1.81		
Pb(II)	2.52	4.0	6.4	8.5
Rare earths *a,e*	1.6–1.9	2.8–3.0	3.3–3.7	
Sr(II)	0.44			
Tl(III)				15.4
UO$_2$(II) *a,e*	2.38	4.36	6.34	
Y(III) *a,e*	1.53	2.65	3.38	
Zn(II)	1.5			
Acetylacetone				
Al(III) *b*	8.6	15.5		
Be(II)	7.8	14.5		
Cd(II)	3.84	6.66		
Ce(III)	5.30	9.27	12.65	
Cr(II)	5.9	11.7		
Co(II)	5.40	9.54		
Cu(II)	8.27	16.34		
Dy(III) *b*	6.03	10.70	14.04	
Er(III) *b*	5.99	10.67	14.09	
Eu(III) *b*	5.87	10.35	13.64	
Fe(II)	5.07	8.67		
Fe(III)	11.4	22.1	26.7	
Ga(III)	9.5	17.9	23.6	
Gd(III) *b*	5.90	10.38	13.79	
Hf(IV)	8.7	15.4	21.8	28.1
Ho(III)	6.05	10.73	14.13	
In(III)	8.0	15.1		
La(III) *b*	5.1	8.90	11.90	
Lu(III) *b*	6.23	11.00	13.63	
Mg(II)	3.65	6.27		
Mn(II)	4.24	7.35		
Mn(III)			3.86	
Nd(III)	5.6	9.9	13.1	
Ni(II) *a*	6.06	10.77	13.09	
Pd(II) *b*	16.2	27.1		
Pr(III) *b*	5.4	9.5	12.5	
Pu(IV) *c*	10.5	19.7	28.1	34.1
Sc(III) *b*	8.0	15.2		

Table 5-15 (*Continued*)
CUMULATIVE FORMATION CONSTANTS FOR METAL COMPLEXES WITH ORGANIC LIGANDS

	$\log K_1$	$\log K_2$	$\log K_3$	$\log K_4$
Sm(III) *b*	5.9	10.4		
Tb(III) *b*	6.02	10.63	14.04	
Th(IV)	8.8	16.2	22.5	26.7
Tm(IV) *b*	6.09	10.85	14.33	
U(IV) *a,c*	8.6	17.0	23.4	29.5
UO_2(II) *b*	7.74	14.19		
VO(II)	8.68	15.79		
V(II)	5.4	10.2	14.7	
Y(III) *b*	6.4	11.1	13.9	
Yb(III) *b*	6.18	11.04	13.64	
Zn(II) *b*	4.98	8.81		
Zr(IV)	8.4	16.0	23.2	30.1
Alizarin red				
Cr(VI)	4.7			
Cu(II)	4.1			
Hf(IV)		10.4		
Mo(VI)		9.6		
Pb(II)	6.0			
Th(IV)		8.24		
UO_2(II)	4.22			
V(V)		8.6		
W(VI)		7.8		
Arsenazo				
Hf(IV)	10.07			
Zr(IV)	12.95			
Aurintricarboxylic acid				
Be(II)	4.54			
Cu(II)	4.1	8.81		
Fe(III)	4.68			
Th(IV)	5.04			
UO_2(II)	4.77			
Benzoylacetone (75% dioxane)				
Ba(II)		9.4		
Be(II)	12.59	24.01		
Cd(II)	7.79	14.36		
Ce(III)	10.09	19.42	27.04	
Co(II)	9.42	17.83		
Cu(II)	12.05	23.01		
La(III)	6.33	11.66	16.78	
Mg(II)	7.69	14.09		
Mn(II)	8.66	15.78		
Ni(II)	9.58	18.00		
Pb(II)	8.84	16.35		
Pr(III)	7.02	13.62	18.74	
UO_2(II)	12.15	23.27		
Y(III)	8.24	14.98	20.57	
Zn(II)	9.62	17.90		
Calmagite				
Ca	6.05			
Mg	8.05			

Table 5-15 (*Continued*)
CUMULATIVE FORMATION CONSTANTS FOR METAL COMPLEXES WITH ORGANIC LIGANDS

	Complex of HL^{2-} Anion		Complex of L^{3-} Anion		Complex of H_2L^-
	$\log K_1$	$\log K_2$	$\log K_1$	$\log K_2$	$\log K_3$
Citric acid					
Ag	7.1				
Al	7.0		20.0		
Ba	2.98				
Be	4.52				
Ca	4.68				
Cd	3.98		11.3		
Ce(III)		6.18		9.65	3.2
Co(II)	4.8		12.5		
Cu(II)	4.35		14.2		
Eu(III)		6.46		9.80	
Fe(II)	3.08		15.5		
Fe(III)	12.5		25.0		
La		6.97		9.45	6.22
Mg	3.29				
Mn(II)	3.67				
Nd(III)		6.32		9.70	
Ni	5.11		14.3		
Pb	6.50				
Pr					3.4
Ra	2.36				
Sr	2.8				
Tl(I)	1.04				
UO_2	8.5	10.8			
Y					3.6
Yb				8	
Zn	4.71		11.4		

	$\log K_1$	$\log K_2$	$\log K_3$		
1,2-Diaminocyclohexane-*N,N,N′,N′*-tetraacetic acid					
Al *c*	17.63				
Ba *c*	8.64				
Ca *c*	12.3				
Cd *c*	19.88				
Ce(III) *c*	16.76				
Co(II) *c*	19.57				
Cu(II) *c*	21.95				
Dy(III) *c*	19.69				
Er(III) *c*	20.20				
Eu(III) *c*	18.77				
Fe(III) *c*	27.48				
Ga *c*	22.91				
Gd *c*	18.80				
Hg(II) *c*	24.4				
Ho *c*	19.89				
La *c*	16.35				
Lu *c*	21.51				
Mg *c*	10.41				
Mn(II) *c*	17.43				
Nd *c*	17.69				
Ni *c*	19.4				
Pb *c*	20.33				
Pr *c*	17.23				

Table 5-15 (*Continued*)
CUMULATIVE FORMATION CONSTANTS FOR METAL COMPLEXES WITH ORGANIC LIGANDS

	$\log K_1$	$\log K_2$	$\log K_3$	$\log K_4$
Sm(III) *c*	18.63			
Sr *c*	8.92			
Tb *c*	19.30			
Tm *c*	20.46			
VO(II) *c*	19.40			
Y *c*	19.41			
Yb *c*	20.80			
Zn *c*	18.6			
Dibenzoylmethane (75% dioxane)				
Ba	6.10	11.50		
Be	13.62	26.03		
Ca	7.17	13.55		
Cd	8.67	16.63		
Ce(III)	10.99	21.53	30.38	
Co(II)	10.35	20.05		
Cu(II)	12.98	24.98		
Cs	3.42			
Fe(II)	11.15	21.50		
K	3.67			
Li	5.95			
Mg	8.54	16.21		
Mn(II)	9.32	17.79		
Na	4.18			
Ni	10.83	20.72		
Pb	9.75	18.79		
Rb	3.52			
Sr	6.40	12.10		
Zn	10.23	19.65		

	$\log K_1$	$\log K_2$	$\log K_3$	$\log K_f$ [MHL]
4,5-Dihydroxybenzene-1,3-disulfonic acid (Tiron)				
Al	19.02	31.10	33.5	
Ba	4.10			14.6
Ca	5.80			14.8
Cd *d*	7.69	13.29		
Ce(III)		3.75		
Co(II) *d*	8.19	14.41		15.7
Cu(II) *d*	12.76	23.73		18.1
Fe(III) *a,c*	20.7	35.9	46.9	22.6
La	12.9			18.6 [La(OH)L]
Mg *a,c*	6.86			14.6
Mn(II) *c*	8.6			
Ni *a,c*	8.56	14.90		15.6
Pb *d*	11.95	18.28		
Sr *c*	4.55			
UO$_2$(II) *c*	15.90			
VO(II)	15.88			
Zn *d*	9.00	16.91		15.9

	$\log K_1$	$\log K_2$	$\log K_f$ [M$_2$L$_3$]
2,3-Dimercaptopropan-1-of (BAL)			
Fe(II)	15.8		
Fe(III)	30.6 [Fe(OH)L]		28
Mn(II)	5.23	10.43	
Ni		22.78	
Zn	13.48	23.3	40.6

Table 5-15 (*Continued*)
CUMULATIVE FORMATION CONSTANTS FOR METAL COMPLEXES WITH ORGANIC LIGANDS

	$\log K_1$	$\log K_2$	$\log K_3$	$\log K_4$
Dimethylglyoxime (50% dioxane)				
Cd	5.7	10.7		
Co(II)	9.80	18.94		
Cu(II)	12.00	33.44		
Fe(II)		7.25		
La	6.6	12.5		
Ni	11.16			
Pb	7.3			
Zn	7.7	13.9		
2,2′-Dipyridyl				
Ag	3.65	7.15		
Cd	4.26	7.81	10.47	
Co(II)	5.73	11.57	17.59	
Cr(II)	4.5	10.5	14.0	
Cu(I)		14.2		
Cu(II)	8.0	13.60	17.08	
Fe(II)	4.36	8.0	17.45	
Hg(II)	9.64	16.74	19.54	
Mg	0.5			
Mn(II) *d*	4.06	7.84	11.47	
Ni	6.80	13.26	18.46	
Pb	3.0			
Ti(III)			25.28	
V(II)	4.9	9.6	13.1	
Zn	5.30	9.83	13.63	
Eriochrome Black T				
Ca	5.4			
Mg	7.0			
Zn	13.5	20.6		
Ethanolamine				
Ag	3.29	6.92		
Cu(II)		6.68		16.48
Hg(II)	8.51	17.32		
Ethylenediamine				
Ag	4.70	7.70		
Cd *a*	5.47	10.09	12.09	
Co(II)	5.91	10.64	13.94	
Co(III)	18.7	34.9	48.69	
Cr(II)	5.15	9.19		
Cu(I)		10.8		
Cu(II)	10.67	20.00	21.0	
Fe(II)	4.34	7.65	9.70	
Hg(II)	14.3	23.3		
Mg	0.37			
Mn(II)	2.73	4.79	5.67	
Ni	7.52	13.84	18.33	
Pd(II)		26.90		
V(II)	4.6	7.5	8.8	
Zn	5.77	10.83	14.11	
Ethylenediamine-*N,N,N′,N′*-tetraacetic acid				
Ag	7.32			
Al	16.11			
Am(III)	18.18			
Ba	7.78			
Be	9.3			
Bi	22.8			
Ca	11.0			
Cd	16.4			
Ce(III)	16.80			

Table 5-15 (*Continued*)
CUMULATIVE FORMATION CONSTANTS FOR METAL COMPLEXES WITH ORGANIC LIGANDS

	$\log K_1$	$\log K_2$	$\log K_3$	$\log K_4$
Cf(III)	19.09			
Cm(III)	18.45			
Co(II)	16.31			
Co(III)	36			
Cr(II)	13.6			
Cr(III)	23			
Cu(II)	18.7			
Dy	18.0			
Er	18.15			
Eu(III)	17.99			
Fe(II)	14.33			
Fe(III)	24.23			
Ga	20.25			
Gd	17.2			
Hg(II)	21.80			
Ho	18.1			
In	24.95			
La	16.34			
Li	2.79			
Lu	19.83			
Mg	8.64			
Mn(II)	13.8			
Mo(V)	6.36			
Na	1.66			
Nd	16.6			
Ni	18.56			
Pb	18.3			
Pd(II)	18.5			
Pm(III)	17.45			
Pr	16.55			
Pu(III)	18.12			
Pu(IV)	17.66			
Pu(VI)	17.66			
Ra	7.4			
Sc	23.1			
Sm	16.43			
Sn(II)	22.1			
Sr	8.80			
Tb	17.6			
Th	23.2			
Ti(III)	21.3			
TiO(II)	17.3			
Tl(III)	22.5			
Tm	19.49			
U(IV)	17.50			
V(II)	12.70			
V(III)	25.9			
VO(II)	18.0			
V(V)	18.05			
Y	18.32			
Yb	18.70			
Zn	16.4			
Zr	19.40			
Glycine				
Ag	3.41	6.89		
Ba	0.77			
Be		4.95		

Table 5-15 (*Continued*)
CUMULATIVE FORMATION CONSTANTS FOR METAL COMPLEXES WITH ORGANIC LIGANDS

	$\log K_1$	$\log K_2$	$\log K_3$	$\log K_4$
Ca	1.38			
Cd	4.74	8.60		
Co(II)	5.23	9.25	10.76	
Cu(II)	8.60	15.54	16.27	
Dy		12.2		
Er		12.7		
Fe(II) *a*	4.3	7.8		
Fe(III) *a,d*	10.0			
Gd		11.9		
Hg(II)	10.3	19.2		
La		11.2		
Mg	3.44	6.46		
Mn(II)	3.6	6.6		
Ni	6.18	11.14	15	
Pb	5.47	8.92		
Pd(II)	9.12	17.55		
Pr		11.5		
Sm		11.7		
Sr	0.91			
Y		12.5		
Yb		13.0		
Zn	5.52	9.96		

N'-(2-Hydroxyethyl)ethylenediamine-*N,N,N'*-triacetic acid

	$\log K_1$	$\log K_2$	$\log K_3$	$\log K_4$
Ba *c*	5.54			
Ca *c*	8.43			
Cd *c*	13.0			
Ce(III) *c*	14.11			
Co(II) *c*	14.4			
Cu(II) *c*	17.40			
Dy *c*	15.30			
Er *c*	15.42			
Eu(III) *c*	15.35			
Fe(II) *c*	11.6			
Fe(III) *c*	19.8			
Gd *c*	15.22			
Hg(II) *c*	20.1			
Ho *c*	15.32			
La *c*	13.46			
Lu *c*	15.88			
Mg *c*	5.78			
Mn(II) *c*	10.7			
Nd *c*	14.86			
Ni *c*	17.0			
Pb *c*	15.5			
Pr *c*	14.61			
Sm *c*	15.28			
Sr *c*	6.92			
Tb *c*	15.32			
Th *c*	18.5			
Tm *c*	15.59			
Y *c*	14.65			
Yb *c*	15.88			
Zn *c*	14.5			

8-Hydroxy-2-methylquinoline (50% dioxane)

	$\log K_1$	$\log K_2$	$\log K_3$	$\log K_4$
Cd	9.00	9.00	16.60	
Ce(III)	7.71			
Co(II)	9.63	18.50		

Table 5-15 (*Continued*)
CUMULATIVE FORMATION CONSTANTS FOR METAL COMPLEXES WITH ORGANIC LIGANDS

	$\log K_1$	$\log K_2$	$\log K_3$	$\log K_4$
Cu(II)	12.48	24.00		
Fe(II)	8.75	17.10		
Mg	5.24	9.64		
Mn(II)	7.44	13.99		
Ni	9.41	17.76		
Pb	10.30	18.50		
UO$_2$(II)	9.4	17		
Zn	9.82	18.72		
8-Hydroxyquinoline-5-sulfonic acid				
Ba	2.31			
Ca	3.52			
Cd	7.70	14.20		
Ce(III)	6.05	11.05	14.95	
Co(II)	8.11	15.05	20.41	
Cu(II)	11.92	21.87		
Er	7.16	13.34	18.56	
Fe(II)	8.4	15.7	21.75	
Fe(III)	11.6	22.0	35.65	
Gd	6.64	12.37	17.27	
La	5.63	10.13	13.83	
Mg	4.79	8.19		
Mn(II)	5.67	10.72		
Nd	6.3	11.6	16.0	
Ni	9.57	18.27	22.9	
Pb	8.53	16.13		
Pr	6.17	11.37	15.67	
Sm	6.58	12.28	17.04	
Sr	2.75			
Th	9.56	18.29	25.92	32.04
UO$_2$(II)	8.52	15.67		
Zn	8.65	16.15		
Lactic acid				
Ba	0.64			
Ca	1.42			
Cd	1.70			
Ce(III) a,c	2.76	4.73	5.96	
Co(II)	1.90			
Cu(II)	3.02	4.85		
Er	2.77	5.11	6.70	
Eu(III)	2.53	4.60	5.88	
Fe(III)	7.1			
Gd	2.53	4.63	5.91	
Ho	2.71	4.97	6.55	
La a,c	2.60	4.34	5.64	
Li	0.20			
Mg	1.37			
Mn(II)	1.43			
Nd	2.47	4.37	5.60	
Ni	2.22			
Pb	2.40	3.80		
Pr a,c	2.85	4.90	6.10	
Rare earths a,c	2.8–3.0	4.9–5.4	6.1–7.8	
Sm	2.56	4.58	5.90	
Sr	0.98			
Tb	2.61	4.73	6.01	
Y	2.53	4.70	6.12	
Yb	2.85	5.27	7.96	
Zn	2.20	3.75		

Table 5-15 (*Continued*)
CUMULATIVE FORMATION CONSTANTS FOR METAL COMPLEXES WITH ORGANIC LIGANDS

	$\log K_1$	$\log K_2$	$\log K_3$	$\log K_4$
Nitrilotriacetic acid				
Al	>10			
Ba *a*	5.88			
Ca	7.60	11.61		
Cd *c*	9.80	15.2		
Ce(III) *c*	10.83	18.67		
Co(II) *c*	10.38	14.5		
Cr(III)	>10			
Cu(II) *c*	13.10			
Dy *c*	11.74	21.15		
Er *c*	12.03	21.29		
Eu(III) *c*	11.52	20.70		
Fe(II) *c*	8.84			
Fe(III) *c*	15.87	24.32		
Gd *c*	11.54	20.80		
Hg(II)	12.7			
Ho *c*	11.90	21.25		
In	15			
La *c*	10.36	17.60		
Li *a*	3.28			
Lu *c*	12.49	21.91		
Mg *c*	5.36	10.2		
Mn(II)	8.60	11.1		
Na	2.15			
Nd *c*	11.26	19.73		
Ni	11.26	16.0		
Pb *a,c*	11.8			
Pr *c*	11.07	19.25		
Sm(III) *c*	11.53	20.53		
Sr	6.73			
Tb *c*	11.59	20.97		
Tl(I)	3.44			
Th *c*	12.4			
Tm *c*	12.22	21.45		
Y *c*	11.48	20.43		
Yb *c*	12.40	21.69		
Zn *c*	10.45	13.45		
Zr *c*	20.8			
1-Nitroso-2-naphthol (75% dioxane)				
Ag	7.74			
Cd	6.18	11.38		
Co(II)	10.67	22.81		
Cu(II)	12.52	23.37		
Mg	6.2	10.60		
Nd	9.5	17.7	25.6	
Ni	10.75	21.29	28.09	
Pb	9.73	17.31		
Pr	9.04	17.06	23.85	
Th *c*	8.50	16.13	24.03	30.29
Y	9.02	17.74	25.04	
Zn	9.32	17.02		
Zr	3.6			
Oxalate				
Ag	2.41			
Al	7.26	13.0	16.3	
Am(III)		9.8		[Am(HL)$_4^-$ 11.0]
Ba	2.31			
Be	4.90			

Table 5-15 (*Continued*)
CUMULATIVE FORMATION CONSTANTS FOR METAL COMPLEXES WITH ORGANIC LIGANDS

	$\log K_1$	$\log K_2$	$\log K_3$	$\log K_4$
Ca	3.0			
Cd	3.52	5.77		
Ce(III)	6.52	10.5	11.3	
Co(II)	4.79	6.7	9.7	
Co(III)			~20	
Cu(II)	6.16	8.5		
Er	4.82	8.21	10.03	
Fe(II)	2.9	4.52	5.22	
Fe(III)	9.4	16.2	20.2	
Gd	7.04			
Hg(II)		6.98		
Mg	3.43	4.38		
Mn(II)	3.97	5.80		
Mn(III) *e*	9.98	16.57	19.42	
Mo(III)	3.38			
Mo(VI)				[$MoO_3(L)^{2-}$ 13.0]
Nd	7.21	11.5	>14	
Ni	5.3	7.64	~8.5	
NpO_2(II)	3.30	7.07		
Pb		6.54		
Pu(III)	9.31	18.70	28	
Pu(IV)	8.74	16.91	23.39	27.50
PuO_2(II)		11.4		
Sr	2.54			
Th				24.48
TiO(II)	2.67			
Tl(I)	2.03			
UO_2(II)		10.57		
VO(II)		9.80		
V(II)	~2.7			
Y	6.52	10.10	11.47	
Yb	7.30	11.7	>14	
Zn	4.89	7.60	8.15	
Zr	9.80	17.14	20.86	21.15
1,10-Phenanthroline				
Ag	5.02	12.07		
Ca	0.7			
Cd	5.93	10.53	14.31	
Co(II)	7.25	13.95	19.90	
Cu(II)	9.08	15.76	20.94	
Fe(II)	5.85	11.45	21.3	
Fe(III)	6.5	11.4	23.5	
Hg(II)		19.65	23.35	
Mg	1.2			
Mn(II)	3.88	7.04	10.11	
Ni	8.80	17.10	24.80	
Pb	4.65	7.5	9	
VO(II)	5.47	9.69		
Zn	6.55	12.35	17.55	
Phthalic acid				
Ba	2.33			
Ca	2.43			
Cd	2.5			
Co(II)	1.81	4.51		
Cu(II)	3.46	4.83		
La		7.74		
Ni	2.14			
Pb *d*	3.4			

Table 5-15 (*Continued*)
CUMULATIVE FORMATION CONSTANTS FOR METAL COMPLEXES WITH ORGANIC LIGANDS

	$\log K_1$	$\log K_2$	$\log K_3$	$\log K_4$
UO$_2$(II)	4.38			
Zn	2.2			
Piperidine				
Ag	3.30	6.48		
Hg(II)	8.70	17.44		
Pt(II)			$\log K_5$ 5.7	$\log K_6$ 8.2
Propylene-1,2-diamine				
Cd *b,c*		9.97	12.12	
Co(II) *d*	5.42	11.47	14.72	
Cu(II) *c*	6.41	20.06		
Hg(II) *c*	10.78	23.53	23.25	
Ni *d*	7.43	13.62	17.89	
Zn *b,c*	5.89	10.87	12.57	
Pyridine				
Ag	1.97	4.35		
Cd	1.40	1.95	2.27	2.50
Co(Ii)	1.14	1.54		
Cu(I)		3.34	4.51	5.44
				$\log K_6$ 6.89
Cu(II)	2.59	4.33	5.93	6.54
			$\log K_5$ 7.00	$\log K_6$ 10.2
Fe(II)	0.71			
Hg(II)	5.1	10.0	10.4	
Mn(II)	1.92	2.77	3.37	3.50
VO(II)	−1.70			
Zn	1.41	1.11	1.61	1.93
Pyridine-2,6-dicarboxylic acid				
Ba *a,d*	3.46			
Ca *a,d*	4.6	7.2		
Cd *a,d*	5.7	10.0		
Ce(III) *a,d*	8.34	14.42	18.80	
Co(II) *a,d*	7.0	12.5		
Cu(II) *a,d*	9.14	16.52		
Dy *a,d*	8.69	16.19	22.14	
Er *a,d*	8.77	16.39	22.14	
Eu(III) *a,d*	8.84	15.98	21.00	
Fe(II) *a,d*	5.71	10.36		
Fe(III) *a,d*	10.91	17.13		
Gd *a,d*	8.74	16.06	21.83	
Ho *a,d*	8.72	16.23	22.08	
La *a,d*	7.98	13.79	18.06	
Lu *a,d*	9.03	16.80	21.48	
Hg(II) *a,d*	20.28			
Mg *a,d*	2.7			
Mn(II) *a,d*	5.01	8.49		
Nd *a,d*	8.78	15.60	20.66	
Ni *a,d*	6.95	13.50		
Pb *a,d*	8.70	10.60		
Pr *a,d*	8.63	15.10	19.94	
Sm *a,d*	8.86	15.88	21.23	
Sr *a,d*	3.89			
Tb *a,d*	8.68	16.11	22.03	
Tm *a,d*	8.83	16.54	22.04	
Y *a,d*	8.46	15.73	21.34	
Yb *a,d*	8.85	16.61	21.83	
Zn *a,d*	6.35	11.88		

Table 5-15 (*Continued*)
CUMULATIVE FORMATION CONSTANTS FOR METAL COMPLEXES WITH ORGANIC LIGANDS

	$\log K_1$	$\log K_2$	$\log K_3$	$\log K_4$
1-(2-Pyridylazo)-2-naphthol (PAN)				
Co(II)	>12			
Cu(II)	16			
Mn(II)	8.5	16.4		
Ni	12.7	25.3		
Tl(III)	2.29			
Zn	11.2	21.7		

	$\log K_f$ [ML]	$\log K_f$ [MHL]	$\log K_f$ [M(HL)$_2$]
4-(2-Pyridylazo)resorcinal (PAR)			
Co(II)		>12	
Cu(II)	10.3		
Mn(II)		9.7	18.9
Ni		13.2	26.0
Sc	4.8		
Tl(III)	4.23		
Zn		12.4	23.5

	$\log K_f$ [ML]	$\log K_f$ [M$_2$L]	$\log K_f$ [MHL]
Pyrocatechol-3,5-disulfonate (Pyrocatechol Violet)			
Al	19.13	4.95	
Bi	27.07	5.25	
Cd	8.13		5.86
Co(II)	9.01		6.53
Cu(II)	16.47		11.18
Ga	22.18	4.65	
In	18.10	4.81	
Mg	4.42	4.6	3.66
Mn(II)	7.13		5.36
Ni	9.35	4.38	6.85
Pb	13.25		10.19
Th	23.36	4.42	
Zn	10.41	6.21	7.21
Zr	27.40	4.18	

	$\log K_1$	$\log K_2$	$\log K_3$	$\log K_4$
8-Quinolinol				
Ba	2.07			
Be	3.36			
Ca (75% dioxane)	7.3	13.2		
Cd	7.2	13.4		
Ce(III) (50% dioxane)	9.15	17.13		
Co(II)	9.1	17.2		
Cu(II)	12.2	23.4		
Fe(II)	8.58	16.93	22.23	
Fe(III)	12.3	23.6	33.9	
La	5.85	16.95		
Mg (50% dioxane)	6.38	11.81		
Mn(II) (50% dioxane)	8.28	15.45		
Ni (50% dioxane)	11.44	21.38		
Pb (50% dioxane)	10.61	18.70		

Table 5-15 (*Continued*)
CUMULATIVE FORMATION CONSTANTS FOR METAL COMPLEXES WITH ORGANIC LIGANDS

	$\log K_1$	$\log K_2$	$\log K_3$	$\log K_4$
Sm	6.84		19.50	
Sr	2.89	6.08		
Th	10.45	20.40	29.85	38.80
UO$_2$(II) (50% dioxane)	11.25	20.89		
V(II)	12.8	23.6		
VO(II)	10.97	20.19		
Y	8.15	14.90	20.25	
Zn (50% dioxane)	9.96	18.86		

	$\log K_f$ [MHL$^+$]	$\log K_f$ [M(HL)$_2$]
Salicylaldoxime		
Ba	0.53	3.72
Be	<7	
Ca	0.92	3.72
Cd	<4.4	
Co(II)		8.13
Cu(II)		8.13
Mg	0.64	4.10
Ni		3.77
Sr		3.77
Zn	<5.2	

	$\log K_1$	$\log K_2$	$\log K_3$	$\log K_4$
Salicylic acid				
Al	14.11			
Be	17.4			
Cd	5.55			
Ce(III)	2.66			
Co(II)	6.72	11.42		
Cr(II)	8.4	15.3		
Cu(II)	10.60	18.45		
Fe(II)	6.55	11.25		
Fe(III) *a,c*	16.48	28.12	36.80	
La	2.64			
Mg (75% dioxane)	4.7			
Mn(II)	5.90	9.80		
Nd	2.70			
Ni	6.95	11.75		
Pr	2.68			
Th	4.25	7.60	10.05	11.60
TiO(II)	6.09			
UO$_2$(II)	13.4			
V(II)	6.3			
Zn	6.85			
Succinic acid				
Ba	2.08			
Be	3.08			
Ca	2.0			
Cd	2.2			
Co(II)	2.22			
Cu(II)	3.33			
Fe(III)	7.49			
Hg(II)		7.28		
La	3.96			

Table 5-15 (*Continued*)
CUMULATIVE FORMATION CONSTANTS FOR METAL COMPLEXES WITH ORGANIC LIGANDS

	$\log K_1$	$\log K_2$	$\log K_3$	$\log K_4$
Mg	1.20			
Mn(II)	2.26			
Nd	8.1			
Ni	2.36			
Pb	2.8			
Ra	1.0			
Sr	1.06			
Zn	1.6			
5-Sulfosalicylic acid				
Al *c*	13.20	22.83	28.89	
Be *c*	11.71	20.81		
Cd *c*	16.68	29.08		
Co(II) *c*	6.13	9.82		
Cr(II) *c*	7.1	12.9		
Cr(III) *c*	9.56			
Cu(II) *c*	9.52	16.45		
Fe(II) *c*	5.90			
Fe(III) *c*	14.64	25.18	32.12	
La *c*	9.11			
Mn(II) *c*	5.24	8.24		
NbO(III) *c*	4.0	7.7		
Ni *c*	6.42	10.24		
UO$_2$(II) *c*	11.14	19.20		
Zn *c*	6.05	10.65		
Tartaric acid				
Ba		1.62		
Bi			8.30	
Ca	2.98	9.01		
Cd	2.8			
Co(II)	2.1			
Cu(II)	3.2	5.11	4.78	6.51
				$\log K_f$ 19.14 [Cu(OH)$_2$L^{2-}]
Eu(III)	4.98	8.11		
Fe(III)	7.49			
La	3.06			
Mg		1.36		
Nd	9.0			
Pb	3.78		4.7	$\log K_f$ 14.1 [Pb(OH)$_2$L^{2-}]
Ra	1.24			
Sr	1.60			
Zn	2.68	8.32		
Thioglycolic acid				
Ce(III) *a,c*	1.99	3.03		
Co(II)	5.84	12.15		
Fe(II)		10.92		
Hg(II)		43.82		
La *a,c*	1.98	2.98		
Mn(II)	4.38	7.56		
Pb	8.5			
Ni	6.98	13.53		
Rare earths *a,c*	1.9–2.1	3.0–3.3		
Y *a,c*	1.91	3.19		
Zn	7.86	15.04		
Thiourea				
Ag	7.4	13.1		
Bi				$\log K_6$ 11.9
Cd	0.6	1.6	2.6	4.6

Table 5-15 (*Continued*)
CUMULATIVE FORMATION CONSTANTS FOR METAL COMPLEXES WITH ORGANIC LIGANDS

	$\log K_1$	$\log K_2$	$\log K_3$	$\log K_4$
Cu(I)			13	15.4
Hg(II)		22.1	24.7	26.8
Pb	1.4	3.1	4.7	8.3
Ru(III)	1.21		0.72	
Thoron				
Th		10.15		
Triethanolamine				
Ag	2.30	3.64		
Co(II)	1.73			
Cu(II)	4.30			
Hg(II)	6.90	13.08		
Ni	2.7			
Zn	2.00			
Triethylenetetramine (Trien)				
Ag	7.7			
Cd	10.75	13.9		
Co(II)	11.0			
Cu(II)	20.4			
Fe(II)	7.8			
Fe(III)	21.9			
Hg(II)	25.26			
Mn(II)	4.9			
Ni	14.0			
Pb	10.4			
Zn	11.9			
1,1,1-Trifluoro-3-2′-Thenoylacetone (TTA)				
Ba		10.6		
Cu(II)	6.55	13.0		
Fe(III)	6.9			
Ni	10.0			
Pr	9.53			
Pu(III)	9.53			
Pu(IV)	8.0			
Th	8.1			
U(IV)	7.2			
Zr	3.03 [as ZrL^{3+}]			
Xylenol orange				
Bi	5.52			
Fe(III)	5.70			
Hf	6.50			
Tl(III)	4.90			
Zn	6.15			
Zr	7.60			
Zincon				
Zn	13.1			

MEASUREMENT OF pH

ELECTROMETRIC MEASUREMENT OF pH

The pH value is defined for an aqueous solution in an operational (arbitrary but reproducible) manner according to the Bates-Guggenheim convention:

$$\mathrm{pH}_x = \mathrm{pH}_s + \frac{E_x - E_s}{2.3026RT/\mathbf{F}}.$$

where R is the gas constant per mole, T is the temperature on the absolute scale, and $\mathbf{F}$ is the faraday. The pH_x of the unknown medium is calculated from that of an accepted standard (pH_s) and the measured difference in the emf (E) of the electrode combination when the standard solution is removed from the cell and replaced by the unknown. The double vertical line marks a liquid junction. Electrodes as fabricated exhibit variations in the reproducibility of the reference electrode, in the liquid-junction potential, and, with glass electrodes, in the asymmetry potential. These differences are all eliminated in the standardizing procedure with standard reference pH buffers. (See R. G. Bates, *Determination of pH, Theory and Practice*, Wiley, New York, 1964.)

Electrode reversible to hydrogen ions	Standard reference buffer or unknown solution	Salt bridge (KCl, 3.5M or saturated)	Reference electrode

An electrometric pH-measurement system consists of (1) pH-responsive electrode, (2) reference electrode, and (3) potential-measuring device—some form of high-impedance electronic voltmeter for glass-electrode combinations and this or a potentiometer arrangement for other pH-responsive electrodes. Electronic pH meters are simply voltmeters with scale divisions in pH units which are equivalent to the values of $2.3026RT/F$ (in mV) per pH unit. Values of this function at several temperatures are given in Table 5-16. There is no compensation incorporated in the meter for the changes in pH of the test solution as a function of temperature. Reliability of an indicator–reference electrode combination must be ascertained by standardization of the pH meter with one standard buffer and checking the pH response by immersing the combination in a second and different reference buffer.

The temperature compensator on a pH meter varies the instrument definition of a pH unit from 54.20 mV at 0° to perhaps 66.10 mV at 60°C. This permits one to measure the pH of the sample (and reference buffer standard) at its actual temperature and thus avoid error due to dissociation equilibria and to junction potentials which have significant temperature coefficients.

Table 5-16
VALUES OF 2.3026RT / F AT SEVERAL TEMPERATURES
In Millivolts

$t\,°C$	Value	$t\,°C$	Value	$t\,°C$	Value	$t\,°C$	Value
0	54.197	25	59.157	50	64.118	80	70.070
5	55.189	30	60.149	55	65.110	85	71.062
10	56.181	35	61.141	60	66.102	90	72.054
15	57.173	38	61.737	65	67.094	95	73.046
18	57.767	40	62.133	70	68.086	100	74.038
20	58.165	45	63.126	75	69.078		

Report of the National Academy of Sciences: National Research Council Committee of Fundamental Constants, 1963.

Hydrogen Electrode. In theory the hydrogen gas electrode is the only perfect hydrogen ion electrode, and corrections must be applied for the imperfect response of other electrodes under specified conditions. The potential of a hydrogen electrode varies with the partial pressure of hydrogen gas in equilibrium with the platinum surface, which is in turn dependent on the barometric pressure and the vapor pressure of water. Customarily the observed emf's of cells with hydrogen electrodes are corrected to a partial pressure of hydrogen of 760 mm of Hg. The barometric corrections, in millivolts to be added to the observed reading, are listed in Table 5-17.

Quinhydrone Electrode. The quinhydrone electrode is formed by dipping a bright platinum or gold electrode, at least 1 cm^2 in area, into a solution saturated with quinhydrone (solubility, 3.63 g per liter), an equimolecular mixture of benzoquinone and hydroquinone (dissociation constant, $K_Q = 0.259$ at 25°C). The metal should be in contact with some of the solid quinhydrone in the cell. Commercial quinhydrone should be recrystallized from water at 70°C and dried at room temperature in subdued light. Standard potentials of the quinhydrone electrode are given in Table 5-18.

Antimony Electrode. This metal–metal oxide electrode consists of high-purity electrolytic antimony cast in stick form. An invisible coating of the very slightly soluble oxide always seems to be present as a surface film. Since the electrode potential differs from one electrode to another, and since the slope of the emf-pH curve is not rectilinear, it is necessary to standardize each electrode with solutions of known pH. It is a rugged electrode; accuracy is ± 0.2 pH unit. The electrode potential is affected by dissolved oxygen, by oxidizing agents, and by reducing agents, and there is a marked sensitivity to complexing agents, notably fluoride ion, certain amino acids, and the anions of hydroxy acids. Ions of metals more electropositive than antimony in the emf series of the elements cannot be tolerated.

Table 5-17
BAROMETRIC PRESSURE CORRECTIONS FOR THE
HYDROGEN ELECTRODE FROM 0 TO 60°C

The first column gives the total barometric pressure. This includes the partial pressure of water vapor above the solution. It has been assumed, in computing the values in the body of the table, that this is equal to the vapor pressure of pure water at the temperatures specified.

Barometric pressure, mm Hg	Correction, mV., at $t =$								
	0°C	10°C	20°C	25°C	30°C	35°C	40°C	50°C	60°C
720	0.71	0.82	0.99	1.13	1.30	1.52	1.81	2.67	4.12
725	0.63	0.73	0.91	1.03	1.20	1.42	1.71	2.56	3.99
730	0.55	0.65	0.82	0.94	1.11	1.32	1.61	2.45	3.87
735	0.47	0.56	0.73	0.85	1.02	1.23	1.51	2.34	3.74
740	0.39	0.48	0.64	0.76	0.92	1.13	1.41	2.23	3.62
745	0.31	0.39	0.55	0.67	0.83	1.04	1.31	2.12	3.50
750	0.23	0.31	0.47	0.58	0.74	0.94	1.21	2.02	3.38
755	0.15	0.23	0.38	0.49	0.64	0.85	1.12	1.91	3.25
760	0.07	0.15	0.30	0.41	0.56	0.76	1.02	1.81	3.14
765	−0.01	0.07	0.21	0.32	0.47	0.67	0.92	1.70	3.02
770	−0.08	−0.01	0.13	0.24	0.38	0.57	0.83	1.60	2.91

Table 5-18
STANDARD POTENTIALS OF THE QUINHYDRONE ELECTRODE

t °C	E°, volts	t °C	E°, volts
0	0.71798	25	0.69976
5	0.71437	30	0.69607
10	0.71073	35	0.69237
15	0.70709	40	0.68865
20	0.70343		

REFERENCE ELECTRODES

To be suitable as a reference electrode, the potential of the half-cell must not be significantly altered if a small current passes through it, the resistance must not be too great, it should be easily assembled, and the components should be stable in contact with the atmosphere. The standard potential of the half-cell includes the liquid junction potential of the cell:

$$\text{Pt; } H_2(g), H^+(a = 1) \parallel \text{KCl solution, reference electrode}$$

Potentials of reference electrodes are given in Table 5-19; the values quoted may vary as much as 2 or 3 mV because of unknown liquid-junction potentials, and the care with which the reference electrodes are prepared and aged.

The calomel electrode is widely used. It comprises a nonattackable element, such as platinum, in contact with mercury and a potassium chloride solution of known concentration that is saturated with respect to mercurous chloride (calomel). Although it is relatively simple to prepare, the rather large temperature coefficient of a saturated calomel electrode (abbreviated SCE) necessitates maintenance of constant temperature in precise work. Electrodes prepared with unsaturated potassium chloride solutions possess smaller temperature coefficients, but these electrodes must be protected from evaporation. Calomel electrodes are not stable above 80°C.

Table 5-19
POTENTIALS OF REFERENCE ELECTRODES IN VOLTS
AS A FUNCTION OF TEMPERATURE
Liquid-junction Potential Included

Temp., °C	0.1M KCl Calomel*	1.0M KCl Calomel*	3.5M KCl Calomel*	Satd. KCl Calomel*	1.0M KCl Ag/AgCl†	1.0M KBr Ag/AgBr‡	1.0M KI Ag/AgI§
0	0.3367	0.2883		0.25918	0.23655	0.08128	−0.14637
5					0.23413	0.07961	−0.14719
10	0.3362	0.2868	0.2556	0.25387	0.23142	0.07773	−0.14822
15	0.3361			0.2511	0.22857	0.07572	−0.14942
20	0.3358	0.2844	0.2520	0.24775	0.22557	0.07349	−0.15081
25	0.3356	0.2830	0.2501	0.24453	0.22234	0.07106	−0.15244
30	0.3354	0.2815	0.2481	0.24118	0.21904	0.06856	−0.15405
35	0.3351			0.2376	0.21565	0.06585	−0.15590
38	0.3350		0.2448	0.2355			
40	0.3345	0.2782	0.2439	0.23449	0.21208	0.06310	−0.15788
45					0.20835	0.06012	−0.15998
50	0.3315	0.2745		0.22737	0.20449	0.05704	−0.16219
55					0.20056		
60	0.3248	0.2702		0.2235	0.19649		
70					0.18782		
80				0.2083	0.1787		
90					0.1695	0.0251	

*Bates et al., *J. Research Natl. Bur. Standards*, **45,** 418 (1950).
†Bates and Bower, *J. Research Natl. Bur. Standards*, **53,** 283 (1954).
‡Hetzer, Robinson and Bates, *J. Phys. Chem.*, **66,** 1423 (1962).
§Hetzer, Robinson and Bates, *J. Phys. Chem.*, **68,** 1929 (1964).

The silver halide electrode consists of metallic silver coated with the silver halide, and immersed in an appropriate halide solution of known concentration. These electrodes find use in systems where the presence of mercury cannot be tolerated, and at temperatures exceeding 80°C. Values at elevated temperatures follow:

Temp., °C	125	150	175	200	225	250	275
1.0M KCl Ag/AgCl*	0.1330	0.1032	0.0708	0.0348	−0.0051	−0.054	−0.090
1.0M KBr Ag/AgBr†	−0.0048	−0.0312	−0.0612	−0.0951			

* Greeley et al., *J. Phys. Chem.*, **64**, 652 (1960).
† Towns et al., *J. Phys. Chem.*, **64**, 1861 (1960).

The values of several additional reference electrodes at 25°C are listed:

Ag/AgCl, satd. KCl	0.198
Ag/AgCl, 0.1M KCl	0.288
Hg/HgO, 1.0M NaOH	0.140
Hg/HgO, 0.1M NaOH	0.165
Hg/Hg$_2$SO$_4$, satd. K$_2$SO$_4$ (22°C)	0.658
Hg/Hg$_2$SO$_4$, satd. KCl	0.655

STANDARD REFERENCE pH BUFFER SOLUTIONS

The assigned values of pH$_s$, according to the Bates-Guggenheim convention [*Pure Applied Chem.* **1**, 163 (1960)], for the primary standard solutions prepared from salts issued by the National Bureau of Standards (U.S.) are given in Table 5-20. These are smoothed values. The ionic strength of these reference solutions is 0.1 or less. Strictly speaking the NBS scale uses a molality concentration system; however, values are given in molarity units for convenience.

As a result of a variable liquid-junction potential, the measured pH may be expected to differ seriously from the pa$_H$ determined from cells without a liquid junction in solutions of high acidity or high alkalinity. Merely to affirm the proper functioning of the glass electrode at the extreme ends of the pH scale, two secondary standards are included in Table 5-20. In addition, values for a 0.1m solution of HCl are given to extend the pH scale up to 275°C [see R. S. Greeley, *Anal. Chem.* **32**, 1717 (1960)]:

t, °C:	25	60	90	125	150	175	200	225–275
pH:	1.10	1.11	1.12	1.13	1.14	1.15	1.16	1.2

Uncertainties in the values are ±0.03 pH unit from 25 to 90°C, ±0.05 pH unit from 125 to 200°C, and ±0.1 pH unit from 225 to 275°C.

Table 5-20
NATIONAL BUREAU OF STANDARDS (U.S.) REFERENCE pH BUFFER SOLUTIONS

Temperature, °C	Secondary Standard, 0.05M K tetroxalate	KH Tartrate (satd. at 25°C)	0.05M KH$_2$ Citrate	0.05M KH Phthalate	0.025M KH$_2$PO$_4$, 0.025M Na$_2$HPO$_4$	0.0087M KH$_2$PO$_4$, 0.0302M Na$_2$HPO$_4$	0.01M Na$_2$B$_4$O$_7$	0.025M NaHCO$_3$, 0.025M Na$_2$CO$_3$	Secondary Standard, Ca(OH)$_2$ (satd. at 25°C)
0	1.666		3.860	4.003	6.984	7.534	9.464	10.317	13.423
5	1.668		3.840	3.999	6.951	7.500	9.395	10.245	13.207
10	1.670		3.820	3.998	6.923	7.472	9.332	10.179	13.003
15	1.672		3.802	3.999	6.900	7.448	9.276	10.118	12.810
20	1.675		3.788	4.002	6.881	7.429	9.225	10.062	12.627
25	1.679	3.557	3.776	4.008	6.865	7.413	9.180	10.012	12.454
30	1.683	3.552	3.766	4.015	6.853	7.400	9.139	9.966	12.289
35	1.688	3.549	3.759	4.024	6.844	7.389	9.102	9.925	12.133
38	1.691	3.548		4.030	6.840	7.384	9.081		12.043
40	1.694	3.547	3.753	4.035	6.838	7.380	9.068	9.889	11.984
45	1.700	3.547		4.047	6.834	7.373	9.038		11.841
50	1.707	3.549	3.749	4.060	6.833	7.367	9.011	9.828	11.705
55	1.715	3.554		4.075	6.834		8.985		11.574
60	1.723	3.560		4.091	6.836		8.962		11.449
70	1.743	3.580		4.126	6.845		8.921		
80	1.766	3.609		4.164	6.859		8.885		
90	1.792	3.650		4.205	6.877		8.850		
95	1.806	3.674		4.227	6.886		8.833		
Dilution value, ΔpH$_{1/2}$	+0.186	+0.049	0.024	+0.052	+0.080	+0.07	+0.01	0.079	−0.28

From R. G. Bates, *J. Research Natl. Bur. Standards (U.S.)*, **66A**, 179 (1962) and B. R. Staples and R. G. Bates, *ibid.*, **73A**, 37 (1969).

The buffer value for the NBS reference pH buffer solutions is given below:

	KH Tartrate	0.05M KH$_2$ Citrate	0.05M KH Phthalate	0.025M KH$_2$PO$_4$, 0.025M Na$_2$HPO$_4$	0.0087M KH$_2$PO$_4$, 0.0302M Na$_2$HPO$_4$	0.01M Na$_2$B$_4$O$_7$	0.025M NaHCO$_3$, 0.025M Na$_2$CO$_3$
Buffer value, β	0.027	0.034	0.016	0.029	0.016	0.020	0.029

For the secondary pH reference standards, the buffer value is 0.070 for potassium tetroxalate and 0.09 for calcium hydroxide.

To prepare the standard pH buffer solutions recommended by the National Bureau of Standards (U.S.), the indicated weights of the pure materials in Table 5-21 should be dissolved in water of specific conductivity not greater than 5 micromhos. The tartrate, phthalate, and phosphates can be dried for 2 hours at 110°C before use. Potassium tetroxalate and calcium hydroxide need not be dried. Fresh-looking crystals of borax should be used. Before use, excess solid potassium hydrogen tartrate and calcium hydroxide must be removed. Buffer solutions pH 6 or above should be stored in plastic containers and should be protected from carbon dioxide with soda-lime traps. The solutions should be replaced within two to three weeks, or sooner if formation of mold is noticed. A crystal of thymol may be added as preservative.

The British Standards Institution scale makes no direct attempt to assign a pH to any buffer solution. A primary standard, consisting of a 0.05M potassium hydrogen phthalate solution, is defined as having a pH of exactly 4.000 at 15°C; all other standard buffer solutions are referred to this (Table 5-22). The defining equation for the pH of the phthalate standard at a series of temperatures is

$$pH = 4.000 + 0.5\left(\frac{t-15}{100}\right)^2$$

for temperatures between 0 and 55°C, and by

$$pH = 4.000 + 0.5\left(\frac{t-15}{100}\right)^2 - \frac{t-55}{500}$$

for temperatures between 55 and 95°C, where t is the temperature in degrees celsius.

The preparation of the buffer solutions of Clark and Lubs plus supplementary systems is given in Table 5-23; these solutions offer a means of preparing aqueous buffers in 0.2 pH intervals.

Table 5-21
COMPOSITIONS OF STANDARD pH BUFFER SOLUTIONS, NATIONAL BUREAU OF STANDARDS (U.S.)
Air Weight of Material per Liter of Buffer Solution

Standard	Weight, g
$KH_3(C_2O_4)_2 \cdot 2H_2O$, 0.05$M$	12.61
Potassium hydrogen tartrate, about 0.034M	Saturated at 25°C
Potassium hydrogen phthalate, 0.05M	10.12
Phosphate:	
KH_2PO_4, 0.025M	3.39
Na_2HPO_4, 0.025M	3.53
Phosphate:	
KH_2PO_4, 0.008665M	1.179
Na_2HPO_4, 0.03032M	4.30
$Na_2B_4O_7 \cdot 10H_2O$, 0.01M	3.80
Carbonate:	
$NaHCO_3$, 0.025M	2.10
Na_2CO_3, 0.025M	2.65
$Ca(OH)_2$, about 0.0203M	Saturated at 25°C

Table 5-22
pH VALUES OF BUFFERS BY THE BRITISH STANDARDS METHOD

The 0.05M KH phthalate solution is defined as having a pH of exactly 4.000 at 15°C, and all other standards are referred to this.

Temperature, °C	0.05M K tetroxalate	KH Tartrate (satd. at 25°C)	0.05M KH Phthalate	0.1M HOAc, 0.1M NaOAc	0.025M KH_2PO_4, 0.025M Na_2HPO_4	0.01M Borax	0.025M $NaHCO_3$, 0.025M Na_2CO_3
0	1.639		4.011	4.684	6.973	9.464	10.284
5	1.642		4.005	4.665	6.953	9.395	10.220
10	1.643		4.001	4.656	6.916	9.333	10.158
15	1.645		4.000	4.660	6.893	9.277	10.101
20	1.646		4.001	4.646	6.873	9.227	10.046
25	1.647	3.556	4.005	4.644	6.856	9.181	9.995
30	1.648	3.550	4.011	4.644	6.844	9.141	9.948
35	1.651	3.547	4.020	4.648	6.843	9.106	9.906
40	1.653	3.546	4.031	4.655	6.827	9.074	9.869
45	1.658	3.549	4.045	4.663	6.825	9.047	9.837
50	1.665	3.554	4.061	4.674	6.826	9.023	9.811
55	1.672	3.563	4.080	4.689	6.830	9.002	
60	1.672	3.564	4.091	4.695	6.827	8.974	

Table 5-23
COMPOSITION AND pH VALUES OF BUFFER SOLUTIONS

Values based on the conventional activity pH scale as defined by the National Bureau of Standards (U.S.) and pertain to a temperature of 25°C. Ref: Bower and Bates, *J. Research Natl. Bur. Standards* (U.S.), **55**, 197 (1955) and Bates and Bower, *Anal. Chem.*, **28**, 1322 (1956). Buffer value is denoted by column headed β.

25 ml 0.2M KCl + x ml 0.2M HCl, Diluted to 100 ml			50 ml 0.1M KH Phthalate + x ml 0.1M HCl, Diluted to 100 ml			50 ml 0.1M KH Phthalate + x ml 0.1M NaOH, Diluted to 100 ml		
pH	x	β	pH	x	β	pH	x	β
1.00	67.0	0.31	2.20	49.5		4.20	3.0	0.017
1.20	42.5	0.34	2.40	42.2	0.036	4.40	6.6	0.020
1.40	26.6	0.19	2.60	35.4	0.033	4.60	11.1	0.025
1.60	16.2	0.077	2.80	28.9	0.032	4.80	16.5	0.029
1.80	10.2	0.049	3.00	22.3	0.030	5.00	22.6	0.031
2.00	6.5	0.030	3.20	15.7	0.026	5.20	28.8	0.030
2.20	3.9	0.022	3.40	10.4	0.023	5.40	34.1	0.025
			3.60	6.3	0.018	5.60	38.8	0.020
			3.80	2.9	0.015	5.80	42.3	0.015

50 ml 0.1M KH₂PO₄ + x ml 0.1M NaOH, Diluted to 100 ml			50 ml 0.1M Tris(hydroxymethyl)aminomethane + x ml of 0.1M HCl, Diluted to 100 ml $\Delta pH/\Delta t \simeq -0.028$ $I = 0.001x$			50 ml of a Mixture 0.1M with Respect to Both KCl and H₃BO₃ + x ml 0.1M NaOH, Diluted to 100 ml		
pH	x	β	pH	x	β	pH	x	β
5.80	3.6		7.00	46.6		8.00	3.9	
6.00	5.6	0.010	7.20	44.7	0.012	8.20	6.0	0.011
6.20	8.1	0.015	7.40	42.0	0.015	8.40	8.6	0.015
6.40	11.6	0.021	7.60	38.5	0.018	8.60	11.8	0.018
6.60	16.4	0.027	7.80	34.5	0.023	8.80	15.8	0.022
6.80	22.4	0.033	8.00	29.2	0.029	9.00	20.8	0.027
7.00	29.1	0.031	8.20	22.9	0.031	9.20	26.4	0.029
7.20	34.7	0.025	8.40	17.2	0.026	9.40	32.1	0.027
7.40	39.1	0.020	8.60	12.4	0.022	9.60	36.9	0.022
7.60	42.4	0.013	8.80	8.5	0.016	9.80	40.6	0.016
7.80	44.5	0.009	9.00	5.7		10.00	43.7	0.014
8.00	46.1					10.20	46.2	

50 ml 0.025M Borax, + x ml 0.1M HCl, Diluted to 100 ml $\Delta pH/\Delta t \simeq -0.008$ $I = 0.025$			50 ml 0.025M Borax + x ml 0.1M NaOH, Diluted to 100 ml $\Delta pH/\Delta t \simeq -0.008$ $I = 0.001(25 + x)$			50 ml 0.05M NaHCO₃ + x ml 0.1M NaOH, Diluted to 100 ml $\Delta pH/\Delta t \simeq -0.009$ $I = 0.001(25 + 2x)$		
pH	x	β	pH	x	β	pH	x	β
8.00	20.5		9.20	0.9		9.60	5.0	
8.20	19.7	0.010	9.40	3.6	0.026	9.80	6.2	0.014
8.40	16.6	0.012	9.60	11.1	0.022	10.00	10.7	0.016
8.60	13.5	0.018	9.80	15.0	0.018	10.20	13.8	0.015

Table 5-23 (*Continued*)
COMPOSITION AND pH VALUES OF BUFFER SOLUTIONS

50 ml 0.025M Borax, + x ml 0.1M HCl, Diluted to 100 ml $\Delta pH/\Delta t \simeq -0.008$ $I = 0.025$			50 ml 0.025M Borax + x ml 0.1M NaOH, Diluted to 100 ml $\Delta pH/\Delta t \simeq -0.008$ $I = 0.001(25 + x)$			50 ml 0.05M NaHCO$_3$ + x ml 0.1M NaOH, Diluted to 100 ml $\Delta pH/\Delta t \simeq -0.009$ $I = 0.001(25 + 2x)$		
pH	x	β	pH	x	β	pH	x	β
8.80	9.4	0.023	10.00	18.3	0.014	10.40	16.5	0.013
9.00	4.6	0.026	10.20	20.5	0.009	10.60	19.1	0.012
9.10	2.0		10.40	22.1	0.007	10.80	21.2	0.009
			10.60	23.3	0.005	11.00	22.7	

50 ml 0.05M Na$_2$HPO$_4$ + x ml 0.1M NaOH, Diluted to 100 ml $\Delta pH/\Delta t \simeq -0.025$ $I = 0.001(77 + 2x)$			25 ml 0.2M KCl + x ml 0.2M NaOH, Diluted to 100 ml $\Delta pH/\Delta t \simeq -0.033$ $I = 0.001(50 + 2x)$		
pH	x	β	pH	x	β
11.00	4.1	0.009	12.00	6.0	0.028
11.20	6.3	0.012	12.20	10.2	0.048
11.40	9.1	0.017	12.40	16.2	0.076
11.60	13.5	0.026	12.60	25.6	0.12
11.80	19.4	0.034	12.80	41.2	0.21
11.90	23.0	0.037	13.00	66.0	0.30

Standards for pH Measurement of Blood and Biological Media.
Blood is a well-buffered medium. In addition to the NBS phosphate standard of 0.025M (pH$_s$ = 6.840 at 38°C), another reference solution containing the same salts, but in the molal ratio 1:4, has an ionic strength of 0.13. It is prepared by dissolving 1.360 g of KH$_2$PO$_4$ and 5.677 g of Na$_2$HPO$_4$ (air weights) in carbon dioxide–free water to make 1 liter of solution. The pH$_s$ is 7.416 ± 0.004 at 37.5 and 38°C.

The compositions and pH$_s$ values of *tris*(hydroxymethyl)aminomethane, covering the pH range 7.0 to 8.9, are listed in Table 7-23.

The phosphate-succinate system gives the values of pH$_s$ shown below:

$\dfrac{\text{Molality}}{\text{KH}_2\text{PO}_4} = \dfrac{\text{Molality}}{\text{Na}_2\text{HC}_6\text{H}_5\text{O}_7}$	pH$_s$	$\Delta(\text{pH}s)/\Delta t$
0.005	6.251	−0.00086 deg^{-1}
0.010	6.197	−0.00071
0.015	6.162	
0.020	6.131	
0.025	6.109	−0.0004

BUFFER SOLUTIONS OTHER THAN STANDARDS

The range of the buffering effect of a single weak acid group is approximately one pH unit on either side of pK_a. The ranges of some useful buffer systems are collected in Table 5-24. After all the components have been brought together, the pH of the resulting solution should be determined with reference to standard reference solutions at the temperature to be employed. Buffer components should be compatible with other components in the system under study; this is particularly significant for buffers employed in biological studies. Check the tables of formation constants to ascertain whether metal-binding character exists.

When there are two or more acid groups per molecule, or a mixture is composed of several overlapping acids, the useful range is larger. Universal buffer solutions consist of a mixture of acid groups which overlap such that successive pK_a values differ by two pH units or less. The Prideaux-Ward mixture comprises phosphate, phenyl acetate, and borate plus HCl, and covers the range from 2 to 12 pH units. McIlvaine's buffer is a mixture of citric acid and Na_2HPO_4 that covers the range from pH 2.2 to 8.0. The Britton-Robinson system consists of acetic acid, phosphoric acid, and boric acid plus NaOH, and covers the range from pH 4.0 to 11.5. A mixture composed of Na_2CO_3, NaH_2PO_4, citric acid, and 2-amino-2-methyl-1,3-propanediol covers the range from pH 2.2 to 11.0.

General directions for the preparation of buffer solutions of varying pH but fixed ionic strength are given by Bates (*Determination of pH, Theory and Practice*, pp. 121–122, Wiley, New York, 1964). Preparation of McIlvaine buffered solutions at ionic strengths of 0.5 and 1.0 and Britton-Robinson solutions of constant ionic strength have been described by Elving, Markowitz, and Rosenthal, *Anal. Chem.*, **28**, 1179 (1956), and Frugoni, *Gazz. Chim. Ital.*, **87**, 403 (1957), respectively.

Table 5-24
pH VALUES OF BUFFER SOLUTIONS FOR CONTROL PURPOSES

Materials	pH range
Glycine and HCl	1.0 to 3.7
Citrate and HCl	1.3 to 4.7
p-Toluenesulfonate and *p*-toluenesulfonic acid	1.1 to 3.3
Formate and HCl	2.8 to 4.6
Succinic acid and borax	3.0 to 5.8
Phenyl acetate and HCl	3.5 to 5.0
Acetate and acetic acid	3.7 to 5.6
Succinate and succinic acid	4.8 to 6.3
2-(*N*-Morpholino)ethanesulfonic acid, "MES," and NaOH	5.2 to 7.1

Table 5-24 (*Continued*)
pH VALUES OF BUFFER SOLUTIONS FOR CONTROL PURPOSES

Materials	pH range
2,2-*Bis*(hydroxymethyl)-2,2',2''-nitrilotriethanol and HCl	5.8 to 7.2
KH$_2$PO$_4$ and borax	5.8 to 9.2
N-*Tris*(hydroxymethyl)methyl-2-aminoethanesulfonic acid, "TES," and NaOH	6.8 to 8.2
KH$_2$PO$_4$ and Na$_2$HPO$_4$	6.1 to 7.5
N-2-Hydroxyethylpiperazine-N'-2-ethanesulfonic acid, "HEPES," and NaOH	6.9 to 8.3
Triethanolamine and HCl	6.9 to 8.5
Diethylbarbiturate (Veronal) and HCl	7.0 to 8.5
Tris(hydroxymethyl)aminomethane, "Tris," and HCl	7.2 to 9.0
N-*Tris*(hydroxymethyl)methylglycine, "Tricine," and HCl	7.2 to 9.0
N,N-*Bis*(2-hydroxyethyl)glycine, "Bicine," and HCl	7.4 to 9.2
Borax and HCl	7.6 to 8.9
Glycine and NaOH	8.2 to 10.1
Ammonia (aqueous) and NH$_4$Cl	8.3 to 9.2
Ethanolamine and HCl	8.6 to 10.4
Borax and NaOH	9.4 to 11.1
Carbonate and hydrogen carbonate	9.2 to 11.0
Na$_2$HPO$_4$ and NaOH	11.0 to 12.0

	x mL of 0.1*M* HCl plus *y* mL of 0.1*M* Glycine (7.505 g Glycine + 5.85 g NaCl per Liter)			*x* mL of 0.1*M* HCl plus *y* mL of 0.1*M* Citrate (21.008 g Citric Acid Monohydrate + 200 ml 1*M* NaOH per Liter)			*x* mL of 0.05*M* Succinic Acid (5.90 g · L^{-1}) plus *y* mL of Borax Solution (19.404 g Na$_2$B$_4$O$_7$ · 10H$_2$O per Liter)	
pH	HCl, mL	Glycine, mL	pH	HCl, mL	Citrate, mL	pH	Succinic Acid, mL	Borax, mL
1.20	84.0	16.0	3.50	52.8	47.2	3.60	90.5	9.5
1.40	71.0	29.0	3.60	51.3	48.7	3.80	86.3	13.7
1.60	61.8	38.2	3.80	48.6	51.4	4.00	82.2	17.8
1.80	55.2	44.8	4.00	43.8	56.2	4.20	77.8	22.2
2.00	49.1	50.9	4.20	38.6	61.4	4.40	73.8	26.2
2.20	42.7	57.3	4.40	34.6	65.4	4.60	70.0	30.0
2.40	36.5	63.5	4.60	24.3	75.7	4.80	66.5	33.5
2.60	30.3	69.7	4.80	11.0	89.0	5.00	63.2	36.8
2.80	24.0	76.0				5.20	60.5	39.5
3.00	17.8	82.2				5.40	57.9	42.1
3.30	10.8	89.2				5.60	55.7	44.3
3.60	6.0	94.0				5.80	54.0	46.0

Table 5-24 (*Continued*)
pH VALUES OF BUFFER SOLUTIONS FOR CONTROL PURPOSES

x mL of 0.2M Sodium Acetate (27.199 g NaOAc · 3H$_2$O per liter) plus y mL of 0.2M Acetic Acid			x mL of 0.1M KH$_2$PO$_4$ (13.617 g · L^{-1}) plus y mL of 0.05M Borax Solution (19.404 g Na$_2$B$_4$O$_7$ · 10H$_2$O per Liter)					
pH	NaOAc, mL	Acetic Acid, mL	pH	KH$_2$PO$_4$, mL	Borax, mL	pH	KH$_2$PO$_4$, mL	Borax, mL
3.60	7.5	92.5	5.80	92.1	7.9	7.60	51.7	48.3
3.80	12.0	88.0	6.00	87.7	12.3	7.80	49.2	50.8
4.00	18.0	82.0	6.200	83.0	17.0	8.00	46.5	53.5
4.20	26.5	73.5	6.40	77.8	22.2	8.20	43.0	57.0
4.40	37.0	63.0	6.60	72.2	27.8	8.40	38.7	61.3
4.60	49.0	51.0	6.80	66.7	33.3	8.60	34.0	66.0
4.80	60.0	40.0	7.00	62.3	37.7	8.80	27.6	72.4
5.00	70.5	29.5	7.20	58.1	41.9	9.00	17.5	82.5
5.20	79.0	21.0	7.40	55.0	45.0	9.20	5.0	95.0
5.40	85.5	14.5						
5.60	90.5	9.5						

x mL of Veronal (20.6 g Na Diethylbarbiturate per Liter) plus y mL of 0.1M HCl			x mL of 0.2M Aqueous NH$_3$ Solution plus y mL of 0.2M NH$_4$Cl (10.699 g · L^{-1})			x mL of 0.1M Citrate (21.0 g Citric Acid Monohydrate + 200 mL 1M NaOH per Liter) plus y mL of 0.1M NaOH		
pH	Veronal, mL	HCl, mL	pH	Aq NH$_3$, mL	NH$_4$Cl, mL	pH	Citrate, mL	NaOH, mL
7.00	53.6	46.4	8.00	5.5	94.5	5.10	90.0	10.0
7.20	55.4	44.6	8.20	8.5	91.5	5.30	80.0	20.0
7.40	58.1	41.9	8.40	12.5	87.5	5.50	71.0	29.0
7.60	61.5	38.5	8.60	18.5	81.5	5.70	67.0	33.0
7.80	66.2	33.8	8.80	26.0	74.0	5.90	62.0	38.0
8.00	71.6	28.4	9.00	36.0	64.0			
8.20	76.9	23.1	9.25	50.0	50.0			
8.40	82.3	17.7	9.40	58.5	41.5			
8.60	87.1	12.9	9.60	69.0	31.0			
8.80	90.8	9.2	9.80	78.0	22.0			
9.00	93.6	6.4	10.00	85.0	15.0			

x mL of 0.2M NaOH Added to 100 mL of Stock Solution (0.04M Acetic Acid, 0.04M H$_3$PO$_4$, and 0.04M Boric Acid)							
pH	NaOH, mL	pH	NaOH, mL	pH	NaOH, mL	pH	NaOH, mL
1.81	0.0	4.10	25.0	6.80	50.0	9.62	75.0
1.89	2.5	4.35	27.5	7.00	52.5	9.91	77.5
1.98	5.0	4.56	30.0	7.24	55.0	10.38	80.0
2.09	7.5	4.78	32.5	7.54	57.5	10.88	82.5
2.21	10.0	5.02	35.0	7.96	60.0	11.20	85.0
2.36	12.5	5.33	37.5	8.36	62.5	11.40	87.5
2.56	15.0	5.72	40.0	8.69	65.0	11.58	90.0
2.87	17.5	6.09	42.5	8.95	67.5	11.70	92.5
3.29	20.0	6.37	45.0	9.15	70.0	11.82	95.0
3.78	22.5	6.59	47.5	9.37	72.5	11.92	97.5

Table 5-24 (*Continued*)
pH VALUES OF BUFFER SOLUTIONS FOR CONTROL PURPOSES

x mL of $0.2M$ $Na_2HPO_4 \cdot 2H_2O$ (35.599 g $\cdot$ L^{-1}) plus
y mL of $0.1M$ Citric Acid (19.213 g $\cdot$ L^{-1})

pH	Na_2HPO_4, mL	Citric Acid, mL	pH	Na_2HPO_4, mL	Citric Acid, mL	pH	Na_2HPO_4, mL	Citric Acid, mL
2.20	2.00	98.00	4.20	41.40	58.60	6.20	66.10	33.90
2.40	6.20	93.80	4.40	44.10	55.90	6.40	69.25	30.75
2.60	10.90	89.10	4.60	46.75	53.25	6.60	72.75	27.25
2.80	15.85	84.15	4.80	49.30	50.70	6.80	77.25	22.75
3.00	20.55	79.45	5.00	51.50	48.50	7.00	82.35	17.65
3.20	24.70	75.30	5.20	53.60	46.40	7.20	86.95	13.05
3.40	28.50	71.50	5.40	55.75	44.25	7.40	90.85	9.15
3.60	32.20	67.80	5.60	58.00	42.00	7.60	93.65	6.35
3.80	35.50	64.50	5.80	60.45	39.55	7.80	95.75	4.25
4.00	38.55	61.45	6.00	63.15	36.85	8.00	97.25	2.75

ACIDITY IN OTHER SOLVENT MEDIA

In nonaqueous and mixed solvents, the operational pH scale is defined as

$$pH_x^* = pH_s^* + \frac{(E_x - E_s)F}{2.3026RT}$$

where standard values of pH_s^* are assigned to buffer solutions, as shown in Table 5-25. Selected values of pH* for buffer solutions in methanol-water and ethanol-water solvents are gathered together in Table 5-26. Meaningful applications of pH* data in cells with liquid junction depends to a very large extent on using reference electrodes in which the salt-bridge solution employs the same solvent as that of the buffer.

Acidity standards for deuterium oxide (heavy water) can be used for measurements of pD. Assignment of pD_s values (Table 5-28) is analogous to those used for assigning pH_s values with the exception that the ultimate reference basis for a scale of deuterium ion activity is the deuterium gas electrode, by convention, taken to be zero at all temperatures. There appears to be a constant difference of 0.45 ± 0.03 between the operational pH, determined with the glass electrode in heavy water solutions with reference to aqueous standards, and the presumed or expected pD in these same solutions.

Table 5-25
STANDARD REFERENCE VALUES pH$_s^*$ FOR THE MEASUREMENT
OF ACIDITY IN 50 WEIGHT PERCENT METHANOL-WATER

Temperature, °C	0.02m HOAc, 0.02m NaOAc, 0.02m NaCl	0.02m NaHSuc, 0.02m NaCl	0.02m KH$_2$PO$_4$, 0.02m Na$_2$HPO$_4$, 0.02m NaCl
10	5.560	5.806	7.937
15	5.549	5.786	7.916
20	5.543	5.770	7.898
25	5.540	5.757	7.884
30	5.540	5.748	7.872
35	5.543	5.743	7.863
40	5.550	5.741	7.858

OAc = acetate   Suc = succinate

Reference: R. G. Bates, *Anal. Chem.*, **40**(6), 35A (1968).

Table 5-26
pH* VALUES FOR BUFFER SOLUTIONS IN
ALCOHOL-WATER SOLVENTS AT 25°C
Liquid-junction Potential not Included

Solvent Composition (weight per cent alcohol)	0.01M H$_2$C$_2$O$_4$, 0.01M NH$_4$HC$_2$O$_4$	0.01M H$_2$Suc, 0.01M LiHSuc	0.01M HSal, 0.01M NaSal
Methanol-Water Solvents			
0	2.15	4.12	
10	2.19	4.30	
20	2.25	4.48	
30	2.30	4.67	
40	2.38	4.87	
50	2.47	5.07	
60	2.58	5.30	
70	2.76	5.57	
80	3.13	6.01	
90	3.73	6.73	
92	3.90	6.92	
94	4.10	7.13	
96	4.39	7.43	
98	4.84	7.89	
99	5.20	8.23	
100	5.79	8.75	7.53
Ethanol-Water Solvents			
0	2.15	4.12	
30	2.32	4.70	
50	2.51	5.07	
71.9	2.98	5.71	
100			8.32

Suc = succinate   Sal = salicylate

Table 5-27
POTENTIALS OF REFERENCE ELECTRODES IN VOLTS AT 25°C FOR WATER–ORGANIC SOLVENT MIXTURES

Electrolyte solution 1 M HCl

Solvent, wt %	Methanol Ag/AgCl	Ethanol Ag/AgCl	2-Propanol Ag/AgCl	Acetone Ag/AgCl	Dioxane Ag/AgCl	Ethylene glycol Ag/AgCl	Methanol Calomel	Dioxane Calomel
5	0.2153	0.2146	0.2180	0.2190		0.2190		
10	0.2090	0.2075	0.2138	0.2156		0.2160	0.255	0.2501
20		0.2003	0.2063	0.2079	0.2031	0.2101		
30	0.1968	0.1945				0.2036		
40				0.1859	0.1635	0.1972	0.243	0.2104
45		0.1859						
50		0.173		0.158				
60	0.1818	0.158				0.1807		
70	0.1492	0.136			0.0659		0.216	0.1126
80								
82				−0.034	−0.0614			−0.0014
90	0.1135	0.196						
94.2	0.0841							
98		0.0215					0.103	
99				−0.53				
100	−0.0099	−0.0081						

Table 5-28
STANDARD REFERENCE VALUES pD$_s$ FOR THE MEASUREMENT OF ACIDITY IN HEAVY WATER

Temperature, °C	0.05M KD$_2$ citrate	0.025M KD$_2$PO$_4$ + 0.025M Na$_2$DPO$_4$	0.025M NaDCO$_3$ + 0.025M Na$_2$CO$_3$
5	4.378	7.539	10.998
10	4.352	7.504	10.924
15	4.329	7.475	10.855
20	4.310	7.449	10.793
25	4.293	7.428	10.736
30	4.279	7.411	10.685
35	4.268	7.397	10.638
40	4.260	7.387	10.597
45	4.253	7.381	10.560
50	4.250	7.377	10.527

References: R. G. Bates, *Anal. Chem.*, **40**(6), 36A (1968); M. Paabo and R. G. Bates, *Anal. Chem.*, **41,** 283 (1969).

Table 5-29
ION ACTIVITY STANDARDS

An approach, similar to the operational definition of pH values, can be applied to the problem of measuring the activities of other ions in solution. Standard reference values for pNa, pCl, pCa, and pF have been suggested for use in standardizing ion-selective electrodes:

Suggested Reference Standard Values, pI ($-\log$ [I]), at 25°C

Material	Molality (mole kg^{-1})	pNa	pCa	pCl	pF
NaCl	0.001	3.015		3.015	
	0.01	2.044		2.044	
	0.1	1.108		1.110	
	1.0	0.160		0.204	
NaF	0.001	3.015			3.015
	0.01	2.044			2.048
	0.1	1.108			1.124
CaCl$_2$	0.000333		3.537	3.191	
	0.00333		2.653	2.220	
	0.0333		1.887	1.286	
	0.333		1.105	0.381	

From R. G. Bates and M. Alfenaar in R. A. Durst (Ed.), "Ion-selective Electrodes," *National Bureau of Standards Spec. Publ.* **314,** Washington, 1969.

Table 5-30
INDICATORS FOR AQUEOUS ACID-BASE TITRATIONS AND COLORIMETRIC DETERMINATION OF pH

This table lists some selected indicators. The pH range or transition interval given in the third column may vary appreciably from one observer to another, and in addition it is affected by ionic strength, temperature, and illumination; consequently only approximate values can be given. They should be considered to refer to solutions having low ionic strengths and a temperature of about 25°C. In the fourth column the $pK_a(-\log K_a)$ of the indicator as determined spectrophotometrically is listed. In the fifth column the wavelength of maximum absorption is given first for the acidic and then for the basic form of the indicator, and the same order is followed in giving the colors in the sixth column. The abbreviations used to describe the colors of the two forms of the indicator are as follows:

B	blue	G	green	P	purple	V	violet
Br	brown						
C	colorless	O	orange	R	red	Y	yellow

Indicator	Chemical Name	pH Range	pK_a	λ_{max}, nm	Color Change
Cresol red (acid range)	o-Cresolsulfonephthalein	0.2 to 1.8			R–Y
Cresol purple (acid range)	m-Cresolsulfonephthalein	1.2 to 2.8	1.51	533, ···	R–Y
Thymol blue (acid range)	Thymolsulfonephthalein	1.2 to 2.8	1.65	544, 430	R–Y
Tropeolin 00	Diphenylamino-p-benzene sodium sulfonate	1.3 to 3.2	2.0	527, ···	R–Y
2,6-Dinitrophenol	2,6-Dinitrophenol	2.4 to 4.0	3.69		C–Y
2,4-Dinitrophenol	2,4-Dinitrophenol	2.5 to 4.3	3.90		C–Y
Methyl yellow	Dimethylaminoazobenzene	2.9 to 4.0	3.3	508, ···	R–Y
Methyl orange	Dimethylaminoazobenzene sodium sulfonate	3.1 to 4.4	3.40	522, 464	R–O
Bromophenol blue	Tetrabromophenolsulfonephthalein	3.0 to 4.6	3.85	436, 592	Y–BV
Bromocresol green	Tetrabromo-m-cresolsulfonephthalein	4.0 to 5.6	4.68	444, 617	Y–B
Methyl red	o-Carboxybenzeneazodimethylaniline	4.4 to 6.2	4.95	530, 427	R–Y
Chlorophenol red	Dichlorophenolsulfonephthalein	5.4 to 6.8	6.0	···, 573	Y–R
Bromocresol purple	Dibromo-o-cresolsulfonephthalein	5.2 to 6.8	6.3	433, 591	Y–P
Bromophenol red	Dibromophenolsulfonephthalein	5.2 to 6.8		···, 574	Y–R
p-Nitrophenol	p-Nitrophenol	5.3 to 7.6	7.15	320, 405	C–Y
Bromothymol blue	Dibromothymolsulfonephthalein	6.2 to 7.6	7.1	433, 617	Y–B
Neutral red	Aminodimethylaminotoluphenazonium chloride	6.8 to 8.0	7.4		R–Y
Phenol red	Phenolsulfonephthalein	6.4 to 8.0	7.9	433, 558	Y–R
m-Nitrophenol	m-Nitrophenol	6.4 to 8.8	8.3	···, 570	C–Y
Cresol red	o-Cresolsulfonephthalein	7.2 to 8.8	8.2	434, 572	Y–R
m-Cresol purple	m-Cresolsulfonephthalein	7.6 to 9.2	8.32	···, 580	Y–P
Thymol blue	Thymolsulfonephthalein	8.0 to 9.6	8.9	430, 596	Y–B
Phenolphthalein	Phenolphthalein	8.0 to 10.0	9.4	···, 553	C–R
α-Naphtholbenzein	α-Naphtholbenzein	9.0 to 11.0			Y–B
Thymolphthalein	Thymolphthalein	9.4 to 10.6	10.0	···, 598	C–B
Alizarin Yellow R	5-(p-Nitrophenylazo)-salicylic acid, Na salt	10.0 to 12.0	11.16		Y–V
Tropeolin 0	p-Sulfobenzeneazoresorcinol	11.0 to 13.0			Y–O Br
Nitramine	2,4,6-Trinitrophenylmethylnitroamine	10.8 to 13.0			C–O Br

Salt Errors of Indicators. The salt error in the colorimetric measurement of pH arises from the fact that the color of the indicator depends on the ratio of the concentrations of the conjugate acid and base forms, whereas it is the ratio of their activities which is fixed by the pH of the solution. Therefore, differences in the activity coefficients of the indicator forms, arising from differences in total ionic strength of the solutions and the indicator charge type, can cause difference in apparent color and pH. The salt error is given by the expression:

$$\text{Salt error} = \log\left(\frac{f_I}{f_{HI}}\right)_x - \log\left(\frac{f_I}{f_{HI}}\right)_s$$

where x and s are the unknown and reference solutions, respectively. An indicator acid can be an uncharged molecule (HI), an anion (HI$^-$), or a protonated cation (HI$^+$). Table 5-31 gives the values of the salt error terms, f_I/f_{HI}, for indicators of the three charge types at various ionic strengths, as calculated from the expanded Debye-Hückel equation.

Table 5-31
SALT ERRORS OF INDICATORS

Ionic strength	$\log(f_I/f_{HI^+})$	$\log(f_{I^-}/f_{HI})$	$\log(f_{I^-}/f_{HI^-})$
0.01	+0.042	−0.042	−0.13
0.02	+0.056	−0.056	−0.17
0.03	+0.064	−0.064	−0.19
0.04	+0.071	−0.071	−0.21
0.05	+0.076	−0.076	−0.23
0.06	+0.080	−0.080	−0.24
0.08	+0.086	−0.086	−0.26
0.10	+0.090	−0.090	−0.27
0.12	+0.093	−0.093	−0.28
0.14	+0.095	−0.095	−0.29
0.16	+0.096	−0.096	−0.29
0.18	+0.097	−0.097	−0.29
0.20	+0.097	−0.097	−0.29
0.25	+0.096	−0.096	−0.29
0.30	+0.094	−0.094	−0.28
0.40	+0.086	−0.086	−0.26
0.50	+0.076	−0.076	−0.23

Table 5-32
MIXED INDICATORS

Mixed indicators give sharp color changes and are especially useful in titrating to a given titration exponent (pI). The information given in this table is from the two-volume work *Volumetric Analysis* by Kolthoff and Stenger, published by Interscience Publishers, Inc., New York, 1942 and 1947, and reproduced with their permission.

Composition of Indicator Solution		pI	Color		Notes
			Acid	Alkaline	
1 part 0.1% methyl yellow in alc. 1 part 0.1% methylene blue in alc.	*	3.25	Blue–violet	Green	Still green at pH 3.4, blue-violet at 3.2†
1 part 0.14% xylene cyanol FF in alc.* 1 part 0.1% methyl orange in aq.	*	3.8	Violet	Green	Color is gray at pH 3.8
1 part 0.1% methyl orange in aq. 1 part 0.25% indigo carmine in aq.	*	4.1	Violet	Green	Good indicator, especially in artificial light
1 part 0.1% methyl orange in aq. 1 part 0.1% aniline blue in aq.		4.3	Violet	Green	
1 part 0.1% bromcresol green sodium salt in aq. 1 part 0.02% methyl orange in aq.		4.3	Orange	Blue–green	Yellow at pH 3.5, greenish yellow at 4.0, weakly green at 4.3
3 parts 0.1% bromcresol green in alc. 1 part 0.2% methyl red in alc.		5.1	Wine–red	Green	Very sharp color change†
1 part 0.2% methyl red in alc. 1 part 0.1% methylene blue in alc.	*	5.4	Red–violet	Green	Color is red-violet at pH 5.2, a dirty blue at 5.4, and a dirty green at 5.6
1 part 0.1% chlorphenol red sodium salt in aq. 1 part 0.1% aniline blue in water		5.8	Green	Violet	Pale violet at pH 5.8

* Keep in a dark bottle.   † Excellent indicator.

Table 5-32 (Continued)
MIXED INDICATORS

Composition of Indicator Solution	pH	Color		Notes
		Acid	Alkaline	
1 part 0.1% bromcresol green sodium salt in aq. 1 part 0.1% chlorphenol red sodium salt in aq.	6.1	Yellow-green	Blue-violet	Blue-green at pH 5.4, blue at 5.8, blue with a touch of violet at 6.0, blue-violet at 6.2
1 part 0.1% bromcresol purple sodium salt in aq. 1 part 0.1% bromthymol blue sodium salt in aq.	6.7	Yellow	Violet-blue	Yellow-violet at pH 6.2, violet at 6.6, blue-violet at 6.8
2 parts 0.1% bromthymol blue sodium salt in aq. 1 part 0.1% azolitmin in aq.	6.9	Violet	Blue	
* 1 part 0.1% neutral red in alc. 1 part 0.1% methylene blue in alc.	7.0	Violet-blue	Green	Violet blue at pH 7.0†
1 part 0.1% neutral red in alc. 1 part 0.1% bromthymol blue in alc.	7.2	Rose	Green	Dirty green at pH 7.4, pale rose at 7.2, clear rose at 7.0
2 parts 0.1% cyanine in 50% alc. 1 part 0.1% phenol red in 50% alc.	7.3	Yellow	Violet	Orange at pH 7.2, beautiful violet at 7.4, color fades on standing
1 part 0.1% bromthymol blue sodium salt in aq. 1 part 0.1% phenol red sodium salt in aq.	7.5	Yellow	Violet	Dirty green at pH 7.2, pale violet at 7.4, strong violet at 7.6†
1 part 0.1% cresol red sodium salt in aq. 3 parts 0.1% thymol blue sodium salt in aq.	8.3	Yellow	Violet	Rose at pH 8.2, distinctly violet at 8.4†
2 parts 0.1% α-naphtholphthalein in alc. 1 part 0.1% cresol red in alc.	8.3	Pale rose	Violet	Pale violet at pH 8.2, strong violet at 8.4

Table 5-32 (*Continued*)
MIXED INDICATORS

Composition of Indicator Solution	pI	Color		Notes
		Acid	Alkaline	
1 part 0.1% α-naphtholphthalein in alc. 3 parts 0.1% phenolphthalein in alc.	8.9	Pale rose	Violet	Pale green at pH 8.6, violet at 9.0
1 part 0.1% phenolphthalein in alc. 2 parts 0.1% methyl green in alc. *	8.9	Green	Violet	Pale blue at pH 8.8, violet at 9.0
1 part 0.1% thymol blue in 50% alc. 3 parts 0.1% phenolphthalein in 50% alc.	9.0	Yellow	Violet	From yellow thru green to violet†
1 part 0.1% phenolphthalein in alc. 1 part 0.1% thymolphthalein in alc.	9.9	Colorless	Violet	Rose at pH 9.6, violet at 10; sharp color change
1 part 0.1% phenolphthalein in alc. 2 parts 0.2% Nile blue in alc.	10.0	Blue	Red	Violet at pH 10†
2 parts 0.1% thymolphthalein in alc. 1 part 0.1% alizarin yellow in alc.	10.2	Yellow	Violet	Sharp color change
2 parts 0.2% Nile blue in aq. 1 part 0.1% alizarin yellow in alc.	10.8	Green	Red-brown	

Table 5-33
FLUORESCENT INDICATORS

Name	pH Range	Color Change Acid to Base	Indicator Solution*
Benzoflavine	−0.3 to 1.7	Yellow to green	1
3,6-Dihydroxyphthalimide	0 to 2.4	Blue to green	1
	6.0 to 8.0	Green to yellow/green	
Eosin (tetrabromofluorescein)	0 to 3.0	Non-fl to green	4, 1%
4-Ethoxyacridone	1.2 to 3.2	Green to blue	1
3,6-Tetramethyldiamino-xanthone	1.2 to 3.4	Green to blue	1
Esculin	1.5 to 2.0	Weak blue to strong blue	
Anthranilic acid	1.5 to 3.0	Non-fl to light blue	2 (50% ethanol)
	4.5 to 6.0	Light blue to dark blue	
	12.5 to 14	Dark blue to non-fl	
3-Amino-1-naphthoic acid	1.5 to 3.0	Non-fl to green	2 (as sulfate in 50% ethanol)
	4.0 to 6.0	Green to blue	
	11.6 to 13.0	Blue to non-fl	
1-Naphthylamino-6-sulfonamide (also the 1-, 7-)	1.9 to 3.9	Non-fl to green	3
	9.6 to 13.0	Green to non-fl	
2-Naphthylamino-6-sulfonamide (also the 2-, 8-)	1.9 to 3.9	Non-fl to dark blue	3
	9.6 to 13.0	Dark blue to non-fl	
1-Naphthylamino-5-sulfonamide	2.0 to 4.0	Non-fl to yellow/orange	3
	9.5 to 13.0	Yellow/orange to non-fl	
1-Naphthoic acid	2.5 to 3.5	Non-fl to blue	4
Salicylic acid	2.5 to 4.0	Non-fl to dark blue	4 (0.5%)
Phloxin BA extra (tetrachloro-tetrabromo-fluorescein)	2.5 to 4.0	Non-fl to dark blue	2
Erythrosin B (tetraiodo-fluorescein)	2.5 to 4.0	Non-fl to light green	4 (0.2%)
2-Naphthylamine	2.8 to 4.4	Non-fl to violet	1
Magdala red	3.0 to 4.0	Non-fl to purple	
p-Aminophenylbenzene-sulfonamide	3.0 to 4.0	Non-fl to light blue	3
2-Hydroxy-3-naphthoic acid	3.0 to 6.8	Blue to green	4 (0.1%)
Chromotropic acid	3.1 to 4.4	Non-fl to light blue	4 (5%)
1-Naphthionic acid	3 to 4	Non-fl to blue	4
	10 to 12	Blue to yellow-green	
1-Naphthylamine	3.4 to 4.8	Non-fl to blue	1
5-Aminosalicylic acid	3.1 to 4.4	Non-fl to light green	1 (0.2% fresh)
Quinine	3.0 to 5.0	Blue to weak violet	1 (0.1%)
	9.5 to 10.0	Weak violet to non-fl	
o-Methoxybenzaldehyde	3.1 to 4.4	Non-fl to green	4 (0.2%)
o-Phenylenediamine	3.1 to 4.4	Green to non-fl	5
p-Phenylenediamine	3.1 to 4.4	Non-fl to orange/yellow	5
Morin (2′,4′,3,5,7-penta-hydroxyflavone)	3.1 to 4.4	Non-fl to green	6 (0.2%)
	8 to 9.8	Green to yellow/green	
Thioflavine S	3.1 to 4.4	Dark blue to light blue	6 (0.2%)
Fluorescein	4.0 to 4.5	Pink/green to green	4 (1%)
Dichlorofluorescein	4.0 to 6.6	Blue green to green	1

Table 5-33 (*Continued*)
FLUORESCENT INDICATORS

Name	pH Range	Color Change Acid to Base	Indicator Solution
β-Methylesculetin	4.0 to 6.2	Non-fl to blue	1
	9.0 to 10.0	Blue to light green	
Quininic acid	4.0 to 5.0	Yellow to blue	6 (satd)
β-Naphthoquinoline	4.4 to 6.3	Blue to non-fl	3
Resorufin (7-oxyphen-oxazone)	4.4 to 6.4	Yellow to orange	
Acridine	5.2 to 6.6	Green to violet	2
3,6-Dihydroxyxanthone	5.4 to 7.6	Non-fl to blue/violet	1
5,7-Dihydroxy-4-methyl-coumarin	5.5 to 5.8	Light blue to dark blue	
3,6-Dihydroxyphthalic acid dinitrile	5.8 to 8.2	Blue to green	1
1,4-Dihydroxybenzene-disulfonic acid	6 to 7	Non-fl to light blue	4 (0.1%)
Luminol	6 to 7	Non-fl to blue	
2-Naphthol-6-sulfonic acid	5-7 to 8-9	Non-fl to blue	4
Quinoline	6.2 to 7.2	Blue to non-fl	6 (satd)
1-Naphthol-5-sulfonic acid	6.5 to 7.5	Non-fl to green	6 (satd)
Umbelliferone	6.5 to 8.0	Non-fl to blue	
Magnesium-8-hydroxy-quinolinate	6.5 to 7.5	Non-fl to yellow	6 (0.1% in 0.01 *M* HCl)
Orcinaurine	6.5 to 8.0	Non-fl to green	6 (0.03%)
Diazo brilliant yellow	6.5 to 7.5	Non-fl to blue	
Coumaric acid	7.2 to 9.0	Non-fl to green	1
β-Methylumbelliferone	>7.0	Non-fl to blue	2 (0.3%)
Harmine	7.2 to 8.9	Blue to yellow	
2-Naphthol-6,8-disulfonic acid	7.5 to 9.1	Blue to light blue	4
Salicylaldehyde semi-carbazone	7.6 to 8.0	Yellow to blue	2
1-Naphthol-2-sulfonic acid	8.0 to 9.0	Dark blue to light blue	4
Salicylaldehyde acetyl-hydrazone	8.3	Non-fl to green/blue	2
Salicylaldehyde thiosemi-carbazone	8.4	Non-fl to blue/green	2
1-Naphthol-4-sulfonic acid	8.2	Dark blue to light blue	4
Naphthol AS	8.2 to 10.3	Non-fl to yellow/green	4
2-Naphthol	8.5 to 9.5	Non-fl to blue	2
Acridine orange	8.4 to 10.4	Non-fl to yellow/green	1
Orcinsulfonephthalein	8.6 to 10.0	Non-fl to yellow	
2-Naphthol-3,6-disulfonic acid	9.0 to 9.5	Dark blue to light blue	4
Ethoxyphenyl-naphtho-stilbazonium chloride	9 to 11	Green to non-fl	1
o-Hydroxyphenylbenzothiazole	9.3	Non-fl to blue green	2
o-Hydroxyphenylbenzoxazole	9.3	Non-fl to blue/violet	2
o-Hydroxyphenylbenzimidazole	9.9	Non-fl to blue/violet	2
Coumarin	9.5 to 10.5	Non-fl to light green	
6,7-Dimethoxyisoquinoline-1-carboxylic acid	9.5 to 11.0	Yellow to blue	0.1% in glycerine/ethanol/water in 2:2:18 ratio
1-Naphthylamino-4-sulfonamide	9.5 to 13.0	Dark blue to white/blue	3

Table 5-33 (*Continued*)
FLUORESCENT INDICATORS

Name	pH Range	Color Change Acid to Base	Indicator Solution
1-Amino-8-hydroxynaphthalene-5,7-disulfonic acid	10 to 12	Violet to yellow/green	4
Cotarnine	>12.5	Yellow to white	
2-Naphthionic acid	12 to 13	Blue to violet	4 (1%)
2-Naphthylamine-6,8-disulfonic acid	12 to 14	Blue to yellow/pink	4 (1%)

* Indicator solutions:
1. 1% solution in ethanol
2. 0.1% solution in ethanol
3. 0.05% solution in 90% ethanol
4. Sodium or potassium salt in distilled water
5. 0.2% solution in 70% ethanol
6. Distilled water

Reference: G. F. Kirkbright, "Fluorescent Indicators," Chapter 9 in *Indicators*, E. Bishop (ed.), Pergamon Press, Oxford, 1972.

CALCULATION OF THE APPROXIMATE pH VALUE OF SOLUTIONS

Strong acid: $\mathrm{pH} = -\log[\text{acid}]$

Strong base: $\mathrm{pH} = 14.00 + \log[\text{base}]$

Weak acid: $\mathrm{pH} = \frac{1}{2}\,pK_a - \frac{1}{2}\log[\text{acid}]$

Weak base: $\mathrm{pH} = 14.00 - \frac{1}{2}\,pK_b + \frac{1}{2}\log[\text{base}]$

Salt formed by a weak acid and a strong base:

$$\mathrm{pH} = 7.00 + \frac{1}{2}\,pK_a + \frac{1}{2}\log[\text{salt}]$$

Acid salts of a dibasic acid:

$$\mathrm{pH} = \frac{1}{2}\,pK_1 + \frac{1}{2}\,pK_2 - \frac{1}{2}\log[\text{salt}] + \frac{1}{2}\log(K_1 + [\text{salt}])$$

Buffer solution consisting of a mixture of a weak acid and its salt:

$$\mathrm{pH} = pK_a + \log\left(\frac{[\text{salt}] + [\mathrm{H_3O^+}] - [\mathrm{OH^-}]}{[\text{acid}] - [\mathrm{H_3O^+}] + [\mathrm{OH^-}]}\right)$$

CALCULATION OF CONCENTRATIONS OF SPECIES PRESENT AT A GIVEN pH

$$\alpha_0 = \frac{[\mathrm{H^+}]^n}{[\mathrm{H^+}]^n + K_1[\mathrm{H^+}]^{n-1} + K_1K_2[\mathrm{H^+}]^{n-2} + \cdots + K_1K_2\cdots K_n} = \frac{[\mathrm{H}_nA]}{C_{\text{acid}}}$$

$$\alpha_1 = \frac{K_1[\mathrm{H^+}]^{n-1}}{[\mathrm{H^+}]^n + K_1[\mathrm{H^+}]^{n-1} + K_1K_2[\mathrm{H^+}]^{n-2} + \cdots + K_1K_2\cdots K_n} = \frac{[\mathrm{H}_{n-1}A^-]}{C_{\text{acid}}}$$

$$\alpha_2 = \frac{K_1K_2[\mathrm{H^+}]^{n-2}}{[\mathrm{H^+}]^n + K_1[\mathrm{H^+}]^{n-1} + K_1K_2[\mathrm{H^+}]^{n-2} + \cdots + K_1K_2\cdots K_n} = \frac{[\mathrm{H}_{n-2}A^{2-}]}{C_{\text{acid}}}$$

$$\vdots$$

$$\alpha_n = \frac{K_1K_2\cdots K_n}{[\mathrm{H^+}]^n + K_1[\mathrm{H^+}]^{n-1} + K_1K_2[\mathrm{H^+}]^{n-2} + \cdots + K_1K_2\cdots K_n} = \frac{[A^{n-}]}{C_{\text{acid}}}$$

SEPARATION METHODS

COLUMN CHROMATOGRAPHY

Retention Behavior

On a chromatogram the distance on the time axis from the point of sample injection to the peak of an eluted component is called the *uncorrected retention time* t_R. The corresponding *retention volume* or peak elution volume is the product of retention time and flow rate, expressed as volume of mobile phase per unit time: $V_R = t_R F_c$. A peak for a nonretained component, such as the air spike in gas chromatography, measures the transit time (t_M) for the mobile phase, and converted to volume (V_M or V_0), it represents the interstitial volume of the column, i.e., the liquid space between the particles of the column packing, plus the effective volume contributions of the injection port and detector. Retention volume measured from the peak of a nonretained substance provides an *adjusted retention volume*: $V'_R = t_R F_c - t_M F_c = V_R - V_M$.

When the mobile phase is compressible, as in gas chromatography, a pressure-gradient correction or compressibility factor, j, must be applied to the adjusted retention volume to give the *net retention volume*: $V_N = j V'_R$. The compressibility factor is given by the expression

$$j = \frac{3[(P_i/P_o)^2 - 1]}{2[(P_i/P_o)^3 - 1]}$$

where P_i is the carrier gas pressure at the inlet to the column, and P_o that at the outlet.

When the solute enters the column, it immediately equilibrates between the stationary phase and the mobile phase. The concentration of solute in each phase is given by the *partition coefficient*: $K_d = C_S/C_M$, where C_S and C_M are the concentrations of solute in the stationary phase and mobile phase, respectively. At the appearance of a peak maximum,

$$V_R - V_M = K_d V_S$$

The capacity factor or *partition ratio* k is a measure of sample retention by the column, in terms of column volumes:

$$k = \frac{V_R - V_M}{V_M} = \frac{C_S V_S}{C_M V_M} = K_d \frac{V_S}{V_M}$$

The volumetric phase ratio is: $\beta = V_M/V_S$.

Column Efficiency

An empirical measure of column efficiency is defined as the theoretical plate number N, in which

$$N = 16 \left(\frac{V'_R}{W}\right)^2$$

Here the peak width at the base, W, is the segment of baseline cut out by the tangents drawn to the inflection points of the peak (Fig. 5-2); it is equivalent to four standard deviations (4σ). An alternative expression is

$$N = 5.54 \left(\frac{V'_R}{W_{1/2}}\right)^2$$

where $W_{1/2}$ is the peak width at one-half the peak height as measured from the baseline.

Resolution

The relative capacity factors or partition ratios of two components are a measure of the column's ability to separate them. This is expressed as the *relative retention*:

$$\alpha = \frac{k_2}{k_1} = \frac{V'_{R,2}}{V'_{R,1}}$$

Two factors influence resolution: the peak-to-peak separation and the average peak width, viz.,

$$R = \frac{V_{R,2} - V_{R,1}}{0.5(W_2 + W_1)}$$

A value of $R = 1.5$ (or 6σ) represents 99.8 percent of baseline resolution, that is, 0.1 percent overlap, whereas in the case of $R = 1.0$ (or

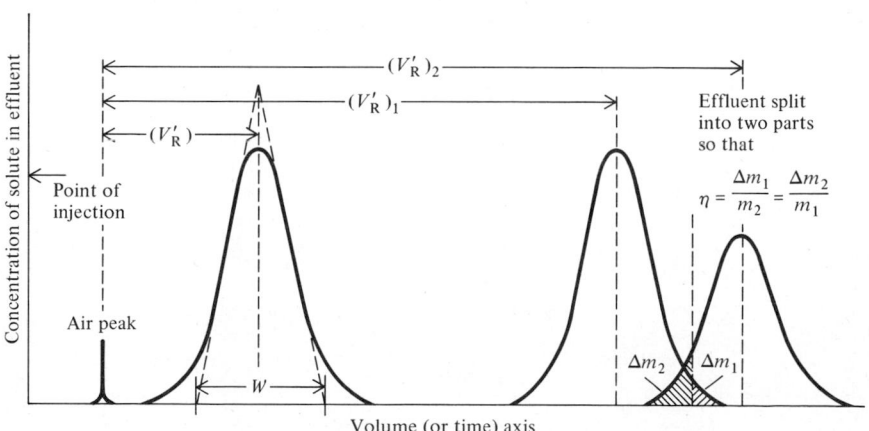

Fig. 5-2. Idealized elution peaks showing method of handling overlapping bands.

4σ), the resolution is about 94 percent complete, which corresponds to a 3 percent overlap of peak areas.

If equal quantities of two solutes in adjacent bands are assumed, resolution can be expressed in terms of k, α, and N:

$$R = \frac{1}{4}\left(\frac{\alpha - 1}{\alpha}\right)\left(\frac{k_2}{1 + k_2}\right)\sqrt{N}$$

Ion-exchange Parameters

At equilibrium the concentration of ions in solution and in the ion-exchange particle is related through the *selectivity coefficient:*

$$E_{\mathrm{H}}^{\mathrm{M}/n} = \frac{[\mathrm{M}^{n+}]_r[\mathrm{H}^+]}{[\mathrm{M}^{n+}][\mathrm{H}^+]_r}$$

where the subscript r refers to the resin phase. The selectivity coefficient expresses the selectivity of the resin for M^{n+} ions in preference to

Table 5-34
SELECTIVITY COEFFICIENTS $E_{\mathrm{H}}^{\mathrm{M}/n}$ FOR SOME
METAL IONS ON DIFFERENTLY
CROSS-LINKED DOWEX 50

	4% DVB	8% DVB	16% DVB
Univalent ions			
Li	0.76	0.79	0.68
H	1.00	1.00	1.00
Na	1.20	1.56	1.61
NH$_4$	1.44	2.01	2.27
K	1.72	2.28	3.06
Rb	1.86	2.49	3.14
Cs	2.02	2.56	3.17
Ag	3.58	6.70	15.6
Bivalent ions			
UO$_2$	0.79	0.85	1.05
Mg	0.99	1.15	1.10
Zn	1.05	1.21	1.18
Co	1.08	1.31	1.19
Cu	1.10	1.35	1.40
Cd	1.13	1.36	1.55
Ni	1.16	1.37	1.27
Mn	1.15	1.43	1.54
Ca	1.39	1.80	2.28
Sr	1.57	2.27	3.16
Pb	2.20	3.46	5.65
Ba	2.50	4.02	6.52
Trivalent ions			
Cr	1.6	2.0	2.5
Ce	1.9	2.8	4.1
La	1.9	2.8	4.1

Based on data of Bonner *et al., J. Phys. Chem.* **61,** 326 (1957); *ibid.* **62,** 250 (1958).

hydrogen ions from a solution containing an equal concentration of each, and for a specified degree of resin loading (usually not exceeding 10 percent). A collection of selectivity coefficients is given in Table 5-34 for cations and in Table 5.35 for anions. Tabulated values refer to equivalents of ion adsorbed from 1 mL of solution per gram of dry resin in the H form for cation exchangers, and in the Cl form for anion exchangers. From the selectivity coefficients listed in the tables, one is able to calculate the selectivity coefficient for the exchange between any pair of ions. For the exchange between a calcium salt solution and a Dowex 50×8 cation-exchange resin loaded with sodium ions:

$$E_{Na}^{Ca/2} = \frac{E_H^{Ca/2}}{E_H^{Na}} = \frac{1.80}{1.56} = 1.15$$

or, expressed on a molar basis to account for the unequal charges of the ions:

$$E_{2Na}^{Ca} = \frac{[Ca^{2+}]_r [Na^+]^2}{[Ca^{2+}][Na^+]_r^2} = (1.15)^2 = 1.32$$

Table 5-35
SELECTIVITY COEFFICIENTS E_{Cl}^X FOR SOME ANIONS ON DOWEX RESINS

	Dowex 1	Dowex 2
Hydroxide	0.09	0.65
Fluoride	0.09	0.13
Aminoacetate	0.10	0.10
Acetate	0.17	0.18
Formate	0.22	0.22
Dihydrogen phosphate	0.25	0.34
Bicarbonate	0.32	0.53
Chloride	1.00	1.00
Bromate	. . .	1.01
Bisulfite	1.3	1.3
Cyanide	1.6	1.3
Bromide	2.8	2.3
Nitrate	3.8	3.3
Benzene sulfonate	. . .	4.0
Bisulfate	4.1	6.1
Phenoxide	5.2	8.7
Iodide	8.7	7.3
p-Toluene sulfonate	. . .	13.7
Thiocyanate	. . .	18.5
Perchlorate	. . .	32
Salicylate	32.2	28

From S. Peterson, *Ann. N. Y. Acad. Sci.* **57,** 144 (1954).

The weight distribution coefficient D_g for a solute M equilibrated with m grams of dry resin contained in a solution volume of V milliliters is given by

$$D_g = \frac{\text{total amount of solute per gram of dry ion exchanger}}{\text{total amount of solute per milliliter in the external solution}}$$

$$= \frac{\text{fraction of solute in resin phase}}{\text{fraction of solute in solution phase}} \cdot \frac{\text{volume of solution phase}}{\text{mass of resin}}$$

For many purposes, the volume distribution coefficient D_v is more convenient in column chromatography:

$$D_v = \frac{\text{amount of sorbed solute in resin per milliliter of resin bed}}{\text{amount of solute per milliliter of external solution}}$$

The two distribution coefficients are related by the expression:

$$D_v = D_g \rho$$

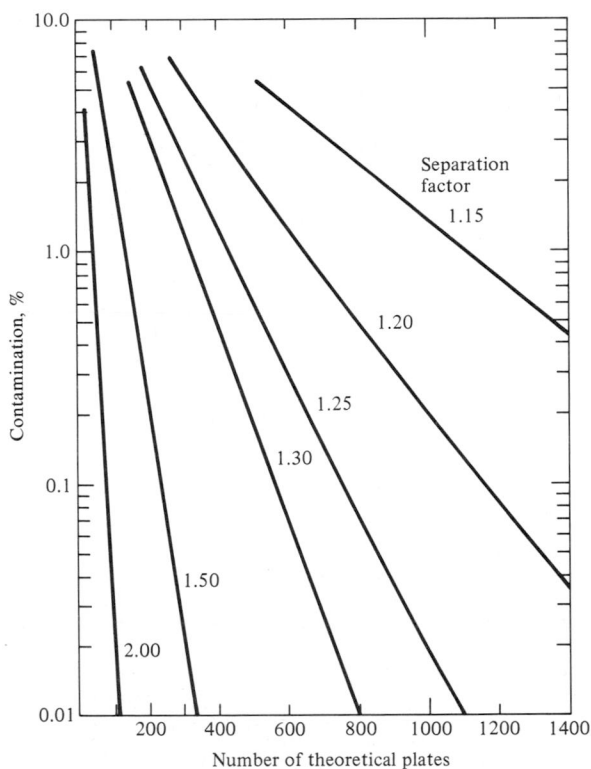

Fig. 5-3. *Cross-contamination, $100\,\Delta m_2/m_1$, for different separation factors, as a function of number of theoretical plates (after Glueckauf), for case that $m_1 = m_2$.*

where ρ is the bed density of a column expressed in the units of mass of dry resin per milliliter of column volume.

D_g (or D_v) can be related to the volume of solution required to remove a solute from a chromatographic column that contains a known amount of exchanger, viz.,

$$V_{\max} = D_g \rho V_b + V_0$$

or

$$V_{\max} = D_v V_b + V_0$$

where V_b is the geometric column volume and V_0 is the interstitial volume. V_0/V_b is approximately 0.4 for most columns, a close approximation to the theoretical value for hexagonal close packing of uniform resin spheres. Thus,

$$D_v = \frac{V_{\max}}{V_b} - \frac{V_0}{V_b} = \frac{V_{\max}}{V_b} - 0.4$$

Theoretical plates needed to achieve desired degrees of purity can be estimated for ion-exchange columns from either Fig. 5-3 or Fig. 5-4.

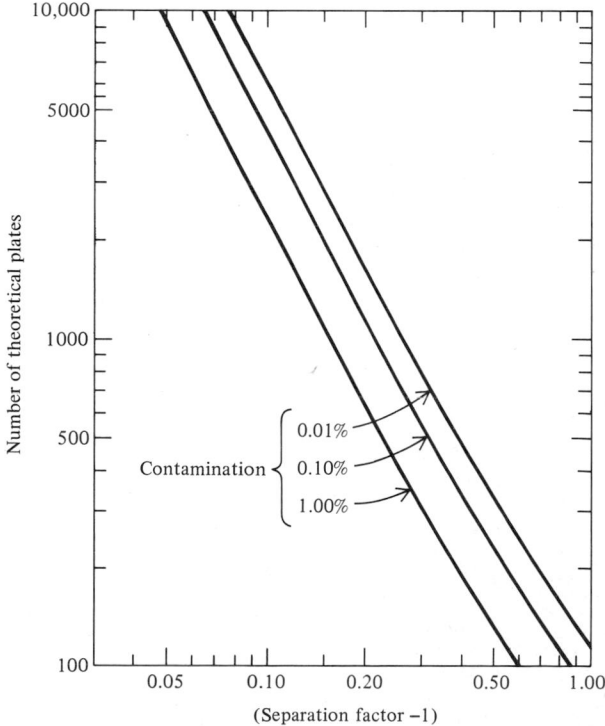

Fig. 5-4. *Theoretical plates needed to achieve desired degrees of purity. (After Glueckauf.)*

Section 6

ELECTROCHEMISTRY

SINGLE ELECTRODE POTENTIALS

It is convenient to consider that the total emf of a cell is the sum of two "single electrode potentials." That is,

$$E_{cell} = E_{ox} + E_{red} \tag{1}$$

where E_{ox} is the single electrode potential of the electrode forming the negative pole of the cell, and E_{red} is the single electrode potential of the electrode forming the positive pole of the cell. An oxidation process always occurs at the negative pole of a primary cell, whereas a reduction process occurs at the positive pole. If a particular substance is more easily oxidized than hydrogen, its E_{ox} is assigned a positive value and its E_{red} a negative value. If a substance is not oxidized as easily as hydrogen, its E_{ox} is negative and E_{red} is positive in sign.

The values given in the table below are E_{red}° values, where E° is the single electrode potential when each substance involved in the electrode reaction is at unit "activity." Values of E_{ox}° may be obtained merely by changing the sign of the E_{red}° values.

If the half-reaction involves H^+ as product or reactant, the solution is acid. Half-reactions involving OH^-, NH_3, CN^-, CO_3^{2-}, or S^{2-} are reactions occurring in an alkaline solution.

An extensive tabulation of E° values is contained in G. Charlot *et al.*, *Selected Constants: Oxidation-Reduction Potentials of Inorganic Substances in Aqueous Solution*, Butterworths, London, 1971.

One substance on the left side of each half-reaction is an oxidizing agent, and one substance on the right side of each half-reaction is a reducing agent. As one proceeds downward in the table, the oxidizing agents decrease in strength and the reducing agents increase in strength. In general, if two half-reactions are represented by the following equations:

$$\text{ox. agent}_1 + ne^- = \text{red. agent}_1$$
$$\text{ox. agent}_2 + ne^- = \text{red. agent}_2$$

and the second equation occurs lower in the table than the first equation, then a reaction may occur between ox. agent$_1$ and red. agent$_2$, whereas no reaction is possible between ox. agent$_2$ and red. agent$_1$. For example, consider the half-reactions

$$Cu^{2+} + I^- + e^- = CuI \qquad (E_{red}^{\circ} = 0.86 \text{ volt})$$
$$I_2 + 2e^- = 2I^- \qquad (E_{red}^{\circ} = 0.535 \text{ volt})$$

from the rule that Cu^{2+} should react with I^-, but no reaction should occur between I_2 and CuI.

To obtain the overall reaction between two half-reactions, the half-reaction which is lower in the table is reversed. Each equation is multiplied through by the proper numbers so that the number of electrons lost in the reversed upper equation is equal to the number of electrons gained in the lower half-reaction. The two half-reactions are then added. For the reaction between Cu^{2+} and I^- the overall reaction is obtained as follows:

$$2I^- = I_2 + 2e^- \qquad \text{(lower half-reaction reversed)}$$
$$\underline{2Cu^{2+} + 2I^- + 2e^- = 2CuI} \qquad \text{(reaction multiplied by 2)}$$
$$2Cu^{2+} + 4I^- = I_2 + 2CuI \qquad \text{(overall reaction)}$$

If two or more reactions between two substances are possible, usually the reaction involving half-reactions which are farthest apart in the table will occur first.

To calculate the single electrode potential, E_{red}, when the concentration of ions involved in the electrode reaction is other than unit activity, the following equation is used:

$$E_{red} = E^\circ_{red} - \frac{0.00019841T}{n} \log_{10} \frac{(a_p)^x}{(a_r)^y} \qquad (2)$$

where n is the number of electrons (e^-) involved in the half-reaction; a_p is the activity of products; a_r is the activity of reactants; x and y are the coefficients of products and reactants, respectively, in the half-reaction *written as a reduction reaction;* and T is the absolute temperature (kelvins). The value of $0.00019841T$ at $25°C$ is 0.05916 (see Table 5-16).

Examples. For the cell:

$$Zn \,|\, Zn^{2+}(a = 1) \,\|\, Cu^{2+}(a = 1) \,|\, Cu$$

As written, the cell diagram specifies that the zinc electrode is the anode. Thus, using E° data from Table 6-1 or Table 6-2,

$$E^\circ_{cell} = 0.337 - (-0.763) = 1.000 \text{ V}$$

The positive sign for the cell potential indicates that the reaction

$$Zn(s) + Cu^{2+} = Zn^{2+} + Cu(s)$$

occurs spontaneously and that this is a galvanic cell.

To calculate the cell emf and overall reaction for the cell:

$$Pt \,|\, Sn^{2+}(a = 0.2),\ Sn^{4+}(a = 0.8) \,\|\, MnO_4^-(a = 0.9),\ Mn^{2+}$$
$$(a = 0.3),\ H^+(a = 0.5) \,|\, Pt$$

use data from Table 6-2 for the half-reaction

$$Sn^{4+} + 2e^- = Sn^{2+}$$

$$E = 0.154 - \frac{0.0592}{2} \log \frac{0.2}{0.8} = 0.154 + 0.018 = 0.172 \text{ V}$$

and for the other half-reaction

$$MnO_4^- + 8H^+ + 5e^- = Mn^{2+} + 4H_2O$$

$$E = 1.51 - \frac{0.0592}{5} \log \frac{0.3}{(0.9)(0.5)^8} = 1.51 - 0.023 = 1.487 \text{ V}$$

$$E_{\text{cell}} = 1.487 - 0.172 = 1.315 \text{ V}$$

As written, the cell describes a spontaneous reaction because the sign for the cell potential is positive. In fact, a spontaneous reaction is indicated whenever two individual half-reactions are added in the manner required to achieve a positive value for E_{cell}, viz.:

$$-5 \ (Sn^{4+} + 2e^- = Sn^{2+}) \qquad\qquad -(E = 0.172)$$
$$\text{(reaction multiplied by 5)}$$
$$2 \ (MnO_4^- + 8H^+ + 5e^- = Mn^{2+} + 4H_2O) \qquad E = 1.315$$
$$\text{(reaction multiplied by 2)}$$

$$\overline{2MnO_4^- + 5Sn^{2+} + 16H^+ = 2Mn^{2+} + 5Sn^{4+} + 8H_2O}$$

Table 6-1
POTENTIALS OF SELECTED HALF-REACTIONS AT 25°C

A summary of oxidation-reduction half-reactions arranged in order of decreasing oxidation strength and useful for selecting reagent systems.

Half-reaction		$E°$, volts
$F_2(g) + 2H^+ + 2e^-$	$= 2HF$	3.06
$O_3 + 2H^+ + 2e^-$	$= O_2 + H_2O$	2.07
$S_2O_8^{2-} + 2e^-$	$= 2SO_4^{2-}$	2.01
$Ag^{2+} + e^-$	$= Ag^+$	2.00
$H_2O_2 + 2H^+ + 2e^-$	$= 2H_2O$	1.77
$MnO_4^- + 4H^+ + 3e^-$	$= MnO_2(s) + 2H_2O$	1.70
$Ce(IV) + e^-$	$= Ce(III)(\text{in } 1M \text{ HClO}_4)$	1.61
$H_5IO_6 + H^+ + 2e^-$	$= IO_3^- + 3H_2O$	1.6
$Bi_2O_4(\text{bismuthate}) + 4H^+ + 2e^-$	$= 2BiO^+ + 2H_2O$	1.59
$BrO_3^- + 6H^+ + 5e^-$	$= \frac{1}{2}Br_2 + 3H_2O$	1.52
$MnO_4^- + 8H^+ + 5e^-$	$= Mn^{2+} + 4H_2O$	1.51
$PbO_2 + 4H^+ + 2e^-$	$= Pb^{2+} + 2H_2O$	1.455
$Cl_2 + 2e^-$	$= 2Cl^-$	1.36
$Cr_2O_7^{2-} + 14H^+ + 6e^-$	$= 2Cr^{3+} + 7H_2O$	1.33
$MnO_2(s) + 4H^+ + 2e^-$	$= Mn^{2+} + 2H_2O$	1.23
$O_2(g) + 4H^+ + 4e^-$	$= 2H_2O$	1.229
$IO_3^- + 6H^+ + 5e^-$	$= \frac{1}{2}I_2 + 3H_2O$	1.20
$Br_2(\text{liq}) + 2e^-$	$= 2Br^-$	1.065
$ICl_2^- + e^-$	$= \frac{1}{2}I_2 + 2Cl^-$	1.06
$VO_2^+ + 2H^+ + e^-$	$= VO^{2+} + H_2O$	1.00
$HNO_2 + H^+ + e^-$	$= NO(g) + H_2O$	1.00
$NO_3^- + 3H^+ + 2e^-$	$= HNO_2 + H_2O$	0.94
$2Hg^{2+} + 2e^-$	$= Hg_2^{2+}$	0.92
$Cu^{2+} + I^- + e^-$	$= CuI$	0.86
$Ag^+ + e^-$	$= Ag$	0.799
$Hg_2^{2+} + 2e^-$	$= 2Hg$	0.79

Table 6-1 (*Continued*)
POTENTIALS OF SELECTED HALF-REACTIONS AT 25°C

Half-reaction		$E°$, volts
$Fe(III) + e^-$	$= Fe^{2+}$	0.771
$O_2(g) + 2H^+ + 2e^-$	$= H_2O_2$	0.682
$2HgCl_2 + 2e^-$	$= Hg_2Cl_2(s) + 2Cl^-$	0.63
$Hg_2SO_4(s) + 2e^-$	$= 2Hg + SO_4^{2-}$	0.615
$H_3AsO_4 + 2H^+ + 2e^-$	$= HAsO_2 + 2H_2O$	0.581
$Sb_2O_5 + 6H^+ + 4e^-$	$= 2SbO^+ + 3H_2O$	0.559
$I_3^- + 2e^-$	$= 3I^-$	0.545
$Cu^+ + e^-$	$= Cu$	0.52
$VO^{2+} + 2H^+ + e^-$	$= V^{3+} + H_2O$	0.337
$Fe(CN)_6^{3-} + e^-$	$= Fe(CN)_6^{4-}$	0.36
$Cu^{2+} + 2e^-$	$= Cu$	0.337
$UO_2^{2+} + 4H^+ + 2e^-$	$= U^{4+} + 2H_2O$	0.334
$BiO^+ + 2H^+ + 3e^-$	$= Bi + H_2O$	0.32
$Hg_2Cl_2(s) + 2e^-$	$= 2Hg + 2Cl^-$	0.2676
$AgCl(s) + e^-$	$= Ag + Cl^-$	0.2223
$SbO^+ + 2H^+ + 3e^-$	$= Sb + H_2O$	0.212
$CuCl_3^{2-} + e^-$	$= Cu + 3Cl^-$	0.178
$SO_4^{2-} + 4H^+ + 2e^-$	$= SO_2(aq) + 2H_2O$	0.17
$Sn^{4+} + 2e^-$	$= Sn^{2+}$	0.154
$S + 2H^+ + 2e^-$	$= H_2S(g)$	0.141
$TiO^{2+} + 2H^+ + e^-$	$= Ti^{3+} + H_2O$	0.10
$S_4O_6^{2-} + 2e^-$	$= 2S_2O_3^{2-}$	0.08
$AgBr(s) + e^-$	$= Ag + Br^-$	0.071
$2H^+ + 2e^-$	$= H_2$	0.0000
$Pb^{2+} + 2e^-$	$= Pb$	−0.126
$Sn^{2+} + 2e^-$	$= Sn$	−0.136
$AgI(s) + e^-$	$= Ag + I^-$	−0.152
$Mo^{3+} + 3e^-$	$= Mo$	ca −0.2
$N_2 + 5H^+ + 4e^-$	$= H_2NNH_3^+$	−0.23
$Ni^{2+} + 2e^-$	$= Ni$	−0.246
$V^{3+} + e^-$	$= V^{2+}$	−0.255
$Co^{2+} + 2e^-$	$= Co$	−0.277
$Ag(CN)_2^- + e^-$	$= Ag + 2CN^-$	−0.31
$Cd^{2+} + 2e^-$	$= Cd$	−0.403
$Cr^{3+} + e^-$	$= Cr^{2+}$	−0.41
$Fe^{2+} + 2e^-$	$= Fe$	−0.440
$2CO_2 + 2H^+ + 2e^-$	$= H_2C_2O_4$	−0.49
$H_3PO_3 + 2H^+ + 2e^-$	$= H_3PO_2 + H_2O$	−0.50
$U^{4+} + e^-$	$= U^{3+}$	−0.61
$Zn^{2+} + 2e^-$	$= Zn$	−0.763
$Cr^{2+} + 2e^-$	$= Cr$	−0.91
$Mn^{2+} + 2e^-$	$= Mn$	−1.18
$Zr^{4+} + 4e^-$	$= Zr$	−1.53
$Ti^{3+} + 3e^-$	$= Ti$	−1.63
$Al^{3+} + 3e^-$	$= Al$	−1.66
$Th^{4+} + 4e^-$	$= Th$	−1.90
$Mg^{2+} + 2e^-$	$= Mg$	−2.37
$La^{3+} + 3e^-$	$= La$	−2.52
$Na^+ + e^-$	$= Na$	−2.714
$Ca^{2+} + 2e^-$	$= Ca$	−2.87
$Sr^{2+} + 2e^-$	$= Sr$	−2.89
$K^+ + e^-$	$= K$	−2.925
$Li^+ + e^-$	$= Li$	−3.045

Table 6-2
POTENTIALS OF THE ELEMENTS AND THEIR COMPOUNDS AT 25°C

Standard potentials are denoted by 0 in the column headed Solution Composition. All other values are formal potentials; concentrations are in mole liter^{-1}.

Half-reaction	Standard or Formal Potential	Solution Composition
Actinium		
$Ac^{3+} + 3e^- = Ac$	-2.4	
Aluminum		
$Al^{3+} + 3e^- = Al$	-1.66	
$AlF_6^{3-} + 3e^- = Al + 6F^-$	-2.07	
$H_2AlO_3^- + H_2O + 3e^- = Al + 4OH^-$	-2.35	
Americium		
$AmO_2^{2+} + 4H^+ + 3e^- = Am^{3+} + 2H_2O$	$+1.75$	
$AmO_2^+ + 4H^+ + 2e^- = Am^{3+} + 2H_2O$	$+1.83$	
$AmO_2^{2+} + e^- = AmO_2^+$	$+1.6$	1 HClO$_4$
$AmO_2^+ + 4H^+ + e^- = Am^{4+} + 2H_2O$	$+1.26$	
$Am^{4+} + e^- = Am^{3+}$	$+2.4$	
$Am(OH)_4 + e^- = Am(OH)_3 + OH^-$	$(+0.5)$	1 NaOH
$Am^{3+} + 3e^- = Am$	-2.38	
Antimony		
$Sb(OH)_6^- + 2e^- = SbO_2^- + 2OH^- + 2H_2O$	(-0.428)	2 KOH
$SbO_3^- + H_2O + 2e^- = SbO_2^- + 2OH^-$	(-0.59)	10 KOH
$Sb(V) + 2e^- = Sb(III)$	$(+0.75)$	3.5 HCl
	$(+0.82)$	6 HCl
	(-0.43)	3 KOH
	(-0.67)	10 KOH
$Sb_2O_5 + 6H^+ + 4e^- = 2SbO^+ + 3H_2O$	$+0.58$	
$Sb_2O_5 + 4H^+ + 4e^- = Sb_2O_3 + 2H_2O$	$+0.69$	
$SbO_2^- + 2H_2O + 3e^- = Sb + 4OH^-$	(-0.67)	10 KOH
$Sb_2O_3 + 6H^+ + 6e^- = 2Sb + 3H_2O$	$+0.15$	
$SbO^+ + 2H^+ + 3e^- = Sb + H_2O$	$+0.21$	
$Sb + 3H^+ + 3e^- = SbH_3$	-0.51	
$Sb + 3H_2O + 3e^- = SbH_3 + 3OH^-$	-1.34	1 KOH
Arsenic		
$H_3AsO_4 + 2H^+ + 2e^- = HAsO_2 + 2H_2O$	$+0.56$	
$AsO_4^{3-} + 2H_2O + 2e^- = AsO_2^- + 4OH^-$	-0.67	
$HAsO_2 + 3H^+ + 3e^- = As + 2H_2O$	$+0.248$	
$AsO_2^- + 2H_2O + 3e^- = As + 4OH^-$	(-0.66)	1 KOH
$As_2O_3 + 6H^+ + 6e^- = 2As + 3H_2O$	$(+0.23)$	0.2–1 HClO$_4$
$As + 3H^+ + 3e^- = AsH_3$	-0.61	
$As + 3H_2O + 3e^- = AsH_3 + 3OH^-$	-1.21	
Astatine		
$H_5AtO_6 + H^+ + 2e^- = AtO_3^- + 3H_2O$	$< +1.6$	
$HAtO_3 + 4H^+ + 4e^- = HAtO + 2H_2O$	$ca. +1.4$	
$HAtO + H^+ + e^- = \frac{1}{2}At_2 + H_2O$	$ca. +0.7$	
$At_2 + 2e^- = 2At^-$	$+0.30$	
Barium		
$Ba^{2+} + 2e^- = Ba$	-2.91	
Berkelium		
$Bk(IV) + e^- = Bk(III)$	$(+1.6)$	1 HClO$_4$
$Bk^{3+} + 3e^- = Bk$	-2.4	
Beryllium		
$Be^{2+} + 2e^- = Be$	-1.85	
$Be_2O_3^{2-} + 3H_2O + 4e^- = 2Be + 6OH^-$	-2.62	

Table 6-2 (*Continued*)
POTENTIALS OF THE ELEMENTS AND THEIR COMPOUNDS AT 25°C

Half-reaction	Standard or Formal Potential	Solution Composition
Bismuth		
Bi_2O_4 ("bismuthate") $+ 4H^+ + 2e^- = 2BiO^+ + 2H_2O$	$+1.59$	
$2BiO_2 + H_2O + 2e^- = Bi_2O_3 + 2OH^-$	$(+0.55)$	1 NaOH
$BiOH^{2+} + H^+ + 3e^- = Bi + H_2O$	$+0.30$	
$BiO^+ + 2H^+ + 3e^- = Bi + H_2O$	$+0.32$	
$Bi^{3+} + 3e^- = Bi$	$+0.293$	
$BiCl_4^- + 3e^- = Bi + 4Cl^-$	$+0.16$	
$BiOCl + 2H^+ + 3e^- = Bi + H_2O + Cl^-$	$+0.16$	
$Bi_2O_3 + 6H^+ + 6e^- = 2Bi + 3H_2O$	-0.46	
$Bi + 3H^+ + 3e^- = BiH_3$	*ca.* -0.8	
Boron		
$H_3BO_3 + 3H^+ + 3e^- = B + 3H_2O$	-0.87	
$BF_4^- + 3e^- = B + 4F^-$	-1.04	
$H_2BO_3^- + H_2O + 3e^- = B + 4OH^-$	-1.79	1 NaOH
Bromine		
$BrO_3^- + 6H^+ + 6e^- = Br^- + 3H_2O$	$+1.44$	
$BrO_3^- + 6H^+ + 5e^- = \frac{1}{2}Br_2 + 3H_2O$	$+1.5$	
$HBrO + H^+ + e^- - \frac{1}{2}Br_2 + H_2O$	$+1.6$	
$BrCl + 2e^- = Br^- + Cl^-$	$+1.35$	
	$(+1.16)$	6 HCl
$BrO^- + H_2O + 2e^- = Br^- + 2OH^-$	$+0.76$	
$Br_3^- + 2e^- = 3Br^-$	$+1.05$	
$Br_2(aq) + 2e^- = 2Br^-$	$+1.08$	
$Br_2(liq) + 2e^- = 2Br^-$	$+1.065$	
Cadmium		
$Cd^{2+} + 2e^- = Cd$	-0.403	
$Cd^{2+} + (Hg) + 2e^- = Cd(Hg)$	-0.352	
$Cd(NH_3)_4^{2+} + 2e^- = Cd + 4NH_3$	-0.61	
$Cd(CN)_4^{2-} + 2e^- = Cd + 4CN^-$	-1.09	
Calcium		
$Ca^{2+} + 2e^- = Ca$	-2.87	
Californium		
$Cf^{3+} + 3e^- = Cf$	-2.1	
Carbon		
$\frac{1}{2}C_2N_2 + H^+ + e^- = HCN$	$+0.37$	
$HCNO + H^+ + e^- = \frac{1}{2}C_2N_2 + H_2O$	$+0.33$	
$CO_2 + 2H^+ + 2e^- = CO + H_2O$	-0.12	
$CO_2 + 2H^+ + 2e^- = HCOOH$	-0.20	
$2CO_2 + 2H^+ + 2e^- = H_2C_2O_4$	-0.49	
$HCHO + 2H_2O + 2e^- = CH_3OH + 2OH^-$	-0.59	
$CNO^- + H_2O + 2e^- = CN^- + 2OH^-$	-0.97	
Cerium		
$Ce(IV) + e^- = Ce(III)$	$(+1.70)$	1 HClO$_4$
	$(+1.60)$	1 HNO$_3$
	$(+1.45)$	0.5 H$_2$SO$_4$
	$(+1.28)$	1 HCl
	$(+0.06)$	2.5 K$_2$CO$_3$
$Ce^{3+} + 3e^- = Ce$	-2.33	
$Ce(OH)^{3+} + H^+ + e^- = Ce^{3+} + H_2O$	$(+1.70)$	1 HClO$_4$
Cesium		
$Cs^+ + e^- = Cs$	-2.92	
$Cs^+ + (Hg) + e^- = Cs(Hg)$	-1.78	

Table 6-2 (*Continued*)
POTENTIALS OF THE ELEMENTS AND THEIR COMPOUNDS AT 25°C

Half-reaction	Standard or Formal Potential	Solution Composition
Chlorine		
$ClO_4^- + 2H^+ + 2e^- = ClO_3^- + H_2O$	+1.19	
$ClO_4^- + 8H^+ + 7e^- = \frac{1}{2}Cl_2 + 4H_2O$	+1.34	
$ClO_3^- + 6H^+ + 5e^- = \frac{1}{2}Cl_2 + 3H_2O$	+1.47	
$ClO_3^- + 6H^+ + 6e^- = Cl^- + 3H_2O$	+1.45	
$ClO_3^- + 2H^+ + e^- = ClO_2 + H_2O$	+1.15	
$HClO_2 + 2H^+ + 2e^- = HClO + H_2O$	+1.64	
$HClO + H^+ + 2e^- = Cl^- + H_2O$	+1.49	
$HClO + H^+ + e^- = \frac{1}{2}Cl_2 + H_2O$	+1.63	
$ClO^- + H_2O + 2e^- = Cl^- + 2OH^-$	+0.89	
$ClO_2(g) + e^- = ClO_2^-$	+0.95	
$ClO_2(g) + H^+ + e^- = HClO_2$	+1.27	
$Cl_2 + 2e^- = 2Cl^-$	+1.39	
Chromium		
$Cr_2O_7^{2-} + 14H^+ + 6e^- = 2Cr^{3+} + 7H_2O$	+1.33	
	(+1.15)	4 H_2SO_4
	(+0.92)	0.1 H_2SO_4
	(+1.03)	1 $HClO_4$
$HCrO_4^- + 7H^+ + 3e^- = Cr^{3+} + 4H_2O$	+1.20	
	(+0.84)	0.1 $HClO_4$
	(+1.08)	3 HCl
	(+0.93)	0.1 HCl
$CrO_4^{2-} + 2H_2O + 3e^- = CrO_2^- + 4OH^-$	−0.12	1 NaOH
$Cr^{3+} + e^- = Cr^{2+}$	(−0.41)	0.0015 H_2SO_4
	(−0.38)	1 HCl
	(−0.51)	1 HF
	(−0.26)	satd. $CaCl_2$
$Cr(CN)_6^{3-} + e^- = Cr(CN)_6^{4-}$	(−1.14)	1 KCN
$Cr^{3+} + 3e^- = Cr$	−0.74	
$Cr^{2+} + 2e^- = Cr$	−0.86	
Cobalt		
$Co(III) + e^- = Co(II)$	(+1.95)	4 $HClO_4$
	(+1.80)	1 HNO_3 or 1 H_2SO_4
$Co(OH)_3 + e^- = Co(OH)_2 + OH^-$	+0.17	
$CoO_2^- + 2H_2O + e^- = Co(OH)_2 + 2OH^-$	(−0.22)	0.01 KOH
$Co(NH_3)_6^{3+} + e^- = Co(NH_3)_6^{2+}$	+0.1	
$Co(NH_3)_5^{3+} + e^- = Co(NH_3)_5^{2+}$	(+0.37)	1 NH_4NO_3
$Co(en)_3^{3+} + e^- = Co(en)_3^{2+}$ [en = ethylenediamine]	(−0.2)	0.1 en + 0.1 KNO_3
$Co(CN)_6^{3-} + e^- = Co(CN)_5^{3-} + CN^-$	< −0.8	0.8 KOH
$Co^{2+} + 2e^- = Co$	−0.29	
$Co(NH_3)_6^{2+} + 2e^- = Co + 6NH_3$	−0.422	
$[Co(CO)_4]_2 + 2e^- = 2Co(CO)_4^-$	−0.40	
Copper		
$Cu_2O_3 + 2H^+ + 2e^- = 2CuO + H_2O$	(*ca.* +1.6)	
$Cu_2O_3 + H_2O + 2e^- = 2CuO + 2OH^-$	(+0.74)	pH 14
$Cu^{2+} + 2e^- = Cu$	+0.34	
$Cu^{2+} + e^- = Cu^+$	+0.17	
$2CuO + 2H^+ + 2e^- = Cu_2O + H_2O$	+0.64	
$Cu^{2+} + 2Br^- + e^- = CuBr_2^-$	(+0.52)	0.7–1 KBr
$Cu^{2+} + I^- + e^- = CuI$	+0.86	
$Cu^{2+} + 2CN^- + e^- = Cu(CN)_2^-$	*ca.* +1.12	
$Cu(CN)_2^{2-} + e^- = Cu + 3CN^-$	(−1.0)	7 KCN
$Cu(NH_3)_4^{2+} + e^- = Cu(NH_3)_2^+ + 2NH_3$	(−0.01)	1 NH_3 + 1 NH_4^+

Table 6-2 (*Continued*)
POTENTIALS OF THE ELEMENTS AND THEIR COMPOUNDS AT 25°C

Half-reaction	Standard or Formal Potential	Solution Composition
Copper (*cont.*)		
$Cu(en)_2^{2+} + e^- = Cu(en)^+ + en$	-0.35	
$CuCl_3^{2-} + e^- = Cu + 3Cl^-$	$(+0.178)$	1 HCl
$Cu(NH_3)_2^+ + e^- = Cu + 2NH_3$	(-0.12)	$1\ NH_3 + 1\ NH_4^+$
$Cu(EDTA)^2 + 2e^- = Cu + EDTA^{4-}$	$+0.13$	0.1 EDTA, pH 4–5
$CuN_3 + e^- = Cu + N_3^-$	$+0.03$	
$Cu^+ + e^- = Cu$	$+0.52$	
$Cu_2O + 2H^+ + 2e^- = 2Cu + H_2O$	-0.36	
Curium		
$Cm(IV) + e^- = Cm(III)$	(*ca.* $+3.2$)	1 $HClO_4$
$Cm^{3+} + 3e^- = Cm$	-2.70	
Dysprosium		
$Dy^{3+} + 3e^- = Dy$	-2.35	
Einsteinium		
$Es^{3+} + 3e^- = Es$	-2.0	
Erbium		
$Er^{3+} + 3e^- = Er$	-2.29	
Europium		
$Eu^{3+} + e^- = Eu^{2+}$	-0.35	
$Eu(EDTA)(III) + e = Eu(EDTA)(II)$	(-0.92)	0.1 EDTA, pH 6–8
$Eu^{3+} + 3e^- = Eu$	-2.40	
Fermium		
$Fm^{3+} + 3e^- = Fm$	-2.1	
Fluorine		
$F_2 + 2H^+ + 2e^- = 2HF$	$+3.06$	
$F_2 + 2e^- = 2F^-$	$+2.87$	
$OF_2 + 2H^+ + 4e^- = H_2O + 2F^-$	$+2.1$	
Francium		
$Fr^+ + e = Fr$	*ca.* -2.9	
Gadolinium		
$Gd^{3+} + 3e^- = Gd$	-2.40	
Gallium		
$Ga^{3+} + 3e^- = Ga$	-0.56	
$Ga(OH)_4^- + 3e^- = Ga + 4OH^-$	-1.3	
Germanium		
$GeO_2(s, hex) + 2H^+ + 2e^- = GeO + H_2O$	-0.118	
$H_2GeO_3 + 4H^+ + 4e^- = Ge + 3H_2O$	$+0.01$	
$GeO_2(s, tetr) + 4H^+ + 2e^- = Ge^{2+} + 2H_2O$	-0.34	
$GeO_2(s, hex) + 4H^+ + 2e^- = Ge^{2+} + 2H_2O$	-0.25	
$Ge^{4+} + 2e^- = Ge^{2+}$	0.0	
$HGeO_3^- + 2H_2O + 4e^- = Ge + 5OH^-$	-1.0	
$Ge^{2+} + 2e^- = Ge$	$+0.23$	
$GeO + 2H^+ + 2e^- = Ge + H_2O$	*ca.* -0.2	
$GeI_2 + 2e^- = Ge + 2I^-$	*ca.* 0.0	
$Ge + 4H^+ + 4e^- = GeH_4$	*ca.* -0.3	
Gold		
$Au(III) + 2e^- = Au(I)$	*ca.* $+1.41$	
$Au(III) + 3e^- = Au$	$+1.50$	
$Au_2O_3 + 6H^+ + 6e^- = 2Au + 3H_2O$	$(+1.36)$	$2\ H_2SO_4 + 2\ Na_2SO_4$
$AuCl_4^- + 2e^- = AuCl_2^- + 2Cl^-$	$(+0.92)$	1 HCl
$AuBr_4^- + 2e^- = AuBr_2^- + 2Br^-$	$(+0.80)$	1 HBr
$Au(SCN)_4^- + 3e^- = Au + 4SCN^-$	$+0.66$	
$Au(SCN)_4^- + 2e^- = Au(SCN)_2^- + 2SCN^-$	$+0.64$	

Table 6-2 (*Continued*)
POTENTIALS OF THE ELEMENTS AND THEIR COMPOUNDS AT 25°C

Half-reaction	Standard or Formal Potential	Solution Composition
Gold (*cont.*)		
$Au(OH)_3 + 3H^+ + 3e^- = Au + 3H_2O$	+1.45	
$AuCl_4^- + 3e^- = Au + 4Cl^-$	+1.00	
$AuBr_4^- + 3e^- = Au + 4Br^-$	+0.85	
$Au(CN)_2^- + e^- = Au + 2CN^-$	−0.6	
$AuCl_2^- + e^- = Au + 2Cl^-$	(+1.15)	1 Cl⁻
$AuBr_2^- + e^- = Au + 2Br^-$	+0.93	
$Au(SCN)_2^- + e^- = Au + 2SCN^-$	+0.69	
Hafnium		
$Hf(IV) + 4e^- = Hf$	−1.70	
$HfO_2 + 4H^+ + 4e^- = Hf + 2H_2O$	−1.57	
Holmium		
$Ho^{3+} + 3e^- = Ho$	−2.32	
Hydrogen		
$2H^+ + 2e^- = H_2$	0.000	
$2D^+ + 2e^- = D_2$	+0.029	
$2H_2O + 2e^- = H_2 + 2OH^-$	−0.828	
$\frac{1}{2}H_2 + e^- = H^-$	−2.25	
Indium		
$In^{3+} + 3e^- = In$	−0.34	
$In(OH)_3 + 3e^- = In + 3OH^-$	−1.0	
$In^{3+} + 2e^- = In^+$	(−0.43)	3 NaClO₄
$In^+ + e^- = In$	(−0.18)	3 NaClO₄
Iodine		
$H_5IO_6 + H^+ + 2e^- = IO_3^- + 3H_2O$	*ca.* +1.6	
$IO_3^- + 5H^+ + 4e^- = HIO + 2H_2O$	+1.14	
$IO_3^- + 6H^+ + 5e^- = \frac{1}{2}I_2 + 3H_2O$	+1.19	
$2ICl_3 + 6e^- = I_2 + 6Cl^-$	+1.28	
$2ICl + 2e^- = I_2 + 2Cl^-$	+1.19	
$2IBr + 2e^- = I_2 + 2Br^-$	+1.02	
$2ICN + 2H^+ + 2e = I_2 + 2HCN$	+0.63	
$2HIO + 2H^+ + 2e^- = I_2 + 2H_2O$	+1.45	
$HIO + H^+ + 2e^- = I^- + H_2O$	+0.99	
$ICl_2^- + e^- = \frac{1}{2}I_2 + 2Cl^-$	+1.06	
$IBr_2^- + e^- = \frac{1}{2}I_2 + 2Br^-$	+0.87	
$I_3^- + 2e^- = 3I^-$	(+0.545)	0.5 H₂SO₄
$I_2(aq) + 2e^- = 2I^-$	+0.621	
$I_2(s) + 2e^- = 2I^-$	+0.535	
Iridium		
$IrCl_6^{2-} + e^- = IrCl_6^{3-}$	+0.87	
	(+0.93)	1 HCl
$IrBr_6^{2-} + e^- = IrBr_6^{3-}$	(+0.95)	1 NaBr
$IrI_6^{2-} + e^- = IrI_6^{3-}$	(+0.48)	1 KI
$IrO_2 + 4H^+ + 4e^- = Ir + 2H_2O$	*ca.* +0.93	
$IrCl_6^{2-} + 4e^- = Ir + 6Cl^-$	+0.835	
$Ir^{3+} + 3e^- = Ir$	+1.15	
$IrCl_6^{3-} + 3e^- = Ir + 6Cl^-$	+0.77	
Iron		
$FeO_4^{2-} + 8H^+ + 3e^- = Fe^{3+} + 4H_2O$	+2.2	
$FeO_4^{2-} + 2H_2O + 3e^- = FeO_2^- + 4OH^-$	(+0.55)	10 NaOH
$Fe(III) + e^- = Fe(II)$	+0.771	
	(+0.75)	1 HClO₄
	(+0.74)	0.2 HNO₃

Table 6-2 (*Continued*)
POTENTIALS OF THE ELEMENTS AND THEIR COMPOUNDS AT 25°C

Half-reaction	Standard or Formal Potential	Solution Composition
Iron (*cont.*)		
$Fe(III) + e^- = Fe(II)$ (*cont.*)	(+0.67)	0.5 H_2SO_4
	(+0.70)	1 HCl
	(+0.53)	10 HCl
	(+0.44)	0.3 H_3PO_4
	(−0.68)	10 NaOH
$Fe(CN)_6^{3-} + e^- = Fe(CN)_6^{4-}$	+0.55	
	(+0.72)	1 $HClO_4$
	(+0.71)	1 HCl
$Fe(EDTA)^- + e^- = Fe(EDTA)^{2-}$	(+0.12)	0.1 EDTA, pH 4–6
$Fe(OH)_3 + 3H^+ + e^- = Fe^{2+} + 3H_2O$	+0.93	
$Fe(OH)_4^- + e^- = Fe(OH)_4^{2-}$	(−0.73)	4 KOH
$Fe_3O_4 + 8H^+ + 2e^- = 3Fe^{2+} + 4H_2O$	+1.23	
$Fe^{2+} + 2e^- = Fe$	−0.44	
$[Fe(CO)_4]_3 + 6e^- = 3Fe(CO)_4^{2-}$	−0.70	
Lanthanum		
$La^{3+} + 3e^- = La$	−2.52	
Lead		
$Pb^{4+} + 2e^- = Pb^{2+}$	(+1.65)	1.1 $HClO_4$
	(+1.8)	1–8 HNO_3
$PbO_2 + SO_4^{2-} + 4H^+ + 2e^- = PbSO_4 + 2H_2O$	+1.69	
$PbO_2 + 4H^+ + 2e^- = Pb^{2+} + 2H_2O$	+1.455	
$PbO_2 + 2H^+ + 2e^- = PbO + H_2O$	+0.28	
$PbO_3^{2-} + 2H_2O + 2e^- = HPbO_2^- + 3OH^-$	(+0.3)	1.7–2.5 NaOH
$PbO_3^{2-} + H_2O + 2e^- = PbO_2^{2-} + 2OH^-$	(+0.2)	8.4 KOH
$Pb^{2+} + 2e^- = Pb$	−0.126	
	(−0.32)	1 NaOAc
$HPbO_2^- + H_2O + 2e^- = Pb + 3OH^-$	−0.54	
$PbO(s, red) + H_2O + 2e^- = Pb + 2OH^-$	(−0.58)	0.075–0.25 $Ba(OH)_2$
$PbSO_4 + 2e^- = Pb + SO_4^{2-}$	−0.356	
$PbF_2 + 2e^- = Pb + 2F^-$	−0.350	
$PbCl_2 + 2e^- = Pb + 2Cl^-$	−0.266	
$PbBr_2 + 2e^- = Pb + 2Br^-$	−0.274	
$PbI_2 + 2e^- = Pb + 2I^-$	−0.364	
$Pb(N_3)_2 + 2e^- = Pb + 2N_3^-$	−0.380	
Lithium		
$Li^+ + e^- = Li$	−3.03	
$Li^+ + (Hg) + e^- = Li(Hg)$	−2.00	
Lutetium		
$Lu^{3+} + 3e^- = Lu$	−2.25	
Magnesium		
$Mg^{2+} + 2e^- = Mg$	−2.37	
$Mg(OH)_2 + 2e^- = Mg + 2OH^-$	−2.69	
Manganese		
$MnO_4^- + e^- = MnO_4^{2-}$	+0.57	
$MnO_4^- + 4H^+ + 3e^- = MnO_2(\beta) + 2H_2O$	+1.68	
	(+1.65)	0.5 H_2SO_4
	(+1.60)	1 HNO_3 or 1 $HClO_4$
$MnO_4^- + 8H^+ + 5e^- = Mn^{2+} + 4H_2O$	+1.51	
$MnO_4^- + 2H_2O + 3e^- = MnO_2 + 4OH^-$	+0.588	
$MnO_4^{2-} + e^- = MnO_4^{3-}$	+0.27	
$MnO_4^{2-} + 2H_2O + 2e^- = MnO_2 + 4OH^-$	*ca.* +0.5	8 KOH

Table 6-2 (*Continued*)
POTENTIALS OF THE ELEMENTS AND THEIR COMPOUNDS AT 25°C

Half-reaction	Standard or Formal Potential	Solution Composition
Manganese (*cont.*)		
$2MnO_2 + 2H^+ + 2e^- = Mn_2O_3 + H_2O$	(+1.04)	5 NH_4Cl
$MnO_2(\beta) + 4H^+ + 2e^- = Mn^{2+} + 2H_2O$	+1.23	
$Mn(III) + e^- = Mn(II)$	(+1.488)	7.5 H_2SO_4
$Mn(H_2P_2O_7)_3^{3-} + 2H^+ + e^- = Mn(H_2P_2O_7)_3^{2-}$		
$\qquad\qquad\qquad + H_4P_2O_7$	(+1.15)	0.4 $H_2P_2O_7^{2-}$
$Mn(CN)_6^{3-} + e^- = Mn(CN)_6^{4-}$	(−0.24)	1.5 NaCN
$Mn(CN)_6^{4-} + e^- = Mn(CN)_6^{5-}$	(−1.05)	1.5–2.5 NaCN
$Mn(OH)_3 + e^- = Mn(OH)_2 + OH^-$	+0.1	
$Mn^{2+} + 2e^- = Mn$	−1.17	
$Mn(OH)_2 + 2e^- = Mn + 2OH^-$	−1.55	
$[Mn(CO)_5]_2 + 2e^- = 2Mn(CO)_5^-$	−0.68	
Mendelevium		
$Md^{3+} + 3e^- = Md$	−2.2	
Mercury		
$2Hg^{2+} + 2e^- = Hg_2^{2+}$	+0.907	
$2HgCl_2 + 2e^- = Hg_2Cl_2 + 2Cl^-$	+0.63	
$Hg^{2+} + 2e^- = Hg$	+0.854	
$HgO + 2H^+ + 2e^- = Hg + H_2O$	+0.926	
$HgO + H_2O + 2e^- = Hg + 2OH^-$	+0.098	
$Hg(CN)_4^{2-} + 2e^- = Hg + 4CN^-$	−0.37	
$HgCl_4^{2-} + 2e^- = Hg + 4Cl^-$	+0.48	
$Hg_2^{2+} + 2e^- = 2Hg$	+0.792	
$Hg_2Cl_2 + 2e^- = 2Hg + 2Cl^-$	+0.268	
$Hg_2Br_2 + 2e^- = 2Hg + 2Br^-$	+0.1392	
$Hg_2I_2 + 2e^- = 2Hg + 2I^-$	−0.040	
$Hg_2SO_4 + 2e^- = 2Hg + SO_4^{2-}$	+0.614	
$Hg_2C_2O_4 + 2e^- = 2Hg + C_2O_4^{2-}$	+0.415	
$Hg_2HPO_4 + 2e^- = 2Hg + HPO_4^{2-}$	+0.64	
$Hg_2(IO_3)_2 + 2e^- = 2Hg + 2IO_3^-$	+0.39	
$Hg_2(N_3)_2 + 2e^- = 2Hg + 2N_3^-$	−0.26	
$(Hg_2)_3[Co(CN)_6]_2 + 6e^- = 6Hg + 2Co(CN)_6^{3-}$	(−0.43)	
Molybdenum		
$H_2MoO_4(aq) + 2H^+ + e^- = MoO_2^+ + 2H_2O$	+0.48	
$MoO_2^{2+} + 2H^+ + e^- = MoO^{3+} + H_2O$	+0.48	
$Mo(VI) + e^- = Mo(V)$	(+0.53)	9.2 H_2SO_4
	(+0.40)	0.5 H_2SO_4
	(+0.70)	8 HCl
	(+0.41)	1 H_3PO_4
	(+0.50)	2 KSCN + 1 HCl
$MoO_2^+ + 4H^+ + 2e^- = Mo^{3+} + 2H_2O$	(−0.01)	0.5 H_2SO_4
$Mo(CN)_8^{3-} + e^- = Mo(CN)_8^{4-}$	(+0.82)	1 H_2SO_4
$Mo(V) + 2e = Mo(III)$ (green)	(+0.25)	2 HCl
(red)	(+0.11)	2 HCl
$Mo(CN)_6^{3-} + e^- = Mo(CN)_6^{4-}$	(+0.73)	0.25 KCl
$Mo(III) + 3e^- = Mo$	ca. −0.2	
$SiMo_{12}O_{40}^{4-} + 4H^+ + 4e^- = H_4SiMo_{12}O_{40}^{4-}$	(0.59)	0.5 H_2SO_4
Neodymium		
$Nd^{3+} + 3e^- = Nd$	−2.45	
Neptunium		
$NpO_2^{2+} + e^- = NpO_2^+$	(+1.14)	1 HNO_3
$NpO_2^+ + 4H^+ + e^- = Np^{4+} + 2H_2O$	(+0.74)	1 HCl

Table 6-2 (*Continued*)
POTENTIALS OF THE ELEMENTS AND THEIR COMPOUNDS AT 25°C

Half-reaction	Standard or Formal Potential	Solution Composition
Neptunium (*cont.*)		
$Np^{4+} + e^- = Np^{3+}$	$(+0.15)$	1 $HClO_4$
	$(+0.14)$	1 HCl
	$(+0.11)$	1 HNO_3
$Np^{3+} + 3e^- = Np$	-1.85	
Nickel		
$Ni(OH)_4 + e^- = Ni(OH)_3 + OH^-$	$>+0.6$	
$NiO_2 + 2H_2O + 2e^- = Ni(OH)_2 + 2OH^-$	$+0.49$	
$NiO_2 + 4H^+ + 2e^- = Ni^{2+} + 2H_2O$	$+1.68$	
$Ni(OH)_3 + 3H^+ + e^- = Ni^{2+} + 3H_2O$	$+2.08$	
$Ni(OH)_3 + e^- = Ni(OH)_2 + OH^-$	$+0.48$	
$Ni(CN)_4^{2-} + e^- = Ni(CN)_4^{3-}$	(-0.82)	1 KCN
$Ni^{2+} + 2e^- = Ni$	-0.25	
$Ni(OH)_2 + 2e^- = Ni + 2OH^-$	-0.72	
$Ni(NH_3)_6^{2+} + 2e^- = Ni + 6NH_3$	-0.48	
Niobium		
$NbO^{3+} + 2H^+ + 2e = Nb^{3+} + H_2O$	(-0.37)	2–6 HCl
$Nb(V) + e^- = Nb(IV)$	(-0.21)	9–12 HCl
$Nb^{3+} + 3e^- = Nb$	*ca.* -1.1	
Nitrogen		
$NO_3^- + 3H^+ + 2e^- = HNO_2 + H_2O$	$+0.94$	
$NO_3^- + 2H^+ + e^- = NO_2 + H_2O$	$+0.80$	
$NO_3^- + NO + e^- = 2NO_2^-$	$+0.49$	
$NO_3^- + 4H^+ + 3e^- = NO + 2H_2O$	$+0.96$	
$NO_3^- + H_2O + 2e^- = NO_2^- + 2OH^-$	$+0.01$	
$NO_2 + H^+ + e^- = HNO_2$	$+1.07$	
$NO_2 + 2H^+ + 2e^- = NO + H_2O$	$+1.03$	
$HNO_2 + H^+ + e^- = NO + H_2O$	$+0.98$	
$2NO + 2H^+ + 2e^- = H_2N_2O_2$	$+0.71$	
$H_2N_2O_2 + 2H^+ + 2e^- = N_2 + 2H_2O$	$+2.65$	
$HONH_3^+ + 2H^+ + 2e^- = NH_4^+ + H_2O$	$+1.35$	
$2HONH_3^+ + H^+ + 2e^- = N_2H_5^+ + 2H_2O$	$+1.42$	
$H_2NNH_3^+ + 3H^+ + 2e^- = 2NH_4^+$	$+1.27$	
$N_2 + 2H_2O + 4H^+ + 2e^- = 2HONH_3^+$	-1.87	
$N_2 + 5H^+ + 4e^- = H_2NNH_3^+$	-0.23	
$3N_2 + 2H^+ + 2e^- = 2HN_3$	-3.1	
Nobelium		
$No^{3+} + 3e^- = No$	-2.5	
Osmium		
$OsO_4 + 4H^+ + 4e^- = OsO_2 + 2H_2O$	$+0.96$	
$HOsO_5^- + 2e^- = OsO_4^{2-} + OH^-$	$+0.3$	
$OsO_4(s, yel) + 8H^+ + 8e^- = Os + 4H_2O$	$+0.85$	
$OsCl_6^{2-} + e^- = OsCl_6^{3-}$	$+0.42$	1 HCl
$OsBr_6^{2-} + e^- = OsBr_6^{3-}$	$+0.35$	2 HBr
$OsCl_6^{3-} + 3e^- = Os + 6Cl^-$	$+0.71$	1 HCl
$Os^{2+} + 2e^- = Os$	$+0.85$	
Oxygen		
$O_3 + 2H^+ + 2e^- = O_2 + H_2O$	$+2.07$	
$O_3 + H_2O + 2e^- = O_2 + 2OH^-$	$+1.24$	
$O_2 + 4H^+ + 4e^- = 2H_2O$	$+1.229$	
$O_2 + 2H^+ + 2e^- = H_2O_2$	$+0.69$	
$O_2 + H_2O + 2e^- = HO_2^- + OH^-$	-0.076	
$H_2O_2 + 2H^+ + 2e^- = 2H_2O$	$+1.77$	

6-13

Table 6-2 (*Continued*)
POTENTIALS OF THE ELEMENTS AND THEIR COMPOUNDS AT 25°C

Half-reaction	Standard or Formal Potential	Solution Composition
Oxygen (*cont.*)		
$H_2O_2 + 2e^- = 2OH^-$	+0.88	
$O_2 + 2H_2O + 4e^- = 4OH^-$	+0.41	1 NaOH
Palladium		
$PdO_3 + 2H^+ + 2e^- = PdO_2 + H_2O$	+1.22	
$PdO_2 + 2H^+ + 2e^- = PdO + H_2O$	+0.95	
$PdCl_6^{2-} + 2e^- = PdCl_4^{2-} + 2Cl^-$	+1.29	
$PdBr_6^{2-} + 2e^- = PdBr_4^{2-} + 2Br^-$	(+0.99)	1 KBr
$PdI_6^{2-} + 2e^- = PdI_4^{2-} + 2I^-$	(+0.48)	1 KI
$Pd^{2+} + 2e^- = Pd$	+0.92	
$PdCl_4^{2-} + 2e^- = Pd + 4Cl^-$	+0.62	
$PdBr_4^{2-} + 2e^- = Pd + 4Br^-$	+0.6	
$Pd(OH)_2 + 2e^- = Pd + 2OH^-$	(−0.19)	0.1 K_2SO_4
$Pd(NH_3)_4^{2+} + 2e^- = Pd + 4NH_3$	(−0.56)	1 NH_3 + 1 NH_4Cl
$Pd(CN)_4^{2-} + 2e^- = Pd + 4CN^-$	(−1.53)	1 KCN
Phosphorus		
$H_3PO_4 + 2H^+ + 2e^- = H_3PO_3 + H_2O$	−0.28	
$H_3PO_3 + 2H^+ + 2e^- = HPH_2O_2 + H_2O$	−0.50	
$HPH_2O_2 + H^+ + e^- = P + 2H_2O$	−0.51	
$4P + 2H^+ + 2e^- = H_2P_4$	−0.35	
$P(white) + 3H^+ + 3e^- = H_3P$	+0.06	
Platinum		
$Pt(CN)_4Cl_2^{2-} + 2e^- = Pt(CN)_4^{2-} + 2Cl^-$	(+0.89)	1 KCl
$PtCl_6^{2-} + 2e^- = PtCl_4^{2-} + 2Cl^-$	+0.73	
$PtBr_6^{2-} + 2e^- = PtBr_4^{2-} + 2Br^-$	(+0.64)	1 KBr
$PtI_6^{2-} + 2e^- = PtI_4^{2-} + 2I^-$	(+0.39)	1 KI
$PtO_4^{2-} + 4H_2O + 2e^- = Pt(OH)_6^{2-} + 2OH^-$	*ca.* +0.4	
$Pt^{2+} + 2e^- = Pt$	*ca.* +1.2	
$Pt(OH)_2 + 2H^+ + 2e^- = Pt + 2H_2O$	+0.98	
$PtCl_4^{2-} + 2e^- = Pt + 4Cl^-$	+0.73	
$PtBr_4^{2-} + 2e^- = Pt + 4Br^-$	+0.58	
$Pt(OH)_2 + 2e^- = Pt + 2OH^-$	(−0.14)	0.1 K_2SO_4
Plutonium		
$Pu(VI) + e^- = Pu(V)$	(+0.92)	1 $HClO_4$ or 0.1 HNO_3
	(+0.91)	1 HCl
$Pu(VI) + 2e^- = Pu(IV)$	(+1.05)	1 HNO_3 or 1 HCl
	(+1.04)	1 $HClO_4$
$Pu^{4+} + e^- = Pu^{3+}$	+1.01	
$Pu(IV) + e^- = Pu(III)$	(+0.97)	1 $HClO_4$
	(*ca.* +0.9)	1 HCl
	(+0.92)	1 HNO_3
	(+0.80)	1 H_3PO_4 + 1 HCl
	(+0.74)	0.5 H_2SO_4
	(+0.50)	1 HF
	(+0.59)	0.6 H_3PO_4 + 1 HCl
	(+0.40)	1 HOAc + 1 NaOAc
$Pu^{3+} + 3e^- = Pu$	−2.03	
Polonium		
$PoO_2 + 4H^+ + 2e^- = Po^{2+} + 2H_2O$	(+0.8)	1 HNO_3
$Po^{4+} + 4e^- = Po$	+0.77	
$Po^{3+} + 3e^- = Po$	+0.56	
$Po^{2+} + 2e^- = Po$	(+0.6–0.7)	HCl
$Po + 2H^+ + 2e^- = H_2Po$	*ca.* −1.0	

Table 6-2 (*Continued*)
POTENTIALS OF THE ELEMENTS AND THEIR COMPOUNDS AT 25°C

Half-reaction	Standard or Formal Potential	Solution Composition
Potassium		
$K^+ + e^- = K$	-2.925	
$K^+ + (Hg) + e^- = K(Hg)$	*ca.* $+1.9$	
Praseodymium		
$Pr^{4+} + e^- = Pr^{3+}$	*ca.* $+2.9$	
$Pr^{3+} + 3e^- = Pr$	-2.47	
Promethium		
$Pm^{3+} + 3e^- = Pm$	-2.42	
Protoactinium		
$PaO_2^+ + 4H^+ + 5e^- = Pa + 2H_2O$	*ca.* -1.0	
$PaF_7^{2-} + 5e^- = Pa + 7F^-$	(-1.03)	
$Pa(V) + e^- = Pa(IV)$	(-0.25)	1 HCl
$PaO_2 + 4H^+ + e^- = Pa^{3+} + 2H_2O$	-0.5	
$Pa^{4+} + e^- = Pa^{3+}$	-1.0	
$Pa^{4+} + 4e^- = Pa$	-1.7	
$Pa^{3+} + 3e^- = Pa$	-1.95	
Radium		
$Ra^{2+} + 2e^- = Ra$	-2.92	
Rhenium		
$ReO_4^- + 2H^+ + e^- = ReO_3 + H_2O$	$+0.77$	
$ReO_4^- + 4H^+ + 3e^- = ReO_2 + 2H_2O$	$+0.51$	
$ReO_4^- + 2H_2O + 3e^- = ReO_2 + 4OH^-$	-0.59	
$ReO_4^- + 8H^+ + 7e^- = Re + 4H_2O$	$+0.37$	
$ReO_3 + 2H^+ + 2e^- = ReO_2 + H_2O$	*ca.* $+0.4$	
$Re(V) + 2e^- = Re(III)$	$(+0.14)$	2 NaCN
$ReO_2 + 4H^+ + 4e^- = Re + 2H_2O$	$+0.26$	
$Re(CN)_6^{3-} + e^- = Re(CN)_6^{4-}$	(-0.72)	pH 0
$ReCl_6^{2-} + e^- = ReCl_4^- + 2Cl^-$	$+0.25$	
$Re^{3+} + 3e^- = Re$	*ca.* $+0.3$	
$Re + e^- = Re^-$	-0.23	
Rhodium		
$Rh(VI) + 2e^- = Rh(IV)$	$(+1.5)$	0.1 H_2SO_4
$Rh(VI) + 3e^- = Rh(III)$	$(+1.5)$	1 $HClO_4$
$RhO^{2+} + 2H^+ + e^- = Rh^{3+} + H_2O$	$(+1.43)$	0.5 H_2SO_4
$RhCl_6^{3-} + 3e^- = Rh + 6Cl^-$	$+0.44$	
Rubidium		
$Rb^+ + e^- = Rb$	-2.93	
$Rb^+ + (Hg) + e^- = Rb(Hg)$	-1.81	
Ruthenium		
$RuO_4 + e^- = RuO_4^-$	$+1.00$	
$Ru(VIII) + 4e^- = Ru(IV)$	$(+1.4)$	1 $HClO_4$
$RuO_4^- + e^- = RuO_4^{2-}$	$+0.59$	
$Ru(IV) + e^- = Ru(III)$	$(+0.86)$	2 HCl
$2Ru(IV) + e^- = Ru(IV) \cdot Ru(III)$	$(+0.56)$	pH 1.15
$Ru(IV) \cdot Ru(III) + e^- = 2Ru(III)$	$(+0.4)$	pH 1.15
	(*ca.* $+1.0$)	1 CF_3COOH
$Ru(CN)_6^{3-} + e^- = Ru(CN)_6^{4-}$	$(+0.8)$	0.05 H_2SO_4
$Ru(III) + e^- = Ru(II)$	(*ca.* 0)	HCl or H_2SO_4 or $HClO_4$
$RuCl_5^{2-} + 3e^- = Ru + 5Cl^-$	$+0.4$	
Samarium		
$Sm^{3+} + 3e^- = Sm$	-2.41	
$Sm^{3+} + e^- = Sm^{2+}$	(-1.56)	0.1 Me_4NI

Table 6-2 (*Continued*)
POTENTIALS OF THE ELEMENTS AND THEIR COMPOUNDS AT 25°C

Half-reaction	Standard or Formal Potential	Solution Composition
Scandium		
$Sc^{3+} + 3e^- = Sc$	-2.1	
Selenium		
$SeO_4^{2-} + 4H^+ + 2e^- = H_2SeO_3 + H_2O$	$+1.15$	
$H_2SeO_3 + 4H^+ + 4e^- = Se + 3H_2O$	$+0.74$	
$Se + 2H^+ + 2e^- = H_2Se$	-0.40	
$Se + 2e^- = Se^{2-}$	-0.92	
Silicon		
$SiO_2 + 4H^+ + 4e^- = Si + 2H_2O$	-0.86	
$SiF_6^{2-} + 4e^- = Se + 6F^-$	-1.2	
$Si + 4H^+ + 4e^- = SiH_4(g)$	$+0.10$	
Silver		
$AgO^+ + 2H^+ + e^- = Ag^{2+} + H_2O$	$(ca. +2.1)$	4 HNO_3
$Ag_2O_3 + 2H^+ + 2e^- = 2AgO + H_2O$	$+1.71$	
$Ag_2O_3 + H_2O + 2e^- = 2AgO + 2OH^-$	$(+0.74)$	1 NaOH
$Ag_2O_3 + 6H^+ + 4e^- = 2Ag^+ + 3H_2O$	$+1.76$	
$Ag^{2+} + e^- = Ag^+$	$(+1.93)$	3 HNO_3
	$(+2.00)$	4 $HClO_4$
$AgO + H^+ + e^- = \frac{1}{2}Ag_2O + \frac{1}{2}H_2O$	$+1.41$	
$Ag_2O + 2H^+ + 2e^- = 2Ag + H_2O$	$+1.17$	
$Ag^+ + e^- = Ag$	$+0.7999$	
$Ag_2SO_4 + 2e^- = 2Ag + SO_4^{2-}$	$+0.653$	
$Ag_2C_2O_4 + 2e^- = 2Ag + C_2O_4^{2-}$	$+0.47$	
$Ag_2CrO_4 + 2e^- = 2Ag + CrO_4^{2-}$	$+0.447$	
$Ag_2O + H_2O + 2e^- = 2Ag + 2OH^-$	$+0.34$	
$Ag(NH_3)_2^+ + e^- = Ag + 2NH_3$	$+0.373$	
$AgCl + e^- = Ag + Cl^-$	$+0.2223$	
$AgBr + e^- = Ag + Br^-$	$+0.071$	
$AgCN + e^- = Ag + CN^-$	-0.017	
$AgI + e^- = Ag + I^-$	-0.152	
$AgSCN + e^- = Ag + SCN^-$	$+0.09$	
$Ag(CN)_2^- + e^- = Ag + 2CN^-$	-0.31	
$Ag_2S + 2e^- = 2Ag + S^{2-}$	-0.71	
$AgN_3 + e^- = Ag + N_3^-$	$+0.29$	
Sodium		
$Na^+ + e^- = Na$	-2.713	
$Na^+ + (Hg) + e^- = Na(Hg)$	-1.84	
Strontium		
$Sr^{2+} + 2e^- = Sr$	-2.89	
Sulfur		
$S_2O_8^{2-} + 2e^- = 2SO_4^{2-}$	$+2.0$	
$2SO_4^{2-} + 4H^+ + 2e^- = S_2O_6^{2-} + 2H_2O$	-0.2	
$SO_4^{2-} + 4H^+ + 2e^- = SO_2(aq) + H_2O$	$+0.17$	
$SO_4^{2-} + H_2O + 2e^- = SO_3^{2-} + 2OH^-$	-0.93	
$S_2O_6^{2-} + 4H^+ + 2e^- = 2H_2S_2O_3$	$+0.6$	
$2SO_3^{2-} + 2H_2O + 2e^- = S_2O_4^{2-} + 4OH^-$	-1.12	
$2SO_2(aq) + H^+ + 2e^- = HS_2O_4^- + H_2O$	-0.08	
$2SO_2(aq) + 2H^+ + 4e^- = S_2O_3^{2-} + H_2O$	$+0.40$	
$2SO_3^{2-} + 3H_2O + 4e^- = S_2O_3^{2-} + 6OH^-$	-0.58	
$SO_3^{2-} + 3H_2O + 4e^- = S + 6OH^-$	-0.66	
$S_4O_6^{2-} + 2e^- = 2S_2O_3^{2-}$	$+0.09$	
$S_2O_3^{2-} + 6H^+ + 4e^- = 2S + 3H_2O$	$+0.5$	
$S + 2H^+ + 2e^- = H_2S(g)$	$+0.14$	

Table 6-2 (*Continued*)
POTENTIALS OF THE ELEMENTS AND THEIR COMPOUNDS AT 25°C

Half-reaction	Standard or Formal Potential	Solution Composition
Sulfur (*cont.*)		
$S + 2e^- = S^{2-}$	-0.48	
$2S + 2e^- = S_2^{2-}$	-0.43	
$3S + 2e^- = S_3^{2-}$	-0.39	
$4S + 2e^- = S_4^{2-}$	-0.36	
$5S + 2e^- = S_5^{2-}$	-0.34	
Tantalum		
$Ta_2O_5 + 10H^+ + 10e^- = 2Ta + 5H_2O$	-0.81	
Technetium		
$TcO_4^- + 4H^+ + 3e^- = TcO_2 + 2H_2O$	$+0.74$	
$TcO_4^- + 2H^+ + e^- = TeO_3 + H_2O$	$+0.7$	
$TcO_4^- + 8H^+ + 7e^- = Tc + 4H_2O$	$+0.47$	
$TcO_3 + 2H^+ + 2e^- = TcO_2 + H_2O$	$+0.8$	
$TcO_2 + 4H^+ + 4e^- = Tc + 2H_2O$	$+0.27$	
$Tc + e^- = Tc^-$	*ca.* -0.5	
Tellurium		
$H_6TeO_6 + 2H^+ + 2e^- = TeO_2 + 4H_2O$	$+1.02$	
$TeCl_6^{2-} + 4e^- = Te + 6Cl^-$	$(+0.63)$	dil HCl
$TeO_2(s) + 4H^+ + 4e^- = Te + 2H_2O$	$+0.59$	
$TeOOH^+ + 3H^+ + 4e^- = Te + 2H_2O$	$+0.559$	
$TeO_3^{2-} + 3H_2O + 4e^- = Te + 6OH^-$	-0.57	
$Te_2 + 2H^+ + 2e^- = H_2Te_2$	-0.36	
$Te_2 + 4H^+ + 4e^- = 2H_2Te(g)$	-0.50	
$Te_2 + 4e^- = Te_2^{2-}$	(-0.84)	0.6–0.8 KOH
Terbium		
$Tb^{3+} + 3e^- = Tb$	-2.39	
Thallium		
$Tl(III) + 2e^- = Tl(I)$	$(+1.26)$	1 HClO$_4$
	$(+1.23)$	0.5–3 HNO$_3$
	$(+1.22)$	0.5–1 H$_2$SO$_4$
	$(+0.77)$	0.5–1 HCl
	$(+0.89)$	0.1 HCl
	$(+0.65)$	1 KBr
$Tl(OH)_3 + 2e^- = Tl^+ + 3OH^-$	-0.05	
$Tl^+ + e^- = Tl$	-0.336	
$TlOH + e^- + Ti + OH^-$	-0.34	
$TlSCN + e^- = Tl + SCN^-$	-0.56	
$TlCl + e^- = Tl + Cl^-$	-0.557	
$TlBr + e^- = Tl + Br^-$	-0.657	
$TlI + e^- = Tl + I^-$	-0.77	
Thorium		
$Th^{4+} + 4e^- = Th$	-1.90	
Thulium		
$Tm^{3+} + 3e^- = Tm$	-2.28	
Tin		
$Sn^{4+} + 2e^- = Sn^{2+}$	$+0.154$	
$SnCl_6^{2-} + 2e^- = SnCl_4^{2-} + 2Cl^-$	$(+0.14)$	1 HCl
$Sn(OH)_6^{2-} + 2e^- = HSnO_2^- + H_2O + 3OH^-$	-0.93	
$Sn^{2+} + 2e^- = Sn$	-0.14	
$SnCl_4^{2-} + 2e^- = Sn + 4Cl^-$	(-0.19)	1 HCl
$HSnO_2^- + H_2O + 2e^- = Sn + 3OH^-$	-0.91	
Titanium		
$TiO^{2+} + 2H^+ + e^- = Ti^{3+} + H_2O$	$+0.1$	

Table 6-2 (*Continued*)
POTENTIALS OF THE ELEMENTS AND THEIR COMPOUNDS AT 25°C

Half-reaction	Standard or Formal Potential	Solution Composition
Titanium (*cont.*)		
$Ti(IV) + e^- = Ti(III)$	(-0.09)	1 HCl
	$(+0.10)$	3 HCl
	$(+0.24)$	6 HCl
	(-0.01)	0.2 H_2SO_4
	$(+0.15)$	5 H_2SO_4
	$(+0.17)$	3 HBr
	(0.0)	1 H_3PO_4
$TiF_6^{2-} + 4e^- = Ti + 6F^-$	-1.24	
$TiO_2 + 4H^+ + 4e^- = Ti + 2H_2O$	-0.86	
$Ti^{3+} + e^- = Ti^{2+}$	-0.37	
$Ti^{2+} + 2e^- = Ti$	-1.63	
Tungsten		
$W(VI) + e^- = W(V)$	$(+0.26)$	12 HCl
	$(+0.22)$	1 H_3PO_4
$2WO_3 + 2H^+ + 2e^- = W_2O_5 + H_2O$	-0.03	
$WO_3 + 6H^+ + 6e^- = W + 3H_2O$	-0.09	
$WO_4^{2-} + 4H_2O + 6e^- = W + 8OH^-$	-1.01	
$W(CN)_8^{3-} + e^- = W(CN)_8^{4-}$	$+0.46$	
$W_2O_5 + 2H^+ + 2e^- = 2WO_2 + H_2O$	-0.04	
$W(V) + 2e^- = W(III)(red)$	(-0.31)	12 HCl
$WO_2 + 4H^+ + 4e^- = W + 2H_2O$	-0.12	
Uranium		
$UO_2^{2+} + e^- = UO_2^+$	$(+0.06)$	0.1 Cl^-
$UO_2^{2+} + 4H^+ + 2e^- = U^{4+} + 2H_2O$	$+0.33$	
$U(VI) + 2e^- = U(IV)$	$(+0.35)$	1 HCl
	$(+0.42)$	0.05–0.5 H_2SO_4
	$(+0.47)$	1 H_3PO_4
$UO_2^+ + 4H^+ + e^- = U^{4+} + 2H_2O$	$(ca.\ +0.55)$	HCl
$U^{4+} + e^- = U^{3+}$	(-0.63)	1 $HClO_4$ or 1 HCl or 0.1 H_2SO_4 + 0.1 KCl
$U^{3+} + 3e^- = U$	-1.80	
Vanadium		
$VO_2^+ + 2H^+ + e^- = VO^{2+} + H_2O$	$+0.999$	
$V(V) + e^- = V(IV)$	$(+1.21)$	12 H_3PO_4
	$(ca.\ +0.94)$	1 H_3PO_4
	(-0.20)	satd. $Na_2P_4O_7$
$VO_2^+ + 4H^+ + 5e^- = V + 4H_2O$	-0.25	
$VO^{2+} + 2H^+ + e^- = V^{3+} + H_2O$	$+0.34$	
$V(IV) + e^- = V(III)$	$(+0.70)$	12 H_3PO_4
	$(+0.39)$	1 H_3PO_4
$V^{3+} + e^- = V^{2+}$	-0.255	
$V(III) + e^- = V(II)$	(-0.27)	1 $HClO_4$ or 1 HCl or 1 H SO
	(-0.22)	0.1–1 NH_4SCN
$V^{2+} + 2e^- = V$	$ca.\ -1.2$	
Xenon		
$H_4XeO_6 + 2H^+ + 2e^- = XeO_3 + 3H_2O$	$ca.\ +3.0$	
$HXeO_6^{3-} + 2H_2O + e^- = HXeO_4 + 4OH^-$	$ca.\ +0.9$	
$HXeO_4 + 3H_2O + 7e^- = Xe + 7OH^-$	$ca.\ +0.9$	
$XeO_3 + 6H^+ + 2F^- + 4e^- = XeF_2 + 3H_2O$	$ca.\ +1.6$	

Table 6-2 (*Continued*)
POTENTIALS OF THE ELEMENTS AND THEIR COMPOUNDS AT 25°C

Half-reaction	Standard or Formal Potential	Solution Composition
Xenon (*cont.*)		
$XeO_3 + 6H^+ + 6e^- = Xe + 3H_2O$	*ca.* +1.8	
$XeF_2 + 2e^- = Xe + 2F^-$	*ca.* +2.2	
Ytterbium		
$Yb^{3+} + e^- = Yb^{2+}$	(−1.17)	0.1 NH_4Cl
$Yb^{3+} + 3e^- = Yb$	−2.25	
Yttrium		
$Y^{3+} + 3e^- = Y$	−2.37	
Zinc		
$Zn^{2+} + 2e^- = Zn$	−0.7628	
$Zn(NH_3)_4^{2+} + 2e^- = Zn + 4NH_3$	−1.04	
$ZnO_2^{2-} + 2H_2O + 2e^- = Zn + 4OH^-$	−1.216	
$Zn(CN)_4^{2-} + 2e^- = Zn + 4CN^-$	−1.26	
Zirconium		
$ZrO_2 + 4H^+ + 4e^- = Zr + 2H_2O$	−1.43	

Table 6-3
SELECTED LIST OF OXIDATION-REDUCTION INDICATORS

Name	Reduction Potential (30°C) in Volts at		Suitable pH Range	Color Change Upon Oxidation
	pH = 0	pH = 7		
Bis(5-bromo-1,10-phenanthroline) ruthenium(II) dinitrate	1.41*			Red to faint blue
Tris(5-nitro-1,10-phenanthroline) iron(II) sulfate	1.25*			Red to faint blue
Iron(II)-2,2′,2″-tripyridine sulfate	1.25*			Pink to faint blue
Tris(4,7-diphenyl-1,10-phenanthroline) iron(II) disulfate	1.13 (4.6 M H_2SO_4)* 0.87(1.0 M H_2SO_4)*			Red to faint blue
o,m'-Diphenylaminedicarboxylic acid	1.12			Colorless to blue-violet
Setopaline	1.06 (*trans*)†			Yellow to orange
p-Nitrodiphenylamine	1.06			Colorless to violet
Tris(1,10-phenanthroline)-iron(II) sulfate	1.06 (1.00 M H_2SO_4)* 1.00 (3.0 M H_2SO_4)* 0.89 (6.0 M H_2SO_4)*			Red to faint blue
Setoglaucine O	1.01 (*trans*)†			Yellow-green to yellow-red
Xylene cyanole FF	1.00 (*trans*)†			Yellow-green to pink
Erioglaucine A	1.00 (*trans*)†			Green-yellow to bluish red
Eriogreen	0.99 (*trans*)†			Green-yellow to orange
Tris(2,2′-bipyridine)-iron(II) hydrochloride	0.97*			Red to faint blue
2-Carboxydiphenylamine [N-phenyl-anthranilic acid]	0.94			Colorless to pink
Benzidine dihydrochloride	0.92			Colorless to blue
o-Toluidine	0.87			Colorless to blue
Bis(1,10-phenanthroline)-osmium(II) perchlorate	0.859 (0.1 M H_2SO_4)			Green to pink
Diphenylamine-4-sulfonate (Na salt)	0.85			Colorless to violet
3,3′-Dimethoxybenzidine dihydrochloride [o-dianisidine]	0.85			Colorless to red
Ferrocyphen	0.81			Yellow to violet

4'-Ethoxy-2,4-diaminoazobenzene	0.76			Red to pale yellow
N,N-Diphenylbenzidine	0.76			Colorless to violet
Diphenylamine	0.76			Colorless to violet
N,N-Dimethyl-p-phenylenediamine	0.76			Colorless to red
Variamine blue B hydrochloride	0.712‡	0.310	1.5–6.3	Colorless to blue
N-Phenyl-1,2,4-benzenetriamine	0.70			Colorless to red
Bindschedler's green	0.680‡	0.224	2–9.5	
2,6-Dichloroindophenol (Na salt)	0.668‡	0.217	6.3–11.4	Colorless to blue
2,6-Dibromophenolindophenol	0.668‡	0.216	7.0–12.3	Colorless to blue
Brilliant cresyl blue [3-amino-9-dimethyl-amino-10-methylphenoxazine chloride]	0.583	0.047	0–11	Colorless to blue
Iron(II)-tetrapyridine chloride	0.59			Red to faint blue
Thionine [Lauth's violet]	0.563‡	0.064	1–13	Colorless to violet
Starch (soluble potato, I_3 present)	0.54			Colorless to blue
Gallocyanine (25°C)		0.021		Colorless to violet-blue
Methylene blue	0.532‡	0.011	1–13	Colorless to blue
Nile blue A [aminonaphthodiethylamino-phenoxazine sulfate]	0.406‡	−0.119	1.4–12.3	Colorless to blue
Indigo-5,5',7,7'-tetrasulfonic acid (Na salt)	0.365‡	−0.046	<9	Colorless to blue
Indigo-5,5',7-trisulfonic acid (Na salt)	0.332‡	−0.081	<9	Colorless to blue
Indigo-5,5'-disulfonic acid (Na salt)	0.291‡	−0.125	<9	Colorless to blue
Phenosafranine	0.280‡	−0.252	1–11	Colorless to violet-blue
Indigo-5-monosulfonic acid (Na salt)	0.262‡	−0.157	<9	Colorless to blue
Safranine T	0.24‡	−0.289	1–12	Colorless to violet-blue
Bis(dimethylglyoximato)-iron(II) chloride	0.155		6–10	Red to colorless
Induline scarlet	0.047‡	−0.299	3–8.6	Colorless to red
Neutral red		−0.323	2–11	Colorless to red-violet

* Transition point is at higher potential than the tabulated formal potential because the molar absorptivity of the reduced form is very much greater than that of the oxidized form.
† Trans = first noticeable color transition; often 60 mV less than $E°$
‡ Values of $E°$ are obtained by extrapolation from measurements in weakly acid or weakly alkaline systems.

Table 6-4
OVERPOTENTIALS FOR COMMON ELECTRODE REACTIONS AT 25°C

The overpotential is defined as the difference between the actual potential of an electrode at a given current density and the reversible electrode potential for the reaction.

Electrode	Current Density, A/cm²					
	0.001	0.01	0.1	0.5	1.0	5.0
	Overpotential, volts					
Liberation of H_2 from $1M$ H_2SO_4						
Ag	0.097	0.13	0.3		0.48	0.69
Al	0.3	0.83	1.00		1.29	
Au	0.017		0.1		0.24	0.33
Bi	0.39	0.4			0.78	0.98
Cd		1.13	1.22		1.25	
Co		0.2				
Cr		0.4				
Cu			0.35		0.48	0.55
Fe		0.56	0.82		1.29	
Graphite	0.002		0.32		0.60	0.73
Hg	0.8	0.93	1.03		1.07	
Ir	0.0026	0.2				
Ni	0.14	0.3			0.56	0.71
Pb	0.40	0.4			0.52	1.06
Pd	0	0.04				
Pt (smooth)	0.0000	0.16	0.29		0.68	
Pt (platinized)	0.0000	0.030	0.041		0.048	0.051
Sb		0.4				
Sn		0.5	1.2			
Ta		0.39	0.4			
Zn	0.48	0.75	1.06		1.23	
Liberation of O_2 from $1M$ KOH						
Ag	0.58	0.73	0.98		1.13	
Au	0.67	0.96	1.24		1.63	
Cu	0.42	0.58	0.66		0.79	
Graphite	0.53	0.90	1.09		1.24	
Ni	0.35	0.52	0.73		0.85	
Pt (smooth)	0.72	0.85	1.28		1.49	
Pt (platinized)	0.40	0.52	0.64		0.77	
Liberation of Cl_2 from saturated NaCl solution						
Graphite			0.25	0.42	0.53	
Platinized Pt	0.006		0.026	0.05		
Smooth Pt	0.008	0.03	0.054	0.161	0.236	
Liberation of Br_2 from saturated NaBr solution						
Graphite		0.002	0.027	0.16	0.33	
Platinized Pt		0.002	0.012	0.069	0.21	
Smooth Pt		0.002	0.006*	0.26	0.38†	
Liberation of I_2 from saturated NaI solution						
Graphite	0.002	0.014	0.097			
Platinized Pt		0.006	0.032		0.196	
Smooth Pt		0.003	0.03	0.12	0.22	

* At 0.23 A/cm².   † At 0.72 A/cm².

The overpotential required for the evolution of O_2 from dilute solutions of $HClO_4$, HNO_3, H_3PO_4 or H_2SO_4 onto smooth platinum electrodes is approximately 0.5 V.

Table 6-5
POLAROGRAPHY

The values in the column headed $E_{1/2}$ are the half-wave potentials in units of volts referred to the saturated calomel electrode. The values in the column headed I_d are for $i_d/C \cdot m_{2/3} \cdot t^{1/6}$, where i_d is the diffusion current in microamperes, C is the concentration in millimoles per liter, m is the rate of flow of mercury in milligrams per second, and t is the drop time in seconds. The term "too positive" means that the diffusion current starts from zero applied emf and that the potential involved is more positive than the oxidation potential of mercury.

Half-Wave Potentials of Inorganic Ions at 25°C

Ion Electrode Reaction	Electrolyte	$E_{1/2}$	I_d
$Ag^+ \rightarrow Ag$	KNO_3	*	.....
$Al^{3+} \rightarrow Al$	$0.05\,N\,BaCl_2$	−1.75	.....
$As^{3+} \rightarrow As$	$1\,N\,H_2SO_4$-0.01% gelatin†	−0.7	8.4
$As \rightarrow AsH_3$		−1.0	
$As^{3+} \rightarrow As^{5+}$	$0.5\,N\,KOH$-0.025% gelatin	−0.26	3.82
$Au^{3+} \rightarrow Au^+$	$0.1\,M\,KCN$†	*	.....
		−1.4	.....
$Ba^{2+} \leftrightarrow Ba(Hg)$	$0.1\,M\,(C_2H_5)_4NI$-50% C_2H_5OH	−1.875	2.91
$Be^{2+} \rightarrow Be$	$0.05\,M\,CH_3CO_2K$	−1.8‡	.....
$Bi^{3+} \rightarrow Bi$	$1\,M\,HCl$-0.01% gelatin	−0.09	.....
$Bi^{3+} \rightarrow Bi$	$0.5\,Na$ tartrate (ph 4.5)	−0.29	.....
$Ca^{2+} \rightarrow Ca(Hg)$	$0.1\,M\,(C_2H_5)_4NI$-50% C_2H_5OH-3.15 × $10^{-5}\,M\,Ba^{2+}$	−2.12	2.87
$Cd^{2+} \rightarrow Cd$	$0.1\,M\,KCl$-0.01% gelatin	−0.599	3.51
$Cd^{2+} \rightarrow Cd$	$1\,M\,HCl$-0.01% gelatin	−0.64	3.58
$Cd^{2+} \rightarrow Cd$	$0.5\,M\,Na$ tartrate-0.01% gelatin (pH 9)	−0.64	2.34
$Cd^{2+} \rightarrow Cd$	$1\,M\,NH_3$-1 $M\,NH_4Cl$-0.01% gelatin	−0.81	3.68
$Cd^{2+} \rightarrow Cd$	$1\,M\,KCN$	−1.18	.....
$Co^{2+} \rightarrow Co$	$0.05\,M\,K_2SO_4$	−1.428	.....
$Co^{3+} \rightarrow Co$	$1\,M\,NH_4Cl$-1 $M\,NH_3$-gelatin†	−0.5	.....
$Co^{2+} \rightarrow Co(Hg)$		−1.3	.....
$Co^{2+} \rightarrow Co$	$0.1\,M$ pyridine-0.1 M pyridinium chloride	−1.07	.....
$Co^{2+} \rightarrow Co$	$1\,M\,KSCN$	−1.03	.....
$Co^{2+} \rightarrow Co^+$	$1\,M\,KCN$	−1.2	.....
$Cr^{3+} \rightarrow Cr^{2+}$	$0.1\,M\,KCl$-0.01% gelatin†	−0.91	.....
$Cr^{2+} \rightarrow Cr$		−1.47	.....
$Cr^{3+} \rightarrow Cr^{2+}$	$0.1\,M$ pyridine-0.1 M pyridinium chloride	−0.95	.....
$Cr^{2+} \rightarrow Cr^{3+}$	$1\,M\,KCl$	−0.40	1.54
$Cr^{2+} \rightarrow Cr^{3+}$	$1\,M\,KSCN$	−0.80	1.64
$Cr^{6+} \rightarrow Cr^{3+}$	$1\,M\,NaOH$	−0.85	.....
$Cr^{6+} \rightarrow Cr^{3+}$	$0.1\,M\,KCl$§	−0.3, −1.0	.....
$Cr^{3+} \rightarrow Cr^{2+}$		−1.5	.....
$Cr^{2+} \rightarrow Cr$		−1.7	.....
$Cs^+ \rightarrow Cs$	$0.1\,M\,(C_2H_5)_4NOH$-50% C_2H_5OH	−2.05	.....
$Cu^{2+} \rightarrow Cu$	$0.1\,M\,KCl\,(HCl)$-0.01% gelatin	+0.04	3.23
$Cu^{2+} \rightarrow Cu$	$1\,M\,NaOH$-0.01% gelatin	−0.42	2.91
$Cu^{2+} \rightarrow Cu$	$0.5\,M\,Na$ tartrate-0.01% gelatin (pH 9)	−0.12	2.24
$Cu^{2+} \rightarrow Cu^+$	$1\,M\,NH_3$-1 $M\,NH_4Cl$†	−0.24	3.75
$Cu^+ \rightarrow Cu$		−0.50	.....
$Cu^{2+} \rightarrow Cu^+$	$1\,M\,K_2C_2O_4$ (pH 5.7 to 10)	−0.27	.....
$Cu^{2+} \leftrightarrow Cu$	$0.1\,M\,Na_4P_2O_7$-0.2 $M\,CH_3CO_2Na$ (pH 4.5)	−0.085	.....
$Eu^{3+} \rightarrow Eu^{2+}$	$0.1\,M\,NH_4Cl$	−0.671	.....

* Too positive.
† One run; two waves.

‡ Poorly defined.
§ One run; three waves.

Table 6-5 (*Continued*)
POLAROGRAPHY

Ion Electrode Reaction	Electrolyte	$E_{1/2}$	I_d
$Fe^{3+} \to Fe^{2+}$	$1\ M\ NH_4ClO_4$†	$\begin{cases}*\end{cases}$	.....
$Fe^{2+} \to Fe$		-1.44^a	.....
$Fe^{3+} \to Fe^{2+}$	$0.5\ M$ Na citrate-0.005% gelatin (pH 6)	-0.222	0.93
$Fe^{3+} \to Fe^{2+}$	$0.5\ M$ Na citrate-0.005% gelatin (pH 5.8)†	-0.21	1.11
$Fe^{2+} \to Fe$		-1.53	.....
$Fe^{3+} \to Fe^{2+}$	$0.2\ M\ Na_2C_2O_4$ (pH 5.25)	-0.245	1.50
$Fe^{2+} \to Fe$	$1\ M$ NaOH	-1.46	.....
$Ga^{3+} \to Ga$	$0.1\ M$ KCl	-1.15^b	.....
$Gd^{3+} \to Gd$	$0.1\ M$ LiCl-0.01% gelatin	-1.77	3.7
$Ge^{2+} \to Ge$	$6\ M$ HCl	-0.45	.....
$Hg^{2+} \to Hg^+$	$0.1\ N$ HNO_3	$*$	.....
$IO_4^- \to IO_3^-$	pH 12	-0.08	.....
$IO_3^- \to I^-$	$0.1\ M$ Na citrate (pH 5.95)	-0.650	.....
$In^{3+} \to In$	$0.1\ M$ KCl-gelatin	-0.561	.....
$K^+ \leftrightarrow K(Hg)$	$0.1\ M\ (C_2H_5)_4NOH$-$50\%\ C_2H_5OH$	-2.10	1.69
$Li^+ \leftrightarrow Li(Hg)$	$0.1\ M\ (C_2H_5)_4NOH$-$50\%\ C_2H_5OH$	-2.31	1.16
$Mg^{2+} \to Mg(Hg)$	$0.1\ M\ (CH_3)_4NCl$	-2.2‡	.....
$Mn^{2+} \to Mn$	$1\ M$ KCl	-1.51^c	.....
$Mn^{2+} \to Mn$	$1\ M\ NH_4Cl$-$1\ M\ NH_3$-0.005% gelatin	-1.65	.....
$Mn^{2+} \to Mn$	$1.5\ M$ KCN	-1.33	.....
$Mn^{2+} \to Mn$	$1\ M$ KSCN-0.01% gelatin	-1.553^c	.....
$Mo^{6+} \to Mo^{5+}$	$0.3\ M$ HCl†	$\begin{cases}-0.26\\-0.63\end{cases}$	.....
$Mo^{5+} \to Mo^{3+}$			.....
$Mo^{6+} \to Mo^{5+}$	$10\ M\ H_2SO_4$†	$\begin{cases}*\\-0.13\end{cases}$	3.99
$Mo^{5+} \to Mo^{3+}$			.....
$Mo^{6+} \to Mo^{5+}$	$0.03\ M\ Na_2HPO_4$-$0.1\ M$ citric acid-$0.1\ M$ KCl†	$\begin{cases}-0.23\\-0.58\end{cases}$	.....
$Mo^{5+} \to Mo^{3+}$			.....
$NO_3^- \to NH_2OH$	$0.01\ M$ LaCl	-1.3 to -1.5	.....
$NO_3^- \to {}_{1/2}N_2$	$0.1\ M$ KCl-$0.01\ M$ HCl-$2 \times 10^{-4}\ M\ UO_2Cl_2$	-0.98^d	13.95
$NO_2^- \to {}_{1/2}N_2$	$0.1\ M$ KCl-$0.01\ M$ HCl-$2 \times 10^{-4}\ M\ UO_2(O_2CCH_3)_2$	-1.0	7.45
$Na^+ \leftrightarrow Na(Hg)$	$0.1\ M\ (C_2H_5)_4NOH$-$50\%\ C_2H_5OH$	-2.07	1.40
$Nb^{5+} \to Nb^{3+}$	$0.9\ M\ HNO_3$	-0.76	.....
$Nd^{3+} \to Nd$	$0.1\ M\ (CH_3)_4NI$-$0.02\ M\ H_2SO_4$-0.1% gelatin	-1.82	4.40
$Ni^{2+} \to Ni$	$0.01\ M$ KCl	-1.1	.....
$Ni^{2+} \to Ni$	$1\ M$ KSCN	-0.70	.....
$Ni^{2+} \to Ni$	$1\ M$ KCl-$0.5\ M$ pyridine	-0.78	.....
$Ni^{2+} \to Ni$	$1\ M\ NH_3$-$1\ M\ NH_4Cl$	-1.09	3.56
$Ni^{2+} \to Ni$	$1\ M$ KCN-0.01% gelatin	-1.36	.....
$Ni^+ \to Ni^{2+}$	$1\ M$ KCN	-0.80	.....
$O_2 \to H_2O_2$	pH 1 to 10 with max. suppressor†	$\begin{cases}-0.05\\-0.94\end{cases}$	12.3
$H_2O_2 \to H_2O$			.....
$Os^{8+} \to Os^{6+}$	Saturated $Ca(OH)_2$§	$\begin{cases}*\\-0.41\\-1.16\end{cases}$	.....
$Os^{6+} \to Os^{5+}$			.....
$Os^{5+} \to Os^{3+}$			.....
$Pb^{2+} \to Pb$	$0.1\ M$ KCl-0.01% gelatin	-0.396	3.80
$Pb^{2+} \to Pb$	$1\ M\ HNO_3$-0.01% gelatin	-0.405	3.67
$Pb^{2+} \to Pb$	$1\ M$ NaOH-0.01% gelatin	-0.755	3.39
$Pb^{2+} \to Pb$	$0.5\ M$ Na tartrate (pH 9)	-0.50	2.30
$Pb^{2+} \to Pb$	$1\ M$ KCN	-0.72	.....
$Pd^{2+} \to Pd$	$1\ M$ KCl-$1\ M$ pyridine	-0.36^f	.....
$Pd^{2+} \to Pd$	$1\ M$ KCN	-1.77	.....

* Too positive. †One run; two waves. ^{a} $0.01\ M$. ^{b} $0.005\ M$.
‡ Poorly defined. ^{c} Two seconds.
§ One run; three waves. ^{d} $4 \times 10^{-4}\ M$. ^{f} $0.005\ M$.

Table 6-5 (*Continued*)
POLAROGRAPHY

Ion Electrode Reaction	Electrolyte	$E_{1/2}$	I_d
$Pd^{2+} \rightarrow Pd$	$1\,M\,NH_3$-$1\,M\,NH_4Cl$	-0.80^f	.....
$Pr^{3+} \rightarrow Pr$	$0.1\,M\,LiCl$-0.01% gelatin	-1.75^g	3.59
$Ra^{2+} \rightarrow Ra$	$0.02\,M\,KCl$	-1.84	.....
$Rb^+ \rightarrow Rb$	$0.1\,M\,(C_2H_5)_4NOH$-$50\%\,C_2H_5OH$	-1.99	.....
$Re^{7+} \rightarrow Re^{4+}$	$4\,M\,HClO_4$	-0.39^a	6.69
$Re^{7+} \rightarrow Re^-$	$2\,M\,KCl$†	$\left\{\begin{array}{l}-1.41\\-1.70\end{array}\right.$	$\begin{array}{l}17.9^b\\.....\end{array}$
Catalytic wave^c			
$Re^{3+} \rightarrow Re^{2+}$	$2\,M\,HClO_4$†	$\left\{\begin{array}{l}-0.28\\-0.46\end{array}\right.$	$\begin{array}{l}.....\\.....\end{array}$
$Re^{2+} \rightarrow Re$			
$Rh^{3+} \rightarrow Rh^{2+}$	$1\,M\,KCN$	-1.47	.....
$Rh^{3+} \rightarrow Rh^{2+}$	$1\,M\,NH_4Cl$	-0.93	.....
$Rh^{3+} \rightarrow Rh^{2+}$	$1\,M\,Pyridine$-$1\,M\,KCl$	-0.41	.....
$Ru^{4+} \rightarrow Ru^{3+}$	$1\,M\,HClO_4$§	$\left\{\begin{array}{l}*, +0.21\\-0.34\end{array}\right.$	$\begin{array}{l}1.53\\1.38\end{array}$
$Ru^{3+} \rightarrow Ru^{2+}$			
$HSO_3^- \rightarrow HSO_2$	pH 6†	$\left\{\begin{array}{l}-0.67\\-1.23\end{array}\right.$	$\begin{array}{l}.....\\.....\end{array}$
$S_2O_4{}^{2-} \rightarrow S_2O_3{}^{2-}$	#		
$H_2SO_3 \leftrightarrow H_2SO_2$	$0.1\,M\,HNO_3$	-0.37	.....
$S_2O_4{}^{2-} \rightarrow 2SO_3{}^{2-}$	$0.5\,M\,(NH_4)_2HPO_4$-$1\,M\,NH_3$-0.01% gelatin	-0.43	4.09
$S \rightarrow H_2S$	CH_3OH-pyridine-pyridinium hydrochloride (pH 6)	-0.50	.....
$Sb^{3+} \rightarrow Sb$	$1\,N\,H_2SO_4$-0.01% gelatin	-0.32	.....
$Sb^{3+} \rightarrow Sb$	$1\,N\,NaOH$-0.01% gelatin	-1.26	.....
$Sb^{3+} \rightarrow Sb$	$0.5\,M\,Na\,tartrate$-$0.1\,M\,NaOH$-0.01% gelatin	-1.32	.....
$Sb^{3+} \rightarrow Sb^{5+}$	$1\,M\,KOH$	-0.45	.....
$Sc^{3+} \rightarrow Sc$	$0.1\,M\,LiCl$-HCl (0.25 of Sc^{3+} concn.)	-1.80	.....
$Se^{4+} \rightarrow Se^{2-}$	$1\,M\,NH_4Cl$-0.1 to $1\,M\,NH_3$-0.003% gelatin (pH 8.0)	-1.44	.....
$Se^{4+} \rightarrow Se^{2-}$	$1\,M\,NH_4Cl$-0.1 to $1\,M\,NH_3$-0.003% gelatin (pH 9.5)	-1.54	.....
$Sm^{3+} \rightarrow Sm^{2+}$	$0.1\,M\,(CH_3)_4NI$-$0.001\,N\,H_2SO_4$-0.01% gelatin†	$\left\{\begin{array}{l}-1.80\\......\end{array}\right.$	$\begin{array}{l}3.85^d\\8.20^d\end{array}$
$Sm^{2+} \rightarrow Sm$			
$Sn^{2+} \rightarrow Sn$	$1\,M\,HCl$-0.01% gelatin	-0.47	4.07
$Sn^{2+} \rightarrow Sn$	$1\,M\,NaOH$-0.01% gelatin†	$\left\{\begin{array}{l}-1.22\\-0.73\end{array}\right.$	$\begin{array}{l}3.45\\3.45\end{array}$
$Sn^{2+} \rightarrow Sn^{4+}$			
$Sn^{2+} \rightarrow Sn$	$0.5\,M\,Na\,tartrate$-0.01% gelatin-$0.1\,M\,NaOH$†	-0.71	2.86
		-1.16	2.86
$Sn^{4+} \rightarrow Sn^{2+}$	$4\,M\,NH_4Cl$-$1\,M\,HCl$-0.005% gelatin†	$\left\{\begin{array}{l}-0.25\\-0.52\end{array}\right.$	$\begin{array}{l}2.84\\3.49\end{array}$
$Sn^{2+} \rightarrow Sn$			
$Sr^{2+} \leftrightarrow Sr(Hg)$	$0.1\,M\,(C_2H_5)_4NI$-$50\%\,C_2H_5OH$	-2.06	3.17
$Te^{4+} \rightarrow Te$	$1\,M\,NH_4Cl$-NH_3-0.003% gelatin (pH 8.4)	-0.63	.....
$Te^{4+} \rightarrow Te$	$1\,M\,NH_4Cl$-NH_3-0.003% gelatin (pH 9.4)	-0.68	.....
$Te^{4+} \rightarrow Te^{2-}$	$1\,M\,NaOH$-0.003% gelatin	-1.19	9.75
$Te^{2-} \leftrightarrow Te$	$1\,M\,HCl$-0.003% gelatin	-0.72	.....
$Ti^{4+} \rightarrow Ti^{2+}$	$0.1\,M\,HCl$-0.005% gelatin	-0.81	1.56
$Ti^{2+} \rightarrow Ti^{4+}$	$0.01\,M\,HCl$	-0.14	.....
$Ti^{4+} \rightarrow Ti^{2+}$	$0.1\,M\,K_2C_2O_4$-$1\,M\,H_2SO_4$-0.005% gelatin	-0.173	1.25
$Ti^{4+} \rightarrow Ti^{2+}$	$0.4\,M\,Na\,tartrate$-0.005% gelatin (pH 6.9)	-1.32	1.22
$Ti^{4+} \rightarrow Ti^{2+}$	$0.4\,M\,Na\,citrate$-0.005% gelatin (pH 5.7)	-0.90	1.02
$Tl^+ \rightarrow Tl$	$0.1\,M\,KNO_3$ or $0.1\,M\,KCl$, etc.	-0.460	2.13
$U^{6+} \leftrightarrow U^{5+}$	$0.1\,M\,KCl$-$0.01\,M\,HCl$-$2 \times 10^{-4}\%$ thymol†	$\left\{\begin{array}{l}-0.18\\-0.92\end{array}\right.$	$\begin{array}{l}1.51\\3.20\end{array}$
$U^{5+} \rightarrow U^{3+}$			
$U^{4+} \rightarrow U^{3+}$	$0.1\,M\,HClO_4$	-0.862	.....

^{f} $0.005\,M$. ^{g} $0.0025\,M$.
^{a} $0.001\,M$.   † One run; two waves. ^{b} $0.0003\,M$.
^{c} $Re^- + 2H^+ \rightarrow Re^+ + H_2$; $Re^+ + 2e \rightarrow Re^-$.
§ One run; three waves.   * Too positive.
# $2HSO_2 \rightarrow 2H^+ + S_2O_4{}^{2-}$.
^{d} $0.001\,M$.

Table 6-5 (*Continued*)
POLAROGRAPHY

Ion Electrode Reaction	Electrolyte	$E_{1/2}$	I_d
$V^{5+} \rightarrow V^{4+}$	$0.05\,M\ H_2SO_4$-0.005% gelatin†	*	1.65
$V^{4+} \rightarrow V^{2+}$		−0.98	3.31
$V^{5+} \rightarrow V^{4+}$	$1\,M\ NH_3$-$1\,M\ NH_4Cl$-0.005% gelatin†	−0.97	4.72
$V^{4+} \rightarrow V^{2+}$		−1.26	.....
$V^{5+} \rightarrow V^{4+}$	$1\,M\ K_2C_2O_4$ (pH 4.6)	*	1.86
$V^{4+} \rightarrow V^{2+}$		−1.33	3.74
$V^{4+} \rightarrow V^{2+}$	$1\,M\ NH_3$-$1\,M\ NH_4Cl$-$0.08\,M\ Na_2SO_3$-0.005%	−1.28	1.82
$V^{4+} \rightarrow V^{5+}$	gelatin†	−0.32	0.94
$V^{4+} \rightarrow V^{5+}$	$1\,M\ NaOH$-$0.08\,M\ Na_2SO_3$	−0.432	1.47
$V^{3+} \leftrightarrow V^{2+}$	$0.5\,M\ H_2SO_4$-0.005% gelatin	−0.55	1.41
$V^{3+} \rightarrow V^{5+}$	$0.5\,M\ KHCO_3$-$0.5\,M\ Na_2CO_3$ (pH 9.4)	−0.337	2.80
$V^{2+} \rightarrow V^{3+}$	$0.5\,M\ H_2SO_4$	−0.508	1.74
$V^{2+} \leftrightarrow V^{3+}$	$1\,M\ K_2C_2O_4$	−1.091	1.43
$V^{2+} \rightarrow V^{3+}$	$1\,M$ Na acetate buffer pH 5.4†	−0.89	1.09
$V^{3+} \rightarrow V^{5+}$		−0.11	3.36
$W^{6+} \rightarrow W^{5+}$	$8\,M\ HCl$†	*	4.70
$W^{5+} \rightarrow W^{3+}$		−0.62	.....
$W^{5+} \rightarrow W^{3+}$	$12\,M\ HCl$	−0.56	2.53
$Yb^{3+} \rightarrow Yb^{2+}$	$0.1\,M\ NH_4Cl$	−1.169	.....
$Yb^{2+} \rightarrow Yb^{+}$	$0.01\,M\ LiCl$	−2.05	.....
$Zn^{2+} \rightarrow Zn$	$0.1\,M\ KCl$-0.01% gelatin	−0.995	3.42
$Zn^{2+} \rightarrow Zn$	$1\,M\ NaOH$-0.01% gelatin	−1.53	3.14
$Zn^{2+} \rightarrow Zn$	$0.5\,M$ Na tartrate-0.01% gelatin (pH 9)	−1.15	2.30
$Zn^{2+} \rightarrow Zn$	$1\,M\ NH_3$-$1\,M\ NH_4Cl$-0.01% gelatin	−1.33	3.82
$Zr^{4+} \rightarrow Zr$	$0.1\,M\ KCl$ (pH 3)	−1.65	.....

* Too positive.   † One run; two waves.

Anodic Depolarization Potentials of Inorganic Anions

Ion	Concn.	Electrolyte	$E_{1/2}$ (S.C.E.)
$2Cl^- + 2Hg \leftrightarrow Hg_2Cl_2$	$0.001\,M$	$0.1\,M\ KNO_3$	+0.25
$2Br^- + 2Hg \leftrightarrow Hg_2Br_2$	$0.001\,M$	$0.1\,M\ KNO_3$	+0.1
$2I^- + 2Hg \leftrightarrow Hg_2I_2$	$0.001\,M$	$0.1\,M\ KNO_3$	−0.05
$S^{2-} + Hg \leftrightarrow HgS$	$0.001\,M$	$0.1\,M\ NaOH$	−0.76
$Se^{2-} + Hg \leftrightarrow HgSe$	$0.00013\,M$	$0.5\,M\ Na_2CO_3$-0.003% gelatin (pH 10.7)	−0.86
$2S_2O_3^{2-} + Hg \leftrightarrow Hg(S_2O_3)_2^{2-}$	$0.001\,M$	$0.1\,M\ KNO_3$	−0.15
$SO_3^{2-} + Hg \leftrightarrow Hg(SO_3)_2^{2-}$	$0.001\,M$	$0.1\,M\ KNO_3$	−0.007
$2CN^- + Hg \leftrightarrow Hg(CN)_2$	$0.0005\,M$	$0.1\,M\ NaOH$	#
$2SCN^- + Hg \leftrightarrow Hg(SCN)_2$	$0.001\,M$	$0.1\,M\ KNO_3$	+0.18
$2OH^- + Hg \leftrightarrow Hg(OH)_2$	$0.001\,M$	$0.1\,M\ KNO_3$	+0.080

Starts at −0.45.

Table 6-5 (*Continued*)
POLAROGRAPHY

Half-Wave Potentials of Organic Compounds at $25°C$

Anodic waves are indicated in the half-wave column. The value in the column headed n indicates the number of electrons involved in the reduction or oxidation. Measurements made at temperatures other than 25°C are indicated by including the temperature value.

Compound	Electrolyte	n	$E_{1/2}$ (S.C.E.)
Aliphatic hydrocarbons:			
Allene	$0.05\,M\,(C_2H_5)_4$ NBr-75% dioxane	2	−2.29
1,3-Butadiene	$0.05\,M\,(C_2H_5)_4$NBr-75% dioxane	2	−2.59
Cyclooctatetraene	$0.1\,M\,(CH_3)_4$NOH-50% C_2H_5OH	2	−1.51
Diacetylene	$0.05\,M\,(C_2H_5)_4$NBr-75% dioxane	2	−2.27
Dimethylfulvene	$0.175\,M\,(C_4H_9)_4$NI-75% dioxane	2	−1.89
Vinylacetylene	$0.05\,M\,(C_2H_5)_4$NBr-75% dioxane	2	−2.40
Aromatic hydrocarbons:			
Acenaphthene	$0.175\,M\,(C_4H_9)_4$NI-75% dioxane	2	−2.58
Anthracene	$0.175\,M\,(C_4H_9)_4$NI-75% dioxane	2	−1.94
1,2-Benzanthracene	$0.175\,M\,(C_4H_9)_4$NI-75% dioxane	2	−2.03
		2	−2.54
Biphenyl	$0.175\,M\,(C_4H_9)_4$NI-75% dioxane	2	−2.70
1,2-Dihydronaphthalone	$0.175\,M\,(C_4H_9)_4$NI-75% dioxane	2	−2.57
Diphenylacetylene	$0.175\,M\,(C_4H_9)_4$NI-75% dioxane	2	−2.20
1,4-Diphenylbuta-1,3-diene	$0.175\,M\,(C_4H_9)_4$NI-75% dioxane	2	−1.98
1,1-Diphenylethylene	$0.175\,M\,(C_4H_9)_4$NI-75% dioxane	2	−2.14
Fluorene	$0.175\,M\,(C_4H_9)_4$NI-75% dioxane	2	−2.65
β-Methylstyrene	$0.175\,M\,(C_4H_9)_4$NI-75% dioxane	2	−2.54
Naphthalene	$0.175\,M\,(C_4H_9)_4$NI-75% dioxane	2	−2.50
Phenanthrene	$0.175\,M\,(C_4H_9)_4$NI-75% dioxane	2	−2.46
		2	−2.71
Phenylacetylene	$0.175\,M\,(C_4H_9)_4$NI-75% dioxane	2	−2.37
3-Phenylindene	$0.175\,M\,(C_4H_9)_4$NI-75% dioxane	2	−2.33
Stilbene	$0.175\,(C_4H_9)_4$NI-75% dioxane	2	−2.26
Styrene	$0.175\,(C_4H_9)4$NI-75% dioxane	2	−2.35
Tetraphenylethylene	$0.175\,M\,(C_4H_9)_4$NI-75% dioxane	2	−2.05
Aldehydes:			
Acetaldehyde	$0.6\,M$ LiOH-$0.07\,M$ LiCl (pH 12.7)	2	−1.89
Acrolein	pH 4.5	...	−1.36
Benzaldehyde	pH 1.2-50% C_2H_5OH	1	−0.94
	pH 11.3-50% C_2H_5OH	2	−1.44
Crotonaldehyde	pH 1.3-50% dioxane	...	−0.92
Formaldehyde	$0.05\,M$ KOH-$0.1\,M$ KCl	2	−1.59
Furfural	pH 3.9	...	−1.06
	pH 7.6	...	−1.38
Glucose ($0.25\,M$)	pH 7	...	−1.55
Glycolaldehyde	$0.1\,M$ NaOH	2	−1.70
Glyoxal	pH 3.4	...	−1.41
p-Hydroxybenzaldehyde	pH 1.81-50% C_2H_5OH	1	−1.16
	pH 11.98-50% C_2H_5OH	2	−1.85
o-Methoxybenzaldehyde	pH 1.81	1	−1.03
	pH 11.98	2	−1.53
p-Methoxybenzaldehyde	pH 1.81	1	−1.07
	pH 11.98	2	−1.60
Methyl glyoxal	pH 4.5	...	−0.83
Propionaldehyde	$0.1\,M$ LiOH	2	−1.93

Table 6-5 (*Continued*)
POLAROGRAPHY

Compound	Electrolyte	n	$E_{1/2}$ (S.C.E.)
Salicylaldehyde	pH 1.81	1	−1.02
	pH 11.98	2	−1.63
Ketones and derivatives:			
Acetone	$0.05\,M\,(C_2H_5)_4NI$-75% dioxane	2	−2.46
	$2.5\,M\,NH_3$-$2.5\,M\,(NH_4)_2SO_4$ (pH 9.3)	...	−1.52
Acetophenone	pH 1.3-50% C_2H_5OH	1	−1.12
	pH 8.6-50% C_2H_5OH	2	−1.62
Aurin (70°C)	pH 7-30% C_2H_5OH	...	−0.76
		...	−1.20
Benzalacetone	pH 1.3-50% C_2H_5OH	1	−0.72
	pH 8.6-50% C_2H_5OH	2	−1.27
Benzalacetophenone	pH 8.6-50% C_2H_5OH	2	−1.10
		2	−1.63
Benzil	pH 1.3-50% C_2H_5OH	2	−0.27
	pH 11.3-50% C_2H_5OH	2	−0.75
Benzoin	pH 1.3-50% C_2H_5OH	...	−0.90
	pH 11.3-50% C_2H_5OH	...	−1.51
Benzophenone	pH 1.3-50% C_2H_5OH	1	−0.94
	pH 11.3-50% C_2H_5OH	2	−1.42
Biacetyl	$0.1\,M$ HCl	...	−0.84
Colchiceine	pH 6.80-50% C_2H_5OH	...	−1.42
Colchicine	pH 6.80-H_2O	...	−1.40
Cyclohexanone	$0.05\,M\,(C_2H_5)_4NI$-75% dioxane	2	−2.45
Dibenzoylethylene (*trans*)	pH 1.3-50% C_2H_5OH	...	−0.12
	pH 11.3-50% C_2H_5OH	...	−0.57
		...	−1.52
Dibenzoylethylene (*cis*)	pH 1.3-50% C_2H_5OH	...	−0.30
	pH 11.3-50% C_2H_5OH	...	−0.62
		...	−1.65
Dibenzoylmethane	pH 1.3-50% C_2H_5OH	...	−0.59
	pH 11.3-50% C_2H_5OH	...	−1.30
		...	−1.62
Fructose	$0.02\,M$ LiCl	...	−1.76
Girard derivatives of aliphatic ketones	pH 8.2	2	−1.52
o-Hydroxyacetophenone	$0.1\,M\,NH_4Cl$-50% C_2H_5OH	...	−1.36
p-Hydroxyacetophenone	$0.1\,M\,NH_4Cl$-50% C_2H_5OH	...	−1.45
Mesityl oxide	pH 1.3-50% C_2H_5OH	...	−1.014
	pH 11.3-50% C_2H_5OH	...	−1.604
Methyl vinyl ketone	$0.1\,M$ KCl	...	−1.42
Santonin	pH 2.5	...	−1.1
	pH 8	...	−1.6
Acids and derivatives:			
Acrylonitrile	$0.05\,M\,(C_2H_5)_4NI$	2	−1.94
Ascorbic acid	pH 3.4 (phosphate buffer-1.5% H_3PO_3)	...	+0.17†
Bromoacetic acid	pH 1.1	2	−0.54
α-Bromopropionic acid	pH 2.0	2	−0.39
Crotonic acid	$0.05\,M\,(C_2H_5)_4NI$-75% dioxane	2	−1.94
Dibromoacetic acid	pH 1.1	2	−0.03
		2	−0.59
Dichloroacetic acid	pH 8.19	2	−1.57
Diethyl fumarate	pH 3.97	2	−0.84
Diethyl maleate	pH 3.98	2	−0.95

† Anodic.

Table 6-5 (*Continued*)
POLAROGRAPHY

Compound	Electrolyte	n	$E_{1/2}$ (S.C.E.)
Ethyl chloroacetate (0°C)	pH 6.8 to 10.4	2	−1.50
Ethyl dichloroacetate (0°C)	pH 6.8 to 10.4	2	−0.86
		2	−1.50
Ethyl trichloroacetate (0°C)	pH 6.8 to 10.4	2	−0.22
		2	−0.86
		2	−1.51
Fumaric acid	pH 8.2	2	−1.60
Iodoacetic acid	pH 1.1	2	−0.16
Maleic acid	pH 8.2	2	−1.36
Methacrylonitrile	0.1 M $(C_2H_5)_4NBr$	2	−2.07
Phenolphthalein	pH 3.5-25% C_2H_5OH	2	−0.11
	pH 10.06-50% C_2H_5OH	...	−1.01
		...	−1.33
Phthalide	0.1 M $(C_4H_9)_4NI$-50% dioxane	2	−2.03
Pyruvic acid	pH 3	...	−0.86
	pH 7	...	−1.30
		...	−1.57
Trichloroacetic acid	pH 8.19	2	−0.84
		2	−1.57
Halogen compounds:			
Allyl bromide	0.05 M $(C_2H_5)_4NBr$-75% dioxane	2	−1.29
Allyl chloride	0.05 M $(C_2H_5)_4NBr$-75% dioxane	2	−1.91
Benzal chloride	0.05 M $(C_2H_5)_4NBr$-75% dioxane	4	−1.81
Benzotrichloride	0.05 M $(C_2H_5)_4NBr$-75% dioxane	2	−0.68
		2	−1.65
		2	−2.00
Benzyl chloride	0.05 M $(C_2H_5)_4NBr$-75% dioxane	2	−1.94
Bromobenzene	0.05 M $(C_2H_5)_4NBr$-75% dioxane	2	−2.32
Bromoform	0.05 M $(C_2H_5)_4NBr$-75% dioxane	2	−0.64
		4	−1.51
n-Butyl bromide	0.05 M $(C_2H_5)_4NBr$-75% dioxane	2	−2.27
Carbon tetrabromide	0.05 M $(C_2H_5)_4NBr$-75% dioxane	2	−0.3
		2	−0.75
		4	−1.49
Carbon tetrachloride	0.05 M $(C_2H_5)_4NBr$-75% dioxane	2	−0.78
		2	−1.71
Chloroform	0.05 M $(C_2H_5)_4NBr$-75% dioxane	2	−1.6
p-Dibromobenzene	0.05 M $(C_2H_5)_4NBr$-75% dioxane	2	−2.10
o-Dichlorobenzene	0.05 M $(C_2H_5)_4NBr$-75% dioxane	2	−2.51
m-Dichlorobenzene	0.05 M $(C_2H_5)_4NBr$-75% dioxane	2	−2.48
p-Dichlorobenzene	0.05 M $(C_2H_5)_4NBr$-75% dioxane	2	−2.49
Ethyl bromide	0.05 M $(C_2H_5)_4NBr$-75% dioxane	2	−2.08
Ethyl iodide	0.05 M $(C_2H_5)_4NBr$-75% dioxane	2	−1.67
γ-Hexachlorocyclohexane	0.1 M $(C_2H_5)_4NI$-80% C_2H_5OH	...	−1.65
	1% KI-50% C_2H_5OH	...	−1.35
Iodobenzene	0.05 M $(C_2H_5)_4NBr$-75% dioxane	2	−1.62
		2	−1.09
		2	−1.50
Methyl bromide	0.05 M $(C_2H_5)_4NBr$-75% dioxane	2	−1.63
Methyl chloride	0.05 M $(C_2H_5)_4NBr$-75% dioxane	2	−2.23
Methyl iodide	0.05 M $(C_2H_5)_4NBr$-75% dioxane	2	−1.63
Methylene bromide	0.05 M $(C_2H_5)_4NBr$-75% dioxane	4	−1.48
Methylene chloride	0.05 M $(C_2H_5)_4NBr$-75% dioxane	4	−1.6
Methylene iodide	0.05 M $(C_2H_5)_4NBr$-75% dioxane	2	−1.12
		2	−1.53

Table 6-5 (*Continued*)
POLAROGRAPHY

Compound	Electrolyte	n	$E_{1/2}$ (S.C.E.)
Nitro compounds and their reduction products:			
Azobenzene (*trans*)	pH 6.26-10% C_2H_5OH	2	−0.33
Azoxybenzene	pH 6.3-20% C_2H_5OH	4	−0.63
Benzenediazonium chloride	0.1 M HCl	1	−0.18
		3	−0.67
m-Dinitrobenzene	pH 1.7-8% C_2H_5OH	4	−0.12
		4	−0.26
Methyl *o*-nitrobenzoate	pH 2-10% C_2H_5OH	4	−0.25
		2	−0.74
Methyl *m*-nitrobenzoate	pH 2-10% C_2H_5OH	4	−0.240
		2	−0.68
Methyl *p*-nitrobenzoate	pH 2-10% C_2H_5OH	4	−0.200
		2	−0.73
p-Nitroaniline	pH 2	6	−0.36
o-Nitroanisole	pH 2-10% C_2H_5OH	4	−0.29
		2	−0.58
	pH 10-10% C_2H_5OH	4	−0.76
m-Nitroanisole	pH 2-10% C_2H_5OH	4	−0.28
		2	−0.69
	pH 10-10% C_2H_5OH	4	−0.71
p-Nitroanisole	pH 2-10% C_2H_5OH	4	−0.35
		2	−0.64
	pH 10-10% C_2H_5OH	4	−0.80
m-Nitrobenzaldehyde	pH 2-10% C_2H_5OH	4	−0.28
		2	−1.20
Nitrobenzene	pH 2-8% C_2H_5OH	4	−0.30
		2	−0.75
o-Nitrobenzoic acid	pH 2-10% C_2H_5OH	4	−0.23
		2	−0.73
m-Nitrobenzoic acid	pH 2-10% C_2H_5OH	4	−0.20
		2	−0.70
p-Nitrobenzoic acid	pH 2-10% C_2H_5OH	4	−0.17
		2	−0.74
Nitromethane	pH 3.3	4	−0.83
	pH 11.9	4	−0.90
o-Nitrophenol	pH 2-10% C_2H_5OH	4	−0.23
	pH 10-8% C_2H_5OH	6	−0.80
m-Nitrophenol	pH 2-8% C_2H_5OH	4	−0.37
		2	−1.10
	pH 10-8% C_2H_5OH	4	−0.76
p-Nitrophenol	pH 2-8% C_2H_5OH	4	−0.35
	pH 10-8% C_2H_5OH	2	−0.92
		4	−1.33
Nitrosobenzene	pH 7-10% C_2H_5OH	2	−0.11
α-Nitroso-β-naphthol	pH 5.9	4	−0.11
N-Nitrosophenylhydroxylamine	pH 2	6	−0.84
o-Nitrotoluene	pH 1-80% dioxane	4	−0.26
		2	−0.64
m-Nitrotoluene	pH 1-80% dioxane	4	−0.22
		2	−0.71
p-Nitrotoluene	pH 1-80% dioxane	4	−0.24
		2	−0.71
Phenylhydroxylamine	pH 7-10% C_2H_5OH	2	−0.09
Tetranitromethane	pH 12	...	−0.41

Table 6-5 (*Continued*)
POLAROGRAPHY

Compound	Electrolyte	n	$E_{1/2}$ (S.C.E.)
Trinitrotoluene	pH 0.5-8% C_2H_5OH	4	−0.01
		4	−0.08
		4	−0.14
Sulfur compounds:			
Cysteine	0.1 M $HClO_4$	1‡	−0.05
Dithiodiglycolic acid	pH 3.00-0.002% gelatin	2	−0.37
Thioglycolic acid	pH 6.75	1‡	−0.38
Peroxides:			
Benzoyl peroxide	0.3 M LiCl-50% CH_3OH-50%C_6H_6	...	0.00
Cumene hydroperoxide	0.3 M LiCl-50% CH_3OH-50% C_6H_6	...	−0.68
Quinones (Reversible system):			
Anthraquinone	pH 7.4-40% dioxane	2	−0.54
Benzoquinone	pH 5.40-50% CH_3OH	2	+0.146
2,3-Dimethylnaphthoquinone	pH 5.40-50% CH_3OH	2	−0.216
Duroquinone	pH 5.40-50% CH_3OH	2	−0.093
2-Methyl-1,4-naphthoquinone	pH 6.24-75% C_2H_5OH	2	−0.17
Toluquinone	pH 5.40-50% CH_3OH	2	+0.090
Heterocyclic compounds containing oxygen:			
Flavanone	pH 7.5-50% $(CH_3)_2CHOH$	...	−1.37
Flavone	pH 7.5-50% $(CH_3)_2CHOH$	...	−1.26
		...	−1.44
2,5,7,8-Tetramethyl-6-hydroxy-chroman	pH 3.56-50% CH_3OH	2	+0.230†
2,4,6,7-Tetramethyl-5-hydroxy-coumaran	pH 3.56-50% CH_3OH	2	+0.219†
Heterocyclic compounds containing nitrogen:			
Acridine	pH 8.29-60% C_2H_5OH	1	−0.80
		1	−1.48
8-Hydroxyquinoline	pH 10	...	−1.39
		...	−1.61
Nicotinamide	pH 8.7	...	−1.56
Iso-Nicotinic acid hydrazide	pH 1.5	...	−0.52
		...	−0.70
Picolinic acid	pH 7	...	−1.17
Pyridine	pH 2.9	...	−1.49
Quinaldinic acid	pH 4-8% C_2H_5OH-gelatin	...	−0.86
		...	−1.19
Quinoline	pH 6.51-50% C_2H_5OH-0.04% gelatin	2	−1.23
Quinoline-8-carboxylic acid	pH 9	...	−1.11
		...	−1.75
Saccharin	pH 7.0	...	−1.77
Thiamin	pH 7.2	...	−1.30

‡HGSR formation.
† Anodic.

Table 6-6
VOLTAIC CELLS AND BATTERIES

Name of Cell	Electrochemical System	Nominal emf (volts)	Remarks
Conventional cells			
Air (oxygen) "depolarized":			
(a) Alkaline	$Zn(Hg)|KOH$ resp $NaOH(aq)|O_2$ [C]	1.4	Wet or "dry"
(b) Leclanché type	$Zn(Hg)|NH_4Cl + ZnCl_2(aq)|O_2$ [C(MnO_2)]	1.4	Wet or "dry"
Alkaline MnO_2	$Zn(Hg)|KOH(aq)|MnO_2$ [C]	1.6	Prim and sec (dry type)
Clark—Standard	$Zn(Hg)|ZnSO_4(aq) \parallel Hg_2SO_4(aq)|Hg$	1.4328[15°]	2-Fluid
Daniell	$Zn(Hg)|ZnSO_4(aq) \parallel CuSO_4(aq)|Cu$	1.08	2-Fluid prototype
Drumm	$Zn(Hg)|KOH + K_2Zn_2O_3(aq)|NiO-OH$	1.8	Secondary
Edison	$Fe|KOH + LiOH(aq)|NiO-OH$	1.5	Secondary
Halogen (chlorine)	$Zn|ZnCl_2(aq)|Cl_2$ [C]	2.05	2.85 v with Mg anode
Lalande-Chaperon	$Zn(Hg)|NaOH(aq)|CuO$ [Fe]	0.95	....
Leclanché wet	$Zn(Hg)|NH_4Cl(aq)|MnO_2$ [C]	1.5	Prototype
Leclanché dry	$Zn(Hg)|NH_4Cl + ZnCl_2(aq)|MnO_2$ [C]	1.6	"Dry battery" of commerce
Magnesium, Leclanché type	$Mg|MgBr_2(aq)|MnO_2$ [C]	1.9	Dry cell
Magnesium, reserve cell:			
(a) AgCl	$Mg|MgCl_2(aq)|AgCl$ [Ag]	1.8	Water-activated
(b) Cu_2Cl_2	$Mg|MgCl_2(aq)|Cu_2Cl_2$ [Cu]	1.75	Water-activated
Main (Reynier)	$Zn(Hg)|H_2SO_4(aq)|PbO_2$ [Pb]	2.5	....
Mercuric oxide (Ruben-Mallory)	$Zn(Hg)|KOH + K_2Zn_2O_3(aq)|HgO$ [C]	1.35	Dry type
Nickel-cadmium	$Cd|KOH(aq)|NiO-OH$	1.3	Sec battery
Perchloric acid	$Pb|HClO_4(aq)|PbO_2$ [Pb]	2.3	Reserve cell
Poggendorff (dichromate)	$Zn(Hg)|K_2Cr_2O_7 + H_2SO_4(aq)|C$	2.0	....
Silver-cadmium	$Cd|KOH(aq)|Ag_2O_2$ [Ag]	1.4	Sec battery

Table 6-6 (*Continued*)
VOLTAIC CELLS AND BATTERIES

Name of Cell	Electrochemical system	Nominal emf (volts)	Remarks
Conventional cells (cont.)			
Silver chloride	Zn(Hg)|NH₄Cl(aq)|AgCl [Ag]	1.02	··········
Silver-zinc	Zn(Hg)|KOH(aq)|Ag₂O₂ [Ag]	1.8	Sec battery
	Zn(Hg)|KOH(aq)|Ag₂O [Ag]	1.6	Prim or sec
Volta	Zn|acid, alk, or salt(aq)|Ag or Cu	0.9–1.0	First battery (wet or dry)
Weston—Standard	Cd(Hg)|CdSO₄(aq) ‖ Hg₂SO₄(aq)|Hg	$1.01827^{20°}$	2-Fluid
Zamboni	Zn|paper|MnO₂ [Ag or Au]	0.7–0.9	High-voltage dry battery of minimal current
Continuous-feed and fuel cells*			
Amalgam	[Fe] Na(Hg)|NaOH(aq)|O₂ [C or Ni(Ag)]	2.05	··········
Hydrogen-oxygen:			
(a) Low temperature	[C or Ni] H₂|KOH(aq)|O₂ [C or Ni(Ag)]	1.1	20–90°C
(b) Intermediate temperature	[Ni] H₂|KOH(aq)|O₂ [Ni]	1.05	200–240°C (>400 psi)
(c) High temperature	[Porous metals] H₂|Fused carbonates|O₂ [Porous metals]	1.0	400–700°C
Ion-exchange membrane	[Pt] H₂|Resin|O₂ [Pt]	1.05	20–80°C

* Cells in early development stages not included; e.g., alcohol, hydrocarbons, redox, etc.

Table 6-7
LIMITING EQUIVALENT IONIC CONDUCTANCES
IN AQUEOUS SOLUTIONS AT 25°C

Except for H^+ (0.0139) and OH^- (0.018), a temperature coefficient of 0.02 deg^{-1} is applicable to the cations and anions.

Ion	Limiting Equivalent Ionic Conductance, mho-cm^2/equivalent	Ion	Limiting Equivalent Ionic Conductance, mho-cm^2/equivalent
Inorganic Cations		**Inorganic Anions**	
Ag^+	61.9	$Au(CN)_2^-$	50
Al^{3+}	61	$Au(CN)_4^-$	36
Ba^{2+}	63.9	$B(C_6H_5)_4^-$	21
Be^{2+}	45	Br^-	78.1
Ca^{2+}	59.5	Br_3^-	43
Cd^{2+}	54	BrO_3^-	55.8
Ce^{3+}	70	Cl^-	76.35
Co^{2+}	53	ClO_2^-	52
$Co(NH_3)_6^{3+}$	100	ClO_3^-	64.6
$Co(en)_3^{3+}$	74.7	ClO_4^-	67.9
Cr^{3+}	67	CN^-	78
Cs^+	77.3	CO_3^{2-}	72
Cu^{2+}	55	$Co(CN)_6^{3-}$	98.9
D^+ (deuterium)	213.7 (18°)	CrO_4^{2-}	85
Dy^{3+}	65.7	F^-	54.4
Er^{3+}	66.0	$Fe(CN)_6^{4-}$	111
Eu^{3+}	67.9	$Fe(CN)_6^{3-}$	101
Fe^{2+}	54	$H_2AsO_4^-$	34
Fe^{3+}	68	HCO_3^-	44.5
Gd^{3+}	67.4	HF_2^-	75
H^+	349.82	HPO_4^{2-}	57
Hg^{2+}	53	$H_2PO_4^-$	33
Ho^{3+}	66.3	$H_2PO_2^-$	46
K^+	73.5	HS^-	65
La^{3+}	69.6	HSO_3^-	50
Li^+	38.69	HSO_4^-	50
Mg^{2+}	53.06	$H_2SbO_4^-$	31
Mn^{2+}	53.5	I^-	76.8
NH_4^+	73.5	IO_3^-	40.5
$N_2H_5^+$	59	IO_4^-	54.5
Na^+	50.11	$N(CN)_2^-$	54.5
Nd^{3+}	69.6	NO_2^-	71.8
Ni^{2+}	50	NO_3^-	71.4
Pb^{2+}	71	$NH_2SO_3^-$	48.6
Pr^{3+}	69.6	N_3^-	69
Ra^{2+}	66.8	OCN^-	64.6
Rb^+	77.8	OH^-	198.6
Sc^{3+}	64.7	PF_6^-	56.9
Sm^{3+}	68.5	PO_3F^{2-}	63.3
Sr^{2+}	59.46	PO_4^{3-}	69.0
Tl^+	76	$P_2O_7^{4-}$	81.4
Tm^{3+}	65.5	$P_3O_9^{3-}$	83.6
UO_2^{2+}	32	$P_3O_{10}^{5-}$	109
Y^{3+}	62	ReO_4^-	54.7
Yb^{3+}	65.2	SCN^-	66
Zn^{2+}	52.8	$SeCN^-$	64.7

Table 6-7 (*Continued*)
LIMITING EQUIVALENT IONIC CONDUCTANCES
IN AQUEOUS SOLUTIONS AT 25°C

Ion	Limiting Equivalent Ionic Conductance, mho-cm^2/equivalent	Ion	Limiting Equivalent Ionic Conductance, mho-cm^2/equivalent
Inorganic Anions		Bromobenzoate	30
SeO_4^{2-}	75.7	n-Butyrate	32.6
SO_3^{2-}	79.9	Chloroacetate	39.7
SO_4^{2-}	80.0	Chlorobenzoate	33
$S_2O_3^{2-}$	85.0	Citrate^{3-}	70.2
$S_2O_4^{2-}$	66.5	α-Crotonate	33.2
$S_2O_6^{2-}$	93	Cyanoacetate	41.8
$S_2O_8^{2-}$	86	Cyclohexane carboxylate	28.7
WO_4^{2-}	69	Cyclopropane-1,1-dicarboxylate^{2-}	53.4
Organic Cations		Decyl sulfonate	26
i-Butylammonium	38	Dichloroacetate	38.3
n-Decylpyridinium	29.5	Diethyl barbiturate^{2-}	26.3
Diethylammonium	42.0	Dihydrogen citrate	30
Dimethylammonium	51.5	Dimethyl malonate^{2-}	49.4
Dipropylammonium	30.1	3,5-Dinitrobenzoate	28.3
n-Dodecylammonium	23.8	Dodecyl sulfonate	24
Ethylammonium	47.2	Ethyl malonate	49.3
Ethyltrimethylammonium	40.5	Ethyl sulfonate	39.6
Methylammonium	58.3	Fluorobenzoate	33
Histidyl	23.0	Formate	54.6
Piperidinium	37.2	$HC_2O_4^-$	40.2
Propylammonium	40.8	Lactate	38.8
Pyrilammonium	24.3	Malonate^{2-}	63.5
Tetra-n-butylammonium	19.1	Methyl sulfonate	48.8
Tetraethylammonium	33.0	$C_2O_4^{2-}$	74.2
Tetramethylammonium	45.3	Octyl sulfonate	29
Tetra-n-propylammonium	23.5	Phenyl acetate	30.6
Triethylammonium	34.3	Picrate	30.2
Triethylsulfonium	36.1	Propionate	35.8
Trimethylammonium	46.6	Propyl sulfonate	37.1
Trimethylsulfonium	51.4	Salicylate	36
Tripropylammonium	26.1	Suberate^{2-}	36
Organic Anions		Succinate^{2-}	58.8
Acetate	40.9	Sulfonate	43.1
p-Anisate	29.0	Tartrate^{2-}	64
Azelate^{2-}	40.6	Trichloroacetate	36.6
Benzoate	32.4		

Table 6-8
STANDARD SOLUTIONS FOR CALIBRATING CONDUCTIVITY VESSELS

The values of conductivity κ are corrected for the conductivity of the water used. The cell constant θ of a conductivity cell can be obtained from the equation

$$\theta = \frac{\kappa R R_{\text{solv}}}{R_{\text{solv}} - R}$$

where R is the resistance measured when the cell is filled with a solution of the composition stated in the table below, and R_{solv} is the resistance when the cell is filled with solvent at the same temperature.

Grams KCl per Kilogram Solution (in vacuo)	Conductivity in $\text{ohm}^{-1} \cdot \text{cm}^{-1}$ at		
	$0°C$	$18°C$	$25°C$
71.135 2	$0.065\ 14_4$	$0.097\ 79_0$	$0.111\ 28_7$
7.419 13	$0.007\ 134_4$	$0.011\ 161_2$	$0.012\ 849_7$
0.745 263*	$0.000\ 773\ 2_6$	$0.001\ 219\ 9_2$	$0.001\ 408\ 0_8$

* Virtually 0.0100 M.
From the data of Jones and Bradshaw, *J. Am. Chem. Soc.*, **55,** 1780 (1933). The original data have been converted from (int. ohm)$^{-1}$ cm^{-1}.

Table 6-9
ELECTRICAL CONDUCTIVITY OF VARIOUS PURE LIQUIDS

Liquid	Temp. °C	mhos/cm or $\text{ohm}^{-1} \cdot \text{cm}^{-1}$	Liquid	Temp. °C	mhos/cm or $\text{ohm}^{-1} \cdot \text{cm}^{-1}$
Acetaldehyde	15	1.7×10^{-6}	Benzylamine	25	$<1.7 \times 10^{-8}$
Acetamide	100	$<4.3 \times 10^{-5}$	Benzyl benzoate	25	$<1 \times 10^{-9}$
Acetic acid	0	5×10^{-9}	Bromine	17.2	1.3×10^{-13}
	25	1.12×10^{-8}	Bromobenzene	25	$<2 \times 10^{-11}$
Acetic anhydride	0	1×10^{-6}	Bromoform	25	$<2 \times 10^{-8}$
	25	4.8×10^{-7}	iso-Butyl alcohol	25	8×10^{-8}
Acetone	18	2×10^{-8}			
	25	6×10^{-8}	Capronitrile	25	3.7×10^{-6}
Acetonitrile	20	7×10^{-6}	Carbon disulfide	1	7.8×10^{-18}
Acetophenone	25	6×10^{-9}	Carbon tetrachloride	18	4×10^{-18}
Acetyl bromide	25	2.4×10^{-6}	Chlorine	-70	$<1 \times 10^{-16}$
Acetyl chloride	25	4×10^{-7}	Chloroacetic acid	60	1.4×10^{-6}
Alizarin	233	1.45×10^{-6}(?)	m-Chloroaniline	25	5×10^{-8}
Allyl alcohol	25	7×10^{-6}	Chloroform	25	$<2 \times 10^{-8}$
Ammonia	-79	1.3×10^{-7}	Chlorohydrin	25	5×10^{-7}
Aniline	25	2.4×10^{-8}	m-Cresol	25	$<1.7 \times 10^{-8}$
Anthracene	230	3×10^{-10}	Cyanogen	$\cdots$	$<7 \times 10^{-9}$
Arsenic tribromide	35	1.5×10^{-6}	Cymene	25	$<2 \times 10^{-8}$
Arsenic trichloride	25	1.2×10^{-6}			
			Dichloroacetic acid	25	7×10^{-8}
			Dichlorohydrin	25	1.2×10^{-5}
Benzaldehyde	25	1.5×10^{-7}	Diethylamine	-33.5	2.2×10^{-9}
Benzene	$\cdots$	7.6×10^{-8}	Diethyl carbonate	25	1.7×10^{-8}
Benzoic acid	125	3×10^{-9}	Diethyl oxalate	25	7.6×10^{-7}
Benzonitrile	25	5×10^{-8}	Diethyl sulfate	25	2.6×10^{-7}
Benzyl alcohol	25	1.8×10^{-6}	Dimethyl sulfate	0	1.6×10^{-7}

Table 6-9 (*Continued*)
ELECTRICAL CONDUCTIVITY OF VARIOUS PURE LIQUIDS

Liquid	Temp. °C	mhos/cm or ohm$^{-1} \cdot$ cm^{-1}	Liquid	Temp. °C	mhos/cm or ohm$^{-1} \cdot$ cm^{-1}
Epichlorohydrin	25	3.4×10^{-8}	Nitromethane	18	6×10^{-7}
Ethyl acetate	25	$<1 \times 10^{-9}$	o- or m-Nitrotoluene	25	$<2 \times 10^{-7}$
Ethyl acetoacetate	25	4×10^{-8}	Nonane	25	$<1.7 \times 10^{-8}$
Ethyl alcohol	25	1.35×10^{-9}	Oleic acid	15	$<2 \times 10^{-10}$
Ethylamine	0	4×10^{-7}			
Ethyl benzoate	25	$<1 \times 10^{-9}$	Pentane	19.5	$<2 \times 10^{-10}$
Ethyl bromide	25	$<2 \times 10^{-8}$	Petroleum	· · ·	3×10^{-13}
Ethylene bromide	19	$<2 \times 10^{-10}$	Phenetole	25	$<1.7 \times 10^{-8}$
Ethylene chloride	25	3×10^{-8}	Phenol	25	$<1.7 \times 10^{-8}$
Ethyl ether	25	$<4 \times 10^{-13}$	Phenyl isothiocyanate	25	1.4×10^{-6}
Ethylidene chloride	25	$<1.7 \times 10^{-8}$	Phosgene	25	7×10^{-9}
Ethyl iodide	25	$<2 \times 10^{-8}$	Phosphorus	25	4×10^{-7}
Ethyl isothiocyanate	25	1.26×10^{-8}	Phosphorus oxychloride	25	2.2×10^{-6}
Ethyl nitrate	25	5.3×10^{-7}			
Ethyl thiocyanate	25	1.2×10^{-6}	Pinene	23	$<2 \times 10^{-10}$
Eugenol	25	$<1.7 \times 10^{-8}$	Piperidine	25	$<2 \times 10^{-7}$
			Propionaldehyde	25	8.5×10^{-7}
Formamide	25	4×10^{-6}	Propionic acid	25	$<1 \times 10^{-9}$
Formic acid	18	5.6×10^{-5}	Propionitrile	25	$<1 \times 10^{-7}$
	25	6.4×10^{-5}	n-Propyl alcohol	18	5×10^{-8}
Furfural	25	1.5×10^{-6}		25	2×10^{-8}
			iso-Propyl alcohol	25	3.5×10^{-6}
Gallium	30	36,800	n-Propyl bromide	25	$<2 \times 10^{-8}$
Glycerol	25	6.4×10^{-8}	Pyridine	18	5.3×10^{-8}
Glycol	25	3×10^{-7}			
Guaiacol	25	2.8×10^{-7}	Quinoline	25	2.2×10^{-8}
Heptane	· · ·	$<1 \times 10^{-13}$	Salicylaldehyde	25	1.6×10^{-7}
Hexane	18	$<1 \times 10^{-18}$	Stearic acid	80	$<4 \times 10^{-13}$
Hydrogen bromide	−80	8×10^{-9}	Sulfonyl chloride, SOCl$_2$	25	2×10^{-6}
Hydrogen chloride	−96	1×10^{-8}			
Hydrogen cyanide	0	3.3×10^{-6}	Sulfur	115	1×10^{-12}
Hydrogen iodide	B.P.	2×10^{-7}		130	5×10^{-11}
Hydrogen sulfide	B.P.	1×10^{-11}		440	1.2×10^{-7}
			Sulfur dioxide	35	1.5×10^{-8}
Iodine	110	1.3×10^{-10}	Sulfuric acid	25	1×10^{-2}
			Sulfuryl chloride, SO$_2$Cl$_2$	25	3×10^{-8}
Kerosene	25	$<1.7 \times 10^{-8}$			
			Toluene	· · ·	$<1 \times 10^{-14}$
Mercury	0	10,629.6	o-Toluidine	25	$<2 \times 10^{-6}$
Methyl acetate	25	3.4×10^{-6}	p-Toluidine	100	6.2×10^{-8}
Methyl alcohol	18	4.4×10^{-7}	Trichloroacetic acid	25	3×10^{-9}
Methyl ethyl ketone	25	1×10^{-7}	Trimethylamine	−33.5	2.2×10^{-10}
Methyl iodide	25	$<2 \times 10^{-8}$	Turpentine	· · ·	2×10^{-13}
Methyl nitrate	25	4.5×10^{-6}			
Methyl thiocyanate	25	1.5×10^{-6}	iso-Valeric acid	80	$<4 \times 10^{-13}$
			Water	18	4×10^{-8}
Naphthalene	82	4×10^{-10}	Xylene	· · ·	$<1 \times 10^{-15}$
Nitrobenzene	0	5×10^{-9}			

Table 6-10
EQUIVALENT CONDUCTIVITIES OF ELECTROLYTES IN AQUEOUS SOLUTIONS AT 18°C

The unit of Λ in the table is $\Omega^{-1} \cdot cm^2 \cdot equiv^{-1}$. The entities to which the equivalent relates are given in the first column.

Electrolyte	Concentration, N										
	0.001	0.005	0.01	0.05	0.1	0.5	1.0	2.0	3.0	4.0	5.0
Acetic acid	41	20.0	14.3	6.48	4.60	2.01	1.32		0.54		0.29
AgNO$_3$	113.2	110.0	107.8	99.5	94.3	77.8	67.8	56.0	48.2	42.1	37.2
$\frac{1}{2}$Ag$_2$SO$_4$	116.3	108.4	102.9								
$\frac{1}{3}$AlBr$_3$ (25°)	132	124	119	103	97						
$\frac{1}{3}$AlCl$_3$	121.1	105.0	93.8			65.0	56.2	44.2	34.7	27.2	
$\frac{1}{3}$AlI$_3$ (25°)	131	124	119	108	88						
$\frac{1}{3}$Al(NO$_3$)$_3$ (25°)	123	115	110	94							
$\frac{1}{6}$Al$_2$(SO$_4$)$_3$ (25°)	107.2	76.8	60.6			43.8	34.3				
$\frac{1}{2}$Ba(OAc)$_2$	85.0	80.4	77.1	65.7	60.2						
$\frac{1}{2}$Ba(BrO$_3$)$_2$ (25°)	113.6	106.8	102.7								
$\frac{1}{2}$BaCl$_2$	115.6	112.3	106.7	96.0	90.8	77.3	70.1	60.3	52.3		
$\frac{1}{2}$Ba(NO$_3$)$_2$	111.7	105.3	101.0	86.8	78.9	56.6	48.4		29.8	23.4	
$\frac{1}{2}$Ba(OH)$_2$	216	213	207	191	180						
Butyric acid						1.66	0.98	0.46	0.26	0.18	0.11
$\frac{1}{2}$Ca(OAc)$_2$	79.6	75.0	71.9	60.3	54.0	36.3	26.3				
$\frac{1}{2}$CaCl$_2$	112.0	106.7	103.4	93.3	88.2	74.9	67.5	58.3	49.7	42.4	35.6
$\frac{1}{2}$Ca(NO$_3$)$_2$	108.5	103.0	99.5	88.4	82.5	65.7	55.9	43.5	35.5	26.0	21.5
$\frac{1}{2}$Ca(OH)$_2$		233	226								
$\frac{1}{2}$CaSO$_4$	104.3	86.3	77.4								
$\frac{1}{2}$CdBr$_2$		86.5	76.3	53.2	44.6	25.3	18.3	12.5	9.1	6.8	5.3
$\frac{1}{2}$CdCl$_2$		91	83	59	50	30.8	22.4	14.4	9.9	7.1	5.4
$\frac{1}{2}$CdI$_2$		76.7	65.6	40.1	31.0	18.3	15.4	12.3	9.7	8.0	
$\frac{1}{2}$Cd(NO$_3$)$_2$		100	96	86.4	80.8	63.9	54.5	41.0	31.4	23.7	17.6
$\frac{1}{2}$CdSO$_4$	97.7	79.7	70.3	49.6	42.2	28.7	23.6	17.7	14.0	11.0	8.35
$\frac{1}{3}$CeCl$_3$ (25°)	137.4		122.1		99.0						
$\frac{1}{6}$Ce$_2$(C$_2$O$_4$)$_3$ (25°)	85.5	54	45.8	29							

Substance										
Chloroacetic acid (25°)	88.4									
Citric acid	54	42.5	22.0	42.9	20.2	13.6	8.1	5.6	4.2	3.3
1/2 CoCl₂	99.3	95.6	82.3	16.1	7.3	5.4	40.3	35.4	30.5	26.4
1/3 CrCl₃				75.0	51.5	45.3	44.8	35.2		
					63.6	56.8				
1/2 CrO₃(H₂CrO₄) (25°)	195	193	191	186						
CsCl	201	130.7	127.5	125.2	113.5	104.3	100.3	95.7	85.1	
1/2 Cu(OAc)₂ (25°)	50.6	55.7	47.2	34.9	28.4					
1/2 CuCl₂	107.9	97.1	93.7	83.7	78.2	67.5	56.8	41.2	31.5	24.5
								45.4	35.3	27.8
1/2 Cu(NO₃)₂ (15°)	98.5	81.0	71.7	53.6	43.8	30.5	25.6	19.7	16.5	16.3
1/2 CuSO₄				207.5	119	82	44.6	26.5		
Dichloroacetic acid (25°)	131	120	103	93						
1/2 FeCl₂ (25°)	82	75	54	44.5	66.5	52.9	37.6	28.1	20.5	15.9
1/3 FeCl₃					30.8	25.8	19.5	15.37		
1/2 FeSO₄	125.6					5.18	3.68	2.93	2.39	1.92
Formic acid	308.2	230.0	187.0	103.4	80.4					
H₃AsO₄ (1 M) (25°)	13.5									
H₃BO₃										
HBr	401	387	373	356	306	282	243	214	179	
HBrO₃ (25°)	377	373	272	156						
HCl		373	370	360	351	327	301	215		
HClO₃				343	317	292	207			
HClO₄ (25°)	413	406	402	392	386	358				
HF		90	60	35.9	31.3	27.0	25.7	24.2		
HI	343.3	332.8	323.9	347	322	297	255	215	179	24.0
HIO₃	375	371	368	253	175	141	106	87	71	
HNO₃	318	279	255	350	324	310		220		
H₃PO₄ (1 M)				357		66		53.1		156
HSCN (25°)	399	394	390	370	205	198		166.8		51.3
1/2 H₂SO₄	361	330	308	225						135.0
1/2 HgCl₂				1.23						
1/3 InBr₃			253	1.85						
KOAc	98.3	95.7	94.0	87.7	53.9	37.0	28.7	19.8	14.4	10.1
KBr	129.4	126.4	124.4	117.8	83.8	71.6	63.4	50.0	40.7	31.4
KBrO₃	109.9	106.9	104.7	97.3	114.2	105.4	102.5	98.0	93.3	87.9
1/3 K₃citrate		109.9	103	87.8	93.0					
KCl	127.3	124.4	122.4	115.8	80.8					
				112.0	102.4	98.3	92.0	88.9		24.5

Table 6-10 (Continued)

EQUIVALENT CONDUCTIVITIES OF ELECTROLYTES IN AQUEOUS SOLUTIONS AT 18°C

Electrolyte	Concentration, N										
	0.001	0.005	0.01	0.05	0.1	0.5	1.0	2.0	3.0	4.0	5.0
KClO$_3$	116.9	113.6	111.6	103.7	99.2	85.3					
KClO$_4$ (25°)	137.9	134.2	131.5	121.6	115.2						
KCN (15°)						104.2	99.7				
$\frac{1}{2}$K$_2$CO$_3$	133.0	121.6	115.5	100.7	94.1	77.8	70.7	65.0	55.6	49.2	42.9
$\frac{1}{2}$K$_2$C$_2$O$_4$	122.4	116.7	112.5	100.8	94.9	80.4	73.7	72.0	59.9		
$\frac{1}{2}$K$_2$CrO$_4$					100.5	86.4	79.5				
$\frac{1}{2}$K$_2$Cr$_2$O$_7$					98.2	85.4					
KF	108.9	106.2	104.3	97.7	94.0	82.6	76.0	63.4	56.5	51.7	46.5
$\frac{1}{3}$K$_3$[Fe(CN)$_6$]	163.1	150.7									
$\frac{1}{4}$K$_4$[Fe(CN)$_6$]	167.2	146.1	134.8	107.7	97.9						
KHCO$_3$ (25°)	115.3	112.2	110.1			86.5	78.9				
KH phthalate	119.3	103.7	99.9	89.3	83.8						
KHS						92.5	91.7	86.4	80.7		69.3
KHSO$_4$						21.0	18.4	15.2			
KH$_2$PO$_4$ (1 M) (25°)	107.1	100.8	98.0	90.7	85.6	$60.0^{18°}$	$45.8^{18°}$				
KI	128.2	125.3	123.4	117.3	114.0	106.2	103.6	101.3	96.4	89.0	81.2
KIO$_3$	96.0	93.2	91.2	84.1	79.7						
KIO$_4$ (25°)	124.9	121.2	118.5	106.7	98.1						
KMnO$_4$ (25°)	133.3		126.5		113						
KNO$_3$	123.6	120.5	118.2	109.9	104.8	89.2	80.5	69.4	61.3		
KOH	234	230	228	219	213	197	184		140.6		105.8
KReO$_4$ (25°)	125.1	121.3	118.5	106.4	97.4						
$\frac{1}{2}$K$_2$S								119.7	108.3	97.2	86.1
KSCN	118.6	115.8	113.9	107.7	104.3	95.7	91.6	86.8	74.6		
$\frac{1}{2}$K$_2$SO$_4$	126.9	120.3	115.8	101.9	94.9	78.5	71.6				
$\frac{1}{3}$LaCl$_3$ (25°)	137.0	127.5	121.8	106.2	99.1						
$\frac{1}{3}$La(NO$_3$)$_3$				86.1	72.1	65.4	54.0	39.1	28.5	19.9	
$\frac{1}{6}$La$_2$(SO$_4$)$_3$				25.7	21.5						
Lactic acid	108.9	53.5	39	18.1	13.2						

Electrolyte											
LiOAc	96.5	93.9		87.9	51.3	37.7	28.9	18.2	11.9	7.2	
LiBr	103.4	100.6		86.1	84.4	73.9	67.2	57.7	45.3	44.2	
LiCl			92.1	88.6	82.4	70.7	63.4	53.1			
LiClO$_4$ (25°)			98.6	92.2							33.3
½Li$_2$CO$_3$				64.2	59.1						
LiI						75.3	69.2	61.0			
LiIO$_3$	65.3	62.9	61.2	55.3	51.5	39.0	31.2	21.4	14.6		
LiNO$_3$	92.9	90.3	88.6	82.7	79.2	68.0	60.8	50.3	34.9	27.3	
LiOH						149.0	134.5	113.5	95.7		
½Li$_2$SO$_4$	96.4		86.9	74.7	68.2	50.5	41.3	30.7	23.3	18.1	13.9
½MgCl$_2$	106.4	101.3	98.1	88.5	83.4	69.6	61.5	52.3	43.3	35.0	28.0
½Mg(NO$_3$)$_2$	102.6	97.7	94.7	85.3	80.5	67.0	59.0	47.0	39.8		
½MgSO$_4$	99.8	84.5	76.2	56.9	49.7	35.4	28.9	23.0	17.3	12.9	9.3
½MnCl$_2$					86.0	68.5	61.0	48.5	38.8	30.2	23.0
½MnSO$_4$						27.6	24.4	18.3	14.0	10.5	7.3
NH$_3$(aq)	28.0	13.2	9.6	4.6	3.3	1.35	0.89		0.36		0.20
NH$_4$OAc		92.9	91.4	84.9		60.5	54.7	42.9	34.0	26.5	
NH$_4$Cl	127.3	124.3	122.1	115.2	110.7	101.4	97.0	92.1	88.2	85.0	80.7
NH$_4$F					90.1	74.5	65.7	55.3	47.9	42.2	
NH$_4$I	124.5		118.0	118.0	115.0	106.0	103.1	100.0		91.4	84.5
NH$_4$NO$_3$				110.0	106.6	94.5	88.8	85.1		71.9	47.6
NH$_4$SCN		120.0	116.5		104.3	94.0	89.9	84.7	79.2	74.0	
½(NH$_4$)$_2$SO$_4$					89.0	79.5	73.0	65.0		55.2	10.5
NaOAc	75.2	72.4	70.2	64.2	61.1	49.4	41.2	29.8	21.5	15.3	
NaBr				99.1	96.0	84.6	78.1	69.1		53.0	
NaBrO$_3$					65.3						
Na n-butyrate (25°)	80.3	77.6	75.8	69.3		61.8	54.5	44.1			
NaCl	106.5	103.8	102.0	95.7	92.0	80.9	74.3	65.0	56.5	49.4	42.7
NaClO$_4$	114.9[25]	111.7[25]	109.6[25]	102.4[25]	98.4[25]	71.7	65.0	55.1	46.0	38.8	
½Na$_2$CO$_3$	112	102.5	96.2	80.3	72.9	54.5	45.5	34.5	27.2		
½Na$_2$CrO$_4$					82.5	66.4	57.7	46.6	38.3		
½Na$_2$Cr$_2$O$_7$ (25°)				98.3	94.9						
NaF	87.8	85.2	83.5	77.0	73.1	60.0	51.9			31.1	
¼Na$_4$[Fe(CN)$_6$] (25°)		129.6	120.0	97.0							
Na formate	88.6				88.2	61.4	53.7	43.1	34.8	28.2	

Table 6-10 (*Continued*)
EQUIVALENT CONDUCTIVITIES OF ELECTROLYTES IN AQUEOUS SOLUTIONS AT 18°C

Electrolyte	Concentration, N										
	0.001	0.005	0.01	0.05	0.1	0.5	1.0	2.0	3.0	4.0	5.0
$NaHCO_3$ (25°)	93.5	90.5	88.4	80.6	76.0						
$\tfrac{1}{3}Na_2HPO_4$	58.4		54.0		44.0	33.5	28.0				
NaH_2PO_4	67.9	65.8	64.4	57.8	54.1						
$\tfrac{1}{4}Na_2H_2P_2O_7$	41.1	39.4	38.2	34.6	32.5						
NaI	124.2	121.2	119.2	112.8	108.8	97.5	89.9	78.6	69.9	62.2	
$NaIO_3$	75.2	72.6	70.9	64.4	60.5						
$\tfrac{1}{2}Na_2MoO_4$	120.8	113	110								
NaN_3 (25°)	117.1	113.8	110.5	101.3	95.7		68.0				
$NaNO_2$ (25°)				91.4	87.2		75.9	63.1	53.6	39.0	39.7
$NaNO_3$	102.9	100.1	98.2			74.1	65.9	54.5	46.0		
$NaOH$	208	203	200	190	183	172	160		108.0		69.0
Na picrate (25°)	78.6	75.7	73.7	66.3	61.8						
$\tfrac{1}{3}Na_3PO_4$	125	122	119	91							
Na propionate (25°)	83.5	80.9	79.1								
$\tfrac{1}{2}Na_2S$						117.0	104.3	85.0	71.0	59.0	47.2
$NaSCN$	144	139	136	124	116	88	72	51	38	27	19
$\tfrac{1}{2}Na_2SiO_3$	106.7	100.8	96.8	83.9	78.4	74.3	68.9	59.8	50.9	43.7	
$\tfrac{1}{2}Na_2SO_4$	120	81.5	74.8	64.3	60.4	59.7	50.8	40.0	33.5		
(mono) Na tartrate	116.1	109.2	104.8	92.2	85.8						
$\tfrac{1}{2}Na_2WO_4$ (25°)											
$\tfrac{1}{2}NiSO_4$	96.3	79.5	70.8	51.0	43.8	30.4	25.1	19.3	15.1		
$\tfrac{1}{2}$Oxalic acid	180.7		158.2	132.9	116.9	75.9	59.4				
$\tfrac{1}{2}Pb(NO_3)_2$	116.1	108.6	103.5	86.3	77.3	53.2	42.0	31.0			
Propionic acid						1.57	1.00	0.54		0.20	
$RbCl$	130.3	127.4	125.3	117.8	113.9		101.9	97.1	92.7	87.2	
$RbOH$				220.6	216.8	204.8	192.0	170.0	148.3		
$\tfrac{1}{4}SnCl_4$	114.5	108.9	105.4	94.4	90.2	75.7	68.5	66.9	47.9	32.7	
$\tfrac{1}{2}SrCl_2$	108.3	102.7	99.0	87.3	80.9	62.7	52.1	58.7	49.9	42.2	
$\tfrac{1}{2}Sr(NO_3)_2$								38.0	29.3	29.3	16.4

Electrolyte											
Tartaric acid (15°)						7.03	4.58	3.32	2.48	1.83	
$\frac{1}{4}$ThCl$_4$	128.2	123.7	120.2	97.4	92.6	61.0	54.0	44.3	36.3	29.8	
TlCl	113.3	108.2	105.4								
TlF	124.7	121.1	118.4	107.9	101.2	78.8	71.5	62.7			
TlNO$_3$	127.4	118.4	112.3	92.7	83.1						
$\frac{1}{2}$Tl$_2$SO$_4$											
Trichloroacetic acid (25°)						273	207	127	79	44	19
$\frac{1}{2}$UO$_2$F$_2$ (25°)	26.10	12.31	9.17	5.43	4.74	3.75	3.22				2.7
$\frac{1}{2}$UO$_2$SO$_4$ (25°)	106.5	63.2	49.2	27.6	22.2	14.4	11.6				
$\frac{1}{3}$YCl$_3$ (25°)	129	122	118	109							
$\frac{1}{2}$Zn(OAc)$_2$ (25°)	83	77	73	58	49						
$\frac{1}{2}$ZnCl$_2$	107	101	98	87	82	65	55	39.6	29.6	23.2	18.5
$\frac{1}{2}$Zn(NO$_3$)$_2$	120	114	111	100							
$\frac{1}{2}$ZnSO$_4$	98.4	82.1	73.2	53.0	45.6	32.3	26.6	20.0	15.9	12.0	9.0

Table 6-11
CONDUCTIVITY OF VERY PURE WATER AT VARIOUS TEMPERATURES

Temperature, °C	0	5	10	15	20	25	30	50
$\kappa \cdot 10^4$, $\Omega^{-1} \cdot cm^{-1}$	1.58	2.22	2.85	3.80	5.00	6.40	8.00	18.9

Ordinary distilled water in equilibrium with air has a conductivity of about 10 $\Omega^{-1} \cdot cm^{-1}$ at room temperature.

COMMON CONDUCTANCE RELATIONS

[SI Units are in Brackets]

Conductivity. The standard unit of conductance is electrolytic conductivity (formerly called specific conductance) κ, which is defined as the reciprocal of the resistance $[\Omega^{-1}]$ of a 1-m cube of liquid at a specified temperature $[\Omega^{-1} \cdot m^{-1}]$. See Table 6-8 and the definition of the cell constant.

In accurate work at low concentrations it is necessary to subtract the conductivity of the pure solvent (Table 6-9) from that of the solution to obtain the conductivity due to the electrolyte.

Resistivity (Specific Resistance):

$$\rho = 1/\kappa \qquad [\Omega \cdot m]$$

Conductance of an Electrolyte Solution:

$$\frac{1}{R} = \kappa \frac{S}{d} \qquad [\Omega^{-1}]$$

where S is the surface area of the electrode, or the mean cross-sectional area of the solution $[m^2]$, and d is the mean distance between the electrodes $[m]$.

Equivalent Conductivity:

$$\Lambda = \kappa/C \qquad [\Omega^{-1} \cdot m^2 \cdot equiv^{-1}]$$

In the older literature, C is the concentration in equivalents per liter. The volume of the solution in cubic centimeters per equivalent is equal to $1000/C$, and $\Lambda = 1000 \, \kappa/C$, the units employed in Table 6-10 $[\Omega^{-1} \cdot cm^2 \cdot equiv^{-1}]$. The formula unit used in expressing the concentration must be specified; for example, NaCl, $\frac{1}{2}K_2SO_4$, $\frac{1}{3}LaCl_3$.

The equivalent conductivity of an electrolyte is the sum of contributions of the individual ions. At infinite dilution: $\Lambda^\circ = \lambda_c^\circ + \lambda_a^\circ$, where λ_c° and λ_a° are the ionic conductances of cations and anions, respectively, at infinite dilution (Table 6-7).

Ionic Mobility and Ionic Equivalent Conductivity:

$$\lambda_c = \mathbf{F}u_c \quad \text{and} \quad \lambda_a = \mathbf{F}u_a \quad [\Omega^{-1} \cdot m^2 \cdot equiv^{-1}]$$

where $\mathbf{F}$ is the Faraday constant, and u_c, u_a are the ionic mobilities $[m^2 \cdot s^{-1} \cdot V^{-1}]$.

$$\Lambda = \alpha\mathbf{F}(u_c + u_a) = \alpha(\lambda_c + \lambda_a)$$

where α is the degree of electrolytic dissociation, Λ/Λ°. The electric mobility u of a species is the magnitude of the velocity in an electric field $[m \cdot s^{-1}]$ divided by the magnitude of the strength of the electric field E $[V \cdot m^{-1}]$.

Ostwald Dilution Law:

$$K_d = \frac{\alpha^2 C}{1 - \alpha}$$

where K_d is the dissociation constant of the weak electrolyte. In general for an electrolyte which yields n ions:

$$K_d = \frac{C^{(n-1)}\Lambda^n}{\Lambda^{\circ(n-1)}(\Lambda^\circ - \Lambda)}$$

Transference Numbers or Hittorf Transport Numbers:

$$T_c = \frac{\lambda_c}{\lambda_c + \lambda_a} \qquad T_a = \frac{\lambda_a}{\lambda_c + \lambda_a} \qquad T_c + T_a = 1$$

$$\frac{T_c}{T_a} = \frac{u_c}{u_a} = \frac{\lambda_c}{\lambda_a}$$

$$\lambda_c = T_c\Lambda \qquad \lambda_a = T_a\Lambda$$

Section 7

ORGANIC CHEMISTRY

NOMENCLATURE

SHORT GLOSSARY OF ORGANIC CHEMICAL NOMENCLATURE

The following rules for naming organic compounds and the examples given in explanation are not intended to cover all of the possible cases. For a more comprehensive and detailed description see Beilstein: *Handbuch der Organischen Chemie*, 4th edition, and Patterson: *Definitive Report of the Commission on the Reform of the Nomenclature of Organic Chemistry, Jour. Am. Chem. Soc. 55*, 3905 (1933); *News Edition, Ind. Eng. Chem. 14, 486* (1936).

HYDROCARBONS—The saturated open-chain hydrocarbons (C_nH_{2n+2}) have names ending in *ane*. The first four are: methane (CH_4); ethane (C_2H_6); propane (C_3H_8); and butane (C_4H_{10}). Beginning with pentane (C_5H_{12}) the prefix (e.g. *penta*) indicates the number of carbon atoms; for example, octane, C_8H_{18}; nonatriacontane, $C_{39}H_{80}$. See special table of *Numerical Prefixes*. Saturated monocyclic hydrocarbons (C_nH_{2n}) have the corresponding names of the open-chain hydrocarbons (C_nH_{2n+2}) preceded by the prefix *cyclo;* for example, cyclopropane (C_3H_6); cyclopentane (C_5H_{10}). The straight (unbranched) chains are termed *normal* (abbreviated: *n*); the grouping ($CH_3)_2CH-$ in a chain is termed *iso;* for example, $CH_3(CH_2)_2CH_3$ is *n*-butane; ($CH_3)_2CHCH_3$ is *iso*-butane. Branched-chain hydrocarbons are often named as derivatives of some relatively simple hydrocarbon by stating what radicals* are attached to the simple hydrocarbon; for example, $CH_3CH_2CH_2CH_3$, which is *n*-butane, may also be named: methyl-ethyl-methane since a methyl radical (CH_3) and an ethyl radical (C_2H_5) are replacing the two hydrogen atoms of the original methane molecule, the carbon atom of which is printed in bold face type. Similarly ($CH_3)_2CHCH_3$, which is *iso*-butane, would be named trimethyl-methane, the carbon atom printed in bold face type being the methane carbon atom corresponding to this name. In this system of naming, a compound might be given any one of several names depending upon which carbon atom is selected; for example, $CH_3CH_2CH_2CH_2CH_3$ might be given any one of the following names: diethyl-methane; methyl-propyl-methane; butyl-methane.

In 1892 an international congress of chemists assembled in Geneva, Switzerland, to devise a uniform and scientific system of nomenclature. This *Geneva system* with modifications proposed at various times is the modern *systematic naming*. According to this system a hydrocarbon is named as a derivative of the normal compound corresponding to the longest chain of carbon atoms in the molecule. The word *normal* is usually omitted, but implied, so that the simple unqualified systematic name always means the compound with the normal constitution; for example, *n*-butane, $CH_3CH_2CH_2CH_3$, is simply butane; iso-butane, ($CH_3)_2CHCH_3$, is 2-methyl-propane. The numeral 2 placed before † the word *methyl* indicates that the methyl radical is attached to the second carbon atom in the longest chain which, in this example, is of three carbons in length and, therefore, a propane. In cases where several different side chains are attached the order of stating them is either according to increasing complexity (e.g., methyl, then ethyl, then propyl, followed by butyl, etc.) or, according to alphabetical order (e.g., butyl, then ethyl, then methyl, followed by propyl, etc.) Where branching occurs in a side chain and for other cases of ambiguity, or if a simpler name would result, then a chain is selected as the fundamental unit which admits of the maximum of substitutions. A detailed discussion of this and other exceptions caused by very complicated structures is beyond the scope of this exposition and the reader is referred to the references given below the title of this section.

UNSATURATED HYDROCARBONS AND UNSATURATION—Hydrocarbons with one double bond (C_nH_{2n}) are named by adding the suffix *ene* to the name of the corresponding alkyl radical (C_nH_{2n+1})*; for example, $CH_2{:}CH_2$ is ethylene; $CH_3CH{:}CH_2$ is propylene. Just as the saturated hydrocarbons may be named as

*See Radicals in this section and also a special section Names and Formulas of Organic Radicals.

† The numbering is such that the lowest numbers are obtained; e.g., $CH_3CH_2CH_2CH(CH_3)_2$ is 2-methylpentane and not 4-methylpentane. Usage varies as to the position in the word for the numerals not only in the case of the hydrocarbons but also with other types such as alcohols, aldehydes, ketones, etc. Some writers place them before, some after, the parts of the name to which they refer. In Beilstein a combination is used wherein numbers placed after are enclosed in parentheses and those placed before are not; e.g., primary isobutyl alcohol is named: 2-methyl-butanol-(1).

ORGANIC CHEMICAL NOMENCLATURE

derivatives of methane, so these C_nH_{2n} hydrocarbons may be named as derivatives of ethylene; for example, $CH_3CH:CHCH_3$ which is a butylene, may also be named *sym*-dimethyl-ethylene since two methyl radicals (CH_3) are replacing two hydrogen atoms of the original ethylene molecule, the carbon atoms of which are printed in bold face type. Similarly $(CH_3)_2C:CH_2$, also a butylene, would be named *uns*-dimethyl-ethylene. The italicized prefixes *sym* and *uns* are abbreviations for *symmetrical* and *unsymmetrical*, respectively, and refer to the symmetrical or unsymmetrical manner in which the two methyl radicals are attached to the ethylene residue.

In the systematic naming, the longest chain is selected as in the case of the saturated hydrocarbons and the compound is named by replacing the *ane* ending of the corresponding saturated hydrocarbon with the ending *ene;* for example, $CH_3CH:CHCH_3$ is 2-butene. The numeral 2 indicates that the double bond is between the second and third carbon atoms. Similarly $(CH_3)_2C:CH_2$ is 2-methyl-1-propene** or 2-methyl-propene-(1), because a methyl group is on the second carbon atom, and the double bond is between the first and second carbon atoms in a chain of three carbon atoms. Where more than one double bond occurs in the molecule a numerical prefix is added to the ending *ene;* for example, isoprene, $CH_2:C(CH_3)CH:CH_2$ is 2-methyl-1,3-butadiene because the longest chain contains four carbon atoms with a methyl radical (CH_3) on the second carbon atom and double bonds between the first and second carbon atoms, and between the third and fourth carbon atoms of the four-carbon chain Similarly benzene would be named cyclohexa-1,3,5-triene.

Hydrocarbons with one triple bond (C_nH_{2n-2}) may be named as derivatives of acetylene ($CH:CH$); for example, $CH_3C:CH$ which is allylene, may also be named methyl-acetylene. In the systematic naming the same scheme is used as with the double bond compounds except that the ending becomes *yne* or *ine*. Thus, allylene is 1-propyne or 1-propine. The Greek letter Δ is often used to indicate unsaturation; the numbers following it indicate the positions of the unsaturated linkings.

Unsaturation in compounds other than the hydrocarbons, such as alcohols, aldehydes, ketones, etc., is indicated in the same general manner by use of *ene, diene, ine*, etc.; for example, allyl alcohol, $CH_2:CHCH_2OH$, is Δ^1-propenol-(3).*

ALCOHOLS—Alcohols are commonly named from the radical to which the hydroxyl radical (OH) is attached; for example, CH_3OH is methyl alcohol; $C_6H_5CH_2OH$ is benzyl alcohol. As in the case of the hydrocarbons, a complex alcohol may be named as the derivative of some relatively simple alcohol. When alcohols are named as derivatives of methyl alcohol the latter is commonly given the name of *carbinol;* for example, $(CH_3)_2CHOH$ is dimethyl carbinol because two of the hydrogen atoms in methyl alcohol have been replaced by two methyl radicals; similarly benzyl alcohol, $C_6H_5CH_2OH$, is phenyl carbinol. The adjectives *primary, secondary*, and *tertiary* are often used in connection with alcohols and refer to alcohol groupings which are of the type $-CH_2OH$, $=CHOH$, $\equiv COH$, respectively; for example, $C_3H_7CH_2OH$ is a primary butyl alcohol; $CH_3CH(OH)C_2H_5$ is a secondary butyl alcohol; $(CH_3)_3COH$ is a tertiary butyl alcohol.

In the systematic naming the hydroxyl radical is indicated by the ending *ol* and its position of attachment by an arabic numeral; for example, $C_2H_5CH_2CH_2OH$ is 1-butanol; $(CH_3)_3COH$ is 2-methyl-2-propanol because a methyl radical and an hydroxyl radical are both attached to the second carbon atom in a chain of three carbon atoms.** Two, three, etc., hydroxyl radicals in a molecule are indicated by the endings *diol, triol*, etc.; for example, glycol, CH_2OHCH_2OH is 1,2-ethandiol.

ETHERS—Ethers are most commonly named by adding the word *ether* to the name of the radicals connected by the ether oxygen atom; for example, $(C_2H_5)_2O$ is ethyl ether or diethyl ether; $CH_3OC_2H_5$ is methyl-ethyl ether. Often the name is such as to designate the presence of an alkoxy (OR) group; for example, $CH_3OC_6H_5$ is methoxy benzene.

ALDEHYDES—Aldehydes are most frequently named by dropping the ending *ic* from the name of the acid to which they are directly oxidizable and adding the suffix *aldehyde;* for example, CH_3CHO is named acetaldehyde because it is directly

---

* In this case the final vowel *e* of the ending *ene* has been dropped from the name because it is followed by the vowel *o* of the ending *ol* which is used to indicate the alcoholic hydroxyl radical.

ORGANIC CHEMICAL NOMENCLATURE

oxidized to acetic acid, CH_3CO_2H; similarly C_6H_5CHO is named benzaldehyde to show its relationship to benzoic acid. Sometimes aldehydes are named in a manner to show a relationship to some simpler aldehyde such as formaldehyde, $HCHO$; for example, CH_3CHO is methyl formaldehyde because the methyl radical has replaced the hydrogen atom in formaldehyde; similarly C_6H_5CHO is named phenyl formaldehyde. In the systematic naming the ending *al* indicates an aldehyde radical (CHO); for example, CH_3CHO is ethanal; $(CH_3)_2CHCHO$ is 2-methyl-1-propanal.**

KETONES—Ketones are most often named by adding the word *ketone* to the names of the radicals connected to the CO radical; for example, $(CH_3)_2CO$ is dimethyl ketone; $CH_3COCH(CH_3)_2$ is methyl-isopropyl ketone. The systematic names of the ketones differ from those of the alcohols and the aldehydes (*q. v.*) only in the termination *one* which is added to the name of the hydrocarbon; for example, $(CH_3)_2CO$ is propanone; $CH_3COCH(CH_3)_2$ is 2-methyl-butanone-(3).**

ACIDS—The common names of the normal carboxylic acids are derived from words having some relation to their natural occurrence or their properties; for example, formic, from ants (*Formicidae*); butyric, from butter; tannic, from the tanning property. Sometimes acids are named in a manner to show a relationship to some simpler acid such as acetic, CH_3CO_2H; for example, pivalic acid, $(CH_3)_3CCO_2H$, is tri-methyl-acetic acid because three methyl radicals (CH_3) are replacing the three hydrogen atoms in acetic acid. The systematic names are derived by adding the suffix *oic* and the word *acid* to the root of the name for the hydrocarbon of the *same number of carbon atoms;* for example, isovaleric acid, $(CH_3)_2CHCH_2CO_2H$, is 2-methyl-butanoic-4-acid; succinic acid, $(CH_2CO_2H)_2$, is butan-1,4-dioic acid. In compounds where such a naming would be awkward the carboxyl radical (CO_2H) is considered as a substituting group and the acid named accordingly; for example, citric acid, $HO_2CCH_2C(OH)(CO_2H)CH_2CO_2H$, is 2-hydroxy-propan-1,2,3-tricarboxylic acid.**

ESTERS AND SALTS—To the name of the alcoholic (or phenolic) radical, or the metal, is added the name of the acid modified to end in *ate;* for example, the ester of ethyl (C_2H_5) alcohol and acetic acid, $CH_3CO_2C_2H_5$, is ethyl acetate; the sodium salt of acetic acid, CH_3CO_2Na, is sodium acetate. To emphasize the ester structure they are often named as follows: acetic acid ethyl ester for ethyl acetate; malonic acid diethyl ester for diethyl malonate, $CH_2(CO_2C_2H_5)_2$.

ACID DERIVATIVES—The most common types in this classification are the anhydrides $(RCO)_2O$; halides $ROCl$; amides $RCONH_2$; amidoximes $RC(:NOH)NH_2$; amidines $RC(:NH)NH_2$; imides $R(CO)_2NH$; nitriles RCN. They are named by adding the endings *anhydride, chloride* (or other *halide*), *amide, amidoxime, amidine, imide, nitrile*, respectively, to the name of the acid or acid radical* (e.g., acetyl, acet, or aceto for the acid radical CH_3CO); for example, acetic anhydride, $(CH_3CO)_2O$; acetyl chloride, CH_3COCl; acetamide, CH_3CONH_2; acetamidoxime, $CH_3C(:NOH)NH_2$; acetamidine, $CH_3C(:NH)NH_2$; acetonitrile, CH_3CN; succinimide, $(CH_2CO)_2NH$. In compounds where such a naming would be awkward the modified carboxyl group is considered as a substituting group and named carbonamide, carbonamidine, carbonitrile, etc.

AMINES—Amines are most often named by adding the suffix *amine* to the name of the radical (or radicals) replacing one (or more) hydrogen atoms in ammonia (NH_3); for example, CH_3NH_2 is methylamine; $CH_3NHC_2H_5$ is methyl-ethylamine. Sometimes amines are named in a manner to show a relationship to some other simpler (or well-known) amine; for example, $C_6H_5NHCH_3$ is methylaniline; $C_6H_5N(CH_3)_2$ is named dimethylaniline to show the relationship to aniline, $C_6H_5NH_2$. In the systematic names the amino (NH_2) or modified amino (NHR, and NR_2) group is considered as a substituting group of the hydrocarbon; for example, $(CH_3)_2CHNH_2$ is 2-aminopropane; $(CH_3)_2CHCH_2NHCH_3$ is 2-methyl-1-methylamino-propane.

RADICALS—Alkyl radicals (C_nH_{2n+1}) are univalent groups derived by removal of one hydrogen atom from saturated hydrocarbons. They are named by replacing the ending *ane* of the corresponding hydrocarbon with the ending *yl;* for example, C_2H_5- is ethyl derived from ethane (C_2H_6); $CH_3CH_2CH_2-$ is propyl; $(CH_3)_2CH-$ is isopropyl.

ORGANIC CHEMICAL NOMENCLATURE

Univalent radicals derived from unsaturated aliphatic hydrocarbons have the endings *enyl; ynyl; dienyl;* etc.; for example, $CH_2:CH-$ commonly called *vinyl* is also ethenyl; $CH:C-$ is ethynyl; $CH_2:CH-CH:CH-$ is buta-1,3-dienyl. In the aromatic series the corresponding *aryl* radicals are named in a similar manner; for example, C_6H_5- is phenyl; $C_6H_4=$ is phenylene; $CH_3C_6H_4-$ is tolyl.

Bivalent radicals derived by removal of two hydrogen atoms from the *same carbon atom* are sometimes named in a manner similar to the corresponding aldehyde; for example, $CH_3CH=$ is acetal; $C_6H_5CH=$ is benzal. In the systematic naming the ending *ylidene* is used to replace the ending *ane* of the corresponding saturated hydrocarbon; for example, $CH_3CH=$ is ethylidene; somewhat analogous is $C_6H_5CH=$ when named benzylidene.

Acid radicals, derived by removal of OH from the carboxyl group ($COOH$), are named by replacing the ending *ic* of the acid name with the ending *yl;* for example, CH_3CO- is acetyl (from acetic) or ethanoyl (from ethanoic).

See also the special table Names and Formulas of Organic Radicals.

RING COMPOUNDS—Positions of substituting groups on the various rings are usually indicated by a system of numbering. In the benzene series of compounds disubstitution positions are often indicated by the terms *ortho* (abbreviated *o*) for the 1,2-positions; *meta* (abbreviated *m*) for the 1,3-positions; *para* (abbreviated *p*) for the 1,4-positions. In cases of trisubstitution the terms *vicinal* (*vic*), *unsymmetrical* (*uns*) and *symmetrical* (*sym*) are often employed to indicate the positions 1,2,3; 1,3,4; and 1,3,5, respectively. In the naphthalene series of compounds, especially in the older literature, disubstitution positions are sometimes indicated by the terms *ortho, meta,* and *para* with the same significance as in the benzene series, and by the terms *ana* for the 1,5-positions; *epi* for the 1,6-positions; *kata* for the 1,7-positions; *peri* for the 1,8-positions; *amphi* for the 2,6-positions; and *pros* for the 2,7-positions. See also the special table Organic Ring Systems.

PREFIXES AND SUFFIXES

-al, a suffix indicating the presence of an *aldehyde group* (-CHO), as in chlor*al,* $Cl_3C\cdot CHO$; or penten*al,* $CH_3(CH_2)_3\cdot CHO$.*

allo-, a prefix indicating the compound to be a close relative, an isomer, or a variety of the compound to the name of which it is prefixed, as *allo*caffeine. In cases of ethylenic isomerism it is prefixed to the name of the more stable form into which the compound can be converted on heating; thus, fumaric acid is *allo*maleic acid since it is the stable isomer formed on heating maleic acid.

alpha-, beta-, gamma-, etc., usually written as letters of the Greek alphabet (α, β, γ, etc.) to indicate the first, second, third, etc., positions of substitution respectively, as in α-aminopropionic acid for $CH_3\cdot CH(NH_2)\cdot CO_2H$ where the carbon atom attached to the CO_2H group is considered the α-carbon atom. See also the definition below for *omega* (ω). See also the numbering in the section Organic Ring Systems.

amphi-, a prefix used with disubstitution products of naphthalene to indicate the 2,6-positions. It is also used in naming certain stereoisomers of dioximes; thus,

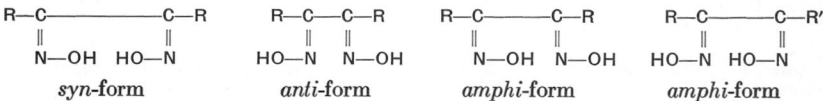

| syn-form | anti-form | amphi-form | amphi-form |

-ane, a suffix usually indicating a saturated (paraffin) hydrocarbon, C_nH_{2n+2}, as in pent*ane,* C_5H_{12}.* It is also used to indicate saturated parent compounds in the cyclic series, as camph*ane,* $C_{10}H_{18}$; or diox*ane,* $C_4H_8O_2$.

anti-, a term used in certain cases of stereoisomerism to indicate an opposed arrangement of groups or atoms, as in *anti*tartaric (mesotartaric) acid. It is often

* See this on preceding pages.

ORGANIC CHEMICAL NOMENCLATURE

employed to distinguish stereoisomers in the case of oximes and hydrazones. See also *amphi* and *syn*.

$$C_6H_5—C—C_6H_4 \cdot CH_3$$
$$\|$$
$$HO—N$$
syn-Phenyl tolyl ketoxime

$$C_6H_5—C—C_6H_4 \cdot CH_3$$
$$\|$$
$$N—OH$$
anti-Phenyl tolyl ketoxime

apo-, a prefix indicating a compound which is related to, or formed from, the compound to the name of which it is prefixed, as *apo*morphine.

-ase, a suffix used in forming the name of an enzyme. Usually the suffix is added to the name, or a part of the name, of a substance upon which the enzyme acts; e.g., malt*ase*.

as-, an abbreviation for *asymmetrical* and used to indicate an asymmetrical arrangement of substitution on a parent compound. See also *Ring Compounds*.

-ate, a suffix used to denote an ester or a salt; e.g., ethyl acet*ate*; or sodium acet*ate*.*

cis, a term used in cases of *cis-trans* (i.e., ethylenic) stereoisomerism to indicate that certain elements or groups are on the *same side* of the molecule; e.g.,

$$H—C—CO_2H$$
$$\|$$
$$H—C—CO_2H$$
cis-form

$$H—C—CO_2H$$
$$\|$$
$$HO_2C—C—H$$
trans-form

d, an abbreviation for *dextro* and usually indicating that a compound is *dextrorotatory*. Exceptions to this meaning are to be noted in certain cases of nomenclature; e.g., *d*-fructose, although levorotatory, is so named because of its relation in configuration to *d*-glucose and *d*-mannose; conversely, *l*-fructose, similarly related to *l*-glucose and *l*-mannose, is dextrorotatory.

-diene, a suffix indicating the *unsaturation* of two ethylenic (i.e., two double bond) linkages; e.g., a diolefin, C_nH_{2n}, as penta*diene*, $CH_3 \cdot CH:CH \cdot CH:CH_2$.*

-diine, or **-diyne,** a suffix indicating the *unsaturation* of two acetylenic (i.e., two triple bond) linkages; e.g., a diacetylene, C_nH_{2n-6}, as buta*diyne*, $CH:C \cdot C:CH$.*

-dioic, a suffix indicating a *dicarboxylic* (two $-CO_2H$) acid; e.g., pentan*dioic* acid, $HO_2C \cdot (CH_2)_3 \cdot CO_2H$.*

-diol, a suffix indicating the presence of *two alcohol* (two hydroxyl) groups; e.g., pentan*diol*, $HO \cdot CH_2 \cdot (CH_2)_3 \cdot CH_2OH$.*

dl, an abbreviation for *dextro-levo* indicating a substance composed of equal parts of the dextro- and levo-optical isomers of a compound.

-ene, a suffix indicating the *unsaturation* of one ethylenic (i.e., one double bond) linkage; e.g., an olefin, C_nH_{2n}, as pent*ene*, $CH_2:CH \cdot C_3H_7$.* It is often used as a suffix in naming aromatic hydrocarbons; e.g., benz*ene*, tolu*ene*, naphthal*ene*, etc.

epi-, a prefix used with disubstitution products of naphthalene to indicate the 1,6-positions. See Ring Compounds preceding. It is also prefixed to the name of an aldose, or related compound, to indicate an isomer that differs from the aldose only in the arrangement of the group about the α-carbon atom (the carbon atom next to the aldehyde group); thus, mannose is also called *epi*glucose. It is also used as a prefix to indicate a *bridge* or intramolecular connection, as in *epi*chlorohydrin, $O \cdot CH_2 \cdot CH \cdot CH_2Cl$.
l_____l

eso-, a prefix indicating that substitution groups are *attached to the ring*, as *eso*-trinitromesitylene for the 2,4,6-trinitro derivative of mesitylene. See also *exo-*.

exo-, a prefix indicating that substitution groups are *attached to the side chain*, as *exo*-nitrotoluene for phenyl-nitromethane, $C_6H_5 \cdot CH_2 \cdot NO_2$. See also *omega-* and *eso-*.

homo-, a prefix indicating a homologue of the compound to the name of which it is prefixed and usually differing by having one more CH_2 in the formula; e.g., *homo*salicylic acid, $CH_3 \cdot C_6H_3(OH)CO_2H$.

i or in, an abbreviation for *inactive* and indicating that the substance exhibits no activity towards polarized light. See also *dl*.

-ic, see **-oic.**

* See this on preceding pages.

ORGANIC CHEMICAL NOMENCLATURE

-in, a suffix used in forming the names of fats (i.e., the glycerides or esters of glycerol), the suffix being used to replace the *ic* ending of the name of the fatty acid forming the glyceride; e.g., tributyr*in*, $(C_3H_7CO)_3C_3H_5O_3$. It is often used as a suffix in the names of proteins and of glucosides; e.g., album*in* and amygdal*in*, resp. See also *i* above.

-ine, a suffix often used in forming the names of basic compounds such as amines, alkaloids, and amino acids; e.g., anil*ine*, quin*ine*, and glyc*ine*, resp. See also *-yne.*

-ite, a suffix used in various ways; thus, for esters of the *ous* acids, e.g., amyl nitr*ite*, $C_5H_{11}O \cdot NO$; for polyhydric alcohols, e.g., mann*ite* (mannitol), $C_6H_8(OH)_6$; for explosives, e.g., cord*ite*, a smokeless powder.

l, an abbreviation for *l(a)evo* and indicating that a compound is levorotatory. See also *d.*

N-, a letter used in the names of certain nitrogen compounds to indicate that the group to which the letter is prefixed has replaced a hydrogen directly attached to a nitrogen atom; e.g., *N*-methylpyrrole, $C_4H_4:N \cdot CH_3$. Sometimes the letter is erroneously used to indicate a *normal* compound, as *N*-butane, $CH_3 \cdot CH_2 \cdot CH_2 \cdot CH_3$, instead of a lower case letter for this meaning, as *n*-butane.

N. F., an abbreviation for the publication *National Formulary* which is issued by the *American Pharmaceutical Association* and quoted in Federal and State laws as having equal authority with the *United States Pharmacopoeia.* See also *U. S. P.*

nor-, a prefix indicating the parent compound from which another may be regarded as derived, as *nor*camphor (of which camphor is the 7-methyl derivative). The prefix has also been used to indicate a *normal* compound which is isomeric with the one to the name of which it is prefixed, as *nor*valine which is α-amino-*n*-valeric acid, whereas valine is α-amino-*iso*-valeric acid.

O-, a letter used in names of certain oxygen compounds to indicate that the group to which the letter is prefixed has replaced a hydrogen directly attached to an oxygen atom; e.g., *O*-ethyl phenol (phenetole), $C_2H_5 \cdot O \cdot C_6H_5$.

-oic, a suffix indicating a carboxylic (CO_2H) *acid;* e.g., pentan*oic* acid.* Often the suffix *-ic* is used in forming the name of an acid; e.g., acet*ic* acid.

-ol, a suffix indicating an *hydroxyl group* (OH) as in an alcohol,* e.g., pentan*ol;* or in a phenol, e.g., resorcin*ol,** $C_6H_4(OH)_2$.

-ole, a suffix used in various ways; thus, for phenolic ethers, e.g., phenet*ole*, $C_2H_5 \cdot O \cdot C_6H_5$; for certain aldehydes, e.g., œnanth*ole*, $CH_3(CH_2)_5 \cdot CHO$; for five-membered rings, usually heterocyclic, e.g., pyrr*ole*, C_4H_5N, or oxaz*ole*, C_3H_3ON.

omega-, a prefix indicating a substitution group *on the last carbon* of a side chain, as *omega*-nitrostyrene for $C_6H_5 \cdot CH:CH \cdot NO_2$. It is most often indicated by the Greek letter omega (ω), as ω-nitrostyrene.

-one, a suffix indicating the presence of a *ketone group* ($:CO$), as in propan*one*,* $CH_3 \cdot CO \cdot CH_3$; or acetophen*one*, $CH_3 \cdot CO \cdot C_6H_5$.

-ose, a suffix usually indicating a *carbohydrate;* e.g., gluc*ose*, $C_6H_{12}O_6$; cellul*ose*, $(C_6H_{10}O_5)_x$. It is also used in naming the primary alteration or hydrolytic products of proteins; e.g., album*ose*.

par(a)-, a prefix or an adjective used to indicate that the substance is related in some way, as a polymer, an isomer, etc., of the compound to whose name it is attached; e.g., *para*ldehyde, *para*lactic acid. See also its significance when used in showing position of substituents in naming *RING COMPOUNDS.**

poly-, a prefix indicating a polymer of the compound to the name of which it is prefixed, as *poly*styrene, $(C_8H_8)_x$.

prim., an abbreviation for *primary* and indicating a replacement to the first degree; e.g., a primary amine ($R \cdot NH_2$) where only one of the three hydrogens in ammonia has been replaced by a group (R) of atoms. It is frequently used in the designation of certain *alcohols.**

pseudo-, a prefix indicating an isomerism with, or some other relation to, the compound to the name of which it is prefixed, as *pseudo*tropine. It is sometimes indicated by the Greek letter psi (ψ); e.g., ψ-tropine.

*See this on preceding pages.

ORGANIC CHEMICAL NOMENCLATURE

S-, a letter used in the names of certain sulfur compounds to indicate that the group to which the letter is prefixed has replaced a hydrogen directly attached to a sulfur atom; e.g., *S*-methyl thioglycolic acid, $CH_3 \cdot S \cdot CH_2 \cdot CO_2H$.

sec., an abbreviation for *secondary* and indicating a replacement to the second degree; e.g., a secondary amine (R_2:NH) where two of the hydrogens in ammonia have been replaced by two groups (R_2) of atoms. It is frequently used in the designation of certain *alcohols.**

s, an abbreviation for *symmetrical* and used to indicate a symmetrical arrangement of substitution on a parent compound. See also *Ring Compounds.**

syn-, a term used in certain cases of isomerism to indicate that certain groups or atoms are on the same side of the molecule. See also *amphi* and *anti.*

C_6H_5—C—H C_6H_5—C—H
‖   ‖
N—OH   HO—N
Benz-*syn*-aldoxime   Benz-*anti*-aldoxime

tert., an abbreviation for *tertiary* and indicating a replacement to the third degree; e.g., a tertiary amine (R_3:N) where three of the hydrogens in ammonia have been replaced by three groups (R_3) of atoms. It is frequently used in the designation of certain *alcohols.**

-thiol, a suffix indicating a mercaptan or thio-alcohol (-SH) group; e.g., pentan-*thiol,* $C_5H_{11} \cdot SH$.

trans, a term used in cases of *cis-trans* stereoisomerism. See also *cis.*

uns., an abbreviation for *unsymmetrical* and used to indicate an unsymmetrical arrangement of substituents on a parent compound. See also *Ring Compounds.**

U. S. P., an abbreviation for *United States Pharmacopoeia* and indicating that a compound conforms to the standards of purity, and other requirements, set by this publication which has been recognized in the United States as a standard in administering laws relating to the Federal Food, Drug and Cosmetic Act (1936, 1939). The Roman numeral often employed with this abbreviation indicates the revision to which reference is being made; e.g., *U. S. P. (XIII)* refers to the 13th revision.

v, an abbreviation for *vicinal.* See under Ring Compounds.*

x-, a symbol for *unknown.* The unknown number of monomers composing a polymeric compound is often indicated by using this symbol as a subscript in the formula; e.g., starch, $(C_5H_{10}O_5)_x$. It is sometimes used to indicate the unknown position of attachment of a substituting group; e.g., in 3,*x*,3′,*x*′-tetranitro-4,4′-diphenol where the positions of substitution of two of the four nitro groups are unknown.

-yl, a suffix used in forming the names of radicals, usually univalent radicals; e.g., alk*yl,* C_nH_{2n+1}; acyl, $R \cdot CO$—.

-ylene, a suffix used in forming the names of unsaturated hydrocarbons *; e.g., eth*ylene,* CH_2:CH_2. It is frequently used in forming the names of bivalent radicals with the free valences on different carbon atoms; e.g., phen*ylene,* -C_6H_4-.

-yne, a suffix indicating *unsaturation* * of one acetylenic (i.e., one triple bond) linkage; e.g., pent*yne,* CH:$CH \cdot C_3H_7$. In this sense modern nomenclature prefers this suffix to that of *-ine.*

* See this on preceding pages.

ORGANIC RING SYSTEMS

The rings given in the following pages are numbered as in Chemical Abstracts beginning with Volume 31 (1937). In all but a few cases (especially anthracene, phenanthrene and purine) the numberings are in accordance with the "Proposed International Rules for Numbering Organic Ring Systems"; cf. Patterson: Jour. Am. Chem. Soc. 47, 543-61 (1925). For a more complete listing of ring systems see the annual indexes of Chemical Abstracts commencing with the year 1916 and each of the decennial indexes; and, Richter: Lexikon der Kohlenstoff-Verbindungen, 3d Edition (1910), Volume I, p. 14, published by Leopold Voss, Hamburg and Leipzig; Patterson and Capell: The Ring Index (1940), Reinhold, N. Y.

When numbering positions in the case of substitution derivatives of benzoic acid, aniline, phenol, toluene, etc., the characteristic radical of each of these substances (i. e., CO_2H, NH_2, OH, CH_3, etc., respectively) is regarded as in position 1.

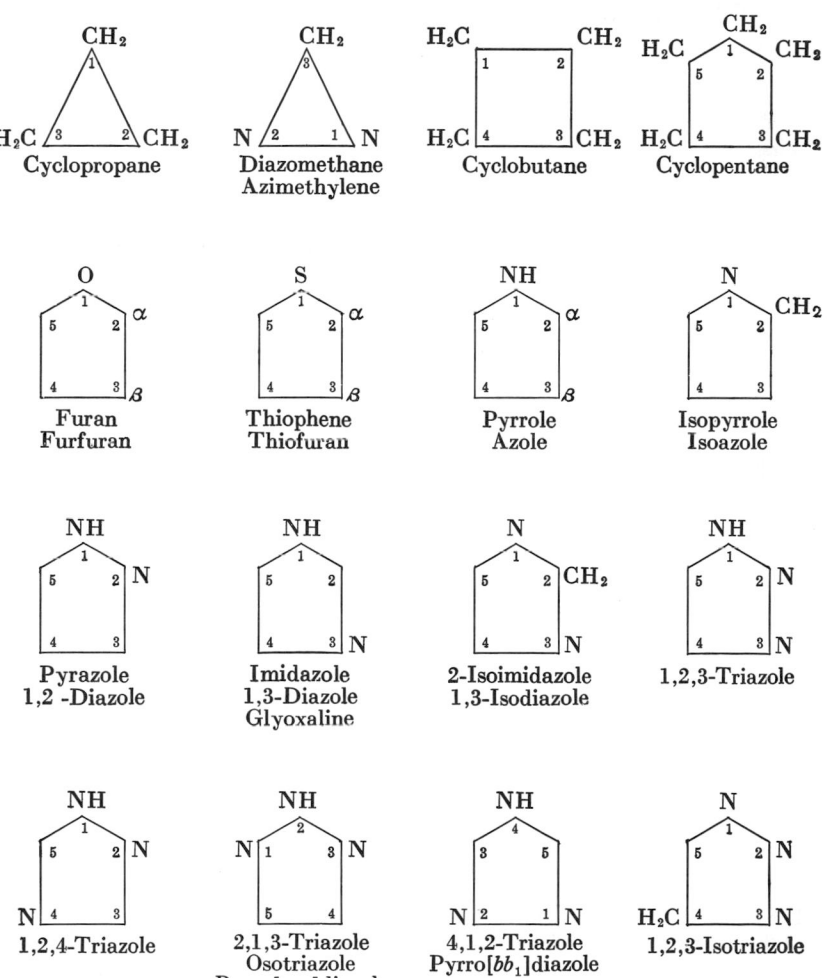

ORGANIC RING SYSTEMS

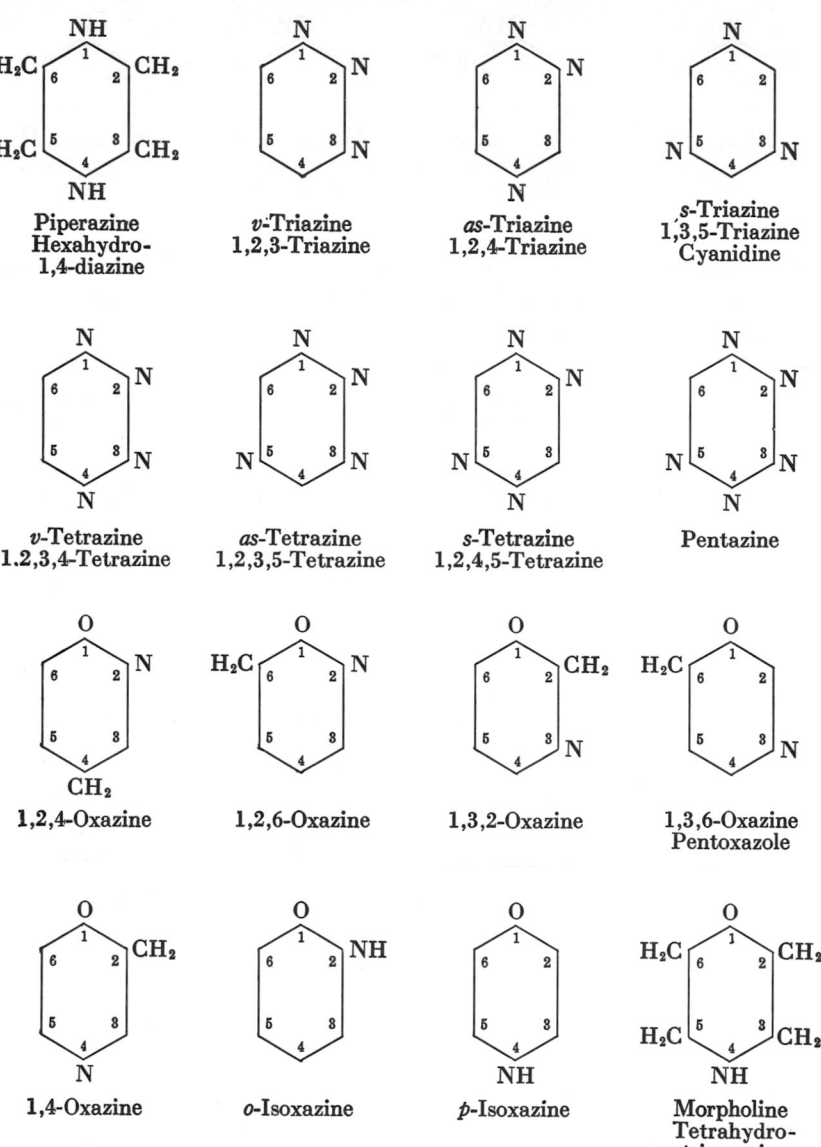

Piperazine
Hexahydro-
1,4-diazine

v-Triazine
1,2,3-Triazine

as-Triazine
1,2,4-Triazine

s-Triazine
1,3,5-Triazine
Cyanidine

v-Tetrazine
1,2,3,4-Tetrazine

as-Tetrazine
1,2,3,5-Tetrazine

s-Tetrazine
1,2,4,5-Tetrazine

Pentazine

1,2,4-Oxazine

1,2,6-Oxazine

1,3,2-Oxazine

1,3,6-Oxazine
Pentoxazole

1,4-Oxazine

o-Isoxazine

p-Isoxazine

Morpholine
Tetrahydro-
p-isoxazine

In a similar manner are derived: oxadiazine; oxatriazine; oxatetrazine; dioxazine; dioxadiazine; dioxatriazine; trioxazine; trioxadiazine; tetroxazine. Sulfur replacing oxygen in these rings forms the corresponding thia derivatives.

ORGANIC RING SYSTEMS

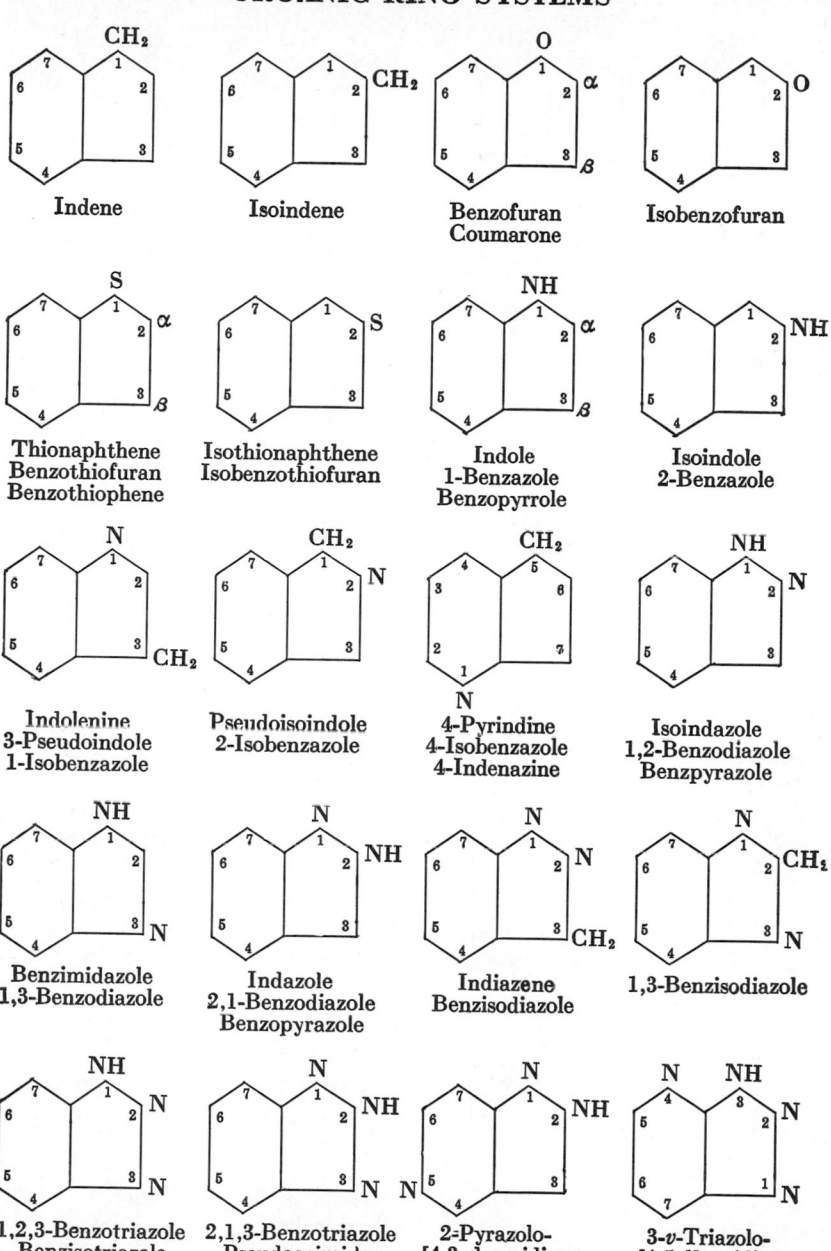

Indene

Isoindene

Benzofuran
Coumarone

Isobenzofuran

Thionaphthene
Benzothiofuran
Benzothiophene

Isothionaphthene
Isobenzothiofuran

Indole
1-Benzazole
Benzopyrrole

Isoindole
2-Benzazole

Indolenine
3-Pseudoindole
1-Isobenzazole

Pseudoisoindole
2-Isobenzazole

4-Pyrindine
4-Isobenzazole
4-Indenazine

Isoindazole
1,2-Benzodiazole
Benzpyrazole

Benzimidazole
1,3-Benzodiazole

Indazole
2,1-Benzodiazole
Benzopyrazole

Indiazene
Benzisodiazole

1,3-Benzisodiazole

1,2,3-Benzotriazole
Benzisotriazole
Azimidobenzene

2,1,3-Benzotriazole
Pseudoazimido-
benzene

2-Pyrazolo-
[4,3-c]-pyridine

3-v-Triazolo-
[4,5-b]pyridine

ORGANIC RING SYSTEMS

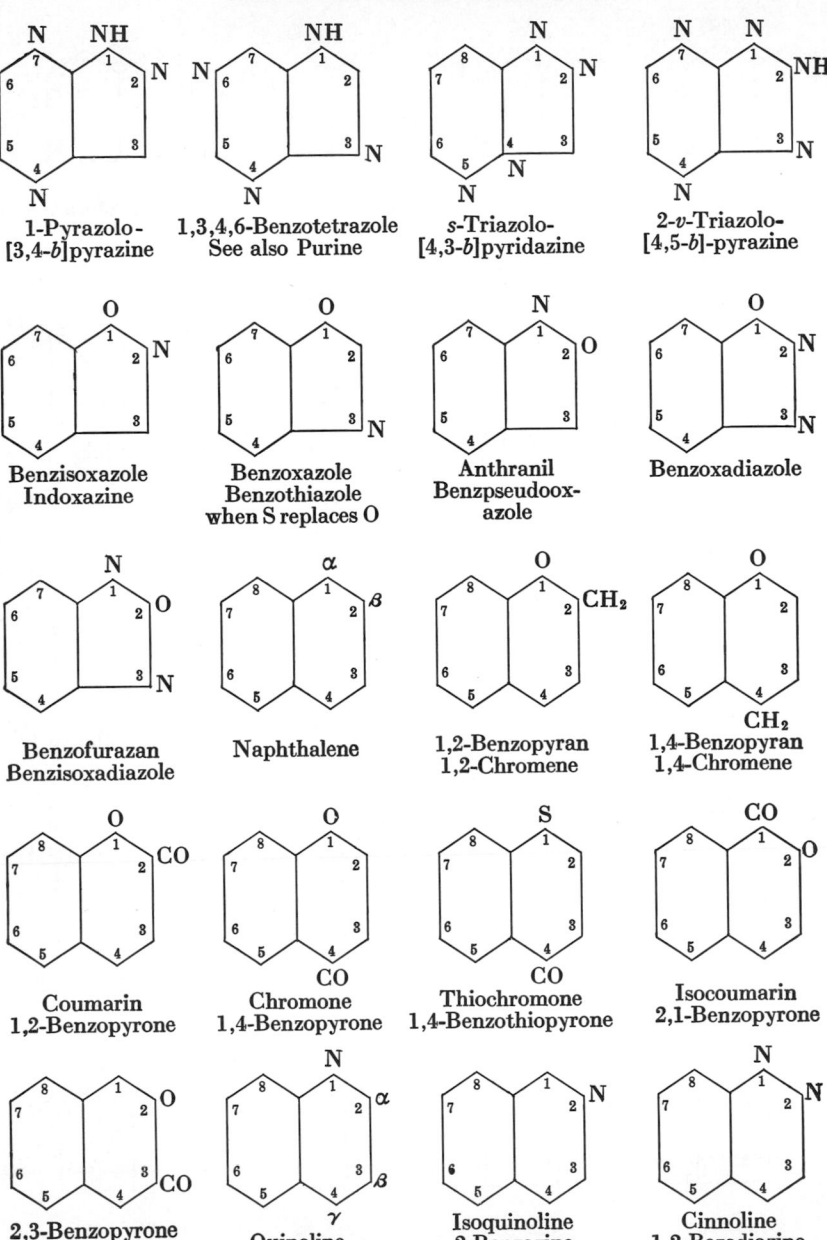

1-Pyrazolo-
[3,4-*b*]pyrazine

1,3,4,6-Benzotetrazole
See also Purine

s-Triazolo-
[4,3-*b*]pyridazine

2-*v*-Triazolo-
[4,5-*b*]-pyrazine

Benzisoxazole
Indoxazine

Benzoxazole
Benzothiazole
when S replaces O

Anthranil
Benzpseudoox-
azole

Benzoxadiazole

Benzofurazan
Benzisoxadiazole

Naphthalene

1,2-Benzopyran
1,2-Chromene

1,4-Benzopyran
1,4-Chromene

Coumarin
1,2-Benzopyrone

Chromone
1,4-Benzopyrone

Thiochromone
1,4-Benzothiopyrone

Isocoumarin
2,1-Benzopyrone

2,3-Benzopyrone

Quinoline
1-Benzazine

Isoquinoline
2-Benzazine
Leucoline

Cinnoline
1,2-Bezodiazine
α-Phenodiazine

ORGANIC RING SYSTEMS

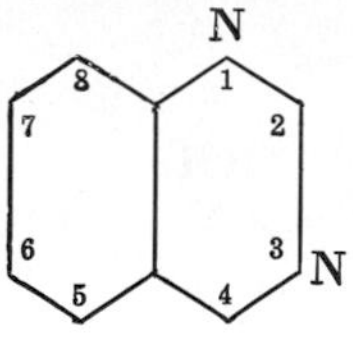

Quinazoline
1,3-Benzodiazine
Phenmiazine

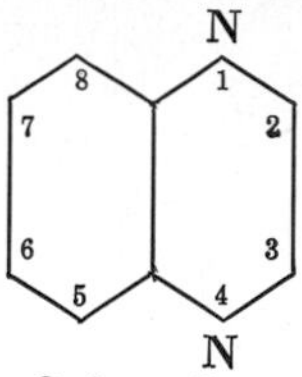

Quinoxaline
1,4-Benzodiazine
Phenpiazine

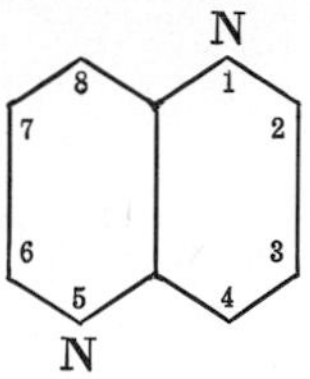

Pyrido[3,2-*b*]-
pyridine

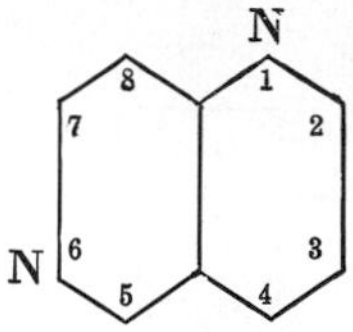

Pyrido[4,3-*b*]-
pyridine

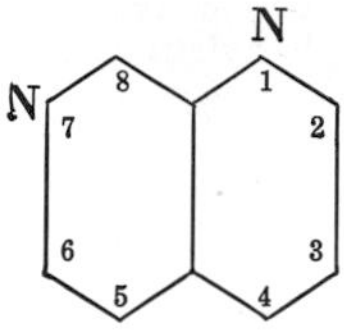

Pyrido[3,4-*b*]-
pyridine

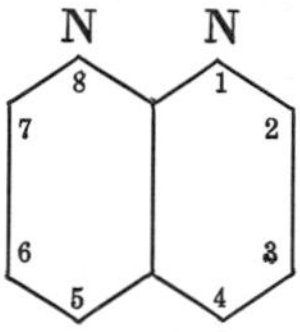

Naphthyridine

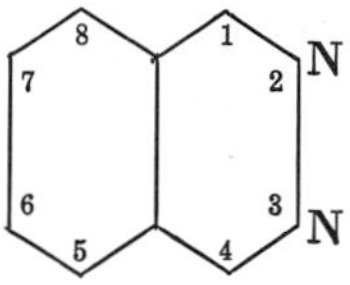

Phthalazine
2,3-Benzodiazine
β-Phenodiazine

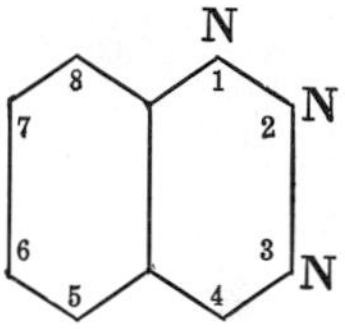

1,2,3-Benzotriazine
Phenotriazine

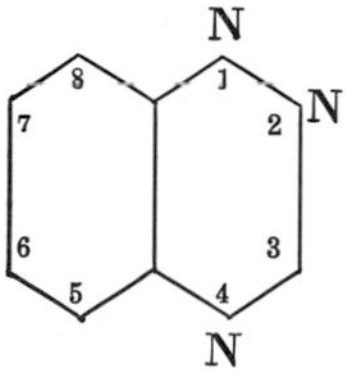

1,2,4-Benzotriazine

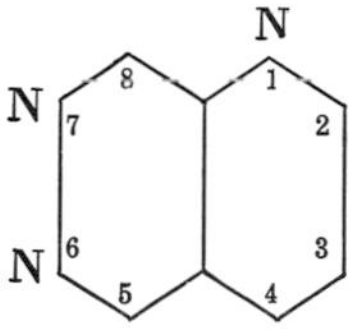

Pyrido[2,3-*d*]-
pyridazine

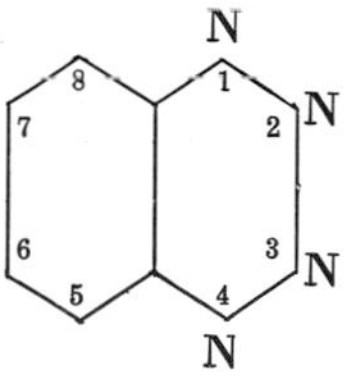

1,2,3,4-Benzotetrazine

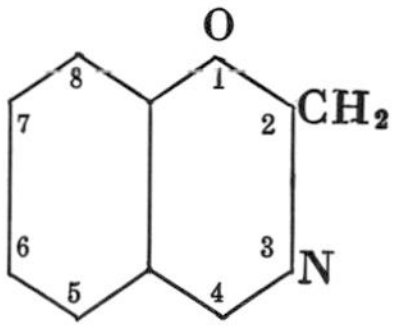

1,3,2-Benzoxazine

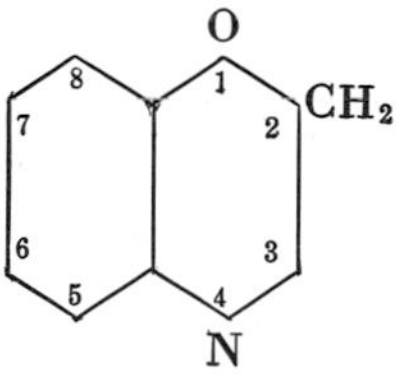

1,4,2-Benzoxazine

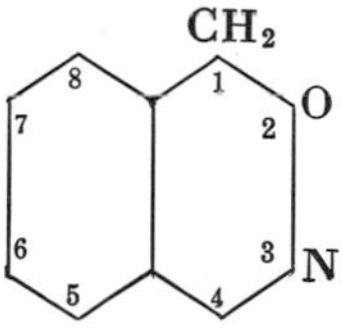

2,3,1-Benzoxazine

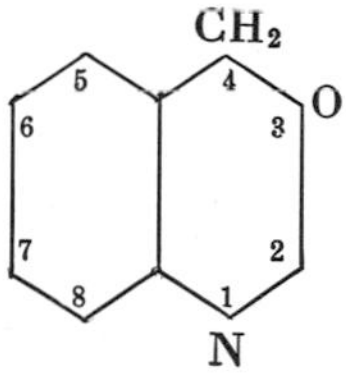

3,1,4-Benzoxazine

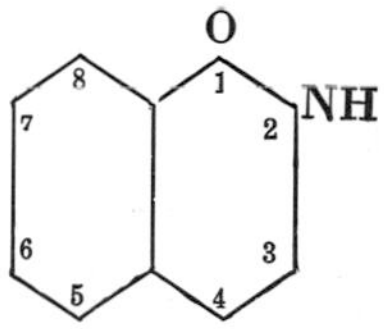

1,2-Benzisoxazine

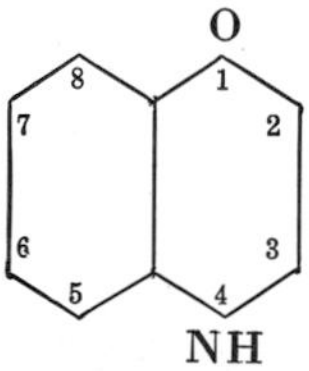

1,4-Benzisoxazine

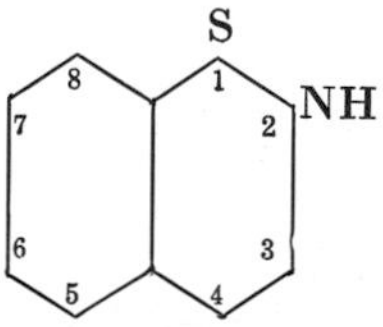

1,2-Benzisothiazine
Also a 1,4-form

In a similar manner are derived: benz-
oxadiazine; benzoxatriazine; etc.; benzo-
dioxazine; etc.; benzotrioxazine; etc.

ORGANIC RING SYSTEMS

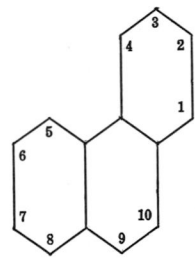

H₂C |1 2| CH₂

8
7
6 5 4 3

Acenaphthene

CH₂
8 9 1
7 2
6 3
5 4

Fluorene
Diphenylenemethane

NH
6 5 4
7 3
8 2
9 1

Carbazole
Dibenzopyrrole
Diphenyleneimide

γ α
8 9 1
7 2 β
6 3
5 10 4

Anthracene

3
4 2
5 1
6
7 10
8 9

Phenanthrene

CH₂
9 10 1
8 2
7 3
6 5 4
O

Xanthene

S
9 10 1
8 2
7 3
6 5 4
S

Thianthrene

N
6 5 4
7 3
8 2
9 10 1

Acridine

N
9 10 1
8 2
7 3
6 5 4
N

Phenazine

ORGANIC RING SYSTEMS

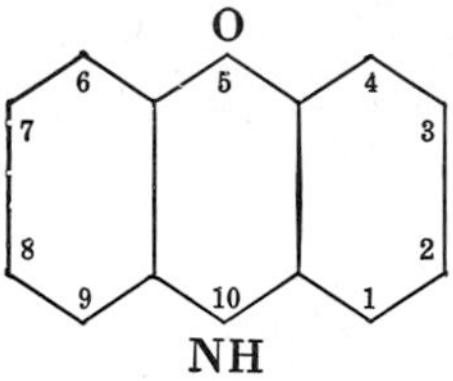

O

$$\begin{array}{c}6\quad5\quad4\\7\qquad\qquad3\\8\qquad\qquad2\\9\quad10\quad1\end{array}$$

NH

Phenoxazine
S replacing O=pheno-
thiazine

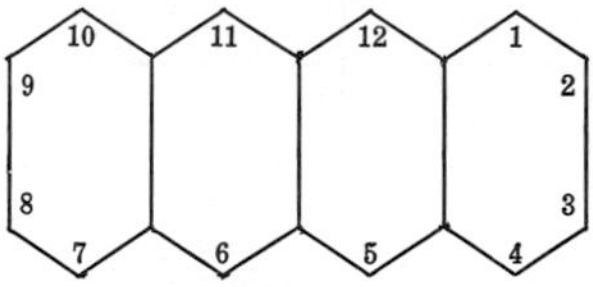

$$\begin{array}{c}10\quad11\quad12\quad1\\9\qquad\qquad\qquad2\\8\qquad\qquad\qquad3\\7\quad6\quad5\quad4\end{array}$$

Naphthacene

H_2C CH_2
   5   3   1
       C
   4   2
H_2C CH_2

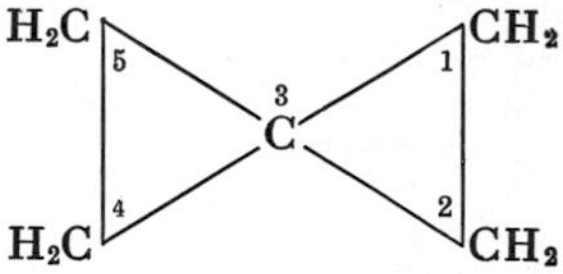

Spiropentane
Rings with but one
atom in common are known
as "spiro compounds".

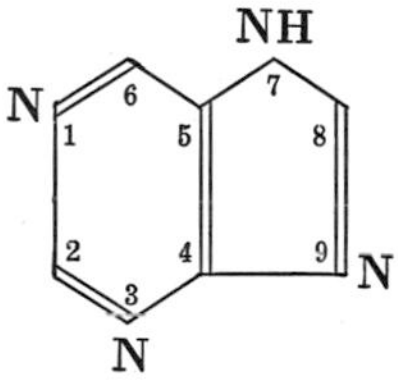

NH

N
   6   7
1   5   8
2   4   9
     3   N

N

Purine

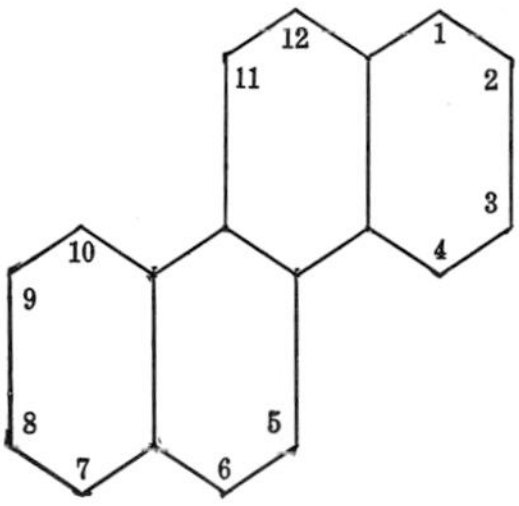

Chrysene

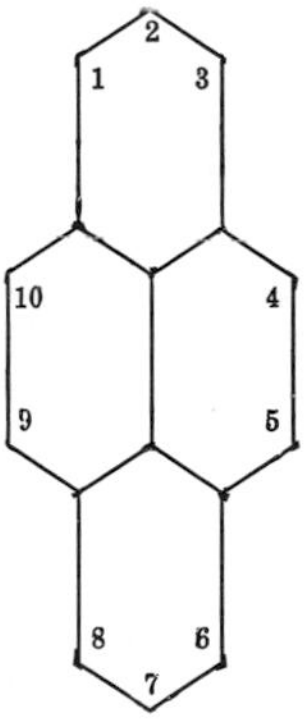

Pyrene

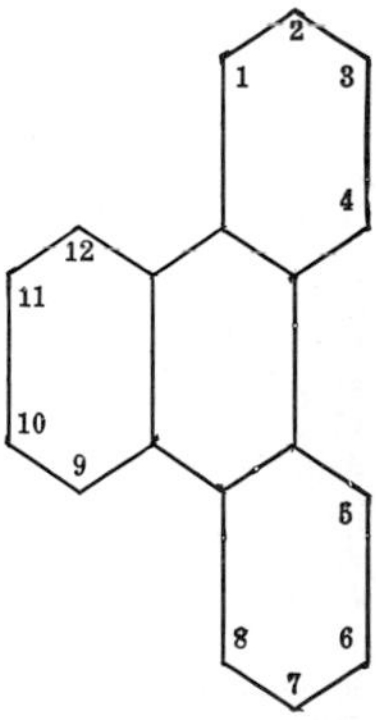

Triphenylene

7-17

Table 7-1
NAMES AND FORMULAS OF ORGANIC RADICALS

For more comprehensive lists see the various lists of radicals given in the subject indexes of the annual and decennial indexes of *Chemical Abstracts*.

Name	Formula	Name	Formula
Acenaphthenyl (from acenaphthene)	$C_{12}H_9-$	Antipyrinyl (from antipyrine)	$CO \cdot N(C_6H_5) \cdot N(CH_3) \cdot$
Acenaphthenylene	$-C_{12}H_8-$		$C(CH_3):C-$
1-Acenaphthenylidene	$C_{12}H_8:$		
Acetamido	$CH_3CO \cdot NH-$	Antipyroyl (from antipyric acid)	$CO \cdot N(C_6H_5) \cdot N(CH_3) \cdot$
Acetenyl	$CH \vdots C-$		$C(CH_3):C \cdot CO-$
Acetimido	$CH_3C(:NH)-$ or		
	$CH_3CON:$	Antipyryl	$CO \cdot N(C_6H_5) \cdot N(CH_3) \cdot$
Acetimidoyl	$CH_3C(:NH)-$		$C(CH_3):C-$
Acetoacetyl	CH_3COCH_2CO-		
Acetonyl	CH_3COCH_2-	Arginyl	$H_2NC(:NH) \cdot NH(CH_2)_3 \cdot$
Acetonylidene	$CH_3COCH:$		$CH(NH_2)CO-$
Acetoxy	$CH_3CO \cdot O-$	Arseno	$-As:As-$
Acetyl	CH_3CO-	Arsenoso	$O:As-$
Acetylene	$:CH \cdot CH:$	Arsinico	$(HO)OAs:$
Acetylimino	$CH_3CON:$	Arsino	H_2As-
Aci-nitro	see nitro	Arso	O_2As-
Acridanyl	$C_{13}H_{10}N-$	Arsono (from arsonic acid)	$(HO)_2OAs-$
Acridinyl (from acridine)	$C_{13}H_8N-$		
Acryloyl	$CH_2:CHCO-$	Arsylene	$HAs:$
Acrylyl	$CH_2:CHCO-$	Asaryl	$(CH_3O)_3C_6H_2-$ (*2,4,5*)
Adipoyl	$-CO(CH_2)_4CO-$	Asparaginyl	$H_2NCOCH_2CH(NH_2) \cdot$
Adipyl	$-CO(CH_2)_4CO-$		$CO-$
Alanyl	$CH_3CH(NH_2)CO-$	Asparagyl	$H_2NCOCH_2CH(NH_2) \cdot$
β-Alanyl	$H_2N(CH_2)_2CO-$		$CO-$
Aldo (in generic sense)	$O:$	Aspartoyl	$-COCH_2CH(NH_2)CO-$
Allyl	$CH_2:CHCH_2-$	Aspartyl	$-COCH_2CH(NH_2)CO-$
β-Allyl	$CH_2:C(CH_3)-$	α-Aspartyl	$HO_2CCH_2CH(NH_2)CO-$
Allylidene	$CH_2:CHCH:$	β-Aspartyl	$HO_2CCH(NH_2)CH_2CO-$
Allyloxy	$CH_2:CHCH_2O-$	Atropoyl	$C_6H_5C(:CH_2)CO-$
Amidino	$H_2N \cdot C(:NH)-$	Auri	$Au \vdots$
Amido	H_2N- (in acid groups)	Auro	$Au-$
Amidoxalyl	$H_2N \cdot CO \cdot CO-$	Azelaoyl	$-CO(CH_2)_7CO-$
Amino	H_2N-	Azido	$N:N \cdot N-$
Amoxy	$CH_3(CH_2)_3CH_2O-$		
Amyl	$CH_3(CH_2)_3CH_2-$	Azimino (azimido)	$-N:N \cdot NH-$
Amylidene	$CH_3(CH_2)_3CH:$	Azino	$:N \cdot N:$
Anilino	C_6H_5NH-	Azo	$-N:N-$
Anisal	$CH_3O \cdot C_6H_4CH:$ (*p*)	Azoxy	$-N(O)N-$
Anisidino	$CH_3O \cdot C_6H_4 \cdot NH_2-$ (*p*)	Benzal	$C_6H_5CH:$
Anisoyl	$CH_3O \cdot C_6H_4CO-$ (*p*)	Benzamido	C_6H_5CONH-
Anisyl	$CH_3O \cdot C_6H_4-$ or	Benzenesulfinyl	C_6H_5SO-
	$CH_3 \cdot C_6H_4CH_2-$	Benzenesulfonamido	$C_6H_4SO_2NH-$
Anisylidene	$CH_3O \cdot C_6H_4CH:$	Benzenesulfonyl	$C_6H_5SO_2-$
Anthraniloyl	$H_2N \cdot C_6H_4CO-$ (*o*)	Benzenyl	$C_6H_5C \vdots$
Anthranoyl	$H_2N \cdot C_6H_4CO-$ (*o*)	Benzhydryl	$(C_6H_5)_2CH-$
Anthraquinonyl (from anthraquinone)	$C_{14}H_7O_2-$ (2 isomers)	Benzhydrylidene	$(C_6H_5)_2C:$
Anthraquinonylene	$-C_{14}H_6O_2-$	Benzidino (from benzidine)	$(p-H_2N \cdot C_6H_4)C_6H_4NH-$ (*p*)
Anthryl (from anthracene)	$C_{14}H_9-$ (5 isomers)	Benziloyl	$(C_6H_5)_2C(OH)CO-$
Anthrylene (from anthracene)	$-C_{14}H_8-$ (11 isomers)	Benzimidazolyl (from benzimidazole)	$C_7H_5N_2-$
Antimono	$-Sb:Sb-$		

Table 7-1 (*Continued*)
NAMES AND FORMULAS OF ORGANIC RADICALS

Name	Formula	Name	Formula
Benzimido	$C_6H_5C(:NH)-$ or	Butyryl	$CH_3(CH_2)_2CO-$
	$C_6H_5CON:$	Cacodyl	$(CH_3)_2As-$
Benzimidoyl	$C_6H_5C(:NH)-$	Camphanyl (from	$C_{10}H_{17}-$ (3 isomers)
Benzofuranyl (from	C_8H_5O-	camphane)	
benzofuran)		Camphoroyl (from	$C_{10}H_{14}O_2:$
Benzofuryl	C_8H_5O-	camphoric acid)	
Benzohydrylidene	$(C_6H_5)_2C:$	Camphoryl (from	$C_{10}H_{15}O-$
Benzopyranyl (from	C_9H_7O- (*2-α*, etc.)	camphor)	
benzopyran)		Camphorylidene (from	$C_{10}H_{14}O:$
Benzoquinonyl	$C_6H_3O_2-$ (*o* or *p*)	camphor)	
Benzoquinonylene	$-C_6H_2O_2-$ (*o* or *p*)	Caprinoyl	$CH_3(CH_2)_8CO-$
Benzothienyl	C_8H_5S-	Caproyl	$CH_3(CH_2)_4CO-$
Benzoxazinyl	C_8H_6NO-	Capryl	$CH_3(CH_2)_8CO-$
Benzoxazolyl (from	C_7H_4NO-	Capryloyl	$CH_3(CH_2)_6CO-$
benzoxazole)		Caprylyl	$CH_3(CH_2)_6CO-$
Benzoxy	$C_6H_5CO\cdot O-$	Carbamido	$H_2N\cdot CO\cdot NH-$
Benzoyl	C_6H_5CO-	Carbamoyl	$H_2N\cdot CO-$
Benzoylene	$-C_6H_4CO-$	Carbamyl	$H_2N\cdot CO-$
Benzoylimino	$C_6H_5CO\cdot N:$	Carbanilino	$C_6H_5NH\cdot CO-$
Benzoyloxy	$C_6H_5CO\cdot O-$	Carbaniloyl	$C_6H_5NH\cdot CO$
Benzyl	$C_6H_5CH_2-$	Carbazolyl	$C_{12}H_8N-$ (5 isomers)
Benzylidene	$C_6H_5CH:$	Carbazyl	$C_{12}H_8N-$ (5 isomers)
Benzylidyne	$C_6H_5C:$	Carbethoxy	$C_2H_5O\cdot CO-$
Benzyloxy	$C_6H_5CH_2\cdot O-$	Carbomethoxy	$CH_3O\cdot CO-$
Benzylthio	$C_6H_5CH_2S-$	Carbonyl	$OC:$
Biphenylene	$-C_6H_4\cdot C_6H_4-$	Carbonyldioxy	$-O\cdot CO\cdot O-$
Biphenylenebisazo	$-N:NC_6H_4\cdot C_6H_4N:N-$	Carboxy	$HO\cdot CO-$
Biphenylyl	$C_6H_5\cdot C_6H_4-$	Carbyl	$-C-$
Biphenylylcarbonyl	$C_6H_5\cdot C_6H_4CO-$	Carvacryl	$(CH_3)[(CH_3)_2CH]:$
Biphenylylene	$-C_6H_4\cdot C_6H_4-$		C_6H_3- (*1,4;2*)
Biphenylylenebisazo	$-N:NC_6H_4\cdot C_6H_4N:N-$	Caryl (from carane)	$C_{10}H_{17}-$
Biphenylyloxy	$C_6H_5\cdot C_6H_4\cdot O-$	Cetyl	$CH_3(CH_2)_{14}CH_2-$
Bismuthino	H_2Bi-	Chaulmoogroyl	$C_6H_7(CH_2)_{12}CO-$
Bornyl (from borneol)	$C_{10}H_{17}-$	Chaulmoogryl	$C_6H_7(CH_2)_{12}CH_2-$
Boryl	$O:B-$	Chloro	$Cl-$
Bromo	$Br-$	Chloroformyl	$Cl\cdot CO-$
Butadienyl	$CH_2:CHCH:CH-$	Chloromercuri	$ClHg-$
	(*1,3*-form shown)	Chromanyl	C_9H_9O-
1-Butenyl	$CH_3CH_2CH:CH-$	Chrysenyl	$C_{18}H_{11}-$
2-Butenyl	$CH_3CH:CHCH_2-$	Cinnamal	$C_6H_5CH:CHCH:$
3-Butenyl	$CH_2:CH(CH_2)_2-$	Cinnamenyl	$C_6H_5CH:CH-$
2-Butenylene	$-CH_2CH:CHCH_2-$	Cinnamoyl	$C_6H_5CH:CHCO-$
Butenylidene	$CH_3CH:CHCH:$	Cinnamyl	$C_6H_5CH:CHCH_2-$
	(*2*-form shown)	Cinnamylidene	$C_6H_5CH:CHCH:$
Butenylidyne	$CH_3CH:CHC:$	Citraconoyl	$-CO\cdot C(CH_3):CH\cdot CO-$
	(*2*-form shown)		(*cis*)
Butoxy	$CH_3(CH_2)_3O-$	Cresotoyl (from cresotinic	$HO(CH_3)C_6H_3CO-$
sec-Butoxy	$C_2H_5CH(CH_3)O-$	acid)	
tert-Butoxy	$(CH_3)_3C\cdot O-$	Cresoxy	$CH_3\cdot C_6H_4\cdot O-$
Butyl	$CH_3(CH_2)_3-$		(*o, m* or *p*)
sec-Butyl	$C_2H_5CH(CH_3)-$	Cresyl	$(HO)(CH_3)C_6H_3-$ or
tert-Butyl	$(CH_3)_3C-$		$CH_3\cdot C_6H_4-$
1,4-Butylene	$-CH_2(CH_2)_2CH_2-$	Cresylene	$-C_6H_3(CH_3)-$
Butylidene	$CH_3(CH_2)_2CH:$		(6 isomers)
sec-Butylidene	$C_2H_5(CH_3)C:$	Crotonoyl	$CH_3CH:CHCO-$
Butylidyne	$CH_3(CH_2)_2C:$	Crotonyl	$CH_3CH:CHCO-$
Butynylene	$-CH_2C:CCH_2-$	Crotyl	$CH_3CH:CHCH_2-$
	(*2*-form shown)		

Table 7-1 (*Continued*)
NAMES AND FORMULAS OF ORGANIC RADICALS

Name	Formula	Name	Formula
Cumal	$(CH_3)_2CH \cdot C_6H_4 \cdot CH:$ (*p*)	Cysteinyl	$HS \cdot CH_2CH(NH_2)CO-$
Cumenyl	$(CH_3)_2CH \cdot C_6H_4-$ (*o, m* or *p*)	Cystyl	$[-COCH(NH_2)CH_2S-]$
		Decanedioyl	$-CO(CH_2)_8CO-$
		Decanoyl	$CH_3(CH_2)_8CO-$
Cumidino	$(CH_3)_2CH \cdot C_6H_4 \cdot NH-$ (*p*)	Decyl	$CH_3(CH_2)_8CH_2-$
		Desyl	$(C_6H_5)(C_6H_5CO)CH-$
Cuminal	$(CH_3)_2CH \cdot C_6H_4 \cdot CH:$ (*p*)	Diacetylamino	$(CH_3CO)_2N-$
		Diazo	$-N:N-$ or $N:N:$
Cuminyl	$(CH_3)_2CH \cdot C_6H_4 \cdot CH_2-$ (*p*)	Diazoamino	$-N:N \cdot NH-$
Cuminylidene	$(CH_3)_2CH \cdot C_6H_4 \cdot CH:$ (*p*)	Diazonium	$\overset{+}{N(N:)}-$
		Dimethyoxyphenethyl	$(CH_3O)_2C_6H_3CH_2CH_2-$
Cumoyl	$(CH_3)_2CH \cdot C_6H_4 \cdot CO-$ (*p*)	Dimethoxyphenylacetyl	$(CH_3O)_2C_6H_3CH_2CO-$
		Dimethylamino	$(CH_3)_2N-$
Cumyl	$(CH_3)_2CH \cdot C_6H_4-$ (*o, m* or *p*)	Dimethylarsino	$(CH_3)_2As-$
		Dimethylbenzoyl	$(CH_3)_2C_6H_3CO-$
Cyanato	$N:CO-$	Dimethylbenzyl	$(CH_3)_2C_6H_3CH_2-$
Cyano	$N:C-$	Diphenylmethyl	$(C_6H_5)_2CH-$
Cyclobutyl	$CH_2CH_2CH_2CH-$ ⌐_____⌐	Diphenylmethylene	$(C_6H_5)_2C:$
		Disilanoxy	$H_3Si \cdot SiH_2 \cdot O-$
Cycloheptyl	$CH_2(CH_2)_5CH-$ ⌐_____⌐	Disilanyl	$H_3Si \cdot SiH_2-$
		Disilanylamino	$H_3Si \cdot SiH_2 \cdot NH-$
Cyclohexadienyl	$CH_2(CH:CH)_2CH-$ ⌐_____⌐ (*2,4*-form shown)	Disilanylene	$-SiH_2 \cdot SiH_2-$
		Disylanylthio	$H_3Si \cdot SiH_2 \cdot S-$
Cyclohexadienylene	$-C_6H_6-$	Disilazanoxy	$H_2Si \cdot NH \cdot SiH_2 \cdot O-$
Cyclohexadienylidene	$CH:CHCH_2CH:CHC:$ ⌐_____⌐ (*2,5*-form shown)	Disilazanyl	$H_2Si \cdot NH \cdot SiH_2-$
		Disilazanylamino	$H_2Si \cdot NH \cdot SiH_2 \cdot NH-$
Cyclohexasilanyl	$SiH_2(SiH_2)_4SiH-$ ⌐_____⌐	Disiloxanoxy	$H_3Si \cdot O \cdot SiH_2 \cdot O-$
		Disiloxanyl	$H_3Si \cdot O \cdot SiH_2-$
Cyclohexenyl (from cyclohexene)	C_6H_9- (3 isomers)	Disiloxanylamino	$H_3Si \cdot O \cdot SiH_2 \cdot NH-$
		Disiloxanylene	$-SiH_2 \cdot O \cdot SiH_2-$
Cyclohexenylene	$-C_6H_8-$	Disiloxanylthio	$H_3Si \cdot O \cdot SiH_2 \cdot S-$
Cyclohexenylidene	$CH_2(CH_2)_2CH:CHC:$ ⌐_____⌐ (*2*-form shown)	Disilthianoxy	$H_3Si \cdot S \cdot SiH_2 \cdot O-$
		Disilthianyl	$H_3Si \cdot S \cdot SiH_2-$
		Disilthianylthio	$H_3Si \cdot S \cdot SiH_2 \cdot S-$
Cyclohexyl (from cyclohexane)	$C_6H_{11}-$	Disilyldisilanyl	$(H_3Si)_3Si-$
		Dithio	$-S \cdot S-$
Cyclohexylene	$-C_6H_{10}-$	Docosyl	$CH_3(CH_2)_{20}CH_2-$
Cyclohexylidene	$CH_2(CH_2)_4C:$ ⌐_____⌐	Dodecanoyl	$CH_3(CH_2)_{10}CO-$
		Dodecyl	$CH_3(CH_2)_{10}CH_2-$
Cyclopentadienyl	C_5H_5-	Dotriacontyl	$CH_3(CH_2)_{30}CH_2-$
Cyclopentadienylidene	$C_5H_4:$	Duryl	$(CH_3)_4C_6H-$ (*2,3,5,6*)
Cyclopentenyl (from cyclopentene)	C_5H_7-	Durylene	$(CH_3)_4C_6:$ (*2,3,5,6*)
Cyclopentylidene	$CH_2CH_2CH:CHC:$ ⌐_____⌐ (*2*-form shown)	Eicosyl	$CH_3(CH_2)_{18}CH_2-$
		Enanthoyl	$CH_3(CH_2)_5CO-$
		Enanthyl	$CH_3(CH_2)_5CO-$
Cyclopentyl (from cyclopentane)	C_5H_9-	Epidioxy	$-O \cdot O-$ (to different atoms or radicals already united in some other way)
Cyclopentylene	$-C_5H_8-$		
Cyclopentylidene	$CH_2(CH_2)_3C:$ ⌐_____⌐	Epithio	$-S-$ (to different atoms or radicals already united in some other way)
Cyclopropyl	CH_2CH_2CH- ⌐_____⌐		
Cymyl (from cymene)	$C_{10}H_{13}-$ (ring attachment)	Epoxy	$-O-$ (to different atoms or radicals already united in some other way)

Table 7-1 (*Continued*)
NAMES AND FORMULAS OF ORGANIC RADICALS

Name	Formula	Name	Formula
Ethanediylidene	$:CH \cdot CH:$	Glutaminyl	$H_2NCOCH_2CH_2CH \cdot$
Ethene	$-CH_2 \cdot CH_2-$		$(NH_2)CO-$
Ethenyl	$CH_2C:$ or $CH_2:CH-$	Glutamoyl	$-OC \cdot CHNH_2 \cdot (CH_2)_2 \cdot CO-$
Ethenylene	$-CH:CH-$	α-Glutamyl	$HO_2C(CH_2)_2CH(NH_2) \cdot$
Ethenylidene	$CH_2:C:$		$CO-$
Ethinyl	$CH:C-$	γ-Glutamyl	$HO_2CCH(NH_2)(CH_2)_2 \cdot$
Ethoxalyl	$C_2H_5O \cdot OC \cdot CO-$		$CO-$
Ethoxy	C_2H_5O-	Glutamyl	see also glutamoyl
Ethoxycarbonyl	$C_2H_5O \cdot CO-$	Glutaryl	$-CO(CH_2)_3CO-$
Ethoxyphenyl	$C_2H_5O \cdot C_6H_4-$	Glyceroyl	$HOCH_2CH(OH)CO-$
Ethyl	CH_3CH_2-	Glyceryl	$-CH_2(CH-)CH_2-$
Ethylamino	C_2H_5NH-	Glycoloyl	$HOCH_2CO-$
Ethylene	$-CH_2 \cdot CH_2-$	Glycolyl	$HOCH_2CO-$
Ethylenedioxy	$-O(CH_2)_2O-$	Glycyl	H_2NCH_2CO-
Ethylidene	$CH_3CH:$	Glyoxalinyl or glyoxalyl	$C_3H_3N_2-$ (4 isomers)
Ethylidyne	$CH_3C:$	Glyoxyloyl	$OCH \cdot CO-$
Ethylthio	C_2H_5S-	Glyoxylyl	$OCH \cdot CO-$
Ethynyl	$CH:C-$	Guaiacyl	$CH_3O \cdot C_6H_4-$ (*o*)
Ethynylene	$-C:C-$	Guanidino	$H_2NC(:NH)NH-$
Fenchyl	$C_{10}H_{17}-$	Guanyl	$H_2NC(:NH)-$
Fluorenyl	$C_{13}H_9-$ (5 isomers)	Hendecyl	$CH_3(CH_2)_9CH_2-$
Fluorenylidene	$C_{13}H_8:$	Heneicosyl	$CH_3(CH_2)_{19}CH_2-$
Fluoro	$F-$	Hentriacontyl	$CH_3(CH_2)_{29}CH_2-$
Formamido	$H \cdot CO \cdot NH-$	Heptacosyl	$CH_3(CH_2)_{25}CH_2-$
Formazyl	$(C_6H_5N:N \cdot)(C_6H_5 \cdot$	Heptadecanoyl	$CH_3(CH_2)_{15}CO-$
	$NHN:)C-$	Heptadecyl	$CH_3(CH_2)_{15}CH_2-$
Formimidoyl	$HC(:NH)-$	Heptanamido	$CH_3(CH_2)_5CONH-$
Formyl	$H \cdot CO-$	Heptanedioyl	$-CO(CH_2)_5CO-$
Formyloxy	$H \cdot CO \cdot O-$	Heptanoyl	$CH_3(CH_2)_5CO-$
Fumaraniloyl	$C_6H_5NHCOCH:CHCO-$	Heptyl	$CH_3(CH_2)_5CH_2-$
	(*trans*)	Hexacontyl	$CH_3(CH_2)_{58}CH_2-$
Fumaroyl	$-COCH:CHCO-$	Hexacosyl	$CH_3(CH_2)_{24}CH_2-$
	(*trans*)	Hexadecanoyl	$CH_3(CH_2)_{14}CO-$
Furfural	$O \cdot CH:CHCH:CCH:$	Hexadecyl	$CH_3(CH_2)_{14}CH_2-$
	⌞_____⌟	Hexamethylene	$-CH_2(CH_2)_4CH_2-$
	(*2*-isomer only)	Hexanedioyl	$-CO(CH_2)_4CO-$
Furfuryl	$O \cdot CH:CH \cdot CH:CCH_2-$	Hexanoyl	$CH_3(CH_2)_4CO-$
	⌞_____⌟	Hexyl	$CH_3(CH_2)_4CH_2-$
	(*2*-isomer only)	Hexylidene	$CH_3(CH_2)_4CH:$
Furfurylidene	$O \cdot CH:CHCH:CCH:$	Hexylidyne	$CH_3(CH_2)_4C:$
	⌞_____⌟	Hippuroyl	$C_6H_5CONHCH_2CO-$
	(*2*-isomer only)	Hippuryl	$C_6H_5CONHCH_2CO-$
Furoyl	$CH:CH \cdot O \cdot CH:CCO-$	Histidyl	$N_2C_3H_3CH_2CH(NH_2) \cdot$
	⌞_____⌟		$CO-$
	(*3*-form shown)	Homopiperonyl	$(CH_2O_2)C_6H_3CH_2CH_2-$
Furyl	C_4H_3O- (2 isomers)		(*3,4*)
Furylidene	$CH:CH \cdot O \cdot CH_2C:$	Homoveratroyl	$(CH_3O)_2C_6H_3CH_2CO-$
	⌞_____⌟		(*3,4*)
	(*3(2H)*-form shown)	Homoveratryl	$(CH_3O)_2C_6H_3CH_2CH_2-$
Furylmethyl	$CH:CH \cdot O \cdot CH:CCH_2-$		(*3,4*)
	⌞_____⌟	Hydantoyl	$H_2NCONHCH_2CO-$
	(*3*)	Hydnocarpoyl	$C_6H_7(CH_2)_{10}CO-$
Galloyl	$(HO)_3C_6H_2CO-$ (*3,4,5*)	Hydnocarpyl	$C_6H_7(CH_2)_{10}CH_2-$
Geranyl (from geraniol)	$C_{10}H_{17}-$	Hydratropoyl	$C_6H_5CH(CH_3)CO-$
Germyl	H_3Ge-	Hydrazi	$-NH \cdot NH-$ (to same
Glucosyl	$C_6H_{11}O_5-$		atom)
Glucosyloxy	$C_6H_{11}O_6-$	Hydrazino	$H_2N \cdot NH-$

Table 7-1 (*Continued*)
NAMES AND FORMULAS OF ORGANIC RADICALS

Name	Formula	Name	Formula
Hydrazo	$-HN \cdot NH-$ (to different atoms)	Isonitro	$(HO)ON:$
		Isonitroso	$HO \cdot N:$
Hydrazono	$H_2N \cdot N:$	l-Isopentenyl	$(CH_3)_2CHCH:CH-$
Hydrocinnamoyl	$C_6H_5CH_2CH_2CO-$	Isopentylidene	$(CH_3)_2CHCH_2CH:$
Hydroperoxy	$HO \cdot O-$	Isopentylidyne	$(CH_3)_2CHCH_2C\vdots$
Hydroxy	$HO-$	Isopentyloxy	$(CH_3)_2CHCH_2CH_2O-$
Hydroxyamino	$HO \cdot NH-$	Isophthalal	$:HC \cdot C_6H_4 \cdot CH:$ (*m*)
Hydroxyimino	$HO \cdot N:$	Isophthaloyl	$-CO \cdot C_6H_4 \cdot CO-$ (*m*)
Hydroxyl	$HO-$	Isophthalylidene	$:HC \cdot C_6H_4 \cdot CH:$ (*m*)
Hydroxymethyl	$HOCH_2-$	Isopropenyl	$CH_2:C(CH_3)-$
Hydroxyphosphinyl	$(HO)HP(:O)-$	Isopropoxy	$(CH_3)_2CH \cdot O-$
ar-Hydroxytolyl	$(HO)(CH_3)C_6H_3-$	Isopropyl	$(CH_3)_2CH-$
-idene (added to any radical usually means a double bond at point of attachment)		Isopropylbenzoyl	$(CH_3)_2CH \cdot C_6H_4 \cdot CO-$
		ar-Isopropylbenzyl	$(CH_3)_2CH \cdot C_6H_4 \cdot CH_2-$
		ar-Isopropylbenzylidene	$(CH_3)_2CH \cdot C_6H_4 \cdot CH:$
		Isopropylidene	$(CH_3)_2C:$
Imidazolidinyl	$C_3H_7N_2$	Isoquinolyl (from isoquinoline)	C_9H_6N- (9 isomers)
Imidazolidyl	$C_3H_7N_2-$		
Imidazolinyl	$C_3H_5N_2-$	Isothiocyanato	$S:C:N-$
Imidazolyl (from imidazole)	$C_3H_3N_2-$ (4 isomers)	Isothiocyano	$S:C:N-$
		Isovaleryl	$(CH_3)_2CHCH_2CO-$
Imido	$HN:$	Isoxazolyl (from isoxazole)	C_3H_2NO- (5 isomers)
Imino	$HN:$		
Indanyl (from indan)	C_9H_9- (4 isomers)	Keto (in specific sense)	$O:$
Indazolyl	$C_7H_5N_2-$	Keto (in generic sense)	$O:$ (to same atom)
Indenyl (from indene)	C_9H_7- (7 isomers)	Lactoyl	$CH_3CH(OH)CO-$
Indolinyl	C_8H_8N-	Lauroyl	$CH_3(CH_2)_{10}CO-$
Indolinylidene	$CH_2 \cdot NH \cdot C_6H_4C:$ (*o*) ⌐_____⌐ (*3*-form shown)	Leucyl	$(CH_3)_2CHCH_2 \cdot CH(NH_2)CO-$
Indolyl (from indole)	C_8H_6N-	Linalyl	$C_{10}H_{17}-$
Indyl	C_8H_6N-	Lysyl	$H_2N(CH_2)_4CH(NH_2)CO-$
Iodo	$I-$	Maleoyl	$-COCH:CHCO-$ (*cis*)
Iodoso	$OI-$	Malonyl	$-CO \cdot CH_2 \cdot CO-$
Iodoxy	O_2I-	Maloyl	$-COCH(OH)CH_2CO-$
Isoallyl	$CH_3CH:CH-$	Mandeloyl	$C_6H_5CH(OH) \cdot CO-$
Isoamoxy	$(CH_3)_2CHCH_2CH_2O-$	Menthyl (from menthane)	$CH_3CH(CH_2)_2CH \cdot$ ⌐_____⌐ $(iso\text{-}C_3H_7)CH_2CH-$
Isoamyl	$(CH_3)_2CHCH_2CH_2-$		
Isoamylidene	$(CH_3)_2CHCH_2CH:$		
Isobornyl	$C_{10}H_{17}-$	Mercapto	$HS-$
Isobutenyl	$(CH_3)_2C:CH-$	Mercuri	$-Hg-$
Isobutoxy	$(CH_3)_2CHCH_2O-$	Mesaconoyl	$-COC(CH_3):CHCO-$ (*trans*)
Isobutyl	$(CH_3)_2CHCH_2-$		
Isobutylidene	$(CH_3)_2CHCH:$	Mesidino	$(CH_3)_3C_6H_2 \cdot NH-$ (*2,4,6*)
Isobutylidyne	$(CH_3)_2CHC\vdots$		
Isobutyryl	$(CH_3)_2CHCO-$	Mesityl	$(CH_3)_3C_6H_2-$ (*2,4,6*)
Isocyanato	$O:C:N-$	α-Mesityl	$(CH_3)_2C_6H_3 \cdot CH_2-$ (*3,5*)
Isocyano	$C:N-$	Mesityloxy	$(CH_3)_3C_6H_2 \cdot O-$ (*2,4,6*)
Isodiazo	$-HN \cdot N:$ (to the same atom)	Mesoxalyl	$-CO \cdot CO \cdot CO-$
		Mesyl	CH_3SO_2-
Isohexyl	$(CH_3)_2CH(CH_2)_2CH_2-$	Metanilamido	$H_2N \cdot C_6H_4 \cdot SO_2NH-$ (*m*)
Isohexylidene	$(CH_3)_2CH(CH_2)_2CH:$	Metanilyl	$H_2N \cdot C_6H_4 \cdot SO_2-$ (*m*)
Isohexylidyne	$(CH_3)_2CH(CH_2)_2C\vdots$	Methacryloyl	$CH_2:C(CH_3) \cdot CO-$
Isoindolinyl	C_8H_8N-	Methallyl	$CH_2:C(CH_3) \cdot CH_2-$
Isoindolyl (from isoindole)	C_8H_6N- (4 isomers)	Methanedisulfonyl	$CH_2(SO_2)_2:$
		Methene	$CH_2:$
Isoleucyl	$C_2H_5CH(CH_3) \cdot CH(NH_2)CO-$	Methenyl	$CH\vdots$

Table 7-1 (*Continued*)
NAMES AND FORMULAS OF ORGANIC RADICALS

Name	Formula	Name	Formula
Methionyl	$CH_3S(CH_2)_2CH(NH_2)\cdot$ $CO-$ or $CH_2(SO_2)_2$:	Nonacosyl	$CH_3(CH_2)_{27}CH_2-$
		Nonadecyl	$CH_3(CH_2)_{17}CH_2-$
Methoxalyl	$CH_3O\cdot CO\cdot CO-$	Nonanedioyl	$-CO(CH_2)_7CO-$
Methoxy	CH_3O-	Nonanoyl	$CH_3(CH_2)_7CO-$
ar-Methoxyanilino	$CH_3O\cdot C_6H_4\cdot NH-$	Nonyl	$CH_3(CH_2)_7CH_2-$
	(*o, m* or *p*)	Norbornyl (from	$C_7H_{11}-$
Methoxybenzoyl	$CH_3O\cdot C_6H_4\cdot CO-$	norbornane)	
	(*o, m* or *p*)	Norcamphanyl	$C_7H_{11}-$
ar-Methoxybenzyl	$CH_3O\cdot C_6H_4\cdot CH_2-$	Norcaryl (from	$C_7H_{11}-$
ar-Methoxybenzylidene	$CH_3O\cdot C_6H_4\cdot CH$:	norcarane)	
Methoxycarbonyl	$CH_3O\cdot CO-$	Norleucyl	$CH_3(CH_2)_3CH(NH_2)CO-$
Methoxyphenyl	$CH_3O\cdot C_6H_4-$	Norpinyl (from	$C_7H_{11}-$
Methyl	CH_3-	norpinane)	
Methylbenzyl	$CH_3\cdot C_6H_4\cdot CH_2-$	Octacosyl	$CH_3(CH_2)_{26}CH_2-$
	(*o, m* or *p*)	Octadecanoyl	$CH_3(CH_2)_{16}CO-$
α-Methylbenzyl	$C_6H_5\cdot CH(CH_3)-$	Octadecyl	$CH_3(CH_2)_{16}CH_2-$
Methylene	CH_2:	Octanedioyl	$-CO(CH_2)_6CO-$
Methylenedioxy	$-O\cdot CH_2\cdot O-$	Octanoyl	$CH_3(CH_2)_6CO-$
Methylidyne	HC:	Octyl	$CH_3(CH_2)_6CH_2-$
Methylol	$HOCH_2-$	Oenanthyl	$CH_3(CH_2)_5CO-$
Methylphenylene	$-C_6H_3(CH_3)-$	Oleoyl	$CH_3(CH_2)_7CH:CH\cdot$
	(6 isomers)		$(CH_2)_7CO-$ (*cis*)
Methylsulfonyl	CH_3SO_2-	Ornithyl	$CH_2(NH_2)(CH_2)_2\cdot$
Methylthio	CH_3S-		$CH(NH_2)CO-$
Morpholino	$CH_2CH_2\cdot O\cdot(CH_2)_2\cdot N-$	Oxalyl	$-OC\cdot CO-$
	(*4*-position only)	Oxamido	$H_2NCO\cdot CONH-$
Morpholinyl	$NH(CH_2)_2OCH_2CH-$	Oxamoyl	$H_2NCO\cdot CO-$
	(*3*-form shown)	Oxamyl	$H_2NCO\cdot CO-$
Myristoyl	$CH_3(CH_2)_{12}CO-$	Oxazinyl	C_4H_4NO-
Naphthal	$C_{10}H_7CH$:	Oxazolidinyl	C_3H_6NO
Naphthalimido (from	$C_{10}H_6(CO)_2N-$ (*1,8*)	Oxazolinyl	C_3H_4NO-
naphthalimide)		Oxazolyl	C_3H_2NO-
Naphthenyl	$C_{10}H_7C$:	Oximido	$HO\cdot N$:
Naphthionyl	$H_2N\cdot C_{10}H_6\cdot SO_2-$ (*4,1*)	Oxo (in specific sense)	O:
Naphthothienyl	$C_{12}H_7S-$	Oxotrimethylene	$-CH_2COCH_2-$ (*2*)
Naphthoxy	$C_{10}H_7O-$	Oxy	$-O-$ (used as a connec-
Naphthoyl	$C_{10}H_7CO-$		tive; see also epoxy and
Naphthoyloxy	$C_{10}H_7CO\cdot O-$		oxo)
Naphthyl	$C_{10}H_7-$	Palmitoyl	$CH_3(CH_2)_{14}CO-$
Naphthylene	$-C_{10}H_6-$	Pelargonoyl	$CH_3(CH_2)_7CO-$
Naphthylidene	$C_6H_4CH_2CH:CH\cdot C$:	Pelargonyl	$CH_3(CH_2)_7CO-$
	(*1(4H)*-form shown)	Pentacontyl	$CH_3(CH_2)_{48}CH_2-$
Naphthylmethylene	$C_{10}H_7CH$:	Pentacosyl	$CH_3(CH_2)_{23}CH_2-$
Naphthylmethylidyne	$C_{10}H_7C$:	Pentadecanoyl	$CH_3(CH_2)_{13}CO-$
Naphthyloxy	$C_{10}H_7O-$	Pentadecyl	$CH_3(CH_2)_{13}CH_2-$
Neopentanetetryl	$C(CH_2-)_4$	Pentamethylene	$-CH_2(CH_2)_3CH_2-$
Neopentyl	$(CH_3)_3CCH_2-$	Pentazolyl	$N:N\cdot:N\cdot N-$
Neryl	$C_{10}H_{17}-$	Pentenyl	$CH_3CH_2CH:CHCH_2-$
Nitramino	$O_2N\cdot NH-$		(*2*-form shown)
Nitrilo	N:	Pentyl	$CH_3(CH_2)_3CH_2-$
Nitro	O_2N-	*tert*-Pentyl	$CH_3CH_2C(CH_3)_2-$
aci-Nitro	$(HO)ON$:	Pentylidene	$CH_3(CH_2)_3CH$:
Nitrosamino	$ON\cdot NH-$	Pentylidyne	$CH_3(CH_2)_3C$:
Nitrosimino	$ON\cdot N$:	Pentyloxy	$CH_3(CH_2)_3CH_2O-$
Nitroso	$ON-$	Perimidinyl (from per-	$C_{11}H_7N_2-$ (8 isomers)
		imidine)	
		Perseleno	$Se:Se$:

Table 7-1 (*Continued*)
NAMES AND FORMULAS OF ORGANIC RADICALS

Name	Formula	Name	Formula
Perthio	S:S: (replacing O only)	Phthalidylidene	$C_6H_4CO \cdot O \cdot C$: (*o*)
Phenacyl	$C_6H_5COCH_2-$		$\lfloor \underline{\qquad} \rfloor$
Phenacylidene	C_6H_5COCH:	Phthalimido	$C_6H_4(CO)_2N-$ (*o*)
Phenanthridinyl	$C_{13}H_8N-$	Phthaloyl	$-OC \cdot C_6H_4 \cdot CO-$ (*o*)
Phenanthryl (from	$C_{14}H_9-$ (5 isomers)	Phthalylidene	$:HCC_6H_4CH$: (*o*)
phenanthrene)		Phytyl	$C_{20}H_{39}-$
Phenanthrylene	$C_{14}H_8$: (several isomers)	Picryl	$(NO_2)_3C_6H_2-$ (*2,4,6*)
Phenazinyl	$C_{12}H_7N_2-$	Pimeloyl	$-CO(CH_2)_5CO-$
Phenenyl	C_6H_3: (*s, as or v*)	Pinanyl (from pinane)	$C_{10}H_{17}-$
Phenethyl	$C_6H_5CH_2CH_2-$	Pinanylene	$-C_{10}H_{16}-$
Phenetidino	$C_2H_5O \cdot C_6H_4NH-$	Pinanylidene	$C_{10}H_{16}$:
	(*o, m or p*)	Piperidino (*1*-position	$CH_2CH_2CH_2CH_2CH_2N-$
Phenetyl	$C_2H_5O \cdot C_6H_4-$	only)	$\lfloor \underline{\qquad} \rfloor$
Phenoxy	C_6H_5O-	Piperidyl	$C_5H_{10}N-$ (*2-, 3- or 4-*)
Phenyl	C_6H_5-	Piperidylidene	C_5H_9N:
N-Phenylacetamido	$CH_3CON(C_6H_5)-$	Piperonyl	$(CH_2O_2):C_6H_3CH_2-$
Phenylacetyl	$C_6H_5CH_2CO-$		(*3,4*)
Phenylazo	$C_6H_5N:N-$	Piperonylidene	$(CH_2O_2):C_6H_3CH$: (*3,4*)
Phenylcarbamido	$C_6H_5NH \cdot CO \cdot NH-$ (*3*)	Piperonyloyl	$(CH_2O_2):C_6H_3CO$ (*3,4*)
Phenylcarbamoyl	C_6H_5NHCO-	Pivaloyl (from pivalic	$(CH_3)_3C \cdot CO-$
Phenylene	$-C_6H_4-$ (*o, m or p*)	acid)	
Phenylenebisazo	$-N:NC_6H_4N:N-$	Pivalyl	$(CH_3)_3C \cdot CO-$
	(*o, m or p*)	Plumbyl	H_3Pb-
Phenylenedimethylene	$-H_2C \cdot C_6H_4 \cdot CH_2-$	Prolyl (from proline)	$HN(CH_2)_3CH \cdot CO-$
Phenylenedimethylidyne	$:HC \cdot C_6H_4 \cdot CH$:		$\lfloor \underline{\qquad} \rfloor$
Phenylethylene	$-CH(C_6H_5)CH_2-$	Propanetriyl	$-CH_2(CH-)CH_2-$
Phenylidene	$CH:CHCH_2CH:CHC$:	Propargyl	$CH \vdots CCH_2-$
	$\lfloor \underline{\qquad} \rfloor$	Propenyl	$CH_3CH:CH-$
	(*2,5*-form shown)	Propenylidene	$CH_3CH:C$:
Phenylimino	C_6H_5N:	Propenylidene (*2*)	$CH_2:CHCH_2-$
α-Phenylphenacyl	$(C_6H_5)(C_6H_5CO)CH-$	Propioloyl	$CH \vdots CCO-$
Phenylpropyl	$C_6H_5CH_2CH_2CH_2-$ (*3*)	Propiolyl	$CH \vdots CCO-$
Phenylsulfamoyl	$C_6H_5NHSO_2-$	Propionamido	CH_3CH_2CONH-
Phenylsulfamyl	$C_6H_5NHSO_2-$	Propionyl	CH_3CH_2CO-
Phenylsulfinyl	C_6H_5SO-	Propionyloxy	$CH_3CH_2CO \cdot O-$
Phenylsulfonamido	$C_6H_4SO_2NH-$	Propoxy	$CH_3CH_2CH_2O-$
Phenylsulfonyl	$C_6H_5SO_2-$	Propyl	$CH_3CH_2CH_2-$ (*n*)
Phenylureido	$C_6H_5NH \cdot CO \cdot NH-$ (*3*)	*sec*-Propyl	$(CH_3)_2CH-$
Phospharseno	$-P:As-$	Propylene	$-CH(CH_3)CH_2-$
Phosphazo	$-P:N-$	Propylidene	CH_3CH_2CH:
Phosphinico	$(HO)OP$: (only as	Propylidyne	$CH_3CH_2C \vdots$
	doubling radical)	Propynyl	$CH_3C \vdots C-$ (*1*)
Phosphinidene	HP:	Propynyl	$CH \vdots CCH_2-$ (*2*)
Phosphinidyne	$P \vdots$	Protocatechuoyl	$(HO)_2C_6H_3CO-$ (*3,4*)
Phosphino	H_2P-	Pseudoallyl	$CH_2:C(CH_3)-$
Phosphinothioyl	$H_2P(:S)-$	Pseudocumidino	$(CH_3)_3C_6H_2NH-$
Phosphinyl	$H_2P(:O)-$		(*2,4,5*)
Phosphinylidene	$HP(:O)$:	*as*-Pseudocumyl	$(CH_3)_3C_6H_2-$ (*2,3,5*)
Phosphinylidyne	$O:P \vdots$	*s*-Pseudocumyl	$(CH_3)_3C_6H_2-$ (*2,4,5*)
Phospho	O_2P-	*v*-Pseudocumyl	$(CH_3)_3C_6H_2-$ (*2,3,6*)
Phosphono	$(HO)_2OP-$	Pseudoindolyl (from	C_8H_6N-
Phosphoro	$-P:P-$	pseudoindole)	
Phosphoroso	$OP-$	Pteridinyl (from	$C_6H_3N_4-$
Phthalal	$:CHC_6H_4CH$: (*o*)	pteridine)	
Phthalamoyl	$H_2NCOC_6H_4CO-$ (*o*)	Pteroyl	$C_{14}H_{11}N_5O_2-$
Phthalazinyl	$C_8H_5N_2-$	Pyranyl	C_5H_7O-
Phthalidyl	$C_6H_4CO \cdot O \cdot CH-$ (*o*)	Pyrazinyl	$C_4H_3N_2-$
	$\lfloor \underline{\qquad} \rfloor$	Pyrazolidinyl	$C_3H_7N_2-$

Table 7-1 (*Continued*)
NAMES AND FORMULAS OF ORGANIC RADICALS

Name	Formula	Name	Formula
Pyrazolidyl	$C_3H_7N_2-$	Stearoyl	$CH_3(CH_2)_{16}CO-$
Pyrazolinyl	$C_3H_5N_2-$	Stibarseno	$-Sb:As-$
Pyrazolyl	$C_3H_3N_2-$ (4 isomers)	Stibinico	$(HO)OSb:$
Pyrenyl	$C_{16}H_9-$	Stibino	H_2Sb-
Pyridazinyl	$C_4H_3N_2-$	Stibo	O_2Sb-
Pyridinediyl	$-C:CHCH:CH\cdot N:C-$	Stibono	$(HO)_2OSb-$
	(*2,3*-form shown)	Stiboso	$OSb-$
		Stibyl	H_2Sb-
Pyridyl (from pyridine)	C_5H_4N- (3 isomers)	Stibylene	$HSb:$
Pyridylidene	$CH:CH\cdot NH\cdot CH:CH\cdot C:$	Styrene	$-CH(C_6H_5)CH_2-$
	(*4(1)*)	Styrolene	$-CH(C_6H_5)CH_2-$
		Styryl	$C_6H_5CH:CH-$
Pyrimidinyl (from pyrimidine)	$C_4H_3N_2-$	Suberoyl	$-CO(CH_2)_6CO-$
		Succinamoyl	$H_2NCOCH_2CH_2CO-$
Pyromucyl	$O\cdot CH:CH\cdot CH:C\cdot CO-$	Succinamyl	$H_2NCOCH_2CH_2CO-$
		Succinimido	$CO\cdot CH_2\cdot CH_2CO\cdot N-$
Pyrrolidinyl (from pyrrolidine)	C_4H_8N- (3 isomers)		
		Succinyl	$-COCH_2CH_2CO-$
Pyrrolidyl	C_4H_8N- (3 isomers)	Sulfamino	HO_3SNH-
Pyrrolinyl	C_4H_6N-	Sulfamoyl	H_2NSO_2-
Pyrrolyl	C_4H_4N-	Sulfamyl	H_2NSO_2-
Pyrrolylcarbonyl	C_4H_3NCO-	Sulfanilamido	$H_2N\cdot C_6H_4\cdot SO_2NH-$ (*p*)
Pyrroyl	C_4H_3NCO-	Sulfanilyl	$H_2N\cdot C_6H_4\cdot SO_2-$ (*p*)
Pyrryl	C_4H_4N-	Sulfhydryl	$HS-$
Pyrruvoyl	$CH_3CO\cdot CO-$	Sulfino	$(HO)OS-$
Quinazolinyl (from quinazoline)	$C_8H_5N_2-$	Sulfinyl	$-SO-$
		Sulfo	$(HO)O_2S-$
Quinolyl (from quinoline)	C_9H_6N-	Sulfoamino	HO_3SNH-
Quinonyl	$C_6H_3O_2-$ (*o* or *p*)	Sulfenamido	$-SO_2NH-$ (only as a
Quinoxalinyl (from quinoxaline)	$C_8H_5N_2-$		doubling radical)
		Sulfonyl	$-SO_2-$
Quinuclidinyl	$C_7H_{12}N-$	Sulfuryl	$-SO_2-$
Salicyl	$HO\cdot C_6H_4\cdot CH_2-$ (*o*)	Tartronoyl	$-COCH(OH)CO-$
Salicylidene	$HO\cdot C_6H_4\cdot CH:$ (*o*)	Tauryl	$H_2NCH_2CH_2SO_2-$
Salicyloyl	$HO\cdot C_6H_4\cdot CO-$ (*o*)	Telluro	$-Te-$
Sabacoyl	$-CO(CH_2)_8CO-$	Terephthalal	$:HCC_6H_4CH:$ (*p*)
Selenino	$(HO)OSe-$	Terephthaloyl	$-COC_6H_4CO-$ (*p*)
Seleninyl	$OSe:$	Terephthalylidene	$:HCC_6H_4CH:$ (*p*)
Seleno	$-Se-$	Terphenylyl (from terphenyl)	$C_{18}H_{13}-$
Selenocyanato	$N:CSe-$		
Selenono	$(HO)O_2Se-$	Terphenylylene (from terphenyl)	$-C_{18}H_{12}-$
Selenonyl	$O_2Se:$		
Selenyl	$HSe-$	Tetracontyl	$CH_3(CH_2)_{38}CH_2-$
Semicarbazido	$H_2NCONHNH-$	Tetracosyl	$CH_3(CH_2)_{22}CH_2-$
Semicarbazono	$H_2NCONH:N:$	Tetradecanoyl	$CH_3(CH_2)_{12}CO-$
Senecicyl (from senecioic acid)	$(CH_3)_2C:CHCO-$	Tetradecyl	$CH_3(CH_2)_{12}CH_2-$
		Tetramethylene	$-CH_2(CH_2)_2CH_2-$
Seryl	$HOCH_2CH(NH_2)CO-$	Tetramethylphenyl	$(CH_3)_4C_6H-$
Siloxy	$H_3Si\cdot O-$	Tetramethylphenylene	$(CH_3)_4C_6:$
Silyl	H_3Si-	*1*-Tetrazeno	$H_2N\cdot NH\cdot N:N-$
Silylamino	$H_3Si\cdot NH-$	Tetrazolyl (from tetrazole)	$N:N\cdot N:N\cdot CH-$
Silyldisilanyl	$(H_3Si)_2:SiH-$		
Silylene	$H_2Si:$		(2 isomers)
Silylidyne	$HSi:$	Thenoyl	$S\cdot CH:CHCH:C\cdot CO-$
Silylthio	$H_3Si\cdot S-$		
Stannono	$(HO)OSn-$		(*2*-form shown)
Stannyl	H_3Sn-	Thenyl	$C_4H_3SCH_2-$
Stannylene	$H_2Sn:$	Thenylidene	$C_4H_3SCH:$

Table 7-1 (*Continued*)
NAMES AND FORMULAS OF ORGANIC RADICALS

Name	Formula	Name	Formula
Thiazinyl	C_4H_4NS-	Triazinyl (from triazine)	$C_3H_2N_3-$
Thiazolidinyl	C_3H_6NS-	Triazolidinyl	$C_2H_6N_3-$
Thiazolidyl	C_3H_6NS-	Triazolyl (from triazole)	$C_2H_2N_3-$
Thiazolinyl	C_3H_4NS-	Tricosyl	$CH_3(CH_2)_{21}CH_2-$
Thiazolyl (from thiazole)	C_3H_2NS- (3 isomers)	Tridecanoyl	$CH_3(CH_2)_{11}CO-$
Thienyl (from thiophene)	C_4H_3S- (2 isomers)	Tridecyl	$CH_3(CH_2)_{11}CH_2-$
Thio	$-S-$	Trimethoxyphenyl	$(CH_3O)_3C_6H_2-$
Thiocarbamoyl	H_2NCS-	Trimethylanilino	$(CH_3)_3C_6H_2NH-$
Thiocarbamyl	H_2NCS-	Trimethylene	$-CH_2CH_2CH_2-$
Thiocarbonyl	$SC:$	Trimethylphenyl	$(CH_3)_3C_6H_2-$
Thiocyanato	$N\!:\!CS-$	Triphenylmethyl	$(C_6H_5)_3C-$
Thiocyano	$N\!:\!CS-$	Triphenylsilyl	$(C_6H_5)_3Si-$
Thioformyl	$S:CH-$	Trisilanyl	$H_3Si\cdot SiH_2\cdot SiH_2-$
Thiohydroxy	$HS-$	Trisilanylene	$-SiH_2\cdot SiH_2\cdot SiH_2-$
Thiomorpholino	$CH_2CH_2\cdot S\cdot CH_2CH_2\cdot N-$	Tritriacontyl	$CH_3(CH_2)_{31}CH_2-$
	$\underline{\qquad\qquad\qquad\qquad}$	Trityl	$(C_6H_5)_3C-$
	(*4*-position only)	Tropoyl	$C_6H_5CH(CH_2OH)CO-$
Thiomorpholinyl	$NH\cdot CH_2CH_2\cdot S\cdot CH_2CH-$	Tryptophyl (from	$C_8H_5NCH_2CH(NH_2)CO-$
(2- or 3-)	$\underline{\qquad\qquad\qquad\qquad}$	tryptophane)	
	(*3*-form shown)	Tyrosyl (from tyrosine)	$HO\cdot C_6H_4\cdot CH_2\cdot CH\cdot$
Thionyl	$-SO-$		$(NH_2)CO-$ (*p*)
Thioxo	$S:$ (replacing 2H in	Undecanoyl	$CH_3(CH_2)_9CO-$
	$:CH_2$)	Undecyl	$CH_3(CH_2)_9CH_2-$
Thiuram	H_2NCS-	Uramino	$H_2NCONH-$
Threonyl	$CH_3CH(OH)CH(NH_2)\cdot$	Ureido	$H_2NCONH-$
	$CO-$	Ureylene	$-NHCONH-$
Thujyl (from thujane)	$C_{10}H_{17}-$	Valeryl	$CH_3(CH_2)_3CO-$
Thymyl (from thymol)	$HC:C(CH_3)\cdot CH:CH\cdot$	Valyl (from valine)	$(CH_3)_2CHCH(NH_2)CO-$
	$\underline{\qquad\qquad\qquad}$	Vanillal	$(CH_3O)(HO)C_6H_3\cdot CH:$
	$C(CH(CH_3)_2):C-$		(*3,4*)
	$\underline{\qquad\qquad\qquad\qquad}$	Vanilloyl	$(CH_3O)(HO)C_6H_3\cdot CO-$
Thyronyl	$(p\text{-}HOC_6H_4O)C_6H_4CH_2\cdot$		(*3,4*)
	$CH(NH_2)CO-$ (*p*)	Vanillyl	$(CH_3O)(HO)C_6H_3\cdot CH_2-$
Toloxy	$CH_3\cdot C_6H_4\cdot O-$		(*3,4*)
	(*o, m* or *p*)	Vanillylidene	$(CH_3O)(HO)C_6H_3\cdot CH:$
Toluenesulfonyl	$C_7H_7SO_2-$		(*3,4*)
Toluidino	$CH_3\cdot C_6H_4\cdot NH-$	Veratral	$(CH_3O)_2C_6H_3\cdot CH:$
	(*o, m* or *p*)		(*3,4*)
Toluoyl	$CH_3\cdot C_6H_4\cdot CO-$	Veratroyl	$(CH_3O)_2C_6H_3\cdot CO-$ (*3,4*)
	(*o, m* or *p*)	*o*-Veratroyl	$(CH_3O)_2C_6H_3\cdot CO-$ (*2,3*)
Toluyl	$CH_3\cdot C_6H_4\cdot CO-$	Veratryl	$(CH_3O)_2C_6H_3\cdot CH_2-$
	(*o, m* or *p*)		(*3,4*)
Tolyl	$CH_3\cdot C_6H_4-$ (*o, m* or *p*)	Veratrylidene	$(CH_3O)_2C_6H_3\cdot CH:$
α-Tolyl	$C_6H_5CH_2-$		(*3,4*)
Tolylene	$-C_6H_3(CH_3)-$	Vinyl	$CH_2:CH-$
	(6 isomers)	Vinylene	$-CH:CH-$
α-Tolylene	$C_6H_5CH:$	Vinylidene	$CH_2:C:$
Tolyloxy	$CH_3\cdot C_6H_4\cdot O-$	Xanthenyl	$C_{13}H_9O-$ (6 isomers)
	(*o, m* or *p*)	Xanthyl (from xanthene)	$C_{13}H_9O-$ (6 isomers)
Tolylsulfonyl	$CH_3\cdot C_6H_4\cdot SO_2-$	Xenyl	$C_6H_5\cdot C_6H_4-$
	(*o, m* or *p*)	Xylidino	$(CH_3)_2C_6H_3NH-$
Tosyl	$CH_3\cdot C_6H_4\cdot SO_2-$	Xyloyl (from xylic acid)	$(CH_3)_2C_6H_3CO-$
	(*o, m* or *p*)		(7 isomers)
Triacontyl	$CH_3(CH_2)_{28}CH_2-$	Xylyl	$(CH_3)_2C_6H_3-$
Triazeno	$H_2N\cdot N:N-$	Xylylene	$-H_2C\cdot C_6H_4\cdot CH_2-$

PHYSICAL PROPERTIES OF PURE SUBSTANCES

Table 7-2
FORMULA INDEX FOR ORGANIC COMPOUNDS

For finding compounds listed below see Table 7-4, Physical Constants of Organic Compounds.

The system used in this formula index is essentially the same as that employed in *Richter's Lexikon der Kohlenstoff Verbindungen*. The succession of the elements combined with carbon is as follows: H, O, N, S, F, Cl, Br, I. All the other elements are placed in alphabetical order according to their symbols. The arrangement depends (1) on the number of carbon atoms, (2) on the number of the other elements, which in addition to carbon, are contained in the compounds, (3) on the kind of elements, which in addition to carbon are contained in the molecule, in accordance with the order as stated above, and (4) on the number of atoms of each element which, in addition to carbon, is contained in the compound Thus 2 III means two carbon atoms combined with three other elements; *e.g.*, $C_2H_2O_2F_2$ 2239, is a compound with 2 carbon atoms and 3 other elements, and it is the 2239th compound in the table to which this formula index is a part.

1 I

CH_4	4061
CO	1239
CO_2	1237
CN	1589
CS_2	1238
CF_4	1244
CCl_4	1243
CBr_4	1242
CI_4	1245

1 II

CHN	3714
CHF_3	3287
$CHCl_3$	1477
$CHBr_3$	1016
CHI_3	3887
CH_2O	3288; 3290
CH_2O_2	3301
CH_2N_2	1577; 1772
CH_2N_4	5870
CH_2S_3	6349
CH_2F_2	4433
CH_2Cl_2	4429
CH_2Br_2	4428
CH_2I_2	4434
CH_3N_3	4134
CH_3F	4240
CH_3Cl	4183
CH_3Br	4147
CH_3I	4287
CH_4O	4100
CH_4N_2	3297
CH_4S	4298
CH_5N	4105
CH_5N_3	3488
CH_5As	4132
CH_5P	4341
CH_6N_2	4269
CH_6N_4	322
COS	1248
$COCl_2$	1247
$COBr_2$	1246
CO_8N_4	5853

$CNCl$	1591
$CNBr$	1590
CNI	1592
$COSCl_2$	5916
$CSCl_4$	5089
$CFCl_3$	3286
CF_2Cl_2	2246
$CClBr_3$	1472
CCl_2Br_2	1983
CCl_2I_2	1986
CCl_3Br	6143
CCl_3I	6157
$CBrI_3$	1004

1 III

$CHON$	1576; 1579
CHO_2Na	3302
CHO_2Tl	3303
CHO_6N_3	6300
$CHNS$	5897
$CHFCl_2$	3271
$CHClBr_2$	1361
$CHClI_2$	1363
$CHCl_2Br$	1970
$CHCl_2I$	2006
$CHBrI_2$	928
$CHBr_2I$	1849
$CH_2O_3N_2$	4323
$CH_2O_4N_2$	2636
CH_2FCl	3267
CH_2ClBr	1344
CH_2ClI	1393
CH_2BrI	950
CH_3ON	3294; 3296
CH_3OCl	4281
CH_3OAs	4131
CH_3O_2N	1228; 4316; 4765
CH_3O_3N	4315
$CH_3O_3N_3$	4863
CH_3NS	5901
CH_3NS_2	2819
CH_3Cl_2As	4214
CH_4ON_2	3298; 3307; 6386
$CH_4O_2N_2$	3842; 4314

$CH_4O_2N_4$	4743
CH_4O_2Si	5615
CH_4O_3S	4382
CH_4O_4S	4384
$CH_4O_6S_2$	4432
CH_4N_2S	5920
CH_4N_2Se	5608
CH_5ON	4279
CH_5ON_3	5609
CH_5O_3As	4130
$CH_5O_4N_3$	6389
CH_5N_3S	5919
CH_6ON_4	1236
$CH_6O_3N_4$	3492
CH_6NCl	4106
CH_6N_3Cl	3491
CO_2NCl_3	4697
CO_2NBr_3	4688
CO_6N_3Br	1005
CO_6N_3I	3908

1 IV

CHO_2NBr_2	1854
$CHFClBr$	3266
CH_2ONCl	1229
CH_2O_2NBr	966
$CH_2O_5S_2Na_2$	3293
CH_3O_2SCl	4383
CH_3O_3SNa	3295
CH_3O_4SNa	3292
CH_5ON_2Cl	6388
CH_6ONCl	4280
CH_6ON_3Cl	5610
$CH_8O_4N_2S$	4270
$CH_8O_4N_4S$	324

2 I

C_2H_2	130
C_2H_4	3192
C_2H_6	2904
C_2N_2	1588
C_2Cl_4	5741
C_2Cl_6	3564
C_2Br_2	1796

Table 7-2 (*Continued*)
FORMULA INDEX FOR ORGANIC COMPOUNDS

C_2Br_4 5721
C_2Br_6 3562
C_2I_2 2382
C_2I_4 5812

2 II

C_2HCl 1305
C_2HCl_3 6155
C_2HCl_5 5054
C_2HBr 891
C_2HBr_3 6105
C_2HBr_5 5050
C_2HI_5 5072
C_2H_2O 3933
$C_2H_2O_2$ 3478
$C_2H_2O_3$ 3480
$C_2H_2O_4$ 5006-7
$C_2H_2N_2$ 3715
$C_2H_2N_4$ 5869
$C_2H_2F_2$ 2250
$C_2H_2Cl_2$ 135-6; 1998
$C_2H_2Cl_4$ 5739-40
$C_2H_2Br_2$ 131-3
$C_2H_2Br_4$ 5719-20
$C_2H_2I_2$ 137-8
C_2H_3N 49; 4195
$C_2H_3N_3$ 6080
C_2H_3F 6439
C_2H_3Cl 6435
$C_2H_3Cl_3$ 6151-2
C_2H_3Br 6434
$C_2H_3Br_3$ 6103
C_2H_3I 6442
$C_2H_3I_3$ 6237
C_2H_4O 5; 3210; 6429
$C_2H_4O_2$ 36; 3470; 4242
$C_2H_4O_3$ 3468
$C_2H_4N_4$ 2080
$C_2H_4N_6$ 1761
$C_2H_4S_2$ 2818
$C_2H_4F_2$ 2249
$C_2H_4Cl_2$ 1993-4
$C_2H_4Br_2$ 1836-7
$C_2H_4I_2$ 2390-1
C_2H_5N 6430
C_2H_5F 3046
C_2H_5Cl 3015
C_2H_5Br 2989
C_2H_5I 3077
C_2H_5Na 5622
C_2H_6O 2478; 2946
$C_2H_6O_2$ 3444
$C_2H_6N_2$ 15
C_2H_6S 2560; 3083
$C_2H_6S_2$ 3209; 4222
C_2H_6Cd 1186
C_2H_6Hg 4044
C_2H_6Se 2553; 3071
C_2H_6Te 2566

C_2H_6Zn 6504
C_2H_7N 2429; 2950
$C_2H_7N_3$ 4258
C_2H_7As 2456; 2970
C_2H_7P 2529; 3135
$C_2H_8N_2$ 2494-5; 3064; 3196-9
C_2OCl_4 6129
$C_2O_2Cl_2$ 5011
$C_2O_2Cl_4$ 6159
$C_2O_4Na_2$ 5008
$C_2O_6N_4$ 6288
C_2N_2S 1593
C_2FCl_3 3285
C_2FCl_5 3276
C_2FBr_5 3275
$C_2F_2Cl_2$ 2245
$C_2F_2Cl_4$ 2253-4
$C_2F_2Br_4$ 2252
C_2F_3Cl 6203
$C_2F_3Cl_3$ 6206
$C_2F_3Br_3$ 6205
C_2ClBr_5 1423
$C_2Cl_2Br_4$ 2073
$C_2Cl_3Br_3$ 6170
$C_2Cl_4Br_2$ 5736-7

2 III

C_2HOCl_3 1281; 1931
C_2HOBr_3 876
C_2HO_2N 5012
$C_2HO_2Cl_3$ 6128
$C_2HO_2Br_3$ 6088
$C_2HO_2I_3$ 6232
C_2HFCl_4 3278
C_2HFBr_4 3277
$C_2HF_2Br_3$ 2255
$C_2HF_3Br_2$ 6204
C_2HClBr_4 1465
$C_2HCl_2Br_3$ 2075-6
$C_2HCl_3Br_2$ 6150
$C_2H_2OCl_2$ 1304; 1926
$C_2H_2OBr_2$ 890
$C_2H_2O_2F_2$ 2239
$C_2H_2O_2Cl_2$ 1928
$C_2H_2O_2Br_2$ 1795
$C_2H_2O_2I_2$ 2381
C_2H_2NCl 1399
C_2H_2NBr 954
C_2H_2NI 3890
$C_2H_2N_2S_3$ 5898
$C_2H_2FCl_3$ 3284
$C_2H_2FBr_3$ 3282-3
$C_2H_2F_2Cl_2$ 2244
$C_2H_2F_2Br_2$ 2242-3
$C_2H_2F_3Cl$ 6202
$C_2H_2ClBr_3$ 1471
$C_2H_2Cl_2Br_2$ 1980-2
$C_2H_2Cl_3Br$ 6142
$C_2H_2Cl_3As$ 1473

C_2H_3ON 3474; 4194
C_2H_3OF 89
C_2H_3OCl 81; 1293
$C_2H_3OCl_3$ 6154
C_2H_3OBr 75
$C_2H_3OBr_3$ 6104
C_2H_3OI 92
$C_2H_3O_2N_3$ 6078
$C_2H_3O_2F$ 3255
$C_2H_3O_2Cl$ 1299; 4187
$C_2H_3O_2Cl_3$ 1286
$C_2H_3O_2Br$ 883
$C_2H_3O_2Br_3$ 877
$C_2H_3O_2I$ 3874
$C_2H_3O_3N$ 5016
$C_2H_3O_4N$ 105; 4592
$C_2H_3O_6N_3$ 6299
C_2H_3NS 4387-8
$C_2H_3FBr_2$ 3268-9
$C_2H_3F_2Cl$ 2241
$C_2H_3F_2Br$ 2240
$C_2H_3F_2I$ 2251
$C_2H_3ClBr_2$ 1359-60
$C_2H_3Cl_2Br$ 1968-9
$C_2H_3Cl_2I$ 2005
C_2H_4OS 5885
$C_2H_4OCl_2$ 1997; 2008
$C_2H_4O_2N_2$ 2256; 3479; **5009**
$C_2H_4O_2N_4$ 603; 2851
$C_2H_4O_2S$ 5905
$C_2H_4O_3N_2$ 3105
$C_2H_4O_4N_2$ 2630; 3463
$C_2H_4O_5S$ 5675
$C_2H_4O_6N_2$ 3462
$C_2H_4N_2S_2$ 2821
$C_2H_4N_6Cl_2$ 1949
C_2H_4ClBr 1342-3
C_2H_4ClI 1390-1
C_2H_4BrI 948-9
C_2H_5ON 10; 14
C_2H_5OCl 1400; 3076; 3194
C_2H_5OBr 3193
C_2H_5OI 3208
$C_2H_5O_2N$ 3097; 3437; 3472; 4175; 4733
$C_2H_5O_2N_3$ 859; 5017
$C_2H_5O_3N$ 3096; 4735
C_2H_5NS 5882
C_2H_5SNa 5624
$C_2H_5Cl_2As$ 3040
$C_2H_6ON_2$ 2513; 3438; 4414
$C_2H_6ON_4$ 2081
C_2H_6OS 2563; 5904
$C_2H_6O_2N_4$ 3689
$C_2H_6O_2S$ 2562; 2905
$C_2H_6O_3S_2$ 2561; 2906
$C_2H_6O_4S$ 2559; 3069
$C_2H_6O_6S_2$ 3206
$C_2H_6N_2S$ 4390-1

Table 7-2 (*Continued*)
FORMULA INDEX FOR ORGANIC COMPOUNDS

$C_2H_6N_4S$ 3494
C_2H_6ClAs 1179
$C_2H_6Cl_3As$ 1181
C_2H_7ON 8; 2907; 3074-5
$C_2H_7O_2As$ 1182
$C_2H_7O_3As$ 2971
$C_2H_7O_4N_3$ 4415
$C_2H_7N_2Cl$ 16
$C_2H_8O_3N_4$ 323; 4259
C_2H_8NCl 2430; 2952
C_2H_8NBr 2951
$C_2H_9N_2Cl$ 2496
$C_2H_{10}ON_2$ 3199
$C_2H_{10}N_2Cl_2$ 3198
$C_2H_{10}N_2Br_2$ 3197
$C_2O_2N_2Ag_2$ 5620
$C_2O_2N_2Hg$ 4040

2 IV

$C_2HFCl_2Br_2$ 3270
$C_2H_2ONCl_3$ 6126
$C_2H_2ONBr_3$ 6086
$C_2H_3ONCl_2$ 1927; 2032
$C_2H_3ONBr_2$ 1793-4
$C_2H_3O_2NCl_2$ 2027
C_2H_4ONCl 1294-5; 1422
$C_2H_4ONCl_3$ 1283
C_2H_4ONBr 879
$C_2H_4O_2NCl$ 1416-7
$C_2H_4O_2NBr$ 965
$C_2H_5O_2SCl$ 3155
$C_2H_5O_3SCl$ 3022
$C_2H_5O_4SNa$ 13
$C_2H_7O_3NS$ 5696
$C_2H_{12}O_4N_6S$ 3493
$C_2H_{14}O_4N_8S$ 325

3 I

C_3H_4 183; 225
C_3H_6 1621; 5459
C_3H_8 5338
C_3O_2 1240
C_3S_2 1241
C_3Cl_8 4936

3 II

C_3HCl_7 3506
C_3H_2O 5349
$C_3H_2O_2$ 5348
$C_3H_2N_2$ 4006
$C_3H_2Br_2$ 1871
C_3H_3N 151
C_3H_3Cl 5343
C_3H_3Br 5342
C_3H_3I 3903; 5344
C_3H_4O 146; 226; 5341
$C_3H_4O_2$ 149
$C_3H_4O_3$ 5525
$C_3H_4O_4$ 4002; 4329

$C_3H_4O_5$ 5695
$C_3H_4O_6$ 4055
$C_3H_4N_2$ 3851; 5478
$C_3H_4Cl_2$ 2058-61
$C_3H_4Br_2$ 1869-70
C_3H_5N 3029; 5363
C_3H_5Cl 202; 5345-6
$C_3H_5Cl_3$ 6166-7
C_3H_5Br 199; 988-90
$C_3H_5Br_3$ 6112
C_3H_5I 208
C_3H_6O 56; 193; 5352; 5468-9
$C_3H_6O_2$ 55; 3048; 3436; 3454; 3942; 4087; 5351
$C_3H_6O_3$ 2292; 2468; 3289; 3380; 3456; 3682; 3939-40; 4063; 4255
$C_3H_6O_4$ 3379
$C_3H_6N_2$ 5479
$C_3H_6N_6$ 4019
C_3H_6S 210
$C_3H_6S_3$ 3309
$C_3H_6Cl_2$ 2050-3
$C_3H_6Br_2$ 1863-6
$C_3H_6I_2$ 2397-9
C_3H_7N 194
C_3H_7F 5406-7
C_3H_7Cl 5400-1
C_3H_7Br 5388-9
C_3H_7I 5417-8
C_3H_8O 4230; 5368-9
$C_3H_8O_2$ 3460; 4424; 5465; 6286
$C_3H_8O_3$ 3381
C_3H_8S 4238; 5424-5
$C_3H_8S_2$ 6284
C_3H_8Se 5448
C_3H_9N 4226; 5371-2; 6248
$C_3H_9N_3$ 2491
C_3H_9Al 231
C_3H_9As 6252
C_3H_9B 872
C_3H_9Bi 856
C_3H_9P 6269
C_3H_9Sb 6274
$C_3H_{10}N_2$ 5464; 6282
$C_3N_3Cl_3$ 6172

3 III

C_3HOBr_5 5048.1
$C_3H_2OCl_4$ 5728-9
$C_3H_2O_3N_2$ 5036
$C_3H_2O_3Br_2$ 1875
$C_3H_2O_4Br_2$ 1851
$C_3H_2O_4Ca$ 4003
C_3H_3ON 5527
$C_3H_3O_2N$ 1583
$C_3H_3O_2Cl$ 1306-7
$C_3H_3O_2Cl_3$ 4403
$C_3H_3O_3N_3$ 1594; 3313

$C_3H_3O_3Cl_3$ 6158
$C_3H_3O_4N$ 4762
$C_3H_3O_4Cl$ 1397
$C_3H_3O_4Br$ 952
C_3H_3NS 5878
$C_3H_3NCl_2$ 2056
$C_3H_3N_3S_3$ 5899
$C_3H_4ON_2$ 1581; 5480
$C_3H_4OCl_2$ 1444; 1929-30
$C_3H_4OBr_2$ 982
$C_3H_4O_2N_2$ 3679
$C_3H_4O_2Cl_2$ 1382; 2054-5
$C_3H_4O_2Br_2$ 1867-8
$C_3H_4O_3N_2$ 182
$C_3H_4O_4N_2$ 5013
$C_3H_4N_2S$ 377
C_3H_5ON 148; 3027-8; 3195; 3944
$C_3H_5ON_5$ 396
C_3H_5OF 5361
C_3H_5OCl 1301; 2879; 5360
$C_3H_5OCl_3$ 6168
C_3H_5OBr 886; 892; 5359
C_3H_5OI 2881; 5362
$C_3H_5O_2N$ 3927; 4829; 4873; 5526
$C_3H_5O_2Cl$ 1398; 1442-3; 3019; 4184
$C_3H_5O_2Br$ 955; 980-1; 4148
$C_3H_5O_2I$ 3901-2
$C_3H_5O_3N_3$ 5014
$C_3H_5O_4N$ 328
$C_3H_5O_6N_3$ 3429
$C_3H_5O_9N_3$ 3428
C_3H_5NS 3158-9
$C_3H_6ON_2$ 3212; 3216
$C_3H_6OS_2$ 6448
$C_3H_6OCl_2$ 3393-4
$C_3H_6OBr_2$ 3388-9
$C_3H_6OI_2$ 3401
$C_3H_6O_2N_2$ 123; 4004
$C_3H_6O_2S$ 5908; 5910
$C_3H_6O_3N_2$ 3678
$C_3H_6O_4N_2$ 2658-9; 4864
$C_3H_6O_7N_2$ 3405-6
C_3H_6ClBr 6279
C_3H_7ON 61; 235; 3047; 4085; 5353; 5356
$C_3H_7ON_3$ 12
C_3H_7OCl 4186; 5461-2; 6280
C_3H_7OBr 5460; 6278
C_3H_7OI 6287
$C_3H_7O_2N$ 158-60; 3291; 3943; 4107; 4827-8; 5428; 5600; 6392
$C_3H_7O_2N_3$ 3442
$C_3H_7O_2Cl$ 3395-6
$C_3H_7O_2I$ 3402
$C_3H_7O_3N$ 5427; 5611-4

Table 7-2 (Continued)
FORMULA INDEX FOR ORGANIC COMPOUNDS

$C_3H_7O_3As$ 197
$C_3H_7O_3Na$ 5623
$C_3H_7O_5N$ 3407-8
C_3H_7NS 5917
$C_3H_7NS_2$ 3157
$C_3H_7N_6Cl$ 4020
$C_3H_8ON_2$ 2572-3; 3183
$C_3H_8O_2N_2$ 5429
$C_3H_8O_2S$ 4239; 5903
$C_3H_8N_2S$ 2568; 3162
C_3H_9ON 4111; 5339; 5416
$C_3H_9O_2N$ 3385
$C_3H_9O_2N_3$ 3489
$C_3H_9O_3As$ 5376
$C_3H_9O_3B$ 6262
$C_3H_9O_4P$ 6268
$C_3H_9O_6P$ 3411
$C_3H_{10}ON_2$ 3384
$C_3H_{10}NCl$ 4227; 6249
$C_3H_{12}O_3N_6$ 3490

3 IV

$C_3H_2ONCl_3$ 1284
$C_3H_3ONS_2$ 5560
$C_3H_3O_2NS$ 2369
$C_3H_4ON_2S$ 5907
$C_3H_4O_2NCl_3$ 1285
$C_3H_5O_2NCl_2$ 2031
$C_3H_5O_3NCl_2$ 2028
$C_3H_6ON_2S$ 122
$C_3H_6O_2NCl$ 1420-1
$C_3H_6NS_2Na$ 2477
C_3H_7ONS 5921; 6450
$C_3H_7O_2NS$ 1625
$C_3H_7O_4SNa$ 64
$C_3H_7O_6PNa_2$ 3412
$C_3H_8O_2NCl$ 5601

3 V

$C_3H_8O_2NSCl$ 1626

4 I

C_4H_2 1679
C_4H_6 1018-9; 1156-7
C_4H_8 1147-9; 1595; 4210
C_4H_{10} 1021-2

4 II

$C_4H_2O_3$ 3996
$C_4H_2O_4$ 134
C_4H_4O 3327
$C_4H_4O_2$ 5871
$C_4H_4O_3$ 5657
$C_4H_4O_4$ 3314; 3475; 3995
$C_4H_4N_2$ 5477; 5482; 5513; 5661
C_4H_4S 5923
C_4H_5N 204-5; 1563; 5521
$C_4H_5N_3$ 3853

C_4H_5Cl 1345
C_4H_6O 1559; 2499; 4058; 4348; 6436
$C_4H_6O_2$ 207; 1557-8; 1667; 3304; 4095-6; 5655; 6426-7
$C_4H_6O_3$ 37-8; 3934; 4366; 6440
$C_4H_6O_4$ 106; 2521; 3066; 3455; 5652-3
$C_4H_6O_5$ 2261; 3473; 3997-8; 4000
$C_4H_6O_6$ 5688; 5690-1
$C_4H_6O_8$ 2367
$C_4H_6N_2$ 4257
C_4H_6S 2852
C_4H_7N 1175-6; 5404-5; 5524
C_4H_7Cl 1566-7; 4060
$C_4H_7Br_3$ 6100-1
C_4H_7I 1568
C_4H_8O 1155; 1160; 1163; 1565; 4059; 4104; 4232; 4422; 5773; 6437-8
$C_4H_8O_2$ 43; 169; 1166-7; 2237; 2935; 3207; 4349; 5408-9; 6285
$C_4H_8O_3$ 2912; 3060; 3397; 3446; 3753-7; 4076; 4229; 4294
$C_4H_8O_4$ 4253
$C_4H_8N_2$ 9; 4256
C_4H_8S 3190
$C_4H_8S_2$ 2228
$C_4H_8Cl_2$ 1971-6
$C_4H_8Br_2$ 1821-5
$C_4H_8I_2$ 2389
C_4H_9N 4102; 5523
C_4H_9F 1085
C_4H_9Cl 1072-5
C_4H_9Br 1057-60
C_4H_9I 1095-8
$C_4H_{10}O$ 1034-7; 2911; 4353-4
$C_4H_{10}O_2$ 1154; 1265; 2196; 2329-32; 2415; 3061; 3459
$C_4H_{10}O_3$ 3404; 3453; 4077
$C_4H_{10}O_4$ 2894
$C_4H_{10}N_2$ 2225-6
$C_4H_{10}S$ 1103-5; 2214
$C_4H_{10}S_2$ 2166
$C_4H_{10}Be$ 845
$C_4H_{10}Cd$ 1185
$C_4H_{10}Hg$ 4043
$C_4H_{10}Se$ 2207
$C_4H_{10}Se_2$ 2208
$C_4H_{10}Sn$ 5949
$C_4H_{10}Te$ 2219
$C_4H_{10}Zn$ 6503
$C_4H_{11}N$ 1039-42; 2109; 2482; 4350

$C_4H_{11}P$ 2197
$C_4H_{12}N_2$ 2178; 5474
$C_4H_{12}As_2$ 1178
$C_4H_{12}Pb$ 3958
$C_4H_{12}Si$ 5617
$C_4H_{12}Sn$ 5951
$C_4H_{13}N_3$ 2238
C_4OCl_{10} 5088

4 III

C_4HNI_4 5817
$C_4H_2O_2Cl_2$ 3316
$C_4H_2O_3Cl_2$ 4442
$C_4H_2O_3Br_2$ 4441
$C_4H_2O_4N_2$ 184-6
$C_4H_2O_4Cl_2$ 2007
$C_4H_2O_4Br_2$ 1840; 1850
$C_4H_2SI_2$ 2400
C_4H_3OBr 942
$C_4H_3O_3N$ 4740
$C_4H_3O_4N_3$ 4862; 6445
$C_4H_3O_4Cl$ 1384; 1396
$C_4H_3O_4Br$ 941; 951
$C_4H_3O_5N_3$ 4632
$C_4H_4O_2N_2$ 6383
$C_4H_4O_2Cl_2$ 5658
$C_4H_4O_3N_2$ 621
$C_4H_4O_3N_4$ 1774
$C_4H_4O_3Cl_2$ 1300
$C_4H_4O_4N_2$ 1686
$C_4H_4O_4Br_2$ 1877-8
$C_4H_4O_4Na_2$ 5654
$C_4H_4O_5N_2$ 186-7
$C_4H_4O_6Ca$ 5689
$C_4H_4O_8Na_2$ 2368
$C_4H_4O_{10}N_2$ 4842
$C_4H_4N_2S_2$ 3211
C_4H_5ON 2880
C_4H_5OCl 1562
$C_4H_5OCl_3$ 1158
$C_4H_5O_2N$ 1587; 3032; 4196; 5660
$C_4H_5O_2N_3$ 386
$C_4H_5O_2Cl$ 1358
$C_4H_5O_2Cl_3$ 3179; 6153
$C_4H_5O_2Br$ 925
$C_4H_5O_2Br_3$ 3178
$C_4H_5O_3N_3$ 6384
$C_4H_5O_4Cl$ 1464
$C_4H_5O_4Br$ 997
C_4H_5NS 219-20; 378
$C_4H_6ON_2$ 4360
$C_4H_6O_2N_2$ 3035; 3315; 3439; 4267-8
$C_4H_6O_2N_4$ 139
$C_4H_6O_2S_2$ 85
$C_4H_6O_2Cl_2$ 1448; 1978-9; 1988; 3039
$C_4H_6O_2Br_2$ 1826-7; 3036

Table 7-2 (*Continued*)
FORMULA INDEX FOR ORGANIC COMPOUNDS

Table 7-2 (*Continued*)
FORMULA INDEX FOR ORGANIC COMPOUNDS

C₅H₅O₃N 6231
C₅H₆OS 3337; 5924
C₅H₆O₂N₂ 4410-3
C₅H₆NCl 5484
C₅H₇ON 3333
C₅H₇O₂N 3030
C₅H₇NCl₂ 5485
C₅H₈O₂N₂ 66
C₅H₈O₂Cl₂ 45
C₅H₈O₃N₂ 1678
C₅H₈O₃Cl₂ 1984
C₅H₉OCl 6404; 6412
C₅H₉O₂N 1674; 4890; 5335
C₅H₉O₂Cl 1077-8;   1445-7;
     3020-1
C₅H₉O₂Br 983-4; 1006-7; 2995-
     6
C₅H₉O₃N 3808-9
C₅H₉O₃Cl 2923
C₅H₉O₄N 3371-2
C₅H₉O₅N 3772
C₅H₉NS 1131-6
C₅H₁₀ON₂ 4898
C₅H₁₀OS₂ 2163-4
C₅H₁₀O₃N₂ 3374-5; 4886
C₅H₁₀N₂S₂ 1249
C₅H₁₁ON 2173; 4356; 4358;
     6244; 6401-2; 6408-9
C₅H₁₁ON₃ 1164; 4234
C₅H₁₁O₂N 387-9; 391-5; 473-4;
     846; 1067-8; 3184;
     4790-3
C₅H₁₁O₃N 472; 2922
C₅H₁₁O₄N 4789
C₅H₁₁NS 6275
C₅H₁₁NS₂ 2162
C₅H₁₁N₃Cl₂ 3668
C₅H₁₂ON₂ 1138-9;   2222-3;
     5845
C₅H₁₂O₂N₂ 5000
C₅H₁₂N₂S 2220
C₅H₁₃ON 4584
C₅H₁₃O₂N 243; 2444
C₅H₁₄N₂S₂ 2449
C₅H₁₅O₂N 1488

5 IV

C₅H₅O₃NS 5505
C₅H₆O₂N₂S 121
C₅H₈O₃NCl₃ 1288
C₅H₁₀O₄NCl 3373
C₅H₁₁O₂NS 4062
C₅H₁₂O₂NCl 390; 847; 2945
C₅H₁₄ONCl 1489

6 I

C₆H₆ 656; 2768
C₆H₈ 1599; 1600

C₆H₁₀ 1609; 2769; 3573-4
C₆H₁₂ 1601; 3655-61; 4209
C₆H₁₄ 3603-7
C₆O₆ 1604
C₆Cl₆ 3563
C₆Br₆ 3561
C₆I₆ 3589

6 II

C₆HCl₅ 5053
C₆HBr₅ 5049
C₆HI₅ 5071
C₆H₂Cl₄ 5733-5
C₆H₂Br₄ 5715-6
C₆H₂I₄ 5809-11
C₆H₃Cl₃ 6134-6
C₆H₃Br₃ 6095-7
C₆H₃I₃ 6233-5
C₆H₄O₂ 5541-2
C₆H₄O₄ 2364
C₆H₄O₆ 5808
C₆H₄Cl₂ 1952-4
C₆H₄Br₂ 1811-3
C₆H₄I₂ 2386-8
C₆H₅N₃ 252; 6079
C₆H₅F 3258
C₆H₅Cl 1324
C₆H₅Br 902
C₆H₅I 3879
C₆H₆O 5118
C₆H₆O₂ 2312-4; 4244
C₆H₆O₃ 150; 4246; 6216-9
C₆H₆O₄ 4443; 5802-3
C₆H₆O₆ 141; 3588
C₆H₆S 5915
C₆H₆S₂ 2820; 2822
C₆H₆Cl₆ 668-9
C₆H₆Br₆ 666
C₆H₆Se 5607
C₆H₇N 513; 5300-2
C₆H₇P 5219
C₆H₈O 2488
C₆H₈O₂ 2285; 5625; 5798
C₆H₈O₄ 209; 2487; 2501; 3945
C₆H₈O₆ 96; 586-7; 6123
C₆H₈O₇ 1515-6; 5578
C₆H₈N₂ 155; 2533; 5196; 5256;
     5258; 5260
C₆H₈N₄ 4425
C₆H₈S 5928-31
C₆H₉N 2535-8; 3149
C₆H₉N₃ 6059-60; 6062
C₆H₉N₁₁ 4018
C₆H₁₀O 191; 206; 1606; 4053;
     4225
C₆H₁₀O₂ 65; 216; 1564; 3024-6;
     3086
C₆H₁₀O₃ 2937; 3417; 3765-6;
     4296; 5357

C₆H₁₀O₄ 153; 2193;   2554-8;
     2940;   3154;   3214;
     3445;   3926;   4235;
     4251-2; 5422-3
C₆H₁₀O₅ 622; 1267; 1661; 2500;
     3443;   3872;   3948;
     3966; 3969; 5630
C₆H₁₀O₆ 2564-5; 3070; 3361;
     3368
C₆H₁₀O₈ 4438-40; 5577
C₆H₁₀N₄ 5079
C₆H₁₀S 1684
C₆H₁₀S₃ 222
C₆H₁₁N 453;   1217-8;   1680;
     2104
C₆H₁₁N₅ 5048
C₆H₁₁Cl 1613
C₆H₁₁Br 1612
C₆H₁₁I 1614
C₆H₁₂O 1212;   1605;   2099;
     3144-5; 3614-5; 4163-6
C₆H₁₂O₂ 454-5; 1029-32; 1208-
     11; 1663; 2101; 2481;
     3006-7; 4404; 4420-1;
     5350; 5441-2
C₆H₁₂O₃ 7; 1266; 3073; 3759-
     60; 5419-20
C₆H₁₂O₄ 2924; 3413
C₆H₁₂O₅ 1608;   3312;   4423;
     5558; 5561
C₆H₁₂O₆ 1602-3; 3616-20
C₆H₁₂O₇ 3359; 3946
C₆H₁₂N₂ 3681; 3932
C₆H₁₂N₄ 3599
C₆H₁₂S₃ 5879-81
C₆H₁₂Cl₂ 2001-2
C₆H₁₂Br₂ 1842-7
C₆H₁₂I₂ 2393
C₆H₁₃N 326; 4344-7
C₆H₁₃Cl 3644-7
C₆H₁₃Br 3641-3
C₆H₁₃I 3650-1
C₆H₁₄O 2783-5; 3000-2; 3624-
     38; 4120-1
C₆H₁₄O₂ 4; 1071; 2139; 2346;
     3598;   3662;   4332;
     5312-3
C₆H₁₄O₃ 1232; 2808; 3471
C₆H₁₄O₄ 6199
C₆H₁₄O₅ 2259; 5557
C₆H₁₄O₆ 3608-9; 3611
C₆H₁₄S 2799-800; 3005; 3652
C₆H₁₄S₂ 2782
C₆H₁₄Hg 4046
C₆H₁₄Zn 6505-6
C₆H₁₅N 2772-3; 3639-40; 6180
C₆H₁₅Al 230
C₆H₁₅As 6183
C₆H₁₅B 871

Table 7-2 (*Continued*)
FORMULA INDEX FOR ORGANIC COMPOUNDS

$C_6H_{15}Bi$ 855
$C_6H_{15}P$ 6189
$C_6H_{15}Sb$ 6198
$C_6H_{16}N_2$ 3596
$C_6H_{16}Si$ 6196
$C_6H_{18}N_4$ 6201
$C_6H_{26}O_8$ 5313
C_6OCl_6 3565-6
$C_6O_2Cl_4$ 5746
$C_6O_2Br_4$ 5726

6 III

C_6HOCl_5 5057
C_6HOBr_5 5051
$C_6HO_2Cl_3$ 6169
$C_6H_2OCl_4$ 5743
$C_6H_2O_2Cl_2$ 2070-1
$C_6H_2O_2Cl_4$ 5742
$C_6H_2O_8N_2$ 4586
$C_6H_2O_9N_4$ 5857
$C_6H_2NCl_5$ 5052
$C_6H_3OCl_3$ 6162-4
$C_6H_3OBr_3$ 6108
$C_6H_3OI_3$ 6238
$C_6H_3O_2Cl_3$ 6156
$C_6H_3O_2Br_3$ 6114
$C_6H_3O_6N_3$ 6292-4
$C_6H_3O_7N_3$ 6307-10
$C_6H_3O_8N_3$ 6312
$C_6H_3NCl_4$ 5730-2
$C_6H_3NBr_4$ 5714
$C_6H_4OCl_2$ 2042
$C_6H_4OBr_2$ 1859-62
$C_6H_4O_2Cl_2$ 2003-4
$C_6H_4O_3N_2$ 4896
$C_6H_4O_4N_2$ 2602-4
$C_6H_4O_5N_2$ 2646-7; 2649-52
$C_6H_4O_6N_2$ 2660-1
$C_6H_4O_6N_4$ 6289
$C_6H_4NCl_3$ 6130-3
$C_6H_4NBr_3$ 6089-93
$C_6H_4N_2Cl_2$ 5544
C_6H_4FCl 3263-5
C_6H_4FBr 3262
C_6H_4FI 3272
C_6H_4ClBr 1340-1
C_6H_4ClI 1387-9
C_6H_4BrI 945-7
C_6H_5ON 4876
C_6H_5OCl 1427-9
C_6H_5OBr 972-4
C_6H_5OI 3898-900; 3910
C_6H_5ONa 5119
$C_6H_5O_2N$ 4646; 5486; 5488; 5496; 5546
$C_6H_5O_2Cl$ 1386; 1460
$C_6H_5O_2Br$ 944
$C_6H_5O_2I$ 3912

$C_6H_5O_3N$ 3791-2; 4799; 4802-3; 4900
$C_6H_5O_3N_3$ 1767
$C_6H_5O_4N$ 4836
$C_6H_5O_4N_3$ 2587-92
$C_6H_5O_5N_3$ 5303
$C_6H_5O_6N_5$ 6311
$C_6H_5NCl_2$ 1932-7
$C_6H_5NBr_2$ 1798-802
$C_6H_5NI_2$ 2385
$C_6H_5N_2Cl$ 1765
$C_6H_5N_2Br_3$ 1768
C_6H_5ClHg 4037
$C_6H_5Cl_2P$ 5270
$C_6H_5Cl_3Si$ 5618
$C_6H_6ON_2$ 4875; 5493
C_6H_6OS 5909
$C_6H_6O_2N_2$ 4618-20; 5212; 5545
$C_6H_6O_2S$ 672
$C_6H_6O_2Si$ 5616
$C_6H_6O_3N_2$ 4610-6
$C_6H_6O_3N_4$ 4417-9
$C_6H_6O_3S$ 675
$C_6H_6O_4N_4$ 2657
$C_6H_6O_4S$ 5123-5
$C_6H_6O_5N_2$ 5490
$C_6H_6O_6S_2$ 664
$C_6H_6O_9S_3$ 686
C_6H_6NF 3256
C_6H_6NCl 1309-11
C_6H_6NBr 893-5
C_6H_6NI 3875-7
$C_6H_6N_2Cl_2$ 2044; 2048
$C_6H_6N_3Cl_9$ 1287
C_6H_7ON 113; 345; 347; 349; 4082; 5200-1
$C_6H_7ON_3$ 4897
$C_6H_7O_2N$ 372
$C_6H_7O_2N_3$ 4813; 4815
$C_6H_7O_3As$ 5158
$C_6H_7O_4As$ 3797
$C_6H_7N_2Cl$ 1433
$C_6H_7N_2Cl_3$ 2049
$C_6H_7N_2Br$ 975
$C_6H_8ON_2$ 1723; 1725-7
$C_6H_8O_2Cl_2$ 154
$C_6H_8O_3N_2$ 521
$C_6H_8O_5Na_2$ 2268
$C_6H_8O_{18}N_6$ 4011
C_6H_8NCl 520
C_6H_8NBr 519
$C_6H_9ON_3$ 6068
$C_6H_9O_2N_3$ 1574; 3669
$C_6H_9O_2Cl_3$ 6144
$C_6H_9O_3N$ 6055
$C_6H_9O_3N_3$ 4197; 5267
$C_6H_9O_3Cl$ 3017-8
$C_6H_9O_3Br$ 2991
$C_6H_9O_4N$ 88

$C_6H_9N_2Cl$ 5197
$C_6H_9N_2Cl_3$ 1434
$C_6H_{10}O_2Br_2$ 3037
$C_6H_{10}O_2S_2$ 2165
$C_6H_{10}O_6Ca$ 3941
$C_6H_{10}N_2Cl_2$ 5257; 5259; 5261
$C_6H_{11}ON$ 1607
$C_6H_{11}OCl$ 1216
$C_6H_{11}O_2N$ 2956
$C_6H_{11}O_2Cl$ 449-50; 1076
$C_6H_{11}O_2Cl_3$ 6124-5
$C_6H_{11}O_2Br$ 920; 2992-3
$C_6H_{11}O_3N$ 1111
$C_6H_{11}O_4N_3$ 1517; 2262
$C_6H_{11}NS$ 490-3
$C_6H_{11}N_3Cl_2$ 6061
$C_6H_{11}N_3Cl_4$ 6067
$C_6H_{12}OCl_2$ 1991-2
$C_6H_{12}O_2N_2$ 152; 2194
$C_6H_{12}O_2Cl_2$ 1925; 6207
$C_6H_{12}O_3N_2$ 3477
$C_6H_{12}O_4N_2$ 2631
$C_6H_{12}N_2S_4$ 5843
$C_6H_{12}N_3Cl_3$ 6063
$C_6H_{13}ON$ 1213; 1664; 2100
$C_6H_{13}ON_3$ 2181; 4359
$C_6H_{13}O_2N$ 282-7; 439-41; 2236; 3496; 3653; 4749-51; 5454
$C_6H_{13}O_2Cl$ 1292
$C_6H_{13}O_3N_3$ 1521
$C_6H_{13}O_5N$ 3311
$C_6H_{13}O_6N$ 3369
$C_6H_{13}NS_2$ 5875
$C_6H_{14}ON_2$ 494-5; 2790-1
$C_6H_{14}O_2N_2$ 3986; 3988
$C_6H_{14}O_2N_4$ 575
$C_6H_{14}O_2S$ 2802-3
$C_6H_{14}O_3S$ 2801
$C_6H_{14}O_4S$ 2798
$C_6H_{15}ON$ 2112
$C_6H_{15}OP$ 6190
$C_6H_{15}O_2N$ 2767
$C_6H_{15}O_3N$ 6178
$C_6H_{15}O_3Al$ 229
$C_6H_{15}O_3B$ 6186
$C_6H_{15}O_3P$ 6192
$C_6H_{15}O_4P$ 6188
$C_6H_{15}SP$ 6191
$C_6H_{16}OSi$ 6194
$C_6H_{16}NCl$ 6182
$C_6H_{16}NBr$ 6181

6 IV

$C_6H_2ONCl_3$ 2072
$C_6H_2O_6N_3Cl$ 5308
$C_6H_3O_2NCl_2$ 2022-6
$C_6H_3O_2SCl_3$ 1956
$C_6H_3O_3NCl_2$ 2029-30

Table 7-2 (*Continued*)
FORMULA INDEX FOR ORGANIC COMPOUNDS

$C_6H_3O_3NBr_2$ 1855
$C_6H_3O_4N_2Cl$ 1364-9
$C_6H_3O_4N_2Br$ 930-1
$C_6H_3O_5N_2Cl$ 1372
$C_6H_3O_5N_2Na$ 2648
C_6H_4ONCl 5543
$C_6H_4O_2NF$ 3274
$C_6H_4O_2NCl$ 1412-4
$C_6H_4O_2NBr$ 960-2
$C_6H_4O_2NI$ 3893-5
$C_6H_4O_2NNa$ 5547
$C_6H_4O_2N_2Cl_2$ 2021
$C_6H_4O_3NCl$ 4698-706
$C_6H_4O_3NK$ 4801
$C_6H_4O_3NNa$ 4800; 4804-5
$C_6H_4O_3N_2S$ 1769-71
$C_6H_4O_3SCl_2$ 1955
$C_6H_4O_4SI_2$ 5626
$C_6H_4O_4S_2Cl_2$ 663
$C_6H_4O_5N_3Na$ 5304
$C_6H_4O_6S_2Na_2$ 665
$C_6H_4O_7N_2S$ 2605
$C_6H_5ONBr_2$ 1797
C_6H_5OClHg 4034-5
$C_6H_5O_2N_2Cl$ 1407-10
$C_6H_5O_2SCl$ 678
$C_6H_5O_2SNa$ 673
$C_6H_5O_3NHg$ 4038
$C_6H_5O_3SCl$ 1326
$C_6H_5O_3SBr$ 903-4
$C_6H_5O_3SNa$ 676
$C_6H_5O_4SCl$ 1430
$C_6H_5O_4SNa$ 5126
$C_6H_5O_5NS$ 4654-5
$C_6H_5O_6NS$ 4806-7
$C_6H_5O_6S_2Cl$ 1325
$C_6H_6O_2NCl$ 5487; 5489
$C_6H_6O_2N_2Cl_2$ 1408; 1411
$C_6H_6O_4AsNa$ 3798
$C_6H_6O_5NAs$ 4810
$C_6H_6O_5N_2S$ 4621-2
$C_6H_7O_2NS$ 677
$C_6H_7O_3NS$ 523-5; 674
$C_6H_7O_4NS$ 353-5
$C_6H_7O_6NS_2$ 517-8
$C_6H_7NSCl_2$ 1308
C_6H_7NClBr 896
C_6H_8ONCl 346; 348; 350
$C_6H_8O_2NCl$ 373
$C_6H_8O_2N_2S$ 527
$C_6H_8O_2N_3Cl$ 4814; 4816
$C_6H_8O_3NAs$ 515
$C_6H_8O_3N_2S$ 5198; 5262
$C_6H_8N_2ClBr$ 976
$C_6H_{10}ON_2Cl_2$ 1724
$C_6H_{10}O_2N_3Cl$ 3671
$C_6H_{10}O_4N_2Cl_2$ 5804
$C_6H_{11}O_2N_2Br$ 1009
$C_6H_{11}O_2N_3Cl_2$ 3670

$C_6H_{12}ONBr$ 2131
$C_6H_{12}O_4N_2S_2$ 1627
$C_6H_{15}O_2N_2Cl$ 2865; 3990
$C_6H_{15}O_2N_4Cl$ 576
$C_6H_{16}O_2N_2Cl_2$ 3987; 3989
$C_6H_{16}O_3NCl$ 6179
$C_6H_{20}O_4N_6S$ 2491

6 V

$C_6H_2ONClBr_2$ 1876
$C_6H_2O_4SI_2Hg$ 5628
$C_6H_3O_4SI_2Na$ 5627
$C_6H_3O_7N_2SNa$ 2606
$C_6H_4O_2SClBr$ 905
$C_6H_4O_4NSCl$ 4657
$C_6H_4O_4SClNa$ 1431
$C_6H_4O_5NSCl$ 1415
$C_6H_4O_5NSBr$ 963
$C_6H_4O_5NSNa$ 4656
$C_6H_5O_3NSCl_2$ 1938
$C_6H_6O_3NSNa$ 526
$C_6H_6O_3N_2SCl_2$ 2045
$C_6H_7O_2NClAs$ 4013
$C_6H_{10}O_3N_2SCl_2$ 5263

7 I

C_7H_8 5960
C_7H_{10} 2286-8
C_7H_{12} 1598; 3510; 3558; 4207-8
C_7H_{14} 1596; 3554-6; 4199
C_7H_{16} 3512-9

7 II

$C_7H_4O_7$ 4016
C_7H_5N 716; 5178
$C_7H_5N_3$ 1766
$C_7H_5Cl_3$ 735
C_7H_6O 642
$C_7H_6O_2$ 711; 3347; 3733-5; 6001
$C_7H_6O_3$ 761; 2310-1; 3350-1; 3742; 3745-6
$C_7H_6O_4$ 2315-20
$C_7H_6O_5$ 6220-3
$C_7H_6N_2$ 267-9; 708; 1578
C_7H_6S 5888-9
$C_7H_6Cl_2$ 635; 1339
$C_7H_6Br_2$ 634; 912; 1880
$C_7H_7N_3$ 263; 788
C_7H_7F 3279-81
C_7H_7Cl 796; 1466-8
C_7H_7Br 793; 999-1001
C_7H_7I 813; 3905-7
C_7H_8O 537; 782; 1545-7
$C_7H_8O_2$ 2370-4; 3750-2; 4079-81; 4998
$C_7H_8O_3$ 3050-1; 3331; 4361; 6396
$C_7H_8N_2$ 649

C_7H_8S 816; 5894-6
C_7H_9N 783; 3146-8; 3981-4; 4125; 5995; 5997; 5999
$C_7H_{10}O_2$ 5772
$C_7H_{10}O_3$ 1668
$C_7H_{10}O_4$ 2469; 2498; 2504; 2939; 5700-1
$C_7H_{10}O_5$ 58; 2195; 3084
$C_7H_{10}N_2$ 808; 1682; 4336; 4338; 4340; 5311; 6003; 6005; 6007; 6022; 6024; 6026
$C_7H_{11}N_3$ 6069
$C_7H_{12}O$ 1597; 4204-6
$C_7H_{12}O_2$ 200; 3042; 3577
$C_7H_{12}O_3$ 3081; 3085; 3587; 5367; 5774
$C_7H_{12}O_4$ 1102; 2184-5; 2489; 3054-5; 4097-9; 4237; 5309-10; 5449; 5466; 6281
$C_7H_{12}O_5$ 3383
$C_7H_{12}O_6$ 2160; 5536
$C_7H_{13}N$ 3648
$C_7H_{14}O$ 2786-7; 3003-4; 3521; 3557; 4122-3; 4200-3
$C_7H_{14}O_2$ 397-403; 1123-5; 3185-6; 3520; 3649; 4173; 5392-5
$C_7H_{14}O_3$ 1099-1100; 2780; 3793
$C_7H_{14}O_4$ 2926; 3392; 4178
$C_7H_{14}O_5$ 2411
$C_7H_{14}O_6$ 4250; 4286; 4297
$C_7H_{14}O_7$ 3358; 4012
$C_7H_{15}Cl$ 3546-7
$C_7H_{15}Br$ 3545
$C_7H_{15}I$ 3549
$C_7H_{16}O$ 2961-2; 3527-43
$C_7H_{16}O_2$ 3511; 5458
$C_7H_{16}O_3$ 3049; 3391
$C_7H_{16}O_7$ 5090
$C_7H_{16}S$ 3525
$C_7H_{17}N$ 3544

7 III

$C_7H_3O_2Cl_3$ 6137-41
$C_7H_3O_2Br_3$ 6098-9
$C_7H_3O_2I_3$ 6236
$C_7H_3O_7N_3$ 6291
$C_7H_3O_8N_3$ 6295
$C_7H_4OCl_2$ 1334-6
$C_7H_4OBr_2$ 910
$C_7H_4OBr_4$ 5717
$C_7H_4O_2N_2$ 4667-9
$C_7H_4O_2Cl_2$ 1959-64
$C_7H_4O_2Br_2$ 1814-9
$C_7H_4O_3N_2$ 1773; 4811
$C_7H_4O_3I_2$ 2394-5
$C_7H_4O_4S$ 5679

Table 7-2 (*Continued*)
FORMULA INDEX FOR ORGANIC COMPOUNDS

$C_7H_4O_5N_2$ 2600-1
$C_7H_4O_5Br_2$ 1841
$C_7H_4O_6N_2$ 2607-11
$C_7H_4O_7N_2$ 2662
C_7H_4NCl 1330
C_7H_4NBr 909
C_7H_5ON 544; 3748; 5177
$C_7H_5ON_3$ 653
C_7H_5OF 757
C_7H_5OCl 755; 1318-20
$C_7H_5OCl_3$ 6146-9
C_7H_5OBr 753
$C_7H_5OBr_3$ 6094
C_7H_5OI 762
$C_7H_5O_2N_3$ 4662
$C_7H_5O_2F$ 3259-61
$C_7H_5O_2Cl$ 1327-9; 1461
$C_7H_5O_2Br$ 906-8
$C_7H_5O_2I$ 3880-2
$C_7H_5O_2Na$ 712
$C_7H_5O_3N$ 4637; 4639-40; 4877-9
$C_7H_5O_3Br$ 993
$C_7H_5O_3I$ 3904; 3911
$C_7H_5O_3Na$ 3744
$C_7H_5O_4N$ 4663-4; 4666; 5497-502
$C_7H_5O_5N$ 4753-60
$C_7H_5O_6N_3$ 2594; 6313-7
$C_7H_5O_7N_3$ 6290; 6298
$C_7H_5O_8N_3$ 6306
$C_7H_5O_8N_5$ 5873
C_7H_5NS 5241
$C_7H_5NS_2$ 4033
$C_7H_5NCl_2$ 5168
$C_7H_6ON_2$ 709
$C_7H_6OBr_2$ 1831
C_7H_6OS 5891
$C_7H_6OS_2$ 2823
$C_7H_6O_2S$ 5918
$C_7H_6O_3N_2$ 4642-4
$C_7H_6O_4N_2$ 2666-71; 4604-9
$C_7H_6O_5N_2$ 2593; 2616-7
$C_7H_6O_5S$ 5676-8
$C_7H_6O_6S$ 5681
$C_7H_6N_2S$ 273
C_7H_6ClBr 913-4; 1337-8
C_7H_7ON 254-6; 643-4; 713; 3299; 4903-5
C_7H_7OF 3257
C_7H_7OCl 1314; 1354-7
C_7H_7OBr 897-8
C_7H_7OI 3878
$C_7H_7O_2N$ 264-6; 710; 3348-9; 3736; 3738-40; 4844-6; 4881; 5167; 5213
$C_7H_7O_2N_3$ 4739
$C_7H_7O_3N$ 374-6; 4623-5; 4678-80; 4710

$C_7H_7O_3N_3$ 4658-60
$C_7H_7O_4N$ 3352; 4742
$C_7H_7O_5As$ 651
$C_7H_7O_5P$ 730
C_7H_7NS 5902
$C_7H_7N_3S$ 734
C_7H_7ClHg 4039
$C_7H_8ON_2$ 257-61; 688; 760; 3308; 4887; 5253
$C_7H_8O_2N_2$ 1700-5; 3684; 4767-9; 4849-58
$C_7H_8O_2S$ 5964
$C_7H_8O_3N$ 352
$C_7H_8O_3N_2$ 4601-3
$C_7H_8O_3N_4$ 1250
$C_7H_8O_3S$ 5966-8
$C_7H_8O_6S_2$ 5962
C_7H_8NCl 1469-70
C_7H_8NBr 1002-3
$C_7H_8N_2S$ 5244
C_7H_9ON 292-3; 295-9; 534-6; 810-1; 4115; 6028-30
$C_7H_9ON_3$ 5236-7
$C_7H_9O_2N_3$ 6064-6
$C_7H_9O_3N$ 1584
$C_7H_9O_3As$ 787
$C_7H_9N_3S$ 5243
$C_7H_{10}ON_2$ 1687
$C_7H_{10}O_4Br_2$ 2155
$C_7H_{10}NCl$ 5996; 5998; 6000
$C_7H_{11}O_4Br$ 2133
$C_7H_{11}O_6N$ 4763
$C_7H_{11}N_2Cl$ 6023; 6025; 6027
$C_7H_{12}O_3Cl_2$ 2057
$C_7H_{12}N_2Cl_2$ 4337; 6004; 6006; 6008
$C_7H_{13}ON$ 112
$C_7H_{13}O_2Br$ 2997-8
$C_7H_{14}N_3Cl_3$ 6070
$C_7H_{15}ON$ 2931; 3522-3; 4124
$C_7H_{15}ON_3$ 5314
$C_7H_{15}O_2N$ 1140-1; 3551; 4744-8
$C_7H_{16}ON_2$ 2806-7
$C_7H_{16}O_4S_2$ 5683
$C_7H_{17}ON$ 2117-8
$C_7H_{17}O_2N$ 2113

7 IV

$C_7H_3O_5N_2Cl$ 2613
$C_7H_3O_6N_2Cl$ 1370-1
$C_7H_3O_6N_2Br$ 932
C_7H_4ONCl 1333
C_7H_4OClBr 911
C_7H_4OClI 3883
$C_7H_4O_3NCl$ 4673-5
$C_7H_4O_4NBr$ 964
$C_7H_4O_4NNa$ 4665
C_7H_4NSCl 1332

$C_7H_5O_2NCl_2$ 4635-6
$C_7H_5O_2NBr_2$ 1804; 4634
$C_7H_5O_2NI_2$ 2383
$C_7H_5O_3NS$ 5579
$C_7H_5O_3NHg$ 4057
$C_7H_5O_4N_2Cl$ 2615
$C_7H_5O_6SNa$ 5682
C_7H_6ONCl 1321-3
C_7H_6ONNa 3300
$C_7H_6O_2NCl$ 1315; 4682-4
$C_7H_6O_2NBr$ 4681
$C_7H_6O_6S_2Na_2$ 5963
$C_7H_7O_2SCl$ 5974-5
$C_7H_7O_2SNa$ 5965
$C_7H_7O_3SNa$ 5969
$C_7H_7O_4NS$ 5671-3
$C_7H_7O_5NS$ 4847
C_7H_8ONCl 1312
$C_7H_9ONCl_2$ 1313
$C_7H_9O_2NS$ 5970-1
$C_7H_9O_3NS$ 379-83
$C_7H_9O_4N_2As$ 1230
$C_7H_{10}O_2N_2Cl_2$ 1705
$C_7H_{10}O_2N_4S$ 5666
$C_7H_{12}ON_2Cl_2$ 1688-9
$C_7H_{12}O_2N_3Cl_3$ 6066
$C_7H_{12}O_4N_2S$ 6009
$C_7H_{14}ONBr$ 4580
$C_7H_{14}O_4N_2S_2$ 2854
$C_7H_{16}ONBr$ 6264
$C_7H_{16}O_2NCl$ 83
$C_7H_{16}O_2NBr$ 82

7 V

$C_7H_4O_3NSNa$ 5580
$C_7H_6O_4NSCl$ 4848
$C_7H_7O_2NSCl_2$ 2074

7 VI

$C_7H_7O_2NSClNa$ 1290

8 I

C_8H_6 5146
C_8H_8 5646
C_8H_{10} 2973; 6454; 6457; 6462
C_8H_{12} 2289-91
C_8H_{14} 1530; 2493; 4991
C_8H_{16} 1617; 1923; 2471-3; 4984-90
C_8H_{18} 4944-58

8 II

$C_8H_4O_3$ 5282
$C_8H_4N_2$ 5288-9
$C_8H_5Cl_5$ 5055
C_8H_6O 1538
$C_8H_6O_2$ 5278-80; 5290
$C_8H_6O_3$ 162-4; 758; 5323
$C_8H_6O_4$ 165-8; 5274; 5276-7; 5327

Table 7-2 (*Continued*)
FORMULA INDEX FOR ORGANIC COMPOUNDS

Table 7-2 (*Continued*)
FORMULA INDEX FOR ORGANIC COMPOUNDS

$C_8H_{11}O_4Br$ 2132
$C_8H_{11}O_6Cl_3$ 1289
$C_8H_{11}N_3S$ 6044
$C_8H_{12}O_3N_2$ 2124
$C_8H_{12}O_4Br_2$ 2156
$C_8H_{14}O_5N_4$ 6208
$C_8H_{15}OCl$ 1225
$C_8H_{15}O_2N$ 79; 4889
$C_8H_{15}O_2Br$ 2994
$C_8H_{15}O_5N$ 1665
$C_8H_{16}O_3N_2$ 3965
$C_8H_{16}O_3Cl_2$ 5771
$C_8H_{16}O_4N_2$ 3215
$C_8H_{17}ON$ 1223
$C_8H_{17}O_2N$ 288; 4437; 4786-8; 4982
$C_8H_{17}O_2Br$ 1026
$C_8H_{17}O_3N$ 4981
$C_8H_{18}O_2S$ 1918
$C_8H_{18}O_3S$ 1917
$C_8H_{18}O_4S$ 1913
$C_8H_{18}O_4S_2$ 6326
$C_8H_{19}O_2N$ 1084
$C_8H_{20}OSi$ 6195
$C_8H_{20}O_4Si$ 3152
$C_8H_{20}NCl$ 5755
$C_8H_{20}NBr$ 5754
$C_8H_{20}NI$ 5757
$C_8H_{21}ON$ 5756
$C_8H_{21}O_5N$ 5753

8 IV

C_8H_4ONCl 3920
$C_8H_4O_2N_2Cl_2$ 1967
$C_8H_4O_2N_2Br_2$ 1820
$C_8H_5ON_2Cl$ 1394
$C_8H_6ONCl_3$ 6127
$C_8H_6ONBr_3$ 6087
$C_8H_6O_2N_2S$ 3325
$C_8H_7O_2NBr_2$ 4212
C_8H_8ONCl 1296-8
C_8H_8ONBr 880-2
C_8H_8ONI 3873
C_8H_8ONNa 5621
$C_8H_8ON_2S$ 774
$C_8H_9O_2SCl$ 6461
$C_8H_9O_3SNa$ 6456; 6459; 6464
$C_8H_{10}ONCl$ 240
$C_8H_{10}O_4NAs$ 577
$C_8H_{10}O_5NAs$ 128
$C_8H_{11}ON_2Cl$ 109
$C_8H_{11}O_2NS$ 5979
$C_8H_{11}O_3NS$ 2451-2; 2966
$C_8H_{11}O_3N_2Na$ 2125
$C_8H_{11}O_4N_2As$ 582
$C_8H_{11}N_2SCl$ 841
$C_8H_{12}ONCl$ 5109
$C_8H_{12}O_2N_2Cl_2$ 1728
$C_8H_{12}O_3NCl$ 5511

$C_8H_{14}O_4N_2S$ 308
$C_8H_{18}O_2NCl$ 100
$C_8H_{18}O_2NBr$ 99
$C_8H_{18}O_2NI$ 101

8 V

$C_8H_8O_4NSNa$ 21
$C_8H_9O_4NAsNa$ 578
$C_8H_9O_4N_2AsNa_2$ 583
$C_8H_{10}O_3NSNa$ 2967

9 I

C_9H_8 3855
C_9H_{10} 3695; 5223-5
C_9H_{12} 3166-8; 5378-9; 6253-5
C_9H_{16} 2492; 4325
C_9H_{18} 3578; 3581; 4930
C_9H_{20} 4910-7

9 II

$C_9H_6O_2$ 1490; 1534-5; 3854; 5220
$C_9H_6O_3$ 6369
$C_9H_6O_4$ 1628; 2898; 6239
$C_9H_6O_5$ 5291
$C_9H_6O_6$ 683-5
$C_9H_6O_7$ 5128
C_9H_7N 5538; 5540
C_9H_8O 1501; 3698-9
$C_9H_8O_2$ 72; 591; 1497-8; 1500
$C_9H_8O_3$ 69; 70-1; 738; 1531-3; 1537
$C_9H_8O_4$ 73; 115; 759; 2333-5; 3674; 6397
$C_9H_8O_5$ 5583
$C_9H_8N_2$ 365-71
C_9H_9N 2284; 3709; 4282-5
$C_9H_{10}O$ 211; 214; 1507; 3129; 3706; 4089; 4146; 6431-3
$C_9H_{10}O_2$ 781; 1552-3; 2458-63; 2914; 2975-8; 3683; 3704; 3810-1; 4065; 4335; 4392-4; 5132; 5221; 6010-2; 6441
$C_9H_{10}O_3$ 2405; 2915-7; 3053; 3072; 3151; 3188; 3481; 3711-3; 3776; 4300-1; 6361; 6424
$C_9H_{10}O_4$ 3052; 3230; 3466; 4078; 6423
$C_9H_{10}O_5$ 3321; 5686
$C_9H_{11}N$ 196; 5797
$C_9H_{11}Br$ 953; 985-6; 5222
$C_9H_{12}O$ 2526; 2986; 3122-3; 3173-5; 3705; 5430-5; 5438; 6265-6
$C_9H_{12}O_2$ 3449; 4054

$C_9H_{12}O_3$ 147; 1090; 3334; 3410; 5266; 5517
$C_9H_{12}N_2$ 62
$C_9H_{12}S$ 3176
$C_9H_{13}N$ 654; 2465; 2569-71; 3170-2; 4228; 5040-1; 5374-5; 6250-1
$C_9H_{14}O$ 5268-9
$C_9H_{14}O_3$ 2948
$C_9H_{14}O_4$ 2142; 2175; 2179; 2186
$C_9H_{14}O_5$ 87; 2103
$C_9H_{14}O_6$ 1201-2; 3418
$C_9H_{14}O_7$ 6263
$C_9H_{14}N_2$ 597
$C_9H_{16}O_2$ 201
$C_9H_{16}O_3$ 1101; 3139-40; 5777
$C_9H_{16}O_4$ 596; 2161; 2169; 2176; 2789
$C_9H_{16}O_6$ 60
$C_9H_{17}N$ 1633; 5047
$C_9H_{18}O$ 1900-1; 4265; 5043
$C_9H_{18}O_2$ 432-5; 1142-5; 3108; 3526; 4174; 4979; 5042; 5396
$C_9H_{18}O_3$ 1892-4
$C_9H_{18}O_5$ 2094
$C_9H_{18}N_2$ 1895
$C_9H_{19}N$ 4114
$C_9H_{19}Cl$ 4928
$C_9H_{19}Br$ 4927
$C_9H_{19}I$ 4929
$C_9H_{20}O$ 4918-25
$C_9H_{20}O_3$ 6193
$C_9H_{20}O_4$ 3012
$C_9H_{21}N$ 4926; 6348

9 III

$C_9H_5O_2Cl$ 1353
$C_9H_5O_4N$ 4817-8
$C_9H_5NCl_2$ 2063-9
$C_9H_6O_2N_2$ 4831-5
$C_9H_6O_2Br_2$ 1829
C_9H_6NCl 1453-9
C_9H_7ON 742; 3820-6
C_9H_7OCl 1505
$C_9H_7O_2N$ 4291-3
$C_9H_7O_2Br$ 921-4
$C_9H_7O_2Na$ 1499
$C_9H_7O_3N$ 3871
$C_9H_7O_4N$ 4707-9
$C_9H_7O_4Br$ 76
$C_9H_7O_4I$ 93-4
$C_9H_7O_4Li$ 117
$C_9H_7O_4Na$ 119
$C_9H_8ON_2$ 1582
$C_9H_8O_2N_2$ 4141-2
$C_9H_8O_2Br_2$ 1830
$C_9H_8O_3I_2$ 3041

Table 7-2 (*Continued*)
FORMULA INDEX FOR ORGANIC COMPOUNDS

$C_9H_8O_5N_2$ 4752
$C_9H_8O_6N_2$ 3043
C_9H_9ON 1502; 3703
C_9H_9OCl 3708
C_9H_9OBr 4333
$C_9H_9O_2N$ 289-91; 4899; 5586
$C_9H_9O_2Cl$ 797
$C_9H_9O_3N$ 125-7; 3664
$C_9H_9O_4N$ 2539; 3099-101; 4677; 5195
$C_9H_9O_6N_3$ 6318-20
$C_9H_9O_7N_3$ 2672
$C_9H_{10}O_3N_2$ 4597; 4766
$C_9H_{10}O_4N_2$ 2635; 4591
$C_9H_{10}O_5N_2$ 4860
$C_9H_{11}ON$ 26-8; 780; 2435-6; 2457; 2972; 3305; 3707; 4086; 5358
$C_9H_{11}OBr$ 987
$C_9H_{11}O_2N$ 18-20; 97-8; 2440-1; 2953-5; 4302; 4711-3; 4764; 5148-50; 5254
$C_9H_{11}O_3N$ 6367
$C_9H_{11}O_4N$ 2361
$C_9H_{12}ON_2$ 330; 3134
$C_9H_{12}O_2N_2$ 5112
$C_9H_{12}O_3S$ 3169
$C_9H_{13}O_2N$ 2882
$C_9H_{13}O_3N$ 157; 5512
$C_9H_{14}O_3N_4$ 1251
$C_9H_{14}NI$ 6267
$C_9H_{16}O_2N_2$ 217; 4906
$C_9H_{17}ON$ 2107; 6056
$C_9H_{17}OCl$ 5046
$C_9H_{17}O_5N$ 5033-4
$C_9H_{17}N_4I$ 3600
$C_9H_{19}ON$ 5044-5; 6414
$C_9H_{19}O_2N$ 4784-5
$C_9H_{20}ON_2$ 1922; 5767
$C_9H_{20}O_4S_2$ 5872
$C_9H_{20}N_2S$ 1921
$C_9H_{21}O_3N$ 6347
$C_9H_{22}N_2S_2$ 2120

9 IV

$C_9H_5ONBr_2$ 1848
$C_9H_7O_4NS$ 3828
$C_9H_8O_3NNa$ 3666
$C_9H_9O_2N_3S_2$ 5670
$C_9H_9O_2NCl_2$ 2077
$C_9H_9O_3NBr_2$ 1881
$C_9H_9O_3NI_2$ 2401-2
$C_9H_{10}ONBr$ 889
$C_9H_{10}O_2N_2S$ 5242
$C_9H_{11}O_3SCl$ 1383
$C_9H_{13}O_2NS$ 5977-8
$C_9H_{14}O_2NCl$ 4582
$C_9H_{14}O_3NCl$ 1523

$C_9H_{15}O_2N_3S$ 2890
$C_9H_{15}O_3N_2Br$ 84
$C_9H_{24}ON_2I_2$ 3591

9 V

C_9H_5ONClI 1392
$C_9H_6O_4NSI$ 3888
$C_9H_9O_2SHgNa$ 4052

10 I

$C_{10}H_8$ 4458
$C_{10}H_{10}$ 2278; 3120
$C_{10}H_{12}$ 2089; 5786
$C_{10}H_{14}$ 1049-52; 1622-4; 2126-8; 2483-6; 3582; 4351-2; 5827-9
$C_{10}H_{16}$ 870; 1189; 3232; 3972-3; 5092-3; 5318; 5576; 5684; 5704-5; 5709
$C_{10}H_{18}$ 1188; 1258; 1630-2; 2514-5; 4023; 5317
$C_{10}H_{20}$ 1650-3; 3579
$C_{10}H_{22}$ 1635-8

10 II

$C_{10}H_5Cl_3$ 6160-1
$C_{10}H_6O_2$ 4520-1; 4523
$C_{10}H_6O_3$ 3786-9
$C_{10}H_6O_4$ 1536; 3328; 4477-8
$C_{10}H_6O_5$ 3343
$C_{10}H_6O_8$ 680-2
$C_{10}H_6Cl_2$ 2009-18
$C_{10}H_6Br_2$ 1852
$C_{10}H_7F$ 3273
$C_{10}H_7Cl$ 1402-3
$C_{10}H_7Br$ 956-7
$C_{10}H_7I$ 3891-2
$C_{10}H_8O$ 4489-90
$C_{10}H_8O_2$ 2348-57; 4192
$C_{10}H_8O_3$ 743; 4406
$C_{10}H_8O_4$ 510; 638; 1504; 3336; 3346; 5604
$C_{10}H_8N_2$ 2810-3; 6051-3
$C_{10}H_8S$ 5913-4
$C_{10}H_8Cl_4$ 4473
$C_{10}H_9N$ 4367; 4369-70; 4372; 4374-5; 4529-30; 5229
$C_{10}H_{10}O$ 626
$C_{10}H_{10}O_2$ 198; 741; 4191; 5175-6; 5581-2
$C_{10}H_{10}O_3$ 637; 771; 2980; 4139; 5094
$C_{10}H_{10}O_4$ 763; 815; 2530-2; 3067; 3236-7; 3716; 4017; 4090; 5239; 5552
$C_{10}H_{10}O_5$ 837; 4993
$C_{10}H_{10}O_6$ 3502
$C_{10}H_{10}N_2$ 2809; 4563; 4565; 4571-9

$C_{10}H_{11}Cl$ 4188
$C_{10}H_{11}Br$ 4152
$C_{10}H_{12}O$ 511; 2901; 2987; **5377**; 5439-40; 5787-90
$C_{10}H_{12}O_2$ 829; 1280; 3118-9; 3163-5; 3218; 3223; 4126; 5185; 5380-5; 5831; 5943; 6256-61; 6481
$C_{10}H_{12}O_3$ 814; 1529; 2968; 5412; 5415; 5446-7; 6419
$C_{10}H_{12}O_4$ 3187; 6226-7
$C_{10}H_{12}O_5$ 3414; 6240-2
$C_{10}H_{12}N_2$ 6363
$C_{10}H_{13}N$ 3930; 5792-5
$C_{10}H_{13}Cl$ 1572
$C_{10}H_{13}Br$ 915; 926
$C_{10}H_{14}O$ 1116; 1119-20; 1255; 1261; 1571; 2466; 5436-7; 5938
$C_{10}H_{14}O_2$ 1070; 2091-3; 5937
$C_{10}H_{14}O_3$ 457; 1200
$C_{10}H_{14}O_4$ 2188; 3400
$C_{10}H_{15}N$ 1046-7; 1256; 2121; 5818-20; 5946
$C_{10}H_{16}O$ 1191; 1257; 1512; 2275; 3233; 5320; 5472; 5935-6
$C_{10}H_{16}O_2$ 585
$C_{10}H_{16}O_4$ 1197-9; 2108; 2788
$C_{10}H_{16}O_5$ 2105
$C_{10}H_{16}O_8$ 1673
$C_{10}H_{16}N_2$ 303; 5606; 5839
$C_{10}H_{16}Cl_2$ 1192-4
$C_{10}H_{17}Cl$ 864; 5319
$C_{10}H_{18}O$ 860; 1260; 1518; 2274; 2516-8; 3217; 3234-5; 3974; 4026; 5471; 5706-8
$C_{10}H_{18}O_2$ 1190
$C_{10}H_{18}O_3$ 2102; 6403; 6410
$C_{10}H_{18}O_4$ 1905-6; 2106; 2187; 2203-4; 2796-7; 3450; 3550; 5605
$C_{10}H_{18}O_6$ 2804-5; 6200
$C_{10}H_{19}N$ 862; 1203-4; 1207
$C_{10}H_{20}O$ 1206; 1259; 1519; 4024-5; 4326; 5414
$C_{10}H_{20}O_2$ 496-7; 1065; 1205; 1219; 3010; 3553; 4331; 4959-60; 5703
$C_{10}H_{20}O_4$ 1069
$C_{10}H_{21}N$ 2145; 4029
$C_{10}H_{21}Cl$ 1646
$C_{10}H_{21}I$ 1647
$C_{10}H_{22}O$ 1640-4; 1740-1; 2519-20; 3106
$C_{10}H_{22}O_2$ 1634; 1654

Table 7-2 (*Continued*)
FORMULA INDEX FOR ORGANIC COMPOUNDS

$C_{10}H_{22}O_3$ 5410; 5703
$C_{10}H_{22}O_5$ 5769
$C_{10}H_{22}S$ 1751-2
$C_{10}H_{22}S_2$ 1753-4
$C_{10}H_{23}N$ 1645; 1734-6

10 III

$C_{10}H_4O_2Cl_2$ 2020
$C_{10}H_4O_8N_4$ 5854-6
$C_{10}H_5OCl_5$ 5056
$C_{10}H_5OBr_3$ 6107
$C_{10}H_5O_6N_3$ 6301-4
$C_{10}H_5O_7N_3$ 6305
$C_{10}H_5O_{10}N$ 5503-4
$C_{10}H_6OCl_2$ 2019
$C_{10}H_6OBr_2$ 1853
$C_{10}H_6O_2N_2$ 5514
$C_{10}H_6O_3S$ 4524
$C_{10}H_6O_4N_2$ 2637-40
$C_{10}H_6O_4Cl_4$ 3068
$C_{10}H_6O_5N_2$ 2641-3
$C_{10}H_6NBr_3$ 6113
$C_{10}H_7OCl$ 1404-6
$C_{10}H_7OBr$ 958-9
$C_{10}H_7O_2N$ 4772-3; 4892-4; 5533
$C_{10}H_7O_3N$ 95; 3937; 4775-9
$C_{10}H_8ON_2$ 4895
$C_{10}H_8O_2N_2$ 4780-3; 4830
$C_{10}H_8O_2S$ 4465-6
$C_{10}H_8O_3S$ 4470-1
$C_{10}H_8O_4N_2$ 3329-30
$C_{10}H_8O_4S$ 4498; 4502-5; 4507; 4510
$C_{10}H_8O_5N_4$ 5305
$C_{10}H_8O_6S_2$ 4461-4
$C_{10}H_8O_7S_2$ 4491
$C_{10}H_8O_8N_2$ 2633
$C_{10}H_8O_8S_2$ 2359
$C_{10}H_8NCl$ 1395
$C_{10}H_8NI$ 3889
$C_{10}H_9ON$ 91; 332-4; 336; 3816-9; 4083; 4179-80; 4567
$C_{10}H_9O_2N$ 3865
$C_{10}H_9O_4N$ 4320-2
$C_{10}H_9O_6N$ 2512
$C_{10}H_{10}ON_2$ 5202
$C_{10}H_{10}O_2N_2$ 2414; 2464; 2981-2
$C_{10}H_{10}O_2Br_2$ 4213
$C_{10}H_{10}O_5N_2$ 4598
$C_{10}H_{10}O_8N_6$ 1541
$C_{10}H_{10}NCl$ 4531-2
$C_{10}H_{11}ON$ 3111; 5134
$C_{10}H_{11}OCl$ 2524
$C_{10}H_{11}O_2N$ 40; 1670
$C_{10}H_{11}O_2N_3$ 4581
$C_{10}H_{11}O_2Br$ 3121
$C_{10}H_{11}O_3N$ 110; 129; 744-5; 1669; 3110

$C_{10}H_{11}N_2Cl$ 4564; 4566
$C_{10}H_{12}ON_2$ 215
$C_{10}H_{12}O_2N_2$ 1676-7
$C_{10}H_{12}O_3N_2$ 1681; 4734; 4871
$C_{10}H_{12}O_4N_2$ 4596
$C_{10}H_{12}O_5N_2$ 2665
$C_{10}H_{13}ON$ 102-4; 1172; 2417-22; 2913; 2934; 5203; 5386; 5874
$C_{10}H_{13}OCl$ 1346; 5939
$C_{10}H_{13}OBr$ 916
$C_{10}H_{13}OI$ 5940
$C_{10}H_{13}O_2N$ 23-5; 2448; 3125; 4714; 4902; 5373
$C_{10}H_{13}O_2Cl$ 5135
$C_{10}H_{13}O_3N$ 4405
$C_{10}H_{13}O_4N_3$ 2619
$C_{10}H_{13}N_2Cl$ 6364
$C_{10}H_{14}ON_2$ 318; 4882
$C_{10}H_{14}OBr_2$ 1828
$C_{10}H_{14}O_2N_2$ 4717-9; 5115
$C_{10}H_{14}O_3N_2$ 5370
$C_{10}H_{14}O_3S$ 5452
$C_{10}H_{14}O_4S$ 5942
$C_{10}H_{14}NBr$ 927
$C_{10}H_{15}ON$ 1262; 2115-6; 2877-8; 3124; 3675
$C_{10}H_{15}OCl$ 1352
$C_{10}H_{15}OBr$ 919
$C_{10}H_{15}O_2N$ 5181; 6476
$C_{10}H_{15}O_3N$ 2345; 4696
$C_{10}H_{16}O_3N_2$ 2776; 2999
$C_{10}H_{16}O_4S$ 1196
$C_{10}H_{16}O_4Br_2$ 2781
$C_{10}H_{17}ON$ 281; 1195
$C_{10}H_{17}N_2Cl$ 304
$C_{10}H_{18}O_5N_4$ 6209
$C_{10}H_{18}N_2Cl_2$ 5840
$C_{10}H_{21}O_2N$ 1649; 4715-6
$C_{10}H_{21}O_3N$ 1648
$C_{10}H_{22}O_2S$ 1756
$C_{10}H_{22}O_3S$ 1755
$C_{10}H_{23}ON$ 1887

10 IV

$C_{10}H_5O_5N_2Na$ 2642
$C_{10}H_5O_5SNa$ 4522
$C_{10}H_6O_2NCl$ 1418-9
$C_{10}H_6O_4SNa$ 4512
$C_{10}H_6O_7S_2Ca$ 4492
$C_{10}H_6O_7S_2K_2$ 4493; 4495
$C_{10}H_6O_7S_2Na_2$ 4494; 4496
$C_{10}H_6O_8N_2S$ 3240
$C_{10}H_7O_2NS$ 4475
$C_{10}H_7O_2SCl$ 4468-9
$C_{10}H_7O_3SNa$ 4472
$C_{10}H_7O_4SK$ 4500; 4509
$C_{10}H_7O_4SNa$ 4501; 4511-2

$C_{10}H_8ON_2S$ 641
$C_{10}H_8O_2NBr$ 940
$C_{10}H_8O_2N_2S$ 773; 5589
$C_{10}H_9O_2NS$ 4467
$C_{10}H_9O_2NCl_2$ 42
$C_{10}H_9O_3NS$ 4541; 4543-53
$C_{10}H_9O_4NS$ 337-40
$C_{10}H_9O_6NS_2$ 4533-5; 4537
$C_{10}H_9O_7NS_2$ 342; 344
$C_{10}H_9O_9NS_3$ 4554
$C_{10}H_{10}ONCl$ 335
$C_{10}H_{10}O_2NCl$ 41
$C_{10}H_{10}O_2NBr$ 884
$C_{10}H_{10}O_2N_4S$ 5665; 5667
$C_{10}H_{10}O_4N_2S$ 5680
$C_{10}H_{13}O_3N_2Br$ 5390
$C_{10}H_{13}O_3SCl$ 1449
$C_{10}H_{15}O_2N_2Cl$ 5116
$C_{10}H_{15}O_3NS$ 2122
$C_{10}H_{16}O_3N_2S$ 853
$C_{10}H_{17}O_6N_3S$ 3378

10 V

$C_{10}H_4O_8NS_3Na_3$ 4476
$C_{10}H_4O_8N_2SNa_2$ 3241
$C_{10}H_5O_8NS_2Na_2$ 4901
$C_{10}H_6O_9NS_3Na_3$ 4555
$C_{10}H_7O_6NS_2Na_2$ 4540
$C_{10}H_7O_7NS_2Na_2$ 343
$C_{10}H_8O_3NSNa$ 4542
$C_{10}H_8O_4NSNa$ 341
$C_{10}H_8O_6NS_2K$ 4536; 4538
$C_{10}H_8O_6NS_2Na$ 4539

11 I

$C_{11}H_{10}$ 4304-5
$C_{11}H_{16}$ 421-4; 2221; 4158-60; 5074
$C_{11}H_{20}$ 6382
$C_{11}H_{22}$ 6373-4
$C_{11}H_{24}$ 6370

11 II

$C_{11}H_6O_{10}$ 671
$C_{11}H_7N$ 4487-8
$C_{11}H_8O$ 4482-3
$C_{11}H_8O_2$ 4306; 4480-1
$C_{11}H_8O_3$ 3778-85; 4275
$C_{11}H_9N$ 5226-8
$C_{11}H_{10}O$ 4311-2
$C_{11}H_{10}O_2$ 3131
$C_{11}H_{10}O_4$ 3696-7; 3971
$C_{11}H_{11}N$ 832; 2540-1; 2544; 2546-8; 4309-10; 6035-7
$C_{11}H_{12}O_2$ 52; 528; 3023; 5151-2
$C_{11}H_{12}O_3$ 739; 6416
$C_{11}H_{12}O_4$ 636; 834; 2941; 3225
$C_{11}H_{14}O$ 1117-8; 6247

Table 7-2 (*Continued*)
FORMULA INDEX FOR ORGANIC COMPOUNDS

$C_{11}H_{14}O_2$ 794; 1053-4; 3065; 3222; 3224; 5347
$C_{11}H_{14}O_3$ 1091-3; 5944; 6507
$C_{11}H_{16}O$ 476; 478-9; 1056; 1079-80; 1115; 5078
$C_{11}H_{16}O_3$ 2231
$C_{11}H_{17}N$ 419-20; 4216-8; 5073
$C_{11}H_{18}O_2$ 3356
$C_{11}H_{18}N_2$ 480
$C_{11}H_{20}O_2$ 6376
$C_{11}H_{20}O_3$ 5778
$C_{11}H_{20}O_4$ 1904; 2134-6; 2158; 2202; 6283
$C_{11}H_{21}N$ 6381
$C_{11}H_{22}O$ 1743-4; 3107; 4324; 6375; 6378
$C_{11}H_{22}O_2$ 436-7; 3113; 4172; 5397; 6377
$C_{11}H_{22}O_3$ 1737-8
$C_{11}H_{22}O_6$ 4386
$C_{11}H_{24}O$ 6371-2
$C_{11}H_{24}O_3$ 3390

11 III

$C_{11}H_7ON$ 4558-9
$C_{11}H_7OCl$ 4486
$C_{11}H_7O_4N$ 4774
$C_{11}H_7NS$ 4570
$C_{11}H_9ON$ 4484-5
$C_{11}H_9O_2N$ 4308; 5528-32
$C_{11}H_9O_3N$ 4278; 5537
$C_{11}H_{10}ON_2$ 3324
$C_{11}H_{11}ON$ 3013
$C_{11}H_{11}O_2N$ 3867
$C_{11}H_{11}O_4N$ 3102-4
$C_{11}H_{12}ON_2$ 565; 2497
$C_{11}H_{12}O_2N_2$ 3126; 6365
$C_{11}H_{12}NI$ 5539
$C_{11}H_{13}O_4N$ 29; 2205
$C_{11}H_{13}O_6N_3$ 6296
$C_{11}H_{15}ON$ 2111; 5364; 6411
$C_{11}H_{15}O_2N$ 1043; 1045
$C_{11}H_{15}O_3N$ 3949; 4064
$C_{11}H_{16}O_3N_2$ 1038
$C_{11}H_{16}O_3S$ 1137
$C_{11}H_{16}O_4N_2$ 571; 5565
$C_{11}H_{18}O_3N_2$ 505
$C_{11}H_{19}ON_3$ 1513
$C_{11}H_{19}N_2Cl$ 302
$C_{11}H_{23}ON$ 6379-80
$C_{11}H_{23}O_2N$ 3038; 4861
$C_{11}H_{25}ON$ 1888-9

11 IV

$C_{11}H_{11}ON_2Br$ 901
$C_{11}H_{11}O_2N_3S$ 5668
$C_{11}H_{15}O_3N_2Br$ 1061
$C_{11}H_{17}O_2NS$ 5973
$C_{11}H_{17}O_3N_2Na$ 506; 4583

$C_{11}H_{18}O_3N_2S$ 854

11 V

$C_{11}H_{10}O_2N_3SNa$ 5669

12 I

$C_{12}H_8$ 3
$C_{12}H_{10}$ 2; 2695
$C_{12}H_{12}$ 2507-9; 3089-90
$C_{12}H_{16}$ 5179
$C_{12}H_{18}$ 2777-8; 3590; 4119; 6184-5
$C_{12}H_{22}$ 2083; 2864
$C_{12}H_{24}$ 2863; 6122
$C_{12}H_{26}$ 2856-7
$C_{12}Cl_{10}$ 1629

12 II

$C_{12}H_6O_3$ 4474
$C_{12}H_6O_{12}$ 667
$C_{12}H_6Cl_4$ 5738
$C_{12}H_8O$ 2763
$C_{12}H_8O_4$ 4459-60; 5037
$C_{12}H_8N_2$ 5101-4
$C_{12}H_8S_2$ 5877
$C_{12}H_8F_2$ 2247-8
$C_{12}H_8Cl_2$ 1989-90
$C_{12}H_8Br_2$ 1833
$C_{12}H_9N$ 1231
$C_{12}H_9Cl$ 1291; 1373-5
$C_{12}H_9Br$ 878; 933-4
$C_{12}H_9I$ 3884-5
$C_{12}H_{10}O$ 2719; 4313; 5216-8
$C_{12}H_{10}O_2$ 48; 2338-43; 3770; 4526-8
$C_{12}H_{10}O_4$ 2817; 5321; 5535
$C_{12}H_{10}N_2$ 599; 2279
$C_{12}H_{10}S$ 2749
$C_{12}H_{10}S_2$ 2715
$C_{12}H_{10}I_2$ 2729
$C_{12}H_{10}Hg$ 4047
$C_{12}H_{10}P_2$ 5271
$C_{12}H_{10}Se$ 2742
$C_{12}H_{11}N$ 830-1; 2698; 5153-5
$C_{12}H_{11}N_3$ 247-9; 1762
$C_{12}H_{12}O$ 3093-4
$C_{12}H_{12}O_2$ 203
$C_{12}H_{12}O_6$ 3775; 5516
$C_{12}H_{12}O_{12}$ 3580
$C_{12}H_{12}N_2$ 313-4; 1710-4; 2725; 2727; 3685
$C_{12}H_{12}N_4$ 1696-9
$C_{12}H_{13}N$ 2505-6; 3091-2; 6270-3
$C_{12}H_{13}N_3$ 262; 1717-8
$C_{12}H_{13}N_5$ 6058
$C_{12}H_{14}O_3$ 3219; 5776
$C_{12}H_{14}O_4$ 569; 1122; 2199-201; 5784

$C_{12}H_{14}N_4$ 1722; 2269-70
$C_{12}H_{16}O$ 481-2; 1615-6; 4181
$C_{12}H_{16}O_2$ 425; 1113-4; 5075; 5945
$C_{12}H_{16}O_3$ 458; 486-7; 584; 1106; 3117; 3487
$C_{12}H_{16}O_8$ 1270
$C_{12}H_{18}O$ 426; 452; 477; 4066
$C_{12}H_{18}O_2$ 1112; 3654
$C_{12}H_{18}O_3$ 5265
$C_{12}H_{18}O_6$ 2146-51
$C_{12}H_{18}O_8$ 2152
$C_{12}H_{19}N$ 2775
$C_{12}H_{20}O_2$ 861; 3354; 3975; 5710
$C_{12}H_{20}O_4$ 1903
$C_{12}H_{20}O_6$ 1666; 2095; 3432
$C_{12}H_{20}O_7$ 6187
$C_{12}H_{22}O_2$ 1520; 4027-8; 4407
$C_{12}H_{22}O_3$ 1214
$C_{12}H_{22}O_4$ 1745; 1910-2; 2171; 2210; 2552; 2771
$C_{12}H_{22}O_5$ 1902
$C_{12}H_{22}O_6$ 1919-20; 2410
$C_{12}H_{22}O_{11}$ 3947; 4007; 5663; 6054
$C_{12}H_{23}N$ 2085; 3956
$C_{12}H_{24}O$ 3952
$C_{12}H_{24}O_2$ 1066; 1639; 3008; 3950
$C_{12}H_{24}O_3$ 1162; 1165
$C_{12}H_{25}Cl$ 2861
$C_{12}H_{25}Br$ 2860
$C_{12}H_{26}O$ 2859
$C_{12}H_{27}N$ 6115-6
$C_{12}H_{30}Sn_2$ 5953

12 III

$C_{12}H_5O_{12}N_7$ 2764
$C_{12}H_6O_8N_4$ 5847-9
$C_{12}H_6O_9N_4$ 5851
$C_{12}H_6O_{10}N_4$ 5846
$C_{12}H_7O_3N$ 4732; 5554
$C_{12}H_7O_4N$ 5551
$C_{12}H_8OS$ 5129
$C_{12}H_8OBr_2$ 1834
$C_{12}H_8O_4N_2$ 2620-3
$C_{12}H_8O_5N_2$ 2627-8
$C_{12}H_8O_5N_4$ 2598
$C_{12}H_8O_9N_4$ 5492
$C_{12}H_9OCl$ 1436-8
$C_{12}H_9OBr$ 885; 935; 977
$C_{12}H_9O_2N$ 3869; 4587; 4723-5
$C_{12}H_9O_3N$ 4729-31
$C_{12}H_9O_4N_3$ 2624-6; 4650-1
$C_{12}H_9O_4N_5$ 2618
$C_{12}H_9O_5N_3$ 2634
$C_{12}H_9O_8N_5$ 5495
$C_{12}H_9NS$ 5138

Table 7-2 (*Continued*)
FORMULA INDEX FOR ORGANIC COMPOUNDS

$C_{12}H_9N_3S$ 5922
$C_{12}H_{10}ON_2$ 615; 2733; 3727-9; 4884
$C_{12}H_{10}OS$ 2752
$C_{12}H_{10}O_2N_2$ 609-11; 662; 4726-7
$C_{12}H_{10}O_2S$ 2750
$C_{12}H_{10}O_2S_2$ 2716
$C_{12}H_{10}O_3N_2$ 4593
$C_{12}H_{10}N_2Cl_2$ 1957
$C_{12}H_{10}ClAs$ 2712
$C_{12}H_{10}Cl_2Se$ 2743
$C_{12}H_{11}ON$ 46-7; 3767-9
$C_{12}H_{11}OI$ 2728
$C_{12}H_{11}O_2N$ 31-2; 3031
$C_{12}H_{11}O_2N_3$ 4661
$C_{12}H_{11}O_4P$ 2735-6
$C_{12}H_{12}ON_2$ 1720
$C_{12}H_{12}O_3N_2$ 5113
$C_{12}H_{12}NCl$ 2699
$C_{12}H_{12}N_2S$ 5886
$C_{12}H_{12}N_2Cl_4$ 1958
$C_{12}H_{12}N_3Cl$ 250
$C_{12}H_{13}O_2N$ 3866
$C_{12}H_{13}O_3N$ 5476
$C_{12}H_{13}O_6N$ 2190
$C_{12}H_{13}N_2Cl$ 315; 2726
$C_{12}H_{13}N_4Cl$ 1699
$C_{12}H_{14}O_2N_2$ 2098; 2129
$C_{12}H_{14}O_8N_4$ 232
$C_{12}H_{14}NI$ 4368; 4371; 4373
$C_{12}H_{14}N_2Cl_2$ 1715
$C_{12}H_{15}ON$ 770
$C_{12}H_{15}O_2N$ 39
$C_{12}H_{15}O_6N_3$ 6297
$C_{12}H_{16}OCl_2$ 1381
$C_{12}H_{16}O_3N_2$ 3231; 5091
$C_{12}H_{16}O_8N_2$ 2528
$C_{12}H_{17}ON$ 77; 1215
$C_{12}H_{17}O_4N$ 2474
$C_{12}H_{17}O_5N$ 2116
$C_{12}H_{17}O_7N$ 5110
$C_{12}H_{18}O_4N_2$ 5559
$C_{12}H_{18}O_5N_2$ 3366-7
$C_{12}H_{19}ON$ 2114
$C_{12}H_{20}O_4Br_2$ 1896
$C_{12}H_{22}O_{14}Ca$ 3360
$C_{12}H_{23}OCl$ 3955
$C_{12}H_{25}ON$ 3953
$C_{12}H_{27}O_2N$ 3951
$C_{12}H_{27}O_3B$ 6117
$C_{12}H_{27}O_4P$ 6121
$C_{12}H_{28}O_4N_2$ 3200
$C_{12}H_{28}NI$ 5866
$C_{12}H_{30}OSi_2$ 6197
$C_{12}H_{30}N_6Cl_2$ 5685

12 IV

$C_{12}H_4O_{12}N_6S$ 2765

$C_{12}H_6O_3SCl_4$ 2043
$C_{12}H_6O_8N_4S_2$ 5850
$C_{12}H_8ON_2Cl_2$ 1950
$C_{12}H_8O_2SBr_2$ 1835
$C_{12}H_8O_3SCl_2$ 5180
$C_{12}H_8O_4N_2S_2$ 2655-6
$C_{12}H_9O_3SCl$ 5170
$C_{12}H_9O_3SNa$ 2751
$C_{12}H_{10}O_4N_2S$ 3730
$C_{12}H_{10}O_5N_2S$ 2307; 4728
$C_{12}H_{11}ONS$ 5906
$C_{12}H_{11}O_3NS$ 2701
$C_{12}H_{11}O_3N_2Na$ 5114
$C_{12}H_{12}ON_2S_2$ 2437
$C_{12}H_{12}O_3N_2S$ 696
$C_{12}H_{12}O_6N_2S_2$ 695
$C_{12}H_{13}O_4NS$ 2700
$C_{12}H_{14}ON_4S$ 5893
$C_{12}H_{14}O_4N_2S$ 1716
$C_{12}H_{16}O_2NCl$ 2413
$C_{12}H_{16}O_4N_2S$ 522
$C_{12}H_{19}O_2N_2Br$ 5470
$C_{12}H_{24}O_6N_6Cu$ 1522

12 V

$C_{12}H_6O_4N_2S_2Cl_2$ 1987
$C_{12}H_6O_8S_2I_4Zn$ 5629
$C_{12}H_9O_4N_2SNa$ 3731
$C_{12}H_9O_5N_2SNa$ 2308
$C_{12}H_{14}O_2N_2Cl_2As_2$ 5592
$C_{12}H_{14}O_2N_5SCl$ 5337
$C_{12}H_{18}ON_4SCl_2$ 5876
$C_{12}H_{19}O_3N_2Na$ 5002

13 I

$C_{13}H_{10}$ 3246
$C_{13}H_{12}$ 2731; 5245-7
$C_{13}H_{26}$ 6175
$C_{13}H_{28}$ 6173

13 II

$C_{13}H_8O$ 3248
$C_{13}H_8O_2$ 6451
$C_{13}H_8O_4$ 3227
$C_{13}H_8O_5$ 2891
$C_{13}H_9N$ 142; 4517-9
$C_{13}H_{10}O$ 724; 3247; 6447
$C_{13}H_{10}O_2$ 3320; 3749; 5159-62; 6453
$C_{13}H_{10}O_3$ 2321-8; 2711; 5133; 5235
$C_{13}H_{10}O_4$ 4276; 6224-5
$C_{13}H_{10}O_6$ 5070
$C_{13}H_{10}N_2$ 690; 1234-5; 4334
$C_{13}H_{10}Cl_2$ 726
$C_{13}H_{11}N$ 632; 2271
$C_{13}H_{12}O$ 691; 823-4; 4073-4
$C_{13}H_{12}O_2$ 2344; 3718
$C_{13}H_{12}N_2$ 646; 2721

$C_{13}H_{13}N$ 693; 786; 4219-20
$C_{13}H_{13}N_3$ 2723; 4211
$C_{13}H_{14}O_3$ 625
$C_{13}H_{14}O_4$ 740
$C_{13}H_{14}O_6$ 5584
$C_{13}H_{14}N_2$ 826; 1721; 4271-3; 4431
$C_{13}H_{15}N$ 5426
$C_{13}H_{16}O_3$ 2983
$C_{13}H_{18}O$ 1611
$C_{13}H_{18}O_3$ 3483
$C_{13}H_{18}O_4$ 1028
$C_{13}H_{20}O_3$ 3663; 3913-5
$C_{13}H_{20}O_8$ 5067
$C_{13}H_{24}O_4$ 2123; 2168
$C_{13}H_{25}N$ 2862
$C_{13}H_{26}O$ 2267; 4408
$C_{13}H_{26}O_2$ 438; 3181; 4295; 6176
$C_{13}H_{28}O$ 6118; 6174
$C_{13}H_{28}O_3$ 1087; 3386
$C_{13}H_{28}O_4$ 5399

13 III

$C_{13}H_7O_3Br_3$ 6110
$C_{13}H_8OCl_2$ 1965-6
$C_{13}H_8O_5N_2$ 2612
$C_{13}H_8O_8N_4$ 5852
$C_{13}H_9ON$ 145; 2714; 3249; 4525
$C_{13}H_9OCl$ 1331
$C_{13}H_9O_2N$ 4738
$C_{13}H_9O_3N$ 4670-2
$C_{13}H_9O_5N$ 4837
$C_{13}H_9O_5N_3$ 4652
$C_{13}H_9NS$ 687; 2755
$C_{13}H_{10}O_3N_2$ 4645
$C_{13}H_{10}O_4N_2$ 2629
$C_{13}H_{10}NCl$ 4519
$C_{13}H_{11}ON$ 270-2; 631; 715; 728; 3306
$C_{13}H_{11}O_2N$ 3741; 5157
$C_{13}H_{11}O_2N_3$ 4638
$C_{13}H_{11}O_4N_3$ 4649
$C_{13}H_{11}NS$ 5890
$C_{13}H_{12}ON_2$ 657-8; 767-8; 828; 1706-8; 2759-60; 3737
$C_{13}H_{12}ON_4$ 2709
$C_{13}H_{12}O_2N_2$ 4771
$C_{13}H_{12}O_2S$ 5252
$C_{13}H_{12}O_3S$ 5248
$C_{13}H_{12}O_5N_6$ 2654
$C_{13}H_{12}N_2S$ 5892
$C_{13}H_{12}N_4S$ 2753
$C_{13}H_{13}ON$ 22; 784; 5791
$C_{13}H_{13}ON_3$ 246; 2744-7
$C_{13}H_{14}ON_4$ 2710
$C_{13}H_{14}O_3N_2$ 5336

Table 7-2 (*Continued*)
FORMULA INDEX FOR ORGANIC COMPOUNDS

$C_{13}H_{14}N_4S$ 2754
$C_{13}H_{15}ON$ 2479
$C_{13}H_{15}N_2Cl$ 827
$C_{13}H_{16}NI$ 2543; 2545
$C_{13}H_{17}O_3$ 363
$C_{13}H_{18}O_3N_4$ 3602
$C_{13}H_{19}ON$ 2774
$C_{13}H_{20}O_2N_2$ 4934-5
$C_{13}H_{25}O_2Br$ 4153
$C_{13}H_{27}ON$ 6177

13 IV

$C_{13}H_8O_5N_3Na$ 4653
$C_{13}H_9O_3SBr_3$ 6111
$C_{13}H_{10}ONCl$ 2708
$C_{13}H_{10}O_3SCl_2$ 2047
$C_{13}H_{11}O_3SCl$ 1439-40
$C_{13}H_{13}O_2NS$ 5972
$C_{13}H_{13}O_4N_3S$ 5328
$C_{13}H_{14}ONCl$ 785
$C_{13}H_{15}O_3N_2Cl_3$ 567
$C_{13}H_{18}O_2NBr$ 1008
$C_{13}H_{21}O_2N_2Cl$ 4934

13 V

$C_{13}H_{16}O_6NHgNa$ 5591

13 VI

$C_{13}H_{13}O_4N_2SAs_2Na$ 5593

14 I

$C_{14}H_{10}$ 538; 5095; 5954
$C_{14}H_{12}$ 2272; 5642-3
$C_{14}H_{14}$ 842-3; 2717-8; 2825-9
$C_{14}H_{16}$ 3576
$C_{14}H_{22}$ 5758-9
$C_{14}H_{26}$ 5752
$C_{14}H_{28}$ 5751
$C_{14}H_{30}$ 5748

14 II

$C_{14}H_6O_8$ 2872
$C_{14}H_8O_2$ 548-50; 5097
$C_{14}H_8O_3$ 3724-6
$C_{14}H_8O_4$ 2297-304; 3747
$C_{14}H_8O_5$ 6212-5
$C_{14}H_8O_6$ 5799-801
$C_{14}H_8O_8$ 5575
$C_{14}H_8Cl_2$ 1939-40
$C_{14}H_8Br_2$ 1803
$C_{14}H_9Br$ 968
$C_{14}H_{10}O$ 545; 561-2; 2730; 5098-100
$C_{14}H_{10}O_2$ 697; 2293-4; 2360; 5019; 5096
$C_{14}H_{10}O_3$ 714; 750-2; 2295-6
$C_{14}H_{10}O_4$ 766; 2688-91
$C_{14}H_{10}O_5$ 5585
$C_{14}H_{10}O_9$ 5687

$C_{14}H_{11}N$ 542-3; 4092-4; 4514-6
$C_{14}H_{12}O$ 1660; 2273; 5144; 5249-51
$C_{14}H_{12}O_2$ 717; 789-92; 2697; 4070; 6013-5
$C_{14}H_{12}O_3$ 694; 699; 809; 833; 3482; 6038-40
$C_{14}H_{12}O_4$ 2347; 2358; 3486
$C_{14}H_{12}N_2$ 633; 5182
$C_{14}H_{13}N$ 3011
$C_{14}H_{14}O$ 806; 2918; 4221; 5164
$C_{14}H_{14}O_2$ 2079; 2693; 3202; 3701-2
$C_{14}H_{14}O_3$ 3931
$C_{14}H_{14}O_8$ 5841
$C_{14}H_{14}N_2$ 54; 612-4; 1729-31; 2696
$C_{14}H_{14}S$ 1791; 2842
$C_{14}H_{14}S_2$ 805; 2836; 3204
$C_{14}H_{14}Hg$ 4048-50
$C_{14}H_{15}N$ 1783; 2830-2; 3044; 4144
$C_{14}H_{15}N_3$ 309-12; 1764; 2431-2
$C_{14}H_{15}P$ 3045
$C_{14}H_{16}N_2$ 1719; 2720; 3692-4; 5955; 5959
$C_{14}H_{17}N$ 2189
$C_{14}H_{18}O$ 451
$C_{14}H_{18}O_4$ 2130; 2794-5; 4231
$C_{14}H_{18}O_6$ 2412
$C_{14}H_{18}O_9$ 1269
$C_{14}H_{20}O_6$ 5781
$C_{14}H_{22}O$ 3918
$C_{14}H_{22}O_4$ 2087
$C_{14}H_{22}O_6$ 5785
$C_{14}H_{22}O_8$ 5764
$C_{14}H_{23}N$ 1891
$C_{14}H_{24}O_2$ 3355
$C_{14}H_{24}N_2$ 300
$C_{14}H_{26}O_3$ 3524
$C_{14}H_{26}O_4$ 1748-50; 1883; 2167; 2177; 2206; 3451
$C_{14}H_{26}O_6$ 1757; 2920
$C_{14}H_{27}N$ 4456
$C_{14}H_{28}O$ 3182; 4451
$C_{14}H_{28}O_2$ 2858: 3080; 3552; 4450
$C_{14}H_{29}Br$ 4457
$C_{14}H_{30}O$ 2264; 5750
$C_{14}H_{30}S$ 2266

14 III

$C_{14}H_4O_{12}N_4$ 1491
$C_{14}H_6O_2Cl_2$ 1941-8
$C_{14}H_6O_2Br_2$ 1805-9
$C_{14}H_6O_6N_2$ 2595-7; 2644
$C_{14}H_7O_2Cl$ 1316-7
$C_{14}H_7O_2Br$ 899-900
$C_{14}H_7O_4N$ 4627-8; 4794

$C_{14}H_7O_6N$ 4599-600
$C_{14}H_8O_5S$ 559
$C_{14}H_8O_7S$ 177
$C_{14}H_8O_8S_2$ 556
$C_{14}H_9O_2N$ 244-5; 4626; **5273**
$C_{14}H_9O_3N$ 172-3; 327
$C_{14}H_9O_4N$ 241-2
$C_{14}H_{10}O_2N_2$ 1690-5; 5163
$C_{14}H_{10}O_2S_2$ 1779
$C_{14}H_{10}O_4N_2$ 600-2; 2663-4
$C_{14}H_{10}O_5N_2$ 616-8
$C_{14}H_{10}O_6Ca$ 3743
$C_{14}H_{11}ON$ 80
$C_{14}H_{11}OCl$ 5215
$C_{14}H_{11}OBr$ 5214
$C_{14}H_{11}O_2N$ 703-5; 1776; 5587
$C_{14}H_{11}O_3N$ 746-8
$C_{14}H_{12}O_2N_2$ 700-2; 5010
$C_{14}H_{12}O_3N_2$ 546
$C_{14}H_{12}O_4N_2$ 3686-8
$C_{14}H_{13}ON$ 84.1; 630; 775-7 5143
$C_{14}H_{13}OCl$ 1376
$C_{14}H_{13}O_2N$ 720-1; 825
$C_{14}H_{14}ON_2$ 533; 4176
$C_{14}H_{14}OS$ 836; 2919
$C_{14}H_{14}O_2N_2$ 598
$C_{14}H_{14}O_2S$ 835; 2843
$C_{14}H_{14}O_3S$ 1554
$C_{14}H_{14}O_4S_2$ 3203
$C_{14}H_{15}ON$ 1786; 5137
$C_{14}H_{16}O_2N_2$ 1759
$C_{14}H_{16}O_6N_2$ 352; 5201
$C_{14}H_{16}N_3Cl$ 2432
$C_{14}H_{17}O_6N$ 3856
$C_{14}H_{18}ON_6$ 1709
$C_{14}H_{18}N_2Cl_2$ 5956
$C_{14}H_{19}O_8N$ 5108
$C_{14}H_{19}O_9Cl$ 44
$C_{14}H_{21}O_4N$ 2159
$C_{14}H_{23}ON_3$ 3916-7
$C_{14}H_{26}N_2Cl_2$ 301
$C_{14}H_{27}OCl$ 4455
$C_{14}H_{29}ON_3$ 4409
$C_{14}H_{29}O_2N$ 4452

14 IV

$C_{14}H_6O_8S_2Na_2$ 553-5; 557
$C_{14}H_7O_5SNa$ 558; 560
$C_{14}H_7O_7NS$ 4630-1
$C_{14}H_7O_7SNa$ 178
$C_{14}H_{10}OClBr$ 1435
$C_{14}H_{13}O_3SCl$ 4189
$C_{14}H_{14}O_6N_2S_2$ 1732
$C_{14}H_{15}O_2NS$ 5980-1
$C_{14}H_{16}O_2N_3As$ 2433
$C_{14}H_{16}O_3N_3As$ 2434
$C_{14}H_{16}O_6N_2S_2$ 5958
$C_{14}H_{17}O_4N_3S_2$ 6368

Table 7-2 (*Continued*)
FORMULA INDEX FOR ORGANIC COMPOUNDS

$C_{14}H_{18}O_4N_2S$ 5957
$C_{14}H_{20}O_6N_2S$ 294; 4116-7
$C_{14}H_{22}O_2NCl$ 5644
$C_{14}H_{22}O_4N_4S$ 4339
$C_{14}H_{23}O_2N_2Cl$ 6366

14 V

$C_{14}H_{14}O_3N_3SNa$ 4328

14 VI

$C_{14}H_{14}O_8N_2S_2As_2Na_2$ 5674

15 I

$C_{15}H_{12}$ 4127-8
$C_{15}H_{16}$ 2985
$C_{15}H_{18}$ 470
$C_{15}H_{24}$ 1263
$C_{15}H_{32}$ 5060

15 II

$C_{15}H_8O_4$ 551-2
$C_{15}H_8O_6$ 176; 2305-6
$C_{15}H_{10}O_2$ 539-41; 640; 3243; 4129
$C_{15}H_{10}O_3$ 2758
$C_{15}H_{10}O_4$ 1494-5; 4101
$C_{15}H_{10}O_5$ 727; 769; 772; 2874
$C_{15}H_{10}O_6$ 5805
$C_{15}H_{10}O_7$ 5806-7
$C_{15}H_{11}N$ 5230-3
$C_{15}H_{12}O$ 627
$C_{15}H_{12}O_2$ 1781; 3763; 5172-4
$C_{15}H_{12}O_3$ 1492; 2258; 4140; 6002
$C_{15}H_{12}O_4$ 120; 4143
$C_{15}H_{13}N$ 2510-1
$C_{15}H_{13}N_3$ 5199
$C_{15}H_{14}O$ 1787; 2838
$C_{15}H_{14}O_2$ 2739; 5190
$C_{15}H_{14}O_3$ 2409; 2833-5
$C_{15}H_{14}O_4$ 4307
$C_{15}H_{14}O_5$ 3484; 5264 5594
$C_{15}H_{16}O$ 1573
$C_{15}H_{16}O_2$ 2694
$C_{15}H_{16}O_3$ 3409
$C_{15}H_{16}O_9$ 2899
$C_{15}H_{17}N$ 2984
$C_{15}H_{17}N_3$ 2837
$C_{15}H_{18}O$ 471
$C_{15}H_{18}O_3$ 5598
$C_{15}H_{20}O_2$ 3497
$C_{15}H_{20}O_4$ 2170; 5597; 5599
$C_{15}H_{20}N_4$ 5713
$C_{15}H_{22}O_{10}$ 4385
$C_{15}H_{24}O$ 5595
$C_{15}H_{24}O_8$ 5765
$C_{15}H_{26}O_2$ 867; 869
$C_{15}H_{26}O_6$ 3421

$C_{15}H_{28}O_2$ 4032
$C_{15}H_{28}O_4$ 2157
$C_{15}H_{30}O$ 2265
$C_{15}H_{30}O_2$ 4303
$C_{15}H_{31}Br$ 5063
$C_{15}H_{32}O$ 5062
$C_{15}H_{33}N$ 6074-6

15 III

$C_{15}H_9O_4N$ 4770
$C_{15}H_{11}O_3N$ 4633
$C_{15}H_{11}O_4N$ 4641
$C_{15}H_{12}ON_2$ 2722
$C_{15}H_{12}OBr_2$ 628-9
$C_{15}H_{12}O_2N_2$ 2724
$C_{15}H_{12}O_3N_2$ 3326
$C_{15}H_{12}O_5N_2$ 3353
$C_{15}H_{12}N_2S$ 317
$C_{15}H_{13}ON$ 30
$C_{15}H_{14}O_2N_2$ 4005
$C_{15}H_{15}ON$ 812; 2442
$C_{15}H_{15}O_2N$ 2761; 5588
$C_{15}H_{15}O_2N_3$ 4376
$C_{15}H_{16}ON_2$ 2467; 2847-9
$C_{15}H_{16}N_2S$ 2844-6
$C_{15}H_{17}N_4Cl$ 4585
$C_{15}H_{25}O_2Br$ 863
$C_{15}H_{31}ON$ 5059
$C_{15}H_{33}O_3B$ 6077

15 IV

$C_{15}H_6O_5NCl$ 4629
$C_{15}H_{11}O_4NI_4$ 5947
$C_{15}H_{11}O_4N_3S$ 679
$C_{15}H_{14}ON_2Cl$ 144
$C_{15}H_{14}O_2N_3Na$ 4377-**8**
$C_{15}H_{20}ONI$ 2480
$C_{15}H_{25}O_2NS$ 5976
$C_{15}H_{25}O_2N_2Cl$ 5032

16 I

$C_{16}H_{10}$ 3245; 5481
$C_{16}H_{12}$ 5204-5
$C_{16}H_{14}$ 2453-5; 2525; 2969
$C_{16}H_{16}$ 2277
$C_{16}H_{26}$ 1739; 5068
$C_{16}H_{30}$ 3571-2
$C_{16}H_{32}$ 3570
$C_{16}H_{34}$ 3569

16 II

$C_{16}H_{10}N_2$ 4513
$C_{16}H_{12}O_3$ 5324
$C_{16}H_{12}O_5$ 874
$C_{16}H_{12}O_6$ 3498
$C_{16}H_{12}O_7$ 5556
$C_{16}H_{13}N$ 5206-7
$C_{16}H_{13}N_3$ 660-1
$C_{16}H_{14}O_2$ 798

$C_{16}H_{14}O_3$ 718; 2257; 2979; 3485; 5991
$C_{16}H_{14}O_4$2748; 3447
$C_{16}H_{14}O_5$ 875
$C_{16}H_{14}O_6$ 3500; 3559; 3672; 5326
$C_{16}H_{14}N_2$ 3242
$C_{16}H_{16}O_2$ 719; 1782
$C_{16}H_{16}O_5$ 180
$C_{16}H_{16}O_6$ 1264
$C_{16}H_{18}O$ 1121
$C_{16}H_{18}O_2$ 3448
$C_{16}H_{18}O_3$ 2692
$C_{16}H_{18}O_4$ 3457
$C_{16}H_{18}O_6$ 3710
$C_{16}H_{18}O_{10}$ 3310
$C_{16}H_{18}N_2$ 2738
$C_{16}H_{19}N$ 2988
$C_{16}H_{20}O_{10}$ 1268
$C_{16}H_{20}N_2$ 5830
$C_{16}H_{21}N_3$ 5833
$C_{16}H_{22}O_4$ 1908
$C_{16}H_{22}O_6$ 2096
$C_{16}H_{22}O_8$ 1528
$C_{16}H_{22}O_{11}$ 3362-3
$C_{16}H_{24}O_4$ 2086
$C_{16}H_{26}O_2$ 1742
$C_{16}H_{28}O_2$ 3680; 5029
$C_{16}H_{30}O$ 4446
$C_{16}H_{30}O_2$ 3849
$C_{16}H_{30}O_3$ 1224
$C_{16}H_{31}N$ 5028
$C_{16}H_{32}O$ 5023
$C_{16}H_{32}O_2$ 2263; 3088; 5022; 5749
$C_{16}H_{32}O_4$ 2234
$C_{16}H_{33}Br$ 1276
$C_{16}H_{33}I$ 1277
$C_{16}H_{34}O$ 1275; 2684
$C_{16}H_{34}O_3$ 456

16 III

$C_{16}H_{10}O_2N_2$ 3857; 3862
$C_{16}H_{10}O_4N_2$ 850
$C_{16}H_{11}O_2N$ 5234
$C_{16}H_{11}O_3N_3$ 4648
$C_{16}H_{12}O_2N_2$ 3863
$C_{16}H_{13}O_{15}N_9$ 1542
$C_{16}H_{14}O_3N_2$ 1509
$C_{16}H_{15}O_3N$ 3848
$C_{16}H_{16}O_2N_2$ 1671; 1780
$C_{16}H_{17}O_3N$ 124
$C_{16}H_{17}O_2N$ 5165
$C_{16}H_{18}O_2N_2$ 606-8
$C_{16}H_{18}O_3N_2$ 851
$C_{16}H_{20}N_3Cl$ 253
$C_{16}H_{27}O_5N_3$ 737
$C_{16}H_{31}OCl$ 5027
$C_{16}H_{33}ON$ 5024-**5**
$C_{16}H_{36}NI$ 5727

Table 7-2 (*Continued*)
FORMULA INDEX FOR ORGANIC COMPOUNDS

16 IV

$C_{16}H_9O_2NI_2$ 2384
$C_{16}H_{10}O_5N_2S$ 3861
$C_{16}H_{10}O_8N_2S_2$ 3860
$C_{16}H_{11}O_{10}N_3S$ 5491
$C_{16}H_{12}O_9N_4S$ 5494
$C_{16}H_{13}O_3NS$ 5208
$C_{16}H_{13}O_3N_3S$ 331
$C_{16}H_{18}O_6S_2Ca$ 6460
$C_{16}H_{18}N_3SCl$ 4427
$C_{16}H_{19}ON_3S$ 4426
$C_{16}H_{24}O_2NCl$ 764; 2119
$C_{16}H_{24}O_6N_2S$ 2447
$C_{16}H_{27}O_2N_2Cl$ 276; 736

16 V

$C_{16}H_8O_8N_2S_2Na_2$ 3858
$C_{16}H_9O_9N_4S_2Na_3$ 5694
$C_{16}H_9O_{10}N_3S_2Na_2$ 4647
$C_{16}H_{11}O_4N_2SNa$ 4994-5
$C_{16}H_{14}O_4N_2As_2Na_2$ 581

17 I

$C_{17}H_{14}$ 817-8
$C_{17}H_{28}$ 3180
$C_{17}H_{36}$ 3508

17 II

$C_{17}H_{11}N$ 547
$C_{17}H_{12}O$ 5210-1
$C_{17}H_{12}O_2$ 4556-7
$C_{17}H_{12}O_3$ 4568-9
$C_{17}H_{14}O$ 819-20; 1506; 1775; 5209
$C_{17}H_{14}N_2$ 689
$C_{17}H_{15}N$ 6031-4
$C_{17}H_{15}N_3$ 5961
$C_{17}H_{16}O_2$ 5192
$C_{17}H_{16}O_3$ 3220; 5387
$C_{17}H_{16}O_6$ 4277
$C_{17}H_{16}O_7$ 3229
$C_{17}H_{19}N_3$ 143
$C_{17}H_{21}N_3$ 592
$C_{17}H_{22}O_3$ 866
$C_{17}H_{22}N_2$ 5834
$C_{17}H_{24}O_2$ 4030; 4977
$C_{17}H_{24}O_3$ 4031
$C_{17}H_{28}O_4$ 868
$C_{17}H_{32}O_3$ 5780
$C_{17}H_{34}O$ 2685; 4015
$C_{17}H_{34}O_2$ 4014; 4330; 5061
$C_{17}H_{36}O$ 3509

17 III

$C_{17}H_9O_4N$ 174
$C_{17}H_{11}O_5N$ 3801
$C_{17}H_{12}O_5N_2$ 2599
$C_{17}H_{13}ON$ 765
$C_{17}H_{20}ON_2$ 2137; 4436

$C_{17}H_{20}O_3N_2$ 2687
$C_{17}H_{20}O_5N_2$ 5117
$C_{17}H_{20}O_6N_4$ 5563
$C_{17}H_{22}ON_2$ 4435

17 IV

$C_{17}H_{19}ON_2Cl$ 5519
$C_{17}H_{24}O_3N_3Cl_3$ 364

17 V

$C_{17}H_{11}O_{10}NS_2Na_2$ 175

18 I

$C_{18}H_{12}$ 650; 1493
$C_{18}H_{14}$ 2703-5
$C_{18}H_{18}$ 5555
$C_{18}H_{20}$ 2476
$C_{18}H_{30}$ 3575; 5867
$C_{18}H_{34}$ 4942
$C_{18}H_{36}$ 4941
$C_{18}H_{38}$ 4938

18 II

$C_{18}H_{10}O_2$ 1496
$C_{18}H_{12}O_4$ 1658
$C_{18}H_{12}N_2$ 2814-6
$C_{18}H_{14}O$ 821
$C_{18}H_{14}O_3$ 1503
$C_{18}H_{14}O_4$ 3790; 5020
$C_{18}H_{15}N$ 6328
$C_{18}H_{15}N_3$ 659
$C_{18}H_{15}As$ 580
$C_{18}H_{15}Bi$ 857
$C_{18}H_{15}P$ 6342
$C_{18}H_{15}Sb$ 563
$C_{18}H_{16}O_2$ 1508
$C_{18}H_{16}O_4$ 1785; 1788; 3924-5
$C_{18}H_{16}O_7$ 6394-5
$C_{18}H_{16}N_2$ 6336
$C_{18}H_{17}N_3$ 2438-9
$C_{18}H_{18}O_2$ 2883-4; 2886-8
$C_{18}H_{18}O_3$ 1055
$C_{18}H_{18}O_4$ 1790; 2921
$C_{18}H_{18}O_6$ 1792
$C_{18}H_{20}O_2$ 2209
$C_{18}H_{22}O_2$ 2903
$C_{18}H_{23}N_3$ 4133
$C_{18}H_{24}O_2$ 2900
$C_{18}H_{24}O_3$ 2902
$C_{18}H_{26}O_4$ 1746-7
$C_{18}H_{26}O_{12}$ 3610; 3612
$C_{18}H_{28}O_4$ 2873
$C_{18}H_{30}O_2$ 3977
$C_{18}H_{30}O_4$ 2084
$C_{18}H_{32}O_2$ 1279; 2871; 3976; 5639
$C_{18}H_{32}O_3$ 5569
$C_{18}H_{32}O_4$ 5641
$C_{18}H_{32}O_6$ 3434-5

$C_{18}H_{32}O_7$ 6119
$C_{18}H_{32}O_{16}$ 5549
$C_{18}H_{34}O_2$ 2870; 3495; 4992
$C_{18}H_{34}O_3$ 5566; 5568
$C_{18}H_{34}O_4$ 1909; 3452
$C_{18}H_{34}O_6$ 2097
$C_{18}H_{35}N$ 5638
$C_{18}H_{36}O$ 5633
$C_{18}H_{36}O_2$ 1274; 2683; 3112; 5631
$C_{18}H_{36}O_3$ 3829-31
$C_{18}H_{36}O_4$ 2365-6
$C_{18}H_{37}I$ 4940
$C_{18}H_{38}O$ 4939
$C_{18}H_{39}N$ 6210

18 III

$C_{18}H_{10}O_6N_2$ 3859
$C_{18}H_{12}ON_2$ 670
$C_{18}H_{12}O_4N_2$ 3205
$C_{18}H_{14}O_2N_4$ 1778
$C_{18}H_{14}O_8Ca$ 116
$C_{18}H_{14}O_8Mg$ 118
$C_{18}H_{15}O_3P$ 6343
$C_{18}H_{15}O_4P$ 6341
$C_{18}H_{15}ClSn$ 5948
$C_{18}H_{15}Cl_2Bi$ 858
$C_{18}H_{15}Cl_2Sb$ 564
$C_{18}H_{16}O_2N_2$ 508; **5171**
$C_{18}H_{17}O_3Na$ 5120
$C_{18}H_{18}O_4N_2$ 5590
$C_{18}H_{22}O_4N_4$ 5193
$C_{18}H_{27}O_3N$ 1227
$C_{18}H_{33}O_3Na$ 5567
$C_{18}H_{35}OCl$ 5637
$C_{18}H_{37}ON$ 5634
$C_{18}H_{39}O_2N$ 5632

18 IV

$C_{18}H_{12}O_4Cl_3P$ 6165
$C_{18}H_{13}O_4Cl_2P$ 2046
$C_{18}H_{14}O_4ClP$ 2713
$C_{18}H_{15}OS_3P$ 6350
$C_{18}H_{15}O_3SP$ 6346
$C_{18}H_{16}O_6N_2S$ 3827
$C_{18}H_{16}O_6N_2Ca$ 3665
$C_{18}H_{23}O_2N_2Cl$ 2910
$C_{18}H_{28}O_4N_2S$ 655
$C_{18}H_{32}O_5N_2S$ 5031
$C_{18}H_{32}O_{10}N_2Ca$ 5035

18 V

$C_{18}H_{14}O_3N_3SNa$ 4056; 4996

19 I

$C_{19}H_{15}$ 6339
$C_{19}H_{16}$ 802-3; 6337
$C_{19}H_{40}$ 4908

Table 7-2 (*Continued*)
FORMULA INDEX FOR ORGANIC COMPOUNDS

19 II

$C_{19}H_{12}O_2$ 4479
$C_{19}H_{13}N$ 5147
$C_{19}H_{14}O$ 2741
$C_{19}H_{14}O_2$ 652; 5255
$C_{19}H_{14}O_3$ 593
$C_{19}H_{14}O_5$ 2766
$C_{19}H_{15}N$ 725
$C_{19}H_{15}Cl$ 6333
$C_{19}H_{15}Br$ 6330
$C_{19}H_{16}O$ 6331
$C_{19}H_{16}O_3$ 3963
$C_{19}H_{16}N_2$ 729
$C_{19}H_{17}N$ 384-5; 804
$C_{19}H_{17}N_3$ 6334-5
$C_{19}H_{18}O$ 2824
$C_{19}H_{18}O_3$ 1758; 3221
$C_{19}H_{18}N_2$ 1733
$C_{19}H_{19}N_3$ 6071-3
$C_{19}H_{22}O_3$ 5004
$C_{19}H_{28}O_2$ 1657; 5711-2
$C_{19}H_{30}O_2$ 509
$C_{19}H_{36}O_2$ 4327
$C_{19}H_{38}O$ 2682; 4261
$C_{19}H_{38}O_2$ 4381; 4909

19 III

$C_{19}H_{13}O_6N_3$ 6322
$C_{19}H_{13}O_7N_3$ 6321
$C_{19}H_{14}O_5S$ 5127
$C_{19}H_{15}O_2N$ 212
$C_{19}H_{17}O_2N$ 3130
$C_{19}H_{18}N_3Cl$ 5039
$C_{19}H_{19}ON_3$ 5038
$C_{19}H_{20}O_4N_2$ 568
$C_{19}H_{21}O_5N$ 6420
$C_{19}H_{24}O_2Br_2$ 865

19 IV

$C_{19}H_{10}O_5SBr_4$ 5724
$C_{19}H_{10}O_5SI_4$ 5815
$C_{19}H_{10}O_{13}N_4S$ 5858
$C_{19}H_{14}ONCl$ 3980
$C_{19}H_{16}O_2NCl$ 213

20 I

$C_{20}H_{14}$ 2574-5; 5156
$C_{20}H_{18}$ 2756-7
$C_{20}H_{38}$ 5293
$C_{20}H_{40}$ 5297
$C_{20}H_{42}$ 1017; 2868; 5294

20 II

$C_{20}H_{12}O_3$ 3244
$C_{20}H_{12}O_5$ 3250; 3720
$C_{20}H_{14}O$ 4560-2
$C_{20}H_{14}O_2$ 2336-7; 5292
$C_{20}H_{14}O_4$ 2737; 5121; 5553
$C_{20}H_{14}O_5$ 3254
$C_{20}H_{14}N_2$ 604-5

$C_{20}H_{14}S$ 2583-4
$C_{20}H_{14}Hg$ 4045
$C_{20}H_{15}N$ 2576
$C_{20}H_{15}N_3$ 251; 1763
$C_{20}H_{16}O_2$ 6327; 6338
$C_{20}H_{16}O_3$ 5573
$C_{20}H_{16}O_4$ 5122
$C_{20}H_{16}N_2$ 3690-1
$C_{20}H_{16}N_4$ 4872
$C_{20}H_{18}O$ 6332
$C_{20}H_{18}O_2$ 529; 3717
$C_{20}H_{19}N$ 1784
$C_{20}H_{20}N_2$ 2839-40
$C_{20}H_{21}N_3$ 3962
$C_{20}H_{24}O_4$ 1555
$C_{20}H_{26}O_3$ 2209
$C_{20}H_{26}O_4$ 2088
$C_{20}H_{30}O_2$ 1
$C_{20}H_{30}O_6$ 1027
$C_{20}H_{32}N_2$ 3961
$C_{20}H_{36}O_2$ 3014
$C_{20}H_{38}O_2$ 3109
$C_{20}H_{40}O$ 5298
$C_{20}H_{40}O_2$ 574; 3153
$C_{20}H_{42}O$ 2869; 5295

20 III

$C_{20}H_8O_5Br_4$ 2875-6
$C_{20}H_8O_5I_4$ 2896-7
$C_{20}H_{10}O_3Cl_2$ 3252
$C_{20}H_{10}O_4Cl_4$ 5744
$C_{20}H_{10}O_4Br_4$ 5722
$C_{20}H_{10}O_4I_4$ 5813
$C_{20}H_{10}O_5Cl_2$ 1999; 2000
$C_{20}H_{10}O_5Br_2$ 1838-9
$C_{20}H_{10}O_5I_2$ 2392
$C_{20}H_{10}O_5Na_2$ 3251
$C_{20}H_{14}ON_2$ 619-20
$C_{20}H_{18}ON_2$ 722-3
$C_{20}H_{20}O_5N_2$ 566
$C_{20}H_{20}N_3Cl$ 5571
$C_{20}H_{21}ON_3$ 5570
$C_{20}H_{22}O_4N_2$ 5307
$C_{20}H_{33}ON$ 4454
$C_{20}H_{41}ONa$ 5296

20 IV

$C_{20}H_6O_5Br_4Na_2$ 2876
$C_{20}H_6O_5I_4Na_2$ 2897
$C_{20}H_8O_4Br_4Na_2$ 5723
$C_{20}H_8O_4I_4Na_2$ 5814
$C_{20}H_{14}O_8S_2Ca$ 4499; 4506; 4508
$C_{20}H_{30}O_2N_3Cl$ 1025
$C_{20}H_{32}O_6N_2S$ 3676-7

20 V

$C_{20}H_8O_6Br_2HgNa_2$ 4041
$C_{20}H_8O_{10}S_2Br_4Na_2$ 998
$C_{20}H_{11}O_{10}N_2S_3Na_3$ 3732

21 I

$C_{21}H_{16}$ 822; 2580-2; 4190
$C_{21}H_{20}$ 5183
$C_{21}H_{42}$ 3504
$C_{21}H_{44}$ 3503

21 II

$C_{21}H_{14}O$ 2577-9
$C_{21}H_{16}N_2$ 3978
$C_{21}H_{18}O$ 2078; 2740
$C_{21}H_{18}N_2$ 233; 3700
$C_{21}H_{20}O_6$ 1575
$C_{21}H_{20}O_9$ 227
$C_{21}H_{21}N$ 6084
$C_{21}H_{21}N_3$ 512
$C_{21}H_{24}O_4$ 2153
$C_{21}H_{30}O_2$ 5333-4
$C_{21}H_{30}N_2$ 5761
$C_{21}H_{32}O_2$ 4084
$C_{21}H_{32}O_{11}$ 3365
$C_{21}H_{36}O_2$ 5329-32
$C_{21}H_{38}O_6$ 3423
$C_{21}H_{40}O_3$ 5782
$C_{21}H_{42}O_4$ 3415-6

21 III

$C_{21}H_{15}ON$ 698
$C_{21}H_{15}O_3N$ 6081
$C_{21}H_{16}ON_2$ 2585
$C_{21}H_{18}O_5S$ 1550-1
$C_{21}H_{20}O_4N_2$ 228
$C_{21}H_{21}O_3P$ 6354-5
$C_{21}H_{21}O_4P$ 6351-3
$C_{21}H_{24}O_5N_2$ 2779
$C_{21}H_{28}ON_2$ 5760

21 IV

$C_{21}H_{14}O_5SBr_4$ 5718
$C_{21}H_{16}O_5SBr_2$ 1832
$C_{21}H_{21}O_3SP$ 6356-8

22 I

$C_{22}H_{14}$ 1777; 5299
$C_{22}H_{46}$ 2855

22 II

$C_{22}H_{15}N_3$ 5572
$C_{22}H_{16}O_3$ 6082-3
$C_{22}H_{16}C_3$ 4999
$C_{22}H_{18}O_4$ 1548; 1789; 2841
$C_{22}H_{20}O_4$ 1549
$C_{22}H_{22}O_8$ 5306
$C_{22}H_{28}O_3$ 5596
$C_{22}H_{32}O_3$ 507
$C_{22}H_{34}O_2$ 2932
$C_{22}H_{40}O_2$ 624
$C_{22}H_{42}O_2$ 873; 1110; 2892
$C_{22}H_{42}O_3$ 1127-8
$C_{22}H_{42}O_4$ 1924; 3201
$C_{22}H_{44}O$ 2893

Table 7-2 (*Continued*)
FORMULA INDEX FOR ORGANIC COMPOUNDS

$C_{22}H_{44}O_2$ 623; 1129-30
$C_{22}H_{44}O_4$ 2309

22 III

$C_{22}H_{16}ON_4$ 5664
$C_{22}H_{21}O_2N$ 5603
$C_{22}H_{23}O_4P$ 2707
$C_{22}H_{40}O_4N_8$ 3601

22 IV

$C_{22}H_{24}O_4N_2S$ 2542

23 I

$C_{23}H_{48}$6171

23 II

$C_{23}H_{17}N$ 50
$C_{23}H_{18}O$ 2732
$C_{23}H_{22}O_6$ 1655; 5574
$C_{23}H_{22}O_7$ 5699
$C_{23}H_{26}N_2$ 3964
$C_{23}H_{27}N_3$ 5844
$C_{23}H_{32}O_6$ 5645
$C_{23}H_{44}O_2$ 475
$C_{23}H_{46}O$ 2850
$C_{23}H_{46}O_2$ 488-9

23 III

$C_{23}H_{23}N_2I$ 2144
$C_{23}H_{25}O_2N_3$ 5838
$C_{23}H_{25}N_2Cl$ (ZnCl$_2$) 3993
$C_{23}H_{26}ON_2$ 5835

23 IV

$C_{23}H_{26}O_3N_3Cl$ 1760

24 I

$C_{24}H_{18}$ 6329
$C_{24}H_{50}$ 5747

24 II

$C_{24}H_{16}O_7$ 3253
$C_{24}H_{20}O_6$ 3420
$C_{24}H_{20}N_2$ 2706; 5863
$C_{24}H_{20}Pb$ 3959
$C_{24}H_{20}Sn$ 5952
$C_{24}H_{28}O_4$ 4931-2
$C_{24}H_{28}N_2$ 5837
$C_{24}H_{40}O_2$ 5238
$C_{24}H_{40}O_4$ 1479
$C_{24}H_{40}O_5$ 1487
$C_{24}H_{44}O_6$ 3425
$C_{24}H_{46}O_3$ 3954
$C_{24}H_{48}O_2$ 3970

24 III

$C_{24}H_{19}O_4P$ 2762
$C_{24}H_{19}O_5N$ 1675
$C_{24}H_{20}BrAs$ 5859

$C_{24}H_{21}O_3N_3$ 801
$C_{24}H_{25}ON_3$ 749
$C_{24}H_{28}ON_2$ 5836
$C_{24}H_{28}N_3Cl$ 5077
$C_{24}H_{29}ON_3$ 5076
$C_{24}H_{41}ON$ 5636

24 IV

$C_{24}H_{14}O_2N_2S_2$ 3868
$C_{24}H_{26}O_4N_4S$ 316

24 V

$C_{24}H_{20}O_6N_2S_2Ba$ 2702

25 I

$C_{25}H_{20}$ 5864
$C_{25}H_{52}$ 5058

25 II

$C_{25}H_{21}N_3$ 5862
$C_{25}H_{22}O_3$ 2885
$C_{25}H_{26}O_9$ 3226
$C_{25}H_{30}O_4$ 4933
$C_{25}H_{31}N_3$ 3595
$C_{25}H_{50}O_2$ 3847

25 III

$C_{25}H_{20}ON_2$ 5865
$C_{25}H_{25}N_2I$ 5316
$C_{25}H_{27}N_2I$ 5003
$C_{25}H_{30}N_3Cl$ 3593
$C_{25}H_{31}ON_3$ 3592

25 IV

$C_{25}H_{29}ON_2I$ 2144

26 I

$C_{26}H_{18}$ 2734
$C_{26}H_{20}$ 5861
$C_{26}H_{22}$ 5860
$C_{26}H_{52}$ 1271
$C_{26}H_{54}$ 3567-8

26 II

$C_{26}H_{20}O$ 731
$C_{26}H_{22}O$ 692
$C_{26}H_{22}O_2$ 732; 2853
$C_{26}H_{22}N_4$ 706-7
$C_{26}H_{42}O_{11}$ 3364
$C_{26}H_{50}O_4$ 3458
$C_{26}H_{52}O_2$ 1272; 4983
$C_{26}H_{54}$ 1273

26 III

$C_{26}H_{31}O_4P$ 1907
$C_{26}H_{33}N_3Cl_2$ 3594
$C_{26}H_{43}O_6N$ 3440

26 IV

$C_{26}H_{42}O_6NNa$ 3441
$C_{26}H_{45}O_7NS$ 5697

26 V

$C_{26}H_{44}O_7NSNa$ 5698

27 I

$C_{27}H_{18}$ 6362
$C_{27}H_{56}$ 3507

27 II

$C_{27}H_{26}O_7$ 6085
$C_{27}H_{36}O_4$ 595
$C_{27}H_{38}O_4$ 594
$C_{27}H_{44}O_3$ 5602
$C_{27}H_{44}O_4$ 1478
$C_{27}H_{46}O$ 1480-2
$C_{27}H_{48}O$ 2276
$C_{27}H_{50}O_6$ 3424

27 III

$C_{27}H_{34}ON_2$ 5762

27 IV

$C_{27}H_{28}O_5SBr_2$ 1879
$C_{27}H_{34}O_4N_2S$ 5763

28 I

$C_{28}H_{58}$ 4937

28 II

$C_{28}H_{18}O_4$ 4497
$C_{28}H_{20}N_2$ 234
$C_{28}H_{30}O_4$ 5941
$C_{28}H_{34}O_{15}$ 3560
$C_{28}H_{44}O$ 1187; 2889; 3979
$C_{28}H_{54}O_3$ 4453
$C_{28}H_{54}O_4$ 2227

28 III

$C_{28}H_{24}O_7N_2$ 4997
$C_{28}H_{33}O_{11}N_5$ 1044

28 IV

$C_{28}H_{31}O_3N_2Cl$ 5766

29 I

$C_{29}H_{60}$ 4907

29 II

$C_{29}H_{48}O_2$ 1483

29 III

$C_{29}H_{35}N_2I$ 1580

30 I

$C_{30}H_{60}$ 4021
$C_{30}H_{62}$ 6057; 4021

Table 7-2 (*Continued*)
FORMULA INDEX FOR ORGANIC COMPOUNDS

30 II

$C_{30}H_{48}O_2$ 4444
$C_{30}H_{48}O_3$ 849
$C_{30}H_{50}O_2$ 848; 1486
$C_{30}H_{58}O_4$ 3461
$C_{30}H_{62}O$ 4449

30 III

$C_{30}H_{23}O_4P$ 5184
$C_{30}H_{39}O_4P$ 6120

31 I

$C_{31}H_{64}$ 3505

31 II

$C_{31}H_{46}O_2$ 4342
$C_{31}H_{62}O$ 5030
$C_{31}H_{62}O_2$ 4022
$C_{31}H_{64}O$ 4449

31 III

$C_{31}H_{43}ON_3$ 3191

32 I

$C_{32}H_{66}$ 2866

32 II

$C_{32}H_{52}O_2$ 2867
$C_{32}H_{62}O_3$ 5026
$C_{32}H_{64}O_2$ 1278

32 III

$C_{32}H_{36}O_6N_4$ 851
$C_{32}H_{36}O_8N_4$ 852

32 V

$C_{32}H_{22}O_6N_6S_2Na_2$ 1527

33 I

$C_{33}H_{68}$ 6359

33 II

$C_{33}H_{62}O_6$ 3422
$C_{33}H_{66}O_2$ 4448

34 I

$C_{34}H_{70}$ 5868

34 II

$C_{34}H_{50}O_2$ 1484-5
$C_{34}H_{66}O_4$ 3464

34 IV

$C_{34}H_{33}O_5N_4Fe$ 3499

34 V

$C_{34}H_{26}O_6N_6S_2Na_2$ 733
$C_{34}H_{26}O_{14}N_6S_4Na_4$ 3228
$C_{34}H_{32}O_4N_4ClFe$ 3501

35 I

$C_{35}H_{72}$ 5080

35 II

$C_{35}H_{28}O_2$ 647-8
$C_{35}H_{40}O_{12}$ 3239
$C_{35}H_{52}O_4$ 4343
$C_{35}H_{70}O$ 5640

36 I

$C_{36}H_{74}$ 3613

36 II

$C_{36}H_{70}O_3$ 5635

36 III

$C_{36}H_{27}O_4P$ 6360
$C_{36}H_{43}O_4P$ 1146

36 IV

$C_{36}H_{62}O_8N_4S$ 1890

37 V

$C_{37}H_{27}O_9N_3S_3Na_2$ 6345

38 II

$C_{38}H_{30}O_2$ 6340
$C_{38}H_{74}O_4$ 3467

38 III

$C_{38}H_{32}N_3Cl$ 516

39 II

$C_{39}H_{74}O_6$ 3426

40 I

$C_{40}H_{56}$ 1252-4; 3985

40 II

$C_{40}H_{50}O_2$ 5562
$C_{40}H_{56}O$ 1570
$C_{40}H_{56}O_2$ 6452

42 IV

$C_{42}H_{84}O_9NP$ 3960

45 II

$C_{45}H_{86}O_6$ 3427

51 II

$C_{51}H_{98}O_6$ 3431

52 III

$C_{52}H_{54}O_{12}N_4$ 3994

57 II

$C_{57}H_{104}O_6$ 3430
$O_{57}H_{110}O_6$ 3433

63 II

$C_{63}H_{122}O_6$ 3419

69 IV

$C_{69}H_{75}N_6Cl_7Zn_2$ 3993

76 IV

$C_{76}H_{64}O_4N_6S$ 6344

Table 7-3
MELTING POINTS OF ORGANIC COMPOUNDS

The compounds below are arranged in the order of the ascending values of the melting points, an arrangement intended to serve as an aid in the identification of organic compounds. The values given may be uncertain by one or more units in the last figure and consequently are not to be used as fixed points for calibration. See special table for *Calibration of Thermometers and Thermocouples.* The numbers listed under a temperature-interval heading refer to the compounds with corresponding numerical listings in the table *Physical Constants of Organic Compounds.* Thus, under the temperature interval 186° to 190° C. is listed *No. 31* which is *acetamino-α-naphthol* m. p. 187°C.

Below −140° C.

498; 499; 872; 1022; 1147; 1149; 1239; 1772; 2246; 2904; 3192; 3574; 3604; 3656; 3933; 4061; 4154; 4209; 5084; 5338; 5459; 6203; 6435; 6439

−130° to −139° C.

183; 202; 500; 501; 502; 1018; 1021; 1072; 1074; 1248; 2478; 3015; 3517; 3655; 3659; 3660; 3661; 3957; 5083; 5346; 5425; 6434

−120° to −129° C.

5; 193; 990; 1148; 1156; 1621; 2783; 2911; 3083; 3271; 3515; 3606; 4155; 4199; 4298; 4957; 4958; 5083; 5368; 5400; 6248

−110° to −119° C.

81; 199; 406; 988; 1035; 1057; 1058; 1059; 1103; 1175; 2245; 2415; 2911; 2946; 2989; 3210; 3280; 3513; 3514; 3516; 3518; 3554; 3556; 3607; 3657; 4154; 4161; 4917; 4946; 4950; 5388; 5401; 5424; 5950; 6180; 6404; 6438

−100° to −109° C.

225; 415; 1019; 1036; 1040; 1095; 1096; 1148.1; 1238; 1240; 1247; 1400; 1609; 1680; 1885; 1916; 1923; 2778; 2782; 3062; 3077; 3096; 3631; 4238; 4424; 4827; 4914; 4916; 4947; 4948; 4949; 4956; 4984; 5086; 5087; 5372; 5468

−90° to −99° C.

56; 75; 208; 427; 442; 455; 871; 1031; 1089; 1097; 1142; 1160; 1174; 1599; 1600; 1619; 1898; 1993; 2208; 2214; 2429; 2499; 2973; 3006; 3167; 3185; 3186; 3288; 3357; 3512; 3578; 3603; 3605; 3973; 4087; 4100; 4105; 4147; 4169; 4183; 4242; 4357; 4420; 4429; 4733; 4828; 4953; 4960; 4986; 5345; 5360; 5363; 5365; 5378; 5379; 5392; 5408; 5417; 5418; 5458; 5960; 6348; 6400; 6405; 6426

−80° to −89° C.

130; 135; 146; 151; 204; 220; 459; 1041; 1049; 1064; 1104; 1123; 1173; 1216; 1246; 1378; 1579; 1618; 1684; 2099; 2127; 2473; 2484; 2493; 2560; 2935; 2950; 3007; 3019; 3142; 3158; 3166; 3279; 3285; 3460; 3558; 4058; 4088; 4164; 4170; 4171; 4232; 4269; 4349; 4422; 4735; 4911; 5352; 5369; 5371; 5389; 5418; 6121

−70° to −79° C.

37; 161; 397; 404; 421; 432; 447; 454; 483; 989; 1029; 1034; 1050; 1104; 1125; 1217; 1265; 1379; 1622; 1636; 1825; 1886; 1914; 2051; 2240; 2472; 2769; 2942; 3048; 3049; 3138; 3279; 3418; 3421; 3449; 3474; 3557; 3658; 4123; 4156; 4355; 4965; 4991; 5300; 5366; 5441; 5455; 6155

−60° to −69° C.

89; 1042; 1083; 1141; 1163; 1266; 1380; 1477; 1559; 1623; 1624; 1740; 1821; 1884; 1980; 2172; 2232; 2483; 2485; 2502; 2569; 2773; 2784; 2965; 3009; 3029; 3108; 3168; 3194; 3266; 3622; 3628; 3648; 3708; 4182; 4262; 4287; 4352; 4697; 4910; 4985; 5046; 5773

−50° to −59° C.

132; 136; 412; 724; 781; 886; 1039; 1051; 1052; 1132; 1171; 1183; 1237; 1281; 1344; 1563; 1595; 1620; 1631; 1638; 1864; 1870; 1970; 2133; 2184; 2242; 2407; 2443; 2471; 2486; 2792; 3269; 3459; 3624; 4053; 4125; 4163; 4166; 4214; 4289; 4387; 4392; 4428; 4944; 5318; 5380; 5710; 6159; 6255; 6399; 6403; 6407; 6448

−40° to −49° C.

49; 319; 994; 1167; 1179; 1180; 1214; 1218; 1301; 1324; 1467; 1606; 1613; 1663; 1735; 1997; 2128; 2138; 2180; 2184; 2193; 2233; 2749; 2777; 2937; 3010; 3113; 3114; 3195; 3257; 3258; 3263; 3521; 3637; 3944; 4174; 4195; 4694; 4980; 5146; 5317; 5357; 5483; 6254; 6255; 6285; 6457; 6504

−30° to −39° C.

537; 796; 843; 902; 1000; 1069; 1098; 1157; 1209; 1466; 1565; 1588; 1632; 1635; 1679; 1824; 1857; 1865; 1883; 1973; 1994; 2053; 2109; 2121; 2126; 2195; 2238; 2772; 2786; 2863; 2975; 3003; 3026; 3047; 3084; 3318; 3406; 3445; 3527; 3529; 3530; 3543; 3637; 4122; 4184; 4383; 4384; 4711; 4737; 4919; 4959; 4962; 5111; 5466; 5646; 5739; 5786; 5923; 5997; 6122; 6152; 6198; 6206; 6207; 6262; 6349; 6382; 6399; 6406; 6427

−20° to −29° C.

68; 91; 211; 306; 470; 642; 735; 800; 916; 926; 933; 948; 986; 999; 1014; 1028; 1053; 1075; 1140; 1152; 1162; 1165; 1233; 1243; 1340; 1346; 1361; 1377; 1381; 1402; 1442; 1459; 1462; 1474; 1475; 1635; 1783; 1823; 1880; 1905; 1910; 1953; 1981; 2106; 2176; 2195; 2202; 2211; 2213; 2416; 2559; 2718; 2771; 2879; 3008; 3016; 3030; 3035; 3057; 3084; 3089; 3265; 3272; 3377; 3393; 3411; 3423; 3424; 3462; 3519; 3544; 3683; 3846; 3879; 3958; 3968; 4025; 4089; 4196; 4201; 4237; 4265; 4282; 4372; 4407; 4430; 4729; 4736; 4765; 4891; 4920; 4964; 5054; 5068; 5105; 5224; 5241; 5351; 5375; 5429; 5828; 5895; 5993; 5995; 6017; 6042; 6103; 6116; 6119; 6143; 6167; 6253; 6282; 6370; 6454; 6503

−10° to −19° C.

55; 57; 137; 190; 230; 409; 493; 535; 544; 635; 716; 782; 845; 915; 936; 1012; 1060; 1256; 1310; 1314; 1343; 1391; 1473; 1538; 1596; 1952; 2101; 2109; 2123; 2183; 2230; 2399; 2481; 2507; 2517; 2636; 2796; 2831; 2906; 3080; 3090; 3109; 3163; 3170; 3223; 3272; 3444; 3453; 3463; 3520; 3547; 3629; 3638; 3639; 3647; 3705; 3714; 3756; 3846; 3955; 4010; 4080; 4126; 4138; 4172; 4181; 4229; 4265; 4266; 4304; 4311; 4403; 4788; 4844; 4961; 4973; 5011; 5085; 5249; 5341; 5538; 5741; 5751; 5885; 5900; 5905; 5992; 5995; 6157; 6166; 6471

0° to −9° C.

65; 68; 133; 231; 321; 482; 513; 735; 753; 755; 793; 917; 946; 953; 980; 996; 1011; 1023; 1166; 1178; 1186; 1208; 1221; 1224; 1255; 1309; 1334; 1363; 1387; 1501; 1591; 1748; 1812; 1900; 1907; 1928; 2046; 2075; 2097; 2174; 2235; 2239; 2468; 2707; 2713; 2762; 2768; 2856; 2864; 3017; 3115; 3159;

Table 7-3 (*Continued*)
MELTING POINTS OF ORGANIC COMPOUNDS

3161; 3256; 3262; 3273; 3344; 3430; 3597; 3733; 3855; 3976; 4026; 4200; 4219; 4257; 4317; 4367; 4380; 4455; 4839; 4844; 4918; 5269; 5322; 5414; 5482; 5719; 5720; 5827; 6150; 6173; 6199; 6375; 6378; 6482

1° to 10° C.

7; 155; 531; 532; 534; 643; 656; 753; 755; 762; 807; 945; 956; 971; 972; 995; 1015; 1016; 1092; 1136; 1184; 1255; 1333; 1387; 1427; 1436; 1468; 1470; 1511; 1543; 1601; 1640; 1811; 1837; 1858; 1928; 2083; 2097; 2174; 2206; 2210; 2237; 2389; 2393; 2396; 2428; 2450; 2468; 2501; 2547; 2742; 2827; 2907; 2914; 2943; 2968; 3067; 3088; 3093; 3095; 3151; 3196; 3199; 3217; 3218; 3233; 3296; 3301; 3424; 3428; 3504; 3570; 3627; 3956; 4063; 4097; 4108; 4144; 4295; 4370; 4434; 4457; 4623; 4646; 4688; 4796; 4868; 5060; 5061; 5107; 5179; 5302; 5568; 5582; 5643; 5748; 5752; 5897; 5910; 5974; 6243

11° to 20° C.

10; 36; 43; 51; 85; 86; 149; 569; 678; 894; 898; 973; 1003; 1005; 1112; 1129; 1136; 1220; 1276; 1302; 1318; 1319; 1336; 1346; 1348; 1349; 1383; 1546; 1558; 1617; 1645; 1743; 1985; 2085; 2089; 2092; 2147; 2153; 2188; 2200; 2217; 2218; 2224; 2234; 2237; 2331; 2428; 2533; 2554; 2825; 2907; 2909; 2914; 2915; 2939; 2953; 2956; 3023; 3088; 3107; 3214; 3336; 3381; 3428; 3478; 3524; 3528; 3569; 3571; 3572; 3940; 3967; 4096; 4241; 4302; 4303; 4324; 4326; 4369; 4382; 4456; 4673; 4679; 4713; 4838; 4845; 4865; 4887; 4941; 4992; 5027; 5042; 5061; 5063; 5136; 5196; 5221; 5283; 5348; 5433; 5439; 5463; 5513; 5525; 5581; 5658; 5749; 5758; 5759; 5797; 5853; 5894; 5908; 5910; 6078; 6112; 6118; 6135; 6154; 6178; 6201; 6246; 6349; 6371; 6372; 6462; 6477; 6484

21° to 30° C.

51; 62; 106; 452; 471; 481; 488; 489; 507; 511; 530; 569; 724; 789; 808; 813; 823; 861; 949; 980; 1001; 1024; 1037; 1079; 1129; 1130; 1274; 1277; 1332; 1339; 1365; 1367; 1413; 1425; 1426; 1469; 1545; 1584; 1605; 1849; 1919; 1932; 1956; 1983; 1989; 2085; 2090; 2147; 2165; 2227; 2254; 2263; 2278; 2386; 2406; 2466; 2526; 2552; 2719; 2731; 2802; 2822; 2859; 2914; 2927; 2944; 3053; 3099; 3112; 3129; 3219; 3274; 3336; 3422; 3452; 3506; 3508; 3511; 3577; 3643; 3883; 3897; 4006; 4071; 4073; 4079; 4083; 4085; 4108; 4146; 4248; 4330; 4395; 4408; 4451; 4679; 4693; 4740; 4840; 4866; 4938; 4942; 5154; 5186; 5196; 5268; 5353; 5431; 5432; 5434; 5474; 5513; 5540; 5548; 5637; 5772; 5797; 5820; 5891; 5901; 5994; 6043; 6052; 6174; 6178; 6200; 6349; 6376; 6377; 6443; 6469; 6485

31° to 40° C.

73; 79; 212; 243; 254; 531; 615; 643; 653; 756; 786; 798; 818; 831; 832; 893; 927; 930; 969; 973; 1024; 1119; 1205; 1350; 1365; 1373; 1412; 1428; 1443; 1453; 1455; 1457; 1458; 1465; 1505; 1507; 1545; 1547; 1670; 1756; 1787; 1859; 1935; 2009; 2022; 2055; 2084; 2089; 2096; 2148; 2190; 2265; 2267; 2332; 2387; 2400; 2498; 2552; 2615; 2683; 2725; 2803; 2868; 2877; 2892; 2893; 2930; 2956; 3050; 3094; 3124; 3153; 3169; 3182; 3201; 3234; 3255; 3319; 3383; 3422; 3426; 3451; 3466; 3488; 3503; 3577; 3843; 3849; 3876; 3883; 3894; 3898; 3899; 3907; 3934; 3936; 3967; 4006; 4015; 4028; 4065; 4150; 4191; 4239; 4305; 4314; 4329; 4330; 4336; 4374; 4381; 4388; 4394; 4487; 4624; 4637; 4674; 4703; 4710; 4723; 4750; 4767; 4797; 4808; 4848; 4908; 4940; 4983; 5023; 5028; 5166; 5312; 5313; 5323; 5353; 5518; 5633; 5706; 5707; 5708; 5759; 5755; 5795; 5909; 5912; 5991; 6016; 6038; 6040;

6056; 6165; 6174; 6245; 6357; 6440; 6450; 6461; 6487; 6495; 6507

41° to 50° C.

10; 147; 274; 307; 329; 563; 626; 632; 636; 675; 714; 724; 761; 770; 799; 866; 883; 887; 888; 911; 914; 918; 960; 1007; 1119; 1121; 1181; 1189; 1229; 1242; 1275; 1282; 1286; 1299; 1300; 1320; 1338; 1355; 1367; 1413; 1429; 1443; 1456; 1457; 1500; 1508; 1535; 1577; 1578; 1615; 1633; 1659; 1668; 1711; 1789; 1790; 1791; 1792; 1795; 1798; 1918; 1930; 1934; 1937; 2013; 2025; 2042; 2055; 2087; 2091; 2111; 2201; 2226; 2387; 2400; 2425; 2457; 2506; 2510; 2564; 2669; 2712; 2716; 2725; 2767; 2791; 2821; 2823; 2834; 2836; 2855; 2871; 2977; 3025; 3026; 3031; 3100; 3102; 3116; 3119; 3133; 3157; 3187; 3235; 3291; 3299; 3384; 3402; 3426; 3427; 3431; 3488; 3565; 3596; 3598; 3698; 3704; 3753; 3754; 3901; 3950; 3951; 3952; 3954; 4024; 4090; 4279; 4301; 4333; 4446; 4486; 4506; 4526; 4529; 4580; 4636; 4637; 4676; 4682; 4683; 4712; 4762; 4764; 4799; 4843; 4905; 5029; 5062; 5094; 5118; 5123; 5134; 5137; 5153; 5155; 5169; 5204; 5212; 5235; 5247; 5265; 5285; 5312; 5313; 5364; 5416; 5485; 5517; 5534; 5638; 5639; 5728; 5733; 5756; 5761; 5874; 5875; 5896; 5917; 5943; 5973; 5999; 6028; 6095; 6128; 6148; 6168; 6171; 6176; 6190; 6244; 6273; 6288; 6341; 6356; 6392; 6424; 6450; 6468; 6472; 6498; 6502; 6507

51° to 55° C.

88; 102; 267; 268; 410; 563; 568; 612; 613; 802; 823; 877; 878; 888; 916; 929; 1067; 1229; 1278; 1286; 1303; 1354; 1357; 1367; 1376; 1472; 1590; 1615; 1764; 1792; 1834; 1860; 1868; 1937; 1954; 2023; 2098; 2253; 2476; 2521; 2527; 2539; 2549; 2659; 2665; 2668; 2685; 2698; 2717; 2736; 2870; 2921; 2934; 3031; 3291; 3307; 3321; 3347; 3408; 3410; 3431; 3433; 3458; 3821; 3864; 3892; 3893; 3908; 3939; 3951; 4030; 4081; 4090; 4140; 4175; 4261; 4299; 4313; 4333; 4453; 4506; 4517; 4556; 4559; 4580; 4625; 4762; 4807; 4846; 4853; 4897; 4904; 4935; 4998; 5055; 5058; 5074; 5155; 5183; 5238; 5251; 5254; 5266; 5325; 5383; 5477; 5515; 5566; 5569; 5661; 5734; 5747; 5756; 5839; 5938; 6014; 6128; 6134; 6162; 6168; 6190; 6346; 6401; 6494

56° to 60° C.

11; 53; 61; 102; 247; 248; 360; 439; 536; 580; 624; 625; 632; 741; 780; 811; 817; 821; 828; 834; 928; 930; 957; 961; 1002; 1043; 1070; 1165; 1271; 1282; 1299; 1354; 1368; 1369; 1376; 1389; 1395; 1403; 1418; 1490; 1500; 1586; 1593; 1611; 1660; 1802; 1827; 1831; 1860; 2160; 2323; 2346; 2348; 2408; 2544; 2669; 2670; 2674; 2682; 2703; 2714; 2755; 2756; 2833; 2842; 2969; 2979; 3054; 3075; 3101; 3125; 3407; 3415; 3427; 3482; 3507; 3522; 3567; 3662; 3680; 3699; 3736; 3759; 3765; 3856; 3875; 3908; 3996; 4014; 4087; 4261; 4272; 4283; 4285; 4313; 4379; 4450; 4556; 4559; 4570; 4639; 4721; 4731; 4772; 4789; 4798; 4811; 4841; 4889; 4890; 4935; 4939; 4998; 5050; 5079; 5129; 5181; 5183; 5192; 5203; 5216; 5251; 5278; 5382; 5398; 5515; 5541; 5702; 5721; 5790; 5824; 5939; 5954; 6011; 6018; 6022; 6051; 6056; 6128; 6299; 6347; 6441

61° to 65° C.

27; 61; 330; 361; 440; 566; 627; 663; 675; 719; 741; 743; 758; 780; 873; 895; 912; 968; 974; 981; 1045; 1068; 1227; 1283; 1284; 1299; 1306; 1404; 1418; 1422; 1433; 1583; 1586; 1768; 1785; 1828; 1867; 1868; 1933; 1978; 2010; 2014; 2026; 2188; 2252; 2348; 2374; 2413; 2426; 2524; 2546; 2564; 2568; 2650; 2666; 2669; 2670; 2694; 2715; 2720; 2756; 2869; 2895; 2922; 2959; 3054; 3289; 3295; 3431; 3433; 3438; 3461; 3468; 3486; 3509; 3568; 3576;

Table 7-3 (*Continued*)
MELTING POINTS OF ORGANIC COMPOUNDS

3699; 3759; 3875; 4014; 4021; 4070; 4098; 4272; 4323; 4343; 4431; 4483; 4635; 4657; 4686; 4712; 4721; 4724; 4864; 4907; 4937; 4943; 5022; 5026; 5044; 5175; 5206; 5229; 5231; 5258; 5290; 5398; 5435; 5541; 5823; 5954; 5978; 6007; 6011; 6026; 6136; 6147; 6161; 6163; 6270; 6272; 6294; 6379; 6419; 6458; 6466; 6497

66° to 70° C.

14; 27; 44; 103; 281; 314; 367; 371; 567; 584; 599; 657; 691; 704; 758; 767; 879; 1017; 1093; 1279; 1291; 1311; 1326; 1341; 1356; 1404; 1406; 1437; 1497; 1534; 1583; 1672; 1687; 1725; 1766; 1784; 1921; 1965; 2011; 2026; 2064; 2072; 2088; 2120; 2131; 2152; 2156; 2292; 2370; 2405; 2424; 2440; 2512; 2531; 2540; 2600; 2613; 2667; 2692; 2695; 2733; 2735; 2752; 2758; 2810; 2813; 2850; 2866; 2869; 2929; 2972; 2976; 3011; 3110; 3204; 3220; 3410; 3433; 3438; 3495; 3498; 3505; 3654; 3751; 3761; 3767; 3769; 3877; 3927; 4323; 4468; 4488; 4510; 4527; 4576; 4654; 4705; 4768; 4873; 4876; 4909; 5088; 5102; 5159; 5164; 5478; 5541; 5631; 5676; 5743; 5788; 5818; 5940; 5975; 6001; 6026; 6029; 6057; 6130; 6149; 6161; 6164; 6250; 6262; 6270; 6290; 6467; 6476

71° to 75° C.

44; 108; 138; 256; 281; 314; 369; 370; 528; 805; 865; 879; 905; 931; 1122; 1170; 1262; 1297; 1311; 1406; 1430; 1437; 1529; 1557; 1573; 1634; 1672; 1781; 1784; 1882; 1920; 1936; 1979; 2024; 2093; 2120; 2216; 2223; 2292; 2414; 2436; 2551; 2676; 2737; 2752; 2757; 2761; 2866; 2871; 2918; 2972; 3055; 3130; 3204; 3220; 3306; 3323; 3343; 3419; 3420; 3433; 3447; 3464; 3496; 3613; 3825; 3943; 4180; 4312; 4320; 4448; 4618; 4673; 4675; 4678; 4684; 4705; 4726; 4719; 4831; 4869; 4903; 4906; 5048.1; 5080; 5159; 5210; 5217; 5290; 5373; 5631; 5632; 5635; 5647; 5819; 5868; 5883; 6015; 6039; 6145; 6265; 6266; 6276; 6359; 6379; 6465; 6470

76° to 80° C.

19; 23; 104; 171; 221; 260; 281; 309; 313; 574; 623; 649; 688; 742; 759; 819; 857; 865; 879; 885; 905; 919; 931; 1112; 1159; 1273; 1331; 1375; 1430; 1438; 1662; 1723; 1781; 1801; 1861; 1979; 2107; 2115; 2137; 2220; 2273; 2382; 2404; 2446; 2448; 2619; 2676; 2711; 2807; 2832; 3103; 3130; 3188; 3400; 3409; 3414; 3420; 3467; 3468; 3496; 3497; 3613; 3757; 3760; 3825; 3831; 3847; 3853; 4051; 4151; 4221; 4251; 4318; 4458; 4469; 4567; 4574; 4595; 4677; 4719; 4726; 4739; 4773; 4888; 5048.1; 5102; 5142; 5210; 5217; 5228; 5286; 5356; 5373; 5764; 5798; 5829; 5868; 5871; 5883; 5979; 6032; 6055; 6104; 6132; 6145; 6146; 6177; 6226; 6263; 6276; 6326; 6342; 6353; 6428; 6492; 6501

81° to 85° C.

14; 19; 40; 148; 170; 171; 223; 281; 364; 366; 385; 465; 659; 672; 718; 742; 754; 803; 824; 885; 940; 959; 1272; 1307; 1312; 1372; 1414; 1418; 1481; 1710; 1781; 1782; 1800; 1801; 1852; 1862; 1876; 1986; 2043; 2066; 2069; 2074; 2086; 2382; 2391; 2404; 2445; 2453; 2496; 2673; 2677; 2708; 2945; 3242; 3248; 3409; 3415; 3484; 3711; 3727; 3768; 3830; 3870; 3874; 3902; 3970; 4135; 4143; 4192; 4221; 4454; 4471; 4515; 4561; 4568; 4595; 4634; 4685; 4739; 4836; 4857; 4882; 5030; 5053; 5067; 5152; 5180; 5200; 5211; 5308; 5347; 5484; 5543; 5675; 5738; 5872; 5881; 5914; 5964; 5977; 6037; 6091; 6109; 6125; 6145; 6177; 6316; 6337; 6418; 6500

86° to 90° C.

18; 258; 269; 271; 281; 377; 441; 573; 658; 754; 784; 795; 804; 812; 881; 934; 963; 1025; 1111; 1242; 1296; 1364; 1366; 1374; 1460; 1607; 1782; 1813; 1826; 2015; 2069; 2150; 2442; 2479; 2489; 2565; 2603; 2616; 2635; 2645; 2679; 2704; 2708; 2763; 2945; 3070; 3105; 3211; 3221; 3322; 3475; 3477; 3484; 3611; 3667; 3715; 3750; 3790; 3844; 3851; 3889; 3926; 4022; 4049; 4074; 4092; 4115; 4127; 4211; 4212; 4220; 4252; 4273; 4449; 4470; 4592; 4680; 4695; 4701; 4739; 4834; 4883; 5024; 5053; 5059; 5170; 5176; 5209; 5230; 5279; 5284; 5347; 5354; 5423; 5543; 5549; 5564; 5640; 5641; 5675; 5696; 5702; 5731; 5732; 5834; 5964; 6003; 6010; 6091; 6094; 6096; 6145; 6271; 6332; 6420; 6463

91° to 95° C.

2; 3; 22; 42; 281; 333; 366; 369; 390; 466; 476; 494; 607; 633; 697; 750; 812; 822; 925; 934; 947; 977; 1172; 1192; 1215; 1242; 1280; 1330; 1352; 1486; 1510; 1544; 1604; 1721; 1872; 1875; 2067; 2248; 2385; 2442; 2581; 2582; 2593; 2623; 2632; 2671; 2678; 2679; 2838; 2955; 3043; 3068; 3070; 3183; 3221; 3303; 3322; 3470; 3523; 3550; 3611; 3616; 3722; 3766; 3774; 3790; 3814; 3829; 3900; 3964; 4022; 4099; 4131; 4189; 4211; 4284; 4471; 4510; 4516; 4518; 4569; 4588; 4597; 4671; 4680; 4695; 4813; 4852; 5101; 5115; 5170; 5182; 5248; 5415; 5449; 5588; 5590; 5603; 5616; 5636; 5678; 5760; 5834; 5837; 5980; 6012; 6030; 6031; 6033; 6084; 6090; 6145; 6191; 6234; 6237; 6337; 6358; 6491

96° to 100° C.

8; 11; 24; 48; 120; 162; 228; 237; 252; 281; 310; 318; 333; 640; 686; 721; 743; 820; 880; 925; 977; 1025; 1120; 1138; 1213; 1280; 1325; 1358; 1461; 1486; 1516; 1580; 1604; 1762; 1794; 1829; 1890; 1927; 2065; 2309; 2338; 2366; 2385; 2421; 2581; 2617; 2632; 2693; 2740; 2820; 2960; 2971; 3111; 3134; 3154; 3202; 3240; 3243; 3276; 3324; 3376; 3412; 3470; 3480; 3523; 3611; 3612; 3616; 3722; 3748; 3772; 3773; 3775; 3962; 3998; 4064; 4067; 4319; 4435; 4465; 4489; 4571; 4597; 4681; 4802; 4849; 4855; 5045; 5081; 5095; 5101; 5115; 5130; 5240; 5415; 5422; 5555; 5609; 5677; 5715; 5760; 5791; 5869; 5989; 6005; 6033; 6053; 6054; 6089; 6108; 6131; 6133; 6145; 6226; 6240; 6296; 6308; 6314; 6446; 6447

101° to 105° C.

11; 48; 54; 81.1; 156; 209; 281; 572; 661; 686; 738; 840; 904; 1102; 1213; 1223; 1287; 1288; 1409; 1506; 1580; 1713; 1733; 1955; 2044; 2063; 2068; 2081; 2276; 2285; 2312; 2340; 2371; 2487; 2508; 2523; 2577; 2649; 2806; 2853; 3206; 3298; 3303; 3348; 3351; 3611; 3616; 3700; 3707; 3931; 3942; 3953; 3964; 4017; 4049; 4086; 4176; 4256; 4271; 4291; 4385; 4414; 4452; 4466; 4562; 4587; 4596; 4670; 4696; 4742; 4777; 4783; 4850; 4945; 5007; 5115; 5131; 5151; 5205; 5256; 5309; 5310; 5358; 5512; 5584; 5594; 5836; 5879; 5890; 5968; 5972; 5982; 6034; 6059; 6060; 6089; 6120; 6258; 6296; 6315; 6380; 6446; 6447; 6496

106° to 110° C.

26; 41; 54; 74; 84; 142; 233; 238; 246; 259; 270; 281; 328; 363; 368; 565; 591; 596; 628; 658; 692; 723; 766; 791; 809; 879; 967; 975; 1020; 1044; 1190; 1223; 1287; 1295; 1386; 1410; 1460; 1656; 1666; 1713; 1733; 1874; 2012; 2019; 2129; 2225; 2269; 2272; 2285; 2313; 2340; 2509; 2534; 2562; 2572; 2580; 2583; 2605; 2841; 2952; 2958; 3180; 3245; 3359; 3566; 3723; 3734; 3770; 3812; 3916; 3991; 4048; 4068; 4113; 4276; 4306; 4359; 4438; 4503;

Table 7-3 (*Continued*)
MELTING POINTS OF ORGANIC COMPOUNDS

4504; 4510; 4557; 4560; 4594; 4613; 4640; 4667; 4806; 4835; 4850; 4854; 4894; 4998; 5020; 5025; 5207; 5232; 5453; 5497; 5634; 5836; 5886; 5921; 5959; 5983; 6046; 6258; 6296; 6297; 6298; 6402

111° to 115° C.

17; 69; 142; 215; 246; 565; 687; 698; 705; 725; 760; 763; 790; 809; 870; 909; 944; 952; 1004; 1020; 1168; 1187; 1396; 1483; 1666; 1713; 1775; 1853; 1874; 2018; 2129; 2222; 2228; 2313; 2367; 2375; 2509; 2627; 2647; 2675; 2811; 2835; 2841; 2978; 3072; 3082; 3162; 3246; 3362; 3587; 3611; 3712; 3728; 3776; 3885; 4011; 4094; 4109; 4137; 4213; 4290; 4307; 4359; 4503; 4510; 4521; 4530; 4613; 4619; 4698; 4725; 4795; 4803; 4899; 5015; 5133; 5160; 5232; 5280; 5292; 5520; 5542; 5831; 5882; 5884; 5906; 5942; 5948; 5983; 6046; 6111; 6114; 6275; 6296; 6301; 6313; 6333; 6350; 6360; 6411; 6421; 6422; 6499

116° to 120° C.

384; 518; 533; 645; 725; 771; 790; 889; 921; 1168; 1187; 1193; 1195; 1285; 1294; 1405; 1407; 1419; 1499; 1514; 1581; 1698; 1713; 1873; 2007; 2016; 2047; 2062; 2071; 2247; 2373; 2375; 2431; 2461; 2586; 2602; 2675; 2878; 3072; 3246; 3259; 3326; 3472; 3494; 3610; 3617; 3675; 3728; 3735; 3887; 3949; 3979; 3992; 4008; 4055; 4094; 4213; 4334; 4390; 4503; 4510; 4521; 4525; 4563; 4572; 4575; 4591; 4601; 4641; 4658; 4661; 4668; 4687; 4699; 4706; 4734; 4858; 5004; 5021; 5099; 5101; 5143; 5201; 5280; 5305; 5385; 5542; 5549; 5553; 5626; 5652; 5657; 5681; 5714; 5730; 5796; 5833; 5867; 5882; 5884; 5942; 5981; 6080; 6092; 6093; 6097; 6218; 6233; 6275; 6291; 6307; 6309; 6361; 6363; 6421; 6499

121° to 125° C.

272; 347; 533; 660; 674; 711; 815; 901; 921; 1164; 1294; 1665; 1730; 1786; 1850; 2029; 2030; 2048; 2071; 2185; 2322; 2342; 2349; 2372; 2467; 2550; 2579; 2601; 2620; 2652; 2681; 2700; 2748; 2828; 2845; 3215; 3225; 3254; 3259; 3260; 3339; 3702; 3716; 3718; 3752; 3815; 3866; 3945; 4047; 4235; 4321; 4409; 4441; 4471; 4490; 4507; 4520; 4565; 4581; 4593; 4603; 4658; 4676; 4753; 5078; 5099; 5144; 5201; 5214; 5237; 5252; 5324; 5334; 5526; 5546; 5557; 5577; 5642; 5652; 5660; 5697; 5880; 5961; 6080; 6086; 6158; 6302; 6310; 6323; 6391; 6453; 6479

126° to 130° C.

20; 111; 197; 249; 273; 311; 312; 365; 542; 619; 644; 657; 674; 713; 750; 874; 943; 951; 958; 962; 1061; 1164; 1169; 1435; 1665; 1709; 1714; 1717; 1758; 2324; 2347; 2369; 2388; 2418; 2449; 2459; 2511; 2558; 2579; 2587; 2750; 2894; 2999; 3200; 3215; 3254; 3473; 3485; 3575; 3694; 3713; 3717; 3777; 3795; 3813; 3854; 3993; 3995; 3997; 4002; 4227; 4361; 4409; 4415; 4425; 4442; 4520; 4581; 4615; 4702; 4756; 4775; 4781; 4847; 4851; 4874; 4896; 5194; 5202; 5215; 5282; 5333; 5376; 5544; 5546; 5558; 5577; 5652; 5660; 5662; 5679; 5683; 5697; 5825; 5862; 5873; 5880; 5926; 5944; 5955; 6113; 6137; 6228; 6256; 6292; 6328; 6416

131° to 135° C.

25; 47; 50; 96; 115; 233; 257; 262; 273; 299; 542; 606; 710; 717; 836; 874; 922; 924; 1061; 1194; 1322; 1397; 1498; 1503; 1513; 1616; 1696; 1706; 1719; 1759; 1779; 1957; 2017; 2159; 2309; 2310; 2347; 2417; 2460; 2474; 2578; 2633; 2651; 2696; 2727; 2916; 3041; 3212; 3340; 3359; 3363; 3619; 3685; 3694; 3741; 3749; 3854; 3867; 3993; 3995; 4002; 4012; 4093; 4133; 4227; 4274; 4292; 4415; 4446;

4523; 4528; 4615; 4700; 4727; 4833; 4856; 5282; 5319; 5493; 5605; 5625; 5652; 5825; 5862; 6088; 6156; 6160; 6216; 6225; 6335; 6386; 6396; 6409

136° to 140° C.

59; 96; 101; 115; 257; 529; 592; 637; 703; 717; 729; 874; 884; 951; 1038; 1139; 1482; 1503; 1657; 1701; 1733; 1829; 1851; 1962; 2119; 2181; 2310; 2339; 2345; 2352; 2354; 2365; 2375; 2419; 2525; 2532; 2578; 2589; 2590; 2633; 2641; 2721; 2727; 2732; 2798; 2839; 2916; 3240; 3346; 3367; 3369; 3370; 3457; 3500; 3601; 3701; 3738; 3842; 3916; 3966; 4460; 4582; 4602; 4610; 4664; 4672; 4771; 4856; 5173; 5220; 5260; 5370; 5384; 5486; 5649; 5735; 5809; 5817; 5857; 5902; 5927; 5937; 5971; 6002; 6217; 6224; 6239; 6317; 6325; 6409

141° to 145° C.

58; 82; 95; 213; 264; 298; 512; 564; 598; 614; 689; 728; 775; 825; 858; 951; 1139; 1321; 1327; 1502; 1701; 1722; 1855; 1962; 1966; 2003; 2257; 2321; 2345; 2360; 2409; 2422; 2458; 2557; 2574; 2628; 2637; 2646; 2721; 2776; 2798; 2873; 3058; 3104; 3162; 3231; 3240; 3308; 3312; 3313; 3341; 3350; 3367; 3380; 3691; 3737; 3794; 3832; 3966; 4000; 4048; 4112; 4113; 4234; 4268; 4270; 4371; 4460; 4513; 4582; 4611; 4633; 4643; 4661; 4664; 4779; 4780; 4782; 4807; 4824; 4884; 5167; 5291; 5326; 5384; 5550; 5561; 5649; 5676; 5677; 5727; 5817; 5841; 5937; 6044; 6049; 6058; 6072; 6126; 6172; 6217; 6224; 6241; 6334; 6336; 6339; 6388; 6493

146° to 150° C.

34; 97; 99; 180; 213; 227; 229; 295; 375; 506; 561; 699; 727; 745; 775; 835; 837; 841; 844; 906; 1287; 1432; 1441; 1480; 1484; 1502; 1592; 1657; 1669; 1776; 1814; 1817; 1841; 1922; 1990; 2078; 2144; 2261; 2318; 2325; 2344; 2470; 2475; 2660; 2680; 2697; 2723; 2754; 2798; 3240; 3290; 3308; 3600; 3618; 3794; 3796; 3802; 3811; 3814; 3917; 3946; 3960; 3971; 4036; 4054; 4234; 4259; 4280; 4460; 4467; 4598; 4620; 4626; 4663; 4669; 4689; 4704; 4753; 4778; 4832; 4895; 4900; 4993; 5018; 5032; 5096; 5253; 5271; 5272; 5481; 5550; 5585; 5659; 5810; 5817; 5843; 5863; 5911; 5988; 6044; 6072; 6073; 6232; 6259; 6304; 6339; 6449; 6493

151° to 155° C.

28; 153; 229; 367; 375; 495; 505; 506; 561; 571; 598; 646; 720; 722; 727; 745; 792; 835; 907; 1008; 1188; 1236; 1287; 1441; 1464; 1480; 1515; 1669; 1678; 1816; 1950; 1961; 2002; 2070; 2144; 2280; 2311; 2318; 2584; 2591; 2739; 2747; 2754; 2798; 3216; 3247; 3290; 3361; 3440; 3673; 3729; 3762; 3763; 3787; 3802; 3839; 3960; 4034; 4423; 4479; 4524; 4589; 4616; 4638; 4645; 4659; 4766; 4769; 4809; 4837; 4895; 5032; 5073; 5100; 5156; 5244; 5355; 5386; 5476; 5532; 5565; 5586; 5695; 5711; 5838; 5844; 5870; 5892; 6045; 6107; 6229; 6261; 6330; 6385; 6389; 6480

156° to 160° C.

46; 93; 229; 289; 292; 362; 389; 495; 505; 510; 545; 570; 608; 629; 646; 650; 668; 677; 722; 777; 792; 801; 923; 997; 1009; 1048; 1328; 1371; 1560; 1718; 1763; 1793; 1950; 2001; 2070; 2256; 2318; 2341; 2344; 2355; 2358; 2574; 2592; 2624; 2653; 2709; 2798; 2843; 2844; 2889; 3290; 3366; 3678; 3729; 3742; 3778; 3848; 3869; 3872; 3960; 4001; 4063; 4267; 4277; 4322; 4423; 4440; 4479; 4480; 4579; 4738; 4815; 4817; 4863; 4881; 4902; 4934; 4936; 5031; 5056; 5113; 5140; 5158; 5161; 5242; 5255; 5314; 5476; 5510; 5533; 5539; 5559; 5651; 5688; 5695; 5700; 5877; 5970; 6238; 6387

Table 7-3 (*Continued*)
MELTING POINTS OF ORGANIC COMPOUNDS

161° to 165° C.

12; 60; 107; 140; 163; 292; 324; 336; 527; 545; 662; 694; 702; 715; 737; 744; 751; 765; 773; 839; 862; 864; 993; 997; 1201; 1316; 1353; 1513; 1542; 1574; 1833; 1960; 2079; 2258; 2270; 2318; 2326; 2341; 2462; 2480; 2579; 2610; 2738; 2746; 2798; 2880; 2889; 3059; 3305; 3328; 3617; 3620; 3678; 3692; 3703; 3740; 3799; 3836; 3848; 3865; 3880; 3928; 3960; 4130; 4250; 4322; 4480; 4614; 4722; 4752; 4776; 4820; 4823; 4830; 4859; 4893; 4902; 5037; 5161; 5174; 5218; 5288; 5480; 5516; 5536; 5574; 5671; 5700; 5803; 5835; 5847; 5918; 6071; 6123; 6138; 6139; 6140; 6306; 6331; 6338

166° to 170° C.

35; 76; 94; 122; 163; 185; 186; 527; 545; 547; 587; 620; 662; 702; 708; 736; 744; 768; 774; 787; 827; 854; 882; 993; 1202; 1235; 1423; 1542; 1570; 1726; 1727; 1877; 1959; 2004; 2209; 2270; 2271; 2314; 2318; 2430; 2432; 2462; 2463; 2464; 2639; 2640; 2741; 2746; 2798; 2867; 3013; 3229; 3230; 3236; 3328; 3329; 3342; 3349; 3562; 3590; 3609; 3617; 3739; 3797; 3833; 3928; 3960; 4004; 4250; 4308; 4439; 4502; 4752; 4759; 4859; 4880; 5098; 5104; 5110; 5113; 5239; 5303; 5560; 5597; 5598; 5601; 5645; 5663; 5671; 5691; 5847; 5888; 6102; 6123; 6169; 6242; 6257; 6264; 6321; 6329; 6390; 6488; 6490

171° to 175° C.

66; 70; 100; 110; 121; 251; 265; 291; 293; 297; 323; 345; 587; 609; 695; 764; 785; 1199; 1230; 1290; 1298; 1479; 1504; 1513; 1569; 1626; 1655; 1658; 1681; 1707; 1815; 1835; 1847; 2209; 2350; 2420; 2430; 2432; 2439; 2576; 2604; 2640; 2662; 2710; 2779; 2814; 2824; 2900; 3230; 3674; 3739; 3797; 3837; 3850; 3895; 3960; 3985; 4136; 4197; 4275; 4373; 4410; 4436; 4439; 4650; 4871; 4875; 5071; 5091; 5103; 5104; 5112; 5113; 5171; 5172; 5236; 5332; 5535; 5597; 5601; 5602; 5610; 5644; 5645; 5663; 5693; 5701; 5852; 6102; 6211; 6230; 6329; 6390; 6425; 6451; 6488; 6490

176° to 180° C.

16; 121; 134; 166; 291; 294; 603; 648, 695; 731; 749; 841; 978; 1191; 1254; 1290; 1298; 1323; 1385; 1392; 1523; 1626; 1729; 1760; 2073; 2281; 2296; 2350; 2353; 2420; 2607; 2608; 2686; 2745; 2814; 2816; 2837; 2846; 3203; 3275; 3305; 3358; 3479; 3595; 3674; 3827; 3834; 3835; 3841; 3850; 3856; 3920; 3960; 4033; 4091; 4110; 4129; 4136; 4190; 4197; 4374; 4444; 4475; 4632; 4642; 4757; 5001; 5332; 5336; 5390; 5470; 5663; 5716; 5920; 5984; 6050; 6102; 6170; 6179; 6227; 6312; 6423

181° to 185° C.

1; 21; 124; 125; 143; 240; 261; 290; 349; 509; 548; 603; 631; 731; 732; 746; 850; 932; 941; 1253; 1528; 1575; 1676; 1742; 1842; 1964; 2573; 2629; 2656; 2729; 2743; 2766; 2840; 3239; 3244; 3261; 3358; 3502; 3835; 3873; 3920; 3960; 4069; 4110; 4376; 4406; 4473; 4481; 4607; 4628; 4732; 4741; 4760; 4818; 5072; 5138; 5147; 5157; 5331; 5578; 5663; 5865; 5919; 5920; 6050; 6235; 6320; 6324; 6340; 6423

186° to 190° C.

29; 31; 63; 124; 125; 240; 266; 349; 521; 548; 603; 604; 683; 732; 899; 941; 964; 1198; 1252; 1289; 1536; 1676; 1700; 1720; 1845; 1942; 2357; 2413; 2575; 2588; 2612; 2760; 2882; 2910; 3330; 3358; 3375; 3378; 3502; 3564; 3608; 3664; 3710; 3782; 3786; 3881; 3960; 4038; 4286; 4293; 4406; 4573; 4760; 4818; 4872; 5006; 5057; 5090; 5108; 5307; 5497; 5570; 5587; 5626; 5653; 5663; 5666; 5812;

5842; 5848; 6047; 6048; 6098; 6110; 6137; 6289; 6305; 6311; 6340; 6452

191° to 195° C.

30; 63; 141; 179; 217; 228; 278; 388; 521; 586; 595; 603; 638; 752; 849; 851; 859; 1196; 1384; 1485; 1495; 1532; 1700; 1797; 1847; 2020; 2021; 2082; 2124; 2301; 2316; 2335; 2363; 2503; 2528; 2598; 2634; 2643; 2655; 2744; 2815; 2917; 3249; 3378; 3583; 3710; 3724; 3782; 3788; 3823; 3960; 3989; 4249; 4297; 4485; 4577; 4818; 4826; 4892; 5005; 5199; 5274; 5307; 5490; 5495; 5522; 5583; 5664; 5668; 5685; 5692; 5830; 5842; 5851; 5854; 6048; 6321; 6368; 6395; 6452

196° to 200° C.

71; 160; 228; 276; 357; 388; 520; 562; 603; 638; 734; 1182; 1196; 1370; 1582; 1702; 1830; 1948; 2299; 2316; 2317; 2319; 2327; 2335; 2528; 2621; 2634; 2657; 2917; 2982; 3126; 3226; 3371; 3490; 3588; 3677; 3696; 3747; 3820; 3861; 3904; 3919; 3960; 4249; 4278; 4644; 4651; 4708; 4933; 5069; 5070; 5243; 5307; 5572; 5608; 5614; 5687; 5699; 5736; 5842; 5851; 5899; 5907; 6395

201° to 205° C.

33; 71; 279; 280; 284; 336; 388; 546; 577; 603; 605; 610; 706; 900; 1197; 1492; 1509; 1521; 1944; 1963; 2299; 2315; 2335; 2402; 2453; 2609; 2611; 2865; 3253; 3373; 3499; 3676; 3686; 3730; 3745; 3747; 3789; 3803; 3821; 3911; 3919; 3947; 4243; 4484; 4604; 4612; 4644; 4662; 4708; 5039; 5193; 5243; 5273; 5307; 5571; 5604; 5608; 5670; 5686; 5690; 5713; 5736; 5757; 5855; 5858; 5907; 6127; 6141; 6394

206° to 210° C.

90; 157; 317; 327; 388; 508; 541; 549; 630; 651; 701; 706; 860; 1308; 1317; 1492; 1517; 1521; 1531; 1533; 1703; 1805; 1940; 1941; 1948; 2080; 2295; 2328; 2334; 2335; 2447; 2556; 2663; 2887; 3780; 3826; 3838; 3840; 3925; 4128; 4438; 4608; 4649; 4660; 4877; 5013; 5075; 5097; 5307; 5487; 5511; 5690; 5717; 5860; 6054; 6081; 6082; 6219; 6220; 6295; 6319; 6417

211° to 215° C.

157; 388; 549; 594; 611; 630; 666; 681; 706; 778; 1315; 1394; 1517; 1521; 1548; 1686; 1948; 1987; 2279; 2282; 2335; 2401; 2522; 2626; 2661; 2663; 2705; 2981; 3238; 3373; 3476; 3492; 3584; 3746; 3784; 3819; 3935; 4438; 4590; 4649; 4774; 4816; 4825; 5016; 5075; 5307; 5487; 5600; 5802; 5860; 6082; 6219; 6260; 6295; 6322; 6366

216° to 220° C.

119; 123; 275; 351; 393; 538; 549; 647; 681; 684; 1521; 1541; 1549; 1585; 1819; 1951; 2337; 2350; 2497; 2522; 2625; 2638; 2663; 2690; 2848; 2981; 3309; 3476; 3584; 3592; 3679; 3684; 3746; 3783; 4360; 4578; 4590; 4825; 5162; 5195; 5234; 5281; 5307; 5321; 5335; 5473; 5487; 5504; 5562; 5712; 5802; 5813; 5846; 5849; 5893; 5996; 6082; 6219; 6221; 6231; 6260; 6295; 6303

221° to 225° C.

232; 263; 358; 576; 681; 707; 1200; 1521; 1778; 1803; 2294; 2542; 2545; 2618; 2625; 2687; 2885; 2887; 2900; 3237; 3372; 3672; 3679; 3684; 3697; 3822; 3852; 3923; 3988; 4005; 4035; 4112; 4179; 4812; 4819; 5017; 5051; 5289; 5335; 5487; 5492; 5579; 5712; 5813; 5846; 5861; 5889; 5893; 5933; 6061; 6106; 6208; 6231; 6236; 6369; 6445

Table 7-3 (*Continued*)
MELTING POINTS OF ORGANIC COMPOUNDS

226° to 230° C.

152; 348; 576; 588; 671; 681; 707; 847; 853; 1250; 1521; 1704; 1731; 1840; 2110; 2305; 2394; 2542; 2543; 2545; 2618; 2687; 2688; 2765; 3237; 3489; 3559; 3563; 3720; 3817; 3852; 3959; 4005; 4106; 4627; 4629; 4754; 4755; 4758; 4812; 4878; 5051; 5306; 5327; 5492; 5500; 5508; 5579; 5612; 5889; 5893; 5932; 5933; 5952; 5998; 6083; 6106; 6231; 6267; 6369

231° to 235° C.

32; 167; 181; 374; 515; 576; 588; 681; 700; 707; 853; 1250; 1496; 1506; 1608; 1704; 1804; 1810; 1818; 2305; 2383; 2394; 2622; 3205; 3437; 3563; 3720; 3726; 3779; 3817; 3818; 3824; 3852; 3990; 4035; 4039; 4368; 4411; 4648; 4754; 4999; 5052; 5109; 5122; 5150; 5274; 5330; 5488; 5528; 5645; 5742; 5821; 5826; 5887; 5947; 6083; 6087; 6099; 6223; 6318

236° to 240° C.

32; 98; 181; 543; 575; 600; 618; 671; 680; 681; 700; 1250; 1496; 1704; 1708; 1804; 2320; 2333; 2437; 2441; 2490; 2599; 2759; 3227; 3437; 3443; 3668; 3779; 3800; 3824; 3852; 3922; 3924; 3990; 4050; 4260; 4609; 4666; 4860; 5122; 5197; 5287; 5329; 5499; 5528; 5545; 5672; 5742; 6239

241° to 245° C.

98; 129; 168; 234; 539; 618; 621; 680; 1231; 1329; 1675; 1697; 1809; 1832; 1881; 2333; 2437; 2455; 2490; 2706; 2764; 3668; 3670; 3800; 3804; 3965; 4045; 4142; 4391; 4459; 4599; 4666; 4707; 5036; 5197; 5507; 5529; 5545; 5623; 5656; 5696; 6000

246° to 250° C.

6; 126; 165; 173; 234; 616; 621; 670; 680; 747; 875; 1489; 1780; 1881; 2077; 2262; 2333; 2437; 2490; 2734; 2764; 3309; 3325; 3498; 3560; 3668; 3671; 3980; 4019; 4101; 4117; 4743; 4761; 4879; 5010; 5337; 5491; 5498; 5506; 5529; 5611; 5613; 5934; 6181

251° to 260° C.

127; 244; 335; 356; 539; 589; 641; 848; 908; 1251; 1493; 1522; 1540; 1602; 1603; 1627; 1628; 1705; 1780; 1878; 1943; 2077; 2333; 2351; 2362; 2437; 2454; 2691; 2847; 2874; 2884; 2886; 2888; 2903; 3252; 3254; 3302; 3325; 3374; 3560; 3671; 3689; 3781; 3791; 3888; 4019; 4037; 4101; 4117; 4405; 4497; 4498; 4532; 4652; 4794; 4821; 4932; 5264; 5337; 5501; 5531; 5589; 5665; 5667; 5678; 5745; 5811; 5856; 5876; 5934; 5941; 6182; 6213; 6364

261° to 270° C.

174; 288; 355; 589; 682; 1478; 1690; 1692; 1777; 1808; 1939; 1946; 2122; 2293; 2298; 2343; 2362; 2437; 2451; 2452; 2594; 2597; 2654; 2849; 2851; 2891; 2898; 2903; 3315; 3442; 3785; 3808; 3809; 3816; 3882; 3888; 4141; 4412; 4544; 4605; 4606; 4770; 4822; 5066; 5121; 5494; 5509; 5514; 5531; 5724; 6209; 6327

271° to 280° C.

109; 174; 286; 287; 355; 359; 525; 540; 582; 589; 590; 748; 1494; 2300; 2343; 2361; 2395; 2614; 2883; 2888; 2898; 2902; 3439; 3442; 3591; 3690; 3828; 3978; 4412; 4474; 4478; 4606; 5148; 5149; 5316; 5489; 5537; 5673; 5722; 5724; 5725; 5801; 5866; 6249; 6398

281° to 290° C.

109; 164; 277; 352; 355; 359; 519; 525; 540; 550; 552; 589; 667; 1555; 1807; 1839; 1947; 2297; 2302; 2491; 2664; 2883; 2890; 2902; 2932; 3314; 3669; 3801; 3806; 3937; 4062; 4600; 4709; 5149; 5163; 5198; 5530; 5563; 5722; 5746; 5806; 5859; 5864; 5866; 6365; 6397; 6398

291° to 300° C.

139; 158; 159; 282; 285; 354; 359; 387; 391; 525; 551; 690; 772; 846; 979; 1539; 1838; 1967; 2336; 2384; 2585; 2664; 2724; 2854; 2890; 2966; 3250; 3586; 4443; 5049; 5475; 5726; 6367; 6397

301° to 310° C.

139; 176; 241; 245; 283; 350; 354; 394; 593; 669; 709; 730; 1677; 1694; 1820; 1838; 2644; 2817; 2854; 3586; 3725; 3792; 3805; 4110; 4931; 5556; 5573; 5744; 6212

311° to 320° C.

392; 395; 617; 669; 709; 1691; 1693; 1694; 2596; 2854; 3561; 4118; 5496; 5669; 5680; 5807; 5816; 6367

321° to 330° C.

602; 1671; 1693; 1695; 2306; 2722; 2854; 4413; 5502; 5816; 6205

331° to 340° C.

601; 602; 1629; 1671; 1695; 2303; 2304; 2595; 2854; 3589; 3807; 6383

341° to 360° C.

145; 601; 779; 2689; 2854; 3589; 4335; 5276; 5956

361° to 380° C.

685; 1594; 4417; 4418; 4562; 5299; 5800; 5805; 6214; 6215; 6362

Above 380° C.

676; 3857; 5009; 5277; 5822; 6384

Table 7-4
PHYSICAL CONSTANTS OF ORGANIC COMPOUNDS

See also special tables of Fats and Oils, Alkaloids, Glucosides, Resins, Nomenclature, Organic Radicals, and Common or Trade Names of Chemicals.

Names of the compounds in the table below are arranged alphabetically. The names were selected in such a manner as to bring isomers and closely related compounds in close sequence to enable a convenient means of comparison. In general, names with prefixes such as *iso* or *tertiary* are placed with the normal compounds; thus, *isopropyl* will be found listed under *propyl*. Indented words are suffixes of the word in bold face type immediately preceding; thus, compound No. 8 is acetaldehyde ammonia. No compound is listed more than once in the table and each compound is given a number.

Synonyms are listed in two ways. In the column headed "Synonym" will be found a different name for the same compound. At the bottom of each page is an alphabetical listing for compounds which are to be found in the main body of the table but under a different name; the number following the name refers to the numerical place of this compound in the table; thus, *acetaldazine* 9 indicates that this compound is the 9th compound in the table where it will be found listed under the name *acetaldehyde azine*. Alkaloids which are listed in another table are indicated by *alkd.*; thus, *quinine, cf. alkd.*, means that this compound does not appear in the table below but will be found in a special table of alkaloids. Similarly, glucosides are indicated by *glcde.*; thus, *absinthin, cf. glcde.* indicates that the compound is listed in the table of glucosides.

Formulas are presented in a semi-structural form. To indicate the position of substituents on aromatic rings, position numbers are given in the same sequence as they appear in the formula; thus, in the formula "$HO \cdot C_6H_3(CHO) \cdot CO_2H$ (2;5,1)", the CO_2H group is at position 1, the CHO at 5 and the OH at 2.

Beil. Ref. In the column so headed will be found the reference to the volume and page numbers of the 4th edition of Beilstein: *Handbuch der Organischen Chemie* (published by Springer, Berlin). Where the Roman numeral, which indicates the volume number, is preceded by an asterisk, reference is made to the first supplementary volumes. Thus, "*I-23", indicates that the compound will be found listed in the supplement to volume 1 and on page 23.

Formula Weights are based upon the International Atomic Weights of 1961 and are computed to the nearest hundredth.

Crystalline Form and Color. In addition to the crystalline form and color, the solvent used in purification is often given; thus, "rhb./al." indicates that rhombic

---

Abbreviations used in the table

a., acid	atm., atmosphere	delq., deliquescent
abs., absolute	Beil. Ref., reference to 4th ed.	dil., dilute
ac., acetic acid	Beilstein; * 1st suppl; ** 2d suppl.	diss., dissociates
Ac, acetyl (i. e. CH_3CO)		*dl*, inactive (i. e. 50% *d* and
act., acetone	b., blue	50% *l*)
act, active	bl., black	off., effloresces
al., ethyl alcohol	b. p., boiling point	et., ether ($C_2H_5OC_2H_5$)
alk., alkali (i. e. aqueous	brn., brown	Et, ethyl (C_2H_5)
NaOH or KOH)	bz., benzene, C_6H_6	expl., explodes
alkd., an alkaloid; see alkaloids	c., cold	fl., flakes
Am, amyl (i. e. C_5H_{11})	carb., carbonate	gly., glycerol
amor., amorphous	chl., chloroform, $CHCl_3$	gn., green
anh., anhydrous	col., colorless or white	gr., gray
anhyd., anhydride	conc., concentrated	h., hot
aq., aqueous; water	cr., crystals or crystalline	hex., hexagonal
art., artificial	*d*, dextrorotatory	hyg., hygroscopic
as, asymmetrical	d., decomposes or decomposed	i., insoluble

No.	Name	Synonym	Formula	Beil. Ref.	Formula Weight
1	**Abietic acid**	sylvic acid	$C_{20}H_{30}O_2$	....	302.46
2	**Acenaphthene**	naphthylene ethylene	$C_{10}H_6(CH_2)_2$	V-586	154.21
3	**Acenaphthylene**	....	$C_{12}H_8$	V-625	152.20

Abalyn 4084   Abietin, cf. glcde.
Abasin 84   Abrastol 4506

Table 7-4 (*Continued*)
PHYSICAL CONSTANTS OF ORGANIC COMPOUNDS

crystals were obtained when the compound was crystallized from alcohol. **Where no color is stated it may generally be inferred that the material is colorless.**

Specific Gravity values are given at room temperatures (15 to 20°C.) unless otherwise indicated by the small figures which follow the value; thus, "$1.069\frac{95}{95}°$" indicates a specific gravity of 1.069 for the substance at 95°C. referred to water at 95°C. Specific gravity of gases with the notation "(A)" or "(D)" indicate the density of the gas referred to (A) air = 1 or (D) hydrogen = 1 respectively. See also special tables.

Melting Point is recorded in a certain case as "82 d." and in some other case as "d. 82", the distinction being made in this manner to indicate that the former is a melting point with decomposition at 82°C., while in the latter decomposition only occurs at 82°C. Where a value such as "$-2H_2O$, 82" is given it indicates loss of two moles of water per formula weight of the compound at a temperature of 82°C.

Boiling Point is given at atmospheric pressure (760 mm of mercury) unless otherwise indicated; thus, "82^{15mm}" indicates the boiling point is 82°C. when the pressure is 15 mm.

Solubility is given in parts by weight (of the formula shown at the extreme left) per 100 parts by weight of the solvent; the small exponent indicates the temperature in degrees C. In the case of gases the solubility is often expressed in some manner as "$5^{10°}$ cc." which indicates that at 10°C., 5 cc. of the gas are soluble in 100 grams of the solvent.

References. The information given in the table has been collected mainly from the following sources:—

Beilstein: *Handbuch der Organischen Chemie*, 4th edition. Published by Springer, Berlin.

The Merck Index, 5th and 7th eds. Published by Merck and Co., Inc., Rahway, N. J. Heilbron: *Dictionary of Organic Compounds*. Published by Oxford Press, New York.

Tables Annuelles Internationales de Constants et Donnes Numeriques. Published by McGraw-Hill Book Co., New York.

International Critical Tables, Vol. 1. Published by McGraw-Hill Book Co., New York.

Seidell: *Solubilities of Inorganic and Organic Compounds*. Published by Van Nostrand and Co., New York.

ign., ignites	*p*, para position	*tert*, tertiary
K$_2$CO$_3$, aqueous potassium carbonate	pa., pale	tet., tetragonal
l, levorotatory	pd., powder	tri., triclinic
lf., leaves or leaflets	pet., petroleum ether	trig., trigonal
lg., ligroin	Ph, phenyl (C$_6$H$_5$)	*v*, vicinal
lq., liquid	pl., plates	v., very
lt., light	pr., prisms	vl., violet
m, meta position	pyr., pyridine	v. s., very soluble
Me, methyl (CH$_3$)	r., red	v. sl. s., very slightly soluble
Me al., methyl alcohol	rhb., rhombic	vac., vacuo or vacuum
met., metallic	s., soluble	wh., white
mn., monoclinic	sc., scales	yel., yellow
m. p., melting point	*sec*, secondary	∞, soluble in all proportions;
n, normal	silv., silvery	i. e., miscible
NaOAc, aqueous sodium acetate	sl., slight or slightly	>, greater than
nd., needles	soln., solution	<, less than
o, ortho position	subl., sublimes	42±, about or near to 42
or., orange	*s*, symmetrical	$-3H_2O$, 100, loses 3 molecules
	syr., syrup	of water at 100° C.

No.	Crystalline Form and Color	Specific Gravity	Melting Point °C.	Boiling Point °C.	Solubility in 100 Parts		
					Water	Alcohol	Ether
1	lf.	....	182	....	i.	v. s.	v. s.
2	rhb./al.	$1.069\frac{95}{95}°$	95	278-9	i.; $26^{18°}$bz.	s. h. $31^{8°}$ac.	s. chl.
3	rhb.	$0.899\frac{16°}{4}$	93.5-4.5	265-75(sl.d.)	i.	v. s.	v. s.

Absinthin, cf. glcde
Accelerine 4883

Acecoline 83

Table 7-4 (*Continued*)
PHYSICAL CONSTANTS OF ORGANIC COMPOUNDS

No.	Name	Synonym	Formula	Beil. Ref.	Formula Weight
4	**Acetal**	acetaldehyde-diethylacetal	$CH_3 \cdot CH(OC_2H_5)_2$	I-603	118.18
5	**Acetaldehyde**	ethanal	$CH_3 \cdot CHO$	I-594	44.05
6	**Acetaldehyde,** meta-	metaldehyde	$(C_2H_4O)_4$	I-602	176.21
7	**Acetaldehyde,** par-	paraldehyde	$(C_2H_4O)_3$	XIX-385	132.16
8	ammonia	. . . .	$(CH_3 \cdot CHOH \cdot NH_2)_3$	XXVI-7	183.25
9	azine	acetaldazine	$CH_3CH:N \cdot N:CHCH_3$	I-609	84.12
10	oxime	acetaldoxime	$CH_3CH:NOH$	I-608	59.07
11	phenylhydrazone	. . . .	$C_6H_5NHN:CHCH_3$	XV-127	134.18
12	semicarbazone	. . . .	$CH_3CH:NNHCONH_2$	III-101	101.11
13	sodium bisulfite	. . . .	$CH_3CHO \cdot NaHSO_3 \cdot$ $\frac{1}{2}H_2O$	I-605	157.12
14	**Acetamide**	ethι namide	$CH_3 \cdot CO \cdot NH_2$	II-175	59.07
15	**Acetamidine**	. . . .	$CH_3 \cdot C(NH)NH_2$	II-185	58.08
16	hydrochloride		$CH_3 \cdot C(NH)NH_2 \cdot HC_l$	II-185	94.54
17	**Acet-**anilide	antifebrin	$C_6H_5NH \cdot COCH_3$	XII-237	135.17
18	anisidide (*o*)	acetyl-*o*-anisidine	$CH_3O \cdot C_6H_4NH \cdot$ $COCH_3$	XIII-371	165.19
19	anisidide (*m*)	acetyl-*m*-anisidine	$CH_3O \cdot C_6H_4NH \cdot$ $COCH_3$	XIII-416	165.19
20	anisidide (*p*)	methacetin; *p*-methoxy-acetanilide	$CH_3O \cdot C_6H_4NH \cdot$ $COCH_3$	XIII-461	165.19
21	metanilic acid (Na)	. . . .	$CH_3CO \cdot NH \cdot C_6H_4 \cdot$ $SO_3Na \cdot 2H_2O$ (1,3)	XIV-691	273.24
22	methyl-*α*-naphthylamide	. . . .	$CH_3CO \cdot (CH_3)N \cdot C_{10}H_7$	XII-1231	199.25
23	*o*-phenetidide	*o*-ethoxyacetanilide	$CH_3CONH \cdot C_6H_4 \cdot$ OC_2H_5	XIII-371	179.22
24	*m*-phenetidide	acetyl-*m*-phenetidine	$CH_3CONH \cdot C_6H_4 \cdot$ OC_2H_5	XIII-416	179.22
25	*p*-phenetidide	phenacetin	$CH_3CONH \cdot C_6H_4 \cdot$ OC_2H_5	XIII-461	179.22
26	toluidide (*o*)	*N*-tolylacetamide	$CH_3C_6H_4NH \cdot COCH_3$	XII-792	149.19
27	toluidide (*m*)	*N*-tolylacetamide	$CH_3C_6H_4NH \cdot COCH_3$	XII-860	149.19
28	toluidide (*p*)	*N*-tolylacetamide	$CH_3C_6H_4NH \cdot COCH_3$	XII-920	149.19
29	**Acetamino-**ethyl-salicylic acid (5;2,1)	benzacetin	$CH_3CONH \cdot C_6H_3 \cdot$ $(OC_2H_5)CO_2H$	XIV-583	223.23
30	fluorene(2)	. . . .	$C_{13}H_9NH \cdot COCH_3$	XII-1331	223.28
31	*α*-naphthol (4,1)	naphthacetol	$CH_3CO \cdot NH \cdot C_{10}H_6OH$	XIII-669	201.23
32	*β*-naphthol (1,2)	. . . .	$CH_3CO \cdot NH \cdot C_{10}H_6OH$	XIII-679	201.23
33	phenol (*o*)	*o*-hydroxyacetanilide	$CH_3CONHC_6H_4OH$	XIII-370	151.17
34	phenol (*m*)	acetyl-aminophenol	$CH_3CONHC_6H_4OH$	XIII-415	151.17
35	phenol (*p*)	. . . .	$CH_3CONHC_6H_4OH$	XIII-460	151.17
36	**Acetic** acid	ethanoic acid	$CH_3 \cdot COOH$	II-96	60.05
37	anhydride	. . . .	$(CH_3CO)_2O$	II-166	102.09
38	**Acetoacetic** acid	acetyl acetic acid	$CH_3CO \cdot CH_2CO_2H$	III-630	102.09
39	*β*-anil	Et *β*-anilinocrotonate	$CH_3C(:NC_6H_5) \cdot CH_2 \cdot$ $CO_2C_2H_5$	XII-518	205.26

Table 7-4 (*Continued*)
PHYSICAL CONSTANTS OF ORGANIC COMPOUNDS

No.	Crystalline Form and Color	Specific Gravity	Melting Point °C.	Boiling Point °C.	Solubility in 100 Parts		
					Water	Alcohol	Ether
4	lq.	$0.821\frac{22}{4}°$	....	102.2	$6^{25°}$	∞	∞
5	col. lq.	$0.783\frac{18}{4}°$	−123.5	20.2	∞	∞	∞
6	nd.	....	246.2*	....	i.; s. chl.	2 h.; s. bz.	$0.5^{35°}$; v. sl. s. h.
7	col. cr.	$0.994\frac{20}{4}°$	12.6	124.4^{752mm}	$12^{13°}$ $6^{100°}$	∞; ∞ chl.	∞
8	col. cr.	....	97	100-10(sl.d.)	v. s.	v. s.	sl. s.
9	lq.	$0.832^{17°}$	....	95-6	....	....	....
10	nd. or lq.	$0.965\frac{20}{4}°$	47(13)	114-5	s.	∞	∞
11	col. nd.	....	(α)98-101 (β)57	$236-7^{20mm}$ $133-6^{20mm}$	....	....	s. pet.
12	nd. / aq. or al.	....	162-3	....	$3^{17°}$	s.	....
13	col. nd.	....	....	....	s.; d. a.	i.	i.
14	col. cr.	1.159	81(69.4)	221.2	s.	s.	v. sl. s.
15	in aq. soln.	....	....	....	d. h.	....	....
16	nd./al.	....	177	....	s.	v. s.	i.
17	rhb./al.	$1.214°$	113-4	305	$0.5^{36°}$; $3.5^{80°}$	$21^{20°}$; $46^{60°}$	$7^{25°}$
18	cr./aq.	....	87-8	303-5	v. s. h.	$55.3^{21°}$	v. s. ac.
19	cr./aq.	....	80-1	....	....	$80^{21°}$	....
20	pl./aq.	....	127	....	$0.2^{15°}$; $8^{100°}$	$12.7^{21°}$; s. act.	s. chl.; s. dil. alk.
21	nd./aq.	....	184-5	....	v. s. h.	v. sl. s.	i.
22	pr./aq.	....	94-5	....	v. sl. s.	s.	s.
23	lf./aq. al.	....	79	>250	i.	s.	....
24	lf./aq.	....	96-7	....	sl. s.	s.	....
25	col. mn.	....	134-5	d.	$0.7^{20°}$	7.4 c.; 40 h.	$1.6^{25°}$; s. gly.
26	rhb.	$1.168^{15°}$	110	296	$0.86^{19°}$	s.; s. bz.	s.; s. chl.
27	mn./aq.	$1.141^{15°}$	65.5	303	$0.44^{13°}$	v. s.	v. s.
28	rhb. or mn.	$1.212^{15°}$	153	306-7	$0.09^{22°}$	$10.2^{25°}$abs.	s.; i. lg.
29	col. nd./aq.	....	189-90	....	sl. s.	s.	....
30	cr./50%ac.	....	194-5	....	....	....	....
31	nd./al.	....	187	....	s. h.	s; s. NH₄OH	s. Na₂CO₃
32	lf./aq. al.	....	235.5	subl. sl. d.	v. s. aq. NaOH	s; s. bz.	s.; s. h. ac.
33	lf./aq. al.	....	203	....	s. h.	s.	s. KOH
34	nd./aq.	....	148-9	....	s.	s.; sl. s. chl.	sl. s.; sl. s. bz.
35	mn./al.	$1.293^{21°}$	168-9	....	v. s. h.	v. s.	sl. s.
36	col. lq.	$1.049\frac{20}{4}°$	16.6	118.1	∞	∞	∞
37	col. lq.	$1.081\frac{20}{4}°$	−73	140.0	12 c.; d. h.	∞; d. h.	∞
38	col. oil	....	36-7	d. <100	∞	∞	s.
39	oil	....	....	d. 240	i.	s.	s.

* In sealed tube; subl. 112-6; partly depolymerized.

Acet-methylamide 4085
Acet-oxime 61
Acet-phenetidide 23-5
Acetamino-, cf. also acetylamino.

Acetic ester 2935
Acetic ether 2935
Acetin 3382
Aceto-acetic ester 2937

Table 7-4 (*Continued*)
PHYSICAL CONSTANTS OF ORGANIC COMPOUNDS

No.	Name	Synonym	Formula	Beil. Ref.	Formula Weight
	Acetoacetic				
40	anilide	acetoacetanilide	$CH_3COCH_2CONH\cdot$ C_6H_5	XII-518	177.20
41	2-chloroanilide	*o*-chloroaceto-acet-anilide	$CH_3CO\cdot CH_2CO\cdot NH\cdot$ C_6H_4Cl	*XII-300	211.65
42	2,5-dichloroanilide	....	$CH_3CO\cdot CH_2CO\cdot NH\cdot$ $C_6H_3Cl_2$	....	246.09
43	**Acetoin**	methylacetyl carbinol	$CH_3CHOH\cdot CO\cdot CH_3$	I-827	88.11
44	**Aceto**-chloroglucose	tetraacetyl-α-gluco-syl chloride	$Cl\cdot CH(CHO_2CCH_3)_3\cdot$ $\underline{CH\cdot (CH_2O_2CCH_3)\cdot O}$	....	366.76
45	dichlorohydrin	dichloropropyl acetate (β,γ)	$CH_3CO_2CH_2CHCl\cdot$ CH_2Cl	II-129	171.02
46	naphthalide (α)	acetyl naphthylamine	$C_{10}H_7NH\cdot COCH_3$	XII-1230	185.23
47	naphthalide (β)	....	$C_{10}H_7NH\cdot COCH_3$	XII-1284	185.23
48	naphthol (1,2)	hydroxy-aceto-naphthone	$HO\cdot C_{10}H_6\cdot COCH_3$	VIII-149	186.21
49	nitrile	methyl cyanide	$CH_3\cdot CN$	II-183	41.05
50	phenine	....	$C_{23}H_{17}N$	....	307.40
51	phenone	methyl-phenyl ketone	$CH_3\cdot CO\cdot C_6H_5$	VII-271	120.15
52	phenone acetone	phenacyl acetone	$C_6H_5\cdot CO\cdot C_2H_4\cdot$ $COCH_3$	VII-687	176.22
53	phenone oxime	....	$C_6H_5(CH_3)C{:}NOH$	VII-278	135.17
54	phenone phenyl-hydrazone	....	$C_6H_5NH\cdot N{:}C(CH_3)\cdot$ C_6H_5	XV-139	210.28
55	**Acetol**	acetyl carbinol	$CH_3CO\cdot CH_2OH$	I-821	74.08
56	**Acetone**	propanone	$CH_3\cdot CO\cdot CH_3$	I-635	58.08
57	cyanohydrin	....	$(CH_3)_2C(OH)\cdot CN$	III-316	85.11
58	diacetic acid	....	$CO{:}(C_2H_4\cdot CO_2H)_2$	III-804	174.15
59	dicarboxylic acid	β-ketoglutaric acid	$CO{:}(CH_2CO_2H)_2$	III-789	146.10
60	glucose, mono	glucose di Me-ketal	$(CH_3)_2C{:}(HCO)_2\cdot$ $\underline{(CHOH)_2\cdot CH\cdot}$ $(CH_2OH)\cdot O$	XXXI-153	220.22
61	oxime	acetoxime	$(CH_3)_2C{:}NOH$	I-649	73.10
62	phenylhydrazone	....	$C_6H_5NH\cdot N{:}C(CH_3)_2$	XV-129	148.21
63	semicarbazone	....	$(CH_3)_2C{:}N\cdot NH\cdot$ $CONH_2$	III-101	115.14
64	sodium bisulfite	....	$(CH_3)_2CO\cdot NaHSO_3$	I-649	162.14
65	**Acetonyl** acetone	....	$(CH_3\cdot CO\cdot CH_2)_2$	I-788	114.15
66	urea	dimethyl hydantoin	$(CH_3)_2{:}$ $\underline{CNHCONHCO}$	XXIV-289	128.13
67	**Acetoxy acetone**	acetol acetate	$CH_3CO_2CH_2COCH_3$	II-155	116.12
68	**Acetyl**-acetone	diacetylmethane	$(CH_3CO)_2CH_2$	I-777	100.12
69	benzoic acid (*o*)	acetophenone car-boxylic acid	$CH_3CO\cdot C_6H_4\cdot CO_2H$	X-690	164.16
70	benzoic acid (*m*)	....	$CH_3CO\cdot C_6H_4\cdot CO_2H$	X-694	164.16
71	benzoic acid (*p*)	....	$CH_3CO\cdot C_6H_4\cdot CO_2H$	X-694	164.16
72	benzoyl	Me-Ph-diketone	$CH_3CO\cdot CO\cdot C_6H_5$	VII-677	148.16

Aceto-acetanilide 40
Aceto-aniside 18-20
Aceto-anisole 5132
Aceto-chloramide 1295
Aceto-cymene 4181

Aceto-mesitylene 6247
Aceto-naphthone 4313
Aceto-phen 115
Aceto-phenone alcohol 754
Aceto-phenone carboxylic acid 69-71

Aceto-propionic acid **3967**
Aceto-pyrine 566
Aceto-sal(in) 115
Aceto-thioamide **5882**
Aceto-vanillone 3776

Table 7-4 (*Continued*)
PHYSICAL CONSTANTS OF ORGANIC COMPOUNDS

No.	Crystalline Form and Color	Specific Gravity	Melting Point °C.	Boiling Point °C.	Solubility in 100 Parts		
					Water	Alcohol	Ether
40	lf.	$1.103^{85°}$	85	d.	sl. s.; s. alk.	s.	s.; s. h. bz.
41	col. nd.	$1.190^{107°}$	107	d.	i.	s.	i.; i. lg.
42	col. nd.	....	94–5	....	i.	sl. s.	i.
43	lq.	$1.011^{\frac{15}{15}°}$	15	148	∞	v. s.	sl. s.; i. lg.
44	col. mn. nd./et.-lg.	....	70–1 (d.)	....	d. h.; i. pet.	s.; s. act.	s.; s. CCl_4
45	col. lq.	$1.168^{15°}$	....	$191{-}2^{755mm}$	....	....	....
46	cr./al.	....	159–60	....	s. h.	$4^{25°}$	....
47	lf./al.	....	134	....	s. h.	s. h.	....
48	yel. nd./lg.	....	103(98)	325(sl. d.)	i.; v. s. bz.	v. sl. s. c.	s. ac.; s. CS_2
49	col. lq.	$0.783^{\frac{20}{4}°}$	−44.9	81.6	∞	∞	∞
50	nd./al.	....	135	....	....	s.	
51	cr. lf.	$1.028^{\frac{20}{4}°}$	19.6	202.0	i.	s.	s.
52	yel. oil	>1	....	162^{12}(sl. d.)	sl. s.	....	i. alk.
53	nd./aq.	....	59	d.	....	v. s.	v. s.
54	col. nd.	....	105–6	....	sl. s.	sl. s. c.	s.
55	col. lq.	$1.082^{\frac{20}{20}°}$	−17	145–6	∞	∞	∞
56	col. lq.	$0.791^{\frac{20}{4}°}$	−94.8	56.2	∞	∞	∞
57	col. lq.	$0.932^{19°}$	−19	82^{23mm}	v. s.	v. s.	v. s.; v. sl. s. pet.
58	rhb./aq.	....	142–3	....	s. h.	s.	sl. s.; i. bz.
59	nd./et.	....	138	....	v. s.	v. s.	sl. s.; i. bz.
60	col. mn. nd./aq.	....	162	....	s.	s.; s. h. EtOAc	i.
61	pr.	$0.97^{\frac{20}{20}°}$	60–1	136.3	v. s.	v. s.	v. s.; s. lg.
62	oil	....	26.6	163^{50mm}	s. c. dil. a.	....	....
63	nd./aq.	....	190–1(d.)	....	s. c.	sl. s. c.	i.
64	col. lf.	....	....	....	s.; d. a.	sl. s.	i.
65	col. lq.	$0.974^{\frac{20}{4}°}$	ca. −6	194^{754mm}	∞; i. KOH	∞	∞
66	tri. pr./al.	....	175	subl.	s.	s.	s.
67	col. oil	$1.075^{\frac{20}{4}°}$	....	174–5	v. s.	v. s.	v. s.
68	lq.	$0.976^{\frac{20}{4}°}$	−23.2(−9)	139^{746mm}	12.5	∞	∞; ∞ chl.
69	cr./aq.	....	114–5	....	s. h.	....	....
70	nd./aq.	....	172	....	s. h.	s.; sl. s. chl.	s.; sl. s. bz.
71	nd./h. aq.	....	200–5(d.)	....	s. h.	sl. s.	sl. s.; i. lg.
72	yel. oil	$1.101^{\frac{20}{4}°}$	....	216–8	$0.3^{20°}$	....	....

Aceto-xylidide 2417-22
Acetozone 73
Acetol acetate 67
Acetol phenyl ether 5132
Acetone azine 3932

Acetone bromoform 6102
Acetone chloride 2053
Acetone chloroform 6145
Acetone chloroform acetate 6144
Acetone diethylsulfone 5683

Acetone oxalic ester 2939
Acetonic acid 3757
Acetonyl-acetophenone 52
Acetonyl-amine 235
Acetonyl anisole 4126

Table 7-4 (*Continued*)
PHYSICAL CONSTANTS OF ORGANIC COMPOUNDS

No.	Name	Synonym	Formula	Beil. Ref.	Formula Weight
	Acetyl				
73	benzoyl peroxide	acetozone	$C_6H_5CO \cdot O_2 \cdot COCH_3$	IX-179	180.16
74	biuret	....	$CH_3CO(NHCO)_2NH_2$	III-72	145.12
75	bromide	ethanoyl bromide	$CH_3CO \cdot Br$	II-174	122.95
76	5-bromosalicylic acid (2;5,1)	5-bromoaspirin	$CH_3CO_2 \cdot C_6H_3(Br) \cdot CO_2H$	X-108	259.06
77	*n*-butylaniline	....	$CH_3CO \cdot N(C_4H_9) \cdot C_6H_5$	XII-247	191.28
78	caproyl	octandione-2,3	$CH_3(CH_2)_4CO \cdot COCH_3$	I-795	142.20
79	caproyl oxime	isonitroso-*n*-amyl- ketone	$C_8H_{15}O_2N$	I-795	157.21
80	carbazole (*N*)	....	$CH_3CO \cdot NC_{12}H_8$	XX-436	209.25
81	chloride	ethanoyl chloride	$CH_3CO \cdot Cl$	II-173	78.50
82	choline bromide	pragmoline	$(CH_3)_3N(Br)(CH_2)_2 \cdot O_2C \cdot CH_3$	*IV-428	226.12
83	choline chloride	acecoline	$(CH_3)_3N(Cl)C_2H_4 \cdot O_2C \cdot CH_3$	IV-281	181.66
84	diethylbromo- acetyl-urea (*N,N'*)	abasin; acetylcar- bromal	$(C_2H_5)_2CBr \cdot CO \cdot NH \cdot CO \cdot NH \cdot COCH_3$	*III-30	279.14
84.1	diphenylamine	*N*-diPh acetamide	$(C_6H_5)_2N \cdot COCH_3$	XII-247	211.27
85	disulfide	diacetyl disulfide	$(CH_3CO)_2S_2$	II-232	150.22
86	ethanolamine (*N*)	....	$HO(CH_2)_2NH \cdot COCH_3$	....	103.12
87	ethylmalonate	....	$CH_3CO \cdot CH \cdot (CO_2C_2H_5)_2$	III-796	202.21
88	ethyloxamate	....	$(CH_3CO)NHC_2O_2 \cdot OC_2H_5$	II-545	159.14
89	fluoride	ethanoyl fluoride	$CH_3CO \cdot F$	II-172	62.04
90	glycine	aceturic acid	$CH_3CO \cdot NH \cdot CH_2 \cdot CO_2H$	IV-354	117.11
91	indole (1)(*N*)	....	$CH_3CO \cdot NC_8H_6$	XX-309	159.19
92	iodide	ethanoyl iodide	$CH_3CO \cdot I$	II-174	169.95
93	iodosalicylic acid	4-iodoaspirin (2;4,1)	$CH_3CO_2C_6H_3(I)CO_2H$	*X-49	306.06
94	iodosalicylic acid	5-iodoaspirin (2;5,1)	$CH_3CO_2C_6H_3(I)CO_2H$	*X-49	306.06
95	isatin (*N*)	....	$CH_3CO \cdot N(CO)_2C_6H_4$	XXI-447	189.17
96	malic acid	....	$C_2H_3O_2 \cdot C_2H_3(CO_2H)_2$	III-429	176.13
97	*o*-methylamino- phenol	....	$CH_3CO \cdot N(CH_3)C_6H_4 \cdot OH$	XIII-372	165.19
98	*p*-methylamino- phenol	....	$CH_3CO \cdot N(CH_3)C_6H_4 \cdot OH$	XIII-466	165.19
99	β-methylcholine bromide	mecholyl bromide	$(CH_3)_3N(Br)CH_2 \cdot CH(CH_3)O_2C \cdot CH_3$	....	240.15
100	β-methylcholine chloride	mecholyl; mecholin	$(CH_3)_3N(Cl)CH_2 \cdot CH(CH_3)O_2C \cdot CH_3$	....	195.69
101	β-methylcholine iodide	mecholyl iodide	$(CH_3)_3N(I)CH_2 \cdot CH(CH_3)O_2C \cdot CH_3$	....	287.14
102	methyl-*o*-toluidine	*N*-Me-acet-*o*- toluidide	$CH_3CO \cdot N(CH_3)C_6H_4 \cdot CH_3$	XII-793	163.22
103	methyl-*m*-toluidine	*N*-Me-acet-*m*- toluidide	$CH_3CO \cdot N(CH_3)C_6H_4 \cdot CH_3$	XII-861	163.22
104	methyl-*p*-toluidine	*N*-Me-acet-*p*- toluidide	$CH_3CO \cdot N(CH_3)C_6H_4 \cdot CH_3$	XII-922	163.22
105	nitrate	....	$CH_3CO_2NO_2$	II-171	105.05
106	peroxide	diacetyl peroxide	$(CH_3CO)_2O_2$	II-170	118.09
107	*p*-phenylenedi- amine	amino-acetanilide (*p*)	$C_2H_3O \cdot NHC_6H_4NH_2$	XIII-94	150.18

Acetoxy-acetophenone 5094
Aceturic acid 90
Acetyl, cf. also aceto or acet.
Acetyl-acetic acid 38

Acetyl-anisidine 18-20
Acetyl-anisole 4065
Acetyl-anthranilic acid (*N*) 125
Acetyl-arsanilic acid 577

Acetyl-benzoin 718
Acetyl-benzoylaconine, cf. **alkd.**
Acetyl-benzylamine 780
Acetyl-carbinol 55

Table 7-4 (*Continued*)
PHYSICAL CONSTANTS OF ORGANIC COMPOUNDS

No.	Crystalline Form and Color	Specific Gravity	Melting Point °C.	Boiling Point °C.	Solubility in 100 Parts		
					Water	Alcohol	Ether
73	col. nd./lg.	....	37-9	130^{19mm}	$0.06^{25°}$ d.	s. oils	s.; s. chl.
74	nd.	....	107	....	s. bz.	v. s.	s.
75	col. lq.	$1.663\frac{16°}{4}$	−96.5	76^{750mm}	d.	d.; ∞ bz.	∞; ∞ chl.
76	cr./al.	....	168	....	i.; i. bz.	$16^{25°}$; i. CCl_4	$10^{25°}$
77	lq.	....	....	$273-5^{718mm}$	....	....	....
78	lq.	....	....	$172-37^{33mm}$	....	....	....
79	cr.	....	39	139^{16mm}	....	....	....
80	nd./aq.	....	69	>360 d.	v. sl. s. h.	v. s.	v. s.; s. bz.
81	col. lq.	$1.105\frac{20°}{4}$	−112.0	51-2	d.	d.; ∞ bz.	∞; ∞ chl.
82	hyg. pr./ abs. al.	....	143	....	v. s.; d. h.	s.	i.
83	hyg. col. pr.	....	....	....	v. s.; d. h.	s.	i.
84	col. cr./aq.	....	108-9	....	sl. s.	s.	s. EtOAc
84, 1	rhb./aq.	....	103	subl.	sl. s.	s.	sl. s.
85	cr.	....	20	d.	i.; s. CS_2	v. s.	v. s.
86	col. syr.	$1.122\frac{20°}{20}$	15.8	d.	sl. s.	s.	s.
87	lq.	$1.083\frac{26°}{5}$	....	120^{17mm}	s. dil. alk.	....	....
88	nd.	....	54	....	d. h.	s.	s.
89	lg. or gas	$0.993^{20°}$	<−60	20.8^{770mm}	5; d.	∞; ∞ bz.	∞
90	cr./aq.	....	206	....	$2.7^{15°}$	s. abs.	i.
91	lq.	$1.071^{0°}$	<−20	239	v. sl. s.	v. s.	v. s.; v. s. bz.
92	col. lq.	$1.98^{17°}$	....	108	d.	d.	s.
93	col. cr.	....	156	....	i.	s.	s.
94	col. cr.	....	166	....	i.	s.	s.
95	yel. nd./bz.	....	141	....	sl. s. c.; d. h.	s.	d. h. HCl
96	col. cr. pd.	....	132-9 d.	....	s.; d. h.	....	i. bz.
97	nd./MeOH	....	150	....	sl. s.	s.	s.
98	sc./aq.	....	240-1	....	v. sl. s.	v. s.	s.
99	col. delq. cr.	....	148-9	....	v. s.	70	i.; d. alk.
100	col. delq. cr./al.	....	172-3	....	v. s.	v. s.	i.; d. alk.
101	col. cr.	....	138-9	....	v. s.	75	i.
102	col. cr.	....	55-6	260	....	s.	....
103	col. cr.	....	66	....	....	s.	....
104	lf./al.-et.	....	80	283	....	s.	s. h. lg.
105	col. lq.	$1.24^{15°}$	expl.	227^{0mm}	....	....	....
106	col. cr.	....	30	$63^{21°}$; expl.	sl. s. c.; d.	....	....
107	nd./aq.	....	162	....	s. h.	v. s.	v. s.

Acetyl-carbromal 84
Acetyl-cyanide 5527
Acetyl-diethylamine 2100
Acetyl-ethylanilide 2934

Acetyl-formic acid 5525
Acetyl-guaiacol 3481
Acetyl-methylurea 4091
Acetyl-naphthalene 4313

Acetyl-naphthylamine 46-7
Acetyl-phenetidine 23-5
Acetyl-phenol 5141

Table 7-4 (*Continued*)
PHYSICAL CONSTANTS OF ORGANIC COMPOUNDS

No.	Name	Synonym	Formula	Beil. Ref.	Formula Weight
	Acetyl				
108	*m*-phenylenedi-amine	aminoacetanilide (*m*)	$C_2H_3O \cdot NHC_6H_4NH_2$	XIII-45	150.18
109	*m*-phenylene diamine HCl	....	$C_2H_3ONHC_6H_4NH_2 \cdot$ HCl	XIII-45	186.64
110	phenylglycine	*N*-Ph-aceturic acid	$C_6H_5N(COCH_3)CH_2 \cdot CO_2H$	XII-476	193.20
111	phenylhydrazine(β)	hydracetin	$CH_3CO \cdot NH \cdot NHC_6H_5$	XV-241	150.18
112	piperidine	....	$CH_3CO \cdot NC_5H_{10}$	XX-45	127.19
113	pyrrol (*N*)	....	$CH_3CO \cdot NC_4H_4$	XX-165	109.13
114	resorcinol (1,3)	euresol; eurisol	$CH_3CO_2 \cdot C_6H_4 \cdot OH$	VI-816	152.15
115	*o*-salicylic acid	aspirin	$CH_3CO_2 \cdot C_6H_4 \cdot CO_2H$	X-67	180.16
116	*o*-salicylic acid (Ca)	aspirin soluble	$(C_9H_7O_4)_2Ca \cdot 2H_2O$	*X-29	434.42
117	*o*-salicylic acid (Li)	litmopyrin	$C_9H_7O_4Li$	X-68	186.09
118	*o*-salicylic acid (Mg)	magnespirin	$(C_9H_7O_4)_2Mg\dagger$	*X-29	382.62
119	*o*-salicylic acid (Na)	hydropyrin	$C_9H_7O_4Na$	X-68	202.14
120	salol	spiroform; vesipyrin	$CH_3CO_2 \cdot C_6H_4 \cdot CO_2C_6H_5$	X-79	256.26
121	2-thiohydantoin	....	$CH_3CO \cdot N \cdot CH_2 \cdot$ | _____ CONH·CS _____|	*XXIV-293	158.18
122	thiourea	....	$CH_3CO \cdot NHCSNH_2$	III-191	118.16
123	urea	....	$CH_3CO \cdot NHCONH_2$	III-61	102.09
124	**Acetylamino-azotoluene**	azodermin (4-NH₂; 3,2'-diMe)	$CH_3 \cdot C_6H_4 \cdot N:N \cdot C_6H_3 \cdot (CH_3) \cdot NH(COCH_3)$	....	267.33
125	benzoic acid (*o*)	*N*-acetyl-anthranilic acid	$CH_3CO \cdot NH \cdot C_6H_4 \cdot CO_2H$	XIV-337	179.18
126	benzoic acid (*m*)	....	$CH_3CO \cdot NH \cdot C_6H_4 \cdot CO_2H$	XIV-396	179.18
127	benzoic acid (*p*)	acetaminobenzoic acid	$CH_3CO \cdot NH \cdot C_6H_4 \cdot CO_2H$	XIV-432	179.18
128	hydroxyphenyl-arsonic acid (3;4,1)	stovarsol; spirocid	$CH_3CONH \cdot C_6H_3(OH) \cdot AsO_3H_2$	....	275.09
129	phenylglycine (*p*)	....	$CH_3CONH \cdot C_6H_4 \cdot CH_2CO_2H$	....	193.20
130	**Acetylene**	ethyne; ethine	$CH\vdots CH$	I-228	26.04
131	dibromide (1,1)	α,α-dibromoethylene	$CH_2:CBr_2$	I-190	185.86
132	dibromide (*cis*)	1,2-dibromoethene	$CHBr:CHBr$	I-190	185.86
133	dibromide (*trans*)	1,2-dibromoethene	$CHBr:CHBr$	I-190	185.86
134	dicarboxylic acid	butynedioic acid	$(C \cdot CO_2H)_2$	II-801	114.06
135	dichloride (*cis*)	1,2-dichloroethene	$CHCl:CHCl$	I-187	96.94
136	dichloride (*trans*)	dioform	$CHCl:CHCl$	I-187	96.94
137	diiodide (*cis*)	1,2-diiodoethene	$CHI:CHI$	I-194	279.85
138	diiodide (*trans*)	1,2-diiodoethene	$CHI:CHI$	I-194	279.85
139	urea	glycoluril	$C_2H_2(CON_2H_2)_2$	XXVI-441	142.12

† Also cr. $+ 3H_2O$, and $+ 4H_2O$.
Acetyl-propiophenone 52
Acetyl-propylaniline 5364

Acetyl-resorcinol 5550
Acetyl-thymol 5945

Table 7-4 (*Continued*)
PHYSICAL CONSTANTS OF ORGANIC COMPOUNDS

No.	Crystalline Form and Color	Specific Gravity	Melting Point °C.	Boiling Point °C.	Solubility in 100 Parts		
					Water	Alcohol	Ether
108	pl./bz.	....	softens >70	d. >87	s.; s. act.	s.; sl. s. bz.	s.; i. lg.
109	pl./al.	....	280±	....	s.; i. bz.	sl. s.	i.; i. lg.
110	lf./aq.	....	190-2	....	v. sl. s.	s.; s. ac.	v. sl. s.
111	pr.	....	130-2	....	s. h.	s.	sl. s.
112	lq.	1.011^{90}	....	226-7	∞	s.	....
113	lq.	....	....	181-2	i.	....	d. HCl
114	yel. oil	....	....	283 d.	i.; s. alk.	∞ ; ∞ bz.	∞ chl.
115	col. nd./aq.	....	135-6	....	$1^{37°}$	v. sl. s. bz.	$5^{20°}$
116	cr./aq. al.	....	....	....	16	1.4	i.
117	wh. hyg. pd.	....	....	....	100	25	i.
118	wh. pd.	....	....	....	s.	sl. s.	i.
119	cr./act.-et.	....	218 sl. d.	....	v. s.	v. s.	sl. s. act.
120	col. cr./ abs. al.	....	97-8	198^{11mm}	i.	s.	s.
121	cr./al.	....	175-6	....	i.	sl. s.	i.
122	pr./aq.	....	166-7	....	s. h.	s.	sl. s.
123	nd./al.	....	218	....	$2^{15°};$ v. s. h.	10 h.	....
124	r. nd./al.	....	185-6	....	i.	s. chl.	s.; s. oils
125	rhb./ac.	....	184-6	....	s. h.	s.	s.; s. bz.
126	nd./al.	....	248-50	subl.	v. sl. s. h.	s. h.	v. sl. s.
127	nd.	....	256.5 d.	....	sl. s.	s.	....
128	wh. pd.	....	....	....	sl. s.	s. alk.	....
129	col. fl.	....	241-2	....	i.	sl. s.	sl. s.
130	col. gas	lq. $0.613^{-80°}$ s. $0.730^{-85°}$ (A) 0.906	-81.5^{891mm}	-84^{760mm}	$100 cc^{18°}$ cf. special table	$600 cc^{18°}$	$2500 cc^{15°}$ act.; 30,000 cc act. 12 atm.
131	col. lq.	$2.178\frac{21°}{4}$	....	$91-2^{754mm}$	i.	v. s.	v. s.
132	col. lq.	$2.285\frac{17.5°}{4}$	-53	110^{754mm}	i.	v. s.	v. s.
133	col. lq.	$2.267\frac{17.5°}{4}$	-6.5	108	i.	v. s.	v. s.
134	pl./et.	....	179-80	....	v. s.	v. s.	v. s.
135	col. lq.	$1.282\frac{20°}{4}$	-80.5	60.3	$0.35^{20°}$	∞	∞
136	col. lq.	$1.255\frac{20°}{4}$	-50	47.7	$0.63^{20°}$	∞	∞
137	col. lq.	$3.063^{20°}$	-13.8	188 d.	i.	s.	s.
138	mn. pr.	$3.303^{21°}$	73	192	i.	s.	s.
139	nd./aq.	....	d. 300±	....	$0.1^{17°};$ $1.5^{100°}$	i.; s. conc. HCl	>1, h. NH_4OH

Acetyl-toluidine 26-8
Acetylamino-phenol 33-5
Acetylene-tetrabromide 5719

Acetylene-tetrachloride 5739
Acetylidene-*o*-aminophenol 2908

Table 7-4 (*Continued*)
PHYSICAL CONSTANTS OF ORGANIC COMPOUNDS

No.	Name	Synonym	Formula	Beil. Ref.	Formula Weight
140	**Aconic acid**	....	CH$_2$·CO·O·CH:C· $\underline{\qquad\qquad}$ CO$_2$H	XVIII-395	128.09
141	**Aconitic acid**	equisetic acid; citridic acid	C$_3$H$_3$(CO$_2$H)$_3$	II-849	174.11
142	**Acridine**	....	C$_6$H$_4$·CH·C$_6$H$_4$N $\underline{\qquad\qquad}$	XX-459	179.22
143	orange (3,6)	....	NC$_{13}$H$_7$[N(CH$_3$)$_2$]$_2$	XXII-487	265.36
144	red (2,7)	....	CH$_3$NH·C$_{13}$H$_7$O: N(Cl)CH$_3$	....	273.74
145	**Acridone**	dihydroketo-acridine	C$_6$H$_4$·CO·C$_6$H$_4$·NH $\underline{\qquad\qquad}$	XXI-335	195.22
146	**Acrolein**	acrylic aldehyde	CH$_2$:CH·CHO	I-725	56.06
147	**Acrolein** (*meta*)	metacrolein	(C$_3$H$_4$O)$_3$	I-727	168.19
148	**Acrylamide**	acrylic amide	CH$_2$:CH·CONH$_2$	II-400	71.08
149	**Acrylic** acid	propenoic acid	CH$_2$:CH·CO$_2$H	II-397	72.06
150	anhydride	....	(CH$_2$:CH·CO)$_2$O	II-400	126.11
151	nitrile	vinyl cyanide	CH$_2$:CH·CN	II-400	53.06
152	**Adipamide**	adipic diamide	(·CH$_2$CH$_2$CONH$_2$)$_2$	II-653	144.17
153	**Adipic** acid	hexandioic acid	(·CH$_2$CH$_2$CO$_2$H)$_2$	II-649	146.14
154	**Adipyl** chloride	....	(·CH$_2$CH$_2$CO·Cl)$_2$	II-653	183.04
155	dinitrile	tetramethylene dicyanide	(·CH$_2$CH$_2$CN)$_2$	II-653	108.14
156	**Adonitol**	adonite	HOCH$_2$(CHOH)$_3$· CH$_2$OH	I-530	152.15
157	**Adrenaline** (*l*)(3,4,1)	*l*-Suprarenin; epinephrine	C$_6$H$_3$(OH)$_2$(CHOH· CH$_2$NHCH$_3$)	XIII-830	183.21
158	**Alanine** (α) (*l* +)	α-aminopropionic acid	H$_2$N·CH(CH$_3$)·CO$_2$H	IV-387	89.09
159	**Alanine** (α) (*dl*)	....	H$_2$N·CH(CH$_3$)·CO$_2$H	IV-387	89.09
160	**Alanine** (β)	β-aminopropionic acid	H$_2$N·CH$_2$·CH$_2$·CO$_2$H	IV-401	89.09
161	**Aldehydin** (2,5)	2-Me-5-Et-pyridine	(CH$_3$)C$_5$H$_3$N(C$_2$H$_5$)	XX-248	121.18
162	**Aldehydo**-benzoic acid (*o*)	phthalic aldehyde	HO$_2$C·C$_6$H$_4$·CHO	X-666	150.14
163	benzoic acid (*m*)	isophthalic aldehyde	HO$_2$C·C$_6$H$_4$·CHO	X-671	150.14
164	benzoic acid (*p*)	terephthalic aldehyde	HO$_2$C·C$_6$H$_4$·CHO	X-671	150 14
165	*o*-hydroxybenzoic acid	(2;5,1)	HO·C$_6$H$_3$(CHO)· CO$_2$H	X-953	166 13
166	*o*-hydroxybenzoic acid	(2;3,1)	HO·C$_6$H$_3$(CHO)· CO$_2$H·1H$_2$O	X-952	184.15
167	*m*-hydroxybenzoic acid	(3;4,1)	HO·C$_6$H$_3$(CHO)· CO$_2$H	X-954	166.13
168	*p*-hydroxybenzoic acid	(4;3,1) formyl sal-icylic acid (5)	HO·C$_6$H$_3$(CHO)· CO$_2$H	X-953	166.13
169	**Aldol**	2-hydroxybutyralde-hyde	CH$_3$·CHOH·CH$_2$· CHO	I-824	88.11
170	**Aldol** par-	paraldol	(C$_4$H$_8$O$_2$)$_2$	I-825	176.21
171	**Aleudrin**	α,α′-diCl-*iso*Pr-carbamate	H$_2$N·CO$_2$·CH(CH$_2$· Cl)$_2$	III-29	172.01

Table 7-4 (*Continued*)
PHYSICAL CONSTANTS OF ORGANIC COMPOUNDS

No.	Crystalline Form and Color	Specific Gravity	Melting Point °C.	Boiling Point °C.	Solubility in 100 Parts		
					Water	Alcohol	Ether
140	rhb./aq.	....	164	....	$18^{15°}$	....	s.
141	lf. or nd./aq.	....	192 d.	....	$33^{15°}$	$50^{12°}$, 88% al.	v. sl. s.
142	rhb./aq. al.	....	110-1	346	sl. s. h.	s.	s.; s. CS_2
143	yel. nd./aq. al.	....	181	....	s.; s. dil. a.	s.; s. act.	v. sl. s. pet.
144	bronze pd.	....	....	....	sl. s.	s.	i.
145	yel. nd./al.	....	354	....	i. aq. KOH; s. h. HCl	s. h.; s. h. ac.	i.; i. bz.; i. chl.
146	col. lq.	$0.841\frac{20°}{4}$	−87.7	52.5	40	s.	s.
147	nd.	....	45-50	sl. diss.	v. sl. s. h.	s.	s.
148	lf./bz.	$1.122^{30°}$	84-5	125^{25mm}	204	74	v. s.
149	col. lq.	$1.062\frac{16°}{4}$	13.0	141-2	∞; ∞ bz.	∞; ∞ chl.	∞; ∞ act.
150	col. lq.	$1.094^{0°}$	....	97^{35mm}	....	....	....
151	col. lq.	$0.806\frac{20°}{4}$	−82	77.3	s.	....	....
152	cr. pd.	....	226-7	....	$0.4^{12°}$	....	....
153	mn. pr.	$1.360\frac{25°}{4}$	151-3	265^{10mm}	$1.4^{15°}$	v. s.	$0.6^{15°}$
154	col. lq.	....	....	$125-8^{11mm}$	d. h.	d. h.	....
155	col. oil	$0.951\frac{19°}{19}$	1	295	v. sl. s.	s.	v. sl. s.
156	pr./aq.	....	102	....	v. s.	s. h.	i.; i. lg.
157	col. cr. pd.	....	d. 207-11	....	$0.027^{20°}$; s. alk.	v. sl. s.; s. a.	i.; i. chl.
158	col. rhb./aq.	....	297 d.	....	$20.5^{45°}$	v. sl. s. abs.	i.
159	col. nd./aq.	....	295 d.	subl. >200	$21.7^{17°}$; $32^{75°}$	0.2, 80% c. al.	i.; i. act.
160	col. rhb.	....	196 d.	....	v. s.	v. sl. s. abs.	i.
161	lq.	$0.918^{23°}$	−70.3	178.3	i.; s. aq. a.	s.	s.;s. H_2SO_4
162	mn.	1.404	97-8	....	v. s.	v. s.	v. s.
163	nd.	....	164-6	....	....	....	....
164	nd./aq.	....	285	....	sl. s. h.	s. chl.	s.
165	nd.	....	248-9	....	$0.04^{25°}$; $0.7^{100°}$	s. h.; s. alk.	s.; i. chl.
166	nd./aq.	....	−H_2O, 100; m. anh. 179	....	$0.06^{25°}$; $6^{100°}$	s.	s. alk.
167	nd.	....	234	....	sl. s. h.	s.	s.
168	pr./aq.	....	243-4	subl.	sl. s.	s.	s.
169	col. lq.	$1.103\frac{20°}{4}$	....	83^{20mm}; d. 85	∞	∞	s.
170	tri.	....	82	in vac. 90-100 (diss.)	v. s.	$25^{25°}$, 99% al.	$5^{23°}$
171	col. cr. pd.	....	80-2	....	sl.s.; s.bz.	s.; s. chl.	s.; s. oil

Table 7-4 (*Continued*)
PHYSICAL CONSTANTS OF ORGANIC COMPOUNDS

No.	Name	Synonym	Formula	Beil. Ref.	Formula Weight
172	**Alizarin** amide (β) (1,2)	....	$C_6H_4(CO)_2C_6H_2\cdot$ $(OH)NH_2$	XIV-267	239.23
173	amide (α) (2,1)	....	$C_6H_4(CO)_2C_6H_2\cdot$ $(OH)NH_2$	XIV-275	239.23
174	blue	diOH-anthraqui- none-quinoline	$C_6H_4(CO)_2C_9H_3N:$ $(OH)_2$	XXI-632	291.27
175	blue S		$C_{17}H_9O_4N\cdot2NaHSO_3$	XXI-633	499.39
176	carboxylic acid (β)	....	$(HO)_2C_6H_2(CO)_2:$ $C_6H_3\cdot CO_2H$	X-1035	284.23
177	3-sulfonic acid (1,2;3)	....	$(HO)_2C_{14}H_5O_2\cdot SO_3H$	XI-355	320.28
178	3-sulfonic acid (Na)	alizarin S; alizarin carmine	$(HO)_2C_{14}H_5O_2\cdot$ $SO_3Na\cdot H_2O$	XI-355	360.28
179	**Aljodan**	iodoethyl allo- phanate	$IC_2H_4\cdot CO_2NHCONH_2$	**III-56	258.02
180	**Alkannin**	....	$C_{16}H_{16}O_5$	**VIII-544	288.30
181	**Allantoin** †	glyoxyldiureide	$C_4H_6O_3N_4$	XXV-474	158.12
182	**Allanturic acid**	lantanuric acid	$NH_2\cdot CO\cdot N:CH\cdot CO_2H$	XXV-475	116.08
183	**Allene**	propadiene	$CH_2:C:CH_2$	I-248	40.07
184	**Alloxan**	mesoxalylurea	$HN\cdot(CO)_3\cdot NH\cdot CO\cdot$ $\overline{}$ $4H_2O$	XXIV-500	214.13
185	**Alloxan**	....	$HN\cdot(CO)_3\cdot NH\cdot CO$ $\overline{}$	XXIV-500	142.07
186	**Alloxan** (mono- hydrate)	....	$CO\cdot NHCONHCO\cdot C:$ $\overline{}$ $(OH)_2$	XXIV-500	160.09
187	**Alloxanic acid**	....	$C_4H_4O_5N_2$	III-772	160.09
188	**Alloxantin**	uroxin	$C_8H_6O_8N_4\cdot2H_2O$	XXVI-556	322.19
189	**Allyl** acetate	....	$CH_3CO_2\cdot C_3H_5$	II-136	100.12
190	acetic acid	pentenoic acid	$CH_2:CH\cdot C_2H_4\cdot CO_2H$	II-425	100.12
191	acetone		$C_3H_5\cdot CH_2\cdot CO\cdot CH_3$	I-734	98.15
192	acetonitrile	....	$C_3H_5\cdot CH_2CN$	II-426	81.12
193	alcohol	propen-1-ol-3	$CH_2:CH\cdot CH_2OH$	I-436	58.08
194	amine	....	$CH_2:CH\cdot CH_2\cdot NH_2$	IV-205	57.10
195	*iso*-amyl ether		$C_3H_5\cdot O\cdot C_5H_{11}$	I-438	128.22
196	aniline	....	$C_3H_5\cdot NH\cdot C_6H_5$	XII-170	133.19
197	arsonic acid	....	$CH_2:CH\cdot CH_2\cdot$ $AsO(OH)_2$	**IV-998	166.01
198	benzoate	....	$C_6H_5\cdot CO_2\cdot C_3H_5$	IX-114	162.19
199	bromide	3-bromo-propene-1	$CH_2:CH\cdot CH_2Br$	I-201	120.98
200	butyrate	....	$CH_3(CH_2)_2CO_2\cdot C_3H_5$	II-272	128.17
201	caproate (*n*)	....	$CH_3(CH_2)_4CO_2\cdot C_3H_5$	**II-285	156.23
202	chloride	3-chloro-propene-1	$CH_2:CH\cdot CH_2Cl$	I-198	76.53
203	cinnamate	....	$C_6H_5\cdot(CH)_2CO_2\cdot C_3H_5$	*IX-230	188.23
204	cyanide	vinyl-acetonitrile	$CH_2:CH\cdot CH_2\cdot CN$	II-408	67.09
205	*iso*-cyanide	allyl carbylamine	$CH_2:CH\cdot CH_2\cdot NC$	IV-208	67.09
206	ether	diallyl ether	$(CH_2:CH\cdot CH_2)_2O$	I-438	98.15
207	formate	....	$HCO_2\cdot CH_2\cdot CH:CH_2$	II-23	86.09
208	iodide	3-iodo-propene-1	$CH_2:CH\cdot CH_2I$	I-202	167.98
209	malonic acid	....	$C_3H_5\cdot CH(CO_2H)_2$	II-776	144.13
210	mercaptan	....	$CH_2:CH\cdot CH_2\cdot SH$	I-440	74.15
211	phenol (*p*)	chavicol	$CH_2:CH\cdot CH_2\cdot C_6H_4\cdot$ OH	VI-571	134.18

† Cf. also Alkaloids table.

Table 7-4 (*Continued*)
PHYSICAL CONSTANTS OF ORGANIC COMPOUNDS

No.	Crystalline Form and Color	Specific Gravity	Melting Point °C.	Boiling Point °C.	Solubility in 100 Parts		
					Water	Alcohol	Ether
172	br. nd./al.	....	....	....	i.	s.	s.; sl. s. NH$_4$OH
173	br. nd./al.	....	250	....	s. alk.	s.	s.; s. alk. carb.
174	b. met. nd./bz.	....	270±	subl.	i.; s. h. bz.	sl. s.; s. ac.	sl. s.
175	r. br. cr.	....	....	....	v. s.; d. h.	sl. s.	i.
176	or. nd./ PhNO$_2$	....	305	subl.	s. aq. NaOAc	sl. s.	v. sl. s.
177	or. yel. cr.	....	....	....	s.	s.	i.
178	or. yel. pd.	....	....	....	s.	s.	....
179	col. pd.	....	192±d.	....	sl. s.	s.; s. act.	v. sl. s. bz.
180	br. met.	....	149	....	s. alk.	v. sl. s.	v. sl. s.
181	nd./h. aq.	....	235-6	....	0.76$^{22°}$; 3.3 h.	sl. s.; s. NaOH	i.
182	hyg. pd.	....	....	....	delq.	i.	d. h. alk.
183	gas	....	−146	−32	....	....	....
184	rhb. pr. /aq.; eff. −3H$_2$O	....	−4H$_2$O, 150	d. 170	s.	s.	....
185	rhb. yel. hyg.	....	d. 170	....	v. s.; s. ac.	v. s.; sl. s. pet.	i.; sl. s. chl.
186	tri./aq.	....	d. 170	....	v. s.	v. s.	i.
187	tri.	....	d.	....	v. s.	16	s.
188	rhb.	....	−2H$_2$O, 120	d. 253-5	0.3$^{25°}$; 6h.	v. sl. s.	v. sl. s.
189	col. lq.	$0.928^{2.0°}_{4}$	....	104^{762mm}	i.	∞	∞
190	col. lq.	$0.984^{1.8°}_{4}$	<−18	187-9	sl. s.	s.	s.
191	col. lq.	$0.847^{1.6°}_{4}$	....	129^{748mm}	i.	....	....
192	lq.	$1.180^{13°}$	....	140	i.	....	....
193	lq.	$0.855^{1.5°}_{4}$	−129	97.1	∞	∞	∞
194	col. lq.	$0.761^{2.2°}_{4}$	....	56.5^{756mm}	∞	∞	∞
195	col. lq.	....	....	120	i.	∞	∞
196	yel. oil	$0.982^{25°}$	....	217-9^{736mm}	i.	s.	s.
197	col. cr./aq.	....	129-30	....	v.s.h.; d.h.a.	s.	....
198	yel. lq.	$1.058^{1.5°}_{15}$	....	230	i.	s.	s.
199	lq.	$1.398^{2.0°}_{4}$	−119.4	70-1^{753mm}	i.	∞	∞
200	col. lq.	....	....	143^{772mm}	i.	s.	s.
201	pa. yel. lq.	....	....	186-8	i.	s.	s.
202	col. lq.	$0.938^{2.0°}_{4}$	−134.5	45.1	<0.1	∞	∞
203	col. lq.	$1.052^{25°}_{25}$	....	150-2^{15mm}	i.	s.	v. s.
204	col. lq.	$0.837^{16°}$	−86.8	118.4	....	s.	....
205	lq.	$0.797^{17°}$	....	106	sl. s.	s.	∞
206	lq.	$0.826^{2.0°}_{4}$	....	94.3	0.3	∞	∞
207	lq.	$0.948^{18°}$	....	83.6^{768mm}	i.	s.	....
208	yel. lq.	$1.848^{1.2°}_{12}$	−99.3	102^{744mm}	i.	∞	∞
209	cr./et.	....	102-5	d.−CO$_2$,180	s.	s.	s.; s. h. bz.
210	lq.	$0.925^{2.8°}_{4}$	....	67-8	....	∞	∞
211	lq.	$1.023^{19.4°}_{4}$	<−25	237	∞ pet.	∞ ; ∞ chl.	∞

Table 7-4 (*Continued*)
PHYSICAL CONSTANTS OF ORGANIC COMPOUNDS

No.	Name	Synonym	Formula	Beil. Ref.	Formula Weight
	Allyl				
212	phenylcinchonate	atoquinol (2,4)	$C_9H_5N(C_6H_5)CO_2\cdot$ C_3H_5	....	289.34
213	phenylcinchonate HCl	....	$C_{19}H_{15}O_2N\cdot HCl$	....	325.80
214	phenyl ether	....	$C_3H_5\cdot O\cdot C_6H_5$	VI-144	134.18
215	phenyl urea (*N,N'*)	....	$C_3H_5NH\cdot CO\cdot$ NHC_6H_5	XII-350	176.22
216	propionate	....	$C_2H_5\cdot CO_2\cdot C_3H_5$	II-241	114.15
217	*iso*-propylacetyl-carbamide	sedormid	$(C_3H_7)(C_3H_5):CH\cdot$ $CO\cdot NHCONH_2$	**III-53	184.24
218	pyridine (2)	....	$C_3H_5\cdot C_5H_4N$	....	119.17
219	thiocyanate	....	$C_3H_5\cdot SCN$	III-177	99.16
220	*iso*-thiocyanate	mustard oil	$C_3H_5\cdot N:C:S$	IV-214	99.16
221	thiourea	thiosinamine	$C_3H_5\cdot NH\cdot CS\cdot NH_2$	IV-211	116.19
222	trisulfide	....	$(C_3H_5)_2S_3$	I-441	178.34
223	urea	....	$C_3H_5\cdot NH\cdot CO\cdot NH_2$	IV-209	100.12
224	*iso*-valerate	....	$(CH_3)_2CH\cdot CH_2\cdot$ $CO_2C_3H_5$	II-313	142.20
225	**Allylene**	propyne; propine	$CH_3\cdot C:CH$	I-246	40.07
226	oxide	....	$CH_3\cdot C:CH\cdot O$	XVII-20	56.06
227	**Aloin**	....	$C_{21}H_{20}O_9$	....	416.39
228	**Alstonine**	....	$C_{21}H_{20}O_4N_2\cdot 3\frac{1}{2}H_2O$	....	427.46
229	**Aluminum** ethoxide	....	$Al(OC_2H_5)_3$	I-313	162.17
230	triethyl	....	$Al(C_2H_5)_3$	IV-643	114.17
231	trimethyl	....	$Al(CH_3)_3$	IV-643	72.09
232	**Amalinic acid**	tetraMe-alloxantin	$(CH_3)_4C_8H_2O_8N_4$	XXVI-559	342.27
233	**Amarine**	triphenyl-imina-zoline	$C_6H_5CHCH(C_6H_5)\cdot$ $N:C(C_6H_5)\cdot NH\cdot$ $\frac{1}{2}H_2O$	XXIII-304	307.40
234	**Amaron**	tetraphenylpyrazine	$C_{28}H_{20}N_2$	XXIII-343	384.49
235	**Amino**-acetone	acetonyl-amine	$NH_2\cdot CH_2\cdot CO\cdot CH_3$	IV-314	73.10
236	acetophenone (*o*)	....	$NH_2\cdot C_6H_4\cdot CO\cdot CH_3$	XIV-41	135.17
237	acetophenone (*m*)	....	$NH_2\cdot C_6H_4\cdot CO\cdot CH_3$	XIV-45	135.17
238	acetophenone (*p*)	....	$NH_2\cdot C_6H_4\cdot CO\cdot CH_3$	XIV-46	135.17
239	acetophenone (*ω*)	phenacylamine	$C_6H_5\cdot CO\cdot CH_2\cdot NH_2$	XIV-49	135.17
240	acetophenone HCl (*ω*)	....	$C_8H_7O\cdot NH_2\cdot HCl$	XIV-49	171.63
241	alizarin (3)	β-aminoalizarin	$C_6H_4(CO)_2C_6H\cdot$ $(OH)_2NH_2$	XIV-285	255.23
242	alizarin (4)	α-aminoalizarin	$C_{14}H_5O_2(OH)_2NH_2$	XIV-286	255.23
243	amylene glycol	2-NH₂-2-Et-1,3-propandiol	$C_2H_5\cdot C(NH_2):$ $(CH_2OH)_2$	....	119.16
244	anthraquinone (1)	....	$C_6H_4(CO)_2C_6H_3NH_2$	XIV-177	223.23
245	anthraquinone (2)	....	$C_6H_4(CO)_2C_6H_3NH_2$	XIV-191	223.23
246	2-azo-5-anisole	benzene-azo-*o*-anis-idine (5;2,1)	$C_6H_5\cdot N:N\cdot C_6H_3(NH_2)\cdot$ OCH_3	XVI-396	227.27
247	azobenzene (*o*)	....	$NH_2\cdot C_6H_4\cdot N_2\cdot C_6H_5$	XVI-303	197.24
248	azobenzene (*m*)	....	$NH_2\cdot C_6H_4\cdot N_2\cdot C_6H_5$	XVI-304	197.24

Table 7-4 (*Continued*)
PHYSICAL CONSTANTS OF ORGANIC COMPOUNDS

No.	Crystalline Form and Color	Specific Gravity	Melting Point °C.	Boiling Point °C.	Solubility in 100 Parts		
					Water	Alcohol	Ether
212	yel. nd./al.	....	36	260^{15mm}	i.; s. oil	8; s. bz.	v. s.
213	yel. nd./al.	....	145-7	....	d.	s. h.	i.
214	col. oil	$0.986\frac{15}{15}$	....	191.7	i.	....	....
215	nd./bz.	....	114.5-5.5	....	s. bz.	....	....
216	col. lq.	....	....	124.5^{774mm}	....	....	....
217	nd./al.	....	194	....	0.03c., 0.5 h.	10	1.3
218	lq.	$0.959^{0°}$	......	189-90	....	....	....
219	oil	$1.056^{15°}$	....	161	v. sl. s.	v. s.	v. s.
220	col. oil	$1.024\frac{15}{4}°$	-102.5	152	0.2	∞	∞
221	col. pr.	$1.219\frac{20}{20}°$	77-8	....	$3^{0°}$; s. h.	s.; i. bz.	v. sl. s.
222	lq.	$1.085^{15°}$	....	$112-22^{16mm}$	v. s.	v. s.	v. sl. s.
223	nd./al.	....	85	....	v. s.	v. s.	v. sl. s.
224	lq.	....	....	155	....	∞	∞
225	gas	$0.660\frac{-1}{4}3°$	-102.7	-23.2	v. sl. s.	v. s.	$3000cc.^{16°}$
226	lq.	....	....	62-3	sl. s.	....	i. aq. K_2CO_3
227	yel. pr./al.	....	147.9	....	sl. s. h.	sl. s. h.	v. s. alk.
228	br. amor.	....	<100; 195 ±(anh.)	....	sl. s.	s.	v. sl. s.
229	pd.	$1.142\frac{20}{0}°$	150-60	$200-5^{10mm}$	d.	i.; sl. s. bz.	v. sl. s.
230	col. lq.	....	<-18	194	d.	d. by air	....
231	col. lq.	....	0	129-30	d.	d. by air	....
232	or./aq.	....	221 d.	....	sl. s. h.	v. sl. s.	s. alk.
233	cr./al.	....	106; 131-3 (anh.)	d. 198 (anh.)	i.	s.	s.
234	red nd./ac.	....	245-6	subl.	i.; v. s. h. bz.	sl. s. h.; s. chl.	sl. s. h.
235	only salts are known	....	....		....	....	....
236	yel. oil	....	....	250 (sl. d.)	i.	s.	s.
237	lf./aq.	....	99.5	289-90	....	....	....
238	cr./aq.	....	106	293-5	s. h.	s.	s.
239	unstable	....	....	....	....	....	....
240	cr.	....	184-8	....	s.	....	i.
241	red pr./ac.	....	>300	subl. d.	s. KOH	sl. s.	sl. s. HCl
242	bl. nd./al.	....	....	....	s. alk.	s.	....
243	col. cr.	$1.099\frac{20}{20}°$	37.5-8.5	$152-3^{10mm}$	∞	∞	v. sl. s.
244	red nd.	....	256	subl.	i.; s. HCl	s.; s. bz.	s.; s. chl.
245	red nd./al.	....	302	subl.	i.	s.	i.; s. bz.
246	red br. pl./ PhMe	....	110.5-1.5	....	s. ac.	s.; s. bz.	s.
247	red pr./al.	....	59	....	i.	v. s.	v. s.
248	or. nd./lg.	....	56-7	....	s. bz.	s.	s.; s. chl.

Table 7-4 (*Continued*)
PHYSICAL CONSTANTS OF ORGANIC COMPOUNDS

No.	Name	Synonym	Formula	Beil. Ref.	Formula Weight
	Amino				
249	azobenzene (*p*)	aniline yellow	$NH_2 \cdot C_6H_4 \cdot N_2 \cdot C_6H_5$	XVI-307	197.24
250	azobenzene HCl (*p*)	....	$C_{12}H_9N_2(NH_2) \cdot HCl$	XVI-310	233.70
251	α-azonaphthalene	azodinaphthylamine	$NH_2 \cdot C_{10}H_6 \cdot N_2 \cdot C_{10}H_7$	XVI-365	297.36
252	azophenylene	benzotriazole	$C_6H_4NH \cdot N : N$	XXVI-38	119.13
253	azoxylene HCl	....	$(CH_3)_4C_{12}H_5(NH_2)N_2 \cdot$ HCl	....	289.81
254	benzaldehyde (*o*)	....	$NH_2 \cdot C_6H_4 \cdot CHO$	XIV-21	121.14
255	benzaldehyde (*m*)	....	$NH_2 \cdot C_6H_4 \cdot CHO$	XIV-28	121.14
256	benzaldehyde (*p*)	....	$NH_2 \cdot C_6H_4 \cdot CHO$	XIV-29	121.14
257	benzaldoxime (*o*)	....	$NH_2 \cdot C_6H_4 \cdot CH : NOH$	XIV-24	136.15
258	benzaldoxime (*m*)	....	$NH_2 \cdot C_6H_4 \cdot CH : NOH$	XIV-28	136.15
259	benzamide (*o*)	anthranilamide	$NH_2 \cdot C_6H_4 \cdot CO \cdot NH_2$	XIV-320	136.15
260	benzamide (*m*)	....	$NH_2 \cdot C_6H_4 \cdot CO \cdot NH_2 \cdot H_2O$	XIV-390	154.17
261	benzamide (*p*)	....	$C_7H_8ON_2 \cdot \frac{1}{4}H_2O$	XIV-425	140.66
262	benzidine (2,4;4′)	triamino-diphenyl	$(NH_2)_2C_6H_3C_6H_4 \cdot NH_2$	XIII-306	199.26
263	benzimidazole (2)	*o*-phenylene-guanidine	$C_6H_4 \cdot NH \cdot C(NH_2) : N$	XXIV-116	133.15
264	benzoic acid (*o*)	anthranilic acid	$NH_2 \cdot C_6H_4 \cdot CO_2H$	XIV-310	137.14
265	benzoic acid (*m*)	....	$NH_2 \cdot C_6H_4 \cdot CO_2H$	XIV-383	137.14
266	benzoic acid (*p*)	aminodracylic acid	$NH_2 \cdot C_6H_4 \cdot CO_2H$	XIV-418	137.14
267	benzonitrile (*o*)	anthranilic nitrile	$NH_2 \cdot C_6H_4 \cdot CN$	XIV-322	118.14
268	benzonitrile (*m*)	....	$NH_2 \cdot C_6H_4 \cdot CN$	XIV-391	118.14
269	benzonitrile (*p*)	....	$NH_2 \cdot C_6H_4 \cdot CN$	XIV-425	118.14
270	benzophenone (*o*)	....	$C_6H_5 \cdot CO \cdot C_6H_4 \cdot NH_2$	XIV-76	197.24
271	benzophenone (*m*)	....	$C_6H_5 \cdot CO \cdot C_6H_4 \cdot NH_2$	XIV-81	197.24
272	benzophenone (*p*)	....	$C_6H_5 \cdot CO \cdot C_6H_4 \cdot NH_2$	XIV-81	197.24
273	benzothiazole (2)	....	$C_6H_4 \cdot N : C(NH_2) \cdot S$	XXVII-182	150.20
274	benzyl-cyanide (*p*)	aminophenyl-aceto-nitrile	$H_2N \cdot C_6H_4 \cdot CH_2 \cdot CN$	XIV-457	132.17
275	benzyl-cyanide HCl (*p*)	....	$H_2N \cdot C_6H_4 \cdot CH_2 \cdot CN \cdot$ HCl	XIV-457	168.63
276	benzoyl-2,2-di-methyl-3-di-ethylaminopro-panol HCl (*p*)	larocaine	$H_2N \cdot C_6H_4 \cdot CO_2 \cdot CH_2 \cdot C(CH_3)_2 \cdot CH_2 \cdot N(C_2H_5)_2 \cdot$ HCl	....	314.86
277	butyric acid (α) (*dl*)	....	$C_2H_5 \cdot CHNH_2 \cdot CO_2H$	IV-408	103.12
278	butyric acid (β) (*dl*)	....	$CH_3 \cdot CHNH_2 \cdot CH_2 \cdot CO_2H$	IV-412	103.12
279	butyric acid (γ)	piperidinic acid	$NH_2 \cdot (CH_2)_3 \cdot CO_2H$	IV-413	103.12
280	*iso*-butyric acid (α)	....	$(CH_3)_2 CNH_2 \cdot CO_2H$	IV-414	103.12
281	camphor (3) (α)	....	$C_{10}H_{15}O \cdot NH_2$	XIV-10	167.25
282	*n*-caproic acid (α) (*dl*)	....	$CH_3(CH_2)_3CHNH_2 \cdot CO_2H$	IV-433	131.18
283	*n*-caproic acid (α)	norleucine (*l* +)	$CH_3(CH_2)_3CHNH_2 \cdot CO_2H$	IV-432	131.18
284	*n*-caproic acid (ε)	....	$H_2N(CH_2)_5CO_2H$	IV-434	131.18

Amino-azotoluene 309-12
Amino-barbituric acid 6384
Amino-benzene 513

Amino-benzene sulfonic acid 523-5
Amino-butyl-guanidine, cf. alko.

Table 7-4 (*Continued*)
PHYSICAL CONSTANTS OF ORGANIC COMPOUNDS

No.	Crystalline Form and Color	Specific Gravity	Melting Point °C.	Boiling Point °C.	Solubility in 100 Parts		
					Water	Alcohol	Ether
249	yel. mn.	....	126-7	225^{120mm}	sl. s. h.	s. h.	s.
250	steel b.*	....	....	....	s.	....	......
251	br. nd.		175	....	sl. s. bz.	sl. s.	sl. s.
252	col. nd./bz.	....	98.5	....	i.	s.	s. bz.
253	br. pd.	....	....	....	i.	s.	sl. s.
254	lf.	....	39-40	d.	v. sl. s.	v. s.	v. s.
255	unstable	....	......	....	....	....	....
256	lf./aq.	....	71.5	....	s.	....	....
257	nd./bz.	....	135-6	subl.	v. sl. s.	s.	s.
258	nd./bz.	....	88	....	v. sl. s. lg.	s.	s.
259	lf./chl.	....	108	300 ± (sl.d.)	v. s. h.	v. s.	sl. s.; i. bz.
260	mn./aq.	....	78-9	$-H_2O > 100$	s.	s.	s.
261	pa. yel. cr.	....	182.9 (anh.)	$-\frac{1}{4}H_2O$, 170	sl. s.	....	....
262	nd.	....	134	....	....	....	....
263	lf./aq.	....	222-4	....	s.; s. alk.; s. dil. a.	s.; s. act.; v. sl. s. bz.	v. sl. s.
264	col. rhb.	....	146-7	subl.	$0.35^{14°}$	$†10.7^{10°}$	$16^{7°}$
265	nd./aq.	$1.511^{4°}$	173-4	....	$†0.6^{15°}$	$†2.2^{10°}$	$1.8^{6°}$
266	mn./pr.	....	187-8	....	$0.3^{13°}$	$†11.3^{10°}$	$8.2^{6°}$
267	nd./CS$_2$	....	51	$267-8^{777mm}$	v. sl. s.	s.	s.
268	nd./aq. al.	....	53	288-90	s. h.	v. s.	v. s.
269	mn. pr.	....	86	d.	v. s. h.	v. s.	v. s.; i. HCl
270	yel. lf./al.	....	106-10	....	....	s.	s.
271	yel. lf./aq.	....	86-7	....	sl. s.	s.	s.
272	lf./aq. al.	....	124	....	v. sl. s. c.	v. s.	$2^{25°}$
273	lf./aq.	....	130-2	d.	v. sl. s.; s. conc. a.	s.; s. chl.	s.
274	lf./aq.	....	46	312	sl. s. h.	s.	s.
275	cr.	....	217-20	....	......	v. sl. s. c.	....
276	col. cr.	....	196-7	....	33.3	10	sl. s. c.
277	lf.	....	285 d.	subl. >300	33 c.	0.2 h.	i.
278	cr.	....	193-4	....	100	i. abs. al.	i. abs. et.
279	lf. or nd.	....	203 d.	....	v. s.	i.	i.
280	pl. or pr./aq.	....	203 d.	subl. 280	v. s.	sl. s.	i.
281	waxy cr.	....	70-110	245-6	i.	s.	s.
282	col. lf./aq.	....	300 ±‡	....	1.25	....	....
283	pl. or lf./aq.	....	301	subl. >275	$1.74^{23°}$	v. sl. s.	....
284	lf./MeOH-et.	....	202-3	softens > 190	v. s.	i.	sl. s. MeOH

* Also a red-black crystalline form.   ‡ Sealed tube.
† In 90% alcohol.

Table 7-4 *(Continued)*
PHYSICAL CONSTANTS OF ORGANIC COMPOUNDS

No.	Name	Synonym	Formula	Beil. Ref.	Formula Weight
285	**Amino** *iso*-caproic (α)	leucine (*l* −)	$(CH_3)_2CHCH_2\cdot$ $CH(NH_2)\cdot CO_2H$	IV-437	131.18
286	caproic acid (α)	isoleucine (*l* +)	$(C_2H_5)(CH_3)CH\cdot$ $CHNH_2CO_2H$	IV-454	131.18
287	caproic acid (α)	allo-isoleucine (*d* −)	$(C_2H_5)(CH_3)CH\cdot$ $CHNH_2CO_2H$	IV-457	131.18
288	caprylic acid (α) (*dl*)	....	$CH_3(CH_2)_5CHNH_2\cdot$ CO_2H	IV-461	159.23
289	cinnamic acid (*o*)	....	$H_2N\cdot C_8H_6\cdot CO_2H$	XIV-517	163.18
290	cinnamic acid (*m*)	....	$H_2N\cdot C_8H_6\cdot CO_2H$	XIV-520	163.18
291	cinnamic acid (*p*)	....	$H_2N\cdot C_8H_6\cdot CO_2H$	XIV-521	163.18
292	*o*-cresol (4;1,2)	amino-hydroxy- toluene	$NH_2\cdot C_6H_3\cdot CH_3(OH)$	XIII-574	123.16
293	*o*-cresol (5;1,2)	....	$NH_2\cdot C_6H_3\cdot CH_3\cdot(OH)$	XIII-576	123.16
294	*o*-cresol H_2SO_4	(5;1,2)	$(C_7H_9ON)_2\cdot H_2SO_4$	XIII-576	344.39
295	*m*-cresol (2;1,3)	....	$NH_2\cdot C_6H_3\cdot CH_3\cdot(OH)$	XIII-589	123.16
296	*m*-cresol (4;1,3)	....	$NH_2\cdot C_6H_3\cdot CH_3\cdot(OH)$	XIII-590	123.16
297	*m*-cresol (6;1,3)	....	$NH_2\cdot C_6H_3\cdot CH_3\cdot(OH)$	XIII-593	123.16
298	*p*-cresol (2;1,4)	....	$NH_2\cdot C_6H_3\cdot CH_3\cdot(OH)$	XIII-598	123.16
299	*p*-cresol (3;1,4)	....	$NH_2\cdot C_6H_3\cdot CH_3\cdot(OH)$	XIII-601	123.16
300	dibutylaniline (*p*)	....	$(C_4H_9)_2N\cdot C_6H_4\cdot NH_2$	*XIII-23	220.36
301	dibutylaniline (*p*)	HCl	$C_{14}H_{24}N_2\cdot 2HCl$	*XIII-23	293.28
302	diethylamino- toluene (mono) HCl	(5;1,2)	$(C_2H_5)_2N\cdot C_6H_3\cdot$ $(CH_3)NH_2\cdot HCl$	....	214.74
303	diethylaniline (*p*)	diEt-phenylene- diNH₂	$(C_2H_5)_2N\cdot C_6H_4\cdot NH_2$	XIII-75	164.25
304	diethylaniline (*p*)	(mono-HCl)	$C_{10}H_{16}N_2\cdot HCl$	....	200.71
305	dimethylaniline (*o*)	diMe-*o*-phenylene- diamine (*N,N*)	$(CH_3)_2N\cdot C_6H_4\cdot NH_2$	XIII-15	136.20
306	dimethylaniline (*m*)	....	$(CH_3)_2N\cdot C_6H_4\cdot NH_2$	XIII-38	136.20
307	dimethylaniline (*p*)	....	$(CH_3)_2N\cdot C_6H_4\cdot NH_2$	XIII-72	136.20
308	dimethylaniline (*p*)	H_2SO_4	$C_8H_{12}N_2\cdot H_2SO_4$	XIII-73	234.28
309	2,3'-dimethyl- azobenzene (4)	amino-azo-toluene (3';1,2,4)	$CH_3\cdot C_6H_4\cdot N_2\cdot C_6H_3\cdot$ $(CH_3)NH_2$	XVI-348	225.30
310	2,3'-dimethyl- azobenzene (4')	(2;1,3',4')	$CH_3\cdot C_6H_4\cdot N_2\cdot C_6H_3\cdot$ $(CH_3)NH_2$	XVI-344	225.30
311	2,4'-dimethyl- azobenzene (4)	(4';1,2,4)	$CH_3\cdot C_6H_4\cdot N_2\cdot C_6H_3\cdot$ $(CH_3)NH_2$	XVI-348	225.30
312	3,4'-dimethyl- azobenzene (4)	(4';1,3,4)	$CH_3\cdot C_6H_4\cdot N_2\cdot C_6H_3\cdot$ $(CH_3)NH_2$	XVI-345	225.30
313	diphenylamine (*o*)	*N*-phenyl-*o*- phenylenediamine	$NH_2\cdot C_6H_4\cdot NH\cdot C_6H_5$	XIII-16	184.24
314	diphenylamine (*p*)	....	$NH_2\cdot C_6H_4\cdot NH\cdot C_6H_5$	XIII-76	184.24
315	diphenylamine (*p*)	HCl	$C_{12}H_{10}N\cdot NH_2\cdot HCl$	*XIII-23	220.70
316	diphenylamine (*p*)	H_2SO_4	$(C_{12}H_{10}N\cdot NH_2)_2\cdot$ H_2SO_4	XIII-78	466.56
317	4-(*p*-diphenyl)- thiazole (2)	....	$C_6H_5\cdot C_6H_4\cdot C:CH\cdot$ $\underset{\rule{1em}{0.4pt}}{\vert}$ $S\cdot C(NH_2):N$	....	252.34
318	ethylacetanilide (*p*)	....	$NH_2C_6H_4N(C_2H_5)\cdot$ $COCH_3$	....	178.24

Table 7-4 (*Continued*)
PHYSICAL CONSTANTS OF ORGANIC COMPOUNDS

No.	Crystalline Form and Color	Specific Gravity	Melting Point °C.	Boiling Point °C.	Solubility in 100 Parts		
					Water	Alcohol	Ether
285	lf./aq. al.	$1.293^{\frac{18}{4}°}$	293–5 d.*	subl.	$2^{17°}$; 6.6 h.	$0.07^{17°}$ abs.	i.; 10.9 ac.
286	rhb./80% al.	....	280 d.*	subl. >280	$3.9^{15.5°}$; $6.1^{75°}$	sl. s. h.; s. h. ac.	i.; s. h. gly.
287	lf. or rods	....	280 d.*	....	$2.6^{20°}$	i. c.	s. h. ac.
288	lf./aq.	....	263–4; subl.	....	sl. s. c.; 0.6 h.	v. sl. s.	v. sl. s.
289	yel. nd.	....	158–9 d.	....	s. h.	s.	s.
290	pa. yel. nd./al.	....	181–2	....	sl. s. h.	s.	s.
291	yel. nd.	....	175–6 d.	....	s. h.	s.	s.
292	col. lf./aq.	....	159–61	subl.	s. h.	v. s.	v. s.
293	lf./bz.	....	174–5	subl.	v. sl. s.	s.	s.; sl. s. bz.
294	....	....	178–9	....	s.	i.	i.
295	lf.	....	148–50	subl.	....	....	....
296	lf.	....	....	....	....	....	....
297	cr./bz.	....	174 d.	....	....	....	....
298	cr./aq.	....	144.5	subl.	sl. s. c.	....	....
299	rhb./bz.	....	135	subl.	i. c.	s.; sl. s. bz.	s.; s. chl.
300	lq.	....	....	....	....	....	....
301	nd./al.	....	....	....	s.	s.	i.
302	col. nd.	....	....	....	s.	sl. s.	i.
303	lq.	....	22–4	260–2	i.	s.	s.
304	col. nd.	....	....	....	s.	s.	i.
305	oil	....	....	$217.5^{751.5mm}$	sl. s.	s.	s.
306	oil	$0.995^{25°}$	<−20	$268\text{–}70^{740mm}$	sl. s.	s.	s.
307	col. nd.	$1.041^{\frac{18}{15}°}$	41	262.3	s. c.	v. s.	s.; s. chl.
308	lf.	....	....	....	v. s.	....	....
309	yel. nd.	....	80	....	i.	s.	....
310	yel. mn./al.	....	100	....	i.	s.	s.; s. chl.
311	yel. lf./al.	....	127	....	i.	s.	sl. s. lg.
312	yel. nd./lg.	....	127–8	....	i.	s.	sl. s. lg.
313	nd./aq.	....	79–80	....	s. chl.	s. act.	s. bz.; sl s. lg.
314	nd./aq. al.	....	66–7†	354 (in H_2)	sl. s.	s. abs. al.	s.
315	grn. pd.	....	....	....	s. h.	s. h.	i.
316	gr. pd.	....	....	....	v. sl. s.	sl. s.	i.
317	col. pd.	....	206–8	....	i.	s. h.	i.
318	col. fl.	....	98–9	....	i.	v. s.	i.

* Sealed tube.
† Cryst./pet. m.p. 75°.
Amino-dracylic acid 266

Amino-ethane 2950
Amino-ethyl alcohol 8, 2907

Table 7-4 (*Continued*)
PHYSICAL CONSTANTS OF ORGANIC COMPOUNDS

No.	Name	Synonym	Formula	Beil. Ref.	Formula Weight
319	**Amino** ethylbenzene (*o*)	*o*-Et-aniline	$NH_2 \cdot C_6H_4 \cdot C_2H_5$	XII-1089	121.18
320	ethylbenzene (*m*)	....	$NH_2 \cdot C_6H_4 \cdot C_2H_5$	XII-1090	121.18
321	ethylbenzene (*p*)	....	$NH_2 \cdot C_6H_4 \cdot C_2H_5$	XII-1090	121.18
322	guanidine	guanyl hydrazine	$H_2N \cdot C(NH) \cdot NH \cdot NH_2$	III-117	74.09
323	guanidine H_2CO_3	(bicarbonate)	$CH_6N_4 \cdot H_2CO_3$	III-118	136.11
324	guanidine H_2SO_4	(acid sulfate)	$CH_6N_4 \cdot H_2SO_4$	III-118	172.16
325	guanidine H_2SO_4	....	$(CH_6N_4)_2 \cdot H_2SO_4 \cdot H_2O$	....	264.26
326	hexahydrobenzene	cyclohexylamine	$H_6C_6H_5 \cdot NH_2$	XII-5	99.19
327	1-hydroxyanthra-quinone (4)	quinizarinamide	$C_6H_4(CO)_2C_6H_2 \cdot (OH)NH_2$	XIV-268	239.23
328	malonic acid	....	$NH_2 \cdot CH(CO_2H)_2$	IV-469	119.08
329	4-methylthiazole (2)	....	$CH_3 \cdot C \cdot CH \cdot S \cdot$ $\overline{\quad\quad\quad\quad}$ $C(NH_2){:}N$ $\underline{\quad\quad\quad\quad}$	XXVII-159	114.17
330	methylacetanilide (*p*)	....	$NH_2 \cdot C_6H_4 \cdot N(CH_3) \cdot COCH_3$	*XIII-30	164.21
331	naphthalene-4-azobenzene-*p*-sulfonic acid (1)	....	$NH_2 \cdot C_{10}H_6 \cdot N{:}N \cdot C_6H_4 \cdot SO_3H$	XVI-367	327.36
332	α-naphthol (4,1)	4-amino-naphthol-1	$NH_2 \cdot C_{10}H_6 \cdot OH$	XIII-667	159.19
333	α-naphthol (8,1)	8-amino-naphthol-1	$NH_2 \cdot C_{10}H_6 \cdot OH$	XIII-672	159.19
334	β-naphthol (1,2)	1-amino-naphthol-2	$NH_2 \cdot C_{10}H_6 \cdot OH$	XIII-676	159.19
335	β-naphthol (1,2)	HCl	$NH_2 \cdot C_{10}H_6 \cdot OH \cdot HCl$	XIII-677	195.65
336	β-naphthol (7,2)	7-amino-naphthol-2	$NH_2 \cdot C_{10}H_6 \cdot OH$	XIII-684	159.19
337	1-naphthol-3-sulfonic acid (6) (6;1,3)	J acid	$NH_2 \cdot C_{10}H_5(OH) \cdot SO_3H$	XIV-823	239.25
338	1-naphthol-3-sulfonic acid (7) (7;1,3)	gamma acid	$NH_2 \cdot C_{10}H_5(OH) \cdot SO_3H$	XIV-828	239.25
339	1-naphthol-5-sulfonic acid (8) (8;1,5)	S acid	$NH_2 \cdot C_{10}H_5(OH) \cdot SO_3H$	XIV-835	239.25
340	2-naphthol-4-sulfonic acid (1) (1;2,4)	1:2:4-acid	$NH_2 \cdot C_{10}H_5(OH) \cdot SO_3H \cdot \frac{1}{2}H_2O$	XIV-846	248.26
341	2-naphthol-6-sulfonic acid (1) Na	eikonogen (1,2,6)	$NH_2 \cdot C_{10}H_5(OH) \cdot SO_3Na \cdot 2\frac{1}{2}H_2O$	XIV-848	306.27
342	1-naphthol-3,6-disulfonic acid (8;1,3,6)	H acid	$NH_2 \cdot C_{10}H_4(OH){:}(SO_3H)_2$	XIV-840	319.31
343	naphthol-disulfonic acid Na	H acid Na salt	$C_{10}H_7O_7NS_2Na_2 \cdot 1\frac{1}{2}H_2O$	XIV-840	390.30
344	1-naphthol-5,7-disulfonic acid (8;1,5,7)	2S acid	$NH_2 \cdot C_{10}H_4(OH){:}(SO_3H)_2$	XIV-845	319.31
345	phenol (*o*)	2-aminophenol	$NH_2 \cdot C_6H_4 \cdot OH$	XIII-354	109.13
346	phenol (*o*) HCl	....	$HO \cdot C_6H_4 \cdot NH_2 \cdot HCl$	XIII-358	145.59
347	phenol (*m*)	3-aminophenol	$NH_2 \cdot C_6H_4 \cdot OH$	XIII-401	109.13
348	phenol (*m*) HCl	....	$HO \cdot C_6H_4 \cdot NH_2 \cdot HCl$	XIII-403	145.59
349	phenol (*p*)	*p*-hydroxyaniline	$NH_2 \cdot C_6H_4 \cdot OH$	XIII-427	109.13
350	phenol (*p*) HCl	....	$HO \cdot C_6H_4 \cdot NH_2 \cdot HCl$	XIII-434	145.59

Amino-ethyl-indole 6363
Amino-ethyl propandiol 247
Amino-form 3599
Amino G acid 4537
Amino-glutaric acid 3371-2

Amino H acid 4555
Amino-hydrocinnamic acid 5149-50
Amino-hydroxybenzoic acid 374-6
Amino-hydroxybutyric acid 5932-4
Amino-hydroxypurine, cf. alkd.

Table 7-4 (*Continued*)
PHYSICAL CONSTANTS OF ORGANIC COMPOUNDS

No.	Crystalline Form and Color	Specific Gravity	Melting Point °C.	Boiling Point °C.	Solubility in 100 Parts — Water	Alcohol	Ether
319	lq.	$0.983^{22°}$	−45	214.3	i.	∞	∞
320	lq.	$0.990^{0°}$	....	$214\text{-}5^{764mm}$	....	....	....
321	col. oil	$0.976^{22°}$	−5	217.4	i.	∞	∞
322	cr.	....	d.	....	s.	s.	i.
323	cr.	....	172 d.	....	i. c.; d. h.	....	....
324	large pl.	....	161	....	s.	....	....
325	....	....	207 ±	....	s.	....	....
326	col. lq.	$0.865^{20°}_{0}$	....	134	s.	s.	....
327	red-violet pd.	....	207-8	....	s. conc. HCl	s.	s. bz.
328	cr.	....	109 d.	....	s. c.	sl. s.	....
329	hyg. cr.	....	42*	231-2 (sl.d.)	v. s.	v. s.	v. s.
330	nd./pet.	....	63	....	....	....	....
331	vl. nd.	....	....	....	i.	v. sl. s.	....
332	nd.	....	....	....	sl. s.	s.	s.
333	cr./bz.-lg.	....	95-7 d.	....	s. h.	s. alk.	s. HCl
334	lf./et.	....	....	....	v. sl. s. h.	....	sl. s.
335	nd.	....	255 d.	....	....	6.6 h.	....
336	nd./al.	....	201 (163)	....	sl. s.	s.	s.
337	....	....	....	....	sl. s.	....	....
338	nd.	....	....	....	0.4 h.	....	....
339	nd.	....	....	....	sl. s.	l.	s. alk.
340	nd.	....	....	....	v. sl. s.	l.	i.; i. bz.
341	col. pd.	....	....	....	s.	i.	i.
342	col. cr.	....	....	....	sl. s.	sl. s.	s. alk.
343	col. cr.	....	....	....	$0.2^{20°}$; $2.4^{60°}$	....	....
344	cr.	....	....	....	s.	....	s. alk.
345	col. nd.	....	173	subl.	$1.7^{0°}$	$4.3^{0°}$	v. s.
346	col. cr.	....	....	208d.	$80^{0°}$	$40^{0°}$	....
347	pr./toluene	....	122-3	....	$2.6^{0°}$	s.	sl. s.
348	pr./aq.	....	229	....	....	....	....
349	lf.	....	184-6 d.	subl. sl. d.	$1.1^{0°}$	$4.6^{0°}$, abs.	i. bz.
350	col. pr.	....	306 d.	....	$71^{0°}$	$10^{0°}$, abs.	....

* Cryst. + 1 HCl, m. 100-2°.

Amino-hydroxytoluene 292-9
Amino-menthane 4029
Amino-methane 4105
Amino-methylaniline 4336

Amino-methylanisole 1544
Amino-naphthalene 4529-30
Amino-nitrophenol 4610-6
Amino-nonane 4926
Amino-octane 4970-2

Amino-oxamide 5017
Amino-pentane, cf. **amylamine.**
Amino-phenetole 5105

Table 7-4 (*Continued*)
PHYSICAL CONSTANTS OF ORGANIC COMPOUNDS

No.	Name	Synonym	Formula	Beil. Ref.	Formula Weight
351	**Amino** phenol (*p*) $H_2C_2O_4$	acid oxalate	$HO \cdot C_6H_4 \cdot NH_2 \cdot H_2C_2O_4$	*XIII-144	199.16
352	phenol (*p*) $H_2C_2O_4$	oxalate	$(HOC_6H_4NH_2)_2 \cdot H_2C_2O_4$	*XIII-144	308.29
353	phenol sulfonic acid	(1,2,4)	$C_6H_3(OH)(NH_2)SO_3H \cdot \frac{1}{2}H_2O$	XIV-814	198.20
354	phenol sulfonic acid	(1,4,2)	$C_6H_3(OH)(NH_2)SO_3H$	XIV-806	189.19
355	phenol sulfonic acid	(1,4,3)	$C_6H_3(OH)(NH_2)SO_3H$	XIV-812	189.19
356	phenylacetic acid (*α*) (*dl*)	....	$C_6H_5 \cdot CH(NH_2) \cdot CO_2H$	XIV-460	151.17
357	phenylacetic acid (*p*)	*p*-amino-toluic acid	$H_2N \cdot C_6H_4 \cdot CH_2CO_2H$	XIV-456	151.17
358	phenylglycine (*p*)	*N*-(4-aminophenyl)-glycine	$H_2N \cdot C_6H_4 \cdot NH \cdot CH_2 \cdot CO_2H$	XIII-105	166.18
359	phthalhydrazide (3)	luminol	$H_2N \cdot C_6H_3 \cdot CO \cdot NH \cdot \overline{NH \cdot CO}$	*XXV-698	177.16
360	pyridine (2) (*α*)	....	$N \cdot CH(CH)_3 \cdot C \cdot NH_2$	XXII-428	94.12
361	pyridine (3) (*β*)	....	$NH_2 \cdot C_5H_4N$	XXII-431	94.12
362	pyridine (4) (*γ*)	....	$NH_2 \cdot C_5H_4N$	XXII-433	94.12
363	pyrine	3-keto-1,5-diMe-4-dimethylamino-2-Ph-2,3-dihydropyrazole	$(CH_3)_2N \cdot C:C(CH_3) \cdot N(CH_3) \cdot N(C_6H_5) \cdot CO$	....	231.30
364	pyrine-butyl-chloral-hydrate	trigemin	$C_{13}H_{17}ON_3 \cdot C_4H_7O_2Cl_3$	XXV-453	424.76
365	quinoline (2) (*α*)	....	$NH_2 \cdot C_9H_6N$	XXII-443	144.18
366	quinoline (3) (*β*)	....	$NH_2 \cdot C_9H_6N$	*XXII-638	144.18
367	quinoline (4) (*γ*)	....	$NH_2 \cdot C_9H_6N \cdot H_2O$	XXII-444	162.19
368	quinoline (5)	....	$NH_2 \cdot C_6H_3 \cdot CH:CH \cdot CH:N$	XXII-445	144.18
369	quinoline (6)	....	$NH_2 \cdot C_9H_6N \cdot 2H_2O$	XXII-447	180.21
370	quinoline (7)	....	$NH_2 \cdot C_9H_6N + aq.$	XXII-450	144.18
371	quinoline (8)	....	$NH_2 \cdot C_9H_6N$	XXII-450	144.18
372	resorcinol (2)	base unknown	....	....	125.13
373	resorcinol (2) HCl	....	$NH_2 \cdot C_6H_3(OH)_2 \cdot HCl$	XIII-782	161.59
374	salicylic acid (3)	3-NH_2-2-OH-benzoic acid	$C_6H_3(CO_2H)(OH) \cdot NH_2$	XIV-577	153.14
375	salicylic acid (4)	....	$C_7H_4O_2(OH)NH_2$	XIV-579	153.14
376	salicylic acid (5)	....	$C_7H_4O_2(OH)NH_2$	XIV-579	153.14
377	thiazole (2)	....	$NH_2 \cdot C:N \cdot CH:CH \cdot S$	XXVII-155	100.14
378	thiophene (*α*)	thiophenine	$NH_2C_4H_3S$	XVII-248	99.16
379	toluene sulfonic acid (1,2,3)	toluidine sulfonic acid	$C_6H_3(CH_3)(NH_2) \cdot SO_3H$	XIV-723	187.22

Table 7-4 (*Continued*)
PHYSICAL CONSTANTS OF ORGANIC COMPOUNDS

No.	Crystalline Form and Color	Specific Gravity	Melting Point °C.	Boiling Point °C.	Solubility in 100 Parts		
					Water	Alcohol	Ether
351	vl. cr./aq.	....	220 d.	....	sl. s.	....	....
352	col. fl.	....	290 d.	....	i.	sl. s. alc.	i.
353	col. cr.	....	....	....	114°	....	....
354	nd. +aq.	....	d. >300	....	$0.07^{14^{\circ}}$	i.	i.
355	nd. +aq.	....	d. >270-85	....	$2^{14^{\circ}}$	i.	i.; s. alk.
356	pr./aq. al.	....	256	subl.256-65	v. sl. s.	v. sl. s.	v. sl. s.; s. alk.
357	col. lf./aq.	....	199-200 d.	....	s. h.	s.	s. alk.
358	lf./aq.	....	222-3 d.	....	sl. s.	....	....
359	yel. nd.	....	280-300	....	i.; s. h. ac.	sl. s.	sl. s.
360	lf./lg.	....	56	204	s.	v. s.	s.; sl. s. lg.
361	lf./bz. lg.	....	64	250-2	v. s.	v. s.	v. s.; i. lg.
362	nd./bz.	....	158	....	s.; s. alk.	s.; sl. s. bz.	sl. s.
363	col. cr.	....	107-9	....	5.5	75	10; 10 bz.
364	col. cr./bz	....	85-6	....	1.5	50	10; sl. s. bz.
365	lf./aq.	....	129	....	v. s. h.	s.; sl. s. bz.	s.; sl. s. lg.
366	cr.	....	94(84)	....	s.	s.	....
367	nd./aq.	....	69-70*	$-H_2O$, 100	s. h.	s.; s. chl.	v. sl. s. CS_2
368	nd./al.	....	109-10	310	sl. s.	s.	s.; i. lg.
369	cr./aq.	....	73.5†	subl.	sl. s.	s.	s.; s. NH_4OH
370	yel. nd./aq.	....	74-5	....	....	....	....
371	yel. nd./al.	....	70	....	s. h.	....	....
372	....	....	....	....	....	....	....
373	gr. pd.	....	....	....	v. s.	s.	i.
374	cr.	....	235 d.	....	....	v. sl. s.	....
375	cr. pd.	....	150-1	....	v. s.	v. s.	sl. s.
376	nd.	....	d. 260-80	....	sl. s. h.	i.; s. HCl	s. CS_2
377	yel. pl./al.	....	90	d.	sl. s.	sl. s.; s. h.	sl. s.
378	oil	....	....	$77-9^{11mm}$	v. s.	v. s.	i.
379	nd.	....	....	....	$0.97^{11^{\circ}}$; v. s. h.	....	....

* Anh. m.p. 154°.
† Anh. m.p. 93-4°.
Amino-purine, cf. alkd.
Amino Schaeffer's acid 4552

Amino-succinic acid 589-90
Amino-thiophenylimine 5922
Amino-toluene (ω) 783

Table 7-4 (*Continued*)
PHYSICAL CONSTANTS OF ORGANIC COMPOUNDS

No.	Name	Synonym	Formula	Beil. Ref.	Formula Weight
380	**Amino** toluene sulfonic acid	(1,2,4)	$C_6H_3(CH_3)(NH_2)\cdot SO_3H\cdot H_2O$	XIV-728	205.23
381	toluene sulfonic acid	(1,4,2)	$C_6H_3(CH_3)(NH_2)\cdot SO_3H\cdot H_2O$	XIV-720	205.23
382	toluene sulfonic acid	(1,4,3)	$C_6H_3(CH_3)(NH_2)\cdot SO_3H\cdot\frac{1}{2}H_2O$	XIV-723	196.23
383	toluene sulfonic acid	(1,2,5)	$C_6H_3(CH_3)(NH_2)\cdot SO_3H\cdot H_2O$	XIV-726	205.23
384	triphenylmethane (*m*)	benzhydryl aniline	$(C_6H_5)_2CH\cdot C_6H_4NH_2$	XII-1342	259.35
385	triphenylmethane	(*p*)	$(C_6H_5)_2CH\cdot C_6H_4NH_2$	XII-1342	259.35
386	uracil (5)	....	$H_2N\cdot C:CH\cdot NH\cdot CO\cdot$ ⌐————⌐ $NH\cdot CO$ ⌐————⌐	XXIV-463	127.10
387	valeric acid (α)	....	$C_2H_5\cdot CH_2\cdot CHNH_2\cdot CO_2H$	IV-416	**117.15**
388	valeric acid (γ) (*dl*)	....	$CH_3\cdot CHNH_2\cdot (CH_2)_2\cdot CO_2H$	IV-418	117.15
389	valeric acid (δ)	....	$NH_2\cdot (CH_2)_4\cdot CO_2H$	IV-418	117.15
390	valeric acid·HCl (δ)	....	$NH_2(CH_2)_4CO_2H\cdot HCl$	IV-419	153.61
391	*iso*-valeric acid (α) (*dl*)	α-valine	$(CH_3)_2CH\cdot CHNH_2\cdot CO_2H$	IV-430	117.15
392	*iso*-valeric acid (α) (*l*+)	valine	$(CH_3)_2CH\cdot CHNH_2\cdot CO_2H$	IV-427	117.15
393	*iso*-valeric acid (β)	β-valine	$(CH_3)_2CNH_2\cdot CH_2\cdot CO_2H$	IV-426	117.15
394	valeric acid (α)	*dl-iso*-valine	$C_2H_5\cdot C(CH_3)(NH_2)\cdot CO_2H$	IV-425	117.15
395	valeric acid (α)	*iso*-valine (*l*+)	$C_2H_5\cdot C(CH_3)(NH_2)\cdot CO_2H\cdot H_2O$	*IV-513	117.15
396	**Ammelin**	triuretdiamidine	$(CN)_3(NH_2)_2OH$	XXVI-244	127.11
397	**Amyl** acetate (*n*)	....	$CH_3\cdot CO_2\cdot C_5H_{11}$	II-131	130.19
398	acetate (*iso*)	common amyl acetate	$CH_3CO_2\cdot CH_2CH_2CH:(CH_3)_2$	II-132	130.19
399	acetate	β-Me-Bu-acetate	$CH_3CO_2\cdot CH_2\cdot CH(CH_3)\cdot C_2H_5$	II-132	130.19
400	acetate (*sec*)	α-Me-Bu-acetate	$CH_3CO_2\cdot CH(CH_3 CH_2\cdot C_2H_5$	II-131	130.19
401	acetate (*sec*)	diEt-carbinol acetate	$CH_3CO_2\cdot CH(C_2H_5)_2$	II-131	130.19
402	acetate (*tert*)	....	$CH_3CO_2\cdot C(CH_3)_2\cdot C_2H_5$	II-132	130.19
403	acetic acid (*iso*)	....	$(CH_3)_2:C_4H_7\cdot CO_2H$	II-342	130.19
404	alcohol (*n*)	pentanol-1	$CH_3\cdot (CH_2)_3\cdot CH_2OH$	I-383	88.15
405	alcohol (*sec. n*)	pentanol-2	$C_2H_5\cdot CH_2\cdot CHOH\cdot CH_3$	I-384	88.15
406	alcohol (*prim. iso*)	2-methyl-butanol-4	$(CH_3)_2CH\cdot CH_2\cdot CH_2OH$	I-392	88.15
407	alcohol (*sec. iso*)	2-methyl-butanol-3	$(CH_3)_2CH\cdot CHOH\cdot CH_3$	I-391	88.15
408	alcohol	pentanol-3	$(C_2H_5)_2:CHOH$	I-385	88.15
409	alcohol (*tert*)	2-methyl-butanol-2	$(CH_3)_2COH\cdot C_2H_5$	I-388	88.15
410	alcohol	2,2-diMe-propanol-1	$(CH_3)_3C\cdot CH_2OH$	I-406	88.15
411	alcohol (*d*)	active amyl alcohol	$C_2H_5\cdot CH(CH_3)\cdot CH_2OH$	I-385	88.15

Amino-toluic acid 357   Amphotropin 3601   Amyl, cf. also diamyl.
Amphetamine 654   Amygdalin, cf. glcde.

Table 7-4 (*Continued*)
PHYSICAL CONSTANTS OF ORGANIC COMPOUNDS

No.	Crystalline Form and Color	Specific Gravity	Melting Point °C.	Boiling Point °C.	Solubility in 100 Parts		
					Water	Alcohol	Ether
380	nd.	....	....	....	$0.9^{11°}$	i.	....
381	mn.	....	d.	....	$0.5^{20°}$	i.	....
382	nd.	....	....	....	0.47	....	....
383	tri./aq.	....	$-H_2O$, 120	....	$2.7^{11°}$ (anh.)	i.	....
384	nd./et.	....	120	....	....	....	....
385	pr./et.	....	83-4	$248^{12mm} \pm$	i.	s. bz.	....
386	nd./aq.	....	d.	subl. d.	0.05 c.; 1.6 h.	s. a.; s. alk.	s. NH_4OH
387	nd./aq.	....	291.5	subl.	$10^{15°}$; v. s. h.	sl. s.	i.
388	cr.	....	193-214 d.	....	v. s.	v. sl. s.	i.; i. bz.
389	lf.	....	157-8 d.	....	v. s.	v. sl. s. abs.	i.
390	pl. or pr.	....	90-2	....	s.	s.	i.
391	lf./al.	....	298 d.	subl.	$7^{25°}$	v. sl. s.	v. sl. s.
392	lf./aq. al.	....	315*	subl. d.	$9.1^{16.5°}$	v. sl. s.	v. sl. s.
393	cr./al.+et.	....	217	subl. >180	v. s.	v. sl. s.	i.
394	rhb.+aq./aq. al.	....	307.5†	subl. $300 \pm$	$36^{20°}$	$0.6^{15°}$	6 h.
395	nd./aq. al.	....	315	subl.	s.	sl. s.	....
396	nd./aq.	....	....	....	$0.008^{23°}$	i.	i.; s. HCl
397	col. lq.	$0.879^{20°}_{20°}$	-70.8	148.4^{737mm}	v. sl. s.	∞	∞
398	col. lq.	$0.876^{15°}_{4}$	....	142^{757mm}	$0.25^{15°}$	∞	∞
399	col. lq.	$0.88^{12.5°}$	....	141-2	v. sl. s.	∞	∞
400	col. lq.	$0.922^{0°}$	....	133.5	sl. s.	∞	∞
401	col. lq.	$0.871^{20°}_{4}$	....	133	sl. s.	∞	∞
402	col. lq.	$0.874^{19°}$	....	124.5^{749mm}	v. sl. s.	∞	∞
403	col. oil	$0.912^{19°}_{19°}$	....	216^{762mm}	sl. s. h.	∞	∞
404	col. lq.	$0.818^{15°}_{4}$	-78.9	138.1	$2.7^{22°}$	∞	∞
405	col. lq.	$0.813^{15°}_{4}$	....	119.9	$4^{20°}$	∞	∞
406	col. lq.	$0.813^{15°}_{4}$	-117.2	132.0	$2^{14°}$	∞	∞
407	col. lq.	$0.825^{15°}_{4}$	....	113-4	$2.8^{30°}$	∞	∞
408	col. lq.	$0.813^{15°}_{4}$	....	116.1	$5.5^{30°}$	∞	∞
409	col. lq.	$0.813^{15°}_{4}$	-9	102.4	sl. s.	s.	s.
410	cr.	....	52-3	113-4	sl. s.	∞	∞
411	col. lq.	$0.816^{20°}_{4}$	....	128	$3.6^{30°}$	∞	∞

* Sealed tube.
† Sealed tube; loses aq. at 100°.

Table 7-4 (Continued)
PHYSICAL CONSTANTS OF ORGANIC COMPOUNDS

No.	Name	Synonym	Formula	Beil. Ref.	Formula Weight
412	**Amyl** amine (*n*)	....	$CH_3(CH_2)_4NH_2$	IV-175	87.17
413	amine (*sec. n*)	....	$(C_3H_7)(CH_3):CHNH_2$	IV-177	87.17
414	amine (*iso*)	....	$(CH_3)_2CH(CH_2)_2NH_2$	IV-180	87.17
415	amine (*tert*)	....	$(C_2H_5)(CH_3)_2C \cdot NH_2$	IV-179	87.17
416	amine	1-NH_2-2-Me-butane	$C_2H_5CH(CH_3) \cdot CH_2NH_2$	IV-178	87.17
417	amine	3-aminopentane	$(C_2H_5)_2:CH \cdot NH_2$	IV-178	87.17
418	amine	3-NH_2-2-Me-butane	$(CH_3)_2CH \cdot CH \cdot (CH_3)NH_2$	IV-179	87.17
419	aniline (*iso*)	....	$C_6H_5 \cdot NH \cdot C_5H_{11}$	XII-169	163.26
420	aniline (*p*) (*tert*)	....	$(C_2H_5)(CH_3)_2C \cdot C_6H_4 \cdot NH_2$	XII-1179	163.26
421	benzene (*n*)	1-phenylpentane	$C_5H_{11} \cdot C_6H_5$	V-434	148.25
422	benzene (*iso*)	....	$C_5H_{11} \cdot C_6H_5$	V-434	148.25
423	benzene (*tert*)	diMe-Et-Ph-methane	$C_5H_{11} \cdot C_6H_5$	V-436	148.25
424	benzene (*sec*)	diEt-Ph-methane	$(C_2H_5)_2CH \cdot C_6H_5$	V-436	148.25
425	benzoate (*iso*)	....	$C_6H_5 \cdot CO_2 \cdot C_5H_{11}$	IX-113	192.26
426	benzyl ether (*iso*)	....	$C_6H_5 \cdot CH_2 \cdot O \cdot C_5H_{11}$	VI-431	178.28
427	bromide (*n*)	1-bromopentane	$CH_3 \cdot (CH_2)_3 \cdot CH_2Br$	I-131	151.05
428	bromide (*iso*)	4-Br-2-Me-butane	$(CH_3)_2CH(CH_2)_2Br$	I-136	151.05
429	bromide (*tert*)	2-Br-2-Me-butane	$(CH_3)_2C(Br) \cdot C_2H_5$	I-136	151.05
430	bromide	1-Br-2, 2-diMe-propane	$(CH_3)_3C \cdot CH_2Br$	I-141	151.05
431	bromide	1-Br-2-Me-butane	$C_2H_5 \cdot CH(CH_3) \cdot CH_2Br$	I-136	151.05
432	*n*-butyrate (*n*)	....	$C_2H_5CH_2CO_2(CH_2)_4 \cdot CH_3$	II-271	158.24
433	*n*-butyrate (*iso*)	....	$C_2H_5CH_2CO_2 \cdot C_5H_{11}$	II-271	158.24
434	*n*-butyrate (*tert*)	....	$C_3H_7CO_2 \cdot C(CH_3)_2 \cdot C_2H_5$	II-271	158.24
435	*iso*-butyrate (*iso*)	....	$(CH_3)_2CHCO_2 \cdot C_5H_{11}$	II-291	158.24
436	*n*-caproate (*n*)	....	$C_5H_{11}CO_2 \cdot C_5H_{11}$	II-323	186.30
437	*n*-caproate (*iso*)	....	$C_5H_{11}CO_2 \cdot (CH_2)_2 \cdot CH:(CH_3)_2$	....	186.30
438	*n*-caprylate (*n*)	....	$C_7H_{15}CO_2(CH_2)_4 \cdot CH_3$	....	214.35
439	carbamate (*n*)	....	$H_2N \cdot CO_2 \cdot C_5H_{11}$	....	131.18
440	carbamate (*iso*)	....	$NH_2 \cdot CO_2 \cdot C_5H_{11}$	III-30	131.18
441	carbamate (*tert*)	Aponal	$H_2N \cdot CO_2 \cdot C(CH_3)_2 \cdot C_2H_5$	*III-14	131.18
442	chloride (*n*)	1-chloropentane	$CH_3 \cdot (CH_2)_3 \cdot CH_2Cl$	I-130	106.60
443	chloride (*sec*)	2-chloropentane	$C_3H_7 \cdot CHCl \cdot CH_3$	I-131	106.60
444	chloride (*sec*)	3-chloropentane	$(C_2H_5)_2CHCl$	I-131	106.60
445	chloride (*iso*)	4-Cl-2-Me-butane	$(CH_3)_2CH(CH_2)_2Cl$	I-135	106.60
446	chloride (*sec. iso*)	3-Cl-2-Me-butane	$(CH_3)_2CH \cdot CHCl \cdot CH_3$	I-135	106.60
447	chloride (*tert*)	2-Cl-2-Me-butane	$(CH_3)_2CCl \cdot C_2H_5$	I-134	106.60
448	chloride	1-Cl-2-Me-butane	$(CH_3)(C_2H_5)CH \cdot CH_2Cl$	I-134	106.60
449	chloroformate (*n*)	*n*-amyl chloro-carbonate	$C_2H_5(CH_2)_3O \cdot COCl$	....	150.61
450	chloroformate (*iso*)	*iso*-amyl chloro-carbonate	$(CH_3)_2CH(CH_2)_2 \cdot O \cdot COCl$	III-12	150.61

Amyl borate 6077   Amyl carbinol (*iso*) 3634

Table 7-4 (*Continued*)
PHYSICAL CONSTANTS OF ORGANIC COMPOUNDS

No.	Crystalline Form and Color	Specific Gravity	Melting Point °C.	Boiling Point °C.	Solubility in 100 Parts		
					Water	Alcohol	Ether
412	col. lq.	$0.766^{19°}$	−55	103-4	s.	s.	s.
413	col. lq.	$0.749_{4}^{20°}$	....	91-2	∞	∞	∞
414	col. lq.	$0.751_{4}^{18°}$	....	95	∞	∞	∞
415	col. lq.	$0.731_{4}^{25°}$	−105	77-8	∞	∞ ; ∞ chl.	∞
416	col. lq.	$0.755^{18°}$	....	95-6	∞	∞	∞
417	col. lq.	$0.749_{4}^{20°}$	....	90-1	∞	∞	∞
418	col. lq.	$0.757^{18.5°}$	....	83-4	∞	∞	∞
419	lq.	$0.928_{4}^{15°}$	....	254.5	....	....	....
420	col. lq.	....	....	259-62	i.	s.	s.
421	lq.	$0.858_{4}^{20°}$	−78.3	205.3	i.	∞	∞
422	lq.	$0.859_{4}^{20°}$	....	198-9	i.	∞	∞
423	lq.	$0.867_{4}^{20°}$	....	189-90	i.	∞	∞
424	col. lq.	$0.876_{4}^{15°}$	....	187^{753mm}	i.	s.	s.
425	col. lq.	$0.992_{14}^{14°}$	....	260.77^{46mm}	i.	∞	∞
426	col. lq.	....	....	237^{748mm}	i.	s.	s.
427	col. lq.	$1.218_{4}^{20°}$	−95	129.7	i.	s.	....
428	col. lq.	$1.203_{4}^{20°}$	−111.9	120.7	i.	s.	s.
429	lq.	$1.216_{0}^{19°}$	....	108^{765mm}	i.	s.	s.
430	col. lq.	$1.260_{4}^{20.4°}$	....	$89\text{-}91^{749mm}$	i.	s.	s.
431	lq.	$1.221_{4}^{20°}$	....	121^{755mm}	i.	s.	s.
432	col. lq.	$0.871_{4}^{15°}$	−73.2	186.4	$0.05^{50°}$	∞	∞
433	col. lq.	$0.866_{18}^{19°}$	....	178.6	i.	∞	∞
434	col. lq.	$0.865_{0}^{14.5°}$	....	164	sl. s.	∞	∞
435	lq.	$0.876_{4}^{0°}$	....	168.8	i.	s.	s.
436	col. lq.	$0.861_{4}^{25°}$	−47	226.2	....	....	....
437	col. lq.	....	....	$94\text{-}6^{10mm}$	i.	s.	....
438	col. lq.	$0.856_{4}^{25°}$	−34.5	260.2	i.	s.	....
439	col. pl.	....	56-7.5	....	i.	s.	s.
440	nd./aq.	$0.944_{4}^{70.6°}$	64	220	s. h.	s.	s.
441	nd./aq. al.	....	86	....	sl. s.	s.	s.; sl. s. pet.
442	col. lq.	$0.878_{4}^{20°}$	−99	108.4	i.	s.	s.
443	lq.	$0.870_{4}^{20°}$	....	96.7	i.	s.	s.
444	col. lq.	$0.895^{21°}$	....	97.3	i.	∞	∞
445	col. lq.	$0.893_{4}^{20°}$	....	99.77^{58mm}	i.	s.	∞
446	lq.	$0.883^{0°}$	....	91^{753mm}	i.	s.	s.
447	lq.	$0.871_{4}^{20°}$	−72.9	85.7	i.	s.	s.
448	lq.	$0.881^{17.5°}$	....	98-9	i.	s.	s.
449	col. lq.	....	....	$43\text{-}57^{mm}$	....	....	....
450	lq.	$1.028_{15}^{18.5°}$	....	155-6	i.	∞	∞

Amyl carbonate 1737-8   Amyl chlorocarbonate 449-50

Table 7-4 (Continued)
PHYSICAL CONSTANTS OF ORGANIC COMPOUNDS

No.	Name	Synonym	Formula	Beil. Ref.	Formula Weight
451	**Amyl** cinnamalde-hyde (n) (α)	....	$C_6H_5 \cdot CH:C(C_5H_{11}) \cdot$ CHO	**VII-310	202.30
452	m-cresol (n)	....	$C_5H_{11} \cdot C_6H_3(OH)CH_3$	....	178.28
453	iso-cyanide (iso)	iso-caproic iso-nitrile	$(CH_3)_2CH(CH_2)_2 \cdot NC$	IV-184	97.16
454	formate (n)	....	$HCO_2 \cdot C_5H_{11}$	II-22	116.16
455	formate (iso)	....	$HCO_2 \cdot C_5H_{11}$	II-22	116.16
456	formate, ortho (n)	....	$HC(OC_5H_{11})_3$	II-22	274.45
457	furoate (iso)	iso-Am pyromucate	$C_4H_3O \cdot CO_2 \cdot C_5H_{11}$	....	182.22
458	β-furylacrylate (n)	....	$C_4H_3O \cdot CH:CH \cdot$ $CO_2C_5H_{11}$	....	208.26
459	iodide (n)	1-iodopentane	$CH_3 \cdot (CH_2)_3 \cdot CH_2I$	I-133	198.05
460	iodide (iso)	4-I-2-Me-butane	$(CH_3)_2CH \cdot CH_2 \cdot CH_2I$	I-138	198.05
461	iodide (sec n)	2-iodopentane	$C_3H_7 \cdot CHI \cdot CH_3$	I-133	198.05
462	iodide (tert)	2-I-2-Me-butane	$(CH_3)_2CI \cdot C_2H_5$	I-138	198.05
463	iodide	1-I-2-Me-butane	$C_2H_5 \cdot CH(CH_3) \cdot CH_2I$	I-138	198.05
464	d-lactate (dl) (−)	....	$C_2H_5OCO_2 \cdot CH_2 \cdot$ $CH(CH_3)C_2H_5$	III-265	160.21
465	malonic acid (n)	....	$C_5H_{11}CH(CO_2H)_2$	II-695	174.20
466	malonic acid (iso)	....	$C_5H_{11}CH(CO_2H)_2$	II-700	174.20
467	mercaptan (n)	pentanthiol-1	$CH_3 \cdot (CH_2)_3 \cdot CH_2 \cdot SH$	I-384	104.22
468	mercaptan (n)	pentanthiol-3	$(C_2H_5)_2CH \cdot SH$	*I-194	104.22
469	mercaptan (iso)	2-Me-butanthiol-4	$(CH_3)_2CH(CH_2)_2SH$	I-405	104.22
470	naphthalene (iso) (β)	....	$C_5H_{11} \cdot C_{10}H_7$	V-574	198.31
471	β-naphthyl ether (iso)	....	$C_5H_{11} \cdot O \cdot C_{10}H_7$	VI-642	214.31
472	nitrate (iso)	....	$(CH_3)_2CH \cdot CH_2 \cdot CH_2 \cdot$ $O \cdot NO_2$	I-403	133.15
473	nitrite (n)	....	$C_5H_{11} \cdot O \cdot NO$	I-384	117.15
474	nitrite (iso)	....	$(CH_3)_2CH \cdot CH_2 \cdot CH_2 \cdot$ $O \cdot NO$	I-402	117.15
475	oleate (iso)	....	$C_8H_{17}CH:CHC_7H_{14} \cdot$ $CO_2 \cdot C_5H_{11}$	II-467	352.61
476	phenol (tert) (p)	Pentaphen	$C_5H_{11} \cdot C_6H_4 \cdot OH$	VI-548	164.25
477	phenol methyl ether (tert) (p)	....	$(C_2H_5)(CH_3)_2C \cdot C_6H_4 \cdot$ $O \cdot CH_3$	VI-549	178.28
478	phenyl ether (n)	....	$CH_3(CH_2)_4 \cdot O \cdot C_6H_5$	*VI-82	164.25
479	phenyl ether (iso)	....	$C_5H_{11} \cdot O \cdot C_6H_5$	VI-143	164.25
480	α-phenylhydrazine	(iso) (α)	$C_6H_5 \cdot N(NH_2)C_5H_{11}$	XV-121	178.28
481	phenyl ketone (n)	caprophenone	$C_5H_{11} \cdot CO \cdot C_6H_5$	VII-333	176.26
482	phenyl ketone (iso)	iso-caprophenone	$C_5H_{11} \cdot CO \cdot C_6H_5$	VII-334	176.26
483	propionate (n)	....	$C_2H_5 \cdot CO_2 \cdot C_5H_{11}$	**II-221	144.22
484	propionate (iso)	....	$C_2H_5 \cdot CO_2 \cdot C_5H_{11}$	II-241	144.22
485	propionate (act)	....	$C_2H_5 \cdot CO_2 \cdot C_5H_{11}$	II-241	144.22
486	salicylate (n)	....	$HO \cdot C_6H_4 \cdot CO_2 \cdot C_5H_{11}$	....	208.26
487	salicylate (iso)	....	$HO \cdot C_6H_4 \cdot CO_2 \cdot C_5H_{11}$	X-76	208.26
488	stearate (iso)	....	$CH_3(CH_2)_{16}CO_2 \cdot$ C_5H_{11}	II-380	354.62
489	stearate	....	$C_{17}H_{35}CO_2 \cdot CH_2 \cdot CH:$ $(CH_3)C_2H_5$	II-380	354.62
490	thiocyanate (iso)	....	$(CH_3)_2CH(CH_2)_2 \cdot$ $S \cdot CN$	III-177	129.23
491	iso-thiocyanate (n)	amyl mustard oil	$CH_3(CH_2)_4 \cdot N:C:S$	IV-176	129.23

Amyl cyanide 1217-8, 2104
Amyl disulfide 1753-4
Amyl ether 1740-1
Amyl-ethyl-barbituric acid 505

Amyl mustard oil 491-3
Amyl oxalate 1745
Amyl phthalate 1746-7

Table 7-4 (*Continued*)
PHYSICAL CONSTANTS OF ORGANIC COMPOUNDS

No.	Crystalline Form and Color	Specific Gravity	Melting Point °C.	Boiling Point °C.	Solubility in 100 Parts		
					Water	Alcohol	Ether
451	yel. lq.	$0.971_{20}^{20°}$	....	174-5^{20mm}	....	....	....
452	col.	....	24-5	137-9^{15mm}	i.	s.	s.
453	lq.	....	....	137-9	i.	s.	s.
454	lq.	$0.902^{0°}$	−73.5	132	v. sl. s.	∞	∞
455	lq.	$0.882_{4}^{20°}$	−93.5	123.5	$0.3^{22°}$	∞	∞
456	lq.	$0.864^{23°}$	....	265-7 dec.	....	....	....
457	col. lq.	....	....	135-7^{25mm}	i.	∞	....
458	col. lq.	$1.032_{4}^{20°}$	....	119^{4mm}	i.	....	....
459	lq.	$1.510_{4}^{20°}$	−86	157.0	i.	s.	∞
460	lq.	$1.515_{4}^{18°}$	....	147^{765mm}	i.	∞	∞
461	lq.	$1.507_{4}^{17°}$	....	144-5	i.	∞	∞
462	lq.	$1.471_{15}^{19°}$	....	127^{765mm}	i.	∞	∞
463	lq.	$1.524_{4}^{20°}$	....	148^{760mm}	i.	∞	∞
464	col. lq.	$0.971_{4}^{20°}$	....	114-5^{36mm}	v. sl. s.	s.	s.
465	col. pr.	....	82	d. 140	v. s.	v. s.	v. s.
466	nd./aq.	....	95d.	d.	v. s.	v. s.	v. s.
467	lq.	$0.857^{20°}$	....	126^{767mm}	i.	∞	∞
468	col. lq.	....	....	105	i.	∞	∞
469	lq.	$0.835_{4}^{20°}$	....	120	i.	∞	∞
470	col. lq.	$0.973^{0°}$	<−21	288-92	i.	s.	s.
471	col. lf.	$1.016_{4}^{12°}$	26.5	323-6	i.	s.	s.
472	lq.	$0.996^{21.7°}$	....	147-8	v. sl. s.	i.	i.
473	pa. yel. lq.	$0.853^{20°}$	....	104^{761mm}	sl. s.	∞	∞
474	lq.	$0.872_{4}^{20°}$	....	99	sl. s.	∞	∞
475	col. lq.	$0.897^{15°}$	....	223-4^{10mm}	i.	s.	v. s.
476	cr.	....	93	265-7	sl. s.	s.	s.
477	col. lq.	....	....	216-7	i.	s.	s.
478	col. lq.	....	....	111^{17mm}	i.	s.	s.
479	col. oil	$0.920_{4}^{22°}$	....	224-5	i.	s.	s.
480	lq.	$0.968^{15°}$	....	260-2	....	....	....
481	lf.	$0.958_{4}^{25°}$	24.7	265.2	....	....	....
482	lq.	$0.962_{5}^{15°}$	−2	240-27^{20mm}	i.	v. s.	v. s.
483	lq.	$0.876_{4}^{15°}$	−73.1	168.7	i.	∞	∞
484	col. lq.	$0.870_{0}^{20°}$	....	160.2	$0.1^{25°}$	∞	∞
485	col. lq.	$0.866_{4}^{20°}$	....	58^{16mm}	v. sl. s.	∞	∞
486	lq.	$1.065^{15°}$	....	265	i.	∞	∞
487	lq.	$1.045_{25}^{25°}$	....	276-7^{743mm}	$0.004^{22°}$	∞	∞ ; ∞ chl.
488	col. pl.	$0.855_{4}^{20°}$	23	185-90^{1mm}	i.	sl. s.	s.
489	col. lq.	$0.855_{4}^{20°}$	21-2	....	i.	s.	s.
490	pa. yel. oil	0.905	....	197	i.	s.	s.
491	lq.	....	....	193.4	v. sl. s.	v. s.	v. s.

Amyl pyromucate 457
Amyl succinate 1748-50
Amyl sulfide (*iso*) 1751-2

Amyl sulfite 1755
Amyl sulfone 1756
Amyl tartrate 1757

Table 7-4 (*Continued*)
PHYSICAL CONSTANTS OF ORGANIC COMPOUNDS

No.	Name	Synonym	Formula	Beil. Ref.	Formula Weight
492	**Amyl** *iso*-thiocyanate (*iso*)	amyl mustard oil	$(CH_3)_2CH(CH_2)_2\cdot$ $N:C:S$	IV-186	129.23
493	*iso*-thiocyanate (*tert*)	amyl mustard oil	$C_5H_{11}N:C:S$	IV-179	129.23
494	urea (*iso*)	....	$C_5H_{11}\cdot NH\cdot CO\cdot NH_2$	IV-185	130.19
495	urea (*tert*)	....	$C_5H_{11}\cdot NH\cdot CO\cdot NH_2$	IV-179	130.19
496	*iso*-valerate (*iso*)	....	$C_4H_9\cdot CO_2\cdot C_5H_{11}$	II-312	172.27
497	*iso*-valerate (*tert*)	valamin	$C_4H_9\cdot CO_2\cdot C_5H_{11}$	II-312	172.27
498	**Amylene** (*n*)	pentene-1	$C_2H_5\cdot CH_2\cdot CH:CH_2$	I-210	70.14
499	**Amylene** (*iso*)	2-methyl-butene-3	$(CH_3)_2CH\cdot CH:CH_2$	I-213	70.14
500	**Amylene** (α)	2-methyl-butene-1	$(C_2H_5)(CH_3)C:CH_2$	I-211	70.14
501	**Amylene** (β)(*cis*)	pentene-2	$C_2H_5\cdot CH:CH\cdot CH_3$	I-210	70.14
501.1	**Amylene** (β)(*trans*)	pentene-2	$C_2H_5\cdot CH:CH\cdot CH_3$	I-210	70.14
502	**Amylene** (β)(*iso*)	2-methyl-butene-2	$(CH_3)_2:C:CH\cdot CH_3$	I-211	70.14
503	glycol (1,4)	γ-pentylene glycol	$C_5H_{10}(OH)_2$	I-480	104.15
504	glycol (2,4)	*iso*-amylene alcohol	$C_5H_{10}(OH)_2$	I-483	104.15
505	**Amytal**	*iso*-Am-Et-bar-bituric acid	$(C_2H_5)(C_5H_{11}):C\cdot CO\cdot$ $\vert\_\_\_$ $NH\cdot CO\cdot NH\cdot CO$ $\_\_\_\_\vert$	....	226.28
506	sodium	....	$C_{11}H_{17}O_3N_2Na$	....	248.26
507	**Anacardic acid**	....	$HO\cdot C_{21}H_{30}\cdot CO_2H$	X-327	344.50
508	**Analgen**	8-EtO-5-benzoyl-amino-quinoline	$C_7H_5ONH\cdot C_6H_2\cdot$ $(OC_2H_5):C_3H_3N$	XXII-503	292.34
509	**Androsterone**	3-*trans*-hydroxy-17-keto-andro-stane	$C_{19}H_{30}O_2$	....	290.45
510	**Anemonin**	pulsatilla camphor	$C_{10}H_8O_4$	....	192.17
511	**Anethole**	*p*-propenyl anisole	$CH_3\cdot CH:CH\cdot C_6H_4\cdot$ OCH_3	VI-566	148.21
512	**Anhydroformald-aniline**	methylene aniline	$(CH_2NC_6H_5)_3$	XXVI-3	315.42
513	**Aniline**	aminobenzene	$C_6H_5\cdot NH_2$	XII-59	93.13
514	acetate	....	$C_6H_5NH_2\cdot C_2H_4O_2$	XII-118	153.18
515	arsonic acid (*p*)	arsanilic acid	$H_2N\cdot C_6H_4\cdot AsO_3H_2$	XVI-878	217.06
516	blue	spirit blue	$C_{38}H_{32}N_3\cdot Cl$	XIII-768	566.15
517	disulfonic acid	(2,5)	$NH_2\cdot C_6H_3:(SO_3H)_2\cdot$ $4H_2O$	XIV-780	325.32
518	disulfonic acid	(2,4)	$NH_2\cdot C_6H_3:(SO_3H)_2\cdot$ $2H_2O$	XIV-778	289.28
519	hydrobromide	....	$C_6H_5NH_2\cdot HBr$	XII-116	174.05
520	hydrochloride	....	$C_6H_5NH_2\cdot HCl$	XII-116	129.59
521	nitrate	....	$C_6H_5NH_2\cdot HNO_3$	XII-116	156.14
522	sulfate	....	$(C_6H_5NH_2)_2\cdot H_2SO_4$	XII-117	284.34
523	*o*-sulfonic acid	orthanilic acid	$NH_2\cdot C_6H_4\cdot SO_3H\dagger$	XIV-681	173.19
524	*m*-sulfonic acid	metanilic acid	$NH_2\cdot C_6H_4\cdot SO_3H\ddagger$	XIV-688	173.19
525	*p*-sulfonic acid	sulfanilic acid	$NH_2\cdot C_6H_4\cdot SO_3H\S$	XIV-695	173.19
526	*p*-sulfonic Na	....	$NH_2\cdot C_6H_4\cdot SO_3Na\cdot$ $2H_2O$	XIV-698	231.20
527	*p*-sulfonic amide	sulfanilamide	$NH_2\cdot C_6H_4\cdot SO_2\cdot NH_2$	XIV-698	172.21

† Also crysts. $+\frac{1}{2}H_2O$.
‡ Also crysts. $+1\frac{1}{2}H_2O$.
§ Also crysts. $+1H_2O$.
Amyl toluene 4119
Amylene alcohol 504, 4103

Amylene chloride 2034
Amylocaine 5644
Anabasine, cf. alkd.
Analgesine 565
Andirine 4405

Anemone camphor 510
Anesthesine 2955
Angelic acid 2425
Angeline 4405
Anhaline, cf. alkd.

Anhalonine, cf. alkd.
Aniline salt 520
Aniline violet 3593
Aniline yellow 249
Anilino-acetic acid 5194

Table 7-4 (*Continued*)
PHYSICAL CONSTANTS OF ORGANIC COMPOUNDS

No.	Crystalline Form and Color	Specific Gravity	Melting Point °C.	Boiling Point °C.	Solubility in 100 Parts		
					Water	Alcohol	Ether
492	yel. lq.	$0.942^{17°}$	....	183-4	v. sl. s.	v. s.	v. s.
493	lq.	....	<−10	166^{770mm}	v. sl. s.	v. s.	v. s.
494	cr.	....	92-3	....	v. sl. s.	....	....
495	mn./aq.	....	152-8	....	$1.3^{27°}$	....	....
496	col. lq.	$0.858^{8\,0°}_{1\,8}$	....	194	v. sl. s.	∞	∞
497	col. lq.	$0.861^{14°}_{0}$	....	173-4	sl. s.	s.	∞ oils
498	lq.	$0.641^{20°}$	−165.2	30.0	i.	∞	∞
499	col. lq.	$0.627^{20°}$	−168.5	20.1	*	∞	∞
500	col. lq.	$0.650^{20°}$	−137.6	31.1	†	∞	∞
501	col. lq.	$0.656^{20°}$	−151.4	37.1	v. sl. s.	∞	∞
501.1	col. lq.	$0.648^{20°}$	−140.2	36.4	v. sl. s.	∞	∞
502	col. lq.	$0.662^{20°}$	−133.8	38.5	†	s.	∞
503	oil	$0.996^{17°}_{4}$	....	$219\text{-}20^{13mm}$	∞	∞	∞ chl.; i. lg.
504	syrup	$0.989^{20°}$	....	202-3	....	....	....
505	col. cr.	....	154-6	....	sl. s.	s.	s.; s. alk.
506	hyg. pd.	....	150-5	....	v. s.	100	i.
507	cr.	....	26	....	i.	s.	s.
508	yel. nd./al.	....	210	....	i.	v. sl. s. c.	s. a.
509	col. nd./ 65% al.	....	184-5	....	i.; s. MeOH	s.; s. CCl_4	s.; s. bz.
510	yel.-wh. cr.	....	157-8	....	sl. s. h.	s. h.	i.
511	lf./al.	$0.991^{20°}_{20}$	22.5	235.3	v. sl. s.	∞ abs. al.	∞
512	pr./al.	....	143	185	i.	sl. s.	s.; s. bz.
513	col. oil	$1.022^{20°}_{4}$	−6.1	184.4	$3.6^{18°}$	∞	∞
514	col. cr.	1.07	....	d., $-H_2O$	∞	∞	....
515	nd./aq.	....	232	d. $-H_2O$, 150	v. s. h.	v. s. h.	i.; i. bz.
516	bl.-brn. cr.	....	....	....	i.	sl. s.	i.
517	cr./aq.	....	....	....	v. s.	v. s.	....
518	nd./aq.	....	d. >120	....	v. s.	v. s.	i.; s. alk.
519	cr. pd.	....	286	....	s.	s.	....
520	lf. or nd.	$1.222^{4°}$	198	245	$88.4^{15°}$; $107^{25°}$	s.	i.
521	rhb.	$1.356^{4°}$	d. 190	....	s.	s.	sl. s.
522	lf./al.	$1.377^{4°}$	d.	....	$5^{14°}$	sl. s.	i.
523	col. cr.	....	d.	....	$1.5^{15°}$	v. sl. s.	v. sl. s.
524	col. nd.	....	d.	....	$2^{15°}$	v. sl. s.	v. sl. s.
525	col. cr.	....	d. 280-300	....	$0.8^{10°}$; $6.6^{100°}$	v. sl. s.	v. sl. s.
526	rhb.	....	....	....	v. s.	s. h.	i.
527	col. lf./ aq. al.	....	165-6	....	0.8 c.; v. s. h.	3 c.	s.; 20 act.

* Insoluble in a mixt. of $2H_2SO_4 + 1H_2O$.
† Very soluble in mixt. of $2H_2SO_4 + 1H_2O$.
Anilino-azobenzene 659
Anilino-benzoic acid 5157
Anilino-ethyl alcohol 2930

Anilino-glyoxylic acid 3921
Anilino-phenol 3767-9
Aniluvitonic acid 5529
Animal starch 3443
Anisal-benzylamine 812

Anise camphor 511
Anisic acid 4069
Anisoyl-anisole 2409
Anisoyl chloride 4071

Table 7-4 (*Continued*)
PHYSICAL CONSTANTS OF ORGANIC COMPOUNDS

No.	Name	Synonym	Formula	Beil. Ref.	Formula Weight
528	**Anis**-alacetone (*p*)	MeO-benzalacetone	$CH_3O \cdot C_6H_4 \cdot CH:CH \cdot CO \cdot CH_3$	VIII-131	176.22
529	alcinnamalacetone	(α,α')	$C_6H_5 \cdot (CH:CH)_2 \cdot CO \cdot CH:CH \cdot C_6H_4 \cdot OCH_3$	VIII-208	290.37
530	alcohol	*p*-MeO-benzyl alcohol	$CH_3O \cdot C_6H_4 \cdot CH_2OH$	VI-897	138.17
531	aldehyde (*o*)	MeO-benzaldehyde	$CH_3O \cdot C_6H_4 \cdot CHO$	VIII-43	136.15
532	aldehyde (*p*)	. . . .	$CH_3O \cdot C_6H_4 \cdot CHO$	VIII-67	136.15
533	aldehyde phenyl-hydrazone (*p*)	. . . .	$C_6H_5NH \cdot N:CH \cdot C_6H_4 \cdot OCH_3$	XV-192	226.28
534	**Anisidine** (*o*)	2-amino-anisole	$NH_2 \cdot C_6H_4 \cdot OCH_3$	XIII-358	123.16
535	**Anisidine** (*m*)	MeO-aniline (*m*)	$NH_2 \cdot C_6H_4 \cdot OCH_3$	XIII-404	123.16
536	**Anisidine** (*p*)	4-amino-anisole	$NH_2 \cdot C_6H_4 \cdot OCH_3$	XIII-435	123.16
537	**Anisole**	methyl-phenyl ether	$CH_3 \cdot O \cdot C_6H_5$	VI-138	108.14
538	**Anthracene**	. . . .	$(C_6H_4CH)_2$	V-657	178.24
539	carboxylic acid (1) (*β*)	. . . .	$C_6H_4:C_2H(CO_2H): C_6H_4$	IX-704	222.25
540	carboxylic acid (2) (*γ*)	. . . .	$C_6H_4:(CH)_2:C_6H_3 \cdot CO_2H$	IX-705	222.25
541	carboxylic acid (9) (*α*)	. . . .	$C_6H_4:(CH)_2:C_6H_3 \cdot CO_2H$	IX-705	222.25
542	**Anthramine** (*α*)	*α*-amino-anthracene	$C_{14}H_9 \cdot NH_2$	XII-1335	193.25
543	**Anthramine** (*β*)	*β*-amino-anthracene	$C_{14}H_9 \cdot NH_2$	XII-1335	193.25
544	**Anthranil**	. . . .	$O \cdot N:C_6H_4:CH$	XXVII-39	119.12
545	**Anthranol**	*γ*-hydroxy anthracene; anthrone	$(C_6H_4:)_2COHCH$ $(C_6H_4:)_2COCH_2$	VII-473	194.24
546	**Anthranoyl anthranilic acid**	. . . .	$NH_2 \cdot C_6H_4 \cdot CO \cdot NHC_6H_4 \cdot CO_2H$	XIV-358	256.26
547	**Anthraquinoline** (*β*)	. . . .	$C_{17}H_{11}N$	XX-506	229.28
548	**Anthraquinone** (1,2)	. . . .	$C_{10}H_6(CO)_2CH:CH$	VII-780	208.22
549	**Anthraquinone** (1,4)	. . . .	$C_{10}H_6(COCH)_2$	VII-781	208.22
550	**Anthraquinone** (9,10)	anthraquinone	$(C_6H_4)_2(CO)_2$	VII-781	208.22
551	carboxylic acid (*α*)	. . . .	$C_6H_4(CO)_2C_6H_3 \cdot CO_2H$	X-834	252.23
552	carboxylic acid (*β*)	. . . .	$C_6H_4(CO)_2C_6H_3 \cdot CO_2H$	X-835	252.23
553	disulfonate Na$_2$ (1,5)	*ρ*-anthraquinone disulfonate	$C_{14}H_6O_2(SO_3Na)_2 \cdot 5H_2O$	XI-340	502.38
554	disulfonate Na$_2$ (1,8)	*χ*-anthraquinone disulfonate	$C_{14}H_6O_2(SO_3Na)_2 \cdot 4H_2O$	XI-341	484.37
555	disulfonate Na$_2$ (2,6)	. . . .	$C_{14}H_6O_2(SO_3Na)_2 \cdot 7H_2O$	*XI-84	538.41
556	disulfonic acid (2,7)	. . . .	$C_{14}H_6O_2(SO_3H)_2$	*XI-84	368.34
557	disulfonate Na$_2$ (2,7)	. . . .	$C_{14}H_6O_2(SO_3Na)_2 \cdot 4H_2O$	*XI-84	484.37
558	sulfonate Na (1)	. . . .	$C_{14}H_7O_2 \cdot SO_3Na$	XI-336	310.26
559	sulfonic acid (2)	*β*-sulfoanthraquinone	$C_{14}H_7O_2 \cdot SO_3H \cdot 3H_2O$	XI-337	342.33

Table 7-4 (*Continued*)
PHYSICAL CONSTANTS OF ORGANIC COMPOUNDS

No.	Crystalline Form and Color	Specific Gravity	Melting Point °C.	Boiling Point °C.	Solubility in 100 Parts		
					Water	Alcohol	Ether
528	lf./et.	....	73–4	....	i.; s. bz.	v. s.	v. s.; s. ac.
529	yel. lf./al. or CS_2	....	138–9	....	i.; s. chl.	sl. s.; s. bz.	sl. s.; sl. s. CCl_4
530	nd.	$1.113\frac{15}{15}°$	25	258.8	i.	s.	s.
531	col. pr.	$1.135\frac{15}{15}°$	35–8 (3)	243–4	i.; sl. s. bz.	sl. s.	v. s.
532	col. oil	$1.123\frac{20}{4}°$	2.5	247–8	v. sl. s.	∞	∞
533	col. nd.	....	120–1	....	s. h. bz.	s. h.	s.
534	col. lq.	$1.098\frac{15}{15}°$	5.2	225	v. sl. s.	∞	∞
535	oil	$1.096\frac{20}{4}°$	<−12	251	v. sl. s.	s.	s.; s. a.
536	pl./aq.	$1.089\frac{55}{55}°$	57.2	243	s. h.	s.	s.
537	col. lq.	$0.990\frac{22}{2}°$	−37.4	153.8	i.	s.	s.
538	col mn.	$1.25\frac{27}{4}°$	217	339.9	i.	$1.9^{20°}$, abs. al.	$12.2^{100°}$, toluene
539	yel. pr./al.	....	245 (260)	subl.	i.	s.	s.; sl. s. bz.
540	yel. lf./al.	....	280±	subl.	i.	sl. s.	i. bz.; i. CS_2
541	lt. yel./al.	....	206 (−CO_2)	....	v. sl. s. h.	s.	....
542	yel. nd./al.	....	130±	....	i. HCl	s.	....
543	yel. lf./al.	....	238	subl. 939^{mm}	i.	sl. s.	sl. s.
544	col. oil	$1.187\frac{15}{4}°$	<−18	d. >215	sl. s. h.	s.	s.; s. c. HCl
545	col. rhb.	....	160–70 d.	....	i.	v. s. h. bz.	s. h. alk.
546	nd./aq. al.	....	202–4	d.	v. sl. s. c.	3, in 50% h. al.	0.3, h. bz.
547	col. lf.	....	170	446	i.	s.	s.
548	or. nd./al.	....	185–90 d.	....	v. sl. s.	3	s. h. chl.
549	yel. nd./al.	....	206–18 d.	....	....	...	....
550	yel. rhb.	$1.438\frac{20}{4}°$	286	379–81	i.	$0.05^{18°}$; 2.25 h.	v. sl. s.
551	yel. nd./aq.	....	293.4	....	sl. s. h.	....	....
552	yel. nd./al.	....	290	....	s. act.	v. sl. s. abs. al.	i.
553	yel. lf.	....	....	....	v. s.	i.	i.
554	yel. pr.	....	....	....	sl. s.	....	....
555	col.cr.darken in light	....	....	....	$3.9^{20°}$; $18^{100°}$	....	....
556	yel. cr.	....	....	....	v. s.	s.	i.; i. bz.
557	cr.	....	....	....	$30.5^{20°}$; $13^{100°}$	v. sl. s.	i.; i. bz.
558	yel. lf.	....	....	....	0.5 c.; s. h.	i.; v. sl. s. act.	i.
559	col. lf.	....	....	....	v. s. c.	v. s.	i.

Table 7-4 (*Continued*)
PHYSICAL CONSTANTS OF ORGANIC COMPOUNDS

No.	Name	Synonym	Formula	Beil. Ref.	Formula Weight
560	**Anthraquinone** sulfonate Na (2)	"silver salt"	$C_{14}H_7O_2 \cdot SO_3Na$	XI-337	310.26
561	**Anthrol** (α)	1-hydroxy-anthra-cene	$C_6H_4(CH)_2C_6H_3OH$	VI-702	194.24
562	**Anthrol** (β)	2-anthrol	$C_6H_4(CH)_2C_6H_3OH$	VI-702	194.24
563	**Antimony triphenyl**	triphenyl stibine	$Sb(C_6H_5)_3$	XVI-891	353.07
564	dichloride	tri-Ph-stibine-di-Cl	$(C_6H_5)_3SbCl_2$	XVI-893	423.98
565	**Antipyrine**	1-Ph-2, 3-diMe-pyrazolone-5	$C_{11}H_{12}ON_2$	XXIV-27	188.23
566	acetylsalicylate	pyrosal	$C_{20}H_{20}O_5N_2$	XXIV-32	368.39
567	chloralhydrate	hypnal; chloral-antipyrine	$Cl_3C \cdot CH(OH)_2 \cdot C_{11}H_{12}ON_2$	XXIV-31	353.64
568	mandelate	tussol	$C_{11}H_{12}ON_2 \cdot C_8H_8O_3$	....	340.38
569	**Apiole**	1-allyl-2, 5-diMeO-3, 4-methylenedi-oxybenzene	$C_{12}H_{14}O_4$	XIX-87	222.24
570	**Arabinose** (α) (d or l)	....	$C_4H_9O_4 \cdot CHO$	I-859	150.13
571	phenylhydrazone (l)	....	$C_6H_5NH \cdot N:C_5H_{10}O_4$	XV-215	240.26
572	**Arabitol** (d)	pentanpentol	$C_5H_7(OH)_5$	I-531	152.15
573	**Arabonic acid**	....	$HO \cdot CH_2(CHOH)_3 \cdot CO_2H$	III-473	166.13
574	**Arachidic acid**	eicosanoic acid	$CH_3(CH_2)_{18} \cdot CO_2H$	II-389	312.54
575	**Arginine** ($l+$)	guanidine-amino-valeric acid	$H_2N \cdot C(:NH) \cdot NH \cdot (CH_2)_3 \cdot CH(NH_2) \cdot CO_2H$	IV-420	174.20
576	hydrochloride (d)	....	$C_6H_{14}O_2N_4 \cdot HCl$	IV-423	210.67
577	**Arsacetin** (p)	acetyl arsanilic acid	$CH_3CO \cdot NH \cdot C_6H_4 \cdot AsO(OH)_2$	XVI-880	259.09
578	Na salt	....	$C_8H_9O_4NAsNa \cdot 4H_2O$	XVI-880	353.14
579	**Arsenic** diethyl	ethyl cacodyl	$[(C_2H_5)_2As]_2$	IV-616	266.09
580	triphenyl	triphenylarsine	$As(C_6H_5)_3$	XVI-828	306.24
581	**Arseno-phenyl-glycine** Na	....	$(As \cdot C_6H_4 \cdot NH \cdot CH_2 \cdot CO_2Na)_2$	....	494.12
582	**Arsono-phenyl-glycinamide** (4)	....	$H_2N \cdot CO \cdot CH_2 \cdot NH \cdot C_6H_4 \cdot AsO(OH)_2$	*XVI-470	274.11
583	Na salt	tryparsamide	$C_2H_5ON_2 \cdot C_6H_4 \cdot AsO(ONa)_2 \cdot 3H_2O$	*XVI-470	372.12
584	**Asarone**	propenyl-2,4,5-tri-MeO-benzene	$CH_3CH:CH \cdot C_6H_2(OCH_3)_3$	VI-1129	208.26
585	**Ascaridole**	....	$CH_3 \cdot C_6H_6O_2 \cdot CH:(CH_3)_2$	XIX-17	168.24
586	**Ascorbic acid** ($l+$)	vitamin C; Redoxon; cevitamic acid	$O \cdot CO \cdot C(OH):C(OH) \cdot CH \cdot CHOH \cdot CH_2OH$	....	176.13
587	*iso-***Ascorbic acid** (d)	....	$C_6H_8O_6$	....	176.13
588	**Asparagine** (l)	....	$HO_2C \cdot CH(NH_2) \cdot CH_2 \cdot CONH_2$	IV-476	132.12
589	**Aspartic acid** ($l-$)	amino-succinic acid	$HO_2C \cdot CH(NH_2) \cdot CH_2 \cdot CO_2H$	IV-472	133.10
590	**Aspartic acid** (dl)	amino-succinic acid	$C_4H_7O_4N$	IV-483	133.10
591	**Atropic acid**	α-phenyl acrylic acid	$C_6H_5C(:CH_2) \cdot CO_2H$	IX-610	148.16

Table 7-4 (*Continued*)
PHYSICAL CONSTANTS OF ORGANIC COMPOUNDS

No.	Crystalline Form and Color	Specific Gravity	Melting Point °C.	Boiling Point °C.	Solubility in 100 Parts		
					Water	Alcohol	Ether
560	silver lf.	....	....	....	0.8 c.; 21 h.	i.	i.
561	nd./al.	....	150-3	....	i.	v. s.	v. s.
562	nd./aq.	....	d. 200	....	i.	v. s.	v. s.
563	tri. pr./pet.	$1.5^{12°}$	53-4.5	>360 d.	i.; v. s. bz.	sl. s.	v. s.
564	col. nd.	....	143	....	v. sl. s.	i. c.	i.; s. bz.
565	mn./aq.	$1.088^{11\frac{3}{4}°}$	113 (109)	319^{174mm}	100	100	sl. s.
566	col. pd.	....	64-5	....	$1^{25°}$; $480°$	$126^{25°}$	$13^{25°}$; s. bz.
567	col. rhb.	....	67-8	....	7.9^{14}	33	sl. s.
568	col. cr.	....	52-5	....	6.6	30	5
569	col. nd.	$1.02^{\frac{20}{4}°}$	30*	294	i.; s. bz.	s.; s. act.	s.; s. oil
570	rhb. pr.	$1.585^{\frac{20}{4}°}$	158.5-9.5	....	$46^{0°}$	$0.5^{9°}$, 90% al.	i.
571	col. nd.	....	151-3	....	$1.2^{15°}$	$1.3^{15°}$, abs.	v. sl. s.
572	pr. or warts	....	102-3	....	v. s.	$2^{12°}$,90% al.	....
573	cr.	....	89	d., $-H_2O$	s.	....	....
574	col. lf.	....	77	328 sl. d.	i.	v. s. h. abs.	v. s.
575	pr./66% al.	....	238 d.	....	$15^{21°}$	sl. s.	i.
576	col. pl.	....	222-35 d.	softens 218	v. s.	....	....
577	lf.	....	>200	....	s. Na_2CO_3	v. sl. s. dil. HCl	....
578	wh. cr. pd.	....	....	....	$10^{20°}$	$33^{50°}$	....
579	lq.	>1	ign. in air	185-90	i.	v. s.	v. s.
580	pl./bz.	1.306	59-60	$>360(CO_2)$	i.; i. HCl	s.; v. s. bz.	v. s.
581	yel. pd.	....	d. in air	....	s.	....	....
582	pl./aq.	....	darkens 280	....	v. s. h.; s. h. ac.	v. sl. s. h.; i. c. chl.	i. c. act.
583	wh. pd.	....	....	....	50	sl. s.	i.; i. chl.
584	mn. nd./aq.	$1.165^{18°}$	67	296	sl. s. h.; s. ac.	$7^{18°}$, 165^{50}, 60% al.	s.; s. chl.
585	col. lq.	$0.999^{\frac{20}{20}°}$	expl. 250	115^{15mm}	i.	s.	expl. with a.
586	col. mn.	....	192	....	33; 1 gly.	4; i. pet.; i. fat	i.; i. bz.
587	col. cr.	....	168-71	....	v. s.	s.; i. act.	v. sl. s.
588	rhb.	$1.543^{\frac{15}{4}°}$	227-35†	235 d.	3.12^{80}	i. c. abs. al.	s. NH^3
589	rhb.	....	251-83 d.	....	$0.5^{16°}$	i.	i.
590	mn. pr.	$1.663^{\frac{13}{13}°}$	271†	d. 290	$2.7^{175°}$	i.	i.
591	mn. nd./aq.	....	106-7	267 d.	0.13 c.	s.; s. chl.	s.; s. CS_2

* Also modifications, m.p. 27.5 and 18-9. † Sealed tube.

Artificial camphor 5319	Asebogenol 5264	Aspirin 115	Atoquinol 212
Asaprol 4506	Asebotin, cf. glcde.	Athenon 3800	Atoxylic acid 515
Asarin 584	Asepsin 882	Aticine, cf. alkd.	Atropine, cf. alkd.
Asaronic acid 6241	Aspidospermine, cf. alkd.	Atisine, cf. alkd.	Atroscine, cf. alkd.
		Atophan 5234	

Table 7-4 (*Continued*)
PHYSICAL CONSTANTS OF ORGANIC COMPOUNDS

No.	Name	Synonym	Formula	Beil. Ref.	Formula Weight
592	**Auramine**	4,4′-dimethylamino-benzophenonimide	$[(CH_3)_2NC_6H_4]_2C{:}NH$	XIV-91	267.38
593	**Aurin**	4′,4″-diOH-fuchsone	$(HOC_6H_4)_2C{:}C_6H_4{:}O$	VIII-361	290.32
594	**Azafrin**	....	$C_{27}H_{38}O_4$	XXX-115	426.60
595	**Azafrinone**	....	$C_{27}H_{36}O_4$	XXX-116	424.59
596	**Azelaic** acid	nonandioic acid	$(CH_2)_7(CO_2H)_2$	II-707	188.23
597	nitrile	....	$(CH_2)_7(CN)_2$	II-709	150.23
598	**Azo**-anisole (2,2′)	diMeO-azobenzene	$(CH_3O{\cdot}C_6H_4N{:})_2$	XVI-92	242.28
599	benzene	diphenyldiimide	$C_6H_5{\cdot}N{:}N{\cdot}C_6H_5$	XVI-8	182.23
600	benzoic acid (2,2′)	....	$(HO_2C{\cdot}C_6H_4{\cdot}N{:})_2$	XVI-228	270.25
601	benzoic acid (3,3′)	....	$(HO_2C{\cdot}C_6H_4{\cdot}N{:})_2{\cdot}$ $\frac{1}{2}H_2O$	XVI-233	279.25
602	benzoic acid (4,4′)	....	$(HO_2C{\cdot}C_6H_4{\cdot}N{:})_2$	XVI-236	270.25
603	dicarbonamide	....	$(NH_2{\cdot}CO{\cdot}N{:})_2$	III-123	116.08
604	naphthalene (α,α')	....	$C_{10}H_7{\cdot}N{:}N{\cdot}C_{10}H_7$	XVI-79	282.35
605	naphthalene (β,β')	....	$C_{10}H_7{\cdot}N{:}N{\cdot}C_{10}H_7$	XVI-80	282.35
606	phenetole (2,2′)	....	$(C_2H_5O{\cdot}C_6H_4{\cdot}N{:})_2$	XVI-92	270.33
607	phenetole (3,3′)	....	$(C_2H_5O{\cdot}C_6H_4{\cdot}N{:})_2$	XVI-95	270.33
608	phenetole (4,4′)	....	$(C_2H_5O{\cdot}C_6H_4{\cdot}N{:})_2$	XVI-113	270.33
609	phenol (2,2′)	dihydroxy-azobenzene	$(HO{\cdot}C_6H_4{\cdot}N{:})_2$	XVI-91	214.23
610	phenol (3,3′)	....	$(HO{\cdot}C_6H_4{\cdot}N{:})_2$	XVI-95	214.23
611	phenol (4,4′)	....	$(HO{\cdot}C_6H_4{\cdot}N{:})_2$	XVI-110	214.23
612	toluene (2,2′)	dimethylazobenzene	$(CH_3{\cdot}C_6H_4{\cdot}N{:})_2$	XVI-61	210.28
613	toluene (3,3′)	....	$(CH_3{\cdot}C_6H_4{\cdot}N{:})_2$	XVI-64	210.28
614	toluene (4,4′)	....	$(CH_3{\cdot}C_6H_4{\cdot}N{:})_2$	XVI-67	210.28
615	**Azoxy**-benzene	....	$C_6H_5{\cdot}N({:}O){:}N{\cdot}C_6H_5$	XVI-622	198.23
616	benzoic acid (2,2′)	....	$(HO_2C{\cdot}C_6H_4)_2{:}N_2O$	XVI-644	286.25
617	benzoic acid (3,3′)	....	$(HO_2C{\cdot}C_6H_4)_2{:}N_2O$	XVI-646	286.25
618	benzoic acid (4,4′)	....	$(HO_2C{\cdot}C_6H_4)_2{:}N_2O$	XVI-647	286.25
619	naphthalene (α,α')	....	$(C_{10}H_7N)_2O$	XVI-633	298.35
620	naphthalene (β,β')	....	$(C_{10}H_7N)_2O$	XVI-633	298.35
621	**Barbituric acid**	malonyl urea	$CH_2CONHCONHCO{\cdot}$ ⎯⎯⎯⎯⎯⎯ $2H_2O$	XXIV-467	164.12
622	**Bassorine**	....	$C_6H_{10}O_5$	....	162.14
623	**Behenic acid**	docosanoic acid	$CH_3{\cdot}(CH_2)_{20}{\cdot}CO_2H$	II-391	340.59
624	**Behenolic acid**	docosinoic acid	$C_8H_{17}C{:}C(CH_2)_{11}{\cdot}$ CO_2H	II-497	336.58
625	**Benzal** acetoacetic ester	....	$C_2H_3O{\cdot}C(C_7H_6){\cdot}$ $CO_2C_2H_5$	X-731	218.25
626	acetone	Me-cinnamyl ketone	$C_6H_5{\cdot}C_2H_2{\cdot}CO{\cdot}CH_3$	VII-364	146.19
627	acetophenone	chalcone	$C_6H_5{\cdot}C_2H_2{\cdot}CO{\cdot}C_6H_5$	VII-479	208.26
628	acetophenone dibromide	....	$C_6H_5{\cdot}(CHBr)_2{\cdot}$ COC_6H_5	VII-445	368.08
629	acetophenone dibromide	high m. p. form	$C_6H_5{\cdot}(CHBr)_2{\cdot}CO{\cdot}$ C_6H_5	VII-445	368.08
630	amino-2-cresol (5)	(5; 1, 2)	$C_6H_5CH{:}N{\cdot}$ $C_6H_3(CH_3){\cdot}OH)$	....	211.27
631	aminophenol (*p*)	....	$C_6H_5CH{:}N{\cdot}C_6H_4{\cdot}OH$	XIII-453	197.24
632	aniline	benzaldehyde anil	$C_6H_5CH{:}N{\cdot}C_6H_5$	XII-195	181.24

Table 7-4 (*Continued*)
PHYSICAL CONSTANTS OF ORGANIC COMPOUNDS

No.	Crystalline Form and Color	Specific Gravity	Melting Point °C.	Boiling Point °C.	Solubility in 100 Parts		
					Water	Alcohol	Ether
592	col./al.	....	136	....	i.	$7^{20°}$	$2.3^{20°}$
593	rhb. red	....	308–10 d.	....	i.; i. bz.	s.	s.; s. alk.
594	or. nd./bz.	....	212–4	....	i.; s. alk.	sl. s.	i.; sl. s. bz.
595	or. cr./act.	....	191	....	....	....	....
596	lf. or nd.	$1.029^{20°}_{4}$	106.5	286.5^{100mm}	0.2 c.; ∞ h.	v. s.	$2.7^{15°}$
597	oil	$0.941^{0°}$	....	$195–6^{19mm}$	i.	v. s.	v. s.
598	or. pr./ MeOH	....	153 (141)	....	i.; s. act.	s.; s. chl.	s.; s. bz.
599	or. mn.	$1.203^{20°}_{4}$	68	297	i.	$4.2^{20°}$	$12^{20°}$ lg.
600	yel. nd./al.	....	237 d.	....	sl. s. h.	v. s. h.	i. bz.
601	yel. nd./ac.	....	340±	d.	v. sl. s.	0.2 h.	sl. s.
602	red nd./ac.	....	d. 330±	....	v. sl. s.	v. sl. s.	v. sl. s.
603	or. red pd.	....	d. 180–200	....	v. sl. s. h.	i.	d. h. HCl
604	red nd./ac.	....	190	subl. >190	i.; s. ac.	v. sl. s.	v. s. bz.
605	red pr./chl.	....	204	subl. 210	i.	sl. s.	s. bz.
606	red pr./al.	....	131	240 d.	i.; s. HCl	s.	s.
607	yel. pr./al.	....	91	d.	i.; i. HCl	s.	s.
608	or. lf.	....	159–60	d.	v. s. chl.	s. h.	v. s.
609	yel. lf.	....	171–2	subl.	i.	0.3 c.	s.; s. KOH
610	brn./aq. al.	....	205	....	v. sl. s.	s. h.	s. Na_2CO_3
611	brn. tri.	....	215 d.	....	sl. s.	s.	s.; s. bz.
612	red mn./et.	....	55	....	i.; s. bz.	$6.03^{15°}$	$147.7^{17°}$
613	or. rhb.	....	54–5	....	i.	s.	s.
614	or. nd./lg.	....	144–5	....	i.	sl. s.	s.; s. lg.
615	yel. rhb.	$1.248^{20°}_{20}$	36	d.	i.	$11.4^{15°}$	$43.5^{15°}$ lg.
616	brn./al.	....	246–8 d.	....	v. sl. s. h.	s. h.	s. h.
617	nd. or lf.	....	320 d.	....	i.	sl. s.	sl. s.
618	yel. armor.	....	d. 240±	....	i.	i.	s. C_5H_5N
619	yel.rhb./al.	....	127	....	i.	s.	....
620	yel.rhb./al.	....	167–8	....	i.	....	....
621	col. pr./aq.	....	d. 245±	....	s. h.	sl. s.	s.; s. HCl
622	wh. amor.	....	....	....	v. sl. s.	i.; d. h. a.	i. $(NH_4)_2S$
623	col. nd.	....	80	306^{60mm}	i.	sl. s.	sl. s.
624	nd.	....	57.5	....	i.	v. s.	v. s.
625	rhb./al.	....	59–60	295–7 sl. d.	v. s. chl.	sl. s. o.	sl. s. c.
626	pl.	$1.035^{20°}_{20}$	41–2	260–2	s. H_2SO_4	s.; s. chl.	s.; s. bz.
627	lt. yel. rhb.	$1.071^{62°}_{4}$	62	345–8 sl. d.	i.	sl. s.	v. s.
628	nd./al.	....	108–9	....	....	$1^{30°}$; s. h.	....
629	....	....	156–7	....	....	$0.16^{30°}$	....
630	brn. fl.	....	210–11	....	i.	sl. s.	i.
631	lf./aq. al.	....	185–6	....	....	v. s.	....
632	yel. cr.	....	56 (48)	300±	i.	s.	s.

No.	Name	Synonym	Formula	Beil. Ref.	Formula Weight
633	**Benzal** azine	dibenzal hydrazine	$(C_6H_5CH:N\cdot)_2$	VII-225	208.27
634	bromide	benzylidene bromide	$C_6H_5\cdot CH:Br_2$	V-308	249.94
635	chloride	benzylidene chloride	$C_6H_5\cdot CH:Cl_2$	V-297	161.03
636	diacetate	benzylidene diacetate	$C_6H_5CH:(O_2CCH_3)_2$	VII-210	208.22
637	lactic acid	phenyl-hydroxy-crotonic acid	$C_6H_5\cdot CH:CH\cdot CHOH\cdot CO_2H$	X-308	178.19
638	malonic acid	....	$C_6H_5\cdot CH:C(CO_2H)_2$	IX-891	192.17
639	methylamine	....	$C_6H_5\cdot CH:N\cdot CH_3$	VII-213	119.17
640	phthalide	benzylidene phthalide	$C_7H_6:C\cdot C_6H_4\cdot CO\cdot O$	XVII-376	222.25
641	2-thiohydantoin (5)	....	$CO\cdot NH\cdot CS\cdot NH\cdot C:C_7H_6$	XXIV-400	204.25
642	**Benzaldehyde**	art. almond oil	$C_6H_5\cdot CH:O$	VII-174	106.13
643	oxime (α; *syn*)	....	$C_6H_5CH:NOH$	VII-218	121.14
644	oxime (β; *anti*)	....	$C_6H_5CH:NOH$	VII-221	121.14
645	oxime carboxylic anhydride (*o*)	benzoxazinone	$C_6H_4\cdot CH:N\cdot O\cdot CO$	XXVII-198	147.13
646	phenylhydrazone	....	$C_6H_5NH\cdot N:CH\cdot C_6H_5$	XV-134	196.25
647	**Benzamaron**	....	$C_6H_5CH[CH(C_6H_5)\cdot COC_6H_5]_2$	VII-849	480.61
648	*iso*-**Benzamaron**	....	$C_{35}H_{28}O_2$	VII-849	480.61
649	**Benzamidine**	benzenylamidine	$C_6H_5C(:NH)\cdot NH_2$	IX-280	120.16
650	**Benzanthracene** (1,2)	naphthanthracene	$C_{18}H_{12}$	V-718	228.30
651	**Benzarsonic acid** (*p*)	....	$(HO)_2OAs\cdot C_6H_4\cdot CO_2H$	XVI-876	246.05
652	**Benzaurine**	....	$C_6H_5C(:C_6H_4O)\cdot C_6H_4OH$	VI-1145	274.32
653	**Benzazide**	benzoyl azide	$C_6H_5CO\cdot N_3$	IX-332	147.14
654	**Benzedrine**	*dl*-desoxy-nor-ephedrine	$C_6H_5CH_2CH(NH_2)\cdot CH_3$	XII-1145	135.21
655	sulfate	....	$(C_9H_{13}N)_2\cdot H_2SO_4$	....	368.50
656	**Benzene**	benzol	C_6H_6	V-179	78.11
657	azo-*o*-cresol (5)	(1;4,3)	$C_6H_5\cdot N_2\cdot C_6H_3(OH)\cdot CH_3$	XVI-130	212.25
658	azo-*m*-cresol (6)	(1;4,2)	$C_6H_5\cdot N_2\cdot C_6H_3(OH)\cdot CH_3$	XVI-134	212.25
659	azodiphenylamine (*p*)	4-anilino-azobenzene	$C_6H_5\cdot N_2\cdot C_6H_4NHC_6H_5$	XVI-314	273.34
660	azo-α-naphthylamine (4)	....	$C_6H_5\cdot N_2\cdot C_{10}H_6\cdot NH_2$	XVI-361	247.30
661	azo-β-naphthylamine (1)	Yellow AB	$C_6H_5\cdot N_2\cdot C_{10}H_6\cdot NH_2$	XVI-369	247.30
662	azoresorcinol	(1;2,4); Sudan G	$C_6H_5\cdot N_2\cdot C_6H_3(OH)_2$	XVI-180	214.23
663	disulfone chloride	(1,3)	$C_6H_4:(SO_2Cl)_2$	XI-200	275.14
664	disulfonic acid (*m*)	....	$C_6H_4\cdot (SO_3H)_2\cdot 2\frac{1}{2}H_2O$	XI-199	283.28
665	disulfonic Na (*m*)	....	$C_6H_4(SO_3Na)_2\cdot 4H_2O$	XI-199	354.27
666	hexabromide (α, *trans*)	hexabromo-cyclohexane	$C_6H_6Br_6$	V-25	557.57
667	hexacarboxylic acid	mellitic acid	$C_6(CO_2H)_6$	IX-1008	342.17
668	hexachloride (α, *trans*)	hexachloro-cyclohexane	$C_6H_6Cl_6$	V-23	290.83

Table 7-4 (*Continued*)
PHYSICAL CONSTANTS OF ORGANIC COMPOUNDS

No.	Crystalline Form and Color	Specific Gravity	Melting Point °C.	Boiling Point °C.	Solubility in 100 Parts		
					Water	Alcohol	Ether
633	yel. pr.	....	93	....	i. c.	s. h.	s.
634	oil	$1.51^{15°}$	....	140^{20mm}	i.	∞	∞
635	col. lq.	$1.295^{16°}$	−16.1	214	i.	∞	∞
636	pl./et.	....	45–6	154^{20mm}	s. dil. alk.	v. s.	v. s.
637	nd./aq.	....	137±	....	s. h.	i. bz., CS_2	sl. s.; i. lg.
638	cr./et.-CS_2	....	195–6 d.	....	v. s. h.	s.	sl. s.
639	col. lq.	....	....	180±	....	....	....
640	mn./al.	....	98–9	....	i. h.	v. s. h.	....
641	yel. nd./al.	....	264–5	....	....	....	....
642	col. lq.	$1.046\frac{20°}{4}$	−26*	179.1	0.3	∞	∞
643	pr.	$1.111\frac{20°}{4}$	35 (5)	117.5^{14mm}	sl. s.	v. s. bz.	v. s.
644	nd./et.	....	128–30	....	s. h.	v. sl. s. bz.	....
645	cr./bz.	....	d. 120	....	....	....	....
646	col.† mn.	....	155–6	....	s. bz.	s. h.	sl. s.
647	cr.	....	218–9	....	$1.6^{12°}$, bz.	....	....
648	cr.	....	179–80	....	$4.1^{12°}$, bz.	....	....
649	cr.	....	80	....	s.	v. s.	sl. s.
650	lf./al.-ac.	....	159–60	subl.	....	sl. s. h.	s. bz.
651	cr./aq.	....	d.-H_2O, 210	....	v. sl. s.	i.	i. ac.
652	r. pd.	....	....	....	i.; s. alk.	s.	s.
653	col. pl./act.	....	32	expl.	i.	sl. s.	s.
654	col. lq.	0.93	....	203	sl. s.; s. a.	s.; s. EtOAc	s.; s. chl.
655	wh. pd.	....	....	....	10	sl. s.	i.
656	col. lq.	$0.879\frac{20°}{4}$	5.5	80.1	$0.07^{22°}$	∞ abs. al.	∞
657	yel. nd./al.	....	129–30	....	v. s. h.; s. alk.	s.; s. chl.	s.; s. bz.
658	yel. nd./lg.	....	109‡	....	....	s.; s. chl.	s.; s. bz.
659	yel. pr./et. al.	....	86–8	....	v. s. lig.	v. s.	v. s.
660	r. nd./al.	....	123	....	s. bz.	s.	s.
661	r. pl./al.	....	102–4	....	i.; s. oil	s.; s. CCl_4	s. ac.
662	r. nd./aq. al.	....	170§	....	i.; s. alk.	v. s.	v. s.; v. s. bz.
663	mn./et.	....	63	210.7^{20mm}	d. 130°	s. h. d.	s.
664	delq. cr.	....	....	....	s.	....	....
665	col. nd.	....	....	....	s.	....	....
666	mn. pr.	....	212	....	....	sl. s.	sl. s.
667	nd./al.	....	286–8¶	d.	v. s.	v. s.	s. h. H_2SO_4
668	mn. pr.	....	157–8	d. >158	$4.4^{15°}$, chl.	s.	$6.5^{18°}$, bz.

* Solidifies −56°.
† Pink on exposure to light.
‡ Crysts. +1H_2O, m. 90°.
§ Cryst. +⅓H_2O, m. 161°.
¶ Sealed tube.

Benzene azonitromethane 4739
Benzene azophenol 3727–9
Benzene diazoanilide 1762
Benzene diazonium, cf. diazobenzene.

Table 7-4 (*Continued*)
PHYSICAL CONSTANTS OF ORGANIC COMPOUNDS

No.	Name	Synonym	Formula	Beil. Ref.	Formula Weight
669	**Benzene** hexachloride (β, *cis*)	....	$C_6H_6Cl_6$	V-23	290.83
670	indone	safranone	$C_{18}H_{12}ON_2$	XXIII-413	272.31
671	pentacarboxylic acid	....	$C_6H(CO_2H)_5·5H_2O$	IX-1006	388.24
672	sulfinic acid	....	$C_6H_5·SO_2H$	XI-2	142.18
673	sulfinic Na	....	$C_6H_5·SO_2Na·2H_2O$	XI-6	200.19
674	sulfohydroxamic acid	Piloty's acid	$C_6H_5SO_2·NHOH$	XI-51	173.19
675	sulfonic acid	....	$C_6H_5·SO_3H$	XI-26	158.18
676	sulfonic Na	....	$C_6H_5SO_3Na·H_2O$	XI-28	198.18
677	sulfonic amide	benzenesulfonamide	$C_6H_5·SO_2·NH_2$	XI-39	157.19
678	sulfonic chloride	benzenesulfonyl chloride	$C_6H_5·SO_2·Cl$	XI-34	176.62
679	1-sulfonic acid-(4-azo-5)-8-hydroxy-quinoline	sulfenazoxine	$HO_3S·C_6H_4·N_2·$ $C_9H_5N(OH)$	XXII-584	329.34
680	tetracarboxylic acid (1,2,3,4)	prehnitic acid	$C_6H_2(CO_2H)_4·2H_2O$	IX-997	290.19
681	tetracarboxylic acid (1,2,3,5)	mellophanic acid	$C_6H_2(CO_2H)_4$	IX-997	254.15
682	tetracarboxylic acid (1,2,4,5)	pyromellitic acid	$C_6H_2(CO_2H)_4·2H_2O$	IX-997	290.19
683	tricarboxylic acid (1,2,3)	hemimellitic acid	$C_6H_3(CO_2H)_3·2H_2O$	IX-976	246.18
684	tricarboxylic acid (1,2,4)	trimellitic acid	$C_6H_3(CO_2H)_3$	IX-977	210.14
685	tricarboxylic acid (1,3,5)	trimesic acid	$C_6H_3(CO_2H)_3$	IX-978	210.14
686	trisulfonic acid	(1,3,5)	$C_6H_3(SO_3H)_3 + aq.$	XI-227	318.30
687	**Benzenyl**-aminothiophenol	μ-phenylbenzothiazole	$C_6H_4·N:C(C_6H_5)·S$	XXVII-74	211.29
688	aminoxime	....	$C_6H_5C(NOH)NH_2$	IX-304	136.15
689	naphthylamidine	α-naphthyl-benz-amidine	$C_6H_5C(:NH)NH·$ $C_{10}H_7$	XII-1233	246.31
690	phenylenediamine (1,2)	α-phenylbenzimid-azole	$C_6H_5C:N·C_6H_4·NH$	XXIII-230	194.24
691	**Benzhydrol**	diphenyl carbinol	$(C_6H_5)_2CHOH$	VI-678	184.24
692	ether	....	$[(C_6H_5)_2CH]_2O$	VI-679	350.46
693	**Benzhydryl** amine	....	$(C_6H_5)_2CH·NH_2$	XII-1323	183.26
694	benzoic acid (*p*)	....	$C_6H_5CH(OH)·$ $C_6H_4CO_2H$	X-346	228.25
695	**Benzidine** disulfonic acid (*o,o'*)	....	$[C_6H_3(NH_2)·SO_3H]_2$	XIV-794	344.37
696	monosulfonic acid	(4;4',3)	$H_2N·C_6H_4·$ $C_6H_3(NH_2)·SO_3H$	XIV-770	264.31
697	**Benzil**	dibenzoyl	$(C_6H_5·CO)_2$	VII-747	210.23
698	**Benzilam**	triphenyloxazole	$C_6H_5C:C(C_6H_5)·O·$ $C(C_6H_5):N$	XXVII-88	297.36
699	**Benzilic acid**	diphenyl glycolic acid	$(C_6H_5)_2C(OH)·CO_2H$	X-342	228.25

Benzene hexahydride 1601
Benzene sulfonamide 677

Benzene sulfonyl chloride 678
Benzene tetrahydride 1609

Table 7-4 (*Continued*)
PHYSICAL CONSTANTS OF ORGANIC COMPOUNDS

No.	Crystalline Form and Color	Specific Gravity	Melting Point °C.	Boiling Point °C.	Solubility in 100 Parts		
					Water	Alcohol	Ether
669	col. cr./ xylene	$1.89^{19°}$	310–2	....	$0.13^{20°}$, chl.	v. sl. s.; $122°$bz.	v. sl. s. ac.
670	brn. met.	....	248–9	....	sl. s.	v. s.	i. alk.
671	rhb.	....	228–30; 238 (anh.)	....	v. s. h.	s.	sl. s.; i. bz.
672	pr./aq.	....	83–4	d. >100	v. s. h.; sl. s. c.	v. s.	v. s.
673	lf./aq.	....	....	....	s.	....	....
674	rhb. pl./aq.	....	126 ±	d. >m. p.	s. h.	s.; v. sl. s. bz.	s.; s. act.
675	col. nd.	....	65–6*	d.	v. s.	v. s.	i.; sl. s. bz.
676	nd./aq. al.	....	450 d.	....	$60^{30°}$	sl. s. h.	....
677	mn. nd./aq.	....	156	....	$0.43^{16°}$	v. s.	v. s.
678	cr.	$1.384^{15°}_{15}$	16.5–17.5	251.5	i.; d. h.	v. s.; d. h.	s.
679	or. nd./aq. al.	....	....	....	....	....	....
680	pr./aq.	....	237–50 d.	....	v. s.	....	sl. s.
681	cr./aq.	....	215–38 d.	....	v. s.	....	....
682	tri./aq.	....	269–71 d.	....	(anh.) $1.4^{16°}$	v. s.	....
683	pl./aq.	....	190 d.	....	$3.21^{9°}$; v. s. h.	....	s.
684	nd./aq.	....	216–8 d.	....	sl. s.	....	sl. s.
685	pr./aq.	....	375–80	subl.	$2.8^{22.5°}$	v. s.	s.
686	hyg. nd.	....	d. >100	....	s.	....	....
687	nd./al.	....	115	360 ± (sl. d.)	i.; s. HCl	sl. s.	s.; s. CS_2
688	mn./aq.	....	79–80	....	sl. s. c.	s.; s. bz.	s.; i. lg.
689	pl./aq.	....	141	....	i.	s.	s.
690	nd./aq.	....	291	....	sl. s.	s.	sl. s. chl., bz.
691	nd./lg.	....	68–9	$297-8^{48mm}$	$0.05^{20°}$	v. s.	v. s.
692	mn./bz.	....	109–10	267^{15mm}	....	sl. s. h.	v. s. bz.
693	lq.	$1.064^{21.5°}_{0}$	....	$301-2^{746mm}$	....	....	....
694	nd./aq.	....	164–5	d.	s. h.	s.	s.; sl. s. chl.
695	pr./aq. ($+3H_2O$)	....	d. >175	....	$0.08^{25°}$	i.	i.
696	lf.	....	....	....	v. sl. s. h.	v. sl. s.	v. sl. s.
697	trig. pr.	$1.23^{15°}$	95	346–8 (sl. d.)	i.	v. s.	v. s.
698	pr./al. et.	....	115	....	v. sl. s.	sl. s. c.	v. s.; s. h. ac.
699	nd./aq.	....	150	....	s. h.	s.	s.; s. H_2SO_4

* Cryst. $+1.5H_2O$, m.p. 43–4.
Benzenyl–amidine 649
Benzhydryl–aniline 384–5

Benzidine 1713–4
Benzidine (β) 1711

Table 7-4 (*Continued*)
PHYSICAL CONSTANTS OF ORGANIC COMPOUNDS

No.	Name	Synonym	Formula	Beil. Ref.	Formula Weight
700	**Benzil**-oxime, di- (*syn*)	α-benzildioxime	$(C_6H_5C:NOH)_2$	VII-760	240.26
701	oxime, di-(*anti*)	β-benzildioxime	$(C_6H_5C:NOH)_2$	VII-761	240.26
702	oxime, di- (*amphi*)	γ-benzildioxime	$(C_6H_5C:NOH)_2 \cdot$ C_2H_5OH	VII-763	286.33
703	oxime, mono- (α)	....	$C_6H_5CO \cdot C(:NOH) \cdot$ C_6H_5	VII-757	225.25
704	oxime, mono- (β)	....	$C_{14}H_{11}O_2N \cdot \frac{1}{2}C_6H_6$	VII-758	264.31
705	oxime, mono- (β)	....	$C_{14}H_{11}O_2N$	VII-758	225.25
706	osazone (α) (*syn*)	....	$(C_6H_5C)_2 \cdot$ $(N \cdot NHC_6H_5)_2$	XV-173	390.49
707	osazone (β) (*anti*)	....	$(C_6H_5C)_2 \cdot$ $(N \cdot NHC_6H_5)_2$	XV-174	390.49
708	**Benzimidazole** (*o*)	....	$C_6H_4 \cdot N:CH \cdot NH$	XXIII-131	118.14
709	**Benzimidazolone** (*o*)	*o*-phenylene urea	$C_6H_4 \cdot NH \cdot CO \cdot NH$	XXIV-116	134.14
710	**Benzohydroxamic acid**	*N*-benzoyl hydroxyl-amine	$C_6H_5CO \cdot NHOH$	IX-301	137.14
711	**Benzoic** acid	....	$C_6H_5 \cdot CO_2H$	IX-92	122.12
712	Na salt	sodium benzoate	$C_6H_5 \cdot CO_2Na \cdot H_2O$	IX-107	162.12
713	amide	benzamide	$C_6H_5CO \cdot NH_2$	IX-195	121.14
714	anhydride	....	$(C_6H_5CO)_2O$	IX-164	226.23
715	anilide	benzanilide; *N*-Ph-benzamide	$C_6H_5NH \cdot COC_6H_5$	XII-262	197.24
716	nitrile	phenyl cyanide	$C_6H_5 \cdot CN$	IX-275	103.12
717	**Benzoin** (*dl*)	....	$C_6H_5 \cdot CHOH \cdot CO \cdot C_6H_5$	VIII-166	212.25
718	acetate	acetyl benzoin	$C_6H_5CO \cdot CH(C_6H_5) \cdot$ $O_2C \cdot CH_3$	VIII-174	254.29
719	ethyl ether	....	$C_6H_5CH(OC_2H_5) \cdot$ $CO \cdot C_6H_5$	VIII-174	240.30
720	oxime (α)	cupron	$C_{13}H_{12}O:C:NOH$	VIII-175	227.27
721	oxime (β)	....	$C_{13}H_{12}O:C:NOH$	VIII-175	227.27
722	phenylhydrazone (α)	....	$C_6H_5NHN:C(C_6H_5) \cdot$ $CHOHC_6H_5$	XV-200	302.38
723	phenylhydrazone (β)	....	$C_6H_5NHN:C(C_6H_5) \cdot$ $CHOHC_6H_5$	XV-200	302.38
724	**Benzophenone**	diphenyl ketone	$(C_6H_5)_2CO$	VII-410	182.22
725	anil	....	$(C_6H_5)_2C:N \cdot C_6H_5$	XII-201	257.34
726	chloride	diPh-diCl-methane	$(C_6H_5)_2CCl_2$	V-590	237.13
727	dicarboxylic acid (2,2′)	....	$(HO_2C \cdot C_6H_4)_2:CO$	X-881	270.24
728	oxime	diphenyl ketoxime	$(C_6H_5)_2C:NOH$	VII-416	197.24
729	phenylhydrazone	....	$C_6H_5NHN:C(C_6H_5)_2$	XV-148	272.35
730	**Benzo**-phosphinic acid (*p*)	....	$HO_2C \cdot C_6H_4 \cdot PO(OH)_2$	XVI-820	202.10

Table 7-4 (*Continued*)
PHYSICAL CONSTANTS OF ORGANIC COMPOUNDS

No.	Crystalline Form and Color	Specific Gravity	Melting Point °C.	Boiling Point °C.	Solubility in 100 Parts		
					Water	Alcohol	Ether
700	lf.	....	235-7 d.	....	i.; s. NaOH	$0.05^{17°}$	v. sl. s.
701	cr.	....	206-7 d.	....	sl. s. h.	$15.3^{17°}$	v. s.
702	nd./al.	....	anh. 164-6	....	i.	$>15.3^{17°}$	v. s.
703	lf./aq. al.	....	138-40	d. 200	i.	s. c.	s.
704	nd./bz.	....	70	....	v. sl. s.	v. s.	v. s.
705	....	....	113-4	....	....	....	....
706	yel. nd.	....	205-15	....	$1.7^{19°}$ act.	sl. s. c.	s.
707	nd.	....	225-35	....	$2.4^{19°}$ act.	sl. s.	sl. s.
708	rhb./al.	....	170	>360	sl. s.; s. a.	v. s.	sl. s.; s. alk.
709	lf./aq.	....	310-2	subl. >270	sl. s. h.	s.; sl. s. bz.	i. aq. a.
710	rhb.	....	131-2	expl.	$2.25^{6°}$	v. s.	sl. s.; i. bz.
711	mn. pr.	$1.316^{28°}_{4}$	122.4; subl. > 100	250.0	$0.21^{17.5°}$; $2.27^{5°}$	$46.6^{15°}$ abs. al.	$66^{15°}$
712	col. cr.	....	$-H_2O$, 120	....	$61^{25°}$; $77^{100°}$	$2.3^{25°}$; $8.3^{78°}$	....
713	col. pr.	1.341	130	290	$1.35^{25°}$	$17.0^{25°}$, abs. al.	sl. s.
714	rhb./et.	$1.199^{15°}_{4}$	42	360	i.; s. act.	s.; s. bz.	s.; s. chl.
715	lf./al.	$1.314^{0°}$	163	$117-9^{10mm}$	i.	$4^{30°}$, abs. al.	sl. s.
716	col. lq.	$1.001^{25°}_{4}$	−13.8	191.1	$1^{100°}$	∞	∞
717	mn.	....	133-7	$343-4^{768mm}$	v. sl. s. h.	s. h.; s. act.	sl. s.
718	pr./et. or al.	....	83	....	....	v. s.	v. s.
719	nd./lg.	....	62-3	....	v. s. bz.	v. s.	v. s.
720	pr./bz.	....	151-2	....	sl. s.	s.	s. NH₄OH
721	pr.	....	99	....	....	....	s.
722	col. nd.	....	155-8	....	i.	$2.2^{20°}$	....
723	col. nd.	....	106	....	....	$8.8^{20°}$	s.
724	col. rhb.	(α) $1.083^{540°}$ (β) $1.108^{230°}$	(α) 48.1 (β) 26.5 (γ) 45.8 (δ) −51	305.9	i.	$6.5^{15°}$; s. chl.	$15^{13°}$
725	yel. rhb./et.	....	117 (112)	356-8	s. bz., CS₂	sl. s.	sl. s.
726	col. lq.	$1.235^{19°}$	....	305 d.	d.	d.	s. bz.
727	cr.	....	150-5 d.	$-H_2O$ >160°	i.	s.	s.
728	cr./lg.	....	143-4	....	v. sl. s.	v. s. act.	v. s.
729	col. nd.	....	137-8	....	.. . ...	sl. s. h.	....
730	nd./aq.	....	>300	d. >300	s.	sl. s.	sl. s. HCl

Benzophenone oxide 6451

Table 7-4 (*Continued*)
PHYSICAL CONSTANTS OF ORGANIC COMPOUNDS

No.	Name	Synonym	Formula	Beil. Ref.	Formula Weight
731	**Benzo**-pinacoline (β)	phenyl-trityl ketone	$(C_6H_5)_3C \cdot CO \cdot C_6H_5$	VII-544	348.45
732	pinacone	....	$[(C_6H_5)_2C \cdot OH]_2$	VI-1058	366.46
733	purpurin 4B	ditolyl-bis-(azo-naphthionic acid)	$[C_{10}H_5(NH_2)(SO_3Na) \cdot N:N \cdot C_6H_3(CH_3) \cdot]_2$	*XVI-342	724.73
734	thiazylhydrazine (2)	....	$C_6H_4N:C(N_2H_3) \cdot S \cdot$ ⌞_____⌟	....	165.22
735	trichloride	phenyl chloroform	$C_6H_5CCl_3$	V-300	195.48
736	**Benzoxyl-2-di-methyl-amino-methyl-1-di-methyl-amino-butane** (2) HCl	alypin hydro-chloride	$[(CH_3)_2N \cdot CH_2]_2:C: (C_2H_5)(O_2C \cdot C_6H_5) \cdot$ HCl	IX-175	314.86
737	nitrate	alypin nitrate	$C_{16}H_{26}O_2N_2 \cdot HNO_3$	*IX-92	341.41
738	**Benzoyl** acetic acid	....	$C_6H_5 \cdot CO \cdot CH_2 \cdot CO_2H$	X-672	164.16
739	acetic ester	ethyl benzoyl-acetate	$C_6H_5 \cdot CO \cdot CH_2 \cdot CO_2 \cdot C_2H_5$	X-674	192.22
740	acetoacetic ester	....	$C_2H_3O \cdot CH(C_7H_5O) \cdot CO_2C_2H_5$	X-817	234.25
741	acetone	....	$C_6H_5 \cdot CO \cdot CH_2 \cdot CO \cdot CH_3$	VII-680	162.19
742	acetonitrile	ω-cyanoacetophenone	$C_6H_5 \cdot CO \cdot CH_2 \cdot CN$	X-680	145.16
743	acrylic acid (β)	....	$C_7H_5O \cdot CH:CH \cdot CO_2H \cdot H_2O$	X-726	194.19
744	alanine (*dl*)	....	$C_6H_5CO \cdot NH \cdot CH(CH_3) \cdot CO_2H$	IX-248	193.20
745	alanine (*l*)	....	$C_{10}H_{11}O_3N$	IX-248	193.20
746	aminobenzoic acid (*o*)	*N*-benzoyl-an-thranilic acid	$C_7H_5O \cdot NH \cdot C_6H_4 \cdot CO_2H$	XIV-340	241.25
747	aminobenzoic acid (*m*)	benzaminobenzoic acid	$C_7H_5O \cdot NH \cdot C_6H_4 \cdot CO_2H$	XIV-397	241.25
748	aminobenzoic acid (*p*)	....	$C_7H_5O \cdot NH \cdot C_6H_4 \cdot CO_2H$	XIV-433	241.25
749	auramine	....	$[(CH_3)_2N \cdot C_6H_4]_2C: N \cdot COC_6H_5$	XIV-95	371.49
750	benzoic acid (*o*)	....	$C_6H_5CO \cdot C_6H_4 \cdot CO_2H \cdot H_2O$	X-747	244.25
751	benzoic acid (*m*)	....	$C_6H_5CO \cdot C_6H_4 \cdot CO_2H$	X-752	226.23
752	benzoic acid (*p*)	....	$C_6H_5CO \cdot C_6H_4 \cdot CO_2H$	X-753	226.23
753	bromide	....	$C_6H_5 \cdot CO \cdot Br$	IX-195	185.03
754	carbinol	phenacyl alcohol	$C_6H_5CO \cdot CH_2OH$	VIII-90	136.15
755	chloride	....	$C_6H_5 \cdot CO \cdot Cl$	IX-182	140.57
756	cyanide	....	$C_6H_5 \cdot CO \cdot CN$	X-659	131.14
757	fluoride	....	$C_6H_5 \cdot CO \cdot F$	IX-181	124.12
758	formic acid	phenylglyoxylic acid	$C_6H_5 \cdot CO \cdot CO_2H$	X-654	150.14
759	glycolic acid	....	$C_7H_5O \cdot OCH_2 \cdot CO_2H$	IX-167	180.16
760	hydrazine	....	$C_6H_5CO \cdot NH \cdot NH_2$	IX-319	136.15
761	hydrogen peroxide	....	$C_6H_5CO_2 \cdot OH$	IX-178	138.12
762	iodide	....	$C_6H_5 \cdot CO \cdot I$	IX-195	232.02
763	lactic acid (*dl*)	....	$CH_3 \cdot CH(O_2C_7H_5) \cdot CO_2H$	IX-167	194.19
764	γ-(2-methylpiper-idino)-propanol HCl	neothesin	$C_{16}H_{23}O_2N \cdot HCl$	....	297.83

Benzo-flavone 4479
Benzo-hydroquinone 2322
Benzo-γ-pyrone 1490
Benzo-pyrrole 3864
Benzo-quinoline 4518

Benzo-quinone 5541-2
Benzo-resorcin 2321
Benzo-salin 4143
Benzosol 3482
Benzo-thiophene 5912

Benzo-triazole 252
Benzozone 73
Benzol 656
Benzoxazinone 645
Benzoyl-acetophenone **1781**

Table 7-4 (*Continued*)
PHYSICAL CONSTANTS OF ORGANIC COMPOUNDS

No.	Crystalline Form and Color	Specific Gravity	Melting Point °C.	Boiling Point °C.	Solubility in 100 Parts		
					Water	Alcohol	Ether
731	nd./al.	....	179–81	....	i.	v. sl. s. c.; 1.2 h.	s.
732	mn.	....	185–6 d.	....	....	2.5 h.	v. s.
733	cr.	....	....	....	s.	....	....
734	pa. yel. cr.	....	197–8	....	i.	s. h.	i.
735	col. lq.	$1.380^{14°}$	−4.75	220.7	i.; d.	s.; s. bz.	s.
736	cr./act.	....	169 d.	....	v. s.; s. act.	s.; s. chl.	i.
737	col. pd.	....	163	....	v. s.	v. s.	v. sl. s.
738	nd./bz.	....	103–4 d.	....	s. h.	s.	s.
739	col. lq.	$1.111^{\frac{25}{4}}$	....	165–7^{720mm}	v. sl. s.	∞	∞
740	oil	....	....	175–6^{12mm} (sl. d.)	....	....	....
741	pr.	$1.090^{\frac{60}{60}}$	60–1	260–2 (sl. d.)	s. h.	v. s.	v. s.
742	pr./aq.	....	80.5	....	sl. s. c.	s.	s.; v. s. bz.
743	cr.	....	65; anh. 99	....	sl. s. c.	s.	s.
744	lf./et.	....	165–6	....	0.4 c.	v. s.	v. sl. s.
745	pl./aq.	....	150–1	....	1.1$^{20°}$	v. s.	v. sl. s.
746	nd./al.	....	181–2	....	i.	s.	s.
747	pr./al.	....	248	....	sl. s.	s.	sl. s.
748	nd./al.	....	278	....	sl. s. h.	s.	s.; s. ac.
749	yel. nd./al.	....	179	....	....	0.16$^{16°}$	0.93$^{16°}$ bz.
750	tri./aq.	....	93–4; anh. 128	....	sl. s.	....	....
751	lf./aq. al.	....	161–2	subl.	sl. s. h.	s.	s.
752	mn.	....	194	subl.	v. sl. s. c.	s.	s.
753	lq.	$1.570^{\frac{20}{4}}$	5.5–7.5	218–9	d.	d. h.	∞
754	hex. pl./al.	1.013	85–6	118–20^{11mm}	sl. s. h.	v. s.	v. s.
755	col. lq.	$1.212^{\frac{20}{4}}$	−0.6	197.9	d.; s. bz.	d. h.	∞; s. CS_2
756	cr. pl.	....	33–4	208–10	i.	....	....
757	lq.	>1	....	161.5^{45mm}	sl. d. h.	d. h.	s.
758	pr.	....	65–6	148^{6mm}	s.	i. CS_2	s.
759	pr.	....	79	....	sl. s. c.; d. h.	s.	s.
760	pl./aq.	....	112.0–2.5	d.	s.	s.; sl. s. bz.	sl. s.
761	lf./lg.	....	41–3	97–100^{13mm}	i.	s.	s.
762	nd.	....	3	135^{25mm}	d. h.	d. h.	s.
763	pl.	....	112	....	0.25 c.; d. h.	v. s.	v. s.
764	wh. cr. pd.	....	171–3	....	100	s.; s. chl.	i.

Benzoyl-anisole 4070
Benzoyl-anthranilic acid (*N*) 746
Benzoyl-azide 653
Penzoyl-ecgonine, cf. alkd.

Benzoyl-eugenol 3220
Benzoyl-formaldoxime 4874
Benzoyl-glucose, cf. glcd.
Benzoyl-glycine 3664

Benzoyl-guaiacol 3482
Benzoyl-hydroquinone 2322
Benzoyl-hydroxylamine (*N*) 710
Benzoyl-naphthalene 5210–1

Table 7-4 (*Continued*)
PHYSICAL CONSTANTS OF ORGANIC COMPOUNDS

No.	Name	Synonym	Formula	Beil. Ref.	Formula Weight
	Benzoyl				
765	α-naphthylamine	benznaphthalide	$C_6H_5CO \cdot NH \cdot C_{10}H_7$	XII-1233	247.30
766	peroxide	....	$(C_6H_5CO)_2O_2$	IX-179	242.23
767	phenylhydrazine (α, α')	....	$C_6H_5N(COC_6H_5) \cdot NH_2$	XV-250	212.25
768	phenylhydrazine	(α, β)	$C_6H_5NH \cdot NHCOC_6H_5$	XV-255	212.25
769	phthalic acid (1;2,3)	....	$C_6H_5O \cdot C_6H_3 \cdot (CO_2H)_2 \cdot H_2O$	X-880	288.26
770	piperidine (*N*)	....	$C_7H_5O \cdot N \cdot CH_2(CH_2)_4$	XX-46	189.26
771	propionic acid (β)	....	$C_7H_5O \cdot CH_2 \cdot CH_2 \cdot CO_2H$	X-696	178.19
772	terephthalic acid	(1;2,5)	$C_{13}H_8O(CO_2H)_2$	X-881	270.24
773	2-thiohydantoin (1)	....	$CH_2 \cdot CO \cdot NH \cdot CS \cdot N \cdot C_7H_5O$	*XXIV-294	220.25
774	thiourea	....	$C_7H_5O \cdot NH \cdot CS \cdot NH_2$	IX-219	180.23
775	toluidide (*o*)	*N*-tolylbenzamide (*o*)	$C_7H_5O \cdot NH \cdot C_6H_4 \cdot CH_3$	XII-795	211.27
776	toluidide (*m*)	....	$C_7H_5O \cdot NH \cdot C_6H_4 \cdot CH_3$	XII-861	211.27
777	toluidide (*p*)	....	$C_7H_5O \cdot NH \cdot C_6H_4 \cdot CH_3$	XII-926	211.27
778	urea (*N*)	....	$C_6H_5CO \cdot NHCONH_2$	IX-215	164.17
779	**Benzoylene urea**	2,4-diketotetrahydro-quinazoline	$C_6H_4 \cdot NH \cdot CO \cdot NH \cdot CO$	XXIV-373	162.15
780	**Benzyl** acetamide (*N*)	acetyl-benzylamine	$CH_3CO \cdot NH \cdot CH_2C_6H_5$	XII-1044	149.19
781	acetate	....	$C_6H_5CH_2O \cdot COCH_3$	VI-435	150.18
782	alcohol	phenyl carbinol	$C_6H_5 \cdot CH_2OH$	VI-428	108.14
783	amine	ω-aminotoluene	$C_6H_5CH_2 \cdot NH_2$	XII-1013	107.16
784	aminophenoı (*p*)	....	$C_6H_5CH_2 \cdot NH \cdot C_6H_4OH$	XIII-448	199.25
785	aminophenol HCl	(*p*)	$C_{13}H_{13}ON \cdot HCl \cdot H_2O$	XIII-448	253.73
786	aniline	phenyl-benzylamine	$C_6H_5CH_2 \cdot NH \cdot C_6H_5$	XII-1023	183.26
787	arsonic acid	....	$C_6H_5CH_2 \cdot AsO(OH)_2$	XVI-872	216.07
788	azide	....	$C_6H_5 \cdot CH_2 \cdot N_3$	V-350	133.15
789	benzoate	....	$C_6H_5CO_2 \cdot CH_2 \cdot C_6H_5$	IX-121	212.25
790	benzoic acid (*o*)	....	$C_6H_5 \cdot CH_2 \cdot C_6H_4 \cdot CO_2H$	IX-676	212.25
791	benzoic acid (*m*)	....	$C_6H_5 \cdot CH_2 \cdot C_6H_4 \cdot CO_2H$	IX-676	212.25
792	benzoic acid (*p*)	....	$C_6H_5 \cdot CH_2 \cdot C_6H_4 \cdot CO_2H$	IX-677	212.25
793	bromide	ω-bromotoluene	$C_6H_5 \cdot CH_2 \cdot Br$	V-306	171.04
794	butyrate	....	$C_2H_5CH_2CO_2CH_2 \cdot C_6H_5$	VI-436	178.23
795	carbamate	....	$NH_2 \cdot CO_2 \cdot CH_2C_6H_5$	VI-437	151.17
796	chloride	ω-chlorotoluene	$C_6H_5 \cdot CH_2 \cdot Cl$	V-292	126.59
797	chloroacetate	....	$ClCH_2CO_2 \cdot CH_2C_6H_5$	VI-435	184.62
798	cinnamate	cinnamein	$C_8H_7 \cdot CO_2 \cdot C_7H_7$	IX-584	238.29
799	cyanamide	....	$C_6H_5 \cdot CH_2 \cdot NH \cdot CN$	XII-1051	132.17
800	cyanide	phenylacetonitrile	$C_6H_5 \cdot CH_2 \cdot CN$	IX-441	117.15
801	cyanurate (*iso*)	....	$(C_6H_5CH_2N \cdot CO)_3$	XXVI-255	399.45
802	diphenyl (*o*)	....	$C_6H_5CH_2 \cdot C_6H_4 \cdot C_6H_5$	V-708	244.34
803	diphenyl (*p*)	....	$C_6H_5CH_2 \cdot C_6H_4 \cdot C_6H_5$	V-708	244.34
804	diphenylamine	diphenyl-benzylamine	$(C_6H_5)_2N \cdot CH_2C_6H_5$	XII-1033	259.35
805	disulfide	dibenzyl-disulfide	$(C_6H_5 \cdot CH_2)_2S_2$	VI-465	246.40

Benzoyl-nitromethane 4594
Benzoyl-persulfide 1779
Benzoyl-pseudotropine, cf. alkd.
Benzoyl-pyrogallol 6224

Benzoyl-resoroin 2321
Benzoyl-salicin, cf. glcde.
Benzoyl-sulfimide 5579
Benzoyl-*pseudo*-tropine, cf. alkd.

Table 7-4 (*Continued*)
PHYSICAL CONSTANTS OF ORGANIC COMPOUNDS

No.	Crystalline Form and Color	Specific Gravity	Melting Point °C.	Boiling Point °C.	Solubility in 100 Parts		
					Water	Alcohol	Ether
765	nd./aq. al.	....	161–2	....	s. ac.	v. sl. s. abs.	....
766	rhb./et.	....	108 d.	expl.	i.	sl. s. c.; s. h.	s.; s. bz.
767	nd./aq.	....	70	....	sl. s. c.	v. s.	v. s.
768	pr./al.	....	168	....	sl. s. h.	s. h.; s. chl.	sl. s.
769	nd./aq.	....	$-H_2O$, 100	$-2H_2O > 100$	s. h.	s.	v. sl. s. bz.
770	col. tri.	....	48	320–1	i.	s.	s.
771	nd./aq.	....	116	sl. d.	s. h.	s.	s.; i. lg.
772	cr./al.	....	>290	....	i.	s.	s.
773	cr./al.	....	165 d.	....	....	....	....
774	pr./aq. al.	....	169–70	....	sl. s. c.	s.	i.
775	rhb.	$1.205^{15°}$	145–6	....	sl. s. h.	s.	....
776	mn./aq. al.	$1.170^{15°}$	....	....	....	$13^{15°}$ abs.	....
777	rhb./al.	$1.202^{15°}$	157–8	232	i.	$4.2^{18°}$ abs.	....
778	nd./al.	....	214–5	....	s. h.	1, c,; 4, h.	i.
779	col. nd.	....	353–4	....	$0.013^{23°}$	v. sl. s.	s. alk.
780	lf./et.	....	60–1	>300	sl. s. pet.	v. s.	v. s.
781	col. lq.	$1.057^{17°}$	−51.5	213.5	i.	∞	∞
782	col. lq.	$1.045\frac{20°}{4}$	−15.2	205.4	$4^{17°}$	∞; ∞ chl.	∞
783	lq.	$0.982\frac{20°}{4}$	....	184.5	∞	∞	∞
784	lf.	....	89	....	v. sl. s.	v. s.	v. s. bz.; s. NaOH
785	pr./aq.	....	172 (anh.)	....	s. h.	s.	i.
786	mn. pr.	$1.065\frac{25°}{25}$	37–8	306^{759mm}	i.	s. h. Me al.	s.; s. chl.
787	nd.	....	167	d.	$0.34^{23°}$; $3.5^{97°}$	$0.87^{23°}$; $5.9^{70°}$	....
788	oil	$1.066^{25°}$	....	108^{23mm}	i.	∞	∞
789	nd. or lf.	1.122 (lq.)	21	323–4	i.; ∞ chl.	∞; s. oil	∞
790	nd./aq. al.	....	114–7	subl.	sl. s.	s.; s. chl.	s.; s. bz.
791	lf./aq. al.	....	107–8	subl.	sl. s.	s.; s. chl.	s.
792	lf./aq. al.	....	155–7	subl.	sl. s.	s.; s. chl.	s.
793	col. lq.	$1.443^{17°}$	−4	198–9	i.; sl. d.	∞	∞
794	col. lq.	$1.016\frac{16°}{18}$	....	238–40	i.	v. s.	v. s.
795	cr.	....	86	d.	v. sl. s.	s.	s.
796	col. lq.	$1.100\frac{20°}{20}$	−39.2	179.4	i.	∞; ∞ chl.	∞
797	col. oil	$1.222\frac{4°}{4}$	....	147.5^{9mm}	....	....	....
798	pr.	....	39	244^{25mm}	i.	s.	s.
799	pl./et.	....	43	....	i.	v. s.	v. s.
800	col. lq.	$1.018\frac{20°}{4}$	−23.8	233–4	i.	∞	∞
801	nd./al.	....	157	>320	i. c.	s.	sl. s.
802	mm.	....	54	$283–7^{110mm}$	v. s. bz.	s.	v. s.
803	lf.	$1.171\frac{0°}{4}$	85	$285–6^{110mm}$	v. s. bz.	sl. s.	v. s.
804	nd.	....	86–7	....	v. sl. s.	s. h.	s.
805	lf./al.	....	71–2	d. >270	v. sl. s.	s. h.	s.; s. bz.

Benzyl, cf. also dibenzyl.
Benzyl-acetic acid 3704
Benzyl-acetoacetic ester 2983
Benzyl-carbamide 844

Benzyl-carbinol 5187
Benzyl-cellosolve 3449
Benzyl citrate 6085

Table 7-4 (*Continued*)
PHYSICAL CONSTANTS OF ORGANIC COMPOUNDS

No.	Name	Synonym	Formula	Beil. Ref.	Formula Weight
806	**Benzyl** ether	dibenzyl ether	$(C_6H_5 \cdot CH_2)_2O$	VI-434	198.27
807	formate	. . . .	$HCO_2 \cdot CH_2 \cdot C_6H_5$	VI-435	136.15
808	hydrazine	. . . .	$C_6H_5 \cdot CH_2 \cdot NH \cdot NH_2$	XV-531	122.17
809	hydroxybenzoate (*p*)	. . . .	$HO \cdot C_6H_4 \cdot CO_2 \cdot C_7H_7$	. . . .	228.25
810	hydroxylamine (*α*)	. . . .	$C_6H_5 \cdot CH_2 \cdot O \cdot NH_2$	VI-440	123.16
811	hydroxylamine (*β*)	. . . .	$C_6H_5 \cdot CH_2 \cdot NH \cdot OH$	XV-17	123.16
812	imino-(4-methoxy-phenyl) methane	anisalbenzylamine	$C_6H_5CH_2N:CH \cdot C_6H_4 \cdot OCH_3$	XII-1043	225.29
813	iodide	*ω*-iodotoluene	$C_6H_5 \cdot CH_2 \cdot I$	V-314	218.04
814	lactate	. . . .	$CH_3CHOHCO_2 \cdot C_7H_7$	**VI-420	180.21
815	malonic acid	. . . .	$C_6H_5CH_2 \cdot CH(CO_2H)_2$	IX-868	194.19
816	mercaptan	. . . .	$C_6H_5 \cdot CH_2 \cdot SH$	VI-453	124.21
817	naphthalene (*α*)	Ph-naphthylmethane	$C_6H_5 \cdot CH_2 \cdot C_{10}H_7$	V-689	218.30
818	naphthalene (*β*)	Ph-naphthylmethane	$C_6H_5 \cdot CH_2 \cdot C_{10}H_7$	V-690	218.30
819	*α*-naphthyl ether	. . . .	$C_6H_5 \cdot CH_2 \cdot O \cdot C_{10}H_7$	. . . .	234.30
820	*β*-naphthyl ether	. . . .	$C_6H_5 \cdot CH_2 \cdot O \cdot C_{10}H_7$	VI-642	234.30
821	*α*-naphthyl ketone	. . . .	$C_6H_5 \cdot CH_2 \cdot CO \cdot C_{10}H_7$	VII-512	246.31
822	phenanthracene (9)	. . . .	$C_7H_7 \cdot C_6H_3 : C_2H_2 : C_6H_4$	*V-359	268.36
823	phenol (*o*)	HO-diPh-methane	$C_6H_5 \cdot CH_2 \cdot C_6H_4 \cdot OH$	VI-675	184.24
824	phenol (*p*)	4-hydroxy-ditane	$C_6H_5 \cdot CH_2 \cdot C_6H_4 \cdot OH$	VI-675	184.24
825	phenyl carbamate (*p*)	Butolan	$C_7H_7 \cdot C_6H_4O_2C \cdot NH_2$	*VI-325	227.27
826	phenylhydrazine (*α,α*)	. . . .	$(C_7H_7)(C_6H_5)N \cdot NH_2$	XV-532	198.27
827	phenylhydrazine HCl	(*α,α*)	$C_{13}H_{12}N \cdot NH_2 \cdot HCl$	XV-533	234.73
828	phenylnitrosamine	nitroso-benzyl-aniline	$C_6H_5CH_2 \cdot N(NO) \cdot C_6H_5$	XII-1071	212.25
829	propionate	. . . .	$C_2H_5CO_2 \cdot CH_2C_6H_5$	VI-436	164.21
830	pyridine (2)(*α*)	. . . .	$C_6H_5 \cdot CH_2 \cdot C_5H_4N$	XX-425	169.23
831	pyridine (3)(*β*)	. . . .	$C_6H_5 \cdot CH_2 \cdot C_5H_4N$	XX-426	169.23
832	pyrrole (*N*)	. . . .	$C_4H_4N \cdot CH_2 \cdot C_6H_5$	XX-164	157.22
833	salicylate	. . . .	$HOC_6H_4CO_2 \cdot CH_2C_6H_5$	X-80	228.25
834	succinate (mono)	. . . .	$C_7H_7O_2C(CH_2)_2 \cdot CO_2H$	VI-436	208.22
835	sulfone	dibenzyl sulfone	$(C_6H_5 \cdot CH_2)_2SO_2$	VI-456	246.33
836	sulfoxide	. . . .	$(C_6H_5 \cdot CH_2)_2SO$	VI-456	230.33
837	tartronic acid	. . . .	$C_7H_7 \cdot C(OH)(CO_2H)_2$	X-515	210.19
838	*iso*-thiocyanate	benzyl mustard oil	$C_6H_5 \cdot CH_2 \cdot N:CS$	XII-1059	149.22
839	thiourea	. . . .	$C_6H_5CH_2 \cdot NH \cdot CS \cdot NH_2$	XII-1051	166.25
840	*iso*-thiourea	. . . .	$C_7H_7 \cdot S \cdot C(:NH) \cdot NH_2$	VI-461	166.25
841	*iso*-thiourea HCl	. . . .	$C_8H_{10}N_2S \cdot HCl$	VI-461	202.71
842	toluene (*m*)	phenyl-tolyl-methane	$C_7H_7 \cdot C_6H_4 \cdot CH_3$	V-607	182.27
843	toluene (*p*)	. . . .	$C_7H_7 \cdot C_6H_4 \cdot CH_3$	V-607	182.27
844	urea	benzyl carbamide	$C_6H_5CH_2 \cdot NH \cdot CO \cdot NH_2$	XII-1050	150.18
845	**Beryllium** diethyl	. . . .	$(C_2H_5)_2Be$	IV-645	67.14
846	**Betaine**†	trimethyl glycine; oxyneurine	$(CH_3)_3N \cdot CH_2 \cdot CO \cdot O$ ⌐⌐	IV-346	117.15
347	hydrochloride	acidol; lycine	$C_5H_{11}O_2N \cdot HCl$	IV-348	153.61
848	**Betulin**	. . . .	$C_{30}H_{50}O_2$	**VI-937	442.73

† See also alkaloids table.
Benzyl-ethyl-aniline 2984
Benzyl-ethyl ether 2986
Benzyl fumarate 1785
Benzyl maleate 1788

Benzyl-methyl ether 4145
Benzyl mustard oil 838
Benzyl phthalate 1789
Benzyl succinate 1790
Benzyl sulfide 1791

Table 7-4 (*Continued*)
PHYSICAL CONSTANTS OF ORGANIC COMPOUNDS

No.	Crystalline Form and Color	Specific Gravity	Melting Point °C.	Boiling Point °C.	Solubility in 100 Parts		
					Water	Alcohol	Ether
806	lq.	$1.036^{16°}$	1.5-3.5	295-8	i.	s. h.	s.
807	col. lq.	$1.081^{23°}$	3.6	$202\text{-}3^{747mm}$	i.	s.	∞
808	col. oil	....	26	135^{29mm}	∞	∞	∞
809	wh. pd.	....	110-2	....	i.	v. s.	s.
810	oil	....	....	$118\text{-}9^{30mm}$	....	....	....
811	nd./pet.	....	57-8	123^{50mm}	s.	....	....
812	col. cr./pet.	....	89-91	....	....	....	....
813	cr.	$1.734^{25°}$	24.1	93^{10mm}	i.	s.	s.
814	col. lq.	....	....	$136\text{-}8^{10mm}$	....	....	....
815	cr./et.	....	121	d. 180	s.	s.	s.; s. li. bz.
816	col. lq.	$1.058^{20°}$	....	194-5	....	....	....
817	lf./al.	$1.165^{0°}$	58-9	350	s. chl., bz.	1.6 c., 3 h.	50 c.
818	mn. pr.	$1.176^{0°}$	35.5	350	....	v. s. h.	v. s. bz.
819	col. cr.	....	76-7	....	i.	s. h.	s.
820	lf./al.	....	98-9	....	s. chl., bz.	s.	s.
821	pl./al.	....	57	....	i.	s.	s.
822	lf./al.	....	91-2	....	i.	....	....
823	col. oil	....	52(21)	312	v. s. h.	s.	s.; s. alk.
824	nd./al.	....	84	320-2	s. h.	s.	s.; s. alk.
825	wh. cr./al.	....	144	....	sl. s.	s. h.	s. bz.
826	col. oil	....	....	$216\text{-}8^{38mm}$	....	....	....
827	nd./aq.	....	167-70	....	....	....	....
828	yel. nd./al	....	58	....	s. chl., lg.	s.	s.
829	lq.	$1.036^{16.5°}_{17.5}$	....	220-2	i.	....	....
830	lq.	$1.054^{20°}_{0}$	....	276^{742mm}	i.	v. s.	v. s.
831	lq.	$1.061^{20°}_{0}$	34	287^{742mm}	i.	v. s.	v. s.
832	....	....	<37.5	246-7	i.	s.	s.
833	col. oil	$1.176^{19.5°}_{15}$	23-5	208^{26mm}	v. sl. s.	∞	∞
834	cr.	....	59	....	i.	s.	s.
835	nd./al. + bz.	....	150-1	290(sl. d.)	i.	sl. s.	s. bz.
836	lf./al.	....	133-4	d. 210	s. h.	v. s.	v. s.
837	pr.	....	147 d.	....	s.	s.	s.
838	lq.	$1.125^{15°}_{4}$	....	243	i.	....	s.
839	pr./aq.	....	162-4	....	i. c.	1.5 c.	....
840	nd./bz.	....	103-4 d.	....	s. chl., bz.	s.	s.
841	pl./HCl	....	176(148)	....	s.	s.	i.
842	lq.	$0.997^{17.5°}$	....	275^{747mm}	....	s.	s.
843	lq.	$0.994^{18°}$	−30	285-6	v. s. chl.	v. s.	v. s.
844	nd./al.	....	147-8	d. 200	$1.74^{5°}$	$3.1^{23°}$, act.	$0.05^{22.5°}$
845	col. lq.	....	−11-3	185-8	d.	d.	....
846	pr./al.	....	293 d.	....	$157^{19°}$	$8.6^{18°}$	i.
847	col. mn./al.	....	236-7 d.	....	60	0.6	i.; i. chl.
848	nd./al.	....	258	subl. sl. d.	i.; i. CS_2	0.7, c.; 4.27, h.	0.4, c.; 3, h.

Table 7-4 (*Continued*)
PHYSICAL CONSTANTS OF ORGANIC COMPOUNDS

No.	Name	Synonym	Formula	Beil. Ref.	Formula Weight
849	**Betulinic acid**	....	$C_{30}H_{48}O_3$	**VI-939	456.72
850	**Bilifuscin**	bile pigment	$C_{16}H_{10}O_4N_2$	....	294.27
851	**Bilirubin**	principal bile pigment	$(C_{16}H_{18}O_3N_2)_2$	....	572.67
852	**Biliverdin**	bile pigment	$C_{32}H_{36}O_8N_4$	....	604.67
853	**Biotin**	vitamin H	$C_{10}H_{16}O_3N_2S$	....	244.31
854	methyl ester	....	$C_{11}H_{18}O_3N_2S$	....	258.34
855	**Bismuth** triethyl	triethyl bismuthine	$(C_2H_5)_3Bi$	IV-622	296.17
856	trimethyl	trimethyl bismuthine	$(CH_3)_3Bi$	IV-622	254.09
857	triphenyl	triphenyl bismuthine	$(C_6H_5)_3Bi$	XVI-898	440.30
858	triphenyl dichloride	triphenylbismuthine dichloride	$(C_6H_5)_3Bi:Cl_2$	XVI-899	511.21
859	**Biuret**	allophanamide	$NH(CONH_2)_2$†	III-70	103.08
860	**Borneol** (*d* or *l*)	....	$C_{10}H_{17}OH$	VI-75	154.25
861	**Bornyl** acetate (*d*)	....	$C_{10}H_{17}O\cdot CO\cdot CH_3$	VI-78	196.29
862	amine (*d*)	....	$C_{10}H_{17}\cdot NH_2$	XII-45	153.27
863	α-bromo-*iso*-valerate (*d*)	brovalol; valisan; eubornyl	$(CH_3)_2CH\cdot CHBr\cdot CO_2C_{10}H_{17}$	VI-79	317.27
864	chloride‡	*iso*-bornyl chloride	$C_{10}H_{17}Cl$	V-97	172.70
865	dibromo-dihydrocinnamate	adamon	$C_6H_5(CHBr)_2CO_2\cdot C_{10}H_{17}$	*IX-202	444.22
866	salicylate (*d*)	salit	$HO\cdot C_6H_4\cdot CO_2\cdot C_{10}H_{17}$	X-76	274.36
867	*iso*-valerate (*d*)	bornyval	$(CH_3)_2CHCH_2CO_2\cdot C_{10}H_{17}$	VI-79	238.37
868	*iso*-valeryl-glycollate	neobornyval	$(CH_3)_2CHCH_2CO_2\cdot CH_2\cdot CO_2\cdot C_{10}H_{17}$	....	296.41
869	*iso*-**Bornyl**-*n*-valerate (*d*)	gynoval	$CH_3(CH_2)_3CO_2\cdot C_{10}H_{17}$	VI-88	238.37
870	**Bornylene** (*l*)	....	$C_{10}H_{16}$	V-155	136.24
871	**Boron** triethyl	....	$(C_2H_5)_3B$	IV-641	98.00
872	trimethyl	....	$(CH_3)_3B$	IV-641	55.92
873	**Brassidic acid**	....	$C_{21}H_{41}\cdot CO_2H$	II-474	338.58
874	**Brazilein**	....	$C_{16}H_{12}O_5\cdot H_2O$	XVIII-194	302.29
875	**Brazilin**	....	$C_{16}H_{14}O_5$	XVII-194	286.29
876	**Bromal**	tribromo-acetaldehyde	$Br_3C\cdot CH:O$	I-626	280.76
877	hydrate	....	$Br_3C\cdot CH(OH)_2$	I-626	298.77
878	**Bromo**-acenapththene (5)	....	$C_{12}H_9Br$	V-587	233.11
879	acetamide (*N*)	acetbromamide	$CH_3\cdot CO\cdot NHBr\cdot H_2O$	II-181	155.98
880	acetanilide (*o*)	....	$Br\cdot C_6H_4\cdot NH\cdot CO\cdot CH_3$	XII-632	214.07
881	acetanilide (*m*)	....	$Br\cdot C_6H_4\cdot NH\cdot CO\cdot CH_3$	XII-634	214.07
882	acetanilide (*p*)	asepsin	$Br\cdot C_6H_4\cdot NH\cdot CO\cdot CH_3$	XII-642	214.07
883	acetic acid	....	$Br\cdot CH_2\cdot CO_2H$	II-213	138.95
884	acetoacetanilide (α)	....	$CH_3CO\cdot CHBr\cdot CO\cdot NH\cdot C_6H_5$	XII-519	256.11
885	aceto-β-naphthone (ω)	....	$C_{10}H_7\cdot CO\cdot CH_2Br$	**VII-338	249.11
886	acetone	....	$Br\cdot CH_2\cdot CO\cdot CH_3$	I-657	136.98
887	acetophenone (ω)	phenacyl bromide	$C_6H_5CO\cdot CH_2\cdot Br$	VII-283	199.05

† Cryst. +$1H_2O$/aq.; −H_2O, 110°. ‡ See also pinene hydrochloride.
Bi-, cf. also di.
Bile acid 1479
Bile pigments 850-2
Bilineurine 1488

Biloptin 2384
Bindschedler's green 5833
Biphenin 1722
Biphenyl 2695

Bismarck brown 6058
Bistriazo-ethane 1761
Biuret-amidine 2081
Biuret base 6209

Bixin 4933
Blue cross gas 2712
Bonoform 5739
Bordeaux DH 3732

Table 7-4 (*Continued*)
PHYSICAL CONSTANTS OF ORGANIC COMPOUNDS

No.	Crystalline Form and Color	Specific Gravity	Melting Point °C.	Boiling Point °C.	Solubility in 100 Parts		
					Water	Alcohol	Ether
849	wh. pd.	....	295-7	....	v. sl. s.	s.	....
850	br. pd.	....	183	....	v. sl. s.	s.; v. s. alk.	v. sl. s.
851	or. pd.	....	darkens with d.	....	i.; s. alk.; s. bz.	v. sl. s.; s. CS_2	v. sl. s.; s. chl.
852	gn. pd.	....	....	....	i.; s. alk.	s.; s. bz.; s. CS_2	v. sl. s.; i. chl.
853	nd.	....	230-2	....	s. NaOH		....
854	cr./al. et.	....	166-7	....	i.	s.; s. Me al.	s.
855	col. oil	1.82	....	*107^{79mm}	i.	v. s.	v. s.
856	col. oil	2.30^{18o}	....	*110	i.	v. s.	v. s.
857	mn./al.	$1.952\frac{15}{4}^{o}$	77-8	242^{14mm}	i.; v. s. chl.	v. sl. s.	s.; s. act.
858	pr./chl. al.	....	141.5	....	v. s. bz.	v. sl. s.	v. sl. s.
859	nd./al.	....	192-3 d.	....	1.25^{0o}; 45^{106o}	s.	....
860	col. cr.	$1.011\frac{20}{4}^{o}$	208.6	215.0	v. sl. s.	v. s.; s. bz.	v. s.
861	rhb./pet.	0.991^{15o}	29	226-7	i.	s.	s.
862	col. cr.	....	163	200	i.	v. s.	v. s.
863	col. oil	1.18	....	163	i.	s.	s.; s. chl.
864	col. cr.	....	161.5	....	i.	v. sl. s. c.	v. s.
865	col. cr./al.	....	75±	....	i.; s. h. chl.	s. h.	s. h.
866	cr.	....	44-5	$230-5^{50mm}$	i.; ∞ oil	∞; ∞ chl.	∞
867	col. lq.	0.951^{20o}	....	255-60	i.	s.	s.
868	col. oil	1.03	....	283-5 d.	i.; s. oil	v. s.; v. s. bz.	v. s.
869	col. lq.	$0.953\frac{18}{4}^{o}$	....	$143-5^{18mm}$	v. sl. s.; s. act.	s.; s. chl.	s.; s. bz.
870	col. cr.	....	113	146^{740mm}	i.	s. Me al.	s. toluene
871	col. lq.	0.696^{23o}	-92.9	95	v. sl. s.	d. by air	....
872	col. gas	....	-160	-20	v. sl. s.	v. s.	v. s.
873	lf./al.	0.859^{57o}	61-2	282^{30mm}	0.7^{25o}	v. sl. s. c.	sl. s.
874	rhd. red	....	softens 130-40	....	sl. s. h.	s. alk.	s. d. H_2SO_4
875	cr./abs. al.	....	250	....	s.; s. alk.	s.	s.
876	yel. lq.	$2.665\frac{25}{4}^{o}$	....	174 d.	forms hydr.	s.	s.
877	cr.	$2.566\frac{40}{4}^{o}$	53.5	....	s.	s.; s. chl.	s.; s. glyc.
878	yel. cr.	1.437^{55o}	51-2	336.4	i.	s.	v. s.
879	pl. + H_2O	....	70-80	m. anh. 108	s.	s.	v. s.
880	nd./al.	....	99	....	i.	s.	s.
881	nd./aq. al.	....	87.5	....	....	v. s.	v. s.
882	mn. pr.	1.717	166-7	....	sl. s. h.	s.; s. bz.	s. chl.
883	pl. or rhb.	$1.934\frac{50}{50}^{o}$	49-50	208	∞^{25o}	∞^{25o}	∞^{25o}
884	col. lf./al.	....	138 d.	....	v. sl. s.; s. alk.	v. s.	sl. s.; sl. s. chl.
885	red fl.	....	84	....	i.	sl. s.	v. s.
886	lq.	1.634^{23o}	-54	136.5^{725mm}	v. sl. s.	v. s.	v. s.
887	rhb.	$1.647\frac{20}{4}^{o}$	50	119	i.; s. bz.	v. s.	v. s.

* Explodes when heated in air.

Table 7-4 (*Continued*)
PHYSICAL CONSTANTS OF ORGANIC COMPOUNDS

No.	Name	Synonym	Formula	Beil. Ref.	Formula Weight
888	**Bromo** acetophe-none (*p*)	Me-*p*-Br-phenyl ketone	$Br \cdot C_6H_4 \cdot CO \cdot CH_3$	VII-283	199.05
889	acetoluidide	3;1,4	$Br \cdot C_6H_3(CH_3) \cdot NH \cdot CO \cdot CH_3$	XII-991	228.10
890	acetyl bromide	....	$Br \cdot CH_2 \cdot CO \cdot Br$	II-215	201.86
891	acetylene	bromo-ethyne	$CH \vdots C \cdot Br$(or $C \vdots CHBr$)	I-245	104.94
892	allyl alcohol (*β*)	....	$CH_2 \vdots CBr \cdot CH_2OH$	I-439	136.98
893	aniline (*o*)	....	$Br \cdot C_6H_4 \cdot NH_2$	XII-631	172.03
894	aniline (*m*)	....	$Br \cdot C_6H_4 \cdot NH_2$	XII-633	172.03
895	aniline (*p*)	....	$Br \cdot C_6H_4 \cdot NH_2$	XII-636	172.03
896	aniline (*p*) HCl	....	$Br \cdot C_6H_4 \cdot NH_2 \cdot HCl$	XII-637	208.49
897	anisole (*o*)	....	$CH_3O \cdot C_6H_4 \cdot Br$	VI-197	187.04
898	anisole (*p*)	....	$CH_3O \cdot C_6H_4 \cdot Br$	VI-199	187.04
899	anthraquinone (1)	....	$C_6H_4 \vdots (CO)_2 \vdots C_6H_3Br$	VII-789	287.12
900	anthraquinone (2)	....	$C_6H_4 \vdots (CO)_2 \vdots C_6H_3Br$	VII-789	287.12
901	antipyrine (*p*)	bromopyrine	$BrC_6H_4 \cdot C_5H_7ON_2$	XXIV-33	267.13
902	benzene	phenyl bromide	C_6H_5Br	V-206	157.02
903	benzene sulfonic acid (*o*)	....	$Br \cdot C_6H_4 \cdot SO_3H$	XI-56	237.08
904	benzene sulfonic acid (*p*)	....	$Br \cdot C_6H_4 \cdot SO_3H$	XI-57	237.08
905	benzene sulfonyl chloride (*p*)	....	$Br \cdot C_6H_4 \cdot SO_2Cl$	XI-57	255.52
906	benzoic acid (*o*)	....	$Br \cdot C_6H_4 \cdot CO_2H$	IX-347	201.03
907	benzoic acid (*m*)	....	$Br \cdot C_6H_4 \cdot CO_2H$	IX-349	201.03
908	benzoic acid (*p*)	....	$Br \cdot C_6H_4 \cdot CO_2H$	IX-351	201.03
909	benzonitrile (*p*)	*p*-Br-phenyl cyanide	$Br \cdot C_6H_4 \cdot CN$	IX-354	182.03
910	benzoyl bromide (*m*)	....	$Br \cdot C_6H_4 \cdot CO \cdot Br$	....	263.93
911	benzoyl chloride (*p*)	....	$Br \cdot C_6H_4 \cdot CO \cdot Cl$	IX-353	219.47
912	benzyl bromide (*p*)	....	$Br \cdot C_6H_4 \cdot CH_2Br$	V-308	249.94
913	benzyl chloride (*o*)	*ω*-Cl-2-Br-toluene	$Br \cdot C_6H_4 \cdot CH_2Cl$	*V-155	205.49
914	benzyl chloride (*p*)	....	$Br \cdot C_6H_4 \cdot CH_2Cl$	V-307	205.49
915	*iso*-butyl benzene (*p*)	....	$Br \cdot C_6H_4 \cdot C_4H_9$	V-415	213.12
916	*tert*-butyl phenol	(4;2,1)	$(CH_3)_3C \cdot C_6H_3(Br)OH$	VI-525	229.12
917	butyric acid (*α*)(*dl*)	....	$C_2H_5 \cdot CHBr \cdot CO_2H$	II-281	167.01
918	*iso*-butyric acid (*α*)	....	$(CH_3)_2CBr \cdot CO_2H$	II-295	167.01
919	camphor (3)(*d*)	*α*-bromocamphor	$C_8H_{14} \cdot CO \cdot CHBr$	VII-120	231.15
920	*n*-caproic acid (*α*)	....	$CH_3(CH_2)_3 \cdot CHBr \cdot CO_2H$	II-325	195.06
921	cinnamic acid (*α*) (*cis*)	bromo-allocin-namic acid	$C_6H_5 \cdot CH \vdots CBr \cdot CO_2H$	IX-600	227.06
922	cinnamic acid (*α*) (*trans*)	....	$C_6H_5 \cdot CH \vdots CBr \cdot CO_2H$	IX-599	227.06
923	cinnamic acid (*β*) (*cis*)	bromo-allocin-namic acid	$C_6H_5 \cdot CBr \vdots CH \cdot CO_2H$	IX-598	227.06

Table 7-4 (Continued)
PHYSICAL CONSTANTS OF ORGANIC COMPOUNDS

No.	Crystalline Form and Color	Specific Gravity	Melting Point °C.	Boiling Point °C.	Solubility in 100 Parts		
					Water	Alcohol	Ether
888	col. lf./al.	....	50–1	255^{736mm}	i.; s. bz.	s.; s. CS_2	v. s.; s. ac.
889	nd./bz.	....	116–7	....	....	....	....
890	lq.	$2.317^{21.5°}_{21.5}$	....	149–50	d.	d.	....
891	poisonous gas	....	....	–2	5000–6000 cc.$^{15°}$, CH_2CBr_2	s.	....
892	lq.	$1.6^{15°}$	....	153–4			
893	cr.	....	31–2	229	i.	v. s.	s.
894	cr.	$1.579^{20-4°}_{4}$	18.5	251	i.	s.	s.
895	rhb.	$1.80^{15-20°}$	63–4	....	i. c.	v. s.	v. s.
896	mn. pr.	....	....	....	s.	s.	i.
897	oil	....	....	218–21	....	s.	....
898	cr.	$1.4949°$	13–14	215	....	s.	....
899	yel. nd./bz.	....	188	subl.	s. H_2SO_4	....	....
900	cr./am. al.	....	204–5	....	....	....	....
901	col. mn./aq.	....	122	$300^{9mm} \pm$	s. h.	s.; s. chl.	sl. s.
902	col. lq.	$1.495^{20°}_{4}$	–30.6	156.2	i.	s.	∞; ∞ chl.
903	delq. nd.	....	....	....	v. s.	v. s.	i.
904	delq. nd.	....	102–3	155^{25mm}	s.	s.	i.
905	tri. pr./et.	....	75–6	153^{15mm}	d. h.	d. h.	....
906	nd./aq.	....	148–50	subl.	$0.18^{25°}$	s.	s.
907	nd.	....	156–8	>280	$0.04^{25°}$	s.	s.
908	mn. pr.	....	251–3	....	sl. s. h.	s.	s.
909	nd./aq. or al.	....	112–3	....	s. h.	s.	v. s.
910	col. lq.	....	....	$118–22^{8mm}$	....	....	....
911	nd./pet.	....	42	245–7 sl. d.	d.; v. s. bz.	v. s. lg.	s.
912	nd./al.	....	61	....	v. sl. s.	s. h.	s.
913	col. lq.	....	....	$124–6^{20mm}$	i.	v. s.	v. s.
914	nd./al.	....	41	236	i.	v. s. h.	v. s.
915	lq.	....	<–18	$232–3^{739mm}$	....	....	....
916	lq.	$1.338^{25°}_{25}$	<–20*	$109–29^{5mm}$	i.	∞; ∞ Me al.	∞ bz., act.
917	col. oil	$1.567^{20°}_{20}$	–4	$127–8^{25mm}$	6.6	s.	s.
918	pl.	$1.523^{60°}_{60}$	48	198–200	forms oil	s.	s.
919	cr.	$1.449^{20°}_{4}$	77–8	274(sl. d.)	i.	$20^{26°}$	v. s.
920	lq.	....	....	$128–31^{10mm}$	s.	s.	s.
921	rhb./aq.	....	120–1	$110^{0.6mm}$	s. h.	s.	s. bz.
922	nd./aq.	....	131–2	$121^{0.6mm}$	v. sl. s. h.	∞	∞
923	mn./al.	....	160	$110^{0.6mm}$	sl. s. h.; sl. d.	sl. s. c.	s.; s. h. bz.

* Crysts. $+1H_2O$, m.p. 51–2°.
Bromo-aminotoluene 1002–3
Bromoaspirin 76

Bromo-butane 1057–60

Table 7-4 (*Continued*)
PHYSICAL CONSTANTS OF ORGANIC COMPOUNDS

No.	Name	Synonym	Formula	Beil. Ref.	Formula Weight
924	**Bromo** cinnamic acid (β) (*trans*)	....	$C_6H_5 \cdot CBr:CH \cdot CO_2H$	IX-597	227.06
925	crotonic acid (β)	....	$CH_3 \cdot C(Br):CH \cdot CO_2H$	II-419	164.99
926	cymene (2)	(4;2,1)	$(CH_3)_2CH \cdot C_6H_3(Br) \cdot CH_3$	V-423	213.12
927	diethylaniline (*p*)	....	$Br \cdot C_6H_4 \cdot N(C_2H_5)_2$	XII-638	228.15
928	diiodo-methane	....	$BrCHI_2$	I-72	346.74
929	dimethylaniline (*p*)	....	$Br \cdot C_6H_4 \cdot N(CH_3)_2$	XII-637	200.08
930	dinitrobenzene	(4;1,2)	$Br \cdot C_6H_3(NO_2)_2$	V-266	247.01
931	dinitrobenzene	(4;1,3)	$Br \cdot C_6H_3(NO_2)_2$	V-266	247.01
932	dinitrobenzoic acid	(4;3,5,1)	$Br \cdot C_6H_2(NO_2)_2CO_2H$	IX-416	291.02
933	diphenyl (*o*)	....	$Br \cdot C_6H_4 \cdot C_6H_5$	V-580	233.11
934	diphenyl (*p*)	....	$Br \cdot C_6H_4 \cdot C_6H_5$	V-580	233.11
935	diphenyl ether (*p*)	....	$Br \cdot C_6H_4 \cdot O \cdot C_6H_5$	*VI-105	249.11
936	ethyl acetate (β)	....	$CH_3CO_2 \cdot CH_2 \cdot CH_2Br$	II-128	167.01
937	ethyl benzene (β)	Ph-Et-bromide	$C_6H_5 \cdot CH_2 \cdot CH_2Br$	V-356	185.07
938	ethyl benzene (α)	Ph-Et-bromide	$C_6H_5 \cdot CH(Br)CH_3$	V-355	185.07
939	ethyl ethyl ether	β-Br-diethyl ether	$BrCH_2CH_2 \cdot O \cdot C_2H_5$	I-338	153.02
940	ethyl phthalimide (β)(*N*)	phthalimino-Et-bromide	$C_6H_4(CO)_2N \cdot CH_2 \cdot CH_2Br$	XXI-461	254.09
941	fumaric acid	....	$CH:CBr(CO_2H)_2$	II-745	194.98
942	furan (2)	....	$Br \cdot C_4H_3O$	XVII-27	146.98
943	furoic acid (3)	....	$Br \cdot C_4H_2O \cdot CO_2H$	XVIII-284	190.99
944	hydroquinone	....	$Br \cdot C_6H_3(OH)_2$	VI-852	189.01
945	iodobenzene (*o*)	....	$Br \cdot C_6H_4 \cdot I$	V-223	282.91
946	iodobenzene (*m*)	....	$Br \cdot C_6H_4 \cdot I$	V-223	282.91
947	iodobenzene (*p*)	....	$Br \cdot C_6H_4 \cdot I$	V-223	282.91
948	iodo-ethane (1,1)	....	$CH_3 \cdot CHBrI$	I-98	234.87
949	iodo-ethane (1,2)	....	$BrCH_2 \cdot CH_2I$	I-98	234.87
950	iodo-methane	....	$BrCH_2I$	I-71	220.84
951	maleic acid	....	$CH:CBr(CO_2H)_2$	II-754	194.98
952	malonic acid	....	$BrCH(CO_2H)_2$	II-594	182.96
953	mesitylene	(2;1,3,5)	$Br \cdot C_6H_2(CH_3)_3$	V-408	199.10
954	methyl cyanide	bromo-acetonitrile	$BrCH_2 \cdot CN$	II-216	119.95
955	methyl acetate	....	$CH_3 \cdot CO_2 \cdot CH_2Br$	II-152	152.98
956	naphthalene (α)	α-naphthyl bromide	$C_{10}H_7Br$	V-547	207.08
957	naphthalene (β)	β-naphthyl bromide	$C_{10}H_7Br$	V-548	207.08
958	α-naphthol (4,1)	....	$Br \cdot C_{10}H_6 \cdot OH$	VI-613	223.08
959	β-naphthol (1,2)	....	$Br \cdot C_{10}H_6 \cdot OH$	VI-650	223.08
960	nitrobenzene (*o*)	....	$Br \cdot C_6H_4 \cdot NO_2$	V-247	202.01
961	nitrobenzene (*m*)	....	$Br \cdot C_6H_4 \cdot NO_2$	V-248	202.01
962	nitrobenzene (*p*)	....	$Br \cdot C_6H_4 \cdot NO_2$	V-248	202.01
963	3-nitrobenzene-1-sulfonic acid (4)	(4;3,1)	$Br \cdot C_6H_3(NO_2) \cdot SO_3H$	XI-74	282.08
964	3-nitrobenzoic acid (2)	(2;3,1)	$Br \cdot C_6H_3(NO_2) \cdot CO_2H$	IX-406	246.02
965	nitroethane (1,1)	....	$CH_3 \cdot CHBr \cdot NO_2$	I-101	153.97
966	nitromethane	....	$Br \cdot CH_2 \cdot NO_2$	I-77	139.94

Table 7-4 (*Continued*)
PHYSICAL CONSTANTS OF ORGANIC COMPOUNDS

No.	Crystalline Form and Color	Specific Gravity	Melting Point °C.	Boiling Point °C.	Solubility in 100 Parts		
					Water	Alcohol	Ether
924	nd./aq.	....	134-5	$122^{0.6mm}$	sl. s. h.	s.	s. h. bz.
925	nd./lg.	....	95-7	....	sl. s. c.	v. s.; v. s. bz.	v. s.; v. s. CS_2
926	lq.	$1.253^{25}_{25}°$	<-20	233-5	i.; ∞^{25} act.	$50^{25°}$ Me al.	∞; ∞ bz.
927	yel. red/ac.	....	33	270	i.	v.s.	v.s.
928	yel. cr.	....	60	110^{25mm}	sl. s. pet.	....	soln. d. in light
929	lf./al.	....	55	264	i.	v. s.	v. s.
930	mn. pr.	$1.801^{60°}$	59.5(34.8)	....	i.	s. h.	s.
931	yel. cr.	....	75.3	....	....	v. s. h.	....
932	pr./aq. H_2SO_4	....	181	....	sl. s.	v. s.	v. s.
933	lq.	....	<-20	296-8	i.	s.	v. s.
934	cr./al.	....	90-1	310	i.; 100^{25} bz.	s.; $325°$ Me al.	$34^{25°}$
935	lq.	$1.449^{13°}$	17-18	305	....	....	....
936	col. lq.	$1.514^{20}_{4}°$	-13.8	162-3	i.	∞	∞
937	lq.	....	....	$217-8^{734mm}$	....	....	....
938	lq.	$1.311^{23°}$	....	200-10 d.	i.	s.	s.
939	col. lq.	$1.357^{20}_{4}°$	....	127-8	sl. s.	∞	∞
940	nd./al.	....	82-3	d.	....	....	....
941	lf./aq.	....	185-6	d.	v. s.	....	....
942	lq.	1.650	....	101-2	i.	s.	....
943	nd./aq.	....	128-9	....	$1.3^{20°}$	s.; v. sl. s. CS_2	s.; v. sl. s. lg.
944	lf./pet.	....	113	subl.	v. s.	....	....
945	col. lq.	$2.257^{25}_{4}°$	5.0(2.1)	257.4^{754mm}	i.	v. sl. s.	v. sl. s. ac.
946	col. lq.	....	-9.3	252^{754mm}	i.	v. sl. s.	v. sl. s. ac.
947	pl./al. et.	....	91-2	251.5^{754mm}	i.	v. sl. s. c.	sl. s.
948	lq.	$2.452^{16°}$	<-20	142-3	....	....	....
949	long nd.	$2.516^{29°}$	28	163	....	v. s. h.	....
950	lq.	$2.926^{17°}$	....	138-40	....	....	....
951	nd. or pr.	....	128*	d.	v. s.	v. s.	v. s.
952	nd./et.	....	112-3 d.	....	....	v. s.	v. s.
953	lq.	$1.319^{10°}$	-1	225-30	....	....	....
954	yel. oil	1.771	....	148-50	....	s.	s.
955	col. lq.	$1.195^{14°}$	....	$130-3^{750mm}$	i.; sl. d.	s.	s.
956	col. oil	$1.482^{20}_{4}°$	6.1	281.2	i.; ∞ bz.	∞ abs. al.	∞
957	lf./al.	$1.605^{0°}$	59	281-2	i.; s. bz.	$6^{20°}$, 92% al.	v. s.; v. s. ehl.
958	nd./aq. al.	....	127-8	....	....	....	....
959	rhb. pr.	....	83-4	d. 130	....	....	....
960	yel. cr.	$1.623^{8}_{4}°$	43	261	i.	v. s.	s.
961	rhb.	$1.704^{20}_{4}°$	56.4	256-7	i.	s.	s.
962	tri.	$1.938^{0}_{4}°$	126-7	255-6	i.	1.4 c.	s.
963	yel. nd.	....	87-8	....	s.	s.	i.
964	cr./aq. al.	....	186-8	....	....	....	....
965	lq.	....	....	146-7	i.	....	....
966	lq.	....	....	152.5^{765mm}	i.	s. alk.	....

* Heated slowly, m.p. 136-8°; rapidly, m.p. 140-1°.

Bromo-methane 4147
Bromo-methylacetophenone 4333
Bromo-methyl-*p*-tolyl ketone 4333
Bromo-nitrotoluene 4681
Bromo-nonane 4927
Bromo-octane 4973-4
Bromo-pentadecane 5063
Bromo-pentane 427-31

Table 7-4 (*Continued*)
PHYSICAL CONSTANTS OF ORGANIC COMPOUNDS

No.	Name	Synonym	Formula	Beil. Ref.	Formula Weight
967	**Bromo** phenacyl bromide (*p*)	diBr–acetophenone	$Br \cdot C_6H_4 \cdot CO \cdot CH_2Br$	VII-285	277.95
968	phenanthrene (9)	. . . .	$C_{14}H_9Br$	V-671	257.14
969	phenetole (*β*)	*β*-PhO-ethyl bromide	$Br \cdot CH_2CH_2 \cdot O \cdot C_6H_5$	VI-142	201.07
970	phenetole (*o*)	. . . .	$Br \cdot C_6H_4 \cdot O \cdot C_2H_5$	VI-197	201.07
971	phenetole (*p*)	. . . .	$Br \cdot C_6H_4 \cdot O \cdot C_2H_5$	VI-199	201.07
972	phenol (*o*)	. . . .	$Br \cdot C_6H_4 \cdot OH$	VI-197	173.02
973	phenol (*m*)	. . . .	$Br \cdot C_6H_4 \cdot OH$	VI-198	173.02
974	phenol (*p*)	. . . .	$Br \cdot C_6H_4 \cdot OH$	VI-198	173.02
975	phenylhydrazine (*p*)	. . . .	$Br \cdot C_6H_4 \cdot NH \cdot NH_2$	XV-434	187.05
976	phenylhydrazine HCl	(*p*)	$Br \cdot C_6H_4 \cdot N_2H_3 \cdot HCl$	XV-435	223.51
977	phenylphenol	(1;3,4)	$C_6H_5 \cdot C_6H_3(Br)OH$	**VI-625	249.11
978	phthalic acid	(3;1,2)	$Br \cdot C_6H_3(CO_2H)_2$	IX-821	245.04
979	(tere)-phthalic acid	(2;1,4)	$Br \cdot C_6H_3(CO_2H)_2$	IX-848	245.04
980	propionic acid (*α*)	(*dl*)	$CH_3 \cdot CHBr \cdot CO_2H$	II-254	152.98
981	propionic acid (*β*)	. . . .	$BrCH_2 \cdot CH_2 \cdot CO_2H$	II-256	152.98
982	propionyl bromide (*α*)	. . . .	$CH_3 \cdot CHBr \cdot CO \cdot Br$	II-256	215.88
983	*n*-propyl acetate (*β*)	. . . .	$CH_3CO_2 \cdot CH_2 \cdot CHBr \cdot CH_3$	. . . .	181.04
984	*n*-propyl acetate (*γ*)	. . . .	$CH_3CO_2 \cdot (CH_2)_3Br$	**II-139'	181.04
985	*n*-propyl benzene (*p*)	. . . .	$Br \cdot C_6H_4 \cdot (CH_2)_2CH_3$	V-391	199.10
986	*iso*-propyl benzene (*p*)	*p*-bromocumene	$Br \cdot C_6H_4 \cdot CH(CH_3)_2$	V-395	199.10
987	*n*-propyl phenyl ether (*γ*)	. . . .	$C_6H_5 \cdot O \cdot (CH_2)_3Br$	VI-142	215.10
988	propylene (1)(*α*)	1-bromo-propene-1	$CH_3 \cdot CH:CHBr$	I-200	120.98
989	propylene (1)(*β*)	1-bromo-propene-1	$CH_3 \cdot CH:CHBr$	I-200	120.98
990	propylene (2)	2-bromo-propene-1	$CH_3 \cdot CBr:CH_2$	I-200	120.98
991	pyridine (2)	*α*-bromo-pyridine	$Br \cdot C_5H_4N$	XX-233	158.01
992	pyridine (3)	*β*-bromo-pyridine	$Br \cdot C_5H_4N$	XX-233	158.01
993	salicylic acid (5)	(5;2,1)	$Br \cdot C_6H_3(OH)CO_2H$	X-107	217.03
994	styrene (*α*)	. . . .	$C_6H_5 \cdot CBr:CH_2$	V-477	183.05
995	styrene (*ω*)	*isomer No. 1*	$C_6H_5 \cdot CH:CHBr$	V-477	183.05
996	styrene (*ω*)	*isomer No. 2*	$C_6H_5 \cdot CH:CHBr$	V-477	183.05
997	succinic acid (*dl*)	. . . .	$(\cdot CH_2CHBr \cdot)(CO_2H)_2$	II-621	196.99
998	sulfalein	di-Na-phenol-tetra-bromo-phthalein-disulfonate	$Br_4C_6 \cdot CO \cdot O \cdot C:$ $\vert$_____$\vert$ $(C_6H_3OHSO_3Na)_2$	. . . .	838.02
999	toluene (*o*)	*o*-tolyl bromide	$Br \cdot C_6H_4 \cdot CH_3$	V-304	171.04
1000	toluene (*m*)	. . . .	$Br \cdot C_6H_4 \cdot CH_3$	V-305	171.04
1001	toluene (*p*)	. . . .	$Br \cdot C_6H_4 \cdot CH_3$	V-305	171.04
1002	*o*-toluidine (5)	(5;2,1)	$Br \cdot C_6H_3(NH_2)CH_3$	XII-838	186.06
1003	*p*-toluidine (3)	(3;4,1)	$Br \cdot C_6H_3(NH_2)CH_3$	XII-991	186.06
1004	triiodomethane	. . . .	$BrCl_3$	I-74	472.63
1005	trinitromethane	. . . .	$BrC(NO_2)_3$	I-79	229.94

Table 7-4 (*Continued*)
PHYSICAL CONSTANTS OF ORGANIC COMPOUNDS

No.	Crystalline Form and Color	Specific Gravity	Melting Point °C.	Boiling Point °C.	Solubility in 100 Parts		
					Water	Alcohol	Ether
967	nd.	....	110-2	....	i.	s. h.	s.
968	pr./al.	1.409$\frac{10°}{4}$	63	>360; subl.	v. s. ac.	....	v. s. CS_2
969	cr.	....	35	240-50 d.	i.	s.	s.
970	lq.	....	....	218-22	....	s.	s.
971	lq.	....	10-12	227-33	....	....	....
972	col. lq.	1.553$^{80°}$	5.6	194-5	s.; s. alk.	s.; ∞ chl.	∞
973	cr.	....	32-3(18)	236-7	s. alk.	s.	s.
974	tet. cr.	1.588$^{80°}$	63.5	238	1.4$^{15°}$	v. s.; 88$^{25°}$ act.	v. s.; 40$^{25°}$ bz.
975	nd./al.	....	106-7	....	s. bz.	s.	s.
976	cr./aq.	....	....	....	s. h.	....	....
977	cr.	....	94-6	....	i.;32$^{25°}$bz.	125$^{25°}$	v. s.
978	nd./aq.	....	178.5 (−H_2O)	....	s.	s.	s.; v. sl. s. chl.
979	nd./al.	....	299	....	0.18$^{24°}$; s. h.	s.	i.; i. bz.
980	pr.	1.700$\frac{20°}{4}$	25.7(−3.9)	205.5 sl. d.	v. s.	v. s.	v. s.
981	pl.	....	62.5	140-2^{45mm}	v. s.	v. s.	v. s.
982	lq.	2.061$\frac{16°}{4}$	....	152-4	....	....	....
983	col. lq.	....	....	51-3^{9mm}	....	....	....
984	col. lq.	....	....	88-90^{22mm}	....	....	....
985	col. lq.	....	....	220	....	....	....
986	lq.	1.365$\frac{22°}{4}$	<−20	216-8	....	....	....
987	col. lq.	1.365$\frac{16°}{16}$	....	211-2^{200mm}	....	....	....
988	lq.	1.434$\frac{15.8°}{4}$	−113	59-60	....	....	....
989	lq.	1.417$\frac{15.8°}{4}$	−76.5	63.3	....	....	....
990	lq.	1.397$\frac{15.8°}{4}$	−125	48.4	....	....	....
991	lq.	1.657$^{15°}$	....	193-4	sl. s.	....	....
992	col. lq.	1.632$^{10°}$	....	173-4^{758mm}	sl. s.	s.	s.
993	nd./aq. al.	....	167-9	subl. >100	0.3$^{80°}$	85$^{25°}$	70$^{25°}$
994	oil	1.406$\frac{20°}{4}$	−43.5	160^{75mm}	....	....	....
995	lq.	1.422$\frac{20°}{4}$	7	221 sl. d.	i.	∞	∞
996	lq.	1.427$\frac{20°}{4}$	−7.5	108^{26mm}	i.	∞	∞
997	col.	2.073	160-1	....	19$^{15°}$	s.	....
998	wh. pd.	....	....	....	s.	i.	i. act.
999	col. lq.	1.422$\frac{20°}{4}$	−28.1	181.5	i.	s.; ∞ bz.	∞$^{25°}$
1000	col. lq.	1.410$\frac{20°}{4}$	−39.8	183.7	i.	s.; ∞ bz.	s.
1001	cr./al.	1.390$\frac{20°}{4}$	26.7	185.0	i.	s.	∞$^{25°}$
1002	rhb./al.	....	58-9	240	v. sl. s.	s.	....
1003	lf.	1.5$^{20°}$	16-8	240	i.	s.	s.
1004	cr.	....	113	....	....	....	....
1005	lq.	2.044$\frac{15°}{4}$	17-8	55-6^{12mm}	0.4$^{25°}$	....	....

Bromo-propylene 199   Bromo-pyrine 901
Bromo-propyne 5342   Bromo-toluene (ω) 793

Table 7-4 (Continued)
PHYSICAL CONSTANTS OF ORGANIC COMPOUNDS

No.	Name	Synonym	Formula	Beil. Ref.	Formula Weight
1006	**Bromo** n-valeric acid (α)	. . . .	$CH_3(CH_2)_2CHBr \cdot CO_2H$	II-302	181.04
1007	*iso*-valeric acid	(α)(*dl*)	$(CH_3)_2CHCHBr \cdot CO_2H$	II-317	181.04
1008	*iso*-valeryl-*p*-phenetidine (α) (*d*)	phenoval	$(CH_3)_2CHCHBr \cdot CO \cdot NH \cdot C_6H_4 \cdot O \cdot C_2H_5$	*XIII-163	300.20
1009	*iso*-valeryl urea (α)	bromural; bromisoval	$(CH_3)_2CHCHBr \cdot CO \cdot NH \cdot CO \cdot NH_2$	III-63	223.08
1010	xylene (3;1,2)	bromo-*o*-xylene (3)	$Br \cdot C_6H_3(CH_3)_2$	V-365	185.07
1011	xylene (4;1,2)	bromo-*o*-xylene (4)	$Br \cdot C_6H_3(CH_3)_2$	V-365	185.07
1012	xylene (2;1,3)	bromo-*m*-xylene (2)	$Br \cdot C_6H_3(CH_3)_2$	V-374	185.07
1013	xylene (4;1,3)	bromo-*m*-xylene (4)	$Br \cdot C_6H_3(CH_3)_2$	V-374	185.07
1014	xylene (5;1,3)	bromo-*m*-xylene (5)	$Br \cdot C_6H_3(CH_3)_2$	V-374	185.07
1015	xylene (2;1,4)	bromo-*p*-xylene (2)	$Br \cdot C_6H_3(CH_3)_2$	V-385	185.07
1016	**Bromoform**	tribromo-methane	$CHBr_3$	I-68	252.75
1017	**Bryonane**	laurane	$C_{20}H_{42}$	I-174	282.56
1018	**Butadiene** (1,2)	methyl-allene	$CH_3 \cdot CH : C : CH_2$	I-249	54.09
1019	**Butadiene** (1,3)	erythrene	$CH_2 : CH \cdot CH : CH_2$	I-249	54.09
1020	**Butandiolamine**	2-NH$_2$-2-Me-propandiol-1,3	$CH_3 \cdot C(NH_2) : (CH_2OH)_2$	IV-303	105.14
1021	**Butane** (n)	diethyl	$CH_3 \cdot CH_2 \cdot CH_2 \cdot CH_3$	I-118	58.12
1022	**Butane** (*iso*)	trimethyl-methane	$(CH_3)_3CH$	I-124	58.12
1023	**Butanolamine**	2-NH$_2$-butanol-1	$C_2H_5 \cdot CH(NH_2) \cdot CH_2OH$	IV-291	89.14
1024	*iso*-**Butanolamine**	2-NH$_2$-2-Me-propanol-1	$(CH_3)_2C(NH_2) \cdot CH_2OH$	. . . .	89.14
1025	**Butoxy**-cinchoninic acid diethyleth-ylene-diamide HCl	nupercaine; percaine	$C_4H_9O \cdot C_9H_5N \cdot CO \cdot NH(CH_2)_2N : (C_2H_5)_2 \cdot HCl$	. . . .	379.93
1026	ethoxyethyl bro-mide (2)(β)	. . . .	$C_4H_9O \cdot (CH_2)_2 \cdot O \cdot (CH_2)_2Br$	. . . .	225.13
1027	ethyl phthalate	. . . .	$C_6H_4(CO_2CH_2CH_2 \cdot O \cdot C_4H_9)_2$	. . . .	366.46
1028	ethyl salicylate (β)(n)	. . . .	$HOC_6H_4CO_2 \cdot CH_2 \cdot CH_2 \cdot O \cdot C_4H_9$	. . . .	238.29
1029	**Butyl** acetate (n)	. . . .	$CH_3CO_2 \cdot CH_2CH_2 \cdot C_2H_5$	II-130	116.16
1030	acetate (*sec*)	. . . .	$CH_3CO_2 \cdot CH(CH_3) \cdot C_2H_5$	II-131	116.16
1031	acetate (*iso*)	. . . .	$CH_3CO_2 \cdot CH_2CH : (CH_3)_2$	II-131	116.16
1032	acetate (*tert*)	. . . .	$CH_3CO_2 \cdot C(CH_3)_3$	II-131	116.16
1033	acetoacetate (n)	. . . .	$CH_3COCH_2CO_2C_4H_9$	. . . .	158.20
1034	alcohol (n)	butanol-1	$C_2H_5 \cdot CH_2 \cdot CH_2OH$	I-367	74.12
1035	alcohol (*sec*)	butanol-2	$C_2H_5 \cdot CHOH \cdot CH_3$	I-371	74.12
1036	alcohol (*iso*)	2-methyl-propanol-1	$(CH_3)_2CH \cdot CH_2OH$	I-373	74.12
1037	alcohol (*tert*)	2-methyl-propanol-2	$(CH_3)_3COH$	I-379	74.12
1038	allylbarbituric acid (*iso*)	sandoptal	$C_4H_9(C_3H_5) : C_4H_2O_3N_2$	. . . .	224.26
1039	amine (n)	. . . .	$C_2H_5 \cdot CH_2 \cdot CH_2 \cdot NH_2$	IV-156	73.14

Table 7-4 (Continued)
PHYSICAL CONSTANTS OF ORGANIC COMPOUNDS

No.	Crystalline Form and Color	Specific Gravity	Melting Point °C.	Boiling Point °C.	Solubility in 100 Parts		
					Water	Alcohol	Ether
1006	lq.	....	....	126-30^{27mm}	sl. s.	v. s.	s.
1007	col. pr./et.	....	44	230 sl. d.	1.4 c.	v. s.	s.
1008	col. cr./aq. al	....	151	....	i.; sl. s. gly.	s. h.; v. sl. s. bz.	sl. s.; sl. s. chl.
1009	lf./toluene	....	160	subl.	2 c.; s. h.	s.	s.; s. alk.
1010	col. lq.	1.365^{20}_{4}	....	213-4	i.	....	....
1011	col. lq.	1.369^{15}_{15}	-0.2	214.5	i.	....	....
1012	col. lq.	....	<-10	206±	i.	....	....
1013	col. lq.	....	....	205-7	i.	....	....
1014	col. lq.	$1.362^{20°}$	<-20	204	....	....	....
1015	lf. or pl.	1.356^{20}_{4}	9-10	205.5^{755mm}	....	....	....
1016	col. lq.	2.890^{20}_{4}	8.3	149.6	0.1 c.	∞; ∞ bz.	∞; ∞ chl.
1017	nd.	....	69	400	v. s. pet.	v. s. h. abs.	v. s. bz.
1018	lq.	$0.652^{20°}$	-136.3	10.3	i.	∞	∞
1019	col. gas	0.621^{20}_{4}	-108.9	-4.41	i.	∞	∞
1020	cr.	....	109-11	151-2^{10mm}	250$^{20°}$	25$^{20°}$	0.1
1021	col. gas	$0.579^{20°}$ (lq.)	-138.3	-0.50	$15\frac{17}{773}°$ cc.	$1883\frac{17}{775}°$ cc.	$2980\frac{18}{773}°$ cc.
1022	col. gas	$0.557^{20°}$ (lq.)	-159.6	-11.7	$13\frac{17}{773}°$ cc.	$1320\frac{17}{775}°$ cc.	$2790\frac{18}{773}°$ cc.
1023	lq.	0.944^{20}_{20}	-2	174-8	∞	∞	∞
1024	col. cr.	0.934^{20}_{20}	30-1	165	∞	∞	∞
1025	wh. cr.	....	90*	....	200; s. chl.	s.; sl. s. bz.	i.; s. act.
1026	col. lq.	....	....	107-9^{12mm}	....	....	....
1027	col. lq.	....	....	189-91^{2mm}	....	....	....
1028	col. lq.	1.077^{25}_{25}	<-20	186-93^{23mm}	i.; ∞ bz.	∞; ∞ chl.	∞
1029	col. lq.	$0.882^{20°}$	-73.5	126.1	0.7	∞	∞
1030	col. lq.	0.865^{25}_{4}	....	111.5-2.5^{744mm}	i.	∞	∞
1031	col. lq.	0.871^{20}_{4}	-98.9	118	0.6$^{25°}$	∞	∞
1032	col. lq.	0.866^{20}_{4}	....	95-6^{750mm}	i.	∞	∞
1033	col. lq.	....	....	98-100^{16mm}	i.	s.	s.
1034	col. lq.	0.810^{20}_{4}	-89.8	118.0	9$^{15°}$	∞	∞
1035	col. lq.	0.808^{20}_{4}	-114.7	99.5	12.5$^{20°}$	∞	∞
1036	col. lq.	0.802^{20}_{4}	-108	108.1	10$^{15°}$	∞	∞
1037	lq. or rhb.	$0.779^{26°}$	25.6	82.6	∞	∞	∞
1038	wh. cr. pd.	....	138-9	....	sl. s.	s.; s. chl.	s.; s. act.
1039	col. lq.	0.739^{25}_{4}	-50	77.8	∞	∞; ∞ gly.	∞

* Free base, m. 97-8°; s. al., aq., bz.; i. et.

Butenoic acid 1557-8, 6427
Butenol 1565, 4422
Butenyl chloride (iso) 4060
Butesin 1043

Butine, cf. butyne
Butoben 1093
Butolan 825
Butoxy-benzene 1116

Butter yellow 2431 or 2438
Butyl, cf. also dibutyl.
Butyl adipate 1883

Table 7-4 (*Continued*)
PHYSICAL CONSTANTS OF ORGANIC COMPOUNDS

No.	Name	Synonym	Formula	Beil. Ref.	Formula Weight
1040	**Butyl** amine (*sec*)	....	$(C_2H_5)(CH_3):CH \cdot NH_2$	IV-160	73.14
1041	amine (*iso*)	....	$(CH_3)_2CH \cdot CH_2 \cdot NH_2$	IV-163	73.14
1042	amine (*tert*)	....	$(CH_3)_3C \cdot NH_2$	IV-173	73.14
1043	p-aminobenzoate (*n*)	butesin	$H_2N \cdot C_6H_4 \cdot CO_2C_4H_9$	....	193.25
1044	aminobenzoate picrate	butesin picrate	$(C_{11}H_{15}O_2N)_2 \cdot C_6H_3O_7N_3$	....	615.60
1045	p-aminobenzoate (*iso*)	cycloform	$H_2N \cdot C_6H_4 \cdot CO_2C_4H_9$	*XIV-567	193.25
1046	aniline (*n*)	....	$C_6H_5 \cdot NH \cdot C_4H_9$	XII-168	149.24
1047	aniline (*iso*)	....	$C_6H_5 \cdot NH \cdot C_4H_9$	XII-168	149.24
1048	arsonic acid (*n*)	....	$C_4H_9 \cdot AsO(OH)_2$	**IV-997	182.05
1049	benzene (*n*)	....	$C_6H_5 \cdot C_4H_9$	V-413	134.22
1050	benzene (*sec*)	....	$C_6H_5 \cdot C_4H_9$	V-414	134.22
1051	benzene (*iso*)	....	$C_6H_5 \cdot C_4H_9$	V-414	134.22
1052	benzene (*tert*)	triMe-Ph-methane	$C_6H_5 \cdot C(CH_3)_3$	V-415	134.22
1053	benzoate (*n*)	....	$C_6H_5CO_2 \cdot C_4H_9$	IX-112	178.23
1054	benzoate (*iso*)	....	$C_6H_5CO_2 \cdot C_4H_9$	IX-113	178.23
1055	o-benzoylbenzo- ate (*n*)	....	$C_6H_5 \cdot CO \cdot C_6H_4 \cdot CO_2 \cdot C_4H_9$	....	282.34
1056	benzyl ether (*n*)	....	$C_4H_9 \cdot O \cdot CH_2C_6H_5$	**VI-410	164.25
1057	bromide (*n*)	1-bromo-butane	$C_2H_5 \cdot CH_2 \cdot CH_2Br$	I-119	137.03
1058	bromide (*sec*)	2-bromo-butane	$C_2H_5 \cdot CHBr \cdot CH_3$	I-119	137.03
1059	bromide (*iso*)	1-Br-2-Me-propane	$(CH_3)_2CH \cdot CH_2Br$	I-126	137.03
1060	bromide (*tert*)	2-Br-2-Me-propane	$(CH_3)_3CBr$	I-127	137.03
1061	β-bromoallyl bar- bituric acid	pernoston	$(CH_2:CBr \cdot CH_2) \cdot (C_4H_9):C_4H_2O_3N_2$	....	303.16
1062	n-butyrate (*n*)	....	$C_3H_7CO_2 \cdot C_4H_9$	II-271	144.22
1063	n-butyrate (*iso*)	....	$C_3H_7CO_2 \cdot C_4H_9$	II-271	144.22
1064	iso-butyrate (*iso*)	....	$(CH_3)_2CHCO_2 \cdot C_4H_9$	II-291	144.22
1065	caproate	....	$C_5H_{11}CO_2 \cdot C_4H_9$	II-323	172.27
1066	caprylate	....	$C_7H_{15}CO_2 \cdot C_4H_9$	II-348	200.32
1067	carbamate (*n*)	....	$NH_2 \cdot CO_2 \cdot C_4H_9$	*III-14	117.15
1068	carbamate (*iso*)	....	$NH_2 \cdot CO_2 \cdot C_4H_9$	III-29	117.15
1069	carbitol acetate	β-BuO-β-EtO- ethyl-acetate	$CH_3CO(OC_2H_4)_2 \cdot O \cdot C_4H_9$	....	204.27
1070	catechol (*p*)(*tert*)	(4;1,2)	$(CH_3)_3C \cdot C_6H_3(OH)_2$	....	166.22
1071	cellosolve (*n*)	2-BuO-ethanol-1	$C_4H_9 \cdot O \cdot CH_2CH_2OH$	**I-519	118.18
1072	chloride (*n*)	1-chloro-butane	$C_2H_5 \cdot CH_2 \cdot CH_2Cl$	I-118	92.57
1073	chloride (*sec*)	2-chloro-butane	$C_2H_5 \cdot CHCl \cdot CH_3$	I-119	92.57
1074	chloride (*iso*)	1-Cl-2-Me-propane	$(CH_3)_2CH \cdot CH_2Cl$	I-124	92.57
1075	chloride (*tert*)	2-Cl-2-Me-propane	$(CH_3)_3C \cdot Cl$	I-125	92.57
1076	chloroacetate (*n*)	....	$Cl \cdot CH_2 \cdot CO_2 \cdot C_4H_9$	II-198	150.61
1077	chloroformate (*n*)	n-Bu-chlorocar- bonate	$Cl \cdot CO_2 \cdot C_4H_9$	**III-11	136.58
1078	chloroformate (*iso*)	iso-Bu-chlorocar- bonate	$Cl \cdot CO_2 \cdot C_4H_9$	III-12	136.58
1079	o-cresol (*p*)(*tert*)	2-Me-4-*tert*-Bu- phenol	$(CH_3)_3C \cdot C_6H_3(CH_3) \cdot OH$	VI-550	164.25
1080	o-cresyl ether (*n*)	n-Bu-o-tolyl ether	$CH_3 \cdot C_6H_4 \cdot O \cdot C_4H_9$	VI-353	164.25
1081	crotonate (*n*)	....	$CH_3CH:CHCO_2 \cdot C_4H_9$	....	142.20
1082	crotonate (*iso*)	....	$CH_3CH:CHCO_2 \cdot C_4H_9$	....	142.20
1083	iso-cyanide (*iso*)	iso-Bu-carbylamine	$(CH_3)_2CH \cdot CH_2 \cdot NC$	IV-167	83.13

Table 7-4 (*Continued*)
PHYSICAL CONSTANTS OF ORGANIC COMPOUNDS

No.	Crystalline Form and Color	Specific Gravity	Melting Point °C.	Boiling Point °C.	Solubility in 100 Parts		
					Water	Alcohol	Ether
1040	col. lq.	$0.724^{20°}_{4}$	−104	66^{772mm}	∞	∞	∞
1041	col. lq.	$0.732^{20°}_{20}$	−85	68–9	∞	∞	∞
1042	col. lq.	$0.698^{18°}_{4}$	−67.5	45.2	....	∞	....
1043	wh. cr. pd.	....	58–9	$173-4^{8mm}$	0.013	s.; s. chl.	s.; s. oil
1044	yel. amor. pd./bz.	....	109–10	....	0.05	s.; s. chl.	s.; s. bz.
1045	nd./lg.	....	65	....	0.022	s.; s. act.	s.; s. bz.
1046	lq.	....	....	235^{720mm}	i.	v. s.	v. s.
1047	oil	$0.940^{20°}_{4}$	....	231–2	$0.01^{15°}$	v. s.	v. s.
1048	col. fl.	....	160–2	....	s.	s.	i.
1049	lq.	$0.860^{20°}$	−88.0	183.3	i.	s.	s.
1050	lq.	$0.862^{20°}$	−75.5	173.3	i.	s.	s.
1051	lq.	$0.853^{20°}$	−51.5	172.8	i.	s.	s.
1052	col. lq.	$0.867^{20°}_{4}$	−57.9	169.1	i.	s.	s.
1053	col. oil	$1.005^{25°}_{25}$	−22	248.5–9.5	i.	s.	s.
1054	col. oil	$0.997^{25°}_{25}$	....	241.5	i.	∞	∞
1055	col. lq.	....	....	$229-32^{15mm}$	....	....	....
1056	col. lq.	$0.931^{10°}_{4}$	....	220.5^{744mm}	i.	∞	∞
1057	lq.	$1.275^{20°}_{4}$	−112.4	101.6	$0.06^{16°}$	∞	∞
1058	lq.	$1.261^{20°}_{4}$	−112.1	91.3	i.	....	....
1059	lq.	$1.264^{20°}_{4}$	−117.4	91.4	$0.06^{18°}$	∞	∞
1060	lq.	$1.220^{20°}_{4}$	−16.2	73.3; d. 210	$0.06^{18°}$	∞	∞
1061	wh. cr. pd.	....	130–3	....	v. sl. s.	s.; s. alk.	s.
1062	col. lq.	$0.872^{20°}_{20}$	....	165.7^{736mm}	i.	∞	∞
1063	col. lq.	$0.863^{18°}_{4}$	....	156.9	i.	∞	∞
1064	col. lq.	$0.875^{0°}_{4}$	−80.7	148.7	i.	∞	∞
1065	col. lq.	$0.862^{25°}_{4}$	−64.3	207.7	i.	....	....
1066	col. lq.	$0.858^{25°}_{4}$	−43.0	245.0	....	....	....
1067	wh. fl.	....	53–4	203–4d.	i.	s.	v. s.
1068	col. lf.	$0.956^{76°}_{4}$	65	206–7	i.	s.	s.
1069	col. lq.	$0.981^{20°}_{20}$	−32.2	246.8	$6.5^{20°}$	....	....
1070	col. cr.	$1.049^{60°}_{25}$	56–7	285	$0.2^{80°}$	s.; v. s. act.	$240^{25°}$
1071	col. lq.	$0.903^{20°}_{4}$	....	171.2	∞	∞	∞
1072	col. lq.	$0.878^{20°}_{4}$	−123.1	78.5	$0.071^{12.5°}$	∞	∞
1073	col. lq.	$0.873^{20°}_{4}$	−131.3	68.3	i.	∞	∞
1074	col. lq.	$0.884^{15°}$	−131.2	68.9	i.	∞	∞
1075	lq.	$0.847^{15°}$	−27.1	50.7	i.	∞	∞
1076	col. lq.	$1.081^{15°}$	....	181–3	....	....	....
1077	col. lq.	$1.074^{25°}_{4}$	....	140–5	d.	d.	∞
1078	lq.	$1.045^{18.5°}_{15}$	....	128.8	slow d.	slow d.; ∞ chl.	∞ ; ∞ bz.
1079	lq.	$0.969^{30°}_{25}$	27	$137-8^{25mm}$	i.; ∞ CCl₄	∞ Me al.	∞ ; ∞ bz.
1080	col. lq.	$0.944^{0°}_{0}$	....	223	....	....	....
1081	lt. yel. lq.	....	....	175–80	....	....	....
1082	lt. yel. oil	....	....	170–2	....	....	....
1083	lq.	$0.787^{4°}$	<−60	110.5	sl. s.	s.	s.

Butyl carbonate 1892–4　　Butyl citrate 6119　　Butyl cyanide (*iso*) 6413
Butyl chlorocarbonate 1077–8　　Butyl cyanide (*n*) 6405　　Butyl cyanide (*tert*) 6246

Table 7-4 (*Continued*)
PHYSICAL CONSTANTS OF ORGANIC COMPOUNDS

No.	Name	Synonym	Formula	Beil. Ref.	Formula Weight
	Butyl				
1084	diethanolamine (*n*)	....	$C_4H_9 \cdot N(CH_2CH_2OH)_2$	IV-285	161.25
1085	fluoride (*iso*)	1-F-2-Me-propane	$(CH_3)_2CH \cdot CH_2F$	I-124	76.11
1086	formate (*n*)	....	$HCO_2 \cdot CH_2CH_2C_2H_5$	II-21	102.13
1087	formate, ortho (*n*)	triBu-orthoformate	$HC(O \cdot CH_2CH_2C_2H_5)_3$	....	232.37
1088	formate (*sec*)	....	$HCO_2CH(CH_3) \cdot C_2H_5$	**II-30	102.13
1089	formate (*iso*)	....	$HCO_2 \cdot CH_2 \cdot CH(CH_3)_2$	II-21	102.13
1090	furoate (*n*)	....	$C_4H_3O \cdot CO_2 \cdot C_4H_9$	....	168.19
1091	β-furylacrylate (*n*)	....	$C_4H_3O \cdot CH{:}CH \cdot CO_2 \cdot C_4H_9$	....	194.23
1092	(*o*)-hydroxy-benzoate	*n*-Bu-salicylate	$HO \cdot C_6H_4 \cdot CO_2 \cdot C_4H_9$	....	194.23
1093	*p*-hydroxy-benzoate (*n*)	butoben	$HO \cdot C_6H_4 \cdot CO_2 \cdot C_4H_9$	....	194.23
1094	α-hydroxy-*iso*-butyrate (*n*)	....	$(CH_3)_2C(OH) \cdot CO_2 \cdot C_4H_9$	....	160.21
1095	iodide (*n*)	1-iodo-butane	$C_2H_5 \cdot CH_2 \cdot CH_2I$	I-123	184.02
1096	iodide (*sec*)	2-iodo-butane	$C_2H_5 \cdot CHI \cdot CH_3$	I-123	184.02
1097	iodide (*iso*)	1-iodo-2-Me-propane	$(CH_3)_2CH \cdot CH_2I$	I-128	184.02
1098	iodide (*tert*)	2-iodo-2-Me-propane	$(CH_3)_3CI$	I-129	184.02
1099	lactate (*n*)	....	$CH_3CHOHCO_2 \cdot C_4H_9$	....	146.19
1100	lactate (*iso*)	....	$CH_3CHOHCO_2 \cdot C_4H_9$	**III-188	146.19
1101	levulinate (*n*)	....	$CH_3CO(CH_2)_2CO_2 \cdot C_4H_9$	**III-207	172.23
1102	malonic acid	....	$C_4H_9 \cdot CH(CO_2H)_2$	II-673	160.17
1103	mercaptan (*n*)	butanthiol-1	$C_2H_5(CH_2)_2 \cdot SH$	I-370	90.19
1104	mercaptan (*iso*)	2-Me-propanthiol-1	$(CH_3)_2CH \cdot CH_2 \cdot SH$	I-378	90.19
1105	mercaptan (*tert*)	....	$(CH_3)_3C \cdot SH$	I-383	90.19
1106	*o*-methoxybenzoate (*n*)	....	$CH_3O \cdot C_6H_4 \cdot CO_2 \cdot C_4H_9$	....	208.26
1107	nitrate (*iso*)	....	$C_4H_9 \cdot O \cdot NO_2$	I-377	119.12
1108	nitrite (*n*)	....	$C_4H_9 \cdot O \cdot NO$	I-369	103.12
1109	nitrite (*iso*)	....	$(CH_3)_2CHCH_2 \cdot O \cdot NO$	I-377	103.12
1110	oleate (*n*)	....	$C_{17}H_{33}CO_2 \cdot C_4H_9$	**II-439	338.58
1111	oxamate (*n*)	....	$H_2N \cdot CO \cdot CO_2 \cdot C_4H_9$	....	145.16
1112	phenoxy-ethanol	(β)(*p-tert*)	$(CH_3)_3C \cdot C_6H_4 \cdot O \cdot CH_2 \cdot CH_2OH$	....	194.28
1113	phenylacetate (*n*)	....	$C_6H_5 \cdot CH_2 \cdot CO_2 \cdot C_4H_9$	....	192.26
1114	phenylacetate (*iso*)	....	$C_6H_5 \cdot CH_2 \cdot CO_2 \cdot C_4H_9$	IX-435	192.26
1115	phenyl carbinol (*n*)	....	$C_6H_5 \cdot CHOH \cdot C_4H_9$	**II-504	164.25
1116	phenyl ether (*n*)	*n*-BuO-benzene	$C_4H_9 \cdot O \cdot C_6H_5$	VI-143	150.22
1117	phenyl ketone (*n*)	valerophenone	$C_4H_9 \cdot CO \cdot C_6H_5$	VII-327	162.23
1118	phenyl ketone (*iso*)	*iso*-valerophenone	$C_4H_9 \cdot CO \cdot C_6H_5$	VII-329	162.23
1119	phenol (*m*)(*tert*)	....	$(CH_3)_3C \cdot C_6H_4 \cdot OH$	....	150.22
1120	phenol (*p*)(*tert*)	....	$(CH_3)_3C \cdot C_6H_4 \cdot OH$	VI-524	150.22
1121	*o*-phenylphenol (*tert*)	(1;2,5)	$C_6H_5 \cdot C_6H_3(OH) \cdot C(CH_3)_3$	....	226.32
1122	phthalate (*n*) (mono)	....	$HO_2C \cdot C_6H_4 \cdot CO_2 \cdot C_4H_9$	**IX-586	222.24
1123	propionate (*n*)	....	$C_2H_5 \cdot CO_2 \cdot C_4H_9$	II-241	130.19
1124	propionate (*sec*)	....	$C_2H_5 \cdot CO_2 \cdot C_4H_9$	II-241	130.19
1125	propionate (*iso*)	....	$C_2H_5 \cdot CO_2 \cdot C_4H_9$	II-241	130.19

Butyl dibromosuccinate 1896   Butyl ether 1898-9   Butyl malate 1902
Butyl disulfide 1897   Butyl ethylene 3655, 3657   Butyl maleate 1903

Table 7-4 (*Continued*)
PHYSICAL CONSTANTS OF ORGANIC COMPOUNDS

No.	Crystalline Form and Color	Specific Gravity	Melting Point °C.	Boiling Point °C.	Solubility in 100 Parts		
					Water	Alcohol	Ether
1084	col. lq.	$0.968^{20°}_{4}$	....	$273\text{-}57^{41mm}$	∞	∞	∞ ; ∞ bz.
1085	gas	....	....	16±	....	....	....
1086	lq.	$0.911^{0°}$	....	106.9	v. sl. s.	∞	∞
1087	lq.	$0.869^{20°}_{4}$	....	245-7	....	....	....
1088	lq.	$0.882^{20°}_{4}$	....	97	sl. s.	∞	∞
1089	lq.	$0.885^{20°}_{4}$	−95.3	98.2	$1.1^{22°}$	∞	∞
1090	col. lq.	$1.056^{20°}_{4}$	....	$118\text{-}20^{25mm}$	i.	∞	∞
1091	col. lq.	$1.048^{20°}_{4}$	....	121^{5mm}	i.	s.	....
1092	col. lq.	....	5.9	259-60	....	....	....
1093	wh. pd.	....	68-9	....	0.02	s.	s.; s. chl.
1094	col. lq.	....	....	$64\text{-}5^{5mm}$	....	....	....
1095	lq.	$1.615^{20°}_{4}$	−103.0	130.4	i.	∞	∞
1096	lq.	$1.597^{20°}$	−104	120.0	i.	∞	∞
1097	lq.	$1.603^{20°}_{4}$	−93.5	121.0	i.	∞	∞
1098	lq.	$1.570^{18.5°}_{15}$	−34	99	i.	∞	∞
1099	col. lq.	0.968	....	$75\text{-}6^{6mm}$	sl. s.	∞	∞
1100	col. lq.	$0.964^{30°}_{4}$	....	$60\text{-}27^{mm}$	....	....	....
1101	lq.	$0.974^{20°}_{4}$	....	237.8	....	....	....
1102	pr./aq.	....	101.5	d. 150	v. s.	v. s.	v. s.
1103	col. lq.	$0.837^{25°}_{4}$	−116	97-8	sl. s.	v. s.	v. s.
1104	lq.	$0.836^{20°}_{4}$	<−79	88	v. sl. s.	s.	s.
1105	lq.	....	....	65-7	....	....	....
1106	col. lq.	....	....	$185\text{-}6^{20mm}$	....	....	....
1107	lq.	$1.015^{20°}_{4}$	....	122.9	i.	∞	∞
1108	lq.	$0.911^{0°}$	....	77-9	....	∞	∞
1109	lq.	$0.870^{20°}_{20}$	....	67-8	sl. s.; d.	∞	....
1110	lq.	$0.868^{25°}$	....	$227\text{-}8^{15mm}$	i.	s.	s.
1111	wh. nd.	....	87-8	....	i.	s. h.	sl. s.
1112	col. lq.	$1.014^{25°}_{25}$	13	$147\text{-}56^{8mm}$	i.; ∞ act.	∞ Me al.	∞ bz.
1113	col. lq.	....	....	$128\text{-}32^{18mm}$	....	....	....
1114	col. lq.	....	....	247	....	....	....
1115	col. lq.	$0.967^{20°}_{20}$	....	$128\text{-}30^{8mm}$	....	....	....
1116	col. lq.	$0.930^{20°}_{4}$	....	210.3	i.	s.	s.
1117	lq.	....	....	248.5	i.	s.	s.
1118	lq.	$0.993^{17.5°}$	....	$227\text{-}8^{720mm}$	i.	∞	∞
1119	col. cr.	....	40.6	240	....	....	....
1120	nd./aq.	$0.908^{113°}_{4}$	99	236-8	s.	s.	s.
1121	col. cr.	$1.022^{25°}_{25}$	50	$196\text{-}9^{25mm}$	i.; v. s. bz.	v. s. Me al.; v. s. act.	v. s.; v. s. CCl₄
1122	col. cr.	....	73-7	....	s. h.	s.	v. s.
1123	col. lq.	$0.883^{15°}$	−89.6	146	i.	∞	∞
1124	col. lq.	$0.866^{20°}_{4}$	....	132.0-2.5	i.	∞	∞
1125	col. lq.	$0.888^{0°}_{4}$	−71	136.8	i.	∞	∞

Butyl malonate 1904
Butyl mustard oil 1133-6

Butyl oxalate 1905-6
Butyl phosphate 6121

Butyl phthalate 1908

Table 7-4 (*Continued*)
PHYSICAL CONSTANTS OF ORGANIC COMPOUNDS

No.	Name	Synonym	Formula	Beil. Ref.	Formula Weight
	Butyl				
1126	pyrrole (*n*)(*N*)	....	$C_4H_4N\cdot C_4H_9$	....	123.20
1127	ricinoleate (*n*)	....	$HO\cdot C_{17}H_{32}CO_2\cdot C_4H_9$	III-388	354.58
1128	ricinoleate (*iso*)	....	$HO\cdot C_{17}H_{32}CO_2\cdot C_4H_9$	III-388	354.58
1129	stearate (*n*)	....	$C_{17}H_{35}CO_2\cdot C_4H_9$	**II-352	340.59
1130	stearate (*iso*)	....	$C_{17}H_{35}CO_2\cdot C_4H_9$	*II-173	340.59
1131	thiocyanate (*n*)	....	$C_2H_5CH_2CH_2\cdot S\cdot CN$	**III-122	115.20
1132	thiocyanate (*iso*)	....	$(CH_3)_2CHCH_2\cdot S\cdot CN$	III-177	115.20
1133	*iso*-thiocyanate (*n*)	butyl mustard oil	$C_2H_5CH_2CH_2\cdot N:CS$	IV-158	115.20
1134	*iso*-thiocyanate	(*sec*)(*d*)	$C_4H_9\cdot N:CS$	IV-161	115.20
1135	*iso*-thiocyanate (*iso*)	*iso*-Bu mustard oil	$(CH_3)_2CHCH_2\cdot N:CS$	IV-171	115.20
1136	*iso*-thiocyanate (*tert*)	....	$(CH_3)_3C\cdot N:CS$	IV-175	115.20
1137	*p*-toluene sulfonate (*n*)	....	$CH_3\cdot C_6H_4\cdot SO_2\cdot O\cdot C_4H_9$	**XI-46	228.31
1138	urea (*n*)(*N*)	*n*-Bu carbamide	$C_4H_9\cdot NH\cdot CO\cdot NH_2$	*IV-371	116.16
1139	urea (*iso*)(*N*)	....	$C_4H_9\cdot NH\cdot CO\cdot NH_2$	IV-168	116.16
1140	urethane (*n*)(*N*)	Et *N-n*-Bu carbamate	$C_4H_9\cdot NH\cdot CO_2\cdot C_2H_5$	IV-158	145.20
1141	urethane (*iso*)	....	$C_4H_9\cdot NH\cdot CO_2\cdot C_2H_5$	IV-168	145.20
1142	*n*-valerate (*n*)	....	$CH_3(CH_2)_3CO_2\cdot C_4H_9$	II-301	158.24
1143	*iso*-valerate (*n*)	....	$(CH_3)_2CHCH_2CO_2\cdot C_4H_9$	**II-275	158.24
1144	*iso*-valerate (*sec*)	....	$(CH_3)_2CHCH_2CO_2\cdot C_4H_9$	II-312	158.24
1145	*iso*-valerate (*iso*)	....	$C_4H_9CO_2\cdot C_4H_9$	II-312	158.24
1146	2-xenyl-di (*p*-*tert*-butyl-phenyl) phosphate (5-*tert*)	phosphen 11	$(CH_3)_3C\cdot C_{12}H_8\cdot O\cdot PO[O\cdot C_6H_4\cdot C(CH_3)_3]_2$	....	570.72
1147	**Butylene** (*α*)	butene-1	$C_2H_5\cdot CH:CH_2$	I-203	56.11
1148	**Butylene** (*β*)(*cis*)	butene-2	$CH_3\cdot CH:CH\cdot CH_3$	I-205	56.11
1148. 1	**Butylene** (*β*)(*trans*)	butene-2	$CH_3\cdot CH:CH\cdot CH_3$	I-205	56.11
1149	**Butylene** (*γ*) (*iso*)	2-methyl-propene-1	$(CH_3)_2C:CH_2$	I-207	56.11
1150	*iso*-butyracetal (*iso*)	....	$(CH_3)_2CH\cdot CH\cdot O\cdot$ ⎤ $C(CH_3)_2\cdot CH_2\cdot O$ ⎦	....	144.22
1151	chlorohydrin (*β, γ*)	3-chloro-butanol-2	$CH_3CHCl\cdot CHOH\cdot CH_3$	I-373	108.57
1152	chlorohydrin (*α*) (*iso*)	1-chloro-2-Me-propanol-2	$(CH_3)_2C(OH)\cdot CH_2Cl$	I-382	108.57
1153	chlorohydrin (*β*) (*iso*)	2-chloro-2-Me-propanol-1	$(CH_3)_2CCl\cdot CH_2OH$	I-378	108.57
1154	glycol (*iso*)	2-Me-propandiol-1,2	$(CH_3)_2COH\cdot CH_2OH$	I-480	90.12
1155	oxide (*iso*)	*α,α* di Me-ethylene oxide	$(CH_3)_2C\cdot CH_2\cdot O$ ⎤⎦	XVII-11	72.11
1156	**Butyne**-1	ethyl-acetylene	$C_2H_5\cdot C:CH$	I-249	54.09
1157	**Butyne**-2	crotonylene	$CH_3\cdot C:C\cdot CH_3$	I-249	54.09
1158	**Butyrchloral**	triCl-butaldehyde	$CH_3\cdot CHCl\cdot CCl_2\cdot CHO$	I-664	175.44
1159	hydrate	butylchloral hydrate	$CH_3\cdot CHCl\cdot CCl_2\cdot CH(OH)_2$	I-664	193.46
1160	**Butyraldehyde** (*n*)	butanal	$C_2H_5\cdot CH_2\cdot CHO$	I-662	72.11
1161	oxime (*n*)	butyraldoxime	$C_2H_5\cdot CH_2\cdot CH:NOH$	I-663	87.12

Butyl salicylate 1092 Butyl sulfate 1913 Butyl sulfone 1918
Butyl sebacate 1909 Butyl sulfide 1914-6 Butyl tartrate 1919-20
Butyl succinate 1910-2 Butyl sulfite 1917 Butyl-toluene 4158-60

Table 7-4 (*Continued*)
PHYSICAL CONSTANTS OF ORGANIC COMPOUNDS

No.	Crystalline Form and Color	Specific Gravity	Melting Point °C.	Boiling Point °C.	Solubility in 100 Parts		
					Water	Alcohol	Ether
1126	yel. lq.	....	....	91-2^{55mm}	....	....	....
1127	lq.	0.906$^{22°}$	....	275^{13mm}	i.	....	s.
1128	lq.	0.903$^{22°}$	....	262^{9mm}	i.	s.	s.
1129	col. lq.	0.855-8$\frac{25}{25}$$^{°}$	27.5(19.5)	220-5^{25mm}	0.3$^{23°}$	s.	s.
1130	waxy	....	25	....	i.	....	....
1131	col. lq.	0.956$^{25°}$	....	185-6	i.	s.	s.
1132	col. lq.	....	−59	175.4	i.	∞	....
1133	lq.	0.956$^{11.2°}$	....	165^{724mm}	i.	s.	s.
1134	lq.	0.943$\frac{20}{4}$$^{°}$	....	159-63	i.	s.	s.
1135	lq.	0.964$\frac{14}{4}$$^{°}$	....	162	i.	s.	s.
1136	lq.	0.919$^{10°}$	10.5	140^{770mm}	i.	s.	s.
1137	lq.	1.120$\frac{20}{4}$$^{°}$	....	174-5^{10mm}	....	....	....
1138	col. nd./bz.	....	96	....	s.	s.	s.
1139	nd./act.	....	140.5-1.5	....	sl. s. act.	sl. s. bz.	v. sl. s.
1140	col. lq.	0.951$^{15°}$	−22	202-3	....	s.	s.
1141	col. lq.	0.943$\frac{20}{4}$$^{°}$	<−65	95-6^{15mm}	i.	s.	....
1142	lq.	0.870$\frac{15}{4}$$^{°}$	−93	186	v. sl. s.	∞	∞
1143	lq.	0.862$\frac{20}{4}$$^{°}$	....	168.8	....	....	....
1144	col. lq.	0.848$\frac{20}{4}$$^{°}$	....	163-4^{752mm}	i.	∞	∞
1145	col. lq.	0.874$\frac{0}{4}$$^{°}$	....	168.7	i.	∞	∞
1146	resin	1.07$\frac{60}{4}$$^{°}$	....	300-25^{5mm}	i.; ∞ CCl$_4$	∞ ; ∞ act.	∞ bz.
1147	col. gas	lq. 0.60$^{20°}$	−185.4	−6.3	i.	v. s.	v. s.
1148	col. gas	lq. 0.62$^{20°}$	−138.9	3.7	....	....	....
1148.1	col. gas	lq. 0.60$^{20°}$	−105.6	0.88			
1149	col. gas	lq. 0.59$^{20°}$	−140.4	−6.9	s. H$_2$SO$_4$	....	....
1150	lq.	0.890$\frac{20}{4}$$^{°}$	....	138	<1	....	....
1151	col. lq.	1.105$^{20°}$	....	136-7.5	6.6$^{20°}$	....	....
1152	col. lq.	1.061$\frac{20}{4}$$^{°}$	−20	127-9	sl. s.	s. HCl	....
1153	col. lq.	....	....	132-3 sl. d.	s. HCl	....	....
1154	lq.	0.994$\frac{20}{4}$$^{°}$	....	178.6	∞	....	....
1155	lq.	0.805$\frac{20}{4}$$^{°}$	....	52.4	5.8	s.	s.
1156	col. lq.	0.65$^{20°}$	−125.8	8.7	i.	s.	s.
1157	col. lq.	0.693$^{20°}$	−32.2	27.0	i.	....	....
1158	lq.	1.396$\frac{20}{4}$$^{°}$	....	164-5^{750mm}	....	....	....
1159	rhb.	1.694^{45}	78 sl. d.	d.	s. h.	v. s.	....
1160	col. lq.	0.817$\frac{20}{4}$$^{°}$	−99	75.7	4	∞	∞
1161	lq.	....	....	152^{715mm}	....	....	....

Table 7-4 (*Continued*)
PHYSICAL CONSTANTS OF ORGANIC COMPOUNDS

No.	Name	Synonym	Formula	Beil. Ref.	Formula Weight
	Butyraldehyde				
1162	trimer	para-butyraldehyde	$(C_4H_8O)_3$	*XIX-807	216.32
1163	**Butyraldehyde** (*iso*)	2-Me-propanal	$(CH_3)_2CH\cdot CHO$	I-671	72.11
1164	semicarbazone (*iso*)	....	$C_3H_7\cdot CH{:}N_3CH_3O$	III-103	129.16
1165	trimer	(para)	$(C_4H_8O)_3$	XIX-390	216.32
1166	**Butyric** acid (*n*)	butanoic acid	$C_2H_5\cdot CH_2\cdot CO_2H$	II-264	88.11
1167	acid (*iso*)	2-Me-propanoic acid	$(CH_3)_2CH\cdot CO_2H$	II-288	88.11
1168	amide (*n*)	*n*-butyramide	$C_2H_5\cdot CH_2\cdot CO\cdot NH_2$	II-275	87.12
1169	amide (*iso*)	*iso*-butyramide	$(CH_3)_2CH\cdot CO\cdot NH_2$	II-293	87.12
1170	anhydride (*n*)	....	$(C_2H_5\cdot CH_2\cdot CO)_2O$	II-274	158.20
1171	anhydride (*iso*)	....	$[(CH_3)_2CH\cdot CO]_2O$	II-292	158.20
1172	anilide (*n*)	*n*-butyranilide	$C_3H_7CO\cdot NHC_6H_5$	XII-252	163.22
1173	chloride (*n*)	butyryl chloride	$C_2H_5\cdot CH_2\cdot CO\cdot Cl$	II-274	106.55
1174	chloride (*iso*)	....	$(CH_3)_2CH\cdot CO\cdot Cl$	II-293	106.55
1175	nitrile (*n*)	*n*-butyronitrile	$C_2H_5\cdot CH_2\cdot CN$	II-275	69.11
1176	nitrile (*iso*)	*iso*-propyl cyanide	$(CH_3)_2CH\cdot CN$	II-294	69.11
1177	**Butyroin**	....	$C_3H_7\cdot CHOH\cdot CO\cdot C_3H_7$	I-840	144.22
1178	**Cacodyl**	....	$(CH_3)_2{:}As\cdot As{:}(CH_3)_2$	IV-615	209.98
1179	chloride	diMe-chloroarsine	$(CH_3)_2AsCl$	IV-607	140.45
1180	sulfide	....	$[(CH_3)_2As]_2S$	IV-608	242.05
1181	trichloride	....	$(CH_3)_2AsCl_3$	IV-612	211.35
1182	**Cacodylic** acid	....	$(CH_3)_2AsO\cdot OH$	IV-610	138.00
1183	oxide	....	$[(CH_3)_2As]_2O$	IV-608	225.98
1184	**Cadaverine**	pentamethylene diamine	$H_2N\cdot (CH_2)_5\cdot NH_2$	IV-266	102.18
1185	**Cadmium** diethyl	....	$(C_2H_5)_2Cd$	IV-677	170.52
1186	dimethyl	....	$(CH_3)_2Cd$	IV-677	142.47
1187	**Calciferol**	vitamin D_2	$C_{28}H_{43}OH$	....	396.66
1188	**Camphane**	....	$C_{10}H_{18}$	V-93	138.25
1189	**Camphene** (*dl*)	....	$C_{10}H_{16}$	V-156	136.24
1190	**Campholic acid**	(*d* or *l*)	$C_{10}H_{18}O_2$	IX-34	170.25
1191	**Camphor** (*d*)	....	$C_{10}H_{16}O$	VII-101	152.24
1192	dichloride (3)(α)	chlorocamphor	$C_{10}H_{16}Cl_2$	VII-117	207.15
1193	dichloride (3)(α')	....	$C_{10}H_{16}Cl_2$	VII-117	207.15
1194	dichloride (3)(β)	....	$C_{10}H_{16}Cl_2$	VII-117	207.15
1195	oxime	....	$C_9H_{16}C{:}NOH$	VII-112	167.25
1196	sulfonic acid (10)	Reychler's acid	$C_{10}H_{15}O\cdot SO_3H$	XI-315	232.30
1197	**Camphoric** acid (*dl*)	paracamphoric acid	$C_8H_{14}(CO_2H)_2$	IX-760	200.24
1198	acid (*d*)	....	$C_8H_{14}(CO_2H)_2$	IX-745	200.24
1199	acid (*iso*)(*l*)	....	$C_8H_{14}(CO_2H)_2$	IX-762	200.24
1200	anhydride (*dl*)	....	$C_{10}H_{14}O_3$	XVII-459	182.22
1201	**Camphoronic acid** (*l*)	....	$C_9H_{14}O_6$	II-837	218.21
1202	acid (*iso*)	....	$C_9H_{14}O_6$	II-835	218.21
1203	**Camphyl** amine (α)	....	$C_8H_{13}\cdot CH_2\cdot CH_2\cdot NH_2$	XII-40	153.27
1204	amine (β)	....	$C_8H_{13}\cdot CH_2\cdot CH_2\cdot NH_2$	XII-40	153.27
1205	**Capric** acid	decanoic acid	$CH_3\cdot (CH_2)_8\cdot CO_2H$	II-355	172.27
1206	aldehyde	decanal	$CH_3\cdot (CH_2)_8\cdot CH{:}O$	I-711	156.27
1207	nitrile	nonyl cyanide	$CH_3\cdot (CH_2)_8\cdot CN$	II-356	153.27
1208	**Caproic** acid (*n*)	hexanoic acid	$CH_3\cdot (CH_2)_4\cdot CO_2H$	II-321	116.16

Butyramido 1168-9
Butyranilide 1172
Butyrolactam 3936

Butyrophenone 5439-40
Butyrone 2786
Butyryl chloride 1173-4

C β acid 4534
Cacodyl hydride 2456
Caffeic acid 2335

Table 7-4 (*Continued*)
PHYSICAL CONSTANTS OF ORGANIC COMPOUNDS

No.	Crystalline Form and Color	Specific Gravity	Melting Point °C.	Boiling Point °C.	Solubility in 100 Parts		
					Water	Alcohol	Ether
1162	oil	0.918	<-20	$115\text{-}7^{17mm}$	....	....	....
1163	col. lq.	0.794^{20}_{4}	-65.9	$63\text{-}47^{57mm}$	112^{00}	∞ ; ∞ chl.	∞; ∞ bz.
1164	lf.	....	125.5-6.0	....	s. bz.	s.	i. pet.
1165	nd./al.	....	59-60	195 sl. d.	i.	s.	v. s.
1166	col. lq.	0.958^{20}_{4}	-5.5	164.1	∞	∞	∞
1167	col. lq.	0.949^{20}_{4}	-47	154.7	20^{200}	∞ ; ∞ chl.	∞
1168	rhb.	1.032	115-6	216	16.3^{150}	s.	sl. s.
1169	mn. pl.	1.013	129-30	216-20	v. s.	s.	sl. s.
1170	col. lq.	0.968^{20}_{20}	-75	199.5^{760mm}	d.	d.	∞
1171	col. lq.	0.950^{25}_{4}	-53.5	181.5^{734mm}	d.	d.	∞
1172	mn. pr.	1.134	92	189^{15mm}	i.	s.	s.
1173	col. lq.	1.028^{20}_{4}	-89	101-2	d.	d.	s.
1174	col. lq.	1.017^{20}_{4}	-90	92	d.	d.	s.
1175	col. lq.	0.795^{15}_{4}	-111.9	117.9	....	....	....
1176	col. lq.	0.775^{15}_{4}	-71.5	103.9	sl. s.	v. s.	v. s.
1177	lq.	$0.911^{16.7}_{4}$	....	180-90	....	....	....
1178	col. oil	1.447^{150}	-5	163	v. sl. s.	s.	s.
1179	col. lq.	1.505^{12}_{4}	<-45	106.5	i.	∞	i.
1180	oil	....	<-40	211	v. sl. s.	s.	s.
1181	cr./et.	....	50 d.	....	d.; s. CS_2	d. abs. al.	s. abs. et.
1182	tri.	....	200	....	82.9^{220}	19.5^{150}	i. abs. et.
1183	col. lq.	1.486^{150}	-57	150	sl. s.	s.	s.
1184	syrup	0.873^{25}_{4}	9	178-80	v. s.	v. s.	sl. s.
1185	col. lq.	1.653^{22}_{4}	-21	$64^{19.5mm}$	....	....	∞
1186	col. lq.	1.985^{17}_{4}	-4.5	105.5^{758mm}	d.	....	∞
1187	cr./act.	....	115-7	....	i.; s. act.	s.; s. chl.	s.
1188	pr.	....	152-3	160^{763mm}	i.	s. h.	s.
1189	cr.	0.822^{780}	50	159-60	i.	s.	s.
1190	mn./80% al.	....	106-7	255-60	0.02^{190}	64.5^{150}, 80% al.	s.
1191	trig.	0.999^{9}_{9}	179.5	207.4	0.1	120^{120}	v. s.
1192	mn.	....	93-4	244-7 sl. d	sl. s. h.	v. s. h.	v. s.
1193	....	....	117	....	220^{180}	....	....
1194	pr./al.	....	132.5	....	v. s. bz.	s.	v. s. ac.
1195	mn./al.	1.011^{160}	119-20	249-54 sl. d.	i.	v. s.	v. s.
1196	pr./ac.	....	195-6	....	v. s.	sl. s. ac.	v. sl. s.
1197	mn.	1.228	202-3	....	0.8^{250}; 10^{1000}	s.	s.
1198	mn.	1.186	187	....	0.6^{120}	s.	i. chl.
1199	cr./aq. al.	1.243	172	....	0.34^{200}	47.5^{200}, abs. al.	....
1200	nd./al.	....	221-3	270	25 chl.	1.5	4
1201	nd./aq.	....	164-5	$195\text{-}210^{13mm}$	12.5^{160}	75.8^{160}, abs. al.	7.4^{160}, abs. et.
1202	tri.	....	170	....	s.	s.	s.
1203	lq.	$0.874^{17.8}_{0}$	....	194-6	....	....	....
1204	lq.	0.870^{20}_{20}	....	205.5-6.5	....	....	....
1205	col. nd.	$0.888^{3.5}_{4}$	31.4	268.70	....	....	....
1206	lq.	0.828^{150}	....	$207\text{-}9^{755mm}$	0.003^{150}	s.; s. chl.	s.; s. bz.
1207	col. lq.	0.823^{15}_{4}	-14.5	235-7	....	....	....
1208	oily lq.	0.931^{15}_{4}	-4.0	205.4	1.10^{200}	s.	s.

Table 7-4 (*Continued*)
PHYSICAL CONSTANTS OF ORGANIC COMPOUNDS

No.	Name	Synonym	Formula	Beil. Ref.	Formula Weight
1209	**Caproic** acid (*iso*)	2-Me-pentanoic-5 acid	$(CH_3)_2CH \cdot (CH_2)_2 \cdot CO_2H$	II-327	116.16
1210	acid	3-Me-pentanoic-1 acid	$C_2H_5 \cdot CH(CH_3) \cdot CH_2 \cdot CO_2H$	II-331	116.16
1211	acid	2-Me-pentanoic-1 acid	$C_2H_5 \cdot CH_2 \cdot CH(CH_3) \cdot CO_2H$	II-326	116.16
1212	aldehyde (*n*)	hexanal	$CH_3 \cdot (CH_2)_4 \cdot CH:O$	I-688	100.16
1213	amide	caproamide	$C_5H_{11} \cdot CO \cdot NH_2$	II-324	115.18
1214	anhydride (*n*)	. . . .	$(C_5H_{11}CO)_2O$	II-324	214.31
1215	anilide (*n*)	*n*-caproanilide	$C_5H_{11}CO \cdot NH \cdot C_6H_5$	XII-255	191.28
1216	chloride (*n*)	*n*-caproyl chloride	$CH_3 \cdot (CH_2)_4 \cdot CO \cdot Cl$	II-324	134.61
1217	nitrile (*n*)	*n*-amyl cyanide	$CH_3 \cdot (CH_2)_4 \cdot CN$	II-324	97.16
1218	nitrile (*iso*)	*iso*-amyl cyanide	$(CH_3)_2CH(CH_2)_2CN$	II-329	97.16
1219	**Capryl** acetate†	(*sec*)-octyl acetate	$CH_3CO_2 \cdot C_8H_{17}$	II-134	172.27
1220	**Caprylic** acid (*n*)	octanoic acid	$CH_3 \cdot (CH_2)_6 \cdot CO_2H$	II-347	144.22
1221	acid	2-Et-hexanoic acid	$CH_3(CH_2)_3CH(C_2H_5) \cdot CO_2H$	II-349	144.22
1222	aldehyde (*n*)	octaldehyde	$CH_3 \cdot (CH_2)_6 \cdot CH:O$	I-704	128.22
1223	amide (*n*)	caprylamide	$C_7H_{15}CO \cdot NH_2$	II-349	143.23
1224	anhydride (*n*)	. . . .	$(C_7H_{15}CO)_2O$	II-348	270.42
1225	chloride (*n*)	octanoyl chloride	$CH_3 \cdot (CH_2)_6 \cdot CO \cdot Cl$	II-348	162.66
1226	nitrile (*n*)	heptyl cyanide	$CH_3 \cdot (CH_2)_6 \cdot CN$	II-349	125.22
1227	**Capsaicin**	(2;1,4)	$CH_3O \cdot C_6H_3(OH) \cdot CH_2 \cdot NH \cdot CO \cdot C_9H_{17}$	*XIII-322	305.42
1228	**Carbamic** acid	known only in derivs.	$H_2N \cdot CO \cdot OH$	III-20	61.04
1229	chloride	urea chloride	$H_2N \cdot CO \cdot Cl$	III-31	79.49
1230	**Carbasone**	*p*-carbamylamino-phenyl-arsonic acid	$H_2NCONH \cdot C_6H_4 \cdot AsO:(OH)_2$	XVI-880	260.08
1231	**Carbazole**	diphenyleneim-ine; dibenzopyrrole	$C_6H_4 \cdot NH \cdot C_6H_4$	XX-433	167.21
1232	**Carbitol**	diethylene glycol mono-Et ether	$HO \cdot (CH_2)_2 \cdot O \cdot (CH_2)_2 \cdot O \cdot C_2H_5$	**I-520	134.18
1233	acetate	. . . .	$CH_3CO_2 \cdot C_6H_{13}O_2$	**II-155	176.21
1234	**Carbodiphenyl-imide**	(α)	$C_6H_5 \cdot N:C:N \cdot C_6H_5$	XII-449	194.24
1235	(β)	. . . .	$C_6H_5 \cdot N:C:N \cdot C_6H_5$	XII-450	194.24
1236	**Carbohydrazide**	carbazide	$(H_2N \cdot NH)_2:CO$	III-121	90.09
1237	**Carbon** dioxide	carbonic anhydride	CO_2	III-4	44.01
1238	disulfide	. . . .	CS_2	III-197	76.14
1239	monoxide	. . . .	CO	I-720	28.01
1240	suboxide	dioxo-allene	$O:C:C:C:O$	I-805	68.03
1241	subsulfide	. . . .	$S:C:C:C:S$	III-207	100.16
1242	tetrabromide	tetrabromo-methane	CBr_4	I-69	331.65
1243	tetrachloride	tetrachloro-methane	CCl_4	I-65	153.82

† See also Nos. 4959-60.

Caproamide 1213		Caprylamide 1223	Caprylone 2265
Caproanilide 1215	Caprone 1743	Caprylamine 4970-2	Carbamic nitrile 1577
Caproic isonitrile 453	Caprophenone (*iso*) 481-2	Caprylene 4984	Carbamide 6386
	Capryl alcohol 4962	Caprylidene 4991	Carbamides, cf. ureas.

Table 7-4 (*Continued*)
PHYSICAL CONSTANTS OF ORGANIC COMPOUNDS

No.	Crystalline Form and Color	Specific Gravity	Melting Point °C.	Boiling Point °C.	Solubility in 100 Parts		
					Water	Alcohol	Ether
1209	col. oil	0.923^{20}_{4}	−33	199.5	v. sl. s.	s.	s.
1210	col. lq.	0.924^{19}	....	197-8	....	....	....
1211	col. lq.	0.928^{18}_{0}	....	193^{748mm}	0.57^{17}	....	....
1212	col. lq.	0.834^{20}_{20}	....	128.6	0.5^{20}	....	....
1213	pl.	0.999	100-1	....	s. h.	s.	s.
1214	col. oil	0.928^{17}	−40.6	241-3 sl. d.	....	s.	....
1215	pr./al.	1.112	95	....	....	v. s.	v. s.
1216	col. lq.	0.981^{15}_{4}	−87.3	152.6	d.	d.; s. chl.	s.
1217	col. lq.	0.809^{15}_{25}	−79.4	164.1	i.	v. s.	v. s.
1218	lq.	0.804^{20}_{4}	−51.1	155.5	i.	∞	∞
1219	col. lq.	0.863^{14}_{4}	....	$194-5^{744mm}$	i.	s.	s.
1220	col. lf.	0.908^{20}_{4}	16.3	239.3	0.25^{100}; 0.07^{15}	s.; s. chl.; s. pet.	s.; s. CS$_2$
1221	lq.	0.908^{20}_{20}	<0	226.9	0.25^{20}	....	....
1222	lq.	0.821^{20}	....	167-70	....	....	....
1223	col. lf.	....	105-10	>200 d.	0.45^{100}	s.	s.
1224	lq.	0.907^{18}_{4}	−1	280-5	....	s.	....
1225	col. lq.	0.953^{15}_{4}	−61.0	195.6	d.	d.	s.
1226	lq.	0.817^{15}_{4}	−45.6	204.6	....	....	....
1227	cr./aq. al.	....	64-5	>210, in vac.	i. c.; s. bz.	s.; s. alk.	s.; s. chl.
1228	....	....	....	....	....	....	....
1229	col. lq.	....	50±	61-2 d.	d.	d.	....
1230	col. cr.	....	174	....	i.	s. alk.	i. HCl
1231	lf. or pl.	....	246-7	354.8	i.; s. bz.; s. chl.	0.92^{14}; 3.88 h.	sl. s.; s. pet.
1232	col. lq.	0.990^{20}_{20}	....	201.9	∞	v. s.	s.
1233	col. lq.	1.011^{20}_{20}	−25±	217.7	∞	∞	∞
1234	syrup	....	....	330-1	v. s. bz.	....	....
1235	cr.	....	168-70	....	v. sl. s. bz.	v. sl. s.	v. sl. s.
1236	nd./aq. al.	....	154	....	i. bz.	i. chl.	i.
1237	col. gas	1.53 (A) $1.101^{-37°}$ (lq.) $1.56^{-79°}$ solid	$-56^{5.2atm.}$	subl. −78.5 cc.	$179.70^{°}$ cc. See special table	s. a.	s. alk.
1238	col. lq.	1.263^{20}_{4} 2.63(A)	−111.5	46.3	0.2^{0}; 0.014^{50}	∞	∞
1239	poison. gas	0.814^{-195}_{4} 0.968(A)	−207	−191.5	0.0044^{0}; 3.5^{0} cc.	s.	s. Cu$_2$Cl$_2$
1240	gas	1.114^{0}	−107	7^{761mm}	d.	....	s.
1241	red lq.	....	....	in vac. 60-70*	v. s. bz.	v. s.	v. s.
1242	col. mn.	3.42	α48.4 β90.1	189.5	i.	s.	s.
1243	col. lq.	1.594^{20}_{4}	−22.96	76.8	0.097^{0}; 0.08^{20}	∞	∞

* Partial polymerization.

Table 7-4 (*Continued*)
PHYSICAL CONSTANTS OF ORGANIC COMPOUNDS

No.	Name	Synonym	Formula	Beil. Ref.	Formula Weight
1244	**Carbon** tetrafluoride	tetrafluoro-methane	CF_4	I-59	88.00
1245	tetraiodide	tetraiodo-methane	CI_4	I-74	519.63
1246	**Carbonyl** bromide	....	$O:C:Br_2$	III-20	187.83
1247	chloride	phosgene	$O:C:Cl_2$	III-13	98.92
1248	sulfide	carbon oxysulfide	$O:C:S$	III-131	60.07
1249	**Carbothialdine**	....	$NH_2CS\cdot S\cdot N(CHCH_3)_2$	XXVI-9	162.28
1250	**Carnine**	....	$C_7H_8O_3N_4\cdot H_2O$	....	214.18
1251	**Carnosine**	$N(\alpha)$-β-alanyl-histidine	$N_2C_3H_3\cdot CH_2\cdot CH\cdot (CO_2H)\cdot NHC_3H_6ON$	XXV-516	226.24
1252	**Carotene** (α)	provitamin A	$C_{40}H_{56}$	XXX-91	536.89
1253	**Carotene** (β)	provitamin A	$C_{40}H_{56}$	XXX-87	536.89
1254	**Carotene** (γ)(*dl*)	provitamin A	$C_{40}H_{56}$	XXX-92	536.89
1255	**Carvacrol**	2-Me-5-*iso*-Pr-phenol	$(CH_3)(C_3H_7)C_6H_3\cdot OH$	VI-527	150.22
1256	**Carvacrylamine** (2;1,4)	2-NH_2-*p*-cymene	$H_2N\cdot C_6H_3(CH_3)\cdot C_3H_7$	XII-1171	149.24
1257	**Carvenone** (*dl*)	....	$C_{10}H_{16}O$	VII-78	152.24
1258	**Carvo**-menthene (*d*)	*p*-menthene-1	$C_{10}H_{18}$	V-84	138.25
1259	menthol	....	$C_{10}H_{19}OH$	VI-26	156.27
1260	menthone	....	$C_{10}H_{18}O$	VII-34	154.25
1261	**Carvone** (*d*)	carvol	$C_{10}H_{14}O$	VII-153	150.22
1262	oxime	carvoxime	$C_{10}H_{14}:NOH$	VII-156	165.24
1263	**Cedrene**	....	$C_{15}H_{24}$	V-461	204.36
1264	**Cedriret**	cörulignone	$[(CH_3O)_2C_6H_2\cdot O\cdot]_2$	VIII-537	304.30
1265	**Cellosolve**	2-ethoxy-ethanol-1	$C_2H_5O(CH_2)_2OH$	I-467	90.12
1266	acetate	....	$CH_3CO_2C_4H_9O$	II-141	132.16
1267	**Cellulose**	....	$(C_6H_{10}O_5)x$	....	162.14
1270	acetate, tri-	....	$C_6H_7O_2(O_2CCH_3)_3$	....	288.26
1271	**Cerotene**	....	$C_{26}H_{52}$	I-227	364.70
1272	**Cerotic acid**	....	$C_{25}H_{51}\cdot CO_2H$	II-394	396.70
1273	**Ceryl alcohol**	....	$C_{26}H_{53}\cdot OH$	I-432	382.72
1274	**Cetyl** acetate	*n*-hexadecyl acetate	$CH_3CO_2\cdot C_{16}H_{33}$	II-136	284.49
1275	alcohol	hexadecanol	$CH_3(CH_2)_{14}CH_2OH$	I-429	242.45
1276	bromide	hexadecyl bromide	$CH_3(CH_2)_{14}CH_2Br$	I-172	305.35
1277	iodide (*n*)	1-iodohexadecane	$CH_3(CH_2)_{14}CH_2I$	I-172	352.35
1278	palmitate	....	$C_{15}H_{31}CO_2\cdot C_{16}H_{33}$	II-373	480.87
1279	**Chaulmoogric acid**	....	$C_5H_7\cdot(CH_2)_{12}\cdot CO_2H$	IX-80	280.45
1280	**Chavibetol** (*iso*) (5;1,2)	propenyl guaiacol	$CH_3\cdot CH:CH\cdot C_6H_3\cdot (OH)OCH_3$	VI-956	164.21
1281	**Chloral**	trichloro-acetalde-hyde	$Cl_3C\cdot CH:O$	I-616	147.39
1282	alcoholate	....	$Cl_3C\cdot CH(OH)\cdot OC_2H_5$	I-621	193.46
1283	ammonia	....	$Cl_3C\cdot CH(OH)NH_2$	I-624	164.42
1284	cyanohydrin	tri-Cl-lactic nitrile	$Cl_3C\cdot CHOH\cdot CN$	III-288	174.41
1285	formamide	chlor(al)amide	$Cl_3C\cdot CHOH\cdot NH\cdot CO\cdot H$	II-27	192.43

Table 7-4 (*Continued*)
PHYSICAL CONSTANTS OF ORGANIC COMPOUNDS

No.	Crystalline Form and Color	Specific Gravity	Melting Point °C.	Boiling Point °C.	Solubility in 100 Parts		
					Water	Alcohol	Ether
1244	gas	$1.96^{-184°}$	−184	−128	sl. s.	....	....
1245	red, cubic	$4.32^{20.2°}$	d.	subl. in vac. 90-100	i.; d. h.	d. h.	s.
1246	col. lq.	$2.45^{15°}$	−80	64-5	v. sl. d.	....	....
1247	poison. gas	$1.371\frac{20°}{4}$	−127.9	7.6	v. sl.; sl. d.	v. s. bz.	v. s. ac.
1248	col. gas	2.105(A) 1.24^{-87}	−138.2	-50.2^{760mm}	$80^{14°}$ cc.	s.	s.
1249	cr./al.	....	....	....	i.	s. h.	i.; s. a.
1250	cr.	....	d. 230-9	....	s. h.	i.	i.
1251	col. nd./al.	....	254 d.	....	$32.3^{25°}$	sl. s.	....
1252	red cr./pet.	$1.000\frac{20°}{20}$	187	....	i.; v. s. chl.	sl. s.; s.bz.	s.; s. CS$_2$
1253	red/bz.-al.	$1.000\frac{20°}{20}$	184	....	i.; v. s. chl.	sl. s.; s. bz.	s.; s. CS$_2$
1254	red pr.	....	177-8	....	i.; s. chl.	s. bz.	s. CS$_2$
1255	col. lq.	$0.977\frac{20°}{4}$	0.5	237.9	v. sl. s.	∞	∞
1256	oil	$0.994^{20°}$	−16	241	v. sl. s.	s.	s.
1257	lq.	$0.926\frac{20°}{4}$	....	233	i.	....	....
1258	lq.	$0.829^{20°}$	....	174-6	....	....	....
1259	oil	$0.908\frac{20°}{4}$	....	220	....	....	....
1260	oil	....	....	222-3	....	....	....
1261	col. lq.	$0.961\frac{20°}{4}$	....	230^{755mm}	i.	∞ ; 20, 60% al.	∞
1262	mn. pl./al.	$1.016^{73°}$	71-2	....	s.	s.	....
1263	col. lq.	$0.930^{15°}$	....	262-3^{750mm}	....	....	....
1264	b. bl. nd.	....	d.	....	i.	s. phenol	s. H$_2$SO$_4$
1265	col. lq.	$0.931\frac{20°}{4}$	−70	135.1	∞	∞ ; ∞ chl.	∞ ; ∞ act.
1266	col. lq.	$0.975\frac{20°}{4}$	−61.7	156.3	22	∞	∞
1267	amor.	1.3-1.4	....	....	i.; *	i.	i.
1270	fl.	....	....	....	i.	s. ac.	i.; s. chl.
1271	cr.	....	57-8	....	....	....	....
1272	col. cr.	$0.836\frac{79}{4}$	82.5	....	i.	v. s. h.	s. h.
1273	cr.	....	80	....	i.	s.	....
1274	nd.	$0.859\frac{25°}{4}$	18.5	200-1^{15mm}	i.	v. sl. s. c.	....
1275	lf.	$0.815\frac{55°}{1}$	49.2	189.5^{15mm}	i.	s.	s.; s. chl.
1276	col. lq.	....	16.5-8	187	....	....	....
1277	lf.	$1.123\frac{20°}{4}$	22-3	211^{15mm}	i.	s.	s.
1278	lf.	$0.832\frac{25°}{4}$	53-4	....	i.; v. s. bz.	$0.05^{22°}$	$21^{22°}$
1279	lf./al.	....	68	247-8^{20mm}	i.	v. sl. s.	s.; s. chl.
1280	cr.	....	95-6	147^{19mm}	....	....	....
1281	col. lq.	$1.505\frac{25°}{4}$	−57 ±	97.6^{768mm}	v. s.	∞	∞
1282	nd.	$1.143^{40°}$	56-7 (50)	115-6	s.	s.	s.
1283	col. nd.	....	63-4	100 ±, d.	v. sl. s.	s.; s. bz.	v. s.
1284	col. cr./aq.	....	61	215-20 sl. d.	s.	s.	s.
1285	col. cr.	....	116-8	d.	5; d. h.	40; d. h.	v. s.

Table 7-4 (Continued)
PHYSICAL CONSTANTS OF ORGANIC COMPOUNDS

No.	Name	Synonym	Formula	Beil. Ref.	Formula Weight
1286	**Chloral** hydrate	....	$Cl_3C \cdot CH(OH)_2$	I-619	165.40
1287	imide (trimer)	tri-Cl-ethylidene-imide	$(Cl_3C \cdot CH:NH)_3$	XXVI-9	439.21
1288	urethane	ural(ine)	$C_2H_2OCl_3 \cdot NHCO_2 \cdot C_2H_5$	III-24	236.48
1289	**Chloralose** (α)	glucochloral	$Cl_3C \cdot CH:O_2:C_6H_7O \cdot (OH)_3$	XXXI-151	309.53
1290	**Chloramine T**	chlorazone (1,4)	$CH_3 \cdot C_6H_4 \cdot S:O \cdot (ONa):NCl \cdot H_2O$	XI-107	245.66
1291	**Chloro-**acenaphthene	(3)	$Cl \cdot C_{10}H_5(CH_2)_2$	*V-276	188.66
1292	acetal	....	$ClCH_2 \cdot CH(OC_2H_5)_2$	I-611	152.62
1293	acetaldehyde	2-chloro-ethanal	$ClCH_2 \cdot CH:O$	I-610	78.50
1294	acetamide	....	$ClCH_2 \cdot CO \cdot NH_2$	II-199	93.51
1295	acetamide (*N*)	acetochloramide	$CH_3 \cdot CO \cdot NHCl$	II-181	93.51
1296	acetanilide (*o*)	....	$Cl \cdot C_6H_4 \cdot NH \cdot COCH_3$	XII-599	169.61
1297	acetanilide (*m*)	....	$Cl \cdot C_6H_4 \cdot NH \cdot COCH_3$	XII-604	169.61
1298	acetanilide (*p*)	....	$Cl \cdot C_6H_4 \cdot NH \cdot COCH_3$	XII-611	169.61
1299	acetic acid	....	$Cl \cdot CH_2 \cdot CO \cdot OH$	II-194	94.50
1300	acetic anhydride	....	$(Cl \cdot CH_2 \cdot CO)_2O$	II-199	170.98
1301	acetone	....	$Cl \cdot CH_2 \cdot CO \cdot CH_3$	I-653	92.53
1302	acetophenone (*p*)	....	$Cl \cdot C_6H_4 \cdot CO \cdot CH_3$	VII-281	154.60
1303	acetophenone (ω)	phenacyl chloride	$C_6H_5 \cdot CO \cdot CH_2Cl$	VII-282	154.60
1304	acetyl chloride	....	$Cl \cdot CH_2 \cdot CO \cdot Cl$	II-199	112.94
1305	acetylene	chloro-ethyne	$CH:C \cdot Cl$ (or $C:CHCl$)	I-244	60.48
1306	acrylic acid (α)	....	$CH_2:CCl \cdot CO_2H$	II-401	106.51
1307	acrylic acid (β)	....	$ClCH:CH \cdot CO_2H$	II-400	106.51
1308	2-amino-thio-phenol hydro-chloride	(4;1,2)	$Cl \cdot C_6H_3(SH) \cdot NH_2 \cdot HCl$	....	196.10
1309	aniline (*o*)	2-chloroaniline	$Cl \cdot C_6H_4 \cdot NH_2$	XII-597	127.57
1310	aniline (*m*)	3-chloroaniline	$Cl \cdot C_6H_4 \cdot NH_2$	XII-602	127.57
1311	aniline (*p*)	4-chloroaniline	$Cl \cdot C_6H_4 \cdot NH_2$	XII-607	127.57
1312	*o*-anisidine	4-Cl-2-NH_2-anisole	$Cl \cdot C_6H_3(OCH_3)NH_2$	XIII-383	157.60
1313	*o*-anisidine HCl	....	$C_7H_8ONCl \cdot HCl$	....	194.06
1314	anisole (*p*)	....	$Cl \cdot C_6H_4 \cdot OCH_3$	VI-186	142.59
1315	anthranilic acid	5-Cl-2-NH_2-benzoic	$Cl \cdot C_6H_3(NH_2)CO_2H$	XIV-365	171.58
1316	anthraquinone (1)	....	$C_6H_4(CO)_2C_6H_3Cl$	VII-787	242.66
1317	anthraquinone (2)	....	$C_6H_4(CO)_2C_6H_3Cl$	VII-787	242.66
1318	benzaldehyde (*o*)	....	$Cl \cdot C_6H_4 \cdot CH:O$	VII-233	140.57
1319	benzaldehyde (*m*)	....	$Cl \cdot C_6H_4 \cdot CH:O$	VII-234	140.57
1320	benzaldehyde (*p*)	....	$Cl \cdot C_6H_4 \cdot CH:O$	VII-235	140.57
1321	benzamide (*o*)	....	$Cl \cdot C_6H_4 \cdot CO \cdot NH_2$	IX-336	155.58
1322	benzamide (*m*)	....	$Cl \cdot C_6H_4 \cdot CO \cdot NH_2$	IX-338	155.58
1323	benzamide (*p*)	....	$Cl \cdot C_6H_4 \cdot CO \cdot NH_2$	IX-341	155.58
1324	benzene	phenyl chloride	$C_6H_5 \cdot Cl$	V-199	112.56
1325	benzene-3,5-di-sulfonic acid	(1;3,5)	$Cl \cdot C_6H_3(SO_3H)_2$	*XI-49	272.68

Table 7-4 (*Continued*)
PHYSICAL CONSTANTS OF ORGANIC COMPOUNDS

No.	Crystalline Form and Color	Specific Gravity	Melting Point °C.	Boiling Point °C.	Solubility in 100 Parts		
					Water	Alcohol	Ether
1286	mn. pr.	1.619^{50}_{4}°	51.7*	96.3^{764} d.	$474^{17°}$	v. s.	s.; s. h. CS_2
1287	rhb./bz.-al.	....	150–5† d.	....	i.; s. bz.	2; s. chl.	s.; s. oil
1288	lf./al.-et.	....	103 sl. d.	....	i.; d. h.	s.	s.
1289	col. nd./ al. or et.	....	187	....	$0.9^{15°}$	$6.6^{21°}$; $0.07^{21°}$ chl.	v. s.
1290	yel. wh. pd.	....	anh. expl. 175–80	....	14c.; 50 h.	d.; i. bz.	i.; i. chl.
1291	nd./al.	1.195^{70}_{4}°	69–70	319^{771mm}	....	....	....
1292	lq.	$1.026^{15°}$	....	157	sl. s.	∞	∞
1293	lq.	....	....	85^{748mm}	....	....	....
1294	mn. pr.	....	120–1	224–5 d.	$10^{24°}$	$9.5^{24°}$ abs.	v. sl. s.
1295	lf.	....	110	....	v. s.	....	s.
1296	nd./aq. ac.	....	88	....	i.	s.	s. bz.
1297	nd./aq. ac.	....	72.5	....	s. CS_2	s.	s. bz.
1298	rhb.	$1.385^{22°}$	176–7	....	sl. s.	s.	v. s.
1299	col. cr.	1.58^{20}_{20}°	α62.8 β56.3 γ50.7 δ43.8(?)	189.4	v. s.; v. s. bz.	s.; s. chl.	s.
1300	pr./bz.	....	46	163^{116mm}	v. sl. s. lg.	sl. s. bz.	v. s.
1301	col. lq.	$1.162^{16°}$	−44.5	121	∞	∞	∞
1302	cr.	$1.188^{20°}$	20	232	i.	∞	∞
1303	rhb.	$1.324^{15°}$	54.0	245–7	i.; s. bz.	v. s.	v. s.
1304	col. lq.	1.498^{20}_{20}°	....	105	d.	d.	....
1305	gas	....	expl.	−30	....	....	....
1306	nd.	....	65	subl.	s.	s.	s.
1307	lf.	....	84–5	....	s.	s.	s.
1308	yel. cr.	....	208 d.	....	s.	s.	i.
1309	lq.	1.213^{20}_{4}°	0	210.5	i.; s. a.	....	s.
1310	lq.	1.216^{20}_{4}°	−10.4	230^{767mm}	i.; s. a.	....	s.
1311	rhb.	$1.427^{19°}$	70–1	230–1	s. h.	s.; s. act.	s.; s. CS_2
1312	nd.	....	83–4	....	i. pet.	s.	s.
1313	col. nd.	....	....	....	s.	sl. s.	i.
1314	lq.	....	<−18	198	i.	s.	s.; s. chl.
1315	nd./al.	....	211–2	....	i.; v. s. act.	$10^{25°}$; s. bz.	i. c.
1316	yel. nd./al.	....	162	subl.	i.; s. ac.	sl. s. h.	s. h. bz.
1317	nd./al.	....	208–9	....	i.	s. h. bz.	s. ac.
1318	nd.	$1.298°$	11	208^{748mm}	v. sl. s.	v. s.	v. s.
1319	pr.	$1.250^{15°}$	17–8	213–4	v. sl. s.	v. s.	v. s.
1320	pr.	$1.196^{61°}$	47.8	213^{748mm}	s. h.	v. s.	v. s.
1321	rhb.	$1.34^{18°}$	142.5	....	s. h.	s.	s.
1322	nd.	....	134.5	....	s. h.	s.	s.
1323	nd./et.	....	179	....	s. h.	v. s.	v. s.
1324	col. lq.	1.106^{20}_{4}°	−45.2	132.0	$0.049^{20°}$	∞; ∞ bz.	∞
1325	hyg. nd.	....	d. 100	....	....	....	....

* Variable m. p., depends upon dissociation.
† Mn. pr./al., m. p. 105–6.
Chloro–acetoacetanilide 41
Chloro–acetonitrile 1399

Chloro–allylene 5343
Chloro–aminoanisole 1312
Chloro–aminotoluene 1469–70

Table 7-4 (*Continued*)
PHYSICAL CONSTANTS OF ORGANIC COMPOUNDS

No.	Name	Synonym	Formula	Beil. Ref.	Formula Weight
1326	**Chloro** benzene sul-fonic acid (1,4)	....	$Cl \cdot C_6H_4 \cdot SO_3H$	XI-54	192.62
1327	benzoic acid (*o*)	....	$Cl \cdot C_6H_4 \cdot CO_2H$	IX-334	156.57
1328	benzoic acid (*m*)	....	$Cl \cdot C_6H_4 \cdot CO_2H$	IX-337	156.57
1329	benzoic acid (*p*)	chlorodracylic acid	$Cl \cdot C_6H_4 \cdot CO_2H$	IX-340	156.57
1330	benzonitrile (*p*)	....	$Cl \cdot C_6H_4 \cdot CN$	IX-341	137.57
1331	benzophenone (*p*)	....	$Cl \cdot C_6H_4 \cdot CO \cdot C_6H_5$	VII-419	216.67
1332	benzothiazole (2)	....	$C_6H_4 \cdot N : C(Cl) \cdot S$	XXVII-44	169.63
1333	benzoxazole (2)	....	$C_6H_4 \cdot N : C(Cl) \cdot O$	XXVII-43	153.57
1334	benzoyl chloride	(1,2)	$Cl \cdot C_6H_4 \cdot CO \cdot Cl$	IX-336	175.02
1335	benzoyl chloride	(1,3)	$Cl \cdot C_6H_4 \cdot CO \cdot Cl$	IX-338	175.02
1336	benzoyl chloride	(1,4)	$Cl \cdot C_6H_4 \cdot CO \cdot Cl$	IX-341	175.02
1337	benzyl bromide	(1,2)	$Cl \cdot C_6H_4 \cdot CH_2Br$	*V-155	205.49
1338	benzyl bromide	(1,4)	$Cl \cdot C_6H_4 \cdot CH_2Br$	V-307	205.49
1339	benzyl chloride	(1,4)	$Cl \cdot C_6H_4 \cdot CH_2Cl$	V-297	161.03
1340	bromobenzene	(1,3)	$Cl \cdot C_6H_4 \cdot Br$	V-209	191.46
1341	bromobenzene	(1,4)	$Cl \cdot C_6H_4 \cdot Br$	V-209	191.46
1342	bromoethane (1,1)	ethylidene chloro-bromide	$CH_3 \cdot CHClBr$	I-89	143.42
1343	bromoethane (1,2)	....	$Cl \cdot CH_2 \cdot CH_2 \cdot Br$	I-89	143.42
1344	bromomethane	....	$Cl \cdot CH_2Br$	I-67	129.39
1345	buta-1,3-diene (2)	chloroprene	$CH_2 : CCl \cdot CH : CH_2$	....	88.54
1346	*tert*-butylphenol	(4;1,2)	$C_4H_9 \cdot C_6H_3(OH)Cl$	....	184.67
1347	butyric acid (*α*)(*dl*)	....	$C_2H_5 \cdot CHCl \cdot CO_2H$	II-276	122.55
1348	butyric acid (*β*)(*dl*)	....	$CH_3 \cdot CHCl \cdot CH_2 \cdot CO_2H$	II-277	122.55
1349	butyric acid (*γ*)	....	$Cl \cdot (CH_2)_3 \cdot CO_2H$	II-278	122.55
1350	*iso*-butyric acid (*α*)	....	$(CH_3)_2 CCl \cdot CO_2H$	II-295	122.55
1351	butyronitrile (*γ*)	....	$Cl \cdot (CH_2)_3 \cdot CN$	II-278	103.55
1352	camphor (*α*)†	3-Cl-*d*-camphor	$Cl \cdot C_{10}H_{15}O$	VII-117	186.68
1353	coumarin (6)	....	$Cl \cdot C_6H_3 \cdot (CH)_2 \cdot CO \cdot O$	XVII-331	180.59
1354	*m*-cresol (*o*)	2-Cl-3-OH-toluene	$CH_3 \cdot C_6H_3(OH)Cl$	**VI-355	142.59
1355	*m*-cresol (*p*)	4-Cl-3-OH-toluene	$CH_3 \cdot C_6H_3(OH)Cl$	*VI-187	142.59
1356	*m*-cresol (6)	6-Cl-3-OH-toluene	$CH_3 \cdot C_6H_3(OH)Cl$	VI-381	142.59
1357	*p*-cresol (*o*)	2-Cl-4-OH-toluene	$CH_3 \cdot C_6H_3(OH)Cl$	VI-402	142.59
1358	crotonic acid (*α*)	....	$CH_3 \cdot CH : CCl \cdot CO_2H$	II-414	120.54
1359	dibromoethane	(1,1,1)	$CH_3 \cdot CClBr_2$	I-92	222.32
1360	dibromoethane	(1,1,2)	$BrCH_2 \cdot CHClBr$	I-92	222.32
1361	dibromomethane	....	$ClCHBr_2$	I-68	208.29
1362	diethyl ether (*β*)	....	$C_2H_5O \cdot CH_2 \cdot CH_2Cl$	I-337	108.57
1363	diiodomethane	....	$ClCHI_2$	I-72	302.28
1364	dinitrobenzene	(3;1,2)	$Cl \cdot C_6H_3(NO_2)_2$	*V-137	202.55
1365	dinitrobenzene (*α*)	(4;1,2)	$Cl \cdot C_6H_3(NO_2)_2$	V-262	202.55
1366	dinitrobenzene	(2;1,3)	$Cl \cdot C_6H_3(NO_2)_2$	V-263	202.55
1367	dinitrobenzene (*α*)	(4;1,3)	$Cl \cdot C_6H_3(NO_2)_2$	V-263	202.55

† See also Nos. 1192-4.
Chloro-biphenyl 1373-5
Chloro-bromopropane 6279
Chloro-bromotoluene 913-4

Chloro-butane 1072-5
Chloro-*iso*-butylene 1567, 4060
Chlorocarbonates, cf. chloroformates.
Chloro-cyanogen 1591

Table 7-4 (Continued)
PHYSICAL CONSTANTS OF ORGANIC COMPOUNDS

No.	Crystalline Form and Color	Specific Gravity	Melting Point °C.	Boiling Point °C.	Solubility in 100 Parts		
					Water	Alcohol	Ether
1326	nd.	....	68	$147\text{-}8^{0.1mm}$	s.	s.	i.; i. bz.
1327	mn./aq.	$1.544^{20°}$	140.2	....	$0.208^{25°}$	s.	s.
1328	pr.	....	154.3	....	$0.041^{25°}$ s.h.	s.	s.
1329	tri.	$1.541^{24°}$	239.7	subl.	$0.008^{25°}$	s.	s.
1330	nd./al.	....	93-4	223^{750mm}	i.; v. s. bz.	v. s.	v. s.
1331	nd./et.-al.	....	77-8	332^{771mm}	i.; $95^{25°}$ bz.	$3^{25°}$ Me al.	$57^{25°}$
1332	lq.	....	24±	248	....	s.	....
1333	cr.	....	7(1)	201-2 d.	i.	s. HCl	....
1334	lq.	....	−4	$229\text{-}30^{773mm}$	d.	d.	....
1335	lq.	....	....	225	d.	d.	....
1336	lq.	1.377	15-6	220-2	d.	d.	....
1337	col. lq.	....	....	$112\text{-}4^{15mm}$	....	....	....
1338	nd./al.	....	49-50	....	v. s. bz.	s.; v. s. ac.	v. s.
1339	nd.	....	29	213-4	i.	v. s. h.	v. s.
1340	col. oil	$1.630^{20°}_{4}$	−21	195-6	i.	v. s.	v. s.
1341	mn. pr.	$1.576^{71°}_{4}$	67.4	196.3	i.	s. h.	s.
1342	lq.	$1.667^{16°}$	....	82.7	....	....	....
1343	lq.	$1.689^{19°}$	−16.6	106.7	$0.693^{0°}$	....	....
1344	col. lq.	$1.991^{19°}$	<−55	68-9	....	....	....
1345	col. lq.	$0.958^{20°}_{25}$	....	59.4	sl. s.	∞	∞
1346	lq.	$1.112^{25°}_{25}$	<−20*	234-51	i.; ∞ chl.	∞ Me al.	∞; ∞ bz.
1347	oily lq.	....	....	101.3^{15mm}	s. h.	∞	∞
1348	col. or.	$1.186^{20°}_{4}$	16.0-16.5	$109\text{-}10^{17mm}$	....	s.	s.
1349	cr.	1.250^{10mm}	16	116^{13mm} sl. d.	....	s.	s.
1350	col. cr.	....	31	118^{50mm}	v. s.	....	....
1351	lq.	$1.162^{10°}$	....	195-7	i.	s.	s.
1352	mn. cr.	....	93-4	244-7 sl. d.	v. sl. s. h.	s. h.; s. bz.	s.; s. chl.
1353	nd./al.	....	161-2	....	i. aq. NH_3	s. h.; v. s. CS_2	s. h.; v. s. bz.
1354	cr./pet.	....	55-6	196	sl. s.	s.	s.
1355	col. pr.	$1.215^{15°}$	46	196	i.	s.	s.
1356	col. cr./lg.	....	66	235	i.; s. bz.	s.; s. chl.	s.; s. act.
1357	col. nd.	....	55	228	sl. s.	s.	s.; s. bz.
1358	nd.	....	99.2-99.5	212	$2^{13°}$	v. s.	v. s.
1359	lq.	$2.134^{16°}$	....	$123\text{-}4^{753mm}$	....	....	....
1360	lq.	$2.268^{16°}$	....	162.5-3.0	....	....	....
1361	col. lq.	$2.445^{15°}$	−22±	120^{730} sl. d.	....	....	....
1362	lq.	$0.989^{20°}_{4}$	....	107-8	....	....	....
1363	col. lq.	3.170°	−4	200±, d.	....	....	....
1364	cr./et.	....	86.8	....	i.	s.	s.
1365	cr./et.	$1.480^{50°}$	α36.3 β37.1 γ38.8 δ28	315 d.	i.	v. s. h.	v. s.
1366	yel. nd./al.	....	87-8	....	i.	s.	s.
1367	rhb./et.	$α1.697^{22°}$ $β1.680^{20°}_{4}$	α53.4 β43 γ27	315 sl. d.	i.; s. bz.	s. h.	s.; s. CS_2

* Crysts. $+H_2O$, m. p. 18°.
Chloro-cyclohexane 1613
Chloro-cymene 1572

Chloro-decane 1646
Chloro-dimethylacetophenone 2524

Table 7-4 (*Continued*)
PHYSICAL CONSTANTS OF ORGANIC COMPOUNDS

No.	Name	Synonym	Formula	Beil. Ref.	Formula Weight
	Chloro				
1368	dinitrobenzene	(5;1,3)	$Cl \cdot C_6H_3(NO_2)_2$	V-264	202.55
1369	dinitrobenzene	(2;1,4)	$Cl \cdot C_6H_3(NO_2)_2$	V-264	202.55
1370	dinitrobenzoic acid	(5;2,3,1)	$Cl \cdot C_6H_2(NO_2)_2CO_2H$	IX-415	246.56
1371	dinitrobenzoic acid	(5;3,4,1)	$Cl \cdot C_6H_2(NO_2)_2CO_2H$	IX-416	246.56
1372	dinitrophenol (α)	(4;2,6,1,)	$Cl \cdot C_6H_2(NO_2)_2OH$	VI-260	218.55
1373	diphenyl (*o*)	chloro-biphenyl	$Cl \cdot C_6H_4 \cdot C_6H_5$	V-579	188.66
1374	diphenyl (*m*)	xenyl chloride	$Cl \cdot C_6H_4 \cdot C_6H_5$	V-579	188.66
1375	diphenyl (*p*)	. . . .	$Cl \cdot C_6H_4 \cdot C_6H_5$	V-579	188.66
1376	ethoxy diphenyl	(2)(*β*)	$C_6H_5 \cdot C_6H_4 \cdot O(CH_2)_2Cl$	. . . .	232.71
1377	ethyl acetate (*β*)	. . . .	$CH_3CO_2 \cdot CH_2 \cdot CH_2Cl$	II-128	122.55
1378	ethylbenzene (*o*)	. . . .	$Cl \cdot C_6H_4 \cdot C_2H_5$	. . . .	140.61
1379	ethylbenzene (*m*)	alkazene 20	$Cl \cdot C_6H_4 \cdot C_2H_5$	. . . .	140.61
1380	ethylbenzene (*p*)	. . . .	$Cl \cdot C_6H_4 \cdot C_2H_5$	V-354	140.61
1381	ethyl 2-chloro-4-*tert*-butyl-phenyl ether (*β*)	(4;2,1)	$(CH_3)_3C \cdot C_6H_3(Cl) \cdot O \cdot CH_2CH_2Cl$	. . . .	247.17
1382	ethyl chloro-formate	β-chloroethyl chlorocarbonate	$Cl \cdot CO \cdot O \cdot CH_2CH_2Cl$	III-11	142.97
1383	ethyl *p*-toluene-sulfonate (*β*)	. . . .	$CH_3 \cdot C_6H_4 \cdot SO_2 \cdot OCH_2CH_2Cl$	**XI-45	234.70
1384	fumaric acid	. . . .	$CH:CCl(CO_2H)_2$	II-744	150.52
1385	furoic acid (5)	5-chloro-pyromucic	$Cl \cdot C_4H_2O \cdot CO_2H$	XVIII-282	146.53
1386	hydroquinone	(2;1,4)	$Cl \cdot C_6H_3(OH)_2$	VI-849	144.56
1387	iodobenzene (*o*)	. . . .	$Cl \cdot C_6H_4 \cdot I$	V-220	238.46
1388	iodobenzene (*m*)	. . . .	$Cl \cdot C_6H_4 \cdot I$	V-220	238.46
1389	iodobenzene (*p*)	. . . .	$Cl \cdot C_6H_4 \cdot I$	V-221	238.46
1390	iodoethane (1,1)	. . . .	$CH_3 \cdot CHClI$	I-98	190.41
1391	iodoethane (2,1)	. . . .	$ClCH_2 \cdot CH_2I$	I-98	190.41
1392	iodo-hydroxy-quinoline	(5,7,8); nioform	$Cl \cdot I \cdot OH \cdot C_6H:C_3H_3N$	XXI-98	305.50
1393	iodomethane	. . . .	$Cl \cdot CH_2 \cdot I$	I-71	176.38
1394	4-ketodihydro-quinazoline (2)	. . . .	$C_6H_4 \cdot N:C(Cl) \cdot NH \cdot CO$	. . . .	180.59
1395	lepidine (α)	2-Cl-4-Me-quinoline	$C_6H_4:C_3HN(CH_3)Cl$	XX-396	177.64
1396	maleic acid	. . . .	$CH:CCl(CO_2H)_2$	II-752	150.52
1397	malonic acid	. . . .	$ClCH(CO_2H)_2$	II-592	138.51
1398	methyl acetate	. . . .	$CH_3CO_2 \cdot CH_2Cl$	II-152	108.53
1399	methyl cyanide	chloro-acetonitrile	$ClCH_2 \cdot CN$	II-201	75.50
1400	methyl ether	. . . .	$ClCH_2 \cdot O \cdot CH_3$	I-580	80.51
1401	methyl-ethyl ketone	1-Cl-2-butanone	$ClCH_2 \cdot CO \cdot C_2H_5$	I-669	106.55
1402	naphthalene (α)	α-naphthyl chloride	$C_{10}H_7Cl$	V-541	162.62
1403	naphthalene (β)	β-naphthyl chloride	$C_{10}H_7Cl$	V-541	162.62
1404	α-naphthol (2,1)	. . . .	$Cl \cdot C_{10}H_6 \cdot OH$	VI-611	178.62
1405	α-naphthol (4,1)	. . . .	$Cl \cdot C_{10}H_6 \cdot OH$	VI-611	178.62
1406	β-naphthol (1,2)	. . . .	$Cl \cdot C_{10}H_6 \cdot OH$	VI-648	178.62
1407	nitroaniline	(4;2,1)	$Cl \cdot C_6H_3(NO_2)NH_2$	XII-729	172.57
1408	nitroaniline HCl	(4;2,1)	$C_6H_5O_2N_2Cl \cdot HCl$	. . . .	209.03
1409	nitroaniline	(4;3,1)	$Cl \cdot C_6H_3(NO_2)NH_2$	XII-731	172.57

Table 7-4 (*Continued*)
PHYSICAL CONSTANTS OF ORGANIC COMPOUNDS

No.	Crystalline Form and Color	Specific Gravity	Melting Point °C.	Boiling Point °C.	Solubility in 100 Parts		
					Water	Alcohol	Ether
1368	col. nd./al.	....	59	vol. in steam	....	v. s.	v. s.
1369	yel. cr./lg.	....	60	....	i.	s.	s.
1370	nd./aq.	....	199-200	expl.	$0.26^{11°}$	....	....
1371	pr./bz.	....	159	....	v. s. act.	s.	v. s.
1372	yel. mn./al.	$1.74^{22°}$	81-2	subl.	v. sl. s. h.	s.	s.; s. chl.
1373	cr.	....	34	267-8	i.	v. s. lg.	....
1374	cr.	....	89	284-5	....	....	....
1375	lf.	....	77.5	282	i.	s. lig.	
1376	col. cr.	$1.142^{60°}_{25}$	54-7	323	v. sl. s.	$12^{25°}$ Me al.	v. s. bz., act.
1377	col. lq.	$1.1522^{25°}_{25}$	<~20	145	$3^{25°}$	∞	∞
1378	lq.	$1.055^{25°}_{25}$	-81	179.2	i.	∞	∞
1379	col. lq.	$1.045^{25°}_{25}$	<-70	183	i.; ∞ bz.	∞ ; ∞ act.	∞ ; ∞ CCl_4
1380	lq.	$1.044^{25°}_{25}$	-62	184.6	i.	∞	∞
1381	col. lq.	$1.149^{25°}_{25}$	<-20	176-80^{18mm}	i.; ∞ bz.	∞ ; ∞ act.	∞ ; ∞ CCl_4
1382	lq.	$1.383^{20°}_{4}$	....	152.5^{752mm}	i.; d.	v. s.	v. s.
1383	col. lq.	....	19-20	210^{21mm}	i.	....	....
1384	pl./ac.	....	191.5-2.5	....	v. s.	v. s.	v. s.
1385	lf.	....	176-7	....	$0.3^{20°}$	s.	s.
1386	mn.	....	106	263 sl. d.	v. s.	v. s.	v. s.
1387	col. oil	$1.952^{25°}_{4}$	0.7	234-5	....	....	....
1388	yel. lq.	....	....	230	....	....	....
1389	col. lf./al.	$1.886^{57°}_{4}$	56-7	227.6^{751mm}	i.	s.	....
1390	lq.	$2.054^{19°}$	....	117-9	.,.....	....	....
1391	lq.	$2.134^{15.3°}_{0}$	-15.6	140.1	....	....	....
1392	yel. nd./ac.	....	177-8	....	i.; s. h. ac.	2.5	s.; s. chl.
1393	oil	$2.49^{20°}$	....	109	....	....	....
1394	nd.	....	212	....	i.; s. NaOH	s.	s. act.
1395	nd./aq. al.	....	59	296	i.	s.	s.; s. chl.
1396	col. cr.	....	114-5	....	....	v. s.	v. s.
1397	pr.	....	133	....	v. s.	v. s.	v. s.
1398	col. lq.	$1.195^{14°}$	....	115-6^{757mm}	d.	....	....
1399	lq.	$1.193^{20°}$	....	124	....	....	....
1400	lq.	$1.070^{20°}_{4}$	-103.5	59.5^{759mm}	d.	d. h.	....
1401	lq.	$1.08^{13°}$	....	139^{755mm}	i.	....	....
1402	col. lq.	$1.194^{20°}_{4}$	-20	259.3	i.	s.; ∞ CS_2	∞ ; ∞ bz.
1403	lf./al.	$1.266^{16°}$	56-7	264-6^{751mm}	i.	v. s.	v. s.
1404	nd./lg.	....	65-70	....	....	v. s.	v. s.
1405	nd./aq. al.	....	117-20	....	....	v. s.	v. s.
1406	cr./aq.	....	70-1	....	....	v. s.	v. s. bz.
1407	or. nd./aq.	....	116-7	....	v. sl. s. lg.	v. s.	v. s.
1408	or. pr.	....	....	....	sl. s.	s.	i.
1409	yel. nd./aq.	....	102-3	....	sl. s. h.	s.; s. chl.	s.

Chloro-hexane 3644-7
Chloro-hydrin 3194, 3395, 6280
Chloro-hydroxydiphenyl 1436-8
Chloro-hydroxytoluene 1354-7
Chloro-isobutylene 1567, 4060

Chloro-mercuri-phenol 4034-5
Chloro-methane 4183
Chloro-methylquinoline 1395

Table 7-4 (*Continued*)
PHYSICAL CONSTANTS OF ORGANIC COMPOUNDS

No.	Name	Synonym	Formula	Beil. Ref.	Formula Weight
1410	**Chloro** nitroaniline	(2;4,1)	$Cl \cdot C_6H_3(NO_2)NH_2$	XII-733	172.57
1411	nitroaniline HCl	(2;4,1)	$C_6H_5O_2N_2Cl \cdot HCl$	XII-733	209.03
1412	nitrobenzene (*o*)	nitrochlorobenzene	$Cl \cdot C_6H_4 \cdot NO_2$	V-241	157.56
1413	nitrobenzene (*m*)	nitrochlorobenzene	$Cl \cdot C_6H_4 \cdot NO_2$	V-243	157.56
1414	nitrobenzene (*p*)	nitrochlorobenzene	$Cl \cdot C_6H_4 \cdot NO_2$	V-243	157.56
1415	nitrobenzene sulfonic acid	(6;3,1)	$Cl \cdot C_6H_3(NO_2)SO_3H \cdot 2H_2O$	XI-73	273.65
1416	nitroethane (1,1)	....	$CH_3 \cdot CHClNO_2$	I-101	109.51
1417	nitroethane (1,2)	....	$ClCH_2 \cdot CH_2NO_2$	I-101	109.51
1418	nitronaphthalene	(1,4)	$Cl \cdot C_{10}H_6 \cdot NO_2$	V-555	207.62
1419	nitronaphthalene	(7,1)	$Cl \cdot C_{10}H_6 \cdot NO_2$	V-556	207.62
1420	nitropropane	(1,1)	$C_2H_5 \cdot CH(Cl)NO_2$	I-116	123.54
1421	nitropropane	(2,2)	$(CH_3)_2C(Cl)NO_2$	**I-79	123.54
1422	nitrosoethane (1,1)	....	$CH_3 \cdot CH(Cl)NO$	I-99	93.51
1423	pentabromoethane	....	$ClCBr_2 \cdot CBr_3$	I-95	459.02
1424	phenetole (*o*)	....	$Cl \cdot C_6H_4 \cdot O \cdot C_2H_5$	VI-184	156.61
1425	phenetole (*p*)	....	$Cl \cdot C_6H_4 \cdot O \cdot C_2H_5$	VI-187	156.61
1426	phenetole (*β*)	....	$C_6H_5 \cdot O \cdot CH_2CH_2Cl$	VI-142	156.61
1427	phenol (*o*)	....	$Cl \cdot C_6H_4 \cdot OH$	VI-183	128.56
1428	phenol (*m*)	....	$Cl \cdot C_6H_4 \cdot OH$	VI-185	128.56
1429	phenol (*p*)	....	$Cl \cdot C_6H_4 \cdot OH$	VI-186	128.56
1430	phenol sulfonic acid	(4;1,2)	$Cl \cdot C_6H_3(OH) \cdot SO_3H \cdot H_2O$	XI-234	226.64
1431	phenol sulfonate Na	(4;1,2)	$C_6H_4OCl(SO_3Na)$	XI-234	230.60
1432	phenoxyacetic acid	(1,2)	$Cl \cdot C_6H_4 \cdot O \cdot CH_2CO_2H$	**VI-172	186.60
1433	phenylenediamine	(2;1,4)	$Cl \cdot C_6H_3(NH_2)_2$	XIII-117	142.59
1434	phenylenediamine	(2;1,4) HCl	$C_6H_7N_2Cl \cdot 2HCl$	XIII-117	215.51
1435	phenylphenacyl bromide	(4;1′,4′)	$Cl \cdot C_6H_4 \cdot C_6H_4 \cdot CO \cdot CH_2Br$	....	309.60
1436	*o*-phenylphenol (2)	(2;6,1)	$C_6H_5 \cdot C_6H_3(Cl)OH$	....	204.66
1437	*o*-phenylphenol (4)	(2;4,1)	$C_6H_5 \cdot C_6H_3(Cl)OH$	....	204.66
1438	*p*-phenylphenol (2)	(4;2,1)	$C_6H_5 \cdot C_6H_3(Cl)OH$	....	204.66
1439	phenyl-4-toluene-sulfonate (2)	(4,1;1′,2′)	$CH_3 \cdot C_6H_4 \cdot SO_3 \cdot C_6H_4 \cdot Cl$	....	282.75
1440	phenyl-4-toluene-sulfonate (4)	(4,1;1′,4′)	$CH_3 \cdot C_6H_4 \cdot SO_3 \cdot C_6H_4 \cdot Cl$	....	282.75
1441	phthalic acid	(4;1,2)	$Cl \cdot C_6H_3(CO_2H)_2$	IX-816	200.58
1442	propionic acid (*α*)	(*dl*)	$CH_3 \cdot CHCl \cdot CO_2H$	II-248	108.53
1443	propionic acid (*β*)	....	$ClCH_2 \cdot CH_2 \cdot CO_2H$	II-249	108.53
1444	propionyl chloride	(*β*)	$ClCH_2 \cdot CH_2 \cdot CO \cdot Cl$	II-250	126.97
1445	propyl acetate (*β*)	....	$CH_3CO_2 \cdot CH_2 \cdot CHCl \cdot CH_3$	II-129	136.58
1446	propyl acetate (*γ*)	....	$CH_3CO_2 \cdot (CH_2)_2CH_2Cl$	*II-58	136.58
1447	*iso*-propyl acetate	(*β*)	$CH_3CO_2 \cdot CH(CH_3) \cdot CH_2Cl$	II-130	136.58
1448	propyl chloroformate (*γ*)	*γ*-chloropropyl chlorocarbonate	$Cl \cdot CO_2 \cdot (CH_2)_2CH_2Cl$	**III-10	157.00
1449	propyl *p*-toluene-sulfonate (*γ*)	....	$CH_3 \cdot C_6H_4 \cdot SO_3 \cdot (CH_2)_2CH_2Cl$	**XI-45	248.73

Table 7-4 (*Continued*)
PHYSICAL CONSTANTS OF ORGANIC COMPOUNDS

No.	Crystalline Form and Color	Specific Gravity	Melting Point °C.	Boiling Point °C.	Solubility in 100 Parts		
					Water	Alcohol	Ether
1410	yel. nd./aq.	....	107-8	....	sl. s. 50% ac.	s.; i. lg.	s.; s. CS_2
1411	pl.	....	....	....	d.	....	....
1412	mn. nd.	$1.305\frac{8}{4}0°$	32.5	245.5^{753}	i.; s. bz.	s. h.	s.
1413	yel. rhb./al.	$1.343\frac{5}{4}0°$	44.5(24.0)	235.6	i.; s. chl.	v. s. h.	v. s.; s. ac.
1414	mn. pr.	$1.298^{91°}$	83-4	242^{761mm}	i.	v. s. h.	v. s.
1415	tri./aq.	....	$-2H_2O$, 110	d. 165-70	v. s.	v. sl. s.	sl. s.
1416	col. lq.	$1.258\frac{20}{20}°$	....	$124-5^{758mm}$	<0.4	s. alk.	....
1417	col. lq.	$1.405^{7°}$	....	173-4	i.	d. HCl	....
1418	yel. nd./al.	....	85(60-1)	....	i.	s.	s.
1419	yel. nd./al.	....	116	....	i.	s.	s.
1420	lq.	$1.209\frac{20}{20}°$	....	140-3	$<0.8^{20°}$	....	....
1421	lq.	$1.193\frac{20}{20}°$	....	132	$<0.5^{20°}$	s.; i. alk.	s.
1422	col. lf. or b. lq.	....	65	....	i. c. NaOH	v. s. Me al.	v. s.; v. s. pet.
1423	cr.	....	170 d.	....			
1424	col. lq.	....	....	208	s. bz.	s.	s.
1425	lq.	....	21	212	....	s.	s.
1426	col. lq.	$1.147\frac{25}{25}°$	27-8	221^{754mm}	v. sl. s.	v. s.; s. bz.	v. s.; s. act.
1427	col. lq.	$1.241\frac{18}{15};2°$	$\alpha 7; \beta 0;$ $\gamma 4.1$	175-6	$2.85^{20°}$	s.	s. alk.
1428	nd.	$1.268^{25°}$	32-3	214	$2.6^{20°}$	s.	s.
1429	nd.	$1.306\frac{20}{4}°$	41-3	217	$2.71^{20°}$	v. s.	v. s.
1430	hyg. pl./aq.	....	75-6	....	v. s.; i. bz.	v. sl. s.; i. chl.	v. sl. s.
1431	nd./aq.	....	....	....	s.	sl. s.	....
1432	col. nd./aq.	....	146-7	....	$0.5^{80°}$	$21^{25°}$	$9^{25°}$; i. bz.
1433	nd./bz. lg.	....	64	....	s.	....	....
1434	nd.	....	....	....	....	....	....
1435	col. nd.	....	126-7	....	i.	s. h.	i.
1436	col. lq.	$1.234\frac{25}{25}°$	6	317-8	i.; s. alk.	∞	∞
1437	yel. nd.	....	67-71	312	i.; s. CCl_4	$85^{25°}$	$250^{25°}$
1438	col. cr.	....	78-80	323	$0.03^{25°}$; s. al.	$>100^{25°}$ act.	$>100^{25°}$
1439	col. rhb.	....	68-70	....			
1440	col. rhb.	....	78-80	....	i.	s. h.	v. s.
1441	nd./al.	....	150.5	$-H_2O$, >150	s.	s.	....
1442	col. lq.	$1.306^{9°}$	<-20	186	∞	∞	∞
1443	col. lf.	....	40-2.5	$203-5^{764mm}$	v. s.	v. s.	v. s.
1444	col. lq.	$1.331^{13°}$	....	$143-5^{763mm}$	i. c.; d. h.	s.; d. h.	v. s.
1445	col. lq.	$1.098^{20°}$	....	$152-3^{750mm}$	i.	s.	s.
1446	lq.	$1.250^{19°}$	....	$163-5^{747mm}$	....	....	....
1447	col. lq.	....	....	149-50	....	....	....
1448	lt. yel. lq.	$1.295\frac{25}{20}°$	....	177	d.	d.	....
1449	col. lq.	$1.267\frac{20}{4}°$	....	$150-1^{2.5mm}$	....	....	....

Table 7-4 (*Continued*)
PHYSICAL CONSTANTS OF ORGANIC COMPOUNDS

No.	Name	Synonym	Formula	Beil. Ref.	Formula Weight
	Chloro				
1450	pyridine (2)(α)	....	$Cl \cdot C_5H_4N$	XX-230	113.55
1451	pyridine (3)(β)	....	$Cl \cdot C_5H_4N$	XX-230	113.55
1452	pyridine (4)(γ)	....	$Cl \cdot C_5H_4N$	XX-231	113.55
1453	quinoline (2)	....	$Cl \cdot C_9H_6N$	XX-359	163.61
1454	quinoline (3)	....	$Cl \cdot C_9H_6N$	XX-359	163.61
1455	quinoline (4)	....	$Cl \cdot C_9H_6N$	XX-360	163.61
1456	quinoline (5)	....	$Cl \cdot C_9H_6N$	XX-360	163.61
1457	quinoline (6)	....	$Cl \cdot C_9H_6N$	XX-360	163.61
1458	quinoline (7)	....	$Cl \cdot C_9H_6N$	XX-361	163.61
1459	quinoline (8)	....	$Cl \cdot C_9H_6N$	XX-361	163.61
1460	resorcinol	(4;1,3)	$Cl \cdot C_6H_3(OH)_2$	**VI-818	144.56
1461	salicylaldehyde	(5;2,1)	$Cl \cdot C_6H_3(OH)CHO$	VIII-53	156.57
1462	styrene (α)	....	$C_6H_5 \cdot CCl:CH_2$	V-476	138.60
1463	styrene (ω)	....	$C_6H_5 \cdot CH:CHCl$	V-476	138.60
1464	succinic acid (*dl*)	....	$(CHCl \cdot CH_2):(CO_2H)_2$	II-619	152.54
1465	tetrabromoethane	(1,1,2,2)	$Br_2CH \cdot CBr_2Cl$	I-95	380.12
1466	toluene (*o*)	tolyl chloride	$Cl \cdot C_6H_4 \cdot CH_3$	V-290	126.59
1467	toluene (*m*)	tolyl chloride	$Cl \cdot C_6H_4 \cdot CH_3$	V-291	126.59
1468	toluene (*p*)	tolyl chloride	$Cl \cdot C_6H_4 \cdot CH_3$	V-292	126.59
1469	*o*-toluidine (*m*)	(5;1,2)	$Cl \cdot C_6H_3(CH_3)NH_2$	XII-835	141.60
1470	*p*-toluidine (*m*)	(3;1,4)	$Cl \cdot C_6H_3(CH_3)NH_2$	XII-989	141.60
1471	tribromoethane	(1,1,2)	$BrCH_2 \cdot CClBr_2$	I-94	301.22
1472	tribromomethane	....	$Cl \cdot CBr_3$	I-68	287.19
1473	vinylarsine dichloride	Lewisite	$ClCH:CH \cdot AsCl_2$	**IV-985	207.32
1474	*o*-xylene (3)	xylyl chloride (*o*)	$Cl \cdot C_6H_3(CH_3)_2$	V-363	140.61
1475	*o*-xylene (4)	(4;1,2)	$Cl \cdot C_6H_3(CH_3)_2$	V-363	140.61
1476	*p*-xylene (2)	(2;1,4)	$Cl \cdot C_6H_3(CH_3)_2$	V-384	140.61
1477	**Chloroform**	trichloromethane	$CHCl_3$	I-61	119.38
1478	**Chlorogenin** (pseudo)	....	$C_{27}H_{44}O_4$	....	432.65
1479	**Choleic acid**	deoxycholic acid	$C_{24}H_{40}O_4$	....	392.58
1480	**Cholesterol** hydrate		$C_{27}H_{45}OH \cdot H_2O$	....	404.68
1481	**Cholesterol**	cholesterin	$C_{27}H_{45}OH$	....	386.64
1482	**Cholesterol** (*iso*)	....	$C_{27}H_{45}OH$	....	386.67
1483	**Cholesteryl** acetate	....	$CH_3CO_2 \cdot C_{27}H_{45}$	....	428.70
1484	benzoate	....	$C_6H_5CO_2 \cdot C_{27}H_{45}$	....	490.78
1485	benzoate (*iso*)	....	$C_6H_5CO_2 \cdot C_{27}H_{45}$	....	490.78
1486	propionate	....	$C_2H_5CO_2 \cdot C_{27}H_{45}$	....	442.73
1487	**Cholic acid**†	cholalic acid	$C_{23}H_{36}(OH)_3CO_2H \cdot H_2O$	....	426.60
1488	**Choline**	bilineurine	$(CH_3)_3N(OH)CH_2 \cdot CH_2OH$	IV-277	121.18
1489	chloride	....	$C_5H_{14}ON \cdot Cl$	IV-280	139.63
1490	**Chromone**	benzo-γ-pyrone	$C_6H_4 \cdot CO \cdot CH:CH \cdot O$	XVII-327	146.15
1491	**Chrysammic acid** (2,4,5,7;1,8)	tetranitro-chrysazine	$(NO_2)_4C_{14}H_2(OH)_2O_2$	VIII-461	420.21
1492	**Chrysarobine**	dihydroxy-methyl-anthranol	$C_{15}H_{12}O_3$	VIII-335	240.26
1493	**Chrysene**	....	$C_{18}H_{12}$	V-718	228.30

† See also No. 3440.
Chrysoidine 1698
Chrysoidine orange 1699
Chrysaphanol 1495
Cicutine, cf. alkd.
Cignolin 2296

Cincholepidine 4370
Cinchomoronic acid 5499, 5501
Cinchonamine, cf. alkd.
Cinchonidine, cf. alkd.
Cinchonine, cf. alkd.

Cinchophen 5234
Cinchotine, cf. alkd.
Cinchovatine, cf. alkd.
Cineole 3217
Cinnamein 798

Table 7-4 (*Continued*)
PHYSICAL CONSTANTS OF ORGANIC COMPOUNDS

No.	Crystalline Form and Color	Specific Gravity	Melting Point °C.	Boiling Point °C.	Solubility in 100 Parts		
					Water	Alcohol	Ether
1450	oil	$1.205^{15°}$	....	1667^{14mm}	s.	....	....
1451	lq.	....	....	1487^{44mm}	s.	....	....
1452	lq.	....	....	147-8	s.	....	....
1453	nd./aq. al.	....	37-8	275^{751mm}	i.; v. s. bz.	s.; v. s. lg.	v. s.
1454	hyg. lq.	....	....	255^{743mm}	....	....	....
1455	cr.	$1.251^{20°}_{4}$	34	$260-1^{744mm}$	s. HCl	v. s.	v. s.
1456	nd.	....	45	256	....	....	....
1457	nd.	....	40-1	$261-2^{740mm}$	....	....	....
1458	nd./al.	....	31-2	267-8	....	....	....
1459	yel. oil	....	<-20	288	....	v. s.	v. s.
1460	wh. amor. pd.	....	105-7	255-6	s.	s.	s.
1461	pl./al.	....	99	....	i.; s. alk.	s.	s.
1462	lq.	$1.102^{18°}_{4}$	-23	199	i.	s.	s.
1463	lq.	$1.110^{18°}_{4}$	....	199	i.	s.	s.
1464	col. cr.	1.679	153-4	....	v. s.	v. s. ac.	v. sl. s. chl.
1465	cr.	$3.366^{16°}$	32-3	$200-5^{285mm}$	v. s. chl.	v. s.	v. s.
1466	col. lq.	$1.082^{20°}_{4}$	-36.5	159.2	i.; ∞ bz.	s.; ∞ chl.	∞
1467	col. lq.	$1.072^{20°}_{4}$	-47.8	161.6	i.; ∞ bz.	s.; ∞ chl.	∞
1468	col. lq.	$1.070^{20°}_{4}$	7.5	162.4	i.; ∞ bz.	s.; ∞ chl.	∞
1469	if./al.	....	29-30	$236-8^{730mm}$	....	....	....
1470	lq.	$1.151^{20°}$	7	222-3	i.	....	....
1471	lq.	$2.602^{16°}$	....	$165-7^{285mm}$	i.	....	....
1472	lf.	$2.71^{15°}$	55	160	i.	....	....
1473	col. lq.	$1.888^{20°}_{4}$	-13	190	i.	v. s.	s.
1474	col. lq.	....	<-20	189.5	i.	∞	∞
1475	col. lq.	$1.069^{15°}_{15}$	<-20	191.5	....	....	....
1476	col. lq.	....	....	186.8	....	∞	∞
1477	col. lq.	$1.489^{20°}$	-63.5	61.2	$0.822^{20°}$	∞	∞
1478	cr.	....	268-70	....	i.	sl. s.	v. sl. s.
1479	col. cr./al.	....	172	....	sl. s. c.	s.; s. alk.	s. ac.; s.act
1480	rhb./al.	1.067	148.5 (anh.)	$subl.^{1.0mm}$	$0.26^{20°}$	$1.1^{17°}$; $117^{8°}$	18; s. bz.
1481	nd./act.	....	148.5	....	i.; s. bz.	s.; s. chl.	v. s.
1482	nd./et.	....	137-8	....	s. h. ac.	s. h.	v. s.
1483	col. nd.	....	113-4	....	i.	sl. s. h.	v. s.
1484	tet. cr.	....	146.6	....	....	v. sl. s. h.	s.
1485	wh. pd.	....	191-5	....	....	sl. s. h.	s.
1486	col. fl.	....	95-7	....	i.	sl. s.	s.
1487	rhb. cr./aq.	....	200-1 (anh.)	....	0.03; s. ac.	5; s. act.	1.4; s. alk.
1488	syrup	....	....	....	v. s.	v. s. abs.	i.
1489	syrup or hyg. cr.	....	247 d.	subl. in vac. 200	v. s.	v. s.	....
1490	nd./pet.	....	58-9	....	i.; s. bz.	s.	s.; s. chl.
1491	yel. mn.	....	d.	expl.	v. sl. s. h.	s.	s.
1492	yel. lf.	....	202-6	....	i. aq. Na_2CO_3	sl. s. c.; s. H_2SO_4	sl. s. c. bz.
1493	col. rhb.	....	253-4	448	i.	$0.1^{16°}$ abs.	v. sl. s.

Table 7-4 (*Continued*)
PHYSICAL CONSTANTS OF ORGANIC COMPOUNDS

No.	Name	Synonym	Formula	Beil. Ref.	Formula Weight
1494	**Chrysin**	5,7-diOH-flavone	$C_{15}H_{10}O_4$	XVIII-124	254.24
1495	**Chrysophanic acid**	2,4-diOH-2-Me-anthraquinone	$C_{14}H_5(OH)_2(CH_3)O_2$	VIII-470	254.24
1496	**Chrysoquinone**	....	$C_{10}H_6(CO)_2C_6H_4$	VII-827	258.28
1497	**Cinnamic** acid (*cis*)	allocinnamic acid	$C_6H_5 \cdot CH{:}CH \cdot CO_2H$	IX-591	148.16
1498	acid (*trans*)	β-phenyl-acrylic acid	$C_6H_5 \cdot CH{:}CH \cdot CO_2H$	IX-572	148.16
1499	Na salt	....	$C_6H_5 \cdot CH{:}CH \cdot CO_2Na$	IX-580	170.14
1500	acid (*iso*)	....	$C_6H_5 \cdot CH{:}CH \cdot CO_2H$	IX-592	148.16
1501	aldehyde	phenyl acrolein	$C_6H_5 \cdot CH{:}CH \cdot CHO$	VII-348	132.16
1502	amide	cinnamide	$C_6H_5 \cdot C_2H_2 \cdot CO \cdot NH_2$	IX-587	147.18
1503	anhydride	....	$(C_6H_5CH{:}CH \cdot CO)_2O$	IX-586	278.31
1504	carboxylic acid (*o*)	....	$(\cdot C_6H_4C_2H_2 \cdot)(CO_2H)_2$	IX-898	192.17
1505	chloride	....	$C_6H_5 \cdot CH{:}CH \cdot CO \cdot Cl$	IX-587	166.61
1506	**Cinnamal acetophenone**	cinnamylidene acetophenone	$C_6H_5 \cdot (CH{:}CH)_2 \cdot CO \cdot C_6H_5$	VII-499	234.30
1507	**Cinnamyl** alcohol	styryl carbinol	$C_6H_5 \cdot CH{:}CH \cdot CH_2OH$	VI-570	134.18
1508	cinnamate	styracin	$C_8H_7CO_2 \cdot C_9H_9$	IX-585	264.33
1509	**Cinnamoyl-*p*-hydroxyphenyl urea**	Elbon®	$C_6H_5 \cdot CH{:}CH \cdot CO_2 \cdot C_6H_4 \cdot NHCONH_2$	*XIII-170	282.30
1510	**Citraconic** acid (*cis*)	methyl-maleic acid	$CH_3 \cdot C(CO_2H){:}CH \cdot CO_2H$	II-768	130.10
1511	anhydride	....	$CH_3 \cdot C{:}CH \cdot CO \cdot O \cdot CO$	XVII-440	112.09
1512	**Citral** (α)	geranial	$C_9H_{15} \cdot CHO$	I-753	152.24
1513	semicarbazone	....	$C_{10}H_{16}{:}N \cdot NHCONH_2$	III-109	209.29
1514	**Citramalic acid** (*dl*)	methyl-malic acid	$CH_3 \cdot C(OH)(CO_2H) \cdot CH_2 \cdot CO_2H$	III-444	148.12
1515	**Citric** acid	....	$HO_2C \cdot CH_2 \cdot C(OH) \cdot (CO_2H) \cdot CH_2 \cdot CO_2H$	III-556	192.13
1516	acid (*iso*)	....	$HO_2C \cdot CHOH \cdot CH \cdot (CO_2H) \cdot CH_2CO_2H$	III-555	192.13
1517	amide	citramide	$C_3H_5O(CONH_2)_3$	III-569	189.17
1518	**Citronellal** (*d*)	rhodinal	$C_9H_{17} \cdot CHO$	I-745	154.25
1519	**Citronellol** (*d*)	2,6-dimethyl-octene-1-ol-8	$C_{10}H_{20}O$	I-451	156.27
1520	**Citronellyl acetate**	....	$C_{12}H_{22}O_2$	II-139	198.31
1521	**Citrulline** (*dl*)	α-NH₂-δ-carbamidovaleric acid	$NH_2CONH(CH_2)_3 \cdot CH \cdot (NH_2) \cdot CO_2H$	....	175.19
1522	Cu salt	....	$(C_6H_{12}O_3N_3)_2Cu$	....	411.90
1523	**Cofebrin** hydrochloride	2,4-diOH-phenyl-propanolamine HCl	$(HO)_2C_6H_3 \cdot CH(OH) \cdot CH(NH_2) \cdot CH_3 \cdot HCl$	....	219.67
1524	**Collidine** (α)	2-Me-4-Et-pyridine	$CH_3 \cdot C_5H_3N \cdot C_2H_5$	XX-248	121.18
1525	**Collidine** (β)	4-Me-3-Et-pyridine	$CH_3 \cdot C_5H_3N \cdot C_2H_5$	XX-250	121.18
1526	**Collidine** (γ)	2,4,6-triMe-pyridine	$(CH_3)_3C_5H_2N$	XX-250	121.18
1527	**Congo red**	diPh-bis-azonaphthionic acid	$[C_{10}H_5(NH_2)(SO_3Na) \cdot N{:}N \cdot C_6H_4]_2$	XVI-410	696.68

Table 7-4 (*Continued*)
PHYSICAL CONSTANTS OF ORGANIC COMPOUNDS

No.	Crystalline Form and Color	Specific Gravity	Melting Point °C.	Boiling Point °C.	Solubility in 100 Parts		
					Water	Alcohol	Ether
1494	yel. pl.	....	275	subl.	i.; s. alk.	0.6 c.; 2.5 h.	sl. s.; i. bz.
1495	yel. lf./al.	....	195	subl. sl. d.	i. c.; s. bz.	s. h.; s. act.	sl. s.; sl. s. pet.
1496	or. nd.	....	235–40	subl.	i.	s. h.	v. sl. s.
1497	mn. pr.	1.2844°	68	125^{19mm}	....	....	s. pet.
1498	mn. pr.	1.245	133	300	$0.04^{18\circ}$	$242^{0\circ}$ abs.	v. s.
1499	wh. cr. pd.	....	d. >115	....	9 c.; 5 h.	0.6	s. gly.
1500	mn. pr.	....	58–9(42)	256 d.	16.6 c. pet.	s.	s.
1501	lq.	$1.110^{20^\circ}_{20}$	–7.5	$252\pm$, sl. d.	v. sl. s.	4, 50% al.	∞
1502	nd./bz.	....	145–6	....	sl. s. h.	s.	s.
1503	nd./al.	....	135–6	....	i.	v. sl. s. c.	v. s. h. bz.
1504	nd./et.	....	173–5 d.	$-H_2O$, d.	v. sl. s.	s.; i. bz.	v. sl. s.
1505	cr.	....	35–6	251–3 sl. d.	d.	s. CCl_4	s. pet.
1506	yel. nd./ al.	....	$\alpha102$; $\beta235$	....	s. H_2SO_4	sl. s.	sl. s.
1507	nd.	$1.040^{35^\circ}_{35}$	33	257.5	sl. s.	v. s.	v. s.
1508	nd. or pr.	$1.085^{16.5\circ}$	44	....	i.	4 c.; 33 h.	33; s. bz.
1509	wh. cr. pd.	....	204	....	i.; i. alk.	s.; s. act.	s. oil
1510	nd.	1.617	92–3	....	$360^{25\circ}$	s.; v. sl. s. chl.	s.; i. bz.; i. CS_2
1511	lq.	$1.25^{15^\circ}_{4}$	7–8	213–4	d.	v. s.	v. s.
1512	col. oil	$0.890^{17^\circ}_{4}$	....	228–9 sl. d.	i.	∞	∞
1513	cr./ac.	....	$\alpha164$ $\beta171$*	....	....	....	....
1514	mn. pr.	....	117–9	....	s.; i. bz.	v. s. abs.	v. s. act.
1515	cr.†	$1.542^{20^\circ}_{4}$	153	d.	133 c.	$76^{15\circ}$, abs.	$2.2^{15\circ}$, abs.
1516	cr.	....	d. 100	....	s.	....	....
1517	cr./aq.	....	210–5 d.	....	$2.7^{18\circ}$; $33.3^{100\circ}$	i.	i.
1518	col. oil	$0.855^{17.5\circ}$	....	204–8	v. sl. s.	∞	∞
1519	col. oil	$0.848^{20^\circ}_{4}$	....	224–5	v. sl. s.	∞	∞
1520	col. lq.	$0.893^{17.5\circ}$	....	$172–3^{34mm}$	....	....	....
1521	col. pr./ aq. Me al.	....	$202–26\pm$	....	s.	i. abs.	i.
1522	cr./aq.	....	257–8 d.	....	i. c.	....	....
1523	wh. cr. pd.	....	178–9 d.	....	66	7	i.
1524	lq.	$0.935^{0\circ}$	....	177.8^{758mm}	sl. s. c.; v. sl. s. h.	s.	s.; s. bz.
1525	lq.	$0.966^{0\circ}$	....	$195–6^{753mm}$	i.	s.	v. s.
1526	lq.	$0.917^{15\circ}$	....	171–2	sl. s. h.	v. s.	....
1527	red	....	....	....	sl. s. h.	sl. s. h. alk.	i.

* Mixture of $\alpha + \beta$, m. p. 135°.
† Crysts./aq. $+1H_2O$; $-H_2O$, 100–32°.

Table 7-4 (*Continued*)
PHYSICAL CONSTANTS OF ORGANIC COMPOUNDS

No.	Name	Synonym	Formula	Beil. Ref.	Formula Weight
1528	**Coniferin**	....	$C_{16}H_{22}O_8 \cdot 2H_2O$	....	378.38
1529	**Coniferyl alcohol**	....	$C_{10}H_{12}O_3$	VI-1131	180.21
1530	**Conylene**	octadiene-x,x	C_8H_{14}	I-258	110.20
1531	**Coumaric** acid (*o*) (*trans*)	*o*-hydroxy-cinnamic acid	$HOC_6H_4 \cdot C_2H_2 \cdot CO_2H$	X-288	164.16
1532	acid (*m*)	*m*-hydroxy-cinnamic	$HOC_6H_4 \cdot C_2H_2 \cdot CO_2H$	X-294	164.16
1533	acid (*p*)	*p*-hydroxy-cinnamic	$HOC_6H_4 \cdot C_2H_2 \cdot CO_2H$	X-297	164.16
1534	**Coumarin**	....	$C_9H_6O_2$	XVII-328	146.15
1535	**Coumarin** (*iso*)	....	$C_6H_4 \cdot CH:CH \cdot O \cdot CO$	XVII-333	146.15
1536	3-carboxylic acid	....	$C_6H_4 \cdot CH:C(CO_2H) \cdot$ $\overline{} \ CO \cdot O \ \overline{}$	XVIII-429	190.16
1537	**Coumarinic acid** (*o*)(*cis*)	*o*-hydroxy-cinnamic acid	$HO \cdot C_6H_4 \cdot CH:CH \cdot$ CO_2H	X-291	164.16
1538	**Coumarone**	....	$C_6H_4 \cdot CH:CH \cdot O$	XVII-54	118.14
1539	**Creatine**	methyl-guanyl-glycine	$HN:C(NH_2)N(CH_3) \cdot$ $CH_2 \cdot CO_2H \cdot H_2O$	IV-363	149.15
1540	**Creatinine**	methyl-glyco-cyamidine	$HN:C.N(CH_3) \cdot CH_2 \cdot$ $\overline{} \ CO \cdot NH \ \overline{}$	XXIV-245	113.12
1541	picrate	....	$C_4H_7ON_3 \cdot C_6H_3O_7N_3$	XXIV-246	342.23
1542	picrate	....	$C_4H_7ON_3 \cdot 2C_6H_3O_7N_3$	XXIV-246	571.33
1543	**Creosol**	(3;1,4)	$CH_3O \cdot C_6H_3(CH_3)OH$	VI-878	138.17
1544	**Cresidine**	2-NH_2-4Me-anisole	$CH_3 \cdot C_6H_3(OCH_3)NH_2$	XIII-602	137.18
1545	**Cresol** (*o*)	*o*-methyl-phenol	$CH_3 \cdot C_6H_4 \cdot OH$	VI-349	108.14
1546	**Cresol** (*m*)	*m*-methyl-phenol	$CH_3 \cdot C_6H_4 \cdot OH$	VI-373	108.14
1547	**Cresol** (*p*)	*p*-methyl-phenol	$CH_3 \cdot C_6H_4 \cdot OH$	VI-389	108.14
1548	phthalein (*o*)	....	$C_6H_4 \cdot CO \cdot O \cdot C:$ $\overline{} \qquad \overline{}$ $(C_6H_3 \cdot CH_3 \cdot OH)_2$	XVIII-153	346.39
1549	phthalin (*o*)	(4',4''-diOH, 2-CO_2H, 3',3''-diMe)	$HO_2C \cdot C_6H_4 \cdot CH:$ $(C_6H_3 \cdot CH_3 \cdot OH)_2$	X-456	348.40
1550	sulfonphthalein (*o*)	cresol red	$C_6H_4 \cdot SO_2 \cdot O \cdot C:$ $\overline{} \qquad \overline{}$ $[C_6H_3 \cdot CH_3 \cdot OH]_2$	XIX-91	382.44
1551	sulfonphthalein (*m*)	metacresol purple	$C_{21}H_{18}O_5S$	....	382.44
1552	**Cresyl**† acetate (*o*)	*o*-tolyl acetate	$CH_3 \cdot C_6H_4 \cdot O_2CCH_3$	VI-355	150.18
1553	acetate (*m*)	cresatin	$CH_3 \cdot C_6H_4 \cdot O_2CCH_3$	VI-379	150.18
1554	*p*-toluenesul-fonate (*o*)	(4;1',2')	$CH_3 \cdot C_6H_4 \cdot SO_3C_6H_4 \cdot$ CH_3	XI-100	262.33
1555	**Crocetin** (*α*)	gardenin	$C_{20}H_{24}O_4$	XXX-106	328.41
1556	**Croconic acid**	crocic acid	$C_5O_3(OH)_2 \cdot 3H_2O$	VIII-488	196.11
1557	**Crotonic** acid (*α*)	butenoic acid	$CH_3 \cdot CH:CH \cdot CO_2H$	II-408	86.09
1558	acid (*β*) (*cis*)	*iso*-crotonic acid	$CH_3 \cdot CH:CH \cdot CO_2H$	II-412	86.09
1559	aldehyde (*α*)	2-butene-1-al	$CH_3 \cdot CH:CH \cdot CHO$	I-728	70.09
1560	amide	crotonamide	$CH_3 \cdot CH:CH \cdot CONH_2$	II-412	85.11
1561	anhydride	....	$(CH_3 \cdot CH:CH \cdot CO)_2O$	II-411	154.17

† See also under tolyl.
Corytuberine, cf. alkd.
Cotarnine, cf. alkd.
Cotoin 2347
Cotton red-4B 733

Cresatin 1553
Cresol purple 1551
Cresol red 1550
Cresorcinol 2371
Cresot(in)ic acid 3836, 3839, 3841

Cresyl-, cf. also tolyl.
Cresyl benzoate 6013-5
Cresyl ethyl ether 3173-5
Cresyl ethyl sulfide 3176
Cresyl methyl ether 4399-4401

Table 7-4 (*Continued*)
PHYSICAL CONSTANTS OF ORGANIC COMPOUNDS

No.	Crystalline Form and Color	Specific Gravity	Melting Point °C.	Boiling Point °C.	Solubility in 100 Parts		
					Water	Alcohol	Ether
1528	nd.	....	185 (anh.)	d.	0.5 c.; v. s. h.	sl. s. abs.	i.
1529	pr.	....	73-4	....	sl. s. h.	s.	s.
1530	lq.	$0.761^{15°}$	....	126^{738mm}	....	s.	....
1531	nd./aq.	....	207-8	subl.	sl. s. c.	s.	v. sl. s.
1532	pr./aq.	....	191	....	s. h.	s.	s.; s. bz.
1533	cr./aq.	....	206-7 d.	....	s. h.	v. s. h.	v. s.; i. lg.
1534	rhb./et.	$0.935^{\frac{20°}{4}}$	70	290-1	0.3 c.; 2 h.	v. s.	s.
1535	pl./bz.	....	47	$285\text{-}6^{719mm}$ sl. d.	i.; s. CS_2	v. s.; v. s. bz.	v. s.
1536	col. nd./ aq.	....	189-90 d.	>290, $-CO^2$	v. sl. s.; $0.7^{25°}$ bz.	$0.9^{25°}$; $0.1^{25°}$ CCl_4	v. sl. s.; v.sl.s.act.
1537	unstable	changes to coumarin	....	....	....	....	....
1538	oil	$1.078^{\frac{15}{15}}$	<-18	173-4	i.	i. alk.	s.
1539	mn./aq.	....	295 (anh.)	$-H_2O$, 100	$1.4^{18°}$	$0.01^{17°}$	i.
1540	mn.	....	260 d.	....	$8.7^{16°}$	$1^{16°}$ abs.	s. h. al.
1541	yel. nd./aq.	....	220	....	....	....	....
1542	yel. pl./al.	....	161-6	....	....	....	....
1543	pr.	$1.092^{\frac{20°}{20}}$	5.5	$221\text{-}2^{765mm}$	v. sl. s.	∞ ; ∞ chl.	∞ ; ∞ bz.
1544	nd./pet.	....	93-4	235	v. sl. s.	s.; s. bz.	s.
1545	cr.	$1.0274^{1°}$	30.9	191.0	2.5	∞ 30°	∞ 30°
1546	lq.	$1.034^{\frac{20°}{4}}$	11.9	202.7	0.5	∞	∞
1547	pr.	$1.018^{41°}$	34.8	201.9	1.8	∞ 36°	∞ 36°
1548	red yel. cr./al.	....	223-5	....	sl. s. h.; s. al. KOH	s.; s. al. NH_3	s.; v. sl. s. bz.
1549	nd.	....	217-8	....	....	....	....
1550	red cr./ac.	....	....	....	sl. s.	s.	s. alk.
1551	gn. fl.	....	....	....	sl. s.	sl. s.	i.
1552	col. lq.	....	....	208	v. sl. s.	v. s.	v. s.
1553	col. lq.	....	....	212	s. h.	v. s.	v. s.
1554	nd.	....	....	53-4	....	....	....
1555	red rhb.	....	283-5 d.	....	s. Ac_2O	s. dil. alk.	i. 10% alk.
1556	yel. lf.	....	$-3H_2O$, 100	....	s.	s.	....
1557	col. mn.	$0.9647^{9.7°}$	72	189	$8.3^{15°}$	s. h. lg.	....
1558	nd.	$1.031^{\frac{15}{4}°}$	15.5	170-1 d.	∞$^{25°}$	s.	....
1559	col. lq.	$0.8532^{\frac{20°}{20}}$	-69	102.2	18	∞	∞
1560	nd./act.	....	159-60	....	2.8 c.	v. s. 0.08 c. bz.	0.09 c.
1561	col. lq.	$1.040^{20°}$	....	$246\text{-}8^{766mm}$	....	....	....

Table 7-4 (*Continued*)
PHYSICAL CONSTANTS OF ORGANIC COMPOUNDS

No.	Name	Synonym	Formula	Beil. Ref.	Formula Weight
1562	**Crotonic** chloride	....	$CH_3 \cdot CH:CH \cdot CO \cdot Cl$	II-411	104.54
1563	nitrile (**trans**)	propenyl cyanide	$CH_3 \cdot CH:CH \cdot CN$	II-412	67.09
1564	**Crotonyl** acetate	crotyl acetate	$CH_3CO_2 \cdot C_4H_7$	II-137	114.15
1565	alcohol	2-butene-1-ol	$CH_3 \cdot CH:CH \cdot CH_2OH$	I-442	72.11
1566	chloride	crotyl chloride	$CH_3 \cdot CH:CH \cdot CH_2Cl$	I-205	90.55
1567	chloride (*iso*)	*iso*-crotyl chloride	$(CH_3)_2C:CH \cdot Cl$	I-209	90.55
1568	iodide	1-iodo-butene-2	$CH_3 \cdot CH:CH \cdot CH_2I$	I-206	182.00
1569	**Cryogenine**	3-semicarbazido-benzamide	$H_2N \cdot CO \cdot NH \cdot NH \cdot C_6H_4 \cdot CO \cdot NH_2$	XV-629	194.19
1570	**Cryptoxanthin**	provitamin A	$C_{40}H_{56}O$	XXX-93	552.89
1571	**Cuminyl** alcohol	cumin alcohol (*p*)	$(CH_3)_2CH \cdot C_6H_4 \cdot CH_2OH$	VI-543	150.22
1572	chloride	1'-chlorocymene	$C_3H_7 \cdot C_6H_4 \cdot CH_2Cl$	V-423	168.67
1573	**Cumyl** phenol (*p*)	(1,4)	$HO \cdot C_6H_4 \cdot C(CH_3)_2 \cdot C_6H_5$	....	212.29
1574	**Cupferon**	....	$C_6H_5 \cdot N(NO) \cdot ONH_4$	XVI-669	155.16
1575	**Curcumin**	....	$C_{21}H_{20}O_6$	VIII-554	368.39
1576	**Cyamelide**	....	$(CNOH)_x$ or $(CNOH)_3$	III-35	$(43.03)_x$
1577	**Cyanamide**	carbamic nitrile	$H_2N \cdot CN$	III-75	42.04
1578	**Cyananilide**	phenyl cyanamide	$C_6H_5 \cdot NH \cdot CN \cdot \frac{1}{2}H_2O$	XII-368	127.15
1579	**Cyanic acid**	....	$HOCN\dagger$	III-33	43.03
1580	**Cyanine**	1,1'-di-*iso*-amyl-4, 4'-quinocyanine iodide	$C_{29}H_{35}N_2I$	XXIII-299	538.52
1581	**Cyano**-acetamide	....	$NC \cdot CH_2 \cdot CO \cdot NH_2$	II-589	84.08
1582	acetanilide	....	$C_6H_5 \cdot NH \cdot CO \cdot CH_2CN$	XII-294	160.18
1583	acetic acid	nitrilomalonic acid	$HO_2C \cdot CH_2 \cdot CN$	II-583	85.06
1584	acetoacetic ester	....	$CH_3CO \cdot CH(CN) \cdot CO_2 \cdot C_2H_5$	III-796	155.15
1585	benzoic acid (*p*)	....	$HO_2C \cdot C_6H_4 \cdot CN$	IX-845	147.13
1586	benzyl chloride (*o*)	....	$ClCH_2 \cdot C_6H_4 \cdot CN$	IX-468	151.60
1587	propionic acid (*α*)	....	$CH_3 \cdot CH(CN) \cdot CO_2H$	II-630	99.09
1588	**Cyanogen**	oxalic nitrile	$NC \cdot CN$	II-549	52.04
1589	**Cyanogen** (para)	....	$(CN)_x$	II-553	$(26.02)_x$
1590	bromide	....	$Br \cdot CN$	III-39	105.93
1591	chloride	....	$Cl \cdot CN$	III-38	61.47
1592	iodide	....	$I \cdot CN$	III-41	152.92
1593	sulfide	....	$NC \cdot S \cdot CN$	III-180	84.10
1594	**Cyanuric acid**	....	$(HNCO)_3 \cdot 2H_2O$	XXVI-239	165.11
1595	**Cyclo**-butane	tetramethylene	$CH_2 \cdot CH_2 \cdot CH_2 \cdot CH_2$	V-17	56.11
1596	heptane	suberane	$CH_2 \cdot (CH_2)_5 \cdot CH_2$	V-29	98.19
1597	heptanone	suberone	$CH_2 \cdot (CH_2)_5 \cdot CO$	VII-13	112.17

† HOCN in aq.; NH:C:O (isocyanic acid) as vapor or in ether soln.
Crotonic acid dibromide 1826-7
Crotonic acid dichloride 1978-9
Crotonyl, cf. also crotyl.
Crotonyl chloride 1562
Crotonylene 1157
Crotyl, cf. also crotonyl.
Crotoyl alcohol 1565
Cry now gas 1303

Cryptopine, cf. alkd.
Crystal violet base 3592
Crystal violet dye salt 3593
Crystal violet leuco base 3595
Cumene 5379
Cumene (pseudo) 6254
Cumenol 5433-5
Cumenol (pseudo) 6265

Cumic acid 5383, 5385
Cumic alcohol 1571
Cumidine 5375
Cumidine (pseudo) **6250**
Cumin alcohol 1571
Cuminic acid 5383, **5385**
Cuminic aldehyde **5377**
Cuminic amide 5386
Cumylic acid 6259

Table 7-4 (*Continued*)
PHYSICAL CONSTANTS OF ORGANIC COMPOUNDS

No.	Crystalline Form and Color	Specific Gravity	Melting Point °C.	Boiling Point °C.	Solubility in 100 Parts Water	Alcohol	Ether
1562	col. lq.	$1.091^{20°}$	....	$124\text{-}5^{759mm}$	....	....	....
1563	col. lq.	$0.826^{15°}_{4}$	−51.5	122	d.	d.	
1564	col. lq.	$0.919^{20°}_{4}$	....	128.5-9.5	2.3	s.	s.
1565	col. lq.	$0.853^{20°}_{4}$	<−30	121-2	16.6	∞	....
1566	col. lq.	$0.934^{20°}_{4}$	....	84	i.	∞	∞
1567	lq.	$0.919^{20°}_{4}$	....	68	<0.1; d. h.	....	....
1568	lq.	$1.682^{20°}$	....	132-3 d.	....	....	....
1569	col. cr.	....	172±	....	1	s.; s. act.	s.; s. chl.
1570	pr./bz.-al.	....	168-9	....	i.	s. CS2	s.; s. chl.
1571	lq.	$0.975^{25°}_{25}$	....	247-8	i.	∞	∞
1572	lq.	$1.020^{22°}_{4}$	....	225-9	....	....	....
1573	col. nd.	....	74-5	187^{10mm}	i.; s. bz.	v. s.	s.; s. act.
1574	col. nd./al.	....	163-4	subl.	v. s.	s. h.	....
1575	red yel. pr.	....	183	....	i.; 0.05 bz.; s. alk.	sl. s. c.; s. ac.	v. sl. s.; i. lg.
1576	wh. amor.	....	....	....	$0.01^{15°}$	i.; sl. s. NH4OH	i.
1577	col. nd.	$1.073^{48°}_{4}$	44-5	140^{19mm}	v. s.	v. s.	v. s.
1578	cr.	....	47	....	sl. s.	v. s.	v. s.
1579	col. gas	$1.140^{0°}$ (lq.)	−80	-64^{0mm}	sl. s.	s. ac.	s.
1580	hyg., mn., gn.	....	100±	d. >150	i.; s. chl.	s. h.	i.; s. act.
1581	nd./al.	....	118-9	d.	15 c.	2 c.	....
1582	cr./al.	....	199-200	....	$0.03^{25°}$	....	....
1583	col. hyg. cr.	....	65-6	$108^{0.15mm}$ d. 165	s.	s.	s.
1584	col. nd.	$1.111^{20°}_{4}$ (lq.)	26	195-7	v. sl. s.	v. s.	v. s.
1585	lf./al.	....	219	....	s. h.	s.	s.; s. h. ac.
1586	mn. pr.	....	60-1.5	252^{759mm}	s. h.	s.	....
1587	oil	$1.14^{20°}$	....	$142\text{-}5^{11mm}$	s.	s.	
1588	col. gas	$0.866^{17.2°}$ 1.804 (A)	−27	−21.2	$450^{20°}$ cc.	$2300^{20°}$ cc.	$500^{20°}$ cc.
1589	brn. bl.	....	(CN)2 at 860	....	i.	i.	sl. s. alk.
1590	nd.	$2.015^{20°}_{4}$	52	61.3^{750mm}	s.	s.	s.
1591	poison. gas	$1.222^{0°}$	−6.9	13.0	$2500^{20°}$ cc.	$10,000^{20°}$ cc.	$5000^{20°}$ cc.
1592	col. nd.	....	146.5*	subl. >m. p.	s. h.	v. s.	v. s.
1593	pl. or lf.	....	60	d.	v. s. d.	s.	s.
1594	mn./aq.	$2.50^{19°}$ anh.	>360	d. to HNCO	$0.27^{17°}$	$0.12^{2°}$	....
1595	col. gas	$0.703^{4°}$	−50	$11\text{-}12^{726mm}$	i.	v. s.	v. s. act.
1596	oil	$0.810^{20°}_{4}$	−12	118-20	i.	....	....
1597	lq.	$0.950^{21°}$	....	179-81	i.	s.	s.

* Sealed tube.

Cupreine, cf. alkd.
Cupron 720
Curangin, cf. glcde.
Cuscohygrine, cf. alkd.
Cuskhygrine, cf. alkd.
Cuspidatin, cf. glcde.
Cyanacetic ester 3030

Cyanethyl carbonate 3032
Cyano-acetophenone (ω) 742
Cyano-benzene 716
Cyano-diallylamine 1682
Cyano-diethylamine 2143
Cyano-furan 3345
Cyano-guanidine 2080
Cyano-phenyl chloride (*iso*) 5168

Cyano-propyl alcohol 3758
Cyano-propylene oxide 2880
Cyanuramide 4019
Cyanuric acid (*iso*) 3313
Cyanuric chloride 6172
Cyclamin, cf. glcde.
Cyclo-form 1045

Table 7-4 (*Continued*)
PHYSICAL CONSTANTS OF ORGANIC COMPOUNDS

No.	Name	Synonym	Formula	Beil. Ref.	Formula Weight
1598	**Cyclo** heptene	suberene	$CH_2(CH_2)_4CH:CH$	V-65	96.17
1599	hexadiene-1,3	*o*-dihydrobenzene	C_6H_8	V-113	80.13
1600	hexadiene-1,4	*p*-dihydrobenzene	C_6H_8	V-113	80.13
1601	hexane	benzene hexa-hydride	$CH_2 \cdot (CH_2)_4 \cdot CH_2$	V-20	84.16
1602	hexanhexol	*d*-inositol	$(CHOH)_6$	VI-1192	180.16
1603	hexanhexol	*i*-inositol; dambose	$(CHOH)_6$	VI-1192	180.16
1604	hexanhexone	triquinoyl	$(CO)_6 \cdot 8H_2O$	VII-907	312.19
1605	hexanol	hexahydro-phenol	$CH_2 \cdot (CH_2)_4 \cdot CHOH$	VI-5	100.16
1606	hexanone	pimelin ketone	$CH_2(CH_2)_4 \cdot CO$	VII-8	98.15
1607	hexanone oxime	....	$CH_2(CH_2)_4C:NOH$	VII-10	113.16
1608	hexanpentol	*d*-quercitol	$CH_2(CHOH)_4 \cdot CHOH$	VI-1186	164.16
1609	hexene	benzene tetra-hydride	$CH_2(CH_2)_3CH:CH$	V-63	82.15
1610	hexyl acetate	....	$CH_3CO_2 \cdot C_6H_{11}$	VI-7	142.20
1611	hexyl anisole (*p*)	....	$CH_3O \cdot C_6H_4 \cdot C_6H_{11}$	**VI-549	190.29
1612	hexyl bromide	....	$C_6H_{11}Br$	V-24	163.06
1613	hexyl chloride	chloro-cyclohexane	$C_6H_{11}Cl$	V-21	118.61
1614	hexyl iodide	....	$C_6H_{11}I$	V-25	210.06
1615	hexyl phenol (*o*)	....	$HO \cdot C_6H_4 \cdot C_6H_{11}$	....	176.26
1616	hexyl phenol (*p*)	....	$HO \cdot C_6H_4 \cdot C_6H_{11}$	VI-583	176.26
1617	octane	octamethylene	$CH_2 \cdot (CH_2)_6 \cdot CH_2$	V-35	112.22
1618	pentadiene-1,3	....	$CH_2 \cdot CH:CH \cdot CH:CH$	V-112	66.10
1619	pentane	pentamethylene	$CH_2 \cdot (CH_2)_3 \cdot CH_2$	V-19	70.14
1620	pentanone	keto-penta-methylene	$CH_2 \cdot (CH_2)_3 \cdot CO$	VII-5	84.12
1621	propane	trimethylene	$CH_2 \cdot CH_2 \cdot CH_2$	V-15	42.08
1622	**Cymene** (*o*)	isopropyl-toluene	$CH_3 \cdot C_6H_4 \cdot CH(CH_3)_2$	V-419	134.22
1623	**Cymene** (*m*)	*iso*-cymene	$CH_3 \cdot C_6H_4 \cdot CH(CH_3)_2$	V-419	134.22
1624	**Cymene** (*p*)	cymene	$CH_3 \cdot C_6H_4 \cdot CH(CH_3)_2$	V-420	134.22
1625	**Cysteine** (*l*)	....	$HS \cdot CH_2 \cdot CH(NH_2) \cdot CO_2H$	IV-506	121.16
1626	hydrochloride	....	$C_3H_7O_2NS \cdot HCl$	IV-506	157.62
1627	**Cystine** (*l*)	....	$[HO_2C \cdot CH(NH_2) \cdot CH_2S \cdot]_2$	IV-507	240.30
1628	**Daphnetin**	7,8-diOH-coumarin	$O \cdot CO \cdot CH:CH \cdot C_6H_2: (OH)_2$	XVIII-100	178.15
1629	**Deca**-chlorodiphenyl	....	$Cl_5C_6 \cdot C_6Cl_5$	V-580	498.66
1630	diene-1,3	....	$CH_3 \cdot (CH_2)_5 \cdot CH:CH \cdot CH:CH_2$	I-260	138.25
1631	hydronaphthalene	"Decalin" (*cis*)	$C_{10}H_{18}$	V-92	138.25
1632	hydronaphthalene	"Decalin" (*trans*)	$C_{10}H_{18}$	V-92	138.25

Cyclo-hexano carboxylic acid 3577
Cyclo-hexylamine 326
Cyclo-hexyl benzene 5179
Cystamin 3599
Cysteine thioformacetal 2854

Cystogen 3599
Cytisine, cf. alkd.
D. A. gas 2712
Dagenan 5668
Dahl's acid 4550-!

Table 7-4 (*Continued*)
PHYSICAL CONSTANTS OF ORGANIC COMPOUNDS

No.	Crystalline Form and Color	Specific Gravity	Melting Point °C.	Boiling Point °C.	Solubility in 100 Parts		
					Water	Alcohol	Ether
1598	oil	$0.823^{20}_{4}°$	....	114-5	i.	s.	s.
1599	lq.*	$0.842^{20}_{20}°$	−98	79-80	i.	s.	s.
1600	lq.*	$0.848^{20}_{4}°$	−90±	85.5	i.	s.	s.
1601	col. lq.	$0.779^{20}_{4}°$	6.5	80.7	i.; ∞ act.	∞ ; ∞ bz.	∞ ; $57^{25°}$ Me al.
1602	mn.	....	253	319 in vac.	$4.5^{15°}$	v. sl. s.	i.
1603	mn./aq.	1.752	253	319^{15mm}	$2^{12°}$	i. abs.	i.
1604	nd./aq. HNO³	....	95-100 d.	....	v. sl. s.; s. alk.	v. sl. s	v. sl. s
1605	col. hyg. nd.	$0.962^{20}_{4}°$	25.5	161.1	$3.6^{20°}$	s.	s.
1606	col. oil	$0.947^{19}_{4}°$	−31.2	155.7	s.	s.	s.
1607	pr./lg.	....	89-90	204 sl. d.	s.; sl. s. lg.	v. s.	v. s.
1608	mn./aq.	$1.585^{13°}$	234-5	....	s.	s. h.	i.
1609	lq.	$0.810^{20}_{4}°$	−103.7	83.3	v. sl. s.	v. s.	v. s.
1610	oil	$0.985^{0}_{4}°$	....	174^{750mm}	i.	∞	∞
1611	yel. rhb.	....	58-9	276	i.	s.	v. s.
1612	col. lq.	$1.324^{20}_{20}°$	....	165^{714mm} sl. d.	i.	s.	s.
1613	col. lq.	$0.977^{18}_{4}°$	−43.9	142	i.	∞ bz.	∞
1614	lq.	$1.626^{15}_{15}°$	....	192^{742mm}	....	....	
1615	cr.	....	50-5	148^{10mm}	v. sl. s.	v. s.	v. s.; s. CCl₄
1616	col. nd./bz.	....	132-3	$155-8^{13mm}$	i.; $4^{25°}$ bz.	$70^{25°}$ act.	$40^{25°}$
1617	cr.	$0.839^{20}_{4}°$	14.4	$150-1^{709mm}$	....	....	....
1618	col. lq.	$0.805^{18}_{4}°$	−85	41-2	i.	∞	∞
1619	col. oil	$0.745^{20}_{4}°$	−93.8	49.3	i.	....	....
1620	oil	$0.948^{20°}$	−51.3	130.7	43.3	....	....
1621	col. gas	$0.720^{-79°}$	−127.4	−32.9	i.	s.	s.
1622	col. lq.	$0.877^{20}_{4}°$	−71.6	178.3	i.	s.	s.
1623	col. lq.	$0.861^{20}_{4}°$	−63.8	175.2	i.	s.	s.
1624	col. lq.	$0.857^{20}_{4}°$	−68.0	177.1	i.	s.	s.
1625	cr. pd.	....	....	....	s.	s. NH₃	s. ac.
1626	col. cr.	....	175-8 d.	....	v. s.	s.	s. act.
1627	pl./aq. HCl	....	d. 258-61	....	$0.01^{19°}$	i.	s. alk.
1628	yel. nd./ aq. al.	....	255-6	subl.	s. h.	v. s. h.; i. chl.	v. sl. s.; i. bz.
1629	rhb.	....	340	....	i.	v. sl. s.	v. sl. s.
1630	lq.	$0.750^{20°}$	....	168-70	....	....	....
1631	lq.	$0.895^{18}_{4}°$	−51	193.3	i.	s.	s.
1632	lq.	$0.872^{20}_{4}°$	−32	185.3	i.	s.	s.

* Probably mixt. of isomers.
Dahlin 3872
Dambose 1603
Daphnin, cf. glcde.

Datiscin, cf. glcde.
Datiscosid, cf. glcde.
Daturine, cf. alkd.

Table 7-4 (*Continued*)
PHYSICAL CONSTANTS OF ORGANIC COMPOUNDS

No.	Name	Synonym	Formula	Beil. Ref.	Formula Weight
1633	**Deca** hydroquinoline	....	$C_9H_{17}N$	XX-156	139.24
1634	methylene glycol	decandiol-1,10	$HOCH_2(CH_2)_8CH_2OH$	I-494	174.29
1635	**Decane** (*n*)	....	$CH_3 \cdot (CH_2)_8 \cdot CH_3$	I-168	142.29
1636	**Decane** (*iso*)	2-methyl-nonane	$(CH_3)_2CH(CH_2)_6CH_3$	I-168	142.29
1637	**Decane**	2,6-diMe-octane	$(CH_3)_2CH(CH_2)_3 \cdot CH(CH_3) \cdot C_2H_5$	I-168	142.29
1638	**Decane** (di-*iso*-amyl)	2,7-diMe-octane	$[(CH_3)_2CH \cdot CH_2CH_2]_2$	I-169	142.29
1639	**Decyl** acetate (*n*)	....	$CH_3CO_2 \cdot C_{10}H_{21}$	II-135	200.32
1640	alcohol (*n*)(*prim*)	decanol-1	$CH_3(CH_2)_8CH_2OH$	I-425	158.29
1641	alcohol (*sec*)	decanol-4	$C_2H_5 \cdot CH_2 \cdot CHOH \cdot (CH_2)_5 \cdot CH_3$	I-426	158.29
1642	alcohol (*tert*)	2-Et-octanol-3	$(C_2H_5)_2COH(CH_2)_4 \cdot CH_3$	I-426	158.29
1643	alcohol (*tert*)	4-Pr-heptanol-4	$(C_2H_5 \cdot CH_2)_3COH$	I-426	158.29
1644	alcohol(*prim*)	2,6-diMe-3-methylolheptane	$(CH_3)_2CH \cdot CH(CH_2 \cdot OH)(CH_2)_2 \cdot CH : (CH_3)_2$	I-427	158.29
1645	amine (*n*)	....	$CH_3(CH_2)_8CH_2NH_2$	IV-199	157.30
1646	chloride (*n*)(*prim*)	1-chlorodecane	$CH_3(CH_2)_8CH_2Cl$	I-168	176.73
1647	iodide (*n*)(*prim*)	1-iododecane	$CH_3(CH_2)_8CH_2I$	I-168	268.18
1648	nitrate (*n*)	....	$C_9H_{19}CH_2 \cdot O \cdot NO_2$	I-425	203.28
1649	nitrite (*n*)	....	$C_9H_{19}CH_2 \cdot O \cdot NO$	I-425	187.28
1650	**Decylene** (*α*)	decene-1	$CH_2 : CH(CH_2)_7CH_3$	I-223	140.27
1651	**Decylene**	2,7-diMe-octene-2	$(CH_3)_2CH(CH_2)_3 \cdot CH : C(CH_3)_2$	I-224	140.27
1652	**Decylene**	2-Me-5-Et-heptene-5	$(CH_3)_2CH(CH_2)_2 \cdot C(C_2H_5) : CHCH_3$	I-224	140.27
1653	**Decylene**	3,3,5-trimethyl-heptene-4	$C_2H_5 \cdot C(CH_3)_2CH : C(CH_3) \cdot C_2H_5$	I-224	140.27
1654	glycol	....	$CH_3 \cdot CHOH \cdot C(C_2H_5) \cdot (C_4H_9)CH_2OH$	....	174.29
1655	**Deguelin**	....	$C_{23}H_{22}O_6$	....	394.43
1656	**Dehydracetic acid**	6-Me-3-aceto-2,4-pyrandione	$C_8H_8O_4$	XVII-559	168.15
1657	**Dehydro**-androsterone	dehydro-*iso*-androsterone	$C_{19}H_{28}O_2$	....	288.43
1658	benzoylacetic acid	....	$C_6H_5CO \cdot CH \cdot CO \cdot CH : C(C_6H_5) \cdot O \cdot CO$ (cyclic)	XVII-575	292.29
1659	**Desoxalic acid**	....	$\cdot CHOH \cdot C(OH) : (CO_2H)_3$	III-586	194.10
1660	**Desoxybenzoin**	phenyl-benzyl-ketone	$C_6H_5 \cdot CO \cdot CH_2 \cdot C_6H_5$	VII-431	196.25
1661	**Dextrin**	starch gum	$(C_6H_{10}O_5)x$	....	(162.14)
1662	**Diacetamide**	....	$(CH_3CO)_2NH$	II-181	101.11
1663	**Diacetone** alcohol	diacetone	$(CH_3)_2COH \cdot CH_2 \cdot CO \cdot CH_3$	I-836	116.16
1664	amine	2-NH_2-2-Me-pentanone-4	$(CH_3)_2C(NH_2)CH_2 \cdot CO \cdot CH_3$	IV-322	115.18
1665	amine acid oxalate	....	$C_6H_{13}ON \cdot C_2H_2O_4 \cdot H_2O$	IV-322	223.23
1666	glucose	glucose bis-diMe-ketal	$C_{12}H_{20}O_6$	XXXI-155	260.29

Table 7-4 (*Continued*)
PHYSICAL CONSTANTS OF ORGANIC COMPOUNDS

No.	Crystalline Form and Color	Specific Gravity	Melting Point °C.	Boiling Point °C.	Solubility in 100 Parts		
					Water	Alcohol	Ether
1633	pr./lg.	....	48.2-8.5	204^{714mm}	s. h.	v. s.	v. s.
1634	nd./aq. al.	....	72-3	179^{15mm}	i. c.	s.; i. pet.	s. h.
1635	col. lq.	0.730$^{20°}$	−29.7	174.1	i.	∞	∞
1636	col. lq.	0.726$^{20°}_{4}$	−74.7	167.0	....	....	....
1637	col. lq.	0.728$^{20°}_{4}$	....	158.5	....	....	....
1638	col. lq.	0.724$^{20°}_{4}$	−54	159.9	....	....	....
1639	col. lq.	....		244	i.	s.	s.; s. bz.
1640	col. oil	0.830$^{20°}_{4}$	7	232.9	....	s.	
1641	col. oil	0.826$^{20°}_{0}$	....	210-1	....		
1642	col. oil	....		199			
1643	col. oil	0.834$^{21°}_{4}$	....	190-2	....		
1644	col. oil	0.849$^{0°}$	....	210-2	....		
1645	cr.	....	17	216-8	i.	....	....
1646	lq.	0.887$^{200°}$	....	180-90^{720mm}	i.		
1647	lq.	1.260$^{16°}_{4}$	....	132^{15mm}	....		
1648	lq.	0.951$^{0°}_{4}$	....	127-8^{11mm}	....		
1649	lq.	....	....	105-8^{12mm}	....		
1650	col. lq.	0.740$^{30°}$	....	172	i.	∞	∞
1651	col. lq.	0.748$^{20°}_{0}$	....	159-62^{650mm}	i.	....	....
1652	col. lq.	0.752$^{11°}_{4}$	....	157-8^{750mm}	i.	....	....
1653	col. lq.	0.773$^{21°}_{21}$	....	157.5^{759mm}			
1654	lq.	0.945$^{20°}_{20}$	....	130^{3mm}	sl. s.	....	....
1655	pa. gn./al.	....	171	....	i.	s.	....
1656	rhb.	....	108.5-9.0	269.9	0.12$^{20°}$; s. h.	v. s. h.	s.
1657	col. cr.	....	148(138)	....	i.	s.	s.
1658	yel. nd./ al.	....	171-2	....	i.; s. chl.; s. NH$_4$OH	sl. s.; s. bz.; s. alk.	s.; sl. s. lg.; s. CS$_2$
1659	hyg. cr.	....	d. 45	....	v. s.	v. s.	....
1660	pl./al.	....	60	320-2	sl. s. h.	v. s.	v. s.
1661	amor.	1.038	....	....	s.	i. abs.	i.
1662	col. nd.	....	78-9	222.5-3.5	v. s.	sl. s. lg.	sl. s.
1663	lq.	0.931$^{25°}$	−47	167.9	∞	∞	∞
1664	col. lq.	....		25$^{0.14mm}$ d. h.	s.	∞	∞
1665	pr./aq.	....	125-6	....	v. s. h.	s. h.	....
1666	col. mn. nd./lg.	....	110-1	subl.	14 h.	s.; s. chl.	s.; s. act.

Table 7-4 (*Continued*)
PHYSICAL CONSTANTS OF ORGANIC COMPOUNDS

No.	Name	Synonym	Formula	Beil. Ref.	Formula Weight
1667	**Diacetyl**	butandione	$CH_3 \cdot CO \cdot CO \cdot CH_3$	I-769	86.09
1668	acetone	....	$(CH_3CO \cdot CH_2)_2CO$	I-808	142.16
1669	*p*-aminophenol	....	$CH_3CO_2 \cdot C_6H_4 \cdot$ $NHCOCH_3$	XIII-464	193.20
1670	aniline (*N*)	*N*-Ph-diacetamide	$C_6H_5N:(COCH_3)_2$	XII-250	177.20
1671	benzidine	....	$(CH_3CO \cdot NH \cdot C_6H_4)_2$	XIII-227	268.32
1672	carbinol	....	$(CH_3CO)_2:CHOH$	I-852	116.12
1673	glucose	....	$(CH_3CO)_2:C_6H_{10}O_6$	....	264.23
1674	monomethoxime	....	$CH_3CO \cdot C(:NOCH_3) \cdot$ CH_3	I-772	115.13
1675	oxyphenyl-isatin	Isacon	$C_6H_4 \cdot N:COH \cdot C:$ $\underset{\underline{\qquad\qquad}}{\mid\qquad\qquad\mid}$ $(C_6H_4O_2C \cdot CH_3)_2$	....	401.42
1676	phenylenediamine	(*o*)	$C_6H_4(NHCOCH_3)_2$	XIII-20	192.22
1677	phenylenediamine	(*p*)	$C_6H_4(NHCOCH_3)_2$	XIII-97	192.22
1678	urea (s)	....	$(CH_3CO \cdot NH)_2CO$	*III-29	144.13
1679	**Diacetylene**	butadiyne	$HC:C \cdot C:CH$	I-266	50.06
1680	**Diallyl** amine	....	$(CH_2:CH \cdot CH_2)_2:NH$	IV-208	97.16
1681	barbituric acid (5,5)	Dial	$CO \cdot NH \cdot CO \cdot NH \cdot$ $\underset{\underline{\qquad\qquad}}{\mid\qquad\qquad}$ $CO \cdot C:(C_3H_5)_2$	*XXIV-422	208.22
1682	cyanamide	....	$(C_3H_5)_2:N \cdot CN$	**IV-666	122.17
1683	oxalate	....	$(CO \cdot O \cdot C_3H_5)_2$	II-540	170.17
1684	sulfide	thioallyl ether	$(CH_2:CH \cdot CH_2)_2S$	I-440	114.21
1685	trisulfide	....	$(C_3H_5)_2S_3$	I-441	178.34
1686	**Dialuric acid**	tartronyl urea	$C_4H_4O_4N_2$	XXV-85	144.09
1687	**Diamino**-anisole	(1;2,4)	$CH_3O \cdot C_6H_3(NH_2)_2$	*XIII-204	138.17
1688	anisole HCl	(1;2,4)	$C_7H_{10}ON_2 \cdot 2HCl$	*XIII-204	211.09
1689	anisole HCl	(1;2,5)	$C_7H_{10}ON_2 \cdot 2HCl$	....	211.09
1690	anthraquinone (1,4)	....	$C_6H_4(CO)_2C_6H_2:$ $(NH_2)_2$	XIV-197	238.25
1691	anthraquinone (1,5)	....	$(H_2N \cdot C_6H_3)_2(CO)_2$	XIV-203	238.25
1692	anthraquinone (1,8)	....	$(H_2N \cdot C_6H_3)_2(CO)_2$	XIV-212	238.25
1693	anthraquinone (2,3)	....	$C_6H_4(CO)_2C_6H_2:$ $(NH_2)_2$	XIV-215	238.25
1694	anthraquinone (2,6)	....	$H_2NC_6H_3(CO)_2C_6H_3 \cdot$ NH_2	XIV-215	238.25
1695	anthraquinone (2,7)	....	$H_2NC_6H_3(CO)_2C_6H_3 \cdot$ NH_2	XIV-216	238.25
1696	azobenzene (2,2′)	....	$(H_2N \cdot C_6H_4 \cdot N:)_2$	XVI-303	212.26
1697	azobenzene (4,4′)	azoaniline	$(H_2N \cdot C_6H_4 \cdot N:)_2$	XVI-334	212.26
1698	azobenzene (2,4)	chrysoidine	$(H_2N)_2C_6H_3 \cdot N_2 \cdot C_6H_5$	XVI-383	212.26
1699	azobenzene HCl	chrysoidine orange	$C_{12}H_{12}N_4 \cdot HCl$	XVI-383	248.72
1700	benzoic acid (2,3)	....	$(H_2N)_2C_6H_3 \cdot CO_2H$	XIV-447	152.15
1701	benzoic acid (2,4)	....	$(H_2N)_2C_6H_3 \cdot CO_2H$	XIV-448	152.15
1702	benzoic acid (2,5)	....	$(H_2N)_2C_6H_3 \cdot CO_2H$	XIV-448	152.15

Diacetoxy-ethane 3214
Diacetyl-, cf. also acetyl.
Diacetyl-dioxime 2490
Diacetyl-disulfide 85
Diacetyl-fluorescein 3253

Diacetyl-hydroquinone 3716
Diacetyl-methane 68
Diacetyl-monoxime 4888
Diacetyl-monoxime methyl ether 1674
Diacetyl-morphine, cf. alkd.

Table 7-4 (*Continued*)
PHYSICAL CONSTANTS OF ORGANIC COMPOUNDS

No.	Crystalline Form and Color	Specific Gravity	Melting Point °C.	Boiling Point °C.	Solubility in 100 Parts		
					Water	Alcohol	Ether
1667	yel. gn. lq.	$0.981\frac{18.5°}{4}$	Ca. −4	87–8	$25^{15°}$	∞	∞
1668	col. lf.	$1.068\frac{40°}{40}$	49	121^{10mm}	s. aq. Na_2CO_3	v. s.	v. s.
1669	lf./aq.	....	150–1	....	....	....	....
1670	pl./lg.	....	37–8	$199-200^{100mm}$	sl. s. c.	s. bz.	s. lg.
1671	nd./ac.	....	330–1	subl. d.	i.	v. sl. s.	v. sl. s.
1672	rhb. nd.	....	67–72	....	v. s. alk.	s. lg.	....
1673	lt. yel. amor.	....	....	....	v. s.	s.	i. bz.
1674	yel. lq.	....	....	125	....	....	....·....
1675	col. cr.	....	241–2	....	i.; i. dil. HCl	sl. s.	i.
1676	nd./aq.	....	185–6	....	s. h.; s. chl.	s.; s. a.	v. sl. s.
1677	cr./ac.	....	310–2	....	v. sl. s.	v. sl. s.	v. sl. s.
1678	nd./50% ac.	....	153.5	d. 170–80	v. sl. s.	s.	....
1679	gas	$0.736\frac{0°}{4}$ ..	−36	9–10	$466^{25°}$cc.	....	....
1680	lq.	$0.789\frac{20°}{20}$	−100	111–2	....	....	....
1681	col. lf./aq.	....	171–3	....	0.33 c.; 2 h.	s.; i. pet.	s.; s. act.
1682	col. lq.	....	....	$140-5^{90mm}$	i.	s.; s. bz. ·	s.
1683	oil	$1.055^{15.5°}$	....	217^{759mm}	i.	s.	....
1684	col. oil	$0.888\frac{27°}{4}$	−83	138.6^{758mm}	sl. s.	∞	∞
1685	lq.	$1.085^{15°}$	....	$112-22^{16mm}$	....	....	....
1686	pr.	....	214 d.	....	sl. s. c.	....	....
1687	nd./et.	....	67–8	....	....	....	....
1688	tan, rhb.	....	....	....	v. s.	i.	i.
1689	wh. pd.	....	....	....	v. s.	s.	i.
1690	cr./al.	....	268	....	s. h.	s.	v. s. bz.
1691	red nd.	....	319	subl.	v. sl. s.	sl. s.	sl. s.
1692	cr./al.	....	262	....	i.	s.	sl. s.
1693	cr./ PhNO_2	....	>320	....	s. H_2SO_4	sl. s. chl.	s. C_5H_5N
1694	red pr./ C_5H_5N	....	310–20 d.	....	s. H_2SO_4	sl. s. h.	i. chl.
1695	or. nd./al.	....	>330	subl.	i.	sl. s.	s. conc. a.
1696	red lf./al.	....	134	....	v. sl. s.	s. h.	v. s.; v. s. act.
1697	yel. nd./ aq. al.	....	241–3	....	v. sl. s.	s.	sl. s. bz., sl. s. lg.
1698	yel. cr.	....	117.5	....	sl. s. h.	s.	s.; v. s. chl.
1699	bl. cr. or red pd.	....	....	....	s.	s.	....
1700	nd.	....	190–1 a.	....	v. sl. s.	....	....
1701	cr.	....	140±	....	v. sl. s.	....	....
1702	pr./aq.	....	d. 200	....	v. sl. s. h.	v. sl. s.	v. sl. s.

Diacetyl-peroxide 106
Diacetyl-resorcinol 5552
Dial 1681
Diallyl-, cf. also allyl.
Diallyl 3574

Dialuramide 6384
Diamine-H 5261
Diamine-P 5260
Diamino-benzene 5256–61
Diamino-benzene sulfonic acid 5262

Table 7-4 (*Continued*)
PHYSICAL CONSTANTS OF ORGANIC COMPOUNDS

No.	Name	Synonym	Formula	Beil. Ref.	Formula Weight
	Diamino				
1703	benzoic acid (3,4)	....	$(H_2N)_2C_6H_3 \cdot CO_2H$	XIV-450	152.15
1704	benzoic acid (3,5)	....	$(H_2N)_2C_6H_3 \cdot CO_2H \cdot$ H_2O	XIV-453	170.17
1705	benzoic HCl (3,5)	....	$(H_2N)_2C_6H_3 \cdot CO_2H \cdot$ $2HCl$	XIV-453	225.08
1706	benzophenone (2,2')	....	$(NH_2 \cdot C_6H_4)_2CO$	XIV-87	212.25
1707	benzophenone (3,3')	....	$(NH_2 \cdot C_6H_4)_2CO$	XIV-88	212.25
1708	benzophenone (4,4')	....	$(NH_2 \cdot C_6H_4)_2CO$	XIV-88	212.25
1709	2-butyloxy-5,5'-azopyridine (2',6')	Niazo; Neotropin	$C_4H_9O \cdot C_5H_3(N) \cdot N_2 \cdot$ $C_5H_2(N)(NH_2)_2$	....	286.34
1710	diphenyl (2,2')	....	$H_2N \cdot C_6H_4 \cdot C_6H_4 \cdot NH_2$	XIII-210	184.24
1711	diphenyl (2,4')	diphenyline	$H_2N \cdot C_6H_4 \cdot C_6H_4 \cdot NH_2$	XIII-211	184.24
1712	diphenyl (3,3')	....	$H_2N \cdot C_6H_4 \cdot C_6H_4 \cdot NH_2$	XIII-213	184.24
1713	diphenyl (4,4')	benzidine	$(H_2N \cdot C_6H_4)_2 \cdot H_2O$ †	XIII-214	202.26
1714	diphenyl (4,4')	benzidine	$(H_2N \cdot C_6H_4)_2$ ‡	XIII-214	184.24
1715	diphenyl (4,4') HCl	benzidine HCl	$(H_2N \cdot C_6H_4)_2 \cdot 2HCl$	XIII-219	257.16
1716	diphenyl H_2SO_4	benzidine sulfate	$(H_2N \cdot C_6H_4)_2 \cdot H_2SO_4$	XIII-219	282.32
1717	diphenylamine (2,4)	....	$C_6H_5 \cdot NH \cdot C_6H_3(NH_2)_2$	XIII-295	199.26
1718	diphenylamine (4,4')	....	$(H_2N \cdot C_6H_4)_2NH$	XIII-110	199.26
1719	diphenylethane (4,4')	diamino-dibenzyl	$(H_2N \cdot C_6H_4 \cdot CH_2)_2$	XIII-248	212.30
1720	diphenyl ether	(4,4')	$H_2N \cdot C_6H_4 \cdot O \cdot C_6H_4 \cdot$ NH_2	XIII-441	200.24
1721	diphenylmethane	4,4'-diaminoditane	$(H_2N \cdot C_6H_4)_2CH_2$	XIII-238	198.27
1722	hydrazobenzene (*p*)	biphenin	$(H_2N \cdot C_6H_4 \cdot NH)_2$	XV-653	214.27
1723	phenol (2,4)	....	$(NH_2)_2C_6H_3 \cdot OH$	XIII-549	124.14
1724	phenol (2,4) HCl	amidol	$(NH_2)_2C_6H_3OH \cdot 2HCl$	XIII-550	197.07
1725	phenol (2,5)	....	$(NH_2)_2C_6H_3 \cdot OH$	XIII-553	124.14
1726	phenol (3,4)	....	$(NH_2)_2C_6H_3 \cdot OH$	XIII-564	124.14
1727	phenol (3,5)	....	$(NH_2)_2C_6H_3 \cdot OH$	XIII-567	124.14
1728	phenylacetic acid HCl	....	$(NH_2)_2C_6H_3CH_2 \cdot$ $CO_2H \cdot 2HCl$	XIV-476	239.10
1729	stilbene (2,2')(α)	....	$(H_2N \cdot C_6H_4 \cdot CH:)_2$	XIII-267	210.28
1730	stilbene (2,2')(β)	....	$(H_2N \cdot C_6H_4 \cdot CH:)_2$	XIII-267	210.28
1731	stilbene (4,4')	....	$(H_2N \cdot C_6H_4 \cdot CH:)_2$	XIII-267	210.28
1732	stilbene-disulfonic acid	(4,4'-diNH₂-2,2'-diS)	$[H_2N \cdot C_6H_3(SO_3H) \cdot$ $CH]_2$	XIV-798	370.41
1733	triphenylmethane	(4,4'); diamino-tritane	$(H_2N \cdot C_6H_4)_2CH \cdot$ C_6H_5	XIII-274	274.37
1734	**Diamyl** amine (*n*)	....	$(C_5H_{11})_2NH$	*IV-378	157.30
1735	amine (*iso*)	....	$(C_5H_{11})_2NH$	IV-182	157.30
1736	amine (*d*)	....	$[C_2H_5CH(CH_3) \cdot$ $CH_2]_2NH$	IV-179	157.30
1737	carbonate (*n*)	amyl carbonate	$(C_5H_{11}O)_2CO$	....	202.30
1738	carbonate (*iso*)	*iso*-amyl carbonate	$(C_5H_{11}O)_2CO$	III-7	202.30
1739	benzene§	....	$(C_5H_{11})_2C_6H_4$	V-470	218.39

† Cryst. from aq. < 60°; ‡ cryst. from aq. > 80°.
§ Mixt. of isomers.
Diamino-butane 5474
Diamino-caproic acid 3986
Diamino-chlorobenzene 1433

Diamino-dibenzyl 1719
Diamino-diphenylsulfide 5886
Diamino-ditane 1721
Diamino-naphthalene 4571-9
Diamino-propane 5464, 6282

Table 7-4 (*Continued*)
PHYSICAL CONSTANTS OF ORGANIC COMPOUNDS

No.	Crystalline Form and Color	Specific Gravity	Melting Point °C.	Boiling Point °C.	Solubility in 100 Parts		
					Water	Alcohol	Ether
1703	lf.	....	210 d.	....	s. h.	....	....
1704	nd./aq.	....	228–36 (anh.)	$-H_2O$, 110	1.18°	s.	s.
1705	nd.	....	253?	....	s.	s.	....
1706	yel. lf./aq. al.	....	134–5	....	i.	s.	....
1707	lt. yel. nd./al.	....	173–4	....	v. sl. s.	s.	s.
1708	yel. nd./ aq. al.	....	237–9	....	v. sl. s. h.	s.	s.
1709	red cr.	....	129	....	sl. s.	s.	s.
1710	nd./al.	....	81	....	....	....	....
1711	nd./aq. al.	....	45	363	v. sl. s.	s.	s.
1712	oil	....	....	....	v. sl. s.	....	s.
1713	cr./aq.	....	105–20	....	....	....	....
1714	cr./aq.	....	127.5–8.7	400–1^{740mm}	1 h.	1 h.	2.2abs.
1715	col. lf.	....	....	....	s., d.	s.; i. HCl	i.
1716	col. pl.	....	....	....	0.0008$^{100°}$	v. sl. s. h.	i.
1717	nd.	....	130	....	....	....	....
1718	lf./aq.	....	158	d.	sl. s.	s.	s.
1719	pl./aq.	....	134–5	subl. sl. d.	sl. s. h.	v. s.	....
1720	cr./al.	....	186–7 d.	....	i.	1.$^{25°}$; 15$^{25°}$ act.	i. bz.; i. CCl$_4$
1721	nd./aq.	....	93–4	249–53^{15mm}	sl. s. c.	s.	s.
1722	yel. cr.	....	145	....	s. h.	s.	s.
1723	lf.	....	78–80 d.	....	s. ac.	s.; s. alk.	sl. s.
1724	nd.	....	....	....	s.	sl. s.	....
1725	cr.	....	68	....	....	....	....
1726	cr.	....	167–8 d.	....	....	....	....
1727	pr.	....	168–70	....	s.	....	sl. s.
1728	tan nd.	....	....	....	v. s.	sl. s.	i.
1729	yel. pr./al.	....	176	....	s. bz.	s.	s.
1730	red nd./aq.	....	123	....	....	....	....
1731	yel. lf./al.	....	227–8	sl. d.	sl. s. h.	s. Me al.	sl. s. bz.
1732	yel. nd.	....	....	....	v. sl. s.	....	....
1733	cr./et.	....	139*	....	v. sl. s.	s.	s.
1734	col. lq.	....	....	202–3^{745mm}	v. sl. s.	v. s.	∞
1735	col. lq.	0.767$^{21°}_{4}$	−44	188–90	sl. s.	s.; s. chl.	∞
1736	col. lq.	0.788$^{0°}$	....	182–4	sl. s.	s.	s.
1737	col. lq.	....	....	130–2^{20mm}	....	....	....
1738	col. lq.	0.912$^{15°}$	....	233	....	....	....
1739	col. lq.	....	....	265±	i.	s.	s.

* Crysts. $+1C_6H_6$/bz., m. p. 104–6°.
Diamino-propanol 3384
Diamino-toluene 6003, 6005, 6007
Diamino-tritane 1733
Diamino-valeric acid 5000

Diamol 1724 or 5260
Diamond green–G base 5672
Diamorphine, cf. alkd.
Di-*iso*-amyl 1638

Table 7-4 (*Continued*)
PHYSICAL CONSTANTS OF ORGANIC COMPOUNDS

No.	Name	Synonym	Formula	Beil. Ref.	Formula Weight
1740	**Diamyl** ether (*n*)	amyl ether	$(C_5H_{11})_2O$	*I-193	158.29
1741	ether (*iso*)	*iso*-amyl ether	$[(CH_3)_2CH(CH_2)_2]_2O$	I-401	158.29
1742	hydroquinone (*tert*)	....	$[C_2H_5C(CH_3)_2]_2:$ $C_6H_2(OH)_2$	VI-952	250.38
1743	ketone (*n*)	caprone	$(C_5H_{11})_2CO$	I-714	170.30
1744	ketone (*iso*)	....	$(C_5H_{11})_2CO$	I-714	170.30
1745	oxalate (*iso*)	....	$(CO_2 \cdot C_5H_{11})_2$	II-540	230.31
1746	phthalate (*n*)	amyl phthalate	$C_6H_4(CO_2 \cdot C_5H_{11})_2$	....	306.41
1747	phthalate (*iso*)	....	$C_6H_4(CO_2 \cdot C_5H_{11})_2$	**IX-587	306.41
1748	succinate (*n*)	amyl succinate	$(CH_2CO_2 \cdot C_5H_{11})_2$	**II-551	258.36
1749	succinate (*iso*)	....	$(CH_2CO_2 \cdot C_5H_{11})_2$	II-611	258.36
1750	succinate (*act*)	....	$(CH_2CO_2 \cdot C_5H_{11})_2$	II-611	258.36
1751	sulfide (*n*)	amyl sulfide	$(C_5H_{11})_2S$	....	174.35
1752	sulfide (*iso*)	....	$(C_5H_{11})_2S$	I-405	174.35
1753	sulfide, di- (*n*)	amyl disulfide	$(C_5H_{11}S)_2$	....	206.41
1754	sulfide, di- (*iso*)	....	$(C_5H_{11}S)_2$	I-406	206.41
1755	sulfite (*n*)	amyl sulfite	$(C_5H_{11}O)_2SO$	....	222.35
1756	sulfone (*iso*)	amyl sulfone	$(C_5H_{11})_2SO_2$	I-406	206.35
1757	tartrate (*iso*)	amyl tartrate	$(HO \cdot CH \cdot CO_2 \cdot C_5H_{11})_2$	*III-179	290.36
1758	**Dianisalacetone**	....	$(CH_3OC_6H_4CH: CH)_2CO$	VIII-354	294.35
1759	**Dianisidine** (*o*)	(3,3'-MeO; 4,4'-NH$_2$)	$(CH_3O \cdot C_6H_3 \cdot NH_2)_2$	XIII-807	244.30
1760	**Dianisyl**-phenetyl-guanidine HCl	guanicaine	$(C_{23}H_{25}O_3N_3 \cdot HCl$	XIII-487	427.93
1761	**Diazidoethane** (1,2)	1,2-bistriazo-ethane	$(CH_2N_3)_2$	I-103	112.09
1762	**Diazo**-aminobenzene	benzene-di-azoanilide	$C_6H_5 \cdot N:N \cdot NHC_6H_5$	XVI-687	197.24
1763	aminonaphthalene	(*β*)	$C_{10}H_7 \cdot N_2 \cdot NH \cdot C_{10}H_7$	....	297.36
1764	aminotoluene	(2,2')	$C_7H_7 \cdot N_2NHC_7H_7$	XVI-703	225.30
1765	benzene chloride	....	$C_6H_5 \cdot N_2Cl$	XVI-431	140.57
1766	benzene cyanide	....	$C_6H_5 \cdot N_2CN$	XVI-432	131.14
1767	benzene nitrate	....	$C_6H_5 \cdot N_2NO_3$	XVI-432	167.13
1768	benzene per-bromide	....	$C_6H_5N(Br_3):N$	XVI-431	344.85
1769	benzene sulfonic acid (*o*)	....	$C_6H_4 \cdot N_2 \cdot O \cdot SO_2$ ⌞_____⌟	XVI-557	184.18
1770	benzene sulfonic acid (*m*)	....	$C_6H_4 \cdot N_2 \cdot O \cdot SO_2$ ⌞_____⌟	XVI-559	184.18
1771	benzene sulfonic acid (*p*)	sulfanilic acid diazide	$C_6H_4 \cdot N_2 \cdot O \cdot SO_2$ ⌞_____⌟	XVI-561	184.18
1772	methane	....	CH_2N_2	XXIII-25	42.04
1773	salicylic acid (5)	....	$HO_2C \cdot C_6H_2(OH)N_2$	XVI-553	164.12
1774	uracil (5)	....	$CO \cdot NH \cdot CO \cdot NH \cdot$ ⌞_____ $CH:C \cdot N:N \cdot OH$ ___⌟	XXV-565	156.10
1775	**Dibenzalacetone**	cinnamone	$(C_6H_5 \cdot CH:CH)_2CO$	VII-500	234.30
1776	**Dibenzamide**	....	$(C_6H_5 \cdot CO)_2NH$	IX-213	225.25
1777	**Dibenzanthracene**	(1,2,5,6)	$C_{22}H_{14}$	*V-369	278.36

Table 7-4 (*Continued*)
PHYSICAL CONSTANTS OF ORGANIC COMPOUNDS

No.	Crystalline Form and Color	Specific Gravity	Melting Point °C.	Boiling Point °C.	Solubility in 100 Parts		
					Water	Alcohol	Ether
1740	lq.	0.787$\frac{15}{4}$°	−69.3	**187.5**	i.	∞	∞
1741	col. lq.	0.777$\frac{20}{4}$°	....	173.4	i.	∞ ; ∞ chl.	∞
1742	cr./bz.	....	185	....	i. aq. NaOH	s.; s. chl.	s.; v. sl. s. lg.
1743	lf.	0.829$\frac{15}{4}$°	14.6	228	i.	v. s.	v. s.
1744	yel. oil	....	....	226	i.	s.	s.
1745	col. lq.	0.968$\frac{11}{11}$°	....	265-7	i.	s.	s.
1746	col. lq.	0.821$\frac{25}{4}$°	....	204-6^{11mm}	....	....	....
1747	col. lq.	1.022$\frac{15.6}{15.6}$°	....	225^{40mm}	i.	s.	s.
1748	col. lq.	0.961$\frac{20}{4}$°	−9	172^{16mm}	....	....	....
1749	col. lq.	0.961$^{13°}$	....	289.9^{728mm}	i.	s.	s.
1750	col. lq.	0.959$\frac{20}{4}$°	....	178-80^{25mm}	i.	s.	s.
1751	col. lq.	....	....	103-5^{12mm}	....	....	....
1752	lq.	0.843$\frac{20}{4}$°	....	216	i.	∞	∞
1753	yel. lq.	....	....	128-30^{12mm}	....	....	....
1754	lq.	0.918$^{18°}$	....	250 sl. d.	....	....	....
1755	col. lq.	....	....	136-8^{15mm}	....	....	....
1756	nd.	....	31	295	v. sl. s.	v. s.	v. s.
1757	lq.	1.063$\frac{14.8}{4}$°	....	195^{16mm}	i.	....	....
1758	yel. lf./EtOAc	....	129-30	....	s. chl., s. bz.	sl. s.	sl. s.
1759	col. lf.	....	131.5	....	i.; s. bz.	s.	s.
1760	col. cr.	....	176	....	6	s.	i. oil
1761	oil	1.170$^{24.9°}$	....	53^{9mm}	....	....	....
1762	yel. lf./al.	....	96-8	d. 150; expl.	i.; v. s. bz.	s. h.	v. s.
1763	red nd./xylene	....	156	....	s. H$_2$SO$_4$	....	....
1764	or. cr.	....	51	....	0.05	....	....
1765	hyg. cr./al. et.	....	expl.	....	v. s.	s. abs.	i.; s. act.
1766	yel.	....	69	....	sl. s.	....	....
1767	nd./al. et.	1.37	expl.	....	v. s.	sl. s.	i.
1768	yel. lf.	....	63.5 d.	....	i. d.	sl. s. d.	i. d.
1769	col.	....	expl.	....	d. h.	d. h.	....
1770	col. nd./aq.	....	expl.	....	s.; d. 60°	d. h.	....
1771	col. nd./aq.	....	expl.	....	s. h.	i.; d. h.	s. dil. alk.
1772	gas	expl. 200°	−145	−23	d.	....	s.
1773	yel. cr.	....	expl. 155	....	s. h.	s. Na$_2$CO$_3$	i.
1774	red or yel. pl./aq.	....	expl.	....	d. h. a.	....	....
1775	lt. yel. mn.	....	112	**d.**	v. s. act.	s. h.	sl. s.
1776	rhb./bz.	....	148	....	0.12$^{15°}$	s.	s.
1777	lf./ac.	....	262	subl.	sl. s. ac.	v. s. bz.	i.

Diazo-ethyl acetate 3035
Diazo-resorcinol 5551

Diazo-resorufine 5554
Dibenzal-hydrazine 633

Table 7-4 (*Continued*)
PHYSICAL CONSTANTS OF ORGANIC COMPOUNDS

No.	Name	Synonym	Formula	Beil. Ref.	Formula Weight
1778	**Dibenzeneazore-sorcinol**	....	$(C_6H_5 \cdot N_2)_2$: $C_6H_2(OH)_2$	XVI-185	318.34
1779	**Dibenzoyl** disulfide	....	$(C_6H_5 \cdot CO)_2S_2$	IX-424	274.36
1780	ethylenediamine	....	$(C_6H_5CO \cdot NHCH_2)_2$	IX-262	268.32
1781	methane	β-hydroxy chalkone	$(C_6H_5 \cdot CO)_2CH_2$	VII-769	224.26
1782	**Dibenzyl**-acetic acid	....	$(C_6H_5CH_2)_2$: $CHCO_2H$	IX-682	240.30
1783	amine	....	$(C_6H_5 \cdot CH_2)_2NH$	XII-1035	197.28
1784	aniline	....	$(C_6H_5 \cdot CH_2)_2N \cdot C_6H_5$	XII-1037	273.38
1785	fumarate	benzyl fumarate	$(:CH \cdot CO_2 \cdot CH_2C_6H_5)_2$	VI-437	296.33
1786	hydroxylamine (β,β)	....	$(C_6H_5CH_2)_2:NOH$	XV-19	213.28
1787	ketone	α,α'-diphenyl acetone	$(C_6H_5CH_2)_2:CO$	VII-445	210.28
1788	maleate	benzyl maleate	$(:CH \cdot CO_2 \cdot CH_2C_6H_5)_2$	VI-437	296.33
1789	phthalate (*o*)	benzyl phthalate	$C_6H_4(CO_2 \cdot C_7H_7)_2$	IX-802	346.39
1790	succinate	....	$(\cdot CH_2CO_2 \cdot C_7H_7)_2$	VI-436	298.34
1791	sulfide	....	$(C_6H_5CH_2)_2S$	VI-455	214.33
1792	tartrate (*d*)	....	$(CHOHCO_2C_7H_7)_2$	*VI-221	330.34
1793	**Dibromo**-acetamide	....	$Br_2CH \cdot CO \cdot NH_2$	II-219	216.87
1794	acetamide (*N*)	acetdibromamide	$CH_3 \cdot CO \cdot NBr_2$	II-182	216.87
1795	acetic acid	....	$Br_2CH \cdot CO_2H$	II-218	217.86
1796	acetylene	dibromoethyne	$BrC:CBr$	I-246	183.84
1797	aminophenol	(4;2,6,1)	$NH_2 \cdot C_6H_2Br_2OH$	XIII-517	266.93
1798	aniline (2,3)	....	$Br_2C_6H_3 \cdot NH_2$	XII-655	250.93
1799	aniline (2,4)	....	$Br_2C_6H_3 \cdot NH_2$	XII-655	250.93
1800	aniline (2,6)	....	$Br_2C_6H_3 \cdot NH_2$	XII-659	250.93
1801	aniline (3,4)	....	$Br_2C_6H_3 \cdot NH_2$	XII-660	250.93
1802	aniline (3,5)	....	$Br_2C_6H_3 \cdot NH_2$	XII-660	250.93
1803	anthracene (9,10)	....	$C_{14}H_8Br_2$	V-665	336.04
1804	anthranilic acid	(3,5)	$Br_2C_6H_2(NH_2)CO_2H$	XIV-371	294.94
1805	anthraquinone (1,3)	....	$Br_2C_{14}H_6O_2$	*VII-414	366.02
1806	anthraquinone (1,5)	....	$Br_2C_{14}H_6O_2$	VII-789	366.02
1807	anthraquinone (2,6)	....	$Br_2C_{14}H_6O_2$	VII-790	366.02
1808	anthraquinone (β)	(2,3)	$Br_2C_6H_2(CO)_2C_6H_4$	VII-790	366.02
1809	anthraquinone (α)	(2,7)	$BrC_6H_3(CO)_2C_6H_3Br$	VII-790	366.02
1810	barbituric acid	(5,5); Dibromin	$Br_2C(CONH)_2CO$	XXIV-472	285.89
1811	benzene (*o*)	....	$C_6H_4Br_2$	V-210	235.92
1812	benzene (*m*)	....	$C_6H_4Br_2$	V-211	235.92
1813	benzene (*p*)	....	$C_6H_4Br_2$	V-211	235.92
1814	benzoic acid (2,3)	....	$Br_2C_6H_3 \cdot CO_2H$	IX-357	279.93
1815	benzoic acid (2,4)	....	$Br_2C_6H_3 \cdot CO_2H$	IX-358	279.93
1816	benzoic acid (2,5)	....	$Br_2C_6H_3 \cdot CO_2H$	IX-358	279.93
1817	benzoic acid (2,6)	....	$Br_2C_6H_3 \cdot CO_2H$	IX-358	279.93
1818	benzoic acid (3,4)	....	$Br_2C_6H_3 \cdot CO_2H$	IX-359	279.93
1819	benzoic acid (3,5)	....	$Br_2C_6H_3 \cdot CO_2H$	IX-359	279.93
1820	benzoylene urea (6,8)	....	$Br_2C_6H_2 \cdot NH \cdot CO \cdot$ $NH \cdot CO$	....	319.95

Dibenzo-furan 2763
Dibenzo-pyranol 6453
Dibenzo-pyrone 6451
Dibenzo-pyrrole 1231
Dibenzo-thioxine 5129

Dibenzoyl 697
Dibenzoyl-resorcinol 5553
Dibenzyl-, cf. also benzyl.
Dibenzyl 2717
Dibenzyl disulfide 805

Table 7-4 (*Continued*)
PHYSICAL CONSTANTS OF ORGANIC COMPOUNDS

No.	Crystalline Form and Color	Specific Gravity	Melting Point °C.	Boiling Point °C.	Solubility in 100 Parts		
					Water	Alcohol	Ether
1778	red nd./ chl.-al.	....	223-4	....	s. h. chl.	i.	v. sl. s. dil. alk.
1779	pr./CS_2	....	133-5	d.	i.; i. NH_4OH	sl. s. h.	sl. s. h.; s. CS_2
1780	nd./al.	....	250-1	d.	i.	$0.08^{22°}$	....
1781	rhb./al.	....	78(72-3)*	$219-21^{18mm}$	i.	$4.4^{19.5°}$	s.
1782	nd./aq.	....	88-9	....	v. sl. s. h.	s.; s. bz.	s.; s. ac.
1783	col. oil	$1.028^{25°}_{25}$	−26	$268-71^{250mm}$	i.	s.	s.
1784	pr./al.	....	70-1	>300 sl. d.	i.	v. s. h.	v. s.
1785	col. pr./lg.	....	64	239^{14mm}	i.	s.	v. sl. s. c.
1786	nd.	....	124	....	sl. s. h.	s.	s.
1787	cr./aq. al.	....	34-5	330.6	....	....	....
1788	oil	....	....	241^{14mm}	....	....	....
1789	pr./al.	....	42-3	274^{12mm}	v. sl. s.	s.	s.
1790	col. lf./al.	....	45-6	238^{14mm}	i.	s.	s.; s. bz.
1791	rhb./et.	$1.071^{0°}_5$	49	....	i.	s.	s.
1792	cr.	$1.204^{72°}$	50±	$250-70^{4mm}$	....	....	....
1793	nd.	....	156	....	....	....	....
1794	yel. nd.	....	100	....	s. h.	s.	s.
1795	col. cr.	....	48-50	232-4 d.	v. s.	v. s.	v. s.
1796	poison. lq.	2±	expl. with trace O_2	76.5 in CO_2	i.	s.	s.
1797	nd./al.	....	192-3	....	i.; s. bz.	s.	sl. s.
1798	pl./aq. al.	....	43	....	v. sl. s.	v. s.	v. s.
1799	rhb.	$2.260^{20°}$	79.5	....	....	s.	s. ac.
1800	nd./al.	....	83-4	262-4	....	v. s. abs.	v. s.
1801	lf./aq. al.	....	80-1	subl. 100	i.	s.	....
1802	nd.	....	56.5	....	....	....	....
1803	yel. nd.	....	221	subl.	s. h. bz.	v. sl. s.	v. sl. s.
1804	nd./al.	....	235-6	....	i.; s. alk.	s.	s.
1805	yel. nd./ac.	....	210	....	s. H_2SO_4	v. sl. s.	v. sl. s.
1806	nd.	....	....	....	....	....	....
1807	yel. cr./ Am. al.	....	289-90	....	....	v. sl. s.	s. h. bz.
1808	yel. nd.	....	269-70	subl.	s. bz.	v. sl. s.	s. chl.
1809	yel. nd./ac.	....	245	....	....	v. sl. s. h.	sl. s. h. ac.
1810	cr./aq. HNO_3	....	240 d.	....	3	s.	s.
1811	col. lq.	$1.956^{20°}_4$	1.8	221-2	i.	s.	s.
1812	col. lq.	$1.952^{20°}_4$	−6.9	219^{755mm}	i.	s.	s.
1813	col. pl./al.	$2.261^{18°}$	87.3	220.3	i.; s. chl.	1.6; s. bz.	$71^{25°}$
1814	nd./aq.	....	147-9	....	s. h. lg.	....	....
1815	lf./aq.	....	171-2	subl.	sl. s. h.	....	....
1816	nd./aq.	....	153	....	v. sl. s. c.	s.	s.
1817	nd./aq.	....	146-7	....	s. h.	v. s.	v. s.; v. s. chl.
1818	nd./aq.	....	232-3	subl.	v. sl. s. c.	s.	s.
1819	nd./aq.	....	219-20	subl.	v. sl. s. c.	v. s.	sl. s. c. bz.
1820	yel. nd./ glycol	....	305-6	....	i.; i. aq. NaOH	i.; s. h. glycol	i.; s. h. $PhNO_2$

* Enol form, two m. p's.; keto, m. p. 81°.
Dibenzyl ether 806
Dibenzyl sulfone 835

Dibromide of cinnamic acid 1830
Dibromin 1810
Dibromo-acetophenone 967

Table 7-4 (*Continued*)
PHYSICAL CONSTANTS OF ORGANIC COMPOUNDS

No.	Name	Synonym	Formula	Beil. Ref.	Formula Weight
	Dibromo				
1821	butane (1,2)	α-butylene bromide	$C_2H_5{\cdot}CHBr{\cdot}CH_2Br$	I-120	215.93
1822	butane (1,3)	....	$CH_3CHBr{\cdot}CH_2CH_2Br$	I-120	215.93
1823	butane (1,4)	....	$Br(CH_2)_4Br$	I-120	215.93
1824	butane (2,3)(*dl*)	β-butylene bromide	$(CH_3{\cdot}CHBr{\cdot})_2$	**II-84	215.93
1825	butane (*iso*)	1,2-diBr-2-Me-propane	$(CH_3)_2CBr{\cdot}CH_2Br$	I-127	215.93
1826	butyric acid (α,β)	crotonic acid diBr	$CH_3(CHBr)_2CO_2H$	II-284	245.91
1827	butyric acid	*iso*-crotonic diBr	$CH_3(CHBr)_2CO_2H$	II-285	245.91
1828	camphor (α,α')	(3,3-diBr-*d*)	$Br_2C_9H_{14}CO$	VII-125	310.04
1829	cinnamic acid (α,β)	"α-acid"	$C_6H_5CBr{:}CBrCO_2H$	IX-601	305.96
1830	cinnamic acid (*dl*)	†	$C_6H_5(CHBr)_2CO_2H$	IX-518	307.98
1831	o-cresol (4,6)	(4,6;2,1)	$Br_2C_6H_2(CH_3)OH$	VI-360	265.94
1832	o-cresolsulfon-phthalein	bromcresol purple	$C_{21}H_{16}O_5Br_2S$	....	540.24
1833	diphenyl (4,4')	....	$BrC_6H_4{\cdot}C_6H_4Br$	V-580	312.02
1834	diphenyl ether (4,4')	....	$(Br{\cdot}C_6H_4)_2O$	VI-200	328.01
1835	diphenylsulfone	(4,4')	$(Br{\cdot}C_6H_4)_2SO_2$	VI-331	376.08
1836	ethane (1,1)	ethylidene dibromide	$CH_3{\cdot}CHBr_2$	I-90	187.87
1837	ethane (1,2)	ethylene bromide	$BrCH_2{\cdot}CH_2Br$	I-90	187.87
1838	fluorescein	(2,4)	$C_{20}H_{10}O_5Br_2$	XIX-228	490.12
1839	fluorescein	(4,5)	$C_{20}H_{10}O_5Br_2$	XIX-228	490.12
1840	fumaric acid	....	$({:}CBr{\cdot}CO_2H)_2$	II-747	273.88
1841	gallic acid	(2,6;3,4,5;1)	$Br_2C_6(OH)_3CO_2H{\cdot}H_2O$	X-490	345.94
1842	hexane (1,2)	α-hexylene di-bromide	$CH_3(CH_2)_3CHBr{\cdot}CH_2Br$	I-144	243.98
1843	hexane (1,5)	....	$CH_3CHBr{\cdot}(CH_2)_3{\cdot}CH_2Br$	I-145	243.98
1844	hexane (1,6)	hexamethylene diBr	$({\cdot}CH_2{\cdot}CH_2{\cdot}CH_2Br)_2$	I-145	243.98
1845	hexane	3,3-dibromo-2,2-dimethyl-butane	$(CH_3)_3C{\cdot}CBr_2{\cdot}CH_3$	I-151	243.98
1846	hexane	3,4-dibromo-2,2-dimethyl-butane	$(CH_3)_3C{\cdot}CHBr{\cdot}CH_2Br$	I-151	243.98
1847	hexane	2,3-dibromo-2,3-dimethyl-butane	$(CH_3)_2CBr{\cdot}CBr(CH_3)_2$	I-152	243.98
1848	8-hydroxy-quinoline (5,7)	....	$Br_2C_9H_4(OH)N$	XXI-97	302.96
1849	iodomethane	....	$ICHBr_2$	I-71	299.74
1850	maleic acid	....	$({:}CBr{\cdot}CO_2H)_2$	II-756	273.88
1851	malonic acid	....	$Br_2C(CO_2H)_2$	II-595	261.86
1852	naphthalene (1,4)	β-dibromonaph-thalene	$C_{10}H_6Br_2$	V-549	285.98
1853	α-naphthol (2,4)	....	$Br_2C_{10}H_5{\cdot}OH$	VI-614	301.98
1854	nitromethane	....	$Br_2CH{\cdot}NO_2$	I-77	218.84
1855	nitrophenol (2,6;4)	....	$Br_2{:}C_6H_2(NO_2)OH$	VI-247	296.91
1856	pentane (1,4)	....	$CH_3{\cdot}CHBr{\cdot}CH_2{\cdot}CH_2{\cdot}CH_2Br$	I-131	229.95
1857	pentane (1,5)	pentamethylene diBr	$Br(CH_2)_5Br$	I-131	229.95

†Dibromide of cinnamic acid.
Dibromo-diphenyl 1833
Dibromo-ethene 131-3

Dibromo-ethylbenzene 5647
Dibromo-ethylene 131-3
Dibromo-ethyne 1796

Table 7-4 (*Continued*)
PHYSICAL CONSTANTS OF ORGANIC COMPOUNDS

No.	Crystalline Form and Color	Specific Gravity	Melting Point °C.	Boiling Point °C.	Solubility in 100 Parts		
					Water	Alcohol	Ether
1821	lq.	$1.820^{20°}_{0°}$	-65	166	i.	∞	....
1822	lq.	$1.807^{18.5°}$	....	174-5	....	....	....
1823	col. lq.	$1.819^{8.5°}_{0°}$	-20	197-8	....	....	....
1824	col. lq.	$1.783^{20°}_{4°}$	-34.5	157-8	i.	....	....
1825	col. lq.	$1.759^{20°}$	-70.3	$148\text{-}9^{737mm}$	i.	....	....
1826	mn. nd.	....	87	....	sl. s. c.	v. s.	v. s.
1827	nd.		58-9	....	sl. s. c.	v. s.	v. s.
1828	col. rhb./lg.	$1.854^{21.6°}_{4°}$	64	subl. >64	i.; s. lg.	$22^{20°}$ abs.	s.; s. bz.
1829	pl./chl. pet.	....	138-9 *	....	i.	s.; s. chl.	s.; sl. s. pet.
1830	mn./chl.	....	199-201	d.	i.; d. h.	s.	
1831	nd.	....	56-7		v. sl. s.	s.; s. alk.	s.; s. bz.
1832	pink pd.	....	241-2	....	v. sl. s.	s.; s. alk.	i.
1833	mn. pr.	1.897	165-7	355-60	i.	v. sl. s. h.	v. s. bz.
1834	cr.	....	58-60	338-40	v. s. bz.	s.	s.
1835	mn. nd./al.	....	172	....	....	sl. s. h.	....
1836	lq.	$2.055^{20°}_{4°}$	....	108-10	i.	v. s.	v. s.
1837	col. lq.	$2.180^{20°}_{4°}$	10	131.7	$0.43^{30°}$	∞	∞
1838	red cr. + EtOH/al.	....	300± (anh.)	....	i.; s. ac.	s. h.	v. sl. s.
1839	red cr./al.	....	285	....	sl. s.; s. ac.	s. h.	v. sl. s.
1840	cr./aq.	....	227-9 d.	....	....	v. s.	v. s.
1841	cr./aq.	....	150	$-H_2O$, 120	$12.4^{15°}$; $200^{100°}$	s.; i. chl.	s.
1842	lq.	$1.596^{13.5°}$	....	87^{16mm}	....	....	....
1843	lq.	$1.599^{20°}_{4°}$	....	$153\text{-}4^{100mm}$	....	....	....
1844	lq.	$1.595^{15°}$	....	243	....	....	....
1845	cr.	....	187†	subl.	....	....	....
1846	col. lq.	$1.616^{0°}$	....	$91\text{-}2^{14mm}$	i.; v. s. bz.	v. s.	v. s.; v. s. chl.
1847	long nd.	$1.811^{15°}$	173**	....	d.	s.	v. s.; s. bz.
1848	nd./al.	....	196	subl.	i. c.	sl. s.; s. a.	s.; s. bz.
1849	tablets	....	22.5	$101\text{-}4^{50mm}$	soln. d. by light	sl. s. pet.	....
1850	nd./et.	....	123.5	....	v. s.	v. s.	v. s.; i. chl.
1851	pr. or nd.	....	136-7 d.	....	v. s.	v. s.	v. s.
1852	nd./al.	....	82-3	310 d.	....	$1.3^{11°}$	v. s.
1853	nd./al.	....	111	....	i.	s.	s.; s. ac.
1854	lq.	....	....	$58.5\text{-}60^{13mm}$	....	....	....
1855	col. pr./al.	....	143-4	d. >145	v. sl. s.; v. sl. s. bz.	s. h.; sl. s. ac.	s. h.; s. h. CS$_2$
1856	lq.	$1.622^{20°}_{4°}$	....	$196\text{-}8^{746}$ d.	....	....	....
1857	lq.	$1.706^{18°}$	-35	221^{763} sl. d.	....	....	....

* β or allo-acid, m.p. 100°
† In sealed tube.
Dibromo-hydrin 3388-9

Dibromo-methane 4428
Dibromo-pene 1869

Table 7-4 (*Continued*)
PHYSICAL CONSTANTS OF ORGANIC COMPOUNDS

No.	Name	Synonym	Formula	Beil. Ref.	Formula Weight
1858	**Dibromo** pentane	2,3-dibromo-2-methyl-butane	$(CH_3)_2CBr \cdot CHBr \cdot CH_3$	I-137	229.95
1859	phenol (2,4)	....	$Br_2C_6H_3 \cdot OH$	VI-202	251.92
1860	phenol (2,6)	....	$Br_2C_6H_3 \cdot OH$	VI-202	251.92
1861	phenol (3,4)	....	$Br_2C_6H_3 \cdot OH$	VI-203	251.92
1862	phenol (3,5)	....	$Br_2C_6H_3 \cdot OH$	VI-203	251.92
1863	propane (1,1)	propylidene diBr	$C_2H_5 \cdot CHBr_2$	I-109	201.90
1864	propane (1,2)	propylene bromide	$CH_3 \cdot CHBr \cdot CH_2Br$	I-109	201.90
1865	propane (1,3)	trimethylene diBr	$BrCH_2 \cdot CH_2 \cdot CH_2Br$	I-110	201.90
1866	propane (2,2)	bromacetol	$(CH_3)_2CBr_2$	I-111	201.90
1867	propionic acid (α,α)	....	$CH_3 \cdot CBr_2 \cdot CO_2H$	II-257	231.88
1868	propionic acid (α,β)	....	$CH_2Br \cdot CHBr \cdot CO_2H$	II-258	231.88
1869	propylene (β,γ)	epidibromohydrin (α)	$CH_2{:}CBr \cdot CH_2Br$	I-201	199.89
1870	propylene (α,γ)	epidibromohydrin (β)	$CH_2Br \cdot CH{:}CHBr$	I-201	199.88
1871	propyne-1 (1,3)	....	$CH_2Br \cdot C{\vdots}CBr$	I-248	197.87
1872	pyridine (2,5)	....	$Br_2C_5H_3N$	XX-233	236.90
1873	pyridine (2,6)	....	$Br_2C_5H_3N$	....	236.90
1874	pyridine (3,5)(α)	....	$Br_2C_5H_3N$	XX-233	236.90
1875	pyruvic acid	....	$CHBr_2 \cdot CO \cdot CO_2H$	III-624	245.87
1876	quinonechloroimide	(2,6)	$Cl \cdot N{:}C_6H_2{:}OBr_2$	VII-640	299.36
1877	succinic acid (*allo*)	....	$Br_2C_2H_2(CO_2H)_2$	II-625	275.89
1878	succinic acid (*meso*)	....	$(CHBr \cdot CO_2H)_2$	II-623	275.89
1879	thymol-sulfon-phthalein	bromthymol blue	$C_{27}H_{28}O_5Br_2S$	*XIX-650	624.40
1880	toluene (2,5)	....	$Br_2C_6H_3 \cdot CH_3$	V-308	249.94
1881	tyrosine (3,5)	*l* (−)	$C_9H_9O_3NBr_2 \cdot 2H_2O$	XIV-619	375.03
1882	*m*-xylene (4,6)	....	$Br_2C_6H_2(CH_3)_2$	V-374	263.97
1883	**Dibutyl** adipate (*n*)	butyl adipate	$(CH_2CH_2CO_2 \cdot C_4H_9)_2$	**II-575	258.36
1884	amine (*n*)	....	$(C_4H_9)_2NH$	IV-157	129.25
1885	amine (*sec*)	....	$(C_4H_9)_2NH$	IV-162	129.25
1886	amine (*iso*)	....	$(C_4H_9)_2NH$	IV-166	129.25
1887	aminoethyl alcohol $(\beta)(n)$	....	$(C_4H_9)_2N \cdot CH_2 \cdot CH_2OH$	....	173.30
1888	aminopropyl alcohol $(\beta)(n)$	....	$(C_4H_9)_2N \cdot CH(CH_3) \cdot CH_2OH$	....	187.33
1889	aminopropyl alcohol $(\gamma)(n)$	....	$(C_4H_9)_2N \cdot CH_2CH_2 \cdot CH_2OH$	....	187.33
1890	aminopropyl-*p*-aminobenzoate H_2SO_4	butyn	$H_2N \cdot C_6H_4 \cdot CO_2 \cdot (CH_2)_3 \cdot N(C_4H_9)_2 \cdot \frac{1}{2}H_2SO_4$	....	355.49
1891	aniline (*n*)(*N*)	....	$C_6H_5N(C_4H_9)_2$	*XII-160	205.35
1892	carbonate (*n*)	butyl carbonate	$CO(OC_4H_9)_2$	III-6	174.24
1893	carbonate (*iso*)	....	$CO(OC_4H_9)_2$	III-6	174.24
1894	carbonate (*sec*)	....	$CO(OC_4H_9)_2$	III-6	174.24
1895	cyanamide (*n*)	....	$(C_4H_9)_2N \cdot CN$	**IV-635	154.26
1896	α,α'-dibromo-succinate (*n*)	....	$(CHBrCO_2C_4H_9)_2$	....	388.11
1897	disulfide (*n*)	....	$C_4H_9 \cdot S \cdot S \cdot C_4H_9$	**I-400	178.36
1898	ether (*n*)	butyl ether	$(C_2H_5 \cdot CH_2 \cdot CH_2)_2O$	I-369	130.23

Dibromo-propanol 3388-9
Dibromo-propyl alcohol 3388-9
Dibromo-sulfobenzide 1835
Dibromo-xylene (ω) 6491-3

Table 7-4 (*Continued*)
PHYSICAL CONSTANTS OF ORGANIC COMPOUNDS

No.	Crystalline Form and Color	Specific Gravity	Melting Point °C.	Boiling Point °C.	Solubility in 100 Parts		
					Water	Alcohol	Ether
1858	lq.	$1.573^{25°}$	7	170-3 sl. d.	....	....	....
1859	cr.	....	40	238-9	$0.2^{15°}$	v. s.	v. s.
1860	nd./aq.	....	55-6	162^{21mm}	....	v. s.	v. s.
1861	nd./aq.	....	79-80	....	....	v. s.	v. s.
1862	cr./lg.	....	81	....	v. sl. s.	v. s.	v. s.
1863	lq.	....	....	130±	....	....	....
1864	col. lq.	$1.933^{\frac{20°}{4}}$	−55.5	141.6	$0.25^{20°}$	s.	v. s.
1865	lq.	$1.987^{\frac{15°}{4}}$	−34.2	167.3	$0.168^{30°}$	s.	s.
1866	lq.	$1.783^{20°}$	....	114.5^{740mm}	....	....	....
1867	rhb.	....	61	200-21 sl. d.	....	....	....
1868	pl.	....	64 (51)	220-40 d.	194^{511mm}	v. s.	$304^{10°}$
1869	lq.	$1.934^{\frac{20°}{4}}$	....	140-2	....	....	....
1870	lq.	$1.995^{\frac{25°}{4}}$	−52	155-6	....	....	....
1871	lq.	$2.137^{0°}$	....	$73-4^{30mm}$	....	....	....
1872	nd./al.	....	94-5	subl.	i.	s.	s.
1873	cr.	....	118-9	....	....	....	....
1874	cr./al.	....	110-2; subl.>100	222	sl. s. h.	s. h.	s.; s. H_2SO_4
1875	nd./et.	....	93	....	s.	....	....
1876	yel. pr./al.	....	83	d. 121	0.0006	sl. s. h.	sl. s. h. ac.
1877	cr.	....	170	d. 180	>217°	v. s.	v. s.
1878	cr.	....	255-6 * d.	subl.>250	$2.0^{17°}$	v. s.; v. sl. s. chl.	v. s.
1879	yel. cr.	....	....	....	v. sl. s.; s. a.	s.	i.; s. alk.
1880	lq.	$1.813^{19°}$	< 20	236	i.	....	....
1881	rhb./aq.	....	245±, d.	−2H_2O, 120	$5^{16°}$; $3.9^{100°}$	v. sl. s.; s. a.	i.; s. alk.
1882	cr.	....	72	255-6	i.	v. s. h.	
1883	col. lq.	$0.965^{\frac{20°}{4}}$	−38	183^{14mm}	i.	∞	∞
1884	col. lq.	$0.768^{\frac{20°}{4}}$	−62	159^{761mm}	v. s.	∞	∞
1885	col. lq.	$0.783^{\frac{0°}{0}}$	−104.5	132^{758mm}	v. s.	....	....
1886	col. lq.	$0.741^{\frac{25°}{4}}$	−70	139-40	v. sl. s.	s.	s.
1887	col. lq.	....	....	$91-67mm$	....	....	....
1888	col. lq.	....	....	$80-6^{6mm}$	....	....	....
1889	col. lq.	....	....	$110-6^{8mm}$	....	....	....
1890	wh. pd.	....	98-100	....	100; sl. s. chl.	s.; s. act.	i.
1891	lq.	....	....	262.8	i.	∞	∞
1892	col. lq.	$0.924^{20°}$	....	207^{740mm}	i.	s.	....
1893	col. lq.	$0.919^{15°}$	....	190	i.	....	....
1894	col. lq.	....	....	178-80	....	....	....
1895	lq.	....	....	$128-30^{20mm}$	i.	s.	s.
1896	col. lq.	....	....	$171-4^{8mm}$	....	....	....
1897	col. lq.	$0.930^{\frac{20°}{4}}$	....	$117-8^{20mm}$	i.	∞	∞
1898	lq.	$0.773^{\frac{15°}{4}}$	−97.9	142.4	<0.05	∞	∞

* In sealed tube.
Dibutyl carbinol (*n*) 4921
Dibutyl carbinol (*iso*) 4925

Table 7-4 (*Continued*)
PHYSICAL CONSTANTS OF ORGANIC COMPOUNDS

No.	Name	Synonym	Formula	Beil. Ref.	Formula Weight
1899	**Dibutyl** ether (*iso*)	*iso*-butyl ether	$[(CH_3)_2CH \cdot CH_2]_2O$	I-376	130.23
1900	ketone (*n*)	nonanone-5	$(C_2H_5 \cdot CH_2 \cdot CH_2)_2CO$	I-709	142.24
1901	ketone (*iso*)	valerone	$(C_4H_9)_2CO$	I-710	142.24
1902	*l*-malate (−)(*n*)	butyl malate	$C_4H_4O_5(C_4H_9)_2$	III-433	246.31
1903	maleate (*n*)	butyl maleate	$(CHCO_2C_4H_9)_2$	....	228.29
1904	malonate (*n*)	butyl malonate	$CH_2(CO_2C_4H_9)_2$	II-581	216.28
1905	oxalate (*n*)	....	$(CO_2C_4H_9)_2$	II-540	202.25
1906	oxalate (*iso*)	....	$(CO_2C_4H_9)_2$	II-540	202.25
1907	phenyl-phenyl-phosphate (*tert*) (*p*)	phosphen-2	$(C_4H_9 \cdot C_6H_4O)_2PO \cdot O \cdot C_6H_5$	....	438.51
1908	*o*-phthalate (*n*)	butyl phthalate	$C_6H_4(CO_2C_4H_9)_2$	**IX-586	278.35
1909	sebacate (*n*)	butyl sebacate	$[(CH_2)_4CO_2C_4H_9]_2$	II-719	314.47
1910	succinate (*n*)	butyl succinate	$(CH_2CO_2C_4H_9)_2$	**II-551	230.31
1911	succinate (*iso*)	....	$(CH_2CO_2C_4H_9)_2$	II-611	230.31
1912	succinate (*sec*)	....	$(CH_2CO_2C_4H_9)_2$	II-611	230.31
1913	sulfate (*n*)	butyl sulfate	$(C_4H_9O)_2SO_2$	....	210.29
1914	sulfide (*n*)	butyl sulfide	$(C_2H_5 \cdot CH_2CH_2)_2S$	I-370	146.30
1915	sulfide (*sec*)	....	$(C_4H_9)_2S$	I-373	146.30
1916	sulfide (*iso*)	....	$[(CH_3)_2CH \cdot CH_2]_2S$	I-379	146.30
1917	sulfite (*n*)	butyl sulfite	$(C_4H_9O)_2SO$	**I-397	194.29
1918	sulfone (*n*)	butyl sulfone	$(C_4H_9)_2SO_2$	I-371	178.30
1919	tartrate (*d*)(*n*)	butyl tartrate	$(CHOH \cdot CO_2C_4H_9)_2$	III-518	262.31
1920	tartrate (*d*)(*iso*)	....	$(CHOH \cdot CO_2C_4H_9)_2$	III-518	262.31
1921	thiourea (*n*)	....	$(C_4H_9 \cdot NH)_2CS$	....	188.34
1922	urea (*n*)(*N,N*)	....	$(C_4H_9)_2N \cdot CO \cdot NH_2$	*IV-372	172.27
1923	**Di-*iso*-butylene** †		$(CH_3)_2C:CH \cdot C(CH_3)_3$	I-222	112.22
1924	**Dicapryl adipate**	(*n*)(*sec*)	$[\cdot (CH_2)_2CO_2CH \cdot (CH_3) \cdot (CH_2)_5 \cdot CH_3]_2$	**II-575	370.58
1925	**Dichloro**-acetal		$Cl_2CH \cdot CH(OC_2H_5)_2$	I-614	187.07
1926	acetaldehyde	2,2-dichloro-ethanal	$Cl_2CH \cdot CHO$	I-613	112.94
1927	acetamide		$Cl_2CH \cdot CO \cdot NH_2$	II-205	127.96
1928	acetic acid	....	$Cl_2CH \cdot CO_2H$	II-202	128.94
1929	acetone (α)(*as*)	....	$Cl_2CH \cdot CO \cdot CH_3$	I-654	126.97
1930	acetone (β)(*s*)	....	$(ClCH_2)_2CO$	I-655	126.97
1931	acetyl chloride	....	$Cl_2CH \cdot CO \cdot Cl$	II-204	147.39
1932	aniline (2,3)	....	$Cl_2C_6H_3 \cdot NH_2$	XII-621	162.02
1933	aniline (2,4)	....	$Cl_2C_6H_3 \cdot NH_2$	XII-621	162.02
1934	aniline (2,5)	....	$Cl_2C_6H_3 \cdot NH_2$	XII-625	162.02
1935	aniline (2,6)	....	$Cl_2C_6H_3 \cdot NH_2$	XII-626	162.02
1936	aniline (3,4)	....	$Cl_2C_6H_3 \cdot NH_2$	XII-626	162.02
1937	aniline (3,5)	....	$Cl_2C_6H_3 \cdot NH_2$	XII-626	162.02
1938	aniline sulfonic	3,6-diCl-sulfanilic	$Cl_2C_6H_2(NH_2)SO_3H$	XIV-707	242.08
1939	anthracene (2,3)	....	$C_{14}H_8Cl_2$	V-664	247.13
1940	anthracene (9,10)	....	$C_{14}H_8Cl_2$	V-664	247.13
1941	anthraquinone (1,3)	....	$C_6H_4(CO)_2C_6H_2Cl_2$	VII-787	277.11
1942	anthraquinone (1,4)	....	$C_6H_4(CO)_2C_6H_2Cl_2$	VII-787	277.11
1943	anthraquinone (1,5)	....	$(ClC_6H_3)_2(CO)_2$	VII-787	277.11
1944	anthraquinone (1,6)	....	$(ClC_6H_3)_2(CO)_2$	*VII-412	277.11
1945	anthraquinone (1,8)	....	$(ClC_6H_3)_2(CO)_2$	VII-788	277.11
1946	anthraquinone (2,3)	....	$C_6H_4(CO)_2C_6H_2Cl_2$	VII-788	277.11
1947	anthraquinone (2,6)	....	$(ClC_6H_3)_2(CO)_2$	VII-788	277.11
1948	anthraquinone (2,7)	....	$(ClC_6H_3)_2(CO)_2$	VII-788	277.11

† See also octylene.
Dibutyl mercury 4042

Dibutyl urethane 3038

Table 7-4 (*Continued*)
PHYSICAL CONSTANTS OF ORGANIC COMPOUNDS

No.	Crystalline Form and Color	Specific Gravity	Melting Point °C.	Boiling Point °C.	Solubility in 100 Parts		
					Water	Alcohol	Ether
1899	lq.	$0.762^{15°}$	....	122-2.5	i.;	∞	∞
1900	lq.	$0.827^{13°}_4$	−5.9	187.7	i.; v. s. CS_2	v. s. chl.	v. s.
1901	oil	$0.806^{20°}_4$	−46.0	168.2	<0.06	∞	∞
1902	lq.	$1.038^{20°}_4$	....	170-1^{13mm}	v. sl. s.	....	....
1903	col. lq.			141-2^{9mm}	....	....	....
1904	lq.	$0.981^{20°}_4$	....	251.5	i.	s.	s.
1905	col. lq.	$0.986^{20°}_4$	−29.6	245.5	i.	s.	s.
1906	col. lq.	$1.002^{14°}$	....	228-9	i.	s.	s.
1907	col. lq.	$1.11^{25°}_{25}$	<0	260-75^{5mm}	i.; ∞ CCl_4	v. s.	∞ bz.
1908	col. lq.	$1.045^{21°}$	....	340	$0.04^{25°}$	∞ ; ∞ bz.	∞ ; ∞ act.
1909	lq.	$0.933^{15°}$	−11	344-5	....	....	....
1910	lq.	$0.965^{20°}$	−29.3	274.5	....	....	....
1911	col. lq.	$0.974^{15°}$	....	265-6	....	....	....
1912	col. lq.	$0.974^{20°}_4$	....	256-7^{750mm}	....	....	....
1913	col. lq.	$1.059^{25°}_{25}$	....	130-2^{11mm}	i.	....	....
1914	lq.	$0.839^{16°}_0$	−79.7	182	i.	....	....
1915	lq.	$0.832^{23°}$	....	165	....	....	....
1916	lq.	$0.836^{10°}$	−105.5	172-3^{747mm}	i.	∞	∞
1917	col. lq.	$1.001^{14°}$	....	108-10^{15mm}	....	....	....
1918	pl./aq.	....	44		v. sl. s.	....	....
1919	pr.	$1.098^{15°}$	22-2.5	200-3^{18mm}	....	....	....
1920	cr.	$1.031^{7.5°}_4$	73-4	323-5	....	....	....
1921	col. nd./al.	....	66-7	....	i.	s.	sl. s.
1922	col. hyg. cr.	....	149-50	118-9^{2mm}	....	s.	s.
1923	col. lq.	$0.721^{20°}_4$	−106.5	104.9	....	....	....
1924	lt. yol. lq.	$0.914^{20°}_4$	....	192-4^{3mm}	....	....	....
1925	lq.	$1.138^{14°}$	....	185	....	....	....
1926	col. lq.*	....	....	88-90	i.	....	....
1927	mn. pr.	....	98	233-47^{45mm}	v. s. h.	v. s.	v. s.
1928	lq.	$1.560^{25°}_{25}$	9.7 (−4)	194.4	∞	∞	∞
1929	lq.	$1.234^{15°}$	....	120	v. sl. s.	s.	s.
1930	pl. or nd.	$1.383^{46°}_4$	45	173.0-3.4	s.	v. s.	v. s.
1931	lq.	....	....	107-8	d.	d.	∞
1932	nd./lg.	....	23-4	252	sl. s. bz.	s.	v. sl. s.
1933	rhb.	$1.567^{20°}_4$	62-3	245^{759mm}	sl. s.	s.	s.
1934	nd./lg.	....	50	251	v. sl. s.	s.	s.
1935	nd./al.	....	39	....	....	....	....
1936	nd./lg.	....	71.5	272	i.; sl. s. bz.	70$^{25°}$	v. s.
1937	nd.	....	50.5	259-60	i.	s.	....
1938	nd./aq.	....	....	....	s. h.	....	....
1939	yel. lf.	....	261	subl. sl. d.	s. ac.	s. h.	....
1940	yel. nd.	....	209-10	....	s. bz.	sl. s.	sl. s.
1941	yel. nd./ac.	....	208-9	....	i.; s. ac.	i.	s. PhNO$_2$
1942	yel. nd./ac.	....	187.5	....	i.; s. h. bz.	v. sl. s.	v. sl. s.
1943	yel. nd./ac.	....	251	....	i.; s. H$_2$SO$_4$	sl. s.	s. PhMe
1944	yel. nd./ac.	....	203-4	....	i.	....	....
1945	pa. yel. nd.	....	202-3	....	s. PhMe	sl. s.	s. PhNO$_2$
1946	yel. nd./ac.	....	268-70	....	i.	sl. s.	s. h. bz.
1947	yel. nd./ac.	....	282	....	i.	....	....
1948	yel. nd.	....	210-1	....	i.	s. PhOMe	....

* Polymerizes on standing.
Dicetyl 2866

Dichloro-acetoacetanilide 42

Table 7-4 (Continued)
PHYSICAL CONSTANTS OF ORGANIC COMPOUNDS

No.	Name	Synonym	Formula	Beil. Ref.	Formula Weight
	Dichloro				
1949	azo-dicarbonamide	azochloramide	$[Cl \cdot N : C(NH_2) \cdot N :]_2$	....	183.00
1950	azoxybenzene	(4,4')	$(ClC_6H_4)_2N_2O$	XVI-625	267.12
1951	barbituric acid (5,5)	dichloromalonyl-urea	$Cl_2C(CONH)_2CO$	XXIV-472	196.98
1952	benzene (o)	....	$C_6H_4Cl_2$	V-201	147.00
1953	benzene (m)	....	$C_6H_4Cl_2$	V-202	147.00
1954	benzene (p)	....	$C_6H_4Cl_2$	V-203	147.00
1955	benzene sulfonic acid (2,5)	....	$Cl_2C_6H_3 \cdot SO_3H$	XI-55	227.08
1956	benzene sulfonyl chloride (3,4)	....	$Cl_2C_6H_3 \cdot SO_2Cl$	*XI-16	245.51
1957	benzidine (3,3')	....	$(H_2N \cdot C_6H_3Cl)_2$	XIII-234	253.13
1958	benzidine HCl	(3,3')	$C_{12}H_{10}N_2Cl_2 \cdot 2HCl$	XIII-234	326.05
1959	benzoic acid (2,3)	....	$Cl_2C_6H_3 \cdot CO_2H$	IX-342	191.01
1960	benzoic acid (2,4)	....	$Cl_2C_6H_3 \cdot CO_2H$	IX-342	191.01
1961	benzoic acid (2,5)	....	$Cl_2C_6H_3 \cdot CO_2H$	IX-342	191.01
1962	benzoic acid (2,6)	....	$Cl_2C_6H_3 \cdot CO_2H$	IX-343	191.01
1963	benzoic acid (3,4)	....	$Cl_2C_6H_3 \cdot CO_2H$	IX-343	191.01
1964	benzoic acid (3,5)	....	$Cl_2C_6H_3 \cdot CO_2H$	IX-344	191.01
1965	benzophenone (2,4')	....	$(Cl \cdot C_6H_4)_2CO$	VII-420	251.11
1966	benzophenone (4,4')	....	$(Cl \cdot C_6H_4)_2CO$	VII-420	251.11
1967	benzoylene urea (6,8)	Sheibley's reagent	$Cl_2C_6H_2 \cdot NH \cdot CO \cdot$ $NH \cdot CO$	....	231.04
1968	1-bromoethane (1,1)	....	$CH_3 \cdot CCl_2Br$	I-90	177.86
1969	1-bromoethane (2,2)	....	$Cl_2CH \cdot CH_2Br$	I-90	177.86
1970	bromomethane	....	Cl_2CHBr	I-67	163.83
1971	butane (1,1)(n)	butylidene di-chloride	$C_2H_5CH_2CHCl_2$	I-119	127.02
1972	butane (1,2)(n)	butylene chloride	$C_2H_5 \cdot CHCl \cdot CH_2Cl$	*I-38	127.02
1973	butane (1,4)(n)	tetramethylene diCl	$ClCH_2(CH_2)_2CH_2Cl$	I-119	127.02
1974	butane (iso)	1,1-dichloro-2-methyl-propane	$(CH_3)_2CH \cdot CHCl_2$	I-126	127.02
1975	butane (iso)	1,3-dichloro-2-methyl-propane	$(ClCH_2)_2CH \cdot CH_3$	**I-88	127.02
1976	butane (iso)	1,2-dichloro-2-methyl-propane	$(CH_3)_2CCl \cdot CH_2Cl$	I-126	127.02
1977	butyl alcohol (tert)	....	$(CH_2Cl)_2COH \cdot CH_3$	I-382	143.01
1978	butyric acid (α,β)	crotonic acid diCl	$CH_3(CHCl)_2CO_2H$	II-279	157.00
1979	butyric acid	iso-crotonic acid diCl	$CH_3(CHCl)_2CO_2H$	II-279	157.00
1980	1,2-dibromoethane (1,1)		$BrCH_2 \cdot CBrCl_2$	I-93	256.76
1981	1,2-dibromoethane (1,2)		$ClBrCH \cdot CHBrCl$	I-93	256.76
1982	1,1-dibromoethane (2,2)		$Br_2CH \cdot CHCl_2$	I-93	256.76
1983	dibromomethane	....	Cl_2CBr_2	I-68	242.74
1984	diethyl carbonate (β,β')		$(ClCH_2CH_2O)_2CO$	**III-5	187.02
1985	diethyl sulfide (β,β')	mustard gas	$(Cl \cdot CH_2CH_2)_2S$	I-349	159.08

Dichloro-iso-butane 1975-6

Table 7-4 (*Continued*)
PHYSICAL CONSTANTS OF ORGANIC COMPOUNDS

No.	Crystalline Form and Color	Specific Gravity	Melting Point °C.	Boiling Point °C.	Solubility in 100 Parts		
					Water	Alcohol	Ether
1949	yel. nd.	....	expl. 155	....	i.	s.	sl. s.
1950	yel. nd./al.	....	155-6	subl.	i.	sl. s.	s.
1951	rhb. pr./aq.	....	219-20 d.	....	sl. s.; v. sl. s. bz.	v. s.; v. sl. s. chl.	v. s.; v. s. ac.
1952	col. lq.	$1.306\frac{20°}{4}$	-17.2	180.4	i.; ∞ bz.	∞	∞
1953	col. lq.	$1.288\frac{20°}{4}$	-24.8	172^{766mm}	i.	s.	s.
1954	col. mn.	$1.458^{21°}$	53.1	174.4	i.; s. bz.	∞ h. abs.	v. s.
1955	nd./aq.	....	>100	....	v. s.	....	v. sl. s.
1956	mn./bz.	....	22.4	$170-4^{20mm}$	d.	d.	....
1957	nd./al.	....	132-3	....	i.	s.; s. bz.	s. ac.
1958	nd.	....	....	....	v. sl. s.	s.	....
1959	nd.	....	166	....	sl. s. h.	s.	s.
1960	nd./aq.	....	164	subl.	s. h.	s.	s.
1961	nd./aq.	....	154	301	$0.09^{14°}$	s.	s. alk.
1962	nd./al.	....	140-3	subl.	s. h.	s. bz.	s. alk.
1963	nd./aq.	....	203-4	....	s. h.	v. s.	s. alk.
1964	nd./al.	....	182-3	subl.	....	v. s.	....
1965	col. mn./al.	$1.393^{14°}$	66-7	$214-5^{22mm}$	i.; s. act.	$190^{25°}$ bz.	$48^{25°}$
1966	col. lf./al.	....	145	353^{757mm}	i.; s. act.	$12^{35°}$ bz.	$2^{25°}$
1967	col. nd./al.	....	296	....	i.; i. aq. NaOH	1 h.; s. aq. KOH	i.; v. sl. s. h. gly.
1968	lq.	$1.752^{16°}$	....	$98-9^{758mm}$	....	....	....
1969	lq.	....	....	138	....	....	....
1970	col. lq.	$2.006\frac{15°}{4}$	-56.9	90.1	....	....	....
1971	oil	....	....	113-5	i.	s.	s.; s. chl.
1972	lq.	....	....	124	....	....	....
1973	lq.	....	-38.7	161-3	....	....	....
1974	lq.	$1.011^{12°}$	....	103-5 d.	....	....	....
1975	col. lq.	$1.138^{20°}$	....	135-9	....	....	....
1976	col. lq.	$1.097^{20°}$	....	107-8	....	....	....
1977	col. lq.	$1.277\frac{20°}{4}$	....	174-5	8.2	s.	s.
1978	pr.	....	62.5-3.0	$124-5^{20mm}$	sl. s.	v. s.	$32^{810.5°}$
1979	long pr.	....	78 (73)	131.5^{20mm}	sl. s.	v. s.	v. s.
1980	lq.	$2.270^{16°}$	-66.9	175	i.	....	....
1981	lq.	....	-26	194-5	i.	....	....
1982	lq.	$2.391^{9°}$	....	195-200	i.	....	....
1983	nd.	$2.42\frac{5°}{0}$	22	135±	i.	....	....
1984	pa. yel. lq.	$1.351\frac{20°}{4}$	8.5	240-1	....	....	....
1985	col. oil	$1.275\frac{20°}{4}$	14.5	217 sl. d.	$0.07^{25°}$	s.; s. bz.; s. chl.	s.; s. ac.

Table 7-4 (*Continued*)
PHYSICAL CONSTANTS OF ORGANIC COMPOUNDS

No.	Name	Synonym	Formula	Beil. Ref.	Formula Weight
	Dichloro				
1986	diiodomethane	....	Cl_2CI_2	I-72	336.73
1987	2,2'-dinitrodi-phenyl disulfide	(4,4')	$(O_2N \cdot C_6H_3Cl \cdot S)_2$	VI-341	377.23
1988	dioxane (2,3)	....	$O \cdot (CHCl)_2O(CH_2)_2$	....	157.00
1989	diphenyl (3,3')	....	$Cl \cdot C_6H_4 \cdot C_6H_4 \cdot Cl$	V-579	223.10
1990	diphenyl (4,4')	....	$Cl \cdot C_6H_4 \cdot C_6H_4 \cdot Cl$	V-579	223.10
1991	dipropyl ether (*n*)	(γ,γ')	$(ClCH_2CH_2CH_2)_2O$	**I-370	171.07
1992	dipropyl ether (*iso*)	(β,β')	$[ClCH_2(CH_3)CH]_2O$	**I-370	171.07
1993	ethane (1,1)	ethylidene dichloride	$CH_3 \cdot CHCl_2$	I-83	98.96
1994	ethane (1,2)	ethylene chloride	$ClCH_2 \cdot CH_2Cl$	I-84	98.96
1995	ether (α,β)	....	$ClCH_2 \cdot CHCl \cdot O \cdot C_2H_5$	I-612	143.01
1996	ether (β,β')(*sym*)	diCl-Et ether	$(ClCH_2 \cdot CH_2)_2O$	**I-335	143.01
1997	ethyl alcohol (β,β)	2,2-dichloro-ethanol-1	$Cl_2CH \cdot CH_2OH$	I-338	114.96
1998	ethylene (α,α)	1,1-dichloro-ethene	$CH_2:CCl_2$	I-186	96.94
1999	fluorescein (2,7)	....	$C_{20}H_{10}O_5Cl_2$	*XIX-722	401.21
2000	fluorescein (3',6')	....	$C_{20}H_{10}O_5Cl_2$	XIX-227	401.21
2001	hexane	2,3-dichloro-2,3-dimethyl-butane	$[(CH_3)_2CCl]_2$	I-152	155.07
2002	hexane	3,3-dichloro-2,2-dimethyl-butane	$(CH_3)_3C \cdot CCl_2 \cdot CH_3$	I-150	155.07
2003	hydroquinone (2,3)	....	$(HO)_2C_6H_2Cl_2$	VI-849	179.00
2004	hydroquinone (2,5)	....	$(HO)_2C_6H_2Cl_2$	VI-850	179.00
2005	1-iodoethane (2,2)	....	$ICH_2 \cdot CHCl_2$	I-98	224.86
2006	iodomethane	....	Cl_2CHI	I-71	210.83
2007	maleic acid	....	$(:CCl \cdot CO_2H)_2$	II-753	184.96
2008	methyl ether (*sym*)	....	$ClCH_2 \cdot O \cdot CH_2Cl$	I-582	114.96
2009	naphthalene (1,2)	....	$C_{10}H_6Cl_2$	V-542	197.07
2010	naphthalene (1,3)	θ-dichloronaphtha-lene	$C_{10}H_6Cl_2$	V-542	197.07
2011	naphthalene (1,4)	β-dichloronaphtha-lene	$C_{10}H_6Cl_2$	V-542	197.07
2012	naphthalene (1,5)	γ-dichloronaphtha-lene	$C_{10}H_6Cl_2$	V-543	197.07
2013	naphthalene (1,6)	η-dichloronaphtha-lene	$C_{10}H_6Cl_2$	V-543	197.07
2014	naphthalene (1,7)	θ'-dichloronaphtha-lene	$C_{10}H_6Cl_2$	V-543	197.07
2015	naphthalene (1,8)	ζ-dichloronaphtha-lene	$C_{10}H_6Cl_2$	V-544	197.07
2016	naphthalene (2,3)	ι-dichloronaphtha-lene	$C_{10}H_6Cl_2$	V-544	197.07
2017	naphthalene (2,6)	ε-dichloronaphtha-lene	$C_{10}H_6Cl_2$	V-544	197.07
2018	naphthalene (2,7)	δ-dichloronaphtha-lene	$C_{10}H_6Cl_2$	V-544	197.07
2019	α-naphthol (2,4)	....	$Cl_2C_{10}H_5OH$	VI-612	213.06
2020	α-naphthoquinone	(2,3)	$O:C_{10}H_4Cl_2:O$	VII-729	227.05

Dichloro-othanal 1926
Dichloro-ethanol 1997

Dichloro-ethene 135-6, 1998
Dichloro-ethyl ether 1996
Dichloro-ethylene 135-6

Table 7-4 (*Continued*)
PHYSICAL CONSTANTS OF ORGANIC COMPOUNDS

No.	Crystalline Form and Color	Specific Gravity	Melting Point °C.	Boiling Point °C.	Solubility in 100 Parts		
					Water	Alcohol	Ether
1986	scales	....	85 d.	....	i.	s. d.	....
1987	yel. cr./ act.-al.	....	213-4	....	v. sl. s. CS_2	v. sl. s.	v. sl. s. lg.; sl. s. bz.
1988	col. lq.	....	....	$88\text{-}9^{19mm}$	....	....	....
1989	nd./al.	....	23	322-4	i.	v. s.	v. s.
1990	pr. or nd.	$1.442^{0°}_{4}$	148	315-9	i.; $14^{25°}$bz.	v. sl. s.	$4^{25°}$
1991	oil	$1.140^{20°}_{20}$	....	215^{745mm}	....	....	....
1992	col. lq.	$1.112^{20°}_{20}$	....	187.3	$0.17^{20°}$	....	....
1993	col. lq.	$1.176^{20°}_{4}$	-97.4	57.3	$0.7°$; $0.5^{30°}$	∞	∞
1994	col. lq.	$1.253^{20°}_{4}$	-35.3	83.5	$0.9°$; $0.9^{30°}$	∞ ; ∞ chl.	∞
1995	lq.	$1.174^{23°}$	....	140-5	....	....	....
1996	lq.	$1.222^{20°}_{20}$	....	178.5	$1.07^{20°}$	s.	s.
1997	lq.	$1.145^{15°}$	-46.1	146	sl. s.	s.	s.
1998	lq.	$1.250^{15°}$	....	37	i.	....	....
1999	or. pd.	....	....	....	sl. s.	sl. s.	sl. s.
2000	or. pd.	....	....	....	v. s. alk.	....	....
2001	cr.	....	159-60	....	i.	....	....
2002	cr.	....	151	subl.	i.	....	
2003	nd./aq.	....	144-5	subl.	i. c. lg.	s.	s.
2004	mn. nd./ act.	$1.815^{24°}$	166-70	subl.	s. h.; s. ac.	v. s.	v. s.
2005	lq.	2.219	....	171-2	i.	....	....
2006	lq.	$2.403^{21.5°}$	....	132	i.	....	....
2007	nd.	....	d. 119-20	....	v. s.; i. chl.	v. s.	v. s.
2008	col. lq.	$1.328^{15°}_{4}$	....	104-5	d.	....	....
2009	nd./al.	$1.315^{48.5°}_{4}$	37	282	....	....	....
2010	nd./al.	....	61.5	291^{775mm}	....	....	....
2011	nd./al.	$1.300^{76°}_{4}$	67-8	$286\text{-}7^{740mm}$	v. s. act.	v. sl. s.	....
2012	lf./al.	....	107	....	i.	s.	s.
2013	nd./al.	....	48-9	....	....	....	....
2014	lf./aq. al.	$1.261^{100°}_{4}$	63-4	285-6	....	s.	s.
2015	cr./al.	$1.292^{100°}_{4}$	88	d.	....	....	....
2016	lf.	....	120	....	....	s. h.	s.
2017	nd./al.	....	135-6	285	....	sl. s.	v. s.
2018	pr.	....	114	....	....	v. s. h.	....
2019	nd./aq. al.	....	106-7	....	s. bz.	s.; s. ac.	s.
2020	yel. nd./al.	....	195-6	....	i.	s. h.	sl. s.

Dichloro-fluoran 3252
Dichloro-hydrin 3393-4

Dichloro-malonyl urea 1951
Dichloro-methane 4429

Table 7-4 (*Continued*)
PHYSICAL CONSTANTS OF ORGANIC COMPOUNDS

No.	Name	Synonym	Formula	Beil. Ref.	Formula Weight
	Dichloro				
2021	4-nitroaniline	(2,6;4,1)	$Cl_2C_6H_2(NO_2)NH_2$	XII-735	207.03
2022	nitrobenzene (2,4)	NO_2-Cl-benzene	$Cl_2C_6H_3 \cdot NO_2$	V-245	192.00
2023	nitrobenzene (2,5)	....	$Cl_2C_6H_3 \cdot NO_2$	V-245	192.00
2024	nitrobenzene (2,6)	....	$Cl_2C_6H_3 \cdot NO_2$	V-246	192.00
2025	nitrobenzene (3,4)	....	$Cl_2C_6H_3 \cdot NO_2$	V-246	192.00
2026	nitrobenzene (3,5)	....	$Cl_2C_6H_3 \cdot NO_2$	V-246	192.00
2027	1-nitroethane (1,1)	....	$Cl_2C(NO_2) \cdot CH_3$	....	143.96
2028	nitrohydrin	diCl-Pr-nitrate	$ClCH_2 \cdot CHCl \cdot CH_2 \cdot NO_3$	I-356	173.98
2029	nitrophenol	(4,6;2,1)	$Cl_2C_6H_2(NO_2)OH$	VI-241	208.00
2030	nitrophenol	(2,6;4,1)	$Cl_2H_6H_2(NO_2)OH$	VI-241	208.00
2031	nitropropane	(1,1,1)	$C_2H_5 \cdot C(NO_2)Cl_2$	....	157.98
2032	1-nitrosoethane (1,1)	....	$CH_3 \cdot CCl_2 \cdot NO$	I-99	127.96
2033	pentane (1,4)	....	$CH_3CHCl(CH_2)_3Cl$	I-131	141.04
2034	pentane (1,5)	amylene chloride	$ClCH_2(CH_2)_3CH_2Cl$	I-131	141.04
2035	pentane (2,3)	....	$C_2H_5(CHCl)_2CH_3$	I-131	141.04
2036	pentane (2,4)	....	$(CH_3CHCl)_2 : CH_2$	*I-43	141.04
2037	pentane	1,4-dichloro-2-methyl-butane	$ClCH_2 \cdot CH(CH_3) \cdot CH_2 \cdot CH_2Cl$	*I-47	141.04
2038	pentane	2,3-dichloro-2-methyl-butane	$(CH_3)_2CCl \cdot CHCl \cdot CH_3$	I-135	141.04
2039	pentane	2,4-dichloro-2-methyl-butane	$(CH_3)_2CCl \cdot CH_2 \cdot CH_2Cl$	I-135	141.04
2040	pentane	3,4-dichloro-2-methyl-butane	$(CH_3)_2CH \cdot CHCl \cdot CH_2Cl$	I-135	141.04
2041	pentane	4,4-dichloro-2-methyl-butane	$(CH_3)_2CH \cdot CH_2 \cdot CHCl_2$	I-135	141.04
2042	phenol (2,4)	....	$Cl_2C_6H_3 \cdot OH$	VI-189	163.00
2043	phenyl-3,4-dichlorobenzene sulfonate (2,4)	....	$Cl_2C_6H_3SO_3 \cdot C_6H_3Cl_2$	....	372.06
2044	phenylhydrazine	(2,5)	$Cl_2C_6H_3 \cdot NH \cdot NH_2$	XV-431	177.03
2045	phenylhydrazine-4-sulfonic acid	(2,5)	$Cl_2C_6H_2(SO_3H) \cdot NH \cdot NH_2$	XV-643	257.10
2046	phenyl-phenyl phosphate (*o,o′*)	phosphen 4	$C_6H_5 \cdot O \cdot PO : (ClC_6H_4 \cdot O)_2$	....	395.18
2047	phenyl-4-toluene sulfonate	....	$CH_3 \cdot C_6H_4 \cdot SO_3 \cdot C_6H_3Cl_2$	....	317.19
2048	*p*-phenylenediamine	(2,6)	$Cl_2C_6H_2(NH_2)_2$	XIII-118	177.03
2049	*p*-phenylenediamine HCl	(2,6)	$Cl_2C_6H_2(NH_2)_2 \cdot HCl$	....	213.50
2050	propane (1,1)	propylidene diCl	$C_2H_5 \cdot CHCl_2$	I-105	112.99
2051	propane (1,2)	propylene chloride	$CH_3 \cdot CHCl \cdot CH_2Cl$	I-105	112.99
2052	propane (1,3)	trimethylene diCl	$ClCH_2 \cdot CH_2 \cdot CH_2Cl$	I-105	112.99
2053	propane (2,2)	acetone chloride	$(CH_3)_2CCl_2$	I-105	112.99
2054	propionic acid (α,α)	....	$CH_3 \cdot CCl_2 \cdot CO_2H$	II-250	142.97
2055	propionic acid (α,β)	....	$ClCH_2 \cdot CHCl \cdot CO_2H$	II-252	142.97
2056	propionitrile (α,α)	....	$CH_3 \cdot CCl_2 \cdot CN$	II-251	123.97
2057	propyl carbonate	(γ,γ')	$[Cl(CH_2)_3O]_2CO$	**III-5	215.08
2058	propylene (α,α)	1,1-diCl-propene-1	$CH_3 \cdot CH : CCl_2$	I-199	110.97
2059	propylene (α,β)	allylene dichloride	$CH_3 \cdot CCl : CHCl$	I-199	110.97

Dichloro-propanol 3393-4
Dichloro-propene 2058-61

Dichloro-propyl acetate 45
Dichloro-propyl carbamate 171

Table 7-4 (*Continued*)
PHYSICAL CONSTANTS OF ORGANIC COMPOUNDS

No.	Crystalline Form and Color	Specific Gravity	Melting Point °C.	Boiling Point °C.	Solubility in 100 Parts		
					Water	Alcohol	Ether
2021	yel. nd./ac.	....	194–5	....	i. h. HCl	s.	....
2022	nd./al.	$1.439^{80°}$	33	258.5	i.	v. s. h.	∞
2023	tri./al.	$1.669^{22°}$	54.6	266	i.	v. s. h.	v. s. bz.
2024	mn. pr.	$1.603^{17°}$	72.5	130^{8mm}	i.	s.	s. CS_2
2025	nd./al.	$1.456_{4}^{7.5°}$	42–3(α)*	255–6	....	....	....
2026	yel. mn.	$1.692^{14°}$	65.4	....	i.	s.	s. ac.
2027	lq.	$1.405_{20}^{20°}$	....	122–5	$<0.5^{20°}$	....	....
2028	lq.	$1.37°$	....	180	....	....	....
2029	yel. mn./al.	$1.822_{4}^{20°}$	122–3	subl. <100	sl. s.	sl. s.; s. bz.	s; s. chl.
2030	yel. pl./et.	....	125 d.	....	i.; sl. s. bz.	s. h.	v. s.
2031	lq.	$1.314_{20}^{20°}$	....	141–4	$<0.5^{20°}$	....	....
2032	b. oil	$1.252^{19°}$	....	68	i.		
2033	col. lq.	....	....	$58-60^{15mm}$	i.	∞	∞
2034	col. lq.	$1.094_{4}^{25°}$	....	180–1	i.; s. CS_2	s.; s. chl.	s.
2035	col. lq.	....	....	138–9	i.	∞	∞
2036	col. lq.	$1.063^{18°}$	....	147–50	i.	∞	∞
2037	col. lq.	$1.103_{4}^{21°}$	....	170–2	i.	∞	∞
2038	lq.	$1.068_{4}^{15°}$	....	130–5	I.	∞	∞
2039	col. lq.	$1.065_{4}^{20°}$	....	152–4 sl. d.	i.	∞	∞
2040	col. lq.	$1.092^{17.5°}$	....	143–5	i.	∞	∞
2041	col. lq.	$1.05^{24°}$	....	130	i.	∞	∞
2042	nd./bz.	$1.383_{25}^{60°}$	45	209–10	$0.45^{20°}$ v. s. chl.	v. s.; $160^{25°}$ CCl_4	v. s.; v. s. bz.
2043	col. cr.	....	81–2	....	i.	s. h.	s.
2044	nd./aq.	....	105	....	sl. s. h.	s.	s.; s. ac.
2045	nd./aq. HCl	....	....	....	....	....	....
2046	col. lq.	$1.34_{25}^{25°}$	<0	$255-75^{mm}$	i.; ∞ bz.	v. s.	∞ CCl_4
2047	col. nd.	....	118–9	....	i.	s. h.	i.
2048	nd./aq. al.	....	123.5	....	....	s.	s.
2049	col. pd.	....	....	....	s.	sl. s.	i.
2050	lq.	$1.143^{10°}$	....	87	v. sl. s.	s.	....
2051	col. lq.	$1.155_{4}^{20°}$	<–70	96.4	$0.27^{20°}$	v. s.	v. s.
2052	lq.	$1.186_{4}^{20°}$	....	123–5	$0.27^{25°}$	s.	s.
2053	lq.	$1.091_{4}^{20°}$	–33.8	70.5	i.	s.	∞ CS_2
2054	col. lq.	$1.389_{4}^{22°}$	....	185–90	v. s.	v. s.	....
2055	nd.	....	50 (36)	210 d.	s.	s.	....
2056	col. lq.	$1.431^{15°}$	....	105	i.	∞	∞
2057	col. lq.	....	....	265–70	....	....	....
2058	lq.	$1.176_{0}^{19.5°}$	....	78	i.	....	....
2059	lq.	....	....	75	i.	....	....

* β modification, liquid, changes to α at 15°.
Dichloro-propyl nitrate 2028

Table 7-4 (*Continued*)
PHYSICAL CONSTANTS OF ORGANIC COMPOUNDS

No.	Name	Synonym	Formula	Beil. Ref.	Formula Weight
	Dichloro				
2060	propylene (β,γ)	epidichlorohydrin (α)	$CH_2:CCl\cdot CH_2Cl$	I-199	110.97
2061	propylene (α,γ)	epidichlorohydrin (β)	$CH_2Cl\cdot CH:CHCl$	I-199	110.97
2062	quinazoline (2,4)	....	$C_6H_4\cdot N:CCl\cdot N:CCl$	XXIII-176	199.04
2063	quinoline (2,3)	....	$C_9H_5NCl_2$	XX-361	198.05
2064	quinoline (2,4)	....	$C_9H_5NCl_2$	XX-361	198.05
2065	quinoline (2,7)	....	$C_9H_5NCl_2$	XX-361	198.05
2066	quinoline (5,6)	....	$C_9H_5NCl_2$	XX-361	198.05
2067	quinoline (5,8)	....	$C_9H_5NCl_2$	XX-362	198.05
2068	quinoline (6,8)	....	$C_9H_5NCl_2$	XX-362	198.05
2069	quinoline (7,8)	....	$C_9H_5NCl_2$	....	198.05
2070	quinone (2,5)	....	$Cl_2C_6H_2O_2$	VII-633	176.99
2071	quinone (2,6)	....	$Cl_2C_6H_2O_2$	VII-633	176.99
2072	quinone-chloro-imide	....	$O:C_6H_2Cl_2:NCl$	VII-634	210.45
2073	tetrabromoethane	(2,2;1,1,1,2)	$Br_3C\cdot CCl_2Br$	I-95	414.56
2074	*p*-toluenesulfon-amide (*N,N*)	dichloramine T	$CH_3\cdot C_6H_4\cdot SO_2NCl_2$	XI-107	240.11
2075	tribromoethane	....	$Br_2CCl\cdot CHBrCl$	I-94	335.66
2076	tribromoethane	....	$Br_2CH\cdot CCl_2Br$	I-94	335.66
2077	tyrosine (3,5)(*dl*)	....	$Cl_2C_6H_2(OH)\cdot CH_2\cdot$ $CHNH_2CO_2H\cdot 2H_2O$	*XIV-670	286.11
2078	**Dicinnamalacetone**	....	$[C_6H_5(CH:CH)_2]_2CO$	VII-524	286.38
2079	**Dicresol** (*o*)	4,4'-diOH;3,3'-diMe	$(CH_3\cdot C_6H_3OH)_2$	VI-1009	214.27
2080	**Dicyandiamide**	param	$H_2N\cdot C(:NH)\cdot NH\cdot CN$	III-91	84.08
2081	**Dicyano**-diamidine	guanyl urea; biuret-amidine	$H_2N\cdot C(:NH)\cdot NH\cdot$ $CONH_2$	III-89	102.10
2082	diamidine H_2SO_4	....	$C_2H_6ON_4\cdot\frac{1}{2}H_2SO_4\cdot$ H_2O	III-90	169.15
2083	**Dicyclohexyl**	decahydro-diPh	$(CH_2(CH_2)_4CH\cdot)_2$	V-108	166.31
2084	adipate	....	$(CH_2CH_2CO_2C_6H_{11})_2$	**VI-11	310.44
2085	amine	....	$(C_6H_{11})_2NH$	XII-6	181.32
2086	maleate	....	$(:CHCO_2C_6H_{11})_2$	....	280.37
2087	oxalate	....	$(CO_2C_6H_{11})_2$	*VI-6	254.33
2088	phthalate	....	$C_6H_4(CO_2C_6H_{11})_2$	IX-799	330.43
2089	**Dicyclopentadiene**	....	$(\cdot CH:CH\cdot CH_2\cdot$ $CH:CH)_2$	V-495	132.21
2090	**Diethanolamine**	iminoethyl alcohol	$HN(CH_2\cdot CH_2OH)_2$	IV-283	105.14
2091	**Diethoxy**-benzene (*o*)	....	$(C_2H_5O)_2C_6H_4$	VI-771	166.22
2092	benzene (*m*)	....	$(C_2H_5O)_2C_6H_4$	VI-814	166.22
2093	benzene (*p*)	....	$(C_2H_5O)_2C_6H_4$	VI-844	166.22
2094	ethyl carbonate	....	$(C_2H_5O\cdot C_2H_4)_2CO_3$	....	206.24
2095	ethyl maleate	....	$(:CHCO_2C_2H_4OC_2H_5)_2$	....	260.29
2096	ethyl phthalate	....	$C_6H_4(CO_2C_2H_4OC_2H_5)_2$	....	310.35
2097	ethyl sebacate	....	$[C_2H_5OC_2H_4CO_2\cdot$ $CH_2(CH_2)_2CH_2\cdot]_2$	....	346.47
2098	quinazoline (2,4)	....	$C_{12}H_{14}O_2N_2$	XXIII-486	218.26

Table 7-4 (*Continued*)
PHYSICAL CONSTANTS OF ORGANIC COMPOUNDS

No.	Crystalline Form and Color	Specific Gravity	Melting Point °C.	Boiling Point °C.	Solubility in 100 Parts		
					Water	Alcohol	Ether
2060	col. lq.	$1.204^{25°}$	....	94	i.	∞	∞
2061	col. lq.	$1.233^{17.5°}$	....	106-9	....	....	....
2062	nd.	....	120	....	i.	s.; i. lg.	s.; s. bz.
2063	cr./aq. al.	....	104-5	....	i.; i. alk.	s.; s. bz.	s.; sl. s. lg.
2064	nd./aq. al.	....	67	280-2	v. sl. s. h.	s.; s. chl.	s.; s. bz.
2065	nd./al.	....	98	....	sl. s.	s.	....
2066	nd.	....	85	....	sl. s.	s.	s. pet.
2067	nd./al.	....	92-3	....	....	s.	s.
2068	nd./al.	....	103-4	....	....	s.	....
2069	nd.	....	85.5	....	....	....	....
2070	lt. yel./bz.	....	155-60	....	sl. s. h.	v. s.	v. s.; sl. s. bz.
2071	rhb./al.	....	120-1	subl. <120	sl. s. h.	v. s. h.	s. chl.
2072	yel. nd./al.	....	67-8	d. 170	v. sl. s.	s. h.	v. s.; s. chl.
2073	cr.	....	180 d.	....	....	....	....
2074	pr./chl.-pet.	....	83	....	v. sl. s.	d. h.; s. chl.	120 bz.
2075	col. lq.	$2.626^{21.5°}$	−5	133^{35mm}	....	....	....
2076	col. lq.	$2.632^{15°}_{4}$	16.8	210	....	....	....
2077	pr./aq.	....	252 ±, d.	....	4; sl. s. bz.	v. sl. s.	v. sl. s.
2078	yel. nd./al.	....	146	....	i.; s. ac.	s. h.	sl. s.
2079	cr./PhMe	....	161	....	s. h.; s. ac.	s.; s. h. bz.	s.
2080	mn. pl.	$1.401^{40°}$	209-11	d.	$2.3^{13°}$	$1.3^{13°}$	$0.01^{13°}$
2081	cr./al.	....	105	d. 160	s. h.	sl. s. c.	i.
2082	col. nd.	....	193-5	−H_2O, 110	5 c.; 33 h.	sl. s.; s. dil. a.	i.
2083	lq.	$0.886^{21°}_{4}$	3.6	236^{758mm}	i.; ∞ act.	$7^{25°}$ MeOH	∞ ; ∞ bz.
2084	col. nd.	....	38-9	$208-12^{9mm}$	i.	s.	s.
2085	col. lq.	$0.925^{18°}$	20±	$254-6^{745mm}$	$0.16^{28°}$	v. s.; s. bz.	∞
2086	col. pl.	....	82-3	....	i.	sl. s.	v. s.
2087	cr./MeOH	....	42-3	$190-1^{73mm}$	....	v. s.	v. s.
2088	pr./al.	....	66	....	i.	s.	....
2089	col. cr.	$0.976^{35°}$	32.9 (19)	170 sl. d.	....	v. s.	v. s.
2090	pr.	$1.097^{20°}_{4}$	28	270^{748mm}	∞ ; i. bz.	∞	v. sl. s.
2091	cr./pet.	....	43-5	....	....	....	....
2092	pr.	....	12.4	$234-5^{756mm}$	i.	s.	s.
2093	mn. lf.	....	71-2	246	i.	s.; s. bz.	s.; s. chl.
2094	col. lq.	....	....	$124-7^{15mm}$	....	....	....
2095	col. lq.	....	....	$174-7^{11mm}$	....	....	....
2096	col. lq.	....	31-3	....	i.	s.	s.
2097	col. lq.	....	−1±	$224-6^{8mm}$	....	....	....
2098	nd.	....	55	....	i.	s.	s.

Diethoxy-ethane 2139
Diethoxy-methane 2172

Table 7-4 (*Continued*)
PHYSICAL CONSTANTS OF ORGANIC COMPOUNDS

No.	Name	Synonym	Formula	Beil. Ref.	Formula Weight
2099	**Diethyl**-acetalde-hyde	2-Et-butyraldehyde	$(C_2H_5)_2CH \cdot CHO$	I-693	100.16
2100	acetamide (*N,N*)	acetyl diEt-amine	$CH_3CO \cdot N(C_2H_5)_2$	IV-110	115.18
2101	acetic acid	....	$(C_2H_5)_2CH \cdot CO_2H$	II-333	116.16
2102	acetoacetic ester	....	$CH_3CO \cdot C(C_2H_5)_2 \cdot CO_2C_2H_5$	III-710	186.25
2103	acetonedicar-boxylate	....	$CO(CH_2CO_2C_2H_5)_2$	III-791	202.21
2104	acetonitrile	3-cyano-pentane	$(C_2H_5)_2CH \cdot CN$	II-334	97.16
2105	acetyl succinate	....	$CH_3CO \cdot C_2H_3 : (CO_2C_2H_5)_2$	III-801	216.24
2106	adipate	....	$[\cdot (CH_2)_2CO_2C_2H_5]_2$	II-652	202.25
2107	allyl-acetamide	novonal	$(C_2H_5)_2C_3H_5 : C \cdot CONH_2$	**II-418	155.24
2108	allyl-malonate	....	$C_3H_5 \cdot CH(CO_2C_2H_5)_2$	II-776	200.24
2109	amine	....	$(C_2H_5)_2NH$	IV-95	73.14
2110	amine hydro-chloride	....	$(C_2H_5)_2NH \cdot HCl$	IV-97	109.60
2111	aminobenzalde-hyde	(*p*)	$(C_2H_5)_2N \cdot C_6H_4 \cdot CHO$	XIV-36	177.25
2112	aminoethanol (*β*)	....	$(C_2H_5)_2N \cdot CH_2CH_2 \cdot OH$	IV-282	117.19
2113	aminoglycerol (*α*)	1-diethylamino-2,3-propandiol	$(C_2H_5)_2N \cdot CH_2 \cdot CHOH \cdot CH_2OH$	IV-302	147.22
2114	aminophenetole (*m*)	diEt-*m*-pheneti-dine	$C_2H_5O \cdot C_6H_4 \cdot N(C_2H_5)_2$	XIII-410	193.29
2115	aminophenol (*m*)	....	$(C_2H_5)_2N \cdot C_6H_4 \cdot OH$	XIII-408	165.24
2116	aminophenol oxalate (*m*)	hydroxy-diEt-aniline oxalate	$[(C_2H_5)_2N \cdot C_6H_4 \cdot OH]_2 : H_2C_2O_4$	XIII-410	420.51
2117	aminopropyl alco-hol (*β*)	....	$(C_2H_5)_2N \cdot CH(CH_3) \cdot CH_2OH$	**IV-734	131.22
2118	aminopropyl alco-hol (*γ*)	....	$(C_2H_5)_2N \cdot CH_2 \cdot CH_2 \cdot CH_2OH$	IV-288	131.22
2119	aminopropyl cin-namate HCl	Apothesine	$(C_2H_5)_2N(CH_2)_3O \cdot CO \cdot C_8H_7 \cdot HCl$	**IX-390	297.83
2120	ammonium diethyl-dithiocarbamate	....	$(C_2H_5)_2N \cdot CS \cdot S \cdot NH_2(C_2H_5)_2$	IV-121	222.42
2121	aniline (*N*)	....	$(C_2H_5)_2N \cdot C_6H_5$	XII-164	149.24
2122	aniline sulfonic acid (*m*)	diethylaminoben-zene sulfonic	$(C_2H_5)_2N \cdot C_6H_4 \cdot SO_3H$	XIV-690	229.30
2123	azelate	ethyl azelate	$(CH_2)_7(CO_2C_2H_5)_2$	II-709	244.33
2124	barbituric acid (5,5)	Veronal; barbital	$C_8H_{12}O_3N_2$	XXIV-485	184.20
2125	barbiturate Na	Medinal	$C_8H_{11}O_3N_2Na$	XXIV-487	206.18
2126	benzene (*o*)	....	$(C_2H_5)_2C_6H_4$	V-426	134.22
2127	benzene (*m*)	....	$(C_2H_5)_2C_6H_4$	V-426	134.22
2128	benzene (*p*)	....	$(C_2H_5)_2C_6H_4$	V-426	134.22
2129	benzoylene urea (1,3)	....	$C_6H_4 \cdot N(C_2H_5) \cdot CO \cdot \overline{\qquad} N(C_2H_5) \cdot CO$	XXIV-376	218.26
2130	benzylmalonate	....	$C_7H_7 \cdot CH(CO_2C_2H_5)_2$	IX-869	250.30
2131	bromoacetamide (*α*)	Neuronal	$(C_2H_5)_2CBr \cdot CO \cdot NH_2$	II-334	194.08
2132	bromomaleate	....	$CH : CBr(CO_2C_2H_5)_2$	II-755	251.08

Diethyl acetal 4
Diethyl-amino-propylene glycol 2113

Diethyl beryllium 845

Table 7-4 (*Continued*)
PHYSICAL CONSTANTS OF ORGANIC COMPOUNDS

No.	Crystalline Form and Color	Specific Gravity	Melting Point °C.	Boiling Point °C.	Solubility in 100 Parts		
					Water	Alcohol	Ether
2099	lq.	$0.816^{20°}_{20°}$	−89	117-8	$0.3^{20°}$	∞	∞
2100	lq.	$0.925^{8.5°}$	....	185-6	....	....	....
2101	col. lq.	$0.920^{18°}_{0°}$	<−15	195-7	sl. s.	s.	s.
2102	col. lq.	$0.965^{25°}_{4}$	....	212	....	....	....
2103	oil	$1.113^{20°}_{4}$	....	250	v. sl. s.	∞	∞
2104	oil	....	....	144-6	....	∞	∞
2105	lq.	$1.080^{25°}_{5}$	....	254-6 d.	i.	s.	....
2106	col. lq.	$1.007^{20°}_{4}$	−19.8	133.8^{15mm}	$0.43^{30°}$	s.	s.
2107	wh. pd.	....	80	155^{16mm}	0.9	s.	s.
2108	col. lq.	$1.006^{25°}_{25}$	....	222-3	i.	s.	s.
2109	col. lq.	$0.709^{15°}_{4}$	−50	55.5^{759mm}	v. s.*	∞	∞
2110	lf./al. et.	$1.048^{21°}_{4}$	228-9	320-30	$232^{25°}$	sl. s. c.	i.
2111	yel. nd./ aq.	....	41	174^{7mm}	....	s.	s.
2112	lq.	$0.885^{20°}_{20°}$	....	162.1	∞	∞	s.
2113	syrup	....	....	233-5	s.	s.	s.; s. chl.
2114	oil	....	....	286	i.	s.	s. ac.
2115	rhb./CS₂- lg.	....	78	276-80	s.; s. chl.	s. CS_2	i. lg.
2116	cr.	....	....	155-6	....	....	....
2117	lq.	....	....	$61-3^{25mm}$	....	....	....
2118	lq.	....	....	189.5	v. s.	....	....
2119	col. cr.	....	136	....	s.	s.; sl. s. act.	sl. s.
2120	pa. yel. pl.	....	82-3	....	v. s.	v. s.	v. sl. s.
2121	oil	$0.935^{20°}_{4}$	−21.3(−34)	217.5	$1.4^{12°}$	s.	s.
2122	cr.	....	270 d.	....	s.	....	....
2123	lq.	$0.973^{20°}_{4}$	−16	291-2	i.	s.	s.
2124	col. cr.	....	191	subl. vac.	0.7 c.; 8 h.	s. h.	s.; s. alk.
2125	col. cr.	....	....	....	20 c.; 40 h.	0.25	i.
2126	col. lq.	$0.881^{20°}_{4}$	−31.4	183.5	i.	s.	s.
2127	col. lq.	$0.864^{20°}_{4}$	−83.9	181.1	i.	s.	s.
2128	col. lq.	$0.862^{20°}_{4}$	−43.2	183.8	i.	s.	s.
2129	nd.	....	110-11	....	i.	s.	....
2130	lq.	$1.077^{15°}_{15}$	....	298-300	i.	....	....
2131	col. cr.	....	66-7	d. 160-70	0.8 c.; s. h.	v. s.; s. bz.	v. s.
2132	col. lq.	$1.410^{17.5°}$	....	256	....	....	....

*Forms a hydrate $+1H_2O$, m. p. −19°.

Table 7-4 (*Continued*)
PHYSICAL CONSTANTS OF ORGANIC COMPOUNDS

No.	Name	Synonym	Formula	Beil. Ref.	Formula Weight
	Diethyl				
2133	bromomalonate	Et bromomalonate	$CHBr(CO_2C_2H_5)_2$	II-594	239.07
2134	*n*-butylmalonate	....	$C_4H_9 \cdot CH(CO_2C_2H_5)_2$	*II-282	216.28
2135	butylmalonate	(*sec*)	$C_5H_{10}(CO_2C_2H_5)_2$	II-679	216.28
2136	butylmalonate	(*iso*)	$C_5H_{10}(CO_2C_2H_5)_2$	II-683	216.28
2137	carbanilide	....	$[(C_2H_5)(C_6H_5)N]_2CO$	XII-422	268.36
2138	carbonate	....	$CO(OC_2H_5)_2$	III-5	118.13
2139	cellosolve	diEtO-ethane	$(C_2H_5O \cdot CH_2)_2$	I-468	118.18
2140	chlorofumarate	....	$CH:CCl(CO_2C_2H_5)_2$	II-745	206.63
2141	chloromaleate	....	$CH:CCl(CO_2C_2H_5)_2$	II-753	206.63
2142	citraconate	....	$CH_3 \cdot C_2H(CO_2C_2H_5)_2$	II-771	186.21
2143	cyanamide	cyano-diEt-amine	$(C_2H_5)_2N \cdot CN$	IV-121	98.15
2144	*iso*-cyanine iodide	ethyl red	$C_{23}H_{23}N_2 \cdot I \cdot C_2H_5OH$	XXIII-298	500.43
2145	cyclohexylamine (*N,N*)	hexahydro-diethyl-aniline	$CH_2(CH_2)_4CH \cdot$ $\vert \underline{\qquad} \vert$ $N(C_2H_5)_2$	XII-6	155.29
2146	diacetosuccinate	....	$(CH_3CO)_2(CHCO_2 \cdot$ $C_2H_5)_2$	III-840	258.27
2147	diacetosuccinate	....	$(CH_3CO)_2(CHCO_2 \cdot$ $C_2H_5)_2$	III-840	258.27
2148	diacetosuccinate	....	$(CH_3CO)_2(CHCO_2 \cdot$ $C_2H_5)_2$	III-840	258.27
2149	diacetosuccinate	....	$(CH_3CO)_2(CHCO_2 \cdot$ $C_2H_5)_2$	III-840	258.27
2150	diacetosuccinate	....	$(CH_3CO)_2(CHCO_2 \cdot$ $C_2H_5)_2$	III-840	258.27
2151	diacetosuccinate	....	$(CH_3CO)_2(CHCO_2 \cdot$ $C_2H_5)_2$	III-840	258.27
2152	diacetyl tartrate	....	$(CH_3CO)_2(O \cdot CH \cdot$ $CO_2C_2H_5)_2$	III-515	290.27
2153	dibenzyl malonate	....	$(C_7H_7)_2C(CO_2C_2H_5)_2$	IX-937	340.42
2154	dibromomaleate	....	$(:CBrCO_2C_2H_5)_2$	II-757	329.97
2155	dibromomalonate	....	$Br_2C(CO_2C_2H_5)_2$	II-595	317.97
2156	dibromosuccinate	(α,α')	$(CHBr \cdot CO_2C_2H_5)_2$	II-624	332.00
2157	dibutylmalonate	(*n*)	$(C_4H_9)_2C(CO_2C_2H_5)_2$	*II-295	272.39
2158	diethylmalonate	....	$(C_2H_5)_2C(CO_2C_2H_5)_2$	II-686	216.28
2159	1,4-dihydrocolli-dine dicar-boxylate	(3,5)	$(CH_3)_3C_5H_2N:$ $(CO_2C_2H_5)_2$	XXII-147	267.33
2160	dihydroxymalonate	diEt-mesoxalate	$(HO)_2C(CO_2C_2H_5)_2$	III-769	192.17
2161	dimethylmalonate	....	$(CH_3)_2C(CO_2C_2H_5)_2$	II-648	188.23
2162	dithiocarbamate †	....	$(C_2H_5)_2N \cdot CS \cdot SH$	IV-121	149.28
2163	dithiocarbonate	carbonyl-disulfethyl	$CO(SC_2H_5)_2$	III-211	150.26
2164	dithiocarbonate	Et-xanthate	$C_2H_5O \cdot CS \cdot SC_2H_5$	III-210	150.26
2165	dithioloxalate	....	$(CO \cdot SC_2H_5)_2$	II-565	178.27
2166	disulfide	....	$C_2H_5S \cdot SC_2H_5$	I-347	122.25
2167	ethyl-*iso*-amyl-malonate	....	$C_5H_{11} \cdot C(C_2H_5):$ $(CO_2C_2H_5)_2$	**II-611	258.36
2168	ethyl-*n*-butyl-malonate	....	$C_4H_9 \cdot C(C_2H_5):$ $(CO_2C_2H_5)_2$	II-712	244.33
2169	ethylmalonate	....	$C_2H_5 \cdot CH(CO_2C_2H_5)_2$	II-644	188.23

† Free acid unstable; Na and K salts s. in aq.   Diethyl *iso*-butyl carbinol 4924
Diethyl cadmium 1185   Diethyl carbinol 408

Table 7-4 (*Continued*)
PHYSICAL CONSTANTS OF ORGANIC COMPOUNDS

No.	Crystalline Form and Color	Specific Gravity	Melting Point °C.	Boiling Point °C.	Solubility in 100 Parts		
					Water	Alcohol	Ether
2133	lq.	$1.402^{25°}_{4}$	<-54	233-5 d.	i.	∞	∞
2134	col. lq.	$0.975^{20°}_{4}$	....	235-40^{760mm}	i.	v. s.	v. s.
2135	col. lq.	$0.988^{15°}$	....	233-4^{774mm}	v. sl. s.	v. s.	v. s.
2136	col. lq.	$0.983^{17°}$	....	225	v. sl. s.	v. s.	v. s.
2137	cr./al.	....	79	....	i.	....	....
2138	col. lq.	$0.975^{20°}_{4}$	-43	126.8	i.	∞	∞
2139	col. lq.	$0.842^{20°}_{4}$	....	124^{759mm}	$21^{20°}$	....	....
2140	col. lq.	$1.189^{20°}_{4}$	....	250 sl. d.	....	....	....
2141	col. lq.	$1.191^{25°}_{4}$	....	235 sl. d.	....	....	....
2142	col. lq.	$1.042^{20°}_{4}$	....	230.3	....	....	....
2143	lq.	$0.854^{20°}_{4}$	....	188-9^{748mm}	i.; d. HCl	s.	s.
2144	hyg. gn. met./al.	....	150-2 d.	....	v. sl. s; v. s. aq. a.	s.; s. act.; s. ac.	i.; i. CS$_2$
2145	lq.	$0.872^{0°}_{0}$	....	193	....	....	....
2146	α_1-oil	....	....	....	v. sl. s.	s.	s.; 10 lg.
2147	α_2-cr.	....	20-2	....	v. sl. s.	v. s.	v. s.
2148	α_3-pr./lg.	....	31-2	....	v. sl. s.	v. s.	v. s.; 35 lg.
2149	α_4-oil	....	....	....	....	....	....
2150	β-stable, mn./al.	$1.209^{20°}_{4}$	89-90	....	....	$6^{20°}$ abs.	2 abs.
2151	meta-stable	$1.176^{20°}_{4}$	....	....	....	....	....
2152	mn.	$1.081^{9°}_{9}$	67-8	288.5^{727mm}	s. h.	v. s. h.	v. s.
2153	oil	$1.093^{20°}_{4}$	13-14	243-6$^{18°}$	....	....	....
2154	col. lq.	$1.698^{25°}_{25}$	....	170-5^{15mm}	....	....	....
2155	lq.	....	....	250-6 sl. d.	....	....	....
2156	rhb. nd.	....	68	d. >130	....	....	....
2157	col. lq.	....	....	153-4^{14mm}	....	....	....
2158	col. lq.	$0.985^{20°}_{4}$	....	230	i.	∞	∞
2159	mn. pr./al.	....	131	>315 d.	v. sl. s.; s. chl.	s. h.; s. bz.	sl. s.; sl. s. CS$_2$
2160	col. pl./bz	....	57	200	$130^{22°}$	v. s. abs.	v. s.
2161	col. lq.	$0.994^{25°}_{25}$	....	196.2-6.7	i.	∞	∞
2162	....	....	....	....	....	....	....
2163	yel. lq.	$1.085^{19°}$	....	198.6-200.1	i.	s.	s.
2164	yel. lq.	$1.085^{19°}$	....	200	i.	s.	s.
2165	yel. nd./et	....	27-8	238-40^{757mm}	....	....	....
2166	oil	$0.993^{20°}_{4}$	....	152.8-3.4	v. sl. s.	....	....
2167	col. lq.	$0.954^{25°}_{25}$	....	150^{20mm}	....	....	....
2168	lq.	$0.965^{25°}_{25}$	....	245-50^{747mm}	....	....	....
2169	col. lq.	$1.004^{20°}_{20}$	....	211^{748mm}	....	....	....

Diethyl carbinol acetate 401
Diethyl carbitol 2233

Diethyl ether 2911

Table 7-4 (*Continued*)
PHYSICAL CONSTANTS OF ORGANIC COMPOUNDS

No.	Name	Synonym	Formula	Beil. Ref.	Formula Weight
2170	**Diethyl** ethyl-phenylmalonate	....	$C_6H_5 \cdot C(C_2H_5):$ $(CO_2C_2H_5)_2$	*IX-384	264.32
2171	ethyl-*iso*-propyl-malonate	....	$C_3H_7 \cdot C(C_2H_5):$ $(CO_2C_2H_5)_2$	II-706	230.31
2172	formal	diEtO—methane	$CH_2(OC_2H_5)_2\dagger$	I-574	104.15
2173	formamide	formyl-diethylamine	$HCON(C_2H_5)_2$	IV-109	101.15
2174	fumarate	....	$(:CH \cdot CO_2C_2H_5)_2$	II-742	172.18
2175	glutaconate	....	$C_3H_4(CO_2C_2H_5)_2$	II-759	186.21
2176	glutarate	....	$(CH_2)_3(CO_2C_2H_5)_2$	II-633	188.23
2177	*n*-heptylmalonate	....	$C_7H_{15}CH(CO_2C_2H_5)_2$	**II-610	258.36
2178	hydrazine (*as*)	....	$(C_2H_5)_2N \cdot NH_2$	IV-550	88.15
2179	itaconate	....	$C_3H_4(CO_2C_2H_5)_2$	II-762	186.21
2180	ketone	pentanone-3	$(C_2H_5)_2CO$	I-679	86.13
2181	ketone semicarbazone	....	$(C_2H_5)_2C:N \cdot NH \cdot CO \cdot$ NH_2	III-103	143.19
2182	malate (*dl*)	....	$(CHOH \cdot CH_2):$ $(CO_2C_2H_5)_2$	III-437	190.20
2183	maleate	....	$(:CH \cdot CO_2C_2H_5)_2$	II-751	172.18
2184	malonate	malonic ester	$CH_2(CO_2C_2H_5)_2$	II-573	160.17
2185	malonic acid	....	$(C_2H_5)_2C(CO_2H)_2$	II-686	160.17
2186	mesaconate	....	$CH_3 \cdot C_2H(CO_2C_2H_5)_2$	II-766	186.21
2187	methyl-ethyl-malonate	....	$C_2H_5 \cdot C(CH_3):$ $(CO_2C_2H_5)_2$	II-664	202.25
2188	muconate	....	$(CH:CH \cdot CO_2C_2H_5)_2$	II-804	198.22
2189	α-naphthylamine (*N*)	....	$C_{10}H_7 \cdot N(C_2H_5)_2$	XII-1223	199.30
2190	4-nitrophthalate	....	$NO_2 \cdot C_6H_3(CO_2C_2H_5)_2$	1X-831	267.24
2191	nitrosamine	nitroso-diethyl-amine	$(C_2H_5)_2N \cdot NO$	IV-129	102.14
2192	oxalacetate	oxalacetic ester	$(CH_2CO)(CO_2C_2H_5)_2$	III-782	188.18
2193	oxalate	ethyl oxalate	$(CO_2C_2H_5)_2$	II-535	146.14
2194	oxamide (*s*)	....	$(CO \cdot NHC_2H_5)_2$	IV-112	144.17
2195	oxomalonate	....	$CO(CO_2C_2H_5)_2$	III-769	174.15
2196	peroxide	....	$C_2H_5O \cdot OC_2H_5$	I-324	90.12
2197	phosphine	....	$(C_2H_5)_2PH$	IV-582	90.11
2198	phosphoric acid	....	$(C_2H_5)_2PO \cdot OH$	I-332	154.10
2199	phthalate (*o*)	ethyl phthalate	$C_6H_4(CO_2C_2H_5)_2$	IX-798	222.24
2200	phthalate (*m*)	ethyl isophthalate	$C_6H_4(CO_2C_2H_5)_2$	IX-834	222.24
2201	phthalate (*p*)	ethyl terephthalate	$C_6H_4(CO_2C_2H_5)_2$	IX-844	222.24
2202	pimelate	....	$(CH_2)_5(CO_2C_2H_5)_2$	II-671	216.28
2203	propylmalonate	....	$C_3H_7 \cdot CH(CO_2C_2H_5)_2$	II-657	202.25
2204	*iso*-propylmalonate	....	$C_3H_7 \cdot CH(CO_2C_2H_5)_2$	II-669	202.25
2205	quinolinate (2,3)	....	$C_5H_3N(CO_2C_2H_5)_2$	XXII-151	223.23
2206	sebacate	ethyl sebacate	$(CH_2)_8(CO_2C_2H_5)_2$	II-719	258.36
2207	selenide, mono-	selenium ethyl	$(C_2H_5)_2Se$	I-349	137.08
2208	selenide, di-	ethyl diselenide	$(C_2H_5Se)_2$	I-349	216.04
2209	stilboestrol (*trans*)	stilbestrol	$[HOC_6H_4(C_2H_5)C:]_2 \cdot$ C_2H_5OH	....	314.43
2210	suberate	....	$(CH_2)_6(CO_2C_2H_5)_2$	II-693	230.31
2211	succinate	ethyl succinate	$(CH_2 \cdot CO_2C_2H_5)_2$	II-609	174.20
2212	*iso*-succinate	....	$CH_3 \cdot CH(CO_2C_2H_5)_2$	II-629	174.20
2213	sulfate	ethyl sulfate	$(C_2H_5O)_2SO_2$	I-327	154.19
2214	sulfide	....	$(C_2H_5)_2S$	I-344	90.19
2215	sulfite	....	$(C_2H_5O)_2SO$	I-325	138.19

† Hydrate $+1H_2O$, s. aq., al., et., chl., bz.
Diethyl mercury 4043

Diethyl mesoxalate 2160, 3084
Diethyl phenetidine 2114

Table 7-4 (*Continued*)
PHYSICAL CONSTANTS OF ORGANIC COMPOUNDS

No.	Crystalline Form and Color	Specific Gravity	Melting Point °C.	Boiling Point °C.	Solubility in 100 Parts		
					Water	Alcohol	Ether
2170	col. lq.	....	....	166^{12mm}	....	....	....
2171	lq.	....	....	232-3	....	....	....
2172	lq.	$0.824_4^{25°}$	−66.5	88.0	$9^{18°};7^{30°}$	∞	∞
2173	col. lq.	$0.908^{19°}$	....	177-8	∞	v. s.	v. s.
2174	col. lq.	$1.052_4^{20°}$	0.6	217.9	....	....	....
2175	col. lq.	$1.050_4^{20°}$	....	236-8	i.	s.	s.
2176	syrup	$1.027_4^{15°}$	−23.8	233.7	$0.882^{20°}$	v. s.	s.
2177	pa. yel. lq.	$0.951^{20°}$	....	$144-6^{8mm}$	....	....	....
2178	hyg. lq.	....	....	96-9	v. s.	v. s.	v. s.
2179	col. lq.	$1.050_{15}^{15°}$	....	228-9	....	....	....
2180	col. lq.	$0.810_4^{25°}$	−39.9	102.0	$4.7^{20°}$	∞	∞
2181	nd./bz.	....	139	....	....	....	s. bz.
2182	col. lq.	$1.124_4^{21°}$	....	253-5	s.	∞	∞
2183	col. lq.	$1.070_{20}^{20°}$	−10.5	225	i.	s.	s.
2184	col. lq.	$1.055_4^{20°}$	−51.5	199.3	$2.08^{20°}$	∞	∞
2185	pr./aq.	....	125	d. 170-80	$65^{16°}$	v. s.	v. s.
2186	col. lq.	$1.047_4^{20°}$	....	229	....	....	....
2187	col. lq.	$0.994_{15}^{15°}$	....	207-8	....	....	....
2188	pr./al.	$0.983_4^{99°}$	63 (13)	200^{12mm}	....	v. s.	....
2189	col. oil	1.005	....	285-90	∞ bz.	∞	∞
2190	pl./al.	....	33-4	$208-10^{15mm}$	i.	s.	s.
2191	yel. oil	$0.943_4^{20°}$	....	176.9	s.	∞	∞
2192	col. oil	$1.131_4^{20°}$	....	$131-2^{24mm}$	i.	∞ ; ∞ bz.	∞
2193	col. lq.	$1.079_4^{20°}$	−40.6	185.4	v. sl. s.	∞	∞
2194	nd./al.	$1.169^{4°}$	....	179	....	....	....
2195	yel. gn. oil	$1.119_{20}^{20°}$	−30±	220±	....	s.	s.
2196	lq.	$0.827_4^{15°}$	....	65	v. sl. s.	∞	∞
2197	col. lq.	<1	....	85	....	....	....
2198	lq.	$1.175^{0°}$	....	203.3	....	....	....
2199	col. lq.	$1.121_{25}^{25°}$	....	289.5	i.	∞	∞
2200	col. lq.	$1.123_4^{25°}$	11.5	302	i.	∞	∞
2201	pr./al.	$1.110_{45}^{45°}$	43-4	302	i.	s.	s.
2202	col. oil	$0.999_4^{15°}$	−23.8	$252-57^{48mm}$	i.	s.	s.
2203	col. lq.	$0.993_{15}^{15°}$	....	$225-67^{71mm}$	i.	s.	s.
2204	col. lq.	$0.993_{15}^{15°}$	....	$215-77^{48mm}$	i.	s.	s.
2205	yel. oil	....	....	280-5 sl. d.	s.	s.	s.; s. bz.
2206	col. lq.	$0.965_4^{20°}$	1.3	306^{773mm}	0.008	s.	s.
2207	lq.	$1.230_4^{28°}$	....	110	i.	....	....
2208	red yel. lq.	$1.696_4^{18°}$	−95	137	....	....	....
2209	col. cr./al.	....	169-71	....	i.; s. alk.	s.	s.
2210	col. lq.	$0.982_4^{20°}$	5.9	282^{763mm}	i.	s.	s.
2211	col. lq.	$1.040_4^{20°}$	−21.3	217.8	i.	∞	∞
2212	col. lq.	$1.021_{15}^{15°}$	....	201.2-1.4	i.	∞	∞
2213	col. lq.	$1.172_4^{25°}$	−25	210.2 sl. d.	i.; sl. d.	s.; d. h.	∞
2214	col. lq.	$0.837_4^{20°}$	−103.3	92.1	$0.313^{20°}$	∞	∞
2215	col. lq.	$1.077_4^{25°}$	....	157.7	s. d.	s.	....

Diethyl phenylenediamine 303
Diethyl *n*-propyl carbinol 4966

Diethyl *iso*-propyl carbinol 4963
Diethyl pyridine 5040-1

Table 7-4 (*Continued*)
PHYSICAL CONSTANTS OF ORGANIC COMPOUNDS

No.	Name	Synonym	Formula	Beil. Ref.	Formula Weight
2216	**Diethyl** sulfone	. . . .	$(C_2H_5)_2SO_2$	I-346	122.19
2217	sulfoxide	. . . .	$(C_2H_5)_2SO$	I-346	106.19
2218	tartrate (*d*)	ethyl tartrate	$(CHOH \cdot CO_2C_2H_5)_2$	III-512	206.20
2219	telluride	tellurium ethyl	$(C_2H_5)_2Te$	I-350	185.72
2220	thiourea (*s*)	. . . .	$(C_2H_5NH)_2CS$	IV-118	132.23
2221	toluene (3,5)	. . . .	$(C_2H_5)_2C_6H_3 \cdot CH_3$	V-441	148.25
2222	urea (*s*)	. . . .	$(C_2H_5NH)_2CO$	IV-115	116.16
2223	urea (*as*)	. . . .	$(C_2H_5)_2N \cdot CO \cdot NH_2$	IV-120	116.16
2224	**Diethylene**-di-acetate	. . . .	$(CH_2CH_2O_2C \cdot CH_3)_2$	II-143	174.20
2225	diamine	piperazine	$NH \cdot C_2H_4 \cdot NH \cdot C_2H_4$	XXIII-4	86.14
2226	diamine hydrate	piperazine hydrate	$C_4H_{10}N_2 \cdot 6H_2O$	XXIII-4	194.23
2227	dilaurate	. . . .	$(C_{11}H_{23}CO_2C_2H_4)_2$	. . . .	454.74
2228	disulfide	dithian-1,4	$S \cdot C_2H_4 \cdot S \cdot C_2H_4$	XIX-3	120.24
2229	glycol diacetate	. . . .	$(CH_3CO_2C_2H_4)_2O$	II-141	190.20
2230	glycol dinitrate	. . . .	$(O_2N \cdot O \cdot C_2H_4)_2O$	**I-521	196.12
2231	glycol monobenzyl ether	. . . .	$HOCH_2CH_2 \cdot O \cdot CH_2 \cdot CH_2 \cdot O \cdot CH_2C_6H_5$	. . . .	196.25
2232	glycol *n*-butyl ether	butyl carbitol	$HOCH_2CH_2 \cdot O \cdot CH_2 \cdot CH_2 \cdot O \cdot C_4H_9$	**I-521	162.23
2233	glycol diethyl ether	diethyl carbitol	$(C_2H_5O \cdot CH_2CH_2)_2O$	**I-520	162.23
2234	glycol monolaurate	glaurin	$C_{11}H_{23}CO_2C_2H_4 \cdot O \cdot C_2H_4OH$	. . . . °	288.43
2235	imide oxide	morpholine	$NH \cdot (CH_2)_2 \cdot O \cdot (CH_2)_2$	XXVII-5	87.12
2236	imide oxide ethanol	morpholine ethanol	$C_4H_8O:N \cdot C_2H_4OH$	. . . .	131.18
2237	oxide	dioxan-1,4	$O:(CH_2)_4:O$	XIX-3	88.11
2238	triamine	. . . .	$(NH_2 \cdot C_2H_4)_2NH$	IV-255	103.17
2239	**Difluoro**-acetic acid	. . . .	$F_2CH \cdot CO_2H$	II-193	96.03
2240	1-bromoethane (2,2)	. . . .	$F_2CH \cdot CH_2Br$	I-89	144.95
2241	1-chloroethane (2,2)	. . . .	$F_2CH \cdot CH_2Cl$	I-83	100.50
2242	1,2-dibromoethane	(2,2)	$F_2CBr \cdot CH_2Br$	I-92	223.85
2243	1,1-dibromoethane	(2,2)	$Br_2CH \cdot CHF_2$	I-92	223.85
2244	1,1-dichloroethane	(2,2)	$F_2CH \cdot CHCl_2$	I-85	134.94
2245	1,2-dichloroethylene (1,2)	. . . .	$FClC:CClF$	. . . .	132.93
2246	dichloromethane	Freon-12	F_2CCl_2	I-61	120.91
2247	diphenyl (2,2')	. . . .	$F \cdot C_6H_4 \cdot C_6H_4 \cdot F$	. . . .	190.19
2248	diphenyl (4,4')	. . . .	$F \cdot C_6H_4 \cdot C_6H_4 \cdot F$	V-579	190.19
2249	ethane (1,2)	ethylene fluoride	$FCH_2 \cdot CH_2F$	I-82	66.05
2250	ethylene (α,α)	1,1-difluoro-ethene	$CH_2:CF_2$	I-186	64.04
2251	1-iodoethane (2,2)	. . . .	$F_2CH \cdot CH_2I$	I-98	191.95
2252	tetrabromoethane	(1,1,2,2;1,2)	$FBr_2C \cdot CFBr_2$	I-95	381.66
2253	tetrachloroethane	(1,1,1,2;2,2)	$Cl_3C \cdot CF_2Cl$	I-86	203.83
2254	tetrachloroethane	(1,1,2,2;1,2)	$FCl_2C \cdot CFCl_2$	. . . .	203.83
2255	tribromoethane	(1,1,2;1,2)	$Br_2FC \cdot CHBrF$	I-94	302.75
2256	**Diformyl hydrazine**	(*s*)	$(\cdot NH \cdot CHO)_2$	II-93	88.07
2257	**Difurfural**-cyclohexanone	. . . .	$(C_4H_3O \cdot CH:)_2 : C_6H_6O$	. . . .	254.29

Diethyl tin 5949
Diethyl toluidine 4216-8

Diethyl zinc 6503
Diethylene glycol 3453

Table 7-4 (*Continued*)
PHYSICAL CONSTANTS OF ORGANIC COMPOUNDS

No.	Crystalline Form and Color	Specific Gravity	Melting Point °C.	Boiling Point °C.	Solubility in 100 Parts		
					Water	Alcohol	Ether
2216	rhb. pl.	$1.357_4^{20°}$	73-4	248	$15.6^{16°}$	v. s. bz.	s. h.
2217	syrup	....	15	$88\text{-}90^{15mm}$	v. s.	s.	s.
2218	lq.	$1.204_4^{20°}$	17	280	sl. s.	∞	∞
2219	red-yel. lq.	$1.599_4^{15°}$	....	137-8	i.	s.	....
2220	cr.	....	77	....	s.	s.	....
2221	lq.	$0.879_4^{20°}$	....	198-200	i.	∞	∞
2222	nd./al.	1.042	112.5	263	v. s.	v. s.	v. s.
2223	nd./et.	....	75	....	v. s.	v. s.	$2.6^{22°}$
2224	lq.	$1.048^{15°}$	12	230^{751mm}	....	....	....
2225	rhb./al.	$1.110_{20}^{24°}$	110	145-6	$15^{20°}$	v. s.	i.
2226	col. cr.	....	44	125-30	v. s.	s.	i.
2227	....	....	28-30	....	....	....	....
2228	mn./et.	....	111-2	199-200	i.; v. s. CS_2	s.	s.
2229	lq.	$1.116_{20}^{20°}$	17-9	250	∞	....	....
2230	lq.	$1.377_4^{25°}$	−11.3	161±	$0.4^{24°}$	v. sl. s.	v. s.
2231	col. lq.	....	....	$176\text{-}9^{18mm}$	....	....	....
2232	col. lq.	$0.956_{20}^{20°}$	−68.1	231	∞ ; ∞ oils	v. s.	v. s.
2233	col. lq.	$0.909_{20}^{20°}$	−44.3	188	∞	v. s.	v. s.
2234	straw colrd. oil	0.960	17-8	>270	i.	s.	s.
2235	col. oil	$1.002_{20}^{20°}$	−3.1	128.9	∞	∞	s.
2236	col. lq.	$1.072_{20}^{20°}$	....	226	∞	....	....
2237	col. lq.	$1.034_4^{20°}$	11.8	101.4	∞	s.	s.
2238	lq.	$0.954_{20}^{20°}$	−39	208 sl. d.	∞	∞	i.
2239	col. lq.	$1.525_4^{20°}$	−0.35	134.2^{766mm}	∞	∞	∞
2240	col. lq.	$1.817_4^{20°}$	−74.5	57.3	v. sl. s.	∞	∞
2241	col. lq.	>1	....	36	....	....	....
2242	col. lq.	$2.242^{12.2°}$	−56.5	93	i.	....	....
2243	col. lq.	$2.312^{20°}$	....	107.5^{760mm}	....	....	....
2244	col. lq.	$1.494^{17°}$	....	60	i.	∞	∞
2245	gas	....	−112	20.9	....	....	....
2246	gas	$1.486^{-30°}$	−155	−29.2	$5.7^{26°}$ cc.	s.	s.
2247	cr.	....	117	....	....	....	....
2248	col. mn./al.	$1.336_4^{25°}$	94-5	254-5	i.; s. oil	s.; s. chl.	s.
2249	gas	....	....	....	....	....	....
2250	col. gas	....	....	....	i.	$150^{18°}$ cc.	$150^{18°}$ cc., chl.
2251	col. lq.	$2.243^{12.2°}$	....	89.5	i.	....	....
2252	col. cr.	....	62.5	186.5^{758mm}	i.	s.	s. ac.
2253	col. cr.	....	52	91	i.	sl. s. c.	v. s.
2254	col. cr.	$1.645_4^{25°}$	24.7	92.8	i.; v. s. bz.	v. s.	v. s.
2255	col. lq.	$2.603^{20°}$	....	146	i.	....	....
2256	pr.	....	159-60	....	v. s.	sl. s.	i.
2257	yel. nd	....	144-5	....	i.	s. h.	sl. s.

Diethylene glycol ethyl ether 1232
Diethylene glycol methyl ether 4177

Difluoro-ethene 2250
Difluoro-methane 4433

Table 7-4 (*Continued*)
PHYSICAL CONSTANTS OF ORGANIC COMPOUNDS

No.	Name	Synonym	Formula	Beil. Ref.	Formula Weight
2258	**Difurfural** cyclopentanone (1,3;5)	pyroxanthin	$(C_4H_3O \cdot CH:)_2:$ C_5H_4O	XIX-140	240.26
2259	**Diglycerol**	....	$[(HO)_2C_3H_5]_2O$	I-513	166.18
2260	**Diglycol chlorohydrin**	Cl-OH-diEt ether	$Cl \cdot CH_2CH_2 \cdot O \cdot$ CH_2CH_2OH	I-467	124.57
2261	**Diglycolic acid**	....	$(HO_2C \cdot CH_2)_2O \cdot H_2O$	III-234	152.10
2262	**Diglycyl glycine**	....	$H_2N(CH_2CONH)_2 \cdot$ CH_2CO_2H	IV-374	189.17
2263	**Diheptyl**-acetic acid (*n*)	....	$(C_7H_{15})_2CH \cdot CO_2H$	II-376	256.43
2264	ether (*n*)	heptyl ether	$(C_7H_{15})_2O$	I-414	214.39
2265	ketone (*n*)	caprylone	$[CH_3(CH_2)_6]_2CO$	I-717	226.41
2266	sulfide (*n*)	heptyl sulfide	$(C_7H_{15})_2S$	I-415	230.46
2267	**Dihexyl ketone** (*n*)	oenanthone	$[CH_3(CH_2)_5]_2CO$	I-715	198.35
2268	**Dihydracrylic acid**	Na salt	$O(CH_2CH_2CO_2Na)_2$	III-297	206.11
2269	**Dihydrazino**-diphenyl	(2,2′)	$(H_2N \cdot NH \cdot C_6H_4)_2$	XV-584	214.27
2270	diphenyl	(4,4′)	$(H_2N \cdot NH \cdot C_6H_4)_2$	XV-585	214.27
2271	**Dihydro**-acridine (9,10)	....	$C_6H_4 \cdot NH \cdot C_6H_4 \cdot CH_2$	XX-443	181.24
2272	anthracene (9,10)	....	$C_6H_4:(CH_2)_2:C_6H_4$	V-641	180.25
2273	anthranol (9,10;9)	....	$C_6H_4 \cdot CH_2 \cdot C_6H_4 \cdot CHOH$	VI-697	196.25
2274	carveol	....	$C_{10}H_{18}O$	VI-63	154.25
2275	carvone (*l*)	....	$C_{10}H_{16}O$	VII-83	152.24
2276	cholesterol	coprosterol	$C_{27}H_{47}OH$	....	388.68
2277	ethylanthracene (9,10;9)	....	$C_6H_4:C_2H_3(C_2H_5):$ C_6H_4	V-649	208.31
2278	naphthalene (1,4)	....	$C_{10}H_{10}$	V-519	130.19
2279	phenazine (9,10)	hydrazophenylene	$NH \cdot C_6H_4 \cdot NH \cdot C_6H_4$	XXIII-209	182.23
2280	*o*-phthalic acid (1,4)	....	$C_6H_6(CO_2H)_2$	IX-781	168.15
2281	*o*-phthalic acid (2,4)	....	$C_6H_6(CO_2H)_2$	IX-781	168.15
2282	*o*-phthalic acid (2,6)	....	$C_6H_6(CO_2H)_2$	IX-782	168.15
2283	*p*-phthalic acid (1,4)	dihydro-terephthalic	$C_6H_6(CO_2H)_2$	IX-785	168.15
2284	quinoline	....	C_9H_9N	....	131.18
2285	resorcinol (*m*)	hydroresorcin	$CO \cdot (CH_2)_3 \cdot COCH_2$	VII-554	112.13
2286	toluene (1,2)	Me-cyclohexadiene	$CH_3 \cdot C_6H_7$	V-115	94.16
2287	toluene (1,3)	....	$CH_3 \cdot C_6H_7$	V-115	94.16
2288	toluene (2,4)	....	$CH_3 \cdot C_6H_7$	V-115	94.16
2289	*o*-xylene (*x,x*)	di Me-cyclohexadiene	C_8H_{12}	V-118	108.18
2290	*m*-xylene (1,5)	....	C_8H_{12}	V-119	108.18
2291	*p*-xylene (1,3)	....	C_8H_{12}	V-120	108.18
2292	**Dihydroxy**-acetone	....	$(HO \cdot CH_2)_2CO$	I-846	90.08
2293	anthracene (*β*)(1,5)	rufol	$C_{14}H_8(OH)_2$	VI-1032	210.23
2294	anthracene (*α*)(1,8)	chrysazol	$C_{14}H_8(OH)_2$	VI-1033	210.23

Table 7-4 (*Continued*)
PHYSICAL CONSTANTS OF ORGANIC COMPOUNDS

No.	Crystalline Form and Color	Specific Gravity	Melting Point °C.	Boiling Point °C.	Solubility in 100 Parts		
					Water	Alcohol	Ether
2258	or. nd./al.; red yel./ bz.	....	162-3	d.	i.	s. h.	v. sl. s.; v. sl. s. CS$_2$
2259	lq.	....	....	220-30^{10mm}	s. h.	....	i.
2260	lq.	$1.170\frac{20°}{20}$	....	196.8	∞	....	....
2261	cr.	....	148	d.	v. s.	v. s.	sl. s.
2262	nd.	....	246 d.	....	v. s. h.	i.	i.
2263	cr.	$0.877\frac{25°}{4}$	26-7	240-50^{85mm}	v. sl. s.	v. s.	v. s.; v. s. bz.
2264	lq.	0.806$^{20°}$	....	261.9	i.	s.	s.
2265	cr./al.	....	39-40	278	....	s.	....
2266	lq.	....	....	298	....	....	....
2267	lf./al.	0.825$^{30°}$	33	264	v. s. chl.	v. s.; s. lg.	v. s.
2268	cr.	....	....	....	i. c. 95% al.	s. h. 90% al.	....
2269	lf./bz.	....	110	....	s. h.; v. s. chl.	v. s.; v. s. h. bz.	i. pet.
2270	lf.	....	165-7 d.	....	sl. s. h.	v. sl. s.	v. sl. s.
2271	col. cr./al.	....	169	subl.; d. 300	i.	s. h.	s.
2272	mn. pr.	$0.897\frac{11°}{4}$	108.5	305	i.; v. s. bz.	v. s.	v. s.
2273	nd./pet.	....	76	....	s. h.	s.	s.; s. bz.
2274	lq.	$0.927\frac{20°}{4}$	....	224-5	....	....	....
2275	oil	$0.925\frac{20°}{4}$	....	221-2	....	....	....
2276	nd.	....	104-5	210-201^{mm}	s. chl., pet.	s. abs.	s.; s. bz.
2277	oil	$1.049\frac{18°}{18}$	....	320-3 sl. d.	i.; ∞ bz.	∞	∞
2278	col. pl.	$0.997\frac{12°}{4}$	25-8	211-2	i.	s.	s.
2279	rhb. lf.	....	212	....	i.	v. sl. s.	i. bz.
2280	mn.	....	153	....	1.6$^{6°}$	....	....
2281	mn. pr.	....	179-80	....	sl. s. h.	s.	....
2282	tri.	....	215	....	0.3$^{25°}$; 6 h.	s.	s. act.
2283	nd./aq.	....	subl. d.	....	0.01 c.	....	....
2284	lq.	....	....	220-6	....	....	....
2285	pr./bz.	....	105-6 sl. d.	....	v. s.; s. act.	v. s.; s. h. bz.	v. sl. s. abs
2286	lq.	....	....	108	....	....	....
2287	lq.	0.835	....	110	....	....	....
2288	lq.	$0.827\frac{20°}{4}$	....	106	....	....	....
2289	col. lq.	....	....	134-5	....	....	s.
2290	col. lq.	$0.823\frac{20°}{4}$	....	129-30^{745mm}	....	....	....
2291	col. lq.	$0.830\frac{20°}{4}$	....	135-8	....	....	....
2292	cr.	....	68-75	....	v. s.; i. lg.	sl. s.	sl. s.
2293	yel./aq. al.	....	265±, d.	....	s. alk.	s.	s.; s. bz.
2294	yel./aq. al.	....	225 d.	....	s. alk.	s.	s.; s. bz.

Dihydro-estrone 2900
Dihydro–ketoacridine 145
Dihydro–morphinone hydrochloride, cf. alkd.
Dihydro–myrcene 2514

Dihydro–naringenin 5264
Dihydro–phytol 5295
Dihydro–pyrrole 5524
Dihydroxy-acetophenone 5550

Table 7-4 (*Continued*)
PHYSICAL CONSTANTS OF ORGANIC COMPOUNDS

No.	Name	Synonym	Formula	Beil. Ref.	Formula Weight
2295	**Dihydroxy** 9-anthranol (3,4)	anthrarobin	$C_{14}H_7(OH)_3$	VIII-330	226.23
2296	*x*-anthranol (1,8)	anthralin	$C_{14}H_7(OH)_3$	VIII-332	226.23
2297	anthraquinone (1,2)	alizarin	$C_6H_4(CO)_2C_6H_2:$ $(OH)_2$	VIII-439	240.22
2298	anthraquinone (1,3)	purpuroxanthin	$C_6H_4(CO)_2C_6H_2:$ $(OH)_2$	VIII-448	240.22
2299	anthraquinone (1,4)	quinizarin	$C_6H_4(CO)_2C_6H_2:$ $(OH)_2$	VIII-450	240.22
2300	anthraquinone (1,5)	anthrarufin	$(HO·C_6H_3)_2(CO)_2$	VIII-453	240.22
2301	anthraquinone (1,8)	chrysazin	$(HO·C_6H_3)_2(CO)_2$	VIII-458	240.22
2302	anthraquinone (2,3)	hystazin; hystazarin	$C_6H_4(CO)_2C_6H_2:$ $(OH)_2$	VIII-462	240.22
2303	anthraquinone (2,6)	anthraflavic acid	$(HO·C_6H_3)_2(CO)_2$	VIII-463	240.22
2304	anthraquinone (2,7)	*iso*-anthraflavic acid	$C_{14}H_8O_4·H_2O$	VIII-466	258.23
2305	anthraquinone carboxylic acid †	(2,4;1 or 1,3;2) munjistin	$C_6H_4(CO)_2C_6H:$ $(OH)_2(CO_2H)$	X-1036	284.23
2306	anthraquinone carboxylic acid	(1;8,3) rheic acid; rhein	$HO·C_6H_3(CO)_2C_6H_2:$ $(OH)(CO_2H)$	X-1033	284.23
2307	azobenzene-4'-sulfonic acid (2,4)	....	$(HO)_2C_6H_3·N:N·$ $C_6H_4·SO_3H$	XVI-275	294.29
2308	azobenzene-4'-sulfonate Na (2,4)	tropeolin O; resorcin yellow	$C_{12}H_9O_5N_2SNa·$ $2\frac{1}{2}H_2O$	XVI-275	361.30
2309	behenic acid	....	$C_{22}H_{44}O_4$	III-410	372.59
2310	benzaldehyde (2,4)	β-resorcyl aldehyde	$(HO)_2C_6H_3·CHO$	VIII-241	138.12
2311	benzaldehyde (3,4)	protocatechuic ald.	$(HO)_2C_6H_3·CHO$	VIII-246	138.12
2312	benzene (*o*)	pyrocatechin	$C_6H_4(OH)_2$	VI-759	110.11
2313	benzene (*m*)	resorcinol	$C_6H_4(OH)_2$	VI-796	110.11
2314	benzene (*p*)	hydroquinone	$C_6H_4(OH)_2$	VI-836	110.11
2315	benzoic acid (2,3)	....	$(HO)_2C_6H_3CO_2H·aq.$	X-375	154.12
2316	benzoic acid (2,4)	resorcylic acid (β)	$(HO)_2C_6H_3CO_2H·aq.$	X-377	154.12
2317	benzoic acid (2,5)	gentisic acid	$(HO)_2C_6H_3·CO_2H$	X-384	154.12
2318	benzoic acid (2,6)	resorcylic acid (γ)	$(HO)_2C_6H_3·CO_2H·$ H_2O	X-388	172.14
2319	benzoic acid (3,4)	protocatechuic acid	$(HO)_2C_6H_3·CO_2H·$ H_2O	X-389	172.14
2320	benzoic acid (3,5)	resorcylic acid (α)	$(HO)_2C_6H_3·CO_2H·$ $1\frac{1}{2}H_2O$	X-404	181.15
2321	benzophenone (2,4)	benzoyl resorcin (4)	$(HO)_2C_6H_3·CO·C_6H_5$	VIII-312	214.22
2322	benzophenone (2,5)	benzoyl hydroquinone (2)	$(HO)_2C_6H_3·CO·C_6H_5$	VIII-312	214.22
2323	benzophenone (2,2')	salicyl-phenol (2)	$(HO·C_6H_4)_2CO$	VIII-313	214.22
2324	benzophenone (2,3')	salicyl-phenol (3)	$(HO·C_6H_4)_2CO$	VIII-315	214.22
2325	benzophenone (2,4')	salicyl-phenol (4)	$(HO·C_6H_4)_2CO$	VIII-315	214.22

† See also No. 176.
Dihydroxy-anthraquinone carboxylic acid 176   Dihydroxy-anthraquinone quinoline 174

Table 7-4 (*Continued*)
PHYSICAL CONSTANTS OF ORGANIC COMPOUNDS

No.	Crystalline Form and Color	Specific Gravity	Melting Point °C.	Boiling Point °C.	Solubility in 100 Parts		
					Water	Alcohol	Ether
2295	yel. nd./ aq. al.	....	208	....	sl. s.; s. ac.	s.; s. alk.	s.; s. act.
2296	yel. pd.	....	178–80	....	i.; s. fat	s. h.; s. bz.	s. dil. alk.
2297	red rhb.	....	289–90	430	$0.03^{100°}$	v. s.	v. s.; s. alk.
2298	yel. lf./bz.	....	263–4	....	i.; s. act.	sl. s.	s. h. ac.
2299	red nd./al.	....	200–2	subl. sl. d.	s. alk.	s. H_2SO_4	s.
2300	yel. lf.	....	280	subl.	i.	sl. s.	s.
2301	red-yel./ al.	....	191	....	s. H_2SO_4	s.; s. alk.	s.; s. ac.
2302	yel. brn. nd.	....	>280	....	s. H_2SO_4	v. sl. s. h.	v. sl. s.
2303	yel. nd./ al.	....	>330	....	s. H_2SO_4	$1.4^{17°}$	i.
2304	yel. nd./ aq. al.	....	>330; subl.	d.–H_2O, 100	s. H_2SO_4	sl. s.	v. sl. s.
2305	yel. lf./ac.	....	230–1	subl. d. > m. p.	s. h.; s. alk.	s. h.; s. H_2SO_4	s.; s. chl.
2306	yel. nd./ Me al.	....	321–2	subl.	i.; s. alk.; s. pyr.	sl. s.; sl. s. bz.	sl. s.; sl. s. chl.
2307	red lf.	....	....	....	$0.2^{19°}$; i. aq. HCl	v. sl. s.	i.
2308	brn. lf.	....	....	....	$0.4^{23°}$; $20^{100°}$	....	....
2309	cr./al.	....	132–3 (99)	....	....	$0.1^{18°}$ abs.	i.
2310	yel. nd./aq.	....	135–6	$220–8^{22mm}$	v. s.	v. s.	v. s.
2311	cr./aq.	....	153–4	....	5 c.; 33 h.	100 h.	v. s.
2312	nd./aq.	$1.344^{0°}$	105	245.6	$45.1^{20°}$	v. s.; s. chl.	v. s.; s. bz.
2313	col. rhb.	$1.272^{15°}$	110(108)	275.9	$147.3^{12.5°}$	v. s.	v. s.
2314	cr.	$1.332^{15°}$	172.3	285^{730mm}	$6^{15°}$	v. s.	v. s.
2315	cr./aq.	....	204(anh.)	–aq. 100	....	....	....
2316	cr.	....	204–6 (anh.) d.	....	$0.26^{17°}$; s. h.	s.	s.
2317	nd./aq.	....	199–200	....	s.; i. CS_2	s.; i. chl.	s.; i. bz.
2318	nd./aq.	....	148–67 (anh.) d.	....	v. s. h.	....	....
2319	nd./aq.	1.542	199 d.	....	$2^{14°}$	v. s.	s.
2320	pr. or nd.	....	237 (anh.)	....	v. s. h.	v. s.	v. s.
2321	nd./h. aq.	....	143–4	....	i. c.	s.	s.; sl. s. bz.
2322	yel. nd./ aq. al.	....	122–4	....	s. bz.	s.	s.
2323	yel. pr./lg.	....	59–60	330–40 sl. d.	i.; v. s. chl.	v. s.	v. s.
2324	yel. cr./et.	....	126	....	....	....	....
2325	lt. yel./aq.	....	147–8	....	sl. s. h.	s. h.; s. bz.	$44^{25°}$

Dihydroxy-azobenzene 609–11, 662   Dihydroxy-benzaldehyde dimethyl ether 2405

Table 7-4 (Continued)
PHYSICAL CONSTANTS OF ORGANIC COMPOUNDS

No.	Name	Synonym	Formula	Beil. Ref.	Formula Weight
2326	**Dihydroxy** benzophenone (3,3')	....	$(HO \cdot C_6 H_4)_2 CO$	VIII-316	214.22
2327	benzophenone (3,4')	....	$(HO \cdot C_6 H_4)_2 CO$	VIII-316	214.22
2328	benzophenone (4,4')	....	$(HO \cdot C_6 H_4)_2 CO$	VIII-316	214.22
2329	butane (1,2)	α-butylene glycol	$C_2 H_5 \cdot CHOH \cdot CH_2 OH$	I-477	90.12
2330	butane (1,3)	β-butylene glycol	$CH_3 \cdot CHOH \cdot CH_2 \cdot CH_2 OH$	I-477	90.12
2331	butane (1,4)	butandiol-1,4	$HO(CH_2)_4 OH$	I-478	90.12
2332	butane (2,3)	butylene glycol (pseudo) (meso)	$CH_3 \cdot (CHOH)_2 \cdot CH_3$	I-479	90.12
2333	cinnamic acid (2,4)	umbellic acid	$(HO)_2 C_6 H_3 \cdot C_2 H_2 \cdot CO_2 H$	X-434	180.16
2334	cinnamic acid (2,5)	....	$(HO)_2 C_6 H_3 \cdot C_2 H_2 \cdot CO_2 H$	X-435	180.16
2335	cinnamic acid (3,4)	caffeic acid	$C_9 H_8 O_4$	X-436	180.16
2336	dinaphthyl (α)	α-dinaphthol	$(HO \cdot C_{10} H_6 \cdot)_2$	VI-1053	286.33
2337	dinaphthyl (2,2', 1,1')	β-dinaphthol	$(HO \cdot C_{10} H_6 \cdot)_2$	VI-1051	286.33
2338	diphenyl (2,5)	Ph-hydroquinone	$(HO)_2 C_6 H_3 \cdot C_6 H_5$	VI-989	186.21
2339	diphenyl (3,4)	Ph-catechol (4)	$(HO)_2 C_6 H_3 \cdot C_6 H_5$	VI-990	186.21
2340	diphenyl (2,2')	o,o'-diphenol	$(HO \cdot C_6 H_4 \cdot)_2$	VI-989	186.21
2341	diphenyl (2,4')	δ-diphenol	$(HO \cdot C_6 H_4 \cdot)_2$	VI-990	186.21
2342	diphenyl (3,3')	....	$(HO \cdot C_6 H_4 \cdot)_2$	VI-991	186.21
2343	diphenyl (4,4')	γ-diphenol	$(HO \cdot C_6 H_4 \cdot)_2$	VI-991	186.21
2344	diphenylmethane (4,4')	(4,4')	$(HO \cdot C_6 H_{\cdot 4})_2 CH_2$	VI-995	200.24
2345	ethylamino-1-hydroxy-benzene	(4)	$HO \cdot C_6 H_4 \cdot N(CH_2 \cdot CH_2 OH)_2$	....	197.24
2346	hexane (2,3)	2,3-hexandiol	$CH_3 (CH_2)_2 (CHOH)_2 \cdot CH_3$	I-484	118.18
2347	4-methoxy-benzophenone (2,6)	cotoin	$C_6 H_5 \cdot CO \cdot C_6 H_2 : (OH)_2 (OCH_3)$	VIII-419	244.25
2348	naphthalene (1,2)	β-naphthohydroquinone	$C_{10} H_6 (OH)_2$	VI-975	160.17
2349	naphthalene (1,3)	naphthoresorcinol	$C_{10} H_6 (OH)_2$	VI-978	160.17
2350	naphthalene (1,4)	α-naphthohydroquinone	$C_{10} H_6 (OH)_2$	VI-979	160.17
2351	naphthalene (1,5)	....	$C_{10} H_6 (OH)_2$	VI-980	160.17
2352	naphthalene (1,6)	....	$C_{10} H_6 (OH)_2$	VI-981	160.17
2353	naphthalene (1,7)	....	$C_{10} H_6 (OH)_2$	VI-981	160.17
2354	naphthalene (1,8)	....	$C_{10} H_6 (OH)_2$	VI-981	160.17
2355	naphthalene (2,3)	....	$C_{10} H_6 (OH)_2$	VI-982	160.17
2356	naphthalene (2,6)	....	$C_{10} H_6 (OH)_2$	VI-984	160.17
2357	naphthalene (2,7)	....	$C_{10} H_6 (OH)_2$	VI-985	160.17
2358	naphthalene diacetate (1,5)	....	$(CH_3 CO_2)_2 C_{10} H_6$	VI-981	244.25
2359	naphthalene-3,6-disulfonic acid	(1,8); chromotropic acid	$(HO)_2 C_{10} H_4 : (SO_3 H)_2 \cdot 2H_2 O$	XI-307	356.33
2360	phenanthrene (3,4)	morphol	$C_{14} H_8 (OH)_2$	VI-1034	210.23
2361	phenyl-α-alanine (3,4)(l-)	hydroxytyrosine; dopa	$(HO)_2 C_6 H_3 \cdot CH_2 \cdot CHNH_2 \cdot CO_2 H$	*XIV-681	197.19

Dihydroxy-chlorobenzene 1460
Dihydroxy-coumarin 1628, 2898
Dihydroxy-diethylsulfide 5900
Dihydroxy-ethylamine 2090

Dihydroxy-ethyl ether 3453
Dihydroxy-flavone 1494
Dihydroxy-fluoran 3253
Dihydroxy-fluorane 3720

Table 7-4 (*Continued*)
PHYSICAL CONSTANTS OF ORGANIC COMPOUNDS

No.	Crystalline Form and Color	Specific Gravity	Melting Point °C.	Boiling Point °C.	Solubility in 100 Parts		
					Water	Alcohol	Ether
2326	nd./aq.	....	163-4	....	sl. s.	s. alk.	....
2327	nd./aq.	....	197-200	....	....	....	....
2328	cr./aq.	....	207-10	....	s. h.	s.; i. CS_2	s.; i. chl.
2329	lq.	$1.006\frac{17.5°}{0}$	....	192-4	v. s.	v. s.	....
2330	oil	$1.006\frac{20°}{20}$	....	206.5	∞	v. s.	i.
2331	oil	$1.020^{20°}$	19	230^{759mm}	∞	....	v. sl. s.
2332	lq.	$1.048^{0°}$	34	181.7^{742mm}	∞	∞	∞
2333	yel. pd.	....	d. 240-60	....	s. h.	s.; i. bz.	i.; i. lg.
2334	cr./aq.	....	207 d.	....	....	....	....
2335	yel. pr./aq.	....	195-213 d.	....	s. h.	s.	sl. s.
2336	pl./al.	....	300	....	i.; s. alk.	s.; sl. s. bz.	v. s.; sl. s. chl.
2337	nd./al.	....	218	subl.	i.; s. alk.	s.	v. s.
2338	nd./aq. al.	....	102-3	....	....	....	....
2339	cr./pet.	....	136-7	>360	s. h.	v. s.; s. CS_2	v. s.; s. chl
2340	pr./toluene	....	103-9	$325-6^{755mm}$	s.	v. s.	v. s.
2341	mn. pr.	....	160-1	324	sl. s. h.	v. s.	v. s.
2342	nd./aq.	....	123	247^{18mm}	s. h.	v. s.	v. s.
2343	rhb./al.	1.25	278-81	....	sl. s.	v. s.	v. s.
2344	lf./aq.	....	158 (148)	....	i. CS_2	s.	v. s.
2345	col. fl.	....	140-1	....	i.	sl. s.	sl. s.
2346	cr.	$0.967^{0°}$	60	206-7	∞	s.	....
2347	yel. pr./al.	....	130-1	....	sl. s. h.; s. alk.	s.; s. bz.; s. CS_2	s.; s. chl.; s. act.
2348	lf.	....	60±	....	s. alk.	....	....
2349	lf.	....	124-5	....	v. s.	v. s.	v. s.
2350	long nd.	....	183-8	....	s. h.	v. s.	v. s.
2351	pr./aq.	....	258-60 d.	....	sl. s.	s.	v. s.
2352	pr./bz.	....	137-8	....	....	sl. s. c.	v. s.
2353	nd./aq.	....	178	....	v. s. h.	v. s.	v. s.
2354	nd. or lf.	....	140	....	sl. s. h.	v. s. bz.	v. s.
2355	mn.	....	159-60	....	sl. s. h.	v. s.	v. s.
2356	rhb./aq.	....	216-8	....	v. s. h.	v. s.	v. s.
2357	nd./aq.	....	190	subl. sl. d.	s. h.	v. s.	v. s.
2358	cr./bz.	....	159-60	....	....	....	....
2359	nd. or lf.	....	....	....	v. s.; i. aq. NaCl	i.	i.
2360	nd.	....	143	....	s.	s. alk.	....
2361	pr. or nd.) aq.; lf./ aq. al.	....	280 d.	....	$0.5^{20°}$; $2.5^{100°}$	i.; i. chl.; v. sl. s. bz.	i.; i. pet.; v. sl. s. CS_2

Dihydroxy-fuchsone 593
Dihydroxy-malonic acid 4055
Dihydroxy-methylanthranol 1492
Dihydroxy-methylanthraquinone 1495

Dihydroxy-methylfuchsone 5573
Dihydroxy-naphthoquinone 4477-8
Dihydroxy-phenylacetic acid 3673

Table 7-4 (*Continued*)
PHYSICAL CONSTANTS OF ORGANIC COMPOUNDS

No.	Name	Synonym	Formula	Beil. Ref.	Formula Weight
	Dihydroxy				
2362	pyridine (2,4)	....	$(HO)_2C_5H_3N$	XXI-160	111.10
2363	pyridine (2,6)	....	$(HO)_2C_5H_3N·\frac{1}{2}H_2O$	XXI-161	120.11
2364	quinone (2,5)	....	$(HO)_2C_6H_2O_2$	VIII-377	140.10
2365	stearic acid (4,9) (*dl*)	....	$C_{18}H_{36}O_4$	III-407	316.49
2366	stearic acid (*dl*)	(9,10)	$C_{18}H_{36}O_4$	III-408	316.49
2367	tartaric acid	....	$[(HO)_2C·CO_2H]_2$	III-830	182.09
2368	tartaric Na salt	....	$[(HO)_2C·CO_2Na]_2·3H_2O$	III-832	280.10
2369	thiazole	mustard oil acetic acid	$HO·C:CH·S·C(OH):N$	XXVII-233	117.13
2370	toluene (2,3)	....	$CH_3·C_6H_3(OH)_2$	VI-872	124.14
2371	toluene (2,4)	4-Me-resorcinol	$CH_3·C_6H_3(OH)_2$	VI-872	124.14
2372	toluene (2,5)	toluhydroquinone	$CH_3·C_6H_3(OH)_2$	VI-874	124.14
2373	toluene (2,6)	2-Me-resorcinol	$CH_3·C_6H_3(OH)_2$	VI-878	124.14
2374	toluene (3,4)	homocatechol	$CH_3·C_6H_3(OH)_2$	VI-878	124.14
2375	*o*-xylene (3,5)	4,5-diMe-resorcinol	$(CH_3)_2C_6H_2(OH)_2$	VI-908	138.17
2376	*o*-xylene (3,6)	2,3-diMe-hydroquinone	$(CH_3)_2C_6H_2(OH)_2$	VI-908	138.17
2377	*m*-xylene (2,4)	2,4-diMe-resorcinol	$(CH_3)_2C_6H_2(OH)_2$	VI-911	138.17
2378	*m*-xylene (2,5)	2,6-diMe-hydroquinone	$(CH_3)_2C_6H_2(OH)_2$	VI-911	138.17
2379	*p*-xylene (2,5)	hydrophloron	$(CH_3)_2C_6H_2(OH)_2$	VI-915	138.17
2380	*p*-xylene (2,6)	β-orcin	$(CH_3)_2C_6H_2(OH)_2$	VI-918	138.17
2381	**Diiodo-**acetic acid	....	$I_2CH·CO_2H$	II-224	311.85
2382	acetylene	diiodo-ethyne	$IC:CI$	I-246	277.83
2383	anthranilic acid	(3,5;2,1)	$I_2C_6H_2(NH_2)CO_2H$	*XIV-554	388.93
2384	atophan	*p*-iodophenyl-6-iodoquinoline-4-carboxylic acid	$C_{16}H_9O_2NI_2$	....	501.06
2385	aniline (2,4)	....	$I_2C_6H_3·NH_2$	XII-675	344.92
2386	benzene (*o*)	*o*-phenylene iodide	$C_6H_4I_2$	V-225	329.91
2387	benzene (*m*)	....	$C_6H_4I_2$	V-225	329.91
2388	benzene (*p*)	....	$C_6H_4I_2$	V-227	329.91
2389	butane (1,4)	tetramethylene di-iodide	$I(CH_2)_4I$	I-123	309.92
2390	ethane (1,1)	ethylidene diiodide	$CH_3·CHI_2$	I-99	281.86
2391	ethane (1,2)	ethylene iodide	$ICH_2·CH_2I$	I-99	281.86
2392	fluorescein	....	$C_{20}H_{10}O_5I_2$	....	584.11
2393	hexane (1,6)	hexamethylene di-iodide	$I(CH_2)_6I$	I-147	337.97
2394	2-hydroxybenzoic acid (3,5)	3,5-diI-salicylic acid	$I_2C_6H_2(OH)CO_2H$	X-113	389.92
2395	4-hydroxybenzoic	(3,5;4,1)	$I_2C_6H_2(OH)CO_2H$	X-180	389.92
2396	pentane (1,5)	pentamethylene di-iodide	$I(CH_2)_5I$	I-133	323.94
2397	propane (1,2)	propylene iodide	$CH_3·CHI·CH_2I$	I-115	295.89
2398	propane (2,2)	iodoacetol	$(CH_3)_2CI_2$	I-115	295.89
2399	propane (1,3)	trimethylene diiodide	$I(CH_2)_3I$	I-115	295.89
2400	thiophene (2,5)	....	$I_2C_4H_2S$	XVII-35	335.93
2401	tyrosine (3,5)(*l*)	iodogorgonic acid	$C_9H_9O_3NI_2$	XIV-619	432.99

Dihydroxy-propane 5465, 6286
Dihydroxy-propionic acid 3379
Dihydroxy- purine, cf. alkd.

Dihydroxy-xanthone 3227
Diiodo-ethene 137-8
Diiodo-ethyne 2382

Table 7-4 (*Continued*)
PHYSICAL CONSTANTS OF ORGANIC COMPOUNDS

No.	Crystalline Form and Color	Specific Gravity	Melting Point °C.	Boiling Point °C.	Solubility in 100 Parts		
					Water	Alcohol	Ether
2362	yel. nd./aq.	....	255–65 d.	....	s. h.	s.	v. sl. s.
2363	nd./aq.	....	195 d.	....	sl. s.	s. h.	i.
2364	yel. nd.	....	subl. 215–20 sl. d.	....	i. c.	s.	i. c.
2365	lf./al.	....	136.5	....	v. s. h. al.	$0.6^{19°}$	$0.2^{18°}$
2366	lf.	....	99–100	....	v. sl. s. c.	$3.6^{18°}$	sl. s.
2367	wh. pd.	....	114–5 d.	....	v. s.; d. h.	....	....
2368	cr.	....	d., $-H_2O$, $-CO^2$	....	$0.04^{0°}$	....	....
2369	rhb.	....	126–8	179^{19mm}; subl.	v. s. h.	s.	s.
2370	lf./bz.	....	68	238–40 sl. d.	s.; v. s. chl.	v. s.	s.; v. s. bz.
2371	cr./bz. pet.	....	104–5	267–70	s.; i. pet.	s.	s.; sl. s. bz.
2372	lf./bz.	....	127–9	....	s.	s.	s.; sl. s. bz.
2373	cr.	....	116	264	v. s.	v. s.	v. s.
2374	pr./bz.	$1.129^{7.4°}_{4}$	65	252	v. s.	v. s.	v. s.
2375	nd./bz.*	....	136–7	subl.	s.; sl. s. lg.	s.	s.
2376	cr./aq.	....	221 sl. d.	....	....	....	....
2377	nd.	....	149–50	....	s.	v. s.	v. s.
2378	nd./ xylene	....	150–1	....	....	....	....
2379	lf./aq.	....	213	subl.	s. h.	v. s.	v. s.; i. bz.
2380	cr./aq. al.	....	163	277–80	s. h.	s.	s.
2381	lt. yel. nd.	....	110 (95–6)	....	s.	s.	s.
2382	col. rhb.	....	78–82	d. 80–100	sl. s. c. lg.	....	....
2383	pr./al.	....	234–5	....	i. h.	sl. s. h. bz.	v. s.
2384	yel. pd.	....	291–2	....	i.	s. h.	v. sl. s.
2385	rhb.	2.75	95–6	....	s. h.	v. s. h.	v. s.
2386	cr./lg.	$2.54^{20°}$	27	286.5	i.	sl. s. c.	....
2387	cr./al. et.	$2.47^{25°}$	40.4	284.7^{757mm}	i.	sl. s.	....
2388	lf./al.	....	129.4	285	i.	sl. s.	....
2389	lq.	$2.307^{18°}$	5.8	$120–5^{12mm}$	....	....	....
2390	lq.	$2.84^{0°}$	....	177–9	i.	v. s.	v. s.
2391	yel. mn.	$2.132^{10°}$	81–2	d.	sl. s.	s.	s.
2392	or.-red pd.	....	....	....	sl. s.	s.	s. alk.
2393	col. nd.	$2.05^{18°}$	9.5	$163^{17.5mm}$	....	....	....
2394	yel.-wh. nd./al.	....	230 $\perp$, d.	....	0.07 c.; 0.15 h.	s.	s.
2395	nd./aq. al.	....	278–9 d.	....	i. h.	v. s.	v. s.
2396	oil	$2.194^{18°}$	9	149^{20mm}	....	....	....
2397	lq.	$2.490^{18.5°}$	....	d.	....	....	....
2398	lq.	$2.15^{0°}$	....	147–8 d.	....	....	....
2399	lq.	$2.576^{15°}$	–13	224	v. sl. s.	s.	s.
2400	col. lf./al.	....	40–1	....	i.	s.	s.; s. chl.
2401	nd./aq. al.	....	213 d.	....	$0.288^{15°}$	d. h. aq.	s. dil. alk.

* Cryst./aq. $+ 1H_2O$, m. p. 115–7°.
Diiodoform 5812
Diiodo-methane 4434

Diiodo-phenol sulfonic acid 5626
Diiodo-propanol 3401
Diiodo-salicylic acid 2394

Table 7-4 (*Continued*)
PHYSICAL CONSTANTS OF ORGANIC COMPOUNDS

No.	Name	Synonym	Formula	Beil. Ref.	Formula Weight
	Diiodo				
2402	tyrosine (3,5)(*dl*)	....	$C_9H_9O_3NI_2$	XIV-622	432.99
2403	**Dimethallyl ether** (*β*)	....	$[CH_2:C(CH_3)\cdot CH_2]_2O$	....	126.20
2404	**Dimethoxy**-aniline	(2,5;1)	$(CH_3O)_2C_6H_3\cdot NH_2$	XIII-738	153.18
2405	benzaldehyde (2,4)	2-MeO-anisalde-hyde	$(CH_3O)_2C_6H_3\cdot CHO$	VIII-242	166.18
2406	benzene (*o*)	veratrole	$(CH_3O)_2C_6H_4$	VI-771	138.17
2407	benzene (*m*)	resorcinol di Me ether	$(CH_3O)_2C_6H_4$	VI-813	138.17
2408	benzene (*p*)	hydroquinone di Me ether	$(CH_3O)_2C_6H_4$	VI-843	138.17
2409	benzophenone (4,4′)	*p*-anisoyl-anisole	$(CH_3O\cdot C_6H_4)_2CO$	VIII-317	242.28
2410	ethyl adipate	....	$[(CH_2)_2\cdot CO_2(CH_2)_2\cdot OCH_3]_2$	....	262.31
2411	ethyl carbonate	....	$[CH_3O(CH_2)_2O]_2CO$	....	178.19
2412	ethyl phthalate	....	$[CH_3O(CH_2)_2O_2C]_2\cdot C_6H_4$	**IX-597	282.30
2413	2-methyl-3,4-dihy-droisoquinoline chloride (6,7)	Iodal	$C_{12}H_{16}O_2N\cdot Cl\cdot 3\frac{1}{2}H_2O$	XXI-170	304.77
2414	quinazoline (2,4)	....	$(CH_3O)_2C_8H_4N_2$	XXIII-486	190.20
2415	**Dimethyl**-acetal	dimethyl aldehyde	$CH_3\cdot CH(OCH_3)_2$	I-603	90.12
2416	acetamide (*N,N*)	acet-dimethylamide	$CH_3CO\cdot N(CH_3)_2$	IV-59	87.12
2417	acetanilide (2,3)	aceto-*o*-xylidide (*v*)	$(CH_3)_2C_6H_3NH\cdot COCH_3$	XII-1101	163.22
2418	acetanilide (2,4)	aceto-*m*-xylidide (*as*)	$(CH_3)_2C_6H_3NH\cdot COCH_3$	XII-1118	163.22
2419	acetanilide (2,5)	aceto-*p*-xylidide	$C_8H_9NH\cdot COCH_3$	XII-1137	163.22
2420	acetanilide (2,6)	aceto-*m*-xylidide (*v*)	$C_8H_9NH\cdot COCH_3$	XII-1109	163.22
2421	acetanilide (3,4)	aceto-*o*-xylidide (*as*)	$C_8H_9NH\cdot COCH_3$	XII-1104	163.22
2422	acetanilide (3,5)	aceto-*m*-xylidide (*s*)	$C_8H_9NH\cdot COCH_3$	XII-1131	163.22
2423	acetoacetic ester	....	$CH_3CO\cdot C(CH_3)_2CO_2\cdot C_2H_5$	III-695	158.20
2424	acrylic acid (*β,β*)	....	$(CH_3)_2C:CH\cdot CO_2H$	II-432	100.12
2425	acrylic acid (*α,β*) (*cis*)	angelic acid	$CH_3\cdot CH:C(CH_3)\cdot CO_2H$	II-428	100.12
2426	acrylic acid (*α,β*) (*trans*)	tiglic acid	$CH_3\cdot CH:C(CH_3)\cdot CO_2H$	II-430	100.12
2427	acrylic aldehyde (*trans*)	tiglic aldehyde	$CH_3\cdot CH:C(CH_3)\cdot CHO$	I-733	84.12
2428	adipate	methyl adipate	$[(CH_2)_2CO_2CH_3]_2$	II-652	174.20
2429	amine	....	$(CH_3)_2NH$	IV-39	45.08
2430	amine hydro-chloride	....	$(CH_3)_2NH\cdot HCl$	IV-41	81.55
2431	amino-azoben-zene (*p*)	benzene-azo-di Me-aniline	$(CH_3)_2N\cdot C_6H_4\cdot N: N\cdot C_6H_5$	XVI-312	225.30
2432	amino-azoben-zene (*p*) HCl	....	$C_{14}H_{15}N_3\cdot HCl$	XVI-312	261.76

Table 7-4 (*Continued*)
PHYSICAL CONSTANTS OF ORGANIC COMPOUNDS

No.	Crystalline Form and Color	Specific Gravity	Melting Point °C.	Boiling Point °C.	Solubility in 100 Parts		
					Water	Alcohol	Ether
2402	pl./aq.	....	>200 d.	....	$0.56^{75°}$	....	....
2403	lq.	$0.816^{20°}_{4}$	....	134	<0.1	....	....
2404	sc./pet.	....	80-1	270 sl. d.	s.	v. s. h.	v. s. h. pet.
2405	nd./aq. al.	....	69-70	165^{10mm}	i.; s. bz.	s.; s. lg.	s.
2406	cr.	$1.091^{15°}_{15}$	22.5(20.9)	206.3	v. sl. s.	s.	s.
2407	lq.	$1.062^{15°}_{15}$	−52	216.5-7.5	v. sl. s.	v. s.	v. s.
2408	lf.	$1.053^{55°}_{55}$	56	212.6	v. sl. s.	v. s.	v. s.
2409	nd./al.	....	144-5	....	s. bz.	s. h.	s. chl.
2410	col. lq.	....	....	$194-6^{20mm}$	....	....	....
2411	col. lq.	....	....	$115-7^{13mm}$	....	....	....
2412	col. lq.	$1.171^{15°}$	....	$178-80^{1mm}$	....	....	....
2413	yel. nd./ aq. act.	....	61-2; 186 d. (anh.)	....	v. s.	v. s.	sl. s. chl.
2414	col. nd.	....	75	....	i.	s.	s.
2415	lq.	$0.850^{20°}_{4}$	−113.2	64^{748mm}	∞	∞	∞
2416	col. lq.	$0.943^{20°}_{4}$	−20	165.5^{754mm}	....	....	....
2417	nd./al.	....	132-3	....	s. h.	s.	s.
2418	nd./aq. al.	....	128-9	....	v. sl. s. o.	s.	....
2419	nd./aq.	....	141-2	....	s. h.	s. h. toluene	....
2420	nd.	....	174-6	....	....	....	....
2421	pr./aq. al.	....	98-9	....	i.	v. s.	....
2422	nd./al.	....	144.5	....	i.	s.	....
2423	oil	$0.977^{20°}_{20}$	....	184-5	v. sl. s.	s.	s.
2424	mn./aq.	$1.006^{24°}$	69-70	194-5	s.	s.	s.
2425	mn. nd.	$0.983^{47°}$	45	185	s. h.	s.	s.
2426	tri. pl.	$0.964^{76°}_{4}$	64.5	198.5	s. h.	s.	s.
2427	lq.	$0.870^{18°}_{4}$	....	115.8^{739mm}	2.5	∞	∞
2428	col. lq.	$1.063^{20°}_{4}$	10-1	115^{13mm}	i.	....	....
2429	col. lq.	$0.680^{9°}_{4}$	−92.2	6.9	v. s.	s.	s.
2430	nd./al.	....	170-1	....	$369^{25°}$	v. s.	i.; $16.9^{25°}$ chl.
2431	yel. lf./al.	....	116-7	d.	i.; s. oil	s.; s. bz.	s.; s. chl.
2432	red-vl. cr.	....	168-74	....	....	....	....

Table 7-4 (Continued)
PHYSICAL CONSTANTS OF ORGANIC COMPOUNDS

No.	Name	Synonym	Formula	Beil. Ref.	Formula Weight
	Dimethyl				
2433	amino-azobenzene-arsinic acid (p)	....	$(CH_3)_2N \cdot C_6H_4 \cdot N:N \cdot C_6H_4 \cdot As(OH)_2$	....	333.22
2434	amino-azo-ben-zene-arsonic acid (p)	....	$(CH_3)_2N \cdot C_6H_4 \cdot N:N \cdot C_6H_4 \cdot AsO(OH)_2$	XVI-885	349.22
2435	amino-benzalde-hyde	(1,2)	$(CH_3)_2N \cdot C_6H_4 \cdot CHO$	XIV-25	149.19
2436	amino-benzalde-hyde (1,4)	Ehrlich's reagent	$(CH_3)_2N \cdot C_6H_4 \cdot CHO$	XIV-31	149.19
2437	amino-benzal-rhodanine-5 (p)	....	$(CH_3)_2N \cdot C_6H_4 \cdot CH: C_3HONS_2$	XXVII-433	264.37
2438	amino-benzene-1-azo-1-naphtha-lene (4)	butter yellow	$(CH_3)_2N \cdot C_6H_4 \cdot N:N \cdot C_{10}H_7$	XVI-321	275.36
2439	amino-benzene-1-azo-2-naphtha-lene (4)	....	$(CH_3)_2N \cdot C_6H_4 \cdot N:N \cdot C_{10}H_7$	XVI-321	275.36
2440	amino-benzoic acid (o)	N,N-diMe-an-thranilic acid	$(CH_3)_2N \cdot C_6H_4 \cdot CO_2H$	XIV-325	165.19
2441	amino-benzoic acid	(p)	$(CH_3)_2N \cdot C_6H_4 \cdot CO_2H$	XIV-426	165.19
2442	amino-benzo-phenone (p)	....	$(CH_3)_2N \cdot C_6H_4 \cdot CO \cdot C_6H_5$	XIV-82	225.29
2443	amino-ethanol	....	$(CH_3)_2N \cdot CH_2 \cdot CH_2OH$	IV-276	89.14
2444	amino-glycerol (α)	γ-dimethylamino-propylene glycol	$(CH_3)_2N \cdot CH_2 \cdot CHOH \cdot CH_2OH$	IV-302	119.16
2445	amino-phenol (m)	....	$(CH_3)_2N \cdot C_6H_4 \cdot OH$	XIII-405	137.18
2446	amino-phenol (p)	....	$(CH_3)_2N \cdot C_6H_4 \cdot OH$	XIII-442	137.18
2447	amino-phenol H_2SO_4	(p)	$C_8H_{11}ON \cdot \frac{1}{2}H_2SO_4$	XIII-442	186.22
2448	amino-phenyl acetate (p)	....	$CH_3CO_2 \cdot C_6H_4 \cdot N(CH_3)_2$	XIII-443	179.22
2449	ammonium-di-methyl-dithiocar-bamate	....	$(CH_3)_2N \cdot CS \cdot S \cdot NH_2(CH_3)_2$	**IV-577	166.31
2450	aniline †	....	$(CH_3)_2N \cdot C_6H_5$	XII-141	121.18
2451	aniline sulfonic acid (m)	....	$(CH_3)_2N \cdot C_6H_4 \cdot SO_3H$	XIV-690	201.25
2452	aniline sulfonic acid (p)	....	$(CH_3)_2N \cdot C_6H_4 \cdot SO_3H \cdot H_2O$	XIV-699	219.26
2453	anthracene (1,3)	....	$(CH_3)_2C_{14}H_8$	V-678	206.29
2454	anthracene (2,3)	....	$(CH_3)_2C_{14}H_8$	V-678	206.29
2455	anthracene (2,6)	....	$(CH_3)_2C_{14}H_8$	V-678	206.29
2456	arsine	cacodyl hydride	$(CH_3)_2AsH$	IV-599	106.00
2457	benzamide (N)	....	$C_6H_5CO \cdot N(CH_3)_2$	IX-201	149.19
2458	benzoic acid (2,3)	hemellitic acid	$(CH_3)_2C_6H_3 \cdot CO_2H$	IX-531	150.18
2459	benzoic acid (2,4)	xylic acid	$(CH_3)_2C_6H_3 \cdot CO_2H$	IX-531	150.18
2460	benzoic acid (2,5)	....	$(CH_3)_2C_6H_3 \cdot CO_2H$	IX-534	150.18
2461	benzoic acid (2,6)	....	$(CH_3)_2C_6H_3 \cdot CO_2H$	IX-531	150.18
2462	benzoic acid (3,4)	....	$(CH_3)_2C_6H_3 \cdot CO_2H$	IX-535	150.18

† See also xylidine.
Dimethyl-amino-azobenzene carboxylic acid 4376
Dimethyl-amino-benzophenonimide 592

Dimethyl-amino-propylene glycol 2444
Dimethyl-n-amyl carbinol 4967
Dimethyl-aniline 6471-5, 6477

Table 7-4 (*Continued*)
PHYSICAL CONSTANTS OF ORGANIC COMPOUNDS

No.	Crystalline Form and Color	Specific Gravity	Melting Point °C.	Boiling Point °C.	Solubility in 100 Parts		
					Water	Alcohol	Ether
2433	vl. cr.	....	....	....	i.	sl. s.	i.
2434	red pd.	....	....	....	s. a.	s. alk.	....
2435	yel. oil	....	....	244	s. dil. ac.	s.	s.
2436	lf./aq.	....	74-5	176-7^{17mm}	i.; s. ac.	s.	s.
2437	red nd./al.	....	240-70 d.	......,....	i.; s. conc. a.	sl. s. h.; v. sl. s. bz.	v. sl. s.; v. sl. s. chl.
2438	red-bl. pr./act.	....	132-4	....	s. conc. H_2SO_4	sl. s.; s. chl.	sl. s.; s. act.
2439	yel. brn. cr./bz.-lg.	....	175	....	....	....	....
2440	nd./et.	....	70	....	s.	s.	sl. s. h.
2441	nd./al.	....	238-9	....	s.	s.	sl. s.
2442	lf./al.	....	90-2	....	i.	v. s. h.	v. s.
2443	col. lq.	$0.887\frac{20°}{4}$	-59	135^{758mm}	∞	....	....
2444	syrup	....	....	220^{749mm}	s.	s.	s.; s. chl.
2445	nd./lg.	....	85	265-8	sl. s. h.	s.	s.
2446	cr./et.-lg.	....	78	....	v. sl. s. lg.	s.	s.
2447	cr.	....	209-10	....	v. s.	sl. s.	....
2448	nd. or pl./aq. al.	....	78-9	....	v. sl. s.; v. sl. s. ac.	s.; v. s. bz.	s.; s. lg.; v. s. chl.
2449	pa. yel. pl.	....	131-3	....	v. s.	v. s.	v. sl. s.
2450	yel. lq.	$0.956\frac{20°}{4}$	2.5	194.2	i.	s.; s. chl.	s.
2451	....	....	d. 266	....	s.	....	....
2452	pr. or lf.	....	270 (anh.) -H$_2$O, 135°	....	s. h.; s. ac.	v. sl. s.	v. sl. s.
2453	lf./et.	....	83	....	....	....	....
2454	lf.	....	252	....	v. s. bz.	....	....
2455	red-yel.; gn.-yel. or silvery	....	243-4	subl.	....	sl. s.	s. bz.
2456	col. lq.	1.213$^{29°}$	....	35.6^{747mm}	....	∞	∞
2457	col. cr.	....	41-2	272-3	v. s.	....	....
2458	pr./al.	....	144	....	v. sl. s. h.	s.	....
2459	mn. or tri.	....	126-7	267^{727mm}	sl. s. h.	s. h.	s. bz., chl.
2460	nd al.	$1.069\frac{20°}{4}$	132	268	v. sl. s. h.	v. s. c.	....
2461	col. nd./lg.	....	116	....	....	sl. s. c. lg.	v. s.
2462	pr./al.	....	165-6	subl.	v. sl. s. h.	s.	....

Dimethyl-anthranilic acid 2440
Dimethyl-azobenzene 612-4
Dimethyl-benzene 6454, 6457, 6462

Dimethyl-benzene sulfonyl chloride 6461
Dimethyl-benzidine 5955, 5959
Dimethyl-benzophenone 2838

Table 7-4 (*Continued*)
PHYSICAL CONSTANTS OF ORGANIC COMPOUNDS

No.	Name	Synonym	Formula	Beil. Ref.	Formula Weight
2463	**Dimethyl** benzoic acid (3,5)	mesitylenic acid	$(CH_3)_2C_6H_3 \cdot CO_2H$	IX-536	150.18
2464	benzoylene urea	....	$C_6H_4 \cdot N(CH_3) \cdot CO \cdot$ $\mid$ $N(CH_3) \cdot CO$ $\mid$	XXIV-375	190.20
2465	benzylamine (*N*)	....	$C_6H_5 \cdot CH_2 \cdot N(CH_3)_2$	XII-1019	135.21
2466	benzyl carbinol	....	$C_6H_5CH_2 \cdot C(CH_3)_2OH$	VI-523	150.22
2467	carbanilide	*N,N'*-diMe-*N,N'*-diPh urea	$[C_6H_5(CH_3)N]_2CO$	XII-418	240.31
2468	carbonate	methyl carbonate	$CO(OCH_3)_2$	III-4	90.08
2469	citraconate	....	$CH_3 \cdot C_2H(CO_2CH_3)_2$	II-770	158.16
2470	cyclohexan-3,5-diol (1,1)	....	$(CH_2 \cdot CHOH)_2CH_2 \cdot C:$ $\mid$ _____ $\mid$ $(CH_3)_2$	*VI-371	144.22
2471	cyclohexane (*o*) *	hexahydro-*o*-xylene	$(CH_3)_2C_6H_{10}$	V-36	112.22
2472	cyclohexane (*m*) *	hexahydro-*m*-xylene	$(CH_3)_2C_6H_{10}$	V-36	112.22
2473	cyclohexane (*p*) *	hexahydro-*p*-xylene	$(CH_3)_2C_6H_{10}$	V-39	112.22
2474	2,4-dicarbethoxy-pyrrole (3,5)	....	$C_{12}H_{17}O_4N$	XXII-133	239.27
2475	dihydroresorcinol (5,5)	1,1-diMe-cyclo-hexan-3,5-dione	$(CH_3)_2C_6H_6O_2$	VII-559	140.18
2476	1,3-diphenylcyclo-butane (1,3)	α-Me-styrene dimer	$[C_6H_5(CH_3)C \cdot CH_2]_2$	V-652	236.36
2477	dithiocarbamate Na	....	$(CH_3)_2N \cdot CS_2Na \cdot$ $2\frac{1}{2}H_2O$	IV-75	188.24
2478	ether	methyl ether	$CH_3 \cdot O \cdot CH_3$	I-281	46.07
2479	6-ethoxyquinoline (2,4)	....	$C_2H_5O \cdot C_9H_4N(CH_3)_2$	....	201.27
2480	6-ethoxyquinoline ethiodide (2,4)	....	$C_{13}H_{15}ON \cdot C_2H_5I$	....	357.24
2481	ethylacetic acid	....	$(CH_3)_2(C_2H_5) \vdots C \cdot$ CO_2H	II-335	116.16
2482	ethylamine	....	$(CH_3)_2N \cdot C_2H_5$	IV-94	73.14
2483	ethylbenzene (3,4)	4-Et-*o*-xylene	$C_2H_5 \cdot C_6H_3(CH_3)_2$	V-427	134.22
2484	ethylbenzene (3,5)	5-Et-*m*-xylene	$C_2H_5 \cdot C_6H_3(CH_3)_2$	V-429	134.22
2485	ethylbenzene (2,4)	6-Et-*m*-xylene	$C_2H_5 \cdot C_6H_3(CH_3)_2$	V-428	134.22
2486	ethylbenzene (2,5)	2-Et-*p*-xylene	$C_2H_5 \cdot C_6H_3(CH_3)_2$	V-428	134.22
2487	fumarate	methyl fumarate	$(\vdots CH \cdot CO_2 \cdot CH_3)_2$	II-741	144.13
2488	furan (2,5)	α,α'-diMe-furan	$(CH_3)_2C_4H_2O$	XVII-41	96.13
2489	glutaric acid (α,α)	....	$(CH_3)_2C_3H_4(CO_2H)_2$	II-676	160.17
2490	glyoxime	diacetyl-dioxime	$(CH_3 \cdot C:NOH)_2$	I-772	116.12
2491	guanidine H_2SO_4 (*uns*)	....	$(CH_3)_2N \cdot C(NH) \cdot$ $NH_2 \cdot \frac{1}{2}H_2SO_4$	**IV-574	136.16
2492	1,3-heptadiene-(2,6)	*iso*-geraniolene	C_9H_{16}	I-260	124.23
2493	1,5-hexadiene (2,5)	....	$[CH_2:C(CH_3) \cdot CH_2 \cdot]_2$	I-259	110.20
2494	hydrazine (*s*)	hydrazomethane	$CH_3NH \cdot NHCH_3$	IV-547	60.10
2495	hydrazine (*as*)	....	$(CH_3)_2N \cdot NH_2$	IV-547	60.10
2496	hydrazine HCl	(*as*)	$(CH_3)_2N \cdot NH_2 \cdot HCl$	IV-547	96.56
2497	2-(2-hydroxy-phenyl)-imida-zole (4,5)	....	$CH_3 \cdot C:C(CH_3) \cdot N:C \cdot$ $\mid$ _____ $\mid$ $(C_6H_4OH) \cdot NH$	XXIII-391	188.23

Dimethyl-benzyl phenol 1573   * cis   Dimethyl-butyl carbinol (*tert*) 3528
Dimethyl-butadiene 2769   Dimethyl cadmium 1186
Dimethyl-butandiol 5312   Dimethyl-chloroarsine 1179
Dimethyl-butyl carbinol (*n*) 3535   Dimethyl-cyclohexadiene 2289-91
Dimethyl-butyl carbinol (*iso*) 3530   Dimethyl-diazine 2533

Table 7-4 (*Continued*)
PHYSICAL CONSTANTS OF ORGANIC COMPOUNDS

No.	Crystalline Form and Color	Specific Gravity	Melting Point °C.	Boiling Point °C.	Solubility in 100 Parts		
					Water	Alcohol	Ether
2463	mn./aq. al.	....	166	subl.	v. sl. s. h.	v. s. c.	....
2464	col. nd.	....	167-8	....	i.	s.	....
2465	lq.	0.915^0	....	$183\text{-}4^{765mm}$	sl. c.; v. sl. h.	∞	∞
2466	nd.	$0.979^{18}_{4}{}^\circ$	24	228	....	....	....
2467	mn. pr./al.	....	121	350	i.	s.	s.; s. bz.
2468	col. lq.	$1.070^{20}_{4}{}^\circ$	0.5	89-90	i.	∞	∞
2469	col. oil	$1.121^{15}_{15}{}^\circ$	....	210.5	$3^{15\circ}$	....	....
2470	pl./act.	....	146-7	....	s.; s. bz.	s.; s. chl.	v. sl. s.
2471	lq.	$0.796^{20}_{4}{}^\circ$	−50.0	129.7	i.	∞	∞
2472	lq.	$0.766^{20}_{4}{}^\circ$	−75.6	120.1	i.	∞	∞
2473	lq.	$0.783^{20}_{4}{}^\circ$	−87.4	124.3	....	....	....
2474	cr./aq. al. or ac.	....	134-5	....	i.; s. chl.; s. H_2SO_4	s.; s. bz.; i. HCl	v. sl. s.; s. ac.
2475	yel. mn. nd./aq.	....	148-9	....	$0.416^{25\circ}$; s. chl.	6.6 h. bz.; s. EtOAc	v. sl. s.
2476	cr./et.	....	52.6	307.5^{760mm}	i.; $125^{25\circ}$ act.	$5^{25\circ}$; $190^{25\circ}$ bz.	$147^{25\circ}$; $104^{25\circ}$ CCl_4
2477	cr.	....	$-2H_2O$, 115	$-2\tfrac{1}{2}H_2O$, 130	....	....	
2478	gas	1.617(A)	−138.5	−23.7	$3700^{18\circ}$ cc.	s.	s.
2479	col. nd.	....	86−8	....	i.	v. s.	v. s.
2480	yel. rhb.	....	163-4	....	i.	s.	i.
2481	lq.	....	−14	187	v. sl. s.	s.	s.
2482	col. lq.	....	....	37.5	....	....	....
2483	col. lq.	$0.875^{20}_{4}{}^\circ$	−67.0	189.8	i.	s.	s.
2484	col. lq.	$0.866^{20}_{4}{}^\circ$	−84.2	183.8	i.	s.	s.
2485	lq.	$0.876^{20}_{4}{}^\circ$	−63.0	188.4	i.	s.	s.
2486	lq.	$0.877^{20}_{4}{}^\circ$	−53.7	186.9	i.	s.	s.
2487	col. tri.	$1.045^{106\circ}$	102	193.3	s. c. chl.	sl. s.	sl. s.
2488	col. lq.	$0.888^{20}_{4}{}^\circ$	....	93−4	i.; i. alk.	∞	∞
2489	nd./HCl	....	90	....	v. s.	v. s.	v. s.
2490	col. cr.	....	240-6	....	$0.06^{20\circ}$	v. s.	v. s.
2491	col. cr./aq.	....	285-7 d.	....	v. s.	i.	i.
2492	lq.	$0.765^{10}_{4}{}^\circ$	....	$143\text{-}5^{755mm}$	i.	....	....
2493	lq.	$0.749^{21}_{21}{}^\circ$	<−80	$113\text{-}4^{755mm}$	i.	....	....
2494	hyg. lq.	$0.827^{20}_{4}{}^\circ$	....	81^{747mm}	∞	∞	∞
2495	hyg. lq.	$0.791^{22\circ}$	....	62.5^{717mm}	v. s.	v. s.	v. s.
2496	col. cr.	....	82-3	....	s.	s.	i.
2497	nd./aq. al.	....	218	....	....	....	....

Dimethyl-diphenyl urea 2467
Dimethyl-disulfide 4222
Dimethyl-ethyl carbinol 409
Dimethyl-ethylene 1148-9
Dimethyl-ethylene glycol 1154

Dimethyl-ethylene oxide 1155
Dimethyl-furfurane carboxylic acid 6396
Dimethyl-hexyl carbinol (*n*) 4922
Dimethyl-hydantoin 66
Dimethyl-hydroquinone 2376

Table 7-4 (*Continued*)
PHYSICAL CONSTANTS OF ORGANIC COMPOUNDS

No.	Name	Synonym	Formula	Beil. Ref.	Formula Weight
2498	**Dimethyl** itaconate	....	$C_3H_4(CO_2CH_3)_2$	II-762	158.16
2499	ketene	....	$(CH_3)_2C{:}CO$	I-731	70.09
2500	malate (*l*)	methyl malate	$C_2H_4O(CO_2CH_3)_2$	III-429	162.14
2501	maleate	methyl maleate	$({:}CH{\cdot}CO_2{\cdot}CH_3)_2$	II-751	144.13
2502	malonate	methyl malonate	$CH_2(CO_2{\cdot}CH_3)_2$	II-572	132.12
2503	malonic acid	....	$(CH_3)_2C(CO_2H)_2$	II-647	132.12
2504	mesaconate	....	$CH_3{\cdot}C_2H(CO_2CH_3)_2$	II-765	158.16
2505	naphthylamine (α)	....	$(CH_3)_2N{\cdot}C_{10}H_7$	XII-1221	171.24
2506	naphthylamine (β)	....	$(CH_3)_2N{\cdot}C_{10}H_7$	XII-1273	171.24
2507	naphthalene (1,4)	....	$(CH_3)_2C_{10}H_6$	V-570	156.23
2508	naphthalene (2,3)	guajen	$(CH_3)_2C_{10}H_6$	*V-268	156.23
2509	naphthalene (2,6)	....	$(CH_3)_2C_{10}H_6$	V-570	156.23
2510	α-naphthoquinoline	(2,4)	$(CH_3)_2C_{13}H_7N$	XX-475	207.28
2511	β-naphthoquinoline	(2,4)	$(CH_3)_2C_{13}H_7N$	XX-476	207.28
2512	4-nitrophthalate	....	$NO_2C_6H_3(CO_2CH_3)_2$	IX-830	239.19
2513	nitrosamine	nitroso-dimethyl-amine	$(CH_3)_2N{\cdot}NO$	IV-84	74.08
2514	2,6-octadiene (2,6)	dihydromyrcene	$C_{10}H_{18}$	I-260	138.25
2515	2,7-octadiene (2,6)	linaloolene	$C_{10}H_{18}$	I-261	138.25
2516	2,6-octadiene-8-ol (2,6)	nerol	$C_{10}H_{18}O$	I-459	154.25
2517	2,6-octadiene-8-ol (2,6)	geraniol	$C_{10}H_{18}O$	I-457	154.25
2518	2,7-octadiene-6-ol (2,6)	*l*-linalool	$C_{10}H_{18}O$	I-460	154.25
2519	octanol-6 (2,6)	linalool tetrahydride	$C_{10}H_{22}O$	I-426	158.29
2520	octanol-8 (2,6)	geraniol tetrahydride	$C_{10}H_{22}O$	I-426	158.29
2521	oxalate	methyl oxalate	$(CO_2{\cdot}CH_3)_2$	II-534	118.09
2522	oxamide (*s*)	....	$(CO{\cdot}NHCH_3)_2$	IV-61	116.12
2523	oxamide (*as*)	....	$(CH_3)_2N{\cdot}(CO)_2NH_2$	IV-61	116.12
2524	phenacyl chloride (2,4)	ω-Cl-2,4-di Me-acetophenone	$(CH_3)_2C_6H_3{\cdot}CO{\cdot}CH_2Cl$	VII-324	182.65
2525	phenanthrene (9,10)	....	$(CH_3)_2C_{14}H_8$	V-680	206.29
2526	phenyl carbinol	....	$C_6H_5{\cdot}C(CH_3)_2{\cdot}OH$	VI-506	136.20
2527	*p*-phenylenedi-amine	(*s*)	$(CH_3NH)_2C_6H_4$	XIII-71	136.20
2528	*p*-phenylenedi-amine oxalate	(*s*)	$(CH_3NH)_2C_6H_4{\cdot}2H_2C_2O_4$	....	316.27
2529	phosphine	....	$(CH_3)_2PH$	IV-580	62.05
2530	*o*-phthalate	methyl phthalate	$C_6H_4(CO_2CH_3)_2$	IX-797	194.19
2531	*m*-phthalate	Me *iso*-phthalate	$C_6H_4(CO_2CH_3)_2$	IX-834	194.19
2532	*p*-phthalate	Me-terephthalate	$C_6H_4(CO_2CH_3)_2$	IX-843	194.19
2533	pyrazine (2,5)	ketine; glycoline	$CH_3C{:}CHN{:}C(CH_3){\cdot} \\ \quad \overline{CH{:}N}_{\_\_\_\!\mid}$	XXIII-96	108.14
2534	pyrazole (3,5)	....	$(CH_3)_2N{:}C{\cdot}CH{:}C{\cdot}NH$	XXIII-74	96.13
2535	pyrrol (1,2)	....	$CH_3{\cdot}C(CH)_3{\cdot}N{\cdot}CH_3$	....	95.15
2536	pyrrol (2,3)	....	$(CH_3)_2C_4H_2NH$	XX-172	95.15
2537	pyrrol (2,4)	....	$(CH_3)_2C_4H_2NH$	XX-172	95.15

Dimethyl-itaconic acid 5700
Dimethyl-ketazine 3932
Dimethy-ketol 43
Dimethyl ketone 56

Dimethyl mercury 4044
Dimethyl-octane 1637-8
Dimethyl-octenol 1519
Dimethyl-pentanol 3529-31

Table 7-4 (*Continued*)
PHYSICAL CONSTANTS OF ORGANIC COMPOUNDS

No.	Crystalline Form and Color	Specific Gravity	Melting Point °C.	Boiling Point °C.	Solubility in 100 Parts		
					Water	Alcohol	Ether
2498	mn./Me al.	$1.124^{18°}_{4}$	38	208	....	....	....
2499	yel. lq.	....	−97.5	34^{750mm}	d.	d.	s.
2500	lq.	$1.233^{20°}_{4}$	....	242	s.	∞	∞
2501	col. lq.	$1.151^{21°}$	7.6	205	....	....	....
2502	col. lq.	$1.154^{20°}_{4}$	−62	180-1	i.	∞	∞
2503	col. pr.	$1.357^{18°}$	192-3 d.	d. > 130	$10^{13°}$	v. s. abs.	v. s.
2504	col. lq.	$1.121^{20°}_{20}$	....	205.5-6.5	....	....	....
2505	col. oil	$1.042^{20°}$	....	274.5^{711mm}	i.	s.	s.
2506	col. cr.	$1.039^{70°}_{70}$	46-7	304-5	i.	s.	s.
2507	lq.	$1.016^{20°}_{4}$	<−18	264-6	i.	....	....
2508	lf./al.	....	104.0-4.5	$265-6^{767mm}$	i.	sl. s.	s. bz.
2509	lf./al.	$1.142^{0°}_{4}$	110-1	$261-2^{762mm}$	i.	sl. s.	
2510	nd./pet.	....	43-4	....	v. s. h. pet.	i. 90% al.	v. s.
2511	nd./et.	....	126-7	>300 d.	v. sl. s. h.	s.; s. ac.	s.; s. act.
2512	cr./aq. al.	....	66-7.5	....	....	....	....
2513	yel. oil	$1.006^{20°}_{4}$	....	153^{774mm}	s.	s.	s.
2514	lq.	$0.775^{21°}_{4}$	....	171.5-3.5	....	....	....
2515	lq.	$0.788^{20°}$	....	165-8	....	....	....
2516	oil	$0.881^{15°}$	....	$224-5^{745mm}$	....	....	....
2517	col. lq.	$0.883^{15°}$	<−15	230	i.	∞	∞
2518	oil	$0.862^{20°}$	....	$197-200^{756mm}$	v. sl. s.	∞	∞
2519	lq.	$0.836^{15°}_{4}$	....	196-7	....	....	....
2520	lq.	$0.849^{0°}_{4}$	....	212-3	....	....	....
2521	col. mn.	$1.148^{54°}$	54	163.3	6	s.	s.
2522	nd./aq.	$1.3^{4°}$	212-7	subl.	$2.5^{9.4°}$; v. s. h.	s.; d. h. alk.	v. sl. s.; s. chl.
2523	pl./bz.	....	104	....	v. s.	v. s.	v. sl.s.
2524	lf.	....	62-3	....	....	s.	....
2525	pr./dil. ac.	....	139	subl.	s. ac.	sl. s.; s. bz.	s. chl.
2526	pr.	$0.972^{19°}_{4}$	23	215-20 sl. d.	....	....	....
2527	cr./pet.	....	53	150^{17mm}	v. sl. s.	s.	s.
2528	col. cr.	....	194-6 d.	....	s.	sl. s.	i.
2529	col. lq.	<1	....	25	i.	....	....
2530	col. lq.	$1.189^{25°}_{25}$	....	280^{734mm}	0.43	....	....
2531	nd./aq. al.	....	67-8	....	i.	....	....
2532	rhb./al.	....	140	>300; subl.	0.3 h.	s. h.	s.
2533	pl.	$0.990^{18°}_{4}$	15	155	∞	∞	∞
2534	pl.	$0.884^{26°}_{4}$	106-7	218^{759mm}	s.; s. bz.	s.; s. chl.	s.
2535	lq.	....	....	65^{14mm}	....	....	....
2536	lq.	....	....	165	....	....	....
2537	lq.	$0.927^{14°}_{4}$	....	171	sl. s.	s.	s.; s. bz.

Dimethyl-phenol 6465-70
Dimethyl-phenyl hydrazine 6500-2
Dimethyl-phenylenediamine 305-7

Dimethyl-propyl carbinol (*n*) 3631
Dimethyl-propyl carbinol (*iso*) 3629
Dimethyl-pyridine 3981-4

Table 7-4 (*Continued*)
PHYSICAL CONSTANTS OF ORGANIC COMPOUNDS

No.	Name	Synonym	Formula	Beil. Ref.	Formula Weight
2538	**Dimethyl** pyrrol (2,5)	....	$(CH_3)_2C_4H_2NH$	XX-172	95.15
2539	quinolinate (2,3)	....	$C_5H_3N(CO_2CH_3)_2$	XXII-151	195.18
2540	quinoline (2,3)	....	$(CH_3)_2C_9H_5N$	XX-406	157.22
2541	quinoline (2,4)	4-Me-quinaldine	$(CH_3)_2C_9H_5N$	XX-407	157.22
2542	quinoline H_2SO_4	(2,4)	$C_{11}H_{11}N \cdot \frac{1}{2}H_2SO_4$	XX-408	206.26
2543	quinoline EtI	(2,4)	$C_{11}H_{11}N \cdot C_2H_5I$	XX-408	313.18
2544	quinoline (2,6)	6-Me-quinaldine	$(CH_3)_2C_9H_5N$	XX-408	157.22
2545	quinoline EtI	(2,6)	$C_{11}H_{11}N \cdot C_2H_5I$	XX-409	313.18
2546	quinoline (3,4)	....	$(CH_3)_2C_9H_5N$	XX-410	157.22
2547	quinoline (5,8)	....	$(CH_3)_2C_9H_5N$	XX-411	157.22
2548	quinoline (6,8)	....	$(CH_3)_2C_9H_5N$	XX-411	157.22
2549	quinone (2,3)	o-xyloquinone	$(CH_3)_2C_6H_2O_2$	VII-656	136.15
2550	quinone (2,5)	p-xyloquinone	$(CH_3)_2C_6H_2O_2$	VII-658	136.15
2551	quinone (2,6)	m-xylo-p-quinone	$(CH_3)_2C_6H_2O_2$	VII-657	136.15
2552	sebacate	methyl sebacate	$(CH_2)_8(CO_2CH_3)_2$	*II-293	230.31
2553	selenide	selenium methyl	$(CH_3)_2Se$	I-291	109.03
2554	succinate	methyl succinate	$(CH_2CO_2 \cdot CH_3)_2$	II-609	146.14
2555	*iso*-succinate	....	$CH_3 \cdot CH(CO_2 \cdot CH_3)_2$	II-628	146.14
2556	succinic acid (*meso*)	....	$(CH_3CH)_2(CO_2H)_2$	II-665	146.14
2557	succinic acid (α,α)	....	$(CH_3)_2C \cdot CH_2(CO_2H)_2$	II-661	146.14
2558	succinic acid (*dl*)	....	$(CH_3CH)_2(CO_2H)_2$	II-667	146.14
2559	sulfate	methyl sulfate	$(CH_3O)_2SO_2$	I-283	126.13
2560	sulfide	methyl sulfide	$(CH_3)_2S$	I-288	62.13
2561	sulfite	....	$(CH_3O)_2SO$	I-282	110.13
2562	sulfone	....	$(CH_3)_2SO_2$	I-289	94.13
2563	sulfoxide	....	$(CH_3)_2SO$	I-289	78.13
2564	tartrate (*d*)	....	$(CHOH \cdot CO_2CH_3)_2$	III-510	178.14
2565	tartrate (racemic)	diMe-racemate	$(CH_3)_2C_4H_4O_6$	III-527	178.14
2566	telluride	tellurium methyl	$(CH_3)_2Te$	I-291	157.67
2567	thetin	....	$HO \cdot S(CH_3)_2 \cdot CH_2 \cdot CO_2H$	III-247	138.19
2568	thiourea (*s*)	....	$(CH_3NH)_2C{:}S$	IV-70	104.17
2569	toluidine (*o*)(*N*)	....	$CH_3 \cdot C_6H_4 \cdot N(CH_3)_2$	XII-785	135.21
2570	toluidine (*m*)(*N*)	....	$CH_3 \cdot C_6H_4 \cdot N(CH_3)_2$	XII-857	135.21
2571	toluidine (*p*)(*N*)	....	$CH_3 \cdot C_6H_4 \cdot N(CH_3)_2$	XII-902	135.21
2572	urea (*s*)	....	$(CH_3NH)_2C{:}O$	IV-65	88.11
2573	urea (*as*)	....	$(CH_3)_2N \cdot CO \cdot NH_2$	IV-73	88.11
2574	**Dinaphthyl** (1,1')	α,α'-binaphthyl	$C_{10}H_7 \cdot C_{10}H_7$	V-726	254.33
2575	**Dinaphthyl** (2,2')	....	$C_{10}H_7 \cdot C_{10}H_7$	V-727	254.33
2576	amine (β,β')	....	$C_{10}H_7 \cdot NH \cdot C_{10}H_7$	XII-1278	269.35
2577	ketone (α,α')	....	$(C_{10}H_7)_2CO$	VII-539	282.35
2578	ketone (α,β')	....	$(C_{10}H_7)_2CO$	VII-539	282.35
2579	ketone (β,β')	....	$(C_{10}H_7)_2CO$	VII-539	282.35
2580	methane (α,α')	....	$(C_{10}H_7)_2CH_2$	V-728	268.36
2581	methane (α,β')	....	$(C_{10}H_7)_2CH_2$	*V-360	268.36
2582	methane (β,β')	....	$(C_{10}H_7)_2CH_2$	V-729	268.36
2583	sulfide (α,α')	....	$(C_{10}H_7)_2S$	VI-623	286.40
2584	sulfide (β,β')	....	$(C_{10}H_7)_2S$	VI-659	286.40
2585	urea (α)(*s*)	α,α dinaphthyl carbamide	$(C_{10}H_7NH)_2CO$	XII-1238	312.38

Dimethyl-resorcinol 2375, 2377, 2380, 6479
Dimethyl-thiophene 5928-31
Dimethyl-xanthine, cf. alkd.

Dimethyl zinc 6504
Dinaphthol 2336-7
Dinaphthyl carbamide 2585

Table 7-4 (*Continued*)
PHYSICAL CONSTANTS OF ORGANIC COMPOUNDS

No.	Crystalline Form and Color	Specific Gravity	Melting Point °C.	Boiling Point °C.	Solubility in 100 Parts		
					Water	Alcohol	Ether
2538	oil	$0.935^{20°}_{4}$	....	169	v. sl. s.	v. s.	v. s.; v. sl. s. alk.
2539	col. pl./bz.	....	53-4	....	s.; s. CS_2	s.; i. lg.	s.; s. bz.
2540	pl.	....	68-9	261^{729mm}	sl. s.	v. s.	s.; s. lg.
2541	lq.	$1.061^{15°}$	....	264-5	i.	s.	s.
2542	nd./al.	....	225-8 d.	....	v. s. h.	v. sl. s.	....
2543	yel. nd./al.	....	226-7	....	....	s. h.	....
2544	cr./et.	....	60	266-7	v. sl. s. h.	s.	s.
2545	cr.	....	225-7	....	....	....	....
2546	cr.	....	65	290^{737mm}	i.	....	....
2547	lq.	$1.070^{21°}$	4-5	265^{736mm}	i.	....	....
2548	lq.	$1.067^{4°}$	....	268-9	i.	....	....
2549	yel. nd.	....	55	subl.	sl. s.	s.	s.
2550	yel. tri.	....	124-5	subl.	sl. s. h.	v. s. h.	s.; s. bz.
2551	yel. nd.	....	72-3	subl.	....	....	....
2552	nd. or pl./et.	$0.988^{8°}_{4}$	38 (26)	293^{754mm}	....	....	....
2553	lq.	$1.408^{15°}_{4}$	....	58.2	i.	....	....
2554	col. cr.	$1.126^{15°}_{18}$	18.2	195.9	0.9	3	....
2555	lq.	$1.028^{25°}_{25}$	....	179	i.	∞	∞
2556	tri.	1.314	209	d.	$<3^{14°}$	v. s.	v. s.
2557	tri.	1.323	142	d. 165	$7.5^{14°}$	v. s.	v. sl. s.
2558	rhb. pr.	1.339	129	d. to anhyd.	$3^{14°}$	v. s.	v. s.
2559	poison. oil	$1.352^{0°}_{4}$	−26.8	188.3-8.6	v. sl. s.	∞	∞
2560	oil	$0.846^{21°}_{4}$	−98.3	37.3	i.	s.	s.
2561	lq.	$1.046^{16.2°}_{4.1}$	....	126.5^{756mm}	d.	s.	s.
2562	pr.	....	109	238	....	....	....
2563	oil	....	....	volt. 100 d.	v. s.	v. s.	v. s.
2564	cr.	$1.328^{20°}_{4}$	61.5 (48)	280	s.*	$200^{15°}$	v. s. bz.*
2565	mn./al.	$1.260^{89.6°}_{4}$	89	282	....	$21^{15°}$	....
2566	lt. yel. oil	<1	....	82	s.	s.	s.
2567	delq. cr.	....	d. −H_2O	....	s.	sl. s.	....
2568	hyg., delq.	....	61-2	....	v. s.	v. s.	sl. s. et.
2569	lq.	$0.929^{20°}_{4}$	−61.3	185-6	i.	s.	s.
2570	yel. oil	$0.941^{20°}_{4}$	....	213-5	i.	s.	s.
2571	lq.	$0.937^{20°}_{4}$	....	210-1	i.	s.	s.
2572	pr./chl. et.	1.142	106	268-70	v. s.	v. s.	v. sl. s.
2573	mn./al.	1.255	182-5	....	s.	sl. s. c.	v. sl. s.
2574	lf./al.	....	160(144-6)	$240-4^{12mm}$	i.; s. bz.	s. h.	s.
2575	pl.	....	187.8	452^{753mm}	i.	v, sl. s. c.	v. sl. s. c.
2576	lf./bz.	....	171-2	471	i.	sl. s. h.	v. s. h. bz.
2577	nd./et.	....	104	....	s. H_2SO_4	s. h.	v. s. bz.
2578	nd./al.	....	135-6	....	s. bz.	$1.3^{14°}$	sl. s. h.
2579	lf. or nd.	....	α125.5 β164.5	....	....	α0.419° β0.0819°	v. sl. s.; v. s. chl.
2580	pr. or nd./al.	....	109	>360	v. s. bz.	0.8 c.; 6.6 h.	v. s.
2581	pr./al.	....	95-6	....	s. bz.	s. EtOAc	....
2582	nd./al.	....	92	....	....	s.	s. bz.
2583	nd./al.	....	110	$289-90^{15mm}$	....	v. sl. s.	v. s. CS_2
2584	lf./al.	....	151	$295-6^{15mm}$	i.	v. sl. s.	v. s. CS_2
2585	nd./ac.	....	297-8	subl.	....	sl. s. h.	....

* Modification m. p. 61.5° less soluble.
Dinaphthyl ether 4560-2
Dinaphthyl hydrazine 3690-1
Dinaphthyl mercury 4045
Dinicotinic acid 5502

Table 7-4 (Continued)
PHYSICAL CONSTANTS OF ORGANIC COMPOUNDS

No.	Name	Synonym	Formula	Beil. Ref.	Formula Weight
2586	**Dinitro**-acetanilide (2,4)	....	$(NO_2)_2C_6H_3\cdot NH\cdot COCH_3$	XII-754	225.16
2587	aniline (2,3)	....	$(NO_2)_2C_6H_3\cdot NH_2$	XII-747	183.12
2588	aniline (2,4)	....	$(NO_2)_2C_6H_3\cdot NH_2$	XII-747	183.12
2589	aniline (2,5)	....	$(NO_2)_2C_6H_3\cdot NH_2$	XII-757	183.12
2590	aniline (2,6)	....	$(NO_2)_2C_6H_3\cdot NH_2$	XII-758	183.12
2591	aniline (3,4)	....	$(NO_2)_2C_6H_3\cdot NH_2$	XII-758	183.12
2592	aniline (3,5)	....	$(NO_2)_2C_6H_3\cdot NH_2$	XII-759	183.12
2593	anisole (2,4)	α-dinitroanisole	$CH_3O\cdot C_6H_3(NO_2)_2$	VI-254	198.14
2594	anthranilic acid (3,5)	....	$C_6H_2(CO_2H)(NO_2)_2\cdot NH_2$	XIV-379	227.13
2595	anthraquinone (1,5)	....	$(NO_2C_6H_3)_2(CO)_2$	VII-793	298.21
2596	anthraquinone (1,8)	....	$(NO_2\cdot C_6H_3)_2(CO)_2$	VII-795	298.21
2597	anthraquinone (2,7)	β-diNO$_2$-anthra-quinone	$(NO_2\cdot C_6H_3)_2(CO)_2$	VII-795	298.21
2598	azoxybenzene (4,4')	....	$(NO_2\cdot C_6H_4)_2N_2O$	XVI-628	288.22
2599	benzalacetone (3,3')	....	$(NO_2C_6H_4CH:CH)_2:CO$	VII-506	324.30
2600	benzaldehyde (2,4)	....	$(NO_2)_2C_6H_3\cdot CHO$	VII-264	196.12
2601	benzaldehyde (2,6)	....	$(NO_2)_2C_6H_3\cdot CHO$	*VII-144	196.12
2602	benzene (o)	....	$(NO_2)_2C_6H_4$	V-257	168.11
2603	benzene (m)	....	$(NO_2)_2C_6H_4$	V-258	168.11
2604	benzene (p)	....	$(NO_2)_2C_6H_4$	V-261	168.11
2605	benzene sulfonic acid	(2,4;1)	$(NO_2)_2C_6H_3\cdot SO_3H\cdot 3H_2O$	XI-78	302.23
2606	benzene sulfonic	Na salt (2,4;1)	$C_6H_3O_7N_2SNa\cdot H_2O$	XI-78	288.17
2607	benzoic acid (2,4)	....	$(NO_2)_2C_6H_3\cdot CO_2H$	IX-411	212.12
2608	benzoic acid (2,5)	....	$(NO_2)_2C_6H_3\cdot CO_2H$	IX-412	212.12
2609	benzoic acid (2,6)	....	$(NO_2)_2C_6H_3\cdot CO_2H$	IX-412	212.12
2610	benzoic acid (3,4)	....	$(NO_2)_2C_6H_3\cdot CO_2H$	IX-413	212.12
2611	benzoic acid (3,5)	....	$(NO_2)_2C_6H_3\cdot CO_2H$	IX-413	212.12
2612	benzophenone (4,4')	....	$(NO_2\cdot C_6H_4)_2CO$	VII-428	272.22
2613	benzoyl chloride (3,5)	(3,5)	$(NO_2)_2C_6H_3\cdot COCl$	IX-414	230.57
2614	benzoylene urea (6,8)	(6,8)	$C_8H_4O_6N_4$	*XXIV-344	252.14
2615	benzyl chloride (2,4)	....	$(NO_2)_2C_6H_3\cdot CH_2Cl$	V-344	216.58
2616	o-cresol (4,6)	3,5-diNO$_2$-2-OH-toluene	$(NO_2)_2C_6H_2(CH_3)\cdot OH$	VI-368	198.14
2617	m-cresol (2,6)	2,4-diNO$_2$-3-OH-toluene	$(NO_2)_2C_6H_2(CH_3)\cdot OH$	VI-387	198.14
2618	diazo-aminoben-zene (4,4')	....	$NO_2C_6H_4\cdot N:N\cdot NH\cdot C_6H_4NO_2$	XVI-700	287.24
2619	diethylaniline	(2,4)	$(NO_2)_2C_6H_3N(C_2H_5)_2$	XII-750	239.23
2620	diphenyl (2,2')	....	$(NO_2\cdot C_6H_4\cdot)_2$	V-583	244.21
2621	diphenyl (3,3')	....	$(NO_2\cdot C_6H_4\cdot)_2$	V-583	244.21
2622	diphenyl (4,4')	....	$(NO_2\cdot C_6H_4\cdot)_2$	V-584	244.21
2623	diphenyl (2,4')	....	$(NO_2\cdot C_6H_4\cdot)_2$	V-584	244.21

Dinitro-aminophenol 5303
Dinitro-bromobenzene 930-1

Dinitro-bromobenzoic acid 932
Dinitro-chlorobenzene 1364-9

Table 7-4 (*Continued*)
PHYSICAL CONSTANTS OF ORGANIC COMPOUNDS

No.	Crystalline Form and Color	Specific Gravity	Melting Point °C.	Boiling Point °C.	Solubility in 100 Parts		
					Water	Alcohol	Ether
2586	nd./al.	....	120	....	i. c.	v. s. h.	s.
2587	or. yel./al.	....	127	....	....	v. s.	s.
2588	yel. or gn. mn.	$1.615^{14°}$	187-8	....	v. sl. s. h.	$0.76^{21°}$	sl. s. h. HCl
2589	or. yel./al.	....	137	....	....	v. s.	....
2590	yel. nd./al.	....	138-40	....	i.	s. h. bz.	i. lg.
2591	yel. nd./aq.	....	154	....	....	v. s.	s.
2592	yel. nd./aq.	....	159-60	....	sl. s. bz.	s.	s.
2593	col. mn.	$1.341^{20°}$	94-5	....	sl. s. h.	$1.5^{20°}$	....
2594	yel. pl./al.	....	268	....	$0.02^{15°}$	sl. s.	....
2595	yel. nd.	....	>330	subl.	i.	v. sl. s.	v. sl. s.
2596	yel. pr.	....	312	....	s. Ac_2O	....	....
2597	nd./ac.	....	262	subl.	i.	sl. s.	sl. s.
2598	yel. nd./ bz.	....	192-3	....	....	....	....
2599	cr./Ac_2O	....	237	....	i.	i.	i
2600	yel. pr.	....	69-70	$190-210^{10mm}$	i.	v. s.	v. s.
2601	lf./ac.	....	123	....	s. h.	s.; sl. s. CS_2	s.; v. sl. s. lg.
2602	col. mn.	$1.59^{18°}$	117-8	319^{774mm}	0.01 c.	$1.9^{21°}$	$5.7^{18°}$ bz.
2603	col. rhb.	$1.575^{20°}_{4}$	89.8	300-2	$0.39^{9°}$	$3.3^{20°}$ abs.	$39.5^{18°}$ bz.
2604	col. mn.	$1.625^{18°}$	173-4	299^{777mm}	$0.18^{100°}$	$0.18^{21°}$	$2.6^{18°}$ bz.
2605	pa. yel., hyg. pr.	....	106-8	$-3H_2O$, >130	s.; s. ac.	s.; i. bz.; i. lg.	v. sl. s.
2606	cr./aq.	....	....	....	v. s.	....	....
2607	cr./aq.	....	179-80	....	$1.85^{25°}$	s.	$0.71^{30°}$ bz.
2608	mn. pr.	....	177-9	....	s. h.	s.	s.
2609	nd./aq.	....	202	....	v. s. h.	....	....
2610	nd.	....	163-4	....	$0.67^{25°}$	s.	s.
2611	mn. pr.	....	204-5	subl.	2 h.	v. s.	sl. s.
2612	col. nd./ac.	....	189	....	i.	....	....
2613	nd./bz.	....	68-9	196^{12mm}	d.	d.	....
2614	gn. yel. pr.	....	274-5	....	$0.016^{23°}$	sl. s. h. s. h. ac.	s. NaOH; sl. s. act.
2615	pl./et.	....	34	....	i.	s.	s.
2616	yel. pr./al.	....	86-7	....	v. sl. s.; v. sl. s. lg.	$10^{15°}$; s. alk.	s.; s. act.
2617	or. red nd.	....	99	....	....	....	....
2618	yel. nd./al.	....	224-6 d.	,........	i.; v. sl. s. chl.	sl. s. h.; v. sl. s. bz.	s.
2619	yel. rhb./ act.	$1.374^{15°}$	80	....	v. sl. s. lg.	v. s. h.	v. s. h.
2620	yel. nd.	1.45	124	....	i.	s. h.	s.
2621	cr.	....	198	....	i.	$0.6^{20°}$	....
2622	nd./al.	1.445	233	....	i.	$1.5^{20°}$	sl. s. bz.
2623	mn.	1.474	93.5	....	....	v. s. h.	....

Dinitro-chlorobenzoic acid 1370-1
Dinitro-chlorophenol 1372

Dinitro-dihydroxyquinone 4586

Table 7-4 (*Continued*)
PHYSICAL CONSTANTS OF ORGANIC COMPOUNDS

No.	Name	Synonym	Formula	Beil. Ref.	Formula Weight
2624	**Dinitro** diphenyl-amine (2,4)	....	$(NO_2)_2C_6H_3 \cdot NH \cdot C_6H_5$	XII-751	259.22
2625	diphenylamine (2,4')	....	$(NO_2 \cdot C_6H_4)_2NH$	XII-715	259.22
2626	diphenylamine (4,4')	....	$(NO_2 \cdot C_6H_4)_2NH$	XII-716	259.22
2627	diphenyl ether (2,2')	....	$(NO_2 \cdot C_6H_4)_2O$	VI-219	260.21
2628	diphenyl ether (4,4')	....	$(NO_2 \cdot C_6H_4)_2O$	VI-232	260.21
2629	diphenylmethane	(4,4')	$(NO_2 \cdot C_6H_4)_2CH_2$	V-595	258.24
2630	ethane (1,1)	....	$CH_3 \cdot CH(NO_2)_2$	I-102	120.07
2631	hexane (1,1)	....	$C_5H_{11} \cdot CH(NO_2)_2$	I-147	176.17
2632	hydroquinone acetate (2,6)	(2,6;1,4)	$(NO_2)_2C_6H_2(OH) \cdot O_2C \cdot CH_3$	VI-858 †	242.15
2633	hydroquinone di-acetate (2,6)	....	$(NO_2)_2C_6H_2: (O_2C \cdot CH_3)_2$	*VI-419	284.18
2634	4'-hydroxydiphen-ylamine (2,4)	....	$(NO_2)_2C_6H_3 \cdot NH \cdot C_6H_4 \cdot OH$	XIII-444	275.22
2635	mesitylene (2,4)	*eso*-dinitro-mesitylene	$(NO_2)_2C_6H(CH_3)_3$	V-411	210.19
2636	methane	....	$(NO_2)_2CH_2$	I-77	106.04
2637	naphthalene (1,3)	γ-dinitronaphthalene	$(NO_2)_2C_{10}H_6$	V-557	218.17
2638	naphthalene (1,5)	α-dinitronaphthalene	$(NO_2)_2C_{10}H_6$	V-558	218.17
2639	naphthalene (1,6)	δ-dinitronaphthalene	$(NO_2)_2C_{10}H_6$	V-559	218.17
2640	naphthalene (1,8)	β-dinitronaphthalene	$(NO_2)_2C_{10}H_6$	V-559	218.17
2641	α-naphthol (2,4)	....	$(NO_2)_2C_{10}H_5 \cdot OH$	VI-617	234.17
2642	α-naphthol (2,4) Na	Martius yellow	$C_{10}H_5O_5N_2Na \cdot H_2O$	VI-618	274.17
2643	β-naphthol (1,6)	....	$(NO_2)_2C_{10}H_5 \cdot OH$	VI-655	234.17
2644	phenanthraquinone	(2,7)	$O:C_{14}H_6(NO_2)_2:O$	VII-807	298.21
2645	phenetole (2,4)	α-dinitrophenetole	$(NO_2)_2C_6H_3 \cdot OC_2H_5$	VI-254	212.16
2646	phenol (2,3)	ϵ-dinitrophenol	$(NO_2)_2C_6H_3 \cdot OH$	VI-251	184.11
2647	phenol (2,4)	α-dinitrophenol	$(NO_2)_2C_6H_3 \cdot OH$	VI-251	184.11
2648	phenol (2,4) Na	....	$C_6H_3O_5N_2Na \cdot H_2O$	*VI-126	224.11
2649	phenol (2,5)	γ-dinitrophenol	$(NO_2)_2C_6H_3 \cdot OH$	VI-256	184.11
2650	phenol (2,6)	β-dinitrophenol	$(NO_2)_2C_6H_3 \cdot OH$	VI-257	184.11
2651	phenol (3,4)	δ-dinitrophenol	$(NO_2)_2C_6H_3 \cdot OH$	VI-257	184.11
2652	phenol (3,5)	θ-dinitrophenol	$(NO_2)_2C_6H_3 \cdot OH$	VI-258	184.11
2653	phenylacetic acid (2,4)	....	$(NO_2)_2C_6H_3 \cdot CH_2 \cdot CO_2H$	IX-459	226.15
2654	phenylcarbazide (4,4')	....	$(NO_2 \cdot C_6H_4 \cdot NH \cdot NH)_2:CO$	....	332.28
2655	phenyl disulfide	(2,2')	$(NO_2 \cdot C_6H_4 \cdot S \cdot)_2$	VI-338	308.34
2656	phenyl disulfide	(4,4')	$(NO_2 \cdot C_6H_4 \cdot S \cdot)_2$	VI-340	308.34
2657	phenylhydrazine	(2,4)	$(NO_2)_2C_6H_3 \cdot NHNH_2$	XV-489	198.14
2658	propane (1,1)	....	$C_2H_5 \cdot CH(NO_2)_2$	I-117	134.09
2659	propane (2,2)	....	$(CH_3)_2C(NO_2)_2$	I-117	134.09
2660	resorcinol (2,4)	....	$(NO_2)_2C_6H_2(OH)_2$	VI-827	200.11
2661	resorcinol (4,6)	....	$(NO_2)_2C_6H_2(OH)_2$	VI-828	200.11
2662	salicylic acid (3,5)	3,5-diNO₂-2-OH-benzoic acid	$(NO_2)_2C_6H_2(OH) \cdot CO_2H \cdot H_2O$	X-122	246.13

† See also correction in *VI-419
Dinitro-diphenyl disulfide 2655-6

Dinitro-glycol 3462
Dinitro-hydroxybenzoic acid 2662

Table 7-4 (*Continued*)
PHYSICAL CONSTANTS OF ORGANIC COMPOUNDS

No.	Crystalline Form and Color	Specific Gravity	Melting Point °C.	Boiling Point °C.	Solubility in 100 Parts		
					Water	Alcohol	Ether
2624	red nd./bz.	....	157	....	s. act.	s. h.	s. chl.
2625	red nd.	....	220-2	....	i.	sl. s.	$0.4^{22°}$ act.
2626	yel. nd.	....	214-5	....	i.	sl. s.	$5.66^{20°}$ act.
2627	nd./al.	....	114.5	....	i.	$0.7^{20°}$	....
2628	nd./al.	....	142-3	....	i.	3 h.	sl. s.
2629	nd./bz.	....	183-5	....	s. h. ac.	i. c.	sl. s.
2630	lq.	$1.350^{23.5°}_{23.5}$	....	185-6	v. sl. s.	s.	s.
2631	yel. oil	>1	....	d.	v. sl. s.	v. s.	v. s.
2632	yel. nd./al.	....	95-6	....	s. chl.	s.	s.
2633	nd./al.	....	135-6	....	i.	s. h.	s. ac.
2634	red nd./ aq. al.	....	195-6	....	s. alk.	s.	v. sl. s. bz.
2635	rhb./al.	....	86	....	i.	s. h.	....
2636	lq.	....	<−15	100 d.	s.	....	....
2637	lt. yel. nd.	....	144-5	subl.	i.	s.	....
2638	nd./ac.	....	216	subl.	i.	s. h. bz.	v. sl. s. CS_2
2639	cr./ac.	....	166	....	i.	....	....
2640	rhb. pl.	....	170-2	d.	i.	$0.19^{19°}$, 88% al.	$0.72^{19°}$ bz.
2641	yel. nd./al.	....	138	....	v. sl. s. h.	sl. s.	sl. s.; s. ac.
2642	yel. red nd.	....	....	....	v. s.	....	....
2643	yel. nd.	....	195	....	v. sl. s. h.	s.	v. s.
2644	yel. cr./ac.	....	303-5	....	....	sl. s.	sl. s. ac.
2645	col. nd./al.	....	86	d.	v. sl. s.	s.	....
2646	yel. mn./al.	$1.681^{20°}$	144-5	....	sl. s.	v. s. h.	v. s.
2647	yel. rhb.	$1.683^{24°}$	112.9	subl.	0.5 c.; 5 h.	$4^{20°}$; s. bz.	v. s. h.
2648	cr.	....	....	....	$4^{25°}$	$2.7^{25°}$ abs.	$1^{25°}$ act.
2649	yel. mn./ aq.	....	104	....	sl. s.	v. s. h.	s.; s. alk.
2650	yel. rhb./ aq.	....	63-4	....	s. h.	s. h.	s.; s. chl.
2651	col. tri./aq.	1.672	134	....	....	v. s.	v. s.
2652	mn./HCl	1.702	123	....	....	v. s.	v. s.
2653	col. nd./ aq.	....	160	d. 179-80	....	....	....
2654	brn. pd.	....	261-3	....	i.	s.	i.
2655	yel. nd./bz.	....	195	d.	v. sl. s. ac.	v. sl. s.	v. sl. s. act.
2656	pl./al.	....	181	....	....	....	....
2657	vl. pr./al.	....	197-8 d.	....	i.	s. dil. a.	i.
2658	acidic oil	$1.258^{22°}$	....	189	s. alk.	....	....
2659	cr.	....	53	185.5	v. sl. s.	i. alk.	....
2660	yel. lf.	....	165 d.	expl.	v. sl. s.	s.	s. alk.
2661	yel. pr.	....	215	subl.	s. h.	s.	s.
2662	pl./aq.	....	173 d.	....	s. c.	v. s.	v. s.

Table 7-4 (*Continued*)
PHYSICAL CONSTANTS OF ORGANIC COMPOUNDS

No.	Name	Synonym	Formula	Beil. Ref.	Formula Weight
2663	**Dinitro** stilbene (4,4′)	low m.p. form	$(NO_2 \cdot C_6H_4 \cdot CH:)_2$	V-637	270.25
2664	stilbene (4,4′)	high m.p. form	$(NO_2 \cdot C_6H_4 \cdot CH:)_2$	V-637	270.25
2665	thymol (2,4)	(2,4;3,1,6)	$(NO_2)_2C_6H(CH_3) \cdot (OH)(C_3H_7)$	VI-543	240.22
2666	toluene (2,3)	....	$(NO_2)_2C_6H_3 \cdot CH_3$	V-339	182.14
2667	toluene (2,4)	....	$(NO_2)_2C_6H_3 \cdot CH_3$	V-339	182.14
2668	toluene (2,5)	....	$(NO_2)_2C_6H_3 \cdot CH_3$	V-341	182.14
2669	toluene (2,6)	....	$(NO_2)_2C_6H_3 \cdot CH_3$	V-341	182.14
2670	toluene (3,4)	....	$(NO_2)_2C_6H_3 \cdot CH_3$	V-341	182.14
2671	toluene (3,5)	....	$(NO_2)_2C_6H_3 \cdot CH_3$	V-341	182.14
2672	tyrosine (3,5)(*l*)	....	$C_9H_9O_7N_3 \cdot H_2O$	*XIV-668	289.20
2673	*o*-xylene (3,4)	....	$(NO_2)_2C_6H_2(CH_3)_2$	V-369	196.16
2674	*o*-xylene (3,6)	....	$(NO_2)_2C_6H_2(CH_3)_2$	V-369	196.16
2675	*o*-xylene (4,5)	....	$(NO_2)_2C_6H_2(CH_3)_2$	V-369	196.16
2676	*o*-xylene (3,5)	....	$(NO_2)_2C_6H_2(CH_3)_2$	V-369	196.16
2677	*m*-xylene (2,4)	....	$(NO_2)_2C_6H_2(CH_3)_2$	V-379	196.16
2678	*m*-xylene (4,6)	....	$(NO_2)_2C_6H_2(CH_3)_2$	V-380	196.16
2679	*p*-xylene (2,3)	β-dinitro-*p*-xylene	$(NO_2)_2C_6H_2(CH_3)_2$	V-387	196.16
2680	*p*-xylene (2,5)	γ-dinitro-*p*-xylene	$(NO_2)_2C_6H_2(CH_3)_2$	V-388	196.16
2681	*p*-xylene (2,6)	α-dinitro-*p*-xylene	$(NO_2)_2C_6H_2(CH_3)_2$	V-388	196.16
2682	**Dinonyl ketone** (*n*)	caprinone	$(C_9H_{19})_2CO$	I-718	282.51
2683	**Dioctyl** acetic acid (*n*)	....	$[CH_3(CH_2)_7]_2CH \cdot CO_2H$	II-388	284.49
2684	ether (*n*)	....	$[CH_3(CH_2)_7]_2O$	I-419	242.45
2685	ketone (*n*)	pelargone	$[CH_3(CH_2)_7]_2CO$	I-718	254.46
2686	**Dioxindole**	oxindole; hydrindic acid	$C_6H_4 \cdot \overline{CHOH \cdot CO \cdot NH}$	XXI-578	149.15
2687	**Diphenetyl urea** (4,4′)	di-(*p*-EtO-phenyl)-urea	$(C_2H_5O \cdot C_6H_4 \cdot NH)_2: CO$	XIII-481	300.36
2688	**Diphenic** acid (2,2′)	diphenic acid (1,10)	$(\cdot C_6H_4 \cdot CO_2H)_2$	IX-922	242.23
2689	acid (3,3′)	diphenic acid (2,9)	$(\cdot C_6H_4 \cdot CO_2H)_2$	IX-927	242.23
2690	acid (2,3′)	diphenic acid (1,9)	$(\cdot C_6H_4 \cdot CO_2H)_2$	IX-926	242.23
2691	acid (2,4′)	diphenic acid (1,8)	$(\cdot C_6H_4 \cdot CO_2H)_2$	IX-926	242.23
2692	**Diphenoxy-**diethyl ether (β,β′)	....	$(C_6H_5OCH_2CH_2)_2O$	*VI-84	258.32
2693	ethane (1,2)	....	$(C_6H_5O \cdot CH_2 \cdot)_2$	VI-146	214.27
2694	propane (1,3)	....	$(C_6H_5O \cdot CH_2)_2CH_2$	VI-147	228.29
2695	**Diphenyl**	biphenyl; xenene	$C_6H_5 \cdot C_6H_5$	V-576	154.21
2696	acetamidine (*N,N′*)	ethenyl-diPh-amidine	$CH_3C(NHC_6H_5): NC_6H_5$	XII-248	210.28
2697	acetic acid	....	$(C_6H_5)_2CH \cdot CO_2H$	IX-673	212.25
2698	amine	....	$(C_6H_5)_2NH$	XII-174	169.23
2699	amine HCl	....	$(C_6H_5)_2NH \cdot HCl$	XII-180	205.69
2700	amine sulfate	....	$(C_6H_5)_2NH \cdot H_2SO_4$	XII-180	267.31
2701	amine sulfonic acid	*N*-Ph-sulfanilic	$C_6H_5 \cdot NH \cdot C_6H_4 \cdot SO_3H$	XIV-699	249.29
2702	amine sulfonic	Ba salt	$(C_{12}H_{10}O_3NS)_2Ba$	XIV-699	633.90
2703	benzene (*o*)	diphenyl-phenylene	$C_6H_5 \cdot C_6H_4 \cdot C_6H_5$	**V-611	230.31
2704	benzene (*m*)	*iso*-diphenyl-benzene	$C_6H_5 \cdot C_6H_4 \cdot C_6H_5$	V-695	230.31
2705	benzene (*p*)	terphenyl	$C_6H_5 \cdot C_6H_4 \cdot C_6H_5$	V-695	230.31

Dioform 136
Diolene 5256
Dionin, cf. alkd.
Dioxane 2237, 6285

Dioxoallene 1240
Dioxolane 3454
Dioxy-purine 6449

Table 7-4 (*Continued*)
PHYSICAL CONSTANTS OF ORGANIC COMPOUNDS

No.	Crystalline Form and Color	Specific Gravity	Melting Point °C.	Boiling Point °C.	Solubility in 100 Parts		
					Water	Alcohol	Ether
2663	red yel./ chl.	....	210-6	....	s. bz., act.	v. sl. s.	v. sl. s.
2664	yel. lf./ac.	....	288-92	....	s. act.	sl. s.	sl. s.
2665	yel. pr./ pet.	....	55	....	v. sl. s.	v. s.	v. s.
2666	nd./pet.	$1.263^{111°}$	61-3	....	....	....	....
2667	nd./CS_2	$1.321^{71°}$	70	300 sl. d.	$0.03^{22°}$	$1.2^{15°}$	$9^{15°}$
2668	nd./al.	$1.282^{111°}$	52.5	....	v. s. bz.	v. s.	v. s. CS_2
2669	rhb.	$1.283^{111°}$	60.5 *	....	i.	s.	....
2670	nd./CS_2	$1.259^{111°}$	60-1	....	i.	$2.2^{17°}$ CS_2	....
2671	mn. pr.	$1.277^{111°}$	92-3	subl.	sl. s.	s. h.	s.
2672	yel. pl.	....	$-H_2O$, 140	d. >220	s. alk.	s. dil. a.	....
2673	nd./al.	....	82	expl. 413	sl. s. pet.	sl. s.	v. s.
2674	col. cr./al.	....	56	....	s. pet.	s.	s.
2675	col. nd./al.	....	115-6	....	sl. s. h.	sl. s. c.	sl. s. c. pet.
2676	yel. nd./al.	....	75-6	expl. 438	v. s. bz.	v. s. act.	v. s. chl.
2677	cr./al.	....	83-4	....	....	....	....
2678	col. pr./al.	....	93-4	....	i.	s. h.	....
2679	mn. pr./al.	....	90-3	....	i.	v. s. h.	....
2680	yel. nd./al.	....	147-8	....	i.	s. h.	s. h.
2681	nd./al.	....	123.5	....	....	....	....
2682	lf./al.	....	58	>350 sl. d.	....	s.	s.
2683	lf./al.	....	39	$270-5^{100mm}$	i.	s. abs.	....
2684	lq.	$0.805\frac{17}{17}°$	....	291.7	sl. s.	s.	s.
2685	pl./Me al.	....	52-3	....	....	sl. s. c.	s. Me al.
2686	rhb./al.	....	180 (violet)	195 d.	8.3 c.; 16.6 h.	6.6 c.; 10 h.	s. alk.
2687	nd./ac.; pr./al.	....	225-6	....	....	....	....
2688	mn. pr.	....	228-9	subl.	sl. s.	s.	s.
2689	lf./al.	....	356-7	....	v. sl. s. h.	s. h.	s.
2690	nd./aq.	....	216	....	sl. s. h.	s.	....
2691	lf./al.	....	251-2	....	....	....	....
2692	col. nd./ aq. al.	....	66-7	....	i.	s. h.	s.
2693	lf./abs. al.	....	97-8	....	i.; $27^{25°}$ bz.	s. h.; $23^{25°}$ act.	$9^{25°}$
2694	lf./al.	....	61	$338-40^{762mm}$	i.	s.	s.
2695	col. mn.	$0.992\frac{73}{4}°$	70.5	256.1	i.	$9.98^{19.5°}$	$6.57^{19.5°}$ Me al.
2696	nd./al.	....	132-3	....	s. a.	s. h.	s.
2697	nd./aq.	....	148	....	s. h.	s.	s.
2698	col. mn.	$1.160\frac{20}{20}°$; $1.054^{61°}$	52.9	302	i.; $58^{19.5°}$ Me al.	$56^{19.5°}$; s. bz.	s.; s. CS_2; s. ac.
2699	col. nd./al.	....	....	....	s.	s.	....
2700	col. cr.	....	123-5	....	i.	s.	s. H_2SO_4
2701	col. lf.	....	....	....	s.	s.	i.
2702	col. lf.	....	....	....	v. sl. s.	....	....
2703	pr./Me al.	....	56-7	332	i.	s.; s. chl.	s.; s. act.
2704	nd./aq. al.	....	86-7	363	i.; s. bz.	s. h.	s.; s. ac.
2705	lf. or nd.	$1.234\frac{0}{4}°$	212-3	376	s. h. bz.	v. sl. s. h.	sl. s.

Dioxy-pyrimidine 6383   *β, m. 65.5°
Diphenol 2340-3   γ, m. 48°
Diphenyl-acetamide 84.1

Diphenyl-acetone 1787
Diphenyl-acetylene 5954
Diphenyl-amine orange 4996

Table 7-4 (Continued)
PHYSICAL CONSTANTS OF ORGANIC COMPOUNDS

No.	Name	Synonym	Formula	Beil. Ref.	Formula Weight
	Diphenyl				
2706	benzidine (N)	....	$(C_6H_5 \cdot NH \cdot C_6H_4)_2$	XIII-223	336.44
2707	butyl-phenyl phosphate (p-tert)	phosphen 1	$C_4H_9 \cdot C_6H_4 \cdot O \cdot PO:$ $(O \cdot C_6H_5)_2$	....	382.40
2708	carbamine chloride	diPh-carbamyl chloride	$(C_6H_5)_2N \cdot CO \cdot Cl$	XII-428	231.68
2709	"carbazone" (s)	....	$C_6H_5 \cdot N:N \cdot CO \cdot NH \cdot$ $NH \cdot C_6H_5$	XVI-24	240.27
2710	carbohydrazide (1,5)	diPh-carbazide	$(C_6H_5NH \cdot NH)_2CO$	XV-292	242.28
2711	carbonate	phenyl carbonate	$(C_6H_5O)_2CO$	VI-158	214.22
2712	chloroarsine	sneezing gas	$(C_6H_5)_2AsCl$	XVI-845	264.59
2713	o-chlorophenyl phosphate	phosphen 3	$(C_6H_5O)_2PO \cdot$ OC_6H_4Cl	....	360.74
2714	iso-cyanate	p-xenylcarbimide	$C_6H_5 \cdot C_6H_4N:CO$	XII-1319	195.22
2715	disulfide	phenyl disulfide	$(C_6H_5S \cdot)_2$	VI-323	218.34
2716	disulfoxide	phenyl oxydisulfide	$(C_6H_5 \cdot SO)_2$	VI-324	250.34
2717	ethane (s)	dibenzyl	$(C_6H_5 \cdot CH_2 \cdot)_2$	V-598	182.27
2718	ethane (as)	....	$(C_6H_5)_2CH \cdot CH_3$	V-605	182.27
2719	ether	phenyl ether	$(C_6H_5)_2O$	VI-146	170.21
2720	ethylenediamine (N)(s)	ethylene-diphenyl-diamine	$(C_6H_5 \cdot NH \cdot CH_2 \cdot)_2$	XII-543	212.30
2721	formamidine (N,N')	methenyl-diphenyl-amine	$HC(:NC_6H_5)NHC_6H_5$	XII-236	196.25
2722	glyoxalone (4,5)	4,5-diPh-imidazolone	$C_6H_5 \cdot C:C(C_6H_5) \cdot$ $NH \cdot CO \cdot NH$	XXIV-211	236.28
2723	guanidine	melaniline	$(C_6H_5NH)_2C:NH$	XII-369	211.27
2724	hydantoin (5,5)	Dilantin	$CO \cdot NH \cdot CO \cdot NH \cdot C:$ $(C_6H_5)_2$	XXIV-410	252.28
2725	hydrazine (α,α)†	....	$(C_6H_5)_2N \cdot NH_2$	XV-122	184.24
2726	hydrazine HCl	(α,α), as, or N,N	$(C_6H_5)_2N \cdot NH_2 \cdot HCl$	XV-123	220.70
2727	hydrazine (p)	4-hydrazino-diPh	$C_6H_5 \cdot C_6H_4 \cdot NH \cdot NH_2$	XV-576	184.24
2728	iodonium hydroxide	known only in soln.	$(C_6H_5)_2I \cdot OH$	V-219	298.13
2729	iodonium iodide	....	$(C_6H_5)_2I \cdot I$	V-219	408.02
2730	ketene	....	$(C_6H_5)_2C:CO$	VII-471	194.24
2731	methane	ditan	$(C_6H_5)_2CH_2$	V-588	168.24
2732	α-naphthyl carbinol	....	$(C_6H_5)_2C(OH) \cdot C_{10}H_7$	VI-729	310.40
2733	nitrosamine	nitroso-diPh-amine	$(C_6H_5)_2N \cdot NO$	XII-580	198.23
2734	phenanthrene	(9,10)	$C_{14}H_8(C_6H_5)_2$	V-747	330.43
2735	phosphate	....	$(C_6H_5O)_2PO \cdot OH$	VI-178	250.19
2736	phosphate	....	$(C_6H_5)_2HPO_4 \cdot 2H_2O$	*VI-95	286.22
2737	phthalate	phenyl phthalate	$C_6H_4(CO_2C_6H_5)_2$	IX-801	318.33
2738	piperazine (N,N')	....	$(C_6H_5N)_2(CH_2)_4$	XXIII-8	238.34
2739	propionic acid (β,β)	β-Ph-hydrocin-namic acid	$(C_6H_5)_2CH \cdot CH_2 \cdot$ CO_2H	IX-680	226.28
2740	propiophenone (β,β)	(β,β)	$(C_6H_5)_2CH \cdot CH_2 \cdot$ $CO \cdot C_6H_5$	VII-524	286.38

† See also No. 3685

Diphenyl-benzylamine 804
Diphenyl-black base-P 314
Diphenyl-carbamyl chloride 2708

Diphenyl-carbazide 2710
Diphenyl carbinol 691
Diphenyl-carbonimide 2714
Diphenyl-carboxylic acid 5160-2

Diphenyl-dichloromethane 726
Diphenyl-diimide 599
Diphenyl-ethyl carbamate 2761
Diphenyl-ethylene 5642-3

Table 7-4 (*Continued*)
PHYSICAL CONSTANTS OF ORGANIC COMPOUNDS

No.	Crystalline Form and Color	Specific Gravity	Melting Point °C.	Boiling Point °C.	Solubility in 100 Parts		
					Water	Alcohol	Ether
2706	lf./PhMe	....	242	....	s. ac.	sl. s.	sl. s. bz.
2707	col. lq.	$1.16^{25°}_{25}$	<0	245-605mm	i.	v. s.; ∞ CCl₄	∞ bz.
2708	lf./al.	....	85-6	....	d. h.	d. alk.	....
2709	or. red nd.	....	157±, d.	....	i.	s.; s. bz.	s. chl.
2710	cr./al.	....	175	....	v. sl. s. h.	s. h.	i.; s. act.
2711	nd./al.	$1.272^{14°}$	80	302-6	i.	v. s.	s.; s. CCl₄
2712	rhb.	$1.583^{40°}$	43-4	333, in CO₂	0.2, d.	20	s.; s. bz.
2713	col. lq.	$1.30^{25°}_{25}$	<0	256-60¹⁵mm	i.	v. s.	∞ bz.; ∞ CCl₄
2714	nd./et.	....	58-60	283 d.	....	....	v. s.
2715	nd./al.	....	61	310	i.	s.	v. s.
2716	mn./al.	....	45	d.	i.; i. alk.	s. h.	s.
2717	col. pr.	$0.978^{50°}_{80}$	52.0	284	i.; i. NH₃	s.; s. SO₂	v. s.
2718	col. oil	$1.004^{20°}$	-21.5	272	i.	∞	∞
2719	col. rhb.	$1.073^{20°}$	26.9	258.3	v. sl. s.; s. ac.	5-10°, 87% al.	∞ ; s. bz.
2720	lt./aq. al.	....	65-7	....	....	v. s.	v. s.
2721	nd./bz.	....	139-43	sl. d.	v. s. chl.	0.05⁹° pet.	s.
2722	nd./al.	....	324-5	....	i.; s. h.	s.; v. sl. ac.	v. sl. s.; v. sl. s. lg.
2723	mn./al.	....	147-8	d. >170	v. sl. s. c.; s. dil. a.	9.1²⁰° 90% al.	sl. s.; s. h. bz.
2724	cr./al.	....	295-8	....	i.; s. alk.; sl. s. bz.	2; s. ac.; sl. s. chl.	sl. s.; 3 act.
2725	tri.	$1.190^{16°}$	44 (36)	220⁵⁰mm	sl. s.	s.	s.
2726	nd./al.-HCl	....	....	....	i.	v. s.	i. HCl
2727	lf./al.	....	135-6 d.	....	v. sl. s.	v. sl. s. lg.	....
2728	....	....	....	....	....	....	....
2729	yel. nd./al.	....	182		....	v. sl. s. h.	....
2730	red-yel. lq.	$1.104^{20°}_{4}$		265-70	....	....	....
2731	col. pr.	$1.001^{26°}_{4}$	26.6	265	i.	v. s.	v. s.
2732	cr./lg.	....	137-8	d.	i.; s. bz.	s. h.	v. s.
2733	yel. mn.	....	66-7	....	v. s. h. bz.	s. h.	....
2734	col. nd./al.	....	249-50	subl.	s. bz.	sl. s.	s.
2735	nd./chl. lg.	....	70		3	s.; s. bz.	s.; s. chl.
2736	ool. or./aq.	$1.242^{60°}_{25}$	51	-2H₂O, 100	3²⁵°	4²⁵° CCl₄	100²⁵°
2737	col. pr./al.	$1.572^{74°}$	73	* 405⁷⁵⁹mm	i.; s. act.	sl. s.	sl. s.
2738	nd./Me al.	....	164	230-5¹²mm	i.; 7²⁵° bz.	4²⁵° act.	2²³°
2739	nd./aq. al.	....	154-5	....	v. sl. s.	s.	....
2740	nd./al.	....	96	....	v. sl. s. lg.; s. bz.	s. h.; s. act.	sl. s.; s. chl.

* Heated rapidly; decomposed when heated slowly.

Diphenyl-glycolic acid 699	Diphenyl-ketoxime 728
Diphenyl-imidazolone 2722	Diphenyl mercury 4047
Diphenyl-ketone 724	Diphenyl-methylenediamine 4431

Diphenyl-oxide 2719
Diphenyl-phenoxy ethane 2853
Diphenyl-phenylene 2703-5
Diphenyl-phthalide 5292

Table 7-4 (*Continued*)
PHYSICAL CONSTANTS OF ORGANIC COMPOUNDS

No.	Name	Synonym	Formula	Beil. Ref.	Formula Weight
2741	**Diphenyl** quino-methane	fuchsone	$(C_6H_5)_2C:C_6H_4:O$	VII-520	258.32
2742	selenium	selenium diphenyl	$(C_6H_5)_2Se$	VI-345	233.17
2743	selenium dichloride	....	$(C_6H_5)_2SeCl_2$	VI-346	304.08
2744	semicarbazide(1,1)	....	$(C_6H_5)_2N\cdot NHCONH_2$	XV-304	227.27
2745	semicarbazide(1,4)	....	$(C_6H_5NH)_2(\cdot NHCO\cdot)$	XV-288	227.27
2746	semicarbazide(2,4)	....	$C_6H_5N(NH_2)\cdot CO\cdot$ $NH\cdot C_6H_5$	XV-277	227.27
2747	semicarbazide(4,4)	....	$(C_6H_5)_2N\cdot CONHNH_2$	*XII-257	227.27
2748	succinate	phenyl succinate	$(CH_2\cdot CO_2C_6H_5)_2$	VI-155	270.29
2749	sulfide	phenyl sulfide	$(C_6H_5)_2S$	VI-299	186.28
2750	sulfone	sulfobenzide	$(C_6H_5)_2SO_2$	VI-300	218.28
2751	4-sulfonic acid	Na salt	$C_6H_5\cdot C_6H_4\cdot SO_3Na$	XI-192	256.26
2752	sulfoxide	....	$(C_6H_5)_2SO$	VI-300	202.28
2753	thiocarbazone	dithizon	$C_6H_5N:N\cdot CS\cdot NH\cdot$ $NH\cdot C_6H_5$	XVI-26	256.33
2754	thiocarbo-hydrazide (1,5)	diPh-thiocarbazide	$(C_6H_5NH\cdot NH)_2CS$	XV-299	258.35
2755	*iso*-thiocyanate	*p*-xenyl mustard oil	$C_{12}H_9N:CS$	XII-1319	211.29
2756	tolyl methane (*m*)	3-methyl tritane	$CH_3\cdot C_6H_4\cdot CH(C_6H_5)_2$	V-710	258.37
2757	tolyl methane (*p*)	4-methyl tritane	$CH_3\cdot C_6H_4\cdot CH(C_6H_5)_2$	V-710	258.37
2758	triketone	....	$(C_6H_5\cdot CO)_2CO$	VII-871	238.25
2759	urea (*s*)	carbanilide	$(C_6H_5\cdot NH)_2CO$	XII-352	212.25
2760	urea (*as*)	....	$(C_6H_5)_2N\cdot CO\cdot NH_2$	XII-429	212.25
2761	urethane	*N*-diPh-ethyl carbamate	$(C_6H_5)_2N\cdot CO_2C_2H_5$	XII-427	241.29
2762	*o*-biphenyl phosphate	phosphen 5	$C_6H_5\cdot C_6H_4O\cdot PO:$ $(O\cdot C_6H_5)_2$	....	402.39
2763	**Diphenylene** oxide	dibenzofurane	$(C_6H_4)_2O$	XVII-70	168.20
2764	**Dipicryl**-amine (2,4,6,2′,4′,6′)	hexaNO$_2$-diPh-amine	$[(NO_2)_3C_6H_2]_2NH$	XII-766	439.21
2765	sulfide (2,4,6,2′,4′,6′)	....	$[(NO_2)_3C_6H_2]_2S$	VI-344	456.26
2766	**Dipiperonalace-tone**	dipiperonylidene-acetone	$(CH_2O_2\cdot C_6H_3\cdot CH:$ $CH)_2CO$	XIX-446	322.32
2767	**Di-*iso*-propanol-amine**	....	$(C_3H_6\cdot OH)_2NH$	**IV-737	133.19
2768	**Dipropargyl**	hexadiyne-1,5	$(CH:C\cdot CH_2\cdot)_2$	I-266	78.11
2769	**Di-*iso*-propenyl**	2,3-dimethyl-buta-diene-1,3	$[CH_2:C(CH_3)]_2$	I-256	82.15
2770	**Dipropyl** acetic acid (*n*)	....	$(C_3H_7)_2CH\cdot CO_2H$	II-350	144.22
2771	adipate (*n*)	*n*-propyl adipate	$[(CH_2)_2CO_2C_3H_7]_2$	**II-574	230.31
2772	amine (*n*)	....	$(C_2H_5\cdot CH_2)_2NH$	IV-138	101.19
2773	amine (*iso*)	....	$(C_3H_7)_2NH$	IV-154	101.19
2774	aminobenzalde-hyde	(*n*)(*p*)	$(C_3H_7)_2N\cdot C_6H_4\cdot CHO$	....	205.30
2775	aniline (*n*)(*N*)	....	$C_6H_5N(C_3H_7)_2$	XII-167	177.29
2776	barbituric acid (5,5)	proponal	$CO\cdot NH\cdot CO\cdot NH\cdot$ $CO\cdot C:(CH_2\cdot C_2H_5)_2$	XXIV-492	212.25
2777	benzene (*iso*)(*o*)	....	$[(CH_3)_2CH]_2C_6H_4$	V-447	162.28
2778	benzene (*iso*)(*m*)	alkazene 12	$[(CH_3)_2CH]_2C_6H_4$	V-447	162.28

Diphenyl selenide 2742
Diphenyl-thiocarbazide 2754
Diphenyl-thiourea 5892
Diphenylene carbinol 3247

Diphenylene disulfide 5877
Diphenylene-imine 1231
Diphenylene ketone 3248
Diphenylene ketone oxide 6451

Table 7-4 (*Continued*)
PHYSICAL CONSTANTS OF ORGANIC COMPOUNDS

No.	Crystalline Form and Color	Specific Gravity	Melting Point °C.	Boiling Point °C.	Water	Alcohol	Ether
					Solubility in 100 Parts		
2741	brn. yel. nd.	....	168-9	....	i.	v. s. act.	sl. s. h.
2742	oil	$1.338\frac{16°}{4}$	2.5	301-2	....	∞	∞
2743	pa. yel. pr.	....	182-5	....	i.	i.	i.
2744	nd./al.	....	195	....	....	s.	s. bz.
2745	nd./al.	....	179-80	....	v. sl. s. h.	s.; s. bz.	i.
2746	col. lf./al.	....	165.5 *	....	v. s. ac.	s.; s. bz.	s.; s. chl.
2747	pr./al.	....	155-6	....	s. h.; s. bz.	s.; s. chl.	v. sl. s.
2748	lf./al.	....	122-3	330	i.; s. bz.	s. CS_2	s.
2749	col. lq.	$1.119\frac{15°}{15}$	<-40	296-7	i.; ∞ bz.	s. h.	∞ ; ∞ CS_2
2750	nd./aq.	$1.248\frac{25°}{4}$	128-9	379	sl. s. h.	s. h.	s. bz.
2751	col. cr.	....	....	....	$1.5^{25°}$	$0.03^{25°}$ CCl_4	$0.03^{25°}$
2752	pr./lg.	....	70.5	340 sl. d.	sl. s. pet.	s.	s.; s. bz.
2753	b.-bl. cr.	....	....	....	i.; s. alk.	v. sl. s.; s. H_2SO_4	sl. s. s.; sl. s. chl.
2754	pr./al.	....	150±	d.	v. sl. s. bz.	v. sl. s.	v. sl. s. ac.
2755	nd./et.	....	58	....	....	....	v. s.
2756	pr./al.	$1.071^{16°}$	60.5-1.5	$353-477^{4mm}$	v. s. bz.	sl. s.	v. s.
2757	pr./Me al.	....	72	>360	sl. s. pet.	v. s. h.	v. s. bz.
2758	yel. nd./lg.	....	69-70	289^{175mm}	i.	sl. s.	s.
2759	rhb.	1.239	241-2	260-2	v. sl. s.	sl. s. h.	s.
2760	rhb.	1.276	189	....	v. sl. s.	s.	s.; s. chl.
2761	pr./lg.	....	72	>360	s.; s. pet.	v. s.; s. bz.	v. s.
2762	col. lq.	$1.20\frac{60°}{4}$	<0	$250-85^{5mm}$	i.; ∞ bz.	v. s.	∞ CCl_4
2763	lf./al.	....	86-7	287-8	i.; s. bz.	s. h.	v. s.
2764	yel. pr./ ac.	....	244-6 d.	....	i.; s. 93% HNO_3	v. sl. s. act.	i.
2765	yel. lf./ac.	....	226-30	expl. 290	i.	v. sl. s.	v. sl. s.
2766	yel. nd./ bz.	....	185	....	i.; i. lg.	sl. s.	s. chl.; s. act.
2767	....	$0.989\frac{45°}{20}$	42	248.7	....	....	....
2768	lq.	$0.805\frac{20°}{4}$	-6	85.4	i.	s.	v. s.
2769	lq.	$0.727\frac{20°}{20}$	-76	68.5	....	....	....
2770	col. lq.	$0.922\frac{0°}{4}$	....	221-2	v. sl. s.	....	....
2771	col. lq.	$0.979\frac{20°}{4}$	-20.3	$143-5^{10mm}$	i.	s.	s.
2772	col. lq.	$0.739\frac{20°}{4}$	-39.6	110-1	s.	∞	∞
2773	col. lq.	$0.722^{22°}$	-61	83.5^{743mm}	s.	s.	....
2774	lt. yel. lq.	....	....	$204-6^{20mm}$	....	....	....
2775	yel. oil	$0.910^{20.4°}$	....	245.4	i.	s.	s.
2776	cr.	....	145	....	v. sl. s. c.; $1.4^{100°}$	v. s.	v. s.
2777	col. lq.	$0.858\frac{25°}{25}$	....	210^{760mm}	i.	∞	∞
2778	col. lq.	$0.860\frac{25°}{25}$	-105	202^{760mm}	i.	∞	∞

* Decomposes to form 1,4-derivative.
Diphenyline 1711
Diphosgene 6159
Diphthalimido-ethane 3205

Dipicolinic acid 5500
Diplosal 5585
Dipropasin 2779
Dipropenyl 3573

Table 7-4 (*Continued*)
PHYSICAL CONSTANTS OF ORGANIC COMPOUNDS

No.	Name	Synonym	Formula	Beil. Ref.	Formula Weight
2779	**Dipropyl** carbanilid-4,4'-dicarboxylate (*n*)	dipropasin	$CO(NH \cdot C_6H_4 \cdot CO_2 \cdot CH_2 \cdot C_2H_5)_2$	XIV-434	384.44
2780	carbonate (*n*)	*n*-propyl carbonate	$(C_2H_5 \cdot CH_2 \cdot O)_2CO$	III-6	146.19
2781	α,α'-dibromo-succinate (*n*)	....	$(CHBr \cdot CO_2C_3H_7)_2$	....	360.05
2782	disulfide (*n*)	....	$(C_2H_5 \cdot CH_2 \cdot S \cdot)_2$	I-360	150.31
2783	ether (*n*)	*n*-propyl ether	$(C_2H_5 \cdot CH_2)_2O$	I-354	102.18
2784	ether (*iso*)	*iso*-propyl ether	$[(CH_3)_2CH]_2O$	I-362	102.18
2785	ether (*n-iso*)	....	$C_2H_5CH_2 \cdot O \cdot CH(CH_3)_2$	I-362	102.18
2786	ketone (*n*)	heptanone-4	$(C_2H_5 \cdot CH_2)_2CO$	I-699	114.19
2787	ketone (*iso*)	2,4-diMe-pen-tanone-3	$[(CH_3)_2CH]_2CO$	I-703	114.19
2788	maleate (*n*)	propyl maleate	$(CH \cdot CO_2C_3H_7)_2$	II-752	200.24
2789	malonate (*n*)	....	$CH_2(CO_2C_3H_7)_2$	II-581	188.23
2790	nitrosamine (*n*)	nitroso-diPr-amine	$(C_3H_7)_2N \cdot NO$	IV-146	130.19
2791	nitrosamine (*iso*)	....	$[(CH_3)_2CH]_2N \cdot NO$	IV-156	130.19
2792	oxalate (*n*)	propyl oxalate	$(CO_2CH_2 \cdot C_2H_5)_2$	II-539	174.20
2793	oxalate (*iso*)	....	$(CO_2 \cdot C_3H_7)_2$	II-539	174.20
2794	phthalate (*n*)	propyl phthalate	$C_6H_4(CO_2 \cdot C_3H_7)_2$	**IX-586	250.30
2795	phthalate (*iso*)	....	$C_6H_4(CO_2 \cdot C_3H_7)_2$	IX-798	250.30
2796	succinate (*n*)	propyl succinate	$(CH_2 \cdot CO_2 \cdot CH_2 \cdot C_2H_5)_2$	II-611	202.25
2797	succinate (*iso*)	....	$(CH_2 \cdot CO_2 \cdot C_3H_7)_2$	II-611	202.25
2798	sulfate (*n*)	*n*-propyl sulfate	$(C_2H_5 \cdot CH_2O)_2SO_2$	I-354	182.24
2799	sulfide (*n*)	*n*-propyl sulfide	$(C_2H_5 \cdot CH_2)_2S$	I-359	118.24
2800	sulfide (*iso*)	....	$[(CH_3)_2CH]_2S$	I-367	118.24
2801	sulfite (*n*)	....	$(C_2H_5 \cdot CH_2 \cdot O)_2SO$	I-354	166.24
2802	sulfone (*n*)	....	$(C_2H_5 \cdot CH_2)_2SO_2$	I-359	150.24
2803	sulfone (*iso*)	....	$[(CH_3)_2CH]_2SO_2$	I-367	150.24
2804	tartrate (*n*)(*d*)	propyl tartrate	$(CHOH \cdot CO_2 \cdot C_3H_7)_2$	III-516	234.25
2805	tartrate (*iso*)(*d*)	....	$(CHOH \cdot CO_2 \cdot C_3H_7)_2$	III-517	234.25
2806	urea (*n*)(*s*)	....	$(C_2H_5 \cdot CH_2 \cdot NH)_2CO$	IV-142	144.22
2807	urea (*n*)(*as*)	....	$(C_3H_7)_2N \cdot CO \cdot NH_2$	IV-143	144.22
2808	**Dipropylene glycol**	....	$(CH_3CHOHCH_2)_2O$	**I-537	134.18
2809	**"Dipyridine"**(?)	nicotyrine	$C_{10}H_{10}N_2$	XXIII-185	158.20
2810	**Dipyridyl** (2,2')	(α,α')	$(C_5H_4N)_2$	XXIII-199	156.19
2811	**Dipyridyl** (4,4')	....	$(C_5H_4N)_2$	XXIII-200	156.19
2812	**Dipyridyl** (2,3')	....	$C_{10}H_8N_2$	XXIII-200	156.19
2813	**Dipyridyl** (3,3')	....	$C_{10}H_8N_2$	XXIII-200	156.19
2814	**Diquinoyl** (2,3')	....	$(C_9H_6N)_2$	XXIII-293	256.31
2815	**Diquinoyl** (3,7')(β)	....	$(C_9H_6N)_2$	XXIII-295	256.31
2816	**Diquinoyl** (6,6')(γ)	....	$(C_9H_6N)_2$	XXIII-295	256.31
2817	**Diresorcinol** (5,5')	....	$[C_6H_3(OH)_2]_2 \cdot 2H_2O$	VI-1164	254.24
2818	**Dithio**-acetic acid	....	$CH_3 \cdot CS \cdot SH$	II-233	92.18
2819	carbamic acid	....	$NH_2 \cdot CS \cdot SH$	III-216	93.17
2820	hydroquinone (*p*)	....	$HS \cdot C_6H_4 \cdot SH$	VI-867	142.24
2821	oxamide	rubeanic acid	$(\cdot C:S \cdot NH_2)_2$	II-565	120.20
2822	resorcinol (*m*)	....	$HS \cdot C_6H_4 \cdot SH$	VI-834	142.24
2823	salicylic acid	(1,2)	$HO \cdot C_6H_4 \cdot CS \cdot SH$	X-134	170.25
2824	**Ditolualacetone** (*p*)	dixylidene acetone	$(CH_3 \cdot C_6H_4 \cdot CH: CH)_2CO$	VII-508	262.35
2825	**Ditolyl** (2,2')	....	$(CH_3 \cdot C_6H_4 \cdot)_2$	V-608	182.27
2826	**Ditolyl** (2,3')	....	$(CH_3 \cdot C_6H_4 \cdot)_2$	V-609	182.27

Dipropyl carbinol (*n*) 3543
Dipropyl carbinol (*iso*) 3531
Dipropyl mercury 4046
Dipropyl zinc 6505-6

Di-*iso*-propylidene acetcne 5268
Disalicylic acid 5585
Disalicylide 3747
Dispermin 2226

Table 7-4 (*Continued*)
PHYSICAL CONSTANTS OF ORGANIC COMPOUNDS

No.	Crystalline Form and Color	Specific Gravity	Melting Point °C.	Boiling Point °C.	Solubility in 100 Parts Water	Alcohol	Ether
2779	wh. pd.	....	171-2	....	i.	s.	....
2780	col. lq.	$0.968^{22°}$	....	168.2	....	....	....
2781	lt. yel. lq.	....	....	147-50^{7mm}	i.	s.	s.
2782	col. lq.	$0.814^{17°}$	-102±	193-5	i.	s.	s.
2783	col. lq.	$0.752^{15°}_{4}$	-122	90.1	sl. s.	∞	∞
2784	col. lq.	$0.725^{20.8°}_{0}$	-60	68.5-9.0	0.2	∞	∞
2785	col. lq.	$0.747^{12.5°}_{4}$	....	82-3	$0.5^{20°}$	....	....
2786	col. lq.	$0.816^{22°}_{4}$	-32.5	143.5	0.43	∞	∞
2787	col. lq.	$0.806^{20°}_{4}$	....	123.7	v. sl. s.	∞	∞ ; s. bz.
2788	col. lq.	$1.030^{18.4°}_{4}$	....	114-7^{6mm}	....	....	....
2789	col lq.	$1.009^{20°}_{4}$	....	228.3	....	....	....
2790	yel. oil	$0.916^{20°}_{4}$	....	205.9	v. sl. s.	....	....
2791	cr./et.	....	46	194.5	v. sl. s.	s.	s.; s. bz.
2792	col. lq.	$1.038^{0°}_{0}$	-51.7	213.5	d. h.	....	....
2793	col. lq.	....	....	190±	....	....	....
2794	col. lq.	....	....	129-32^{1mm}	i.	s.	s.
2795	col. lq.	....	....	160-19^{mm}	i.	s.	s.
2796	col. lq.	$1.001^{20°}_{4}$	-10.4	250.8	....	....	....
2797	col. lq.	$1.019^{0°}_{0}$	....	247.1	....	....	....
2798	oil	$1.106^{20°}_{4}$	d. 140-70	120^{20mm}	....	....	v. s. pet.
2799	lq.	$0.839^{20°}_{4}$	....	142-377^{2mm}	i.	s.	s.
2800	lq.	....	....	120.5^{763mm}	....	....	....
2801	col. lq.	$1.030^{20°}_{4}$	....	194	i.	s.	s.
2802	scales	$1.028^{50°}_{4}$	29-30	....	....	....	....
2803	cr.	....	36	....	v. s.	....	....
2804	lq.	$1.139^{20°}_{4}$	....	303	sl. s.	s.	s.
2805	col. lq.	$1.130^{20°}$	....	275	....	....	....
2806	nd./aq.	....	105	255	sl. s. c.	v. s.	v. s.
2807	nd./pet.	....	76	....	v. s.	....	....
2808	col. lq.	$1.025^{20°}_{20}$	....	231.8	∞	....	....
2809	lq.	$1.124^{13°}$	....	280-1	v. sl. s. h.	s.	s.
2810	pr./pet.	....	69.5	272.5	0.5	s.; s. bz.	s.; s. chl.
2811	pl.	....	111-2	304.8	s. h.	v. s.	v. s.
2812	lq.	....	....	296	v. sl. s.	....	s.
2813	hyg. nd.	$1.164^{20°}$	68	295.5-6.5	∞	∞	sl. s.
2814	yel. nd.	....	175.5-6.0	>400 sl. d.	i. h.	s. h.	s.
2815	pl./al.	....	192.5	subl.	i.	s. h.	sl. s.
2816	mn./al.	....	178	....	v. sl. s. h.	sl. s.	sl. s.; s. bz.
2817	pl. or nd.	....	310 (anh.,	-2H$_2$O, 100	s. h.	sl. s.	i. ac.
2818	red-yel. oil	$1.24^{20°}$	....	37^{15mm}	i. c.	v. s.	v. s.; s. bz.
2819	col. nd.	....	....	....	v. s. d.	v. s.; d. h.	v. s.
2820	lf./aq. al.	....	98	....	s. bz.	s.; s. lg.	s. ac.
2821	or. red cr.	....	d. 190-200	subl.	v. sl. o. o.	e.; s. alk.	i.
2822	cr.	....	27	243	....	....	....
2823	yel. nd./ pet.	....	48-50	....	sl. s.	s.	s.; s. bz.
2824	yel. nd./al.	....	175	....	....	....	....
2825	cr./al.	$0.955^{10°}$	17.8	258^{738mm}	v. s. bz.	v. s.	v. s.
2826	lq.	$0.998^{22°}$	....	273-4	....	v. s.	v. s.

Table 7-4 (*Continued*)
PHYSICAL CONSTANTS OF ORGANIC COMPOUNDS

No.	Name	Synonym	Formula	Beil. Ref.	Formula Weight
2827	**Ditolyl** (3,3′)	....	$(CH_3 \cdot C_6H_4 \cdot)_2$	V-609	182.27
2828	**Ditolyl** (4,4′)	....	$(CH_3 \cdot C_6H_4 \cdot)_2$	V-610	182.27
2829	**Ditolyl** (2,4′)	....	$(CH_3 \cdot C_6H_4 \cdot)_2$	V-609	182.27
2830	amine (2,2′)	....	$(CH_3 \cdot C_6H_4)_2NH$	XII-787	197.28
2831	amine (3,3′)	....	$(CH_3 \cdot C_6H_4)_2NH$	XII-858	197.28
2832	amine (4,4′)	....	$(CH_3 \cdot C_6H_4)_2NH$	XII-907	197.28
2833	carbonate (*o*)	dicresyl carbonate	$(CH_3 \cdot C_6H_4 \cdot O)_2CO$	VI-356	242.28
2834	carbonate (*m*)	....	$(CH_3 \cdot C_6H_4 \cdot O)_2CO$	VI-379	242.28
2835	carbonate (*p*)	....	$(CH_3 \cdot C_6H_4 \cdot O)_2CO$	VI-398	242.28
2836	disulfide (*p*)	....	$(CH_3 \cdot C_6H_4 \cdot S)_2$	VI-425	246.40
2837	guanidine (*o*)	....	$(C_7H_7 \cdot NH)_2C:NH$	XII-803	239.32
2838	ketone (*p*)	diMe-benzophenone	$(CH_3 \cdot C_6H_4)_2CO$	VII-451	210.28
2839	*m*-phenylene-diamine (*p*)	....	$(CH_3 \cdot C_6H_4NH)_2C_6H_4$	XIII-42	288.41
2840	*p*-phenylene-diamine (*p*)	....	$(CH_3 \cdot C_6H_4NH)_2C_6H_4$	XIII-81	288.41
2841	phthalate (*o*)	*o*-cresyl phthalate	$C_6H_4(CO_2 \cdot C_7H_7)_2$	....	346.39
2842	sulfide (*p*)	diMe-diPh sulfide	$(CH_3 \cdot C_6H_4)_2S$	VI-419	214.33
2843	sulfone (*p*)	....	$(CH_3 \cdot C_6H_4)_2SO_2$	VI-419	246.33
2844	thiourea (*o*)(*s*)	....	$(CH_3 \cdot C_6H_4 \cdot NH)_2CS$	XII-807	256.37
2845	thiourea (*m*)(*s*)	....	$(CH_3 \cdot C_6H_4 \cdot NH)_2CS$	XII-864	256.37
2846	thiourea (*p*)(*s*)	....	$(CH_3 \cdot C_6H_4 \cdot NH)_2CS$	XII-948	256.37
2847	urea (*o*)(*s*)	....	$(CH_3 \cdot C_6H_4 \cdot NH)_2CO$	XII-801	240.31
2848	urea (*m*)(*s*)	....	$(CH_3 \cdot C_6H_4 \cdot NH)_2CO$	XII-863	240.31
2849	urea (*p*)(*s*)	....	$(CH_3 \cdot C_6H_4 \cdot NH)_2CO$	XII-941	240.31
2850	**Diundecyl ketone**	laurone	$(C_{11}H_{23})_2CO$	I-719	338.62
2851	**Diurea**	*p*-urazine	$CO:(NH \cdot NH)_2:CO$	XXVI-204	116.08
2852	**Divinyl sulfide**	....	$(CH_2:CH)_2S$	I-434	86.16
2853	**Dixenoxy ethane** (*o*)	diPh-phenoxy ethane	$(C_6H_5 \cdot C_6H_4 \cdot O \cdot CH_2)_2$	....	366.46
2854	**Djenkolic acid** (*l*)(−)	cysteine thioform-acetal	$[HO_2C \cdot CH(NH_2) \cdot CH_2S]_2CH_2$	....	254.33
2855	**Docosane** (*n*)	....	$CH_3(CH_2)_{20}CH_3$	I-174	310.61
2856	**Dodecane** (*n*)	dihexyl	$CH_3(CH_2)_{10}CH_3$	I-171	170.34
2857	**Dodecane**	2,4,5,7-tetra-methyl-octane	$C_{12}H_{26}$	I-171	170.34
2858	**Dodecyl** acetate (*n*)	....	$CH_3 \cdot CO_2 \cdot C_{12}H_{25}$	II-136	228.38
2859	alcohol (*n*)	dodecanol-1	$CH_3(CH_2)_{10}CH_2OH$	I-428	186.34
2860	bromide (*n*)	lauryl bromide	$CH_3(CH_2)_{10}CH_2Br$	**I-133	249.24
2861	chloride (*n*)	lauryl chloride	$CH_3(CH_2)_{10}CH_2Cl$	**I-133	204.79
2862	cyanide (*n*)	....	$CH_3(CH_2)_{10}CH_2CN$	II-364	195.35
2863	**Dodecylene** (*α*)	dodecene-1	$CH_3(CH_2)_9CH:CH_2$	I-225	168.33
2864	**Dodecyne**-2	....	$CH_3(CH_2)_8C:C \cdot CH_3$	I-261	166.31
2865	**Doryl**	carbamylcholine chloride	$NH_2 \cdot CO_2 \cdot CH_2 \cdot CH_2 \cdot N(CH_3)_3Cl$	....	182.65
2866	**Dotriacontane** (*n*)	dicetyl	$CH_3(CH_2)_{30}CH_3$	I-177	450.88
2867	**Echitin**	....	$C_{32}H_{52}O_2$	....	468.77
2868	**Eicosane** (*n*)	....	$CH_3(CH_2)_{18}CH_3$	I-174	282.56
2869	**Eicosyl alcohol**	eicosanol-1	$CH_3(CH_2)_{18}CH_2OH$	I-431	298.56
2870	**Elaidic acid**	....	$C_{17}H_{33} \cdot CO_2H$	II-469	282.47
2871	**Eleostearic acid**	....	$C_{18}H_{32}O_2$	II-497	280.45

Table 7-4 (*Continued*)
PHYSICAL CONSTANTS OF ORGANIC COMPOUNDS

No.	Crystalline Form and Color	Specific Gravity	Melting Point °C.	Boiling Point °C.	Solubility in 100 Parts		
					Water	Alcohol	Ether
2827	oil	$0.999\frac{16}{4}°$	5-7	287^{13mm}	i.; s. bz.	s.	s.
2828	mn./et.	....	121-2	295	i.; s. bz.	v. sl. s. c.	s.
2829	lq.	....	....	273-6	i.; s. bz.	s.	s.
2830	lq.	....	....	$312^{727.5mm}$	v. sl. s.	....	....
2831	lq.	....	<-12	320-4	v. sl. s.	s.	s.
2832	nd.	....	79	330.5	v. sl. s.	....	....
2833	nd./al.	....	60	....	s. ac.	....	....
2834	col. cr.	....	47-9	....	i.	s. h.	s.
2835	nd./al.	....	111-3	....	v. sl. s.	sl. s. c.	....
2836	nd./al.	....	46	$210-5^{20mm}$	....	s.	v. s.
2837	cr./aq. al.	$1.10\frac{20}{4}°$	178-9	....	v. sl. s.	s. h.	s.
2838	rhb./al.	....	95	333^{725mm}	i.; s. CS_2	v. s. abs.	v. s.
2839	nd./al.	....	138-9	d.	sl. s. bz.; sl. s. ac.	sl. s. c.	sl. s.
2840	lf.	....	182	....	sl. s. bz.; sl. s. ac.	v. sl. s. c.	v. sl. s. pet.
2841	col. cr.	....	110-2	....	i.	s. h.	i.
2842	nd./al.	....	56-7	>300	i.; s. bz.	v. s. h.	v. s.
2843	pr./bz.	....	158-9	405^{714mm}	s. CS_2	s. h.	s. chl.
2844	nd./al.	....	158	216-8	i.	v. s. h.	i.
2845	nd.	....	122	....	v. sl. s. h.	s.	s. bz.
2846	rhb.	....	178	....	i.	v. sl. s. h.	i.
2847	nd./ac.	....	255-6	....	i.	sl. s. h.	s. h. ac.
2848	nd./al.	....	223-5	....	i.	s. h.	....
2849	cr./al.	....	266-8	....	i.	sl. s. c.	sl. s.
2850	pl.	$0.809\frac{69}{4}°$	69-70	....	....	i. c.	....
2851	mn./aq.	....	267-70	....	sl. s. c.	sl. s.	sl. s. h. ac.
2852	oil	$0.917\frac{15}{4}°$	....	85-6	v. sl. s.	∞	∞
2853	col. cr.	....	101-2	....	$0.02^{25°}$	$19^{25°}$ act.	$55^{25°}$ bz.
2854	nd.	....	300-50	....	....	....	....
2855	cr.	$0.778\frac{44}{4}°$	44.5	224.5^{15mm}	i.	4 h.	v. s.
2856	lq.	$0.751\frac{20}{4}°$	-9.6	214.5	i.	v. s.	v. s.
2857	oily lq.	....	....	208-10	s. lg.	....	....
2858	col. lq.	....	....	$151-2^{15mm}$	....	....	....
2859	lf.	$0.831\frac{24}{4}°$	24	255-9	i.	s.	s.
2860	col. lq.	....	ca -10	$175-80^{45mm}$	i.	s.	s.
2861	col. lq.	....	....	$128-30^{11mm}$	i.	....	s.
2862	col. lq.	....	8-9	275	....	v. s.	v. s.
2863	col. lq.	$0.762\frac{15}{4}°$	-31.5	96^{15mm}	i.	v. s.	v. s.
2864	lq.	$0.792\frac{15}{4}°$	-9	105^{15mm}	....	v. s.	v. s.
2865	col. cr.	....	203-4 d.	....	100	2	i.
2866	cr./chl.	$0.78^{70.1°}$	70-1	310^{15mm}	s. h. ac.	v. sl. s. c.	s. h.
2867	lf.	....	170	....	v. s. chl.	$0.07^{15°}$, 80% al.	sl. s.
2868	cr.	$0.778\frac{86-7}{4}°$	36.5	205^{15mm}	i.	....	∞
2869	col. wax	....	65-6	220^{3mm}	s. h. pet.	v. s. c.	s. h. bz.
2870	lf./al.	$0.851\frac{79-4}{4}°$	51-2	288^{100mm}	i.	v. s.	v. s.
2871	α lf./al. βnd./al.	....	α48-9 β72	$235^{12mm}±$, sl. d.	αv. s. et. βv. s. CS_2	β s. h. ac.	....

Table 7-4 (*Continued*)
PHYSICAL CONSTANTS OF ORGANIC COMPOUNDS

No.	Name	Synonym	Formula	Beil. Ref.	Formula Weight
2872	**Ellagic acid**	dilactone	$C_{14}H_2O_4(OH)_4\cdot 2H_2O$	XIX-261	338.23
2873	**Embelin**	embelic acid	$C_{18}H_{26}O_2(OH)_2$	....	308.42
2874	**Emodin**	4,5,7-triOH-2-Me-anthraquinone	$C_{15}H_7O_2(OH)_3$	VIII-520	270.24
2875	**Eosine**	tetraBr-fluorescein	$C_{20}H_8O_5Br_4$	XIX-228	647.92
2876	salt	soluble eosine	$C_{20}H_6O_5Br_4Na_2$	XIX-230	691.88
2877	**Ephedrine** (*l*) †	....	$C_6H_5\cdot CHOH\cdot CH\cdot (CH_3)\cdot NHCH_3$	XIII-636	165.24
2878	*pseudo-***Ephedrine** (*d*) †	*d*-isoephedrine	$C_6H_5\cdot CHOH\cdot CH\cdot (CH_3)\cdot NHCH_3$	XIII-637	165.24
2879	**Epichlorohydrin** (*α*)	chloropropylene oxide	$O\cdot CH_2\cdot CH\cdot CH_2\cdot Cl$	XVII-6	92.53
2880	**Epicyanohydrin**	cyanopropylene oxide	$O\cdot CH_2\cdot CH\cdot CH_2CN$	XVIII-261	83.09
2881	**Epiiodohydrin** (*α*)	iodopropylene oxide	$O\cdot CH_2\cdot CH\cdot CH_2I$	XVII-10	183.98
2882	**Epinine**	3,4-diOH-Ph-EtMe-amine	$(HO)_2C_6H_3\cdot CH_2\cdot CH_2\cdot NHCH_3$	*XIII-325	167.21
2883	**Equilenin** (*dl*)	(synthetic)	$C_{18}H_{18}O_2$	....	266.34
2884	**Equilenin** (*d*)	(from mares)	$C_{18}H_{18}O_2$	....	266.34
2885	benzoate	....	$C_{18}H_{17}O\cdot O_2C\cdot C_6H_5$	....	370.45
2886	**Equilenin** (*l*)	(synthetic)	$C_{18}H_{18}O_2$	....	266.34
2887	*iso-***Equilenin** (*dl*)	....	$C_{18}H_{18}O_2$	....	266.34
2888	*iso-***Equilenin** (*d*)	14-epi-equilenin	$C_{18}H_{18}O_2$	....	266.34
2889	**Ergosterol**	ergosterin	$C_{28}H_{44}O$	....	396.66
2890	**Ergothioneine** (*d*)	thiohistidine betaine	$C_9H_{15}O_2N_3S\cdot 2H_2O$	XXV-521	265.33
2891	**Eriodictyol**‡	5,7,3′,4′-tetraOH-flavanone	$(HO)_2C_7H_2O_2\cdot C_6H_3(OH)$	VIII-543	244.21
2892	**Erucic acid**	docosenoic acid	$CH_3(CH_2)_7CH:CH\cdot (CH_2)_{11}CO_2H$	II-472	338.58
2893	**Erucyl alcohol**	docosene-9-ol-22	$C_{22}H_{44}O$	I-453	324.60
2894	**Erythritol** (*dl*)	butantetrol-1,2,3,4	$(CHOH\cdot CH_2OH)_2$	I-525	122.12
2895	tetranitrate	nitroerythrite	$C_4H_6(ONO_2)_4$	I-527	302.11
2896	**Erythrosin**	tetraiodo-fluorescein	$C_{20}H_8O_5I_4$	XIX-231	835.90
2897	salt	....	$C_{20}H_6O_5I_4Na_2$	....	879.87
2898	**Esculetin**	6,7-diOH-coumarin	$C_9H_6O_4\cdot H_2O$	XVIII-98	196.16
2899	**Esculin**	....	$C_{15}H_{16}O_9\cdot 1\frac{1}{2}H_2O$	....	367.31
2900	**Estradiol** (*β*)	dihydroestrone	$CH_3\cdot C_{17}H_{19}(OH)_2$	....	272.39
2901	**Estragole**	*p*-methoxy-allyl-phenol	$CH_2:CH\cdot CH_2\cdot C_6H_4\cdot OCH_3$	VI-571	148.21
2902	**Estriol**	estrone hydrate	$C_{18}H_{24}O_3$	....	288.39
2903	**Estrone**	ketohydroxy-estrin	$CH_3\cdot C_{17}H_{18}O(OH)$	....	270.37
2904	**Ethane**	....	$CH_3\cdot CH_3$	I-80	30.07
2905	sulfinic acid	ethyl sulfinic acid	$C_2H_5\cdot SO\cdot OH$	IV-1	94.13
2906	sulfonic acid	ethyl sulfonic acid	$C_2H_5\cdot SO_2\cdot OH$	IV-5	110.13
2907	**Ethanolamine**	amino-ethyl alcohol	$NH_2\cdot CH_2\cdot CH_2OH$	IV-274	61.08

† See also Alkaloid Table.
‡ Beilstein shows structure as a chalcone.
Elon 4117
Embelic acid 2873
Embutal 4583

Emerald green base 5762
Emetine, cf. alkd.
Empirin 115
Enanthaldehyde 3521
Enanthaldoxime 3522

Enanthic acid 3520
Endoiodin 3591
Ephedrine (iso) 2878
Epiallocholesterol 1481
Epicarin 3790

Table 7-4 (Continued)
PHYSICAL CONSTANTS OF ORGANIC COMPOUNDS

No.	Crystalline Form and Color	Specific Gravity	Melting Point °C.	Boiling Point °C.	Solubility in 100 Parts		
					Water	Alcohol	Ether
2872	yel. pd.	$1.667^{18°}$	d.	$-2H_2O >$ 120	v. sl. s. h.	sl. s.	i.
2873	or. lf.	....	143	subl.	i.; s. alk.	s.	sl. s. pet.
2874	red nd./ac.	....	255-7	....	i.; s. alk.	s.; s. bz.; s. chl.	s.; s. ac.
2875	col. cr./ac.	....	....	....	i.	s.	sl. s. h. ac.
2876	red brn. pd.	....	....	....	s.	s.	....
2877	cr./et.	....	40	255	5	500	s.; s. chl.
2878	rhb. pl.	....	117	....	sl. s. c.	s.	s.
2879	lq.	$1.183\frac{25°}{25}$	-57.2	117^{756mm}	<5	∞	∞
2880	pr.	....	162	....	s. h.	s.	d. a.
2881	lq.	$2.03^{13°}$	....	160-80	i.	....	....
2882	col. cr./al.	....	188-9	....	sl. s.	sl. s. h.	....
2883	lf./act. al.	....	288 (278)	....	v. sl. s.	....	....
2884	nd./aq. al.	....	251 (vac.)	....	v. sl. s.	s.	s.
2885	cr.	....	223 (vac.)	....	....	....	....
2886	col. cr.	....	251 (vac.)	....	v. sl. s.	s.	s.
2887	col. cr.	....	223 (206)	....	v. sl. s.	s.	s.
2888	col. cr.	....	273 (258)	....	v. sl. s.	s.	s.
2889	cr.	1.04	160-3	....	i.; s. chl.	s.; s. bz.	s.
2890	col. mn./aq.	....	$290 \pm$, d.	....	$11.6^{20°}$	v. sl. s.	i.; i. chl.
2891	col. pl./al.	....	267	....	sl. s. h.	sl. s. h.	s. alk.
2892	nd./al.	$0.860\frac{5.5°}{4}$	33-4	281^{30mm}	i.	v. s.	v. s.
2893	cr.	....	34.5	$241-2^{10mm}$	v. s. ac.	v. s.	v. s. bz.
2894	tet. pr.	$1.451\frac{20°}{4}$	121.5	329-31	60	sl. s. c.	i.
2895	lf./al.	....	61	expl.	i. c.	s.; s. glyc.	s.
2896	or. cr./et.	....	....	....	i.; i. bz.	sl. s.; i. chl.	i. abs.
2897	brn. pd.	....	....	....	s.	s.	....
2898	nd.	....	>270 d. (anh.)	....	s. h.	s.	v. sl. s.
2899	pr.	....	160 d. (anh.)	$-H_2O$, 120-30	$0.15^{11°}$	4 h.; s. ac.	i. abs.
2900	nd./bz. act.	....	223 (175)	d.	s. dioxane	s.	s.
2901	lq.	$0.965^{21°}$	....	$214-6^{764mm}$	....	....	....
2902	cr./al. EtOAc	....	283 (275)	....	0.003	s.	sl. s.
2903	col. cr./al.	....	259-61 *	....	i.; s. alk.	s.; s. bz.	s.; s. chl.
2904	col. gas	$0.546^{-88°}$ 1.049(A)	-183.2	-88.6	$4.7^{20°}$ cc.; $1.8^{80°}$cc.	150 cc., abs.	....
2905	syrup	....	....	....	s. alk.	....	....
2906	hyg. cr.	$1.334^{25°}$	-17	....	s.	s.	s. alk.
2907	col. oil	$1.022^{20°}$	10.5	171^{757mm}	∞ ; sl. s. bz.	∞ ; s. chl.	1

---

* Two other forms, m. p. 254° and 256°.
Epidibromohydrin 1869-70
Epidichlorohydrin 2060-1
Epihydrin alcohol 3436
Epinephrine 157

Equisetic acid 141
Ergosterin 2889
Ergotinine, cf. alkd.
Ergotoxine, cf. alkd.
Ericin 4078

Eriodictyol (homo) 3672
Eriodictyonone 3672
Erythrene 1019
Esculetin methyl ether 5604
Esculin, cf. glcde.

Table 7-4 (Continued)
PHYSICAL CONSTANTS OF ORGANIC COMPOUNDS

No.	Name	Synonym	Formula	Beil. Ref.	Formula Weight
2908	**Ethenyl-**aminophenol	2-methyl-benzoxazole	$C_6H_4N{:}C(CH_3)O$ |_____|	XXVII-46	133.15
2909	aminothiophenol	2-methyl-benzothiazole	$C_6H_4N{:}C(CH_3)S$ |_____|	XXVII-46	149.22
2910	*p,p'*-diethoxy-diphenylamidine HCl	holocaine-HCl; phenacaine-HCl	$C_2H_5O{\cdot}C_6H_4{\cdot}NH{\cdot}C{:}$ $(CH_3)N{\cdot}C_6H_4{\cdot}$ $OC_2H_5{\cdot}HCl{\cdot}H_2O$	XIII-468	352.86
2911	**Ether**	(di)ethyl ether	$(C_2H_5)_2O$	I-314	74.12
2912	**Ethoxy-**acetic acid	glycolic ethyl ether	$C_2H_5O{\cdot}CH_2{\cdot}CO_2H$	III-233	104.11
2913	acetoanil (*p*)	*N*-ethylidene-*p*-phenetidine	$C_2H_5O{\cdot}C_6H_4{\cdot}N{:}$ $CH{\cdot}CH_3$	. . . .	163.22
2914	benzaldehyde (*o*)	. . . .	$C_2H_5O{\cdot}C_6H_4{\cdot}CHO$	VIII-43	150.18
2915	benzoic acid (*o*)	. . . .	$C_2H_5O{\cdot}C_6H_4{\cdot}CO_2H$	X-64	166.18
2916	benzoic acid (*m*)	. . . .	$C_2H_5O{\cdot}C_6H_4{\cdot}CO_2H$	X-138	166.18
2917	benzoic acid (*p*)	. . . .	$C_2H_5O{\cdot}C_6H_4{\cdot}CO_2H$	X-156	166.18
2918	diphenyl (*p*)	. . . .	$C_2H_5O{\cdot}C_6H_4{\cdot}C_6H_5$	. . . .	198.27
2919	diphenyl sulfide	(*p*)	$C_2H_5O{\cdot}C_6H_4{\cdot}S{\cdot}C_6H_5$	. . . .	230.33
2920	ethyl adipate (*β*)	. . . .	$(CH_2)_4(CO_2{\cdot}CH_2{\cdot}$ $CH_2{\cdot}O{\cdot}C_2H_5)_2$	. . . .	290.36
2921	ethyl *o*-benzoyl-benzoate (*β*)	. . . .	$C_6H_5{\cdot}CO{\cdot}C_6H_4{\cdot}CO_2{\cdot}$ $CH_2CH_2{\cdot}OC_2H_5$	. . . .	298.34
2922	ethyl carbamate	(*β*)	$NH_2{\cdot}CO_2CH_2CH_2$ OC_2H_5	. . . .	133.15
2923	ethyl chloroform-ate (*β*)	*β*-EtO-ethyl chloro-carbonate	$Cl{\cdot}CO_2CH_2{\cdot}CH_2{\cdot}$ OC_2H_5	. . . .	152.58
2924	ethyl glycolate	(*β*)	$HOCH_2CO_2{\cdot}$ $(CH_2)_2OC_2H_5$	. . . .	148.16
2925	ethyl *α*-hydroxy-*iso*-butyrate	. . . .	$(CH_3)_2C(OH)CO_2{\cdot}$ $CH_2CH_2{\cdot}OC_2H_5$	. . . .	176.21
2926	ethyl lactate (*β*)	. . . .	$CH_3{\cdot}CHOH{\cdot}CO_2CH_2{\cdot}$ $CH_2{\cdot}OC_2H_5$	. . . .	162.19
2927	phenol (*o*)	guaethol	$C_2H_5O{\cdot}C_6H_4{\cdot}OH$	VI-771	138.17
2928	phenol (*m*)	resorcinol ethyl ether	$C_2H_5O{\cdot}C_6H_4{\cdot}OH$	VI-814	138.17
2929	phenol (*p*)	hydroquinone ethyl ether	$C_2H_5O{\cdot}C_6H_4{\cdot}OH$	VI-843	138.17
2930	**Ethoxyl-**aniline	*β*-anilino-ethanol	$C_6H_5{\cdot}NH(C_2H_4OH)$	XII-182	137.18
2931	piperidine	*γ*-pipecolyl carbinol	$C_5H_{10}N{\cdot}C_2H_4OH$	XXI-4	129.20
2932	**Ethyl** abietate	. . . .	$C_{19}H_{29}CO_2{\cdot}C_2H_5$	. . . .	330.52
2933	acetamide (*N*)	. . . .	$CH_3CO{\cdot}NH{\cdot}C_2H_5$	IV-109	87.12
2934	acetanilide	acetethylanilide	$CH_3CO{\cdot}N(C_2H_5)C_6H_5$	XII-246	163.22
2935	acetate	acetic ether	$CH_3CO_2{\cdot}C_2H_5$	II-125	88.11
2936	acetate, ortho	ethenyl triEt-ether	$CH_3C(OC_2H_5)_3$	II-129	162.23
2937	acetoacetate	acetoacetic ester	$CH_3CO{\cdot}CH_2{\cdot}CO_2{\cdot}$ C_2H_5	III-632	130.14
2938	acetoacetic ester	. . . .	$CH_3CO{\cdot}CH(C_2H_5){\cdot}$ $CO_2{\cdot}C_2H_5$	III-691	158.20
2939	acetopyruvate	oxalacetone	$CH_3CO{\cdot}CH_2{\cdot}CO{\cdot}$ $CO_2{\cdot}C_2H_5$	III-747	158.16
2940	acetylglycolate	. . . .	$CH_3CO_2{\cdot}CH_2{\cdot}CO_2{\cdot}$ C_2H_5	III-237	146.14
2941	acetylsalicylate	. . . .	$CH_3CO_2{\cdot}C_6H_4{\cdot}CO_2{\cdot}$ C_2H_5	X-75	208.22

Table 7-4 (*Continued*)
PHYSICAL CONSTANTS OF ORGANIC COMPOUNDS

No.	Crystalline Form and Color	Specific Gravity	Melting Point °C.	Boiling Point °C.	Solubility in 100 Parts		
					Water	Alcohol	Ether
2908	col. lq.	1.137^0	9–10	200–1	i.	s.	∞
2909	lq.	....	12–4	238	i.	s.	s. HCl
2910	col. cr./aq.	....	189 (anh.)	....	2	s.	i.; s. chl.
2911	col. lq.	$0.708\frac{25}{4}$	α –116.3 β –123.3	34.6	7.5^{20}	∞	∞ chl.
2912	col. lq.	$1.102\frac{20}{4}$	....	206–7 sl. d.	s.	s.	s.
2913	yel. lq.	....	....	$160–27^{mm}$	....	....	....
2914	col.	....	20–2(6)	247–9	....	∞	∞
2915	oil	....	21–3	d. 300±	v. sl. s.	....	....
2916	nd./aq.	....	135–7	subl.	v. sl. s. h.	s.	s.
2917	nd.	....	196–8	....	v. sl. s. h.	....	....
2918	col. fl.	....	71–2	....	i.	s. h.	s.
2919	yel. lq.	....	....	$186–9^{10mm}$	....	....	....
2920	col. lq.	....	....	$164–6^{4mm}$	....	....	....
2921	col. pl.	....	51–3	....	i.	s.	v. s.
2922	col. cr.	....	61–2	....	s.	s.	sl. s.
2923	lt. yel. lq.	....	....	$55–65^{15mm}$	....	....	....
2924	col. lq.	....	....	$98–100^{7mm}$	....	....	....
2925	col. lq.	....	....	204–6	....	....	....
2926	col. lq.	....	....	$99–101^{10mm}$	....	....	....
2927	oil	....	28	216–7	sl. s.	∞	∞
2928	yel. lq.	....	....	246–7	v. sl. s.	s.	s.; s. bz.
2929	lf./aq.	....	66–7	246–7	v. s. h.	v. s.	v. s.
2930	lq.	$1.097\frac{20}{20}$	35±	286	4.6^{20}	s.	s.; s. chl.
2931	lq.	$1.006\frac{15}{4}$	....	227–8	∞	∞	....
2932	lq.	$1.020\frac{20}{20}$	d. >280	200^{4mm}	i.	....	....
2933	oily lq.	$0.942^{4.5^0}$	....	205	∞	∞	i. aq. alk.
2934	rhb./aq.	$0.994\frac{60}{4}$	53–4	258	i.	s.	v. s.
2935	col. lq.	$0.901\frac{20}{4}$	–83.6	77.2	8.5^{15^0}	∞	∞
2936	col. lq.	$0.885\frac{25}{4}$	....	144–6	i.c.	∞ ; ∞ chl.	∞ act.
2937	col. lq.	$1.025\frac{20}{4}$	–45	180^{755mm}	13^{17^0}; ∞	∞ ; ∞ chl.	∞
2938	oil	$0.986\frac{20}{4}$	....	198	sl. s.	∞	∞
2939	cr.	$1.125\frac{20}{4}$	18	213–5	....	....	....
2940	lq.	1.099^{17^0}	....	180	v. sl. s.	....	....
2941	col. lq.	1.157^{15^0}	....	272 sl. d.	i.	s.	s.

Table 7-4 (*Continued*)
PHYSICAL CONSTANTS OF ORGANIC COMPOUNDS

No.	Name	Synonym	Formula	Beil. Ref.	Formula Weight
2942	**Ethyl** acrylate	(polymerizes readily)	$CH_2:CH \cdot CO_2 \cdot C_2H_5$	II-399	100.12
2943	acrylic acid (β)	pentenoic acid	$C_2H_5 \cdot CH:CH \cdot CO_2H$	II-426	100.12
2944	adipate (mono)	Et hydrogen adipate	$C_2H_5O_2C \cdot (CH_2)_4 \cdot CO_2H$	* II-277	174.20
2945	alaninate HCl (*dl*)	alanine ethyl ester HCl	$CH_3CH(NH_2)CO_2 \cdot C_2H_5 \cdot HCl$	IV-390	153.61
2946	alcohol	ethanol; alcohol	$CH_3 \cdot CH_2OH$	I-292	46.07
2947	allophanate	....	$NH_2CO \cdot NHCO_2C_2H_5$	III-69	132.12
2948	allylacetoacetate	....	$CH_3CO \cdot CH(C_3H_5) \cdot CO_2C_2H_5$	III-738	170.21
2949	allyl ether	....	$CH_2:CH \cdot CH_2 \cdot O \cdot C_2H_5$	I-438	86.13
2950	amine	amino-ethane	$C_2H_5 \cdot NH_2$	IV-87	45.08
2951	amine hydro-bromide	....	$C_2H_5NH_2 \cdot HBr$	IV-91	126.00
2952	amine hydro-chloride	....	$C_2H_5NH_2 \cdot HCl$	IV-91	81.55
2953	aminobenzoate (*o*)	ethyl anthranilate	$NH_2 \cdot C_6H_4 \cdot CO_2C_2H_5$	XIV-319	165.19
2954	aminobenzoate (*m*)	....	$NH_2 \cdot C_6H_4 \cdot CO_2C_2H_5$	XIV-389	165.19
2955	aminobenzoate (*p*)	Anesthesin	$NH_2 \cdot C_6H_4 \cdot CO_2C_2H_5$	XIV-422	165.19
2956	aminocrotonate (β)	....	$CH_3C(NH_2):CH \cdot CO_2C_2H_5$	III-654	129.16
2957	amino ethanol (β)	....	$C_2H_5NH \cdot CH_2 \cdot CH_2OH$	IV-282	89.14
2958	amino phenol (*o*)	....	$C_2H_5NH \cdot C_6H_4 \cdot OH$	XIII-364	137.18
2959	amino phenol (*m*)	....	$C_2H_5NH \cdot C_6H_4 \cdot OH$	XIII-408	137.18
2960	amino phenol (*p*)	....	$C_2H_5NH \cdot C_6H_4 \cdot OH$	XIII-443	137.18
2961	amyl ether (*act.*)	....	$C_2H_5 \cdot O \cdot C_5H_{11}$	I-387	116.20
2962	amyl ether (*iso*)	....	$C_2H_5 \cdot O \cdot C_5H_{11}$	I-401	116.20
2963	*n*-amyl ketone	octanone-3	$C_2H_5 \cdot CO \cdot C_5H_{11}$	I-706	128.22
2964	*iso*-amyl ketone	....	$C_2H_5 \cdot CO \cdot C_5H_{11}$	I-706	128.22
2965	aniline †	Et-Ph-amine	$C_6H_5 \cdot NH \cdot C_2H_5$	XII-159	121.18
2966	aniline sulfonic acid (*m*)	*N*-ethyl-metanilic acid	$C_2H_5NH \cdot C_6H_4 \cdot SO_3H$	XIV-690	201.25
2967	aniline sulfonate	Na salt	$C_8H_{10}O_3NSNa \cdot 2H_2O$	XIV-690	259.26
2968	anisate (*p*)	....	$CH_3O \cdot C_6H_4 \cdot CO_2 \cdot C_2H_5$	X-159	180.21
2969	anthracene (9)	....	$(C_6H_4)_2C_2H \cdot C_2H_5$	V-678	206.29
2970	arsine	....	$C_2H_5 \cdot AsH_2$	IV-601	106.00
2971	arsonic acid	....	$C_2H_5 \cdot AsO(OH)_2$	IV-614	154.00
2972	benzamide (*N*)	....	$C_6H_5CO \cdot NH \cdot C_2H_5$	IX-202	149.19
2973	benzene	phenylethane	$C_6H_5 \cdot C_2H_5$	V-351	106.17
2974	benzene sulfonate	....	$C_6H_5 \cdot SO_3 \cdot C_2H_5$	XI-30	186.23
2975	benzoate	....	$C_6H_5 \cdot CO_2 \cdot C_2H_5$	IX-110	150.18
2976	benzoic acid (*o*)	....	$C_2H_5 \cdot C_6H_4 \cdot CO_2H$	IX-526	150.18
2977	benzoic acid (*m*)	....	$C_2H_5 \cdot C_6H_4 \cdot CO_2H$	IX-528	150.18
2978	benzoic acid (*p*)	....	$C_2H_5 \cdot C_6H_4 \cdot CO_2H$	IX-529	150.18
2979	*o*-benzoylbenzoate	....	$C_6H_5CO \cdot C_6H_4 \cdot CO_2 \cdot C_2H_5$	X-749	254.29
2980	benzoylformate	....	$C_6H_5CO \cdot CO_2C_2H_5$	X-657	178.19
2981	benzoylene urea (1)	....	$C_{10}H_{10}O_2N_2$	....	190.20
2982	benzoylene urea (3)	....	$C_{10}H_{10}O_2N_2$	XXIV-375	190.20
2983	benzylacetoacetate	....	$CH_3CO \cdot CH(C_7H_7) \cdot CO_2C_2H_5$	X-710	220.27
2984	benzylaniline	benzyl-Et-aniline	$C_6H_5N(C_2H_5)C_7H_7$	XII-1026	211.31

† See also Nos. 319-21.
Ethyl adipate 2944
Ethyl aminobenzene sulfonic acid 2966

Ethyl anilinocrotonate 39
Ethyl anthranilate 2953

Table 7-4 (*Continued*)
PHYSICAL CONSTANTS OF ORGANIC COMPOUNDS

No.	Crystalline Form and Color	Specific Gravity	Melting Point °C.	Boiling Point °C.	Solubility in 100 Parts		
					Water	Alcohol	Ether
2942	col. lq.	$0.925^{15°}$	−72	100-1	sl. s.	....	....
2943	col. lq.	$0.992^{15°}$	9-10	197-200	$6.3^{20°}$	....	....
2944	hyg. cr./ et.-pet.	....	29	180^{19mm}	....	....	....
2945	col. hyg. cr.	....	85-7	....	v. s.	s.	i.
2946	col. lq.	$*0.789^{20°}_{4}$	−114.5	78.4	∞	∞ chl.	∞
2947	cr./bz.	....	197-8	d.	s. h.	$0.5^{21°}$	$0.1^{20°}$
2948	lq.	$0.992^{17.6°}_{4}$	....	206 sl. d.	....	....	....
2949	lq.	$0.765^{20°}_{4}$	....	$66-7^{742.9mm}$	i.	∞	∞
2950	col. lq.	$0.689^{15°}_{15}$	−80.6	16.6	∞	∞	∞
2951	mn. nd./al.	1.741	....	....	s. act.	v. s.	i. chl.
2952	mn.	1.216	108-9	....	$240^{17°}$	v. s.	i.
2953	cr.	$1.117^{20°}_{4}$	13	266-8	....	s.	s.
2954	oil	....	....	294	sl. s. h.	∞	∞
2955	cr./al.	....	91-2	....	i.	s.	s.
2956	mn. pr.	$1.021^{20°}_{4}$	33.9(20)	210-5 d.	i.	s.	s.; s. bz.
2957	oil	$0.914^{20°}_{4}$	....	$167-9^{751mm}$	v. s.	v. s.	v. s.
2958	pl.	....	108-9	....	i.; sl. s. CS_2	v. s.; s. h. bz.	sl. s.
2959	cr./bz. lg.	....	62	176^{12mm}	s. h.	s.; sl. s. lg.	s.; v. s. chl.
2960	nd./aq.	....	100	....	s. h.	s.	s.
2961	lq.	$0.759^{18°}_{4}$	....	$108-9^{736mm}$	i.	∞	∞
2962	lq.	$0.764^{18°}$	....	112	i.	∞	∞
2963	lq.	$0.850^{0°}$	....	$169-70^{738mm}$	i.	∞	∞
2964	lq.	$0.830^{20°}$	....	163.5	i.	∞	∞
2965	lq.	$0.963^{20°}_{4}$	−65.8	205.5	i.	∞	∞
2966	nd./aq.	....	d. 294	....	$2.15^{15°}$	....	....
2967	lf./aq. al.	....	....	....	....	....	....
2968	lq.	$1.103^{25°}_{25}$	7-8	269-70	i.	s.	s.
2969	lf./al.	$1.041^{99°}$	59-60	....	i.	s.	....
2970	col. lq.	$1.217^{22°}$	....	36	$0.011^{9°}$	....	....
2971	cr./al.	....	99.5	....	$70^{27°}$	$39.4^{25°}$	....
2972	nd./aq.	....	70-1	298-300	sl. s. h.	....	....
2973	col. lq.	$0.867^{20°}_{4}$	−95.0	136.2	$0.011^{15°}$	∞; i. NH_3	∞; s. SO_2
2974	col. lq.	$1.219^{17°}_{?}$	....	156^{15mm}	d. h.	∞; bz.	∞; ∞ chl.
2975	col. lq.	$1.052^{15°}_{15}$	−34.7	212.4	i.; ∞ pet.	∞; ∞ chl.	s.
2976	nd./h. aq.	....	68	259	v. sl. s.	s.	s.
2977	nd./aq. al.	$1.042^{100°}_{4}$	47	....	i. c.	s.	....
2978	pr./al.	....	112-3	....	s. h.	s.	s.
2979	col. rhb. pr.	$1.122^{64.4°}_{4}$	58	....	....	v. s.	v. s.
2980	lq.	$1.122^{25.1°}_{4}$	....	265^{766mm}	....	....	....
2981	cr.	....	215-7	....	i.; s. NaOH	sl. s.	s. H_2SO_4
2982	nd.	....	197.5-8.5	....	i.	s.	s. NaOH
2983	col. lq.	$1.036^{15.5°}_{16.5}$	....	283-4	i.	∞	∞
2984	lt. yel. oil	$1.034^{18.5°}$	....	$165-79^{mm}$	i.; ∞ chl.	18	∞

* See also special table of specific gravities.   "Ethyl" base 5762
Ethyl azelate 2123   Ethyl benzoylacetate 739

Table 7-4 (*Continued*)
PHYSICAL CONSTANTS OF ORGANIC COMPOUNDS

No.	Name	Synonym	Formula	Beil. Ref.	Formula Weight
	Ethyl				
2985	benzylbenzene (*p*)	....	$C_2H_5 \cdot C_6H_4 \cdot CH_2 \cdot C_6H_5$	V-614	196.29
2986	benzyl ether	....	$C_2H_5O \cdot CH_2 \cdot C_6H_5$	VI-431	136.20
2987	benzyl ketone		$C_2H_5 \cdot CO \cdot CH_2 \cdot C_6H_5$	VII-314	148.21
2988	benzyl-*o*-toluidine	Et *o*-tolyl-benzyl-amine	$CH_3 \cdot C_6H_4 \cdot N(C_2H_5) \cdot CH_2 \cdot C_6H_5$	XII-1033	225.34
2989	bromide	bromoethane	$C_2H_5 \cdot Br$	I-88	108.97
2990	bromoacetate	....	$Br \cdot CH_2 \cdot CO_2 \cdot C_2H_5$	II-214	167.01
2991	α-bromoaceto-acetate	....	$CH_3CO \cdot CHBr \cdot CO_2 \cdot C_2H_5$	III-664	209.05
2992	bromobutyrate	(α)(*n*)	$C_2H_5 \cdot CHBr \cdot CO_2C_2H_5$	II-282	195.06
2993	bromo-*iso*-butyrate	(α)	$(CH_3)_2CBr \cdot CO_2C_2H_5$	II-296	195.06
2994	α-bromo-*n*-caproate	....	$CH_3(CH_2)_3CHBr \cdot CO_2C_2H_5$	II-325	223.12
2995	α-bromopropionate	....	$CH_3 \cdot CHBr \cdot CO_2C_2H_5$	II-255	181.04
2996	β-bromopropionate	....	$Br(CH_2)_2CO_2 \cdot C_2H_5$	II-256	181.04
2997	α-bromovalerate (*n*)	....	$CH_3(CH_2)_2CHBr \cdot CO_2C_2H_5$	II-302	209.09
2998	α-bromo-*iso*-valerate	....	$(CH_3)_2CH \cdot CHBr \cdot CO_2C_2H_5$	II-317	209.09
2999	*n*-butyl-barbituric acid (5,5)	Soneryl; Neonal	$(CONH)_2CO \cdot C:$ $\overline{\rule{1.5em}{0pt}}\ \overline{\rule{1.5em}{0pt}}$ $(C_2H_5)(C_4H_9)$	....	212.25
3000	*n*-butyl ether	....	$C_2H_5 \cdot O \cdot C_2H_4 \cdot C_2H_5$	I-369	102.18
3001	*iso*-butyl ether	....	$C_2H_5 \cdot O \cdot CH_2 \cdot CH: (CH_3)_2$	I-376	102.18
3002	*tert*-butyl ether	....	$C_2H_5 \cdot O \cdot C(CH_3)_3$	I-381	102.18
3003	*n*-butyl ketone	heptanone-3	$C_2H_5 \cdot CO \cdot C_4H_9$	I-699	114.19
3004	*iso*-butyl ketone	....	$C_2H_5 \cdot CO \cdot C_4H_9$	I-700	114.19
3005	*n*-butyl sulfide	....	$C_2H_5 \cdot S \cdot C_4H_9$	**I-399	118.24
3006	*n*-butyrate	....	$C_2H_5 \cdot CH_2 \cdot CO_2 \cdot C_2H_5$	II-270	116.16
3007	*iso*-butyrate	....	$(CH_3)_2CH \cdot CO_2 \cdot C_2H_5$	II-291	116.16
3008	*n*-caprate	....	$CH_3(CH_2)_8 \cdot CO_2C_2H_5$	II-356	200.32
3009	*n*-caproate	....	$CH_3(CH_2)_4 \cdot CO_2C_2H_5$	II-323	144.22
3010	*n*-caprylate	....	$CH_3(CH_2)_6 \cdot CO_2C_2H_5$	II-348	172.27
3011	carbazole (*N*)	....	$C_{12}H_8N \cdot C_2H_5$	XX-436	195.27
3012	carbonate, ortho †	....	$C(OC_2H_5)_4$	III-5	192.26
3013	carbostyril	....	$C_6H_4CH:C(C_2H_5) \cdot$ $\overline{\rule{1em}{0pt}}$ $NH \cdot CO$ $\overline{\rule{1em}{0pt}}$	XXI-115	173.22
3014	chaulmoograte	Chaulmestrol; Moogrol	$C_5H_7(CH_2)_{12}CO_2C_2H_5$	IX-80	308.51
3015	chloride	chloroethane	$C_2H_5 \cdot Cl$	I-82	64.52
3016	chloroacetate	....	$Cl \cdot CH_2 \cdot CO_2 \cdot C_2H_5$	II-197	122.55
3017	chloroacetoacetate	....	$Cl \cdot CH_2 \cdot CO \cdot CH_2 \cdot CO_2 \cdot C_2H_5$	III-663	164.59
3018	α-chloroaceto-acetate	....	$CH_3CO \cdot CHCl \cdot CO_2 \cdot C_2H_5$	III-662	164.59
3019	chloroformate	ethyl chlorocar-bonate	$Cl \cdot CO_2 \cdot C_2H_5$	III-10	108.53
3020	α-chloropropionate	....	$CH_3 \cdot CHCl \cdot CO_2C_2H_5$	II-249	136.58

† See also No. 2138.
Ethyl borate 6186
Ethyl bromomalonate 2133
Ethyl-butyl-acetaldehyde 3062
Ethyl-butyl alcohol 3638

Ethyl-butylamine 3640
Ethyl-butyl carbamate 1140-1
Ethyl-butyl carbinol (*iso*) 3537
Ethyl-butyl carbinol (*sec*) 3541
Ethyl butylmalonate 2134-6

Table 7-4 (*Continued*)
PHYSICAL CONSTANTS OF ORGANIC COMPOUNDS

No.	Crystalline Form and Color	Specific Gravity	Melting Point °C.	Boiling Point °C.	Solubility in 100 Parts		
					Water	Alcohol	Ether
2985	lq.	$0.985^{19°}$	....	294-5	s. chl.	s.	s.
2986	oil	$0.949^{20°}_{4}$	....	187-97^{32mm}	i.	∞	∞
2987	lq.	$0.998^{17°}$	....	230^{755mm}	i.	s.	∞
2988	yel. oil	....	....	153-71^{0mm}	....	....	....
2989	col. lq.	$1.460^{20°}_{4}$	−118.9	38.4	1.06°°; 0.93°°	∞	∞
2990	lq.	$1.506^{20°}_{20}$	....	168	i.	∞	∞
2991	lq.	$1.429^{14°}_{4}$	....	210-5 d.	....	....	....
2992	col. lq.	$1.330^{20°}_{20}$	....	177.5^{765mm} sl. d.	i.	∞	∞
2993	col. lq.	$1.329^{20°}_{20}$	....	163.6^{762mm}	i.	∞	∞
2994	lq.	....	....	205-10	....	....	....
2995	col. lq.	$1.394^{20°}_{4}$	....	160-5 sl. d.	i.	∞	∞
2996	lq.	$1.412^{18°}_{4}$	....	72-5^{15mm}	....	....	....
2997	lq.	$1.226^{18°}_{4}$	....	190-2	i.	∞	∞
2998	lq.	$1.278^{12°}_{12}$	....	186	....	....	....
2999	col. cr./ aq. al.	....	127-8	....	sl. s.; s. dil. alk.	20	11
3000	col. lq.	$0.752^{20°}_{20}$	....	91.4	i.	∞	∞
3001	col. lq.	0.751	....	78-80	i.	∞	∞
3002	col. lq.	$0.752^{20°}$	....	70^{758mm}	i.	∞	∞
3003	lq.	$0.816^{20°}_{20}$	−39	149-50	i.	∞	∞
3004	lq.	$0.815^{17°}_{4}$	....	135^{735mm}	i.	∞	∞
3005	yel. lq.	$0.857^{25°}_{25}$	....	143-5	....	....	....
3006	col. lq.	$0.874^{25°}_{4}$	−100.8	121.6	0.68$^{25°}$	∞	∞
3007	col. lq.	$0.866^{20°}_{4}$	−88.2	110.0	sl. s.	∞	∞
3008	lq.	$0.856^{25°}_{4}$	−18.0(−20)	244.9	i.; ∞ chl.	∞	∞
3009	col. lq.	$0.859^{20°}_{20}$	−67.5	167.9	i.	∞	∞
3010	col. lq.	$0.871^{18°}_{4}$	−43.2(−59)	208.5	i.	∞	m
3011	lf./et.	....	67-8	....	i.	v. s. h.	v. s.
3012	col. lq.	$0.919^{18°}_{4}$	....	158-9	....	....	....
3013	cr./aq. HCl	....	168	....	....	....	....
3014	pa. yel. lq.	$0.906^{15°}_{4}$	....	230^{20mm}	i.	∞	∞ chl.
3015	col. lq.	$0.903^{10°}$	−138	12.3	0.45$^{0°}$	∞	∞
3016	col. lq.	$1.159^{20°}_{4}$	−26	144	i.	∞	∞
3017	lq.	$1.218^{17°}_{4}$	−8±	220^{756mm}	v. sl. s.	∞	∞
3018	lq.	$1.19^{14°}_{18}$	....	193 d.	v. sl. s.	s.	s.
3019	col. lq.	$1.138^{20°}_{4}$	−80.6	94-5	d.	∞; ∞ bz.	∞; ∞ chl.
3020	col. lq.	$1.087^{20°}_{4}$	....	147-8	i.	∞	∞

Table 7-4 (*Continued*)
PHYSICAL CONSTANTS OF ORGANIC COMPOUNDS

No.	Name	Synonym	Formula	Beil. Ref.	Formula Weight
	Ethyl				
3021	β-chloropropionate	. . . .	$Cl·CH_2·CH_2·CO_2C_2H_5$	II-250	136.58
3022	chlorosulfate	ethyl chlorosulfonate	$C_2H_5O·SO_2Cl$	I-327	144.58
3023	cinnamate (*trans*)	. . . .	$C_6H_5·C_2H_2·CO_2C_2H_5$	IX-581	176.22
3024	crotonate (α)	. . . .	$C_3H_5·CO_2·C_2H_5$	II-411	114.15
3025	crotonic acid (α)	. . . .	$CH_3·CH:C(C_2H_5)·$ CO_2H	II-440	114.15
3026	crotonic acid (α)	. . . .	$CH_3·CH:C(C_2H_5)·$ CO_2H	II-440 *II-408	114.15 114.15
3027	cyanate	. . . .	$NCO(C_2H_5)$	. . . .	71.08
3028	*iso*-cyanate	. . . .	$C_2H_5·N:CO$	IV-122	71.08
3029	*iso*-cyanide	ethyl carbylamine	$C_2H_5·NC$	IV-107	55.08
3030	cyanoacetate	cyanacetic ester	$NC·CH_2·CO_2·C_2H_5$	II-585	113.12
3031	α-cyanocinnamate	Et β-Ph-α-cyanoacrylate	$C_6H_5·CH:C(CN)·$ $CO_2·C_2H_5$	IX-894	201.23
3032	cyanoformate	cyanethyl carbonate	$NC·CO_2·C_2H_5$	II-547	99.09
3033	cyclopentan-1-one-2-carboxylate	. . . .	$CO(CH_2)_3CH·CO_2·$ $\underline{\quad\quad\quad}$ C_2H_5	X-597	156.18
3034	diacetoacetate	diacetoethyl acetate	$(C_2H_3O)_2CH·CO_2C_2H_5$	III-751	172.18
3035	diazoacetate	diazoethyl acetate	$N_2CH·CO_2·C_2H_5$	* III-211	114.10
3036	dibromoacetate	. . . .	$Br_2CH·CO_2·C_2H_5$	II-219	245.91
3037	α,β-dibromobutyrate	. . . .	$CH_3(CHBr)_2CO_2C_2H_5$	II-284	273.96
3038	di-*n*-butylcarbamate	(*N,N*)	$(C_4H_9)_2N·CO_2C_2H_5$	. . . .	201.31
3039	dichloroacetate	. . . .	$Cl_2CH·CO_2·C_2H_5$	II-203	157.00
3040	dichloroarsine	. . . .	$C_2H_5·AsCl_2$	IV-603	174.89
3041	3,5-diiodosalicylate	(3,5;2,1)	$I_2C_6H_2(OH)CO_2C_2H_5$	X-114	417.97
3042	β,β-dimethylacrylate	. . . .	$(CH_3)_2C:CHCO_2C_2H_5$	II-433	128.17
3043	3,5-dinitrobenzoate	. . . .	$(NO_2)_2C_6H_3·CO_2C_2H_5$	IX-414	240.17
3044	diphenylamine	. . . .	$C_2H_5·N:(C_6H_5)_2$	XII-181	197.28
3045	diphenylphosphine	. . . .	$C_2H_5·P:(C_6H_5)_2$	XVI-759	214.25
3046	fluoride	fluoroethane	$CH_3·CH_2F$	I-82	48.06
3047	formamide (*N*)	. . . .	$H·CO·NHC_2H_5$	IV-108	73.10
3048	formate	. . . .	$H·CO_2·C_2H_5$	II-19	74.08
3049	formate, ortho	aethon	$H·C(OC_2H_5)_3$	II-20	148.20
3050	furoate (α)	ethyl pyromucate	$C_4H_3O·CO_2·C_2H_5$	XVIII-275	140.14
3051	furoate (β)	. . . .	$C_4H_3O·CO_2·C_2H_5$	. . . .	140.14
3052	furoylacetate (α)	. . . .	$C_4H_3O·CO·CH_2·$ $CO_2·C_2H_5$	XVIII-408	182.18
3053	β-furylacrylate	. . . .	$C_4H_3O·CH:CH·CO_2·$ C_2H_5	XVIII-300	166.18
3054	glutaric acid (α)	. . . .	$C_2H_5·C_3H_5(CO_2H)_2$	II-676	160.17
3055	glutaric acid (β)	. . . .	$C_2H_5·CH(CH_2CO_2H)_2$	II-676	160.17
3056	glycerate	. . . .	$(HO)_2C_3H_3·CO_2C_2H_5$	III-397	134.13
3057	glycinate	. . . .	$H_2N·CH_2·CO_2·C_2H_5$	IV-340	103.12
3058	glycinate HCl	. . . .	$C_4H_9O_2N·HCl$	IV-342	139.58
3059	glycine (*N*)	. . . .	$C_2H_5NH·CH_2·CO_2H$	IV-349	103.12
3060	glycolate	. . . .	$HO·CH_2·CO_2·C_2H_5$	III-236	104.11

Table 7-4 (*Continued*)
PHYSICAL CONSTANTS OF ORGANIC COMPOUNDS

No.	Crystalline Form and Color	Specific Gravity	Melting Point °C.	Boiling Point °C.	Solubility in 100 Parts Water	Alcohol	Ether
3021	col. lq.	$1.109^{20°}_{4}$	....	$162\text{-}37^{65mm}$	....	....	....
3022	lq.	$1.263^{18°}$	....	58^{20mm}	s. chl.	s. bz.	s.
3023	col. lq.	$1.049^{20°}_{4}$	12(7.5)	271	i.	∞	∞
3024	col. lq.	$0.924^{15°}_{4}$	....	138^{748mm}	i.	s.	s.
3025	col. pr.	....	41.5	209	v. sl. s.	v. s.	v. s.
3026	oily lq.*;	....	−35	199.5^{750mm}	i.	∞	∞
	nd./al.†	$0.958^{50°}_{4}$	41-2	$204\text{-}77^{55mm}$			
3027	lq.	$1.127^{15°}$	....	d.	i.	∞	∞
3028	lq.	$0.907^{16°}_{4}$	....	60	d.	....	....
3029	col. lq.	$0.738^{25°}$	<−66	78-9	sl. s.		s.
3030	col. lq.	$1.062^{20°}_{4}$	−22.5	208^{753mm}	$22^{5°}$; $9^{80°}$	∞	∞
3031	nd./al.**	....	50-1	360 sl. d.	s. chl.; s. ac.	12 c.	s.; s. bz.
3032	lq.	$1\ 003^{20°}_{4}$	....	115-6	....	....	....
3033	col. lq.	$1.098^{0°}$	....	218 sl. d.	....	....	....
3034	col. lq.	$1.089^{25°}_{25}$	....	209-11 sl. d.	sl. s.	v. s.	v. s.
3035	yel. oil	$1.085^{18°}_{4}$	−22	$140\text{-}17^{20mm}$	sl. s.	∞	∞
3036	oil	$1.903^{20°}_{20}$	....	192-4	i.	∞	∞
3037	lq.	....	....	$123\text{-}4^{30mm}$	....	....	....
3038	col. lq.	....	....	$101\text{-}3^{6mm}$	....	....	....
3039	col. lq.	$1.282^{20°}_{4}$	....	158	i.	∞	∞
3040	col. lq.	$1.742^{15°}_{4}$	....	156	sl. s.	∞	m
3041	col. lf./al.	....	133	d. >200°	i. c.	s. h.	sl. s.
3042	lq.	$0.922^{21°}_{21}$	....	154-5	v. s. CS₂; v. s. lg.	v. s.; v. s. chl.	v. s.; v. s. bz.
3043	nd./al.	$1.295^{111°}$	91-2	....	....	$0.6^{13°}$	s. h. al.
3044	lq.	....	....	295-7	i.	s.	....
3045	oil	....	....	293	....	s.	s. bz.
3046	gas	1.7(A)	....	−32	$190^{14°}$cc.	v. s.	....
3047	lq.	$0.952^{21°}$	<−30	197-9	∞	∞	∞
3048	col. lq.	$0.923^{20°}_{4}$	−79.4	54.2	$1^{118°}$	∞	∞
3049	lq.	$0.939^{25°}_{25}$	−76	145-6	v. sl. s.	∞	∞
3050	lf.	$1.117^{20.8°}_{8}$	34	195^{755mm}	i.	∞	∞
3051	lq.	$1.038^{20°}_{20}$	....	$65\text{-}71^{4mm}$	....	....	....
3052	lt. yel. oil	$1.165^{17°}$	....	$143\text{-}5^{10mm}$	i.; v. s. NH₄OH	s.	s.
3053	yel. cr.	$1.09^{20°}_{4}$	24.5	232-3	i.	∞	∞
3054	col. or.	....	60.5	250-60 sl. d.	v. s.	v. s.	v. s.
3055	pr./chl.	....	73	....	v. s.	v. s.	v. s.
3056	lq.	$1.191^{15°}_{15}$	....	121^{14mm}	s.	s.	....
3057	col. oil	$1.028^{20°}_{4}$	<−20	148^{748mm} sl. d.	∞	∞	∞
3058	nd.	....	144	subl.	v. s.	v. s.	d. alk.
3059	lf./al.	....	>160 d.	....	s.	s.	....
3060	col. lq.	$1.087^{15°}_{4}$	....	160	....	v. s.	v. s.

* Easily converted to solid modification.
† Also a liquid form, d. on boiling & forms some solid modification.
Ethyl dithiooxalate 2165
Ethyl ether 2911

Ethyl ethyl-acetoacetate 2938
Ethyl ethyl-amyl-malonate 2167
Ethyl ethyl-butyl-malonate 2168
Ethyl *N*-ethyl-carbamate 3184
Ethyl-ethylene 1147

Table 7-4 (*Continued*)
PHYSICAL CONSTANTS OF ORGANIC COMPOUNDS

No.	Name	Synonym	Formula	Beil. Ref.	Formula Weight
3061	**Ethyl** glycol ether	....	$HO \cdot CH_2 \cdot CH_2 \cdot O \cdot C_2H_5$	I-467	90.12
3062	hexaldehyde (2)	Et-Bu-acetaldehyde	$C_4H_9 \cdot CH(C_2H_5) \cdot CHO$	I-707	128.22
3063	hydracrylate	....	$HO \cdot CH_2 \cdot CH_2 \cdot CO_2C_2H_5$	III-297	118.13
3064	hydrazine		$C_2H_5 \cdot NH \cdot NH_2$	IV-550	60.10
3065	hydrocinnamate	....	$C_6H_5(CH_2)_2CO_2C_2H_5$	IX-511	178.23
3066	hydrogen oxalate	ethyl oxalic acid	$HO_2C \cdot CO_2 \cdot C_2H_5$	II-535	118.09
3067	hydrogen phthalate (*o*)	....	$C_6H_4(CO_2C_2H_5) \cdot CO_2H$	IX-797	194.19
3068	hydrogen tetra-chlorophthalate	....	$HO_2C \cdot C_6Cl_4 \cdot CO_2 \cdot C_2H_5$	IX-820	331.97
3069	hydrogen sulfate	ethyl sulfuric acid	$C_2H_5O \cdot SO_2 \cdot OH$	I-325	126.13
3070	hydrogen tartrate (*d*)	....	$HO_2C \cdot (CHOH)_2 \cdot CO_2C_2H_5$	III-512	178.14
3071	hydroselenide		$C_2H_5 \cdot SeH$	I-349	109.03
3072	*p*-hydroxybenzoate †	....	$HO \cdot C_6H_4 \cdot CO_2C_2H_5$	X-159	166.18
3073	α-hydroxy-*iso*-butyrate	....	$(CH_3)_2C(OH)CO_2 \cdot C_2H_5$	III-315	132.16
3074	hydroxylamine (α)	....	$C_2H_5O \cdot NH_2$	I-336	61.08
3075	hydroxylamine (β)	....	$C_2H_5 \cdot NH \cdot OH$	IV-535	61.08
3076	hypochlorite	....	$C_2H_5 \cdot O \cdot Cl$	I-324	80.51
3077	iodide	iodoethane	$CH_3 \cdot CH_2I$	I-96	155.97
3078	iodoacetate	....	$ICH_2 \cdot CO_2 \cdot C_2H_5$	II-222	214.00
3079	lactate	....	$CH_3 \cdot CHOH \cdot CO_2C_2H_5$	III-280	118.13
3080	laurate	....	$CH_3(CH_2)_{10}CO_2C_2H_5$	II-361	228.38
3081	levulinate	....	$CH_3 \cdot CO(CH_2)_2 \cdot CO_2 \cdot C_2H_5$	III-675	144.17
3082	malonic acid	....	$C_2H_5 \cdot CH(CO_2H)_2$	II-643	132.12
3083	mercaptan	ethanthiol	$C_2H_5 \cdot SH$	I-340	62.13
3084	mesoxalate	....	$CO(CO_2C_2H_5)_2$	III-769	174.15
3085	methylaceto-acetate	Me-acetoacetic ester	$CH_3CO \cdot CH(CH_3) \cdot CO_2C_2H_5$	III-679	144.17
3086	methyl acrylate	....	$C_3H_5 \cdot CO_2 \cdot C_2H_5$	II-423	114.15
3087	*S*-methylxanthate	....	$C_2H_5O \cdot CS \cdot SCH_3$	III-210	136.24
3088	myristate	....	$C_{13}H_{27}CO_2 \cdot C_2H_5$	II-365	256.43
3089	naphthalene (α)	....	$C_2H_5 \cdot C_{10}H_7$	V-569	156.23
3090	naphthalene (β)	....	$C_2H_5 \cdot C_{10}H_7$	V-569	156.23
3091	α-naphthylamine	....	$C_2H_5 \cdot NH \cdot C_{10}H_7$	XII-1222	171.24
3092	β-naphthylamine	....	$C_2H_5 \cdot NH \cdot C_{10}H_7$	XII-1274	171.24
3093	α-naphthyl ether	....	$C_{10}H_7 \cdot O \cdot C_2H_5$	VI-606	172.23
3094	β-naphthyl ether	neroline; bromelia	$C_{10}H_7 \cdot O \cdot C_2H_5$	VI-641	172.23
3095	nicotinate	....	$C_5H_4N \cdot CO_2C_2H_5$	XXII-39	151.17
3096	nitrate	nitric ether	$C_2H_5 \cdot O \cdot NO_2$	I-329	91.07
3097	nitrite	nitrous ether	$C_2H_5 \cdot O \cdot NO$	I-329	75.07
3098	nitroacetate	....	$NO_2 \cdot CH_2 \cdot CO_2 \cdot C_2H_5$	II-225	133.10
3099	nitrobenzoate (*o*)	....	$NO_2 \cdot C_6H_4 \cdot CO_2C_2H_5$	IX-372	195.18
3100	nitrobenzoate (*m*)	....	$NO_2 \cdot C_6H_4 \cdot CO_2C_2H_5$	IX-378	195.18
3101	nitrobenzoate (*p*)	....	$NO_2 \cdot C_6H_4 \cdot CO_2C_2H_5$	IX-390	195.18
3102	nitrocinnamate (*o*)	....	$NO_2 \cdot C_6H_4 \cdot CH:CH \cdot CO_2 \cdot C_2H_5$	IX-605	221.21
3103	nitrocinnamate (*m*) (*trans*)	....	$NO_2 \cdot C_6H_4 \cdot CH:CH \cdot CO_2 \cdot C_2H_5$	IX-606	221.21
3104	nitrocinnamate (*p*)	....	$NO_2 \cdot C_6H_4 \cdot CH:CH \cdot CO_2 \cdot C_2H_5$	IX-607	221.21

† See also ethyl salicylate.

Ethyl ethyl-malonate 2169	Ethyl fumarate 2174	Ethyl heptyl-malonate **2177**
Ethyl ethyl-phenyl-malonate 2170	Ethyl glutaconate 2175	Ethyl hexenal 3141
Ethyl ethyl-propyl-malonate 2171	Ethyl glutarate 2176	Ethyl-hexyl carbinol 4920
Ethyl *S*-ethylxanthate 2164	Ethyl glycolic acid 2912	Ethyl itaconate 2179
	Ethyl *n*-heptylate 3108	Ethyl malate 2182

Table 7-4 (*Continued*)
PHYSICAL CONSTANTS OF ORGANIC COMPOUNDS

No.	Crystalline Form and Color	Specific Gravity	Melting Point °C.	Boiling Point °C.	Solubility in 100 Parts		
					Water	Alcohol	Ether
3061	lq.	$0.935^{15°}_{15°}$	....	$134\text{-}5^{748mm}$	....	....	s. lq. NH_3
3062	col. lq.	$0.820^{20°}_{20°}$	<-100	163-4	$0.07^{20°}$	....	....
3063	col. lq.	$1.064^{25°}$	....	185-90	∞	∞	∞
3064	hyg. lq.	....	....	99.5^{709mm}	v. s.	v. s.	v. s.
3065	lq.	$1.015^{20°}_{4}$	....	249	i.	s.	s.
3066	col. lq.	$1.218^{20°}_{4}$	....	117^{15mm}	....	....	....
3067	oil	....	2	d.	sl. s.	s.	s.
3068	cr.	....	94-5	d. 150	i.; s. aq. alk. carb.	s.	s.
3069	syrup	$1.316^{17°}$	....	d.	∞; d. h.	∞; d. h.	∞
3070	col. hyg. pr.	....	90±	....	s.; d. h.	s.	i.
3071	lq.	$1.395^{24°}_{4}$	....	53.5	i.	....	....
3072	col. cr.	....	116-8	297-8	$1^{80°}$	$72^{25°}$ act.	$45^{25°}$
3073	col. lq.	....	....	149-50	d. h.	....	....
3074	col. lq.	$0.883^{7.5°}$	....	68	∞	∞	∞
3075	nd./lg.	$0.908^{20°}_{4}$	59 d.	....	v. s.	v. s.	sl. s.
3076	yel. lq.	$1.013^{-4°}_{4}$	expl.	36^{752mm}	∞ bz.	∞ chl.	∞
3077	col. lq.	$1.933^{20°}_{4}$	−110.9	72.4	$0.4^{20°}$	∞	∞
3078	col. oil	$1.817^{12.7°}$	....	178-80	....	....	....
3079	oil	$1.030^{25°}_{4}$	....	155	∞	∞	∞
3080	oil	$0.868^{13°}_{4}$	−10.7	269	i.	s.	∞
3081	col. lq.	$1.016^{20°}_{20}$	....	205.2^{756mm}	v. s.	∞	
3082	col. pr.	....	111.5	d. 160	v. s.	v. s.	v. s.
3083	lq.	$0.839^{20°}_{4}$	−147	35.1	1.5; s. alk.	s.	s.
3084	yel. gn. lq.	$1.119^{20°}_{20}$	−30±	220±	s.	....	s.
3085	col. lq.	$1.019^{20°}_{4}$	....	186.8	i.	s.	s.
3086	col. lq.	$0.913^{15.6}$	....	118	i.	s.	s.
3087	lq.	$1.119^{25°}_{4}$	....	183-4	i.	s.	s.
3088	col. cr.	$0.856^{20°}_{4}$	10.5-1.5	295	i.	sl. s.	sl. s.
3089	lq.	$0.990^{25°}_{25}$	−27	258^{758mm} sl. d.	i.	∞	∞
3090	lq.	$1.002^{25°}_{0}$	−19	251	i.	∞	∞
3091	oil	$1.060^{20°}_{4}$	....	303^{723mm}	i.	s.	s.
3092	oil	$1.057^{20°}_{4}$	....	316-7	i.	s.	s.
3093	cr.	$1.061^{20°}_{20}$	5.5	276.4	i.	s.	s.
3094	pl.	$1.064^{20°}_{20}$	37.5	282	i.; s. chl.	s.; s. bz.	s.; s. pet.
3095	col. oil	....	8-9	225	sl. s.	s.	s.; s. bz.
3096	col. lq.	$1.100^{25°}_{4}$	−102	87-8	$1.3^{55°}$	∞	∞
3097	lq.	$0.900^{15.5°}$	....	17	v. sl. s.	∞	∞
3098	col. lq.	$1.199^{20°}_{4}$	....	$105\text{-}7^{25mm}$	v. sl. s.	∞	....
3099	tri.	....	30	149^{10mm}	i.	s.	s.
3100	mn. pr.	....	41	298±	i.	s.	s.
3101	tri./al.	....	57	....	i.	s.	s.
3102	yel. rhb.	....	44	....	v. s. bz.	v. s. h.	v. s.
3103	col. mn./al.	....	78-9	....	i.	sl. s.	sl. s.
3104	wh. tri.	....	141-2	....	i.	v. sl. s. c.	s. ac.

Table 7-4 (*Continued*)
PHYSICAL CONSTANTS OF ORGANIC COMPOUNDS

No.	Name	Synonym	Formula	Beil. Ref.	Formula Weight
3105	**Ethyl** nitrolic acid	....	$CH_3 \cdot C(NOH) \cdot NO_2$	II-189	104.07
3106	octyl ether (*n*)	....	$C_2H_5 \cdot O \cdot C_8H_{17}$	I-419	158.29
3107	octyl ketone (*n*)	undecanone-3	$C_2H_5 \cdot CO \cdot C_8H_{17}$	I-713	170.30
3108	oenanthylate	ethyl *n*-heptylate	$C_6H_{13}CO_2 \cdot C_2H_5$	II-340	158.24
3109	oleate	....	$C_{17}H_{33}CO_2 \cdot C_2H_5$	II-467	310.52
3110	oxanilate	....	$C_6H_5 \cdot NH \cdot CO \cdot CO_2 \cdot C_2H_5$	XII-282	193.20
3111	oxindole (1)	N-Et-oxindole	$C_8H_6ON \cdot C_2H_5$	XXI-283	161.21
3112	palmitate	....	$C_{15}H_{31}CO_2 \cdot C_2H_5$	II-372	284.49
3113	pelargonate	....	$C_8H_{17}CO_2 \cdot C_2H_5$	II-353	186.30
3114	phenol (*o*)	phlorol	$C_2H_5 \cdot C_6H_4 \cdot OH$	VI-470	122.17
3115	phenol (*m*)	....	$C_2H_5 \cdot C_6H_4 \cdot OH$	VI-471	122.17
3116	phenol (*p*)	....	$C_2H_5 \cdot C_6H_4 \cdot OH$	VI-472	122.17
3117	γ-phenoxybutyrate	....	$C_6H_5O \cdot CH_2(CH_2)_2 \cdot CO_2 \cdot C_2H_5$	**VI-159	208.26
3118	phenylacetate	....	$C_6H_5 \cdot CH_2 \cdot CO_2 \cdot C_2H_5$	IX-434	164.21
3119	phenylacetic acid	....	$C_6H_5 \cdot CH(C_2H_5)CO_2H$	IX-541	164.21
3120	phenylacetylene	1-phenyl-butyne-1	$C_6H_5 \cdot C \vdots C \cdot C_2H_5$	V-517	130.19
3121	phenylbromo-acetate	(*dl*)	$C_6H_5 \cdot CHBrCO_2C_2H_5$	IX-452	243.11
3122	phenyl carbinol (*dl*)	sec-Ph-Pr-alcohol	$C_6H_5 \cdot CHOH \cdot C_2H_5$	VI-502	136.20
3123	phenyl carbinol (*l*)	....	$C_6H_5 \cdot CHOH \cdot C_2H_5$	VI-502	136.20
3124	phenyl-ethanol-amine	β-ethylanilino-ethyl alcohol	$C_6H_5N(C_2H_5)CH_2 \cdot CH_2OH$	XII-183	165.24
3125	phenylglycinate	N-Ph-glycine Et ester	$C_6H_5NH \cdot CH_2CO_2 \cdot C_2H_5$	XII-470	179.22
3126	phenylhydantoin (*dl*) (5,5)	Nirvanol	$\underset{(C_2H_5)(C_6H_5)}{CO \cdot NH \cdot CO \cdot NH \cdot C :}$	*XXIV-348	204.23
3127	phenylhydrazine	(α,α)	$C_6H_5N(C_2H_5) \cdot NH_2$	XV-119	136.20
3128	phenylhydrazine	(α,β)	$C_6H_5 \cdot NH \cdot NH \cdot C_2H_5$	XV-120	136.20
3129	phenyl ketone	propiophenone	$C_2H_5 \cdot CO \cdot C_6H_5$	VII-300	134.18
3130	2-phenyl-6-methyl-cinchoninate	neocincophen; neo-quinophan; Novatophan	$CH_3 \cdot C_9H_4N(C_6H_5) \cdot CO_2 \cdot C_2H_5$	*XXII-520	291.35
3131	phenylpropiolate	....	$C_6H_5 \cdot C \vdots C \cdot CO_2 \cdot C_2H_5$	IX-634	174.20
3132	phenyl sulfide	thiophenetole	$C_6H_5 \cdot S \cdot C_2H_5$	VI-297	138.23
3133	phenylsulfone	....	$C_6H_5 \cdot SO_2 \cdot C_2H_5$	VI-297	170.23
3134	phenyl urea (*N,N'*)	....	$C_2H_5NH \cdot CO \cdot NHC_6H_5$	XII-348	164.21
3135	phosphine	....	$C_2H_5 \cdot PH_2$	IV-581	62.05
3136	propargyl ether	....	$CH \vdots C \cdot CH_2 \cdot O \cdot C_2H_5$	I-454	84.12
3137	propiolate	....	$CH \vdots C \cdot CO_2 \cdot C_2H_5$	II-477	98.10
3138	propionate	....	$C_2H_5 \cdot CO_2 \cdot C_2H_5$	II-240	102.13
3139	*n*-propylaceto-acetate	....	$CH_3CO \cdot CH(C_3H_7) \cdot CO_2 \cdot C_2H_5$	III-700	172.23
3140	*iso*-propylaceto-acetate	....	$CH_3CO \cdot CH(C_3H_7) \cdot CO_2 \cdot C_2H_5$	III-702	172.23
3141	β-*n*-propyl-acrolein (α)	2-Et-hexenal	$C_3H_7 \cdot CH \vdots C(C_2H_5) \cdot CHO$	I-744	126.20
3142	*n*-propyl ether	....	$C_2H_5 \cdot O \cdot CH_2 \cdot C_2H_5$	I-354	88.15
3143	*iso*-propyl ether	....	$C_2H_5 \cdot O \cdot CH(CH_3)_2$	I-362	88.15
3144	*n*-propyl ketone	hexanone-3	$C_2H_5 \cdot CO \cdot CH_2 \cdot C_2H_5$	I-690	100.16
3145	*iso*-propyl ketone	2-Me-pentanone-3	$C_2H_5 \cdot CO \cdot CH(CH_3)_2$	I-691	100.16
3146	pyridine (2)	....	$C_2H_5 \cdot C_5H_4N$	XX-241	107.16

Ethyl nitrophthalate 2190
Ethyl nitroso-ethylcarbamate 4886
Ethyl nitroso-methylcarbamate 4891
Ethyl orthoformate 3049
Ethyl orthopropionate 6193

Ethyl oxalate 2193
Ethyl oxalic acid 3066
Ethyl oxamate 5015
Ethyl-phenylamine 2965
Ethyl-phenyl-barbituric acid 5113

Table 7-4 (*Continued*)
PHYSICAL CONSTANTS OF ORGANIC COMPOUNDS

No.	Crystalline Form and Color	Specific Gravity	Melting Point °C.	Boiling Point °C.	Solubility in 100 Parts		
					Water	Alcohol	Ether
3105	rhb.	....	88 d.	....	s.	s.	s.
3106	lq.	$0.801^{0°}$	....	189	i.	s.	s.
3107	lq.	$0.827^{20°}_{4}$	12.5	227	i.	s.	s.
3108	col. lq.	$0.872^{20°}_{20}$	−66.1	187-8	$0.029^{20°}$	∞	∞ ; ∞ chl.
3109	oil	$0.869^{20°}_{4}$	<−15	216-8^{15mm}	i.	∞	∞
3110	pl. or pr./al.	....	66-7	260-300 (sl. d.)	sl. s. h.	s.	s.
3111	nd./act.	....	97	....	sl. s. h.	....	....
3112	col. nd.	$0.858^{25°}_{4}$	24-5	191^{10mm}	i.	s.	s.
3113	col. lq.	$0.866^{17.5°}$	−36.8(−44)	227.0	i.	∞	∞
3114	col. lq.	$1.018^{25°}_{25}$	−45	207-8^{756mm}	v. sl. s.	∞ ; s. bz.	∞
3115	col. lq.	$1.001^{25°}_{25}$	−4	214^{752mm}	v. sl. s.	∞	∞
3116	nd.	....	46-7	218.5-9.5	v. sl. s.	1200$^{25°}$	710$^{25°}$
3117	lq.	$1.048^{23°}_{25}$	....	156-62^{20mm}	....	....	....
3118	col. lq.	$1.033^{20°}_{4}$	....	227	i.	∞	∞
3119	pl./et.	....	42	270	....	....	....
3120	lq.	$0.923^{21°}$	....	201-3	i.	s.	s.
3121	oil	$1.415^{20°}_{4}$	....	145^{15mm}	....	....	....
3122	col. lq.	$0.994^{23°}_{0}$	....	219-20 sl. d.	i.	s.	s.
3123	col. lq.	$0.998^{13.8°}_{4}$	....	....	i.	s.	s.
3124	lq.	$1.04^{20°}_{20}$	37.2	268^{740mm}	$0.5^{20°}$	....	....
3125	lf.	....	57-8	273 sl. d.	v. sl. s. h.	v. s. h.	v. s.
3126	col. nd./al.	....	199-200	....	0.06 c.; 0.9 h.	6; i. bz.; s. alk.	0.6; s. ac.
3127	oil	$1.018^{15°}$	....	237	....	....	....
3128	oil	$1.004^{15°}_{15}$	....	*237-40	sl. s.	v. s.	v. s.
3129	pl.	$1.012^{20°}_{20}$	21	218	i.	s.	s.
3130	yel. cr./al.	....	75-6	....	i.; s. a.	s. h.	s.; s. chl.
3131	oil	$1.063^{13°}_{4}$	....	260-70 sl. d.	....	....	....
3132	lq.	$1.024^{15°}_{4}$	....	205-6	....	....	....
3133	mn.	$1.010^{22°}$	42	>300	s. h.	s.	s.; s. a.
3134	nd./aq. al.	....	99	....	....	s.	....
3135	col. lq.	<1	....	25	....	∞	....
3136	lq.	$0.833^{20°}_{4}$	....	80-2	sl. s.	∞	....
3137	col. lq.	$0.968^{15°}_{15}$	....	119^{745mm}	i.; v. s. chl.	v. s.	v. s.
3138	col. lq.	$0.896^{15°}_{4}$	−73.9	99.1	$2.4^{20°}$	∞	∞
3139	lq.	$0.948^{25°}_{4}$	....	223.6	....	....	....
3140	lq.	$0.960^{25°}_{25}$	....	205 d.	v. sl. s.	∞	∞
3141	col. lq.	$0.848^{20°}_{4}$	....	174.5^{748mm}	$0.07^{20°}$	s.	∞ ; ∞ bz.
3142	col. lq.	$0.739^{20°}_{4}$	<−79	62-3	sl. s.	∞	∞
3143	col. lq.	$0.745^{0°}$	....	54	sl. s.	∞	∞
3144	col. lq.	$0.813^{21.8°}_{4}$	....	123-4	v. sl. s.	∞	∞
3145	col. lq.	$0.814^{18°}_{0}$	....	114^{745mm}	v. sl. s.	v. s.	∞
3146	lq.	$0.950^{0°}$	....	148.5^{753mm}	sl. s.	∞	v. s.

*In nitrogen.

Ethyl phenyl-benzyl-carbamate 5165
Ethyl phenyl-carbamate 5254
Ethyl phenyl-cyanoacrylate 3031
Ethyl-phenyl ether 5111

Ethyl-phenylnitrosoamine 4885
Ethyl phosphate 6188
Ethyl phosphite 6192
Ethyl phthalate 2199, 3067
Ethyl pimelate 2202

Ethyl-propyl carbinol (*n*) 3626
Ethyl-propyl carbinol (*iso*) 3632
Ethyl propyl-malonate 2203-4

Table 7-4 (Continued)
PHYSICAL CONSTANTS OF ORGANIC COMPOUNDS

No.	Name	Synonym	Formula	Beil. Ref.	Formula Weight
3147	**Ethyl** pyridine (3)	β-lutidine	$C_2H_5 \cdot C_5H_4N$	XX-242	107.16
3148	pyridine (4)	....	$C_2H_5 \cdot C_5H_4N$	XX-243	107.16
3149	pyrrole (N)	Et-pyrrolylamine	$C_4H_4N \cdot C_2H_5$	XX-163	95.15
3150	pyruvate	....	$CH_3CO \cdot CO_2 \cdot C_2H_5$	III-616	116.12
3151	salicylate (o)	Et hydroxybenzoic †	$HO \cdot C_6H_4 \cdot CO_2 \cdot C_2H_5$	X-73	166.18
3152	silicate, ortho	....	$Si(OC_2H_5)_4$	I-334	208.33
3153	stearate	....	$C_{17}H_{35}CO_2 \cdot C_2H_5$	II-379	312.54
3154	succinic acid	....	$HO_2C \cdot CH(C_2H_5) \cdot CH_2 \cdot CO_2H$	II-660	146.14
3155	sulfone chloride	Et-sulfonyl chloride	$C_2H_5 \cdot SO_2 \cdot Cl$	IV-6	128.58
3156	thioacetate (S)	....	$CH_3CO \cdot S \cdot C_2H_5$	II-232	104.17
3157	thiocarbamate	dithiourethane	$NH_2 \cdot CS \cdot S \cdot C_2H_5$	III-218	121.22
3158	thiocyanate	....	$C_2H_5 \cdot S \cdot CN$	III-175	87.14
3159	iso-thiocyanate	ethyl mustard oil	$C_2H_5 \cdot N:C:S$	IV-123	87.14
3160	thioglycolate	....	$HS \cdot CH_2CO_2 \cdot C_2H_5$	III-255	120.17
3161	thioglycolic acid	(S)	$C_2H_5S \cdot CH_2 \cdot CO_2H$	III-248	120.17
3162	thiourea	....	$C_2H_5 \cdot NH \cdot CS \cdot NH_2$	IV-117	104.17
3163	toluate (o)	....	$CH_3 \cdot C_6H_4 \cdot CO_2 \cdot C_2H_5$	IX-463	164.21
3164	toluate (m)	....	$CH_3 \cdot C_6H_4 \cdot CO_2 \cdot C_2H_5$	IX-476	164.21
3165	toluate (p)	....	$CH_3 \cdot C_6H_4 \cdot CO_2 \cdot C_2H_5$	IX-484	164.21
3166	toluene (o)	methyl-ethyl-benzene	$C_2H_5 \cdot C_6H_4 \cdot CH_3$	V-396	120.20
3167	toluene (m)	....	$C_2H_5 \cdot C_6H_4 \cdot CH_3$	V-396	120.20
3168	toluene (p)	....	$C_2H_5 \cdot C_6H_4 \cdot CH_3$	V-397	120.20
3169	p-toluene sulfonate	....	$CH_3 \cdot C_6H_4 \cdot SO_3 \cdot C_2H_5$	XI-99	200.26
3170	o-toluidine	....	$CH_3 \cdot C_6H_4 \cdot NHC_2H_5$	XII-786	135.21
3171	m-toluidine	....	$CH_3 \cdot C_6H_4 \cdot NHC_2H_5$	XII-857	135.21
3172	p-toluidine	....	$CH_3 \cdot C_6H_4 \cdot NHC_2H_5$	XII-904	135.21
3173	o-tolyl ether	Et o-cresyl ether	$CH_3 \cdot C_6H_4 \cdot O \cdot C_2H_5$	VI-352	136.20
3174	m-tolyl ether	....	$CH_3 \cdot C_6H_4 \cdot O \cdot C_2H_5$	VI-376	136.20
3175	p-tolyl ether	....	$CH_3 \cdot C_6H_4 \cdot O \cdot C_2H_5$	VI-393	136.20
3176	p-tolyl sulfide	Et p-cresyl sulfide	$CH_3 \cdot C_6H_4 \cdot S \cdot C_2H_5$	VI-417	152.26
3177	triazoacetate	....	$N_3CH_2 \cdot CO_2 \cdot C_2H_5$	II-229	129.12
3178	tribromoacetate	....	$Br_3C \cdot CO_2 \cdot C_2H_5$	II-221	324.81
3179	trichloroacetate	....	$Cl_3C \cdot CO_2 \cdot C_2H_5$	II-209	191.44
3180	tripropylbenzene	(iso)	$C_2H_5 \cdot C_6H_2(C_3H_7)_3$	....	232.41
3181	undecylate (n)	....	$C_{10}H_{21}CO_2 \cdot C_2H_5$	II-358	214.35
3182	n-undecyl ketone	tetradecanone-3	$C_2H_5 \cdot CO \cdot C_{11}H_{23}$	I-716	212.38
3183	urea	....	$C_2H_5 \cdot NH \cdot CO \cdot NH_2$	IV-115	88.11
3184	urethane	Et N-Et carbamate	$C_2H_5 \cdot NHCO_2C_2H_5$	IV-114	117.15
3185	valerate (n)	....	$CH_3(CH_2)_3CO_2C_2H_5$	II-301	130.19
3186	iso-valerate	....	$(CH_3)_2CH \cdot CH_2 \cdot CO_2 \cdot C_2H_5$	II-312	130.19
3187	vanillate (4,3;1)	....	$HO(CH_3O)C_6H_3 \cdot CO_2 \cdot C_2H_5$	X-397	196.20
3188	vanillin	3-EtO-4-OH-ben-zaldehyde; ethovan	$HO(C_2H_5O)C_6H_3 \cdot CHO$	VIII-256	166.18
3189	vinyl carbinol	pentene-1-ol-3	$CH_2:CH \cdot CHOH \cdot C_2H_5$	I-443	86.13
3190	vinyl sulfide	....	$CH_2:CH \cdot S \cdot C_2H_5$	I-434	88.17
3191	violet base	hexaEt-4,4′,4″-tri-NH_2-triPh-carbinol	$[(C_2H_5)_2N \cdot C_6H_4]_3 C \cdot OH$	XIII-759	473.71

† See also No. 3072
Ethyl pyromucate 3050
Ethyl-pyrrolylamine 3149
Ethyl red 2144
Ethyl sebacate 2206

Ethyl selenomercaptan 3071
Ethyl suberate 2210
Ethyl succinate 2211-2
Ethyl sulfate 2213
Ethyl sulfinic acid 2905

Ethyl sulfonic acid 2906
Ethyl sulfonyl chloride 3155
Ethyl sulfuric acid 3069
Ethyl tartrate 2218, 3070
Ethyl thioncarbamate 6450

Table 7-4 (*Continued*)
PHYSICAL CONSTANTS OF ORGANIC COMPOUNDS

No.	Crystalline Form and Color	Specific Gravity	Melting Point °C.	Boiling Point °C.	Solubility in 100 Parts		
					Water	Alcohol	Ether
3147	lq.	0.959$\frac{9}{4}$°	....	165.3	sl. s. c.; v. sl. s. h.	s.	s.
3148	lq.	0.936$^{20°}$	....	166	s. aq. a.	....	....
3149	lq.	0.888$^{16°}$	....	130-1	i. c.; s. a.	∞	∞
3150	col. lq.	1.060$\frac{16°}{4}$	....	155	i.	∞	∞
3151	col. lq.	1.136$\frac{15°}{4}$	1.3	233-4	i.	∞	∞
3152	lq.	0.936$\frac{20°}{20}$	−82.5	168.6	sl. d.	∞	....
3153	col. cr.	0.848$^{36.3°}$	33.8(31.1)	199-201^{10mm}	i.	s.	s.
3154	col. pr.	....	98	....	v. s.	v. s.	v. s.
3155	lq.	1.357$^{22.5°}$	....	177.5	sl. d.	sl. d.	v. s.
3156	lq.	0.976$\frac{28°}{4}$	....	116-7	i.	v. s.	v. s.
3157	lf./et.	....	41-2	d.	i.	v. s.	v. s.
3158	lq.	0.996$\frac{25°}{4}$	−85.5	145^{765mm}	i.	∞	∞
3159	col. lq.	1.004$\frac{15°}{4}$	−5.9	131-2	i.	∞	∞
3160	lq.	1.096$^{15°}$	....	156-8	....	....	....
3161	oil	1.150$\frac{20°}{4}$	−8.7	123-4^{15mm}	∞	s.	s.
3162	nd.	....	114 (113)	....	v. s.	v. s.	....
3163	lq.	1.032$\frac{25°}{25}$	<−10	227	i.	∞	∞
3164	lq.	1.030$\frac{20°}{20}$	....	231^{750mm}	i.	∞	∞
3165	lq.	1.024$\frac{25°}{25}$	....	235.5	i.	∞	∞
3166	lq.	0.881$\frac{20°}{4}$	−80.8	165.2	i.	∞	∞
3167	lq.	0.865$\frac{20°}{4}$	−95.6	161.3	i.	∞	∞
3168	lq.	0.861$\frac{20°}{4}$	−62.4	162.0	i.	∞	∞
3169	mn. pr./al.	1.166$\frac{48°}{4}$	33-4	221.3	i.	s.	s.
3170	lq.	0.948$\frac{25°}{4}$	<−15	215-6	i.	....	....
3171	lq.	....	....	221-2	....	....	....
3172	lq.	0.942$\frac{25°}{4}$	....	217	i.	....	....
3173	lq.	0.959$\frac{13.3°}{4}$	....	184	....	....	....
3174	lq.	0.956$\frac{0°}{0}$	....	192	....	....	....
3175	lq.	0.966$\frac{0°}{0}$	....	188-9	....	....	....
3176	lq.	1.002$^{17.5°}$	....	220-1	....	....	....
3177	col. oil	1.127$\frac{20°}{20}$	....	70^{20mm}	....	....	....
3178	col. lq.	2.230$\frac{20°}{20}$	....	225	i.	∞	∞
3179	col. lq.	1.383$\frac{20°}{4}$	....	167-8	i.	∞	∞
3180	col. cr.	....	106.9	260	i.	1.5$^{25°}$	105$^{25°}$
3181	lq.	....	....	140^{20mm}	i.	....	....
3182	cr./Me al.	....	34	152^{16mm}	....	....	....
3183	nd.	1.213$^{18°}$	92	....	v. s.	80	i.
3184	col. lq.	0.981$\frac{20°}{4}$	....	174-6	63$^{15°}$	d. h. alk.	....
3185	col. lq.	0.877$^{20°}$	−91.2	145.5	0.24$^{25°}$	∞	∞
3186	col. lq.	0.867$\frac{20°}{4}$	−99.3	135	0.17$^{20°}$	∞ ; ∞ bz.	∞
3187	nd.	....	44	291-3	i.; s. alk.	v. s.	v. s.
3188	col. pl./aq.	....	77-8	....	v. sl. s.	s.	s.
3189	lq.	0.840$\frac{19.5°}{0}$	....	114.5-4.7	....	....	....
3190	col. lq.	0.887$^{14°}$	....	90.5-1.5	....	....	....
3191	cr./lg.	....	....	....	....	sl. d.	....

Table 7-4 (*Continued*)
PHYSICAL CONSTANTS OF ORGANIC COMPOUNDS

No.	Name	Synonym	Formula	Beil. Ref.	Formula Weight
3192	**Ethylene**	ethene	$CH_2:CH_2$	I-180	28.05
3193	bromohydrin	glycol bromohydrin	$Br·CH_2·CH_2OH$	I-338	124.97
3194	chlorohydrin	2-chloro-ethanol-1	$ClCH_2·CH_2OH$	I-337	80.51
3195	cyanohydrin	β-OH-propionitrile	$HO·CH_2·CH_2·CN$	III-298	71.08
3196	diamine	....	$NH_2·CH_2·CH_2·NH_2$	IV-230	60.10
3197	diamine HBr	....	$(CH_2NH_2)_2·2HBr$	IV-232	221.93
3198	diamine HCl	....	$(CH_2NH_2)_2·2HCl$	IV-232	133.02
3199	diamine hydrate	....	$(CH_2NH_2)_2·H_2O$	IV-230	78.11
3200	diamine *iso*-valerate	*iso*-valeryl-ethyl-ene-diamine	$C_{12}H_{28}O_4N_2$	....	264.37
3201	dicaprate	....	$(C_9H_{19}CO_2)_2C_2H_4$	....	370.58
3202	diphenyl ether	....	$(C_6H_5·O·CH_2)_2$	VI-146	214.27
3203	diphenylsulfone	....	$(C_6H_5·SO_2·CH_2)_2$	VI-302	310.39
3204	diphenylthioether	....	$(C_6H_5·S·CH_2)_2$	VI-301	246.40
3205	diphthalimide (*N,N'*)	....	$[C_6H_4(CO)_2:N·CH_2]_2$	XXI-492	320.31
3206	disulfonic acid	....	$(HO_3S·CH_2)_2$	IV-11	190.19
3207	ethylidene oxide	glycol-ethylidene-diacetal	$CH_3·CH:(OCH_2·)_2$	XIX-8	88.11
3208	iodohydrin	2-iodo-ethanol-1	$ICH_2·CH_2OH$	I-339	171.97
3209	mercaptan	ethandithiol-1,2	$HS·CH_2·CH_2·SH$	I-471	94.20
3210	oxide	....	$CH_2·CH_2·O$	XVII-4	44.05
3211	thiocyanate	....	$(CH_2·SCN)_2$	III-178	144.22
3212	urea	....	$(CH_2NH)_2CO$	XXIV-2	86.09
3213	**Ethylidene** acetone	Me-propenyl ketone	$CH_3·CH:CH·CO·CH_3$	I-732	84.12
3214	diacetate	diacetoxyethane (1,1)	$CH_3·CH(O_2C·CH_3)_2$	II-152	146.14
3215	diurethane	....	$CH_3·CH:(NH·CO_2·C_2H_5)_2$	III-24	204.23
3216	urea	....	$CH_3·CH:(NHCONH)$	III-60	86.09
3217	**Eucalyptol**	cineol	$C_{10}H_{18}O$	XVII-24	154.25
3218	**Eugenol** (1,3,4)	allyl guaiacol	$C_6H_3(C_3H_5)(OCH_3)·OH$	VI-961	164.21
3219	acetate	....	$CH_3·CO_2·C_{10}H_{11}O$	VI-965	206.24
3220	benzoate	benzoyl eugenol	$C_6H_5CO_2C_{10}H_{11}O$	IX-135	268.32
3221	cinnamate	....	$C_8H_7CO_2C_{10}H_{11}O$	IX-586	294.35
3222	methyl ether	....	$C_3H_5·C_6H_3:(OCH_3)_2$	VI-963	178.23
3223	*iso*-**Eugenol** (1,3,4)	propenyl guaiacol	$C_6H_3(C_3H_5)·(OCH_3)·OH$	VI-955	164.21
3224	methyl ether	Me isoeugenol	$C_3H_5·C_6H_3:(OCH_3)_2$	VI-956	178.23
3225	**Eugetinic acid**	(5;6,3,1)	$CH_3O·C_6H_2(OH)·(C_3H_5)CO_2H$	X-441	208.22
3226	**Eupittonic acid**	....	$C_{19}H_8(OCH_3)_6O_3$	VIII-574	470.48
3227	**Euxanthone**	1,7-diOH-xanthone	$CO:(C_6H_3OH)_2O$	XVIII-113	228.21
3228	**Evans blue**	....	$C_{34}H_{24}O_{14}N_6S_4Na_4$	....	960.82
3229	**Evernic acid**	....	$C_{16}H_{13}O_6·OCH_3$	X-416	332.31
3230	**Everninic acid**	(6,4;2,1)	$HO(CH_3O)C_6H_2·(CH_3)CO_2H$	X-413	182.18
3231	**Evipan** (1,5;5)	*N*-Me-cyclohexenyl-methylbarbituric	$C_4HO_3N_2(CH_3)_2·C_6H_9$	....	236.27
3232	**Fenchene** (*dl*)	....	$C_{10}H_{16}$	V-163	136.24
3233	**Fenchone** (*d*)	....	$C_{10}H_{16}O$	VII-96	152.24

Table 7-4 (*Continued*)
PHYSICAL CONSTANTS OF ORGANIC COMPOUNDS

No.	Crystalline Form and Color	Specific Gravity	Melting Point °C.	Boiling Point °C.	Solubility in 100 Parts		
					Water	Alcohol	Ether
3192	col. gas	0.975(A) 0.566$^{-192°}_{4}$	−169.2	−103.7	25.6$^{0°}$ cc.	360 cc.	s.
3193	col. lq.	1.772$^{20°}_{4}$	....	150 sl. d.	s.	s.	i. pet.
3194	col. lq.	1.202$^{20°}_{4}$	−67.5	128.6*	∞	∞	∞
3195	lq.	1.059$^{0°}$	−46.2	220-275^{4mm}	∞	∞	2.3$^{15°}$
3196	col. lq.	0.900$^{20°}_{20}$	8.5	117.2	∞	∞; i. bz.	0.3
3197	col. pr.	....	....	....	s.	i.	i.
3198	mn. pr.	....	subl.	....	s.	i.	i.
3199	col. lq.	0.963$^{21°}_{4}$	10	118	∞	....	....
3200	wh. pd.	....	129	....	s.	s.	....
3201	col. pl.	....	37-8	....	i.	s.	v. s.
3202	lf./abs. al.	....	97-8	....	i.	s. h.	s.; s. chl.
3203	nd./al.	....	179-80	....	v. sl. s. h.	s. h.; s. bz.	v. s. h. ac.
3204	nd.	....	70.0-0.5	....	i.	....	....
3205	nd./ac.	....	233-4	....	i.	s. h.	i.; s. bz.
3206	nd.	....	104	....	v. s.	v. s.	....
3207	lq.	0.987$^{15°}_{4}$	....	82.5^{766mm}	66.6; i. aq. CaCl$_2$	∞	∞
3208	col. lq.	2.197$^{20°}_{4}$	....	85^{25mm}	s.	s.	....
3209	lq.	1.123$^{23.5°}$	....	146	v. s. alk.	v. s.	s. NH$_4$OH
3210	lq.	0.887$^{7°}_{4}$	−111.7	10.7	∞	∞	v. s.
3211	pl. or nd.	....	90	d.	sl. s.	s.	s.
3212	nd.	....	131	....	s.	s. h.	v. sl. s.
3213	col. lq.	0.856$^{20°}$	....	122-4^{745mm}	s.	....	....
3214	col. lq.	1.061$^{12°}$	18.85	168^{740mm}	sl. s.	∞	d. alk.
3215	nd.	....	125-6	170-80^{20mm}	s. h.	s.	s.
3216	nd.	....	154	d. 160	i.	sl. s.	i.
3217	col. oil	0.927$^{200°}$	1.5	176-7	1.9$^{15°}$	∞; s. oils	∞; s. ac.
3218	oil	1.066$^{20°}_{4}$	−9.2	254.8	v. sl. s.	∞; ∞ chl.	∞
3219	pl./al.	1.087$^{15°}_{15}$	29-30	281-275^{2mm}	i.	s.	s.
3220	col. cr.	....	70.5	360	i.; s. act.	s. h.	s.; s. chl.
3221	col. nd.	....	90-1	....	i.; s. act.	s. h.	s.; s. chl.
3222	lq.	1.055$^{15°}$	....	248-9	i.	25, 60% al.	∞
3223	oil	1.091$^{15°}_{15}$	16-9	267.5	v. sl. s.	∞	∞
3224	col. lq.	1.055$^{22°}$	....	263-4	i.	s.	s.
3225	pr./aq.	....	124	d.	v. sl. s. c.	s.; s. (NH$_4$)$_2$CO$_3$	s.
3226	or. nd./al.	....	200 d.	....	s. alk.	sl. s. h. abs.	s. ac.
3227	yel. nd.	....	240	subl. sl. d.	i.; s. alk.	s. h.	sl. s.
3228	b. pd.	....	....	....	s.	i.	i.
3229	pr./al.	....	168-9 d.	....	v. sl. s. h.	s. h.	v. sl. s.
3230	cr./aq.	....	170-1 d.	....	v. sl. s. h.; s. EtOAc	s. h.; s. dil. NaOH	v. sl. s.
3231	col. cr. pd.	....	143-5	....	sl. s.	s. h.	sl. s.
3232	lq.	0.868$^{19°}$	....	155-6	....	....	....
3233	oil	0.948$^{18°}$	5-6	193-5	i.	v. s.	v. s.

* Constant b. p. mixt., 42.5% aq. 95.8°.

Ethylene tribromide 6105	Ethylidene phenetidine 2913	Evipal 3231
Ethylene trichloride 6155	Ethyne 130	Evodiamine, cf. alkd.
Ethylidene chlorobromide 1342	Eubornyl 863	Ewer and Pick's acid 4462
Ethylidene dibromide 1836	Eudermol, cf. alkd.	Exalgin 4086
Ethylidene dichloride 1993	Eupyrin 6420	F acid 4510
Ethylidene diiodide 2390	Euresol 114	Fantan 5171
Ethylidene dimethyl ether 2415	Eurisol 114	Fast green-J base 5672
	Euxanthic acid, cf. glcde.	Fast red-D 3732

Table 7-4 (*Continued*)
PHYSICAL CONSTANTS OF ORGANIC COMPOUNDS

No.	Name	Synonym	Formula	Beil. Ref.	Formula Weight
3234	**Fenchyl** alcohol (*dl*)	....	$C_{10}H_{17}OH$	VI-71	154.25
3235	alcohol (*d*)(*α*)	fenchol	$C_{10}H_{17}OH$	VI-70	154.25
3236	**Ferulic acid** (3;4,1)	MeO-OH-cinnamic acid	$CH_3O \cdot C_6H_3(OH) \cdot CH:CHCO_2H$	X-436	194.19
3237	*iso*-**Ferulic acid** (4;3,1)	hesperetinic acid	$CH_3O \cdot C_6H_3(OH) \cdot CH:CHCO_2H$	X-437	194.19
3238	**Filicic acid**	filicin (1,1;2,4,6)	$(CH_3)_2C_6H_4(:O)_3$	VII-856	154.17
3239	**Filixic acid**	....	$C_{35}H_{40}O_{12}$	VIII-576	652.70
3240	**Flavianic** acid	2,4-diNO₂-1-naph-thol-7-sulfonic	$(NO_2)_2C_{10}H_4(OH) \cdot SO_3H \cdot 3H_2O$	XI-275	368.28
3241	sodium salt	naphthol yellow S	$C_{10}H_4O_8N_2SNa_2 \cdot 3H_2O$	XI-275	412.24
3242	**Flavaniline** (*α*)	*p*-aminophenyl-lepidine	$NH_2 \cdot C_6H_4 \cdot C_9H_5N \cdot CH_3$	XXII-469	234.30
3243	**Flavone**	3-phenyl-benzo-4-pyrone	$C_6H_4 \cdot OC(C_6H_5):$ ⎯⎯⎯ $CH \cdot CO$ ⎯⎯	XVII-373	222.25
3244	**Fluoran**	....	$C_{20}H_{12}O_3$	XIX-146	300.32
3245	**Fluoranthene**	....	$C_{16}H_{10}$	V-685	202.26
3246	**Fluorene**	....	$C_6H_4 \cdot CH_2 \cdot C_6H_4$	V-625	166.22
3247	**Fluorenol**	diphenylene carbinol	$C_6H_4 \cdot C_6H_4 \cdot CHOH$	VI-691	182.22
3248	**Fluorenone**	diphenylene ketone	$C_6H_4 \cdot CO \cdot C_6H_4$	VII-465	180.21
3249	oxime	....	$C_{12}H_8:C:NOH$	VII-467	195.22
3250	**Fluorescein**	resorcinol phthalein	$C_{20}H_{12}O_5$	XIX-222	332.32
3251	sodium salt	uranin	$C_{20}H_{10}O_5Na_2$	XIX-225	376.28
3252	chloride	3,6-diCl-fluoran	$C_{20}H_{10}O_3Cl_2$	XIX-147	369.21
3253	diacetate	3,6-diAcO-fluoran	$C_{24}H_{16}O_7$	XIX-227	416.39
3254	**Fluorescin**	resorcinol phthalin	$C_{20}H_{14}O_5$	XVIII-358	334.33
3255	**Fluoro**-acetic acid	....	$FCH_2 \cdot CO_2H$	II-193	78.04
3256	aniline (*p*)	....	$F \cdot C_6H_4 \cdot NH_2$	XII-597	111.12
3257	anisole (*p*)	....	$F \cdot C_6H_4 \cdot OCH_3$	*VI-98	126.13
3258	benzene	phenyl fluoride	$C_6H_5 \cdot F$	V-198	96.11
3259	benzoic acid (*o*)	....	$F \cdot C_6H_4 \cdot CO_2H$	IX-333	140.12
3260	benzoic acid (*m*)	....	$F \cdot C_6H_4 \cdot CO_2H$	IX-333	140.12
3261	benzoic acid (*p*)	....	$F \cdot C_6H_4 \cdot CO_2H$	IX-333	140.12
3262	bromobenzene (*p*)	....	$F \cdot C_6H_4 \cdot Br$	V-209	175.01
3263	chlorobenzene (*o*)	....	$F \cdot C_6H_4 \cdot Cl$	*V-110	130.55
3264	chlorobenzene (*m*)	....	$F \cdot C_6H_4 \cdot Cl$	....	130.55
3265	chlorobenzene (*p*)	....	$F \cdot C_6H_4 \cdot Cl$	V-201	130.55
3266	chlorobromo-methane	....	$CHClBrF$	I-67	147.38
3267	chloromethane	....	CH_2ClF	I-60	68.48
3268	dibromoethane	(2;1,1)	$FCH_2 \cdot CHBr_2$	I-92	205.86
3269	dibromoethane	(2;1,2)	$FCHBr \cdot CH_2Br$	I-92	205.86
3270	1,1-dichloro-1,2-di-bromoethane (2)	....	$Cl_2CBr \cdot CHFBr$	I-93	274.75
3271	dichloromethane	Freon-21	$FCHCl_2$	I-61	102.92
3272	iodobenzene (*p*)	....	$F \cdot C_6H_4 \cdot I$	V-220	222.00
3273	naphthalene (*α*)	naphthyl fluoride	$C_{10}H_7F$	V-540	146.17
3274	nitrobenzene (*p*)	....	$NO_2 \cdot C_6H_4 \cdot F$	V-241	141.10

Table 7-4 (*Continued*)
PHYSICAL CONSTANTS OF ORGANIC COMPOUNDS

No.	Crystalline Form and Color	Specific Gravity	Melting Point °C.	Boiling Point °C.	Solubility in 100 Parts		
					Water	Alcohol	Ether
3234	col. cr.	$0.935^{40°}$	33-5*	201	sl. s.	....	....
3235	col. pr.	$0.964^{20°}_4$	42	201-2	sl. s.	s.	s.; s. pet.
3236	pr./aq.	....	169-70	d.	s. h.	s.; sl. s. bz.	sl. s.
3237	nd. or pr.	....	225-8	....	sl. s. h.	s.	s.; i. lg.
3238	col. cr./al.	....	213-5 d.	subl. sl. d.	1.4 h.	10 h.	sl. s.
3239	cr.	....	184 d.	....	i.; s. chl.	i. abs.	sl. s.; s. bz.
3240	pa. yel. nd./HCl	....	100; 140-50(anh.)	d. >175	v. s.	v. s.	....
3241	yel. pd.	....	....	....	s.	....	....
3242	yel. nd./al.	....	83-4	....	i.	s.	s.; s. bz.
3243	col./lg.	....	97	....	i.	s.	s.
3244	nd.	....	184	....	s. H_2SO_4	s. HNO_3	....
3245	nd./al.	$1.252^{0°}_4$	109-10	$250\text{-}160^{mm}$	i.	v. s. h.	v. s.
3246	col. cr./al.	$1.203^{0°}_4$	115-6	293-5	i.; i.NH_3; s. SO_2	s. h.	s.
3247	nd./aq.	....	153	....		s.	s.
3248	yel. rhb.	....	83-4	341.5	i.	v. s.	v. s.
3249	nd.	....	192-3	....	s. chl.	i. pet.	....
3250	yel. red pd.	....	d. >290	....	v. sl. s. h.	s. h.; s. alk.	s. h. ac.
3251	or. red pd.	....	....	....	s.	sl. s.	
3252	col. cr.	....	258-60	....	i.; s. chl.	v. sl. s.	s. h. bz.
3253	col. cr./et. bz.	....	201-5	....	i. c. alk.	v. sl. s.	s. ac.
3254	col. nd./ac.	....	125-7†	....	i.; s. alk.	s.	s.
3255	cr.	....	33	165	....	....	....
3256	oil	$1.152^{25°}_4$	-0.8	187.4	v. sl. s.	....	....
3257	col. lq.	....	-43.5	156-7	....	∞	∞
3258	col. lq.	$1.024^{20°}_4$	-41.9	84.8	i.	∞	∞
3259	nd./aq.	....	120-2	....	s. h.	s.	s.
3260	lf./aq.	....	124	....	sl. s. h.		
3261	mn./aq.	....	184-6	....	sl. s. h.	s.	s.
3262	col. lq.	$1.597^{20°}_4$	-8	152^{755mm}	i.	v. s.	v. s.
3263	lq.	....	-42.5	138^{774mm}	....	....	....
3264	col. lq.	....		123-5			
3265	col. lq.	$1.226^{20.5°}_4$	-26.9	130^{756mm}			
3266	col. lq.	$1.906^{16°}$	<-65	38	....		
3267	gas	....	....	....	d.		
3268	lq.	....	....	117.5			
3269	col. lq.	$2.257^{17°}$	-54	122.5^{761mm}			
3270	col. lq.	$2.130^{23°}$	....	163.5			
3271	gas	$1.426^{0°}$	-135	8.9	i.	s.	s.
3272	col. lq.	$1.925^{15°}$	-18 (-27)	183.2^{760mm}	i.	s.	s.
3273	lq.	$1.133^{19.5°}_4$	-8	212-6	s. bz.	s.; s. ac.	s. chl.
3274	pa. yel. cr.	$1.320^{0°}_4$	27(21.6)	206.7	....	....	....

* Rapid heating, m. p. 37-8°.
† Solvent-free crysts. m. p. 253-4°.
Fluoro-ethane 3046

Fluoro-ethylene 6439
Fluoro-methane 4240
Fluoro-octane 4978

Table 7-4 (Continued)
PHYSICAL CONSTANTS OF ORGANIC COMPOUNDS

No.	Name	Synonym	Formula	Beil. Ref.	Formula Weight
	Fluoro				
3275	pentabromoethane	....	$Br_3C \cdot CBr_2F$	I-95	442.57
3276	pentachloroethane	....	$Cl_3C \cdot CCl_2F$	....	220.29
3277	tetrabromoethane	....	$Br_3C \cdot CHBrF$	I-95	363.66
3278	tetrachloroethane	....	$Cl_3C \cdot CHClF$	I-86	185.84
3279	toluene (*o*)	....	$CH_3 \cdot C_6H_4 \cdot F$	V-290	110.13
3280	toluene (*m*)	....	$CH_3 \cdot C_6H_4 \cdot F$	V-290	110.13
3281	toluene (*p*)	....	$CH_3 \cdot C_6H_4 \cdot F$	V-290	110.13
3282	tribromoethane	....	$BrCH_2 \cdot CFBr_2$	I-93	284.76
3283	tribromoethane	....	$Br_2CH \cdot CHBrF$	I-93	284.76
3284	trichloroethane	....	$Cl_2CH \cdot CHClF$	I-85	151.40
3285	trichloroethylene	....	$FClC:CCl_2$	....	149.38
3286	trichloromethane	Freon-11	Cl_3CF	I-64	137.37
3287	**Fluoroform**	trifluoromethane	CHF_3	I-59	70.01
3288	**Formaldehyde**	methanal	$H \cdot CH:O$	I-558	30.03
3289	**Formaldehyde** (meta)	α-trioxymethylene	$(CH_2O)_3$	XIX-381	90.08
3290	**Formaldehyde** (para)	....	$(CH_2O)_x \cdot xH_2O$	I-566	(30.03)
3291	acetamide	formicin †	$CH_3CONH \cdot CH_2OH$	II-178	89.09
3292	bisulfite Na	....	$CH_2OH \cdot SO_3Na \cdot H_2O$	I-578	152.10
3293	hydrosulfite	hydrosulfite N. F.	$CH_2O \cdot Na_2S_2O_4 \cdot H_2O$	I-578	222.15
3294	oxime	formaldoxime	$H_2C:NOH$	I-590	45.04
3295	sulfoxylate Na	....	$CH_2OH \cdot SO_2Na \cdot 2H_2O$	I-577	154.12
3296	**Formamide**	....	$H \cdot CO \cdot NH_2$	II-26	45.04
3297	**Formamidine**	salts only known	$HN:CH \cdot NH_2$	II-90	44.06
3298	**Formamidoxime**	....	$H_2N \cdot CH:NOH$	II-91	60.06
3299	**Formanilide**	formylaniline	$C_6H_5 \cdot NH \cdot CHO$	XII-230	121.14
3300	sodium salt	....	$C_6H_5 \cdot N(Na) \cdot CHO$	XII-233	143.13
3301	**Formic acid**	methanoic acid	$H \cdot CO_2H$	II-8	46.03
3302	sodium salt	See Section 4	....	....	....
3303	thallium salt	See Section 4	....	....	....
3304	**Formyl** acetone	....	$HO \cdot CH:CH \cdot CO \cdot CH_3$	I-767	86.09
3305	amino-1,3-di-methylbenzene (2)	form-*m*-xylidide (*v*)	$(CH_3)_2C_6H_3NH \cdot OCH$	XII-1109	149.19
3306	diphenylamine	*N*-Ph-formanilide	$H \cdot CO \cdot N(C_6H_5)_2$	XII-235	197.24
3307	hydrazine	formhydrazide	$H_2N \cdot NH \cdot CHO$	II-93	60.06
3308	phenylhydrazine (β)	....	$C_6H_5NH \cdot NH \cdot CHO$	XV-233	136.15
3309	thioaldehyde	....	$CH_2 \cdot (S \cdot CH_2)_2 \cdot S$	XIX-382	138.27
3310	**Fraxin** ‡	....	$C_{16}H_{18}O_{10}$ + aq.	XXXI-249	370.32
3311	**Fructosamine** (*d*)	*iso*-glucosamine	$C_6H_{11}O_5(NH_2)$	IV-332	179.17
3312	**Fucose**	....	$C_5H_{11}O_4 \cdot CHO$	I-876	164.16
3313	**Fulminuric acid**	*iso*-cyanuric acid	$NC \cdot CH(NO_2) \cdot CONH_2$	II-598	129.08
3314	**Fumaric** acid (*trans*)	butendioic acid	$(:CH \cdot CO_2H)_2$	II-737	116.07
3315	amide	....	$(:CH \cdot CO \cdot NH_2)_2$	II-743	114.10
3316	chloride	fumaryl chloride	$(:CH \cdot CO \cdot Cl)_2$	II-743	152.97
3317	**Furfural** (3)	....	$C_4H_3O \cdot CHO$	....	96.09
3318	**Furfural** (2)	furfurol	$C_4H_3O \cdot CHO$	XVII-272	96.09

† Formicin is a commercial syrupy product.   ‡ See also Glucoside table.

Fluoro-propane 5406-7	Formaldoxime 3294	Formosul 3295
Folliculin 2903	Formamine 3599	Formyl-aniline 3299
Formaldehyde diacetate 4430	Formhydrazide 3307	Formyl-diethylamine 2173
Formaldehyde diethylacetal 2172	Formicin 3291	Formyl-salicylic acid 168
Formaldehyde dimethylacetal 4424	Formin 3599	Formyl-xylidine 3305
Formaldehyde dipropylacetal 5458	Formonitrile 3714	Form-xylidide 3305
Formaldehyde trimethyleneacetal 6285	Formopan 3295	Forsling's acid-I 4551

Table 7-4 (*Continued*)
PHYSICAL CONSTANTS OF ORGANIC COMPOUNDS

No.	Crystalline Form and Color	Specific Gravity	Melting Point °C.	Boiling Point °C.	Solubility in 100 Parts		
					Water	Alcohol	Ether
3275	col. cr.	....	176 d.	subl. 120	s. bz., chl.	sl. s. c.	s.
3276	col. cr.	$1.74^{25°}$	100	136.8	i.	....	....
3277	lq.	$1.939^{16°}$	....	103.5^{23mm}	....	....	....
3278	col. lq.	$1.63^{116.7°}$	....	116.5	....	....	....
3279	col. lq.	$1.004^{13.2°}$	−80±	113–4	i.	s.	s.
3280	col. lq.	$0.997^{13.4°}$	−110.8	115–6	i.	s.	s.
3281	col. lq.	$1.001^{\frac{16°}{4}}$	....	116–7	i.	s.	s.
3282	col. lq.	$2.605^{17.5°}$	....	162.7^{757mm}	....	....	....
3283	col. lq.	$2.674^{18°}$	....	178	....	....	....
3284	col. lq.	$1.539^{\frac{20°}{4}}$	....	103	i.	∞	∞
3285	col. lq.	$1.530^{25°}$	−82	71.0	d.	....	....
3286	col. lq.	$1.494^{17.2°}$	....	24.9	i.	∞	∞
3287	gas	lq. $1.47^{-84°}$	....	$20^{40atm.}$	75 cc.	500 cc.	sl. s. chl.
3288	gas	$0.815^{-20°}$	−92±	−21	v. s.	v. s.	v. s.
3289	wh. solid	$1.17^{65°}$	64; subl. 46	114.5^{759mm}	$21^{25°}$; ∞ h.	s.	s.; sl. s. pet.
3290	wh. amor.	....	150–60	subl. 120±	$20–30^{18°}$	i.	i.
3291	hyg. cr.	1.2	50–2±	d.	v. s.	v. s.	i.
3292	nd./aq.	....	....	....	s.; d. alk.	sl. s.; d. a.	s. Me al.
3293	nd.	....	....	....	s.	....	....
3294	col. lq.	....	....	84	10–20	....	....
3295	hyg. nd./aq.	....	63–4	d. >125	50–60	i. abs.	i.; i. bz.
3296	hyg. lq.	$1.133^{\frac{20°}{4}}$	2.5	$109^{15 mm}$	∞	∞	v. sl. s.
3297	....	....	....	....	....	....	....
3298	rhb.	....	104–5	....	v. s.; i. bz.	v. sl. s.	v. sl. s.
3299	mn.	$1.147^{\frac{15°}{15}}$	47	216^{120mm}	sl. s.	v. s.	s.
3300	....	....	....	....	d.	v. sl. s.	....
3301	col. lq.	$1.220^{\frac{20°}{4}}$	8.40	100.8	∞	∞	....
3302	....	....	....	....	....	....	....
3303	....	....	....	....	....	....	....
3304	exists only as salts	....	....	....	....	....	....
3305	nd./ai.	....	164–5*	....	....	s.	....
3306	rhb./al.	$1.230^{\frac{20°}{4}}$	73–4	$189–90^{13mm}$	i.	s.	s.; s. bz.
3307	pl. or nd.	....	54	....	v. s. chl.	v. s.; s. bz.	v. s.
3308	lf./al.	....	145–6	....	s. h.	s.	sl. s. bz.
3309	pr./chl.	....	247(218)	subl. 150	sl. s. h.	sl. s.	sl. s.; s. bz.
3310	nd./al.	....	−H_2O, 110	d. 201–5	s. h.	s. h.	i.
3311	syrup	....	....	....	....	s.	i.
3312	nd.	....	145	....	v. s.	v. sl. s.	....
3313	pr./ai.	....	145 d.	....	v. s.; i. bz.	v. s.	v. sl. s.
3314	col. pr.	$1.635^{\frac{20°}{4}}$	286–7†; subl. 7^{200}	290	$0.717°$; $9.8^{100°}$	$5.75^{29.7°}$	$0.72^{5°}$
3315	pr./aq.	....	d. 265–70	....	s. h.	sl. s. h.	i. ac.; i. chl.
3316	col. lq.	$1.415^{\frac{20°}{20}}$	....	161–4	d.	d.	....
3317	lq.	$1.111^{\frac{20°}{20}}$	....	144^{732mm}	....	....	....
3318	lq.	$1.159^{\frac{20°}{4}}$	....	161.7^{760mm}	$9.11^{3°}$	∞	∞

* When heated rapidly, m. p. 176–7°.   † In sealed tube.

Forsling's acid-II 4550	Fructose 3616	Fumarine, cf. alkd.
Frangulin, cf. glcde.	Fructosone 3368	Fumaryl chloride 3316
Frangulinic acid 2874	Fruit sugar 3616	Furacrolein 3347
Franguloside, cf. glcde.	Fuchsine 5571	Furan 3327
Fraxin, cf. glcde.	Fuchsine (para) 5039	Furan carboxylic acid 3339–40
Fraxinin, cf. glcde.	Fuchsine carbinol base 5570	Furan tetrahydride 5773
Freon 2246, 3271, 3286, 6206	Fuchsone 2741	Furfuracrolein 3347
		Furfuraldehyde 3318

Table 7-4 (*Continued*)
PHYSICAL CONSTANTS OF ORGANIC COMPOUNDS

No.	Name	Synonym	Formula	Beil. Ref.	Formula Weight
3319	**Furfural** acetone	....	$C_4H_3O \cdot CH:CH \cdot CO \cdot CH_3$	XVII-306	136.15
3320	acetophenone (ω)	....	$C_4H_3O \cdot CH:CH \cdot CO \cdot C_6H_5$	XVII-353	198.22
3321	diacetate	....	$C_4H_3O \cdot CH(O_2C \cdot CH_3)_2$	XVII-278	198.18
3322	oxime (α)(*syn*)	furfuraldoxime	$C_4H_3O \cdot CH:NOH$	XVII-281	111.10
3323	oxime (β)(*anti*)	....	$C_4H_3O \cdot CH:NOH$	XVII-281	111.10
3324	phenylhydrazone	....	$C_4H_3O \cdot CH:N \cdot NH \cdot C_6H_5$	XVII-282	186.22
3325	2-thiohydantoin (5)	....	CO·NH·CS·NH·C: |_____| CH·C₄H₃O	....	194.21
3326	**Furfuramide**	....	$(C_5H_4O)_3N_2$	XVII-281	268.27
3327	**Furfuran**	furan	CH:CH·CH:CH·O |_____|	XVII-27	68.08
3328	**Furil** (α,α)	difurfuroyl	$(C_4H_3O \cdot CO)_2$	XIX-166	190.16
3329	dioxime (*syn*)(α)	....	$(C_4H_3O \cdot C:NOH)_2 \cdot H_2O$	XIX-166	238.20
3330	dioxime (β)	....	$(C_4H_3O \cdot C:NOH)_2$	XIX-166	220.19
3331	**Furfuryl** acetate	....	$C_4H_3O \cdot CH_2 \cdot O_2C \cdot CH_3$	XVII-112	140.14
3332	alcohol	α-furyl carbinol	$C_4H_3O \cdot CH_2OH$	XVII-112	98.10
3333	amine	....	$C_4H_3O \cdot CH_2NH_2$	XVIII-584	97.12
3334	butyrate (α)	....	$C_3H_7CO_2 \cdot CH_2 \cdot C_4H_3O$	....	168.19
3335	chloride (α)	....	$C_4H_3O \cdot CH_2Cl$	....	116.55
3336	furoate (α)	furfuryl pyromucate	$C_4H_3O \cdot CO_2 \cdot CH_2 \cdot C_4H_3O$	....	192.17
3337	mercaptan (α)	....	$C_4H_3O \cdot CH_2 \cdot SH$	....	114.17
3338	propionate (α)	....	$C_2H_5CO_2 \cdot CH_2 \cdot C_4H_3O$	....	154.17
3339	**Furoic** acid (β)(3)	....	$C_4H_3O \cdot CO_2H$	*XVIII-439	112.09
3340	acid (α)(2)	pyromucic acid	$C_4H_3O \cdot CO_2H$	XVIII-272	112.09
3341	amide (α)	pyromucamide	$C_4H_3O \cdot CO \cdot NH_2$	XVIII-276	111.10
3342	amide (β)	....	$C_4H_3O \cdot CO \cdot NH_2$	....	111.10
3343	anhydride (α)	pyromucic anhydride	$(C_4H_3O \cdot CO)_2O$	XVIII-276	206.16
3344	chloride (α)	furoyl chloride	$C_4H_3O \cdot COCl$	XVIII-276	130.53
3345	nitrile (α)	2-cyanofuran	$C_4H_3O \cdot CN$	XVIII-278	93.09
3346	**Furoin** (α,α)	....	$C_4H_3O \cdot CHOH \cdot CO \cdot C_4H_3O$	XIX-204	192.17
3347	**Furyl**-acrolein	fur (fur)acrolein	$C_4H_3O \cdot CH:CH \cdot CHO$	XVII-305	122.12
3348	acrolein oxime	....	$C_4H_3O \cdot (CH)_3NOH$	....	137.14
3349	acrylamide	....	$C_4H_3O \cdot CH:CH \cdot CO \cdot NH_2$	XVIII-300	137.14
3350	acrylic acid	(stable form)	$C_4H_3O \cdot CH:CH \cdot CO_2H$	XVIII-300	138.12
3351	acrylic acid	(labile form)	$C_6H_5O \cdot CO_2H$	XVIII-301	138.12
3352	**Gallamide** (3,4,5;1)	gallamic acid	$(HO)_3C_6H_2 \cdot CONH_2 \cdot 1\frac{1}{2}H_2O$	X-487	196.16
3353	**Gallocyanine**	solid violet	$C_{15}H_{12}O_5N_2$	XXVII-438	300.27
3354	**Geranyl** acetate	geraniol acetate	$CH_3CO_2 \cdot C_{10}H_{17}$	II-140	196.29
3355	*n*-butyrate	....	$C_3H_7 \cdot CO_2 \cdot C_{10}H_{17}$	II-272	224.35
3356	formate	....	$H \cdot CO_2 \cdot C_{10}H_{17}$	II-23	182.26
3357	**Germanium tetra-ethyl**	tetraEt germanium	$(C_2H_5)_4Ge$	IV-631	188.84

Table 7-4 (*Continued*)
PHYSICAL CONSTANTS OF ORGANIC COMPOUNDS

No.	Crystalline Form and Color	Specific Gravity	Melting Point °C.	Boiling Point °C.	Solubility in 100 Parts		
					Water	Alcohol	Ether
3319	col. nd.	....	39–40	135-7^{33mm}	i.	s.	s.; s. chl.
3320	oil	1.114$^{20°}$	....	317	....	....	....
3321	col. cr./pet.	....	52–3	220	i.; s. bz.	s.	v. s.
3322	nd./lg.	....	90–1	201-8 sl. d.	v. sl. s. c.	s.; s. bz.	s.; s. ac.
3323	nd./lg.	....	74–5	....	sl. s.	v. s.; s. bz.	v. s.
3324	yel. lf./al.	....	97–8	....	i.; i. lg.	s.	s.
3325	gn. cr.	....	250–2	....	i.	sl. s.	i.
3326	nd./al.	....	117–21	250 d.	i.; d. a.	s.	s.
3327	col. lq.	0.937$^{20°}_{4}$	....	31-2^{756mm}	i.	s.	s.
3328	yel. nd./bz.	....	165–6	....	i.; s. chl.	sl. s.	sl. s.
3329	col. nd./aq.	....	166–8	....	v. sl. s. bz.	v. s.	v. s.
3330	cr./lg.-et.	....	188–90 d.	....	i.	....	v. sl. s.
3331	col. oil	1.118$^{20°}_{4}$	....	175–7	i.	s.	s.
3332	oil	1.129$^{25°}_{4}$	....	169.5^{752mm}	∞; slow d.	s.	s.
3333	lq.	<1	....	145^{754mm}	∞	s.	s.
3334	col. lq.	1.053$^{20°}_{4}$	....	212–3	v. sl. s.	s.	∞
3335	col. lq.	1.178$^{20°}_{4}$	....	49^{26mm}	i.	s.	s.
3336	col. lq.	α1.33 β1.40	α19.5 β27.5	>350^{760} d.	i.; i. pet.	∞; ∞ bz.	∞; ∞ chl.
3337	col. oil	1.132$^{20°}_{4}$	....	84^{65mm}	i.	....	....
3338	col. lq.	1.109$^{20°}_{4}$	....	195–6	v. sl. s.	s.	∞
3339	col. nd./aq.	....	121–2	105-10^{12mm}	s. h.	s.	v. s.
3340	mn. pr.	....	133–4 subl.>100	230–2 d. >250	3.6$^{15°}$; 25 h.	s.	s.
3341	cr.	....	142–3	subl. >100	....	....	....
3342	cr.	....	169	....	....	....	....
3343	nd./al.	....	73	325 d.	....	s.	s.
3344	col. lq.	....	−2 to 0	173–6	sl. d. h.	sl. d. h.	s.
3345	col. lq.	....	....	140-8	v. sl. s.	∞	∞
3346	pa. brn. nd./Me al.	....	138–9	d.	v. sl. s. h.	sl. s. h.; s. h. toluene	sl. s.
3347	yel. nd./lg.	....	54	>200 d.	i.	s.	s.
3348	yel. nd.	....	103–5	....	i.	s.	s.
3349	sc./aq.	....	168–9	....	v. sl. s. c.	....	....
3350	nd./aq.	....	141	286 d.; in vac. subl.	0.2 c.; s. h.	s.; v. s. ac.; 1.5$^{19°}$ bz.	v. s.; i. lg.; i. CS$_2$
3351	pr. or pl.	....	103–4	....	s. h.	7.5$^{19°}$ bz.	....
3352	lf./aq.	....	....	244 d. (anh.)	s. h.	....	....
3353	gn. cr.	....	....	....	v. sl. s. h.	s.; s. act.	s. alk.
3354	col. lq.	0.917$^{15°}$	....	242-5^{764} d.	v. sl. s.	s.	∞
3355	col. lq.	0.901$^{17°}_{4}$	....	151-31^{8mm}	i.	s.	s.
3356	col. lq.	0.927$^{20°}_{4}$	....	113-41^{5mm}	i.	s.	s.
3357	col. lq.	0.991$^{24.5°}_{24.5}$	−90	163.5	i.	s.	s.

Table 7-4 (*Continued*)
PHYSICAL CONSTANTS OF ORGANIC COMPOUNDS

No.	Name	Synonym	Formula	Beil. Ref.	Formula Weight
3358	**Gluco-α-heptose** (*d*)	....	$C_7H_{14}O_7$	I-934	210.19
3359	**Gluconic** acid	....	$CH_2OH(CHOH)_4\cdot CO_2H$	III-542	196.16
3360	Ca salt	Ca gluconate	$C_{12}H_{22}O_{14}Ca\cdot H_2O$	III-544	448.40
3361	δ-lactone	glucono-δ-lactone	$C_6H_{10}O_6$	*XVIII-405	178.14
3362	**Glucose** (α) penta-acetate (*d*)	penta-acetyl glucose	$C_6H_7O_6(COCH_3)_5$	XXXI-119	390.35
3363	(β)pentacetate (*d*)	....	$C_6H_7O_6(COCH_3)_5$	XXXI-120	390.35
3364	pentabutyrate	....	$C_6H_7O_6(COC_3H_7)_5$	....	530.62
3365	pentapropionate	....	$C_6H_7O_6(COC_2H_5)_5$	....	460.48
3366	phenylhydrazone	(*d*)(α)	$C_6H_5NH\cdot N:C_6H_{12}O_5$	XV-221	270.29
3367	phenylhydrazone (β)	....	$C_6H_5NH\cdot N:C_6H_{12}O_5$	XV-221	270.29
3368	**Glucosone** (*d*)	fructosone	$C_4H_9O_4\cdot CO\cdot CHO$	I-932	178.14
3369	**Glucosoxime** (*d*)	....	$C_6H_{12}O_5:NOH$	I-902	195.17
3370	**Glutaconic acid** (trans)	pentendioic acid	$HO_2C\cdot CH_2\cdot CH:CH\cdot CO_2H$	II-758	130.10
3371	**Glutamic** acid (*dl*)	glutaminic acid	$HO_2C\cdot CHNH_2\cdot (CH_2)_2\cdot CO_2H$	IV-493	147.13
3372	acid (*l*)(+)	α-NH₂-glutaric	$C_5H_9O_4N$	IV-488	147.13
3373	acid HCl (*d*)	Acidulin	$C_5H_9O_4N\cdot HCl$	IV-491	183.59
3374	amide (*dl*)	glutamine	$HO_2C\cdot C_3H_5NH_2\cdot CO\cdot NH_2$	IV-491	146.15
3375	amide (*l*)(+)	*l*-glutamine	$C_5H_{10}O_3N_2$	IV-491	146.15
3376	**Glutaric** acid	pentandioic acid	$CH_2(CH_2\cdot CO_2H)_2$	II-631	132.12
3377	nitrile	....	$CH_2(CH_2\cdot CN)_2$	II-635	94.12
3378	**Glutathione**	....	$C_{10}H_{17}O_6N_3S$	**IV-931	307.33
3379	**Glyceric** acid (*dl*)	dihydroxy-propionic	$HOCH_2CHOH\cdot CO_2H$	III-392	106.08
3380	aldehyde (*dl*)	....	$HOCH_2CHOH\cdot CHO$	I-845	90.08
3381	**Glycerol**	glycerin	$CHOH(CH_2OH)_2$	I-502	92.10
3382	acetate, mono	monacetin †	$C_5H_{10}O_4$	II-146	134.13
3383	acetate, di	diacetin	$HO\cdot C_3H_5(O_2C\cdot CH_3)_2$	II-147	176.17
3384	amine, di (αγ)	1,3-diNH₂-2-propanol	$CHOH(CH_2NH_2)_2$	IV-290	90.13
3385	amine, mono (α)	1-NH₂-2,3-propandiol	$CH_2OH\cdot CHOH\cdot CH_2\cdot NH_2$	IV-301	91.11
3386	*iso*-amyl ether	(di)(αγ)	$(C_5H_{11}OCH_2)_2:CHOH$	I-513	232.37
3387	*iso*-amyl ether	(mono)(α)	$CH_2OH\cdot CHOH\cdot CH_2\cdot O\cdot C_5H_{11}$	I-513	162.23
3388	bromohydrin, di (α)	1,3-dibromo-2-propanol	$(CH_2Br)_2:CHOH$	I-365	217.90
3389	bromohydrin, di (β)	2,3-dibromo-1-propanol	$Br\cdot CH_2\cdot CHBr\cdot CH_2OH$	I-357	217.90
3390	*n*-butyl ether	(di)(αγ)	$(C_4H_9OCH_2)_2:CHOH$	....	204.31
3391	*n*-butyl ether	(mono)(α)	$CH_2OH\cdot CHOH\cdot CH_2\cdot O\cdot C_4H_9$	....	148.20
3392	*n*-butyrate, mono	α-monobutyrin	$C_3H_7CO_2\cdot CH_2\cdot CHOH\cdot CH_2OH$	II-273	162.19
3393	chlorohydrin, di (α)	1,3-dichloro-2-propanol	$(CH_2Cl)_2:CHOH$	I-364	128.99
3394	chlorohydrin, di (β)	2,3-dichloro-1-propanol	$Cl\cdot CH_2\cdot CHCl\cdot CH_2OH$	I-356	128.99

† Mixture of both isomers.
Gluco-chloral 1289
Gluco-gallin, cf. glcde.
Glucosamine (*iso*) 3311
Glucose 3618

Glucose dimethyl-ketal 60
Glucosido-methyl salicylate, cf. glcde.
Glucosido-salicylaldehyde, cf. glcde.
Gluside 5579
Glutamine 3374-5

Table 7-4 (*Continued*)
PHYSICAL CONSTANTS OF ORGANIC COMPOUNDS

No.	Crystalline Form and Color	Specific Gravity	Melting Point °C.	Boiling Point °C.	Solubility in 100 Parts Water	Alcohol	Ether
3358	rhb.	....	180-90 d.	....	$10^{14°}$; v. s. h.	v. sl. s. abs. al.	....
3359	nd./al. et.	....	softens 110; 131	....	s.	i. abs. al.	i.
3360	col. nd./aq.	....	....	....	$3.8^{16.5°}$	i.	20 h. aq.
3361	nd./al.	....	153	....	d.	....	....
3362	mn. nd./al.	....	112-3	....	$0.15^{18.5°}$; i. pet.	$1.3^{19°}$ abs. al.	$2.8^{15°}$
3363	mn. nd./al.	....	132-4	subl. vac.	$0.1^{18°}$; i. pet.	$0.8^{19°}$ abs. al.	$2.1^{15°}$
3364	pa. yel. lq.	$1.094^{25°}$	....	$228^{1.5mm}$	i.	s.	s.; s. chl.
3365	pa. yel. lq.	$1.151^{25°}$	....	....	i.	s.	s.; s. chl.
3366	col. cr.	....	159-60	....	s.	sl. s. c.	sl. s. c.
3367	col. cr.	....	140-1	....	sl. s.	sl. s. c.	v. sl. s. c.
3368	amor.	....	....	....	s.	s. h.	i.
3369	nd./Me al.	....	137-8	....	v. s.	v. sl. s.	i.
3370	pr./et.	....	137-8	....	v. s.	v. s.	v. s.
3371	cr./aq.	1.460	199 d.	....	$1.5^{20°}$	v. sl. s.	v. sl. s.
3372	rhb.	$1.538^{\frac{20}{4}}$	224-5 d.	....	$0.7^{20°}$	i. abs.	i.
3373	rhb.	....	202-4*	....	33.3 c.*	i. HCl	....
3374	nd./aq.	....	256	....	$3.6^{18°}$	v. sl. s.	....
3375	nd./aq. al.	....	186	....	$3.6^{18°}$	$0.0005^{25°}$	v. sl. s.
3376	col. cr.	$1.429^{15°}$	97.5	200^{20mm}	$63.9^{20°}$	v. s. abs.	v. s.; s. bz.
3377	col. lq.	$0.991^{\frac{15}{4}}$	-29.5(-32)	285-7	s.	s.	i.
3378	col. pr./al.	....	190-2 d.	....	$10^{0°}$; d.h.	i. abs.	i.; i. ac.
3379	syrup	....	....	....	∞	∞	i.
3380	nd. or pr.	$1.455^{\frac{18}{18}}$	142	$140-50^{0.8mm}$	$3^{18°}$	sl. s.	sl. s.; i. bz.
3381	col. lq.	$1.261^{\frac{20}{4}}$	18.2†	290	∞; i. bz.	∞; i. pet.	i.; i. chl.
3382	col. oil	$1.20^{\frac{20}{4}}$	....	158^{165mm}	v. s.	v. s.	sl. s.; i. bz.
3383	col. lq.	$1.178^{\frac{15}{5}}$	40	$175-6^{40mm}$	s.	s.	sl. s.
3384	col. hyg.	$1.096^{\frac{25}{25}}$	42	130^{10mm} ±	∞; d. act	∞; i. bz.	i.; i. CCl_4
3385	oil	$1.185^{\frac{25}{25}}$	....	264 sl. d.	s.	s.	i.; i. bz.
3386	col. lq.	$0.903^{\frac{25}{25}}$	....	270-2	i.	∞	∞
3387	col. lq.	$0.987^{\frac{25}{25}}$	....	260-2	s.	∞	∞
3388	col. lq.	$2.135^{\frac{21}{21}}$	....	214.8	i.	....	....
3389	col. lq.	$2.120^{\frac{20}{4}}$	....	219 sl. d.	i.	∞; ∞ act.	∞; ∞ bz.
3390	col. lq.	....	....	$121-5^{13mm}$	i.	∞	∞
3391	col. lq.	$0.945^{\frac{25}{25}}$	....	$133-7^{18mm}$	s.	∞	∞
3392	col. lq.	$1.129^{18°}$	....	269-71	∞	....	....
3393	col. lq.	$1.367^{\frac{20}{4}}$	<-20	174.3^{760mm}	$12^{20°}$, $19^{70°}$	∞	∞
3394	col. lq.	$1.362^{\frac{20}{4}}$	....	182-3	14.5	∞	∞

* Heated rapidly, m. p. 213°; d. − HCl, in aq.
† Solidifies at much lower temp.
Glutaminic acid 3371

Glycerin(e) 3381
Glycerol, cf. also glyceryl.

Table 7-4 (*Continued*)
PHYSICAL CONSTANTS OF ORGANIC COMPOUNDS

No.	Name	Synonym	Formula	Beil. Ref.	Formula Weight
	Glycerol				
3395	chlorohydrin, mono	α-chlorohydrin	$CH_2Cl \cdot CHOH \cdot CH_2OH$	I-473	110.54
3396	chlorohydrin, mono	β-chlorohydrin	$CHCl(CH_2OH)_2$	I-476	110.54
3397	formal†	....	$H_2C:C_3H_6O_3$	....	104.11
3398	formate, di†	....	$(HCO_2)_2C_3H_6O$	II-24	148.12
3399	furfural†	....	$C_5H_4O:C_3H_6O_3$	....	170.17
3400	guaiacol ether, mono	guaiamar; oresol; oreson	$CH_3O \cdot C_6H_4 \cdot O \cdot$ $C_3H_5(OH)_2$	....	198.22
3401	iodohydrin, di (α)	1,3-diiodo-2-pro-panol	$(CH_2I)_2:CHOH$	I-366	311.89
3402	iodohydrin, mono (α)	3-iodo-1,2-propan-diol; Alival	$CH_2I \cdot CHOH \cdot CH_2OH$	I-475	201.99
3403	methyl ether, di	(αγ)	$(CH_3OCH_2)_2CHOH$	I-512	120.15
3404	methyl ether, mono	(α)	$CH_3O \cdot C_3H_5(OH)_2$	I-512	106.12
3405	nitrate, di (1,2)	....	$C_3H_5(OH)(O \cdot NO_2)_2$	I-515	182.09
3406	nitrate, di (1,3)	....	$CHOH(CH_2NO_3)_2$	I-515	182.09
3407	nitrate, mono (α)	....	$(HO)_2C_3H_5 \cdot O \cdot NO_2$	I-514	137.09
3408	nitrate, mono (β)	....	$(CH_2OH)_2CHO \cdot NO_2$	I-515	137.09
3409	phenyl ether, di	(αγ)	$(C_6H_5OCH_2)_2CHOH$	VI-149	244.29
3410	phenyl ether, mono (α)	antodyne	$C_6H_5 \cdot O \cdot CH_2 \cdot CHOH \cdot$ CH_2OH	VI-149	168.19
3411	phosphoric acid	....	$(HO)_2C_3H_5 \cdot OPO_3H_2$	I-517	172.08
3412	phosphate, sodium	(β)	$(HOCH_2)_2CH \cdot O \cdot$ $PO(ONa)_2 \cdot 5H_2O$	*I-517	306.12
3413	propionate, mono	α-monopropionin	$C_2H_5CO_2 \cdot CH_2 \cdot$ $CHOH \cdot CH_2OH$	*II-107	148.16
3414	salicylate, mono (α or β)	monosalicylin	$HO \cdot C_6H_4CO_2 \cdot$ $C_3H_5(OH)_2$	X-82	212.20
3415	stearate, mono	α-monostearin	$C_{17}H_{35}CO_2 \cdot CH_2 \cdot$ $CHOH \cdot CH_2OH$	II-380	358.57
3416	stearate, mono	β-monostearin	$C_{17}H_{35}CO_2 \cdot CH:$ $(CH_2OH)_2$	II-380	358.57
3417	**Glyceryl** ether	....	$C_3H_5O_3:C_3H_5$	XIX-393	130.14
3418	triacetate	triacetin	$(CH_3CO_2)_3C_3H_5$	II-147	218.21
3419	triarachidate	triarachin	$(C_{19}H_{39}CO_2)_3C_3H_5$	II-390	975.67
3420	tribenzoate	tribenzoin	$(C_6H_5CO_2)_3C_3H_5$	IX-140	404.42
3421	tributyrate	tributyrin	$(C_3H_7CO_2)_3C_3H_5$	II-273	302.37
3422	tricaprate	tricaprin	$(C_9H_{19}CO_2)_3C_3H_5$	II-356	554.86
3423	tricaproate	tricaproin	$(C_5H_{11}CO_2)_3C_3H_5$	II-324	386.53
3424	tricaprylate	tricaprylin	$(C_7H_{15}CO_2)_3C_3H_5$	II-348	470.70
3425	triheptylate	triheptylin	$(C_6H_{13}CO_2)_3C_3H_5$	**II-295	428.61
3426	trilaurate	trilaurin	$(C_{11}H_{23}CO_2)_3C_3H_5$	II-362	639.02
3427	trimyristate	trimyristin	$(C_{13}H_{27}CO_2)_3C_3H_5$	II-367	723.18
3428	trinitrate	nitroglycerin	$(O_2N \cdot O)_3C_3H_5$	I-516	227.09
3429	trinitrite	....	$(ON \cdot O)_3C_3H_5$	I-514	179.09
3430	trioleate	triolein	$(C_{17}H_{33}CO_2)_3C_3H_5$	II-468	885.46
3431	tripalmitate	tripalmitin	$(C_{15}H_{31}CO_2)_3C_3H_5$	II-373	807.35
3432	tripropionate	tripropionin	$(C_2H_5CO_2)_3C_3H_5$	*II-107	260.29
3433	tristearate	tristearin	$(C_{17}H_{35}CO_2)_3C_3H_5$	II-383	891.51
3434	trivalerate	trivalerin	$(C_4H_9CO_2)_3C_3H_5$	....	344.45

† Mixture of both isomers.
Glyceryl, cf. also glycerol.

Glyceryl-amine 3384-5
Glyceryl diacetate 3383

Table 7-4 (*Continued*)
PHYSICAL CONSTANTS OF ORGANIC COMPOUNDS

No.	Crystalline Form and Color	Specific Gravity	Melting Point °C.	Boiling Point °C.	Solubility in 100 Parts		
					Water	Alcohol	Ether
3395	col. lq.	1.318_{25}^{25}°	....	213	∞	∞	s.
3396	col. lq.	1.321_{4}^{20}°	....	146^{18mm}	s.	∞ ; s. bz.	∞ ; ∞ act
3397	col. lq.	1.220_{25}^{25}°	....	192-5	∞	∞	∞
3398	col. lq.	1.304^{15}°	....	$163\text{-}6^{30mm}$	i. CS_2	....	....
3399	yel. lq.	1.267_{25}^{25}°	....	$163\text{-}7^{22mm}$	s.	∞	sl. s.
3400	col. cr. pd.	....	78-9	....	5	s.	s.; s. chl.
3401	yel. oil	2.4^{15}°	freezes −16 to −20	d.	1.3; d. alk.	∞ ; ∞ bz.	∞ ; ∞ chl.
3402	col. cr.	2.03^{13}°	49±	....	v. s.	v. s.	....
3403	col. lq.	1.004_{4}^{25}°	....	169	∞	∞	∞
3404	col. lq.	1.111_{4}^{25}°	....	220	∞	∞	s.
3405	oil	....	expl.	....	d. by a.	d. by alk.	....
3406	oil	1.471^{15}°	<−30; expl.	$146\text{-}8^{15mm}$ sl. d.	d. by a., alk.	v. s.	v. s.
3407	col. pr.	1.40	58-9	155-60*	70^{15}°	v. s.	v. sl. s.
3408	lf.	1.40	54	155-60	....	v. s.	sl. s.
3409	col. hex.	1.179_{4}^{24}°	80-1	$200\text{-}10^{10mm}$	i.; s. bz.	s. h.	s.; s. chl.
3410	col. nd./abs. et.	1.225_{4}^{30}°	69-70 (53-4)	200^{22mm}	∞ ; v. sl. s. lg.	∞ ; v. s. bz.	∞
3411	syrup	1.50^{14}°	−20	....	∞	∞ abs.	....
3412	pl./aq.	....	98-100	d. >130	66 c.	....	....
3413	col. lq.	1.154_{4}^{20}°	....	$141\text{-}4^{6mm}$	....	....	....
3414	col. nd./et.	....	76	....	1 c.; v. s. h.	s.; ∞ gly.	sl. s.; s. h. bz.
3415	col. nd./Me al.	0.984_{4}^{20}°	82(57)	....	i.	s. h.	s.
3416	nd. or wax	....	....	....	i.	s. h.	s.
3417	lq.	1.091^{118}°	....	171-2	∞	∞	∞
3418	col. lq.	1.161_{4}^{17}°	−78	258-9	7.17^{15}°	∞ ; ∞ chl.	∞ ; ∞ bz.
3419	col. cr.	....	72.2	....	i.	s. h.	s. h.
3420	nd./Me al.	1.228^{12}°	75-6	d.	i.; s. bz.	s. h.	s.; s. chl.
3421	col. lq.	1.032_{4}^{20}°	<−75	305-9	i.	s.	s.
3422	col. cr.	0.921_{4}^{40}°	31(25)	....	i.; s. bz.	s. h.	v. s.
3423	col. lq.	0.987_{4}^{20}°	−25	....	i.	s. 85% al.	s.
3424	col. lq.	0.954_{4}^{20}°	8.3(−21)	....	i.	s. 85% al.	s.
3425	col. lq.	0.969^{20}°	....	$223\text{-}5^{3mm}$	i.	s.	s.
3426	col. nd.	0.894_{4}^{60}°	46.4(36)†	....	v. s. bz.	sl. s. c.	v. s.
3427	lf.	0.885_{4}^{60}°	56.5(49)†	....	i.; v. s. bz.	v. s.	v. s. chl.
3428	col. or yel. oil	1.594_{4}^{20}°	13.3(2.0)	160^{15mm}; expl. 270	0.18^{20}°	54^{20}° abs.	∞
3429	yel. lq.	1.291_{18}^{10}°	....	150 sl. d.	d.; i. CS_2	d.; s. chl.	s.; s. bz.
3430	col. oil	0.915^{150}°	−4	240^{18mm} sl. d.	i.	sl. s.	v. s.
3431	col. nd.	0.866_{4}^{80}°	65.1 (55; 45)	$310\text{-}20^{0.1mm}$	i.	0.004^{21}° abs. al.	v. s.
3432	lq.	1.100_{18}^{20}°	....	$177\text{-}82^{20mm}$	....	....	....
3433	col pr.	0.862_{4}^{80}°	70.8(54.5; 64.5)	....	i.; s. bz.	s. h.	s. h.; sl. s. pet.
3434	pa. yel. lq	1.030^{20}°	....	$152\text{-}5^{1mm}$	i.	s.	s.

* Not explosive.
† Probably several other modifications.

Glyceryl tribromohydrin 6112
Glyceryl trichlorohydrin 6166

Table 7-4 (*Continued*)
PHYSICAL CONSTANTS OF ORGANIC COMPOUNDS

No.	Name	Synonym	Formula	Beil. Ref.	Formula Weight
	Glyceryl				
3435	trivalerate (*iso*)	tri-*iso*-valerin	$(C_4H_9CO_2)_3C_3H_5$	II-314	344.45
3436	**Glycide**	glycidol; epihydrin alcohol	$CH_2 \cdot O \cdot CH \cdot CH_2OH$	XVII-104	74.08
3437	**Glycine**	amino-acetic acid	$NH_2 \cdot CH_2 \cdot CO_2H$	IV-333	75.07
3438	amide	....	$NH_2 \cdot CH_2 \cdot CO \cdot NH_2$	IV-343	74.08
3439	anhydride	diketopiperazine	$C_4H_6O_2N_2$	XXIV-264	114.10
3440	**Glycocholic** acid	cholylglycine; "cholic acid"	$C_{24}H_{39}O_4 \cdot NH \cdot CH_2 \cdot CO_2H$	....	465.64
3441	sodium salt	Na glycocholate	$C_{26}H_{42}O_6NNa$	....	487.62
3442	**Glycocyamine**	guanyl glycine	$HN:C(NH_2) \cdot NH \cdot CH_2CO_2H$	IV-359	117.11
3443	**Glycogen**	animal starch	$(C_6H_{10}O_5)_x$	....	(162.14)
3444	**Glycol**	ethandiol-1,2	$HOCH_2 \cdot CH_2OH$	I-465	62.07
3445	acetate, di	ethylene acetate	$(CH_3CO_2 \cdot CH_2 \cdot)_2$	II-142	146.14
3446	acetate, mono	....	$CH_3CO_2 \cdot CH_2CH_2OH$	II-141	104.11
3447	benzoate, di	ethylene benzoate	$(C_6H_5CO_2 \cdot CH_2)_2$	IX-129	270.29
3448	benzyl ether, di	....	$(C_6H_5CH_2 \cdot O \cdot CH_2)_2$	....	242.32
3449	benzyl ether, mono	benzyl cellosolve	$C_7H_7O \cdot C_2H_4OH$	....	152.19
3450	butyrate, di	ethylene dibutyrate	$(C_3H_7CO_2 \cdot CH_2)_2$	II-272	202.25
3451	capr(o)ate, di	ethylene dicaprate	$(C_5H_{11}CO_2 \cdot CH_2)_2$	....	258.36
3452	caprylate, di	ethylene dicaprylate	$(C_7H_{15}CO_2 \cdot CH_2)_2$	**II-303	314.47
3453	ether	diethylene glycol	$(HO \cdot CH_2 \cdot CH_2)_2O$	I-468	106.12
3454	formal	dioxolane; glycol-methylene ether	$O \cdot CH_2 \cdot CH_2 \cdot O \cdot CH_2$	XIX-2	74.08
3455	formate, di	ethylene formate	$(HCO_2 \cdot CH_2)_2$	II-23	118.09
3456	formate, mono	....	$HCO_2 \cdot CH_2 \cdot CH_2OH$	II-23	90.08
3457	guaiacyl ether, di	....	$(CH_3OC_6H_4O \cdot CH_2)_2$	....	274.32
3458	laurate, di	ethylene dilaurate	$(C_{11}H_{23}CO_2 \cdot CH_2)_2$	II-361	426.69
3459	methyl ether, di	diMeO-ethane	$(CH_3 \cdot O \cdot CH_2)_2$	I-467	90.12
3460	methyl ether, mono	methyl Cellosolve	$CH_3O(CH_2)_2OH$	I-467	76.10
3461	myristate, di	ethylene dimyristate	$(C_{13}H_{27}CO_2 \cdot CH_2)_2$	II-366	482.79
3462	nitrate, di	dinitroglycol	$(O_2N \cdot O \cdot CH_2)_2$	I-469	152.06
3463	nitrite, di	ethylene nitrite	$(ON \cdot O \cdot CH_2)_2$	I-469	120.07
3464	palmitate, di	ethylene dipalmitate	$(C_{15}H_{31}CO_2 \cdot CH_2)_2$	II-373	538.90
3465	propionate, di	ethylene dipropionate	$(C_2H_5CO_2 \cdot CH_2)_2$	II-242	174.20
3466	salicylate, mono	glysal; Spirosal	$C_7H_5O_3 \cdot C_2H_4OH$	X-81	182.18
3467	stearate, di	ethylene distearate	$(C_{17}H_{35}CO_2 \cdot CH_2)_2$	II-380	595.01
3468	**Glycolic** acid	hydroxy-acetic acid	$HO \cdot CH_2 \cdot CO_2H$	III-228	76.05
3469	Ca salt	calcium glycolate	$C_4H_6O_6Ca \cdot 4H_2O$	III-232	262.23
3470	aldehyde	hydroxy-acetaldehyde	$HO \cdot CH_2 \cdot CHO$	I-817	60.05
3471	aldehyde-diethyl-acetal	....	$HO \cdot CH_2 \cdot CH:(OC_2H_5)_2$	I-818	134.18
3472	amide	hydroxy-acetamide	$HO \cdot CH_2 \cdot CO \cdot NH_2$	III-240	75.07
3473	anhydride	....	$(HO \cdot CH_2 \cdot CO)_2O$	III-239	134.09
3474	nitrile	....	$HO \cdot CH_2 \cdot CN$	III-242	57.05
3475	**Glycolide**	glycolid	$CH_2 \cdot CO_2 \cdot CH_2 \cdot CO \cdot O$	XIX-153	116.07
3476	**Glycyl** glycine	....	$H_2N \cdot CH_2 \cdot CO \cdot NH \cdot CH_2 \cdot CO_2H$	IV-371	132.12
3477	glycine ester	....	$C_3H_7ON_2 \cdot CO_2C_2H_5$	IV-373	160.17
3478	**Glyoxal**	ethandial	$O:CH \cdot CH:O$	I-759	58.04

Table 7-4 (*Continued*)
PHYSICAL CONSTANTS OF ORGANIC COMPOUNDS

No.	Crystalline Form and Color	Specific Gravity	Melting Point °C.	Boiling Point °C.	Solubility in 100 Parts		
					Water	Alcohol	Ether
3435	oil	$0.998^{20°}$	....	153-6^{2mm}	....	....	....
3436	col. lq.	$1.114\frac{16}{16}°$	....	166-7 sl. d.	∞; sl. s. pet.	∞; sl. s. xylene	∞; s. bz.
3437	mn.	$1.575^{50°}$	232-6 d.	....	23 c.	0.1 c.	i.
3438	hyg. nd.	....	65-7	....	v. s.	v. s.	v. sl. s.
3439	pl./aq.	....	275 d.	subl. 260	s. h.	s.; d. a.	d. alk.
3440	nd.	....	154-5 d.	....	$3.3^{20°}$; $8.5^{100°}$	v. s.; v. sl. s. bz.	$0.09^{20°}$; v. sl. s. chl.
3441	hyg. pd.	....	....	....	s.	s.	....
3442	nd./aq.	....	270-80	....	$0.45^{15°}$	v. sl. s.	v. sl. s.
3443	wh. pd.	....	240	....	s.	i.	i.
3444	col. lq.	$1.113\frac{19}{4}°$	-15.6	197.9	∞	∞	1.0
3445	col. lq.	$1.109\frac{14}{4}°$	-31	190.5	$14.3^{22°}$	∞	∞
3446	col. lq.	1.108	....	182	∞	∞	
3447	rhb./et.	....	73-4	>360	i.	....	s.
3448	col. lq.	....	....	131-6^{2mm}	....	....	
3449	col. lq.	1.07	<-75	256	0.4	....	
3450	col. lq.	1.024^{00}	....	240	i.	v. s.	v. s.
3451	....	....	37-8	....	....	....	
3452	pa. yel. lq.	....	22	....	i.	....	
3453	lq.	$1.118\frac{20}{20}°$	-10.5	244.8	∞; i. bz.	∞; i. chl.	i.
3454	lq.	$1.060\frac{20}{4}°$	....	75-6	∞		
3455	lq.	....	....	174	....	s.	s.
3456	lq.	$1.199\frac{15}{4}°$	....	180	∞	....	
3457	col. nd.	....	138-9	....	v. sl. s.	s. h.	....
3458	col. amor.	....	52-4	188^{20mm}	i.	v. s.	v. s.
3459	lq.	$0.863\frac{20}{4}°$	-58	84-5	∞		
3460	col. lq.	$0.965\frac{20}{4}°$	-85.1	124-5	∞	∞	∞; ∞ bz.
3461	....	....	63-4	$208^{0.1mm}$			
3462	yel. lq.	$1.488\frac{20}{4}°$	-22.3	expl. 114	i.; d. alk.	s.; ∞ bz.	∞; ∞ act.
3463	lq.	1.216^{00}	<-15	96-8	i.; d. alk.	s. d.	s.
3464	nd./al. chl.	....	71-2	$226^{0.1mm}$	i.	s.	s.
3465	lq.	$1.045^{25°}$	....	211-2	sl. s.	∞	∞
3466	col. cr.	....	37	169-70^{12mm}	$0.97^{22°}$	v. s.; s. bz.	v. s.
3467	lf.	....	76-7	$241^{0.1mm}$	i.	$0.12^{40°}$	v. s.
3468	nd./aq.	....	79(63)	d.	v. s.	$90^{25°}$	v. s.
3469	nd.	....	$-4H_2O,110°$	....	$1.3^{15°}$	$5^{100°}$ aq.	....
3470	col. pl.	$1.366^{100°}$	95-7	....	v. s.	v. s. h.	v. sl. s.
3471	lq.	$0.888^{24°}$	....	167	d. h.	....	....
3472	rhb.	....	120	....	v. s.	sl. s.	....
3473	cr. pd.	....	128-30	....	i. c.; d. h.	i.	i.
3474	col. oil	$1.104^{19°}$	<-72	183 sl. d.	v. s.	v. s.	v. s.; i. bz.
3475	lf./al.	....	86-7	....	s. h.; s. ac.	s. h.	sl. s. c.; s. h. chl.
3476	lf./aq.	....	d. 215-20	....	v. s. h.	sl. s.	i.
3477	nd.	....	88-9	....	v. s.	v. s.	sl. s.
3478	yel. cr.	1.142^{00}	15	51^{776mm}	....	v. s. abs.	v. s. abs.

Glycol iodohydrin 3208
Glycol methylene ether 3454
Glycol phenyl ether 5136
Glycolic ethyl ether 2912
Glycolic methyl ether 4063

Glycoline 2533
Glycoluril 139
Glycolyl thiourea 5907
Glycolyl urea 3679

Glycovanillin, cf. glcde.
Glycophillin, cf. glcde.
Glyoxalic acid 3480
Glyoxalin 3851

Table 7-4 (*Continued*)
PHYSICAL CONSTANTS OF ORGANIC COMPOUNDS

No.	Name	Synonym	Formula	Beil. Ref.	Formula Weight
3479	**Glyoxime**	....	$HON:CH\cdot CH:NOH$	I-761	88.07
3480	**Glyoxylic acid**	glyoxalic acid	$HO_2C\cdot CH:O\cdot H_2O$	III-594	92.05
3481	**Guaiacol** acetate	....	$CH_3OC_6H_4\cdot O_2CCH_3$	VI-774	166.18
3482	benzoate	Benzosol	$CH_3OC_6H_4O_2C_7H_5$	IX-130	228.25
3483	*n*-caproate	....	$CH_3OC_6H_4O_2C_6H_{11}$	....	222.29
3484	carbonate	duotal	$(CH_3O\cdot C_6H_4O)_2CO$	VI-776	274.28
3485	cinnamate	Styracol	$CH_3OC_6H_4O_2C_9H_7$	IX-585	254.29
3486	salicylate	guaiacol salol	$C_{14}H_{12}O_4$	X-81	244.25
3487	valerate	....	$CH_3OC_6H_4O_2C_5H_9$	....	208.26
3488	**Guanidine**	imino-urea	$(H_2N)_2C:NH$	III-82	59.07
3489	acetate	....	$CH_5N_3\cdot HC_2H_3O_2$	III-86	119.12
3490	carbonate	....	$(CH_5N_3)_2\cdot H_2CO_3$	III-86	180.17
3491	hydrochloride	....	$CH_5N_3\cdot HCl$	III-86	95.54
3492	nitrate	....	$CH_5N_3\cdot HNO_3$	III-86	122.08
3493	sulfate	....	$(CH_5N_3)_2\cdot H_2SO_4\cdot \frac{1}{2}H_2O$	III-86	225.23
3494	thiocyanate	....	$CH_5N_3\cdot HCNS$	III-169	118.16
3495	**Gynocardic acid**	....	$C_{17}H_{33}CO_2H$	*IX-45	282.47
3496	**Hedonal**	Me-*n*-Pr-carbinol-urethane	$H_2N\cdot CO_2\cdot CH(CH_3)\cdot (CH_2CH_2CH_3)$	III-29	131.18
3497	**Helenin**	alantolactone	$C_{15}H_{20}O_2$	XVII-327	232.33
3498	**Hematein**	....	$C_{16}H_{12}O_6$	XVIII-227	300.27
3499	**Hematin**	phenodin	$C_{34}H_{33}O_5N_4Fe$	....	633.51
3500	**Hematoxylin**	hydroxybrasilin	$C_{16}H_{14}O_6\cdot 3H_2O$	XVII-219	356.33
3501	**Hemin**	....	$C_{34}H_{32}O_4N_4ClFe$	....	651.96
3502	**Hemipinic acid** (3,4;1,2)	3,4-di MeO-phthalic acid	$(CH_3O)_2C_6H_2: (CO_2H)_2 + aq.$	X-543	226.19
3503	**Heneicosane** (*n*)	....	$CH_3(CH_2)_{19}CH_3$	I-174	296.58
3504	**Heneicosene**-9	....	$CH_3(CH_2)_7CH: CH(CH_2)_{10}CH_3$	I-227	294.57
3505	**Hentriacontane** (*n*)	....	$CH_3(CH_2)_{29}CH_3$	I-177	436.86
3506	**Hepta**-chloropropane	(1,1,1,2,2,3,3)	$Cl_3C\cdot CCl_2\cdot CHCl_2$	I-108	285.21
3507	cosane (*n*)	....	$CH_3(CH_2)_{25}CH_3$	I-176	380.75
3508	decane (*n*)	....	$CH_3(CH_2)_{15}CH_3$	I-173	240.48
3509	decyl alcohol	heptadecanol-9	$[CH_3(CH_2)_7]_2CHOH$	I-430	256.48
3510	diene-2,4	....	$CH_3\cdot (CH:CH)_2\cdot C_2H_5$	I-257	96.17
3511	**Heptandiol**-1,7	....	$HO\cdot (CH_2)_7\cdot OH$	I-489	132.20
3512	**Heptane** (*n*)	....	$CH_3\cdot (CH_2)_5\cdot CH_3$	I-154	100.21
3513	**Heptane** (*iso*)	2-methyl-hexane	$(CH_3)_2CH\cdot C_4H_9$	I-156	100.21
3514	**Heptane**	3-methyl-hexane	$C_3H_7\cdot CH(CH_3)\cdot C_2H_5$	I-157	100.21
3515	**Heptane**	2,2-dimethyl-pentane	$(CH_3)_3C\cdot CH_2\cdot C_2H_5$	I-157	100.21
3516	**Heptane**	2,4-dimethyl-pentane	$[(CH_3)_2CH]_2CH_2$	I-158	100.21
3517	**Heptane**	3,3-dimethyl-pentane	$(CH_3)_2C(C_2H_5)_2$	I-158	100.21
3518	**Heptane**	3-ethyl-pentane	$(C_2H_5)_3CH$	I-157	100.21
3519	**Heptane**	2,2,3-tri Me-butane	$(CH_3)_3C\cdot CH(CH_3)_2$	**I-121	100.21
3520	**Heptanoic** acid	oenanthylic acid	$CH_3(CH_2)_5CO_2H$	II-338	130.19
3521	aldehyde	oenanthal	$CH_3(CH_2)_5\cdot CHO$	I-695	114.19
3522	aldehyde oxime	enanthaldoxime	$CH_3(CH_2)_5\cdot CH:NOH$	I-698	129.20
3523	amide	*n*-heptamide	$CH_3(CH_2)_5CO\cdot NH_2$	II-340	129.20

Table 7-4 (*Continued*)
PHYSICAL CONSTANTS OF ORGANIC COMPOUNDS

No.	Crystalline Form and Color	Specific Gravity	Melting Point °C.	Boiling Point °C.	Solubility in 100 Parts		
					Water	Alcohol	Ether
3479	pr./aq.	....	178	....	v. s. h.	v. s.	v. s.
3480	mn./aq.	....	98	....	v. s.	v. sl. s.	v. sl. s.
3481	col. lq.	$1.156^{0°}$	....	240-1	....	∞	∞
3482	col. cr. pd.	....	57-8	....	v. sl. s.	s. h.	s.; s. chl.
3483	col. lq.	....	....	$168-70^{15mm}$	i.	s.	s.
3484	cr.	....	86-8	....	i.	2; s. h.	7; s. bz.
3485	nd./al.	....	130	....	i.	s.; s. bz.	s.; s. chl.
3486	col. cr./al.	....	65	....	i.	s.	s.; s. chl.
3487	yel. lq.	1.05	....	265	i.	s.; s. bz.	s.; s. chl.
3488	hyg. col. cr.	....	$50\pm$	....	v. s.	s.	....
3489	nd.	....	229-30	....	v. s.	v. s.	i.
3490	col. cr.	1.25	197 d.	....	$50^{24°}$	$0.224°$	i. NH_3
3491	col. cr.	....	....	....	v. s.	s.	....
3492	col. cr.	....	214	....	$14^{25°}$	$1.6^{25°}$	$0.6^{25°}$ act.
3493	col. cr.	....	....	....	v. s.	i.	....
3494	col. lf.	....	118	....	$73^{0°}$	$135^{15°}$ aq.	....
3495	col. lf./al.	....	67.5	....	i.	s.	s.
3496	col. nd.	....	74-6	$215\pm$	$0.8^{35°}$	s.	s.; s. chl.
3497	nd./aq. al.	....	76	275	v. sl. s. h.	s.; s. bz.	s.; s. chl.
3498	silv. nd.	....	250 d.	....	$0.06^{20°}$; s. NH_4OH	sl. s.; i. chl.	$0.013^{20°}$; i. bz.
3499	brn. pd.	....	>200	....	i.; s. alk.	i.; i. chl.	i.
3500	cr./aq. $(NH_4)_2SO_3$	....	140(anh.)	$-H_2O$, 100-20	s. h.	s.; s. aq. borax	s.
3501	h. bl. rhb.	....	....	....	i.; s. ac.	i.	i.; i. chl.
3502	mn.	....	182-6 d.	....	sl. s. c.	s.	0.09
3503	col. cr.	$0.778^{40.4°}_{4}$	40.4	215^{15mm}	i.	....	....
3504	col. lq.	$0.802^{20°}_{4}$	3	$201-2^{11mm}$	i.	....	....
3505	col. cr.	$0.781^{68.1°}_{4}$	68.1	302^{15mm}	i.	$0.7^{15°}$ chl.	sl. s.
3506	col. cr.	$1.805^{34°}_{4}$	30	247-8	....	....	....
3507	col. cr.	$0.780^{59.5°}_{4}$	59.5	270^{15mm}	i.	....	....
3508	col. cr.	$0.775^{20°}_{4}$	22.5	303	i.	sl. s.	s.
3509	col. pl.	....	61	....	....	sl. s. aq. al.	....
3510	col. lq.	$0.733^{21.5°}_{4}$	....	107	....	....	....
3511	col. cr.	....	22.5	262	v. s.	v. s.	i.
3512	col. lq.	$0.684^{20°}_{4}$	−90.6	98.4^{760mm}	$0.00521^{8°}$	sl. s.	∞; ∞ chl.
3513	col. lq.	$0.679^{20°}_{4}$	−118.3	90.1	i.	s.	∞
3514	col. lq.	$0.687^{20°}_{4}$	−119.4	92.0	i.	s.	∞
3515	col. lq.	$0.674^{20°}_{4}$	−123.8	79.2	i.	s.	∞
3516	col. lq.	$0.673^{20°}_{4}$	−119.2	80.5	i.	s.	∞
3517	col. lq.	$0.693^{20°}_{4}$	−134.5	86.1	i.	s.	∞
3518	col. lq.	$0.698^{20°}_{4}$	−118.6	93.5	i.	s.	∞
3519	col. lq.	$0.690^{20°}_{4}$	−25.0	80.9	i.	s.	∞
3520	col. lq.	$0.922^{15°}_{4}$	−7.5	223.0	$0.25^{15°}$	s.	s.
3521	col. lq.	$0.822^{15°}_{4}$	−43.3	152.8	$0.02^{20°}$	∞	∞
3522	pl./al.	$0.858^{54.7°}_{4}$	56	195	sl. s. c.	s.	s.
3523	nd./al.	$0.849^{11.2°}_{4}$	95-6	250-8	v. s.	v. s.	v. s.

Guvacine, cf. alkd.
Gynesin, cf. alkd.
Gynoval 869
H acid 342
Harmaline, cf. alkd.
Harmine, cf. alkd.
Hasting's naphtha 4100
Hederin (α), cf. glcde.

Helianthin-B 4328
Helicin, cf. glcde.
Helicon 115
Heliotropine 5323
Helixin, cf. glcde.
Helleborein, cf. glcde.
Helleborin, cf. glcde.
Hemellitic acid 2458

Hemimellitic acid 683
Hemimellitene 6253
Hemisine 157
Hendecenoic acid 6376
Hepta-decanal 4015
Hepta-decanoic acid 4014
Hepta-decanol 3509
Hepta-decyl cyanide 5638

Hepta-methylene 1596
Hepta-methylene dicyanide 597
Heptamide 3523
Heptandioic acid 5309

Table 7-4 (*Continued*)
PHYSICAL CONSTANTS OF ORGANIC COMPOUNDS

No.	Name	Synonym	Formula	Beil. Ref.	Formula Weight
	Heptanoic				
3524	anhydride (*n*)	heptylic anhydride	$(C_6H_{13}CO)_2O$	II-340	242.36
3525	**Heptanthiol-2**	heptyl mercaptan	$CH_3 \cdot CH(SH) \cdot C_5H_{11}$	I-415	132.27
3526	**Heptyl** acetate (*n*)	....	$CH_3CO_2 \cdot C_7H_{15}$	II-134	158.24
3527	alcohol (*n*)	heptanol-1	$CH_3(CH_2)_5CH_2OH$	I-414	116.20
3528	alcohol	2,2,3-trimethyl-butanol-3	$(CH_3)_3C \cdot C(CH_3)_2 \cdot OH$	I-418	116.20
3529	alcohol	2,3-dimethyl-pentanol-3	$(CH_3)_2CH \cdot C(OH) \cdot (CH_3) \cdot C_2H_5$	I-417	116.20
3530	alcohol	2,4-dimethyl-pentanol-2	$(CH_3)_2C(OH) \cdot CH_2 \cdot CH(CH_3)_2$	I-417	116.20
3531	alcohol	2,4-diMe-pentanol-3	$[(CH_3)_2CH]_2CHOH$	I-417	116.20
3532	alcohol	3-ethyl-pentanol-2	$(C_2H_5)_2CH \cdot CHOH \cdot CH_3$	I-416	116.20
3533	alcohol	3-ethyl-pentanol-3	$(C_2H_5)_3COH$	I-417	116.20
3534	alcohol	2-methyl-hexanol-1	$CH_3(CH_2)_3CH(CH_3) \cdot CH_2OH$	I-415	116.20
3535	alcohol	2-methyl-hexanol-2	$(CH_3)_2C(OH)(CH_2)_3 \cdot CH_3$	I-415	116.20
3536	alcohol	2-methyl-hexanol-3	$(CH_3)_2CH \cdot CHOH \cdot (CH_2)_2CH_3$	I-416	116.20
3537	alcohol	2-methyl-hexanol-4	$(CH_3)_2CH \cdot CH_2 \cdot CHOH \cdot C_2H_5$	I-416	116.20
3538	alcohol	2-methyl-hexanol-5	$(CH_3)_2CH(CH_2)_2 \cdot CHOH \cdot CH_3$	I-416	116.20
3539	alcohol (*iso*)	2-methyl-hexanol-6	$(CH_3)_2CH(CH_2)_3 \cdot CH_2OH$	I-416	116.20
3540	alcohol	3-methyl-hexanol-3	$C_2H_5 \cdot CH_2 \cdot C(CH_3) \cdot (OH) \cdot C_2H_5$	I-416	116.20
3541	alcohol	3-methyl-hexanol-4	$C_2H_5 \cdot CH(CH_3) \cdot CHOH \cdot C_2H_5$	I-416	116.20
3542	alcohol	heptanol-2	$CH_3(CH_2)_4 \cdot CHOH \cdot CH_3$	I-415	116.20
3543	alcohol	heptanol-4	$(CH_3 \cdot CH_2 \cdot CH_2)_2 : CHOH$	I-415	116.20
3544	amine (*n*)	....	$C_7H_{15} \cdot NH_2$	IV-193	115.22
3545	bromide (*n*)	1-bromoheptane	$CH_3(CH_2)_5 \cdot CH_2Br$	I-155	179.11
3546	chloride (*n*)	1-chloroheptane	$CH_3(CH_2)_5 \cdot CH_2Cl$	I-154	134.65
3547	chloride	3-chloro-2,3-di-methyl-petane	$(CH_3)_2CH \cdot CCl(CH_3) \cdot C_2H_5$	I-157	134.65
3548	formate (*n*)	....	$HCO_2 \cdot C_7H_{15}$	II-22	144.22
3549	iodide	1-iodo-2-methyl-hexane	$ICH_2 \cdot CH(CH_3) \cdot (CH_2)_3 \cdot CH_3$	I-157	226.10
3550	malonic acid (*n*)	....	$C_7H_{15}CH(CO_2H)_2$	II-721	202.25
3551	nitrite (*n*)	....	$CH_3(CH_2)_6 \cdot O \cdot NO$	I-415	145.20
3552	oenanthylate	*n*-heptyl *n*-heptylate	$C_6H_{13}CO_2 \cdot C_7H_{15}$	II-340	228.38
3553	propionate (*n*)	....	$C_2H_5CO_2 \cdot C_7H_{15}$	II-241	172.27
3554	**Heptylene** (*α*)	heptene-1	$CH_3(CH_2)_4 \cdot CH : CH_2$	I-219	98.19
3555	**Heptylene**	2,4-dimethyl-pentene-2	$(CH_3)_2C : CH \cdot CH : (CH_3)_2$	I-220	98.19
3556	**Heptylene**	2,3,3-triMe-butene-1	$(CH_3)_3C \cdot C(CH_3) : CH_2$	**I-199	98.19
3557	alcohol	2-Me-penten-4-ol-2	$C_3H_5 \cdot CH_2 \cdot C(CH_3)_2 \cdot OH$	I-445	114.19
3558	**Heptyne-1**	oenanthylidene	$CH_3(CH_2)_4 \cdot C : CH$	I-256	96.17

Heptanol 3527-43
Heptanone 2786, 3003, 4122
Heptene 3554-6
Heptyl, cf. also diheptyl.
Heptyl aldehyde 3521

Heptyl carbinol 4961
Heptyl cyanide 1226
Heptyl ether 2264
Heptyl heptylate 3552
Heptyl mercaptan 3525

Table 7-4 (*Continued*)
PHYSICAL CONSTANTS OF ORGANIC COMPOUNDS

No.	Crystalline Form and Color	Specific Gravity	Melting Point °C.	Boiling Point °C.	Solubility in 100 Parts		
					Water	Alcohol	Ether
3524	lq.	$0.932^{21\circ}$	17	258-68	....	....	....
3525	lq.	$0.835^{20\circ}$	....	174-5^{765mm}	i.	....	....
3526	col. lq.	$0.875^{15\circ}_{4}$	−50.2	192.5	i.	s.	s.
3527	col. lq.	$0.824^{20\circ}_{4}$	.34.6	175^{756mm}	$0.18^{25\circ}$	∞	∞
3528	col. lq.	....	15-7	131-2	hydrate +1H_2O	∞	∞
3529	lq.	$0.840^{20\circ}_{4}$	<−30	138-40^{750mm}	....	∞	∞
3530	lq.	$0.816^{20\circ}_{4}$	<−20	132-3^{760mm}	i.	∞	∞
3531	col. lq.	$0.829^{20\circ}_{4}$	....	140	v. sl. s.	∞	∞
3532	lq.	$0.853^{0\circ}$	....	151^{743mm}	....	∞	∞
3533	col. oil	$0.839^{20\circ}_{4}$	....	142^{764mm}	sl. s.	∞	∞
3534	lq.	$0.831^{13\circ}_{4}$	....	162-4^{750mm}	....	∞	∞
3535	col. lq.	$0.815^{20\circ}_{20}$	....	141-2^{755mm}	i.	∞	∞
3536	col. oil	$0.827^{16.7\circ}_{4}$	....	145-6	i.	∞	∞
3537	lq.	....	....	147-8^{756mm}	....	∞	∞
3538	lq.	$0.819^{17.5\circ}$	....	148-50	....	∞	∞
3539	lq.	$0.825^{11.5\circ}_{4}$	....	170.5^{755mm}	v. sl. s.	∞	∞
3540	lq.	$0.828^{18\circ}_{4}$	....	140.3^{745mm}	....	∞	∞
3541	lq.	$0.852^{0\circ}$	....	149-50	i.	s.	s.
3542	lq.	$0.819^{20\circ}_{4}$	....	160.4	$0.4^{20\circ}$	s.	s.
3543	lq.	$0.820^{20\circ}_{4}$	−37	156	i.	s.	s.
3544	col. lq.	$0.777^{20\circ}$	−23	155	v. sl. s.	∞	∞
3545	lq.	$1.145^{15\circ}_{4}$	−58.1	180.0	i.	∞	∞
3546	lq.	$0.881^{16\circ}$	....	159.2^{750mm}	i.	∞	∞
3547	lq.	$0.884^{22\circ}$	<−15	135-8^{757mm}	i.	∞	∞
3548	lq.	$0.883^{15\circ}_{4}$	−46.4	178.1	i.	s.	s.
3549	lq.	$1.366^{21\circ}_{4}$	....	78-9^{19mm}	i.	s.	s.
3550	cr./bz.	....	95 d.	....	i.	s.	s.; v. s. act.
3551	lq.	$0.894^{0\circ}$	....	155	i.	....	s.
3552	col. lq.	$0.864^{15\circ}_{4}$	−33.3	277.2	i.	s.	s.
3553	lq.	$0.872^{15\circ}_{4}$	−50.9	210.0	i.	s.	s.
3554	col. lq.	$0.697^{20\circ}_{4}$	−119.2	93.3	i.	s.	s.
3555	col. lq.	$0.696^{20\circ}_{4}$	....	82.4	i.	s.	s.
3556	lq.	$0.705^{20\circ}$	−111.4	77.9	....	....	....
3557	lq.	$0.831^{18\circ}_{0}$	−73	119.5	....	....	....
3558	lq.	$0.732^{20\circ}_{4}$	−81	99	....	....	....

Table 7-4 (*Continued*)
PHYSICAL CONSTANTS OF ORGANIC COMPOUNDS

No.	Name	Synonym	Formula	Beil. Ref.	Formula Weight
3559	**Hesperetin**	5,7,3'-triOH-4'-MeO-flavanone	$C_{16}H_{14}O_6$	VIII-544	302.29
3560	**Hesperidin**	....	$C_{28}H_{34}O_{15}$	....	610.57
3561	**Hexabromo-**benzene	perbromobenzene	C_6Br_6	V-215	551.52
3562	ethane	perbromoethane	$Br_3C \cdot CBr_3$	I-96	503.48
3563	**Hexachloro-**benzene	perchlorobenzene	C_6Cl_6	V-205	284.78
3564	ethane	carbon hexachloride	$Cl_3C \cdot CCl_3$	I-87	236.74
3565	chlorophenol	(low melting)	C_6OCl_6	VI-194	300.78
3566	chlorophenol	(high melting)	C_6OCl_6	VII-144	300.78
3567	**Hexacosane** (*n*)	....	$CH_3(CH_2)_{24} \cdot CH_3$	I-175	366.72
3568	**Hexacosane** (*iso*) (impure?)	cerane	$(CH_3)_2CH \cdot (CH_2)_{22} \cdot CH_3$	*I-70	366.72
3569	**Hexadecane** (*n*)	cetane	$CH_3(CH_2)_{14} \cdot CH_3$	I-172	226.45
3570	**Hexadecylene** (*α*)	hexadecene-1	$CH_3(CH_2)_{13} \cdot CH:CH_2$	I-226	224.43
3571	**Hexadecyne**-1	....	$CH_3(CH_2)_{13} \cdot C:CH$	I-262	222.42
3572	**Hexadecyne**-2	cetylene	$CH_3(CH_2)_{12} \cdot C:C \cdot CH_3$	I-262	222.42
3573	**Hexadiene**-2,4	dipropenyl	$(CH_3 \cdot CH:CH \cdot)_2$	I-254	82.15
3574	**Hexadiene**-1,5	diallyl	$(CH_2:CH \cdot CH_2 \cdot)_2$	I-253	82.15
3575	**Hexaethylbenzene**	"alkazene 6"	$(C_2H_5)_6C_6$	V-471	246.44
3576	**Hexahydro-**anthracene	anthracene hexahydride	$C_{14}H_{16}$	V-573	184.28
3577	benzoic acid	cyclohexanecarboxylic acid	$CH_2(CH_2)_4 \cdot CH \cdot CO_2H$ ⌊⎯⎯⎯⎯⎯⎯⌋	IX-7	128.17
3578	cumene	....	$C_3H_7 \cdot C_6H_{11}$	V-41	126.24
3579	cymene (*p*)	menthane; terpane	$CH_3 \cdot C_6H_{10} \cdot C_3H_7$	V-47	140.27
3580	mellitic acid	....	$C_6H_6(CO_2H)_6$	IX-1007	348.22
3581	mesitylene	(1,3,5)	$C_6H_9(CH_3)_3$	V-45	126.24
3582	naphthalene	....	$C_{10}H_{14}$	V-433	134.22
3583	*o*-phthalic acid (*cis*)	....	$C_6H_{10}(CO_2H)_2$	IX-730	172.18
3584	*o*-phthalic acid (*dl*)	(*trans*)	$C_6H_{10}(CO_2H)_2$	IX-730	172.18
3585	*p*-phthalic acid	(*cis*)	$C_6H_{10}(CO_2H)_2$	IX-733	172.18
3586	*p*-phthalic acid	(*trans*)	$C_6H_{10}(CO_2H)_2$	IX-733	172.18
3587	salicylic acid (*o*)	....	$HO \cdot C_6H_{10} \cdot CO_2H$	X-5	144.17
3588	**Hexahydroxy-**benzene	....	$C_6(OH)_6$	VI-1198	174.11
3589	**Hexaiodobenzene**	periodobenzene	C_6I_6	V-230	833.49
3590	**Hexamethyl-**benzene	....	$C_6(CH_3)_6$	V-450	162.28
3591	diamino-*iso*-propanol-diiodide	endoiodin	$[I(CH_3)_3N \cdot CH_2]_2:$ CHOH	....	430.11
3592	pararosaniline base	4,4',4''-tris-diMe-amino-triPh-carbinol	$[(CH_3)_2N \cdot C_6H_4]_3$ $C \cdot OH$	XIII-755	389.55
3593	pararosaniline chloride	(dye salt) ‡	$C_{25}H_{30}N_3Cl \cdot 9H_2O$	XIII-756	570.13
3594	pararosaniline-hydroxymethylate	(dye salt) §	$C_{26}H_{33}N_3Cl_2$	XIII-758	458.48
3595	4,4',4''-triamino-triphenyl-methane	(leuco base)	$[(CH_3)_2N \cdot C_6H_4]_3CH$	XIII-315	373.55

‡ Commercial product usually mixed with pentamethyl deriv.
§ Base not isolated; comes into commerce as $ZnCl_2$ double salt.
Hexadecyl acetylene 4942
Hexadecyl bromide 1276
Hexadienoic acid 5625
Hexadiyne 2768

Hexaethyl-pararosaniline 3191
Hexahydro-aniline 326
Hexahydro-benzene 1601
Hexahydro-cresol 4200-1
Hexahydro-diethylaniline 2145

Table 7-4 (*Continued*)
PHYSICAL CONSTANTS OF ORGANIC COMPOUNDS

No.	Crystalline Form and Color	Specific Gravity	Melting Point °C.	Boiling Point °C.	Solubility in 100 Parts		
					Water	Alcohol	Ether
3559	yel. wh. cr./ EtOAc	....	226-7 d.	....	i.; v. sl. s. alk.	s.; sl. s. bz.	s.; sl. s. chl.
3560	hyg. nd.	....	251 ±, d.	....	0.02 h.	sl. s.	i.; i. bz.
3561	nd./bz.	....	316	....	i. h.	$0.01^{20°}$	v. sl. s.
3562	rhb.	3.823	170 d.	....	v. s. CS_2	v. sl. s. h.	v. sl. s.
3563	mn.	$2.044^{24°}$	228-31	309^{742mm}	i.; s. bz.	v. sl. s. h.	s. h.
3564	rhb.	$2.091^{20°}_{4}$	186.9-7.4*	184.4	$0.005^{22°}$	v. s.	v. s.
3565	yel. cr.	....	46	....	....	....	....
3566	yel. cr.	....	106	d. 210	i.	s.; s. chl.	s. pet.
3567	cr.	$0.779^{5.7°}_{4}$	56.6	262^{15mm}	s. bz.	v. sl. s.	$3.5^{15°}$chl.
3568	pl./et.	....	61	$207^{0.7mm}$	i.	s.	s.
3569	lf.	$0.774^{20°}_{4}$	18.5(16.2)	287.5	i.	∞	∞
3570	lq.	$0.784^{15°}_{4}$	4	274	i.	....	....
3571	cr.	$0.797^{20°}$	15	155^{15mm}	i.	....	....
3572	pl.	$0.804^{20°}_{4}$	20	160^{15mm}	i.	....	....
3573	oil	$0.720^{20°}$	....	80	i.	....	....
3574	lq.	$0.691^{20°}$	141	59.6	i.	....	....
3575	pr./al.	$0.831^{13°}_{0}$	130	298.3	i.; v. s. bz.	$0.75^{25°}$	$8^{25°}$
3576	lf.	....	63	290	i.; v. s. bz.	v. s.	v. s.
3577	pr.	$1.034^{22°}_{4}$	30-1	232-4	sl. s.	v. s.; s. chl.	v. s.; s. pet.
3578	lq.	$0.790^{20°}_{0}$	−90.6	154.7	i.	v. s.	v. s.
3579	lq.	$0.793^{20°}_{0}$	....	169-70	i.	v. s.	v. s.
3580	syrup	....	d., $-H_2O$	....	v. s.	v. s.	sl. s.
3581	lq.	$0.787^{4°}$	....	135-8	....	....	....
3582	lq.	$0.934^{23°}_{3}$	....	200	....	....	....
3583	tri./aq.	....	192	d., $-H_2O$ >192	$>0.2^{20°}$	s.	....
3584	mn./aq.	....	215-21	....	$0.2^{20°}$	s. act.	....
3585	lf./aq.	....	....	....	$>0.08^{17°}$	s.	s.; s. chl.
3586	mn./aq.	....	300 ±	subl.	$0.08^{17°}$ 1.3 h.	s.; s. act.	sl. s.; i. chl.
3587	cr./aq.	....	111	....	s.	s.	s.; sl. s. bz.
3588	nd./aq. HCl	....	d. 200	....	sl. s. c.	sl. s.	sl. s.; sl. s. bz.
3589	red nd./bz.	....	340-50 d.	....	i.	i.	i.
3590	pl./al.	....	165.5	265	i.	$0.2^{0°}$; 4.6 h.	v. s.; v. s. bz.
3591	col. cr. pd.	....	275 d.	....	s.	sl. s.	i.; i. act.
3592	col. cr./bz.	....	219	....	i.; s. bz.; s. pet.	sl. s.; s. CS_2	s.; s. chl.
3593	gn. brn. met. cr./aq.	....	$-8H_2O$, 70-80°	....	s.	s.	v. s. chl.
3594	gn. pd.	....	....	....	s.	sl. s.	....
3595	lf./al.	....	178-9	....	i. c.; s. bz.	s. h.; s. ac.	s.; s. chl.

* Sealed tube.
Hexahydro-phenol 1605
Hexahydro-pyridine 5322
Hexahydro-toluene 4199
Hexahydro-triphenyltriazine 512
Hexahydro-xylene 2471-3
Hexahydroxy-anthraquinone 5575
Hexahydroxy-hexahydrobenzene 1602-3
Hexalin 1605
Hexamethyl-ethane 4945

Table 7-4 (*Continued*)
PHYSICAL CONSTANTS OF ORGANIC COMPOUNDS

No.	Name	Synonym	Formula	Beil. Ref.	Formula Weight
3596	**Hexamethylene-** diamine	....	$NH_2 \cdot (CH_2)_6 \cdot NH_2$	IV-269	116.21
3597	dicyanide	....	$NC \cdot (CH_2)_6 \cdot CN$	II-694	136.20
3598	glycol	hexandiol-1,6	$HO \cdot (CH_2)_6 \cdot OH$	I-484	118.18
3599	tetramine	Urotropin; Formin; methenamine	$(CH_2)_6N_4$	I-583	140.19
3600	tetramine-alliodide	allyliodourotropin	$C_6H_{12}N_4 \cdot C_3H_5I$	I-588	308.17
3601	camphorate	Amphotropin	$C_{12}H_{24}N_8 \cdot C_{10}H_{16}O_4$	*I-314	480.62
3602	salicylate	saliformin	$C_6H_{12}N_4 \cdot C_7H_6O_3$	....	278.31
3603	**Hexane** (*n*)	....	$CH_3 \cdot (CH_2)_4 \cdot CH_3$	I-142	86.18
3604	**Hexane** (*iso*)	2-methyl-pentane	$(CH_3)_2CH \cdot (CH_2)_2 \cdot CH_3$	I-148	86.18
3605	**Hexane**	2,2-dimethyl-butane	$(CH_3)_3C \cdot C_2H_5$	I-150	86.18
3606	**Hexane**	2,3-dimethyl-butane	$[(CH_3)_2CH]_2$	I-151	86.18
3607	**Hexane**	3-methyl-pentane	$(C_2H_5)_2CH \cdot CH_3$	I-149	86.18
3608	**Hexanhexol**	dulcitol	$HOCH_2(CHOH)_4 \cdot$ CH_2OH	I-544	182.17
3609	**Hexanhexol**	*d*-mannitol; mannite	$HOCH_2(CHOH)_4 \cdot$ CH_2OH	I-534	182.17
3610	hexaacetate	mannitol hexa- acetate	$C_6H_8O_6(COCH_3)_6$	II-150	434.40
3611	**Hexanhexol**	*d*-sorbitol; sorbite	$HOCH_2(CHOH)_4 \cdot$ CH_2OH	I-533	182.17
3612	hexaacetate	sorbitol hexaacetate	$C_6H_8O_6(COCH_3)_6$	II-150	434.40
3613	**Hexatriacontane**	....	$C_{36}H_{74}$	I-178	506.99
3614	**Hexenyl** alcohol	....	$C_6H_{11}OH$	I-446	100.16
3615	alcohol	1-hexen-5-ol	$CH_2:CH(CH_2)_2 \cdot$ $CHOH \cdot CH_3$	I-444	100.16
3616	**Hexose**	*d*-fructose; levulose	$C_5H_{12}O_5:CO$	I-918	180.16
3617	**Hexose**	*d*-galactose	$C_5H_{11}O_5 \cdot CHO$	I-911	180.16
3618	**Hexose**	*d*-glucose; dextrose	$C_5H_{11}O_5 \cdot CHO$	I-879	180.16
3619	**Hexose**	*d*-mannose; semi- nose	$C_5H_{11}O_5 \cdot CHO$	I-905	180.16
3620	**Hexose**	*d*-sorbose; sorbinose	$C_6H_{12}O_6$	I-927	180.16
3621	**Hexyl** acetate (*n*)	....	$CH_3CO_2 \cdot C_6H_{13}$	II-132	144.22
3622	acetate	Me-*iso*-Bu carbinol acetate	$CH_3CO_2 \cdot CH(CH_3) \cdot$ $CH_2CH(CH_3)_2$	II-133	144.22
3623	acetate (*iso*)	....	$CH_3CO_2 \cdot C_4H_7(CH_3)_2$	II-133	144.22
3624	alcohol (*n*)	hexanol-1	$CH_3(CH_2)_4CH_2OH$	I-407	102.18
3625	alcohol (*n*)	hexanol-2	$CH_3 \cdot CHOH \cdot C_4H_9$	I-408	102.18
3626	alcohol (*n*)	hexanol-3	$C_2H_5 \cdot CHOH \cdot C_3H_7$	I-408	102.18
3627	alcohol	2,2-diMe-butanol-3	$(CH_3)_3C \cdot CHOH \cdot CH_3$	I-412	102.18
3628	alcohol	2,2-diMe-butanol-4	$(CH_3)_3CCH_2CH_2OH$	I-412	102.18
3629	alcohol	2,3-diMe-butanol-2	$(CH_3)_2CH \cdot C(CH_3)_2 \cdot$ OH	I-413	102.18
3630	alcohol	2-Me-pentanol-1	$C_2H_5 \cdot CH_2 \cdot CH(CH_3) \cdot$ CH_2OH	I-409	102.18
3631	alcohol	2-Me-pentanol-2	$(CH_3)_2COH \cdot CH_2\ C_2H_5$	I-409	102.18
3632	alcohol	2-Me-pentanol-3	$(CH_3)_2CH \cdot CHOH \cdot$ C_2H_5	I-410	102.18
3633	alcohol	2-Me-pentanol-4	$CH_3 \cdot CHOH \cdot CH_2 \cdot$ $CH(CH_3)_2$	I-410	102.18

Table 7-4 (*Continued*)
PHYSICAL CONSTANTS OF ORGANIC COMPOUNDS

No.	Crystalline Form and Color	Specific Gravity	Melting Point °C.	Boiling Point °C.	Solubility in 100 Parts		
					Water	Alcohol	Ether
3596	lf.	....	42	204–5	v. s.	s.	....
3597	col. lq.	$0.954^{18°}$	−3.5	185^{15mm}	....	....	....
3598	nd./aq.	....	42	250	s.	s.	sl. s. h.
3599	col. rhb.	....	....	subl. vac. 230–70	$81^{12°}$	$31^{2°}$ abs.	v. sl. s.
3600	col. cr.	....	148±, d.	....	v. s.	i. chl.	i.
3601	col. cr.	....	....	....	10	s.; s. chl.	i.; i. bz.
3602	col. cr.	....	....	....	s.	s.	....
3603	col. lq.	$0.659\frac{20°}{4}$	−95.3	68.7	i.; ∞ chl.	$50^{33°}$	∞
3604	lq.	$0.654\frac{20°}{4}$	−153.7	60.3	i.	....	s.
3605	lq.	$0.649\frac{20°}{20}$	−99.7	49.7	i.	....	s.
3606	lq.	$0.662\frac{20°}{4}$	−128.4	58.0	i.	....	s.
3607	lq.	$0.664\frac{20°}{4}$	−118	63.3	i.	....	s.
3608	mn.	$1.466^{15°}$	189	$290-5^{3mm}$	$3.2^{15°}$; v. s. h.	v. sl. s.	i.
3609	col. rhb.	$1.489\frac{20°}{4}$	167–9	$290-5^{3mm}$	$13^{14°}$	$0.01^{14°}$ abs. al.	i.
3610	col. cr.	....	122–4	....	i.	sl. s. h.	i.; s. ac.
3611	cr. + ½ or 1 H₂O	....	110–2 (anh.)*	....	v. s.	v. s. h.	....
3612	col. cr./al.	....	99–100	....	sl. s.	sl. s.	i.
3613	cr.	$0.782^{76°}$	75.8	$265^{1.0mm}$	sl. s. chl.	v. sl. s.	sl. s.
3614	lq.	$0.891^{10°}$	....	137^{765mm}	$10^{10°}$	∞	∞
3615	lq.	$0.842\frac{16.2°}{17.5}$	....	140^{759mm}	v. sl. s.	....	....
3616	nd./aq.	$1.669^{17.5°}$	95–105	....	v. s.	$8.5^{18°}$ abs. al.	s. act.
3617	pr.	....	165.5 (anh.)†	....	$10.3^{0°}$ $68.3^{25°}$	$0.6^{38.5°}$ 85% al.	s. pyr.
3618	rhb.‡	$1.544^{25°}$	146(anh.)	....	$82^{17.5°}$	sl. s.	i.
3619	rhb.	$1.539\frac{20°}{4}$	132	....	$248^{17°}$	v. sl. s. abs.	i.
3620	rhb.	$1.654^{15°}$	165	....	$55^{17°}$	v. sl. s. abs.	sl. s. Me al.
3621	col. lq.	$0.878\frac{15°}{4}$	−60.9	171.5	i.	v. s.	v. s.
3622	col. lq.	$0.860\frac{20°}{20}$	−63.8	146–7	$0.13^{20°}$	....	....
3623	col. lq.	....	....	159^{755mm}	i.	v. s.	v. s.
3624	col. lq.	$0.822\frac{15°}{4}$	−51.6	157.5	$0.6^{20°}$	∞	∞
3625	col. lq.	$0.818\frac{16.8°}{4}$	....	139–40	sl. s.	∞	∞
3626	col. lq.	$0.818\frac{24°}{4}$	....	135	sl. s.	∞	∞
3627	lq. or nd.	$0.812^{25°}$	5.5	120–1	v. sl. s.	∞	∞
3628	oily lq.	....	−60	....	v. sl. s.	∞	∞
3629	lq.	$0.821\frac{20°}{0}$	−14	120–1	v. sl. s.	∞	∞
3630	lq.	$0.826\frac{20°}{4}$	....	148^{762mm}	v. sl. s.	∞	∞
3631	lq.	$0.809\frac{20°}{4}$	−107	122.5– 3.5^{762mm}	v. sl. s.	∞	∞
3632	lq.	$0.826\frac{18°}{4}$	....	127.5^{721mm}	v. sl. s.	∞	∞
3633	lq.	$0.813\frac{20°}{4}$	....	131.8	$1.6^{20°}$	∞	∞

* Crysts./al., m. 87–95°.
† Crysts. + 1H₂O, m. 118–20°.
‡ Crysts. + 1H₂O, mn. pl.
Hexene 3655–61

Hexophan 3801
Hexone 4164
Hexyl, cf. also dihexyl.

Table 7-4 (*Continued*)
PHYSICAL CONSTANTS OF ORGANIC COMPOUNDS

No.	Name	Synonym	Formula	Beil. Ref.	Formula Weight
3634	**Hexyl** alcohol	2-Me-pentanol-5	$(CH_3)_2CH \cdot CH_2 \cdot CH_2 \cdot CH_2OH$	I-411	102.18
3635	alcohol (*act.*)	3-Me-pentanol-1	$C_2H_5 \cdot CH(CH_3) \cdot CH_2 \cdot CH_2OH$	I-411	102.18
3636	alcohol (*act.*)	3-Me-pentanol-2	$CH_3 \cdot CHOH \cdot CH \cdot (CH_3) \cdot C_2H_5$	I-411	102.18
3637	alcohol	3-Me-pentanol-3	$C_2H_5 \cdot C(CH_3) \cdot (OH) \cdot C_2H_5$	I-411	102.18
3638	alcohol	3-methylol-pentane	$(C_2H_5)_2CH \cdot CH_2OH$	I-412	102.18
3639	amine (*n*)	....	$CH_3(CH_2)_5 \cdot NH_2$	IV-188	101.19
3640	amine	2-Et Bu-amine	$(C_2H_5)_2CHCH_2 \cdot NH_2$	IV-192	101.19
3641	bromide (*n*)	1-bromohexane	$CH_3(CH_2)_4CH_2Br$	I-144	165.08
3642	bromide (*iso*)	5-bromo-2-methyl-pentane	$(CH_3)_2CH \cdot (CH_2)_2 \cdot CH_2Br$	I-148	165.08
3643	bromide	2-bromo-2,3-diMe-butane	$(CH_3)_2CH \cdot CBr: (CH_3)_2$	I-152	165.08
3644	chloride (*n*)	1-chlorohexane	$CH_3 \cdot (CH_2)_4 \cdot CH_2Cl$	I-143	120.62
3645	chloride	2-chlorohexane	$CH_3(CH_2)_3CHCl \cdot CH_3$	I-144	120.62
3646	chloride	1-chloro-2,3-diMe-butane	$(CH_3)_2CH \cdot CH: (CH_2Cl)CH_3$	I-151	120.62
3647	chloride	2-chloro-2,3-diMe-butane	$(CH_3)_2CH \cdot CCl(CH_3)_2$	I-151	120.62
3648	cyanide (*n*)	oenanthylic nitrile	$CH_3(CH_2)_5 \cdot CN$	II-341	111.19
3649	formate (*n*)	....	$HCO_2 \cdot C_6H_{13}$	II-22	130.19
3650	iodide (*n*)	1-iodohexane	$CH_3 \cdot (CH_2)_4 \cdot CH_2I$	I-146	212.07
3651	iodide	2-iodo-2,3-dimethyl-butane	$(CH_3)_2CI \cdot CH(CH_3)_2$	I-153	212.07
3652	mercaptan (*n*)	hexanthiol-1	$CH_3(CH_2)_4 \cdot CH_2 \cdot SH$	I-408	118.24
3653	nitrite (*n*)	....	$C_6H_{13} \cdot O \cdot NO$	I-407	131.18
3654	resorcinol (*n*)	(1;2,4)	$C_6H_{13} \cdot C_6H_3(OH)_2$	**VI-904	194.28
3655	**Hexylene** (α)	1-hexene	$CH_3(CH_2)_3CH:CH_2$	I-215	84.16
3656	**Hexylene**	2,3-diMe-butene-1	$(CH_3)_2CH \cdot C(CH_3): CH_2$	I-218	84.16
3657	**Hexylene**	3,3-diMe-butene-1	$(CH_3)_3C \cdot CH:CH_2$	....	84.16
3658	**Hexylene**	2,3-diMe-butene-2	$(CH_3)_2C:C(CH_3)_2$	I-218	84.16
3659	**Hexylene**	2-methyl-pentene-2	$(CH_3)_2C:CH \cdot C_2H_5$	I-217	84.16
3660	**Hexylene** (*cis?*)	3-methyl-pentene-2	$(C_2H_5)(CH_3)C:CH \cdot CH_3$	I-217	84.16
3661	*trans?*	3-methyl-pentene-2	C_6H_{12}	I-217	84.16
3662	glycol (2,3)	hexandiol-2,3	$C_3H_7 \cdot (CHOH)_2 \cdot CH_3$	I-484	118.18
3663	**Hexylphenyl-carbinol** (*n*)	α-Ph-*n*-heptyl alcohol	$C_6H_{13} \cdot CHOH \cdot C_6H_5$	*VI-272	192.30
3664	**Hippuric** acid	benzoyl glycine	$C_6H_5CO \cdot NHCH_2 \cdot CO_2H$	IX-225	179.18
3665	calcium salt	Ca hippurate	$(C_9H_8O_3N)_2Ca \cdot 3H_2O$	IX-229	450.46
3666	sodium salt	Na hippurate	$C_9H_8O_3NNa$	IX-229	201.16
3667	**Histamine**	β-imidazolyl-4-ethylamine	$C_3H_3N_2 \cdot C_2H_4 \cdot NH_2$	*XXV-629	111.15
3668	hydrochloride, di	....	$C_5H_9N_3 \cdot 2HCl$	*XXV-630	184.08
3669	**Histidine** (*l*)(−)	β-imidazolyl-α-alanine	$C_3H_3N_2 \cdot CH_2 \cdot CH \cdot (NH_2) \cdot CO_2H$	XXV-513	155.16
3670	hydrochloride, di	....	$C_6H_9O_2N_3 \cdot 2HCl$	XXV-515	228.08

Hexylene bromide 1842
Hexylene iodide 2393
Holocaine HCl 2910

Homoatropine, cf. alkd.
Homocatechol 2374
Homogentisic acid 3673

Table 7-4 (*Continued*)
PHYSICAL CONSTANTS OF ORGANIC COMPOUNDS

No.	Crystalline Form and Color	Specific Gravity	Melting Point °C.	Boiling Point °C.	Solubility in 100 Parts		
					Water	Alcohol	Ether
3634	lq.	0.816^{20}_{4}	....	$151\text{-}2^{747mm}$	v. sl. s.	∞	∞
3635	lq.	$0.826^{20°}$	....	152-3	i.	s.	s.
3636	oil	$0.831^{18°}_{4}$	....	135^{762mm}	....	....	....
3637	lq.	0.824^{20}_{0}	<−38	122.5^{758mm}	i.	....	....
3638	lq.	0.833^{20}_{0}	<−15	148.9	0.43	....	....
3639	col. lq.	0.763^{25}_{4}	−19	$129\text{-}30^{742mm}$	v. sl. s.	∞	∞
3640	col. lq.	....	....	125.3	s.	∞	∞
3641	lq.	$1.173^{20°}$	....	155.5^{744mm}	....	....	....
3642	col. lq.	$1.168^{20°}$	....	146-7	....	....	....
3643	col. cr.	1.177^{20}_{0}	24-5	$132\text{-}37^{42mm}$	....	....	....
3644	lq.	0.876^{20}_{4}	....	132.9^{765mm}	i.	....	....
3645	lq.	0.869^{21}_{4}	....	122.5^{754mm}	....	....	....
3646	lq.	$0.887^{22°}$	....	122	....	....	....
3647	lq.	$0.878^{19°}$	−10.4	112	....	....	....
3648	col. lq.	0.813^{15}_{4}	−64	184.6	i.	....	....
3649	lq.	0.886^{20}_{4}	−62.7	155.5	....	∞	∞
3650	lq.	1.439^{20}_{0}	....	180^{763mm}	....	....	....
3651	nd.	1.446^{20}_{0}	....	140^{749mm}	....	....	....
3652	col. lq.	$0.849^{20°}$	....	$149\text{-}50^{768mm}$	i.	....	....
3653	yel. lq.	$0.885^{20°}$	....	$129\text{-}30^{774mm}$	i.	s.	s.
3654	col. nd.	....	68-70	179^{7mm}	0.05	v. s.; s. bz.	s.; s. act.
3655	col. lq.	0.673^{20}_{4}	−139	63.6	i.	∞	∞
3656	col. lq.	$0.678^{20°}$	−140.0	55.6	i.; ∞ CS_2	∞	∞
3657	col. lq.	$0.653^{20°}$	−115.5	41.2	i.	....	....
3658	lq.	$0.708^{20°}$	−75.4	73.2^{760mm}	....	s. act.	s.
3659	lq.	$0.686^{20°}$	....	67.2	s. 2 vol. H_2SO_4 + 1 aq.	....	....
3660	lq.	$0.699^{20°}$	−138.4	70.5	....	....	....
3661	lq.	$0.694^{20°}$	−135.2	67.8	....	....	....
3662	cr.	$0.967^{0°}$	60	206-7	∞	s.	....
3663	lq.	0.946	....	275	....	....	....
3664	rhb.	1.371^{20}_{4}	189-90	d.	$0.4^{20°}$; i. pet.	s. h.; i. bz.	$0.25^{18°}$
3665	col. pr.	1.32	....	....	5 c.; 16 h.	....	....
3666	wh. pd.	....	....	....	s.	s.	....
3667	col. hyg. cr./aq.	....	86*	$209\text{-}10^{18mm}$	v. s.; s. h. chl.	s.	i.
3668	pr./aq. al.	....	239-46 (d.)	....	s.	sl. s.	i.
3669	lf./aq.	....	d. 287-8	....	s.	v. sl. s.	i.
3670	rhb.	....	245	....	s. d.	i.	i.

* Sealed tube.
Homohydroquinone 2372
Homopyrocatechin 2374

Homosalicylaldehyde 4379
Homosalicylic acid 3833, 3836, 3839, 3841
Hot stuff gas 1985

Table 7-4 (Continued)
PHYSICAL CONSTANTS OF ORGANIC COMPOUNDS

No.	Name	Synonym	Formula	Beil. Ref.	Formula Weight
3671	**Histidine** hydrochloride, mono	. . . .	$C_6H_9O_2N_3 \cdot HCl \cdot H_2O$	XXV-515	209.63
3672	**Homo**-eriodictyol	5,7,4'-triOH-3'-methoxy-flavanone	$C_{16}H_{14}O_6$†	. . . .	302.29
3673	gentisinic acid	2,5-diOH-phenyl-acetic acid	$(HO)_2C_6H_3CH_2 \cdot CO_2H \cdot H_2O$	X-407	186.17
3674	phthalic acid (*o*)	. . . .	$HO_2C \cdot C_6H_4 \cdot CH_2 \cdot CO_2H$	IX-857	180.16
3675	**Hordenine** (*p*)‡	. . . .	$HO \cdot C_6H_4 \cdot CH_2 \cdot CH_2 \cdot N(CH_3)_2$	XIII-626	165.24
3676	sulfate	. . . .	$(C_{10}H_{15}ON)_2 \cdot H_2SO_4 \cdot H_2O$	XIII-626	446.57
3677	sulfate	. . . .	$(C_{10}H_{15}ON)_2 \cdot H_2SO_4 \cdot 2H_2O$	*XIII-236	464.58
3678	**Hydantoic acid**	. . . .	$H_2N \cdot CO \cdot NH \cdot CH_2 \cdot CO_2H$	IV-359	118.09
3679	**Hydantoin**	glycolyl urea	$HN \cdot CH_2 \cdot CO \cdot NH \cdot CO$ $\underline{\qquad\qquad}$	XXIV-242	100.08
3680	**Hydnocarpic acid**	. . . .	$C_5H_7(CH_2)_{10}CO_2H$	IX-79	252.40
3681	**Hydracetamide**	. . . .	$(CH_3 \cdot CH)_3N_2$	I-608	112.18
3682	**Hydracrylic acid**	ethylene lactic acid	$HO \cdot CH_2CH_2 \cdot CO_2H$	III-295	90.08
3683	**Hydratropic acid**	α-Ph-propionic acid	$C_6H_5CH(CH_3)CO_2H$	IX-524	150.18
3684	**Hydrazinobenzoic acid** (*p*)	. . . .	$H_2N \cdot NH \cdot C_6H_4 \cdot CO_2H$	XV-631	152.15
3685	**Hydrazo**-benzene	*N,N'*-diphenyl-hydrazine	$C_6H_5NH \cdot NHC_6H_5$	XV-123	184.24
3686	benzoic acid (*o*)	. . . .	$(HO_2C \cdot C_6H_4 \cdot NH \cdot)_2$	XV-626	272.26
3687	benzoic acid (*m*)	. . . .	$(HO_2C \cdot C_6H_4 \cdot NH \cdot)_2$	XV-629	272.26
3688	benzoic acid (*p*)	. . . .	$(HO_2C \cdot C_6H_4 \cdot NH \cdot)_2$	XV-632	272.26
3689	dicarbonamide	. . . .	$(H_2N \cdot CO \cdot NH \cdot)_2$	III-116	118.10
3690	naphthalene (α,α')	. . . .	$C_{10}H_7NH \cdot NHC_{10}H_7$	XV-562	284.36
3691	naphthalene (β,β')	. . . .	$C_{10}H_7NH \cdot NHC_{10}H_7$	XV-569	284.36
3692	toluene (*o*)	. . . .	$(CH_3 \cdot C_6H_4 \cdot NH \cdot)_2$	XV-497	212.30
3693	toluene (*m*)	. . . .	$(CH_3 \cdot C_6H_4 \cdot NH \cdot)_2$	XV-506	212.30
3694	toluene (*p*)	. . . .	$(CH_3 \cdot C_6H_4 \cdot NH \cdot)_2$	XV-511	212.30
3695	**Hydrindene** (1,2)	indane	$C_6H_4 \cdot CH_2 \cdot CH_2 \cdot CH_2$ $\underline{\qquad\qquad}$	V-486	118.16
3696	dicarboxylic acid	(β,β)	$C_9H_8(CO_2H)_2$	IX-904	206.20
3697	dicarboxylic acid	(α,β)	$C_9H_8(CO_2H)_2$	*IX-391	206.20
3698	**Hydrindone** (α)	indanone-1	$C_6H_4 \cdot CH_2 \cdot CH_2 \cdot CO$ $\underline{\qquad\qquad}$	VII-360	132.16
3699	**Hydrindone** (β)	indanone-2	$C_6H_4 \cdot CH_2 \cdot CO \cdot CH_2$ $\underline{\qquad\qquad}$	VII-363	132.16
3700	**Hydro**-benzamide	tribenzaldiamine	$C_6H_5CH:$ $(N:CHC_6H_5)_2$	VII-215	298.39
3701	benzoin	. . . .	$(C_6H_5 \cdot CHOH \cdot)_2$	VI-1003	214.27
3702	benzoin (*iso*)	. . . .	$C_{14}H_{12}(OH)_2$	VI-1004	214.27
3703	carbostyril	. . . .	C_9H_9ON	XXI-288	147.18
3704	cinnamic acid	β-Ph-propionic acid	$C_6H_5CH_2CH_2CO_2H$	IX-508	150.18
3705	cinnamic alcohol	γ-Ph-Pr-alcohol	$C_6H_5(CH_2)_3OH$	VI-503	136.20
3706	cinnamic aldehyde	. . . .	$C_6H_5CH_2CH_2 \cdot CHO$	VII-304	134.18

† Beilstein gives structure as a chalcone.
‡ See also Alkaloid table.
HS (hot stuff) gas 1985
Hydracetin 111

Hydraergotocin, cf. alkd.
Hydramine, mixt. of 2314 and 5260
Hydrastine, cf. alkd.
Hydrastinine, cf. alkd.

Table 7-4 (*Continued*)
PHYSICAL CONSTANTS OF ORGANIC COMPOUNDS

No.	Crystalline Form and Color	Specific Gravity	Melting Point °C.	Boiling Point °C.	Solubility in 100 Parts		
					Water	Alcohol	Ether
3671	col. rhb./aq.	....	250-5	$-H_2O$, 100	s.	i.	i.
3672	col. nd./ac.	....	224-5	....	i.; i. chl.	s.; i. bz.	sl. s. EtOAc
3673	pr./aq.	....	152(anh.)	....	v. s.	v. s.	v. s.; i. bz.
3674	cr./aq.	....	175-80	....	s. h.	v. s.	sl. s.
3675	col. rhb.	....	117.8 subl. > 140	173-4^{11mm}	s.; i. pet.; s. chl.	v. s.; s. dil. a.	v. s.; s. alk.
3676	pr. nd.	....	205(anh.)	....	s.	v. sl. s.	....
3677	col. cr.	....	197	....	s.	sl. s.	i.
3678	mn.	....	160-1	....	$3^{20°}$; s. h.	$0.5^{20°}$; s. h.	....
3679	nd.	....	220-1	....	s. h.; s. alk.	1.7 h.	i.
3680	lf./al.	....	59-60	....	i.	v. sl. s.	v. sl. s.
3681	yel. pd.	....	....	d.	v. s. d.	v. s.	....
3682	syrup	....	....	d.	....	....	....
3683	col. lq.	1.1$^{0°}$	<−20	266-7	v. sl. s.	....	....
3684	nd. or pl./aq.	....	220-5 d.	....	sl. s. h.	....	....
3685	lt. yel./al.	1.158$^{16°}$	131	d.	v. sl. s.	5$^{16°}$	i. ac.
3686	lf./al.	....	205	....	i.	s. h.	s. alk.
3687	lt. yel./al.	....	....	....	i.	sl. s. h.	....
3688	nd./al.	....	....	....	v. sl. s.	sl. s. h.	....
3689	pl./aq.	....	254-9	....	0.02$^{16°}$; sl. s. h.	i.; s. in 30%NaOH	i.
3690	lf./bz.	....	274(271)	....	s. bz.	s.	s.
3691	lt. red lf.	....	140-1	....	i.	v. s.	v. s.
3692	lf./al.	....	165	....	s. bz.	v. sl. s. aq. al.	s.
3693	col. oil	....	....	....	....	v. s.	....
3694	mn.	0.957	133-4(126)	d.	v. s. bz.	v. s.	v. s.
3695	col. lq.	0.963$^{10°}_{4}$	....	177	i.	∞	∞
3696	lf./aq.	....	199	$-CO_2$ > 200	....	....	....
3697	cr.	....	222	....	v. sl. s. bz.	s.; s. act.	sl. s.
3698	nd./aq.	1.099$^{42°}$	41-2	243-5	sl. s.	v. s.	v. s.
3699	nd./al.	1.071$^{67°}$	58-61	250-5 sl. d.	i.	v. s.	v. s.
3700	cr./al.	....	101-2	....	i.	v. s.	v. s.
3701	mn.	0.927$^{134°}$	138-9	>300	0.25$^{15°}$	v. s. h.	....
3702	mn./al.	....	121-2	133$^{0.02mm}$	0.19$^{15°}$; 1.25 h.	s.	s.
3703	rhb./al.	....	163	....	i.; s. h. HCl	s.	s.
3704	mn.	1.071$^{48.7°}_{4}$	48.5	279.8	0.6$^{20°}$	s.; s. chl.	s.; s. bz.
3705	oil	1.008$^{20°}_{4}$	<−18	235-7	s.	∞	∞
3706	lq.	....	....	221-4^{744mm}	i.	16.7	....

Table 7-4 (*Continued*)
PHYSICAL CONSTANTS OF ORGANIC COMPOUNDS

No.	Name	Synonym	Formula	Beil. Ref.	Formula Weight
	Hydro				
3707	cinnamic amide	β-Ph-propionamide	$C_6H_5CH_2CH_2CONH_2$	IX-511	149.19
3708	cinnamic chloride	....	$C_6H_5CH_2CH_2COCl$	IX-511	168.62
3709	cinnamic nitrile	β-Ph-propionitrile	$C_6H_5CH_2CH_2CN$	IX-512	131.18
3710	corulignone (4;3,5)	....	$[HOC_6H_2(OCH_3)_2]_2$	VI-1200	306.32
3711	coumaric acid (*o*)	melilotic acid	$HOC_6H_4(CH_2)_2CO_2H$	X-241	166.18
3712	coumaric acid (*m*)	phenol propionic acid	$HOC_6H_4(CH_2)_2CO_2H$	X-244	166.18
3713	coumaric acid (*p*)	phlorentinic acid	$HOC_6H_4(CH_2)_2CO_2H$	X-244	166.18
3714	cyanic acid	formonitrile	HCN	II-29	27.03
3715	cyanic acid (dimolecular)	prussic acid	$HN:CH \cdot NC$	II-28	54.05
3716	**Hydroquinone** acetate, di	diacetyl-hydro-quinone (*p*)	$(CH_3CO_2)_2:C_6H_4$	VI-846	194.19
3717	benzyl ether, di	....	$(C_6H_5CH_2O)_2C_6H_4$	VI-845	290.37
3718	benzyl ether, mono	....	$C_6H_5CH_2O \cdot C_6H_4OH$	VI-845	200.24
3719	dicarboxylic acid	(2,5;1,4)	$(HO)_2C_6H_2(CO_2H)_2$	X-554	198.13
3720	**Hydroquino-phthalein**	2,7-dihydroxy-fluorane	$C_{20}H_{12}O_5$	XIX-219	332.32
3721	**Hydroxy**-aceto-phenone (*o*)†	....	$HO \cdot C_6H_4 \cdot CO \cdot CH_3$	VIII-85	136.15
3722	acetophenone (*m*)	....	$HO \cdot C_6H_4 \cdot CO \cdot CH_3$	VIII-86	136.15
3723	acetophenone (*p*)	....	$HO \cdot C_6H_4 \cdot CO \cdot CH_3$	VIII-87	136.15
3724	anthraquinone (1)	....	$C_6H_4(CO)_2C_6H_3 \cdot OH$	VIII-338	224.22
3725	anthraquinone (2)	....	$C_6H_4(CO)_2C_6H_3 \cdot OH$	VIII-342	224.22
3726	anthraquinone (2)	2-hydroxy-anthraquinone-1,4	$C_{10}H_6(CO)_2CH:COH$	VIII-337	224.22
3727	azobenzene (*o*)	....	$HO \cdot C_6H_4 \cdot N:N \cdot C_6H_5$	XVI-90	198.23
3728	azobenzene (*m*)	....	$HO \cdot C_6H_4 \cdot N:N \cdot C_6H_5$	XVI-94	198.23
3729	azobenzene (*p*)	....	$HO \cdot C_6H_4 \cdot N:N \cdot C_6H_5$	XVI-96	198.23
3730	azobenzene sul-fonic acid (4,4')	....	$HO \cdot C_6H_4 \cdot N:N \cdot C_6H_4 \cdot SO_3H$	XVI-272	278.29
3731	azobenzene sul-fonic Na	(4,4')	$HO \cdot C_6H_4 \cdot N:N \cdot C_6H_4 \cdot SO_3Na \cdot 2H_2O$	XVI-272	336.29
3732	1,1'-azonaphtha-lene-3,6,4'-tri-sulfonic acid (2) Na	bordeaux S; fast red D	$NaO_3S \cdot C_{10}H_6N:N \cdot C_{10}H_4(OH)(SO_3Na)_2$	*XVI-305	604.48
3733	benzaldehyde (*o*)	salicylaldehyde	$HO \cdot C_6H_4 \cdot CHO$	VIII-31	122.12
3734	benzaldehyde (*m*)	....	$HO \cdot C_6H_4 \cdot CHO$	VIII-58	122.12
3735	benzaldehyde (*p*)	....	$HO \cdot C_6H_4 \cdot CHO$	VIII-64	122.12
3736	benzaldehyde oxime (*o*)	salicylaldoxime	$HO \cdot C_6H_4 \cdot CH:NOH$	VIII-49	137.14
3737	benzaldehyde phenylhydrazone	(*o*)	$HO \cdot C_6H_4 \cdot CH:N \cdot NH \cdot C_6H_5$	XV-188	212.25
3738	benzamide (*o*)	salicylamide	$HO \cdot C_6H_4 \cdot CO \cdot NH_2$	X-87	137.14
3739	benzamide (*m*)	....	$HO \cdot C_6H_4 \cdot CO \cdot NH_2$	X-140	137.14
3740	benzamide (*p*)	....	$HOC_6H_4CONH_2 \cdot H_2O$	X-164	155.15
3741	benzanilide (*o*)	salicylanilide	$HO \cdot C_6H_4CO \cdot NHC_6H_5$	XII-500	213.24
3742	benzoic acid (*o*)	salicylic acid	$HO \cdot C_6H_4 \cdot CO_2H$	X-43	138.12

† See also No. 754.
Hydro-cotarnine, cf. alkd.
Hydro-naphthoquinone 2348, 2350
Hydro-phloron 2379
Hydro-quinine, cf. alkd.

Hydro-quinone 2314
Hydro-quinone diethyl ether 2093
Hydro-quinone dimethyl ether 2408
Hydro-quinone ethyl ether 2929
Hydro-quinone methyl ether 4081

Hydro-resorcin 2285
Hydro-sulfite N.F. 3293
Hydrol 4435
Hydrolit 3295
Hydropyrin 119

Table 7-4 (*Continued*)
PHYSICAL CONSTANTS OF ORGANIC COMPOUNDS

No.	Crystalline Form and Color	Specific Gravity	Melting Point °C.	Boiling Point °C.	Solubility in 100 Parts		
					Water	Alcohol	Ether
3707	nd./aq.	....	104-5	....	....	s.	s.
3708	lq.	$1.135^{21°}$	<−60	225 d.	....	....	....
3709	lq.	$1.001^{18°}$	....	261	....	....	....
3710	mn./al.	....	190±	d.	v. sl. s.	i. CS_2	v. sl. s.
3711	cr./aq.	....	82-3	....	$5^{18°}$; $108^{40°}$	v. s.	v. s.
3712	mn./bz. lg.	....	111	....	i. lg.	s.; s. bz.	s.
3713	mn./et.	....	128-30	....	v. s. h.	v. s.	v. s.; i. CS_2
3714	poison. lq.	$0.688^{\frac{20°}{4}}$	−13.3	25.7	∞	∞	∞
3715	rhb.	....	87	120-5	v. s.	sl. s.	sl. s.
3716	pl./al.	....	123-4	....	sl. s. h.	s.	s.; s. chl.
3717	pl./al.	....	128	....	v. sl. s. ac.	2.5 h.	v. sl. s.
3718	pl./aq.	....	122-3	....	s. h.	s.; s. alk.	s.; s. bz.
3719	yel. cr./al.	....	d.	....	sl. s. h.	sl. s. h.	sl. s.
3720	nd./et.	....	227-32	....	v. sl. s. h.	s.; s. alk.	s.; i. lg.
3721	oil	$1.131^{\frac{21°}{4}}$	4-6	213^{717mm}	sl. s.	∞	∞; ∞ ac.
3722	nd. or lf.	$1.099^{109°}$	95-6	296^{756mm}	s. h.	s.	s.; s. bz.
3723	nd./aq. al.	1.109	109	$147-8^{3mm}$	$12^{2°}$; $7.1^{100°}$	s.	s.
3724	or. red/al.	....	194-5	subl.	....	s.	v. s.
3725	yel. nd./al.	....	306-8	subl.	v. sl. s. c.	s.	s.
3726	yel. nd./al.	....	d. 235	subl.	s. alk.	s. alk. carb.	....
3727	or. nd./et.	....	82.5-3	....	sl. s.	s.	s.
3728	yel. pr./bz.	....	114-6	....	0.08 h.	s.	s.
3729	or. rhb./al.	....	155-6	$220-30^{20mm}$ sl. d.	$0.002^{25°}$; 0.08 h.	$31^{25°}$; $0.6^{25°}CCl_4$	v. s.; $2^{25°}$ bz.
3730	yel. red pr./aq.	....	d. > 200	....	sl. s.; i. HCl	s.	i.
3731	red yel. pl.	....	....	....	$0.71^{5°}$; v. s. h.	....	....
3732	red brn. pd.	....	....	....	s.	sl. s.	....
3733	col. oil	$1.153^{\frac{25°}{4}}$	1-2	196.5	sl. s.	∞	∞
3734	nd./aq.	....	106-8	240±	s. h.	s.	s.
3735	nd./aq.	$1.129^{130°}$	116-7	subl.	$1.38^{31°}$	$70^{25°}$ act.	$4^{25°}$ bz.
3736	col. pr./bz.-pet.	....	57-9	d.	v. sl. s. c.	s.; s. dil. HCl	s.; i. lg.; s. bz.
3737	col. nd.	....	142-3	234^{28mm}	sl. s. h.	s. h.	s.
3738	lf./aq.	....	140	d. 270	sl. s.	s.; s. chl.	s.
3739	lf./aq.	....	170.5	....	s. h.	s.; i. chl.	s.; i. CS_2
3740	nd./aq.	....	162(anh.)	−H_2O, 100	s. h.	s.; i. chl.	s.; i. CS_2
3741	pr./al.	....	135	d.	v. sl. s. h.	s.; s. bz.; s. chl.	s.; sl. s. CS_2
3742	mn.	$1.443^{\frac{20°}{4}}$	158.3; subl. 76	$211^{20mm}±$	$0.16^{4°}$; $2.6^{75°}$	$49.6^{15°}$ abs. al.	$50.5^{15°}$

Table 7-4 (*Continued*)
PHYSICAL CONSTANTS OF ORGANIC COMPOUNDS

No.	Name	Synonym	Formula	Beil. Ref.	Formula Weight
	Hydroxy				
3743	benzoic Ca (*o*)	calcium salicylate	$(C_7H_5O_3)_2Ca \cdot H_2O$†	X-60	332.33
3744	benzoic Na (*o*)	sodium salicylate	$HO \cdot C_6H_4 \cdot CO_2Na$	X-59	160.11
3745	benzoic acid (*m*)	....	$HO \cdot C_6H_4 \cdot CO_2H$	X-134	138.12
3746	benzoic acid (*p*)	....	$HO \cdot C_6H_4 \cdot CO_2H$	X-149	138.12
3747	benzoic anhydride (*o*)	salicylide; β-disalicylide	$(C_6H_4)_2O_2(CO)_2$	XIX-171	240.22
3748	benzonitrile (*o*)	salicylic nitrile	$HO \cdot C_6H_4 \cdot CN$	X-96	119.12
3749	benzophenone (*p*)	....	$C_6H_5 \cdot CO \cdot C_6H_4 \cdot OH$	VIII-158	198.22
3750	benzyl alcohol (*o*)	salicyl alcohol	$HO \cdot C_6H_4 \cdot CH_2OH$	VI-891	124.14
3751	benzyl alcohol (*m*)	....	$HO \cdot C_6H_4 \cdot CH_2OH$	VI-896	124.14
3752	benzyl alcohol (*p*)	....	$HO \cdot C_6H_4 \cdot CH_2OH$	VI-897	124.14
3753	butyric acid (α) (*dl*)	....	$C_2H_5 \cdot CHOH \cdot CO_2H$	III-303	104.11
3754	butyric acid (β)(*l*)	....	$CH_3 \cdot CHOH \cdot CH_2 \cdot CO_2H$	III-307	104.11
3755	butyric acid (β)(*dl*)	....	$HO \cdot C_3H_6 \cdot CO_2H$	III-308	104.11
3756	butyric acid (γ)	....	$HO(CH_2)_3CO_2H$	III-311	104.11
3757	butyric acid (*iso*)(α)	acetonic acid	$(CH_3)_2COH \cdot CO_2H$	III-313	104.11
3758	butyric nitrile (γ)	(*dl*)	$HO \cdot (CH_2)_3 \cdot CN$	III-311	85.11
3759	caproic acid (α)(*dl*)	oxycaproic acid	$CH_3(CH_2)_3CHOH \cdot CO_2H$	III-332	132.16
3760	caproic acid(*iso*)(α)	leucic acid (*dl*)	$HO \cdot C_5H_{10} \cdot CO_2H$	III-336	132.16
3761	caprylic acid (α)	....	$C_6H_{13} \cdot CHOH \cdot CO_2H$	III-348	160.21
3762	caprylic acid (*iso*) (α)	tetramethyl-hydracrylic acid	$[(CH_3)_2CH]_2 : C(OH) \cdot CO_2H$	III-348	160.21
3763	chalcone (2)‡	o-OH-benzalacetophenone	$C_6H_5 \cdot CO \cdot CH : CH \cdot C_6H_4 \cdot OH$	VIII-191	224.26
3764	γ-chloro-*n*-butyronitrile (β)(*dl*)	....	$ClCH_2 \cdot CHOH \cdot CH_2 \cdot CN$	III-310	119.55
3765	β,β-dimethyl-γ-butyrolactone (α)	(*dl*)§	$\begin{array}{c} CHOH \cdot CO \cdot O \cdot \\ \vert \qquad\qquad \vert \\ CH_2 \cdot C : (CH_3)_2 \end{array}$	XVIII-3	130.14
3766	β,β-dimethyl-γ-butyrolactone (α)	(*l*)§	$C_6H_{10}O_3$	....	130.14
3767	diphenylamine (*o*)	anilino-phenol	$HO \cdot C_6H_4 \cdot NH \cdot C_6H_5$	XIII-365	185.23
3768	diphenylamine (*m*)	anilino-phenol	$HO \cdot C_6H_4 \cdot NH \cdot C_6H_5$	XIII-410	185.23
3769	diphenylamine (*p*)	anilino-phenol	$HO \cdot C_6H_4 \cdot NH \cdot C_6H_5$	XIII-444	185.23
3770	diphenyl ether (*o*)	phenoxy-phenol	$HO \cdot C_6H_4 \cdot O \cdot C_6H_5$	VI-772	186.21
3771	ethyl ethylenediamine	....	$HO \cdot CH_2 \cdot CH_2 \cdot NH \cdot CH_2 \cdot CH_2 \cdot NH_2$	IV-286	104.15
3772	glutamic acid (β)	(*l*+)	$HO_2C \cdot CH_2 \cdot CHOH \cdot CH(NH_2) \cdot CO_2H$	*IV-550	163.13
3773	glutaric acid (α) (*dl*)	....	$HO_2C \cdot CHOH \cdot CH_2 \cdot CH_2 \cdot CO_2H$	III-442	148.12
3774	glutaric acid (β)	....	$HOCH : (CH_2CO_2H)_2$	III-443	148.12

† Crysts.+ $2H_2O$, $2.8^{16°}$ aq., $-H_2O$ 100°; + $3H_2O$, $2.3^{15°}$, $35.8^{100°}$ aq.
‡ See also No. 1781.
§ Lactone portion of pantothenic acid is the *l*-form.
Hydroxy-benzoquinone-oxime 4900
Hydroxy-brasilin 3500
Hydroxy-butene 4422
Hydroxy-butyraldehyde 169
Hydroxy-chlorobenzaldehyde 1461

Hydroxy-cinchonine, cf. alkd.
Hydroxy-cinnamic acid 1531-3, 1537
Hydroxy-coniine, cf. alkd.
Hydroxy-coumarin 6369
Hydroxy-cumene 5433-5
Hydroxy-deguelin 5699

Table 7-4 (*Continued*)
PHYSICAL CONSTANTS OF ORGANIC COMPOUNDS

No.	Crystalline Form and Color	Specific Gravity	Melting Point °C.	Boiling Point °C.	Solubility in 100 Parts		
					Water	Alcohol	Ether
3743	col. nd.	....	....	....	1.4 c.; 4 h.	s.	s.
3744	col. pl./al.	....	....	....	$125^{25°}$	$171^{5°}$	25 gly.
3745	rhb./aq.	1.473	201	....	$0.8^{19°}$; s. h.	s. h.	$10^{10°}$
3746	mn./aq. al.	1.4684^{0}	214.5-5.5	....	$0.17^{0°}$; $2.75^{5°}$	v. s.	$23^{25°}$; i. CS_2
3747	nd./chl.	....	200-1	d.	i.	s. h.	....
3748	pr./bz.	....	97-8	149^{14mm}	v. sl. s.	v. s.	v. s.
3749	rhb.	....	134-5	....	s. h.	$25^{25°}$ act.	$8^{25°}$
3750	rhb./aq.	$1.161^{25°}$	86-7	subl.	$6.6^{15°}$	v. s.	v. s.; s. bz.
3751	cr./bz.	....	67	300±, d.	v. s. h.	v. s.	v. s.
3752	pr./aq.	....	125	252	v. s.	v. s.	v. s.
3753	hyg. cr.	$1.125^{20°}$	43-4	subl. > 60; 225-60 d.	s.	s.	s.
3754	mn.	....	48-50	....	v. s.	v. s.	v. s.; i. bz.
3755	hyg. syrup	....	....	130^{12mm} ±	....	....	....
3756	lq.	....	<−17	d. slowly room temp.	....	....	....
3757	hyg. pr.	....	79; subl. 50	212	v. s.	v. s.	v. s.; v. sl. s. bz.
3758	lq.	$1.029^{8°}$	....	238-40	s.	s.; i. CS_2	s.; s. chl.
3759	nd.	....	60-2	subl. 100 sl. d.	v. s.	v. s.	v. s.
3760	pl./et. + pet.	....	76-7	....	s.	s.	s.
3761	pl.	....	69.5	....	v. sl. s.	v. s.	v. s.
3762	cr./et.	....	152-3 d.	192-3 d.	v. s.	v. s.	v. s.
3763	yel. lf./ aq. al.	....	154-5 d.	....	v. sl. s. CS_2	v. s.	sl. s. chl.
3764	yel. lq.	....	....	250±, d.	s.	s.	s.
3765	col. hyg. nd.	....	56-8	$119-21^{15mm}$	s.; s. bz.	s.; s. chl.; s, CS_2	s.; sl. s. pot.
3766	cr./bz. pet.	....	91-2	....	s.; sl. s. pet.	s.	s.
3767	pr./aq.	....	69-70	$180-9^{20mm}$	sl. s. h.	s.	s.; sl. s. bz.
3768	lf./aq.	....	81.5-2.0	340	sl. s. h.; s. alk.	s.: s. dil. a.	s.; sl. s. lg.
3769	lf.	....	70	330	v. sl. s. c.; s. alk.	s.; s. dil. a.	s.; s. chl.
3770	nd./aq.	....	106-7	$151-5^{11mm}$	sl. s. h.	v. s. h.	v. s. h.
3771	lq.	....	....	238-40	v. s.	v. s.	i.
3772	pr./aq.	....	softens 100	d. > 100	v. s.	i.	i.; v. s. ac.
3773	col. cr.	....	98-100 d.	....	s.	....	....
3774	nd./aq.	....	95	....	v. s.	v. s.	sl. s.

Table 7-4 (*Continued*)
PHYSICAL CONSTANTS OF ORGANIC COMPOUNDS

No.	Name	Synonym	Formula	Beil. Ref.	Formula Weight
	Hydroxy				
3775	hydroquinone triacetate (1,2,4)	triacetyl-hydroxy-hydroquinone	$(CH_3CO_2)_3C_6H_3$	VI-1089	252.23
3776	3-methoxyacetophenone (4)	acetovanillone; apocynin	$CH_3CO \cdot C_6H_3(OH) \cdot OCH_3$	VIII-272	166.18
3777	methyl benzoic acid	(*o*)	$HO \cdot CH_2 \cdot C_6H_4 \cdot CO_2H$	X-218	152.15
3778	α-naphthoic acid	(2,1)	$HO \cdot C_{10}H_6 \cdot CO_2H$	X-328	188.18
3779	α-naphthoic acid	(5,1)	$HO \cdot C_{10}H_6 \cdot CO_2H$	X-330	188.18
3780	α-naphthoic acid	(6,1)	$HO \cdot C_{10}H_6 \cdot CO_2H$	X-330	188.18
3781	α-naphthoic acid	(7,1)	$HO \cdot C_{10}H_6 \cdot CO_2H$	X-330	188.18
3782	β-naphthoic acid	(1,2)	$HO \cdot C_{10}H_6 \cdot CO_2H$	X-331	188.18
3783	β-naphthoic acid	(3,2)	$HO \cdot C_{10}H_6 \cdot CO_2H$	X-333	188.18
3784	β-naphthoic acid	(5,2)	$HO \cdot C_{10}H_6 \cdot CO_2H$	X-337	188.18
3785	β-naphthoic acid	(7,2)	$HO \cdot C_{10}H_6 \cdot CO_2H$	X-337	188.18
3786	α-naphthoquinone	(2)	$C_{10}H_5O_2 \cdot OH$	VIII-300	174.16
3787	α-naphthoquinone	(5); juglone	$C_{10}H_5O_2 \cdot OH$	VIII-308	174.16
3788	β-naphthoquinone	(7)	$C_{10}H_5O_2 \cdot OH$	VIII-299	174.16
3789	β-naphthoquinone	(6)	$C_{10}H_5O_2 \cdot OH$	*VIII-638	174.16
3790	naphthyl-*o*-hydroxy-*m*-toluic acid (β)	Epicarine	$HO \cdot C_{10}H_6 \cdot CH_2 \cdot C_6H_3(OH) \cdot CO_2H$	....	294.31
3791	nicotinic acid (α)	....	$HO \cdot C_5H_3N \cdot CO_2H$	XXII-214	139.11
3792	nicotinic acid (*p*)	....	$HO \cdot C_5H_3N \cdot CO_2H$	XXII-215	139.11
3793	oenanthylic acid (α)	....	$C_5H_{11} \cdot CHOH \cdot CO_2H$	III-342	146.19
3794	phenylacetic acid (*o*)	....	$HO \cdot C_6H_4 \cdot CH_2 \cdot CO_2H$	X-187	152.15
3795	phenylacetic acid (*m*)	....	$HO \cdot C_6H_4 \cdot CH_2 \cdot CO_2H$	X-189	152.15
3796	phenylacetic acid (*p*)	....	$HO \cdot C_6H_4 \cdot CH_2 \cdot CO_2H$	X-190	152.15
3797	phenylarsonic acid	(*p*)	$HO \cdot C_6H_4 \cdot AsO(OH)_2$	XVI-874	218.04
3798	phenylarsonate Na	(*p*)	$HO \cdot C_6H_4 \cdot As:O (OH)ONa \cdot 2\frac{1}{2}H_2O$	XVI-874	285.06
3799	phenylethylamine (*p*)	tyramine	$HO \cdot C_6H_4 \cdot CH_2 \cdot CH_2 \cdot NH_2$	XIII-625	137.18
3800	phenylglycine (*p*)	photo-glycin	$HO \cdot C_6H_4 \cdot NH \cdot CH_2 \cdot CO_2H$	XIII-488	167.17
3801	phenyl-quinoline-dicarboxylic acid	2-(4-OH-3-CO_2H-Ph)cinchoninic acid	$HO_2C \cdot C_9H_5N \cdot C_6H_3(OH)CO_2H$	*XXII-567	309.28
3802	*o*-phthalic acid	(3;1,2)	$HO \cdot C_6H_3(CO_2H)_2$	X-498	182.13
3803	*o*-phthalic acid	(4;1,2)	$HO \cdot C_6H_3(CO_2H)_2$	X-499	182.13
3804	*m*-phthalic acid	(2;1,3)	$HO \cdot C_6H_3(CO_2H)_2 \cdot H_2O$	X-501	200.15
3805	*m*-phthalic acid	(4;1,3)	$HO \cdot C_6H_3(CO_2H)_2$	X-502	182.13
3806	*m*-phthalic acid	(5;1,3)	$HO \cdot C_6H_3(CO_2H)_2 \cdot 2H_2O$	X-504	218.16
3807	*p*-phthalic acid	(2;1,4)	$HO \cdot C_6H_3(CO_2H)_2$	X-505	182.13
3808	proline (4)(*dl*)	4-OH-pyrrolidine-2-carboxylic acid	$HN:C_4H_5(OH)CO_2H$	XXII-190	131.13
3809	proline (4)(*l*−)	....	$C_5H_9O_3N$	XXII-191	131.13
3810	propiophenone (*o*)	*o*-propionyl-phenol	$HOC_6H_4COC_2H_5$	VIII-102	150.18

Hydroxy-methoxycoumarin 5604
Hydroxy-methyl-azobenzene 657-8
Hydroxy-methylbenzaldehyde 4379
Hydroxy-methylcoumarin 4406
Hydroxy-methylene acetone 3304
Hydroxy-naphthalene 4489-90
Hydroxy-phenetole 5136
Hydroxy-phenylalanine 6367

Hydroxy-phenyl-mercurichloride 4034-5
Hydroxy-propionitrile 3195
Hydroxy-purine, cf. alkd.
Hydroxy-quinol 6217
Hydroxy-quinoline carboxylic acid 3937
Hydroxy-stachydrine, cf. alkd.
Hydroxy-styrene 6443-4
Hydroxy-succinic acid 3997-8

Table 7-4 (*Continued*)
PHYSICAL CONSTANTS OF ORGANIC COMPOUNDS

No.	Crystalline Form and Color	Specific Gravity	Melting Point °C.	Boiling Point °C.	Solubility in 100 Parts		
					Water	Alcohol	Ether
3775	nd./abs. al.	....	96-7	>300 sl. d.	d. a.	d. alk.	....
3776	col. pr./ aq.	....	115	295-300	v. s. h.; i. pet.	7.79°	s.; s. bz.
3777	nd.	....	128 d.	....	0.420°	s.	s.
3778	nd./aq. al.	....	156-7 d.	....	v. sl. s.	v. s. abs.	s.; s. bz.
3779	nd./h. aq.	....	234-7	subl.	s. h.	v. s.	s.; s. ac.
3780	nd./aq.	....	208-9	....	sl. s. h.	v. s.	....
3781	nd./aq.	....	253-4	....	s. h.	s.	....
3782	nd./al.	....	188-91	....	v. sl. s. h.	s.; s. alk.	s.; s. bz.
3783	yel. lf./aq.	....	216	....	sl. s. h.	s.; s. bz.	s.; s. chl.
3784	nd./al.	....	211-2	....	sl. s. h.	s.	....
3785	lf.	....	262	....	s.	s.	s.
3786	red cr./ac.	....	190d.	subl.	sl. s. h.	s.	s.
3787	yel. red/ bz.	....	153-4	d.	i.; s. h. ac.	sl. s. c.	sl. s.
3788	brn. nd.	....	194	....	i. bz.	s.; s. ac.	i.
3789	lt. yel./aq. al.	....	>200	d. > 220	s. h.	s.	....
3790	wh. pd.	....	90±, d.	....	i.	s.; i. chl.	s.; s. act.
3791	nd./aq.	....	256	....	v. sl. s.	....	....
3792	nd./aq.	....	301-2 d.	subl.	sl. s. h.	v. sl. s.	v. sl. s.
3793	pr.	....	65	d.	v. sl. s. c.	....	....
3794	nd./ct.	....	145-7	240-3 d.	s.	sl. s. c. chl.	s.
3795	nd./bz. lg.	....	129	190^{11mm}	v. s.	v. s.	v. s.
3796	nd./aq.	....	148	....	v. s. h.	v. s.	v. s.
3797	wh. pd.	....	175-80 d.	....	v. s.	s.	....
3798	nd./aq. al.	....	-2$\frac{1}{2}$H$_2$O, 100	....	s.	....	....
3799	lf./bz.	....	161	175-81^{8mm}	<10 h.	10± h.; s. bz.	sl. s. h. xylene
3800	lf./aq.	....	240-1 d.	....	sl. s.; s. alk.	sl. s.	i.
3801	yel. pd.	....	283-4 d.	....	i.; s. alk.	sl. s.; sl. s. EtOAc	i. lg.
3802	nd./aq.	....	150±, d.	....	s.	s.	s.
3803	cr./aq.	....	204-5 d.	....	v. sl. s. bz.	s.	s.; sl. s. pet.
3804	nd./aq.	....	243-4 (anh.)	-H$_2$O, 100	0.14$^{24°}$; 2.6$^{100°}$	s.	s.; sl. s. chl.
3805	nd./aq.	....	310	....	0.02$^{10°}$; 0.7$^{100°}$	s.; s. h. ac.	s.; i. chl.
3806	nd./aq.	....	288(anh.)	-2H$_2$O, 100	0.035°; 18$^{99°}$	s.	s.; s. bz.
3807	pd./aq.	....	>330	subl. sl. d.	v. sl. s.	s.; s. Me al.	sl. s.
3808	col. pl./ Me al.	....	261-2 d.	....	v. s.	v. sl. s.	....
3809	lf./aq. al.	....	270	....	v. s.	v. sl. s.	....
3810	lq.	....	....	115^{15mm}	v. sl. s.	s.; s. alk.	s.

Table 7-4 (*Continued*)
PHYSICAL CONSTANTS OF ORGANIC COMPOUNDS

No.	Name	Synonym	Formula	Beil. Ref.	Formula Weight
	Hydroxy				
3811	propiophenone (*p*)	*p*-propionyl-phenol	$HOC_6H_4COC_2H_5$	VIII-102	150.18
3812	pyridine (2)(α)	α-pyridone	$HO \cdot C_5H_4N$	XXI-43	95.10
3813	pyridine (3)	β-pyridone	$HO \cdot C_5H_4N$	XXI-46	95.10
3814	pyridine (γ)	γ-pyridone	$HO \cdot C_5H_4N \cdot H_2O$	XXI-48	113.12
3815	pyrotartaric acid (α)	....	$HO_2C \cdot CH(CH_3) \cdot CHOH \cdot CO_2H$	III-445	148.12
3816	quinaldine (3)	....	$C_{10}H_9ON$	XXI-103	159.19
3817	quinaldine (4)	....	$C_{10}H_9ON$	XXI-104	159.19
3818	quinaldine (5)	....	$C_{10}H_9ON$	XXI-106	159.19
3819	quinaldine (6)	....	$C_{10}H_9ON$	XXI-106	159.19
3820	quinoline (2)(α)	carbostyril	$HO \cdot C_9H_6N$	XXI-77	145.16
3821	quinoline (4)(γ)	....	$HO \cdot C_9H_6N \cdot 3H_2O$	XXI-83	199.21
3822	quinoline (5)(*ana*)	....	$HO \cdot C_9H_6N$	XXI-84	145.16
3823	quinoline (6)(*p*)	....	$HO \cdot C_9H_6N$	XXI-85	145.16
3824	quinoline (7)(*m*)	....	$HO \cdot C_9H_6N$	XXI-91	145.16
3825	quinoline (8)(*o*)	oxine	$HO \cdot C_9H_6N$	XXI-91	145.16
3826	*iso*-quinoline (1)	*iso*-carbostyril	$C_6H_4CO \cdot NH \cdot CH:CH$	XXI-100	145.16
3827	quinoline (8) sulfate	Chinosol; quinosol	$(HO \cdot C_9H_6N)_2 \cdot H_2SO_4$	XXI-92	388.40
3828	quinoline-5-sulfonic acid (8)	....	$HOC_9H_5N \cdot SO_3H \cdot 2H_2O$	XXII-407	261.26
3829	stearic acid (α)	....	$C_{16}H_{33}CHOH \cdot CO_2H$	III-364	300.49
3830	stearic acid (10)	....	$CH_3(CH_2)_7 \cdot CHOH \cdot (CH_2)_8CO_2H$	III-365	300.49
3831	stearic acid (λ)	....	$CH_3(CH_2)_5 \cdot CHOH \cdot (CH_2)_{10}CO_2H$	III-366	300.49
3832	*o*-toluic acid (1;2,6)	....	$CH_3C_6H_3(OH)CO_2H$	X-214	152.15
3833	*o*-toluic acid (1;3,2)	Me-salicylic acid (6)	$CH_3C_6H_3(OH)CO_2H$	X-217	152.15
3834	*o*-toluic acid (1;3,6)	....	$CH_3C_6H_3(OH)CO_2H \cdot \frac{1}{2}H_2O$	X-214	161.16
3835	*o*-toluic acid (1;4,2)	....	$CH_3C_6H_3(OH)CO_2H$	X-215	152.15
3836	*m*-toluic acid (1;2,3)	β-cresotinic acid	$CH_3C_6H_3(OH)CO_2H$	X-220	152.15
3837	*m*-toluic acid (1;2,5)	....	$CH_3C_6H_3(OH)CO_2H \cdot \frac{1}{2}H_2O$	X-225	161.16
3838	*m*-toluic acid (1;3,5)	....	$CH_3C_6H_3(OH)CO_2H$	X-227	152.15
3839	*m*-toluic acid (1;4,3)	α-cresotinic acid	$CH_3C_6H_3(OH)CO_2H$	X-227	152.15
3840	*p*-toluic acid (1;2,4)	....	$CH_3C_6H_3(OH)CO_2H$	X-237	152.15
3841	*p*-toluic acid (1;3,4)	γ-cresotinic acid	$CH_3C_6H_3(OH)CO_2H$	X-233	152.15
3842	urea	....	$NH_2 \cdot CO \cdot NHOH$	III-95	76.06
3843	valeric acid (α)(*n*)	....	$C_3H_7 \cdot CHOH \cdot CO_2H$	III-320	118.13
3844	*iso*-valeric acid (α)	(*dl*)	$(CH_3)_2CH \cdot CHOH \cdot CO_2H$	III-328	118.13
3845	*iso*-valeric acid (β)	....	$(CH_3)_2COH \cdot CH_2 \cdot CO_2H$	III-327	118.13

Table 7-4 (*Continued*)
PHYSICAL CONSTANTS OF ORGANIC COMPOUNDS

No.	Crystalline Form and Color	Specific Gravity	Melting Point °C.	Boiling Point °C.	Solubility in 100 Parts		
					Water	Alcohol	Ether
3811	nd./aq.	....	149-50	....	$0.04^{15°}$	$3.3^{100°}$ aq.	$42^{5°}$
3812	nd./bz.	....	106-7	280-1	v. s.	v. s.	s.; sl. s. lg.
3813	nd./bz.	....	129	subl.	s.	s.	sl. s. bz.
3814	mn.	....	92; 148.5 (anh.)	>350	$100^{15°}$	s.; v. sl. s. chl.	i.; i. bz.
3815	mn.	....	123	d.	....	....	....
3816	nd./al.	....	265 d.	....	sl. s.	s.	....
3817	cr.	....	230-1	>360 d.	1 c.; 10 h.	s.; v. sl. s. bz.	v. sl. s.
3818	lf./al.	....	232-4	sl. d.	i.	sl. s. c.	s.
3819	cr./aq.	....	213	sl. d.	sl. s.	s.	s.
3820	pr./al.	....	199-200	subl.	s. h.	v. s.	v. s.
3821	nd./aq.	....	52*	>300 d.	$0.47^{15°}$	v. s. h.	v. sl. s.
3822	nd./al.	....	224 d.	subl.	v. s. h. Na_2CO_3	s.	sl. s.; i. lg.
3823	pr./abs. al.	....	193	>360	v. sl. s. c.	sl. s.; s. alk.	v. sl. s. ;s. a.
3824	pr./abs. al.	....	235-8 d.	....	v. sl. s.	v. s.	s. alk.
3825	pr./aq. al.	....	75-6	266.6^{752mm}	v. sl. s. o.	s.; s. dil. alk.	sl. s.
3826	mn./bz.	....	208-9	subl.	sl. s.	s.; s. chl.	sl. s.; sl. s. bz.
3827	yel. cr. pd.	....	177-8	....	v. s.	sl. s.	i.; s. gly.
3828	pa. yel. cr./HCl	....	275±, d.	....	sl. s. h.	sl. s.	v. sl. s.
3829	nd./chl.	....	92-3	....	v. s. h. bz.	$0.6^{20°}$	s.
3830	tab./al.	....	83-5	....	....	$9.7^{20°}$	$2.4^{20°}$
3831	cr./al.	....	78	....	s. chl.; i. lg.	$13^{20°}$	$5.4^{19°}$
3832	cr./aq.	....	142-5	....	....	....	....
3833	nd./aq.	....	167-8	....	$0.14^{25°}$	v. s.	v. s.
3834	cr./aq.	....	178(anh.)	$-H_2O$, 100	s. h.	v. s.	v. s.; i. chl.
3835	nd./aq.	....	180-2	subl.	s.	v. s.	v. s.
3836	nd./h. aq.	....	163-4	....	s. h.	s.	s.; s. chl.
3837	nd./aq.	....	173-4 (anh.)	$-H_2O$, 100	s. h.	s.	s.; i. CS_2
3838	nd./aq.	....	208-10	subl.	s.	....	....
3839	nd./aq.	....	152-3	subl. sl. d.	s. h.	s.	s.; s. chl.
3840	nd./aq.	....	206-7	subl.	s. h.	s.	s.; i. chl.
3841	nd./aq.	....	177-8	subl.	s.	s.	s. chl.
3842	nd./al.	....	139-40	d.	v. s.	s. h.	....
3843	hyg. pl.	....	34	subl.	v. s.	v. s.	v. s.
3844	rhb.	....	86	....	v. s.	v. s.	v. s.
3845	syrup	....	....	....	v. s.	v. s.	v. s.

* Loses $3H_2O$, 110°; m. p. anh. 201°

Iodo-gorgonic acid 2401-2
Iodo-heptane 3549
Iodo-hexadecane 1277
Iodo-hexahydrobenzene 1614
Iodo-hexane 3650-1
Iodo-hydrin 3208, 6287
Iodo-methane 4287

Iodo-nonane 4929
Iodo-octadecane 4940
Iodo-octane 4980
Iodo-pentane 459-63
Iodo-propene 208
Iodo-phthalein 5813
Iodo-propane 5417-8
Iodo-propyl alcohol 6287

Iodo-propylene 208
Iodo-propylene glycol 3402
Iodo-propylene oxide 2881
Iodo-quinine sulfate, cf. alkd.
Iodo-thymol 5940
Iodo-toluene (ω) 813
Iodol 5817
Iodophen 5813

Table 7-4 (*Continued*)
PHYSICAL CONSTANTS OF ORGANIC COMPOUNDS

No.	Name	Synonym	Formula	Beil. Ref.	Formula Weight
3846	**Hydroxy** valeric lactone (γ)	....	$CH_3 \cdot CHCH_2CH_2CO \cdot O$ ⌐———————⌐	XVII-235	100.12
3847	**Hyenic acid**	....	$CH_3(CH_2)_{23} \cdot CO_2H$	II-394	382.68
3848	**Hypnoacetin**	....	$C_{16}H_{15}O_3N$	XIII-464	269.30
3849	**Hypogaeic acid**	....	$C_{15}H_{29} \cdot CO_2H$	II-461	254.42
3850	**Imesatin**	....	$C_6H_4 \cdot C(NH)CO \cdot NH$ ⌐—————————⌐	XXI-440	146.15
3851	**Imidazol**	glyoxalin	$C_3H_4N_2$	XXIII-45	68.08
3852	**Imino-**diacetic acid	....	$HN:(CH_2 \cdot CO_2H)_2$	IV-365	133.10
3853	diaceto-dinitrile	....	$HN:(CH_2 \cdot CN)_2$	IV-367	95.10
3854	**Indandione** (1,3)	α,γ-diketo-hydrindene	$C_6H_4 \cdot CO \cdot CH_2 \cdot CO$ ⌐————————⌐	VII-694	146.15
3855	**Indene**	....	$C_6H_4 \cdot CH_2 \cdot CH:CH$ ⌐——————⌐	V-515	116.16
3856	**Indican** (β)	indoxyl-β-glucoside	$C_{14}H_{17}O_6N \cdot 3H_2O$	....	349.34
3857	**Indigo**	indigotin	$C_{16}H_{10}O_2N_2$	XXIV-417	262.27
3858	carmine	soluble indigo	$C_{16}H_8O_2N_2(SO_3Na)_2$	XXV-304	466.36
3859	dicarboxylic acid	....	$C_{18}H_{10}O_6N_2$	XXV-273	350.29
3860	disulfonic acid	....	$C_{16}H_8O_2N_2(SO_3H)_2$	XXV-304	422.39
3861	monosulfonic acid	....	$C_{16}H_9O_2N_2(SO_3H)$	XXV-303	342.33
3862	purpurin	indirubin	$(NH \cdot C_6H_4CO \cdot C:)_2$	XXIV-430	262.27
3863	white	....	$(C_6H_4C(OH):C \cdot NH)_2$ ⌐—————————⌐	XXIII-538	264.29
3864	**Indole**	benzopyrrole	C_8H_7N	XX-304	117.15
3865	**Indolyl-**acetic acid (3)	skatole carboxylic acid (ω)	$C_8H_6N \cdot CH_2CO_2H$	XXII-66	175.19
3866	butyric acid (3)(γ)	....	$C_8H_6N \cdot (CH_2)_3CO_2H$	....	203.24
3867	propionic acid (3)(β)	skatole-ω-acetic	$C_8H_6N \cdot (CH_2)_2CO_2H$	XXII-69	189.22
3868	**Indophenin**	....	$(C_{12}H_7NOS)_2$	XXI-438	426.52
3869	**Indophenol**	....	$HO \cdot C_6H_4 \cdot N:C_6H_4:O$	....	199.21
3870	**Indoxyl**	....	$HN \cdot CH:C(OH) \cdot C_6H_4$ ⌐——————————⌐	XXI-69	133.15
3871	**Indoxylic acid**	....	$C_6H_4 \cdot COH \cdot C(CO_2H):$ ⌐———————————— NH ——⌐	XXII-226	177.16
3872	**Inulin**	dahlin; alantin; alant starch	$(C_6H_{10}O_5)_6 \cdot H_2O$	....	990.88
3873	**Iodo-**acetanilide (p)	....	$I \cdot C_6H_4 \cdot NH \cdot COCH_3$	XII-671	261.06
3874	acetic acid	....	$I \cdot CH_2 \cdot CO_2H$	II-222	185.95
3875	aniline (o)	....	$I \cdot C_6H_4 \cdot NH_2$	XII-669	219.03
3876	aniline (m)	....	$I \cdot C_6H_4 \cdot NH_2$	XII-670	219.03
3877	aniline (p)	....	$I \cdot C_6H_4 \cdot NH_2$	XII-670	219.03
3878	anisole (o)	....	$CH_3O \cdot C_6H_4 \cdot I$	VI-207	234.04
3879	benzene	phenyl iodide	$C_6H_5 \cdot I$	V-215	204.01
3880	benzoic acid (o)	....	$I \cdot C_6H_4 \cdot CO_2H$	IX-363	248.02
3881	benzoic acid (m)	....	$I \cdot C_6H_4 \cdot CO_2H$	IX-365	248.02
3882	benzoic acid (p)	....	$I \cdot C_6H_4 \cdot CO_2H$	IX-366	248.02
3883	benzoyl chloride (o)	....	$I \cdot C_6H_4 \cdot CO \cdot Cl$	IX-364	266.47

Table 7-4 (*Continued*)
PHYSICAL CONSTANTS OF ORGANIC COMPOUNDS

No.	Crystalline Form and Color	Specific Gravity	Melting Point °C.	Boiling Point °C.	Solubility in 100 Parts		
					Water	Alcohol	Ether
3846	lq.	$1.050\frac{22}{4}°$	<−18	207–8	∞; i. aq. K_2CO_3	....	....
3847	cr./et.	....	77–8	....	....	v. sl. s. c.	v. s.
3848	lf./al.	....	160±	....	i.	$0.323°$	i.
3849	col. nd.	....	33	236^{15mm}	i.	v. s.	....
3850	yel. pr.	....	175–6	....	i.	s. h.	sl. s.
3851	pr.	....	89–90	255–6	s.	v. s.	sl. s.
3852	rhb.	....	d. 225–36	....	$2.45°$	i.	i.
3853	lf./et.	....	78	....	s.	s.	sl. s.
3854	cr./lg.	....	129–31 d.	....	v. sl. s. c.	s. h.	s. bz.
3855	col. lq.	$0.991\frac{25}{25}°$	−2	181–2	i.	s.	∞
3856	brn. rhb.	....	57; 176–8 (anh.)	....	v. s.	v. s.; sl. s. bz.	sl. s.
3857	b. cr./ aniline	1.35	390–2	subl.	i.	i.	i.; s. h. act.
3858	b. amor.	....	....	....	sl. s. c.	i.	i. aq. NaCl
3859	b. bl. ppt.	....	....	....	s. H_2SO_4	i.	i.; i. chl.
3860	b. amor.	....	....	....	v. s.	v. s.	....
3861	b. amor.	....	d. 200	....	v. s.	v. s.	....
3862	brn. nd.	....	....	subl.	i.	sl. s.	s. ac.
3863	col.–gray	....	....	....	i.; s. alk.	s.	s.
3864	lf./aq.	....	52	253–4	s. h.	s. h.	s.; s. bz.
3865	col. lf./bz.	....	165–8	....	v. sl. s. c.	s.; s. act.	s.; i. chl.
3866	rhb./bz. pet.	....	124–5	....	i.	s.; s. act.	s.; i. chl.
3867	cr./aq.	....	133–4	....	v. sl. s. c.	s.; s. ac.	s.
3868	b. pd.	....	d.	....	i.; s. H_2SO_4	v. sl. s.	v. sl. s.; i. bz.
3869	pl./act. pet.	....	160	....	s.; i. pet.	s.; s. bz.	s.; s. chl.
3870	yel. pr.	....	85	110	s.; s. act.	s.	s.; v. sl. s. pet.
3871	cr.	....	subl. 122–3	d.	v. sl. s; d. h.	....	....
3872	hyg. pd.	1.4(anh.)	d. 160	....	$0.01°$; $37^{100°}$	$0.02^{16°}$	....
3873	mn.	$1.989^{15-20°}$	183–4	....	s. h.	$6.4^{21°}$	v. s. ac.
3874	col. pl.	....	82–3	....	s.	s.	sl. s.
3875	nd.	....	60–1	....	v. sl. s.	v. s.	v. s.
3876	lf. or nd.	....	33	....	i.	s.	
3877	nd./aq.	....	67–8	....	sl. s.	s.; s. chl.	s.
3878	yel. lq.	$1.8^{20°}$	....	240–1	i.; s. bz.	∞; ∞ chl.	∞
3879	col. lq.	$1.824\frac{25}{4}°$	−31.3	188.5	i.; ∞ chl.	s.	∞
3880	nd./aq.	2.25	162	....	sl. s. h.	v. s.	v. s.
3881	cr./act.	....	187–8	subl.	sl. s.	s.	....
3882	lf.	....	269–70	subl.	sl. s. h.	s.	....
3883	cr.	....	30–1	159^{27mm}	d.	d.	....

Table 7-4 (*Continued*)
PHYSICAL CONSTANTS OF ORGANIC COMPOUNDS

No.	Name	Synonym	Formula	Beil. Ref.	Formula Weight
3884	**Iodo** diphenyl (*o*)	....	$I \cdot C_6H_4 \cdot C_6H_5$	**V-486	280.11
3885	diphenyl (*p*)	....	$I \cdot C_6H_4 \cdot C_6H_5$	V-581	280.11
3886	ethylacetate (*β*)	....	$CH_3CO_2 \cdot CH_2 \cdot CH_2I$	II-129	214.00
3887	form	triiodomethane	HCI_3	I-73	393.73
3888	8-hydroxyquino- line-5 sulfonic acid (7)	loretin; ferron	$IC_9H_4N(OH) \cdot SO_3H$	XXII-408	351.12
3889	lepidine (2)	2-I-4-Me-quinoline	$C_{10}H_8NI$	XX-397	269.09
3890	methyl cyanide	iodoacetonitrile	$I \cdot CH_2 \cdot CN$	II-223	166.95
3891	naphthalene (*α*)	*α*-naphthyl iodide	$C_{10}H_7I$	V-550	254.07
3892	naphthalene (*β*)	*β*-naphthyl iodide	$C_{10}H_7I$	V-552	254.07
3893	nitrobenzene (*o*)	....	$I \cdot C_6H_4 \cdot NO_2$	V-252	249.01
3894	nitrobenzene (*m*)	....	$I \cdot C_6H_4 \cdot NO_2$	V-253	249.01
3895	nitrobenzene (*p*)	....	$I \cdot C_6H_4 \cdot NO_2$	V-253	249.01
3896	phenetole (*o*)	....	$I \cdot C_6H_4 \cdot O \cdot C_2H_5$	VI-207	248.07
3897	phenetole (*p*)	....	$I \cdot C_6H_4 \cdot O \cdot C_2H_5$	VI-208	248.07
3898	phenol (*o*)	....	$I \cdot C_6H_4 \cdot OH$	VI-207	220.01
3899	phenol (*m*)	....	$I \cdot C_6H_4 \cdot OH$	VI-207	220.01
3900	phenol (*p*)	....	$I \cdot C_6H_4 \cdot OH$	VI-209	220.01
3901	propionic acid (*α*)	....	$CH_3 \cdot CHI \cdot CO_2H$	II-261	199.98
3902	propionic acid (*β*)	....	$ICH_2 \cdot CH_2 \cdot CO_2H$	II-261	199.98
3903	propyne-1 (1)†	....	$CH_3 \cdot C \vdots CI$	I-248	165.96
3904	salicylic acid (3)	*m*-I-salicylic acid	$I \cdot C_6H_3(OH)CO_2H$	X-112	264.02
3905	toluene (*o*)	*o*-tolyl iodide	$I \cdot C_6H_4 \cdot CH_3$	V-310	218.04
3906	toluene (*m*)	*m*-tolyl iodide	$I \cdot C_6H_4 \cdot CH_3$	V-311	218.04
3907	toluene (*p*)	*p*-tolyl iodide	$I \cdot C_6H_4 \cdot CH_3$	V-312	218.04
3908	trinitromethane	....	$IC(NO_2)_3$	I-79	276.93
3909	*m*-xylene (2)	2-I-1,3-diMe-ben- zene	$I \cdot C_6H_3(CH_3)_2$	V-375	232.07
3910	**Iodoso**-benzene	....	C_6H_5IO	V-217	220.01
3911	benzoic acid (*o*)	....	$OI \cdot C_6H_4 \cdot CO_2H$	IX-363	264.02
3912	**Iodoxybenzene**	....	$C_6H_5IO_2$	V-218	236.01
3913	**Ionone** (*α*)	....	$C_{10}H_{16} \vdots CH \cdot CO \cdot CH_3$	VII-168	192.30
3914	**Ionone** (*β*)	....	$C_{10}H_{16} \vdots CH \cdot CO \cdot CH_3$	VII-167	192.30
3915	**Ionone** (*pseudo*)	citrylidene acetone	$C_{13}H_{20}O$	I-757	192.30
3916	semicarbazone (*α*)	....	$C_{13}H_{20} \vdots NNHCONH_2$	VII-169	249.36
3917	semicarbazone (*β*)	....	$C_{13}H_{20} \vdots NNHCONH_2$	VII-168	249.36
3918	**Irone** (*β*)	irone	$C_{14}H_{22}O$	VII-169	206.33
3919	**Isatin**	....	$C_6H_4 \cdot CO \cdot COH \vdots N$ $\lfloor \qquad \qquad \rfloor$	XXI-432	147.13
3920	chloride	....	$C_6H_4 \cdot CO \cdot CCI \vdots N$ $\lfloor \qquad \qquad \rfloor$	XXI-302	165.58
3921	**Isatinic acid**	anilino-glyoxylic acid	$NH_2 \cdot C_6H_4 \cdot CO \cdot CO_2H$	XIV-648	165.15
3922	**Isatoic anhydride**	....	$C_6H_4 \cdot CO_2 \cdot CO \cdot NH$	XXVII-264	163.13
3923	**Isatoxime** (*β*)	nitroso-oxindol	$C_6H_4 \cdot N \vdots C(OH)C \vdots NOH$ $\lfloor \qquad \qquad \rfloor$	XXI-443	162.15
3924	**Isatropic acid** (*α*)	1-Ph-1,2,3,4-tetra- hydronaphthalene- 1,4-dicarboxylic acid	$(C_9H_8O_2)_2$	IX-957	296.33

† See also No. 5344.
Kyanol 513
L acid 4503
Laburnine, cf. alkd.

Lactic anhydride 3948
Lactophenin 3949
Lactyl urea 4268
Lambda acid 4544
Lantanuric acid 182

Table 7-4 (*Continued*)
PHYSICAL CONSTANTS OF ORGANIC COMPOUNDS

No.	Crystalline Form and Color	Specific Gravity	Melting Point °C.	Boiling Point °C.	Solubility in 100 Parts		
					Water	Alcohol	Ether
3884	lq.	$1.604\frac{25°}{25}$	....	$189-92^{36mm}$	....	....	....
3885	col. cr./al.	....	112-3	320 sl. d.	i.; s. bz.	s. h.	s.; s. ac.
3886	col. lq.	$2.441^{20°}$	....	184^{743mm}	....	....	....
3887	yel. hex.	$4.008^{17°}$	119	subl.	$0.01^{25°}$	$1.5^{17°}$; 11 h.	$13.6^{25°}$
3888	yel. lf.	....	d. 260±	....	0.2 c.; 0.6 h.	sl. s.; s. H_2SO_4	i.; i. bz.; i. chl.
3889	nd./lg.	....	90	....	....	s.	s.; s. lg.
3890	oil	2.307	....	$182-4^{720}$ d.	s. d.	....	....
3891	oil	$1.734^{15°}$	....	305	i.	∞	∞
3892	lf.	$1.632\frac{0.9°}{4}$	54-5	308-10	i.	v. s.	v. s.
3893	yel. nd.	$1.883\frac{100°}{4}$	52-4	$288-9^{729mm}$	i.	s. h.	v. s.
3894	mn.	$1.878\frac{100°}{4}$	37-8	280±	....	....	....
3895	yel. nd./al.	2.273 (solid)	171.5	287^{726mm}	....	....	....
3896	lq.	....	....	245^{736mm}	i.	s.	s.
3897	cr.	....	29	249^{729mm}	i.	s.	s.; s. chl.
3898	nd. or pl.	$1.876^{80°}$	40.4	$186-7^{160mm}$	s. h.	v. s.	v. s.
3899	nd./lg.	....	40	d.	....	s.	s.
3900	nd./aq.	$1.857^{112°}$	93-4	d.	sl. s.	v. s.	v. s.
3901	nd.	....	44.5-5.5	$105^{0.3mm}$	v. sl. s.	s.	s.
3902	lf.	....	82	....	s. h.	v. s.	v. s.
3903	col. oil	$2.08^{22°}$	....	110.2^{757mm}	....	....	....
3904	col. nd./aq.	....	199	....	s. h.	s.	s. alk.
3905	lq.	$1.698^{20°}$	....	211-2	i.	∞	∞
3906	lq.	$1.698^{20°}$	....	213	i.	∞	∞
3907	lf.	$1.678^{40°}$	35-6	211.5	i.	v. s.	v. s.
3908	unstable, yel. pr.	....	55-6	48^{13mm}	i.	s. h.	s. bz., lg.
3909	lq.	....	....	228-30	....	....	....
3910	yel. amor.	....	expl. 210	....	s. h.	s. h.	v. sl. s.
3911	col. lf./aq.	....	>200 d.	....	s. h.	s. h.	v. sl. s.
3912	nd./aq.	....	expl. 236-7	....	s. h.	i.; i. bz.	s. h. ac.
3913	col. oil	$0.930^{20°}$	....	136.1^{17mm}	sl. s.	∞	∞
3914	col. oil	$0.944^{20°}$	....	140^{18mm}	sl. s.	∞	∞
3915	oil	$0.898^{20°}$	....	$143-5^{12mm}$	....	....	....
3916	cr./bz. lg.	....	107-8 (137-8)	....	....	s.	....
3917	nd./al.	....	148-9	....	i.	s.	s.; s. bz.
3918	col. oil	$0.939^{20°}$	....	144^{16mm}	v. sl. s.	v. s.	v. s.
3919	yel. red mn.	....	200-1	....	s. h.	v. s. h.	sl. s.; s. alk.
3920	brn. nd.	....	180±, d.	....	i.	s.	s.; blue
3921	wh. pd.	....	d.	....	s.	....	....
3922	mn./act.	....	240 d.	....	d. h.	$38°$	sl. s.; 1.3 h. act.
3923	yel. nd.	....	225 d.	....	v. sl. s; s. a.	sl. s.	s. KOH; i. bz.
3924	cr.	....	237	....	sl. s. h.	sl. s.; i. CS_2	i.; i. bz.

Laricin, cf. glcde.
Larocaine 276
Laudanine, cf. alkd.
Laudanosine, cf. alkd.
Laurane 1017

Laurent's acid 4548
Laurone 2850
Lauroyl chloride 3955
Lauryl alcohol 2859
Lauryl bromide 2860

Table 7-4 (*Continued*)
PHYSICAL CONSTANTS OF ORGANIC COMPOUNDS

No.	Name	Synonym	Formula	Beil. Ref.	Formula Weight
3925	**Isatropic acid** (β)	....	$(C_9H_8O_2)_2$	IX-957	296.33
3926	**Isomannide**	....	$C_6H_{10}O_4$	I-540	146.14
3927	**Isonitrosoacetone**	....	$CH_3\cdot CO\cdot CH:NOH$	I-763	87.08
3928	**Itaconic acid**	methylene succinic acid	$CH_2:C(CO_2H)\cdot CH_2\cdot CO_2H$	II-760	130.10
3929	**Itamalic acid**	free acid non-existent	$CH_2OH\cdot CH(CO_2H)\cdot CH_2\cdot CO_2H$	III-446	148.12
3930	**Kairoline**	N-methyl-tetrahydroquinoline	$C_9H_{10}N\cdot CH_3$	XX-264	147.22
3931	**Kawain**	....	$CH_3O\cdot C_{13}H_{11}O_2$	*XIX-418	230.27
3932	**Ketazine**	acetone azine	$[(CH_3)_2C:N\cdot]_2$	I-651	112.18
3933	**Ketene**	....	$H_2C:C:O$	I-724	42.04
3934	**Keto-**butyric acid (α)	....	$CH_3\cdot CH_2\cdot CO\cdot CO_2H$	III-629	102.09
3935	dihydroquinazoline (4)	....	$C_6H_4\cdot N:CH\cdot NH\cdot CO$	XXIV-143	146.15
3936	pyrrolidine	α,γ-butyrolactam	$NH\cdot (CH_2)_3\cdot CO\cdot H_2O$	XXI-236	103.12
3937	**Kynurenic acid**	4-hydroxyquinoline-3-carboxylic acid	$C_9H_5N(OH)CO_2H\cdot H_2O$	XXII-230	207.19
3938	**Lacmoid**	resorcinol blue	†	*VI-399	......
3939	**Lactic** acid (l) (+)	paralactic acid	$CH_3\cdot CHOH\cdot CO_2H$	III-261	90.08
3940	acid (dl)	....	$CH_3\cdot CHOH\cdot CO_2H$	III-268	90.08
3941	Ca salt	calcium lactate	$(C_3H_5O_3)_2Ca\cdot 5H_2O$	III-277	308.30
3942	aldehyde	....	$CH_3\cdot CHOH\cdot CHO$	I-819	74.08
3943	amide	....	$CH_3\cdot CHOH\cdot CO\cdot NH_2$	III-283	89.09
3944	nitrile	aldehyde-cyanohydrin	$CH_3\cdot CHOH\cdot CN$	III-284	71.08
3945	**Lactide** (dl)	dilactide	$C_6H_8O_4$	XIX-154	144.13
3946	**Lactoic acid** (d)	d-galactonic acid	$CH_2OH\cdot (CHOH)_4\cdot CO_2H$	III-549	196.16
3947	**Lactose**	milk sugar	$C_{12}H_{22}O_{11}\cdot H_2O$	XXXI-407	360.32
3948	**Lactyl** lactic acid	"lactic anhydride"	$C_2H_5O\cdot CO_2\cdot C_2H_4\cdot CO_2H$	III-282	162.14
3949	phenetidine (p)	lactophenin	$C_2H_5O\cdot C_6H_4\cdot NH\cdot CO\cdot CHOH\cdot CH_3$	XIII-491	209.25
3950	**Lauric** acid	dodecanoic acid	$CH_3(CH_2)_{10}\cdot CO_2H$	II-359	200.32
3951	ammonium salt	ammonium laurate	$C_{11}H_{23}CO_2NH_4$	*II-156	217.35
3952	aldehyde	dodecanal	$CH_3(CH_2)_{10}\cdot CHO$	I-714	184.32
3953	amide	....	$C_{11}H_{23}\cdot CO\cdot NH_2$	II-363	199.34
3954	anhydride	....	$(C_{11}H_{23}CO)_2O$	II-362	382.63
3955	chloride	lauroyl chloride	$CH_3(CH_2)_{10}\cdot CO\cdot Cl$	II-363	218.77
3956	nitrile	undecyl cyanide	$C_{11}H_{23}\cdot CN$	II-363	181.32
3957	**Lead** tetraethyl	tetraethyl lead	$(C_2H_5)_4Pb$	IV-639	323.44
3958	tetramethyl	tetramethyl lead	$(CH_3)_4Pb$	IV-639	267.33
3959	tetraphenyl	tetraphenyl lead	$(C_6H_5)_4Pb$	XVI-917	515.62
3960	**Lecithin**	protagon	$C_{42}H_{84}O_9PN$	....	778.11
3961	**Lepamine**	....	$C_{20}H_{32}N_2$	....	300.49
3962	**Leuco-**aniline	3-Me; $NH_2,4,4',4''$	$CH_3\cdot C_6H_3(NH_2)CH:(C_6H_4\cdot NH_2)_2$	XIII-321	303.41
3963	aurine (4,4',4'')	....	$(HO\cdot C_6H_4)_3CH$	VI-1143	292.34
3964	malachite green	$Me_2N, 4,4'$	$[(CH_3)_2N\cdot C_6H_4]_2:CH\cdot C_6H_5$	XIII-275	330.45

† Mixed dye; no formula.
Lauryl chloride 2861
Lauth's violet 5922
Lemonflavin, of. quercitrin, glcde.

Lenigallol 5516
Lepidine 4370
Lepidine ethiodide 4371
Lepidone 4179

Table 7-4 (*Continued*)
PHYSICAL CONSTANTS OF ORGANIC COMPOUNDS

No.	Crystalline Form and Color	Specific Gravity	Melting Point °C.	Boiling Point °C.	Solubility in 100 Parts		
					Water	Alcohol	Ether
3925	pl./aq.	....	206	d. to α, 220	....	....	....
3926	hyg., mn.	....	87	176^{30mm}	v. s.	s.; sl. s. chl.	i.; i. bz.
3927	lf./et.	$1.074^{67.5°}$	69	subl.	v. s.	v. sl. s. pet.	v. s.
3928	rhb.	1.63	165-6 sl. d.	....	$8.3^{20°}$; v. sl. s. bz.	$25^{15°}$, 88% al.	v. sl. s.; v. sl. s. chl.
3929	....	....	....	....	....	....	....
3930	lq.	$1.022^{20°}_{4}$	....	$247-50^{758mm}$	....	v. s.	sl. s.
3931	col. cr.	....	105-6	$195-70^{0.1mm}$	i.	s.	s.; s. act.
3932	lq.	$0.843^{20°}_{4}$	....	131	∞	∞	∞
3933	col. gas	....	−151	−56	d.	d.	s.; s. act.
3934	hyg. pl.	....	31.5-2.0	85^{21mm}	v. s.	v. s.	v. sl. s.
3935	col. nd.	....	213-4	....	s. h.	s.	s. NaOH
3936	cr.	$1.120^{20°}_{4}$	35	251*	v. s.	....	....
3937	nd.	....	290(anh.)	$-H_2O$, 140-5	$0.09^{100°}$	s. h.	i.
3938	vl. pd.	....	....	....	sl. s.	s.	i.; s. act.
3939	delq. pr.	....	52.8	....	∞	∞	∞
3940	hyg.	$1.249^{15°}_{4}$	16.8	122^{14mm}	∞	∞	∞
3941	ool. cr.	....	$-5H_2O$, 100	....	10 c.	3	∞ h. aq.
3942	nd.	....	101-5	....	s.; v. s. ac.	s.; s. act.	i.; i. bz.
3943	cr.	$1.138^{80°}_{4}$	74	....	v. s.	v. s.	....
3944	col. lq.	$0.992^{18°}_{4}$	−40	182-4 sl. d.	∞	∞	i. pet.
3945	tri./al.	$0.862^{10°}_{4}$	124.5	255^{757mm}	v. sl. s. c.; d. h.	v. sl. s. c. abs. al.	....
3946	nd./aq.	....	147.5	....	s.	....	....
3947	col. rhb.	$1.525^{20°}$	202(anh.)	d.	17 c.; 40 h.	i.	i.
3948	lt. yel. oil.	....	....	d. 250-60	v. sl. s.	s.	s.
3949	col. nd./aq.	....	118	....	0.3 c.; 1.8 h.	13; sl. s. lg.	sl. s.; s. h. bz.
3950	col. nd.	$0.871^{50°}_{4}$	44.1	225^{100mm}	i.; s. bz.	s.	s.; s.pet.
3951	waxy	0.88	48-55	$-NH_3 > 50$	s. h.	s.; i. chl.	i.; i. bz.
3952	col. lf.	....	44.5	$184-5^{100mm}$	i.	s.	s.
3953	col. nd.	....	102	$200^{12.5mm}$	i.	v. s.	....
3954	col. cr.	$0.855^{70°}_{4}$	41	$166^{0.1mm}$	....	....	....
3955	col. lq.	....	−17	145^{18mm}	d.	d.	s.
3956	oil	$0.827^{15°}$	4	198^{100mm}	....	....	∞
3957	col. lq.	$1.659^{18°}_{4}$	−136	152^{291mm}	i.; s. bz.	sl. s.	∞
3958	col. lq.	$1.995^{20°}_{4}$	−27.5	110^{760mm}	i.	∞	∞
3959	tet./bz.	$1.530^{20°}$	228-9	d. 270	i.	$0.1^{30°}$	$1.7^{30°}$ bz.
3960	waxy	....	150-200 d.	....	i.; s. chl.	s. h.	s. h.
3961	lq.	....	....	275	....	....	....
3962	cr./aq.	....	100	....	sl. s. h.	v. s.	sl. s.
3963	col. pr./ac.	....	....	....	sl. s.	s.; s. alk.	s. ac.
3964	nd./bz.	....	102(94)	d.	i.	s.; s. bz.	s.; sl. s. lg.

* B. p. of anh. compound.
Leucic acid 3760
Leucine 283, 285-7
Leuco- aniline 6071-3

Leuco-alizarin 2295
Leuco-crystal violet 3595
Leuco-dimethylphenylene green 5833
Leucoline 5538

Table 7-4 *(Continued)*
PHYSICAL CONSTANTS OF ORGANIC COMPOUNDS

No.	Name	Synonym	Formula	Beil. Ref.	Formula Weight
3965	**Leucyl glycine** *(dl)*	....	$C_5H_{12}NCO \cdot NHCH_2 \cdot$ CO_2H	IV-448	188.23
3966	**Levulin**	....	$C_6H_{10}O_5$	I-925	162.14
3967	**Levulinic** acid	acetopropionic acid	$CH_3CO(CH_2)_2CO_2H$	III-671	116.12
3968	aldehyde	....	$CH_3CO(CH_2)_2CHO$	I-774	100.12
3969	**Lichenin**	moss starch	$(C_6H_{10}O_5)x$	....	(162.14)
3970	**Lignoceric acid**	....	$C_{24}H_{48}O_2$	II-393	368.65
3971	**Limettin**	5,7-di MeO-coumarin	$(CH_3O)_2C_9H_4O_2$	XVIII-97	206.20
3972	**Limonene** *(dl)*	dipentene	$C_{10}H_{16}$	V-137	136.24
3973	**Limonene** *(d* or *l)*	p-menthadiene-1,8(9)	$C_{10}H_{16}$	V-133	136.24
3974	**Linalool** *(d)*†	coriandrol	$C_{10}H_{18}O$	I-461	154.25
3975	**Linalyl acetate**	bergamol	$CH_3CO_2 \cdot C_{10}H_{17}$	II-141	196.29
3976	**Linoleic acid**	octadecadienoic acid	$C_{18}H_{32}O_2$	II-496	280.45
3977	**Linolenic acid**	octadecatrienoic acid	$C_{18}H_{30}O_2$	II-499	278.44
3978	**Lophine**	triphenyl-imidazole	$(C_6H_5)_3C \cdot C \cdot N : C \cdot NH$	XXIII-318	296.38
3979	**Lumisterol**	....	$C_{28}H_{43}OH$	....	396.66
3980	**Luteol**	oxychloro-diphenyl-quinoxaline	$C_{19}H_{14}ONCl$	....	307.79
3981	**Lutidine** (2,5)‡	dimethyl-pyridine	$(CH_3)_2C_5H_3N$	XX-244	107.16
3982	**Lutidine** (2,4)(α,γ)	dimethyl-pyridine	$(CH_3)_2C_5H_3N$	XX-244	107.16
3983	**Lutidine** (2,6)(α,α')	dimethyl-pyridine	$(CH_3)_2C_5H_3N$	XX-244	107.16
3984	**Lutidine** (3,4)(β,γ)	dimethyl-pyridine	$(CH_3)_2C_5H_3N$	XX-246	107.16
3985	**Lycopene**	lycopin	$C_{40}H_{56}$	XXX-81	536.89
3986	**Lysine** *(dl)*	α,ϵ-diaminocaproic acid	$NH_2 \cdot (CH_2)_4 \cdot CHNH_2 \cdot$ CO_2H	IV-436	146.19
3987	dihydrochloride	....	$C_6H_{14}O_2N_2 \cdot 2HCl$	IV-437	219.11
3988	**Lysine** *(l+)*	....	$(NH_2)_2C_5H_9 \cdot CO_2H$	IV-435	146.19
3989	hydrochloride *(d)*	....	$C_6H_{14}O_2N_2 \cdot 2HCl$	IV-436	219.11
3990	monohydrochloride	*(d)*	$C_6H_{14}O_2N_2 \cdot HCl$	....	182.65
3991	**Lyxose** $(\alpha)(d)$	....	$CH_2(CHOH)_4 \cdot O$	XXXI-56	150.13
3992	**Lyxose** $(\beta)(d)$	....	$CH_2(CHOH)_4 \cdot O$	XXXI-56	150.13
3993	**Malachite** green§	benzaldehyde green; (zinc salt)	$3C_{23}H_{25}N_2Cl \cdot 2ZnCl_2 \cdot$ $2H_2O$	XIII-745	1403.35
3994	green	(oxalate salt)	$2C_{23}H_{25}N_2 \cdot C_2HO_4 \cdot$ $H_2C_2O_4$	XIII-745	927.03
3995	**Maleic** acid *(cis)*	butendioic acid; toxilic acid	$(:CH \cdot CO_2H)_2$	II-748	116.07
3996	anhydride	....	$(:CH \cdot CO)_2O$	XVII-432	98.06
3997	**Malic** acid *(dl)*	hydroxysuccinic acid	$HO_2C \cdot CHOH \cdot CH_2 \cdot$ CO_2H	III-435	134.09
3998	acid *(d* or *l)*	....	$HO \cdot C_2H_3(CO_2H)_2$	III-417	134.09
3999	Ca acid salt *(l)*	Ca bimalate	$Ca(HC_4H_4O_5)_2 \cdot 6H_2O$	....	414.33
4000	acid (α), *iso-*	methyl tartronic acid	$CH_3 \cdot C(OH)(CO_2H)_2$	III-440	134.09
4001	amide *(l)*	*l*-malamide	$HO \cdot C_2H_3(CO \cdot NH_2)_2$	III-435	132.12
4002	**Malonic** acid	propandioic acid	$CH_2:(CO_2H)_2$	II-566	104.06
4003	Ca salt	calcium malonate	$CaC_3H_2O_4 \cdot 4H_2O$	II-570	214.19
4004	amide	malonamide	$CH_2:(CO \cdot NH_2)_2$	II-582	102.09

† See also No. 2518.
‡ See also No. 3147.
§ In medicinal use without ZnCl₂

Levulose 3616
Lewisite 1473

Light green 3594
Lilacin, cf. glcde.
Linalool tetrahydride 2519
Linaloolene 2515
Linamarin, cf. glcde.
Lindol 6351

Litmopyrin 117
Lobeline, cf. alkd.
Lodal 2413
Loretin 3888
Luminal 5113
Luminol 359

Lupanine, cf. alkd.
Lupinidine, cf. alkd.
Lupinine, cf. alkd.
Lutein 6452
Luteosterone 5334
Lutidinic acid 5498

Table 7-4 (*Continued*)
PHYSICAL CONSTANTS OF ORGANIC COMPOUNDS

No.	Crystalline Form and Color	Specific Gravity	Melting Point °C.	Boiling Point °C.	Solubility in 100 Parts		
					Water	Alcohol	Ether
3965	cr./aq.	....	243 d.	....	6.6 h.	v. sl. s.	v. sl. s.
3966	delq. amor.	....	d. 140-5	....	v. s.	v. sl. s.	i.
3967	lf.	$1.140\frac{20}{20}°$	33.5(18-9)	245-6 sl. d.	v. s.	v. s.	v. s.
3968	col. lq.	$1.018\frac{20}{4}°$	<-21	186-8 sl. d.	∞	∞	s.
3969	col. amor.	....	....	....	s. h.	s. HCl	i.
3970	col. nd.	....	81-2	....	s. bz.	s.; s. ac.	s.; s. CS_2
3971	col. cr./al.	....	147.5	200 sl. d.	i.; s. chl.	s. h.	v. sl. s.
3972	col. lq.	$0.844^{20°}$	....	178	i.	∞	∞
3973	lq.	$0.842\frac{20}{4}°$	-96.9	177	i.	∞	∞
3974	col. oil	$0.868^{20°}$	....	198-200	v. sl. s.	10, 50% al.	∞
3975	col. lq.	$0.895^{20°}$	....	$220^{762}±$, d.	v. sl. s.	∞	∞
3976	lt. yel. oil	$0.903\frac{18}{4}°$	-9.5	$229-30^{16mm}$	i.	∞	∞
3977	oil	$0.914\frac{18}{4}°$	....	$230-2^{17mm}$	i.	s.	v. s.
3978	nd.	....	275	....	i.	2.8 h., abs.	$0.3^{20°}$
3979	col. nd.	....	118	....	i.; s. act.	s.; s. chl.	v. s.
3980	yel. nd.	....	246	....	i.	s.	s.
3981	lq.	$0.938^{0°}$	....	156.5	25 c.; sl. s. h.	∞	∞
3982	lq.	$0.949\frac{0}{4}°$	....	157-9	20 c.; sl. s. h.	s.	s.
3983	lq.	$0.923^{25°}$	-6.6	142-3	∞ c.	sl. s. h. aq.	....
3984	lq.	....	....	163.5-4.5	....	....	......
3985	red pr./pet.	....	174-5	....	i.; s. h. bz.	v. sl. s. h.	0.03 h.
3986	syrup	....	....	....	s.	....	i.
3987	cr.	....	192-3	....	v. s.	....	....
3988	nd./aq.	....	224-5 d.	....	v. s.	i.	....
3989	cr./HCl	....	193	....	....	....	....
3990	wh. pd.	....	235-6	....	s.	i.	i.
3991	cr./al. et.	....	106-7	....	v. s.	$2.5^{17°}$	....
3992	nd./al.	....	117-8	....	v. s.	sl. s.	....
3993	gn. pr./ aq. al.	....	130±	....	v. s.	v. s.	....
3994	gn. pr.	....	....	....	v. s. h.	s.	....
3995	mn.	1.609	130.5	138±; d., anhyd.	$79^{25°}$; $393^{98°}$	$70^{30°}$	$8^{25°}$
3996	cr./chl.	1.5	52.8	202; subl.	$16.3^{30°}$	v. sl. s. CCl_4	....
3997	col. cr.	$1.601\frac{20}{4}°$	128-9	150 d.	$144^{26°}$; $411^{79°}$	v. s.	v. s.
3998	col. cr.	$1.595\frac{20}{4}°$	99-100	140±, d.	v. s.	v. s.	$8.4^{15°}$
3999	col. rhb.	....	....	....	sl. s. c.	sl. s. ac.	s. a.
4000	col. cr.	....	142 d.	d. 170 ±	v. s.	v. s.	v. s.
4001	pr./aq.	....	156-8 d.	....	s.	....	....
4002	col. tri.	$1.631^{15°}$	130-5 d.	....	$138^{16°}$	$42^{25°}$	$8^{15°}$ abs.
4003	col. nd.	....	$-4H_2O,180$	$-3H_2O, 100$	$0.4^{0°}$	$0.7^{100°}$ aq.	....
4004	tet. or mn.	....	170	....	$8.3^{8°}$	i. abs.	i.

Table 7-4 (*Continued*)
PHYSICAL CONSTANTS OF ORGANIC COMPOUNDS

No.	Name	Synonym	Formula	Beil. Ref.	Formula Weight
4005	**Malonic** anilide	malonanilide	$CH_2(CONH \cdot C_6H_5)_2$	XII-293	254.29
4006	nitrile	methylene dicyanide	$CH_2(CN)_2$	II-589	66.06
4007	**Maltose**	malt sugar	$C_{12}H_{22}O_{11} \cdot H_2O$	XXXI-386	360.32
4008	**Mandelic** acid (*dl*)	....	$C_6H_5 \cdot CHOH \cdot CO_2H$	X-197	152.15
4009	sodium salt	sodium mandelate	$C_6H_5 \cdot CHOH \cdot CO_2Na$	....	174.13
4010	nitrile (*dl*)	....	$C_6H_5 \cdot CHOH \cdot CN$	X-206	133.15
4011	**Mannitol hexani-** trate	....	$C_6H_8(O \cdot NO_2)_6$	I-543	452.17
4012	**Mannoheptose** (*d*)	....	$C_7H_{14}O_7$	I-935	210.19
4013	**Mapharsen**	3-NH₂-4-OH- phenyl-arsinoxide HCl	$O:As \cdot C_6H_3(OH)NH_2 \cdot HCl \cdot \frac{1}{2}C_2H_5OH$	*XVI-447	258.54
4014	**Margaric** acid	heptadecanoic acid	$CH_3 \cdot (CH_2)_{15} \cdot CO_2H$	II-376	270.46
4015	aldehyde	heptadecanal	$CH_3 \cdot (CH_2)_{15} \cdot CHO$	I-717	254.46
4016	**Meconic** acid	....	$C_7H_4O_7 \cdot 3H_2O$	XVIII-503	254.15
4017	lactone	6,7-diMeO-phthal-ide; meconine; opianyl	$CO \cdot O \cdot CH_2 \cdot C_6H_2:$ $\mid_____\mid$ $(OCH_3)_2$	XVIII-89	194.19
4018	**Melam**	....	$C_6H_9N_{11}$	III-169	235.21
4019	**Melamine**	cyanuramide	$C_3N_3(NH_2)_3$	XXVI-245	126.12
4020	hydrochloride	....	$C_3H_6N_6 \cdot HCl \cdot \frac{1}{2}H_2O$	XXVI-246	171.59
4021	**Melene**	....	$C_{30}H_{60}$ (or $C_{30}H_{62}$)?	I-227	420.81 422.83
4022	**Melissic acid**	....	$C_{30}H_{61} \cdot CO_2H$	II-396	466.84
4023	**Menthene** (*d*)	....	$C_{10}H_{18}$	V-87	138.25
4024	**Menthol** (*α*)(*l*)	hexahydrothymol	$C_{10}H_{19}OH$	VI-28	156.27
4025	**Menthol** (*d*)(*neo*)	*p*-menthanol-3	$C_{10}H_{19}OH$	VI-28	156.27
4026	**Menthone** (*l*)	....	$C_{10}H_{18}O$	VII-38	154.25
4027	**Menthyl** acetate (*l*)	....	$CH_3 \cdot CO_2 \cdot C_{10}H_{19}$	VI-32	198.31
4028	acetate (*d*)(*neo*)	....	$CH_3 \cdot CO_2 \cdot C_{10}H_{19}$	....	198.31
4029	amine (*l*)	aminomenthane	$C_{10}H_{19} \cdot NH_2$	XII-26	155.29
4030	benzoate (*α*)	....	$C_6H_5 \cdot CO_2 \cdot C_{10}H_{19}$	IX-115	260.38
4031	salicylate	....	$HO \cdot C_6H_4 \cdot CO_2 \cdot C_{10}H_{19}$	X-76	276.38
4032	*iso*-valerate (*l*)	Validol	$C_4H_9 \cdot CO_2 \cdot C_{10}H_{19}$	VI-33	240.39
4033	**Mercaptobenzo-** thiazole (2)	....	$C_6H_4 \cdot N:C(SH) \cdot S$ $\mid_____\mid$	XXVII-185	167.25
4034	**Mercuri-**hydroxy-phenyl chloride (*o*)	*o*-Cl-mercuriphenol	$HO \cdot C_6H_4 \cdot HgCl$	XVI-959	329.15
4035	hydroxyphenyl chloride (*p*)	*p*-Cl-mercuriphenol	$HO \cdot C_6H_4 \cdot HgCl$	XVI-961	329.15
4036	phenyl acetate	Ph-mercuric acetate	$C_6H_5 \cdot Hg \cdot O_2CCH_3$	XVI-954	336.74
4037	phenyl chloride	Ph-mercuric chloride	$C_6H_5 \cdot HgCl$	XVI-953	313.15
4038	phenyl nitrate	Ph-mercuric nitrate	$C_6H_5 \cdot Hg \cdot O \cdot NO_2$	XVI-953	339.70
4039	tolyl chloride (*p*)	....	$CH_3 \cdot C_6H_4 \cdot HgCl$	XVI-956	327.18
4040	**Mercuric fulminate**	....	$Hg(ONC)_2 \cdot \frac{1}{2}H_2O$	....	293.63
4041	**Mercurochrome**	diNa-diBr-hydroxy-mercury-fluorescein	$C_{20}H_7O_5Br_2Na_2HgOH \cdot 3H_2O$	....	804.72
4042	**Mercury** dibutyl	di-*n*-butyl mercury	$(C_4H_9)_2Hg$	**IV-1049	314.82
4043	diethyl	diethyl mercury	$(C_2H_5)_2Hg$	IV-679	258.71
4044	dimethyl	dimethyl mercury	$(CH_3)_2Hg$	IV-678	230.66

Malonyl thiourea 5887
Malonyl urea 621
Malt sugar 4007
Mandelonitrile glucoside, cf. glcde.
Mannite 3609
Mannitol 3609
Mannitol hexaacetate 3610

Mannoheptitol 5090
Mannose 3619
Marsh gas 4061
Martius yellow 2642
Matecite 4286
Matezite 4286
Mecholin 100

Mecholyl 100
Mecholyl bromide 99
Mecholyl iodide 101
Meconine 4017
Medinal 2125
Melaniline 2723
Meletin 5807

Table 7-4 (*Continued*)
PHYSICAL CONSTANTS OF ORGANIC COMPOUNDS

No.	Crystalline Form and Color	Specific Gravity	Melting Point °C.	Boiling Point °C.	Solubility in 100 Parts		
					Water	Alcohol	Ether
4005	nd./al.	....	229-31	....	i.; s. ac.	v. s. h.	i.
4006	col. cr.	$1.049^{34°}_4$	31.7	223-4	13	40	20
4007	col. nd.	$1.540^{17°}$	d.	....	v. s.	v. sl. s. c.	i.
4008	rhb./aq.	$1.300^{20°}_4$	118.1	d.	$16^{20°}$	s.	s.
4009	wh. cr. pd.	....	....	....	99	2	....
4010	oil	1.124	−10	d. 170	i.	s.	s.; s. chl.
4011	nd.	$1.604^{0°}$	112-3	expl.	i.	$2.9^{13°}$	$4^{9°}$
4012	nd.	....	134-5	....	v. s.	sl. s. abs.	....
4013	wh. pd.	....	....	....	v. s.; s. alk.	v. s.; v. sl. s. act.	v. sl. s.
4014	col. pl.	$0.853^{60°}$	60.9	227^{100mm}	i.	$32^{28°}$ abs.	v. s.
4015	cr.	....	36	$203-4^{26mm}$	sl. s. c. al.	v. s. h.	v. s.
4016	rhb.	....	$-3H_2O$, 100	d.	d. h.	s.; d. h. a.	sl. s.
4017	col. nd./aq.	....	102-2.5	subl.	$0.14^{15°}$; $4.5^{100°}$	s.; s. chl.	s.; s. bz.
4018	pd.	....	d.	....	i.	sl. s. a.	s. h. KOH
4019	mn.	$1.573^{250°}$	360 d.	subl.	s. h.	sl. s. h.	i.
4020	nd.	....	....	....	....	i.	....
4021	cr.	$0.913^{25°}$	62-3	$218^{0.5mm}$	i.; v. sl. s. bz.	3.6 h., abs.	v. sl. s.
4022	nd./al.	....	90-1	....	i.	v. s. h.	v. sl. s.
4023	lq.	$0.814^{18°}_4$	....	168-9	i.	v. s. h.	....
4024	col. cr.	$0.890^{15°}_{15}$	42.5*	216.3	0.04 c.; s. pet.	v. s.; v. s. chl.	v. s.; v. s. ac.
4025	col. lq.	0.9±	−22	98^{16mm}	....	....	....
4026	oil	$0.896^{20°}_{20}$	−6.6	207	sl. s.	∞	∞
4027	col. lq.	$0.919^{20°}_4$	....	227	sl. s.	∞	∞
4028	col. cr.	....	37-8	....	i.	s.	v. s.
4029	oil	$0.861^{20°}_4$	....	209 ±	s. c.	v. s.	....
4030	rhb./al.	†$1.002^{20°}_4$	54-5	301^{748mm}	....	....	....
4031	col. oil	$1.045^{25°}_{25}$	....	$203-5^{20mm}$	i.	s.	s.
4032	col. lq.	$0.907^{15°}_4$	....	129^{9mm}	i.	s.	s.; s. chl.
4033	col. nd./ aq. al.	$1.42^{20°}_4$	179	d.	i.; s. alk.; s. ac.	s.; s. alk. carb.	sl. s.
4034	col. cr./ aq. al.	....	152.5	....	sl. s.	s.; v. sl. s. chl.	s. h. bz.
4035	lf./act.	....	225-6	....	....	sl. s. h.	....
4036	col. pr./bz.	....	149	....	s. h.	s.; s. ac.	s. bz.
4037	col. lf./bz.	....	251	subl.	i.; s. pyr.	sl. s. h.	s.; s. bz.
4038	pl./al.	....	188-9 d.	....	v. sl. s. h.	s. h.	s. bz.
4039	col. lf./bz.	....	232-3	....	i.	v. sl. s. h.	i.
4040	cr./aq.	4.42(anh.)	expl.	....	$0.07^{12°}$; $0.17^{49°}$	s.; d. h. aq. KOH	s. NH$_4$OH
4041	gn. scales	....	....	....	s.	0.02	i.; i. chl.
4042	col. lq.	$1.778^{20°}_4$	....	$120-3^{23mm}$	i.	sl. s.	v. s.
4043	col. lq.	$2.423^{23°}_4$	....	159	i.	sl. s.	v. s.
4044	col. lq.	$2.954^{22°}_4$	....	95-6	v. sl. s.	v. s.	v. s.

* M. p. β 35.5°; γ 33.5°; δ 31.5°.
† Supercooled liquid.
Melilotic acid 3711
Melin, cf. glcde.
Melissyl alcohol 4449
Melitose 5549

Mellitic acid 667
Mellophanic acid 681
Menthadiene 3973, 5092-3, 5704-5, 5709
Menthane 3579
Menthanol 4025

Menthene 1258
Mercapto-, cf. also thio- and sulfo-.
Mercapto-ethanol 5904
Mercapto-succinic acid 5911

Table 7-4 (*Continued*)
PHYSICAL CONSTANTS OF ORGANIC COMPOUNDS

No.	Name	Synonym	Formula	Beil. Ref.	Formula Weight
4045	**Mercury** dinaphthyl (α)	. . . .	$(C_{10}H_7)_2Hg$	XVI-949	454.92
4046	dipropyl	dipropyl mercury	$(C_2H_5 \cdot CH_2)_2Hg$	IV-679	286.77
4047	diphenyl	diphenyl mercury	$(C_6H_5)_2Hg$	XVI-946	354.80
4048	ditolyl (*o*)	ditolyl mercury	$(CH_3 \cdot C_6H_4)_2Hg$	XVI-947	382.86
4049	ditolyl (*m*)	ditolyl mercury	$(CH_3 \cdot C_6H_4)_2Hg$	XVI-947	382.86
4050	ditolyl (*p*)	ditolyl mercury	$(CH_3 \cdot C_6H_4)_2Hg$	XVI-947	382.86
4051	mercaptide	. . . .	$(C_2H_5 \cdot S)_2Hg$	I-342	322.84
4052	**Merthiolate**	Na ethylmercuri-thiosalicylate	$C_2H_5 \cdot Hg \cdot S \cdot C_6H_4 \cdot CO_2Na$	. . . .	404.81
4053	**Mesityl oxide**	. . . .	$(CH_3)_2C{:}CH \cdot CO \cdot CH_3$	I-736	98.15
4054	**Mesorcin** (1,3,5;2,4)	. . . .	$(CH_3)_3C_6H(OH)_2$	VI-939	152.19
4055	**Mesoxalic acid**	dihydroxy malonic	$(HO)_2C(CO_2H)_2$	III-766	136.06
4056	**Metanil yellow**	Na *m*-sulfonate-azo-diphenylamine	$C_6H_5 \cdot NH \cdot C_6H_4 \cdot N{:}N \cdot C_6H_4 \cdot SO_3Na$	XVI-330	375.38
4057	**Metaphen**®	4-NO_2-3-OH-mercuri-*o*-cresol	$Hg \cdot C_7H_5O_3N$	. . . .	351.71
4058	**Methacrolein**	α-Me-acrolein	$CH_2{:}C(CH_3) \cdot CHO$	I-731	70.09
4059	**Methallyl** alcohol	isopropenyl carbinol	$CH_2{:}C(CH_3) \cdot CH_2OH$	I-443	72.11
4060	chloride (β)	γ-Cl-isobutylene	$CH_2{:}C(CH_3) \cdot CH_2Cl$	I-209	90.55
4061	**Methane**	marsh gas	CH_4	I-56	16.04
4062	**Methionine** (*l*–)	α-NH_2-γ-methyl-thiolbutyric acid	$CH_3 \cdot S \cdot (CH_2)_2 \cdot CHNH_2 \cdot CO_2H$	**IV-938	149.21
4063	**Methoxy**-acetic acid	glycolic methyl ether	$CH_3O \cdot CH_2 \cdot CO_2H$	III-232	90.08
4064	acetophenetidine	kryofine	$C_2H_5O \cdot C_6H_4 \cdot NH \cdot CO \cdot CH_2 \cdot OCH_3$	XIII-489	209.25
4065	acetophenone (*p*)	*p*-acetyl-anisole	$CH_3O \cdot C_6H_4 \cdot COCH_3$	VIII-87	150.18
4066	*tert*-amylbenzene	(*p*)	$CH_3O \cdot C_6H_4 \cdot C_5H_{11}$	VI-549	178.28
4067	benzoic acid (*o*)	salicylic methyl ether	$CH_3O \cdot C_6H_4 \cdot CO_2H$	X-64	152.15
4068	benzoic acid (*m*)	. . . .	$CH_3O \cdot C_6H_4 \cdot CO_2H$	X-137	152.15
4069	benzoic acid (*p*)	*p*-anisic acid	$CH_3O \cdot C_6H_4 \cdot CO_2H$	X-154	152.15
4070	benzophenone (*p*)	*p*-benzoyl anisole	$CH_3O \cdot C_6H_4COC_6H_5$	VIII-159	212.25
4071	benzoyl chloride (*p*)	anisoyl chloride	$CH_3O \cdot C_6H_4 \cdot COCl$	X-163	170.60
4072	benzyl alcohol (*o*)	saligenin-2-Me-ether	$CH_3O \cdot C_6H_4 \cdot CH_2OH$	VI-893	138.17
4073	diphenyl (*o*)	Ph-phenol-Me-ether	$CH_3O \cdot C_6H_4 \cdot C_6H_5$	VI-672	184.24
4074	diphenyl (*p*)	Me Ph-phenol ether	$CH_3O \cdot C_6H_4 \cdot C_6H_5$	VI-674	184.24
4075	ethyl chloroformate (β)	β-methoxyethyl chlorocarbonate	$Cl \cdot CO_2 \cdot CH_2 \cdot CH_2 \cdot OCH_3$	. . . .	138.55
4076	ethyl formate (β)	. . . .	$HCO_2 \cdot CH_2 \cdot CH_2 \cdot OCH_3$	*II-19	104.11
4077	methylal	. . . .	$(CH_3O \cdot CH_2)_2O$	I-576	106.12
4078	methyl salicylate	mesotan; salmester	$HO \cdot C_6H_4 \cdot CO_2CH_2 \cdot OCH_3$	X-83	182.18
4079	phenol (*o*)	guaiacol	$CH_3O \cdot C_6H_4 \cdot OH$	VI-768	124.14
4080	phenol (*m*)	resorcinol Me ether	$CH_3O \cdot C_6H_4 \cdot OH$	VI-813	124.14
4081	phenol (*p*)	hydroquinone methyl ether	$CH_3O \cdot C_6H_4 \cdot OH$	VI-843	124.14
4082	pyridine (*p*)(γ)	. . . .	$CH_3O \cdot C_5H_4N$	XXI-49	109.13
4083	quinoline (*p*)	quinanisole	$CH_3O \cdot C_9H_6N$	XXI-85	159.19
4084	**Methyl** abietate	Abalyn	$C_{19}H_{29}CO_2 \cdot CH_3$	. . . .	316.49
4085	acetamide (*N*)	acet-methylamide	$CH_3 \cdot CO \cdot NH \cdot CH_3$	IV-58	73.10
4086	acetanilide (*N*)	exalgin	$CH_3CO \cdot N(CH_3)C_6H_5$	XII-245	149.19
4087	acetate	. . . .	$CH_3CO_2 \cdot CH_3$	II-124	74.08
4088	acetoacetate	. . . .	$CH_3CO \cdot CH_2 \cdot CO_2 \cdot CH_3$	III-632	116.12

Meritol, mixt. of 2312 and 5260.
Mersalyl 5591
Mesaconic acid 4243
Mescaline, cf. alkd.
Mesidine 6251

Mesitol 6266
Mesitylene 6255
Mesitylenic acid 2463
Mesotan 4078
Mesoxalylurea 184-6

Metacresol purple **1551**
Metacrolein 147
Metaldehyde 6
Metanilic acid 524
Methacetin 20

Table 7-4 (*Continued*)
PHYSICAL CONSTANTS OF ORGANIC COMPOUNDS

No.	Crystalline Form and Color	Specific Gravity	Melting Point °C.	Boiling Point °C.	Solubility in 100 Parts		
					Water	Alcohol	Ether
4045	cr./bz.	1.929	243	d.	i.; s. h. chl.	sl. s. h.	s. CS_2
4046	col. lq.	$2.124^{16°}$	....	189-91	i.	sl. s.	v. s.
4047	nd./bz.	$2.318^{4°}$	124-5	>300 d.	i.; s. chl.	sl. s. h.	sl. s.
4048	tri./bz.	....	141(107)	219^{14mm}	....	....	....
4049	nd.	....	102(89)	....	sl. s. chl.	sl. s.	sl. s.
4050	nd./bz.	....	244-6	....	s. h. bz.	sl. s. h.	s. CS_2
4051	lf./al.	....	76-7	....	....	7 h.	....
4052	pd.	....	....	....	100	14	i.; i. bz.
4053	lq.	$0.858\frac{20}{4}°$	−59	$129-30^{750mm}$	$32^{20°}$	∞	∞
4054	lf.	....	149-50	274.5-5.5	v. sl. s.	s.	s.
4055	hyg. cr.	....	119-20 sl. d.	....	v. s.	s.	s.
4056	brn. yel. pd.	....	....	....	s.	....	....
4057	yel. pd.	....	....	....	i.; s. h. ac.	i.; s. dil. NaOH	i.; s. NH_4OH
4058	lq.	$0.837\frac{20}{4}°$	−81	68.4	6.4	....	....
4059	lq.	$0.852\frac{20}{4}°$	....	114.5	25	....	....
4060	col. lq.	$0.926\frac{20}{4}°$	....	72-3	<0.1	....	....
4061	col. gas	$(A)0.554\frac{[0°}{760}$	−182.5	−161.5	$3.3^{20°}$ cc.	$47.1^{20°}$ cc.	$104^{10°}$ cc.
4062	hex. pl./ 65% al.	....	283 d.	....	s. c.	s.; i. abs.	i.; i. bz.; i. act.
4063	hyg. lq.	$1.177\frac{20}{4}°$	8	203-4	∞	∞	∞
4064	col. nd.	....	98-9	....	0.17 c.; 2 h.	s.	s.
4065	pl./et.	$1.082\frac{41}{4}°$	38-9	258	v. sl. s.	v. s.	v. s.
4066	lq.	....	....	$113-4^{13mm}$	....	....	....
4067	pl./aq.	1.180	98.5-9.0	200	$0.5^{30°}$	v. s.	v. s.
4068	nd./aq.	....	107-9	$170-2^{10mm}$	s. h.	s.	s.
4069	mn./aq.	$1.385^{4°}$	184.2	275-80	$0.03^{19°}$	v. s.	v. s.
4070	pr./et.	....	61-2	$354-5^{729mm}$	....	v. s.	v. s.
4071	col. nd.	....	22-3(18)	263 sl. d.	i. d.	s. d.	s. bz.
4072	col. lq.	$1.043\frac{25}{25}°$	....	248-50	v. sl. s.	s.	∞
4073	pr./pet.	....	29-30	274	....	....	....
4074	lf./al.	....	90	....	....	s. h.	....
4075	col. lq.	....	....	$54-60^{13mm}$	....	....	....
4076	col. lq.	$1.048\frac{15}{4}°$	....	131-1.5	....	....	....
4077	col. lq.	$0.959\frac{20}{20}°$	....	106-8	....	....	....
4078	col. oil	$1.2^{15°}$	....	162^{42mm}	v. sl. s.	∞; ∞ bz.	∞; ∞ chl.
4079	pr.	$1.140\frac{15}{15}°$	28.3	205	$1.7^{15°}$	v. s.	v. s.
4080	lq.	>1	<−17.5	243.3-4.3	sl. s.	∞; s. alk.	∞
4081	lf./aq.	....	53	243	....	....	v. s. bz.
4082	lq.	....	....	$190-1^{738mm}$	∞	....	....
4083	lq.	$1.154^{20°}$	26-8	254^{310mm}	....	s.	....
4084	pa. yel. lq.	$1.040\frac{20}{20}°$	....	360-5 d.	i.	∞	∞
4085	col. nd.	....	28	206	v. s.	v. s.	v. s.; i. lg.
4086	rhb./al.	$1.004\frac{105}{4}°$	102-4	253^{712mm}	1.7 c.	2	14
4087	col. lq.	$0.933\frac{20}{4}°$	−98.7	57.3	$33^{22°}$	∞	∞
4088	col. lq.	$1.077\frac{20}{4}°$	−80	169-70 sl. d.	38	∞	∞

Table 7-4 (*Continued*)
PHYSICAL CONSTANTS OF ORGANIC COMPOUNDS

No.	Name	Synonym	Formula	Beil. Ref.	Formula Weight
4089	**Methyl** acetophenone (*p*)	methyl-*p*-tolyl ketone	$CH_3 \cdot C_6H_4 \cdot CO \cdot CH_3$	VII-307	134.18
4090	acetylsalicylate	methyl aspirin; methyl Rhodine	$CH_3CO_2 \cdot C_6H_4 \cdot CO_2 \cdot CH_3$	X-73	194.19
4091	acetylurea	....	$CH_3 \cdot NH \cdot CO \cdot NH \cdot CO \cdot CH_3$	IV-66	116.12
4092	acridine (4)	....	$CH_3 \cdot C_{13}H_8N$	XX-470	193.25
4093	acridine (2)	....	$CH_3 \cdot C_{13}H_8N$	XX-470	193.25
4094	acridine (9)	....	$CH_3 \cdot C_{13}H_8N$	XX-470	193.25
4095	acrylate	....	$CH_2 : CH \cdot CO_2 \cdot CH_3$	II-399	86.09
4096	acrylic acid (α)	....	$CH_2 : C(CH_3) \cdot CO_2H$	II-421	86.09
4097	hydrogen adipate	Me adipate (mono)	$CH_3O_2C(CH_2)_4 \cdot CO_2H$	II-652	160.17
4098	adipic acid (α)	....	$HO_2C \cdot CH(CH_3) \cdot (CH_2)_3 \cdot CO_2H$	II-672	160.17
4099	adipic acid (β)(*d*)	....	$HO_2C \cdot (CH_2)_2 \cdot CH \cdot (CH_3) \cdot CH_2 \cdot CO_2H$	II-673	160.17
4100	alcohol	wood alcohol	$CH_3 \cdot OH$	I-273	32.04
4101	alizarin (2;3,4)	3-methyl-alizarin	$CH_3(OH)_2C_6H(CO)_2 : C_6H_4$	VIII-469	254.24
4102	allylamine	....	$CH_3 \cdot NH \cdot C_3H_5$	IV-206	71.12
4103	allyl carbinol	amylene alcohol	$(CH_3)(C_3H_5) : CHOH$	I-443	86.13
4104	allyl ether	....	$CH_3 \cdot O \cdot CH_2 \cdot CH : CH_2$	I-437	72.11
4105	amine	amino-methane	$CH_3 \cdot NH_2$	IV-32	31.06
4106	amine hydrochloride	....	$CH_3NH_2 \cdot HCl$	IV-36	67.52
4107	aminoacetate	methyl glycinate	$NH_2 \cdot CH_2 \cdot CO_2 \cdot CH_3$	IV-340	89.09
4108	aminobenzoate (*o*)	methyl anthranilate	$NH_2 \cdot C_6H_4 \cdot CO_2 \cdot CH_3$	XIV-317	151.17
4109	aminobenzoate (*p*)	....	$NH_2 \cdot C_6H_4 \cdot CO_2 \cdot CH_3$	XIV-422	151.17
4110	aminobenzoic acid (*o*)	*N*-methyl anthranilic acid	$CH_3NH \cdot C_6H_4 \cdot CO_2H$	XIV-323	151.17
4111	amino ethanol (β)	....	$CH_3NH \cdot CH_2 \cdot CH_2OH$	IV-276	75.11
4112	3-amino-4-hydroxy benzoate	orthoform new	$NH_2(OH)C_6H_3 \cdot CO_2 \cdot CH_3$	XIV-593	167.17
4113	amino-*p*-hydroxybenzoic acid (3)	....	$CH_3NH(OH)C_6H_3 \cdot CO_2H$	XIV-593	167.17
4114	amino-2-methylheptene (6)	Octin	$C_8H_{15} \cdot NHCH_3$	....	141.26
4115	amino-phenol (*o*)	....	$CH_3NH \cdot C_6H_4 \cdot OH$	XIII-362	123.16
4116	amino-phenol H_2SO_4	(*o*)	$(C_7H_9ON)_2 \cdot H_2SO_4$	XIII-362	344.39
4117	amino-phenol H_2SO_4	(*p*); Metol; Elon	$(C_7H_9ON)_2 \cdot H_2SO_4$	XIII-442	344.39
4118	amino-propionic acid (α)(*dl*)	....	$CH_3 \cdot CH(NHCH_3) \cdot CO_2H$	IV-391	103.12
4119	amylbenzene (*m*)	amyl toluene	$CH_3 \cdot C_6H_4 \cdot C_5H_{11}$	V-446	162.28
4120	*n*-amyl ether	....	$CH_3 \cdot O \cdot C_5H_{11}$	*I-193	102.18
4121	*iso*-amyl ether	....	$CH_3 \cdot O \cdot C_5H_{11}$	I-400	102.18
4122	*n*-amyl ketone	heptanone-2	$CH_3(CH_2)_4 \cdot CO \cdot CH_3$	I-699	114.19
4123	*iso*-amyl ketone	2-Me-hexanone-5	$CH_3 \cdot CO \cdot C_5H_{11}$	I-701	114.19
4124	*iso*-amyl ketoxime	....	$CH_3 \cdot C(:NOH) \cdot C_5H_{11}$	I-701	129.20
4125	aniline†	Me-Ph-amine	$C_6H_5 \cdot NH \cdot CH_3$	XII-135	107.16

† See also Nos. 5995-6000.
Methoxy-benzaldehyde 531-2
Methoxy-benzene 537
Methoxy-benzyl alcohol 530
Methoxy-cinchonine, cf. alkd.
Methoxy-cinchoninic acid 5537
Methoxy-coniferin, cf. glcde.
Methoxy-hydroxycinnamic acid 3236-7

Methoxy-styrene 6431-3
Methoxy-styryl ketone 528
Methoxy-tetrahydroquinoline 5874
Methoxylamine 4280
Methyl, cf. also dimethyl.
Methyl-acetanilide 26-8
Methyl-acetoacetic ester 3085
Methyl-acetopyrandione 1656

Table 7-4 (*Continued*)
PHYSICAL CONSTANTS OF ORGANIC COMPOUNDS

No.	Crystalline Form and Color	Specific Gravity	Melting Point °C.	Boiling Point °C.	Solubility in 100 Parts		
					Water	Alcohol	Ether
4089	nd.	$0.9892^{22°}$	−23	228^{759mm}	i.	v. s.	v. s.
4090	col. cr./al.	....	50-1	$134-6^{9mm}$	i.	s.; s. gly.	s.; s. chl.
4091	mn./aq.	....	180	....	v. s. h.	s. h.	v. sl. s. h.
4092	nd./al.	....	88		v. sl. s. h.	s.	....
4093	yel. nd./ aq. al.	....	134	....	s. bz.	s.	s.
4094	pl./lg.	....	114-7	360^{740mm}	....	....	....
4095	col. lq.	$0.974^{0°}$	....	80.3*	....	s.	s.
4096	pr.	$1.015\frac{20°}{4}$	15-16	161-3	s. h.	∞	∞
4097	lq.	....	8-9	162^{10mm}	....	....	....
4098	col. cr.	....	64	$216-20^{28mm}$	v. s.	v. s.	v. s.
4099	col. cr.	....	93-4.5	230^{30mm}	v. s.	v. s.	v. s.
4100	col. lq.	$†0.792\frac{20°}{4}$	−97.8	64.7	∞	∞	∞
4101	yel. nd.	....	250-2	....	s. alk.	s.	s.
4102	lq.	....	....	64-6	s.	....	....
4103	lq.	$0.834\frac{20°}{0}$	....	$115-6^{750mm}$	12.5	∞	∞
4104	lq.	$0.771^{11°}$	....	$42-3^{757mm}$	v. sl. s.	∞	∞
4105	col. gas	$0.699^{-11°}$	−92.5	-6.7^{758mm}	$959^{25°}$ cc./cc.	s.	....
4106	pl./al.	1.23	226-8	$225-30^{15mm}$	v. s.	23 h. abs.	i.; i. chl.
4107	lq.	....	....	130 d.	....	....	....
4108	col. lq.	$1.168\frac{18.6°}{4}$	24(8.2)	135.5^{15mm}	sl. s.	s.	s.
4109	lf./al.	....	112	....			
4110	lf./al.	....	179-82	subl. d.	0.2 c.; 0.4 h.	s.	s.
4111	col. oil	$0.937^{20°}$	....	159^{747mm}	∞	∞	∞
4112	nd./bz. cr./chl.	....	142‡ 110-1	....	v. sl. s.; i. pet.	18; s. act.	2; s. alk ; s. bz.
4113	cr.	....	142(110)	....	s. h.	s.	s.
4114	col. oil	0.795	....	176-8	i.	s.	s.; s. a.
4115	pl./bz. pet.	....	86-7	....	i.	s.	s. bz.
4116	ool. or.	....	....	....	s.	i.	i.
4117	col. nd./aq.	....	250-60 d.	....	$4^{25°}$	sl. s.	i.
4118	rhb./al.	....	307-17 d.	subl. 292 sl. d.	s.	10 h.; i. c. abs.	....
4119	lq.	$0.868^{22°}$	....	207-8	....	....	....
4120	col. lq.	0.75	....	99-100	....	....	....
4121	lq.	$0.687\frac{91°}{4}$	....	91^{765mm}			
4122	col. lq.	$0.820\frac{15°}{4}$	−35.5	151.5	$0.4^{20°}$	s.	s.
4123	col. lq.	$0.813^{20°}$	−73.9	144	v. sl. s.	∞	∞
4124	oil	$0.888\frac{20°}{4}$	....	$195-6^{761mm}$	....	....	....
4125	lq.	$0.986\frac{20°}{4}$	−57	196.3	$0.01^{25°}$	s.	∞

* Polymerizes readily.
† See also special table of specific gravities.
‡ HCl salt, m. p. 225°; s. 10/100 aq.
Methyl acet-toluidide 102-4
Methyl-acetyl carbinol 43
Methyl-acetylene 225
Methyl-acrolein 4058
Methyl adipate 2428, 4097

Methyl-allene 1018
Methyl-amyl alcohol 3630-7
Methyl-amyl carbinol (*n*) 3542
Methyl-amyl carbinol (*iso*) 3538
Methyl-aniline violet 5076
Methyl anisate 4301

Table 7-4 (*Continued*)
PHYSICAL CONSTANTS OF ORGANIC COMPOUNDS

No.	Name	Synonym	Formula	Beil. Ref.	Formula Weight
4126	**Methyl** *p*-anisyl ketone	*p*-acetonyl-anisole	$CH_3O \cdot C_6H_4 \cdot CH_2 \cdot CO \cdot CH_3$	VIII-106	164.21
4127	anthracene (α)	....	$CH_3 \cdot C_{14}H_9$	V-674	192.26
4128	anthracene (β)	....	$CH_3 \cdot C_{14}H_9$	V-674	192.26
4129	anthraquinone (2)	....	$CH_3 \cdot C_6H_3(CO)_2C_6H_4$	VII-809	222.25
4130	arsenic acid	methyl arsonic acid	$CH_3 \cdot AsO(OH)_2$	IV-613	139.97
4131	arsenious oxide	....	$CH_3 \cdot As{:}O$	IV-610	105.96
4132	arsine	....	$CH_3 \cdot AsH_2$	IV-599	91.97
4133	auramine	....	$[(CH_3)_2N \cdot C_6H_4]_2C{:}N \cdot CH_3$	XIV-93	281.40
4134	azide	....	$CH_3 \cdot N(N)_2$	I-80	57.06
4135	benzamide (*N*)	....	$C_6H_5 \cdot CO \cdot NH \cdot CH_3$	IX-201	135.17
4136	benzimidazole (2)	....	$C_6H_4 \cdot N{:}C(CH_3)NH$	XXIII-145	132.17
4137	benzimidazole (5)	*m*-toliminazole	$CH_3 \cdot C_6H_3 \cdot N{:}CHNH$	XXIII-151	132.17
4138	benzoate	oil of niobe	$C_6H_5 \cdot CO_2 \cdot CH_3$	IX-109	136.15
4139	benzoylacetate	....	$C_6H_5CO \cdot CH_2 \cdot CO_2CH_3$	X-673	178.19
4140	*o*-benzoylbenzoate	....	$C_6H_5CO \cdot C_6H_4 \cdot CO_2 \cdot CH_3$	X-748	240.26
4141	benzoylene urea (1)	....	$C_6H_4N(CH_3)CONHCO$	XXIV-375	176.18
4142	benzoylene urea (3)	....	$C_6H_4NHCON(CH_3)CO$	XXIV-375	176.18
4143	benzoylsalicylate	Benzosalin	$C_6H_5CO_2 \cdot C_6H_4 \cdot CO_2 \cdot CH_3$	X-73	256.26
4144	benzylaniline (*N*)	....	$C_6H_5 \cdot CH_2 \cdot N(CH_3) \cdot C_6H_5$	XII-1024	197.28
4145	benzyl ether	....	$CH_3 \cdot O \cdot CH_2 \cdot C_6H_5$	VI-431	122.17
4146	benzyl ketone	phenyl acetone	$CH_3 \cdot CO \cdot CH_2 \cdot C_6H_5$	VII-303	134.18
4147	bromide	bromomethane	$CH_3 \cdot Br$	I-67	94.94
4148	bromoacetate	....	$Br \cdot CH_2 \cdot CO_2 \cdot CH_3$	II-213	152.98
4149	*o*-bromobenzoate	....	$Br \cdot C_6H_4 \cdot CO_2 \cdot CH_3$	IX-348	215.05
4150	*m*-bromobenzoate	....	$Br \cdot C_6H_4 \cdot CO_2 \cdot CH_3$	IX-350	215.05
4151	*p*-bromobenzoate	....	$Br \cdot C_6H_4 \cdot CO_2 \cdot CH_3$	IX-352	215.05
4152	4-bromohydrindene (7)	....	$CH_3(Br)C_6H_2(CH_2)_3$	....	211.11
4153	α-bromolaurate	....	$C_{10}H_{21}CHBrCO_2CH_3$	....	293.25
4154	butadiene-1,3 (2)	isoprene	$CH_2{:}CH \cdot C(CH_3){:}CH_2$	I-252	68.12
4155	butadiene-2,3 (2)	dimethyl-allene (*as*)	$(CH_3)_2C{:}C{:}CH_2$	I-252	68.12
4156	butylamine (*n*)	....	$CH_3 \cdot NH \cdot C_4H_9$	IV-157	87.17
4157	*iso*-butylamine	....	$CH_3 \cdot NH \cdot C_4H_9$	IV-164	87.17
4158	*n*-butylbenzene (*o*)	butyl-toluene	$CH_3 \cdot C_6H_4 \cdot C_4H_9$	V-437	148.25
4159	*n*-butylbenzene (*m*)	....	$CH_3 \cdot C_6H_4 \cdot C_4H_9$	V-437	148.25
4160	*n*-butylbenzene (*p*)	....	$CH_3 \cdot C_6H_4 \cdot C_4H_9$	V-437	148.25
4161	*n*-butyl ether	....	$CH_3 \cdot O \cdot C_4H_9$	I-369	88.15
4162	*iso*-butyl ether	....	$CH_3 \cdot O \cdot C_4H_9$	I-376	88.15
4163	*n*-butyl ketone	hexanone-2	$CH_3 \cdot CO \cdot C_4H_9$	I-689	100.16
4164	*iso*-butyl ketone	2-Me-pentanone-4	$CH_3 \cdot CO \cdot C_4H_9$	I-691	100.16
4165	butyl ketone (*sec*)	3-Me-pentanone-2	$CH_3 \cdot CO \cdot C_4H_9$	I-693	100.16
4166	butyl ketone (*tert*)	pinacolin	$CH_3 \cdot CO \cdot C(CH_3)_3$	I-694	100.16
4167	*n*-butyl sulfide	....	$CH_3 \cdot S \cdot C_4H_9$	....	104.22

Table 7-4 (*Continued*)
PHYSICAL CONSTANTS OF ORGANIC COMPOUNDS

No.	Crystalline Form and Color	Specific Gravity	Melting Point °C.	Boiling Point °C.	Solubility in 100 Parts		
					Water	Alcohol	Ether
4126	oil	$1.071\frac{17}{17}°$	<−15	267-9	i.	s.	s.
4127	lf./al.	$1.047^{99.4°}$	86	....	s. bz.	....	....
4128	col. lf.	$1.181\frac{0}{4}°$	207	....	i.; s. bz.	v. sl. s.	v. sl. s.
4129	col. nd./al.	....	176-7	subl.	s. H_2SO_4	s.	s.; v. s. bz.
4130	mn.	....	161	....	v. s.	s.	....
4131	cr./CS_2	....	95	275±, d.	....	....	....
4132	col. lq.	....	....	2	0.01	∞	∞
4133	yel. cr./al.	....	133	....	v. sl. s.	s.	v. s. ac.
4134	col. lq.	$0.869\frac{8}{15}°$	expl. > 500	20-1	....	....	....
4135	pl./al.	....	82	291^{765mm}	s.	sl. s. lg.	i. alk.
4136	nd./aq.	....	175-6	....	s. h.; s. NaOH	sl. s.	sl. s.
4137	cr./aq.	....	114	....	....	....	....
4138	col. lq.	$1.087\frac{25}{0}°$	−12.4(−14)	199.5	$0.016^{30°}$	∞	∞
4139	lt. yel. lq.	$1.173\frac{0}{0}°$	....	152^{15mm}	i.	∞	∞
4140	mn. pr.	$1.190\frac{19.4°}{4}°$	52	350-2	....	....	....
4141	col. nd.	....	264-5	....	s. h.	sl. s.	s. aq. NaOH
4142	col. nd.	....	242	....	i.	s.	....
4143	pr./al.	....	84-5	$270\text{-}80^{120mm}$	i. c.	3; s. bz.	s.; s. chi.
4144	lq.	....	9.2	305-6	i.	s.	s.
4145	lq.	$0.971\frac{15}{15}°$	....	174	i.	s.	s.
4146	cr.	$1.003^{20°}$	27	210-2	i.	s.	s.
4147	col. gas	$1.732\frac{0}{0}°$	−93.7	3.5	v. sl. s.	s.; s. CS_2	s.; s. chi.
4148	lq.	....	....	$51\text{-}2^{15mm}$	....	....	....
4149	lq.	....	....	244-6	....	....	....
4150	pl.	....	31-2	122.5^{15mm}	....	....	....
4151	rhb./aq. al.	1.689	79-80	....	....	s.	s.
4152	lt. yel. lq.	....	....	$120\text{-}2^{10mm}$	i.	sl. s.	s.
4153	col. lq.	....	....	$145\text{-}8^{8mm}$	....	....	....
4154	col. lq.	$0.681\frac{20}{4}°$	−146.0	34.1	i.	∞	∞
4155	lq.	$0.680^{20°}$	−120	40	....	....	....
4156	lq.	$0.736\frac{18}{4}°$	−75	91-2	....	....	....
4157	lq.	$0.722^{18°}$	....	76-8	....	....	....
4158	oil	$0.870\frac{18}{4}°$	....	200-1	i.	sl. s.	s.
4159	oil	$0.861\frac{20}{4}°$	....	197.5	i.	sl. s.	s.
4160	oil	$0.864\frac{20}{4}°$	....	197-8	i.	sl. s.	s.
4161	lq.	$0.744\frac{20}{4}°$	−115.5	71	i.	∞	∞
4162	lq.	$0.731\frac{20}{4}°$	....	59^{741mm}	i.	∞	∞
4163	col. lq.	$0.816\frac{15}{4}°$	−56.9	127.2	v. sl. s.	∞	∞
4164	col. lq.	$0.801\frac{20}{4}°$	−84.7	117-9	$2^{20°}$	∞	∞ ; ∞ bz.
4165	col. lq.	$0.815\frac{18}{4}°$	....	117.8	....	....	....
4166	col. lq.	$0.800^{16°}$	−52.5	106.2	$2.5^{15°}$	s.; v. s. act.	s.
4167	col. lq.	....	....	122-4	....	....	....

Table 7-4 (*Continued*)
PHYSICAL CONSTANTS OF ORGANIC COMPOUNDS

No.	Name	Synonym	Formula	Beil. Ref.	Formula Weight
	Methyl				
4168	butyne-3 (2)	*iso*-propyl acetylene	$(CH_3)_2CH \cdot C \vdots CH$	I-251	68.12
4169	butyrate (*n*)	....	$C_2H_5 \cdot CH_2 \cdot CO_2 \cdot CH_3$	II-270	102.13
4170	*iso*-butyrate	....	$(CH_3)_2CH \cdot CO_2 \cdot CH_3$	II-290	102.13
4171	butyric acid†(α)(*dl*)	Me-Et-acetic acid	$C_2H_5 \cdot CH(CH_3)CO_2H$	II-304	102.13
4172	caprate	....	$C_9H_{19} \cdot CO_2 \cdot CH_3$	II-356	186.30
4173	caproate (*n*)	....	$CH_3(CH_2)_4CO_2 \cdot CH_3$	II-323	130.19
4174	caprylate	....	$CH_3(CH_2)_6CO_2 \cdot CH_3$	II-348	158.24
4175	carbamate	"methyl urethane"	$NH_2 \cdot CO_2 \cdot CH_3$	III-21	75.07
4176	carbanilide	....	$C_6H_5 \cdot N(CH_3) \cdot CO \cdot NH \cdot C_6H_5$	XII-418	226.27
4177	carbitol	diethylene-glycol (mono)methyl ether	$CH_3O \cdot CH_2 \cdot CH_2 \cdot O \cdot CH_2 \cdot CH_2 \cdot OH$	....	120.15
4178	carbitol acetate	....	$CH_3CO_2 \cdot C_5H_{11}O_2$	....	162.19
4179	carbostyril (γ)	lepidone	$CH_3 \cdot C_9H_6ON$	XXI-107	159.19
4180	carbostyril (*N*)	*N*-Me-α-quinolone	$C_6H_4CH:CH \cdot CO \cdot N \cdot \overline{} CH_3$	XXI-304	159.19
4181	carvacryl ketone	2-aceto-*p*-cymene	$CH_3 \cdot CO \cdot C_{10}H_{13}$	VII-336	176.26
4182	Cellosolve acetate	....	$CH_3CO_2 \cdot CH_2 \cdot CH_2 \cdot OCH_3$	II-141	118.13
4183	chloride	chloromethane	$CH_3 \cdot Cl$	I-59	50.49
4184	chloroacetate	....	$ClCH_2 \cdot CO_2 \cdot CH_3$	II-197	108.53
4185	*o*-chlorobenzoate	....	$Cl \cdot C_6H_4 \cdot CO_2 \cdot CH_3$	IX-336	170.60
4186	β-chloroethyl ether	....	$CH_3 \cdot O \cdot CH_2 \cdot CH_2Cl$	I-337	94.54
4187	chloroformate	Me-chlorocarbonate	$Cl \cdot CO_2 \cdot CH_3$	III-9	94.50
4188	4-chloro-hydrindene (7)	....	$CH_3(Cl)C_6H_2(CH_2)_3$	....	166.65
4189	4-chlorophenyl-4-toluene sulfonate	(3,4′-diMe;4-Cl)	$CH_3 \cdot C_6H_4 \cdot SO_3 \cdot C_6H_3(Cl) \cdot CH_3$	....	296.77
4190	cholanthrene	....	$C_{20}H_{13} \cdot CH_3$	....	268.36
4191	cinnamate	....	$C_6H_5 \cdot C_2H_2 \cdot CO_2 \cdot CH_3$	IX-581	162.19
4192	coumarin (β)	....	$C_6H_4 \cdot C(CH_3):CH \cdot CO_2 \overline{}$	XVII-336	160.17
4193	crotonate (α)	....	$C_3H_5 \cdot CO_2 \cdot CH_3$	II-410	100.12
4194	*iso*-cyanate	methyl-carbonimide	$CH_3 \cdot N:CO$	IV-77	57.05
4195	*iso*-cyanide	methyl-carbylamine	$CH_3 \cdot NC$	IV-56	41.05
4196	cyanoacetate	....	$CH_3O \cdot CO \cdot CH_2 \cdot CN$	II-584	99.09
4197	*iso*-cyanurate	....	$(CH_3)_3(CON)_3$	XXVI-249	171.16
4198	cyclobutane	Me-tetramethylene	$CH_3 \cdot C_4H_7$	V-20	70.14
4199	cyclohexane	toluene hexahydride	$CH_3(CHC_5H_{10})$	V-29	98.19
4200	cyclohexanol (*o*)	hexahydro-*o*-cresol (*cis*) (*dl*)	$CH_3 \cdot C_6H_{10} \cdot OH$	VI-11	114.19
4201	cyclohexanol (*o*)	(*trans*)	$CH_3 \cdot C_6H_{10} \cdot OH$	**VI-18	114.19
4202	cyclohexanol (*p*)	4-methylhexalin (*cis*) (*dl*)	$CH_3 \cdot C_6H_{10} \cdot OH$	VI-14	114.19
4203	cyclohexanol (*p*)	(*trans*)	$CH_3 \cdot C_6H_{10} \cdot OH$	....	114.19
4204	cyclohexanone (*o*)	....	$CH_3 \cdot C_6H_9:O$	VII-14	112.17
4205	cyclohexanone (*m*)	(*dl*)	$CH_3 \cdot C_6H_9:O$	VII-17	112.17
4206	cyclohexanone (*p*)	....	$CH_3 \cdot C_6H_9:O$	VII-18	112.17
4207	cyclohexene (Δ¹)	toluene tetrahydride	$CH_3 \cdot C_6H_9$	V-66	96.17
4208	cyclohexene (Δ³)	....	$CH_2CH:CH(CH_2)_2CH \cdot \overline{} CH_3$	V-67	96.17

† See also No. 6406.
Methyl carbonate 2468
Methyl-carbonimide 4194
Methyl-carbylamine 4195

Methyl cellosolve 3460
Methyl chlorocarbonate 4187
Methyl-chloroform 6151
Methyl-cinnamyl ketone 626

Table 7-4 (*Continued*)
PHYSICAL CONSTANTS OF ORGANIC COMPOUNDS

No.	Crystalline Form and Color	Specific Gravity	Melting Point °C.	Boiling Point °C.	Solubility in 100 Parts		
					Water	Alcohol	Ether
4168	col. lq.	$0.665^{20°}$	....	28	i.	∞	∞
4169	col. lq.	$0.903^{\frac{16}{4}°}$	−84.8	102.8	1.7	∞	∞
4170	col. lq.	$0.889^{\frac{20}{4}°}$	−87.7	92.3	v. sl. s.	∞	∞
4171	col. lq.	$0.938^{\frac{20}{20}°}$	<−80	174-7	sl. s.	∞	∞
4172	lq.	....	−18	223-4	i.	∞	∞
4173	col. lq.	$0.889^{\frac{15}{4}°}$	−71.0	151.3	i.	∞	∞
4174	col. lq.	$0.887^{18°}$	−40	192-4	i.	∞	∞
4175	col. pl.	$1.136^{\frac{5.6}{4}°}$	54.2	177	$217^{11°}$	$73^{15°}$	s.
4176	nd./al.	....	104	203-5	sl. s. h.	sl. s. c.	v. s.; v. s. bz.
4177	col. lq.	$1.035^{\frac{20}{20}°}$	....	193.2	∞	....	....
4178	col. lq.	$1.040^{\frac{20}{20}°}$	....	209.1	∞	....	....
4179	nd./aq.	....	223.7	270^{17mm}	sl. s. h.; s. a.	s. h.; i. alk.	sl. s.; sl. s. bz.
4180	col. nd./lg.	....	74	324^{728mm} sl. d.	1.4 c.	s.; s. chl.; sl. s. lg.	s. act.; v. s. bz.
4181	lq.	$0.956^{\frac{20}{4}°}$	<−10	240-4	....	....	....
4182	col. lq.	$1.007^{\frac{20}{4}°}$	−65.1	144.5	∞	....	....
4183	col. gas	(A)1.785	−97.7	−24	$280^{16°}$ cc.	$350^{20°}$ cc.	4000 cc., ac.
4184	col. lq.	$1.236^{\frac{20}{4}°}$	−32.7	130^{740mm}	v. sl. s.	∞	∞
4185	col. lq.	....	....	234-5			
4186	col. lq.	$1.031^{\frac{20}{4}°}$	....	90.5	8 c.	∞	∞
4187	col. lq.	$1.236^{15°}$	....	71-2	d.	∞; ∞ chl.	∞; ∞ bz.
4188	yel. lq.	....	....	$111-4^{10mm}$	....		
4189	col. cr.	....	94-5	....	i.	sl. s.	v. s.
4190	yel. nd./bz.	....	176.5-7.5	....			
4191	cr.	$1.042^{\frac{8.6}{0}°}$	33.4	263	i.	v. s.	v. s.
4192	nd./bz.	....	82	....	....	s.	s. bz.
4193	lq.	$0.981^{4°}$	....	128	i.	∞	∞
4194	col. lq.	$0.967^{\frac{16}{4}°}$	....	43-5	....		
4195	col. lq.	$0.734^{\frac{18}{4}°}$	−45	59.6	$10^{15°}$	∞	∞
4196	lq.	$1.123^{\frac{15}{4}°}$	−22.5	203	i.	∞	∞
4197	mn.	....	175-6	274	sl. s. h.	s.	....
4198	lq.	$0.694^{\frac{20}{4}°}$	....	39-42	i.	....	s.
4199	col. lq.	$0.769^{\frac{20}{4}°}$	−126.6	100.9	i.	s.	s.
4200	col. lq.	$0.934^{\frac{20}{4}°}$	−9.5	165	v. sl. s.	∞	∞
4201	col. lq.	$0.924^{\frac{20}{4}°}$	−21	166.5	....		
4202	lq.	$0.913^{\frac{21}{4}°}$	....	$173-4^{750mm}$	v. sl. s.	∞	∞
4203	lq.	$0.912^{\frac{21}{4}°}$	....	$173-4^{745mm}$	....		
4204	lq.	$0.925^{\frac{20}{4}°}$	−14	165.1	i.	s.	s.
4205	lq.	$0.915^{\frac{20}{4}°}$	−73.5	169.6	i.	s.	s.
4206	lq.	$0.916^{\frac{20}{4}°}$	−40.6	171.3	i.	s.	s.
4207	lq.	$0.809^{\frac{20}{4}°}$	....	110-1	i.	s.	s.
4208	lq.	$0.800^{\frac{20}{20}°}$	....	103	i.	s.	s.

Methyl citraconate 2469
Methyl citrate 6263
Methyl-cresyl ether 4399-4401

Methyl-cresyl sulfide 4402
Methyl cyanide 49
Methyl-cyclohexadiene 2286-8

Table 7-4 (*Continued*)
PHYSICAL CONSTANTS OF ORGANIC COMPOUNDS

No.	Name	Synonym	Formula	Beil. Ref.	Formula Weight
4209	**Methyl** cyclopentane	....	$CH_3(CH \cdot C_4H_8)$	V-27	84.16
4210	cyclopropane	methyl-trimethylene	$CH_3 \cdot CH \cdot CH_2 \cdot CH_2$ $\lfloor\underline{\quad\quad}\rfloor$	V-18	56.11
4211	diazoaminoben-zene (4)	....	$CH_3 \cdot C_6H_4 \cdot N:N \cdot$ $NH \cdot C_6H_5$	XVI-705	211.27
4212	dibromoanthranil-ate	(3,5)	$Br_2C_6H_2(NH_2) \cdot$ CO_2CH_3	*XIV-553	308.97
4213	dibromocinnamate (α,β)(*dl*)	Me β-Ph-α,β-diBr-propionate	$C_6H_5 \cdot CHBr \cdot CHBr \cdot$ CO_2CH_3	IX-518	322.01
4214	dichloroarsine	....	$CH_3 \cdot AsCl_2$	IV-601	160.86
4215	diethylamine	....	$CH_3 \cdot N(C_2H_5)_2$	IV-99	87.17
4216	diethylaminoben-zene (*o*)	*o*-diethyltoluidine	$CH_3 \cdot C_6H_4 \cdot N(C_2H_5)_2$	XII-786	163.26
4217	diethylaminoben-zene (*m*)	*m*-diethyltoluidine	$CH_3 \cdot C_6H_4 \cdot N(C_2H_5)_2$	XII-857	163.26
4218	diethylaminoben-zene (*p*)	*p*-diethyltoluidine	$CH_3 \cdot C_6H_4 \cdot N(C_2H_5)_2$	XII-904	163.26
4219	diphenylamine (*N*)	....	$CH_3 \cdot N(C_6H_5)_2$	XII-180	183.26
4220	diphenylamine (*p*)	Ph-*p*-tolylamine	$CH_3 \cdot C_6H_4 \cdot NH \cdot C_6H_5$	XII-905	183.26
4221	diphenyl carbinol	....	$(C_6H_5)_2C(OH) \cdot CH_3$	VI-685	198.27
4222	disulfide	dimethyl-disulfide	$CH_3 \cdot S \cdot S \cdot CH_3$	I-291	94.20
4223	epichlorohydrin(β)	....	$O \cdot CH_2 \cdot C(CH_3) \cdot CH_2Cl$ $\lfloor\underline{\quad\quad}\rfloor$	....	106.55
4224	ethyl-acetaldehyde	2-methyl-butanal-1	$C_2H_5 \cdot CH(CH_3) \cdot CHO$	I-682	86.13
4225	β-ethylacrolein(α)	2-Me-pentene-2-al-1	$C_2H_5 \cdot CH:C(CH_3)CHO$	I-735	98.15
4226	ethylamine	....	$CH_3 \cdot NH \cdot C_2H_5$	IV-94	59.11
4227	ethylamine HCl	....	$C_2H_5(CH_3)NH \cdot HCl$	IV-94	95.57
4228	ethylaniline	....	$C_6H_5 \cdot N(CH_3)C_2H_5$	XII-162	135.21
4229	ethyl carbonate	....	$CH_3O \cdot CO \cdot OC_2H_5$	III-4	104.11
4230	ethyl ether	....	$CH_3 \cdot O \cdot C_2H_5$	I-314	60.10
4231	ethyl ethyl-phenyl-malonate	....	$CH_3O_2C \cdot C(C_6H_5):$ $(C_2H_5)CO_2C_2H_5$	....	250.30
4232	ethyl ketone	butanone	$CH_3 \cdot CO \cdot C_2H_5$	I-666	72.11
4233	ethyl ketoxime	butanoxime	$CH_3 \cdot C(:NOH) \cdot C_2H_5$	I-668	87.12
4234	ethyl ketone semi-carbazone	....	$C_3H_8:C:N \cdot NHCONH_2$	III-102	129.16
4235	ethyl malonic acid	....	$C_2H_5(CH_3)C:(CO_2H)_2$	II-664	146.14
4236	ethyl oxalate	....	$CH_3O \cdot (CO)_2 \cdot OC_2H_5$	II-535	132.12
4237	ethyl succinate	....	$C_7H_{12}O_4$	II-609	160.17
4238	ethyl sulfide	....	$CH_3 \cdot S \cdot C_2H_5$	I-343	76.16
4239	ethyl sulfone	....	$CH_3 \cdot SO_2 \cdot C_2H_5$	I-343	108.16
4240	fluoride	fluoromethane	$CH_3 \cdot F$	I-59	34.03
4241	formanilide	....	$C_6H_5(CH_3)N \cdot CHO$	XII-234	135.17
4242	formate	....	$HCO_2 \cdot CH_3$	II-18	60.05
4243	fumaric acid (*trans*)	mesaconic acid	$CH_3 \cdot C(CO_2H):CH \cdot$ CO_2H	II-763	130.10
4244	furfural (2,5)	....	$CH_3 \cdot C_4H_2O \cdot CHO$	XVII-289	110.11
4245	furfuran (2)	sylvane	$CH_3 \cdot C_4H_3O$	XVII-36	82.10
4246	furoate	....	$C_4H_3O \cdot CO_2CH_3$	XVIII-274	126.11
4247	furoylacetate (α)	....	$C_4H_3O \cdot CO \cdot CH_2 \cdot$ CO_2CH_3	....	168.15
4248	furylacrylate (β)	....	$C_4H_3O \cdot CH:CH \cdot$ CO_2CH_3	XVIII-300	152.15
4249	gallate	gallicin	$(HO)_3C_6H_2 \cdot CO_2CH_3$	X-483	184.15
4250	glucoside (α)(*d*)	....	$CH_3O \cdot C_6H_{11}O_5$	XXXI-179	194.19

Table 7-4 (*Continued*)
PHYSICAL CONSTANTS OF ORGANIC COMPOUNDS

No.	Crystalline Form and Color	Specific Gravity	Melting Point °C.	Boiling Point °C.	Solubility in 100 Parts		
					Water	Alcohol	Ether
4209	col. lq.	$0.749^{20°}_{4}$	-142.4	71.8	i.	....	s.
4210	col. gas	$0.691^{-20°}_{0}$	....	4-5	i.	....	s.
4211	yel. lf.	....	90-1	d.	i.	....	....
4212	nd.	....	87-8	....	....	s.	....
4213	mn. pl./CS_2	....	115-6	....	....	....	....
4214	col. lq.	$1.838^{20°}_{4}$	-59	133-6	sl. s.	v. s.	s.
4215	col. lq.	....	....	65-7	v. s.	s.	s.
4216	lq.	....	....	$208-9^{755mm}$	i.	s.	s.
4217	lq.	....	....	231-2	i.	s.	s.
4218	lq.	$0.924^{15.5°}$	....	228-9	....	....	....
4219	col. lq.	$1.048^{20°}_{4}$	-7.6	295-6	i.	s.	s.
4220	cr.	....	87-9	$317-8^{728mm}$	i.	s.	s.
4221	pr./et.	....	80-1	$175-80^{20mm}$	v. s. bz.	....	v. s.
4222	lq.	$1.057^{16°}_{4}$	....	116-8	....	....	....
4223	lq.	$1.103^{20°}_{4}$	....	122.3	3.1	....	....
4224	lq.	$0.807^{20°}$	....	90-2	i.	....	....
4225	lq.	$0.854^{25°}_{4}$	....	137.3^{759mm}	i.	....	....
4226	col. lq.	....	....	34-5	....	....	....
4227	lf./al. et.	....	130-3	....	v. s.	v. s.	i.; s. chl.
4228	lq.	....	....	201	i.	∞	∞
4229	lq.	$1.002^{27°}$	-14.5	109.2	i.	∞	∞
4230	col. lq.	$0.697^{21.1°}_{4}$	....	7.6	s.	∞	∞
4231	yel. lq.	....	....	$165-70^{20mm}$	i.	s.	s.
4232	col. lq.	$0.805^{20°}_{4}$	-86.9	79.6	37	∞	∞; ∞ bz.
4233	col. oil	$0.923^{20°}_{4}$	-29.5	152-3	10	∞	∞
4234	lf./aq.	....	148; slowly 143	....	v. s. h.	v. s.	....
4235	pr. or nd.	....	122	d. 180	v. s.	v. s.	v. s.
4236	lq.	$1.156^{0°}_{0}$	....	173.7	i.	v. s.	v. s.
4237	lq.	$1.093^{0°}$	<-20	208.2	i.	v. s.	v. s.
4238	oil	$0.837^{20°}$	-104.8	66.9	i.	∞	∞
4239	nd.	....	36	....	v. s.	v. s.	v. sl. s. c.
4240	gas	....	....	-78.4	166^{15c} cc.	....	....
4241	lq.	$1.095^{20°}_{4}$	14-5	253^{716mm}	v. sl. s.	v. s.	....
4242	lq.	$0.974^{20°}_{4}$	-99.0	31.8	$30^{20°}$	∞	s. Me al.
4243	cr. pd.	1.466	202	250 d.	$2.7^{18°}$; $118^{100°}$	$30.6^{17°}$, 90% al.	v. sl. s. chl.
4244	oil	$1.109^{18°}$	....	186-7	3.3	v. s.	....
4245	col. lq.	0.916	....	65^{759mm}	....	....	....
4246	col. lq.	$1.179^{21.4°}_{4}$	....	181.3	i.	∞	∞
4247	yel. lq.	....	....	$148-52^{13mm}$	i.	s.	s.
4248	col. lq.	....	27.5	$227-8^{774mm}$	i.; s. lg.	s.; s. bz.	s.
4249	mn./Me al.	....	194-8	....	$1.072^{23°}$	v. s.	s.
4250	col. nd./al.	....	168-9	$200^{0.2mm}$	63	1.6	i.

Methyl-ethylacetic acid 4171
Methyl-ethyl acetone 4165
Methyl-ethyl-acetylene 5087
Methyl-ethylbenzene 3166-8
Methyl-ethyl-*n*-butyl carbinol 4968

Methyl-ethyl carbinol 1035
Methyl-ethyl ketone diethylsulfone 6326
Methyl-ethyl-*n*-propyl carbinol 3540
Methyl-ethyl-*iso*-propyl carbinol 3529
Methyl-ethyl-protocatechuic aldehyde 6419

Table 7-4 (Continued)
PHYSICAL CONSTANTS OF ORGANIC COMPOUNDS

No.	Name	Synonym	Formula	Beil. Ref.	Formula Weight
4251	**Methyl** glutaric acid (α)	2-methyl-pen-tandioic acid	$CH_3 \cdot CH(CO_2H) \cdot CH_2 \cdot CH_2 \cdot CO_2H$	II-655	146.14
4252	glutaric acid (β)	3-Me-pentandioic	$CH_3 \cdot CH(CH_2CO_2H)_2$	II-659	146.14
4253	glycerinate (d)	....	$CH_2OH \cdot CHOH \cdot CO_2 \cdot CH_3$	III-397	120.11
4254	glycerine chloro-hydrin (β),mono	....	$ClCH_2 \cdot C(CH_3)OH \cdot CH_2OH$	....	124.57
4255	glycolate	....	$HO \cdot CH_2 \cdot CO_2 \cdot CH_3$	III-236	90.08
4256	glyoxalidine	2-Me-Δ²-imidazoline	$CH_3 \cdot C{:}N(CH_2)_2NH$	XXIII-31	84.12
4257	glyoxaline (1)	1-methyl-imidazole	$CH_3 \cdot N \cdot CH{:}N \cdot CH{:}CH$	XXIII-46	82.11
4258	guanidine	....	$CH_3NH \cdot CH_3N_2$	IV-68	73.10
4259	guanidine nitrate	....	$C_2H_7N_3 \cdot HNO_3$	IV-69	138.11
4260	guanidine sulfate		$(C_2H_7N_3)_2 \cdot H_2SO_4$	....	244.27
4261	heptadecyl ketone	nonadecanone-2	$CH_3(CH_2)_{16} \cdot CO \cdot CH_3$	I-718	282.51
4262	heptenone	....	$(CH_3)_2C{:}CH(CH_2)_2 \cdot CO \cdot CH_3$	I-741	126.20
4263	n-heptylate	Me oenanthylate	$CH_3(CH_2)_5CO_2CH_3$	II-339	144.22
4264	n-heptyl ether	....	$CH_3 \cdot O \cdot (CH_2)_6CH_3$	I-414	130.23
4265	heptyl ketone	nonanone-2	$CH_3 \cdot CO \cdot (CH_2)_6CH_3$	I-709	142.24
4266	hexyl ketone	octanone-2	$CH_3 \cdot CO \cdot (CH_2)_5CH_3$	I-704	128.22
4267	hydantoin (β)	....	$C_4H_6O_2N_2$	XXIV-244	114.10
4268	hydantoin (γ)	lactyl urea	$C_4H_6O_2N_2 \cdot H_2O$	XXIV-279	132.12
4269	hydrazine	....	$CH_3 \cdot NH \cdot NH_2$	IV-546	46.07
4270	hydrazine sulfate	....	$CH_3 \cdot NH \cdot NH_2 \cdot H_2SO_4$	IV-547	144.15
4271	hydrazobenzene (o)	β-Ph-o-tolylhydra-zine	$CH_3 \cdot C_6H_4 \cdot NH \cdot NH \cdot C_6H_5$	XV-497	198.27
4272	hydrazobenzene (m)	....	$CH_3 \cdot C_6H_4(NH)_2C_6H_5$	XV-506	198.27
4273	hydrazobenzene (p)	....	$CH_3 \cdot C_6H_4(NH)_2C_6H_5$	XV-511	198.27
4274	hydroxybenzoate (p)		$HO \cdot C_6H_4 \cdot CO_2CH_3$	X-158	152.15
4275	3-hydroxy-1,4-naphthoquinone (2)	phthiocol	$CH_3(OH)C_{10}H_4O_2$	....	188.18
4276	hydroxynaphtho-quinone acetate	phthiocol mono-acetate	$C_{13}H_{10}O_4$	....	230.22
4277	hydroxynaphtho-quinone acetate	phthiocol tri-acetate	$C_{17}H_{16}O_6$	....	316.31
4278	hydroxynaphtho-quinone oxime	phthiocol monoxime	$C_{11}H_8O_2{:}NOH$	....	203.20
4279	hydroxylamine (β)	....	$CH_3 \cdot NH \cdot OH$	IV-534	47.06
4280	hydroxylamine HCl	(α)	$CH_3O \cdot NH_2 \cdot HCl$	I-288	83.52
4281	hypochlorite		$CH_3 \cdot O \cdot Cl$	I-282	66.49
4282	indole (1)	....	C_9H_9N	XX-308	131.18
4283	indole (2)	methyl ketole	$C_6H_4 \cdot CH{:}C(CH_3) \cdot NH$	XX-311	131.18
4284	indole (3)	skatole	$C_6H_4C(CH_3){:}CH \cdot NH$	XX-316	131.18
4285	indole (5)	....	$CH_3 \cdot C_6H_3 \cdot CH{:}CH \cdot NH$	XX-317	131.18
4286	inositol	d-inositol Me ether	$(HO)_5C_6H_6 \cdot OCH_3$	VI-1193	194.19

Methyl-ethyl-pyridine 161, 1524–5
Methyl-exaltone 4446
Methyl fumarate 2487
Methyl glycinate 4107
Methyl-glycine 5600

Methyl-glycocyamidine 1540
Methyl-granatonine, cf. alkd.
Methyl green 3594
Methyl-guanylglycine 1539
Methyl-heptyl carbinol (n) 4919

Table 7-4 (*Continued*)
PHYSICAL CONSTANTS OF ORGANIC COMPOUNDS

No.	Crystalline Form and Color	Specific Gravity	Melting Point °C.	Boiling Point °C.	Solubility in 100 Parts		
					Water	Alcohol	Ether
4251	pr./aq.	....	78-80	$214\text{-}5^{22mm}$	v. s.	v. s.	v. s.
4252	pr. or pl.	....	86-7	....	v. s.	v. s.	v. s.
4253	lq.	$1.281\frac{15°}{15}$	....	$119\text{-}20^{14mm}$	∞	∞	v. sl. s.
4254	lq.	$1.237\frac{20°}{4}$	....	$95\text{-}67^{mm}$	∞	....	....
4255	lq.	$1.168^{18°}$	....	151.2	....	....	....
4256	col. hyg. cr.	....	105	195-8	s.	s.	i.; s. chl.
4257	lq.	$1.036^{10°}$	-6	197-9	∞	....	....
4258	col. delq.	....	d.	....	v. s.	s.	....
4259	col. cr./al.	....	149-50	....	s.	....	....
4260	cr./aq.	....	239-40	....	....	....	....
4261	lf.	$0.811^{55.5°}$	55-6	266.5^{110mm}	v. s. chl.	v. s.	v. s.
4262	lq.	$0.860^{20°}$	-67.1	173-4	i.	∞	∞
4263	lq.	$0.885\frac{15°}{4}$	-55.8	173.8	i.	....	....
4264	lq.	$0.795\frac{20°}{0}$	....	149.8	i.	∞	∞
4265	lq.	$0.825\frac{15°}{4}$	-7.8	194-6	i.	s.	s.
4266	col. lq.	$0.819\frac{20°}{4}$	-21	172.9	i.	∞	∞
4267	pr.	....	156-7	subl.	s.	s.	3
4268	rhb.	....	145(anh.)	....	s.	s.	sl. s.
4269	hyg. lq.	....	<-80	87^{745mm}	s.	∞	∞; i. lg.
4270	cr./Me al.	....	142	....	v. s.	v. sl. s.	....
4271	lf./al.	....	101-2	....	i.	sl. s. c.	s.
4272	lt. yel. cr.	....	59-61	....	i.; s. bz.	v. s.	sl. s.
4273	pl./lg.	....	86-7	....	v. s. bz.	v. s.	v. s.
4274	nd./aq. al.	....	131	270-80 d.	0.25	s.; s. act.	s.
4275	yel. pr./aq. Me al.	....	172-3	vol. in steam	s.; s. dil. alk.	s.	s.
4276	yel. nd.	....	106-7	....	i.; s. alk.	s. Me al.	s. pyr.
4277	col. pr.	....	158-9	....	i.	s. Me al.	....
4278	yel. nd.	....	199-200 d.	....	i.	s.	s. pyr.
4279	hyg. pr.	$1.000\frac{20°}{4}$	42	62.5^{15mm}	v. s.	v. s.	sl. s.
4280	pr./al. et.	....	149	....	s.	s.	i.
4281	gas	....	....	12^{726mm}	....	....	....
4282	lq.	$1.071^{0°}$	< 20	242.4	i.	v. s.	v. s.
4283	nd.	1.07	59-60	272^{750mm}	s. h.; s. HCl	v. s.	v. s.
4284	lf./lg.	....	95	$265\text{-}6^{755mm}$	0.05 c.	s.; s. chl.	s.; s. bz.
4285	nd./aq.	....	58.5	....	s. h.	s.	s.; s. bz.
4286	cr./aq.	1.52	186-7	....	v. s.	i. abs.	i.; i. chl.

Methyl-hexalin 4200-3
Methyl-hexyl carbinol 4962
Methyl-hydrocupreine, cf. alkd.
Methyl hydrogen sulfate 4384

Methyl-hydroquinone 2372
Methyl-imidazole 4257
Methyl-imidazoline 4256

Table 7-4 (Continued)
PHYSICAL CONSTANTS OF ORGANIC COMPOUNDS

No.	Name	Synonym	Formula	Beil. Ref.	Formula Weight
4287	**Methyl** iodide	iodomethane	CH_3I	I-70	141.94
4288	iodobenzoate (o)	....	$I \cdot C_6H_4 \cdot CO_2CH_3$	IX-364	262.05
4289	iodobenzoate (m)	....	$I \cdot C_6H_4 \cdot CO_2CH_3$	IX-365	262.05
4290	iodobenzoate (p)	....	$I \cdot C_6H_4 \cdot CO_2CH_3$	IX-367	262.05
4291	isatin (O)	....	$C_6H_4 \cdot N{:}C(OCH_3)CO$	XXI-583	161.16
4292	isatin (N)	....	$C_6H_4N(CH_3) \cdot CO \cdot CO$	XXI-446	161.16
4293	isatin (5)	p-methyl isatin	$CH_3 \cdot C_6H_3 \cdot CO \cdot CO \cdot NH$	XXI-509	161.16
4294	lactate	....	$CH_3 \cdot CHOH \cdot CO_2CH_3$	III-280	104.11
4295	laurate	....	$CH_3(CH_2)_{10}CO_2CH_3$	II-361	214.35
4296	levulinate	....	$CH_3CO(CH_2)_2CO_2CH_3$	III-675	130.14
4297	d-mannoside (α)	....	$CH_3O \cdot C_6H_{11}O_5$	I-907	194.19
4298	mercaptan	methanthiol	$CH_3 \cdot SH$	I-288	48.11
4299	mercaptobenzo-thiazole (2)	....	$C_6H_4 \cdot N{:}C(SCH_3) \cdot S$	XXVII-109	181.28
4300	methoxybenzoate (o)	....	$CH_3O \cdot C_6H_4CO_2CH_3$	X-71	166.18
4301	methoxybenzoate (p)	methyl anisate	$CH_3O \cdot C_6H_4CO_2CH_3$	X-159	166.18
4302	methylanthra-nilate (N)	methylamino-methyl benzoate	$CH_3NH \cdot C_6H_4 \cdot CO_2 \cdot CH_3$	XIV-324	165.19
4303	myristate	....	$CH_3(CH_2)_{12}CO_2CH_3$	II-365	242.41
4304	naphthalene (α)	....	$C_{10}H_7 \cdot CH_3$	V-566	142.20
4305	naphthalene (β)	....	$C_{10}H_7 \cdot CH_3$	V-567	142.20
4306	1,4-naphthoquin-one (2)	....	$CH_3 \cdot C_{10}H_5O_2$	**VII-656	172.19
4307	naphthohydro-quinonediacetate	(2;1,4)	$CH_3 \cdot C_{10}H_5{:}$ $(O_2C \cdot CH_3)_2$	**VI-958	258.28
4308	naphthoquinone monoxime	(2;4,1)	$CH_3 \cdot C_{10}H_5O{:}NOH$	....	187.20
4309	naphthylamine (α)	....	$C_{10}H_7 \cdot NHCH_3$	XII-1221	157.22
4310	naphthylamine (β)	....	$C_{10}H_7 \cdot NHCH_3$	XII-1273	157.22
4311	naphthyl ether (α)	....	$CH_3 \cdot O \cdot C_{10}H_7$	VI-606	158.20
4312	naphthyl ether (β)	nerolin; yara yara	$CH_3 \cdot O \cdot C_{10}H_7$	VI-640	158.20
4313	naphthyl ketone (β)	β-acetonaphthone	$CH_3 \cdot CO \cdot C_{10}H_7$	VII-402	170.21
4314	nitramine	nitraminomethane	$CH_3 \cdot NH \cdot NO_2$	IV-567	76.06
4315	nitrate	....	$CH_3O \cdot NO_2$	I-284	77.04
4316	nitrite	....	$CH_3O \cdot NO$	I-284	61.04
4317	nitrobenzoate (o)	....	$NO_2 \cdot C_6H_4 \cdot CO_2CH_3$	IX-372	181.15
4318	nitrobenzoate (m)	....	$NO_2 \cdot C_6H_4 \cdot CO_2CH_3$	IX-378	181.15
4319	nitrobenzoate (p)	....	$NO_2 \cdot C_6H_4 \cdot CO_2CH_3$	IX-390	181.15
4320	nitrocinnamate (o)	....	$NO_2 \cdot C_6H_4 \cdot CH{:}CH \cdot CO_2CH_3$	IX-605	207.19
4321	nitrocinnamate (m)(trans)	....	$NO_2 \cdot C_6H_4 \cdot CH{:}CH \cdot CO_2CH_3$	IX-606	207.19
4322	nitrocinnamate (p)(trans)	....	$NO_2 \cdot C_6H_4 \cdot CH{:}CH \cdot CO_2CH_3$	IX-607	207.19
4323	nitrolic acid	....	$HC({:}NOH)NO_2$	II-92	90.04
4324	n-nonyl ketone	undecanone-2	$CH_3(CH_2)_8 \cdot CO \cdot CH_3$	I-713	170.30
4325	octadiene-4,6 (2)	....	$CH_3 \cdot CH{:}(CH)_2{:}CH \cdot CH_2CH(CH_3)_2$	I-260	124.23
4326	n-octyl ketone	decanone-2	$CH_3 \cdot CO \cdot C_8H_{17}$	I-711	156.27
4327	oleate	....	$C_{17}H_{33}CO_2CH_3$	II-467	296.50

Methyl-iodoform 6237
Methyl-isoeugenol 3224
Methyl itaconate 2498
Methyl-ketole 4283
Methyl malate 2500

Methyl maleate 2501
Methyl-maleic acid 1510
Methyl-malic acid 1514
Methyl malonate 2502
Methyl-malonic acid 5652

Methyl mesaconate 2504
Methyl-morphine, cf. alkd.
Methyl mustard oil 4388
Methyl nonylate 4331
Methyl oenanthylate 4263

Table 7-4 (*Continued*)
PHYSICAL CONSTANTS OF ORGANIC COMPOUNDS

No.	Crystalline Form and Color	Specific Gravity	Melting Point °C.	Boiling Point °C.	Solubility in 100 Parts		
					Water	Alcohol	Ether
4287	col. lq.	$2.279\frac{20}{4}°$	−66.5	42.4	$1.8^{15°}$	∞	∞
4288	lq.	....	....	$277\text{-}8^{729mm}$	....	....	....
4289	nd.	....	50	$276\text{-}7^{739mm}$	i.; i. lg.	s. h.	s. h.
4290	rhb./et. al.	$2.020^{10°}$	114	subl.	....	....	....
4291	red lf./aq.	....	101-2	....	sl. s. c.	s. alk.	s. h. HCl
4292	red nd./ aq.	....	134	....	s. aq. NaOH	....	....
4293	red nd./ aq. al.	....	187	....	v. sl. s. h.;s.alk.	s.; s. h. HCl	sl. s.
4294	lq.	$1.090^{19°}$	....	144.8	∞ ; d.	s.	s.
4295	lq.	....	5	148^{18mm}	i.	....	....
4296	lq.	$1.047\frac{20}{4}°$	....	196	....	....	....
4297	col. nd./al.	$1.473\frac{7}{4}°$	193-4	....	$30.7^{15°}$	3	i.
4298	lq. or gas	$0.896^{0°}$	−121	5.96	s.	v. s.	v. s.
4299	pr./aq. al.	....	52	....	s. conc. a.	....	....
4300	lq.	$1.157\frac{19}{4}°$	....	245-6	....	s.	....
4301	pl./al.	....	48-9	255-6	i.	s.	s.
4302	cr./pet.	$1.120^{15°}$	18-9	256	i.	s.	s.
4303	cr./al.	....	18-9	295^{751mm}	i.	....	....
4304	oil	$1.025\frac{14}{4}°$	−30.8	244.8	i.	v. s.	v. s.
4305	mn.	$0.990\frac{40}{4}°$	34.4	241.1	i.	v. s.	v. s.
4306	yel. nd./ al.	....	106	volt. in steam	i.; sl. s. pet.	s.; s. bz.	s.
4307	col. pr./aq. Me al.	....	114	....	i.	s.	v. s. act.
4308	hex. pl./aq. Me al.	....	166-8	....	i.	s.	s. pyr.
4309	oil	....	....	293-5	s. CS_2	s.	s.
4310	oil	....	....	$308\text{-}10^{761mm}$	....	....	....
4311	col. oil	$1.096\frac{14}{4}°$	<−10	265-9	i.	s.	s.; s. bz.
4312	lf./et.	....	73-5	274	sl. s. CS_2	sl. s.	s.; s. bz.
4313	nd./lg.	....	55-6	301-3	....	....	....
4314	nd./et.	$1.243\frac{49}{4}°$	38	....	v. s.	v. s.	sl. s.
4315	lq.	$1.203^{25°}$	expl.	65	sl. s.	s.	s.
4316	gas	lq. $0.991^{15°}$	....	−12	....	s.	s.
4317	yel. lq.	$1.286^{20°}$	−8	275	i. pet.	s.	s.
4318	col. nd.	....	78.5	279	....	sl. s.	sl. s.
4319	mn.	....	96	....	i.	s.	s.
4320	col. nd./ aq.	....	72-3	$187\text{-}9^{15mm}$	v. sl. s. h.	s. h.	....
4321	yel. pr./ Me al.	....	123-4	d.	i.; v. sl. s. CS_2	v. sl. s.; s. chl.	v. sl. s.; s. bz.
4322	col. nd./al.	....	160-1	281-6	....	sl. s.	i.
4323	nd./et.	....	64-8 d.	....	s.	s.	s.
4324	col. oil	$0.829\frac{15}{4}°$	12.8	223	i.	s.	s.
4325	lq.	$0.752\frac{18}{4}°$	....	149	i.	....	....
4326	lq.	$0.828\frac{15}{4}°$	13.1	209^{750mm}	i.	s.	s.
4327	oil	$0.874\frac{20}{4}°$	....	$190\text{-}1^{10mm}$	i	∞	∞

Methyl oxalate 4329
Methyl pentenal 4225
Methyl phenol 1545-7
Methyl-phenylamine 4125
Methyl-phenyl carbinol 5186

Methyl phenyl-dibromopropionate 4213
Methyl-phenyl diketone 72
Methyl-phenyl ether 537
Methyl-phenyl glyoxal 72
Methyl-phenyl ketone 51

Table 7-4 (*Continued*)
PHYSICAL CONSTANTS OF ORGANIC COMPOUNDS

No.	Name	Synonym	Formula	Beil. Ref.	Formula Weight
4328	**Methyl** orange	tropeolin D; orange III	$(CH_3)_2N \cdot C_6H_4 \cdot N : N \cdot C_6H_4 \cdot SO_3Na$	XVI-331	327.34
4329	oxalic acid	Me acid oxalate	$CH_3O_2C \cdot CO_2H$	II-534	104.06
4330	palmitate	....	$C_{15}H_{31}CO_2CH_3$	II-372	270.46
4331	pelargonate	methyl nonylate	$CH_3(CH_2)_7CO_2CH_3$	II-353	172.27
4332	2,4-pentandiol (2)	....	$(CH_3)_2C(OH) \cdot CH_2 \cdot CHOH \cdot CH_3$	I-486	118.18
4333	phenacyl bromide (*p*)	ω-Br-*p*-methyl-acetophenone	$CH_3 \cdot C_6H_4 \cdot CO \cdot CH_2Br$	VII-309	213.08
4334	phenazine (2)	tolazine	$CH_3 \cdot C_6H_3 : N_2 : C_6H_4$	XXIII-237	194.24
4335	phenylacetate	....	$C_6H_5 \cdot CH_2 \cdot CO_2CH_3$	IX-434	150.18
4336	phenylenediamine	(*N*)(*p*)	$CH_3NH \cdot C_6H_4 \cdot NH_2$	XIII-71	122.17
4337	phenylenediamine HCl (*N*)(*p*)	*p*-amino-methyl-aniline di-HCl	$C_7H_{10}N_2 \cdot 2HCl$	....	195.09
4338	phenylhydrazine	(α,α)	$C_6H_5 \cdot N(CH_3) \cdot NH_2$	XV-117	122.17
4339	phenylhydrazine sulfate (α,α)	....	$(C_7H_{10}N_2)_2 \cdot H_2SO_4$	XV-118	342.42
4340	phenylhydrazine	(α,β)	$C_6H_5 \cdot NH \cdot NH \cdot CH_3$	XV-118	122.17
4341	phosphine	....	$CH_3 \cdot PH_2$	IV-580	48.02
4342	3-phytyl-1,4-naph-thoquinone (2)	vitamin K-1	$C_{31}H_{46}O_2$	....	450.71
4343	3-phytyl-1,4-naph-thoquinone di-acetate (2)	dihydrovitamin K-1 diacetate	$C_{35}H_{52}O_4$	....	536.80
4344	piperidine (*N*)	....	$C_5H_{10}N \cdot CH_3$	XX-16	99.19
4345	piperidine (2)	α-pipecoline	$CH_3 \cdot CH(CH_2)_4 \cdot NH$	XX-95	99.19
4346	piperidine (3)	β-pipecoline	$CH_2 \cdot NH \cdot (CH_2)_3 \cdot CH \cdot CH_3$	XX-100	99.19
4347	piperidine (4)	γ-pipecoline	$(CH_2)_2 \cdot NH \cdot (CH_2)_2 \cdot CH \cdot CH_3$	XX-101	99.19
4348	propargyl ether	....	$CH_3 \cdot O \cdot CH_2 \cdot C \vdots CH$	I-454	70.09
4349	propionate	....	$C_2H_5 \cdot CO_2CH_3$	II-239	88.11
4350	*n*-propylamine	....	$CH_3 \cdot NH \cdot CH_2 \cdot C_2H_5$	IV-137	73.14
4351	*n*-propylbenzene (*m*)	*m*-propyl toluene	$CH_3 \cdot C_6H_4 \cdot CH_2 \cdot C_2H_5$	V-418	134.22
4352	*n*-propylbenzene (*p*)	*p*-propyl toluene	$CH_3 \cdot C_6H_4 \cdot CH_2 \cdot C_2H_5$	V-419	134.22
4353	*n*-propyl ether	....	$CH_3 \cdot O \cdot CH_2 \cdot C_2H_5$	I-354	74.12
4354	*iso*-propyl ether	....	$CH_3 \cdot O \cdot CH(CH_3)_2$	I-362	74.12
4355	*n*-propyl ketone	pentanone-2	$CH_3 \cdot CO \cdot CH_2 \cdot C_2H_5$	I-676	86.13
4356	propyl ketone oxime	(*n*)	$CH_3 \cdot C(:NOH) \cdot C_3H_7$	I-677	101.15
4357	*iso*-propyl ketone	2-methyl-butanone-3	$CH_3 \cdot CO \cdot CH(CH_3)_2$	I-682	86.13
4358	*iso*-propyl ketoxime	....	$CH_3 \cdot C(:NOH) \cdot C_3H_7$	I-683	101.15
4359	*iso*-propyl ketone semicarbazone	....	$C_3H_7(CH_3)C : N \cdot NH \cdot CO \cdot NH_2$	III-103	143.19
4360	5-pyrazolone (3)	....	$CH_2 \cdot CO \cdot NH \cdot N : C \cdot CH_3$	XXIV-19	98.11

Methyl-phenyl-nitrosoamine 4887
Methyl-phenylphenol ether 4073-4
Methyl phosphate 6268
Methyl phthalate 2530-2
Methyl *iso*-phthalic acid 6397

Methyl picrate 6290
Methyl-picrylnitramine 5873
Methyl pivalate 4404
Methyl-proline-methylbetaine, cf. alkd.

Table 7-4 (*Continued*)
PHYSICAL CONSTANTS OF ORGANIC COMPOUNDS

No.	Crystalline Form and Color	Specific Gravity	Melting Point °C.	Boiling Point °C.	Solubility in 100 Parts		
					Water	Alcohol	Ether
4328	red pd.	....	....	....	0.2 c.	i.	....
4329	cr.	....	37±	108-9^{12mm}	....	....	....
4330	col. cr.	....	29.5-30.5	196^{15mm}	i.	s.	s.
4331	lq.	0.877$^{17.5°}$	....	213-47^{57mm}	i.	s.	s.
4332	lq.	0.924$^{17°}_{4}$	....	196	s.	s.	s.
4333	lf./al.	....	50-1	....	i. d.	v. s. d.	v. s.
4334	nd.	....	117	350±, sl. d.	sl. s. h.	s.; s. H_2SO_4	s.; s. chl.
4335	lq.	1.044$^{16°}$	d. 360	220	i.	∞	∞
4336	col. lf.	....	35.5	257-9.5	s.	s.	s.
4337	yel. pd.	....	....	....	v. s.	sl. s.	i.
4338	yel. lq.	1.038$^{21.5·6°}_{4}$	....	131^{35mm}	sl. s. h.	∞; ∞ chl.	∞; ∞ bz.
4339	lf.	....	....	....	v. s.	v. sl. s. c.	....
4340	oil	1.04$^{15°}_{15}$	....	110-2$^{12-15mm}$	....	s.	s.
4341	col. gas	....	....	-14^{759mm}	i.	sl. s.	7000$^{0°}$ cc.
4342	pa. yel. oil	....	....	....	i.	s. abs.	s. act.; s. pet.
4343	col. nd./al.	....	62-3	....	i.	v. s.; v. s. bz.	v. s. act.
4344	lq.	0.821$^{15°}$	....	107	....	....	....
4345	lq.	0.862$^{0°}$	....	118-9^{753mm}	s.	i. aq. KOH	....
4346	lq.	0.864$^{0°}_{4}$	....	125-6	s.	....	....
4347	lq.	0.867$^{0°}$	....	127-9	s.	....	....
4348	lq.	0.83$^{12.5°}$	....	63	v. sl. s.	∞	∞
4349	col. lq.	0.915$^{20°}_{4}$	-87.5	79.7	0.5$^{20°}$	∞	∞
4350	lq.	0.720$^{17°}$	....	62-4	s.	s.	∞
4351	lq.	0.863$^{16°}$	....	182	i.	s. abs.	∞
4352	lq.	0.864$^{15.4°}_{4}$	-63	183.5	i.	s. abs.	∞
4353	col. lq.	0.738$^{20°}$	....	39.1	sl. s.	∞	∞
4354	col. lq.	0.735$^{20°}_{20}$	....	32.5^{777mm}	v. sl. s.	∞	∞
4355	col. lq.	0.812$^{15°}_{15}$	-77.8	102.4	v. sl. s.	∞	∞
4356	oil	0.910$^{20°}_{4}$	....	168^{748mm}	s.	∞	∞
4357	col. lq.	0.803$^{20°}_{0}$	-92	95	v. sl. s.	∞	∞
4358	lq.	....	....	157-8	s.	∞	∞
4359	cr./et.	....	110-4	....	s.	v. s.	v. s.
4360	pr./aq.; nd./al.	....	216-7	subl.	v. s. h.; sl. s. Na_2CO_3	sl. s. h.	sl. s. aq. a.

Methyl-propenyl ketone 3213
Methyl-*iso*-propyl-acetophenone 4181
Methyl-*iso*-propylbenzene 1622-4
Methyl-propyl carbinol (*n*) 405

Methyl-propyl carbinol (*iso*) 407
Methyl-*iso*-propyl-phenanthrene 5555
Methyl-*iso*-propyl-phenol 1255, 5938
Methyl-pyridine 5300-2

Table 7-4 (*Continued*)
PHYSICAL CONSTANTS OF ORGANIC COMPOUNDS

No.	Name	Synonym	Formula	Beil. Ref.	Formula Weight
4361	**Methyl** pyrogallol (1;3,4,5)	trihydroxy-toluene	$CH_3 \cdot C_6H_2(OH)_3$	VI-1112	140.14
4362	pyrrole (1)	*N*-methyl pyrrole	$C_4H_4N \cdot CH_3$	XX-163	81.12
4363	pyrrole (2)(α)	. . . .	$CH_3 \cdot C_4H_4N$	XX-170	81.12
4364	pyrrole (3)(β)	. . . .	$CH_3 \cdot C_4H_4N$	XX-171	81.12
4365	pyrroline (*N*)	1-Me-3-pyrroline	$CH_3 \cdot N \cdot CH_2(CH)_2CH_2$	XX-133	83.13
4366	pyruviate	. . . .	$C_3H_3O_3 \cdot CH_3$	III-616	102.09
4367	quinoline (2)	quinaldine	$CH_3 \cdot C_9H_6N$	XX-388	143.19
4368	quinoline ethiodide	quinaldine ethiodide	$CH_3 \cdot C_9H_6N : (I)C_2H_5$	XX-392	299.16
4369	quinoline (3)	*Py*-3 (β)	$CH_3 \cdot C_9H_6N$	XX-394	143.19
4370	quinoline (4)	cincholepidine; lepidine (4)	$C_6H_4N : CH \cdot CH : C \cdot CH_3$	XX-395	143.19
4371	quinoline ethiodide	lepidine ethiodide	$CH_3 \cdot C_9H_6N : (I)C_2H_5$	XX-396	299.16
4372	quinoline (6)	*p*-toluquinoline	$CH_3 \cdot C_9H_6N$	XX-398	143.19
4373	quinoline ethiodide	*p*-toluquinoline EtI	$CH_3 \cdot C_9H_6N : (I)C_2H_5$	XX-398	299.16
4374	quinoline (7)	*m*-toluquinoline	$CH_3 \cdot C_9H_6N$	XX-400	143.19
4375	quinoline (8)	*o*-toluquinoline	$CH_3 \cdot C_9H_6N$	XX-401	143.19
4376	red	*p*-diMe-amino-azobenzene-*o'*-carboxylic	$HO_2C \cdot C_6H_4 \cdot N : N \cdot C_6H_4 \cdot N(CH_3)_2$	XVI-329	269.31
4377	red, Na salt	. . . .	$C_{15}H_{14}O_2N_3Na$	*XVI-316	291.29
4378	red, para	(4',4)	$NaO_2C \cdot C_6H_4 \cdot N : N \cdot C_6H_4 \cdot N(CH_3)_2$	XVI-329	291.29
4379	salicylaldehyde (3)	(3;6,1)	$CH_3 \cdot C_6H_3(OH)CHO$	VIII-100	136.15
4380	salicylate (*o*)	oil of wintergreen	$HO \cdot C_6H_4 \cdot CO_2CH_3$	X-70	152.15
4381	stearate	. . . .	$C_{17}H_{35}CO_2CH_3$	II-379	298.51
4382	sulfonic acid	methane sulfonic acid	$CH_3 \cdot SO_3H$	IV-4	96.11
4383	sulfonic chloride	Me-sulfone chloride	$CH_3 \cdot SO_2Cl$	IV-5	114.55
4384	sulfuric acid	Me-hydrogen sulfate	$CH_3O \cdot SO_2 \cdot OH$	I-283	112.10
4385	tetraacetyl-*d*-glucoside (α)	. . . .	$CH_3O \cdot C_{14}H_{19}O_9$	XXXI-180	362.34
4386	tetramethyl-*d*-glucoside (α)	. . . .	$(CH_3O)_5C_6H_7O$	XXXI-180	250.29
4387	thiocyanate	. . . .	$CH_3 \cdot S \cdot CN$	III-175	73.12
4388	*iso*-thiocyanate	methyl mustard oil	$CH_3 \cdot N : C : S$	IV-77	73.12
4389	thiophene (β)	β-thiotolene	$CH_3 \cdot C_4H_3S$	XVII-38	98.17
4390	thiourea	Me-thiocarbamide	$NH_2 \cdot CS \cdot NH \cdot CH_3$	IV-70	90.15
4391	*iso*-thiourea sulfate	. . . .	$(C_2H_6N_2S)_2 \cdot H_2SO_4$	*III-78	278.37
4392	toluate (*o*)	. . . .	$CH_3 \cdot C_6H_4 \cdot CO_2CH_3$	IX-463	150.18
4393	toluate (*m*)	. . . .	$CH_3 \cdot C_6H_4 \cdot CO_2CH_3$	IX-475	150.18
4394	toluate (*p*)	. . . .	$CH_3 \cdot C_6H_4 \cdot CO_2CH_3$	IX-484	150.18
4395	toluene sulfonate	(1,4)	$CH_3 \cdot C_6H_4 \cdot SO_3CH_3$	XI-99	186.23
4396	toluidine (*o*)	. . . .	$CH_3 \cdot C_6H_4 \cdot NHCH_3$	XII-784	121.18
4397	toluidine (*m*)	. . . .	$CH_3 \cdot C_6H_4 \cdot NHCH_3$	XII-856	121.18
4398	toluidine (*p*)	. . . .	$CH_3 \cdot C_6H_4 \cdot NHCH_3$	XII-902	121.18
4399	*o*-tolyl ether	Me *o*-cresyl ether	$CH_3 \cdot O \cdot C_6H_4 \cdot CH_3$	VI-352	122.17
4400	*m*-tolyl ether	*m*-cresyl Me ether	$CH_3 \cdot O \cdot C_6H_4 \cdot CH_3$	VI-376	122.17
4401	*p*-tolyl ether	. . . .	$CH_3 \cdot O \cdot C_6H_4 \cdot CH_3$	VI-392	122.17
4402	*p*-tolyl sulfide	thiocresol Me ether	$CH_3 \cdot S \cdot C_6H_4 \cdot CH_3$	VI-417	138.23
4403	trichloroacetate	. . . .	$Cl_3C \cdot CO_2CH_3$	II-208	177.42
4404	trimethylacetate	methyl pivalate	$(CH_3)_3C \cdot CO_2CH_3$	II-320	116.16

Methyl-quercetin 5556
Methyl-quinaldine 2541, 2544
Methyl quinolinate 2539
Methyl-quinoline carboxylic acid 5528-32
Methyl-quinolone 4180
Methyl-resorcinol 2371, 2373, 4998
Methyl-rhodin 4090
Methyl rosaniline 3593

Methyl-salicylic acid 3833, 4067
Methyl sebacate 2551
Methyl-styrene (α) 5224
Methyl-styrene (dimer) 2476
Methyl-styryl ketone 626
Methyl succinate 2554-5
Methyl-succinic acid 5520
Methyl sulfate 2559

Table 7-4 (*Continued*)
PHYSICAL CONSTANTS OF ORGANIC COMPOUNDS

No.	Crystalline Form and Color	Specific Gravity	Melting Point °C.	Boiling Point °C.	Solubility in 100 Parts		
					Water	Alcohol	Ether
4361	nd./bz.	....	129	....	d. fused alk.	....	....
4362	lq.	$0.920^{10°}$	....	$114\text{-}5^{748mm}$	i.	∞	∞
4363	lq.	0.945	....	$147\text{-}8^{750mm}$	sl. d. a.	....	....
4364	lq.	....	....	$142\text{-}3^{743mm}$	sl. d. a.	....	....
4365	col. lq.	....	....	79-80	∞	s.	s.; s. chl.
4366	lq.	$1.154^{0°}$	....	134-7	....	....	....
4367	lq.	$1.059^{\frac{20°}{4}}$	−1	$244\text{-}5^{750mm}$	v. sl. s.	s. chl.	s.
4368	yel. nd./al.	....	233-4	....	s.	v. sl. s.	i.
4369	cr.	$1.067^{\frac{20°}{4}}$	16	250^{710mm}	i.	s.	s.
4370	lq.	$1.086^{20°}$	9-10	261-3	sl. s.	∞; ∞ bz.	∞; ∞ lg.
4371	yel. cr./al.	....	142-3	....	s.	s.	i.
4372	lq.	$1.068^{20°}$	−22	258-9	v. sl. s.	s.	s.
4373	cr.	....	171-3	....	....	....	....
4374	yel. oil	$1.061^{\frac{20°}{4}}$	39	257.6^{750mm}	v. sl. s.	s.	s.
4375	lq.	$1.073^{20°}$	....	247.8^{760mm}	v. sl. s.	∞	∞
4376	vl. nd. or red pd.	....	179-83	....	v. sl. s.	s.	s. ac.
4377	red pl.	....	....	....	s.	....	....
4378	or. pd.	....	....	....	....	....	....
4379	lf./aq.al.	$1.091^{\frac{5.8°}{4}}$	56	217-8	v. sl. s.	s.	s.; s. chl.
4380	col. lq.	$1.182^{\frac{25°}{25}}$	−8.6	222.9	$0.073^{30°}$	∞; s. chl.	∞
4381	col. cr.	....	38-9	$214\text{-}5^{15mm}$	i.	s.	s.
4382	col. lq.	$1.481^{\frac{18°}{4}}$	20	167^{10mm}	∞	....	....
4383	yel. lq.	$1.481^{\frac{18°}{4}}$	−32	161.5^{730mm}	i. c.; d. h.	....	s.
4384	oil	....	<−30	d.	v. s.	s.	∞ abs.
4385	col. rhb./al.	....	101	....	i.; s. chl.	s.; s. act.	s.; s. pet.
4386	col. lq.	$1.108^{20°}$	....	$145\text{-}50^{13mm}$	sl. s.	s.; s. chl.	sl. s.
4387	lq.	$1.069^{\frac{24°}{4}}$	−51	130-3	v. sl. s.	∞	∞
4388	col. cr.	$1.069^{\frac{37°}{4}}$	35-6	119^{759mm}	v. sl. s.	∞	∞
4389	oil	$1.022^{\frac{20°}{4}}$	−68.9(−74)	115.4	....	....	....
4390	pr.	....	118-9	....	v. s.	v. s.	sl. s.
4391	cr. pd.	....	241-2 d.	....	s. h.	i.	....
4392	col. lq.	$1.073^{15°}$	<−50	213	i.	∞	∞
4393	col. lq.	$1.066^{15°}$	....	215	i.	∞	∞
4394	cr./pet.	....	33-4	217	i.	v. s.	v. s.
4395	cr./et. lg.	....	28	....	i.; s. bz.	v. s.	v. s.
4396	lq.	$0.973^{15°}$	....	206-7	i.	∞	∞
4397	lq.	....	....	206-7	i.	∞	∞
4398	lq.	$0.935^{\frac{5.8°}{5}}$	....	$209\text{-}11^{761mm}$	i.	∞	∞
4399	lq.	$0.985^{\frac{15°}{15}}$	....	171-2	i.	v. s.	v. s.
4400	lq.	$0.977^{\frac{15°}{15}}$	....	177	....	....	....
4401	lq.	$0.976^{\frac{15°}{15}}$	....	176	....	....	....
4402	lq.	$1.030^{\frac{18°}{6}}$	....	209^{747mm}	....	....	....
4403	lq.	$1.489^{\frac{13.2°}{3.2}}$	−17.5	$152\text{-}3^{765mm}$	d.	d.	s.
4404	lq.	$0.891^{\frac{0°}{4}}$	....	100-2	....	....	....

Table 7-4 (*Continued*)
PHYSICAL CONSTANTS OF ORGANIC COMPOUNDS

No.	Name	Synonym	Formula	Beil. Ref.	Formula Weight
4405	**Methyl** (tyrosine N)($l+$)	ratanhine; andirine	$CH_3NH \cdot C_9H_9O_3$	XIV-612	195.22
4406	umbelliferone (β)	7-OH-4-Me coumarin	$C_{10}H_8O_3$	XVIII-31	176.17
4407	undecylenate	....	$CH_2{:}CH(CH_2)_8CO_2 \cdot CH_3$	II-459	198.31
4408	n-undecyl ketone	tridecanone-2	$CH_3 \cdot CO(CH_2)_{10}CH_3$	I-715	198.35
4409	n-undecyl ketone semicarbazone	....	$C_{11}H_{23}(CH_3)C{:}N \cdot NH \cdot CO \cdot NH_2$	*I-371	255.41

4410	uracil (1)	....	$CO \cdot (CH)_2NH \cdot CO \cdot N \cdot$
			    |_____|
			      CH_3
				XXIV-316	126.12

4411	uracil (3)	....	$CO \cdot NH \cdot CO(CH)_2 \cdot N \cdot$
			   |_____|
			      CH_3
				XXIV-316	126.12

4412	uracil (4)	....	$(NH \cdot CO)_2CH{:}C \cdot CH_3$ |_____|	XXIV-342	126.12
4413	uracil (5)	thymine	$(CO \cdot NH)_2CH{:}C \cdot CH_3$ $_7$ |_____|	XXIV-353	126.12
4414	urea (N)	....	$CH_3 \cdot NH \cdot CO \cdot NH_2$	IV-64	74.08
4415	urea nitrate	....	$C_2H_6ON_2 \cdot HNO_3$	IV-65	137.10
4416	urethane	*cf. also* Me-carbamate	$CH_3 \cdot NH \cdot CO_2C_2H_5$	IV-64	103.12
4417	uric acid (1)	....	$C_6H_6O_3N_4$	XXVI-524	182.14
4418	uric acid (3)	....	$C_6H_6O_3N_4 \cdot \frac{1}{2}H_2O$	XXVI-524	191.15
4419	uric acid (7)	....	$C_6H_6O_3N_4 \cdot H_2O$	XXVI-525	200.16
4420	n-valerate	....	$CH_3 \cdot (CH_2)_3 \cdot CO_2CH_3$	II-301	116.16
4421	*iso*-valerate	....	$(CH_3)_2C_2H_3 \cdot CO_2CH_3$	II-311	116.16
4422	vinyl carbinol	buten-1-ol-3	$CH_2{:}CH \cdot CHOH \cdot CH_3$	I-441	72.11
4423	d-xyloside (β)	....	$CH_3O \cdot C_5H_9O_4$	XXXI-54	164.16
4424	**Methylal**	dimethoxy-methane	$CH_2(OCH_3)_2$	I-574	76.10
4425	**Methylene**-amino-acetonitrile	(dimolecular)	$(CH_2{:}N \cdot CH_2 \cdot CN)_2$	II-89	136.16
4426	blue (base)	N,N,N',N',-tetra-Me-thionine	$[(CH_3)_2N]_2C_{12}H_6 \cdot NS(OH)$	XXVII-393	301.41
4427	blue chloride	Me-thionine chloride	$C_{16}H_{18}N_3S(Cl) \cdot 3H_2O$	XXVII-395	373.90
4428	bromide	dibromomethane	$CH_2{:}Br_2$	I-67	173.85
4429	chloride	dichloromethane	$CH_2{:}Cl_2$	I-60	84.93
4430	diacetate	....	$(CH_3CO_2)_2CH_2$	II-152	132.12
4431	dianiline	....	$(C_6H_5 \cdot NH)_2CH_2$	XII-184	198.27
4432	disulfonic acid	methionic acid	$CH_2{:}(SO_2OH)_2$	I-579	176.17
4433	fluoride	difluoromethane	$CH_2{:}F_2$	I-59	52.02
4434	iodide	diiodomethane	$CH_2{:}I_2$	I-71	267.84
4435	**Michler's** hydrol (p,p')	....	$[(CH_3)_2N \cdot C_6H_4]_2{:}CHOH$	XIII-698	270.38
4436	ketone (p,p')	....	$[(CH_3)_2N \cdot C_6H_4]_2CO$	XIV-89	268.36
4437	**Morpholine-ethan-ol ethyl ether**	....	$O{:}(CH_2CH_2)_2{:}N \cdot C_2H_4 \cdot O \cdot C_2H_5$	....	159.23
4438	**Mucic** acid	....	$(CHOH)_4(CO_2H)_2$	III-581	210.14
4439	acid, allo-	....	$(CHOH)_4(CO_2H)_2$	III-576	210.14

Table 7-4 (*Continued*)
PHYSICAL CONSTANTS OF ORGANIC COMPOUNDS

No.	Crystalline Form and Color	Specific Gravity	Melting Point °C.	Boiling Point °C.	Solubility in 100 Parts		
					Water	Alcohol	Ether
4405	col. nd.	....	257	d. 280	$0.14^{20°}$; $0.5^{100°}$	0.007 abs.	i.; s. NH$_4$OH
4406	nd./al.*	....	188-9	d.	v. sl. s. h.; s. alk.	s.; s. ac.; i. Na$_2$CO$_3$	v. sl. s.; sl. s. chl.
4407	lq.	$0.889^{15°}$	−27.5	248-9	....	....	....
4408	cr.	$0.823^{28°}$	28-9	260-5	i.	v. s.	v. s.
4409	cr.	....	122-6	....	....	....	....
4410	pr./aq. al.	....	174-5	....	v. s.	v. s.	i. bz.
4411	pr./al.	....	232	....	s.; i. dil. HCl	s. h.	s. aq. NaOH
4412	nd./al.	....	d. 270-80	....	$0.7^{22°}$; s. alk.	sl. s.; s. H$_2$SO$_4$	v. sl. s.; s. NH$_4$OH
4413	pl./aq.	....	326 d.	....	$0.4^{25°}$; s. alk.	sl. s.; s. H$_2$SO$_4$	v. sl. s.
4414	pr./aq.	1.204	101-2	d.	v. s.	v. s.	i.
4415	col. cr.	....	128-32	....	....	....	....
4416	col. lq.	$1.009\frac{19}{4}°$	....	170	$69^{16°}$	s.	....
4417	cr.	....	>360 d.	....	0.05 h.	v. sl. s.	....
4418	cr.	....	>360 d.	....	0.2 h.	....	s. alk.
4419	cr.	....	>370 d.	....	$1.3^{100°}$	....	s. alk.
4420	lq.	$0.895\frac{15}{4}°$	−91	127.3	v. sl. s.	∞	∞
4421	col. lq.	$0.881\frac{20}{4}°$	....	116-7^{764mm}	v. sl. s.	∞	∞
4422	lq.	$0.831\frac{20}{4}°$	<−80	97	∞		
4423	cr./EtOAc	....	155-6	....	v. s.	s. h.	5 h. act.
4424	col. lq.	$0.866\frac{15}{4}°$	−105	42.3	33	∞	∞
4425	col. pr./ aq.	....	129.5	....	s. h.	s. h.	i.; i. bz.
4426	dark, amor.	....	....	....	v. s.	v. s.	i.
4427	gn. b. met. /aq. HCl	....	−2H$_2$O, 100	−3H$_2$O, 150	4	2	i.; s. chl.
4428	col. lq.	$2.495\frac{20}{4}°$	−52.7	97.0	$1.17^{30°}$; $1.15^{20°}$	∞ ; ∞ act.	∞
4429	col. lq.	$1.336\frac{20}{4}°$	−96.7	40.2	$2^{20°}$	∞	∞
4430	lq.	$1.132\frac{20}{20}°$	−23	164-5	sl. s.	v. s.	∞
4431	pl./et. pet.	....	65	208-9 d.	i.; i. pet.	s.	s.
4432	hyg. nd.	....	....	....	$245.8^{25°}$	s.	....
4433	gas	....	....	....	....	....	....
4434	col. lq.	$3.325\frac{20}{4}°$	6.1(5.6)	180 d.	$1.6^{0°}$; $1.4^{20°}$	∞	∞
4435	gn. lf./bz.	....	96-7	....	i.; s. bz.	s. h.	s.; s. ac.
4436	lf./al.	....	174	>360 d.	i.; v. s. bz.	sl. s.	v. sl. s.
4437	col. lq.	$0.965\frac{20}{20}°$	....	206.2	∞	....	....
4438	pd.	....	206-14 d. (108)	....	$0.33^{14°}$; $1.6^{100°}$	i.	i.; s. alk.
4439	nd./aq.	....	166-71 d.	....	9 h.	sl. s.	

*Crysts. + 1H$_2$O/aq.; − H$_2$O, 110°.

Monacetin 3382
Monazol 3800
Monobutyrin 3392
Monochlorohydrin 3395-6
Monoethanolamine 2907
Monopropionin 3413

Monosalicylin 3414
Monostearin 3415-6
Monosulfonic acid-F 4510
Monotol 4117
Monotropitoside, cf. glcde.
Moogrol 3014
Morin 5806

Mori(n)tannic acid 5070
Morphine, cf. alkd.
Morphol 2360
Morpholine 2235
Morpholine-ethanol 2236
Morpholine-ethanol ethyl ether 4437
Moss starch 3969

Table 7-4 (*Continued*)
PHYSICAL CONSTANTS OF ORGANIC COMPOUNDS

No.	Name	Synonym	Formula	Beil. Ref.	Formula Weight
4440	**Mucic** acid, talo-	(*d* or *l*)	$(CHOH)_4(CO_2H)_2$	III-577	210.14
4441	**Mucobromic acid**	. . . .	$HO_2C \cdot CBr:CBr \cdot CHO$	III-728	257.88
4442	**Mucochloric acid**	. . . .	$HO_2C \cdot CCl:CCl \cdot CHO$	III-727	168.96
4443	**Muconic acid**	. . . .	$(\cdot CH:CH \cdot CO_2H)_2$	II-803	142.11
4444	**Mudarol**	mudarin	$C_{30}H_{47}O(OH)$	. . . .	440.72
4445	**Murexide**	ammonium salt purpuric acid	$C_8H_4O_6N_5 \cdot NH_4 \cdot H_2O$	XXV-499	302.20
4446	**Muscone** (*l*)	3-Me-cyclopenta-decanone	$CH_3 \cdot C_{15}H_{27}:O$	**VII-51	238.42
4447	**Myochrysine**	Na aurothiomalate	$NaO_2C \cdot CH_2 \cdot CH:$ $(SAu)CO_2Na$	. . . .	390.08
4448	**Myricyl** acetate	. . . .	$CH_3CO_2 \cdot C_{31}H_{63}$	*II-63	494.89
4449	alcohol	melissyl alcohol	$C_{31}H_{63}OH$ $C_{30}H_{61}OH(?)$	*I-222	452.86 438.83
4450	**Myristic** acid	tetradecanoic acid	$CH_3(CH_2)_{12} \cdot CO_2H$	II-365	228.38
4451	aldehyde	tetradecanal	$CH_3(CH_2)_{12} \cdot CHO$	I-716	212.38
4452	amide	myristamide	$C_{13}H_{27} \cdot CO_2NH_2$	II-368	243.39
4453	anhydride	. . . .	$(C_{13}H_{27}CO)_2O$	II-367	438.74
4454	anilide	myristanilide	$C_{13}H_{27}CO \cdot NHC_6H_5$	XII-257	303.49
4455	chloride	myristoyl chloride	$C_{13}H_{27} \cdot COCl$	II-368	246.82
4456	nitrile	tridecyl cyanide	$C_{13}H_{27} \cdot CN$	II-368	209.38
4457	**Myristyl bromide**	. . . .	$CH_3(CH_2)_{12}CH_2Br$	**I-136	277.30
4458	**Naphthalene**	. . . .	$C_{10}H_8$	V-531	128.18
4459	dicarboxylic acid	naphthalic acid (1,4)	$C_{10}H_6(CO_2H)_2$	IX-917	216.20
4460	dicarboxylic acid	naphthalic acid (1,8)	$C_{10}H_6(CO_2H)_2$	IX-918	216.20
4461	disulfonic acid (1,5)	Armstrong's acid	$C_{10}H_6(SO_3H)_2$	XI-212	288.30
4462	disulfonic acid (1,6)	(δ)	$C_{10}H_6(SO_3H)_2$	XI-213	288.30
4463	disulfonic acid (2,6)	. . . .	$C_{10}H_6(SO_3H)_2$	XI-215	288.30
4464	disulfonic acid (2,7)	(α)	$C_{10}H_6(SO_3H)_2$	XI-216	288.30
4465	sulfinic acid (α)	. . . .	$C_{10}H_7 \cdot SO_2H$	XI-15	192.24
4466	sulfinic acid (β)	. . . .	$C_{10}H_7 \cdot SO_2H$	XI-16	192.24
4467	sulfone amide (α)	. . . .	$C_{10}H_7 \cdot SO_2NH_2$	XI-157	207.25
4468	sulfone chloride (α)	. . . .	$C_{10}H_7 \cdot SO_2Cl$	XI-157	226.68
4469	sulfone chloride (β)	. . . .	$C_{10}H_7 \cdot SO_2Cl$	XI-173	226.68
4470	sulfonic acid (α)	. . . .	$C_{10}H_7 \cdot SO_3H \cdot 2H_2O$	XI-155	244.27
4471	sulfonic acid (β)	. . . .	$C_{10}H_7 \cdot SO_3H \cdot H_2O$	XI-171	226.25
4472	sulfonic Na salt	(β)	$C_{10}H_7O_3SNa$	XI-171	230.22
4473	tetrachloride	(1,2,3,4)	$C_{10}H_8Cl_4$	V-492	269.99
4474	**Naphthalic anhydride**	(1,8)	$C_{10}H_6:(CO)_2O$	XVII-521	198.18
4475	**Naphthasultam**	(1,8)	$C_{10}H_7O_2NS$	XXVII-59	205.24
4476	disulfonate Na	(2,4)	$C_{10}H_4O_8NS_3Na_3 \cdot$ $8\frac{1}{2}H_2O$	XXVII-356	584.44
4477	**Naphthazarin** (5,8;1,4)	5,8-dihydroxy-α-naphthoquinone	$(HO)_2C_{10}H_4O_2$	VIII-412	190.16
4478	*iso*-**Naphthazarin**	(2,3;1,4)	$(HO)_2C_{10}H_4O_2$	VIII-411	190.16
4479	**Naphthoflavone** (α)	7,8-benzoflavone	$C_{19}H_{12}O_2$	XVII-390	272.31
4480	**Naphthoic** acid (α)	. . . .	$C_{10}H_7 \cdot CO_2H$	IX-647	172.19
4481	acid (β)	isonaphthoic acid	$C_{10}H_7 \cdot CO_2H$	IX-656	172.19
4482	aldehyde (α)	. . . .	$C_{10}H_7 \cdot CHO$	VII-400	156.19
4483	aldehyde (β)	. . . .	$C_{10}H_7 \cdot CHO$	VII-401	156.19
4484	amide (α)	. . . .	$C_{10}H_7 \cdot CONH_2$	IX-648	171.20

Table 7-4 (*Continued*)
PHYSICAL CONSTANTS OF ORGANIC COMPOUNDS

No.	Crystalline Form and Color	Specific Gravity	Melting Point °C.	Boiling Point °C.	Solubility in 100 Parts		
					Water	Alcohol	Ether
4440	lf./act.	....	158 ± d.	....	v. s.	v. s. h.	i.
4441	pl./et. lg.	....	122-5	....	v. s. h.	v. s.	v. s.
4442	pl./aq.	....	127	....	s. h.	s.	s.; i. lg.
4443	nd./aq.	....	298 d.	....	0.002 c.	s. h.	s. h. ac.
4444	hex./al. et.	....	176	....	i.	s.	s.
4445	gn. red pd.	....	....	....	v. sl. s. c.	i.	i.
4446	col. oil	$0.922\frac{17°}{4}$	*	328	v. sl. s.	∞	....
4447	lt. yel. pd.	....	....	....	v. s.	....	....
4448	nd.	....	73-5	$311\text{-}3^{8mm}$	....	....	....
4449	nd. or lf.	$0.777^{95°}$	88	....	i.	v. sl. s. c.; v. sl. s. pet.	v. s.; v. s. bz.
4450	col. lf.	$0.853\frac{70°}{4}$	54.2	250.5^{100mm}	i.; v. s. bz.	v. s. abs.	v. s.
4451	cr.	....	23.5	166^{24mm}			
4452	nd.	....	102	217^{12mm}	i.	s.	sl. s.
4453	col. cr.	$0.850\frac{20°}{4}$	53.5	$198^{0.1mm}$	....	....	....
4454	nd./al.	....	84	113^{10mm}	i.; s. chl.	v. s. bz.	v. s.
4455	col. lq.	....	1-3	168^{15mm}	d.	d.	s.
4456	cr.	$0.828\frac{19°}{4}$	19	226.5^{100mm}	....	s.	
4457	col. lq.	....	3-5	....	i.	....	s.
4458	col. pl./al.	$1.145\frac{20°}{4}$	80.2	218.0	$0.003^{25°}$; v. s. CCl_4	$9.5^{19.5°}$; v. s. CS_2	v. s.; $46^{16°}$ bz.
4459	nd.	....	>240	....	i. h.	s.	s.
4460	nd./al.	....	d. >140	....	v. sl. s.	s. h.	sl. s.
4461	lf.	....	d.	....	$102^{20°}$	s.	i.
4462	cr.	....	....	d. 125	$164^{20°}$	s.	i.
4463	hyg. lf.	....	....	....	s.	....	
4464	hyg. nd.	....	....	....	s.	sl. c. HCl	....
4465	nd./aq.	....	98-9	....	s.	s.	sl. s.
4466	nd.	....	105	....	s.	s.	s.
4467	cr./al.	....	150	....	....	v. s.	v. s.
4468	lf./et.	....	68	195^{13mm}	i.	v. s.	v. s.
4469	lf.	....	79	201^{13mm}	i.	sl. s. pet.	s. bz.
4470	cr.	....	90	....	v. s.	v. s.	sl. s.
4471	hyg. cr.	....	125 †	....	$77^{30°}$	0.2 h. bz.	....
4472	cr./aq.	....	....	d.	$6^{23.9°}$	....	
4473	mn./chl.	....	187-9	....	i.	v. sl. s. h.	sl. s. h.
4474	nd./al.	....	273-4	....	sl. s. ac.	v. sl. s.	v. sl. s.
4475	nd./bz.	....	177-8	....	s. h.	sl. s.	s.
4476	yel. lf./aq. al.	....	−8H_2O, 160	....	v. s.	sl. s.	....
4477	brn. nd., al.	....	subl. vac.	....	sl. s. h.	sl. s.; s. alk.	sl. s.
4478	brn. lf./ac.	....	276-80	subl.	sl. s. h.	sl. s.	v. sl. s.
4479	cr./aq. al.	....	155-6	....	....	....	....
4480	nd./aq. al.	....	161-3	300	v. sl. s. h.	s. h.	s.
4481	mn.	$1.077\frac{100°}{4}$	184	>300	$0.007^{25°}$	s.	s.
4482	lq.	$1.148\frac{20°}{4}$	....	291.6	....	....	....
4483	lf./h. aq.	$1.078^{99.4°}$	60.5-1.0	....	s. h.	v. s.	v. s.
4484	nd./al.	....	202	....	....	v. sl. s.	....

* Oxime, m. p. 45°; semicarbazone, m. p. 134°.
† Crysts. + 3H_2O, m. p. 83°; anh., m. p. 91°.
Myristamide 4452
Myristanilide 4454
Myristoyl chloride 4455

Myristyl alcohol 5750
Nandinine, cf. alkd.
Napelline, cf. alkd.

Naphthacetol 31
Naphthalene acetic acid 4528
Naphthalene decahydride 1631-2
Naphthalic acid 4459-60
Naphthalidine 4529

Table 7-4 (*Continued*)
PHYSICAL CONSTANTS OF ORGANIC COMPOUNDS

No.	Name	Synonym	Formula	Beil. Ref.	Formula Weight
	Naphthoic				
4485	amide (β)	....	$C_{10}H_7 \cdot CONH_2$	IX-657	171.20
4486	chloride (β)	β-naphthoyl chloride	$C_{10}H_7 \cdot COCl$	IX-657	190.63
4487	nitrile (α)	α-naphthyl cyanide	$C_{10}H_7 \cdot CN$	IX-649	153.19
4488	nitrile (β)	β-naphthyl cyanide	$C_{10}H_7 \cdot CN$	IX-659	153.19
4489	**Naphthol** (α)	α-hydroxy naphthalene	$C_{10}H_7 \cdot OH$	VI-596	144.17
4490	**Naphthol** (β)	β-hydroxy naphthalene	$C_{10}H_7 \cdot OH$	VI-627	144.17
4491	disulfonic acid (β)	(2;3,6); R-acid	$HO \cdot C_{10}H_5(SO_3H)_2$	XI-288	304.30
4492	disulfonic Ca	R-acid Ca salt	$HO \cdot C_{10}H_5(SO_3)_2Ca$	....	342.36
4493	disulfonic K	R-acid K salt	$HO \cdot C_{10}H_5(SO_3K)_2$	....	380.49
4494	disulfonic Na	R-acid Na salt	$HO \cdot C_{10}H_5(SO_3Na)_2$	XI-289	348.26
4495	disulfonic K (β)	(2;6,8); G-acid K	$HO \cdot C_{10}H_5(SO_3K)_2$	....	380.49
4496	disulfonic Na (β)	G-acid Na salt	$HO \cdot C_{10}H_5(SO_3Na)_2$	XI-290	348.26
4497	phthalein (α)		$C_8H_4O_2(C_{10}H_6OH)_2$	XVIII-157	418.45
4498	sulfonic acid (α)	(1,2); Schaeffer's (α)	$HO \cdot C_{10}H_6 \cdot SO_3H$	XI-269	224.24
4499	sulfonic Ca (α)	Schaeffer's Ca (α)	$(C_{10}H_7O_4S)_2Ca \cdot H_2O$	XI-270	504.55
4500	sulfonic K (α)	Schaeffer's K	$C_{10}H_7O_4SK \cdot \frac{1}{2}H_2O$	XI-270	271.34
4501	sulfonic Na (α)	Schaeffer's Na	$C_{10}H_7O_4SNa$	XI-269	246.22
4502	sulfonic acid (α)	(1,4);Neville-Winther	$HO \cdot C_{10}H_6 \cdot SO_3H$	XI-271	224.24
4503	sulfonic acid (α)	(1-OH; 5-SO$_3$H)	$HO \cdot C_{10}H_6 \cdot SO_3H$	XI-273	224.24
4504	sulfonic acid (α)	(1,8)	$HO \cdot C_{10}H_6 \cdot SO_3H \cdot H_2O$	XI-275	242.25
4505	sulfonic acid (β)	(2-OH; 1-SO$_3$H)	$HO \cdot C_{10}H_6 \cdot SO_3H$	XI-281	224.24
4506	sulfonic Ca (β)	(2,1); asaprol	$(C_{10}H_7O_4S)_2Ca \cdot 3H_2O$	....	540.58
4507	sulfonic acid (β)	(2,6); Schaeffer's (β)	$HO \cdot C_{10}H_6 \cdot SO_3H$	XI-282	224.24
4508	sulfonic Ca (β)	Schaeffer's Ca (β)	$(C_{10}H_7O_4S)_2Ca \cdot 5H_2O$	XI-282	576.61
4509	sulfonic K (β)	Schaeffer's K (β)	$C_{10}H_7O_4SK \cdot xH_2O$	XI-283	(262.33)
4510	sulfonic acid (β)	(2-OH; 7-SO$_3$H)	$HO \cdot C_{10}H_6 \cdot SO_3H$	XI-285	224.24
4511	sulfonic Na (β)	(2,7); F-acid Na	$C_{10}H_7O_4SNa \cdot 2\frac{1}{2}H_2O$	XI-286	291.26
4512	sulfonic Na (β)	(2,8) Bayer-acid Na	$C_{10}H_7O_4SNa$	XI-286	246.22
4513	**Naphtho-**phenazine	(α,β)	$C_{10}H_6 : N_2 : C_6H_4$	XXIII-276	230.27
4514	quinaldine (α)	....	$C_{13}H_8NCH_3$	XX-471	193.25
4515	quinaldine (β)	....	$C_{13}H_8NCH_3$	XX-471	193.25
4516	quinaldine (γ)	....	$C_{13}H_8NCH_3$	....	193.25
4517	quinoline (α)	....	$C_{13}H_9N$	XX-463	179.22
4518	quinoline (β)	5,6-benzoquinoline	$C_{13}H_9N$	XX-464	179.22
4519	quinoline HCl (β)	....	$C_{13}H_9N \cdot HCl \cdot 2H_2O$	XX-465	251.72
4520	quinone (α)(1,4)	....	$C_{10}H_6O_2$	VII-724	158.16
4521	quinone (β)(1,2)	....	$C_{10}H_6O_2$	VII-709	158.16
4522	quinone-4-sulfonate Na (β)	....	$C_{10}H_5O_2 \cdot SO_3Na$	XI-330	260.20
4523	quinone (amphi)	(2,6)	$C_{10}H_6O_2$	VII-733	158.16
4524	sulfone (1,8)	naphthosultone	$C_{10}H_6O \cdot SO_2$ $\underline{\quad\quad}$	XIX-43	206.22
4525	**Naphthoylacetonitrile** (β)	....	$C_{10}H_7 \cdot CO \cdot CH_2 \cdot CN$	....	195.22
4526	**Naphthyl** acetate (α)	....	$CH_3CO_2 \cdot C_{10}H_7$	VI-608	186.21
4527	acetate (β)	....	$CH_3CO_2 \cdot C_{10}H_7$	VI-644	186.21
4528	acetic acid (α)	....	$C_{10}H_7 \cdot CH_2CO_2H$	IX-666	186.21
4529	amine (α)	1-NH$_2$-naphthalene	$C_{10}H_7 \cdot NH_2$	XII-1212	143.19

Naphthane 1631-2
Naphthanthracene 650
Naphthenic acid 3577
Naphthionic acid 4541, 4544

Naphthol yellow 2642
Naphthol yellow-S 3241
Naphtho-hydroquinone 2348, 2350

Naphtho-nitrile 4487-8
Naphtho-picric acid 6305
Naphtho-quinone oxime 4892

Table 7-4 (*Continued*)
PHYSICAL CONSTANTS OF ORGANIC COMPOUNDS

No.	Crystalline Form and Color	Specific Gravity	Melting Point °C.	Boiling Point °C.	Solubility in 100 Parts Water	Alcohol	Ether
4485	pl./al.	....	192	....	s. bz.	s. h.	s.; s. chl.
4486	cr.	....	43	304–6	d.	s. h. bz.	s. h.
4487	nd./lg.	1.111^{25}_{25}	36–7	299	....	v. s.	....
4488	lf./lg.	1.094^{60}_{60}	66	305–6	v. sl. s.	s.	s.
4489	mn.	1.224^{40}	96; subl.	278–80	sl. s. h.; s. alk.	v. s.; s. bz.	v. s.; s. chl.
4490	mn.	1.217^{40}	122–3	285–6	0.1c.; 1.25 h.	v. s.	v. s.; chl.
4491	delq. nd.	....	....	....	v. s.	v. s.	i.
4492	cr.	....	....	....	$30.6^{25\circ}$	$61^{90\circ}$ aq.	....
4493	cr.	....	....	....	$29.5^{25\circ}$	$58^{100\circ}$ aq.	....
4494	col. nd.	....	....	....	$25.2^{25\circ}$	v. sl. s.	$41^{90\circ}$ aq.
4495	col. cr.	....	....	....	$8.01^{25\circ}$	....	$32^{90\circ}$ aq.
4496	col. cr.	....	....	....	$34.2^{20\circ}$	....	$63^{90\circ}$ aq.
4497	cr./bz.*	....	253–5*	....	i.	s.	s. alk.
4498	pl./aq.	....	>250	....	v. s. h.	....	i.
4499	cr.	....	....	....	v. sl. s.	....	....
4500	pr./aq.	....	....	....	$2.8^{18\circ}$	v. sl. s.	i. KCl
4501	cr./al.	....	....	....	$6.29^{25\circ}$	$23^{90\circ}$ aq.	....
4502	pl./aq.	....	170 d.	....	v. s.	....	....
4503	hyg. cr.	....	110–20	....	v. s.	....	....
4504	cr.	....	106–7	$-H_2O$, 180	v. s.	....	....
4505	cr.	....	....	....	v. s.	....	....
4506	wh. pd.	....	50±	....	6.6	4	....
4507	lf.	....	125	....	v. s.	v. s.	....
4508	col. lf.	....	....	....	$4.76^{20\circ}$	s.	$16^{90\circ}$ aq.
4509	nd. or lf.	....	....	....	v. s. h.	i.	i.
4510	nd./HCl	....	115–6†	d. 150	s.	s.	i.; i. bz.
4511	lf.	....	....	....	$8^{15\circ}$	....	....
4512	lf.	....	....	....	v. s.	v. sl. s.	....
4513	yel. nd./bz.	....	142.5	>360	sl. s. bz.	v. sl. s.	v. sl. s.
4514	lq.	....	....	>300	....	....	....
4515	nd./aq. al.	....	82	>300	sl. s.	s.	s.
4516	cr.	....	91–2	....	....	s.	s.; s. bz.
4517	mn./et.	....	52	223^{347mm}	v. sl. s.	s.	s.; s. bz.
4518	pl./h. aq.	....	93.5	350^{721mm}	sl. s. h.; s. aq. a.	v. s.	v. s.; v. s. bz.
4519	nd.	....	....	....	v. s.	i.	....
4520	yel. trl.	....	125–6	subl. 100	v. sl. s. c.	v. s. h.	s.
4521	red nd./et.	....	d. 115–20	....	s.; s. H_2SO_4	s. bz.	s.
4522	cr./50% al.	....	....	....	v. s.	i.	....
4523	red pr.	....	135	....	i. pet.	s. d.	v. sl. s.
4524	pr./bz.	....	154	>360 sl. d.	sl. s. CS_2	sl. s.; v. s. chl.	s. h. bz.
4525	yel. nd.	....	118–20	....	i.	s.	sl. s.
4526	nd./al.	....	46–9	....	....	s.	s.
4527	nd./al.	....	69–70	....	s. chl.	s.	s.
4528	col. nd./aq.	....	133	....	s. h.	3.3	s.; s. bz.
4529	rhb./aq. al.	1.123^{25}_{25}	50; subl.	300.8	0.17 c.	v. s.	v. s.

* Crysts. $+ C_6H_6$; loses C_6H_6 > 100°; m. anh. 253–5°.
† Crysts. $+ 1H_2O$, m. p. 108–9; $+ 2H_2O$, m. p. 95°; $+ 4H_2O$, m. p. 67°.

Naphtho-resorcinol 2349
Naphtho-salol 4569

Naphtho-sultone 4524
Naphthoyl chloride 4486

Table 7-4 (*Continued*)
PHYSICAL CONSTANTS OF ORGANIC COMPOUNDS

No.	Name	Synonym	Formula	Beil. Ref.	Formula Weight
4530	**Naphthyl** amine (β)	amino-naphthalene	$C_{10}H_7 \cdot NH_2$	XII-1265	143.19
4531	amine HCl (α)	....	$C_{10}H_7 \cdot NH_2 \cdot HCl$	XII-1220	179.65
4532	amine HCl (β)	....	$C_{10}H_7 \cdot NH_2 \cdot HCl$	XII-1272	179.65
4533	amine-4,8-disulfonic acid (α)	α,S-acid; δ-acid	$NH_2 \cdot C_{10}H_5(SO_3H)_2$	XIV-787	303.31
4534	amine-4,8-disulfonic acid (β)	β,C-acid	$NH_2 \cdot C_{10}H_5(SO_3H)_2$	XIV-786	303.31
4535	amine-5,7-disulfonic acid (β)	....	$C_{10}H_9O_6NS_2 \cdot 5H_2O$	XIV-783	393.39
4536	amine-5,7-disulfonic K salt (β)	....	$C_{10}H_8O_6NS_2K$	....	341.41
4537	amine-6,8-disulfonic acid (β)	β,γ-acid; amido-G acid	$NH_2 \cdot C_{10}H_5(SO_3H)_2 \cdot 4H_2O$	XIV-784	375.38
4538	amine-6,8-disulfonic K salt (β)	....	$C_{10}H_8O_6NS_2K$	....	341.41
4539	amine-6,8-disulfonic Na salt (β)	....	$C_{10}H_8O_6NS_2Na$	....	325.30
4540	amine-6,8-disulfonic Na salt (β)	....	$C_{10}H_7O_6NS_2Na_2$	....	348.28
4541	amine-*p*-sulfonic acid (α)	naphthionic acid; (1,4)	$NH_2 \cdot C_{10}H_6 \cdot SO_3H$	XIV-739	223.25
4542	amine-*p*-sulfonic Na (α)	naphthionic Na salt	$C_{10}H_8O_3NSNa \cdot 4H_2O$	XIV-739	317.30
4543	amine sulfonic acid (α)	Schollkopf's acid; peri acid; (1,8)	$NH_2 \cdot C_{10}H_6 \cdot SO_3H$	XIV-752	223.25
4544	amine sulfonic acid	(α)(1,2)	$NH_2 \cdot C_{10}H_6 \cdot SO_3H$	XIV-757	223.25
4545	amine sulfonic acid (α)(4,2)	Cleve's γ-acid	$NH_2 \cdot C_{10}H_6 \cdot SO_3H$	XIV-757	223.25
4546	amine sulfonic (α)	(5,2); Cleve's β-acid	$NH_2 \cdot C_{10}H_6 \cdot SO_3H$	XIV-758	223.25
4547	amine sulfonic (α)	(8,2); Cleve's θ-acid	$NH_2 \cdot C_{10}H_6 \cdot SO_3H$	XIV-765	223.25
4548	amine sulfonic (α)	(5,1); Laurent's acid	$NH_2 \cdot C_{10}H_6 \cdot SO_3H$	XIV-744	223.25
4549	amine sulfonic (β)	(2,1); Tobias' acid	$NH_2 \cdot C_{10}H_6 \cdot SO_3H$	XIV-738	223.25
4550	amine sulfonic (β)	(6,1); Dahl's acid	$NH_2 \cdot C_{10}H_6 \cdot SO_3H$	XIV-748	223.25
4551	amine sulfonic (β)	(7,1); Dahl's acid	$NH_2 \cdot C_{10}H_6 \cdot SO_3H$	XIV-750	223.25
4552	amine sulfonic (β)	(6,2); Bronner's acid	$NH_2 \cdot C_{10}H_6 \cdot SO_3H$	XIV-760	223.25
4553	amine sulfonic (β)	(7,2); δ-acid	$NH_2 \cdot C_{10}H_6 \cdot SO_3H$	XIV-763	223.25
4554	amine-3,6,8-trisulfonic acid (α)	Koch acid; amino-H acid	$NH_2 \cdot C_{10}H_4(SO_3H)_3 \cdot 6H_2O$	XIV-801	491.47
4555	amine-3,6,8-trisulfonic Na (β)	Koch acid Na; amino-H acid Na	$C_{10}H_6O_9NS_3Na_3$	....	449.32
4556	benzoate (α)	....	$C_6H_5CO_2 \cdot C_{10}H_7$	IX-125	248.28
4557	benzoate (β)	....	$C_6H_5CO_2 \cdot C_{10}H_7$	IX-125	248.28
4558	*iso*-cyanate (α)	....	$C_{10}H_7 \cdot N{:}CO$	XII-1244	169.18
4559	*iso*-cyanate (β)	....	$C_{10}H_7 \cdot N{:}CO$	XII-1297	169.18
4560	ether (α)	dinaphthyl ether	$C_{10}H_7 \cdot O \cdot C_{10}H_7$	VI-607	270.33
4561	ether (α,β')	dinaphthyl ether	$C_{10}H_7 \cdot O \cdot C_{10}H_7$	VI-642	270.33
4562	ether (β)	dinaphthyl ether	$C_{10}H_7 \cdot O \cdot C_{10}H_7$	VI-642	270.33
4563	hydrazine (α)	....	$C_{10}H_7 \cdot NH \cdot NH_2$	XV-561	158.20

Naphthylamine red-G 3732
Naphthyl-benzamidine 689
Naphthyl bromide 956-7
Naphthyl chloride 1402-3
Naphthyl cyanide 4487-8

Naphthyl-ethyl ether 3093-4
Naphthyl fluoride 3273
Naphthyl iodide 3891-2
Naphthyl-mercaptan 5913-4
Naphthyl-nitrosohydroxylamine 4581

Table 7-4 (*Continued*)
PHYSICAL CONSTANTS OF ORGANIC COMPOUNDS

No.	Crystalline Form and Color	Specific Gravity	Melting Point °C.	Boiling Point °C.	Solubility in 100 Parts		
					Water	Alcohol	Ether
4530	lf./aq.	$1.061\frac{98}{4}°$	111–2	306.1	v. s. h.	s.	s.
4531	nd.	....	....	subl.	$3.8^{20°}$	s.	s.
4532	lf.	....	254	....	v. s.	v. s.	....
4533	cr.	....	....	....	s. alk.	....	....
4534	pr.	....	....	....	s. alk.	v. sl. s. H_2SO_4	....
4535	rhb./aq.	....	....	....	$23^{20°}$	$10^{20°}$ aq.*	$245^{4°}$ aq.*
4536	cr.	....	....	....	$3.4^{18°}$	$67^{1°}$ aq.	....
4537	nd./aq.	....	....	....	$9^{20°}$	sl. s.	....
4538	cr.	....	....	....	$12.8^{20°}$	$347^{8°}$ aq.	....
4539	cr.	....	....	....	$2.7^{18°}$	$127^{8°}$ aq.	....
4540	cr.	....	....	....	v. sl. s.	....	....
4541	nd. + $\frac{1}{2}H_2O$	....	d.	....	$0.03^{10°}$ $0.2^{100°}$	i.; s. alk.	i.
4542	mn. or rhb.	....	$-3\frac{1}{2}H_2O$, 80	$-4H_2O$, 130	s.; v. sl. s. aq. alk.	sl. s.	i.
4543	nd. + $1H_2O$	....	d.	....	$0.02^{21°}$; $0.4^{100°}$	s. ac.	....
4544	nd./aq.	....	262–5 d.	....	$0.24^{0°}$; $3.1^{100°}$	i.	i. bz.
4545	nd.	....	....	....	sl. s.	....	....
4546	cr./aq.	....	....	....	$0.1^{16°}$	i.	i.
4547	nd. + $1H_2O$	....	....	....	$0.46^{25°}$	v. sl. s.	v. sl. s.
4548	cr. + $1H_2O$	....	d.	$-H_2O$, 110	0.1 c.	i.	i.
4549	cr./HCl	....	....	....	d. h.	....	....
4550	pl./aq.	....	....	....	$0.03^{20°}$	i.	i.
4551	pr./aq.	....	....	....	$0.06^{20°}$	i.	....
4552	cr. + $1H_2O$	....	....	....	$0.01^{20°}$; $0.16^{100°}$	....	....
4553	cr. + $1H_2O$	....	....	....	$0.02^{20°}$; $0.3^{100°}$	....	....
4554	cr.	....	....	....	$200^{18°}$	$12.5^{18°}$	....
4555	cr.	....	....	....	$7.21^{20°}$	$16^{54°}$ aq.	....
4556	cr./et. al.	....	56	....	....	....	v. s.
4557	nd./al.	....	107–8	....	i.; s. chl.	s. h.	sl. s.
4558	col. lq.	1.18	....	269–70	d.	s.; s. pet.	s.; s. chl.
4559	lf.	....	55–6	....	d.	v. s.	s.; s. bz.
4560	lf./al.	....	110	>360	i.; s. bz.	sl. s. c.	s.
4561	nd./et. al.	....	81	264^{15mm}	s. bz.	....	s.
4562	nd./al.	....	105; d. 380	250^{19mm} sl. d.	v. s. bz.	s. h.	v. s.; sl. s. c. ac.
4563	col. cr.	....	116–7	203^{20mm}	v. sl. s. c.	v. s. h.	sl. s.

* Solubility of anhy. compd.
Naphthylene ethylene 2
Narceine, cf. alkd.
Narcissine, cf. alkd.
Narcosine, cf. alkd.

Narcotine, cf. alkd.
Naringin, cf. glcde.
Neoarsphenamine 5593
Neobornyval 868
Neocincophen 3130

Neol 376
Neomenthol 4025
Neomenthyl acetate 4028
Neonal 2999
Neoquinophan 3130

Table 7-4 (Continued)
PHYSICAL CONSTANTS OF ORGANIC COMPOUNDS

No.	Name	Synonym	Formula	Beil. Ref.	Formula Weight
	Naphthyl				
4564	hydrazine HCl (α)	....	$C_{10}H_7 \cdot NH \cdot NH_2 \cdot HCl$	XV-562	194.67
4565	hydrazine (β)	....	$C_{10}H_7 \cdot NH \cdot NH_2$	XV-568	158.20
4566	hydrazine HCl (β)	....	$C_{10}H_7 \cdot NH \cdot NH_2 \cdot HCl$	XV-568	194.67
4567	hydroxylamine (α)	....	$C_{10}H_7 \cdot NHOH$	XV-32	159.19
4568	salicylate (α)	alphol	$HOC_6H_4CO_2C_{10}H_7$	X-80	264.28
4569	salicylate (β)	betol	$HOC_6H_4CO_2C_{10}H_7$	X-80	264.28
4570	*iso*-thiocyanate (α)	....	$C_{10}H_7 \cdot N:CS$	XII-1244	185.25
4571	**Naphthylene** di- amine (1,2)	diamino- naphthalene	$C_{10}H_6(NH_2)_2$	XIII-196	158.20
4572	diamine (1,4)	....	$C_{10}H_6(NH_2)_2$	XIII-201	158.20
4573	diamine (1,5)	....	$C_{10}H_6(NH_2)_2$	XIII-203	158.20
4574	diamine (1,6)	....	$C_{10}H_6(NH_2)_2$	XIII-204	158.20
4575	diamine (1,7)	....	$C_{10}H_6(NH_2)_2$	XIII-205	158.20
4576	diamine (1,8)	....	$C_{10}H_6(NH_2)_2$	XIII-205	158.20
4577	diamine (2,3)	....	$C_{10}H_6(NH_2)_2$	XIII-207	158.20
4578	diamine (2,6)	....	$C_{10}H_6(NH_2)_2$	XIII-208	158.20
4579	diamine (2,7)	....	$C_{10}H_6(NH_2)_2$	XIII-208	158.20
4580	**Neo**-dorm(e)	α-*iso*-Pr-α-Br- butyramide	$(CH_3)_2CH \cdot C(C_2H_5) \cdot$ $(Br)CONH_2$	**II-299	208.10
4581	cupfer(r)on	α-naphthyl-nitroso- hydroxylamine NH_4	$C_{10}H_7N(NO)ONH_4$	*XVI-396	205.22
4582	synephrine HCl	α-OH-β-methyl- aminoethyl-3-OH benzene HCl	$HO \cdot C_6H_4 \cdot CHOH \cdot$ $CH_2 \cdot NHCH_3 \cdot HCl$	....	203.67
4583	**Nembutal**®	pentobarbital Na	$C_2H_5 \cdot C_4HN_2O_3Na \cdot$ $CH(CH_3) \cdot C_3H_7$	....	248.26
4584	**Neurine**	trimethyl vinyl am- monium hydroxide	$CH_2:CH \cdot N(CH_3)_3 \cdot$ OH	IV-203	103.17
4585	**Neutral red**	toluylene red	$C_{15}H_{16}N_4 \cdot HCl$	XXV-401	288.78
4586	**Nitranilic acid**	3,6-diNO_2-2,5- diOH-quinone	$(NO_2)_2C_6(OH)_2O_2 \cdot$ aq.	VIII-384	230.09
4587	**Nitro**-acenaphthene	(5-nitro)	$(NO_2)C_{10}H_5:(CH_2)_2$	V-588	199.21
4588	acetanilide (*o*)	....	$NO_2 \cdot C_6H_4 \cdot NHCOCH_3$	XII-691	180.16
4589	acetanilide (*m*)	....	$NO_2 \cdot C_6H_4 \cdot NHCOCH_3$	XII-703	180.16
4590	acetanilide (*p*)	....	$NO_2 \cdot C_6H_4 \cdot NHCOCH_3$	XII-719	180.16
4591	*p*-acetanisidide (3)	(3;1,4)	$NO_2 \cdot C_6H_3(OCH_3) \cdot$ $NHCOCH_3$	XIII-522	210.19
4592	acetic acid	....	$NO_2 \cdot CH_2 \cdot CO_2H$	II-225	105.05
4593	2-acetnaphthalide (1)	1-NO_2-2-acetyl- naphthylamine	$NO_2 \cdot C_{10}H_6 \cdot$ $NHCOCH_3$	XII-1313	230.23
4594	acetophenone (ω)	benzoyl-nitro- methane	$C_6H_5 \cdot CO \cdot CH_2 \cdot NO_2$	VII-289	165.15
4595	acetophenone (*m*)	....	$NO_2 \cdot C_6H_4 \cdot CO \cdot CH_3$	VII-288	165.15
4596	*p*-acetphenetidide	(3;1,4)	$NO_2 \cdot C_6H_3(OC_2H_5) \cdot$ $NHCOCH_3$	XIII-522	224.22
4597	acet-*p*-toluidide	3-NO_2-4-acetyl- aminotoluene	$NO_2 \cdot C_6H_3(CH_3) \cdot$ $NHCOCH_3$	XII-1002	194.19
4598	4-acetylamino- phenylacetate	(3;4,1)	$NO_2 \cdot C_6H_3(NHCO \cdot$ $CH_3) \cdot O_2CCH_3$	....	238.20
4599	alizarin (3)(β)	alizarin orange	$(HO)_2C_{14}H_5O_2 \cdot NO_2$	VIII-447	285.21
4600	alizarin (4)(α)	....	$(HO)_2C_{14}H_5O_2 \cdot NO_2$	VIII-447	285.21

Table 7-4 (*Continued*)
PHYSICAL CONSTANTS OF ORGANIC COMPOUNDS

No.	Crystalline Form and Color	Specific Gravity	Melting Point °C.	Boiling Point °C.	Solubility in 100 Parts		
					Water	Alcohol	Ether
4564	pl./HCl	....	....	....	s. c.	....	....
4565	col. lf./aq.	....	124–5	....	sl. s. h.	s. h.	sl. s.
4566	nd./aq.	....	233 d.	....	s. h.	....	....
4567	cr./aq.	....	79	....	s.	s.	s.
4568	col. cr.	....	83	....	i.; s. oil	s.	s.
4569	col. cr./al.	....	95	....	i.	s. h.	s.; s. bz.
4570	nd./al.	....	58	....	i.	s. h.	s.
4571	lf./aq.	....	96–8	150–1$^{0.5mm}$	sl. s. h.	s.	s.
4572	pr./h. aq.	....	120	....	sl. s. h.	v. s.	v. s.
4573	pr./et.	....	189.5	....	s. h.	s. h.	s.
4574	nd./aq.	1.147$\frac{99.4°}{4}$	77.5	....	s. h.	s. h.	s. h.
4575	nd./aq.	....	117.5	....	sl. s.	s.	v. sl. s.
4576	cr./aq. al.	1.127$\frac{99.4°}{4}$	66.5	205^{12mm}	sl. s. h.	∞	∞
4577	lf./et.	....	191–3	....	....	s.	s.
4578	nd./aq.	....	216–8	....	v. sl. s. h.	sl. s.	sl. s.
4579	lf./aq.	....	159	....	....	....	....
4580	col. cr.	....	50–1	subl.	0.7 c.; d. h.	s.; s. pet.	s.; s. bz.
4581	lf./al. NH₃	....	125–6	....	s.	s. Me al.	i.
4582	ool. cr.	....	139–41	....	s.	s.	....
4583	wh. cr. pd.	....	....	....	s.	s.	i.
4584	syrup, poison.	....	....	....	s.	d. aq. NH₄Cl	....
4585	gn. pd.	....	....	....	d., –HCl	s.	....
4586	yel. pl.	....	–H₂O, 100	expl. 170	v. s.	v. s.	i.
4587	yel. nd./lg.	....	101–2	....	s. h.	s.	s.
4588	yel. mn.	1.419$^{15°}$	93–4	....	s. h.; s. chl.	s.; v. s. 10% KOH	s.
4589	col. lf.	....	154	....	sl. s. h.	s.	i.; s. chl.
4590	rhb.	....	215–6	....	s. h.	s.; s. KOH	s.
4591	yel. nd./al.	....	117–8	....	s. h.	s.; s. bz.	s.; s. ac.
4592	nd./chl.	....	87–9 d.	....	d.	v. s.	v. s.; i. pet.
4593	yel. rhb. nd./al.	....	123.5	....	sl. s. h.; v. sl. s. lg.	s.; s. bz.; s. ac.	v. sl. s.
4594	lf./aq. al.	....	106–8	....	i. c.	v. s.	v. s.
4595	nd.	....	80–1	202	i.	s.	....
4596	yel. nd./aq.	....	103–4	....	....	s. abs.	s.; s. chl.
4597	yel. nd./pet.	....	96; softens 94	....	sl. s. h.	s.	....
4598	yel. pr./aq. al.	....	147–8	....	....	....	....
4599	yel. lf./al.	....	244 d.	subl. sl. d.	sl. s.	s. chl., bz.	s. aq. alk.
4600	yel. nd./al.	....	289 d.	....	v. sl. s.; s. H₂SO₄	s. chl., bz.	s. aq. alk.

Table 7-4 (*Continued*)
PHYSICAL CONSTANTS OF ORGANIC COMPOUNDS

No.	Name	Synonym	Formula	Beil. Ref.	Formula Weight
	Nitro				
4601	2-aminoanisole (4)	(4;1,2)	$NO_2 \cdot C_6H_3(OCH_3)NH_2$	XIII-389	168.15
4602	2-aminoanisole (5)	(5;1,2)	$NO_2 \cdot C_6H_3(OCH_3)NH_2$	XIII-390	168.15
4603	4-aminoanisole (3)	(3;1,4)	$NO_2 \cdot C_6H_3(OCH_3)NH_2$	XIII-521	168.15
4604	*o*-aminobenzoic acid (3)	(3;1,2) nitro-anthranilic acid	$NO_2 \cdot C_6H_3(CO_2H) \cdot NH_2$	XIV-373	182.14
4605	*o*-aminobenzoic acid (4)	(4;1,2)	$NO_2 \cdot C_6H_3(CO_2H) \cdot NH_2$	XIV-374	182.14
4606	*o*-aminobenzoic acid (5)	(5;1,2)	$NO_2 \cdot C_6H_3(CO_2H) \cdot NH_2$	XIV-375	182.14
4607	*o*-aminobenzoic acid (6)	(6;1,2)	$NO_2 \cdot C_6H_3(CO_2H) \cdot NH_2$	XIV-378	182.14
4608	*m*-aminobenzoic acid (5)	(5;1,3)	$NO_2 \cdot C_6H_3(CO_2H) \cdot NH_2$	XIV-415	182.14
4609	*p*-aminobenzoic acid (2)	(2;1,4)	$NO_2 \cdot C_6H_3(CO_2H) \cdot NH_2$	XIV-439	182.14
4610	*o*-aminophenol (3)	(3;2,1)	$NO_2 \cdot C_6H_3(NH_2)OH$	....	154.13
4611	*o*-aminophenol (4)	(4;2,1)	$NO_2 \cdot C_6H_3(NH_2)OH$	XIII-388	154.13
4612	*o*-aminophenol (5)	(5;2,1)	$NO_2 \cdot C_6H_3(NH_2)OH$	XIII-390	154.13
4613	*o*-aminophenol (6)	(6;2,1)	$NO_2 \cdot C_6H_3(NH_2)OH$	XIII-391	154.13
4614	*m*-aminophenol (5)	(5;3,1)	$NO_2 \cdot C_6H_3(NH_2)OH$	XIII-422	154.13
4615	*p*-aminophenol (2)	(2;4,1)	$NO_2 \cdot C_6H_3(NH_2)OH$	XIII-520	154.13
4616	*p*-aminophenol (3)	(3;4,1)	$NO_2 \cdot C_6H_3(NH_2)OH$	XIII-521	154.13
4618	aniline (*o*)	*o*-nitraniline	$NO_2 \cdot C_6H_4 \cdot NH_2$	XII-687	138.13
4619	aniline (*m*)	*m*-nitraniline	$NO_2 \cdot C_6H_4 \cdot NH_2$	XII-698	138.13
4620	aniline (*p*)	*p*-nitraniline	$NO_2 \cdot C_6H_4 \cdot NH_2$	XII-711	138.13
4621	aniline-4-sulfonic acid (2)	nitrosulfanilic acid	$NO_2 \cdot C_6H_3(NH_2)SO_3H$	XIV-708	218.19
4622	aniline-2-sulfonic acid (4)	(4;1,2)	$NO_2 \cdot C_6H_3(NH_2)SO_3H$	XIV-686	218.19
4623	anisole (*o*)	....	$CH_3O \cdot C_6H_4 \cdot NO_2$	VI-217	153.14
4624	anisole (*m*)	....	$CH_3O \cdot C_6H_4 \cdot NO_2$	VI-224	153.14
4625	anisole (*p*)	....	$CH_3O \cdot C_6H_4 \cdot NO_2$	VI-230	153.14
4626	anthracene (9)	nitrosoanthron	$C_{14}H_9 \cdot NO_2$	V-666	223.23
4627	anthraquinone (1)	....	$C_6H_4(CO)_2C_6H_3 \cdot NO_2$	VII-791	253.22
4628	anthraquinone (2)	....	$O_2C_{14}H_7 \cdot NO_2$	VII-792	253.22
4629	anthraquinone-2-carbonyl chloride	(1-NO₂)	$O_2C_{14}H_6(NO_2)COCl$	....	315.67
4630	anthraquinone-5-sulfonic acid (1)	(1-NO₂)	$O_2C_{14}H_6(NO_2)SO_3H$	XI-336	333.28
4631	anthraquinone-8-sulfonic acid (1)	(1-NO₂)	$O_2C_{14}H_6(NO_2)SO_3H$	XI-337	333.28
4632	barbituric acid (5)	dilituric acid	$NO_2 \cdot C_4H_3O_3N_2 \cdot 3H_2O$	XXIV-474	227.13
4633	benzaceto-phenone	(1,3)	$NO_2 \cdot C_6H_4 \cdot CH{:}CH \cdot CO \cdot C_6H_5$	VII-482	253.26
4634	benzal bromide (*p*)	....	$NO_2 \cdot C_6H_4 \cdot CH{:}Br_2$	V-336	294.94
4635	benzal chloride (*m*)	....	$NO_2 \cdot C_6H_4 \cdot CH \cdot Cl_2$	V-332	206.03
4636	benzal chloride (*p*)	....	$NO_2 \cdot C_6H_4 \cdot CH{:}Cl_2$	V-332	206.03

Table 7-4 (*Continued*)
PHYSICAL CONSTANTS OF ORGANIC COMPOUNDS

No.	Crystalline Form and Color	Specific Gravity	Melting Point °C.	Boiling Point °C.	Solubility in 100 Parts		
					Water	Alcohol	Ether
4601	red nd./al.	$1.207^{156°}$	118	....	s. h. bz.	s.; s. ac.	v. sl. s. lg.
4602	pa. yel. nd.	$1.211^{156°}$	139–40	....	....	....	....
4603	red/aq. al.	....	123	....	sl. s.	s.	s.
4604	yel. mn.	$1.558^{15°}$	204	....	i.	v. s.	v. s.
4605	yel. red nd.	....	269.5	....	v. sl. s. h.	s. xylene	....
4606	lt. yel. nd.	....	270–80 d.	....	s. h.	s.	s.
4607	yel. lf./aq.	....	183–4 d.	....	s. h.	v. s.	v. s.
4608	yel. pr./aq.	....	208	....	sl. s.	s. h.	sl. s.; v. s. h. ac.
4609	red nd./ aq.	....	240	....	s. h.	v. s.	v. s. ac.
4610	cr.	....	136	....	....	....	....
4611	or. pr.	....	142–3	....	sl. s. c.	v. s.	v. s.
4612	brn.nd./aq.	....	201–2	....	....	s. h.	....
4613	red nd./ aq. al.	....	110–1	....	v. sl. s. c.	s.; v. s. chl.	v. s.; v. s. bz.
4614	yel. cr.	....	165	....	v. sl. s. bz., chl.	v. s.	v. s.
4615	red nd./al.	....	128–31	....	....	....	....
4616	red pr./et. al.	....	154	....	s.	s.	s.
4618	yel. rhb.	$1.442^{15°}$	71.5	284.1	s. h.	v. s.	v. s.
4619	yel. rhb.	1.43	114	306.4	$0.11^{20°}$	$7.1^{20°}$	$7.9^{20°}$
4620	yel. mn.	$1.437^{14°}$	147.5	331.7	$0.08^{18.5°}$; 2.2 h.	$5.8^{20°}$	$6.1^{20°}$
4621	yel. nd.	....	....	....	v. s.; s. aq. H_2SO_4	sl. s.	s. conc. HCl
4622	yel. cr.	....	....	....	....	....	....
4623	col. cr.	$1.254^{20°}_{4}$	9.5–10.5	272–3	$0.17^{30°}$	∞	∞
4624	nd./al.	$1.373^{18°}$	38	258	i.	s.	....
4625	pr./al.	$1.233^{20°}$	54	274	$0.06^{30°}$	v. s.	v. s.
4626	yel. nd./al.	....	146	>360	i. aq. alk.	sl. s.	v. s. bz.
4627	nd./ac.	....	230	270^{7mm}	i.	sl. s.	v. sl. s.
4628	yel. nd./al.	....	184.5–5.0	$270-17^{mm}$	s. H_2SO_4	v. sl. s. c.	sl. s.; v. s. chl.
4629	tan flakes	....	230 d.	....	i.	i.	i.
4630	yel. cr./aq.	....	....	....	s.	i.	i.
4631	yel. cr./aq.	....	....	....	sl. s.	i.	i.
4632	pr./aq.	....	181–3 (anh.)	....	$0.09^{25°}$	s.; s. alk.	i.
4633	yel. nd./ al. or bz.	....	144–5	....	i. lg.; v. s. bz.	s.; s. chl.	i.; s. ac.
4634	nd./al.	....	82.0–2.5	....	i.	s.	s.
4635	mn.	....	65	....	i.	v. s. h.	v. s.
4636	pr./al.	....	46	....	i.	v. s.	v. s.

Table 7-4 (*Continued*)
PHYSICAL CONSTANTS OF ORGANIC COMPOUNDS

No.	Name	Synonym	Formula	Beil. Ref.	Formula Weight
4637	**Nitro** benzaldehyde (*o*)	....	$NO_2 \cdot C_6H_4 \cdot CH:O$	VII-243	151.12
4638	benzaldehyde phenylhydrazone (*o*)	....	$NO_2 \cdot C_6H_4 \cdot CH:N \cdot NH \cdot C_6H_5$	XV-136	241.25
4639	benzaldehyde (*m*)	....	$NO_2 \cdot C_6H_4 \cdot CH:O$	VII-250	151.12
4640	benzaldehyde (*p*)	....	$NO_2 \cdot C_6H_4 \cdot CH:O$	VII-256	151.12
4641	benzalfurfural-acetone (*m*)	....	$NO_2 \cdot C_6H_4 \cdot CH:CH \cdot CO \cdot CH:CH \cdot C_4H_3O$	....	269.26
4642	benzamide (*o*)	....	$NO_2 \cdot C_6H_4 \cdot CO \cdot NH_2$	IX-373	166.14
4643	benzamide (*m*)	....	$NO_2 \cdot C_6H_4 \cdot CO \cdot NH_2$	IX-381	166.14
4644	benzamide (*p*)	....	$NO_2 \cdot C_6H_4 \cdot CO \cdot NH_2$	IX-394	166.14
4645	benzanilide (*m*)	....	$NO_2C_6H_4CONHC_6H_5$	XII-267	242.24
4646	benzene	oil of mirbane	$C_6H_5 \cdot NO_2$	V-233	123.11
4647	benzeneazo-chromotropic Na salt	*p*-nitrobenzeneazo-1,8-diOH-naphthalene-3,6-disulfonic Na	$NO_2 \cdot C_6H_4 \cdot N:N \cdot C_{10}H_3(OH)_2(SO_3Na)_2$	....	513.37
4648	benzeneazo-α-naphthol (*p*)	....	$NO_2 \cdot C_6H_4 \cdot N:N \cdot C_{10}H_6(OH)$	XVI-151	293.28
4649	benzeneazo-orcinol (*p*)	(4′;4,6,2)	$NO_2 \cdot C_6H_4 \cdot N:N \cdot C_6H_2(OH)_2CH_3$	....	273.25
4650	benzeneazo-resorcinol (*m*)	(3′;2,4)	$NO_2 \cdot C_6H_4 \cdot N:N \cdot C_6H_3(OH)_2$	....	259.22
4651	benzeneazo-resorcinol (*p*)	(4′;2,4)	$NO_2 \cdot C_6H_4 \cdot N:N \cdot C_6H_3(OH)_2$	XVI-181	259.22
4652	benzeneazo-salicylic acid (*p*)	(4′;4,3)	$NO_2 \cdot C_6H_4 \cdot N:N \cdot C_6H_3(OH)CO_2H$	XVI-247	287.23
4653	benzeneazo-salicylic Na (*p*)	alizarin yellow R	$C_{13}H_8O_5N_3Na$	....	309.22
4654	benzene sulfonic acid (*o*)	....	$NO_2 \cdot C_6H_4 \cdot SO_3H$	XI-67	203.17
4655	benzene sulfonic acid (*m*)	....	$NO_2 \cdot C_6H_4 \cdot SO_3H$	XI-68	203.17
4656	benzene sulfonic Na	(*m*)	$NO_2 \cdot C_6H_4 \cdot SO_3Na$	XI-68	225.16
4657	benzene sulfonic chloride (*m*)	....	$NO_2 \cdot C_6H_4 \cdot SO_2Cl$	XI-69	221.62
4658	benzhydrazide (*o*)	....	$NO_2 \cdot C_6H_4CO \cdot NH \cdot NH_2$	IX-375	181.15
4659	benzhydrazide (*m*)	....	$NO_2 \cdot C_6H_4CO \cdot NH \cdot NH_2$	IX-388	181.15
4660	benzhydrazide (*p*)	....	$NO_2 \cdot C_6H_4CO \cdot NH \cdot NH_2$	IX-399	181.15
4661	benzidine (2)	....	$NO_2(NH_2)C_6H_3 \cdot C_6H_4 \cdot NH_2$	XIII-235	229.24
4662	benzimidazole (6)	....	$HN \cdot CH:N \cdot C_6H_3 \cdot NO_2$	XXIII-135	163.14
4663	benzoic acid (*o*)	....	$NO_2 \cdot C_6H_4 \cdot CO_2H$	IX-370	167.12
4664	benzoic acid (*m*)	....	$NO_2 \cdot C_6H_4 \cdot CO_2H$	IX-376	167.12
4665	benzoic Na (*m*)	....	$C_7H_4O_4NNa \cdot 3H_2O$	IX-377	243.15
4666	benzoic acid (*p*)	....	$NO_2 \cdot C_6H_4 \cdot CO_2H$	IX-389	167.12
4667	benzonitrile (*o*)	....	$NO_2 \cdot C_6H_4 \cdot CN$	IX-374	148.12
4668	benzonitrile (*m*)	....	$NO_2 \cdot C_6H_4 \cdot CN$	IX-385	148.12

Table 7-4 (*Continued*)
PHYSICAL CONSTANTS OF ORGANIC COMPOUNDS

No.	Crystalline Form and Color	Specific Gravity	Melting Point °C.	Boiling Point °C.	Solubility in 100 Parts		
					Water	Alcohol	Ether
4637	yel. nd./aq.	....	42-3.5 (37.9)	153^{23mm}	v. sl. s.	v. s.	v. s.; s. bz.
4638	red nd.	....	154-5	....	v. s. act.; i. lg.	sl. s.	sl. s.
4639	nd./aq.	....	58	164^{23mm}	0.16$^{25°}$	v. s. h.	s.; s. chl.
4640	pr./aq.	....	106.5	....	v. sl. s.	v. s	sl. s.
4641	yel. pd.	....	120-4	....	i.	sl. s. h.	i.
4642	nd./aq. al.	$1.462^{\frac{32}{4}°}$	176.6	317	s. h.	s. h.	s.
4643	yel. mn./aq.	....	142-3	310-5	v. sl. s.	s.	s.
4644	nd./aq.	....	200-1.4	....	sl. s.	s.	s.
4645	lf./al.	....	153-4	subl.	sl. s.	s.	s.; s. bz.
4646	lt. yel. lq.	$1.203^{\frac{20}{4}°}$	5.7	210.9	0.19$^{20°}$	v. s.	∞; ∞ bz.
4647	red brn. pd.	....	....	....	s.	i.	....
4648	rod gn. nd.	....	234-5	d. 255-60	s. am. al.; v. sl. s. chl.	v. sl. s.; s. xylene	v. sl. s.; v. sl. s. bz.
4649	dark red pd.	....	210-2 d.	....	i.	s.	sl. s.
4650	brn. pd.	....	174-5 d.	....	i.	s.	sl. s.
4651	red pd./Me al.	....	199-200	....	s. alk.; v. sl. s. ac.	v. sl. s. h.	v. sl. s. toluene
4652	or. brn. nd./aq. ac.	....	254-7 d.	....	s. ac.	s.	sl. s. h. toluene
4653	brn. yel. pd.	....	....	....	s.	....	....
4654	hyg. lf.	....	70	d.	v. s.	s.; s. alk.	i.
4655	hyg. lf.	....	....	....	....	s. h.	....
4656	pl./aq.	....	....	....	v. s.	....	....
4657	mn. pr./et.; nd./lg.	....	63-4	....	i.; d. h.	s. h.	....
4658	yel. brn./aq.	....	120-1	....	s.	s.; i. chl.	i.; i. bz.
4659	nd./aq.	....	152	....	sl. s.	sl. s.	i.; i. bz.
4660	yel. nd./aq.	....	210	....	v. sl. s.	v. sl. s.	i.; i. bz.
4661	red nd./aq.	....	143(117)	....	sl. s. h.	....	....
4662	nd./aq.	....	204	....	sl. s.; s. alk. carb.	s.; s. a.; sl. s. chl.	sl. s.; sl. s. bz.
4663	tri./aq.	1.575	147.5	....	0.65$^{20°}$	28$^{11°}$, 90%	22$^{11°}$
4664	mn.	1.494	141-2	....	0.24$^{16.5°}$	31$^{11.7°}$	25$^{10.2°}$
4665	mn.	....	....	....	s. h.	....	....
4666	pa. yel. mn.	$1.550^{\frac{32}{4}°}$	240-2	subl.	0.02$^{15°}$	0.9$^{10°}$, 90%	2.2$^{12.5°}$
4667	nd./aq.	....	109-10	....	s. h.	s.	s. ac.
4668	nd./aq.	....	117-8	subl.	s. h.	s.	v. s.

Table 7-4 (*Continued*)
PHYSICAL CONSTANTS OF ORGANIC COMPOUNDS

No.	Name	Synonym	Formula	Beil. Ref.	Formula Weight
4669	**Nitro** benzonitrile (*p*)	....	$NO_2 \cdot C_6H_4 \cdot CN$	IX-397	148.12
4670	benzophenone (*o*)	....	$NO_2 \cdot C_6H_4 \cdot CO \cdot C_6H_5$	VII-425	227.22
4671	benzophenone (*m*)	....	$NO_2 \cdot C_6H_4 \cdot CO \cdot C_6H_5$	VII-425	227.22
4672	benzophenone (*p*)	....	$NO_2 \cdot C_6H_4 \cdot CO \cdot C_6H_5$	VII-426	227.22
4673	benzoyl chloride (*o*)	....	$NO_2 \cdot C_6H_4 \cdot COCl$	IX-373	185.57
4674	benzoyl chloride(*m*)	....	$NO_2 \cdot C_6H_4 \cdot COCl$	IX-381	185.57
4675	benzoyl chloride (*p*)	....	$NO_2 \cdot C_6H_4 \cdot COCl$	IX-394	185.57
4676	benzoyl formic acid (*o*)	....	$NO_2 \cdot C_6H_4 \cdot CO \cdot CO_2H \cdot$ aq.	X-664	195.13
4677	benzyl acetate (*p*)	....	$NO_2 \cdot C_6H_4 \cdot CH_2 \cdot O_2C \cdot CH_3$	VI-451	195.18
4678	benzyl alcohol (*o*)	....	$NO_2 \cdot C_6H_4 \cdot CH_2OH$	VI-447	153.14
4679	benzyl alcohol (*m*)	....	$NO_2 \cdot C_6H_4 \cdot CH_2OH$	VI-449	153.14
4680	benzyl alcohol (*p*)	....	$NO_2 \cdot C_6H_4 \cdot CH_2OH$	VI-450	153.14
4681	benzyl bromide (*p*)	ω-Br-*p*-nitrotoluene	$NO_2 \cdot C_6H_4 \cdot CH_2Br$	V-334	216.04
4682	benzyl chloride (*o*)	ω-Cl-*o*-nitrotoluene	$NO_2 \cdot C_6H_4 \cdot CH_2Cl$	V-327	171.58
4683	benzyl chloride (*m*)	....	$NO_2 \cdot C_6H_4 \cdot CH_2Cl$	V-329	171.58
4684	benzyl chloride (*p*)	....	$NO_2 \cdot C_6H_4 \cdot CH_2Cl$	V-329	171.58
4685	benzyl cyanide (*o*)	nitro-α-toluic nitrile	$NO_2 \cdot C_6H_4 \cdot CH_2 \cdot CN$	IX-455	162.15
4686	benzyl cyanide (*m*)	....	$NO_2 \cdot C_6H_4 \cdot CH_2 \cdot CN$	IX-455	162.15
4687	benzyl cyanide (*p*)	....	$NO_2 \cdot C_6H_4 \cdot CH_2 \cdot CN$	IX-456	162.15
4688	bromoform	bromopicrin	$NO_2 \cdot CBr_3$	I-77	297.74
4689	butandiol	2-nitro-2-methyl-1, 3-propandiol	$CH_3 \cdot C \cdot NO_2 \cdot (CH_2OH)_2$	I-480	135.12
4690	butane (1)	1-nitrobutane	$C_2H_5 \cdot CH_2 \cdot CH_2 \cdot NO_2$	I-123	103.12
4691	butane (2)	2-nitrobutane	$C_2H_5 \cdot CHNO_2 \cdot CH_3$	I-123	103.12
4692	butane (α)(*iso*)	1-nitro-2-Me-propane	$(CH_3)_2CH \cdot CH_2 \cdot NO_2$	I-129	103.12
4693	butane (*tert*)	2-nitro-2-Me-propane	$(CH_3)_3C \cdot NO_2$	I-129	103.12
4694	1-butanol (2)	β-NO_2-butyl alcohol	$C_2H_5 \cdot CHNO_2 \cdot CH_2OH$	I-370	119.12
4695	butanol	2NO_2-2-Me-propanol	$(CH_3)_2CNO_2 \cdot CH_2OH$	I-378	119.12
4696	camphor (α)(3)	....	$C_8H_{14} \cdot CO \cdot CH \cdot NO_2$	VII-129	197.24
4697	chloroform	chloropicrin	$NO_2 \cdot CCl_3$	I-76	164.38
4698	chlorophenol	(1;2,4)	$HO \cdot C_6H_3(Cl) \cdot NO_2$	VI-240	173.57
4699	chlorophenol	(1;2,3)	$HO \cdot C_6H_3(Cl) \cdot NO_2$	VI-239	173.57
4700	chlorophenol	(1;3,4)	$HO \cdot C_6H_3(Cl) \cdot NO_2$	VI-240	173.57
4701	chlorophenol	(1;4,2)	$HO \cdot C_6H_3(Cl) \cdot NO_2$	VI-238	173.57
4702	chlorophenol	(1;4,3)	$HO \cdot C_6H_3(Cl) \cdot NO_2$	VI-239	173.57
4703	chlorophenol	(1;5,2)	$HO \cdot C_6H_3(Cl) \cdot NO_2$	VI-238	173.57
4704	chlorophenol	(1;5,3)	$HO \cdot C_6H_3(Cl) \cdot NO_2$	VI-239	173.57
4705	chlorophenol	(1;6,2)	$HO \cdot C_6H_3(Cl) \cdot NO_2$	VI-239	173.57
4706	chlorophenol	(1;6,3)	$HO \cdot C_6H_3(Cl) \cdot NO_2$	VI-240	173.57
4707	cinnamic acid (*o*)	....	$NO_2 \cdot C_6H_4 \cdot CH:CH \cdot CO_2H$	IX-604	193.16
4708	cinnamic acid (*m*)	....	$NO_2 \cdot C_6H_4 \cdot CH:CH \cdot CO_2H$	IX-605	193.16
4709	cinnamic acid (*p*)	....	$NO_2 \cdot C_6H_4 \cdot CH:CH \cdot CO_2H$	IX-606	193.16
4710	*p*-cresol (2)	3-NO_2-4-OH-toluene	$NO_2 \cdot C_6H_3(CH_3)OH$	VI-412	153.14

Nitro-benzoylhydrazine 4658-60
Nitro-benzylidene bromide 4634
Nitro-benzylidene chloride 4635-6
Nitro-bromobenzene 960-2

Nitro-bromobenzoic acid 964
Nitro-butyl alcohol 4694-5
Nitro-carb(in)ol 4765

Table 7-4 (*Continued*)
PHYSICAL CONSTANTS OF ORGANIC COMPOUNDS

No.	Crystalline Form and Color	Specific Gravity	Melting Point °C.	Boiling Point °C.	Solubility in 100 Parts		
					Water	Alcohol	Ether
4669	yel. lf./al.	....	147-9	....	sl. s.	s. h.	s. ac.
4670	mn./al.	....	105	....	....	sl. s. abs.	....
4671	nd./al.	....	94-5	234^{18mm}	....	s. h.	....
4672	lf./abs. al.	....	138	....	v. sl. s.	s. h.	sl. s. CS_2
4673	col. cr.	....	75(20)	205^{105mm}	d.	d.	s.
4674	cr.	....	34-5	275-8 sl. d.	d.	d.	v. s.
4675	nd./lg.	....	72	154^{15mm}	d.	d.	s.
4676	pr./aq.	....	46-7; 122 d. (anh.)	....	∞ h.	....	....
4677	yel. nd./al.	....	78	....	....	s. h.	....
4678	nd./aq.	....	74	270 sl. d.	sl. s.	v.s.	v. s.
4679	cr.	....	27(13-5)	175-80^{3mm}	....	....	v. s.
4680	nd./aq.	....	93(87-9)	185^{12mm}	s. h.	v. s.	v. s.
4681	nd./al.	....	99-100	....	i.	2$^{19°}$	v. s.
4682	cr.	....	48-9	....	i.	26.3$^{30°}$	v. s. h.
4683	yel. nd.	....	45-7	173-83^{30mm}	i.	30.4$^{30°}$	v. s.
4684	nd./al.	....	71	....	i.	8.2$^{30°}$	s.
4685	nd./aq.	....	115-6	....	v. s. h.	s.	s.
4686	cr.	....	61-2	....	v. sl. s.	s.	s.
4687	lf./al.	....	116-7	....	i.	s.	....
4688	pr.	2.811$^{12.5°}$	10.3	*127^{118mm}	i.	s.	s.
4689	mn. cr.	....	147-9	d.	80$^{20°}$	45$^{20°}$	4$^{20°}$
4690	lq.	....	....	151-2	....	....	....
4691	lq.	0.988$^{0°}$	....	138-9^{747mm}	....	....	....
4692	col. lq.	0.9877$^{0°}$	....	158-9^{755mm}	i.	s.	s.
4693	cr.	....	24	126^{748mm}	i. alk.	∞	∞ ; ∞ bz.
4694	lq.	1.134$^{20°}_{20}$	−47	105^{10mm}	20$^{20°}$	∞$^{20°}$	∞$^{20°}$
4695	cr./Me al.	....	90.5	95.5^{10mm}	350$^{20°}$	490$^{20°}$	120$^{20°}$
4696	mn./bz.	....	102-3	....	i.	s.	s.; v. s. bz.
4697	lq.	1.651$^{22.8°}_{4}$	−64	112.3^{766mm}	0.17$^{18°}$	37 cc., 80% al.	s.
4698	col. nd./aq.	....	111	....	v. s. chl.	v. s.	v. s.
4699	cr./aq.	....	120	..... ,...	....	....	....
4700	nd./bz.	....	133	....	....	....	....
4701	yel. mn.	....	87	....	v. sl. s.	s.	v. s.
4702	nd./aq.	....	126-7	....	....	....	....
4703	yel. pr./aq.	....	38.9 (32.7)	....	sl. s.	s.	s.; s. ac.
4704	cr.	....	147	....	....	....	....
4705	yel. nd./aq.	....	70-1	....	v. sl. s.	v. s. chl.	....
4706	nd./aq.	....	118-9	....	....	....	....
4707	nd./al.	....	243-5	subl.	i.	0.2$^{25°}$ abs.	....
4708	col./al.	....	203-5	....	....	1$^{25°}$ abs.	....
4709	lt. yel./al.	....	286-8	....	v. sl. s. h.	0.01$^{25°}$	v. sl. s.; i. CS_2
4710	yel./aq. al.	1.240$^{38.6°}_{4}$	32	125^{22mm}	v. sl. s.	v. s.	v. s.

* Explosive.
Nitro-chalcone 4633
Nitro-chloroaniline 1407-11
Nitro-chlorobenzene 1412-4

Nitro-chlorobenzene sulfonic acid 1415
Nitro-chloroethane 1416-7
Nitro-chloronaphthalene 1418-9
Nitro-chloropropane 1420-1

Table 7-4 (*Continued*)
PHYSICAL CONSTANTS OF ORGANIC COMPOUNDS

No.	Name	Synonym	Formula	Beil. Ref.	Formula Weight
	Nitro				
4711	cumene (*o* and *p*)	. . . .	$NO_2 \cdot C_6H_4 \cdot CH(CH_3)_2$	**V-307	165.19
4712	cumene, pseudo-	(5;1,2,4)	$NO_2 \cdot C_6H_2(CH_3)_3$	V-404	165.19
4713	cumene, pseudo-	(6;1,2,4)	$NO_2 \cdot C_6H_2(CH_3)_3$	V-404	165.19
4714	*p*-cymene (2;1,4)	2-nitrocymene	$NO_2 \cdot C_6H_3(CH_3) \cdot C_3H_7$	V-424	179.22
4715	decane	1-nitro-2,7-dimethyl-octane	$(CH_3)_2CH(CH_2)_4 \cdot CH(CH_3)CH_2NO_2$	I-169	187.28
4716	decane	2-nitro-2,7-dimethyl-octane	$(CH_3)_2C(NO_2)(CH_2)_4 \cdot CH(CH_3)_2$	I-169	187.28
4717	diethylaniline (*o*)	. . . .	$NO_2 \cdot C_6H_4 \cdot N(C_2H_5)_2$	*XII-341	194.24
4718	diethylaniline (*m*)	. . . .	$NO_2 \cdot C_6H_4 \cdot N(C_2H_5)_2$	XII-702	194.24
4719	diethylaniline (*p*)	. . . .	$NO_2 \cdot C_6H_4 \cdot N(C_2H_5)_2$	XII-715	194.24
4720	dimethylaniline (*o*)	. . . .	$NO_2 \cdot C_6H_4 \cdot N(CH_3)_2$	XII-690	166.18
4721	dimethylaniline (*m*)	. . . .	$NO_2 \cdot C_6H_4 \cdot N(CH_3)_2$	XII-701	166.18
4722	dimethylaniline (*p*)	. . . .	$NO_2 \cdot C_6H_4 \cdot N(CH_3)_2$	XII-714	166.18
4723	diphenyl (2)	. . . .	$C_6H_5 \cdot C_6H_4 \cdot NO_2$	V-582	199.21
4724	diphenyl (3)	. . . .	$C_6H_5 \cdot C_6H_4 \cdot NO_2$	V-582	199.21
4725	diphenyl (4)	. . . .	$C_6H_5 \cdot C_6H_4 \cdot NO_2$	V-583	199.21
4726	diphenylamine (*o*)	. . . .	$C_6H_5 \cdot NH \cdot C_6H_4 \cdot NO_2$	XII-690	214.23
4727	diphenylamine (*p*)	. . . .	$C_6H_5 \cdot NH \cdot C_6H_4 \cdot NO_2$	XII-715	214.23
4728	diphenylamine-2-sulfonic acid (4)	. . . .	$C_6H_5 \cdot NH \cdot C_6H_3(NO_2) \cdot SO_3H$	XIV-686	294.29
4729	diphenyl ether (2)	. . . .	$C_6H_5 \cdot O \cdot C_6H_4 \cdot NO_2$	VI-218	215.21
4730	diphenyl ether (3)	. . . .	$C_6H_5 \cdot O \cdot C_6H_4 \cdot NO_2$	VI-224	215.21
4731	diphenyl ether (4)	. . . .	$C_6H_5 \cdot O \cdot C_6H_4 \cdot NO_2$	VI-232	215.21
4732	diphenylene oxide	2-NO_2-dibenzfuran	$NO_2 \cdot C_{12}H_7 \cdot O$	XVII-72	213.19
4733	ethane	. . . .	$CH_3 \cdot CH_2 \cdot NO_2$	I-99	75.07
4734	ethylacetanilide	(*p*)	$NO_2 \cdot C_6H_4 \cdot N(C_2H_5) \cdot COCH_3$	XII-720	208.22
4735	ethyl alcohol	2-nitro-ethanol-1	$HO \cdot CH_2 \cdot CH_2 \cdot NO_2$	I-339	91.07
4736	ethylbenzene (*o*)	. . . .	$NO_2 \cdot C_6H_4 \cdot C_2H_5$	V-358	151.17
4737	ethylbenzene (*p*)	. . . .	$NO_2 \cdot C_6H_4 \cdot C_2H_5$	V-358	151.17
4738	fluorene (2)	. . . .	$NO_2 \cdot C_{12}H_7 : CH_2$	V-628	211.22
4739	formaldehyde-phenylhydrazone	benzene-azo-nitro-methane	$C_6H_5NH \cdot N : CH \cdot NO_2$	XV-235	165.15
4740	furan (2)	. . . .	$NO_2 \cdot C_4H_3O$	XVII-28	113.07
4741	furoic acid (5)	nitro-pyromucic acid (4;2,1)	$NO_2 \cdot C_4H_2O \cdot CO_2H$	XVIII-287	157.08
4742	guaiacol (4)		$NO_2 \cdot C_6H_3(OCH_3)OH$	VI-788	169.14
4743	guanidine	. . . .	$NO_2 \cdot NH \cdot C(NH) \cdot NH_2$	III-126	104.07
4744	heptane (1)	. . . .	$C_6H_{13} \cdot CH_2 \cdot NO_2$	I-155	145.20
4745	heptane (2)	. . . .	$C_5H_{11} \cdot CHNO_2 \cdot CH_3$	I-156	145.20
4746	heptane	2-NO_2-2,4-dimethylpentane	$(CH_3)_2CNO_2 \cdot CH_2 \cdot CH(CH_3)_2$	I-158	145.20
4747	heptane	3-NO_2-2,2-dimethylpentane	$(CH_3)_3C \cdot CH(NO_2) \cdot C_2H_5$	I-157	145.20
4748	heptane	3-NO_2-3-Et-pentane	$(C_2H_5)_3C \cdot NO_2$	I-157	145.20
4749	hexane (1)	. . . .	$CH_3(CH_2)_4CH_2NO_2$	I-147	131.18

Nitro-dibenzfuran 4732
Nitro-dichloroaniline 2021
Nitro-dichlorobenzene 2022-6

Nitro-dichlorophenol 2029-30
Nitro-dimethylbenzene 4865-70

Table 7-4 (*Continued*)
PHYSICAL CONSTANTS OF ORGANIC COMPOUNDS

No.	Crystalline Form and Color	Specific Gravity	Melting Point °C.	Boiling Point °C.	Solubility in 100 Parts		
					Water	Alcohol	Ether
4711	yel. oil	$1.11^{12°}$	−35	224 d.	....	....	....
4712	col. or gn.	....	65(45-6)	265	col.: v. sl. s. pet.	s. h.	gn.: v. s. pet.
4713	pr.	....	20	....	....	....	....
4714	oil	$1.067^{20°}_{4}$	....	152^{15mm}	i.	....	....
4715	lq.	$0.925^{21°}_{0}$	....	235-7 d.	....	....	....
4716	lq.	$0.909^{20°}_{4}$	....	235-7749^{mm} d.	....	....	....
4717	yel. oil	....	....	$153\text{-}5^{20mm}$	sl. s.	s.	s.
4718	yel. oil	....	....	288-90	....	....	....
4719	yel. mn./al.	1.225	77-8	....	....	v. s. h.	sl. s. lg.
4720	yel. oil	$1.179^{20°}_{4}$	....	$151\text{-}3^{30mm}$	sl. s.	v. s.	v. s.
4721	red mn.	$1.313^{17°}$	60-1	280-5	i.	s.	s.
4722	yel. nd.	....	163-4	....	i.	s. h.	s. h. ac.
4723	rhb.	1.44	37	320	i.	s.	v. s.
4724	yel. lf.	....	61	....	i.	v. s.	v. s. ac.
4725	nd./al.	....	113-4	340	i.	sl. s. c.	v. s.
4726	or./aq. al.	....	75-6	....	....	....	....
4727	yel. nd.	....	132-3	....	i. dil. a.	s.	s. ac.
4728	gn. lf./HCl	....	....	....	v. s.	v. s.	....
4729	yel. oil	$1.258^{15°}$	<−20	235^{60mm}	i.	s. abs.	s. bz.
4730	yel. oil	$1.245^{15°}$	....	$202\text{-}4^{14mm}$	....	....	....
4731	tab.	....	56-7	320±	i.	sl. s. c.	v. s.
4732	yel. nd./ac.	....	181-2	....	s. h. ac.	sl. s. h.	sl. s.
4733	lq.	$1.052^{20°}_{20}$	−90	114.8^{761mm}	$4.5^{20°}$; s. a., alk.	∞; ∞ chl.	∞
4734	lf.	....	118-9	....	v. sl. s.; s. bz.	s.; i. lg.	v. sl. s.; s. CS$_2$
4735	col. lq.	$1.270^{15°}$	<−80	194^{765mm}	v. s.	v. s.	v. s.
4736	col. oil	$1.126^{24.5°}$	−23	227-8	i.	v. s.	s.
4737	col. oil	$1.124^{25°}$	−32	245-6	i.	v. s.	v. s.
4738	nd./50% ac.	....	157-8	....	....	....	....
4739	or. pr.	....	85-6 (75-6)	....	s. lg.	s. chl.	s. bz.
4740	lt. yel./pet.	....	28	....	v. sl. s.	s. alk.	s.
4741	lt. yel./aq.	....	185	subl.	v. sl. s.	s.	s.; i. chl.
4742	yel. nd./aq.	....	104-5	....	....	....	....
4743	nd./aq.	....	246-7	....	$9^{100°}$	sl. s.	v. sl. s.
4744	lt. yel. oil	$0.948^{17°}$	....	193-5	i.	v. s.	v. s.
4745	lq.	$0.947^{0°}$	....	194-8 sl. d.	s. conc. alk.	....	....
4746	lq.	$0.956^{0°}_{0}$	....	$181\text{-}2^{742mm}$	....	....	....
4747	lq.	$0.940^{20°}_{20}$	....	$89\text{-}90^{40mm}$	....	....	....
4748	lq.	$0.955^{0°}$	....	185-90	....	....	....
4749	col. lq.	$0.949^{20°}$	....	$193\text{-}4^{765mm}$	i.	v. s.	v. s. al. alk.

Nitroerythrite 2895
Nitro-fluorobenzene 3274

Nitroform 6300
Nitro-glycerin(e) 3428

Table 7-4 (*Continued*)
PHYSICAL CONSTANTS OF ORGANIC COMPOUNDS

No.	Name	Synonym	Formula	Beil. Ref.	Formula Weight
4750	**Nitro** hexane	3-NO$_2$-2,2-di-methylbutane	(CH$_3$)$_3$C·CH(NO$_2$)·CH$_3$	I-151	131.18
4751	hexane	2-NO$_2$-2-Me-pentane	(CH$_3$)$_2$C(NO$_2$)·C$_3$H$_7$	I-149	131.18
4752	hippuric acid (*m*)	....	NO$_2$·C$_6$H$_4$·CO·NH·CH$_2$·CO$_2$H	IX-383	224.17
4753	*o*-hydroxybenzoic acid (3;2,1)	3-nitrosalicylic acid	NO$_2$·C$_6$H$_3$(OH)·CO$_2$H·H$_2$O	X-114	201.14
4754	*o*-hydroxybenzoic acid (4;2,1)	4-nitrosalicylic acid	NO$_2$·C$_6$H$_3$(OH)CO$_2$H	X-116	183.12
4755	*o*-hydroxybenzoic acid (5;2,1)	5-nitrosalicylic acid	NO$_2$·C$_6$H$_3$(OH)CO$_2$H	X-116	183.12
4756	*o*-hydroxybenzoic acid (6;2,1)	6-nitrosalicylic acid	NO$_2$·C$_6$H$_3$(OH)CO$_2$H	....	183.12
4757	*m*-hydroxybenzoic acid (2;3,1)	....	NO$_2$C$_6$H$_3$(OH)CO$_2$H·H$_2$O	X-146	201.14
4758	*m*-hydroxybenzoic acid (4;3,1)	....	NO$_2$·C$_6$H$_3$(OH)CO$_2$H	X-146	183.12
4759	*m*-hydroxybenzoic acid (6;3,1)	....	NO$_2$C$_6$H$_3$(OH)CO$_2$H·H$_2$O	X-147	201.14
4760	*p*-hydroxybenzoic acid (3;4,1)	....	NO$_2$·C$_6$H$_3$(OH)CO$_2$H	X-181	183.12
4761	isatin (5)	....	NO$_2$·C$_6$H$_3$·NH·CO·CO$\underline{}$	XXI-456	192.13
4762	malonic dialdehyde	2-nitro-propandial	NO$_2$·CH(CHO)$_2$	I-766	117.06
4763	malonic ester	....	NO$_2$·CH(CO$_2$C$_2$H$_5$)$_2$	II-596	205.17
4764	mesitylene (1,3,5;2)	....	(CH$_3$)$_3$C$_6$H$_2$·NO$_2$	V-410	165.19
4765	methane	....	CH$_3$·NO$_2$	I-74	61.04
4766	*N*-methylacetanilide (*p*)	....	NO$_2$·C$_6$H$_4$·N(CH$_3$)·COCH$_3$	XII-719	194.19
4767	*N*-methylaniline (*o*)	....	NO$_2$·C$_6$H$_4$·NHCH$_3$	XII-689	152.15
4768	*N*-methylaniline (*m*)	....	NO$_2$·C$_6$H$_4$·NHCH$_3$	XII-700	152.15
4769	*N*-methylaniline (*p*)	....	NO$_2$·C$_6$H$_4$·NHCH$_3$	XII-714	152.15
4770	2-methylanthra-quinone (1)	....	NO$_2$(CH$_3$)C$_{14}$H$_6$O$_2$	VII-811	267.24
4771	4'-methyldiphenyl-amine (4)	....	NO$_2$·C$_6$H$_4$·NH·C$_7$H$_7$	XII-906	228.25
4772	naphthalene (α)	....	NO$_2$·C$_{10}$H$_7$	V-553	173.17
4773	naphthalene (β)	....	NO$_2$·C$_{10}$H$_7$	V-555	173.17
4774	naphthoic acid (8,1)	....	NO$_2$·C$_{10}$H$_6$·CO$_2$H	IX-653	217.18
4775	α-naphthol (2,1)	....	NO$_2$·C$_{10}$H$_6$·OH	VI-615	189.17
4776	α-naphthol (4,1)	....	NO$_2$·C$_{10}$H$_6$·OH	VI-615	189.17
4777	β-naphthol (1,2)	....	NO$_2$·C$_{10}$H$_6$·OH	VI-653	189.17
4778	β-naphthol (5,2)	....	NO$_2$·C$_{10}$H$_6$·OH	VI-654	189.17
4779	β-naphthol (8,2)	....	NO$_2$·C$_{10}$H$_6$·OH	VI-655	189.17
4780	α-naphthylamine	(2,1)	NO$_2$·C$_{10}$H$_6$·NH$_2$	XII-1258	188.19
4781	β-naphthylamine	(1,2)	NO$_2$·C$_{10}$H$_6$·NH$_2$	XII-1313	188.19
4782	β-naphthylamine	(5,2)	NO$_2$·C$_{10}$H$_6$·NH$_2$	XII-1314	188.19
4783	β-naphthylamine	(8,2)	NO$_2$·C$_{10}$H$_6$·NH$_2$	XII-1315	188.19

Table 7-4 (*Continued*)
PHYSICAL CONSTANTS OF ORGANIC COMPOUNDS

No.	Crystalline Form and Color	Specific Gravity	Melting Point °C.	Boiling Point °C.	Solubility in 100 Parts		
					Water	Alcohol	Ether
4750	col. pr.	....	40	167.5-8748mm	v. s. pet.	v. s.	v. s.
4751	lq.	0.949^{0}_{0}	....	172-6756mm	i. alk.	....	....
4752	nd./aq.	....	165-7	....	$0.4^{23°}$	s.	s.
4753	rhb./aq.	....	123-5; 148-9(anh.)	....	$0.13^{15.5°}$	v. s.; s. bz.	v. s.; s. chl.
4754	col. nd./aq.	....	226-35	....	s. h.; s. chl.	s.; i. lg.	sl. s. bz.
4755	nd./aq.	$1.650^{20°}$	229-32	....	$0.18^{22°}$	v. s.; s. act.	v. s.
4756	cr.	....	130	....	s. act.	sl. s.	v. s.
4757	pl./aq.	....	178	....	v. sl. s.	s.	s.
4758	yel. lf./aq.	....	230	....	v. sl. s.	....	....
4759	yel. nd./aq.	....	169	....	v. s.	v. s.	v. s.
4760	nd./aq.	....	185-6	....	v. s. h.	v. s.	s.
4761	nd./al.	....	248-50	....	sl. s.	s.	s. alk.
4762	pr.	....	50-1	....	d.; v.s.chl.	v. s.	v. s.
4763	col. oil	$1.199^{20°}_{4}$	....	152-3³⁸mm	i.	....	....
4764	rhb./al.	....	44	255	....	v. s. h.	....
4765	oil	$1.131^{25°}_{4}$	−28.6	101.3	$9.5^{20°}$	s.; s. alk.	s.
4766	lf./aq.	....	152-3	....	....	s.	s.
4767	red nd./pet.	....	36-7	....	v. sl. s. c.	s.	s.
4768	red yel./al.	....	67-8	....	s. h.	s.	s.
4769	br. yel./al.	....	151-2	....	s. bz.	s.	v. sl. s. lg.
4770	pa. yel. nd./ac.	....	269-70	subl.; d. 330-2	sl. s. bz.; s. $C_6H_5NO_2$	v. sl. s.; sl. s. chl.	v. sl. s.
4771	yel. nd./al. or bz.	....	138-9	....	v. s. ac.; v.sl.s.bz.	v. s. h.	v. sl. s.
4772	yel. nd./al.	$1.223^{61.5°}$	59-60	304	i.; s. CS_2	s.; s. chl.	s.
4773	col./al.	....	79	165¹⁵mm	i.	v. s.	v. s.
4774	pr./al.	....	215	....	0.04	4.7	sl. s.
4775	yel. nd./al.	....	128	....	v. sl. s.	s.	....
4776	col. cr.	....	164	....	s. h.	v. s.	v. s. ac.
4777	yel. nd./al.	....	103	....	i.; s. alk.	s.; s. ac.	v. s.
4778	lt. yel./aq.	....	147	....	v. s. h.	v. s.	v. s.
4779	yel. nd./aq.	....	144-5	....	s. bz.	s.; s. chl.	s.
4780	red yel. mn.	....	144	....	....	s.	....
4781	red yel./al.	....	126-7	....	s. h.	s.	s. ac.
4782	red nd./al.	....	143.5	....	s. bz.	s.	i. lg.
4783	red nd.	....	104-5	....	i. lg.	s.	s.

Table 7-4 *(Continued)*

PHYSICAL CONSTANTS OF ORGANIC COMPOUNDS

No.	Name	Synonym	Formula	Beil. Ref.	Formula Weight
4784	**Nitro** nonane (1)	....	$CH_3(CH_2)_7CH_2NO_2$	I-166	173.26
4785	nonane	2-NO_2-2,6-di-methylheptane	$(CH_3)_2C(NO_2)\cdot(CH_2)_3CH(CH_3)_2$	I-167	173.26
4786	octane (1)	....	$CH_3(CH_2)_6CH_2NO_2$	I-161	159.23
4787	octane	1-NO_2-2,5-di-methylhexane	$(CH_3)_2CH(CH_2)_2\cdot CH(CH_3)CH_2NO_2$	I-163	159.23
4788	octane	2-NO_2-2,5-di-methylhexane	$(CH_3)_2C(NO_2)\cdot(CH_2)_2CH(CH_3)_2$	I-163	159.23
4789	pentandiol	2-NO_2-2-Et-1,3-propandiol	$C_2H_5\cdot C(NO_2)\cdot(CH_2OH)_2$	I-483	149.15
4790	pentane (1)	....	$CH_3(CH_2)_3CH_2NO_2$	I-133	117.15
4791	pentane (3)	....	$(C_2H_5)_2CH\cdot NO_2$	I-133	117.15
4792	pentane	2-NO_2-2-Me-butane	$(CH_3)_2C(NO_2)C_2H_5$	I-140	117.15
4793	pentane	4-NO_2-2-Me-butane	$(CH_3)_2CH(CH_2)_2NO_2$	I-140	117.15
4794	phenanthraquinone	(2)	$NO_2\cdot C_6H_3(CO)_2C_6H_4$	VII-806	253.22
4795	*p*-phenetidine (3)	3-NO_2-4-NH_2-phenetole	$C_6H_3(NH_2)OC_2H_5\cdot NO_2$	XIII-521	182.18
4796	phenetole (*o*)	....	$C_2H_5O\cdot C_6H_4\cdot NO_2$	VI-218	167.17
4797	phenetole (*m*)	....	$C_2H_5O\cdot C_6H_4\cdot NO_2$	VI-224	167.17
4798	phenetole (*p*)	....	$C_2H_5O\cdot C_6H_4\cdot NO_2$	VI-231	167.17
4799	phenol (*o*)	....	$HO\cdot C_6H_4\cdot NO_2$	VI-213	139.11
4800	phenol Na (*o*)	Na nitrophenoxide	$NaO\cdot C_6H_4\cdot NO_2$	VI-217	161.10
4801	phenol K (*o*)	K nitrophenoxide	$KO\cdot C_6H_4\cdot NO_2\cdot\frac{1}{2}H_2O$	VI-217	186.22
4802	phenol (*m*)	....	$HO\cdot C_6H_4\cdot NO_2$	VI-222	139.11
4803	phenol (*p*)	....	$HO\cdot C_6H_4\cdot NO_2$	VI-226	139.11
4804	phenol Na (*p*)	....	$NaC_6H_4O_3N\cdot 2H_2O$	VI-230	197.13
4805	phenol Na (*p*)	....	$NaC_6H_4O_3N\cdot 4H_2O$	VI-230	233.16
4806	*o*-phenol sulfonic acid	(1;4,2)	$HOC_6H_3(NO_2)SO_3H\cdot 3H_2O$	XI-237	273.22
4807	*p*-phenol sulfonic acid	(1;2,4)	$HOC_6H_3(NO_2)SO_3H\cdot 3H_2O$	XI-246	273.22
4808	phenylacetate (*o*)	....	$NO_2\cdot C_6H_4\cdot O_2CCH_3$	VI-219	181.15
4809	phenylacetic acid (*p*)	*p*-NO_2-α-toluic acid	$NO_2\cdot C_6H_4\cdot CH_2CO_2H$	IX-455	181.15
4810	phenylarsonic acid	(*m*)	$NO_2\cdot C_6H_4\cdot AsO(OH)_2$	XVI-869	247.04
4811	phenyl *iso*-cyanate	(*p*)	$NO_2\cdot C_6H_4\cdot N{:}CO$	XII-725	164.12
4812	phenylglycine (*p*)	....	$NO_2\cdot C_6H_4\cdot NH\cdot CH_2\cdot CO_2H$	XII-725	196.16
4813	phenylhydrazine (*m*)	....	$NO_2\cdot C_6H_4\cdot NH\cdot NH_2$	XV-460	153.14
4814	phenylhydrazine (*m*)	HCl salt	$NO_2\cdot C_6H_4\cdot N_2H_3\cdot HCl$	XV-460	189.60
4815	phenylhydrazine (*p*)	....	$NO_2\cdot C_6H_4\cdot NH\cdot NH_2$	XV-468	153.14
4816	phenylhydrazine (*p*)	HCl salt	$NO_2\cdot C_6H_4\cdot N_2H_3\cdot HCl$	XV-468	189.60
4817	phenylpropiolic acid	(*o*)	$NO_2\cdot C_6H_4\cdot C{:}C\cdot CO_2H$	IX-636	191.14
4818	phenylpropiolic acid	(*p*)	$NO_2\cdot C_6H_4\cdot C{:}C\cdot CO_2H$	IX-637	191.14
4819	*o*-phthalic acid (3)	....	$NO_2\cdot C_6H_3(CO_2H)_2$	IX-823	211.13
4820	*o*-phthalic acid (4)	....	$NO_2\cdot C_6H_3(CO_2H)_2$	IX-828	211.13

Nitro-orthanilic acid 4622   Nitro-phenylacetonitrile 4685-7

Table 7-4 (*Continued*)
PHYSICAL CONSTANTS OF ORGANIC COMPOUNDS

No.	Crystalline Form and Color	Specific Gravity	Melting Point °C.	Boiling Point °C.	Solubility in 100 Parts		
					Water	Alcohol	Ether
4784	lt. yel. lq.	$0.923^{17°}$	....	215-8 d.	....	....	....
4785	lq.	$0.915^{1.8°}_{0}$	....	113-4^{25mm}	....	....	....
4786	lt. yel. lq.	$0.935^{20°}$	....	206-10 sl. d.	....	....	....
4787	lq.	....	....	100-5^{20mm}	....	....	....
4788	lq.	$0.921^{20°}_{0}$	−18	201-2^{755mm}	s. alk.	....	....
4789	nd./aq.	....	57-8	d.	v. s.	s.	s.
4790	col. lq.	$0.948^{20°}$	....	172-3	....	....	....
4791	lq.	$0.958^{0°}$	....	152-5^{746mm}	i.	s.	s.
4792	lq.	$0.97^{0°}$	....	149-51^{748mm}	....	....	....
4793	lq.	$0.960^{20.6°}_{4}$	....	164^{756mm}	....	....	....
4794	yel. lf./ac.	....	257-8	....	....	v. sl. s.	sl. s. ac.
4795	red pr./al.	....	113	....	....	s. h.	s.; s. chl.
4796	oil	$1.190^{15°}$	5-6	275	v. sl. s.	v. s.	v. s.
4797	yel. cr.	....	34	169^{70mm}	v. sl. s.	v. s. h.	v. s.
4798	col. mn.	$1.181^{5°}$	59-60	283^{758mm}	v. sl. s.	v. s. h.	v. s.
4799	yel. mn.	$1.295^{45°}$	45.0	217.2	$0.21^{20°}$	v. s.	v. s.; s. bz.
4800	red lf./al.	....	....	....	v. s.	v. sl. s. NaOH	....
4801	or. nd./al.	$1.682^{20°}$	....	....	166°	$21^{15°}$ aq.	....
4802	col. mn.	$1.485^{20°}$	96-7	194^{70mm}	$1.35^{20°}$	v. s.	s.; i. pet.
4803	yel. pr.	$1.48^{20°}$	114.0	subl.	$1.6^{25°}$	v. s.	v. s.
4804	or. cr./aq.	....	−2H$_2$O, 110	....	6.5 c.	....	....
4805	yol. mn./aq.	....	....	....	s.	....	....
4806	nd.	....	d. 110	....	v. s.	v. s.	sl. s.
4807	nd./aq.	....	51.5; 141-2 (anh.)	....	v. s.	v. s.	v. s. chl.
4808	col. nd./lg.	....	39-40	253 d.	v. s. bz.	v. s.	v. s.
4809	nd./aq.	....	152-3	....	v. sl. s. c.	s.	s.; s. chl.
4810	lf./aq.	....	d.	....	$2^{18°}$	sl. s. chl.	i. lg.
4811	nd.	....	56-7	....	s. h. lg.	s. chl.	s.; s. bz.
4812	yel. cr./aq.	....	225-30 d.	....	v. sl. s.	v. s. h.	sl. s.
4813	yel. nd./al.	....	93	....	v. sl. s. h.	v. sl. s. bz.	s. ac.
4814	yel. pl.	....	....	....	v. sl. s. c.	v. sl. s. c.	sl. s. HCl
4815	or. red nd.	....	157	....	s. h.	v. s.	s.
4816	or. red pl.	....	214-5 d.	....	s.	i.	i.
4817	nd./aq.	....	157 d.	expl. 155-6	s. h.	s.; i. CS$_2$	v. sl. s. chl.
4818	nd./al.	....	181-93 d.	....	sl. s.	s. h.	s.; i. pet.
4819	lt. yel./aq.	....	222*	....	$2^{25°}$	v. s. h.	sl. s.
4820	lt. yel. cr.	....	164-5	....	v. s.	v. s.	s.

* Sealed tube.

Table 7-4 (*Continued*)
PHYSICAL CONSTANTS OF ORGANIC COMPOUNDS

No.	Name	Synonym	Formula	Beil. Ref.	Formula Weight
4821	**Nitro** *m*-phthalic acid (5)	....	$NO_2 \cdot C_6H_3(CO_2H)_2$	IX-840	211.13
4822	*p*-phthalic acid (2)	....	$NO_2 \cdot C_6H_3(CO_2H)_2$	IX-851	211.13
4823	phthalic anhydride	(3)	$NO_2 \cdot C_6H_3(CO)_2O$	XVII-486	193.12
4824	phthalide (6)	....	$NO_2 \cdot C_6H_3 \cdot CH_2 \cdot O \cdot CO$	XVII-313	179.13
4825	phthalimide (3)	....	$NO_2 \cdot C_6H_3(CO)_2NH$	XXI-505	192.13
4826	phthalimide (4)	....	$NO_2 \cdot C_6H_3(CO)_2NH$	XXI-506	192.13
4827	propane (1)	....	$CH_3 \cdot CH_2 \cdot CH_2 \cdot NO_2$	I-115	89.09
4828	propane (2)	*β*-nitropropane	$(CH_3)_2CH \cdot NO_2$	I-116	89.09
4829	propene (3,1)	nitroallyl	$CH_2{:}CH \cdot CH_2 \cdot NO_2$	I-203	87.08
4830	quinaldine (6)	6-NO_2-2-Me-quinoline	$NO_2 \cdot C_9H_5N \cdot CH_3$	XX-394	188.19
4831	quinoline (5)(*ana*)	....	$NO_2 \cdot C_9H_6N$	XX-371	174.16
4832	quinoline (6)(*p*)	....	$NO_2 \cdot C_9H_6N$	XX-372	174.16
4833	quinoline (7)(*m*)	....	$NO_2 \cdot C_9H_6N$	XX-372	174.16
4834	quinoline (8)(*o*)	....	$NO_2 \cdot C_9H_6N$	XX-373	174.16
4835	*iso*-quinoline (5 or 8)	....	$NO_2 \cdot C_6H_3{:}C_3H_3N$	XX-386	174.16
4836	resorcinol (2)	(2;1,3)	$NO_2 \cdot C_6H_3(OH)_2$	VI-823	155.11
4837	salol (*α*)	5-nitrosalol	$NO_2 \cdot C_6H_3(OH) \cdot CO_2 \cdot C_6H_5$	X-118	259.22
4838	styrene (*o*)	....	$NO_2 \cdot C_6H_4 \cdot CH{:}CH_2$	V-478	149.15
4839	styrene (*m*)	....	$NO_2 \cdot C_6H_4 \cdot CH{:}CH_2$	V-478	149.15
4840	styrene (*p*)	....	$NO_2 \cdot C_6H_4 \cdot CH{:}CH_2$	V-478	149.15
4841	styrene (*ω*)(*β*)	....	$C_6H_5 \cdot CH{:}CH \cdot NO_2$	V-478	149.15
4842	tartaric acid	....	$(NO_2 \cdot O \cdot CH \cdot CO_2H)_2$	III-509	240.08
4843	thiophene (2)	....	$NO_2 \cdot C_4H_3S$	XVII-35	129.14
4844	toluene† (*o*)	....	$CH_3 \cdot C_6H_4 \cdot NO_2$	V-318	137.14
4845	toluene (*m*)	....	$CH_3 \cdot C_6H_4 \cdot NO_2$	V-321	137.14
4846	toluene (*p*)	....	$CH_3 \cdot C_6H_4 \cdot NO_2$	V-323	137.14
4847	*o*-toluene sulfonic acid	(1;4,2)	$CH_3 \cdot C_6H_3(NO_2) \cdot SO_3H \cdot 2H_2O$	XI-90	253.23
4848	*p*-toluene sulfonyl chloride	(1;2,4)	$CH_3 \cdot C_6H_3(NO_2) \cdot SO_2Cl$	XI-111	235.65
4849	*o*-toluidine (3;1,2)	....	$NO_2 \cdot C_6H_3(CH_3)NH_2$	XII-843	152.15
4850	*o*-toluidine (4;1,2)	....	$NO_2 \cdot C_6H_3(CH_3)NH_2$	XII-844	152.15
4851	*o*-toluidine (5;1,2)	....	$NO_2 \cdot C_6H_3(CH_3)NH_2$	XII-846	152.15
4852	*o*-toluidine (6;1,2)	....	$NO_2 \cdot C_6H_3(CH_3)NH_2$	XII-848	152.15
4853	*m*-toluidine (2;1,3)	....	$NO_2 \cdot C_6H_3(CH_3)NH_2$	XII-876	152.15
4854	*m*-toluidine (4;1,3)	....	$NO_2 \cdot C_6H_3(CH_3)NH_2$	XII-876	152.15
4855	*m*-toluidine (5;1,3)	....	$NO_2 \cdot C_6H_3(CH_3)NH_2$	XII-877	152.15
4856	*m*-toluidine (6;1,3)	....	$NO_2 \cdot C_6H_3(CH_3)NH_2$	XII-877	152.15
4857	*p*-toluidine (2;1,4)	....	$NO_2 \cdot C_6H_3(CH_3)NH_2$	XII-996	152.15
4858	*p*-toluidine (3;1,4)	....	$NO_2 \cdot C_6H_3(CH_3)NH_2$	XII-1000	152.15
4859	trimethylol-methane	2-NO_2-2-methylol-1,3-propandiol	$NO_2 \cdot C(CH_2OH)_3$	I-520	151.12
4860	tyrosine (3)(*l*)	....	$NO_2 \cdot C_6H_3(OH) \cdot CH_2 \cdot CH(NH_2) \cdot CO_2H$	XIV-620	226.19
4861	undecane (1)	....	$CH_3 \cdot (CH_2)_9CH_2 \cdot NO_2$	I-170	201.31
4862	uracil (5)	....	$NO_2 \cdot C{:}CH \cdot (NHCO)_2$	XXIV-320	157.09

† See also No. 5213.
Nitro-propandial 4762
Nitro-*iso*-propyl-methyl-benzene 4714
Nitro-pyromucic acid 4741

Nitro-salicylic acid 4753-6
Nitro-sulfanilic acid 4621
Nitro-toluic acid (*α*) 4809
Nitro-toluic nitrile (*α*) 4685-**7**

Table 7-4 (*Continued*)
PHYSICAL CONSTANTS OF ORGANIC COMPOUNDS

No.	Crystalline Form and Color	Specific Gravity	Melting Point °C.	Boiling Point °C.	Solubility in 100 Parts		
					Water	Alcohol	Ether
4821	col. cr.	....	255 sl. d.	....	$0.15^{15°}$; $81^{99°}$	v. s.	v. s.
4822	nd./h. aq.	....	263-70	....	s. h.	s. h.	....
4823	col. nd./ac.	....	162-3	....	s. act.	s. h.	v. sl. s. bz.
4824	nd.	....	141	....	i. c.; v. s. h. chl.	sl. s. c.	sl. s.; i. alk. carb.
4825	yel. lf./al.	....	215-6	subl.	i.; s. ac.	s. h.	i.; i. lg.
4826	yel. lf.	....	199-201	subl.	v. sl. s. h.	s.; s. ac.	s. act.
4827	oil	$1.003^{20°}_{20}$	-108	131.6	$1.4^{20°}$	∞	∞
4828	lq.	$1.024^{0°}$	-93	120.3	$1.7^{20°}$	$∞^{20°}$	$∞^{20°}$
4829	col. lq.	$1.051^{21°}$	....	125-30	i.	s.	s.
4830	yel. nd./aq.	....	163-4	....	v. sl. s.	s.	i.
4831	nd./aq.	....	72	subl.	v. sl. s. h.	s. h.	s. bz.
4832	nd.	....	149-50	subl.	sl. s. c.	sl. s. c.	sl. s.; s. bz.
4833	nd./al.	....	132-3	....	....	v. sl. s. c.	s.
4834	mn./al.	....	88-9	....	s. h.	s.	s.; s. bz.
4835	nd./aq.	....	110	subl.	s. h.	s.	s.; s. bz.
4836	or./aq. al.	....	83-5	d. > 180°	....	....	....
4837	nd./al.	....	151-2	....	....	s.	s. ac.
4838	oil	....	12-13.5	....	s. H_2SO_4	....	....
4839	yel. oil	....	-5	....	v. s. lg.	v. s. abs.	v. s.
4840	pr./lg.	....	29	d.	v. sl. s. c. lg.	s. h.	v. s. h.
4841	yel. pr./al.	....	58	250-60 d.	v. sl. s. h.	s.	v. s.
4842	silky nd.	....	d.	....	d. 0^{v}	v. s. c.; d. h.	v. s.; l. bz.
4843	mn./al.	....	46	224-5	i. alk.	s.	s.
4844	yel. lq.	$1.163^{20°}_{4}$	(α) -9.3 (β) -3.2	221.7	$0.065^{30°}$; s. pet.	∞; ∞ bz.; s. SO_2	∞; sl. s. NH_3
4845	lq.	$1.157^{20°}_{4}$	16.1	232.6	$0.050^{30°}$	$8.6^{15°}$	∞
4846	rhb.	$1.123^{55°}$	51.7	238.5	$0.004^{15°}$	s.; s. bz.	$80.8^{15°}$
4847	pl./aq.	....	130 (anh.)	....	$47.7^{23°}$	v. s.	v. s.; s. chl.
4848	pl./et.	....	33-4	....	d.	d.	....
4849	or./45% al.	....	97	....	sl. s.	s.; s. chl.	s.; s. bz.
4850	yel. mn.	$1.365^{15°}$	105-7	....	v. sl. s.	s.	s.
4851	yel. mn.	$1.366^{15°}$	129-30	....	v. sl. s. h.	s.	....
4852	yel. rhb.	$1.378^{15°}$	91-2	305 d.	1.3 h.	v. s.	v. s.; v. s. bz.
4853	yel. nd.	....	53	....	sl. s. c.	s.	....
4854	yel. lf./aq.	....	109-10	....	s. bz.	s.	s.; s. chl.
4855	yel. brn. nd.	....	98	....	sl. s. c.	s.	v. s.; s. bz.
4856	yel. nd./aq.	....	135-8	....	s. h.	s.	s.
4857	yel. nd./aq.	....	81.5	....	sl. s. h.	v. s. h.	s.
4858	red mn.	$1.312^{17°}$	116-7	....	sl. s. h.	s.	....
4859	nd. or pr.	....	165-70 d.	....	$220^{20°}$	$45^{20°}$	$1^{20°}$
4860	pa. yel. nd./aq.	....	237	....	v. sl. s. h.	i.; s. alk.	i.; s. aq. a.
4861	yel. lq.	$0.900^{15°}$	....	d.	....	v. s.	v. s.
4862	col. nd.	....	expl.	....	sl. s. c.	s.	....

Nitroso–amyl ketone 79
Nitroso–anthron 4626
Nitroso–barbituric acid 6445
Nitroso–benzylaniline 828

Nitroso–diethylamine 2191
Nitroso–dimethylamine 2513
Nitroso–diphenylamine 2733
Nitroso–dipropylamine 2790-1

Table 7-4 (*Continued*)
PHYSICAL CONSTANTS OF ORGANIC COMPOUNDS

No.	Name	Synonym	Formula	Beil. Ref.	Formula Weight
4863	**Nitro** urea	....	$NO_2 \cdot NH \cdot CO \cdot NH_2$	III-125	105.05
4864	urethane	....	$NO_2 \cdot NH \cdot CO_2C_2H_5$	III-125	134.09
4865	*o*-xylene (3;1,2)	....	$NO_2 \cdot C_6H_3(CH_3)_2$	V-367	151.17
4866	*o*-xylene (4;1,2)	....	$NO_2 \cdot C_6H_3(CH_3)_2$	V-368	151.17
4867	*m*-xylene (2;1,3)	....	$NO_2 \cdot C_6H_3(CH_3)_2$	V-378	151.17
4868	*m*-xylene (4;1,3)	....	$NO_2 \cdot C_6H_3(CH_3)_2$	V-378	151.17
4869	*m*-xylene (5;1,3)	....	$NO_2 \cdot C_6H_3(CH_3)_2$	V-378	151.17
4870	*p*-xylene (2;1,4)	....	$NO_2 \cdot C_6H_3(CH_3)_2$	V-387	151.17
4871	*m*-xylidine (5), aceto-	(5;1,3,4)	$NO_2 \cdot C_6H_2(CH_3)_2 \cdot NH \cdot COCH_3$	XII-1128	208.22
4872	**Nitron**	4,5-dihydro-1,4-diPh-3,5-phenyl-imino-1,2,4-tri-azole	$C_{20}H_{16}N_4$	XXVI-349	312.38
4873	**Nitroso**-acetone (*iso*)	pyroracemic aldoxime	$CH_3 \cdot CO \cdot CH:NOH$	I-763	87.08
4874	acetophenone (*iso*)	Ph-glyoxaloxime	$C_6H_5 \cdot CO \cdot CH:NOH$	VII-671	149.15
4875	aniline (*p*)	*p*-quinone-imide-oxime tautomer	$ON \cdot C_6H_4 \cdot NH_2$ or $HN:C_6H_4:NOH$	VII-625	122.13
4876	benzene	....	$C_6H_5 \cdot NO$	V-230	107.11
4877	benzoic acid (*o*)	....	$ON \cdot C_6H_4 \cdot CO_2H$	IX-368	151.12
4878	benzoic acid (*m*)	....	$ON \cdot C_6H_4 \cdot CO_2H$	IX-369	151.12
4879	benzoic acid (*p*)	....	$ON \cdot C_6H_4 \cdot CO_2H$	IX-369	151.12
4880	*n*-butyric acid (*iso*)	(*α*)	$C_2H_5 \cdot C(:NOH) \cdot CO_2H$	III-629	117.11
4881	*m*-cresol (4)	(4;3,1)	$ON \cdot C_6H_3(CH_3)OH$	VII-648	137.14
4882	diethylaniline (*p*)	....	$ON \cdot C_6H_4 \cdot N(C_2H_5)_2$	XII-684	178.24
4883	dimethylaniline (*p*)	....	$ON \cdot C_6H_4 \cdot N(CH_3)_2$	XII-677	150.18
4884	diphenylamine (*p*)	*p*-NO-Ph aniline	$ON \cdot C_6H_4 \cdot NH \cdot C_6H_5$	XII-207	198.23
4885	ethylaniline (*N*)	Et-Ph-nitrosoamine	$C_6H_5 \cdot N(NO)C_2H_5$	XII-580	150.18
4886	*N*-ethyl-urethane	*N*-NO-*N*-Et car-bamate	$C_2H_5N(NO) \cdot CO_2C_2H_5$	IV-129	146.15
4887	methylaniline (*N*)	....	$C_6H_5 \cdot N(NO)CH_3$	XII-579	136.15
4888	methyl-ethyl ketone (*iso*)	diacetyl monoxime	$CH_3COC(:NOH) \cdot CH_3$	I-772	101.11
4889	methyl-*n*-hexyl ketone	(*iso*)	$CH_3 \cdot CO \cdot C(:NOH) \cdot (CH_2)_4CH_3$	I-795	157.21
4890	methyl-propyl ketone (*iso*)	....	$CH_3 \cdot CO \cdot C(:NOH) \cdot C_2H_5$	I-776	115.13
4891	*N*-methyl-urethane	....	$CH_3N(NO) \cdot CO_2C_2H_5$	IV-85	132.12
4892	*α*-naphthol (1,4)	naphthoquinone oxime tautomer	$HO \cdot C_{10}H_6 \cdot NO$ or $O:]C_{10}H_6:NOH$	VII-727	173.17
4893	*α*-naphthol (1,2)	2-nitroso-naphthol-1	$HO \cdot C_{10}H_6 \cdot NO$	VII-715	173.17
4894	*β*-naphthol (2,1)	1-nitroso-naphthol-2	$HO \cdot C_{10}H_6 \cdot NO$	VII-712	173.17
4895	*β*-naphthylamine	(2,1)	$ON \cdot C_{10}H_6 \cdot NH_2$	VII-717	172.19
4896	nitrobenzene (*o*)	....	$ON \cdot C_6H_4 \cdot NO_2$	V-256	152.11
4897	phenylhydrazine (*α*)	....	$C_6H_5 \cdot N(NO) \cdot NH_2$	XV-416	137.14
4898	piperidine (*N*)	....	$C_5H_{10}N \cdot NO$	XX-83	114.15
4899	propiophenone (*α*)	(*iso*); methyl-benzoyl ketoxime (*β*)	$C_6H_5 \cdot CO \cdot C(:NOH) \cdot CH_3$	VII-677	163.18
4900	resorcinol (4)	2-OH-*p*-benzoqui-none-1-oxime tautomer	$ON \cdot C_6H_3(OH)_2$ or $O:C_6H_3(:NOH)OH$	VIII-235	139.11

Table 7-4 (*Continued*)
PHYSICAL CONSTANTS OF ORGANIC COMPOUNDS

No.	Crystalline Form and Color	Specific Gravity	Melting Point °C.	Boiling Point °C.	Solubility in 100 Parts		
					Water	Alcohol	Ether
4863	wh. pd./aq.	....	158-9 d.	....	s. h.	s.; sl. s. bz	v. sl. s.
4864	lf./lg.	....	64	....	s.	v. s.	v. s.
4865	yel. oil	$1.147^{15°}$	15	240-5	....	....	....
4866	yel. pr./al.	$1.139^{30°}_{30}$	29-30	258 sl. d.	i.	$\infty^{30°}$	v. s.
4867	lq.	$1.112^{15°}$	13-5	225^{744mm}	....	....	....
4868	yel. lq.	$1.135^{15°}$	2	244	i.	s.	s.
4869	col. nd./al.	....	74-5	273^{739mm}	....	....	....
4870	yel. lq.	$1.132^{15°}$	....	238.9^{739mm}	....	....	....
4871	yel. nd./aq.	....	172-3	....			
4872	yel. lf./al.	....	189-90 d.	....	i.; s. act.	s. h.; s. bz.	v. sl. s.; s. chl.
4873	lf./et.	$1.074^{67.5°}$	69	subl.	v. s.	v. sl. s. pet.	v. s.
4874	mn./aq. al.	....	126-8	....	s. h.	s. alk.	....
4875	b. nd./bz.	....	173-4	....	s.	s. bz.	....
4876	col. rhb.*	....	67.5-8.0	$57-9^{18mm}$	i.; i.NH_3	s.	sl. s. lg.
4877	cr./al.	....	210 d.	....	s. ac.	s. h.	v. sl. s.
4878	col. cr.	....	d. 230	....	....	....	....
4879	yel. pd.	....	d. 250	....	sl. s. bz.	sl. s. h.	sl. s. ac.
4880	nd./aq.	....	169-70	....	sl. s.	s.	sl. s.
4881	nd./aq.	....	158-60	....	sl. s. h.	s.; s. bz.	v. sl. s.
4882	gn. mn.	$1.24^{15°}$	84	....	v. sl. s.	s.	s.
4883	gn. tri.	....	86-7	....	i.	s.	s.
4884	gn. pl./bz.	....	144.6	....	sl. s.; v. s. chl.	v. s.; sl. s. lg.	v. s.; s. bz.
4885	yel. oil	$1.087^{20°}_{4}$	....	$119-20^{15mm}$	i.	....	....
4886	red oil	$1.071^{20°}_{4}$	....	86^{36mm}	....	....	....
4887	yel. oil	$1.124^{20°}_{4}$	14-5	128^{19mm}	....	s.	s.
4888	pr./chl.	....	76	185-6	sl. s.	v. s.; s. alk.	v. s.; s. chl.
4889	cr./lg.	....	58-9	133^{11mm}	....	....	v. s.
4890	lf./lg.	....	58-9	183-7 sl. d.	sl. s. c.	v. s.; v. s. chl.	v. s.
4891	yel. red lq.	$1.122^{20°}_{4}$	<-20	65^{13mm}	sl. s. h.	∞	∞; ∞ bz.
4892	nd./aq. al.	....	193-4	....	i.	v. s.	v. s.
4893	yel. nd./aq.	....	162-4 d.	....	s. h.	v. s.	sl. s.
4894	brn. pr./al.	....	109.5	....	$0.1^{20°}$	$2.4^{13°}$	v. sl. s. pet.
4895	gn./aq. al.	....	150-2	....	sl. s. h.	v. s. h.	s. h.
4896	yel. cr./act.	....	126†	....	i.; i. pet.	s. h.	v. s. chl.
4897	lt. yel. lf.	....	51	....	s.	poisonous	s.
4898	lt. yel. oil	$1.063^{18.5°}_{4}$	....	217-8	s.	v. s. aq. a.	....
4899	nd./aq.	....	113-4	....	s. alk.	....	....
4900	yel. nd. + 1H_2O/aq.	....	$-H_2O$, 105; d. 148	....	s.; s. chl.	v. s.; s. ac.	s.; i. bz.; i. CS_2

* Green as liquid or in solution.
† Green when liquid.
Nonene 4930
Nonyl cyanide 1207
Nonylic acid 5042

Nor-arecaidine, cf. alkd.
Nosophen 5813
Nostal 5390
Novacetyl 118
Novarsenobenzol 5593

Novatophan 3130
Novonal 2107
Nucin 3787
Numal 5370
Nupercaine 1025

Table 7-4 (*Continued*)
PHYSICAL CONSTANTS OF ORGANIC COMPOUNDS

No.	Name	Synonym	Formula	Beil. Ref.	Formula Weight
4901	**Nitroso** R salt	(1;3,6;2)	$ON \cdot C_{10}H_4(SO_3Na)_2 \cdot OH$	....	377.26
4902	thymol (*p*)	(quinoneoxime tautomer)	$ON \cdot C_{10}H_{12} \cdot OH$ or $O:C_{10}H_{12}:NOH$	VII-664	179.22
4903	toluene (*o*)	....	$CH_3 \cdot C_6H_4 \cdot NO$	V-317	121.14
4904	toluene (*m*)	....	$CH_3 \cdot C_6H_4 \cdot NO$	V-318	121.14
4905	toluene (*p*)	....	$CH_3 \cdot C_6H_4 \cdot NO$	V-318	121.14
4906	triacetonamine (*N*)	....	$C_9H_{16}ON \cdot NO$	XXI-251	184.24
4907	**Nonacosane**	....	$C_{29}H_{60}$	I-176	408.80
4908	**Nonadecane** (*n*)	....	$CH_3 \cdot (CH_2)_{17} \cdot CH_3$	I-174	268.53
4909	**Nonadecylic acid**	....	$CH_3 \cdot (CH_2)_{17} \cdot CO_2H$	II-389	298.51
4910	**Nonane** (*n*)	....	$CH_3 \cdot (CH_2)_7 \cdot CH_3$	I-165	128.26
4911	**Nonane** (*iso*)	2-methyl-octane	$(CH_3)_2CH \cdot C_6H_{13}$	I-166	128.26
4912	**Nonane**	2,4-di Me-heptane	C_9H_{20}	*I-64	128.26
4913	**Nonane**	2,5-di Me-heptane	C_9H_{20}	I-167	128.26
4914	**Nonane**	2,6-di Me-heptane	$[(CH_3)_2CH \cdot CH_2]_2CH_2$	I-167	128.26
4915	**Nonane**	4-ethyl-heptane	$C_2H_5 \cdot CH(CH_2C_2H_5)_2$	I-167	128.26
4916	**Nonane**	3-methyl-octane	$(C_2H_5)(CH_3)CH \cdot (CH_2)_4CH_3$	I-166	128.26
4917	**Nonane**	4-methyl-octane	C_9H_{20}	*I-63	128.26
4918	**Nonyl** alcohol (*n*)	nonanol-1	$CH_3(CH_2)_7CH_2OH$	I-423	144.26
4919	alcohol (*n*)	nonanol-2	$C_7H_{15} \cdot CHOH \cdot CH_3$	I-423	144.26
4920	alcohol (*n*)	nonanol-3	$C_6H_{13} \cdot CHOH \cdot C_2H_5$	I-424	144.26
4921	alcohol (*n*)	nonanol-5	$(C_4H_9)_2CHOH$	I-424	144.26
4922	alcohol	2-methyl-octanol-2	$(CH_3)_2C(OH)C_6H_{13}$	I-424	144.26
4923	alcohol	4-ethyl-heptanol-4	$C_2H_5 \cdot C(OH):(CH_2 \cdot CH_2 \cdot CH_3)_2$	I-424	144.26
4924	alcohol	2-Me-4-Et-hexanol-4	$(C_2H_5)_2C(OH) \cdot CH_2 \cdot CH(CH_3)_2$	I-425	144.26
4925	alcohol	2,6-di Me-heptanol-4	$[(CH_3)_2CH \cdot CH_2]_2:CHOH$	I-425	144.26
4926	amine (*n*)	1-amino-nonane	$CH_3(CH_2)_7CH_2NH_2$	IV-198	143.27
4927	bromide (2)	2-bromo-nonane	$C_7H_{15} \cdot CHBr \cdot CH_3$	I-166	207.16
4928	chloride (2)	2-chloro-nonane	$C_7H_{15} \cdot CHCl \cdot CH_3$	I-166	162.70
4929	iodide (*n*)	1-iodo-nonane	$CH_3(CH_2)_7 \cdot CH_2I$	I-166	254.16
4930	**Nonylene** (*β*)	nonene-2	$C_6H_{13} \cdot CH:CH \cdot CH_3$	I-223	126.24
4931	**Norbixin** (*trans*)	(stable)(*β*)	$C_{24}H_{28}O_4$	XXX-109	380.49
4932	**Norbixin** (*cis*)	(labile)	$C_{24}H_{28}O_4$	XXX-110	380.49
4933	methyl ester (*cis*)	(labile)-bixin	$C_{25}H_{30}O_4$	XXX-112	394.52
4934	**Novocaine**®	Ethocaine; procaine HCl	$C_{13}H_{20}O_2N_2 \cdot HCl$	XIV-424	272.78
4935	base	procaine	$NH_2 \cdot C_6H_4CO_2(CH_2)_2 \cdot N(C_2H_5)_2$	XIV-424	236.32
4936	**Octa**-chloropropane	perchloro-propane	C_3Cl_8	I-108	319.66
4937	cosane	....	$C_{28}H_{58}$	I-176	394.77
4938	decane	*n*-octadecane	$CH_3 \cdot (CH_2)_{16} \cdot CH_3$	I-173	254.50
4939	decyl alcohol (*n*)	octadecanol-1	$CH_3(CH_2)_{16}CH_2OH$	I-431	270.50
4940	decyl iodide (*n*)	1-iodo-octadecane	$CH_3(CH_2)_{16}CH_2I$	I-173	380.40
4941	decylene (*α*)	octadecene-1	$CH_3(CH_2)_{15}CH:CH_2$	I-226	252.49
4942	decyne-1	hexadecyl acetylene	$CH_3(CH_2)_{15}C:CH$	I-262	250.47
4943	**Octandiol** (1,8)	....	$HO \cdot (CH_2)_8 \cdot OH$	I-490	146.23
4944	**Octane** (*n*)	....	$CH_3 \cdot (CH_2)_6 \cdot CH_3$	I-159	114.23
4945	**Octane**	2,2,3,3-tetramethyl-butane	$[(CH_3)_3C]_2$	I-165	114.23

Table 7-4 (*Continued*)
PHYSICAL CONSTANTS OF ORGANIC COMPOUNDS

No.	Crystalline Form and Color	Specific Gravity	Melting Point °C.	Boiling Point °C.	Solubility in 100 Parts		
					Water	Alcohol	Ether
4901	yel. cr.	....	....	....	2.5	sl. s.	....
4902	lt. yel. nd./chl.	....	160-4 sl. d.	....	v. sl. s. h.; s. alk.	s.	s.; s. chl.
4903	nd. or pr.	....	72.5	....	v. s. chl.	v. s.	v. s.
4904	nd.	....	53.5	....	i.	....	s.
4905	col.nd./lg.*	....	48.5	....	v. sl. s.	v.s. Me al.	v. s. bz.
4906	nd./aq. al.	1.14	72-3	subl.	....	v. s.	v. s.
4907	cr.	lq. $0.780^{63.8°}$	63.8	$346-8^{40mm}$	$1.1^{15°}$ chl.	v. s. h.	v. s.
4908	cr.	$0.777^{32°}_4$	32	330	i.	sl. s.	s.
4909	lf./al.	....	66.5	$297-8^{100mm}$	i.	....	....
4910	col. lq.	$0.718^{20°}_4$	-53.6	150.8	i.	s. abs.	s.
4911	lq.	$0.713^{20°}_4$	-80.4	143.3	....	....	....
4912	col. lq.	$0.715^{20°}_4$	....	132.9	i.	∞	∞
4913	col. lq.	$0.717^{20°}_4$	....	136	....	....	....
4914	lq.	$0.709^{20°}_4$	-102.9	135.2	i.	....	s.
4915	lq.	$0.728^{20°}_4$	....	141.2	i.	....	s.
4916	lq.	$0.725^{20°}_4$	-107.6	144.2	i.	....	s.
4917	col. lq.	$0.720^{20°}_4$	-113.2	142.4	i.	....	....
4918	lq.	$0.828^{20°}_4$	-5	213.5	i.	∞	∞
4919	lq.	$0.823^{20°}_4$	-35	193-4	i.	s.	s.
4920	lq.	$0.825^{20°}_4$	-22	194.5^{750mm}	i.	v. s.	v. s.
4921	oil	$0.823^{20°}$	....	193^{766mm}	i.	∞	∞
4922	lq.	$0.823^{18°}_{19}$	....	178	i.	s.	s.
4923	lq.	$0.835^{20°}$	....	179.5	i.	s.	s.
4924	lq.	$0.840^{22°}$	....	172	i.	s.	s.
4925	lq.	$0.816^{12°}_4$	....	$172-4^{750mm}$	i.	s.	s.
4926	lq.	....	....	201	....	....	....
4927	lq.	$1.081^{20°}$	....	$208-9^{767}$ d.	....	....	....
4928	col. lq.	$0.856^{20°}$	....	190^{764mm}	....	....	....
4929	lq.	$1.287^{16°}_4$	....	117^{15mm}	....	....	....
4930	lq.	$0.754^{15°}_{15}$	....	148-9	i.		
4931	b. red cr.	....	>300	....	i.	i.	s. h. pyr.
4932	red nd./ac.	....	254-5	....	i.; s. pyr.	s. alk.	l.
4933	vl. cr./ac.	....	198	....	i.; s. pyr.	sl. s. h.	$0.3^{20°}$ chl.
4934	nd./abs. al.	....	156	....	v. s.	....	l.
4935	cr./lg.	....	59-60†	....	sl. s.	s.	s. chl.
4936	pl.	....	160	$268-9^{734mm}$	s. lg.	s.	s.
4937	cr.	lq. $0.779^{61.6°}$	61.6	$316-8^{40mm}$	$2^{15°}$ chl.	....	....
4938	cr.	$0.775^{28°}_4$	28.0(27.4)	317	i.	sl. s.	s.; s. act.
4939	lf.	$0.812^{59°}_4$	58.5	210.5^{15mm}	i.	s.	s.
4940	cr. lf.	....	33.5-4.0	$170^{0.5mm}$	....	v. sl. s. c.	....
4941	cr.	$0.791^{18°}_4$	18	179^{15mm}	....	....	....
4942	cr.	$0.796^{30°}$	26	180^{15mm}	....	....	....
4943	nd.	....	63	172^{20mm}	sl. s.	v. s.	sl. s.
4944	col. lq.	$0.703^{20°}_4$	-56.8	125.7	$0.002^{16°}$	sl. s.	s.
4945	lf.	....	100.7	106.3	i.	sl. s.	s.

* Green as liquid or in solution.
† Needles + $2H_2O$/aq. al., m. p. 51°.
Octandioic acid 5649
Octandione 78
Octanoic acid 1220

Octanol 4961-9
Octanone 2963, 4266
Octanoyl chloride 1225
Octene 4984-90
Octin 4114

Table 7-4 (*Continued*)
PHYSICAL CONSTANTS OF ORGANIC COMPOUNDS

No.	Name	Synonym	Formula	Beil. Ref.	Formula Weight
4946	**Octane**	2,2,3-triMe-pentane	$(CH_3)_3C \cdot CH(CH_3) \cdot C_2H_5$	*I-62	114.23
4947	**Octane** ("iso")	2,2,4-triMe-pentane	$(CH_3)_3C \cdot CH_2 \cdot CH: (CH_3)_2$	**I-127	114.23
4948	**Octane**	2,3,3-triMe-pentane	$C_2H_5 \cdot C(CH_3)_2 \cdot CH: (CH_3)_2$	....	114.23
4949	**Octane**	2,3,4-triMe-pentane	$[(CH_3)_2CH]_2CH \cdot CH_3$	....	114.23
4950	**Octane**	2-Me-3-Et-pentane	$(C_2H_5)_2CH \cdot CH: (CH_3)_2$	I-164	114.23
4951	**Octane**	2,3-dimethyl-hexane	$(CH_3)_2CH \cdot CH(CH_3) \cdot CH_2 \cdot C_2H_5$	*I-62	114.23
4952	**Octane**	2,4-dimethyl-hexane	$C_2H_5 \cdot CH(CH_3) \cdot CH_2 \cdot CH(CH_3)_2$	I-162	114.23
4953	**Octane**	2,5-dimethyl-hexane	$[(CH_3)_2CH \cdot CH_2 \cdot]_2$	I-162	114.23
4954	**Octane**	3,4-dimethyl-hexane	$[C_2H_5 \cdot CH(CH_3) \cdot]_2$	I-163	114.23
4955	**Octane**	3-ethyl-hexane	$(C_2H_5)_2CH \cdot CH_2 \cdot C_2H_5$	*I-62	114.23
4956	**Octane** (*iso*)	2-methyl-heptane	$(CH_3)_2CH(CH_2)_4CH_3$	I-161	114.23
4957	**Octane**	3-methyl-heptane	$C_2H_5 \cdot CH(CH_3) \cdot (CH_2)_3 \cdot CH_3$	I-162	114.23
4958	**Octane**	4-methyl-heptane	$(C_2H_5CH_2)_2CH \cdot CH \cdot CH_3$	I-162	114.23
4959	**Octyl** acetate (*n*)	(*See also* No. 1219)	$CH_3 \cdot CO_2C_8H_{17}$	II-134	172.27
4960	acetate	2-Et-hexyl acetate	$CH_3 \cdot CO_2CH_2 \cdot CH: (C_2H_5)(C_4H_9)$	....	172.27
4961	alcohol (*n*)	octanol-1	$CH_3(CH_2)_6CH_2OH$	I-418	130.23
4962	alcohol (capryl alcohol)	octanol-2	$CH_3(CH_2)_5CHOH \cdot CH_3$	I-419	130.23
4963	alcohol	2-Me-3-Et-pentanol-3	$(C_2H_5)_2C(OH) \cdot CH(CH_3)_2$	I-423	130.23
4964	alcohol	2,2,4-trimethyl-pentanol-4	$(CH_3)_3CCH_2C(OH): (CH_3)_2$	I-423	130.23
4965	alcohol	2-Et-hexanol-1	$C_4H_9(C_2H_5)CH \cdot CH_2OH$	**I-453	130.23
4966	alcohol	3-Et-hexanol-3	$(C_2H_5)_2C(OH) \cdot C_3H_7$	I-421	130.23
4967	alcohol	2-methyl-heptanol-2	$(CH_3)_2C(OH) \cdot CH_2 \cdot (CH_2)_3CH_3$	I-420	130.23
4968	alcohol (*dl*)	3-methyl-heptanol-3	$C_2H_5(CH_3)C(OH) \cdot CH_2 \cdot CH_2 \cdot C_2H_5$	I-421	130.23
4969	alcohol (*dl*)	4-methyl-heptanol-4	$(C_2H_5CH_2)_2C(OH) \cdot CH_3$	I-421	130.23
4970	amine (*n*)	1-amino-octane	$CH_3(CH_2)_7NH_2$	IV-196	129.25
4971	amine (*n*)(*sec*)	2-amino-octane	$C_6H_{13} \cdot CH(NH_2) \cdot CH_3$	IV-196	129.25
4972	amine	2-Et-1-amino-hexane	$C_4H_9(C_2H_5)CH \cdot CH_2 \cdot NH_2$	....	129.25
4973	bromide (*n*)	1-bromo-octane	$CH_3(CH_2)_6CH_2Br$	I-160	193.13
4974	bromide (*n*)(*sec*)	2-bromo-octane	$C_6H_{13} \cdot CHBr \cdot CH_3$	I-160	193.13
4975	chloride (*n*)	1-chloro-octane	$CH_3(CH_2)_6CH_2Cl$	I-159	148.68
4976	chloride (*n*)(*sec*)	2-chloro-octane	$C_6H_{13} \cdot CHCl \cdot CH_3$	I-160	148.68
4977	cinnamate (*n*)(*sec*)	....	$C_6H_5 \cdot CH:CH \cdot CO_2 \cdot CH(CH_3)C_6H_{13}$	*IX-230	260.38
4978	fluoride (*n*)	1-fluoro-octane	$CH_3(CH_2)_6CH_2F$	I-159	132.22
4979	formate (*d*)	β-octyl formate	$HCO_2 \cdot C_8H_{17}$	II-22	158.24
4980	iodide (*n*)	1-iodo-octane	$CH_3(CH_2)_6CH_2I$	I-160	240.13
4981	nitrate (*n*)	....	$CH_3(CH_2)_7O \cdot NO_2$	I-419	175.23
4982	nitrite (*n*)	....	$CH_3(CH_2)_7O \cdot NO$	I-419	159.23

Table 7-4 (*Continued*)
PHYSICAL CONSTANTS OF ORGANIC COMPOUNDS

No.	Crystalline Form and Color	Specific Gravity	Melting Point °C.	Boiling Point °C.	Solubility in 100 Parts		
					Water	Alcohol	Ether
4946	col. lq.	$0.716^{20°}_{4}$	−112.3	109.8^{760mm}	i.	sl. s.	s.
4947	col. lq.	$0.692^{20°}_{4}$	−107.4	99.2	i.	sl. s.	s.
4948	col. lq.	$0.726^{20°}_{4}$	−100.7	114.8	i.	sl. s.	s.
4949	col. lq.	$0.719^{20°}_{4}$	−109.2	113.5	i.	sl. s.	s.
4950	lq.	$0.719^{20°}_{4}$	−115.0	115.7	i.	sl. s.	s.
4951	lq.	$0.712^{20°}_{4}$	....	115.6	i.	sl. s.	s.
4952	lq.	$0.700^{20°}_{4}$	....	109.4	i.	sl. s.	s.
4953	col. lq.	$0.694^{20°}_{4}$	−91.2	109.1	i.	sl. s.	s.
4954	lq.	$0.719^{20°}_{4}$	....	117.7	i.	sl. s.	s.
4955	lq.	$0.714^{20°}_{4}$	....	118.5	i.	sl. s.	s.
4956	col. lq.	$0.698^{20°}_{4}$	−109.0	117.7	i.	sl. s.	s.
4957	lq.	$0.706^{20°}_{4}$	−120.5	118.9	i.	sl. s.	s.
4958	lq.	$0.705^{20°}_{4}$	−121.0	117.7	i.	sl. s.	s.
4959	col. lq.	$0.885^{0°}_{4}$	−38.5	210	i.	s.	s.
4960	col. lq.	$0.873^{20°}_{20}$	−93	199	$<0.03^{20°}$	∞	∞
4961	col. lq.	$0.829^{20°}_{4}$	−16.7	194.5	$0.054^{20°}$	∞	∞ ; ∞ chl.
4962	col. lq.	$0.822^{20°}_{4}$	−38.6	179−80	$0.096^{25°}$	∞	∞
4963	lq.	$0.830^{20°}$	....	$160-1^{750mm}$	....	s.	s.
4964	lq.	$0.842^{0°}$	−20	146.5-7.5	i.	sl. s.	s.
4965	col. lq.	$0.834^{20°}_{20}$	< −76	184	$0.1^{20°}$	s.	s.
4966	lq.	$0.838^{20°}_{4}$	....	160.5		∞	∞
4967	lq.	$0.879^{20°}$	....	162	....	....	....
4968	lq.	$0.820^{21°}$	....	162	i.	....	....
4969	lq.	$0.825^{20°}$	....	161.5	i.	....	....
4970	col. lq.	$0.777^{27°}$	....	$175-7^{745mm}$	v. sl. s.	s.	s.
4971	lq.	$0.771^{25°}_{4}$	....	164−5	....	....	....
4972	lq.	$0.792^{20°}_{20}$	....	167−8	....	....	....
4973	col. lq.	$1.118^{15°}$	−55	201.5	i.	∞	∞
4974	lq.	$1.099^{22°}$	....	188−9	....	....	....
4975	lq.	$0.892^{0°}_{4}$	....	182.5-3.5	i.	....	....
4976	lq.	$0.871^{15°}$	....	171−3	i.	....	s.
4977	lq.	$0.972^{17°}_{4}$	....	240^{60mm}	....	....	....
4978	col. lq.	$0.804^{21°}$	....	142.8	i.	....	....
4979	col. lq.	$0.872^{12.5°}$	−39.1	198.8	i.	....	....
4980	lq.	$1.336^{15°}_{4}$	−45.7	225.5	....	....	....
4981	lq.	$0.975^{0°}_{4}$	....	$110-2^{20mm}$	....	....	....
4982	gn. lq.	$0.862^{17°}$	....	175−7	....	....	....

Table 7-4 (*Continued*)
PHYSICAL CONSTANTS OF ORGANIC COMPOUNDS

No.	Name	Synonym	Formula	Beil. Ref.	Formula Weight
4983	**Octyl** stearate (*n*) (*sec*)	....	$C_{17}H_{35}CO_2 \cdot CH(CH_3) \cdot$ C_6H_{13}	*II-173	396.70
4984	**Octylene** (α)†	octene-1	$CH_3(CH_2)_5CH{:}CH_2$	I-221	112.22
4985	**Octylene**	2,3,3-trimethyl-pentene-1	$C_2H_5 \cdot C(CH_3)_2 \cdot$ $C(CH_3){:}CH_2$	....	112.22
4986	**Octylene**	2,4,4-trimethyl-pentene-1	$(CH_3)_3C \cdot CH_2 \cdot$ $C(CH_3){:}CH_2$	I-222	112.22
4987	**Octylene**	2,3,4-trimethyl-pentene-2	$(CH_3)_2CH \cdot C(CH_3){:}$ $C(CH_3)_2$	....	112.22
4988	**Octylene**	3,4,4-trimethyl-pentene-2	$(CH_3)_3C \cdot C(CH_3){:}$ $CH \cdot CH_3$	....	112.22
4989	**Octylene**	3-ethyl-hexene-2	$CH_3 \cdot CH{:}C(C_2H_5) \cdot$ $CH_2 \cdot C_2H_5$	I-222	112.22
4990	**Octylene**	2-methyl-heptene-2	$(CH_3)_2C{:}CH \cdot CH_2 \cdot$ $CH_2 \cdot C_2H_5$	I-222	112.22
4991	**Octyne-1**	caprylidene	$CH_3 \cdot (CH_2)_5 \cdot C{:}CH$	I-258	110.20
4992	**Oleic acid**	....	$CH_3(CH_2)_7CH{:}CH \cdot$ $(CH_2)_7CO_2H$	II-463	282.47
4993	**Opianic acid**	(5,6;2,1)	$(CH_3O)_2C_6H_2{:}$ $(CHO)CO_2H$	X-990	210.19
4994	**Orange I**	*p*-sulfobenzene-azo-α-naphthol Na	$NaO_3S \cdot C_6H_4 \cdot N{:}$ $N \cdot C_{10}H_6 \cdot OH$	*XVI-296	350.33
4995	**Orange II**	*p*-sulfobenzene-azo-β-naphthol Na	$NaO_3S \cdot C_6H_4 \cdot N{:}$ $N \cdot C_{10}H_6 \cdot OH$	*XVI-296	350.33
4996	**Orange IV**	4'-anilino-azo-ben-zene-4-sulfonic Na	$NaO_3S \cdot C_6H_4 \cdot N{:}N \cdot$ $C_6H_4 \cdot NH \cdot C_6H_5$	*XVI-319	375.38
4997	**Orcein**	....	$C_{28}H_{24}O_7N_2$	I-886	500.51
4998	**Orcinol**	5-methyl resorcinol	$CH_3 \cdot C_6H_3(OH)_2$	VI-882	124.14
4999	**Orcin-phthalein**	....	$C_{22}H_{16}O_5$	XIX-236	360.37
5000	**Ornithine**	α,δ-diaminovaleric acid	$H_2N \cdot (CH_2)_3 \cdot CH{:}$ $(NH_2)CO_2H$	IV-420	132.16
5001	**Orsellinic acid** (4,6;2,1)	....	$(HO)_2C_6H_2(CH_3) \cdot$ $CO_2H \cdot aq.$	X-412	168.15
5002	**Ortal sodium**	*n*-hexyl-ethyl-barbituric Na	$C_6H_{13}(C_2H_5){:}$ $C_4HO_2N_2ONa$	....	262.29
5003	**Orthochrom T**	....	$C_{23}H_{22}N_2 \cdot (C_2H_5I)$	XXIII-310	482.41
5004	**Ostruthin**	....	$C_{19}H_{22}O_3$	....	298.39
5005	**Oxalenediur-amidoxime**	oxaldiureide-dioxime	$[H_2N \cdot CO \cdot NH \cdot C \cdot$ $({:}NCH)]_2$	III-65	204.15
5006	**Oxalic** acid	ethandioic acid	$HO_2C \cdot CO_2H$	II-502	90.04
5007	acid	ethandioic acid	$(CO_2H)_2 \cdot 2H_2O$	II-502	126.07
5008	Na salt	sodium oxalate	$(CO_2Na)_2$	II-513	134.00
5009	amide	oxamide	$(CONH_2)_2$	II-545	88.07
5010	anilide	oxanilide	$(CO \cdot NH \cdot C_6H_5)_2$	XII-284	240.26
5011	chloride	oxalyl chloride	$Cl \cdot CO \cdot CO \cdot Cl$	II-542	126.93
5012	imide	oximide	$CO \cdot CO \cdot NH$ $\mid\!\!\underline{\qquad}\!\!\mid$	XXI-368	71.04
5013	**Oxaluric acid**	....	$NH_2 \cdot CO \cdot NH \cdot CO \cdot$ CO_2H	III-64	132.08
5014	amide	oxalan	$C_3H_5O_3N_3$	III-65	131.09
5015	**Oxamaethane**	ethyl oxamate	$NH_2 \cdot CO \cdot CO_2C_2H_5$	II-544	117.11

† See also di-iso-butylene.
Orange-III 4328
Orange-GS 4996
Orange-N 4996
Orcin (β) 2380

Oreson 3400
Orexin 5182
Orizabin, cf. glcde.
Orthamine 5256
Orthanilic acid 523

Orthoform new 4112
Ortol, mixt. of 2314 and 4116
Ouabain, cf. glcde.
Oxalacetic ester 2192
Oxalacetone 2939

Table 7-4 (*Continued*)
PHYSICAL CONSTANTS OF ORGANIC COMPOUNDS

No.	Crystalline Form and Color	Specific Gravity	Melting Point °C.	Boiling Point °C.	Solubility in 100 Parts		
					Water	Alcohol	Ether
4983	col. cr.	$0.841\frac{3.8}{4}°$	34	235^{6mm}	....	....	....
4984	lq.	$0.716^{20°}$	−102.4	121.3	i.	∞	∞
4985	col. lq.	$0.737^{20°}$	−69	108.2	....	....	....
4986	col. lq.	$0.715^{20°}$	−93.5	101.4	....	....	....
4987	col. lq.	$0.743^{20°}$	....	116.3	....	....	....
4988	col. lq.	$0.739^{20°}$	....	112	....	....	....
4989	lq.	$0.737^{20°}$	....	121	i.	....	....
4990	col. lq.	$0.725\frac{20}{0}°$	....	122	i.	....	....
4991	lq.	$0.743\frac{25}{4}°$	−79	131–2^{762mm}	....	....	....
4992	col. nd.	$0.891\frac{20}{4}°$	16	285–6^{100mm}	i.	∞	∞
4993	nd./aq.	....	150	....	0.25 c.; 1.7 h.	s.	s.
4994	or. red pd.	....	....	....	s.	....	....
4995	or. red pd.	....	....	....	s.	....	....
4996	or. yel. pd.	....	....	....	s.	....	....
4997	brn. cr.	....	....	....	i. chl., CS_2	s.; s. ac.	i.; i. bz.
4998	pr./bz.	1.290^{40}	107–8*	287–90	v. s.	v. s.	v. s.
4999	pr./act.	....	d. >230	....	i.; s. h. ac.	s.; s. alk.	i.; i. bz.
5000	syrup	....	....	....	v. s.	v. s.	sl. s.
5001	cr.	....	176 d. (anh.)	−H_2O, 100	s. gly.	v. s.	22$^{20°}$
5002	wh. pd.	....	....	....	v. s.	s.	i.
5003	gn. pr./al.	....	....	....	sl. s. h.	s. h.	....
5004	cr./aq. al.	....	117–9	....	i.; i. pet.	s. h.	s. chl.
5005	nd./aq. al.	....	191–2 d.	....	i. c.; i. bz.; s. a.	s.; i. lg.; s. alk.	i.; i. chl.
5006	col. rhb.	1.90	186–7 d.	subl. > 100	10$^{20°}$; 120$^{100°}$	24$^{15°}$ abs.	1.3$^{15°}$ abs.
5007	col. mn.	$1.653\frac{19}{4}°$	101.5	−2H_2O, 100	....	....	....
5008	cr. pd.	....	....	....	3.2$^{16°}$	i.	6.6 h. aq.
5009	mn.	1.667	417–9 d.	....	0.047.3^{0}; 0.6$^{100°}$	i.	i.
5010	pl./bz.	....	251–3	>320	i. h.	sl. s. h.	sl. s. h.
5011	col. lq.	$1.488\frac{13}{4}°$	−12	63–4^{763mm}	d.	d.	s.
5012	pr.	....	....	....	v. sl. s.; d. h.	sl. s. NH_4OH	....
5013	cr./aq.	....	d. 208–10	....	v. sl. s.; d. h.	i.	i.
5014	cr. ppt.	....	d.	....	i. c.	s. H_2SO_4	s. d. KOH
5015	rhb.	....	114–5	....	s.	v. sl. s. bz.	s.

* Crysts. + 1H_2O/aq., m. p. 56° ±.

Table 7-4 (*Continued*)
PHYSICAL CONSTANTS OF ORGANIC COMPOUNDS

No.	Name	Synonym	Formula	Beil. Ref.	Formula Weight
5016	**Oxamic** acid	....	$NH_2 \cdot CO \cdot CO_2H$	II-543	89.05
5017	hydrazide	amino-oxamide	$NH_2 \cdot CO \cdot CO \cdot NH \cdot NH_2$	II-559	103.08
5018	**Oxanilic acid**	....	$C_6H_5NH \cdot CO \cdot CO_2H$	XII-281	165.15
5019	**Oxanthranol**	anthrahydroquinone	$C_6H_4(COCHOH)C_6H_4$	VIII-190	210.23
5020	diacetate	....	$C_{14}H_8(O_2C \cdot CH_3)_2$	VI-1034	294.31
5021	**Oxindol** †	....	$C_6H_4 \cdot CH_2 \cdot CO \cdot NH$ |_____|	XXI-282	133.15
5022	**Palmitic** acid	hexadecanoic acid	$CH_3(CH_2)_{14} \cdot CO_2H$	II-370	256.43
5023	aldehyde	hexadecanal	$CH_3(CH_2)_{14} \cdot CHO$	I-717	240.43
5024	aldehyde oxime	....	$C_{15}H_{31} \cdot CH:NOH$	I-717	255.45
5025	amide	....	$C_{15}H_{31} \cdot CO \cdot NH_2$	II-374	255.45
5026	anhydride	....	$(C_{15}H_{31}CO)_2O$	II-374	494.85
5027	chloride	palmityl chloride	$C_{15}H_{31} \cdot CO \cdot Cl$	II-374	274.88
5028	nitrile	pentadecyl cyanide	$C_{15}H_{31} \cdot CN$	II-375	237.43
5029	**Palmitolic acid**	....	$CH_3(CH_2)_7C:C \cdot (CH_2)_5 \cdot CO_2H$	II-494	252.40
5030	**Palmitone**	....	$(C_{15}H_{31})_2CO$	I-719	450.84
5031	**Panthesin**	....	$C_{18}H_{32}O_5N_2S$	....	388.53
5032	**Pantocaine**®	*p*-butylaminobenzoyl-dimethyl-aminoethanol	$C_4H_9NH \cdot C_6H_4CO_2 \cdot (CH_2)_2N(CH_3)_2HCl$	....	300.83
5033	**Pantothenic** acid (*dl*)	....	$HOCH_2C(CH_3)_2CH \cdot (OH)CONH \cdot (CH_2)_2 \cdot CO_2H$	....	219.24
5034	acid (*d*)	....	$C_9H_{17}O_5N$	....	219.24
5035	Ca salt (*d*)	....	$(C_9H_{16}O_5N)_2Ca$	....	476.54
5036	**Parabanic acid**	oxalyl urea	$C_3H_2O_3N_2$	XXIV-449	114.06
5037	**Paracotoin**	....	$CH_2O_2:C_{11}H_6O_2$	....	216.20
5038	**Pararosaniline** base (4,4′,4″)	triamino-triphenyl-carbinol	$HO \cdot C(C_6H_4 \cdot NH_2)_3$	XIII-750	305.38
5039	hydrochloride	parafuchsin	$C_{19}H_{18}N_3Cl \cdot H_2O$	XIII-752	341.84
5040	**Parvoline** (α)	2,4-diEt-pyridine	$C_9H_{13}N$	XX-253	135.21
5041	**Parvoline** (β)	3,4-diEt-pyridine	$C_9H_{13}N$	XX-253	135.21
5042	**Pelargonic** acid	nonanoic acid	$CH_3(CH_2)_7 \cdot CO_2H$	II-352	158.24
5043	aldehyde	nonanal	$CH_3(CH_2)_7 \cdot CHO$	I-708	142.24
5044	aldehyde oxime	....	$C_8H_{17} \cdot CH:NOH$	I-708	157.26
5045	amide	nonanamide	$CH_3(CH_2)_7CO \cdot NH_2$	II-353	157.26
5046	chloride	pelargonyl chloride	$CH_3(CH_2)_7 \cdot COCl$	II-353	176.69
5047	nitrile	octyl cyanide	$CH_3(CH_2)_7 \cdot CN$	II-354	139.24
5048	**Penta**-aminobenzene	....	$(NH_2)_5C_6H$	XIII-346	153.19
5048.1	bromoacetone	....	$Br_3C \cdot CO \cdot CHBr_2$	I-659	452.59
5049	bromobenzene	....	Br_5C_6H	V-215	472.62
5050	bromoethane	....	$CHBr_2 \cdot CBr_3$	I-95	424.58
5051	bromophenol	....	$Br_5C_6 \cdot OH$	VI-206	488.62
5052	chloroaniline	....	$Cl_5C_6 \cdot NH_2$	XII-631	265.35
5053	chlorobenzene	....	Cl_5C_6H	V-205	250.34
5054	chloroethane	pentalin	$CHCl_2 \cdot CCl_3$	I-87	202.30
5055	chloroethylbenzene	alkazene-32	$Cl_5C_6 \cdot C_2H_5$	V-355	278.39
5056	chloro-1-keto-1,2,3,4-tetrahydro-naphthalene	2,2,3,4,4-penta-Cl	$C_{10}H_5OCl_5$	VII-370	318.42
5057	chlorophenol	....	$Cl_5C_6 \cdot OH$	VI-194	266.34
5058	cosane	....	$C_{25}H_{52}$	I-175	352.69

† See also No. 2686.
Oxychalcone 3763
Oxydimorphine, cf. alkd.
Oxyjuglon(e) 4477
Oxymethyl-acetamide 3291
Oxyneurine 846
Oxyneurine HCl, cf. alkd.

Oxypurine, cf. alkd.
Palmitin 3431
Palmityl chloride 5027
Papaveraldine, cf. alkd.
Papaverine, cf. alkd.
Parabutyraldehyde 1162, 1165
Paracamphoric acid 1197

Paraconine, cf. alkd.
Paraformaldehyde 3290
Parafuchsine 5039
Parafuchsine carbinol base **5038**
Paralactic acid 3939
Paraldehyde 7
Paraldol 170

Table 7-4 (*Continued*)
PHYSICAL CONSTANTS OF ORGANIC COMPOUNDS

No.	Crystalline Form and Color	Specific Gravity	Melting Point °C.	Boiling Point °C.	Solubility in 100 Parts		
					Water	Alcohol	Ether
5016	cr. pd.	....	214 d.	....	v. sl. s.	i. abs.	i.
5017	col. lf./aq.	....	221-3 d.	....	$0.32^{20°}$	i.; s. a.	i.; s. alk.
5018	nd./bz.	....	150	....	sl. s. h.	v. s.	v. s.
5019	yel. nd.	....	....	....	....	s.	s. alk.
5020	nd./ac.	....	108-9	....	....	s.	s. ac.
5021	col. nd./aq.	....	120	227^{73mm}	s. h.	s.	s.; s. alk.
5022	col. pl.	$0.849\frac{7.0°}{4}$	62.8	271.5^{100mm}	i.; v. sl. s. pet.	$9.3^{20°}$ abs.	s.; s. chl.
5023	wh. wax	....	34	$200\text{-}2^{29mm}$	i.	s.	s.
5024	nd./aq. al.	....	88	....	sl. s. bz.	sl. s. pet.	s.; s. chl.
5025	cr. lf.	....	106-7	$235\text{-}6^{12mm}$	i.	v. s.	v. sl. s.
5026	col. cr.	$0.847\frac{7.0°}{4}$	64	....	....	$0.2^{20°}$	s.
5027	col. cr.	....	12	$194\text{-}5^{17mm}$	d.	d.	sl. s.
5028	pl.	$0.822\frac{31.1°}{4}$	31	251.5^{100mm}	i.	s.	s.
5029	nd./aq.	....	42	240^{15mm}	i.	v. s.	v. s.
5030	lf./al.	$0.800\frac{8.3°}{4}$	82.8	....	i.	....	....
5031	wh. pd.	....	157-9	....	33	s.	s.
5032	wh. pd.	....	148-50*	....	14	s.	i.
5033	col. syrup	....	....	....	s.	s.	s. act.
5034	syrup	....	....	....	s.	s.	s. act.
5035	col. cr.	....	....	....	s.	s. Me al.	sl. s. act.
5036	mn.	....	243 d.	subl. 100	$4.7^{8°}$	s.	i.
5037	yel. cr.	....	162	....	....	s.	s.; s. chl.
5038	col. lf.	....	205 d.	....	v. sl. s.	s.	i.
5039	gn. met.	....	d. 250	....	$0.31^{22°}$	....	....
5040	lq.	$0.934^{0°}$	....	188	sl. s.	s.	....
5041	lq.	0.916	....	209^{710mm}	....	....	....
5042	col. oil	$0.906\frac{2.0°}{4}$	12.5	253-4	v. sl. s.	s.; s. chl.	s.
5043	lq.	$0.827\frac{19°}{19}$	....	185	....	....	....
5044	lf./aq. al.	....	63-4	....	i.	s.	s.
5045	col. cr.	....	99-100	....	i. c.	sl. s.	sl. s.
5046	col. lq.	$0.946\frac{15°}{4}$	−60.5	215.4^{760mm}	d.	d.	s.
5047	col. lq.	$0.821\frac{15°}{4}$	−34.2	224.0	i.	s.	s.
5048	salts only	....	....	....	....	....	....
5048.1	rhb. nd./al.	....	79-80(72)	subl.	i. c.	v. s.	v. s.
5049	nd./ac.	....	293	....	i.; s. bz.	sl. s.	sl. s.
5050	mn. pr.	3.312	56-7	210^{300mm} d.	i.	s.	v. s.
5051	mn./al.	....	225-6	subl.	i.	sl. s.	sl. s.
5052	nd./al.	....	232	....	sl. s. lg.	s.	s.
5053	nd./al.	$1.834^{17°}$	85-6	275-7	i.	s. h.	v. s.
5054	col. lq.	$1.671\frac{25°}{4}$	−29	162	$0.05^{20°}$	∞	∞
5055	cr./al. bz.	$1.552\frac{60°}{25}$	53	305±	i.; v. s. bz.	$2^{25°}$ Me al.	$166^{25°}$
5056	mn./bz.	....	156-7	d. >200, −HCl	s. h. bz.	v. s.	i.
5057	mn.	$1.978^{22°}$	188-9	309^{754mm} d.	$0.003^{50°}$	v. s.	$148^{25°}$
5058	cr.	$0.791^{60°}$	53.3	$282\text{-}4^{40mm}$	....	s. abs.	$5.4^{15°}$ chl.

Table 7-4 (*Continued*)
PHYSICAL CONSTANTS OF ORGANIC COMPOUNDS

No.	Name	Synonym	Formula	Beil. Ref.	Formula Weight
5059	**Penta** decanaldoxime	....	$C_{14}H_{29} \cdot CH:NOH$	I-716	241.42
5060	decane (*n*)	....	$CH_3(CH_2)_{13} \cdot CH_3$	I-172	212.42
5061	decyl acetate (*n*)	....	$CH_3CO_2 \cdot C_{15}H_{31}$	II-136	270.46
5062	decyl alcohol (*n*)	pentadecanol-1	$CH_3(CH_2)_{13} \cdot CH_2OH$	I-429	228.42
5063	decyl bromide	1-bromo-pentadecane	$CH_3(CH_2)_{13} \cdot CH_2Br$	I-172	291.32
5064	diene-1,3	piperylene	$CH_3CH:CH \cdot CH:CH_2$	I-251	68.12
5065	diene-2,3	....	$CH_3CH:C:CHCH_3$	I-251	68.12
5066	erythritol	pentaerythrite	$C(CH_2OH)_4$	I-528	136.15
5067	erythritol tetra-acetate	....	$C(CH_2O_2CCH_3)_4$	II-150	304.30
5068	ethylbenzene	....	$(C_2H_5)_5C_6H$	V-471	218.39
5069	glycerol	....	$CH_3 \cdot C(CH_2OH)_3$	I-520	120.15
5070	hydroxy benzo-phenone	mori(n)tannic acid; maclurin	$(HO)_2C_6H_3 \cdot CO \cdot C_6H_2(OH)_3 \cdot H_2O$	VIII-538	280.24
5071	iodobenzene	....	I_5C_6H	V-229	707.60
5072	iodoethane	....	$CHI_2 \cdot CI_3$	*I-31	659.55
5073	methyl-amino-benzene	....	$(CH_3)_5C_6 \cdot NH_2$	XII-1182	163.26
5074	methyl benzene	....	$(CH_3)_5C_6H$	V-443	148.25
5075	methyl benzoic acid	....	$(CH_3)_5C_6 \cdot CO_2H$	IX-569	192.26
5076	methyl para-rosaniline	methyl violet base	$C_{24}H_{29}ON_3$	XIII-755	375.52
5077	methyl pararosani-line chloride	methyl violet dye salt †	$C_{24}H_{28}N_3 \cdot Cl$	XIII-755	393.96
5078	methyl phenol	....	$(CH_3)_5C_6 \cdot OH$	VI-551	164.25
5079	methylene-tetrazol	Metrazol; Cardiazol	$\cdot N(CH_2)_5C:(N_3)$	....	138.17
5080	triacontane (*n*)	....	$CH_3(CH_2)_{33} \cdot CH_3$	I-177	492.96
5081	**Pentaldol**	2,2-diMe-propan-3-ol-1-al	$(CH_3)_2C(CHO) \cdot CH_2OH$	I-833	102.13
5083	**Pentane**	*n*-pentane	$(C_2H_5)_2CH_2$	I-130	72.15
5084	**Pentane** (*iso*)	2-methyl-butane	$(CH_3)_2CH \cdot C_2H_5$	I-134	72.15
5085	**Pentane** (*neo*)	2,2-dimethyl-propane	$(CH_3)_4C$	I-141	72.15
5086	**Pentyne**-1	propyl acetylene	$C_2H_5 \cdot CH_2 \cdot C:CH$	I-250	68.12
5087	**Pentyne**-2	valerylene	$C_2H_5 \cdot C:C \cdot CH_3$	I-250	68.12
5088	**Perchloro**-ether	....	$Cl_5C_2 \cdot O \cdot C_2Cl_5$	II-210	418.57
5089	methyl mercaptan	....	$Cl_3C \cdot S \cdot Cl$	III-135	185.89
5090	**Perseitol**	mannoheptitol	$C_7H_{16}O_7$	I-548	212.20
5091	**Phanodorn**®	Et-cyclohexenyl-barbituric acid	$(C_2H_5)(C_6H_9): C_4H_2O_3N_2$	....	236.27
5092	**Phellandrene** (α)(*d*)	*p*-menthadiene-1,5	$C_{10}H_{16}$	V-129	136.24
5093	**Phellandrene** (β)	*p*-menthadiene-2,1(7)	$C_{10}H_{16}$	V-132	136.24
5094	**Phenacyl** acetate	ω-acetoxy-acetophenone	$C_6H_5 \cdot CO \cdot CH_2 \cdot O \cdot CO \cdot CH_3$	VIII-92	178.19
5095	**Phenanthrene**	....	$(C_6H_4CH)_2$	V-667	178.24
5096	hydroquinone	(9,10)	$C_{14}H_8(OH)_2$	VI-1035	210.23
5097	**Phenanthraqui-none**	phenanthrene quinone	$C_6H_4(CO)_2C_6H_4$	VII-796	208.22
5098	**Phenanthrol** (2)	....	$C_{14}H_9OH$	VI-704	194.24
5099	**Phenanthrol** (3)	....	$C_{14}H_9OH$	VI-705	194.24

† Commercial dye mostly penta- with some hexa-methyl derivative.

Table 7-4 (*Continued*)
PHYSICAL CONSTANTS OF ORGANIC COMPOUNDS

No.	Crystalline Form and Color	Specific Gravity	Melting Point °C.	Boiling Point °C.	Solubility in 100 Parts		
					Water	Alcohol	Ether
5059	nd./aq. al.	....	86	....	sl. s. c. pet.	sl. s. c.	v. s.; sl. s. bz.
5060	col. lq.	0.770^{20}_{4}	10	270.5	i.	v. s.	v. s.
5061	wax	....	10–1	230^{70mm}	....	....	....
5062	cr.	....	45–6	....	....	....	....
5063	cr.	....	14–5	....	....	....	....
5064	lq.	0.676^{200}	....	42.3	....	....	....
5065	lq.	0.66^{200}	....	40	....	....	....
5066	cr.	....	262	276^{30mm}	$5.6^{15°}$	v. sl.s.	i.
5067	nd./aq. or bz.	1.273^{18}_{4}	83–4	....	....	....	....
5068	lq.	0.896^{20}_{4}	<−20	277	i.	....	....
5069	nd./abs. al.	....	199	subl.	v. s.	v. s.	i.
5070	yel. pr./aq.	....	200(anh.)	−H_2O, 130–40	$0.5^{15°}$	s.	s.
5071	cr./al.	....	172	....	s. h. ac.	sl. s. c.	sl. s.
5072	col./ac.	....	182–4	....	s. ac.	s.	s. bz.
5073	mn./al.	....	151–2	278–80	i. h.	s.	s.
5074	pr./aq. al.	0.847^{107}_{4}	54.3	230–1	i.	v. s.	v. s. bz.
5075	nd./aq. al.	....	210.5	subl.	v. sl. s.	v. s. h.	....
5076	salts only	....	....	....	....	....	....
5077	red or b. vl.	....	....	....	s.	....	....
5078	nd./al.	....	125	267	0.15 h.	s. h. aq. NaOH	....
5079	wh. cr. pd.	....	57–8	....	s.	s.	s.
5080	cr.	0.782^{74}_{47}	α75; β72	331^{15mm}	....	....	....
5081	col. nd./al.	....	96–7	172–37^{47mm} d.	5 c.	2.5 c.	s.
5083	col. lq.	0.626^{200}	−129.7	36.1	$0.036^{16°}$	∞	∞
5084	col. lq.	0.621^{190}	−159.9	27.9	i.	∞	∞
5085	lq.	0.613^{20}_{4}	−16.6	9.5	i.	s.	s.
5086	col. lq.	0.691^{200}	−106	40.2	i.	....	s.
5087	lq.	0.711^{200}	−109.3	56.1	i.	....	....
5088	tet.	$1.900^{14.5°}$	69	d.	....	....	....
5089	yel. lq.	$1.695^{17.5°}$	....	149 sl. d.	....	....	....
5090	nd.	1.485	188	....	0.918^{0}	s. h.	447$^{4°}$ aq.
5091	wh. cr. pd.	....	173±	....	sl. s.	22	6
5092	lq.	0.845^{200}	....	175	....	i.	s.
5093	lq.	0.852^{20}_{4}	....	171–2	i.	i.	s.
5094	rhb. pl./et. or lg.	....	48–9	270	i.; s. chl.	s.; sl. s. bz.	s.; ol. o. lg.
5095	pl./al.	$1.179^{25°}$	100.5	340	i.; s. CS_2	$2^{14°}$; 10 h.	v. s.; s. bz.
5096	col. nd.	....	147–8	....	s. h.	v. s.	v. s.; v.s.bz.
5097	or. nd.	$1.405^{4°}$	206.5–7.5	>300; subl.	v. sl. s.	s. h.; s. bz.	sl. s.
5098	lf./aq. al.	....	168–9	....	....	v. s.	v. s.
5099	nd./aq. al.	....	119–22	....	sl. s. h.	v. s.	v. s.

Table 7-4 (*Continued*)
PHYSICAL CONSTANTS OF ORGANIC COMPOUNDS

No.	Name	Synonym	Formula	Beil. Ref.	Formula Weight
5100	**Phenanthrol** (9)	phenanthrone	$C_{14}H_9OH$	VI-706	194.24
5101	**Phenanthroline** (*o*)	(1,10)	$C_{12}H_8N_2$	XXIII-227	180.21
5102	**Phenanthroline** (*m*)	(1,5)	$C_{12}H_8N_2 \cdot 2H_2O$	XXIII-227	216.24
5103	**Phenanthroline** (*p*)	(1,8)	$C_{12}H_8N_2 \cdot 4H_2O$	XXIII-228	252.27
5104	**Phenazine**	azophenylene	$C_6H_4{:}N_2{:}C_6H_4$	XXIII-223	180.21
5105	**Phenetidine** (*o*)	*o*-aminophenetole	$C_2H_5O \cdot C_6H_4 \cdot NH_2$	XIII-359	137.18
5106	**Phenetidine** (*m*)	....	$C_2H_5O \cdot C_6H_4 \cdot NH_2$	XIII-404	137.18
5107	**Phenetidine** (*p*)	....	$C_2H_5O \cdot C_6H_4 \cdot NH_2$	XIII-436	137.18
5108	citrate	Citrophen	$C_8H_{11}ON \cdot C_6H_8O_7$	XIII-437	329.31
5109	hydrochloride	....	$C_8H_{11}ON \cdot HCl$	XIII-437	173.64
5110	tartrate	Vinopyrin	$C_8H_{11}ON \cdot C_4H_6O_6$	XIII-437	287.27
5111	**Phenetole**	ethyl-phenyl ether	$C_2H_5 \cdot O \cdot C_6H_5$	VI-140	122.17
5112	**Phenetyl urea** (*p*)	dulcin; Sucrol; Valzin	$C_2H_5O \cdot C_6H_4 \cdot NH \cdot CO \cdot NH_2$	XIII-480	180.21
5113	**Phenobarbital**	Et-Ph-barbituric acid; Luminal; Gardenal	$(C_2H_5)(C_6H_5){:}C_4H_2O_3N_2$	*XXIV-423	232.24
5114	Na salt	"soluble"	$C_{12}H_{11}O_3N_2Na$	....	254.22
5115	**Phenocoll**	glycine-*p*-phenetidide	$C_2H_5O \cdot C_6H_4 \cdot NH \cdot CO \cdot CH_2 \cdot NH_2 \cdot H_2O$	XIII-506	212.25
5116	hydrochloride	phenamine-HCl	$C_{10}H_{14}O_2N_2 \cdot HCl$	XIII-506	230.70
5117	salicylate	salocoll	$C_{10}H_{14}O_2N_2 \cdot C_7H_6O_3$	....	332.36
5118	**Phenol**	carbolic acid	$C_6H_5 \cdot OH$	VI-113	94.11
5119	Na salt	sodium phenoxide	$C_6H_5ONa \cdot 3H_2O$	VI-136	170.14
5120	Na salt	....	$C_6H_5ONa \cdot 2C_6H_5OH$	*VI-78	304.32
5121	phthalein	....	$C_{20}H_{14}O_4$	XVIII-143	318.33
5122	phthalin	phthalin	$C_{20}H_{16}O_4$	X-455	320.35
5123	sulfonic acid (*o*)	....	$HO \cdot C_6H_4 \cdot SO_3H \cdot \frac{3}{4}H_2O$	XI-234	187.69
5124	sulfonic acid (*m*)	....	$HO \cdot C_6H_4 \cdot SO_3H \cdot 2H_2O$	XI-239	210.21
5125	sulfonic acid (*p*)	....	$HO \cdot C_6H_4 \cdot SO_3H$	XI-241	174.18
5126	sulfonic Na (*p*)	Na sulfocarbolate	$C_6H_5O_4SNa \cdot 2H_2O$	XI-241	232.19
5127	sulfonphthalein	phenol red	$C_{19}H_{14}O_5S$	XIX-91	354.38
5128	tricarboxylic acid	(2;1,3,5)	$HOC_6H_2(CO_2H)_3 \cdot aq.$	X-580	226.14
5129	**Phenoxthine**	phenothioxine; dibenzothioxine (1,4)	$C_6H_4 \cdot O \cdot C_6H_4 \cdot S \cdot$ ‖	XIX-45	200.26
5130	**Phenoxy**-acetic acid	....	$C_6H_5O \cdot CH_2 \cdot CO_2H$	VI-161	152.15
5131	acetic amide	phenoxyacetamide	$C_6H_5O \cdot CH_2 \cdot CONH_2$	VI-162	151.17
5132	acetone	ω-acetoanisole	$C_6H_5O \cdot CH_2 \cdot COCH_3$	VI-151	150.18
5133	benzoic acid (*o*)	salicylic phenyl ether	$C_6H_5O \cdot C_6H_4 \cdot CO_2H$	X-65	214.22
5134	butyronitrile (γ)	....	$C_6H_5O \cdot (CH_2)_3 \cdot CN$	VI-164	161.21
5135	β'-chloroethyl ether (β)	(β)	$C_6H_5O \cdot (CH_2)_2 \cdot O \cdot CH_2 \cdot CH_2Cl$	**VI-150	200.67
5136	ethyl alcohol	phenyl cellosolve	$C_6H_5O \cdot (CH_2)_2 \cdot OH$	VI-146	138.17
5137	ethyl aniline	(N)(β)	$C_6H_5O \cdot (CH_2)_2 \cdot NH \cdot C_6H_5$	**XII-107	213.28
5138	**Phenthiazine**	thiodiphenylamine	$C_6H_4 \cdot NH \cdot C_6H_4 \cdot S \cdot$ ‖	XXVII-63	199.27
5139	**Phenyl** acetaldehyde	α-toluylaldehyde	$C_6H_5 \cdot CH_2 \cdot CHO$	VII-292	120.15
5140	acetamide	....	$C_6H_5 \cdot CH_2 \cdot CO \cdot NH_2$	IX-437	135.17

Table 7-4 (*Continued*)
PHYSICAL CONSTANTS OF ORGANIC COMPOUNDS

No.	Crystalline Form and Color	Specific Gravity	Melting Point °C.	Boiling Point °C.	Solubility in 100 Parts		
					Water	Alcohol	Ether
5100	col. nd.	....	152-3	....	v. sl. s.	v. s.	v. s.
5101	cr./bz.	....	98-100*	>300	0.3	s.	s.
5102	pl./aq.	....	66; anh. 78	>360	s. h.; i. bz.	v. s.	v. sl. s.
5103	nd./aq.	....	173(anh.)	subl. > 100	s. h.	s.	sl. s.
5104	yel. nd.	....	170-1	>360	v. sl. s.	2 c.; s. h.	s.
5105	oil	....	<-21	228-9	i.	s.	s.
5106	lq.	....	....	180-205^{100mm}	i.	s.	s.
5107	lq.	1.061$^{15°}$	3-4	254-5	i.	s.	s.
5108	wh. cr. pd.	....	186-8		2.5	sl. s.	....
5109	col. pl.	....	233-4	subl.	v. s.	....	....
5110	col. lf.	....	168±	d. 150-5	v. s.	sl. s.	i.
5111	col. lq.	0.965$^{20°}_4$	-29.5	170	i.	∞	∞
5112	lf./aq. al.	....	173-4	....	0.12$^{15°}$; 0.5$^{100°}$	4	s.
5113	lf./aq.	....	174 (156; 166)	....	0.1 c.; s. h.	14; s. alk.	10
5114	wh. cr. pd.	....	....	....	100	11	i.; i. chl.
5115	nd.	....	95±;100.5 (anh.)	....	sl. s.	s.	....
5116	wh. cr. pd.	....	....	....	5	s.	sl. s.
5117	wh. nd.	....	....	....	0.5 c.	5 h. aq.	....
5118	col. nd.	1.071$^{25°}_4$ 1.054$^{45°}$	40.9	181.8	8.2$^{15°}$; ∞$^{65.3°}$	∞	∞
5119	col. cr.	....	....	d. by CO_2	23.8 c.	7.4$^{100°}$	0.5$^{25°}$
5120	cr.	....	....	....	80$^{25°}$±		
5121	col. rhb.	1.299$^{25°}_4$	261-2	....	0.2$^{20°}$; s. h.	10$^{25°}$; s. alk.	5.9 c.; v. sl. s. chl.
5122	nd./aq. al.	....	237-8	....	0.02$^{20°}$	....	....
5123	cr.	....	50 d.	....	v. s.	v. s.	
5124	nd.	....	-1.5H_2O, 100	-2H_2O, 140	....		
5125	delq. nd.	....	....	....	s.	s.	....
5126	col. mn.	....	....	....	24 c.; 125 h.	0.8 c.; 8 h.	
5127	red nd./ac.	....	....	....	0.08	0.3	i.; s. alk.
5128	cr./aq.	....	-H_2O, 120	d. 180	0.5$^{100°}$	s. h.	sl. s.; i. chl
5129	nd./al.	1.226$^{60°}_{25}$	56-7	185-7^{23mm}	i.; 165$^{25°}$ bz.	7$^{25°}$ Me al. 200$^{25°}$act.	165$^{25°}$
5130	nd./aq.	....	98-9	285 sl. d.	1.2$^{10°}$; v. s. h.	s.; s. ac.	29$^{25°}$; 3$^{25°}$ bz.
5131	nd./aq.	....	101-2	....	v. sl. s. h.	s. h.	....
5132	col. oil	....	....	229-30			
5133	lf./aq. al.	....	113-14.5	355 sl. d.	sl. s. h.	v. s.	v. s.
5134	nd.	....	45-6	287-9^{765mm}	....	....	....
5135	pa. yel. lq.	1.149$^{15°}_{15}$	....	138-43^{8mm}	i.	s.	s.
5136	oil	1.109$^{20°}_{20}$	14.0	244.7	2$^{25°}$	∞; s. KOH	∞; ∞ bz.
5137	cr.	1.070$^{60°}_{25}$	44-9	202^{10mm}	i.; v. s. bz.	v. s.; v. s. act.	v. s.; v. s. CCl_4
5138	yel. lf./al.	....	184-5	subl. 371	i.; 3$^{25°}$ bz.	2$^{25°}$; 27$^{25°}$ act.	7$^{25°}$
5139	lq.	1.025$^{20°}$	....	193-4	v. sl. s.	∞	∞
5140	lf.	....	156-7	d. 280-90	v. sl. s. c.	s.	v. sl. s.

* Crysts. + 1H_2O/aq., m. p. 92-3°; anh., m. p. 117°.

Phenol red 5127
Phenol-tetrachlorophthalein 5744
Phenoquin 5234
Phenothiazine 5138
Phenothioxine 5129

Phenoval 1008
Phenoxy-acetamide 5131
Phenoxy-ethyl bromide 969
Phenoxy-phenol 3770
Phenyl-, cf. also diphenyl-.

Table 7-4 (Continued)
PHYSICAL CONSTANTS OF ORGANIC COMPOUNDS

No.	Name	Synonym	Formula	Beil. Ref.	Formula Weight
5141	**Phenyl** acetate	acetyl phenol	$C_6H_5 \cdot O \cdot CO \cdot CH_3$	VI-152	136.15
5142	acetic acid	α-toluic acid	$C_6H_5 \cdot CH_2 \cdot CO_2H$	IX-431	136.15
5143	acetanilide	phenacetyl-anilide	$C_6H_5CH_2CO \cdot NHC_6H_5$	XII-275	211.27
5144	acetophenone (*p*)	Me-diphenylyl ketone	$C_{12}H_9 \cdot CO \cdot CH_3$	VII-443	196.25
5145	acetyl chloride	....	$C_6H_5 \cdot CH_2 \cdot CO \cdot Cl$	IX-436	154.60
5146	acetylene	....	$C_6H_5 \cdot C \vdots CH$	V-511	102.13
5147	acridine (9)	....	$C_6H_5 \cdot C_{13}H_8N$	XX-514	255.32
5148	alanine (*dl*)	α-amino-β-phenyl-propionic acid	$C_6H_5CH_2 \cdot CH(NH_2) \cdot CO_2H$	XIV-498	165.19
5149	alanine (β)(*l*)	α-amino-hydrocin-namic acid	$C_6H_5CH_2 \cdot CH(NH_2) \cdot CO_2H$	XIV-495	165.19
5150	alanine (β,β)	β-amino-hydrocin-namic acid	$C_6H_5 \cdot CH(NH_2) \cdot CH_2 \cdot CO_2H$	XIV-493	165.19
5151	angelic acid	α-ethyl-cinnamic acid	$C_6H_5 \cdot CH \vdots C(C_2H_5) \cdot CO_2H$	IX-623	176.22
5152	angelic acid (*stereo isomer*)	....	$C_6H_5 \cdot CH \vdots C(C_2H_5) \cdot CO_2H$	IX-623	176.22
5153	aniline (*o*)	2-amino-diphenyl	$C_6H_5 \cdot C_6H_4 \cdot NH_2$	XII-1317	169.23
5154	aniline (*m*)	3-amino-diphenyl	$C_6H_5 \cdot C_6H_4 \cdot NH_2$	XII-1318	169.23
5155	aniline (*p*)	4-amino-diphenyl	$C_6H_5 \cdot C_6H_4 \cdot NH_2$	XII-1318	169.23
5156	anthracene (9)	....	$C_6H_5 \cdot C_{14}H_9$	V-725	254.33
5157	anthranilic acid (*N*)(*o*)	*o*-anilino-benzoic acid	$C_6H_5NH \cdot C_6H_4 \cdot CO_2H$	XIV-327	213.24
5158	arsonic acid	Ph-arsinic acid	$C_6H_5 \cdot AsO(OH)_2$	XVI-868	202.04
5159	benzoate	....	$C_6H_5 \cdot CO_2 \cdot C_6H_5$	IX-116	198.22
5160	benzoic acid (*o*)	diPh-carboxylic acid	$C_6H_5 \cdot C_6H_4 \cdot CO_2H$	IX-669	198.22
5161	benzoic acid (*m*)	....	$C_6H_5 \cdot C_6H_4 \cdot CO_2H$	IX-671	198.22
5162	benzoic acid (*p*)	....	$C_6H_5 \cdot C_6H_4 \cdot CO_2H$	IX-671	198.22
5163	benzoylene urea (3)	....	$C_6H_4 \cdot NH \cdot CO \cdot N \cdot \underline{\qquad} (C_6H_5) \cdot CO \underline{\qquad}$	XXIV-376	238.25
5164	benzylcarbinol	....	$C_6H_5 \cdot CHOH \cdot CH_2 \cdot C_6H_5$	VI-683	198.27
5165	benzylurethane	ethyl *N*-Ph-*N*-benzyl-carbamate	$C_6H_5CH_2(C_6H_5)N \cdot CO_2C_2H_5$	**XII-565	255.32
5166	bromoacetate	....	$Br \cdot CH_2 \cdot CO_2C_6H_5$	VI-154	215.05
5167	carbamate	"phenyl urethane"	$NH_2 \cdot CO_2 \cdot C_6H_5$	VI-159	137.14
5168	carbylamine chloride	*iso*-cyanophenyl dichloride	$C_6H_5 \cdot N \vdots CCl_2$	XII-447	174.03
5169	chloroacetate	....	$Cl \cdot CH_2 \cdot CO_2C_6H_5$	VI-153	170.60
5170	4-chlorobenzene-sulfonate	....	$Cl \cdot C_6H_4 \cdot SO_3 \cdot C_6H_5$	....	268.72
5171	cinchonoyl-urethane	fantan	$C_6H_5 \cdot C_9H_5N \cdot NH \cdot CO_2C_2H_5$	....	292.34
5172	cinnamic acid (α)(*trans*)	....	$C_6H_5 \cdot CH \vdots C(C_6H_5) \cdot CO_2H$	IX-691	224.26
5173	cinnamic acid (α)(*cis*)	....	$C_6H_5 \cdot CH \vdots C(C_6H_5) \cdot CO_2H$	IX-693	224.26
5174	cinnamic acid (β)	....	$(C_6H_5)_2C \vdots CH \cdot CO_2H$	IX-699	224.26
5175	crotonic acid (γ)(3)	....	$C_{10}H_{10}O_2$	IX-614	162.19

Table 7-4 (*Continued*)
PHYSICAL CONSTANTS OF ORGANIC COMPOUNDS

No.	Crystalline Form and Color	Specific Gravity	Melting Point °C.	Boiling Point °C.	Solubility in 100 Parts		
					Water	Alcohol	Ether
5141	col. lq.	$1.073\frac{25°}{25}$	....	195.8^{756mm}	v. sl. s.	∞ ; ∞ chl.	∞
5142	lf.	$1.081\frac{80°}{4}$	76–7	265.5	s. h.	v. s.	v. s.
5143	pr./al.	....	117–8	....	i. aq. KOH	s.; i. H_2SO_4	s.
5144	pr./act.	....	121	325–7	....	s.	s. act.
5145	col. lq.	$1.168\frac{20°}{4}$	....	170^{250mm}	d.	d.	....
5146	col. lq.	$0.930\frac{20°}{4}$	–43	142–3	i.	∞	∞
5147	yel. pr.	....	181	403–4	i.; s. bz.	s. h.	s.
5148	lf./al.	....	271–3 d.	....	sl. s. c.	v. sl. s. h.	i.
5149	cr./aq.	....	d. 275–83	....	$3.1^{25°}$	v. sl. s.	v. sl. s.
5150	cr./aq.	....	231 d.	....	s. h.; s. alk.	s. h.; s. aq. a.	sl. s.
5151	nd./aq.	....	104–5	....	$0.01^{25°}$; s. h.	v. s.	sl. s. pet.
5152	nd./aq.	....	82	....	sl. s. h.	s.	s.; s. bz.
5153	cr./aq. al.	....	50–2	299^{760mm}	sl. s.	s.	....
5154	nd.	....	30	254^{135mm}	sl. s.	s.	s.
5155	lf./aq. al.	....	53–4	302	s. h.	s.	s.
5156	lf./al.	....	152–3	417	s. h. bz.	s. h.	s. h.
5157	nd./al.	....	185–7	d. > 184	v. sl. s. h.	s. h.	v. sl. s.
5158	cr./aq.	1.760	157–8	$-H_2O > 158$	$3.3^{28°}$	$15.5^{26°}$	i. chl.
5159	mn.	$1.235^{31°}$	70–1	314	i.	v. s. h.	v. s. h.
5160	cr./aq. al.	....	113–4	343–4	i. c.	s.	s. bz.
5161	lf./al.	....	160–1	....	sl. s.	s.	s.
5162	nd./al.	....	225–6.5	subl.	sl. s. h.	s.	s.
5163	nd.	....	282	....	i.	sl. s.	s. NaOH
5164	nd./aq. al.	....	67–8	$167-70^{10mm}$	0.06 h.	$420^{7°}$	v. s.
5165	col. lq.	$1.076\frac{5.5°}{4}$	....	$180-2^{12mm}$	....	....	....
5166	lf./al.	....	32	140^{20mm}	....	....	....
5167	nd./aq.	$1.078^{63.8°}$	141–3	....	sl. s.	s.	....
5168	lq.	....	....	209–10	sl. d. h.	d.	....
5169	nd./al.	....	44–5	230–5	i.	s.	s.
5170	col. cr.	....	90–1	....	i.	s. h.	s.
5171	yel. wh. cr.	....	173–4	....	i.; s. chl.	sl. s.; sl. s. bz.	sl. s.
5172	wh. nd./aq. al.	....	172	subl.	s. h.	s.	s.
5173	nd.	....	137–8	....	s. h.	....	....
5174	lf./al.	....	162	....	sl. s. h.	s.	s.
5175	pl./bz.	....	65	....	s. bz.	s.; s. CS_2	s.

Table 7-4 (*Continued*)
PHYSICAL CONSTANTS OF ORGANIC COMPOUNDS

No.	Name	Synonym	Formula	Beil. Ref.	Formula Weight
5176	**Phenyl** crotonic acid (*iso*)	. . . .	$C_6H_5 \cdot CH:CH \cdot CH_2 \cdot$ CO_2H	IX-612	162.19
5177	*iso*-cyanate	phenyl carbonimide	$C_6H_5 \cdot N:CO$	XII-437	119.12
5178	*iso*-cyanide	phenyl carbylamine	$C_6H_5 \cdot N:C$	XII-191	103.12
5179	cyclohexane	cyclohexylbenzene	$C_6H_5 \cdot CH(CH_2)_4 \cdot CH_2 \cdot$	V-503	160.26
5180	3,4-dichloroben- zene sulfonate	. . . .	$Cl_2C_6H_3 \cdot SO_3 \cdot C_6H_5$	. . . .	303.17
5181	diethanolamine	. . . .	$(HO \cdot CH_2 \cdot CH_2)_2N \cdot$ C_6H_5	XII-183	181.24
5182	dihydroquinazo- line	Orexin	$C_6H_4 \cdot CH_2 \cdot N(C_6H_5) \cdot$ $CH:N$	XXIII-137	208.27
5183	ditolylmethane (4,4′)	. . . .	$C_6H_5 \cdot CH(C_6H_4 \cdot CH_3)_2$	V-712	272.39
5184	di-*o*-xenyl phosphate	phosphen 6	$(C_{12}H_9O)_2:PO \cdot$ OC_6H_5	. . . .	478.49
5185	ethyl acetate (β)	. . . .	$C_6H_5 \cdot C_2H_4 \cdot O_2CCH_3$	VI-479	164.21
5186	ethyl alcohol (α)	Me-Ph-carbinol	$C_6H_5 \cdot CHOH \cdot CH_3$	VI-475	122.17
5187	ethyl alcohol (β)	benzyl carbinol	$C_6H_5 \cdot CH_2 \cdot CH_2OH$	VI-478	122.17
5188	ethylamine (α)(*dl*)	. . . .	$C_6H_5 \cdot CH(NH_2) \cdot CH_3$	XII-1094	121.18
5189	ethylamine (β)	ω-phenyl-ethylamine	$C_6H_5 \cdot CH_2 \cdot CH_2 \cdot NH_2$	XII-1096	121.18
5190	ethyl benzoate (β)	. . . .	$C_6H_5CO_2 \cdot C_2H_4 \cdot C_6H_5$	. . . .	226.28
5191	ethyl chloride (β)	β-Cl-ethyl benzene	$C_6H_5 \cdot CH_2CH_2Cl$	V-354	140.61
5192	ethyl cinnamate (β)	. . . .	$C_6H_5 \cdot CH:CH \cdot CO_2 \cdot$ $CH_2 \cdot CH_2 \cdot C_6H_5$	. . . .	252.32
5193	*d*-glucosazone	Ph-*d*-fructosazone	$(C_6H_5NHN)_2C_6H_{10}O_4$	XV-225	358.40
5194	glycine (*N*)	anilino-acetic acid	$C_6H_5 \cdot NH \cdot CH_2 \cdot CO_2H$	XII-468	151.17
5195	glycine *o*-car- boxylic acid	. . . .	$HO_2C \cdot C_6H_4 \cdot NHCH_2 \cdot$ CO_2H	XIV-348	195.18
5196	hydrazine	. . . .	$C_6H_5 \cdot NH \cdot NH_2$	XV-67	108.14
5197	hydrazine HCl	. . . .	$C_6H_5N_2H_3 \cdot HCl$	XV-108	144.61
5198	hydrazine-*p*-sul- fonic acid	. . . .	$H_2N \cdot NH \cdot C_6H_4 \cdot SO_3H$	XV-639	188.21
5199	hydrazoquinoline	(α)	$C_6H_5(NH)_2C_9H_6N$	XXII-564	235.29
5200	hydroxylamine (β)	. . . .	$C_6H_5 \cdot NH \cdot OH$	XV-2	109.13
5201	hydroxylamine (β)	(oxalate salt)	$2C_6H_5NHOH \cdot H_2C_2O_4$	. . . .	308.29
5202	3-methylpyrazolone	(*N*)	$C_4H_5ON_2 \cdot C_6H_5$	XXIV-20	174.20
5203	morpholine (4)	. . . .	$(CH_2)_2 \cdot O \cdot (CH_2)_2 \cdot N \cdot$ C_6H_5	XXVII-6	163.22
5204	naphthalene (α)	. . . .	$C_6H_5 \cdot C_{10}H_7$	V-687	204.27
5205	naphthalene (β)	. . . .	$C_6H_5 \cdot C_{10}H_7$	V-687	204.27
5206	α-naphthylamine	(*N*)	$C_6H_5 \cdot NH \cdot C_{10}H_7$	XII-1224	219.29
5207	β-naphthylamine	(*N*)	$C_6H_5 \cdot NH \cdot C_{10}H_7$	XII-1275	219.29
5208	1-naphthylamine- 8-sulfonic acid	(*N*); phenyl (peri) acid	$C_6H_5 \cdot NH \cdot C_{10}H_6 \cdot$ SO_3H	XIV-753	299.35
5209	α-naphthyl carbinol	. . . .	$C_6H_5 \cdot CHOH \cdot C_{10}H_7$	VI-710	234.30
5210	α-naphthyl ketone	benzoyl-naphthalene	$C_6H_5 \cdot CO \cdot C_{10}H_7$	VII-510	232.28
5211	β-naphthyl ketone	. . . .	$C_6H_5 \cdot CO \cdot C_{10}H_7$	VII-511	232.28
5212	nitroamine	. . . .	$C_6H_5 \cdot NH \cdot NO_2$	. . . .	138.13
5213	nitromethane	ω-nitrotoluene	$C_6H_5 \cdot CH_2 \cdot NO_2$	V-325	137.14

Phenyl-ethyl-hydantoin 3126 Phenyl fluoride 3258 Phenyl-glyoxylic acid 758
Phenyl-ethyl ketone 3129 Phenyl-formanilide 3306 Phenyl-hydrocinnamic acid 2739
Phenyl-ethyl sulfide 3132 Phenyl-fructosazone 5193 Phenyl-hydroquinone 2338
Phenyl-ethylene 5646 Phenyl-glycine ethyl ester 3125 Phenyl-hydroxycrotonic acid **637**
Phenyl-ethylene oxide 5648 Phenyl-glyoxaloxime 4874 Phenyl iodide 3879

Table 7-4 (*Continued*)
PHYSICAL CONSTANTS OF ORGANIC COMPOUNDS

No.	Crystalline Form and Color	Specific Gravity	Melting Point °C.	Boiling Point °C.	Solubility in 100 Parts		
					Water	Alcohol	Ether
5176	nd./aq.	....	88	302 sl. d.	sl. s. h.	v. s.	v. s.
5177	lq.	1.096^{20}_{4}	....	166^{769mm}	d.	d.	v. s.
5178	col. lq.	$0.978^{15°}$	....	78^{40mm}	d.	d.	s.
5179	oil	0.944^{20}_{0}	7-8	239^{745mm}	....	....	....
5180	col. nd.	....	82-5	....	i.	s. h.	v. s.
5181	cr.	1.120^{60}_{20}	58-9	>350 sl. d.	$5^{25°}$;$23^{25°}$ bz.	v. s.; v. s. act.	$29^{25°}$
5182	pl./et. lg.	$1.290^{4°}$	95	d.	i.	s.	s.
5183	nd./Me al.	....	55-6	....	v. s. bz.	sl. s. c.	v. s.
5184	col. lq.	1.20^{60}_{4}	....	285-330^{5mm}	i.	v. s.	∞ bz.; ∞ CCl₄
5185	lq.	$1.051^{22.5°}$	....	232	....	....	
5186	lq.	1.019^{13}_{4}	21.4	203.6^{745mm}	i.	∞	∞
5187	col. oil	1.023^{13}_{4}	−27	219-21^{750mm}	$1.6^{20°}$	s. aq. al.	∞
5188	oil	$0.940^{15°}$	ca. −65	187.5^{741mm}	$4.2^{20°}$	∞	∞
5189	lq.	0.958^{24}_{4}	....	198	s.	∞	∞
5190	yel. lq.	....	....	204-6^{25mm}	i.	s.	s.
5191	oil	1.069^{25}_{4}	....	190-200 sl. d.	....	....	....
5192	col. cr.	....	57-8	....	i.	sl. s.	v. s.
5193	yel. nd.	....	205 d.	....	i.	0.66 abs.	25 c. pyr.
5194	or.	....	127	....	s.	s.	sl. s.
5195	nd./Me al.	....	218-20	....	sl. s.	s.	s.; i. bz.
5196	lt. yel. oil	$1.097^{22.7}_{4}$	19.6*	243.5	sl. s. h.	∞; ∞ chl.	∞; ∞ bz.
5197	lf./al.	....	249 d.	....	v. s.	s.	i.
5198	cr./al.	....	286	....	$0.6^{12°}$; 3 h.	sl. s.	
5199	nd./al.	....	191	....	s. chl.	sl. s.	i.; s. ac.
5200	col. nd.	....	81-2	....	2 c.; 10 h.	v. s.	v. s.; v. sl. s. lg.
5201	col. nd.	....	120-2 d.	....	sl. s.	s.	i.
5202	pr./aq.	....	128	191^{17mm}	$120°$	v. s. h.	v. sl. s.
5203	cr./al. et.	$1.058^{270°}$	57	269.9	$1.0^{20°}$	s.	s.
5204	waxy	....	45±	336-7	v. s. bz.	v. s.	v. s.
5205	lf./al.	....	108-9	345-6	v. s. bz.	sl. s.	sl. s.
5206	pr./al.	....	62	335^{528mm}	s. bz.	s.; s. ac.	s.; s. chl.
5207	rhb./Me al.	....	107-8	395.5	i.; s. h. bz.	v. s. h.	v. s. h.
5208	lf.	....			v. sl. s.		
5209	cr./al.	....	88	>360	i.	v. s.	v. s.
5210	rhb./al.	....	75.5	385	i.	$2.4^{12°}$ abs.	
5211	rhb./al.	....	82	398^{754mm}	i.	$2^{12°}$ abs.	....
5212	lf./lg.	....	46	expl. 98	sl. s. c.	v. s.	sl. s. lg.
5213	yel. lq.	1.160^{20}_{0}	....	141-2^{35} sl. d.	....	....	....

* Crysts. $+ \frac{1}{3}H_2O$, m. 24° ±.

Phenyl mercaptan 5915
Phenyl-mercuric acetate 4036
Phenyl-mercuric chloride 4037
Phenyl-mercuric nitrate 4038

Phenyl-methyl-, cf. also methyl-phenyl-.
Phenyl-methylamino-propanol, cf. alkd.
Phenyl-methyl carbinol 5186
Phenyl-methyl-hydrazine 4338-40

Phenyl-methyl ketone 51
Phenyl mustard oil 5241
Phenyl-naphthyl-methane 817-8

Table 7-4 (*Continued*)
PHYSICAL CONSTANTS OF ORGANIC COMPOUNDS

No.	Name	Synonym	Formula	Beil. Ref.	Formula Weight
5214	**Phenyl** phenacyl bromide	ω-Br-*p*-phenyl-acetophenone	$C_6H_5 \cdot C_6H_4 \cdot CO \cdot CH_2 \cdot Br$	....	275.15
5215	phenacyl chloride	ω-Cl-*p*-phenyl-acetophenone	$C_6H_5 \cdot C_6H_4 \cdot CO \cdot CH_2 \cdot Cl$	VII-443	230.70
5216	phenol (*o*)	2-hydroxy-diphenyl	$C_6H_5 \cdot C_6H_4 \cdot OH$	VI-672	170.21
5217	phenol (*m*)	3-hydroxy-diphenyl	$C_6H_5 \cdot C_6H_4 \cdot OH$	VI-673	170.21
5218	phenol (*p*)	4-hydroxy-diphenyl	$C_6H_5 \cdot C_6H_4 \cdot OH$	VI-674	170.21
5219	phosphine	....	$C_6H_5 \cdot PH_2$	XVI-757	110.10
5220	propiolic acid	....	$C_6H_5 \cdot C \vdots C \cdot CO_2H$	IX-633	146.15
5221	propionate	....	$C_2H_5 \cdot CO_2 \cdot C_6H_5$	VI-154	150.18
5222	*n*-propyl bromide	(γ)	$C_6H_5 \cdot (CH_2)_3 \cdot Br$	V-391	199.10
5223	propylene (α,α)	*iso*-allylbenzene	$C_6H_5 \cdot CH:CH \cdot CH_3$	V-481	118.18
5224	propylene (β)	α-methyl-styrene	$C_6H_5 \cdot C(CH_3):CH_2$	V-484	118.18
5225	propylene (γ)	....	$C_6H_5 \cdot CH_2 \cdot CH:CH_2$	V-484	118.18
5226	pyridine (2)(α)	....	$C_6H_5 \cdot C_5H_4N$	XX-424	155.20
5227	pyridine (3)(β)	....	$C_6H_5 \cdot C_5H_4N$	XX-424	155.20
5228	pyridine (4)(γ)	....	$C_6H_5 \cdot C_5H_4N$	XX-424	155.20
5229	pyrrole (*N*)	....	$C_4H_4N \cdot C_6H_5$	XX-164	143.19
5230	quinoline (2)(α)	....	$C_6H_5 \cdot C_9H_6N$	XX-481	205.26
5231	quinoline (4)	....	$C_6H_5 \cdot C_9H_6N$	XX-483	205.26
5232	quinoline (6)(*p*)	....	$C_6H_5 \cdot C_9H_6N$	XX-483	205.26
5233	quinoline (8)(*o*)	....	$C_6H_5 \cdot C_9H_6N$	XX-484	205.26
5234	quinoline-4-carboxylic acid (2)	2-Ph-cinchonic acid; Atophan	$C_6H_5 \cdot C_9H_5N \cdot CO_2H$	XXII-103	249.27
5235	salicylate	salol	$HO \cdot C_6H_4 \cdot CO_2 \cdot C_6H_5$	X-76	214.22
5236	semicarbazide (1)	....	$C_6H_5NH \cdot NH \cdot CO \cdot NH_2$	XV-287	151.17
5237	semicarbazide (4)	....	$C_6H_5 \cdot NH \cdot CO \cdot N_2H_3$	XII-378	151.17
5238	stearate	....	$C_{17}H_{35}CO_2 \cdot C_6H_5$	VI-155	360.59
5239	succinic acid (α)	(*dl*)	$C_6H_5 \cdot C_2H_3(CO_2H)_2$	IX-865	194.19
5240	thioacetamide	thio-Ph-acetamide	$C_6H_5 \cdot CH_2 \cdot CS \cdot NH_2$	IX-460	151.23
5241	*iso*-thiocyanate	phenyl mustard oil	$C_6H_5 \cdot N:C:S$	XII-453	135.19
5242	thiohydantoic acid	phenyl-pseudothio-hydantoic acid	$C_6H_5N:C(NH_2) \cdot S \cdot CH_2 \cdot CO_2H$	XII-411	210.26
5243	thiosemicarbazide	(1)	$C_6H_5NH \cdot NH \cdot CS \cdot NH_2$	XV-294	167.23
5244	thiourea	Ph thiocarbamide	$C_6H_5 \cdot NH \cdot CS \cdot NH_2$	XII-388	152.22
5245	toluene (*o*)	....	$C_6H_5 \cdot C_6H_4 \cdot CH_3$	V-596	168.24
5246	toluene (*m*)	....	$C_6H_5 \cdot C_6H_4 \cdot CH_3$	V-596	168.24
5247	toluene (*p*)	....	$C_6H_5 \cdot C_6H_4 \cdot CH_3$	V-597	168.24
5248	*p*-toluene sulfonate	....	$CH_3 \cdot C_6H_4 \cdot SO_3 \cdot C_6H_5$	XI-99	248.30
5249	*o*-tolyl ketone	Me benzophenone	$C_6H_5 \cdot CO \cdot C_6H_4 \cdot CH_3$	VII-439	196.25
5250	*m*-tolyl ketone	....	$C_6H_5 \cdot CO \cdot C_6H_4 \cdot CH_3$	VII-440	196.25
5251	*p*-tolyl ketone	....	$C_6H_5 \cdot CO \cdot C_6H_4 \cdot CH_3$	VII-440	196.25
5252	*p*-tolyl sulfone	Me diphenyl sulfone	$C_6H_5 \cdot SO_2 \cdot C_6H_4 \cdot CH_3$	VI-418	232.30
5253	urea	phenyl carbamide	$C_6H_5 \cdot NH \cdot CO \cdot NH_2$	XII-346	136.15
5254	urethane†(*N*)	Et-Ph-carbamate	$C_6H_5 \cdot NH \cdot CO_2C_2H_5$	XII-320	165.19
5255	xanthydrol (9)	....	$C_6H_5 \cdot C_{13}H_8O(OH)$	XVII-138	274.32
5256	**Phenylene** diamine	1,2-diaminobenzene	$NH_2 \cdot C_6H_4 \cdot NH_2$	XIII-6	108.14
5257	diamine HCl (*o*)	....	$C_6H_4(NH_2)_2 \cdot 2HCl$	XIII-14	181.07
5258	diamine (*m*)	1,3-diaminobenzene	$NH_2 \cdot C_6H_4 \cdot NH_2$	XIII-33	108.14
5259	diamine HCl (*m*)	....	$C_6H_4(NH_2)_2 \cdot 2HCl$	XIII-38	181.07
5260	diamine (*p*)	1,4-diaminobenzene	$NH_2 \cdot C_6H_4 \cdot NH_2$	XIII-61	108.14
5261	diamine HCl (*p*)	....	$C_6H_4(NH_2)_2 \cdot 2HCl$	....	181.07
5262	diamine sulfonic acid (*p*)	2,5-diaminobenzene 1-sulfonic acid	$(NH_2)_2C_6H_3 \cdot SO_3H \cdot 2H_2O$	XIV-712	224.24

† See also No. 5167.

Phenyl-nitrosohydroxylamine 1574
Phenyl-oxydisulfide 2716
Phenyl-pentane 421-4
Phenyl peri acid 5208
Phenyl-phenol methyl ether 4073-4

Phenyl-phenylenediamine 313-4
Phenyl phosphate 6341
Phenyl phosphite 6343
Phenyl phthalate 2737
Phenyl phthalimide 5273
Phenyl-propionaldehyde 3706
Phenyl-propionamide 3707
Phenyl-propionic acid 3683, 3704

Phenyl-propionitrile 3709
Phenyl-propionyl chloride 3708
Phenyl-propyl alcohol 3122, 3705
Phenyl-propyl carbinol 5436-7
Phenyl-propyl ketone 5439-40
Phenyl-pyrocatechol 2339
Phenyl succinate 2748
Phenyl-sulfanilic acid 2701

Table 7-4 (*Continued*)
PHYSICAL CONSTANTS OF ORGANIC COMPOUNDS

No.	Crystalline Form and Color	Specific Gravity	Melting Point °C.	Boiling Point °C.	Solubility in 100 Parts		
					Water	Alcohol	Ether
5214	col. nd.	....	124–5	....	....	$1.3^{25°}$	$6.778°$ al.
5215	yel. cr. pd.	....	126–7	....	....	v. s. h.	....
5216	nd./pet.	....	56–7	275	i.; s. lg.	s.; s. alk.	s.
5217	nd./aq.	....	75–8	>300	sl. s.	v. s.	s. KOH
5218	nd./aq. al.	....	164–5	305–8	i.; s. alk.	v. s.	v. s.
5219	lq.	$1.001^{15°}$		160–1	i.	v. s.	v. s.
5220	nd./aq.	....	136–7	subl.	v. sl. s.	v. s.	v. s.
5221	pr.	$1.047\frac{25°}{25}$	20	211^{760mm}	....	....	....
5222	col. lq.	....	....	$121–2^{20mm}$	....	....	s.
5223	col. lq.	$0.914\frac{20°}{4}$	....	176–7	i.	s.	s.
5224	col. lq.	$0.911^{20°}$	−23.2	165.4	i.	∞	∞
5225	col. lq.	$0.893\frac{20°}{0}$	....	156–7	i.	s.	s.
5226	lq.	>1	....	$269–70^{749mm}$	i.	s.	s.
5227	oil	>1	....	$269–70^{749mm}$	i.	s.	s.
5228	lf./aq.	....	77–8	274–5	s. h.	s.; s. chl.	s.; s. bz.
5229	pl.	....	62	234	i.; s. pet.	s.; s. chl.	s.; s. bz.
5230	nd./aq. al.	....	86	363	sl. s.	s. h.	s.
5231	nd./et.	....	61–2	....	i.	v. s.	v. s.
5232	cr./al.	$1.195^{20°}$	110–1	260^{77mm}	v. sl. s.	s.	s.
5233	lq.	....	....	283^{187mm}	s. bz.	s.	s.
5234	nd./Me al.	....	217–9	....	i.; s. alk.; 4h. Me al.	5 h. abs.; 3 h. act.	s.; s. a.; 0.4 h. bz.
5235	rhb./al.	$1.250\frac{20°}{4}$	**41.4**	$172–3^{12mm}$	0.015	v. s. h.	s.; 80 bz.
5236	lf./aq.	....	174–6	....	s. h.	s.; s. act.	sl. s.
5237	rhb./aq.	....	122	d. 140–70	sl. s. h.	s.	i.
5238	cr.	....	52	267^{15mm}	i.	....	....
5239	cr./aq.	....	168	$-H_2O > 168$	s. h.; s. ac.	s.; i. lg.	s.; i. bz.
5240	rhb./al.	....	97–8	d.	v. sl. s. h.	s.	s.
5241	col. lq.	$1.138\frac{15°}{15}$	−21	219–20	i.	s.	s.
5242	col. nd.	....	157–8	....	v. sl. s.; s. a.; s. alk.	v. sl. s.; sl. s. CS_2	v. sl. s.; v. sl. s. bz.
5243	pr./al.	....	200–1 d.	....	sl. s.	s. h.	sl. s.
5244	nd./aq.	1.3	154	....	$0.26^{18°}$	s.	$5.9^{100°}$ aq.
5245	lq.	$1.010\frac{20°}{4}$	....	261–4	i.	s. abs.	s.
5246	lq.	$1.031^{0°}$	....	272–7	i.	s. abs.	s.
5247	lf./lg.	$1.015^{27°}$	47–8	267	i.	s. abs.	s.
5248	rhb. nd./al.	....	94–5	....	i.; v. s. bz.	v. s.	v. s.
5249	col. lq.	....	<−18	$312–5^{735mm}$	i.	v. s. 80% al.	v. s.
5250	lq.	$1.088^{17.5°}$	....	321^{761mm}	ω chl.	∞	∞
5251	mn.	....	59–60(55)	326.5	i.; s. bz.	s.	s.
5252	pl./al.	....	124.5	....	sl. s. ac.	$1.62^{20°}$	sl. s. bz.
5253	mn.	1.302	147	d. 160	s. h.	s.	sl. s.
5254	pl./al.	$1.106\frac{30°}{4}$	52–3	237 sl. d.	i. c.	s.	s. h.; s. bz.
5255	pr./bz. lg.	....	159–60	subl.	sl. s. ac.	s. h.; s. bz.	s.; i. lg.
5256	lf./aq.	....	103–4	266–8	$1.23^{35°}$	v. s.	v. s.
5257	nd.	....	....	....	v. s.	....	....
5258	rhb.	$1.139\frac{15°}{15}$	62.8	284–7	v. s.	v. s.	s.
5259	col. nd.	....	....	....	v. s.	s.	....
5260	mn.	....	140	267	$3.82^{4°}$	s.	s.; s. chl.
5261	col. tri.	....	....	....	v. s.	sl. s.	i. HCl
5262	pl./aq.	....	....	....	s.	i.	i.; i. bz.

Table 7-4 (*Continued*)
PHYSICAL CONSTANTS OF ORGANIC COMPOUNDS

No.	Name	Synonym	Formula	Beil. Ref.	Formula Weight
	Phenylene				
5263	diamine sulfonic acid (*p*) HCl	....	$C_6H_8O_3N_2S \cdot 2HCl$	....	261.13
5264	**Phloretin**	dihydronaringenin	$C_{15}H_{14}O_5$	VIII-498	274.28
5265	**Phloroglucinol** tri-ethyl ether	(1,3,5)	$C_6H_3(OC_2H_5)_3$	VI-1103	210.28
5266	trimethyl ether	(1,3,5)	$C_6H_3(OCH_3)_3$	VI-1101	168.19
5267	trioxime (1,3,5)	....	$C_6H_6(:NOH)_3$	XV-34	171.16
5268	**Phorone**	*sym*-diisopropyli-dene acetone	$[(CH_3)_2C:CH]_2CO$	I-751	138.21
5269	*iso*-**Phorone**	1,1,3-triMe-cyclo-hexene-3-one-5	$(CH_3)_3C_6H_5O$	VII-65	138.21
5270	**Phosphenyl chloride**	Ph-diCl-phosphine	$C_6H_5 \cdot PCl_2$	XVI-763	178.99
5271	**Phosphobenzene**	....	$C_6H_5 \cdot P:P \cdot C_6H_5$	XVI-824	216.16
5272	**Phthalamidic acid** (*o*)	*o*-phthalamic acid	$NH_2 \cdot CO \cdot C_6H_4 \cdot CO_2H$	IX-809	165.15
5273	**Phthalanil**	*N*-Ph-phthalimide	$C_6H_5 \cdot N:C_8H_4O_2$	XXI-464	223.23
5274	**Phthalic** acid (*o*)	phthalic acid	$C_6H_4(CO_2H)_2$	IX-791	166.13
5275	acid Na	(mono Na salt)	$C_8H_5O_4Na \cdot 2H_2O$	IX-796	224.15
5276	acid (*m*)	isophthalic acid	$C_6H_4(CO_2H)_2$	IX-832	166.13
5277	acid (*p*)	terephthalic acid	$C_6H_4(CO_2H)_2$	IX-841	166.13
5278	aldehyde†(*o*)	....	$C_6H_4(CHO)_2$	VII-674	134.14
5279	aldehyde (*m*)	....	$C_6H_4(CHO)_2$	VII-675	134.14
5280	aldehyde (*p*)	....	$C_6H_4(CHO)_2$	VII-675	134.14
5281	amide (*o*)	phthaldiamide	$C_6H_4(CO \cdot NH_2)_2$	IX-814	164.17
5282	anhydride (*o*)	....	$C_6H_4(CO)_2O$	XVII-469	148.12
5283	dichloride (*o*) (*s*)	phthalyl dichloride	$C_6H_4(COCl)_2$	IX-805	203.03
5284	dichloride (*o*) (*as*)	phthalyl dichloride	$C_6H_4 \cdot CO \cdot O \cdot C:Cl_2$	IX-805	203.03
5285	dichloride (*m*)	isophthalyldichlo-ride	$C_6H_4(COCl)_2$	IX-834	203.03
5286	dichloride (*p*)	terephthalyl diCl	$C_6H_4(COCl)_2$	IX-844	203.03
5287	imide (*o*)	*o*-phthalimide	$C_6H_4(CO)_2NH$	XXI-458	147.13
5288	nitrile (*m*)	isophthalic nitrile	$C_6H_4(CN)_2$	IX-836	128.13
5289	nitrile (*p*)	terephthalic nitrile	$C_6H_4(CN)_2$	IX-846	128.13
5290	**Phthalide**	....	$C_6H_4 \cdot CH_2 \cdot O \cdot CO$	XVII-310	134.14
5291	**Phthalonic acid**	(1,2)	$C_6H_4(CO_2H) \cdot CO \cdot CO_2H$	X-857	194.15
5292	**Phthalophenone**	triPh-carbinol-*o*-carboxylic anhyd.	$(C_6H_5)_2C \cdot C_6H_4 \cdot CO \cdot O$	XVII-391	286.33
5293	**Phytadiene**	....	$C_{20}H_{38}$	I-263	278.53
5294	**Phytane**	....	$C_{20}H_{42}$	I-174	282.56
5295	**Phytanol**	dihydrophytol	$C_{20}H_{41}OH$	I-431	298.56
5296	sodium	....	$C_{20}H_{41}ONa$	....	320.54
5297	**Phytene**	....	$C_{20}H_{40}$	I-227	280.54
5298	**Phytol**	....	$C_{20}H_{39}OH$	I-453	296.54
5299	**Picene**	....	$C_{22}H_{14}$	V-735	278.36
5300	**Picoline** (α)	2-methyl-pyridine	$CH_3 \cdot C_5H_4N$	XX-234	93.13
5301	**Picoline** (β)	3-methyl-pyridine	$CH_3 \cdot C_5H_4N$	XX-239	93.13

† See also Nos. 162-4.
Phosgene 1247
Phosphen-1 2707
Phosphen-2 1907
Phosphen-3 2713

Phosphen-4 2046
Phosphen-5 2762
Phosphen-7 6120
Phosphen-8 6165
Phosphen-9 6360

Phosphen-11 1146
Photo-glycin 3800
Photol 4117
Phthal alcohol 6497-9
Phthalamic acid 5272

Table 7-4 (*Continued*)
PHYSICAL CONSTANTS OF ORGANIC COMPOUNDS

No.	Crystalline Form and Color	Specific Gravity	Melting Point °C.	Boiling Point °C.	Solubility in 100 Parts		
					Water	Alcohol	Ether
5263	tan pd.	....	....	....	s.	sl. s.	i.
5264	lf.	....	253-5 d.	....	v. sl. s. h.	∞	$0.35^{16°}$ abs.
5265	cr.	....	43	175^{24mm}	i.	v. s.	v. s.
5266	pr./al.	....	52.5	255.5	i.	v. s.	v. s.
5267	pd.	....	expl. 155	....	v. sl. s.	v. sl. s.	sl. s. chl.
5268	yel. gn. pr.	$0.885^{20/4°}$	28	197.2^{743mm}	$0.1^{50°}$	s.	s.
5269	lq.	$0.923^{20/20°}$	−8.1	215-6	i.	....	....
5270	fum. lq.	$1.319^{20°}$	....	224.6	d.	∞ bz.	∞ CS_2
5271	lt. yel. pd.	....	149-50	....	i. h.	i.	i.; s. h. bz.
5272	pr.	....	148-9	155 d.	s.	s.	sl. s.
5273	nd./al.	....	208-10	subl.	i.	s.	....
5274	mn./aq.	$1.593^{20/4°}$	191* 231†	....	$0.54^{14°}$; $18^{99°}$	$11.7^{18°}$ abs.	$0.68^{15°}$
5275	pr./aq.	....	−H_2O, 100	....	$10^{25°}$	....	....
5276	nd./h. aq.	....	347-8	subl.	$0.01^{25°}$ 0.2 h.	s.; s. ac.	i. bz.; i. pet.
5277	cr. or amor.	....	425*	subl. >300	0.001 c.	sl. s. h.	i.; s. alk.
5278	lt. yel. nd.	....	56	....	1.4 h.	v. s.	v. s.
5279	nd.	....	89-90	....	sl. s.	v. s.	i. pet.
5280	nd./aq.	....	115-6	$245-8^{771mm}$	1.7 h.	v. s.	s.
5281	cr.	....	221-3 d.	....	v. sl. s. o.	v. sl. s.	i.
5282	rhb.	$1.5274°$	131.5-2.0	284.5	v. sl. s.	s.	sl. s.
5283	col. lq.	$1.414^{25/25°}$	16	281.1	d.	d.	s.
5284	cr.	....	88-9	275.2^{720mm}	....	....	....
5285	cr.	....	41	276	d.	d.	....
5286	nd.	....	78-9	259	....	....	....
5287	cr./et.	....	238	subl.	$0.04^{25°}$	5; s. alk.	s. h.; i. bz.
5288	nd.	....	161.5	subl.	s. h.; i. lg.	v. s. h.	v. s. h.
5289	nd./bz.	....	224-6	....	i.; s. h. ac.	sl. s. h.	v. sl. s. h.
5290	nd./aq.	$1.164^{99/4°}$	73(65)	290	v. sl. s.	s.	....
5291	pr./bz. al.	....	144.5	....	$115^{15°}$	s.	s.; sl. s. chl.
5292	lf./al.	....	115	419-28 sl. d.	d. h.	s.; s. H_2SO_4	i. c. alk.
5293	lq.	$0.826^{0°}$	....	$185-8^{22mm}$	∞ pet.	∞ Me al.	∞ ac.
5294	lq.	$0.803^{0/4°}$	....	$169.5^{9.5mm}$	v. sl. s. ac.	v. sl. s.	....
5295	oil	$0.840^{20/4°}$	....	202^{10mm}	i.	s.	s.
5296	oil	....	....	....	....	s. pet.	s.
5297	col. oil	$0.817^{0/4°}$	....	$177-8^{10.5mm}$	v. s. pet.	sl. s. h.	sl. s.
5298	col. oil	$0.852^{20/4°}$	....	204^{10mm}	i.	∞ Me al.	∞
5299	col. lf.	....	364	518-20	i.	v. sl. s.	v. sl. s.
5300	col. lq.	$0.944^{20/4°}$	−66.6	129.4	v. s.	∞	∞
5301	col. lq.	$0.961^{15/4°}$		143.5	∞	∞	∞

* Sealed tube.
† Heated rapidly.
Phthaldiamide 5281
Phthalimide 5287
Phthalimino-ethyl bromide 940

Phthalin 5122
Phthalyl dichloride 5283-4
Phthiocol 4275
Phthiocol acetate 4276-7
Phthiocol oxime 4278

Phyllyrin, cf. glcde.
Physostigmine, cf. alkd.
Picein, cf. glcde.
Piceoside, cf. glcde.
Picoline dicarboxylic acid 6398

Table 7-4 (*Continued*)
PHYSICAL CONSTANTS OF ORGANIC COMPOUNDS

No.	Name	Synonym	Formula	Beil. Ref.	Formula Weight
5302	**Picoline** (γ)	4-methyl-pyridine	$CH_3 \cdot C_5H_4N$	XX-240	93.13
5303	**Picramic** acid	4,6-diNO$_2$-2-NH$_2$-phenol	$HO \cdot C_6H_2(NH_2)$: (NO$_2$)$_2$	XIII-394	199.12
5304	sodium salt	sodium picramate	$C_6H_4O_5N_3Na \cdot H_2O$	XIII-395	239.13
5305	**Picrolonic acid**	4-NO$_2$-3-Me-1-*p*-NO$_2$-phenylpyrazolone-5	$C_{10}H_8O_5N_4$	XXIV-51	264.20
5306	**Picropodophyllin**	....	$C_{22}H_{22}O_8$	XIX-424	414.42
5307	**Picrorocellin**	....	$C_{20}H_{22}O_4N_2$	XXV-93	354.41
5308	**Picryl chloride** (1;2,4,6)	....	$Cl \cdot C_6H_2(NO_2)_3$	V-273	247.55
5309	**Pimelic** acid	heptandioic acid	$(CH_2)_5(CO_2H)_2$	II-670	160.17
5310	acid (*iso*)	....	$(C_2H_5)(CH_3)C \cdot CH_2(CO_2H)_2$	II-685	160.17
5311	nitrile	....	$(CH_2)_5(CN)_2$	II-671	122.17
5312	**Pinacol**	2,3-diMe-butandiol-2,3	$[(CH_3)_2C \cdot OH]_2$	I-487	118.18
5313	hydrate	pinacone hydrate	$C_6H_{14}O_2 \cdot 6H_2O$	I-488	226.27
5314	**Pinacoline semi-carbazone**	....	$C_5H_{12}C:N \cdot NH \cdot CO \cdot NH_2$	III-104	157.22
5315	**Pinacolyl acetate**	....	$CH_3CO_2 \cdot C_6H_{13}$	II-133	144.22
5316	**Pinacyanole**	sensitol red	$C_{23}H_{20}N_2 \cdot C_2H_5I$	XXIII-320	480.40
5317	**Pinane** (*d*)	....	$C_{10}H_{18}$	V-93	138.25
5318	**Pinene** (α)(*dl*)	....	$C_{10}H_{16}$	V-144	136.24
5319	hydrochloride	artificial camphor	$C_{10}H_{16} \cdot HCl$	V-94	172.70
5320	**Pinol** (*dl*)	sobrerone	$C_{10}H_{16}O$	XVII-45	152.24
5321	**Piperic acid**	....	$CH_2O_2:C_6H_3 \cdot C_5H_5O_2$	XIX-281	218.21
5322	**Piperidine**	hexazane	$CH_2(CH_2)_4NH$	XX-6	85.15
5323	**Piperonal**	heliotropine	$CH_2O_2:C_6H_3 \cdot CHO$	XIX-115	150.14
5324	acetophenone	....	$C_8H_6O_2:CH \cdot COC_6H_5$	XIX-141	252.27
5325	**Piperonyl** alcohol	....	$CH_2O_2:C_6H_3 \cdot CH_2OH$	XIX-67	152.15
5326	phloroglucinoldimethyl ether	protocotoin; (2,4-diMeO)	$(CH_3O)_2C_{14}H_8O_4$	XIX-242	302.29
5327	**Piperonylic acid**	....	$CH_2O_2:C_6H_3 \cdot CO_2H$	XIX-269	166.13
5328	**Proflavine**	3,6-diNH$_2$-acridine sulfate; trypaflavine	$C_{13}H_{11}N_3 \cdot H_2SO_4 \cdot H_2O$	*XXII-650	325.35
5329	**Pregnan(e)diol**	23(α),20(α)	$C_{21}H_{36}O_2$	....	320.52
5330	**Pregnan(e)diol**	3(α),20(β)	$C_{21}H_{36}O_2$	....	320.52
5331	**Pregnan(e)diol**	3(β),20(α)	$C_{21}H_{36}O_2$	....	320.52
5332	**Pregnan(e)diol**	3(β),20(β)	$C_{21}H_{36}O_2$	....	320.52
5333	**Progesterone**(α)	....	$C_{21}H_{30}O_2$	....	314.47
5334	**Progesterone**(β)	progestin; corporin	$C_{21}H_{30}O_2$	....	314.47
5335	**Proline** (*l*-)	pyrrolidine-2-carboxylic acid	$HN \cdot (CH_2)_3 \cdot CH \cdot CO_2H$	XXII-2	115.13
5336	**Prominal**	1-Me-5-Et-5-Ph-barbituric acid	$(C_6H_5)(C_2H_5)(CH_3) \cdot C_4HO_3N_2$	....	246.27
5337	**Prontosil**®† (4,1;1',2',4')	prontosil red; rubiazol; streptozon	$H_2NSO_2 \cdot C_6H_4 \cdot N:N \cdot C_6H_3(NH_2)_2 \cdot HCl$	....	327.79
5338	**Propane**	....	$CH_3 \cdot CH_2 \cdot CH_3$	I-104	44.10
5339	**iso-Propanolamine**	α-NH$_2$-*iso*Pr alcohol	$NH_2 \cdot CH_2CHOH \cdot CH_3$	IV-289	75.11

† *p*-Aminobenzene-sulfonamide, better known by its trade name, "Prontosil."

Table 7-4 (*Continued*)
PHYSICAL CONSTANTS OF ORGANIC COMPOUNDS

No.	Crystalline Form and Color	Specific Gravity	Melting Point °C.	Boiling Point °C.	Solubility in 100 Parts Water	Alcohol	Ether
5302	lq.	$0.957\frac{15}{4}°$	3.7	143.1	∞	∞	∞
5303	red nd./al.	....	169	....	$0.14^{22°}$; s. ac.	s.; sl. s. chl.	sl. s.; s. bz.
5304	red crust	....	....	....	$2^{16°}$	....	....
5305	yel. nd./al.	....	121-2	d. 125	$0.91^{17°}$; 0.93 h.	$4.8^{17°}$; 9 hot	$0.51^{17°}$; $0.6^{17°}$ Me al.
5306	col. nd./bz.	....	228	....	i.; s. alk.	s.	s. chl.
5307	col. pr./al.	....	190-220	....	i.; s. chl.	s. h.	sl. s.
5308	yel. mn.	$1.797^{20°}$	83	....	i.; d. aq. alk.; s.bz.	$4.8^{17°}$; s. h. chl.	$7^{17°}$; sl. s. pet.
5309	mn./aq.	$1.291\frac{25}{4}°$	103-5	272^{100mm}	$2.5^{14°}$	v. s.	v. s.
5310	rhb./aq.	....	103-4	d. 135	$15.4^{15°}$	v. s.	v. s. h.
5311	col. lq.	$0.949^{18°}$	....	$175-6^{14mm}$	i.	∞	∞
5312	col. nd.	lq. $0.967^{15°}$	41-3(38)*	$171-2^{739mm}$	sl. s. c.; s. h.	v. s.	v. s.; sl. s. CS_2
5313	cr./aq.	....	46-7†	d. $-H_2O$	sl. s. c.	s.	s.
5314	nd./aq.	....	157	....	sl. s. c.	s.	v. s.
5315	col. lq.	....	....	143^{757mm}	....	....	....
5316	b. gn./al.	....	276-8	....	sl. s.	s.	s. pyr.
5317	col. lq.	$0.839\frac{20}{4}°$	−45	169.4	....	....	....
5318	col. lq.	$0.878\frac{20}{4}°$	−55	154-6	v. sl. s.	∞ abs.	∞
5319	cr. lf.	....	131-2	207-8	i.	33	s.
5320	lq.	$0.953\frac{20}{20}°$	....	183-4	s. bz.	s.	s.
5321	yel. nd./al.	....	216-7	subl. sl. d.	i.; 2 h.	3.6 c.	s.; sl. s. bz.
5322	lq.	$0.860\frac{20}{4}°$	−10.5	106.4	∞	∞	....
5323	cr./aq.	....	37	263	$0.1^{0°}$; $0.66^{78°}$	$15^{0°}$; $550^{78°}$	∞
5324	yel. nd./al.	....	122	....	....	....	....
5325	nd.	....	52-3	d.	s. h.	∞	∞
5326	lt. yel. mn./al.	....	141-2	....	i.; s. alk.; s. ac.	s.; i. alk. carb.	s.; s. bz.; s. chl.
5327	nd./al.	....	228	subl.	v. sl. s. h.	sl. s. h.	sl. s.
5328	red nd.	....	....	....	1 c.	s.	i.; i. chl.
5329	col. cr./act.	....	237-9	....	sl. s.	s.	s.
5330	col. cr./al.	....	232-4	....	sl. s.	s.	s.
5331	col. cr./et.	....	182-4	....	sl. s.	s.	s.
5332	cr./aq. al.	....	174-6	....	sl. s.	s.	s.
5333	pr.	....	128.5	....	i.	s.	s.
5334	nd./pet.	....	121	....	i.	s.	s.
5335	hyg. nd./al. et.	....	220-2 d.	....	v. s.	$1.6^{19°}$	i.
5336	col. cr.	....	176	....	s. h.	s.	....
5337	red cr. pd.	....	247-51	....	0.25	....	....
5338	col. gas	(A)1.562 $0.585^{-\frac{44.5}{4}°}$	−187.7	−42.1	$6.5^{17.8°}$ cc.	$926^{16.6°}$ cc.	$1299^{16.6°}$ cc.
5339	lq.	$0.973^{18°}$	0	160-1	v. s.	s.	i.

* Crysts. + $6H_2O$/aq., m. p. 47°.
† Anhydrous crysts./et., m. p. 38°; b. p. 172-5°.

Planocaine 4935
Poirrier's blue 6344
Polygonin, cf. glcde.
Populin, cf. glcde.
Porphyroxine, cf. alkd.
Potassium myronate, cf. glcde.

Pragmoline 82
Prehnitic acid 680
Prehnitol 5827
Prehnitylic acid 6257
Primulin, cf. glcde.
Procaine 4935

Progestin(e) 5334
Proline-betaine, cf. alkd.
Prontosil album 527
Propadiene 183
Propanal 5352
Propandioic acid 4002
Propandiol 5465, 6286
Propandithiol 6284

Table 7-4 (*Continued*)
PHYSICAL CONSTANTS OF ORGANIC COMPOUNDS

No.	Name	Synonym	Formula	Beil. Ref.	Formula Weight
5340	**Propargyl** acetate	....	$CH_3CO_2 \cdot CH_2 \cdot C \vdots CH$	II-140	98.10
5341	alcohol	propyne-1-ol-3	$HC \vdots C \cdot CH_2OH$	I-454	56.06
5342	bromide	3-bromo-propyne-1	$HC \vdots C \cdot CH_2Br$	I-248	118.97
5343	chloride	chloro-allylene	$HC \vdots C \cdot CH_2Cl$	I-248	74.51
5344	iodide	3-iodo-propyne-1	$HC \vdots C \cdot CH_2I$	I-248	165.96
5345	**Propenyl** chloride†	1-chloro-propene-1	$CH_3 \cdot CH \vdots CHCl$	I-198	76.53
5346	chloride (*iso*)	2-chloro-propene-1	$CH_3 \cdot CCl \vdots CH_2$	I-198	76.53
5347	guaethol	(4;1,2)	$CH_3 \cdot CH \vdots CH \cdot C_6H_3 \vdots$ $(OH)(OC_2H_5)$	....	178.23
5348	**Propiolic** acid	propynoic acid	$HC \vdots C \cdot CO_2H$	II-477	70.05
5349	aldehyde	propargyl aldehyde	$HC \vdots C \cdot CHO$	I-750	54.05
5350	**Propionaldol**	....	$C_2H_5 \cdot CHOH \cdot CH \vdots$ $(CH_3)CHO$	I-836	116.16
5351	**Propionic** acid	propanoic acid	$CH_3 \cdot CH_2 \cdot CO_2H$	II-234	74.08
5352	aldehyde	propanal	$CH_3 \cdot CH_2 \cdot CHO$	I-629	58.08
5353	aldehyde oxime	propionaldoxime	$C_2H_5 \cdot CH \vdots NOH$	I-631	73.10
5354	aldehyde semi-carbazone (*α*)	....	$C_2H_5 \cdot CH \vdots N \cdot NH \cdot$ $CO \cdot NH_2$	III-101	115.14
5355	aldehyde semi-carbazone	....	$C_2H_5 \cdot CH \vdots N \cdot NH \cdot$ $CO \cdot NH_2$	III-101	115.14
5356	amide	propionamide	$C_2H_5 \cdot CO \cdot NH_2$	II-243	73.10
5357	anhydride	....	$(C_2H_5 \cdot CO)_2O$	II-242	130.14
5358	anilide	propionanilide	$C_2H_5 \cdot CO \cdot NH \cdot C_6H_5$	XII-250	149.19
5359	bromide	propionyl bromide	$C_2H_5 \cdot CO \cdot Br$	II-243	136.98
5360	chloride	propionyl chloride	$C_2H_5 \cdot CO \cdot Cl$	II-243	92.53
5361	fluoride	propionyl fluoride	$C_2H_5 \cdot CO \cdot F$	II-243	76.07
5362	iodide	propionyl iodide	$C_2H_5 \cdot CO \cdot I$	II-243	183.98
5363	nitrile	ethyl cyanide	$CH_3 \cdot CH_2 \cdot CN$	II-245	55.08
5364	**Propyl** acetanilide (*N*)	acetyl *n*-propyl-aniline	$C_6H_5 \cdot N(COCH_3) \cdot$ $CH_2 \cdot C_2H_5$	XII-246	177.25
5365	acetate (*n*)	............··..	$CH_3CO_2 \cdot CH_2 \cdot C_2H_5$	II-129	102.13
5366	acetate (*iso*)	....	$CH_3CO_2 \cdot CH(CH_3)_2$	II-130	102.13
5367	acetoacetate (*iso*)	....	$CH_3COCH_2CO_2C_3H_7$	III-659	144.17
5368	alcohol (*n*)	propanol-1	$CH_3 \cdot CH_2 \cdot CH_2OH$	I-350	60.10
5369	alcohol (*iso*)	propanol-2	$(CH_3)_2CHOH$	I-360	60.10
5370	allyl barbituric acid (*iso*)	Numal; Alurate	$(C_3H_7)(C_3H_5) \vdots$ $C_4H_2O_3N_2$	....	210.23
5371	amine (*n*)	1-aminopropane	$CH_3 \cdot CH_2 \cdot CH_2 \cdot NH_2$	IV-136	59.11
5372	amine (*iso*)	2-aminopropane	$(CH_3)_2CH \cdot NH_2$	IV-152	59.11
5373	*p*-aminobenzoate	Propaesin	$NH_2 \cdot C_6H_4 \cdot CO_2C_3H_7$	XIV-423	179.22
5374	aniline (*n*)(*N*)	....	$C_6H_5 \cdot NH \cdot CH_2 \cdot C_2H_5$	XII-166	135.21
5375	aniline (*iso*)(*p*)	cumidine	$(CH_3)_2CH \cdot C_6H_4 \cdot NH_2$	XII-1147	135.21
5376	arsonic acid (*n*)	....	$C_3H_7 \cdot AsO(OH)_2$	IV-615	168.02
5377	benzaldehyde (*iso*) (*p*)	*p*-cuminic aldehyde	$(CH_3)_2CH \cdot C_6H_4 \cdot CHO$	VII-318	148.21
5378	benzene (*n*)	....	$C_6H_5 \cdot CH_2 \cdot C_2H_5$	V-390	120.20
5379	benzene (*iso*)	cumene	$C_6H_5 \cdot CH(CH_3)_2$	V-393	120.20
5380	benzoate (*n*)	....	$C_6H_5 \cdot CO_2 \cdot CH_2 \cdot C_2H_5$	IX-112	164.21
5381	benzoate (*iso*)	....	$C_6H_5 \cdot CO_2 \cdot CH(CH_3)_2$	IX-112	164.21
5382	benzoic acid (*n*)(*o*)	....	$C_3H_7 \cdot C_6H_4 \cdot CO_2H$	IX-544	164.21
5383	benzoic acid (*iso*) (*o*)	*o*-cuminic acid	$(CH_3)_2CH \cdot C_6H_4 \cdot$ CO_2H	IX-546	164.21
5384	benzoic acid (*n*)(*p*)	....	$C_3H_7 \cdot C_6H_4 \cdot CO_2H$	IX-545	164.21

† Properties for trans-; cis, m. p. −134.8°, b. p. 32.8°.

Propanoic acid 5351	Propene 5459	Propenyl cyanide 1563
Propanol 5368-9	Propenoic acid 149	Propenyl guaiacol 1280, 3223
Propanone 56	Propenol 193	Propenyl trimethoxybenzene 584
Propanselenol 5448	Propenyl anisole 511	Propesine 5373
Propanthiol 5424-5	Propenyl bromide (*iso*) 990	Propine 225
Propargyl aldehyde 5349	Propenyl carbinol (*iso*) 4059	Propionamide 5356

Table 7-4 (*Continued*)
PHYSICAL CONSTANTS OF ORGANIC COMPOUNDS

No.	Crystalline Form and Color	Specific Gravity	Melting Point °C.	Boiling Point °C.	Solubility in 100 Parts		
					Water	Alcohol	Ether
5340	col. lq.	$1.005^{20°}_{4}$	....	124-5	....	s.	s.
5341	col. lq.	$0.972^{20°}_{4}$	−17	114-5	s.	∞	∞
5342	lq.	$1.58^{20°}$	....	82	....	....	....
5343	lq.	$1.045^{5°}$	....	56	....	....	....
5344	lq.	$2.018^{0°}$	....	115	....	....	s.
5345	col. lq.	$0.935^{15°}_{5}$	−99	37	i.	∞; ∞ act.	∞; ∞ bz.
5346	lq.	$0.9189^{0°}$	−137.4	22.65	i.	....	....
5347	col. pd.	....	85-7	....	i.	s.	s.
5348	col. lq.	$1.139^{15°}_{5}$	18	102^{200mm}	s.	s.	s.
5349	oil	....	....	59-61	v. s.	....	....
5350	col. lq.	$0.986^{25°}_{4}$	....	$84-6^{11mm}$	s.	....	....
5351	col. lq.	$0.993^{20°}_{4}$	−20.8	141.4	∞	∞	∞; ∞ chl.
5352	col. lq.	$0.807^{20°}_{4}$	−81	49.5^{740mm}	$20^{20°}$	∞	∞
5353	lq.	$0.926^{20°}_{4}$	21(40?)	130-2	....	....	....
5354	nd./bz. + lg.	....	88-90	....	v. s.	v. s. h. bz.	....
5355	pl./aq.	....	154	less s. bz. than α	....	....	....
5356	rhb. pl.	$1.042^{20°}_{4}$	79-80	222.2	v. s.	v. s.	v. s.
5357	col. lq.	$1.012^{20°}_{4}$	−45	167	d.	d.	....
5358	lf./al.	1.175	105-7	....	$0.4^{24°}$	s.; s. h. aq.	s.
5359	lq.	$1.521^{16.4°}_{4}$	....	103-4	d.	d.	s.
5360	col. lq.	$1.065^{20°}_{4}$	−94	80	d.	d.	s.
5361	col. lq.	$0.972^{15°}$	....	44	d.	d.	....
5362	lq.	....	....	127-8	d.	d.	....
5363	col. lq.	$0.782^{20°}_{4}$	−91.9	97.2	s.	∞	∞
5364	mn. pl./et. or lg.	....	47-8	266^{712mm}	i.	s.	s.
5365	col. lq.	$0.886^{20°}_{4}$	−95	101.6	$1.6^{16°}$	∞	∞
5366	col. lq.	$0.874^{20°}_{20}$	−73.4	88.4	$3^{20°}$	∞	∞
5367	col. lq.	....	....	185-7	v. sl. s.	s.	s.
5368	col. lq.	$0.804^{20°}_{4}$	−127	97.2	∞	∞	∞
5369	col. lq.	$0.785^{20°}_{4}$	−89.5	82.4	∞	∞	∞
5370	col. cr./aq.	....	137-8	....	sl. s.	s.	s.
5371	col. lq.	$0.718^{20°}_{20}$	−83	$49-50^{761mm}$	∞'	∞	∞
5372	col. lq.	$0.694^{15°}_{4}$	−101	33-4	∞	∞	∞
5373	col. pr.	....	74-6	....	v. sl. s.	s.; s. chl.	s.; s. bz.
5374	lq.	$0.9491^{8°}$	....	???	i.	v. s.	v. s.
5375	lq.	0.953	<−20	225^{761mm}	i.	....	....
5376	col. nd./al.	....	126-7	....	v. s.	v. s.	i.
5377	lq.	$0.978^{20°}_{4}$	....	236-7	i.	s.	s.
5378	col. lq.	$0.862^{20°}_{4}$	−99.5	159.2	i.	∞; s. SO_2	∞; i. NH_3
5379	col. lq.	$0.862^{20°}_{4}$	−96.0	152.4	i.	∞	∞
5380	col. lq.	$1.021^{25°}_{25}$	−51.6	231	i.	s.	s.
5381	col. lq.	$1.010^{25°}_{25}$	....	218.5	i.	s.	s.
5382	lf./aq. al.	....	58	272^{739mm}	sl. s. h.	s.	....
5383	pr./aq.	....	51	....	sl. s. h.	s.	s.; s. bz.
5384	pr./h. aq.	....	140-1	....	sl. s. h.	s.	s.; s. bz.

Table 7-4 (*Continued*)
PHYSICAL CONSTANTS OF ORGANIC COMPOUNDS

No.	Name	Synonym	Formula	Beil. Ref.	Formula Weight
5385	**Propyl** benzoic acid (*iso*) (*p*)	cuminic acid	$(CH_3)_2CH \cdot C_6H_4 \cdot CO_2H$	IX-546	164.21
5386	benzoic amide (*iso*)(*p*)	cuminic amide	$(CH_3)_2CH \cdot C_6H_4 \cdot CO \cdot NH_2$	IX-547	163.22
5387	*o*-benzoylbenzoate (*n*)	....	$C_6H_5 \cdot CO \cdot C_6H_4 \cdot CO_2 \cdot CH_2 \cdot C_2H_5$	....	268.32
5388	bromide (*n*)	1-bromopropane	$CH_3 \cdot CH_2 \cdot CH_2Br$	I-108	123.00
5389	bromide (*iso*)	2-bromopropane	$CH_3 \cdot CHBr \cdot CH_3$	I-108	123.00
5390	bromopropenyl barbituric acid	Nostral; Noctal	$(C_3H_7)(CH_3 \cdot CBr : CH) : C_4H_2O_3N_2$	....	289.14
5391	*iso*-butyl ketone (*n*)	....	$C_2H_5 \cdot CH_2 \cdot CO \cdot C_4H_9$	I-706	128.22
5392	*n*-butyrate (*n*)	....	$C_3H_7 \cdot CO_2 \cdot C_3H_7$	II-271	130.19
5393	*iso*-butyrate (*n*)	....	$(CH_3)_2CHCO_2C_3H_7$	II-291	130.19
5394	*n*-butyrate (*iso*)	....	$C_2H_5 \cdot CH_2 \cdot CO_2 \cdot CH : (CH_3)_2$	II-271	130.19
5395	*iso*-butyrate (*iso*)	....	$C_3H_7 \cdot CO_2 \cdot C_3H_7$	II-291	130.19
5396	caproate (*n*)	....	$C_5H_{11} \cdot CO_2 \cdot CH_2 \cdot C_2H_5$	II-323	158.24
5397	caprylate (*n*)	....	$C_7H_{15} \cdot CO_2 \cdot CH_2 \cdot C_2H_5$	II-348	186.30
5398	carbamate (*n*)	....	$NH_2 \cdot CO_2 \cdot CH_2 \cdot C_2H_5$	III-28	103.12
5399	carbonate, ortho-	....	$C(OCH_2 \cdot C_2H_5)_4$	III-6	248.37
5400	chloride (*n*)	1-chloropropane	$CH_3 \cdot CH_2 \cdot CH_2Cl$	I-104	78.54
5401	chloride (*iso*)	2-chloropropane	$CH_3 \cdot CHCl \cdot CH_3$	I-105	78.54
5402	chloroformate (*n*)	*n*-Pr chloro-carbonate	$Cl \cdot CO_2 \cdot CH_2 \cdot C_2H_5$	III-11	122.55
5403	chloroformate (*iso*)	*iso*-Pr chlorocarb.	$Cl \cdot CO_2 \cdot CH(CH_3)_2$	III-12	122.55
5404	*iso*-cyanide (*n*)	*n*-propyl carbylamine	$C_2H_5 \cdot CH_2 \cdot N : C$	IV-141	69.11
5405	*iso*-cyanide (*iso*)	*iso*-Pr carbylamine	$(CH_3)_2CH \cdot N : C$	IV-154	69.11
5406	fluoride (*n*)	1-fluoropropane	$CH_3 \cdot CH_2 \cdot CH_2F$	I-104	62.09
5407	fluoride (*iso*)	2-fluoropropane	$CH_3 \cdot CHF \cdot CH_3$	I-104	62.09
5408	formate (*n*)	....	$H \cdot CO_2 \cdot CH_2 \cdot C_2H_5$	II-21	88.11
5409	formate (*iso*)	....	$H \cdot CO_2 \cdot CH(CH_3)_2$	II-21	88.11
5410	formate, ortho-(*n*)	....	$H \cdot C(O \cdot CH_2 \cdot C_2H_5)_3$	II-21	190.29
5411	furoate (*n*)	....	$C_4H_3O \cdot CO_2 \cdot C_3H_7$	XVIII-275	154.17
5412	β-furylacrylate (*n*)	....	$C_4H_3O \cdot CH : CH \cdot CO_2 \cdot CH_2 \cdot C_2H_5$	....	180.21
5413	glycolate (*iso*)	....	$HOCH_2 \cdot CO_2 \cdot C_3H_7$	**III-172	118.13
5414	*n*-hexyl ketone (*n*)	decanone-4	$C_3H_7 \cdot CO \cdot C_6H_{13}$	I-711	156.27
5415	*p*-hydroxybenzoate (*n*)	(*n*); nipasol	$HOC_6H_4CO_2 \cdot C_3H_7$	X-160	180.21
5416	hydroxylamine (β)	....	$C_2H_5 \cdot CH_2 \cdot NHOH$	IV-537	75.11
5417	iodide (*n*)	1-iodopropane	$CH_3 \cdot CH_2 \cdot CH_2I$	I-113	169.99
5418	iodide (*iso*)	2-iodopropane	$CH_3 \cdot CHI \cdot CH_3$	I-114	169.99
5419	lactate (*n*)	....	$C_2H_4(OH)CO_2C_3H_7$	III-265	132.16
5420	lactate (*iso*)	....	$C_2H_4(OH)CO_2C_3H_7$	III-282	132.16
5421	levulinate (*n*)	....	$C_4H_7O \cdot CO_2C_3H_7$	III-675	158.20
5422	malonic acid (*n*)	....	$C_3H_7 \cdot CH(CO_2H)_2$	II-657	146.14
5423	malonic acid (*iso*)	....	$C_3H_7 \cdot CH(CO_2H)_2$	II-669	146.14
5424	mercaptan (*n*)	propanthiol-1	$CH_3 \cdot CH_2 \cdot CH_2 \cdot SH$	I-359	76.16
5425	mercaptan (*iso*)	propanthiol-2	$(CH_3)_2CH \cdot SH$	I-367	76.16
5426	α-naphthylamine (*n*)	....	$C_{10}H_7 \cdot NH \cdot CH_2 \cdot C_2H_5$	XII-1224	185.27
5427	nitrate (*n*)	....	$CH_3 \cdot CH_2 \cdot CH_2 \cdot O \cdot NO_2$	I-355	105.09
5428	nitrite (*n*)	....	$CH_3 \cdot CH_2 \cdot CH_2 \cdot O \cdot NO$	I-355	89.09

Propyl-benzyl alcohol (*iso*) 1571
Propyl-bromobutyramide 4580
Propyl carbinol (*n*) 1034
Propyl carbinol (*iso*) 1036
Propyl carbonate 2780

Propyl carbylamine 5404-5
Propyl chlorocarbonate 5402-3
Propyl cyanide 1175-6
Propyl dibromosuccinate 2781
Propyl disulfide 2782

Propyl ether 2783-5
Propyl-ethylene 498-9
Propyl maleate 2788
Propyl malonate 2789
Propyl mustard oil **5450**

Table 7-4 (*Continued*)
PHYSICAL CONSTANTS OF ORGANIC COMPOUNDS

No.	Crystalline Form and Color	Specific Gravity	Melting Point °C.	Boiling Point °C.	Solubility in 100 Parts		
					Water	Alcohol	Ether
5385	tri.	1.162	116-7	subl.	$0.015^{25°}$	s.; s. conc. H_2SO_4	s.
5386	nd. or pl.	....	155	....	sl. s. h.	∞	sl. s.
5387	col. lq.	....	....	$223\text{-}5^{15mm}$	i.	s.	s.
5388	col. lq.	$1.353^{20°}_{4}$	−109.9	71.0	$0.25^{20°}$	∞	∞
5389	col. lq.	$1.310^{20°}_{4}$	−90.0	59.4	$0.32^{20°}$	∞ ; ∞ chl.	∞ ; ∞ bz.
5390	col. cr.	....	178	....	s^{l}. s.	s.	sl. s.
5391	lq.	$0.813^{22°}_{0}$	....	155^{750mm}	i.; s. alk.	s.; s. act.	s.; sl. s. chl.
5392	col. lq.	$0.879^{15°}$	−95.2	142.7	$0.17^{17°}$	∞	∞
5393	col. lq.	$0.884^{°}_{4}$	....	134-5	....	....	....
5394	col. lq.	$0.865^{13°}$	....	128	....	....	....
5395	col. lq.	$0.847^{21°}_{4}$	....	120.8	....	....	....
5396	col. lq.	$0.863^{25°}_{4}$	−68.7	187.2	....	....	....
5397	col. lq.	$0.880^{0°}_{0}$	−43.0	226.4	....	....	....
5398	pr.	....	60-1	200	v. s.	v. s.	v. s.
5399	col. lq.	$0.911^{8°}$	....	224.2			
5400	col. lq.	$0.890^{20°}_{4}$	−122.8	46.7	$0.27^{20°}$	∞	∞
5401	col. lq.	$0.859^{20°}$	−117	34.8	$0.31^{20°}$	∞	∞
5402	col. lq.	$1.090^{20°}_{4}$	....	$114\text{-}5^{768mm}$	i.; sl. d.	i.; sl. d.	∞ ; ∞ bz.
5403	col. lq.	....	....	$103\text{-}5^{721mm}$	i.	v. s.	v. s.
5404	lq.	$0.753^{25°}$	....	99.5	i.	∞	∞
5405	lq.	$0.760^{0°}$	....	87	i.	∞	∞
5406	gas	....	....	−3	....	....	....
5407	gas	....	....	−11	....	....	....
5408	col. lq.	$0.901^{20°}_{4}$	−92.9	80.9	$2.2^{22°}$	∞	∞
5409	lq.	$0.873^{20°}_{4}$	....	$68\text{-}71^{751mm}$	$2.1^{22°}$		
5410	col. lq.	$0.881^{20°}_{4}$	....	196-8	$2.1^{22°}$		
5411	col. lq.	$1.075^{26°}_{4}$	....	211	v. sl. s.	s.	∞
5412	col. lq.	$1.074^{20°}_{4}$	....	119^{7mm}	i.	s.	....
5413	col. lq.	$1.043^{18°}_{4}$	....	164	s.	s.	s.
5414	lq.	$0.824^{21°}_{0}$	−9	206-7	v. sl. s.	∞	∞
5415	col. pr./et.	....	95-6	....	$0.2^{80°}$	$100^{25°}$ Me al.	$50^{25°}$
5416	nd./et.	....	46±	....	v. s.	v. s.	sl. s. lg.
5417	lq.	$1.743^{20°}_{4}$	−98.7	102.5	$0.11^{20°}$	∞	∞
5418	lq.	$1.714^{15°}_{4}$	−90.0	89.5	$0.14^{20°}$	∞ ; ∞ chl.	∞ ; ∞ bz.
5419	col. lq.	....	....	$122\text{-}3^{150mm}$	s.	s.	s.
5420	col. lq.	....	....	167.5	s.	s.	s.
5421	lq.	$0.990^{20°}_{4}$	....	221	....	....	....
5422	pl./bz.	....	96	d.	v. s.	v. s.	....
5423	col. pr.	....	87	d. 180	....	v. s.	v. s.
5424	lq.	$0.836^{25°}_{4}$	−112	67-8	v. sl. s.	s.	s.
5425	lq.	$0.809^{25°}_{4}$	−130.7	58-60	sl. s.	∞	∞
5426	lt. yel. oil	....	....	$316\text{-}8^{771mm}$	i.	....	....
5427	lq.	$1.058^{20°}_{4}$	....	110.5	....	s.	s.
5428	lq.	$0.935^{20°}_{4}$	....	57	....	s.	s.

Table 7-4 (*Continued*)
PHYSICAL CONSTANTS OF ORGANIC COMPOUNDS

No.	Name	Synonym	Formula	Beil. Ref.	Formula Weight
	Propyl				
5429	nitroamine (*n*)	....	$C_2H_5 \cdot CH_2 \cdot NH \cdot NO_2$	IV-570	104.11
5430	phenol (*o*)(*n*)	....	$C_2H_5 \cdot CH_2 \cdot C_6H_4 \cdot OH$	VI-499	136.20
5431	phenol (*m*)(*n*)	....	$C_2H_5 \cdot CH_2 \cdot C_6H_4 \cdot OH$	VI-499	136.20
5432	phenol (*p*)(*n*)	....	$C_2H_5 \cdot CH_2 \cdot C_6H_4 \cdot OH$	VI-500	136.20
5433	phenol (*o*)(*iso*)	*o*-cumenol	$(CH_3)_2CH \cdot C_6H_4 \cdot OH$	VI-504	136.20
5434	phenol (*m*)(*iso*)	*m*-hydroxy-cumene	$(CH_3)_2CH \cdot C_6H_4 \cdot OH$	VI-505	136.20
5435	phenol (*p*)(*iso*)	*p*-hydroxy-cumene	$(CH_3)_2CH \cdot C_6H_4 \cdot OH$	VI-505	136.20
5436	phenyl carbinol (*n*)	....	$C_3H_7 \cdot CHOH \cdot C_6H_5$	VI-522	150.22
5437	phenyl carbinol (*iso*)	....	$C_3H_7 \cdot CHOH \cdot C_6H_5$	VI-523	150.22
5438	phenyl ether (*n*)	....	$C_2H_5 \cdot CH_2 \cdot O \cdot C_6H_5$	VI-142	136.20
5439	phenyl ketone (*n*)	butyrophenone	$C_2H_5 \cdot CH_2 \cdot CO \cdot C_6H_5$	VII-313	148.21
5440	phenyl ketone (*iso*)	*iso*-butyrophenone	$(CH_3)_2CH \cdot CO \cdot C_6H_5$	VII-316	148.21
5441	propionate (*n*)	....	$C_2H_5 \cdot CO_2 \cdot CH_2 \cdot C_2H_5$	II-240	116.16
5442	propionate (*iso*)	....	$C_2H_5 \cdot CO_2 \cdot CH(CH_3)_2$	II-241	116.16
5443	pyridine (2)(*α*)(*n*)	conyrine	$C_2H_5 \cdot CH_2 \cdot C_5H_4N$	XX-247	121.18
5444	pyridine (2)(*α*)(*iso*)	....	$(CH_3)_2CH \cdot C_5H_4N$	XX-247	121.18
5445	pyridine (4)(*γ*)(*iso*)	....	$(CH_3)_2CH \cdot C_5H_4N$	XX-248	121.18
5446	salicylate (*o*)(*n*)	....	$HO \cdot C_6H_4 \cdot CO_2 \cdot C_3H_7$	X-75	180.21
5447	salicylate (*o*)(*iso*)	....	$HO \cdot C_6H_4 \cdot CO_2 \cdot C_3H_7$	....	180.21
5448	selenomercaptan	propanselenol-1	$CH_3 \cdot CH_2 \cdot CH_2 \cdot SeH$	I-360	123.06
5449	succinic acid (*n*)	....	$C_3H_7 \cdot C_2H_3(CO_2H)_2$	II-675	160.17
5450	*iso*-thiocyanate (*n*)	*n*-propyl mustard oil	$CH_3 \cdot CH_2 \cdot CH_2 \cdot N:CS$	IV-145	101.17
5451	thiocyanate (*iso*)	....	$(CH_3)_2CH \cdot S \cdot CN$	III-177	101.17
5452	*p*-toluene sulfonate	(*n*)	$CH_3C_6H_4SO_3C_3H_7$	**XI-45	214.29
5453	urea (*n*)	....	$C_3H_7 \cdot NH \cdot CO \cdot NH_2$	IV-142	102.14
5454	urethane (*n*)	....	$C_3H_7 \cdot NH \cdot CO_2 \cdot C_2H_5$	IV-143	131.18
5455	*n*-valerate (*n*)	....	$C_4H_9CO_2 \cdot C_3H_7$	II-301	144.22
5456	*iso*-valerate (*n*)	....	$C_4H_9CO_2 \cdot CH_2 \cdot C_2H_5$	II-312	144.22
5457	*iso*-valerate (*iso*)	....	$C_4H_9CO_2 \cdot CH(CH_3)_2$	II-312	144.22
5458	**Propylal** (*n*)	methylene diPr ether	$CH_2(OCH_2 \cdot C_2H_5)_2$	I-575	132.20
5459	**Propylene**	propene	$CH_3 \cdot CH:CH_2$	I-196	42.08
5460	bromohydrin (*prim*)	*β*-Br-propyl alcohol	$CH_3 \cdot CHBr \cdot CH_2OH$	*I-181	139.00
5461	chlorohydrin (*prim*)	2-chloro-propanol-1	$CH_3 \cdot CHCl \cdot CH_2OH$	I-356	94.54
5462	chlorohydrin (*sec*)	1-chloro-propanol-2	$CH_3 \cdot CHOH \cdot CH_2Cl$	I-363	94.54
5463	cyanide	....	$CH_3 \cdot CH(CN) \cdot CH_2 \cdot CN$	II-640	94.12
5464	diamine (*dl*)	1,2-diamino-propane	$CH_3 \cdot CH(NH_2) \cdot CH_2 \cdot NH_2$	IV-257	74.13
5465	glycol (*α*)	propandiol-1,2	$CH_3 \cdot CHOH \cdot CH_2OH$	I-472	76.10
5466	glycol acetate, di-	....	$(CH_3CO_2)_2CH_2 \cdot CH(CH_3)$	II-142	160.17
5467	glycol acetate, mono-(*α*)	....	$CH_3CO_2 \cdot CH_2 \cdot CHOH \cdot CH_3$	II-142	118.13
5468	oxide (1,2)	....	$CH_3 \cdot CH \cdot CH_2 \cdot O$ (ring)	XVII-6	58.08
5469	oxide (1,3)	trimethylene oxide	$CH_2 \cdot CH_2 \cdot CH_2 \cdot O$ (ring)	XVII-6	58.08
5470	**Prostigmine bromide**	(1,3)	$Br(CH_3)_3N \cdot C_6H_4 \cdot O_2C \cdot N(CH_3)_2$	....	303.21

Table 7-4 (*Continued*)
PHYSICAL CONSTANTS OF ORGANIC COMPOUNDS

No.	Crystalline Form and Color	Specific Gravity	Melting Point °C.	Boiling Point °C.	Solubility in 100 Parts		
					Water	Alcohol	Ether
5429	col. lq.	$1.104^{15°}$	−22	128–94^{0mm}	sl. s.	∞	∞
5430	lq.	$1.015^{0°}$	....	221–6	v. sl. s.	s.	s.
5431	lq.	....	26	228	v. sl. s.	s.	....
5432	cr.	$1.009^{0°}$	21–2	230–2	v. sl. s.	s.	....
5433	lq.	$1.012^{20°}$	15–6	214–5	v. sl. s.	∞	∞
5434	cr.	....	26	228	v. sl. s.	....	....
5435	nd.	$0.990^{20°}$	61	228.2–9.2	v. sl. s.	$316^{25°}$	$350^{25°}$
5436	oil	$1.021\frac{18°}{4}$	....	168–70^{100mm}	....	....	....
5437	oil	$0.987\frac{13.7°}{4}$	....	218–21	....	....	....
5438	col. lq.	$0.953\frac{15°}{15}$	....	189–90	....	....	....
5439	lq.	$0.990\frac{18°}{4}$	11	231^{727mm}	i.	∞	∞
5440	lq.	$0.985\frac{20°}{20}$	....	220^{746mm}	i.	s.	s.
5441	col. lq.	$0.883\frac{20°}{4}$	−75	122.6	$0.56^{25°}$	∞	∞
5442	col. lq.	$0.893^{0°}$	....	109–11^{750mm}	$0.6^{25°}$	....	....
5443	lq.	<1	....	165–8	....	....	....
5444	lq.	$0.934^{0°}$	....	158–9	sl. s.	....	....
5445	lq.	$0.944^{0°}$	....	177–8	....	sl. s.	....
5446	col. lq.	$1.099^{15°}$	....	238–40	v. sl. s.	∞	∞
5447	col. lq.	$1.010^{25°}$	....	120–21^{8mm}	....	....	....
5448	oil	$1.302\frac{20°}{4}$	....	84	....	....	....
5449	col. cr.	....	92–3	....	s.	2.8 c. chl.	....
5450	lq.	$0.978\frac{18°}{4}$	....	152.7^{743mm}	....	....	....
5451	lq.	$0.963^{20°}$	....	152–37^{54mm}	i.	∞	∞
5452	col. lq.	$1.144\frac{20°}{4}$	<−20	164–61^{0mm}	i.	s.	s.
5453	cr.	....	107	....	s.	....	....
5454	lq.	....	....	192–3	....	....	....
5455	lq.	$0.874^{15°}$	−70.7	167.5	i.	∞	∞
5456	col. lq.	$0.863\frac{20°}{4}$	....	155.9	i.	∞ ; ∞ chl.	∞
5457	col. lq.	$0.854^{17°}$	....	142^{756mm}	....	....	....
5458	lq.	$0.834^{20°}$	−97	137–8	....	....	....
5459	col. gas	(A)1.498 0.609–$\frac{47°}{4}$	−185.3	−47.7	44.6 cc.	1200 cc.	500 cc. ac.
5460	lq.	....	....	52–3^{16mm}	....	....	....
5461	col. lq.	$1.103^{20°}$	....	133–4	s.	s.	s.
5462	col. lq.	$1.115\frac{20°}{20}$	....	126–77^{62mm}	∞	∞	....
5463	col. lq.	....	12	252–4	....	....	....
5464	col. lq.	$0.878^{15°}$	....	119–20	∞	....	....
5465	col. oil	$1.040^{19.4°}$	....	188–9	∞	∞	8
5466	col. lq.	....	−31	190.2	10	....	....
5467	col. lq.	$1.055^{20°}$	....	182–3	s.	....	....
5468	col. lq.	$0.831\frac{20°}{20}$	−104.4	35	$33^{30°}$	∞	∞
5469	lq.	....	....	50	∞	....	....
5470	wh. cr. pd.	....	176 d.	....	v. s.	....	....

PS (puking stuff) gas 4697
Pseudoaconitine, cf. alkd.
Pseudochlorogenin 1478
Pseudocinchonine, cf. alkd.
Pseudocumene 6254

Pseudocumidine 6250
Pseudocumene 6254
Pseudocumenol 6265
Pseudoephedrine 2878
Pseudomorphine, cf. alkd.

Pseudopelletierine, cf. alkd.
Pseudopunicine, cf. alkd.
Pseudosarsasapogenin 5602
Pseudotropine, cf. alkd.
Pukateine, cf. alkd.

Table 7-4 (*Continued*)
PHYSICAL CONSTANTS OF ORGANIC COMPOUNDS

No.	Name	Synonym	Formula	Beil. Ref.	Formula Weight
5471	**Pulegol** (*iso*)(*d*)	....	$C_{10}H_{17}OH$	VI-65	154.25
5472	**Pulegone**	....	$C_{10}H_{16}O$	VII-81	152.24
5473	**Purine**	....	$C_5H_4N_4$	XXVI-354	120.11
5474	**Putrescine**	....	$NH_2 \cdot (CH_2)_4 \cdot NH_2$	IV-264	88.15
5475	hydrochloride	....	$C_4H_{12}N_2 \cdot 2HCl$	IV-264	161.08
5476	**Pyrantin**	*p*-ethoxyphenyl-succinimide	$C_2H_5O \cdot C_6H_4 \cdot$ $N(COCH_2)_2$	XXI-377	219.24
5477	**Pyrazine**	1,4-diazine	N:CH·CH:N·CH:CH	XXIII-91	80.09
5478	**Pyrazole**	....	N:CH·CH:CH·NH	XXIII-39	68.08
5479	**Pyrazoline**	....	N:CH·CH₂·CH₂·NH	XXIII-28	70.09
5480	**Pyrazolone** (5)	1,2-pentadiazenone	CO·CH₂·CH:N·NH	XXIV-13	84.08
5481	**Pyrene**	....	$C_{16}H_{10}$	V-693	202.26
5482	**Pyridazine**	1,2-diazine	N:CH·CH:CH·CH:N	XXIII-89	80.09
5483	**Pyridine**	....	C_5H_5N	XX-181	79.10
5484	hydrochloride	....	$C_5H_5N \cdot HCl$	XX-189	115.57
5485	hydrochloride	....	$C_5H_5N \cdot 2HCl$	XX-189	152.02
5486	carboxylic acid (2)	picolinic acid (2)(α)	$C_5H_4N \cdot CO_2H$	XXII-33	123.11
5487	carboxylic HCl (2)	α-picolinic HCl	$C_5H_4N \cdot CO_2H \cdot HCl$	XXII-34	159.57
5488	carboxylic acid (3)	nicotinic acid (3)(β)	$C_5H_4N \cdot CO_2H$	XXII-38	123.11
5489	carboxylic HCl (3)	nicotinic HCl	$C_5H_4N \cdot CO_2H \cdot HCl$	XXII-39	159.57
5490	carboxylic HNO₃	nicotinic nitrate	$C_5H_4N \cdot CO_2H \cdot HNO_3 \cdot$ H_2O	XXII-39	204.14
5491	3-carboxylic acid flavianate	nicotinic flavianate	$C_6H_5O_2N \cdot$ $C_{10}H_6O_8N_2S$	....	437.34
5492	3-carboxylic acid picrate	nicotinic acid picrate	$C_6H_5O_2N \cdot$ $C_6H_3O_7N_3$	*XXII-503	352.22
5493	3-carboxylic amide	nicotin(ic)amide	$C_5H_4N \cdot CONH_2$	XXII-40	122.13
5494	3-carboxylic amide flavianate	nicotinamide flavianate	$C_6H_6ON_2 \cdot$ $C_{10}H_6O_8N_2S$	....	436.36
5495	3-carboxylic amide picrate	nicotinamide picrate	$C_6H_6ON_2 \cdot C_6H_3O_7N_3$	....	351.23
5496	carboxylic acid (4)	*iso*-nicotinic acid (4)	$C_5H_4N \cdot CO_2H$	XXII-45	123.11
5497	dicarboxylic acid (2,3)	quinolinic acid	$C_5H_3N(CO_2H)_2$	XXII-150	167.12
5498	dicarboxylic acid (2,4)	lutidinic acid (2,4)	$C_5H_3N(CO_2H)_2 \cdot H_2O$	XXII-153	185.14
5499	dicarboxylic acid (2,5)	*iso*-cinchomeronic acid	$C_5H_3N(CO_2H)_2 \cdot H_2O$	XXII-153	185.14
5500	dicarboxylic acid (2,6)	dipicolinic acid	$C_5H_3N(CO_2H)_2 \cdot$ $1\frac{1}{2}H_2O$	XXII-154	194.15
5501	dicarboxylic acid (3,4)	cinchomeronic acid	$C_5H_3N(CO_2H)_2$	XXII-155	167.12
5502	dicarboxylic acid (3,5)	dinicotinic acid	$C_5H_3N(CO_2H)_2$	XXII-160	167.12
5503	pentacarboxylic acid	....	$C_5N(CO_2H)_5 \cdot 2H_2O$	XXII-190	335.18
5504	pentacarboxylic acid	....	$C_5N(CO_2H)_5 \cdot 3H_2O$	XXII-190	353.20
5505	sulfonic acid (3)	....	$C_5H_4N \cdot SO_3H$	XXII-387	159.16

Table 7-4 (*Continued*)
PHYSICAL CONSTANTS OF ORGANIC COMPOUNDS

No.	Crystalline Form and Color	Specific Gravity	Melting Point °C.	Boiling Point °C.	Solubility in 100 Parts		
					Water	Alcohol	Ether
5471	col. lq.	0.911$\frac{20°}{4}$	....	86-9^{10mm}	v. sl. s.	....	....
5472	col. lq.	0.932$\frac{20°}{20}$	....	224^{754mm}	i.	∞	∞
5473	cr./al.	....	217	d.	s.	s. h.	s. toluene
5474	cr.	....	27-8	158-60	v. s.	....	....
5475	nd./aq.	0.877$\frac{25°}{4}$	>290	....	v. s.	i. Me al.	....
5476	pr./al.	....	155-8	....	0.08$^{17°}$; 1.2$^{100°}$	s. h.	i.
5477	pr./aq.	1.031$\frac{61°}{4}$	52-3	118^{768mm}	∞	s.	s.; s. HCl
5478	nd./et.	....	70	186-8	s.	s.	s.; s. bz.
5479	lq.	....	....	144	∞	∞	sl. s.
5480	nd./toluene	....	165	subl. d.	s.	v. s.	v. sl. s.
5481	lt. yel. pr.	1.277$\frac{0°}{4}$	150.4	>360	i.	3.1 h. abs.	v. s.
5482	lq.	1.107$\frac{20°}{4}$	-8	208	∞ ; s. HCl	s.; i. lg.	s.; s. bz.
5483	col. lq.	0.983$\frac{20°}{4}$	-41.5	115.5*	∞	∞	s.
5484	hyg. pl./al.	....	82	218-9	s.	s.	i.; s. chl.
5485	pr.	....	46-7	d. > 55	....	....	i.
5486	nd./al.	....	137-9	d. -CO$_2$	v. s.	v. s.	v. sl. s.
5487	rhb.	....	210-25 d.	....	....	....	....
5488	nd./al.	....	235.2	subl.	s. h.	s. h.	v. sl. s.
5489	pr./aq.	....	274-5	....	....	s.	....
5490	lf. or pr./aq.	....	192-4	....	....	....	....
5491	yel. pl./al.	....	249-50	....	s. h.	sl. s. abs.	i.
5492	pa. yel. pl. or nd./aq.	....	225-7 d.	....	s. h.	sl. s. abs.	i.
5493	col. nd./bz.	....	133	....	100	66.6	v. sl. s.
5494	pa. yel. cr./75% al.	....	269-70 d.	....	s.	sl. s. abs.	i.
5495	yel. pl. or nd.	....	193	....	s. h.	sl. s. abs.	i.
5496	nd./aq.	....	317†	d.	s. h.	sl. s. h.	v. sl. s.
5497	col. mn./aq.	....	d. 110 (slow)	d. > 190 (rapid)	0.67°; s. h.	sl. s.; s. alk.	0.02; i. bz.
5498	pl./aq.	0.942	248-50	....	v. s. h.	s. h.	i.
5499	cr./aq.	....	236-7 (anh.)	subl. d.	sl. s. h.; s. h. HCl	v. sl. s.	v. sl. s.
5500	nd./aq.	....	226 d. (anh.)	....	sl. s. c.	v. sl. s.	v. sl. s.
5501	cr./HCl	....	258-9 d.	subl. d.	sl. s. h.	sl. s.	i.; i. chl.
5502	cr.	....	323 d.	subl. sl. d.	v. sl. s.	v. sl. s.	v. sl. s.; s. HCl
5503	cr./et.	....	-H$_2$O, 100	d.	s.	....	v. sl. s.
5504	cr./aq.	....	d. 220 (anh.)	....	s.	....	v. sl. s.
5505	nd. or lf.	....	d.	....	v. s.	v. s.	i.

* Liquid + 3H$_2$O, b. p. 92-3°.
† Sealed tube.

Pyrocatechin 2312
Pyrocatechol 2312
Pyrocatechol ethyl ether 2927

Pyrocatechuic acid 2312
Pyrodin 111
Pyrogallic acid 6216
Pyrogallol 6216
Pyrogallol carboxylic acid 6220, 6223

Pyroligneous spirit 4100
Pyromellitic acid 682
Pyromucic acid 3340
Pyromuc(ic)amide 3341
Pyromucic anhydride 3343

Table 7-4 (*Continued*)
PHYSICAL CONSTANTS OF ORGANIC COMPOUNDS

No.	Name	Synonym	Formula	Beil. Ref.	Formula Weight
5506	**Pyridine** tricarboxylic acid (2,3,4)	α-carbocinchomeronic acid	$C_5H_2N(CO_2H)_3 \cdot 1\frac{1}{2}H_2O$	XXII-182	238.15
5507	tricarboxylic acid (2,4,5)	berberonic acid	$C_5H_2N(CO_2H)_3 \cdot 2H_2O$	XXII-185	247.16
5508	tricarboxylic acid (2,4,6)	trimesitinic acid	$C_5H_2N(CO_2H)_3 \cdot 2H_2O$	XXII-185	247.16
5509	tricarboxylic acid (3,4,5)	β-carbocinchomeronic acid	$C_5H_2N(CO_2H)_3 \cdot 3H_2O$	XXII-186	265.17
5510	**Pyridoxin**	2-Me-3-OH-4,5-di-(hydroxymethyl)-pyridine	$C_8H_{11}O_3N$	....	169.18
5511	hydrochloride	....	$C_8H_{11}O_3N \cdot HCl$	....	205.64
5512	3-methyl ether	....	$CH_3O \cdot C_8H_{10}O_2N$	....	183.21
5513	**Pyrimidine**	*m*-diazine (1,3)	$CH:CH \cdot CH:N \cdot CH:N$	XXIII-89	80.09
5514	**Pyrocoll**	....	$C_4H_3 \cdot N:(CO)_2:N \cdot C_4H_3$	XXIV-403	186.17
5515	**Pyrogallol** dimethyl ether (1,3;2)	2,6-diMeO-phenol	$(CH_3O)_2C_6H_3 \cdot OH$	VI-1081	154.17
5516	triacetate (1,2,3)	lenigallol	$C_6H_3(O_2CCH_3)_3$	VI-1083	252.23
5517	trimethyl ether	(1,2,3)	$C_6H_3(OCH_3)_3$	VI-1081	168.19
5518	**Pyrone** (1,4)	....	$C_5H_4O_2$	XVII-271	96.09
5519	**Pyronine**	....	$C_{17}H_{19}ON_2 \cdot Cl$	XVIII-596	302.81
5520	**Pyrotartaric acid** (*dl*)	methyl succinic acid	$HO_2C(CH_3)CH \cdot CH_2 \cdot CO_2H$	II-636	132.12
5521	**Pyrrole**	....	$(\cdot CH:CH)_2:NH$	XX-159	67.09
5522	carboxylic acid (α)	(2)	$C_4H_3NH(CO_2H)$	XXII-22	111.10
5523	**Pyrrolidine**	tetrahydropyrrole	$(CH_2)_4:NH$	XX-4	71.12
5524	**Pyrroline**	dihydropyrrole	$CH_2 \cdot CH_2 \cdot NH \cdot CH:CH$	XX-133	69.11
5525	**Pyruvic** acid	pyro-racemic acid	$CH_3 \cdot CO \cdot CO_2H$	III-608	88.06
5526	amide	....	$CH_3 \cdot CO \cdot CO \cdot NH_2$	III-620	87.08
5527	nitrile	acetyl cyanide	$CH_3 \cdot CO \cdot CN$	III-620	69.06
5528	**Quinaldine** carboxylic acid (3)	2-Me-quinoline-3-carboxylic acid	$CH_3 \cdot C_9H_5N \cdot CO_2H$	XXII-83	187.20
5529	4-carboxylic acid	aniluvitonic acid	$CH_3 \cdot C_9H_5N \cdot CO_2H \cdot$ aq.	XXII-85	187.20
5530	5-carboxylic acid	(2;5)	$CH_3 \cdot C_9H_5N \cdot CO_2H$	XXII-86	187.20
5531	6-carboxylic acid	(2;6)	$CH_3 \cdot C_9H_5N \cdot CO_2H$	XXII-87	187.20
5532	8-carboxylic acid	(2;8	$CH_3 \cdot C_9H_5N \cdot CO_2H \cdot 1\frac{1}{2}H_2O$	XXII-87	214.22
5533	**Quinaldinic acid**	quinoline-2-carboxylic acid	$C_9H_6N \cdot CO_2H \cdot 2H_2O$	XXII-71	209.20
5534	**Quinazoline**	....	$C_6H_4 \cdot CH:N \cdot CH:N$	XXIII-175	130.15
5535	**Quinhydrone**	....	$C_6H_4O_2 \cdot C_6H_4(OH)_2$	VII-617	218.21
5536	**Quinic acid** (*l*)	....	$(HO)_4C_6H_7 \cdot CO_2H$	X-535	192.17
5537	**Quininic acid** (6;4)	6-MeO-cinchoninic acid	$CH_3O \cdot C_9H_5N \cdot CO_2H$	XXII-234	203.20
5538	**Quinoline**	leucoline	C_9H_7N	XX-339	129.16
5539	ethiodide	....	$C_9H_7N \cdot (C_2H_5I)$	XX-353	285.13
5540	**Quinoline** (*iso*)	....	C_9H_7N	XX-380	129.16

Table 7-4 (*Continued*)
PHYSICAL CONSTANTS OF ORGANIC COMPOUNDS

No.	Crystalline Form and Color	Specific Gravity	Melting Point °C.	Boiling Point °C.	Solubility in 100 Parts		
					Water	Alcohol	Ether
5506	rhb./aq.	....	249–50 d. (anh.)	$-H_2O$, 115–20	18°; s. h.	sl. s.	i.; i. bz.
5507	tri.	....	243	....	s. aq. a.	sl. s. h.	i.; i. bz.
5508	pl./aq. H_2SO_4	....	227 d. (anh.)	subl. d.	v. sl. s.	sl. s.	sl. s.
5509	pl. or lf.	....	d. 261 (anh.)	$-H_2O$, 115	s. h.	....	....
5510	col. nd.	....	160 d.	subl. in vac.	s.; s. act.	s.; sl. s. $CHCl_3$	sl. s.
5511	col. pr.	....	209–10 d.	subl. in vac.	22	1.1	sl. s.; sl. s. act.
5512	cr./chl. lg.	....	103–5	....	s.	s.	s.
5513	cr.	....	20–2	123–4	∞	s.	s.
5514	yel. mn.	....	268–9*	subl.	i.; s. c. H_2SO_4	v. sl. s. c.	v. sl. s.; s. ac.
5515	mn./aq.	....	55–6	262.7	1.8$^{13°}$	s.	s.
5516	wh. cr. pd.	....	164–5	....	i.	s.	d. aq. alk.
5517	rhb.	1.099$\frac{75°}{75}$	47	241	s. bz.	v. s.	v. s.
5518	cr.	1.190$^{40.3°}$	32.5	215–7	v. sl. s.	v. s.	v. s.
5519	gꞧ. pd.	....	....	....	s.	sl. s.	i.
5520	tri.	1.411	111–2	....	66$^{20°}$	v. s.	v. s.
5521	lq.	0.970$\frac{20°}{4}$	−24	131	8$^{25°}$	s.; s. bz.	s.; i. alk.
5522	mn.	....	192 sl. d.	d. 208.5	s.	s.	s.
6523	lq.	0.852$^{22.5°}$	....	87.5–8.5	∞	∞; ∞ chl.	∞
5524	lq.	0.910$\frac{20°}{4}$	....	90–1	v. s.	∞	∞
5525	col. lq.	1.267$\frac{20°}{4}$	13.6±	165 sl. d.	∞	∞	∞
5526	pl./al.	....	124–5	subl. 100	v. s.	s.	sl. s. c. bz.
5527	lq.	0.975$\frac{20°}{4}$	....	93	....	....	....
5528	nd./al. or bz.	....	235–8 d.	....	i.	v. sl. s.	v. sl. s.
5529	yel. nd./aq.	....	244–6 d.	subl. d.	sl. s.; s. h. ac.	sl. s.; i. h. chl.; s. a.	sl. s.; i. pet.
5530	nd./al.	....	285 d.	subl. d.	v. sl. s.	s.; s. h. a.	i.; i. bz.
5531	nd./al.	....	259–61	subl. d.	v. sl. s. h.	s. h.	....
5532	nd./aq.	....	151	....	v. s. h.	s.; s. a.	s. alk.
5533	nd./aq.	....	156 (anh.)	$-H_2O$, 100	s. h.; s. alk.	s.	s. h. bz.
5534	pl./pet.	....	48.0–8.5	243^{773mm}	v. s.	s.	s.
5535	red brn. rhb.	1.401$^{20°}$	171	subl. sl. d.	s. h.; s. NH_4OH	s.	s.; d. chl.
5536	mn./aq.	1.637	162–3	d.	40$^{9°}$	s.; s. ac.	v. sl. s.
5537	yel. pr./aq. HCl	....	280 d.	....	v. sl. s. c.; s. alk.	1.2 h. abs.; s. a.	v. sl. s.; v. sl. s. bz.
5538	hyg. lq.	1.095$^{20°}$	−15.6	237.1^{747mm}	0.6	∞	∞; s. CS_2
5539	yel. mn./al.	....	159–60	....	301$^{25°}$	1.8$^{25°}$ chl.	i.; s. al.
5540	pl.	1.091$\frac{30°}{4}$	26.5	243.3	sl. s.	....	s. a.

* Sealed tube.

Quercitrinic acid, cf. glcde.	Quinalizarin 5801	Quinine, cf. alkd.	Quinoline blue 1580
Quinaldic acid 5533	Quinanisole 4083	Quinizarin 2299	Quinoline carboxylic
Quinaldine 4367	Quinazine 5548	Quinizarin amide 327	acid 5533
Quinaldine ethiodide 4368	Quinicine, cf. alkd.	Quinoform, cf. alkd.	Quinolinic acid 5497
	Quinidine, cf. alkd.	Quinol 2314	

Table 7-4 (*Continued*)
PHYSICAL CONSTANTS OF ORGANIC COMPOUNDS

No.	Name	Synonym	Formula	Beil. Ref.	Formula Weight
5541	**Quinone** (*o*)	*o*-benzoquinone	$CH\cdot(CH)_3\cdot CO\cdot CO$	VII-600	108.10
5542	**Quinone** (*p*)	quinone	$CO:(CH:CH)_2:CO$	VII-609	108.10
5543	chlorimide (1,4)	....	$O:C_6H_4:N\cdot Cl$	VII-619	141.56
5544	dichlordiimide (*p*)	....	$Cl\cdot N:C_6H_4:N\cdot Cl$	VII-621	175.02
5545	dioxime (*p*)	....	$HO\cdot N:C_6H_4:N\cdot OH$	VII-627	138.13
5546	monoxime (*p*)	*p*-nitrosophenol tautomer	$O:C_6H_4:N\cdot OH$ or $HO\cdot C_6H_4\cdot NO$	VII-622	123.11
5547	monoxime Na	(sodium salt)	$NaO\cdot C_6H_4\cdot NO\cdot 2H_2O$	VII-624	181.12
5548	**Quinoxaline**	quinazine	$C_6H_4\cdot N:CH\cdot CH:N$	XXIII-176	130.15
5549	**Raffinose**	melitose	$C_{18}H_{32}O_{16}\cdot 5H_2O$	XXXI-462	594.52
5550	**Resacetophenone**	2,4-diOH-acetophenone	$(HO)_2C_6H_3\cdot COCH_3$	VIII-266	152.15
5551	**Resazurin**	"azoresorcin"	$C_{12}H_7O_4N$	XXVII-128	229.19
5552	**Resorcinol** diacetate	(1,3)	$C_6H_4(O_2C\cdot CH_3)_2$	VI-816	194.19
5553	dibenzoate	(1,3)	$C_6H_4(O_2C\cdot C_6H_5)_2$	IX-131	318.33
5554	**Resorufin**	9-OH-isophenoxazone	$HC\cdot C_{12}H_6ON:O$	XXVII-128	213.19
5555	**Retene**	1-Me-7-isoPr-phenanthrene	$C_{18}H_{18}$	V-683	234.34
5556	**Rhamnetin**	quercetin-7-Me-ether	$C_{16}H_{12}O_7$	XVIII-245	316.27
5557	**Rhamnitol**	rhamnite	$CH_3\cdot C_5H_{11}O_5$	I-532	166.18
5558	**Rhamnose** (*β*)	isodulcit	$C_6H_{12}O_5\cdot H_2O$	I-870	182.17
5559	phenylhydrazone	....	$C_6H_{12}O_4:N\cdot NHC_6H_5$	XV-216	254.29
5560	**Rhodanine**	rhodanic acid	$S\cdot CH_2\cdot CO\cdot NH\cdot CS$	XXVII-242	133.19
5561	**Rhodeose**	....	$C_5H_{11}O_4\cdot CHO$	I-876	164.16
5562	**Rhodoxanthin**	....	$C_{40}H_{50}O_2$	XXX-101	562.84
5563	**Riboflavin**	vitamin B_2 or G	$C_{17}H_{20}O_6N_4$	....	376.37
5564	**Ribose** (*d*) (−)	....	$C_4H_9O_4\cdot CHO$	I-859	150.13
5565	phenylhydrazone (*l*)	....	$C_5H_{10}O_4:N\cdot NHC_6H_5$	XV-215	240.26
5566	**Ricinelaidic** acid	....	$C_{18}H_{34}O_3$	III-388	298.47
5567	sodium salt†	....	$C_{17}H_{33}O\cdot CO_2Na$	....	320.45
5568	**Ricinoleic acid**	....	$C_{18}H_{34}O_3$	III-385	298.47
5569	**Ricinstearolic acid**	....	$C_6H_{13}CHOHCH_2C\vdots C\cdot C_7H_{14}CO_2H$	III-391	296.45
5570	**Rosaniline** (3-Me; 4,4′,4″-NH_2)	fuchsine carbinol base	$(CH_3\cdot C_6H_3NH_2)\cdot COH(C_6H_4NH_2)_2$	XIII-763	319.41
5571	hydrochloride	fuchsine	$C_{20}H_{20}N_3Cl\cdot aq.$	XIII-765	337.86
5572	**Rosinduline**	....	$HN:C_{10}H_5:NC_6H_4N\cdot$ C_6H_5	XXV-348	321.38
5573	**Rosolic acid**‡	4,4′-diOH-3-Me-fuchsone	$C_{20}H_{16}O_3$	VIII-365	304.35

† Usually a mixture of salts of fatty acids from castor oil. ‡ See also No. 593.

Quinone-imide-oxime 4875
Quinophan 5234
Quinophenol 3825
Quinosol 3827
Quinotol 1386
Quinotoxine, cf. alkd.
R acid 4491
Racemic acid 5690
Ratanhin 4405
Redoxon 586
Resazoin 5551
Resorcin yellow 2308
Resorcine 2313
Resorcinol 2313
Resorcinol blue 3938
Resorcinol diethyl ether 2092
Resorcinol dimethyl ether 2407
Resorcinol ethyl ether 2928
Resorcinol methyl ether 4080
Resorcinol monacetate 114
Resorcinol phthalein 3250
Resorcinol phthalin 3254
Resorcyl aldehyde (*β*) 2310
Resorcylic acid 2316, 2318, 2320

Table 7-4 (*Continued*)
PHYSICAL CONSTANTS OF ORGANIC COMPOUNDS

No.	Crystalline Form and Color	Specific Gravity	Melting Point °C.	Boiling Point °C.	Solubility in 100 Parts		
					Water	Alcohol	Ether
5541	red pr.	....	d. 60-70	....	v. s. act.	i. pet.	sl. s.
5542	yel. mn./aq.	$1.318\frac{20}{4}°$	112.9	subl.	v. sl. s. c.	s.; s. h. lg.	s.
5543	yel. lq.	....	85-6	expl.	s. h.	s.	s.; s. a.
5544	nd./aq.	....	126 d.	....	v. sl. s. h.	v. s. h.	v. s.
5545	col. or yel.	....	d. 240±	....	v. sl. s. aq. NH_4OH	s. conc. NH_4OH	....
5546	lt. yel. rhb.	....	124-6	d. 144	s. h.	v. s.	v. s.
5547	red nd./al.	....	$-2H_2O$, 100	....	v. s.	s.; s. act.	I.
5548	cr.	$1.133\frac{48}{4}°$	29-30	225-6	s. c.; sl. s. h.	∞	∞; ∞ bz.
5549	cr./aq.	$1.465^{0}°$	80 partly; 118-9 (anh.)	$-H_2O$, 110	$14.3^{20°}$; ∞ h.	$0.1^{20°}$	....
5550	col. cr.	$1.18^{141°}$	144-6	d.	i.; s. pyr.	s. h.; s. ac.	i.; i. bz.
5551	gn. red/ac.	....	d.	....	i.	sl. s.	i.; i. alk.
5552	lq.	....	....	278 sl. d.	....	....	....
5553	pl./al.	....	117	....	....	$5^{18°}$ abs.	....
5554	brn./HCl	....	....	....	i.; s. alk.	v. sl. s.	i.
5555	lf./al.	$1.13^{16°}$	98-9	390-4	i.; s. bz.	69 h.; s. CS_2	v. s. h.
5556	yel. n.d/al.	....	>300	....	v. sl. s. h.	s. h.	s. alk.
5557	tri.	....	121	....	v. s.; sl. s. act.	v. s.; sl. s. chl.	v. sl. s.
5558	col. mn.	$1.471\frac{20}{4}°$	126	....	$60.8^{21°}$	54 Me al.	i.
5559	col. lf.	....	159	....	$1.25^{15°}$	....	i.
5560	pa. yel. pr./al.	....	168-70 d.	....	$0.2^{25°}$; s. h.; s.alk.	v. s.; s. NH_4OH	v. s.
5561	nd.	....	144	....	v. s.	v. sl. s.	....
5562	b. bl. lf./bz. Me al.	....	219	....	i.; sl. s. bz.	v. sl. s.; s. chl.	s. pyr.; i. pet.
5563	or. yel. nd.	....	290 d.		$0.3^{25°}$	i. abs.	i.; s. pyr.
5564	pl./abs. al.	....	87	....	s.	sl. s.	
5565	col. cr.	....	154-5 sl. d.	....	v. s.	....	....
5566	nd./al.	....	53	$240-2^{10mm}$	....	v. s.	v. sl. s. pet.
5567	wh. yel. pd.	....	....	....	s.	s.	....
5568	lq.	$0.954^{16°}$	4-5	$226-8^{10mm}$	i.	∞; ∞ chl.	∞
5569	nd./al.	....	53	260^{10mm}	....	s.	s.
5570	col. nd./aq.	....	186 d.	....	v. sl. s.; s. a.	sl. s.	i.
5571	gn. red	l.22	d. > 200	....	0.3	s.	I.; s. HCl
5572	brn. nd./et.	....	198-9	....	i.	s.	s.; s. bz.
5573	red lf.	....	308-10 d.	d.	$0.12^{25°}$	v. s. h.	sl. s.; s. alk.

Table 7-4 (*Continued*)
PHYSICAL CONSTANTS OF ORGANIC COMPOUNDS

No.	Name	Synonym	Formula	Beil. Ref.	Formula Weight
5574	**Rotenone**	....	$C_{23}H_{22}O_6$	....	394.43
5575	**Rufigallic acid**	1,2,3,5,6,7-hexa-OH-anthraquinone	$C_{14}H_2O_2(OH)_6$	VIII-567	304.22
5576	**Sabinene**	....	$C_{10}H_{16}$	V-143	136.24
5577	**Saccharic acid** (*d* or *l*)	....	$(CHOH)_4(CO_2H)_2$	III-577	210.14
5578	*iso*-**Saccharic acid**	....	$(CHOH \cdot CH \cdot CO_2H)_2O$	XVIII-364	192.13
5579	**Saccharin**	*o*-benzoyl sulfimide	$C_6H_4 \cdot CO \cdot NH \cdot SO_2$ ⌐___⌐	XXVII-168	183.19
5580	soluble	(sodium salt)†	$C_7H_4O_3NSNa \cdot 2H_2O$	XXVII-170	241.20
5581	**Safrole**	3,4-methylenedioxy-allylbenzene	$CH_2O_2 : C_6H_3 \cdot CH_2 \cdot CH : CH_2$	XIX-39	162.19
5582	*iso*-**Safrole**	3,4-methylenedioxy-propenylbenzene	$CH_2O_2 : C_6H_3 \cdot CH : CH \cdot CH_3$	XIX-35	162.19
5583	**Salicyl**-acetic acid	....	$C_6H_4(OCH_2CO_2H) \cdot CO_2H$	X-69	196.16
5584	aldehyde triacetate	acetylsalicylalde-hyde diacetate	$C_6H_4CH : (O_2CCH_3)_2 \cdot O_2C \cdot CH_3$	VIII-45	266.25
5585	salicylic acid	disalicylic acid; salysal; diplosal	$HO \cdot C_6H_4 \cdot CO_2 \cdot C_6H_4 \cdot CO_2H$	X-84	258.23
5586	**Salicylidene**-acetamide	....	$HO \cdot C_6H_4 \cdot CH : N \cdot CO \cdot CH_3$	VIII-47	163.18
5587	benzamide	....	$HO \cdot C_6H_4 \cdot CH : N \cdot CO \cdot C_6H_5$	IX-212	225.25
5588	*p*-phenetidine	malakin	$HO \cdot C_6H_4 \cdot CH : N \cdot C_6H_4 \cdot OC_2H_5$	XIII-458	241.29
5589	2-thiohydantoin	5-(2-hydroxyben-zal)-2-thio-hydantoin	$HO \cdot C_6H_4 \cdot CH : C_3H_2ON_2S$	*XXV-502	220.25
5590	**Salipyrine**	antipyrene salicylate	$C_{18}H_{18}O_4N_2$	XXIV-32	326.36
5591	**Salyrgan**	mersalyl	$C_{13}H_{16}O_6NHgNa$	....	505.86
5592	**Salvarsan**® (1-As;3-NH₂; 4-OH)	arsphenamine; "606"	$(:As \cdot C_6H_3(OH)NH_2 \cdot HCl)_2 \cdot 2H_2O$	*XVI-507	475.04
5593	**Salvarsan** (neo)	neoarsphenamine; novarsenobenzol	$NH_2(OH)C_6H_3As : As \cdot C_6H_3(OH)NH \cdot CH_2OSONa‡$	*XVI-508	466.16
5594	**Santalic acid**	....	$C_{15}H_{14}O_5$	....	274.28
5595	**Santalol** (α)	arheol	$C_{15}H_{24}O$	VI-558	220.36
5596	**Santylyl salicylate**	santyl	$HO \cdot C_6H_4 \cdot CO_2 \cdot C_{15}H_{23}$	X-80	340.47
5597	**Santonic acid**	....	$C_{15}H_{20}O_4$	X-804	264.32
5598	**Santonin**	santoninic anhydride	$C_{15}H_{18}O_3$	XVII-499	246.31
5599	**Santoninic acid**	....	$C_{15}H_{20}O_4$	X-962	264.32
5600	**Sarcosine**	*N*-methyl glycine	$CH_3NH \cdot CH_2 \cdot CO_2H$	IV-345	89.09
5601	hydrochloride	....	$C_3H_7O_2N \cdot HCl$	IV-345	125.56
5602	**Sarsasapogenin,** pseudo	pseudo-smilagenin	$C_{27}H_{44}O_3$	....	416.65
5603	**Schonberg's reagent**	benzylimido-di-(4-methoxyphenyl) methane	$(CH_3O \cdot C_6H_4)_2C : N \cdot CH_2 \cdot C_6H_5$	....	331.42

† Usually with 5% H_2O. ‡ Contains also solvent and inorganic salts.

Rubiazol 5337	S acid 339, 344	Salicin, cf. glcde.
Rufiopin 5799	Sα acid 4533	Salicoyl-phenol 2323-5
Rufol 2293	Sacchar(in)ol 5579	Salicoyl-resorcinol 6225
Rumpff acid 4512	Saccharinose 5579	Salicyl-acetophenone 3763
Rutaecarpine, cf. alkd.	Safranone 670	Salicyl alcohol 3750
Rutin, cf. glcde.	Salamid 3738	Salicyl aldehyde 3733
Rutylidene 6382	Salazolon 5590	Salicyl-aldehyde ethyl ether 2914

Table 7-4 (*Continued*)
PHYSICAL CONSTANTS OF ORGANIC COMPOUNDS

No.	Crystalline Form and Color	Specific Gravity	Melting Point °C.	Boiling Point °C.	Solubility in 100 Parts		
					Water	Alcohol	Ether
5574	pl./al.	....	163	....	i.; s. chl.	sl. s.; s. act.	sl. s.; v. s. pet.
5575	red cr.	....	subl. sl. d.	....	i.; s. H_2SO_4	v. sl. s.; s. alk.	v. sl. s.
5576	lq.	$0.848\frac{15}{15}°$	....	164-5	i.	∞	∞
5577	nd./al.	....	125-6	d. lactone	v. s.	v. s.	i.
5578	rhb.	....	185	d.	s.	s.	v. sl. s.
5579	mn./act.	....	225-8 sl. d.	subl. 300 in vac.	$0.4^{25°}$; sl. s. h.	3.1 c.; sl. s. chl.	1.05 c.; s. alk. carb.
5580	col. pd.	....	....	....	83	2.2	....
5581	col. mn.	$1.100\frac{20°}{4}$	11.2	233-4	i.	s.	∞; ∞ chl.
5582	col. lq.	$1.122\frac{20°}{4}$	6-7	252-3	i.	∞	∞; ∞ bz.
5583	nd./aq.	....	191-2	....	s. h.	s.; s. act.	s.; s. ac.
5584	nd. or pl./al.	....	103-4	....	i.	s. h.; s. CCl_4	s.; s. bz.
5585	col. cr./bz. or chl.	....	148-9	d.	i.; $1^{25°}$ bz.	$45^{25°}$; $66^{25°}$ act.	$28^{25°}$; v. s. CCl_4
5586	yel. pd.	....	d. > 150	....	i.; i. alk. carb.	i. alk.	s. conc. H_2SO_4
5587	yel. pd.	....	d. 190	....	v. s. act.*	s.; s. alk.	i.; v. sl. s. bz.
5588	pa. yel. cr./al.	....	91-2	....	i.	s. h.	s.; s. bz.
5589	nd./ac.	....	251-3	....	....	....	....
5590	cr. pd.	....	91-2	....	$0.5^{15°}$; $4^{100°}$	s.; v. s. chl.	sl. s.
5591	delq. wh.	....	....	....	100	35	i.
5592	yel. pd.	....	....	....	s.	sl. s.	v. sl. s.
5593	yel. pd.	....	....	....	s.; d. by heat, air, light.	sl. s.; v. s. gly.	i.; i. chl.; sl. s. act.
5594	red pr.	....	104	....	i.; s. alk.	∞ abs.	s.
5595	col. oil	$0.977\frac{25°}{25}$	....	300±	i.	s.	....
5596	yel. oil	1.07	....	126.6^{20mm}	i.	s.	s.
5597	col. rhb./aq.	$1.251\frac{26°}{4}$	170-2	285^{15mm}	$0.56^{17°}$	s.	s.; s. chl.
5598	col. pr.	1.187	169-70	....	$0.02^{17.5°}$; $0.4^{100°}$	$2.3^{22.5°}$; $37^{80°}$	$1.3^{17.5°}$; $2.4^{40°}$
5599	rhb. pl./al.	....	$-H_2O$, 120	....	s. h.	s.; s. chl.	sl. s.
5600	rhb.	....	210-5 d.	....	v. s.	sl. s.	....
5601	nd./al.	....	170-2	....	v. s.	v. sl. s.	v. sl. s.
5602	nd./act.	....	171-3	....	sl. s.	s.	s.; s. ac.
5603	pa. yel. cr./abs. al.	....	93	....	i.; sl. s. pet.	s. h.	sl. s.

* Also a modification insoluble in act.

Salicyl-aldehyde glucose, cf. glcde.	Salicylic nitrile 3748	Saligenol 3750
Salicyl-aldoxime 3736	Salicylic phenyl ether 5133	Salimenthol 4031
Salicyl-amide 3738	Salicylic sulfonic acid 5681	Salinigrin, cf. glcde.
Salicyl-anilide 3741	Salicylide 3747	Salipyrazolon 5590
Salicyl-phenol 2323-5	Salicyl(oyl)-resorcinol 6225	Salit 866
Salicylic acid 3742	Saligenin 3750	Salmester 4078
Salicylic methyl ether 4067	Saligenin methyl ether 4072	Salocoll 5117
		Salol 5235

Table 7-4 (*Continued*)
PHYSICAL CONSTANTS OF ORGANIC COMPOUNDS

No.	Name	Synonym	Formula	Beil. Ref.	Formula Weight
5604	**Scopoletin**	7-OH-6-MeO-coumarin	$C_{10}H_8O_4$	XVIII-99	192.17
5605	**Sebacic** acid	decandioic acid	$(CH_2)_8(CO_2H)_2$	II-718	202.25
5606	nitrile	....	$(CH_2)_8(CN)_2$	II-720	164.25
5607	**Seleno**-phenol	....	$C_6H_5 \cdot SeH$	VI-345	157.07
5608	urea	....	$NH_2 \cdot CSe \cdot NH_2$	III-227	123.02
5609	**Semicarbazide**	....	$NH_2 \cdot NH \cdot CO \cdot NH_2$	III-98	75.07
5610	hydrochloride	....	$NH_2CON_2H_3 \cdot HCl$	III-100	111.53
5611	**Serine** (*dl*)	....	$HO \cdot CH_2 \cdot CHNH_2 \cdot CO_2H$	IV-512	105.09
5612	**Serine** (*d* or *l*)	....	$C_3H_7O_3N$	IV-505	105.09
5613	*iso*-**Serine** (*dl*)	....	$NH_2 \cdot CH_2 \cdot CHOH \cdot CO_2H$	IV-503	105.09
5614	*iso*-**Serine** (*d* or *l*)	....	$C_3H_7O_3N$	IV-503	105.09
5615	**Silico**-acetic acid	....	$CH_3 \cdot SiO \cdot OH$	IV-629	76.13
5616	benzoic acid	....	$C_6H_5 \cdot SiO \cdot OH$	XVI-911	138.20
5617	**Silicon** methyl	....	$(CH_3)_4Si$	IV-625	88.23
5618	phenyl trichloride	....	$C_6H_5 \cdot SiCl_3$	XVI-911	211.55
5619	tetraethyl	silicononane	$(C_2H_5)_4Si$	IV-625	144.33
5620	**Silver**† fulminate	....	$Ag_2C_2O_2N_2$	I-722	299.77
5621	**Sodium**† acetanilide	....	$CH_3CON(Na) \cdot C_6H_5$	XII-237	157.15
5622	ethyl	....	$C_2H_5 \cdot Na$	*IV-618	52.05
5623	glycerolate	....	$NaC_3H_7O_3$	I-511	114.08
5624	mercaptide	....	$C_2H_5 \cdot S \cdot Na$	I-341	84.12
5625	**Sorbic acid**	hexadienoic acid	$CH_3(CH:CH)_2CO_2H$	II-483	112.13
5626	**Sozoiodolic** acid	2,6-diiodophenol-4-sulfonic acid	$I_2(OH)C_6H_2 \cdot SO_3H \cdot 3H_2O$	XI-245	480.03
5627	Na salt	sozoiodol-sodium	$C_6H_3O_4I_2SNa \cdot 2H_2O$	XI-245	483.98
5628	Hg salt	....	$C_6H_2O_4I_2SHg$	*XI-56	624.54
5629	Zn salt	....	$(C_6H_3O_4I_2S)_2Zn \cdot 6H_2O$	XI-245	1023.38
5630	**Starch**‡	....	$(C_6H_{10}O_5)_x$	....	(162.14)
5631	**Stearic** acid	octadecanoic acid	$CH_3(CH_2)_{16}CO_2H$	II-377	284.47
5632	ammonium salt	amm. stearate	$C_{17}H_{35}CO_2NH_4$	*II-171	301.52
5633	aldehyde	octadecanal	$CH_3(CH_2)_{16}CHO$	I-718	268.49
5634	amide	stearamide	$C_{17}H_{35} \cdot CO \cdot NH_2$	II-384	283.50
5635	anhydride	....	$(C_{17}H_{35}CO)_2O$	II-384	550.96
5636	anilide	stearanilide	$C_{17}H_{35}CO \cdot NHC_6H_5$	XII-257	359.60
5637	chloride	stearyl chloride	$C_{17}H_{35} \cdot CO \cdot Cl$	II-384	302.93
5638	nitrile	heptadecyl cyanide	$C_{17}H_{35} \cdot CN$	II-384	265.49
5639	**Stearolic acid**	octadecynoic acid	$C_8H_{17}C \vdots C(CH_2)_7 \cdot CO_2H$	II-495	280.45
5640	**Stearone**	....	$C_{35}H_{70}O$	I-720	506.95
5641	**Stearoxylic acid**	....	$CH_3(CH_2)_7(CO)_2 \cdot (CH_2)_7CO_2H$	III-761	312.45
5642	**Stilbene** (*trans*)	diPh-ethylene	$C_6H_5 \cdot CH:CH \cdot C_6H_5$	V-630	180.25
5643	**Stilbene** (*cis*)	diPh-ethylene	$C_6H_5 \cdot CH:CH \cdot C_6H_5$	V-633	180.25
5644	**Stovaine**§	benzoyl-EtdiMe-amino-*iso*-propanol HCl	$C_6H_5CO_2C(C_2H_5) \cdot (CH_3)CH_2N(CH_3)_2 \cdot HCl$	IX-175	271.79
5645	**Strophanthidin**	....	$C_{23}H_{32}O_6 \cdot \frac{1}{2}H_2O$	....	413.52
5646	**Styrene**	phenyl-ethylene	$C_6H_5 \cdot CH:CH_2$	V-474	104.15

† See also inorganic compounds.
‡ See also No. 3443.
§ Free base, liquid, b. p. 149^{25mm}, s. organic solvents.
Salt of amber 5653
Salysal 5585
Sandoptal 1038
Santoninic anhydride 5598
Santyl 5596
Saponin. cf. glcde.

Saporubrin, cf. glcde.
Sapotoxin, cf. glcde.
Sarcine, cf. alkd.
Sarcolactic acid 3939
Sarsasaponin, cf. glcde.
Satrapol 4117

Saxin 5579
Scammonin, cf. glcde.
Schaeffer acid 4498, 4507
Schollkopf's acid 4543
Scopolamine, cf. alkd.
Secaline, cf. alkd.
Sedatine 565
Sedormid 217
Selenium diphenyl 2742

Table 7-4 (*Continued*)
PHYSICAL CONSTANTS OF ORGANIC COMPOUNDS

No.	Crystalline Form and Color	Specific Gravity	Melting Point °C.	Boiling Point °C.	Solubility in 100 Parts		
					Water	Alcohol	Ether
5604	nd. or pr.	....	204	subl.	v. sl. s.	s. h.; i. bz.	s. h. ac.
5605	pr./HNO₃	$1.207^{25°}_{4}$	134.5	294.5¹⁰⁰mm	0.1 c.; 2 h.	v. s.	v. s.
5606	col. lq.	....	7-9	199-200¹⁵mm	....	....	....
5607	oil	$1.487^{15°}$	....	183.6	v. sl. s.	v. s. CCl₄	v. s.
5608	pr./aq.	....	200± d.	....	10¹⁹°	3¹⁸°	0.6¹⁸°
5609	pr./al.	....	96	....	v. s.	v. s.	i.
5610	pr./aq. al.	....	173 d.	....	v. s.	i. abs.	i.
5611	pr./aq.	....	246 d.	....	4²⁰°	i.	i.
5612	col. pr.	....	228 d.	....	33²⁵°	i.	....
5613	mn.	....	248	....	1.5²⁰° s. h. aq.	v. sl. s.	v. sl. s.
5614	col. cr./aq.	....	199-200 d.	....	250°	....	....
5615	amor. pd.	....	....	....	i.	i. h. aq. Na₂CO₃	s. conc. KOH
5616	glass/et.	....	-99(-102)	....	i.	s. KOH	s.
5617	lq.	$0.648^{19°}_{1}$	ign. in air	26.6	i. H₂SO₄	....	....
5618	lq.	$1.326^{18.8°}_{4}$	....	201.5	d.	d.; s. chl.	s.
5619	col. lq.	$0.768^{22°}_{4}$	....	154.7	i.	....	....
5620	wh. nd./aq.	....	....	expl.	0.02³⁰°	s.	i. HNO₃
5621	cr. pd.	....	....	....	d. h.	....	....
5622	wh. pd.	....	ign. in air	....	d.	i. bz.	....
5623	wh. pd.	....	d. 245	....	d.	s.	i.; i. CS₂
5624	cr./al.	....	....	....	v. s.; d. h.	v. s.; d. h.	....
5625	nd./aq.	....	134.5	228 d.	s. h.	v. s.	v. s.
5626	col. mn. pr.	....	120 (anh.); d. 190	-3H₂O, 100	s.	s.	s.
5627	col. cr.	....	....	....	16	6	i.
5628	or. pd.	....	....	....	0.05²⁰°	i.; 13.3 aq. 15%NaCl	i.; s. aq. KI
5629	col. nd.	....	....	....	5	30	i.
5630	wh. amor.	$1.50^{21°}$	d.	....	i.	i.	i.
5631	mn./CS₂	$0.847^{69.3°}$	69.6	291¹¹⁰mm	0.03²⁵°; s. bz.; s.act.	22⁰°; 100⁵⁰°	6¹⁵°; s. CCl₄
5632	waxy	0.89	73-5	d.	v. sl. s.	s. h.	i.; i. bz.
5633	lf.	....	38	251-2¹⁰⁰mm	....	....	s.
5634	col. cr.	....	108-9	250-1¹² sl. d.	i.	s. h.	s. h.
5635	col. cr.	$0.855^{8.0°}_{4}$	72	....	i.; s. chl.	0.02²⁰°abs.	0.2¹⁵°
5636	nd./al.	....	94-5	153.5¹⁰mm	i.; s. act.	s.; s. chl.	v. s.; s. bz.
5637	col. cr.	....	23	215¹⁵ sl. d.	i.	....	....
5638	col. cr.	$0.818^{41°}_{4}$	41	274.5¹⁰⁰mm	i.	....	....
5639	pr./al.	....	48	260	i.	s. h.	s.
5640	lf./lg.	$0.798^{8.9°}_{4}$	88.4	345¹²mm	i.	sl. s. h.	sl. s. h.
5641	yel. pl.	....	86	....	sl. s. lg.	s. h.	s.
5642	mn./al.	$0.970^{12.5°}_{13}$	124	306-7	i.	0.9¹⁷° abs.	7.9¹⁴°
5643	yel. oil	....	1	145¹³mm	....	....	....
5644	nd./al.	....	175	....	50; 2 chl.	22 abs.	i.; i. act.
5645	lf./aq. al.	....	170-5*	....	sl. s.; s. bz.	s.; s. chl.	s.; i. pet.
5646	col. lq.	$0.906^{20°}$	-30.6	145.2	v. sl. s.	∞	∞

* Anhyd., m. p. 235°.

Table 7-4 (*Continued*)
PHYSICAL CONSTANTS OF ORGANIC COMPOUNDS

No.	Name	Synonym	Formula	Beil. Ref.	Formula Weight
5647	**Styrene** dibromide	α,β-diBr-ethyl-benzene	$C_6H_5CHBr\cdot CH_2Br$	V-356	263.97
5648	oxide	phenylethylene oxide	$C_6H_5\cdot CH\cdot O\cdot CH_2$ $\lvert\_\_\_\_\rvert$	XVII-49	120.15
5649	**Suberic** acid	octandioic acid	$(CH_2)_6(CO_2H)_2$	II-691	174.20
5650	aldehyde	octandial	$(CH_2)_6(CHO)_2$	I-795	142.20
5651	**Succinamic acid**	. . . .	$NH_2\cdot CO\cdot C_2H_4\cdot CO_2H$	II-613	117.11
5652	*iso*-**Succinic acid**	methyl-malonic acid	$CH_3\cdot CH(CO_2H)_2$	II-627	118.09
5653	**Succinic** acid	butandioic acid	$(CH_2\cdot CO_2H)_2$	II-601	118.09
5654	sodium salt	sodium succinate	$C_4H_4O_4Na_2\cdot 6H_2O$	*II-262	270.15
5655	aldehyde	succinic dialdehyde	$(CH_2\cdot CHO)_2$	I-767	86.09
5656	amide	succinamide	$(CH_2\cdot CO\cdot NH_2)_2$	II-614	116.12
5657	anhydride	. . . .	$(CH_2\cdot CO)_2O$	XVII-407	100.07
5658	chloride	succinyl chloride	$(CH_2\cdot COCl)_2$	II-613	154.98
5659	*N*-chloroimide	succinchlorimide	$(CH_2\cdot CO)_2N\cdot Cl$	XXI-380	133.53
5660	imide	succinimide	$(CH_2\cdot CO)_2NH$	XXI-369	99.09
5661	nitrile	ethylene dicyanide	$(CH_2\cdot CN)_2$	II-615	80.09
5662	peroxide	succinyl peroxide	$(HO_2C\cdot C_2H_4\cdot CO)_2O_2$	II-613	234.16
5663	**Sucrose**	cane sugar; beet sugar	$C_{12}H_{22}O_{11}$	XXXI-424	342.30
5664	**Sudan III**	benzene-azo-*p*-benzene-azo-β-naphthol	$C_6H_5\cdot N\!:\!N\cdot C_6H_4\cdot N\!:$ $N\cdot C_{10}H_6OH$	XVI-171	352.40
5665	**Sulfa**-diazine	2-sulfanilamido-pyrimidine	$H_2N\cdot C_6H_4\cdot SO_2\cdot NH\cdot$ $C_4H_3N_2$	. . . .	250.28
5666	guanidine	sulfanilyl-guanidine	$H_2N\cdot C_6H_4\cdot SO_2\cdot NH\cdot$ $C(:NH)\cdot NH_2$	. . . .	214.25
5667	pyrazine	2-sulfanilamido-pyrazine	$H_2N\cdot C_6H_4\cdot SO_2\cdot NH\cdot$ $C_4H_3N_2$	. . . .	250.28
5668	pyridine	2-sulfanilamido-pyridine	$H_2N\cdot C_6H_4\cdot SO_2\cdot NH\cdot$ C_5H_4N	. . . .	249.29
5669	pyridine sodium	. . . .	$C_{11}H_{10}O_2N_3SNa$	. . . .	271.28
5670	thiazole	2-sulfanilamido-thiazole	$H_2N\cdot C_6H_4\cdot SO_2\cdot NH\cdot$ C_3H_2NS	. . . .	255.32
5671	**Sulfamino**-benzoic acid (*o*)	benzoic acid-(*o*)-sulfamide	$NH_2\cdot SO_2\cdot C_6H_4\cdot CO_2H$	XI-376	201.20
5672	benzoic acid (*m*)	. . . .	$NH_2\cdot SO_2\cdot C_6H_4\cdot CO_2H$	XI-386	201.20
5673	benzoic acid (*p*)	. . . .	$NH_2\cdot SO_2\cdot C_6H_4\cdot CO_2H$	XI-390	201.20
5674	**Sulfarsphenamine**	diNa-3,3'-diNH$_2$-4, 4'-diOH-arseno-benzene-*N*-di-methylenesulfonate	$[NaSO_3\cdot CH_2\cdot$ $HN(OH)\cdot C_6H_3\cdot$ $As:]_2$	*XVI-509	598.23
5675	**Sulfo**-acetic acid	. . . .	$HO_3S\cdot CH_2CO_2H\cdot H_2O$	IV-21	158.13
5676	benzoic acid (*o*)	. . . .	$HO_3S\cdot C_6H_4\cdot CO_2H\cdot$ $3H_2O$	XI-369	256.23
5677	benzoic acid (*m*)	. . . .	$HO_3S\cdot C_6H_4\cdot CO_2H\cdot$ $2H_2O$	XI-384	238.22
5678	benzoic acid (*p*)	. . . .	$HO_3S\cdot C_6H_4\cdot CO_2H\cdot$ $3H_2O$	XI-389	256.23
5679	benzoic anhydride	(1,2)	$C_6H_4(CO)(SO_2):O$	XIX-110	184.17
5680	phenyl-3-methyl-5-pyrazolone	(1-*p*-sulfophenyl)	$HO_3S\cdot C_6H_4\cdot N\cdot$ $C_4H_5ON\cdot H_2O$	XXIV-44	272.28

Table 7-4 (*Continued*)
PHYSICAL CONSTANTS OF ORGANIC COMPOUNDS

No.	Crystalline Form and Color	Specific Gravity	Melting Point °C.	Boiling Point °C.	Solubility in 100 Parts		
					Water	Alcohol	Ether
5647	cr./al.	....	74–5	139–41^{15mm}	i.; s. lg.	s.; s. bz.	v. s.
5648	lq.	$1.052\frac{16°}{4}$	....	191–2	....	....	s.
5649	nd./aq.	$1.266\frac{25°}{4}$	140–4	279^{100mm}	0.14$^{16°}$	s.; i. chl.	0.8$^{15°}$
5650	oil	....	....	230–40 sl. d.	v. s.	....	....
5651	col. nd.	....	157	....	sl. s.	v. sl. s. abs.	i. bz.
5652	col. nd.	1.455	d. 120–35	....	66$^{20°}$	v. s.	v. s.
5653	col. mn.	$1.572\frac{25°}{4}$	189–90	235(–H_2O)	6.8$^{20°}$; 121$^{100°}$	9.9$^{15°}$	1.2$^{15°}$
5654	wh. pd.	....	–6H_2O, 120	....	2 c.	v. sl. s.	87$^{75°}$ aq.
5655	lq.	$1.069\frac{18°}{4}$	....	169–70 sl. d.	s.	s.	s.
5656	col. nd.	....	242–3	....	0.5$^{15°}$; 11$^{100°}$	i. abs.	i.
5657	col. cr.	1.503	119.6	261	v. sl. s.	v. sl. s. pet.	sl. s.
5658	col. cr.	$1.377\frac{20°}{4}$	16.7	192–3	d.	d.; s. bz.	i. pet.
5659	rhb./bz.	1.65	148	....	s. d.	0.9	sl. s.
5660	cr./act.	1.412$^{16°}$	125–6	287–8	v. s.	s.; i. chl.	v. sl. s.
5661	col.	$0.985\frac{63°}{4}$	57.2	265–7	v. s.	v. s.	sl. s.
5662	col. pl.	....	128 d.	....	33	s.; s. act.	sl. s.; i. bz.
5663	col. mn,	1.588$^{15°}$	170–86 d.	....	179$^{0°}$	0.9	i.
5664	brn. lf./ac.	....	195	....	i.; i. alk.; s. xylene	sl. s.; s. H_2SO_4	s.; s. oil; s. chl.
5665	col. mn./ aq.	....	255–6	....	0.012$^{37°}$	sl. s.	i.
5666	col. mn./ aq.	....	189–90	....	0.19$^{37°}$	sl. s.	l.
5667	col. cr.	....	255–7 d.	....	†0.005$^{17°}$	....	....
5668	col. pr./aq.	....	191–2	....	<0.03 c.; 0.05$^{37°}$	25; v. s. aq. HCl	sl. s.; v. s. alk.
5669	col. cr./al.	....	317 d.	....	63$^{25°}$	11	....
5670	brn./75% al.	....	201–2	....	0.09$^{37°}$	....	....
5671	rhb./al.	....	165–7	....	s.	s.	s.
5672	pl./aq.	....	237–8	....	v. sl. s. c.	s.	sl. s.
5673	pr./aq.	....	d. 280	....	sl. s. h.	v. sl. s. bz.	....
5674	yel. pd.	....	....	....	v. s.	v. sl. s.	....
5675	hyg. cr.	....	84–6	245 d.	s.	s.; i. chl.	i. abs.
5676	cr./aq.	....	68–9‡	–3H_2O, 105	v. s.	v. s.	i.
5677	delq. cr.	....	98‡	....	s.	s.; i. bz.	anh. s.
5678	nd./aq.	....	94	m. anh. 260	s.	s.	anh. s.
5679	col. cr./bz.	....	128–9	184–6^{18mm}	s. h.	s. chl.	s.; s. bz.
5680	nd./aq.	....	d. 320	–H_2O, 120	0.5$^{20°}$	v. sl. s. abs.	i.; v. s. alk.

† Sodium salt very soluble in aq.
‡ Anhydrous crysts., m. p. 141°.
Strophanthin, cf. glcde.
Strychnine, cf. alkd.
Styphnic acid 6312
Stypticin, cf. alkd.
Styptol, cf. alkd.

Styracin 1508
Styracol 3485
Styrolene 5646
Styron 5646
Styrone 1507
Styryl carbinol 1507
Suberane 1596

Suberene 1598
Suberone 1597
Succinamide 5656
Succinimide 5660
Succin-chlorimide 5659
Succinyl chloride 5658
Succinyl peroxide 5662
Sucrol 5112

Table 7-4 (*Continued*)
PHYSICAL CONSTANTS OF ORGANIC COMPOUNDS

No.	Name	Synonym	Formula	Beil. Ref.	Formula Weight
	Sulfo				
5681	salicylic acid (5)	. . . .	$HO_3S \cdot C_6H_3(OH)CO_2H$	XI-411	218.19
5682	salicylic Na	(Na acid salt)	$C_7H_5O_6SNa \cdot 2H_2O$	XI-412	276.20
5683	**Sulfonal**	acetone diethyl-sulfone	$(CH_3)_2C(SO_2 \cdot C_2H_5)_2$	I-662	228.33
5684	**Sylvestrine** (*d* or *l*)	. . . .	$C_{10}H_{16}$	V-125	136.24
5685	**Synthalin**	decamethylene-di-guanidine di-HCl	$[H_2N \cdot C(:NH)NH \cdot (CH_2)_5]_2 \cdot 2HCl$	**IV-712	329.32
5686	**Syringic acid**	gallic acid-3,5-dimethyl ether	$HO(CH_3O)_2 \cdot C_6H_2 \cdot CO_2H$	X-480	198.18
5687	**Tannin**	digallic acid	$(HO)_3C_6H_2 \cdot CO_2 \cdot C_6H_2(OH)_2CO_2H$	. . . .	322.23
5688	**Tartaric** acid (meso)	*i*-tartaric acid	$(CHOH \cdot CO_2H)_2$	III-528	150.09
5689	Ca salt (meso)	Ca meso-tartrate	$C_4H_4O_6Ca \cdot 3H_2O$	III-529	242.20
5690	acid (*racemic*) (*dl*)	"Traubensaüre"	$(CHOH \cdot CO_2H)_2 \cdot H_2O$	III-522	168.10
5691	acid (*d* or *l*)	"Weinsaüre"	$(CHOH \cdot CO_2H)_2$	III-481	150.09
5692	amide (*d*)	tartramide	$(CHOH \cdot CONH_2)_2$	III-520	148.12
5693	**Tartramidic acid**	. . . .	$NH_2 \cdot CO \cdot (CHOH)_2 \cdot CO_2H$	III-520	149.10
5694	**Tartrazine**	hydrazine yellow	$C_{16}H_9O_9N_4S_2Na_3$	XXV-252	534.37
5695	**Tartronic acid**	hydroxymalonic acid	$HO \cdot CH(CO_2H)_2 \cdot \frac{1}{2}H_2O$	III-415	129.07
5696	**Taurine**	. . . .	$NH_2 \cdot CH_2 \cdot CH_2 \cdot SO_3H$	IV-528	125.15
5697	**Taurocholic acid**	cholaic acid	$C_{26}H_{45}O_7NS \cdot H_2O$	. . . .	533.75
5698	Na salt	. . . .	$C_{26}H_{44}O_7NSNa$	. . . .	537.70
5699	**Tephrosin**	hydroxy-deguelin	$C_{23}H_{22}O_7$	. . . .	410.43
5700	**Teraconic acid**	γ-di Me-itaconic acid	$(CH_3)_2C:C(CO_2H) \cdot CH_2 \cdot CO_2H$	II-786	158.16
5701	**Terebic acid**	terebinic acid	$(CH_3)_2C \cdot CH(CO_2H) \cdot$ $\underset{\rule{2.5cm}{0.4pt}}{\overline{CH_2 \cdot CO_2}}$	XVIII-377	158.16
5702	**Terpenylic acid** (*dl*)	terpenolic acid	$C_8H_{12}O_4 \cdot H_2O$	XVIII-385	190.20
5703	**Terpin hydrate** (*cis*)	. . . .	$C_{10}H_{20}O_2 \cdot H_2O$	VI-745	190.29
5704	**Terpinene** (α)	*p*-menthadiene (1,3)	$C_{10}H_{16}$	V-126	136.24
5705	**Terpinene** (β)	*p*-menthadiene (3,1)(7)	$C_{10}H_{16}$	V-132	136.24
5706	**Terpinenol**-4 (*d*)	terpenol	$C_{10}H_{18}O$	VI-55	154.25
5707	**Terpineol** (α)(*d* or *l*)	. . . .	$C_{10}H_{18}O$	VI-56	154.25
5708	**Terpineol** (α)(*dl*)	. . . .	$C_{10}H_{18}O$	VI-58	154.25
5709	**Terpinolene**	*p*-menthadiene (1,4)(8)	$C_{10}H_{16}$	V-133	136.24
5710	**Terpinyl acetate** (α)	(*dl*)	$C_{10}H_{17} \cdot O_2C \cdot CH_3$	VI-60	196.29
5711	**Testerone** (*trans*)	. . . .	$C_{19}H_{28}O_2$	. . . .	288.43
5712	**Testerone** (*cis*)	. . . .	$C_{19}H_{28}O_2$	. . . .	288.43
5713	**Tetraamino**-3,3′-dimethyl-di-phenylmethane	(4,6,4′,6′-tetraNH_2)	$[CH_3(NH_2)_2C_6H_2]_2:CH_2$	XIII-342	256.35
5714	**Tetrabromo**-aniline	(2,3,4,6)	$Br_4C_6H \cdot NH_2$	XII-668	408.73
5715	benzene (1,2,3,5)	. . . .	$Br_4C_6H_2$	V-214	393.72

Table 7-4 (*Continued*)
PHYSICAL CONSTANTS OF ORGANIC COMPOUNDS

No.	Crystalline Form and Color	Specific Gravity	Melting Point °C.	Boiling Point °C.	Solubility in 100 Parts		
					Water	Alcohol	Ether
5681	hyg.nd./aq.	....	120d.(anh.)	....	s.	s.	s.
5682	col. cr.	....	....	....	3.3	i.	....
5683	pr./al.	$1.183\frac{13\,2°}{4}$	127-8	300 sl. d.	$0.2^{15°}$	$1.5^{15°}$; s. bz.	$0.7^{15°}$; s. chl.
5684	lq.	$0.863\frac{20°}{4}$	....	176-7	....	....	....
5685	cr./al. et.	....	193	....	s.	....	....
5686	nd./aq. or et.	....	205-7	....	v. sl. s.	s.; s. chl.	s.
5687	amor. pd.	....	200 d.	....	s.	sl. s. abs.	i. abs.
5688	cr.	**1.737**	159-60	....	$120^{15°}$	....	....
5689	cr./aq.	....	$-3H_2O$, 170	....	0.17 h.	0.03^{18} ac.	$0.09^{100°}$ ac.
5690	tri.	$1.697\frac{20°}{4}$	205-6	$-H_2O$, 100	$20.6^{20°}$; $185^{100°}$	20^{0}	0.09
5691	mn.	$1.760\frac{20°}{4}$	168-70	d.	$139^{20°}$; $343^{100°}$	$25^{15°}$ abs.	$0.4^{15°}$
5692	rhb.	....	195 d.	....	i. bz.	$0.04^{12°}$	i.
5693	rhb./aq.	....	171-2	....	....	....	....
5694	or. yel. pd.	....	....	....	v. s.	i.	....
5695	col. pr./aq.	....	d. 155-8; $-\frac{1}{2}H_2O$,60	subl. 110	v. s.	v. s.	i.; anh. s.
5696	....	....	>240 d.*	....	$6.4^{12°}$	$0.004^{17°}$	i.
5697	delq. nd.	....	125±, d.	....	d. h.	s.	v. sl. s.
5698	yel. pd.	....	....	....	v. s.	v. s. h.	sl. s.
5699	col. pr.	....	198	....	i.	s. act.	s.; s. chl.
5700	tri./et.	....	160-1 d.	....	v. s. h.	v. s.	sl. s.; v. sl. s. bz.
5701	mn./al.	$0.815\frac{24°}{4}$	174-5	....	s. h.	s.	$1.7^{10°}$
5702	col. cr./aq.	....	56; 90 (anh.)	subl. 130	s. h.	....	....
5703	rhb.	....	$-H_2O$ > 117	....	$0.4^{15°}$; 3.3 h.	$10^{15°}$; 50 h.	$1^{15°}$; $0.5^{15°}$ chl.
5704	lq.	$0.834\frac{20°}{4}$	....	181.5	i.	∞	∞
5705	lq.	$0.838^{22°}$	....	173-4	i.	∞	∞
5706	cr.	$0.936^{15°}$	38-40	$219\text{-}21^{760mm}$	i.	v. s.	v. s.
5707	col. cr.	$0.935^{15°}$	38-40	219-21	i.	v. s.	v. s.
5708	col. cr.	$0.935\frac{20°}{20}$	35	$218\text{-}9^{752mm}$	i.	v. s.	v. s.
5709	lq.	$0.862\frac{20°}{4}$	....	186-7	i.	∞	∞
5710	lq.	$0.966\frac{20°}{4}$	<-50	220 d.	i.	20	....
5711	col.nd./act.	....	154-5	....	i.	s.	s.
5712	cr.	....	220-1	....	i.	s.	s.
5713	lf./aq.	....	203-4	....	v. sl. s.	v. sl. s.	....
5714	nd.	....	116-7	....	....	v. s.	v. s.
5715	nd./al.	....	98.5	329	i.	s. h.	v.s.; v.s.bz.

* The sulfonimide, m. p. 88°.

Sulfanilyl-guanidine 5666	Sulfhydryl-benzoic acid 5918	Sulfon-methane 5683
Sulfarsenol 5674	Sulfo-anthraquinone 559	Sulfur yellow-S 3241
Sulfenazoxine 679	Sulfo-benzide 2750	Sulfuric ether 2911
Sulfenthal 5127	Sulfo-cyanic acid 5897	Superpalite 6159
	Sulfon-ethylmethane 6326	Suprarenine 157

Table 7-4 (*Continued*)
PHYSICAL CONSTANTS OF ORGANIC COMPOUNDS

No.	Name	Synonym	Formula	Beil. Ref.	Formula Weight
	Tetrabromo				
5716	benzene (1,2,4,5)	....	$Br_4C_6H_2$	V-214	393.72
5717	*o*-cresol	....	$Br_4C_6(OH)\cdot CH_3$	VI-362	423.75
5718	*m*-cresol-sulfon-phthalein	brom cresol green	$C_{21}H_{14}O_5Br_4S$	....	698.04
5719	ethane (1,1,2,2) (*sym*)	acetylene tetra-bromide	$Br_2CH\cdot CHBr_2$	I-94	345.67
5720	ethane (1,1,1,2)	(*uns*)	$Br_3C\cdot CH_2Br$	I-94	345.67
5721	ethylene	ethylene tetraBr	$Br_2C:CBr_2$	I-192	343.66
5722	phenolphthalein	(3',5',3'',5'')	$C_{20}H_{10}O_4Br_4$	XVIII-149	633.94
5723	phenolphthalein	(sodium salt)	$C_{20}H_8O_4Br_4Na_2$	....	677.90
5724	phenolsulfon-phthalein	brom phenol blue	$C_{19}H_{10}O_5Br_4S$	*XIX-649	669.99
5725	phthalic anhydride	....	$C_6Br_4\cdot CO\cdot O\cdot CO$	XVII-485	463.72
5726	quinone	bromanil	$O:C_6Br_4:O$	VII-642	423.70
5727	**Tetrabutylammonium iodide** (*n*)	....	$(C_4H_9)_4N\cdot I$	IV-157	369.37
5728	**Tetrachloro**-acetone	(*s*)	$(Cl_2CH)_2CO\cdot 4H_2O$	I-656	267.92
5729	acetone (*s*)	....	$(Cl_2CH)_2CO$	I-656	195.86
5730	aniline (2,3,4,5)	....	$Cl_4C_6H\cdot NH_2$	XII-630	230.91
5731	aniline (2,3,5,6)	....	$Cl_4C_6H\cdot NH_2$	**XII-340	230.91
5732	aniline (2,3,4,6)	....	$Cl_4C_6H\cdot NH_2$	XII-630	230.91
5733	benzene (1,2,3,4)	....	$Cl_4C_6H_2$	V-204	215.89
5734	benzene (1,2,3,5)	....	$Cl_4C_6H_2$	V-204	215.89
5735	benzene (1,2,4,5)	....	$Cl_4C_6H_2$	V-205	215.89
5736	1,2-dibromo-ethane	(1,1,2,2)	$Cl_2BrC\cdot CBrCl_2$	I-93	325.65
5737	1,1-dibromo-ethane	(1,2,2,2)	$Cl_3C\cdot CClBr_2$	I-93	325.65
5738	diphenyl (2,4,2',4')	....	$(C_6H_3Cl_2)_2$	V-579	291.99
5739	ethane (*s*)	acetylene tetraCl	$Cl_2CH\cdot CHCl_2$	I-86	167.85
5740	ethane (*as*)	(1,1,1,2-tetraCl)	$Cl_3C\cdot CH_2Cl$	I-86	167.85
5741	ethylene	ethylene tetraCl	$Cl_2C:CCl_2$	I-187	165.83
5742	hydroquinone	....	$(HO)_2C_6Cl_4$	VI-851	247.89
5743	phenol (2,3,4,6)	....	$HO\cdot C_6HCl_4$	VI-193	231.89
5744	phenolphthalein	(4,5,6,7)	$C_{20}H_{10}O_4Cl_4$	XVIII-148	456.11
5745	*o*-phthalic acid	....	$Cl_4C_6(CO_2H)_2\cdot \frac{1}{2}H_2O$	IX-819	312.92
5746	quinone	chloranil	$O:C_6Cl_4:O$	VII-636	245.88
5747	**Tetracosane** (*n*)	....	$CH_3\cdot (CH_2)_{22}\cdot CH_3$	I-175	338.67
5748	**Tetradecane** (*n*)	....	$CH_3\cdot (CH_2)_{12}\cdot CH_3$	I-171	198.40
5749	**Tetradecyl** acetate	(*n*)	$CH_3CO_2\cdot C_{14}H_{29}$	II-136	256.43
5750	alcohol (*n*)	tetradecanol-1	$CH_3(CH_2)_{12}CH_2OH$	I-428	214.39
5751	**Tetradecylene** (*α*)	tetradecene-1	$CH_3(CH_2)_{11}CH:CH_2$	I-226	196.38
5752	**Tetradecyne**-2	....	$CH_3(CH_2)_{10}C:C\cdot CH_3$	I-262	194.36
5753	**Tetraethanolam-monium hydroxide**	....	$(HO\cdot CH_2CH_2)_4NOH$	IV-285	211.26
5754	**Tetraethylam-monium bromide**	....	$(C_2H_5)_4NBr$	IV-104	210.16
5755	ohloride	....	$(C_2H_5)_4NCl\cdot 4H_2O$	IV-104	237.77

Table 7-4 (*Continued*)
PHYSICAL CONSTANTS OF ORGANIC COMPOUNDS

No.	Crystalline Form and Color	Specific Gravity	Melting Point °C.	Boiling Point °C.	Solubility in 100 Parts		
					Water	Alcohol	Ether
5716	mn./CS_2	$3.027^{20°}$	178-80	....	i.	....	....
5717	nd./ac.	....	207-8	....	sl. s. ac.	s.; s. alk.	s.
5718	lt. yel. pd.	....	....	....	sl. s.	s.; s. alk.	i.
5719	col. lq.	$2.964^{20°}_{4}$	0-2; d. >190	151^{54mm}	i.; ∞ ac.	∞; ∞ chl.	∞; ∞ aniline
5720	col. lq.	$2.875^{20°}_{4}$	0; d. 175	$103.5^{13.5mm}$	....	s.	....
5721	cr. pl.	....	56-7	226-7	....	....	....
5722	col. nd./al.	....	294	....	i.; s. alk.	v. sl. s.	s.
5723	b. pd.	....	....	....	v. s.	sl. s.	i.
5724	cr./ac. act.	....	270-1 d.	....	0.07	sl. s.	i.; s. alk.
5725	cr./ac. + xylene	....	275-80	....	v. sl. s. ac.	v. sl. s. bz.	s. $PhNO_2$
5726	yel. mn./bz.	....	300	subl.	i.	s. h.	sl. s.
5727	lf./bz.	....	144-5	....	sl. s.	s.	s.
5728	tri.	....	48-9	....	....	....	....
5729	lq.	....	....	150-2 sl. d.	v. s. bz.	v. s.	v. s.
5730	nd./al.	....	118-20	....	s. bz.	s.; s. ac.	s.
5731	cr.	....	110	....	i.	s.	....
5732	nd./lg.	....	88	....	s. CS_2	s.	s. lg.
5733	nd.	....	46-7	254^{761mm}	i.	sl. s.	v. s.
5734	nd./al.	....	54-5	246	i.	sl. s. c.	v. s. CS_2
5735	nd./et.	$1.858^{22°}$	138-40	240-6	i.	s. c.	s. bz.
5736	rhb.	2.713	200-5 d.	....	....	....	....
5737	rhb.	2.794	subl.	....	....	s. h.	s.
5738	cr.	....	83	....	$0.29^{20°}$	v. s. h.	sl. s. lg.
5739	col. lq.	$1.600^{20°}_{4}$	−36	146.2	$0.29^{20°}$	∞	∞
5740	lq.	$1.588^{20°}_{4}$	....	129-30	$0.02^{20°}$	∞	∞
5741	col. lq.	$1.631^{15°}_{4}$	−22.4.	121.2	i.	∞; ∞ chl.	∞; ∞ bz.
5742	mn.	....	238-40	subl.	i.; s. act.	$20^{25°}$	$20^{25°}$
5743	nd./lg.	$1.6^{60°}_{4}$	69-70	164^{23mm}	v. sl. s.	v. s.	v. s.
5744	pl./Me al.	....	>300	....	i.; s. alk.	s.; i. ohl.	s.; i. bz.
5745	cr./aq.	....	d. −1½H_2O^*	....	$0.6^{14°}$; $3.0^{99°}$	s.	s.; v. s. act.
5746	yel.mn./bz.	....	290†	subl. 80°	i. $bz.^{25°}$	i. c.; sl. s. h.	i. c.; sl. s. CCl_4
5747	cr.	$0.779^{51°}_{4}$	51.1	324	....	$8.4^{15°}$ chl.	s.
5748	col. lq.	$0.765^{20°}_{4}$	5.5	252.5	i.	v. s.	v. s.
5749	col. cr.	....	12-3	$176-7^{15mm}$	....	....	....
5750	cr.	$0.824^{38°}_{4}$	38	$167-70^{15mm}$	<0.02	sl. s.	s.
5751	lq.	$0.775^{15°}_{4}$	−12	127^{15mm}	....	....	....
5752	cr.	$0.800^{15°}_{4}$	6.5	134^{15mm}	....	....	....
5753	....	....	....	....	∞		
5754	cr./abs. al.	....	....	....	....	v. s.	s. chl.
5755	mn.	$1.112^{25°}_{4}$	37.5	....	$141^{25°}$	$8.2^{25°}$ chl.	....

* Anhydride, m. p. 255°.
† Sealed tube.
Terpenolic acid 5702
Terphenyl 2705
Tethrothalein 5722

Tetraacetyl-glucosyl chloride 44
Tetraanhydro-berberine, cf. alkd.
Tetrabase 5834
Tetrabromo-fluorescein 2875-6
Tetrabromo-methane 1242

Tetracaine 5032
Tetrachloro-methane 1243
Tetradecanal 4451
Tetradecanoic acid 4450
Tetradecanol 5750

Table 7-4 (*Continued*)
PHYSICAL CONSTANTS OF ORGANIC COMPOUNDS

No.	Name	Synonym	Formula	Beil. Ref.	Formula Weight
5756	**Tetraethylammo-nium** hydroxide	....	$(C_2H_5)_4NOH$	IV-103	147.26
5757	iodide	....	$(C_2H_5)_4NI$	IV-104	257.16
5758	**Tetraethyl-**benzene	(1,2,4,5)	$(C_2H_5)_4C_6H_2$	V-455	190.33
5759	benzene (1,2,3,4)	....	$(C_2H_5)_4C_6H_2$	V-455	190.33
5760	diaminoben-zophenone	(4,4')	$[(C_2H_5)_2N\cdot C_6H_4]_2CO$	XIV-98	324.47
5761	diaminodiphenyl-methane	(4,4')	$[(C_2H_5)_2N\cdot C_6H_4]_2CH_2$	XIII-242	310.49
5762	diaminotriphenyl carbinol	brilliant green base	$[(C_2H_5)_2N\cdot C_6H_4]_2\cdot$ $C(OH)\cdot C_6H_5$	XIII-746	402.58
5763	diaminotriphenyl carbinol sulfate	brilliant green dye salt	$(C_{27}H_{33}N_2)\cdot HSO_4$	XIII-746	482.65
5764	ethanetetracarb-oxylate (*sym*)	....	$[(C_2H_5O_2C)_2CH]_2$	II-858	318.33
5765	propanetetra-carboxylate	$(\alpha,\alpha,\beta,\gamma)$	$(CH_2\cdot CH\cdot CH)\cdot$ $(CO_2\cdot C_2H_5)_4$	II-859	332.35
5766	rhodamine	rhodamine B	$C_{28}H_{31}O_3N_2Cl$	XIX-346	479.02
5767	urea	....	$[(C_2H_5)_2N]_2CO$	IV-120	172.27
5768	**Tetraethylene** glycol	....	$(\cdot CH_2OCH_2\cdot)_3$: $(\cdot CH_2OH)_2$	I-468	194.23
5769	glycol dimethyl ether	diMeO-tetraglycol	$(CH_3O\cdot CH_2\cdot CH_2\cdot$ $O\cdot CH_2\cdot CH_2)_2O$	....	222.28
5770	pentamine	....	$NH_2\cdot (CH_2CH_2NH)_3\cdot$ $CH_2\cdot CH_2\cdot NH_2$	....	189.31
5771	**Tetraglycol dichloride**	....	$(Cl\cdot CH_2\cdot CH_2\cdot O\cdot$ $CH_2\cdot CH_2)_2O$	....	231.12
5772	**Tetrahydro-**ben-zoic acid (Δ_1)	....	$CH_2(CH_2)_3CH:C\cdot$ $\overline{}$ CO_2H	IX-41	126.16
5773	furan	tetramethylene oxide	$CH_2(CH_2)_2CH_2\cdot O$ $\overline{}$	XVII-10	72.11
5774	furfuryl acetate	....	$CH_3CO_2\cdot CH_2\cdot C_4H_7O$	**XVII-107	144.17
5775	furfuryl alcohol	....	$C_4H_7O\cdot CH_2OH$	**XVII-106	102.13
5776	furfuryl benzoate	....	$C_6H_5CO_2\cdot CH_2\cdot C_4H_7O$	....	206.24
5777	furfuryl butyrate	....	$C_3H_7CO_2\cdot CH_2\cdot C_4H_7O$	....	172.23
5778	furfuryl *n*-cap-roate	....	$C_5H_{11}CO_2\cdot CH_2\cdot$ C_4H_7O	....	200.28
5779	furfuryl lactate	....	$CH_3\cdot CHOH\cdot CO_2\cdot$ $CH_2\cdot C_4H_7O$	....	174.20
5780	furfuryl laurate	....	$C_{11}H_{23}CO_2\cdot CH_2\cdot$ C_4H_7O	....	284.44
5781	furfuryl maleate	....	$(:CH\cdot CO_2\cdot CH_2\cdot$ $C_4H_7O)_2$	....	284.31
5782	furfuryl palmitate	....	$C_{15}H_{31}CO_2\cdot CH_2\cdot$ C_4H_7O	....	340.55
5783	furfuryl propionate	....	$C_2H_5CO_2\cdot CH_2\cdot C_4H_7O$	....	158.20
5784	furfuryl salicylate	....	$HO\cdot C_6H_4CO_2\cdot$ $CH_2\cdot C_4H_7O$	....	222.24
5785	furfuryl succinate	....	$(\cdot CH_2\cdot CO_2\cdot CH_2\cdot$ $C_4H_7O)_2$	....	286.33
5786	naphthalene (*trans*) (1,2,3,4)	"Tetralin"	$C_6H_4CH_2(CH_2)_2CH_2$ $\overline{}$	V-491	132.21

Table 7-4 (Continued)
PHYSICAL CONSTANTS OF ORGANIC COMPOUNDS

No.	Crystalline Form and Color	Specific Gravity	Melting Point °C.	Boiling Point °C.	Solubility in 100 Parts		
					Water	Alcohol	Ether
5756	only in solns.*	....	*	d.	s.	....	....
5757	col. cr./aq.	$1.5594°$	>200	....	$45^{25°}$	$1.6^{25°}$ chl.	i.; s. al.
5758	lq.	$0.888\frac{16°}{4}$	13	250	i.	s. abs.	v. s.
5759	lq.	$0.887\frac{20°}{4}$	11.6	248	i.	s. abs.	v. s.
5760	lf./al.	....	95-6	....	....	....	....
5761	cr./al.	....	41-2	253^{10mm}	....	....	....
5762	red brn.	....	....	....	v. sl. s.	s.	s. aq. a.
5763	rhb. gold nd.	....	....	....	v. s.	v. s.	....
5764	tet. pr.	$1.064^{79.5°}$	76	305 d.	i. pet. ..	s.	s.
5765	oil	$1.118\frac{20°}{4}$	....	$200\text{-}1^{14mm}$	....	....	....
5766	lf./HCl	....	....	....	v. s.	v. s.	sl. s. alk.
5767	lq.	$0.886\frac{20°}{4}$	....	210-5	i.	s. a.	i. alk.
5768	lq.	$1.125\frac{20°}{20}$	....	327-8	∞	....	....
5769	lq.	$1.013\frac{20°}{20}$	....	275.8	∞	....	....
5770	lq.	$0.999\frac{20°}{20}$	....	333	∞	....	....
5771	lq.	$1.186\frac{20°}{20}$	....	114^{2mm}	sl. s.	....	....
5772	cr.	$1.072\frac{47°}{4}$	29	240-3	$0.7^{20°}$	....	....
5773	col. lq.	$0.888\frac{21°}{4}$	−108.5	65-6	s.	s.	s.
5774	col. lq.	$1.062\frac{25°}{4}$	....	$192\text{-}4^{740mm}$	∞	∞	∞; ∞ chl.
5775	col. lq.	$1.050\frac{20°}{4}$	....	$177\text{-}8^{743mm}$	∞	∞	∞
5776	lq.	$1.137\frac{20°}{0}$	....	$300\text{-}2^{750mm}$	i.	∞	∞; ∞ chl.
5777	lq.	$1.012\frac{20°}{0}$	....	225-7	i.	∞	∞; ∞ chl.
5778	ool. lq.	....	....	$141\text{-}3^{19mm}$	i.	s.	s.
5779	yel. lq.	....	....	$146\text{-}9^{18mm}$	s.	s.	s.
5780	pa. yel. lq.	....	....	$184\text{-}6^{6mm}$	i.	s.	s.
5781	yel. viscous lq.	....	....	$190\text{-}3^{2mm}$	sl. s.	s.	s.
5782	pa. yel. lq.	....	20-2	$195\text{-}81.5mm$	i.	s.	s.
5783	lq.	$1.044\frac{20°}{4}$	....	$204\text{-}7^{756mm}$	....	....	....
5784	pa. yel. lq.	....	....	$131\text{-}3^{2mm}$	i.	s.	s.
5785	pa. yel. lq.	....	....	$219\text{-}21^{9mm}$	i.	s.	s.
5786	col. lq.	$0.970\frac{20°}{4}$	−31.5	194	i.; ∞ bz.	s.; ∞ act.	s.

* Crysts. $+ 4H_2O$, m. p. 49-50°; $+ 6H_2O$, m. p. 55°.
Tetrahydro-nicotinic acid, cf. alkd.
Tetrahydro-pyrrole 5523
Tetrahydro-toluene 4207-8
Tetrahydroxy-flavanone 2891
Tetraiodo-fluorescein 2896
Tetraiodo-methane 1245

Table 7-4 (*Continued*)
PHYSICAL CONSTANTS OF ORGANIC COMPOUNDS

No.	Name	Synonym	Formula	Beil. Ref.	Formula Weight
	Tetrahydro				
5787	α-naphthol (*ac.*)	....	$C_6H_4\cdot CHOH\cdot$ $\mid$ _____ $(CH_2)_2\cdot CH_2$ $\mid$ ____	**VI-541	148.21
5788	α-naphthol (*ar.*)	....	$(CH_2)_4:C_6H_3OH$	VI-578	148.21
5789	β-naphthol (*ac.*)	....	$C_6H_4(CH_2)_2CHOHCH_2$ $\mid$ _____ $\mid$	VI-579	148.21
5790	β-naphthol (*ar.*)	....	$(CH_2)_4:C_6H_3OH$	VI-579	148.21
5791	β-naphthoylaceto- nitrile (*ar.*)	....	$(CH_2)_4:C_6H_3\cdot CO\cdot$ $CH_2\cdot CN$	....	199.25
5792	α-naphthylamine (*ac.*)	....	$C_6H_4\cdot CH(NH_2)\cdot$ $\mid$ _____ $(CH_2)_2\cdot CH_2$ $\mid$ ____	XII-1200	147.22
5793	α-naphthylamine	(*ar.*)	$(CH_2)_4:C_6H_3\cdot NH_2$	XII-1197	147.22
5794	β-naphthylamine (*ac.*)(*dl*)	....	$C_6H_4CH_2CH(NH_2)\cdot$ $\mid$ _____ $CH_2\cdot CH_2$ ____$\mid$	XII-1200	147.22
5795	β-naphthylamine	(*ar.*)	$(CH_2)_4:C_6H_3\cdot NH_2$	XII-1198	147.22
5796	o-phthalic acid (Δ_1)	....	$C_8H_{10}O_4$	IX-770	170.17
5797	quinoline (1,2,3,4)	....	$C_9H_{10}NH$	XX-262	133.19
5798	quinone (*p*)	....	$OC:(CH_2)_4:CO$	VII-556	112.13
5799	**Tetrahydroxy-** anthraquinone	rufiopin (1,2,5,6)	$C_{14}H_4O_2(OH)_4$	VIII-549	272.22
5800	anthraquinone (1,3,5,7)	anthrachryson	$C_{14}H_4O_2(OH)_4\cdot 2H_2O$	VIII-551	308.25
5801	anthraquinone (1,2,5,8)	quinalizarin; alizarin bordeaux	$C_{14}H_4O_2(OH)_4$	VIII-549	272.22
5802	benzene (1,2,4,5)	....	$(HO)_4C_6H_2$	VI-1155	142.11
5803	benzene (1,2,3,5)	....	$(HO)_4C_6H_2$	VI-1154	142.11
5804	3,6-diaminoben- zene HCl	(1,2,4,5)	$(HO)_4C_6(NH_2)_2\cdot 2HCl$	XIII-842	245.06
5805	flavone (3,7,3′,4′)	fisetin	$(HO)_2C_9H_3O_2\cdot C_6H_3:$ $(OH)_2$	XVIII-221	286.24
5806	flavanol (5,7,2′,4′)	morin; 3,5,7,2′,4′- pentaOH-flavone	$C_{15}H_{10}O_7\cdot 2H_2O$	XVIII-239	338.27
5807	flavanol (5,7,3′,4′)	quercetin	$C_{15}H_{10}O_7\cdot 2H_2O$	XVIII-242	338.27
5808	quinone	....	$(HO)_4C_6O_2$	VIII-534	172.10
5809	**Tetraiodo-**benzene	(1,2,3,4)	$I_4C_6H_2$	V-229	581.70
5810	benzene (1,2,3,5)	....	$I_4C_6H_2$	V-229	581.70
5811	benzene (1,2,4,5)	....	$I_4C_6H_2$	V-229	581.70
5812	ethylene	periodo-ethene	$I_2C:CI_2$	I-195	531.64
5813	phenolphthalein	nosophen; Iodophene	$C_{20}H_{10}O_4I_4$	XVIII-151	821.92
5814	phenolphthalein	(Na salt); Iodeikon	$C_{20}H_8O_4I_4Na_2\cdot 3H_2O$	XVIII-151	919.93
5815	phenolsulfon- phthalein	....	$C_{19}H_{10}O_5I_4S$	....	857.97
5816	phthalic anhydride	....	$I_4C_6:(CO)_2O$	XVII-486	651.71
5817	pyrrole	Iodol	I_4C_4NH	XX-168	570.68

Table 7-4 (Continued)
PHYSICAL CONSTANTS OF ORGANIC COMPOUNDS

No.	Crystalline Form and Color	Specific Gravity	Melting Point °C.	Boiling Point °C.	Solubility in 100 Parts		
					Water	Alcohol	Ether
5787	col. lq.	$1.090\frac{17}{4}°$	....	140^{17mm}	....	....	....
5788	mn.	... .	68.5-9.0	$264-5^{705mm}$	sl. s. h.	v. s.	v. s.
5789	oil	$1.071\frac{20}{4}°$	....	264^{716mm}	v. sl. s.	v. s.	v. s.
5790	nd./lg.	....	58-9	275-6	v. sl. s.	v. s.	v. s.
5791	pa. yel. cr.	....	98-100	....	i.	sl. s.	s.
5792	oil	....	....	246.5^{714mm}	s. h.	s.	s.
5793	oil	$1.067\frac{15}{15}°$	....	275^{12mm}	v. sl. s.	s.	s.
5794	lq.	$1.034\frac{15}{15}°$	....	250^{710} sl. d.	sl. s. h.	s.	s.
5795	nd./lg.	$1.029\frac{22}{4}°$	38	$275-7^{713mm}$	....	s.	s.
5796	lf.	....	120(−H$_2$O)	....	v. s.	....	....
5797	cr.	$1.070^{4°}$	15-6	251	s.	∞	∞
5798	mn./aq.	....	78	subl. 100	s.	s.	s.
5799	yel. red	....	subl. d.	....	sl. s. h.; s.H$_2$SO$_4$	s.; s. h. ac.	v. sl. s.
5800	yel. nd.	....	>360	−H$_2$O, 150	i.; i. CS$_2$	sl. s.; s. ac.	v. sl. s.
5801	red nd./ PhNO$_2$	....	>275	subl.	s. alk.; s. H$_2$SO$_4$	....	....
5802	lf./ac.	....	215-20	....	v. s.; sl. s. HCl	v. s.	v. s.
5803	nd./aq.	....	165	....	v. s.	v. s.	i. chl., bz.
5804	nd.	....	....	....	v. s.	i. HCl	....
5805	yel. nd.	....	>360 d.	....	i. c.; s. alk.	s.; s. act.; sl. s. pet.	sl. s.; sl. s. bz., chl.
5806	yel. nd./al.	....	285 (anh.)	....	$0.03^{20°}$; $0.1^{100°}$	s.; i. CS$_2$	sl. s.; s. alk.
5807	yel. nd.	....	313-4 (anh.)	....	i. c.; v. sl. s. h.	0.4 c.; 5.5 h.	v. sl. s.; s. alk.
5808	b. bl.	....	....	....	v. s. h.	v. s.	sl. s.
5809	pr./CS$_2$	... .	136	subl.	s. chl.	s.	s.
5810	pr./et.	....	148	subl.	v. s. h. ac.	sl. s.	sl. s.
5811	nd./et.	....	254	subl. in vac.	v. s. CS$_2$	v. sl. s.	v. sl. s.
5812	yel. mn.	$2.983^{20°}$	187-92	subl. in vac.	v. s. CS$_2$	v. sl. s. c.	s. ac., bz.
5813	amor. pd.	....	d. 220±	....	i.; s. alk.	sl. s.	s.; s. chl.
5814	pa. b. hyg.	....	d. by CO$_2$	....	14	sl. s.	....
5815	col. amor.	....	....	....	i.	i.	i.
5816	yel. cr./ac.	....	329-31	subl.	s. PhOH	s. PhNO$_2$	i.
5817	yel./aq. al.	....	d. 140-50	....	0.02	$5.8^{15°}$; s. h.	50

Table 7-4 *(Continued)*
PHYSICAL CONSTANTS OF ORGANIC COMPOUNDS

No.	Name	Synonym	Formula	Beil. Ref.	Formula Weight
5818	**Tetramethyl-** aminobenzene	2,3,4,5-tetramethyl- aniline	$(CH_3)_4C_6H \cdot NH_2$	XII-1175	149.24
5819	aminobenzene	(2,3,5,6); duridine	$(CH_3)_4C_6H \cdot NH_2$	XII-1177	149 24
5820	aminobenzene	(2,3,4,6); *iso-* duridine	$(CH_3)_4C_6H \cdot NH_2$	XII-1175	149 24
5821	ammonium bromide	....	$(CH_3)_4N \cdot Br$	IV-51	154.06
5822	ammonium chloride	....	$(CH_3)_4N \cdot Cl$	IV-51	109.60
5823	ammonium hydroxide	....	$(CH_3)_4NOH \cdot 5H_2O$	IV-50	181.23
5824	amm. hydroxide	....	$(CH_3)_4NOH \cdot 3H_2O$	IV-50	145.19
5825	amm. hydroxide	....	$(CH_3)_4NOH \cdot H_2O$	IV-50	109.17
5826	ammonium iodide	....	$(CH_3)_4N \cdot I$	IV-51	201.05
5827	benzene (1,2,3,4)	prehnitol	$(CH_3)_4C_6H_2$	V-430	134.22
5828	benzene (1,2,3,5)	isodurene	$(CH_3)_4C_6H_2$	V-430	134.22
5829	benzene (1,2,4,5)	durene	$(CH_3)_4C_6H_2$	V-431	134.22
5830	benzidine	....	$[(CH_3)_2N \cdot C_6H_4 \cdot]_2$	XIII-221	240.35
5831	benzoquinone	duroquinone	$(CH_3)_4C_6O_2$	VII-669	164.21
5832	diaminobutane(α,δ)	....	$[(CH_3)_2N \cdot CH_2CH_2 \cdot]_2$	IV-265	144.26
5833	diaminodiphenyl- amine (*p*)	Bindschedler's green leuco base	$[(CH_3)_2N \cdot C_6H_4]_2NH$	XIII-112	255.37
5834	diaminodiphenyl methane (*p,p'*)	Michler's hydride	$[(CH_3)_2N \cdot C_6H_4]_2CH_2$	XIII-239	254.38
5835	diamino-4''-hy- droxy-triphenyl- methane	(4,4')	$[(CH_3)_2N \cdot C_6H_4]_2CH \cdot C_6H_4 \cdot OH$	XIII-737	346.48
5836	diamino-4''-di- methoxy-tri- phenylmethane	(4,4')	$[(CH_3)_2N \cdot C_6H_4]_2CH \cdot C_6H_4 \cdot OCH_3$	XIII-737	360.50
5837	diamino-4''- methyl-tri- phenylmethane	(4,4')	$[(CH_3)_2N \cdot C_6H_4]_2CH \cdot C_6H_4 \cdot CH_3$	XIII-282	344.50
5838	diamino-3''-nitro- triphenylmethane	(4,4')	$[(CH_3)_2N \cdot C_6H_4]_2CH \cdot C_6H_4 \cdot NO_2$	XIII-279	375.47
5839	*p*-phenylene- diamine	....	$(CH_3)_2N \cdot C_6H_4 \cdot N(CH_3)_2$	XIII-74	164.25
5840	*p*-phenylene- diamine HCl	Wurster's reagent	$C_{10}H_{16}N_2 \cdot 2HCl$	XIII-74	237.17
5841	pyromellitate	(1,2,4,5)	$C_6H_2(CO_2CH_3)_4$	IX-998	310.26
5842	succinic acid	....	$[(CH_3)_2C \cdot CO_2H]_2$	II-706	174.20
5843	thiouram disulfide	....	$[(CH_3)_2N \cdot CS \cdot S \cdot]_2$	IV-76	240.43
5844	triaminotriphenyl- methane	tetramethyl-*p*- leucaniline	$[(CH_3)_2N \cdot C_6H_4]_2CH \cdot C_6H_4 \cdot NH_2$	XIII-314	345.49
5845	urea	....	$[(CH_3)_2N]_2CO$	IV-74	116.16
5846	**Tetranitro-**diphenol	(3,x,3',x;4,4')	$[(NO_2)_2C_6H_2 \cdot OH]_2$	VI-992	366.20
5847	diphenyl (2,4,2',4')	....	$C_{12}H_6(NO_2)_4$	V-585	334.20
5848	diphenyl (3,4,3',4')	....	$C_{12}H_6(NO_2)_4$	V-585	334.20
5849	diphenyl (2,6,2',6')	....	$C_{12}H_6(NO_2)_4$	*V-274	334.20
5850	diphenyldisulfide	(2,2',4,4')	$[(NO_2)_2C_6H_3 \cdot S \cdot]_2$	VI-344	398.33
5851	diphenyl ether	(2,4,2',4')	$[C_6H_3(NO_2)_2]_2O$	VI-255	350.20
5852	diphenylmethane	(2,4,2',4')	$C_{13}H_8(NO_2)_4$	V-596	348.23
5853	methane	....	$C(NO_2)_4$	I-80	196.03

Tetramethyl thionine 4426
Tetramethyl tin 5951
Tetramethylene 1595
Tetramethylene-diamine 5474

Tetramethylene dibromide 1823
Tetramethylene dichloride 1973
Tetramethylene dicyanide 155
Tetramethylene diiodide 2389

Table 7-4 (*Continued*)
PHYSICAL CONSTANTS OF ORGANIC COMPOUNDS

No.	Crystalline Form and Color	Specific Gravity	Melting Point °C.	Boiling Point °C.	Solubility in 100 Parts		
					Water	Alcohol	Ether
5818	lf./bz.	....	70	259-60	sl. s. h.	s.	s.; s. pet.
5819	pr./h. aq.	....	75	261-2	v. sl. s. c.	s.	s.
5820	cr.	$0.978^{24°}$	23-4	253-5	....	....	....
5821	cr.	1.56	d. > 230	subl. > 360	$55^{15°}$	sl. s. abs.	i.; s. SO_2
5822	col. cr.	1.169	425±*	subl. >300	s.	s. h.	i.; i. chl.
5823	hyg. nd.	....	62-3	d.	$151^{0°}$	$\infty^{63°}$ aq.	....
5824	....	....	59-60	....	....	....	....
5825	....	....	d. 130-5	....	....	....	....
5826	col. pr.	1.84	d. > 230	....	sl. s. c.	0.1 h.	i.; s. SO_2
5827	lq.	$0.905^{20°}_{4}$	−6.3	205.0	....	....	....
5828	lq.	$0.890^{20°}_{4}$	−24	197.9	i.	s.	....
5829	mn.	$0.838^{81°}_{4}$	79.3	196	i.; s. bz.	s.	s.
5830	nd./bz. pet.	....	193-4	>360	s. h. bz.	v. sl. s.	sl. s.
5831	yel. nd./lg.	....	111	subl. 100	i.; s. bz.	v. s.	v. s.
5832	col. lq.	$0.804^{19°}_{4}$	....	169	∞	s.	s.
5833	pl.	....	119	....	....	s.	....
5834	lf./al.	....	90-1; subl.	300	i.; s. bz.	s. h.; s. ac.	s.; s. chl.
5835	cr./al.	....	163-5	....	i.; s. dil. alk.	v. sl. s. lg.	s. bz.
5836	nd./al.	....	105-6	....	v. s. lg.; v. s. chl.	s.; v. s. CS_2	v. s.
5837	nd./al.	....	94-5	....	i.	s.	s.
5838	yel. pr./al.	....	152	....	i.; s. bz.	sl. s.; sl. s. lg.	sl. s.
5839	lf./aq. al.	....	51	260	sl. s. h.	v. s.; v. s. lg.	v. s.; v. s. chl.
5840	cr.	....	....	....	s.	....	....
5841	lf./Me al.	....	141.5	....	....	sl. s. h.	....
5842	cr.	....	190-200 d.	....	$0.48^{13.5°}$	s.	s.; i. lg.
5843	cr./al. chl.	$1.29^{20°}_{4}$	146	....	i.; s. ohl.	v. sl. s.	v. sl. s.
5844	cr./al.	....	151-2	....	....	sl. s.	....
5845	lq.	$0.972^{15°}$	....	177.5	....	v. s.	v. s.
5846	yel. nd.	....	220-5	....	i.	s.	....
5847	yel. pr./bz.	....	165-6	d.	s. bz.; s. ac.	sl. s.	sl. s.
5848	yel. pr.	....	186	....	s. bz.	v. sl. s. lg.	s. ac.
5849	yel. nd./ac.	....	217-8	....	....	....	....
5850	yel. nd.	....	expl. > 280	....	i.; s. pyr.	i.; s. $PhNO_2$	i.; s. $PhNH_2$
5851	cr.	....	195±	....	i.; s. h. bz.	v. sl. s.	sl. s.
5852	lt. yel./ac.	....	172	....	sl. s. bz.	i.	i.
5853	col. lq.	$1.639^{20°}_{4}$	13.8	125.7 sl. d.	i.	v. s.	v. s.

*Sealed tube.

Table 7-4 (*Continued*)
PHYSICAL CONSTANTS OF ORGANIC COMPOUNDS

No.	Name	Synonym	Formula	Beil. Ref.	Formula Weight
	Tetranitro				
5854	naphthalene (γ)	(1,3,5,8)	$C_{10}H_4(NO_2)_4$	V-564	308.17
5855	naphthalene (β)	(1,3,6,8)	$C_{10}H_4(NO_2)_4$	V-564	308.17
5856	naphthalene (α)	(1,5,x,x)	$C_{10}H_4(NO_2)_4$	V-564	308.17
5857	phenol (2,3,4,6)	....	$HO \cdot C_6H(NO_2)_4$	VI-292	274.10
5858	phenolsulfon-phthalein	....	$C_{19}H_{10}O_{13}N_4S$	*XIX-650	534.37
5859	**Tetraphenyl-arsonium bromide**	....	$(C_6H_5)_4AsBr \cdot 2H_2O$	....	499.29
5860	ethane (*s*)	....	$[(C_6H_5)_2CH \cdot]_2$	V-738	334.47
5861	ethylene	....	$(C_6H_5)_2C \colon C(C_6H_5)_2$	V-743	332.45
5862	guanidine	....	$HN \colon C[N(C_6H_5)_2]_2$	XII-430	363.47
5863	hydrazine	....	$(C_6H_5)_2N \cdot N(C_6H_5)_2$	XV-125	336.44
5864	methane	....	$(C_6H_5)_4C$	V-739	320.44
5865	urea	....	$[(C_6H_5)_2N]_2CO$	XII-429	364.45
5866	**Tetrapropyl**-ammonium iodide (*n*)	....	$(C_2H_5 \cdot CH_2)_4NI$	IV-140	313.27
5867	benzene (*iso*)	(1,2,4,5)	$[(CH_3)_2CH]_4C_6H_2$	**V-358	246.44
5868	**Tetratriacontane** (*n*)	....	$CH_3(CH_2)_{32}CH_3$	I-177	478.94
5869	**Tetrazine** (1,2,4,5)	....	$(\colon N \cdot CH \cdot N)_2$	XXVI-353	82.07
5870	**Tetrazole**	....	$CH \colon N \cdot NH \cdot N \colon N$	XXVI-346	70.05
5871	**Tetrolic acid**	butynoic acid	$CH_3 \cdot C \colon C \cdot CO_2H$	II-479	84.08
5872	**Tetronal**	pentane-3,3-di-ethyl-sulfone	$(C_2H_5)_2C(SO_2C_2H_5)_2$	I-681	256.39
5873	**Tetryl** (2,4,6)	tri-NO$_2$-phenyl-methylnitramine	$(NO_2)_3C_6H_2 \cdot N(CH_3)NO_2$	XII-770	287.15
5874	**Thallin**	6-methoxy-tetra-hydroquinoline	$CH_3O \cdot C_9H_{10}N$	XXI-61	163.22
5875	**Thialdine**	....	$(CH_3)_3C_3H_4S_2N$	XXVII-461	163.31
5876	**Thiamine chloride**	vitamin B$_1$	$C_{12}H_{17}ON_4ClS \cdot HCl$	....	337.27
5877	**Thianthrene**	diphenylene disul-fide	$(C_6H_4)_2S_2$	XIX-45	216.33
5878	**Thiazole**	....	C_3H_3NS	XXVII-15	85.13
5879	**Thio**-acetaldehyde (α)	sulfaldehyde	$(CH_3 \cdot CHS)_3$	XIX-387	180.35
5880	acetaldehyde (β)	sulfaldehyde	$(CH_3 \cdot CHS)_3$	XIX-387	180.35
5881	acetaldehyde (γ)	sulfaldehyde	$(CH_3 \cdot CHS)_3$	....	180.35
5882	acetamide	aceto-thioamide	$CH_3 \cdot CS \cdot NH_2$	II-232	75.13
5883	acetanilide	....	$C_6H_5 \cdot NH \cdot CS \cdot CH_3$	XII-245	151.23
5884	acetdimethylamide	....	$CH_3 \cdot CS \cdot N(CH_3)_2$	*IV-329	103.19
5885	acetic acid	....	$CH_3 \cdot CO \cdot SH$	II-230	76.12
5886	aniline (4,4')	diNH$_2$-diPH-sulfide	$(NH_2 \cdot C_6H_4)_2S$	XIII-535	216.31
5887	barbituric acid	malonyl thiourea	$C_4H_4O_2N_2S$	XXIV-476	144.15
5888	benzaldehyde (α)	....	$(C_6H_5 \cdot CHS)_3$	XIX-396	366.57
5889	benzaldehyde (β)	....	$(C_6H_5 \cdot CHS)_3$	XIX-397	366.57
5890	benzanilide	....	$C_6H_5 \cdot NH \cdot CS \cdot C_6H_5$	XII-269	213.30
5891	benzoic acid	....	$C_6H_5 \cdot CO \cdot SH$	IX-419	138.19
5892	carbanilide	diphenyl thiourea	$(C_6H_5 \cdot NH)_2CS$	XII-394	228.32
5893	chrome	....	$C_{12}H_{14}ON_4S$	....	262.34
5894	cresol (*o*)	*o*-tolyl mercaptan	$CH_3 \cdot C_6H_4 \cdot SH$	VI-370	124.21
5895	cresol (*m*)	*m*-tolyl mercaptan	$CH_3 \cdot C_6H_4 \cdot SH$	VI-388	124.21
5896	cresol (*p*)	*p*-tolyl mercaptan	$CH_3 \cdot C_6H_4 \cdot SH$	VI-416	124.21

Theelol 2902
Theine, cf. alkd.
Theobromine, cf. alkd.
Theocine, cf. alkd.

Theophylline, cf. alkd.
Thevetin, cf. glcde.
Thienyl alcohol 5924
Thio-, cf. also sulfo-

Table 7-4 (Continued)
PHYSICAL CONSTANTS OF ORGANIC COMPOUNDS

No.	Crystalline Form and Color	Specific Gravity	Melting Point °C.	Boiling Point °C.	Solubility in 100 Parts		
					Water	Alcohol	Ether
5854	yel./act.	....	194-5	....	s. HNO_3	sl. s.	s. act.
5855	nd./al.	....	203	expl.	i.	....	....
5856	lt. yel./chl.	....	259	expl.	i.	v. sl. s.	....
5857	lt. yel./chl.	....	140	expl.	d. h.	v. sl. s. bz.	v. sl. s. lg.
5858	yel. nd./ ac. act.	....	>200	....	s.	s.	i.
5859	col. cr.	....	281-4 (anh.)	$-2H_2O >$ 100	1.6	s.	sl. s. act.
5860	rhb.	1.182	209-11	379-83	14 h. bz.	0.8 h.	5 h. ac.
5861	tri. or mn.	$1.155_4^{0°}$ ...	221-3	415-25	s. h. bz.	sl. s.	sl. s.
5862	rhb./lg.	....	130-1	....	i.; v. s. bz.	v. s.	v. s.
5863	rhb./al. chl.	....	147-9	....	s. bz.	v. sl. s. h.	s. act.
5864	rhb./bz.	....	285	431	i. lg.	i.; s. h. bz.	i.; i. ac.
5865	rhb.	1.222	183	....	i.	s. h.	
5866	rhb.	$1.314_4^{2.5°}$	d. 280 ±	....	$18.6^{25°}$	$55^{25°}$ chl.	....
5867	cr./al.	....	117	260^{775mm}	i.	$1^{25°}$	$85^{25°}$
5868	cr.	lq. $0.781^{73°}$	α73;β77(?)	$255^{1.0mm}$	....	....	....
5869	red pr.	....	99	....	s.	s.	s.
5870	lf./al.	....	155	subl.	s.; s. ac.	s.; sl. s. bz.	sl. s.
5871	pl./et.	....	77-8	203	v. s.	v. s.	v. s.
5872	pl./aq.	....	85	....	0.22 c.	$5.4^{15°}$ abs.	$10^{15°}$
5873	yel. mn./al.	$1.57^{19°}$	129	expl. 187	i.; s. bz.	s. h.	s.; s. ac.
5874	pr.	....	42-3	283^{735mm}	sl. s. c.	v. s.; v. s. bz.	v. s.
5875	mn.	$1.191^{18°}$	43	d.	v. sl. s.	s.	v. s.
5876	hyg. pr.	....	252 sl. d.	....	100	0.1	i.; i. bz.
5877	mn./al.	$1.706_4^{18°}$	158-60	364-6	i.	0.25 c.	s. h.
5878	lq.	$1.200_4^{17°}$	....	116.8	sl. s.	s.	s.
5879	cr./al.	....	101	246-7	i.; s. bz.	$3.86^{25°}$	$15.6^{25°}$
5880	cr./al.	....	125-6	246-7	i.; s. chl.	$3.97^{25°}$	$13.7^{25°}$
5881	cr.	....	81	100	....	....	....
5882	mn.	....	115-6	....	v. s.	sl. s.	sl. s.
5883	nd./aq.	....	75-6	d.	i.	i. a.	s. alk.
5884	col. cr.	....	114-6	....	sl. s.	s.	v. s.
5885	yel. lq.	$1.074^{10°}$	<-17	93	s.	∞	∞
5886	nd./aq.	....	108	....	sl. s. h.	s.	s.; s. h. bz.
5887	lf./aq.	....	235 d.	....	sl. s.	s.; s. alk.	s. alk. carb.
5888	nd./bz. al.	....	166-7	....	i.	$0.2^{25°}$	$1.1^{25°}$
5889	nd./bz.	....	225-6 d.	....	s. h. ac.	$0.04^{25°}$	$0.4^{25°}$
5890	yel. pr./al.	....	101-2	....	i.	s.	v. s.
5891	yel. oil	....	24	d.	i.	v. s.	∞
5892	rhb./al.	$1.32^{40°}$	154	d.	i.	v. s.	v. s.
5893	yel. pr./chl.	....	227-8*	....	s.	sl. s.	sl. s.
5894	lf.	....	15	194.3	i.	s.	s.
5895	lq.	$1.052_4^{12°}$	<-20	195.4	i.	s.	s.
5896	lf./al.	....	43-4	195	i.	sl. s.	v. s.

* HCl salt, m. p. 217-21°.
Thio-allyl ether 1684
Thio-benzyl alcohol 816

Thio-carbamide 5920
Thio-carbonyl chloride 5916
Thio-cresol methyl ether 4402

Table 7-4 (*Continued*)
PHYSICAL CONSTANTS OF ORGANIC COMPOUNDS

No.	Name	Synonym	Formula	Beil. Ref.	Formula Weight
5897	**Thio** cyanic acid	sulfocyanic acid	HS·CN	III-143	59.09
5898	cyanic acid, per-	3,5-dimercapto-1,2,4-thiodiazol	CS·NH·CS·NH·S ⌞_____⌟	XXVII-665	150.24
5899	cyanuric acid	....	$C_3H_3N_3S_3$	XXVI-259	177.27
5900	diglycol (β)	β,β′diOH-diEt sulfide	$(HO·CH_2CH_2)_2S$	I-470	122.19
5901	formamide	....	$H·CS·NH_2$	II-95	61.11
5902	formanilide	....	$C_6H_5·NH·CH:S$	XII-233	137.20
5903	glycerol (1)	....	$HS·CH_2·C_2H_3(OH)_2$	I-519	108.16
5904	glycol, mono-	ethanol-1-thiol-2	$HO·CH_2·CH_2·SH$	I-470	78.13
5905	glycolic acid	....	$HS·CH_2·CO_2H$	III-245	92.12
5906	glycolic-β-amino-naphthalide	thionalide	$C_{10}H_7·NH·CO·CH_2·SH$	....	217.29
5907	hydantoin	glycolyl thiourea	$C_3H_4ON_2S$	XXIV-260	116.14
5908	hydracrylic acid	....	$HS·CH_2CH_2CO_2H$	III-299	106.14
5909	hydroquinone, mono-	(1,4)	$HS·C_6H_4·OH$	VI-859	126.18
5910	lactic acid	....	$CH_3·CH(SH)·CO_2H$	III-289	106.14
5911	malic acid (*dl*)	mercaptosuccinic	$(HS·C_2H_3)(CO_2H)_2$	III-439	150.15
5912	naphthene	benzothiophene	C_8H_6S	XVII-59	134.20
5913	naphthol (α)	α-naphthyl mercaptan	$C_{10}H_7·SH$	VI-621	160.24
5914	naphthol (β)	β-naphthyl mercaptan	$C_{10}H_7·SH$	VI-657	160.24
5915	phenol	phenyl mercaptan	$C_6H_5·SH$	VI-294	110.18
5916	phosgene	thiocarbonyl chloride	$Cl_2C:S$	III-134	114.98
5917	propionamide	....	$C_2H_5·CS·NH_2$	II-264	89.16
5918	salicylic acid (*o*)	sulfhydryl benzoic	$HS·C_6H_4·CO_2H$	X-125	154.19
5919	semicarbazide	....	$NH_2·CS·NH·NH_2$	III-195	91.14
5920	urea	thiocarbamide	$NH_2·CS·NH_2$	III-180	76.12
5921	urethane	ethyl thiourethane	$NH_2·CO·SC_2H_5$	III-138	105.16
5922	**Thionin**(e)	Lauth's violet	$C_{12}H_9N_3S$	XXVII-391	227.28
5923	**Thiophene**	....	C_4H_4S	XVII-29	84.14
5924	alcohol (α)	thienyl carbinol	$C_4H_3S·CH_2OH$	XVII-113	114.17
5925	aldehyde (α)(2)	....	$C_4H_3S·CHO$	XVII-285	112.15
5926	carboxylic acid (α)	(1,2)	$C_4H_3S·CO_2H$	XVIII-289	128.15
5927	carboxylic acid (β)	(1,3)	$C_4H_3S·CO_2H$	XVIII-292	128.15
5928	**Thioxene** (*o*)	2,5-diMe-thiophene	$(CH_3)_2C_4H_2S$	XVII-41	112.19
5929	**Thioxene** (*o*)	2,3-diMe-thiophene	$(CH_3)_2C_4H_2S$	XVII-40	112.19
5930	**Thioxene** (*m*)	2,4-diMe-thiophene	$(CH_3)_2C_4H_2S$	XVII-41	112.19
5931	**Thioxene** (*m*)	3,4-diMe-thiophene	$(CH_3)_2C_4H_2S$	XVII-42	112.19
5932	**Threonine** (*dl*)	α-amino-β-hydroxy-butyric acid	$CH_3·CHOH·CH:(NH_2)CO_2H$	IV-514	119.12
5933	**Threonine** (*d*-)	....	$C_4H_9O_3N$	IV-514	119.12
5934	**Threonine,** allo-	(*dl*)	$C_4H_9O_3N$	....	119.12
5935	**Thujone** (α)	....	$C_{10}H_{16}O$	VII-93	152.24
5936	**Thujone** (β)	tanacetone	$C_{10}H_{16}O$	VII-93	152.24
5937	**Thymohydro-quinone**	....	$C_{10}H_{14}O_2$	VI-945	166.22
5938	**Thymol**	5-Me-2-isoPr-phenol	$(CH_3)(C_3H_7)C_6H_3OH$	VI-532	150.22
5939	chloride	(3,6;1,4)	$(CH_3)(C_3H_7)C_6H_2:(OH)Cl$	VI-539	184.67
5940	iodide	4-iodothymol	$C_{10}H_{13}OI$	VI-541	276.12
5941	phthalein	....	$C_{28}H_{30}O_4$	*XVIII-381	430.55

Thio-diphenylamine 5138
Thio-histidine betaine 2890
Thio-phenetole 3132
Thio-phenine 378

Thio-phenylacetamide 5240
Thio-sinamine 221
Thio-trimethylacetamide 6275
Thiols, cf. mercaptans.

Table 7-4 (*Continued*)
PHYSICAL CONSTANTS OF ORGANIC COMPOUNDS

No.	Crystalline Form and Color	Specific Gravity	Melting Point °C.	Boiling Point °C.	Solubility in 100 Parts		
					Water	Alcohol	Ether
5897	col. lq.	....	5±	d.	∞, d.	v. s.	v. s.
5898	not known; salts only	....	....	....	....	....	....
5899	cr.	....	d. 200	....	sl. s. h.	sl. s.	sl. s.
5900	syrup	1.221^{20}_{20}	−10	164-6^{20mm}	∞	s. chl.	....
5901	col. cr.	....	28-9	....	s.; i. bz.	s.; s. act.	s.; s. CS_2
5902	nd./aq.	....	138	....	i. aq. HCl	s.	v. s.
5903	yel. lq.	$1.295^{14.4°}$	....	d.	v. sl. s.	∞	i.
5904	lq.	1.114^{20}_{4}	....	153-7	v. s.	v. s.	v.s.; ∞ bz.
5905	col. lq.	$1.325^{20°}$	−16.5	123^{29mm}	....	....	....
5906	col. nd.	....	111-2	....	....	s.	s.
5907	nd./aq.	....	d. 200±	....	sl. s.	i.	i.
5908	col. cr.	$1.218^{21°}$	16.8	111-2^{15mm}	∞	∞	∞
5909	cr.	....	32-4	166-8^{45mm}	s.	s. H_2SO_4	....
5910	oil	....	10±	98-9^{14mm}	∞	∞	∞
5911	col. cr.	....	149-50	....	s.; s. act.	s.; sl. s. bz.	sl. s.
5912	lf.	1.165^{20}_{4}	31-2	220-1	....	....	....
5913	lq.	1.155^{23}_{4}	....	208.5^{200mm}	sl. s.	v. s.	v. s.
5914	cr./al.	....	81	286-8	v. sl. s.	v. s.	v. s.
5915	col. lq.	1.074^{23}_{4}	−14.9	169.5	v. sl. s.	v. s.; ∞ bz.	∞; s. CS_2
5916	red lq.	$1.509^{15°}$	....	73.5	d.	d.	s.
5917	lf./bz.	....	41-3	....	v. sl. s.	v. sl. s.	s.; s. bz.
5918	lt. yel. nd.	....	164	subl.	sl. s. h.	s.	s. ac.
5919	nd./aq.	....	181-3 d.	....	s.	s.	....
5920	rhb./al.	1.405^{90}_{4}	180-2	d.	9.2$^{13°}$	s.	sl. s.
5921	lf.	....	108-9	subl. sl. d.	s. h.	s.	sl. s.
5922	brn. bl. lf.	....	....	....	i. c.; i. lg.	sl. s.; s. chl.	sl. s.
5923	col. lq.	1.070^{15}_{4}	−38.3	84	i.; s. bz.	s.	s. H_2SO_4
5924	lq.	....	....	207	....	....	....
5925	oil	$1.215^{21°}$	....	198	....	....	s.
5926	nd./aq.	....	126.5	260 sl. d.	v. s. h.	v. s.	s.; sl. s. lg.
5927	mn. nd.	....	136	subl.	0.43$^{25°}$	....	....
5928	lq.	$0.976^{17.5°}$	....	136.5-7.5	i.	s.	s.
5929	lq.	$0.994^{21°}$	....	136-7	....	....	....
5930	lq.	$0.996^{20°}$	....	137-8	i.	s.	s.
5931	lq.	1.008^{23}_{22}	....	144-6	....	....	....
5932	col. cr.	....	235-8 d.	....	20$^{25°}$; v. s. h.	i; i. chl.	i.
5933	hex. pl.	....	225-7 d.	....	s.	....	....
5934	col. cr.	....	250-2	....	s. h.	....	....
5935	col. lq.	0.912	....	200-1	i.	s.	....
5936	col. lq.	0.916^{20}_{4}	....	201-3	i.	s.	....
5937	pr.	....	140-3	290	s. h.	v. s.	v. s.
5938	cr./act.	0.972^{25}_{25}	49.6	232.9	0.09$^{19°}$; 0.11$^{100°}$	v. s.; v. s. chl.	v. s.; s. alk.
5939	col. pl./lg.	....	59-61	....	0.1; s. dil. alk.	222; 60 bz.	83
5940	nd.	....	68-9	....	sl. s. h.	sl. s.	s.; s. chl.
5941	col. nd./al.	....	252-3	....	i.; s. H_2SO_4	s.; s. act.	s. dil. alk.

Thionalid 5906
Thiotolene 4389
Thujin, cf. glcde.
Thymine 4413

Thymiode 5940
Thymiodol 5940
Thymodin 5940
Thymol-hexahydride 4024

Table 7-4 (*Continued*)
PHYSICAL CONSTANTS OF ORGANIC COMPOUNDS

No.	Name	Synonym	Formula	Beil. Ref.	Formula Weight
5942	**Thymol** sulfonic acid (α)	(3,6;1,4)	$(CH_3)(C_3H_7)C_6H_2$: $(OH)SO_3H \cdot H_2O$	XI-267	248.30
5943	**Thymoquinone**	....	$O:C_6H_2(CH_3) \cdot (C_3H_7):O$	VII-662	164.21
5944	**Thymotic acid** (*o*)	(1,4;3,2)	$(CH_3)(C_3H_7)C_6H_2$: $(OH)CO_2H$	X-280	194.23
5945	**Thymyl**-acetate	acetyl thymol	$C_{10}H_{13} \cdot O_2C \cdot CH_3$	VI-537	192.26
5946	amine (1,3;4)	....	$CH_3(NH_2)C_6H_3 \cdot C_3H_7$	XII-1171	149.24
5947	**Thyroxine** (*l*)	....	$C_{15}H_{11}O_4NI_4$	*XIV-671	776.88
5948	**Tin** chloride triphenyl	triphenyltin chloride	$(C_6H_5)_3SnCl$	*XVI-540	385.46
5949	diethyl	diethyl tin	$(C_2H_5)_2Sn$	IV-631	176.81
5950	tetraethyl	tetraethyl tin	$(C_2H_5)_4Sn$	IV-632	234.94
5951	tetramethyl	tetramethyl tin	$(CH_3)_4Sn$	IV-632	178.83
5952	tetraphenyl	tetraphenyl tin	$(C_6H_5)_4Sn$	XVI-914	427.12
5953	triethyl	triethyl tin	$[(C_2H_5)_3Sn \cdot]_2$	IV-638	411.75
5954	**Tolane**	diphenyl-acetylene	$C_6H_5 \cdot C : C \cdot C_6H_5$	V-656	178.24
5955	**Tolidine** (*o*)	3,3′-diMe-benzidine	$(CH_3 \cdot C_6H_3 \cdot NH_2)_2$	XIII-256	212.30
5956	hydrochloride	....	$C_{14}H_{16}N_2 \cdot 2HCl$	XIII-257	285.22
5957	sulfate	....	$C_{14}H_{16}N_2 \cdot H_2SO_4$	XIII-257	310.37
5958	disulfonic acid	(3,4;6)	$[(CH_3)(NH_2)C_6H_2 \cdot SO_3H]_2 \cdot 1\frac{1}{2}H_2O$	XIV-796	399.44
5959	**Tolidine** (*m*)	2,2′-diMe-benzidine	$(CH_3 \cdot C_6H_3 \cdot NH_2)_2$	XIII-255	212.30
5960	**Toluene**	methyl-benzene	$CH_3 \cdot C_6H_5$	V-280	92.14
5961	azo-1-naphthyl-amine-2 (*o*)	(2,1;1′,2′); yellow OB	$CH_3 \cdot C_6H_4 \cdot N:N \cdot C_{10}H_6 \cdot NH_2$	XVI-373	261.33
5962	disulfonic acid	(2,4)	$CH_3 \cdot C_6H_3(SO_3H)_2$	XI-204	252.27
5963	disulfonic Na	(2,4); (Na salt)	$C_7H_6O_6S_2Na_2 \cdot 7H_2O$	XI-204	422.34
5964	sulfinic acid (*p*)	....	$CH_3 \cdot C_6H_4 \cdot SO_2H$	XI-9	156.20
5965	sulfinic Na (*p*)	(sodium salt)	$C_7H_7O_2SNa \cdot 2H_2O$	XI-9	214.22
5966	sulfonic acid (*o*)	....	$CH_3 \cdot C_6H_4 \cdot SO_3H \cdot 2H_2O$	XI-83	208.23
5967	sulfonic acid (*m*)	....	$CH_3 \cdot C_6H_4 \cdot SO_3H \cdot aq.$	XI-94	172.20
5968	sulfonic acid (*p*)	....	$CH_3 \cdot C_6H_4 \cdot SO_3H \cdot H_2O$	XI-97	190.22
5969	sulfonic Na (*p*)	(sodium salt)	$C_7H_7O_3SNa \cdot 2H_2O$	XI-97	230.22
5970	**Toluenesulfonyl**-amide (*o*)	toluenesulfonamide	$CH_3 \cdot C_6H_4 \cdot SO_2 \cdot NH_2$	XI-86	171.22
5971	amide (*p*)	....	$CH_3 \cdot C_6H_4 \cdot SO_2 \cdot NH_2$	XI-104	171.22
5972	anilide (*p*)	toluene-sul-fonanilide	$CH_3 \cdot C_6H_4 \cdot SO_2 \cdot NH \cdot C_6H_5$	XII-567	247.32
5973	*n*-butylamide (*p*)	....	$CH_3 \cdot C_6H_4 \cdot SO_2 \cdot NH \cdot C_4H_9$	....	227.33
5974	chloride (*o*)	toluene sulfon-chloride	$CH_3 \cdot C_6H_4 \cdot SO_2Cl$	XI-86	190.65
5975	chloride (*p*)	....	$CH_3 \cdot C_6H_4 \cdot SO_2Cl$	XI-103	190.65
5976	di-*n*-butylamide	(*p*)	$CH_3 \cdot C_6H_4 \cdot SO_2 \cdot N: (C_4H_9)_2$	....	283.44
5977	dimethylamide (*p*)	....	$CH_3 \cdot C_6H_4 \cdot SO_2 \cdot N: (CH_3)_2$	**XI-56	199.27
5978	ethylamide (*p*)	....	$CH_3 \cdot C_6H_4 \cdot SO_2 \cdot NH \cdot C_2H_5$	XI-105	199.27
5979	methylamide	....	$CH_3 \cdot C_6H_4 \cdot SO_2 \cdot NH \cdot CH_3$	XI-105	185.25
5980	methylanilide	....	$CH_3 \cdot C_6H_4 \cdot SO_2 \cdot N: (CH_3)(C_6H_5)$	XII-575	261.35

Table 7-4 (*Continued*)
PHYSICAL CONSTANTS OF ORGANIC COMPOUNDS

No.	Crystalline Form and Color	Specific Gravity	Melting Point °C.	Boiling Point °C.	Solubility in 100 Parts		
					Water	Alcohol	Ether
5942	col. pl.	....	115-6	....	v. s.	....	....
5943	yel. tri.	....	46-7	232	v. sl. s.	s.	s.
5944	mn./aq.	....	127	subl.	0.01 c.	s.	s.; s. bz.
5945	lq.	$1.009^{0°}$	....	245^{757mm}	i.; ∞ chl.	∞ ; ∞ bz.	∞
5946	oil	....	....	230	v. sl. s.	s.	s.
5947	col. nd.	....	232 d.	....	i.; s. alk.	i.	i.
5948	col. cr./al.	....	112-3	$240^{13.5mm}$	i.	s.	s.
5949	yel. oil	$1.558^{15°}$	....	d.	i.	s.	s.
5950	col. lq.	$1.199^{20°}_{4}$	-112	181	i.	....	s.
5951	lq.	$1.291^{26°}_{4}$	....	78	i.	....	....
5952	tet. pr./chl.	$1.490^{0°}_{4}$	226	>420	i.; s. pyr.	v. sl. s.	v. sl. s.
5953	lq.	$1.412^{0°}$	....	265-70	i.	i. aq. al.	s. bz.
5954	mn. pr.	$0.966^{100°}_{4}$	60-2	300	i.	v. s. h.	v. s.
5955	lf.	....	129-31	....	v. sl. s.	s.	s.; s. ac.
5956	scales	....	d. > 340	....	$0.9^{12°}$	....	....
5957	gray	....	d.	....	0.1c.; d.h.	v. sl. s.	s. dil. a.
5958	nd.	....	-H$_2$O, 150	....	$0.2^{18°}$ (anh.)	i.; i. ac.	i.
5959	pr./aq.	....	107-8	....	sl. s. h.	v. s.	v. s.
5960	col. lq.	$0.866^{20°}_{4}$	-95	110.6	i.; s. act.	∞ abs.	∞
5961	yel. pd.	....	122-5	....	i.; s. oil	s.; s. CCl$_4$	s.; s. bz.
5962	syrup	....	....	....	s.	....	....
5963	pr.	....	....	....	s.	....	....
5964	cr./aq.	....	85-90	....	sl. s.	v. s.; s. h. bz.	v. s.
5965	cr.	....	....	....	s.	....	....
5966	delq. cr.	....	*	$128.8^{0.1mm}$	v. s.	s.	....
5967	syrup	....	....	....	v. s.	s.	....
5968	mn.	....	104-5	$146-7^{0.1mm}$	v. s.	s.	....
5969	lf.	....	....	....	s.	....	....
5970	tet. pr.	....	156	....	$0.11^{9°}$	$3.575^{0°}$	....
5971	mn.	....	137	....	$0.2^{9°}$	$7.45^{0°}$	....
5972	tri./aq. al. or bz.	....	103	....	....	v. s.	....
5973	col. rhb.	....	41-2.5	....	i.	s.	v. s.
5974	oil	$1.344^{17°}$	10	126^{10mm}	i.	....	....
5975	tri.	....	69	134.5^{10mm}	i.	s.	s.; s. bz.
5976	pa. yel. lq.	....	....	$233-4^{20mm}$	i.	s.	s.
5977	nd./pet.	....	86-7	....	i.	sl. s.; v. s. act.	v. s.; v. s. bz.
5978	cr./lg.; pl./aq. al.	....	63-4.5	....	....	....	....
5979	pl./aq. al.	....	77-8	....	v. sl. s.	v. s.	....
5980	mn. pl./ EtOAc	....	93-4	....	....	v. s.	v. s.

* Stable < 100°; forms *p*-isomer 140-50°.
Tolubenzyl alcohol 6016-8
Tolubenzyl-amine 6482-4
Toluene-azo-toluidine 309-12

Tolubenzyl acetate 6481
Toluene hexahydride 4199
Toluene tetrahydride 4207-8
Toluhydroquinone 2372

Table 7-4 (*Continued*)
PHYSICAL CONSTANTS OF ORGANIC COMPOUNDS

No.	Name	Synonym	Formula	Beil. Ref.	Formula Weight
5981	**Toluenesulfonyl** *p*-toluidide (*p*)	....	$CH_3 \cdot C_6H_4 \cdot SO_2 \cdot NH \cdot$ $C_6H_4 \cdot CH_3$	XII-981	261.35
5982	**Toluic** acid†(*o*)	2-Me-benzoic acid	$CH_3 \cdot C_6H_4 \cdot CO_2H$	IX-462	136.15
5983	acid (*m*)	3-Me-benzoic acid	$CH_3 \cdot C_6H_4 \cdot CO_2H$	IX-475	136.15
5984	acid (*p*)	4-Me-benzoic acid	$CH_3 \cdot C_6H_4 \cdot CO_2H$	IX-483	136.15
5985	aldehyde (*o*)	*o*-toluylaldehyde	$CH_3 \cdot C_6H_4 \cdot CHO$	VII-295	120.15
5986	aldehyde (*m*)	*m*-Me-benzaldehyde	$CH_3 \cdot C_6H_4 \cdot CHO$	VII-296	120.15
5987	aldehyde (*p*)	4-Me-benzaldehyde	$CH_3 \cdot C_6H_4 \cdot CHO$	VII-297	120.15
5988	amide (*o*)	*o*-toluamide	$CH_3 \cdot C_6H_4 \cdot CO \cdot NH_2$	IX-465	135.17
5989	amide (*m*)	*m*-toluamide	$CH_3 \cdot C_6H_4 \cdot CO \cdot NH_2$	IX-477	135.17
5990	amide (*p*)	4-toluamide	$CH_3 \cdot C_6H_4 \cdot CO \cdot NH_2$	IX-486	135.17
5991	anhydride (*o*)	....	$(CH_3 \cdot C_6H_4 \cdot CO)_2O$	IX-464	254.29
5992	nitrile (*o*)	*o*-tolunitrile	$CH_3 \cdot C_6H_4 \cdot CN$	IX-466	117.15
5993	nitrile (*m*)	3-tolyl cyanide	$CH_3 \cdot C_6H_4 \cdot CN$	IX-477	117.15
5994	nitrile (*p*)	4-tolunitrile	$CH_3 \cdot C_6H_4 \cdot CN$	IX-489	117.15
5995	**Toluidine** (*o*)	2-methyl aniline	$CH_3 \cdot C_6H_4 \cdot NH_2$	XII-772	107.16
5996	hydrochloride	....	$CH_3 \cdot C_6H_4 \cdot NH_2 \cdot HCl$	XII-782	143.62
5997	**Toluidine** (*m*)	3-methyl aniline	$CH_3 \cdot C_6H_4 \cdot NH_2$	XII-853	107.16
5998	hydrochloride	....	$CH_3 \cdot C_6H_4 \cdot NH_2 \cdot HCl$	XII-856	143.62
5999	**Toluidine** (*p*)	4-methyl aniline	$CH_3 \cdot C_6H_4 \cdot NH_2$	XII-880	107.16
6000	hydrochloride	....	$CH_3 \cdot C_6H_4 \cdot NH_2 \cdot HCl$	XII-896	143.62
6001	**Toluquinone**	*o*-Me-*p*-benzo-quinone	$O{:}C_6H_3(CH_3){:}O$	VII-645	122.12
6002	**Toluyl-*o*-benzoic** acid (*p*)	(1,4;1′,2′)	$CH_3 \cdot C_6H_4 \cdot CO \cdot$ $C_6H_4 \cdot CO_2H$	X-759	240.26
6003	**Toluylene** diamine	(1;3,4)(*o*)(*uns*)	$CH_3 \cdot C_6H_3(NH_2)_2$	XIII-148	122.17
6004	diamine HCl	(1;3,4)	$C_7H_{10}N_2 \cdot 2HCl$	XIII-148	195.09
6005	diamine (1;2,4)	(*m*)(*uns*)	$CH_3 \cdot C_6H_3(NH_2)_2$	XIII-124	122.17
6006	diamine HCl	(1;2,4)	$C_7H_{10}N_2 \cdot 2HCl$	XIII-129	195.09
6007	diamine (1;2,5)	(*p*)	$CH_3 \cdot C_6H_3(NH_2)_2$	XIII-144	122.17
6008	diamine HCl	(1;2,5)	$C_7H_{10}N_2 \cdot 2HCl$	XIII-144	195.09
6009	diamine H_2SO_4	(1;2,5)	$C_7H_{10}N_2 \cdot H_2SO_4$	XIII-144	220.25
6010	**Tolyl** acetic acid (*o*)	....	$CH_3 \cdot C_6H_4 \cdot CH_2 \cdot CO_2H$	IX-527	150.18
6011	acetic acid (*m*)	....	$CH_3 \cdot C_6H_4 \cdot CH_2 \cdot CO_2H$	IX-528	150.18
6012	acetic acid (*p*)	....	$CH_3 \cdot C_6H_4 \cdot CH_2 \cdot CO_2H$	IX-530	150.18
6013	benzoate (*o*)	*o*-cresyl benzoate	$C_6H_5CO_2 \cdot C_6H_4 \cdot CH_3$	IX-119	212.25
6014	benzoate (*m*)	*m*-cresyl benzoate	$C_6H_5CO_2 \cdot C_6H_4 \cdot CH_3$	IX-120	212.25
6015	benzoate (*p*)	*p*-cresyl benzoate	$C_6H_5CO_2 \cdot C_6H_4 \cdot CH_3$	IX-120	212.25
6016	carbinol (*o*)	*o*-tolubenzyl alcohol	$CH_3 \cdot C_6H_4 \cdot CH_2OH$	VI-484	122.17
6017	carbinol (*m*)	*m*-xylyl alcohol	$CH_3 \cdot C_6H_4 \cdot CH_2OH$	VI-494	122.17
6018	carbinol (*p*)	....	$CH_3 \cdot C_6H_4 \cdot CH_2OH$	VI-498	122.17
6019	*iso*-cyanate (*o*)	*o*-tolylcarbonimide	$CH_3 \cdot C_6H_4 \cdot N{:}CO$	XII-812	133.15
6020	*iso*-cyanate (*m*)	*m*-tolylcarbonimide	$CH_3 \cdot C_6H_4 \cdot N{:}CO$	XII-864	133.15
6021	*iso*-cyanate (*p*)	*p*-tolylcarbonimide	$CH_3 \cdot C_6H_4 \cdot N{:}CO$	XII-955	133.15
6022	hydrazine (*o*)	....	$CH_3 \cdot C_6H_4 \cdot NH \cdot NH_2$	XV-496	122.17
6023	hydrazine HCl (*o*)	....	$C_7H_{10}N_2 \cdot HCl \cdot H_2O$	XV-496	176.65
6024	hydrazine (*m*)	....	$CH_3 \cdot C_6H_4 \cdot NH \cdot NH_2$	XV-506	122.17
6025	hydrazine HCl (*m*)	....	$C_7H_{10}N_2 \cdot HCl$	XV-506	158.63
6026	hydrazine (*p*)	....	$CH_3 \cdot C_6H_4 \cdot NH \cdot NH_2$	XV-510	122.17
6027	hydrazine HCl (*p*)	....	$C_7H_{10}N_2 \cdot HCl$	....	158.63
6028	hydroxylamine	(*o*)(*β*)	$CH_3 \cdot C_6H_4 \cdot NHOH$	XV-13	123.16

† See also No. 5142.
Toluidine sulfonic acid 379-83
Tolunitrile 5992-4
Toluol 5960
Toluphenazine 4334

Toluquinoline 4372, 4374-5
Toluquinone oxime 4881
Toluylaldehyde 5139, 5985-7
Toluylene red 4585

Table 7-4 (*Continued*)
PHYSICAL CONSTANTS OF ORGANIC COMPOUNDS

No.	Crystalline Form and Color	Specific Gravity	Melting Point °C.	Boiling Point °C.	Solubility in 100 Parts		
					Water	Alcohol	Ether
5981	col. tri./ac.	....	117-8	....	....	v. s. h.	....
5982	cr./aq.	$1.062\frac{115}{4}$	104-5	259^{751mm}	$2.17^{100°}$	v. s.	s. chl.
5983	pr./aq.	$1.054\frac{112}{4}$	110-1	263	$0.09^{15°}$; $1.6^{100°}$	v. s.	v. s.
5984	cr./aq.	....	179-80	274-5	$1.26^{100°}$	v. s.	v. s.
5985	lq.	$1.039\frac{20}{4}$	....	196-9	sl. s.	∞	∞
5986	lq.	$1.019\frac{20}{4}$	....	199	sl. s.	∞	∞
5987	lq.	$1.019\frac{16.7}{4}$	....	204-5	sl. s.	∞	∞
5988	nd./aq.	....	147	....	v. s. h.	v. s.	s.
5989	nd./et.	....	97	....	v. sl. s. bz.	s.	v. sl. s.
5990	nd./aq.	....	159-60	....	s. h.	s.	s.
5991	cr./et.	....	39	>325	....	....	....
5992	col. lq.	$0.998\frac{15}{15}$	-13	205.2	i.	∞	∞
5993	col. lq.	$0.976^{15°}$	-23	210^{773mm}	0.09 c.	1.7 h. aq.	....
5994	nd./al.	$0.981\frac{30}{30}$	29.5	217.6	i.	v. s.	v. s.
5995	col. lq.	$0.999\frac{20}{4}$	-16.4 (-24.4)	200.4	$1.5^{25°}$; s. dil. a.	∞	∞
5996	mn. pr.	....	218-20	242	s.	sl. s.	....
5997	col. lq.	$0.989\frac{20}{4}$	-31.3	203.4	sl. s.	∞	∞
5998	lf./aq.	....	228	250	$96.3^{12°}$	$61.9^{9°}$	....
5999	cr.	0.962^{2500} $1.046\frac{20}{4}$	43.8	200.6	$0.74^{21°}$; $1.1^{32°}$	v. s.; s. dil. a.	v. s.; s. CS_2
6000	nd./ac. et.	....	243	257.5	$22.9^{11°}$	$25^{17°}$	i.; i. bz.
6001	yel. nd.	....	68-9	subl.	sl. s. c.	v. s.	v. s.
6002	nd./ toluene	....	139-40	d.	v. sl. s. h.; s. act.	v. s.	v. s.
6003	lf./lg.	....	89-90	265	s. c.	....	....
6004	nd.	....	....	....	v. s.	....	....
6005	rhb.	....	99	283-5	s. h.	s.	s.
6006	nd.	....	....	....	s.	....	....
6007	pl./bz.	....	64	273-4	s.; s. h. bz.	s.	s.
6008	lf.	....	....	....	s.	....	....
6009	pd.	....	....	....	$0.8^{11°}$	....	....
6010	nd./aq.	....	88-9	....	s. h.	....	....
6011	nd.	....	60.5-1.5	....	s. h.	....	....
6012	nd./aq.	....	92-4	265-7	s. h.	s.	s.; s. bz.
6013	lq.	....	....	307	i.	....	s.
6014	cr.	....	54-5	313-4	i.	....	....
6015	pl./et. al.	....	71-2	315-6	i.	....	....
6016	nd.	$1.023^{40°}$	34-6	223^{750mm}	1 c.	v. s. abs.	v. s.
6017	lq.	$0.916^{17°}$	<-20	217	5 c.	s.	s.
6018	nd.	....	60°	217	sl. s.	s.	s.
6019	lq.	....	....	184-7	i.; d. h.	d. h.	s.
6020	pa. yel. lq.	....	....	195-8	i.	s.	s.
6021	lq.	....	....	187^{751mm}	....	....	....
6022	nd.	....	56-9	....	sl. s. lg.	v. s.	v. s.
6023	pl.	....	....	....	s.	s.	s.
6024	oil	....	....	240-4	i.	s.	s.; s. chl.
6025	nd.	....	....	....	s.	s.	....
6026	rhb.	....	65-6	240-4 sl. d.	sl. s.	v. s.; s. bz.	v. s.
6027	tan pl.	....	....	....	s.	sl.	i.
6028	col./bz. et.	....	44	....	sl. s. lg.	s.	s.

Table 7-4 (*Continued*)
PHYSICAL CONSTANTS OF ORGANIC COMPOUNDS

No.	Name	Synonym	Formula	Beil. Ref.	Formula Weight
6029	**Tolyl** hydroxylamine	$(m)(\beta)$	$CH_3 \cdot C_6H_4 \cdot NHOH$	XV-14	123.16
6030	hydroxylamine	$(p)(\beta)$	$CH_3 \cdot C_6H_4 \cdot NHOH$	XV-15	123.16
6031	α-naphthylamine	$(o)(N)$	$C_{10}H_7 \cdot NH \cdot C_6H_4 \cdot CH_3$	XII-1225	233.32
6032	α-naphthylamine	$(p)(N)$	$C_{10}H_7 \cdot NH \cdot C_6H_4 \cdot CH_3$	XII-1225	233.32
6033	β-naphthylamine	$(o)(N)$	$C_{10}H_7 \cdot NH \cdot C_6H_4 \cdot CH_3$	XII-1277	233.32
6034	β-naphthylamine	$(p)(N)$	$C_{10}H_7 \cdot NH \cdot C_6H_4 \cdot CH_3$	XII-1277	233.32
6035	pyrrole $(o)(N)$	....	$C_4H_4N \cdot C_6H_4 \cdot CH_3$	XX-164	157.22
6036	pyrrole $(m)(N)$	....	$C_4H_4N \cdot C_6H_4 \cdot CH_3$	....	157.22
6037	pyrrole $(p)(N)$	....	$C_4H_4N \cdot C_6H_4 \cdot CH_3$	XX-164	157.22
6038	salicylate (o)	$(1,2;1',2')$	$HO \cdot C_6H_4 \cdot CO_2 \cdot C_6H_4 \cdot CH_3$	X-80	228.25
6039	salicylate (m)	$(1,2;1',3')$	$HO \cdot C_6H_4 \cdot CO_2 \cdot C_6H_4 \cdot CH_3$	X-80	228.25
6040	salicylate (p)	....	$HO \cdot C_6H_4 \cdot CO_2 \cdot C_6H_4 \cdot CH_3$	X-80	228.25
6041	*iso*-thiocyanate (o)	o-tolyl mustard oil	$CH_3 \cdot C_6H_4 \cdot N:CS$	XII-813	149.22
6042	*iso*-thiocyanate (m)	m-tolyl mustard oil	$CH_3 \cdot C_6H_4 \cdot N:CS$	XII-865	149.22
6043	*iso*-thiocyanate (p)	p-tolyl mustard oil	$CH_3 \cdot C_6H_4 \cdot N:CS$	XII-956	149.22
6044	thiosemicarbazide	(o)	$CH_3 \cdot C_6H_4 \cdot NH \cdot CS \cdot NH \cdot NH_2$	XII-952	181.26
6045	thiourea (o)	....	$CH_3 \cdot C_6H_4 \cdot NH \cdot CS \cdot NH_2$	XII-806	166.25
6046	thiourea (m)	....	$CH_3 \cdot C_6H_4 \cdot NH \cdot CS \cdot NH_2$	XII-863	166.25
6047	thiourea (p)	....	$CH_3 \cdot C_6H_4 \cdot NH \cdot CS \cdot NH_2$	XII-947	**166.25**
6048	urea (o)	....	$CH_3 \cdot C_6H_4 \cdot NH \cdot CO \cdot NH_2$	XII-801	150.18
6049	urea (m)	....	$CH_3 \cdot C_6H_4 \cdot NH \cdot CO \cdot NH_2$	XII-862	150.18
6050	urea (p)	....	$CH_3 \cdot C_6H_4 \cdot NH \cdot CO \cdot NH_2$	XII-940	150.18
6051	**Tolylene** cyanide (o)	xylylene dicyanide	$C_6H_4(CH_2 \cdot CN)_2$	IX-874	156.19
6052	cyanide (m)	xylylene dicyanide	$C_6H_4(CH_2 \cdot CN)_2$	IX-875	156.19
6053	cyanide (p)	xylylene dicyanide	$C_6H_4(CH_2 \cdot CN)_2$	IX-875	156.19
6054	**Trehalose**	mycose	$C_{12}H_{22}O_{11} \cdot 2H_2O$	XXXI-378	378.33
6055	**Triacetamide**	....	$(CH_3CO)_3N$	II-181	143.14
6056	**Triacetonamine hydrate**	tetramethyl-piperidone (γ)	$C_9H_{17}ON \cdot H_2O$	XXI-249	173.26
6057	**Triacontane** (n)	....	$CH_3(CH_2)_{28}CH_3$	I-176	422.83
6058	**Triamino-azobenzene**	$(2,4,3')$; Bismarck brown	$NH_2 \cdot C_6H_4 \cdot N:N \cdot C_6H_3(NH_2)_2$	XVI-386	227.27
6059	benzene (1,2,3)	....	$(NH_2)_3C_6H_3$	XIII-294	123.16
6060	benzene (1,2,4)	....	$(NH_2)_3C_6H_3$	XIII-294	123.16
6061	benzene HCl	$(1,2,4)$	$(NH_2)_3C_6H_3 \cdot 2HCl$	XIII-295	196.08
6062	benzene (1,3,5)	(free base unknown)	$(NH_2)_3C_6H_3$	XIII-299	123.16
6063	benzene HCl	$(1,3,5)$	$(NH_2)_3C_6H_3 \cdot 3HCl$	XIII-299	232.54
6064	benzoic acid (2,3,5)	....	$(NH_2)_3C_6H_2 \cdot CO_2H$	XIV-455	167.17

Tolyl iodide 3905-7
Tolyl-mercaptan 5894-6
Tolyl-mercuric chloride 4039
Tolyl-mustard oil 6041-3

Tolyl phosphate 6351-3
Tolyl phosphite 6354-5
Tolyl thiophosphate 6356-8
Tolylene alcohol 6497-9

Table 7-4 (*Continued*)
PHYSICAL CONSTANTS OF ORGANIC COMPOUNDS

No.	Crystalline Form and Color	Specific Gravity	Melting Point °C.	Boiling Point °C.	Solubility in 100 Parts		
					Water	Alcohol	Ether
6029	lf./bz. pet.	....	68.5	....	sl. s. h.	s.	s.; sl. s. lg.
6030	lf./bz.	....	93-4	....	1 c.; 5 h.	v. s.	s.; sl. s. bz.
6031	nd./lg.	....	94-5	....	i.; v. s. bz.	v. s.	v. s.
6032	pr./al.	....	78-9	236^{15mm}	s. bz.	s. h.	s.; sl. s. h. pet.
6033	cr./lg.	....	95-6	400-5	v. s. bz.; v. s. lg.	v. s.; v. s. chl.	v. s.; v. s. act.
6034	lf./al.	....	102-3	....	sl. s. lg.	sl. s. c.	s. bz.
6035	col. oil	....	....	246	v. sl. s. h.	v. s.	v. s. bz.
6036	yel. lq.	....	....	$149-51^{41mm}$	i.	s.	s.
6037	lf./aq. al.	....	82	252^{729mm}	v. sl. s. h.	v. s.	v. s. pet.
6038	col. cr.	....	35	....	i.	s.	s.
6039	col. cr.	....	74	....	i.	s.	s.
6040	col. cr.	....	39	....	i.	s.	s.
6041	col. lq.	$1.104\frac{25°}{25}$	....	239	i.	v. s.	∞
6042	lq.	....	<−20	244^{732mm}	....	....	....
6043	nd./et.	$1.087\frac{25°}{25}$	26	237	d. h.	d. h.	s.
6044	lf./bz.	....	145-6	....	i.	s.	s.; i. lg.
6045	cr./aq.	....	151-2	....	v. s. h.	v. s.	v. sl. s.
6046	pr./al.	....	110-1	....	s. h.	s.	s.
6047	pl./al.	....	188	....	v. sl. s. c.	s. h.	....
6048	lf./al.	....	190-2	....	$0.25^{45°}$; s. h.	s.	s.
6049	lf./aq.	....	142-3	....	....	....	....
6050	nd./aq.	....	180-1	....	$0.3^{45°}$	s.	$0.06^{23°}$
6051	cr./et.	....	59-60	....	....	s.	s.
6052	cr.	....	28-9	$305-10^{300mm}$ sl. d.	i.	s.	s.; s. chl.
6053	nd./aq.	....	98	....	sl. s. h.	s.	s.; s. chl.
6054	rhb./al.	....	97*	$-2H_2O > 130^\circ$	s. h.	sl. s. h.	i.
6055	nd./et.	....	78-9	....	....	....	s.
6056	pl./aq.	....	59; 35 (anh.)	205 sl. d.	s.	s.	s.
6057	cr.	lq. $0.780^{65.9°}$	65.9	$235^{1.0mm}$	s. bz.	sl. s.	s.
6058	or. mn./aq.	....	143.5	....	i.	s.	s.
6059	cr.	....	103±	336	v. s.	v. s.	v. s.
6060	lf./chl.	....	<100	340±	v. s.	v. s.	sl. s.; sl. s. chl.
6061	nd.	....	225	....	s.	v. sl. s.	sl. s. HCl
6062	....	....	....	....	....	....	....
6063	col. cr.	....	....	....	s.; d. h.	sl. s.	i.
6064	cr./aq.	....	....	....	s. h.	v. sl. s. h.	i.

* Anhydrous, m. p. 210°.
Toxilic acid 3995
"Traubensäure" 5690

Triacetin 3418
Triacetyl-hydroquinone 3775
Triacetyl-pyrogallol 5516

Table 7-4 (*Continued*)
PHYSICAL CONSTANTS OF ORGANIC COMPOUNDS

No.	Name	Synonym	Formula	Beil. Ref.	Formula Weight
6065	**Triamino** benzoic acid (3,4,5)	....	$(NH_2)_3C_6H_2 \cdot CO_2H \cdot$ $\frac{1}{2}H_2O$	XIV-455	176.18
6066	benzoic acid HCl	(2,4,6)	$(NH_2)_3C_6H_2 \cdot CO_2H \cdot$ 3HCl	XIV-455	276.55
6067	chlorobenzene HCl	(2,4,6)	$(NH_2)_3C_6H_2Cl \cdot 3HCl$	....	266.99
6068	phenol (2,4,6)	....	$(NH_2)_3C_6H_2 \cdot OH$	XIII-569	139.16
6069	toluene (2,4,6)	....	$(NH_2)_3C_6H_2 \cdot CH_3$	XIII-303	137.19
6070	toluene HCl	(2,4,6)	$(NH_2)_3C_6H_2CH_3 \cdot 3HCl$	XIII-303	246.57
6071	triphenylmethane (*o,p′,p″*)	*o*-leucoaniline	$(NH_2 \cdot C_6H_4)_3CH$	XIII-311	289.38
6072	triphenylmethane (*m,p′,p″*)	pseudo-leucoaniline	$(NH_2 \cdot C_6H_4)_3CH$	XIII-312	289.38
6073	triphenylmethane (*p,p′,p″*)	*p*-leucoaniline	$(NH_2 \cdot C_6H_4)_3CH$	XIII-313	289.38
6074	**Triamyl**-amine (*n*)	....	$(C_5H_{11})_3N$	*IV-378	227.44
6075	amine (*iso*)	....	$(C_5H_{11})_3N$	IV-183	227.44
6076	amine (*d*)	....	$[C_2H_5 \cdot CH(CH_3) \cdot$ $CH_2]_3N$	IV-179	227.44
6077	borate (*n*)	*n*-amyl borate	$(C_5H_{11})_3BO_3$	....	272.24
6078	**Triazo**-acetic acid	....	$(N:N):N \cdot CH_2 \cdot CO_2H$	II-229	101.07
6079	benzene	phenyl azide	$C_6H_5 \cdot N_3$	V-276	119.13
6080	**Triazole**	pyrrodiazole	$CH:N \cdot NH \cdot CH:N$	XXVI-13	69.07
6081	**Tribenzamide**	....	$(C_6H_5CO)_3N$	IX-214	329.36
6082	**Tribenzoyl** methane (*α*)	....	$(C_6H_5CO)_2C:$ $C(OH) \cdot C_6H_5$	VII-877	328.37
6083	methane (*β*)	....	$(C_6H_5CO)_3CH$	VII-877	328.37
6084	**Tribenzyl**-amine	....	$(C_6H_5 \cdot CH_2)_3N$	XII-1038	287.41
6085	citrate	benzyl citrate	$(C_6H_5CH_2)_3C_6H_5O_7$	XXVI-421	462.50
6086	**Tribromo**-acetamide	....	$Br_3C \cdot CO \cdot NH_2$	II-221	295.77
6087	acetanilide (2,4,6)	....	$Br_3C_6H_2 \cdot NH \cdot COCH_3$	XII-665	371.87
6088	acetic acid	....	$Br_3C \cdot CO_2H$	II-220	296.76
6089	aniline (2,3,4)	....	$Br_3C_6H_2 \cdot NH_2$	XII-662	329.83
6090	aniline (2,3,5)	....	$Br_3C_6H_2 \cdot NH_2$	XII-662	329.83
6091	aniline (2,4,5)	....	$Br_3C_6H_2 \cdot NH_2$	XII-662	329.83
6092	aniline (2,4,6)	....	$Br_3C_6H_2 \cdot NH_2$	XII-663	329.83
6093	aniline (3,4,5)	....	$Br_3C_6H_2 \cdot NH_2$	XII-668	329.83
6094	anisole (2,4,6)	....	$Br_3C_6H_2 \cdot OCH_3$	VI-205	344.84
6095	benzene (1,2,4)	(*as*)	$Br_3C_6H_3$	V-213	314.82
6096	benzene (1,2,3)	(*v*)	$Br_3C_6H_3$	V-213	314.82
6097	benzene (1,3,5)	(*s*)	$Br_3C_6H_3$	V-213	314.82
6098	benzoic acid (2,4,6)	....	$Br_3C_6H_2 \cdot CO_2H$	IX-360	358.83
6099	benzoic acid (3,4,5)	....	$Br_3C_6H_2 \cdot CO_2H$	IX-361	358.83
6100	butane (1,2,3)	....	$CH_3(CHBr)_2CH_2Br$	I-121	294.83
6101	butane	1,2,3-triBr-2-Me-propane	$(CH_2Br)_2:CBr \cdot CH_3$	I-128	294.83
6102	*tert*-butyl alcohol	acetone-bromoform	$(CH_3)_2C(OH) \cdot CBr_3$	*I-193	310.83
6103	ethane (1,1,2)	vinyl tribromide	$Br \cdot CH_2 \cdot CHBr_2$	I-93	266.77
6104	ethyl alcohol	Avertin	$Br_3C \cdot CH_2OH$	**I-338 ..	282.77
6105	ethylene	ethylene tribromide	$Br \cdot CH:CBr_2$	I-191	264.76
6106	mesitylene	(2,4,6;1,3,5)	$Br_3C_6(CH_3)_3$	V-409	356.90
6107	*β*-naphthol	(1,3,6;2)	$Br_3C_{10}H_4 \cdot OH$	VI-652	380.88
6108	phenol (2,4,6)	(*s*)	$Br_3C_6H_2 \cdot OH$	VI-203	330.82

Triamino-diphenyl 262
Triamino-triazine 4020
Triamino-triphenyl carbinol 5038

Triarachin 3419
Tribenzal-diamine 3700
Tribenzoin 3420

Table 7-4 (*Continued*)
PHYSICAL CONSTANTS OF ORGANIC COMPOUNDS

No.	Crystalline Form and Color	Specific Gravity	Melting Point °C.	Boiling Point °C.	Solubility in 100 Parts		
					Water	Alcohol	Ether
6065	nd./aq.	....	$-H_2O, >$ 100	....	s. h.	i.	i.
6066	pr./aq.	....	....	....	s.	sl. s.	i.
6067	brn. pd.	....	....	....	s.	sl. s.	i.
6068	unstable	....	....	257	....	....	....
6069	oil	....	....	....	....	....	....
6070	nd.	....	....	....	s.; d. h.	....	....
6071	cr./al.	....	165	....	....	....	....
6072	nd./et. lg.*	....	150*	....	v. sl. s. lg.	s.	sl. s.
6073	lf./aq.	....	148	....	sl. s. c.	s. abs.	s. bz.
6074	lq.	....	....	240-5	....	....	....
6075	col. lq.	$0.786\frac{20°}{4}$	....	235	....	....	....
6076	col. lq.	$0.796^{13°}$	....	230-7	i.	v. s.	∞
6077	col. lq.	....	....	$130\text{-}2^{6mm}$	....	....	....
6078	col. hyg.	$1.354^{33°}$	16 ± ; expl.	93^{3mm}	....	....	....
6079	yel. oil	$1.088\frac{20°}{20}$	expl.	73.5^{24mm}	i.	sl. s.	sl. s.
6080	nd./et.	....	120-1	260	v. s.	v. s.	sl. s.
6081	nd./al.	....	207-8	subl.	s. h. bz.	i. c.	v. sl. s.
6082	cr. ppt.	....	210-20	subl.	0.5/3 cc. chl.	s.	s. act.
6083	nd./al.	....	226-31	....	0.5 c. act. 0.2 chl.	0.01 c.	i. 10% Na$_2$CO$_3$
6084	mn./et.	$0.991\frac{9.5°}{4}$	92-3	380-90	v. sl. s.	s. h.	s.
6085	col. pl.	....	51	....	i.	sl. s.	i.
6086	mn.	....	121-2	....	v. sl. s. bz.	s. h.	s. h.
6087	nd./al.	....	232	....	....	....	....
6088	cr.	....	135	245 d.	v. s.; d. h.	v. s.	v. s.
6089	lf./aq. al.	....	100.6	....	sl. s.	v. s.	v. s. bz.
6090	nd./al.	....	91	....	....	....	....
6091	nd./al.	....	85-6	....	v. s. bz.	v. s.	v. s.
6092	rhb./bz.	$2.35\frac{20°}{20}$	119-20	300	i.; s. chl.	v. s. h.	v. s.
6093	nd.	....	118-9	....	i.	s.	s.
6094	nd./al.	....	87-8	....	....	115°	....
6095	nd./al.	....	44-5	275-6	i.; v. s. bz.	v. s. h.	v. s.
6096	col./al.	2.658	87-8	....	....	....	....
6097	nd./al.	....	119-20	271^{765mm}	i.	sl. s. h.	s.
6098	pr./aq.	....	187	....	$0.35^{15°}$	$0.55^{100°}$ aq.	....
6099	nd./aq. al.	....	234-5	....	v. sl. s. h.	s.	s. h.
6100	col. lq.	$2.190\frac{16°}{4}$	-19 ±	$110\text{-}3^{19mm}$	....	....	....
6101	col. lq.	$2.211\frac{14°}{4}$	....	222-5	i.	....	....
6102	cr./aq. al.	....	167-76	....	sl. s.	s.	s.
6103	lq.	$2.579\frac{20°}{4}$	-26	$187\text{-}8^{752mm}$	....	....	....
6104	col. cr.	....	79-80	$92\text{-}3^{10mm}$	$2.54^{0°}$	s.	s.; s. bz.
6105	lq.	$2.708^{20.5°}$	....	163-4	....	....	....
6106	tri./al.	....	223-6	....	i.	v. sl. s. h.	s. bz.
6107	nd./ac.	....	155	....	....	....	....
6108	nd./aq.	$2.55\frac{20°}{20}$	96	subl.	$0.01^{15°}$; $50^{25°}$ bz.	v. s.; v. s. act.	s.; $12^{25°}$ CCl$_4$

* Crysts./bz. + 1C$_6$H$_6$, m. p. 145°.
Tribenzylene-benzene 6362
Tribromo-acetaldehyde 876

Tribromo-hydrin 6112
Tribromo-methane 1016
Tribromo-nitromethane 4688

Table 7-4 (*Continued*)
PHYSICAL CONSTANTS OF ORGANIC COMPOUNDS

No.	Name	Synonym	Formula	Beil. Ref.	Formula Weight
	Tribromo				
6109	phenyl acetate	(1;2,4,6)	$CH_3CO_2 \cdot C_6H_2Br_3$	VI-205	372.85
6110	phenyl salicylate	(2,1;2′,4′,6′); tribromosalol	$HO \cdot C_6H_4 \cdot CO_2 \cdot$ $C_6H_2Br_3$	X-78	450.93
6111	phenyl-*p*-toluene-sulfonate (2,4,6)	....	$CH_3 \cdot C_6H_4 \cdot SO_3 \cdot$ $C_6H_2Br_3$	....	485.01
6112	propane (1,2,3)	glyceryl tribromo-hydrin	$(CH_2Br)_2{:}CHBr$	I-112	280.80
6113	quinaldine (ω)	....	$C_9H_6N \cdot CBr_3$	....	379.89
6114	resorcinol (2,4,6)	....	$Br_3C_6H(OH)_2$	VI-822	346.82
6115	**Tributyl**-amine (*n*)	....	$(C_4H_9)_3N$	IV-157	185.36
6116	amine (*iso*)	....	$(C_4H_9)_3N$	IV-166	185.36
6117	borate (*n*)	*n*-butyl borate	$B(OC_4H_9)_3$	**I-398	230.16
6118	carbinol (*n*)	....	$(C_4H_9)_3C \cdot OH$	**I-464	200.37
6119	citrate (*n*)	*n*-butyl citrate	$(C_4H_9)_3C_6H_5O_7$	**III-371	360.45
6120	phenyl phosphate (*p; tert*)	phosphen 7	$[(CH_3)_3C \cdot C_6H_4 \cdot O]_3PO$	....	494.62
6121	phosphate (*n*)	*n*-butyl phosphate	$(C_4H_9O)_3PO$	**I-397	266.32
6122	**Tri-*iso*-butylene**	....	$(CH_3)_2C{:}C[C(CH_3)_3]_2$	I-225	168.33
6123	**Tricarballylic acid**	....	$(HO_2C \cdot CH_2)_2{:}CH \cdot$ CO_2H	II-815	176.13
6124	**Trichloro**-acetal	....	$Cl_3C \cdot CH(OC_2H_5)_2$	I-621	221.51
6125	acetal	....	$Cl_3C \cdot CH(OC_2H_5)_2$	I-621	221.51
6126	acetamide	....	$Cl_3C \cdot CO \cdot NH_2$	II-211	162.40
6127	acetanilide (2,4,6)	....	$CH_3 \cdot CO \cdot NH \cdot$ $C_6H_2Cl_3$	XII-628	238.50
6128	acetic acid	....	$Cl_3C \cdot CO_2H$	II-206	163.39
6129	acetyl chloride	....	$Cl_3C \cdot CO \cdot Cl$	II-210	181.83
6130	aniline (2,3,4)	....	$Cl_3C_6H_2 \cdot NH_2$	XII-626	196.46
6131	aniline (2,4,5)	....	$Cl_3C_6H_2 \cdot NH_2$	XII-627	196.46
6132	aniline (2,4,6)	....	$Cl_3C_6H_2 \cdot NH_2$	XII-627	196.46
6133	aniline (3,4,5)	....	$Cl_3C_6H_2 \cdot NH_2$	XII-630	196.46
6134	benzene (1,2,3)	(*v*)	$Cl_3C_6H_3$	V-203	181.45
6135	benzene (1,2,4)	(*as*)	$Cl_3C_6H_3$	V-203	181.45
6136	benzene (1,3,5)	(*s*)	$Cl_3C_6H_3$	V-203	181.45
6137	benzoic acid (2,3,4)	....	$Cl_3C_6H_2 \cdot CO_2H$	IX-345	225.46
6138	benzoic acid (2,3,5)	....	$Cl_3C_6H_2 \cdot CO_2H$	IX-345	225.46
6139	benzoic acid (2,4,5)	....	$Cl_3C_6H_2 \cdot CO_2H$	IX-345	225.46
6140	benzoic acid (2,4,6)	....	$Cl_3C_6H_2 \cdot CO_2H$	IX-345	225.46
6141	benzoic acid (3,4,5)	....	$Cl_3C_6H_2 \cdot CO_2H$	IX-346	225.46
6142	bromoethane (1;2,2,2)		$BrCH_2 \cdot CCl_3$	I-90	212.31
6143	bromomethane		$Br \cdot CCl_3$	I-67	198.28
6144	*tert*-butyl acetate	(*β,β,β*)	$CH_3CO_2 \cdot C \cdot$ $(CH_3)_2CCl_3$	II-131	219.50
6145	*tert*-butyl alcohol	acetone chloroform	$Cl_3C \cdot COH{:}(CH_3)_2$	I-382	177.46
6146	*o*-cresol (4,5,6)	(4,5,6;2,1)	$Cl_3C_6H(OH) \cdot CH_3$	*VI-175	211.48
6147	*o*-cresol (3,5,6)	(3,5,6;2,1)	$Cl_3C_6H(OH) \cdot CH_3$	*VI-175	211.48
6148	*m*-cresol (2,4,6)	(2,4,6;3,1)	$Cl_3C_6H(OH) \cdot CH_3$	*VI-189	211.48
6149	*p*-cresol (2,3,6)	(2,3,6;4,1)	$Cl_3C_6H(OH) \cdot CH_3$	VI-404	211.48
6150	dibromoethane (1,1;2,2,2)		$Br_2CH \cdot CCl_3$	I-93	291.21
6151	ethane (1,1,1)	methyl chloroform	$CH_3 \cdot CCl_3$	I-85	133.41
6152	ethane (1,1,2)	vinyl trichloride	$ClCH_2 \cdot CHCl_2$	I-85	133.41
6153	ethyl acetate	....	$CH_3CO_2 \cdot CH_2 \cdot CCl_3$	II-128	191.44
6154	ethyl alcohol	....	$Cl_3C \cdot CH_2OH$	I-338	149.40

Table 7-4 (Continued)
PHYSICAL CONSTANTS OF ORGANIC COMPOUNDS

No.	Crystalline Form and Color	Specific Gravity	Melting Point °C.	Boiling Point °C.	Solubility in 100 Parts		
					Water	Alcohol	Ether
6109	nd./al.	....	82-3	290-300	....	....	....
6110	wh. pd.	....	189	....	i.; s. chl.	sl. s.; s. act.	s. bz.; s. ac.
6111	col. cr.	....	113-4		i.	s.	s.
6112	cr.	$2.436^{23°}$	16-7	219-21	i.	s.	s.
6113	tan pl.	....	127-8	....	i.	i.	i.
6114	nd./aq.	....	111	....	sl. s. c.	s. h.	s.
6115	col. lq.	0.778^{20}_{20}	....	216.5^{761mm}	i.	s.	∞
6116	col. lq.	0.764^{25}_{4}	−22	191.5	i.	s.	∞
6117	col. lq.	....	....	$103-6^{8mm}$	....		....
6118	col. lq.	$0.844^{18°}_{4}$	20	$117-20^{10mm}$	i.	s.	s.
6119	col. lq.	1.045^{20}_{20}	−20	233^{17mm}	i.	∞	∞
6120	col. cr.	....	102-5	300^{5mm}	i.; ∞ bz.	$2^{25°}$	∞ CCl₄
6121	col. lq.	0.976^{25}_{25}	<−80	289 d.	0.6	∞	∞
6122	col. liq.†	0.759^{20}_{4}	<−30‡	$178-80^{752mm}$	i.	....	....
6123	pr./aq.	....	165-6	d.	$41^{15°}$	v. s.	$1^{18°}$
6124	lq.	$1.266^{15°}$	....	197	0.5	∞	∞
6125	solid	....	83	230 d.	....	....	....
6126	mn.	....	141	$238-9^{746mm}$	v. sl. s.	v. s.	v. s.
6127	nd.	....	203-4	....	s. 50% ac.	s.; v. sl. s. CS₂	v. sl. s.; v. sl. s. lg.
6128	cr.	1.617^{46}_{15}	58(50±)	195.5^{754mm}	$120^{25°}$	s.	s.
6129	col. lq.	$1.629^{16.2°}$	....	118	d.	d.	....
6130	nd./lg.	....	67.5	292^{774mm}	s. lg.	v. s.	....
6131	nd./lg.	....	96	270±	sl. s. lg.	s.; s. CS₂	s. 50% ac.
6132	nd./lg.	....	77.5-8.5	262^{746mm}	i. H₃PO₄	s.	s.
6133	nd./aq. al.	....	100	....			
6134	cr./al.	....	52-3	218-9	i.	sl. s.	....
6135	col. cr.	$1.446^{26°}$	17	213	i.	s.	....
6136	nd.	....	63.5	208.5^{764mm}	i.	sl. s.	....
6137	nd.	....	186(129)	....	sl. s.	s.	s.
6138	nd./aq.	....	163	....	v. sl. s. c.	s.	s.
6139	nd./aq.	....	163-4	subl.	v. sl. s. c.	s. c. abs.	....
6140	or./aq.	....	164		sl. s.	s.	s.
6141	nd./aq. al.	....	203	subl.	v. sl. s. c.	s.	s.
6142	lq.	1.884	....	151-3	....	....	....
6143	col. lq.	2.055^{0}_{4}	−21	104.1			
6144	lq.	....	....	191	i.; v. s. act.	v. s.; v. s. chl.	v. s.; v. s. bz.
6145	col. cr.	....	97*	167	0.8 c.	111	s.; s. chl.
6146	col./pet.	....	77	....	sl. s.	s.; s. alk.	s.
6147	nd./ac.	....	62	....	i. pet.	s.; s. alk.	s.
6148	col. nd./aq.	....	47	265	sl. s.	s.; s. alk.	s.; s. lg.
6149	nd./pet.	....	66-7	....	sl. s.	s.; s. alk.	s.
6150	lq.	2.317^{0}_{4}	−4.5	$93-5^{14mm}$	....	....	....
6151	lq.	1.346^{15}_{4}	−32.7	74.0	i.	∞	∞
6152	col. lq.	$1.441^{25.5°}_{4}$	−36.7	113.5	$0.44^{20°}$	∞	∞
6153	col. oil	$1.189^{15°}$	....	170^{747mm}	....	....	....
6154	rhb. pl.	$1.550^{23.3°}$	17.8	151^{737mm}	v. sl. s.	∞	∞

* Forms solid solution with H₂O, m. p. 75-97°.   † High-boiling form, ‡I-180, d. 0.776, b. p. 195-6°
Trichloro-butaldehyde 1158

Table 7-4 (*Continued*)
PHYSICAL CONSTANTS OF ORGANIC COMPOUNDS

No.	Name	Synonym	Formula	Beil. Ref.	Formula Weight
6155	**Trichloro** ethylene	ethylene trichloride	$ClCH:CCl_2$	I-187	131.39
6156	hydroquinone	(2,3,5)	$(HO)_2C_6HCl_3$	VI-850	213.45
6157	iodomethane	....	$Cl_3C \cdot I$	I-71	245.27
6158	lactic acid	....	$Cl_3C \cdot CHOH \cdot CO_2H$	III-286	193.41
6159	methylchloro-formate	diphosgene	$Cl \cdot CO_2 \cdot CCl_3$	III-18	197.83
6160	naphthalene (1,4,5)	(δ)	$C_{10}H_5Cl_3$	V-545	231.51
6161	naphthalene (1,4,6)	(ϵ)	$C_{10}H_5Cl_3$	V-546	231.51
6162	phenol (2,3,5)	....	$Cl_3C_6H_2 \cdot OH$	VI-190	197.45
6163	phenol (2,4,5)	....	$Cl_3C_6H_2 \cdot OH$	**VI-180	197.45
6164	phenol (2,4,6)	omal	$Cl_3C_6H_2 \cdot OH$	VI-190	197.45
6165	phenyl phosphate	(*o*); phosphen 8	$(Cl \cdot C_6H_4 \cdot O)_3PO$	....	429.63
6166	propane (1,2,3)	glyceryl trichloro-hydrin	$(ClCH_2)_2:CHCl$	I-106	147.43
6167	propane (1,1,2)	....	$CH_3 \cdot CHCl \cdot CHCl_2$	I-106	147.43
6168	*iso*-propyl alcohol	(1,1,1); isopral	$Cl_3C \cdot CHOH \cdot CH_3$	I-365	163.43
6169	quinone	....	$O:C_6HCl_3:O$	VII-634	211.43
6170	tribromoethane	(1,2,2;1,1,2)	$Cl_2CBr \cdot CClBr_2$	I-94	370.11
6171	**Tricosane** (*n*)	....	$CH_3(CH_2)_{21}CH_3$	I-175	324.64
6172	**Tricyanogen chloride**	cyanuric chloride	$C_3N_3Cl_3$	XXVI-35	184.41
6173	**Tridecane** (*n*)	....	$CH_3(CH_2)_{11}CH_3$	I-171	184.37
6174	**Tridecyl alcohol** (*n*)	tridecanol-1	$CH_3(CH_2)_{11}CH_2OH$	I-428	200.37
6175	**Tridecylene**	....	$C_{13}H_{26}$	I-225	182.35
6176	**Tridecylic** acid	tridecanoic acid	$CH_3(CH_2)_{11} \cdot CO_2H$	II-364	214.35
6177	aldoxime	....	$C_{12}H_{25} \cdot CH:NOH$	I-715	213.37
6178	**Triethanol**-amine	....	$N(CH_2 \cdot CH_2 \cdot OH)_3$	IV-285	149.19
6179	amine HCl	....	$C_6H_{15}O_3N \cdot HCl$	IV-285	185.65
6180	**Triethylamine**	....	$(C_2H_5)_3N$	IV-99	101.19
6181	hydrobromide	....	$(C_2H_5)_3N \cdot HBr$	IV-101	182.11
6182	hydrochloride	....	$(C_2H_5)_3N \cdot HCl$	IV-101	137.65
6183	**Triethyl** arsine	arsenic triethyl	$(C_2H_5)_3As$	IV-602	162.11
6184	benzene (1,3,5)	(*s*)	$(C_2H_5)_3C_6H_3$	V-449	162.28
6185	benzene (1,2,4)	(*as*)	$(C_2H_5)_3C_6H_3$	V-448	162.28
6186	borate	ethyl borate	$(C_2H_5O)_3B$	I-335	146.00
6187	citrate	ethyl citrate	$C_3H_5O(CO_2C_2H_5)_3$	III-568	276.29
6188	phosphate	ethyl phosphate	$(C_2H_5O)_3PO$	I-332	182.16
6189	phosphine	....	$(C_2H_5)_3P$	IV-582	118.16
6190	phosphine oxide	....	$(C_2H_5)_3PO$	IV-592	134.16
6191	phosphine sulfide	....	$(C_2H_5)_3PS$	IV-592	150.22
6192	phosphite	ethyl phosphite	$(C_2H_5O)_3P$	I-330	166.16
6193	propionate, ortho-	Et orthopropionate	$C_2H_5 \cdot C(OC_2H_5)_3$	II-240	176.26
6194	silicol	....	$(C_2H_5)_3Si \cdot OH$	IV-627	132.28
6195	silicol ethyl ether	....	$(C_2H_5)_3Si \cdot OC_2H_5$	IV-627	160.33
6196	silicon hydride	....	$(C_2H_5)_3Si \cdot H$	IV-625	116.28
6197	silicon oxide	....	$[(C_2H_5)_3Si]_2O$	IV-627	246.54
6198	stibine	antimony triethyl	$(C_2H_5)_3Sb$	IV-618	208.94
6199	**Triethylene** glycol	....	$(HOCH_2CH_2OCH_2)_2$	I-468	150.18
6200	glycol diacetate	....	$(CH_3CO_2 \cdot CH_2 \cdot CH_2 \cdot O \cdot CH_2)_2$	II-141	234.25
6201	tetramine	....	$(H_2N \cdot CH_2 \cdot CH_2 \cdot NH \cdot CH_2)_2$	IV-255	146.24
6202	**Trifluoro**-1-chloro-ethane	(1,2,2)	$F_2CH \cdot CHFCl$	I-83	118.49

Trichloro–hydroxy toluene 6146-9
Trichloro–lactic nitrile 1284
Trichloro–methane 1477
Trichloro–nitromethane 4697

Tricresol, mixt. of 1545-7
Tridecanoic acid 6176
Tridecanol 6174
Tridecanone 4408

Table 7-4 (*Continued*)
PHYSICAL CONSTANTS OF ORGANIC COMPOUNDS

No.	Crystalline Form and Color	Specific Gravity	Melting Point °C.	Boiling Point °C.	Solubility in 100 Parts		
					Water	Alcohol	Ether
6155	col. lq.	$1.466^{20°}_{20°}$	−73	87.2	$0.12^{25°}$	∞	∞
6156	pr./aq.	....	134	subl.	$0.61^{15°}$	v. s.	v. s.
6157	lq.	$2.36^{17°}$	−19	42 d.	....	....	....
6158	cr./et.	....	124	$140\text{-}70^{45mm}$	v. s.	v. s.	s.; s. chl.
6159	lq.	$1.653^{14°}$	−57	127.5	**d. ?**	....	....
6160	nd./al.	....	133	....	....	v. s. h.	....
6161	nd./al.	....	65-6	....	....	sl. s. h.	....
6162	nd./aq. al.	....	55	249-50	sl. s. h.	s.	s.; s. lg.
6163	col./pet.	....	61-3	252	i.; s. CCl_4	s.; s. bz.	s.
6164	nd.	$1.490^{7.5°}_{4}$	68-9	246	$0.09^{25°}$	v. s.	v. s.
6165	col. cr.	$1.38^{60°}_{4}$	35	$255\text{-}65^{5mm}$	i.; ∞ bz.	v. s.	∞ CCl_4
6166	lq.	$1.391^{20°}_{4}$	−14.7	156.6	<0.1	∞	∞
6167	oil	$1.372^{25°}$	<−20	140	i.	∞	∞
6168	col. mn.	....	50-1	161.8^{773mm}	3	s.; s. KOH	s.
6169	yel. lf./aq.	....	168-9	subl.	i. c.	v. s. h.	v. s.
6170	pr.	$2.44^{18°}$	178-80 d.	....	....	....	....
6171	lf.	$0.779^{47.7°}_{4}$	47.7	234^{15mm}	i.	....	....
6172	mn./et.	1.32	145	190	d. h.; s. ac.	v. s. chl.	s. h. abs.
6173	col. lq.	$0.757^{20°}_{4}$	−6.2	234	i.	v. s.	v. s.
6174	cr.	$0.822^{31°}_{4}$	30.5	$155\text{-}6^{15mm}$	....	....	....
6175	lq.	$0.845^{0°}$	....	232.7	i.	s.	v. s.
6176	cr./al.	....	41	$199\text{-}200^{24mm}$	i.	v. s.	v. s.
6177	nd./aq. al.	....	80.5	....	sl. s. bz.	sl. s.	v. s.; s. chl.
6178	col. lq.	$1.126^{20°}_{20°}$	20-1	$277\text{-}9^{150mm}$	∞	∞; s. chl.	sl. s.
6179	cr./al.	....	177	....	v. s.	v. sl. s.	....
6180	col. oil	$0.729^{20°}_{20°}$	−114.7	89.4	∞ < 19°	∞	∞
6181	hex./chl.	1.322	248	subl. > 225	$151^{125°}$	$23^{25°}$ chl.	i.; s. al.
6182	cr./al.	$1.069^{21°}_{4}$	253-4	subl. > 245	$150^{28°}$	s.; s. chl.	i.
6183	col. lq.	$1.150^{20°}_{4}$	....	140^{736} sl. d.	i.	∞	∞
6184	lq.	$0.861^{20°}_{4}$	....	215	i.	s. abs.	s.
6185	lq.	$0.882^{17°}_{4}$	....	$217\text{-}8^{755mm}$	i.	s. abs.	s.
6186	lq.	$0.864^{20°}_{4}$	....	120	d.	....	....
6187	oil	$1.137^{20°}_{4}$	....	294 ±	i.	∞	∞
6188	col. lq.	$1.068^{20°}_{4}$	....	215-6	$100^{25°}$	s.	s.
6189	col. lq.	$0.800^{15°}_{4}$	....	127.5^{744mm}	i.	∞	∞
6190	col. nd.	....	52-3(44)	242.8-3.0	∞; i. KOH	∞	sl. s.
6191	hex. pr.	....	94	ign. 70	....	....	....
6192	lq.	$0.969^{20°}_{4}$	....	$155\text{-}7^{760mm}$	i.	v. s.	v. s.
6193	col. lq.	$0.887^{20°}$	....	161^{766mm}	....	....	....
6194	lq.	$0.871^{0°}$	....	154	i.	....	....
6195	lq.	$0.840^{0°}$	....	153	i.	∞	∞
6196	lq.	$0.721^{15°}_{4}$	....	95-6	i.	i. H_2SO_4	....
6197	lq.	$0.859^{0°}$	....	231	i.	s. H_2SO_4	....
6198	col. lq.	$1.324^{16°}$	< 29	158-9	....	v. s.	v. s.
6199	col. hyg.	$1.125^{20°}_{20°}$	−5	290	∞; i. pet.	∞; ∞ bz.	v. sl. s.
6200	col. lq.	....	....	300	∞	∞	∞
6201	lq.	$0.982^{15°}$	12	266-7	v. s.	v. s.	....
6202	col. lq.	$1.365^{0°}$	....	17	....	v. s.	....

Tridecyl cyanide 4456
Triethyl aluminum 230
Triethyl bismuthine 855

Triethyl boron 871
Triethyl carbinol 3533
Triethyl tin 5953

Table 7-4 *(Continued)*
PHYSICAL CONSTANTS OF ORGANIC COMPOUNDS

No.	Name	Synonym	Formula	Beil. Ref.	Formula Weight
	Trifluoro				
6203	2-chloroethylene	(1,1,2)	$F_2C:CFCl$	....	116.47
6204	1,2-dibromoethane	(1,1,2)	$F_2BrC \cdot CHBrF$	I-92	241.84
6205	1,1,2-tribromo-ethane	(1,2,2)	$Br_2CF \cdot CBrF_2$	I-94	320.74
6206	1,1,2-trichloro-ethane	(1,2,2) Freon-113	$Cl_2CF \cdot CClF_2$	....	187.38
6207	**Triglycol dichloride**	....	$Cl(CH_2 \cdot CH_2 \cdot O)_2 \cdot CH_2 \cdot CH_2 \cdot Cl$	....	187.07
6208	**Triglycyl** glycine	....	$NH_2(CH_2CONH)_3 \cdot CH_2 \cdot CO_2H$	IV-377	246.22
6209	glycine ester	biuret base	$C_7H_{13}O_3N_4 \cdot CO_2C_2H_5$	IV-377	274.28
6210	**Trihexylamine** (*n*)	....	$(C_6H_{13})_3N$	IV-188	269.52
6211	**Trihydroxy**-aceto-phenone (2,3,4)	gallacetophenone; alizarin yellow C	$(HO)_3C_6H_2 \cdot COCH_3$	VIII-393	168.15
6212	anthraquinone	anthragallol (1,2,3)	$C_6H_4(CO)_2C_6H(OH)_3$	VIII-505	256.22
6213	anthraquinone	(1,2,4); purpurin	$C_6H_4(CO)_2C_6H(OH)_3$	VIII-509	256.22
6214	anthraquinone (1,2,6)	flavopurpurin	$HO \cdot C_6H_3(CO)_2C_6H_2:(OH)_2$	VIII-513	256.22
6215	anthraquinone (1,2,7)	anthrapurpurin	$HO \cdot C_6H_3(CO)_2C_6H_2:(OH)_2$	VIII-516	256.22
6216	benzene (1,2,3)(*v*)	pyrogallol	$(HO)_3C_6H_3$	VI-1071	126.11
6217	benzene (1,2,4) (*uns*)	hydroxy-hydro-quinone	$(HO)_3C_6H_3$	VI-1087	126.11
6218	benzene (1,3,5) (*sym*)	phloroglucinol	$(HO)_3C_6H_3 \cdot 2H_2O$	VI-1092	162.14
6219	benzene (1,3,5)	phloroglucinol	$(HO)_3C_6H_3$	VI-1092	126.11
6220	benzoic acid (2,3,4)	pyrogallol carboxylic acid	$(HO)_3C_6H_2 \cdot CO_2H \cdot xH_2O$	X-464	170.12
6221	benzoic acid (2,4,5)	....	$(HO)_3C_6H_2 \cdot CO_2H \cdot \frac{1}{2}H_2O$	X-468	179.13
6222	benzoic acid (2,4,6)	phloroglucinol car-boxylic acid	$(HO)_3C_6H_2 \cdot CO_2H \cdot H_2O$	X-468	188.14
6223	benzoic acid (3,4,5)	gallic acid	$(HO)_3C_6H_2 \cdot CO_2H \cdot H_2O$	X-470	188.14
6224	benzophenone (2,3,4)	alizarin yellow A	$C_6H_5 \cdot CO \cdot C_6H_2(OH)_3 \cdot H_2O$	VIII-417	248.24
6225	benzophenone (2,6,2')	salicyloyl-resorcinol	$(HO)_2C_6H_3 \cdot CO \cdot C_6H_4OH$	VIII-422	230.22
6226	butyrophenone (2,3,4)	*n*-butyropyrogallol	$(HO)_3C_6H_2 \cdot COC_3H_7 \cdot H_2O$	VIII-399	214.22
6227	butyrophenone (*n*)	(2,4,6)	$(HO)_3C_6H_2 \cdot COC_3H_7 \cdot H_2O$	*VIII-691	214.22
6228	glutaric acid	(*d* or *l*)	$(CHOH)_3(CO_2H)_2$	III-553	180.12
6229	glutaric acid (*dl*)	....	$(CHOH)_3(CO_2H)_2$	III-553	180.12
6230	methyl-amino-methane	....	$H_2N \cdot C(CH_2OH)_3$	IV-303	121.14
6231	pyridine (*sym*)	(2,4,6)	$(HO)_3C_5H_2N$	XXI-197	127.10
6232	**Triiodo-**acetic acid	....	$I_3C \cdot CO_2H$	II-225	437.74
6233	benzene (1,2,3)	(*v*)	$I_3C_6H_3$	V-228	455.80
6234	benzene (1,2,4)	(*as*)	$I_3C_6H_3$	V-228	455.80
6235	benzene (1,3,5)	(*s*)	$I_3C_6H_3$	V-228	455.80
6236	benzoic acid	(2,3,5)	$I_3C_6H_2 \cdot CO_2H$	*IX-150	499.81

Trifluoro-methane 3287
Trigemin 364
Trigenolline, cf. alkd.

Trigonelline, cf. alkd.
Triheptylin 3425
Trihydroxy-ethylamine 6178

Table 7-4 (*Continued*)
PHYSICAL CONSTANTS OF ORGANIC COMPOUNDS

No.	Crystalline Form and Color	Specific Gravity	Melting Point °C.	Boiling Point °C.	Solubility in 100 Parts		
					Water	Alcohol	Ether
6203	gas	....	−157.5	−27.9	d.	....	....
6204	col. lq.	$2.254^{14°}$	....	76.5	....	....	....
6205	lq.	$2.5677°$	d. 330	117	....	....	....
6206	lq.	$1.576^{20°}_{4}$	−35	47.6	i.	∞	∞; ∞ bz.
6207	lq.	$1.197^{20°}_{20}$	−31.5	241	$1.9^{20°}$	....	....
6208	col. pd.	....	d. > 220	....	$2^{15°}$; $4^{100°}$	v. sl. s.	....
6209	pl./aq.	....	d. 270	....	v. s. c.	v. sl. s.	v. sl. s.
6210	col. lq.	....	....	263–5	v. sl. s.	v. s.	v. s.
6211	col. cr./aq.	....	173	....	0.2 c.; s. h.	s.	v. sl. s. bz.
6212	cr./al. ac.	....	310 d.	subl. 290±	v. sl. s.	s.; s. H_2SO_4	s.
6213	red nd./al.	....	256–7	....	sl. s. h.	s.	s.
6214	yel. nd./al.	....	>360	459 sl. d.	v. sl. s. h.	s.	sl. s.
6215	or. nd./al.	....	369	462 sl. d.	sl. s. h.	v. s. h.	sl. s.
6216	nd.	$1.453^{4°}$	133–4	309	$40^{13°}$	s.; sl. s. bz.	s.
6217	mn./aq.	....	140.5	....	v. s.	v. s.	v. s.; sl. s. bz.
6218	rhb.	....	117; −2H_2O, 110	subl. sl. d.	$1.13^{25°}$	v. s.	v. s.
6219	cr.	....	209–19	subl.	....	....	....
6220	nd./aq.	....	206 d. (anh.)	subl. in CO_2	$0.13^{12.5°}$	s.	sl. s.
6221	nd./aq.	....	217–8 d. (anh.)	$-\frac{1}{2}H_2O$, 105	s. h.	s.	....
6222	....	....	....	....	sl. s. c.	s.	v. s.
6223	mn./aq.	$1.694^{4°}_{4}$	235 d. (anh.)	−H_2O, 100	$1^{13°}$; $3.3^{100°}$	$28^{15°}$ abs.	$2.5^{15°}$; 20 act.
6224	yel. nd./ aq. al.	....	140–1 (anh.)	....	v. sl. s. c.	s.; s. alk.; sl. s. bz.	s.; s. H_2SO_4
6225	yel. pl./al.	....	133–4	....	v. sl. s. h.	s.; s. alk.	s. bz.
6226	yel. nd./aq.	....	76–80; 100 (anh.)	....	....	....	....
6227	nd./aq.	....	179–80 (anh.)	−H_2O, 110	sl. s.	v. s.	v. s.
6228	lf./al.	....	128	....	v. s.	s.	s. act.
6229	pl./act.	....	154–5 d.	....	v. s.	v. s.	s. act.
6230	nd./al.	....	171–2	219–20^{10mm}	v. s.	$0.4^{20°}$	i.; v. sl. s. act.(d.)
6231	nd. or pd.	....	220–30 d.	....	sl. s.; d. h.	i.	i.
6232	yel. lf.	....	150 d.	....	s.	....	....
6233	nd./al.	....	116	....	i.	v. s.	v. s.
6234	nd./al.	....	91.4	....	i.	s.	s. chl.
6235	nd./ac.	....	182–4	....	i.	sl. s.	sl. s.
6236	pr./al.	....	223–4	....	i.; v. sl. s. bz.	s. h.	i.

Trihydroxy-methylanthraquinone 2874
Trihydroxy-toluene 4361
Trihydroxy-triphenylmethane 3963

Table 7-4 (*Continued*)
PHYSICAL CONSTANTS OF ORGANIC COMPOUNDS

No.	Name	Synonym	Formula	Beil. Ref.	Formula Weight
6237	**Triiodo** ethane (1,1,1)	methyl iodoform	$CH_3 \cdot CI_3$	I-99	407.76
6238	phenol (2,4,6)	. . . .	$I_3C_6H_2 \cdot OH$	VI-211	471.80
6239	**Triketohydrindene hydrate**	ninhydrin	$C_6H_4(CO)_2 : C(OH)_2$	*VII-475	178.15
6240	**Trimethoxy** benzoic acid (2,3,4)	. . . .	$(CH_3O)_3C_6H_2 \cdot CO_2H$	X-465	212.20
6241	benzoic acid (2,4,5)	asaronic acid	$(CH_3O)_3C_6H_2 \cdot CO_2H$	X-468	212.20
6242	benzoic acid (3,4,5)	gallic acid trimethyl ether	$(CH_3O)_3C_6H_2 \cdot CO_2H$	X-481	212.20
6243	**Trimethyl-** acetaldehyde	pivalic aldehyde	$(CH_3)_3C \cdot CHO$	I-688	86.13
6244	acetaldehyde oxime	. . . .	$(CH_3)_3C \cdot CH : NOH$	*I-354	101.15
6245	acetic acid	pivalic acid	$(CH_3)_3C \cdot CO_2H$	II-319	102.13
6246	acetonitrile	*tert*-butyl cyanide	$(CH_3)_3C \cdot CN$	II-320	83.13
6247	acetophenone (2,4,6)	acetomesitylene	$(CH_3)_3C_6H_2 \cdot COCH_3$	VII-332	162.23
6248	amine	. . . .	$(CH_3)_3N$	IV-43	59.11
6249	amine HCl	. . . .	$(CH_3)_3N \cdot HCl$	IV-46	95.57
6250	aniline (2,4,5)	pseudocumidine	$(CH_3)_3C_6H_2 \cdot NH_2$	XII-1150	135.21
6251	aniline (2,4,6)	mesidine	$(CH_3)_3C_6H_2 \cdot NH_2$	XII-1160	135.21
6252	arsine	arsenic trimethyl	$(CH_3)_3As$	IV-600	120.03
6253	benzene (1,2,3)	hemimellitene	$(CH_3)_3C_6H_3$	V-399	120.20
6254	benzene (1,2,4)	pseudocumene	$(CH_3)_3C_6H_3$	V-400	120.20
6255	benzene (1,3,5)	mesitylene	$(CH_3)_3C_6H_3$	V-406	120.20
6256	benzoic acid (2,3,5)	γ-*iso*-durylic acid	$(CH_3)_3C_6H_2 \cdot CO_2H$	IX-552	164.21
6257	benzoic acid (2,3,4)	prehnitylic acid	$(CH_3)_3C_6H_2 \cdot CO_2H$	IX-552	164.21
6258	benzoic acid (2,3,6)	. . . .	$(CH_3)_3C_6H_2 \cdot CO_2H$	IX-552	164.21
6259	benzoic acid (2,4,5)	durylic acid	$(CH_3)_3C_6H_2 \cdot CO_2H$	IX-554	164.21
6260	benzoic acid (3,4,5)	α-*iso*-durylic acid	$(CH_3)_3C_6H_2 \cdot CO_2H$	IX-554	164.21
6261	benzoic acid (2,4,6)	β-*iso*-durylic acid	$(CH_3)_3C_6H_2 \cdot CO_2H$	IX-553	164.21
6262	borate	methyl borate	$(CH_3O)_3B$	I-287	103.91
6263	citrate	. . . .	$C_3H_5O(CO_2CH_3)_3$	III-567	234.21
6264	methoxy-propenyl ammonium bromide	esmodil	$(CH_3)_3N(Br) \cdot CH_2 \cdot C(:CH_2) \cdot OCH_3$	. . . .	210.12
6265	phenol (2,4,5)	pseudocumenol	$(CH_3)_3C_6H_2 \cdot OH$	VI-509	136.20
6266	phenol (2,4,6)	mesitol	$(CH_3)_3C_6H_2 \cdot OH$	VI-518	136.20
6267	phenyl ammonium iodide	. . . .	$(CH_3)_3C_6H_5N \cdot I$	XII-159	263.12
6268	phosphate	methyl phosphate	$(CH_3O)_3PO$	I-286	140.08
6269	phosphine	. . . .	$(CH_3)_3P$	IV-580	76.08
6270	quinoline (2,3,4)	. . . .	$(CH_3)_3C_9H_4N$	XX-414	171.24
6271	quinoline (2,3,6)	. . . .	$(CH_3)_3C_9H_4N$	XX-414	171.24
6272	quinoline (2,4,6)	. . . .	$(CH_3)_3C_9H_4N \cdot aq.$	XX-414	171.24
6273	quinoline (2,6,8)	. . . .	$(CH_3)_3C_9H_4N$	XX-415	171.24
6274	stibine	antimony trimethyl	$(CH_3)_3Sb$	IV-617	166.86
6275	thioacetamide	. . . .	$(CH_3)_3C \cdot CS \cdot NH_2$	. . . .	117.21
6276	urea	. . . .	$(CH_3)_2N \cdot CO \cdot NHCH_3$	IV-74	102.14
6277	**Trimethylene-** acetal	. . . .	$CH_3 \cdot CH \cdot O(CH_2)_3 \cdot O$ ⌞_____⌟	. . . .	102.13
6278	bromohydrin	α-bromohydrin	$Br \cdot (CH_2)_3 \cdot OH$	I-356	139.00

Table 7-4 (*Continued*)
PHYSICAL CONSTANTS OF ORGANIC COMPOUNDS

No.	Crystalline Form and Color	Specific Gravity	Melting Point °C.	Boiling Point °C.	Solubility in 100 Parts		
					Water	Alcohol	Ether
6237	yel. octahed.	....	95 d.	....	v. s. CS_2; v. s. bz.	v. sl. s.	v. s.; sl. s. lg.
6238	nd./aq. al.	....	157-8	d.	s. act.	2	s.
6239	pr./aq.	....	239-40 d.; sl. d. 139	....	s. h.	s. alk.	v. sl. s.
6240	cr./et.	....	97-9	....	s.	....	....
6241	nd./al.	....	144	300±	s. h.	s.; s. lg.	s. bz.
6242	mn./aq.	....	169-70	225-7^{10mm}	v. sl. s.	v. s.	v. s.; s. chl.
6243	lq.	0.793$^{17°}$	3	74-5	....	....	....
6244	cr.	....	41	65^{20mm}	....	....	....
6245	nd.	0.905$^{50°}$	35.5	163.8	2.1$^{20°}$	v. s.	v. s.
6246	cr.	....	15-6	105-6	....	....	....
6247	lq.	0.975$^{20°}_{4}$	....	240.5^{735mm}			
6248	col. gas	0.662$^{-5°}$	−117.1	2.9	41$^{19°}$	s.	s.
6249	cr./al.	....	271-8 d.	....	s.	s.	i.
6250	nd./aq.	....	66-8	234-5	0.12$^{19°}$	....	....
6251	lq.	0.963	....	229-30	....		
6252	col. lq.	1.124$^{22°}$	....	52.8	sl. s.	....	
6253	col. lq.	0.894$^{20°}_{4}$	−25.4	176.1	i.	s.	s.
6254	col. lq.	0.876$^{20°}_{4}$	−43.8	169.4	i.	s.; s. SO_2	s.; s. bz.
6255	col. lq.	0.865$^{20°}_{4}$	−44.7(−52)	164.7	i.	s.; ∞ bz.	∞; s. SO_2
6256	pl./lg.	....	127	....	....	....	
6257	pr./al.	....	167.5	....	....		
6258	nd./aq.	....	105-6	....	....		
6259	nd./bz.	....	149-50	....	v. sl. s. h.	v. s.	v. s.; s. bz.
6260	nd./aq.	....	215-6	....	v. sl. s. h.	s.	s.
6261	cr./al.	....	152-5	....	s. chl.	s.	s.
6262	lq. burns with	0.920$^{23°}_{4}$	−29 green flame	68.7	d.	∞	∞
6263	tri.	....	78-9	283-7 sl. d.	....	....	....
6264	wh. cr. pd.	....	169	....	s.	s.	....
6265	nd.	....	71-2	231-4	v. sl. s. c.	v. s.	v. s.
6266	nd.	....	72	221	v. sl. s.	v. s.	v. s.
6267	lf./al.	....	228-30 d.	subl.	s.	2$^{8°}$	i. chl.
6268	lq.	1.197$^{19.5°}_{0}$	....	197.2	100$^{25°}$	s.	s.
6269	col. lq.	<1	....	40-2	i.	....	s.
6270	cr.	....	65±	285	....	....	s.
6271	cr./lg.	....	86-7	285	sl. s. bz.	s.	v. s.
6272	hyg. cr.	....	63-4	277-8	sl. s.	sl. s.	v. s.
6273	mn./lg.	....	46	260^{719mm}	i.	v. s.	v. s.
6274	col. lq.	1.523$^{15°}$	....	80.6	v. sl. s.	s.	s.
6275	cr.	....	114-6	....	....	....	
6276	mn.	1.19	75.5	232.5^{765mm}	v. s.	v. s.	s.
6277	col. lq.	....	....	109-11	s.	s.	s.
6278	lq.	1.571$^{20°}_{4}$	....	98-112^{185mm}	16.6 c.	....	....

Trimethyl–glycocoll, cf. alkd.
Trimethyl–methane 1022
Trimethyl–pentane 4946
Trimethyl–phenylmethane 1052
Trimethyl–pyridine 1526

Trimethyl–trimethylene glycol 4332
Trimethyl–trithiane 5879-81
Trimethyl–vinyl–ammonium hydroxide 4584
Trimethyl–xanthine, cf. alkd.
Trimethylene 1621

Table 7-4 (*Continued*)
PHYSICAL CONSTANTS OF ORGANIC COMPOUNDS

No.	Name	Synonym	Formula	Beil. Ref.	Formula Weight
	Trimethylene				
6279	chlorobromide	3-Cl-1-Br-propane	$Cl \cdot (CH_2)_3 \cdot Br$	I-109	157.44
6280	chlorohydrin	α-chlorohydrin	$Cl \cdot (CH_2)_3 \cdot OH$	I-356	94.54
6281	diacetate	....	$CH_2(CH_2O_2C \cdot CH_3)_2$	II-143	160.17
6282	diamine	....	$NH_2(CH_2)_3NH_2$	IV-261	74.13
6283	dibutyrate	....	$CH_2(CH_2 \cdot O_2C \cdot C_3H_7)_2$	....	216.28
6284	dimercaptan	propandithiol-1,3	$HS \cdot (CH_2)_3 \cdot SH$	I-476	108.23
6285	formal	1,3-dioxane	$CH_2 \cdot O \cdot (CH_2)_3 \cdot O \vert _____ \vert$	XIX-2	88.11
6286	glycol	propandiol-1,3	$HO \cdot (CH_2)_3 \cdot OH$	I-475	76.10
6287	iodohydrin	α-iodohydrin	$I \cdot (CH_2)_3 \cdot OH$	I-358	185.99
6288	**Trinitro**-acetonitrile	....	$(NO_2)_3C \cdot CN$	II-229	176.05
6289	aniline (2,4,6)	picramide	$(NO_2)_3C_6H_2 \cdot NH_2$	XII-763	228.12
6290	anisole (2,4,6)	methyl picrate	$(NO_2)_3C_6H_2 \cdot OCH_3$	VI-288	243.13
6291	benzaldehyde (2,4,6)	....	$(NO_2)_3C_6H_2 \cdot CHO$	VII-265	241.12
6292	benzene (1,2,3)	(*v*)	$(NO_2)_3C_6H_3$	*V-140	213.11
6293	benzene (1,3,5)	(*s*)	$(NO_2)_3C_6H_3$	V-271	213.11
6294	benzene (1,2,4)	(*as*)	$(NO_2)_3C_6H_3$	V-271	213.11
6295	benzoic acid (2,4,6)	....	$(NO_2)_3C_6H_2 \cdot CO_2H$	IX-417	257.12
6296	*tert*-butyltoluene (2,4,6;1,3)	artificial musk	$(NO_2)_3C_6H:(CH_3) \cdot C(CH_3)_3$	V-439	283.24
6297	*tert*-butylxylene (2,4,6;1,3;5)	musk xylene	$(NO_2)_3C_6(CH_3)_2 \cdot C(CH_3)_3$	V-448	297.27
6298	*m*-cresol	(2,4,6;1,3)	$(NO_2)_3C_6H(CH_3) \cdot OH$	VI-387	243.13
6299	ethane (1,1,1)	....	$CH_3 \cdot C(NO_2)_3$	I-103	165.06
6300	methane	nitroform	$(NO_2)_3CH$	I-79	151.04
6301	naphthalene (1,2,5)	δ-trinitronaphthalene	$(NO_2)_3C_{10}H_5$	V-563	263.17
6302	naphthalene (1,3,5)	α-trinitronaphthalene	$(NO_2)_3C_{10}H_5$	V-563	263.17
6303	naphthalene (1,3,8)	β-trinitronaphthalene	$(NO_2)_3C_{10}H_5$	V-563	263.17
6304	naphthalene (1,4,5)	γ-trinitronaphthalene	$(NO_2)_3C_{10}H_5$	V-563	263.17
6305	α-naphthol (2,4,5)	naphthopicric acid	$(NO_2)_3C_{10}H_4 \cdot OH$	VI-619	279.17
6306	orcinol	(2,4,6;1.3,5)	$(NO_2)_3C_6(OH)_2CH_3$	VI-890	259.13
6307	phenol (γ)	(2,3,6)	$(NO_2)_3C_6H_2 \cdot OH$	VI-265	229.11
6308	phenol (β)	(2,4,5)	$(NO_2)_3C_6H_2 \cdot OH$	VI-265	229.11
6309	phenol (2,3,5)	....	$(NO_2)_3C_6H_2 \cdot OH$	VI-264	229.11
6310	phenol (2,4,6)	picric acid	$(NO_2)_3C_6H_2 \cdot OH$	VI-265	229.11
6311	phenylhydrazine	(2,4,6)	$(NO_2)_3C_6H_2 \cdot N_2H_3$	XV-493	243.14
6312	resorcinol (2,4,6)	styphnic acid	$(NO_2)_3C_6H(OH)_2$	VI-830	245.11
6313	toluene (β)	(2,3,4)	$(NO_2)_3C_6H_2 \cdot CH_3$	*V-172	227.13
6314	toluene (ϵ)	(2,3,5)	$(NO_2)_3C_6H_2 \cdot CH_3$	*V-172	227.13
6315	toluene (γ)	(2,4,5)	$(NO_2)_3C_6H_2 \cdot CH_3$	V-347	227.13
6316	toluene (α)	(2,4,6); TNT	$(NO_2)_3C_6H_2 \cdot CH_3$	V-347	227.13
6317	toluene (δ)	(3,4,5)	$(NO_2)_3C_6H_2 \cdot CH_3$	*V-173	227.13
6318	trimethyl benzene (2,4,6; 1,3,5)	*eso*-trinitro-mesitylene	$(NO_2)_3C_6(CH_3)_3$	V-412	255.19
6319	trimethyl benzene (4,5,6;1,2,3)	trinitro-hemi-mellitene	$(NO_2)_3C_6(CH_3)_3$	V-400	255.19
6320	trimethyl benzene (3,4,6;1,2,4)	*eso*-trinitro-pseudocumene	$(NO_2)_3C_6(CH_3)_3$	V-405	255.19

Table 7-4 (*Continued*)
PHYSICAL CONSTANTS OF ORGANIC COMPOUNDS

No.	Crystalline Form and Color	Specific Gravity	Melting Point °C.	Boiling Point °C.	Solubility in 100 Parts		
					Water	Alcohol	Ether
6279	col. lq.	$1.638^{8°}$	....	142-3	i.	....	....
6280	lq.	$1.131\frac{20}{4}°$	....	160-2	50 c.	s.	s.
6281	col. lq.	$1.070^{19°}$	....	209-10	10	....	....
6282	col. lq.	$0.884\frac{25}{4}°$	−23.5	$135-67^{38mm}$	....	∞	∞
6283	col. lq.	....	....	$125-30^{8mm}$	i.	s.	s.
6284	oil	....	....	169-70	v. sl. s.	∞ ; ∞ chl.	∞ ; ∞ bz.
6285	col. lq.	$1.034\frac{20}{4}°$	−42	105-6	∞	∞	∞
6286	oil	$1.060\frac{20}{4}°$	....	214	∞	∞ ; i. chl.	i. bz.
6287	lq.	$1.998\frac{20}{4}°$	....	225^{748mm}	v. sl. s.	s.	s.
6288	waxy	....	41.5	expl. 220	d.	d.	s.
6289	yel. mn.	....	188-90	expl.	i.	sl. s.	s. h. act.
6290	col. mn.	$1.408^{20°}$	68.4	....	s. bz.	s. ac.	....
6291	pl./bz.	....	119	....	....	....	....
6292	lt. gn./al.	....	127.5	....	i.	10 h.	....
6293	col. rhb.	$1.688\frac{20}{4}°$	121(61)	d.	0.04 c.	$1.9^{17.5°}$	$1.5^{17.5°}$
6294	col. cr.	$1.73^{16°}$	61-2	....	....	$5.5^{15.5°}$	$7.1^{15.5°}$
6295	rhb./aq.	....	210-20 d.	....	$2.05^{24°}$	....	....
6296	lt. yel./al.	....	112-3; (105-6)	....	i.	s.	s.; s. bz.
6297	nd./al.	....	110	....••....	i.	sl. s.	s.
6298	yel. nd./aq.	....	109.5	expl. 150	$0.22^{20°}$; $0.81^{100°}$	v. s.	v. s.
6299	cr.	....	56	....	v. sl. s.	s.; sl. s. lg.	s.
6300	col. cr.	$1.597\frac{24}{4}°$	23; expl.	$45-7^{22mm}$	s.	....	....
6301	nd./al.	....	112-3	....	....	s.	....
6302	rhb./chl.	....	122-3	....	s. ac.	s.	s. chl.
6303	cr./al.	....	218-9	....	$0.02^{100°}$	$*0.05^{23°}$	$0.13^{15°}$
6304	yel./chl.	....	148-9	..... ••	$1.1^{18°}$ bz.	$*0.11^{19°}$	$0.4^{19°}$
6305	yel. nd./aq.	....	190	expl.	sl. s. h.	sl. s.	0.3 c. ac.
6306	yel. nd.	....	162-3	expl.	s. h.	i. aq. a.	v. s. h. bz.
6307	wh. nd.	....	117-8	....	s. h.	v. s.	v. s.
6308	wh. nd./aq.	....	96	....	s. h.	v. s.	v. s.
6309	yel. nd./aq.	....	119-20	....	s. ac.	s.	s. bz.
6310	yel. rhb.	$1.763\frac{20}{4}°$	121.8	expl. > 300	$1.23^{20°}$	$6.23^{20°}$ abs.	$1.08^{13°}$ abs.
6311	yel. nd./al.	....	186	d.	i.; s. ac.	s. h.; i. bz.	i.; i. chl.
6312	yel./act.	1.829	180	....	$0.6^{14°}$	v. s.	v. s.
6313	cr.	$1.620\frac{20}{4}°$	112	expl. 290-310	i.	sl. s. c.	s.
6314	yel. rhb.	....	97.2	d. 335	....	s.	i.
6315	yel. pl/act.	$1.620\frac{20}{4}°$	104	expl. 290	i.	s. h.	v. s.
6316	cr./al.	1.654	80.1	expl. 280	0.15 h.	$1.5^{22°}$	$5^{33°}$
6317	cr.	....	137.5	d. 313	....	$1^{15°}$	....
6318	tri./al.	1.48	232	expl. 415	sl. s. act.	sl. s. h.	sl. s. h.
6319	pl./al.	....	209	....	....	....	....
6320	pr.	....	185	....	....	v. sl. s. h.	v. s. h. bz.

* From 85% alcohol.
Trimyristin 3427
Trinitro-chlorobenzene 5308
Trinitro-cyanomethane 6288

Trinitro-hemimellitene 6319
Trinitro-mesitylene 6318
Trinitro-phenyl-methylnitramine 5873
Trinitro-pseudocumene 6320

Table 7-4 *(Continued)*
PHYSICAL CONSTANTS OF ORGANIC COMPOUNDS

No.	Name	Synonym	Formula	Beil. Ref.	Formula Weight
6321	**Trinitro** triphenyl carbinol	(4,4',4'')	$(NO_2 \cdot C_6H_4)_3C \cdot OH$	VI-720	395.33
6322	triphenyl methane	(4,4',4'')	$(NO_2 \cdot C_6H_4)_3CH$	V-707	379.33
6323	*m*-xylene (4,5,6)	....	$(NO_2)_3C_6H(CH_3)_2$	V-381	241.16
6324	*m*-xylene (2,4,6)	....	$(NO_2)_3C_6H(CH_3)_2$	V-381	241.16
6325	*p*-xylene (2,3,6)	eso-tri-NO_2-*p*-xylene	$(NO_2)_3C_6H(CH_3)_2$	V-389	241.16
6326	**Trional**	Me-Et ketone di-Et sulfone	$(C_2H_5)(CH_3)C:$ $(SO_2C_2H_5)_2$	I-671	242.36
6327	**Triphenyl** acetic acid	....	$(C_6H_5)_3C \cdot CO_2H$	IX-712	288.35
6328	amine	....	$(C_6H_5)_3N$	XII-181	245.33
6329	benzene (1,3,5)	....	$(C_6H_5)_3C_6H_3$	V-737	306.41
6330	bromomethane	....	$(C_6H_5)_3C \cdot Br$	V-704	323.24
6331	carbinol	tritanol	$(C_6H_5)_3C \cdot OH$	VI-713	260.34
6332	carbinol methyl ether	methyl-trityl ether	$(C_6H_5)_3C \cdot OCH_3$	VI-716	274.37
6333	chloromethane	....	$(C_6H_5)_3C \cdot Cl$	V-700	278.78
6334	guanidine (α)	....	$C_6H_5N:C(NHC_6H_5)_2$	XII-451	287.37
6335	guanidine (β)	....	$HN:C(NHC_6H_5)N:$ $(C_6H_5)_2$	XII-430	287.37
6336	hydrazine	....	$(C_6H_5)_2N \cdot NHC_6H_5$	XV-125	260.34
6337	methane	tritane	$(C_6H_5)_3C \cdot H$	V-698	244.34
6338	methane *o*-carboxylic acid	....	$(C_6H_5)_2CH \cdot C_6H_4 \cdot$ CO_2H	IX-714	288.35
6339	methyl	trityl	$(C_6H_5)_3C \cdots$	V-715	243.33
6340	methyl peroxide	....	$[(C_6H_5)_3C \cdot O \cdot]_2$	VI-716	518.66
6341	phosphate	phenyl phosphate	$(C_6H_5O)_3PO$	VI-179	326.29
6342	phosphine	....	$(C_6H_5)_3P$	XVI-759	262.29
6343	phosphite	phenyl phosphite	$(C_6H_5O)_3P$	VI-177	310.29
6344	rosaniline sulfate	Poirrier's blue	$(C_{38}H_{32}N_3)_2SO_4$	XIII-768	1157.46
6345	*p*-rosaniline trisulfonic Na	methyl blue; brilliant cotton blue	$C_{37}H_{27}O_9N_3S_3Na_2$	....	799.81
6346	thiophosphate	....	$(C_6H_5O)_3PS$	VI-181	342.36
6347	**Tri-*iso*-propanolamine**	....	$N(CH_2 \cdot CHOH \cdot CH_3)_3$	....	191.27
6348	**Tripropylamine** (*n*)	....	$(C_2H_5 \cdot CH_2)_3N$	IV-139	143.27
6349	**Trithio-**carbonic acid	....	$HS \cdot CS \cdot SH$	III-221	110.22
6350	phenyl phosphate	....	$(C_6H_5S)_3PO$	VI-182	374.49
6351	**Tritolyl** phosphate (*o*)	tricresyl phosphate	$(CH_3 \cdot C_6H_4 \cdot O)_3PO$	VI-358	368.37
6352	phosphate (*m*)	tricresyl phosphate	$(CH_3 \cdot C_6H_4 \cdot O)_3PO$	....	368.37
6353	phosphate (*p*)	tricresyl phosphate	$(CH_3 \cdot C_6H_4 \cdot O)_3PO$	VI-401	368.37
6354	phosphite (*m*)	tricresyl phosphite	$(CH_3 \cdot C_6H_4 \cdot O)_3P$	VI-381	352.37
6355	phosphite (*p*)	tricresyl phosphite	$(CH_3 \cdot C_6H_4 \cdot O)_3P$	VI-401	352.37
6356	thiophosphate (*o*)	tricresyl thiophosphate	$(CH_3 \cdot C_6H_4 \cdot O)_3PS$	*VI-173	384.44
6357	thiophosphate (*m*)	....	$(CH_3 \cdot C_6H_4 \cdot O)_3PS$	*VI-203	384.44
6358	thiophosphate (*p*)	....	$(CH_3 \cdot C_6H_4 \cdot O)_3PS$	*VI-203	384.44
6359	**Tritriacontane**	....	$C_{33}H_{68}$	I-177	464.91
6360	**Tri-*o*-xenyl phosphate**	tri-*o*-phenylphenyl phosphate	$(C_6H_5 \cdot C_6H_4 \cdot O)_3PO$	....	554.59
6361	**Tropic acid** (*dl*)	....	$C_6H_5 \cdot CH(CH_2OH) \cdot$ CO_2H	X-261	166.18
6362	**Truxene** (α)	tribenzylene benzene	$C_{27}H_{18}$	V-752	342.44

Table 7-4 (*Continued*)
PHYSICAL CONSTANTS OF ORGANIC COMPOUNDS

No.	Crystalline Form and Color	Specific Gravity	Melting Point °C.	Boiling Point °C.	Solubility in 100 Parts		
					Water	Alcohol	Ether
6321	mn. or rhb.	....	193(167)	....	s. bz.; s. ac.	sl. s. h.	sl. s.
6322	cr./bz.	....	212.5	....	v. sl. s. bz.	v. sl. s. ac.	....
6323	pr./al.	$1.494^{19°}$	125	....	i.	$1.2^{20°}$	sl. s.
6324	yel./al. bz.	$1.604^{19°}$	182	....	i.	$0.04^{20°}$	sl. s.
6325	col. mn./al.	$1.59^{19°}$	139–40	expl. 410	....	....	....
6326	pl./al.	$1.199^{8.5°}_{4}$	76	....	$0.3^{15°}$	$5.8^{5°}$ abs.	$6.6^{15°}$
6327	mn.	....	264–5 sl. d.	....	sl. s.	s.	v. sl. s. bz.
6328	mn.	$0.774^{0°}_{0}$	126.5	365	i.; s. act.	sl. s.	s.
6329	rhb.	1.205	170–1	....	v. s. bz.	s. abs.	s.
6330	lt. yel. cr.	1.55	152	230^{15mm}	d.	v. s. CS_2	v. s. bz.
6331	cr./bz.	$1.188^{2.0°}_{4}$	162.5	>360	v. s. bz.	v. s.	v. s.
6332	tri./Me al.	....	87–8	....	....	....	....
6333	col. cr.	....	112–3	$230–5^{20mm}$	d.	v. s. CS_2	v. s. bz.
6334	rhb./al.	1.13	145–7	d.	i.	$4.6^{0°}$ abs.	....
6335	pl.	....	131	....	sl. s. bz.	s.	s.
6336	nd./bz. pet.	....	142	....	i.	s.	v. s. bz.
6337	cr.	$1.014^{2.0°}_{0}$	93.4(81)	$358–9^{754mm}$	i.; 44° bz.	v. s. h.	v. s.
6338	nd./al.	....	162	subl.	i.	s.	s.
6339	col. cr.	....	145–7	d.	i.	sl. s. h.	v. s. chl.
6340	cr./CS_2	....	185–6	....	i.	i.	i.
6341	pr./al.	$1.206^{5.8°}_{4}$	49–50	245^{11mm}	i.	$155^{25°}$	v. s.
6342	mn./et.	1.194	79	>360*	i.; s. HCl	sl. s.	v. s.; s. bz.
6343	lq.	$1.184^{18°}_{18}$	22–4	360	i.; s. chl.	s.; s. bz.	s.
6344	b. pd.	....	....	....		....	....
6345	b. pd.	....	....	....	s.	....	....
6346	pr./al.	$†1.23^{20°}_{4}$	52–3	>360 d.	i.; s. chl.	s.; s. bz.	s.; s. act.
6347	col. pl.	$0.991^{6.0°}_{20}$	58	306.5	s.	s.	s.
6348	col. lq.	$0.757^{2.0°}_{4}$	−93.5	156.5	v. sl. s.	∞	∞
6349	red oil	$1.47^{17°}_{4}$	−30	d. 20–30	i. d.	sl. s.	sl. s.
6350	mn./et.	....	114–5	....	i.	s.	s.; s. chl.
6351	lq.	....	....	410 sl. d.	i.; s. bz.	v. s.	v. s.
6352	col. lq.	....	....	$273–5^{17mm}$	i.	s.	s.
6353	nd./aq.	....	77–8	....	v. s. bz.	v. s.	v. s.
6354	col. lq.	....	....	$235–87^{mm}$	....	....	....
6355	pa. yel. lq.	....	....	$236–97^{mm}$	....	....	....
6356	col. nd./al.	....	45–6	....	i.	sl. s.	v. s.
6357	rhb. cr.	....	33–4	....	i.	sl. s.	v. s.
6358	nd./al.	....	93–4	....	i.; v. s. bz.	v. s. chl.	sl. s. lg.
6359	cr.	lq. $0.780^{71.8°}$	71.8	328^{15mm}	....	....	....
6360	col. rhb.	....	112–3	....	i.	$i.^{25°}$	$8^{25°}$ bz.; 125° CCl_4
6361	nd./aq.	....	117–8	d.	$2^{15°}$; s. h.	s.; sl. s. bz.	s.; i. CS_2
6362	pl./xylene	....	365–8	....	i.; s. aniline	i.; s. $PhNO_2$	i.

* In hydrogen atm.
† Supercooled liquid.
Triphenyl-dihydroglyoxalin 233
Triphenyl-imidazole 3978
Triphenyl-imidazoline 233

Triphenyl-methyl chloride 6333
Triphenyl-oxazole 698
Triphenyl-stibine 563
Triphenyl-stibine dichloride 564
Triphenyl-tin chloride 5948

Tripropin 3432
Tripropionin 3432
Triptane 3519
Triquinoyl 1604
Tristearin 3433

Table 7-4 (*Continued*)
PHYSICAL CONSTANTS OF ORGANIC COMPOUNDS

No.	Name	Synonym	Formula	Beil. Ref.	Formula Weight
6363	**Tryptamine**	aminoethyl-indole (ω)(2)	$NH \cdot C_6H_4 \cdot CH:C\cdot$ $\underline{\hspace{1.2cm}}$ $CH_2 \cdot CH_2 \cdot NH_2$	....	160.22
6364	hydrochloride	....	$C_{10}H_{12}N_2 \cdot HCl$	....	196.68
6365	**Tryptophan** (*l*)	β-indolyl-α-alanine	$C_6H_4 \cdot NH \cdot CH:C\cdot$ $\underline{\hspace{1.2cm}}$ $C_2H_3(NH_2)CO_2H$	XXII-546	204.23
6366	**Tutocaine**	*p*-aminobenzoyl-di-methylamino-1,2-dimethyl propanol HCl	$NH_2 \cdot C_6H_4 \cdot CO_2CH:$ $(CH_3) \cdot CH(CH_3)\cdot$ $CH_2 \cdot N(CH_3)_2 \cdot HCl$	....	286.80
6367	**Tyrosine** (*l*) (−)	β-(*p*-hydroxy-phenyl)-alanine	$HO \cdot C_6H_4 \cdot C_2H_3(NH_2)\cdot$ CO_2H	XIV-605	181.19
6368	**Uliron**	4-(4'-aminophenyl-sulfonamido)-phenylsulfondi-methylamide	$NH_2 \cdot C_6H_4 \cdot SO_2 \cdot NH\cdot$ $C_6H_4 \cdot SO_2N(CH_3)_2$	....	355.44
6369	**Umbelliferone**	7-hydroxy-coumarin	$HO \cdot C_6H_3 \cdot CH:CH \cdot CO \cdot O$ $\underline{\hspace{1.8cm}}$	XVIII-27	162.15
6370	**Undecane** (*n*)	....	$CH_3 \cdot (CH_2)_9 \cdot CH_3$	I-170	156.31
6371	**Undecyl** alcohol (*n*)	undecanol-1	$CH_3(CH_2)_9CH_2OH$	I-427	172.31
6372	alcohol (*n*)(*sec*)	undecanol-2	$C_9H_{19} \cdot CHOH \cdot CH_3$	I-427	172.31
6373	**Undecylene** (α)	undecene-1	$CH_3(CH_2)_8CH:CH_2$	I-225	154.30
6374	**Undecylene** (β)	undecene-2	$C_8H_{17} \cdot CH:CH \cdot CH_3$	I-225	154.30
6375	alcohol	undecene-1-ol-11	$CH_2:CH(CH_2)_9 \cdot OH$	I-452	170.30
6376	**Undecylenic acid**	10-hendecenoic acid	$CH_2:CH(CH_2)_8CO_2H$	II-458	184.28
6377	**Undecylic** acid	undecanoic acid	$CH_3 \cdot (CH_2)_9 \cdot CO_2H$	II-358	186.30
6378	aldehyde	undecanal	$CH_3 \cdot (CH_2)_9 \cdot CHO$	I-712	170.30
6379	aldehyde oxime	....	$CH_3(CH_2)_9CH:NOH$	I-713	185.31
6380	amide	....	$CH_3(CH_2)_9CO \cdot NH_2$	II-358	185.31
6381	nitrile	decyl cyanide	$CH_3 \cdot (CH_2)_9 \cdot CN$	II-358	167.30
6382	**Undecyne**-1	rutylidene	$CH_3 \cdot (CH_2)_8 \cdot C:CH$	I-261	152.28
6383	**Uracil** (2,4)	2,6-dioxy-pyrimi-dine	$CH:CH \cdot CO \cdot NH\cdot$ $\underline{\hspace{1cm}}$ $CO \cdot NH$ $\underline{\hspace{0.6cm}}$	XXIV-312	112.09
6384	**Uramil**	5-amino-barbituric acid	$(CONH)_2COCH \cdot NH_2$ $\underline{\hspace{1.6cm}}$	XXV-492	143.10
6385	**Uraminobenzoic acid** (*o*)	....	$NH_2 \cdot CO \cdot NH \cdot C_6H_4\cdot$ CO_2H	....	180.16
6386	**Urea**	carbamide	$NH_2 \cdot CO \cdot NH_2$	III-42	60.06
6387	calcium chloride	afenil	$CaCl_2 \cdot 4CH_4ON_2$	*III-26	351.21
6388	hydrochloride	....	$CO(NH_2)_2 \cdot HCl$	III-54	96.52
6389	nitrate	acidogen nitrate	$CO(NH_2)_2 \cdot HNO_3$	III-54	123.07
6390	oxalate	....	$2CH_4ON_2 \cdot C_2H_2O_4$	III-55	210.15
6391	oxalate	....	$2CH_4ON_2 \cdot C_2H_2O_4\cdot$ $2H_2O$	III-55	246.18
6392	**Urethane**	ethyl carbamate	$NH_2 \cdot CO_2 \cdot C_2H_5$	III-22	89.09
6393	**Uric acid**	2,6,8-trioxy-purine	$C_5H_4O_3N_4$	XXVI-513	168.11
6394	**Usnic acid** (*d*)	....	$C_{18}H_{16}O_7$	XIX-316	344.32
6395	**Usnic acid** (*dl*)	....	$C_{18}H_{16}O_7$	XIX-316	344.32

Table 7-4 (*Continued*)
PHYSICAL CONSTANTS OF ORGANIC COMPOUNDS

No.	Crystalline Form and Color	Specific Gravity	Melting Point °C.	Boiling Point °C.	Solubility in 100 Parts		
					Water	Alcohol	Ether
6363	cr.	....	120	....	....	....	....
6364	col. cr.	....	255-8	....	s.	sl. s.	i.
6365	hex., rhb. nd.	....	289	....	sl. s. c.; v. s. h.	i.; i. alk.	i.; i. chl.
6366	col. cr. pd.	....	212-5	....	25	2.5	....
6367	nd./aq.	$1.456^{20°}$	>290 d. *314 d.	....	$0.04^{17°}$; $0.65^{100°}$	$0.01^{17°}$; s. alk.	i.; s. a.
6368	col. cr. pd.	....	193-5	....	sl. s.; v. s. alk.	s.	s. act.
6369	nd./aq.	....	224-7	subl.	1 h.	s.; s. HCl	sl. s.; s. ac.
6370	col. lq.	$0.741\frac{20°}{4}$	-25.6	194.5	i.	∞	∞
6371	lq.	$0.822\frac{35°}{4}$	19	131^{15mm}	<0.02	s.	....
6372	lq.	$0.827\frac{20°}{4}$	12	228-9	i.	s.	....
6373	col. lq.	$0.763\frac{20°}{4}$	....	188-90	i.	∞	∞
6374	lq.	$0.774\frac{15°}{15}$	....	192-3	i.	∞	∞
6375	lq.	$0.850^{15°}$	-2	245^{756mm}	....	....	....
6376	cr.	$0.907\frac{24°}{4}$	24.5	295 d.	i.	s.	s.; s. chl.
6377	col. cr.	$0.891^{30°}$	29-30	228^{160mm}	i.	s.	s.
6378	lq.	$0.862\frac{15°}{15}$	-4	$116\text{-}7^{18mm}$	....	....	....
6379	nd./Me al.	....	72(61)	....	....	v. s.	v. s.
6380	col. cr.	....	103	....	....	....	....
6381	col. lq.	....	....	253-4	....	s.	s.
6382	lq.	$0.867\frac{25°}{4}$	-33	210-5	i.	s.	s.
6383	nd./aq.	....	338 d.	....	s. h.; s. NH_4OH	i.	i.
6384	nd.	....	>400	....	sl. s. h.; s. alk.	s. c. H_2SO_4	i.; s. NH_3
6385	nd.	....	152	d. 171-2	d. h.	s.; s. Me al.	s. act.
6386	col. pr.	$1.335\frac{20°}{4}$	132.7	d.	$100^{17°}$; ∞ h.	$20^{20°}$	sl. s.
6387	wh. hyg. pd.	....	158-60	....	v. s.	sl. s.	i. Me al.
6388	wh. hyg. lf.	....	d. 145	....	s.	....	....
6389	col. mn. pr.	....	152 d.	....	v. s. h.	s.	i. HNO_3
6390	mn. pr.	....	170-1	d.	$4.4^{16°}$	$1.6^{16°}$	i.
6391	....	....	d. > 120	$-H_2O$, 120	....	....	....
6392	col. lf.	$1.11\frac{20°}{20°}$	49-50	184	v. s.	v. s.	v. s.
6393	cr.	$1.893^{20°}$	d.	....	0.06 h.	i.; s. aq. LiOH	i.
6394	cr.	....	203	d.	i.	v. sl. s.	sl. s.
6395	yel. pr.	....	195-6	....	i.	v. sl. s. c.	$0.3^{20°}$

* Heated rapidly.
Trypaflavine 5328
Tryparsamide 583
Turicine, cf. alkd.
Tussol 568

Tylcalsin 116
Tyllithin 117
Tylnatrin 119
Tyramine 3799
Ulexine, cf. alkd.

Ultraquinine, cf. alkd.
Umbellic acid 2333
Undecanal 6378
Undecanoic acid 6377
Undecanol 6371-2

Undecanone 3107, 4324
Undecene 6373-4
Undecenol 6375
Undecyl cyanide 3956
Ural 1288

Table 7-4 (*Continued*)
PHYSICAL CONSTANTS OF ORGANIC COMPOUNDS

No.	Name	Synonym	Formula	Beil. Ref.	Formula Weight
6396	**Uvinic acid**	2,5-diMe-furfurane-3-carboxylic acid	$(CH_3)_2C_4HO \cdot CO_2H$	XVIII-297	140.14
6397	**Uvitic acid** (5;1,3)	5-methyl-isophthalic acid	$CH_3 \cdot C_6H_3(CO_2H)_2$	IX-864	180.16
6398	**Uvitonic acid** (2;4,6)	o-picoline-o,p-dicarboxylic acid	$CH_3 \cdot C_5H_2N(CO_2H)_2$	XXII-161	181.15
6399	**Valeric** acid (*n*)	pentanoic acid	$C_2H_5 \cdot CH_2 \cdot CH_2 \cdot CO_2H$	II-299	102.13
6400	aldehyde (*n*)	pentanal	$C_2H_5 \cdot CH_2 \cdot CH_2 \cdot CHO$	I-676	86.13
6401	aldehyde oxime	....	$C_4H_9 \cdot CH:NOH$	I-676	101.15
6402	amide (*n*)	....	$CH_3(CH_2)_3CONH_2$	II-301	101.15
6403	anhydride (*n*)	....	$(C_4H_9CO)_2O$	II-301	186.25
6404	chloride (*n*)	n-valeryl chloride	$CH_3 \cdot (CH_2)_3 \cdot COCl$	II-301	120.58
6405	nitrile (*n*)	n-butyl cyanide	$CH_3(CH_2)_3 \cdot CN$	II-301	83.13
6406	*iso*-**Valeric** acid	β-Me-butyric acid	$(CH_3)_2CH \cdot CH_2 \cdot CO_2H$	II-309	102.13
6407	aldehyde	2-methyl-butanal-4	$(CH_3)_2CH \cdot CH_2 \cdot CHO$	I-684	86.13
6408	aldehyde oxime	....	$C_4H_9 \cdot CH:NOH$	I-686	101.15
6409	amide	....	$C_4H_9 \cdot CO \cdot NH_2$	II-315	101.15
6410	anhydride	....	$(C_4H_9CO)_2O$	II-314	186.25
6411	anilide	....	$C_4H_9 \cdot CO \cdot NHC_6H_5$	XII-254	177.25
6412	chloride	*iso*-valeryl chloride	$(CH_3)_2CH \cdot CH_2COCl$	II-315	120.58
6413	nitrile	*iso*-butyl cyanide	$(CH_3)_2CH \cdot CH_2 \cdot CN$	II-315	83.13
6414	**Valeryl diethyl-amide** (*iso*)	Valyl	$(CH_3)_2CH \cdot CH_2 \cdot CO \cdot N(C_2H_5)_2$	....	157.26
6415	**Valylene**	....	$CH_2:C(CH_3):C:CH$	I-263	66.10
6416	**Vanillalacetone** (1;3,4)	ferulic methyl ketone	$CH_3 \cdot CO \cdot CH:CH \cdot C_6H_3(OCH_3)OH$	VIII-291	192.22
6417	**Vanillic acid** (3;4,1)	4-OH-3-MeO-benzoic acid	$(CH_3OC_6H_3(OH) \cdot CO_2H$	X-392	168.15
6418	**Vanillin** (3;4,1)	4-OH-3-MeO-benzaldehyde	$(CH_3O)C_6H_3(OH) \cdot CHO$	VIII-247	152.15
6419	ethyl ether	4-EtO-3-MeO-benzaldehyde	$(CH_3O)C_6H_3:(OC_2H_5)CHO$	VIII-256	180.21
6420	*p*-phenetidine ethyl carboxylate	(4,1;1′,3′,4′); Eupyrin	$C_2H_5O \cdot C_6H_4 \cdot N:CH \cdot C_6H_3(OCH_3)O \cdot CO_2 \cdot C_2H_5$	....	343.38
6421	*iso*-**Vanillin**	3-OH-4-MeO-benzaldehyde	$(CH_3O)C_6H_3(OH) \cdot CHO$	VIII-254	152.15
6422	**Vanillyl alcohol**	4-OH-3-MeO-benzyl alcohol	$(CH_3O)C_6H_3(OH) \cdot CH_2OH$	VI-1113	154.17
6423	**Veratric** acid	3,4-dimethoxy-benzoic acid	$(CH_3O)_2C_6H_3 \cdot CO_2H$	X-393	182.18
6424	aldehyde (3,4;1)	vanillin methyl ether	$(CH_3O)_2C_6H_3 \cdot CHO$	VIII-255	166.18
6425	**Vinaconic acid**	ethylene malonic acid	$CH_2 \cdot CH_2 \cdot C(CO_2H)_2$	IX-722	130.10
6426	**Vinyl** acetate	....	$CH_3CO_2 \cdot CH:CH_2$	*II-63	86.09
6427	acetic acid	butenoic acid	$CH_2:CH \cdot CH_2 \cdot CO_2H$	II-407	86.09
6428	acrylic acid (β)	....	$CH_2:(CH)_2:CH \cdot CO_2H$	II-481	98.10
6429	alcohol	ethenol	$CH_2:CH \cdot OH$	I-601	44.05
6430	amine†	....	$CH_2:CH \cdot NH_2$	IV-203	43.07
6431	anisole (*o*)	o-methoxy styrene	$CH_2:CH \cdot C_6H_4 \cdot OCH_3$	VI-560	134.18
6432	anisole (*m*)	m-methoxy styrene	$CH_2:CH \cdot C_6H_4 \cdot OCH_3$	VI-561	134.18
6433	anisole (*p*)	p-methoxy styrene	$CH_2:CH \cdot C_6H_4 \cdot OCH_3$	VI-561	134.18
6434	bromide	bromo-ethylene	$CH_2:CH \cdot Br$	I-188	106.96

† This product is actually ethylene imine, $CH_2 \cdot CH_2 \cdot NH$.

Uraline 1288
Uralium 1288
Uranin 3251
Urazine 2851

Urea chloride 1229
Ureido-hydantoin, cf. alkd.
Ureous acid, cf. alkd.
Urotropine 3599

Uroxin 188
Ursin, cf. glcde.
Ursol-D 5260
Ursol-P 349
Vacciniin, cf. glcde.

Table 7-4 (*Continued*)
PHYSICAL CONSTANTS OF ORGANIC COMPOUNDS

No.	Crystalline Form and Color	Specific Gravity	Melting Point °C.	Boiling Point °C.	Solubility in 100 Parts		
					Water	Alcohol	Ether
6396	cr./aq.	....	135	subl.	0.25 h.	s.	v. s.
6397	nd./aq.	....	290-1	subl.	sl. s. h.	s.	s.
6398	cr. pd.	....	274-82 d.	....	i. c.; s. a.	s. h. aniline	v. sl. s. h. bz.
6399	col. lq.	$0.940_4^{20°}$	−34.5(−59)	186.4	$3.3^{16°}$	∞	∞
6400	lq.	$0.819^{11°}$	−92	103.4	v. sl. s.	s.	s.
6401	cr.	....	52	....	....	....	....
6402	mn. pl.	1.023	106	....	v. s.	v. s.	v. s.
6403	lq.	$0.922_4^{17°}$	−56.1	227.5	d. h.	....	....
6404	col. lq.	$1.016^{15°}$	−110	127-8	d.	d.	....
6405	col. lq.	$0.804_4^{15°}$	−96	140.8	i.	s.	s.
6406	col. lq.	$0.925_4^{20°}$	−29.3	176.5	$4.2^{20°}$	∞ ; ∞ chl.	∞
6407	col. lq.	$0.803^{17°}$	−51	92.5	sl. s.	s.	s.
6408	oil	$0.893_4^{20°}$	....	161.3^{759mm}	....	s.	s.
6409	mn.	$0.965_4^{20°}$	135-7	232	s.	s.	s.
6410	col. lq.	$0.929_4^{27°}$	....	215	....	....	....
6411	nd./lg.	1.078	113-4	....	sl. s. h.	s.	s.
6412	col. lq.	$0.989_4^{20°}$	....	116-7.5	d.	d.	s.
6413	col. lq.	$0.795_4^{15°}$	−100.9	130.3	....	....	....
6414	col. lq.	....	....	210	4	s.	s.
6415	col. lq.	....	....	50	....	....	....
6416	yel. nd./ aq. al.	....	129-30	....	v. sl. s.	s.; s. conc. H₂SO₄	s.; s. bz.
6417	nd./aq.	....	207	subl.	$0.12^{14°}$; $2.5^{100°}$	v. s.	v. s.
6418	mn.	1.056	82-3.5	285(in CO₂)	$114°$; $575°$	v. s.; v. s. chl.	v. s.; v. s. CS₂
6419	mn.	....	64-5	subl.	v. sl. s. h.	s.	s.
6420	yel. cr.	....	87-8	....	sl. s.	s. h.	s.; s. chl.
6421	mn.	$1.196_4^{20°}$	115-7	subl. sl. d.	s. h.	s.; sl. s. CS₂	s.
6422	mn./aq.	....	115	d.	v. s. h.	v. s.	v. s.
6423	cr./aq.*	....	180-1	subl.	$0.05^{14°}$; $0.6^{100°}$	v. s.	v. s.
6424	nd./et.	....	44-7	280-5	sl. s. h.	s.	s.
6425	tri./et.	....	175	210^{30mm}	v. s.	s.	....
6426	col. lq.†	$0.932_4^{20°}$	−92.8	72-3	$2^{20°}$	∞	∞
6427	col. lq.	$1.013_{15}^{15°}$	−39	163	s.	∞	∞
6428	hyg. pr./ et.	....	80	d. 110-5	s. h.	s.	s.; sl. s. pet.
6429	not known;	acetalde-	hyde	des-	motrope	....	....
6430	lq.	0.832	....	56	....	s.	....
6431	lq.	$1.005_4^{17.2°}$	....	195-200	i.	s.	s.
6432	lq.	....	....	$89-90^{14mm}$	i.	s.	s.
6433	lq.	$1.000_4^{13°}$	....	$204-5^{756mm}$	i.	s.	s.
6434	lq.	$1.529_4^{11°}$	−137.8	15.8	i.	∞	∞

* Crysts. + 1H₂O < 50°.
† Polymerizes in light.
Valamin 497
Valdivin, cf. glcde.
Valerone 1901

Valerophenone 1117-8
Valeryl chloride 6404, 6412
Valerylene 5087
Validol 4032
Valine 391-5

Valisan 863
Valyl 6414
Valzin 5112
Vanillal 3188
Vanillin methyl ether 6424

Table 7-4 (*Continued*)
PHYSICAL CONSTANTS OF ORGANIC COMPOUNDS

No.	Name	Synonym	Formula	Beil. Ref.	Formula Weight
6435	**Vinyl** chloride	chloro-ethylene	$CH_2:CH\cdot Cl$	I-186	62.50
6436	ether	....	$(CH_2:CH)_2O$	I-433	70.09
6437	ethyl alcohol	allyl carbinol	$C_3H_5\cdot CH_2OH$	I-441	72.11
6438	ethyl ether	ethyl-vinyl ether	$CH_2:CH\cdot O\cdot C_2H_5$	I-433	72.11
6439	fluoride	fluoro ethylene	$CH_2:CH\cdot F$	I-186	46.04
6440	glycolic acid	....	$CH_2:CH\cdot CHOH\cdot CO_2H$	III-370	102.09
6441	guaiacol (1;3,4)	hesperetole	$CH_2:CH\cdot C_6H_3(OH)\cdot$ (OCH_3)	VI-954	150.18
6442	iodide	iodoethylene	$CH_2:CH\cdot I$	I-192	153.95
6443	phenol (*o*)	*o*-hydroxy styrene	$CH_2:CH\cdot C_6H_4\cdot OH$	VI-560	120.15
6444	phenol (*m*)	*m*-hydroxy styrene	$CH_2:CH\cdot C_6H_4\cdot OH$	VI-561	120.15
6445	**Violuric** acid	nitroso-barbituric acid	$CO\cdot(NHCO)_2\cdot C:NOH$	XXIV-506	157.09
6446	sodium salt	....	$C_4H_2O_4N_3Na$	XXIV-507	179.07
6447	**Xanthene**	2,2'-methylene-diphenyl ether	$C_6H_4\cdot CH_2\cdot C_6H_4\cdot O$	XVII-73	182.22
6448	**Xanthic acid***	xanthogenic acid	$C_2H_5O\cdot CS\cdot SH$	III-209	122.21
6449	**Xanthine†**	2,6-dioxy-purine	$C_5H_4O_2N_4$	XXVI-447	152.11
6450	**Xanthogenamide**	Et-thioncarbamate	$C_2H_5O\cdot CS\cdot NH_2$	III-137	105.16
6451	**Xanthone**	benzophenone oxide	$OC:(C_6H_4)_2O$	XVII-354	196.21
6452	**Xanthophyll**	lutein	$C_{40}H_{56}O_2$	XXX-95	568.89
6453	**Xanthydrol**	9-OH-xanthene	$HOCH:(C_6H_4)_2O$	XVII-129	198.22
6454	**Xylene** (*o*)	1,2-dimethylbenzene	$C_6H_4(CH_3)_2$	V-362	106.17
6455	sulfonic acid	(1,2;4)	$(CH_3)_2C_6H_3\cdot SO_3H\cdot$ $2H_2O$	XI-121	222.26
6456	sulfonic Na	(1,2;4)	$(CH_3)_2C_6H_3\cdot SO_3Na\cdot$ $5H_2O$	XI-121	298.29
6457	**Xylene** (*m*)	1,3-dimethylbenzene	$C_6H_4(CH_3)_2$	V-370	106.17
6458	sulfonic acid	(1,3;4)	$(CH_3)_2C_6H_3\cdot SO_3H\cdot$ $2H_2O$	XI-123	222.26
6459	sulfonic Na	(1,3;4)	$(CH_3)_2C_6H_3\cdot SO_3Na\cdot$ H_2O	XI-123	226.23
6460	sulfonic Ca	(1,3;4)	$[(CH_3)_2C_6H_3\cdot SO_3]_2Ca$	....	410.53
6461	sulfonyl chloride	(1,3;4)	$(CH_3)_2C_6H_3\cdot SO_2Cl$	XI-123	204.68
6462	**Xylene** (*p*)	1,4-dimethylbenzene	$C_6H_4(CH_3)_2$	V-382	106.17
6463	sulfonic acid	(1,4;2)	$(CH_3)_2C_6H_3\cdot SO_3H\cdot$ $2H_2O$	XI-127	222.26
6464	sulfonic Na	(1,4;2)	$(CH_3)_2C_6H_3\cdot SO_3Na\cdot$ H_2O	XI-127	226.23
6465	**Xylenol** (*v*)(*o*)	2,3-dimethyl phenol	$(CH_3)_2C_6H_3\cdot OH$	VI-480	122.17
6466	**Xylenol** (*as*)(*o*)	3,4-dimethyl phenol	$(CH_3)_2C_6H_3\cdot OH$	VI-480	122.17
6467	**Xylenol** (*s*)(*m*)	3,5-dimethyl phenol	$(CH_3)_2C_6H_3\cdot OH$	VI-492	122.17
6468	**Xylenol** (*v*)(*m*)	2,6-dimethyl phenol	$(CH_3)_2C_6H_3\cdot OH$	VI-485	122.17
6469	**Xylenol** (*as*)(*m*)	2,4-dimethyl phenol	$(CH_3)_2C_6H_3\cdot OH$	VI-486	122.17
6470	**Xylenol** (*p*)	2,5-dimethyl phenol	$(CH_3)_2C_6H_3\cdot OH$	VI-494	122.17
6471	**Xylidine** (*v*)(*o*)	2,3-dimethyl aniline	$(CH_3)_2C_6H_3\cdot NH_2$	XII-1101	121.18
6472	**Xylidine** (*as*)(*o*)	3,4-dimethyl aniline	$(CH_3)_2C_6H_3\cdot NH_2$	XII-1103	121.18
6473	**Xylidine** (*s*)(*m*)	3,5-dimethyl aniline	$(CH_3)_2C_6H_3\cdot NH_2$	XII-1131	121.18
6474	**Xylidine** (*v*)(*m*)	2,6-dimethyl aniline	$(CH_3)_2C_6H_3\cdot NH_2$	XII-1107	121.18
6475	**Xylidine** (*as*)(*m*)	2,4-dimethyl aniline	$(CH_3)_2C_6H_3\cdot NH_2$	XII-1111	121.18
6476	acetate	....	$C_8H_{11}N\cdot HC_2H_3O_2$	....	181.24
6477	**Xylidine** (*p*)	2,5-dimethyl aniline	$(CH_3)_2C_6H_3\cdot NH_2$	XII-1135	121.18

† See also Alkaloid table.
Vanirome 3188
Vasicine, cf. alkd.
Veratridine, cf. alkd.
Veratrine, cf. alkd.

Veratrole 2406
Veratroylaconine, cf. alkd.
Veritol 4117
Veronal 2124
Vesipyrin 120

Vicianin, cf. glcde.
Vicine, cf. alkd.
Victoria yellow 4056
Vinetine, cf. alkd.
Vinopyrin 5110

* Really (new nomenclature) ethyl xanthic acid. Xanthic acid (hypothetical) is HOCSSH.

Table 7-4 (*Continued*)
PHYSICAL CONSTANTS OF ORGANIC COMPOUNDS

No.	Crystalline Form and Color	Specific Gravity	Melting Point °C.	Boiling Point °C.	Solubility in 100 Parts		
					Water	Alcohol	Ether
6435	gas	$0.908\frac{25}{25}°$	−160	−13.9	sl. s.	s.	v. s.
6436	col. lq.	$0.773\frac{20}{20}°$	....	28.3	v. sl. s.	∞	∞
6437	lq.	$0.838\frac{17.5°}{4}$	....	113.5^{748mm}	s.	....	....
6438	lq.	$0.763\frac{14.5°}{17.5}$	−115.3	35.5	v. sl. s.	s.	....
6439	col. gas	$0.853^{-26°}$	−160.5	−72.2	i.	$400^{20°}$ cc.	$550^{20°}$ cc. act.
6440	hyg. nd.	....	33-40	$129\text{-}30^{12mm}$	v. s.	s.	s.; i. CS_2
6441	cr.	....	57	....	sl. s.	s.	s.
6442	lq.	$2.08^{0°}$	....	56	....	....	....
6443	nd.	$1.061\frac{19.2°}{4}$	29	108^{15mm}	s. alk.	s.	s.
6444	oil	....	....	$114\text{-}6^{17mm}$	....	....	....
6445	rhb.	....	224 d.	$-H_2O$, 100	s. h.	s.	....
6446	lf./al.	....	100.5	315	sl. s.	....	....
6447	lf./al.	....	100.5	315	sl. s.; s. H_2SO_4	sl. s. c.	s.; s. bz.
6448	oil	>1	−53	d. 24	v. sl. s.	....	....
6449	pd.	....	d. > 150	....	$0.26^{17°}$	$0.03^{17°}$	s. KOH
6450	mn.	....	40-1	....	$2.3^{20°}$	∞	....
6451	nd./al.	....	173-4	$349\text{-}50^{730mm}$	sl. s. h.; s. chl.	0.7 c.; 8.5 h.	sl. s.; sl. s. lg.; s. bz.
6452	brn. red	....	190-3	....	i.; s. chl.	sl. s. h.	sl. s.
6453	col./aq. al.	....	122-3 d.	....	v. sl. s.	s.	s. chl.
6454	col. lq.	$0.880\frac{20°}{4}$	−25.2	144.4	i.	∞ abs.	∞
6455	pl./aq. H_2SO_4	....	d.	....	s.	....	....
6456	pr.	....	....	....	s.		
6457	col. lq.	$0.864\frac{20°}{4}$	−47.9	139.1	i.	∞ abs.	∞
6458	cr./aq. H_2SO_4	....	63-4	....	s.	....	....
6459	pl.	....	....	....	s.		
6460	col. cr.	....	....	....	s.	i.	i.
6461	cr.	....	34	$135\text{-}6^{11mm}$	d. h.	d. h.	v. s.
6462	pl.	$0.861\frac{20°}{4}$	13.3	138.4	i.	s.	v. s.
6463	col. lf./aq.	....	86	$149^{0.1mm}$	s.	s. chl.	....
6464	mn.pr./aq.	$1.522^{15°}$	....	....	s.	....	....
6465	nd./aq.	....	75	218	s.	s.	....
6466	cr./10% al.	$1.023\frac{17}{15}°$	64.5-6.5	225^{757mm}	sl. s.	s.	∞
6467	nd./aq.	....	68	219.5	sl. s.	s.	....
6468	lf.	....	48-9	212	s. h.	s.	....
6469	nd.	$1.036\frac{20°}{4}$	25-6	211.5^{766mm}	....	....	....
6470	mn.	....	74.5	211.5-3.5	....	....	....
6471	lq.	$0.991^{15°}$	<−15	223	v. sl. s.	s.	s.
6472	pr. lg.	$1.076^{17.5°}$	49-50	224-6	v. sl. s. c.	s. pet.	....
6473	oil	$0.972\frac{20°}{4}$	....	221-2	....	....	....
6474	lq.	$0.980^{15°}$	10-2	216-7	....	....	....
6475	lq.	$0.978\frac{19.6°}{4}$	....	213-4	v. sl. s.	....	....
6476	cr.	....	70	....	....	....	....
6477	oil	$0.979\frac{21°}{4}$	15.5	215^{739mm}	v. sl. s.	....	....

Table 7-4 (*Continued*)

PHYSICAL CONSTANTS OF ORGANIC COMPOUNDS

No.	Name	Synonym	Formula	Beil. Ref.	Formula Weight
6478	**Xylitol**	pentanpentol	$C_5H_{12}O_5$	I-531	152.15
6479	**Xylorcinol** (*m*)	4,6-diMe-resorcinol	$(CH_3)_2C_6H_2(OH)_2$	VI-912	138.17
6480	*l*-**Xylose** (+)	wood sugar	$C_4H_9O_4 \cdot CHO$	I-865	150.13
6481	**Xylyl** acetate (*p*)	*p*-tolubenzyl acetate	$CH_3C_6H_4CH_2 \cdot O_2C \cdot CH_3$	VI-498	164.21
6482	amine (*o*)	*o*-tolubenzylamine	$CH_3 \cdot C_6H_4 \cdot CH_2 \cdot NH_2$	XII-1106	121.18
6483	amine (*m*)	....	$CH_3 \cdot C_6H_4 \cdot CH_2 \cdot NH_2$	XII-1134	121.18
6484	amine (*p*)	....	$CH_3 \cdot C_6H_4 \cdot CH_2 \cdot NH_2$	XII-1141	121.18
6485	bromide (*o*)	ω-bromo-*o*-xylene	$CH_3 \cdot C_6H_4 \cdot CH_2 \cdot Br$	V-365	185.07
6486	bromide (*m*)	ω-bromo-*m*-xylene	$CH_3 \cdot C_6H_4 \cdot CH_2 \cdot Br$	V-374	185.07
6487	bromide (*p*)	ω-bromo-*p*-xylene	$CH_3 \cdot C_6H_4 \cdot CH_2 \cdot Br$	V-385	185.07
6488	chloride† (*o*)	ω-chloro-*o*-xylene	$CH_3 \cdot C_6H_4 \cdot CH_2 \cdot Cl$	V-364	140.61
6489	chloride (*m*)	....	$CH_3 \cdot C_6H_4 \cdot CH_2 \cdot Cl$	V-373	140.61
6490	chloride (*p*)	....	$CH_3 \cdot C_6H_4 \cdot CH_2 \cdot Cl$	V-384	140.61
6491	**Xylylene** dibromide (*o*)	ω,ω'-diBr-*o*-xylene	$C_6H_4(CH_2Br)_2$	V-366	263.97
6492	dibromide (*m*)	....	$C_6H_4(CH_2Br)_2$	V-374	263.97
6493	dibromide (*p*)	....	$C_6H_4(CH_2Br)_2$	V-385	263.97
6494	dichloride (*o*)	ω,ω' diCl-*o*-xylene	$C_6H_4(CH_2Cl)_2$	V-364	175.06
6495	dichloride (*m*)	....	$C_6H_4(CH_2Cl)_2$	V-373	175.06
6496	dichloride (*p*)	....	$C_6H_4(CH_2Cl)_2$	V-384	175.06
6497	glycol (*o*)	phthal alcohol	$C_6H_4(CH_2OH)_2$	VI-910	138.17
6498	glycol (*m*)	....	$C_6H_4(CH_2OH)_2$	VI-914	138.17
6499	glycol (*p*)	....	$C_6H_4(CH_2OH)_2$	VI-919	138.17
6500	**Xylyl** hydrazine	(2,4;1)	$(CH_3)_2C_6H_3 \cdot NHNH_2$	XV-549	136.20
6501	hydrazine (2,5;1)	diMe-Ph-hydrazine	$(CH_3)_2C_6H_3 \cdot NHNH_2$	XV-552	136.20
6502	hydrazine (2,6;1)	diMe-Ph-hydrazine	$(CH_3)_2C_6H_3 \cdot NHNH_2$	XV-548	136.20
6503	**Zinc** diethyl	zinc ethide	$(C_2H_5)_2Zn$	IV-672	123.49
6504	dimethyl	dimethyl zinc	$(CH_3)_2Zn$	IV-671	95.44
6505	dipropyl (*n*)	*n*-dipropyl zinc	$(C_2H_5 \cdot CH_2)_2Zn$	IV-675	151.55
6506	di-*iso*-propyl	....	$[(CH_3)_2CH]_2Zn$	IV-675	151.55
6507	**Zingerone**	3-MeO-4-OH-benzylacetone	$HO(CH_3O)C_6H_3 \cdot CH_2 \cdot CH_2 \cdot COCH_3$	*VIII-623	194.23

† See also Nos. 1474-6.
Vitamin-B6 5510
Vitamin-C 586
Vitamin-D2 1187
Vitamin-G 5563
Vitamin-H 853
Vitamin-K 4342
Vomicine, cf. alkd.
Waldivin, cf. glcde.

War gases 1247; 1303; 1473; 1985; 4697
"Weinsaüre" 5691
Westrosol 6155
White damp 1239
White tar 4458
Wintergreen oil 4380
Wood alcohol 4100
Wood naphtha 4100

Wood spirit 4100
Wood sugar 6480
Wrightine, cf. alkd.
Wurster's reagent 5840
Xanthaline, cf. alkd.
Xanthenol 6453
Xanthogenic acid 6448
Xanthopuccine, cf. alkd.
Xanthopurpurin 2298

Table 7-4 (*Continued*)
PHYSICAL CONSTANTS OF ORGANIC COMPOUNDS

No.	Crystalline Form and Color	Specific Gravity	Melting Point °C.	Boiling Point °C.	Solubility in 100 Parts		
					Water	Alcohol	Ether
6478	syrup	....	....	....	s.	....	....
6479	lf./chl.	....	124-5	276-9	v. s.	v. s.	v. s.
6480	nd.	$1.535^{0\circ}$	153-4	....	$117^{20\circ}$	v. sl. s. c.	i.
6481	lq.	....	....	227	....	....	....
6482	oil	$0.977\frac{19}{0}^{\circ}$	0	205.6^{745mm}	....	....	....
6483	oil	$0.965\frac{20}{0}^{\circ}$	....	205^{750mm}	i.	s.	s.
6484	lq.	$0.952\frac{20}{0}^{\circ}$	12.6-13.2	204^{739mm}	v. sl. s.	....	....
6485	pr.	$1.381^{23\circ}$	21	223-4	i.	s.	s.
6486	col. lq.	$1.371^{23\circ}$	....	212-5 sl. d.	i.	s.	s.
6487	nd./al.	1.324	38	$218-20^{740mm}$	i.	v. s. chl.	v. s. h.
6488	col. lq.	....	d. 170 ±	195-203	i.	∞ abs.	∞
6489	lq.	$1.064^{20\circ}$	d. 170 ±	195-6	i.	∞ abs.	∞
6490	oil	....	d. 170 ±	200-2	i.	∞ abs.	∞
6491	rhb.	$1.988^{0\circ}$	94.5	d.	16.6 pet.	s.	20
6492	mn./chl.	$1.959^{0\circ}$	76-7	$135-40^{20mm}$	33 pet.	v. s. chl.	v. s.
6493	mn./bz.	$2.012^{0\circ}$	145-7	245	d. >80°	v. s. h. chl.	$2.7^{20\circ}$
6494	mn.?	$1.393^{0\circ}$	55	239-41	v. s. chl.	v. s.	v. s.
6495	cr.	$1.302^{0\circ}$	34.2	250-5	i.	....	....
6496	mn.	$1.417^{0\circ}$	100.5	240-5 d.	i.; s. act.	s.; s. chl.	v. sl. s.
6497	pl./et.	....	64.2-4.8	....	>25$^{18\circ}$	>25$^{18\circ}$	25$^{18\circ}$
6498	cr./bz.	lq. $1.161^{18\circ}$	46-7	$154-9^{13mm}$	v. s.	....	s.
6499	nd.	....	115-6	....	sl. s.	v. s.	v. s.
6500	nd./et.	....	85	d.	v. sl. s.	v. s.	s.
6501	col. nd.	....	78	....	i.	s.	s.
6502	nd./pet.	....	46 ±	....	s. lg.	....	....
6503	col. lq.	$1.182^{18\circ}$	−28	118	d.	d.	....
6504	col. lq.	$1.386^{11\circ}$	−40	46	d.	d.	....
6505	col. lq.	$1.072\frac{21}{21}^{\circ}$	....	158-60	d.	d.	....
6506	col. lq.	....	....	$94-8^{40mm}$	d.	d.	....
6507	col. cr./ et. pet.	....	40-1	....	sl. s.	s. dil. alk.	s.; sl. s. pet.

Table 7-5
PHYSICAL CONSTANTS OF ALKALOIDS
Compiled by F. E. SHEIBLEY, Ph.D.

Names of the compounds in the table below are arranged alphabetically. No compound is listed more than once in the table and each compound is given a number.

Synonyms. At the bottom of each page is an alphabetical listing of names for compounds which are to be found in the main body of the table but under a different name; the number following the name refers to the numerical place of this compound in the table; thus, *acetyl-benzoyl-aconine 3* indicates that this compound is the 3d compound in the table where it will be found listed under the name *aconitine*.

Appearance. In addition to the crystalline form and color, the solvent used in purification is often given; thus, "rhomb./al." indicates that rhombic crystals were obtained when the compound was crystallized from alcohol.

Optical Properties are given in the column headed "[α]"; only the directions of the specific rotations have been indicated, since the factors affecting the determination of this property are too variable to come within the scope of the table.

Color Reactions with Sulfuric Acid, where available, have been included as an

Abbreviations used in the table

a., acid
abs., absolute
ac. a., acetic acid
act., acetone
al., alcohol
alk., alkali (i. e. aqueous NaOH or KOH)
am. al., amyl alcohol
amor., amorphous
anh., anhydrous
aq., aqueous; water
bl., blue
B. P., boiling point
br., brown
bz., benzene, C_6H_6
c., cold
chl., chloroform, $CHCl_3$
colorl., colorless
cryst., crystals or crystalline

d, dextro-rotatory
d., decomposes or decomposition
deliq., deliquescent
dil., dilute
dk., dark
diss., dissociates
effl., efflorescent
et., ether $(C_2H_5)_2O$
Et, ethyl (C_2H_5)
EtOAc, ethyl acetate
gly., glycerol
grn., green
h., hot
hex., hexagonal
hyg., hygroscopic
i., insoluble
in., inactive
l, levo-rotatory
leaf., leaflets or leaves

No.	Name	[α]	Formula	Appearance	Melting Point °C.
1	**Aconine**	*d*	$C_{25}H_{41}O_9N$	hyg., amor.	132
2	salts	*l*	. . . .	hyg.	. . . .
3	**Aconitine**	*d*	$C_{34}H_{47}O_{11}N$	rhomb. pr./chl.	204
4	hydrobromide	*l*	$C_{34}H_{47}O_{11}N \cdot HBr \cdot 2\frac{1}{2}H_2O$	hex. tab./aq.	sint. 160
5	hydrobromide	. . .	$C_{34}H_{47}O_{11}N \cdot HBr \cdot \frac{1}{2}H_2O$	need./al.	206–7
6	hydrochloride	*l*	$C_{34}H_{47}O_{11}N \cdot HCl \cdot 3H_2O$	cryst.	149; 170*
7	**Adenine**	. . .	$C_5H_5N_5$	need. $+ 3H_2O$/aq.	360–5 d. subl. 220
8	**Agmatine**	. . .	$HN:C(NH_2) \cdot NH \cdot (CH_2)_4NH_2$	. . . .	. . . .

* Melting point of the anhydrous compound.

Acetyl-benzoyl-aconine 3   Acidol 34

Table 7-5 (*Continued*)
PHYSICAL CONSTANTS OF ALKALOIDS

aid in the making of rapid preliminary examinations.

Solubilities expressed by numbers are given in parts by weight of solvent required to dissolve one part of the alkaloid at a temperature of approximately 25°C. Because of the wide discrepancies existing between many of these figures as found in the literature no claim to accuracy can be made, and the values stated are perhaps best considered as upper limits.

References. The information given in the table has been collected mainly from the following sources: Henry: **Plant Alkaloids**, 4th edition, published by The Blakiston Co., Philadelphia-Toronto (1949); Manske and Holmes, **The Alkaloids**, published by Academic Press, New York (1950–1960); **The Merck Index**, 7th edition, published by Merck and Co., Inc., Rahway, N.J. (1977); Heilbron: **Dictionary of Organic Compounds**, published by Oxford University Press, New York (1934); Beilstein, **Handbuch der Organischen Chemie**, 3d edition.

---

lig., ligroin	r., red
liq., liquid	rhomb., rhombic
lt., light	s., soluble
lustr., lustrous	sint., sinters
Me, methyl (CH_3)	sl., slight or slightly
MeOH, methyl alcohol	subl., sublimes
met., metallic	tab., tabular
mon., monoclinic	tricl., triclinic
need., needles	trim., trimetric
octahedrl., octahedral	v., very
org., orange	v. s., very soluble
orthorhomb., orthorhombic	v. sl. s., very slightly soluble
pa., pale	wh., white
pet., petroleum ether	yel., yellow
powd., powder	∞, soluble in all proportions;
pr., prisms	i.e., miscible
pyr., pyridine	>, greater than

No.	Reaction with H_2SO_4	Solubility Expressed in Parts of Solvent Required to Dissolve 1 Part Alkaloid					
		Water	Alcohol	Ether	Chloroform	Benzene	Others
1	....	v. s.	v. s.	i.	s.	....	i. pet.
2	....	....	....	....	....	....	....
3	colorl. when pure	3300	23	47	3	6.2	....
4	....	s.	s.	....	....	....	....
5	....	....	....	....	....	....	....
6	....	s.	s.	....	....	....	....
7	....	1086 c.; 40 h.	sl. s. h.	i.	i.	s. a.	s. h. NH_4OH
8	....	....	....	....	....	....	....

ψ-Aconitine 185   Acraconitine 185

Table 7-5 (*Continued*)
PHYSICAL CONSTANTS OF ALKALOIDS

No.	Name	[α]	Formula	Appearance	Melting Point °C.
9	sulfate	...	$C_5H_{14}N_4 \cdot H_2SO_4$	colorl. cryst./aq. MeOH	226–9
10	**Allantoin†**	*in*	$C_4H_6O_3N_4$	need./h. aq.	235–6
11	**Anabasine**	*l*	$C_{10}H_{14}N_2$	colorl. liq.	B.P. 276
12	**Anhalonine**	*l*	$C_{12}H_{15}O_3N$	wh. need.	85
13	**Apomorphine**	...	$C_{17}H_{17}O_2N$	pr. + $1Et_2O$/et.	170 d.
14	hydrochloride	*l*	$C_{17}H_{17}O_2N \cdot HCl \cdot \frac{1}{2}H_2O$	pr./aq.	....
15	**Apoquinine**	*l*	$C_{19}H_{22}O_2N_2$	need./et.	180–90 d.
16	**Arecoline**	*in*	$C_8H_{13}O_2N$	very alkaline oil	B.P. 209
17	hydrobromide	*in*	$C_8H_{13}O_2N \cdot HBr$	pr./al.	169–71
18	hydrochloride	...	$C_8H_{13}O_2N \cdot HCl$	cryst.	158
19	**Aspidospermine**	*l*	$C_{22}H_{30}O_2N_2$	need./al. or pet.	208
20	**Atisine**	*l*	$C_{22}H_{33}O_2N$	wh., amor.	indefinite
21	hydrochloride	*d*	$C_{22}H_{33}O_2N \cdot HCl$	prisms	296
22	**Atropine**	*in*	$C_{17}H_{23}O_3N$	colorl. pr.	118, subl.
23	sulfate	*in*	$(C_{17}H_{23}O_3N)_2 \cdot H_2SO_4 \cdot H_2O$	need.	194*
24	**Bebeerine, α**	*l*	$C_{18}H_{19}O_3N$	pr./MeOH	214
25	**Bebeerine, β**	*d*	$C_{18}H_{19}O_3N$	yel., amor.	142–50
26	hydrochloride	...	$C_{18}H_{19}O_3N \cdot HCl$	need. or scales	259–60
27	**Benzoylecgonine**	*l*	$C_{16}H_{19}O_4N \cdot 4H_2O$	lustr. need./aq.	90–2; 193–5*
28	**Berberine**	*in*	$C_{20}H_{19}O_5N \cdot 6H_2O$	red-yel. need./aq.	145d.
29	bisulfate	...	$C_{20}H_{17}O_4N \cdot H_2SO_4$	yel. need.	....
30	chloroform	...	$C_{20}H_{19}O_5N \cdot CHCl_3$	tricl. tab./chl.	179
31	hydrochloride	...	$C_{20}H_{17}O_4N \cdot HCl \cdot 2H_2O$	org. need. or yel. powd.	....
32	nitrate	...	$C_{20}H_{17}O_4N \cdot HNO_3$	yel. need.	....
33	**Betaine†**	*in*	$C_5H_{11}O_2N \cdot H_2O$	sweet deliq. cryst.; anh. at 100°	293*
34	hydrochloride	...	$C_5H_{11}O_2N \cdot HCl$	mon. cryst.	227–8 d.
35	**Brucine**	*l*	$C_{23}H_{26}O_4N_2 \cdot 4H_2O$	mon. pr./al.	105; 178*
36	hydrochloride	...	$C_{23}H_{26}O_4N_2 \cdot HCl$	wh. need.	....
37	nitrate	...	$C_{23}H_{26}O_4N_2 \cdot HNO_3 \cdot 2H_2O$	wh. pr.	230 d.*
38	sulfate	...	$(C_{23}H_{26}O_4N_2)_2 \cdot H_2SO_4 \cdot 7H_2O$	long need.	....
39	**Caffeine**	*in*	$C_8H_{10}O_2N_4 \cdot H_2O$	need./al.; anh. 100°	235*; subl. 178
40	citrate (true)	...	$C_8H_{10}O_2N_4 \cdot C_6H_8O_7$	mon.	....
41	hydrochloride	...	$C_8H_{10}O_2N_4 \cdot HCl \cdot 2H_2O$	mon.	d. 80–100
42	mercurichloride	...	$C_8H_{10}O_2N_4 \cdot HgCl_2$	colorl. need.	246
43	sulfate	...	$C_8H_{10}O_2N_4 \cdot H_2SO_4$	wh. need.	....
44	triiodide	...	$C_8H_{10}O_2N_4I_2 \cdot HI \cdot 1\frac{1}{2}H_2O$	long grn. met. pr.	171
45	**Canadine**	*l*	$C_{20}H_{21}O_4N$	silky need./al.	133–4
46	**Carpaine**	*d*	$C_{14}H_{25}O_2N$	pr./al.	121
47	**Carpiline**	*d*	$C_{16}H_{18}O_3N_2$	wh. cryst.	187
48	**Cephaeline**	*l*	$C_{28}H_{38}O_4N_2$	fine need./et.	107–8; 120–30*

* Melting point of the anhydrous compound.
† See also listing in the table Physical Constants of Organic Compounds.
Amino-butyl-guanidine 8   6-Amino-purine 7
2-Amino-6-hydroxy-purine 115   Anhaline 121

Table 7-5 (*Continued*)
PHYSICAL CONSTANTS OF ALKALOIDS

No.	Reaction with H_2SO_4	Solubility Expressed in Parts of Solvent Required to Dissolve 1 Part Alkaloid					
		Water	Alcohol	Ether	Chloroform	Benzene	Others
9	....	s.	v. sl. s.	....	....	....	....
10	....	132 c.; 30 h.	5000 abs.	i.	....	....	s. NaOH
11	....	s.	s.	s.	....	s.	....
12	....	s.	s.	s.	s.		....
13	colorl.	sl. s.	s.	sl. s.	v. s.	sl. s.	s. alk.
14	....	39.5	38.2	1864	sl. s.	....	....
15	fluorescent dil.	s. h.	s.	sl. s.	v. s.	v. s.	s. KOH
16	....	s.	s.	s.	s.	....	....
17	....	1	8 c.; 2 h.	sl. s.	sl. s.	....	....
18	....	s.	s.	....	....	....	....
19	colorl.	6000	48	106	s.	s.	s. dil. a.
20	....	sl. s.	s.	s.	s.	s.	s. dil. a.
21	....	v. s.	v. s.	i.	....	....	....
22	colorl.	300	1.46	16.6	1.56	s.	s. dil. a.
23	colorl.	0.38	3.7	2140	620	....	....
24	br.→r. h.	i.	sl. s.	sl. s.	s.	....	s. a., act.
25	....	i.	sl. s.	....	s.	s.	s. dil. a.
26	....	s.	s.	....	....	....	....
27	....	s. h.	s.	i.	....	s. dil. a.	s. alk.
28	grn.→yel.	22	100	v. sl. s.	sl. s.	sl. s.	unstable
29	....	100; s. h.	sl. s.	....	....	....	....
30	....	....	....	....	....	....	....
31	....	400; s. h.	s. h.	i.	i.	....	....
32	....	sl. s.	....	....	....	....	....
33	....	s.	s.	sl. s.	....	....	....
34	....	1.7	15	i.	i.	....	....
35	dk. yel. h.; HNO_3→r.	320 c.; 150 h.	1.1	133	7.5	88	i. alk.
36	....	s.	s.	....	....	....	....
37	....	s.	s.	....	....	....	....
38	....	75 c.; 10 h.	84	....	254	....	....
39	+$K_2Cr_2O_7$ →grn.	45.6	53.2	375 c.; 339 h.	8 c.; 6.4 h.	88 c.; 18.9 h.	s. EtOAc sl. s. pet.
40	....	s. d.	s. d.	....	....	....	....
41	....	s. d.	s. d.	....	....	....	....
42	....	260	....	....	....	....	....
43	....	s. d.	s. d.	....	....	....	....
44	....	i.	s.	....	sl. s.	....	....
45	....	i.	s.	v. s.	v. s.	v. s.	....
46	....	sl. s.	9	33	s.	5.5	s. dil. a.
47	....	s. h.	....	sl. s.	s.	s.	....
48	....	i.	s.	sl. s.	s.	s.	s. alk.; s. dil. a.

Table 7-5 (*Continued*)
PHYSICAL CONSTANTS OF ALKALOIDS

No.	Name	[α]	Formula	Appearance	Melting Point °C.
49	**Cevadine**	d	$C_{32}H_{49}O_9N \cdot 2C_2H_5OH$	rhomb.; becomes anh. 130–40	205*
50	**Chelerythrine**	in	$C_{21}H_{19}O_5N \cdot C_2H_5OH$	pr. leaf./al.	207
51	**Chelidonine**	d	$C_{20}H_{19}O_5N \cdot H_2O$	mon. tab./dil. HCl	135–6*
52	**Cinchonamine**	d	$C_{19}H_{24}ON_2$	orthorhomb.need./al.	185
53	**Cinchonidine**	l	$C_{19}H_{22}ON_2$	trim. pr./al.	210.5
54	hydrochloride	l	$C_{19}H_{22}ON_2 \cdot HCl \cdot 2H_2O$	pyramids or pr.	242*
55	sulfate	l	$(C_{19}H_{22}ON_2)_2 \cdot H_2SO_4 \cdot 3H_2O$	mon. pr.	240 d.*
56	**Cinchonine**	d	$C_{19}H_{22}ON_2$	rhomb. pr./al.	264
57	bisulfate	...	$C_{19}H_{22}ON_2 \cdot H_2SO_4 \cdot 4H_2O$	octahedrl.	....
58	hydrochloride	d	$C_{19}H_{22}ON_2 \cdot HCl \cdot 2H_2O$	mon.	217–8 d.*
59	sulfate	d	$(C_{19}H_{22}ON_2)_2 \cdot H_2SO_4 \cdot 2H_2O$	rhomb.	198.5*
60	**Cinchotine**	d	$C_{19}H_{24}ON_2$	pr. or scales	268–9
61	**Cinnamylcocaine**	l	$C_{19}H_{23}O_4N$	need./bz.	121
62	**Cocaine**	l	$C_{17}H_{21}O_4N$	mon.pr./al.;need. /aq.	98
63	chromate	...	$C_{17}H_{21}O_4N \cdot H_2CrO_4 \cdot H_2O$	org. yel. leaf.	127
64	hydrochloride	l	$C_{17}H_{21}O_4N \cdot HCl$	short pr./al.	195
65	**Coclaurine**	l	$C_{17}H_{19}O_3N$	bitter need.	221
66	**Codamine**	...	$C_{20}H_{25}O_4N$	pr./al.	121
67	**Codeine**	l	$C_{18}H_{21}O_3N \cdot H_2O$	rhomb. pr./aq.	155*
68	hydrochloride	l	$C_{18}H_{21}O_3N \cdot HCl \cdot 2H_2O$	need., pr./aq.	280 d.
69	phosphate	l	$C_{18}H_{21}O_3N \cdot H_3PO_4 \cdot 2H_2O$	need. or pr.	235 d.
70	sulfate	l	$(C_{18}H_{21}O_3N)_2 \cdot H_2SO_4 \cdot 5H_2O$	rhomb. pr.	278 d.
71	**Colchicine**	l	$C_{22}H_{25}O_6N$	yel. varnish; yel. need./EtOAc	143–7*; 155–7
72	chloroform	...	$C_{22}H_{25}O_6N \cdot CHCl_3$	need./chl.	d. 60–70
73	**Columbamine**	...	$C_{20}H_{21}O_5N$	free base unknown	....
74	chloride	...	$C_{20}H_{20}O_4NCl \cdot 2\frac{1}{2}H_2O$	yel. need.	194
75	chloride	...	$C_{20}H_{20}O_4NCl \cdot 4H_2O$	br. pr.	184
76	**Conessine**	d	$C_{24}H_{40}N_2$	leaf. or need./act.	123–5
77	**Conhydrine**	d	$C_8H_{17}ON$	wh. cryst./et.	121; B.P. 226
78	**Coniine**	d	$C_3H_7 \cdot C_5H_{10}N$	colorl. liq.	−2; B.P. 166–7
79	hydrochloride	d	$C_8H_{17}N \cdot HCl$	rhombs/aq.	220
80	picrate	...	$C_8H_{17}N \cdot C_6H_3O_7N_3$	yel. need./h. aq.	75
81	**Corybulbine**	l	$C_{21}H_{25}O_4N$	light–sensitive crysts.	238
82	**Corycavine**	in	$C_{21}H_{21}O_5N$	rhomb. tab./al.	218–9

* Melting point of the anhydrous compound.

Chinicine 193	α–Chondodendrine 24
Chinidine 195	Cicutine 78
Chinotine 195	Cinchovatine 53
Choline sinapate 215	Cinnamoylcocaine 61

Table 7-5 (*Continued*)
PHYSICAL CONSTANTS OF ALKALOIDS

No.	Reaction with H₂SO₄	Solubility Expressed in Parts of Solvent Required to Dissolve 1 Part Alkaloid					
		Water	Alcohol	Ether	Chloroform	Benzene	Others
49	yel.→r.	sl. s.	10	12	s.	....	s. CS₂
50	grn.→yel.	i.	sl. s.	sl. s.	v. s.	s. dil. a.	sl. s. act.
51	crimson with guaiacum	i.	v. s.	v. s.	s.	....	s. am. al.
52	....	v. sl. s.	30	100	s. h.	s. h.	s. dil. a.
53	no fluorescence	5000	20	200	s.	....	s. dil. a.
54	....	20	s.	300	v. s.	....	....
55	....	63 c.; 20 h.	72 c.; 32 h.	v. sl. s.	923	....	....
56	....	3670	48 c.; 20 h.	370	165	....	s. am. al.
57	....	0.4	0.8	....	....	....	....
58	....	22 c.; 3.5 h.	1 c.	275	22	....	....
59	....	60 c.; 30 h.	10 c.; 6 h.	3230	70	....	....
60	....	1300	sl. s.	534	v. sl. s.	....	....
61	....	i.	s.	s.	s.	s.	....
62	colorl.	600	5	2.5	1.1	s.	s. act.
63	....	sl. s.	....	....	....	....	....
64	....	0.4	2.6	i.	19	....	s. act.
65	....	sl. s.	v. s. h.	sl. s.	sl. s.	i.	s. alk.; s. dil. a.
66	....	s. h.	s.	s.	s.	s.	s. dil. a.
67	h.→bl.	120	1.6 c.; 1 h.	12.5	0.66	10.4	68 NH₄OH
68	....	20 c.; 1 h.	145	....	....	....	....
69	....	2.25	261	1310	6700	....	....
70	....	30 c.; 6.3 h.	1200	i.	i.	....	....
71	yol.→r. h.	22	s.	157	s.	88	i. pet.
72	....	d. h.	....	....	....	....	....
73	....	....	....	....	....	....	....
74	....	s.	s.	....	....	....	....
75	....	s.	s.	....	....	....	....
76	....	v. sl. s.	s.	s.	s.	....	....
77	....	sl. s.	s.	s.	s.	....	....
78	colorl.	100	v. s.	v. s.	sl. s.	s.	sl. s. CS₂
79	....	2	s.	....	s.	....	....
80	....	....	s.	s.	....	....	....
81	....	v. sl. s.	s. h.	v. sl. s.	s.	s.	....
82	....	i.	v. sl. s.	....	s.	s. dil. a.	i. alk.

Cinnamoylecgonine methyl ester 61
Coffearin 234
Conchinine 195
Conicine 78

Conquinine 195
Conydrine 77
Cordianine 10
Cornutine 107

Table 7-5 (*Continued*)
PHYSICAL CONSTANTS OF ALKALOIDS

No.	Name	[α]	Formula	Appearance	Melting Point °C.
83	**Corydaline**	d	$C_{22}H_{27}O_4N$	colorl. pr./al.	135
84	**Corytuberine**	d	$C_{19}H_{21}O_4N$	silky need./et.	240
85	**Cotarnine**	...	$C_{12}H_{15}O_4N$	need./bz.	132–3 d.
86	hydrochloride	...	$C_{12}H_{14}O_3NCl \cdot 2H_2O$	pa. yel. silky need.	197 d.
87	phthalate	...	$(C_{12}H_{14}O_3N)_2 \cdot C_6H_4(CO_2)_2$	yel. cryst. or powd.	103
88	**Cryptopine**	in	$C_{21}H_{23}O_5N$	pr./al. or bz.	220–1
89	**Cupreine**	l	$C_{19}H_{22}O_2N_2 \cdot 2H_2O$	pr./et.	198*
90	**Cuscohygrine**	in	$C_{13}H_{24}ON_2$	oil	B.P. 215^{50}
91	hydrate	...	$C_{13}H_{24}ON_2 \cdot 3\frac{1}{2}H_2O$	need.	40–1; 120–30†
92	**Cytisine**	l	$C_{11}H_{14}ON_2$	large rhomb. cryst.	152–3
93	**Delphinine**	d	$C_{33}H_{45}O_9N$	plates/al.	198–200
94	hydrochloride	...	$C_{33}H_{45}O_9N \cdot HCl$	need./MeOH + et.	208–10
95	**Diacetylmorphine**	...	$C_{21}H_{23}O_5N$	bitter cryst./MeOH	172
96	hydrochloride	l	$C_{21}H_{23}O_5N \cdot HCl \cdot H_2O$	cryst. powd.	230 d.
97	**Dilaudid**	...	$C_{17}H_{19}O_3N \cdot HCl$	cryst. powd.	....
98	**Dionin**		$C_{19}H_{23}O_3N \cdot HCl \cdot H_2O$	wh. cryst. powd.	123 d.; 170 d.*
99	**Ecgonine**	l	$C_9H_{15}O_3N \cdot H_2O$	mon. pr./al.	198; 205*
100	hydrochloride	l	$C_9H_{15}O_3N \cdot HCl$	rhomb. or tricl. tab.	246
101	**Emetine**	l	$C_{29}H_{40}O_4N_2$	plates/al. or et.	74
102	hydrochloride	d	$C_{29}H_{40}O_4N_2 \cdot 2HCl \cdot 7H_2O$	woolly need./h. aq.; thick pr./c. satd. soln.	235–55; dry, d.
103	**Ephedrine‡**	l	$C_{10}H_{15}ON$	unctuous, colorl. cryst.	40; B.P. 255
104	hydrochloride	l	$C_{10}H_{15}ON \cdot HCl$	need.	216 d.
105	sulfate	l	$(C_{10}H_{15}ON)_2 \cdot H_2SO_4$	wh. odorless cryst.	245 d.
106	**Ergotinine**	d	$C_{35}H_{39}O_5N_5$	long need./al.	239 d.
107	**Ergotoxine**	l	$C_{35}H_{41}O_6N_5$	pr./bz.	190–200
108	**Evodiamine**	d	$C_{19}H_{17}ON_3$	yel. leaf./al.	278
109	hydrate	in	$C_{19}H_{19}O_2N_3$	rhomb. leaf.	146–7
110	**Gelsemine**	d	$C_{20}H_{22}O_2N_2$	wh. cryst.	178
111	acetone	...	$C_{20}H_{22}O_2N_2 \cdot (CH_3)_2CO$	pr./act.	−act. at 120
112	hydrochloride	d	$C_{20}H_{22}O_2N_2 \cdot HCl$	pr./aq. al. or aq.	300
113	**Glaucine**	d	$C_{21}H_{25}O_4N$	yel. rhomb. pr.	119–20
114	**Gnoscopine**	in	$C_{22}H_{23}O_7N$	long need./MeOH	232 d.
115	**Guanine**	in	$C_5H_5ON_5$	wh. cryst. powd.	>360 d.
116	**Guvacine**	...	$C_6H_9O_2N \cdot H_2O$	short rods/dil. al.	285 d.

* Melting point of the anhydrous compound.
† Becomes anhydrous at 120–30°
‡ See also listing in table Physical Properties of Organic Compounds.

Corynine 244	Diamorphine 95	
Cuskhygrine 90	Dihydromorphinone hydrochloride **97**	
Daturine 22	2,6-Dihydroxy-purine 243	1,3-Dimethyl-xanthine 233
Dehydromorphine 188	Dimethoxy-strychnine 35	3,7-Dimethyl-xanthine 232

Table 7-5 (*Continued*)
PHYSICAL CONSTANTS OF ALKALOIDS

No.	Reaction with H$_2$SO$_4$	Solubility Expressed in Parts of Solvent Required to Dissolve 1 Part Alkaloid					
		Water	Alcohol	Ether	Chloroform	Benzene	Others
83	....	i.	s. h.	v. s.	v. s.	s.	i. alk.
84	....	s. h.	s.	i.	i.	i.	....
85	....	s. h.	s.	s.	....	s. dil. a.	s. NH$_4$OH
86	....	1	4	....	....	....	....
87	....	v. s.	....	....	....	....	....
88	violet→ grn.→yel.	v. sl. s.	100 h.	v. sl. s.	sl. s.	v. sl. s.	s. h. pyr.
89	dil.→no fluorescence	i.	s.	sl. s.	sl. s.	sl. s.	s. alk.; i. NH$_4$OH
90	....	∞	s.	s.	....	s.	....
91	....	s.	s.	s.†	....	s.*	....
92	....	s.	s.	i.	s.	s.	i. pet.
93	+malic a. →org.→bl.	50,000	20	10	15	....	....
94	....	....	....	....	....	....	....
95	....	1700	24	70	2.2	s.	s. dil. a.; s. alk.
96	....	2	s.	i.	i.	....	....
97	....	s.	s.	i.	....	....	....
98	....	7	1.4	i.	i.	....	....
99	....	4	50	i.	i.	i.	s. EtOAc
100	....	s.	sl. s.	....	....	....	....
101	....	1000	s.	s.	s.	sl. s.	....
102	....	4	s.	s.	....	....	....
103	....	20	0.2	s.	s.	....	s. oils
104	....	3	12	i.	....	....	....
105	....	1.2	76	i.	....	....	....
106	with et.→ org.→bl.	sl. s.	200 c.; 52 h.	1020	s.	77 h.	26 act.
107	ditto	i.	s.	sl. s.	s.	s. h.	s. NaOH
108	....	i.	sl. s.	sl. s.	sl. s.	i.	i. dil. a.
109	....	....	....	....	....	....	....
110	colorl.→yel. br.→yel. grn.	sl. s.	s.	s.	s.	s.	s. dil. a.
111	....	....	....	....	....	....	....
112	....	s.	sl. s.	....	....	....	....
113	colorl.→bl. in time	s. h.	v. s.	s.	v. s.	sl. s.	sl. s. pet.
114	....	i.	1500	....	s. h.	sl. s.	i. alk.
115	....	i.	v. sl. s.	v. sl. s.	s. a.	v. sl. s. NH$_4$OH	s. KOH
116	....	s.	....	i.	....	....	....

* With separation of droplets of water.

Eserine 174
Ethyl-morphine-HCl 98
Eudermol 163
Feraconitine 185

Fumarine 183
Gelseminine 110
Glyoxyl-diureide 10
Guaranine 39

Gynesin 234

Table 7-5 (*Continued*)
PHYSICAL CONSTANTS OF ALKALOIDS

No.	Name	[α]	Formula	Appearance	Melting Point °C.
117	**Harmaline**	*in*	$C_{13}H_{14}ON_2$	pr./al. + bz.	250 d.
118	**Harmine**	*in*	$C_{13}H_{12}ON_2$	rhomb. pr./al.	257–9 d.
119	**Homoatropine**	...	$C_{16}H_{21}O_3N$	deliq. pr./et.	99–100
120	hydrobromide	...	$C_{16}H_{21}O_3N \cdot HBr$	rhomb.	217–8 d.
121	**Hordenine**	*in*	$C_{10}H_{15}ON$	orthorhomb. pr.	117.8; subl. 140–50
122	sulfate	...	$(C_{10}H_{15}ON)_2 \cdot H_2SO_4 \cdot 2H_2O$	colorl. cryst.	208–10*
123	**Hydrastine**	*l*	$C_{21}H_{21}O_6N$	colorl. rhomb. pr./al.	132
124	hydrochloride	*d*	$C_{21}H_{21}O_6N \cdot HCl$	hyg. powd.	116
125	**Hydrastinine**	*in*	$C_{11}H_{13}O_3N$	need./lig.	116–7
126	hydrochloride	*in*	$C_{11}H_{12}O_2NCl$	yel. need.	212 d.
127	**Hydrocotarnine**	*in*	$C_{12}H_{15}O_3N \cdot \frac{1}{2}H_2O$	mon. pr./al.	55–6
128	**Hydroquinine**	*l*	$C_{20}H_{26}O_2N_2 \cdot 2H_2O$	need./chl. or et.	172*
129	**Hyoscine**	*l*	$C_{17}H_{21}O_4N$	syrup or cryst./et.	59
130	hydrobromide	*l*	$C_{17}H_{21}O_4N \cdot HBr \cdot 3H_2O$	rhomb. tab. or need./aq.	194–7*
131	**Hyoscyamine**	*l*	$C_{17}H_{23}O_3N$	silky need./aq. al.	106–8
132	hydrobromide	*l*	$C_{17}H_{23}O_3N \cdot HBr$	deliq. pr.	152
133	hydrochloride	...	$C_{17}H_{23}O_3N \cdot HCl$	wh. cryst.	149–51
134	sulfate	*l*	$(C_{17}H_{23}O_3N)_2 \cdot H_2SO_4 \cdot 2H_2O$	need./al.	206*
135	**Hypaphorine**	*d*	$C_{14}H_{18}O_2N_2 \cdot 2H_2O$	large mon. cryst./aq.	255*
136	**Hypoxanthine**	*in*	$C_5H_4ON_4$	minute need.	d. 150
137	**Japaconitine**	*d*	$C_{34}H_{47}O_{11}N$	need./al., et. or chl.	202–9
138	**Jervine**	*l*	$C_{26}H_{37}O_3N \cdot 2H_2O$	long grouped pr.	238–42
139	**Laudanine**	*in*	$C_{20}H_{25}O_4N$	pr./aq. al.	166–7
140	**Laudanosine**	*d*	$C_{21}H_{27}O_4N$	need./bz.	90
141	**Lobeline**	*l*	$C_{22}H_{27}O_2N$	broad colorl. need.	130–1
142	hydrochloride	*l*	$C_{22}H_{27}O_2N \cdot HCl$	wh. granular powd.	180
143	**Lupanine**	*d*	$C_{15}H_{24}ON_2$	need./pet.; very alkaline	40
144	hydrochloride	*d*	$C_{15}H_{24}ON_2 \cdot HCl \cdot 2H_2O$	....	127–8; 250–2*
145	**Lupinine**	*l*	$C_{10}H_{19}ON$	rhomb. cryst./pet.	69–71
146	hydrochloride	*l*	$C_{10}H_{19}ON \cdot HCl$	rhomb. pr./aq. al.	212–3
147	**Lycorine**	*l*	$C_{16}H_{17}O_4N$	colorl. pr./al.	280 d.
148	**Mezcaline**	*in*	$C_{11}H_{17}O_3N$	colorl. alk. oil or cryst.	B.P. 180^{12}; 35–6
149	**Morphine**	*l*	$C_{17}H_{19}O_3N \cdot H_2O$	trim. pr./al.	254 d.*
150	acetate	*l*	$C_{17}H_{19}O_3N \cdot CH_3CO_2H \cdot 3H_2O$	cryst. powd./al.	200 d.

* Melting point of the anhydrous compound.

Herapathite 207	Hydro-berberine 45	Hydroxy-coniine 77	*dl*-Hyoscyamine 22
Heroin 95	Hydro-cinchonine 60	6-Hydroxy-purine 136	Iodoquinine sulfate 207
Hydra-ergotocin 244	Hydroxy-cinchonine 89	Hydroxy-stachydrine 238	Laburnine 92

Table 7-5 (*Continued*)
PHYSICAL CONSTANTS OF ALKALOIDS

No.	Reaction with H_2SO_4	Solubility Expressed in Parts of Solvent Required to Dissolve 1 Part Alkaloid					
		Water	Alcohol	Ether	Chloroform	Benzene	Others
117	....	v. sl. s.	s. h.	sl. s.	....	....	s. dil. a.
118	yel. with grn. fluorescence	v. sl. s.	sl. s.	sl. s.	s.	....	s. dil. a.
119	....	sl. s.	s.	s.	s.	s.	s. dil. a.
120	....	6	32	i.	625	....	....
121	....	s.	v. s.	v. s.	s.	sl. s.	s. dil. a.; s. alk.
122	....	s.	sl. s.	i.	....	....	....
123	olive grn. with $(NH_4)_2MoO_4$	i.	170	175	1.4	15	....
124	....	s.	s.	v. sl. s.	sl. s.	....	....
125	....	s. h.	s.	s.	s.	d.	s. a.
126	....	v. s.	v. s.	300	286	....	....
127	yel.→r.h.	i.	v. s.	v. s.	v. s.	v. s.	i. alk.
128	dil.→fluorescence	v. sl. s.	s.	s.	s.	s. act.; i. pet.	s. NH_4OH
129	h.→bl.	sl. s.	s.	s.	s.	sl. s.	sl. s. pet.
130	....	1.5	16	i.	750	....	....
131	colorl.	281	s.	49	1.5	132	s. dil. a.
132	....	v. s.	2.5	1610	1.7	....	....
133	....	s.	s.	....	....	....	....
134	....	0.5	4.5	v. sl. s.	v. sl. s.	....	....
135	....	v. s.	v. s.	....	....	....	i.
136	....	1370 c.; 69.5^{100}*	900 h.	i.	....	s. a.	s. alk.
137	....	i.	s.	s.	s.	....	i. pet.
138	yel.→grn. h.	i.	s.	sl. s.	s.	sl. s.	s. act.
139	red	v. sl. s.	s.	600	v. s.	v. s.	s. alk. carb.
140	rose red→red violet 150°	i.	s.	19	s.	s. h.	i. alk.
141	red-br.	v. sl. s.; d. h.	s. h.	s.	s.	s.	v. sl. s. pet.
142	....	40	10	....	v. s.	....	....
143	....	s.	s.	s.	s.	....	....
144	....	....	....	....	....	....	....
145	....	s.	s.	s.	s.	s.	sl. s. pet.
146	....	s.	....	....	....	....	....
147	$+MoO_3$→grn.→bl.	i.	sl. s.	sl. s.	sl. s.	sl. s. EtOAc	s. a.
148	yel.→violet	s.	s.	i.	s.	s.	i. pet.
149	pink→grn. h.→br.	3533 c.; 1075 h.	170 c.; 80 h.	4450	1525	9000	475 EtOAc
150	....	2.25	17.3	i.	710	....	6.5 gly.

Table 7-5 (*Continued*)
PHYSICAL CONSTANTS OF ALKALOIDS

No.	Name	[α]	Formula	Appearance	Melting Point °C.
151	hydrochloride	*l*	$C_{17}H_{19}ON \cdot HCl \cdot 3H_2O$	silky need./aq.	200 d.
152	sulfate	*l*	$(C_{17}H_{19}O_3N)_2 \cdot H_2SO_4 \cdot 5H_2O$	silky or cubic cryst./aq.	250 d.*
153	**Muscarine**	*d*	$C_8H_{19}O_3N$	deliq. cryst.	. . . .
154	**Nandinine**	*d*	$C_{19}H_{19}O_4N$	leaf.	145–6
155	**Narceine**	*in*	$C_{23}H_{27}O_8N \cdot 3H_2O$	need. or pr./aq.	145*
156	bisulfate	. . .	$C_{23}H_{27}O_8N \cdot H_2SO_4 \cdot 10H_2O$	cryst. powd.	d.→yel.
157	hydrochloride	. . .	$C_{23}H_{27}O_8N \cdot HCl \cdot 3H_2O$	cryst./HCl	192*
158	**Narcotine**	*l*	$C_{22}H_{23}O_7N$	long need./h. al.	176
159	hydrochloride	. . .	$C_{22}H_{23}O_7N \cdot HCl \cdot H_2O$	lustrous cryst.	197–8
160	**Nicotine**	*l*	$C_{10}H_{14}N_2$	colorl. oil	B.P. 246[730]
161	hydrochloride	*d*	$C_{10}H_{14}N_2 \cdot 2HCl$	deliq. cryst.	. . . .
162	picrate	. . .	$C_{10}H_{14}N_2 \cdot 2C_6H_3O_7N_3$	yel. need. or pr./al.	218
163	salicylate	*d*	$C_{10}H_{14}N_2 \cdot C_7H_6O_3$	wh. plates	117–8
164	tartrate	*d*	$C_{10}H_{14}N_2 \cdot 2C_4H_6O_6 \cdot 2H_2O$	reddish-wh. cryst.	88–90
165	**Oxyacanthine**	*d*	$C_{38}H_{38}O_6N_2$	need./al. or et.	216–7
166	nitrate	. . .	$C_{38}H_{38}O_6N_2 \cdot 2HNO_3 \cdot 4H_2O$	need.	195–200
167	**Papaveraldine**	. . .	$C_{20}H_{19}O_5N$	cryst./bz. or pet.	210
168	**Papaverine**	*in*	$C_{20}H_{21}O_4N$	rhomb. pr. or need./al.-et.	147–8
169	hydrochloride	. . .	$C_{20}H_{21}O_4N \cdot HCl$	mon. pl./aq.	220–1 d.
170	**Paraconiine**	*in*	$C_8H_{15}N$	yel. liq.	B.P. 168–70
171	**Pelletierine**	*in*	$C_8H_{15}ON$	colorl. oil	B.P. 106[21]; 195[760]
172	**Pellotine**	. . .	$C_{13}H_{19}O_3N$	plates/al.	110–2
173	hydrochloride	. . .	$C_{13}H_{19}O_3N \cdot HCl$	wh. cryst.	. . . .
174	**Physostigmine**	*l*	$C_{15}H_{21}O_2N_3$	hyg. cryst. (2 forms)	86–7; 105–6
175	**Pilocarpidine**	*d*	$C_{10}H_{14}O_2N_2$	viscid oil	. . . .
176	nitrate	*d*	$C_{10}H_{14}O_2N_2 \cdot HNO_3$	pr./aq.	137
177	**Pilocarpine**	*d*	$C_{11}H_{16}O_2N_2$	colorl. oil or need.	34
178	hydrochloride	*d*	$C_{11}H_{16}O_2N_2 \cdot HCl$	pr. or need.	204–5
179	nitrate	*d*	$C_{11}H_{16}O_2N_2 \cdot HNO_3$	pr./al. or aq.	176–8
180	**Piperine**	*in*	$C_{17}H_{19}O_3N$	mon. need./al.	129–30
181	periodide	. . .	$(C_{17}H_{19}O_3N)_2 \cdot HI \cdot I_2$	steel bl. need.	145
182	**Porphyroxine**	*l*	$C_{19}H_{23}O_4N$	pr./lig.	134–5
183	**Protopine**	*in*	$C_{20}H_{19}O_5N$	mon. cryst./al.	207–8
184	**Protoveratrine**	. . .	$C_{32}H_{51}O_{11}N$	rectangular tab.	245–50 d.
185	**Pseudoaconitine**	*d*	$C_{36}H_{51}O_{12}N$	rhombs/chl. + et.	212–4
186	**Pseudoephedrine**	*d*	$C_{10}H_{15}ON$	rhomb. tab./et.	118–9
187	hydrochloride	. . .	$C_{10}H_{15}ON \cdot HCl$	need.	176
188	**Pseudomorphine**	*l*	$C_{34}H_{36}O_6N_2 \cdot 3H_2O$	crusts or need.	d. 327

* Melting point of the anhydrous compound.
Napelline 3
Narcissine 147
Narcosine 158
dl-Narcotine 114
Nepaline 185

Neriine 76
Nor-arecaidine 116
Opianin 158
Opin 182
Oxy-dimorphine 188

Table 7-5 (*Continued*)
PHYSICAL CONSTANTS OF ALKALOIDS

No.	Reaction with H_2SO_4	Solubility Expressed in Parts of Solvent Required to Dissolve 1 Part Alkaloid					
		Water	Alcohol	Ether	Chloroform	Benzene	Others
151	....	17.2 c.; 0.5 h.	42	i.	i.	....	19 gly.
152	....	15.3	452	i.	i.	....	....
153	....	v. s.	v. s.	sl. s.	sl. s.	....	d.a.; stable alk.
154	....	sl. s.	s.	s.	s.	s.	s. dil. a.
155	br.→r. h.	769 c.; 220 h.	945 c.; s. h.	i.	v. sl. s.	i.; s. alk.	s. NH₄OH; s. dil. a.
156	....	s.→basic salt	s. h.	s.	s.	....	
157	....	s. h.	s. h.	....	....	....	s. MeOH
158	yel. grn.→ r. h.	3300	100	166	3	22	s. h. alk.
159	....	diss.	s.	....	s.	....	....
160	colorl.	s.	∞	∞	∞	....	s. pet.
161	....	s.	s.	....	....	....	....
162	....	....	....	....	....	....	low solubility
163	....	s.	s.	....	....	....	....
164	....	v. s.	v. s.	....	....	....	....
165	colorl.	i.	s.	s.	s.	s.	s. dil. a.
166	....	sl. s.	....	....	....	....	....
167	....	i.	sl. s.	sl. s.	s.	s.	s. a.; sl. s. pet.
168	colorl.→ rose r. h.	i.	45 c.; 4 h.	250	s. h.	s. h.	13 pyr.
169	....	37	s.	i.	s.	....	....
170	....	v. sl. s.	∞	∞	....	....	....
171	grn. with $K_2Cr_2O_7$	20	s.	s.	s.	s.	....
172	yel.→r. with HNO_3	v. sl. s.	s.	s.	s.	....	sl. s. pet.
173	....	s.	....	....	....	....	....
174	colorl.→yel.	sl. s.	s.	s.	s.	s.	....
175	....	s.	s.	....	s.	....	....
176	....	2	82	....	....	....	....
177	colorl.	v. s.	v. s.	sl. s.	v. s.	v. sl. s.	i. pet.; s. alk.
178	....	0.3	3	i.	545	....	....
179	....	4	60	i.	i.	....	....
180	dk. r.→ br. blk.	sl. s.	12	26	2.6	s.	i. pet.
181	....	....	s.	....	v. s.	s.	....
182	red	sl. s.	s.	s.	s.	s.	s. dil. a.
183	yel.→ violet→grn.	i.; sl.s. act.	900	1000	15	v. sl. s.	sl. s. NH₄OH
184	grn.→bl.→ violet	i.	s. h.	sl. s.	s.	i.	i. pet.
185	....	v. sl. s.	s.	sl. s.	s.	....	*l* salts
186	....	sl. s.	s.	s.	s.	....	....
187	....	s.	s.	....	....	....	....
188	+sucrose→ dk. grn.→ br.	i.	i.	i.	i.	s. alk.	s. h. NH₄OH; s. pyr.

Oxy-neurine HCl 34
6-Oxy-purine 136
Paramorphine 228
Peganine 239
1-Phenyl-2-methylamino-propanol 103

Pilosine 47
Pitayine 195
Proline-betaine 222
2-Propyl-piperidine 78
Pseudocinchonine 60

Table 7-5 (*Continued*)
PHYSICAL CONSTANTS OF ALKALOIDS

No.	Name	[α]	Formula	Appearance	Melting Point °C.
189	hydrochloride	*l*	$C_{34}H_{36}O_6N_2 \cdot 2HCl \cdot 2H_2O$	cryst. powd.	
190	**Pseudopelletierine**	*in*	$C_9H_{15}ON$	anh. prism. tab.	48; B.P. 246
191	**Pseudotropine**	*in*	$C_8H_{15}ON$	tab. or pr./et.	108
192	**Pukateine**	*l*	$C_{18}H_{17}O_3N$	cryst./et.	200
193	**Quinicine**	*d*	$C_{20}H_{24}O_2N_2$	yel. oil; hardens on standing	(60)
194	oxalate	*d*	$(C_{20}H_{24}O_2N_2)_2 \cdot H_2C_2O_4 \cdot 9H_2O$	pr./chl. or need./al.	149
195	**Quinidine**	*d*	$C_{20}H_{24}O_2N_2$	pr.+al./al.; tab.+ et./et.	174–5*
196	bisulfate	...	$C_{20}H_{24}O_2N_2 \cdot H_2SO_4 \cdot 4H_2O$	hair-like need. or pr.	....
197	hydrochloride	*d*	$C_{20}H_{24}O_2N_2 \cdot HCl \cdot H_2O$	asbestos-like pr.	258–9 d.*
198	sulfate	*d*	$(C_{20}H_{24}O_2N_2)_2 \cdot H_2SO_4 \cdot 2H_2O$	pr. or need./h. aq.	....
199	**Quinine**	*l*	$C_{20}H_{24}O_2N_2$	wh. need. or powd.	175
200	arsenate	...	$3(C_{20}H_{24}O_2N_2) \cdot 2H_3AsO_4 \cdot 5H_2O$	wh. effl. cryst.	....
201	bisulfate	*l*	$C_{20}H_{24}O_2N_2 \cdot H_2SO_4 \cdot 7H_2O$	pr./aq. or al.	160 d.*
202	formate	*l*	$C_{20}H_{24}O_2N_2 \cdot HCO_2H$	cryst. powd.; need.	109 d.
203	hydrate	*l*	$C_{20}H_{24}O_2N_2 \cdot 3H_2O$	efflorescent; anh. at 100°	57
204	hydrobromide	...	$C_{20}H_{24}O_2N_2 \cdot HBr \cdot H_2O$	hyg., silky need.	152–200
205	hydrochloride	*l*	$C_{20}H_{24}O_2N_2 \cdot HCl \cdot 2H_2O$	effl., silky need.	158–60*
206	hydrochloride, di-	*l*	$C_{20}H_{24}O_2N_2 \cdot 2HCl$	wh. powd. or need.	180–5
207	iodosulfate	...	$4C_{20}H_{24}O_2N_2 \cdot 3H_2SO_4 \cdot 2HI \cdot I_4 \cdot 6H_2O$	pl./al.; r. or grn. by reflected or transmitted light	$-H_2O$, 100
208	salicylate	...	$C_{20}H_{24}O_2N_2 \cdot C_7H_6O_3 \cdot H_2O$	need./aq.	195
209	sulfate	*l*	$(C_{20}H_{24}O_2N_2)_2 \cdot H_2SO_4 \cdot 7H_2O$	efflorescent need.	235*
210	sulfate	*l*	$(C_{20}H_{24}O_2N_2)_2 \cdot H_2SO_4 \cdot 2H_2O$	by drying in air	205
211	valerate	...	$C_{20}H_{24}O_2N_2 \cdot C_5H_{10}O_2 \cdot H_2O$	cryst. powd.	95
212	**Rhoeadine**	...	$C_{21}H_{21}O_6N$	small pr. or need.	245–7 d.
213	**Ricinine**	...	$C_8H_8O_2N_2$	pr. or tab./al. or aq.	201, subl.
214	**Rutecarpine**	...	$C_{18}H_{13}ON_3$	yel. pl.; need./EtOAc	260–2
215	**Sinapine**	...	$C_{16}H_{25}O_6N$	free base unknown	....
216	bisulfate	...	$C_{16}H_{24}O_5NHSO_4 \cdot 3H_2O$	leaf./al.	127*
217	thiocyanate	...	$C_{16}H_{24}O_5NSCN \cdot H_2O$	pale yel. need./aq.	178
218	**Solanidine**	*l*	$C_{26}H_{41}ON$	need./et. or al.	219
219	**Solanine**	*l*	$C_{44}H_{71}O_{15}N$	slender need./al.	244–54 d.

* Melting point of the anhydrous compound.
Pseudo-punicine 190
Punicine 171
Pyridyl-*N*-methyl-pyrrolidine 160
2-(3-Pyridyl)-piperidine 11

Quebrachine 244
α-Quinidine 53
β-Quinine 195

Table 7-5 (*Continued*)
PHYSICAL CONSTANTS OF ALKALOIDS

No.	Reaction with H_2SO_4	Solubility Expressed in Parts of Solvent Required to Dissolve 1 Part Alkaloid					
		Water	Alcohol	Ether	Chloroform	Benzene	Others
189	....	70	....	....	....	....	....
190	+CrO₃→ grn.	s.	s.	s.	s.	....	sl. s. pet.
191	....	v. s.	v. s.	sl. s.	s.	....	very alkaline
192	org. →r. and violet h.	i.	s.	161	s.	s. pyr.	s. alk
193	no fluor- escence	v. sl. s.	s.	s.	s.	....	....
194	....	s. h.	s.	....	s.	....	....
195	dil.→bl. fluores- cence	2000 c.; 800 h.	26	22	2.3	s.	sl. s. lig.
196	....	8; fluores- cence	....	....	....	....	....
197	....	60 c.; v. s. h.	s.	sl. s.	s.	....	....
198	....	100 c.; 15 h.	8	v. sl. s.	15	i.	....
199	colorl.→lt. yel.→br. h.	1750	0.6	22.6	1.9	166 c.; 30 h.	s. CS₂; s. NH₄OH
200	....	650 c.; 120 h.	200 c.; 50 h.	....	....	....	s. dil. a.
201	....	fluorescence; 9	19	1770	920	....	18 gly.
202	....	19	s.	v. sl. s.	s.	....	....
203	....	1560 c.; 800 h.	0.6	1.4	1.6	70	212 gly.
204	dil.→fluor- escence	40 o.; 3 h.	1	23	1	....	9 gly.
205	ditto	16 c.; 0.5 h.	0.6	340	1	....	9 gly.
206	....	0.6	5	v. sl. s.	7	....	....
207	....	1000 h.	800 c.; 50 h.	....	....	....	60 h. ac. a.
208	....	1500	14	114	38	....	18 gly.
209	dil.→fluor- escence	725 c.; 30 h.	60	sl. s.	1000	....	24 gly.
210	....	810 c.; 30 h.	96	sl. s.	sl. s.	....	....
211	....	70 c.; 40 h.	2	10	....	....	....
212	purple r.	1200	700	800	v. sl. s.	i.	s. a., d.
213	....	s. h.; i. pet.	sl. s. c.; s. h.	sl. s.	s.	sl. s.	forms no salts
214	bright yel.	i.	sl. s.	s.	s.	s.	....
215	....	....	....	....	....	....	....
216	....	s.	s. h.	i.	....	....	....
217	....	sl. s.	sl. s.	....	....	....	....
218	....	v. sl. s. h.	s. h.	sl. s.	s.	....	....
219	yel.→rose →r.	i.	s. h.	i.	i.	i.	....

Table 7-5 (*Continued*)
PHYSICAL CONSTANTS OF ALKALOIDS

No.	Name	[a]	Formula	Appearance	Melting Point °C.
220	**Sparteine**	*l*	$C_{15}H_{26}N_2$	colorl. oil	B.P. 325^{754} in H_2
221	bisulfate	...	$C_{15}H_{26}N_2 \cdot H_2SO_4 \cdot 5H_2O$	transparent cryst.	150–2*
222	**Stachydrine**	...	$C_7H_{13}O_2N \cdot H_2O$	deliq. cryst.	235 d*
223	oxalate	...	$C_7H_{13}O_2N \cdot H_2C_2O_4$	need.	105–7
224	**Strychnine**	*l*	$C_{21}H_{22}O_2N_2$	colorl. rhombs/al.	286–8; B.P. 270^5
225	hydrochloride	...	$C_{21}H_{22}O_2N_2 \cdot HCl \cdot 2H_2O$	efflorescent pr.	....
226	nitrate	*l*	$C_{21}H_{22}O_2N_2 \cdot HNO_3$	shining need.	....
227	sulfate	...	$(C_{21}H_{22}O_2N_2)_2 \cdot H_2SO_4 \cdot 5H_2O$	effl., mon. pr.	200 d.*
228	**Thebaine**	*l*	$C_{19}H_{21}O_3N$	leaf. or pr./al.	193
229	hydrochloride	*l*	$C_{19}H_{21}O_3N \cdot HCl \cdot H_2O$	large rhombs or yel. powd.	....
230	*iso*-**Thebaine**	*d*	$C_{19}H_{21}O_3N$	rhomb./al. or et.	203–4
231	sulfate	...	$(C_{19}H_{21}O_3N)_2 \cdot H_2SO_4$	....	120–1 d.
232	**Theobromine**	*in*	$C_7H_8O_2N_4$	minute rhomb. cryst.; mon./h. aq.	330†; subl. 290
233	**Theophylline**	*in*	$C_7H_8O_2N_4 \cdot H_2O$	mon. tab. or need./h. aq.	269–72
234	**Trigonelline**	...	$C_7H_7O_2N \cdot H_2O$	hyg. pr./al.	218 d.*
235	**Tropacocaine**	*in*	$C_{15}H_{19}O_2N$	need. or plates	49
236	hydrochloride	...	$C_{15}H_{19}O_2N \cdot HCl$	need.; pl./aq. al.	283 d.
237	**Tropine**	*in*	$C_8H_{15}ON$	hyg. tab./abs. et.	63
238	**Turicine**	*d*	$C_7H_{13}O_3N \cdot H_2O$	sweet, effl. pr. or need./aq. al.	260 d.*
239	**Vasicine**	*l*	$C_{11}H_{12}ON_2$	need./al.	211–2
240	**Veratridine**	*in*	$C_{36}H_{51}O_{11}N$	amor.; yel.	180
241	**Vicine**	*l*	$C_{10}H_{16}O_7N_4 \cdot 2H_2O$	need.	239–42
242	**Vomicine**	*d*	$C_{22}H_{24}O_4N_2$	need./aq. al.	278–80
243	**Xanthine**	*in*	$C_5H_4O_2N_4 \cdot H_2O$	small pl.; anh. at 125°	>150 d.
244	**Yohimbine**	*d*	$C_{21}H_{26}O_3N_2$	need./aq. al.	247–8
245	hydrochloride	*d*	$C_{21}H_{26}O_3N_2 \cdot HCl$	plates	302
246	nitrate	...	....	colorl. pr.	276
247	thiocyanate	...	....	rectangular pr./h. aq.	233–4
248	**Zygadenine**	*l*	$C_{39}H_{63}O_{10}N$	need./bz.	200–1

* Melting point of the anhydrous compound.
† Sealed tube.

Stypticin 86	Tetrahydro-nicotinic acid 116	Trimethyl-glycocoll 33
Styptol 87	Theine 39	1,3,7-Trimethyl-xanthine 39
Telepathine 118	Theocine 233	Tropine mandelate 119
Tetraanhydro-berberine 45	Trigenolline 234	Ulexine 92

Table 7-5 (*Continued*)
PHYSICAL CONSTANTS OF ALKALOIDS

No.	Reaction with H_2SO_4	Solubility Expressed in Parts of Solvent Required to Dissolve 1 Part Alkaloid					
		Water	Alcohol	Ether	Chloroform	Benzene	Others
220	colorl.	328	s.	s.	s.	i.	....
221	colorl.	1.1	2.4	i.	i.	....	....
222	....	s.	s.	i.	i.	s. dil. a.	d. in air
223	....	....	i. c.	....	....	....	....
224	colorl.	6400 c.; 3100 h.	110 c.; 28 h.	v. sl. s.	6	150	173 PhMe
225	....	35	60	i.	....	....	i. HCl
226	....	42 c.; 10 h.	120	i.	156	....	60 gly.
227	....	31 c.; 7 h.	65	i.	325	....	7 gly.
228	blood r.	i.	10	135	19	18	i. alk.
229	....	12	s.	....	....	....	....
230	....	....	s.	sl. s.	s.	....	....
231	....	....	....	....	....	....	....
232	....	2000 c.; 150 h.	1775 c.; 260 h.	3125 h.	157 c.; 100 h.	100,000	s. alk.; 4700 h. CCl_4
233	....	120 c.; s. h.	64	sl. s.	164	s. alk.	s. NH_4OH
234	....	v. s.	s.	v. sl. s.	v. sl. s.	i.	neutral reaction
235	....	i.	s.	s.	v. s.	v. s.	s. dil. NH_4OH
236	....	s.	sl. s.	....	....	....	....
237	....	v. s.	v. s.	s.	s.	s.	....
238	....	v. s.	sl. s.	....	....	....	....
239	colorl.	sl. s.	s.	sl. s.	s.	sl. s.	i. pet.; s. dil. a.
240	yel.	s.	....	sl. s.	....	....	....
241	yel.	sl. s.	i. abs.	....	....	s. dil. a.	s. MeOH
242	+CrO_3→ deep r.	....	s. h.	sl. s.	s.	....	s. act.
243	....	14,400 c.; 1500 h.	2400	3000	....	....	s. alk.; s. a.
244	colorl. + $K_2Cr_2O_7$→ dirty grn.	v. sl. s.	s.	sl. s.	s.	s. h.	....
245	....	120	400	....	....	....	....
246	....	....	....	....	....	....	....
247	....	....	....	....	....	....	....
248	or. → cherry r.	....	s.	....	s.	s.	....

Ultraquinine 89
5–Ureido–hydantoin 10
Ureous acid 243
Veratrine, a mixture of 49,240 *et al.*
Veratrine (crystallized) 49

Veratroylaconine 185
Vinetine 165
Viridine 138
Wrightine 76

Xanthaline 167
Xanthopuccine 45
Yageine 118
Yajeine 118

Table 7-6
PHYSICAL CONSTANTS OF GLUCOSIDES
Compiled by F. E. SHEIBLEY, Ph.D.

Names of the compounds in the table below are arranged alphabetically.

Appearance. In addition to the crystalline form and color, the solvent used in purification is often given; thus, "rhomb./al." indicates that rhombic crystals were obtained when the compound was crystallized from alcohol.

Optical Properties. These are indicated by the symbols *d*, *l*, or *in.* following the name. Most of the optically active glucosides are levo-rotatory.

Solubilities. Most glucosides are soluble in cold or hot water. The solubilities given for alcohol refer to the ordinary alcohol of approximately 95% concentration. Solubilities expressed by numbers are given in parts by weight of solvent required to dissolve one part of the glucoside. Because of discrepancies existing between many of these figures as found in the literature no claim to accuracy can be made,

Abbreviations used in the table

a., acid	colorl., colorless
abs., absolute	conc., concentrated
ac., a., acetic acid	cryst., crystals or crystalline
act., acetone	crystn., crystallization
al., alcohol	d., decomposes or decomposition
alk., alkali (i. e. aqueous NaOH or KOH)	*d*, dextrorotatory
amor., amorphous	dil., dilute
anh., anhydrous	et., ether $(C_2H_5)_2O$
aq., aqueous; water	EtOAc, ethyl acetate
br., brown	h., hot
bz., benzene, C_6H_6	hyd., hydrate
c., cold	hyg., hygroscopic
Chl., chloroform, $CHCl_3$	

No.	Name	Formula	Appearance	Melting Point °C.
1	Absinthin	$C_{30}H_{40}O_8$	glossy need. or yel. amor. powd.	68
2	Aesculin (*l*)	$C_{15}H_{16}O_9 \cdot 2H_2O$	pr./aq. or dil. al.	205 d.
3	Amygdalin (*l*)	$C_{20}H_{27}O_{11}N \cdot 3H_2O$	orthorhomb. pr./aq.; glossy scales $(+2H_2O)/$ 80% al.	214–6*
4	Antiarin	$C_{27}H_{42}O_{10} \cdot 4H_2O$	plates/aq.	220–5
5	Apiin (*l*)	$C_{26}H_{28}O_{14} \cdot H_2O$	glossy need. or yel. cryst. powd.	228
6	Arbutin (*l*)	$C_{12}H_{16}O_7 \cdot H_2O$	long silky need./aq.	195–200*
7	Baptisin (*l*)	$C_{28}H_{30}O_{14} \cdot 3H_2O$	thin wh. need./al. slowly becomes anh.	sint. 150; m. 249–51
8	Bryonin (*d*)	$C_{34}H_{50}O_9$	amor. bright yel. powd.	softens 208
9	Carminic acid	$C_{22}H_{20}O_{13}$	purplish-br. mass or bright r. powd.; r. pr.	d. 136

* Melting point of the anhydrous compound.

Abietin 15	Asebotin 55
Acocantherin 51	Aurantiin 49
Arthanitin 22	

Table 7-6 (*Continued*)
PHYSICAL CONSTANTS OF GLUCOSIDES

and the numerical values stated are perhaps best considered as upper limits.

Hydrolysis. On being heated with aqueous mineral acids, usually dilute hydrochloric acid or sulfuric acid, glucosides undergo hydrolysis to yield the original cyclic hemiacetal (parent saccharide) and the reactant alcohol originally linked to the noncyclic oxygen on the anomeric carbon.

References. The information given in the table has been collected mainly from the following sources:—

J. J. L. van Rijn: **Die Glykoside.** Published by Borntraeger, Berlin (1900). **The Merck Index,** 7th edition. Published by Merck and Co., Inc., Rahway, N. J. Beilstein: **Handbuch der Organischen Chemie,** 3d edition.

i., insoluble	pet., petroleum ether
in, inactive	powd., powder
K$_2$CO$_3$, aq. soln. of	pr., prisms
KOH, aq. soln. of	pyr., pyridine
l, levo-rotatory	r., red
leaf., leaflets or leaves	rhomb., rhombic
m-, meta	s., soluble
m., melts	sint., sinters
m.p., M.P., melting point	sl., slight or slightly
mon., monoclinic	v., very
need., needles	wh., white
org., orange	yel., yellow
orthorhomb., orthorhombic	>, greater than

No.	Solubility Expressed in Parts of Solvent Required to Dissolve 1 Part Glucoside					Hydrolytic Products		
	Water	Alcohol	Ether	Chloro-form	Others	Principle	M.P. °C.	Sugar
1	sl. s.	s.	s.	s.	s. bz.; s. NaOH	resinous sub-stance	. . . . . .	glucose
2	576 c.; s. h.*	24 h.	sl. s.	s. h.	v. s. dil. alk.	aesculetin	270 d.	glucose
3	12 c.; v. s. h.	720 c.; 9 h.	i.	. . . . . .	. . . .	hydrocyanic acid benzaldehyde	−12 −26	glucose
4	s.	s.	sl. s.	. . . . . .	. . . .	antiarigenin	180	antiarose
5	sl. s. c.; s. h.	s. h.	i.	. . . . . .	. . . .	apigenin	subl. 292−5	glucose + apiose
6	8 c.; 1 h.	13	i.	i.	i. CS$_2$	hydroquinone	170.3	glucose
7	sl. s. c.; s. h.	sl. s. dil.	v. sl. s.	v. sl. s.	s. glacia. ac. a.	baptigenin	296−8	glucose + rhamnose
8	s.	s	i.	i.	. . . .	bryogenin (resinous)	. . . . . .	glucose
9	s.	s.	v. sl. s.	i.	s. alk.; s. conc. H$_2$SO$_4$	carmine red	. . . . . .	sugar

* Aqueous solutions fluoresce faint blue.

Avenein 35	6-Benzoyl-*d*-glucose 79
Avornin 30	6-Benzoyl-salicin 59

Table 7-6 (*Continued*)
PHYSICAL CONSTANTS OF GLUCOSIDES

No.	Name	Formula	Appearance	Melting Point °C.
10	Cerberin (*l*)	$C_{27}H_{40}O_8$	glossy cryst./et.	191–2
11	α-Chinovin (*d*)	$C_{30}H_{48}O_8$	rosettes of small need./al.	. . . .
12	β-Chinovin (*d*)	$C_{30}H_{48}O_8$	scales/dil. al.	235 d.
13	Clavicepsin (*d*)	$C_{18}H_{34}O_{16} \cdot H_2O$	wh. cryst.	91; 198*
14	Colocynthin	$C_{56}H_{84}O_{23}$	microscopic pr.	. . . .
15	Coniferin (*l*)	$C_{16}H_{22}O_8 \cdot 2H_2O$	wh. pointed satiny need.	185
16	Convallamarin (*l*)	$C_{23}H_{44}O_{12}$, mixture	wh., cryst. powd.	. . . .
17	Convallarin (*l*)	$C_{34}H_{62}O_{11}$, mixture	rectangular pr.	. . . .
18	Convolvulin (*l*)	$C_{54}H_{96}O_{27}$	wh. amor. powd.	155–8
19	Coriamyrtin (*d*)	$C_{15}H_{18}O_5$	mon. pr.	228–30
20	Crocin	$C_{44}H_{64}O_{26} \cdot H_2O$	br. red cryst.	186 d.
21	Curangin (*d*)	$C_{48}H_{77}O_{20}$	amor.	172
22	Cyclamin (*l*)	$C_{27}H_{38}O_{13}$	wh. amor. powd.; microscopic cryst.	236
23	Daphnin (*l*)	$C_{15}H_{16}O_9 \cdot 2H_2O$	pr. or need./aq.	215 d.
24	Datiscin (*l*)	$C_{27}H_{30}O_{15} \cdot 4H_2O$	glossy need. or pl./aq.	192
25	Digitalin	$C_{36}H_{56}O_{14}$	wh. cryst. powd.	229
26	Digitonin (*l*)	$C_{55}H_{90}O_{29}$	wh. cryst. powd.	sint. 225; d. 235
27	Digitoxin†	$C_{41}H_{64}O_{13}$	wh. leaf.	255–6*
28	Digitoxin hydrate	$C_{41}H_{64}O_{13} \cdot 6H_2O$	leaf./al.	145
29	Euxanthic acid	$C_{19}H_{16}O_{10} \cdot H_2O$	straw yel. need.	155–8 d.
30	Frangulin (*l*)	$C_{21}H_{20}O_9 \cdot H_2O$	org. need./aq. pyr.	246–9
31	Fraxin	$C_{16}H_{18}O_{10}$	need./al.	205
32	Fustin	$C_{36}H_{26}O_{14}$	wh. glossy need./aq.	217 d.
33	Gaultherin (*l*)	$C_{14}H_{18}O_8 \cdot H_2O$	need. or pr./al.	179–80
34	Glucogallin (*l*)	$C_{13}H_{16}O_{10}$	wh.-yel. cryst.	193 d.
35	Glycovanillin (*l*)	$C_{14}H_{18}O_8 \cdot 2H_2O$	wh. need./dil. al.	192
36	Glycyphyllin	$C_{21}H_{24}O_9 \cdot 3H_2O$	long, thin, glossy pr./et.	175–80 d.
37	Gratiolin	$C_{43}H_{70}O_{15}$	fine, glossy need.	235–7 d.
38	α-Hederin (*l*)	$C_{41}H_{64}O_{11}$	wh. need.	256–7
39	Helicin ‡ (*l*)	$C_{13}H_{16}O_7 \cdot \frac{3}{4}H_2O$	fine, radiating need./aq.	175 *; —aq. 100
40	Helleborein (*l*)	$C_{37}H_{56}O_{18}$	warts of fine need./al.	270

* Melting point of the anhydrous compound.
† Crystalline digitalin.
‡ Does not occur naturally but is obtained by oxidizing salicin.

Cuspidatin 58	Digitin 26
Datiscosid 24	Esculin 2
Digitalin, crystalline 27	Esculinic acid 2

Table 7-6 (*Continued*)
PHYSICAL CONSTANTS OF GLUCOSIDES

No.	Solubility Expressed in Parts of Solvent Required to Dissolve 1 Part Glucoside					Hydrolytic Products		
	Water	Alcohol	Ether	Chloro-form	Others	Principle	M.P. °C.	Sugar
10	sl. s.	12	sl. s.	9	sl. s. CCl₄	cerberetin	85.5	glucose
11	i.	s. dil.	v. sl. s.	v. sl. s.	s. alk.	chinovic acid	d. 295	chinovose
12	....	v. s.*	i.	......	i. EtOAc	chinovic acid	d. 295	chinovose
13	s.	sl. s.	i.	i.	i. bz.	d-mannitol	166	glucose
14	s.	s. h.	i.	......	....	colocynthein (resinous)	......	glucose
15	200 c.; v. s. h.	sl. s.	i.	......	....	coniferyl alcohol†	73–4	glucose
16	s.	s. dil.	sl. s.	i.	....	convallamaretin	......	sugar
17	sl. s.	s.	i.	......	....	convallaretin	......	sugar
18	sl. s.	s.	i.	sl. s.	s. EtOAc	methylethyl-acetic and other acids	......	glucose, rhodeose
19	sl. s. c.	s. h.	s.	s.	....	indefinite	......	sugar
20	sl. s. c.; s. h.	sl. s. abs.	i.	......	....	crocetin	285	gentiobiose
21	v. sl. s.	s.	sl. s.	......	s. aq. act.	curangaegenin	132	rhamnose + a little glucose
22	sl. s.	57	i.	i.	i. bz.	cyclamiretin	198	fructose, cyclose
23	sl. s. c.; s. h.	v. s. h.	i.	......	s. alk.	daphnetin	253–6 d.	glucose
24	s.	s.	i.	......	....	datiscetin	276	rhamnose
25	1000	12	sl. s.	sl. s.	s. MeOH	digitaligenin	210–2	glucose, digitalose
26	sl. s.	s.	v. sl. s.	v. sl. s.	s. MeOH	digitogenin	softens 250	glucose, galactose
27	v. sl. s.	s.	v. sl. s.	s.	....	digitoxigenin	230	digitoxose
28	....	....	......	......	....	....	......	....
29	s. h.	s. h.	v. sl. s.	......	s. alk.	euxanthone	240	glycuronic acid
30	i.	sl. s. h.	i.	s. alk.	s. h. bz.	frangula-emodin	256–7	rhamnose
31	sl. s. c.; s. h.	s. h.	i.	......	‡	fraxetin	227	glucose
32	v. s. h.	v. s.	sl. s.	......	s. alk.	fisetin	>360	rhamnose
33	slowly s.	s.	v. sl. s.	v. sl. s.	v. sl. s. act.	methyl salicylate	−8.3	glucose
34	s.	s.	sl. s.	i.	s. alk.	gallic acid	235 d.	glucose
35	s.	sl. s.	i.	......	....	vanillin	80	glucose
36	s. h.	s.	s.	i.	i. bz.	phloretin	180	isodulcite
37	s. h.	s.	i.	......	....	gratiogenin	198	glucose
38	i.	s.	i.	......	s. ac. a.	hederagenin	331	arabinose, rhamnose
39	60 c.; v. s. h.	s.	sl. s.	......	....	salicylaldehyde	−7	glucose
40	v. s.	sl. s.	i.	......	....	helleboretin	>200	glucose, arabinose, ac. a.

*Heat evolved; forms penta-alcoholate, m.p. 70–80°, which separates.
†By the action of emulsin; heating with dilute acids gives resinous material.
‡Solutions fluoresce blue.

Franguloside 30
Fraxinin 31
Glucosido-methyl salicylate 33

Glucosido-salicylaldehyde 39
Gratus strophanthin 51
Helixin 38

Table 7-6 (*Continued*)
PHYSICAL CONSTANTS OF GLUCOSIDES

No.	Name	Formula	Appearance	Melting Point °C.
41	Helleborin	$C_{28}H_{36}O_6$	lustrous need.	>250
42	Hesperidin	$C_{28}H_{34}O_{15}$	wh., micro. need./aq. MeOH	251–2; d. 254
43	Indican	$C_{14}H_{17}O_6N \cdot 3H_2O$	need./aq.	57–8; 176–8 *
44	Iridin	$C_{24}H_{26}O_{13}$	wh. need.→yel. in air	208
45	Jalapin (*l*)	$C_{34}H_{56}O_{16}$	colorl. amor. mass	131–50
46	Linamarin (*l*)	$C_{10}H_{17}O_6N$	need.	142–3
47	Maclayin	$C_{17}H_{32}O_{10}$	deliquescent cryst.	158–65
48	Murrayin	$C_{18}H_{22}O_{10}$	microscopic need.	170
49	Naringin (*l*)	$C_{27}H_{32}O_{14} \cdot 8H_2O$	need/aq.	82
50	Ononin	$C_{25}H_{26}O_{11}$	small pr., need. or plates	210
51	Ouabain (*l*)	$C_{29}H_{44}O_{12} \cdot 7H_2O$	transparent plates	185 *
52	Parillin (*l*)	$C_{26}H_{44}O_{10} \cdot 2\frac{1}{2}H_2O$	fine plates	177
53	Periplocin (*d*)	$C_{30}H_{48}O_{12}$	long, thin need.	205
54	Phillyrin	$C_{27}H_{34}O_{11}$	need. or plates	162
55	Phloridzin (*l*)	$C_{21}H_{24}O_{10} \cdot 2H_2O$	small, wh., silky need.	108†
56	Picein (*l*)	$C_{14}H_{18}O_7 \cdot H_2O$	need./aq.	194 *
57	Picrocrocin (*l*)	$C_{16}H_{26}O_7$	pr.-et.-chl.-MeOH mixt.	154–6
58	Polygonin	$C_{21}H_{20}O_{10}$	glossy, yel. need.	202–3
59	Populin (*l*)	$C_{20}H_{22}O_8 \cdot 2H_2O$	wh. cryst. powd. or very fine need./aq.	180 *
60	Prulaurasin (*l*)	$C_{14}H_{17}O_6N$	bitter need.	120–2
61	Quercitrin	$C_{21}H_{20}O_{11} \cdot 2H_2O$	pale yel. need. or plates	182–5; 250–2 *
62	Robinin	$C_{32}H_{40}O_{19} \cdot 7\frac{1}{2}H_2O$	yellowish need./aq.	195 *
63	Ruberythric acid	$C_{25}H_{26}O_{13}$	small, citron yel. need.; yel. pr./aq.	258–60
64	Rubiadin glucoside	$C_{21}H_{20}O_9$	yel. need./glacial ac. a.	270 d.
65	Rutin	$C_{27}H_{30}O_{16} \cdot 2H_2O$	bright yel. need./aq.	188–90
66	Salicin (*l*)	$C_{13}H_{18}O_7$	glossy need., plates or rhomb. pr./aq.	199–201, then re-melts 230–40
67	Saponin	$C_{32}H_{52}O_{17}$	wh. amor. powd.	d. 195
68	Saporubrin (*l*)	$(C_{18}H_{28}O_{10})_4$	wh. amor. powd.	....
69	Sapotoxin	$C_{17}H_{26}O_{10}$	wh. amor. powd.	....
70	Sarsasaponin (*l*)	$C_{44}H_{76}O_{20} \cdot 7H_2O$	long need./al	sint. 200; m. 248

* Melting point of the anhydrous compound.
† Melts at 108°, then solidifies at 130° and remelts at 170–1 d.

iso-Hesperidin 49	*dl*-Mandelonitrile-glucoside 60
Indoxyl-β-glucoside 43	Melin 65
Kalmin 55	Methoxy-coniferin 74
Laricin 15	Monotropitoside 33
Lilacin 74	Orizabin 45
Macleyin 47	Paviin 31

Table 7-6 (*Continued*)
PHYSICAL CONSTANTS OF GLUCOSIDES

No.	Solubility Expressed in Parts of Solvent Required to Dissolve 1 Part Glucoside					Hydrolytic Products		
	Water	Alcohol	Ether	Chloro-form	Others	Principle	M.P. °C.	Sugar
41	i.	sl. s.	sl. s.	s.	....	helleboresin	d. >140	glucose
42	v. sl. s.	sl. s.	i.	i.	v. s. dil. alk.	hesperetin	226	glucose, rhamnose
43	v. s.	v. s.	sl. s.	sl. s.	sl. s. bz.	indigo	390–2 d.	glucose
44	v. sl. s.	s. h.	i.	i.	sl. s. act.	irigenin	186	glucose
45	sl. s	s.	s. h.	s.	d. alk.	jalapinolic acid	67–9	sugars
46	v. s.	sl. s.	sl. s.	sl. s.	s. h. act.	hydrocyanic acid	−12	glucose, act.
47	....	sl. s.	i.	i.	....	maclayetin	209–10	glucose
48	s. h.	s.	i.	......	s. alk.	murrayetin	110	glucose
49	sl. s. c.; s. h.	v. s.	i.	i.	v. s. h. ac. a.	naringenin	251	glucose, rhamnose
50	sl. s. h.	s. h.	i.	......	s. h. KOH	formononetin	265	glucose
51	100 c.; 5 h.	20 c.; 8 h.	i.	i.	i. EtOAc	acocanthic acid lactone	......	rhamnose
52	v. sl. s. c.; 20 h.	s.	i.	s.	i. pet.	parigenin	......	sugars
53	125 c.	s.	v. sl. s.	v. sl. s.	i. bz.	periplogenin	185	glucose
54	s. h.	s.	i.	s. h.	....	phillygenin	......	glucose
55	1000 c.; s. h.	4	sl. s.	i.	....	phloretin	262–4	glucose
56	sl. s.	s.	s.	i.	s. ac. a.	*p*-hydroxy-acetophenone	109	glucose
57	v. s.	v. s.	sl. s.	sl. s.	i. bz.	safranol	......	glucose
58	sl. s. h.	sl. s. h.	i.	......	....	emodin	254	glucose
59	i. c.; sl. s. h.	sl. s. c.; s. h.	i.	s. dil. a.	s. dil. alk.	saligenin	87	glucose, benzoic a.
60	s.	s.	i.	......	....	*dl*-mandelonitrile	−10	glucose
61	i. c.; sl. s. h.	s.	sl. s.	......	s. alk.	quercetin	313–4 *	rhamnose
62	s. h.	s. h.	i.	......	s. alk.	caempferol	271	rhamnose
63	sl. s. c.; s. h.	v. sl. s. abs.	v. sl. s.	s. alk. →r.	i. bz.	alizarin	290	glucose; xylose
64	v. sl. s. h.	s.	s.	......	i. K₂CO₃	rubiadin	290	glucose
65	s. h.	s. h.	i.	i.	s. alk.	quercetin	313–4 *	glucose, rhamnose
66	23 c.; 3 h.	72	i.	i.	s. alk.	saligenin	87	glucose
67	s.*	l.	i.	i.	i. bz.	sapogenin	257–60	sugar
68	v. s.*	s. dil.	i.	i.	i. bz.	sapogenin	257–60	glucose
69	s.*	s. dil.	i.	i.	s. alk.	sapotoxin-sapogenin	......	sugar
70	v. s.*	s. h.	v. sl. s.	......	....	sarsasapogenin	197–8	glucose

* Aqueous solutions foam on shaking.

Phaseolunatin 46
Phillyroside 54
Phlorizin 55
Phyllyrin 54
Piceoside 56
Potassium myronate 72
Primulin 22

Quercimelin 61
Quercitrinic acid 61
Rhamnin 83
Rhamnoxanthin 30
Rhodeoretin 18
Rubianic acid 63
Salinigrin 56
Scammonin 45

Table 7-6 (*Continued*)
PHYSICAL CONSTANTS OF GLUCOSIDES

No.	Name	Formula	Appearance	Melting Point °C.
71	Sinalbine (*l*)	$C_{30}H_{42}O_{15}N_2S_2 \cdot 5H_2O$	pale yel. need.	83–4; 139 *
72	Sinigrin (*l*)	$C_{10}H_{16}O_9NS_2K \cdot H_2O$	rhomb. pr./aq.; need./al.	127–9; 179 *
73	Strophanthin (*in*)	$C_{31}H_{48}O_{12}$	microcryst., hyg.	179
74	Syringin (*l*)	$C_{17}H_{24}O_9 \cdot H_2O$	rosettes of long need./aq.	192
75	Tampicin	$C_{34}H_{54}O_{14}$	amor., colorl. to yel.	130
76	Tannic acid (*d*)	$C_{76}H_{52}O_{46}$	yel. to br. amor. bulky powd. or spongy masses; shining scales	d. 210–5
77	Thevetin (*l*)	$C_{42}H_{66}O_{18} \cdot 3H_2O$	need./al.	210
78	Thujin	$C_{20}H_{22}O_{12}$	yel. microscopic tablets	. . . .
79	Vacciniin (*d*)	$C_6H_{11}O_6COC_6H_5 \cdot H_2O$	cryst./aq. act.	104–6
80	Valdivin (*in*)	$C_{36}H_{48}O_{20} \cdot 2H_2O$	hexagonal pr.	230 d.*
81	Vicianin (*l*)	$C_{19}H_{25}O_{10}N$	wh. need.	147–8
82	Violutin (*l*)	$C_{19}H_{26}O_{12}$	wh. cryst.	169–72
83	Xanthorhamnin (*d*)	$C_{34}H_{42}O_{20}$	yel. micro need. + EtOH/al.	—EtOH, at 120

* Melting point of anhydrous compound.
Sinigroside 72
Smilacin 52

Sophorin 65
Strophanthin Thoms 51

Table 7-6 (*Continued*)
PHYSICAL CONSTANTS OF GLUCOSIDES

No.	Solubility Expressed in Parts of Solvent Required to Dissolve 1 Part Glucoside					Hydrolytic Products		
	Water	Alcohol	Ether	Chloroform	Others	Principle	M.P. °C.	Sugar
71	s.	sl. s.	i.	......	i. CS$_2$	sinapine sulfate, and higher mustard oils	127 *	glucose
72	s.	sl. s. c.; i. abs.	i.	i.	i. bz.	allyl mustard oil	−80	glucose
73	43	s.	i.	i. CS$_2$	i. bz.	strophanthidin	170; 235 *	sugars; no glucose
74	s. h.	s. h.	i.	......	....	syringenin	......	glucose
75	i.	s.	s.	......	....	tampicolic acid	......	sugar
76	v. s.	sl. s.	v. sl. s.	v. sl. s.	v. sl. s. act.	gallic acid	235 d.*	glucose
77	s. h.	s.	sl. s.	sl. s.	s. EtOAc	thevetigenin	140	glucose
78	s. h.	s.	......	......	....	thujetin	......	glucose
79	s.	s.	i.	sl. s.	sl. s. bz.	benzoic acid	121.7	glucose
80	s. h.	s. dil.	i.	v. s.	....	?	......	sugar
81	s. h.	sl. s.	......	i.	i. bz.	benzaldehyde hydrocyanic acid	−26 −12	vicianose
82	s.	s.	i.	......	i. act.	methy.salicylate	−8.3	glucose; arabinose
83	v. s.; d.	s.	i.	i.	i. bz.	rhamnetin	>300	rhamnose

* Melting point of anhydrous compound.　　Violutiside 82
Ursin 6　　　　　　　　　　　　　　　　Waldivin 80

PROPERTIES OF HORMONES

JOHN B. STANBURY, M.D., and EDWIN D. BRANSOME, JR., M.D.

Unit of Experimental Medicine, Department of Nutrition
and Food Science, Massachusetts Institute of Technology,
Cambridge, Mass.

HORMONAL STEROIDS

Steroids are non-saponifiable lipids (they are not rendered water soluble by treatment with alkali) which are secondary alcohols and have a perhydrocyclopentanophenanthrene nucleus in common. The structure consisting of three hexane (A, B, and C) and one pentane (D) nuclei, has a conventional order of numbering of its 17 carbon atoms: illustrated in the two-dimensional structure below.

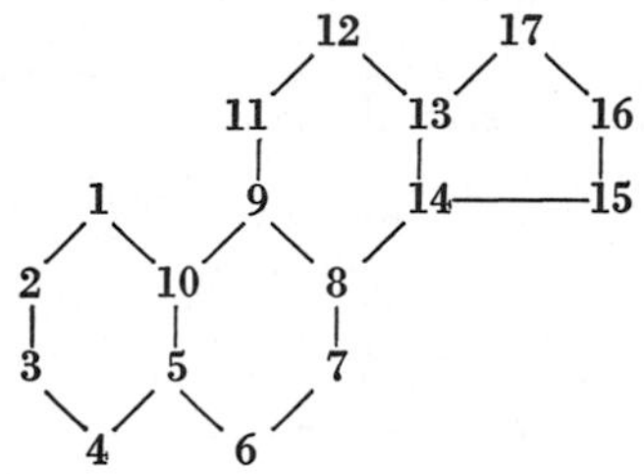

Steroid hormones are synthesized by the adrenal cortex, ovary, testis, and placenta. They have a number of biological activities: effects on intermediary metabolism (glucocorticoids), upon salt retention (mineralocorticoids), upon sexual characteristics and function (male androgens, female estrogens, female reproductive progestogens). Several hundred natural steroids have been isolated and characterized so far; most do not have the biologic activity of hormones and are either biosynthetic precursors of steroid hormones or metabolic products. Only hormonal steroids and some of their synthetic analogs will be briefly mentioned here. Further information may be discovered through perusal of the appropriate references on p. 7-427.

Nomenclature. Steroid hormones differ in respect to which —H, —OH, or =O radicals, and —CH₃ or carbon side chains are attached to the basic carbon skeleton. They may also differ in the presence and position of double bonds between skeleton carbon atoms. Small differences exert great effects on the physical and biological behavior of the compounds. Four types of names for steroids are used: (1) a trade name if the compound is sold as a drug, (2) a common or trivial name, (3) a modification of a trivial name, (4) a modification of the parent hydrocarbon. An example may be provided by the adrenal cortex hormone, hydrocortisone:

(1) Cortef, Cort-Dome, etc.; (2) Hydrocortisone, cortisol, Kendall's compound F; (3) 17-hydroxycorticosterone; (4) 11β,17,21-trihydroxy pregn-4-ene-3,20-dione.

The parent compounds named according to substitution of the 17 carbon skeleton are: estrane (one —CH₃ group, C-18, attached to C-13); androstane (an additional —CH₃, C-19 at C-10), and pregnane (androstane plus a side chain, C-20 and C-21, from C-17). Further modifications and their descriptions follow.

(a) *Double bond:* This is indicated in the spelling of the hydrocarbon and the position, e.g., pregnane (0); pregn-4-ene (1); pregn-1,4-diene (2); pregn-1,4,9-triene (3).

(b) *Oxygen substitution:* This may be a hydroxyl group (suffix **ol** preceded by carbon atom number or prefix **hydroxy**) or a ketone group (suffix **one;** prefix **keto** or **oxo**).

(c) *Isomerism:* Hydroxyl groups or hydrogens at locations of steroisomerism may be designated as α if they extend below the relatively flat plane of the ring and are represented by a dotted line; if above the plane of the ring, they are β and are indicated by a solid line.

(d) *Other changes:* **nor** means shortening of a side chain or elimination of a skeleton carbon. **Deoxy:** substitution of —H for —OH; **Dehydro:** loss of an —H; **Dihydro:** substitution of 2H for a double bond. (Please see References 1 and 2 for further information.)

PROPERTIES OF COMMON STEROIDS

In their free alcohol forms, these compounds (except for the estrogens) have little or no solubility in water. Identification is accomplished by chromatographic (paper, column, thin layer, gas-liquid) separation and subsequent purification to constant specific radioactivity after addition of labeled standard. Physicochemical characteristics which are useful include infrared absorption spectra (cf. standards), O.R.D. (optical rotatory dispersion), fluorescence and precipitability with digitonin.

Abbreviations used in the tables

a. acid	aq., water	i., insoluble	s., soluble
alk., alkali	But., butanol	IEP, isoelectric point	solv., solvent
al., alcohol	dil., dilute	ORD, optical rotatory dispersion	vol., volatile

Table 7-7
STEROID HORMONES

No.	Name		Biology	
	Trivial	Chemical	Main source	Action
1	**Hydrocortisone** (cortisol)	$11\beta,17\alpha,21$-trihydroxy-pregn-4-ene-3,20-dione	Adrenal cortex	Glucocorticoid. Anti-inflammatory. Rx as ester.
2	**Cortisone** 11-dehydrohydro-cortisone	$17\alpha,21$-dihydroxy-pregn-4-ene-3,11,20-trione	Metabolism of (1) synthetic	Glucocorticoid. Anti-inflammatory. Rx as ester. 0.7 times potency of hydrocortisone in rats and humans.
3	**Corticosterone**	$11\beta,21$-dihydroxy-pregn-4-ene-3,20-dione	Adrenal cortex	Replaces (1) in some species (rats, mice, birds).
4	**Aldosterone**	$11\beta,21$-dihydroxy-18-formyl-pregn-4-ene-3,20-dione	Adrenal cortex (synthetic)	Mineralocorticoid. Potency: about 25 times deoxycorticosterone re NaCl retention.
5	**Deoxycorticosterone** (DOC)	21-hydroxy-pregn-4-ene-3,20-dione	Adrenal cortex (synthetic)	Mineralocorticoid. Rx as esters.
6	**9α fluorohydrocortisone** (fluorinef)	9α-fluoro-$11\beta,17\alpha$,21-trihydroxy-pregn-4-ene-3,20-dione	Synthetic only	Mineralocorticoid (and glucocorticoid). NaCl retaining potency equal to aldosterone.
7	**Prednisolone** (Δ-1 hydrocortisone)	$11\beta,17\alpha,21$-trihydroxy-pregn-1,4-diene-3,20-dione	Synthetic only	Glucocorticoid. Anti-inflammatory potency: 4 times hydrocortisone in man; 3.5 times in rats.
8	**Methyl prednisolone** (6α methyl Δ-1 hydro-cortisone)	$11\beta,17\alpha,21$-trihydroxy-6α methyl-pregn-1,4-diene-3,20-dione	Synthetic only	Same. Rx as alcohol or ester. Anti-inflammatory potency: 5 times hydrocortisone in man and rats.
9	**Dexamethasone** (16α methyl-9α-fluoro Δ-1 hydrocortisone)	9α fluoro-$11\beta,17\alpha,21$-trihydroxy-16α-methyl-pregn-1,4-diene-3,20-diene	Synthetic only	Same. Anti-inflammatory potency: 30 times hydrocortisone in man; 150 times in rats.
10	**Triamcinolone** (16α-hydroxy-9α fluoro Δ-1 hydrocortisone	9α fluoro-11β, 16α-17α 21-tetrahydroxy-pregn-1,4-diene-3,20-dione	Synthetic only	Same. Anti-inflammatory potency: 4 times hydrocortisone in man; 3 times in rats.
11	**Estradiol** 17β	$3\beta,17\beta$ dihydroxy-estr-1,3,5-triene	Ovary	Estrogenic: the principal biological estrogen. Rx as esters.

Table 7-7 (*Continued*)
STEROID HORMONES

No.	Chemistry				
	Formula	Mol. wt.	Melting point, °C	Identification (O.R.D., etc.)	Abs. max.
1	$C_{21}H_{30}O_5$	362.47	217–220	$[\alpha]^{25}D +150-156°$ (dioxane) green fluorescence in H_2SO_4	242 mμ (methanol)
2	$C_{21}H_{28}O_5$	360.44	220–224	$[\alpha]^{25}D +209°$ (ethanol) green fluorescence in H_2SO_4	237 mμ
3	$C_{21}H_{30}O_4$	346.45	180–182	$[\alpha]^{55}D +223°$ (ethanol) green fluorescence in H_2SO_4	240 mμ
4	$C_{21}H_{28}O_5$	360.45	108–112 (hydrate)	$[\alpha]^{25}D +161°$ (chloroform)	240 mμ
5	$C_{21}H_{30}O_3$	330.45	141–142	$[\alpha]^{22}D +178°$ (ethanol) no H_2SO_4 fluorescence	240 mμ
6	$C_{21}H_{29}FO_5$	380.46	260–262*	$[\alpha]^{23}D +139°$ (ethanol)	239 mμ
7	$C_{21}H_{28}O_5$	360.44	240–241	$[\alpha]^{25}D +102°$ (dioxane)	242 mμ
8	$C_{22}H_{30}O_5$	374.46	228–237	$[\alpha]^{20}D +83°$ (dioxane)	243 mμ
9	$C_{22}H_{29}FO_5$	392.45	262–264	$[\alpha]^{25}D +73°$ (chloroform) for 21-acetate	239 mμ
10	$C_{25}H_{31}FO_8$	478.52	186–188	$[\alpha]^{25}D +22°$ (chloroform)	239 mμ
11	$C_{18}H_{24}O_2$	272.37	173–179	$[\alpha]^{25}D +76-83°$ (dioxane)	225, 280 mμ

Table 7-7 (*Continued*)
STEROID HORMONES

No.	Name		Biology	
	Trivial	Chemical	Main source	Action
12	**Estrone** (theelin)	3β hydroxy-estr-1,3,5-triene-17-one	Ovary, placenta	Estrogenic (weak).
13	**Estriol**	3β,16α,17β,tri-hydroxy-estr-1,3,5-triene	Ovary	Estrogenic (weak).
14	**Progesterone**	Pregn-4-ene-3,20-dione	Ovary, placenta	Progestogen. Maintains secretory endometrium.
15	**Pregnanediol**	3α,20α-dihydroxy-pregnane	Metabolism of (14)	Metabolite of progesterone; presence in urine used in pregnancy test.
16	**Testosterone**	17β-hydroxy-4-androsten-3-one	Testis, synthetic	Androgen; the principal biologic androgen. Rx as esters.

* Decomposition point.

Table 7-7 (*Continued*)
STEROID HORMONES

No.	Chemistry				
	Formula	Mol. wt.	Melting point, °C	Identification (O.R.D., etc.)	Abs. max.
12	$C_{18}H_{22}O_2$	270.36	258–262	$[\alpha]^{25}D$ +158–168° (dioxane); precipitated by digitonin	283, 285 mμ
13	$C_{18}H_{24}O_2$	288.37	282	$[\alpha]^{25}D$ +58° (dioxane); precipitated by digitonin	280 mμ
14	$C_{21}H_{30}O_2$	314.45	121	$[\alpha]^{20}D$ +172–182° (dioxane)	240 mμ
15	$C_{21}H_{36}O_2$	320.50	238	$[\alpha]^{20}D$ +27.4° (ethanol) water soluble not precipitated by digitonin	
16	$C_{19}H_{28}O_2$	288.41	155	$[\alpha]^{24}D$ +109° (ethanol)	238 mμ

Table 7-8

HORMONAL AMINO ACIDS AND POLYPEPTIDES

These include water-soluble hormones from the anterior and posterior pituitary, thyroid, and parathyroid glands; the adrenal medulla; the pancreas; and the gastrointestinal tract.

Name	Action	Source	Formula; mol. wt.	Method of testing	Use and method of administration
Adrenocortico-tropin (ACTH, corticotropin).	Maintenance and stimulation of adrenal corticosteroid biosynthesis.	Anterior pituitary.	Polypeptide, 39 residues. Sol.: aq.-s.	Bioassay—adrenal ascorbic acid depletion, adrenal vein corticosterone concentration: immunoassay.	Parenterally, supplanted by corticosteroids in rheumatoid arthritis, asthma, and other allergic conditions. Used to maintain the integrity of the adrenal cortex.
Chorionic gonadotropin (HCG, HPL).	Stimulates rupture of ovarian follicles and formation of corpora lutea; stimulates testosterone production in males and seminiferous tubule growth in males.	Urine of pregnant women, placenta.	Glyco-polypeptide, about 30,000. Sol.: aq.-s., glycerol-s.	Immunoassay; mouse ovarian weight increase; ovulation in mouse or rabbit; spermatozoa discharge in frog.	Cryptorchism; male hyogonadism; some cases of male and female sterility.
Calcitonin (Thyrocalcitonin).	Regulation of plasma calcium levels: hypocalcemic.	Bovine and porcine thyroid glands.	Polypeptide.	Plasma calcium lowering in rats.	Commercial preparations for human use not available.
Follicle stimulating hormone (FSH, perganol (commercial preparation).	Stimulates graafian follicle maturation in the ovary; stimulates spermatogenesis and seminiferous tubule growth in the ovary.	Anterior pituitary; postmenopausal urine.	Polypeptide; 30,000 (human). Sol.: aq.-s.; 50% al.-s.	Bioassay, mouse ovarian weight increase; radioimmunoassay.	Induction of ovulation (in combination with chorionic gonadotropin), parenterally.
Glucagon.	Transient elevation of blood sugar; inhibition of intestinal motility.	Pancreas—α cells.	Polypeptide 29 residues, 3,485. Sol.: dil. a.-s.; dil. alk-s.	Bioassay (glycogenolysis); immunoassay	For diagnosis in glycogen deposition disease parenterally; rarely for hypoglycemia.
Growth hormone (Somatotropin, GH, STH).	Species specific promotion of somatic growth; effects on intermediary metabolism.	Anterior pituitary, bovine, porcine.	Polypeptide, about 45,000. Sol.: aq.-s.	Radioimmunoassay; bioassay for growth in hypophysectomized rats.	Occasionally in hypopituitary dwarfism, intramuscularly.
Insulin.	Promotes uptake of blood glucose by liver, muscle; other metabolic effects.	Pancreas—β cells.	Polypeptide; α chain 21 residues; β chain 30 residues, 5,700. Sol. aq.-i.; dil. a. + alk-s.	Bioassay; hypoglycemic effect. Radioimmunoassay.	In diabetes mellitus subcutaneously; occasionally intramuscularly; by subcutaneous injection in combination with zinc, protamine or globin for prolonged effect.

Name	Action	Source	Formula: mol. wt.	Method of testing	Use and method of administration
L-triiodothyronine β-[4-(4-hydroxy-3-iodophenoxy)-3,5-diiodophenyl] alanine (liothyronine).	More rapid but the same as thyroxine; present in very low concentration.	Thyroid gland; synthetic.	$C_{15}H_{12}I_3NO_4$, 651.01. Sol. aq.-i.; al.-i.; dil. alk.-s.	Iodometric; bioassay, chromatography.	In hypothyroid states orally and occasionally parenterally.
Luteinizing hormone (L.H.; ICSH).	Stimulates graafian follicle rupture and corpus luteum formation; with FSH stimulates estrogen synthesis; with prolactin, progesterone synthesis.	Pituitary, ovine, porcine.	Polypeptide 26,000 (human). Sol.: aq.-s.	Rat ventral prostate hypertrophy radioimmunoassay.	Pure preparation not commercially available.
L-thyroxine—$\beta l(\beta,5$-diiodo-4-hydroxyphenoxy)-3,5-diiodophenyl] alanine; 3,5,3'5'-tetraiodothyronine.	Control of metabolic rate; also promotion of growth.	Thyroid gland; synthetic.	$C_{15}H_{11}O_4N I_4$; 776.93. Sol.: aq.-i.; al.-i.; alk.-s.; al. $+$ a.-s.; vol. solv.-i.	Iodometric; bioassay; chromatography.	In hypothyroid states orally and occasionally parenterally.
Vasopressin (antidiuretic hormone; ADH; Petressin).	Antidiuretic action on kidneys.	Bovine neurohypophysis; synthetic.	Octapeptide—lysine vasopressin (swine); mol. wt. 1056; arginine vasopressin mol wt. 1084. IEP: pH 10.9. Sol.: aq.-s.	Blood pressure rise in dogs. Radioimmunoassay.	In diabetes insipidus as tannate; in oil intramuscularly crude posterior pituitary powder; by nasal insufflation.
Melanocyte stimulating hormone (α-MSH; β-MSH), (Intermedian, MSH, melanotropic hormone).	Increase of skin pigmentation.	Neurohypophysis.	Polypeptides; α-MSH: 14 residues; β-MSH: 22 residues (human). Shares same amino acid sequence with ACTH.	Darkening of frog skin.	Preparations for human use not available.
Melatonin, N-acetyl-5-methoxytryptamine.	Decreases skin pigmentation.	Pineal gland; synthetic.	$C_{13}H_{16}O_2N_2$, 231.	Blanching of pigmented frog skin.	Not used in man.
Norepinephrine (noradrenaline)	Neuro-humoral transmitter and regulates blood pressure by stimulating vasoconstriction.	Sympathetic nervous system postganglionic fibers, adrenal medulla.	$C_8H_{11}O_3N$, 169.18. Benzene ring, ethylamine side chain with $+OH$ groups at ring C_3, C_4 and the side chain βC. Sol.: aq.-s.	Bioassay, (blood pressure rise); colorimetry, fluorescence on oxidation, chromatography.	Support of blood pressure in shock (intravenous).

Table 7-8 (*Continued*)
HORMONAL AMINO ACIDS AND POLYPEPTIDES

Name	Action	Source	Formula: mol. wt.	Method of testing	Use and method of administration
Oxytocin (pitocin).	Promotes uterine contraction, milk ejection.	Bovine neurohypophysis; synthetic.	Octapeptide, 1007. IEP: pH 7.7. Sol.: aq.-s.; But.-s.	Bioassay—isolated uterus contraction. Avian vasodepression.	To induce labor and to promote involution of postpartum uterus; promotion of milk ejection; by intranasal spray.
Parathyroid hormone (parathormone).	Regulation of plasma calcium; hypercalcemic; stimulation of calcium absorption by intestine, breakdown of bone structure.	Bovine parathyroid glands.	Polypeptide, about 9,000. Sol.: dil. a.-s.	Increase in plasma calcium in dog. Radioimmunoassay.	Occasionally in initial control of parathyroprivic tetany.
Prolactin (lactogenic hormone; luteotropin).	Stimulates lactation; with L.H., promotes progesterone synthesis by ovary.	Ovine pituitary.	Polypeptide, 23,500. Sol.: aq.-i.; saline-s.; al. + a.-s.	Pigeon crop sac hypertrophy.	No preparation available for human use.
Thyrotropin (TSH).	Maintenance of thyroid function.	Anterior pituitary; bovine, porcine.	Glyco-polypeptide, about 10,000. Sol.: aq.-s.	Thyroid iodine depletion—murine or chick.	Diagnostic for distinguishing primary and secondary myxedema.
Epinephrine (adrenaline).	Support of blood pressure by increasing cardiac output; other metabolic effects including increased oxygen consumption.	Adrenal medulla.	$C_9H_{13}O_3N$, 183.21. Structure of norepinephrine with methyl substitution of the amine. Sol.: aq.-s.	Cf. norepinephrine.	Intravenously or subcutaneously in allergic reactions.

HORMONES OF THE GASTROINTESTINAL TRACT

A number of polypeptides are secreted by mucosal cells of the gastrointestinal tract in response to chemical or mechanical stimuli. Their mode of action is not yet very well understood.

Hormone	Origin	Target
Secretin	Duodenum	Pancreatic secretion; bile flow
Gastrin	Pylorus	Gastric secretion
Cholecystokinin	Small intestine	Contraction of gall bladder
Enterogastrone	Small intestine	Inhibits gastric secretion, contraction
Pancreozymin	Upper intestine	Pancreatic secretion

REFERENCES

1. L. F. Fieser and M. Fieser, "Steroids," Reinhold Publ. Co., New York (1959).

2. R. I. Dorfman and F. Ungar, "Metabolism of Steroid Hormones," Academic Press, New York (1965).

3. A. B. Eisenstein (ed.), "The Adrenal Cortex," Little, Brown & Co., Boston (1967).

4. P. H. Katzman and W. H. Elliott, Chemistry of Estrogens (review), in M. Florkin and E. H. Statz (eds.), "Comprehensive Biochemistry," p. 47, Elsevier, Amsterdam (1963).

5. G. I. Fujimoto and R. W. Ledeen, Chemistry of Androgens and Other C^{19} Steroids (review), in M. Florkin and E. H. Statz (eds.), "Comprehensive Biochemistry," p. 33, Elsevier, Amsterdam (1963).

6. E. Diczfalusy and P. Troen, Placental Hormones (review), *Vitamins, Hormones* 19, 229 (1961).

7. P. G. Stecher, M. J. Finkel, O. H. Siegmund, and B. M. Szatranski (eds.), "The Merck Index of Chemicals and Drugs," Rahway, N.J. (1960).

8. "The United States Pharmacopeia," 17th revision, Mack Co., Easton, Pa., 1965.

COMMERCIAL ORGANIC MATERIALS

Table 7-9
CONSTANTS OF FATS, OILS, AND WAXES

Classification: Class SV, semi-drying vegetable oil; NVO, non-drying vegetable oil of the olive oil type; AF, animal fat; AW, animal wax; IW, insect wax; NVR, non-drying vegetable oil of the rape oil

No.	Name	Class	Specific Gravity $\frac{15°}{15}$C.
1	Acorn, *Quercus agrifolia*	SV	0.916
2	Almond, *Prunus amygdalus*	NVO	0.914-0.921
3	Apricot kernel, *Prunus Armeniaca*	NVO	0.915-0.926
4	Beef marrow, *Bos taurus*	AF	0.931-0.938
5	Beef tallow, *Bos taurus*	AF	0.895
6	Beechnut, *Fagus sylvatica F. Americana*	SV	0.922
7	Beeswax (ordinary), *Apis mellifera*	IW	0.953-0.970
	Indian	IW	0.953-0.970
8	Black mustard, *Sinapis nigra*	NVR	0.915-0.919
9	Black walnut, *Juglans nigra*; (*cf. Walnut*)	DV	0.918-0.921
10	Bone fat, *Sevum ossis*	AF	0.914-0.916
11	Brazil nut, *Bertholletia excelsis*	SV	0.917-0.918
12	Butter fat, *Vaccae lactis adeps*	AF	$0.907\text{-}0.912\frac{40°}{15}$
13	Candelilla, *Euphorbia cerifera*	VW	0.981-0.994
14	Candlenut, *Aleurites moluccana*	DV	0.925
15	Candlenut, *Aleurites triloba*	DV	0.927
16	Carnauba wax, *Corypha cerifera*		
	No. 1 Yellow	VW	0.990-0.996
	No. 3 Crude	VW	0.994-1.010
	No. 3 Refined	VW	0.990-0.996
17	Castor, *Ricinus communis*	NVC	0.960-0.967
18	Ceresine	MW	0.900-0.920
19	Chaulmoogra, U. S. P. X Revision	SV	$0.950^{25°}$
20	Chaulmoogra, *Taraktogenos Kurzii*	SV	0.943-0.954
21	Cherry kernel, *Prunus cerasus*	NVO	0.918-0.929
22	Chicken fat, *Gallus domesticus*	AF	0.924
23	Chinese insect wax, *Coccus cerifera*	IW	0.950-0.970
24	Chinese vegetable tallow, *Stillingia sebifera*	VF	0.918-0.922
25	Coconut, *Cocos butyracea; C. nucifera*	VF	0.926
26	Cocoa (Cacao) butter, *Theobroma cacao*	VF	0.964-0.974
27	Cod liver, *Gadus morrhua*	MA	0.922-0.931
28	Corn (Maize), *Zea Mays*	SV	0.921-0.928
29	Cottonseed, *Gossypium* Species	SV	$0.917\text{-}0.918\frac{25°}{25}$
30	Cottonseed stearin, *Gossypium*	VF	$0.867\text{-}0.868^{100°}$
31	Croton, *Croton tiglium*	SV	0.942-0.944
32	Date kernel, *Phoenix dactylifera*	NVO	—
33	Deer fat, *Cervus elephus*	AF	0.962-0.967
34	Dolphin, *Delphinus globiceps*	MA	0.908-0.930
35	Esparto, *Stipa tenacissima*	VW	0.985-0.995
36	Goat's butter, *Capellae lactis adeps*	AF	$0.917\text{-}0.935^{37.7°}$
37	Goose fat, *Anser cinereus*	AF	$0.923\text{-}0.930^{37.7}$
38	Grape seed, *Vitis vinefera*	NVC	0.917-0.933
39	Hazelnut, *Corylus avellana*	NVO	0.917
40	Hemp seed, *Cannibis sativa*	DV	0.928-0.934
41	Herring, *Clupea harengus*	MA	0.920-0.939
42	Horse fat, *Equus caballus*	AF	0.919-0.933
43	Human fat	AF	0.9033
44	Japan wax, *Rhus succedaneum*	VW	0.970-0.998
			$0.875^{100°}$

Table 7-9 (*Continued*)
CONSTANTS OF FATS, OILS, AND WAXES

type; DV, drying vegetable oil; VW, vegetable wax; NVC, non-drying vegetable oil of the castor oil type; VF, vegetable fat; MA, marine animal or fish oil; MW, mineral wax; NA, non-drying animal oil; SMW, semi-mineral wax; Sp., sperm oil.

No.	Solidification point °C.	Acid value	Saponification value	Iodine value	Reichert-Meissl value
1	−10	....	199.3	100.0	....
2	−15 to −20	0.5-3.5	183.3-207.6	93-103.4	0.5
3	−17	3.5	191.4-198.2	100-108.7	0.2
4	29 to 31	1.6	196-199	39-55.4	2
5	31 to 38	0.25	196-200	35.4-42.3	0.25
6	−17	....	191-196	97-111	....
7	62 to 66	17.0-21.0	88-100	8-11	....
	61 to 67	5.0-10.5	87-117	4-10.5	....
8	16	5.7-7.3	173-175	99-110	....
9	turbid −12	8.6-9.0	190.1-191.5	141-142.7	....
10	15 to 17	29.6-53	185-198	46-55.8	0.2-1.7
11	0 to 3	1.4	193	90-106	....
12	20 to 23	0.45-35.4	210-230	26-38	17.0-34.5
13	73 to 77	18.6-23.9	55.0-64.2	....	....
14	<−18	2	189-195	163-164	1.2
15	....	....	202-204	139-143.8	....
16					
	86 to 88	1.5-2.5	75-86	....	....
	86 to 90	3.0-8.5	75-89	....	....
	86 to 89	3.0-5.0	76-85	7.0-14.5	....
17	turbid −12	0.12-0.8	175-183	84	1.4
	solid −17 to −18				
18	56 to 82	0	0	4.0-8.0	....
19	<25	....	196-213	98-104	....
20	20 to 25	0.79-21.5	196-213	97.6-110.4	....
21	−19 to −20	1.1	193.3-195	110-114.3	....
22	21 to 27	1.2	193-204.6	66-71.5	1.8
23	80 to 85	1.9-8.9	78-93	1.0-2.5	....
24	24 to 34	2.4	179-206	23-40.5	0.2-0.9
25	14 to 22	2.5-10	253.4-262	6.2-10	6.6-7.5
26	21.5-23	1.1-1.9	192.8-195	32.8-41.7	0.3-1
27	−3	5.6	171-189	137-166	0.2
28	−10 to −20	1.37-2.02	187-193	111-128	4.3
29	+12 to −13	0.6-0.9	194-196	103-111.3	0.95
30	16 to 22	4-10	195	88.7-93.6	0.22
31	−8 to −18	27-30.9	193-215	108-109	12-13.6
32	18.1	....	211	52.3	0.88
33	....	0.8-5.3	194.5-200	26-36	0.68
34	+5 to −3	2-12	body 203.4 jaw 290	126.9 body 32.8 jaw	body 46.9 jaw 65.9
35	75 to 79	22.0-27.0	58.0-72.5	7.0-15.0	....
36	....	....	233-236	25-37	20.8-27.7
37	22 to 24	0.59	191-193	58-67	0.2-0.98
38	−10 to −17	0.75	171-191	94.3-135	0.46
39	−17 to −18	....	191-197	87	0.99
40	−15 to −28	0.45	190-195	145-161.7	....
41	....	1.8-44	170-194	102-149	....
42	20 to 45	0-2.4	195-200	75-86	1.6-2.1
43	15	....	193-200	57-73	....
44	49 to 56	4.0-15.0	210-235	4.0-15.0	....

Table 7-9 (*Continued*)
CONSTANTS OF FATS, OILS, AND WAXES

No.	Name	Class	Specific Gravity $\frac{15°}{15}$C.
45	Lard oil, *Sus scrofa*	NA	0.913-0.915
46	Lard oil (fatty tissue), *Sus scrofa*	AF	0.934-0.938 $0.861^{\frac{100°}{15.5}}$
47	Laurel (bayberry), *Laurus nobilis*	VF	$0.880^{100°}$
48	Linseed, *Linum usitatissimum*	DV	0.930-0.938
49	Menhaden, *Alosa manhaden—Brevortia tyrannus*	MA	0.923-0.933
50	Montan:		
	Reibeck	SMW	0.995-1.040
	Domestic	SMW	1.020-1.040
51	Mutton tallow, *Ovis aries*	AF	0.937-0.953 $0.858^{\frac{100°}{15.5}}$
52	Myrtle wax, *Myrica cerifera—M. Carolinensis*	VF	$0.995-0.875^{100°}$
53	Neat's foot, *Bos taurus*	NA	0.913-0.918
54	Nutmeg (mace) butter, *Myristica officinalis*	VF	0.945-0.966
55	Olive, *Olea Europaea sativa*	NVO	0.9140-0.918
56	Ouricury, *Syagrus coronata* (Mart.)	VW	0.990-1.010
57	Ozokerite	MW	0.900-0.996
58	Palm, *Elaeis guineensis* (W. Africa)	VF	$0.924; 0.858^{100°}$
59	Palm kernel, *Elaeis guineensis* (S. America)	VF	....
60	Palm kernel, *Elaeis guineensis* (W. Africa)	VF	$0.866-0.873^{100°}$
61	Peach kernel, *Amygdalus Persica*	NVO	0.918-0.925
62	Peanut, *Arachis hypogaea*	NVO	0.917-0.926
63	Perilla, *Perilla ocimoides*	DV	0.930-0.937
64	Pistachio nut, *Pistacia vera*	NVO	0.913-0.919
65	Plum kernel, *Prunus domestica* and *P. damascena*	NVO	0.912-0.913
66	Poppy seed, *Papaver somniferum*	DV	0.924-0.926
67	Porpoise—body oil, *Delphinus phocaena*	MA	0.926
68	Pumpkin seed, *Cucurbita pepo*	SV	0.923-0.925
69	Rabbit fat, *Lepus cuniculus*	AF	0.934-0.936 $0.861^{\frac{100°}{15}}$
70	Rape seed, *Brassica campestris*	NVR	0.913-0.917
71	Safflower, *Carthamus tinctorius*	DV	0.925-0.928
72	Sardine, *Clupea pilchardus* and *C. scombrinus*	MA	0.920-0.934
73	Seal, *Phoca* species	MA	0.915-0.926
74	Sesame, *Sesamum indicum*	SV	$0.919^{\frac{25°}{25}}$
75	Shark, *Selache* (*cetorhinus*) *maxima et al.*	MA	0.916-0.919
76	Soya, soy or soja bean, *Soja hispida; Dolichos hispida*	SV	0.924-0.927
77	Sperm, *Physeter macrocephalus*	Sp	0.878-0.884
78	Spermaceti, *Cetaceum*	AW	0.905-0.960
79	Spermaceti, *Physeter macrocephalus*	AW	0.905-0.945
79.1	Sugar cane wax, *Saccharum officinarum*	VW	0.970-0.980
80	Sunflower, *Helianthus annuus*	DV	0.924-0.926
81	Tallow, *Oleum adipis bovis*	NA	0.914-0.919
82	Tallow, *Sevum*	AF	0.925-0.950
83	Tung,—China wood, *Aleurites fordii*	DV	0.939-0.949
84	Tung,—China wood, *Aleurites montana*	DV	0.939-0.949
85	Walnut, *Juglans regia* (cf. Black walnut)	DV	0.925-0.927
86	Whale, *Balaena mysticetus*	MA	0.917-0.924
87	White mustard seed, *Sinapis alba*	NVR	0.912-0.916
88	Wool fat, *Ovis aries*	AW	0.970-0.973

Table 7-9 (*Continued*)
CONSTANTS OF FATS, OILS, AND WAXES

No.	Solidification point °C.	Acid value	Saponification value	Iodine value	Reichert-Meissl value
45	+4 to −2	0.1-2.5	193-198	62.5-79	0-0.2
46	27.1 to 29.9	0.5-0.8	195-203	47-66.5	0.5-0.8
47	25	26.3	198-199	68-80	1.6
48	−19 to −27	1-3.5	188-195	175-202	0.95
49	−5	3-11.6	189-192.9	148-185	1.2
50					
	83 to 89	35.0-45.0	80-99	....	....
	80 to 86	35.0-45.0	100-115	....	....
51	32 to 41	1.7-14	195-196	48-61	....
52	39 to 43	3-4.4	205.5-211.7	3.9-9.5	0.5
53	+10 to −2	0.1-0.6	193-199	57.5-75	0.9-1.2
54	41 to 42	17.2	154-178	40-81	1.1-4.2
55	turbid +2, ppt. −6	0.3-1.0	185-196	79-88	0.6-1.5
56	86 to 89	12.0-18.8	88.0-95.8	....	....
57	56 to 82	0	0	4.0-8.0	
58	35 to 42	10	200-205	49.2-58.9	0.9-1.9
59	27.4	0.33-0.55	220.2-231.4	25.5-31.6	....
60	....	....	243-255	10.5-17.5	5-6.8
61	−20	1-1.5	191-193	92-99.7	
62	3	0.8	186-194	88-98	0.4
63	....	....	188-194	185-206	....
64	−5 to −10	....	191	83-87	
65	−5 to −8	0.55	191-193	100-103.6	....
66	−16 to −18	2.5	193-195	128-141	0.6
67	−16	body 1.2 jaw 5.0	body 203.4 jaw 253-272	body 126.9 jaw 30.9-49.6	body 46.9 jaw 132
68	−15	....	188-193	121-130	4.45
69	17 to 23	1.4-7.2	199-203	70-99.8	0.7-2.8
70	−10	0.36-1.0	168-179	94-105	0.0-7.9
71	−13 to −18	0.6	188-203	122-141	0.0-0.2
72	20 to 22	4-25	187.7-196	150-193	0.5-1
73	3	1.9-40	187.5-196.2	130-152	0.2
74	−4 to −6	9.8	188-193	103-117	1.1-1.2
75	....	....	157-164	115-139	....
76	−10 to −16	0.3-1.8	189-193.5	122-134	0.5-2.8
77	15.5	13.2	120-137	80-84	0.6
78	41 to 49	0.5-3.0	121-135	2.5-8.5	
79	42 to 47	0.5-2.8	126-135	3.8-9.5	
79.1	76-82	20-30	55-70	....	....
80	−17	11.2	188-193	129-136	0.5
81	2 to 7.5	0.2-0.25	193.5-199	56-60.5	0.3
82	....	....	193-198	35-45	0.5-1.0
83	<17	2	190-197	163-171	1.10
84	<17	2	190-197	163-171	0.35
85	−15 to −27	2.5	190.1-197	139-150	0.92
86	−2 to 0	1.9	160-202	90-146	14
87	−8 to −16	5.4	171-174	94-98.4	....
88	38 to 40	59.8	82-130	17-29	5-8

Table 7-9 (*Continued*)
CONSTANTS OF FATS, OILS, AND WAXES

No.	Name	Index of Refraction at 25° C.	Hehner value	Acetyl value	Unsaponifiable matter
1	Acorn	....	....	....	....
2	Almond	1.4593-1.4646*	96.0	9.6	0.75
3	Apricot kernel	1.4636-1.4705	....	12.2	....
4	Beef marrow	....	....	....	....
5	Beef tallow	1.4552-1.4587*	96-96.5	2.7-8.6	....
6	Beechnut	1.4698	95-96	....	....
7	Beeswax (ordinary)	1.4445-1.4473†	....	13.0-16.0	52-55
	Indian	1.4380-1.4420‡	....	....	52-55
8	Black mustard	1.4718	96	....	3.3
9	Black walnut	....	95.8	....	....
10	Bone fat	....	91-95	11.3	0.5-1.8
11	Brazil nut	1.4671	....	....	....
12	Butter fat	1.4555-1.4578*	87.6-89.6	1.9-8.6	0.3-0.6
13	Candelilla	1.4565-1.4610‡	....	....	....
14	Candlenut *Al. mol.*	1.4760-1.4790	95-96	9.8	0.5-9
15	Candlenut *Al. tri.*	1.4760-1.4790	....	....	....
16	Carnauba wax, yellow	1.4490-1.4525§	....	....	50-55
	crude	1.4460-1.4520§	....	....	50-55
	refined	1.4465-1.4500§	....	54.0-56.0	50-55
17	Castor	1.4771	....	146-150.5	0.6
18	Ceresine	1.4320-1.4370‡	....	....	100
19	Chaulmoogra, U.S.P. X Revision	....	....	....	....
20	Chaulmoogra, *Taraktogenos Kursii*	1.4777-1.4779	....	....	2.4-2.6
21	Cherry kernel	1.4635	....	....	....
22	Chicken fat	1.4580*	94.6	45	....
23	Chinese insect wax	1.4566*	....	....	49-55
24	Chinese vegetable tallow	1.4470-1.4579*	95.3	....	....
25	Coconut	1.453	82.3-90.5	2.3-6.9	....
26	Cocoa (Cacao) butter	1.4537-1.4580*	94-95	1.97	....
27	Cod liver	1.4758-1.4783	95.3	1.15	0.54-2.68
28	Corn	1.4733	93-95	7.5-11.5	1.5-2.8
29	Cottonseed	1.4743-1.4752	95.7	21-25	1.1
30	Cottonseed stearin	1.4700-1.4725	96.5	....	....
31	Croton	1.4710*	89.0	19.8-38.6	0.6
32	Date kernel	1.4535-1.4633	95.2	....	....
33	Deer fat	....	95.8	....	0.52
34	Dolphin	1.4665	93.1	....	2
35	Esparto	1.4550-1.4590‡	....	....	15-20
36	Goat's butter	1.4499-1.4551*	....	....	....
37	Goose fat	1.4583-1.4626	94.5-95.3	....	....
38	Grape seed	1.4713-1.4725	92	13.5-14.5	1.6
39	Hazelnut	1.4667	95.5	3.2	0.5
40	Hemp seed	1.4740-1.4745*	....	....	1.08
41	Herring	1.4665-1.4729	95-96	....	1-2
42	Horse fat	1.4658-1.4702	95-98	....	....
43	Human fat	1.4593-1.4607*	94-96	....	....
44	Japan wax	1.4520-1.4585‡	89-91	17.0-25.0	1.0-3
45	Lard oil	1.4607*	97	2.6	0.6
46	Lard oil (fatty tissue)	1.4609-1.4620	93-95	2.6	....
47	Laurel (bayberry)	1.4783	....	....	....
48	Linseed	1.4797-1.4802	94.5-95.5	....	0.4-1.2
49	Menhaden	1.4787	....	....	0.6-1.43
50	Montan, Reibeck	....	....	....	30-45
	Domestic	....	....	....	....
51	Mutton tallow	1.4545-1.4585*	95.5	....	....
52	Myrtle wax	1.4511*	92-94	....	....

* At 40° C; † at 65° C; ‡ at 80° C; § at 95° C.

Table 7-9 (*Continued*)
CONSTANTS OF FATS, OILS, AND WAXES

No.	Maumené number	Insoluble fatty acids			
		Melting point °C.	Solidification point or Titer	Acid value	Iodine value
1	....	25	....	....	....
2	51-54	13-14	9.5-11.5	196-207	93-96.5
3	42.5	2.3-4.5	....	197	99.4-108
4	....	44-46	....	204.5	44-56
5	....	42.5-44	37.9-46.2	197-202	41.3
6	64	23-24	17	....	114
7	....	....	....	....	....
8	43	16-17	13.4-13.7	187.1	109.6
9	....	0	....	....	....
10	....	41-43	39-43	200	48-57
11	....	28-30	31.1-32.2	....	....
12	....	38-41	33-39	210-233	36-52
13	....	....	....	....	....
14	....	20-21	....	....	....
15	....	....	17.8	....	....
16	....	....	....	....	....
17	46-7	13	3	192	87-93
18	....	....	....	....	....
19	....	....	....	....	....
20	....	44-45	39.6	215	103
21	45	19-21	13-15	....	104-114
22	....	38-40	32-34	200.8	64.6
23	....	92	....	....	....
24	....	39-57	45.2-47.2	202-208	30-55
25	21	24-27	21.2-25.2	258-273	8.4-9
26	....	48-53	47.2-49.2	190-198	33-39
27	102-113	21.8-38	17.5-24.3	204-207	130-170
28	74-86	17-20	14-16	198.4	113-126
29	75-81	34.5	....	202-208	111-115
30	....	27-45	39.9-51	....	94
31	....	17-19	17-19	201	112
32	....	....	....	....	....
33	....	50-64	46-50	....	....
34	....	....	....	....	....
35	....	....	....	....	....
36	....	....	....	....	....
37	....	36.6-40	31-34	202	65.3
38	53	23-25	18-20	187.4	99
39	36	22-25	19-20	201	91-98
40	97	17-21	15.6-16.6	....	141
41	....	30-32	....	179	....
42	46-54	31.3-53.4	33.7-45	203	72-87
43	....	35.5	....	....	64
44	....	53-56.5	54-55	214	....
45	40-47	33-38.4	27-33	....	....
46	24-28	37-46.6	36-42.4	202	64.2
47	116	....	15.1	....	82
48	103-126	20-24	16-20.6	197	179-192
49	123-128	....	....	....	....
50	....	....	....	....	....
51	....	33.5-49	40-48.5	210	34.8
52	....	47-48	....	231	....

Table 7-9 (*Continued*)
CONSTANTS OF FATS, OILS, AND WAXES

No.	Name	Index of Refraction at 25° C.	Hehner value	Acetyl value	Unsaponifiable matter
53	Neat's foot	1.4643-1.4685	94.8-95.9	7.7-9.3	0.12-0.65
54	Nutmeg	1.4700-1.4812*	....	....	....
55	Olive	1.4657-1.4667	95	10.5	0.4-1.0
56	Ouricury	1.4530-1.4555§	....	....	....
57	Ozokerite	1.4320-1.4370‡	....	....	100
58	Palm	1.4603-1.4639*	94.5-97	15.7	....
59	Palm kernel (S. America)	....	....	....	....
60	Palm kernel (W. Africa)	1.4492-1.4543*	91-91.5	7.6	....
61	Peach kernel	1.4682-1.4701	94-96	6.5	....
62	Peanut	1.4620-1.4653*	95	3.5	0.5-0.9
63	Perilla	1.4753*	95.8	....	....
64	Pistachio nut	1.4672	96	....	....
65	Plum kernel	1.4679-1.4702	95.2	....	....
66	Poppy seed	1.4739-1.4742	95.4	....	0.43
67	Porpoise—body oil	1.4622-1.4625	body 85.5 jaw 68-72	....	body 3.7 jaw 16-17
68	Pumpkin seed	1.4724-1.4739	96	....	....
69	Rabbit fat	1.4586*	99.5	....	....
70	Rape seed	1.4649-1.4659	94.5-96.3	14.75	1.48
71	Safflower	1.4769	95	16.1	....
72	Sardine	1.4763-1.4852	93.3-96	21-22	0.98
73	Seal	1.4742-1.4762	93-96	33-34	0.3-1.0
74	Sesame	1.4704-1.4717	95	....	0.9-1.3
75	Shark	1.4825	87-97	11.9	2.8-15.2
76	Soya	1.4723-1.4756	93-94.5	4.9	1.27-1.54
77	Sperm	1.4573	....	4.5-6.4	37-41
78	Spermaceti (*Cet.*)	....	....	2.0-3.0	50-55
79	Spermaceti (*Phys.*)	....	....	2.6	51.5
80	Sunflower	1.4659-1.4721	95	....	0.31
81	Tallow (oil)	....	....	....	....
82	Tallow (sevum)	....	95-96	....	....
83	Tung (*Al. ford.*)	1.515-1.520	96	....	0.4-0.8
84	Tung (*Al. mont.*)	1.515-1.520	....	....	0.4-0.8
85	Walnut	1.4770	93.4-95.4	....	....
86	Whale	1.4679-1.4724	93-95	11-23	1-4
87	White mustard seed	1.4649*	96-97	....	....
88	Wool fat	1.4784-1.4822	91	23	39-44

* At 40° C; † at 65° C; ‡ at 80° C; § at 95° C.

Synthetic Waxes

No.	Name	Melting point °C.	Color
1	Acrawax (Glyco)	95-97	tan
2	Acrawax B (Glyco)	86-90	light brown
3	Acrawax C (Glyco)	140-142	light brown
4	Armowax (Armour Chemical Div.)	132	light tan
5	Carbowax 1500 (Carbide & Carbon)	30-40	white
6	Carbowax 4000	54-57	white
7	Castor Wax (Bakers Castor Oil Co	84-87	white
8	Chlorowax (Diamond Alkali Co.)	100	cream
9	Gersthofen Wax (Formerly IG Wax)	79-82	yellow
10	Opal Wax (Du Pont)	77-81	white
11	Santowax (Monsanto)	56-104	yellow

Table 7-9 (*Continued*)
CONSTANTS OF FATS, OILS, AND WAXES

No.	Maumené number	Insoluble fatty acids			
		Melting point °C.	Solidification point or Titer	Acid value	Iodine value
53	47-58.5	29-41	16-26.5	200.6	62-76
54	....	42.5	36	....	....
55	35-52	26-30	16.9-26.4	193-198	86-90
56	....	....	....	....	....
57	....	....	....	....	....
58	....	50	42.5-45.5	204-207	53
59	....	....	....	....	....
60	....	25-28.5	20-25.5	258-264	12
61	42.5	10-18	13-13.5	201-205	94-102
62	44-67	26-36	30.5-39	202	96-103
63	124	−5	....	....	200-211
64	44.5-45	17-20	13-14	....	89-96
65	45	12.4-18	....	....	95.7-104
66	71-88	20.5	17-19	199	139
67	body 50	....	....	207	126
68	....	26	26-28	197	134
69	....	39-50	35-41	218	64
70	50-67	18.5-20	11.7-13.6	185	99-106
71	....	11-17	7-12	199	148
72	....	30-34.8	28.2	177-185	....
73	....	22-23	13-17	193	186-201
74	61-68.5	25-35	24	197-201	110-116
75	....	21-22	....	....	....
76	59-61	26.2-27.5	24	198	115-140
77	51	13.4	11.9	23.6	83-86
78	....	....	....	....	....
79	....	....	....	....	....
80	60-75	22-24	18-19.8	202	124-134
81	....	43-46	35-37.5	197-202	25-40
82	....	....	40-50	....	....
83	....	31-44	....	189-198	144-159
84	....	....	....	....	....
85	103	15-20	14.3	200	150
86	85-92	14-27	10-24	....	131
87	44-49	15-16	9-10	185.8	95
88	....	41.8	....	....	17

Synthetic Waxes (*Continued*)

No.	Specific gravity $\frac{15°}{15°}$ C.	Flash point °C.	Acid value	Saponification value	Iodine value
1	1.04	230	....	....	....
2	0.97	235	2.0	....	....
3	0.97	285	10.0	....	....
4	....	250	12.0	17.0	....
5	1.151	430	....	....	....
6	1.204	535	....	....	....
7	0.98-1.00	....	2.0	175-185	3-6
8	1.62-1.70				
9	1.01-1.02	....	17-25	158-178	....
10	0.98-1.00	....	2.0	175-185	2-5
11	1.097	191	....	....	....

Table 7-10
PHYSICAL AND CHEMICAL PROPERTIES OF NATURAL RESINS

For a review of modified natural resins see *Bull. 738* (May, 1950) of Scientific Section of Natl. Paint, Varnish & Lacquer Assoc., Inc.—Resin Index of 1950; and, for resins of interest to the plastics trade, the latest *Modern Plastics Encyclopedia* published by Plastics Catalogue Corp.

Classification. The 3d column headed "Class" places each resin into one of the following groups:

I. Natural gum-resins of vegetable origin, which contain some resinous constituents in admixture with carbohydrate bodies, so that the resulting complex will yield some water-soluble constituents.

II. Natural resins of animal origin.

III. Natural resins of vegetable origin—largely mixtures of resin acids, alcohols and esters, with some resenes or hydrocarbon bodies. This class is divided into the following groups:

    A. Oleoresins and balsams—very fresh exudations containing volatile solvents which act as carriers.

    B. New resins—same as **IIIA**—with the solvents gone completely.

Name	Source	Class	Sp. Gr.
Gum Accroides Used to color spirit varnishes and nitrocellulose lacquers, and in sealing wax. Composition is 85% *p*-coumaric ester of xanthoresino-tannol.	A species of Yellow *Xanthorroea* in variety Australia Red variety	III. B III. B	....
Accra Copal	W. Africa	III. C	1.033
Amber Used in jewelry; it is the hardest and most highly fossilized resin. Composition is 28% succino-abietic acid, 70% esters of succinic acid and succino-resinol.	Baltic coast and Burma. Fossil resin from extinct conifers	III. D	1.052 (1.05–1.11)
Gum Ammoniacum Composition:—20-26% gum, 65-75% resin, balance water, ash, etc.	*Dorema Ammoniacum,* from Persia and Africa	I.	1.19–1.21
Red Angola Copal Commonly used in Europe; less in in U. S. A.	Semi-fossil from Angola, W. Africa	III. C	1.066– 1.068
White Angola Copal	W. Africa	III. C	1.055
Gum Arabic (Acacia) Used in adhesives and emulsions. Varieties:—gum senegal, kordo-fan.	Dried juice from bark of *Acacia senegal*	True gum	....
Asafoetida Composition:—25% gum, 60% resin ester, 7% essential oil. U. S. P. requires ash to be under 10%. Used in treatment of heaves.	Species of *Ferula*	I.	....
Batu An important variety of East India resin (*q.v.*).			

Table 7-10 (*Continued*)
PHYSICAL AND CHEMICAL PROPERTIES OF NATURAL RESINS

C. Semi-fossil resins—the products of the action of the elements on new resins over a relatively short time. These resins are not entirely soluble unless melted and partially depolymerized.

D. Fossil resins—those exposed for long periods of time. They are substantially insoluble.

IV. Natural hydrocarbon resins.

The 4th column gives the **specific gravity**, the 5th gives the **softening point** (**S. P.**), and the 6th gives the **melting point** (**M. P.**). The next three columns give the **acid, saponification,** and **iodine numbers** in that order. Figures enclosed in parentheses indicate the extreme values reported by different investigators. The 10th and last column gives the solubility of the resins in various solvents. The figures 1 to 12 refer to the following solvents: 1-ethyl alcohol; 2-methyl alcohol; 3-ether; 4-benzene; 5-acetone; 6-amyl alcohol; 7-chloroform; 8-aniline; 9-benzaldehyde; 10-carbon tetrachloride; 11-turpentine; 12-amyl acetate. The letter "s." (soluble), indicates that over 90% of the resin is soluble in the solvents numbered thereafter; "p. s." (partly soluble), indicates that from 41–90% of the resin will dissolve, while "i." (insoluble), indicates that 40% or less dissolves. The **ester number** is indicated by **E**, and the methoxyl number by **M**.

S. P., °C.	M. P., °C.	Acid No.	Sap. No.	Iod. No.	Solubility
....	....	64–88	98–176	156–192	s. 1
....	....	64–106	18–25	....	....
75	120 (106–156)	98 (46–129)	140 (133–168)	58 (58–62)	s. 6, 8, 12; p. s. 1, 3, 5, 7, 9; i. 2, 4, 10, 11.
175	300 (280–315)	15 (15–34)	115 (86–145)	62 E.=71–91	v. sl. s. or i. 1, 2, 3, 4, 5, 6, 7, 8, 9, 10, 11, 12.
....	....	Resin:- 57–135	Resin:- E.=19–98	M.=8.6–11.0	....
90	>300 (305)	128 (128–143)	132 (132–162)	63–137 E.=58–62	s. 5, 6, 8, 9, 12; p. s. 1, 3; i. 2, 4, 7, 10, 11.
45	95 (95–125)	127 (57–127)	160 (132–160)	130	s. 5, 6, 8, 9, 12; p. s. 1, 2, 3, 4, 7; i. 10, 12.
....	....	2–8	56–90	0.5	s. water.
....	....	Resin.- 11–82	Resin:- E.=82– 214	M.=7–18	s. in 90% alcohol over 50%

Table 7-10 (*Continued*)
PHYSICAL AND CHEMICAL PROPERTIES OF NATURAL RESINS

Name	Source	Class	Sp. Gr.
Benguela Copal An uncommon resin.	W. Africa	III. C	1.058 (1.035-1.062)
Gum Benzoin Composition:—69% cinnamic acid esters, 30% cinnamic acid, 1% or less of vanillin.	Sumatra and Siam *Styrax Benzoin*	III. A, B	1.063-1.092
Brazil Copal	S. America *Hymenae courbaril*	III. C	1.053
Cameron Copal	W. Africa	III. C	1.052
Canada Balsam Used in optometric work.	*Abies balsamea*	III. A	0.90
Coal Resin A dark colored natural hydro-carbon resin	Utah coal	IV.	1.03
Colombia Copal	S. America	III. C	1.054
Congo Copal Composition:—89% resin acids, 8.5% resenes. It is the commonest fossil varnish resin, widely used because of its low price. Loango copal is a variety of Congo.	Belgian Congo	III. C	1.061
Copaiba Balsam Oleo resin, named, as are many others, from port of shipment. Used in medicine.	S. American species of *Copaiba*	III. A	0.916-0.995
Dammar A common resin, used in varnish and lacquer. It is dissolved in coal tar solvents and alcohol, which precipitate incompatible ingredients.	East Indian trees of *Shorea* variety	III. B	1.031 (1.031-1.123)
Demerara Copal Semi fossil gum. Not in common use.	British Guiana	III. D	1.047
Dragon's Blood Used as red coloring agent.	E. Indies (Sumatra) *Calamus draco*	III. A, B	1.25
East India A very common resin, widely used in the varnish industry. Composition:—12% resin acids, 78% resenes.	Species of *Dipterocarpus* in East Indies	III. B, C	....
Gum Elemi Contains some essential oils, mixed with resins. Used in lacquers and spirit varnishes.	Mainly from the Philippines; *Canarium commune*	III. A, B	1.018-1.083

Table 7-10 (*Continued*)
PHYSICAL AND CHEMICAL PROPERTIES OF NATURAL RESINS

S. P., °C.	M. P., °C.	Acid No.	Sap. No.	Iod. No.	Solubility
65 (65-95)	165 (140-215)	123 (123-137)	157 (135-168)	61-85 E. = 50-64	s. 6, 8, 12; p. s. 1, 2, 3, 5, 7, 9; i. 4, 10, 11.
....	< 100	127-142 (98-142)	190-207 (148-207)	57-76 M. = 13-44	s. completely in 1.
50	100	123 (123-130)	133 (133-143)	123-134 E. = 1-5	s. 6, 8, 12; p. s. 1, 2, 3, 4, 5, 7, 9, 10, 11.
100 (96-110)	150 (110-150)	160 (129-160)	70 (70-168)	65-70	s. 8; p. s. 3, 6, 9, 12; i, 1, 2, 4, 5, 7, 10, 11.
....	Liquid	88-106	105-116 E. = 4-10	Refr. Index = 1.532; M. = 0	....
....	160-165	very low	0	....	s. 3, 4, 7, 10, 11; i. 1, 2, 5, 6, 12
90	>300	119	156	....	s. 6, 8, 12; p. s. 1, 3, 5, 7, 9; i. 2, 4, 10, 11.
90 (90-95)	195 (115-195)	132 (132-151)	132-179	58-59	s. 6, 8, 12; p. s. 1, 2, 3, 5, 9; i. 4, 7, 10, 11.
....	....	34-98	E. = 2-33	....	....
75	100 E. = 4-20 M. = 0	35 (21-35)	39 (31-47)	64-142	s. 3, 4, 7, 8, 9, 10, 11, 12; p. s. 1, 2, 5, 6.
90	180	98	102	....	p. s. 3, 7, 9, 12; i. 1, 2, 4, 5, 8, 10, 11.
....	100 E. = 142	11 M. = 25-34	153	54-98	s. 1, 3; p. s. 7, CS_2, petr. ether, ethyl acetate.
....	....	21	35	....	s. coal tar hydrocarbons; i. 1, esters.
....	soft	15-22 E. = 3-24	24-28 M. = 0-2.5	81-175	....

Table 7-10 (*Continued*)
PHYSICAL AND CHEMICAL PROPERTIES OF NATURAL RESINS

Name	Source	Class	Sp. Gr.
Galbanum Composition:—64% alcohol-soluble resin, 27% gum, 9% essential oil. Used in medicine.	*Ferula galbaniflua*	I.	1.109-1.133
Gamboge Composition:—65-75% resin, 20-25% gum. Used as a yellow coloring agent and as a purgative.	Siam. *Garcinia Hanburii*	I.	. . . .
Gilsonite A natural, dark, asphaltic resin.	Mineral deposits	IV.	1.015
Guaiacum Used in medicine.	From wood of *Guajacum officinale* and *G. sanctum*	III. B	. . . .
Gurjun Balsam	Java and Cochin, China *Dipterocarpus*	III. A	0.960–0.966
Jalap, Resin of Used in medicine; purgative.	*Exogonium jalapa*	III. B	. . . .
Karaya Gum (Indian gum) Resembles tragacanth. Used in hair "wave set" and in foods.	India; Africa *Sterculia urens*	I.	1.461–1.480
Kauri, Bush Composition:—64.5% resin acids, 12% resenes.	A recent to semi-fossil variety of Kauri. It is the softest grade of Kauri on the market.	III. B, C	1.030–1.038
Kauri, Fossil Composition:—69% resin acids, 9.5% resenes. A common resin used for high grade varnishes.	*Agathis australis* (the New Zealand Kauri Pine)	III. C, D	1.053
Kissel Copal Said to make poor varnishes.	Africa	III. C	1.066
Gum Lac (Shellac) (Data given is for orange shellac.)	Secretion of *Coccus lacca*, the lac insect	II.	1.182 (1.113-1.214)
Madagascar Copal A hard fossil resin, used little because of its high price.	Fossil gum from Madagascar	III. D	1.056
Manila, Soft Used as a cheap shellac substitute in spirit varnishes.	E. Indies. Trees of *Hymenea* group	III. B	1.060
Manila, Hard The hardest grade used in the varnishes. A very common resin.	E. Indies	III. C	1.065

Table 7-10 (*Continued*)
PHYSICAL AND CHEMICAL PROPERTIES OF NATURAL RESINS

S. P., °C.	M. P., °C.	Acid No.	Sap. No.	Iod. No.	Solubility
....	....	Resin:- 19-40	Resin:- E.=55- 91	....	....
....	....	79-100 E.=57- 67	148 M.=0-2.4	71 (116)	s. 1; emulsifies in water.
115-120	130-140	very low	0	....	s. 3, 4, 7, 10, 11; i. 1, 2, 5, 6, 12.
....	....	23-44	M.=74- 84	....	54-74% soluble in ether.
....	Sticky liquid	....	....	....	s. 4, 7, CS$_2$; p. s. 1.
....	....	12-13	E.=120	M.=0	....
....	....	13-23	....	....	....
50	125	82 (74-82)	87 (79-102)	74-170	s. 1, 6, 8, 9, 12; p. s. 2, 3, 4, 5, 7; i. 10, 11.
90	185	79 (63-79)	90 E.=26- 36	120 (120-164)	s. 6, 8, 9, 12; p. s. 1, 5, 7; i. 2, 3, 4, 10, 11.
65	110	70	118	....	s. 8; p. s. 1, 3, 5, 7, 9, 12; i. 2, 4, 6, 10, 11.
95	150	61 (55-65) E.=150	201 (200-212) Unsap.= 3.5	18 (15-18)	s. alcohols, borax solutions; p. s. 3, 5, 7, CS$_2$, ethyl acetate; i. 4, toluene, petr. ether.
130	300	66	78.5	126	p. s. 6, 8, 9, 12; i. 1, 2, 3, 4, 5, 7, 10, 11.
45	120	145 (136-150)	185 (185-196)	91-111	s. 1, 2, 5, 6, 8, 9, 12; p. s. 3, 4, 7; i. 10, 11.
80	190	73 E.=44-50	87	90 (86-91)	s. 6, 8, 9; p. s. 1, 3, 5, 7; i. 2, 4, 10, 11.

Table 7-10 (*Continued*)
PHYSICAL AND CHEMICAL PROPERTIES OF NATURAL RESINS

Name	Source	Class	Sp. Gr.
Mastic Formerly much used in lacquers and spirit varnishes. Not so popular now.	Evergreen shrubs in Mediterranean regions. *Pistacia lentiscus*	III. B	1.057 (1.04–1.07)
Mecca Balsam (Balm of Gilead) Used as a tonic in the Orient.	*Balsamodendron gileadense* in Arabia	III. A	....
Myrrh (Bdellium is a variety.) Has 50–60% gum. Used in medicine.	Turkey, India, Arabia, Somaliland. Species of *Comniphora*	I.	....
Peru Balsam Has 55–66% cinnamein. Used in medicine and perfumery.	Central America. Black liquid from *Toluifera pereirae*	III. A	1.14–1.151
Olibanum or **Frankincense** Used in perfumery.	*Boswellia Carterii*, etc. Somaliland and S. Australia	I.	....
Pontianak Composition:—84.5% resin acids, 4% resenes. A recent to semi-fossil resin. Similar to, but harder and less odorous than soft manila. Used in spirit varnishes.	Borneo	III. B, C	1.037
Rosin or **Colophony** Largely abietic acid, $C_{20}H_{30}O_2$. Commonest and cheapest resin. Graded according to color. Used in paints, soap, paper, etc.	*Pinus* species. Mostly *Pinus palustris*. Gum obtained on distillation	III. B	1.045– 1.086
Gum Sandarac Composition:—85% sandaracolic acid, 10% callitrolic acid. Common resin for use in spirit varnishes.	N. E. Africa. Small trees: *Callitris quadrivalvis*	III. B	1.073
Sierra Leone Copal The hardest of the recent resins.	W. Africa. *Guibourtia Copalifera*	III. B	1.072
Balsam Storax Used in medicine.	Levant and Greece. *Styrax officinalis*	III. A	....
Balsam Tolu	S. America. *Toluifera balsamum*	III. A	....
Gum Thus A soft, fresh oleoresin from slashed trees, yielding turpentine and rosin on distillation.	*Pinus* species	III. A	....
Gum Tragacanth Swells in water—used in adhesives.	Asia Minor and Persia. *Astragalus gummifer*	Gum	....

Table 7-10 (*Continued*)
PHYSICAL AND CHEMICAL PROPERTIES OF NATURAL RESINS

S. P., °C.	M. P., °C.	Acid No.	Sap. No.	Iod. No.	Solubility
80-93	95	63 (50-71)	79 (70-194)	64-159 E. = 23-29	s. 3, 4, 6, 7, 8, 9, 10, 11, 12; p. s. 1, 2, 5.
....	....	....	....	....	....
....	M. = ca. 13	Resin:- 42-70	Resin:- 159-216	Resin:- E. = 95-145	....
....	....	....	....	....	s. 1, 3.
....	E. = 7- 131	42-50	M. = 5-7	....	....
55	135	134	186	119-142	s. 1, 5, 6, 8, 9, 12; p. s. 2, 3, 7; i. 4, 10, 11.
70-80	120-135	155-175 E. = 8-23	167-194 M. = 0	80-220	s. nearly all organic solvents; p. s. 3, petr. ether.
....	145	140 (154)	154 (142-174)	66-160 E. = 1-11 M. = 0	s. 1, 3, 6, 8, 12; p. s. 2, 5, 7, 9; i. 4, 10, 11.
60	130 (130-200)	110 (73-130)	123 (123-158)	63-133	s. 6, 8, 9, 12; p. s. 2, 3, 4, 5, 7; i. 1, 10, 11.
....	....	128-131	191-206	65	s. 3, 1 (hot).
....	....	....	....	....	....
....	....	108-145	140-161	E. = 3-60	....
....	....	....	....	....	....

Table 7-10 (*Continued*)
PHYSICAL AND CHEMICAL PROPERTIES OF NATURAL RESINS

Name	Source	Class	Sp. Gr.
Venice Turpentine	European larch. *Larix europaea*	III. A	1.094– 1.190
Zanzibar Copal Composition:—85% resin acids, 10% essential oils, 5% resenes. A common resin, little used only because of its high price. Soluble in oils after depolymerization.	Zanzibar and E. Africa *Trachylobium verrucosum*	III. D	1.054– 1.063

GLYCERIDE CONTENT OF DRYING OILS

The values in the table give the percentage content of various fatty acid esters present in the glyceride structure of the common drying oils.

The composition of different samples of the same drying oil may vary considerably. These variations are often reflected in changes in the constants of the oil, particularly the iodine value.

Oil	Saturated	Oleate	Linoleate	Linolenate	Eleostearate
Cottonseed	25	40	35	..	..
Soybean	14	26	52	8	..
Dehydrated Castor	5	10	85	..	..
Linseed	10	18	17	55	..
Perilla	7	14	16	63	..
Tung	5	7	3	..	85
Oiticica	10	6	10	..	74*

* Licanate, $(C_{17}H_{27}OCO_2)_3C_3H_5$

Table 7-10 (*Continued*)
PHYSICAL AND CHEMICAL PROPERTIES OF NATURAL RESINS

S. P., °C.	M. P. ,°C.	Acid No.	Sap. No.	Iod. No.	Solubility
E. = 30-56	M. = 0	67-101	81-127	144	s. 1, 3, 4, 7, petr. ether.
150	300	93 (87-93)	93 (70-93)	115-123	p. s. 8, 12; i. 1, 2, 3, 4, 5, 6, 7, 9, 10, 11.

Table 7-11
THE VITAMINS

In compiling data for this revision certain facts became evident in making additions to the previous table. Some of the new factors which have now been added have not as yet been shown to be indispensable to humans. The importance of some are now being recognized, and clinical experimentation is practically just beginning. However, it is known that the animal body does not function properly where the intake of these factors is low. Some of these factors have been shown to be synthesized in the intestines of some animals by the intestinal flora and consequently it is very difficult to produce a recognizable deficiency. Almost invariably a deficiency of any factor in humans is accompanied by a deficiency of other factors; this gives the chief impetus to preparation of polyvitamin products. Thus far no factor essential for lower animals, bacteria, etc., has been found which is not also essential for humans. Some signs of deficiencies are apparently attributable to more than one vitamin. Some deficiencies are only partially cured by one or another vitamin, while complete cure is obtained only by administration of all vitamins concerned. Most of the vitamins are dependent upon the presence of one another for proper functioning.

INTERNATIONAL VITAMIN STANDARDS

1 Unit vitamin A ⇌ 0.344 microgram (μg) pure vitamin A acetate ⇌ 0.3 microgram pure vitamin A alcohol; 1 unit provitamin A = 0.6 microgram β-carotene.
1 Unit vitamin B$_1$ ⇌ 3.3 micrograms thiamine chloride.
1 Unit vitamin C ⇌ 50 micrograms of l-ascorbic acid.
1 Unit vitamin D ⇌ 1 milligram of standard solution of irradiated ergosterol (corresponding to 0.1 microgram of the ergosterol used in its preparation or 0.025 microgram crystalline vitamin D$_2$ (viosterol or calciferol).

Name and Synonym	Physiological Effect	Concentrates for Medicinal Use	Plant	Animal
			Sources	
Vitamin A Axerophthol Provitamins A (α-, β-, and γ-carotenes, cryptoxanthin, aphanin, aphanicin, echinenone, myxoxanthin, leprotene (and probably other carotinoids) Anti-infective Anti-ophthalmic (Fat soluble)	Increases resistance to infection of eyes, sinuses, ears, kidneys, skin, and mucous membranes. Prevents xerophthalmia, hyperkeratosis of skin and mucous membranes. Promotes appetite and growth. Essential for reproduction. The vitamin is stored in the body and is required at all ages. Deficiency results in night blindness and injury to the nervous system. Adult human requirements approximately 3000 to 5000 units daily.	Carotene, obtained from carrots or vegetables, is converted by the body into vitamin A. Unsaponifiable fractions of halibut liver oil, cod liver oil and other fish liver oils.	All yellow and green vegetables. Raw carrots, yellow turnips, yellow corn, spinach, squash, canned and fresh tomatoes, sweet potatoes, oranges, prunes, apricots and pimento peppers.	Halibut liver oil, cod liver oil and other fish liver oils, fresh milk, cream, butter, cheddar and cream cheese, egg yolk, liver and kidney.
Thiamine chloride Vitamin B$_1$ Anti-neuritic	Promotes appetite and digestion by stimulating metabolism. Vital to health at all ages. Required by the mother for reproduction and	Synthetic thiamine chloride. From yeast and wheat germ.	Whole grain cereals, green leafy vegetables, water cress, carrots,	Milk, cheese, raw oysters, brain, kidney and liver.

Name / Properties	Function	Source (synthetic / origin)	Food Source	Distribution
Anti-beriberi (Soluble in water and in 90% alcohol; heat labile; acid stable; adsorbed from acid solution)	lactation. Prevents the diseases beriberi and polyneuritis. Promotes normal oxidation of carbohydrates in the body and formation of fat from carbohydrates. Adult human requirements approximately 1.5 to 2.0 milligrams daily.		canned tomatoes, green corn, baked potatoes, dried lima beans, turnips and nuts.	
p-Aminobenzoic acid PAB Chromotrichia factor Anti-gray-hair factor	Influences enzyme activity. In experimental rats it apparently prevents achromotrichia. Inhibits oxidation of adrenaline. Nutrilite for certain bacteria. Antagonizes sulfanilamide and sulfapyridine.	Synthetic. Extracts of yeast.	Yeast.	Widely distributed over entire plant and animal kingdoms.
Choline (Choline is replaceable by methionine or betaine with estimated efficacy of 30%; also replaceable by dimethyl-thetin or dimethyl-β-propiothetin)	Maintains integrity of tissues. In chickens it is necessary for normal nutrition and egg production. Deficiency results in marked deposition of liver fat. Acts as transmethylating agent. Deficiency in young rats produces severe renal lesions. Important in lipid metabolism. Deficiency causes a type of cirrhosis of the liver, hemorrhagic degeneration of kidneys, necrosis of kidney cortex, enlargement of spleen. Deficiency results in anemia in the rat. Necessary for normal growth and lactation of fat.	Synthetic choline chloride. As lecithin in phospholipids.	Rice bran, yeast.	Liver. Present in phospholipids of practically all animal and plant cells.
Ascorbic acid Vitamin C Cevitamic acid Antiscorbutic (Soluble in water; heat labile at alkaline reaction)	Prevents and cures scurvy. Important in formation and maintenance of connective tissue in the body; essential for normal tooth and bone formation and strength of capillary walls. Inhibits dental caries, pyorrhea, certain gum infections, anorexia, anemia, failure of wound healing and susceptibility to infection. Important in carbohydrate metabolism and in controlling infective processes. Body stores readily depleted. Adult human requirements 75 to 100 milligrams daily.	Synthetic l-ascorbic acid.	Oranges, lemons, grapefruit, raw and leafy vegetables, raw apples, tomatoes, carrots and paprika.	Adrenal cortex, pituitary, ovary, thymus, intestinal wall, and milk.
Vitamin D Anti-rachitic (Fat soluble)	Prevents and cures rickets in young and osteoporosis in adults. Promotes absorption of calcium and phosphorus, and highly important for normal metabolism of these mineral elements in the young. Prevents spasmophilia and osteomalacia. Limited storage in the body. Approximate human requirements: for premature babies 600 to 800 units with milk; for full term infants 300 to 400 units with milk; for children 300 to 400 units	The effect of this vitamin is produced by the exposure of the body to direct sunlight or ultraviolet light. Ergosterol and foods exposed to (or irradiated with) ultraviolet. Unsaponifiable fractions of liver oils of halibut, cod, and many	Practically none.	Cod, halibut and other fish liver oils, butter, egg yolk, and fish.

Table 7-11 (*Continued*)
THE VITAMINS

Name and Synonym	Physiological Effect	Sources		
		Concentrates for Medicinal Use	Plant	Animal
	and for pregnant and nursing women 800 units daily.	other fish. Cholesterol and several of its derivatives after exposure to ultraviolet light.		
Pantothenic acid Pantothen Filtrate factor Factor II Chick anti-dermatitis factor	Prevents dermatitis in chicks. Deficiency may produce lesions in nervous system or endocrine systems, or may involve skin or hair. May play a part in maintaining integrity of nervous tissue. Necessary in treatment of peripheral neuritis along with other B-vitamins. Prevents hemorrhagic adrenals and achromotrichia in rats. Prevents duodenal ulcers and atrophy of duodenal mucosa. Essential for growth in various experimental animals and plants. Needed for reproduction in chicks and increases egg hatchability. Is necessary for acetylations in the rat. It is a constituent of coenzyme A. As coenzyme A it yields citrate from acetate and oxaloacetate.	Synthetic calcium *d*-pantothenate. Liver and yeast extracts.	Rice bran, wheat bran, brewer's yeast, Irish potato, taro root, alfalfa, wheat germ, molasses, peanuts, cabbage, cucumbers, and rolled oats.	Mammalian liver, egg yolk, eggs, canned salmon, whole milk, kidney, heart, spleen, brain, pancreas, tongue, lung, muscle, beef.
Pyridoxin Pyridoxal Pyridoxamine Vitamin B$_6$ Adermin Factor Y Anti-dermatitis Anti-acrodynia factor (Soluble in water and in dilute alcohol; heat and acid stable)	Deficiency of this vitamin causes dermatitis in rats; anemia in dogs and pigs, and is presumably essential for hemoglobin formation. Controls both iron and copper absorption in rats. In experimental animals deficiency produces convulsions resembling human epileptic fits. Requirements proportional to metabolism. Apparently concerned with metabolism of the amino acids to carbohydrates. Deficiency impairs liver function, produces erythrodermic dermatosis, impairs growth.	Synthetic pyridoxin. From the germ and bran of wheat and rice; yeast, and liver extracts.	Grains, chiefly in the germ, yeast, maize, and molasses, seeds, legumes, spinach, and lettuce.	Liver, heart muscles, milk, and egg yolk.
Riboflavin Vitamin B$_2$ Vitamin G (Soluble in water and in dilute alcohol; heat and acid stable)	Plays an important role in vision mechanism. In fowls important in production and hatchability of eggs. Deficiency produces cheilosis, glossitis, seborrheic lesions about eyes and nose, vascularizing keratitis, and mild photophobia. Is a component of several enzymes. Plays part in metabolism of amino acids and carbohydrates. Deficiency produces loss of	Synthetic *d*-riboflavin. From milk whey, liver and yeast. Fermentation residues.	Fruits, leafy vegetables, soy beans, wheat germ, yeast.	Kidney, liver, milk, eggs.

Vitamin and factor	Function	Synthetic and source derivatives	Plant sources	Animal sources
	hair, injury to nervous system, failure to grow, yellow liver, anemia and death. Adult human requirements approximately 2 to 3 milligrams daily.			
Vitamin E α-, β-, and γ-Tocoph-erols Anti-sterility (Fat soluble)	Absence of this vitamin causes sterility in either sex. It is stored in the body to a certain degree. Exerts favorable influence on growth. Limited value in certain cases of muscular dystrophy and amyotropic lateral sclerosis. Aids in resisting protein deficiency and prevents nutritional liver damage.	Synthetic dl-α-tocopherol. Unsaponifiable fraction of wheat germ oil.	Vegetable oils, lettuce, beans, whole wheat, rice, barley, corn, and water cress.	Meat, liver, milk, and eggs.
Nicotinic acid Niacin Nicotinic acid amide Niacin amide Anti-pellagra Pellagra preventive (P-P) factor	Prevents and cures black tongue in dogs and pellagra in man. Is a component of several enzymes. Influences carbohydrate, protein, water, and heavy metals metabolism. Deficiency produces a certain type of anemia. Adult human requirement approximately 10 to 20 milligrams daily.	Synthetic nicotinic acid, its amide, and all hydrolyzable nicotinoyl derivatives. From liver and yeast.	Yeast, wheat germ, peas, beans, green leafy vegetables.	Liver, lean meat, milk, egg yolk.
Vitamin K Anti-hemorrhagic Coagulation factor Prothrombin factor Phylloquinones Menadione (2-methyl-1, 4-naphthoquinone) (Fat soluble, heat stable, alkali labile)	Promotes normal blood coagulation time. In chicks it prevents intestinal, subcutaneous and intramuscular hemorrhage. Deficiency reduces prothrombin formation, hemorrhagic diathesis in newborn, certain types of hepatic and biliary diseases.	Synthetic K-vitamers; i.e., many related 1,4-naphthoquinones and compounds readily converted to them, most important is 2-methyl-1, 4-naphthoquinone, which replaces commercially the vitamin, and is known commercially as menadine.	All green vegetables and green leaves, tomatoes, oat sprouts, dried carrot tops, hemp-seed oil, cotton-seed oil, soy-bean oil, and bacterial action on rice bran.	Hog-liver fat, egg yolk, bacterial action on casein and fish meal. K_2 from putrefied fish meal.
Biotin Vitamin H Coenzyme R Anti-egg-white factor Curative factor of egg-white injury Bios II Bios II$_b$ Factor X Factor W	Essential in vital economy of animals and in normal pigment metabolism. Takes part in carbon dioxide fixation; deaminase activity stimulates lipid utilization. Important in intermediary metabolism. Prevents dermatitis. Pervents perosis in chicks and fatty livers. Deficiency results in seborrheic desquamative dermatitis, emaciation and death. Nutrilite for certain bacteria and for yeast. It is antagonized by avidin which occurs in egg white.	Synthetic. Extracts of egg yolk, liver, milk-sugar residues, fermentation residues.	Widely distributed in nature. Yeast, seeds, nuts, tomatoes, carrots, corn and other vegetables.	Liver, kidney, pancreas, egg yolk, pork, and milk. Occurs in small amounts in all mammals and birds; tissue of eye only exception found so far.
Inositol (Meso-inositol) Bios I Anti-alopecia factor	Promotes growth, cures baldness in mice, important in fat metabolism, prevents fatty liver, favors growth of yeast and other microorganisms.	Liver extracts. Wheat bran extracts in the form of the hexa-phosphate ester (phytic acid) which on hydrolysis yields inositol.	Citrus fruits, nuts, yeast, various whole cereals and grains.	Muscle, kidney, liver, brain, milk, and eggs.

Table 7-11 (*Continued*)
THE VITAMINS

Name and Synonym	Physiological Effect	Concentrates for Medicinal Use	Sources	
			Plant	Animal
Folic acid Pteroylglutamic acid (L. casei factor)	Nutrilite for yeast and certain bacteria. Anti-anemia factor. Appears to be a growth factor for rats. Deficiency in mice is accompanied by reduction in cellular elements of blood and arrest of bone marrow maturation.	Extracts of yeast, milk sugar residue, spinach. Synthetic pteroylglutamic acid.	Yeast, mushrooms. Leaves of many plants.	Milk. Present in all animal tissues.
Citrin Vitamin P Permeability vitamin	Promotes tissue and capillary permeability. Prolongs life of scorbutic guinea pigs. Deficiency causes nutritional purpura.	Extracts of citrus fruits.	Citrus fruit juices and the leaves of several plants.	None.
Vitamin B₁₂ Antipernicious factor Animal protein factor	Prevents: pernicious anemia; megaloblastic anemia of infancy; nutritional macrocytic anemia; sprue; nutritional glossitis. Required for gestation and lactation.	Liver extracts.		Liver, milk, fish meal.

Table 7-12
FORMULAS OF THERMOPLASTIC AND THERMOSETTING MATERIALS AND SYNTHETIC RUBBERS

A. Thermoplastic Materials

(1) Polyethylene

$CH_2:CH_2 \longrightarrow [—CH_2\cdot CH_2—]n$

Ethylene   Polyethylene
            Polythene

(2) Polystyrene

$CH(C_6H_5):CH_2 \longrightarrow [—CH(C_6H_5)\cdot CH_2—]n$

Styrene   Styron,
            Lustron

(3) Polyvinylchloride

$CH_2:CHCl \longrightarrow [—CH_2\cdot CHCl—]n$

Vinyl   Geon,
chloride   Koroseal

(4) Polyvinylchloride Acetate

$CH_2:CHCl + CH_2:CH(OAc) \longrightarrow [—CH_2\cdot CHCl\cdot CH_2\cdot CH(OAc)\cdot CH_2\cdot CHCl—]n$

Vinyl   Vinyl   Vinylite V
chloride   acetate

(5) Polyvinylidene Chloride

$CH_2:CHCl + CH_2:CCl_2 \longrightarrow [—CH_2\cdot CCl_2\cdot CH_2\cdot CHCl—]n$

Vinyl   Vinylidene   Saran,
chloride   chloride   Geon, Velon

(6) Polytetrafluoroethylene

$CF_2:CF_2 \longrightarrow [—CF_2\cdot CF_2—]n$

Tetra-   Teflon
fluoro-
ethylene

(7) Polyvinyl Butyral

$[—CH_2\cdot CH(OAc)—]n \longrightarrow [—CH_2\cdot CH(OH)—]n + C_3H_7\cdot CHO \longrightarrow$

Polyvinyl   Polyvinyl   Butyr-
acetate   alcohol   aldehyde

$$\left[\begin{array}{c} —CH_2\cdot CH\cdot CH_2\cdot CH— \\ O—CH—O \\ C_3H_7 \end{array} \right]n$$

Saflex, Butacite,
Vinylite X

(8) Polymethyl Methacrylate

$CH_2:C(CH_3)\cdot CO\cdot OCH_3 \longrightarrow [—CH_2\cdot C(CH_3)(CO\cdot OCH_3)—]n$

Methyl methacrylate   Plexiglas, Lucite

(9) Nylon

$C_6H_{11}OH \quad N:C(CH_2)_4C:N \longleftarrow [(CH_2)_2CO_2H]_2 + H_2N(CH_2)_6NH_2$

Cyclohexanol
          Adiponitrile   Adipic acid   Hexamethylene
                                            diamine

$\longrightarrow [—CO\cdot(CH_2)_4\cdot CO\cdot NH\cdot(CH_2)_6\cdot NH—]n$

Nylon

(10) Cellulose Nitrate

$C_6H_{10}O_5 + HNO_3 \longrightarrow [—C_6H_7O_2(OH)(ONO_2)_2—]n$

Cellulose   Nitric   Celluloid
          acid

(11) Cellulose Acetate

$C_6H_{10}O_5 + Ac_2O \longrightarrow [—C_6H_7O_2(OH)(OAc)_2—]n$

Cellulose   Acetic   Lumarith,
          anhydride   Tenite

(12) Cellulose Acetate Butyrate

$C_6H_{10}O_5 + Ac\cdot O\cdot COC_3H_7 \longrightarrow [—C_6H_7O_2(OH)(OAc)(O_2CC_3H_7)—]n$

Cellulose   Acetic-butyric   Tenite II
          anhydride

(13) Ethyl Cellulose

$C_6H_{10}O_5 + C_2H_5Cl \longrightarrow [—C_6H_7O_2(OH)(OC_2H_5)_2—]n$

Cellulose   Ethyl   Ethocel
          chloride

Table 7-12 (*Continued*)

FORMULAS OF THERMOPLASTIC AND THERMOSETTING MATERIALS AND SYNTHETIC RUBBERS

B. Thermosetting Materials

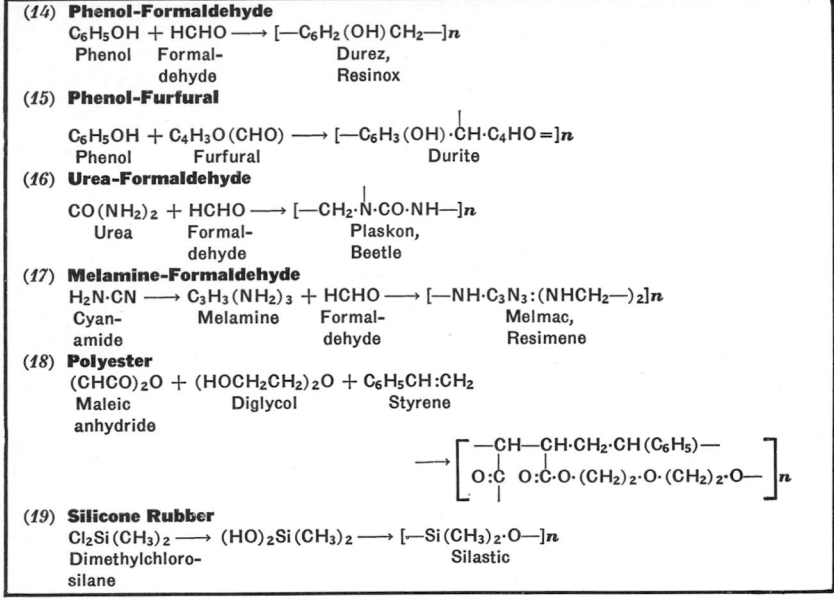

(14) Phenol-Formaldehyde

$C_6H_5OH + HCHO \longrightarrow [—C_6H_2(OH)CH_2—]n$

Phenol · Formal- · Durez,
dehyde · Resinox

(15) Phenol-Furfural

$C_6H_5OH + C_4H_3O(CHO) \longrightarrow [—C_6H_3(OH)·\overset{|}{C}H·C_4HO=]n$

Phenol · Furfural · Durite

(16) Urea-Formaldehyde

$CO(NH_2)_2 + HCHO \longrightarrow [—CH_2·\overset{|}{N}·CO·NH—]n$

Urea · Formal- · Plaskon,
dehyde · Beetle

(17) Melamine-Formaldehyde

$H_2N·CN \longrightarrow C_3H_3(NH_2)_3 + HCHO \longrightarrow [—NH·C_3N_3:(NHCH_2—)_2]n$

Cyan- · Melamine · Formal- · Melmac,
amide · dehyde · Resimene

(18) Polyester

$(CHCO)_2O + (HOCH_2CH_2)_2O + C_6H_5CH:CH_2$

Maleic · Diglycol · Styrene
anhydride

$$\longrightarrow \left[\begin{array}{l} —CH—CH·CH_2·CH(C_6H_5)— \\ O\!:\!C \ \ O\!:\!C·O·(CH_2)_2·O·(CH_2)_2·O— \end{array} \right]n$$

(19) Silicone Rubber

$Cl_2Si(CH_3)_2 \longrightarrow (HO)_2Si(CH_3)_2 \longrightarrow [—Si(CH_3)_2·O—]n$

Dimethylchloro- · Silastic
silane

C. Synthetic Rubbers

Approximate proportions in parts by weight.

(a) GR-S

$CH_2:CH·CH:CH_2 + CH(C_6H_5):CH_2 \longrightarrow [—CH_2·CH:CH·CH_2·CH(C_6H_5)·CH_2—]n$

1,3-Butadiene · Styrene · GR-S
75 parts · 25 parts

(b) Butene-Diene

$CH_2:C(CH_3)_2 + CH_2:C(CH_3)·CH:CH_2 \longrightarrow \left[\left(CH_2·\overset{\overset{CH_3}{|}}{\underset{\underset{CH_3}{|}}{C}} \right)_{90} \overset{\overset{H\ CH_3}{}}{\underset{\underset{H}{}}{—C·C}} :CH·CH— \right]n$

Isobutene · Diene · Butyl
98 parts · 2 parts

(c) Neoprene

$CH_2:C(Cl)·CH:CH_2 \longrightarrow [—CH_2·C(Cl):CH·CH_2—]n$

Chloroprene · Neoprene

(d) Acrylonitrile copolymers

$CH_2:CH·CN + CH_2:CH·CH:CH_2 \longrightarrow [—CH_2·CH:CH·CH_2·CH(CN)—]n$

Acrylonitrile · Butadiene · Hycar,
Low 10-20 parts · 60-80 parts · Chemigum,
Medium 20-35 parts · · Butaprene,
High 35 parts · · Paracril

(e) Thiokol

$Cl·R·Cl + Na_2S_4 \longrightarrow [—R—S_4—]n$

Dichloride · Sodium · Thiokol
polysulfide

where R — $CH_2·CH_2—$
or — $CH_2·CH_2·O·CH_2·CH_2—$
or — $CH_2·CH_2·O·CH_2·O·CH_2·CH_2—$

Table 7-13
PROPERTIES OF THERMOPLASTIC AND THERMOSETTING MATERIALS

A. Thermoplastic Materials*

Property	(1) Poly-ethylene	(2) Poly-styrene	(3) Polyvinyl Chloride Non-Rigid	(4) Polyvinylchloride Acetate — Rigid	(4) Polyvinylchloride Acetate — Non-Rigid	(5) Poly-vinylidene Chloride	(6) Poly-tetrafluoro-ethylene
Fabrication							
Bulk factor	2.2-3.6	2.0-2.3	......	2	......	2	......
Injectn. moldg. temp., °F	325-375	325-500	330-375	280-300	300-330	300-400	......
Injectn. moldg. press., psi $\times 10^{-3}$	4.5-30	10-30	15-25	18-30	7-20	10-30	......
Mold shrinkage, mils/in.	20-50	2-8	17	1	20-100	5-15	......
Physical							
Specific gravity	0.92	1.06	1.18-1.65	1.35-1.45	1.15-1.45	1.65-1.72	2.1-2.3
Specific volume, in.^{3}/lb	30	26	16-20	19-21	17-23	16-17	12-13
Coefficient thermal expansion, linear $\times 10^5$	18	6-8	......	6.9	7-25	19	9
Specific heat, cal/g	0.52	0.32	0.36	0.24	0.3-0.5	0.32	0.25
Thermal conductivity $\times 10^4$	8	1.9	3.9	4	3.9	2.2	6
Heat distortion temperature, °F	115-122	165-190	......	125-135	......	150-180	266
Heat resistance—continuous, °F	......	150-170	125	130	150	160-200	550
Flammability, in./min	1.1-1.4	2.0	0-0.3	None	0-0.7	None	None
Water absorption, %	0.01-0.03	0.04-0.06	0.4-1.3	0.15	0.1-1.0	<0.1	0
Mechanical							
Impact strength, Izod ft-lbs/in.	No break	0.25-0.6	......	0.25-0.50	1-3.2	0.3-1.0	4
Tensile strength, psi $\times 10^{-3}$	1.4-2.4	3-8.5	1-2.6	4.5-8	......	4-7	2-4.5
Elongation at break, %	200-600	1.5-3.5	200-390	7.5-13	225-370	10-40	200-300
Flexural strength, psi $\times 10^{-3}$	1.5-1.7	4.8-19	......	9-10	......	15-17	2.0
Compressive strength, psi $\times 10^{-3}$	......	11.5-17	......			4.5-5.5	1.7
Electrical							
Dielec. strength, short, v/mil	400-475	500-700	325-425	400-425	150-400	350-400	450
Vol. resistivity, ohms-cm	10^7-10^9	10^{17}-10^{19}	10^{11}-10^{15}	>10^{14}	10^7	10^{14}-10^{16}	10^{16}
Dielec. constant, 60 cycles	2.3	2.5-2.7	6.2-6.4	3.2-3.3	8.1-8.3	3-5	2
" " , 1,000 cycles	2.3	2.5-2.7	4.2-4.9	3.1-3.2	6.9-8.4	3-5	2
" " , 10^6 cycles	2.3	2.5-2.7		3.0-3.1		3-5	2
Power factor, 60 cycles	0.3-0.5	0.06-0.5	97-150	7-10	50-100	30-80	0.2
" , 1,000 cycles	0.3-0.5	0.05-0.5	109-115	11-13	85-108	30-150	0.2
" , 10^6 cycles	0.3-0.5	0.10-0.5		18-19		30-50	0.2

* The values were chosen as the most representative for each type. Considerable variation may occur since the properties of a molded article depend not only on the plastic used but on many other factors including conditions of forming and design of the molded part itself. Improvement of specific properties may often be accomplished by changes in formulation. Often improvement in one property may be accomplished at the expense of other properties.

Table 7-13 (Continued)

PROPERTIES OF THERMOPLASTIC AND THERMOSETTING MATERIALS

A. Thermoplastic Materials

Property	(7) Polyvinyl Butyral Plasticized	(8) Polymethyl Methacrylate	(9) Nylon	(10) Cellulose Nitrate	(11) Cellulose Acetate	(12) Cellulose Acetate Butyrate	(13) Ethyl Cellulose
Fabrication							
Bulk factor	....	1.7-2.5	2.3	....	2.0-2.6	2.0-2.4	2.0-2.5
Injectn. moldg. temp., °F	250-340	325-480	510-600	....	300-500	330-430	350-500
Injectn. moldg. press., psi $\times 10^{-3}$	15-30	10-30	Low	....	8-32	8-32	3-30
Mold shrinkage, mils/in.	4-20	3-6	12-15	....	1-7	1-5	1-10
Physical							
Specific gravity	1.05-1.50	1.18-1.20	1.14-1.16	1.35-1.57	1.26-1.5	1.1-1.23	1.07-1.18
Specific volume, in.^{3}/lb	18-26	18-20	24	20	20-22	22	23-26
Coefficient thermal expansion, linear $\times 10^5$	0.4	7-9	10.3	9-16	8-16	11-17	9-16
Specific heat, cal/g	....	0.35	0.43	0.34-0.38	0.4	0.4	0.36
Thermal conductivity $\times 10^4$	....	4-6	6	3.1-5.1	4-8	4-8	3-5
Heat distortion point, °F	....	135-200	165-170	110-150	100-235	103-182	100-180
Heat resistance—continuous, °F	....	120-140	....	Poor	140-220	140-220	140-220
Flammability, in./min	....	0.5-1.0	Self ext.	V. rapid	2-self ext.	1.5-self ext.	2-self ext.
Water absorption, %	1-2	0.3-0.6	1.5	0.7-4.0	2-6	1.3-2.4	0.7-4.0
Mechanical							
Impact strength, Izod ft-lbs/in.	0.5-3.5	0.2-0.4	0.6-0.9	0.2-8.0	0.5-5.7	0.4-9.4	3-11
Tensile strength, psi $\times 10^{-3}$	....	4-6	9-10.5	3-10	1.5-8	1.4-6.5	2.5-8
Elongation at yield, %	150-450	1-10	45-55	25-50	10-75	35-100	2-100
Flexural strength, psi $\times 10^{-3}$	....	10-20	11-13	6-15	2-16	1.6-11	....
Compressive strength, psi $\times 10^{-3}$	....	9-15	14-16	20-30	10-37	....	8-20
Electrical							
Dielec. strength, short, v/mil	175-375	500	350-400	250-365	250-265	250-400	470-550
Vol. resistivity, ohms-cm	10^{10}	10^{15}-10^{19}	10^{11}-10^{13}	10^{11}	10^{10}-10^{12}	10^{10}-10^{12}	10
Dielec. constant, 60 cycles	5.6	3.4-3.6	4-5	6.7-7.3	3.7-7.5	3.5-6.4	3.2-4
" , 1,000 cycles	4-13	3.3-3.5	4.5-5	....	3.5-7.0	....	3.3-3.8
" , 10^6 cycles	4	2.8-3.3	3.4-4.0	....	3.2-7.0	....	3.3-3.7
Power factor, 60 cycles	115	50-60	14-50	60-150	10-60	10-40	8-30
" , 1,000 cycles	10-100	40-60	20-50	....	10-60	....	8-15
" , 10^6 cycles	20-100	28-33	40-70	....	10-100	....	17-36

B. Thermosetting Materials*

(14) Phenol-Formaldehyde

Property	No Filler	Wood Flour	Mica	Asbestos	Fabric	Sisal Felt	No Filler Cast
Fabrication							
Bulk factor	2.3	2.2	2.5	2-12	4-18	2-5	......
Molding temp., °F	300	300-350	350	350	350	300	200
Molding press., psi $\times 10^{-3}$	3	4	5	2-6	8	3	None
Colors	Dark			Limited			......
Mold shrinkage, mils/in.	10	7	3	3	6	20	......
Molding qualities	Fair	Excellent	Fair	Good-Fair	Fair	Fair	......
Physical							
Specific gravity	1.27	1.37	1.6-2.0	1.80	1.40	0.7-1.4	1.30
Specific volume, in.^{3}/lb	22	20	14-17	1.6	19-21	20-40	22
Coefficient thermal expansion, linear $\times 10^5$	2.5-6	3-4.4	2	2	1-3	7-20	5-15
Specific heat, cal/g	0.40	0.4	0.28-0.32	0.28-0.32	0.30-0.35	0.3-0.4	0.4
Thermal conductivity $\times 10^4$	3-6	4-7	10-14	8-16	4-7	2-8	3-5
Heat distortion point, °F	240-260	260-300	210-320	285-350	250-300	320	100-170
Heat resistance—continuous, °F	250	300	250-300	350-400	250	250	180
Water absorption, %	0.1-0.2	0.8	0.02-0.1	0.1-0.3	0.5-1.8	0.5-15	0.2-2.0
Mechanical							
Impact strength, Izod ft-lbs/in.	0.4-0.5	0.2-0.4	4.5-7	4-8	1-8	3-16	0.3-0.4
Tensile strength, psi $\times 10^{-3}$	7-10	6.5-9.5	......	......	4-8	7-12	2-9
Elongation at yield, %	1-1.5	0.4-0.8	......	......	0.7	......	......
Flexural strength, psi $\times 10^{-3}$	12-17	9-12	7-13	7-13	9-13	16-20	6-9
Compressive strength, psi $\times 10^{-3}$	10-30	24-32	15-25	15-25	15-30	10-35	9-25
Electrical							
Dielec. strength, short, v/mil	300-400	300-375	325-500	200-350	200-325	25-400	75-450
Vol. resistivity, ohms-cm	10^{12}	10^{11}	10^{11}-10^{14}	10^9-10^{12}	10^{10}-10^{12}	10^{11}-10^{12}	10^{11}-10^{12}
Dielec. constant, 60 cycles	5-6	5.5-7	5-6	12-50	6-10	5-10	6.5-30
" , 1,000 cycles	4-5	4.8-6	4.5-5.5	18-50	5.5-9	......	5-18
" , 10^6 cycles	4.5-5	4.5-5.5	4.5-5.2	5.5-8.0	4.5-6	3-5	5-11
Power factor, 60 cyc es	0.3-0.65	0.16-0.84	0.01-0.06	0.15-0.4	0.08-0.3	0.1-0.3	0.4-4.0
" , 1,000 cycles	0.14-0.45	0.14-0.36	0.01-0.04	0.1-0.4	0.04-0.18	......	1.2-3.6
" , 10^6 cycles	0.07-0.15	0.13-0.28	0.05-0.03	0.05-0.2	0.03-0.06	0.3-0.5	0.05-1.1
Arc resistance, secs	......	......	......	......	......	......	120-250

* The values were chosen as the most representative for each type. Considerable variation may occur since the properties of a molded article depend not only on the plastic used but on many other factors including conditions of forming and design of the molded part itself. Improvement of specific properties may often be accomplished by changes in formulation. Often improvement in one property may be accomplished at the expense of other properties.

Table 7-13 (Continued)
PROPERTIES OF THERMOPLASTIC AND THERMOSETTING MATERIALS

B. Thermosetting Materials

Property	(15) Phenol-Furfural Wood Flour	(15) Phenol-Furfural Fabric	(16) Urea-Formaldehyde Cellulose	(17) Melamine-Formaldehyde Cellulose	(17) Melamine-Formaldehyde Asbestos	(17) Melamine-Formaldehyde Fabric	(18) Polyester None	(19) Silicone Rubber Mineral
Fabrication								
Bulk factor	2.2-3	2.5	2.2-3	2.2-2.7	2.1-2.5	5-10	Cast	1-1.5
Molding temp., °F	280-400	270-360	275-370	275-370	275-370	275-340	......	300
Molding press., psi $\times 10^{-3}$	1-5	1-8	1.56	1-6	1-7	3-5	......	0.1-0.8
Colors	Limited	Limited	Unltd.	Unltd.	Gray	......	......	Unltd.
Mold shrinkage, mils/in.	5-10	3-7.0	5-11	6-12	3-5	3-5	......	4-6
Molding qualities	Excellent	Good	Excellent	Excellent	Good	Good	Excellent	Good
Physical								
Specific gravity	1.32-1.42	1.3-1.45	1.45-1.55	1.40-1.55	1.7-2.00	1.5	1.32-1.40	1.4-2.0
Specific volume, in.^{3}/lb	20	20	18-19	18-19	14-16	18	20-21	14-20
Coefficient thermal expansion, linear $\times 10^5$	3-7	2-16	2.5-3	2.5-4.5	2-4.5	......	5.5-10	......
Specific heat, cal/g	0.3-0.4	0.3-0.4	0.4	......	......	......	0.3-0.5	......
Thermal conductivity $\times 10^4$	3-12	3-8	7-10	10	13.7	11	5	......
Heat distortion point, °F	240-300	240-285	260-280	285-300	265	310	140-200	......
Heat resistance—continuous, °F	300	240-300	180	210	250-400	250	220	500
Water absorption, %	0.2-0.6	0.5-2.5	0.75-2.0	0.08-1.7	0.08-1.4	0.3-0.6	0.2	0.25-1.0
Mechanical								
Impact strength, Izod ft-lbs/in.	0.2-0.4	0.6-4.8	0.24-0.36	0.24-0.35	0.3-0.4	1.0	0.3-0.4	0.200-0.60
Tensile strength, psi $\times 10^{-3}$	5-8.5	5.5-8	6-13	7-13	6-7	7-8	5-6	100-225
Elongation at yield, %								
Flexural strength, psi $\times 10^{-3}$	8-15	8-13	10-18	9-16	9-11	10-15	8-14	......
Compressive strength, psi $\times 10^{-3}$	16-36	20-32	27-43	30	21-32	30	19-23	......
Electrical								
Dielec. strength, short, v/mil	300-550	150-450	300-400	300-400	350-400	250-350	570-1275	290-450
Vol. resistivity, ohms-cm	10^9-10^{12}	10^9-10^{11}	10^{12}-10^{13}	10^{10}-10^{13}	10^{11}	......	10^{14}	10^{11}
Dielec. constant, 60 cycles	5-12	5-10	7-9.5	8.0-9.5	6.4-9.9	7.7-8.3	3.45	......
" , 1,000 cycles	4-10	4-9	6.4-9	7.6-8.7	......	7.2-7.6	3.50	......
" , 10^6 cycles	4-9	4-8	6.3-7.5	7.2-8.0	6.7	6.7	3.6	3.2-7.4
Power factor, 60 cycles	0.04-6.3	0.06-0.3	0.035-0.045	0.025-0.075	0.07-0.17	0.075-0.12	0.2	......
" , 1,000 cycles	0.04-0.8	0.06-0.2	0.027-0.055	0.015-0.035	......	0.035-0.05	0.014	......
" , 10^6 cycles	0.03-0.1	0.03-0.1	0.027-0.04	0.025-0.45	0.041	0.035-0.036	0.06	0.003
Arc resistance, secs	......	......	100-140	125	120-140	......	126	......

Table 7-14
PROPERTIES OF NATURAL AND SYNTHETIC RUBBERS*

	Natural Rubber	(a) Styrene-Butadiene Copolymer, GR-S	(b) Butene-Diene Copolymer	(c) Neoprene Polychloroprene	(d) Butadiene-Acrylonitrile Copolymers Acrylonitrile Content			(e) Polysulfide Thiokol
					Low	Medium	High	
Density	0.92	0.94	0.915	1.23	0.96	0.98	1.00	1.35
Refractive index	1.52	1.53	1.51	1.56	1.54	1.52	1.54	1.65
Specific heat, cal/g	0.452	0.454	0.464	......	......	......	......	......
Gum Stocks:								
Tensile strength, psi	3,100	300	2,000	2,800	500	700	900	1,000
Elongation, %	775	380	800	600	400	400	500	200
Black-loaded Stocks:								
Tensile strength, psi	3,900	3,000	2,200	3,600	3,000	3,500	4,000	1,000
Elongation, %	780	650	600	350	400	450	400	200
Stress 300%, psi	1,400	1,200	8,000	2,000	1,500	1,500	1,500	......
Swelling, % by volume, in:								
Kerosene at 25°C	200	100	300	60	10	8	4	4
Benzene at 25°C	200	200	300	150	......	120	......	50
Acetone at 25°C	25	30	30	20	60	100	150	25
Mineral oil at 70°C	120	150	130	10	10	5	2	1
Brittle point, °C	−56	−60	−45	−40	−40	−30	−1	−35
Rel. permeability to H_2	50	......	......	12	......	20	......	4
Rel. permeability to air	11	......	1	......	......	......	......	......
Insulation resist., ohms/cm	10^{17}	10^{15}	10^{16}	10^{10}	10^{10}	10^{10}	10^{10}	10^{15}
Resilience, %	90	75	50	75	......	74	63	62
Tear resistance, psi	1,640	550	1,000	1,100	......	......	......	......
Creep, 70°C	26	14.6	......	62	......	17	......	......

* The values were chosen as the most representative for each type. Considerable variation may occur since properties of a rubber compound can be varied greatly by choice of fillers, softeners, resins, accelerators, condition of cure, and numerous other factors.

Section 8

SPECTROSCOPY

X-RAY METHODS

An X-ray tube operating at a voltage V (in keV) emits a continuous X-ray spectrum, the minimum wavelength of which is given by $\lambda_{min} = 12.398/V$ with the wavelength expressed in angstroms. For expressing the wavelength in kX units, divide by the factor 1.00202. Tables 8-1 and 8-2 are based on the K and L wavelength values as published by Y. Cauchois and H. Hulubei (*Tables de Constantes et Données Numériques*, I. *Longueurs d'Onde des Émissions X et des Discontinuités d'Absorption X*, Hermann, Paris, 1947) and by the International Union of Crystallography (*International Tables for X-ray Crystallography*, Kynoch Press, Birmingham, England, 1962). Wavelength accuracy is only to about 1 in 25,000 except for the lines employed in X-ray diffraction work.

Use of energy-proportional detectors for X-rays creates a need for energy values of K and L absorption edges (Table 8-3) and emission series (Table 8-4). These values were obtained by a conversion to keV of tabulated experimental wavelength values and smoothed by a fit to Moseley's law. Although values are listed to 1 eV, chemical form may shift absorption edges and emission lines as much as 10 to 20 eV. S. Fine and C. F. Hendee [*Nucleonics*, **13**(3), 36 (1955)] also give values for $K\beta_2$, $L\gamma_1$, and $L\beta_2$ lines.

The relative intensities of X-ray emission lines from targets varies for different elements. However, one can assume a ratio of $K\alpha_1/K\alpha_2 = 2$ for the commonly used targets. The ratio of $K\alpha_2/K\beta_1$ from these targets varies from 6 to 3.5. The intensities of $K\beta_2$ radiations amount to about one per cent of that of the corresponding $K\alpha_1$ radiation. In practical applications these ratios have to be corrected for differential absorption in the window of the tube and air path, the ratio of scattering factors for and differential absorption in the crystal, and for sensitivity characteristics of the detector. Generalizing, the intensities of radiations from the K and L series are as follows:

Emission line	$K\alpha_1$	$K\alpha_2$	$K\beta_1$	$K\beta_2$	$L\alpha_1$	$L\alpha_2$	$L\beta_1$	$L\beta_2$	$L\gamma_1$
Relative intensity	500	250	80–150	5	100	10	30	60	40

For angles at which the $K\alpha_1$, $K\alpha_2$ doublet is not resolved, a mean wavelength $[K\overline{\alpha} = (2K\alpha_1 + K\alpha_2)/3]$ can be used.

Table 8-1
WAVELENGTHS OF X-RAY EMISSION SPECTRA
IN ANGSTROMS

Atomic No.	Element	$K\alpha_2$	$K\alpha_1$	$K\beta_1$	$L\alpha_1$	$L\beta_1$
3	Li		240			
4	Be		113			
5	B		67			
6	C		44			
7	N		31.60			
8	O		23.71			
9	F		18.31			
10	Ne		14.616	14.464		
11	Na		11.909	11.617	407.6	
12	Mg		9.889	9.558	251.0	
13	Al	8.3392	8.3367	7.981	169.8	
14	Si	7.1277	7.1253	6.7681	123	
15	P		6.1549	5.8038		
16	S	5.3747	5.3720	5.0317		
17	Cl	4.7305	4.7276	4.4031		
18	Ar	4.1946	4.1916	3.8848		
19	K	3.7446	3.7412	3.4538	42.7	
20	Ca	3.3616	3.3583	3.0896	36.32	35.95
21	Sc	3.0345	3.0311	2.7795	31.33	31.01
22	Ti	2.75207	2.7484	2.5138	27.39	27.02
23	V	2.5073	2.5035	2.2843	24.26	23.85
24	Cr	2.29351	2.28962	2.08480	21.67	21.28
25	Mn	2.1057	2.1018	1.9102	19.45	19.12
26	Fe	1.93991	1.93597	1.75653	17.567	17.255
27	Co	1.79278	1.78892	1.62075	15.968	15.667
28	Ni	1.66169	1.65784	1.50010	14.566	14.279
29	Cu	1.54433	1.54051	1.39217	13.330	13.053
30	Zn	1.4389	1.4351	1.2952	12.257	11.985
31	Ga	1.3439	1.3400	1.20784	11.290	11.023
32	Ge	1.2580	1.2540	1.1289	10.435	10.174
33	As	1.1798	1.1758	1.0573	9.671	9.414
34	Se	1.1088	1.1047	0.9921	8.990	8.736
35	Br	1.0438	1.0397	0.9327	8.375	8.125
36	Kr	0.9841	0.9801	0.8785	7.822	7.574
37	Rb	0.9296	0.9255	0.8286	7.3181	7.076
38	Sr	0.8794	0.8752	0.7829	6.8625	6.6237
39	Y	0.8330	0.8279	0.7407	6.4485	6.2117
40	Zr	0.7901	0.7859	0.7017	6.0702	5.8358
41	Nb	0.7504	0.7462	0.6657	5.7240	5.4921
42	Mo	0.713543	0.70926	0.632253	5.4063	5.1768
43	Tc	0.6793	0.6749	0.6014	5.1126	4.8782
44	Ru	0.6474	0.6430	0.5725	4.8455	4.6204
45	Rh	0.6176	0.6132	0.5456	4.5973	4.3739
46	Pd	0.5898	0.5854	0.5205	4.3676	4.1460
47	Ag	0.563775	0.559363	0.49701	4.1541	3.9344

Table 8-1 (*Continued*)
WAVELENGTHS OF X-RAY EMISSION SPECTRA
IN ANGSTROMS

Atomic No.	Element	$K\alpha_2$	$K\alpha_1$	$K\beta_1$	$L\alpha_1$	$L\beta_1$
48	Cd	0.5394	0.5350	0.4751	3.9563	3.7381
49	In	0.5165	0.5121	0.4545	3.7719	3.5552
50	Sn	0.4950	0.4906	0.4352	3.5999	3.3848
51	Sb	0.4748	0.4703	0.4171	3.4392	3.2256
52	Te	0.4558	0.4513	0.4000	3.2891	3.0767
53	I	0.4378	0.4333	0.3839	3.1485	2.9373
54	Xe	0.4204	0.4160	0.3685	3.016	2.807
55	Cs	0.4048	0.4003	0.3543	2.9016	2.8920
56	Ba	0.3896	0.3851	0.3408	2.7752	2.5674
57	La	0.3753	0.3707	0.3280	2.6651	2.4583
58	Ce	0.3617	0.3571	0.3158	2.5612	2.3558
59	Pr	0.3487	0.3441	0.3042	2.4627	2.2584
60	Nd	0.3565	0.3318	0.2933	2.3701	2.1666
61	Pm	0.3249	0.3207	0.2821	2.282	2.0796
62	Sm	0.3137	0.3190	0.2731	2.1994	1.9976
63	Eu	0.3133	0.2985	0.2636	2.1206	1.9202
64	Gd	0.2932	0.2884	0.2544	2.0460	1.8462
65	Tb	0.2834	0.2788	0.2460	1.9755	1.7763
66	Dy	0.2743	0.2696	0.2376	1.9088	1.7100
67	Ho	0.2655	0.2608	0.2302	1.8447	1.6468
68	Er	0.2572	0.2525	0.2226	1.7843	1.5873
69	Tm	0.2491	0.2444	0.2153	1.7263	1.5299
70	Yb	0.2415	0.2368	0.2088	1.6719	1.4756
71	Lu	0.2341	0.2293	0.2021	1.6194	1.4235
72	Hf	0.2270	0.2222	0.1955	1.5696	1.3740
73	Ta	0.2203	0.2155	0.1901	1.5219	1.3270
74	W	0.213813	0.208992	0.184363	1.4764	1.2818
75	Re	0.2076	0.2028	0.1789	1.4329	1.2385
76	Os	0.2016	0.1968	0.1736	1.3911	1.1972
77	Ir	0.1959	0.1910	0.1685	1.3513	1.1578
78	Pt	0.1904	0.1855	0.1637	1.3130	1.1198
79	Au	0.1851	0.1802	0.1590	1.2764	1.0836
80	Hg	0.1799	0.1750	0.1544	1.2411	1.0486
81	Tl	0.1750	0.1701	0.1501	1.2074	1.0152
82	Pb	0.1703	0.1654	0.1460	1.1750	0.9822
83	Bi	0.1657	0.1608	0.1419	1.1439	0.9520
84	Po	0.1608	0.1559	0.1382	1.1138	0.9222
85	At	0.1570	0.1521	0.1343	1.0850	0.8936
86	Rn	0.1529	0.1479	0.1307	1.0572	0.8659
87	Fr	0.1489	0.1440	0.1272	1.030	0.840
88	Ra	0.1450	0.1401	0.1237	1.0047	0.8137
89	Ac	0.1414	0.1364	0.1205	0.9799	0.7890
90	Th	0.1378	0.1328	0.1174	0.9560	0.7652
91	Pa	0.1344	0.1294	0.1143	0.9328	0.7422
92	U	0.1310	0.1259	0.1114	0.9105	0.7200

Table 8-1 (*Continued*)
WAVELENGTHS OF X-RAY EMISSION SPECTRA
IN ANGSTROMS

Atomic No.	Element	$K\alpha_2$	$K\alpha_1$	$K\beta_1$	$L\alpha_1$	$L\beta_1$
93	Np	0.1278	0.1226	0.1085	0.8893	0.6984
94	Pu	0.1246	0.1195	0.1058	0.8682	0.6777
95	Am	0.1215	0.1165	0.1031	0.8481	0.6576
96	Cm	0.1186	0.1135	0.1005	0.8287	0.6388
97	Bk	0.1157	0.1107	0.0980	0.8098	0.6203
98	Cf	0.1130	0.1079	0.0956	0.7917	0.6023
99	Es	0.1103	0.1052	0.0933	0.7740	0.5850
100	Fm	0.1077	0.1026	0.0910	0.7570	0.5682

Table 8-2
WAVELENGTHS OF ABSORPTION EDGES IN ANGSTROMS

Atomic No.	Element	K	L_I	L_{II}	L_{III}
3	Li	226.5			
4	Be	110.68			
5	B	66.289			
6	C	43.68			
7	N	30.99			
8	O	23.32			
9	F	17.913			
10	Ne	14.183			
11	Na	11.478		400	
12	Mg	9.512	197.4	247.92	
13	Al	7.951	142.5	170	
14	Si	6.745	105.1	126.48	
15	P	5.787	81.0	96.84	
16	S	5.018	64.23	76.05	
17	Cl	4.397	52.08	61.37	62.93
18	Ar	3.871	43.19	50.39	50.60
19	K	3.436	36.35	42.02	42.17
20	Ca	3.070	31.07	35.20	35.49
21	Sc	2.757	26.83	30.16	30.53
22	Ti	2.497	23.39	26.83	27.37
23	V	2.269	20.52	23.70	24.26
24	Cr	2.07012	16.7	17.9	20.7
25	Mn	1.896	16.27	18.90	19.40
26	Fe	1.74334	14.60	17.17	17.53
27	Co	1.60811	13.34	15.53	15.93
28	Ni	1.48802	12.27	14.13	14.58
29	Cu	1.38043	11.27	13.01	13.29
30	Zn	1.283	10.33	11.86	12.13
31	Ga	1.195	9.54	10.61	11.15
32	Ge	1.116	8.73	9.97	10.23

Table 8-2 (*Continued*)
WAVELENGTHS OF ABSORPTION EDGES IN ANGSTROMS

Atomic No.	Element	K	L_{I}	L_{II}	L_{III}
33	As	1.044	8.108	9.124	9.367
34	Se	0.9800	7.505	8.417	8.646
35	Br	0.9199	6.925	7.752	7.989
36	Kr	0.8655	6.456	7.165	7.395
37	Rb	0.8155	5.997	6.643	6.863
38	Sr	0.7697	5.582	6.172	6.387
39	Y	0.7276	5.233	5.756	5.962
40	Zr	0.6888	4.867	5.378	5.583
41	Nb	0.6529	4.581	5.025	5.223
42	Mo	0.61977	4.299	4.719	4.912
43	Tc	0.5888	4.064	4.427	4.629
44	Ru	0.5605	3.841	4.179	4.369
45	Rh	0.5338	3.626	3.942	4.130
46	Pd	0.5092	3.428	3.724	3.908
47	Ag	0.48582	3.254	3.514	3.698
48	Cd	0.4641	3.084	3.326	3.504
49	In	0.4439	2.926	3.147	3.324
50	Sn	0.4247	2.778	2.982	3.156
51	Sb	0.4066	2.639	2.830	3.000
52	Te	0.3897	2.510	2.687	2.855
53	I	0.3738	2.390	2.553	2.719
54	Xe	0.3585	2.274	2.429	2.592
55	Cs	0.3447	2.167	2.314	2.474
56	Ba	0.3314	2.068	2.204	2.363
57	La	0.3184	1.973	2.103	2.258
58	Ce	0.3065	1.891	2.009	2.164
59	Pr	0.2952	1.811	1.924	2.077
60	Nd	0.2845	1.735	1.843	1.995
61	Pm	0.2743	1.668	1.766	1.918
62	Sm	0.2646	1.598	1.702	1.845
63	Eu	0.2555	1.536	1.626	1.775
64	Gd	0.2468	1.477	1.561	1.709
65	Tb	0.2384	1.421	1.501	1.649
66	Dy	0.2305	1.365	1.438	1.579
67	Ho	0.2229	1.319	1.390	1.535
68	Er	0.2157	1.269	1.339	1.483
69	Tm	0.2089	1.222	1.288	1.433
70	Yb	0.2022	1.181	1.243	1.386
71	Lu	0.1958	1.140	1.198	1.341
72	Hf	0.1898	1.099	1.154	1.297
73	Ta	0.1839	1.061	1.113	1.255
74	W	0.17837	1.025	1.074	1.215
75	Re	0.1731	0.9901	1.036	1.177
76	Os	0.1678	0.9557	1.001	1.140
77	Ir	0.1629	0.9243	0.9670	1.106

Table 8-2 (*Continued*)
WAVELENGTHS OF ABSORPTION EDGES IN ANGSTROMS

Atomic No.	Element	K	L_I	L_{II}	L_{III}
78	Pt	0.1582	0.8914	0.9348	1.072
79	Au	0.1534	0.8638	0.9028	1.040
80	Hg	0.1492	0.8353	0.8779	1.009
81	Tl	0.1447	0.8079	0.8436	0.9793
82	Pb	0.1408	0.7815	0.8155	0.9503
83	Bi	0.1371	0.7565	0.7891	0.9234
84	Po	0.1332	0.7322	0.7638	0.8970
85	At	0.1295	0.7092	0.7387	0.8720
86	Rn	0.1260	0.6868	0.7153	0.8479
87	Fr	0.1225	0.6654	0.6929	0.8248
88	Ra	0.1192	0.6446	0.6711	0.8027
89	Ac	0.1161	0.6248	0.6500	0.7813
90	Th	0.1129	0.6061	0.6301	0.7606
91	Pa	0.1101	0.5875	0.6106	0.7411
92	U	0.1068	0.5697	0.5919	0.7233
93	Np	0.1045	0.5531	0.5742	0.7042
94	Pu	0.1018	0.5366	0.5571	0.6867
95	Am	0.0992	0.5208	0.5404	0.6700
96	Cm	0.0967	0.5060	0.5246	0.6532
97	Bk	0.0943	0.4913	0.5093	0.6375
98	Cf	0.0920	0.4771	0.4945	0.6223
99	Es	0.0897	0.4636	0.4801	0.6076
100	Fm	0.0875	0.4506	0.4665	0.5935

Table 8-3
CRITICAL X-RAY ABSORPTION ENERGIES IN keV

Atomic No.	Element	K	L_I	L_{II}	L_{III}
1	H	0.0136			
2	He	0.0246			
3	Li	0.0547			
4	Be	0.112			
5	B	0.187			
6	C	0.284			
7	N	0.400			
8	O	0.532			
9	F	0.692			
10	Ne	0.874	0.048	0.022	
11	Na	1.08	0.055	0.034	
12	Mg	1.30	0.0628	0.0502	
13	Al	1.559	0.0870	0.0720	
14	Si	1.838	0.118	0.0977	
15	P	2.142	0.153	0.128	

Table 8-3 (*Continued*)
CRITICAL X-RAY ABSORPTION ENERGIES IN keV

Atomic No.	Element	K	L_I	L_{II}	L_{III}
16	S	2.469	0.193	0.163	0.162
17	Cl	2.822	0.238	0.202	0.201
18	Ar	3.200	0.287	0.246	0.244
19	K	3.606	0.341	0.295	0.292
20	Ca	4.038	0.399	0.350	0.346
21	Sc	4.496	0.462	0.411	0.407
22	Ti	4.966	0.530	0.462	0.456
23	V	5.467	0.604	0.523	0.515
24	Cr	5.988	0.679	0.584	0.574
25	Mn	6.542	0.762	0.656	0.644
26	Fe	7.113	0.849	0.722	0.709
27	Co	7.713	0.929	0.798	0.783
28	Ni	8.337	1.02	0.877	0.858
29	Cu	8.982	1.10	0.954	0.935
30	Zn	9.662	1.20	1.05	1.02
31	Ga	10.39	1.30	1.17	1.14
32	Ge	11.10	1.42	1.24	1.21
33	As	11.87	1.529	1.358	1.32
34	Se	12.65	1.66	1.472	1.431
35	Br	13.48	1.791	1.599	1.552
36	Kr	14.32	1.92	1.729	1.674
37	Rb	15.197	2.064	1.863	1.803
38	Sr	16.101	2.212	2.004	1.937
39	Y	17.053	2.387	2.171	2.096
40	Zr	17.998	2.533	2.308	2.224
41	Nb	18.986	2.700	2.467	2.372
42	Mo	20.003	2.869	2.630	2.525
43	Tc	21.050	3.045	2.796	2.680
44	Ru	22.117	3.227	2.968	2.839
45	Rh	23.210	3.404	3.139	2.995
46	Pd	24.356	3.614	3.338	3.181
47	Ag	25.535	3.828	3.547	3.375
48	Cd	26.712	4.019	3.731	3.541
49	In	27.929	4.226	3.929	3.732
50	Sn	29.182	4.445	4.139	3.911
51	Sb	30.497	4.708	4.391	4.137
52	Te	31.817	4.953	4.621	4.347
53	I	33.164	5.187	4.855	4.559
54	Xe	34.551	5.448	5.103	4.783
55	Cs	35.974	5.706	5.360	5.014
56	Ba	37.432	5.995	5.629	5.250
57	La	38.923	6.264	5.902	5.490
58	Ce	40.43	6.556	6.169	5.728
59	Pr	41.99	6.837	6.446	5.968
60	Nd	43.57	7.134	6.728	6.215

Table 8-3 (*Continued*)
CRITICAL X-RAY ABSORPTION ENERGIES IN keV

Atomic No.	Element	K	L_I	L_{II}	L_{III}
61	Pm	45.19	7.431	7.022	6.462
62	Sm	46.85	7.742	7.316	6.720
63	Eu	48.51	8.059	7.624	6.984
64	Gd	50.23	8.383	7.942	7.251
65	Tb	52.00	8.713	8.258	7.520
66	Dy	53.77	9.053	8.587	7.795
67	Ho	55.61	9.395	8.918	8.074
68	Er	57.47	9.754	9.270	8.362
69	Tm	59.38	10.12	9.622	8.656
70	Yb	61.31	10.49	9.985	8.949
71	Lu	63.32	10.87	10.35	9.248
72	Hf	65.37	11.28	10.75	9.567
73	Ta	67.46	11.68	11.14	9.883
74	W	69.51	12.09	11.54	10.20
75	Re	71.67	12.52	11.96	10.53
76	Os	73.87	12.97	12.38	10.86
77	Ir	76.11	13.41	12.82	11.21
78	Pt	78.35	13.865	13.26	11.55
79	Au	80.67	14.351	13.731	11.92
80	Hg	83.08	14.838	14.205	12.278
81	Tl	85.52	15.344	14.695	12.65
82	Pb	87.95	15.861	15.200	13.03
83	Bi	90.54	16.386	15.709	13.42
84	Po	93.16	16.925	16.233	13.81
85	At	95.73	17.481	16.777	14.21
86	Rn	98.45	18.054	17.331	14.61
87	Fa	101.1	18.628	17.893	15.02
88	Ra	103.9	19.228	18.473	15.44
89	Ac	107.7	19.829	19.071	15.86
90	Th	109.8	20.452	19.673	16.278
91	Pa	112.4	21.096	20.295	16.720
92	U	115.0	21.757	20.944	17.163
93	Np	118.2	22.411	21.585	17.606
94	Pu	121.2	23.117	22.250	18.062
95	Am	124.3	23.795	22.935	18.524
96	Cm	127.2	24.502	23.629	18.992
97	Bk	131.3	25.231	24.344	19.466
98	Cf	133.6	26.010	25.070	19.954
99	Es	138.1	26.729	25.824	20.422
100	Fm	141.5	27.503	26.584	20.912

Table 8-4
X-RAY EMISSION ENERGIES IN keV

Atomic No.	Element	$K\beta_1$	$K\alpha_1$	$L\beta_1$	$L\alpha_1$
3	Li		0.052		
4	Be		0.110		
5	B		0.185		
6	C		0.282		
7	N		0.392		
8	O		0.523		
9	F		0.677		
10	Ne		0.851		
11	Na	1.067	1.041		
12	Mg	1.297	1.254		
13	Al	1.553	1.487		
14	Si	1.832	1.740		
15	P	2.136	2.015		
16	S	2.464	2.308		
17	Cl	2.815	2.622		
18	Ar	3.192	2.957		
19	K	3.589	3.313		
20	Ca	4.012	3.691	0.344	0.341
21	Sc	4.460	4.090	0.399	0.395
22	Ti	4.931	4.510	0.458	0.452
23	V	5.427	4.952	0.519	0.512
24	Cr	5.946	5.414	0.581	0.571
25	Mn	6.490	5.898	0.647	0.636
26	Fe	7.057	6.403	0.717	0.704
27	Co	7.649	6.930	0.790	0.775
28	Ni	8.264	7.477	0.866	0.849
29	Cu	8.904	8.047	0.948	0.928
30	Zn	9.571	8.638	1.032	1.009
31	Ga	10.263	9.251	1.122	1.096
32	Ge	10.981	9.885	1.216	1.186
33	As	11.725	10.543	1.317	1.282
34	Se	12.495	11.221	1.419	1.379
35	Br	13.290	11.923	1.526	1.480
36	Kr	14.112	12.649	1.638	1.587
37	Rb	14.960	13.394	1.752	1.694
38	Sr	15.834	14.164	1.872	1.806
39	Y	16.736	14.957	1.996	1.922
40	Zr	17.666	15.774	2.124	2.042
41	Nb	18.621	16.614	2.257	2.166
42	Mo	19.607	17.478	2.395	2.293
43	Tc	20.612	18.370	2.538	2.424
44	Ru	21.655	19.278	2.683	2.558
45	Rh	22.721	20.214	2.834	2.696
46	Pd	23.816	21.175	2.990	2.838
47	Ag	24.942	22.162	3.151	2.984

Table 8-4 (*Continued*)
X-RAY EMISSION ENERGIES IN keV

Atomic No.	Element	$K\beta_1$	$K\alpha_1$	$L\beta_1$	$L\alpha_1$
48	Cd	26.093	23.172	3.316	3.133
49	In	27.274	24.207	3.487	3.287
50	Sn	28.483	25.270	3.662	3.444
51	Sb	29.723	26.357	3.843	3.605
52	Te	30.993	27.471	4.029	3.769
53	I	32.292	28.610	4.220	3.937
54	Xe	33.644	29.779	4.422	4.111
55	Cs	34.984	30.970	4.620	4.286
56	Ba	36.376	32.191	4.828	4.467
57	La	37.799	33.440	5.043	4.651
58	Ce	39.255	34.717	5.262	4.840
59	Pr	40.746	36.023	5.489	5.034
60	Nd	42.269	37.359	5.722	5.230
61	Pm	43.811	38.726	5.956	5.431
62	Sm	45.400	40.124	6.206	5.636
63	Eu	47.027	41.529	6.456	5.846
64	Gd	48.718	42.983	6.714	6.059
65	Tb	50.391	44.470	6.979	6.275
66	Dy	52.178	45.985	7.249	6.495
67	Ho	53.934	47.528	7.528	6.720
68	Er	55.690	49.099	7.810	6.948
69	Tm	57.487	50.730	8.103	7.181
70	Yb	59.352	52.360	8.401	7.414
71	Lu	61.282	54.063	8.708	7.654
72	Hf	63.209	55.757	9.021	7.898
73	Ta	65.210	57.524	9.341	8.145
74	W	67.233	59.310	9.670	8.396
75	Re	69.298	61.131	10.008	8.651
76	Os	71.404	62.991	10.354	8.910
77	Ir	73.549	64.886	10.706	9.173
78	Pt	75.736	66.820	11.069	9.441
79	Au	77.968	68.794	11.439	9.711
80	Hg	80.258	70.821	11.823	9.987
81	Tl	82.558	72.860	12.210	10.266
82	Pb	84.922	74.957	12.611	10.549
83	Bi	87.335	77.097	13.021	10.836
84	Po	89.809	79.296	13.441	11.128
85	At	92.319	81.525	13.873	11.424
86	Rn	94.877	83.800	14.316	11.724
87	Fr	97.483	86.119	14.770	12.029
88	Ra	100.136	88.485	15.233	12.338
89	Ac	102.846	90.894	15.712	12.650
90	Th	105.592	93.334	16.200	12.966
91	Pa	108.408	95.851	16.700	13.291
92	U	111.289	98.428	17.218	13.613

Table 8-4 (*Continued*)
X-RAY EMISSION ENERGIES IN keV

Atomic No.	Element	$K\beta_1$	$K\alpha_1$	$L\beta_1$	$L\alpha_1$
93	Np	114.181	101.005	17.740	13.945
94	Pu	117.146	103.653	18.278	14.279
95	Am	120.163	106.351	18.829	14.618
96	Cm	123.235	109.098	19.393	14.961
97	Bk	126.362	111.896	19.971	15.309
98	Cf	129.544	114.745	20.562	15.661
99	Es	132.781	117.646	21.166	16.018
100	Fm	136.075	120.598	21.785	16.379

Filters. The K spectra of the light metals, often used as target material in the production of X-rays for diffraction studies, contain three strong lines, α_1, α_2 and β_1, of which the α lines form a doublet with a narrow wavelength separation. The $K\beta$ radiation can be eliminated by using a thin foil filter, usually of the element of next lower atomic number to that of the target element; the $K\alpha$ lines are transmitted with a relatively small loss of intensity. Table 8-5, restricted to the K wavelengths of target elements in common use, lists the calculated thicknesses of β filters required to reduce the $K\beta_1/K\alpha_1$ integrated intensity ratio to $1/100$.

Table 8-5
β FILTERS FOR COMMON TARGET ELEMENTS

Target Element	$K\bar{\alpha}$, Å	Excitation Voltage, keV	$K\beta_1/K\alpha_1 = 1/100$			% Loss $K\alpha_1$
			Absorber	Thickness, mm	g/cm^2	
Ag	0.560834	25.52	Pd	0.062	0.074	60
Mo	0.71069	20.00	Zr	0.081	0.053	57
Cu	1.54178	8.981	Ni	0.015	0.013	45
Ni	1.65912	8.331	Co	0.013	0.011	42
Co	1.79021	7.709	Fe	0.012	0.009	39
Fe	1.93728	7.111	Mn	0.011	0.008	38
			MnO$_2$	0.026	0.013	45
Cr	2.29092	5.989	V	0.011	0.007	37
			V$_2$O$_5$	0.036	0.012	48
	$L\alpha_1$		$L\beta_1/L\alpha_1 = 1/100$			% Loss $L\alpha_1$
W	1.4763	10.200	Cu	0.035		77

Interplanar Spacings. Diffractometer alignment procedures require the use of a well-prepared polycrystalline specimen. Two standard samples found to be suitable are silicon and α-quartz (including Novaculite). The 2θ values of several of the most intense reflections for these materials are listed in Table 8-6 (*Tables of Interplanar Spacings d vs Diffraction Angle 2θ for Selected Targets*, Picker Nuclear, White Plains, N.Y., 1966). To convert to d for $K\bar{\alpha}$ or to d for $K\alpha_2$, multiply the tabulated d value (Table 8-6) for $K\alpha_1$ by the factor given below:

Element	$K\bar{\alpha}$	$K\alpha_2$
W	1.00769	1.02307
Ag	1.00263	1.00789
Mo	1.00202	1.00604
Cu	1.00082	1.00248
Ni	1.00077	1.00232
Co	1.00072	1.00216
Fe	1.00067	1.00204
Cr	1.00057	1.00170

Table 8-6
INTERPLANAR SPACINGS FOR $K\alpha_1$ RADIATION, d vs 2θ

α-quartz (Including Novaculite)

hkl d(Å)	100 4.260	101 3.343	110 2.458	102 2.282	200 2.128	112 1.817	202 1.672	211 1.541	203 1.375	301 1.372
W $K\alpha_1$: 2θ	2.81	3.58	4.87	5.25	5.63	6.59	7.17	7.78	8.72	8.74
Ag $K\alpha_1$: 2θ	7.53	9.60	13.07	14.08	15.10	17.71	19.26	20.91	23.47	23.52
Mo $K\alpha_1$: 2θ	9.55	12.18	16.59	17.88	19.19	22.51	24.49	26.61	29.89	29.96
Cu $K\alpha_1$: 2θ	20.83	26.64	36.52	39.45	42.44	50.16	54.86	59.98	68.14	68.31
Ni $K\alpha_1$: 2θ	22.44	28.71	39.42	42.60	45.85	54.28	59.44	65.08	74.15	74.34
Co $K\alpha_1$: 2θ	24.24	31.04	42.68	46.15	49.71	58.98	64.68	70.96	81.16	81.38
Fe $K\alpha_1$: 2θ	26.27	33.66	46.38	50.20	54.11	64.38	70.75	77.83	89.50	89.74
Cr $K\alpha_1$: 2θ	31.18	40.05	55.52	60.22	65.09	78.11	86.42	95.96	112.73	113.11

Silicon

hkl d(Å)	111 3.1353	220 1.91997	311 1.63736	400 1.357630	331 1.24584	422 1.1085	511,333 1.0451	440 0.959986	531 0.917922	620 0.858637
W $K\alpha_1$: 2θ	3.82	6.24	7.32	8.83	9.62	10.82	11.48	12.50	13.07	13.98
Ag $K\alpha_1$: 2θ	10.24	16.75	19.67	23.78	25.95	29.23	31.04	33.88	35.48	38.02
Mo $K\alpha_1$: 2θ	12.99	21.29	25.02	30.28	33.08	37.32	39.67	43.36	45.45	48.79
Cu $K\alpha_1$: 2θ	28.44	47.30	56.12	69.13	76.38	88.03	94.96	106.71	114.10	127.55
Ni $K\alpha_1$: 2θ	30.66	51.16	60.83	75.26	83.42	96.80	104.96	119.42	129.12	149.76
Co $K\alpha_1$: 2θ	33.15	55.53	66.22	82.42	91.77	107.59	117.71	137.42	154.04	
Fe $K\alpha_1$: 2θ	35.97	60.55	72.48	90.96	101.97	121.67	135.70			
Cr $K\alpha_1$: 2θ	42.83	73.21	88.72	114.97	133.53					

Analyzing Crystals. The range of wavelengths usable with various analyzing crystals are governed by the d spacings of the crystal planes and by the geometric limits to which the goniometer can be rotated. The d value should be small enough to make the angle 2θ greater than approximately 10 or 15 deg, even at the shortest wavelength used; otherwise excessively long analyzing crystals would be needed to prevent the direct fluorescent beam from entering the detector. A small d value is also favorable for producing a large dispersion of the spectrum to give good separation of adjacent lines. On the other hand, a small d value imposes an upper limit to the range of wavelengths that can be analyzed. Actually the goniometer is limited mechanically to about 150 deg for a 2θ value. A final requirement is the reflection efficiency and minimization of higher-order reflections. Table 8-7 gives a list of crystals commonly used for X-ray spectroscopy.

The long-wavelength analyzers are prepared by dipping an optical flat into the film of the metal fatty acid about 50 times to produce a layer 180 molecules in thickness.

Lithium fluoride is the optimum crystal for all wavelengths less than 3 Å. Pentaerythritol (PET) and potassium hydrogen phthalate (KAP) are usually the crystals of choice for wavelengths from 3 to 20 Å. Two crystals suppress even-ordered reflections: silicon (111) and calcium fluoride (111).

Table 8-7
ANALYZING CRYSTALS FOR X-RAY SPECTROSCOPY

Crystal	Reflecting Plane	$2d$ Spacing, Å	Reflectivity
Quartz	$50\overline{5}2$	1.624	Low
Aluminum	111	2.338	High
Topaz	303	2.712	Medium
Quartz	$20\overline{2}3$	2.750	Low
Lithium fluoride	220	2.848	High
Silicon	111	3.135	High
Quartz	112	3.636	Medium
Lithium fluoride	200	4.028	High
Sodium chloride	200	5.639	High
Calcium fluoride	111	6.32	High
Quartz	$10\overline{1}1$	6.686	High
Quartz	$10\overline{1}0$	8.50	Medium
Pentaerythritol (PET)	002	8.742	High
Ethylenediamine tartrate (EDT)	020	8.808	Medium
Ammonium dihydrogen phosphate (ADP)	110	10.648	Low
Gypsum	020	15.185	Medium
Mica	002	19.92	Low
Potassium hydrogen phthalate (KAP)	$10\overline{1}1$	26.4	Medium
Lead palmitate		45.6	
Strontium behenate		61.3	
Lead stearate		100.4	Medium

Mass Absorption Coefficients. Radiation traversing a layer of substance is diminished in intensity by a constant fraction per centimeter thickness x of material. The emergent radiant power P, in terms of incident radiant power P_0, is given by

$$P = P_0 \exp(-\mu x)$$

which defines the total linear absorption coefficient μ. Since the reduction of intensity is determined by the quantity of matter traversed by the primary beam, the absorber thickness is best expressed on a mass basis, in g/cm^2. The mass absorption coefficient μ/ρ, expressed in units cm^2/g, where ρ is the density of the material, is approximately independent of the physical state of the material and, to a good approximation, is additive with respect to the elements composing a substance.

Table 8-8 contains values of μ/ρ for the common target elements employed in X-ray work. A more extensive set of mass absorption coefficients for K, L, and M emission lines within the wavelength range from 0.7 to 12 Å is contained in Heinrich's paper in T. D. McKinley, K. F. J. Heinrich, and D. B. Wittry (eds.), "The Electron Microprobe," pp. 351–377, Wiley, New York, 1966. This article should be consulted to ascertain the probable accuracy of the values and for a compilation of coefficients and exponents employed in the computations.

Table 8-8
MASS ABSORPTION COEFFICIENTS FOR $K\alpha_1$ LINES AND W $L\alpha_1$ LINE

Emitter Wavelength, Å	Ag $K\alpha_1$ 0.559	Mo $K\alpha_1$ 0.709	Cu $K\alpha_1$ 1.541	Ni $K\alpha_1$ 1.658	Co $K\alpha_1$ 1.789	Fe $K\alpha_1$ 1.936	Cr $K\alpha_1$ 2.290	W $L\alpha_1$ 1.476
Absorber								
1 H	0.37	0.38	0.43	0.4	0.4	0.5	0.5	0.4
2 He	0.16	0.18	0.37	0.4	0.4	0.5	0.7	0.3
3 Li	0.18	0.22	0.50	0.6	0.7	0.9	1.5	0.4
4 Be	0.22	0.30	1.2	1.5	1.9	2.3	3.7	1.1
5 B	0.30	0.45	2.5	3.1	3.9	4.9	7.9	2.2
6 C	0.42	0.50	4.6	5.7	7.1	8.8	14.2	4.1
7 N	0.60	0.83	7.5	9.3	11.5	14.4	23.1	6.7
8 O	0.80	1.45	12.9	15.8	19.5	24.5	39.4	11.4
9 F	1.00	1.9	16.5	20.3	25.2	31.4	50.3	14.6
10 Ne	1.41	2.6	22.8	27.9	34.6	43.1	69.0	20.1
11 Na	1.75	3.5	30.3	37.2	45.9	57.2	91.4	26.8
12 Mg	2.27	4.6	39.5	48.4	59.8	74.6	119.1	34.9
13 Al	2.74	5.8	49.6	60.7	75.0	93.4	149.0	43.9
14 Si	3.44	7.3	61.4	75.2	92.8	115.5	183.8	54.4
15 P	4.20	8.8	74.7	91.4	112.9	140.5	223.6	66.2
16 S	5.15	10.6	89.2	109.2	134.7	167.4	266.1	79.1
17 Cl	5.86	12.4	104.8	128.2	158.1	196.6	312.4	92.8

Table 8-8 (*Continued*)
MASS ABSORPTION COEFFICIENTS FOR $K\alpha_1$ LINES
AND W $L\alpha_1$ LINE

Emitter Wavelength, Å Absorber	Ag $K\alpha_1$ 0.559	Mo $K\alpha_1$ 0.709	Cu $K\alpha_1$ 1.541	Ni $K\alpha_1$ 1.658	Co $K\alpha_1$ 1.789	Fe $K\alpha_1$ 1.936	Cr $K\alpha_1$ 2.290	W $L\alpha_1$ 1.476
18 Ar	6.40	14.5	121.4	148.5	183.0	227.3	360.7	107.6
19 K	8.0	16.7	139.8	171	211	262	415	124
20 Ca	9.7	18.9	158.6	194	239	296	469	141
21 Sc	10.5	21.8	180.5	221	272	337	534	160
22 Ti	11.8	25.3	203	247	304	378	597	180
23 V	13.3	27.7	228	278	342	424	77	202
24 Cr	15.7	31.0	254	311	382	474	88	226
25 Mn	17.4	34.5	282	344	423	63.5	101	250
26 Fe	19.9	38.1	311	380	57.6	71.4	113	276
27 Co	21.8	42.1	341	52.8	64.9	80.6	127	303
28 Ni	25.0	46.4	48.3	58.9	72.5	90.0	142	333
29 Cu	26.4	50.7	53.7	65.5	80.6	100.0	158	47.6
30 Zn	28.2	55.4	59.5	72.7	89.4	110.9	175	52.8
31 Ga	30.8	60.1	65.9	80.5	99.0	122.8	194	58.5
32 Ge	33.5	65.2	72.3	88.2	108.6	134.7	213	64.1
33 As	36.5	70.5	79.1	96.6	118.9	147	233	70.2
34 Se	38.5	76.0	86.1	105.1	129.4	161	254	76.4
35 Br	42.3	82.5	93.9	114.7	141.2	175	277	83.4
36 Kr	45.0	88.3	101.9	124.5	153.2	190	300	90.5
37 Rb	48	95	84	103	127	158	252	98
38 Sr	52	102	90	110	137	170	271	106
39 Y	56	109	97	119	147	183	292	114
40 Zr	61	17	104	128	158	197	314	122
41 Nb	66	18	112	138	170	212	338	132
42 Mo	71	19	119	146	180	225	358	140
43 Tc	76	20	128	157	194	241	384	150
44 Ru	12	22	137	168	207	258	410	160
45 Rh	13	23	146	179	221	275	438	171
46 Pd	14	24	155	190	235	292	466	182
47 Ag	15	26	165	202	249	310	493	193
48 Cd	15	28	174	213	263	327	520	204
49 In	16	30	185	227	280	347	553	217
50 Sn	17	32	195	239	295	367	583	229
51 Sb	19	34	206	252	310	386	612	241
52 Te	19	36	216	265	326	405	644	253
53 I	21	37	230	281	346	431	684	269
54 Xe	22	39	239	293	361	448	710	280
55 Cs	24	42	332	404	495	612	822	295
56 Ba	25	44	349	425	522	645	622	311
57 La	26	46	365	444	545	673	647	325
58 Ce	28	48	383	466	571	603	216	341
59 Pr	29	51	401	487	597	453	229	356
60 Nd	31	54	420	510	534	473	241	373

Table 8-8 (*Continued*)
MASS ABSORPTION COEFFICIENTS FOR $K\alpha_1$ LINES AND W $L\alpha_1$ LINE

Emitter Wavelength, Å	Ag $K\alpha_1$ 0.559	Mo $K\alpha_1$ 0.709	Cu $K\alpha_1$ 1.541	Ni $K\alpha_1$ 1.658	Co $K\alpha_1$ 1.789	Fe $K\alpha_1$ 1.936	Cr $K\alpha_1$ 2.290	W $L\alpha_1$ 1.476
Absorber								
61 Pm	32	56	440	535		164	254	392
62 Sm	33	59	L_I 456	473	417	173	268	406
63 Eu	35	61	405	354	148	182	282	423
64 Gd	36	64	L_{II} 424	370	156	191	296	
65 Tb	38	67	316	135	164	201	311	393 L_I
66 Dy	39	70	L_{III} 329	141	172	211	327	293 L_{II}
67 Ho	41	72	123	148	181	222	343	304
68 Er	43	75	129	156	189	233	360	316 L_{III}
69 Tm	45	79	135	163	199	244	377	120
70 Yb	46	82	141	171	208	256	395	126
71 Lu	48	84	148	179	218	267	414	132
72 Hf	51	88	155	187	228	280	433	138
73 Ta	52	91	162	196	238	293	453	144
74 W	55	95	169	204	249	306	473	151
75 Re	57	98	176	213	260	319	494	157
76 Os	59	102	184	223	271	333	515	164
77 Ir	61	106	192	232	283	347	538	171
78 Pt	64	109	200	242	295	362	560	179
79 Au	67	113	209	252	307	377	584	186
80 Hg	69	117	218	263	321	394	609	194
81 Tl	72	121	227	275	334	411	635	203
82 Pb	74	125	236	286	348	428	662	211
83 Bi	78	129	247	298	363	446	690	220
84 Po		131	258	311	380	466	721	230
85 At			269	325	397	487	753	240
86 Rn	85		281	340	414	509	787	251
87 Fr		89	294	356	433	532	823	262
88 Ra	91		307	372	453	556	861	274
89 Ac			322	389	474	582	900	287
90 Th	97		337	408	497	610	944	301
91 Pa			353	427	520	639	988	315
92 U	104		372	450	548	673	898	332
93 Np			392	474	578	709	945	350
94 Pu		54	418	505	615	755	835	373

ELECTRONIC EMISSION AND ABSORPTION SPECTROSCOPY

The tables of emission and absorption lines are presented in two parts. In Table 8-9 the data are arranged by element in alphabetical order of chemical symbol, whereas in Table 8-10 the sensitive lines of the elements are arranged in order of decreasing wavelengths.

The wavelengths in column 2 of Table 8-9 are all normal air wavelengths and are given to the nearest 0.01 Å, except for band systems. A Roman numeral II following the wavelength signifies a singly ionized atom, the letter d an unresolved double line, and t a triplet line.

The relative intensity numbers in column 3 of Table 8-9 are taken from W. F. Meggers, C. H. Corliss, and B. F. Scribner, "Tables of Spectral-line Intensities, Part I," National Bureau of Standards Monograph 32, U.S. Government Printing Office, Washington, D.C., 1961. All emission lines are assigned relative intensities proportional to their limiting detectabilities. In a fully exposed spectrogram of copper containing 0.1 atomic per cent of another element any faint but unmistakable line at a given wavelength is assigned unit intensity. For example, a line of intensity 10 should show plainly at 0.01 atomic per cent, while one of intensity 1000 should be easily seen at 0.0001 atomic per cent (one in a million). Actual values of detection limits reported in the literature for a dc arc and a spark-porous cup are included in columns 4 and 5, respectively.

The flame emission detection limit is quite dependent on instrument and operating variables, particularly the detector, the fuel and oxidant gases, and the slit width. Many of the data are for a Beckman model DU spectrophotometer or a 0.5-meter Jarrell-Ash monochromator, equipped with a 1P28 multiplier phototube and a sprayer-burner combination. For example, a value of 0.6 μg/ml per 0.1 mV for silver implies that the balancing potentiometer moved through one division out of 100 total divisions (0.1 mV for a 10-mV recorder) when a solution containing 0.6 μg/ml of silver was sprayed into the flame.

The atomic absorption sensitivity values are also dependent on instrument and operating variables, particularly lamp current, slit width, burner type, path length, and type of flame. Sensitivity is defined as the concentration of test element required to cause an absorption of 1 per cent (0.004 absorbance unit). A table for conversion of per cent absorption values into absorbance units is in Section 2. The concentration range in which the stated sensitivity is valid is given in parentheses for many wavelengths. I am indebted to Mr. T. C. Rains, Analytical Chemistry Division, National Bureau of Standards, Washington, D.C., for many of the atomic absorption data.

Abbreviations in the table

AA, air-acetylene flame
AH, air-hydrogen flame
AP, air-propane flame
NA, nitrous oxide-acetylene flame
(N-A)A, (nitrous oxide-air)-acetylene flame
OA, oxygen-acetylene flame
OH, oxygen-hydrogen flame
n, organic solvent aspirated directly into flame
r, fuel-rich flame
w, lean flame
z, emission from inner-conal gases

Table 8-9
EMISSION AND ABSORPTION LINES 1900 to 9000 Å

Element	Wavelength, Å	Relative Intensity	Detection limit: DC Arc, μg/g	Detection limit: Spark-porous Cup, μg/ml	Detection limit: Flame Emission, μg/ml/0.1 mV	Atomic Absorption: Sensitivity, μg/ml/1% Abs (conc'n range for which valid)
Ag	3280.68	5500	0.1	0.02	1.0 OH	0.13 AA (1–10)
	3382.89	2800			0.6 OH	0.22 AA (1–50)
Al	2269.09	2				9 NAr (10–1000)
	2367.06	18				8 NAr (10–1000)
	2373.13 *d*	36				5 NAr (100–1000)
	2567.99	24				16 NAr (100–1000)
	2575.10 *d*	48				11 NAr (100–1000)
	3082.16	320				4 NAr (10–500)
	3092.71 *d*	650	1.0			2.2 NAr (5–100)
	3944.03	450			0.3 OAn	5 NAr (10–200)
	3961.53	900	1.0	0.3	0.5 OAn	3 NAr (10–200)
AlO	4842				0.5 OAn	
As	1890					1.7 AA; 2.8 AH
	1936.96	17				1.3 AAw (3–50) 1.2 AH (3–100)
	1971.97	28				1.8 AAw (3–100) 2.3 AH (3–100)
	2288.12	44		3	4.0 OAnz	
	2349.84	85	30	3	2.2 OAnz	
	2780.22	140	50	10	13.0 OAnz	
Au	2427.95	200	2	30		0.3 AA and NA
	2675.95	340	2	20	5 OA	0.6 AA
B	2088.93	7				50 NAr (50–1000)
	2089.59	11				45 NAr (50–1000)
	2496.78	240	2	2	7 OAnrz	63 NAr (200–1000)
	2497.73	480	2	1		35 NAr (200–1000)
BO_2	4530				30 OAn (50%)	
	4715				10 OAn (50%)	
	4920				5 OAn (50%)	
	5180				3 OAn (50%)	
	5476				3 OAn (50%)	
	5790				6 OAn (50%)	

Table 8-9 (*Continued*)
EMISSION AND ABSORPTION LINES 1900 to 9000 Å

Element	Wavelength, Å	Relative Intensity	DC Arc, μg/g	Spark-porous Cup, μg/ml	Flame Emission, μg/ml/0.1 mV	Atomic Absorption: Sensitivity, μg/ml/1% Abs (conc'n range for which valid)
Ba	2304.24 II	28		0.5		
	2335.27 II	55		0.5		
	3071.58	18	100			45 (N–A) Ar (100–1000)
	3501.11	50				92 (N–A) Ar (200–1000)
	4130.66 II	150		4		
	4554.03 II	6500	0.1	0.1	0.06 OA	
	4934.09 II	2000			0.08 OA	
	5535.48	650	0.1		0.03 OA	3 (N–A) Ar (10–500)
Be	2348.61	300	0.1	0.02	1.0 OAnr	0.3 NAr
	3130.42 II	480	0.1	0.02		
	3131.07 II	320		0.003		
	3321.34 t	100	1			
Bi	2061.70	55				5 AA (5–500)
	2110.26	10				17 AA (50–500)
	2228.25	3			11.5 OAn	1.6 AA (5–100)
	2230.61	14				0.7 AA (5–100)
	2276.58	5			6.4 OAn	9 AA (10–1000)
	2897.98	400	10			
	3067.72	3600	1	1		2 AA (5–500)
Br(InBr)	3758				1.6 AH	
C	2478.57	10				
Ca	2398.56	4			0.001 NA	
	3933.67 II	4200	0.1	0.01	0.005 OAz	
	3968.47 II	2200			0.01 OAz	
	4226.73	1100	0.1		0.07 OA	0.08 AA (1–50)
	4454.78	140	1.0			
CaOH	5540				0.25 OH	
	6220				1.6 OH	
Cd	2265.02 II	110		0.2		
	2288.02	1500	10	0.2	10 OH; 4 OAn	0.07 AA (1–30)
	3261.06	32	10		5 OH; 2 OAn	38 AA (50–1000)
Ce	3801.53 II	200	40			
	3942.75 II	190		25		
	3999.24 II	200	60			
	4012.38 II	190		10		
	4040.76 II	150	10			
	4186.60 II	250		3		
	5200.12 d	6				30 NA
	5223.49	28				30 NA
	5697.00	32				39 NA
	5699.23	40			16 NAn	
Cl (CuCl)	4354				10 OH	
(InCl)	3599				0.7 AH	
CN	3883				20 OHn	
	3851				60 OHn	
Co	2174.60	2				3.5 AA (5–100)
	2286.16 II	26		0.5		

Table 8-9 (*Continued*)
EMISSION AND ABSORPTION LINES 1900 to 9000 Å

Element	Wavelength, Å	Relative Intensity	Detection limit			Atomic Absorption: Sensitivity, μg/ml/1% Abs (conc'n range for which valid)
			DC Arc, μg/g	Spark-porous Cup, μg/ml	Flame Emission, μg/ml/0.1 mV	
	2309.02	24				7 AA (10–500)
	2363.79 II	30		0.5		
	2407.25	140				0.02 AA (1–50)
	2424.93	130			1.7 OAr	0.2 AA (1–50)
	2435.83	25				1.4 AA (5–100)
	2521.36	180	3			0.8 AA (3–100)
	2987.16	36				12 AA (50–1000)
	2989.59	36				12 AA (50–1000)
	3044.00	160				2.3 AA (5–500)
	3405.12	700			6.2 OA	
	3412.63	140			5.2 OA	0.7 AA (3–100)
	3431.58	160			10 OA	
	3443.64	550			14 OA	
	3453.50	1300	1	2.0	3.4 OA; 0.8 OAn	
	3465.80	320			6.4 OA	6 AA (10–1000)
	3474.02	500			8 OA	10 AA (50–1000)
	3506.32	440			6 OA	
	3512.48	240			10 OA	
	3526.85	400			4 OA	4 AA (10–1000)
	3873.12	240			6 OA	
	4121.32	190			15 OA	
Cr	2364.71	3				40 AAr (50–500)
	2677.16 II	200		0.1		
	2835.63 II	280	10	0.3		
	2843.25 II	190		0.3		
	3578.69	2400			0.2 OAn	0.22 AAr (1–50)
	3593.49	2100			0.25 OAn	0.29 AAr (1–50)
	3605.33	1600			0.33 OAn	0.35 AAr (1–30)
	3615.64	11				23 AAr
	4254.35	1700	1		0.10 OAn	0.6 AAr (1–50)
	4274.80	1300			0.13 OAn	0.8 AAr (1–100)
	4289.72	850			0.17 OAn	1.1 AAr (1–100)
	5204.52	440				49 A
	5208.44	900				19 AA
Cs	4555.36	40	10		2 OH	65 AA (100–500)
	4593.18	20			8 OH	200 AA (500–1000)
	8521.10	1500	10		0.5 OH	0.8 AA (10–100)
	8943.50	800			0.5 OH	13 AA (50–100)
Cu	2024.34	2				0.6 AA (5–50)
	2178.94	8				0.5 AA (10–50)
	2181.72	6				0.7 AA (10–50)
	2225.70	4				1.2 AA (5–500)
	2441.64	4				34 AA (100–500)
	2492.15	36				7 AA (10–100)
	3247.54	5000	0.2	0.2	0.6 OA	0.1 AA (1–10)
	3273.96	2500		0.05	0.8 OA	0.2 AA (1–10)
Dy	3407.79 II	480		2		
	3531.70 II	7000	1			
	4000.48 II	650	10			
	4045.99	1000			0.07 NA	0.75 NA
	4186.78	950				0.9 NA
	4191.60	180				16 NA

Table 8-9 (*Continued*)
EMISSION AND ABSORPTION LINES 1900 to 9000 Å

Element	Wavelength, Å	Relative Intensity	DC Arc, μg/g	Spark-porous Cup, μg/ml	Flame Emission, μg/ml/0.1 mV	Atomic Absorption: Sensitivity, μg/ml/1% Abs (conc'n range for which valid)
					Detection limit	
	4194.85	550				1.4 NA
	4211.72	1300			0.5 OA	0.7 NA
	4225.14	220				26 NA
DyO	5280				0.11 OAn; 1 OH	
	5400				0.20 OAn	
	5490				0.14 OAn	
	5730				0.08 OAn	
	5830				0.1 OAn; 5 OH	
Er	3372.76 II	750	5	2		
	3499.11 II	650	10			
	3692.64 II	700	10			
	3892.69	340				3.7 NA
	3896.25 II	420	10			
	3906.34 II	850	4			
	4007.97	1100	60		2 OA	0.9 NA
	4087.65	280				0.9 NA; 6.4 AA
	4151.10	550	300			1.3 NA
ErO	5040				0.07 OAn	
	5520				0.1 OAn; 2 OH	
	5650				3 OH	
Eu	3819.67 II	3400	10			
	4205.05 II	4000		0.5		
	4435.56 II	900	10			
	4594.03	750			0.05 OAn	0.8 NAr
	4627.22	650			0.06 OAn	0.9 NA
	4661.88	550			0.2 OAn	1.1 NA
F(CaF)	5291		100		300 OA	
(MgF)	3594		325			
(SrF)	6633		225			
Fe	2166.77	15				0.7 AAr (1–100)
	2382.04	60	0.5			
	2395.62	60	0.5			
	2483.27	280				0.15 AAr (1–30)
	2484.19	90				0.18 AAr (1–50)
	2487.97	4				0.3 AAr (1–30)
	2489.75 / 2490.64 } d	180				0.6 AAr (3–50)
	2510.83	90				1.6 AAr (5–500)
	2524.39	50				4 AAr (10–100)
	2527.43	140				0.8 AAr (5–100)
	2599.40 II	200	0.2			
	2719.02	260				0.5 AAr (1–100)
	2744.07	30				4 AAr (30–400)
	3020.64	280			0.2 OAn	0.5 AAr (5–50)
	3440.61	400			0.44 OAn	2.4 AAr (10–500)
	3581.20	600			0.8 OAn; 15 OA	
	3719.95	600	0.2		0.12 OAn; 3 OA	1 AAr (5–100)
	3734.87	700			0.16 OAn	
	3748.26	140			0.23 OAn	4 AAr (30–1000)
	3859.91	420			0.14 OAn; 3 OA	2.1 AAr (30–500)
	3886.28	180				6 AAr (50–1000)
	3920.26	36			3.5 OAn	40 AAr (300–1000)
	3927.92	70			10 OA	24 AAr (100–1000)

Table 8-9 (*Continued*)
EMISSION AND ABSORPTION LINES 1900 to 9000 Å

Element	Wavelength, Å	Relative Intensity	DC Arc, µg/g	Spark-porous Cup, µg/ml	Flame Emission, µg/ml/0.1 mV	Atomic Absorption: Sensitivity, µg/ml/1% Abs (conc'n range for which valid)
Ga	2874.24	500				2.3 AA
	2943.64	950	10	0.5		2.4 AA
	4032.98	1000		10	1.2 OA	6.2 AA
	4172.06	2000	1		0.5 OA	3.7 AA
Gd	3422.47 II	700		4		
	3684.13	200				17 NA
	3768.39 II	850	4			
	3783.05	280				17 NA
	3850.97	500		0.5		
	4058.22	240				19 NA
	4078.70	260				17 NA
	4190.78	200				40 NA
	4251.73 II	160	10			
	4346.46 d	200			25 OA	27 NA
	4401.86	130			2 NA	
GdO	4640				0.1 OAn; 1 OA	
	5680				0.2 OAn; 2 OA	
	6010				0.05 OAn	
	6220				0.03 OAn	
Ge	2592.54	500				4 NA
	2651.18 d	1200	1.0	0.5	2 OAn	2 NA
	2691.34	500				10 NA
	2709.63	850				5 NA
	2754.59	650				5 NA
	3039.06	750	1.0	10		40 NA
Hf	2641.41 II	120		5		
	2773.36 II	110	100			
	2820.22 II	140		4		
	2866.37	240				18 NA
	2898.26	200				40 NA
	2916.48	220	100			
	3072.88	240				14 NA
	3682.24	220			75 OAnz	45 NA
Hg	2536.52	1500	5	10	2.5 OAn	2 AA
	3650.15	280	100			
	4046.56	180				
	4358.35	400				
	5460.74	320				
Ho	3398.98 II	900		0.5		
	3891.02 II	1500	30			
	4053.93	900				1.9 NA
	4103.84	1000			0.5 OA	1.4 NA
	4163.03	900				2.4 NA
HoO	5160				0.1 OAn	
	5320				0.2 OAn	
	5660				0.05 OAn; 2 OA	
I(Inl)	4099				2.0 AH	
In	2710.26	160				8.8 AA
	3039.36	800			8.0 OH	0.9 AA
	3256.09	1300	20	10	2.2 OH	1.0 AA
	3258.56	300				11 AA
	4101.76	1700		3	0.14 OH	2.6 AA
	4511.31	1800	2		0.07 OH; 0.3 OA	2.8 AA

Table 8-9 (*Continued*)
EMISSION AND ABSORPTION LINES 1900 to 9000 Å

| Element | Wavelength, Å | Relative Intensity | Detection limit | | | Atomic Absorption: Sensitivity, μg/ml/1% Abs (conc'n range for which valid) |
			DC Arc, μg/g	Spark-porous Cup, μg/ml	Flame Emission, μg/ml/0.1 mV	
Ir	2088.82	13				8 AA
	2543.97	380				20 AA
	2639.71	170				13 AA
	2664.79	200				15 AA
	2849.72	280	10			18 AA
	3220.78	500	10	10		
K	4044.14	32	100	200	1.7 OH	6 AA (10–1000)
	4047.20	16				13 AA (50–1000)
	7664.91	1800	1		0.2 OH	0.06 AA (2–10)
	7698.98	900				0.1 AA (2–10)
La	3337.49 II	200		0.3		
	3949.10 II	900	10	1.0		
	4086.72 II	550	5			
	4187.32	28			40 OA	49 NA
	5501.34	36			8 NA	34 NA
	5791.34	34			2 NA	
LaO	4371/6				0.06 OAn	
	4418.24				0.06 OAn	
	5406/8				1.4 OAn	
	5430				1.4 OAn	
	5600				0.18 OAn	
	7410				0.005 OAn	
	7910				0.005 OAn	
Li	2741.20	5				15 AA (100–1000)
	3232.61	17	10	10	46 OA	18 AA (50–1000)
	4602.86	13			13 OA	
	6103.64	320			4 OA	192 AP
	6707.84	3600	0.1	0.1	0.07 OA	0.04 AA (0.1–10)
Lu	2615.42 II	1200	0.5			
	2911.39 II	600	10	0.5		
	3312.11	360			6 OA	21 NA
	3359.56	440				12 NA
	3567.84	280				27 NA
LuO	4680				0.05 OAn; 3 OH	
	5170				0.05 OAn; 4 OH	
Mg	2025.82					2.0 AA
	2795.53 II	1000	0.2	0.01		
	2802.70 II	600		0.003		
	2852.13	6000	0.2		0.2 OAr	0.008 AAw
MgOH	3702				1.4 OH	
	3810/30				1.6 OH	
Mn	2576.10 II	1200	1	0.02		
	2593.73 II	800		0.05		
	2213.85	4				1.3 AA (2–500)
	2794.82	800				0.06 AA (1–20)
	2798.27	650				0.08 AA (1–10)
	2801.06	480	1			0.12 AA (1–50)
	4030.76	2000	1		0.1 OA	0.6 AA (2–100)
	4033.07	1400				0.8 AA (2–200)
	4034.49	800				1.0 AA (2–200)
Mo	2816.15 II	220		1.0		

Table 8-9 (*Continued*)
EMISSION AND ABSORPTION LINES 1900 to 9000 Å

| Element | Wavelength, Å | Relative Intensity | Detection limit | | | Atomic Absorption: Sensitivity, µg/ml/1% Abs (conc'n range for which valid) |
			DC Arc, µg/g	Spark-porous Cup, µg/ml	Flame Emission, µg/ml/0.1 mV	
	3112.12	170			48 OAn	20 (N–A) Ar (100–1000)
	3132.59	1800	10	0.3	2 OAn	0.8 (N–A) Ar (2–200)
	3158.16	750			10 OAn	4 (N–A) Ar (10–1000)
	3170.35	1100	1		4 OAn	2 (N–A) Ar (5–500)
	3193.97	950			4 OAn	3.5 (N–A) Ar (5–200)
	3208.83	380			19 OAn	14 (N–A) Ar (100–1000)
	3798.25	3200	1		0.5 OAn	2 (N–A) Ar (5–200)
	3864.11	2800			0.6 OAn	4 (N–A) Ar (5–500)
	3902.96	1800			0.7 OAn	4 (N–A) Ar (10–500)
	5506.49	480			9 OAn	
Na	3302.32/.99	30	10	35	12.5 OH	2.8 AA (10–500)
	5889.95	2000	0.1		0.001 OH	0.016 AA (0.1–5)
	5895.92	1000				0.03 AA (0.1–10)
Nb	2950.88 II	180		2		
	3349.06	200			72 OAn	27 NA
	3580.27	800			43 OAn	27 NA
	3713.01	340			56 OAn	
	3742.39	180			31 OAn	
	4058.94	1700			13 OAn	36 NA
	4079.73	1200			17 OAn	32 NA
	4100.92	700			19 OAn	42 NA
	4123.81	550			28 OAn	40 NA
Nd	4012.25 II	220		5		
	4061.09 II	280	20			
	4303.58	320	10			
	4634.24	30				10 NA
	4896.93	24				14 NA
	4924.53	40			5 OAn	8 NA
NdO	5550				0.2 OAn	
	6630				0.4 OAn; 1 OH	
	7020				0.2 OAn; 1 OH	
	7120				0.4 OAn; 1 OH	
Ni	2289.98	18				0.6 AA (5–50)
	2310.96	30				0.22 AA (3-50)
	2320.03	44				0.15 AA (1–30)
	2345.54	26				0.5 AA (5–100)
	3002.49	320	3			0.7 AA (3–300)
	3037.94	140				2 AA (10–300)
	3050.82	280				0.7 AA (3–300)
	3232.96	100				5 AA (10–500)
	3369.57	260				3 AA (3–300)
	3391.05	120				6 AA (10–500)
	3392.99	300			10 OA	2.4 AA (3–100)
	3414.76	750	3	0.8	4 OA	
	3417.8					0.6 AA (1–100)
	3433.56	240			10 OA	2.5 AA (5–500)
	3446.26	440			10 OA	
	3458.47	460			5 OA	
	3461.65	460			11 OA	1.1 AA (1–500)
	3492.96	500			6 OA	

Table 8-9 (*Continued*)
EMISSION AND ABSORPTION LINES 1900 to 9000 Å

Element	Wavelength, Å	Relative Intensity	DC Arc, µg/g	Spark-porous Cup, µg/ml	Flame Emission, µg/ml/0.1 mV	Atomic Absorption: Sensitivity, µg/ml/1% Abs (conc'n range for which valid)
	3515.05	600			5 OA	
	3524.54	750			2 OA; 0.2 OAn	0.6 AA (1–50)
	3619.39	600			7 OA	
NO	2363				10 OH; 14 AH (sheathed)	
Os	2637.13	360				1.8 NA
	2644.11	180				4.8 NA
	2714.64	280				4.2 NA
	2806.91	260				4.6 NA
	2909.06	900	10			1.0 NA
	3018.04	460				3.2 NA
	3058.66	900		15		1.6 NA
	3301.56	800				3.6 NA
	4260.85	440				30 NA; 21 AA
	4420.47	440			10 OAnz	19 NA
P	2135.47 ⌉					240 AA or NA
	2136.20 ⌋					
	2534.01	70		5	230 OAnr	
	2535.65	60	30		50 OAnr	
	2553.28	38	100		120 OAnr	
HPO	5100				13 AH reversed	
	5249				6 AH reversed	
	5600				8 AH reversed	
PO	2464.2				3 OAnr; 19 OHn	
Pb	2022.02	5				7 AA (20–1000)
	2053.27	8				6 AA (20–1000)
	2169.99	22			550 OHn	0.23 AA (1–50)
	2614.18	700			13 OHn	
	2801.99	1000			10 OHn	
	2833.06	950	1	4	6 OHn	0.6 AA (1–100)
	3639.58	550			4 OHn	
	3683.48	1400			2 OHn; 21 OA	
	4057.83	3400	1		2 OHn; 14 OA	
Pd	2447.91	65				0.3 AA
	2476.42	100		2		0.3 AA
	2763.09	160				1.0 AA
	3404.58	2600	1	2	0.1 OHn; 1 OAn	1.2 AA
	3421.24	1400			1.0 OHn; 5 OAn	
	3516.94	1300			0.3 OHn; 3 OAn	
	3609.55	2200			0.2 OHn; 2 OAn	
	3634.70	2200			0.1 OHn; 1 OAn	
Pr	3908.41 II	320	25			
	4100.75 II	260		2		
	4225.33 II	340	10			
	4914.03	12				19 NA
	4951.36	34			0.4 OAn; 15 OA	13 NA
	5133.42	24			0.4 OAn	23 NA
Pt	2144.23	6				7.3 AA
	2174.67	7				3.3 AA
	2487.17	100			200 OAn	
	2628.03	110			100 OAn	5.3 AA
	2659.45	280		1	15 OAn; 13 OHn	2.2 AA

Table 8-9 (*Continued*)
EMISSION AND ABSORPTION LINES 1900 to 9000 Å

Element	Wavelength, Å	Relative Intensity	Detection limit DC Arc, μg/g	Detection limit Spark-porous Cup, μg/ml	Detection limit Flame Emission, μg/ml/0.1 mV	Atomic Absorption: Sensitivity, μg/ml/1% Abs (conc'n range for which valid)
	2830.30	140				7.4 AA
	3064.71	320	2		15 OAn; 10 OHn	4.6 AA
Rb	4201.85	32	100		4 OH	12 AA (25–100)
	4215.56	16			15 OH	24 AA (50–100)
	7800.23	3000	1		0.6 OH	0.2 AA (0.5–10)
	7947.60	1500			0.7 OH	0.35 AA (1–10)
Re	2274.62	24				29 NA
	2287.51	40				20 NA
	2294.49	160				25 NA
	3451.88	1600	10		11 OAn	29 NA
	3460.46	5500	1	10	3 OAn	11 NA
	3464.73	4000		5	5 OAn	19 NA
	4889.14	220			8 OAn	
	5275.56	160			12 OAn	
Rh	3396.85	480				0.8 AA
	3434.89	700	1	0.7	3 OAn; 2 OHn	0.3 AA
	3502.52	500			6 OAn; 2 OHn	1.3 AA
	3507.32	240				1.3 AA
	3528.02	750			4 OAn; 5 OHn	
	3657.99	700			4 OAn; 3 OHn	1.7 AA
	3692.36	800			1 OAn; 1 OHn	0.6 AA
Ru	3436.74	650	10			
	3498.94	850		2	2 OHn; 3 OAn	1.3 AA
	3728.03	1000			0.5 OHn	
	3799.35	700			0.4 OAn	
	3925.92	300				14 AA
S₂	3645				5 AH (reversed shielded)	
	3740				4 AH	
	3837				3 AH	
	3940				3 AH	
	4050				4 AH	
	4150				4 AH	
SO₂	2070				10 AH	
Sb	2068.33	55				0.8 AA (2–100)
	2127.39	5				20 AA (100–1000)
	2175.81	38				0.6 AA (1–100)
	2179.19	7				1.5 AA
	2311.47	45			0.5 AH	1.5 AA (5–500)
	2528.52				1 OAn	
	2598.05	600	7	2	0.6 OAn	
	2877.92	140	10			
Sc	3269.91	400				1.6 NA
	3273.63	500				5 NA
	3353.73 II	900	2			
	3613.84 II	2500	0.5	0.05		
	3630.75 II	1800	1			
	3642.79 II	1200	1			
	3907.49	1800	10		1 OAn	0.5 NA
	3911.81	2100			0.7 OAnr	0.5 Na
	4020.40	1800			0.1 OAnr	0.9 NA
	4023.69	1800				0.7 NA

Table 8-9 (*Continued*)
EMISSION AND ABSORPTION LINES 1900 to 9000 Å

Element	Wavelength, Å	Relative Intensity	DC Arc, µg/g	Spark-porous Cup, µg/ml	Flame Emission, µg/ml/0.1 mV	Atomic Absorption: Sensitivity, µg/ml/1% Abs (conc'n range for which valid)
			Detection limit			
	4054.55	500				1.4 NA
	4246.83 II	1400	1			
ScO	6700				0.01 OAn	
Se	1960.26	34		70		0.7 AAw (1–100)
	2039.85	40			3 OHn	11 AA
	2062.79	15				43 AA
	2074.79	3				50 AAw (100–500)
Si	2216.67	3				9 NA
	2506.90	170			12 OAnr	6 NA
	2514.32	160			14 OAnr	7 NA
	2516.11	360	1	1	4 OAnr; 10 OAr	2 NA
	2519.21	120			16 OAnr	11 NA
	2524.11	240			14 OAnr	8 NA
	2528.51	400			12 OAnr	7 NA
Sm	3609.49 II	280	20			
	3634.29 II	280	40	3		
	3885.29 II	280	20			
	4296.74	110				8.5 NA
	4424.34 II	200	10			
	4760.27	75				24 NA
	4783.10	60			5 OA	
	5200.59	34				13 NA
SmO	6140				0.1 OAn	
	6240				0.1 OAn	
	6400				0.1 OAn	
	6520				0.1 OAn; 3 OHn	
Sn	2246.02	45		2		
	2334.80	38			17 OAn	5.5 AH
	2429.49	420			2 OAnr	
	2661.24	140				19 AH
	2839.99	1400	3		0.3 NA	
	2863.33	1000				10 AH (50–1000)
	3034.12	850			9 OAn	
	3175.02	550	1			
Sr	3464.46 II	65	10		26 OAn	
	4077.71 II	4600	0.1	0.006		
	4607.33	650	0.1		0.09 OH; 0.06 OA	0.06 AA or NA
SrOH	6060				0.3 OH; 0.6 OA	
Ta	2608.63	160				21 NA
	2685.15 II	180		2		
	2714.67	300				11 NA
	2775.88	90				21 NA
	3012.54 II	240	100			
	3311.16	140	100			
	4812.75	20			18 OAnz	
Tb	3509.17 II	600	15	3		
	3901.35	150			4 OAnrz	12 NA
	4061.59	120				13 NA
	4278.52 II	70	10			
	4318.85	200			10 OAn	9 NA
	4326.47	280			0.5 NA	8 NA
	4338.45	160				14 NA

Table 8-9 (*Continued*)
EMISSION AND ABSORPTION LINES 1900 to 9000 Å

| Element | Wavelength, Å | Relative Intensity | Detection limit | | | Atomic Absorption: Sensitivity, μg/ml/1% Abs (conc'n range for which valid) |
			DC Arc, μg/g	Spark-porous Cup, μg/ml	Flame Emission, μg/ml/0.1 mV	
TbO	5340				2 OHn; 0.1 OAn	
	5730				0.1 OAn	
	6080				0.2 OAn	
	6120				0.3 OAn	
Te	2142.75	55			7 OAn	0.5 AA (10–100)
	2259.04	6			7 OAn	4 AA (10–500)
	2383.25	55	100		4 OAn	67 AA
	2385.76	70	70	10	2 OAn; 380 OA	43 AA
Th	3244.46	20				850 NA
	3392.03 II	90	10			
	3539.59	48	35			
Ti	3186.51	200				3 NA
	3234.52 II	550		0.1		
	3241.99 II	220	10			
	3341.88	480			16 OAn	
	3349.04 II	1000		3		
	3354.64	340				2.9 NA
	3371.45	360			16 OAn	2.0 NA
	3372.80 II	480	1			
	3635.46	400			8 OAn	4.4 NA
	3642.68	550			7 OAn	2.2 NA
	3653.50	600	1		6 OAn	2.5 NA
	3741.06	280				2.6 NA
	3752.86	440			6 OAn	2.5 NA
	3948.67	380			13 OAn	5.0 NA
	3958.21	440			6 OAn	5.0 NA
	3989.76	480			6 OAn	5.2 NA
	3998.64	650			5 OAn	16 OAn
Tl	2767.87	440	20	3		0.3 AA
	2580.14	70				30 AA
	3519.24	2000	10			
	3775.72	1200			0.6 OH	1 AA
	5350.46	1800	1		1.2 OH	
Tm	3131.26 II	700		2		
	3462.20 II	800	5			
	3717.92	650				0.5 NA
	4094.19	750				1.2 NA
	4105.84	700				1.0 NA
	4203.73	440				1.2 NA
	4359.93	200				3.6 NA
TmO	4850				0.12 OAn	
	4900				0.16 OAn	
	5350				0.16 OAn; 3 OHn	
	5570				0.10 OAn	
U	3566.60	95				180 NA
	3584.88	130				120 NA
	3670.07 II	160		100		
	4241.67 II	75	100			
	5915.40	20			10 OAnz	
V	3033.82 II	38		0.3		
	3066.38	320			9 OAn	7 NA (20–200)
	3093.11 II	500		1		
	3183.98	700	1		5 OAn	2 NA

Table 8-9 (*Continued*)
EMISSION AND ABSORPTION LINES 1900 to 9000 Å

| Element | Wavelength, Å | Relative Intensity | Detection limit | | | Atomic Absorption: Sensitivity, $\mu g/ml/1\%$ Abs (conc'n range for which valid) |
			DC Arc, $\mu g/g$	Spark-porous Cup, $\mu g/ml$	Flame Emission, $\mu g/ml/0.1$ mV	
	3185.40	500	10		6 OAn	1.9 NA (5–100)
	3271.12 II	500		1		
	3840.75	280			6 OAn	
	3855.84	320			5 OAn	
	4379.24	950	1		3 OAn	9 NA
	4384.72	550			5 OAn	
	4389.97	380			6 OAn	9 NA
W	2397.09 II	34		3		
	2551.35	280				5 NA
	2658.04 II	50		10		
	2946.98	300	100			
	4008.75	950		30	4 OAnz	20 NA
	4294.61	450	100			
Y	3242.28 II	800	10	0.2		
	3710.30 II	1500		0.1		
	4077.38	950				5.7 NA
	4102.38	1000			3 OAn	5 NA
	4128.31	900				5.4 NA
	4142.85	750				11 NA
	4374.94	1200	1			
YO	4280				0.1 OAn	
	5980				0.02 OAn	
	6140				0.02 OAn	
Yb	2464.49	65				1.5 NA
	2671.98	55				14 NA
	3289.37 II	2600	10	0.04		
	3464.36	340				0.8 NA
	3694.19 II	3200	0.5			
	3987.98	1900			0.02 OAn	0.3 NA
	5556.48	140			0.2 OH; 0.06 OAn	
Zn	2138.56	1000	10		77 OAn	0.02 AA (0.1–10)
	3075.90	26	20			60 AA (300–1000)
	3345.02	140	10	4		
	4810.53	110	100			
Zr	3273.05 II	160		2		
	3391.98 II	900	1	0.2		
	3438.23 II	750	10			
	3519.60	320			52 OAnz	20 NA
	3601.19	550			75 OAnz	15 NA

Table 8-10
SENSITIVE LINES OF THE ELEMENTS

In this table the sensitive lines of the elements are arranged in order of decreasing wavelengths. In the column headed Sensitivity the most sensitive line of the neutral or un-ionized atom is indicated by U1, and other lines by U2, U3, etc., in order of decreasing sensitivity. For the singly ionized atom the corresponding designations are V1, V2, V3, etc. Where U1 is not given, the most sensitive lines lie outside the range of 10,000 to 2,000 Å.

The table is taken by permission of George R. Harrison and the Technology Press from *M.I.T. Wavelength Tables*, 1939 ed., John Wiley & Sons, New York.

The *abbreviations* used in this table are those employed in the *M.I.T. Tables*, and are as follows:

bh, band head
d, double line
h, hazy, diffuse
l, shaded, or displaced to longer wavelengths
r, narrow self-reversal
R, wide self-reversal
s, shaded, or displaced to shorter wavelengths

w, wide or complex
W, very wide or complex
I, line classified as being emitted by the un-ionized atom
II, line classified as being emitted by the singly ionized atom
?, line listed in M.I.T. Tables, but not found on a plate made from pure sample of the element in question

Wavelength	Element	Intensity		Sensitivity	Wavelength	Element	Intensity		Sensitivity
		Arc	Spk [Dis]				Arc	Spk [Dis]	
9237.49	S I	....	[200]	U6	5777.665	Ba I	500R	100R	U2
9228.11	S I	....	[200]	U5	5688.224	Na I	300	....	..
9212.91	S I	....	[200]	U4	5682.657	Na I	80	....	..
8943.50	Cs I	2000R	....	U2	5679.56	N II	....	[500]	V2
8521.10	Cs I	5000R	....	U1	5676.02	N II	....	[100]	V4
8115.311	Ar I	....	[5000]	U2	5666.64	N II	....	[300]	V3
7947.60	Rb I	5000R	....	U2	5608.8	Pb II	....	[40]	V2
7800.227	Rb I	9000R	....	U1	5570.2895	Kr I	....	[2000]	U3
7775.433	O I	....	[100]	U4	5535.551	Ba I	1000R	200R	U1
7774.138	O I	....	[300]	U3	5519.115	Ba I	200R	60R	U3
7771.928	O I	....	[1000]	U2	5465.487	Ag I	1000R	500R	U4
7698.979	K I	5000R	....	U2	5464.61	I II	....	[900]	..
7664.907	K I	9000R	....	U1	5460.740	Hg I	....	[2000]	..
7503.867	Ar I	....	[700]	U4	5455.146	La I	200	1	U3
7450.00	Rn I	....	[600]	U2	5424.616	Ba I	100R	30R	U4
7067.217	Ar I	....	[400]	U3	5400.562	Ne I	....	2000	..
7055.42	Rn I	....	[400]	U3	5350.46	Tl I	5000R	2000R	U1
6965.430	Ar I	....	[400]	U3	5291.0	bhCaF	200	....	..
6902.46	F I	....	[500]	U3	5218.202	Cu I	700	....	U3
6856.02	F I	....	[1000]	U2	5209.067	Ag I	1500R	1000R	U3
6707.844	Li I	3000R	200	U1	5208.436	Cr I	500R	100	U4
6562.79	H I	....	[3000]	U2	5206.039	Cr I	500R	200	U5
6438.4696	Cd I	2000	1000	..	5204.518	Cr I	400R	100	U6
6402.246	Ne I	....	[2000]	..	5183.618	Mg I	500wh	300	..
6362.347	Zn I	1000Wh	500	..	5172.699	Mg I	200wh	100wh	..
6249.929	La I	300	....	U1	5167.343	Mg I	100wh	50	..
6243.36	Al II	....	100	V3	5161.188	I II	....	[300]	..
6231.76	Al II	....	30	..	5153.235	Cu I	600	....	U4
6103.642	Li I	2000R	300	U3	5105.541	Cu I	500	....	U5
5930.648	La I	250	....	U2	5007.213	Ti I	200	40	..
5895.923	Na I	5000R	500R	U2	4999.510	Ti I	200	80	..
5889.953	Na I	9000R	1000R	U1	4991.066	Ti I	200	100	..
5875.618	He I	....	[1000]	U3	4981.733	Ti I	300	125	U1
5870.9158	Kr I	....	[3000]	U2	4962.263	Sr I	40	....	U4
5852.488	Ne I	....	[2000]	..	4934.086	Ba II	400h	400h	V2

Table 8-10 (*Continued*)
SENSITIVE LINES OF THE ELEMENTS

Wave-length	Element		Intensity		Sensi-tivity	Wave-length	Element		Intensity		Sensi-tivity
			Arc	Spk [Dis]					Arc	Spk [Dis]	
4889.17	Re	I	2000w	....	U2	4420.468	Os	I	400R	100	..
4872.493	Sr	I	25	....	U3	4390.865	Sm	II	150	150	..
4861.327	H	I	....	[500]	U3	4389.974	V	I	80R	60R	..
4832.075	Sr	I	200	8	U2	4384.722	V	I	125R	125R	..
4825.91	Ra	I	....	[800]	U1	4379.238	V	I	200R	200R	U1
4819.46	Cl	II	....	[200]	V4	4358.35	Hg	I	3000w	500	..
4816.71	Br	II	....	[300]	V3	4305.447	Sr	II	40	....	..
4810.534	Zn	I	400w	300h	..	4303.573	Nd		100	40	..
4810.06	Cl	II	....	[200]	V3	4302.108	W	I	60	60	U1
4794.54	Cl	II	....	[250]	V2	4294.614	W	I	50	50	U2
4785.50	Br	II	....	[400]	V2	4289.721	Cr	I	3000R	8000r	U3
4772.312	Zr	I	100	....	..	4274.803	Cr	I	4000R	800r	U2
4742.25	Se	I	....	[500]	U6	4267.27	C	II	....	500	V2
4739.478	Zr	I	100	....	..	4267.02	C	II	....	350	V3
4739.03	Se	I	....	[800]	U5	4254.346	Cr	I	5000R	1000	U1
4730.78	Se	I	....	[1000]	U4	4241.669	U		40	50	..
4722.552	Bi	I	1000	100	..	4226.728	Ca	I	500R	50W	U1
4722.159	Zn	I	400w	300h	..	4226.570	Ge	I	200	50	..
4710.075	Zr	I	60	....	..	4225.327	Pr		50	40	..
4704.86	Br	II	....	[250]	V1	4215.556	Rb	I	1000R	300	U4
4696.25	S	I	....	[15]	U9	4215.524	Sr	II	300r	400W	V2
4695.45	S	I	....	[30]	U8	4211.719	Dy		200	15	..
4694.13	S	I	....	[500]	U7	4205.046	Eu	II	200R	50	..
4687.803	Zr	I	125	....	U4	4201.851	Rb	I	2000R	500	U3
4685.75	He	II	....	[300]	..	4189.518	Pr		100	50	..
4682.28	Ra	II	....	[800]	V2	4186.599	Ce	II	80	25	..
4680.138	Zn	I	300w	200h	..	4179 422	Pr		200	40	..
4674.848	Y	I	80	100	U1	4177.321	Nd		15	25	..
4671.226	Xe	I	....	[2000]	U2	4172.056	Ga	I	2000R	1000R	U1
4643.695	Y	I	50	100	U2	4167.966	Dy		50	12	..
4624.276	Xe	I	....	[1000]	U3	4165.606	Ce	II	40	6	..
4607.331	Sr	I	1000R	50R	U1	4137.095	Cb	I	100	60	U5
4603.00	Li	I	800	....	U4	4130.664	Ba	II	50r	60Wh	V3
4593.177	Cs	I	1000R	50	U4	4129.737	Eu	II	150R	50R	..
4555.355	Cs	I	2000R	100	U3	4123.810	Cb	I	200	125	U4
4554.042	Ba	II	1000R	200	V1	4123.228	La	II	500	500	V4
4524.741	Sn		500wh	50	..	4109.98	N	I	....	[1000]	U2
4518.57	Lu		300	40	..	4103.37	N	III	....	[80]	..
4511.323	In	I	5000R	4000R	U1	4101.773	In	I	2000R	1000R	U2
4500.977	Xe	I	....	[500]	U4	4100.923	Cb	I	300w	200w	U3
4454.781	Ca	I	200	....	U2	4099.94	N	I	....	[150]	U3
4434.960	Ca	I	150	....	U3	4097.31	N	III	....	[100]	..
4434.321	Sm	II	200	200	V2	4093.161	Hf	II	25	20	..
4425.441	Ca	I	100	....	U4	4079.729	Cb	I	500w	200w	U2
4424.342	Sm	II	300	300	V1	4077.974	Dy		150r	100	..

Table 8-10 (*Continued*)
SENSITIVE LINES OF THE ELEMENTS

Wave-length	Element	Intensity		Sensi-tivity	Wave-length	Element	Intensity		Sensi-tivity
		Arc	Spk [Dis]				Arc	Spk [Dis]	
4077.714	Sr II	400r	500W	V1	3774.332	Y II	12	100	..
4077.340	La II	600	400	V3	3768.405	Gd	20	20	..
4062.817	Pr	150	50	..	3761.917	Tm	200	120	..
4058.938	Cb I	1000w	400w	U1	3761.333	Tm	250	150	..
4057.820	Pb I	2000R	300R	U1	3748.264	Fe I	500	200	U4
4047.201	K I	400	200	U4	3748.17	Ho	60	40	..
4046.561	Hg I	200	300	..	3745.903	Fe I	150	100	U5
4045.983	Dy	150	12	..	3745.564	Fe I	500	500	U3
4044.140	K I	800	400	U3	3737.133	Fe I	1000r	600	U2
4040.762	Ce II	70	5	..	3719.935	Fe I	1000R	700	U1
4034.490	Mn I	250r	20	U3	3710.290	Y II	80	150	V1
4033.073	Mn I	400r	20	U2	3694.203	Yb	500R	1000R	..
4032.982	Ga I	1000R	500R	U2	3692.652	Er	20	12	..
4030.755	Mn I	500r	20	U1	3692.357	Rh I	500hd	150wd	..
4023.688	Sc I	100	25	U3	3683.471	Pb I	300	50	U2
4020.399	Sc I	50	20	U4	3672.579	U	8	15	..
4019.137	Th	8	8	..	3663.276	Hg I	500	400	U5
4012.388	Ce I, II	60	20	..	3657.987	Rh I	500W	200W	..
4008.753	W I	45	45	U3	3654.833	Hg I	....	[200]	U4
4000.454	Dy	400	300	..	3653.496	Ti I	500	200	U2
3987.994	Yb	1000R	500R	..	3650.146	Hg I	200	500	U3
3968 468	Ca II	500R	500R	V2	3646.196	Gd	200w	150	..
3961.527	Al I	3000	2000	U1	3642.785	Sc II	60	50	V3
3951.154	Nd	40	30	..	3642.675	Ti I	300	125	..
3949.106	La II	1000	800	V2	3639.580	Pb I	300	50h	..
3944.032	Al I	2000	1000	U2	3635.463	Ti I	200	100	..
3933.666	Ca II	600R	600R	V1	3634.695	Pd	2000R	1000R	U3
3911.810	Sc I	150	30	U1	3633.123	Y II	50	100	..
3907.476	Sc I	125	25	U2	3630.740	Sc II	50	70	V2
3906.316	Er	25	12	..	3613.836	So II	40	70	V1
3905.528	Si I	20	15W	..	3613.790	W II	10	30	..
3902.963	Mo I	1000R	500R	U3	3610.510	Cd I	1000	500	..
3891.785	Ba II	18	25	V4	3609.548	Pd I	1000R	700R	..
3891.02	Ho	200	40	..	3601.193	Zr I	400	15	U1
3888.646	He I	....	[1000]	U2	3601.040	Th	8	10	..
3874.18	Tb	200	200	..	3600.734	Y II	100	300	..
3864.110	Mo I	1000R	500R	U2	3596.179	Ru I	30	100	U3
3848.75	Tb	100	200	..	3572.473	Zr II	60	80	V4
3838.258	Mg I	300	200	U2	3561.74	Tb	200	200	..
3832.306	Mg I	250	200	U3	3554.43	Lu	50	150	..
3829.350	Mg I	100w	150	U4	3552.172	U	8	12	..
3814.42	Ra II	....	[2000]	V1	3547.682	Zr I	200	12	U2
3798.252	Mo I	1000R	1000R	U1	3538.75	Th	....	50	..
3788.697	Y II	30	30	..	3529.813	Co I	1000R	30	U3
3775.72	Tl I	3000R	1000R	U2	3524.541	Ni I	1000R	100wh	..

Table 8-10 (*Continued*)
SENSITIVE LINES OF THE ELEMENTS

Wave-length	Element		Intensity		Sensi-tivity	Wave-length	Element		Intensity		Sensi-tivity
			Arc	Spk [Dis]					Arc	Spk [Dis]	
3519.605	Zr	I	100	10	U3	3302.988	Na	I	300R	150R	U4
3519.24	Tl	I	2000R	1000R	U3	3302.588	Zn	I	800	300	U3
3516.943	Pd	I	1000R	500R	..	3302.323	Na	I	600R	300R	U3
3515.054	Ni	I	1000R	50h	..	3290.59	Th		....	40h	..
3513.645	Ir	I	100h	100	U2	3289.37	Yb		500R	1000R	..
3509.17	Tb		200	200	..	3282.333	Zn	I	500R	300	U4
3499.104	Er		18	15	..	3280.683	Ag	I	2000R	1000R	U1
3498.942	Ru	I	500R	200	U1	3273.962	Cu	I	3000R	1500R	U2
3496.210	Zr	II	100	100	V3	3269.494	Ge	I	300	300	U3
3492.956	Ni	I	1000R	100h	U2	3267.945	Os	I	400R	30	..
3474.887	Sr	II	80	50	..	3267.502	Sb	I	150	150Wh	..
3472.48	Lu		50	150	..	3262.328	Sn	I	400h	300h	U3
3466.201	Cd	I	1000	500	..	3262.290	Os	I	500R	50	..
3465.800	Co	I	2000R	25	U2	3261.057	Cd	I	300	300	..
3464.57	Sr	II	200	200	..	3258.564	In	I	500R	300R	U5
3462.21	Tm		200	100	..	3256.090	In	I	1500R	600R	U3
3460.47	Re	I	1000W	....	U1	3247.540	Cu	I	5000R	2000R	U1
3453.505	Co	I	3000R	200	U1	3242.280	Y	II	60	100	..
3451.41	B	II	5	30	V2	3232.61	Li	I	1000R	500	U2
3438.230	Zr	II	250	200	V2	3232.499	Sb	I	150	250wh	..
3437.015	Ir	I	20	15	..	3229.75	Tl	I	2000	800	..
3436.737	Ru	I	300R	150	U2	3225.479	Cb	II	150w	800wr	..
3434.893	Rh		1000R	200r	U1	3220.780	Ir	I	100	30	U1
3421.24	Pd	I	2000R	1000R	U2	3215.560	W	I	10	9	..
3414.765	Ni	I	1000R	50wh	U1	3194.977	Cb	II	30	300	..
3406.664	Ta		70w	18s	..	3185.396	V	I	500R	400R	U2
3405.120	Co	I	2000R	150	..	3183.982	V	I	500R	400R	..
3404.580	Pd	I	2000R	1000R	U1	3183.406	V	I	200R	100R	..
3403.653	Cd	I	800	500h	..	3179.332	Ca	II	100	400w	V3
3397.07	Lu		50	20r	..	3175.019	Sn	I	500h	400hr	..
3396.85	Rh	I	1000w	500	..	3163.402	Cb	II	15	8	..
3391.975	Zr	II	300	400	V1	3158.869	Ca	II	100	300w	V4
3383.761	Ti	II	70	300R	..	3134.718	Hf		80	125	..
3382.891	Ag	I	1000R	700R	U2	3131.072	Be	II	200	150	V2
3380.711	Sr	II	150	200	..	3130.786	Cb	II	100	100	..
3372.800	Ti	II	80	400R	V3	3130.416	Be	II	200	200	V1
3361.213	Ti	II	100	600R	V2	3125.284	V	II	80	200R	..
3349.035	Ti	II	125	800R	V1	3118.383	V	II	70	200R	V4
3345.020	Zn	I	800	300	U2	3110.706	V	II	70	300R	V3
3323.092	Rh	I	1000	200	..	3102.299	V	II	70	300R	V2
3321.343	Be	I	1000r	30	U2	3094.183	Cb	II	100	1000	V1
3321.086	Be	I	100	....	U3	3093.108	V	II	100R	400R	V1
3321.013	Be	I	50	....	U4	3092.713	Al	I	1000	1000	U3
3318.840	Ta		125	35	..	3082.155	Al	I	800	800	U4
3311.162	Ta		300w	70w	U1	3072.877	Hf	I	80	18	..

Table 8-10 (*Continued*)
SENSITIVE LINES OF THE ELEMENTS

Wave-length	Element		Intensity		Sensi-tivity	Wave-length	Element		Intensity		Sensi-tivity
			Arc	Spk [Dis]					Arc	Spk [Dis]	
3071.591	Ba	I	100R	50R	U5	2837.602	C	II	....	40	V5
3067.716	Bi	I	3000hR	2000wh	U1	2836.710	C	II	....	200	V4
3064.712	Pt	I	2000R	300R	U1	2835.633	Cr	II	100	400r	V1
3058.66	Os	I	500R	500	..	2833.069	Pb	I	500R	80R	..
3039.356	In	I	1000R	500R	U4	2830.295	Pt	I	1000R	600r	..
3039.064	Ge	I	1000	1000	U2	2820.224	Hf	II	40	100	..
3034.121	Sn	I	200wh	150wh	..	2816.179	Al	II	10	100	V2
3009.147	Sn	I	300h	200h	..	2816.154	Mo	II	200	300h	V1
2997.967	Pt	I	1000R	200r	..	2809.625	Bi	I	200w	100	..
2989.029	Bi	I	250wh	100wh	..	2802.695	Mg	II	150	300	V2
2976.586	Ru		60	200	..	2802.19	Au		....	200	..
2965.546	Ru		60	200	..	2795.53	Mg	II	150	300	V1
2945.668	Ru		60	300	..	2780.521	Bi	I	200w	100	..
2943.637	Ga	I	10	20r	U3	2780.197	As	I	75R	75	U5
2940.772	Hf	I	60	12	..	2773.357	Hf	II	25	60	..
2938.298	Bi	I	300w	300w	..	2769.67	Te	I	....	[30]	..
2936.77	Ho		....	1000R	..	2767.87	Tl	I	400R	300R	..
2929.794	Pt	I	800R	200w	..	2748.58	Cd	II	5	200	..
2924.792	Ir	I	25wh	15	..	2712.410	Ru		80	300	..
2918.32	Tl	I	400R	200R	..	2709.626	Ge	I	30	20	..
2916.481	Hf	I	50	15	..	2692.065	Ru		8	200	..
2911.39	Lu		100	300	..	2678.758	Ru		100	300	..
2909.116	Mo	II	25	40h	V5	2675.95	Au	I	250R	100	U2
2909.061	Os	I	500R	400	U1	2669.166	Al	II	3	100	V1
2904.408	Hf	I	30	6	..	2659.454	Pt	I	2000R	500R	U2
2898.71	As	I	25r	40	..	2658.722	Pd	II	20	300	..
2898.259	Hf	I	50	12	..	2651.575	Ge	I	30	20	..
2897.975	Bi	I	500WR	500WR	U2	2651.178	Ge	I	40	20	..
2894.84	Lu		60	200	..	2650.781	Be	I	25	....	U5
2890.994	Mo	II	30	50h	V4	2641.406	Hf	II	40	125	..
2881.578	Si	I	500	400	U1	2631.553	Al	II	....	40	..
2877.915	Sb	I	250W	150	..	2614.178	Pb	I	200r	80	..
2874.244	Ga	I	10	15r	U4	2605.688	Mn	II	100R	500R	V3
2871.508	Mo	II	100	100h	V3	2598.062	Sb	I	200	100	..
2863.327	Sn	I	300R	300R	U2	2593.729	Mn	II	200R	1000R	V2
2860.934	Cr	II	60	100	V5	2589.167	W	II	15d	25	..
2860.452	As	I	50r	50	..	2576.104	Mn	II	300R	2000R	V1
2855.676	Cr	II	60	200Wh	V4	2573.09	Cd	II	3	150	..
2854.581	Pd	II	4	500h	..	2557.958	Zn	II	10	300	V3
2852.129	Mg	I	300R	100R	U1	2554.93	P	I	60	[20]	..
2849.838	Cr	II	80	150r	V3	2553.28	P	I	80	[20]	U3
2849.725	Ir	I	40h	20h	..	2536.519	Hg	I	2000R	1000R	U2
2848.232	Mo	II	125	200h	V2	2535.65	P	I	100	[30]	U2
2843.252	Cr	II	125	400r	V2	2534.01	P	I	50	[20]	..
2839.989	Sn	I	300R	300R	U1	2530.70	Te	I	....	[30]	..

Table 8-10 (*Continued*)
SENSITIVE LINES OF THE ELEMENTS

Wave-length	Element		Intensity		Sensi-tivity	Wave-length	Element		Intensity		Sensi-tivity
			Arc	Spk [Dis]					Arc	Spk [Dis]	
2528.535	Sb	I	300R	200	..	2307.857	Co	II	25	50w	..
2528.516	Si	I	400	500	U2	2304.235	Ba	II	60R	80R	..
2519.822	Co	II	40	200	..	2296.89	C	III	....	200	..
2516.881	Hf	II	35	100	..	2288.12	As	I	250R	5	U3
2516.123	Si	I	500	500	U3	2288.018	Cd	I	1500R	300R	U1
2513.028	Hf	II	25	70	..	2287.084	Ni	II	100	500	V1
2506.899	Si	I	300	200	U4	2286.156	Co	II	40	300*l*	V1
2505.739	Pd	II	3	30	..	2276.578	Bi	I	100R	40	..
2502.001	Zn	II	20	400w	V4	2270.213	Ni	II	100	400	V2
2498.784	Pd	II	4	150h	..	2265.017	Cd	II	25d	300	V2
2497.733	B	I	500	400	U1	2264.457	Ni	II	150	400	V3
2496.778	B	I	300	300	U2	2253.86	Ni	II	100	300	V4
2488.921	Pd	II	10	30	..	2246.995	Cu	II	30	500	V3
2478.573	C	I	400	[400]	U2	2246.412	Ag	II	25	300hs	V3
2456.53	As	I	100r	8	U4	2203.505	Pb	II	50W	5000R	V1
2437.791	Ag	II	60	500wh	V2	2192.260	Cu	II	25	500h	V2
2427.95	Au	I	400R	100	U1	2175.890	Sb	I	300	40	U2
2413.309	Fe	II	60	100h	V5	2169.994	Pb	I	1000R	1000R	..
2410.517	Fe	II	50	70h	V4	2144.382	Cd	II	50	200R	V1
2404.882	Fe	II	50	100wh	V3	2142.75	Te	I	60R	....	..
2397.091	W	II	18	30	..	2138.56	Zn	I	800R	500	U1
2395.625	Fe	II	50	100wh	V2	2135.976	Cu	II	25	500w	V1
2388.918	Co	II	10	35	..	2068.38	Sb	I	300R	3	U1
2385.76	Te	I	600	[300]	U2	2062.788	Se	I	....	[800]	U3
2383.25	Te	I	500	[300]	U3	2062.38	I		....	[900]	..
2382.039	Fe	II	40r	100R	V1	2061.91	Zn	II	100	100	V2
2378.622	Co	II	25	50w	..	2061.70	Bi	I	300R	100	..
2370.77	As	I	50r	3	..	2039.851	Se	I	....	[1000]	U2
2369.67	As	I	40r	....	..	2025.51	Zn	II	200	200	V1
2363.787	Co	II	25	50	..						
2349.84	As	I	250R	18	U3						
2348.610	Be	I	2000R	50	U1						
2335.269	Ba	II	60R	100R	..						
2312.84	Cd	II	1	200	..						
2311.469	Sb	I	150R	50	..						

SOME COMMON SPECTROSCOPIC RELATIONSHIPS

Electromagnetic Radiation

Electromagnetic radiation travels in straight lines in a uniform medium, has a velocity of 299 792 500 m $\cdot$ s^{-1} in a vacuum, and possesses properties of both a wave motion and a particle (photon). *Wavelength* λ is the distance from crest to crest; *frequency* ν is the number of waves passing a fixed point in a unit length of time. Wavelength and frequency are related by the relation

$$c = \lambda\nu$$

where c is the velocity of light (in a vacuum). In any material medium the speed of propagation is smaller than this and is given by the product nc, where n is the refractive index of the medium.

Radiation is absorbed or emitted only in discrete packets called photons or quanta:

$$E = h\nu$$

where E is the energy of the quantum and h is Planck's constant.

The relation between energy and mass is given by the *Einstein equation:*

$$\Delta E = \Delta mc^2$$

where ΔE is the energy release and Δm is the loss of mass. Strictly, the mass of a particle depends on its velocity, but here the masses are equated to their rest masses (at zero velocity).

The *Wien displacement law* states that the wavelength of maximum emission, λ_m, of a blackbody varies inversely with absolute temperature; the product $\lambda_m T$ remains constant. When λ_m is expressed in micrometers, the law becomes

$$\lambda_m T = 2898$$

In terms of σ_m, the wavenumber of maximum emission:

$$\sigma_m = 3.48T$$

Another useful version is: $h\nu_m = 5kT$, where k is the Boltzmann constant.

Stefan's law states that the total energy J radiated by a blackbody per unit time and area (power per unit area) varies as the fourth power of the absolute temperature:

$$J = aT^{-4}$$

where a is a constant whose value is 5.67×10^{-8} W $\cdot$ m^{-2} $\cdot$ K^{-4}.

The relationship between the voltage of an X-ray tube (or other energy source), in volts, and the wavelength is given by the *Duane-Hunt equation:*

$$\lambda = \frac{hc}{eV} = \frac{12\ 398}{V}$$

where the wavelength is expressed in angstrom units.

Laws of Photometry

The time rate at which energy is transported in a beam of radiant energy is denoted by the symbol P_0 for the incident beam, and by P for the quantity remaining unabsorbed after passage through a sample or container. The ratio of radiant power transmitted by the sample to the radiant power incident on the sample is the *transmittance T:*

$$T = \frac{P}{P_0}$$

The logarithm (base 10) of the reciprocal of the transmittance is the *absorbance A:*

$$A = -\log T = \log\left(\frac{1}{T}\right)$$

When a beam of monochromatic light, previously rendered plane parallel, enters an absorbing medium at right angles to the plane-parallel surfaces of the medium, the rate of decrease in radiant power with the length of light path (cuvette interior) b, or with the concentration of absorbing material C (in grams per liter) will follow the exponential progression, often referred to as *Beer's law:*

$$T = 10^{-abC} \qquad \text{or} \qquad A = abC$$

where a is the absorptivity of the component of interest in the solution. When C is expressed in moles per liter,

$$T = 10^{-\varepsilon bC} \qquad \text{or} \qquad A = \varepsilon bC$$

where ε is the molar absorptivity.

The total fluorescence (or phosphorescence) intensity is proportional to the quanta of light absorbed, $P_0 - P$, and to the efficiency ϕ, which is the ratio of quanta absorbed to quanta emitted:

$$F = (P_0 - P)\phi = P_0\phi(1 - e^{-\varepsilon bC})$$

When the term εbC is not greater than 0.05 (or 0.01 in phosphorescence),

$$F = k\phi P_0 \varepsilon bC$$

where the term k has been introduced to handle instrumental artifacts and the geometry factor because fluorescence (and phosphorescence) is emitted in all directions but is viewed only through a limited aperture.

The thickness of a transparent film or the path length of infrared absorption cells b, in centimeters, is given by

$$b = \frac{1}{2n_{\mathrm{D}}} \left(\frac{n}{\bar{\nu}_1 - \bar{\nu}_2} \right)$$

where n is the number of fringes (peaks or troughs) between two wavenumbers $\bar{\nu}_1$ and $\bar{\nu}_2$, and n_D is the refractive index of the sample material (unity for the air path of an empty cuvette). If measurements are made in wavelength, as micrometers, the expression is

$$b = \frac{1}{2n_{\mathrm{D}}} \left(\frac{n\lambda_1\lambda_2}{\lambda_2 - \lambda_1} \right)$$

Grating Equation

The light incident on each groove is diffracted or spread out over a range of angles, and in certain directions reinforcement or constructive interference occurs, as stated in the grating formula:

$$m\lambda = b(\sin i \pm \sin r)$$

where b is the distance between adjacent grooves, i is the angle of incidence, r is the angle of reflection (both angles relative to the grating normal), and m is the order number. A positive sign applies where incoming and emergent beams are on the same side of the grating normal.

The *blaze wavelength* is that wavelength for which the angle of reflectance from the groove face and the angle of reflection (usually the angle of incidence) from the grating are identical.

The *Bragg equation:*

$$m\lambda = 2d \sin \theta$$

states the condition for reinforcement of reflection from a crystal lattice, where d is the distance between each set of atomic planes and θ is the angle of reflection.

Ionization of Metals in a Plasma

A loss in spectrochemical sensitivity results when a free metal atom is split into a positive ion and an electron:

$$\mathrm{M} = \mathrm{M}^+ + e^-$$

The degree of ionization, α_i, is defined as

$$\alpha_i = \frac{[M^+]}{[M^+] + [M]}$$

At equilibrium, when the ionization and recombination rates are balanced, the ionization constant K_i (in atm) is given by

$$K_i = \frac{[M^+][e^-]}{[M]} = \left(\frac{\alpha_i^2}{1 - \alpha_i^2}\right) p_{\Sigma M}$$

where $p_{\Sigma M}$ (in atm) is the total atom concentration of metal in all forms in the plasma.

The ionization constant can be calculated from the *Saha equation:*

$$\log K_i = -5040\,\frac{E_i}{T} + \frac{5}{2}\log T - 6.49 + \log\frac{g_{M^+}g_{e^-}}{g_M}$$

where E_i is the ionization potential of the metal in eV (Table 3-2), T is the absolute temperature of the plasma (in kelvins), and the g terms are the statistical weights of the ionized atom, the electron, and the neutral atom. For the alkali metals the final term is zero; for the alkaline earth metals, it is 0.6.

To suppress the ionization of a metal, another easily ionized metal (denoted a *deionizer* or *radiation buffer*) is added to the sample. To ensure that ionization is suppressed for the test element, the product $(K_i)_M p_M$ of the deionizer must exceed the similar product for the test element one hundred-fold (for 1 percent residual ionization of the test element).

NUCLEAR MAGNETIC RESONANCE

Table 8-11
PROTON CHEMICAL SHIFTS

Values are given on the officially approved δ scale; $\tau = 10.00 - \delta$.

Substituent Group	Methyl Protons	Methylene Protons	Methine Proton
HC—C—CH₂	0.95	1.20	1.55
HC—C—NR₂	1.05	1.45	1.70
HC—C—C=C	1.00	1.35	1.70
HC—C—C=O	1.05	1.55	1.95
HC—C—NRAr	1.10	1.50	1.80
HC—C—NH(C=O)R	1.10	1.50	1.90
HC—C—(C=O)NR₂	1.10	1.50	1.80
HC—C—(C=O)Ar	1.15	1.55	1.90
HC—C—(C=O)OR	1.15	1.70	1.90
HC—C—Ar	1.15	1.55	1.80
HC—C—OH	1.20	1.50	1.75
HC—C—OR	1.20	1.50	1.75
HC—C—C≡CR	1.20	1.50	1.80
HC—C—C≡N	1.25	1.65	2.00
HC—C—SR	1.25	1.60	1.90
HC—C—OAr	1.30	1.55	2.00
HC—C—O(C=O)R	1.30	1.60	1.80
HC—C—SH	1.30	1.60	1.65
HC—C—(S=O)R and —SO₂R	1.35	1.70	
HC—C—NR₃⁺	1.40	1.75	2.05
HC—C—O—N=O	1.40		
HC—C—O(C=O)CF₃	1.40	1.65	
HC—C—Cl	1.55	1.80	1.95
HC—C—F	1.55	1.85	2.15
HC—C—NO₂	1.60	2.05	2.50
HC—C—O(C=O)Ar	1.65	1.75	1.85
HC—C—I	1.75	1.80	2.10
HC—C—Br	1.80	1.85	1.90
HC—CH₂	0.90	1.30	1.50
HC—C=C	1.60	2.05	
HC—C≡C	1.70	2.20	2.80
HC—(C=O)OR	2.00	2.25	2.50
HC—(C=O)NR₂	2.00	2.25	2.40
HC—SR	2.05	2.55	3.00
HC—O—O	2.10	2.30	2.55
HC—(C=O)R	2.10	2.35	2.65
HC—C≡N	2.15	2.45	2.90
HC—I	2.15	3.15	4.25
HC—CHO	2.20	2.40	
HC—Ar	2.25	2.45	2.85
HC—NR₂	2.25	2.40	2.80
HC—SSR	2.35	2.70	
HC—(C=O)Ar	2.40	2.70	3.40
HC—SAr	2.40		
HC—NRAr	2.60	3.10	3.60
HC—SO₂R and —(SO)R	2.60	3.05	
HC—Br	2.70	3.40	4.10
HC—NR₃⁺	2.95	3.10	3.60
HC—NH(C=O)R	2.95	3.35	3.85

R = alkyl group.   Ar = aryl group.

8-41

Table 8-11 (*Continued*)
PROTON CHEMICAL SHIFTS

Substituent Group	Methyl Protons	Methylene Protons	Methine Proton
HC—SO$_3$R	2.95		
HC—Cl	3.05	3.45	4.05
HC—OH and —OR	3.20	3.40	3.60
HC—PAr$_3$	3.20	3.40	
HC—NH$_2$	3.50	3.75	4.05
HC—O(C=O)R	3.65	4.10	4.95
HC—OAr	3.80	4.00	4.60
HC—O(C=O)Ar	3.80	4.20	5.05
HC—O(C=O)CF$_3$	3.95	4.30	
HC—F	4.25	4.50	4.80
HC—NO$_2$	4.30	4.35	4.60
Cyclopropane		0.20	0.40
Cyclobutane		2.45	
Cyclopentane		1.65	
Cyclohexane		1.50	1.80
Cycloheptane		1.25	

Substituent Group	Proton Shift	Substituent Group	Proton Shift
HC≡CH	2.35	HO—C=O	10–12
HC≡CAr	2.90	HO—SO$_2$	11–12
HC≡C—C=C	2.75	HO—Ar	4.5–6.5
HAr	7.20	HO—R	0.5–4.5
HCO—O	8.1	HS—Ar	2.8–3.6
HCO—R	9.4–10.0	HS—R	1–2
HCO—Ar	9.7–10.5	HN—Ar	3–6
HO—N=C (oxime)	9–12	HN—R	0.5–5

Saturated Heterocyclic Ring Systems

Table 8-11 (*Continued*)
PROTON CHEMICAL SHIFTS

Unsaturated Cyclic Systems

$$\delta_{CH_2} = 0.23 + C_1 + C_2 \qquad \delta_{CH} = 0.23 + C_1 + C_2 + C_3$$

Table 8-12
ESTIMATION OF CHEMICAL SHIFT FOR PROTONS OF
$-CH_2-$ AND $>CH-$ GROUPS

X	C	X	C	X	C
$-CH_3$	0.5	$-SR$	1.6	$-OR$	2.4
$-CF_3$	1.1	$-C\equiv C-Ar$	1.7	$-Cl$	2.5
$>C=C<$	1.3	$-CN$	1.7	$-OH$	2.6
$-C\equiv C-R$	1.4	$-CO-R$	1.7	$-N=C=S$	2.9
$-COOR$	1.5	$-I$	1.8	$-OCOR$	3.1
$-NR_2$	1.6	$-Ph$	1.8	$-OPh$	3.2
$-CONR_2$	1.6	$-Br$	2.3		

Table 8-13
ESTIMATION OF CHEMICAL SHIFT OF PROTON
ATTACHED TO A DOUBLE BOND

Positive Z values indicate a downfield shift and an arrow indicates the point of attachment of the substituent group to the double bond.

$$\delta_{C=C} = 5.25 + Z_{gem} + Z_{cis} + Z_{trans}$$

$$\begin{array}{c} R_{cis} \quad\quad H \\ C=C \\ R_{trans} \quad R_{gem} \end{array}$$

R	Z_i for R (ppm)		
	Z_{gem}	Z_{cis}	Z_{trans}
→H	0	0	0
→alkyl	0.45	−0.22	−0.28
→alkyl—ring (5- or 6-member)	0.69	−0.25	−0.28
→CH$_2$O—	0.64	−0.01	−0.02
→CH$_2$S—	0.71	−0.13	−0.22
→CH$_2$X(X:F. Cl. Br)	0.70	0.11	−0.04
→CH$_2$N⟨	0.58	−0.10	−0.08
⟩C=C⟨ (isolated)	1.00	−0.09	−0.23
⟩C=C⟨ (conjugated)	1.24	0.02	−0.05
→C≡N	0.27	0.75	0.55
→C≡C—	0.47	0.38	0.12
⟩C=O (isolated)	1.10	1.12	0.87
⟩C=O (conjugated)	1.06	0.91	0.74
→COOH (isolated)	0.97	1.41	0.71
→COOH (conjugated)	0.80	0.98	0.32
→COOR (isolated)	0.80	1.18	0.55
→COOR (conjugated)	0.78	1.01	0.46
→C=O with H—N⟨	1.02	0.95	1.17
→C=O with Cl	1.37	0.98	0.46
→C=O with Cl	1.11	1.46	1.01
→OR (R: aliphatic)	1.22	−1.07	−1.21
→OR (R: conjugated)	1.21	−0.60	−1.00
→OCOR	2.11	−0.35	−0.64
→CH$_2$—C=O; →CH$_2$—C≡N	0.69	−0.08	−0.06
→CH$_2$—aromatic ring	1.05	−0.29	−0.32
→Cl	1.08	0.18	0.13
→Br	1.07	0.45	0.55
→I	1.14	0.81	0.88
→N—R (R: aliphatic)	0.80	−1.26	−1.21

Table 8-13 (*Continued*)
ESTIMATION OF CHEMICAL SHIFT OF PROTON
ATTACHED TO A DOUBLE BOND

R	Z_i for R (ppm)		
	Z_{gem}	Z_{cis}	Z_{trans}
→N—R (R: conjugated)	1.17	−0.53	−0.99
→N—C=O	2.08	−0.57	−0.72
→aromatic	1.38	0.36	−0.07
→SR	1.11	−0.29	−0.13
→SO₂	1.55	1.16	0.93

Table 8-14
CHEMICAL SHIFTS IN MONOSUBSTITUTED BENZENE

$$\delta = 7.27 + \Delta_i$$

Substituent	Δ_{ortho}	Δ_{meta}	Δ_{para}
NO_2	0.94	0.18	0.39
CHO	0.58	0.20	0.26
COOH	0.80	0.16	0.25
$COOCH_3$	0.71	0.08	0.20
COCl	0.82	0.21	0.35
CCl_3	0.8	0.2	0.2
$COCH_3$	0.62	0.10	0.25
CN	0.26	0.18	0.30
$CONH_2$	0.65	0.20	0.22
NH_3^+	0.4	0.2	0.2
CH_2X *	0.0–0.1	0.0–0.1	0.0–0.1
CH_3	−0.16	−0.09	−0.17
CH_2CH_3	−0.15	−0.06	−0.18
$CH(CH_3)_2$	−0.14	−0.09	−0.18
$C(CH_3)_3$	−0.09	0.05	−0.23
F	−0.30	−0.02	−0.23
Cl	0.01	−0.06	−0.08
Br	0.19	−0.12	−0.05
I	0.39	−0.25	−0.02
NH_2	−0.76	−0.25	−0.63
OCH_3	−0.46	−0.10	−0.41
OH	−0.49	−0.13	−0.2
OCOR	−0.2	0.1	−0.2
$NHCH_3$	−0.8	−0.3	−0.6
$N(CH_3)_2$	−0.60	−0.10	−0.62

*X = Cl, alkyl, OH, or NH_2.

Table 8-15
PROTON SPIN COUPLING CONSTANTS

Structure	J, Hz	Structure	J, Hz
$\diagdown$C$\diagdown^{\text{H}}_{\text{H}}$	12–15	H_a (a-a) H_e (a-e) H_e (e-e) H_a	8–10 2–3 2–3
$\diagdown$CH—CH$\diagup$ (free rotation)	6–8		
$\diagdown$CH—OH (no exchange)	5	Cyclopentane (cis)	4–6
(—NH)		(trans)	4–6
$\diagdown$CH—C$\overset{\text{H}}{=}$O	1–3	Cyclobutane (cis)	8
		(trans)	8
$H_t\diagdown$C=C$\diagup^{H_g}$ (gem) $H_c\diagup$ $\diagdown^H$ (cis) (trans)	0–3 6–14 11–18	Cyclopropane (cis) (trans) (hetero)	9–11 6–8 4–6
$H_c\diagdown$C=C$\diagup^{\text{CH}\diagdown}$ (cis) $H_t\diagup$ $\diagdown^{H_g}$ (trans) (gem)	0.5–3 0.5–3 4–10	$\langle\text{ring}\rangle$—H (o) (m) (p)	6–10 1–3 0–1
$\diagdown$C=CH—CH=C$\diagup$	10–13	(2–3)	5–6
=CH—C$\overset{|}{\underset{H}{=}}$O	6	(3–4) (2–4) (3–5) (2–5) (2–6)	7–9 1–2 1–2 0–1 0–1
—CH$_2$—C≡C—C$\overset{|}{\diagdown}$H	0–3	(1–2) (1–3) (2–3) (3–4) (2–4) (2–5)	2–3 2–3 2–3 3–4 1–2 1–3
$\diagdown$CH—C≡CH	0–3		
$H\diagdown$C=C$\diagup^H$ (3-member) (ring) (4-member) (5-member) (6-member) (7-member)	0–2 2–4 5–7 6–9 10–13	$\diagdown$C$\overset{\text{H}}{\underset{\text{F}}{\diagdown}}$	45–52
(2–3) (3–4) (2–4) (2–5)	1.8 3.5 0–1 1–2	$\diagdown$CH—CF$\diagdown$ (gauche) (trans)	0–12 10–45
(2–3) (3–4) (2–4) (2–5)	5–6 3.5–5.0 1.5 3.4	$H_t\diagdown$C=C$\diagup^{H_g}$ (gem) $H_c\diagup$ $\diagdown^F$ (cis) (trans)	72–90 1–8 12–40
$\langle\rangle$—F (o) (m) (p)	6–10 5–6 0–2	$\diagdown$C=C$\diagup^{\diagdown}_{\text{CH}_3}$ F$\diagup$	2–4
$\langle\rangle$—CH$_3$ (o) (m) F (p)	2.5 1.5 0	$\diagdown$C=C$\diagup^{\text{CF}}$ H$\diagup$	0–6

Table 8-16
SOLVENT POSITIONS OF RESIDUAL PROTONS IN
INCOMPLETELY DEUTERATED SOLVENTS

Solvent	Group	δ (ppm)
d_4-Acetic acid	Methyl	2.05
	Hydroxyl	11.5*
d_6-Acetone	Methyl	2.05
d_3-Acetonitrile	Methyl	1.95
d_6-Benzene	Methine	7.3
d_1-t-Butanol (OD)	Methyl	1.28
d_1-Chloroform	Methine	7.25
d_{12}-Cyclohexane	Methylene	1.40
Deuterium oxide	Hydroxyl	4.7*
d_7-Dimethylformamide	Methyl	2.75
	Methyl	2.95
	Formyl	8.05
d_6-Dimethylsulfoxide	Methyl	2.5
	Absorbed water	3.3*
d_8-Dioxan	Methylene	3.55
d_{18}-Hexamethylphosphoramide	Methyl	2.60, d, $J = 9$ Hz
d_4-Methanol	Methyl	3.35
	Hydroxyl	4.8*
d_2-Methylene chloride	Methylene	5.35
d_5-Pyridine	C—2 Methine	8.5
	C—3 Methine	7.0
	C—4 Methine	7.35
d_8-Toluene	Methyl	2.3
	Methine	7.2
d_1-Trifluoroacetic acid	Hydroxyl	11.3*

* These values may vary greatly, depending upon the solute and its concentration.

INFRARED SPECTROSCOPY

Table 8-17
ABSORPTION FREQUENCIES OF SINGLE BONDS TO HYDROGEN

From Williams and Fleming, *Spectroscopic Methods in Organic Chemistry*, 2d ed., McGraw-Hill Book Company, Ltd.; by permission.

Saturated C—H and C—C

Group	Band	Remarks
$>CH_2$ ⎫ $-CH_3$ ⎬	2960–2850(s)	Two or three bands usually; $>C-H$ stretching
$>CH$	2890–2880(w)	
$>CH_2$ $-CH_3$	1470–1430(m)	$>C-H$ deformations
$-CH_3$	1390–1370(m)	$-CH_3$ symmetrical deformation
$>CH_2$	ca. 720(w)	$>CH_2$ rocking

Miscellaneous C—H

Group	Band	Remarks
Cyclopropane C—H ⎫ Epoxide C—H ⎬ $-CCH_2$—halogen ⎭	ca. 3050(w)	C—H stretching; cf. alkenes
$-CO-CH_3$	3100–2900(w)	Often very weak
$-CHO$	2900–2700(w)	Usually two bands, one near 2720 cm^{-1}
$-O-CH_3$	2850–2810(m)	
$-O-CH_2-O-$	2790–2770(m)	
$N-CH_3$ and $N-CH_2-$	2820–2780(m)	
$-C(CH_3)_3$	1395–1385(m) 1365(s)	
$>C(CH_3)_2$	ca. 1380(m)	A roughly symmetrical doublet
$-O-CO-CH_3$ $-CO-CH_3$	1385–1365(s) 1360–1355(s)	The high intensity of these bands often dominates this region of the spectrum

Table 8-17 (*Continued*)
ABSORPTION FREQUENCIES OF SINGLE BONDS TO HYDROGEN

Alkene and Aromatic C—H

See also Table 8-21 for the corresponding double bond absorptions, and Table 8-22 for the aromatic C—H out-of-plane bending vibrations.

Group	Band	Remarks
—C≡C—H	ca. 3300(s)	
$\diagup C = C \diagdown$ (H, H)	3095–3075(m)	C—H stretching; sometimes obscured by the much stronger bands of saturated C—H groups which occur below 3000 cm^{-1}
$\diagup C = C \diagdown$ (H)	3040–3010(m)	
Aryl—H	3040–3010(w)	Often obscured
$\diagup C = C \diagdown$ (R, H / H, R)	970–960(s)	C—H out-of-plane deformation. When the double bond is conjugated with, for example, a C=O group this band is shifted towards 990 cm^{-1}
RCH=CH$_2$	995–985(s) and 940–900(s)	
R$_2$C=CH$_2$	895–885(s)	
R$_2$C=C (H, R)	840–790(m)	
$\diagup C = C \diagdown$ (H, H / R, R)	730–675(m)	

Alcohol and Phenol —O—H

Group	Band	Remarks
Water in solution	3710	
Free —OH	3650–3590(v)	Sharp; O—H stretching
H-bonded —OH (solid, liquid, and dilute solution)	3600–3200(s)	Often broad but may be sharp for some intramolecular single bridge H-bonds; the lower the frequency, the stronger the H-bond
Intramolecular H-bonded —OH in chelate form (see also Table 8-20, carboxylic acids)	3200–2500(v)	Broad; the lower the frequency, the stronger the H-bond; sometimes so broad as to be overlooked

Table 8-17 (*Continued*)
ABSORPTION FREQUENCIES OF SINGLE BONDS TO HYDROGEN

Alcohol and Phenol —O—H (*continued*)

Group	Band	Remarks
Water of crystal- lization (solid state spectra)	3600–3100(w)	Usually a weak band at 1640– 1615 cm^{-1} also; water in trace amounts in KBr discs shows a broad band at 3450 cm^{-1}
—O—H	1410–1260(s)	O—H bending
⟩C—OH	1150–1040(s)	C—O stretching

Amine, Imine, Ammonium, and Amide N—H
N—H Stretching

Much is known of amide N—H absorptions, the appearance of two bands being ascribed to forms I and II. The carbonyl region of many amides (Table 8-20) also shows two bands.

Group	Band	Remarks
Amine and imine ⟩N—H =N—H	3500–3300(m)	Primary amines show two bands in this range: the unsymmetrical and symmetrical stretching. Secondary amines absorb weak- ly. The pyrrole and indole N—H band is sharp
—NH$_3^+$ Amino acids	3130–3030(m)	Values for solid state; broad; bands also (but not always) near 2500 and 2000 cm^{-1}
Amino salts	ca. 3000(m)	
⟩NH$_2^+$ ⟩NH$^+$ =NH̟	2700–2250(m)	Values for solid state; broad, due to the presence of overtone bands, etc.
Primary amide —CONH$_2$	ca. 3500(m) ca. 3400(m)	Lowered ~150 cm^{-1} in the solid state and on H-bonding; often several bands 3200–3050 cm^{-1}
Secondary amide —CONH—	3460–3400(m)	Two bands; lowered on H- bonding and in the solid state. Only one band with lactams
	3100–3070(w)	A weak extra band with bonded and solid state samples

Table 8-17 *(Continued)*
ABSORPTION FREQUENCIES OF SINGLE BONDS TO HYDROGEN

N—H Bending

See also Table 8-20 for amide absorptions in this region.

Group	Band	Remarks
—NH$_2$	1650–1560(m)	
>NH	1580–1490(w)	Often too weak to be noticed
—NH$_3^+$	1600(s) 1500(s)	Secondary amine salts have the 1600 cm^{-1} band

Miscellaneous R—H

Group	Band	Remarks
—S—H	2600–2550(w)	Weaker than O—H and less affected by H-bonding
P—H	2440–2350(m)	Sharp
P(=O)(OH)	2700–2560(m)	Associated OH
R—D	1/1.37 times the corresponding R—H frequency	Useful when assigning R—H bands, deuteration leading to a known shift to lower frequency

Table 8-18
TRIPLE BONDS

Group	Band	Remarks
—C≡C—H	3300(m) 2140–2100(w)	C—H stretching C≡C stretching
—C≡C—	2260–2150(v)	*†‡
—C≡N	2260–2200(v)	C≡N stretching; stronger and to the lower end of the range when conjugated; occasionally very weak or absent; for example, some cyanohydrins show no C≡N absorption
Diazonium salts R—N≡N (R—N⁺≡N)	ca. 2260	
Thiocyanates R—S—C≡N	2175–2140(s)	Aryl thiocyanates at upper end of the range, alkyl at the lower end

*Conjugation with olefinic or acetylenic groups lowers the frequency and raises the intensity. Conjugation with carbonyl groups usually has little effect on the position of absorption.

†Symmetrical and nearly symmetrical substitution makes the C≡C stretching frequency inactive in the infrared. It is, however, seen clearly in the Raman spectrum.

‡When more than one acetylenic linkage is present, and sometimes when there is only one, there are frequently more absorption bands in this region than there are triple bonds to account for them.

Table 8-19
CUMULATED DOUBLE BONDS

Group	Band	Remarks
Carbon dioxide O=C=O	2349(s)	Appears in many spectra due to inequalities in path length
Isocyanates —N=C=O	2275–2250(s)	Very high intensity; position un-affected by conjugation
Azides —N$_3$	2160–2120(s)	
Carbodiimides —N=C=N—	2155–2130(s)	Very high intensity; split into an unsymmetrical doublet by conjugation with aryl groups
Ketenes >C=C=O	ca. 2150(s)	
Isothiocyanates —N=C=S	2140–1990(s)	Broad and very intense
Diazoalkanes R$_2$C=$\overset{+}{N}$=$\overset{-}{N}$	ca. 2100(s)	
Ketenimines C=C—N—	ca. 2000(s)	
Allenes C=C=C	ca. 1950(m)	Two bands when terminal allene or when bonded to electron attracting groups, e.g., —CO$_2$H

Table 8-20
CARBONYL ABSORPTION BANDS

All bands quoted are strong.

Groups	Band	Remarks
Acid anhydrides		
—CO—O—CO—		
Saturated	1850–1800	Two bands usually separated by about 60 cm^{-1}. The higher frequency band is more intense in acyclic anhydrides and the lower frequency band is more intense in cyclic anhydrides
	1790–1740	
Aryl and $\alpha\beta$-unsaturated	1830–1780	
	1700–1710	
Saturated five-ring	1870–1820	
	1800–1750	
All classes	1300–1050	One or two strong bands due to C—O stretching
Acid chlorides —COCl		
Saturated	1815–1790	Acid fluorides higher, bromides and iodides lower
Aryl and $\alpha\beta$-unsaturated	1790–1750	
Acid peroxides		
—CO—O—O—CO—		
Saturated	1820–1810	
	1800–1780	
Aryl and $\alpha\beta$-unsaturated	1805–1780	
	1785–1755	
Esters and lactones		
—CO—O—		
Saturated	1750–1735	
Aryl and $\alpha\beta$-unsaturated	1730–1715	
Aryl and vinyl esters		
C=C—O—CO—alkyl	1800–1750	The C=C stretching band also shifts to higher frequency
Esters with electronegative α-substituents; e.g.,		
$>$CCl—CO—O—	1770–1745	
α-Keto esters	1755–1740	
Six-ring and larger lactones	Similar values to the corresponding open chain esters	
Five-ring lactone	1780–1760	
$\alpha\beta$-Unsaturated five-ring lactone	1770–1740	When α-C—H present there are two bands, the relative intensity depending on the solvent
$\beta\gamma$-Unsaturated five-ring lactone; i.e., vinyl ester type	~1800	
Four-ring lactone	~1820	
β-Keto ester in H-bonding enol form	~1650	Keto from normal; chelate type H-bond causes shift to lower frequency than the normal ester. The C=C is usually near 1630(s) cm^{-1}
All classes	1300–1050	Usually two strong bands due to C—O stretching

Table 8-20 (*Continued*)
CARBONYL ABSORPTION BANDS

Groups	Band	Remarks
Aldehydes —CHO		
(See also Table 8-17 for C—H.) All values given below are lowered in liquid film or solid state spectra by about 10–20 cm^{-1}. Vapor phase spectra have values raised about 20 cm^{-1}		
Saturated	1740–1720	
Aryl	1715–1695	Orthohydroxy or amino groups shift this value to 1655–1625 cm^{-1} due to intramolecular H-bonding
$\alpha\beta$-Unsaturated	1705–1680	
$\alpha\beta,\gamma\delta$-Unsaturated	1680–1660	
β-Ketoaldehyde in enol form	1670–1645	Lowering caused by chelate type H-bonding
Ketones >C=O		
All values given below are lowered in liquid film or solid state spectra by about 10–20 cm^{-1}. Vapor phase spectra have values raised about 20 cm^{-1}		
Saturated	1725–1705	
Aryl	1700–1680	
$\alpha\beta$-Unsaturated	1685–1665	
$\alpha\beta,\alpha'\beta'$-Unsaturated and diaryl	1670–1660	
Cyclopropyl	1705–1685	
Six-ring ketones and larger	Similar values to the corresponding open chain ketones	
Five-ring ketones	1750–1740	$\alpha\beta$-Unsaturation, etc., has a similar effect on these values as on those of open chain ketones
Four-ring ketones	~1780	
α-Halo ketones	1745–1725	Affected by conformation; highest values are obtained when both halogens are in the same plane as the C=O
α,α'-Dihalo ketones	1765–1745	
1,2-Diketones s-*trans* (i.e., open chains)	1730–1710	Antisymmetrical stretching frequency of both C=O's. The symmetrical stretching is inactive in the infrared but active in the Raman
1,2-Diketones s-*cis*, six-ring	1760 and 1730	
1,2-Diketones s-*cis*, five-ring	1775 and 1760	
o-Amino- or o-hydroxy-aryl ketones	1655–1635	Low due to intramolecular H-bonding. Other substituents and steric hindrance, etc., affect the position of the band
Quinones	1690–1660	C=C usually near 1600(s) cm^{-1}
Extended quinones	1655–1635	
Tropone	1650	Near 1600 cm^{-1} when lowered by H-bonding as in tropolones

8-55

Table 8-20 (*Continued*)
CARBONYL ABSORPTION BANDS

Groups	Band	Remarks
Carboxylic acids —CO₂H		
All types	3000–2500	O—H stretching; a characteristic group of small bands due to combination bands, etc.
Saturated	1725–1700	The monomer is near 1760 cm⁻¹, but is rarely observed. Occasionally both bands, the free monomer, and the H-bonded dimer can be seen in solution spectra. Ether solvents give one band near 1730 cm⁻¹
αβ-Unsaturated	1715–1690	
Aryl	1700–1680	
α-Halo-	1740–1720	
Carboxylate ions —CO₂⁻		
Most types	1610–1550 1420–1300	Antisymmetrical and symmetrical stretching, respectively
Amides —CO—N<		
(See also Table 8-17 for N—H stretching and bending.)		
Primary —CONH₂		
In solution	~1690	Amide I; C=O stretching
Solid state	~1650	
In solution	~1600	Amide II; mostly N—H bending
Solid state	~1640	
		Amide I is generally more intense than amide II. (In the solid state, amides I and II may overlap.)
Secondary —CONH—		
In solution	1700–1670	Amide I
Solid state	1680–1630	
In solution	1550–1510	Amide II; found in open chain amides only
Solid state	1570–1515	
		Amide I is generally more intense than amide II
Tertiary	1670–1630	Since H-bonding is absent, solid and solution spectra are much the same
Lactams		
Six- and larger rings	~1670	
Five-ring	~1700	Shifted to higher frequency when the N atom is in a bridged system
Four-ring	~1745	
R—CO—N—C=C		Shifted +15 cm⁻¹ by the additional double bond
C=C—CO—N		Shifted by up to +15 cm⁻¹ by the additional double bond. This is an unusual effect for αβ-unsaturation. It is said to be due to the inductive effect of the C=C on the well conjugated CO—N system, the usual conjugation effect being less important in such a system

Table 8-20 (*Continued*)
CARBONYL ABSORPTION BANDS

Groups	Band	Remarks
Imides —CO—N—CO—		
Cyclic six-ring	ca. 1710 and ca. 1700	Shift of $+15$ cm^{-1} with $\alpha\beta$-unsaturation
Cyclic five-ring	ca. 1770 and ca. 1700	
Ureas N—CO—N		
RNHCONHR	ca. 1660	
Six-ring	ca. 1640	
Five-ring	ca. 1720	
Urethanes		
R—O—CO—N	1740–1690	Also shows amide II band when non- or monosubstituted on N
Thioesters and Acids RCO—S—R'		
RCOSH	ca. 1720	$\alpha\beta$-Unsaturated or aryl acid or ester shifted ca. -25 cm^{-1}
RCOS—alkyl	ca. 1690	
RCOS—aryl	ca. 1710	

Intensities of Carbonyl Bands. Acids generally absorb more strongly than esters, and esters more strongly than ketones or aldehydes. Amide absorption is usually similar in intensity to that of ketones but is subject to much greater variations.

Position of Carbonyl Absorption. The general trends of structural variation on the position of C=O stretching frequencies may be summarized as follows:

(1) The more electronegative the group X in the system R—CO—X—, the higher is the frequency.

(2) $\alpha\beta$-Unsaturation causes a lowering of frequency of 15–40 cm^{-1}, except in amides, where little shift is observed and that usually to higher frequency.

(3) Further conjugation has relatively little effect.

(4) Ring strain in cyclic compounds causes a relatively large shift to higher frequency. This phenomenon provides a remarkably reliable test of ring size, distinguishing clearly between four, five, and larger membered ring ketones, lactones, and lactams. Six-ring and larger ketones, etc., show the normal frequency found for the open chain compounds.

(5) Hydrogen bonding to a carbonyl group causes a shift to lower frequency of 40–60 cm^{-1}. Acids, amides, enolized β-keto carbonyl systems, and *o*-hydroxy- and *o*-aminophenyl carbonyl compounds show this effect. All carbonyl compounds tend to give slightly lower values for the carbonyl stretching frequency in the solid state compared with the value for dilute solutions.

(6) Where more than one of the structural influences on a particular carbonyl group is operating, the net effect is usually close to additive.

Table 8-21
OTHER DOUBLE BOND ABSORPTION BANDS

Alkenes $>$C$=$C$<$

See also Table 8-17 for the $=$C$-$H absorptions of alkenes.

Group	Band	Remarks
Nonconjugated		
$>$C$=$C$<$	1680–1620(v)	May be very weak if more or less symmetrically substituted
Conjugated with aromatic ring	ca. 1625(m)	More intense than with unconjugated double bonds
Dienes, trienes, etc.	1650(s) and 1600(s)	Lower-frequency band usually more intense and may hide or overlap the higher-frequency band
$\alpha\beta$-Unsaturated carbonyl compounds	1640–1590(s)	Usually much weaker than the C$=$O band
Enol esters, enol ethers, and enamines	1690–1650(s)	

Imines, Oximes, etc. $>$C$=$N$--$

Group	Band	Remarks
$>$C$=$N$-$H	3400–3300(m)	N$-$H stretching; lowered on H-bonding
$>$C$=$N$-$	1690–1640(v)	Difficult to identify due to large variations in intensity and the
$\alpha\beta$-Unsaturated	1600–1630(v)	closeness to C$=$C stretching region. Oximes usually give very
Conjugated cyclic systems	1660–1480(v)	weak bands

Azo Compounds $-$N$=$N$-$

Group	Band	Remarks
$-$N$=$N$-$	ca. 1575(v)	Very weak or inactive in infrared. Sometimes seen in Raman
$-\overset{+}{\text{N}}=$N$-$ $\underset{\text{O}^-}{|}$	ca. 1570	

Nitro, Nitroso, etc. N$=$O

Group	Band	Remarks
C$-$NO$_2$	ca. 1560(s) ca. 1350(s)	Lowered $\sim$30 cm^{-1} when conjugated. The two bands are due to asymmetrical and symmetrical stretching of the NO bonds
Nitrates O$-$NO$_2$	1650–1600(s) 1270–1250(s)	

Table 8-21 (*Continued*)
OTHER DOUBLE BOND ABSORPTION BANDS

Nitro, Nitroso, etc. N=O (*Continued*)

Group	Band	Remarks
Nitramines N—NO$_2$	1630–1550(s) 1300–1250(s)	
C—N=O	1600–1500(s)	
O—N=O	1680–1610(s)	Two bands
N—N=O	1500–1430(s)	
$\overset{+}{\geqslant}$N—$\overset{-}{\text{O}}$		
Aromatic Aliphatic	1300–1200(s) 970–950(s)	Very strong bands
NO$_3^-$	1410–1340 860–800	

Table 8-22
AROMATIC ABSORPTION BANDS

Aromatic Compounds

See also Table 8-17 for aryl—H vibration frequencies.

Group	Band	Remarks
Aromatic rings	ca. 1600(m) ca. 1580(m) ca. 1500(m)	 Stronger when the ring is further conjugated This is usually the strongest of the two or three bands

Substitution Patterns of the Benzene Ring

Group	Band	Remarks
Five adjacent H	770–730(s) and 720–680(s)	Monosubstituted
Four adjacent H	770–735(s)	Ortho-disubstituted
Three adjacent H	810–750(s)	Meta-disubstituted, etc., and 1,2,3-trisubstituted
Two adjacent H	860–800(s)	Para-disubstituted, etc.
Isolated H	900–800(w)	Meta-disubstituted, etc.; usually not strong enough to be useful

Values hold reasonably well for condensed ring systems and pyridines.

Table 8-23
MISCELLANEOUS ABSORPTION BANDS
Sulfur Compounds

Group	Band	Remarks
—S—H	2600–2550(w)	S—H stretching; weaker than O—H and less affected by H-bonding. This absorption is strong in the Raman
>C=S	1200–1050(s)	
>C—N< (with S below C)	ca. 3400	N—H stretching; lowered to ~3150 cm^{-1} in the solid state
	1500–1460(s)	Amide II
	1300–1100(s)	Amide I
>S=O	1060–1040(s)	
>SO$_2$	1350–1310(s)	
	1160–1120(s)	
—SO$_2$—N<	1370–1330(s)	
	1180–1160(s)	
—SO$_2$—O—	1420–1330(s)	
	1200–1145(s)	

Phosphorus Compounds

Group	Band	Remarks
P—H	2440–2350(s)	Sharp
P—Ph	1440(s)	Sharp
P—O—alkyl	1050–1030(s)	
P—O—aryl	1240–1190(s)	
P=O	1300–1250(s)	
P—O—P	970–910	Broad
P(=O)(OH)	2700–2560	H-bonded O—H
	1240–1180(s)	P=O stretching

Ethers

Group	Band	Remarks
>C—O—C<	1150–1070(s)	C—O stretching
=C—O—C<	1275–1200(s)	
	1075–1020(s)	

Table 8-23 (*Continued*)
MISCELLANEOUS ABSORPTION BANDS

Ethers (*Continued*)

Group	Band	Remarks
C—O—CH$_3$	2850–2810(m)	C—H stretching; aryl ethers at higher end of the range
>C—C< over O (epoxide)	ca. 1250 ca. 900 ca. 800	

Halogen Compounds

Group	Band	Remarks
C—F	1400–1000(s)	
C—Cl	800–600(s)	
C—Br	750–500(s)	
C—I	~500(s)	

Inorganic Ions

Group	Band	Remarks
Ammonium	3300–3030	All bands strong
Cyanide, thiocyanate, cyanate	2200–2000	
Carbonate	1450–1410	
Sulfate	1130–1080	
Nitrate	1380–1350	
Nitrite	1250–1230	
Phosphates	1100–1000	

Section 9
THERMODYNAMIC PROPERTIES

ENTHALPIES AND GIBBS (FREE) ENERGIES OF FORMATION, ENTROPIES, AND HEAT CAPACITIES OF ELEMENTS AND COMPOUNDS

The tables contain values of the enthalpy and Gibbs (formerly free) energy of formation, entropy, and heat capacity at 298.15K (25°C). No values are given in these tables for metal alloys or other solid solutions, fused salts, or for substances of undefined chemical composition.

For a more complete listing of compounds see the tables of "Selected Values of Chemical Thermodynamic Properties," *National Bureau of Standards Technical Notes* 270-3, 270-4, 270-5, 270-6, and 270-7 by D. D. Wagman et al., Washington, D.C.; "JANAF Thermochemical Tables," by D. R. Stull and H. Prophet, *National Bureau of Standards Publication* 37, Washington, D.C.; supplements to JANAF appearing in *J. Phys. Chem. Reference Data*; D. R. Stull, E. F. Westrum, Jr., and G. C. Sinke, *The Chemical Thermodynamics of Organic Compounds*, Wiley-Interscience, New York, 1969; and I. Barin and O. Knacke, *Thermochemical Properties of Inorganic Substances*, Springer-Verlag, Berlin, 1973.

The physical state of each substance is indicated in the column headed State as crystalline solid (c), liquid (liq), gaseous (g), or amorphous (amorp). Solutions in water are listed as aqueous (aq).

The values of the thermodynamic properties of the pure substances given in these tables are, for the substances in their standard states, defined as follows: For a pure solid or liquid, the standard state is the substance in the condensed phase under a pressure of one atmosphere. For a gas the standard state is the hypothetical ideal gas at unit fugacity, in which state the enthalpy is that of the real gas at the same temperature and at zero pressure.

The values of $\Delta Hf°$ and $\Delta Gf°$ given in the tables represent the change in the appropriate thermodynamic quantity when one gram-formula weight of the substance in its standard state is formed, isothermally at the indicated temperature, from the elements, each in its appropriate standard reference state. The standard reference state at 25°C for each element has been chosen to be the standard state that is thermodynamically stable at 25°C and one atmosphere pressure. The standard reference states are indicated in the tables by the fact that the values of $\Delta Hf°$ and $\Delta Gf°$ are exactly zero.

The values of $S°$ represent the virtual or "thermal" entropy of the substance in the standard state at 298.15 K, omitting contributions from nuclear spins. Isotope mixing effects are also excluded except in the case of the (^{1}H-^{2}H) system.

THERMODYNAMIC PROPERTIES

Solutions in water are designated as aqueous, and the concentration of the solution is expressed in terms of the number of moles of solvent associated with one mole of the solute. If no concentration is indicated, the solution is assumed to be dilute. The standard state for a solute in aqueous solution is taken as the hypothetical ideal solution of unit molality (indicated as std. state, $m = 1$). In this state the partial molal enthalpy and the heat capacity of the solute are the same as in the infinitely dilute real solution (aq, ∞).

The value of ΔHf° given for a solute in its standard state is the apparent molal enthalpy of formation of the substance in the infinitely dilute real solution. The experimental value for a heat of dilution is obtained directly as the difference between the two values of ΔHf° at the corresponding concentrations.

Values of ΔHf° and ΔGf° (or ΔFf°) in the tables are expressed in kilocalories per mole; values of S° and C_p° are expressed in calories per degree per mole.

Table 9-1
ELEMENTS AND INORGANIC COMPOUNDS

Formula and Description	State	$\Delta Hf°$	$\Delta Gf°$	$S°$	$C_p°$
Actinium					
Ac	c	0	0	15.00	6.50
	g	94.0			
AcBr$_3$	c	−220			
AcCl$_3$	c	−271			
AcF$_3$	c	−477			
Ac$_2$O$_3$		−444	−425		
AcOF	c	−265			
Aluminum					
Al	c	0	0	6.77	5.81
	liq	2.07	1.58	8.42	7.59
	g	78.00	68.30	39.30	5.11
Al^{3+} std. state, $m = 1$	aq	−127	−116	−76.9	
AlAs	c	−27.8			
Al$_6$BeO$_{10}$	c	−1344.2	−1270.9	41.96	63.38
	liq	−1266.6	−1203.2	75.26	63.38
Al(BH$_4$)$_3$	liq	−3.9	34.6	69.1	46.5
	g	3	35	90.6	
AlBO$_2$	g	−129.4	−131.6	64.4	12.63
AlBr	g	(3.8)	5.8	57.25	8.51
AlBr$_3$	c	−122.16	−116.71	43.08	24.04
	liq	−119.79	−116.22	49.35	29.87
	g	−98.2	−104.8	83.43	18.03
std. state, $m = 1$	aq	−214	−191	−17.8	
Al$_2$Br$_6$	g	−244.0	−226.4	130.8	39.88
AlC	g	164.8	151.3	53.38	7.71
Al$_4$C$_3$	c	−49.54	−56.99	25	27.74
	g	−51.6	−48.6	21.3	27.91
Al(CH$_3$)$_3$	liq	−32.6	−2.4	50.05	37.19
	g	−17.7			
Al$_2$(CH$_3$)$_6$	g	−55.19	−2.34	125.4	
Al(OAc)$_3$	c	−452.3			
AlCl	g	−12.3	−18.6	54.46	8.28
AlCl$_2$	g	−69.0	−71.6	68.9	12.51
AlCl$_3$	c	−168.65	−150.59	26.12	21.78
	liq	−161.28	−147.75	41.33	30.00
	g	−139.70	−136.25	75.12	17.18
std. state, $m = 1$	aq	−247	−210	−36.4	
AlCl$_3 \cdot$ 6H$_2$O	c	−643.3	−542.4	90	
Al$_2$Cl$_6$	g	−309.6	−291.8	113.65	37.74
AlF	g	−63.4	−69.5	51.4	7.63
AlF$_2$	g	−175	−177	62.9	10.97
AlF$_3$	c	−361.0	−342.0	15.89	17.96
	g	−289.0	−285.1	66.15	14.95
AlF$_3 \cdot$ 3H$_2$O	c	−549.1	−490.4	50	
Al$_2$F$_6$	g	−629	−607	92.5	31.96
AlH	g	61.96	55.25	44.88	7.02
AlH$_3$	c	−11			
AlI$_3$	c	−74	−73	45.3	23.64
	liq	−71	−72	52.5	29.00
	g	−49	−60	86.8	18.28
std. state, $m = 1$	aq	−167	−153	2.9	
Al$_2$I$_6$	g	−121	−134	139.6	40.89
AlN	c	−76.0	−68.6	4.82	7.20
	g	104	98	50.6	7.44
Al(NO$_3$)$_3$ std. state, $m = 1$	aq	−276	−196	28.1	
Al(NO$_3$)$_3 \cdot$ 6H$_2$O	c	−681.28	−526.74	111.8	103.5
Al(NO$_3$)$_3 \cdot$ 9H$_2$O	c	−897.96	−700.2	136	
AlO	g	20	13.8	52.17	7.38
AlO$_2^-$ std. state, $m = 1$	aq	−219.6	−196.8	−5	
Al$_2$O	g	−31.4	−38.5	62.0	10.92
Al$_2$O$_3$ α, corundum	c	−400.4	−378.1	12.17	18.89
γ	c	−396.0	−373.5	14.3	18.89
	liq	−377.90	−358.33	21.41	18.89

Table 9-1 (*Continued*)
ELEMENTS AND INORGANIC COMPOUNDS

Formula and Description	State	$\Delta Hf°$	$\Delta Gf°$	$S°$	$C_p°$
$Al_2O_3 \cdot H_2O$ boehmite	c	−472.0	−436.3	23.15	31.37
diaspore	c	−478	−440	16.86	25.22
$Al_2O_3 \cdot 3H_2O$ gibbsite	c	−612.5	−546.7	33.51	44.49
bayerite	c	−610.1			
AlOCl	c	−189.60	−176.21	13.00	13.60
	g	−83.2	−83.7	59.47	11.97
AlOF	g	−140.2	−140.3	55.99	10.69
AlOH	g	−43	−44	51.7	7.63
$Al(OH)^{2+}$ std. state, $m = 1$	aq		−165.9		
$Al(OH)_3$	c	−307	−312.1	17	22.26
$Al(OH)_4^-$ std. state, $m = 1$	aq	−356.2	−310.2	28	
AlP	c	−39.8			
$AlPO_4$ berlinite	c	−404.4	−382.7	21.70	22.27
AlS	g	48	35.9	55.09	7.98
Al_2S_3	c	−173		(23)	
$Al_2(SO_4)_3$	c	−821.0	−838.1	57.2	62.00
std. state, $m = 1$	aq	−906	−766	−139.4	
Al_2Se_3	c	−135			
Al_2SiO_5 andalusite	c	−619.5	−584.3	22.3	29.34
kyanite	c	−620.5	−584.1	20.03	29.10
sillimanite	c	−619.8	−583.8	22.99	29.21
$Al_2Si_2O_7 \cdot 2H_2O$ halloysite	c	−975.1	−898.5	48.6	58.86
kaolinite	c	−979.6	−903.0	48.5	58.62
$Al_6Si_2O_{13}$ mullite	c	−1630	−1540	65.7	77.83
Al_2Te_3	c	−78			
Americium					
Am	c	0	0	15.0	
Am^{3+} std. state, $m = 1$	aq	−163.2	−160.5	−38.0	
Am^{4+} std. state, $m = 1$	aq	−122.3	−110.2	−89	
$AmCl_3$	c	−251	−234		
AmF_3	c	−394			
	aq			17.6	
AmF_4	c	−400			
Am_2O_3	c	−420	−401	36.98	
AmO_2	c	−240.2	−227.1	20.0	
AmOCl	c	−228			
$Am(OH)_3$	c		−300.0		
$Am(OH)_4$	c		−347.0		
Ammonium					
NH_3	g	−11.02	−3.94	45.97	8.38
undissoc; std. state, $m = 1$	aq	−19.19	−6.35	26.6	
	aq, 1	−18.011	−18.011		
	aq, 10	−19.074	−19.074		
	aq, 100	−19.167	−19.167		
NH_4^+ std. state, $m = 1$	aq	−31.67	−18.97	27.1	19.1
NH_4OH	liq	−86.33	−60.74	39.57	37.02
undissoc; std. state, $m = 1$	aq	−87.505	−63.04	43.3	
ionized; std. state, $m = 1$	aq	−86.64	−56.56	24.5	−16.4
	aq, 1	−86.875			
	aq, 2	−87.078			
	aq, 10	−87.396			
	aq, 100	−87.483			
$NH_4Al(SO_4)_2$	c	−562.2	−487.2	51.7	54.12
std. state, $m = 1$	aq	−593	−491	−40.2	
NH_4AsO_2 std. state, $m = 1$	aq	−134.21	−102.63	37.0	
$NH_4H_2AsO_3$ std. state, $m = 1$	aq	−202.51	−159.32	53.5	
$NH_4H_2AsO_4$	c	−523.3	−199.1	41.12	36.13
std. state, $m = 1$	aq	−249.06	−199.01	55.1	
$(NH_4)_2HAsO_4$	c	−282.4			
std. state, $m = 1$	aq	−279.9	−208.7	53.8	
$(NH_4)_3AsO_4$	c	−307.4			
std. state, $m = 1$	aq	−307.28	−211.91	42.4	

Table 9-1 (*Continued*)
ELEMENTS AND INORGANIC COMPOUNDS

Formula and Description	State	$\Delta Hf°$	$\Delta Gf°$	$S°$	$C_p°$
NH_4BO_2 std. state, $m = 1$	aq	−216.27	−181.24	18.2	
NH_4Br	c	−64.73	−41.9	27.00	21.19
std. state, $m = 1$	aq	−60.72	−43.82	46.8	−14.8
	aq, 100	−60.614			
	aq, 1000	−60.650			
NH_4BrO	aq	−54.2	−27.0	37	
NH_4BrO_3	aq	−51.7	−18.6	66.1	
$(NH_4)_2CO_3$ std. state, $m = 1$	aq	−225.18	−164.11	40.6	
NH_4HCO_3	c	−203.0	−159.2	28.9	
std. state, $m = 1$	aq	−197.06	−159.23	48.9	
NH_4 carbamate	c	−154.17	−107.09	31.9	
NH_4CN	c	0.10			32
std. state, $m = 1$	aq	4.3	22.2	49.6	
NH_4CNO cyanate	c	−72.75			
std. state, $m = 1$	aq	−66.6	−42.3	52.6	
NH_4CNS thiocyanate	c	−18.8			
std. state, $m = 1$	aq	−13.40	3.18	61.6	9.5
NH_4 formate	c	−135.63			
std. state, $m = 1$	aq	−133.38	−102.9	49	−1.9
NH_4 acetate	c	−147.26			
std. state, $m = 1$	aq	−147.83	−107.26	47.8	17.6
NH_4 chloroacetate	c	−153.7			
NH_4 trichloroacetate	c	−156.7			
$(NH_4)_2C_2O_4$	c	−268.72			
	aq, 2100	−260.6	−196.2		
NH_4 dithiocarbamate	c	−30.3			
NH_4Cl	c	−75.15	−48.51	22.6	20.1
std. state, $m = 1$	aq	−71.62	−50.34	40.6	−13.5
	aq, 10	−71.567			
	aq, 100	−71.487			
NH_4ClO std. state, $m = 1$	aq	−57.3	−27.8	37	
NH_4ClO_2 std. state, $m = 1$	aq	−47.6	−14.9	51.3	
NH_4ClO_3 std. state, $m = 1$	aq	−55.4	−19.8	65.9	
NH_4ClO_4	c	−70.58	−21.25	44.02	30.61
std. state, $m = 1$	aq	−62.58	−21.03	70.6	
NH_4HCrO_4 std. state, $m = 1$	aq	−241.6	−201.8	71.1	
$(NH_4)_2CrO_4$	c	−279.0			
std. state, $m = 1$	aq	−273.5	−211.90	66.2	
$(NH_4)_2Cr_2O_7$	c	−431.8			
std. state, $m = 1$	aq	−419.5	−348.9	116.8	
$NH_4Cr(SO_4)_2 \cdot 12H_2O$	c			170.9	168.5
NH_4F	c	−110.89	−83.36	17.20	15.60
std. state, $m = 1$	aq	−111.17	−85.61	23.8	−6.4
NH_4HF_2	c	−191.9	−155.6	27.61	25.50
std. state, $m = 1$	aq	−187.01	−157.15	49.2	
NH_4I	c	−48.14	−26.9	28	19.54
std. state, $m = 1$	aq	−44.86	−31.30	53.7	−14.9
	aq, 100	−44.784			
NH_4IO std. state, $m = 1$	aq	−57.4	−28.2	25.8	
NH_4IO_3	c	−92.2			
std. state, $m = 1$	aq	−84.6	−49.6	55.4	
NH_4IO_4	aq	−66.9			
NH_4N_3 azide	c	27.6	65.5	26.9	
	aq	34.1	64.3	52.7	
NH_4NO_2	c	−61.3			
std. state, $m = 1$	aq	−56.7	−27.9	60.6	−4.2
NH_4NO_3	c	−87.37	−43.98	36.11	33.3
std. state, $m = 1$	aq	−81.23	−45.58	62.1	−1.6
	aq, 10	−82.470			
	aq, 100	−81.340			
$(NH_4)_2O$	liq	−102.94	−63.84	63.94	59.08
NH_4PO_3	aq	−265.2			

Table 9-1 (*Continued*)
ELEMENTS AND INORGANIC COMPOUNDS

Formula and Description	State	$\Delta Hf°$	$\Delta Gf°$	$S°$	$C_p°$
NH₄H₂PO₂ hypophosphite	c	−180.0			
NH₄H₂PO₄	c	−345.38	−289.33	36.32	34.00
std. state, $m = 1$	aq	−341.49	−289.14	48.7	
(NH₄)₂HPO₄	c	−374.50			45
std. state, $m = 1$	aq	−372.17	−298.28	46.2	
(NH₄)₃PO₄	c	−399.6			
std. state, $m = 1$	aq	−400.3	−300.4	28	
(NH₄)₄P₂O₇ std. state, $m = 1$	aq	−669.5	−534.6	80	
(NH₄)₂PoCl₆ std. state, $m = 1$	aq		−176		
(NH₄)₂PtCl₆	c	−192.0			56.8
NH₄ReO₄	c	−226.0	−185.2	55.6	
NH₄HS	c	−37.5	−12.1	23.3	
std. state, $m = 1$	aq	−35.9	−16.09	42.1	
(NH₄)₂S std. state, $m = 1$	aq	−55.4	−17.4	50.7	
(NH₄)₂S₂ std. state, $m = 1$	aq	−56.1	−18.9	61.0	
NH₄HSO₃	c	−183.7			
std. state, $m = 1$	aq	−181.34	−145.12	60.5	
NH₄HSO₄	c	−245.45			
std. state, $m = 1$	aq	−243.75	−199.66	58.6	−0.9
	aq, 200	−245.65			
(NH₄)₂SO₃	c	−211.6			
std. state, $m = 1$	aq	−215.2	−154.2	47.2	
(NH₄)₂SO₄	c	−282.23	−215.56	52.6	44.81
std. state, $m = 1$	aq	−280.66	−215.77	58.6	−31.8
	aq, 100	−280.407			
(NH₄)₂S₂O₃	aq	−219.2			
(NH₄)₂S₂O₄ std. state, $m = 1$	aq	−243.4	−181.4	76	
(NH₄)₂S₂O₆	aq	−349.7			
(NH₄)₂S₂O₇	aq	−398.2			
(NH₄)₂S₂O₈	c	−392.5			
std. state, $m = 1$	aq	−383.3	−303.3	113.5	
(NH₄)₂S₄O₆	aq	−355.92			
(NH₄)₂Sb₂S₄ std. state, $m = 1$	aq	−115.7	−61.7	41.7	
NH₄HSe	c	−31.8	−5.6	23.1	
std. state, $m = 1$	aq	−27.9	−8.5	46	
(NH₄)₂Se std. state, $m = 1$	aq		−7.0		
NH₄HSeO₃ std. state, $m = 1$	aq	−154.65	−117.33	60.2	
NH₄HSeO₄ std. state, $m = 1$	aq	−170.7	−127.1	62.8	
(NH₄)₂SeO₃ std. state, $m = 1$	aq	−185.0	−126.3	57.3	
(NH₄)₂SeO₄	c	−209.0			
std. state, $m = 1$	aq	−206.5	−143.4	67.1	
(NH₄)₂SiF₆ hexagonal	c	−640.94	−565.38	66.98	54.52
(NH₄)₂SnCl₆	c	−295.6			
NH₄HTe	c	0.3			
NH₄H₅TeO₆	aq	−333.2			
(NH₄)₂TeO₃	aq	−205.9			
NH₄VO₃	c	−251.7	−212.3	33.6	30.91
Antimony					
Sb III	c	0	0	10.92	6.03
IV explosive	amorp	2.54			
	g	62.7	53.1	43.06	4.97
Sb₂	g	56.3	44.7	60.90	8.70
Sb₄	g	49.0	33.8	84	
SbBr₃	c	−62.0	−57.2	43.0	25.94
SbCl	g	−6.22			8.49
SbCl₂	g	−18.5			
SbCl₃	c	−91.34	−77.37	44.0	25.8
	g	−75.0	−72.0	80.71	18.33
SbCl₅	liq	−105.2	−83.7	72	
	g	−94.25	−79.91	96.04	28.95
SbF	g	−11.29			7.97
SbF₃	c	−218.8			

9-7

Table 9-1 (*Continued*)
ELEMENTS AND INORGANIC COMPOUNDS

Formula and Description	State	$\Delta Hf°$	$\Delta Gf°$	$S°$	$C_p°$
SbH_3	g	34.681	35.31	55.61	9.81
Sb_2H_4	g	57.2			
SbI_3	c	−24.0		51.5	23.32
SbN	g	63.66			7.41
SbO	g	47.67			
SbO^+ std. state, $m = 1$	aq		−42.33		
SbO_2^- std. state, $m = 1$	aq		−81.32		
Sb_2O_3	c	−164.9		29.40	24.20
Sb_2O_4	c	−216.9	−190.2	30.4	27.39
Sb_2O_5	c	−232.3	−198.2	29.9	28.11
Sb_4O_6 II, cubic	c	−344.3	−303.1	52.8	
I, orthorhombic	c	−338.7	−299.5	58.8	48.46
$Sb(OH)_3$	c		−163.8		
undissoc; std. state, $m = 1$	aq	−184.9	−154.1	27.8	
$HSb(OH)_6$	aq	−353.4			
SbOCl	c	−89.4			
SbOF undissoc; std. state, $m = 1$	aq		−116.5		
Sb_2S_3 black	c	−41.8	−41.5	43.5	28.65
orange	amorp	−35.2			
$Sb_2S_4^{2-}$ std. state, $m = 1$	aq	−52.4	−23.8	−12.5	
$Sb_2(SO_4)_3$	c	−574.2			
Sb_2Te_3	c	−13.5	−13.2	56	
Argon					
Ar	g	0	0	36.982	4.968
std. state, $m = 1$	aq	−2.9	3.9	14.2	
Arsenic					
As α, gray	c	0	0	8.4	5.89
γ, yellow, cubic	c	3.5			
β	amorp	1.0			
As_2	g	53.1	41.1	57.2	8.366
As_4	g	34.4	22.1	75	
$AsBr_3$	c	−47.2		(53)	
	liq	−43.1			
	g	−31	−38	86.94	18.92
$AsCl_3$	liq	−72.9	−62.0	51.7	31.90
	g	−62.5	−59.5	78.17	18.10
AsF_3	liq	(−226)	−198	(43)	
AsH_3	g	15.88	16.47	53.22	9.10
As_2H_4	g	35.2			
AsI_3	c	−13.9	−14.2	50.92	25.28
	g			92.79	19.27
AsN	g	46.91	40.15	53.9	7.27
AsO	g	16.72			
AsO_2^- std. state, $m = 1$	aq	−102.54	−83.66	9.9	
AsO_4^{3-} std. state, $m = 1$	aq	−212.27	−55.00	−38.9	
As_2O_3	c	−189.72		(36)	22.86
As_2O_5	c	−221.05	−187.0	25.2	27.85
As_4O_6 octahedral	c	−314.04	−275.46	51.2	45.72
monoclinic	c	−313.0	−275.82	56	
	g	−289.0	−262.4	91	
As_2S_2	c	−34.1			
As_2S_3	c	−40.4	−40.3	39.1	27.8
Astatine					
At	c	0	0	29.0	
Barium					
Ba	c	0	0	14.9	6.72
	liq	1.19	0.92	15.95	6.72
	g	42.8	35.1	40.63	4.97
Ba^{2+} std. state, $m = 1$	aq	−128.50	−134.02	2.3	
$Ba(acetate)_2$	c	−354.8			
std. state, $m = 1$	aq	−360.82	−310.60	43.7	
$Ba_3(AsO_4)_2$	c	−819			

Table 9-1 (*Continued*)
ELEMENTS AND INORGANIC COMPOUNDS

Formula and Description	State	$\Delta Hf°$	$\Delta Gf°$	$S°$	$C_p°$
BaBr$_2$	c	−181.0	−176.1	35	17.81
	g	−105	−113	79	14.7
std. state, m = 1	aq	−186.60	−183.72	41.7	
BaBr$_2$ · 2H$_2$O	c	−326.5	−294.1	54	
Ba(BrO$_3$)$_2$	c	−179.89	−138.0	58	
Ba(BrO$_3$)$_2$ · H$_2$O	c	−252.1	−197.09	69.9	52.90
BaC$_2$	c	−18			
Ba(CN)$_2$	c	−52.2			
	aq	−55.0			
Ba(CNO)$_2$ iso	c	−213			
BaCO$_3$ witherite	c	−290.7	−271.9	26.8	20.40
std. state, m = 1	aq	−290.34	−260.19	−11.3	
BaC$_2$O$_4$	c	−327.1			
BaC$_2$O$_4$ · 2H$_2$O	c	−471.1			
BaCl$_2$	c	−205.1	−193.7	29.56	17.96
	liq	−198.96	−188.85	34.30	17.96
	g	−119.2	−122.06	77.83	13.42
std. state, m = 1	aq	−208.40			
BaCl$_2$ · 2H$_2$O	c	−348.98	−309.86	48.5	38.71
Ba(ClO$_2$)$_2$	c	−162.6	−127.0	47	
Ba(ClO$_3$)$_2$	c	−182.3			
	aq, 400	−176.2			
Ba(ClO$_4$)$_2$	c	−191.2			
Ba(ClO$_4$)$_2$ · 3H$_2$O	c	−404.3	−303.7	94	
BaCrO$_4$	c	−345.6	−321.53	37.9	
BaF$_2$	c	−288.9	−276.9	23.04	17.26
	liq	−279.95	−269.69	28.98	17.26
	g	−192.1	−194.67	71.98	12.85
std. state, m = 1	aq	−287.50	−267.30	−4.3	
Ba$_3$[Fe(CN)$_6$]$_2$ std. state, m = 1	aq	−116.9	−53.5	136.1	
Ba(formate)$_2$	c	−333.6			
	aq, 400	−331.6			
Ba(glycolate)$_2$	c	−444.7			
	aq	−439.6			
BaHAsO$_4$ · H$_2$O	c	−412.6			
Ba(H$_2$AsO$_4$)$_2$ · 2H$_2$O	c	−696.9			
Ba(HCO$_3$)$_2$ std. state, m = 1	aq	−459.28	−414.54	45.9	
BaHPO$_4$	c	−433.7			
Ba(H$_2$PO$_2$)$_2$	c	−421.2			
	aq, inf	−421.9			
Ba(H$_2$PO$_4$)$_2$	c	−747			
BaI$_2$	c	−144.7	−143.74	39.47	18.52
	liq	−140.03	−140.39	43.90	18.52
	g	−72.4	−84.47	83.2	13.73
std. state, m = 1	aq	−154.88	−158.68	55.5	
Ba(IO$_3$)$_2$	c	−245.5	−206.7	59.6	44.8
std. state, m = 1	aq	−243.3	−195.2	58.9	
Ba(IO$_3$)$_2$ · H$_2$O	c	−316.0	−263.9	71	
Ba$_5$(IO$_6$)$_2$	c	−944.7			
BaMoO$_4$	c	−370	−344.1	33	33.6
Ba(N$_3$)$_2$	c	−0.8			
Ba(N$_3$)$_2$ · H$_2$O	c	−73.7	−25.1	45	
Ba(NH$_2$)$_2$	c	−78.5			
Ba(NO$_2$)$_2$	c	−183.6			
	aq	−178.5			
Ba(NO$_3$)$_2$	c	−237.11	−190.42	51.1	36.18
std. state, m = 1	aq	−227.62	−187.24	72.3	
BaO	c	−131.0	−124.38	17.23	11.30
	liq	−117.50	−112.63	23.08	11.30
	g	−29.60	−34.61	56.25	7.86
BaO$_2$	c	−151.6			
BaO$_2$ · 8H$_2$O	c	−718.6			

Table 9-1 (*Continued*)
ELEMENTS AND INORGANIC COMPOUNDS

Formula and Description	State	$\Delta Hf°$	$\Delta Gf°$	$S°$	$C_p°$
$Ba(OCl)_2$	aq	−179.3			
BaOH	g	−52			
$Ba(OH)_2$	c	−225.8			
	g	−140			
	aq, 500	−237.9			
$Ba(OH)_2 \cdot 8H_2O$	c	−798.8	−667.6	102	
$Ba_3(PO_4)_2$ colloidal	c	−978			
$BaPtCl_6$	c	−283.2			
$Ba(ReO_4)_2 \cdot 4H_2O$	c	−805	−697.5	90	
BaS	c	−110	−109	18.7	11.80
	g	+12			
	aq	−117.9			
$BaSO_3$	c	−281.9			
$BaSO_4$	c	−352.1	−325.6	31.6	24.32
std. state, $m = 1$	aq	−345.82	−311.99	7.1	
BaS_2O_6	aq	−415.4			
$BaS_2O_8 \cdot 4H_2O$	c	−738.4			
BaSe	c	−89			
$BaSeO_3$	c	−248.7	−231.4	40	
$BaSeO_4$	c	−274.0	−249.7	42	
$BaSiF_6$	c	−705.6	−667.8	39	
$BaSiO_3$	c	−388.05	−368.13	26.2	21.51
$BaTiO_3$	c	−396.7	−375.8	25.8	24.49
Ba_2TiO_4	c	−536.1	−509.8	47.0	36.48
$BaWO_4$	c	−407			
$BaZrO_3$	c	−425.3	−405.0	29.8	24.31
Beryllium					
Be	c	0	0	2.28	3.93
	liq	2.88	2.38	3.95	3.81
	g	77.5	68.5	32.55	4.97
Be^{2+} std. state, $m = 1$	aq	−91.5	−90.75	−31.0	
$BeAl_2O_4$	c	−549.9	−520.7	15.84	25.18
$3BeO \cdot B_2O_3$	c	−742.0	−702.4	24.00	33.40
$BeBr_2$	c	−88.4	−84.4	25.40	16.50
	aq	−142	−127.9		
Be_2C	c	−28.0	−21	3.90	10.34
BeC_2	g	135	121	52.2	9.87
$BeCl_2$ β	c	−118.6	−107.44	18.12	14.92
	aq	−171.1			
$BeCl_2 \cdot 4H_2O$	c	−436.8			
BeF_2 α	c	−245.4	−234.1	12.75	12.39
	aq	−251.4			
BeH	g	78.1	71.3	40.84	6.96
BeH_2	c	−4.60			
BeI_2	c	−46.0	−50	28.8	17.0
	aq	−112	−103.4		
Be_3N_2 α, cubic	c	−140.6	−127.4	8.16	15.38
$Be(NO_3)_2$	aq	−191.0			
BeO α	c	−145.4	−138.4	3.29	6.11
	g	31	24.9	47.21	7.05
BeO_2^{2-} std. state, $m = 1$	aq	−189.0	−153.0	38.0	
$Be(OH)_2$ β	c	−216.5	−195.2	11.00	15.63
BeS	c	−56.0			
Be_2SiO_4	c	−513.7	−485.8	15.37	22.84
$BeSO_4$ α	c	−288.05	−261.44	18.64	20.84
std. state, $m = 1$	aq	−308.8	−268.7	−26.2	
$BeSO_4 \cdot 4H_2O$	c	−579.29	−497.29	55.68	51.77
$BeWO_4$	c	−362	−336	21.12	23.25
Bismuth					
Bi	c	0	0	13.56	6.10
	g	49.5	40.2	44.67	4.97
$BiBr_3$	c	63	56	54	26

Table 9-1 (*Continued*)
ELEMENTS AND INORGANIC COMPOUNDS

Formula and Description	State	$\Delta Hf°$	$\Delta Gf°$	$S°$	$C_p°$
$BiCl_3$	c	−90.6	−75.3	42.3	25
	g	−63.5	−61.2	85.74	19.04
BiH_3	g	66.4			
BiI_3	c	−24.0	−41.9		
BiO^+ std. state, $m = 1$	aq		−35.0		
Bi_2O_3	c	−137.16	−118.0	36.2	27.13
$BiO(OH)$	c		−88.0		
$Bi(OH)_3$	c	−170.0			
$BiOBr$	c		−71.0		
$BiOCl$	c	−87.7	−77.0	28.8	
$BiONO_3$	c		−67.0		
BiS	g	43	29	68	
Bi_2S_3	c	−34.2	−33.6	47.9	29.2
$Bi_2(SO_4)_3$	c	−608.1			
$BiSe$	g	42.0			
$BiTe$	g	42.8			
Bi_2Te_3	c	−18.5	−18.4	62.36	28.8
Boron					
B	c	0	0	1.40	2.65
	amorp	0.9		1.56	2.86
	g	134.5	124.0	36.65	4.971
B_2	g	198.5	185.0	48.23	7.30
BBr	g	56.9	46.7	53.75	7.87
BBr_3	liq	−57.3	−57.0	54.9	
	g	−49.15	−55.56	77.47	16.20
B_4C	c	−17	−17	6.48	12.62
$B(CH_3)_3$	liq	−34.2	−7.7	57.1	
	g	−29.7	−8.6	75.2	21.15
BCl	g	35.73	28.90	50.94	7.57
BCl_3	liq	−102.1	−92.6	49.3	25.5
	g	−96.50	−92.91	69.31	14.99
B_2Cl_4	liq	−125.0	−111.1	62.7	32.9
$BOCl$	g	−75.6		56.7	10.73
$(BOCl)_3$	g	−390.4	−370.5	91	
$BClF_2$	g	−212.8	−209.4	65	
BCl_2F	g	−154.2	−150.9	68	
BF	g	−29.2	−35.8	47.89	7.07
BF_3	g	−271.75	−267.77	60.71	12.06
BF_4^- std. state, $m = 1$	aq	−376.4	−355.4	43	
B_2F_4	g	−145			
BOF	g	−145			
HBF_4	aq	−375.5			
BH	g	107.46	100.29	41.05	6.97
BH_3	g	24			
BH_4^- std. state, $m = 1$	aq	11.51	27.31	26.4	
B_2H_6	g	8.5	20.7	55.45	13.60
B_4H_{10}	g	15.8			
B_5H_9	liq	10.20	41.03	44.03	36.12
BI_3	g	17.00	4.96	83.43	16.92
BN	c	−60.8	−54.6	3.54	4.71
	g	154.75	146.87	50.71	7.04
$B_3N_3H_6$	liq	−129.3	−93.88	47.7	
BO	g	6	−1	48.62	6.98
BO_2	g	−71.8	−73.1	54.84	10.28
BO_2^- std. state, $m = 1$	aq	−184.60	−162.27	−8.9	
B_2O_2	g	−108.7	−110.5	57.93	13.69
B_2O_3	c	−304.20	−285.30	12.90	15.04
	amorp	−299.84	−282.6	18.6	14.6
	g	−201.67	−198.85	66.85	15.98
$B_4O_7^{2-}$ std. state, $m = 1$	aq		−622.6		
$B(OH)_4^-$ std. state, $m = 1$	aq	−321.23	−275.65	24.5	
BP cubic	c	−19			

Table 9-1 (*Continued*)
ELEMENTS AND INORGANIC COMPOUNDS

Formula and Description	State	$\Delta Hf°$	$\Delta Gf°$	$S°$	$C_p°$
BS	b	81.74	69.02	51.65	7.18
B_2S_3	c	−57.5			
Bromine					
Br	g	26.741	19.701	41.805	4.968
Br⁻	g	−55.9			
Br⁻ std. state, $m = 1$	aq	−29.05	−24.85	19.7	−33.9
Br_2	liq	0	0	36.384	18.090
	g	7.387	0.751	58.641	8.61
std. state, $m = 1$	aq	−0.62	0.94	31.2	
	CCl_4	0.71	0.36	37.6	
Br_3^- std. state, $m = 1$	aq	−31.17	−25.59	51.5	
BrCl	g	3.50	−0.23	57.36	8.36
Br_2Cl^- std. state, $m = 1$	aq	−40.7	−30.7	45.1	
BrF	g	−22.43	−26.09	54.70	7.88
BrF_3	liq	−71.9	−57.5	42.6	29.78
	g	−61.09	−54.84	69.89	15.92
BrF_5	liq	−109.6	−84.1	53.8	
BrO	g	30.06	25.87	56.75	7.67
BrO⁻ std. state, $m = 1$	aq	−22.5	−8.0	10	
BrO_2	c	11.6			
BrO_3^- std. state, $m = 1$	aq	−20.0	0.4	39.0	
Cadmium					
Cd γ	c	0	0	12.37	6.21
α	c	−0.14	−0.14	12.37	
	g	26.77	18.51	40.066	4.968
$CdAs_2$	c	−4.2			
Cd_3As_2	c	−10.0			
$Cd_3(AsO_4)_2$	c		−410.2		
$Cd(BO_2)_2$	c		−354.87		
$CdBr_2$	c	−75.57	−70.82	32.8	18.32
std. state, $m = 1$	aq	−76.24	−68.24	21.9	
$CdBr_2 \cdot 4H_2O$	c	−356.73	−298.287	75.6	
$CdCl_2$	c	−93.57	−82.21	27.55	17.85
std. state, $m = 1$	aq	−98.04	−81.286	9.5	
$CdCl_2 \cdot {}^5\!/_2 H_2O$	c	−270.54	−225.644	54.3	
$CdCl_3^-$ std. state, $m = 1$	aq	−134.1	−116.4	48.5	
$Cd(ClO_4)_2$ std. state, $m = 1$	aq	−79.96	−22.66	69.5	
$Cd(ClO_4)_2 \cdot 6H_2O$	c	−490.6			
$Cd(CN)_2$	c	38.8			
std. state, $m = 1$	aq	53.9	63.9	27.5	
$Cd(CN)_4^{2-}$ std. state, $m = 1$	aq	102.3	121.3	77	
$Cd(CNS)_2$ thiocyanate	c	12.43			
std. state, $m = 1$	aq	18.40	25.76	51.4	
$CdCO_3$	c	−179.4	−160.0	22.1	
CdC_2O_4	c	−218.1			
std. state, $m = 1$	aq	−215.3	−179.6	−6.6	
$Cd(acetate)_2$ std. state, $m = 1$	aq	−250.46	−195.12	23.9	
$Cd(formate)_2$ std. state, $m = 1$	aq	−221.56	−186.23	26	
Cd fulminate	c	90			
CdF_2	c	−167.4	−154.8	18.5	
std. state, $m = 1$	aq	−177.14	−151.82	−24.1	
CdI_2	c	−48.6	−48.13	38.5	19.11
std. state, $m = 1$	aq	−44.52	−43.20	35.7	
CdI_3^- std. state, $m = 1$	aq		−62.0		
CdI_4^{2-} std. state, $m = 1$	aq	−81.7	−75.5	78	
$Cd(IO_3)_2$	c		−90.13		
std. state, $m = 1$	aq	−123.9	−79.7	39.1	
$Cd(N_3)_2$	c	108			
std. state, $m = 1$	aq	113.38	147.9	34.1	
Cd_3N_2	c	38.7			
$Cd(NH_3)_4^{2+}$ std. state, $m = 1$	aq	−107.6	−54.1	80.4	
$Cd(NO_3)_2$	c	−109.06			
std. state, $m = 1$	aq	−117.26	−71.76	52.5	

Table 9-1 (*Continued*)
ELEMENTS AND INORGANIC COMPOUNDS

Formula and Description	State	$\Delta Hf°$	$\Delta Gf°$	$S°$	$C_p°$
$Cd(NO_3)_2 \cdot 4H_2O$	c	−394.11			
Cd_3P_2	c	−27.4			
$Cd_3(PO_4)_2$	c		−587.1		
CdO	c	−61.7	−54.6	13.1	10.38
CdO_2^{2-} std. state, $m = 1$	aq		−68.0		
$CdOH^+$ std. state, $m = 1$	aq		−62.4		
$HCdO_2^-$ std. state, $m = 1$	aq		−86.9		
$Cd(OH)_2$ pptd	c	−134.0	−113.2	23	
std. state, $m = 1$	aq	−128.08	−93.73	−22.6	
$Cd(OH)_3^-$ std. state, $m = 1$	aq		−143.6		
$Cd(OH)_4^{2-}$ std. state, $m = 1$	aq		−181.3		
CdS	c	−38.7	−37.4	15.5	13.17
$CdSO_4$	c	−223.06	−196.65	29.407	23.80
std. state, $m = 1$	aq	−235.46	−196.51	−12.7	
$CdSO_4 \cdot H_2O$	c	−296.26	−255.46	36.814	32.16
$CdSO_4 \cdot \frac{8}{3}H_2O$	c	−413.33	−350.224	54.883	50.97
CdSb	c	−3.44	−3.11	22.2	
Cd_3Sb_2	c	−13.9			
$CdSeO_3$	c	−137.5	−119.0	34.0	
std. state, $m = 1$	aq	−139.8	−106.9	−14.4	
$CdSeO_4$	c	−151.3	−127.1	39.3	
std. state, $m = 1$	aq	−161.3	−124.0	−4.6	
$CdSiO_3$	c	−284.20	−264.20	23.3	21.17
CdTe	c	−22.1	−22.0	24	
CdSb	c	−3.6		22	10.92
Calcium					
Ca	c	0	0	9.90	6.05
	liq	2.61	1.96	12.11	7.20
	g	42.85	34.78	36.99	4.97
Ca^+	g	185.3	175.3	38.369	4.968
Ca^{2+} std. state, $m = 1$	aq	−129.74	−132.30	−12.7	
$Ca(acetate)_2$	c	−353.6			
std. state, $m = 1$	aq	−362.06	−308.88	28.7	
$CaO \cdot Al_2O_3$	c	−556.0	−527.9	27.30	28.87
glassy	amorp	−550			
$CaO \cdot 2Al_2O_3$	c	−950.7	−901.2	42.50	48.00
$3CaO \cdot Al_2O_3$	c	−857.5	−815.5	49.2	50.16
glassy	amorp	−849			
$Ca_3(AsO_4)_2$	c	−788.4	−732.1	54	
$CaO \cdot B_2O_3$	c	−485.41	−459.87	25.06	24.85
$CaO \cdot 2B_2O_3$	c	−803.12	−756.96	32.2	37.75
glassy	amorp	−790.48			
$2CaO \cdot B_2O_3$	c	−653.54	−620.62	34.68	35.16
$3CaO \cdot B_2O_3$	c	−819.57	−779.14	43.9	44.90
$CaBr_2$	c	−163.3	−158.73	31.00	17.94
	liq	−158.46	−155.19	35.34	17.94
	g	−92.0	−100.61	75.20	14.43
std. state, $m = 1$	aq	−187.84	−182.00	26.7	
$CaBr_2 \cdot 6H_2O$	c	−599.0	−514.6	98	
$Ca(BrO_3)_2$	c	−171.8			
	aq, 1000	−170.4			
CaC_2	c	−14.3	−15.5	16.72	14.99
$Ca(CN)_2$	c	−44.1			
	aq	−56.9			
$CaCN_2$ cyanamide	c	−83.8			
$CaCO_3$ aragonite	c	−288.51	−269.55	21.2	19.42
calcite	c	−288.46	−269.80	22.2	19.57
std. state, $m = 1$	aq	−291.58	−258.47	−26.3	
CaC_2O_4	c	−325.2			
std. state, $m = 1$	aq	−326.9	−293.37	−1.8	
$CaC_2O_4 \cdot H_2O$	c	−400.30	−361.85	37.4	36.52
$CaCl_2$	c	−190.2	−178.8	25.0	17.41

Table 9-1 (*Continued*)
ELEMENTS AND INORGANIC COMPOUNDS

Formula and Description	State	$\Delta Hf°$	$\Delta Gf°$	$S°$	$C_p°$
	liq	−185.0	−175.0	29.6	17.41
	g	−112.7	−114.5	69.3	14.18
std. state, $m = 1$	aq	−209.64	−195.04	14.3	
$CaCl_2 \cdot H_2O$	c	−265.1			
$CaCl_2 \cdot 2H_2O$	c	−335.3			
$CaCl_2 \cdot 4H_2O$	c	−480.3			
$CaCl_2 \cdot 6H_2O$	c	−623.3			
$Ca(ClO_2)_2$	c	−162.1			
$Ca(ClO_4)_2$	c	−176.09			
std. state, $m = 1$	aq	−191.56	−136.42	74.3	
$Ca(ClO_4)_2 \cdot 4H_2O$	c	−465.8	−352.97	103.6	
$CaCrO_4$	c	−329.6	−305.3	32	
	aq	−340.9			
CaF_2	c	−291.5	−279.0	16.46	16.02
	liq	−283	−273	22.13	16.39
	g	−187	−190	65.4	12.25
std. state, $m = 1$	aq	−288.74	−265.58	−19.3	
$Ca_3[Fe(CN)_6]_2$ std. state, $m = 1$	aq	−120.6	−48.3	91.1	
$Ca_2[Fe(CN)_6]$ std. state, $m = 1$	aq	−150.6	−98.51	−2.7	
$CaO \cdot Fe_2O_3$	c	−363.37	−337.67	34.74	36.71
$2CaO \cdot Fe_2O_3$	c	−511.30	−478.44	45.12	46.19
$Ca(formate)_2$	c	−331.4			
	aq, 400	−332.5			
CaH_2	c	−44.5	−35.2	10	
$CaHPO_4$	c	−433.65	−401.83	26.62	26.30
std. state, $m = 1$	aq	−438.57	−392.64	45.28	47.10
$CaHPO_4 \cdot 2H_2O$	c	−574.47	−515.00	45.28	47.10
$Ca(H_2PO_2)_2$	c	−418.9			
std. state, $m = 1$	aq	−423.1			
$Ca(H_2PO_4)_2$	c	−742.04			
std. state, $m = 1$	aq	−749.38	−672.64	30.5	
$Ca(H_2PO_4)_2 \cdot H_2O$	c	−814.93	−730.98	62.1	61.86
CaI_2	c	−128.30	−127.42	34.72	18.44
	liq	−119.54	−121.06	42.77	18.44
	g	−61.70	−73.80	78.26	18.44
std. state, $m = 1$	aq	−156.12	−156.96	40.5	
$CaI_2 \cdot 8H_2O$	c	−700.2			
$Ca(IO_3)_2$	c	−239.6	−200.6	55	
$Ca(IO_3)_2 \cdot 6H_2O$	c	−664.6	−542.0	108	
$Ca_5(IO_6)_2$	c	−915		108.4	
$Ca[Mg(CO_3)_2]$ dolomite	c	−556.0	−517.1	37.09	37.65
$CaO \cdot MgO \cdot SiO_2$ monticellite	c	−540.89			
$CaO \cdot MgO \cdot 2SiO_2$ diopside	c	−766.3	−724.7	34.16	39.80
$2CaO \cdot MgO \cdot 2SiO_2$ akermanite	c	−926.67	−879.53	50.0	50.67
$3CaO \cdot MgO \cdot 2SiO_2$ merwinite	c	−1091.7	−1037.4	60.5	60.29
$CaMoO_4$	c	−368.4	−342.9	29.3	27.32
std. state, $m = 1$	aq	−368.2	−332.2	−6.2	
$Ca(N_3)_2$	c	+3.5			
$Ca(NO_2)_2$	c	−177.2			
	aq, 800	−179.5			
$Ca(NO_3)_2$	c	−224.28	−177.63	46.2	35.70
std. state, $m = 1$	aq	−228.86	−185.52	57.3	
$Ca(NO_3)_2 \cdot 2H_2O$	c	−368.25	−293.82	64.4	
$Ca(NO_3)_2 \cdot 3H_2O$	c	−439.3	−351.8	76.3	
$Ca(NO_3)_2 \cdot 4H_2O$	c	−509.64	−409.53	89.7	
CaO	c	−151.80	−144.25	9.13	10.07
	liq	−133.21	−127.38	14.89	10.07
$CaOCl_2$	c	−178.4			
	aq	−188.9			
$Ca(OCl)_2$ hypochlorite	aq	−180.3			
$CaOH$	g	−48			
$Ca(OH)_2$	c	−235.70	−214.75	19.93	20.91
std. state, $m = 1$	aq	−239.68	−207.49	−17.8	

Table 9-1 (*Continued*)
ELEMENTS AND INORGANIC COMPOUNDS

Formula and Description	State	$\Delta Hf°$	$\Delta Gf°$	$S°$	$C_p°$
Ca_3P_2	c	−121			
$Ca(PO_3)_2$ β	c			35.05	34.68
glassy	amorp		−587.0		
$Ca_3(PO_4)_2$ β	c	−984.9	−928.5	56.4	54.45
α	c	−982.3	−926.3	57.58	55.35
std. state, $m = 1$	aq	−999.8	−883.9	−144	
$Ca_2P_2O_7$ β	c	−798.0	−748.6	45.23	44.89
$Ca_{10}(PO_4)_6F_2$ fluorapatite	c	−3285	−3103	185.4	179.7
$Ca_{10}(PO_4)_6(OH)_2$ hydroxyapatite	c	−3221	−3030	186.6	184.0
std. state, $m = 1$	aq	−3239.1	−2859.2	−450	
CaS	c	−113.5	−112.3	13.5	11.34
$CaSO_3$	c			24.23	12.92
$CaSO_3 \cdot 2H_2O$	c	−418.9	−371.7	44	42.7
$CaSO_4$ anhydrite, insoluble	c	−342.76	−315.93	25.5	23.82
α, soluble	c	−340.64	−313.93	25.9	23.95
β, soluble	c	−339.58	−312.87	25.9	23.67
std. state, $m = 1$	aq	−347.06	−310.27	−7.9	
$CaSO_4 \cdot \frac{1}{2}H_2O$ α, macro	c	−376.85	−343.41	31.2	28.54
β, micro	c	−376.35	−343.18	32.1	29.69
$CaSO_4 \cdot 2H_2O$	c	−483.42	−429.60	46.4	44.46
CaS_2O_3	aq, inf	−285.6			
$CaS_2O_6 \cdot 4H_2O$	c	−697			
CaSe	c	−88.0	−86.8	16	
$CaSeO_4 \cdot 2H_2O$	c	−407.9	−355.4	53	
$CaSiO_3$ wollastonite	c	−390.76	−370.39	19.58	20.38
pseudowollastonite	c	−389.2	−369.2	20.88	20.67
glassy	amorp	−382.65			
Ca_2SiO_4 β	c	−551.5	−524.1	30.53	30.78
γ	c	−554.0	−526.1	28.87	30.27
$3CaO \cdot SiO_2$	c	−700.1	−665.4	40.3	41.08
$3CaO \cdot 2SiO_2$	c	−946.7	−899.0	50.38	51.24
$CaTiO_3$ perovskite	c	−396.9	−376.5	22.38	23.34
$3CaO \cdot 2TiO_2$	c	−944.2	−896.6	56.1	57.20
$CaTiSiO_5$ sphene	c	−622.2	−588.4	30.88	33.21
$CaWO_4$	c	−393.20	−367.71	30.21	27.28
std. state, $m = 1$	aq	−386.8			
$CaO \cdot V_2O_5$	c	−556.71	−518.57	42.8	39.86
$2CaO \cdot V_2O_5$	c	−736.94	−691.49	52.7	50.08
$3CaO \cdot V_2O_5$	c	−902.95	−851.12	65.7	60.29
$CaZrO_3$	c	−422.3	−401.8	23.92	23.88
Carbon*					
C graphite	c	0	0	1.361	2.066
diamond	c	0.4533	0.6930	0.568	1.4617
	g	171.291	160.442	37.7597	4.9805
C^-	g	140.5	131.6	36.16	4.968
C_2	g	200.2	186.8	47.63	10.31
C_2^-	g	106	94	46.96	6.99
C_3	g	196.0	180.3	56.7	9.02
CBr	g	122	111	55.8	8.61
CCl	g	120	112	53.61	7.71
CF	g	61	53	50.89	7.18
CF^+	g	274.7	266.5	48.1	7.084
CF_2	g	−43.5	−45.8	57.53	9.31
CF_2^+	g	225.1	220.9	58.95	9.26
CN	g	104.0	96.8	48.41	6.97
CN^-	g	14.5	9.26	46.8	6.96
std. state, $m = 1$	aq	36.0	41.2	22.5	
CN^+	g	430.9	421.4	50.99	7.03
CN_2	g	139	137	55.35	10.10
C_4N_2	g	127.5	122.1	69.31	20.53
$(CN)_2$ cyanogen	g	73.84	71.03	57.90	13.60

*For the values of organic compounds see Table 9-2.

Table 9-1 (*Continued*)
ELEMENTS AND INORGANIC COMPOUNDS

Formula and Description	State	$\Delta Hf°$	$\Delta Gf°$	$S°$	$C_p°$
CNBr	c	32.5			
	g	43.35	38.39	59.07	11.12
CNCl	g	31.60	29.99	56.28	10.69
CNF	g			53.67	9.99
CNI	c	38.3	40.48	30.80	
	g	53.80	46.88	61.33	11.55
CN·N$_3$ cyanogen azide	c	92.6			
CO	g	−26.42	−32.81	47.30	6.97
CO$_2$	g	−94.05	−94.26	51.07	8.87
undissoc; std. state, $m = 1$	aq	−98.90	−92.26	28.1	
CO$_3^{2-}$ std. state, $m = 1$	aq	−161.84	−126.17	−13.6	
C$_3$O$_2$	liq	−28.03	−25.10	43.28	25.8
	g	−22.40	−26.25	66.05	16.01
COBr$_2$	g	−23.0	−26.5	73.85	14.78
COCl$_2$	g	−52.80	−49.42	67.82	13.79
COF$_2$	g	−153.00	−149.28	61.84	11.29
COS	g	−33.08	−39.59	55.32	9.92
CS	g	56	44	50.30	7.12
CS$_2$	liq	21.44	15.60	36.17	18.1
	g	27.98	15.99	56.83	10.87
Cerium					
Ce	c	0	0	17.2	6.44
	g	101	92	45.81	5.52
Ce^{3+} std. state, $m = 1$	aq	−166.4	−160.6	−49	
Ce^{4+} std. state, $m = 1$	aq	−128.4	−120.4	−72	
Ce(acetate)$_3$					
undissoc; std. state, $m = 1$	aq	−509.27	−429.83	44.9	
CeC$_2$	c	−15	−15.2	20	
	g	136.2	122.9	64	10.5
CeC$_4$	g	168	152	73	17.3
Ce$_2$(C$_2$O$_4$)$_3$·10H$_2$O	c	−1621	−1411	191	
CeCl$_3$	c	−251.8	−233.7	36	20.9
	g	−174			
std. state, $m = 1$	aq	−286.2	−254.7	−9	
CeCrO$_3$	c	−368	−347	25	
CeF$_3$	c	−391.0	(−372)	27.5	22.3
CeH$_2$	c	−49		13.3	9.78
CeI$_3$	c	−155.3	(−161)	(50)	
std. state, $m = 1$	aq	−213.9	−207.5	34	
Ce(IO$_3$)$_3$	c	−332			
CeN	c	−78.3	−70.8		
	g	89.3			7.70
Ce(NO$_3$)$_3$	c	−293.0			
	aq, 2600	−314.7			
Ce(NO$_3$)$_3$·6H$_2$O	c	−729.14			
CeO$_2$	c	−260.2	−244.9	14.89	14.73
Ce$_2$O$_3$	c	−429.3	−407.8	36.0	27.4
CeOCl	c	−239			
CeS	c	−109.8	−107.9	18.7	11.94
	g	31.4	20.2	62.2	8.3
CeS$_2$	c	−146.3			
	g	2.4	−8.8	70	13.8
Ce$_2$S$_3$	c	−284			
Ce$_3$S$_4$	c	−397			
Ce(SO$_4$)$_2$	c	−560			
Ce$_2$(SO$_4$)$_3$	c	−945.1			
std. state, $m = 1$	aq	−998.3	−873.0	−76	
Ce$_2$(SO$_4$)$_3$·8H$_2$O	c	−1320.6	−1340.2		
Cesium					
Cs	c	0	0	20.35	7.70
	liq	0.499	0.006	22.01	7.75
	g	18.3	11.9	41.94	4.97

Table 9-1 (*Continued*)
ELEMENTS AND INORGANIC COMPOUNDS

Formula and Description	State	$\Delta Hf°$	$\Delta Gf°$	$S°$	$C_p°$
Cs^+	g	109.6	102.1	40.565	4.968
$CsAl(SO_4)_2 \cdot 12H_2O$	c	−1449.5	−1218.5	164	
CsBr	c	−96.96	−92	27.1	12.37
std. state, $m = 1$	aq	−88.1	−91.98	51.1	
$CsBrO_3$	c	−83.33		38.8	
CsCl	c	−105.84	−99.0	24.183	12.53
	liq	−103.8	−97.1	24.31	18.50
	g	−57.4	−61.6	61.18	8.83
std. state, $m = 1$	aq	−99.2	−88.76	45.0	
$CsClO_3$	c	−94.6		40	
$CsClO_4$	c	−105.81	(−73)	41.89	
std. state, $m = 1$	aq	−90.6	−69.98	75.3	
CsF	c	−132.6	−125.6	21.1	12.42
	liq	−129.98	−123.11	21.53	17.70
	g	−85.2	−89.2	58.10	8.57
std. state, $m = 1$	aq	−135.9	−133.49	29.5	
CsH	g	29.0	24.3	51.25	
$CsHF_2$	c	−216.1			
CsI	c	−80.5	−79.76	30.0	12.32
std. state, $m = 1$	aq	−72.6	−79.76	57.9	
$CsIO_3$	c	−125.70		40	
$CsNH_2$	c	−25.4			
$CsNO_2$	c	−85			
$CsNO_3$	c	−118.11			
std. state, $m = 1$	aq	−108.6	−93.82	86.8	
Cs_2O	c	−75.9			
	g	−22	−25	76	13.24
Cs_2O_2	c	−96.2			
CsOH	c	−99.6	−86.6	23.6	16.22
	liq	−97.04	−87.45	28.31	19.50
	g	−62.0	−62.1	60.98	11.88
std. state, $m = 1$	aq	−114.2	−105.00	29.3	
$CsReO_4$	c	−257.2			
Cs_2S	c	−81.1			
Cs_2SO_4	c	−349.8			
std. state, $m = 1$	aq	−335.3	−312.16	67.7	
Cs_2SiF_6	c	−669.5			
Chlorine					
Cl	g	28.99	25.17	39.45	5.22
Cl^-	g	−58.8			
std. state, $m = 1$	aq	−39.95	−31.37	13.5	−32.6
Cl_2	g	0	0	53.29	8.115
ClF	g	−13.02	−13.37	52.05	7.66
ClF_3	liq	−45.3			
	g	−38.0	−28.4	67.28	15.26
ClF_5	g	−57	−35	74.24	23.22
$ClF_3 \cdot HF$	g	−107.7	−91.8	86	
Cl_2F_6	g	−81.1	−56.7	117	
ClO	g	24.19	23.30	54.15	7.52
ClO^- std. state, $m = 1$	aq	−25.6	−8.8	10	
ClO_2	g	24.5	28.76	61.37	10.04
ClO_2^- std. state, $m = 1$	aq	−15.9	4.1	24.2	
$(ClO_3)_2$	g	37			
ClO_3^- std. state, $m = 1$	aq	−23.7	−0.8	38.8	
ClO_4^- std. state, $m = 1$	aq	−30.91	−2.06	43.5	
Cl_2O	g	19.20	23.30	64.02	11.43
Cl_2O_7	liq	56.9			
ClO_3F	g	−6.49	10.72	66.65	15.52
Chromium					
Cr	c	0	0	5.645	5.601
	liq	6.239	5.340	8.660	5.601
	g	95.00	84.27	41.64	4.968

Table 9-1 (*Continued*)
ELEMENTS AND INORGANIC COMPOUNDS

Formula and Description	State	$\Delta Hf°$	$\Delta Gf°$	$S°$	$C_p°$
Cr^{2+} std. state, $m = 1$	aq	−34.3			
CrBr$_2$	c	−72.2			
	g	−17			
[Cr(H$_2$O)$_6$]Br$_3$ purple	c	−550.4			
Cr$_3$C$_2$	c	−20.40	−20.63	20.42	23.53
Cr$_7$C$_3$	c	−38.7	−39.9	48.0	49.92
Cr$_{23}$C$_6$	c	−87.2	−89.3	145.8	149.2
CrCl$_2$	c	−94.5	−85.1	27.56	17.01
	g	−30.7			
	aq	−114.2			
CrCl$_2 \cdot$ 2H$_2$O	c	−237.1			
CrCl$_2 \cdot$ 3H$_2$O	c	−308.9			
CrCl$_2 \cdot$ 4H$_2$O	c	−384.4			
CrCl$_3$	c	−133.0	−116.2	29.4	21.94
CrCl$_4$	g	−102	(−91.6)		
Cr(CO)$_6$	c	−257.41		70.03	54.07
CrO$_2$Cl$_2$	liq	−138.5	−122.1	53.0	
	g	−128.6	−119.9	78.8	20.2
[Cr(H$_2$O)$_6$]Cl$_3$ violet	c	−586.2			
CrF$_2$	c	−186	(−172)		
	g	−99			
CrF$_3$	c	−277	−260	22.44	18.82
CrF$_4$	c	−298			
Cr$_7$H$_2$	c	−3.8			
CrI$_2$	c	−37.5			
	g	24			
	aq	−60.1			
CrI$_3$	c	−49.0			
CrN	c	−28.00	−22.18	9.01	12.59
	g	120.70	112.79	55.08	7.35
Cr$_2$N	c	−30.0	−24.43	15.5	15.79
CrO	g	45.00	36.95	57.16	7.486
CrO$_2$	g	−18.00	−20.88	64.32	10.37
CrO$_3$	g	−70.00	−65.36	63.59	13.39
Cr$_2$O$_3$	c	−271.20	−251.70	19.40	28.77
	liq	−243.40	−227.07	30.02	24.32
Cr$_3$O$_4$	c	−366			
CrO$_4^{2-}$ std. state, $m = 1$	aq	−210.60	−173.96	12.00	
Cr$_2$O$_7^{2-}$ std. state, $m = 1$	aq	−356.2	−311.0	62.6	
HCrO$_4^-$ std. state, $m = 1$	aq	−209.9	−182.8	44.0	
CrO$_2$(OH)$_2$	g	−174			
Cr(OH)$_3$	c	−254.3			
[Cr(H$_2$O)$_6$]$^{3+}$	aq	−477.8			
Cr$_2$(SO$_4$)$_3$	c	−145.7		64.5	72.31
Cr$_2$(SO$_4$)$_3 \cdot$ 18H$_2$O	c			223	
CrSb	c				12.7
CrSb$_2$	c				19.7
CrSi	c	−15			9.2
CrSi$_2$	c	−23			12.7
Cr$_3$Si	c	−30			19.3
Cr$_5$Si$_3$	c	−64			34.9
Cobalt					
Co hexagonal	c	0	0	7.18	5.93
face-centered cubic	c	0.11	0.06	7.34	
Co^{2+} std. state, $m = 1$	aq	−13.9	−13.0	−27	
Co^{3+} std. state, $m = 1$	aq	22	32	−73	
Co$_3$(AsO$_4$)$_2$	c		−387.4		
Co(BO$_2$)$_2$	c		−325.8		
CoBr$_2$	c	−52.8			
std. state, $m = 1$	aq	−72.0	−62.7	12	
CoBr$_2 \cdot$ 6H$_2$O	c	−482.8			
CoCO$_3$	c	−170.4			

Table 9-1 (*Continued*)
ELEMENTS AND INORGANIC COMPOUNDS

Formula and Description	State	$\Delta Hf°$	$\Delta Gf°$	$S°$	$C_p°$
$CoCl_2$	c	−74.7	−64.5	26.09	18.76
std. state, $m = 1$	aq	−93.8	−75.7	0	
$CoCl_2 \cdot H_2O$	c	−147			
$CoCl_2 \cdot 2H_2O$	c	−220.6	−182.8	45	
$CoCl_2 \cdot 6H_2O$	c	−505.6	−412.4	82	
$CoCl_3$	g	−39.10	−36.93	79.85	18.22
$Co(ClO_4)_2$ std. state, $m = 1$	aq	−75.7	−17.1	60	
$Co(ClO_4)_2 \cdot 6H_2O$	c	−487.2			
$Co(CNO)_2$ cyanate	c	−51.8			
$Co(CNS)_2$ thiocyanate	c	24.2			
$Co(formate)_2$	c	−208.7			
CoC_2O_4	c	−203.5			
std. state, $m = 1$	aq	−211.1	−174.1	−16	
CoF_2	c	−165.4	−154.7	19.59	16.44
CoF_3	c	−189	−172	22.6	21.94
CoI_2	c	−21.2			
std. state, $m = 1$	aq	−40.3	−37.7	26	
$Co(IO_3)_2$ std. state, $m = 1$	aq	−119.7	−74.2	30	
$Co(IO_3)_2 \cdot 2H_2O$	c	−258.6	−190.2	64	
$Co(NH_3)_6^{3+}$ std. state, $m = 1$	aq	−139.8	−37.6	35	
$Co(NH_3)_6^{2+}$ std. state, $m = 1$	aq		−45.3		
$[Co(NH_3)_6]Br_2$	c	−216.4			
$[Co(NH_3)_6]Br_3$	c	−239.7	−119.8	77.7	78.1
std. state, $m = 1$	aq	−227.0	−112.2	94	
$[Co(NH_3)_6]Cl_2$	c	−238.0			
$[Co(NH_3)_6]Cl_3$	c	−268.7			76.6
std. state, $m = 1$	aq	−259.7	−131.7	75	
$[Co(NH_3)_5Cl]Cl_2$	c	−243.1	−139.3	87.5	57.2
$[Co(NH_3)_6](ClO_4)_3$	c	−247.3	−53.0	147	
std. state, $m = 1$	aq	−232.5	−43.8	166	
$[Co(NH_3)_6]I_2$	c	−189.9			69.1
$[Co(NH_3)_6]I_3$	c	−104.8			74.3
$[Co(NH_3)_5NO_2]^{2+}$ std. state, $m = 1$	aq	−146.6	−41.3	43	
$[Co(NH_3)_5NO_2](NO_3)_2$	c	−260.2	−100.0	83	
	aq	−245.7	−94.5	113	
$[Co(NH_3)_6](NO_3)_3$ std. state, $m = 1$	aq	−288.5	−117.4	140	
	c	−306.4	−123.5	107	
$Co(NO_3)_2$	c	−100.5			
std. state, $m = 1$	aq	−113.0	−66.2	43	
$Co(NO_3)_2 \cdot 2H_2O$	c	−244.2			
$Co(NO_3)_2 \cdot 6H_2O$	c	−528.49			
CoO	c	−56.87	−51.20	12.66	13.20
Co_3O_4	c	−217.5	−190.0	27.3	29.5
$Co(OH)_2$ blue	c		−107.6		
pink	c	−129.0	−108.6	19	
std. state, $m = 1$	aq	−123.8	−88.2	−32	
$Co(OH)_3$	c	−171.3			
Co_2P	c	−45			
$Co_3(PO_4)_2$	c		−573.3		
$CoHPO_4$	c		−282.5		
CoS	c	−19.8			
Co_2S_3 pptd	c	−35.2			
$CoSO_4$	c	−212.3	−187.0	28.2	24.67
std. state, $m = 1$	aq	−231.2	−191.0	−22	
$CoSO_4 \cdot 6H_2O$	c	−641.4	−534.35	87.86	84.46
$CoSO_4 \cdot 7H_2O$	c	−712.22	−591.26	97.05	93.33
$CoSe$	c	−14.6			
$CoSeO_3 \cdot 2H_2O$	c	−266.5			
$CoSi$	c	−24.0	−23.6	10.3	10.6
Co_2SiO_4	c	−353			
$CoTe_2$	c	−31			
Co_3Te_4	c	−77			

Table 9-1 (*Continued*)
ELEMENTS AND INORGANIC COMPOUNDS

Formula and Description	State	$\Delta Hf°$	$\Delta Gf°$	$S°$	$C_p°$
Copper					
Cu	c	0	0	7.923	5.840
	g	80.86	71.37	39.74	4.968
Cu⁺ std. state, $m = 1$	aq	17.13	11.95	9.7	
Cu²⁺ std. state, $m = 1$	aq	15.48	15.66	−23.8	
Cu_2	g	115.72	103.24	57.71	8.75
Cu_3As	c	−2.8			
$Cu_3(AsO_4)_2$	c		−310.9		
std. state, $m = 1$	aq	−378.10	−263.02	−149.2	
CuBr	c	−25.0	−24.1	22.97	13.08
$CuBr_2$	c	−33.9			
$CuBr_2 \cdot 4H_2O$	c	−317.0			
$CuCO_3 \cdot Cu(OH)_2$ malachite	c	−251.3	−213.6	44.5	
$2CuCO_3 \cdot Cu(OH)_2$ azurite	c	−390.1			
CuCl	c	−32.8	−28.65	20.6	11.6
$CuCl_2^-$ std. state, $m = 1$	aq		−57.4		
$CuCl_3^{2-}$ std. state, $m = 1$	aq		−90		
$2CuCl \cdot CO \cdot 2H_2O$	c	−225			
$2CuCl \cdot C_2H_2$	c	−23.3	−7.63	50.7	
$3CuCl \cdot C_2H_2$	c	−56.4	−36.52	71.0	
$CuCl_2$	c	−49.2	−38.7	25.83	17.18
undissoc; std. state, $m = 1$	aq		−47.3		
$CuCl_2 \cdot 2H_2O$	c	−196.3	−156.8	40	
$Cu(ClO_4)_2$ std. state, $m = 1$	aq	−46.34	11.54	63.2	
$Cu(ClO_4)_2 \cdot 6H_2O$	c	−460.9			
CuCN	c	22.7	25.9	21.51	14.59
$Cu(CN)_2^-$ std. state, $m = 1$	aq		61.6		
$Cu(CN)_3^{2-}$ std. state, $m = 1$	aq		96.5		
$Cu(CN)_4^{3-}$ std. state, $m = 1$	aq		135.4		
CuCNS thiocyanate	c		16.7		
std. state, $m = 1$	aq	35.40	34.10	44.2	
$Cu(CNS)_2$ std. state, $m = 1$	aq	52.02	59.96	45.2	
$Cu(acetate)_2$	c	−213.5			
std. state, $m = 1$	aq	−216.84	−160.92	17.6	
$Cu(formate)_2$	c	−186.7			
std. state, $m = 1$	aq	−187.94	−152.1	20	
CuONC fulminate	c	26.3			
CuC_2O_4	c		−158.2		
std. state, $m = 1$	aq	−181.7	−145.5	−12.9	
$Cu(C_2O_4)_2^{2-}$ std. state, $m = 1$	aq	−380.5	−319.3	35	
CuF	c	−46	−41	15.5	10.72
CuF_2	c	−131.2	−119.3	16.4	16.77
$CuF_2 \cdot 2H_2O$	c		−234.6		
$CuFeO_2$	c	−127.3	−114.7	21.2	19.13
$CuFe_2O_4$	c	−230.69	−205.26	33.7	35.52
CuH	c	5.1			
	g	70			
CuI	c	−16.2	−16.6	23.1	12.92
$Cu(IO_3)_2$ std. state, $m = 1$	aq	−90.3	−45.5	32.8	
$Cu(IO_3)_2 \cdot H_2O$	c	−165.4	−112.0	59.1	
$CuMoO_4$	c	−225			
CuN_3	c	66.7	82.4	24	
$Cu(N_3)_2$	c	143.0			
Cu_3N	c	17.8			22
$Cu(NH_3)^{2+}$ std. state, $m = 1$	aq	−9.3	3.72	2.9	
$Cu(NH_3)_2^{2+}$ std. state, $m = 1$	aq	−34.0	−7.28	26.6	
$Cu(NH_3)_3^{2+}$ std. state, $m = 1$	aq	−58.7	−17.48	47.7	
$Cu(NH_3)_4^{2+}$ std. state, $m = 1$	aq	−83.3	−26.60	65.4	
$Cu(NO_3)_2$	c	−72.4			
std. state, $m = 1$	aq	−83.64	−37.56	46.2	
$Cu(NO_3)_2 \cdot 6H_2O$	c	−504.5			
CuO	c	−37.6	−31.0	10.19	10.11

Table 9-1 (*Continued*)
ELEMENTS AND INORGANIC COMPOUNDS

Formula and Description	State	$\Delta Hf°$	$\Delta Gf°$	$S°$	$C_p°$
Cu_2O	c	−40.3	−34.9	22.26	15.21
$Cu(OH)_2$	c	−107.6	−89.1	25.9	22.75
std. state, $m = 1$	aq	−94.46	−59.53	−28.9	
CuP_2	c	−29			
Cu_3P	c	−36.2			
$Cu_2P_2O_7$	c		−448.0		
std. state, $m = 1$	aq	−511.8	−427.4	−76	
$Cu_3(PO_4)_2$	c		−490.3		
CuS	c	−12.7	−12.8	15.9	11.43
Cu_2S α	c	−19.0	−20.6	28.9	18.24
$CuSO_4$	c	−184.36	−158.2	26	23.9
std. state, $m = 1$	aq	−201.84	−162.31	−19.0	
	aq, 100	−200.374			
$CuSO_4 \cdot H_2O$	c	−259.52	−219.46	34.9	32
$CuSO_4 \cdot 3H_2O$	c	−402.56	−334.65	52.9	49
$CuSO_4 \cdot 5H_2O$	c	−544.85	−449.344	71.8	67
Cu_2SO_3 std. state, $m = 1$	aq	−117.6	−92.4	12	
Cu_2SO_4	c	−179.6			
Cu_2Sb	c	−2.75		30.4	18.35
CuSe	c	−9.45			
$CuSe_2$	c	−10.3			
Cu_2Se	c	−14.2		37.6	21.2
$CuSeO_3$	c		−83.2		
std. state, $m = 1$	aq	−106.2	−72.7	−20.7	
$CuSeO_4$	c	−114.36			
$CuSeO_4 \cdot 5H_2O$	c	−466.96			
$CuWO_4$	c	−250.0			
Curium					
Cm	c	0	0		
$CmCl_3$	c	−226.4			
CmF_3	c			28.1	
Dysprosium					
Dy	c	0	0	18.0	6.49
	g	69.4	60.8	46.97	4.97
Dy^{3+} std. state, $m = 1$	aq	−167	−159	−55.2	5
$Dy(acetate)_3$	c	−476			
undissoc; std. state, $m = 1$	aq	−511.17	−429.11	38.9	
$DyBr_3$	c	−209			
$Dy(BrO_3)_3 \cdot 9H_2O$	c	−856.1			
DyC_2	c	206.1	193.2	64	10.5
$DyCl_3$ β	c	−239			
γ	c	−236			
std. state, $m = 1$	aq	−286	−253	−14.8	−93
$DyCl_3 \cdot 6H_2O$	c	−686	−586	96.00	82.7
DyI_3	c	−145			
std. state, $m = 1$	aq	−206.1	−198.2		
$Dy(IO_3)_3$	c	−329			
Dy_2O_3	c	−445.3	−423.4	35.8	27.79
$Dy(OH)_3$	c		−305.8		
$Dy_2(SO_4)_3$ std. state, $m = 1$	aq	−982.7	−854.4		
$Dy_2(SO_4)_3 \cdot 8H_2O$	c	−1322.0			
Erbium					
Er	c	0	0	17.49	6.72
	g	75.8	67.2	46.72	4.97
Er^{3+} std. state, $m = 1$	aq	−168.6	−159.9	−58.4	5
$Er(acetate)_3$					
undissoc; std. state, $m = 1$	aq	−511.91	−429.84	38.0	
$ErBr_3$	c	(−205)			
$Er(BrO_3)_3 \cdot 9H_2O$	c	−856.8			
ErC_2	g	138.2	125.4	63	10.5
$Er_2(C_2O_4)_3 \cdot 6H_2O$	c	−1183			
$ErCl_3$	c	−238.7		24	
std. state, $m = 1$	aq	−288.5	−254.0	−18.0	−93

Table 9-1 (*Continued*)
ELEMENTS AND INORGANIC COMPOUNDS

Formula and Description	State	$\Delta Hf°$	$\Delta Gf°$	$S°$	$C_p°$
ErCl$_3$ · 6H$_2$O	c	−687.0	−586.6	95.3	82.0
ErF$_3$	c	−409			
ErH$_2$	c	−49.0			
ErI$_3$	c	−146.5			
std. state, $m = 1$	aq	−202.4	−194.5		
Er$_2$O$_3$	c	−453.6	−432.3	37.2	25.93
Er$_2$(SO$_4$)$_3$ std. state, $m = 1$	aq	−975.3	−847.0		
Europium					
Eu	c	0	0	18.59	6.61
	g	41.9	34.0	45.10	4.97
Eu^{2+} std. state, $m = 1$	aq	−125	−129.1	1	
Eu^{3+} std. state, $m = 1$	aq	−144.6	−137.2	−53	2
EuBr$_3$	c	(−202)			
Eu(BrO$_3$)$_3$ · 9H$_2$O	c	−835.3			
EuC$_2$	c	−15	−16	24	
EuCl$_2$	c	−194			
EuCl$_3$	c	−223.7			
std. state, $m = 1$	aq	−264.4	−231.3	−13	−96
EuCl$_3$ · 6H$_2$O	c	−665.6	−565.5	97.3	87.7
EuF$_3$	g	−322			
EuF$_3$ · ½H$_2$O	c	−413.7			
Eu(IO$_3$)$_3$	c	−308.4			
Eu(NO$_3$)$_3$ · 6H$_2$O	c	−706.9			
EuO	c	−141.5	−133.1	15	
Eu$_2$O$_3$ cubic	c	−397.4			29.6
monoclinic	c	−394.7	−372.1	35	29.2
Eu$_3$O$_4$	c	−543	−512	49	
Eu(OH)$_3$	c	−285.5			
Eu$_2$(SO$_4$)$_3$ std. state, $m = 1$	aq	−989.3	−862.2		
Fluorine					
F	g	18.88	14.80	37.92	5.436
F$^-$	g	−61.1	−62.7	34.77	4.968
F$_2$	g	0	0	48.45	7.49
FNO$_3$	g	2.5	17.6	70	15.59
FO	g	41			
F$_2$O	g	−4.39		59.12	10.30
Francium					
Fr	c	0	0	22.50	7.60
	g	17.40	11.15	43.48	
FrBr	c	−95		30	11.8
	g	44.4		65.59	
FrCl	c	−105		26.7	11.80
	g	55.8		62.8	
FrF	c	−123		23.6	11.6
	g	82.6		59.9	
FrI	c	−82		32.2	
	g	33.9		67.4	
Fr$_2$O	c	−81	−71.5	37.5	
Gadolinium					
Gd	c	0	0	16.27	8.85
	g	95.0	86.0	46.42	6.58
Gd^{3+} std. state, $m = 1$	aq	−164	−158	−49.2	0
Gd(acetate)$_3$ std. state, $m = 1$	aq	−508.94	−428.0	42.0	
Gd(BrO$_3$)$_3$ · 9H$_2$O	c	−854.8			
GdC$_2$	c	−25			
Gd$_2$(C$_2$O$_4$)$_3$ · 10H$_2$O	c		−1407		
GdCl$_3$	c	−241			
std. state, $m = 1$	aq	−284	−253	−8.8	−98
GdCl$_3$ · 6H$_2$O	c	−685	−586	97.56	83.0
GdF$_3$ · ½H$_2$O	c	−433			
GdH$_2$	c	−45.5			
GdI$_3$	c	−142			
std. state, $m = 1$	aq	−208.9	−201.6	−31.3	

Table 9-1 (*Continued*)
ELEMENTS AND INORGANIC COMPOUNDS

Formula and Description	State	$\Delta Hf°$	$\Delta Gf°$	$S°$	$C_p°$
Gd(IO$_3$)$_3$	c	−327			
Gd$_2$O$_3$ monoclinic	c	−434.9			25.5
cubic	c			36.0	25.22
Gd$_2$(SO$_4$)$_3$ std. state, $m = 1$	aq	−988.3	−861.2	−81.9	
Gd$_2$(SO$_4$)$_3 \cdot$ 8H$_2$O	c	−1513	−1322	155.8	140.5
Gallium					
Ga	c	0	0	9.77	6.18
	liq	1.33			
	g	66.2	57.1	40.38	6.06
Ga$_2$	g	104.8			
GaAs	c	−17	−16.2	15.34	11.05
Ga$_2$C$_2$	g	134			
Ga(CH$_3$)$_3$	liq	−18.7			
	g	−10.8			
GaBr	g	−11.9	−21.5	60.2	8.70
GaBr$_3$	c	−92.4	−86.0	43	
	g	−70			
GaBr$_4^-$ std. state, $m = 1$	aq	−158.2	−131.5	8.6	
GaCl	g	−19.1	−25.4	57.4	8.50
GaCl$_3$	c	−125.4	−108.7	34	
	g	−107.0			
GaF	g	−60.2			7.95
GaF$_3$	c	−278	−259.4	20	
GaH	g	52.7	46.3	46.69	7.00
GaI	g	6.9			8.76
GaI$_3$	c	−57.1			
GaN	c	−26.4			
	g	42	36	54	
GaO	g	66.8	60.6	55.2	7.66
Ga$_2$O$_3$ rhombic	c	−260.3	−238.6	20.31	19.28
GaOH	g	−27.4			
Ga(OH)$^{2+}$ std. state, $m = 1$	aq		−90.9		
Ga(OH)$_2^+$ std. state, $m = 1$	aq		−142.8		
GaO$_3^{3-}$ std. state, $m = 1$	aq		−148		
H$_2$GaO$_3^-$ std. state, $m = 1$	aq		−178		
Ga(OH)$_3$	c	−230.5	−198.7	24	
GaP	c	−21			
GaPO$_4$	c		−310.1		
Ga$_2$(SO$_4$)$_3$	c				62.4
GaSb	c	−10.0	−9.3	18.18	11.60
Ga$_2$Te$_3$	c				41.2
Germanium					
Ge	c	0	0	7.43	5.580
	g	90.0	80.3	40.103	7.345
Ge$_2$	g	113.08	99.5	60.4	8.5
GeBr	g	56.32			8.87
GeBr$_2$	g	−15.0	−25.5	79.1	
GeBr$_4$	liq	−83.1	−79.2	67.1	
	g	−71.7	−76.0	94.66	24.34
GeH$_3$Br	g			65.66	13.47
GeC	g	151			
Ge(C$_2$H$_5$)$_4$	liq	−49.5			
GeCl	g	37	30	58.8	8.81
GeCl$_4$	liq	−127.1	−110.6	58.7	
	g	−118.5	−109.3	83.08	22.97
GeH$_3$Cl	g			63.00	13.08
GeHCl$_3$	liq			53.6	
GeF	g	−7.97			8.30
GeF$_4$	g	−284.4		72.36	19.56
GeH$_4$	g	21.7	27.1	51.87	10.76
Ge$_2$H$_6$	liq	32.82			
	g	38.8			

Table 9-1 (*Continued*)
ELEMENTS AND INORGANIC COMPOUNDS

Formula and Description	State	$\Delta Hf°$	$\Delta Gf°$	$S°$	$C_p°$
Ge_3H_8	liq	46.3			
	g	54.2			
GeI_2	c	−21	−20	32	
	g	11.2	−1.0	76	
GeI_4	c	−33.9	−34.5	64.8	
	g	−13.6	−25.4	102.49	24.89
GeH_3I	g			67.65	13.75
Ge_3N_4	c	−15.1			
GeO brown	c	−50.7	−56.7	12	
yellow	c		−49.5		
	g	−11.04	−17.49	53.58	7.39
GeO_2 hexagonal	c	−131.7	−118.8	13.21	12.45
tetragonal	c	−138.7			
Ge_2O_2	g	−112			
Ge_2O_3	g	−212			
H_2GeO_3	aq	−195.73			
GeP	c	−5	−4	15	
GeS	c	−16.5	−17.1	17	
	g	22	10	56	8.05
GeS_2	c	−45.3			
GeSe	c	−22.0			
	g	22.84			8.42
GeSi	g	127			
GeTe	c	−6			
	g	42			8.59
$GeTe_2$	g	44			
Gold					
Au	c	0	0	11.33	6.075
	g	87.5	78.0	43.115	4.968
Au_2	g	123.1			8.808
AuBr	c	−3.34		27.0	11.96
$AuBr_3$	c	−12.73			
$AuBr_4^-$ std. state, $m = 1$	aq	−45.8	−40.0	80.3	
$Au(CN)_2^-$ std. state, $m = 1$	aq	57.9	68.3	41	
AuCl	c	−8.6		22.2	11.65
$AuCl_3$	c	−28.1		35.4	22.66
$AuCl_4^-$ std. state, $m = 1$	aq	−77.0	−56.72	63.8	
$HAuCl_4$ std. state, $m = 1$	aq	−77.0	−56.22	63.8	
AuF_3	c	−86.9		27.3	21.82
AuH	g	70.5	63.5	50.441	6.968
AuI	c	0.2		28.5	12.39
AuO_3^{3-} std. state, $m = 1$	aq		−12.4		
$Au(OH)_3$ pptd	c	−101.5	−75.77	45.3	
Au_2P_3	g	−23.8			
$AuSb_2$	c	−4.65		28.5	18.50
Hafnium					
Hf hexagonal	c	0	0	10.41	6.15
	g	148.0	137.8	44.642	4.972
HfB	c	−47			
HfB_2	c	−80.3	−79.4	10.2	11.89
HfC	c	−60.1		9.85	8.23
$HfCl_4$	c	−236.70	−215.42	45.6	28.80
	g	−211.4			
	aq	−302.7			
HfF_4 monoclinic	c	−461.4	−437.5	27	
	g	−399.1			
HfN	c	−88.3			
HfO	g	12			
HfO_2	c	−273.6	−245.5	14.18	14.40
hydrous ppt		−269.0			
$HfOOH^+$	aq	−279.5			
Helium					
He	g	0	0	30.124	4.9679

Table 9-1 (*Continued*)
ELEMENTS AND INORGANIC COMPOUNDS

Formula and Description	State	$\Delta Hf°$	$\Delta Gf°$	$S°$	$C_p°$
Holmium					
Ho	c	0	0	18.0	6.49
	g	71.9	63.3	46.72	4.97
Ho^{3+} std. state, $m = 1$	aq	−168.5	−161.0	−54.2	4
$Ho(acetate)_3$					
undissoc; std. state, $m = 1$	aq	−512.45	−431.06	40.5	
$HoBr_3$	c	(−225)			
$Ho(BrO_3)_3 \cdot 9H_2O$	c	−857.5			
HoC_2	c	−26	−26.7	23	
	g	135.1			10.5
Ho_2C_3	c	−56			
$HoCl_3$	c	−240.3			21
std. state, $m = 1$	aq	−288.4	−255.1	−13.8	−94
$HoCl_3 \cdot 6H_2O$	c	−687.9	−588.0	97.08	83.0
HoF_3	c	−408			
	g	−294			
HoH_2	c	−51.7			
HoI_3	c	−149.0			
std. state, $m = 1$	aq	−203.8	−196.2		
$Ho(IO_3)_3$	c	−330			
Ho_2O_3	c	−449.5	−428.1	37.8	27.48
$Ho_2(SO_4)_3 \cdot 8H_2O$ std. state, $m = 1$	aq	−978.1	−850.4		
Hydrogen					
^1H	g	52.095	48.581	27.391	4.968
^2H	g	52.981	49.360	29.455	4.968
H^+ std. state, $m = 1$	aq	0	0	0	0
H_2	g	0	0	31.211	6.892
$^2H_2$	g	0	0	34.620	6.978
$^1H^2H$	g	0.076	−0.350	34.343	6.978
H_2 std. state, $m = 1$	aq	−1.0	4.2	13.8	
$HAsO_2$ undissoc; std. state, $m = 1$	aq	−109.1	−96.25	30.1	
H_2AsO_3 undissoc; std. state, $m = 1$	aq	−170.84	−140.35	26.4	
H_3AsO_3 undissoc; std. state, $m = 1$	aq	−177.4	−152.94	46.6	
$HAsO_4^{2-}$ undissoc; std. state, $m = 1$	aq	−216.62	−170.82	−0.4	
$H_2AsO_4^-$ undissoc; std. state, $m = 1$	aq	−217.39	−180.04	28	
H_3AsO_4	c	−216.6			
undissoc; std. state, $m = 1$	aq	−215.7	−183.1	44	
HBO_2 cubic	c	−192.17			
monoclinic	c	−189.83	−172.9	9	
orthorhombic	c	−188.52	−172.5	12	
H_3BO_3	c	−261.55	−231.60	21.23	19.45
	g	−237.6			
undissoc; std. state, $m = 1$	aq	−256.29	−231.56	38.8	
HBr	g	−8.71	−12.79	47.47	6.97
std. state, $m = 1$	aq	−29.05	−24.85	19.7	−33.9
HBrO undissoc; std. state, $m = 1$	aq	−27.0	−19.7	34	
$HBrO_3$ std. state, $m = 1$	aq	−20.0	0.4	39.0	
HCl	g	−22.06	−22.78	44.64	6.96
std. state, $m = 1$	aq	−39.952	−31.37	13.5	−32.6
^2HCl	g	−22.40	−23.01	46.02	6.97
HClO	g	−22.0	−19.0	56.55	8.88
undissoc; std. state, $m = 1$	aq	−28.9	−19.1	34	
$HClO_2$ undissoc; std. state, $m = 1$	aq	−12.4	1.4	45.0	
$HClO_3$ std. state, $m = 1$	aq	−23.7	−0.8	38.8	
$HClO_4$	liq	−9.70			
std. state, $m = 1$	aq	−30.91	−2.06	43.5	
$HClO_4 \cdot H_2O$	c	−91.35			
$HClO_4 \cdot 2H_2O$	liq	−162.04			
HCN	liq	26.02	29.86	26.97	16.88
	g	32.3	29.8	48.20	8.57
std. state, $m = 1$	aq	36.0	41.2	22.5	
undissoc; std. state, $m = 1$	aq	25.6	28.6	29.8	

Table 9-1 (*Continued*)
ELEMENTS AND INORGANIC COMPOUNDS

Formula and Description	State	$\Delta Hf°$	$\Delta Gf°$	$S°$	$C_p°$
H_2CO_3 std. state, $m = 1$	aq	−167.22	−148.94	44.8	
HCO_3^- std. state, $m = 1$	aq	−165.39	−140.26	21.8	
H_2CS_3 trithiocarbonic acid	liq	6.0	6.65	53.3	35.8
HF	liq	−71.65		18.02	12.35
	g	−64.80	−65.30	41.51	6.96
ionized; std. state, $m = 1$	aq	−79.50			
undissoc; std. state, $m = 1$	aq	−76.50	−70.95	21.2	
HF_2^- std. state, $m = 1$	aq	−155.34	−138.18	22.1	
^2HF	g	−65.94	−66.35	42.90	6.96
HI	g	6.33	0.41	49.35	6.97
std. state, $m = 1$	aq	−13.19	−12.33	26.6	−34.0
HIO undissoc; std. state, $m = 1$	aq	−33.0	−23.7	22.8	
HIO_3	c	−55.0			
	aq, inf	−52.2			
H_5IO_6	aq	−180.4			
H_2MoO_4	c	−250.0			
	g	−203.4	−188.2	85	24.89
	aq	−240.8			
HNCO isocyanic acid	g	−27.90	−25.66	56.91	10.72
HNCS isothiocyanic acid	g	30.5	27.0	59.2	11.2
HNO_2 *cis*	g	−18.3	−10.0	59.59	10.84
trans	g	−18.8	−10.5	59.55	11.00
undissoc; std. state, $m = 1$	aq	−28.5	−13.3	36.5	
HNO_3	liq	−41.40	−19.10	37.19	
	g	−32.28	−17.87	63.64	12.75
std. state, $m = 1$	aq	−49.56	−26.61	35.0	−20.7
$H_2N_2O_2$ hyponitrous acid	aq	−13.7	8.6	52	
OH	g	9.49	8.37	43.88	7.17
OH^+	g	314.8	312.2	43.66	6.97
OH^-	g	−34.3	−33.2	41.2	6.97
std. state, $m = 1$	aq	−54.97	−37.59	−2.57	−35.5
HO_2 hydroperoxyl	g	(5)	8	54.38	8.34
HO_2^- std. state, $m = 1$	aq	−38.32	−16.1	5.7	
H_2O	liq	−68.315	−56.687	16.71	17.995
	g	−57.796	−54.634	45.104	8.025
^2H_2O	liq	−70.411	−58.195	18.15	20.16
	g	−59.560	−56.059	47.378	8.19
$^1H^2HO$	liq	−69.285	−57.817	18.95	
H_3O^+	g	234.3			
H_2O_2	liq	−44.88	−28.78	26.2	21.3
	g	−32.53	−25.21	55.66	10.31
undissoc; std. state, $m = 1$	aq	−45.69	−32.05	34.4	
$H_2O_2^+$	g	220.7			
HOCN ionized; std. state, $m = 1$	aq	−34.90	−23.3	25.5	
undissoc; std. state, $m = 1$	aq	−36.90	−28.0	34.6	
OCN^- cyanate ion; std. state, $m = 1$	aq	−34.9	−23.3	25.5	
HOF	g	−23.5	−20.47	54.17	8.59
HPO_3	c	−226.7			
	aq	−233.5			
HPO_4^{2-} std. state, $m = 1$	aq	−308.83	−260.34	−8.0	
$H_2PO_4^-$ std. state, $m = 1$	aq	−309.82	−270.17	21.6	
H_3PO_3	c	−230.5			
H_3PO_2 hypophosphorous acid	c	−144.5			
H_3PO_4	c	−302.8	−265.9	26.42	25.35
	liq	−299.8	−265.7	36.0	34.67
ionized; std. state, $m = 1$	aq	−305.3	−243.5	−53	
undissoc; std. state, $m = 1$	aq	−307.92	−273.10	37.8	
$H_4P_2O_7$	c	−535.6			
undissoc; std. state, $m = 1$	aq	−542.2	−485.7	64	
HS	g	34.10	27.08	46.74	7.72
HS^- std. state, $m = 1$	aq	−4.2	2.88	15.0	
H_2S	g	−4.82	−7.90	49.18	8.17
std. state, $m = 1$	aq	−9.5	−6.66	29	

Table 9-1 (*Continued*)
ELEMENTS AND INORGANIC COMPOUNDS

Formula and Description	State	$\Delta Hf°$	$\Delta Gf°$	$S°$	$C_p°$
H_2S_2	g	3.71			
H_2S_3	g	7.29			
H_2S_4	g	10.57			
H_2S_5	g	13.84			
$HSbO_2$ undissoc; std. state, $m = 1$	aq	−116.6	−97.4	11.1	
H_3SbO_4	aq	−216.8			
HSCN undissoc; std. state, $m = 1$	aq		23.31		
SCN^- std. state, $m = 1$	aq	18.27	22.15	34.5	−9.6
H_2SO_3 undissoc; std. state, $m = 1$	aq	−145.51	−128.56	55.5	
HSO_3^- std. state, $m = 1$	aq	−149.67	−126.15	33.4	
H_2SO_4	liq	−194.55	−164.93	37.50	33.20
	g	−177.0	−156.8	69.1	19.27
std. state, $m = 1$	aq	−217.32	−177.97	4.8	−70
$H_2SO_4 \cdot H_2O$	liq	−269.51	−227.18	50.56	51.35
$H_2SO_4 \cdot 2H_2O$	liq	−341.09	−286.77	66.06	62.34
$H_2S_2O_4$ std. state, $m = 1$	aq		−147.4		
$H_2S_2O_6$	aq	−286.4			
$H_2S_2O_7$	c	−304.4			
$H_2S_2O_8$ std. state, $m = 1$	aq	−320.0	−265.4	59.3	
HSO_3Cl	liq	−143.7			
HSO_3F	liq	−186			
	g	−180	−165	71.0	17.98
H_2Se	g	7.1	3.8	52.32	8.30
std. state, $m = 1$	aq	4.6	5.3	39.1	
HSe^- std. state, $m = 1$	aq	3.8	10.5	19	
H_2SeO_3	c	−125.35			
undissoc; std. state, $m = 1$	aq	−121.29	−101.87	49.7	
$HSeO_3^-$ std. state, $m = 1$	aq	−122.98	−98.36	32.3	
H_2SeO_4	c	−126.7			
	aq, inf	−140.3			
$HSeO_4^-$ std. state, $m = 1$	aq	−139.0	−108.1	35.7	
H_2SiO_3	c	−284.1	−261.1	32	
undissoc; std. state, $m = 1$	aq	−282.7	−258.0	26	
H_4SiO_4	c	−354.0	−318.6	46	
undissoc; std. state, $m = 1$	aq	−351.0	−314.7	43	
H_2Te		23.8		54.7	8.50
H_2TeO_3 std. state, $m = 1$	aq		−76.2		
H_6TeO_6	c	−310.4			
H_2TiF_6	aq	−573.7			
H_2WO_4	c	−270.5	−240.0	35	26.92
	g	−216.4	−200.7	84	24.50
Indium					
In	c	0	0	13.82	6.39
	g	58.15	49.89	41.51	4.98
In_2	g	91.04			
InAs	c	−14.0	−12.8	18.1	11.42
InBr	c	−41.9	−40.4	27	
	g	−13.6	−22.54	61.99	8.76
$InBr_3$	c	−102.5			
	g	−67.4			
InCl II	c	−44.5			
	g	−18			
$InCl_3$	c	−128.4			
	g	−89.4			
In_2Cl_3	g	−103.6			
InF	g	−48.61			
InH	g	51.5	45.49	49.60	7.07
InI	g	1.8	−9.0	63.87	8.80
	c	−27.8	−28.8	31	
InI_3	c	−57			
InN	c	−4.2			
InO	g	92.5	87.1	56.5	7.78

Table 9-1 (*Continued*)
ELEMENTS AND INORGANIC COMPOUNDS

Formula and Description	State	$\Delta Hf°$	$\Delta Gf°$	$S°$	$C_p°$
In_2O_3	c	−221.27	−198.55	24.9	22
InOH	g	−19			
$In(OH)^{2+}$ std. state, $m = 1$	aq	−88.5	−74.8	−21	
$In(OH)_2^+$ std. state, $m = 1$	aq	−148	−125.5	6	
InP	c	−21.2	−18.4	14.3	10.86
InS	c	−33.0	−31.5	16	
	g	90			
In_2S	g	15	3.1	76	
In_2S_3	c	−102	−98.6	39.1	28.20
$In_2(SO_4)_3$	c	−666	−583	65	67
InSe	c	−28			
In_2Se_3	c	−82			
InSb	c	−7.3	−6.1	20.6	11.82
	g	82.3			
InTe	c	−23			
In_2Te_3	c	−47			
Iodine					
I	g	25.535	16.798	43.184	4.968
I^-	g	−47.0			
std. state, $m = 1$	aq	−13.19	−12.33	26.6	−34.0
I_2	c	0	0	27.757	13.011
	g	14.923	4.627	62.28	8.82
std. state, $m = 1$	aq	5.4	3.92	32.8	
std. state, $m = 1$	CCl_4	6.0	2.66	39.0	
I_3^- std. state, $m = 1$	aq	−12.3	−12.3	57.2	
IBr	g	9.76	0.89	61.822	8.71
IBr_2^- std. state, $m = 1$	aq		−29.4		
BrI_2^- std. state, $m = 1$	aq	−30.6	−26.3	47.2	
$IBrCl^-$ std. state, $m = 1$	aq		−35.0		
ICl	liq	−5.71	−3.25	32.3	
	g	4.25	−1.30	59.140	8.50
ICl_2^- std. state, $m = 1$	aq		−38.5		
ICl_3	c	−21.4	−5.34	40.0	
I_2Cl^- std. state, $m = 1$	aq		−32.9	52.9	
IF	g	−22.86	−28.32	56.42	7.99
IF_5	liq	−206.7			
	g	−200.8	−184.4	80.0	24.59
IF_7	g	−225.6	−195.6	82.8	32.6
IO	g	41.84	35.80	58.65	7.86
IO^- std. state, $m = 1$	aq	−25.7	−9.2	−1.3	
IO_3^- std. state, $m = 1$	aq	−52.9	−30.6	28.3	
IO_4^-	aq	−35.2			
I_2O_5	c	−37.78			
Iridium					
Ir	c	0	0	8.48	6.00
	g	159.0	147.7	46.240	4.968
$IrCl_3$	c	−58.7	−43	27	
	g	25	24	90	
$IrCl_6^{2-}$	aq	−148.1	(−129)	(50)	
$IrCl_6^{3-}$	aq	−179.5	(−109)	(70)	
IrF_6	c	−138.54	−110.34	59.2	
	g	−130	−110	85.5	28.94
IrO_2	c	−65.5	(−42)	13.7	13.70
IrO_3	g	1.9	(7)	(69)	
IrS_2	c	−33	(−32)	(15)	
Ir_2S_3	c	−56	(−53)	(23)	
Iron					
Fe α	c	0	0	6.52	6.00
	liq	3.138	2.641	8.195	5.99
Fe^{2+} std. state, $m = 1$	aq	−21.3	−18.85	−32.9	
Fe^{3+} std. state, $m = 1$	aq	−11.6	−1.1	−75.5	

Table 9-1 (*Continued*)
ELEMENTS AND INORGANIC COMPOUNDS

Formula and Description	State	$\Delta Hf°$	$\Delta Gf°$	$S°$	$C_p°$
$FeAl_2O_4$	c	-470	-442	25.4	29.53
FeAsS	c	-10	-12	29	
$FeBr_2$	c	-59.7	-56.7	33.62	19.18
std. state, $m = 1$	aq	-79.4	-68.55	6.5	
$FeBr_3$	c	-64.1			
	g	-29.6			
std. state, $m = 1$	aq	-98.8	-75.7	-16.4	
Fe_3C α-cementite	c	6.0	4.8	25.0	25.3
$Fe(CN)_6^{3-}$ std. state, $m = 1$	aq	134.3	174.3	64.6	
$Fe(CN)_6^{4-}$ std. state, $m = 1$	aq	108.9	166.09	22.7	
$H_2Fe(CN)_6^{2-}$ std. state, $m = 1$	aq	108.9	157.37	52	
$FeCO_3$ siderite	c	-177.00	-159.35	22.2	19.63
$Fe(CO)_5$	liq	-185.0	-168.6	80.8	57.5
	g	-175.4	-166.65	106.4	
$Fe_2(C_2O_4)_3$ std. state, $m = 1$	aq	-614.8	-485.5	-118.3	
$Fe(acetate)_3$ std. state, $m = 1$	aq	-360.1	-266.0	-13.4	
$FeCNS^{2+}$ std. state, $m = 1$	aq	5.6	17.0	-31	
$FeCl^{2+}$ std. state, $m = 1$	aq	-43.1	-34.4	-27	
$FeCl_2$	c	-81.69	-72.26	28.19	18.32
std. state, $m = 1$	aq	-101.2	-81.59	-5.9	
	aq, 100	-99.88			
$FeCl_2 \cdot 2H_2O$	c	-227.8			
$FeCl_2 \cdot 4H_2O$	c	-370.3			
$FeCl_3$	c	-95.48	-79.84	34.0	23.10
std. state, $m = 1$	aq	-131.5	-95.2	-35.0	
$FeCl_3 \cdot 6H_2O$	c	-531.5			
Fe_2Cl_6	g	-156.5			
$Fe(ClO_4)_2$ std. state, $m = 1$	aq	-83.1	-22.97	54.1	
$Fe(ClO_4)_2 \cdot 6H_2O$	c	-494.4			
$FeCr_2O_4$	c	-345.3	-321.2	34.9	31.94
FeF_2	c	-168	-158	20.79	16.28
std. state, $m = 1$	aq	-180.3	-152.13	-39.5	
FeF_3	c	-249	-232	23.5	21.75
std. state, $m = 1$	aq	-250.1	-201.0	-85.4	
FeI_2	c	-25	-27	40	20.00
std. state, $m = 1$	aq	-47.7	-43.51	20.3	
FeI_3	g	17			
std. state, $m = 1$	aq	-51.2	-38.1	4.3	
$FeMoO_4$	c	-257	-233	30.9	28.31
$Fe_2(MoO_4)_3$	c	-702			
Fe_4N	c	-2.5	0.9	37.3	29.27
$Fe(NO_3)_3$ std. state, $m = 1$	aq	-160.3	-80.9	29.5	
$Fe(NO_3)_3 \cdot 9H_2O$	c	-785.2			
$Fe_{0.947}O$ wistite	c	-63.64	-58.59	13.74	11.50
FeO	c	-65.0	-60.10	14.52	11.93
Fe_2O_3 hematite	c	-197.0	-177.4	20.89	24.82
Fe_3O_4 magnetite	c	-267.3	-242.7	35.0	35.19
$FeOH^+$ std. state, $m = 1$	aq	-77.6	-66.3	-7	
$Fe(OH)^{2+}$ std. state, $m = 1$	aq	-69.5	-54.83	-34	
$Fe(OH)_2$ pptd	c	-136.0	-116.3	21	23.20
$Fe(OH)_2^+$ std. state, $m = 1$	aq		-104.7		
$Fe(OH)_3$ pptd	c	-196.7	-166.5	25.5	24.30
FeP	c	-30			
FeP_2	c	-46			
Fe_2P	c	-39			
Fe_3P	c	-39			
$FePO_4$	c	-310.1			
$FePO_4 \cdot 2H_2O$ strengite	c	-451.3	-396.2	40.93	43.15
FeS iron-rich pyrrhotite	c	-23.9	-24.0	14.41	12.08
FeS_2 pyrite	c	-42.6	-39.9	12.65	14.86
Fe_7S_8 sulfur-rich pyrrhotite	c	-176.0	-178.9	116.1	95.26
$FeSO_4$	c	-221.9	-197.2	28.9	24.04
std. state, $m = 1$	aq	-238.6	-196.82	-28.1	

Table 9-1 (*Continued*)
ELEMENTS AND INORGANIC COMPOUNDS

Formula and Description	State	$\Delta Hf°$	$\Delta Gf°$	$S°$	$C_p°$
$FeSO_4 \cdot 7H_2O$	c	−720.50	−599.97	97.8	94.28
$Fe_2(SO_4)_3$	c	−617.0	−540.9	73.5	63.27
std. state, $m = 1$	aq	−675.2	−536.1	−136.4	
FeSe	c	−18.0			
$FeSe_2$	c			20.75	17.42
FeSi	c	−17.6	−17.6	11.0	11.4
$FeSi_2$ β-lebanite	c	−19.4	−18.7	13.3	15.79
Fe_3Si	c	−22.4	−22.6	24.8	23.50
Fe_2SiO_4 fayalite	c	−353.7	−329.6	34.7	31.76
FeTe	c	−15.0			
$FeTiO_3$ ilmenite	c	−297.9		25.3	23.78
$FeWO_4$	c	−276	−252	31.5	27.39
$Fe_2(WO_4)_3 \cdot 8H_2O$	c	−1355			
Krypton					
Kr	g	0	0	39.191	4.968
KrF_2	c	14.4			
Lanthanum					
La	c	0	0	13.6	6.48
	g	103.0	94.07	43.56	5.44
La^{3+} std. state, $m = 1$	aq	−169.0	−163.4	−52.0	−3
$La(acetate)_3$					
undissoc; std. state, $m = 1$	aq	−512.86	−432.41	39.3	
$LaBr_3$	c	(−233)			
$La(BrO_3)_3 \cdot 9H_2O$	c	−857.9			
LaC_2	c	−17	−17.3	17	
	g	140.4	127.1	61	10.6
$La_2(CO_3)_3$	c	−750.9			
$La_2(C_2O_4)_3 \cdot 10H_2O$	c	−1414			
$LaCl_3$	c	−256.0		34.5	24.76
std. state, $m = 1$	aq	−288.9	−257,5	−12	−101
$LaCl_3 \cdot 7H_2O$	c	−759.7	−648.5	110.6	103.0
LaH_2	c	−48.3			
LaI_3	c	−159.4			
std. state, $m = 1$	aq	−216.3	−209.9	−34	
$La(IO_3)_3$	c	−334	−270.4	62	
LaN	c	−72.1	−64.6		
$La(NO_3)_3$	c	−299.8			
	aq, inf	−317.7			
La_2O_3	c	−428.7	−407.7	30.43	26.00
$La(OH)_3$	c	−337.0	−312.8		
LaS	c	−109	−107.9	17.5	14
La_2S_3	c	−289			
$La_2(SO_4)_3$	c	−942.0		67	
std. state, $m = 1$	aq	−1003.1	−877.8	−76	
$La_2(SO_4)_3 \cdot 9H_2O$	c	−1589		152	
$La_2(SeO_4)_3$	c	−688.20	−629.5	81	
La_2Te_3	c	−173	−170.8	55.36	31.58
Lead					
Pb	c	0	0	15.48	6.41
	liq	1.025	0.531	17.14	6.41
	g	46.75	38.87	41.89	4.97
$Pb(acetate)_2$	c	−230.5			
PbB_2O_4	c	−372.0	−346.6	31.20	25.60
PbB_4O_7	c	−683.0	−637.5	39.9	40.20
PbBr	g	17	7.6	65.10	8.82
$PbBr_2$	c	−66.30	−62.32	38.51	19.02
	liq	−63.91	−60.84	41.56	19.02
	g	−24.95	−33.66	81.09	13.60
$PbBr_4$	g	−109.07	−113.12	101.84	25.01
PbCl	g	3.6	−2.3	62.02	8.66
$PbCl_2$	c	−85.90	−75.09	32.50	18.42
	liq	−82.28	−72.71	36.61	18.42
	g	−41.60	−43.69	75.79	13.19

Table 9-1 (*Continued*)
ELEMENTS AND INORGANIC COMPOUNDS

Formula and Description	State	$\Delta Hf°$	$\Delta Gf°$	$S°$	$C_p°$
PbCl$_4$	g	−132.03	−122.82	91.19	24.03
PbClF	c	−127.8	−116.7	29.1	
Pb(ClO$_4$)$_2$	c	−52.0			
Pb(CH$_3$)$_4$	liq	23.4			
	g	32.48			
Pb(C$_2$H$_5$)$_4$	liq	12.6			
	g	26.19			
PbCO$_3$	c	−167.1	−149.5	31.3	20.89
PbC$_2$O$_4$	c	−203.5	−179.3	34.9	25.2
PbCrO$_4$	c	−222.5			
PbF	g	−19.2	−25.1	59.71	8.22
PbF$_2$ α	c	−161.8	−150.8	27.00	17.27
β	c	−161.6	−150.65	27.35	17.75
PbF$_4$	g	−270.90	−261.16	79.71	21.73
PbI$_2$	c	−41.92	−41.49	41.79	18.54
	liq	−37.69	−38.97	47.54	18.54
PbI$_4$	g	−53.65	−65.70	111.41	25.41
Pb(IO$_3$)$_2$	c	−118.4	−84.0	74.8	
PbMoO$_4$	c	−251.4	−227.4	39.7	28.61
Pb(N$_3$)$_2$ monoclinic	c	114.3	149.3	35.4	
orthorhombic	c	113.8	148.7	35.7	
Pb(NO$_3$)$_2$	c	−108.0			
PbO red	c	−52.34	−45.23	15.9	10.94
yellow	c	−52.12	−45.09	16.42	10.94
PbO$_2$	c	−65.6	−51.5	17.16	14.62
Pb$_2$O$_3$	c			36.3	25.74
Pb$_3$O$_4$	c	−171.8	−143.8	50.7	37.03
PbO · PbCO$_3$	c	−219.5	−195.2	48.8	
Pb(OH)$_2$	c		−108.1	21	
pptd	c	−123.3			
Pb$_3$(PO$_4$)$_2$	c	−620.3	−581.4	84.4	61.25
Pb(ReO$_4$)$_2$ · 2H$_2$O	c	−534	−455	74	
PbS	c	−23.50	−23.12	21.83	11.82
PbSO$_3$	c	−160.1			
PbSO$_4$	c	−219.87	−194.36	35.51	24.67
PbSe	c	−24.6	−24.3	24.5	12.0
PbSeO$_3$	c	−128.5			
PbSeO$_4$	c	−145.6	−120.7	40.1	
PbSiO$_3$	c	−273.7	−253.6	26.3	21.52
Pb$_2$SiO$_4$	c	−329	−303	44.7	32.74
PbSiO$_4$	c	−483.7	−456.4	20.08	23.58
PbTe	c	−16.9	−16.6	26.3	12.08
Lithium					
Li	c	0	0	38.4	5.90
	liq	0.569	0.223	8.11	7.48
	g	38.4	30.6	33.14	4.97
LiAlF$_4$	g	−443	−433	78	24.10
Li$_3$AlF$_6$	c	−808.7	−770.5	44.9	48.40
LiAlH$_4$	c	−28	−11.6	21	20.65
LiAlO$_2$	c	−284.3	−269.4	12.75	16.21
LiBeF$_3$	c	−394.8	−376.7	21.3	21.95
Li$_2$BeF$_4$	c	−543.4	−519.0	31.2	32.33
LiBH$_4$	c	−45.52	−29.82	18.12	19.73
LiBO$_2$	c	−243.6	−230.2	12.36	14.43
Li$_2$B$_4$O$_7$	c	−804	−758	37.20	43.75
LiBr	c	−83.87	−81.65	17.70	11.69
std. state, $m = 1$	aq	−95.45	−94.79	22.7	
LiBrO$_3$	c	−76.49			
	aq	−77.9	−65.70		
Li$_2$C$_2$	c	−14.2			
LiCl	c	−97.58	−91.79	14.17	11.48
	aq, inf	−106.58	−101.57	16.6	

Table 9-1 (*Continued*)
ELEMENTS AND INORGANIC COMPOUNDS

Formula and Description	State	$\Delta Hf°$	$\Delta Gf°$	$S°$	$C_p°$
LiCl · H₂O	c	−170.31	−151.2	24.8	
LiClO₃	c	−70		18	
LiClO₄	c	−91.0	−60.7	30	25.10
	aq	−106.3	−81.4		
Li₂CO₃	c	−290.64	−270.60	21.55	23.00
	aq, inf	−294.74	−266.66	−5.9	
LiF	c	−147.45	−140.70	8.52	9.99
	aq, inf	−145.21	−136.30	1.1	
LiH	c	−21.66	−16.36	4.79	6.68
LiHCO₃	aq, inf	−231.73	−210.53	29.5	
LiI	c	−64.55	−64.45	20.5	11.97
	aq, inf	−79.92	−82.57	29.5	
LiIO₃	c	−122.35			
Li₅(IO₆)₂	c	−488.5			
Li₃N	c	−47.2	−36.8	9	18.48
LiNO₂	c	−96.6			
LiNO₃	c	−115.1			
	aq, inf	−115.93	−96.63	38.4	
LiO	g	20.0	14.45	50.40	7.75
Li₂O	c	−143.1	−134.3	9.06	12.93
Li₂O₂	c	−151.2	−136.5	13.5	16.88
LiOH	c	−115.9	−104.9	10.23	11.85
	aq, inf	−121.51	−107.82	0.9	
LiOH · H₂O	c	−188.77	−164.8	22	
LiReO₄	c	−253.4			
Li₂SO₄	c	−342.83	−314.66		
	aq, inf	−350.01	−317.78	10.9	
Li₂SO₄ · H₂O	c	−414.20	−375.07		
Li₂Se	c	−91.1			
Li₂SiF₄	c	−688.9			
Li₂SiO₃	c	−394.2	−372.6	19.2	24.04
Li₂Si₂O₅	c	−612.1	−577.7	30	33.00
Li₂TiO₃	c	−399.3	−377.6	21.93	26.26
Lutetium					
Lu	c	0	0	12.18	6.42
	g	102.2	96.7	44.14	4.99
Lu³⁺ std. state, *m* = 1	aq	−159	−150	−63	6
LuCl₃	c	−226.0			
std. state, *m* = 1	aq	−279	−244	−23	−92
LuCl₃ · 6H₂O	c	−676.6	−576.3	89.9	82.0
LuI₃	c	−131			
	aq, inf	−200.2	−192.0		
Lu(NO₃)₃ · 5H₂O	c	−646.26			
Lu₂O₃	c	−448.9	−427.6	26.28	24.32
Lu₂(SO₄)₃	aq, inf	−970.9	−842.0		
Magnesium					
Mg	c	0	0	7.814	5.95
	liq	2.16	1.46	10.16	5.95
	g	35.28	27.03	35.50	4.97
Mg⁺	g	213.1	202.9	36.88	4.97
Mg²⁺ std. state, *m* = 1	aq	−111.58	−108.7	−33.0	
MgAl₂O₄	c	−552.8	−523.5	21.20	27.71
Mg₂Al₄Si₅O₁₈ cordierite	c	−2177	−2055	97.3	108.1
Mg₃As₂	c	−88.8			
Mg₃(AsO₄)₂	c	−739.2			
MgBr₂	c	−125.3	−120.4	28.0	17.49
std. state, *m* = 1	aq	−169.68	−158.4	6.4	
MgBr₂ · 6H₂O	c	−576.0	−491.4	95	
MgCl₂	c	−153.35	−141.52	21.42	17.06
std. state, *m* = 1	aq	−191.48	−171.4	−6.0	
MgCl₂ · 2H₂O	c	−305.86	−267.24	43.0	38.05
MgCl₂ · 6H₂O	c	−597.28	−505.49	87.5	75.30

Table 9-1 (*Continued*)
ELEMENTS AND INORGANIC COMPOUNDS

Formula and Description	State	$\Delta Hf°$	$\Delta Gf°$	$S°$	$C_p°$
$Mg(ClO_4)_2$	c	−135.97			
std. state, $m = 1$	aq	−173.40	−112.8	54.0	
$Mg(ClO_4)_2 \cdot 6H_2O$	c	−584.5	−445.3	124.5	
$MgCO_3$ magnesite	c	−261.9	−241.9	15.7	18.05
MgC_2O_4	c	−303.3			
std. state, $m = 1$	aq	−308.8	−269.8	−22.1	
$MgCrO_4$	c	−321.1			
$MgCr_2O_4$	c	−426.3	−398.9	25.34	30.30
MgF_2	c	−268.7	−256.0	13.68	14.72
$MgFe_2O_4$	c	−341.4	−314.8	29.6	34.35
Mg_2Ge	c	−26.0	−25.3	20.67	16.62
MgH_2	c	−18.0	−8.6	7.43	8.45
MgI_2	c	−87.0	−85.6	31.0	
std. state, $m = 1$	aq	−137.96	−133.4	20.2	
$MgMoO_4$	c	−334.81	−309.69	28.4	26.57
Mg_3N_2	c	−110.1	−95.8	21.0	24.98
$MgNH_4PO_4 \cdot 6H_2O$	c	−880.0			
$Mg(NO_3)_2$	c	−188.97	−140.9	39.2	33.92
std. state, $m = 1$	aq	−210.70	−161.9	37.0	
$Mg(NO_3)_2 \cdot 6H_2O$	c	−624.59	−497.3	108	
MgO microcrystal	c	−142.92	−135.27	6.67	9.00
periclase	c	−143.8	−136.0	6.44	8.88
$Mg(OH)_2$	c	−221.0	−199.3	15.10	18.44
std. state, $m = 1$	aq	−221.52	−183.9	−35.6	
$Mg_3(PO_4)_2$	c	−903.6	−845.8	45.22	51.02
Mg_3Sb_2 α	c	−56			
$MgSeO_3$	c	−215.15			
$MgSeO_4$	c	−231.48			
std. state, $m = 1$	aq	−254.8	−214.2	−20.1	
Mg_2Si	c	−18.6	−18.0	16.0	16.21
$MgSiO_3$ clinoenstatite	c	−370.22	−349.46	16.20	19.45
Mg_2SiO_4 forsterite	c	−519.6	−491.2	22.74	28.37
$Mg_3Si_2O_5(OH)_4$ chrysotile	c	−1043.4	−965.1	52.9	65.41
$Mg_3Si_4O_{10}(OH)_2$ talc	c	−1415.5	−1324.8	62.3	76.9
MgS	c	−82.7	−81.7	12.03	10.90
$MgSO_3$	c	−241.0			
$MgSO_4$	c	−307.1	−279.8	21.9	23.06
std. state, $m = 1$	aq	−328.90	−286.7	−28.2	
$MgSO_4 \cdot H_2O$	c	−382.9	−341.5	30.2	
$MgSO_4 \cdot 6H_2O$	c	−737.8	−629.1	83.2	
$MgSO_4 \cdot 7H_2O$	c	−809.92	−686.4	89	
$MgTe$	c	−50			
$MgTiO_3$	c	−375.9	−354.7	17.82	21.96
Mg_2TiO_4	c	−517.3	−489.4	26.13	30.72
$MgTi_2O_5$	c	−599.8	−565.7	30.42	35.10
$Mg(VO_3)_2$	c	−526.19	−487.43	38.4	39.47
$Mg_2V_2O_7$	c	−677.80	−632.24	47.9	48.63
$MgWO_4$	c	−366.3	−339.6	24.18	26.14
Manganese					
Mn α	c	0	0	7.65	6.29
β	c			8.22	6.34
γ	c	0.37	0.34	7.75	9.59
	g	67.1	57.0	41.49	4.97
$MnAs$	c	−14			
$Mn_3(AsO_4)_2$	c	−512.8			
$MnBr_2$	c	−92.0	−89	(33)	
std. state, $m = 1$	aq	−110.9	−97.8		
$MnBr_2 \cdot H_2O$	c	−168.5			
$MnBr_2 \cdot 4H_2O$	c	−380.1			
Mn_3C	c	1.1	1.3	23.6	22.33
Mn_7C_3	c	−10			
$MnCO_3$ natural	c	−213.7	−195.2	20.5	19.48
pptd	c	−211	−194	27.0	

Table 9-1 (*Continued*)
ELEMENTS AND INORGANIC COMPOUNDS

Formula and Description	State	$\Delta Hf°$	$\Delta Gf°$	$S°$	$C_p°$
MnC_2O_4	c	-246.2			
undissoc; std. state, $m = 1$	aq	-248.5	-221.0	16.1	
$MnC_2O_4 \cdot 2H_2O$	c	-389.2	-338.2	48	
$MnC_2O_4 \cdot 3H_2O$	c	-459.1			
$Mn(acetate)_2$	c	-274.4			
$Mn(acetate)_2 \cdot 4H_2O$	c	-558.8			
$Mn_2(CO)_{10}$	c	-401.0			
	g	-386.0			
MnCl	g	10.1			8.05
$MnCl_2$	c	-115.03	-105.29	28.26	17.43
std. state, $m = 1$	aq	-132.66	-117.3	9.3	-53
$MnCl_2 \cdot H_2O$	c	-188.8	-166.4	41.6	
$MnCl_2 \cdot 2H_2O$	c	-261.0	-225.2	52.3	
$MnCl_2 \cdot 4H_2O$	c	-403.3	-340.3	72.5	
Mn^{2+} std. state, $m = 1$	aq	-52.76	-54.5	-17.6	12
MnF_2	c	-189	-179	22.05	15.96
MnI_2	c	-59.3	(-65)	36.0	18.01
$MnI_2 \cdot 2H_2O$	c	-201.4			
$MnI_2 \cdot 4H_2O$	c	-343.9			
$Mn(IO_3)_2$	c	-160	-124.4	63	
$MnMoO_4$	c	-284.8	-261	(32)	
MnN_6 azide	c	92.2			
Mn_5N_2	c	-48.8			
$Mn(NO_3)_2$	c	-137.73			
std. state, $m = 1$	aq	-151.9	-107.8	52	-29
$Mn(NO_3)_2 \cdot 6H_2O$ glassy	amorp	-566.9			
	liq	-557.27			
MnO	c	-92.07	-86.74	14.27	10.86
	g	29.6			
MnO_2	c	-124.29	-111.18	12.68	12.94
pptd	amorp	-120.1			
Mn_2O_3	c	-229.2	-210.6	26.4	23.60
Mn_3O_4	c	-331.7	-306.7	37.2	33.38
MnO_4^- std. state, $m = 1$	aq	-129.4	-106.9	45.7	
MnO_4^{2-} std. state, $m = 1$	aq	-156	-119.7	14	
$Mn(OH)_2$ pptd	amorp	-166.2	-147.0	23.7	
MnP	c	-27			
MnP_3	c	-51			
$Mn_3(PO_4)_2$	c	-744.9			
$MnHPO_4$	c		-332.5		
MnS green	c	-51.2	-52.2	18.7	11.94
pink pptd	amorp	-51.1			
$MnSO_4$	c	-254.60	-228.83	26.8	24.02
std. state, $m = 1$	aq	-270.1	-232.5		
$MnSO_4 \cdot H_2O$ α	c	-329.0	-289.9	(37)	
β	c	-322.2			
$MnSO_4 \cdot 4H_2O$	c	-539.7			
$MnSO_4 \cdot 5H_2O$	c	-610.2			78
$MnSO_4 \cdot 7H_2O$	c	-750.3			
$Mn_2(SO_4)_3$	c	-666.9			
$MnS_2O_6 \cdot 6H_2O$	c	-751.0			
MnSb	c	-12			
MnSc	c	-25.5	-26.7	21.7	12.20
$MnSiO_3$	c	-315.7	-296.5	21.3	20.66
Mn_2SiO_4	c	-413.6	-390.1	39.0	31.04
$MnTiO_3$	c	-324.0		25.3	23.86
$MnWO_4$	c	-311.9		33.6	30.45
Mercury					
Hg	liq	0	0	18.17	6.688
	g	14.655	7.613	41.79	4.968
$HgBr_2$	c	-40.8	-36.6	40.71	18.00
undissoc; std. state, $m = 1$	aq	-38.4	-34.2	41	

Table 9-1 (*Continued*)
ELEMENTS AND INORGANIC COMPOUNDS

Formula and Description	State	$\Delta Hf°$	$\Delta Gf°$	$S°$	$C_p°$
HgBr$_3^-$ std. state, $m = 1$	aq	−70.1	−62.0	62	
HgBr$_4^{2-}$ std. state, $m = 1$	aq	−103.0	−88.7	74	
Hg$_2$Br$_2$	c	−49.45	−43.278	52.28	25.00
HgBrCl std. state, $m = 1$	aq	−45.5	−38.7	40	
HgBrI undissoc; std. state, $m = 1$	aq	−28.1	−26.7	46	
Hg(CH$_3$)$_2$	liq	14.3	33.5	50	
	g	22.56	34.9	73	
Hg(C$_2$H$_5$)$_2$	liq	7.2			
	g	18.8			
Hg(C$_6$H$_5$)$_2$	c	67.6			
Hg(CN)$_2$	c	63.0			
	g	91			
undissoc; std. state, $m = 1$	aq	66.5	74.6	37.3	
Hg(CN)$_3^-$ std. state, $m = 1$	aq	94.9	110.7	51.4	
Hg(CN)$_4^{2-}$ std. state, $m = 1$	aq	125.8	147.8	71	
Hg(II) fulminate	c	64			
Hg(I) acetate	c	−199.6	−153.3	78	
Hg$_2$CO$_3$	c	−132.3	−111.9	43	
HgC$_2$O$_4$	c	−162.1			
Hg$_2$C$_2$O$_4$	c		−141.8		
Hg(CNS)$_2$ undissoc; std. state	aq	46.9	60.1	37.6	
Hg(CNS)$_4^{2-}$ std. state, $m = 1$	aq	78.0	98.3	109	
Hg$_2$(CNS)$_2$ thiocyanate	c		54.4		
HgCl	g	20.1	15.0	62.09	8.68
HgCl$_2$	c	−53.6	−42.7	34.9	17.66
undissoc; std. state, $m = 1$	aq	−51.7	−41.4	37	
HgCl$_3^-$ std. state, $m = 1$	aq	−92.9	−73.9	50	
HgCl$_4^{2-}$ std. state, $m = 1$	aq	−132.4	−106.8	70	
Hg$_2$Cl$_2$	c	−63.39	−50.377	46.0	
HgF$_2$	c	−101.0	−89	27.8	17.89
Hg$_2$F$_2$	c	−116	−102	38	24.00
HgH	g	57.36	51.63	52.46	7.16
HgI	g	31.64	21.14	67.26	8.99
HgI$_2$ red	c	−25.2	−24.3	43.3	18.58
yellow	c	−24.6			
	g	−4.1	−14.3	80.31	14.60
undissoc; std. state, $m = 1$	aq	−19.0	−18.0	42	
HgI$_3^-$ std. state, $m = 1$	aq	−36.5	−35.5	72	
HgI$_4^{2-}$ std. state, $m = 1$	aq	−56.2	−50.6	86	
Hg$_2$I$_2$	c	−29.00	−26.53	58	25.3
Hg$_2$(N$_3$)$_2$	c	142.0	178.4	49	
Hg(NO$_3$)$_2$ · ½H$_2$O	c	−93.8			
Hg$_2$(NO$_3$)$_2$ · 2H$_2$O	c	−207.5			
HgO red, orthorhombic	c	−21.71	−13.995	16.80	10.53
yellow	c	−21.62	−13.964	17.0	
red, hexagonal	c	−21.4	−13.92	17.0	
Hg(OH)$_2$ undissoc; std. state	aq	−84.9	−65.7	34	
HgS red	c	−13.9	−12.1	19.7	11.57
black	c	−12.8	−11.4	21.1	
HgSO$_4$	c	−169.1	−142	(34)	
Hg$_2$SO$_4$	c	−177.61	−149.589	47.96	31.54
Hg(HS)$_2$ undissoc; std. state, $m = 1$	aq		−6.4		
HgSe	c	−11	−9.1	22.5	
	g	18.1	7.5	63.82	8.8
HgSeO$_3$	c		−68.0		
Hg$_2$SeO$_3$	c		−71.1		
HgTe	c	−8.1	−6.7	25.5	
	g			65.73	
Molybdenum					
Mo	c	0	0	6.85	5.75
	g	157.3	146.4	43.461	4.968

Table 9-1 (*Continued*)
ELEMENTS AND INORGANIC COMPOUNDS

Formula and Description	State	$\Delta Hf°$	$\Delta Gf°$	$S°$	$C_p°$
MoBr$_2$	c	−62.4	−53		18.3
MoBr$_4$	c	−76.8			
MoO$_2$Br$_2$	c	−150.4			
MoB	c	−21			
MoC	c	−2.4		8.2	8.99
Mo$_2$C	c	−10.9		15.7	14.10
Mo(CO)$_6$	c	−234.9	−209.8	77.9	57.90
	g	−218.0	−204.6	117	49
MoCl$_2$	c	−67.4	(−35)		
MoCl$_3$	c	−92.5		(32.6)	
MoCl$_4$	c	−114.8	−96	53.5	31.00
MoCl$_5$	c	−126.0	−101	57	37.20
MoCl$_6$	c	−125	−93	61	41.90
MoOCl$_2$	c	−126			
MoO$_2$Cl$_2$	c	−171.4			
	g	−151.6	−143	80.7	20.37
	aq	−190.4			
MoO$_2$Cl$_2$ · H$_2$O	c	−245.4			
MoOCl$_4$	c	−153.0			
MoF$_6$	liq	−378.95	−352.08	62.06	40.58
	g	−372.29	−351.88	83.75	28.82
MoOF$_4$	g	−300	−285	79.0	23.12
MoI$_2$	g	32			
Mo$_2$N	c	−16.6		21.0	15.30
MoO	g	92.5		56.9	7.31
MoO$_2$	c	−140.76	−127.40	11.06	13.38
MoO$_3$	c	−178.08	−159.66	18.58	17.92
	g	−86	−82	67	14.35
	aq	−172.5			
MoO$_4^{2-}$ std. state	aq	−238.5	−199.9	6.5	
MoS$_2$	c	−56.2	−54.0	14.96	15.19
MoS$_3$	c	−61.48		15.9	16.19
Mo$_2$S$_3$	c	−92.5		28.5	28.06
MoSi$_2$	c	−28			
Mo$_3$Si	c	−23	−23	25.4	22.23
Mo$_5$Si$_3$	c	−68			
Neodymium					
Nd	c	0	0	17.1	6.56
	g	78.3	69.9	45.24	5.28
Nd^{3+} std. state, $m = 1$	aq	−166.4	−160.5	−49.4	−5
Nd(acetate)$_3$					
undissoc; std. state, $m = 1$	aq	−510.52	−430.17	43.4	
Nd(BrO$_3$)$_3$ · 9H$_2$O	c	−856.7			
NdC$_2$	g	130.75	117.9	6.3	10.6
NdCl$_3$	c	−248.8			27
std. state, $m = 1$	aq	−286.3	−254.7	−9.0	−103
NdCl$_3$ · 6H$_2$O	c	−687.0	−588.1	99.7	86.25
Nd$_2$(CO$_3$)$_3$	c		−744.5		
Nd$_2$(C$_2$O$_4$)$_3$ · 10H$_2$O	c	−1621	−1412	−191	
NdF$_3$	c	−396			
NdH$_2$	c	−46			
NdI$_3$	c	−152.8			
	aq, inf	−211.3	−204.6		
Nd(NO$_3$)$_3$	c	−294.2			
Nd$_2$O$_3$ hexagonal	c	−432.1	−411.3	36.9	26.60
Nd$_2$S$_3$	c	−284	−280.2	44.28	29.28
Nd$_2$(SO$_4$)$_3$	c	−948.1			
	aq, inf	−993.1	−867.2		
Nd$_2$(SO$_4$)$_3$ · 8H$_2$O	c	−1513.1		160.9	144.9
Nd$_2$Se$_3$	c			53.6	31.1
Neon					
Ne	g	0	0	34.95	4.968

Table 9-1 (*Continued*)
ELEMENTS AND INORGANIC COMPOUNDS

Formula and Description	State	$\Delta Hf°$	$\Delta Gf°$	$S°$	$C_p°$
Neptunium					
Np	c	0	0		7.04
NpBr$_3$	c	−174			
NpBr$_4$	c	−183			
NpCl$_3$	c	−216			
NpCl$_4$	c	−237			
NpCl$_5$	c	−246			
NpF$_3$	c	−360			
NpF$_4$	c	−428			
NpI$_3$	c	−120			
NpO$_2$	c	−246	−234	19.2	15.8
Nickel					
Ni	c	0	0	7.14	6.23
	g	102.7	91.9	43.519	5.583
Ni^{2+} std. state, $m = 1$	aq	−12.9	−10.9	−30.8	
Ni$_3$(AsO$_4$)$_2$	c		−377.5		
Ni(BO$_2$)$_2$	c		−347.3		
NiBr$_2$	c	−50.7			
std. state, $m = 1$	aq	−71.0	−60.6	8.6	
Ni$_3$C	c	16.1			
Ni(CN)$_2$ pptd	c	30.5			
Ni(CN)$_4^{2-}$ std. state, $m = 1$	aq	87.9	112.8	52	
Ni(CNO)$_2$ cyanate	c	−54.4			
Ni(CNS)$_2$ thiocyanate	c	22.8			
Ni(CO)$_4$	liq	−151.3	−140.6	74.9	48.9
	g	−144.10	−140.36	98.1	34.70
Ni(acetate)$_2$ std. state, $m = 1$	aq	−245.2	−187.5	10.6	
NiCO$_3$	c		−146.4		
NiC$_2$O$_4$	c	−204.8			
std. state, $m = 1$	aq	−210.1	−172.0	−19.9	
NiCl$_2$	c	−72.976	−61.918	23.34	17.13
std. state, $m = 1$	aq	−92.8	−73.6	−3.6	
NiCl$_2 \cdot$ 2H$_2$O	c	−220.4	−181.7	42	
NiCl$_2 \cdot$ 4H$_2$O	c	−362.5	−295.2	58	
NiCl$_2 \cdot$ 6H$_2$O	c	−502.67	−409.54	82.3	
Ni(ClO$_4$)$_2$ std. state, $m = 1$	aq	−74.7	−15.0	56.2	
Ni(ClO$_4$)$_2 \cdot$ 6H$_2$O	c	−486.6			
NiF$_2$	c	−155.7	−144.4	17.59	18.00
NiI$_2$	c	−18.7			
std. state, $m = 1$	aq	−39.3	−35.6	22.4	
Ni(IO$_3$)$_2$	c	−116.9	−78.0	51	
Ni(NO$_3$)$_2$	c	−99.2			
std. state, $m = 1$	aq	−112.0	−64.2	39.2	
Ni(NO$_3$)$_2 \cdot$ 6H$_2$O	c	−528.6			111
NiO	c	−57.3	−50.6	9.08	10.59
Ni$_2$O$_3$	c	−117.0			
Ni(OH)$_2$	c	−126.6	−106.9	21	
Ni(OH)$_3$ pptd	c	−160			
Ni$_3$P	c	−50.2			
Ni$_5$P$_2$	c	−97.7			
Ni$_2$P$_2$O$_7$	c		−497.9		
Ni$_3$(PO$_4$)$_2$	c		−562.4		
NiS	c	−19.6	−19.0	12.66	13.06
pptd	c	−18.5			
Ni$_3$S$_2$	c	−48.5	−47.1	32.0	28.12
NiSO$_4$	c	−208.63	−181.6	23.2	33.05
std. state, $m = 1$	aq	−230.2	−188.9	−26.0	
NiSO$_4 \cdot$ 6H$_2$O tetrahedral green	c	−641.21	−531.78	79.391	78.36
NiSO$_4 \cdot$ 7H$_2$O	c	−711.36	−588.49	90.57	87.14
NiSe	c	−14.1			
NiSeO$_3 \cdot$ 2H$_2$O	c	−271.11			
NiSi	c	−20.6			10.9

Table 9-1 (*Continued*)
ELEMENTS AND INORGANIC COMPOUNDS

Formula and Description	State	$\Delta Hf°$	$\Delta Gf°$	$S°$	$C_p°$
Ni_2Si	c	−33.6			16.8
NiTe	c	−12.8			12.5
Ni_4W	c	−43			29.0
$NiWO_4$	c	−270.9		28.2	32.5
Nioblum					
Nb	c	0	0	8.70	5.88
	g	173.5	162.8	44.490	7.208
$NbBr_5$	c	−132.9		73.0	37.21
	g	−104.8			
NbC	c	−33.2	−32.7	8.46	8.81
Nb_2C	c	−45.4	−44.4	15.3	14.48
$NbCl_3$	g	−86		35.2	22.25
$NbCl_4$	c	−166.0		44.0	28.64
	g	−134			
$NbCl_5$	c	−190.6	−163.3	50.3	35.4
	g	−168.2	−154.4	95.71	28.88
$NbCo_2$	c	−13.7	−13.2	22	
$NbCo_3$	c	−14.1	−13.7	29	
$NbCr_2$	c	−5.0	−5.0	19.97	17.45
NbF_5	c	−433.5	−406.1	38.3	32.2
	g	−415.8	−401.1	76.9	23.2
$NbFe_2$	c	−11.1	−11.8	24	13.0
$NbGe_2$	c	−20.8			
NbI_5	c	−64.2		82	37.2
NbN	c	−56.2	−49.2	8.25	9.32
Nb_2N	c	−59.9		19.0	16.13
NbO	c	−97.0	−90.5	11.5	9.86
	g	51	44	57.09	7.36
NbO_2	c	−190.3	−177.0	13.03	13.74
	g	−51.3	−52.3	61.0	
NbO_3^- ionic strength = 1	aq		−222.8		
Nb_2O_5 high-temperature form	c	−454.0	−422.1	32.80	31.57
$Nb(OH)_5$ undissoc; ionic strength = 1	aq		−346.2		
$NbOCl_2$	c	−185.1		29.0	22.28
$NbOCl_3$	c	−210.2	−187	38	28.64
	g	−179.8	−171.6	85.6	22.0
Nitrogen					
N	g	112.979	108.886	36.613	4.968
N_2	g	0	0	45.77	6.961
N_3^- std. state, m = 1	aq	65.76	83.2	25.8	
HN_3	g	70.3	78.4	57.09	10.44
undissoc; std. state, m = 1	aq	62.16	76.9	34.9	
NCl_3	liq	55			
NCN radical	g	113.0	110.9	54.04	10.08
NCO radical	g	38.1	36.1	55.48	9.57
NF_3	g	−31.4	−21.5	62.30	12.76
N_2F_2 cis	g	16	26	62.1	11.94
trans	g	19.4	28.8	62.75	12.78
NH imidogen	g	90.16	88.73	43.29	6.97
NH_2 amidogen	g	45.50	47.76	46.51	8.02
N_2H_2 cis, diimide	g	51	58.1	52.2	8.74
trans	g	43.6	50.9	52.61	8.93
N_2H_4	liq	12.10	35.67	28.97	23.63
	g	22.80	38.07	56.97	11.85
undissoc; std. state, m = 1	aq	8.20	30.6	33	
$N_2H_5^+$ std. state, m = 1	aq	−1.8	19.7	36	16.8
N_2H_5Br	c	−37.2			
std. state, m = 1	aq	−30.8	−5.2	55.7	−17.1
$N_2H_5Br \cdot HBr$	c	−64.8			

Table 9-1 (*Continued*)
ELEMENTS AND INORGANIC COMPOUNDS

Formula and Description	State	$\Delta Hf°$	$\Delta Gf°$	$S°$	$C_p°$
N_2H_5Cl	c	−47.0			
std. state, $m = 1$	aq	−41.8	−11.7	49.5	−15.8
$N_2H_5Cl \cdot HCl$	c	−87.8			
$N_2H_5ClO_4$	c	−42.2			
std. state, $m = 1$	aq	−32.7	17.6	79.7	
N_2H_5OH	liq	−58.01			
	g	−49.0	−18.9	63	
undissoc; std. state, $m = 1$	aq	−60.11	−26.1	49.7	17.5
$N_2H_5NO_3$	c	−60.13			
std. state, $m = 1$	aq	−51.41	−6.91	71	
$(N_2H_5)_2SO_4$	c	−229.2			
std. state, $m = 1$	aq	−221.0	−138.6	77	−36
HNF_2	g			60.40	10.37
NO	g	21.57	20.69	50.347	7.133
NO_2	g	7.93	12.26	57.35	8.89
NO_2^- std. state, $m = 1$	aq	−25.0	−8.9	33.5	−23.3
NO_3	g	16.95	27.36	60.36	11.22
NO_3^- nitrate; std. state, $m = 1$	aq	−49.56	−26.61	35.0	−20.7
peroxynitrite	aq	−10.7			
N_2O	g	19.61	24.90	52.52	9.19
N_2O_2	g	40.72	48.49	68.72	15.18
$N_2O_2^{2-}$ hyponitrite	aq	−4.1	33.2	6.6	
$HN_2O_2^-$	aq	−9.4	18.2	34	
N_2O_3	g	20.01	33.32	74.61	15.68
N_2O_4	c	−8.37	23.79	35.92	29.18
	liq	−4.68	23.28	50.01	34.06
	g	2.19	23.38	72.70	18.47
N_2O_5	g	2.70	28.13	82.98	22.87
NOBr	g	19.64	19.70	65.38	10.87
NOCl	g	12.36	15.79	62.52	10.68
$NOClO_4$	c	−36.9			
NOF	g	−15.70	−12.02	59.27	9.88
NOF_3	g	−39	−23	66.54	16.22
NOSCN std. state, $m = 1$	aq	55.2	63.5	51.2	
NO_2Cl	g	3.0	13.0	65.02	12.71
NO_2ClO_4	c	8.7			
NO_2F	g	−19.00	−8.90	62.2	11.92
$N_2O_3(SO_3)_2$	c	−253			
N_4S_4	c	128.0			
NSe	c	42.3			
NSF	g			62.07	10.55
NSF_3	g			68.48	17.18
Osmium					
Os	c	0	0	7.8	5.9
	g	189	178	46.000	4.968
$OsCl_3$	c	−45.5	−29	31	
$OsCl_4$	c	−60.9	−38	37	
	g	−19			
OsF_6 cubic	c			58.8	
	g			85.56	28.88
OsO_2	c		−46		
OsO_3	g	−67.8			
OsO_4 yellow	c	−94.2	−72.9	34.4	
white	c	−92.2	−72.6	40.1	
	g	−80.6	−70.0	70.2	17.7
$Os(OH)_4$	amorp		−161.0		
OsS_2	c	−34.9	−32	13	
Oxygen					
O	g	59.553	55.389	38.467	5.237
O_2	g	0	0	49.003	7.016

Table 9-1 (*Continued*)
ELEMENTS AND INORGANIC COMPOUNDS

Formula and Description	State	$\Delta H f°$	$\Delta G f°$	$S°$	$C_p°$
O_3	g	34.1	39.0	57.08	9.37
OF	g	29.70	28.71	52.04	7.32
OF_2	g	5.86	9.98	59.11	10.35
O_2F_2	g	4.73	14.68	64.08	12.92
O_2F_3	g	3.8			
Palladium					
Pd	c	0	0	9.04	6.21
	g	90.4	81.2	39.90	4.968
Pd^{2+}	aq, ∞	40.5	42.2	−28	
$PdBr_2$	c	−24.9			
$PdBr_4^{2-}$	aq, ∞	−88.8	−76.0	70	
$Pd(CN)_2$	c	56.9			
$PdCl_2$	c	−41.0	−29.9	25	
$PdCl_4^{2-}$ 1M HCl	aq, ∞	−124.8	−99.6	62	
$PdCl_6^{2-}$ 1M HCl	aq, ∞	−143	−102.8	65	
$Pd(CNS)_2$	c		56.0		
Pd_2H	c	−4.7	−1.2	21.9	
PdI_2	c	−15.2	−15.0	36	
PdO	c	−20.4		13.4	7.53
$Pd(OH)_2$ pptd	c	−89.0	−72		
$Pd(OH)_4$ pptd	c	−156.0	−115	(35)	
PdS	c	−18	−16	11	
PdS_2	c	−19.4	−17.8	19	
PdTe	c			21.42	12.23
$PdTe_2$	c			30.25	18.31
Phosphorus					
P red, V	c	0	0	5.45	5.07
	liq	4.32	2.89	10.25	6.29
	g	79.80	69.80	38.98	4.97
α, white	c	4.17	2.87	9.82	5.70
black	c	−9.4			
P_2	g	34.94	24.8	52.11	7.66
P_4	g	30.8	17.3	30.8	16.05
PBr_3	liq	−44.1	−42.0	57.4	
	g	−33.3	−38.9	83.17	18.16
PBr_5	c	−64.5			
PCl_3	liq	−76.4	−65.1	51.9	
	g	−68.6	−64.0	74.49	17.17
PCl_5	c	−106.0			
	g	−81.9	−66.5	87.11	26.96
PF	g			53.74	7.56
PF_3	g	−219.6	−214.5	65.28	14.03
PF_5	g	−376.9	−360.6	71.9	20.28
PH	g	61	53	46.9	6.97
PH_3	g	5.5	6.1	50.24	8.07
std. state, $m = 1$	aq	−2.27	6.05	28.7	
PH_4^+ std. state, $m = 1$	aq		22.0		
P_2H_4	liq	−1.2	16.0	40	
	g	5.0			
PH_4Br	c	−30.5	−11.4	26.3	
PH_4Cl	c	−34.7			
PH_4I	c	−16.7	0.2	29.4	26.2
PH_4OH std. state, $m = 1$	aq	−70.48	−50.64	45.4	
PI_3	c	−10.9			
	g			89.45	18.73
PN	g	7.76	2.47	50.45	7.10
P_3N_5	c	−71.4			
$P_3N_3Cl_6$	c	−194.1			
	g	−175.9			
$P_4N_4Cl_8$	c	−259.2			
	g	−236.1			
PO	g	−2.90	−9.84	53.221	7.59

Table 9-1 (*Continued*)
ELEMENTS AND INORGANIC COMPOUNDS

Formula and Description	State	$\Delta Hf°$	$\Delta Gf°$	$S°$	$C_p°$
PO_3^-	aq	−233.5			
PO_4^{3-} std. state, $m=1$	aq	−305.3	−243.5	−53	
$P_2O_7^{4-}$ std. state, $m=1$	aq	−542.8	−458.7	−28	
P_4O_6	c	−392.0			
P_4O_{10} hexagonal	c	−702.7	−639.4	54.70	50.60
orthorhombic	c	−713.2		54.70	48.94
$POBr_3$	c	−109.6			
	g	−93.00	−93.43	85.97	21.48
$POCl_3$	liq	−142.7	−124.5	53.17	33.17
	g	−129.6	−120.1	77.76	20.30
$POClF_2$	g	−228	−218	72.08	16.45
$POCl_2F$	g	−179	−170	76.55	18.96
POF_3	g	−295.6	−285.3	68.19	16.45
P_2S_3	c	−19.2			
P_4S_3	c	−37	−38	48	35.00
	liq	−36.1	−37.5	49.5	44.00
	g	−19.4	−28.8	76.28	37
$PSBr_3$	g	−63	−69	89.08	22.69
$PSCl_3$	g	−86.8	−83.1	80.60	21.47
PSF_3	g	−237	−232.7	71.23	17.82
Platinum					
Pt	c	0	0	9.95	6.20
	g	135.1	124.4	45.960	6.102
PtBr	c	−11	(−7)	(28)	
$PtBr_2$	c	−23.1		(44)	
$PtBr_3$	c	−30.9		(56)	
$PtBr_4$	c	−38.0		(68)	
$PtBr_4^{2-}$	aq	−89	(−69)	(47)	
$PtBr_6^{2-}$	aq	−114	(−89)	(67)	
PtCl	c	−9		(27)	
$PtCl_2$	c	−26.5		(28)	
std. state, $m=1$	aq		−18.3		
$PtCl_3$	c	−41.6	(−32)	(36)	
$PtCl_4$	c	−56.5	(−41)	(42)	
	aq	−76.2			
$PtCl_4^{2-}$ std. state, $m=1$	aq	−120.3	−88.1	40	
$PtCl_6^{2-}$ std. state, $m=1$	aq	−161	−117	52.6	
PtF_6 cubic	c			56.3	
	g			83.23	29.35
PtI_4	c	−17.4			
PtO_2	g	41.0	40.1	62	
Pt_3O_4	c	−39			
$Pt(OH)_2$	c	−84.1	(−66)	(30)	
PtS	c	−19.5	−18.2	13.16	10.37
PtS_2	c	−26.0	−23.8	17.85	15.75
PtTe	c			19.41	11.93
$PtTe_2$	c			30.25	18.31
Plutonium					
Pu	c	0	0	12.3	8.48
	g		74.9	42.90	
Pu^{3+}	aq	−138.6	−140.5	−39	
Pu^{4+}	aq	−138.6	−356		
$PuBr_3$	c	−198.8	−192.3	46.10	25.78
Pu_2C_3	c	−1.70	−4.70	40.50	22.93
$PuCl_3$	c	−229.8	−213.37	38.00	24.58
$PuCl_4$	c	−330			
PuF_3	c	−371	−353.45	27.00	23.14
PuF_4	c	−414	−393.1	38.7	28.88
PuF_6	c	6.09	6.5	53.20	40.00
PuH_2	c	−33.3	−24.3	14.3	9.33
PuH_3	c	−33	−19.71	15.5	10.33
PuI_3	c	−155.0	−153.92	51.20	26.73

Table 9-1 (*Continued*)
ELEMENTS AND INORGANIC COMPOUNDS

Formula and Description	State	$\Delta Hf°$	$\Delta Gf°$	$S°$	$C_p°$
PuN	c	−75.70	−69.18	14.20	12.75
PuO	c	−135	−128.8	16.9	12.25
PuO$_2$	c	−252.9	−240.4	19.7	16.4
Pu$_2$O$_3$ α	c	−430	−411.2	33.2	31.8
β	c	−410.00	−390.12	36.4	31.3
PuOBr	c	−212.40	−204.24	28.5	20.99
PuOCl	c	−222.7	−211.27	26.00	19.99
PuOF	c	−269.80	−257.87	21.90	18.99
PuOI	c	−197.80	−191.43	30.20	21.99
Pu(SO$_4$)$_2$	c	−526.00	−470.71	39.00	43.49
PuS	c	−105.00	−104.37	18.70	12.90
Pu$_2$S$_3$	c	−236.50	−235.53	46.00	30.99
Polonium					
Po	c	0	0	15.0	6.3
	g	34.8	25.8	45.13	
Po^{2+} std. state, $m = 1$	aq		17		
Po^{4+} std. state, $m = 1$	aq		70		
PoCl$_6^{2-}$ std. state, $m = 1$	aq		−138		
Po(OH)$_4$	c		−130		
Po(OH)$_2^{4+}$ std. state, $m = 1$	aq		−113		
PoO$_2$	c	−60	47	17	14.7
PoO$_3$	c		−33		
PoS	c		52		
Potassium					
K	c	0	0	15.46	7.05
	liq	0.546	0.063	17.08	7.82
	g	21.3	14.5	21.52	4.97
KAlCl$_4$	c	−286	−262	47	37.4
K$_3$AlCl$_6$	c	−500.0	−463.3	90.0	59.49
K$_3$AlF$_6$	c	−795		68	52.84
K$_2$O · Al$_2$O$_3$ · 4SiO$_2$ leucite	c	−1406.4			
K$_2$O · Al$_2$O$_3$ · 6SiO$_2$ microcline	c	−1816			
(adularia)	c	−1842			
KAl(SO$_4$)$_2$	c	−589.24	−534.29	48.9	46.11
	aq	−616.9			
KAl(SO$_4$)$_2$ · H$_2$O	c	−667.5			
KAl(SO$_4$)$_2$ · 2H$_2$O	c	−742			
KAl(SO$_4$)$_2$ · 3H$_2$O	c	−814			
KAl(SO$_4$)$_2$ · 12H$_2$O	c	−1447.74	−1227.8	164.3	155.80
K$_3$AsO$_3$	aq	−323.0			
K$_3$AsO$_4$	aq	−390.3	−355.7		
KH$_2$AsO$_4$	c	−271.5	−237.0	37.08	
	aq	−276.2			
KBF$_4$	c	−451.0	−426.6	32.0	27.36
KBH$_4$	c	−54.2	−38.2	25.48	23.08
KBO$_2$	c	−237.8	−223.9	19.12	16.02
K$_2$B$_4$O$_7$	c	−796.9	−749.7	49.80	40.75
KBr	c	−93.73	−90.63	23.05	12.50
	aq, 10	−89.77			
	aq, 50	−89.07			
	aq, 500	−88.87	−89		
	aq, 1000	−88.88			
	aq, ∞	−88.94	−92.04	43.8	
KBrO	aq	−82.0			
KBrO$_3$	c	−79.4	−58.2	35.65	
	aq, ∞	−69.6	−56.6	63.4	
K$_2$CO$_3$	c	−274.9	−254.4	37.17	27.35
	aq, 50	−281.02			
	aq, 1000	−281.56	−264		
K$_2$CO$_3$ · ½H$_2$O	c	−210.43			
K$_2$CO$_3$ · 1½H$_2$O	c	−283.40			

Table 9-1 (*Continued*)
ELEMENTS AND INORGANIC COMPOUNDS

Formula and Description	State	$\Delta Hf°$	$\Delta Gf°$	$S°$	$C_p°$
KHCO$_3$	c	−229.3			
	aq, 1500	−224.5	−207.7		
K(HCO$_2$) formate	c	−158.0			
	aq	−157.7			
KC$_2$H$_3$O$_2$	c	−173.2			
	aq, 10	−175.6			
	aq, ∞	−176.88	−156.7		
K$_2$C$_2$O$_4$	c	−320.8			
	aq, 100	−316.88			
	aq, 5000	−316.90	−393.1		
	aq, ∞	−317.1			
K$_2$C$_2$O$_4$ · H$_2$O	c	−392.17			
KCN	c	−27.12	−24.39	30.54	15.87
	aq	−24.1	−28		
KCNO cyanate	c	−98.5			
	aq	−93.5	−90.85		
KCNS thiocyanate	c	−48.62			
	aq, 2	−45.46			
	aq, 8	−44.32			
	aq, 50	−43.11			
	aq, ∞	−42.8	−44		
KCl	g	−51.6	−56.2	57.24	
	c	−104.18	−97.59	19.76	12.26
	aq, 50	−100.11			
	aq, ∞	−100.06	−98.82	37.7	
KClO	aq	−85.4			
KClO$_3$	c	−93.50	−69.29	34.17	
	aq, 100	−84.09			
	aq, ∞	−83.54	−68.09	63.5	
KClO$_4$	c	−102.8	−71.8	36.1	26.87
	aq, 500	−91.58			
	aq, ∞	−91.45	−70.04	68.0	
K$_2$CrO$_4$	c	−330.49			
	aq, 17.2	−328.2			
	aq, 54	−327.0			
	aq, ∞	−326.0	−306		
K$_2$Cr$_2$O$_7$	c	−485.90			
	aq, 135	−470.41			
	aq, 800	−469.50	−441		
	aq, 2000	−468.7			
KF	c	−135.9	−128.8	15.91	11.71
	aq, 4	−137.35			
	aq, 50	−138.51			
	aq, ∞	−138.70	−133		
KF · 2H$_2$O	c	−277.0	−242.7	36	
KF · 4H$_2$O	c	−418.0			
KH	c	−13.82	−8.14	12	9.06
KHF$_2$	c	−222.6	−206.3	24.92	18.39
	aq, 25	−213.73			
	aq, 200	−213.4			
KF · 2HF	c	−296.7			
KF · 3HF	c	−373.0			
KI	c	−78.31	−77.03	25.43	12.61
	aq, 6	−74.94			
	aq, ∞	−73.41	−79.82	50.6	
KIO$_3$	c	−121.5	−101.7	36.20	
	aq, 200	−115.16			
	aq, ∞	−115.0	−99.9	52.2	
KIO$_4$	c	−112.8			
KMnO$_4$	c	−194.4	−170.6	41.04	
	aq, 140	−184.44			
	aq, 400	−183.9	−168.0		

Table 9-1 (*Continued*)
ELEMENTS AND INORGANIC COMPOUNDS

Formula and Description	State	$\Delta Hf°$	$\Delta Gf°$	$S°$	$C_p°$
K_2MoO_4	aq, 880	−374.1			
KNH_2	c	−28.3			
KNO_2	c	−88.5			
	aq, 400	−85.2	−76		
KNO_3	c	−117.76	−93.96	31.77	23.01
	aq, 400	−109.48	−93.68		
	aq, ∞	−109.41	−93.88	69.5	
K_2O	c	−86.8	−77.0	22.5	20.00
K_2O_2	c	−118.5	−102.7	27.0	23.94
KO_2	c	−68.0	−57.5	29.3	18.53
KOH	c	−101.78	−90.6	18.85	15.51
	aq, 3	−111.77			
	aq, 5	−113.31			
	aq, 25	−114.70			
	aq, 500	−114.86	−105		
	aq, ∞	−115.00	−105.06	22.0	
$KOH \cdot \frac{3}{4}H_2O$	c	−161.7			
$KOH \cdot H_2O$	c	−179.6			
$KOH \cdot 2H_2O$	c	−251.2			
K_2OsCl_6	c	−280			
K_3PO_3	aq	−397.5			
K_3PO_4	aq	−478.7	−443.3		
KH_2PO_4	c	−374.9	−326		
	aq, 35	−370.70			
	aq, 755	−370.4			
$KReO_4$	c	−264.02			
K_3RhCl_6	c	−343			
K_2S	c	−100			
	aq, 7	−107.04			
	aq, 10	−108.95			
	aq, 20	−110.17			
	aq, 25	−110.27			
	aq, 50	−110.17			
	aq, 200	−109.94	−111.4		
$K_2S \cdot 2H_2O$	c	−243.0			
$K_2S \cdot 5H_2O$	c	−456.7			
KHS	c	−63.2			
	aq, 10	−64.35			
	aq, 200	−64.1			
$KHS \cdot \frac{1}{4}H_2O$	c	−80.1			
K_2S_4	c	−113.0			
	aq	−114.6			
$K_2S_4 \cdot \frac{1}{2}H_2O$	c	−150.5			
$K_2S_4 \cdot 2H_2O$	c	−258.7			
K_2SO_3	c	−266.9			
	aq	−269.1	−251.3		
$KHSO_3$	aq, 385	−209.7			
K_2SO_4	c	−342.66	−314.62	42.0	31.36
	aq, 1000	−336.75	−310		
	aq, 5000	−336.81			
$KHSO_4$	c	−276.8			
	aq, 20	−273.38			
	aq, 100	−273.44			
	aq, 800	−274.3			
$K_2S_2O_5$	c	−362.6			
	aq	−351.9			
$K_2S_2O_5 \cdot 1\frac{1}{2}H_2O$	c	−397.0			
$K_2S_2O_6$	c	−413.6			
	aq, 400	−401.0			
$K_2S_2O_8$	c	−458.3			
	aq	−445.3			
$K_2S_4O_6$	c	−422			
	aq	−410			

Table 9-1 (*Continued*)
ELEMENTS AND INORGANIC COMPOUNDS

Formula and Description	State	$\Delta Hf°$	$\Delta Gf°$	$S°$	$C_p°$
$K_2S_5O_6 \cdot 1\frac{1}{2}H_2O$	c	−515.4			
K_2Se	c	−79.3			
	aq	−88.5	−99		
$K_2Se \cdot 9H_2O$	c	−722.0			
$K_2Se \cdot 14H_2O$	c	−1066			
$K_2Se \cdot 19H_2O$	c	−1417			
KHSe	c	−35.9			
	aq	−35.4			
K_2SeO_4	aq, 440	−265.3	−240		
$KHSeO_4$	aq, 220	−202.8			
K_2TeO_3	aq	−261.6			
K_2TeO_4	aq	−290.3			
KVO_3	aq	−284.0			
KVO_5	aq	−254.9			
K_2SiF_6	c	−671			
K_2SiO_3	c	−370.0	−347.9	34.93	28.30
Praseodymium					
Pr	c	0	0	17.5	6.50
	g	85.0	76.7	45.34	5.11
Pr^{3+} std. state, $m = 1$	aq	−168.4	−162.3	−50	−7
$Pr(acetate)_3$ std. state, $m = 1$	aq	−513.27	−431.54	39.4	
$Pr(BrO_3)_3 \cdot 9H_2O$	c	−858.5			
PrC_2	g	131.3	118.7	62.6	10.6
$PrCl_3$	c	−252.6			24
std. state, $m = 1$	aq	−288.3	−256.4	−10	−105
$Pr_2(CO_3)_3$	c	−768			
$Pr_2(C_2O_4)_3 \cdot 10H_2O$	c	−1415			
$PrF_3 \cdot \frac{1}{2}H_2O$	c	−439.4			
PrH_2	c	−47.4	−36.9	13.6	9.8
PrI_3	c	−156.4			
	aq, inf	−212.8	−206.1		
$Pr(IO_3)_3$	c	−333.8			
$Pr(NO_3)_3$	c	−293.8			
$Pr(NO_3)_3 \cdot 6H_2O$	c	−731.05			
Pr_2O_3 hexagonal	c	−432.5		28.06	
cubic	c	−432.5			
$Pr(OH)_3$	c		−307.1		
$PrSO_4^+$ std. state, $m = 1$	aq	−382.3	−345.1	−17	
$Pr(SO_4)_2^-$ std. state, $m = 1$	aq	−598.4	−525.6	0	
Promethium-147					
Pm	c	0	0		
	g			44.69	5.80
$PmCl_3$	c	−251.9			
	aq, inf	−290.5			
Radium					
Ra	c	0	0	17	
	g	38	31	42.15	4.97
Ra^{2+} std. state, $m = 1$	aq	−126.1	−134.2	13	
$RaBr_2$	c	−195			
$RaCl_2$	c			32	
std. state, $m = 1$	aq	−206.0	−196.9	40	
$RaCl_2 \cdot 2H_2O$	c	−350	−311.4	51	
$Ra(IO_3)_2$	c	−245.4	−207.6	65	
$Ra(NO_3)_2$	c	−237	−190.3	53	
std. state, $m = 1$	aq	−225.2	−187.4	83	
RaO	c	−125			
$RaSO_4$	c	−351.6	−326.4	33	
std. state, $m = 1$	aq	−343.4	−312.2	18	
Radon					
Rn	g	0	0	42.09	4.968
Rhenium					
Re	c	0	0	8.81	6.16
	g	184.0	173.2	45.131	4.968

Table 9-1 (*Continued*)
ELEMENTS AND INORGANIC COMPOUNDS

Formula and Description	State	$\Delta Hf°$	$\Delta Gf°$	$S°$	$C_p°$
Re⁻ std. state, $m = 1$	aq	11	2.4	55	
ReAs₂	c	1.3			
ReBr₃	c	−40			
Re₃Br₉	g	−69			
ReCl₃	c	−63	−45	29.6	22.08
ReCl₅	c	−89			
Re₃Cl₉	g	−137			
ReCl₆²⁻ std. state, $m = 1$	aq	−182	−141	60	
H₂ReCl₄	c	−152			
ReF₆	g	−273			
ReO₂	c	−101	−88	41	
ReO₂ · 2H₂O pptd	c	−236			
ReO₃	c	−144.6	−127	61.5	
Re₂O₇	c	−296.4	−254.8	49.5	39.7
	g	−263	−237.6	108	
HReO₄	c	−182.2	−156.9	37.8	
	g	−159			
std. state, $m = 1$	aq	−188.2	−166.0	48.1	−3.2
ReS₂	c	−43			
Re₂S₇	c	−107			
Re₃Si	c	0			
Rhodium					
Rh	c	0	0	7.56	5.95
	g	133.1	122.1	44.383	5.022
RhCl₂	g	30.3			
RhCl₃	c	−71.5	(−54)	(21)	
	g	16			
Rh₂O₃	c	−68.3	−65	26.5	24.85
Rh(OH)₃	c		(−112)		
Rubidium					
Rb	g	20.51	13.35	40.63	4.97
	c	0	0	16.6	7.36
RbAl(SO₄)₂	c	−569			
RbAl(SO₄)₂ · H₂O	c	−647			
RbAl(SO₄)₂ · 2H₂O	c	−722			
RbAl(SO₄)₂ · 3H₂O	c	−797			
RbAl(SO₄)₂ · 12H₂O	c	−1448.0			
RbBr	c	−93.03	−90.38	25.88	12.36
	aq, ∞	−87.8	−92.02	49.0	
RbBrO₃	c	−81.32			
Rb₂CO₃	c	−269.6		37.9	
	aq, 200	−279.4	−263.8		
	aq, 2000	−278.0			
Rb₂CO₃ · H₂O	c	−344.2			
Rb₂CO₃ · 1½H₂O	c	−381.5			
Rb₂CO₃ · 3½H₂O	c	−521.7			
RbHCO₃	c	−228.5			
	aq	−224.1	−209.1		
RbCN	aq	−25.9			
RbSCN thiocyanate	c	−54			
	aq	−41.7			
RbCl	c	−103.98	−98.48	22.6	12.24
	aq, ∞	−98.9	−98.80	42.9	
RbClO₃	c	−93.8	−69.8	36.3	
	aq, ∞	−82.4	−68.07	68.7	
RbClO₄	c	−103.87	−73.19	38.4	
	aq, ∞	−90.3	−70.02	73.2	
RbF	c	−131.28		18.0	12.07
	aq, ∞	−137.6	−133.53	27.4	
RbF · ⅓H₂O	c	−156.12			
RbF · 1½H₂O	c	−240.33			
RbHF₂	c	−217.3			
	aq	−212.5			

Table 9-1 (*Continued*)
ELEMENTS AND INORGANIC COMPOUNDS

Formula and Description	State	$\Delta Hf°$	$\Delta Gf°$	$S°$	$C_p°$
RbI	c	−78.5	−77.8	28.21	12.38
	aq, ∞	−72.3	−79.80	55.8	
Rb$_2$IrCl$_6$	c	−290.9			
RbNO$_3$	c	−117.04			
	aq, ∞	−108.5	−93.86	64.7	
RbNH$_2$	c	−25.7			
Rb$_2$O	c	−78.9			
Rb$_2$O$_2$	c	−101.7			
RbOH	c	−98.9			
	aq, ∞	−113.9	−105.05	27.2	
RbOH · H$_2$O	c	−177.8			
RbOH · 2H$_2$O	c	−250.8			
RbReO$_4$	c	−256.9			
	aq	−249.2			
Rb$_2$S	c	−83.2			
	aq, 500	−107.8			
RbHS	c	−62.4			
	aq	−63.1			
RbHSO$_4$	c	−273.7			
	aq, 300	−270.4			
Rb$_2$SO$_4$	c	−340.50			
	aq, ∞	−334.7	−312.24	33.8	
RbHSe	c	−35.5			
	aq	−34.3			
Rb$_2$SiF$_6$	c	−678.4			
Ruthenium					
Ru	c	0	0	6.82	5.75
	g	153.6	142.4	44.550	5.144
RuBr$_3$	c	−33			
RuCl$_3$ black	c	−49			
	g	−0.3			
RuCl$_4$	g	−12.4			
RuCl$_5$(OH)$^{2-}$	aq, ∞		−168.7		
RuF$_5$	c	−213.4			
	g	−189			
RuI$_3$	c	−15.7			
RuO$_2$	c	−72.9			
hydrated	amorp		−51.3		
RuO$_3$	g	−18.7			
RuO$_4$	c	−57.2	−36.4	35.0	
	liq	−54.6	−36.4	43.8	
	g	−44.0	−33.4	69.3	18.14
RuO$_4^-$	aq, ∞		−58.7		
RuO$_4^{2-}$	aq, ∞		−72.6		
RuS$_2$	c	−47	−44		
Samarium					
Sm	c	0	0	16.63	7.06
	g	49.4	41.3	43.72	7.26
Sm^{2+} std. state, $m = 1$	aq		−118.9		
Sm^{3+} std. state, $m = 1$	aq	−165.3	−159.3	−50.6	−5
Sm(acetate)$_3$ std. state, $m = 1$	aq	−510.01	−429.49	42.0	
Sm(BrO$_3$)$_3$ · 9H$_2$O	c	−856.1			
SmC$_2$	c	−17	−18.1	23	
SmCl$_2$	c	−194.9			
SmCl$_3$	c	−245.2			
std. state, $m = 1$	aq	−285.2	−253.4	−10.2	−103
SmCl$_3$ · 6H$_2$O	c	−686.0	−587.1	99	86.3
Sm$_2$(CO$_3$)$_3$	c		−741.4		
SmF$_3$ · ½H$_2$O	c	−436.2			
SmI$_3$	c	−148.2			
Sm(IO$_3$)$_3$	c	−330			
Sm(NO$_3$)$_3$	c	−289.7			

Table 9-1 (*Continued*)
ELEMENTS AND INORGANIC COMPOUNDS

Formula and Description	State	$\Delta Hf°$	$\Delta Gf°$	$S°$	$C_p°$
Sm_2O_3 monoclinic	c	−435.7	−414.6	36.1	27.37
$Sm_2(SO_4)_3$	c	−931.9			
$Sm_2(SO_4)_3 \cdot 8H_2O$	c	−1513.1	−1323.8	160.7	145.0
Scandium					
Sc	c	0	0	8.28	6.10
	g	90.3	80.32	41.75	5.28
Sc_2	g	154.9	141.6	61	8.7
Sc^{3+} std. state, $m = 1$	aq	−146.8	−140.2	−61	
$ScBr_2$	g			77.6	13.0
$ScBr_3$	c	−177.6			
$ScCl$	g	26.9	20.6	56.00	8.40
$ScCl_2$	g			72.5	12.6
$ScCl_3$	c	−221.1		29.0	22.38
	aq, 5500	−268.8			
$ScCl_3 \cdot 6H_2O$	c	−671.6			
$Sc(CNS)^{2+}$ std. state, $m = 1$	aq		−119.6		
ScF	g	−33.2	−39.3	53.11	7.74
ScF_2	g	−153.5	−156.6	67.0	11.5
ScF_3	c	−389.4	−371.8	22	
	g	−298	−295	71.8	16.2
ScI_2	g			81.6	13.3
ScO	g	−13.68	−19.90	53.65	7.38
Sc_2O	g	−6.9			11.2
Sc_2O_3	c	−456.22	−434.85	18.4	24.85
$Sc(OH)^{2+}$ std. state, $m = 1$	aq	−205.9	−191.5	−32	
$Sc(OH)_3$	c	−325.9	−294.8	24	
$Sc(OH)_2Cl$	c	−303	−276.3	26	
ScS	g	41.8	29.7	56.3	8.0
$Sc(SO_4)^+$ std. state, $m = 1$	aq		−321.7		
$Sc(SO_4)_2^-$ std. state, $m = 1$	aq		−501.5		
$Sc_2(SeO_3)_3 \cdot 10H_2O$	c	−1326.5			
Selenium					
Se hexagonal, black	c	0	0	10.144	6.062
monoclinic, red	c	1.6			
	g	54.27	44.71	42.21	4.978
glassy	amorp	1.2			
Se_2	g	34.9	23.0	60.2	8.46
Se^{2-} std. state, $m = 1$	aq		30.9		
Se_6	g	39.2			
$SeBr_2$	g	−5			
Se_2Br_2	g	7			
$SeCl_2$	g	−7.6			
$SeCl_4$	c	−43.8			
Se_2Cl_2	liq	−19.7			
	g	4			
SeF_6	g	−267	−243	74.99	26.4
HSe^- std. state, $m = 1$	aq	3.8	10.5	19	
H_2Se	g	7.1	3.8	52.32	8.30
std. state, $m = 1$	aq	4.6	5.3	39.1	
SeO	g	12.75	6.41	55.9	7.47
SeO_2	c	−53.86			
SeO_3	c	−39.9			
SeO_3^{2-} std. state, $m = 1$	aq	−121.7	−88.4	3	
SeO_4^{2-} std. state, $m = 1$	aq	−143.2	−105.5	12.9	
Se_2O_5	c	−97.6			
$SeOCl_2$	g	−6			
$HSeO_3^-$ std. state, $m = 1$	aq	−122.98	−98.36	32.3	
$HSeO_4^-$ std. state, $m = 1$	aq	−139.0	−108.1	35.7	
H_2SeO_3	c	−125.35			
undissoc; std. state, $m = 1$	aq	−121.29	−101.87	49.7	
H_2SeO_4	c	−126.7			
	aq, ∞	−140.3			

Table 9-1 (*Continued*)
ELEMENTS AND INORGANIC COMPOUNDS

Formula and Description	State	$\Delta Hf°$	$\Delta Gf°$	$S°$	$C_p°$
Silicon					
Si	c	0	0	4.50	4.78
	amorp	1.0			
	g	108.9	98.3	40.12	5.318
Si_2	g	142	128	54.92	8.22
SiBr	g	50			9.23
$SiBr_4$	liq	−109.3	−106.1	66.4	
	g	−99.3	−103.2	90.29	23.21
SiC β, cubic	c	−17.5	−16.9	3.97	6.42
α, hexagonal	c	−17.1	−16.5	3.94	6.38
	g	147	132	56.55	10.90
SiCl	g	45.39		56.82	8.81
$SiCl_2$	g	−39.59	−42.35	67.0	12.16
$SiCl_4$	liq	−164.2	148.16	57.3	34.73
	g	−158.4	−148.8	79.02	21.57
SiH_3Cl	g	−48	−43	59.88	12.20
SiH_2Cl_2	g			68.26	14.45
$SiHCl_3$	liq	−128.9	−115.34	54.4	
SiF	g	1.7	−5.8	53.94	7.80
SiF_2	g	−140.5	−143.0	61.38	10.49
SiF_4	g	−385.98	−375.88	67.49	17.60
SiF_6^{2-} std. state, m = 1	aq	−571.0	−525.7	29.2	
SiH_3F	g	−105	−100	56.95	11.33
$SiHF_3$	g			64.96	14.47
SiH	g	83.3		47.42	6.98
SiH_4	g	7.3	13.6	48.88	10.24
Si_2H_6	g	19.2	30.4	65.14	19.31
Si_3H_8	liq	22.1			
	g	28.9			
SiI_4	c	−45.3			
SiN	g	116.28	109.01	51.78	7.21
Si_3N_4	c	−177.7	−139	27	28.85
SiO_2 quartz	c	−217.72	−204.75	10.00	10.62
cristobalite	c	−217.37	−204.56	10.20	10.56
tridymite	c	−217.27	−204.42	10.4	10.66
	amorp	−215.94	−203.33	11.2	10.6
$SiOF_2$	g	−231	−227	64.81	12.83
SiS	g	26.88	14.56	53.43	7.71
SiS_2	c	−51.0	−50.8	19.2	18.52
SiSe	g	23.78			8.04
$SiSe_2$	c	−7			
SiTe	g	30.99			8.31
Silver					
Ag	c	0	0	10.17	6.059
	g	68.01	58.72	41.321	4.968
Ag_2	g	97.99	85.75	61.43	8.84
Ag^+ std. state, m = 1	aq	25.234	18.433	17.37	5.2
Ag^{2+} in 4M $HClO_4$, std. state	aq	64.2	64.3	−21	
Ag_3AsO_4	c		−129.7		
AgAt	c	−10.8		31.8	13.3
AgBr	c	−23.99	−23.16	25.6	12.52
std. state, m = 1	aq	−3.82	−6.42	37.1	−28.7
$AgBrO_3$	c	−6.5	13.0	36.5	
AgCl	c	−30.370	−26.244	23.0	12.14
	g			58.75	8.57
std. state, m = 1	aq	−14.718	−12.939	30.9	−27.4
$AgClO_2$	c	2.10	18.1	32.16	20.87
std. state, m = 1	aq	9.3	22.5	41.6	
$AgClO_3$	c	−6.1			
std. state, m = 1	aq	1.5	17.6	56.2	

Table 9-1 (*Continued*)
ELEMENTS AND INORGANIC COMPOUNDS

Formula and Description	State	$\Delta Hf°$	$\Delta Gf°$	$S°$	$C_p°$
AgClO$_4$	c	−7.44			
std. state, $m = 1$	aq	−5.68	16.37	60.9	
Ag$_2$CrO$_4$	c	−174.89	−153.40	52.0	34.00
AgCN	c	34.9	37.5	25.62	15.95
Ag(CN)$_2^-$ std. state, $m = 1$	aq	64.6	73.0	46	
AgCN$_2$ cyanamide	c	56.2			
AgONC fulminate	c	43			
AgOCN cyanate	c	−22.8	−13.9	29	
AgSCN thiocyanate	c	21.0	24.23	31.3	15
Ag$_2$CO$_3$	c	−120.9	−104.4	40.0	26.83
Ag$_2$C$_2$O$_4$	c	−160.9	−139.6	50	
Ag acetate	c	−95.3	−73.56	35.8	
AgF	c	−48.9		20.0	12.41
std. state, $m = 1$	aq	−54.27	−48.21	14.1	−20.3
AgF · 2H$_2$O	c	−191.4	−160.4	41.8	31
AgF$_2$	c	−87.3			
AgI	c	−14.78	−15.82	27.6	13.58
std. state, $m = 1$	aq	12.04	6.10	44.0	−28.8
AgIO$_3$	c	−40.9	−22.4	35.7	24.60
std. state, $m = 1$	aq	−27.7	−12.2	45.7	
Ag$_2$MoO$_4$	c	−200.9	−178.8	51	
AgN$_3$	c	73.8	89.9	24.9	
Ag$_3$N	c	47.6			
Ag(NH$_3$)$_2^+$ std. state, $m = 1$	aq	−26.60	−4.12	58.6	
AgNO$_2$	c	−10.77	4.56	30.64	19.17
AgNO$_3$	c	−29.73	−8.00	33.68	22.24
Ag$_2$O	c	−7.42	−2.68	29.0	15.74
AgO	c	−2.73	3.40	13.810	10.76
Ag$_2$O$_2$	c	−5.8	6.6	28	21
Ag$_2$O$_3$	c	8.1	29.0	24	
AgOH std. state, $m = 1$	aq	−29.736	−19.161	14.80	−30.3
AgP$_2$	c	−11.0			
AgP$_3$	c	−16.6			
Ag$_3$PO$_4$	c		−210		
Ag$_4$P$_2$O$_7$	c	−453			
AgReO$_4$	c	−176	−151.9	36.6	
Ag$_2$S α, orthorhombic	c	−7.79	−9.72	34.42	18.29
β	c	−7.03	−9.43	36.0	
Ag$_2$SO$_3$	c	−117.3	−98.3	37.8	
std. state, $m = 1$	aq	−101.4	−79.4	27.8	
Ag$_2$SO$_4$	c	−171.10	−147.82	47.9	31.40
std. state, $m = 1$	aq	−166.85	−141.10	39.6	−60
Ag$_2$Se	c	−9	−10.6	36.02	19.54
Ag$_2$SeO$_3$	c	−87.3	−72.7	55.0	
Ag$_2$SeO$_4$	c	−100.5	−79.9	59.4	
Ag$_2$Te	c	−8.9	10.3	37.0	20.9
Ag$_2$WO$_4$	c	−221.2		49.0	35.47
Sodium					
Na	c	0	0	12.30	6.73
	liq	0.575	0.119	13.83	7.82
	g	25.75	18.48	36.71	4.97
NaAlCl$_4$	c	−273.0	−249.0	45.0	37.04
Na$_3$AlCl$_6$	c	−473	−437	83.0	58.35
Na$_3$AlF$_5$ cryolite	c	−791	−751	57	51.56
	liq	−774.0	−738.1	68.5	51.56
NaAlO$_2$	c	−270.8	−255.6	16.83	17.61
Na$_3$AsO$_4$	c	−365			
NaBH$_4$	c	−45.85	−30.38	24.23	20.67
NaBO$_2$	c	−233.2	−219.7	17.57	15.76
Na$_2$B$_4$O$_7$	c	−783	−737	45.3	44.64
Na$_2$B$_4$O$_7$ · 10H$_2$O	c	−1497.2			
Na$_3$BiO$_4$	c	−288			

Table 9-1 (*Continued*)
ELEMENTS AND INORGANIC COMPOUNDS

Formula and Description	State	$\Delta Hf°$	$\Delta Gf°$	$S°$	$C_p°$
NaBr	c	−86.38	−83.48	20.75	12.29
	aq, inf	−86.18	87.16	33.7	
NaBrO$_3$	c	−73.46			31.2
	aq, 400	−68.89	−57.59		
NaCN	c	−21.46		28.32	16.40
NaOCN cyanate	c	−95.6			
NaCNS thiocyanate	c	−41.73			
	aq, ∞	−40.1			
Na$_2$CO$_3$	c	−270.3	−250.4	32.5	26.53
	aq, 400	−275.9	−251.4		
Na$_2$CO$_3 \cdot$ H$_2$O	c	−341.8			
Na$_2$CO$_3 \cdot$ 7H$_2$O	c	−765.1			
Na$_2$CO$_3 \cdot$ 10H$_2$O	c	−975.6			
NaHCO$_3$	c	−226.5	−203.6	24.4	20.98
	aq	−222.5	−203.9		
Na$_2$CO$_3 \cdot$ NaHCO$_3 \cdot$ 2H$_2$O	c	−641.2			
NaHCO$_2$ formate	c	−155.03			
	aq, 400	−154.92			
NaHCO$_2 \cdot$ 2H$_2$O	c	−296.6			
NaHCO$_2 \cdot$ 3H$_2$O	c	−364.2			
NaC$_2$H$_3$O$_2$	c	−169.8			
	aq, 6400	−174.07	−152.3		
	aq, ∞	−174.12			
NaC$_2$H$_3$O$_2 \cdot$ 3H$_2$O	c	−383.50			
Na$_2$C$_2$O$_4$	c	−314.3			
	aq, 600	−310.5	−283.4		
NaHC$_2$O$_4$	c	−257.8			
	aq, 400	−252.7			
NaHC$_2$O$_4 \cdot$ H$_2$O	c	−330.2			
NaCl	g	−43.50		54.9	8.55
	c	−98.23	−91.79	17.30	12.07
	aq, ∞	−97.30	−93.94	27.6	
NaClO	aq	−82.7			
NaClO$_2$	c	−73.85			
	aq, 1000	−73.80			
NaClO$_3$	c	−85.73			30.2
	aq, ∞	−80.78	−63.21	53.4	
NaClO$_4$	c	−91.48	−60.79	34.00	26.60
	aq, ∞	−88.69	−65.16	57.9	
Na$_2$CrO$_4$	c	−317.6			
	aq, 600	−320.6	−296		
Na$_2$CrO$_4 \cdot$ 4H$_2$O	c	−601.3			
Na$_2$Cr$_2$O$_7$	aq, 600	−463.4	−431		
NaF	c	−137.5	−130.3	12.24	11.20
	aq, ∞	−135.94	−128.67	12.1	
NaHF$_2$	c	−216.6			
	aq	−211.2			
NaH	g	29.88	24.78	44.93	7.24
	c	−13.49	−8.02	9.56	8.70
NaI	c	−68.84	−68.0	23.5	12.48
	aq, ∞	−70.65	−70.94	40.5	
NaI $\cdot$ 2H$_2$O	c	−211.05			
NaIO$_3$	c	−117.28			32.3
	aq, 500	−112.13	−94.8		
	aq, ∞	−112.2			
Na$_2$IrCl$_6$	c	−233.4			
Na$_2$MoO$_4$	c	−368			
	aq, 800	−368.6	−333		
NaNH$_2$	c	−28.4			
NaNO$_2$	c	−85.9			
	aq	−82.6	−71.0		

Table 9-1 (*Continued*)
ELEMENTS AND INORGANIC COMPOUNDS

Formula and Description	State	$\Delta Hf°$	$\Delta Gf°$	$S°$	$C_p°$
NaNO$_3$	c	−111.54	−87.45	27.8	22.24
	aq, ∞	−106.65	−89.00	49.4	
Na$_2$O	c	−99.4	−90.0	17.4	16.52
Na$_2$O$_2$	c	−122.7	−107.5	22.66	21.34
NaOH	c	−101.99	−90.60	15.40	14.23
	liq	−99.64	−89.42	18.13	20.99
	aq, ∞	−112.24	−100.18	11.9	
NaOH · H$_2$O	c	−175.17	−149.00	20.2	
NaPO$_3$	c	−288.6			
	aq, 600	−292.8			
Na$_3$PO$_3$	aq, 1000	−389.1			
NaH$_2$PO$_3$	c	−289.4			
	aq, 600	−290.5			
NaH$_2$PO$_3$ · 2½H$_2$O	c	−454.8			
Na$_2$HPO$_3$	c	−338.0			
	aq, 800	−347.5			
Na$_2$HPO$_3$ · 5H$_2$O	c	−684.2			
NaH$_2$PO$_4$	aq, 300	−367.7			
Na$_2$HPO$_4$	c	−417.4			
	aq, 200	−423.89			
	aq, 1000	−423.44			
Na$_2$HPO$_4$ · 2H$_2$O	c	−560.2			
Na$_2$HPO$_4$ · 7H$_2$O	c	−913.3			
Na$_2$HPO$_4$ · 12H$_2$O	c	−1266.4			
Na$_3$PO$_4$	c	−460			
	aq, 300	−475.0	−428.7		
	aq, 1000	−473.9			
Na$_3$PO$_4$ · 12H$_2$O	c	−1309.0			
NaNH$_4$HPO$_4$	aq, 500	−398.8			
NaNH$_4$HPO$_4$ · 4H$_2$O	c	−682.7			
NaH$_3$P$_2$O$_7$	c	−602.7			
	aq, 120	−603.9			
NaH$_3$P$_2$O$_7$ · H$_2$O	c	−670.6			
Na$_2$H$_2$P$_2$O$_7$	c	−663.4			
	aq, 1500	−661.6			
Na$_2$H$_2$P$_2$O$_7$ · 6H$_2$O	c	−1085.5			
Na$_3$HP$_2$O$_7$	c	−711.4			
	aq, 1500	−718.5			
Na$_3$HP$_2$O$_7$ · H$_2$O	c	−788.2			
Na$_3$HP$_2$O$_7$ · 6H$_2$O	c	−1135.7			
Na$_4$P$_2$O$_7$	c	−760.8			
	aq, 1500	−772.9			
Na$_4$P$_2$O$_7$ · 10H$_2$O	c	−1468.2			
Na$_2$PbO$_3$	c	−205			
Na$_2$PtBr$_6$	c	−221.8			
	aq	−231.7			
Na$_2$PtBr$_6$ · 6H$_2$O	c	−650.1			
Na$_2$PtCl$_4$	aq	−240.0	−216.8		
Na$_2$PtCl$_6$	c	−273.6			
	aq	−282.0			
Na$_2$PtCl$_6$ · 2H$_2$O	c	−418.5			
Na$_2$PtCl$_6$ · 6H$_2$O	c	−702.5			
Na$_2$PtI$_6$	aq	−170.1			
NaReO$_4$	c	−249.4			
	aq	−247.6			
Na$_2$S	c	−89.2	−86	23.4	18.99
	aq, 800	−104.36	−101.8		
Na$_2$S · 4½H$_2$O	c	−416.9			
Na$_2$S · 5H$_2$O	c	−452.7			
Na$_2$S · 9H$_2$O	c	−736.7			
NaHS	c	−56.5			
	aq, 800	−61.25			

Table 9-1 (*Continued*)
ELEMENTS AND INORGANIC COMPOUNDS

Formula and Description	State	$\Delta Hf°$	$\Delta Gf°$	$S°$	$C_p°$
NaHS · 2H$_2$O	c	−199.27			
Na$_2$S$_2$	aq	−104.6			
Na$_2$S$_3$	aq	−106.5			
Na$_2$S$_4$	c	−98.4			
Na$_2$SO$_3$	c	−260.6	−239.5	34.9	
	aq	−263.8			
Na$_2$SO$_3$ · 7H$_2$O	c	−753.4			
NaHSO$_3$	aq	−206.6			
Na$_2$SO$_4$ III	c	−330.90	−302.78	35.73	31.79
	aq, ∞	−331.46	−302.52	32.9	
Na$_2$SO$_4$ · 10H$_2$O	c	−1033.48	−870.93	141.7	
NaHSO$_4$	c	−269.2			
	aq, 200	−270.6			
NaHSO$_4$ · H$_2$O	c	−339.2			
Na$_2$SO$_4$ · (NH$_4$)$_2$SO$_4$ · H$_2$O	c	−691.5			
Na$_2$S$_2$O$_3$	c	−267.0			
	aq	−269			
Na$_2$S$_2$O$_3$ · 5H$_2$O I	c	−621.89			
II	c	−620.60			
Na$_2$S$_2$O$_5$	c	−349.1			
	aq	−345			
Na$_2$S$_2$O$_6$	c	−399.9			
	aq	−394.6			
Na$_2$S$_2$O$_6$ · 2H$_2$O	c	−542.5			
Na$_2$S$_3$O$_6$	aq	−409			
Na$_2$S$_3$O$_6$ · 3H$_2$O	c	−623.0			
Na$_2$S$_4$O$_6$	aq	−405			
Na$_2$S$_4$O$_6$ · 2H$_2$O	c	−550.0			
Na$_3$SbO$_4$	c	−352			
Na$_2$Se	c	−63.0			
	aq	−82.9	−89.4		
Na$_2$Se · 4½H$_2$O	c	−398.2			
Na$_2$Se · 9H$_2$O	c	−709.1			
Na$_2$Se · 16H$_2$O	c	−1199.4			
NaHSe	c	−27.8			
	aq	−32.7			
Na$_2$SeO$_3$	aq	−236.6			
NaHSeO$_3$	aq	−180.7			
Na$_2$SeO$_4$	c	−258			
	aq	−259.7	−230.3		
NaHSeO$_4$	aq	−201.1			
Na$_2$SiO$_3$	c	−363	−341	27.2	26.74
Na$_2$SiO$_3$ · 5H$_2$O	c	−720.0			
Na$_2$SiO$_3$ · 9H$_2$O	c	−1002.0			
Na$_2$Si$_2$O$_5$	c	−595.4		39.21	37.40
Na$_2$SiF$_6$	c	−677			
	aq, 600	−671.2			
NaHSiF$_6$	aq, 400	−614.1			
Na$_2$SnO$_3$	c	−276			
Na$_4$SnO$_4$	aq, 1200	−455.5			
Na$_2$Te	c	−84.0			
Na$_2$Te$_2$	c	−101.5			
Na$_2$TeO$_4$	c	−313			
Na$_2$UO$_4$	c	−501			
Na$_2$U$_2$O$_7$ · 1½H$_2$O	c	−880			
(Na$_2$O$_2$)$_2$ · UO$_4$	aq	−596			
(Na$_2$O$_2$)$_2$ · UO$_4$ · 9H$_2$O	c	−1225			
Na$_3$VO$_4$	c	−420			
Na$_2$WO$_4$	c	−369	−342	38.3	33.41
	aq, 220	−380.9	−345		
Na$_2$ZnO$_2$	c	−188			

Table 9-1 (*Continued*)
ELEMENTS AND INORGANIC COMPOUNDS

Formula and Description	State	$\Delta Hf°$	$\Delta Gf°$	$S°$	$C_p°$
Strontium					
Sr	c	0	0	12.5	6.39
	liq	1.82	1.47	13.66	8.40
	g	39.2	31.2	39.32	4.97
Sr^{2+} std. state, $m = 1$	aq	−130.45	−133.71	−7.8	
$Sr(acetate)_2$	c	−355.5			
Sr_3As_2	c	−147.3			
$Sr_3(AsO_4)_2$	c	−792.8	−736.2	61	
$Sr(BF_4)_2$	c	−921			
$SrBr_2$	c	−171.60	−167.25	34.28	18.37
std. state, $m = 1$	aq	−188.55	−183.41	31.6	
$SrBr_2 \cdot 6H_2O$	c	−605.0	−519.7	97	82.1
$Sr(BrO_3)_2 \cdot H_2O$	c	−264	−189.1	67	
$SrCl_2$	c	−198.10	−186.67	27.45	18.07
std. state, $m = 1$	aq	−210.35	−196.45	19.2	
$SrCl_2 \cdot 2H_2O$	c	−343.7	−306.4	52	38.3
$SrCl_2 \cdot 6H_2O$	c	−627.1	−535.67	93.4	
$Sr(ClO_4)_2$	c	−182.31			
std. state, $m = 1$	aq	−192.27	−137.83	79.2	
$Sr(CN)_2 \cdot 4H_2O$	c	−333.7			
$SrCO_3$ strontianite	c	−291.6	−272.5	23.2	19.46
std. state, $m = 1$	aq	−292.29	−259.88	−21.4	
SrC_2O_4	c	−327.6			
std. state, $m = 1$	aq	−327.6	−294.8	2.1	
SrF_2	c	−290.90	−278.58	19.63	16.73
$Sr_2[Fe(CN)_6]$ std. state, $m = 1$	aq	−152.0	−101.33	7.1	
$Sr_3[Fe(CN)_6]_2$ std. state, $m = 1$	aq	−122.8	−52.5	105.8	
$Sr(formate)_2$	c	−333.0			
SrH_2	c	−43.1			
$Sr(HCO_3)_2$	aq, inf	−460.7	−413.8	36.0	
$SrHPO_4$	c	−435.4	−403.6	29	
$Sr(H_2PO_4)_2$	c	−749.2			
SrI_2	c	−134.20	−133.54	38.03	18.63
std. state, $m = 1$	aq	−156.83	−158.37	45.4	
$SrI_2 \cdot 6H_2O$	c	−570.9			84.9
$Sr(IO_3)_2$	c	−243.6	−204.4	56	
$Sr(IO_3)_2 \cdot 6H_2O$	c	−666.8	−543.7	109	
$Sr_5(IO_6)_2$	c	−950.1			
$SrMoO_4$	c	−370		30.8	27.98
Sr_3N_2	c	−93.4	−76.5		
$Sr(N_3)_2$ azide	c	48.9			
$Sr(NO_2)_2$	c	−182.2			
$Sr(NO_3)_2$	c	−233.80	−186.46	46.50	35.82
std. state, $m = 1$	aq	−229.57	−186.93	62.2	
$Sr(NO_3)_2 \cdot 4H_2O$	c	−515.0	−413.65	88.2	
SrO	c	−141.50	−134.42	13.27	10.85
SrO_2	c	−151.4		13	19.0
$Sr(OH)_2$	c	−229.2			
	aq, inf	−240.29	−208.2	−14.4	
Sr_3P_2	c	−152			
$Sr_3(PO_4)_2$	c	−985.4			
SrS	c	−108.3	−107.2	16.3	11.64
$SrSe$	c	−92.2			
$SrSeO_3$	c	−250.4			
$SrSeO_4$	c	−273.1			
$SrSiO_3$	c	−390.5	−370.4	23.1	21.16
Sr_2SiO_4	c	−550.8	−523.7	36.6	32.09
$SrSO_3$	c	−281.3			
$SrSO_4$	c	−347.3	−320.5	28.1	25.76
std. state, $m = 1$	aq	−347.77	−311.68	−3.0	
$SrS_2O_6 \cdot 4H_2O$	c	−699.1			
$SrTiO_3$	c	−399.71	−379.64	26.0	23.51

Table 9-1 (*Continued*)
ELEMENTS AND INORGANIC COMPOUNDS

Formula and Description	State	$\Delta Hf°$	$\Delta Gf°$	$S°$	$C_p°$
Sr_2TiO_4	c	−546.7	−520.7	38.0	34.34
$SrWO_4$	c	−391.9	−366	33	
$SrZrO_3$	c	−422.4	−402.2	27.5	24.71
Sulfur					
S rhombic	c	0	0	7.63	5.40
	liq	0.34	0.09	8.4	7.58
	g	66.29	56.61	40.09	5.66
S_2	g	30.84	19.14	54.51	7.76
S_8	g	24.20	11.75	102.82	37.30
S^{2-} std. state, $m = 1$	aq	7.9	20.5	−3.5	
S_2Br_2	liq	−3			
SCN^-	aq	18.27	22.15	34.5	−9.6
SCl_2	liq	−12			
	g	−4.7			
S_2Cl_2	g	−4.66	−6.99	76.35	17.41
S_3Cl_2	liq	−12.4			
S_4Cl_2	liq	−10.2			
S_5Cl_2	liq	−8.8			
SCl_4	liq	−13.7			
SF_4	g	−174.10	−163.68	69.58	17.21
SF_6	g	−291.8	−266.7	69.72	23.25
std. state, $m = 1$	aq	−293.0	−259.3	39.8	
S_2F_{10}	liq	(−485)			
SF_2Cl	g	−250.5	−226.9	76.26	24.9
SO	g	1.167	−5.06	53.02	7.21
SO_2	liq	−76.6			
	g	−70.944	−71.748	59.30	9.53
undissoc; std. state, $m = 1$	aq	−77.194	−71.871	38.7	
	aq	−80.584			
SO_3 β	c	−108.63	−88.19	12.5	
	liq	−105.41	−88.04	22.85	
	g	−94.58	−88.69	61.34	12.11
SO_3^{2-} std. state, $m = 1$	aq	−151.9	−116.3	−7	
SO_4^{2-} std. state, $m = 1$	aq	−217.32	−177.97	4.8	−70
$S_2O_3^{2-}$	aq	−155.9			
$S_2O_4^{2-}$ std. state, $m = 1$	aq	−180.1	−143.5	22	
$S_2O_6^{2-}$	aq	−286.4			
$S_2O_7^{2-}$	aq	−334.9			
$S_2O_8^{2-}$ std. state, $m = 1$	aq	−320.0	−265.4	59.3	
$S_3O_6^{2-}$	aq	−286.7			
$S_4O_6^{2-}$	aq	−292.58			
$SOBr_2$	g	−21.8			
$SOCl_2$	liq	−58.7			
	g	−50.8	−47.4	74.01	15.9
SO_2Cl_2	liq	−94.2			
	g	−84.8	−74.2	74.53	18.4
$S_2O_5Cl_2$	liq	−168.7			56
SOF_2	g	(−130)		66.58	13.58
SO_2F_2	g	−181.3	−170.2	67.86	15.78
SO_3F^-	aq	−193.0			
$SO_2(NH_2)_2$ sulfamide	c	−129.3			
Tantalum					
Ta	c	0	0	9.92	6.06
	g	186.9	176.7	44.241	4.985
TaB_2	c	−46		10.6	11.5
TaC	c	−35.0	−34.6	10.11	8.79
Ta_2C	c	−51.0	−50.8	20.7	14.6
$TaBr_5$	c	−143.0		73.0	37.22
	g	−115.6			
$TaCl_3$	c	−132.2		37.0	22.25
$TaCl_4$	c	−167.7		46.0	28.64
	g	−134.0		92.0	23.55
$TaCl_5$	c	−205.3		59.4	37.21
	g	−181.3			

Table 9-1 (*Continued*)
ELEMENTS AND INORGANIC COMPOUNDS

Formula and Description	State	$\Delta Hf°$	$\Delta Gf°$	$S°$	$C_p°$
TaF$_5$	c	−454.97		46.6	31.18
std. state, undissoc	aq		−137.6		
TaF$_6^-$ std. state	aq		−209.2		
TaF$_7^{2-}$ std. state	aq		−280.2		
Ta$_2$H	c	−7.8	−16.5	18.9	21.7
TaI$_5$	c	−117		82	37.2
TaN	c	−60.1		12.1	9.7
Ta$_2$N	c	−65		22.0	20.00
TaO	g	60	53	57.6	7.31
TaO$_2$	g	−48	−50	67	10.52
Ta$_2$O$_5$ β	c	−489.0	−456.8	34.2	32.30
	aq	−496.7			
TaOCl$_3$	g	−186.6		86.4	23.55
TaS$_2$	c	−111			
TaSi$_2$	c	−28			
Ta$_5$Si$_3$	c	−76			
Technetium					
Tc	c	0	0	8.00	5.80
	g	162		43.25	4.97
Tc$_2$O$_7$	c	−266			
HTcO$_4$	c	−167			
std. state, $m = 1$	aq	−173			
Tellurium					
Te	c	0	0	11.88	6.15
	g	47.02	37.55	43.65	4.968
	amorp	2.7			
Te$_2$	g	40.2	28.2	64.06	8.78
TeBr$_4$	c	−45.5			
TeCl$_4$	c	−78.0		50	33.1
TeF$_6$	g	−315		80.25	27.94
TeO	g	15.6	9.2	57.7	7.19
TeO$_2$	c	−77.1	−64.6	19.0	15.27
TeO$_3^{2-}$	aq	−142.6			
Te(OH)$_3^+$ std. state, $m = 1$	aq	−145.4	−118.6	26.7	
TeSe	g	38.0	26.0	63.5	
Terbium					
Tb	c	0	0	17.50	6.91
	g	92.9	83.6	48.63	5.87
Tb^{3+} std. state, $m = 1$	aq	−163.2	−155.8	−54	4
TbC$_2$	c	211.7	198.7	64	10.5
TbCl$_3$	c	−238.3			
std. state, $m = 1$	aq	−283.0	−249.9	−14	−94
TbCl$_3 \cdot 6H_2O$	c	−683.4	−583.4	96.4	
Tb$_2$(CO$_3$)$_3$	c	−795.7			
Tb$_2$O$_3$	c	−445.8			27.7
TbO$_2$	c	−232.2			
Tb$_2$(SO$_4$)$_3$	aq, inf	−987.5	−859.8		
Thallium					
Tl	c	0	0	15.34	6.29
	g	43.55	35.24	43.225	4.968
TlBr	c	−41.4	−40.00	28.8	12.07
std. state, $m = 1$	aq	−27.77	−32.59	49.7	
TlBr^{2+} std. state, $m = 1$	aq	9.0	13.5	−13	
TlBr$_2^+$ std. state, $m = 1$	aq	−26.1	−21.4	20	
TlBr$_2^-$ std. state, $m = 1$	aq	−53.8	−58.8	84	
TlBr$_3$ undissoc; std. state	aq	−59.7	−53.7	49	
std. state, $m = 1$	aq	−40.2	−23.2	13	
TlBr$_4^-$ std. state, $m = 1$	aq	−90.9	−84.2	80	
TlBrO$_3$	c	−32.6	−12.70	40.3	
std. state, $m = 1$	aq	−18.7	−7.3	69.0	
TlCl	c	−48.79	−44.20	26.59	12.60
std. state, $m = 1$	aq	−36.67	−39.11	43.5	
undissoc; std. state, $m = 1$	aq	−41.10	−39.91	41.3	

Table 9-1 (*Continued*)
ELEMENTS AND INORGANIC COMPOUNDS

Formula and Description	State	$\Delta Hf°$	$\Delta Gf°$	$S°$	$C_p°$
TICl²⁺ std. state, $m = 1$	aq	1.0	9.7	−19	
TICl₂⁺ std. state, $m = 1$	aq	−43.0	−29.6	7	
TICl₂⁻ std. state, $m = 1$	aq		−70.7		
TICl₃	c	−75.3			
std. state, $m = 1$	aq	−72.9	−42.8	−5.5	
undissoc; std. state, $m = 1$	aq	−84.0	−65.6	32	
TICl₄⁻ std. state, $m = 1$	aq	−124.1	−100.8	58	
TIClO₃	aq	−22.4	−8.5	68.0	
Tl₂CO₃	c	−167.3	−146.9	37.1	
Tl(I) acetate	c	−126.1			
Tl(I) fulminate	c	27.6			
TICNS thiocyanate	c	6.8	9.21	39	
std. state, $m = 1$	aq	19.55	14.41	64.5	
Tl₂CrO₄	c	−225.8	−205.9	67.5	
TIF	c	−77.6		19.9	13.09
std. state, $m = 1$	aq	−78.22	−74.38	26.7	
TIHF₂	c			34.92	21.35
TII	c	−29.6	−29.97	30.5	12.55
	g	1.7			
std. state, $m = 1$	aq	−11.91	−20.07	56.6	
TII₂⁻ std. state, $m = 1$	aq		−35.1		
TII₄⁻ std. state, $m = 1$	aq		−39.3		
TIIO₃	c	−63.9	−45.86	42.2	
std. state, $m = 1$	aq	−51.6	−38.3	58.3	
TIN₃	c	55.8	70.38	35.1	
TINO₃	c	−58.30	−36.44	38.4	23.78
std. state, $m = 1$	aq	−48.28	−34.35	65.0	
Tl₂O	c	−42.7	−35.2	30	
Tl₂O₃	c	−74.5			
Tl₂O₄	c	−83.0			
TIOH	c	−57.1	−46.8	21	
std. state, $m = 1$	aq	−53.69	−45.33	27.4	
Tl(OH)₃	c		−121.2		
Tl₂S	c	−23.2	−22.4	36	
Tl₂SO₄	c	−222.7	−198.49	55.1	
Tl₂Se	c	−14	−14.1	41	
Tl₂SeO₄	c	−151	−126.4	56	
Tl₂Te	c	−22			
Thorium					
Th	c	0	0	13.6	6.53
ThBr₄	c	−227.1			
	aq	−298.6			
ThBr₄ · 7H₂O	c	−753.7			
ThBr₄ · 10H₂O	c	−971.6			
ThBr₄ · 12H₂O	c	−1116.0			
ThOBr₂	c	−252.8			
ThC₂	c	−29.9		16.4	13.55
ThCl₄	c	−285		44.05	28.86
	aq	−343.0			
ThCl₄ · 2H₂O	c	−437.4			
ThCl₄ · 4H₂O	c	−589.2			
ThCl₄ · 7H₂O	c	−805.9			
ThCl₄ · 8H₂O	c	−877.6			
ThCl₄ · NH₄Cl	c	−373.6			
ThCl₄ · 2NH₄Cl · 10H₂O	c	−1172.6			
ThOCl₂	c	−274.8		27.15	21.81
Th(OH)Cl₃ · H₂O	c	−377.4			
ThF₄	c	−477			
ThH₄	c	−43			
ThI₄	c	−131			
ThOI₂	c	−228.2			
ThOI₂ · 3½H₂O	c	−479.3			

Table 9-1 (*Continued*)
ELEMENTS AND INORGANIC COMPOUNDS

Formula and Description	State	$\Delta Hf°$	$\Delta Gf°$	$S°$	$C_p°$
Th(OH)I$_3$ · 10H$_2$O	c	−952.4			
Th$_3$N$_4$	c	−308	−282	42.7	37.26
Th(NO$_3$)$_4$	aq, 20	−379.10			
	aq, 50	−380.36			
	aq, 100	−380.48			
	aq, 500	−380.5			
ThO$_2$	c	−292	−280.1	15.6	14.76
Th(OH)$_4$	c*	−421.5			
Th$_2$S$_3$	c	−262.0			
Th(SO$_4$)$_2$	c	−602		35.4	41.46
	aq	−616.8			
Th(SO$_4$)$_2$ · 4H$_2$O	c	−882.7			
Th(SO$_4$)$_2$ · 8H$_2$O	c	−1168.6			
ThOSO$_4$	c	−487			
Thulium					
Tm	c	0	0	17.69	6.46
	g	55.5	47.2	45.41	4.97
Tm^{3+} std. state, $m = 1$	aq	−166.8	−158.2	−58	6
TmCl$_3$	c	−235.8			
std. state, $m = 1$	aq	−286.6	−252.3	−18	−92
TmI$_3$	c	−143.8			
std. state, $m = 1$	aq	−201.4	−193.5		
Tm$_2$O$_3$	c	−451.4	−428.9	33.4	27.9
Tin					
Sn I, white	c	0	0	12.32	6.45
II, gray	c	−0.50	0.03	10.55	6.16
	g	72.2	63.9	40.243	5.081
SnBr$_2$	c	−58.2			
SnBr$_4$	c	−90.2	−83.7	63.2	32.61
	g	−75.2	−79.2	98.43	24.71
SnCl$_2$	c	−79.3		31	18.96
std. state, $m = 1$	aq	−78.8	−71.6	41	
SnCl$_2$ · 2H$_2$O	c	−220.2			
SnCl$_3^-$ in aq HCl, std. state	aq	−116.4	−102.8	62	
SnCl$_4$	liq	−122.2	−105.2	61.8	39.5
SnH$_4$	g	38.9	45.0	54.39	11.70
Sn$_2$H$_6$	g	65.6			
SnI$_2$	c	−34.3			
SnI$_4$	c				20.3
SnO	c	−68.3	−61.4	13.5	10.59
	g			55.45	7.55
SnO$_2$	c	−138.8	−124.2	12.5	12.57
Sn(OH)$_2$ pptd	c	−134.1	−117.5	37	
Sn(OH)$_4$ pptd	c	−265.3			
	g			106.6	25.2
SnS	c	−24	−23.5	18.4	11.77
	g	28.5			
SnS$_2$	c	−40.0		20.9	16.76
Sn(SO$_4$)$_2$	c	−389.4			
std. state, $m = 1$	aq		−354.2		
SnSe	c	−21.7			
	g	30.8			
SnTe	c	−14.6			
	g	38.4			
Titanium					
Ti α	c	0	0	7.33	5.99
β	c	1.433	1.026	8.691	6.21
TiAs	c	−35.8			
TiB	c	−38.3	−38.2	8.3	7.09
TiB$_2$	c	−67	−65	6.80	10.58
TiBr$_2$	c	−97	−91.6	25.90	18.81
TiBr$_3$	c	−131.5	−125.6	42.17	24.31

Table 9-1 (*Continued*)
ELEMENTS AND INORGANIC COMPOUNDS

Formula and Description	State	$\Delta Hf°$	$\Delta Gf°$	$S°$	$C_p°$
TiBr$_4$	c	−147.7	−141.2	58.23	31.43
	liq	−144.7	−141.0	67.92	36.30
	g	−131.5	−136.0	95.35	24.07
TiC	c	−44.0	−43.1	5.79	8.08
TiCl$_2$	c	−122.8	−111.0	20.9	16.69
TiCl$_3$	c	−172.3	−156.2	33.4	23.22
TiCl$_4$	c	−194.8	−175.7	49.9	30.94
	liq	−192.2	−176.2	60.31	34.70
	g	−182.4	−173.7	84.8	22.8
TiF$_2$	g	−164	−166	61	12.64
TiF$_3$	c	−343	−325.6	21	21.99
TiF$_4$	c	−394.2	−372.7	32.02	27.31
	g	−370.8	−362.2	75.2	20.36
TiH$_2$	c	−34.5	−25.1	7.1	7.2
TiI$_2$	c	−64	−62	29	20.61
TiI$_3$	c	−77	−76	46	27.91
TiI$_4$	c	−89.8	−88.6	58.8	30.03
	liq	−83.25	−86.74	74.53	37.40
TiN	c	−80.8	−74.0	7.23	8.86
TiO α	c	−129.7	−122.7	8.31	9.55
TiO^{2+} in HClO$_4$ medium	aq	−164.9			
TiO$_2$ anatase	c	−224.36	−211.12	11.93	13.21
brookite	c	−225.1			
rutile	c	−225.8	−212.6	12.03	13.15
Ti$_2$O$_3$	c	−363.5	−342.8	18.83	23.27
Ti$_3$O$_5$ α	c	−587.8	−553.9	30.9	37.00
Ti$_4$O$_7$	c	−813.7	−768.0	47.5	49.83
TiOCl	c	−180			
TiOCl$_2$	g	−130.4	−127.9	76.70	17.20
TiOF$_2$	g	−221	−217	68.02	15.00
TiP	c	−67.6			
TiS	c	−57			
TiS$_2$	c	−80.0		18.73	16.23
TiSi	c	−31			
TiSi$_2$	c	−32			
Ti$_5$Si$_3$	c	−138			
Tungsten					
W	c	0	0	7.81	5.81
	liq	11.22	9.66	10.92	5.81
	g	203.4	193.3	41.55	5.09
WBr$_5$	c	−74.5	−64.4	65.0	37.16
WBr$_6$	c	−82	−69	75	43.35
WCl$_2$	c	−61.5	−52.6	31.2	18.60
WCl$_4$	c	−106	−86	47.4	31.00
WCl$_5$	c	−123	−96	52.0	37.20
WCl$_6$	c	−142	−109	57.0	41.93
WC	c	−9.69		7.74	8.45
W$_2$C	c	−6.3		19.5	18.31
W(CO)$_6$	c	−227.9		79.3	57.96
WF$_6$	c	−418.2			
	liq	−417.9	−390.0	59.6	40.50
	g	−229.7	−199.8	83.1	32.13
WO$_2$	c	−140.94	−127.60	12.08	13.32
WO$_3$	c	−201.46	−182.63	18.14	17.48
WO$_4^{2-}$ std. state, $m = 1$	aq	−257.1			
W$_3$O$_8$	g	−409	−378	118.0	49.30
WOBr$_4$	c	−130.1			
WOCl$_4$	c	−160.4	−131.3	41.3	34.95
	g	−137	−122	90.1	25.36
WOF$_4$	c	(−333)	−310	42.0	31.93
	g	−319	−305	80.0	22.91
WO$_2$Br$_2$	c	−170.3			

Table 9-1 (*Continued*)
ELEMENTS AND INORGANIC COMPOUNDS

Formula and Description	State	$\Delta Hf°$	$\Delta Gf°$	$S°$	$C_p°$
WO_2Cl_2	c	−186.5	−168.0	48.0	24.95
WS_2	c	−50			
WSi_2	c	−22			
Uranium					
U	g	125			
	c	0	0	12.03	6.60
UBr_3	c	−170.1	−164.7	49	
UBr_4	c	−196.6	−188.5	56.0	29.32
UC	c	−23.2		14.15	11.98
$UC_{1.9}$	c	−23.0		16.3	25.15
UCl_3	c	−213.0	−196.9	37.99	24.29
UCl_4	c	−251.2	−230.0	47.4	28.87
UCl_5	c	−262.1	−237.4	58.0	34.55
UCl_6	c	−272.4	−241.5	68.3	41.96
UF_3	c	−357	−339	28.0	23.02
UF_4	c	−443	−421	36.1	27.72
UF_5	c	−488	−461	47.3	31.63
UF_6	g	−505	−485	54.4	40.03
UH_3	c	−30.4	−17.35	15.27	11.78
UI_3	c	−114.7	−115.3	56	
UI_4	c	−127.0	−126.1	67.0	31.00
$UIBr_3$	c	−177.1			
$UICl_3$	c	−219.9	−204.4	54	
UN	c	−80	−75	14.9	11.31
U_2N_3	c	−213	−194	29	
UO_2	c	−270	−257	18.6	15.23
UO_3	c	−302	−283	23.57	19.51
$UO_3 \cdot H_2O$	c	−375.4			
$UO_3 \cdot 2H_2O$	c	−446.2			
$UO_4 \cdot 2H_2O$	c	−436			
U_3O_8	c	−854.4		67.5	56.7
$UOBr_2$	c	−246.0		37.7	23.41
UO_2Br_2	aq	−308.2			
$UO_2(C_2H_3O_2)_2$	aq	−484.0			
$UO_2(C_2H_3O_2)_2 \cdot 2H_2O$	c	−624.9			
$UO_2(C_2H_3O_2)_2 \cdot$					
$NH_4C_2H_3O_2 \cdot 6H_2O$	c	−1045.8			
UOCl	c	−226.4		24.6	17.44
$UOCl_2$	c	−260.0		33.1	22.64
$UOCl_3$	c	−284.2		42.0	27.95
UO_2Cl_2	c	−302.9		36.0	25.73
UO_2F_2	c	−391.3		32.40	24.63
UO_2CrO_4	aq	−456.7			
$UO_2CrO_4 \cdot 5\frac{1}{2}H_2O$	c	−838.8			
$UO_2(NO_3)_2$	c	−329.2	−273.1	66	
	aq, ∞	−349.1	−289.2	53	
$UO_2(NO_3)_2 \cdot H_2O$	c	−404.8	−335.3	76	
$UO_2(NO_3)_2 \cdot 2H_2O$	c	−480.0	−396.6	85	
$UO_2(NO_3)_2 \cdot 3H_2O$	c	−552.2	−454.7	94	
$UO_2(NO_3)_2 \cdot 6H_2O$	c	−764.3	−625.0	120.85	
UO_2SO_4	aq, ∞	−467.3	−413.7	−13	
$UO_2SO_4 \cdot 3H_2O$	c	−666.8	−586.0	63	
US	c	−73.2		18.6	12.10
US_2	c	−120.0		26.0	17.79
$U(SO_4)_2$	c	−563			
USi_2	c	−31.0		19.6	20.00
USi_3	c	−31.2		25.4	24.00
Vanadium					
V	c	0	0	6.91	5.95
	g	122.90	108.32	43.544	6.217
VBr_2	c	−87.3			
	g	−37.1			

Table 9-1 (*Continued*)
ELEMENTS AND INORGANIC COMPOUNDS

Formula and Description	State	$\Delta Hf°$	$\Delta Gf°$	$S°$	$C_p°$
VBr_3	c	−103.6			
	g	−56.8			
VBr_4	g	−80.5			
VCl_2	c	−108	−97	23.2	17.26
VCl_3	c	−138.8	−122.2	31.3	22.27
VCl_4	liq	−136.1	−120.4	61	
	g	−125.6	−117.6	86.6	23.0
$V(CO)_6$	g	−236			
VF_3	c			23.18	21.62
VF_4	c	−335.4			
VF_5	liq	−353.8	−328.2	42.0	
	g	−342.7	−327.4	76.67	23.56
VI_2	c	−60.1		35.0	17.89
VI_3	c	−64.7		48.5	23.84
VI_4	g	−29.3			
VN	c	−51.9	−45.7	8.91	9.08
VO	c	−103.2	−96.6	9.3	10.86
	g	25	18	55.8	7.3
VO^{2+} std. state	aq	−116.3	−106.7	−32.0	
VO_2	c	−171.5	12.3	14.96	
VO_2^+ std. state	aq	−155.3	−140.3	−10.1	
VO_3^- std. state	aq	−212.3	−187.3	12	
V_2O_3	c	−293.5	−272.3	23.5	24.67
V_2O_4 α	c	−341.1	−315.1	24.5	27.96
V_2O_5	c	−370.6	−339.3	31.3	30.51
V_3O_5	c	−465	−434	39	
V_4O_7	c	−635	−591	52	
V_6O_{13}	c	−1062			
HVO_4^{2-} std. state	aq	−277.0	−233.0	4	
$H_2VO_4^-$ std. state	aq	−280.6	−244.0	29	
$[VO_2H_2O_2]^+$ std. state	aq		−178.4		
VOCl	c	−138.4			
VO_2Cl	c	−185.6			
$VOCl_2$	c	−165.0			
$VOCl_3$	liq	−175.6	−159.8	49	36.0
	g	−166.25	−157.58	82.26	21.49
$VOSO_4$	c	−312.9	−279.6	26.0	
std. state, undissoc	aq		−282.6		
$VOSCN^+$ std. state	aq	−98	−86	8	
V_2S_3	c	−227			
VSi_2	c	−73			
V_2Si	c	−37			
V_3Si	c	−26			
V_5Si_3	c	−94			
Xenon					
Xe	g	0	0	40.529	4.968
XeF_4	c	−62.5	−29.4		
	g	−51.5	−50		28.334
XeF_2	c	−39.2			
	g	−25.9	−37		
XeF_6	s	−86			
	g	−71			
XeO_3	c	96			
$XeOF_4$	liq	35			
XeO_2F_2	c	35			
Ytterbium					
Yb	c	0	0	14.31	6.39
	g	36.4	28.3	41.35	4.97
Yb^{2+} std. state, $m = 1$	aq	−126			
Yb^{3+} std. state, $m = 1$	aq	−161.2	−153.9	57	6
$Yb(acetate)_3$					
undissoc; std. state, $m = 1$	aq	−503.1	−423.72	43.8	

Table 9-1 (*Continued*)
ELEMENTS AND INORGANIC COMPOUNDS

Formula and Description	State	$\Delta Hf°$	$\Delta Gf°$	$S°$	$C_p°$
Yb(acetate)$_2^+$ std. state, $m = 1$	aq	−387.46	−334.45	18.1	
Yb(acetate)$^{2+}$ std. state, $m = 1$	aq	−273.85	−244.49	−16.8	
YbC$_2$	c	−17.9	−18.5	19	
YbCl$_2$	c	−191.1			
YbCl$_3$	c	−229.4			
std. state, $m = 1$	aq	−281.1	−248.0	−17	−92
YbCl$_3 \cdot$ 6H$_2$O	c	−680.2	−580.6	94.6	81.6
YbH$_2$	c	−42.1			
Yb(IO$_3$)$_3$	c	−322			
Yb(NO$_3$)$_3$	aq, inf	−309.9			
Yb$_2$O$_3$	c	−433.7	−412.7	31.8	27.57
Yb$_2$(SO$_4$)$_3$	aq	−971.9	−843.0		
Yttrium					
Y	c	0	0	10.62	6.34
	g	100.7	91.1	42.87	6.18
Y$_2$	g	163.5	150.7	64	8.7
Y^{3+} std. state, $m = 1$	aq	−172.9	−156.8	−60	
YBr^{2+} std. state, $m = 1$	aq	−203.8	−191.6	−43	
YC$_2$	c	−26	−26	13	
	g	142.6	128.4	61	10.7
Y$_2$(CO$_3$)$_3$			−752.4		
Y$_2$(C$_2$O$_4$)$_3 \cdot$ 9H$_2$O	c		−1363.8		
Y(acetate)$_3$	aq	−516.1			
Y(SCN)$^{2+}$ std. state, $m = 1$	aq		−144.7		
YCl	g	47.8	41.5	58.33	8.56
YCl^{2+} std. state, $m = 1$	aq	−214.0	−198.7	−46	
YCl$_3$	c	−239.0		32.7	23.41
	g	−179.3			18
	aq, 4000	−291.96			
YCl$_3 \cdot$ 6H$_2$O	c	−691.3	−592.1	92	
YF	g	−33	−39	55.38	7.92
YF$_3$	c	−410.8	−393.1	24	
	g	−308.0	−305.4	74.5	16.8
YH$_2$	c	−37.5	−27.8	9.17	8.24
YH$_3$	c	−47.3	−33.2	10.02	10.36
YI$_3$	c	−147.4			
Y(IO$_3$)$_3$	c		−271.2		
YO	g	−9.3	−15.5	55.88	7.53
Y$_2$O$_2$	g	−127.4			15.8
Y$_2$O$_3$	c	−455.38	−434.19	23.68	24.50
Y(OH)$^{2+}$ std. state, $m = 1$	aq		−210.1		
Y(OH)$_3$	c		−308.6		
Y(OH)$_2$Cl	c		−297.9		
Y(ReO$_4$)$_3$	c	−701.9	−629.4	88	
YS	g	41.7	29.7	58	8.2
Zinc					
Zn	c	0	0	9.95	6.07
	g	31.245	22.748	38.450	4.968
ZnAs$_2$	c	−10.0			
Zn$_3$As$_2$	c	−7.7			
Zn(BO$_2$)$_2$	c		−373.58		
ZnBr$_2$	c	−78.55	−74.60	33.1	15.70
std. state, $m = 1$	aq	−94.88	−84.84	12.6	57
ZnBr$_2 \cdot$ 2H$_2$O	c	−224.0	−191.1	47.5	
ZnCO$_3$	c	−94.26	−174.85	19.7	19.05
ZnC$_2$O$_4$ std. state, $m = 1$	aq	−234.0	−196.2	−15.9	
ZnC$_2$O$_4 \cdot$ 2H$_2$O	c	−374.0	−321.7	46.7	
Zn(C$_2$O$_4$)$_2^{2-}$ std. state, $m = 1$	aq	−430.7	−367.5	31	
Zn(formate)$_2$	c	−235.8			
std. state, $m = 1$	aq	−240.20	−202.9	17	
Zn(acetate)$_2$	c	−257.8			
std. state, $m = 1$	aq	−269.10	−211.72	14.6	

Table 9-1 (*Continued*)
ELEMENTS AND INORGANIC COMPOUNDS

Formula and Description	State	$\Delta Hf°$	$\Delta Gf°$	$S°$	$C_p°$
Zn(CN)$_2$	c	22.9			
Zn(CN)$_4^{2-}$ std. state, $m = 1$	aq	81.8	106.8	54	
ZnCl$_2$	c	−99.20	−88.296	25.9	16.14
	g	−63.6			
std. state, $m = 1$	aq	−116.68	−97.88	0.2	−54
ZnCl$_3^-$ std. state, $m = 1$	aq		−129.2		
ZnCl$_4^{2-}$ std. state, $m = 1$	aq		−159.2		
Zn(ClO$_4$)$_2$ std. state, $m = 1$	aq	−98.60	−39.26	60.2	
Zn(ClO$_4$)$_2 \cdot$6H$_2$O	c	−509.89	−371.8	130.4	
ZnF$_2$	c	−182.7	−170.5	17.61	15.69
std. state, $m = 1$	aq	−195.78	−168.42	−33.4	−40
ZnI$_2$	c	−49.72	−49.94	38.5	15.70
std. state, $m = 1$	aq	−63.16	−59.80	25.2	−57
Zn(IO$_3$)$_2$	c		−103.68		
std. state, $m = 1$	aq	−142.6	−96.3	29.8	
Zn(N$_3$)$_2$	c	52			
Zn$_3$N$_2$	c	−5.4			26
Zn(NO$_3$)$_2$	c	−115.6			
ionized, std. state	aq	−135.90	−88.36	43.2	−30
Zn(NO$_3$)$_2 \cdot$6H$_2$O	c	−551.30	−423.79	109.2	77.2
Zn(NH$_3$)$_4^{2+}$ std. state, $m = 1$	aq	−127.5	−72.2	72	
ZnO	c	−83.24	−76.08	10.43	9.62
ZnO$_2^{2-}$ std. state, $m = 1$	aq		−91.85		
ZnO$_2 \cdot$2H$_2$O	c	−207.6			
Zn(OH)$^+$ std. state, $m = 1$	aq		−78.9		
HZnO$_2^-$ std. state, $m = 1$	aq		−109.26		
Zn(OH)$_2$ β	c	−153.42	−132.31	19.4	
ϵ	c	−152.74	−132.68	19.5	17.3
std. state, $m = 1$	aq	−146.72	−110.33	−31.9	−60
Zn(OH)$_3^-$ std. state, $m = 1$	aq		−165.95		
Zn(OH)$_4^{2-}$ std. state, $m = 1$	aq		−205.23		
Zn$_3$P$_2$	c	−113			
Zn(PO$_3$)$_2$	c	−497.9			
Zn$_2$P$_2$O$_7$	c	−600.0			
Zn$_3$(PO$_4$)$_2$	c	−691.3			
ZnS wurtzite	c	−46.04			
sphalerite	c	−49.23	−48.11	13.8	11.0
ZnSO$_4$	c	−234.9	−209.0	30.6	27.33
std. state, $m = 1$	aq	−254.10	−213.11	−22.0	−59
ZnSO$_4 \cdot$H$_2$O	c	−311.78	−270.58	33.1	
ZnSO$_4 \cdot$6H$_2$O	c	−663.83	−555.64	86.9	85.49
ZnSO$_4 \cdot$7H$_2$O	c	−735.60	−612.59	92.9	91.64
ZnS$_2$O$_6$	aq	−323.2			
ZnS$_2$O$_6 \cdot$6H$_2$O	c	−735.2			
ZnSb	c	−3.5			
ZnSe	c	39	39	20	
ZnSeO$_3$ std. state, $m = 1$	aq	−158.5	−123.5	−23.7	
ZnSeO$_3 \cdot$H$_2$O	c	−222.5	−189.5	39	
ZnSeO$_4$	c	−158.8			
std. state, $m = 1$	aq	−180.0	−140.6	−13.9	
ZnSeO$_4 \cdot$6H$_2$O	c	−587.5			
ZnSiO$_3$	c	−301.2		21.4	20.26
Zn$_2$SiO$_4$	c	−391.19	−364.06	31.4	29.48
ZnTe	c	−28.1			
ZnWO$_4$	c	−293		34.5	29.03
Zirconium					
Zr α, hexagonal	c	0	0	9.32	6.06
β	c	1.71	1.16	11.15	5.89
ZrB$_2$	c	−77.1	−76.0	8.59	11.53
ZrBr$_2$	c	−97	−91	28	20.73
ZrBr$_3$	c	−152	−145	41.14	23.78
ZrBr$_4$	c	−181.6	−173.1	53.70	29.83

Table 9-1 (*Continued*)
ELEMENTS AND INORGANIC COMPOUNDS

Formula and Description	State	$\Delta Hf°$	$\Delta Gf°$	$S°$	$C_p°$
ZrC	c	−47.0	−46.2	7.96	9.06
ZrCl$_2$	c	−103	−92	26	17.35
ZrCl$_3$	c	−171	−154	34.8	22.99
ZrCl$_4$	c	−234.35	−212.7	43.4	28.63
ZrF$_2$	c	−230	−218	18	15.76
ZrF$_3$	c	−335	−317	21	20.0
ZrF$_4$	c	−456.8	−432.6	25.02	24.76
ZrH$_2$	c	−40.4	−30.8	8.37	7.40
ZrI$_2$	c	−62	−61.7	35.9	22.50
ZrI$_3$	c	−95	−94.3	48.90	24.81
ZrI$_4$	c	−115.9	−114.9	61.41	29.54
ZrN	c	−87.2	−80.4	9.29	9.66
ZrO$_2$ monoclinic	c	−262.3	−248.5	12.04	13.43
ZrSiO$_4$	c	−483.7		20.08	23.41

Table 9-2
ORGANIC COMPOUNDS

Substance	State	$\Delta Hf°$	$\Delta Gf°$	$S°$	$C_p°$
Acenaphthene	c	16.8			
Acenaphthylene	c	44.7			
Acetaldehyde	liq	−45.96	−30.64	38.3	65.6
	g	−39.76	−31.86	63.15	13.06
Acetaldoxime	c	−18.6			
	liq	−19.5			
Acetamide	c	−76.0			
Acetamidoguanidine nitrate	c	−118.1			
1-Acetamido-2-nitroguanidine	c	−46.3			
5-Acetamidotetrazole	c	−1.2			
Acetanilide	c	−50.3			
Acetic acid	liq	−115.71	−93.2	38.2	29.7
	g	−103.93	−90.03	67.52	15.90
ionized; std. state, $m = 1$	aq	−116.16	−88.29	20.7	−1.5
nonionized; std. state, $m = 1$	aq	−116.70	−94.78	42.7	
Acetic anhydride	liq	−149.14	−116.83	64.2	
	g	−137.60	−113.93	93.20	23.78
Acetone	liq	−59.18	−37.22	47.9	30.22
	g	−51.78	−36.58	70.49	17.90
Acetone glyceraldehyde	liq	−180			
Acetonitrile	liq	12.8	23.7	35.76	21.86
	g	21.00	25.24	58.19	12.48
Acetophenone	liq	−34.07	−4.06	59.62	
	g	−20.76	0.44	89.12	
Acetyl radical	g	−4.0			
N-Acetylbenzidine	c	−38.0			
Acetyl bromide	liq	−53.5			
Acetyl chloride	liq	−65.44	−49.73	48.0	28
	g	−58.30	−49.29	70.47	16.21
Acetylene	g	54.19	50.00	48.00	10.50
std. state, $m = 1$	aq	50.54	51.88	29.5	
Acetylenedicarbonitrile	liq	119.6			
	g	127.50	122.10	69.31	20.53
Acetylene dicarboxylic acid	c	−138.1			
Acetyl fluoride	liq	−112.4			
N-Acetylhydrazobenzene	c	−2.0			
o-Acetylhydroxybenzoic acid	c	−194.93			
N-Acetylimidazole	c	−28.6			
Acetyl iodide	liq	−39.3			
4-Acetylresorcinol	c	−137.1			
N-Acetyltetrazole	c	19.49			
Acridine	c	44.8			
Acrolein	liq	−27.97	−16.17		
	g	−20.50	−15.45		
Acrylic acid	liq	−91.8			
	g	−80.36	−68.37	75.29	18.59
Acrylonitrile	liq	36.1			
	g	44.20	46.68	65.47	15.24
Adenine	c	23.21	71.58	36.1	
Adipic acid	c	−237.60			
	liq	−235.51	−177.17		
Aetioporphyrin I	c	−6.0			
Aetioporphyrin II	c	0.4			
α-Alanine d	c	−134.03	−88.23	31.6	
l	c	−133.96	−88.49	30.88	
dl	c	−134.55	−88.92	31.6	

Table 9-2 (*Continued*)
ORGANIC COMPOUNDS

Substance	State	$\Delta Hf°$	$\Delta Gf°$	$S°$	$C_p°$
Alanine anhydride	c	−128.0			
α-Alanylglycine *dl*	c	−185.64	−117.00	51.0	
l	c	−197.52	−127.30	46.62	
Alanylphenylalanine *dl*	c	−170.2			
Alanylphenylalanyl anhydride	c	−89.3			
Allantoin (5-ureidohydantoin)	c	−171.50	−106.65	46.6	
Allomucic acid	c	−412			
Alloxan monohydrate	c	−239.08	−182.08	44.6	
Alloxantin dihydrate	c	−510.3			
Allyl radical	g	38			
1-Allyl-5-allylaminotetrazole	c	83.7			
1-Allyl-5-aminotetrazole	c	63.4			
2-Allyl-5-aminotetrazole	c	67.6			
Allylcyclopentane	liq	−15.74			
Allyl ethyl sulfoxide	liq	−41.83			
Allyl trichloroacetate	liq	−94.5			
Amalic acid	c	−367.0			
Amarine	c	63			
p-Aminoacetophenone	c	70.2			
3-Aminoacridine	c	39.8			
5-Aminoacridine	c	38.1			
2-Aminobenzoic acid	c	−95.8			
3-Aminobenzoic acid	c	−98.2			
4-Aminobenzoic acid	c	−98.8			
2-Aminobiphenyl	c	26.8			
4-Aminobiphenyl	c	19.4			
1-Aminobutane (butylamine)	liq	−30.52			
	g	−22.00	11.76	86.76	28.33
2-Aminobutane (*sec*-butylamine)	g	−24.90	9.71	83.90	27.99
4-Aminobutanoic acid	c	−138.1			
2-Aminoethanesulfonic acid	c	−187.7	−134.3	36.8	33.6
ionized; std. state, $m = 1$	aq	−171.92	−121.76	47.8	
nonionized; std. state, $m = 1$	aq	−181.92	−134.12	55.7	
2-Aminohexanoic acid (norleucine)	c	−152.7			
4-Aminohexanoic acid	c	−154.5			
5-Aminohexanoic acid	c	−153.7			
6-Aminohexanoic acid	c	−152.7			
3-Amino-2-methylpropane	liq	−31.68			
(2-butylamine)					
5-Aminopentanoic acid	c	−144.5			
5-Aminotetrazole	c	49.7			
5-Aminotetrazole nitrate	c	−6.6			
3-Amino-1,2,4-triazole	c	18.4			
Amygdalin	c	−455			
1,2-Anhydroglucose-3,5,6-triacetate	c	−411.7			
Aniline	liq	7.55	35.63	45.72	45.90
	g	20.76	39.84	76.28	25.91
Anisine	c	−51			
Anisoyl glycine	c	−180.9			
Anthracene	c	29.0	68.30	49.58	49.7
9,10-Anthracenedione	c	−49.6			
β-D-Arabinose	c	−252.84			
β-L-Arabinose	c	−252.84			
D-Arabonic acid-γ-lactone	c	−238.2			
L-Arginine	c	−148.66			

Table 9-2 (*Continued*)
ORGANIC COMPOUNDS

Substance	State	$\Delta Hf°$	$\Delta Gf°$	$S°$	$C_p°$
D-Arginine	c	−149.05	−57.43	59.9	
L-Ascorbic acid (vitamin C)	c	−278.34			
L-Asparagine	c	−188.50	−126.73	41.7	
L-Aspartic acid	c	−232.47	−174.53	40.66	
Azobenzene cis	c	86.7			
trans	c	76.6			
Azodicarbamide	c	−69.90			
Azulene	g	66.90	84.10	80.75	30.69
Barbituric acid	c	−152.2			
Benzaldehyde	liq	−21.23	2.24		
	g	−9.57	5.85		
Benzamide	c	−48.42			
Benzanilide	c	−22.3			
1,2-Benzanthracene	c	41			
2,3-Benzanthracene	c	38.3	85.79	51.48	
1,2-Benzanthra-9,10-quinone	c	−55.4			
Benzene	liq	11.71	29.72	41.41	19.52
	g	19.82	30.99	64.34	
Benzenethiol (thiophenol)	liq	15.32	32.02	53.25	41.40
	g	26.66	35.28	80.51	25.07
Benzidine	c	16.9			
Benzil	c	−36.8			
Benzoic acid	c	−92.03	−58.62	40.05	34.97
Benzoic anhydride	c	−103.0			
Benzonitrile	g	52.30	62.33	76.73	26.07
Benzophenone	c	−8.0	33.5	58.6	
p-Benzoquinone	c	−44.33			
Benzotriazole	c	59.74			
DL-Benzoylalanine	c	−147.9			
Benzoyl bromide	liq	−25.58			
Benzoyl chloride	liq	−39.17			
Benzoyl iodide	liq	−12.31			
Benzoylphenylalanine	c	−129.6			
Benzoyl sarcosine	c	−135.7			
3,4-Benzphenanthrene	c	44.2			
Benzyl radical	g	45			
Benzyl alcohol	liq	−38.49	−6.57	51.8	
Benzyl bromide (2-bromotoluene)	liq	5.6			
Benzyl chloride	liq	−7.8			
N-Benzyldiphenylamine	c	44.2			
Benzyl ethyl sulfide	liq	−1.17			
Benzyl iodide	liq	13.8			
Benzyl mercaptan	liq	10.4			
Benzyl methyl ketone	liq	−36.30			
Benzyl methyl sulfide	liq	6.27			
Bicyclo[4,1,0]heptane	g	0.33			
Bicyclo[3,1,0]hexane	g	9.09			
Bicyclo[4,2,0]octane	g	−6.39			
Bicyclo[5,1,0]octane	g	−3.85			
Bicyclopropyl	g	30.9			
Biphenyl	c	24.02	60.75	49.2	38.80
	liq	28.5	62.07	59.8	
Biphenylene	liq	84.4			
N,N'-Bisuccinimide	c	−169.5			
Brassidic acid	c	−214			

Table 9-2 (*Continued*)
ORGANIC COMPOUNDS

Substance	State	$\Delta Hf°$	$\Delta Gf°$	$S°$	$C_p°$
Bromal	liq	−31.13			
Bromal hydrate	c	−112			
Bromobenzene	liq	14.5	30.12	52.0	37.17
4-Bromobenzoic acid	c	−90.4			
1-Bromobutane	g	−25.65	−3.08	88.39	26.13
2-Bromobutane	liq	−37.2	−4.60		
	g	−28.70	−6.16	88.50	26.48
Bromochlorodifluoromethane	g	−112.7	−107.18	76.14	
Bromochlorofluoromethane	g	−70.5	−66.58	72.88	
Bromochloromethane	g	−12.0	−9.39	68.67	
Bromodichlorofluoromethane	g	−64.4	−58.98	78.87	
Bromodichloromethane	g	−14.0	−10.16	75.56	
Bromodifluoromethane	g	−110.8	−106.90	70.51	
Bromoethane	liq	−21.99	−6.64	47.5	24.1
	g	−15.30	−6.29	68.71	15.45
Bromoethene (vinyl bromide)	g	18.73	19.30	65.83	13.26
Bromofluoromethane	g	−60.4	−57.71	65.97	
1-Bromoheptane	liq	−52.21			
1-Bromohexane	liq	−46.42			
Bromoiodomethane	g	12.0	9.36	73.49	
Bromomethane	g	−9.02	−6.75	58.76	10.15
2-Bromo-2-methylpropane	liq	−39.3			
	g	−32.00	−6.73	79.34	27.85
1-Bromooctane	liq	−58.57			
1-Bromopentane	liq	−40.68			
	g	−30.87	−1.37	97.70	31.60
1-Bromopropane	g	−21.00	−5.37	79.08	20.66
2-Bromopropane	g	−23.20	−6.51	75.53	21.37
N-Bromosuccinimide	c	−80.35			
Bromotrichloromethane	g	−8.9	−2.96	79.55	
Bromotrifluoromethane	g	−155.1	−148.8	71.16	16.57
Brucine	c	−118.6			
1,2-Butadiene	g	38.77	47.43	70.03	19.15
1,3-Butadiene	g	26.33	36.01	66.62	19.01
Butadiyne (biacetylene)	g	113.00	106.11	59.76	17.60
n-Butane	liq	−35.29	−3.60	55.2	
	g	−30.15	−4.10	74.12	23.29
1,2-Butanediamine	liq	−28.74			
2,3-Butanedione (diacetyl)	liq	−87.44			
1,4-Butanedithiol	liq	−25.11			
1-Butanethiol (n-butyl mercaptan)	liq	−29.79	0.97	65.96	
	g	−21.05	2.64	89.68	28.24
2-Butanethiol	liq	−31.13	−0.04	64.87	
	g	−23.00	1.29	87.65	28.51
1-Butanol	liq	−78.18	−38.84	54.1	42.31
	g	−65.65	−36.04	86.7	26.29
2-Butanol	liq	−81.88	−42.31	53.8	47.5
	g	−69.94	−40.06	85.8	27.08
2-Butanone (methyl ethyl ketone)	liq	−65.29	−36.18	57.08	37.98
	g	−56.26	−34.91	80.81	24.59
1-Butene	g	−0.03	17.04	73.04	20.47
2-Butene cis	g	−1.67	15.74	71.90	18.86
trans	g	−2.67	15.05	70.86	20.99
1-Buten-3-yne	g	72.80	73.13	66.77	17.49
t-Butoxy radical	g	−24.7			

Table 9-2 (*Continued*)
ORGANIC COMPOUNDS

Substance	State	$\Delta H f°$	$\Delta G f°$	$S°$	$C_p°$
t-Butyl radical	g	6.7			
N-Butylacetamide	liq	−91.02			
Butyl acetate	liq	−126.52			
t-Butylamine	liq	−35.97			
	g	−28.65	6.90	80.76	28.67
n-Butylbenzene	liq	−18.67[18°]	27.50		
	g	−3.30	34.58	105.04	41.85
sec-Butylbenzene	liq	−15.87			
t-Butylbenzene	liq	−16.90			
sec-Butyl butyrate	liq	−141.6			
n-Butyl chloroacetate	liq	−128.7			
n-Butyl 2-chlorobutyrate	liq	−156.6			
n-Butyl 3-chlorobutyrate	liq	−146.0			
n-Butyl 4-chlorobutyrate	liq	−147.7			
n-Butyl 2-chloropropionate	liq	−136.7			
n-Butyl 3-chloropropionate	liq	−133.4			
n-Butyl crotonate	liq	−111.8			
n-Butylcyclohexane	g	−50.95	13.49	109.58	49.50
n-Butylcyclopentane	g	−40.22	14.67	109.04	42.42
n-Butyl dichloroacetate	liq	−131.5			
n-Butyl ether	liq	−156.1			
	g	−87.2	114.96	48.82	
t-Butyl hydroperoxide	liq	−70.2			
n-Butyl lithium	liq	−31.6			
n-Butyl trichloroacetate	liq	−130.6			
1-Butyne (ethyl acetylene)	g	39.48	48.30	69.51	19.46
2-Butyne (dimethyl acetylene)	g	34.97	44.32	67.71	18.63
n-Butyraldehyde	g	−49.00	−27.43	82.44	24.52
n-Butyramide	c	−87.5			
n-Butyric acid	liq	−127.59	−90.27	54.1	42.1
n-Butyronitrile	g	8.14	25.97	77.78	23.19
Caffeine (methyl theobromine)	c	−76.2			
Capric acid (decanoic acid)	c	−170.59			
Caproic acid (hexanoic acid)	liq	−139.71			
ε-Caprolactam	c	−78.54	−22.72	40.3	
Caprylic acid (octanoic acid)	liq	−151.93			
Carbazole	c	30.3			
Carboxyl radical	g	−54			
CCH radical	g	114	105	49.6	8.87
Cellobiose	c	−532.5			
Chloroacetamide	c	−80.9			
Chloroacetic acid	c, l	−122.3			
ionized	aq	−119.81			
nonionized, std. state, $m = 1$	aq	−118.92			
Chloroacetyl chloride	liq	−68.0			
2-Chlorobenzaldehyde	liq	−28.4			
3-Chlorobenzaldehyde	liq	−30.2			
4-Chlorobenzaldehyde	c	−35.1			
Chlorobenzene	liq	2.58	21.32	50.0	35.9
2-Chlorobenzoic acid	c	−95.3			
3-Chlorobenzoic acid	c	−101.2			
4-Chlorobenzoic acid	c	−102.19			
Chlorobenzoquinone	c	−52.7			
1-Chlorobutane	g	−35.20	−9.27	85.58	25.71
2-Chlorobutane	g	−38.60	−12.78	85.94	25.93

Table 9-2 (*Continued*)
ORGANIC COMPOUNDS

Substance	State	$\Delta Hf°$	$\Delta Gf°$	$S°$	$C_p°$
2-Chlorobutyric acid	liq	−137.6			
3-Chlorobutyric acid	liq	−133.0			
4-Chlorobutyric acid	liq	−135.4			
Chlorocyclohexane	liq	−49.54			
2-Chloro-1,1-difluoroethylene	g	−79.2	−72.90	72.28	
Chlorodifluoromethane	g	−115.6	−108.1	67.12	13.35
Chloroethane (ethyl chloride)	g	−26.83	−14.46	65.91	14.97
Chloroethylene (vinyl chloride)	g	8.40	12.31	63.08	12.84
Chloroethyne	g	51	47	57.81	12.98
Chlorofluoromethane	g	−63.2	−57.11	63.16	11.24
Chloroform	liq	−31.6	−17.17	48.5	
	g	−24.60	−16.76	70.63	15.63
Chloroiodomethane	g	3.0	3.69	70.78	
Chloromethane (methyl chloride)	g	−19.59	−13.97	55.97	9.74
Chloromethyloxirane	liq	−35.48			
1-Chloro-2-methylpropane	g	−38.10	−11.87	84.56	25.93
2-Chloro-2-methylpropane	g	−43.80	−15.32	77.00	27.30
1-Chloronaphthalene	liq	13.0			
2-Chloronaphthalene	c	13.2			
1-Chloropentane	g	−41.80	−8.94	94.89	31.18
3-Chlorophenol	c	−49.4			
4-Chlorophenol	c	−47.3			
1-Chloropropan-2,3-diol	liq	−125.58			
2-Chloropropan-1,3-diol	liq	−123.71			
1-Chloropropane	g	−31.10	−12.11	76.27	20.23
2-Chloropropane	g	−35.00	−14.94	72.70	20.87
3-Chloro-1-propene (allyl chloride)	g	−0.15	10.42	73.29	18.01
2-Chloropropionic acid	liq	−125.0			
3-Chloropropionic acid	c	−131.4			
N-Chlorosuccinimide	c	−85.58			
Chlorotrifluoromethane	g	−169.20	−159.38	68.16	15.98
Chlorotrinitromethane	liq	−6.54			
Chrysene	c	34.7			
Cinchonamine	c	−10.4			
Cinchonidine	c	7.1			
Cinchonine	c	7.4			
Cinnamic acid *cis*	c	−72.0			
trans	c	−80.53			
Cinnamic anhydride	c	−83.1			
Citraconic acid	c	−197.04			
Citric acid	c	−369.0	−295.5	39.73	
Citric acid monohydrate	c	−439.4	−352.0	67.74	1.276
Codeine monohydrate	c	−151.2			
Coniine	liq	−57.6			
Creatine	c	−128.16	−63.32	45.3	
Creatine hydrate	c	−199.1			
Creatinine	c	−56.77	−6.97	40.10	
o-Cresol	g	−30.74	−8.86	85.47	31.15
m-Cresol	g	−31.63	−9.69	85.27	29.27
p-Cresol	g	−29.97	−7.38	83.09	29.75
m-Cresol acetate	liq	−89.41			
Crotonic acid *cis*	liq	−83			
trans	c	−102.9			
Crotononitrile *trans*	g	35.77	46.22	71.31	19.62
Cyanamide	c	14.05			

Table 9-2 (*Continued*)
ORGANIC COMPOUNDS

Substance	State	$\Delta Hf°$	$\Delta Gf°$	$S°$	$C_p°$
1-Cyanoguanidine	c	5.4	42.9	30.90	28.40
3-Cyanopyridine	c	46.23			
5-Cyanotetrazole	c	96.1			
4-Cyanothiazole	c	52.63			
Cyclobutane	g	6.37	26.30	63.43	17.26
Cyclobutene	g	31.00	41.76	62.98	16.03
Cyclododecane	c	−73.29			
Cycloheptane	liq	−37.47	12.92	57.97	29.42
Cycloheptanone	liq	−71.5			
1,3,5-Cycloheptatriene	liq	34.22	58.09	51.30	38.90
1,3-Cyclohexadien-5-yl radical	g	49.4			
Cyclohexane	liq	−37.34	6.37	48.84	37.4
	g	−29.43	7.59	71.28	25.40
Cyclohexane-1,2-dicarboxylic					
acid *cis*	c	−229.7			
trans	c	−232.0			
Cyclohexanethiol	g	−22.80			
Cyclohexanol	liq	−83.22	−31.87	47.7	
Cyclohexanone	g	−55.00	−21.69	77.00	26.21
Cyclohexene	liq	−9.28	24.28	51.67	34.9
	g	−1.28	25.54	74.27	25.10
Cyclohexen-3-yl radical	g	29			
1-Cyclohexenylmethanol	liq	−91.4			
Cyclohexyl radical	g	13			
Cyclooctane	liq	−40.58	18.60	62.62	
Cyclooctanone	liq	−77.9			
1,3,5,7-Cyclooctatetraene	liq	60.93	85.70	52.65	
Cyclopentadiene	g	32.00	42.86	64.00	
Cyclopentane	liq	−25.28	8.70	48.82	30.80
	g	−18.46	9.23	70.00	19.84
Cyclopentanediol-1,2 *cis*	c	−115.9			
trans	c	−117.1			
Cyclopentanethiol	g	−11.45	13.63	86.38	25.79
Cyclopentanol	liq	−71.74	−30.55	49.2	
Cyclopentanone	liq	−56.24			
	g	−46.03			
Cyclopentene	liq	1.02	25.93	48.10	29.24
	g	7.87	26.48	69.23	17.95
1-Cyclopentenylmethanol	liq	8.2			
Cyclopentyl-1-thiaethane	g	−15.41			
Cyclopropane	g	12.74	24.95	56.75	13.37
Cyclopropene	g	66.0	68.42	58.38	
Cyclopropyl radical	g	55			
L-Cysteine	c	−124.5			
L-Cystine	c	−245.7			
Decahydronaphthalene *cis*	liq	−52.45	16.47	63.34	55.45
trans	liq	−55.14	13.79	63.32	54.61
Decanal	g	−79.09	−15.90	138.28	57.29
n-Decane	liq	−71.95	−4.19	101.70	75.16
1,10-Decanediol	c	−165.74			
1-Decanethiol	liq	−66.07			
	g	−50.54	14.68	145.82	61.08
1-Decanoic acid	c	−170.59			
1-Decanol	liq	−114.6	−31.6	102.9	
	g	−96.0	−24.9	142.8	59.1

Table 9-2 (*Continued*)
ORGANIC COMPOUNDS

Substance	State	$\Delta Hf°$	$\Delta Gf°$	$S°$	$C_p°$
1-Decene	liq	−41.73	25.10	101.58	
1-Decyne	g	9.85	60.28	125.36	52.51
Deoxybenzoin	c	−16.96			
Desoxyamalic acid	c	−285.7			
Diacetamide	c	−117			
Diacetyl peroxide	liq	−127.9			
o-Diallyl phthalate	liq	−131.6			
Dialuric acid	c	−314.4			
2,6-Diaminopyridine	c	−1.56			
Diamylose	c	−850			
Diazomethane	g	46.0	52.06	58.02	12.55
Dibenzoylethane	c	−61.1			
Dibenzoylethylene	c	−27.4			
Dibenzoylmethane	c	−53.6			
Dibenzoyl peroxide	c	−100			
Dibenzyl	c	10.53	62.15	64.4	61.0
Dibenzyl ketone	c	−20.1			
Dibenzyl sulfide	c	23.74			
Dibenzyl sulfone	c	−42.1			
1,2-Dibromobutane	g	−23.70	−3.14	97.70	30.38
Dibromochlorofluoromethane	g	−55.4	−53.40	81.89	
Dibromochloromethane	g	−5.0	−4.50	78.31	
1,2-Dibromocycloheptane	liq	−37.67			
1,2-Dibromocyclohexane	liq	−38.8			
1,2-Dibromocyclooctane	liq	−41.41			
Dibromodichloromethane	g	−7.0	−4.67	83.23	
Dibromodifluoromethane	g	−102.7	−100.16	77.66	
1,2-Dibromoethane	liq	−19.4	−5.0	53.37	32.51
Dibromofluoromethane	g	−53.4	−52.84	75.70	
Dibromomethane	g	−3.53	−3.87	70.10	13.04
1,2-Dibromopropane	g	−17.40	−4.22	89.90	24.57
Di-n-butylborinic acid	liq	−146.3			
Di-n-butyl ether	g	−79.80	−21.16	119.60	48.76
Di-n-butyl mercury	liq	−23.4			
Di-t-butyl peroxide	liq	−91.0			
Di-n-butyl-o-phthalate	c	−201			
Di-n-butyl sulfate	liq	−216.1			
Di-n-butyl sulfite	liq	−165.6			
Di-n-butyl sulfone	c	−145.76			
Dichloroacetic acid	liq	−119.0			
ionized	aq	−122.4			
nonionized	aq	−120.4			
Dichloroacetylene	g	50	47	65	15.67
1,2-Dichlorobenzene	g	7.16	19.76	81.61	27.12
1,3-Dichlorobenzene	g	6.32	18.78	82.09	27.20
1,4-Dichlorobenzene	g	5.50	18.44	80.47	27.22
Dichlorodifluoromethane	g	−117.90	−108.51	71.91	17.31
1,1-Dichloroethane	liq	−38.3	−18.1	50.61	30.18
	g	−31.10	−17.52	72.91	18.25
1,2-Dichloroethane	liq	−39.49	−19.03	49.84	30.9
	g	−31.00	−17.65	73.66	18.80
1,1-Dichloroethylene	liq	−5.8	5.85	48.17	26.60
	g	0.30	5.78	68.85	16.02
1,2-Dichloroethylene *cis*	liq	−6.6	5.27	47.42	27
	g	0.45	5.82	69.20	15.55

Table 9-2 (*Continued*)
ORGANIC COMPOUNDS

Substance	State	$\Delta Hf°$	$\Delta Gf°$	$S°$	$C_p°$
1,2-Dichloroethylene *trans*	g	1.00	6.35	69.29	15.93
Dichlorofluoromethane	g	−68.10	−60.77	70.04	14.58
Dichloromethane	liq	−29.7	−16.83	42.7	
	g	−22.80	−16.46	64.61	12.16
1,2-Dichloropropane	g	−39.60	−19.86	84.80	23.47
1,3-Dichloropropane	g	−38.60	−19.74	87.76	23.81
2,2-Dichloropropane	g	−42.00	−20.21	77.92	25.30
Dicyanoacetylene	liq	119.6			
1,4-Dicyanobutyne-2	c	87.6			
Dicyclohexadiene	liq	6.3			
Dicyclopentadiene	c	27.9			
Dicyclopentyl	liq	−41.8			
2,2-Diethoxypropane	liq	−128.83			
Diethylamine	g	−17.30	17.23	84.18	27.66
Diethylbarbituric acid (veronal)	c	−178.7			
1,2-Diethylbenzene	g	−4.53	33.72	103.81	43.63
1,3-Diethylbenzene	g	−5.22	32.67	104.99	42.27
1,4-Diethylbenzene	g	−5.32	32.95	103.73	42.10
Diethylenediamine	c	−3.2	57.4	20.5	
Diethylene glycol	liq	−150.2			
	g	−136.5		105.4	32.3
Diethyl ether (ethyl ether)	liq	−65.30	−27.88	60.5	40.8
	g	−60.26	−29.24	81.90	26.89
Diethyl mercury	liq	7.1			
Diethylmethyl phosphonate	liq	−245.3			
Diethylnitramine	liq	−25.4			
Diethyl oxalate	liq	−192.51			
Diethyl peroxide	liq	−53.4			
Diethyl *o*-phthalate	liq	−186			
Diethyl selenide	liq	−23.0			
Diethyl sulfate	liq	−194.28			
Diethyl sulfite	liq	−143.50			
Diethyl sulfone	c	−123.13			
Diethyl sulfoxide	liq	−63.97			
Diethyl zinc	liq	4.0			
1,2-Difluorobenzene	liq	−79.04	−59.41	53.20	38.01
1,3-Difluorobenzene	g	−74.09	−61.43	76.57	25.40
1,4-Difluorobenzene	g	−73.43	−60.43	75.43	25.55
2,2′-Difluorobiphenyl	c	−70.73			
4,4-Difluorobiphenyl	c	−70.91			
2,2-Difluorochloroethylene	g	−75.4	−69.1	72.39	17.23
1,1-Difluoroethane	g	−119.70	−105.87	67.50	16.24
1,1-Difluoroethylene	g	−82.50	−76.84	63.38	14.14
Difluoromethane	g	−108.24	−101.66	58.94	10.25
Diglycylglycine	c	−230.8			
9,10-Dihydroanthracene	c	15.87			
1,2-Dihydronaphthalene	liq	18.0			
1,4-Dihydronaphthalene	liq	21.0			
Dihydropyran, 4H	liq	−37.5			
5,12-Dihydrotetracene	c	25.44			
2,3-Dihydrothiophene	liq	12.73			
2,5-Dihydrothiophene	liq	11.31			
1,2-Dihydroxybenzene	c	−86.3	−50.20	35.9	31.6
1,3-Dihydroxybenzene	c	−87.95	−50.00	35.3	31.3

Table 9-2 (*Continued*)
ORGANIC COMPOUNDS

Substance	State	$\Delta Hf°$	$\Delta Gf°$	$S°$	$C_p°$
1,2-Diiodobenzene	c	41.2			
1,3-Diiodobenzene	c	44.7			
1,4-Diiodobenzene	c	38.4			
1,2-Diiodoethane	g	15.90	18.76	83.30	19.67
Diiodomethane	g	28.30	24.24	73.95	13.80
Diisopropyl ether	liq	−83.94	−21.1	70.4	
	g	−76.20	−29.13	93.27	37.83
Diisopropyl ketone	g	−74.40			
Diisopropyl mercury	liq	−3.1			
1,2-Dimethoxybenzene	liq	−69.4			
Dimethoxyborane	liq	−144.5			
1,2-Dimethoxyethane	liq	−90.02			
2,2-Dimethoxypropane	liq	−108.92			
α,β-Dimethylacrylic acid *cis*	c	−117.3			
Dimethyl adipate	liq	−211.9			
Dimethylamine	g	−4.50	16.25	65.24	16.50
std. state, $m = 1$	aq	−16.88	13.85	31.8	
$(CH_3)_2NH_2^+$; std. state, $m = 1$	aq	−28.74	−0.80	41.2	
Dimethylaminotrimethylsilane	liq	−66.8			
N,N-Dimethylaniline	liq	8.2			
2,2-Dimethylbutane	g	−44.35	−2.30	85.62	33.91
2,3-Dimethylbutane	g	−42.49	−0.98	87.42	33.59
2,3-Dimethyl-1-butene	g	−13.32	18.89	87.39	34.29
2,3-Dimethyl-2-butene	g	−14.15	18.18	87.15	29.54
3,3-Dimethyl-1-butene	g	−10.31	23.46	82.16	30.23
2,3-Dimethyl-2-butenoic acid	c	−108.9			
Dimethyl cadmium	g	9.528		72.40	31.5
Dimethylchlorosilane	liq	−79.8			
1,1-Dimethylcyclohexane	liq	−52.31	6.34	63.87	
	g	−43.26	8.42	87.24	36.90
1,2-Dimethylcyclohexane *cis*	g	−41.15	9.85	89.51	37.40
trans	g	−43.02	8.24	88.65	38.00
1,3-Dimethylcyclohexane *cis*	g	−44.16	7.13	88.54	37.60
trans	g	−42.20	8.68	89.92	37.60
1,4-Dimethylcyclohexane *cis*	g	−42.22	9.07	88.54	37.60
trans	g	−44.12	7.58	87.19	37.70
1,1-Dimethylcyclopentane	g	−33.05	9.33	85.87	31.86
1,2-Dimethylcyclopentane *cis*	g	−30.96	10.93	87.51	32.06
trans	g	−32.67	9.17	87.67	32.14
1,3-Dimethylcyclopentane *cis*	g	−32.47	9.37	87.67	32.14
trans	g	−31.93	9.91	87.67	32.14
Dimethyldichlorosilane	g	−110.2		80.16	24.17
2,4-Dimethyl-1,3-dioxane *cis*	liq	−111.79			
4,5-Dimethyl-1,3-dioxane	liq	−108.32			
5,5-Dimethyl-1,3-dioxane	liq	−110.53			
4,4′-Dimethyldiphenylamine	c	−2.8			
Dimethyl ether	g	−43.99	−26.99	63.83	15.73
N,N-Dimethylformamide	liq	−57.2		28.5	37.45
Dimethylfulvene	liq	21.5			
Dimethyl fumarate	liq	−174.3			
Dimethyl glutarate	liq	−205.9			
Dimethylglyoxime	c	−42.51			
2,2-Dimethylhexane	liq	−62.63	0.71	79.33	
	g	−53.71	2.56	103.06	
2,3-Dimethylhexane	liq	−60.40	2.17	81.91	

Table 9-2 (*Continued*)
ORGANIC COMPOUNDS

Substance	State	$\Delta Hf°$	$\Delta Gf°$	$S°$	$C_p°$
2,3-Dimethylhexane (*Continued*)	g	−51.13	4.23	106.11	
2,4-Dimethylhexane	liq	−61.47	0.89	82.62	
	g	−52.44	2.80	106.51	
2,5-Dimethylhexane	liq	−62.26	0.60	80.96	
	g	−53.21	2.50	104.93	
3,3-Dimethylhexane	liq	−61.58	1.23	81.12	
	g	−52.61	3.17	104.70	
3,4-Dimethylhexane	liq	−60.23	2.03	82.97	
	g	−50.91	4.14	107.15	
2,2-Dimethyl-3-hexene *cis*	liq	−30.22			
trans	liq	−34.64			
5,5-Dimethyl hydantoin	c	−126.4			
1,1-Dimethylhydrazine	liq	11.8	49.4	47.32	39.21
1,2-Dimethylhydrazine	liq	13.3	50.8	47.60	40.88
Dimethyl ketone radical	g	−11			
Dimethyl maleate	liq	−168.2			
Dimethylmaleic anhydride	c	−139.0			
Dimethyl malonate	liq	−190.2			
Dimethyl mercury	liq	14.0			
Dimethylnitramine	c	−16.9			
Dimethyl oxalate	liq	−181.0			
2,2-Dimethylpentane	g	−49.27	0.02	93.90	39.67
2,3-Dimethylpentane	g	−47.62	0.16	98.96	39.67
2,4-Dimethylpentane	g	−48.28	0.74	94.80	39.67
3,3-Dimethylpentane	g	−48.17	0.63	95.53	39.67
2,7-Dimethylphenanthrene	c	8.70			
4,5-Dimethylphenanthrene	c	21.26			
9,10-Dimethylphenanthrene	c	11.4			
Dimethyl *m*-phthalate	c	−171			
Dimethyl *o*-phthalate	liq	−162			
Dimethyl *p*-phthalate	c	−170			
2,2-Dimethylpropane	g	−39.67	−0.364	73.23	29.07
2,3-Dimethylpyridine	liq	4.62			
2,4-Dimethylpyridine	liq	3.85			
2,5-Dimethylpyridine	liq	4.45			
2,6-Dimethylpyridine	liq	3.02			
3,4-Dimethylpyridine	liq	4.36			
3,5-Dimethylpyridine	liq	5.36			
Dimethyl succinate	liq	−199.6			
1,1-Dimethylsuccinic acid	c	−236.08			
1,2-Dimethylsuccinic acid *cis*	c	−233.6			
trans	c	−235.1			
Dimethyl sulfate	liq	−175.23			
Dimethyl sulfite	liq	−125.07			
Dimethyl sulfone	c	−107.8	−72.3	34.77	
Dimethyl sulfoxide	liq	−48.6	−23.7	45.0	35.2
3,3-Dimethyl-2-thiabutane	liq	−37.49			
2,2-Dimethylthiacyclopropane	liq	−5.78			
2,2-Dimethyl-3-thiapentane	liq	−44.7			
2,4-Dimethyl-3-thiapentane	g	−33.76	6.48	99.30	40.45
2,3-Dinitroaniline	c	−2.8			
2,4-Dinitroaniline	c	−16.3			
2,5-Dinitroaniline	c	−10.6			
2,6-Dinitroaniline	c	−12.1			
3,4-Dinitroaniline	c	−7.8			

Table 9-2 (*Continued*)
ORGANIC COMPOUNDS

Substance	State	$\Delta Hf°$	$\Delta Gf°$	$S°$	$C_p°$
3,5-Dinitroaniline	c	−9.3			
2,4-Dinitroanisole	c	−44.6			
2,6-Dinitroanisole	c	−45.2			
1,2-Dinitrobenzene	c	2.06	50.56	51.7	
1,3-Dinitrobenzene	c	−4.04	44.13	52.8	
2,4-Dinitrophenol	c	−55.6			
2,6-Dinitrophenol	c	−50.2			
2,4-Dinitroresorcinol	c	−99.3			
4,6-Dinitroresorcinol	c	−105.1			
2,4-Dinitrotoluene	c	−17.1			
2,6-Dinitrotoluene	c	−12.2			
1,4-Dioxane	liq	−84.47	−44.96	46.67	
	g	−75.30	−43.21	71.65	22.48
1,3-Dioxane	liq	−89.99			
1,4-Dioxatetralin	liq	−60.9			
Dioxindole	c	−76.9			
1,3-Dioxolan	g	−71.1			
Dipentene	liq	−12.1			
N,N-Diphenylacetamide	c	−10.3			
Diphenylamine	c	31.07			
1,4-Diphenylbutadiene *cis, cis*	c	47.51			
trans, trans	c	42.73			
Diphenylbutadiyne	c	123.91			
1,4-Diphenylbutane	c	−2.36			
1,4-Diphenyl-1,4-butanedione	c	−61.24	1.87	77.6	
1,4-Diphenyl-2-butene-1,4-dione	c	−27.55	26.64	76.3	
Diphenylcarbinol	c	−25.04			
Diphenyl carbonate	c	−95.93	−42.05	66.54	
Diphenyldichlorosilane	liq	−66.5			
Diphenyl disulfide	c	35.8			
Diphenyl disulfone	c	−153.59			
1,1-Diphenylethane	liq	11.7	58.58	80.28	
1,2-Diphenylethane	liq	12.31	63.87	64.6	
1,1-Diphenylethene	liq	41.21			
Diphenyl ether	liq	−3.48	34.47	69.62	
Diphenylethyne	c	74.66			
Diphenylfulvene	c	7.1			
Diphenyl mercury	c	66.8			
Diphenylmethane	liq	21.25	66.19	57.2	55.7
Diphenyl sulfide	liq	39.1			
Diphenyl sulfone	c	−53.71			
Diphenyl sulfoxide	c	2.40			
Di-n-propyl ether	g	−70.00	−25.23	100.98	37.83
Di-n-propyl mercury	liq	−5.0			
Di-n-propyl sulfate	liq	−205.22			
Di-n-propyl sulfite	liq	−154.52			
Di-n-propyl sulfone	liq	−130.94			
Di-n-propyl sulfoxide	liq	−78.65			
2,3-Dithiabutane	liq	−14.82	1.67	56.26	34.92
5,6-Dithiadecane	g	−37.86	12.87	136.91	55.23
3,4-Dithiahexane	liq	−28.69	2.28	72.90	
1,3-Dithian-2-thione	c	−3.1			
4,5-Dithiaoctane	liq	−40.95	4.56	89.28	
N,N-Dithiodiethylamine	liq	−29.1			
1,3-Dithiolan-2-thione	c	3.1			

Table 9-2 (*Continued*)
ORGANIC COMPOUNDS

Substance	State	$\Delta Hf°$	$\Delta Gf°$	$S°$	$C_p°$
Di-*p*-tolyl sulfone	c	−74.32			
Divinyl ether	g	−9.53			
Divinyl sulfone	liq	−49.5			
Dodecane	liq	−84.16	6.71	117.26	89.86
1-Dodecene	g	−39.52	32.96	147.78	64.43
1-Dodecyne	g	−0.01	64.22	143.98	63.44
Dulcitol	c	−321.9			
Eicosane	g	−108.93	28.04	223.26	110.73
Eicosanoic acid (arachidic acid)	c	−241.9			
1-Eicosene	g	−78.93	49.03	222.26	108.15
Ergosterol	c	−188.8			
Erythritol *meso*	c	−127.56	−152.12	39.9	
Ethane	g	−20.24	−7.84	54.76	12.54
1,2-Ethanedithiol	liq	−12.83			
Ethanethiol	g	−11.02	−1.12	70.77	17.37
Ethanol	liq	−66.20	−41.63	38.49	26.76
	g	−56.03	−40.13	67.54	15.64
Ethoxy radical	g	−6			
Ethyl radical	g	26.0	31	59.2	
Ethyl acetate	liq	−114.49	−79.52	62.0	
	g	−105.86	−78.25	86.70	27.16
Ethyl allyl sulfone	liq	−96.95			
Ethylamine	g	−11.00	8.91	68.08	17.36
N-Ethylaniline	liq	0.9	45.10	57.2	
Ethylbenzene	liq	−2.98	28.61	60.99	
	g	7.12	31.21	86.15	30.69
2-Ethyl-1-butene	g	−12.32	19.11	90.01	31.92
Ethyl carbamate (urethane)	c	−124.4			
Ethyl crotonate	liq	−100.4			
Ethylcyclohexane	liq	−50.72	6.95	67.14	
1-Ethylcyclohexene	liq	−25.50			
Ethylcyclopentane	liq	−39.08	8.92	67.00	
	g	−30.37	10.65	90.42	31.49
Ethyldiethylcarbamate	liq	−141.6			
Ethylene	g	12.50	16.31	52.39	10.24
Ethylene carbonate	c	−138.9			
Ethylone chlorohydrin	liq	−70.6			
1,2-Ethylenediamine	liq	−15.06		50	
	aq, 200	−13.32			
Ethylenediaminetetraacetic acid	c	−420.5			
Ethylenediammonium chloride	c	−122.7			
	aq, 5000	−115.92			
Ethylene glycol (1,2-ethanediol)	liq	−108.70	−77.25	39.9	35.8
	g	−93.05	−72.77	77.33	23.20
	aq, 1	−109.01			
Ethyleneimine (azirane)	g	29.50	42.54	59.90	12.55
Ethylene oxide	g	−12.58	−3.13	57.94	11.54
2-Ethyl-1-hexanal	liq	−83.30			
2-Ethyl-2-hexanal	liq	−62.44			
3-Ethylhexane	liq	−59.88	1.79	84.95	
Ethylidenecyclohexane	liq	−21.19			
Ethyl isovalerate	liq	−136.5			
Ethyl lithium	c	−14.0			
Ethylmercury bromide	c	−25.7			
Ethylmercury chloride	c	−33.7			

Table 9-2 (*Continued*)
ORGANIC COMPOUNDS

Substance	State	$\Delta Hf°$	$\Delta Gf°$	$S°$	$C_p°$
Ethylmercury iodide	c	−15.7			
Ethyl methyl ether	g	−51.73	−28.12	74.24	21.45
Ethyl nitrate	g	−36.80	−8.81	83.25	23.27
Ethyl nitrite	g	−24.9		24.74	23.71
3-Ethylpentane	g	−45.33	2.63	98.35	39.67
Ethyl pentanoate	liq	−132.2			
Ethyl peroxyl radical	g	(−2)			
2-Ethylphenol	c	−49.91			
3-Ethylphenol	c	−51.21			
4-Ethylphenol	c	−53.63			
Ethylphosphonic acid	c	−251.3			
Ethyl *n*-propanoate	liq	−122.16	−79.16		
2-Ethylpyridine	liq	−1.2			
Ethylsuccinic acid	c	−236.4			
Ethyl thiolacetate	liq	−64.01			
Ethyl *β*-vinylacrylate	liq	−80.8			
Ethyl vinyl ether	g	−33.63			
Ethynylbenzene (phenylacetylene)	g	78.22	86.46	76.88	27.46
Fluoranthene	c	45.75	82.60	55.09	
Fluoroacetamide	c	−118.7			
Fluoroacetic acid	c	−164.5			
Fluorobenzene	g	−27.86	−16.50	72.33	22.57
2-Fluorobenzoic acid	c	−135.67			
3-Fluorobenzoic acid	c	−139.13			
4-Fluorobenzoic acid	c	−140.00			
Fluoroethane	g	−62.90	−50.44	63.34	14.21
2-Fluoroethanol	liq	−111.3			
Fluoromethane	g	−56.80	−51.09	53.25	8.96
1-Fluoropropane	g	−67.20	−47.87	72.71	19.75
2-Fluoropropane	g	−69.00	−48.81	69.82	19.60
4-Fluorotoluene	liq	−44.80	−19.06	56.67	
Fluorotrinitromethane	liq	−52.8			
Formaldehyde	g	−27.70	−26.27	52.29	8.46
unhydrolyzed	aq	−35.9	−31.02		
Formamide	liq	−60.7			
	g	−44.5	−33.71	59.41	10.84
Formanilide	c	−36.2			
Formic acid	liq	−101.51	−86.38	30.82	23.67
	g	−90.49	−83.89	59.45	10.81
ionized; std. state, *m* = 1	aq	−101.71	−83.9	22	−21.0
nonionized; std. state, *m* = 1	aq	−101.68	−89.0	39	
dimer	g	−195.08			
Formyl HCO	g	10.4	6.76	53.66	8.27
HCO+	g	204	201	48.3	8.62
Formyl fluoride	g	−90	−88	59.0	9.66
N-Formyl-DL-leucine	c	−222.1			
Formyl urea	c	−118			
β-D-Fructose	c	−302.2			
D-Fucose	c	−262.7			
Fumaric acid	c	−193.84	−156.70	39.7	
Fumaronitrile	c	64.11			
Furan	g	−8.23	0.21	63.86	15.64
Furfural	liq	−47.8			
Furfuryl alcohol	liq	−66.05	−36.85	51.50	

Table 9-2 (*Continued*)
ORGANIC COMPOUNDS

Substance	State	$\Delta Hf°$	$\Delta Gf°$	$S°$	$C_p°$
2-Furoic acid (pyromucic acid)	c	−119.12			
Furylacrylic acid	c	−109.7			
Furylethylene	liq	−2.5			
D-Galactonic acid	c	−384.8			
D-Galactose	c	−304.1	−219.60	49.1	
D-Glucaric acid-1,4-lactone	c	−343.2			
D-Glucaric acid-3,6-lactone	c	−343.6			
D-Gluconic acid	c	−379.3			
D-Gluconic acid-δ-lactone	c	−300.3			
D-Glucose α	c	−304.26	−217.6	50.7	
β	c	−302.76			
D-Glutamic acid	c	−240.19	−173.87	45.7	
L-Glutamic acid	c	−241.32	−174.78	44.98	
L-Glutamine	c	−197.3			
Glutaric acid	c	−229.44			
Glyceraldehyde	liq	−143			
Glycerol	liq	−159.76	−114.01	48.87	35.9
Glyceryl-1-acetate	liq	−217.5			
Glyceryl-1-benzoate	c	−185.80			
Glyceryl-2-benzoate	c	−184.71			
Glyceryl-1-caprate	c	−265.05			
Glyceryl-2-caprate	c	−261.90			
Glyceryl-1,3-diacetate	liq	−268.2			
Glyceryl-1-laurate	c	−277.46			
Glyceryl-2-laurate	c	−275.48			
Glyceryl-1-myristate	c	−292.31			
Glyceryl-1-palmitate	c	−306.28			
Glyceryl-1-stearate	c	−319.64			
Glyceryl triacetate	liq	−318.3			
Glyceryl trilaurate	c	−489			
Glyceryl trimyristate	c	−520			
Glyceryl trinitrate	liq	−88.6			
Glycine	c	−126.22	−88.09	24.74	23.71
ionized; std. state, $m = 1$	aq	−112.28	−75.28	26.54	
nonionized; std. state, $m = 1$	aq	−122.85	−88.62	37.84	
$NH_3^+CH_2COOH$; std. state, $m = 1$	aq	−123.78	−91.82	45.46	
Glycol acetal	liq	91.1			
Glycolic acid	c	−158.6			
Glycylglycine	c	−178.51	−117.25	45.4	
Glycylphenylalanine	c	−163.9			
Glycylvaline	c	−200.0			
Glyoxal	g	−50.66			
Glyoxime	c	−21.63			
Glyoxylic acid	c	−199.7			
Guanidine	c	−13.39			
Guanidine carbonate	c	−232.10	−133.23	70.6	61.87
Guanidine nitrate	c	−92.5			
Guanidine sulfate	c	−288.0			
Guanine	c	−43.72	11.33	38.3	
Guanylurea nitrate	c	−102.1			
Heptadecane	g	−94.15	22.01	195.33	94.33
Heptadecanoic acid	c	−220.9			
1-Heptadecene	g	−64.15	43.00	194.33	91.76
1-Heptanal	g	−63.10	−20.71	110.34	40.89
n-Heptane	liq	−53.63	0.42	77.92	53.76

Table 9-2 (*Continued*)
ORGANIC COMPOUNDS

Substance	State	$\Delta Hf°$	$\Delta Gf°$	$S°$	$C_p°$
n-Heptane (*Continued*)	g	−44.88	1.91	102.27	39.67
1-Heptanethiol	g	−35.76	8.65	117.89	44.68
Heptanoic acid (enanthic acid)	liq	−145.75			
1-Heptanol	liq	−95.8	−34.0	76.5	66.5
	g	−79.3	−28.9	114.8	42.7
1-Heptene	liq	−23.41	21.22	78.31	50.62
	g	−14.89	22.90	101.24	37.10
1-Heptyne	g	24.62	54.18	97.44	36.11
Hexachlorobenzene	c	−31.30	0.25	62.20	48.11
	g	−8.10	10.56	105.45	41.40
Hexachloroethane	g	−33.20	−13.13	95.30	32.68
Hexadecafluoroethylcyclohexane	liq	−799.1			
Hexadecafluoroheptane	liq	−817.6	−739.24	134.28	
	g	−808.9	−737.87	158.88	
Hexadecane	g	−89.23	20.00	186.02	88.86
Hexadecanoic acid (palmitic acid)	c	−213.3	−75.54	108.12	
1-Hexadecanol (cetyl alcohol)	c, II	−163.4	−23.6	108.0	104.8
	liq	−151.86	−23.08	145.0	
1-Hexadecene	g	−59.23	40.99	185.02	86.29
Hexafluorobenzene	liq	−237.25	−211.43	66.90	52.96
	g	−228.64	−210.18	91.59	37.43
Hexafluoroethane	g	−320.90	−300.15	79.30	25.43
Hexahydroindane *cis*	g	−30.4			
trans	g	−31.4			
Hexamethylbenzene	c	−39.19	28.06	71.66	61.5
Hexamethyldisiloxane	liq	−194.7	−129.5	103.69	74.42
Hexamethylenetetramine	c	30.0	103.92	39.05	
Hexanal	g	−59.37	−23.93	101.07	35.43
Hexanamide	c	−101.48			
Hexane	liq	−47.52	−0.91	70.76	45.2
	g	−39.96	−0.06	92.83	34.20
1-Hexanethiol	g	−30.83	6.65	108.58	39.21
1-Hexanol	liq	−90.7	−36.4	69.2	56.6
	g	−75.9	−32.4	105.5	37.2
1-Hexene	liq	−17.30	19.93	70.55	43.81
	g	−9.96	20.90	91.93	31.63
2-Hexene *cis*	g	−12.51	18.22	92.37	30.04
trans	g	−12.27	18.27	90.97	31.64
3-Hexene *cis*	g	−11.38	19.84	90.73	29.55
trans	g	−13.01	18.55	89.59	31.75
1-Hexyne	g	29.55	52.24	88.13	30.65
Hippuric acid (benzoylglycine)	c	−145.63	−88.33	57.2	
Hydantoic acid	c	−179			
Hydantoin	c	−107.2			
Hydrazobenzene	c	52.9			
Hydroquinone	c	−87.08	−49.48	33.5	33.9
Hydrosorbic acid	llq	−110.2			
o-Hydroxybenzoic acid	c	−140.64	−100.7	42.6	38.03
m-Hydroxybenzoic acid	c	−139.8	−99.74	42.3	37.59
p-Hydroxybenzoic acid	c	−139.7	−99.55	42.0	37.08
β-Hydroxybutyric acid	liq	−162.3			
Hydroxyisobutyric acid	c	−177.9			
L-Hydroxyproline	c	−158.1			
8-Hydroxyquinoline	c	−19.9			
Hypoxanthene (6-oxypurine)	c	−26.47	18.39	34.8	

Table 9-2 (*Continued*)
ORGANIC COMPOUNDS

Substance	State	$\Delta Hf°$	$\Delta Gf°$	$S°$	$C_p°$
Imidazole	c	14.5			
Indane	liq	2.56	36.04	56.01	45.47
Indene	liq	26.39	52.00	51.19	44.68
Indole	c	29.8			
Iodobenzene	g	38.85	44.88	79.84	24.08
2-Iodobenzoic acid	c	−72.2			
3-Iodobenzoic acid	c	−75.7			
4-Iodobenzoic acid	c	−75.5			
Iodocyclohexane	liq	−23.5			
Iodoethane	liq	−9.6	3.5	50.6	27.5
	g	−2.00	5.10	70.82	15.76
Iodomethane	liq	−3.29	3.61	38.9	
	g	3.29	3.72	60.64	10.54
2-Iodo-2-methylpropane	g	−17.60	5.65	81.79	28.27
1-Iodonaphthalene	liq	38.6			
2-Iodonaphthalene	c	34.5			
2-Iodophenol	c	−22.9			
3-Iodophenol	c	−22.6			
4-Iodophenol	c	−22.8			
1-Iodopropane	g	−7.30	6.68	80.32	21.48
2-Iodopropane	g	−10.00	4.80	77.55	21.53
3-Iodopropene (allyl iodide)	liq	13.7			
3-Iodopropionic acid	c	−109.9			
2-Iodotoluene	liq	18.7			
3-Iodotoluene	liq	18.9			
4-Iodotoluene	liq	16.1			
Isatin	c	−62.7			
Isobutylbenzene	liq	−16.68			
Isobutyl dichloroacetate	liq	−132.4			
Isobutyl phenyl ketone	liq	−52.63			
Isobutyl trichloroacetate	liq	−132.4			
Isobutyronitrile	g	6.07	24.76	74.88	23.04
L-Isoleucine	c	−151.8	−82.97	49.71	45.00
Isopropenyl acetate	liq	−92.31			
Isopropyl radical	g	17.6			
Isopropyl acetate	liq	−124.01			
Isopropylbenzene (cumene)	liq	−9.85	29.70	66.87	
	g	0.94	32.74	92.87	36.26
Isopropyl nitrate	g	−45.65	−9.72	89.20	28.84
Isopropyl thiolacetate	liq	−71.26			
Isopropyl trichloroacetate	liq	−128.2			
Isoquinoline	c	37.9			
L-Isoserine	c	−177.8			
Isothiocyanic acid	g	30.50	26.98	59.28	11.09
Itaconic acid	c	−201.06			
Ketene	g	−14.60	−14.41	57.79	12.37
α-Ketoglutaric acid	c	−245.35			
D-Lactic acid	c	−165.88		34.3	
L-Lactic acid	c	−165.89	−124.98	34.00	
	liq	−161.2	−123.84	45.9	
β-Lactose	c	−534.1	−374.52	92.3	
Lauric acid (dodecanoic acid)	c	−185.14			
D-Leucine	c	−152.36	−82.97	49.71	
L-Leucine	c	−154.6	−82.76	50.62	48.03
DL-Leucine	c	−153.14	−83.54	49.5	

Table 9-2 (*Continued*)
ORGANIC COMPOUNDS

Substance	State	$\Delta Hf°$	$\Delta Gf°$	$S°$	$C_p°$
DL-Leucylglycine	c	−205.7	−112.14	67.2	
Leucylglycylglycine	c	−259.6			
Levulinic acid	c	−166.6			
Levulinic lactone	liq	−76.2			
Limonene (+)	liq	−13.0			
DL-Lysine	c	−162.2			
Maleic acid	c	−188.94	−149.40	38.1	32.36
Maleic anhydride	c	−112.08			
L-Malic acid	c	−263.78	−211.45		
DL-Malic acid	c	−264.27			
Malonamide	c	−130.5			
Malonic acid	c	−212.96			
Malonic diamide	c	−130.52			
Malononitrile	c	44.6			
Maltose	c	−530.8	−412.60		
L-Mandelic acid	c	−138.8			
D-Mannitol	c	−139.61	−225.20	57.0	
D-Mannose	c	−301.9			
Melamine (triaminotriazine)	c	−17.3	44.10	35.63	
Melezitose	c	−815			
2-Mercaptopropionic acid	liq	−111.9	−82.19	54.70	
Mesaconic acid	c	−197			
Mesoxalic acid	c	−290.7			
2,2-Metacyclophane	g	40.8			
Methane	g	−17.89	−12.15	44.52	8.54
Methanethiol (methyl mercaptan)	g	−5.49	−2.37	60.96	12.01
Methanol	liq	−57.13	−39.87	30.41	19.40
	g	−48.06	−38.82	57.29	10.49
std. state, $m = 1$	aq	−58.78			
L-Methionine	c	−180.4	−120.88	55.32	
Methoxyl radical	g	(2)			
2-Methoxybenzaldehyde	c	−63.7			
3-Methoxybenzaldehyde	liq	−66.0			
4-Methoxybenzaldehyde	liq	−63.9			
Methoxybenzene (anisole)	g	−17.3			
Methoxymethyl radical	g	(−4)			
2-Methoxytetrahydropyran	liq	−105.7			
5-Methoxytetrazole	c	16.6			
Methyl CH_3	g	34.82	35.35	46.38	9.25
Methyl acetate	liq	−106.4			
Methyl acrylate	g	−70.10	−56.78		
Methyl allantoin (pyvurile)	c	−177.0			
Methyl allyl sulfone	liq	−91.95			
Methylamido radical CH_3NH	g	35			
Methylamine	g	−5.50	7.71	57.98	11.97
std. state, $m = 1$	aq	−16.77	4.94	29.5	
Methylaminolithium	c	−22.92			
N-Methylaniline	liq	7.7			
Methyl benzoate	liq	−79.8			
Methyl benzyl sulfone	c	−88.65			
2-Methylbiphenyl	liq	25.8			
3-Methylbiphenyl	liq	20.4			
4-Methylbiphenyl	c	13.2			
2-Methyl-1,3-butadiene (isoprene)	g	18.10	34.86	75.44	25.00
3-Methyl-1,2-butadiene	g	31.00	47.47	76.40	25.20

Table 9-2 (*Continued*)
ORGANIC COMPOUNDS

Substance		State	$\Delta Hf°$	$\Delta Gf°$	$S°$	$C_p°$
2-Methylbutane		g	−36.92	−3.54	82.12	28.39
3-Methyl-1-butanethiol		g	−27.44			
2-Methyl-2-butanethiol		liq	−38.90	0.56	69.34	
		g	−30.36	2.20	92.48	34.30
2-Methyl-1-butanol		liq	−85.2			52.6
3-Methyl-1-butanol		liq	−85.2			50.3
2-Methyl-2-butanol		liq	−90.7	−41.9	54.8	59.2
		g	−78.8	−39.5	86.7	
3-Methyl-2-butanol		liq	−87.5			55.5
2-Methyl-1-butene		g	−8.68	15.68	81.15	26.28
3-Methyl-1-butene		g	−6.92	17.87	79.70	28.35
2-Methyl-2-butene		g	−10.17	14.26	80.92	25.10
Methyl *n*-butyl sulfone		liq	−128.00			
Methyl *t*-butyl sulfone		c	−132.8			
3-Methyl-1-butyne		g	32.60	49.12	76.23	25.02
Methyl caprate		liq	−153.07			
Methyl caproate		liq	−129.10			
N-Methylcaprolactam		liq	−73.3			
5-Methylcaprolactam		c	−86.9			
7-Methylcaprolactam		c	−86.5			
Methyl caprylate		liq	−141.07			
Methyl crotonate		liq	−91.5			
Methylcyclohexane		liq	−45.45	4.86	59.26	
		g	−36.99	6.52	82.06	32.27
2-Methylcyclohexanol	*cis*	liq	−93.3			
	trans	liq	−99.4			
3-Methylcyclohexanol	*cis*	liq	−99.5			
	trans	liq	−94.3			
4-Methylcyclohexanol	*cis*	liq	−98.8			
	trans	liq	−103.6			
2-Methylcyclohexanone		liq	−68.8			
Methylcyclopentane		g	−25.50	8.55	81.24	26.24
1-Methylcyclopentanol		liq	−82.3			
2-Methylcyclopentanone		liq	−63.4			
1-Methylcyclopentene		g	−1.30	24.41	78.00	24.10
3-Methylcyclopentene		g	2.07	27.48	79.00	23.90
4-Methylcyclopentene		g	3.53	29.06	78.60	23.90
Methyldichlorosilane		liq	−105.9			
2-Methyl-1,3-dioxane		liq	−104.60			
4-Methyl-1,3-dioxane		liq	−99.80			
N-Methyldiphenylamine		liq	28.8			
4-Methyldiphenylamine		c	11.7			
Methylene		g	92.35	88.25	46.32	8.27
2-Methylenecyclohexanol		liq	−66.3			
2-Methylenecyclopentanol		liq	11.2			
β-Methylene-*β*-propiolactone (diketene)		liq	−55.72			
Methylene sulfate		c	−164.6			
1-Methyl-2-ethylbenzene		g	0.29	31.33	95.42	37.74
1-Methyl-3-ethylbenzene		g	−0.46	30.22	96.60	36.38
1-Methyl-4-ethylbenzene		g	−0.78	30.28	95.34	36.22
2-Methyl-3-ethylpentane		liq	−59.69	3.03	81.41	
		g	−50.48	5.08	105.43	
3-Methyl-3-ethylpentane		liq	−60.46	2.69	79.97	
		g	−51.38	4.76	103.48	

Table 9-2 (*Continued*)
ORGANIC COMPOUNDS

Substance	State	$\Delta Hf°$	$\Delta Gf°$	$S°$	$C_p°$
2-Methyl-3-ethyl-1-pentene	g	−23.97			
Methyl ethyl sulfite	liq	−135.55			
Methyl ethyl sulfone	c	−116.17			
Methyl formate	liq	−90.60	−71.53	29	
	g	−83.70	−71.03	72.00	15.90
Methylglyoxal	g	−64.8			
Methylglyoxime	c	−30.3			
2-Methylheptane	liq	−60.98	0.92	84.16	
	g	−51.50	3.05	108.81	
3-Methylheptane	liq	−60.34	1.12	85.66	
	g	−50.82	3.28	110.32	
4-Methylheptane	liq	−60.17	1.87	83.72	
	g	−50.69	4.00	108.35	
Methyl heptanoate	liq	−135.54			
2-Methylhexane	liq	−54.93	−0.69	77.28	53.28
	g	−46.59	0.77	100.38	39.67
3-Methylhexane	liq	−54.35	−0.39	78.23	
	g	−45.96	1.10	101.37	39.67
Methyl hexanoate	liq	−129.11			
5-Methylhydantoin	c	−116.3			
Methylhydrazine	liq	12.9	43.0	39.66	32.25
	g	22.55	44.66	66.61	17.0
Methylidyne CH	g	142.00	134.02	43.72	6.97
CH+	g	388.8	380.1	41.00	6.97
α-Methylindole	c	14.5			
Methyl isocyanide	g	35.6	39.6	58.99	12.65
1-Methyl-2-isopropylbenzene (*o*-cymene)	liq	−18.19			
1-Methyl-3-isopropylbenzene	liq	−18.69			
1-Methyl-4-isopropylbenzene	liq	−18.7	28.65	73.28	
Methyl isopropyl ether	g	−60.24	−28.89	80.86	26.55
Methyl isopropyl ketone	g	−62.76			
Methyl isopropyl sulfone	liq	−120.44			
Methyl isothiocyanate CH_3NCS	g	31.3	34.5	69.29	15.65
3-Methylisoxazole	liq	−5.0			
5-Methylisoxazole	liq	−6.4			
Methyl laurate	liq	−165.66			
Methylmercury bromide	c	−20.6			
Methylmercury chloride	c	−27.8			
Methylmercury iodide	c	−10.4			
Methyl myristate	liq	−177.80			
1-Methylnaphthalene	liq	13.43	46.26	60.90	53.63
2-Methylnaphthalene	c	10.72	46.03	52.58	46.84
Methyl nitrate	liq	−38.0	−10.4	51.9	37.6
	g	−29.8	−9.4	76.1	
Methyl nitrite	g	−15.30	0.24	67.95	15.11
Methyl oleate	liq	−174.2			
Methyl pelargonate	liq	−147.29			
2-Methylpentane	g	−41.66	−1.20	90.95	34.46
3-Methylpentane	g	−41.02	−0.51	90.77	34.20
Methyl pentanoate	liq	−122.90			
2-Methyl-1-pentene	g	−12.49	18.55	91.34	32.41
3-Methyl-1-pentene	g	−10.76	20.66	90.06	34.04
4-Methyl-1-pentene	g	−10.54	21.52	87.89	30.23
2-Methyl-2-pentene	g	−14.28	17.02	90.45	30.26

Table 9-2 (*Continued*)
ORGANIC COMPOUNDS

Substance	State	$\Delta Hf°$	$\Delta Gf°$	$S°$	$C_p°$
3-Methyl-2-pentene *cis*	g	−13.80	17.50	90.45	30.26
trans	g	−14.02	17.04	91.26	30.26
4-Methyl-2-pentene *cis*	g	−12.03	19.63	89.23	31.92
trans	g	−12.99	19.03	88.02	33.80
Methyl phenyl sulfone	c	−82.49			
Methylphosphonic acid	c	−252			
2-Methylpropanal	g	−52.25			
2-Methylpropane	g	−32.15	−4.99	70.42	23.14
2-Methyl-1,2-propanediamine	liq	−32.00			
2-Methyl-1-propanethiol	g	−23.24	1.33	86.73	28.28
2-Methyl-2-propanethiol	g	−26.17	0.17	80.79	28.91
2-Methyl-1-propanol	g	−67.69	−39.99	85.81	26.6
2-Methyl-2-propanol	liq	−85.86	−44.14	46.10	52.61
	g	−74.67	−42.46	77.98	27.10
2-Methylpropene	g	−4.04	13.88	70.17	21.30
Methyl propyl ether	g	−56.82	−26.27	83.52	26.89
7-Methylpurine	c	51.3			
2-Methylpyridine (2-picoline)	liq	13.83	39.80	52.07	37.86
	g	24.05	42.32	77.68	23.90
3-Methylpyridine	liq	15.57	41.16	51.70	37.93
4-Methylpyridine	liq	13.58			
N-Methylpyrrolidone	liq	−62.64			
Methyl salicylate	liq	−127.1			
α-Methylstyrene	liq	16.8			
	g	27.00	49.84	91.70	34.70
β-Methylstyrene *cis*	g	29.00	51.84	91.70	34.70
trans	g	28.00	51.08	90.90	34.90
Methylsuccinic acid	c	−229.02			
3-Methyl-2-thiabutane	g	−21.61	3.21	85.87	28.00
2-Methylthiacyclopentane	g	−15.12			
2-Methyl-3-thiapentane	liq	−37.3			
4-Methylthiazole	liq	16.31			
2-Methylthiophene	liq	10.75	27.35	52.22	29.43
3-Methylthiophene	liq	10.38	27.00	52.19	29.38
4-Methyluracil	c	−109.2			
Methyl valerate	liq	−122.89			
Morphine monohydrate	c	−170.1			
Mucic acid	c	−423			
Murexide	c	−289.7			
Myrcene	liq	3.5			
Myristic acid (tetradecanoic acid)	c	−199.21			
Naphthalene	c	18.0	48.05	39.89	
	g	35.6	53.44	80.22	31.68
1-Naphthol	g	−5.1			
2-Naphthol	g	−10.1			
1,4-Naphthoquinone	c	−43.83			
1-Naphthyl acetate	c	−68.89			
2-Naphthyl acetate	c	−72.72			
1-Naphthylamine	c	16.2			
2-Naphthylamine	c	14.4			
Narceine dihydrate	c	−421.2			
Narcotine	c	−210.9			
Nicotine	liq	9.4			
Nitrilotriacetic acid	c		−312.5		
2-Nitroaniline	c	−3.45	42.60	42.1	39.3

Table 9-2 (*Continued*)
ORGANIC COMPOUNDS

Substance	State	$\Delta Hf°$	$\Delta Gf°$	$S°$	$C_p°$
3-Nitroaniline	c	−4.46	41.60	42.1	40.2
4-Nitroaniline	c	−9.91	36.10	42.1	40.4
Nitrobenzene	liq	3.80	34.95	53.6	44.4
2-Nitrobenzoic acid	c	−94.25	−46.95	49.8	
3-Nitrobenzoic acid	c	−100.25	−52.71	49.0	
4-Nitrobenzoic acid	c	−101.25	−53.07	50.2	43.3
3-Nitrobiphenyl	c	15.6			
4-Nitrobiphenyl	c	9.7			
1-Nitrobutane	g	−34.40	2.42	94.28	29.85
2-Nitrobutane	g	−39.10	−1.49	91.62	29.51
3-Nitro-2-butanol	liq	−93.2			
2-Nitrodiphenylamine	c	15.4			
Nitroethane	g	−24.4	−1.17	75.39	18.69
aci form	aq	−30.7			
nitro form	aq	−32			
2-Nitroethanol	liq	−83.8			
Nitroguanidine	c	−22.1			
Nitromethane	liq	−27.03	−3.47	41.05	25.33
	g	−17.86	−1.66	65.73	13.70
1-Nitronaphthalene	c	10.2			
1-Nitropropane	g	−30.00	0.08	85.00	24.41
2-Nitropropane	g	−33.21	−3.06	83.10	24.26
4-Nitrosodiphenylamine	c	50.9			
Nonadecane	g	−104.00	26.03	213.95	105.26
1-Nonadecene	g	−74.00	47.02	212.95	102.69
1-Nonanal	g	−74.16	−17.91	128.97	51.82
Nonane	liq	−65.84	2.81	94.09	
	g	−54.74	5.93	120.86	50.60
1-Nonanethiol	g	−45.61	12.67	136.51	55.61
Nonanoic acid	liq	−157.68			
1-Nonanol	liq	−109.2	−32.4	91.3	67.50
1-Nonene	g	−24.74	26.93	119.86	48.03
Octadecane	g	−99.08	24.02	204.64	99.80
Octadecanoic acid	c	−226.5			
1-Octadecene	g	−69.08	45.01	203.64	97.22
Octafluorocyclobutane	g	−365.20	−334.33	95.69	37.32
1-Octanal	g	−69.23	−19.91	119.66	46.36
Octanamide	c	−113.1			
Octane	liq	−59.74	1.77	85.50	45.14
	g	−49.82	3.92	111.55	45.14
1-Octanethiol	g	−40.68	10.67	127.20	50.14
Octanoic acid	liq	−151.93			
1-Octanol	liq	−101.6	−34.2	90.2	77.7
2-Octanone	liq	−91.9	−33.54	89.35	65.31
1-Octene	liq	−29.52	22.49	86.15	57.65
	g	−19.82	24.91	110.55	42.56
1-Octyne	g	10.70	56.26	106.75	41.58
Oleic acid	c	−187.2			
DL-Ornithine	c	−156.0			
Oxacyclobutane (trimethylene oxide)	g	−19.25	−2.33	65.46	
Oxalic acid	c	−197.7	−166.8	28.7	
std. state, $m = 1$	aq	−197.2	−161.1	10.9	
Oxalic acid dihydrate	c	−341.0			
Oxalyl chloride	liq	−85.6			
Oxamic acid	c	−160.4			

Table 9-2 (*Continued*)
ORGANIC COMPOUNDS

Substance	State	$\Delta Hf°$	$\Delta Gf°$	$S°$	$C_p°$
Oxamide	c	−123.0	−81.9	28.2	
Oxindole	c	−41.2			
8-Oxypurine	c	−15.4			
Palmitic acid	c	−213.10			
Papaverine	c	−120.2			
Parabanic acid	c	−138.0			
[1,8]-Paracyclophane	c	−19.6			
[2,2]-Paracyclophane	g	59.9			
[6,6]-Paracyclophane	c	−46.1			
Paraldehyde	liq	−164.2			
Pentachloroethane	g	−34.8	−16.79	91.17	28.22
Pentachlorofluoroethane	g	−75.8	−55.93	93.54	
Pentachlorophenol	c	−70.6	−34.44	60.21	48.27
Pentadecane	g	−84.31	17.98	176.71	83.40
1-Pentadecene	g	−54.31	38.97	175.71	80.82
1-Pentadecyne	g	−14.78	70.25	171.91	79.84
1,2-Pentadiene	g	34.80	50.29	79.70	25.20
1,3-Pentadiene *cis*	g	18.70	34.84	77.50	22.60
trans	g	18.60	35.07	76.40	24.70
1,4-Pentadiene	g	25.20	40.69	79.70	25.10
2,3-Pentadiene	g	33.10	49.21	77.60	24.20
Pentaerythritol	c	−220.0	−146.73	47.34	45.51
Pentaerythritol tetranitrate	c	−128.8			
Pentafluorobenzoic acid	c	−296.34			
Pentafluoroethane	g	−264.00	−246.00	79.76	22.88
Pentafluorophenol	c	−244.86			
Pentamethylbenzene	liq	−32.33	25.64	70.22	51.74
Pentamethylbenzoic acid	c	−128.13			
1-Pentanal	g	−54.45	−25.88	91.53	29.96
Pentanamide	c	−90.70			
Pentan-2,4-dione (acetylacetone)	liq	−101.33			
	g	−90.47		95.1	28.7
Pentan-1,5-dithiol	liq	−30.99			
Pentane	g	−35.00	−2.00	83.40	28.73
1-Pentanethiol	liq	−35.72	2.28	74.18	
Pentanoic acid	liq	−133.71	−89.10	62.10	50.48
1-Pentanol	liq	−85.0	−38.3	62.0	49.8
2-Pentanol	liq	−87.7			
3-Pentanol	liq	−88.5	−40.4	57.4	60.0
2-Pentanone	g	−61.82	−32.76	89.91	28.91
3-Pentanone	liq	−70.87			
1-Pentene	g	−5.00	18.91	82.65	26.19
2-Pentene *cis*	g	−6.71	17.17	82.76	24.32
trans	g	−7.59	16.71	81.36	25.92
2-Pentenoic acid	liq	−106.7			
3-Pentenoic acid	liq	−103.9			
4-Pentenoic acid	liq	−102.9			
1-Pentyne	g	34.50	50.25	78.82	25.50
2-Pentyne	g	30.80	46.41	79.30	23.59
Perfluoropiperidine	liq	−482.9	−422.67	94.02	70.93
Perylene	c	43.69			
α-Phellandrene	liq	−14.3			
Phenacetin	c	−101.1			
9,10-Phenanthraquinone	c	−55.18			
Phenanthrene	c	27.3	64.12	50.6	

Table 9-2 (*Continued*)
ORGANIC COMPOUNDS

Substance	State	$\Delta Hf°$	$\Delta Gf°$	$S°$	$C_p°$
Phenazine	c	56.4			
Phenol	c	−39.44	−12.05	34.42	32.2
	liq	−37.80	−11.02		30.46
	g	−23.03	−7.86	75.43	24.75
Phenoxy radical	g	10			
Phenoxyacetic acid	c	−122.8			
Phenyl radical	g	71			
Phenyl acetate	liq	−80.02			
Phenylacetic acid	c	−95.3			
β-Phenyl-1-alanine DL- and L-	c	−111.9	−50.6	51.06	48.52
Phenyl benzoate	c	−57.7			
2-Phenylbenzoic acid	c	−83.4			
Phenylboronic acid	c	−172.0			
1-Phenylcyclohexene	liq	−4.0			
Phenylcyclopropane	liq	24.7			
N-Phenyldiacetimide	c	−86.63			
p-Phenylenediamine	c	0.73			
Phenyl ethyl sulfide	liq	5.29			
DL-Phenylglyceric acid	c	−178.5			
N-Phenylglycine	c	−96.2			
α-Phenylglycine	c	−103.2			
Phenylglyoxime α	c	−4.9			
β	c	10.1			
Phenylglyoxylic acid	c	−115.3			
Phenylhydrazine	liq	34.03			
Phenyl methyl sulfide	liq	11.5			
N-Phenyl-2-naphthylamine	c	38.2			
N-Phenylpyrrole	c	38.1			
2-Phenylpyrrole	c	34.5			
Phenyl salicylate	c	−104.3			
Phenyl thiolacetate	liq	−29.16			
Phosgene	g	−52.80	−49.42	67.82	13.79
Phthalamide	c	−104.4			
m-Phthalic acid	c	−191.91			
o-Phthalic acid	c	−186.91	−141.39	49.7	45.0
p-Phthalic acid	c	−195.05			
Phthalic anhydride	c	−110.1	−79.12	42.9	38.5
Phthalonitrile	c	65.82			
Pimelic acid	c	−241.25			
Pinene α	liq	−3.9			
β	liq	−1.8			
Piperazine	c	−10.90			
Piperidine	liq	−21.05			
α-Piperidone	c	−73.3	−26.79	39.4	
DL-Proline	c	−125.7			
Propadiene	g	45.92	48.37	58.30	14.10
Propane	g	−24.82	−5.63	64.58	17.59
1,2-Propanediamine	liq	−23.38			
1,2-Propanediol	liq	−119.6			
1,3-Propanediol	liq	−124.4			
1,3-Propanedithiol	liq	−18.83			
2,3-Propanedithiol	liq	−18.82			
1-Propanethiol	g	−16.22	0.52	80.40	22.65
2-Propanethiol	g	−18.22	−0.61	77.51	22.94
1-Propanol	liq	−72.66	−40.78	46.5	33.7

Table 9-2 (*Continued*)
ORGANIC COMPOUNDS

Substance	State	$\Delta Hf°$	$\Delta Gf°$	$S°$	$C_p°$
1-Propanol (*Continued*)	g	−61.28	−38.67	77.61	20.82
2-Propanol	liq	−75.97	−43.09	43.16	36.06
	g	−65.11	−41.44	74.07	21.21
1,2,3-Propenetricarboxylic acid					
cis	c	−292.7			
trans	c	−294.7			
2-Propen-1-ol (allyl alcohol)	g	−31.55	−17.03	73.51	18.17
Propionaldehyde	g	−45.90	−31.18	72.83	18.80
Propionamide	c	−81.7			
Propionic acid	liq	−122.07	−91.65		
Propionic anhydride	liq	−161.53	−113.66		
Propionitrile	liq	3.5	21.31	45.25	
	g	12.10	22.98	68.50	17.46
1-Propylamine	g	−17.30	9.51	77.48	22.89
2-Propylamine	liq	−26.83			
n-Propylbenzene	g	1.87	32.80	95.76	36.41
n-Propylcarbamate	c	−132.07			
n-Propyl chloroacetate	liq	−123.3			
n-Propylcyclohexane	g	−46.20	11.31	100.27	44.03
n-Propylcyclopentane	g	−35.39	12.57	99.73	36.96
Propylene (propene)	g	4.88	15.02	63.72	15.37
Propylene oxide	g	−22.17	−6.16	68.53	17.29
n-Propyl nitrate	g	−41.60	−6.53	92.10	28.99
n-Propyl phenyl ketone	liq	−45.14			
n-Propyl thiolacetate	liq	−70.29			
n-Propyl trichloroacetate	liq	−122.7			
Propyne (methyl acetylene)	g	44.32	46.47	59.30	14.50
Pyrazine	c	33.41			
Pyrazole	c	28.3			
Pyrene	c	27.44	64.40	53.75	56.4
Pyridazine	liq	53.74			
Pyridine	liq	23.96	43.34	42.52	31.72
	g	33.61	45.46	67.59	18.67
Pyrimidine	liq	35.04			
Pyrrole	liq	15.08			
Pyrrole-2-aldehyde	c	−24.8			
Pyrrole-2-aldoxime	c	2.9			
Pyrrolidine	liq	−9.84	25.94	48.76	
	g	−0.86	27.41	73.97	19.39
2-Pyrrolidone	c	−68.3			
Pyruvic acid	liq	−139.7	−110.75	42.9	
Quinaldine	c	39.3			
Quinhydrone	c	−19.79	−77.19	77.9	66.2
Quinidine	c	−38.3			
Quinine	c	−37.1			
Quinoline	liq	37.33	65.90	51.9	
p-Quinone	c	−44.10	−20.0	38.9	
Raffinose	c	−761			
L-Rhamnose	c	−256.5			
Rhamnose triacetate	c	−455.4			
D-Ribose	c	−251.16			
Saccharinic acid lactone	c	−249.6			
Salicylaldehyde	liq	−66.9			
Salicylaldoxime	c	−43.91			
Salicylic acid	c	−140.9	−99.93	42.6	

Table 9-2 (*Continued*)
ORGANIC COMPOUNDS

Substance	State	$\Delta Hf°$	$\Delta Gf°$	$S°$	$C_p°$
Sarcosine	c	−121.2			
Sebacic acid	c	−258.8			
Semicarbazide std. state, $m = 1$	aq	−39.9	−9.7	71.2	
L-Serine	c	−173.6			
Serylserine	c	−281.8			
Sorbic acid	c	−93.4			
L-Sorbose	c	−303.68	−217.10	52.8	
5,5′-Spiro-bis(1,3-dioxane)	c	−167.8			
Spiropentane	g	44.27	63.41	67.45	21.06
Stearic acid	c	−226.5			
Stilbene cis	liq	43.81			
trans	c	32.27	75.90	60.0	
Strychnine	c	−41.0			
Styrene	liq	24.83	48.37	56.78	43.64
	g	35.22	51.10	82.48	29.18
Suberic acid	c	−248.1			
Succinamide	c	−138.9			
Succinic acid	c	−224.79	−178.64	42.0	35.8
Sucrose	c	−531.9	−369.18	86.1	
L-Tartaric acid	c	−306.5			
DL-Tartaric acid	c	−308.5			
Tartaric acid meso	c	−305.9			
Tetrabromomethane	g	19.00	15.61	85.53	21.78
Tetracene	c	37.95			
Tetrachlorobenzoquinone	c	−69.0			
1,1,1,2-Tetrachlorodifluoroethane	g	−117.1	−97.3	91.5	29.5
1,1,1,2-Tetrachloroethane	g	−35.7	−19.2	85.05	24.67
1,1,2,2-Tetrachloroethane	liq	−47.0	−22.7	59.0	39.6
	g	−36.50	−20.45	86.69	24.09
Tetrachloroethylene	g	−3.40	4.90	81.46	22.69
Tetrachloromethane	liq	−31.75	−14.97	51.67	
	g	−22.90	−12.80	74.07	19.94
1,1,2,2-Tetracyanocyclopropane	c	141			
Tetracyanoethylene	c	149.1			
Tetradecane	g	−79.38	15.97	167.40	77.93
Tetradecanoic acid	c	−199.2			
1-Tetradecene	g	−49.36	36.99	166.40	75.36
Tetraethylene glycol	liq	−234.6			
Tetraethyl lead	liq	12.7	80.4	112.92	
	g	26.3			
1,1,1,2-Tetrafluoroethane	g	−214.10	−197.46	75.58	20.62
Tetrafluoroethylene	g	−157.40	−149.07	71.69	19.24
Tetrafluoromethane	g	−223.0	−212.3	62.45	14.59
Tetrahydrofuran	liq	−51.67			
Tetrahydrofurfuryl alcohol	liq	−104.1			
1,2,3,4-Tetrahydronaphthalene	liq	−6.1			
Tetrahydropyran	liq	−61.1			
1,2,5,6-Tetrahydropyridine	liq	8.0			
Tetraiodomethane	g	62.84	51.89	93.60	22.91
1,2,3,4-Tetramethylbenzene	liq	−23.0	25.49	69.45	
1,2,3,5-Tetramethylbenzene	liq	−23.54	23.58	99.55	57.5
1,2,4,5-Tetramethylbenzene	liq	−29.48	24.20	71.83	51.6
2,2,3,3-Tetramethylbutane	g	−53.99	5.26	93.06	
Tetramethyl lead	liq	23.5	62.8	76.5	
	g	32.6	64.7	100.5	34.42

Table 9-2 (*Continued*)
ORGANIC COMPOUNDS

Substance	State	$\Delta Hf°$	$\Delta Gf°$	$S°$	$C_p°$
Tetramethylsilane	g	−68.50	−35.46	86.30	31.12
Tetramethylsuccinic acid	c	−242.0			
Tetramethylthiacyclopropane	c	−19.84			
Tetranitromethane	liq	8.9			
1,1,1,2-Tetraphenylethane	c	53.31			
1,1,2,2-Tetraphenylethane	c	51.63			
Tetraphenylethene	c	74.46			
Tetraphenylhydrazine	c	109.4			
Tetraphenylmethane	c	59.1	137.20		
Tetrazole	c	56.7			
Thebaine	c	−63.0			
Theobromine	c	−86.4			
Thiaadamantane	c	−34.22			
2-Thiabutane	liq	−21.89	1.79	57.14	
	g	−14.25	2.73	79.62	22.73
Thiacyclobutane	g	14.61	25.69	68.17	16.57
Thiacycloheptane	g	−14.66	20.09	86.50	29.78
Thiacyclohexane	liq	−25.32	9.96	52.16	
	g	−15.12	12.68	77.26	25.86
Thiacyclopentane	liq	−17.39	8.97	49.67	
	g	−8.08	11.00	73.94	21.72
Thiacyclopropane	liq	12.41	22.52	38.84	
	g	19.65	23.16	61.01	12.83
4-Thia-5,5-dimethylhex-1-ene	liq	−21.68			
2-Thiaheptane	g	−29.34	8.39	107.73	39.10
3-Thiaheptane	g	−29.92	7.65	108.27	38.71
4-Thiaheptane	liq	−40.62	5.12	80.85	
	g	−29.96	7.94	107.16	38.53
2-Thiahexane	liq	−34.15	4.08	73.49	
	g	−24.42	6.37	98.43	33.64
3-Thiahexane	liq	−34.58	3.50	73.98	
	g	−25.00	5.63	98.97	33.25
5-Thianonane	liq	−52.74	7.66	96.82	
	g	−39.99	11.76	125.76	49.46
2-Thiapentane	liq	−28.21	2.79	65.14	
	g	−19.54	4.40	88.84	28.05
3-Thiapentane	liq	−28.43	2.81	64.36	40.97
	g	−19.95	4.25	87.96	27.97
2-Thiapropane	g	−8.97	1.66	68.32	17.71
6-Thiaundecane	liq	−63.61			
Thioacetic acid	g	−43.49	−36.81	74.86	19.33
Thiohydantoic acid	c	−132.6			
Thiohydantoin	c	−59.5			
Thiolacetic acid	liq	−52.39			
β-Thiolactic acid	liq	−111.6			
Thiophene	liq	19.24	28.97	43.30	
	g	27.66	30.30	66.65	17.42
Thiosemicarbazide	c	6.0			
Thiourea	c	−21.13	5.2	27.7	
	aq, 100	−15.6			
L-Threonine (also DL-)	c	−181.4			
Thymine	c	−111.9			
Thymol	c	−74.0			
Tiglic acid	c	−117.3			
Toluene	liq	2.87	27.19	52.81	37.58

Table 9-2 (*Continued*)
ORGANIC COMPOUNDS

Substance	State	$\Delta Hf°$	$\Delta Gf°$	$S°$	$C_p°$
Toluene (*Continued*)	g	11.95	29.16	76.64	24.77
2-Toluenethiol	liq	10.57			
m-Toluic acid	c	−101.85			
o-Toluic acid	c	−99.55			
p-Toluic acid	c	−102.59			
o-Toluic anhydride	c	−127.5			
p-Toluic anhydride	c	−124.5			
Trehalose	c	−531.3			
2,4,6-Triamino-1,3,5-triazine	g	−17.13	42.33	74.10	20.93
2-Triazoethanol	liq	22.6			
Tribenzylamine	c	33.6			
Tribromochloromethane	g	3.0	2.17	85.36	
Tribromofluoromethane	g	−45.4	−46.14	82.65	
Tribromomethane	g	4.00	1.78	79.01	16.96
Tri-*n*-butylamine	liq	−67.32			
Tri-*n*-butyl borate	liq	−286.7			
Tri-*n*-butylboron	liq	−83.4			
Tri-*n*-butyl phosphate	liq	−348			
Tri-*n*-butyl phosphine oxide	c	−110			
Trichloroacetaldehyde	liq	−56.1			
Trichloroacetamide	c	−85.6			
Trichloroacetic acid	c	−120.7			
ionized	aq	−123.4			
Trichloroacetyl chloride	liq	−66.4			
Trichlorobenzoquinone	c	−64.5			
1,1,1-Trichloroethane	g	−34.01	−18.21	76.49	22.07
1,1,2-Trichloroethane	g	−33.10	−18.52	80.57	21.47
Trichloroethylene	g	−1.40	4.75	77.63	19.17
Trichlorofluoromethane	g	−68.10	−58.68	74.06	18.66
Trichloromethyl	g	19	22	70.9	15.21
1,2,3-Trichloropropane	g	−44.40	−23.37	91.52	26.82
1,1,1-Tricyanoethane	c	83.9			
Tricyanoethylene	c	105.0			
Tridecane	g	−74.45	13.97	158.09	72.47
Tridecanoic acid	c	−192.8			
1-Tridecene	g	−44.45	34.96	157.09	69.89
Triethyl aluminum	liq	−56.6			
Triethylamine	g	−23.80	26.36	96.90	38.46
Triethylaminoborane	liq	−47.47			
Triethyl arsenite	liq	−168.9			
Triethylarsine	liq	3.1			
Triethyl borate	liq	−250.4			
Triethylenediamine	c	−3.4	57.28	37.67	
Triethylene glycol	liq	−192.2			
Triethyl phosphate	liq	−297			
Triethylphosphine	liq	−21.3			
Triethyl phosphite	liq	−205.9			
Triethylstibine	liq	1.2			
Triethylsuccinic acid	c	−254.9			
Triethyl thionophosphate	liq	−232.5			
Trifluoroacetic acid	liq	−255.4			
Trifluoroacetonitrile	g	−118.4	−110.4	71.3	18.70
1,1,1-Trifluoroethane	g	−178.20	−162.11	68.67	18.76
2,2,2-Trifluoroethanol	liq	−207.4			
Trifluoroethylene	g	−118.50	−112.22	69.94	16.54

Table 9-2 (*Continued*)
ORGANIC COMPOUNDS

Substance	State	$\Delta Hf°$	$\Delta Gf°$	$S°$	$C_p°$
Trifluoroiodomethane	g	−141.0	−136.70	73.50	
Trifluoromethane	g	−165.71	−157.48	62.04	12.22
Trifluoromethyl CF_3	g	−112.4	−109.2	63.3	11.90
CF_3^+	g	100.6	103.1	60.8	11.87
Trifluoromethylbenzene	liq	−152.40	−123.98	64.89	
	g	−143.42	−122.20	89.05	31.17
Trifluoromethylhypofluorite					
CF_3OF	g	−183	−169	77.06	18.97
DL-Trihydroxyglutaric acid	c	−356			
Triiodomethane	g	50.40	42.54	84.97	17.94
Trimethylacetic acid	liq	−134.9			
Trimethylacetic anhydride	liq	−186.4			
2,4,5-Trimethylacetophenone	liq	−60.3			
2,4,6-Trimethylacetophenone	liq	−63.9			
Trimethyl aluminum	liq	−36.1		50.05	37.19
Trimethylamine	g	−5.70	23.64	69.02	21.93
std. state, $m = 1$	aq	−18.17	22.22	31.9	
Trimethylamine aluminum					
chloride adduct	c	−210.1			
Trimethylammonium ion					
std. state, $m = 1$	aq	−26.99	8.90	47.0	
Trimethyl arsenite	liq	−141.2			
Trimethylarsine	liq	−3.9			
1,2,3-Trimethylbenzene	liq	−14.01	25.68	66.40	
1,2,4-Trimethylbenzene	liq	−14.79	24.46	67.93	
1,3,5-Trimethylbenzene	liq	−15.18	24.83	65.38	
Trimethyl borate	liq	−222.9			
Trimethylboron	liq	−34.1			
2,2,3-Trimethylbutane	g	−48.95	1.02	91.61	39.33
Trimethylchlorosilane	liq	−91.8			
1,3,5-Trimethylcyclohexane					
cis, cis	g	−51.48	8.10	93.30	42.93
2,2,3-Trimethylpentane	liq	−61.44	2.21	78.30	
	g	−52.61	4.09	101.62	
2,2,4-Trimethylpentane	liq	−61.97	1.65	78.40	
	g	−53.57	3.27	101.15	
2,3,3-Trimethylpentane	liq	−60.63	2.54	79.93	
	g	−51.73	4.52	103.14	
2,3,4-Trimethylpentane	liq	−60.98	2.55	78.71	
	g	−51.97	4.52	102.31	
2,4,4-Trimethyl-1-pentene	liq	−35.21	20.66	73.2	
2,4,4-Trimethyl-2-pentene	liq	−34.44	21.04	74.5	
Trimethylphosphine	liq	−29.2			
Trimethylphosphine-N-ethylimine	liq	−35.8			
Trimethylphosphine oxide	c	−114.2			
Trimethyl phosphite	liq	−177.1			
Trimethylsilanol	liq	−130.3			
Trimethylstibine	liq	0.2			
Trimethylsuccinic acid	c	−239.2			
Trimethylsuccinic anhydride	c	−164.5			
Trimethylthiacyclopropane	liq	−14.47			
Trimethylurea	c	−79.0			
2,4,6-Trinitroanisole	c	−37.6			
1,3,5-Trinitrobenzene	c	−10.40			
Trinitromethane	c	−11.50			
1,4,5-Trinitronaphthalene	c	8.7			

Table 9-2 (*Continued*)
ORGANIC COMPOUNDS

Substance	State	$\Delta Hf°$	$\Delta Gf°$	$S°$	$C_p°$
1,3,8-Trinitronaphthalene	c	5.8			
2,4,6-Trinitrophenetole	c	−48.9			
2,4,6-Trinitrophenol	c	−51.23			
2,4,6-Trinitrophenylhydrazine	c	8.8			
2,4,6-Trinitrotoluene	c	−16.0			
2,4,6-Trinitro-*m*-xylene	c	−24.5			
Triphenylamine	c	58.70[18°]	120.50		
Triphenylarsine	c	74.1			
Triphenylcarbinol	c	−0.80	65.2	78.7	
Triphenylene	c	33.72	78.68	60.87	
1,1,1-Triphenylethane	c	37.56			
1,1,2-Triphenylethane	c	31.11			
Triphenylethylene	c	55.8	123.00		
Triphenylmethane	c	38.71	98.60	74.6	70.5
Triphenyl phosphate	c	−181			
Triphenylphosphine	c	55.5			
Triphenylphosphine oxide	c	−14.4			
Tri-*n*-propylamine	liq	−49.51			
Tris(acetylacetonato)-chromium	c	−366.4			
1,1,1-Tris(hydroxymethyl)ethane	c	−177.96			
Tropolone	c	−57.18			
L-Tryptophan	c	−99.8	−28.54	60.00	56.92
L-Tyrosine	c	−163.4	−92.18	51.15	51.73
Undecane	liq	−78.05	5.44	109.49	
	g	−64.60	9.94	139.48	61.53
1-Undecene	g	−34.60	30.94	138.48	58.96
Urea	c	−79.71	−47.04	25.00	22.26
std. state, *m* = 1	aq	−75.95			
Urea nitrate	c	−134.8			
Urea oxalate	c	−365.3			
Uric acid	c	−147.73	−85.75	41.4	
L-Valine (also DL-)	c	−148.2	−85.80	42.75	40.35
Valylphenylalanine	c	−183.5			
Vinyl radical	g	63			
Vinyl bromide	g	18.7	19.3	65.90	13.27
Vinyl chloride	g	8.5	12.4	63.07	12.84
Vinylcyclohexane	liq	−21.19			
Vinylcyclopropane	liq	29.3			
2-Vinylpyridine	liq	37.2			
Xanthine	c	−90.49	−39.64	38.5	
o-Xylene	liq	−5.84	26.37	58.91	44.9
	g	4.54	29.18	84.31	31.85
m-Xylene	liq	−6.08	25.73	60.27	43.8
	g	4.12	28.41	85.49	30.49
p-Xylene	liq	−5.84	26.31	59.12	
	g	4.29	28.95	84.23	30.32
2,3-Xylenol	g	−37.57			
2,4-Xylenol	g	−38.93			
2,5-Xylenol	g	−38.63			
2,6-Xylenol	g	−38.66			
3,4-Xylenol	g	−37.42			
3,5-Xylenol	g	−38.61			
Xylitol	c	−267.32			
D-Xylose	c	−252.8			

SOME THERMODYNAMIC RELATIONS

Enthalpy

The heat attending the formation of a chemical compound by direct (or hypothetical) union of its elements at constant pressure is termed the enthalpy of formation ΔHf.

In a reaction at constant temperature and pressure, the enthalpy

$$\Delta H = H_{\text{products}} - H_{\text{reactants}} \tag{9-1}$$

is equal to the heat absorbed from the surroundings.

Enthalpy, as defined by the equation

$$H = E + PV \tag{9-2}$$

is particularly convenient for dealing with constant-pressure processes. For any infinitesimal change in the system

$$dH = dE + P\,dV + V\,dP \tag{9-3}$$

For a constant-pressure process, $dP = 0$, and

$$dH = dE + P\,dV = dq_{\text{rev}} \tag{9-4}$$

where q_{rev} is the reversible heat of the reaction.

For solids and liquids, changes in enthalpy are usually almost equal to the changes in internal energy, dE, because for these states the volumes are relatively small and changes in the PV term are often negligible.

Enthalpy of Formation

Since methods are available for obtaining only enthalpy differences, it is customary to choose some state for each element and to calculate the enthalpy of that element in terms of that state (see page 9-1). Once standard enthalpies are assigned to the elements, it is possible to determine standard enthalpies for compounds. These enthalpies are called standard enthalpies of formation and designated $\Delta Hf°$. For the reaction

$$\text{C(graphite)} + O_2(g) \rightarrow CO_2(g) \qquad \Delta H^\circ_{298} = -94.052 \text{ kcal}$$

Since the elements are in their standard states, the enthalpy change for the reaction is equal to the standard enthalpy of CO_2 less the standard enthalpies of C and O_2, which are zero in each instance. Thus,

$$\Delta Hf° = -94.052 - 0 - 0 = -94.052 \text{ kcal}$$

Tables of enthalpies, such as Tables 9-1 and 9-2, can be used to determine the enthalpy for any reaction at 1 atm and 298 K involving the elements and any of the compounds appearing in the tables.

Enthalpies of Formation of Ions

The solution of 1 mole of HCl gas in a large amount of water (infinitely dilute real solution) is represented by

$$HCl(g) + \inf H_2O \rightarrow H^+(aq) + Cl^-(aq)$$

The heat evolved in the reaction is $\Delta H° = -17.892$ kcal. With the value of $\Delta Hf°$ for HCl from Table 9-1, one has for the reaction

$$\Delta H° = -17.960 = \Delta Hf°[H^+(aq)] + \Delta Hf°[Cl^-(aq)]$$
$$-\Delta Hf°[HCl(g)]$$
$$= -17.960 + (-22.060) = -39.952 \text{ kcal}$$

for the standard enthalpy of formation of the pair of ions H^+ and Cl^- in aqueous solution (standard state, $m = 1$).

To obtain the $\Delta Hf°$ values for individual ions, the enthalpy of formation of $H^+(aq)$ is arbitrarily assigned the value zero at 25°C. Thus, $\Delta Hf°[Cl^-(aq)] = -39.952$ kcal. With similar data from Tables 9-1 and 9-2, one can calculate enthalpies of formation of other ions. Thus, for the $\Delta Hf°[KCl(aq, \infty)]$ of -100.060 kcal and the foregoing,

$$\Delta Hf°[K^+(aq, \infty)] = \Delta Hf°[KCl(aq, \infty)] - \Delta Hf°[Cl^-(aq, \infty)]$$
$$= -100.060 - (-39.952) = -60.108 \text{ kcal}$$

Enthalpy of Vaporization (or Sublimation)

When the pressure of the vapor in equilibrium with a liquid reaches 1 atm, the liquid boils and is completely converted to vapor on absorption of the enthalpy of vaporization ΔHv at the normal boiling point T_b. A rough empirical relationship between the normal boiling point and the enthalpy of vaporization (*Trouton's rule*) is

$$\frac{\Delta Hv}{T_b} = 21 \text{ cal} \cdot \text{mol}^{-1} \cdot \text{K}^{-1}$$

It is best applied to nonpolar liquids which form unassociated vapors.

To a first approximation, the enthalpy of sublimation ΔHs at constant temperature is

$$\Delta Hs = \Delta Hm + \Delta Hv \tag{9-5}$$

where ΔHm is the enthalpy of melting.

The *Clapeyron equation* expresses the dynamic equilibrium existing between the vapor and the condensed phase of a pure substance:

$$\frac{dP}{dT} = \frac{\Delta Hv}{T \, \Delta V} \tag{9-6}$$

where ΔV is the volume increment between the vapor phase and the condensed phase. If the condensed phase is solid, the enthalpy increment is that of sublimation.

Substitution of $V = RT/P$ into Eq. 9-6 and rearranging gives the *Clausius-Clapeyron equation*,

$$\frac{dP}{P\,dT} = \frac{\Delta Hv}{RT^2} \qquad (9\text{-}7)$$

or

$$\Delta Hv = -R\,\frac{d(\ln P)}{d(1/T)} \qquad (9\text{-}8)$$

which may be used for calculating the enthalpy of vaporization of any compound provided its boiling point at any pressure is known. If an Antoine equation is available (such as Eq. 1, page 10–29), differentiation and insertion into Eq. 9-8 gives

$$\Delta Hv = \frac{4.5757\,T^2 B}{(T + C - 273.15)^2} \qquad (9\text{-}9)$$

Inclusion of a compressibility factor into Eq. 9-7, as suggested by the *Haggenmacher equation*, improves the estimate of ΔHv:

$$\Delta Hv = \frac{RT^2}{P}\left(\frac{dP}{dT}\right)\left[1 - \frac{T_c^3 P}{T^3 P_c}\right]^{1/2} \qquad (9\text{-}10)$$

where T_c and P_c are critical constants (Table 9-7). Although critical constants may be unknown, the compressibility factor is very nearly constant for all compounds belonging to the same family, and an estimate can be deduced from a related compound whose critical constants are available.

Heat Capacity (or Specific Heat)

Heat capacity is defined as the heat absorbed per degree rise in temperature for a system. Symbolically it is given by the limit:

$$\lim_{\Delta T \to 0} \frac{q}{\Delta T}$$

If the process takes place at constant volume, no expansion work is done and

$$\left(\frac{\partial q}{\partial T}\right)_v = C_v \qquad (9\text{-}11)$$

the heat capacity at constant volume. For heat addition at constant pressure, $dH = dq$, and

$$\left(\frac{\partial H}{\partial T}\right)_p = C_p \tag{9-12}$$

For an ideal gas,

$$C_p = C_v + R \tag{9-13}$$

and for nonideal situations, the exact difference is

$$C_p - C_v = \left[P + \left(\frac{\partial E}{\partial V}\right)_T\right]\left(\frac{\partial V}{\partial T}\right)_P \tag{9-14}$$

The temperature dependence of the heat capacity is complex. If the temperature range is restricted, the heat capacity of any phase may be represented adequately by an expression such as

$$C_p = a + bT + cT^2 \tag{9-15}$$

in which a, b, and c are empirical constants. These constants may be evaluated by taking three pieces of data: $(T_1, C_{p,1})$, $(T_2, C_{p,2})$ and $(T_3, C_{p,3})$, and substituting in the following expressions:

$$\frac{C_{p,1}}{(T_1 - T_2)(T_1 - T_3)} + \frac{C_{p,2}}{(T_2 - T_1)(T_2 - T_3)}$$
$$+ \frac{C_{p,3}}{(T_3 - T_2)(T_3 - T_1)} = c \tag{9-16}$$

$$\left(\frac{C_{p,1} - C_{p,2}}{T_1 - T_2}\right) - [(T_1 - T_2)c] = b \tag{9-17}$$

$$(C_{p,1} - bT_1) - cT_1^2 = a \tag{9-18}$$

Smoothed data presented at rounded temperatures, such as are available in Tables 9-3 and 9-4, plus the C_p° values at 298 K listed in Tables 9-1 and 9-2, are especially suitable for substitution in the foregoing parabolic equations. The use of such a parabolic fit is appropriate for interpolation, but data extrapolated outside the original temperature range should not be sought.

Enthalpy of a System

The enthalpy increment of a system over the interval of temperature from T_1 to T_2, under constraint of constant pressure, is given by the integration of Eq. 9-12:

$$H_2 - H_1 = \int_{T_1}^{T_2} C_p \, dT \tag{9-19}$$

The enthalpy over a temperature range that includes phase transitions, melting, and vaporization, is represented by

$$H_2 - H_1 = \int_{T_1}^{T_t} C_p(\text{s,II})\, dT + \Delta Ht + \int_{T_t}^{T_m} C_p(\text{s,I})\, dT + \Delta Hm$$

$$+ \int_{T_m}^{T_v} C_p(\text{liq})\, dT + \Delta Hv + \int_{T_b}^{T_2} C_p(\text{g})\, dT \qquad (9\text{-}20)$$

Integration of heat capacities, expressed as Eq. 9-15, leads to

$$\Delta H = a(T_2 - T_1) + \frac{b(T_2^2 - T_1^2)}{2} + \frac{c(T_2^3 - T_1^3)}{3} \qquad (9\text{-}21)$$

The integration can also be performed graphically.

Entropy

Every substance can be assigned a standard entropy $S°$, based on the third law of thermodynamics; that is, a pure crystalline substance has zero entropy at zero on the thermodynamic temperature scale (0 K). The difference in entropy between $0\,\text{K}$ and a temperature T can be deduced from the defining equation for entropy:

$$S_T - S_0 = \int_0^T \frac{dq_{\text{rev}}}{T} \qquad (9\text{-}22)$$

In the physical change of state, $\Delta Sm = \Delta Hm/T_m$ is the entropy of melting (or fusion), $\Delta Sv = \Delta Hv/T_b$ is the entropy of vaporization, and $\Delta Ss = \Delta Hs/T_s$ is the entropy of sublimation.

A general expression for entropy of a system, involving phase transitions, is

$$S_2 - S_1 = \int_{T_1}^{T_t} \frac{C_p(\text{s,II})\, dT}{T} + \frac{\Delta Ht}{T_t} + \int_{T_t}^{T_m} \frac{C_p(\text{s,I})\, dT}{T} + \frac{\Delta Hm}{T_m}$$

$$+ \int_{T_m}^{T_b} \frac{C_p(\text{liq})\, dT}{T} + \frac{\Delta Hv}{T_b} + \int_{T_b}^{T_2} \frac{C_p(\text{g})\, dT}{T} \qquad (9\text{-}23)$$

If C_p is independent of temperature,

$$\Delta S = C_p(\ln T_2 - \ln T_1) = 2.303\, C_p \log\left(\frac{T_2}{T_1}\right) \qquad (9\text{-}24)$$

If the heat capacities change with temperature, an empirical equation like Eq. 9-15 may be inserted in Eq. 9-23 before integration. Usually the integration is performed graphically from a plot of either C_p/T versus T or C_p versus $\ln T$.

Gibbs Free Energy

Gibbs free energy is given by

$$G = E + PV - TS = H - TS \qquad (9\text{-}25)$$

Under conditions of constant temperature and pressure, a process can occur only in the direction of decreasing Gibbs free energy, that is,

$$\Delta G = \Delta H - T\,\Delta S \qquad (9\text{-}26)$$

A spontaneous process leads to the minimum possible value of ΔG, and the sign for ΔG must be negative. When $\Delta G = 0$, the system is at equilibrium. A positive value for ΔG indicates that the process tends to proceed spontaneously in the opposite direction.

The standard Gibbs free energy of formation, $\Delta Gf°$ (at 298.15 K), is defined as the change in Gibbs free energy corresponding to the formation of the substance in its standard state from its elements in their standard states.

The change in Gibbs free energy with temperature is given by the *Gibbs-Helmholtz equation:*

$$\left[\frac{\partial(G/T)}{\partial T}\right]_P = \frac{-H}{T^2} \qquad (9\text{-}27)$$

or

$$\left[\frac{\partial(\Delta G/T)}{\partial(1/T)}\right]_P = \Delta H \qquad (9\text{-}28)$$

With the information available in Tables 9-1 to 9-4, it is possible to calculate $\Delta G°$ at any temperature in the range where the C_p data are valid using Eq. 9-19 and substituting this value of ΔH into Eq. 9-28 and integrating.

An equilibrium constant may be calculated from the expression

$$\Delta G° = -RT \ln K° \qquad (9\text{-}29)$$

where $K°$ is the equilibrium constant in terms of partial pressures $(K_p°)$ or concentrations $(K_c°)$, both expressed in standard states.

The variation of an equilibrium constant with temperature is given by the *van't Hoff equation:*

$$\ln K = -\frac{\Delta H°}{RT} + \frac{\Delta S°}{R} \qquad (9\text{-}30)$$

when $\Delta H°$ is relatively independent of temperature and, when this is the case, $\Delta S°$ is also relatively independent of temperature. For many purposes it is satisfactory to assume a constant value of $\overline{\Delta H}$ that corresponds to an average value over the temperature range of the equilibrium-constant measurements. Then,

$$\frac{d(\ln K)}{dT} = \frac{\Delta H^\circ}{RT^2} \tag{9-31}$$

Integration between the limits T_1 and T_2 gives:

$$\ln \frac{K_2}{K_1} = \frac{\Delta H^\circ}{R} \left(\frac{1}{T_1} - \frac{1}{T_2} \right) \tag{9-32}$$

It is sometimes more convenient to write Eq. 9-31 in the form

$$\frac{d(\ln K)}{d(1/T)} = - \frac{\Delta H^\circ}{R} \tag{9-33}$$

The integrated form of the foregoing equation, and using Briggsian logarithms, is

$$\log K = - \frac{\Delta H^\circ}{2.303R} \left(\frac{1}{T} \right) + \text{constant} \tag{9-34}$$

Table 9-3
HEATS OF MELTING AND VAPORIZATION (OR SUBLIMATION) AND SPECIFIC HEAT AT VARIOUS TEMPERATURES OF THE ELEMENTS AND INORGANIC COMPOUNDS
Abbreviations Used in the Table

ΔHm, enthalpy of melting at the melting point in $kcal \cdot mol^{-1}$

ΔHv, enthalpy of vaporization at the boiling point in $kcal \cdot mol^{-1}$

ΔHs, enthalpy of sublimation at 25°C in $kcal \cdot mol^{-1}$

C_p, specific heat at temperature specified (K) for physical state in existence at the particular temperature, expressed in $cal \cdot K^{-1} \cdot mol^{-1}$

dec, decomposition

t, transition point at temperature specified (°C) with enthalpy of transition in $kcal \cdot mol^{-1}$

Substance	ΔHm	ΔHv	ΔHs	C_p (at specified temperature, K)			
				400	600	800	1000
Actinium							
Ac	(3.4)		94				
Aluminum							
Al	2.56	69.50	78.00	6.16	6.72	7.37	8.43
Al_6BeO_{10}	96			77.50	90.96	97.47	101.62
$AlBO_2$				14.29	16.29	17.43	18.13
$AlBr_3$	2.69	5.69*	24				
Al_4C_3				33.45	38.31	40.90	42.84
$AlCl_3$	8.45		27.66*	23.00	23.26	23.55	24.10
AlF_3 ΔHt, 0.1346[455°]	67.0		72.0	20.62			
			93*				
AlI_3	3.8		26.8*	25.94			
AlN				8.70	10.44	11.26	11.66

Al$_2$O$_3$ α	28.0			22.96	26.90	28.71	29.82
γ	22.5			22.96	26.90	28.71	29.82
Al$_2$O$_3$·H$_2$O				31.20	33.28		
Al$_2$O$_3$·3H$_2$O				54.71	74.75		
Al(OH)$_3$				27.37	37.40		
AlOCl				15.37	17.36	18.37	18.94
AlOF				12.02	13.38	13.99	14.30
Al$_2$(SO$_4$)$_3$				76.86	89.11	95.35	99.84
Al$_2$O$_3$·SiO$_2$ andalusite				36.12	41.69	44.40	46.29
kyanite				36.14	42.02	44.85	46.80
sillimanite				35.70	41.25	44.11	46.18
Al$_2$O$_3$·2SiO$_2$·2H$_2$O				66.67	76.46	84.48	
3Al$_2$O$_3$·2SiO$_2$ mullite				96.50	112.08	119.64	124.56
Al$_2$O$_3$·TiO$_2$				38.74	43.70	46.12	47.81
Ammonium							
NH$_3$	1.351	5.581		9.24	10.81	12.23	13.47
NH$_4$Br ΔHt, $0.77^{138°}$				24.61	24.30		
NH$_4$Cl ΔHt, $0.25^{-30.6°}$					24.70	28.13	
ΔHt, $0.944^{184.6°}$							
NH$_4$I ΔHt, $0.70^{-13°}$	5.0		$40.28^{525°}$	21.27			
Antimony							
Sb	4.74	46.23		6.20	6.62	7.05	7.50
SbBr$_3$	3.51	16.7		30.00	19.51	19.65	19.72
SbCl$_3$	3.1	10.4		29.50	19.51	19.65	19.72
SbCl$_5$	2.40	11.57				19.65	
SbI$_3$	5.45	14.6		25.48	34.30		19.72
Sb$_2$O$_3$ ΔHt, $1.7^{573°}$	13.0			25.94	29.36	32.78	36.00
Sb$_2$S$_3$				29.48	32.12	34.76	
Sb$_2$Se$_3$			29.5				
Argon							
Ar	0.2808	1.558					
Arsenic							
As	5.100		7.630	6.12	6.56	7.01	
AsBr$_3$	4.10	10.0		31.90	21.10	21.10	
AsCl$_3$	2.42	7.57					

* To dimer.

Table 9-3 (*Continued*)
HEATS OF MELTING AND VAPORIZATION (OR SUBLIMATION) AND SPECIFIC HEAT AT VARIOUS TEMPERATURES OF THE ELEMENTS AND INORGANIC COMPOUNDS

Substance	ΔH_m	ΔH_v	ΔH_s	C_p (at specified temperature, K)			
				400	600	800	1000
AsF_3	12.47	7.10					
AsF_5	2.80	4.98					
AsH_3	4.34						
As_4O_6 monoclinic	8.00	14.30		10.86	12.71	14.06	15.27
ΔHt, 8.22 (rh → mn)							
As_2O_3 arsenolite	15.86			27.81			
As_2O_3 ΔHt, $0.67^{-33°}$	5.07			27.81			
(arsenolite to claudetite)	4.4						
As_2O_5				27.85	27.85		
Barium							
Ba	1.92	33.52	42.8	7.94	12.11	11.13	10.35
$BaBr_2$ ΔHt, $4.04^{925°}$	7.5	58.90		18.45	19.69	20.94	22.18
$BaCl_2$ ΔHt, $4.5^{806°}$; $0.7^{968°}$	3.82			18.48	19.22	20.15	21.38
$BaCO_3$	5.58	68.2	96.8	18.15	19.20	20.30	22.60
BaF_2 ΔHt, $0.0^{967°}$ ($\alpha \to \beta$)				19.01	19.96	20.92	21.88
ΔHt, $0.64^{1207°}$ ($\beta \to \gamma$)							
BaI_2	6.34	43.88	72.3	11.93	12.72	13.24	13.68
BaO	14	70.7	101.4				
$Ba_3(PO_4)_2$	18.60			28.53	31.46	32.38	32.96
$BaSO_4$	9.70			24.73	28.20	30.62	32.73
$BaSiO_3$				18.10	19.35	27.00	
Beryllium							
Be	2.8	71.137	77.5	4.77	5.58	6.07	6.52
$BeAl_2O_4$	41.6			31.21	37.03	39.98	41.64
$BeBr_2$	4.5	23.9	31			27.00	27.00

Compound							
Be₂C	13.92	12.90	11.88	10.35			18
BeCl₂ ΔHt, 1.63⁴⁰³°	29.02	29.02	19.14	17.39	32.5	25	2.07
BeF₂ α	20.47	18.80	16.12	14.94	55.1	47.65	1.14
BeI₂	27.00	27.00	19.65	18.50	26	23	5
Be₃N₂	29.54	28.10	25.46	20.18			30.9
BeO ΔHt, 1.6²⁰⁹⁹° α	11.79	11.15	10.12	8.07			14.1
β	11.79	11.15	10.12	8.07			15.7
Be₂SiO₄	41.60	39.68	35.65	28.86			
BeSO₄ ΔHt, 0.266⁵⁹⁰°	41.69	35.8	30.31	24.83			
ΔHt, 4.673⁶³⁵°							
BeWO₄	36.57	34.15	31.39	27.00			
Bismuth							
Bi	7.60	7.60	7.60	6.45			
BiBr₃						36.20	2.60
BiCl₃						18.02	5.19
BiF₅						17.35	2.60
BiI₃						14.9	
Bi₂O₃ ΔHt, 27.9⁷¹⁷°	32.74	31.14	29.54	27.94		5.00	6.80
Bi₂S₅		45.95	42.96	39.27			8.90
Bi₂Te₃							28.8
Boron							
B	5.95	5.56	4.99	3.72	132	121	5.39
BBr₃	19.35	19.07	18.55	17.36		7.30	
B₄C	27.24	25.59	23.24	18.45			25
BCl₃	19.08	18.68	17.94	16.35		5.70	
B₂Cl₄					18	8.03	2.58
BF₃	18.12	17.34	16.03	13.75		4.62	0.48
BH₃	13.96	12.51	10.86	9.30			
B₂H₆	32.60	29.08	24.22	17.75		3.45	1.06
B₄H₁₀						6.47	
B₅H₉						6.79	1.47
B₅H₁₁						7.7	
B₆H₁₀	60.81	54.34	44.84	31.12		9.16	
B₁₀H₁₄	110.03	99.71	84.05	59.74		10.3	
B₂H₅Br						6.23	5.25

Table 9-3 (Continued)

HEATS OF MELTING AND VAPORIZATION (OR SUBLIMATION) AND SPECIFIC HEAT AT VARIOUS TEMPERATURES OF THE ELEMENTS AND INORGANIC COMPOUNDS

Substance	ΔH_m	ΔH_v	ΔH_s	C_p (at specified temperature, K)			
				400	600	800	1000
BI_3	(1)	10.1		17.92	18.88	19.28	19.49
BN			174	6.28	8.42	9.67	10.60
$B_3N_3H_6$		7.67		30.32	40.49	47.14	51.76
BO_2				11.35	12.73	13.50	13.94
B_2O_3	5.75	86.4		18.63	23.45	31.00	31.00
BOCl			104.2	11.72	12.71	13.31	13.74
$(BOF)_3$			21				
$B_2(OH)_4$			30.1	29.86	36.76	41.02	43.96
$B_3O_3H_3$			10.7	28.70	38.90	46.50	51.20
Bromine							
Br			4.97	4.98	5.03	5.11	
Br_2	2.53	7.06		8.78	8.91	8.97	9.01
BrCl	2.4	8.3					
BrCN			43.3				
BrF		6.0					
BrF_3	2.88	10.24		17.35	18.63	19.15	19.40
BrF_5	1.36	7.31		27.01	29.44	30.42	30.90
Cadmium							
Cd	1.45	23.87		6.49	7.10	7.10	7.10
$CdBr_2$	5.00	27.00					
$CdCl_2$	7.2	29.86		19.08	20.62	22.16	25.00
CdF_2	5.40	52.0					
CdI_2	3.66	25.40					
$Cd(NO_3)_2 \cdot 4H_2O$	7.80						
CdO			53.82	10.48	10.90	11.31	11.73
CdS			50.10	13.26	13.44	13.62	13.80

$CdSb$	4.79			12.45	15.45	14.30	14.30
$CdSO_4$				25.88	29.58	33.28	36.98
Calcium							
Ca ΔH_t, $0.22^{448°}$				6.43	6.52	8.25	9.31
$CaBr_2$	2.04	36.97	42.85	18.64	19.24	19.95	21.18
CaC_2	6.95	48.0	71.3	16.24	17.53	17.00	17.40
$CaCl_2$	6.8	56.2	77.5	18.08	18.69	19.34	20.50
$CaCO_3$ calcite				23.20	26.40	28.20	29.60
$CaCO_3 \cdot MgCO_3$ dolomite				41.80	47.44	51.92	56.05
CaF_2 ΔH_t, $1.15^{1151°}$	7.10	73.8	104.5	17.65	18.77	20.06	21.54
CaI_2	10.00	42.88	66.6	18.92	19.87	20.81	21.75
$Ca(NO_3)_2$	5.12			41.51	50.31	58.17	
CaO	19			11.14	12.07	12.52	12.84
$Ca(OH)_2$ ΔH_{dec}, $23.7^{521.6°}$				23.52	25.80		
$Ca_3(PO_4)_2$ ΔH_t, $3.7^{1100°}$				60.98	70.65	79.19	87.41
CaS				11.80	12.20	12.60	13.00
$CaSiO_3$ ΔH_t, $1.7^{1190°}$	13.40			24.01	26.99	28.50	29.59
Ca_2SiO_4 ΔH_t, $1.06^{675°}$; $0.78^{1420°}$				35.00	38.92	42.84	43.98
$3CaO \cdot SiO_2$				46.95	52.20	55.16	57.46
$3CaO \cdot 2SiO_2$				57.25	64.82	68.65	71.39
$CaSO_4$	6.70			26.22	30.94	35.66	40.38
$CaSO_4 \cdot \frac{1}{2}H_2O$				35.24	39.96	44.68	49.40
$CaSO_4 \cdot 2H_2O$				62.28	67.00	71.72	76.44
$CaTiO_3$ ΔH_t, $0.55^{1257°}$				26.83	29.43	30.51	31.16
Carbon							
C graphite	25.00		171.29	2.85	4.03	4.75	5.14
C_2				9.48	8.61	8.51	8.60
C_3				8.96	9.49	10.15	10.71
CF^+				7.33	7.86	8.24	8.48
CF_2				10.30	11.72	12.50	12.95
CN				7.03	7.33	7.69	8.00
CN^-				7.00	7.24	7.58	7.88
CN^+				7.24	7.89	8.23	8.52
CN_2				10.80	11.75	12.42	12.93

Table 9-3 (*Continued*)

HEATS OF MELTING AND VAPORIZATION (OR SUBLIMATION) AND SPECIFIC HEAT AT VARIOUS TEMPERATURES OF THE ELEMENTS AND INORGANIC COMPOUNDS

Substance	ΔH_m	ΔH_v	ΔH_s	C_p (at specified temperature, K)			
				400	600	800	1000
C_4N_2	1.94			22.66	25.37	27.26	28.62
$(CN)_2$ cyanogen		5.58	4.96	14.80	16.33	17.43	18.25
CNBr			10.85	11.85	12.68	13.24	13.64
CNCl	2.72	5.78		11.54	12.48	13.12	13.56
CNF							
CNI		13.98	15.5	12.15	12.85	13.34	13.72
CO ΔH_t, $0.151^{-211.62°}$	0.200	1.444		7.01	7.28	7.62	7.93
CO_2	1.99		6.03	9.87	11.31	12.29	12.97
C_3O_2	1.29	5.832	5.65	17.92	20.43	22.15	23.34
$COCl_2$ I	1.372			15.28	16.98	17.92	18.49
II	1.335						
III	1.131						
COF_2	1.130	3.86		13.09	15.51	16.93	17.79
COS		4.423		10.96	12.25	13.07	13.62
CS	1.05			7.40	7.94	8.29	8.51
CS_2			6.61	11.82	12.97	13.64	14.03
Cerium							
Ce ΔH_t, $0.72^{730°}$	1.238		78	7.32	7.36	7.68	8.07
$CeCl_3$	12.4	40.8					
CeI_3							
CeO_2				16.00	16.50	17.00	17.50
Cesium							
Cs	0.499	16.198	18.3	7.75	7.75	7.75	8.50
CsBr	5.65	36		12.64	13.15	13.67	18.50
CsCl ΔH_t, $0.90^{470°}$	3.8	27.52	48.4	13.07	14.13	15.22	18.50
CsF	5.19	27.6	47.4	12.85	13.70	14.55	17.70

CsI	5.7	35.9		12.41	13.81	15.65	16.21
CsIO₃ ΔHt, 0.38¹⁵°	3.11						
CsIO₄ ΔHt, 0.17¹⁵⁰°							
CsOH ΔHt, 0.31¹³⁷° ΔHt, 1.45²²⁰°	1.09	28.6	37.6	17.78	19.50	19.50	19.50
Chlorine							
Cl₂	1.531	4.878		8.44	8.74	8.88	8.96
ClF				8.07	8.51	8.73	8.85
ClF₃	1.82	6.58		16.88	18.37	18.98	19.29
ClF₅		5.48		26.30	29.06	30.19	30.75
ClO				7.94	8.43	8.68	8.82
ClO₂		6.52		11.01	12.28	12.96	13.35
ClO₃							
Cl₂O		6.28	12.3				
Cl₂O₇		8.29		18.15			
ClO₃F	0.916	4.619		12.28	13.09	13.42	13.59
Chromium							
Cr	4.9	81.14	95.00	6.03	21.32	22.97	23.88
Cr₃C₂				27.60	6.63	7.04	7.62
Cr(CO)₆			16.9	55.90	31.43	33.25	34.73
CrCl₂	7.72	47.50	56.8	17.35	18.41	19.47	20.53
CrCl₃			86.8	22.25	23.66	25.06	26.47
CrF₂			60				
CrF₃							
CrN ΔHdec, 26.7¹²⁸²°				11.73	12.05	12.36	12.86
Cr₂N ΔHdec, 27.7¹⁵³⁷°				17.55	19.00	20.31	21.63
CrO				7.88	8.38	8.63	8.78
CrO₂				11.24	12.36	12.94	13.26
CrO₃	3.77			15.26	17.34	18.32	18.84
Cr₂O₃	31.0			26.93	28.81	29.71	30.36
CrO₂Cl₂		8.25					
CrO₂F₂		8.2					
Cr₂(SO₄)₃	5.6			75.75	82.51	89.26	96.02

Table 9-3 (Continued)
HEATS OF MELTING AND VAPORIZATION (OR SUBLIMATION) AND SPECIFIC HEAT AT VARIOUS TEMPERATURES OF THE ELEMENTS AND INORGANIC COMPOUNDS

Substance	ΔH_m	ΔH_v	ΔH_s	C_p (at specified temperature, K)			
				400	600	800	1000
Cobalt							
Co ΔH_t, $0.108^{427°}$	3.87	89.22	101.5	6.34	7.14	7.85	8.95
$CoCl_2$	10.7	34.8	52.3	19.53	20.22	23.70	23.70
$CoCl_3$	14	48.3	75	19.38	20.04	20.13	20.13
CoF_2				23.30	24.01	24.37	24.83
CoF_3				12.65	12.98	13.10	13.39
CoO				34.10	38.91	44.28	50.26
Co_3O_4				28.51	33.62	36.34	37.79
$CoSO_4$ ΔH_t, $0.51^{691°}$							
Copper							
Cu	3.17	72.74	81.0	6.01	6.31	6.61	6.91
CuBr ΔH_t, $1.4^{380°}$; 0.7^{465}	2.3	5.18	57.8	13.50	14.29	16.00	16.00
CuCl	2.45			12.49	14.16	16.00	16.00
CuCN	(3)			15.95	17.48	18.65	
CuF_2	9.4	43.98	62.5	17.47	18.67	19.87	21.07
CuI	2.6			13.24	13.82	14.39	16.00
Cu_2O	13.58			16.32	17.72	19.12	20.52
CuO	2.82			11.20	12.49	13.46	14.34
CuS				11.66	12.18	12.71	13.24
Cu_2S ΔH_t, $0.92^{103°}$; $0.2^{350°}$	5.50			23.25	23.25	20.32	20.32
Cu_2Se ΔH_t, $1.16^{110°}$				21.72	21.92	22.12	22.32
$CuSO_4$				27.58	32.49	35.36	36.67
Dysprosium							
Dy	(4.1)	60.0	69.4				
Erbium							
Er	(4.1)		75.8				

Europium							
Eu	(2.5)						
Fluorine							
F ΔH_t, $0.17^{-227.6°}$	0.122	1.562	41.9	7.90	8.43	8.71	8.89
Francium							
Fr			17.40				
Gadolinium							
Gd	3.7	72.0	95.0	8.74	8.49	8.25	8.01
Gd$_2$O$_3$				27.11	28.70	29.72	30.57
Gallium							
Ga	1.336			6.65	6.65	6.65	6.65
GaCl$_3$	2.55						
Ga$_2$O$_3$				21.85	26.89	31.92	
Germanium							
Ge	8.8	78.3		5.96	6.32	6.55	6.74
GeBr$_4$		8.56					
GeCl$_4$		7.03					
GeF$_2$			27.0				
GeH$_4$		3.61					
Ge$_2$H$_6$		5.99					
Ge$_3$H$_8$		7.55					
GeHCl$_3$		8.00					
Ge(CH$_3$)$_4$		6.46					
GeO$_2$	10.5			14.07	15.50	16.94	18.37
Gold							
Au	2.955	80.08		6.17	6.40	6.64	6.89
AuSb$_2$				18.98	19.90		
Hafnium							
Hf ΔH_t, $1.65^{1750°}$	5.75	136.4		6.34	6.70	7.06	7.43
Hf(BH$_4$)$_4$	2.85						
HfCl$_4$	18		23.8	29.98	25.28	25.50	25.61
HfF$_4$			63.0				
HfI$_4$			48.9				
HfO$_2$ ΔH_t, $2.5^{1700°}$	25.0			16.17	17.67	18.48	19.09
Helium							
He	0.0033	0.0194					

Table 9-3 (Continued)
HEATS OF MELTING AND VAPORIZATION (OR SUBLIMATION) AND SPECIFIC HEAT AT VARIOUS TEMPERATURES OF THE ELEMENTS AND INORGANIC COMPOUNDS

Substance	ΔH_m	ΔH_v	ΔH_s	C_p (at specified temperature, K)			
				400	600	800	1000
Holmium							
Ho	(4.1)	(60)	71.9	6.974	7.008	7.078	7.217
Hydrogen							
H_2	0.028	0.216					
$^2H_2$	0.057	0.293					
$^3H_2$		0.322					
HBO_2			57.9	14.70	13.40	14.85	15.90
HBr	0.575	4.210		6.98	7.12	7.38	7.68
HCl ΔH_t, $0.284^{-174.77°}$	0.476	3.86		6.97	7.07	7.29	7.56
HClO				9.55	10.51	11.13	11.59
HCN	2.009	6.027		9.42	10.56	11.45	12.18
HF	0.939	1.8		6.97	6.99	7.06	7.21
HI	0.686	4.724		7.01	7.25	7.60	7.92
H_2MoO_4				27.58	30.78	32.58	33.74
HN_3		7.29					
HNCO				12.09	13.93	15.19	16.13
HNCS				12.71	14.57	15.74	16.57
HNO_2 cis				10.84	14.32	15.64	16.55
trans				12.44	14.41	15.68	16.55
HNO_3	2.503	9.43		15.21	18.57	20.64	21.95
HO				7.09	7.06	7.15	7.33
HO^+				6.977	7.066	7.279	7.550
HO^-				6.97	7.00	7.11	7.30
HO_2				8.91	9.98	10.77	11.37
H_2O	1.436	9.717		8.19	8.68	9.25	9.85
^2H_2O	1.501	9.945					

H_2O_2	2.987	10.53	12.34	11.58	13.31	14.30	15.93
HOF				9.22	10.24	10.93	11.46
HPH_2O_2	2.31						
H_3PO_3	3.07			42.00	56.40	70.80	85.20
H_3PO_4	3.2			8.51	9.32	10.18	10.97
$H_4P_2O_7$	8.30	4.463					
H_2S ΔHt, $0.366^{-169.61°}$	0.568						
ΔHt, $0.108^{-146.91°}$							
H_2S_2	1.805	7.497					
H_2S_3		9.327					
H_2S_4		11.261					
H_2S_5		13.34					
H_2Se	3.45	4.75		8.71	9.61	10.48	11.35
H_2SeO_4	2.56	13.39		36.70	27.34	29.74	31.49
H_2SO_4		8.4		20.91	24.53	26.54	27.79
HSO_3F		5.7					
H_2Te				8.99			
Indium							
In	0.781	55.41		6.82	7.19	7.19	7.19
$In(CH_3)_3$	3.77						
Iodine							
I_2	3.709	10.026	14.924	19.28	8.98	9.03	9.06
ICl	2.773		12.65				
IF_5	3.84						
IF_7			7.46				
Iridium							
Ir	6.3	146.3		6.13	6.41	6.68	6.96
IrO_2				15.25	18.29	21.33	24.37
Iron							
Fe ΔHt, $0.225^{911°}$	3.630	83.68	99.5	6.50	7.58	9.17	13.56
ΔHt, $0.260^{1392°}$							
$FeBr_2$ ΔHt, $0.1^{377°}$	12			19.72	20.78	21.85	25.50
Fe_3C ΔHt, $0.180^{190°}$	12.33		49.6	27.64	27.42	28.02	28.62
$FeCl_2$	10.28	26.3	48.0	19.04	19.86	20.44	24.42
Fe_2Cl_4			63.3	30.80	31.34	31.57	31.75

Table 9-3 (Continued)

HEATS OF MELTING AND VAPORIZATION (OR SUBLIMATION) AND SPECIFIC HEAT AT VARIOUS TEMPERATURES OF THE ELEMENTS AND INORGANIC COMPOUNDS

Substance	ΔH_m	ΔH_v	ΔH_s	C_p (at specified temperature, K)			
				400	600	800	1000
$FeCl_3$	10.30	10.46*		25.50	32.00	19.68	19.48
Fe_2Cl_6		9.00		42.45	43.14	43.39	43.51
$Fe(CO)_5$	3.25		75				
FeF_2	12.4	53.64	52.8	17.20	18.44	19.18	19.63
FeF_3			46	23.03	23.14	23.74	24.34
FeI_2 $\Delta H_t, 0.2^{377°}$	10.7	25.0	52	20.06	20.18	26.50	27.00
Fe_2I_4			125	31.56	31.70	31.80	31.93
FeO	5.75			12.39	13.12	13.70	14.19
Fe_2O_3 $\Delta H_t, 0.16^{677°}$				28.71	33.74	37.82	36.00
Fe_3O_4	33.0			40.90	50.80	60.44	48.00
$Fe(OH)_2$			58.2	24.40	26.60	28.43	29.49
$Fe(OH)_3$				28.20	33.60	37.00	39.40
FeS $\Delta H_t, 0.57^{138°}; 0.12^{325°}$	7.73			15.75	13.63	14.10	14.58
FeS_2				16.50	17.83	18.46	18.90
$FeSO_4$				27.88	32.94	35.71	
$Fe_2(SO_4)_3$				73.38	86.83	93.99	97.81
$FeSiO_3$				24.10	27.31	29.76	31.99
Fe_2SiO_4	22			36.07	40.27	42.95	45.20
$FeTiO_3$	21.7			26.62	29.16	30.61	31.75
$FeWO_4$				32.24	34.23	36.07	37.86
Krypton							
Kr	0.3907	2.158					
Lanthanum							
La $\Delta H_t, 0.68^{868°}$	2.03	96.1		6.81	7.13	7.45	7.77
$LaCl_3$	13.0	45.9		25.28	26.31	27.33	28.36

Compound							
La₂O₃	1.141			28.04	29.80	30.81	31.61
Lead							
Pb		42.53	46.75	6.63	7.03	7.17	7.03
PbB₂O₄				30.98	38.83	44.14	47.42
PbB₄O₇				49.36	63.47	72.78	78.88
PbBr				8.91	8.98	9.02	9.06
PbBr₂	3.93	28.23	41.35	19.42	21.22	26.80	26.80
PbBr₄				25.36	25.62	25.71	25.76
PbCl₂	5.23	30.2	44.3	19.14	20.53	26.65	26.65
PbCl₄				24.78	25.35	25.56	25.65
PbCO₃ ΔH_t, $0.35^{310°}$	3.52			23.83	29.55	35.27	
PbF₂		37.5		18.18	21.78	25.32	22.57
PbF₄				23.25	24.57	25.10	25.35
PbI₂	5.6	28.34	41.16	18.84	20.08	25.95	25.95
PbI₄				25.60	25.73	25.77	25.79
PbO yellow ΔH_t, $0.04^{488°}$ (to yellow)	6.1	(49.5)	68.9	11.60	12.45	13.15	13.81
red							
PbO₂	4.5			16.16	17.81	18.54	18.95
Pb₃O₄				41.36	45.60	47.60	49.60
PbS	6.22		55.0	12.06	12.52	12.98	13.43
PbSiO₃ ΔH_t, $4.1^{866°}$	9.60			24.26	27.14	30.02	33.07
PbSO₄				25.99	30.73	36.42	42.38
Lithium							
Li	0.717	35.16		6.60	7.06	6.92	6.89
Li₃AlF₆ ΔH_t, $0.5^{475°}$ ΔH_t, $0.3^{575°}$ ΔH_t, $0.1^{705°}$	20.6			54.94	62.69	68.00	73.00
LiAlO₂	(21)			19.49	22.15	23.46	24.38
LiBeF₃	6.5			25.0	31.0	38.0	38.0
Li₂BeF₄	10.5			35.95	43.08	55.47	55.47
LiBH₄				21.75	25.13		
LiBO₂	8.08	63.5		17.00	20.33	23.17	25.90
Li₂B₄O₇	29			47.24	57.63	65.59	71.80

*To dimer.

Table 9-3 (Continued)
HEATS OF MELTING AND VAPORIZATION (OR SUBLIMATION) AND SPECIFIC HEAT AT VARIOUS TEMPERATURES OF THE ELEMENTS AND INORGANIC COMPOUNDS

Substance	ΔH_m	ΔH_v	ΔH_s	C_p (at specified temperature, K)			
				400	600	800	1000
LiBr ΔH_v, 25.6 to equilibrium mixture	4.22	35.2 to monomer	47	12.25	13.41	15.41	15.60
LiCl	4.74		52	12.18	13.27	14.33	15.28
LiClO$_4$	5.00			31.00	38.50	38.50	38.50
Li$_2$CO$_3$ ΔHt, 0.134$^{350°}$ ΔHt, 0.535$^{410°}$	10.7			27.12	35.60	37.99	44.32
LiF	6.474	35.08 to equil. mixture	66.0 to monomer	11.12	12.33	13.32	14.24
LiH	5.40	10.9	55.27	8.32	11.09	13.70	14.00
Li$_2$HfCl$_6$	8.80						
LiI	3.50	23.3	86.3	12.77	14.02	15.10	15.10
LiIO$_3$ ΔHt, 0.53$^{260°}$	4.20						
Li$_2$MoO$_4$							
Li$_3$N	14			20.99	25.66	29.64	31.90
Li$_2$O				15.30	17.64	19.26	20.61
Li$_2$O$_2$				19.76	19.17	19.46	19.61
LiOH	4.99	44.9	59.9	13.88	16.30	20.81	20.81
Li$_2$SiO$_3$	6.7			28.40	32.10	34.52	36.40
Li$_2$Si$_2$O$_5$ ΔHt, 0.225$^{936°}$	12.86			41.80	49.16	53.20	56.25
Li$_2$SO$_4$	3.04						
Li$_2$TiO$_3$ ΔHt, 2.75$^{1212°}$	26.33			30.44	33.82	35.62	36.78
Lutetium							
Lu	(4.6)		102.2				
Magnesium							
Mg	2.14	30.5	35.28	6.29	6.80	7.42	7.88

MgAl$_2$O$_4$	47			33.25	37.92	40.39	42.22
MgBr$_2$	9.4	35.6	53	18.47	19.46	20.21	25.00
MgCl$_2$	10.30	37.34	57.5	18.18	19.18	19.71	22.10
MgCO$_3$				21.54	25.74	29.01	
MgF$_2$	13.90	63.48	95.5	16.39	17.83	18.60	19.23
MgH$_2$				10.35	$11.95^{500°}$		
MgI$_2$			46				
Mg$_3$N$_2$ ΔHt, $0.110^{550°}$ ΔHt, $0.220^{788°}$				25.73	27.19	28.65	29.60
Mg(NO$_3$)$_2$	18.5	113.4	145	40.28	53.90		
MgO macrocrystal	29			10.17	11.29	11.86	12.23
Mg$_3$(PO$_4$)$_2$	20.5			57.40	67.45	76.63	84.02
Mg$_2$Si	18			17.63	19.08	20.05	20.89
MgSiO$_3$ ΔHt, $0.16^{630°}$ ΔHt, $0.39^{985°}$				22.51	25.57	27.67	28.76
Mg$_2$SiO$_4$	17			32.88	37.39	39.93	41.73
MgS	3.5			11.43	11.90	12.30	12.70
MgSO$_4$	21.6			26.29	30.50	33.58	36.25
MgTiO$_3$	31			25.15	28.31	29.97	31.06
Mg$_2$TiO$_4$	35			34.96	39.12	41.76	43.90
MgTi$_2$O$_5$				39.72	44.15	46.90	49.15
MgWO$_4$				29.50	32.75	34.93	37.00
Manganese							
Mn ΔHt, $0.535^{727°}$ ΔHt, $0.545^{1101°}$ ΔHt, $0.430^{1137°}$	3.50	52.52		6.82	7.62	8.34	8.99
MnBr$_2$	8			18.60	19.79	20.97	24.00
Mn$_3$C ΔHt, $3.57^{1037°}$		27		24.96	27.49	29.10	30.45
MnCl$_2$	8.97	35.6		18.45	19.55	20.35	23.00
Mn$_2$(CO)$_{10}$			15.0				
MnF$_2$	5.5		76.1	16.87	18.08	19.30	20.52
MnI$_2$	10			18.67	19.98	21.28	26.00
MnO	13.00			11.34	12.03	12.52	12.96
MnO$_2$				15.15	16.99	17.95	
Mn$_2$O$_3$				26.06	28.86	30.93	32.79

Table 9-3 (Continued)

HEATS OF MELTING AND VAPORIZATION (OR SUBLIMATION) AND SPECIFIC HEAT AT VARIOUS TEMPERATURES OF THE ELEMENTS AND INORGANIC COMPOUNDS

Substance	ΔH_m	ΔH_v	ΔH_s	C_p (at specified temperature, K)			
				400	600	800	1000
Mn_3O_4 $\Delta Ht,$ 4.97[172°]				37.59	40.52	42.95	45.24
MnO_3F		8.1					
MnS	6.3			12.12	12.48	12.84	13.20
$MnSiO_3$	16.0			24.12	27.04	28.56	29.68
Mn_2SiO_4	22.6			35.48	39.45	42.41	45.07
$MnSO_4$				28.43	32.66	35.30	
$MnTiO_3$				26.70	28.96	30.04	30.78
$MnWO_4$				31.78	33.83	35.69	37.48
Mercury							
Hg	0.5486	14.13	14.65	6.55	6.49	4.97	4.97
Hg_2Br_2				26.20	27.63		
$HgBr_2$	4.28	14.15	20	18.71	24.00	24.40	24.40
Hg_2Cl_2				25.34	26.80		
$HgCl_2$	4.64	14.08	20.0	18.39	14.63	14.74	14.79
HgF_2	5.5	22	31	18.40	19.40	20.40	24.40
Hg_2I_2	6.5			26.39	32.60		
HgI_2 $\Delta Ht.$ 0.60[129°]	4.53	14.1	21.3	19.60	24.40	14.86	14.88
HgO				11.55	12.94	14.28	
HgS $\Delta Ht.$ 1.0[386°]				11.46	12.19	12.92	
Molybdenum							
Mo	6.6	141	157.3	6.05	6.38	6.55	6.70
$MoCl_2$	6.0						
$MoCl_3$			52.0				
$MoCl_4$	4	14.7	22	32.3	35.00	35.00	35.00
$MoCl_5$	4.5	12	19	40.0	42.00	42.00	42.00
$MoCl_6$			20	46.0	36.86	37.24	37.43

Mo(CO)₆				31.86	34.75	35.98	36.59
MoF₄	1.02	12.09		15.19	17.01	18.28	19.46
MoF₅	1.034	12.37		19.86	21.94	23.91	26.06
MoF₆		6.512	16.7				
MoO₂				25.93	26.72	29.95	30.58
MoO₃			92	22.03	23.77	24.58	25.00
MoOCl₃	11.60	16.19	25				
MoOCl₄				16.60	19.30	26.24	30.24
MoOF₄							
MoO₂Cl₂				18.23	22.23	25.49	26.46
MoS₂				30.23	34.49		
MoS₃							
Mo₂S₃				23.19	24.45		
Mo₃Si							
Neodymium							
Nd ΔHt, $0.713^{862°}$	1.705	60		6.76	7.66	8.82	10.04
Nd₂O₃				28.74	31.15	32.92	34.51
Neon							
Ne	0.0801	0.422					
Neptunium							
Np ΔHt, $2.00^{280°}$	2.6			8.32			
Nickel							
Ni ΔHt, $0.14^{357°}$	4.210	89.60		6.17	7.17	7.55	7.88
NiBr₂	18.47	54.84					
NiCl₂	3.3	53.81		18.02	19.07	19.84	20.54
Ni(CO)₄		7.0	79.4	37.52	42.88	48.24	
NiF₂				18.25	18.75	19.24	19.73
NiO				12.47	12.39	12.80	13.20
NiS				14.37	16.93		
Ni₂S	2.98						
Ni₃S₂	5.80						
NiSO₄				34.07	36.05	38.04	40.02
Niobium							
Nb	6.43	164.93	175.2	6.07	6.30	6.50	6.69
NbB	20.0			13.80	16.29	18.07	19.54

Table 9-3 (Continued)
HEATS OF MELTING AND VAPORIZATION (OR SUBLIMATION) AND SPECIFIC HEAT AT VARIOUS TEMPERATURES OF THE ELEMENTS AND INORGANIC COMPOUNDS

Substance	ΔH_m	ΔH_v	ΔH_s	C_p (at specified temperature, K)			
				400	600	800	1000
NbBr$_5$	8.5	12.0		40.39			12.41
NbC			31	10.01	11.32	11.98	25.60
NbCl$_4$	8.8	13.1		30.09	31.09	25.49	31.23
NbCl$_5$	4.2	11.0	22.8	40.79	30.57	31.02	
NbF$_5$	9.0			43.50	44.00		
NbI$_5$	11.0	14.0		39.33			
NbN ΔH_t, $1.0^{1370°}$			148	10.85	11.93	12.33	12.72
NbO	20.4			10.52	11.28	11.82	12.30
NbO$_2$ ΔH_t, $0.82^{817°}$	22			15.19	17.13	19.00	20.91
Nb$_2$O$_5$	24.9		142.8	34.65	38.42	40.64	41.95
Nitrogen							
N$_2$ ΔH_t, $0.055^{-237.53°}$	0.172	1.333					
N$_3$				10.68	12.05	12.93	13.50
NCN radical				11.04	12.42	13.25	13.75
NCO radical				10.48	11.84	12.75	13.35
NF$_3$	3.670	2.769		14.79	17.07	18.16	18.73
N$_2$F$_2$ cis	3.400	21.9		13.90	16.32	17.60	18.31
trans		21.0		14.39	16.47	17.64	18.32
N$_2$F$_4$	3.17	15.9					
NClF$_2$		4.35					
NH				6.97	7.04	7.22	7.47
NH$_2$				8.22	8.80	9.49	10.18
NH$_3$	1.351	5.581		9.24	10.81	12.23	13.47
N$_2$H$_2$ cis				9.78	11.88	13.55	14.83
N$_2$H$_4$	3.025	9.70		15.10	18.30	20.60	22.30
NHF$_2$		5.94					

NO	0.550	3.292		7.16	7.47	7.83	8.12
NO₂⁻				9.93	11.39	12.26	12.77
NO₃⁻				13.37	16.11	17.51	18.28
N₂O	1.563	3.956		10.20	11.56	12.48	13.11
N₂O⁺		9.40		10.98	12.27	13.10	13.62
N₂O₃		9.11 equil. mixture		17.39	19.82	21.39	22.38
N₂O₄	3.502			9.64	11.02	12.00	12.63
N₂O₅		6.14	14.9	26.49	30.68	32.74	33.81
NOBr		4.61		11.58	12.26	12.61	12.86
NOCl				11.27	12.11	12.71	13.12
NOF		5.0		10.65	11.69	12.35	12.78
NOF₃		4.31		18.82	21.72	23.19	24.02
NO₂Cl		13.2		14.26	16.29	17.49	18.20
NO₂F		7.6		13.64	15.88	17.19	17.98
N₃P₃Cl₆	5.0						
N₃P₃F₆	5.3						
Osmium							
Os	7.59			6.00	6.18	6.37	6.55
OsF₆	4.06	6.84					
OsO₄ yellow	2.34	9.45					
white							
Oxygen							
O	0.106	1.630		5.13	5.05	5.02	5.00
O₂ ΔHt, 0.022⁻²⁴⁹·⁴⁹°				7.19	7.67	8.06	8.34
ΔHt, 0.178⁻²²⁹·³⁸°							
O₃		2.59		10.43	11.87	12.62	13.04
OF₂		2.65		11.37	12.52	13.07	13.35
O₂F₂		4.583					
O₃F₂		4.58					
Palladium							
Pd	4.2	86.4		6.34	6.61	6.89	7.16
PdCl₂	9.58			8.98	11.82	14.66	
PdO							

Table 9-3 (Continued)
HEATS OF MELTING AND VAPORIZATION (OR SUBLIMATION) AND SPECIFIC HEAT AT VARIOUS TEMPERATURES OF THE ELEMENTS AND INORGANIC COMPOUNDS

Substance	ΔHm	ΔHv	ΔHs	C_p (at specified temperature, K)			
				400	600	800	1000
Phosphorus							
P red, V	4.5	$3.37^{25°}$	7.69	5.54	6.17	6.90	
white ΔHt, $0.125^{-77.8}$	0.157						
black			8.21				
P_2				8.05	8.49	→P	
P_4				17.51	18.74	19.21	19.45
PBr_3		9.5		18.86	19.40	19.60	19.70
PBr_5		4.2					
$PBrF_2$		5.45					
PBr_2F		7.34					
PCl_3	1.08	7.3		18.17	19.06	19.40	19.57
PCl_5		5.95	15.5	28.70	30.31	30.94	31.24
$PClF_2$		5.45					
PCl_2F		3.49					
PF_3		4.11		15.79	17.68	18.54	18.99
PF_5				23.72	27.42	29.13	30.02
PH				7.01	7.24	7.59	7.91
PH_3	0.270	3.489		9.19	10.35	11.13	11.80
P_2H_4		6.89					
$P(NCO)_3$		11.9					
PO	3.360			7.73	8.13	8.42	8.60
P_2O_6		10.38		41.75	47.89	50.71	52.40
P_4O_{10}			25.34	62.20	80.30	94.80	105.80
$POBr_3$		10.9		22.77	24.11	24.75	25.10
$POCl_3$	8.2	8.4		21.98	23.69	24.50	25.94
$POClF_2$		6.08		18.96	21.89	23.36	24.16

$POCl_2F$		7.40		20.95	23.08	24.12	24.67
POF_3	3.60	5.55		18.90	21.81	23.29	24.10
$PO(NCO)_3$		13.41					
$PO(NCS)_3$		14.82					
P_4S_3	2.2	14.3		44.0	44.0	37.00	37.00
$PSBr_3$				23.86	24.87	25.27	25.47
$PSCl_3$		5.70		23.07	24.47	25.04	25.31
$PSClF_2$		6.89					
$PSCl_2F$		4.68					
PSF_3				15.79	17.68	18.54	18.99
$PS(NCS)_3$		14.82					
Platinum							
Pt	4.70	121.8		6.31	6.57	6.82	7.08
Plutonium							
Pu $\Delta Ht,\ 3.2^{122°}$	16.0	79.71		9.45	11.20	9.70	9.70
$\Delta Ht,\ 0.7^{206°}$							
$\Delta Ht,\ 0.8^{319°}$							
$\Delta Ht,\ 16.0^{480°}$							
$PuBr_3$	13.4	56.5	69.9				
$PuCl_3$	15.2	57.6	72.8				
PuF_3	13		89.6				
PuF_4	10.2	7.15	71.6				
PuF_6	4.456		11.6				
PuI_3	12.0						
PuO_2		133.8					
Polonium							
Po		24.60	80				
Potassium							
K	0.558	18.38		7.53	7.20	7.11	7.26
K_3AlCl_6				61.94	66.80	71.60	72.19
K_3AlF_6				58.43	64.38	68.54	
KBH_4				24.12	25.34		
KBF_4 $\Delta Ht,\ 3.36^{283°}$	4.22	57.1		31.27	33.96	36.08	39.94
KBO_2	7.5			18.32	21.47	23.54	24.97
$K_2B_4O_7$	25			49.30	59.88	64.80	67.70

Table 9-3 (Continued)

HEATS OF MELTING AND VAPORIZATION (OR SUBLIMATION) AND SPECIFIC HEAT AT VARIOUS TEMPERATURES OF THE ELEMENTS AND INORGANIC COMPOUNDS

Substance	ΔH_m	ΔH_v	ΔH_s	C_p (at specified temperature, K)			
				400	600	800	1000
KBr ΔH_s, 59.04*	6.1	35.65	51.08§	12.86	13.47	14.44	16.26
KCl ΔH_v, 37.14§	6.282	29.7†		12.70	13.46	14.36	16.01
$KClO_4$ ΔH_t, $3.29^{299.6°}$				33.10	39.45	43.56	
KCN ΔH_t, $0.279^{-104.9°}$	3.5	37.55	46§	15.85	15.87	15.89	18.00
K_2CO_3	6.6			30.63	36.01	40.64	45.16
K_2CrO_4	6.92						
$K_2Cr_2O_7$	8.77						
KF	6.50	33.9†	57.8§	12.20	12.97	13.72	14.62
KH				10.55	12.41	8.68	8.87
KHF_2 ΔH_t, $2.68^{196.70°}$	1.582			20.57	25.00	25.00	25.00
KI	5.74	45.62		12.89	13.70	14.97	17.30
KNO_3 ΔH_t, $1.22^{128°}$	2.30			25.91	28.80	25.19	27.13
K_2O				21.88	23.28		
K_2O_2 ΔH_t, $0.072^{-79.7°}$	6.10			25.60	28.85		
KO_2 ΔH_t, $0.0375^{-42.3°}$	3.0			20.05	21.55		
K_2O_3	7.03	34.1	45.9				
KOH ΔH_t, $1.54^{243°}$	2.06			17.13	18.80	19.86	19.86
KOH^+				12.28	12.65	12.88	13.11
KPO_3	2.11						
K_3PO_4	8.90						
$K_4P_2O_7$	14.00						
$KReO_4$	20.40						
K_2SiO_3	(12)			32.40	37.70	40.80	42.80
K_2SO_4 ΔH_t, $1.79^{584°}$	8.48			35.35	40.76	50.44	45.79

K_2WO_4	4.65						
K_2ZrCl_6	5.50						
Praseodymium							
Pr	2.70	79.5					
Protactinium							
Pa	(3.5)						
$PaCl_5$	22.2	14.65					
Radium							
Ra	(2.0)						
Radon							
Rn	(0.69)	4.01					
Rhenium							
Re	8.0	170.8		6.22	6.44	6.69	6.95
ReF_5	1.107	13.88					
ReF_6		6.867					
ReF_7	1.799	9.154					
ReO_2	5.20	17.7	65.64				
ReO_3	15.8		49.8				
Re_2O_7	3.80						
Re_2O_8							
$ReOCl_4$	3.23	10.9					
$ReOF_4$		14.59					
$ReOF_5$		7.72	8.94				
ReO_2F_3		15.7					
ReS_2			22.66				
Rhodium							
Rh	5.15	118.4		6.21	6.69	7.17	7.65
Rubidium							
Rb	0.525	18.11		7.50	7.50	7.50	
RbBr	3.70	37.12		12.62	13.13	13.64	16.00
RbCl	4.40	36.92		12.50	12.99	13.49	15.30
$RbClO_4$ ΔH_t, $3.01^{284°}$							
RbF	6.3	42.5		12.41	13.83	15.52	17.29
RbI	2.99	35.96		12.65	13.18	13.70	16.00
$RbNO_3$	1.34						

*To dimer.   †To equilibrium mixture.   §To monomer.

Table 9-3 (Continued)

HEATS OF MELTING AND VAPORIZATION (OR SUBLIMATION) AND SPECIFIC HEAT AT VARIOUS TEMPERATURES OF THE ELEMENTS AND INORGANIC COMPOUNDS

Substance	ΔH_m	ΔH_v	ΔH_s	C_p (at specified temperature, K)			
				400	600	800	1000
RbOH	1.62						
Ruthenium							
Ru ΔHt, $0.23^{1500°}$	6.21	141.4		5.85	6.15	6.45	6.75
Samarium							
Sm ΔHt, $0.744^{917°}$	2.13	39.38	49.5	7.96	9.34	10.58	11.78
Sm$_2$O$_3$ ΔHt, $0.25^{922°}$				29.92	32.34	33.79	34.96
Scandium							
Sc	(3.85)	(72.85)					
Sc$_2$O$_3$	1.30			25.43	26.55	27.68	28.81
Selenium							
Se ΔHt, $0.18^{150°}$				6.71	8.40	8.40	
Se$_2$		25.49		8.66	8.81	8.86	8.89
Se$_6$		20.60					
SeF$_4$		11.2					
SeF$_6$	2.0	22.58	6.4	30.56	33.78	35.16	36.01
SeO$_2$		10.2					
SeOCl$_2$	1.01	6.77					
SeO$_2$F$_2$							
Silicon							
Si	12.0	85.8	107.7	5.29	5.77	6.06	6.30
SiBr$_4$		9.05					
SiC				8.15	9.97	10.94	11.55
SiCl$_4$	1.845	6.86		23.13	24.51	25.06	25.33
SiF$_4$			6.13	23.13	24.51	25.06	25.33
SiH$_4$		2.96		12.30	15.75	18.34	20.21
Si$_2$H$_6$		5.11					

Si₃H₈	6.78			14.49	17.69	19.87	21.38
SiH₃Cl	6.36			21.10	23.31	24.31	24.82
SiClF₃				26.45	31.00	34.85	37.80
Si₃N₄		1.84		12.77	15.40	17.62	16.48
SiO₂ quartz ΔHt, 0.17$^{575°}$				11.32	14.83	16.09	16.71
ΔHt, 0.48$^{806°}$							
cristobalite, low							
ΔHt, 0.32$^{270°}$							
(to high cristobalite)							
cristobalite, high							
SiOF₂		2.3		14.66	16.82	17.93	18.55
SiS₂		5.00		18.79	19.53	19.87	20.41
Silver							
Ag	61.68	2.855		6.10	6.41	6.78	7.17
AgBr	45.9	2.18		14.09	17.17	14.90	14.90
AgCl	42.5	3.155		13.59	14.73	16.00	16.00
AgF	42.8	4.0		12.93	13.96		
AgI ΔHt, 1.47$^{150°}$	34.45	2.25		15.46	13.50	13.50	14.00
AgNO₃ ΔHt, 0.6$^{160°}$		2.755		26.88	30.60		
Ag₂S ΔHt, 1.4$^{176°}$; 1.4$^{586°}$		2.7		20.69	21.64	21.64	21.64
Ag₃Sb				25.93	29.13	32.33	
Ag₂SO₄ ΔHt, 3.75$^{430°}$		4.3		34.26	39.84	45.42	49.00
Ag₂Te ΔHt, 0.17$^{137°}$				23.10	22.20	22.20	22.20
Sodium							
Na ΔHs, 32.87 (→Na₂)	23.285	0.622	25.75	7.53	7.12	6.92	6.92
Na₂				9.04	9.15	9.24	9.32
NaAlCl₄				19.94	22.53	23.60	24.45
Na₃AlCl₆				60.80	65.25	69.00	71.85
Na₃AlF₆ ΔHt, 2.0$^{565°}$		25.64		56.08	62.58	70.94	67.40
ΔHt, 0.1$^{880°}$							
NaAlO₂ ΔHt, 0.310$^{467°}$		8.0		19.94	22.53	23.60	24.45
NaBH₄ ΔHt, 0.239$^{-83.3°}$				22.60	25.96		
NaBO₂	57.3	19.4	77.0	18.02	21.17	23.24	24.67
Na₂B₄O₇				51.10	57.72	62.65	66.85
NaBr	38.42	6.24	51.98	12.78	13.40	14.00	14.60

Table 9-3 (Continued)

HEATS OF MELTING AND VAPORIZATION (OR SUBLIMATION) AND SPECIFIC HEAT AT VARIOUS TEMPERATURES OF THE ELEMENTS AND INORGANIC COMPOUNDS

Substance	ΔH_m	ΔH_v	ΔH_s	C_p (at specified temperature, K)			
				400	600	800	1000
$NaCl$	6.73			12.51	13.26	14.18	15.50
$NaClO_3$	5.40						
$NaClO_4$ ΔH_t, $0.60^{309°}$				32.50	38.30		19.00
$NaCN$	2.1	35.39	41.3	16.41	16.46	16.50	
Na_2CO_3 ΔH_t, $0.165^{450°}$	7.090			29.90	39.03	36.65	42.83
NaF	7.97	42.1	68.1	11.85	12.59	13.32	14.22
$NaFeCl_4$	4.30						
NaH				10.15	12.12	13.98	15.50
NaI	5.64			12.86	13.44		
$NaIO_3$ ΔH_t, $8.39^{422°}$							
Na_2MoO_4	3.60						
$NaNO_3$	3.49			27.72	37.19	22.14	
NaO_2 ΔH_t, $0.350^{-76.7°}$; ΔH_t, $0.370^{-49.9°}$				18.23	20.19	22.14	
Na_2O ΔH_t, $0.42^{750.1°}$; ΔH_t, $2.85^{970.1°}$	11.4	41.9	54.5	18.11	20.49	21.82	22.69
Na_2O_2 ΔH_t, $1.37^{512°}$	1.58			23.36	25.92	27.15	20.01
$NaOH$ ΔH_t, $1.72^{299°}$	1.6			15.52	20.57	20.29	21.1
Na_2S				19.3	19.9	20.5	
Na_2SiO_3 ΔH_t, $0.16^{78°}$	12.38			30.54	35.16	38.18	40.49
$Na_2Si_2O_5$ ΔH_t, $0.15^{707°}$	8.5			43.83	52.00	56.21	70.00
Na_2SO_4 ΔH_t, $2.584^{248.6°}$ (V $\to$ l); ΔH_t, $1.63^{248.6°}$ (III $\to$ l); ΔH_t, $0.08^{707°}$ (l $\to$ δ)	5.50			35.76	41.92	44.80	48.25
Na_2TiO_3	16.8						

Compound				37.13	42.60	47.50	50.00
Na$_2$WO$_4$ ΔH_t, $7.373^{587.7°}$ ΔH_t, $0.983^{588.9°}$	5.688						
Na$_2$ZrCl$_6$	4.0						
Strontium							
Sr ΔH_t, $0.18^{555°}$	1.96	32.73	39.2	6.79	7.65	8.80	9.00
SrBr$_2$ ΔH_t, $2.92^{645°}$	2.42	46.4	74	18.89	19.76	20.93	27.82
SrCl$_2$ ΔH_t, $1.44^{727°}$	3.88	59.31	83.5	18.86	20.00	21.70	25.29
SrCO$_3$ ΔH_t, $4.7^{924°}$	7.09	76.4	107.8	22.72	25.61	27.74	29.64
SrF$_2$ ΔH_t, $0.01^{1148°}$ ΔH_t, $0.01^{1211°}$				17.86	18.84	19.35	20.50
SrI$_2$	4.70	45.35	68.5	19.30	20.62	21.93	26.30
SrO	18			11.59	12.44	12.98	13.40
Sr$_3$(PO$_4$)$_2$	18.50						
SrSO$_4$				27.12	29.78	32.44	35.10
Sulfur							
S rhombic ΔH_t, $0.096^{95.31°}$ (rh → mn)	0.4105	2.30	66.7	7.73	8.20	4.37	4.42
S$_2$				8.14	8.55	8.74	8.84
S$_8$				39.84	41.89	42.67	43.04
S$_2$Cl$_2$		8.61		18.37	19.16	19.46	19.61
SF$_4$		6.32		19.99	22.78	24.01	24.63
SF$_6$		4.08		27.85	32.54	34.62	35.69
S$_2$F$_{10}$		6.15	5.60				
SH				7.56	7.47	7.62	7.86
SO				7.54	8.09	8.41	8.62
SO$_2$ (mp 16.9°) III	1.769	5.955		10.39	11.72	12.53	13.02
SO$_3$ (mp 32.6°) II	0.47	9.99		14.06	16.90	18.61	19.76
(mp 62.3°) I	2.47	10.40					
	6.09						
SOBr$_2$	7.41						
SOCl$_2$		5.90		17.03	18.25	18.85	19.18
SOClF		5.18					
SOF$_2$		5.1		15.36	17.30	18.26	18.78
SOF$_4$							

Table 9-3 (Continued)
HEATS OF MELTING AND VAPORIZATION (OR SUBLIMATION) AND SPECIFIC HEAT AT VARIOUS TEMPERATURES OF THE ELEMENTS AND INORGANIC COMPOUNDS

Substance	ΔH_m	ΔH_v	ΔH_s	C_p (at specified temperature, K)			
				400	600	800	1000
SO_2Cl_2		7.50		20.37	22.59	23.75	24.41
SO_2F_2		4.79		18.29	21.35	22.96	23.87
SO_2ClF				19.38	22.01	23.39	24.16
SO_3BrF		8.67					
SO_3F_2 (FSO_2OF)		5.35					
$S_2O_5Cl_2$		13.2					
S_2O_5ClF		8.1					
$S_2O_5F_2$		7.6					
Tantalum							
Ta	8.7	176.15		6.18	6.42	6.56	6.68
$TaBr_5$	9.0	14.0		40.20			
TaC	25			9.96	11.10	11.73	12.21
$TaCl_5$	8.9	13.6		40.56	31.07	31.30	31.41
TaF_5	4.5	11.0		43.50			
TaI_5	10.0	15.5		39.33	43.50	28.67	28.83
TaO_2				11.41	12.50	13.04	13.32
Ta_2O_5	28.7			35.26	39.30	41.87	43.68
Technetium							
Tc	5.50	138.0		6.00	6.40	6.80	7.20
TcF_6 cubic	1.128	7.427					
TcO_3F	5.377	9.453					
Tellurium							
Te	4.180	12.10		6.68	7.73	9.00	9.00
$TeCl_4$	4.510	16.8		33.20	53.2	26.00	26.00
TeF_4		8.2					
TeF_6			6.74	31.60	34.36	35.53	36.25
Te_2F_{10}		9.44					

TeH_2		5.7	59				
TeO_2	7.00	51.7	92.9	16.22	17.33	18.18	18.94
Terbium							
Tb	(3.9)	(70)					
Thallium							
Tl ΔHt, $0.09^{234°}$	0.975	39.41		6.57	7.20	7.20	7.20
TlBr	3.9	24.7		12.79	14.21	18.02	16.21
TlCl	3.8	24.75		12.80	13.20	14.20	14.20
Tl_2CO_3	4.40						
TlF	3.3	27.7		12.40	16.08	16.08	16.08
TlI	3.5	24.8		12.89	14.48	17.20	17.20
$TlNO_3$	2.29						
Tl_2S	3.00						
Tl_2SO_4	5.50						
Thorium							
Th ΔHt, $0.653^{1630°}$	3.85	122.8		6.79	7.30	7.81	8.23
ThC_2 ΔHt, $1.5^{1415°}$ ΔHt, $1.5^{1500°}$				14.95	16.29	17.14	17.84
$ThCl_4$ ΔHt, $1.2^{406°}$	10.50	36.5		30.28	31.69	32.60	33.72
ThI_4	8.00	31.5		40.50	46.86	53.22	
Th_3N_4							
ThO_2	291.1			16.10	17.31	18.00	18.56
Thulium							
Tm	(4.4)	(51)	55.5				
Tin							
Sn ΔHt, $0.50^{13°}$	1.67	70.78		6.90	6.90	6.85	6.85
$SnBr_2$	1.72	32.50					
$SnBr_4$	2.85	9.8		37.75	25.52	25.64	25.69
$SnCl_2$	3.05	19.50		19.90	22.00	22.00	14.00
$SnCl_4$	2.19	8.1		24.48	25.17	25.44	25.58
$Sn(CH_3)_4$		7.32					
SnH_4		4.42					
SnI_2		24.00					
SnO				10.95	11.65	12.35	13.05
SnO_2 ΔHt, $0.45^{410°}$ ΔHt, $0.3^{540°}$				15.40	17.67	18.77	19.54

Table 9-3 (*Continued*)

HEATS OF MELTING AND VAPORIZATION (OR SUBLIMATION) AND SPECIFIC HEAT AT VARIOUS TEMPERATURES OF THE ELEMENTS AND INORGANIC COMPOUNDS

Substance	ΔHm	ΔHv	ΔHs	C_p (at specified temperature, K)			
				400	600	800	1000
SnS ΔHt, $0.16^{602°}$	7.55			12.08	13.27	14.66	13.52
SnS$_2$				17.19	18.03	18.87	19.71
Titanium							
Ti $\alpha \rightarrow \beta$ ΔHt, $0.99^{882°}$	4.45	101.63	111.57	6.34	6.76	7.23	7.77
β							
TiB				9.64	11.61	12.17	12.41
TiB$_2$	24			13.12	15.81	17.23	18.38
TiBr$_2$			49.29	19.06	19.63	20.18	20.73
TiBr$_3$ ΔHt, $0^{93.3°}$	3.08	10.80	33.17	25.29	30.00	35.20	37.45
TiBr$_4$	17			36.30	24.34	25.56	25.66
TiC				9.72	11.39	11.93	12.23
TiCl$_2$			50.7	17.54	18.73	19.64	20.54
TiCl$_3$ ΔHt, $0.0^{53.1°}$			39.71	23.65	24.39	24.96	25.50
TiCl$_4$	2.382	8.55		34.94	24.96	25.33	25.51
TiF$_2$			52.97	13.49	14.21	14.50	14.66
TiF$_3$			23.4	22.30	23.54	24.75	26.04
TiF$_4$				30.27	23.91	24.68	25.08
TiH$_2$			51.8	9.40	12.86	15.09	16.38
TiI$_2$			41	20.79	21.13	21.48	21.83
TiI$_3$				28.09	28.43	28.78	20.63
TiI$_4$ ΔHt, $2.37^{106°}$	4.74	13.5		35.40	37.40	25.70	25.75
TiN	16		143	10.47	11.65	12.10	12.46
TiO ΔHt, $1^{992°}$	10			10.75	12.15	13.19	14.12
TiO$_2$ anatase	13.86		161	15.20	16.94	17.65	18.01
rutile	16		163.8	14.62	15.99	16.82	17.50
Ti$_2$O$_3$ ΔHt, $0.272^{197°}$	25			28.09	32.61	34.17	34.98
Ti$_3$O$_5$ ΔHt, $3.2^{177°}$	41			43.70	45.20	47.60	50.00
Ti$_4$O$_7$	54			57.34	64.81	68.35	70.46

$TiOCl_2$	9.5	192.8	203.4	18.10	18.96	19.33	19.52
$TiOF_2$	4.10	19.48		16.63	18.20	18.87	19.21
Tungsten							
W				5.95	6.16	6.37	6.59
WBr_5				39.60	43.50	31.59	31.67
WBr_6				46.00	37.35	37.53	37.61
WCl_2				19.13	20.18	21.22	
WCl_4				32.34	34.95	31.31	31.49
WCl_5 ΔH_t, 1.0$^{177°}$	4.9	16.266	24	40.02	30.95	37.24	37.42
WCl_6 ΔH_t, 3.77$^{230°}$	1.60	14.32	23.9	46.00	48.00		
WC			17.2	9.76	10.90	11.51	11.96
$W(CO)_6$				64.34			
WF_6	0.980	6.47		31.65	34.66	35.93	36.56
WO_2			159.24	15.16	17.04	18.04	18.68
WO_3 ΔH_t, 0.355$^{777°}$	17.55	18.3	131.46	19.65	22.26	23.48	24.30
W_3O_8				53.45	57.40	59.10	59.96
$WOCl_4$	10.85	10.5		37.50	29.44	30.36	30.84
WOF_4		13.4		43.50	28.62	29.89	30.53
WO_2Cl_2	1.22		26	27.50	32.40	23.92	25.05
Uranium							
U ΔH_t, 0.700$^{672°}$ ΔH_t, 1.145$^{772°}$	3.00	99.7		6.92	8.31	9.95	10.00
UBr_4	13.2	28.5		31.40	33.48	39.00	39.00
UC	11.1			13.06	13.93	14.42	14.86
UCl_3	10.70	46.13		24.56	25.74	27.16	28.66
UCl_4	8.50	33.8		30.13	32.12	33.93	38.83
UCl_5	5.0	18.0		36.06	44.62	32.15	34.20
UCl_6	10.2	12.0		43.70	51.14	37.96	40.16
UF_4	8.0	53.0		28.46	29.88	31.29	32.69
UF_5				32.59	34.19	39.82	
UF_6	4.59	31.21	11.52	33.58	35.53	36.38	36.91
UI_4	16.9			33.60	35.72	39.60	39.60
UN				12.48	13.45	13.94	14.30
UO_2				17.37	19.07	19.88	20.42

Table 9-3 (Continued)

HEATS OF MELTING AND VAPORIZATION (OR SUBLIMATION) AND SPECIFIC HEAT AT VARIOUS TEMPERATURES OF THE ELEMENTS AND INORGANIC COMPOUNDS

Substance	ΔH_m	ΔH_v	ΔH_s	C_p (at specified temperature, K)			
				400	600	800	1000
UO_3				21.25	22.79	23.66	
U_3O_8				63.57	69.48	72.70	
$UOBr_2$				25.52	27.40	28.49	
$UOCl$				18.10	19.77	21.59	
$UOCl_2$				24.35	26.19	27.50	
$UOCl_3$				28.95	30.87	32.78	
UO_2Cl_2				28.22	30.17	31.07	
UO_2F_2				27.22	29.28	30.27	30.95
US				12.20	12.40	12.59	12.79
US_2				18.68	20.42		
Vanadium							
V	5.5	106.77	123.2	6.27	6.57	6.85	7.19
VBr_3			43.3				
VCl_2			44.8	17.89	18.69	19.32	19.90
VCl_3		9.1	44.7	23.51	24.88	25.86	
VCl_4	2.3			38.65	23.93	24.52	25.02
VF_5	11.94	10.62		18.09	18.49	18.89	19.29
VI_2			44	10.35	11.52	12.24	12.83
VN ΔH_{dec}, $54.4^{2346°}$			177				
VO	13.0			11.85	12.79	13.65	14.45
V_2O_3 ΔH_t, $0.388\text{-}104.3°$	28.0			28.09	30.43	31.70	32.97
VO_2 ΔH_t, $1.037^{2°}$	13.6			16.06	17.77	18.59	19.16
V_2O_4 ΔH_t, $2.156^{7°}$	26.79	63.00		32.33	35.48	37.16	38.40
V_2O_5	15.56	8.0		36.08	40.23	42.36	45.60
$VOCl_3$				36.00	25.24	25.49	25.60
Xenon							
Xe	0.5495	3.020					

Substance	ΔH fus	ΔH vap	ΔH				
XeF₂			12.3	6.53	6.81	7.16	7.52
XeF₄			15.3	27.09	28.99	29.81	30.32
XeF₆			15.3				
Yttrium							
Y ΔHt, 1.189$^{1485°}$	2.73	87.13		6.31			
Y₂O₃ ΔHt, 0.31$^{1057°}$							
Zinc							
Zn	1.765	27.62		6.31	7.50	7.50	7.50
ZnBr₂	3.74	23.5		16.76	18.84	27.20	14.70
ZnCl₂	2.45	28.5		16.70	18.84	24.10	24.10
ZnF₂		44.00		16.76			
Zn(C₂H₅)₂		9.6					
ZnI₂	4.50	23.00		10.84	11.84	12.35	12.71
ZnO	4.47			11.81	12.53	12.94	13.26
ZnS ΔHt, 3.2$^{1020°}$					32.82	36.46	40.10
ZnSO₄				29.18			
Zirconium							
Zr ΔHt, 0.96$^{862°}$	5.0	141.13	148.3	6.37	6.92	7.48	8.03
ZrB₂	25	31.5	55	13.74	15.73	16.66	17.25
ZrBr₂	5.5		49	21.01	21.55	22.10	21.75
ZrBr₃			28	24.50	25.00	25.20	25.35
ZrBr₄				30.90	31.90	25.63	25.71
ZrC	19	45.0		10.42	11.80	12.45	12.76
ZrCl₂	6.4		45.4	18.17	19.13	19.87	20.54
ZrCl₃			26.4	24.07	25.24	26.06	26.79
ZrCl₄	12	69.0	96.6	29.97	31.34	25.46	25.59
ZrF₂			70.1	16.70	18.20	19.27	20.08
ZrF₃	9.0		56.80	21.95	24.00	25.10	25.75
ZrF₄		51.64		27.12	29.63	30.93	32.05
ZrI₂	15.35	27		22.71	23.09	25.37	29.54
ZrI₃	6		42	25.30	25.50	25.60	25.77
ZrI₄			31.0	30.17	30.81	25.73	19.82
ZrN	16.1			10.70	11.63	12.18	12.61
ZrO₂ ΔHt, 1.42$^{1205°}$	20.8	149.2		15.26	16.79	17.56	18.10
ZrSiO₄				27.85	31.81	33.84	35.09

Table 9-4

HEATS OF MELTING AND VAPORIZATION (OR SUBLIMATION) AND SPECIFIC HEAT AT VARIOUS TEMPERATURES OF ORGANIC COMPOUNDS

See Table 9-3 for abbreviations used in the table.

Substance	ΔH_m	ΔH_v	ΔH_s	C_p (at specified temperature, K)			
				400	600	800	1000
Acenaphthene			20.6				
Acenaphthylene			17.0				
Acetaldehyde	0.770	6.24		15.73	20.52	24.20	29.96
Acetanilide			19.3				
Acetic acid	2.80	5.663		19.52	25.15	29.08	31.99
Acetic anhydride	2.51	9.85	11.54	30.86	41.62	48.91	54.11
Acetone	1.366	6.952		22.00	29.34	34.93	39.15
Acetonitrile	1.952	7.3	7.94	14.62	18.35	21.26	23.50
ΔHt, 0.215$^{-56°}$							
Acetophenone		9.275	13.4				
Acetyl bromide			7.9				
Acetyl chloride			7.2	18.86	23.18	26.30	28.60
Acetylene	0.900	4.05	5.1	11.97	13.73	14.93	15.92
Acetylenedicarbonitrile			6.88	22.66	25.37	27.26	28.62
Acetyl fluoride			6.0				
Acetyl iodide			7.9				
Acrylic acid		11.21	12.98	22.94	29.50	33.93	37.12
Acrylonitrile		7.8		18.36	23.11	26.43	28.88
Adenine			25.8				
Adipic acid			30.8				
α-Alanine			33.0				
Allyl ethyl sulfoxide			17.1				
Allyl trichloroacetate			12.5				
1-Aminobutane		8.50		35.44	47.30	56.01	62.54
2-Aminobutane			7.5	35.40	47.55	56.42	62.54

Aniline	2.519	10.643	13.325	34.17	46.09	53.79	59.18
Anthracene		13.5	24.7				
9,10-Anthracenedione			26.8				
Azoisopropane			8.5				
Azulene	2.89	13.26	22.8	42.15	59.32	70.59	78.24
Benzaldehyde			12				
1,2-Benzanthra-9,10-quinone			19.8				
Benzene	2.358	7.352	8.090	26.74	37.73	45.06	50.16
Benzenethiol	2.736	9.53	11.64	32.76	44.13	51.59	56.79
Benzil			23.5				
Benzilidene anil			20.5				
Benzoic acid	4.32	12.10	22.70				
Benzoic anhydride			23				
Benzonitrile	2.60	11.0	13.26	33.65	44.80	52.08	57.08
Benzophenone			22.5				
1,4-Benzoquinone			15.00				
Benzoyl bromide			14.0				
Benzoyl chloride			13.1				
Benzoyl iodide			14.8				
3,4-Benzphenanthrene			25.4				
Benzyl bromide			11.3				
Benzyl chloride			12.3				
Benzyl ethyl sulfide			13.6				
Benzyl iodide			11.3				
Benzyl methyl ketone			12.78				
Benzyl methyl sulfide			12.8				
Bicyclo[4.1.0]heptane			9.14				
Bicyclo[3.1.0]hexane			7.85				
Bicyclo[4.2.0]octane			9.85				
Bicyclo[5.1.0]octane			10.42				
Bicyclopropyl			8.0				
Biphenyl	4.44	10.9		52.83	73.54	86.92	96.00
Biphenylene			30.8				
Bromobenzene	2.54	9.05	10.62	30.44	40.99	47.78	52.40
4-Bromobenzoic acid			21.0				

Table 9-4 (Continued)
HEATS OF MELTING AND VAPORIZATION (OR SUBLIMATION) AND SPECIFIC HEAT AT VARIOUS TEMPERATURES OF ORGANIC COMPOUNDS

Substance	ΔH_m	ΔH_v	ΔH_s	C_p (at specified temperature, K) 400	600	800	1000
1-Bromobutane	1.6	7.78		32.64	43.00	50.48	56.03
2-Bromobutane			8.45	33.09	43.76	51.31	56.93
Bromoethane	1.4	6.41	6.57	18.93	24.56	28.58	31.59
Bromoethene			12.05	15.91	19.83	22.50	24.46
1-Bromoheptane			10.91				
1-Bromohexane							
Bromomethane							
$\quad$ΔHt, $0.113^{-99.4°}$	1.429	5.715		11.94	14.98	17.26	19.01
2-Bromo-2-methylpropane			7.4	34.93	45.58	52.65	57.74
$\quad$ΔHt, $1.35^{-64.5°}$							
$\quad$ΔHt, $0.25^{-41.6°}$	0.47						
1-Bromooctane	2.74	8.24	13.14				
1-Bromopentane	1.56	7.14		39.58	52.34	61.55	68.36
1-Bromopropane		6.79		25.70	33.66	39.41	43.70
2-Bromopropane				26.34	34.42	40.09	44.26
1,2-Butadiene	1.665	5.82	5.71	23.54	30.72	36.01	40.02
1,3-Butadiene	1.908	5.42	5.03	24.29	31.84	36.84	40.52
α-Butadiene sulfone			14.7				
Butadiyne				20.17	23.14	25.11	26.61
Butane		5.352		29.60	40.30	48.23	54.22
$\quad$ΔHt, $0.494^{-165.60°}$	1.114		5.035				
2,3-Butanedione			9.25				
1,4-Butanedithiol			13.22				
1-Butanethiol	2.500	7.702	8.73	34.95	46.54	55.68	62.95
2-Butanethiol	1.548	7.312	8.14	35.38	46.42	54.29	60.02
1-Butanol	2.24	10.31	12.52	32.80	43.90	52.11	58.26
2-Butanol		9.75	11.87	33.70	44.72	52.68	58.62
2-Butanone	2.017	7.475	8.34	29.81	39.09	46.08	51.33

Compound							
1-Butene	0.920	5.238	4.81	26.04	35.14	41.80	46.82
2-Butene cis	1.747	5.580	5.29	24.33	33.80	40.87	46.15
trans	2.332	5.439	5.10	26.02	34.80	44.20	46.58
1-Buten-3-yne				21.26	26.67	30.40	33.16
N-Butylacetamide			18.2				
n-Butyl acetate			10.42				
t-Butylamine		8.58	7.10	36.46	48.87	57.49	63.79
Butylbenzene stable(I)	2.682(I)	9.38		54.75	75.20	89.37	99.49
metastable(II)	2.691(II)		11.98				
sec-Butylbenzene			11.72				
t-Butylbenzene			11.50				
sec-Butyl butyrate			11.3				
n-Butyl chloroacetate			12.2				
n-Butyl 2-chlorobutyrate			12.6				
n-Butyl 3-chlorobutyrate			12.7				
n-Butyl 4-chlorobutyrate			13.0				
n-Butyl 2-chloropropionate			13.0				
n-Butyl 3-chloropropionate			13.3				
n-Butyl crotonate			12.4				
sec-Butyl crotonate			11.8				
n-Butylcyclohexane	3.384	9.20	11.96	66.00	93.10	112.30	125.70
n-Butylcyclopentane	2.704	8.69	11.00	57.77	80.38	97.35	114.80
N-Butyldiacetimide			15.4				
n-Butyl dichloroacetate			12.5				
t-Butyl hydroperoxide			11.41				
n-Butylisobutylamine			10.73				
n-Butyl lithium			25.6				
n-Butyl trichloroacetate			12.8				
1-Butyne	1.441	5.861	5.67	23.87	30.83	35.95	39.84
2-Butyne	2.207	6.340	6.38	22.62	29.68	35.14	39.29
n-Butyraldehyde	2.654		8.05	30.20	39.60	46.60	51.70
Butyric acid	2.50		15.2				
Butyronitrile	1.2	10.04	9.53	28.39	37.07	43.48	48.22
D-Camphor	1.635	8.13					
ε-Caprolactam		14.22	19.9				
Carbazole			20.2				

Table 9-4 (Continued)
HEATS OF MELTING AND VAPORIZATION (OR SUBLIMATION) AND SPECIFIC HEAT AT VARIOUS TEMPERATURES OF ORGANIC COMPOUNDS

Substance	ΔH_m	ΔH_v	ΔH_s	C_p (at specified temperature, K)			
				400	600	800	1000
Carbon disulfide	1.049	6.401					
Chloroacetic acid			18				
Chloroacetyl chloride			9.3				
2-Chlorobenzaldehyde			13.3				
Chlorobenzene	2.28	8.73	9.81	30.62	41.16	47.89	52.48
2-Chlorobenzoic acid			19.0				
3-Chlorobenzoic acid			19.6				
4-Chlorobenzoic acid			21.0				
Chlorobenzoquinone			16.5				
1-Chlorobutane		7.38	8.0	32.30	42.77	50.31	55.92
2-Chlorobutane		6.98	7.60	32.52	43.18	50.84	56.60
Chlorocyclohexane			10.4				
Chlorodifluoromethane	0.985	4.833		15.63	18.87	20.84	22.10
Chloroethane	1.064	5.892		18.54	24.28	28.39	31.48
1-Chloro-2-ethylbenzene			11.3				
1-Chloro-4-ethylbenzene			11.5				
Chloroethylene				15.56	19.61	22.35	24.35
Chloroethyne				14.39	15.97	16.98	17.75
Chlorofluoromethane				13.29	16.57	18.81	20.39
Chloroform	2.28	7.08	7.48	17.75	20.38	21.87	22.83
Chloromethane	1.537	5.147		11.52	14.66	17.04	18.86
Chloromethyloxirane			9.7				
1-Chloro-2-methylpropane			7.57	32.52	43.18	50.84	56.60
2-Chloro-2-methylpropane	0.48	6.6		34.00	44.20	51.50	57.00

$\Delta H t$, $0.41^{-90.1°}$
$\Delta H t$, $1.39^{-53.6°}$

1-Chloronaphthalene			15.6				
2-Chloronaphthalene			19.6				
1-Chloropentane		7.93	9.1	39.24	52.11	61.38	68.25
3-Chlorophenol			12.7				
4-Chlorophenol			12.4				
1-Chloropropane		6.62	6.9	25.36	33.43	39.24	43.59
2-Chloropropane		6.34	6.47	25.99	34.20	39.94	44.16
3-Chloro-1-propene				22.12	28.43	32.93	36.30
Chlorotrifluoromethane				18.53	21.60	23.17	24.03
Chlorotrinitromethane			10.86				
Chrysene			28.1				
o-Cresol		10.20	18.17	39.74	52.77	61.55	68.82
m-Cresol		10.32	14.75	38.74	52.26	61.27	68.50
p-Cresol		10.32	17.67	38.65	52.10	61.11	68.48
m-Cresyl acetate			14.51				
Cubane			19.2				
4-Cyanothiazole			17.67				
Cyclobutane	0.260	5.781	5.65	23.89	34.76	42.42	47.96
$\Delta H t$, 1.38$^{-126.79°}$							
Cyclobutene				21.59	30.30	36.26	40.53
Cyclododecane			18.26				
Cycloheptane	0.450	7.93	9.21	41.82	62.42	77.03	87.40
$\Delta H t$, 1.187$^{-138.4°}$							
$\Delta H t$, 0.069$^{-75.0°}$							
$\Delta H t$, 0.108$^{-60.8°}$							
Cycloheptanone	0.277	9.250	12.4	37.13	50.07	58.58	64.58
1,3,5-Cycloheptatriene							
$\Delta H t$, 0.561$^{-119.19°}$							
Cyclohexane	0.640	7.160	7.896	35.82	53.83	66.76	75.80
$\Delta H t$, 1.611$^{-87°}$							
Cyclohexanol	0.406	10.875	12.820	41.14	59.29	72.18	81.13
$\Delta H t$, 1.96$^{-9.7°}$							
Cyclohexanone		9.00	10.77	36.00	52.90	65.00	73.00
Cyclohexene	0.787	7.285	8.00	34.64	49.45	59.49	66.62
$\Delta H t$, 1.016$^{-134.4°}$							
Cyclooctane	0.576	8.58	10.36	47.82	71.00	87.30	99.01
$\Delta H t$, 1.507$^{-106.7°}$							
$\Delta H t$, 0.114$^{-89.35°}$							

Table 9-4 (Continued)
HEATS OF MELTING AND VAPORIZATION (OR SUBLIMATION) AND SPECIFIC HEAT AT VARIOUS TEMPERATURES OF ORGANIC COMPOUNDS

Substance	ΔH_m	ΔH_v	ΔH_s	C_p (at specified temperature, K)			
				400	600	800	1000
Cyclooctanone	2.695		13.0	38.45	52.77	62.23	68.88
1,3,5,7-Cyclooctatetraene		8.700	10.30				
Cyclopentadiene			6.78				
Cyclopentane	0.1455	6.524	6.818	28.38	42.57	52.60	59.84
ΔH_t, 1.167$^{-150.76°}$							
ΔH_t, 0.0823$^{-135.08°}$							
Cyclopentanethiol	1.872	8.443	9.93	34.53	48.65	58.61	65.84
Cyclopentanol			13.74				
Cyclopentanone			10.21				
Cyclopentene	0.804		6.71	25.08	37.19	45.78	51.94
ΔH_t, 0.115$^{-186.08°}$							
Cyclopropane	1.301	4.793		18.31	26.15	33.57	35.39
Decahydronaphthalene							
cis ΔH_t, 0.511$^{-57.1°}$	2.268	9.940	12.0	56.64	84.14	103.36	116.91
trans	3.455	9.260	11.6	56.78	84.20	103.40	116.93
Decanal		9.388	12.277	71.80	95.70	113.00	125.70
Decane	6.863			71.24	96.36	114.92	128.20
1-Decanethiol	7.4	11.1	15.5	76.63	102.63	122.10	136.98
Decanoic acid	7.0		28.4				
1-Decanol	9.0	11.9	18.6	74.44	99.94	118.53	132.24
1-Decene	3.300	9.24	12.06	67.79	91.27	108.28	120.90
ΔH_t, 1.90$^{-74.8°}$							
1-Decyne				65.64	86.96	102.42	113.90
Deoxybenzoin			22.3				
Dibenzilidene azine			22.3				
Dibenzyl ketone			21.3				

Compound							
Dibenzyl sulfide			22.3				
Dibenzyl sulfone			27.8				
1,2-Dibromobutane			10.8	36.77	46.70	53.60	58.50
1,2-Dibromocycloheptane			12.43				
1,2-Dibromocyclohexane			12.07				
1,2-Dibromocyclooctane			13.04				
1,2-Dibromoethane	2.62	8.69	9.86	23.83	29.24	32.94	35.80
1,2-Dibromoheptane			13.01				
1,2-Dibromopropane				29.74	37.63	42.91	46.74
Di-n-butylborinic acid			15				
Di-n-butyl ether		8.83	10.5	60.78	81.29	96.52	107.86
Di-n-butyl mercury			15.6				
Di-t-butyl peroxide			7.6				
Di-n-butyl o-phthalate			21.9				
Di-n-butyl sulfate			18.1				
Di-n-butyl sulfite			16.2				
Di-n-butyl sulfone			24.0				
Dichloroacetyl chloride		9.7	9.4				
1,2-Dichlorobenzene	3.19	9.7	11.56	34.12	44.07	50.28	54.42
1,3-Dichlorobenzene			11.44	34.18	44.09	50.29	54.42
1,4-Dichlorobenzene	4.34	9.5	15.5	34.24	44.16	50.35	54.46
2,6-Dichlorobenzoquinone			16.7				
2,2'-Dichlorobiphenyl			23.0				
4,4'-Dichlorobiphenyl			24.8				
Dichlorodifluoromethane				19.69	22.37	23.69	24.39
1,1-Dichloroethane	1.881	6.97	7.36	21.85	27.18	30.79	33.40
1,2-Dichloroethane	2.112	7.65	8.47	22.00	26.90	30.40	33.00
1,1-Dichloroethylene	1.557	6.26	6.328	18.80	22.44	24.71	26.29
1,2-Dichloroethylene cis	1.72	7.08	7.43	18.41	22.23	24.60	26.23
trans	1.72	6.65	6.92	18.58	22.28	24.62	26.24
Dichlorofluoromethane				16.78	19.70	21.41	22.51
Dichloromethane	1.1	6.74	6.94	14.24	17.30	19.32	20.76
1,2-Dichloropropane		7.59	8.68	28.60	36.47	41.97	46.08
1,3-Dichloropropane		8.10	9.66	28.69	36.22	41.56	45.50
2,2-Dichloropropane		7.0	7.8	30.56	38.06	43.00	46.56
Dicyanoacetylene			6.88				

Table 9-4 (Continued)
HEATS OF MELTING AND VAPORIZATION (OR SUBLIMATION) AND SPECIFIC HEAT AT VARIOUS TEMPERATURES OF ORGANIC COMPOUNDS

Substance	ΔH_m	ΔH_v	ΔH_s	C_p (at specified temperature, K)			
				400	600	800	1000
2,2-Diethoxypropane			7.61				
Diethylamine	4.01		7.6	34.88	47.14	56.16	62.91
1,2-Diethylbenzene	2.62	9.42	12.61	56.01	75.66	89.54	99.49
1,3-Diethylbenzene	2.53	9.41	12.55	55.01	75.19	89.31	99.37
1,4-Diethylbenzene		9.41	12.54	54.68	74.84	89.04	99.16
Diethylene glycol		12.50	13.7				
Diethyl ether	1.745	6.38	6.516	33.01	43.92	52.26	58.51
Diethyl mercury			10.7				
Diethylmethyl phosphonate			13.5				
Diethylnitramine		10.04	12.7				
Diethyl oxalate			15.2				
Diethyl peroxide			7.3				
Diethyl o-phthalate			21.1				
Diethyl selenide			9.3				
Diethyl sulfate			13.6				
Diethyl sulfite			11.6				
Diethyl sulfone			20.6				
Diethyl sulfoxide			14.9				
1,2-Difluorobenzene	2.640	7.699	8.65	32.76	43.33	50.12	54.72
1,3-Difluorobenzene			8.29	32.72	43.13	49.67	53.93
1,4-Difluorobenzene			8.51	32.84	43.20	49.68	53.99
2,2'-Difluorobiphenyl			22.7				
4,4'-Difluorobiphenyl			21.8				
1,1-Difluoroethane		5.1		19.93	25.70	29.70	32.57
1,1-Difluoroethylene				17.16	21.32	23.95	25.74

Compound							
Difluoromethane			22.3	12.22	15.72	18.22	19.98
9,10-Dihydroanthracene			7.7				
Dihydropyran, 4H			27.7				
5,12-Dihydrotetracene			9.02				
2,3-Dihydrothiophene			9.55				
2,5-Dihydrothiophene			15.5				
1,2-Diiodobenzene			15.7	22.94	27.92	31.37	33.84
1,2-Diiodoethane			12.2	15.74	18.37	20.06	21.29
Diiodomethane	3.02(I) 2.88(II)	6.95					
Diisopropyl ether	2.635		7.75	46.90	62.61	74.39	83.17
Diisopropyl ketone			9.93				
Diisopropyl mercury			12.8				
1,2-Dimethoxybenzene			16.0				
Dimethoxyborane			6.14				
2,2-Dimethoxypropane	1.420		7.03				
Dimethylamine		6.330	6.07	20.89	28.41	33.94	38.19
Dimethylaminotrimethyl-silane		6.287	7.6	43.70	60.00	71.40	79.70
2,2-Dimethylbutane ΔH_t, $1.289^{-147.34°}$ ΔH_t, $0.068^{-132.28°}$	0.138	6.519	6.618	43.30	59.20	75.20	79.10
2,3-Dimethylbutane ΔH_t, $1.552^{-137.08°}$	0.194	6.55	6.96	42.60	55.40	65.00	72.20
2,3-Dimethyl-1-butene			6.97				
2,3-Dimethyl-2-butene ΔH_t, $0.844^{-76.34°}$	1.542	7.083	7.776	37.48	51.78	62.78	71.14
3,3-Dimethyl-1-butene ΔH_t, $1.037^{-148.3°}$	0.261	6.13	6.36	38.90	53.40	63.60	71.00
Dimethyl cadmium			9.07				
1,1-Dimethylcyclohexane ΔH_t, $1.430^{-120.01°}$	0.495	7.79	9.043	50.70	74.10	90.70	102.20
1,2-Dimethylcyclohexane cis ΔH_t, $1.974^{-100.6°}$	0.393	8.04	9.492	51.10	74.00	90.10	101.40
trans	2.491(I) 2.508(II)	7.86	9.168	51.90	74.60	90.50	101.70

Table 9-4 (Continued)

HEATS OF MELTING AND VAPORIZATION (OR SUBLIMATION) AND SPECIFIC HEAT AT VARIOUS TEMPERATURES OF ORGANIC COMPOUNDS

Substance		ΔH_m	ΔH_v	ΔH_s	C_p (at specified temperature, K)			
					400	600	800	1000
1,3-Dimethylcyclohexane	cis	2.586	7.84	9.137	51.20	74.20	90.50	102.00
	trans	2.358	8.09	9.369	51.10	73.80	89.80	101.10
1,4-Dimethylcyclohexane	cis	2.225	8.07	9.329	51.10	73.80	89.80	101.10
	trans	2.947	7.79	9.053	51.60	74.60	90.60	101.90
1,1-Dimethylcyclopentane ΔHt, $1.551^{-126.36°}$		0.258	7.239	8.079	43.55	62.78	76.18	85.83
1,2-Dimethylcyclopentane cis ΔHt, $1.594^{-131.66°}$		0.396	7.576	8.549	43.67	62.72	75.98	85.57
	trans	1.713	7.375	8.259	43.71	62.66	75.84	85.43
1,3-Dimethylcyclopentane	cis	1.761	7.265	8.200	43.71	62.66	75.84	85.43
	trans	1.738	7.361	8.248	43.71	62.66	75.84	85.43
Dimethyldichlorosilane				8.2				
2,4-Dimethyl-1,3-dioxane	cis			9.53				
4,5-Dimethyl-1,3-dioxane				10.16				
5,5-Dimethyl-1,3-dioxane				9.86				
Dimethyl ether		1.180	5.141		19.02	25.16	30.04	33.79
N,N-Dimethylformamide				11.4				
Dimethylfulvene				10.6				
Dimethylglyoxime				23.2				
2,2-Dimethylhexane		1.62	7.71	8.91				
2,3-Dimethylhexane			7.94	9.27				
2,4-Dimethylhexane			7.79	9.03				
2,5-Dimethylhexane		3.096	7.80	9.05				
3,3-Dimethylhexane		1.7	7.76	8.97				
3,4-Dimethylhexane			7.95	9.32				

Compound							
2,2-Dimethyl-3-hexene cis			8.88				
trans			8.91				
1,1-Dimethylhydrazine			8.37				
1,2-Dimethylhydrazine			9.40				
Dimethyl mercury			8.26				
Dimethylnitramine			16.7				
2,2-Dimethylpentane	1.392	6.97	7.75	50.42	68.33	81.43	91.20
2,3-Dimethylpentane		7.26	8.19	50.42	68.33	81.43	91.20
2,4-Dimethylpentane	1.636	7.05	7.86	50.42	68.33	81.43	91.20
3,3-Dimethylpentane	1.689	7.09	7.89	50.42	68.33	81.43	91.20
2,7-Dimethylphenanthrene			25.5				
4,5-Dimethylphenanthrene			25.0				
9,10-Dimethylphenanthrene			28.6				
2,2-Dimethylpropane	0.752	5.438	5.205	37.55	51.21	60.78	67.80
ΔHt, $0.616^{-133.14°}$							
2,3-Dimethylpyridine			11.70				
2,4-Dimethylpyridine			11.42				
2,5-Dimethylpyridine			11.43				
2,6-Dimethylpyridine			11.01				
3,4-Dimethylpyridine			12.38				
3,5-Dimethylpyridine			12.04				
Dimethyl sulfate			11.6				
Dimethyl sulfite			9.6				
Dimethyl sulfone			18.4				
Dimethyl sulfoxide	1.56	12.66	12.64				
3,3-Dimethyl-2-thiabutane	2.011(I)	7.523	8.57				
	1.83(II)						
2,2-Dimethylthiacyclopropane	1.69	8.00	8.55				
2,2-Dimethyl-3-thiapentane	2.49	8.04	9.4	50.64	66.22	77.12	85.24
2,4-Dimethyl-3-thiapentane			9.44				
1,3-Dinitrobenzene			14.3				
2,4-Dinitrophenol			25				
2,6-Dinitrophenol			26.8				
1,1-Dinitropropane			14.93				
1,4-Dioxane	3.07		9.20	30.23	43.44	52.15	58.05
ΔHt, $0.562^{-0.3°}$							

Table 9-4 (Continued)

HEATS OF MELTING AND VAPORIZATION (OR SUBLIMATION) AND SPECIFIC HEAT AT VARIOUS TEMPERATURES OF ORGANIC COMPOUNDS

Substance	ΔH_m	ΔH_v	ΔH_s	C_p (at specified temperature, K)			
				400	600	800	1000
1,3-Dioxolan			8.5				
Dipentene			11.5				
Diphenylamine			23.1				
Diphenylchlorosilane			16.6				
Diphenyl disulfide			22.7				
Diphenyl disulfone			38.7				
1,2-Diphenylethane		12.3	20.1				
1,1-Diphenylethene			17.5				
Diphenyl ether	4.115	15.5$^{25°}$	19.6				
Diphenylfulvene			25				
Diphenyl mercury			26.95				
Diphenylmethane			19.7				
Diphenyl sulfide			16.2				
Diphenyl sulfone			25.4				
Diphenyl sulfoxide			23.2				
Di-n-propyl ether			8.6	46.90	62.61	74.39	83.17
Di-n-propyl mercury			13.2				
Di-n-propyl sulfate			16.0				
Di-n-propyl sulfite			14.0				
Di-n-propyl sulfone			19.1				
Di-n-propyl sulfoxide			17.8				
2,3-Dithiabutane	2.197	8.05	9.17	26.36	32.83	37.66	41.31
5,6-Dithiadecane		11.2	15.2	68.38	89.98	105.83	117.86
3,4-Dithiahexane	2.248	9.01	10.89	40.90	52.24	60.19	65.97
1,3-Dithian-2-thione			21.85				
4,5-Dithiaoctane	3.30	10.02	12.55	44.50	71.30	83.70	93.20

Compound							
_N,N_-Dithiodiethylamine							
1,3-Dithiolan-2-thione							
Di-_p_-tolyl sulfone			12.6				
Divinyl ether			19.56				
Divinyl sulfone			26.2				
Dodecane	8.57	10.43	6.26	85.13	115.04	136.76	152.90
Dodecanedioic acid			13.5				
1-Dodecene	4.76	10.27	14.65	8.68	109.95	130.41	145.50
ΔH_t, 1.088 $^{-60.2°}$			36.6				
Eicosane	16.70	13.74	14.42	140.65	189.78	225.28	251.60
Eicosanoic acid	17.2		24.1	137.20	184.69	218.93	244.20
1-Eicosene	8.2	13.35	48				
Erythritol _meso_			23.86				
Ethane	0.683	3.517	32.3	15.65	21.35	25.81	29.30
1,2-Ethanedithiol			1.200				
Ethanethiol	1.189	6.401	10.68	21.08	27.21	31.83	35.38
Ethanol	1.198	9.255	6.526	19.36	25.69	30.33	33.83
Ethyl acetate	2.505	7.720	10.11	32.84	43.65	51.01	56.05
Ethyl allyl sulfone			8.63				
Ethylamine			20.0	21.65	28.68	33.89	37.88
_N_-Ethylaniline	2.195	6.7	6.7				
Ethylbenzene		8.50	12.5	40.76	56.44	67.15	74.77
3-Ethyl-1-butene		6.88	10.10	40.70	54.50	64.40	71.90
Ethyl crotonate			7.41				
Ethylcyclohexane	1.992	8.20	10.6	51.60	74.10	90.10	101.30
1-Ethylcyclohexene			9.67				
Ethylcyclopentane	1.642(I) 1.889(II)	7.715	10.34	43.89	61.70	75.22	85.16
Ethylene	0.801	3.237	8.72	12.67	17.87	20.03	22.43
Ethylene carbonate	2.41						
Ethylene glycol	2.78	11.86	17.5	27.06	32.72	36.90	39.88
Ethyleneimine		7.24	15.68	16.83	23.56	28.14	31.45
Ethylene oxide	1.236	6.101	7.55	14.95	20.62	24.60	27.47
Ethyl formate	2.20	7.201	5.96				
2-Ethyl-1-hexanal			11.70				
3-Ethylhexane	8.03		9.48				

Table 9-4. (Continued)

HEATS OF MELTING AND VAPORIZATION (OR SUBLIMATION) AND SPECIFIC HEAT AT VARIOUS TEMPERATURES OF ORGANIC COMPOUNDS

Substance	ΔH_m	ΔH_v	ΔH_s	C_p (at specified temperature, K)			
				400	600	800	1000
Ethyl isovalerate			10.5				
Ethyl lithium			27.9				
Ethylmercury bromide			18.3				
Ethylmercury chloride			18.2				
Ethylmercury iodide			19.0				
Ethyl methyl ether		7.92		26.08	34.58	41.19	46.18
Ethyl nitrate	2.04	7.40	8.67	28.73	37.07	42.72	46.69
3-Ethylpentane	2.282		8.42	50.42	68.33	81.43	91.20
Ethyl pentanoate			11.0				
2-Ethylphenol			15.20				
3-Ethylphenol			16.30				
4-Ethylphenol			19.20				
Ethylphosphonic acid			12.1				
Ethyl n-propanoate		8.178	9.0				
Ethyl β-vinylacrylate			11.6				
Ethyl vinyl ether			6.35				
Ethynylbenzene			24.65	35.95	48.01	55.79	61.17
Fluoranthrene	2.702		8.27	29.99	40.86	47.83	52.58
Fluorobenzene		7.457	21.8				
4-Fluorobenzoic acid				17.71	23.56	27.82	31.00
Fluoroethane				10.56	13.83	16.45	18.44
Fluoromethane				24.55	32.82	38.88	43.37
1-Fluoropropane				24.72	33.14	39.14	43.55
2-Fluoropropane	2.235	8.144	9.42	36.43	49.70	58.60	64.84
4-Fluorotoluene			8.3				
Fluorotrinitromethane							

Compound							
Formaldehyde	3.035	5.85		9.38	11.52	13.37	14.81
Formic acid		5.24	11.03	12.85	16.02	18.35	19.95
Formyl HCO				8.73	9.79	10.75	11.49
HCO$^+$				9.39	10.39	11.14	11.78
Fumaric acid			32.5				
Fumaronitrile			17.2				
Furan ΔH_t, 0.489$^{-123.2°}$	0.909	6.474	6.61	21.20	29.31	34.41	37.89
Furfuryl alcohol	3.12		15.4				
2-Furoic acid			25.92				
Furylethylene	4.416		9.1				
Glycerol			20.5				
Glyceryl triacetate			19.6				
Glyceryl trinitrate			23.9				
Heptadecane ΔH_t, 2.62$^{11.1°}$	9.67	12.64	20.6	119.83	161.75	192.08	214.60
Heptadecanoic acid	12.3			116.38	156.66	185.74	207.20
1-Heptadecene	7.5	12.39	20.32				
1-Heptanal	5.637	7.575	11.40	51.00	67.70	79.80	88.70
Heptane	3.359	9.5	8.74	50.42	68.33	81.43	91.20
1-Heptanethiol	6.067		12.06	55.81	74.60	88.91	99.98
Heptanoic acid			18.0	53.62	71.92	85.32	95.25
1-Heptanol	3.16	11.5	16.5	46.97	63.24	75.08	83.90
1-Heptene ΔH_t, 0.07$^{-136°}$	2.964(I) 3.021(II)	7.43	8.52	48.08	55.78	59.96	62.34
Hexachlorobenzene	6.1		23.2	36.21	39.82	41.48	42.38
Hexachloroethane	2.33	12.2	16.5				
$\quad\Delta H_t$, 1.97$^{1.3°}$			9.20				
Hexadecafluoroethylcyclohexane							
Hexadecafluoroheptane	12.39		8.7				
Hexadecane	12.8	12.24	19.38	112.89	152.41	181.02	202.20
Hexadecanoic acid			36.9	116.09	156.00	184.90	206.30
1-Hexadecanol	7.8		40.5	109.44	147.32	174.67	194.80
$\quad\Delta H_t$, 4.8$^{44.0°}$, 5.7$^{49.1°}$							
1-Hexadecene	7.216	12.05	19.14				
Hexafluorobenzene	2.770	7.571	8.61	43.88	52.55	57.62	60.63
Hexafluoroethane	0.642	3.860		30.01	35.60	38.40	39.87
$\quad\Delta H_t$, 0.893$^{-169.17°}$							

Table 9-4 (Continued)
HEATS OF MELTING AND VAPORIZATION (OR SUBLIMATION) AND SPECIFIC HEAT AT VARIOUS TEMPERATURES OF ORGANIC COMPOUNDS

Substance	ΔH_m	ΔH_v	ΔH_s	C_p (at specified temperature, K)			
				400	600	800	1000
Hexahydroindane cis			11.0				
trans			10.7				
Hexamethylbenzene	4.93		17.9	74.18	97.13	113.51	125.55
ΔH_t, $0.269^{-156.67°}$							
ΔH_t, $0.422^{110.7°}$							
Hexamethyldisiloxane			8.9				
Hexanal				44.00	58.30	68.70	76.40
Hexanamide	3.126		22.72				
Hexane		6.896	7.54	43.47	58.99	70.36	78.89
1-Hexanethiol	4.305	8.9	11.14	48.87	65.26	77.84	87.65
Hexanoic acid	6.98	15.45	17.3				
1-Hexanol	3.68	11.6	14.8	46.68	62.58	74.25	82.92
1-Hexene	2.234	6.76	7.32	40.03	53.90	64.02	71.54
2-Hexene cis		6.96	7.52	38.60	53.00	63.40	71.20
trans		6.91	7.54	39.70	53.40	63.60	71.20
3-Hexene cis		6.86	7.47	38.50	53.20	63.50	71.20
trans		6.92	7.54	40.20	53.90	63.90	71.40
1-Hexyne			23.7	37.87	49.59	58.16	64.56
Hydroquinone			26.0				
8-Hydroxyquinoline			11.8				
Indane			12.64				
Indene			16.7				
Iodobenzene	2.33	9.44	11.85	31.10	41.43	48.07	52.60
4-Iodobenzoic acid			21.0				
Iodocyclohexane			11.3				

Iodoethane		7.115	7.7	19.18	24.64	28.65	31.65
Iodomethane	3.47	6.52	6.63	12.33	15.28	17.47	19.17
2-Iodo-2-methylpropane			8.46	35.27	45.82	52.85	57.91
1-Iodonaphthalene			17.3				
2-Iodonaphthalene			21.7				
1-Iodopropane			8.6	26.27	34.11	39.80	44.03
2-Iodopropane			8.14	26.59	34.58	40.21	44.34
3-Iodopropene			9.1				
2-Iodotoluene (also 3- and 4-)			13.0				
Isobutylbenzene			11.54				
Isobutyl dichloroacetate			12.5				
Isobutyl phenyl ketone			14.22				
Isobutyl trichloroacetate			12.7				
Isobutyronitrile	1.86	7.754	8.99	28.56	37.39	43.74	48.40
Isopropyl acetate		8.97	8.89	48.00	66.20	78.60	87.30
Isopropylbenzene		8.35	10.79	35.96	46.81	54.13	59.26
Isopropyl nitrate			9.27	12.71	14.57	15.74	16.57
Isopropyl trichloroacetate			12.4				
Isothiocyanic acid			4.18	14.22	16.89	18.80	20.25
Ketene			31.7				
Lauric acid	8.8		36.0				
Leucine			11.5				
Limonene (+)			26.3				
Maleic acid			17.1				
Maleic anhydride			18.9				
Malononitrile			29.7				
D-Mannitol	5.39		22.0				
Melamine							
2,2-Metacyclophane							
Methane ΔH_t, 0.0187$^{-248\ \text{to}\ -252.7°}$	0.225	1.953		9.71	12.55	15.18	17.40
Methanethiol ΔH_t, 0.0525$^{-135.6°}$	1.411	5.872	5.7	14.04	17.57	20.32	22.48
Methanol ΔH_t, 0.152$^{-115.8°}$	0.768	8.24	8.94	12.29	16.02	19.04	21.38
4-Methoxybenzaldehyde			15.42				

Table 9-4 (Continued)
HEATS OF MELTING AND VAPORIZATION (OR SUBLIMATION) AND SPECIFIC HEAT AT VARIOUS TEMPERATURES OF ORGANIC COMPOUNDS

Substance	ΔH_m	ΔH_v	ΔH_s	C_p (at specified temperature, K)			
				400	600	800	1000
Methoxybenzene			11.18				
2-Methoxytetrahydropyran			10.2				
Methyl CH$_3$			19.0	10.05	11.54	12.89	14.09
Methyl allyl sulfone			5.80				
Methylamine	1.466	6.169	23.7	14.38	18.86	22.44	25.26
Methyl benzyl sulfone							
2-Methyl-1,3-butadiene	1.155	6.191	6.32	31.80	41.40	48.00	52.90
3-Methyl-1,2-butadiene		6.51	6.68	31.00	40.30	47.20	52.40
2-Methylbutane	1.231	5.901	5.94	36.49	49.89	59.71	67.12
2-Methyl-1-butanethiol	1.78	8.0					
3-Methyl-1-butanethiol							
2-Methyl-2-butanethiol	0.1454	7.50	8.51	42.79	56.58	66.28	73.30
$\Delta Ht,\ 1.907^{-114.0°}$	1.750						
3-Methylbutanoic acid		10.32	12.9				
2-Methyl-1-butanol		10.5	13.0				
3-Methyl-1-butanol		10.54					
2-Methyl-2-butanol	1.06	9.6	11.9				
$\Delta Ht,\ 0.47^{-127.2°}$		9.9	12.4				
3-Methyl-2-butanol							
2-Methyl-1-butanol	1.891	6.094	6.181	33.20	44.72	53.15	59.43
3-Methyl-1-butene	1.281	5.750	5.70	35.26	45.90	53.85	59.83
2-Methyl-2-butene	1.816	6.287	6.468	31.93	43.42	52.05	58.55
Methyl-n-butyl sulfone			18.2				
Methyl t-butyl sulfone			19.7				
3-Methyl-1-butyne		6.25	6.16	31.10	40.60	47.40	52.40
Methyl crotonate			9.8				

Compound							
Methylcyclohexane	1.614	7.44	8.45	44.35	64.46	78.74	88.79
2-Methylcyclohexanol *cis (and trans)*			15.1				
3-Methylcyclohexanol *cis*			15.6				
trans			15.7				
4-Methylcyclohexanol *cis*			15.7				
trans			15.8				
Methylcyclopentane	1.656	6.95	7.55	36.11	52.43	64.00	72.44
1-Methylcyclopentene			7.55	32.50	46.80	57.00	64.30
3-Methylcyclopentene			7.7	32.60	47.10	57.20	64.50
4-Methylcyclopentene			7.7	32.60	47.00	57.10	64.40
Methyldichlorosilane			6.7				
2-Methyl-1,3-dioxane			9.23				
4-Methyl-1,3-dioxane			9.36				
Methylene CH_2				8.64	9.37	10.14	10.89
1-Methyl-2-ethylbenzene	2.38(I) 2.28(II)	9.29	11.40	48.50	65.80	78.10	86.90
1-Methyl-3-ethylbenzene	1.82(I) 1.79(II)	9.21	11.21	47.50	65.40	77.80	86.80
1-Methyl-4-ethylbenzene	3.19	9.18	11.14	47.20	65.00	77.60	86.60
2-Methyl-3-ethylpentane	2.71	7.88	9.20				
3-Methyl-3-ethylpentane	2.59	7.84	9.08				
2-Methyl-3-ethyl-1-pentene			8.98				
Methyl ethyl sulfite			10.4				
Methyl ethyl sulfone			18.6				
Methyl formate	1.800	6.75		19.50	25.20	29.10	32.00
Methylglyoxal			9.1				
2-Methylheptane	2.839	8.08	9.48				
3-Methylheptane	2.779	8.10	9.52				
4-Methylheptane	2.59	8.10	9.48				
Methyl heptanoate			12.0				
2-Methylhexane	2.195	7.33	8.32	50.42	68.33	81.43	91.20
3-Methylhexane		7.36	8.39	50.42	68.33	81.43	91.20
Methyl hexanoate			11.1				
Methylhydrazine			9.65				

Table 9-4 (*Continued*)
HEATS OF MELTING AND VAPORIZATION (OR SUBLIMATION) AND SPECIFIC HEAT AT VARIOUS TEMPERATURES OF ORGANIC COMPOUNDS

Substance	ΔH_m	ΔH_v	ΔH_s	C_p (at specified temperature, K)			
				400	600	800	1000
Methylidyne CH				6.98	7.11	7.40	7.78
CH+				6.98	7.10	7.36	7.65
1-Methyl-2-isopropyl-benzene	2.39	9.17	12.10				
1-Methyl-3-isopropyl-benzene	3.27	9.11	11.94				
1-Methyl-4-isopropyl-benzene	2.31	9.12	12.02				
Methyl isopropyl ether			6.27	32.97	44.17	52.67	59.08
Methyl isopropyl ketone			8.82				
Methyl isopropyl sulfone			16.8				
3-Methylisoxazole			9.8				
5-Methylisoxazole			10.0				
Methylmercury bromide			16.2				
Methylmercury chloride			15.5				
Methylmercury iodide			15.6				
1-Methylnaphthalene	1.160	11.0		50.74	69.79	82.48	91.21
ΔH_t, $1.190^{-32.37°}$							
2-Methylnaphthalene	2.808	11.0		50.50	69.31	82.03	90.86
ΔH_t, $1.341^{15.4°}$							
Methyl nitrate	1.97	7.54	8.1	21.87	27.54	31.47	34.19
Methyl nitrite		5.0	5.4	18.24	23.35	26.97	29.52
2-Methylpentane	1.498	6.643	7.138	44.00	59.60	70.80	79.20
3-Methylpentane		6.711	7.236	43.47	59.00	70.40	78.90
Methyl pentanoate			10.2				
2-Methyl-1-pentene		6.71	7.29	40.80	54.40	64.40	71.80

3-Methyl-1-pentene		6.43	6.83	42.50	55.60	65.20	72.30
4-Methyl-1-pentene		6.47	6.86	38.90	52.90	63.10	70.70
2-Methyl-2-pentene		6.93	7.55	39.00	53.20	58.60	71.10
3-Methyl-2-pentene *cis*		6.89	7.49	39.00	53.20	63.40	71.10
trans		7.00	7.67	39.00	53.20	63.40	71.10
4-Methyl-2-pentene *cis*		6.59	7.04	40.05	54.10	64.00	71.50
trans		6.68	7.16	41.90	54.80	64.50	71.80
Methyl phenyl sulfone			22.0				
Methylphosphonic acid			11.5				
2-Methylpropanal			7.5				
2-Methylpropane	1.085	5.089	4.57	29.77	40.62	48.49	54.40
2-Methyl-1-propanethiol	1.191	7.412	8.28	35.31	46.26	53.77	59.17
2-Methyl-2-propanethiol	0.593	6.80	7.36	36.13	47.60	55.53	61.24
ΔHt, $0.972^{-121.6°}$							
ΔHt, $0.155^{-116.2°}$							
ΔHt, $0.232^{-73.8°}$							
2-Methyl-1-propanol	1.602	9.80	12.04	34.16	45.37	53.28	59.16
2-Methyl-2-propanol		9.33	12.73				
ΔHt, $0.20^{12.99°}$							
2-Methylpropene	1.418	5.286	4.92	26.57	35.30	41.86	46.85
Methyl propyl ether			6.6	33.01	43.92	52.26	58.51
2-Methylpyridine	2.324	8.654	10.15	31.92	44.55	53.21	59.34
3-Methylpyridine	3.389	8.932	10.62	31.82	44.47	53.12	59.23
α-Methylstyrene				44.80	60.70	71.80	79.80
β-Methylstyrene *cis*				44.80	60.70	71.80	79.80
trans				45.20	61.20	72.20	80.00
3-Methyl-2-thiabutane	2.236	7.338	8.15	34.69	46.01	54.95	62.29
2-Methylthiacyclopentane	2.08	8.7	10.0				
2-Methyl-3-thiapentane			9.2				
4-Methylthiazole			10.48				
2-Methylthiophene	2.263	8.103	9.26	29.43	39.57	46.43	51.30
3-Methylthiophene	2.518	8.186	9.44	29.38	39.34	45.95	50.59
Naphthalene	4.536	10.34	17.6	42.83	59.67	70.77	78.38
1-Naphthol			21.9				
2-Naphthol			19.8				

Table 9-4 (Continued)
HEATS OF MELTING AND VAPORIZATION (OR SUBLIMATION) AND SPECIFIC HEAT AT VARIOUS TEMPERATURES OF ORGANIC COMPOUNDS

Substance	ΔH_m	ΔH_v	ΔH_s	C_p (at specified temperature, K)			
				400	600	800	1000
1,4-Naphthoquinone			17.3				
1-Naphthylamine			21.5				
2-Naphthylamine			21.1				
p-Nitroaniline	5.04		26				
Nitrobenzene	2.78	9.744					
1-Nitrobutane		9.3	11.6	37.65	50.21	59.03	65.39
2-Nitrobutane		8.8	10.48	37.61	50.46	59.44	65.96
Nitroethane		8.4	9.9	23.66	31.45	36.81	40.67
Nitromethane	2.319	8.12	9.17	16.80	21.92	25.56	28.17
1-Nitronaphthalene			25.6				
1-Nitropropane		8.8	10.37	30.72	40.87	47.96	53.06
2-Nitropropane		8.4	9.88	30.89	41.19	48.22	53.24
Nonadecane							
ΔH_t, $3.30^{22.8°}$	10.95	13.39	22.9	133.71	180.43	214.21	239.20
1-Nonadecene	8.0	13.06	22.68	130.26	175.35	207.86	231.80
1-Nonanal			17.28	64.80	86.40	101.90	113.40
Nonane							
ΔH_t, $1.50^{-55.97°}$	3.72	8.82	11.10	64.30	87.01	103.56	115.90
1-Nonanethiol	8.0	10.6		69.69	93.28	111.04	124.65
Nonanoic acid			19.7				
1-Nonanol	4.3	13.0	18.6	67.50	90.60	107.46	119.91
1-Nonene		8.68	10.88	60.85	81.93	97.22	108.50
Octadecane	14.81	13.02	21.7	126.77	171.09	203.15	226.90
Octadecanoic acid	15.1		39.8				
1-Octadecene	7.8	12.74	21.50	123.32	166.00	196.80	219.50
Octafluorocyclobutane	0.662	5.58		44.50	53.85	58.65	61.50
1-Octanal			16.28	57.90	77.00	90.90	101.00

Octanamide			26.4				
Octane	4.957	8.225	9.916	57.35	77.67	92.50	103.60
1-Octanethiol	5.8	10.1	19.2	62.75	83.94	99.97	112.31
Octanoic acid	3.30	16.73	15.6				
1-Octanol	10.1	11.2	9.70	60.56	81.26	96.39	107.58
1-Octene				53.91	72.58	86.15	96.20
1-Octyne	3.660	8.07	23.4	51.75	68.28	80.30	89.20
Oxalic acid							
ΔHt, 0.3 (α → β)							
Oxalyl chloride			7.6				
Oxamide			26.8				
Palmitic acid	10.30		37				
[1,8]-Paracyclophane			26.5				
[2,2]-Paracyclophane			23.0				
[6,6]-Paracyclophane			27.5				
Paraldehyde			9.9				
Pentachloroethane	2.7	8.9	10.9	31.96	36.35	38.71	40.17
Pentachlorofluoroethane	0.449						
Pentachlorophenol			16.1				
Pentadecane	8.31	11.82	18.20	105.95	143.07	169.95	189.90
ΔHt, 2.19$^{-2.25°}$							
1-Pentadecene	6.9	11.63	17.96	102.50	137.98	163.60	182.50
1,2-Pentadiene		6.59	6.85	31.40	40.80	47.70	52.80
1,3-Pentadiene cis		6.60	6.77	29.50	39.90	47.00	52.20
trans		6.46	6.64	31.20	40.90	47.70	52.60
1,4-Pentadiene	1.468	6.01	6.01	31.30	40.80	47.60	52.70
2,3-Pentadiene		6.75	7.05	29.90	39.40	46.60	52.00
Pentaerythritol			34.4				
Pentaerythritol tetranitrate			36.3				
Pentafluorobenzoic acid			21.9				
Pentafluoroethane				27.20	32.94	36.12	37.98
Pentafluorophenol	2.95		16.1				
Pentamethylbenzene				65.00	86.08	101.29	112.33
ΔHt, 0.473$^{23.7°}$							
1-Pentanal				37.10	49.00	57.70	64.00
Pentanamide			21.34				

Table 9-4 (Continued)
HEATS OF MELTING AND VAPORIZATION (OR SUBLIMATION) AND SPECIFIC HEAT AT VARIOUS TEMPERATURES OF ORGANIC COMPOUNDS

Substance	ΔH_m	ΔH_v	ΔH_s	C_p (at specified temperature, K)			
				400	600	800	1000
Pentan-2,4-dione			10.82				
Pentane	2.008	6.16	6.32	36.53	49.64	59.30	66.55
Pentan-1,5-dithiol			14.17				
Pentanenitrile	1.130	7.98					
1-Pentanethiol	4.19	8.34	9.83	41.93	55.92	66.78	75.32
Pentanoic acid	3.850	10.53	16.6				
1-Pentanol	2.34	10.6	13.61	39.74	53.24	63.18	70.59
2-Pentanol		10.3	12.7				
3-Pentanol		10.1	12.8				
2-Pentanone	1.388	7.98	9.89	36.42	48.32	57.13	63.61
1-Pentene	1.700	6.02	6.09	33.10	44.56	52.95	59.21
2-Pentene cis		6.24	6.41	31.57	43.62	52.29	58.78
trans	1.996	6.23	6.38	32.67	44.02	52.45	58.81
1-Pentyne		6.63	6.79	31.10	40.40	47.10	52.20
2-Pentyne		6.99	7.35	29.20	38.70	45.90	51.40
Perylene			30.0				
α-Phellandrene			12.1				
9,10-Phenanthraquinone			21.9				
Phenanthrene		13.3	21.1				
Phenol	2.752	9.73	16.41	32.45	43.54	50.62	55.49
Phenyl acetate			13.0				
β-Phenyl-1-alanine DL- and L-			36.8				
Phenyl benzoate			23.0				
N-Phenyldiacetimide			21.5				
Phenyl ethyl sulfide			13.2				
Phenylhydrazine			14.69				

1-Phenyl-2-methylpropane	2.99	9.04	11.82	15.28	16.98	17.92	18.49
Phenyl methyl sulfide			12.1				
Phenyl salicylate		5.832	22.0				
Phosgene I	1.372						
II	1.335						
III	1.131						
m-Phthalic acid			25.5				
p-Phthalic acid			23.5				
Phthalic anhydride			21.19				
α-Pinene			10.7				
β-Pinene			11.1				
Propadiene		4.45		17.21	22.00	25.42	28.00
1-Propanal			7.09	23.09	30.22	35.45	39.27
Propane	0.842	4.487	3.605	22.47	30.76	36.99	41.73
Propane-2,3-dithiol			11.87				
1-Propanethiol	1.309	7.059	7.62	27.86	36.72	43.60	49.01
ΔH_t, $0.949^{-131.06°}$							
2-Propanethiol	1.371	6.670	7.039	28.35	37.02	43.26	47.92
ΔH_t, $0.013^{-160.6°}$							
1-Propanol	1.242	9.982	11.36	25.86	34.56	41.04	45.93
2-Propanol	1.293	9.510	10.85	26.78	35.76	42.13	46.82
2-Propen-1-ol	1.800	7.716	11.3	22.81	30.11	35.28	39.06
Propionic acid			13.7				
Propionic anhydride			12.6				
Propionitrile	1.202	7.353	8.632	21.18	27.42	32.14	35.70
ΔH_t, $0.408^{-96.19°}$							
1-Propylamine			7.46	28.51	37.99	44.94	50.21
n-Propylbenzene I	2.215	9.14	11.05	47.82	65.86	78.30	87.16
II	2.03						
n-Propyl carbamate			19.4				
n-Propyl chloroacetate	2.479	8.62	11.6				
n-Propylcyclohexane	2.398	8.15	10.78	59.10	83.80	101.20	113.40
n-Propylcyclopentane			9.82	50.83	71.04	86.28	97.50
Propylene	0.718	4.40		19.23	25.81	30.77	34.52
Propylene oxide	1.561	6.87	6.67	22.16	30.07	35.68	39.79

Table 9-4 (*Continued*)
HEATS OF MELTING AND VAPORIZATION (OR SUBLIMATION) AND SPECIFIC HEAT AT VARIOUS TEMPERATURES OF ORGANIC COMPOUNDS

Substance	ΔH_m	ΔH_v	ΔH_s	C_p (at specified temperature, K)			
				400	600	800	1000
n-Propyl nitrate		8.58	9.70	35.79	46.49	53.87	59.08
n-Propyl phenyl ketone			14.51				
n-Propyl trichloroacetate		5.29	12.7	17.33	21.80	25.14	27.71
Propyne			13.45				
Pyrazine			22.5				
Pyrene			12.78				
Pyridazine	1.979	8.39	9.61	25.42	35.72	42.49	47.17
Pyrimidine			11.95				
Pyrrole			10.80				
Pyrrolidine	2.050	7.89	8.98	27.33	40.31	49.35	55.84
ΔH_t, 0.129−66.01°							
Salicylic acid			22.74				
Sebacic acid			38.4				
5,5'-Spiro-bis(1,3-dioxane)			17.4				
Spiropentane	1.538	6.39	6.58	28.55	40.10	47.91	53.51
Stilbene *cis*			16.5				
Styrene	2.617	8.85	10.50	38.32	52.14	61.40	67.92
Suberic acid			34.2				
Succinic acid			28.1				
Tetrabromomethane				23.20	24.51	25.51	25.32
Tetracene			30				
Tetrachlorobenzoquinone			23.6				
1,1,1,2-Tetrachloroethane		9.24	10.7	28.36	33.28	36.24	38.17
1,1,2,2-Tetrachloroethane				27.90	32.91	35.85	37.76
Tetrachloroethylene	2.5	8.3	9.4	25.10	27.86	29.29	30.07

Compound							
Tetrachloromethane	0.601	7.16	7.79	21.92	23.82	24.64	25.05
$\quad\Delta H_t$, 1.095$^{-47.9°}$							
Tetracyanoethylene			19.4				
Tetradecane	10.90	11.38	17.01	99.01	133.72	158.89	177.60
Tetradecanoic acid			33.4				
1-Tetradecene	6.6	11.21	16.78	95.56	128.64	152.54	170.20
Tetraethylene glycol			24				
Tetraethyl lead			13.6				
1,1,2-Tetrafluoroethane	1.844	4.02		24.90	30.76	34.20	36.36
Tetrafluoroethylene	0.167			21.97	25.53	27.61	28.86
Tetrafluoromethane				17.30	20.74	22.58	23.61
$\quad\Delta H_t$, 0.353$^{-196.92°}$							
Tetrahydrofuran			7.65				
Tetrahydrofurfuryl alcohol			15.9				
1,2,3,4-Tetrahydronaphthalene			13.4				
Tetrahydropyran			8.35				
Tetraiodomethane	2.684	10.76	13.66	24.00	24.94	25.31	25.49
1,2,3,4-Tetramethylbenzene	2.561	10.47	13.34	56.81	75.68	89.42	99.47
1,2,3,5-Tetramethylbenzene	5.02	10.88	18	55.76	74.81	88.79	99.01
1,2,4,5-Tetramethylbenzene	1.802	7.51	10.24	55.50	74.38	88.41	98.71
2,2,3,3-Tetramethylbutane							
$\quad\Delta H_t$, 0.478$^{-120.66°}$							
Tetramethyl lead			9.1				
Tetranitromethane			10.3				
Tetrazole			23				
2-Thiabutane	2.333	7.06	7.61	27.81	36.41	42.93	47.94
Thiacyclobutane	1.971	7.7	8.56	21.89	30.45	36.40	40.67
$\quad\Delta H_t$, 0.160$^{-96.45°}$							
Thiacycloheptane			11.30	42.0	65.0	79.0	88.0
Thiacyclohexane	0.585	8.60	10.22	35.71	52.37	64.00	72.34
$\quad\Delta H_t$, 0.262$^{-71.75°}$							
$\quad\Delta H_t$, 1.858$^{-33.14°}$							
Thiacyclopentane	1.757	8.28	9.28	28.95	40.04	47.66	53.14
Thiacyclopropane		6.98	7.24	16.53	21.99	25.61	28.21
4-Thia-5,5'-dimethyl-1-hexene			10.6				

Table 9-4 (Continued)

HEATS OF MELTING AND VAPORIZATION (OR SUBLIMATION) AND SPECIFIC HEAT AT VARIOUS TEMPERATURES OF ORGANIC COMPOUNDS

Substance	ΔH_m	ΔH_v	ΔH_s	C_p (at specified temperature, K)			
				400	600	800	1000
2-Thiaheptane	2.96	8.78	10.88	48.67	65.02	77.59	87.41
3-Thiaheptane	2.90	8.76	10.74	48.37	64.96	77.74	87.75
4-Thiaheptane	2.90		10.64	48.21	65.13	78.45	89.05
2-Thiahexane	2.976	8.2	9.8	41.73	55.68	66.53	75.08
3-Thiahexane	2.529	8.3	9.58	41.43	55.62	66.68	75.42
5-Thianonane	4.64		12.75	62.09	83.81	100.58	113.71
2-Thiapentane	2.369	7.67	8.65	34.64	45.86	54.45	61.14
3-Thiapentane	2.845	7.59	8.55	34.65	46.11	54.91	61.79
2-Thiapropane	1.908	6.45	6.61	21.12	27.01	31.58	35.17
6-Thiaundecane			14.7				
Thioacetic acid	1.216		8.27	22.25	26.72	30.41	32.62
Thiophene ΔH_t, 0.152$^{-101.6°}$		7.52		23.02	30.95	36.01	39.54
Thymol			21.8				
Toluene	1.586	7.93	9.08	33.48	47.20	56.61	63.32
2-Toluenethiol			12.3				
2,4,6-Triamino-1,3,5-triazine			29.5				
Tribromomethane			17.2	18.80	21.03	22.29	23.12
Tri-n-butyl phosphate			9.8				
Trichloroacetyl chloride			21.2				
Trichlorobenzoquinone							
1,1,1-Trichloroethane ΔH_t, 1.79$^{-48.95°}$	0.45	7.96	7.76	25.72	30.68	33.73	35.81
1,1,2-Trichloroethane	2.7	8.3	9.4	25.03	30.13	33.28	35.42
Trichloroethylene		7.52	8.2	21.80	25.06	26.94	28.15

Trichlorofluoromethane				20.84	23.13	24.19	24.74
Trichloromethyl CCl₃				16.66	18.16	18.83	19.18
1,2,3-Trichloropropane		8.87	11.22	31.71	38.87	43.79	47.34
Tricyanoethylene			19.4				
Tridecane ΔHt, 1.831⁻¹⁸·²°	6.81	10.91	15.83	92.07	124.38	147.82	165.20
Tridecanoic acid	8.2		35.0				
1-Tridecene	6.2	10.75	15.60	88.62	119.29	141.48	157.80
Triethyl aluminum			17.5				
Triethylamine			8.29	48.70	66.10	78.56	87.80
Triethylaminoborane			14.5				
Triethyl arsenite			12.1				
Triethylarsine			10.3				
Triethyl borate			10.5				
Triethylenediamine ΔHt, 2.30⁷·⁹·⁸°	1.45		14.8				
Triethylene glycol		17.07	18.9				
Triethyl phosphate			13.7				
Triethylphosphine			9.5				
Triethyl phosphite			10.0				
Triethylstibine			10.4				
1,1,1-Trifluoroethane	1.480	4.58		22.75	28.38	31.98	34.44
Trifluoroethylene				19.39	23.30	25.69	27.23
Trifluoromethane	0.970	3.99		14.61	18.16	20.35	21.76
Trifluoromethyl CF₃			8.98	13.74	16.17	17.50	18.25
CF₃‡				13.62	16.00	17.35	18.13
Trifluoromethylbenzene	3.29			40.59	54.20	62.75	68.45
Triiodomethane	3.9	7.80	16.7	19.60	21.52	22.64	23.38
2,4,5-Trimethylaceto-phenone			15.1				
2,4,6-Trimethylaceto-phenone			15.1				
Trimethyl aluminum	1.564	5.48	14.9	28.08	38.34	45.62	50.98
Trimethylamine			5.26				
Trimethyl arsenite			10.1				

Table 9-4 (Continued)

HEATS OF MELTING AND VAPORIZATION (OR SUBLIMATION) AND SPECIFIC HEAT AT VARIOUS TEMPERATURES OF ORGANIC COMPOUNDS

Substance	ΔHm	ΔHv	ΔHs	C_p (at specified temperature, K)			
				400	600	800	1000
Trimethylarsine			6.9				
1,2,3-Trimethylbenzene	1.955	9.57	11.73	46.90	64.00	76.70	85.90
ΔHt, $0.157^{-54.46°}$							
ΔHt, $0.319^{-42.89°}$							
1,2,4-Trimethylbenzene I	3.153	9.38	11.46	46.96	64.29	76.93	86.10
1,3,5-Trimethylbenzene I	2.274	9.33	11.35	46.41	64.08	76.84	86.07
$=$	1.932						
$\equiv$	1.892						
Trimethyl borate			8.3				
Trimethylboron	0.540	6.92	4.83				
2,2,3-Trimethylbutane			7.66	50.83	69.61	82.73	92.32
ΔHt, $0.586^{-151.8°}$							
Trimethylchlorosilane			7.2				
1,3,5-Trimethylcyclo-hexane cis, cis				58.05	83.94	102.20	115.21
2,2,3-Trimethylpentane	2.06	7.65	8.82				
2,2,4-Trimethylpentane	2.20	7.41	8.40				
2,3,3-Trimethylpentane	0.205	7.73	8.90				
ΔHt, $1.850^{-109.01°}$							
2,3,4-Trimethylpentane	2.215	7.82	9.01				
2,4,4-Trimethyl-1-pentene		7.5	8.5				
2,4,4-Trimethyl-2-pentene		7.8	8.9				
Trimethylphosphine			6.7				
Trimethylphosphine oxide			12.0				
Trimethyl phosphite			8.8				
Trimethylsilanol			10.9				

Compound							
Trimethylstibine			7.5				
Trimethylsuccinic anhydride			17.7				
Trimethylthiacyclopropane			9.40				
2,4,6-Trinitroanisole			31.8				
1,3,5-Trinitrobenzene			23.8				
Trinitromethane			11.15				
2,4,6-Trinitrophenetole			28.8				
2,4,6-Trinitrotoluene			28.3				
Triphenylarsine			23.5				
Triphenylene			28.2				
Triphenylmethane			23.9				
Triphenylphosphine			23				
Tropolone			20.0				
Undecane	5.28	9.92	13.47	78.18	105.80	125.69	140.60
$\Delta H t$, $1.64^{-36.55°}$							
Undecanoic acid	6.2		29.0				
1-Undecene	4.06	9.77	13.24	74.74	100.61	119.34	133.20
$\Delta H t$, $2.202^{-55.8°}$							
Urea			21.0				
o-Xylene	3.25	8.80	10.38	41.03	55.98	66.64	74.35
m-Xylene	2.765	8.69	10.20	40.03	55.51	66.41	74.23
p-Xylene	4.09	8.60	10.13	39.70	55.16	66.14	74.02
2,3-Xylenol			20.1				
2,4-Xylenol			15.74				
2,5-Xylenol			20.31				
2,6-Xylenol			18.07				
3,4-Xylenol			20.49				
3,5-Xylenol			19.80				

Table 9-5
HEAT CAPACITY STANDARDS

In the following tables for water, mercury, and aluminum oxide the units of temperature are based on the International Temperature Scale of 1948 using 0°C = 273.16°K. Values expressed in calories are in units of the "defined" calorie. This calorie, which is independent of the properties of water, is by definition equal to 4.1833 international joules. Redefined in January 1948, this calorie is equal to 4.1840 absolute joules. One absolute joule = 0.999835 international joule = 0.239006 calorie.

A. Water

The values in the following table were determined by Ginnings and Furukawa of the National Bureau of Standards. Published in the *J. Am. Chem. Soc.*, **75**, 522 (1953), they are reproduced here by permission of the American Chemical Society. For international use in calorimetry, the International Committee on Weights and Measures has approved a heat capacity table (Procès-Verbaux of the International Committee on Weights and Measures, Session of 1950, p. 92) which differs from this table by less than 0.005 joule per degree per gram-mole.

Values in the columns headed *C-satd.* and C_p give the heat capacities (specific heats) at saturation pressure and at 1 atmosphere (760 mm Hg) pressure, respectively. The units in the columns headed J are in absolute joules per degree per gram-mole (mol. wt. = 18.016); the units in the columns headed **cal.** are in calories per degree per gram.

	C-satd.		C_p			*C-satd.*		C_p	
°C	J	cal.	J	cal.	°C	J	cal.	J	cal.
0	75.993	1.00814	75.985	1.00804	55	75.350	0.99962	75.348	0.99959
5	75.714	1.00445	75.706	1.00434	60	75.385	1.00008	75.385	1.00008
10	75.532	1.00203	75.525	1.00194	65	75.428	1.00065	75.428	1.00065
15	75.417	1.00050	75.410	1.00040	70	75.426	1.00129	75.478	1.00132
20	75.345	0.99955	75.339	0.99947	75	75.532	1.00203	75.536	1.00208
25	75.303	0.99899	75.298	0.99893	80	75.594	1.00285	75.601	1.00295
30	75.282	0.99871	75.278	0.99866	85	75.667	1.00382	75.675	1.00391
35	75.277	0.99865	75.273	0.99860	90	75.746	1.00489	75.757	1.00502
40	75.283	0.99873	75.280	0.99869	95	75.835	1.00605	75.850	1.00625
45	75.298	0.99893	75.295	0.99889	100	75.934	1.00736	75.954	1.00763
50	75.320	0.99922	75.318	0.99919					

B. Mercury

The values in the following table were determined by Douglas, Ball and Ginnings, and published in *J. Research Natl. Bur. Standards*, **46**, 334 (1951). Values in the columns headed *C-satd.*, C_p, and C_v are those for liquid-vapor equilibrium, maintenance of constant pressure, and maintenance of constant volume, respectively. The units are in "defined calories." (1 calorie = 4.1840 absolute joules = 4.1833 international joules.)

	Liquid			Vapor
°C	*C-satd.*	C_p	C_v	C_p
−38.88*	0.033686	0.033686	0.02975	0.02476
−20	0.033534	0.033534	0.02941	0.02476
0	0.033817	0.033817	0.02920	0.02476
20	0.033240	0.033240	0.02876	0.02476
25	0.033206	0.033206	0.02867	0.02476
40	0.033110	0.033110	0.02845	0.02476
60	0.032987	0.032987	0.02816	0.02476
80	0.032877	0.032877	0.02789	0.02476
100	0.032776	0.032776	0.02764	0.02476
120	0.032686	0.032686	0.02739	0.02476
140	0.032606	0.032606	0.02716	0.02476

* Triple point.

Table 9-5 (*Continued*)
HEAT CAPACITY STANDARDS

B. Mercury (*Continued*)

| °C | Liquid | | | Vapor |
	C-satd.	C_p	C_v	C_p
160	0.032536	0.032536	0.02693	0.02476
180	0.032476	0.032476	0.02674	0.02477
200	0.032426	0.032426	0.02659	0.02477
220	0.032386	0.032386	....	0.02477
240	0.032356	0.032356	....	0.02477
260	0.032335	0.032336	....	0.02478
280	0.032324	0.032325	....	0.02479
300	0.032321	0.032323	....	0.02480
320	0.032328	0.032330	....	0.02481
340	0.032343	0.032346	....	0.02482
356.58†	0.032362	0.032366	....	0.02484
360	0.032367	0.032371	....	0.02484
380	0.032398	0.032404	....	0.02486
400	0.032437	0.032445	....	0.02489
420	0.032483	0.032494	....	0.02492
440	0.032536	0.032550	....	0.02495
460	0.032596	0.032614	....	0.02499
480	0.032661	0.032684	....	0.02503
500	0.032733	0.032762	....	0.02507

C. Aluminum Oxide

The values in the following table were determined by Ginnings and Furukawa of the National Bureau of Standards on aluminum oxide in the form of synthetic sapphire (corundum) in pieces that passed a #10 and were retained by a #40 sieve and with impurities between 0.01 and 0.02% by weight. Heat capacity values below the experimental range were obtained by extrapolating a Debye equation fitted to the experimental values at the lowest temperature. The units in the columns headed J are in absolute joules per degree per gram-mole (mol. wt. = 101.96) at constant pressure of 1 atmosphere (760 mm Hg); the units in the columns headed **cal.** are in calories per degree per gram at constant pressure of 1 atmosphere. Published in the *J. Am. Chem. Soc.*, **75**, 522 (1953), they are reproduced here by permission of the American Chemical Society.

| | C_p | | | C_p | |
K	J	cal.	K	J	cal.
0	0.0	0.0	65	3.620	0.008486
5	0.0012	0.000003	70	4.582	0.01074
10	0.0094	0.000022	75	5.668	0.01329
15	0.0316	0.000074	80	6.895	0.01616
20	0.0759	0.000178	85	8.247	0.01933
25	0.1417	0.000332	90	9.692	0.02272
30	0.2627	0.000616	95	11.223	0.02631
35	0.4377	0.001026	100	12.84	0.03010
40	0.6907	0.001619	110	16.31	0.03823
45	1.039	0.002436	120	20.05	0.04700
50	1.492	0.003497	130	23.96	0.05617
55	2.069	0.004850	140	27.96	0.06554
60	2.780	0.006517	150	31.99	0.07499

† Boiling point.

Table 9-5 (*Continued*)
HEAT CAPACITY STANDARDS

C. Aluminum Oxide (*Continued*)

	C_p			C_p	
K	J	cal.	K	J	cal.
160	35.99	0.08436	560	110.21	0.25835
170	39.94	0.09362	570	110.82	0.25978
180	43.79	0.10263	580	111.40	0.26114
190	47.53	0.11147	590	111.97	0.26247
200	51.14	0.11988	600	112.51	0.26374
210	54.60	0.12799	610	113.03	0.26495
220	57.92	0.13577	620	113.54	0.26615
230	61.09	0.14320	630	114.03	0.26730
240	64.12	0.15030	640	114.50	0.26841
250	67.01	0.15708	650	114.95	0.26946
260	69.75	0.16350	660	115.40	0.27051
270	72.36	0.16962	670	115.82	0.27150
280	74.84	0.17543	680	116.24	0.27248
290	77.19	0.18094	690	116.64	0.27342
298.16	79.01	0.18521	700	117.03	0.27433
300	79.41	0.18615	720	117.77	0.27607
310	81.52	0.19109	740	118.46	0.27768
320	83.50	0.19574	760	119.12	0.27923
330	85.39	0.20017	780	119.74	0.28069
340	87.18	0.20436	800	120.33	0.28207
350	88.88	0.20834	820	120.88	0.28336
360	90.52	0.21219	840	121.40	0.28458
370	92.06	0.21580	860	121.90	0.28575
380	93.51	0.21920	880	122.37	0.28685
390	94.88	0.22241	900	122.82	0.28791
400	96.18	0.22545	920	123.24	0.28889
410	97.40	0.22832	940	123.65	0.28985
420	98.55	0.23102	960	124.03	0.29074
430	99.64	0.23357	980	124.40	0.29161
440	100.69	0.23603	1000	124.74	0.29241
450	101.68	0.23835	1020	125.08	0.29321
460	102.64	0.24061	1040	125.39	0.29393
470	103.54	0.24271	1060	125.69	0.29464
480	104.42	0.24477	1080	125.98	0.29531
490	105.25	0.24672	1100	126.26	0.29597
500	106.05	0.24860	1120	126.52	0.29658
510	106.81	0.25038	1140	126.77	0.29718
520	107.55	0.25211	1160	127.01	0.29773
530	108.26	0.25378	1180	127.24	0.29827
540	108.93	0.25535	1200	127.47	0.29881
550	109.59	0.25690			

KOPP'S RULE

This rule, which should be used only where experimental values are lacking, states that the specific heat of a compound is approximately equal to the sum of the heat capacities of the constituent elements and that an *approximate* value expressed in gram calories per gram formula weight can be calculated by assigning the following atomic heat capacities to the elements:

For solids: C, 1.8; H, 2.3; O, 4.0; S, 5.4; P, 5.4; F, 5.0; Si, 3.8; B, 2.7; all other elements, 6.2.

For liquids: C, 2.8; H, 4.8; O, 6.0; S, 7.4; P, 7.4; F, 7.0; Si, 5.8; B, 4.7; all other elements, 8.0.

Example: For $BaCO_3$: $6.2 + 1.8 + (3 \times 4.0) = 20.0$ cal per gram-formula weight; or, $20.0 \div 197.37 = 0.1013$ cal $\cdot$ g^{-1}. Found by experiment, 0.0999.

ESTIMATION OF GIBBS FREE ENERGY

When the Gibbs free energy of a compound is not available, it may be estimated by summation of the contributions of the component groups, as formulated by van Krevelen and Chermin [*Chem. Eng. Sci.*, **1**, 66 (1951)].

$$\Delta Gf = \Sigma \text{ contributions of composing groups} + RT \ln \sigma$$

in which $RT \ln \sigma$ represents the correction for the symmetry, that is, the number of indistinguishable positions in space the molecule may be made to assume by a simple rigid rotation. The group contributions are represented as a linear function of the temperature in the ranges from 300 to 600 K and from 600 to 1500 K.

$$\Delta Gf = A + \frac{B}{100} T$$

In comparison to the expression: $\Delta Gf = \Delta H - T \Delta S$, it follows that the A term has the dimension of enthalpy of formation and the term $B/100$ has the dimension of an entropy of formation, both at 298.15 K. The estimated values of ΔGf are in kilocalories per mole, and are for the gaseous state. A large number of symmetry numbers may be found in G. Herzberg, *Infrared and Raman Spectra of Polyatomic Molecules*, Van Nostrand, New York, 1945.

Table 9-6
FREE ENERGY OF COMPONENT GROUPS

	300 to 600 K		600 to 1500 K	
Group	A	B	A	B
Alkane				
$-\overset{\textstyle\mid}{\underset{\textstyle\mid}{C}}-$	1.958	3.735	4.385	3.350
$-\overset{\textstyle\mid}{C}H$	−0.705	2.910	−0.705	2.910
$-CH_2-$	−5.193	2.430	−5.830	2.544
$-CH_3$	−10.943	2.215	−12.310	2.436
Alkene				
$H_2C{=}CH-$	13.737	1.655	12.465	1.762
$H_2C{=}C\big\langle$	16.467	1.915	16.255	1.966
$\overset{H}{}{>}C{=}C{<}\overset{H}{}$	17.663	1.965	16.180	2.116
$\overset{H}{}{>}C{=}C\big\langle{}_H$	17.187	1.915	15.815	2.062
$\overset{H}{}{>}C{=}C\big\langle$	20.217	2.295	19.584	2.354
${>}C{=}C\big\langle$	25.135	2.573	25.135	2.573
$H_2C{=}C{=}C\big\langle{}^H$	49.377	1.035	4.170	1.208
$H_2C{=}C{=}C\big\langle$	51.084	1.474	51.084	1.474
$\overset{H}{}{>}C{=}C{=}C\big\langle{}^H$	52.460	1.483	52.460	1.483
Conjugated Alkene				
$H_2C \leftrightarrow$	5.437	0.675	4.500	0.832
$\overset{H}{}{>}C \leftrightarrow$	7.407	1.035	6.980	1.088
${>}C \leftrightarrow$	9.152	1.505	10.370	1.308
Alkyne				
$HC{\equiv}$	27.048	−0.765	26.700	−0.704
$-C{\equiv}$	26.938	−0.525	26.555	−0.550
Aryl				
$HC{\big\langle}$	3.047	0.615	2.505	0.706
$-C{\big\langle}$	4.675	1.150	5.010	0.988
Cycloalkane Rings				
3-membered ring	23.458	−3.045	22.915	−2.966

Table 9-6 (*Continued*)
FREE ENERGY OF COMPONENT GROUPS

Group	300 to 600 K		600 to 1500 K	
	A	B	A	B
4-membered ring	10.73	−2.65	10.60	−2.50
5-membered ring	4.275	−2.350	2.665	−2.182
6-membered ring	−1.128	−1.635	−1.930	−1.504
5-membered ring	−3.657	−2.395	−3.915	−2.250
6-membered ring	−9.102	−2.045	−8.810	−2.071
Branching in Alkane Chains				
Side chain with 2 or more C-atoms	1.31	0	1.31	0
3 adjacent —CH groups	2.12	0	2.12	0
Adjacent —CH and —C— groups	1.80	0	1.80	0
2 adjacent —C— groups	2.58	0	2.58	0
Branching in Cycloalkane 5-membered Ring				
Single branching	−1.04	0	−1.69	0
Double branching:				
1,1 position	−1.85	0	−1.190	−0.160
1,2 position *cis*	−0.38	0	−0.38	0
trans	−2.55	0	−0.945	−0.266
1,3 position *cis*	−1.20	0	−0.370	−0.166
trans	−2.35	0	−0.800	−0.264
Branching in Cycloalkane 6-membered Ring				
Single branching	−0.93	0	0.230	−0.192
Double branching:				
1,1 position	0.835	−0.367	1.745	−0.556
1,2 position *cis*	−0.19	0	1.470	−0.276
trans	−2.41	0	0.045	−0.398
1,3 position *cis*	−2.70	0	−1.647	−0.185
trans	−1.60	0	0.260	−0.290
1,4 position *cis*	−1.11	0	−1.11	0
trans	−2.80	0	−0.995	−0.245
Branching in Aromatic Rings				
Double branching				
1,2 position	1.02	0	1.02	0
1,3 position	−0.31	0	−0.31	0
1,4 position	0.93	0	0.93	0
Triple branching				
1,2,3 position	1.91	0	2.10	0
1,2,4 position	1.10	0	1.10	0
1,3,5 position	0	0	0	0
Oxygen-containing Groups				
—OH	−41.56	1.28	−41.56	1.28
—O—	−15.79	−0.85		
O (ring)	−18.37	0.80	−16.07	0.40
H—C=O	−29.28	0.77	−30.15	0.83
C=O	−28.08	0.91	−28.08	0.91

Table 9-6 (*Continued*)
FREE ENERGY OF COMPONENT GROUPS

Group	300 to 600 K		600 to 1500 K	
	A	B	A	B
—C(=O)OH (carboxyl group)	−98.39	2.86	−98.83	2.93
—C(=O)O— (ester group)	−92.62	2.61	−92.62	2.61
H—C=C=O (ketene group with H)	−12.86	0.46	−12.86	0.46
C=C=O	−9.62	0.72	−9.38	0.73
epoxide (C—O—C ring)	12.86	−0.63	12.86	−0.63
dioxolane-type ring	−5.82	0.25	−3.53	−0.16
Nitrogen-containing Groups				
—C≡N	30.75	−0.72	30.75	−0.72
—N=C	46.32	−0.89	46.32	−0.89
—NH₂	2.82	2.71	−6.78	3.98
NH	12.93	3.16	12.93	3.16
N—	19.46	3.82	19.46	3.82
N (aromatic)	11.32	1.11	12.26	0.96
—NO₂	−9.0	3.70	−14.19	4.38
Sulfur-containing Groups				
—SH	5.12	1.07	5.12	1.07
—S—	12.58	1.42	12.29	1.44
S (aromatic)	14.83	0.51	15.15	0.44
SO	−14.39	3.39	−14.39	3.39
SO₂	5.22	5.58	7.11	5.26
Halogen-containing Groups				
—F	−45.10	0.20		
—Cl	−8.25	0	−8.25	0
—Br	2.21	−0.26	2.21	−0.26
—I	15.26	0	15.26	0

↔ Indicates resonant bond in aromatic ring.

CRITICAL PHENOMENA

The *critical temperature* T_c (or t_c) of a gas is the temperature above which the gas cannot be liquefied no matter how high the pressure.

The *critical pressure* P_c is the lowest pressure which will liquefy the gas at its critical temperature.

The *critical molar volume* V_c is the volume of one mole at the critical temperature and the critical pressure. It can be computed from the critical density ρ_c as follows:

$$\frac{\text{Molecular weight in g} \cdot \text{mol}^{-1}}{(\rho_c \text{ in g} \cdot \text{cm}^{-3})(1000 \text{ cm}^3 \cdot \text{L}^{-1})} = V_c \text{ in L} \cdot \text{mol}^{-1}$$

The critical pressure, critical molar volume, and critical temperature are the values of the pressure, molar volume, and thermodynamic temperature at which the densities of coexisting liquid and gaseous phases just become identical.

Table 9-7
CRITICAL PROPERTIES

Substance	T_c, °C	P_c, atm	ρ_c, g/cm³
Acetaldehyde (ethanal)	188		
Acetic acid	321.3	57.1	0.351
Acetic anhydride	296	46.2	
Acetone	236.5	47.2	0.278
Acetonitrile	274.7	47.7	0.237
Acetylene (ethyne)	35.18	60.59	0.231
Air	−140.6	37.2	0.313
Allene	120		
Allyl alcohol	272		
Allyl sulfide	380		
Ammonia	132.4	111.3	0.235
Aniline	426	52.4	0.34
Anisole	368	41.2	
Antimony tribromide	904.5	56	
Argon	−122.44	48.00	0.5307
Benzaldehyde	352	21.5	
Benzene	288.94	48.34	0.302
Benzonitrile	426.2	41.6	
Biphenyl	516	38	0.307
Boron pentafluoride	197		
Boron tribromide	300		
Boron trichloride	178.8	38.2	
Boron trifluoride	−12.3	49.2	
Bromine	311	102	1.18
Bromobenzene	397	44.6	0.485
Bromoethane (ethyl bromide)	230.7	61.5	0.507
Bromomethane (methyl bromide)	191		
Bromopentafluorobenzene	397	44.6	
Bromotrifluoromethane	67.0	39.2	0.76
1,3-Butadiene	152	42.7	0.245
n-Butane	152.01	37.7	0.228
Butane nitrile (butyronitrile)	309.1	37.4	
Butanoic acid (n-butyric acid)	355	52	0.304
1-Butanol (n-butyl alcohol)	289.78	43.55	0.270
2-Butanol (sec-butyl alcohol)	262.80	41.39	0.276
2-Butanone (ethyl methyl ketone)	262.4	41.0	0.270
1-Butene	146.4	39.7	0.234
2-Butene cis	162.40	41.5	0.240
trans	155.46	40.5	0.236
n-Butyl acetate	306		
n-Butylamine	251	41	
n-Butylbenzene	387.3	28.49	0.270
1-Butyne (ethylacetylene)	190.5		
2-Butyne (dimethylacetylene)	215.5		
Carbon dioxide	31.04	72.85	0.468
Carbon disulfide	279	78	0.44
Carbon monoxide	−140.23	34.53	0.301
Carbon tetrachloride	283.15	44.97	0.558
Carbonyl chloride (phosgene)	182	56	0.52
Carbonyl sulfide	105	61	
Chlorine	144.0	76.1	0.573

Table 9-7 (*Continued*)
CRITICAL PROPERTIES

Substance	T_c, °C	P_c, atm	ρ_c, g/cm^3
Chlorine pentafluoride	142.6	51.9	0.565
Chlorine trifluoride	153.5		
Chlorobenzene	359.2	44.6	0.365
1-Chloro-1,1-difluoroethane	137.1	40.7	0.435
2-Chloro-1,1-difluoroethylene	127.4	44.0	0.499
Chlorodifluoromethane (freon 22)	96.0	49.12	0.525
Chloroethane (ethyl chloride)	187.2	52	
Chloroform	263.4	54	0.50
Chloromethane (methyl chloride)	143.1	65.92	0.353
Chloropentafluoroethane	80.0	31.16	0.613
1-Chloropropane (*n*-propyl chloride)	230	45.2	
3-Chloropropene (allyl chloride)	241		
Chlorotrifluoromethane (freon 13)	28.9	38.7	0.579
o-Cresol	424.4	49.4	0.384
m-Cresol	432.6	45.0	0.346
p-Cresol	431.4	50.8	0.391
Cyanogen	128	59	
Cyclohexene	280.3	40.2	0.273
Chclohexene	287.26		
Cyclopentane	238.5	44.49	0.27
Cyclopentene	232.9		
Cyclopropane	124.65	54.23	
Cymene	385		
Decalin *cis*	429.0		
trans	413.8		
1-Decanol	427		0.264
Decanenitrile (caprylonitrile)	348.8	32.1	
Deuterium (equilibrium)	−234.90	16.28	0.0668
(normal)	−234.81	16.43	
Deuterium bromide (DBr)	88.8		
Deuterium chloride	50.3		
Deuterium hydride (DH)	−237.25	14.64	0.0481
Deuterium iodide	148.6		
Diborane (B_2H_6)	16.7	39.53	
Dibromomethane (methylene bromide)	310	71	
1,2-Dibromoethane	309.8	70.6	
Dichlorodifluoromethane (freon 12)	111.80	40.71	0.558
1,1-Dichloroethane	250	50	0.42
1,2-Dichloroethane	288	53	0.44
1,1-Dichloroethylene	271.0		
1,2-Dichloroethylene	243.3	54.4	
Dichlorofluoromethane (freon 21)	178.5	51.0	0.522
Dichloromethane (methylene chloride)	237	60	
1,1-Dichloro-1,2,2,2-tetrafluoroethane	145.5	32.6	0.582
1,2-Dichloro-1,1,2,2-tetrafluoroethane (freon 114)	145.7	32.2	0.582
Dideuterium oxide (D_2O)	371.0	215.7	0.363
Diethylamine	223.5	36.6	0.243
1,4-Diethylbenzene	384.73	27.66	
Diethyl ether	193.55	35.9	0.265
Difluoroamine (HNF_2)	130	93	
Difluorochloromethane	96.4	48.5	0.525
Difluorodiazine *cis* (N_2F_2)	−1	70	
trans	−13	55	

Table 9-7 (*Continued*)
CRITICAL PROPERTIES

Substance	T_c, °C	P_c, atm	ρ_c, g/cm^3
Difluorodichloromethane (freon 12)	111.8	40.71	0.558
1,1-Difluoroethane	113.5	44.37	0.365
1,1-Difluoroethylene	30.1	43.75	0.417
Dihydrogen disulfide	299	58.3	
Dihydrogen trisulfide	465	50.6	
Dihydrogen tetrasulfide	582	43.1	
Dihydrogen pentasulfide	657	38.4	
Dihydrogen hexasulfide	707	36	
Dihydrogen heptasulfide	742	33	
Dihydrogen octasulfide	767	32	
Dimethylamine	164.5	52.4	
N,N-Dimethylaniline	414	35.8	
2,2-Dimethylbutane	215.58	30.40	0.240
2,3-Dimethylbutane	226.78	30.86	0.241
Dimethyl ether	126.9	53	0.242
2,2-Dimethylhexane	276.65	24.96	0.239
2,3-Dimethylhexane	290.27	25.94	0.244
2,4-Dimethylhexane	280.30	25.23	0.242
2,5-Dimethylhexane	276.84	24.54	0.237
3,3-Dimethylhexane	288.80	26.19	0.258
3,4-Dimethylhexane	295.63	26.57	0.245
Dimethyl oxalate	355	39.3	
2,2-Dimethylpentane	247.29	27.37	0.241
2,3-Dimethylpentane	264.14	28.70	0.255
2,4-Dimethylpentane	246.58	27.01	0.240
3,3-Dimethylpentane	263.19	29.07	0.242
2,2-Dimethylpropane (neopentane)	160.60	31.57	0.238
2,3-Dimethylpyridine (2,3-lutidine)	382.3		
2,4-Dimethylpyridine	374		
2,5-Dimethylpyridine	371.0		
2,6-Dimethylpyridine	350.6		
3,4-Dimethylpyridine	410.6		
3,5-Dimethylpyridine	394.1		
N,N-Dimethyl-*o*-toluidine	395	30.8	
Dioxane	314	51.4	0.370
Di-*n*-propylamine	277	31	
3-4-Dithiahexane (ethyl disulfide)	369		
Ethane	32.28	48.16	0.203
Ethanethiol (ethyl mercaptan)	226	54.2	0.300
Ethanol (ethyl alcohol)	243.1	62.96	0.276
Ethyl acetate	250.1	37.99	0.308
Ethylacetylene	190.5		
Ethylal (propionaldehyde)	254		
Ethyl allyl ether	245		
Ethylamine	183	55.5	
Ethylbenzene	343.94	35.62	0.284
Ethyl bromide	230.7	61.5	0.507
Ethyl *n*-butanoate (ethyl *n*-butyrate)	293	30	0.28
Ethyl chloride	187.2	52	
Ethyl crotonate	326		
Ethylcyclopentane	296.3	33.5	0.262
Ethyl disulfide	369		

Table 9-7 (*Continued*)
CRITICAL PROPERTIES

Substance	T_c, °C	P_c, atm	ρ_c, g/cm^3
Ethyl ether, *see* Diethyl ether			
Ethylene (ethene)	9.21	49.66	0.218
Ethylene oxide	196	71.0	0.314
Ethyl fluoride	102.16	49.62	
Ethyl formate	235.3	46.76	0.323
3-Ethylhexane	292.27	25.74	0.251
Ethyl mercaptan	226	54.2	0.300
Ethyl-2-methylpropanoate (ethyl isobutyrate)	280	30	0.28
3-Ethylpentane	267.42	28.53	0.241
Ethyl propanoate (ethyl propionate)	272.9	33.18	0.296
Ethyl *n*-propyl ether	226.4	32.1	0.361
Ethyl sulfide	284	39.1	0.279
o-Ethyltoluene	380	31	0.28
m-Ethyltoluene	363	31	0.28
p-Ethyltoluene	363	31	0.28
Fluorine	−129.0	55	0.63
Fluorobenzene	286.6	44.6	0.354
Fluorodichloromethane	178.5	51.0	0.522
Fluoroethane	102.16	49.62	
Fluorotrichloromethane (freon 11)	198.0	43.2	0.554
Fluoromethane	44.55	58.0	0.300
Germanium tetrachloride	276.9	38	
Helium	−267.96	2.261	0.06930
Helium-3	−269.81	1.167	0.041
n-Heptane	267.0	27.00	0.232
1-Heptanol	360		0.267
1-Heptene	264.08		
1,5-Hexadiene	234.4		
Hexamethylbenzene	494		
n-Hexane	234.2	29.3	0.233
1-Hexanol	337		0.268
1-Hexene	230.83		
Hydrazine	380	145	
Hydrogen (equilibrium)	−240.17	12.77	0.0308
(normal)	−239.91	12.80	0.0310
Hydrogen bromide	89.8	84	
Hydrogen chloride	51.40	81.5	0.42
Hydrogen cyanide	183.5	53.2	0.195
Hydrogen deuteride, *see* Deuterium hydride			
Hydrogen fluoride	188	64	0.29
Hydrogen iodide	150.7	81	
Hydrogen selenide	137	88	
Hydrogen sulfide	100.4	88.9	0.31
Iodine	535		
Iodobenzene	448	44.6	0.581
Iodomethane (methyl iodide)	255		
Isobutyl acetate	288		
Isobutylbenzene	377	31	
Isobutyl butanoate	338		

Table 9-7 (*Continued*)
CRITICAL PROPERTIES

Substance	T_c, °C	P_c, atm	ρ_c, g/cm³
Isobutyl formate	278	38.3	0.29
Isobutyl-3-methyl butanoate	348		
Isobutyl propanoate	319		
Isopropylbenzene	357.9	31.67	0.28
Isopropyl ether	226.9	28.4	0.265
Isoquinoline	530		
Isoxazole	278.9		
Krypton	−63.75	54.20	0.9085
Mercury	900	180	
Methane	−82.60	45.44	0.162
Methanethiol (methyl mercaptan)	196.8	71.4	0.332
Methanol (methyl alcohol)	239.43	79.9	0.272
Methoxybenzene (anisole)	368	41.2	
Methyl acetate	233.7	46.33	0.325
Methylamine	156.9	73.6	
N-Methylaniline	428	51.3	
2-Methylbutane (isopentane)	187.24	33.37	0.236
3-Methylbutanoic acid (isovaleric acid)	361		
Methyl butanoate	281.3	34.28	0.300
3-Methyl-1-butanol	306.25		
2-Methyl-2-butanol	272		
3-Methyl-2-butanone (methyl isopropyl ketone)	280.2	38.0	0.278
2-Methyl-1-butene	192	34	
2-Methyl-2-butene	197	34	
Methylcyclohexane	299.1	34.32	0.285
Methylcyclopentane	259.6	37.4	0.264
Methyl ether, *see* Dimethyl ether			
Methylethylamine	223.5	36.6	0.243
Methyl ethyl ether	164.7	43.4	0.272
2-Methyl-3-ethylpentane	293.87	26.65	0.258
3-Methyl-3-ethylpentane	303.36	27.71	0.251
Methyl ethyl sulfide	260	42	
Methyl formate	214.0	59.25	0.349
2-Methylheptane	286.42	24.52	0.234
3-Methylheptane	290.45	25.13	0.246
4-Methylheptane	288.52	25.09	0.240
2-Methylhexane	257.16	26.98	0.238
3-Methylhexane	262.04	27.77	0.248
Methylhydrazine	292	79.3	0.170
Methyl-2-methyl propanoate	267.55	33.87	0.301
1-Methylnaphthalene	499		
2-Methylnaphthalene	488		
Methyl oxalate	260	9.48	
2-Methylpentane	224.30	29.71	0.235
3-Methylpentane	231.20	30.83	0.235
4-Methyl-2-pentanone (methyl isobutyl ketone)	298.3	32.3	
2-Methylpropane (isobutane)	134.98	36.00	0.221
Methyl propanoate	257.4	39.52	0.312
2-Methylpropanoic acid (isobutyric acid)	336	40	0.302
2-Methyl-1-propanol (isobutyl alcohol)	274.58	42.39	0.272
2-Methyl-2-propanol (*t*-butyl alcohol)	233.0	39.20	0.270

Table 9-7 (*Continued*)
CRITICAL PROPERTIES

Substance	T_c, °C	P_c, atm	ρ_c, g/cm³
2-Methylpropene	144.73	39.48	0.235
2-Methylpyridine (α-picoline)	348		
3-Methylpyridine (β-picoline)	372		
4-Methylpyridine (γ-picoline)	373		
Methyl sulfide	229.9	54.6	0.309
Naphthalene	475.2	39.98	0.31
Neon	−228.71	26.86	0.4835
Niobium pentachloride	534	0.46	0.68
Nitric oxide	−92.9	64.6	0.52
Nitrogen-14	−146.89	33.54	0.3110
Nitrogen-15	−146.8	33.5	0.332
Nitrogen dioxide (equilibrium)	158.2	100	0.557
Nitrogen trifluoride	−39.3	44.7	
Nitromethane	315	62.3	0.352
Nitrous oxide	36.434	71.596	0.4525
Nitryl fluoride	76.3		
n-Nonane	321.41	22.8	
1-Nonanol	404		0.264
n-Octane	295.61	24.54	0.232
1-Octanol	385		0.266
2-Octanol	364		
1-Octene	293.4		
Oxygen	−118.38	50.14	0.419
Oxygen difluoride (fluorine oxide)	−58.0	48.9	0.553
Ozone	−12.10	54.6	0.436
Paraldehyde	290		
Pentafluorochloroacetone	137.5	28.4	
Pentafluorobenzene	258.8	34.7	
1,1,2-H-Pentafluoropropane (refrigerant 245)	106.96	30.96	0.491
n-Pentane	196.5	33.35	0.237
Pentanoic acid (*n*-valeric acid)	378		
1-Pentanol	313		0.270
2-Pentanone (methyl *n*-propyl ketone)	290.8	38.4	0.286
3-Pentanone (diethyl ketone)	287.8	36.9	0.256
1-Pentene	191.59	40	
2-Pentene cis	203	36	
trans	202	36	
n-Pentyl formate	303		
1-Pentyne (propylacetylene)	220.3		
Perchloryl fluoride	95.2	53.0	0.64
Perfluoroacetone (hexafluoroacetone)	84.1	28.0	
Perfluorobenzene	243.57	32.61	
Perfluoro-*n*-butane	113.2	22.93	0.629
Perfluoro-(2-butyltetrahydrofuran)	227.1	15.86	0.707
Perfluorocyclobutane	115.22	27.41	0.616
Perfluorocyclohexane	184.0	24	
Perfluorocyclohexene	188.6		
Perfluoro-*n*-decane	269.2	14.3	
Perfluoroethane	19.7		0.617
Perfluoroethene (tetrafluoroethylene)	33.3	38.92	0.58
Perfluoro-*n*-heptane	201.6	16.0	0.584

Table 9-7 (*Continued*)
CRITICAL PROPERTIES

Substance	T_c, °C	P_c, atm	ρ_c, g/cm³
Perfluoro-1-heptene	205.0		
Perfluoro-*n*-hexane	174.5	18.8	
Perfluoro-1-hexene	181.2		
Perfluoromethane (tetrafluoromethane)	−45.6	36.9	0.630
Perfluoro-(methylcyclohexane)	213.6	23	
Perfluoronaphthalene	399.9		
Perfluoro-*n*-nonane	250.8	15.4	
Perfluoro-*n*-octane	229	16.4	
Perfluoro-*n*-pentane	149	20.1	
Perfluoro-*n*-propane	71.9	26.45	0.628
Phenetole	374	33.8	
Phenol	421.1	60.5	0.41
Phosphine	51.3	64.5	
Phosphonium chloride	49.1	72.7	
Phosphorus bromide difluoride (PBrF$_2$)	113		
Phosphorus chloride difluoride	89.17	44.61	
Phosphorus dibromide fluoride	254		
Phosphorus dichloride fluoride	189.84	49.3	
Phosphorus pentachloride	372		
Phosphorus trichloride	285.5		
Phosphorus trifluoride	−2.05	42.69	
Phosphoryl chloride difluoride (POClF$_2$)	150.6	43.4	
Phosphoryl trichloride	329		
Phosphoryl trifluoride	73.3	41.8	
Piperidine	320.9		
Propane	96.67	41.94	0.217
Propanenitrile (propionitrile)	291.2	41.3	0.240
Propanoic acid (propionic acid)	339	53	0.32
1-Propanol (*n*-propyl alcohol)	263.56	51.02	0.275
2-Propanol (isopropyl alcohol)	235.16	47.02	0.273
Propene (propylene)	91.8	45.6	0.233
n-Propyl acetate	276.2	33.19	0.269
n-Propylamine	233.8	46.8	
n-Propylbenzene	365.15	31.58	0.273
n-Propyl formate	264.9	40.08	0.309
n-Propyl propanoate	305		
Propyne (methylacetylene)	129.23	55.54	0.245
Pyridine	346.8	55.6	0.312
Pyrrole	366.6		
Pyrrolidine	295.4	55.4	0.286
Quinoline	509		
Radon	103.84	62	1.6
Silicon chloride trifluoride (SiClF$_3$)	34.5	34.2	
Silicon hydride (silane)	−3.5	47.8	
Silicon tetrachloride	233		
Silicon tetrafluoride	−14.1	36.7	
Silicon trichlorofluoride	165.3	35.3	
Sulfane di- (H$_2$S$_2$)	572	58.3	
tri- (H$_2$S$_3$)	738	50.6	
tetra- (H$_2$S$_4$)	855	43.1	
penta- (H$_2$S$_5$)	930	38.4	

Table 9-7 (*Continued*)
CRITICAL PROPERTIES

Substance	T_c, °C	P_c, atm	ρ_c, g/cm³
Sulfur	1040	116	
Sulfur dioxide	157.50	77.9	0.5240
Sulfur hexafluoride	45.55	37.11	0.734
Sulfur tetrafluoride	91		
Sulfur trioxide	218.2	83.8	0.633
Tantalum pentachloride	494	0.43	0.89
1,1,2,2-Tetrachloro-1,2-difluoroethane	278		
1,1,2,2-Tetrachloroethane	388.0		
Tetrachloroethene	347.1		
Tetrafluorohydrazine (F_2NNF_2)	36	77	
Tetrahydrothiophene	358.8		
1,2,4,5-Tetramethylbenzene	402	29	
2,2,3,3-Tetramethylbutane	294.7	28.3	0.248
o-Terphenyl	617.8	38.5	0.306
m-Terphenyl	651.7	34.6	0.300
p-Terphenyl	652.8	32.8	0.302
2-Thiabutane (methyl ethyl sulfide)	260	42	
4-Thia-1,5-heptadiene (diallyl sulfide)	380		
3-Thiapentane (diethyl sulfide)	284	39.1	0.284
2-Thiapropane (dimethyl sulfide)	229.9	54.6	0.309
Thiophene	307	56.2	0.385
Thymol	425		
Tin(IV) chloride	318.7	37.0	0.742
Titanium tetrachloride	355		
Toluene	318.57	40.55	0.292
Toluonitrile	450		
Trichloroethene	271.0	49.5	
Trichlorofluoromethane	198.0	43.5	0.554
1,2,2-Trichloro-1,1,2-trifluoroethane (freon 113)	214.1	33.7	0.576
1H-Tridecafluorohexane	198.6		
Triethylamine	262	30	0.26
Trifluoroacetic acid	218.1	32.15	0.559
1,1,1-Trifluoroethane	73.1	37.09	0.434
Trifluoromethane	25.74	47.73	0.525
Trimethylamine	160.1	40.2	0.233
1,2,3-Trimethylbenzene	391.3	34.09	
1,2,4-Trimethylbenzene	375.90	31.90	
1,3,5-Trimethylbenzene	364.13	30.86	
2,2,3-Trimethylbutane	257.96	29.15	0.252
2,2,3-Trimethylpentane	290.28	26.94	0.262
2,2,4-Trimethylpentane	270.74	25.34	0.244
2,3,3-Trimethylpentane	300.34	27.83	0.251
2,3,4-Trimethylpentane	293.19	26.94	0.248
2,4,6-Trimethyl-s-trioxane (paraldehyde)	290		
1H-Undecafluoropentane	170.8		
Uranium hexafluoride	230.2	45.5	
Vinyl fluoride (fluoroethene)	54.7	51.7	0.320

Table 9-7 (*Continued*)
CRITICAL PROPERTIES

Substance	T_c, °C	P_c, atm	ρ_c, g/cm^3
Water	374.2	218.3	0.325
Water, heavy, *see* Dideuterium oxide			
Xenon	16.59	57.62	1.105
o-Xylene	357.1	36.84	0.288
m-Xylene	343.82	34.95	0.282
p-Xylene	343.0	34.65	0.280
2,3-Xylenol	449.7	48	0.26
2,4-Xylenol	434.4	43	0.24
2,5-Xylenol	449.9	48	0.26
2,6-Xylenol	427.8	42	0.24
3,4-Xylenol	456.7	49	0.27
3,5-Xylenol	442.4	36	0.20

Section 10

PHYSICAL PROPERTIES

SOLUBILITIES

Table 10-1
SOLUBILITY OF GASES IN WATER

The column (or line entry) headed "α" gives the volume of gas (in ml) measured at standard conditions (0°C and 760 mm or 101.325 kN $\cdot$ m^{-2}) dissolved in 1 ml of water at the temperature stated (in °C) and when the pressure of the gas without that of the water vapor is 760 mm. The line entry "A" indicates the same quantity except that the gas itself is at the uniform pressure of 760 mm when in equilibrium with water.

The column headed "l" gives the volume of the gas (in ml) dissolved in 1 ml of water when the pressure of the gas plus that of the water vapor is 760 mm.

The column headed "q" gives the weight of gas in grams dissolved in 100 grams of water when the pressure of the gas plus that of the water vapor is 760 mm.

Temp. °C	ACETYLENE		AIR*		AMMONIA		BROMINE	
	α	q	$\alpha(\times 10^3)$	%Oxygen in dissd. air.	α	q	α	q
0	1.73	0.200	29.18	34.91	1130	89.5	60.5	42.9
1	1.68	0.194	28.42	34.87	....	....	....	....
2	1.63	0.188	27.69	34.82	....	....	54.1	38.3
3	1.58	0.182	26.99	34.78	....	....	....	....
4	1.53	0.176	26.32	34.74	1047	79.6	48.3	34.2
5	1.49	0.171	25.68	34.69	....	....	....	....
6	1.45	0.167	25.06	34.65	....	....	43.3	30.6
7	1.41	0.162	24.47	34.60	....	....	....	....
8	1.37	0.157	23.90	34.56	947	72.0	38.9	27.5
9	1.34	0.154	23.36	34.52	....	....	....	....
10	1.31	0.150	22.84	34.47	870	68.4	35.1	24.8
11	1.27	0.146	22.34	34.43	....	....	....	....
12	1.24	0.142	21.87	34.38	857	65.1	31.5	22.2
13	1.21	0.138	21.41	34.34	837	63.6	....	....
14	1.18	0.135	20.97	34.30	....	....	28.4	20.0
15	1.15	0.131	20.55	34.25	770	....	....	....
16	1.13	0.129	20.14	34.21	775	58.7	25.7	18.0
17	1.10	0.125	19.75	34.17	....	....	....	....
18	1.08	0.123	19.38	34.12	....	....	23.4	16.4
19	1.05	0.119	19.02	34.08	....	....	....	....
20	1.03	0.117	18.68	34.03	680	52.9	21.3	14.9
21	1.01	0.115	18.34	33.99	....	....	....	....
22	0.99	0.112	18.01	33.95	....	....	19.4	13.5
23	0.97	0.110	17.69	33.90	....	....	....	....
24	0.95	0.107	17.38	33.86	639	48.2	17.7	12.3
25	0.93	0.105	17.08	33.82	....	....	....	....
26	0.91	0.102	16.79	33.77	....	....	16.3	11.3
27	0.89	0.100	16.50	33.73	....	....	....	....
28	0.87	0.098	16.21	33.68	586	44.0	15.0	10.3
29	0.85	0.095	15.92	33.64	....	....	....	....
30	0.84	0.094	15.64	33.60	530	41.0	13.8	9.5
35	....	.....	....	.....	....	....	....	....
40	....	.....	14.18	.....	400	31.6	9.4	6.3
45	....	.....	....	.....	....	....	....	....
50	....	.....	12.97	.....	290	23.5	6.5	4.1
60	....	.....	12.16	.....	200	16.8	4.9	2.9
70	....	.....	....	.....	....	11.1	3.8	1.9
80	....	.....	11.26	.....	....	6.5	3.0	1.2
90	....	.....	....	.....	....	3.0	....	....
100	....	.....	11.05	.....	....	0.0	....	....

* Free from NH$_3$ and CO$_2$; total pressure of air + aq. tension is 760 mm.

Table 10-1 (*Continued*)
SOLUBILITY OF GASES IN WATER

Temp. °C.	CARBON DIOXIDE α	q	CARBON MONOXIDE α	q	CHLORINE l	q	ETHANE α	q	ETHYLENE α	q	HYDROGEN α	q
0	1.713	0.3346	0.03537	0.004397	.....	.....	0.09874	0.01317	0.226	0.0281	0.02148	0.0001922
1	1.646	0.3213	0.03455	0.004293	.....	.....	0.09476	0.01263	0.219	0.0272	0.02126	0.0001901
2	1.584	0.3091	0.03375	0.004191	.....	.....	0.09093	0.01212	0.211	0.0262	0.02105	0.0001881
3	1.527	0.2978	0.03297	0.004092	.....	.....	0.08725	0.01162	0.204	0.0253	0.02084	0.0001862
4	1.473	0.2871	0.03222	0.003996	.....	.....	0.08372	0.01114	0.197	0.0244	0.02064	0.0001843
5	1.424	0.2774	0.03149	0.003903	.....	.....	0.08033	0.01069	0.191	0.0237	0.02044	0.0001824
6	1.377	0.2681	0.03078	0.003813	.....	.....	0.07709	0.01025	0.184	0.0228	0.02025	0.0001806
7	1.331	0.2589	0.03009	0.003725	.....	.....	0.07406	0.00983	0.178	0.0220	0.02007	0.0001789
8	1.282	0.2492	0.02942	0.003640	.....	.....	0.07106	0.00943	0.173	0.0214	0.01989	0.0001772
9	1.237	0.2403	0.02878	0.003559	.....	.....	0.06826	0.00906	0.167	0.0207	0.01972	0.0001756
10	1.194	0.2318	0.02816	0.003479	3.148	0.9972	0.06561	0.00870	0.162	0.0200	0.01955	0.0001740
11	1.154	0.2239	0.02757	0.003405	3.047	0.9654	0.06328	0.00838	0.157	0.0194	0.01940	0.0001725
12	1.117	0.2165	0.02701	0.003332	2.950	0.9346	0.06106	0.00808	0.152	0.0188	0.01925	0.0001710
13	1.083	0.2098	0.02646	0.003261	2.856	0.9050	0.05894	0.00780	0.148	0.0183	0.01911	0.0001696
14	1.050	0.2032	0.02593	0.003194	2.767	0.8768	0.05694	0.00753	0.143	0.0176	0.01897	0.0001682
15	1.019	0.1970	0.02543	0.003130	2.680	0.8495	0.05504	0.00727	0.139	0.0171	0.01883	0.0001668
16	0.985	0.1903	0.02494	0.003066	2.597	0.8232	0.05326	0.00703	0.136	0.0167	0.01869	0.0001654
17	0.956	0.1845	0.02448	0.003007	2.517	0.7979	0.05159	0.00680	0.132	0.0162	0.01856	0.0001641
18	0.928	0.1789	0.02402	0.002947	2.440	0.7738	0.05003	0.00659	0.129	0.0158	0.01844	0.0001628
19	0.902	0.1737	0.02360	0.002891	2.368	0.7510	0.04858	0.00639	0.125	0.0153	0.01831	0.0001616
20	0.878	0.1688	0.02319	0.002838	2.299	0.7293	0.04724	0.00620	0.122	0.0149	0.01819	0.0001603
21	0.854	0.1640	0.02281	0.002789	2.238	0.7100	0.04589	0.00602	0.119	0.0146	0.01805	0.0001588
22	0.829	0.1590	0.02244	0.002739	2.180	0.6918	0.04459	0.00584	0.116	0.0142	0.01792	0.0001575
23	0.804	0.1540	0.02208	0.002691	2.123	0.6739	0.04335	0.00567	0.114	0.0139	0.01779	0.0001561
24	0.781	0.1493	0.02174	0.002646	2.070	0.6572	0.04217	0.00551	0.111	0.0135	0.01766	0.0001548
25	0.759	0.1449	0.02142	0.002603	2.019	0.6413	0.04104	0.00535	0.108	0.0131	0.01754	0.0001535
26	0.738	0.1406	0.02110	0.002560	1.970	0.6259	0.03997	0.00520	0.106	0.0129	0.01742	0.0001522
27	0.718	0.1366	0.02080	0.002519	1.923	0.6112	0.03895	0.00506	0.104	0.0126	0.01731	0.0001509
28	0.699	0.1327	0.02051	0.002479	1.880	0.5975	0.03799	0.00493	0.102	0.0123	0.01720	0.0001496
29	0.682	0.1292	0.02024	0.002442	1.839	0.5847	0.03709	0.00480	0.100	0.0121	0.01709	0.0001484
30	0.665	0.1257	0.01998	0.002405	1.799	0.5723	0.03624	0.00468	0.098	0.0118	0.01699	0.0001474
35	0.592	0.1105	0.01877	0.002231	1.602	0.5104	0.03230	0.00412	.....	.....	0.01666	0.0001425
40	0.530	0.0973	0.01775	0.002075	1.438	0.4590	0.02915	0.00366	.....	.....	0.01644	0.0001384
45	0.479	0.0860	0.01690	0.001933	1.322	0.4228	0.02660	0.00327	.....	.....	0.01624	0.0001341
50	0.436	0.0761	0.01615	0.001797	1.225	0.3925	0.02459	0.00294	.....	.....	0.01608	0.0001287
60	0.359	0.0576	0.01488	0.001522	1.023	0.3295	0.02177	0.00239	.....	.....	0.01600	0.0001178
70	.....	.....	0.01440	0.001276	0.862	0.2793	0.01948	0.00185	.....	.....	0.0160	0.000102
80	.....	.....	0.01430	0.000980	0.683	0.2227	0.01826	0.00134	.....	.....	0.0160	0.000079
90	.....	.....	0.0142	0.00057	0.39	0.127	0.0176	0.0008	.....	.....	0.0160	0.000046
100	.....	.....	0.0141	0.00000	0.00	0.000	0.0172	0.0000	.....	.....	0.0160	0.000000

Table 10-1 (Continued)
SOLUBILITY OF GASES IN WATER

Temp. °C.	HYDROGEN SULFIDE		METHANE		NITRIC OXIDE		NITROGEN*		OXYGEN		SULFUR DIOXIDE	
	α	q	α	q	α	q	α	q	α	q	l	q
0	4.670	0.7066	0.05563	0.003959	0.07381	0.009833	0.02354	0.002942	0.04889	0.006945	79.789	22.83
1	4.522	0.6839	0.05401	0.003842	0.07184	0.009564	0.02297	0.002869	0.04758	0.006756	77.210	22.09
2	4.379	0.6619	0.05244	0.003728	0.06993	0.009305	0.02241	0.002798	0.04633	0.006574	74.691	21.37
3	4.241	0.6407	0.05093	0.003619	0.06809	0.009057	0.02187	0.002730	0.04512	0.006400	72.230	20.66
4	4.107	0.6201	0.04946	0.003513	0.06632	0.008816	0.02135	0.002663	0.04397	0.006232	69.828	19.98
5	3.977	0.6001	0.04805	0.003410	0.06461	0.008584	0.02086	0.002600	0.04287	0.006072	67.485	19.31
6	3.852	0.5809	0.04669	0.003312	0.06298	0.008361	0.02037	0.002537	0.04180	0.005918	65.200	18.65
7	3.732	0.5624	0.04539	0.003217	0.06140	0.008147	0.01990	0.002477	0.04080	0.005773	62.973	18.02
8	3.616	0.5446	0.04413	0.003127	0.05990	0.007943	0.01945	0.002419	0.03983	0.005632	60.805	17.40
9	3.505	0.5276	0.04292	0.003039	0.05846	0.007747	0.01902	0.002365	0.03891	0.005498	58.697	16.80
10	3.399	0.5112	0.04177	0.002955	0.05709	0.007560	0.01861	0.002312	0.03802	0.005368	56.647	16.21
11	3.300	0.4960	0.04072	0.002879	0.05587	0.007393	0.01823	0.002263	0.03718	0.005246	54.655	15.64
12	3.206	0.4814	0.03970	0.002805	0.05470	0.007233	0.01786	0.002216	0.03637	0.005128	52.723	15.09
13	3.115	0.4674	0.03872	0.002733	0.05357	0.007078	0.01750	0.002170	0.03559	0.005014	50.849	14.56
14	3.028	0.4540	0.03779	0.002665	0.05250	0.006930	0.01717	0.002126	0.03486	0.004906	49.033	14.04
15	2.945	0.4411	0.03690	0.002599	0.05147	0.006788	0.01685	0.002085	0.03415	0.004802	47.276	13.54
16	2.865	0.4287	0.03606	0.002538	0.05049	0.006652	0.01654	0.002045	0.03348	0.004703	45.578	13.05
17	2.789	0.4169	0.03525	0.002478	0.04956	0.006524	0.01625	0.002006	0.03283	0.004606	43.939	12.59
18	2.717	0.4056	0.03448	0.002422	0.04868	0.006400	0.01597	0.001970	0.03220	0.004514	42.360	12.14
19	2.647	0.3948	0.03376	0.002369	0.04785	0.006283	0.01570	0.001935	0.03161	0.004426	40.838	11.70
20	2.582	0.3846	0.03308	0.002319	0.04706	0.006173	0.01545	0.001901	0.03102	0.004339	39.374	11.28
21	2.517	0.3745	0.03243	0.002270	0.04625	0.006059	0.01522	0.001869	0.03044	0.004252	37.970	10.88
22	2.456	0.3648	0.03180	0.002222	0.04545	0.005947	0.01498	0.001838	0.02988	0.004169	36.617	10.50
23	2.396	0.3554	0.03119	0.002177	0.04469	0.005838	0.01475	0.001809	0.02934	0.004087	35.302	10.12
24	2.338	0.3463	0.03061	0.002133	0.04395	0.005733	0.01454	0.001780	0.02881	0.004007	34.026	9.76
25	2.282	0.3375	0.03006	0.002091	0.04323	0.005630	0.01434	0.001751	0.02831	0.003931	32.786	9.41
26	2.229	0.3290	0.02952	0.002050	0.04254	0.005530	0.01413	0.001724	0.02783	0.003857	31.584	9.06
27	2.177	0.3208	0.02901	0.002011	0.04188	0.005435	0.01394	0.001698	0.02736	0.003787	30.422	8.73
28	2.128	0.3130	0.02852	0.001974	0.04124	0.005342	0.01376	0.001672	0.02691	0.003718	29.314	8.42
29	2.081	0.3055	0.02806	0.001938	0.04063	0.005252	0.01358	0.001647	0.02649	0.003651	28.210	8.10
30	2.037	0.2983	0.02762	0.001904	0.04004	0.005165	0.01342	0.001624	0.02608	0.003588	27.161	7.80
35	1.831	0.2648	0.02546	0.001733	0.03734	0.004757	0.01256	0.001501	0.02440	0.003315	22.489	6.47
40	1.660	0.2361	0.02369	0.001586	0.03507	0.004394	0.01184	0.001391	0.02306	0.003082	18.766	5.41
45	1.516	0.2110	0.02238	0.001466	0.03311	0.004059	0.01130	0.001300	0.02187	0.002858	.....	....
50	1.392	0.1883	0.02134	0.001359	0.03152	0.003758	0.01088	0.001216	0.02090	0.002657	.....	....
60	1.190	0.1480	0.01954	0.001144	0.02954	0.003237	0.01023	0.001052	0.01946	0.002274	.....	....
70	1.022	0.1101	0.01825	0.000926	0.02810	0.002668	0.00977	0.000851	0.01833	0.001856	.....	....
80	0.917	0.0765	0.01770	0.000695	0.02700	0.001984	0.00958	0.000660	0.01761	0.001381	.....	....
90	0.84	0.041	0.01735	0.00040	0.0265	0.00113	0.0095	0.00038	0.0172	0.00079	.....	....
100	0.81	0.000	0.0170	0.00000	0.0263	0.00000	0.0095	0.00000	0.0170	0.00000	.....	....

*Atmospheric Nitrogen containing 98.815% N₂ by volume + 1.185% inert gases.

Table 10-1 (Continued)
SOLUBILITY OF GASES IN WATER

Substance		0°	10°	20°	30°	40°	60°	80°
Argon	α	0.052 8	0.041 3	0.033 7	0.028 8	0.025 1	0.020 9	0.018 4
Helium	A	0.009 8	0.009 11	0.008 6	0.008 39	0.008 41	0.009 02	$0.009\ 42^{70°}$
Hydrogen chloride	α	512	475	442	412	385	339	
Krypton	α	0.110 5	0.081 0	0.062 6	0.051 1	0.043 3	0.035 7	
Neon	A		$0.011\ 79°$	0.010 6	0.010 0	$0.009\ 48^{42°}$		$0.009\ 84^{73°}$
Nitrous oxide	A		0.88	0.63				
Ozone	g · liter^{-1}	0.039 4	$0.029\ 9^{12°}$	$0.021\ 0^{19°}$	$0.013\ 9^{27°}$	0.004 2	0	
Radon	α	0.510	0.326	0.222	0.162	0.126	0.085	
Xenon	α	0.242	0.174	0.123	0.098	0.082		

Table 10-2

SOLUBILITIES OF INORGANIC COMPOUNDS AND METAL SALTS OF ORGANIC ACIDS IN WATER AT VARIOUS TEMPERATURES

Solubilities are expressed as the number of grams of substance of stated molecular formula which when dissolved in 100 g of water make a saturated solution at the temperature stated (°C).

Substance	Formula	0°	10°	20°	30°	40°	60°	80°	90°	100°
Aluminum chloride	$AlCl_3$	43.9	44.9	45.8	46.6	47.3	48.1	48.6		49.0
fluoride	AlF_3	0.56	0.56	0.67	0.78	0.91	1.1	1.32		1.72
nitrate	$Al(NO_3)_3$	60.0	66.7	73.9	81.8	88.7	106	132	153	160
perchlorate	$Al(ClO_4)_3$	122	128	133						182
sulfate	$Al_2(SO_4)_3$	31.2	33.5	36.4	40.4	45.8	59.2	73.0	80.8	89.0
thallium(I) sulfate	$Al_2Tl_2(SO_4)_4$	3.15	4.60	6.39	9.37	14.39	35.35			
Ammonium aluminum sulfate	$NH_4Al(SO_4)_2$	2.10	5.00	7.74	10.9	14.9	26.7			
azide	NH_4N_3	16.0		25.3		37.1				
bromide	NH_4Br	60.5	68.1	76.4	83.2	91.2	108	125	135	145
chloride	NH_4Cl	29.4	33.2	37.2	41.4	45.8	55.3	65.6	71.2	77.3
chloroiridate(IV)	$(NH_4)_2IrCl_6$	0.56	0.71	0.95	1.20	1.56	2.45	4.38		
chloroplatinate(IV)	$(NH_4)_2PtCl_6$	0.289	0.374	0.499	0.637	0.815	1.44	2.16	2.61	3.36
chromate	$(NH_4)_2CrO_4$	25.0	29.2	34.0	39.3	45.3	59.0	76.1		
chromium(III) sulfate	$(NH_4)Cr(SO_4)_2$	3.95			18.8	32.6				
cobalt(II) sulfate	$(NH_4)_2Co(SO_4)_2$	6.0	9.5	13.0	17.0	22.0	33.5	49.0	58.0	75.1
dichromate	$(NH_4)_2Cr_2O_7$	18.2	25.5	35.6	46.5	58.5	86.0	115		156
dihydrogen arsenate	$NH_4H_2AsO_4$	33.7		48.7		63.8	83.0	107	122	
dihydrogen phosphate	$NH_4H_2PO_4$	22.7	29.5	37.4	46.4	56.7	82.5	118		173
dithionate	$(NH_4)_2S_2O_6$	133	151	166	179	204	311	533		
formate	NH_4CHO_2	102		143						
hydrogen carbonate	NH_4HCO_3	11.9	16.1	21.7	28.4	36.6	59.2	109	170	354
hydrogen phosphate	$(NH_4)_2HPO_4$	42.9	62.9	68.9	75.1	81.8	97.2			
hydrogen tartrate	$NH_4C_4H_5O_6$	1.00	1.88	2.70						
iodide	NH_4I	155	163	172	182	191	209	229		250
iron(II) sulfate	$(NH_4)_2Fe(SO_4)_2$	12.5	17.2	26.4	33	46				

Table 10-2 (Continued)
SOLUBILITIES OF INORGANIC COMPOUNDS AND METAL SALTS OF ORGANIC ACIDS IN WATER AT VARIOUS TEMPERATURES

Substance	Formula	0°	10°	20°	30°	40°	60°	80°	90°	100°
Ammonium magnesium										
sulfate	$(NH_4)_2Mg(SO_4)_2$	11.8	14.6	18.0	21.7	25.8	35.1	48.3		65.7
nickel sulfate	$(NH_4)_2Ni(SO_4)_2$	1.00	4.00	6.50	9.20	12.0	17.0			
nitrate	NH_4NO_3	118	150	192	242	297	421	580	740	871
oxalate	$(NH_4)_2C_2O_4$	2.2	3.21	4.45	6.09	8.18	14.0	22.4	27.9	34.7
perchlorate	NH_4ClO_4	12.0	16.4	21.7	27.7	34.6	49.9	68.9		
selenite	$(NH_4)_2SeO_3$	96	105	115	126	143	192			
sulfate	$(NH_4)_2SO_4$	70.6	73.0	75.4	78.0	81	88	95		103
sulfite	$(NH_4)_2SO_3$	47.9	54.0	60.8	68.8	78.4	104	144	150	153
tartrate	$(NH_4)_2C_4H_4O_6$	45.0	55.0	63.0	70.5	76.5	86.9			
thioantimonate(V)	$(NH_4)_3SbS_4$	71.2		91.2	120					
thiocyanate	NH_4SCN	120	144	170	208	234	346			
vanadate	NH_4VO_3			0.48	0.84	1.32	2.42			
zinc sulfate	$(NH_4)_2Zn(SO_4)_2$	7.0	9.5	12.5	16.0	20.0	30.0	46.6	58.0	72.4
Antimony(III) chloride	$SbCl_3$	602		910	1087	1368	[completely miscible at 72°]			
fluoride	SbF_3	385		444	562					
Arsenic hydride										
(760 mm), cc	AsH_3	42	30	28						
oxide (pent-)	As_2O_5	59.5	62.1	65.8	69.8	71.2	73.0	75.1		76.7
oxide (tri-)	As_2O_3	1.20	1.49	1.82	2.31	2.93	4.31	6.11		8.2
Barium acetate	$Ba(C_2H_3O_2)_2 \cdot 3H_2O$	58.8	62	72	75	78.5	75.0	74.0		74.8
azide	$Ba(N_3)_2$	12.5	16.1	$17.4^{17°}$						
bromate	$Ba(BrO_3)_2 \cdot H_2O$	0.29	0.44	0.65	0.95	1.31	2.27	3.52	4.26	5.39
bromide	$BaBr_2 \cdot 2H_2O$	98	101	104	109	114	123	135		149
n-butyrate	$Ba(C_4H_7O_2)_2$	37.0	36.1	35.4	34.9	35.2	37.2	41.7	45.5	$48.1^{95°}$
caproate	$Ba(C_6H_{11}O_2)_2 \cdot 3.5H_2O$	11.71	8.38	6.89	5.87	5.79	8.39	14.71	19.28	
chlorate	$Ba(ClO_3)_2 \cdot H_2O$	20.3	26.9	33.9	41.6	49.7	66.7	84.8		105
chloride	$BaCl_2 \cdot 2H_2O$	31.2	33.5	35.8	38.1	40.8	46.2	52.5	55.8	59.4
chlorite	$Ba(ClO_2)_2$	43.9	44.6	45.4		47.9	53.8	66.6		80.8
fluoride	BaF_2		0.159	0.160	0.162					

formate	Ba(CHO₂)₂	26.2	28.0	29.9	31.9	34.0	38.6	44.2	47.6	51.3
hydroxide	Ba(OH)₂	1.67	2.48	3.89	5.59	8.22	20.94	101.4		
iodate	Ba(IO₃)₂			0.035	0.046	0.057				
iodide	BaI₂·2H₂O	182	201	223	250	264		291		301
nitrate	Ba(NO₃)₂	4.95	6.67	9.02	11.48	14.1	20.4	27.2		34.4
nitrite	Ba(NO₂)₂·H₂O	50.3	60	72.8		102	151	222	261	325
perchlorate	Ba(ClO₄)₂·3H₂O	239		336		416	495	575		653
propionate	Ba(C₃H₅O₂)₂·H₂O	57.2	56.8		57.5	59.0	62.0	67.8	73.0	82.7
isosuccinate	BaC₄H₄O₄	0.421	0.432	0.418	0.393	0.366	0.306	0.237		
sulfamate	Ba(SO₃NH₂)₂	18.3	22.3	26.8	32.5	38.5	49.6	61.5	67.34	73.5
sulfide	BaS	2.88	4.89	7.86	10.38	14.89	27.69	49.91		60.29
tartrate	Ba(C₂H₂O₃)₂	0.021	0.024	0.028	0.032	0.035	0.044	0.053		
Beryllium nitrate	Be(NO₃)₂	97	102	108	113	125	178			
sulfate	BeSO₄	37.0	37.6	39.1	41.4	45.8	53.1	67.2		82.8
Boric acid	H₃BO₃	2.67	3.73	5.04	6.72	8.72	14.81	23.62	30.38	40.25
Cadmium bromide	CdBr₂	56.3	75.4	98.8	129	152	153	156		160
chlorate	Cd(ClO₃)₂	299	308	322	348	376	455			
chloride	CdCl₂·2.5H₂O	90	100	113	132	135	136	140		147
	CdCl₂·H₂O		135	135	135	135				
formate	Cd(CHO₂)₂	8.3	11.1	14.4	18.6	25.3	59.5	80.5	85.2	94.6
iodide	CdI₂	78.7	84.7	87.9	92.1	100	111			125
nitrate	Cd(NO₃)₂	122	136	150	167	194	203	221		272
perchlorate	Cd(ClO₄)₂·6H₂O		180	188	195	203	221	243		
selenate	CdSeO₄	72.5	68.4	64.0	58.9	55.0	44.2	32.5	27.2	22.0
sulfate	CdSO₄	75.4	76.0	76.6		78.5	81.8	66.7	63.1	60.8
Calcium acetate	Ca(OAc)₂·2H₂O	37.4	36.0	34.7	33.8	33.2	32.7	33.5	31.1	29.7
benzoate	Ca(OBz)₂·3H₂O	2.32	2.45	2.72	3.02	3.42	4.71	6.87	8.55	8.70
bromide	CaBr₂·6H₂O	125	132	143	$185^{34°}$	213	278	295		$312^{105°}$
butyrate	Ca(C₄H₇O₂)₂	20.31	19.15	18.20	17.25	16.40	15.15	14.95		15.85
cacodylate	Ca(C₂H₇O₂)₂	48	52	59	71	100				
	Ca(C₂H₆AsO₂)₂·9H₂O	59.5								
chloride	CaCl₂·6H₂O	59.5	64.7	74.5	100	128	137	147	154	159
chromate	CaCrO₄	4.5		2.25	1.83	1.49	0.83			
(mn)	CaCrO₄·2H₂O	17.3		16.6	16.1	17.05				
formate	Ca(CHO₂)₂	16.15		16.60		17.05	17.50	17.95		18.40
gluconate	Ca(C₆H₁₁O₇)₂·H₂O			3.72		5.29	12.11		36.80	$57.29^{6°}$
hydrogen carbonate	Ca(HCO₃)₂	16.15		16.60		17.05	17.50	17.95		18.40
hydroxide	Ca(OH)₂	0.189	0.182	0.173	0.160	0.141	0.121		0.086	0.076

Table 10-2 (Continued)
SOLUBILITIES OF INORGANIC COMPOUNDS AND METAL SALTS OF ORGANIC ACIDS IN WATER AT VARIOUS TEMPERATURES

Substance	Formula	0°	10°	20°	30°	40°	60°	80°	90°	100°
Calcium iodate	$Ca(IO_3)_2 \cdot 6H_2O$	0.090		0.24	0.38	0.52	0.65	0.66	0.67	
iodide	CaI_2	64.6	66.0	67.6	69.0	70.8	74	78		81
lactate	$Ca(C_3H_5O_3)_2 \cdot 5H_2O$	3.1		$5.41^{15°}$	7.9		$88.7^{55°}$			
levulinate	$Ca(C_{10}H_{14}O_6) \cdot 2H_2O$	38.1		$45.1^{16°}$	55.0	$70.3^{45°}$				
malonate	$Ca(C_3H_2O_4)$	0.29	0.33	0.36	0.40	0.42	0.46	0.48		
nitrate	$Ca(NO_3)_2 \cdot 4H_2O$	102	115	129	152	191		358	363	
nitrite	$Ca(NO_2)_2 \cdot 4H_2O$	63.9		$84.5^{18°}$	104		134	151	166	178
propionate	$Ca(C_3H_5O_2)_2 \cdot H_2O$	42.80		39.85			38.25	39.85	42.15	48.44
selenate	$CaSeO_4 \cdot 2H_2O$	9.73	9.77	9.22	8.79	7.14				
succinate	$Ca(C_3H_2O_2)_2 \cdot 3H_2O$	1.127	1.22	1.28		1.18	0.89	0.68		0.66
sulfamate	$Ca(SO_3NH_2)_2$	56.5	62.8	72.3	84.5	100.1	150.0	215.2	$242^{95°}$	
sulfate	$CaSO_4 \cdot {}^{1}/_{2}H_2O$			0.32	$0.29^{25°}$	$0.26^{35°}$	$0.21^{45°}$	$0.145^{65°}$	$0.125^{75°}$	0.071
	$CaSO_4 \cdot 2H_2O$	0.223	0.244	$0.255^{18°}$	0.264	0.265	$0.244^{65°}$	$0.234^{75°}$		0.205
tartrate	$CaC_4H_4O_6 \cdot 4H_2O$	0.026	0.029	0.034	0.046	0.063	0.091	0.130		
uranyl carbonate	$Ca_2UO_2(CO_3)_3 \cdot 10H_2O$	0.1		$0.4^{23°}$		0.8	$1.5^{55°}$			
valerate	$Ca(C_5H_9O_2)_2$	9.82	9.25	8.80	8.40	8.05	7.78	7.95	8.20	8.78
*iso*valerate	$Ca(C_5H_9O_2)_2 \cdot 3H_2O$	26.05	22.70	21.80	21.68	22.00	18.38	16.88	16.65	16.55
Carbon disulfide	CS_2	0.204	0.194	0.179	0.155	0.111				
oxide sulfide (STP) mL/100 mL	COS	133.3	83.6	56.1	40.3					
tetrafluoride (STP) mL/100 g	CF_4		0.595	0.490	0.415	0.366				
Cerium(III) ammonium nitrate	$Ce(NH_4)_2(NO_3)_5$		242	276	318	376	681			
(IV) ammonium nitrate	$Ce(NH_4)_2(NO_3)_6$			135	150	169	213			
(III) ammonium sulfate	$Ce(NH_4)(SO_4)_2$			5.53	4.49	3.48	2.02	1.33		
(III) selenate	$Ce_2(SeO_3)_3$	39.5	37.2	35.2	33.2	32.6	13.7	4.6	2.1	

Solubility table (values in g per 100 mL / 100 g water at successive temperatures; temperature-column headings are cut off at the top of the page). Superscripts indicate the actual measurement temperature for off-grid entries.

Compound	Formula									
(III) sulfate	$Ce_2(SO_4)_3 \cdot 9H_2O$	21.4		9.84	7.24	5.63	3.87	5.40	10.5	22.7
	$Ce_2(SO_4)_3 \cdot 8H_2O$			9.43	7.10	5.70	4.04			
Cesium aluminum sulfate	$Cs_2Al_2(SO_4)_4$	18.8								
bromate	$CsBrO_3$	0.21	0.30	0.40	0.61	0.85	2.00			
chlorate	$CsClO_3$	2.46	3.8	6.2	9.5	13.8	26.2	45.0	58.0	79.0
chloride	$CsCl$	161	175	187	197	208	230	250	260	271
chloroaurate(III)	$CsAuCl_4$		0.5	0.8	1.7	3.3	8.9	19.5	27.7	37.9
chloroplatinate(IV)	Cs_2PtCl_6	0.0047	0.0064	0.0087	0.0119	0.0158	0.0290	0.0525	0.0675	0.0914
formate	$CsCHO_2$	335	381	450	533	694				
iodide	CsI	44.1	58.5	76.5	96	$124^{45°}$	150	190	205	
nitrate	$CsNO_3$	9.33	14.9	23.0	33.9	47.2	83.8	134	163	197
perchlorate	$CsClO_4$	0.8	1.0	1.6	2.6	4.0	7.3	14.4	20.5	30.0
sulfate	Cs_2SO_4	167	173	179	184	190	200	210	215	220
Chlorine dioxide	ClO_2	2.76	6.00	$8.70^{15°}$						
Chromium(III) nitrate	$Cr(NO_3)_3$	$108^{5°}$	$124^{15°}$	$130^{25°}$	$152^{35°}$					
(VI) oxide	CrO_3	164.8	167.2			172.5	183.9	191.6		206.8
(III) perchlorate	$Cr(ClO_4)_3$	104	130							
Cobalt(II) bromide	$CoBr_2$	91.9	112		128	163	227	241		257
chlorate	$Co(ClO_3)_2$	135	180		195	214	316			
chloride	$CoCl_2$	43.5	47.7	52.9	59.7	69.5	93.8	97.6	101	106
iodate	$Co(IO_3)_2$			1.02	0.90	0.88	0.82	0.73		0.70
nitrate	$Co(NO_3)_2$	84.0	89.6	97.4	111	125	174	204		
nitrite	$Co(NO_2)_2$	0.076	0.24	0.40	0.61	0.85				
sulfate	$CoSO_4$	25.5	30.5	36.1	42.0	48.8	55.0	53.8	45.3	38.9
	$CoSO_4 \cdot 7H_2O$	44.8	56.3	65.4	73.0	88.1	101			
Copper(II) ammonium chloride	$CuCl_2 \cdot 2NH_4Cl$	28.2	$32.0^{12°}$	35.0	38.3	43.8	56.6	76.5	76.5	107
ammonium sulfate	$CuSO_4 \cdot (NH_4)_2SO_4$	11.5	15.1	19.4	24.4	30.5	46.3	69.7	86.1	
bromide	$CuBr_2$	107	116	126	128					
chloride	$CuCl_2$	68.6	70.9	73.0	77.3	87.6	96.5	104	108	120
fluorosilicate	$CuSiF_6$	73.5	76.5	81.6	$84.1^{25°}$	$91.2^{50°}$		$93.2^{75°}$		
nitrate	$Cu(NO_3)_2$	83.5	100	125	156	163	182	208	222	247
potassium sulfate	$CuSO_4 \cdot K_2SO_4$	5.1	7.2	10.0	13.6	18.2	36.50	53.68		
selenate	$CuSeO_4$	12.04	14.53	17.51	21.04	25.22				
sulfate	$CuSO_4 \cdot 5H_2O$	23.1	27.5	32.0	37.8	44.6	61.8	83.8		
tartrate	$CuC_4H_4O_6 \cdot 3H_2O$	$0.020^{15°}$		0.042	0.089	0.142	0.197	0.144		
Gadolinium bromate	$Gd(BrO_3)_3 \cdot 9H_2O$	50.2	70.1	95.6	126	166				114
sulfate	$Gd_2(SO_4)_3$	3.98	3.30	2.60	2.32					

Table 10-2 (Continued)
SOLUBILITIES OF INORGANIC COMPOUNDS AND METAL SALTS OF ORGANIC ACIDS IN WATER AT VARIOUS TEMPERATURES

Substance	Formula	0°	10°	20°	30°	40°	60°	80°	90°	100°
Germanium(IV) oxide	GeO_2		0.49	0.43	0.50	0.61				
Holmium sulfate	$Ho_2(SO_4)_3 \cdot 8H_2O$			8.18	6.71^{25}	4.52				
Hydrazinium (1+) nitrate	$N_2H_5NO_3$		175	266	402	607	2127			
(2+) sulfate	$N_2H_6SO_4$			2.87	3.89	4.15	9.08	14.39		
(1+) sulfate	$(N_2H_5)_2SO_4$				221	300	554			
Hydrogen bromide	HBr	221.2	210.3	204.0^{15}		171.5^{50}		150.5^{75}		130.0
chloride	HCl	82.3	78.0^{8}	71.9	67.3	63.3	56.1			
selenide, mL at STP	H_2Se	386	351	289						
Iodine	I_2	0.014	0.020	0.029	0.039	0.052	0.100	0.225	0.315	0.445
Iridium(IV) ammonium chloride	$(NH_4)_2IrCl_6$	0.556	0.706	0.77	1.21	1.57	2.46	4.38	dec	
sodium chloride	Na_2IrCl_6		34.46^{15}		56.17	96.00	191.2	279.3		
Iron(II) ammonium sulfate	$FeSO_4 \cdot (NH_4)_2SO_4 \cdot 6H_2O$	17.23	31.0	36.47	45.0					
(II) bromide	$FeBr_2$	101	109	117	124	133	144	168	176	184
(II) chloride	$FeCl_2$	49.7	59.0	62.5	66.7	70.0	78.3	88.7	92.3	94.9
(III) chloride	$FeCl_3 \cdot 6H_2O$	74.4		91.8	106.8					
(II) fluoro-silicate	$FeSiF_6 \cdot 6H_2O$	72.1	74.4		77.0^{25}		83.7^{50}	88.1^{75}		100.1^{106}
(II) nitrate	$Fe(NO_3)_2 \cdot 6H_2O$	113	134				266			
(III) nitrate	$Fe(NO_3)_3 \cdot 9H_2O$	112.0		137.7		175.0				
(III) perchlorate	$Fe(ClO_4)_3$	289		368	422	478	772			
(II) sulfate	$FeSO_4 \cdot 7H_2O$	28.8	40.0	48.0	60.0	73.3	100.7			
Lanthanum bromate	$La(BrO_3)_3$	98	120	149	200					
nitrate	$La(NO_3)_3$	100		136		168	247			
selenate	$La_2(SeO_3)_3$	50.5	45	45	45	45	18.5	5.4	2.2	
sulfate	$La_2(SO_4)_3$	3.00	2.72	2.33	1.90	1.67	1.26	0.91	0.79	0.68
Lead(II) acetate	$Pb(C_2H_3O_2)_2$	19.8	29.5	44.3	69.8	116				
bromide	$PbBr_2$	0.45	0.63	0.86	1.12	1.50	2.29	3.23	3.86	4.55
chloride	$PbCl_2$	0.67	0.82	1.00	1.20	1.42	1.94	2.54	2.88	3.20
fluorosilicate	$PbSiF_6$	190		222			403	428		463

Substance	Formula	0.044	0.056	0.069	0.090	0.124	0.193	0.294		0.42
iodide	PbI_2	37.5	46.2	54.3	63.4	72.1	91.6	111		133
nitrate	$Pb(NO_3)_2$	31.2	35.1	40.8	50.6	68.6	56.5			
Lithium acetate	$LiC_2H_3O_2$		55.2		55.9	56.1	86.6			100
ammonium sulfate	$LiNH_4SO_4$									
azide	LiN_3	61.3	64.2	67.2	71.2	75.4				
benzoate	$LiC_7H_5O_2$	38.9	41.6	44.7	53.8					
borate (meta-)	$LiBO_2$	0.90	1.3	2.7	5.7	10.9				
bromate	$LiBrO_3$	154	166	179	198	221	269	308	329	355
bromide	$LiBr$	143	147	160	183	211	223	245	239	266
carbonate	Li_2CO_3	1.54	1.43	1.33	1.26	1.17	1.01	0.85		0.72
chlorate	$LiClO_3$	241	283	372	488	604	777			
chloride	$LiCl$	69.2	74.5	83.5	86.2	89.8	98.4	112	121	128
chloroaurate(III)	$LiAuCl_4$	105	113	136	167	206	324	599		138
cyanoplatinate(II)	$Li_2Pt(CN)_4$			141	153	160	178	216	239	
formate	$LiCHO_2$	32.3	35.7	39.3	44.1	49.5	64.7	92.7	116	
hydrogen phosphite	Li_2HPO_3	9.97			7.61	7.11	6.03			4.43
hydroxide	$LiOH$	11.91	12.11	12.35	12.70	13.22	14.63	16.56		19.12
iodide	LiI	151	157	165	171	179	202	435	440	481
molybdate	Li_2MoO_4	82.6		79.5	79.4	78.0	175			73.9
nitrate	$LiNO_3$	53.4	60.8	70.1	138	152	177	233	272	
nitrite	$LiNO_2$	70.9	82.5	96.8	114	133			151	324
perchlorate	$LiClO_4$	42.7	49.0	56.1	63.6	72.3	92.3	128		
phosphate (meta-)	$LiPO_3$	0.101		$0.058^{25°}$		0.048				
selenite	Li_2SeO_3	25.0	23.3	21.5	19.6	17.9		11.9	11.1	9.9
sulfate	Li_2SO_4	36.1	35.5	34.8	34.2	33.7		31.4	30.9	
tartrate (*d-*)	$Li_2C_4H_4O_6$	42.0	31.8	27.1	26.6	27.2				
thiocyanate	$LiSCN$			114	131	153				
vanadate	Li_3VO_4	2.50		4.82	6.28	4.38	2.67			
Magnesium acetate	$Mg(C_2H_3O_2)_2$	56.7	59.7	53.4	68.6	75.7	118			125
bromide	$MgBr_2$	98	99	101	104	106	112		268	
chlorate	$Mg(ClO_3)_2$	114	123	135	155	178	242			
chloride	$MgCl_2$	52.9	53.6	54.6	55.8	57.5	61.0	66.1	69.5	73.3
fluorosilicate	$MgSiF_6$	26.3		30.8		34.9	44.4			
formate	$Mg(CHO_2)_2$	14.0	14.2	14.4	14.9	15.9	17.9	20.5	22.2	
iodate	$Mg(IO_3)_2$		7.2	8.6	10.0	11.7	15.2	15.5	15.6	23.9
iodide	MgI_2	120	140	140	173			186		

Table 10-2 (*Continued*)

SOLUBILITIES OF INORGANIC COMPOUNDS AND METAL SALTS OF ORGANIC ACIDS IN WATER AT VARIOUS TEMPERATURES

Substance	Formula	0°	10°	20°	30°	40°	60°	80°	90°	100°
Magnesium nitrate	$Mg(NO_3)_2$	62.1	66.0	69.5	73.6	78.9	78.9	91.6	106	
selenate	$MgSeO_4$	20.0	30.4	38.3	44.3	48.6	55.8			
sulfate	$MgSO_4$	22.0	28.2	33.7	38.9	44.5	54.6	55.8	52.9	50.4
sulfite	$MgSO_3$	0.339	0.446	0.573	0.751	0.959	0.779	0.642	0.622	
tartrate	$MgC_4H_4O_6$	0.54	0.78	1.06		1.02				
Manganese bromide	$MnBr_2$	127	136	147	157	169	197	225	226	228
chloride	$MnCl_2$	63.4	68.1	73.9	80.8	88.5	109	113	114	115
fluoride	MnF_2			1.06		0.67	0.44			0.48
nitrate	$Mn(NO_3)_2$	102	118	139	206					
oxalate	MnC_2O_4	0.020	0.024	0.028	0.033					
sulfate	$MnSO_4$	52.9	59.7	62.9	62.9	60.0	53.6	45.6	40.9	35.3
Mercury(II) bromide	$HgBr_2$	0.30	0.40	0.56	0.66	0.91	1.68	2.77		4.9
(II) chloride	$HgCl_2$	3.63	4.82	6.57	8.34	10.2	16.3	30.0		61.3
(I) perchlorate	$Hg_2(ClO_4)_2$	282	325	367	407	455	499	541		580
Molybdenum trioxide	MoO_3			0.134	0.285	0.454	1.08	1.74		
Neodymium bromate	$Nd(BrO_3)_3$	43.9	59.2	75.6	95.2	116				
chloride	$NdCl_3$		96.7	98.0	99.6	102	105			
nitrate	$Nd(NO_3)_3$	127	133	142	145	159	211			
selenate	$Nd_2(SeO_3)_3$	46.2	44.6	41.8	39.9	39.9	43.9		3.3	
sulfate	$Nd_2(SO_4)_3$	13.0	9.7	7.1	5.3	4.1	2.8		1.2	
Nickel bromide	$NiBr_2$	113	122	131	138	144	153	154		155
chlorate	$Ni(ClO_3)_2$	111	120	133	155	181	221	308		
chloride	$NiCl_2$	53.4	56.3	60.8	70.6	73.2	81.2	86.6		87.6
fluoride	NiF_2		2.55	2.56						
iodate	$Ni(IO_3)_2$				1.15		2.56		2.59	
	$Ni(IO_3)_2 \cdot 4H_2O$	0.74		1.09	1.43		1.06		1.00	
iodide	NiI_2	124	135	148	161	174	184	187	188	
nitrate	$Ni(NO_3)_2$	79.2		94.2	105	119	158	187	188	
perchlorate	$Ni(ClO_4)_2$	105	107	110	113	117				

Solubility data (continued). Values in g/100 g water.

Compound	Formula									
Nickel sulfate	$NiSO_4 \cdot 6H_2O$ (pale blue)	26.2		40.1	43.6	47.6	55.6	64.5	70.1	76.7
	$NiSO_4 \cdot 7H_2O$ (green)		32.4	37.7	44.4	46.6	49.2	50.4		
Osmium tetroxide	OsO_4	5.26	5.75	6.43						
Oxalic acid	$H_2C_2O_4$	3.54	6.08	9.52	14.23	21.52	44.32	84.5	120	
Potassium acetate	$KC_2H_3O_2$	216	233	256	283	324	350	381	398	
aluminum sulfate	$KAl(SO_4)_2$	3.00	3.99	5.90	8.39	11.7	24.8	71.0	109	
azide	KN_3	41.4	46.2	50.8	55.8	61.0				106
benzoate	$KC_7H_5O_2$		65.8	70.7	76.7	82.1	85.5			
bromate	$KBrO_3$	3.09	4.72	6.91	9.64	13.1	22.7	34.1	46.0	49.9
bromide	KBr	53.6	59.5	65.3	70.7	75.4	85.5	94.9	99.2	104
cadmium bromide	$KCdBr_3$	116	133	150	170	191	233	276	298	325
cadmium chloride	$KCdCl_3$	26.6	32.3	38.9	45.6	53.1	67.5	83.5		101
carbonate	K_2CO_3	105	108	111	114	117	127	140	148	156
chlorate	$KClO_3$	3.3	5.2	7.3	10.1	13.9	23.8	37.6	46.0	56.3
chloride	KCl	28.0	31.2	34.2	37.2	40.1	45.8	51.3	53.9	56.3
chloroaurate(III)	$KAuCl_4$		38.3	61.8	94.9	145	405			
chloroplatinate(IV)	K_2PtCl_6	0.48	0.60	0.78	1.00	1.36	2.45	3.71		5.03
chromate	K_2CrO_4	56.3	60.0	63.7	66.7	67.8	70.1		74.5	
citrate	$K_3C_6H_5O_7$		153	172	194					
cobalt(II) sulfate	$K_2Co(SO_4)_2$	8.5	11.7	15.5	19.3	23.3	32.5	47.7		
copper(II) sulfate	$K_2Cu(SO_4)_2$	5.1	7.2	10.0	13.6	18.2				
cyanoplatinate(II)	$K_2Pt(CN)_4$	11.6	19.8	33.9	52.0	78.3	139	177	194	
dichromate	$K_2Cr_2O_7$	4.7	7.0	12.3	18.1	26.3	45.6	73.0	83.5	
dihydrogen phosphate	KH_2PO_4	14.8	18.3	22.6	28.0	33.5	50.2	70.4	71.5	
dithionate	$K_2S_2O_6$	2.6	4.2	6.6	9.3					
ferricyanide	$K_3Fe(CN)_6$	30.2	38	46	53	59.3	70			91
ferrocyanide	$K_4Fe(CN)_6$	14.3	21.1	28.2	35.1	41.4	54.8	66.9	71.5	74.2
fluoride	KF	44.7	53.5	94.9	108	138	142	150		
fluorogermanate(IV)	K_2GeF_6	0.25	0.36	0.50	0.66	0.96				
fluorosilicate	K_2SiF_6	0.077	0.102	0.151	0.202	0.253				
fluorotitanate(IV)	K_2TiF_6	0.55	0.91	1.28						
formate	$KCHO_2$		313	337	361	398	471	580	658	
hydrogen carbonate	$KHCO_3$	22.5	27.4	33.7	39.9	47.5	65.6			

Table 10-2 (Continued)
SOLUBILITIES OF INORGANIC COMPOUNDS AND METAL SALTS OF ORGANIC ACIDS IN WATER AT VARIOUS TEMPERATURES

Substance	Formula	0°	10°	20°	30°	40°	60°	80°	90°	100°
Potassium hydrogen fluoride	KHF_2	24.5	30.1	39.2	46.8	56.5	78.8	114		
hydrogen selenite	$KH_3(SeO_3)_2$	115	162	215	300	408	900			
hydrogen sulfate	$KHSO_4$	36.2		48.6	54.3	61.0	76.4	96.1		122
hydrogen tartrate	$KC_4H_5O_6$	0.231	0.358	0.523	0.762					
hydroxide	KOH	95.7	103	112	126	134	154			178
iodate	KIO_3	4.60	6.27	8.08	10.3	12.6	18.3	24.8		32.3
iodide	KI	128	136	144	153	162	176	192	198	206
iron(II) sulfate	$K_2Fe(SO_4)_2$	19.6	24.5	32.1	39.1	44.9	57.2	63.4		
magnesium sulfate	$K_2Mg(SO_4)_2$	14.0	19.5	25.0	30.4	36.6	50.2			
nickel sulfate	$K_2Ni(SO_4)_2$	3.37	4.50	5.94	7.72	9.85	15.4	23.0	27.8	33.4
nitrate	KNO_3	13.9	21.2	31.6	45.3	61.3	106	167	203	245
nitrite	KNO_2	279	292	306	320	329	348	376	390	410
oxalate	$K_2C_2O_4$	25.5	31.9	36.4	39.9	43.8	53.2	63.6	69.2	75.3
perchlorate	$KClO_4$	0.76	1.06	1.68	2.56	3.73	7.3	13.4	17.7	22.3
periodate	KIO_4	0.17	0.28	0.42	0.65	1.0	2.1	4.4	5.9	
permanganate	$KMnO_4$	2.83	4.31	6.34	9.03	12.6	22.1			
peroxodisulfate	$K_2S_2O_8$	1.65	2.67	4.70	7.75	11.0				
perrhenate	$KReO_4$	0.34	0.63	0.99	1.47	2.2	4.58	8.7		
phosphate	K_3PO_4		81.5	92.3	108	133				
salicylate	$KC_7H_5O_3$	21.2	32.4	47.1	61.3	78.6	116	156		
selenate	K_2SeO_4	107	109	111	113	115	119	121		122
selenite	K_2SeO_3	169	186	203	217	217	220			217
sulfate	K_2SO_4	7.4	9.3	11.1	13.0	14.8	18.2	21.4	22.9	24.1
sulfite	K_2SO_3	106		106	107	107	108			112
tellurate	K_2TeO_4	8.8		27.5	50.4					
thioantimonate(V)	K_3SbS_4	306	320		302	315		381		
thiocyanate	$KSCN$	177	198	224	255	289	372	492	571	675
thiosulfate	$K_2S_2O_3$	96		155	175	205	238	293	312	
zinc sulfate	$K_2Zn(SO_4)_2 \cdot 6H_2O$	13.0	18.9	25.9	35.0	44.9	72.1			

Compound	Formula									
Praseodymium bromate	Pr(BrO₃)₃	55.9	73.0	91.8	114	144				
nitrate	Pr(NO₃)₃			112	162	178				
selenate	Pr₂(SeO₃)₃	36.2			32.4	31.2	30.4	5.43	3.6	
sulfate	Pr₂(SO₄)₃	19.8	15.6	12.6	9.89	2.56	5.04	3.5	1.1	0.91
Rubidium aluminum sulfate	Rb₂Al₂(SO₄)₄	0.72	1.05	1.50	2.20	3.25	7.40	21.6		
bromate	RbBrO₃				3.6	5.1				
bromide	RbBr	90	99	108	119	132	158			
chlorate	RbClO₃	2.1	3.4	5.4	8.0	11.6	22	38	49	63
chloride	RbCl	77	84	91	98	104	115	127	133	143
chloroaurate(III)	RbAuCl₄		4.8	9.9	15.5	21.5	36.2	54.6	65.8	79.2
chloroplatinate(IV)	Rb₂PtCl₆	0.014	0.020	0.028	0.040	0.056	0.090	0.182	0.247	0.333
chromate	Rb₂CrO₄	62.0	67.5	73.6	78.9	85.6	95.7			
cobalt sulfate	Rb₂Co(SO₄)₂	5.10	7.47	10.8	14.5	18.2	30.2	44.9	55.0	70.1
dichromate (mn)	Rb₂Cr₂O₇			5.9	10.0	15.2	32.3			
(tric)				5.8	9.5	14.8	32.4			
formate	RbCHO₂		443	554	614	694	900			
iron(III) sulfate	RbFe(SO₄)₂·12H₂O		8.0	20	35	52	200			
nitrate	RbNO₃	19.5	33.0	52.9	81.2	117		310	374	452
perchlorate	RbClO₄	1.09	1.19	1.55	2.20	3.26	6.27	11.0	15.5	22.0
salicylate	RbC₇H₅O₃	187	187	212	238	268	324			
sulfate	Rb₂SO₄	37.5	42.6	48.1	53.6	58.5	67.5	75.1	78.6	81.8
Samarium bromate	Sm(BrO₃)₃	34.2	47.6	62.5	79.0	98.5				
chloride	SmCl₃		92.4	93.4	94.6	96.9				
Selenic acid	H₂SeO₄	426	122.2	567	1328	344.4	383.1	383.1	385.4	
Selenious acid	H₂SeO₃	90.1	222	166.7	235.6	335	440			
Selenium dioxide	SeO₂			257	291					
Silver acetate	AgC₂H₃O₂	0.73	0.89	1.05	1.23	1.43	1.93	2.59	1.33	
bromate	AgBrO₃		0.11	0.16	0.23	0.32	0.57	0.94		
chlorate	AgClO₃		10.4	15.3	20.9	26.8				
fluoride	AgF	85.9	120	172	190	203		585	652	733
nitrate	AgNO₃	122	167	216	265	311	440			793
nitrite	AgNO₂	0.16	0.22	0.34	0.51	0.73	1.39			
perchlorate	AgClO₄	455	484	525	594	635				
sulfamate	AgNH₂SO₃	2.30	4.82	7.53	10.3	15.3	28.5			
sulfate	Ag₂SO₄	0.57	0.70	0.80	0.89	0.98	1.15	1.30	1.36	1.41

Table 10-2 (Continued)
SOLUBILITIES OF INORGANIC COMPOUNDS AND METAL SALTS OF ORGANIC ACIDS IN WATER AT VARIOUS TEMPERATURES

Substance	Formula	0°	10°	20°	30°	40°	60°	80°	90°	100°
Sodium acetate	$NaC_2H_3O_2$	36.2	40.8	46.4	54.6	65.6	139	153	161	170
aluminum sulfate	$Na_2Al_2(SO_4)_4$	37.4	39.3	39.7	41.7	43.8				
azide	NaN_3	38.9	39.9	40.8						55.3
benzoate	$NaC_7H_5O_2$	62.6	62.8	62.8	62.9	63.1	64.5	68.6	70.6	73.3
borate (penta-)	$NaB_{10}O_{16}$	6.4	8.6	12.0	16.4	22.0	37.9	63.4	83.5	108
borate (tetra-)	$Na_2B_4O_7$	1.11	1.60	2.56	3.86	6.67	19.0	31.4	41.0	52.5
bromate	$NaBrO_3$	24.2	30.3	36.4	42.6	48.8	62.6	75.7		90.8
bromide	$NaBr$	80.2	85.2	90.8	98.4	107	118	120	121	121
carbonate	Na_2CO_3	7.00	12.5	21.5	39.7	49.0	46.0	43.9	43.9	
chlorate	$NaClO_3$	79.6	87.6	95.9	105	115	137	167	184	204
chloride	$NaCl$	35.7	35.8	35.9	36.1	36.4	37.1	38.0	38.5	39.2
chloroaurate(III)	$NaAuCl_4$		139	151	178	227	900	279		
chloroiridate(IV)	Na_2IrCl_6		31.6	39.3	56.2	96.1	192	125		
chromate	Na_2CrO_4	31.7	50.1	84.0	88.0	96.0	115			126
cyanide	$NaCN$	40.8	48.1	58.7	71.2					
dichromate	$Na_2Cr_2O_7$	163	172	183	198	215	269	376	405	415
diethyl barbiturate	$NaC_8H_{11}N_2O_3$		12.7	21.5	24.7				48.0	
dihydrogen phosphate (ortho-)	NaH_2PO_4	56.5	69.8	86.9	107	133	172	211	234	
dihydrogen phosphate (pyro-)	$Na_2H_2P_2O_7$	4.47	6.95	12.0	17.1	18.4				
dithionate	$Na_2S_2O_6$	6.3	11.1	15.1	19.6	24.7	36.1	49.3	56.3	64.7
dodecanesulfonate	$NaC_{12}H_{25}SO_3$			0.13	0.25	6.54				
dodecanoate	$NaC_{12}H_{23}O_2$				4.58	22.7	105	170		
EDTA (Y)*	$Na_2H_2Y \cdot 2H_2O$	10.6		11.1	12.8	14.2	17.0	22.2	24.3	27.09[98]
ferrocyanide	$Na_4Fe(CN)_6$	11.2	14.8	18.8	23.8	29.9	43.7	62.1		
fluoride	NaF	3.66		4.06	4.22	4.40	4.68	4.89		5.08
fluoroberyllate	Na_2BeF_4	1.33		1.44	1.92	1.92	2.24	2.62	2.73	
fluorogermanate	Na_2GeF_6	1.52	1.68		2.25	2.83		3.36		
fluorosilicate	Na_2SiF_6	4.35	5.7	7.2	8.6	10.3	14.3	18.7	21.5	24.5

	T1	T2	T3	T4	T5	T6	T7	T8	T9
formate $NaCHO_2$	43.9	62.5	81.2	102	108	122	138	147	160
germanate Na_2GeO_3	14.4	18.8	23.8	28.7	37.2	65.0	116	188	198
hydrogen arsenate Na_2HAsO_4	5.9	13.0	33.9	49.3	69.5	144	186		
hydrogen carbonate $NaHCO_3$	7.0	8.1	9.6	11.1	12.7	16.0			
hydrogen phosphate Na_2HPO_4	1.68	3.53	7.83	22.0	55.3	82.8	92.3	102	104
hydrogen phosphite Na_2HPO_3	418	424	429	566					
hydrogen succinate $NaC_4H_5O_4$	17.5	25.3	34.8	47.7	61.6	74.5	90.1		
hydroxide $NaOH$	46.0	98	109	119	129	174			
hydroxostannate(IV) $Na_2Sn(OH)_6$	29.4	36.4	43.7	42.7	38.9				
hypochlorite $NaClO$			53.4	100	110				
iodate $NaIO_3$	2.48	4.59	8.08	10.7	13.3	19.8	26.6	29.5	33.0
iodide NaI	159	167	178	191	205	257	295		302
molybdate Na_2MoO_4	44.1	64.7	65.3	66.9	68.6	71.8		68.1	
nitrate $NaNO_3$	73.0	80.8	87.6	94.9	102	122	148		180
nitrite $NaNO_2$	71.2	75.1	80.8	87.6	94.9	111	133		160
oxalate $Na_2C_2O_4$	2.69	3.05	3.41	3.81	4.18	4.93	5.71		6.50
perchlorate $NaClO_4$	167	183	201	222	245	288	306		329
periodate $NaIO_4$	1.83	5.6	10.3	19.9	30.4	29.9			
phosphate Na_3PO_4	4.5	8.2	12.1	16.3	20.2		60.0		77.0
potassium tartrate $NaKC_4H_4O_6$	31.9	46.6	67.8	102	117	130	144		
salicylate $NaC_7H_5O_3$		44.7	95.3	111					
selenate Na_2SeO_4	13.3	25.2	26.9	77.0	81.8	78.6	74.8	73.0	72.7
selenite Na_2SeO_3	78.6	81.2	86.2	94.2	96.5	91.6	86.6	84.5	82.5
sulfate Na_2SO_4	4.9	9.1	19.5	40.8	48.8	45.3	43.7	42.7	42.5
$Na_2SO_4 \cdot 7H_2O$	19.5	30.0	44.1	20.5	26.6				
sulfide Na_2S	9.6	12.1	15.7	35.5	37.2	39.1	55.0	65.3	
sulfite Na_2SO_3	14.4	19.5	26.3	37.2	49.3	32.6	29.4	27.9	
thioantimonate(V) Na_3SbS_4	13.4	20.0	27.9			53.8	88.3		
thiocyanate $NaSCN$	111	111	134	164	176	192	210	218	
thiosulfate $Na_2S_2O_3 \cdot 5H_2O$	50.2	59.7	70.1	83.2	104				
tungstate Na_2WO_4	71.5		73.0		77.6		90.8		97.2
vanadate $NaVO_3$			19.3	22.5	26.3	33.0	40.8		
Strontium acetate $Sr(C_2H_3O_2)_2$	37.0	42.9	41.1	39.5	38.3	36.8	36.1	36.2	36.4
bromide $SrBr_2$	85.2	93.4	102	112	123	150	182		223
chloride $SrCl_2$	43.5	47.7	52.9	58.7	65.3	81.8	90.5		101
chromate $SrCrO_4$	0.085	0.085	0.090				0.058		

*Properly called dihydrogen ethylenediaminetetraacetate ($Na_2H_2EDTA \cdot 2H_2O$).

Table 10-2 (Continued)
SOLUBILITIES OF INORGANIC COMPOUNDS AND METAL SALTS OF ORGANIC ACIDS IN WATER AT VARIOUS TEMPERATURES

Substance	Formula	0°	10°	20°	30°	40°	60°	80°	90°	100°
Strontium fluoride	SrF_2	0.0113		0.0117	0.0119					
formate	$Sr(CHO_2)_2$	9.1	10.6	12.7	15.2	17.8	25.0	31.9	32.9	34.4
hydroxide	$Sr(OH)_2$	0.91	1.25	1.77	2.64	3.95	8.42	20.2	44.5	91.2
iodide	SrI_2	165		178		192	218	270	365	383
nitrate	$Sr(NO_3)_2$	39.5	52.9	69.5	88.7	89.4	93.4	96.9	98.4	
nitrite	$Sr(NO_2)_2$			65	72	79	97	130	134	
oxide	SrO				1.03	1.05	3.40	9.15	13.13	12.15
sulfate	$SrSO_4$	0.0113	0.0129	0.0132	0.0138	0.0141	0.0131	0.0016	0.0015	
Sulfamic acid	H_2NSO_3H	14.7	18.6	21.3	26.1	29.5	37.1	47.1		
Telluric acid	H_2TeO_4	16.2	33.8	41.6	50.0	57.2	77.5	106		155
Terbium bromate	$Tb(BrO_3)_3 \cdot 9H_2O$	66.4	89.7	117	152	198				
Thallium(I) azide	TlN_3	0.171	0.236	0.364						27.2
bromide	$TlBr$	0.022	0.032	0.048	0.068	0.097	0.177			
carbonate	Tl_2CO_3	2.00		5.3		$12.7^{50°}$	12.2	36.6		57.3
chlorate	$TlClO_3$			3.92						
chloride	$TlCl$	0.21	0.25	0.33	0.42	0.52	0.80	1.20		1.80
hydroxide	$TlOH$	25.4	29.6	35.0	40.4	49.4	73.3	106	126	150
iodide	TlI	0.002		0.006		0.015	0.035	0.070		0.120
nitrate	$TlNO_3$	3.90	6.22	9.55	14.3	21.0	46.1	110	200	414
nitrite	$TlNO_2$	17.9	28.9	40.3	53.2	83.6	216	1150	750	
perchlorate	$TlClO_4$	6.00	8.04	13.1	19.7	28.3	50.8	81.5		
picrate	$TlOC_6H_2(NO_2)_3$	0.135		0.40	0.57	0.83	1.73			
selenate	Tl_2SeO_4		2.17	2.80				8.50		10.8
sulfate	Tl_2SO_4	2.73	3.70	4.87	6.16	7.53	11.0	14.6	16.5	18.4
Thorium nitrate	$Th(NO_3)_4$	186	187	191						
sulfate	$Th(SO_4)_2 \cdot 4H_2O$	0.74	0.99	1.38	1.99	4.04	1.63	3.04	3.58	
sulfate	$Th(SO_4)_2 \cdot 9H_2O$			0.99	1.17	3.00	2.11			
Tin(II) iodide	SnI_2					1.42				
Uranium(IV) sulfate	$U(SO_4)_2 \cdot 4H_2O$			11.9	10.1	9.0	7.7			4.20
sulfate	$U(SO_4)_2 \cdot 8H_2O$				17.9	29.2	55.8			

		98	107	122	141	167	317	388	426	474
Uranyl nitrate	$UO_2(NO_3)_2$	98	107	122	141	167	317	388	426	474
oxalate	$UO_2C_2O_4$		0.45	0.50	0.61	0.80	1.22	1.94		3.16
Ytterbium sulfate	$Yb_2(SO_4)_3$	44.2	37.5		22.2	17.2	10.4	6.4	5.8	4.7
Yttrium bromide	YBr_3	63.9	78.1	75.1	79.6	87.3	101	116	123	
chloride	YCl_3	77.3		78.8		80.8				
nitrate	$Y(NO_3)_3$	93.1	106	123	143	163	200			
sulfate	$Y_2(SO_4)_3$	8.05	7.67	7.30	6.78	6.09	4.44	2.89	2.2	
Zinc bromide	$ZnBr_2$	389		446	528	591	618	645		672
chlorate	$Zn(ClO_3)_2$	145	152	200	209	223				
chloride	$ZnCl_2$	342	363	395	437	452	488	541		614
formate	$Zn(CHO_2)_2$	3.70	4.30	5.20	6.10	7.40	11.8	21.2	28.8	38.0
iodide	ZnI_2	430		432		445	467	490		510
nitrate	$Zn(NO_3)_2$	98			138	211				
sulfate (rh)	$ZnSO_4$	41.6	47.2	53.8	61.3	70.5	75.4	71.1		60.5
sulfate (mn)			54.4	60.0	65.5					
tartrate	$ZnC_4H_4O_6$			0.022	0.041	0.060	0.104	0.059		

VAPOR PRESSURES

Table 10-3
VAPOR PRESSURE OF MERCURY

Temp. °C.	mm of Hg	Temp. °C.	mm of Hg	Temp. °C.	mm of Hg
−38	0.0_5145	88	0.1413	214	26.826
−36	0.0_5197	90	0.1582	216	28.504
−34	0.0_5266	92	0.1769	218	30.271
−32	0.0_5359	94	0.1976	220	32.133
−30	0.0_5478	96	0.2202	222	34.092
−28	0.0_5630	98	0.2453	224	36.153
−26	0.0_5828	100	0.2729	226	38.318
−24	0.0_4108	102	0.3032	228	40.595
−22	0.0_4140	104	0.3366	230	42.989
−20	0.0_4181	106	0.3731	232	45.503
−18	0.0_4232	108	0.4132	234	48.141
−16	0.0_4298	110	0.4572	236	50.909
−14	0.0_4380	112	0.5052	238	53.812
−12	0.0_4481	114	0.5576	240	56.855
−10	0.0_4606	116	0.6150	242	60.044
− 8	0.0_4762	118	0.6776	244	63.384
− 6	0.0_4954	120	0.7457	246	66.882
− 4	0.0_3119	122	0.8198	248	70.543
− 2	0.0_3149	124	0.9004	250	74.375
0	0.0_3185	126	0.9882	252	78.381
+ 2	0.0_3228	128	1.084	254	82.568
4	0.0_3276	130	1.186	256	86.944
6	0.0_3335	132	1.298	258	91.518
8	0.0_3406	134	1.419	260	96.296
10	0.0_3490	136	1.551	262	101.28
12	0.0_3588	138	1.692	264	106.48
14	0.0_3706	140	1.845	266	111.91
16	0.0_3846	142	2.010	268	117.57
18	0.001009	144	2.188	270	123.47
20	0.001201	146	2.379	272	129.62
22	0.001426	148	2.585	274	136.02
24	0.001691	150	2.807	276	142.69
26	0.002000	152	3.046	278	149.64
28	0.002359	154	3.303	280	156.87
30	0.002777	156	3.578	282	164.39
32	0.003261	158	3.873	284	172.21
34	0.003823	160	4.189	286	180.34
36	0.004471	162	4.528	288	188.79
38	0.005219	164	4.890	290	197.57
40	0.006079	166	5.277	292	206.70
42	0.007067	168	5.689	294	216.17
44	0.008200	170	6.128	296	226.00
46	0.009497	172	6.596	298	236.21
48	0.01098	174	7.095	300	246.80
50	0.01267	176	7.626	302	257.78
52	0.01459	178	8.193	304	269.17
54	0.01677	180	8.796	306	280.98
56	0.01925	182	9.436	308	293.21
58	0.02206	184	10.116	310	305.89
60	0.02524	186	10.839	312	319.02
62	0.02883	188	11.607	314	332.62
64	0.03287	190	12.423	316	346.70
66	0.03740	192	13.287	318	361.26
68	0.04251	194	14.203	320	376.33
70	0.04825	196	15.173	322	391.92
72	0.05469	198	16.200	324	408.04
74	0.06189	200	17.287	326	424.71
76	0.06993	202	18.437	328	441.94
78	0.07889	204	19.652	330	459.74
80	0.08880	206	20.936	332	478.13
82	0.1000	208	22.292	334	497.12
84	0.1124	210	23.723	336	516.74
86	0.1261	212	25.233	338	537.00

Table 10-3 (*Continued*)
VAPOR PRESSURE OF MERCURY

Temp. °C.	mm of Hg	Temp. °C.	mm of Hg	Temp. °C.	mm of Hg
340	557.90	374	1028.9	520	7691
342	579.45	376	1064.4	550	10650
344	601.69	378	1100.9	600	22.87atm.
346	624.64	380	1138.4	650	35.49atm.
348	648.30	382	1177.0	700	52.51atm.
350	672.69	384	1216.6	750	74.86atm.
352	697.83	386	1257.3	800	103.31atm.
354	723.73	388	1299.1	850	138.42atm.
356	750.43	390	1341.9	900*	180.92atm.
358	777.92	392	1386.1	950	226.58atm.
360	806.23	394	1431.3	1000	290.5atm.
362	835.38	396	1477.7	1050	358.1atm.
364	865.36	398	1525.2	1100	437.3atm.
366	896.23	400	1574.1	1150	521.3atm.
368	928.02	430	2464	1200	616.8atm.
370	960.66	460	3715	1250	721.4atm.
372	994.34	490	5420	1300	835.9atm.

* Critical point.

Table 10-4
VAPOR PRESSURE OF LIQUID AMMONIA, NH_3

t°C.	p in atm	t°C.	p in atm	t°C.	p in atm
−78	0.0582	−6	3.3677	66	29.784
−76	0.0683	−4	3.6405	68	31.211
−74	0.0797	-2	3.9303	70	32.687
−72	0.0929	0	4.2380	72	34.227
−70	0.1078	+2	4.5640	74	35.813
−68	0.1246	4	4.9090	76	37.453
−66	0.1437	6	5.2750	78	39.149
−64	0.1651	8	5.6610	80	40.902
−62	0.1891	10	6.0685	82	42.712
−60	0.2161	12	6.4985	84	44.582
−58	0.2461	14	6.9520	86	46.511
−56	0.2796	16	7.4290	88	48.503
−54	0.3167	18	7.9310	90	50.558
−52	0.3578	20	8.4585	92	52.677
−50	0.4034	22	9.0125	94	54.860
−48	0.4536	24	9.5940	96	57.111
−46	0.5087	26	10.2040	98	59.429
−44	0.5693	28	10.8430	100	61.816
−42	0.6357	30	11.512	102	64.274
−40	0.7083	32	12.212	104	66.804
−38	0.7875	34	12.943	106	69.406
−36	0.8738	36	13.708	108	72.084
−34	0.9676	38	14.507	110	74.837
−32	1.0695	40	15.339	112	77.668
−30	1.1799	42	16.209	114	80.578
−28	1.2992	44	17.113	116	83.570
−26	1.4281	46	18.056	118	86.644
−24	1.5671	48	19.038	120	89.802
−22	1.7166	50	20.059	122	93.045
−20	1.8774	52	21.121	124	96.376
−18	2.0499	54	22.224	126	99.796
−16	2.2349	56	23.372	128	103.309
−14	2.4328	58	24.562	130	106.913
−12	2.6443	60	25.797	132	110.613
−10	2.8703	62	27.079	132.3	111.3(c.p.)
− 8	3.1112	64	28.407		

Table 10-5
VAPOR PRESSURE OF ICE IN MILLIMETERS OF MERCURY
For Temperatures from −99 to 0°C

The values in the table are for ice in contact with its own vapor. Where the ice is in contact with air at a temperature t°C, this correction must be added: Correction = $20p/(100)(t + 273)$.

t, °C	p, mm Hg	t, °C	p, mm Hg	t, °C	p, mm Hg
−99	0.000 012	−49	0.033 4	−14.5	1.300
−98	0.000 015	−48	0.037 8	−14.0	1.361
−97	0.000 018	−47	0.042 6	−13.5	1.424
−96	0.000 022	−46	0.048 1	−13.0	1.490
−95	0.000 027	−45	0.054 1	−12.5	1.559
−94	0.000 033	−44	0.060 9	−12.0	1.632
−93	0.000 040	−43	0.068 4	−11.5	1.707
−92	0.000 048	−42	0.076 8	−11.0	1.785
−91	0.000 058	−41	0.086 2	−10.5	1.866
−90	0.000 070	−40	0.096 6	−10.0	1.950
−89	0.000 084	−39	0.108 1	−9.8	1.985
−88	0.000 10	−38	0.120 9	−9.6	2.021
−87	0.000 12	−37	0.135 1	−9.4	2.057
−86	0.000 14	−36	0.150 7	−9.2	2.093
−85	0.000 17	−35	0.168 1	−9.0	2.131
−84	0.000 20	−34	0.187 3	−8.8	2.168
−83	0.000 24	−33	0.208 4	−8.6	2.207
−82	0.000 29	−32	0.231 8	−8.4	2.246
−81	0.000 34	−31	0.257 5	−8.2	2.285
−80	0.000 40	−30.0	0.285 9	−8.0	2.326
−79	0.000 47	−29.5	0.301	−7.8	2.367
−78	0.000 56	−29.0	0.317	−7.6	2.408
−77	0.000 66	−28.5	0.334	−7.4	2.450
−76	0.000 77	−28.0	0.351	−7.2	2.493
−75	0.000 90	−27.5	0.370	−7.0	2.537
−74	0.001 05	−27.0	0.389	−6.8	2.581
−73	0.001 23	−26.5	0.409	−6.6	2.626
−72	0.001 43	−26.0	0.430	−6.4	2.672
−71	0.001 67	−25.5	0.453	−6.2	2.718
−70	0.001 94	−25.0	0.476	−6.0	2.765
−69	0.002 25	−24.5	0.500	−5.8	2.813
−68	0.002 61	−24.0	0.526	−5.6	2.862
−67	0.003 02	−23.5	0.552	−5.4	2.912
−66	0.003 49	−23.0	0.580	−5.2	2.962
−65	0.004 03	−22.5	0.609	−5.0	3.013
−64	0.004 64	−22.0	0.640	−4.8	3.065
−63	0.005 34	−21.5	0.672	−4.6	3.117
−62	0.006 14	−21.0	0.705	−4.4	3.171
−61	0.007 03	−20.5	0.740	−4.2	3.225
−60	0.008 08	−20.0	0.776	−4.0	3.280
−59	0.009 25	−19.5	0.814	−3.8	3.336
−58	0.010 6	−19.0	0.854	−3.6	3.393
−57	0.012 1	−18.5	0.895	−3.4	3.451
−56	0.013 8	−18.0	0.939	−3.2	3.509
−55	0.015 7	−17.5	0.984	−3.0	3.568
−54	0.017 8	−17.0	1.031	−2.8	3.360
−53	0.020 3	−16.5	1.080	−2.6	3.691
−52	0.023 0	−16.0	1.132	−2.4	3.753

Table 10-5 (*Continued*)
VAPOR PRESSURE OF ICE IN MILLIMETERS OF MERCURY

t, °C	p, mm Hg	t, °C	p, mm Hg	t, °C	p, mm Hg
−51	0.026 1	−15.5	1.186	−2.2	3.816
−50	0.029 6	−15.0	0.241	−2.0	3.880
−1.8	3.946	−1.0	4.217	−0.2	4.504
−1.6	4.012	−0.8	4.287	0.0	4.579
−1.4	4.079	−0.6	4.359		
−1.2	4.147	−0.4	4.431		

Table 10-6
VAPOR PRESSURE OF WATER IN MILLIMETERS OF MERCURY
For Temperatures from −10 to 120°C

The values in the table are for water in contact with its own vapor. Where the water is in contact with air at a temperature t °C, the following correction must be added: Correction (for temperatures up to 40°C) $= p(0.775 - 0.000\ 313\ t)/100$; correction (for temperatures above 50°C) $= p(0.0652 - 0.000\ 087\ 5\ t)/100$.

t, °C	p, mm Hg	t, °C	p, mm Hg	t, °C	p, mm Hg	t, °C	p, mm Hg
−10.0	2.149	14.5	12.382	24.6	23.198	34.4	40.796
−9.5	2.236	15.0	12.788	24.8	23.476	34.6	41.251
−9.0	2.326	15.2	12.953	25.0	23.756	34.8	41.710
−8.5	2.418	15.4	13.121	25.2	24.039	35.0	42.175
−8.0	2.514	15.6	13.290	25.4	24.326	35.2	42.644
−7.5	2.613	15.8	13.461	25.6	24.617	35.4	43.117
−7.0	2.715	16.0	13.634	25.8	24.912	35.6	43.595
−6.5	2.822	16.2	13.809	26.0	25.209	35.8	44.078
−6.0	2.931	16.4	13.987	26.2	25.509	36.0	44.563
−5.5	3.046	16.6	14.166	26.4	25.812	36.2	45.054
−5.0	3.163	16.8	13.347	26.6	26.117	36.4	45.549
−4.5	3.284	17.0	14.530	26.8	26.426	36.6	46.050
−4.0	3.410	17.2	14.715	27.0	26.739	36.8	46.556
−3.5	3.540	17.4	14.903	27.2	27.055	37.0	47.067
−3.0	3.673	17.6	15.092	27.4	27.374	37.2	47.582
−2.5	3.813	17.8	15.284	27.6	27.696	37.4	48.102
−2.0	3.956	18.0	15.477	27.8	28.021	37.6	48.627
−1.5	4.105	18.2	15.673	28.0	28.349	37.8	49.157
−1.0	4.258	18.4	15.871	28.2	28.680	38.0	49.692
−0.5	4.416	18.6	16.071	28.4	29.015	38.2	50.231
0.0	4.579	18.8	16.272	28.6	29.354	38.4	50.774
0.5	4.750	19.0	16.477	28.8	29.697	38.6	51.323
1.0	4.926	19.2	16.685	29.0	30.043	38.8	51.879
1.5	5.107	19.4	16.894	29.2	30.392	39.0	52.442
2.0	5.294	19.6	17.105	29.4	30.745	39.2	53.009
2.5	5.486	19.8	17.319	29.6	31.102	39.4	53.580
3.0	5.685	20.0	17.535	29.8	31.461	39.6	54.156
3.5	5.889	20.2	17.753	30.0	31.824	39.8	54.737
4.0	6.101	20.4	17.974	30.2	32.191	40.0	55.324
4.5	6.318	20.6	18.197	30.4	32.561	40.5	56.81
5.0	6.543	20.8	18.422	30.6	32.934	41.0	58.34
5.5	6.775	21.0	18.650	30.8	33.312	41.5	59.90
6.0	7.013	21.2	18.880	31.0	33.695	42.0	61.50
6.5	7.259	21.4	19.113	31.2	34.082	42.5	63.13
7.0	7.513	21.6	19.349	31.4	34.471	43.0	64.80
7.5	7.775	21.8	19.587	31.6	34.864	43.5	66.51
8.0	8.045	22.0	19.827	31.8	35.261	44.0	68.26
8.5	8.323	22.2	20.070	32.0	35.663	44.5	70.05
9.0	8.609	22.4	20.316	32.2	36.068	45.0	71.88
9.5	8.905	22.6	20.565	32.4	36.477	45.5	73.74
10.0	9.209	22.8	20.815	32.6	36.891	46.0	75.65
10.5	9.521	23.0	21.068	32.8	37.308	46.5	77.61
11.0	9.844	23.2	21.324	33.0	37.729	47.0	79.60
11.5	10.176	23.4	21.583	33.2	38.155	47.5	81.64
12.0	10.518	23.6	21.845	33.4	38.584	48.0	83.71
12.5	10.870	23.8	22.110	33.6	39.018	48.5	85.85
13.0	11.231	24.0	22.387	33.8	39.457	49.0	88.02
13.5	11.604	24.2	22.648	34.0	39.898	49.5	90.24
14.0	11.987	24.4	22.922	34.2	40.344	50.0	92.51

Table 10-6 (*Continued*)
VAPOR PRESSURE OF WATER IN MILLIMETERS OF MERCURY

t, °C	p, mm Hg	t, °C	p, mm Hg	t, °C	p, mm Hg	t, °C	p, mm Hg
50.5	94.86	67.5	209.57	84.5	425.2	97.6	697.10
51.0	97.20	68.0	214.17	85.0	433.6	97.8	702.17
51.5	99.65	68.5	218.95	85.5	442.3	98.0	707.27
52.0	102.09	69.0	223.73	86.0	450.9	98.2	712.40
52.5	104.65	69.5	228.72	86.5	459.8	98.4	717.56
53.0	107.20	70.0	233.7	87.0	468.7	98.6	722.75
53.5	109.86	70.5	238.8	87.5	477.9	98.8	727.98
54.0	112.51	71.0	243.9	88.0	487.1	99.0	733.24
54.5	115.28	71.5	249.3	88.5	496.6	99.2	738.53
55.0	118.04	72.0	254.6	89.0	506.1	99.4	743.85
55.5	120.92	72.5	260.2	89.5	515.9	99.6	749.20
56.0	123.80	73.0	265.7	90.0	525.76	99.8	754.58
56.5	126.81	73.5	271.5	90.5	535.83	100.0	760.00
57.0	129.82	74.0	277.2	91.0	546.05	101.0	787.57
57.5	132.95	74.5	283.2	91.5	556.44	102.0	815.86
58.0	136.08	75.0	289.1	92.0	566.99	103.0	845.12
58.5	139.34	75.5	295.3	92.5	577.71	104.0	875.06
59.0	142.60	76.0	301.4	93.0	588.60	105.0	906.07
59.5	145.99	76.5	307.7	93.5	599.66	106.0	937.92
60.0	149.38	77.0	314.1	94.0	610.90	107.0	970.60
60.5	152.91	77.5	320.7	94.5	622.31	108.0	1004.42
61.0	156.43	78.0	327.3	95.0	633.90	109.0	1038.92
61.5	160.10	78.5	334.2	95.2	638.59	110.0	1074.56
62.0	163.77	79.0	341.0	95.4	643.30	111.0	1111.20
62.5	167.58	79.5	348.1	95.6	648.05	112.0	1148.74
63.0	171.38	80.0	355.1	95.8	652.82	113.0	1187.42
63.5	175.35	80.5	362.4	96.0	657.62	114.0	1227.25
64.0	179.31	81.0	369.7	96.2	662.45	115.0	1267.98
64.5	183.43	81.5	377.3	96.4	667.31	116.0	1309.94
65.0	187.54	82.0	384.9	96.6	672.20	117.0	1352.95
65.5	191.82	82.5	392.8	96.8	677.12	118.0	1397.18
66.0	196.09	83.0	400.6	97.0	682.07	119.0	1442.63
66.5	200.53	83.5	408.7	97.2	687.04	120.0	1489.14
67.0	204.96	84.0	416.8	97.4	692.05		

VAPOR-PRESSURE EQUATIONS

Numerous mathematical formulas relating the temperature and pressure of the gas phase in equilibrium with the condensed phase have been proposed. The Antoine equation (Eq. 1) gives good correlation with experimental values. Equation 2 is simpler and is often suitable over restricted temperature ranges. In these equations, and the derived differential coefficients for use in the Haggenmacher and Clausius-Clapeyron equations, the p term is the vapor pressure of the compound in millimeters of mercury (torr), the t term is the temperature in degrees Celsius, and the T term is the absolute temperature in kelvins ($t°C + 273.15$).

Eq.	Vapor-pressure Equation	(dp/dT)	$-[d(\ln p)/d(1/T)]$
1	$\log p = A - \dfrac{B}{t + C}$	$\dfrac{2.303pB}{(t + C)^2}$	$\dfrac{2.303BT^2}{(t + C)^2}$
2	$\log p = A - \dfrac{B}{T}$	$\dfrac{2.303pB}{T^2}$	$2.303B$
3	$\log p = A - \dfrac{B}{T} - C \log T$	$p\left(\dfrac{2.303B}{T^2} - \dfrac{C}{T}\right)$	$2.303B - CT$

Equations 1 and 2 are easily rearranged to calculate the temperature of the normal boiling point:

$$\text{(Eq. 1)} \quad t = \frac{B}{A - \log p} - C \qquad \text{(Eq. 2)} \quad T = \frac{B}{A - \log p}$$

The constants in the Antoine equation may be estimated by selecting three widely spaced data points and substituting in the following equations in sequence:

$$\left(\frac{y_3 - y_2}{y_2 - y_1}\right)\left(\frac{t_2 - t_1}{t_3 - t_2}\right) = 1 - \left(\frac{t_3 - t_1}{t_3 + C}\right)$$

$$B = \left(\frac{y_3 - y_1}{t_3 - t_1}\right)(t_1 + C)(t_3 + C)$$

$$A = y_2 + \left(\frac{B}{t_2 + C}\right)$$

In these equations, $y_i = \log p_i$.

Table 10-7
VAPOR PRESSURES OF VARIOUS INORGANIC COMPOUNDS

Substance	State	Eq.	Range, °C	A	B	C
Aluminum						
$AlCl_3$		2	70–190	16.24	6 006	
Al_2O_3		2	1840–2000	14.22	28 200	
Ammonium						
NH_3	c*	1		9.963 82	1 617.907	272.55
	liq	1		7.360 50	926.132	240.17
NH_4Br	subl c	1		9.220 0	3 947	227.0
NH_4Cl	subl c	1		9.355 7	3 703.7	232.0
NH_4I	subl c	1		9.147 0	3 858	226.0
NH_4N_3	c	1		10.433 4	2 821.0	240.0
Antimony						
Sb	c	2	1070–1325	9.051	9 871	
$SbBr_3$		2	235–324	8.005	2 873	
$SbCl_3$		2	170–253	8.090	2 582.3	
SbI_3		2	330–445	7.831	3 350.55	
Sb_2Se_3	subl c	2		8.790 6	6 432.3	
Argon						
Ar	c	1		7.505 81	399.085	272.63
	liq	1		6.616 51	304.227	267.32
Arsenic						
As		2	440–815	10.800	6 947	
		2	800–860	6.692	2 460	
$AsCl_3$		2	50–100	7.953	2 042.7	
As_2O_3		2	100–310	12.127	5 815.81	
		2	315–490	6.513	2 722.2	
Barium						
Ba		2	930–1130	15.765	18 280	
BaH_2 [97% pure]		2	500–1000	6.86	4 000	
Bismuth						
Bi		2	1210–1420	8.876	10 446	
$BiCl_3$		2	91–213	2.681	685.519	
Boron						
BBr_3		2	−40 to 90	7.655	1 740.3	
BCl_3		1		6.188 11	756.89	214.0
$B(CH_3)_3$		2	−118 to −20	7.459 5	1 157.99	
B_2H_6	liq	1		6.366 38	521.490	241.98
B_5H_{11}	liq	2	−43 to 8.4	7.901	1 690.3	
Bromine						
Br_2	c	1		9.7209	2 041.3	260.1
	liq	1		6.877 80	1 119.68	221.38
BrF_3	liq	1		7.729 74	1 673.95	219.48
BrF_5	liq	1		7.273 68	1 219.28	236.40
BrO_2F	liq	1		7.436 51	1 195.8	260.1
Cadmium						
Cd		2	150–321	8.564	5 693	
		2	500–840	7.897	5 218	
CdI_2		2	385–450	9.269	6 383	
Calcium						
Ca		2	500–700	9.697	10 185	
		2	960–1100	16.240	19 325	
Carbon						
C [as C(g)]	liq	1		11.042 8	37 736	302.2
[as $C_2(g)$]	liq	1		12.583 2	43 281	318.3
[all species]	liq	1		9.381 3	27 240	264.0

* Crystalline solid.

Table 10-7 (*Continued*)
VAPOR PRESSURES OF VARIOUS INORGANIC COMPOUNDS

Substance	State	Eq.	Range, °C	A	B	C
Carbon						
CNBr	subl c	1		9.488 9	2 041.8	251.70
CNF		1	−76 to −47	6.778 9	697.61	224.95
CO	c l	1		7.414 8	342.50	269.0
	liq	1		6.694 22	291.743	267.99
CO₂	c	1		9.810 66	1 347.786	273.00
C₃O₂	liq	1		7.188 99	1 100.94	249.15
COCl₂	liq	1		6.971 33	998.770	236.68
COF₂		1	−109 to −84	6.885 5	576.70	228.58
COS		1	−111 to −49	6.907 23	804.48	250.0
CS₂		1	3–80	6.942 79	1 169.11	241.59
CSe₂		1	0–50	6.776 73	1 353.20	219.95
CSeS		1	to 84	6.699 6	1 161.97	219.59
Cesium						
Cs		2	200–350	6.949	3 833.7	
CsBr		2	978–1305	7.990	8 022.53	
CsCl		2	986–1295	8.340	8 523.94	
CsF		2	1033–1255	7.703	7 359.21	
CsH		2	245–378	11.79	5 900	
		2	340–440	9.25	4 410	
CsI		2	1052–1280	9.124	9 699.11	
Chlorine						
Cl₂	c	1		9.705 12	1 444.19	267.13
	liq	1		6.937 90	861.34	246.33
ClF	liq	1		6.989	682.1	256
ClF₃	liq	1		7.366 85	1 096.28	232.63
ClF₅		1		6.269 33	653.06	206.6
ClO₂	liq	1		6.036 11	590.09	176.15
Cl₂O	liq	1		7.132 68	1 021.56	238.16
ClOClO₃	liq	1		7.538 67	1 404.18	257.00
Cl₂O₇	liq	1		6.869 29	1 214.00	220.79
ClO₂F	liq	1		6.677 15	809.78	218.96
ClO₃F	liq	1		6.895 19	791.73	243.88
Copper						
CuBr		2	997–1351	5.460	4 173.2	
CuCl		2	878–1369	5.454	4 215.0	
CuI		2	991–1154	5.570	4 215.0	
Fluorine						
F₂	liq	1		6.765 88	304.35	266.54
FNO₃	liq	1		6.658 6	769.5	248.0
Germanium						
GeCl₄		2	10.4–86	7.340	2 010.9	
Helium						
³He	liq	1	−271.13 to −270.86	4.272 7	5.594	273.840
	liq	1	−271.13 to −269.92	5.100 0	11.062	274.950
⁴He		1	−271.4 to −270.1	4.558 7	8.1548	273.710
		1	−271.4 to −268.9	5.320 75	14.6515	274.950
		1	−271.4 to −268.1	6.004 60	24.0668	276.650
Hydrogen						
¹H₂ normal, 25% para	c	1		6.043 86	66.507	274.630
	liq	1		5.824 38	67.5078	275.700
equilibrium	c	1		6.042 07	65.961	274.60
	liq	1		5.814 64	66.7945	275.650
¹H²H (DH)	c	1		6.960 08	99.968	276.590
	liq	1		6.016 12	77.1349	275.620
²H₂ (D₂) normal,	c	1		7.726 05	135.461	278.550
66.7% ortho	liq	1		6.128 25	83.5251	275.216

Table 10-7 (*Continued*)
VAPOR PRESSURES OF VARIOUS INORGANIC COMPOUNDS

Substance	State	Eq.	Range, °C	A	B	C
$^2H_2$ equilibrium,	c	1		7.751 10	135.58	278.50
97.8% ortho	liq	1		6.044 68	79.5888	274.680
$^3H_2$ (T_2) normal, 25%	c	1		6.184 03	76.7445	271.850
para	liq	1		6.089 21	81.8971	273.650
^1HBr	c	1		7.667 61	878.57	253.2
	liq	1		6.287 53	540.82	225.44
^2HBr (DBr)	c	1		7.500 93	820.68	247.3
	liq	1		6.162 38	505.68	220.6
^1HCl	c	1		8.134 73	941.57	268.06
	liq	1		7.170 00	745.80	258.88
^2HCl (DCl)	c	1		7.850 47	843.32	258.32
	liq	1		6.935 96	668.20	249.50
HCN	liq	1	−16 to 46	7.528 2	1329.5	260.4
^1HF	liq	1		7.680 98	1475.60	287.88
^2HF (DF)	liq	1		7.217 04	1268.37	273.87
^1HI	c	1		7.315 6	894.32	239.6
	liq	1		5.608 9	416.04	188.1
^2HI (DI)	c	1		7.314 9	889.52	238.8
	liq	1		5.601 8	413.98	187.8
HN_3	liq	1		6.857	1 066	232
HNO_3	liq	1		7.511 9	1 406	221.0
^1H_2O			[See Tables 10-8 and 10-9]			
^2H_2O (D_2O)		1	0–60	8.143 71	1 746.15	230.59
		1	60–120	7.918 37	1 616.76	219.54
$H_2{}^{18}O$		1	0–60	8.133 2	1 762.39	235.660
		1	60–120	7.972 08	1 668.84	227.700
H_2O_2	liq	1		7.969 17	1 886.76	220.6
HPO_2F	liq	1		6.735 3	1 342.9	232.0
H_2S	c	1		7.614 18	885.319	250.25
	liq	1		6.993 92	768.130	249.09
H_2S_2	liq	1		6.974	1 232	225
H_2S_3	liq	1		6.807	1 488	209
H_2S_4	liq	1		6.945	1 772	196
H_2S_5	liq	1		7.320	2 104	189
HSO_3Cl	liq	1		7.049	1 480	201
HSO_3F	liq	1		7.399 5	1 521	174.0
H_2Se	c	1		7.635 4	927.6	240.0
	liq	1		6.966 0	787.67	235.0
H_2Te	liq	1		7.000	935	229
Iodine						
I_2	c	1		9.810 9	2 901.0	256.00
	liq	1		7.018 1	1 610.9	205.0
ICl	liq	1		7.702 1	1 517.9	217.0
IF_5	c	1		10.964	2 538	245
	liq	1		7.464 8	1 460	216.0
IF_7	c	1		7.998	1 340	256
Iridium						
IrF_6	c	2	0.4–44	8.618	1 868	
	liq	2	44–54	7.952	1 657	
Iron						
$FeCl_2$	liq	2	708–834	9.794	7 455	
	liq	2	700–930	8.33	7 061	
$FeCl_3$	c	2	160–304	15.11	7 142	
FeI_2		2	517–577	13.183	10 778	
		2	601–686	9.674	7 716	

Table 10-7 (*Continued*)
VAPOR PRESSURES OF VARIOUS INORGANIC COMPOUNDS

Substance	State	Eq.	Range, °C	A	B	C
Krypton						
Kr	c	1		7.539 55	539.48	269.8
	liq	1		6.630 70	416.38	264.45
Lead						
Pb		2	525–1325	7.827	9 845.4	
PbBr$_2$		2	735–918	8.064	6 163.1	
PbCl$_2$		2	500–950	8.961	7 411.4	
PbF$_2$		2	1078–1289	8.391	8 623.2	
Lithium						
LiBr		2	1010–1265	8.068	7 975.5	
LiCl		2	1045–1325	7.939	8 142.7	
LiF		2	1398–1666	8.753	11 407	
LiH		2	500–650	11.227	9 600	
		2	700–800	9.926	8 204	
LiI		2	940–1140	8.011	7 500	
Magnesium						
Mg		2	900–1070	12.993	13 579.8	
MgH$_2$		2	337–415	9.78	3 857	
Mercury						
Hg			[See Table 10-5]			
HgBr$_2$		2	130–270	10.094	4 168.0	
HgCl$_2$		2	130–270	10.094	4 118.34	
		2	275–309	8.409	3 187.1	
Hg$_2$Cl$_2$		1		8.521 51	3 110.96	168.0
HgI$_2$		2	266–360	8.115	3 278.5	
Neon						
Ne	c	1		7.065 16	110.61	272.00
	liq	1		6.084 44	78.380	270.550
Neptunium						
NpF$_6$	liq	3	55.1–76.8	0.010 23	1 191.1	−2.582 5
Nickel						
Ni(CO)$_4$		2	2–40	7.780	1 556.5	
Niobium						
NbBr$_5$	liq	2		8.92	3 850	
NbCl$_5$	liq	2	210–254	8.37	2 827	
NbF$_5$	liq	2		8.439	2 824	
Nitrogen						
N$_2$ natural	c	1		7.345 12	322.222	269.980
	liq	1		6.494 57	255.680	266.550
^{15}N$_2$	c	1		7.363 96	323.17	269.88
	liq	1		6.494 14	255.535	266.451
NCl$_3$		1		6.956	1 190	221
NF$_3$	liq	1		6.779 66	501.913	257.79
NH$_3$			[See Table 10-6]			
N$_2$H$_4$	liq	1		7.801 9	1 679.07	227.7
NO natural	c	1		9.628 26	758.736	266.00
	liq	1		8.743 00	682.938	268.27
N$_2$O	c	1		9.437 00	1 174.020	268.22
	liq	1		7.003 94	654.260	247.16
N$_2$O$_4$ equilibrium	c	1		10.736 31	2 075.53	252.80
mixture	liq	1		8.917 12	1 798.54	276.80
N$_2$O$_5$	c	1		11.644 5	2 510	253.0
NOCl	c	1		8.540 8	1 397.3	261.0
	liq	1		7.361 54	1 094.73	249.70
N$_2$O$_3$		2	−25 to 0	10.30	2 057.9	
NOF	liq	1		6.443 5	556.13	216.0

Table 10-7 (*Continued*)
VAPOR PRESSURES OF VARIOUS INORGANIC COMPOUNDS

Substance	State	Eq.	Range, °C	A	B	C
Nitrogen						
NO₂Cl	liq	1		5.372 3	395.40	174.0
NO₂F	liq	1		6.833 4	654.55	238.0
Osmium						
OsF₅		2	75–180	9.75	3 429	
OsF₆		2	34–48	7.470	1 473	
OsF₈		2	38–47	7.650	1 525	
OsO₄		2	−38 to 40	10.710 0	2 951.00	
OsO₃F₂		2	59–105	7.994	1 911	
Oxygen						
O₂	liq	1		6.691 44	319.013	266.697
O₃	liq	1		6.837	552.5	251.0
OF₂	liq	1		7.236 19	545.05	269.91
O₂F₂	liq	1		6.779 02	756.39	250.16
O₃F₂		2	79–114	6.134 3	675.57	
Palladium						
PdCl₂		2	680–857	6.32	5 032	
Phosphorus						
P red, V	subl c	1		11.060	5 323	220
white	subl c	1		6.936 9	1 907.6	190.0
P₄ black, o-rh		1		12.405	6 671	247
PBr₃	liq	1	−40 to 173	6.915 5	1 590.5	221.0
PBr₅	liq	1	to 104	6.948	1 320	214
PBrF₂	liq	1	−133 to −16	6.904 2	885.12	236.0
PBr₂F	liq	1	−115 to 78	6.858 0	1 210.3	226.0
PCl₃	liq	1	−92 to 76	6.826 7	1 196	227.0
PCl₅	c	1	to 160	10.206 8	2 903.1	237.0
	liq	1		7.033	1 490	200.0
PClF₂	liq	1	−165 to −47	6.639 6	780.88	255.0
PCl₂F	liq	1	−144 to 14	6.796 56	982.332	237.00
P(OCN)₃	liq	2	−2 to 169	8.745 5	2 595	
PF₃	liq	1	−152 to −101	6.860 4	620.22	257.0
PF₅	liq	1	−93.8 to −84.5	6.914 4	647.21	245.0
PH₃	c	1		7.482 35	794.496	265.20
	liq	1		6.715 59	645.512	256.066
P₂H₄	liq	1		6.862 8	1 137	227.0
P₄O₆	liq	1	24–175	6.716 37	1 412.8	193.0
P₄O₁₀	c III	1		9.707 0	3 822	201.0
	c I	1		10.843 2	6 424	213
	liq	1		6.935 2	3 069	152
POBr₃	liq	1	51–192	7.007 8	1 609.2	198.0
POBrCl₂	liq	1	31–165	6.924	1 411	213
POBrClF	liq	1		6.914	1 214	222
POBrF₂	liq	1	−85 to 32	7.101 9	1 118.9	233.0
POBr₂F	liq	1	−117 to 110	6.721 2	1 328.9	236.0
POCl₃	liq	1	1.2–105	6.865 8	1 297.2	220.0
POClF₂	liq	1	−96 to 3	6.926 6	946.96	231.0
POCl₂F	liq	1	−80 to 53	7.084 65	1 201.86	233.00
POF₃	c	1		10.930 5	1 783	261.0
	liq	1		7.115 5	810.1	231.0
PO(OCN)₃		2	5–193	9.168 2	2 931	
PO(SCN)₃		2	14–300	8.533 0	3 240	
P₄S₁₀		2		9.17	4 940	
PSBr₃	c	2		10.105	3 196.2	
	liq	2		8.338 3	2 641.9	
PS(OCN)₃		2		10.032	3 492	

Table 10-7 (*Continued*)
VAPOR PRESSURES OF VARIOUS INORGANIC COMPOUNDS

Substance	State	Eq.	Range, °C	A	B	C
Platinum						
Pt		2	1425–1765	7.786	25 384	
PtF$_6$	liq	1	61.3–81.7	89.15	5 686	27.49
Polonium						
Po	liq	1		7.041 4	5 017.6	241.0
PoCl$_4$	liq	1		7.554	2 360	115
Potassium						
K		2	260–760	7.183	4 434.33	
KBr		2	1095–1375	7.936	8 555.3	
KCl		2	1116–1418	8.130	8 863.4	
KF		2	1278–1500	9.000	10 838	
KOH		2	1170–1327	7.330	7 103.3	
KI		2	1063–1333	7.949	8 132.2	
Protactinium	liq	2		17.27	7 377	
Radon						
Rn	c	1		7.495 5	884.41	255.0
	liq	1		6.701 5	718.25	250.0
Rhenium						
ReF$_5$	c	2		9.024	3 037	
ReF$_6$	c	3	−3.45 to 18.5	9.123 0	1 765.4	0.1790
	liq	3	18.5–48	18.208 1	1 956.7	3.599
ReF$_7$	c	3	−14.5 to 48.3	13.043 2	2 205.8	1.470 3
	liq	3	48.3–74.6	−21.583 5	244.28	−9.908 3
ReO$_2$	c	2	650–785	11.65	14 437	
	liq	2	480–660	5.345	4 742	
ReO$_3$	c	2	325–420	15.16	10 882	
	liq	2	300–480	7.745	4 966	
Re$_2$O$_7$	liq	2	230–360	8.98	3 868	
ReOF$_4$	liq	2	108–172	10.09	3 206	
ReOF$_5$	liq	2	41–73	7.727	1 679	
ReS$_2$	c	2	500–700	3.214	4 976	
Re$_2$S$_7$	c	2	260–410	8.86	4 800	
Rubidium						
Rb		2	250–370	6.976	3 969.5	
RbCl		2	1142–1395	9.111	10 373	
RbF		2	1142–1400	8.570	9 568.4	
Ruthenium						
RuOF$_4$		2	120–160	8.60	2 616	
Selenium						
Se	liq	1		7.631 6	4 213.0	202.0
SeCl$_4$	c	1		10.250 9	3 068.8	225.0
SeF$_4$	liq	1		7.888 7	1 603.0	215.0
SeF$_6$	c	1		8.385 4	1 121.4	250.0
SeO$_2$		1		6.577 81	1 879.81	179.0
SeOCl$_2$	liq	1		6.257 3	970.87	112.0
SeOF$_2$	liq	1		7.420	1 380	178
Silicon						
SiCl$_4$	liq	1	0–53	6.857 26	1 138.92	228.88
SiH$_4$		2	−160 to −112	6.881	645.9	
Si$_2$H$_6$		2	−115 to −14.6	7.258	1 133.4	
Si$_3$H$_8$		2	−70 to 52	7.676	1 559.1	
Silver						
AgCl		2	1255–1442	8.179	9 688.7	
Sodium						
Na		2	180–883	7.553	5 395.4	
NaCl		2	976–1155	8.329 7	9 417.07	

Table 10-7 (*Continued*)
VAPOR PRESSURES OF VARIOUS INORGANIC COMPOUNDS

Substance	State	Eq.	Range, °C	A	B	C
NaCl		2	1156–1430	8.548	9 704.3	
NaCN		2	800–1360	7.472	8 122.81	
NaF		2	1562–1701	8.640	11 396.6	
NaI		2	1063–1307	8.371	8 623.2	
NaOH		2	1010–1402	7.030	6 894	
Strontium						
Sr		2	940–1140	16.056	18 802.8	
Sulfur						
S equilibrium	liq	1		6.843 59	2 500.12	186.30
S_2Br_2	liq	1		7.177	1 660	185
SCl_2	liq	1		8.454	1 594	227
S_2Cl_2	liq	1		6.783 6	1 341	206.0
S_2F_2	liq	1		6.684	628	256
SF_4	liq	1		6.839 5	823.4	248.0
SF_6	c	1		8.416 0	1 096.5	262.0
S_2F_{10}	liq	1		7.067 6	1 100.6	234.0
SO_2	c	1		9.754 3	1 553.8	225.0
	liq	1		7.282 28	999.900	237.190
SO_3 "icelike"	c III	1		10.565 7	2 273.8	255.0
"woollike"	c II	1		11.590 1	2 665.6	264.0
	c I	1		14.255 9	3 692.1	273.0
	liq	1		9.050 85	1 735.31	236.50
$SOBr_2$	liq	1		7.056	1 445	206
$SOCl_2$	liq	1		7.287 45	1 446.7	252.7
SOClF	liq	1		7.173 1	1 100.1	244.00
SOF_2	liq	1		6.959 06	775.48	234.00
SOF_4	liq	1		7.071 8	840.3	249.0
$S_2O_2F_{10}$	liq	1		6.874	1 110	229
$S_2O_5Cl_2$	liq	1		7.019	1 460	202
S_2O_5ClF	liq	1		7.015 6	1 257.4	204.0
$S_2O_5F_2$	liq	1		6.881	1 120	229
$S_2O_5F_4$	liq	1		6.885	1 140	227
SO_2BrF	liq	1		7.142 8	1 155	231.0
SO_2Cl_2	liq	1		7.001 7	1 209	224.0
SO_2ClF	liq	1		6.521 5	793.73	210.70
SO_2F_2	liq	1		6.907 0	784.3	250
Tantalum						
$TaBr_5$	liq	2		8.11	3 260	
$TaCl_5$	liq	2	220–240	8.68	2 970	
TaF_5	liq	2		8.524	2 834	
TaI_5	liq	2		7.67	3 950	
Technetium						
TcF_6	liq	3	37.4–51.7	24.808 7	2 405	5.803 6
TcO_3F	liq	2	18.3–51.8	8.417	2 065	
Tc_2O_7	c	2		18.279	7 205	
	liq	2		8.999	3 571	
Tellurium						
Te	liq	1		7.301 0	5 370.6	221
$TeCl_4$	liq	1		7.558 6	2 355	115
TeF_6	liq	1		6.748 8	807.0	247.0
Te_2F_{10}	liq	1		6.901 8	1 150	227.0
TeO_2		2	450–733	12.328 4	13 222	
Thallium						
Tl		2	950–1200	6.1240	6 268	
TlF		2	282–298	12.52	5 484	

Table 10-7 (*Continued*)
VAPOR PRESSURES OF VARIOUS INORGANIC COMPOUNDS

Substance	State	Eq.	Range, °C	A	B	C
Thorium						
ThF_4	liq	2		10.821	15 270	
ThH_2		2	up to 883	9.50	7 650	
Tin						
$SnCl_4$		2	−52 to −38	9.824	2 441.23	
SnH_4		2	−148 to −49	7.400	999.68	
Titanium						
$TiCl_2$	subl c	2		9.30	8 500	
$TiCl_3$	subl c	2	455–550	10.401	8 296	
$TiCl_4$	liq	2	−23 to 136	7.683	1 964	
TiI_4	liq	2	160–360	7.577	3 054	
Tungsten						
W		2	2230–2770	9.920	46 850	
Uranium						
UF_6	liq	1	64–116	6.994 64	1 126.288	221.963
	liq	1	116–230	7.690 69	1 683.165	302.148
UH_3 dissociation		2	200–430	9.39	4 590	
U^2H_3 (UD_3)		2		9.43	4 500	
U^3H_3 (UT_3)		2		9.46	4 471	
Vanadium						
VBr_2	c	2	541–716	9.08	10 460	
	subl c	2	800–905	5.9	9 830	
VBr_3		2	314–427	11.12	7 470	
VCl_2	subl c	2	910–1100	5.725	9 721	
VCl_3		2	352–567	11.20	9 777	
VCl_4	liq	2	30–153	7.62	2 020	
VF_3	subl c	2	650–920	12.357	15 603	
VF_5	subl c	2	−20 to 19.5	8.168	2 608	
	liq	2	19.5–45.5	7.549	2 423	
VI_2	subl c	2	850–1016	2.56	5 600	
$VOCl_3$	liq	2	15.4–125	7.69	1 920	
Xenon						
Xe	c	1		7.484 5	714.896	264.0
	liq	1		6.642 89	566.282	258.660
XeF_2	subl c	1		10.019 47	2 683.96	261.68
XeF_4	subl c	1		10.913 87	3 095.06	269.56
Zinc						
Zn	c	2	250–419	9.200	6 946.6	

Table 10-8
VAPOR PRESSURES OF VARIOUS ORGANIC COMPOUNDS

Substance	Eq.	Range, °C	A	B	C
Acenaphthene	1	147–187	7.728 19	2 534.234	245.576
	2	147–288	8.033	2 834.99	
Acetaldehyde	1	liq	8.005 52	1 600.017	291.809
Acetic acid	1	liq	7.387 82	1 533.313	222.309
Acetic anhydride	1	liq	7.149 48	1 444.718	199.817
Acetone	1	liq	7.117 14	1 210.595	229.664
Acetonitrile	1	liq	7.119 88	1 314.4	230
Acetophenone	2	30–100	9.135 2	2 878.8	
Acetyl bromide	1	liq	5.197 02	545.784	150.396
Acetyl chloride	1	liq	6.948 87	1 115.954	223.554
Acetylene	1	−130 to −83	9.140 2	1 232.6	280.9
	1	−82 to −72	7.099 9	711.0	253.4
Acetyl iodide	1	liq	4.181 44	355.452	108.160
Acrylic acid	1	20–70	5.652 04	648.629	154.683
Acrylonitrile	1	−20 to 140	7.038 55	1 232.53	222.47
Allyl isothiocyanate	1	10–50	5.126 58	791.434	154.019
m-Aminobenzotrifluoride	1	0–96	7.651 86	1 940.6	218.0
		96–300	7.170 30	1 650.21	193.58
p-Aminophenol	1	130–185	−3.357 50	699.157	−331.343
Aniline	1	102–185	7.320 10	1 731.515	206.049
Anthracene	2	100–160	8.91	3 761	
	1	176–380	7.674 01	2 819.63	247.02
9,10-Anthracenedione	2	224–286	12.305	5 747.9	
	2	285–370	8.002	3 341.94	
Benzene	1	−12 to 3	9.106 4	1 885.9	244.2
	1	8–103	6.905 65	1 211.033	220.790
Benzenethiol	1	52–198	6.990 19	1 529.454	203.048
Benzoic acid	2	60–110	9.033	3 333.3	
Benzonitrile	1	liq	6.746 31	1 436.72	181.0
Benzophenone	1	48–202	7.349 66	2 331.4	195.0
	1	200–306	7.162 94	2 051.855	173.074
Benzotrifluoride	1	−20 to 180	7.007 08	1 331.30	220.58
Benzoyl chloride	2	140–200	7.924 5	2 372.1	
Benzyl acetate	1	46–156	8.457 05	2 623.206	259.067
Benzyl alcohol	1	122–205	7.198 17	1 632.593	172.790
Biphenyl	1	69–271	7.245 41	1 998.725	202.733
2-(2-Biphenylyloxy)ethanol	1	240–300	8.005 87	2 776.761	206.914
Bromobenzene	1	56–154	6.860 64	1 438.817	205.441
2-Bromobenzyl cyanide	1	85–152	5.044 59	734.821	59.273
1-Bromobutane	1	−78 to 23	5.281 38	685.001	160.880
Bromochloromethane	1	16–68	6.496 06	942.267	192.587
Bromochlorodifluoromethane	1	−95 to 10	6.839 98	935.632	240.330
2-Bromo-2-chloro-1,1,1-tri- fluoroethane	1	−51 to 55	6.945 02	1 127.856	227.341
Bromocyclohexane	1	68–260	6.979 80	1 572.19	217.38
p-Bromodiphenyl ether	1	25–190	7.009 3	1 902.7	153.3
	1	190–400	6.681 43	1 683.84	132.90
Bromoethane	1	28–75	6.988 6	1 121.9	234.7
Bromoethene	1	−88 to 16	6.997 4	1 009.9	251.6
2-Bromoethylbenzene	1	127–217	7.800	2 235.4	238.7
4-Bromoethylbenzene	1	liq	6.982 09	1 632.60	193
2-Bromo-2-methylpropane	1	0–72.8	7.395 9	1 512.7	262.2
1-Bromonaphthalene	1	liq	7.003 50	1 927.05	186.0
o-Bromostyrene	1	liq	6.910 38	1 631.2	195

Table 10-8 (*Continued*)
VAPOR PRESSURES OF VARIOUS ORGANIC COMPOUNDS

Substance	Eq.	Range, °C	A	B	C
p-Bromostyrene	1		7.228 38	1 743.67	218.0
4-Bromotoluene	1	85–280	7.007 62	1 612.35	206.36
2-Bromovinylbenzene	1	110–129	0.564 97	82.913	−191.71
4-Bromovinylbenzene	1	119–147	12.504 2	7 349.00	559.02
1,2-Butadiene	1	−69 to −34	7.398 22	1 219.877	259.776
	1	−26 to 30	6.993 83	1 041.117	242.274
1,3-Butadiene	1	−80 to −62	7.035 55	998.106	245.233
	1	−58 to 15	6.849 99	930.546	238.854
n-Butane	1	−77 to 19	6.808 96	935.86	238.73
1-Butanethiol	1	−2 to 123	6.927 54	1 281.018	218.100
2-Butanethiol	1	−13 to 110	6.886 98	1 229.904	222.021
1-Butanol	1	15–131	7.476 80	1 362.39	178.77
2-Butanol	1	25–120	7.474 31	1 314.19	186.55
2-Butanone	1	43–88	7.063 56	1 261.34	221.97
1-Butene	1	−82 to 13	6.792 90	908.80	238.54
2-Butene cis	1	−73 to 23	6.884 68	967.32	237.87
trans	1	−76 to 20	6.883 37	967.50	240.84
Butyl acetate	1	60–126	7.127 12	1 430.418	210.745
n-Butylamine trimethylboron	1	0–99	8.465 21	1 980.98	193.60
n-Butylbenzene	1	62–213	6.983 17	1 577.965	201.378
sec-Butylbenzene	1	87–174	6.942 19	1 533.95	204.39
t-Butylbenzene	1	84–170	6.922 55	1 505.987	203.490
n-Butyl borate	1	117–218	7.406 87	1 905.035	186.134
n-Butyl-t-butyl ether	1	83–124	6.955 56	1 348.702	206.303
Butyl carbitol	1	50–153	7.741 14	2 056.904	195.655
Butyl cellosolve	1	93–170	6.956 59	1 399.903	172.154
sec-Butyl chloroacetate	1	30–172	7.933 38	2 103.30	249.29
n-Butylcyclohexane	1	60–211	6.910 30	1 538.518	200.833
sec-Butylcyclohexane	1	91–180	6.890 96	1 530.70	202.373
t-Butylcyclohexane	1	84–173	6.856 80	1 501.724	206.108
n-Butylcyclopentane	1	41–185	6.899 35	1 457.08	205.99
n-Butyl formate	1	29–112	7.693 6	1 698.7	247.4
sec-Butyl formate	1	30–100	6.493	972.9	176.0
n-Butyl-α-hydroxyisobutyrate	1	112–185	8.421 7	2 617.32	287.09
1-n-Butylnaphthalene	1	25–170	7.434 47	2 227.7	202.2
	1	170–345	7.081 4	1 971.5	180
2-n-Butylnaphthalene	1	25–170	7.438 08	2 242.2	202.3
	1	170–345	7.084 8	1 984.3	180
n-Butyl nitrate	1	0–70	8.054 27	1 992.83	254.30
1-Butyl pentafluoropropionate	1	82–116	6.651 00	1 108.02	177.04
2-sec-Butylphenol	1	179–240	6.951 93	1 593.74	163.79
2-t-Butylphenol	1	135–225	7.217 56	1 822.81	196.23
4-t-Butylphenol	1	198–252	7.000 38	1 627.51	155.24
Butyl phenyl ether	1	119–210	7.299 7	1 882.70	215.82
n-Butyl propionate	1	32–93	9.484 89	2 852.58	296.98
n-Butyl trifluoroacetate	1	71–104	8.567 94	2 305.22	301.06
1-Butyl trimethylsilyl ether	1	71–124	7.763 00	1 884.68	261.31
1-Butyne	1	−68 to 27	6.981 98	988.75	233.01
2-Butyne	1	−51 to −34	7.037 91	896.91	199.06
	1	−31 to 47	7.073 38	1 101.71	235.81
n-Butyraldehyde	1	31–74	6.385 44	913.59	185.48
Butyric acid	1	90–163	7.739 9	1 764.7	199.9
Camphor	2	0–180	8.799	2 797.39	
	1	178–232	6.106	1 043.6	116.4
Capric acid	1	153–187	6.255 3	1 106.3	57.96

Table 10-8 (*Continued*)
VAPOR PRESSURES OF VARIOUS ORGANIC COMPOUNDS

Substance	Eq.	Range, °C	A	B	C
Caproic acid	1	98–179	6.924 9	1 340.8	126.6
Capronitrile	1	92–164	7.123 1	1 597.2	212.8
Caprylic acid	1	130–206	7.770 64	1 933.05	159.36
Carbazole	1	253–358	7.086 3	2 179.4	163.5
Carbitol	1	40–151	7.640 81	1 801.31	183.97
Chloroacetic acid	1	104–190	7.550 16	1 723.365	179.98
4-Chloroacetophenone	1	122–212	7.084 57	1 693.63	190.95
Chloroacetyl chloride	1	28–107	7.149 77	1 340.79	208.70
N-Chloroaniline	1	61–125	3.037 67	171.35	−14.99
2-Chloroaniline	1	20–108	7.562 65	1 998.6	220.0
	1	108–300	7.192 40	1 762.74	200.0
3-Chloroaniline	1	15–125	7.559 39	2 073.75	215
	1	125–310	7.236 03	1 857.75	196.64
o-Chloroanisole	1	115–186	7.121 36	1 655.80	188.77
Chlorobenzene	1	62–131.7	6.978 08	1 431.05	217.55
o-Chlorobenzotrichloride	1	30–150	7.504 30	2 228.07	220.0
	1	150–350	7.117 94	1 951.37	196.27
1-Chloro-4-bromobenzene	2	23–63	11.629	3 643.30	
1-Chlorobutane	1	−17 to 78.6	6.836 94	1 173.79	218.13
2-Chlorobutane	1	0–40	6.799 23	1 149.12	224.68
1-Chlorodecane	1	86–225.9	6.939 86	1 639.06	177.94
1-Chlorododecane	1	116–246	6.834 08	1 654.82	155.09
Chloroethane	1	−56 to 12.2	6.986 47	1 030.01	238.61
2-Chloroethylbenzene	1		6.981 69	1 556.0	201.0
3-Chloroethylbenzene	1		6.990 82	1 577.3	200
4-Chloroethylbenzene	1		6.983 09	1 577.0	200
Chloroethylene	1	−65 to −13	6.891 17	905.01	239.48
Chloroform	1	−35 to 61	6.493 4	929.44	196.03
1-Chloroheptane	1	34–160	6.916 70	1 453.96	199.83
1-Chlorohexadecane	1	166–327	7.282 03	2 152.61	162.73
1-Chlorohexane	1	15–136	7.051 36	1 461.72	215.57
Chlorohexylisocyanate	1	90–180	7.740 95	2 340.50	241.90
Chloromethane	1	−75 to −5	7.093 49	948.58	249.34
Chloromethoxytrichlorosilane	1	0–50	7.312 92	1 545.71	226.10
2-Chloro-2-methylpropane	1	22–47	4.896	334.99	114.0
1-Chlorononane	1	69–205	7.046 54	1 655.57	192.26
1-Chlorooctane	1	54–184	7.051 52	1 600.24	200.28
Chloropentafluorobenzene	1	36–140	7.068 83	1 389.19	213.75
p-Chlorophenetole	1	122–212	7.084 57	1 693.63	190.95
2-Chlorophenol	1	80–200	6.877 31	1 471.61	193.17
β-Chloro-β-phenylethyl alcohol	1	166–259	6.917 33	1 635.63	145.87
1-Chlorophenylisocyanate	1	50–160	12.265 9	6 532.55	499.59
m-Chlorophenylisocyanate	1	71–158	6.797 29	1 512.43	180.90
Chloroprene	1	20–60	6.161 50	783.45	179.7
1-Chloropropane	1	−25 to 47	6.926 48	1 110.19	227.94
2-Chloropropane	1	0–30	7.771	1 582	288
3-Chloro-1-propene	1	13–44	5.297 16	418.375	128.168
2-Chloropropionitrile	1	0–84	7.329 73	1 732.55	211.79
	1	84–240	7.200 85	1 657.25	205.3
γ-Chloropropyltrichlorosilane	1	87–179	7.156 4	1 679.07	210.38
1-Chlorotetradecane	1	142–296.8	7.200 7	2 018.9	170.6
o-Chlorotoluene	1	0–65	7.367 97	1 735.8	230.0
	1	65–220	6.947 63	1 497.2	209.0
1-Chloro-2,4,6-trinitrobenzene	1	200–270	3.080 9	184.93	−117.9
1-Chloroundecane	1	101–245	6.967 6	1 709.4	172.9

Table 10-8 (*Continued*)
VAPOR PRESSURES OF VARIOUS ORGANIC COMPOUNDS

Substance		Eq.	Range, °C	A	B	C
o-Chlorovinylbenzene		1	98–155	6.956 6	1 602.2	204.5
p-Chlorovinylbenzene		1	100–127	9.969 1	4 093.5	392.4
2-Chlorovinyldichloroarsine	cis	1	68–109	5.487 9	785.09	115.61
	trans	1	50–150	6.814 0	1 465.07	178.53
3-Chlorovinyldichloroarsine		1	66–110	2.810 5	97.17	−27.51
o-Cresol		1	120–191	6.911 7	1 435.50	165.16
m-Cresol		1	150–201	7.508 0	1 856.36	199.07
p-Cresol		1	128–202	7.035 08	1 511.08	161.85
Cyanic acid		1	−76 to −6	7.568 59	1 251.86	243.79
Cyclobutane		1	−60 to 12	6.916 31	1 054.54	241.37
Cyclobutanone		1	−24 to 25	6.116 68	933.95	183.19
Cyclobutene		1	−77 to 2	7.305 7	1 166.0	261.06
Cycloheptane		1	68–159	6.853 95	1 331.57	216.35
1,3,5-Cycloheptatriene		1	0–65	6.974 33	1 376.84	220.75
Cyclohexane		1	20–81	6.841 30	1 201.53	222.65
Cyclohexanethiol		1	84–203	6.886 73	1 476.70	209.83
Cyclohexanol		1	94–161	6.255 3	912.87	109.13
Cyclohexene		1		6.886 17	1 229.973	224.10
Cyclohexyl acetate		1	95–172	7.975 86	2 167.99	252.30
Cyclohexylamine		1	61–128	6.689 54	1 229.42	188.80
1-Cyclohexylamino-2-propanol		1	150–238	7.011 56	1 655.02	162.59
Cyclohexylpentafluoropropionate		1	82–155	7.725 5	1 844.73	224.89
Cyclohexyltrifluoroacetate		1	72–147	7.802 35	1 954.66	249.33
Cyclohexyltrimethylsilyl ether		1	91–168	8.090 52	2 276.62	267.94
Cyclooctane		1	97–194	6.861 87	1 437.79	210.02
1,3,5,7-Cyclooctatetraene		1	0–75	7.006 69	1 472.11	215.84
Cyclopentane		1	−40 to 72	6.886 76	1 124.162	231.36
Cyclopentanethiol		1	81–173	6.914 97	1 388.63	212.05
Cyclopentanone		1	0–26	2.902 47	162.90	63.22
Cyclopentene		1		6.920 66	1 121.818	223.45
Cyclopentyl-1-thiaethane		1	83–199	6.940 83	1 480.70	208.47
Cyclopropane		1	−90 to −32	6.887 88	856.01	246.50
o-Cymene		1	81–180	7.266 10	1 768.45	224.95
m-Cymene		1	79–176	7.123 74	1 644.95	212.76
p-Cymene		1	107–178	7.050 74	1 608.91	208.72
Decahydronaphthalene	cis	1	68–228	6.875 29	1 594.460	203.39
	trans	1	61–219	6.856 81	1 564.683	206.26
Decane		1	58–203	6.943 65	1 495.17	193.86
1-Decanethiol		1	109–271	6.998 1	1 713.6	177.0
1-Decanol		1	25–52	11.560	4 055	273.2
		1	103–230	6.922 44	1 472.01	133.98
1-Decene		1	54–199	6.934 77	1 484.98	195.707
Decylbenzene		1	203–298	7.035 96	1 903.98	160.33
Decylcyclohexane		1	197–298	7.019 37	1 899.33	161.35
Decylcyclopentane		1	182–279	6.999 12	1 822.05	163.05
Deuterodiborane		1	−155 to −94	6.480 83	545.20	244.73
Diacetone alcohol		1	28–115	8.502 42	2 400.56	263.79
1,3-Diacetylbenzene		1	50–145	0.056 24	64.188	−196.97
1,4-Diacetylbenzene		1	116–157	2.803 71	177.25	−46.43
Diacetylene		1	−78 to 0	4.990 79	356.36	143.22
Diallyl sulfide		1	10–40	4.829 30	643.18	142.34
4,4'-Diaminodiphenylmethane		1	198–272	3.172 31	210.49	−137.41
Diamyl ether		1	105–187	7.067 10	1 604.77	196.58
Dibenzyl ketone		2	285–325	8.257	3 244.42	

Table 10-8 (*Continued*)
VAPOR PRESSURES OF VARIOUS ORGANIC COMPOUNDS

Substance	Eq.	Range, °C	A	B	C
1,2-Dibromobenzene	1	20–117	7.501 28	2 093.7	230
	1	117–300	7.102 65	1 825.77	207.0
Dibromodichloroethane	1	25–130	5.197 53	763.44	110.81
Dibromodifluoromethane	1	−26 to 23	7.152 22	1 181.612	253.85
1,2-Dibromoethane	1	52–131	6.721 48	1 280.82	201.75
1,2-Dibromoethylene *cis*	1	26–78	7.038 74	1 349.84	209.26
trans	1	4–71	4.581 11	393.641	103.56
1,2-Dibromopropane	1	0–50	7.303 98	1 644.4	232.0
	1	50–250	6.891 05	1 419.60	212.0
1,3-Dibromopropane	1	0–71	7.549 84	1 890.56	240.0
	1	71–275	7.198 74	1 678.26	222.0
Di-*n*-butyl ether	1	89–140	6.796 3	1 297.29	191.03
Di-*t*-butyl ether	1	4–109	6.932 9	1 348.53	233.79
Di-*n*-butyl phthalate	1	126–202	6.639 80	1 744.20	113.69
Di-*n*-butyl sebacate	1	128–208	7.587 66	2 364.89	147.54
Di-*n*-butyl sulfide	1	10–40	6.769 3	1 208.80	217.51
1,2-Dichlorobenzene	1	131–181	7.143 78	1 704.49	219.42
1,3-Dichlorobenzene	1	91–173	7.040 1	1 607.05	213.38
1,4-Dichlorobenzene	1	95–174	7.020 8	1 590.9	210.2
Dichlorobenzotrichloride	1	20–167	7.439 54	2 190.0	200
	1	167–340	6.985 24	1 868.91	172.00
Dichlorobenzyl chloride	1	20–138	7.504 57	2 125.9	213.8
	1	138–350	7.147 35	1 881.38	192.93
1,1-Dichloroethane	1	−39 to 18	6.977 0	1 174.02	229.06
1,2-Dichloroethane	1	−31 to 99	7.025 3	1 271.3	222.9
1,1-Dichloroethylene	1	−28 to 32	6.972 2	1 099.4	237.2
1,2-Dichloroethylene *cis*	1	0–84	7.022 3	1 205.4	230.6
trans	1	−38 to 85	6.965 1	1 141.9	231.9
2,2′-Dichloroethyl sulfide	1	15–76	8.587 41	2 588.23	246.06
1,2-Dichloroethyltrichloro-silane	1	102–181	7.826	2 144.9	253.1
Dichloromethane	1	−40 to 40	7.409 2	1 325.9	252.6
2-(2,4-Dichlorophenoxy)-ethanol	1	212–286	7.240 09	2 004.31	157.25
3,4-Dichlorophenylisocyanate	1	60–190	8.679 3	3 312.3	333.9
1,2-Dichloropropane	1	45–96	6.980 7	1 308.1	222.8
3,4-Dichlorotoluene	1	0–105	7.343 94	1 882.5	215.0
	1	105–330	6.979 25	1 655.44	195.0
Diethanolamine	1	194–241	8.138 8	2 327.9	174.4
1,1-Diethoxyethane	1	0–70	6.757 63	1 191.60	203.12
Diethoxymethane	1	0–75	6.908 41	1 229.52	217.01
Diethylaluminum chloride	1	44–125	8.229 70	2 484.53	255.45
Diethylamine	1	31–61	5.801 6	583.30	144.1
N,N-Diethylaniline	1	50–218	7.466 0	1 993.57	218.5
1,2-Diethylbenzene	1	liq	6.987 80	1 576.940	200.51
1,3-Diethylbenzene	1	liq	7.003 60	1 575.310	200.96
1,4-Diethylbenzene	1	liq	6.998 20	1 588.310	201.97
Diethyldichlorosilane	1	48–128	6.862 9	1 346.3	207.7
Diethyl disulfide	1	15–61	7.349 89	1 695.00	227.29
	1	61–230	6.975 07	1 485.970	208.96
Diethylene glycol	1	130–243	7.636 7	1 939.4	162.7
Diethyl ether	1	−61 to 20	6.920 32	1 064.07	228.80
Diethyl ethylphosphate	1	76–134	4.101 6	315.17	15.50
N,N-Diethylformamide	1	30–90	6.395 4	1 203.8	165.6
Diethyl ketone	1		6.857 91	1 216.3	204
3,3-Diethylpentane	1	63–147	6.896 03	1 453.48	215.83

Table 10-8 (*Continued*)
VAPOR PRESSURES OF VARIOUS ORGANIC COMPOUNDS

Substance		Eq.	Range, °C	A	B	C
3,5-Diethylphenol		1	114–248	7.651 3	2 228	218.5
Diethylpropylphosphonate		1	87–134	4.558 1	446.50	26.17
Diethyl sulfide		1	0–150	6.928 36	1 257.83	218.66
1,2-*bis*-Difluoroamino-4-methylpentane		1	−20 to 20	8.009 11	1 944.92	245.44
Difluoromethane		1	−82 to −32	7.138 9	821.7	244.7
1,2-Dihydroxybenzene		1	118–246	7.577	2 054	187
1,3-Dihydroxybenzene		1	151–276	7.889	2 231	169
1,2-Diiodoethylene	*cis*	1	29–152	5.522	797.8	106.4
	trans	1	77–130	6.093 1	1 197.0	172.3
Diisoamyl sulfide		1	10–80	−1.959 8	390.61	−219.33
p-Diisopropylbenzene		1	120–211	6.993 3	1 663.88	194.41
Diisopropyl ether		1	23–67	6.849 5	1 139.34	218.7
2,4-Diisopropylphenol		1	122–255	6.714	1 506	138
1,2-Dimethoxyethane		1	0–60	6.718 9	1 050.5	209.2
N,N-Dimethylacetamide		1	30–90	9.720 9	3 273.8	334.5
Dimethylamine		1	−72 to 6.9	7.082 12	960.242	221.67
bis-Dimethylaminoborane		1	−25 to 62.5	5.584 52	774.371	170.64
N-Dimethylaminodiborane		1	−38 to 14	8.340 1	1 917.35	302.73
bis-Dimethylaminodifluorosilane		1	24–88	5.952	748.7	146.9
N,N-Dimethylaniline		1	71–197	7.367 7	1 857.08	220.36
Dimethyl beryllium		1	100–180	19.089 9	11 535.45	496.64
1,4-Dimethyl-bicyclo(2,2,1)-heptane		1	56–119	6.761 96	1 342.66	213.53
2,3-Dimethyl-bicyclo(2,2,1)-heptane	*trans*	1	72–138	6.868 15	1 420.32	212.94
2,3-Dimethyl-1,3-butadiene		1	0–68.5	7.119 7	1 299.69	238.09
2,2-Dimethylbutane		1	−42 to 73	6.754 83	1 081.176	229.34
2,3-Dimethylbutane		1	−35 to 81	6.809 83	1 127.187	228.90
2,3-Dimethyl-2-butanethiol		1	56–167	6.839 56	1 354.24	215.96
2,3-Dimethyl-1-butene		1	−36 to 78	6.862 36	1 134.675	229.37
2,3-Dimethyl-2-butene		1	−21 to 97	6.950 58	1 215.428	225.44
3,3-Dimethyl-1-butene		1	−47 to 64	6.677 51	1 010.516	224.91
Dimethyl cadmium		1	−2 to 23	6.490 55	1 126.36	201.07
1,1-Dimethylcyclohexane		1	10–147	6.798 21	1 321.705	217.85
1,2-Dimethylcyclohexane	*cis*	1	18–158	6.837 46	1 367.311	215.84
	trans	1	13–151	6.833 08	1 353.881	219.13
1,3-Dimethylcyclohexane	*cis*	1	11–147	6.838 83	1 338.473	218.07
	trans	1	15–152	6.834 55	1 343.687	215.39
1,4-Dimethylcyclohexane	*cis*	1	15–152	6.832 87	1 345.613	216.15
	trans	1	10–147	6.817 73	1 330.437	218.58
1,1-Dimethylcyclopentane		1	−12 to 113	6.817 24	1 219.474	221.95
1,2-Dimethylcyclopentane	*cis*	1	−3 to 125	6.850 08	1 269.140	220.21
	trans	1	−9 to 117	6.844 22	1 242.748	221.69
1,3-Dimethylcyclopentane	*cis*	1	−10 to 116	6.837 15	1 237.456	222.01
	trans	1	−9 to 117	6.838 17	1 240.023	221.62
Dimethyldichlorosilane		1	28–72	7.062 1	1 280.29	235.65
1,2-Dimethyldisilane		1	−46 to 0	4.024 3	255.4	129.2
Dimethyl ether		1	−71 to −25	6.976 03	889.264	241.96
N,N-Dimethylformamide		1	30–90	6.928 0	1 400.87	196.43
2,2-Dimethylhexane		1		6.837 15	1 273.59	215.07
2,3-Dimethylhexane		1		6.870 04	1 315.50	214.16
2,4-Dimethylhexane		1		6.853 05	1 287.88	214.79
2,5-Dimethylhexane		1		6.859 84	1 287.27	214.41
3,3-Dimethylhexane		1		6.851 21	1 307.88	217.44

Table 10-8 (*Continued*)
VAPOR PRESSURES OF VARIOUS ORGANIC COMPOUNDS

Substance	Eq.	Range, °C	A	B	C
3,4-Dimethylhexane	1		6.879 86	1 330.04	214.86
1,1-Dimethylhydrazine	1	−35 to 20	7.408 13	1 305.91	225.53
1,2-Dimethylhydrazine	1	1–25	5.611 9	633.59	143.17
N,N-Dimethylhydroxylamine	1	17–90	7.565 8	1 415.96	201.93
O,N-Dimethylhydroxylamine	1	−45 to 42.2	7.405 4	1 245.58	233.06
Dimethylmalononitrile	1	49–140	7.035 5	1 546.99	202.00
1,3-Dimethylnaphthalene	1	20–148	7.634 7	2 295.4	232.4
	1	148–310	7.269 8	2 076.0	210
1,4-Dimethylnaphthalene	1	20–148	7.634 7	2 345.8	232.6
(same for 1,6- and 1,7-)	1	148–310	7.269 8	2 076.0	210
1,8-Dimethylnaphthalene	1	25–150	7.407 89	2 123.2	201.2
	1	150–320	7.056 4	1 879	180
2,3-Dimethylnaphthalene	1	20–155	7.403 96	2 111.9	201.1
	1	155–315	7.052 7	1 869	180
2,6-Dimethylnaphthalene	1	20–150	7.396 8	2 080.3	200.8
	1	150–310	7.046 0	1 841	180
2,7-Dimethylnaphthalene	1	25–150	7.398 75	2 085.9	200.9
	1	150–310	7.047 8	1 846	180
2,2-Dimethylpentane	1	−19 to 103	6.814 80	1 190.033	223.30
2,3-Dimethylpentane	1	−10 to 115	6.853 82	1 238.017	221.82
2,4-Dimethylpentane	1	−17 to 105	6.826 21	1 192.04	225.32
3,3-Dimethylpentane	1	−14 to 112	6.826 67	1 228.663	225.32
2,4-Dimethyl-3-pentanone	1	48–125	6.968 53	1 382.84	213.06
Dimethyl-o-phthalate	1	82–151	4.522 32	700.31	51.42
2,2-Dimethylpropane	1	−14 to 29	6.604 27	883.42	227.78
2,2-Dimethyl-1-propanol	1	55–115	7.875 3	1 604.7	208.2
2,5-Dimethylpyrrole	1	100–199	7.203 06	1 509.60	181.76
2,4-Dimethylquinoline	1	185–269	7.025 4	1 830.29	174.44
2,6-Dimethylquinoline	1	188–267	6.931 12	1 748.73	166.37
Dimethyl sulfide	1	−22 to 20	7.150 9	1 195.58	242.68
3,3-Dimethyl-2-thiabutane	1	liq	6.847 09	1 259.648	218.69
2,2-Dimethyl-3-thiapentane	1	liq	6.850 86	1 323.24	212.89
2,4-Dimethyl-3-thiapentane	1	liq	6.871 18	1 327.12	212.55
2,3-Dimethylthiophene	1	50–205	6.924 9	1 430.0	212
2,4-Dimethylthiophene	1	50–205	6.993 9	1 450.7	212.0
2,5-Dimethylthiophene	1	47–200	6.961 1	1 427.7	213.2
3,4-Dimethylthiophene	1	54–205	6.996 1	1 467.1	211.5
1,3-Dinitrobenzene	1	252–292	4.337	229.2	−137
2,4-Dinitrotoluene	1	200–299	5.798	1 118	61.8
2,6-Dinitrotoluene	1	150–260	4.372	380	−43.6
3,5-Dinitrotoluene	1	220–270	1.556	30.59	−302
1,4-Dioxane	1	20–105	7.431 55	1 554.68	240.34
Dipentene	1	21–170	7.111 6	1 613.42	207.8
2,2′-Diphenol	1	171–325	8.193 5	3 067.6	253.1
Diphenyldichlorosilane	1	192–281	6.999 03	1 918.20	161.41
Diphenyl ether	1	204–271	7.011 04	1 799.71	177.74
Diphenylmethane	1	217–282	6.291	1 261	105
Di-n-propyl ether	1	26–89	6.947 6	1 256.5	219.0
Disilanyl chloride	1	−46 to 18	7.104 8	1 211.8	245.2
2,3-Dithiabutane	1	6–135	6.977 92	1 346.342	218.86
5,6-Dithiadecane	1	101–263	6.963 8	1 684.1	181.3
3,4-Dithiahexane	1	40–182	6.975 07	1 485.970	208.96
4,5-Dithiaoctane	1	72–226	6.975 29	1 603.793	195.85
Dodecane	1	91–247	6.997 95	1 639.27	181.84
1-Dodecanethiol	1		7.024 4	1 817.8	164.1

Table 10-8 (*Continued*)
VAPOR PRESSURES OF VARIOUS ORGANIC COMPOUNDS

Substance	Eq.	Range, °C	A	B	C
Dodecanoic acid	1	106–176	7.860 8	2 159.1	143.2
1-Dodecanol	1	138–214	7.539 86	2 003.29	168.13
1-Dodecene	1	89–244	6.976 07	1 621.11	182.45
Durenol	1	108–249	7.758	2 432	250
Eicosane	1	198–379	7.152 2	2 032.7	132.1
1-Eicosanethiol	1		7.114	2 125	119
1-Eicosene	1	liq	7.135 1	2 043.0	137.9
Ethane	1	−142 to −75	6.829 15	663.72	256.68
Ethanethiol	1	−49 to 56	6.952 06	1 084.531	231.39
Ethanol	1	−2 to 100	8.321 09	1 718.10	237.52
Ethanolamine	1	65–171	7.456 8	1 577.67	173.37
Ethyl acetate	1	15–76	7.101 79	1 244.95	217.88
m-Ethylacetophenone	1	19–143	3.767 2	708.05	182.6
p-Ethylacetophenone	1	21–94	4.274 6	629.34	120.9
Ethylamine	1	−20 to 90	7.054 13	987.31	220.0
N-Ethylaniline	1	50–207	7.422 8	1 903.4	214.3
Ethylbenzene	1	26–164	6.957 19	1 424.255	213.21
2-Ethyl-1-butene	1	−28 to 88	6.997 12	1 218.352	231.30
Ethyl butyl ether	1	38–92	6.944 4	1 256.4	216.9
Ethyl chloroacetate	1	25–146	6.967	1 355.9	188.2
p-Ethylchlorobenzene	1	109–184	6.951 1	1 557.1	198.1
Ethylcyclohexane	1	20–160	6.867 28	1 382.466	214.99
Ethylcyclopentane	1	−0.1 to 129	6.887 09	1 298.599	220.68
Ethylene	1	−153 to −91	6.744 19	594.99	256.16
Ethylene glycol	1	50–200	8.090 8	2 088.9	203.5
Ethylene glycol monoethyl ether	1	63–134	7.874 6	1 843.5	234.2
Ethylene glycol monomethyl ether	1	56–124	7.849 8	1 793.9	236.9
Ethylene oxide	1	−49 to 12	7.128 43	1 054.54	237.76
Ethyl formate	1	4–54	7.009 0	1 123.94	218.2
3-Ethylhexane	1		6.890 98	1 327.88	212.60
2-Ethyl-1-hexanol	1	74–184	6.914 7	1 339.7	147.8
2-Ethyl-2-hexenal	1	54–175	6.861 3	1 457.4	190.6
Ethyl iodoacetate	1	29–89	4.073 7	374.64	54.8
Ethyl isothiocyanate	1	10–50	7.106 0	1 567.5	234.2
Ethyl methyl ether	1	5–7.7	5.518	434.5	158
Ethyl methyl ketone	1		6.974 21	1 209.6	216
3-Ethyl-5-methylphenol	1	195–247	7.040 83	1 615.44	152.6
2-Ethyl-4-methyl-1-pentanol	1	70–176	6.582 6	1 134.6	129.2
Ethyl nitrate	1	0–60	7.163 7	1 338.8	224.9
3-Ethylpentane	1	−7 to 119	6.875 64	1 251.827	219.89
2-Ethylphenol	1	86–208	7.800 3	2 140.4	227
3-Ethylphenol	1	97–218	7.468	1 856	187
4-Ethylphenol	1	101–218	8.291	2 423	229
Ethyl phenyl ether	1	117–181	7.021 38	1 508.39	194.49
Ethyl *n*-propanoate	1	34–98	6.994 9	1 260.6	207.4
Ethyl *n*-propyl ether	1	20–63	6.985 1	1 188.5	226.4
Ethyl *n*-propyl ketone	1	75–133	7.000 82	1 365.79	208.01
m-Ethylstyrene	1		7.039 28	1 614.0	198
p-Ethylstyrene	1		6.900 71	1 570.9	198
Ethyl trichloroacetate	1	44–95	7.725 4	1 927.0	233.7
Ethyl trichlorosilane	1	28–96	6.606	1 118	201
Ethyl triexthoxysilane	1	64–153	6.886 8	1 377.9	183.0
Ethyl vinyldichlorosilane	1	45–122	6.859	1 331	210.8

Table 10-8 (*Continued*)
VAPOR PRESSURES OF VARIOUS ORGANIC COMPOUNDS

Substance	Eq.	Range, °C	A	B	C
Fenchyl alcohol	1	59–200	5.693	797.6	84.6
Fluoranthene	1	197–384	6.373	1 756	118
Fluorene	1	161–300	7.761 8	2 637.1	243.2
Fluorobenzene	1	−18 to 84	7.187 0	1 381.8	235.6
m-Fluorobenzotrifluoride	1	40–137	7.006 59	1 304.35	215.67
bis-(Fluorocarbonyl)-peroxide	1	−47 to −7	9.608 4	2 247.64	319.83
p-Fluorotoluene	1	68–155	6.994 26	1 374.055	217.40
Formaldehyde	1	−109 to −22	7.195 8	970.6	244.1
Formic acid	1	37–101	7.581 8	1 699.2	260.7
Formyl fluoride	1	−95 to −61	5.270	362	175
Furan	1	2–61	6.975 27	1 060.87	227.74
2-Furfuraldehyde	1	56–161	6.575 9	1 198.7	162.8
Glycerol	1	183–260	6.165	1 036	28
Glyceryl-1,3-diacetate	1	100–190	6.407 3	1 092.0	119.3
Guaiacol	1	82–205	6.161	1 051	116
Hemellitenol	1	123–248	6.972	1 563	134
Heptadecane	1	161–337	7.014 3	1 865.1	149.20
1-Heptadecene	1		7.008 67	1 868.9	152.50
Heptane	1	−2 to 124	6.896 77	1 264.90	216.54
1-Heptanethiol	1	58–206	6.952 49	1 525.311	197.70
Heptanoic acid	1	112–150	5.287 4	665.54	42.07
1-Heptanol	1	60–176	6.647 67	1 140.64	126.56
1-Heptene	1	−6 to 118	6.901 87	1 258.345	219.30
Hexadecane	1	149–321	7.028 67	1 830.51	154.45
1-Hexadecanethiol	1		7.075	1 990	140
1-Hexadecanol	1	50–103	7.281 7	1 909.7	128.1
	1	145–190	6.158 6	1 380.0	91
1-Hexadecene	1		7.040 11	1 840.52	157.57
1,5-Hexadiene	1	0–59	6.574 1	1 013.5	214.8
Hexafluoroacetone	1	−79 to −27	6.650 2	725.90	219.9
Hexafluorobenzene	1	5–114	7.032 95	1 227.98	215.49
Hexafluorodisiloxane	1	−39 to −23	7.471 2	1 169.3	278.1
Hexafluoroethane	1	−93 to −78	6.793 35	657.06	246.2
Hexahydroindane cis	1	77–168	6.868 22	1 497.33	207.67
trans	1	71–161	6.861 19	1 475.70	209.66
Hexamethyldisiloxane	1	36–138	6.773 79	1 202.03	208.25
Hexane	1	−25 to 92	6.876 01	1 171.17	224.41
1-Hexanethiol	1	40–181	6.946 64	1 454.004	204.95
1-Hexanol	1	35–157	7.860 45	1 761.26	196.66
2-Hexanol	1	25–142	7.261 0	1 371.7	173.2
3-Hexanol	1	25–138	7.689	1 670.0	211.8
1-Hexene	1	16–64	6.857 70	1 148.62	225.35
3-Hexyne	1	−20 to 24	5.895	863.3	194
Hydroquinone	1	159–286	8.137	2 461	183
3-Hydroxy-3-methyl-2-butanone	1	45–146	7.340 9	1 653.6	227.5
Iodobenzene	1	20–188	7.011 9	1 640.1	208.8
Iodoethane	1	30–60	6.959	1 232	229
Isoamyl acetate	1	41–95	7.436	1 606.6	216
Isobutylbenzene	1	86–174	6.935 56	1 530.05	204.59
Isobutyl borate	1	99–200	7.197	1 745.8	193
Isobutyl cellosolve	1	71–159	7.694 8	1 825.9	219.6
Isobutylcyclohexane	1	85–172	6.867 97	1 493.10	203.16
Isobutyl nitrate	1	0–70	8.164 3	2 022.7	262.4
Isobutyraldehyde	1	13–63	6.735 1	1 053.2	209.1
Isobutyric acid	1	58–152	4.894	382.6	38

Table 10-8 (*Continued*)
VAPOR PRESSURES OF VARIOUS ORGANIC COMPOUNDS

Substance	Eq.	Range, °C	A	B	C
Isocaproic acid	1	96–133	6.258	1 038.6	103
Isopropylbenzene	1	39–181	6.936 66	1 460.793	207.78
Isopropyl borate	1	65–139	8.070	2 120	269
o-Isopropylbromobenzene	1	132–210	6.717 8	1 462.7	170.9
Isopropyl caprate	1	90–178	9.959	4 013.9	326.5
Isopropyl caprylate	1	65–146	8.032 2	2 213.6	220.9
Isopropyl cellosolve	1	67–140	7.500 0	1 639.2	213.3
Isopropyl chloroacetate	1	35–153	8.382	2 328	275
Isopropylcyclohexane	1	71–155	6.873 14	1 453.20	209.44
Isopropylcyclopentane	1	47–127	6.887 36	1 380.12	218.05
Isopropyl laurate	1	117–196	8.532 6	2 951.6	240.7
Isopropyl myristate	1	140–193	10.418 0	4 866.48	314.17
Isopropyl nitrate	1	0–70	7.266 6	1 434.4	225.2
Isopropyl palmitate	1	160–197	10.916 4	5 572.0	364.8
o-Isopropylphenol	1	97–215	8.167	2 343	229
p-Isopropylphenol	1	108–228	8.666	2 810	258
Isopropyl phenyl ether	1	72–175	6.517 6	1 238.0	163.0
Isopropyl stearate	1	182–207	0.079 3	10.41	−221
Isopseudocumenol	1	106–233	5.602	768	49
Isoquinoline	1	167–244	6.912 2	1 723.4	184.3
Isovaleric acid	1	86–104	3.946 55	255.41	11.3
Ketene	1	−88 to −49	7.615	1 036	269
Lauric acid	1	106–176	7.860 8	2 159.1	143.2
Lepidine	1	199–266	7.271 2	1 946.14	177.64
2,3-Lutidine	1	155–162	7.447 8	1 832.6	240.1
2,4-Lutidine	1	150–160	7.339 0	1 733.4	230.4
2,5-Lutidine	1	85–157	7.081 0	1 539.6	209.6
2,6-Lutidine	1	79–144	7.056 7	1 470.2	208.0
3,4-Lutidine	1	172–180	7.362 0	1 840.1	231.5
3,5-Lutidine	1	163–173	7.333 1	1 783.6	228.7
Mesitol	1	94–221	6.659	1 392	148
Mesityl oxide	1	14–130	6.635 8	1 186.1	186.0
Methacrylonitrile	1		6.980 2	1 274.96	220.7
Methane c	1	−195 to −183	7.193 09	451.64	268.49
liq	1	−181 to −152	6.695 61	405.42	267.78
Methanol	1	−14 to 65	7.897 50	1 474.08	229.13
	1	64–110	7.973 28	1 515.14	232.85
Methoxybenzene	1	110–164	7.052 69	1 489.99	203.57
N-Methylacetamide	1	40–90	2.631 1	121.7	−9.3
Methyl acetate	1	1–56	7.065 2	1 157.63	219.73
Methylal	1	0–35	6.872 2	1 049.2	220.6
Methylamine	1	−83 to −6	7.336 9	1 011.5	233.3
N-Methylaniline	1	50–200	7.081 9	1 631.3	192.4
Methyl benzoate	1	111–199	7.273	1 847	221
Methyl borate	1	31–68	7.646 0	1 491.5	245.5
Methyl boric anhydride	1	0–55	8.004 1	1 726.1	257.9
2-Methyl-1,3-butadiene	1	−52 to −24	7.011 87	1 126.159	238.88
	1	−19 to 55	6.885 64	1 071.578	233.51
3-Methyl-1,2-butadiene	1	−45 to −20	7.151 95	1 194.537	239.47
	1	−20 to 62	6.943 50	1 103.901	230.89
2-Methylbutane	1	−57 to 49	6.833 15	1 040.73	235.45
2-Methyl-1-butanethiol	1	liq	6.913 85	1 347.317	215.07
3-Methyl-1-butanethiol	1	liq	6.914 91	1 342.509	214.45
2-Methyl-2-butanethiol	1	liq	6.828 37	1 254.885	218.76
2-Methyl-1-butanol	1	34–129	7.067 30	1 195.26	156.83

Table 10-8 (*Continued*)
VAPOR PRESSURES OF VARIOUS ORGANIC COMPOUNDS

Substance	Eq.	Range, °C	A	B	C
3-Methyl-1-butanol	1	25–153	7.258 21	1 314.36	169.36
2-Methyl-2-butanol	1	25–102	6.519 3	863.4	135.3
3-Methyl-2-butanol	1	25–111	6.942 1	1 090.9	157.2
2-Methyl-1-butene	1	−53 to 52	6.846 37	1 039.69	236.65
3-Methyl-1-butene	1	−63 to 41	6.824 55	1 012.37	236.65
2-Methyl-2-butene	1	−48 to 60	6.966 59	1 124.33	236.63
Methyl butyl ether	1	23–69	6.887 1	1 162.1	219.9
3-Methyl-1-butyne	1	−55 to 47	6.884 80	1 014.81	227.11
2-Methyl-3-butyn-2-ol	1	21–106	6.657 5	976.5	154.1
Methyl *n*-butyrate	1		6.972 11	1 272.73	208.5
Methyl caprate	1	107–188	7.190 0	1 783.8	181.6
Methyl caproate	1	44–105	7.409 3	1 672.74	218.98
Methyl caprylate	1	100–146	6.916 5	1 496.3	176.5
Methyl carbitol	1	112–193	7.424	1 751	192
Methyl cellosolve acetate	1	70–144	7.125 1	1 447.0	196.1
Methyl chloroacetate	1	45–130	7.004 4	1 306.3	187.3
Methylcyclohexane	1	−3 to 127	6.823 00	1 270.763	221.42
Methylcyclopentane	1	−24 to 96	6.862 83	1 186.059	226.04
Methyldichlorosilane	1	1–41	7.027 8	1 167.8	240.7
1-Methyl-2-ethylbenzene	1	48–194	7.003 14	1 535.374	207.30
1-Methyl-3-ethylbenzene	1	46–190	7.015 82	1 529.184	208.51
1-Methyl-4-ethylbenzene	1	46–191	6.998 02	1 527.113	208.92
1-Methyl-1-ethylcyclopentane	1	43–122	6.859 20	1 347.602	217.21
1-Methyl-2-ethylcyclo-pentane *cis*	1	49–129	6.905 88	1 388.412	216.89
2-Methyl-3-ethylpentane	1		6.867 31	1 318.12	215.31
3-Methyl-3-ethylpentane	1		6.867 31	1 347	219.68
3-Methyl-5-ethylphenol	1	111–233	7.958	2 236	208
2-Methyl-5-ethylpyridine	1	52–177	5.050	517	59
N-Methylformamide	1	96–200	7.497 4	1 849.4	201.1
Methyl formate	1	21–32	3.027	3.02	−11.9
2-Methylheptane	1	42–119	6.917 35	1 337.47	213.69
3-Methylheptane	1	43–120	6.899 44	1 331.53	212.41
4-Methylheptane	1		6.900 65	1 327.66	212.57
2-Methylhexane	1	−9 to 115	6.873 18	1 236.026	219.55
3-Methylhexane	1	−8 to 117	6.867 64	1 240.196	219.22
Methylhydrazine	1	2–25	6.576 2	1 007.5	181.4
N-Methylhydroxylamine	1	40–65	7.045 6	1 223.3	172.1
O-Methylhydroxylamine	1	−63 to 48	7.363 9	1 225.3	225.2
Methyl isobutyl ketone	1	22–116	6.672 7	1 168.4	191.9
1-Methyl-2-isopropylbenzene	1	liq	6.940 4	1 548.05	203.15
1-Methyl-3-isopropylbenzene	1	liq	6.940 5	1 539.05	203.93
1-Methyl-4-isopropylbenzene	1	liq	6.923 7	1 537.06	203.05
3-Methylisoquinoline	1	176–225	6.969 2	1 717.3	166.9
Methyl isothiocyanate	1	10–50	2.896 8	103.6	45.4
Methyl laurate	1	158–212	6.767 1	1 589.72	140.5
Methyl linolate	1	166–206	6.111 1	1 660.1	118.8
Methyl methacrylate	1	39–89	8.409 2	2 050.5	274.4
Methyl myristate	1	166–238	7.622 3	2 283.93	184.8
1-Methylnaphthalene	1	108–278	7.035 92	1 826.948	195.00
2-Methylnaphthalene	1	105–274	7.068 50	1 840.268	198.40
Methyl oleate	1	166–205	7.544 1	2 656.9	200.7
Methyl palmitate	1	148–202	9.594 4	4 146.43	297.76
2-Methylpentane	1	−32 to 83	6.839 10	1 135.410	226.57
3-Methylpentane	1	−30 to 87	6.848 87	1 152.368	227.13

Table 10-8 (*Continued*)
VAPOR PRESSURES OF VARIOUS ORGANIC COMPOUNDS

Substance		Eq.	Range, °C	A	B	C
2-Methyl-2-pentanethiol		1	56–165	6.858 5	1 343.79	212.8
2-Methyl-1-pentanol		1	25–150	7.520 1	1 564.7	189.2
2-Methyl-4-pentanol		1	25–133	8.467 1	2 174.9	257.8
2-Methyl-1-pentene		1	−30 to 85	6.850 30	1 138.516	224.70
3-Methyl-1-pentene		1	−38 to 77	6.755 23	1 086.316	226.20
4-Methyl-1-pentene		1	−38 to 77	6.835 29	1 121.302	229.687
2-Methyl-2-pentene		1	−26 to 90	6.923 67	1 183.837	225.51
3-Methyl-2-pentene	*cis*	1	−26 to 91	6.910 73	1 186.402	226.70
	trans	1	−23 to 94	6.926 34	1 194.527	224.83
4-Methyl-2-pentene	*cis*	1	−35 to 79	6.841 29	1 120.707	226.59
	trans	1	−33 to 81	6.880 30	1 142.874	227.14
Methyl phenyl ether		1	110–164	7.052 69	1 489.99	203.57
2-Methylpiperidine		1	51–158	6.818 59	1 274.61	205.40
2-Methylpropane		1	−87 to 7	6.910 48	946.35	246.68
2-Methyl-1-propanethiol		1	−10 to 113	6.887 46	1 237.282	220.31
2-Methyl-2-propanethiol		1	1–88	6.787 81	1 115.565	221.31
2-Methyl-1-propanol		1	20–115	7.327 05	1 248.48	172.92
2-Methyl-2-propanol		1	20–103	7.319 94	1 154.48	117.65
2-Methylpropene		1	−82 to 12	6.684 66	866.25	234.64
N-Methylpropionamide		1	30–90	−0.9103	119.4	−148.0
Methyl propionate		1	21–79	6.942 4	1 170.2	208.8
2-Methyl-2-propylamine		1	19–75	6.783 2	993.33	210.50
Methyl propyl ether		1	0–39	6.118 6	708.69	179.9
2-Methylpyridine		1	80–168	7.032 4	1 415.73	211.63
3-Methylpyridine		1	74–185	7.050 21	1 481.78	211.25
4-Methylpyridine		1	75–186	7.041 77	1 480.68	210.50
1-Methylpyrrole		1	49–149	7.085 0	1 368.66	212.80
6-Methylquinoline		1	187–266	6.927 2	1 746.08	166.46
7-Methylquinoline		1	238–258	7.597 7	2 229.4	214.9
Methyl salicylate		1	79–220	7.083 3	1 712.8	187.1
Methyl stearate		1	204–240	2.357 0	68.92	−156.5
o-Methylstyrene		1	32–112	7.212 9	1 664.08	214.59
		1	75–255	6.884 61	1 485.41	200.0
m-Methylstyrene		1	10–72	7.275 34	1 695.4	220.0
		1	72–250	6.879 28	1 471.44	200.0
p-Methylstyrene		1	68–170	7.011 2	1 535.1	200.7
α-Methylstyrene		1		6.923 66	1 486.88	202.4
β-Methylstyrene		1		6.923 39	1 499.80	201.0
Methyl sulfoxide		1	20–50	7.763 7	2 048.7	231.6
3-Methyl-2-thiabutane		1	−13 to 109	6.901 96	1 232.170	221.67
2-Methylthiacyclopentane		1	liq	6.944 12	1 409.503	214.41
3-Methylthiacyclopentane		1	67–179	6.949 1	1 431.8	213.6
2-Methyl-3-thiapentane		1	liq	6.891 30	1 293.05	215.04
Methyl-2-thiazole		1	80–128	7.042 1	1 407.05	209.33
2-Methylthiophene		1	9–138	6.938 97	1 326.48	214.31
3-Methylthiophene		1	11–141	6.986 11	1 363.83	216.78
Methyl trichlorosilane		1	13–64	7.088 2	1 289.2	239.9
2-Methyl-5-vinylpyridine		1	69–183	6.156	1 023	129
Morpholine		1	0–44	7.718 13	1 745.8	235.0
		1	44–170	7.160 30	1 447.70	210.0
Naphthalene	c	1	86–250	7.010 65	1 733.71	201.86
	liq	1	125–218	6.818 1	1 585.86	184.82
1-Naphthol		1	141–282	7.284 21	2 077.56	184.0
2-Naphthol		1	144–288	7.347 14	2 135.00	183.0
Nicotine		1	134–246	6.789	1 650	176

Table 10-8 (*Continued*)
VAPOR PRESSURES OF VARIOUS ORGANIC COMPOUNDS

Substance	Eq.	Range, °C	A	B	C
o-Nitroaniline	2	150–260	8.868 4	3 336.50	
m-Nitroaniline	2	170–260	8.818 8	3 440.9	
p-Nitroaniline	2	190–260	9.559 5	4 039.73	
Nitrobenzene	1	134–211	7.115 6	1 746.6	201.8
m-Nitrobenzotrifluoride	1	10–105	7.653 15	2 006.1	220.0
	1	104–280	7.180 25	1 710.60	195.12
Nitromethane	1	56–136	7.281 66	1 446.94	227.60
1-Nitropropane	1	59–131	7.114 6	1 467.45	215.23
o-Nitrotoluene	1	129–222	5.851	946	96
p-Nitrotoluene	1	148–233	6.994 8	1 720.39	184.9
Nonadecane	1	184–366	7.015 3	1 932.8	137.6
1-Nonadecene	1	liq	7.115 1	1 997.4	142.7
Nonafluorocyclopentane	1	17–75	6.945 3	1 051.7	220.1
Nonane	1	39–179	6.938 93	1 431.82	202.01
1-Nonanethiol	1	93–251	6.983 9	1 655.6	183.7
Nonanoic acid	1	137–177	3.235 9	143.97	−75.6
1-Nonanol	1	94–214	7.827 8	1 953.8	181.9
1-Nonene	1	35–175	6.954 30	1 436.20	205.69
Octadecane	1	172–352	7.002 2	1 894.3	143.30
1-Octadecanethiol	1	liq	7.096	2 061	129
1-Octadecanol	1	120–218	6.461 6	1 599	90
1-Octadecene	1		7.060 65	1 997.4	147.50
Octane	1	19–152	6.918 68	1 351.99	209.15
1-Octanethiol	1	76–229	6.969 09	1 593.0	190.61
1-Octanol	1	0–80	12.070 1	4 506.8	319.9
	1	70–195	6.837 90	1 310.62	136.05
2-Octanol	1	72–180	6.388 8	1 060.4	122.5
3-Octanol	1	76–176	5.221 5	560.3	64.7
4-Octanol	1	71–176	5.739 6	760.5	89.5
1-Octene	1	15–147	6.934 95	1 355.46	213.05
5-Oxyhydrindene	1	120–251	9.213 7	3 665.8	326.4
Pentachloroethane	1	25–162	6.740	1 378	197
Pentadecane	1	136–304	7.023 59	1 789.95	161.38
1-Pentadecene	1		7.022 91	1 788.58	163.347
1,2-Pentadiene	1	−42 to −26	7.259 90	1 250.293	241.96
	1	−21 to 67	6.918 20	1 104.991	228.85
1,3-Pentadiene cis	1	−43 to −22	7.193 87	1 223.602	240.62
	1	−18 to 66	6.910 89	1 101.923	229.37
trans	1	−45 to −20	7.102 12	1 185.389	239.41
	1	−18 to 64	6.913 17	1 103.840	231.72
1,4-Pentadiene	1	−57 to −37	7.174 01	1 155.378	244.30
	1	−33 to 47	6.835 43	1 017.995	231.46
2,3-Pentadiene	1	−39 to −18	7.202 53	1 231.768	237.56
	1	−14 to 70	6.962 16	1 126.837	227.84
Pentafluorobenzene	1	49–94	7.036 65	1 254.07	216.02
Pentafluorochloroacetone	1	−40 to 32	6.848 4	925.3	225.4
Pentafluorochloroethane	1	−95 to −39	6.833 34	802.97	242.27
Pentafluorophenol	1	105–155	7.066 0	1 379.15	183.91
2,2,3,3,3-Pentafluoropropanol	1	0–23	6.308 7	830.56	153.8
Pentafluorotoluene	1	39–138	7.084 78	1 392.20	213.67
bis-Pentamethyldisilanoxydi- silane	1	169–201	8.556 64	3 051.316	258.85
bis-Pentamethyldisilanyl ether	1	88–183	8.161 44	2 575.250	273.32
Pentane	1	−50 to 58	6.852 96	1 064.84	233.01
Pentanenitrile	1	69–141	7.104 9	1 519.4	218.4

Table 10-8 (*Continued*)
VAPOR PRESSURES OF VARIOUS ORGANIC COMPOUNDS

Substance	Eq.	Range, °C	A	B	C
1-Pentanethiol	1	19–153	6.933 11	1 369.479	211.31
Pentanoic acid	1	72–174	5.412	591	60
1-Pentanol	1	37–138	7.177 58	1 314.56	168.11
2-Pentanol	1	25–120	7.275 75	1 271.92	170.37
3-Pentanol	1	21–116	7.414 93	1 354.42	183.41
2-Pentanone	1	56–111	7.021 93	1 313.85	215.01
3-Pentanone	1	56–111	7.025 29	1 310.28	214.19
1-Pentene	1	−55 to 51	6.844 24	1 044.01	233.50
2-Pentene *cis*	1	−49 to 58	6.843 08	1 052.44	228.69
trans	1	−49 to 58	6.899 83	1 080.76	232.57
1-Pentyne	1	−44 to 61	6.967 34	1 092.52	227.18
2-Pentyne	1	−33 to 78	7.046 14	1 189.87	229.60
Perdeuterobenzene	1	10–82	6.892 35	1 198.39	219.43
Perdeuterocyclohexane	1	10–80	6.837 86	1 190.38	222.40
Perfluorobutane	1	−39 to −4	7.035 1	990.27	240.4
Perfluorobutene	1	−28 to 20	9.222	2 401.6	382
Perfluorocyclobutane	1	−32 to 0	6.815 29	862.49	225.19
Perfluorocyclohexane	1	19–65	6.04	597	136
Perfluorocyclopentane	1	17–56	7.039 6	1 069.3	234.6
Perfluoroheptane	1	−2 to 106	6.937 72	1 181.14	208.66
Perfluorohexane	1	30–57	6.875 2	1 080.8	213.4
Perfluoromethylcyclohexane	1	33–111	6.824 06	1 133.76	211.22
Perfluorooctane	1	37–105	5.902 5	1 225.93	198.99
Perfluoropentane	1	9–65	7.017 9	1 072.9	230.0
Perfluoropiperidine	1	29–81	6.853 4	1 059.95	217.2
Perfluoropropane	1	−79 to −36	6.919 4	825.8	241.2
Perfluoropropene	1	−41 to 20	7.355	1 012.1	257
Phenanthrene	1	176–379	7.260 82	2 379.04	203.76
Phenol	1	107–182	7.133 0	1 516.79	174.95
β-Phenylethyl acetate	1	149–233	6.834 3	1 555.2	160.8
α-Phenylethyl alcohol	1	82–190	1.508	91	−263
o-Phenylethylphenol	1	169–250	4.506 0	516.8	−32.1
p-Phenylethylphenol	1	174–251	4.304 1	459.3	−52.4
Phenylisocyanate	1	10–80	−0.708 0	106.4	−146.6
4-Phenylphenol	1	177–308	8.657 5	3 022.8	216.1
Phosgene	1	−68 to 68	6.842 97	941.25	230
Phthalic anhydride	2	160–285	8.022	2 868.5	
α-Pinene	1	19–156	6.852 5	1 446.4	208.0
β-Pinene	1	19–166	6.898 4	1 511.7	210.2
Piperidine	1	42–144	6.855 69	1 238.80	205.43
Propadiene	1	−99 to −16	5.713 7	458.06	196.07
Propane	1	−108 to −25	6.803 38	804.00	247.04
1-Propanethiol	1	−25 to 91	6.928 46	1 183.307	224.62
2-Propanethiol	1	−37 to 75	6.877 34	1 113.895	226.16
1-Propanol	1	2–120	7.847 67	1 499.21	204.64
2-Propanol	1	0–101	8.117 78	1 580.92	219.61
2-Propen-1-ol	1	21–97	11.187 0	4 068.5	392.7
Propionic acid	1	56–139.5	6.403	950.2	130.3
Propionic anhydride	1	67–167	5.819 5	810.3	108.7
Propionitrile	1	−84 to 22	5.278 2	665.52	159.10
Propiophenone	1	132–201	7.370	1 894	205
Propyl acetate	1	39–101	7.016 15	1 282.28	208.60
1-Propylamine	1	23–77	6.926 51	1 044.05	210.84
2-Propylamine	1	4–61	6.890 25	985.69	214.07
n-Propylbenzene	1	43–188	6.951 42	1 491.297	207.14

Table 10-8 (*Continued*)
VAPOR PRESSURES OF VARIOUS ORGANIC COMPOUNDS

Substance	Eq.	Range, °C	A	B	C
n-Propyl borate	1	85–179	7.399 8	1 741	206
n-Propyl caprate	1	97–186	8.701 22	2 945.99	253.63
n-Propyl caproate	1	43–120	8.667 1	2 556.0	262.9
n-Propyl caprylate	1	70–153	8.516 7	2 599.5	246.2
n-Propyl cellosolve	1	77–149	7.146 4	1 440.6	187.7
n-Propylcyclohexane	1	40–186	6.886 46	1 460.800	207.94
n-Propylcyclopentane	1	21–158	6.903 92	1 384.386	213.16
Propylene	1	−112 to −32	6.778 11	770.85	245.51
1,2-Propylene oxide	1	−35 to 130	7.064 92	1 113.6	232
n-Propyl formate	1	26–82	6.848	1 127	203
n-Propyl laurate	1	124–205	8.068 9	2 692.4	222.5
n-Propyl myristate	1	147–200	9.216 8	3 744.68	272.87
n-Propyl nitrate	1	0–70	6.954 9	1 294.4	206.7
n-Propyl palmitate	1	166–204	14.129 2	9 759.2	539.7
o-(n-Propyl)phenol	1	104–222	9.215	3 254	292
p-(n-Propyl)phenol	1	0–234	8.329 6	2 661	254
n-Propyl phenyl ether	1	101–190	7.734 3	2 146.2	252.3
Propyne	1	−90 to −6	6.784 85	803.73	229.08
Pseudocumenol	1	107–232	6.915	1 547	152
Pyrene	1	200–395	5.618 4	1 122.0	15.2
Pyridine	1	67–153	7.041 15	1 373.80	214.98
Pyrogallol	1	177–309	6.092	1 031	12
Pyrrole	1	66–166	7.294 70	1 501.56	210.42
Quinaldine	1	178–248	7.179 00	1 857.84	184.50
Quinoline	1	164–238	6.817 59	1 668.73	186.26
Spiropentane	1	3–71	6.917 00	1 090.08	231.10
Styrene	1	32–82	7.140 16	1 574.51	224.09
Terpenyl acetate	1	37–150	6.443 46	1 377.27	143.85
α-Terpineol	1	84–217	8.141 2	2 479.4	253.7
Terpinolene	1	40–179	7.169	1 706	211
Tetrabutyl tin	1	100–300	6.545	1 649	148
1,1,2,2-Tetrachloro-1,2-difluoroethane	1	10–91.5	10.995	4 437.1	455.2
1,1,1,2-Tetrachloroethane	1	59–130	6.898 75	1 365.88	209.74
1,1,2,2-Tetrachloroethane	1	25–130	6.631 7	1 228.1	179.9
Tetrachloroethylene	1	37–120	6.976 83	1 386.92	217.53
Tetrachloromethane	1		8.879 26	1 212.021	226.41
Tetradecane	1	122–286	7.013 00	1 740.88	167.72
1-Tetradecanethiol	1		7.048 5	1 909.2	151.9
1-Tetradecanol	1	130–264	6.674 1	1 204.5	54.0
1-Tetradecene	1	119–283	7.030 65	1 754.09	171.52
1,2,3,4-Tetrafluorobenzene	1	6–50	7.084 6	1 339.23	223.49
1,2,3,5-Tetrafluorobenzene	1	6–50	6.986 17	1 245.20	218.35
Tetrafluoroethylene	1	−131 to −65	6.896 59	683.84	245.93
Tetrafluoromethane	1		6.972 31	540.50	260.10
Tetrahydrofuran	1	23–100	6.995 15	1 202.29	226.25
Tetraiodothiophene	1	−65 to 24	5.585 44	871.25	175.59
Tetralin	1	94–206	7.070 55	1 741.30	208.26
1,2,3,4-Tetramethylbenzene	1	80–217	7.059 4	1 690.54	199.48
1,2,3,5-Tetramethylbenzene	1	75–228	7.077 9	1 675.43	201.14
1,2,4,5-Tetramethylbenzene	1	74–227	7.080 0	1 672.43	201.43
2,2,3,3-Tetramethylbutane	1	0–65	6.876 65	1 329.93	226.36
Tetramethyl lead	1	0–60	6.937 7	1 335.3	219.1
2,2,3,3-Tetramethylpentane	1	57–141	6.830 60	1 398.67	213.84

Table 10-8 (*Continued*)
VAPOR PRESSURES OF VARIOUS ORGANIC COMPOUNDS

Substance	Eq.	Range, °C	A	B	C
2,2,3,4-Tetramethylpentane	1	52–134	6.834 18	1 375.59	214.94
2,2,4,4-Tetramethylpentane	1	43–123	6.796 20	1 324.59	216.02
Tetramethylsilane	1	−64 to 21	6.822 39	1 033.72	235.62
2-Thiabutane	1	−26 to 90	6.938 49	1 182.562	224.78
Thiacyclobutane	1	−5 to 120	7.016 67	1 321.331	224.51
Thiacyclohexane	1	29–170	6.905 18	1 422.47	211.72
Thiacyclopentane	1	14–148	6.995 40	1 401.939	219.61
Thiacyclopropane	1	−35 to 77	7.037 25	1 194.37	232.42
3-Thiaheptane	1	33–172	6.941 02	1 421.32	205.81
4-Thiaheptane	1	32–170	6.935 77	1 413.44	205.73
2-Thiahexane	1	17–150	6.945 83	1 363.808	212.07
3-Thiahexane	1	14–144	6.933 80	1 341.57	212.51
2-Thiapentane	1	−4 to 120	6.955 45	1 284.32	219.66
3-Thiapentane	1	−13 to 109	6.928 36	1 257.833	218.66
2-Thiapropane	1	−47 to 58	6.948 79	1 090.755	230.80
Thiazole	1	63–118	7.142 01	1 425.35	216.26
Thiophene	1	−12 to 108	6.959 26	1 246.02	221.35
Toluene	1	6–137	6.954 64	1 344.800	219.48
o-Toluidine	1	118–200	7.082 03	1 627.72	187.13
m-Toluidine	1	122–203	7.093 67	1 631.43	183.91
p-Toluidine	1		7.260 22	1 758.55	201.0
m-Tolyl pentafluoro-propionate	1	98–174	7.427 20	1 707.59	201.70
p-Tolyl pentafluoro-propionate	1	99–176	8.078 6	2 223.8	252.1
m-Tolyl trifluoroacetate	1	91–166	7.681 0	1 874.84	223.48
p-Tolyl trifluoroacetate	1	92–169	7.913 8	2 055.41	238.99
Tribromomethane	1	30–101	6.821 8	1 376.7	201.0
1,2,3-Tribromopropane	1	128–205	7.037 2	1 735.32	195.42
Trichloroacetic acid	1	112–198	7.273 0	1 594.3	165.4
Trichloroacetonitrile	1	17–83	7.183 5	1 368.3	232.5
Trichloroacetyl chloride	1	32–119	6.990 75	1 390.47	220.11
1,1,1-Trichloroethane	1	−6 to 17	8.643 4	2 136.6	302.8
1,1,2-Trichloroethane	1	50–114	6.951 85	1 314.41	209.20
Trichloroethylene	1	18–86	6.518 3	1 018.6	192.7
Trichlorofluoromethane	1		6.884 28	1 043.004	236.88
Trichlorosilane	1	2–32	6.773 9	1 009.0	227.2
bis-Trichlorosilylethane	1	91–160	7.835 11	2 241.769	249.84
1,1,1-Trichloro-2,2,2-trifluoroethane	1	14–36	4.437 3	204.1	83.9
1,1,2-Trichloro-1,2,2-trifluoroethane	1	−25 to 83	6.880 3	1 099.9	227.5
Tridecane	1	107–267	7.007 56	1 690.67	174.22
1-Tridecene	1	105–264	6.981 02	1 672.00	174.95
Triethanolamine	1	252–305	10.067 5	4 542.78	297.76
Triethyl aluminum	1	57–126	11.646 1	4 466.59	322.87
Triethylamine	1	50–95	5.858 8	695.7	144.8
Triethyl borate	1	29–109	7.511 1	1 641.7	236.3
Triethylsilanol	1	24–140	7.793 7	1 756.1	202.4
Trifluoroacetic acid	1	12–72	6.147 76	1 228.60	216.09
Trifluoroacetic anhydride	1	−2 to 39	6.135 8	1 026.1	202.0
Trifluoroacetonitrile	1	−132 to −68	7.127 6	773.82	249.9
1,3,5-Trifluorobenzene	1	6–50	6.919 8	1 197.13	219.12
Trifluorochloroethylene	1	−67 to −11	6.896 16	848.33	239.64
1,1,1-Trifluoroethane	1	−110 to −48	6.903 78	788.20	243.23

Table 10-8 (*Continued*)
VAPOR PRESSURES OF VARIOUS ORGANIC COMPOUNDS

Substance	Eq.	Range, °C	A	B	C
2,2,2-Trifluoroethanol	1	−0.5 to 25	6.788 2	978.13	173.06
Trifluoromethane	1	−128 to −82	7.088 6	705.33	249.78
bis-(Trifluoromethyl)-acetoxyphosphine	1	0–40	7.391 31	1 426.254	220.37
2,2,2-Trifluoro-1-methyl-benzene	1	55–139	6.970 45	1 306.35	217.38
bis-(Trifluoromethyl)-chlorophosphine	1	−80 to 0	7.661 06	1 386.652	267.14
Trifluoromethylhypofluorite	1	145–189	6.950 6	650.1	−18.4
bis-(Trifluoromethyl)-iodophosphine	1	0–47	6.901 39	1 180.723	222.95
Triisobutylene	1	56–179	7.002 1	1 613.47	212.5
Trimethyl aluminum	1	64–127	7.570 29	1 734.72	242.78
Trimethylamine	1	−80 to 3	6.857 55	955.94	237.52
1,2,3-Trimethylbenzene	1	57–205	7.040 82	1 593.958	207.08
1,2,4-Trimethylbenzene	1	52–198	7.043 83	1 573.267	208.56
1,3,5-Trimethylbenzene	1	49–193	7.074 36	1 569.622	209.58
2,2,3-Trimethylbutane	1	−19 to 106	6.792 30	1 200.563	226.05
Trimethylchlorosilane	1	2–55	7.055 8	1 245.5	240.7
1,1,3-Trimethylcyclohexane	1	55–137	6.839 51	1 394.88	215.73
1,1,2-Trimethylcyclopentane	1	36–115	6.822 38	1 309.81	218.58
1,1,3-Trimethylcyclopentane	1	29–106	6.809 31	1 275.92	219.89
1,2,4-Trimethylcyclopentane					
cis, cis, trans	1	39–118	6.857 38	1 335.69	219.16
cis, trans, cis	1	33–110	6.851 3	1 307.10	219.92
1,3,5-Trimethyl-2-ethyl-benzene	1	88–210	6.790 8	1 505.8	174.7
1,4,5-Trimethyl-2-ethyl-benzene	1	87–132	3.029 3	116.4	−34.6
2,2,5-Trimethylhexane	1	46–125	6.837 75	1 325.54	210.91
2,4,4-Trimethylhexane	1	51–131	6.856 54	1 371.81	214.40
Trimethylhydrazine	1	−16 to 14	7.106 80	1 189.88	222.06
O,N,N-Trimethylhydroxylamine	1	−79 to 23	6.765 8	979.55	222.2
2,2,3-Trimethylpentane	1		6.825 46	1 294.88	218.42
2,2,4-Trimethylpentane	1	24–100	6.811 89	1 257.84	220.74
2,3,3-Trimethylpentane	1		6.843 53	1 328.05	220.38
2,3,4-Trimethylpentane	1	36–114	6.853 96	1 315.08	217.53
2,4,4-Trimethyl-1-pentene	1	−3 to 128	6.834 57	1 273.416	220.62
2,4,4-Trimethyl-2-pentene	1	2–131	6.859 22	1 272.717	214.99
2,3,5-Trimethylphenol	1	186–247	7.080 12	1 685.90	166.14
Trimethylsilanol	1	18–85	8.126 6	1 657.6	219.2
2,4,5-Trimethylstyrene	1	79–216	7.331 5	1 880.7	205.7
2,4,6-Trimethylstyrene	1	90–208	7.089 1	1 702.61	195.93
1,2,4-Trinitrobenzene	1	250–300	3.194	87	−199
1,3,5-Trinitrobenzene	1	202–312	5.534 5	993.6	11.2
2,4,6-Trinitrobenzene	1	249–342	9.621 1	4 987.9	329.9
2,4,6-Trinitrotoluene	1	230–250	7.671 52	2 669.4	205.6
α-Trioxane	1	56–114	7.818 6	1 783.3	247.1
Trivinylarsine	1	22–66	7.894 1	2 115.6	293.9
Trivinyl bismuth	1	20–74	7.237 2	1 667.0	215.1
Trivinylphosphine	1	16–61	7.928 4	2 110.0	301.3
Trivinylstibine	1	20–70	8.322 1	2 446.3	303.8
Undecane	1	75–226	6.972 20	1 569.57	187.70
1-Undecanethiol	1		7.012 2	1 767.4	170.4
1-Undecene	1	72–222	6.966 77	1 563.21	189.87

Table 10-8 (*Continued*)
VAPOR PRESSURES OF VARIOUS ORGANIC COMPOUNDS

Substance	Eq.	Range, °C	A	B	C
Urethane	1		7.421 64	1 758.21	205.0
Vinyl acetate	1	22–72	7.210 1	1 296.13	226.66
o-Xylene	1	32–172	6.998 91	1 474.679	213.69
m-Xylene	1	28–166	7.009 08	1 462.266	215.11
p-Xylene	1	27–166	6.990 52	1 453.430	215.31
2,3-Xylenol	1	149–218	7.053 97	1 617.57	170.74
2,4-Xylenol	1	144–212	7.055 39	1 587.46	169.34
2,5-Xylenol	1	144–212	7.051 56	1 592.70	170.74
2,6-Xylenol	1	145–204	7.070 70	1 628.32	187.60
3,4-Xylenol	1	172–229	7.079 19	1 621.45	159.26
3,5-Xylenol	1	155–223	7.130 76	1 639.86	164.16

BOILING POINTS

Table 10-9
BOILING POINTS OF WATER

A. Barometric Pressures at Various Temperatures

Temp. °C.	0.0° mm of Hg	0.2° mm of Hg	0.4° mm of Hg	0.6° mm of Hg	0.8° mm of Hg
80	355.40	358.28	361.19	364.11	367.06
81	370.03	373.01	376.02	379.05	382.09
82	385.16	388.25	391.36	394.49	397.64
83	400.81	404.00	407.22	410.45	413.71
84	416.99	420.29	423.61	426.95	430.32
85	433.71	437.12	440.55	444.01	447.49
86	450.99	454.51	458.06	461.63	465.22
87	468.84	472.48	476.14	479.83	483.54
88	487.28	491.04	494.82	498.63	502.46
89	506.32	510.20	514.11	518.04	521.99
90	525.97	529.98	534.01	538.07	542.15
91	546.26	550.40	554.56	558.75	562.96
92	567.20	571.47	575.76	580.08	584.43
93	588.80	593.20	597.63	602.09	606.57
94	611.08	615.62	620.19	624.79	629.41
95	634.06	638.74	643.45	648.19	652.96
96	657.75	662.58	667.43	672.32	677.23
97	682.18	687.15	692.15	697.19	702.25
98	707.35	712.47	717.63	722.81	728.03
99	733.28	738.56	743.87	749.22	754.59
100	760.00	765.44	770.91	776.42	781.95

B. Boiling Points of Water at Various Pressures

Pressure, atm.	Boiling Point, °C.	Pressure, atm.	Boiling Point, °C.	Pressure, atm.	Boiling Point, °C.	Pressure, atm.	Boiling Point, °C.
0.5	80.9	7	164.2	14	194.1	21	213.9
1	100.0	8	169.6	15	197.4	22	216.2
2	119.6	9	174.5	16	200.4	23	218.5
3	132.9	10	179.0	17	203.4	24	220.8
4	142.9	11	183.2	18	206.1	25	222.9
5	151.1	12	187.1	19	208.8	26	225.0
6	158.1	13	190.7	20	211.4	27	227.0

Table 10-10
CORRECTION OF BOILING POINT TEMPERATURE AT
VARIOUS PRESSURES TO NORMAL PRESSURE (760 mm. Hg)

To correct for small differences in barometric pressure the following formula is employed:

$$T_c = T_o + C(760 - P_{mm})*$$

where T_c = corrected boiling point.
T_o = observed boiling point.
P = atmospheric pressure in millimeters of mercury.
C = a constant having the value of 0.037 at $25 - 40°C$; 0.043 at $41 - 75°C$; 0.044 at $76 - 100°C$; 0.046 at $101 - 120°C$; 0.048 at $121 - 140°C$; 0.051 at $141 - 155°C$; 0.055 at $156 - 220°C$; 0.057 at $221 - 300°C$; 0.064 at $310 - 325°C$.

A. S. T. M. METHOD

In the *A. S. T. M. Standards on Petroleum Products and Lubricants*, September 1937, page 96, the Sydney Young equation is used for calculating the proper correction to the observed boiling point. In this equation the temperature range as well as the pressure is considered. The Sydney Young equation is as follows:

For Centigrade scale: $C_c = 0.00012 (760 - P) (273 + t_c)$,
For Fahrenheit scale: $C_f = 0.00012 (760 - P) (460 + t_f)$

in which C_c and C_f are, respectively, corrections to be made on the observed temperatures t_c or t_f, and P is the actual barometric pressure in millimeters of mercury.

The table which follows is a convenient approximation of the corrections as calculated from the equations above. These approximations will suffice for ordinary work. For more precise work, the correction should be calculated using the Sydney Young equation and the temperature when 50% of the material has boiled. A new correction should be made for each variation of more than 10°C. or 18°F.

As pointed out by Hoyt, *Jour. Chem. Ed.* 11, *405 (1934)*, the Sydney Young equation is applicable only to non-polar substances with low dielectric constants such as the hydrocarbons for which it was proposed by the *A. S. T. M.* For other types of compounds a more exact value can be obtained by use of the following equations and constants:

For Centigrade scale: $C_c = K (760 - P) (273 + t_c)$,
For Fahrenheit scale: $C_f = K (760 - P) (460 + t_f)$,

where the value of the constant K is for:

Hydrocarbons	0.000125	Esters	0.000121
Halogen derivatives	0.000125	Ketones	0.000121
Ethers	0.000125	Amines	0.000118
Aldehydes	0.000125	Alcohols	0.000100

The variation in the case of acids is too wide to permit of a single value for K.

* This formula is applicable only to liquids of particular types such as a hydrocarbon, an ether, an aldehyde, a ketone, an ester, or a halogen derivative.

Table 10-10 (*Continued*)
CORRECTION OF BOILING POINT TEMPERATURES

Boiling Point		Correction for Each One mm. Difference in Pressure*	
t_c in °C.	t_f in °F.	C_c in °C.	C_f in °F.
10–30	50–86	0.035	0.063
30–50	86–122	.038	.068
50–70	122–158	.040	.072
70–90	158–194	.042	.076
90–110	194–230	.045	.081
110–130	230–266	.047	.085
130–150	266–302	.050	.089
150–170	302–338	.052	.094
170–190	338–374	.054	.098
190–210	374–410	.057	.102
210–230	410–446	.059	.106
230–250	446–482	.062	.111
250–270	482–518	.064	.115
270–290	518–554	.066	.119
290–310	554–590	.069	.124
310–330	590–626	.071	.128
330–350	626–662	.074	.132
350–370	662–698	.076	.137
370–390	698–734	.078	.141
390–410	734–770	.081	.145

* To be added in case barometric pressure is below 760 mm.; to be subtracted in case the pressure is above 760 mm.

DREISBACH METHOD

A formula (2) has been developed from the Antoine equation (1) which gives the rate of change of boiling point with pressure:

$$(1) \quad \log p = A - [B \div (t^0\mathrm{C} + 230)]$$
$$(2) \quad dt/dp = B \div [2.3026p(A - \log p)^2]$$

This equation will give an accurate value for dt/dp at any pressure but cannot be used to correct boiling points at various barometric readings to boiling points at 760 mm pressure since the dt/dp value varies with pressure. For equations to correct for these variations and for the numerical values of the constants A and B for innumerable compounds belonging to 23 classes of organic compounds (hydrocarbons, alcohols, aldehydes, ketones, acids, esters, amines, etc.) see Dreisbach's *P-V-T Relationships of Organic Compounds*, published by Handbook Publishers, Inc., Sandusky, Ohio, U. S. A. (1952).

The use of this book of tables is simple and does not involve complicated mathematical operations. All of the computations have been done by the author. Its use involves merely the reading of temperature values in the body of the table corresponding to pressure values, ranging from 760 mm to 0.1 mm, which are shown across the top of the page. Values for latent heats of vaporization, vapor and liquid densities, index of refraction and flash points are also given.

This book of tables, which has been in constant use in the laboratories and plants of The Dow Chemical Co., will be of value wherever such data are required for not only compounds for which data are recorded in the literature but also for those compounds for which experimental values have not as yet been determined.

Table 10-11
CALCULATION OF BOILING POINTS OF ORGANIC COMPOUNDS

Cf. Kinney, *Jour. Am. Chem. Soc. 60, 3032 (1938); Ind. Eng. Chem. 32, 559 (1940).*

Boiling points of organic compounds may be calculated from their structures as follows: Obtain the molecular *B.P.N.* by adding together the appropriate boiling point numbers (*b.p.n.*) for the various atoms and structural groupings given in Table A. From the *B.P.N.* obtain the B.P. (boiling point in degrees Celsius) using Table B.

Certain rules must be followed in applying the data:

1. The *b.p.n.'s* of 0.8 and 1.0 are applied to the carbon and hydrogen atoms, respectively, of the longest aliphatic chain in the molecule. It is essential that the longest aliphatic chain be used as the base because the boiling point is dependent, in part, upon this factor.
2. All groups or atoms attached to this chain are assigned *b.p.n.'s* as indicated in Table A.
3. With many atoms and groups the position of the group, or the presence of other groups, alters the *b.p.n.* These, as far as are known, are given in Table A.

Example: To calculate the boiling point of 1-methyl-4-(1-methylethyl)-1-cyclopentene, $(CH_3)_2$: $CH \cdot CH \cdot CH_2 \cdot C(CH_3):CH \cdot CH_2$.

Carbon in the longest aliphatic chain (2×0.8)	1.6
Hydrogen in the longest aliphatic chain (4×1.0)	4.0
Methyl radical attached to the aliphatic chain	3.05
Carbon in the cyclopentene ring (5×0.8)	4.0
Hydrogen in the cyclopentene ring (6×1.0)	6.0
B.P.N. for the 5-membered ring	2.5
B.P.N. for the ethylenic linkage, type $R_2C:CHR$	2.3
Methyl attached to the cyclopentene ring	3.05
Calculated *B.P.N.*, sum of the items above	26.50
Calculated B.P., from Table B	142.5°C.
Observed B.P.	143.1°C.

Example: To calculate the boiling point of 2,8-dimethyl-5-nonanone, $[(CH_3)_2CH \cdot CH_2 \cdot CH_2]_2CO$.

Carbon in the longest aliphatic chain (9×0.8)	7.2
Hydrogen in the longest aliphatic chain (16×1.0)	16.0
Two methyl radicals (2×3.05)	6.1
One carbonyl oxygen, type RCH_2COCH_2R	7.5
Calculated *B.P.N.*, sum of the items above	36.8
Calculated B.P., from Table B	222.5°C.
Observed B.P.	226.0°C.

A. Atomic and Group Boiling Point Numbers

Carbon, in the main chain	0.8	Type of olefinic linkage		
Hydrogen, attached to the main		$CH_2{=}CH_2$	1.2	
chain	1.0	$RCH{=}CH_2$	1.5	
		$RCH{=}CHR$	1.9	
Radicals, saturated, attached to the		$R_2C{=}CHR$	2.3	
main chain or to cyclic rings		$R_2C{=}CR_2$	2.8	
Methyl	3.05			
Ethyl	5.5	Radicals, unsaturated, attached to		
Propyl	7.0	main chain		
Butyl	9.7	Methylene	4.4	
2,2-Dimethyl grouping	−0.4	Ethylidene	7.0	
		Vinyl	5.4	
		Propylidene	9.0	
Two or three alkyls attached to ad-		Butylidene	10.4	
jacent carbons of saturated main				
chains of 6 carbons or less	+0.5	Type of acetylenic linkage		
		$HC{\equiv}CH$	4.0	
		$RC{\equiv}CH$	4.4	
Four or more alkyls attached to ad-		$RC{\equiv}CCH_3$	5.4	
jacent carbons of saturated main		$RC{\equiv}CR$	4.8	
chains of 6 carbons or less	+1.0			

Table 10-11 (*Continued*)
CALCULATION OF BOILING POINTS OF ORGANIC COMPOUNDS

A. Atomic and Group Boiling Point Numbers

Type of diolefin			$R_2CHOCHR_2$, R_3COCHR_2	1.1
Allenes	4.8		R_3COCR_3	(0.2?)
Conjugated, normal values of double bonds plus	0.8		**Aldehyde** $=O$	
Not conjugated, normal values of double bonds only			RCH_2CHO	8.2
			R_2CHCHO	7.6
Type of triolefin			R_3CCHO	7.0
All bonds conjugated, normal values of double bonds plus	2.4		**Ketone** $=O$	
Two bonds conjugated, normal values of double bonds plus	0.8		RCH_2COCH_3	8.0
			$R_2CHCOCH_3$, RCH_2COCH_2R	7.5
No conjugation, normal values of double bonds only			R_3CCOCH_3, $R_2CHCOCH_2R$	7.0
			R_3CCOCH_2R, $R_2CHCOCHR_2$	6.5
			R_3CCOCH_2R	(6.0?)
Type of diacetylene			R_3CCOCR_3	(5.5?)
1,3-Diacetylenes, normal values of triple bonds only			**Ester** $-OO-$	
All other conjugated, normal values of triple bonds plus	3.0		RCH_2COOCH_3, CH_3COOCH_2R	8.5
No conjugation, normal values of triple bonds only			$R_2CHCOOCH_3$, RCH_2COOCH_2R, $HCOOCHR_2$, $CH_3COOCHR_2$	7.6
Type of enyne			$R_3CCOOCH_3$, $R_2CHCOOCH_2R$, $RCH_2COOCHR_2$, $HCOOCR_3$, CH_3COOCR_3	6.7
Conjugated, normal values of bonds plus	0.8		$R_3CCOOCH_2R$, $R_2CHCOOCHR_2$, RCH_2COOCR_3	5.8
No conjugation, normal values of bonds only			$R_3CCOOCHR_2$, $R_2CHCOOCR_3$	4.9
Type of dienyne			$R_3CCOOCR_3$	4.0
Conjugated, normal values of bonds plus	2.4		**Acid** $-COOH$	
No conjugation, normal values of bonds only			RCH_2COOH	19.3
			$R_2CHCOOH$	18.6
			R_3CCOOH	17.9
Cyclic radicals: Add 0.8 for each carbon, 1.0 for each hydrogen, the normal values of any unsaturated linkages, and the following values for the ring:			**Amine, primary** $-NH_2$	
			RCH_2NH_2	7.3
			R_2CHNH_2	6.2
			R_3CNH_2	5.1
Cyclo-propyl, -butyl, -pentyl, -hexyl, respectively	2.1, 2.3, 2.5, 2.7		**Amine, secondary** $-NH-$	
Cyclo-heptyl, -octyl, etc.,	3.4, 3.9, etc.		RCH_2NHCH_3	5.0
(add 0.5 for each additional CH_2 in the ring)			$R_2CHNHCH_3$, RCH_2NHCH_2R	4.0
			R_3CNHCH_3, $R_2CHNHCH_2R$	3.5
			R_3CNHCH_2R, $R_2CHNHCHR_2$	3.0
Alcohol $-OH$			**Amine, tertiary** $=N-$	
RCH_2OH	10.8		$RCH_2N(CH_3)_2$	2.0
R_2CHOH	8.8		$R_2CHN(CH_3)_2$, $(RCH_2)_2NCH_3$	1.5
R_3COH	6.8		$R_3CN(CH_3)_2$, $(R_2CH)_2NCH_3$, $(RCH_2)_3N$	1.25
Ether $-O-$			**Cyanide** $-CN$	
RCH_2OCH_3, R_2CHOCH_3, R_3COCH_3	2.9		RCH_2CN	14.0
			R_2CHCN	12.8
RCH_2OCH_2R, R_2CHOCH_2R, R_3COCH_2R	2.0		R_3CCN	11.6
			Isocyanide $-NC$	
			RCH_2NC	12.2
			R_2CHNC	11.1
			R_3CNC	10.0

Table 10-11 (*Continued*)
CALCULATION OF BOILING POINTS OF ORGANIC COMPOUNDS

B. Molecular Boiling Point Numbers and Their Boiling Points

Calculated from the formula: $B.P. = 230.14 \sqrt[3]{B.P.N.} - 543$

B. P. N.	B. P., ° C.	Av. Increase per 0.1 Unit, ° C.	B. P. N.	B. P., ° C.	Av. Increase per 0.1 Unit, ° C
5	−149.5	2.50	39	237.5	0.66
6	−124.5	2.18	40	244.1	0.65
7	−102.7	2.00	41	250.6	0.64
8	−82.7	1.84	42	257.0	0.63
9	−64.3	1.71	43	263.3	0.62
10	−47.2	1.60	44	269.5	0.61
11	−31.2	1.51	45	275.6	0.60
12	−16.1	1.43	46	281.6	0.59
13	−1.8	1.35	47	287.5	0.59
14	+11.7	1.29	48	293.4	0.58
15	24.6	1.23	49	299.2	0.57
16	36.9	1.19	50	304.9	0.56
17	48.8	1.14	51	310.5	0.55
18	60.2	1.10	52	316.0	0.54
19	71.1	1.06	53	321.5	0.55
20	81.7	1.03	54	326.9	0.53
21	92.0	0.99	55	332.2	0.53
22	101.9	0.96	56	337.5	0.52
23	111.5	0.94	57	342.7	0.52
24	120.9	0.90	58	347.9	0.50
25	129.9	0.89	59	352.9	0.51
26	138.8	0.86	60	358.0	0.50
27	147.4	0.85	61	363.0	0.49
28	155.9	0.82	62	367.9	0.49
29	164.1	0.80	63	372.8	0.48
30	172.1	0.79	64	377.6	0.47
31	180.0	0.76	65	382.3	0.47
32	187.6	0.76	66	387.0	0.48
33	195.2	0.74	67	391.8	0.45
34	202.6	0.73	68	396.3	0.47
35	209.9	0.70	69	401.0	0.45
36	216.9	0.70	70	405.5	0.45
37	223.9	0.68	71	410.0	
38	230.7	0.68			

Table 10-12
ORGANIC SOLVENTS ARRANGED BY BOILING POINTS

For methods of testing the purity, the purification, and the drying of many of the solvents listed below, the reader is referred to Riddick and Bunger, "Organic Solvents," in *Techniques of Chemistry*, 3d ed., A. Weissberger (ed.), Vol. II, Wiley-Interscience, New York, 1970.

Name	B. P. 760 mm °C	Name	B. P. 760 mm °C
Ethyl chloride	13	Trichloroethylene	87.2
Ethylene oxide	14	Isopropyl acetate	88.4
Furan	31–2	Isobutyl bromide	91.5
Methyl formate	32	2,5-Dimethyl-furan	93–4
Diethyl ether	34.6	Ethyl chloroformate	94–5
Propylene oxide	35	Allyl alcohol	96.6
n-Pentane	36.1	1,2-Dichloropropane	96.8
Ethyl bromide	38.4	n-Propyl alcohol	97.8
Methylene chloride	40	n-Heptane	98.4
Methylal	42.3	Ethyl propionate	99.1
Carbon disulfide	46.3	sec-Butyl alcohol	99.5
Ethyl formate	54	Isoamyl chloride	99.7
Acetone	56.5	Ligroin (high boiling)	100–150
Methyl acetate	57.1	Formic acid	100.8
Ethylidene dichloride	57.3	Methyl-cyclohexane	100.9
Acetylene dichloride*	59–61	Dioxane (1,4)	101.1
Ligroin (low boiling)	60–120	Nitromethane	101.5
Chloroform	61.2	n-Propyl acetate	101.6
Methyl alcohol	64.7	Diethyl ketone	101.7
Tetrahydrofuran	65–6	tert-Amyl alcohol	102
Di-isopropyl ether	68.5–69.0	Acetal	102.2
n-Hexane	68.7	n-Butyl formate	106.9
Isobutyl chloride	68.9	Isobutyl alcohol	107–8
Trichloroethane (1,1,1)	74.1	Acetylene dibromide*	110
Dioxolane	75–6	Toluene	110.6
Carbon tetrachloride	76.8	sec-Butyl acetate	112–3
Ethyl acetate	77.1	Trichloroethane (1,1,2)	113.5
n-Butyl chloride	77.9	Nitroethane	114.8
Ethyl alcohol	78.4	Pyridine	115–6
Methyl-ethyl ketone	79.6	Pentan-3-ol	115.6
2-Methyl-tetrahydrofuran	80.0	Epichlorohydrin	117
Benzene	80.1	n-Butyl alcohol	117
Cyclohexane	80.7	Isobutyl acetate	118
n-Propyl formate	81.3	Methyl isobutyl ketone	118
Acetonitrile	82	Acetic acid	118.1
Isopropyl alcohol	82.5	Propylene glycol monomethyl ether	119
tert-Butyl alcohol	82.9	Ethyl n-butyrate	120–1
Cyclohexene	83.3	2-Nitropropane	120.3
Ethylene chloride	83.7	Isoamyl bromide	120.4
Thiophene	84	Tetrachloroethylene	120.8

* Mixture of isomers.

Table 10-12 (*Continued*)
ORGANIC SOLVENTS ARRANGED BY BOILING POINTS

Name	B. P. 760 mm °C	Name	B. P. 760 mm °C
Di-isopropyl ketone	123.7	Bromobenzene	156.2
Ethylene glycol diethyl ether		Monoethyl-glycol acetate	156.5
(Diethyl Cellosolve)	124	Hexan-1-ol	157.2
Ethylene glycol monomethyl		Trichloropropane (1,2,3)	158
ether (Methyl Cellosolve)	124–5	Ethylene glycol mono-isobutyl	
n-Octane	125.7	ether	158.8
Diethyl carbonate	126		
		Cyclohexanol	160–1
n-Butyl acetate	126	Isoamyl propionate	160.2
sec-Propylene chlorohydrin	127	Heptan-2-ol	160.4
Ethylene chlorohydrin	128.8	Furfural	161.7
Mesityl oxide	130.1	Pentachloroethane	162
Ethylene bromide	131.5		
		Diacetone alcohol	167.9
1-Nitropropane	131.6	Di-isobutyl ketone	168.1
2-Methyl-pentan-4-ol	131.8	Methyl acetoacetate†	169–70
Isobutyl carbinol	132.0	Furfuryl alcohol	170
Chlorobenzene	132.1	Methyl o-tolyl ether	171–2
Xylene*	*ca* 133		
		Ethylene glycol mono-n-butyl-	
Cyclohexylamine	134	ether	171.2
Ethylene glycol monoethyl		Phenetole	172
ether (Cellosolve)	135.1	Di-isoamyl ether	173.4
Ethylbenzene	136.2	n-Decane	174.0
n-Amyl alcohol (*prim.*)	138	Glycol diformate	174
Acetic anhydride	139.6		
		Cyclohexyl acetate	174–5
Di-isopropyl carbinol	140	2,6-Dimethylheptan-4-ol	174–5
Acetylacetone	140.5	α,α'-Dichlorohydrin	174.3
Isoamyl acetate	142	Furfuryl acetate	175–7
Di-n-butyl ether	142.4	Methyl p-tolyl ether	176
Ethylene glycol mono-iso-			
propyl ether	144	Eucalyptole	176–7
		Methyl m-tolyl ether	177
Monomethylglycol acetate		p-Cymene	177.1
(Methyl-cellosolve Acetate)	144.5	Dichloroethyl ether (*sym*)	178.5
Acetylene tetrachloride	146.3	iso-Amyl n-butyrate	178.6
2-Methylpentan-1-ol	148		
3-Methylol-pentane	148.9	o-Dichlorobenzene	179
n-Amyl acetate	149	Octan-2-ol	179–80
		Ethyl acetoacetate	180
Ethyl-n-butyl ketone	149–50	Ethylene glycol mono-isoamyl	
Bromoform	150.5	ether	181
n-Nonane	150.8	Phenol	181.4
Methyl-n-amyl ketone	151.5		
Isopropylbenzene	152.4	2-Ethylhexan-2-ol	184
		Aniline	184.4
Anisole	154–5	Diethyl oxalate	186
Ethyl lactate	155	Diethylene glycol diethyl ether	188
Heptan-4-ol	156	α-Propylene glycol	188–9
Cyclohexanone	155–6		
Heptan-4-ol	156	Ethyl benzyl ether	190
		Glycol diacetate	190.5
		Benzonitrile	190.7

† Slight decompn.

Table 10-12 (*Continued*)
ORGANIC SOLVENTS ARRANGED BY BOILING POINTS

Name	B. P. 760 mm °C	Name	B. P. 760 mm °C
Decalin*	191.7	n-Propyl benzoate	231
Dimethylaniline	193	Tributyl borate	231
		Decan-1-ol	232.9
Isoamyl isovalerate	194	Benzyl cyanide	233–4
Octan-1-ol	194–5	Quinoline	238
Acetonyl acetone	194.1		
Diethylene glycol monomethyl ether	194.2	Ethylene glycol monophenyl ether (Phenyl Cellosolve)	244.7
Diethylene glycol monoethyl ether	195	Diethylene glycol	244.8
		n-Dibutyl oxalate	245.5
Glycol	197.4	Diethylene glycol monobutyl ether acetate	246.4
Methyl benzoate	198–9	Triethylene glycol monoethyl ether	248
Diethyl malonate	198.9		
o-Toluidine	199.7		
"Carbitol"	200.1	n-Butyl benzoate	250
		Diethylene glycol mono-n-hexyl ether	252
p-Toluidine	200.3	Diethylene glycol di-n-butyl ether (Dibutyl Carbitol)	255
Acetophenone	203	Triacetin	258–9
Ethylene glycol dibutyl ether	203	α-Chloronaphthalene	259.3
Methyl-phenyl carbinol	203.9		
Benzyl alcohol	204.7		
Tetralin	206–7	Isoamyl benzoate	262
Butan-1,3-diol	206.5	α-Monobutyrin	269–71
γ-Valerolactone	207–8	Ethyl cinnamate	271
Ethylene glycol monomethyl ether acetal	207.2	o-Nitroanisole	272–3
Camphor	209.1	Tetraethylene glycol dimethyl ether	275.8
Diethylene glycol monomethyl ether acetate	209.1	Isoamyl salicylate	277–8
o-Chloroaniline	210.5	α-Bromonaphthalene	281.1
Nitrobenzene	210.9	Dimethyl phthalate	282
Ethyl benzoate	211–2	Glycerol	290
Isophorone	215–6	Triethylene glycol	290
Diethylene glycol monoethyl ether acetate	217.7	Diethyl phthalate	298–9
Naphthalene	217.9	Benzyl benzoate	323–4
Acetamide	222	Tetraethylene glycol dibutyl ether	330
Methyl salicylate	222.2	n-Dibutyl phthalate	340
Diethyl maleate	225		

* Mixture of isomers.

AZEOTROPIC MIXTURES

These tables give the data for various binary and ternary azeotropic or constant boiling mixtures.

For a very complete listing of azeotropes see Horsley, *Ind. Eng. Chem., Analytical Edition, 19, 508 (1947)*.

For a thorough discussion of Azeotropy, see Lecat, *Traite de Chimie Organique*, V. Grignard, Mason et Cie, Paris, 1935, Tome I, pp. 121 to 267; or Young, *Distillation Principles and Processes*, Macmillan, London, 1922.

The boiling points listed are in °C. at 760 mm.

These data are taken from Lecat, *"L'Azeotropisme"*, Brussels, 1918; *Ann. soc. sci. Bruxelles 45B, 169–76, 284–94 (1926); 47B, i, 21–7, 63–71, 108–14, 149–58 (1927); 48B, i, 13–22, 54–62, 113–26; ii, 1–18 (1928); 49B, ii, 17–47, 109–43 (1929); 50B, 21–33 (1930); 55B, 43–7, 253–65 (1935); 56B, 41–54, 221–34 (1936); Rec. trav. chim. Pays-Bas 45, 620–7 (1926); 46, 240–7 (1927); 47, 13–18 (1928)*, unless otherwise designated as follows:

(1) Young and Fortey, *Trans. Chem. Soc., 81, 717, 739, 752 (1902); 83, 45 (1903)*.
(2) Roscoe, *Quart. Jour. Chem. Soc., 13, 146 (1861); 15, 270 (1862)*.
(3) Bogin, *U. S. P. Re. 17, 157*.
(4) Young, *"Distillation Principles and Processes"*, Macmillan, London, 1922.
(5) Miller and Bliss, *Ind. Eng. Chem., 32, 123 (1940)*.
(6) Hannotte, *Bull. Soc. Chim. Belg., 35, 85–109 (1926)*.
(7) Wuyts, *ibid., 33, 167–192 (1924)*.
(8) Beduwe, *ibid., 34, 41–55 (1925)*.
(9) Atkins, *Trans. Chem. Soc., 117, 218–20 (1920)*.
(10) Booklet by Shell Chemical Co., "Organic Chemicals Manufactured by Shell Chemical Company", 1939.
(11) Azeotropic Data, Advances in Chemistry Series (6), American Chemical Society (1952).

Table 10-13
BINARY AZEOTROPES CONTAINING WATER

	B.P. 760 mm.		% By weight	
	Other component	Azeotrope	Water	Other component
A—Alcohols				
Ethyl alcohol (1)	78.4	78.1	4.5	95.5
n-Propyl alcohol (1)	97.2	87.7	28.3	71.7
Isopropyl alcohol (1)	82.5	80.4	12.1	87.9
n-Butyl alcohol	117.8	92.4	38	62
Isobutyl alcohol	108.0	90.0	33.2	66.8
sec-Butyl alcohol	99.5	88.5	32.1	67.9
tert-Butyl alcohol	82.8	79.9	11.7	88.3
n-Amyl alcohol (1-Pentanol)	137.8	96.0	54.0	46.0
prim-Isoamyl alcohol (2-Methyl-1-butanol)	131.4	95.2	49.6	50.4
tert-Amyl alcohol (2-Methyl-2-butanol)	102.3	87.4	27.5	72.5
sec-Amyl alcohol (3-Pentanol)	115.4	91.7	36.0	64.0
act-sec-Amyl alcohol (2-Pentanol)	119.3	92.5	38.5	61.5
n-Hexyl alcohol	157.9	97.8	75	25
n-Heptyl alcohol	176.2	98.7	83	17
n-Octyl alcohol	195.2	99.4	90	10

Table 10-13 (*Continued*)
BINARY AZEOTROPES CONTAINING WATER

	B.P. 760 mm.		% By weight	
	Other component	Azeotrope	Water	Other component
Allyl alcohol	97.0	88.2	27.1	72.9
Benzyl alcohol	205.2	99.9	91	9
Furfuryl alcohol	169.4	98.5	80	20
B—Hydrocarbons				
Benzene	80.2	69.3	8.9	91.1
Toluene	110.8	84.1	19.6	80.4
C—Substituted Hydrocarbons				
Ethylene chloride	83.7	72	8.3	91.7
Propylene chloride	96.8	78	12	88
D—Ethers				
Diethyl ether	34.5	34.2	1.3	98.7
Di-isopropyl ether	68.4	62.2	4.5	95.5
Ethyl *n*-propyl ether	63.6	59.5	4	96
Di-isobutyl ether	122.2	88.6	23	77
Di-isoamyl ether	172.6	97.4	54	46
Diphenyl ether	259.3	99.3	96.8	3.2
Phenetole	170.4	97.3	59	41
Anisole	153.9	95.5	40.5	59.5
Resorcinol diethyl ether	235.0	99.7	91	9
E—Esters				
n-Propyl formate	80.9	71.9	3.6	96.4
n-Butyl formate	106.8	83.8	15	85
Isobutyl formate	98.4	80.4	7.8	92.2
n-Amyl formate	132.0	91.6	28.4	71.6
Isoamyl formate	123.9	89.7	23.5	76.5
Benzyl formate	202.3	99.2	80	20
Ethyl acetate	77.1	70.4	6.1	93.9
n-Propyl acetate	101.6	82.4	14	86
Isopropyl acetate	91.0	77.4	6.2	93.8
n-Butyl acetate	126.2	90.2	28.7	71.3
Isobutyl acetate	117.2	87.5	19.5	80.5
n-Amyl acetate	148.8	95.2	41	59
Isoamyl acetate	142.1	93.8	36.2	63.8
Benzyl acetate	214.9	99.6	87.5	12.5
Phenyl acetate	195.7	98.9	75.1	24.9
Methyl propionate	79.9	71.4	3.9	96.1
Ethyl propionate	99.2	81.2	10	90
n-Propyl propionate	122.1	88.9	23	77
Isobutyl propionate	136.9	92.8	32.2	67.8
Isoamyl propionate	160.3	96.6	48.5	51.5
Methyl butyrate	102.7	82.7	11.5	88.5
Ethyl butyrate	120.1	87.9	21.5	78.5
n-Propyl butyrate	142.8	94.1	36.4	63.6
n-Butyl butyrate	165.7	97.2	53	47
Isobutyl butyrate	156.8	96.3	46	54
Isoamyl butyrate	178.5	98.1	63.5	36.5
Methyl Isobutyrate	92.3	77.7	6.8	93.2
Ethyl isobuyrate	110.1	85.2	15.2	84.8
n-Propyl isobutyrate	133.9	92.2	30.8	69.2

Table 10-13 (*Continued*)
BINARY AZEOTROPES CONTAINING WATER

	B.P. 760 mm.		% By weight	
	Other component	Azeotrope	Water	Other component
iso-Butyl isobutyrate	147.3	95.5	39.4	60.6
iso-Amyl isobutyrate	168.9	97.4	56.0	44.0
Methyl isovalerate	116.3	87.2	19.2	80.8
Ethyl isovalerate	134.7	92.2	30.2	69.8
n-Propyl isovalerate	155.8	96.2	45.2	54.8
iso-Butyl isovalerate	168.7	97.4	55.8	44.2
iso-Amyl isovalerate	193.5	98.8	74.1	25.9
Ethyl caproate	166.8	97.2	54	46
Methyl cinnamate	261.9	99.9	95.5	4.5
Methyl benzoate	199.5	99.1	79.2	20.8
Ethyl benzoate	212.4	99.4	84.0	16.0
n-Propyl benzoate	230.9	99.7	90.9	9.1
n-Butyl benzoate	249.8	99.9	94	6
Isobutyl benzoate	242.2	99.8	92.6	7.4
Isoamyl benzoate	262.3	99.9	95.6	4.4
Ethyl phenylacetate	228.8	99.7	91.3	8.7
Ethyl nitrate	87.7	74.4	22	78
n-Propyl nitrate	110.5	84.8	20	80
Isobutyl nitrate	122.9	89.0	25	75
F—Organic Acids				
Formic acid (Max.)	100.8	107.3	22.5	77.5
Acetic acid	118.1	No	Azeotrope	
Propionic acid	141.1	99.98	82.3	17.7
Butyric acid	163.5	99.4	81.6	18.4
Isobutyrate acid	154.5	99.3	79	21
G—Inorganic Acids (2)				
Nitric acid (Max.)	86.0	120.5	32	68
Perchloric acid (Max.)	110.0	203	28.4	71.6
Hydrofluoric acid (Max.)	19.4	120	63	37
Hydrochloric acid (Max.)	−84	110	79.76	20.24
Hydrobromic acid (Max.)	−73	126	52.5	47.5
Hydriodic acid (Max.)	−34	127	43	57
H—Ketones				
Methyl ethyl ketone	79.6	73.5	11	89
Methyl n-propyl ketone (10)	102.0	83.3	19.5	80.5
Methyl isobutyl ketone (10)	115.9	87.9	24.3	75.7
Mesityl oxide (10)	129.5	91.8	34.8	65.2
Diacetone alcohol (10)	166	98.8	87.3	12.7
I—Aldehydes				
Furfural	161.5	97.5	65	35
Butyraldehyde (3)	75.7	68	6	94
J—Amines				
Pyridine	115.5	92.6	43	57

Table 10-14
BINARY AZEOTROPES CONTAINING ALCOHOLS

	B.P. 760 mm.		% By weight	
	Other component	Azeotrope	Alcohol	Other component
A—Methyl Alcohol (B.P. 64.7°)				
Methylal	42.3	41.9	8.2	91.8
Dimethyl sulfide	37.3	34.0	15	85
Methyl borate	68.7	54.6	32	68
Dimethyl carbonate	90.4	62.7	70	30
Methyl acetate	57.0	53.8	18.7	81.3
Methyl propionate	79.8	62.5	47.5	52.5
Ethyl formate	54.1	51.0	16	84
Ethyl acetate	77.1	62.3	44	56
Isopropyl acetate	91.0	64.5	80	20
n-Pentane	36.2	30.8	9	91
n-Hexane	68.9	50.6	28	72
n-Heptane	98.5	59.1	51.5	48.5
Benzene	80.2	58.3	39.6	60.4
Toluene	110.8	63.8	69	31
Cyclohexane	80.8	54.2	37.2	62.8
Nitromethane	101.2	64.6	91	9
Methyl iodide	42.6	38.0	6.5	93.5
Chloroform	61.1	53.5	12.6	87.4
Carbon tetrachloride (4)	76.8	55.7	20.6	79.4
Ethyl bromide	38.4	35.0	4.5	95.5
Ethylene chloride	83.7	61.0	32	68
n-Propyl chloride	46.6	40.5	9.5	90.5
n-Propyl bromide	71.0	54.5	21	79
Isopropyl chloride	36.3	33.4	6	94
Isopropyl bromide	59.8	48.6	15.0	85.0
Isopropyl iodide	89.4	61.0	38	62
n-Butyl chloride	78.1	57.0	27	73
Isobutyl bromide	91.0	61.3	41.7	58.3
Di-n-propyl ether	90.4	63.8	72	28
Acetone	56.5	55.7	12.1	87.9
B—Ethyl Alcohol (B.P. 78.3°)				
Methyl acetate	57.0	56.9	3	97
Methyl propionate	79.7	72.0	33	67
Ethyl nitrate	87.7	71.9	44	56
Ethyl acetate	77.1	71.8	30.8	69.2
Ethyl propionate	99.2	78.0	75	25
n-Propyl formate	80.8	71.8	38	62
n-Propyl acetate	101.6	78.2	85	15
Isopropyl acetate	91.0	76.8	57	43
Benzene	80.2	68.2	32.4	67.6
Toluene	110.8	76.7	68	32
n-Pentane	36.2	34.3	5	95
n-Hexane	68.9	58.7	21	79
n-Heptane	98.5	70.9	49	51
n-Octane	125.6	77.0	78	22
Methyl iodide	42.6	41.2	3.2	96.8
Ethylene chloride	83.7	70.5	37	63
Carbon tetrachloride	76.8	65.1	15.8	84.2
Allyl chloride	45.7	44	5	95

Table 10-14 (*Continued*)
BINARY AZEOTROPES CONTAINING ALCOHOLS

	B.P. 760 mm.		% By weight	
	Other component	Azeotrope	Alcohol	Other component
Chloroform	61.1	59.4	7	93
n-Propyl chloride	46.7	45.0	6	94
n-Propyl bromide	71.0	62.8	20.5	79.5
n-Propyl iodide	102.4	75.4	44	56
Isopropyl chloride	36.3	35.6	2.8	97.2
Isopropyl bromide	59.8	55.6	10.5	89.5
Isopropyl iodide	89.4	71.5	27	73
n-Butyl chloride	78.1	65.7	20.3	79.7
n-Butyl bromide	100.3	75.0	43	57
Isobutyl bromide	91.0	72.5	31	69
Acetal	103.6	78.0	76	24
Di-n-propyl ether	90.4	74.5	44	56
Methyl ethyl ketone	79.6	74.8	40	60
C—Isopropyl Alcohol (B.P. 82.5°)				
Methyl propionate	79.8	76.4	37	63
Ethyl acetate	77.1	75.3	25	75
Isopropyl acetate	91.0	81.3	60	40
Benzene	80.2	71.9	33.3	66.7
Toluene	110.8	81.3	79	21
n-Pentane	36.2	35.5	6	94
n-Hexane	68.9	62.7	23	77
n-Heptane	98.5	76.3	54	46
Carbon tetrachloride	76.8	69.0	18	82
Chloroform	61.1	60.8	4.2	95.8
Allyl bromide	70.8	66.5	20	80
Ethyl iodide	72.3	67.1	15	85
Ethylene chloride	83.7	74.7	43.5	56.5
Isopropyl bromide	59.8	57.8	12	88
Isopropyl iodide	89.4	76.0	32	68
n-Propyl chloride	46.7	46.4	2.8	97.2
n-Propyl bromide	71.0	66.8	20.5	79.5
n-Propyl iodide	102.4	79.8	42	58
n-Butyl chloride	78.1	70.8	23	77
Acetal	103.6	81.3	63	37
Ethyl n-propyl ether	63.6	62.0	10	90
Di-isopropyl ether (5)	82.3	66.2	14.1	85.9
Methyl ethyl ketone	79.0	77.5	32	68
D—n-Propyl Alcohol (B.P. 97.2°)				
Methyl butyrate	102.7	94.4	49	51
Ethyl propionate	99.2	93.4	48	52
n-Propyl formate	80.8	80.65	3	97
n-Propyl acetate	101.6	94.7	51	49
Benzene	80.2	77.1	16.9	83.1
Toluene	110.8	92.4	52.5	47.5
n-Hexane	68.9	65.7	4	96
Carbon tetrachloride	76.8	73.1	11.5	88.5
Chlorobenzene	132.0	96.9	83	17
Ethylene chloride	83.7	80.7	19	81
n-Propyl bromide	71.0	69.7	9	91
n-Butyl chloride	78.1	74.8	18	82
Acetal	103.6	92.4	37	63
Di-n-propyl ether	90.4	85.7	30	70

Table 10-14 (*Continued*)
BINARY AZEOTROPES CONTAINING ALCOHOLS

	B.P. 760 mm.		% By weight	
	Other component	Azeotrope	Alcohol	Other component
E—*Isobutyl Alcohol* (B.P. 107.9°)				
Methyl butyrate	102.7	101.3	25	75
Isobutyl formate	97.9	97.4	12	88
Isobutyl acetate	117.5	107.6	92	8
Benzene	80.2	79.8	9.3	90.7
Toluene	110.8	100.9	44.5	55.5
m-Xylene	139.0	107.7	87	13
Cyclohexane	80.8	78.1	14	86
n-Hexane	68.9	68.3	2.5	97.5
Chlorobenzene	132.0	107.1	63	37
Ethylene chloride	83.7	83.5	6.5	93.5
n-Butyl chloride	78.1	77.7	4	96
n-Butyl bromide	100.3	95.0	21	79
Isobutyl bromide	91.0	88.8	12	88
Isobutyl iodide	120.4	104.0	36	64
Isoamyl chloride	99.8	94.5	22	78
Acetal	103.6	98.2	20	80
F—*n-Butyl Alcohol* (B.P. 117.8°)				
Ethyl butyrate	120.0	115.7	64	36
n-Butyl formate	106.6	105.8	23.7	76.3
n-Butyl acetate	126.2	117.2	47	53
Toluene	110.8	105.7	27	73
m-Xylene	139.0	116.0	80	20
Cyclohexane	80.8	79.8	4	96
n-Heptane	98.5	94.4	18	82
Carbon tetrachloride	76.8	76.6	2.5	97.5
Chlorobenzene	132.0	115.3	56	44
Isobutyl bromide	91.0	90.2	7	93
Isobutyl iodide	120.4	110.5	30	70
Acetal	103.6	101	13	87
G—*Isoamyl Alcohol* (B.P. 131.4°)				
Isoamyl formate	123.8	123.7	10	90
Isoamyl acetate	142.1	131.3	98.5	1.5
Toluene	110.8	110.0	14	86
m-Xylene	139.0	127.0	53.3	46.7
Chlorobenzene	132.0	124.3	35	65
Isoamyl bromide	120.3	116.8	21	79
Isoamyl iodide	147.7	129.2	54	46
Paraldehyde	124.0	122.9	22	78
H—*n-Amyl Alcohol* (B.P. 137.8°)				
n-Amyl formate	132.0	130.4	43	57
Di-*n*-butyl ether	142.1	134.0	52	48
I—*Cyclohexyl Alcohol* (B.P. 160.7°)				
o-Xylene	143.6	143.0	14	86
Isoamyl iodide	147.5	147.0	10	90
Furfural	161.4	155.6	45	55
Di-isoamyl ether	172.6	158.8	78	22
Phenetole	170.4	159.2	72	28

Table 10-14 (*Continued*)
BINARY AZEOTROPES CONTAINING ALCOHOLS

	B.P. 760 mm.		% By weight	
	Other component	Azeotrope	Alcohol	Other component
J—Allyl Alcohol (B.P. 97.0°)				
Methyl butyrate	102.7	93.8	55	45
n-Propyl acetate	101.6	94.2	53	47
Benzene	80.2	76.8	17.4	82.6
Toluene	110.8	92.4	50	50
Cyclohexane	80.8	74	20	80
Allyl iodide	102.0	89.4	28	72
Chlorobenzene	132.0	96.2	85	15
Ethylene chloride	83.7	79.9	18	82
K—Benzyl Alcohol (B.P. 205.2°)				
Nitrobenzene	210.8	204.0	58	42
Iodobenzene	188.6	187.8	12	88
o-Bromotoluene	181.4	181.25	7	93
Naphthalene	218.1	204.1	60	40
m-Cresol (Max.)	202.2	207.1	61	39
Dimethyl aniline	194.1	193.9	6.5	93.5
L—Ethylene Glycol (B.P. 197.4°)				
Ethyl benzoate	212.6	186.1	46.5	53.5
Isoamyl acetate	142.1	141.95	3	97
Diphenyl	254.9	192.0	64	36
Mesitylene	164.6	156.0	13	87
Naphthalene	218.1	183.9	51	49
Toluene	110.8	110.2	6.5	93.5
m-Xylene	139.0	135.6	15	85
Ethylene bromide	131.7	129.8	4	96
Nitrobenzene	210.9	185.9	59	41
Chlorobenzene	132.0	130.1	94.4	5.6
Benzyl chloride	179.3	167.0	30	70
Benzyl alcohol	205.1	193.1	56	44
Di-n-butyl ether	142.1	140.0	10	90
Diphenyl ether	259.3	193.1	60	40
Anisole	153.9	150.5	10.5	89.5
β-Phenylethyl alcohol	219.4	194.4	69	31
Acetophenone	202.1	185.7	52	48
Aniline	184.4	180.6	24	76
Dimethyl aniline	194.1	175.9	33.5	66.5
o-Cresol	191.1	189.6	27	73
M—Glycerol (B.P. 291.0°)				
n-Propyl benzoate	230.9	228.8	8	92
n-Butyl benzoate	249.8	243.0	17	83
Diphenyl	254.9	243.8	55	45
Naphthalene	218.1	215.2	10	90
p-Dibromobenzene	220.3	217.1	10	90
Diphenyl ether	257.7	246.3	22	78

Table 10-15
BINARY AZEOTROPES CONTAINING ORGANIC ACIDS

	B.P. 760 mm.		% By weight	
	Other component	Azeotrope	Acid	Other component
A—*Formic Acid* (B.P. 100.8°)				
Benzene	80.2	71.7	31	69
Toluene	110.8	85.8	50	50
m-Xylene	139.0	94.2	70.2	29.8
n-Pentane	36.2	34.2	10	90
n-Hexane	68.9	60.6	28	72
n-Heptane	98.5	78.2	43.5	56.5
n-Octane	125.8	90.5	63	37
Chloroform	61.2	59.2	15	85
Carbon tetrachloride	76.8	66.7	18.5	81.5
Carbon disulfide	46.3	42.6	17	83
Methyl iodide	42.6	42.1	6	94
Ethyl bromide	38.4	38.2	3	97
Ethylene chloride	83.6	77.4	14	86
Ethylene bromide	131.7	94.7	51.5	48.5
n-Propyl chloride	46.7	45.6	8	92
n-Propyl bromide	71.0	64.7	27	73
Isopropyl chloride	34.8	34.7	1.5	98.5
Isopropyl bromide	59.4	56.0	14	86
Isobutyl chloride	68.9	63.0	19	81
Isoamyl chloride	99.8	80.0	33.5	66.5
Chlorobenzene	132.0	95.0	55	45
Diethyl ketone (Max.)	102.2	105.4	33	67
Methyl *n*-propyl ketone (Max.)	102.3	105.3	32	68
B—*Acetic Acid* (B.P. 118.5°)				
Benzene	80.2	80.05	2	98
Toluene	110.8	105.0	34	66
m-Xylene	139.0	115.4	72.5	27.5
n-Heptane	98.5	92.3	30	70
n-Octane	125.8	109.0	50	50
Carbon tetrachloride	76.8	76.6	3	97
Ethylene bromide	131.7	114.4	55	45
Isopropyl iodide	89.2	88.3	9	91
n-Butyl bromide	100.4	97.6	18	82
Isobutyl bromide	91.3	90.2	12	88
Isoamyl chloride	99.8	97.2	18.5	81.5
Chlorobenzene	132.0	114.7	58.5	41.5
C—*Propionic Acid* (B.P. 140.9°)				
m-Xylene	139.0	132.7	35.5	64.5
Ethylene bromide	131.7	127.8	17.5	82.5
Isobutyl iodide	120.4	119.5	9	91
Isoamyl bromide	120.3	119.2	10	90
Chlorobenzene	132.0	128.9	18	82
Anisole	153.9	140.8	96	4
D—*Butyric Acid* (B.P. 162.5°)				
m-Xylene	139.0	138.3	6	94
Ethylene bromide	131.7	131.1	3.5	96.5
Chlorobenzene	132.0	131.8	2.8	97.2

Table 10-15 (*Continued*)
BINARY AZEOTROPES CONTAINING ORGANIC ACIDS

	B.P. 760 mm.		% By weight	
	Other component	Azeotrope	Acid	Other component
Benzyl chloride	179.3	160.8	65	35
Anisole	153.9	152.9	12	88
Furfural	161.5	159.4	42.5	57.5
E—Isobutyric Acid (B.P. 154.4°)				
m-Xylene	139.0	136.8	14	86
Ethylene bromide	131.7	130.5	6.5	93.5
Chlorobenzene	132.0	131.2	8	92
Benzyl chloride	179.3	153.5	80	20
Anisole	153.9	148.5	42	58
F—Isovaleric Acid (B.P. 176.5°)				
Mesitylene	164.6	162.8	20	80
Benzyl chloride	179.3	171.2	36	64
Benzaldehyde	179.2	174.5	68	32
Isoamyl butyrate	178.5	176.1	70	30
Phenetole	170.5	168.5	20	80
G—Caproic Acid (B.P. 205.2°)				
Naphthalene	218.1	202.0	70	30
Benzyl chloride	179.3	179.0	3	97
Nitrobenzene	210.8	202.0	70	30
H—Caprylic Acid (B.P. 237.5°)				
Naphthalene	218.1	216.2	6	94
p-Dibromobenzene	220.3	218.8	10	90
I—Chloroacetic Acid (B.P. 189.4°)				
Mesitylene	164.6	162	17	83
Naphthalene	218.1	187.1	78	22
Benzyl chloride	179.3	173.8	25	75
p-Dichlorobenzene	174.1	167.6	24.5	75.5
o-Cresol	191.1	187.5	54	46
J—Phenylacetic Acid (B.P. 266.5°)				
α-Chloronaphthalene	262.7	255.9	30	70
Diphenyl	255.9	252.2	23.3	76.7
Isoamyl benzoate	262.0	259.9	26	74
Methyl cinnamate	261.9	261.8	3	97
Diphenyl ether	259.3	255.4	27.8	72.2
K—Benzoic Acid (B.P. 250.5°)				
Diphenyl	255.9	245.9	50.5	49.5
Naphthalene	218.1	217.7	5	95
p-Dibromobenzene	220.3	219.5	3.8	96.2
p-Nitrotoluene	239.0	237.4	11	89
Ethyl salicylate	234.0	233.85	6	94
Isobutyl benzoate	241.9	241.2	12	88
Diphenyl ether	259.3	247.0	59	41

Table 10-16
TERNARY AZEOTROPES CONTAINING WATER AND ALCOHOLS

	B.P. 760 mm.		% By weight		
Other component	Other component	Azeotrope	Water	Alcohol	Other component
A—Ethyl Alcohol (B.P. 78.3°)					
Ethyl acetate (6)	77.1	70.3	7.8	9.0	83.2
Diethyl formal (7)	87.5	73.2	12.1	18.4	69.5
Diethyl acetal (8)	103.6	77.8	11.4	27.6	61.0
Cyclohexane	80.8	62.1	7	17	76
Benzene (4)	80.2	64.9	7.4	18.5	74.1
Chloroform	61.2	55.5	3.5	4.0	92.5
Carbon tetrachloride	76.8	61.8	4.3	9.7	86.0
Ethyl iodide	72.3	61	5	9	86
Ethylene chloride	83.7	66.7	5	17	78
B—n-Propyl Alcohol (B.P. 97.2°)					
n-Propyl formate (6)	80.9	70.8	13	5	82
n-Propyl acetate (6)	101.6	82.2	21.0	19.5	59.5
Di-*n*-propyl formal (7)	137.4	86.4	8.0	44.8	47.2
Di-*n*-propyl acetal (8)	147.7	87.6	27.4	51.6	21.0
Di-*n*-propyl ether (7)	91.0	74.8	11.7	20.2	68.1
Cyclohexane	80.8	66.6	8.5	10.0	81.5
Benzene (4)	80.2	68.5	8.6	9.0	82.4
Carbon tetrachloride	76.8	65.4	5	11	84
Diethyl ketone	102.2	81.2	20	20	60
C—Isopropyl Alcohol (B.P. 82.5°)					
Cyclohexane	80.8	64.3	7.5	18.5	74.0
Benzene (4)	80.2	66.5	7.5	18.7	73.8
D—n-Butyl Alcohol (B.P. 117.8°)					
n-Butyl formate (6)	106.6	83.6	21.3	10.0	68.7
n-Butyl acetate (6)	126.2	89.4	37.3	27.4	35.3
Di-*n*-butyl ether (7)	141.9	91	29.3	42.9	27.7
E—Isobutyl Alcohol (B.P. 108.0°)					
Isobutyl formate (6)	94.4	80.2	17.3	6.7	76.0
Isobutyl acetate	117.2	86.8	30.4	23.1	46.5
F—tert-Butyl Alcohol (B.P. 82.6°)					
Benzene (4)	80.2	67.3	8.1	21.4	70.5
Carbon tetrachloride (9)	76.8	64.7	3.1	11.9	85.0
G—n-Amyl Alcohol (B.P. 137.8°)					
n-Amyl formate (6)	131.0	91.4	37.6	21.2	41.2
n-Amyl acetate (6)	148.8	94.8	56.2	33.3	10.5
H—Isoamyl Alcohol (B.P. 131.4°)					
Isoamyl formate (6)	124.2	89.8	32.4	19.6	48.0
Isoamyl acetate (6)	142.0	93.6	44.8	31.2	24.0
I—Allyl Alcohol (B.P. 97.0°)					
n-Hexane	69.0	59.7	5	5	90
Cyclohexane	80.8	66.2	8	11	81
Benzene	80.2	68.2	8.6	9.2	82.2
Carbon tetrachloride	76.8	65.2	5	11	84

Table 10-17
MOLECULAR ELEVATION OF THE BOILING POINT
(EBULLIOSCOPIC CONSTANTS)

Molecular weights can be determined with the relation

$$M = K_b \frac{1000 w_2}{w_1 \Delta T_b}$$

where ΔT_b is the elevation of the boiling point brought about by the addition of w_2 grams of solute to w_1 grams of solvent and K_b is the ebullioscopic constant. In the column headed Barometric Correction is given the number of degrees for each millimeter of difference between the barometric reading and 760 mm Hg to be subtracted from K_b if the pressure is lower, or added if higher, than 760 mm. In general, the effect is within experimental error if the pressure is within 10 mm of 760 mm.

K_b	Boiling Point, °C	Barometric Correction	Compound
0.515	100.00	0.0008	Water
0.785	64.70	0.0002	Methanol
1.160	78.29	0.0003	Ethanol
1.59	97.20		1-Propanol
1.71	56.29	0.0004	Acetone
1.824	34.55	0.0005	Diethyl ether
2.061	56.323	0.0005	Methyl acetate
2.125	42.30		Dimethoxymethane
2.28	79.64		2-Butanone
2.35	46.225	0.0006	Carbon disulfide
2.53	80.100	0.0007	Benzene
2.530	117.9	0.0008	Acetic acid
2.583	77.114	0.0007	Ethyl acetate
2.710	115.256		Pyridine
2.75	80.725	0.0007	Cyclohexane
2.84	106.40		Piperidine
3.22	184.40	0.0009	Aniline
3.22	166.1		N,N-Dimethylacetamide
3.270	101.320		1,4-Dioxane
3.29	110.625	0.0008	Toluene
3.43	98.427	0.0008	Heptane
3.51	140.83		Propionic acid
3.60	181.839	0.0009	Phenol
3.62	61.152	0.0009	Chloroform
4.15	131.687	0.0011	Chlorobenzene
4.48	76.75	0.0013	Carbon tetrachloride
5.611	207.42	0.0015	Camphor
5.80	217.955	0.0014	Naphthalene
6.26	155.908	0.0016	Bromobenzene
6.608	131.36	0.0016	1,2-Dibromoethane

MELTING POINTS

Table 10-18
FREEZING MIXTURES

The table below gives the percent of the anhydrous material as shown in the column on the left in the eutectic mixture with ice. The eutectic temperature is the lowest temperature which can be obtained from a mixture of the substance with ice. To obtain the maximum cooling effect, the freezing mixture should be prepared with ice rather than with water and the other ingredient should be cooled to 0°C. For most purposes sodium chloride and ice, or calcium chloride and ice, mixtures are most common providing temperatures of -21.2°C. and -55°C. respectively.

Formula of Substance	%	Eutectic Temp. °C.	Formula of Substance	%	Eutectic Temp. °C.
$BaCl_2$	22.5	−7.8	$MnSO_4$	32.2	−10.5
$CaCl_2$	29.8	−55	NH_4Cl	18.6	−15.8
$Ca(NO_3)_2$	35	−16	NH_4NO_3	41.2	−17.35
$CuCl_2$	36	−40	$(NH_4)_2SO_4$	38.3	−19.05
$Cu(NO_3)_2$	36	−24	$NaBr$	40.3	−28
$CuSO_4$	11.9	−1.6	Na_2CO_3	5.9	−2.1
$FeCl_3$	33.1	−55	$NaCl$	23.3	−21.13
$FeSO_4$	13.04	−1.824	$NaOH$	19	−28
HCl	24.8	−86	$NaNO_3$	37	−18.5
HNO_3	32.7	−43	Na_2SO_4	12.7	−3.55
K_2CO_3	39.5	−36.5	$Na_2S_2O_3$	30	−11
KCl	19.75	−11.1	$NiSO_4$	20.6	−4.15
K_2CrO_4	36.6	−11.3	SO_3	32	−75
KOH	31.5	−65	$SrCl_2$	26	−18.7
KNO_3	10.9	−2.9	$Sr(NO_3)_2$	24.5	−5.75
$MgCl_2$	21.6	−33.6	$ZnCl_2$	51	−62
$Mg(NO_3)_2$	34.6	−29	$Zn(NO_3)_2$	39.4	−29
$MgSO_4$	19	−3.9	$ZnSO_4$	27.2	−6.55

Table 10-19
NONAQUEOUS COOLING BATHS

Low temperatures may be produced by mixtures of various substances with carbon dioxide snow or by evaporation of low boiling liquids.*

Substance	Deg. C.	Substance	Deg. C.
Alcohol-carbon dioxide	−72	Nitrogen, boiling point	−196
Ammonia, boiling point	−33.4	Oxygen, boiling point	−183
Chloroform-carbon dioxide	−77	Sulfur dioxide, boiling point	−10
Ether-carbon dioxide	ca −78		
Ethyl chloride, boiling point	+12.5		
Liquid air, boiling point	−190		

* The assertion that certain baths with CO_2 snow, especially the acetone bath, give a temperature materially lower than pure dry CO_2 snow is erroneous. Cf. Thiel and Caspar, *Z. physik. Chem.* **86,** 257–93 (1914); *C. A.* **8,** 1229 (1914). Liquid nitrogen (but not liquid air, which is hazardous) poured into ethyl alcohol until a slush is formed gives a cooling bath −115° to −125°C.

According to Kolb, *Chemist-Analyst,* 1-methoxy-2-propanol is a very good liquid for dry ice cooling baths. This solvent is available from the Dow Chemical Co., Midland, Mich., under the trade name DOWANOL 33-B.

Table 10-20
COMPOSITIONS OF AQUEOUS ANTIFREEZE SOLUTIONS

FREEZING POINT OF ETHYL ALCOHOL-WATER MIXTURES*

Specific Gravity 20°/4°C. (68°F.)	% alcohol by weight	% alcohol by volume	Freezing Point °C.	Freezing Point °F.
0.99363	2.5	3.13	−1.0	30.2
0.98971	4.8	6.00	−2.0	28.4
0.98658	6.8	8.47	−3.0	26.6
0.98006	11.3	14.0	−5.0	23.0
0.97670	13.8	17.0	−6.1	21.0
0.97336	16.4	20.2	−7.5	18.5
0.97194	17.5	21.5	−8.7	16.3
0.97024	18.8	23.1	−9.4	15.1
0.96823	20.3	24.8	−10.6	12.9
0.96578	22.1	27.0	−12.2	10.0
0.96283	24.2	29.5	−14.0	6.8
0.95914	26.7	32.4	−16.0	3.2
0.95400	29.9	36.1	−18.9	−2.0
0.94715	33.8	40.5	−23.6	−10.5
0.93720	39.0	46.3	−28.7	−19.7
0.92193	46.3	53.8	−33.9	−29.0
0.90008	56.1	63.6	−41.0	−41.8
0.86311	71.9	78.2	−51.3	−60.3

FREEZING POINT OF METHYL (WOOD) ALCOHOL-WATER MIXTURES*

Specific Gravity 15.6°C. (60°F.)	% alcohol by weight	% alcohol by volume	Freezing Point °C.	Freezing Point °F.
0.993	3.9	5	−2.2	28
0.986	8.1	10	−5.0	23
0.980	12.2	15	−8.3	17
0.974	16.4	20	−11.7	11
0.968	20.6	25	−15.6	4
0.963	24.9	30	−20.0	−4
0.956	29.2	35	−25.0	−13
0.949	33.6	40	−30.0	−22
0.942	38.0	45	−35.6	−32

FREEZING POINT OF PRESTONE—WATER MIXTURES†

% Prestone By Weight	% Prestone By Volume	Specific Gravity 15°/15C. (59°F.)	Freezing Point °C.	Freezing Point °F.
10	9.2	1.013	−3.6	25.6
15	13.8	1.019	−5.6	22.0
20	18.3	1.026	−7.9	17.8
25	23.0	1.033	−10.7	12.8
30	28.0	1.040	−14.0	6.8
40	37.8	1.053	−22.3	−8.2
50	47.8	1.067	−33.8	−28.8
60	58.1	1.079	−49.3	−56.7

* Values are for pure alcohol. Since some commercial antifreezes contain small amounts of water, slightly higher volume concentrations than those given in the table may be required. Antifreezes also contain corrosion inhibitors and other additives to make them function properly as cooling liquids. These affect freezing point slightly and specific gravity to a greater degree. If a protection table is furnished by the manufacturer it should be used in preference to the values given above for the pure substance.

† Eveready Prestone (manufactured by the National Carbon Co.), marketed for antifreeze purposes, is 97% ethylene glycol containing fractional percentages of soluble and insoluble ingredients to prevent foaming, creepage and water corrosion in automobile cooling systems.

Table 10-20 (*Continued*)
COMPOSITIONS OF AQUEOUS ANTIFREEZE SOLUTIONS
FREEZING POINT OF ETHYL ALCOHOL–WATER MIXTURES

Specific Gravity 15.6°C. (60°F.)	% alcohol by volume	Freezing Point	
		°C.	°F.
0.990	5	−1.7	29
0.984	10	−3.3	26
0.978	15	−6.1	21
0.972	20	−8.3	17
0.964	25	−11.1	12
0.955	30	−14.4	6
0.945	35	−17.8	0
0.933	40	−18.3	−1
0.922	45	−18.9	−2
0.910	50	−20.0	−4
0.899	55	−21.7	−7
0.887	60	−23.3	−10
0.875	65	−24.4	−12
0.864	70	−26.7	−16
0.852	75	−32.2	−26
0.840	80	−41.7	−43

FREEZING POINT OF PROPYLENE GLYCOL-WATER MIXTURES*

Specific Gravity 15.6°C. (60°F.)	% glycol by volume	Freezing Point	
		°C.	°F.
1.004	5	−1.1	30
1.006	10	−2.2	28
1.012	15	−3.9	25
1.017	20	−6.7	20
1.020	25	−8.9	16
1.024	30	−12.8	9
1.028	35	−16.1	3
1.032	40	−20.6	−5
1.037	45	−26.7	−16
1.040	50	−33.3	−28

FREEZING POINT OF GLYCEROL (GLYCERINE)—WATER MIXTURES†

% Glycerol by Weight	Specific Gravity 15°/15°C. (59°F.)	Specific Gravity 20°/20°C. (68°F.)	Freezing Point	
			°C.	°F.
10	1.02415	1.02395	−1.6	29.1
20	1.04935	1.04880	−4.8	23.4
30	1.07560	1.07470	−9.5	14.9
40	1.10255	1.10135	−15.5	4.3
50	1.12985	1.12845	−22.0	−7.4
60	1.15770	1.15605	−33.6	−28.5
70	1.18540	1.18355	−37.8	−36.0
80	1.21290	1.21090	−19.2	−2.3
90	1.23950	1.23755	−1.6	29.1
100	1.26557	1.26362	17.0	62.6

* See footnote on preceding page.

† The values are those reported by Bosart and Snoddy (*Jour. Ind. Eng. Chem.*, **19**, 506 (1927)), and Lane (*Jour. Ind. Eng. Chem.*, **17**, 924 (1925)) but modified by adding 2°F to all temperatures below 0°F in accordance with the suggestion of the Procter and Gamble Co.

Table 10-20 (*Continued*)
COMPOSITIONS OF AQUEOUS ANTIFREEZE SOLUTIONS
FREEZING POINT OF MAGNESIUM CHLORIDE BRINES

% MgCl₂ by weight	Spec. Grav. 15.6°C. (60°F.)	Freezing Point °C.	Freezing Point °F.	% MgCl₂ by weight	Spec. Grav. 15.6°C. (60°F.)	Freezing Point °C.	Freezing Point °F.
5	1.043	−3.11	26.4	18	1.161	−22.1	−7.7
6	1.051	−3.89	25.0	19	1.170	−25.6	−12.2
7	1.060	−4.72	23.5	20	1.180	−27.4	−17.3
8	1.069	−5.67	21.8	21	1.190	−30.6	−23.0
9	1.078	−6.67	20.0	22	1.200	−32.8	−27.0
10	1.086	−7.83	17.9	23	1.210	−28.9	−20.0
11	1.096	−9.05	15.7	24	1.220	−25.6	−14.0
12	1.105	−10.5	13.1	25	1.230	−23.3	−10.0
13	1.114	−12.1	10.3	26	1.241	−21.1	−6.0
14	1.123	−13.7	7.3	27	1.251	−19.4	−3.0
15	1.132	−15.6	4.0	28	1.262	−18.3	−1.0
16	1.142	−17.6	0.4	29	1.273	−17.2	+1.0
17	1.151	−19.7	−3.5	30	1.283	−16.7	2.0

FREEZING POINT OF SODIUM CHLORIDE BRINES

Compiled in collaboration with C. D. Looker, Ph.D., International Salt Co., Inc.

% NaCl by weight	Spec. Grav. 15°C. (59°F.)	Freezing Point °C.	Freezing Point °F.	% NaCl by weight	Spec. Grav. 15°C. (59°F.)	Freezing Point °C.	Freezing Point °F.
0	1.000	0.00	32.0	15	1.112	−10.88	12.4
1	1.007	−0.58	31.0	16	1.119	−11.90	10.6
2	1.014	−1.13	30.0	17	1.127	−12.93	8.7
3	1.021	−1.72	28.9	18	1.135	−14.03	6.7
4	1.028	−2.35	27.8	19	1.143	−15.21	4.6
5	1.036	−2.97	26.7	20	1.152	−16.46	2.4
6	1.043	−3.63	25.5	21	1.159	−17.78	+0.0
7	1.051	−4.32	24.2	22	1.168	−19.19	−2.5
8	1.059	−5.03	22.9	23	1.176	−20.69	−5.2
9	1.067	−5.77	21.6	23.3 (*E*)	1.179	−21.13	−6.0
10	1.074	−6.54	20.2	24	1.184	−17.0*	+1.4*
11	1.082	−7.34	18.8	25	1.193	−10.4*	13.3*
12	1.089	−8.17	17.3	26	1.201	−2.3*	27.9*
13	1.097	−9.03	15.7	26.3	1.203	0.0*	32.0*
14	1.104	−9.94	14.1				

* Saturation temperatures of sodium chloride dihydrate; at these temperatures NaCl·2H₂O separates leaving the brine of the eutectic composition (*E*).

PROPYLENE GLYCOL-GLYCEROL

Propylene glycol, a satisfactory antifreeze with the advantage of being non-toxic, can be combined with glycerol, also an efficient non-toxic antifreeze, to give a mixture that can be tested for freezing point with an ethylene glycol (Prestone) hydrometer. A mixture of 70% propylene glycol and 30% glycerol (% by weight of water-free materials), when diluted, can be tested on the standard instrument used for ethylene glycol solutions.

Table 10-20 (Continued)
COMPOSITIONS OF AQUEOUS ANTIFREEZE SOLUTIONS
FREEZING POINT OF CALCIUM CHLORIDE BRINES

Reproduced from "Dow Calcium Chloride for Refrigeration," copyright 1929 by the Dow Chemical Co., by permission.

Density of Brine @ 60°F.	Specific Baume Gravity Scale	Freezing Point °F.	Freezing Point °C.	Ammonia Suction Pressure lbs. gauge Corresponding to Freezing Point	Calcium Chloride Content of Brine in Per Cent — Anhydrous CaCl2	Calcium Chloride Content of Brine in Per Cent — Dow 73-75% CaCl2	Anhydrous CaCl2 Content of Brine† — Pounds Per Gallon	Anhydrous CaCl2 Content of Brine† — Pounds Per Cu. Ft.	Weight of Brine @ 60°F. — Pounds Per Gallon	Weight of Brine @ 60°F. — Pounds Per Cu. Ft.	Specific Heat −20	Specific Heat −10	Specific Heat 0	Specific Heat +10	Specific Heat +20	Heat Capacity −20	Heat Capacity −10	Heat Capacity 0	Heat Capacity +10	Heat Capacity +20
1.00	0.0	+32.0	0.0	47.6	0.0	0.0	0.0	0.0	8.34	62.4										
1.01	1.4	+31.1	−0.5	46.4	1.1	1.48	0.093	0.69	8.41	63.1										
1.02	2.8	+30.2	−1.0	45.3	2.3	3.11	0.196	1.46	8.50	63.7										
1.03	4.2	+29.1	−1.8	43.9	3.5	4.73	0.301	2.25	8.59	64.3										
1.04	5.6	+28.0	−2.2	42.6	4.7	6.35	0.407	3.04	8.67	64.9										
1.05	6.9	+27.0	−2.8	41.4	5.8	7.99	0.508	3.80	8.75	65.5										
1.06	8.2	+25.9	−3.2	40.1	7.0	9.46	0.619	4.62	8.84	66.2										
1.07	9.5	+24.6	−4.1	38.6	8.1	10.93	0.723	5.40	8.92	66.8										
1.08	10.7	+23.4	−4.8	37.2	9.2	12.41	0.829	6.19	9.00	67.4										
1.09	12.0	+21.7	−5.9	35.4	10.4	14.08	0.945	7.06	9.09	68.1										
1.10	13.2	+20.3	−6.5	33.8	11.4	15.40	1.05	7.81	9.17	68.7					0.829					7.60
1.11	14.4	+18.5	−7.5	31.9	12.5	16.90	1.16	8.64	9.25	69.3					0.815					7.55
1.12	15.5	+16.5	−8.6	29.9	13.5	18.25	1.26	9.42	9.34	69.9					0.802					7.49
1.13	16.7	+14.4	−9.9	27.9	14.6	19.75	1.38	10.28	9.42	70.6					0.789					7.44
1.14	17.8	+12.0	−11.1	25.6	15.6	21.05	1.48	11.08	9.50	71.2					0.776					7.38
1.15	18.9	+9.7	−12.4	23.5	16.6	22.41	1.59	11.89	9.58	71.8				0.759	0.764				7.28	7.33
1.16	20.0	+7.0	−13.9	21.2	17.6	23.80	1.70	12.71	9.67	72.4				0.748	0.753				7.24	7.29
1.17	21.1	+4.1	−15.6	18.9	18.6	25.15	1.81	13.56	9.75	73.1				0.737	0.742				7.19	7.24
1.18	22.1	+1.4	−17.0	16.8	19.5	26.38	1.92	14.34	9.83	73.7				0.727	0.732				7.15	7.20
1.19	23.2	−2.2	−19.0	14.2	20.5	27.75	2.02	15.20	9.92	74.3			0.712	0.717	0.722			7.07	7.12	7.17
1.20	24.2	−5.8	−21.0	11.7	21.5	29.06	2.13	16.07	10.01	74.9			0.703	0.707	0.712			7.03	7.08	7.13
1.21	25.2	−9.4	−23.0	9.4	22.4	30.30	2.25	16.89	10.08	75.5			0.694	0.699	0.704			7.00	7.05	7.10
1.22	26.2	−13.2	−25.1	7.3	23.3	31.50	2.37	17.71	10.17	76.2		0.681	0.686	0.691	0.696		6.93	6.98	7.03	7.08
1.23	27.1	−17.1	−27.2	5.1	24.2	32.70	2.48	18.54	10.25	76.8		0.674	0.679	0.683	0.688		6.92	6.96	6.99	7.06
1.24	28.1	−21.3	−29.6	3.1	25.1	33.97	2.60	19.39	10.33	77.4	0.663	0.667	0.672	0.676	0.681	6.84	6.89	6.94	6.99	7.04
1.25	29.0	−25.8	−32.1	1.3	26.0	35.13	2.71	20.25	10.41	78.1	0.657	0.661	0.665	0.669	0.674	6.84	6.89	6.93	6.96	7.02
1.26	29.9	−30.8	−34.9	* 1.7	26.9	36.40	2.83	21.12	10.50	78.7	0.651	0.655	0.659	0.663	0.667	6.84	6.88	6.92	6.96	7.00
1.27	30.8	−36.4	−38.0	* 6.1	27.8	37.60	2.94	21.93	10.58	79.3	0.646	0.650	0.653	0.657	0.660	6.84	6.87	6.91	6.95	6.98
1.28	31.7	−44.1	−42.3	*11.2	28.7	38.80	3.06	22.89	10.66	79.9	0.642	0.645	0.647	0.650	0.654	6.84	6.88	6.90	6.93	6.98
1.29	32.6	−59.8	−51.0	*18.5	29.6	40.00	3.18	23.79	10.75	80.5	0.637	0.640	0.642	0.645	0.648	6.85	6.88	6.90	6.93	6.96
1.30	33.5	−41.8	−41.0	* 9.9	30.5	41.20	3.30	24.70	10.84	81.2	0.633	0.635	0.637	0.640	0.643	6.86	6.88	6.91	6.94	6.97

* Inches of mercury below one standard atmosphere (29.92 in.)

* To find equivalent weight of Dow 73-75% Solid multiply by 1.35 and for weight of Dow Flake 77-80% Calcium Chloride multiply by 1.28.

Table 10-21

PHYSICAL PROPERTIES OF REFRIGERANTS

Compiled from information contained in a private communication from KINETIC CHEMICALS, INC., Wilmington, Del.; U.S.A. and from Cir. No. 2, *Am. Soc. Refrigeration Engineers* (1926).

Refrigerant	Ammonia	Carbon Dioxide	Ethyl Chloride	Methyl Chloride	Sulfur Dioxide	Freon*	F—11*	F—21*	F—114*
Chemical symbol	NH_3	CO_2	C_2H_5Cl	CH_3Cl	SO_2	CCl_2F_2	CCl_3F	$CHCl_2F$	$C_2Cl_2F_4$
Molecular weight	17.032	44.00	64.50	50.48	64.06	120.914	137.371	102.922	170.914
Color of liquid	colorless	colorless	colorless	colorless	colorless	colorless	colorless	colorless	colorless
Odor	pungent aromatic	odorless	pungent etherial odor; sweetish taste	similar to chloroform, but less sweet	pungent characteristic	etherial odor	etherial odor	etherial odor	etherial odor
Density of liquid (Water=1)	0.6818	See Note X	0.9232	0.998	1.4601	1.445	1.568	1.446	1.570
at, deg. C.	−33.35		0.0	−24.09	−10.0	−15.0	−15.0	−15.0	−15.0
Density of gas, grams 1 liter (See Note Y)	0.7708	1.9768	2.31	2.3045	2.9267	5.44	6.20	4.60	7.75
(Air=1)	0.5962	1.5290		1.7824	2.2636	4.21	4.80	3.56	6.0
Boiling point at 1 atm., deg. C.	−33.35	See Note X	13.1	−24.09	−10.0	−29.8	23.7	8.9	3.5
deg. F.	−28.03		55.6	−11.36	14.0	−21.7	74.67	48.0	38.4
Melting point, deg. C.	−77.70	−78.52	−138.7	−91.5	−75.2	−155.0	−88.0	−127.0	−105.5
deg. F.	−107.86	−109.34	−217.7	−132.7	−103.4	−247.0	−126.4	−196.6	−158.0
Critical temperature, deg. C.	132.9	31.00	182.8	143.12	157.12	111.7	196.6		146.1
deg. F.	271.2	87.80	361.0	289.6	314.82	233.0	386.0		295.0
Critical pressure, atm. abs.	112.3	72.85	53.3	65.93	77.65	39.4	41.7		35.5
lbs. per sq. in. abs.	1651	1071	784	969.2	1141.55	580	612		550
Specific heat of constant pressure (C_p)	0.5202	0.2025	0.273	0.24	0.1511	0.1476	0.147	0.18	
Specific heat of constant volume (C_v)	0.4011	0.1558		0.20		0.1297	0.1296	0.1607	
Ratio of specific heats (C_p/C_v) at deg. C.	1.2969	1.3003	1.1257	1.1991	1.256	1.138	1.135	1.12	1.106
(at deg. C.)		0.0	73	66—86	16—34				
Latent heat of vaporization at 1 atm., BTU per lb.	589.4	256.3 See Note X	168.6	180.6	172.3	71.95	78.8	101.8	58.3
Suction pressure at 5°F., lb. per sq. in. abs.	34.5	331.8	4.65	20.89	11.82	26.51	2.96	5.5	7.2
Head pressure at 86°F., lb. per sq. in. abs.	168.5	1039.6	27.1	95.53	65.9	107.9	18.3	30.5	35.8

*Trade names for dichlorodifluoromethane, trichloromonofluoromethane, dichloromonofluoromethane, and dichlorotetrafluoroethane.
Note X—Carbon Dioxide not a liquid at atmospheric pressure.
Note Y—Density of gas at 0°C. (32°F.) and 760 mm. (1 atm.), except for Ethyl Chloride for which no temperature has been given.

Table 10-22
CRYOSCOPIC CONSTANTS

The cryoscopic constant K_f gives the depression of the melting point, ΔT (in °C), produced when one mole of solute is dissolved in 1000 g of a solvent. It is applicable only to dilute solutions for which the number of moles of solute is negligible in comparison with the number of moles of solvent. It is often used for molecular weight determinations,

$$M_2 = \frac{1000 w_2 K_f}{w_1 \, \Delta T}$$

where w_1 is the weight of solvent and w_2 is weight of solute whose molecular weight is M_2.

K_f	Melting Point, °C	Compound	K_f	Melting Point, °C	Compound
1.853	0.0	Water	7.40	40.90	Phenol
2.77	8.27	Formic acid	7.88	26.87	Diphenyl ether
3.85	2.55	Formamide	8.37	25.82	2-Methyl-2-propanol
3.90	16.66	Acetic acid	11.0	52	2,2-Dimethyl-1-propanol
4.04	80.00	Acetamide	12.5	9.79	1,2-Dibromoethane
4.07	18.54	Dimethyl sulfoxide	14.4	8.05	Bromoform
4.3	13.263	p-Xylene	14.52	3.63	Bicyclohexane
4.63	11.8	1,4-Dioxane	17.95	73.4	Antimony(III) chloride
5.12	5.533	Benzene	18.26	57.88	Succinonitrile
5.60	30.944	o-Cresol	20.0	6.54	Cyclohexane
6.27	3.60	Dibenzyl ether	21.7	0.0	1,1,2,2-Tetrabromoethane
6.65	30.55	N-Methylacetamide	29.8	−22.95	Carbon tetrachloride
6.852	5.76	Nitrobenzene	37.7	178.75	Camphor
6.94	80.290	Naphthalene	39.3	6.544	Cyclohexanol
6.96	34.739	p-Cresol	64.1	−30.9	Sulfolane

HUMIDIFICATION AND DRYING

Table 10-23
MASS OF WATER VAPOR IN SATURATED AIR

The values in the table are grams of water contained in a cubic meter (m^3) of saturated air at a total pressure of 101 325 Pa (1 atm).

°C	$g \cdot m^{-3}$	°C	$g \cdot m^{-3}$	°C	$g \cdot m^{-3}$
−30	0.341	12	10.65	53	95.56
−29	0.375	13	11.35	54	100.0
−28	0.413	14	12.05	55	104.5
−27	0.456	15	12.80	56	109.1
−26	0.504	16	13.60	57	114.1
−25	0.554	17	14.45	58	119.2
−24	0.607	18	15.35	59	124.7
−23	0.667	19	16.30	60	130.2
−22	0.733	20	17.30	61	136.0
−21	0.804	21	18.35	62	142.1
−20	0.883	22	19.40	63	148.4
−19	0.968	23	20.55	64	154.9
−18	1.063	24	21.75	65	161.3
−17	1.164	25	23.05	66	167.9
−16	1.273	26	24.35	67	175.1
−15	1.375	27	25.75	68	182.6
−14	1.510	28	27.20	69	190.3
−13	1.650	29	28.75	70	198.2
−12	1.800	30	30.35	71	206.5
−11	1.965	31	32.05	72	215.1
−10	2.140	32	33.80	73	223.7
−9	2.331	33	35.60	74	233.0
−8	2.539	34	37.55	75	242.0
−7	2.761	35	39.55	76	251.2
−6	3.003	36	41.65	77	261.1
−5	3.250	37	43.90	78	271.6
−4	3.512	38	46.20	79	282.3
−3	3.810	39	48.60	80	293.4
−2	4.131	40	51.21	81	304.8
−1	4.473	41	53.86	82	316.6
0	4.849	42	56.61	83	328.7
1	5.199	43	59.51	84	341.2
2	5.569	44	62.53	85	353.6
3	5.947	45	65.52	86	366.2
4	6.35	46	68.61	87	379.9
5	6.80	47	72.00	88	394.1
6	7.25	48	75.56	89	408.6
7	7.75	49	79.24	90	423.5
8	8.25	50	83.05	91	439.0
9	8.80	51	87.04	92	454.8
10	9.40	52	91.22	93	471.2
11	10.00				

Table 10-24
HUMIDITY AND DEW POINT FROM WET AND DRY BULB READINGS

In the table below $d-w$ is the difference between the dry and the wet bulb thermometer readings, $r.h.$ is the relative humidity, and $a.h.$ is the absolute humidity.

	d−w =0°C.			d−w =1°C.			d−w =2°C.			d−w =3°C.		
Temp. °C.	r. h. %	dew point °C.	a. h. mm of Hg	r. h. %	dew point °C.	a. h. mm of Hg	r. h. %	dew point °C.	a. h. mm of Hg	r. h. %	dew point °C.	a. h. mm of Hg
−20	100	−20	0.8	..	......	....	..	......	....	..	......	....
−10	100	−10	1.9	66	−14.6	1.3	32	−22.1	0.6	..	......	....
− 5	100	− 5	3.0	75	− 8.3	2.3	51	−12.7	1.5	27	−19.4	0.8
0	100	0	4.6	81	− 2.5	3.7	63	− 5.5	2.9	45	− 9.3	2.1
+ 5	100	+ 5	6.5	86	+ 2.8	5.6	72	+ 0.3	4.7	58	− 2.3	3.8
10	100	10	9.2	88	8.1	8.1	76	6.1	7.0	65	+ 3.8	6.0
15	100	15	12.8	90	13.4	11.5	80	11.6	10.2	71	9.7	9.0
16	100	16	13.6	90	14.4	12.3	81	12.7	11.0	71	10.8	9.7
17	100	17	14.5	90	15.4	13.1	81	13.7	11.8	72	12.0	10.5
18	100	18	15.5	91	16.5	14.0	82	14.8	12.6	73	13.1	11.3
19	100	19	16.5	91	17.5	15.0	82	15.9	13.5	74	14.2	12.1
20	100	20	17.5	91	18.5	16.0	83	16.9	14.5	74	15.3	13.0
21	100	21	18.7	91	19.5	17.0	83	18.0	15.5	75	16.4	14.0
22	100	22	19.8	92	20.6	18.2	83	19.1	16.5	76	17.5	15.0
23	100	23	21.1	92	21.6	19.3	84	20.1	17.6	76	18.6	16.0
25	100	25	23.8	92	23.6	21.9	84	22.2	20.1	77	20.7	18.3
30	100	30	31.8	93	28.7	29.5	86	27.4	27.3	79	26.0	25.2

	d−w =4°C.			d−w =5°C.			d−w =6°C.			d−w =7°C.		
Temp. C.	r. h. %	dew point °C	a. h. mm of Hg	r. h. %	dew point °C.	a. h. mm of Hg	r. h. %	dew point °C.	a. h. mm of Hg	r. h. %	dew point °C.	a. h. mm of Hg
0	28	−14.6	1.3	11	−24.2	0.5	..	......	....	..	......	....
5	45	− 5.3	2.9	32	− 9.3	2.1	19	−15.2	1.2	6	−27.1	0.4
10	54	+ 1.2	5.0	44	− 1.5	4.0	34	− 4.6	3.1	24	− 8.7	2.2
12	57	3.9	6.0	48	+ 1.2	5.0	38	− 1.6	4.0	29	− 4.9	3.0
14	60	6.4	7.2	51	4.0	6.1	42	+ 1.3	5.0	34	− 1.6	4.0
15	61	7.6	7.8	52	5.4	6.7	44	2.8	5.6	36	− 0.1	4.5
16	62	8.8	8.5	54	6.7	7.3	46	4.3	6.2	37	+ 1.5	5.1
17	64	10.0	9.2	55	8.0	8.0	47	5.6	6.8	39	3.1	5.7
18	65	11.2	10.0	56	9.2	8.7	49	7.0	7.5	41	4.6	6.3
19	65	12.4	10.8	58	10.5	9.5	50	8.3	8.2	43	6.0	7.0
20	66	13.5	11.6	59	11.7	10.3	51	9.6	9.0	44	7.4	7.7
21	67	14.7	12.5	60	12.9	11.1	52	10.9	9.8	46	8.8	8.5
22	68	15.8	13.5	61	14.1	12.0	54	12.2	10.6	47	10.1	9.3
23	69	16.9	14.5	61	15.2	13.0	55	13.4	11.5	48	11.4	10.1
25	70	19.2	16.7	63	17.5	15.0	57	15.8	13.5	50	14.0	12.0
30	73	24.6	23.3	67	23.2	21.3	61	21.6	19.4	55	20.0	17.6

Table 10-24 (*Continued*)
HUMIDITY AND DEW POINT FROM WET AND DRY BULB READINGS

Temp. °C.	d−w=8°C. r.h. %	d−w=8°C. dew point °C.	d−w=8°C. a.h. mm of Hg	d−w=9°C. r.h. %	d−w=9°C. dew point °C.	d−w=9°C. a.h. mm of Hg	d−w=10°C. r.h. %	d−w=10°C. dew point °C.	d−w=10°C. a.h. mm of Hg	d−w=11°C. r.h. %	d−w=11°C. dew point °C.	d−w=11°C. a.h. mm of Hg
8	7	−22.9	0.6	..	.....	....	..	.....	....	..	.....	....
10	14	−14.5	1.3	5	−26.0	0.4	..	.....	....	..	.....	....
12	20	− 9.1	2.1	11	−15.5	1.2	..	.....	....	..	.....	....
14	25	− 5.0	3.0	17	− 9.5	2.0	9	−16.3	1.1	..	.....	....
15	27	− 3.2	3.5	20	− 7.1	2.5	12	−12.6	1.5	5	−22.6	0.6
16	30	− 1.5	4.0	22	− 5.0	3.0	15	− 9.6	2.0	8	−16.8	1.0
17	32	+ 0.1	4.6	24	− 3.1	3.5	17	− 7.1	2.5	10	−12.8	1.5
18	34	1.8	5.2	27	− 1.3	4.1	20	− 4.9	3.0	13	− 9.6	2.0
19	35	3.4	5.8	29	+ 0.4	4.7	22	− 2.9	3.6	15	− 6.9	2.5
20	37	5.0	6.5	30	2.1	5.3	24	− 1.0	4.2	18	− 4.6	3.1
21	39	6.4	7.2	32	3.8	6.0	26	+ 0.8	4.8	20	− 2.5	3.7
22	40	7.9	8.0	34	5.4	6.7	28	2.6	5.5	22	− 0.6	4.3
23	42	9.3	8.8	36	6.9	7.5	30	4.3	6.2	24	+ 1.3	5.0
24	43	10.7	9.6	37	8.4	8.3	31	5.9	7.0	26	3.1	5.7
25	44	12.0	10.5	38	9.9	9.1	33	7.5	7.8	27	4.9	6.5
26	46	13.3	11.5	40	11.3	10.0	34	9.0	8.6	29	6.6	7.3
28	48	15.9	13.5	42	14.0	12.0	37	11.9	10.5	32	9.7	9.0
30	50	18.4	15.8	44	16.6	14.1	39	14.7	12.5	34	12.7	11.0

Table 10-25
RELATIVE HUMIDITIES AND AQUEOUS TENSIONS OF AQUEOUS SOLUTIONS OF H_2SO_4, NaOH AND $CaCl_2$ AT 25°C

Concentrations are expressed in percentage of anhydrous solute by weight. Stokes and Robinson, *Ind. Eng. Chem.*, 41, 2013 (1949).

% Humidity	Aqueous Tension	% H_2SO_4	% NaOH	% $CaCl_2$	% Humidity	Aqueous Tension	% H_2SO_4	% NaOII	% $CaCl_2$
100	23.756	0.00	0.00	0.00	50	11.88	43.10	28.15	35.64
95	22.57	11.02	5.54	9.33	45	10.69	45.41	29.86	37.61
90	21.38	17.91	9.83	14.95	40	9.50	47.71	31.58	39.62
85	20.19	22.88	13.32	19.03	35	8.31	50.04	33.38	41.83
80	19.00	26.79	16.10	22.25	30	7.13	52.45	35.29	44.36
75	17.82	30.14	18.60	24.95	25	5.94	55.01	37.45	.....
70	16.63	33.09	20.80	27.40	20	4.75	57.76	40.00	.....
65	15.44	35.80	22.80	29.64	15	3.56	60.80	43.32	.....
60	14.25	38.35	24.66	31.73	10	2.38	64.45	47.97	.....
55	13.07	40.75	26.42	33.71	5	1.19	69.44	.....	.....

Table 10-26
SOLUTIONS FOR MAINTAINING CONSTANT HUMIDITY

A saturated aqueous solution in contact with an excess of a definite solid phase at a given temperature will maintain constant humidity in an enclosed space. Table 10-25 and this table give a number of salts suitable for this purpose. The aqueous tension (in mm Hg) of a solution at a given temperature is found by multiplying the decimal fraction of the humidity by the aqueous tension at 100 per cent humidity for the specific temperature. For example, the aqueous tension of a saturated solution of NaCl at 20°C is $0.757 \times 17.54 = 13.28$ mm Hg, and at 80°C is $0.764 \times 355.1 = 271.3$ mm Hg.

Solid Phase	% Humidity at Specified Temperatures (°C)						
	10	20	25	30	40	60	80
$K_2Cr_2O_7$			98.0				
K_2SO_4	98	97	97	96	96	96	
KNO_3	95	93	92.5	91	88	82	
KCl	88	85.0	84.3	84	81.7	80.7	79.5
KBr		84	80.7		79.6	79.0	79.3
NaCl	76	75.7	75.3	74.9	74.7	74.9	76.4
$NaNO_3$			73.8	72.8	71.5	67.5	65.5
$NaNO_2$		66	65	63.0	61.5	59.3	58.9
$NaBr \cdot 2H_2O$		57.9	57.7		52.4	49.9	50.0
$Na_2Cr_2O_7 \cdot 2H_2O$	58	55	54		53.6	55.2	56.0
$Mg(NO_3)_2 \cdot 6H_2O$	57	55	52.9	52	49	43	
$K_2CO_3 \cdot 2H_2O$	47	44	42.8		42		
$MgCl_2 \cdot 6H_2O$	34	33	33.0	33	32	30	
$KF \cdot 2H_2O$				27.4	22.8	21.0	22.8
$KC_2H_3O_2 \cdot 1.5H_2O$	24	23	22.5	22	20		
$LiCl \cdot H_2O$	13	12	10.2	12	11	11	
KOH	13	9	8	7	6	5	
100% Humidity: Aqueous Tension (mm Hg)	9.21	17.54	23.76	31.82	55.32	149.4	355.1

Table 10-27
DRYING AGENTS

Drying Agent	Most Useful For	Residual Water, mg H_2O per Liter of Dry Air (25°C)	Grams Water Removed per Gram of Desiccant	Regeneration, °C
Al_2O_3	Hydrocarbons	0.002–0.005	0.2	175 (24 h)
$Ba(ClO_4)_2$[a]	Inert gas streams	0.6–0.8	0.17	140
BaO	Basic gases: hydrocarbons, aldehydes, alcohols	0.0007–0.003	0.12	1000
CaC_2[b]	Ethers		0.56	Impossible
$CaCl_2$[c]	Inert organics	0.1–0.2	0.15 (1 H_2O) 0.30 (2 H_2O)	250
CaH_2[d]	Hydrocarbons, ethers, amines, esters, higher alcohols	1×10^{-5}	0.85	Impossible
CaO	Ethers, esters, alcohols, amines	0.01–0.003	0.31	Difficult, 1000
$CaSO_4$	Most organic substances	0.005–0.07	0.07	225
Dow Desiccant 812[e]	Most materials	(5–200 ppm)		No
K_2CO_3	Most materials except acids and phenols		0.16	158
KOH	Amines	0.01–0.9		Impossible
$LiAlH_4$[f]	Hydrocarbons		1.9	Impossible
$Mg(ClO_4)_2$[a]	Gas streams	0.0005–0.002	0.24	250 (high vacuum)
MgO	All but acidic compounds	0.008	0.45	800
$MgSO_4$	Most organic compounds	1–12	0.15–0.75	Not feasible
Molecular sieves: 4X	Molecules with effective diameter > 4Å	0.001	0.18	250
5X	Molecules with effective diameter > 5Å	0.001	0.18	250
9.5% Na-Pb alloy[d]	Hydrocarbons, ethers	(For solvents only)	0.08	Impossible
Na_2SO_4	Ketones, acids, alkyl and aryl halides	12	1.25	150
P_2O_5	Gas streams; not suitable for alcohols, amines, ketones, or amines	2×10^{-5}	0.5	Not feasible
Silica gel	Most organic amines	0.002–0.07	0.2	200–350
Sulfuric acid	Air and inert gas streams	0.003–0.008	Indefinite	Not feasible

[a] May form explosive mixtures when contacting organic material.   [b] Explosive C_2H_2 formed.
[c] Slow in drying action.   [d] H_2 formed.   [e] Used as column drying of organic liquids.   [f] Strong reductant.

DENSITY AND SPECIFIC GRAVITY

HYDROMETERS

Various hydrometers and the relation between the various scales.

Alcoholometer. This hydrometer is used in determining the density of aqueous ethyl alcohol solutions; the reading in degrees is numerically the same as the percentage of alcohol by volume. The scale known as Tralle gives the percentage by volume. Wine and Must hydrometer relations are given below.

Ammoniameter. This hydrometer, employed in finding the density of aqueous ammonia solutions, has a scale graduated in equal divisions from 0° to 40°. To convert the reading to specific gravity multiply by 3 and subtract the resulting number from 1000.

Balling Hydrometer. See under Saccharometer.

Barkometer or **Barktrometer.** This hydrometer, which is used in determining the density of tanning liquors, has a scale from 0° to 80° Bk; the number to the right of the decimal point of a specific gravity reading is the corresponding Bk degree; thus, a specific gravity of 1.015 is 15°Bk.

Baumé Hydrometers. For liquids heavier than water.—This hydrometer was originally based on the density of a 10% sodium chloride solution, which was given the value of 10°, and the density of pure water, which was given the value of 0°; the interval between these two values was divided into 10 equal parts. Other reference points have been taken with the result that so much confusion exists that there are about 36 different scales in use, many of which are incorrect. In general a Baumé hydrometer should have inscribed on it the temperature at which it was calibrated and also the temperature of the water used in relating the density to a specific gravity.

The following expression gives the relation between the specific gravity and several of the Baumé scales:—

$$\text{Specific gravity} = \frac{m}{m - \text{Baumé}}$$

$m = 145$ at 60°/60°F (15.56°C) for the American Scale

$m = 144$ for the old scale used in Holland

$m = 146.3$ at 15°C for the Gerlach Scale

$m = 144.3$ at 15°C for the Rational Scale generally used in Germany

For liquids lighter than water.— Originally the density of a solution of 1 gram of sodium chloride in 9 grams of water at 12.5°C was given a value of 0°Bé and pure water a value of 10°Bé. The scale between these points was divided into ten equal parts and these divisions were repeated throughout the scale giving a relation which could be expressed by the formula: Specific gravity $= 145.88/(135.88 + \text{Bé})$, which is approximately equal to $146/(136 + \text{Bé}.)$ Other scales have since come into more general use such as that of the Bu. of Standards in which the specific gravity at 60°/60°F $= 140/(130 + \text{Bé}.)$ and that of the American Petroleum Institute (A.P.I. Scale) in which the specific gravity at 60°/60°F $= 141.5/(131.5 + \text{API}°.)$

See also special table for conversion to density and Twaddell scale.

Beck's Hydrometer. This hydrometer is graduated to show a reading of 0° in pure water and a reading of 30° in a solution with a specific gravity of 0.850, with equal scale divisions above and below these two points.

Brix Hydrometer. See under Saccharometers.

Cartier's Hydrometer. This hydrometer shows a reading of 22° when immersed in a solution having a density of 22° Baumé but the scale divisions are smaller than on the Baumé hydrometer in the ratio of 16 Cartier to 15 Baumé.

Fatty Oil Hydrometer. The graduations on this hydrometer are in specific gravity within the range 0.908 to 0.938. The letters on the scale correspond to the specific gravity of the various common oils as follows: *R*, rape; *O*, olive; *A*, almond; *S*, sesame; *HL*, hoof oil; *HP*, hemp; *C*, cotton seed; *L*, linseed. See also Oleometer below.

Lactometers. These hydrometers are used in determining the density of milk. The various scales in common use are the following:

New York Board of Health has a scale graduated into 120 equal parts, 0° being equal to the specific gravity of water and 100° being equal to a specific gravity of 1.029.

Quevenne lactometer is graduated from 15° to 40° corresponding to specific gravities from 1.015 to 1.040.

Soxhlet lactometer has a scale from 25° to 35° corresponding to specific gravities from 1.025 to 1.035 respectively.

Oleometer. A hydrometer for determining the density of vegetable and sperm oils with a scale from 50° to 0° corresponding to specific gravities from 0.870 to 0.970. See also Fatty Oil hydrometer above.

Saccharometers. These hydrometers are used in determining the density of sugar solutions. Solutions of the same concentration but of different carbohydrates have very nearly the same specific gravity and in general a concentration of 10 grams of carbohydrate per 100 cc of solution shows a specific gravity of 1.0386. Thus, the wt. of sugar in 1000 ml soln. is, (a) For conc. <12g/100 ml: (wt. of 1000 ml soln. −1000) ÷0.386; (b) for conc. >12g/100 ml: (wt. of 1000 ml soln. − 1000) ÷0.385.

Brix hydrometer is graduated so that the number of degrees is identical with the percentage by weight of cane sugar and is used at the temperature indicated on the hydrometer.

Balling's saccharometer is used in Europe and is practically identical with the Brix hydrometer.

Bates brewers' saccharometer which is used in determining the density of malt worts is graduated so that the divisions express pounds per barrel (32 gallons). The relation between degrees Bates (=b) and degrees Balling (=B) is shown by the following formula: B= 260b/(360+b).

See also below under Wine and Must.

Salinometer. This hydrometer, which is used in the pickling and meat packing plants, is graduated to show percentage of saturation of a sodium chloride solution. An aqueous solution is completely saturated when it contains 26.4% pure sodium chloride. The range from 0% to 26.4% is divided into 100 parts, each division therefore representing 1% of saturation. In another type of salinometer, the degrees correspond to percentages of sodium chloride expressed in grams of sodium chloride per 100 cc of water.

Sprayometer (Parrot and Stewart).— This hydrometer which is used in determining the density of *lime sulfur* solutions has two scales; one scale is graduated from 0° to 38° Baumé and the other scale is from 1.000 to 1.350 specific gravity.

Tralle Hydrometer. See Alcoholometer above.

Twaddell Hydrometer. This hydrometer, which is used only for liquids heavier than water, has a scale such that when the reading is multiplied by 5 and added to 1000 the resulting number is the specific gravity with reference to water as 1000. To convert specific gravity at 60°/60°F to Twaddell degrees, take the decimal portion of the specific gravity value and multiply it by 200; thus a specific gravity of 1.032=0.032 × 200=6.4° Tw. See also special table for conversion to density and Baumé scale.

Wine and Must Hydrometer. This instrument has three scales. One scale shows readings of 0° to 30° Brix for sugar (see Brix hydrometer above); another scale from 0° to 15° Tralle is used for sweet wines to indicate the percentage of alcohol by volume; and a third scale from 0° to 20° Tralle is used for tart wines to indicate the percentage of alcohol by volume.

SPECIFIC GRAVITY CORRECTIONS FOR THE BUOYANT EFFECT OF THE AIR

Determinations made with a pyknometer.

$$D_{vac} = \frac{W_2}{W_1}d - 0.0012\left(\frac{W_2 d}{W_1} - 1\right)$$

$$S_{vac} = \frac{W_2}{W_1} - 0.0012\left(\frac{W_2}{W_1} - 1\right)$$

Where

D_{vac} = density of the liquid in grams per milliliter at t°C corrected for the buoyant effect of air.

W_1 = weight in air of the water required to fill the pyknometer at t°C.

W_2 = weight in air of the liquid required to fill the pyknometer at t°C.

d = density of water in grams per milliliter at t°C.

S_{vac} = specific gravity of the liquid at t°C. referred to water at t°C corrected for the buoyant effect of air.

When the weight of the water is determined at a temperature of t°C, and that of the liquid at a different temperature t′, the equations above are modified as follows:

$$D_{vac} = \frac{W_2}{W_1}d - 0.0012\left(\frac{W_2}{W_1}d - 1\right) + 0.000026\,(t' - t°)\left(\frac{W_2}{W_1}d\right)$$

$$S_{vac} = \frac{W_2}{W_1} - 0.0012\left(\frac{W_2}{W_1} - 1\right) + 0.000026\,(t' - t°)\left(\frac{W_2}{W_1}\right)$$

Determinations made with a plummet or sinker.

The equations above may also be used when the density is determined with plummet or sinker, but in this case

W_1 = weight of the plummet in air minus its weight in water.

W_2 = weight of the plummet in air minus its weight in the liquid.

CONVERSION OF SPECIFIC GRAVITY AT 25°/25° C. TO DENSITY AT ANY TEMPERATURE FROM 0° TO 40° C.

Cf. Dreisbach, *Ind. Eng. Chem., Anal. Ed. 12, 160 (1940).*

Liquids change volume with change in temperature, but the amount of this change, β (coefficient of cubical expansion), varies widely with different liquids, and to some extent for the same liquid at different temperatures. See special table, *Coefficients of Cubical Expansion for Various Liquids and Aqueous Solutions.*

The table below, which is calculated from the relationship:

$$F\beta_t = \frac{\text{density of water at } 25°C. \ (=0.99705)}{[1-\beta\ (25-t)]} \quad (1)$$

may be used to find d^t, the density (weight of 1 cc.), of a liquid at any temperature (t) between 0° and 40°C. if the specific gravity at 25°/25°C. (S) and the coefficient of cubical expansion (β) are known. Substitutions are made in the equations:

$$d^t = SF\beta_t \quad (2)$$
$$S = \frac{d^t}{F\beta_t} \quad (3)$$

TABLE—FACTORS ($F\beta_t$)

Density $t°C. =$ Sp. Gr. $25°/25° \times F\beta_t$

$*\beta \times 10^3$ \ °C.	0	5	10	15	20	25	30	35	40
1.3	1.0306	1.0237	1.0169	1.0102	1.0036	0.99705	0.99065	0.9843	0.9780
1.2	1.0279	1.0216	1.0154	1.0092	1.0031	0.99705	0.9911	0.9853	0.9794
1.1	1.0253	1.0195	1.0138	1.0082	1.0026	0.99705	0.9916	0.9963	0.9809
1.0	1.0227	1.0174	1.0123	1.0072	1.0021	0.99705	0.9921	0.9872	0.98234
0.9	1.0200	1.0153	1.0107	1.0060	1.0016	0.99705	0.99262	0.9882	0.9638
0.8	1.0174	1.0133	1.0092	1.0051	1.0011	0.99705	0.9931	0.98918	0.9851
0.7	1.0148	1.0113	1.0077	1.0041	1.0006	0.99705	0.9936	0.99015	0.98672
0.6	1.0122	1.0092	1.0061	1.0031	1.0001	0.99705	0.9941	0.9911	0.9882
0.5	1.0097	1.0072	1.0046	1.0021	0.99958	0.99705	0.9944	0.9921	0.9897
0.	1.0071	1.0051	1.0031	1.0011	0.99908	0.99705	0.9951	0.9931	0.9911

$*\beta =$ coefficient of cubical expansion.

EXAMPLES

All examples are based upon an assumed coefficient of cubical expansion, β, of 1.3×10^{-3}.

Example 1. To find the density of a liquid at 20°C., d^{20}, which has a specific gravity (S) of $1.2500_{25}^{25°}$.

From the table above $F\beta_t$ at 20°C. $= 1.0036$.

$$d^{20} = d^t = SF\beta_t = 1.2500 \times 1.0036 = 1.2545$$

CONVERSIONS OF DENSITY—SPECIFIC GRAVITY

Example 2. To find the density at 20°C. (d^{20}) of a liquid which has a specific gravity of $1.2500^{17°}_{4°}$.

Since the density of water at 4°C. is equal to 1, specific gravity at $17°/4° = d^{17} = 1.2500$.

Substitution in equation *3* with $F_{\beta t}$ at 17°C., by interpolation from the table, equal to 1.00756, gives

$$Sp.\ gr.\ 25°/25° = S = 1.2500 \div 1.00756$$

Substitution of this value for S in equation *2* with $F_{\beta t}$ at 20°C., from the table, equal to 1.0036, gives

$$d^{20} = d^t = (1.2500 \div 1.00756) \times 1.0036 = 1.2451$$

Example 3. To find the specific gravity at 20°/4°C. of a liquid which has a specific gravity of $1.2500^{25°}_{4°}$.

Since the density of water at 4°C. is equal to 1, specific gravity $25°/4° = d^{25} = 1.2500$; and, specific gravity $20°/4° = d^{20}$.

Substitution in equation *3*, with $d^t = 1.2500$; and, with $F_{\beta t}$ at 25°C., from the table, equal to 0.99705, gives

$$Sp.\ gr.\ 25°/25° = S = 1.2500 \div 0.99705$$

Substitution of this value for S in equation *2*, with $F_{\beta t}$ at 20°C., from the table, equal to 1.0036, gives

$$Sp.\ gr.\ 20°/4° = d^{20} = (1.2500 \div 0.99705) \times 1.0036 = 1.2582$$

Example 4. To find the density at 25°C. of a liquid which has a specific gravity of $1.2500^{15°}_{15°}$.

Since the density of water (see special table, *Absolute Density of Water*) at 15°C. = 0.99910,

$$d^{15} = sp.\ gr.\ 15°/15° \times 0.99910 = 1.2500 \times 0.99910$$

Substitution in equation *3*, with $F_{\beta t}$ at 15°C., from the table, equal to 1.0102, gives

$$Sp.\ gr.\ 25°/25° = S = (1.2500 \times 0.99910) \div 1.0102$$

Substitution of this value for S in equation *2*, with $F_{\beta t}$ at 25°, from the table, equal to 0.99705, gives

$$d^{25} = d^t = (1.2500 \times 0.99910 \div 1.0102) \times 0.99705 = 1.2326$$

Table 10-28
DENSITY OF MERCURY AND WATER

The density of mercury and pure air-free water under a pressure of 101 325 Pa (1 atm) is given in units of grams per cubic centimeter ($g \cdot cm^{-3}$). For mercury, the values are based on the density at 20°C being 13.545 884 $g \cdot cm^{-3}$. Water attains its maximum density of 0.999 973 $g \cdot cm^{-3}$ at 3.98°C. For water, the temperature (t_m, °C) of maximum density at different pressures (p) in atmospheres is given by

$$t_m = 3.98 - 0.0225(p - 1)$$

Density of Water	Temp., °C	Density of Mercury	Density of Water	Temp., °C	Density of Mercury
	−20	13.644 59	0.987 12	52	13.467 68
	−18	13.639 62	0.986 18	54	13.462 82
	−16	13.634 66	0.985 21	56	13.457 96
	−14	13.629 70	0.984 22	58	13.453 09
	−12	13.624 75	0.983 20	60	13.448 23
	−10	13.619 79	0.982 16	62	13.443 37
	−8	13.614 85	0.981 09	64	13.438 52
	−6	13.609 90	0.980 01	66	13.433 67
	−4	13.604 96	0.978 90	68	13.428 82
	−2	13.600 02	0.977 77	70	13.423 97
0.999 84	0	13.595 08	0.976 61	72	13.419 13
0.999 94	2	13.590 15	0.975 44	74	13.414 28
0.999 97	4	13.585 22	0.974 24	76	13.409 43
0.999 94	6	13.580 29	0.973 03	78	13.404 60
0.999 85	8	13.575 36	0.971 79	80	13.399 77
0.999 70	10	13.570 44	0.970 53	82	13.394 92
0.999 50	12	13.565 52	0.969 26	84	13.390 09
0.999 24	14	13.560 60	0.967 96	86	13.385 26
0.998 94	16	13.555 70	0.966 65	88	13.380 42
0.998 60	18	13.550 79	0.965 31	90	13.375 60
0.998 20	20	13.545 88	0.963 96	92	13.370 77
0.997 77	22	13.540 97	0.962 59	94	13.365 94
0.997 30	24	13.536 06	0.961 20	96	13.361 12
0.996 78	26	13.531 17	0.959 79	98	13.356 30
0.996 23	28	13.526 26	0.958 36	100	13.351 48
0.995 65	30	13.521 37		120	13.303 4
0.995 03	32	13.516 47		140	13.255 4
0.994 37	34	13.511 58		160	13.207 6
0.993 69	36	13.506 70		180	13.159 8
0.992 97	38	13.501 82		200	13.112 0
0.992 22	40	13.496 93		220	13.064 5
0.991 44	42	13.492 07		240	13.016 9
0.990 63	44	13.487 18		260	12.969 2
0.989 79	46	13.482 29		280	12.921 5
0.988 93	48	13.477 42		300	12.873 7
0.988 04	50	13.472 56			

Table 10-29
SPECIFIC GRAVITY OF AIR AT VARIOUS TEMPERATURES

The table below gives the weight in grams $\times 10^4$ of one milliliter of air at 760 mm of mercury pressure and at the temperature indicated. Density in grams per milliliter is the same as the specific gravity referred to water at 4°C. as unity. To convert to density referred to air at 70°F. as unity, divide the values below by 12.00.

t°C.	Sp.Gr. $\times 10^4$	t°C.	Sp.Gr. $\times 10^4$	t°C.	Sp.Gr. $\times 10^4$	t°C.	Sp.Gr. $\times 10^4$
−25	14.240	15	12.255	60	10.596	140	8.541
−24	14.182	16	12.213	62	10.532	142	8.500
−23	14.125	17	12.170	64	10.470	144	8.459
−22	14.069	18	12.129	66	10.408	146	8.419
−21	14.013	19	12.087	68	10.347	148	8.379
−20	13.957	20	12.046	70	10.286	150	8.339
−19	13.902	21	12.004	72	10.227	155	8.242
−18	13.847	22	11.964	74	10.168	160	8.147
−17	13.793	23	11.923	76	10.109	165	8.054
−16	13.739	24	11.883	78	10.052	170	7.963
−15	13.685	25	11.843	80	9.995	175	7.874
−14	13.632	26	11.803	82	9.938	180	7.787
−13	13.580	27	11.764	84	9.882	185	7.702
−12	13.527	28	11.725	86	9.828	190	7.619
−11	13.476	29	11.686	88	9.773	195	7.537
−10	13.424	30	11.647	90	9.719	200	7.457
−9	13.373	31	11.609	92	9.666	205	7.379
−8	13.322	32	11.570	94	9.613	210	7.303
−7	13.272	33	11.533	96	9.561	215	7.228
−6	13.222	34	11.495	98	9.509	220	7.155
−5	13.173	35	11.458	100	9.458	230	7.013
−4	13.124	36	11.420	102	9.408	240	6.881
−3	13.075	37	11.383	104	9.358	250	6.753
−2	13.026	38	11.347	106	9.308	260	6.624
−1	12.978	39	11.310	108	9.259	270	6.504
0	12.931	40	11.274	110	9.211	280	6.389
+1	12.883	41	11.238	112	9.163	290	6.277
2	12.836	42	11.202	114	9.116	300	6.166
3	12.790	43	11.167	116	9.069	310	6.062
4	12.743	44	11.132	118	9.022	320	5.942
5	12.697	45	11.097	120	8.976	330	5.847
6	12.652	46	11.062	122	8.931	340	5.755
7	12.606	47	11.027	124	8.886	350	5.664
8	12.561	48	10.993	126	8.841	360	5.578
9	12.517	49	10.958	128	8.797	370	5.493
10	12.472	50	10.924	130	8.753	380	5.407
11	12.428	52	10.857	132	8.710	400	5.248
12	12.385	54	10.791	134	8.667	420	5.101
13	12.341	56	10.725	136	8.625	440	4.952
14	12.298	58	10.660	138	8.583	460	4.812

DENSITY OF MOIST AIR

The density of moist air depends upon the temperature, the humidity, and the barometric pressure. It is expressed by the equation:

$$d_t = D_t \times \frac{P - 0.3783e}{760}$$

where d_t is the density of the moist air at the temperature t; D_t is the density of dry air at the temperature t (see the table *Specific Gravity of Air*); P is the height of the barometer after correction and reduction to standard conditions, and is expressed in millimeters of mercury (see the section on *Barometry*); e is the vapor pressure of water at the temperature of the dew point and is expressed in millimeters of mercury (see the table *Vapor Pressure of Water*).

Example. To find the density of moist air at a temperature of 20°C, with a dew point of 10°C, and a corrected barometric pressure of 750 mm.

Reference to the table of *Specific Gravity of Air* shows that D at 20°C is equal to 0.0012046 g/ml. Reference to the table of *Vapor Pressure of Water* shows that at 10°C (the temperature of the dew point) e is equal to 9.209 mm. Therefore,

$$d = 0.0012046 \times \frac{750 - (0.3783 \times 9.209)}{760}$$

$$d = 0.0011832 \text{ g/ml} = 1.1832 \text{ g/l}$$

REFRACTIVE INDEX, DIPOLE MOMENT, DIELECTRIC CONSTANT, VISCOSITY, AND SURFACE TENSION

REFRACTIVE INDEX

The refractive index n is the ratio of the velocity of light in a particular substance to the velocity of light in vacuum. Values reported refer to the ratio of the velocity in air to that in the substance saturated with air. Usually the yellow sodium doublet lines are used; they have a weighted mean of 589.26 nm and are symbolized by D. When only a single refractive index is available, as in Tables 10-34 and 10-35, approximate values over a small temperature range may be calculated using a mean value of 0.000 45 per degree for dn/dt, and remembering that n_D decreases with an increase in temperature. If a transition point lies within the temperature range, extrapolation is not reliable.

The *specific refraction* r_D is given by the Lorentz and Lorenz equation,

$$r_D = \frac{n_D^2 - 1}{n_D^2 + 2} \cdot \frac{1}{\rho}$$

where ρ is the density at the same temperature as the refractive index, and is independent of temperature and pressure. The molar refraction is equal to the specific refraction multiplied by the molecular weight. It is a more or less additive property of the groups or elements comprising the compound. A set of atomic refractions is given in Table 10-30; an extensive discussion will be found in Bauer, Fajans, and Lewin, in *Physical Methods of Organic Chemistry*, 3d ed., A. Weissberger (ed.), Vol. 1, Part II, Chapter 28, Wiley-Interscience, New York, 1960.

The empirical Eykman equation

$$\frac{n_D^2 - 1}{n_D + 0.4} \cdot \frac{1}{\rho} = \text{constant}$$

offers a more accurate means for checking the accuracy of experimental densities and refractive indices, and for calculating one from the other, than does the Lorentz and Lorenz equation.

The refractive index of moist air can be calculated from the expression

$$(n - 1) \times 10^6 = \frac{103.49}{T} p_1 + \frac{177.4}{T} p_2 + \frac{86.26}{T} \left(1 + \frac{5748}{T}\right) p_3$$

where p_1 is the partial pressure of dry air (in mm Hg), p_2 is the partial pressure of carbon dioxide (in mm Hg), p_3 is the partial pressure of water vapor (in mm Hg), and T is the temperature (in kelvins).

Table 10-30
ATOMIC AND GROUP REFRACTIONS

Group	Mr_D	Group	Mr_D
H	1.100	N (primary aliphatic amine)	2.322
C	2.418	N (sec-aliphatic amine)	2.499
Double bond (C=C)	1.733	N (tert-aliphatic amine)	2.840
Triple bond (C≡C)	2.398	N (primary aromatic amine)	3.21
Phenyl (C_6H_5)	25.463	N (sec-aromatic amine)	3.59
Naphthyl ($C_{10}H_7$)	43.00	N (tert-aromatic amine)	4.36
O (carbonyl) (C=O)	2.211	N (primary amide)	2.65
O (hydroxyl) (O—H)	1.525	N (sec amide)	2.27
O (ether, ester) (C—O—)	1.643	N (tert amide)	2.71
F (one fluoride)	0.95	N (imidine)	3.776
(polyfluorides)	1.1	N (oximido)	3.901
Cl	5.967	N (carbimido)	4.10
Br	8.865	N (hydrazone)	3.46
I	13.900	N (hydroxylamine)	2.48
S (thiocarbonyl) (C=S)	7.97	N (hydrazine)	2.47
S (thiol) (S—H)	7.69	N (aliphatic cyanide) (C≡N)	3.05
S (dithia) (—S—S—)	8.11	N (aromatic cyanide)	3.79
Se (alkyl selenides)	11.17	N (aliphatic oxime)	3.93
3-membered ring	0.71	NO (nitroso)	5.91
4-membered ring	0.48	NO (nitrosoamine)	5.37
		NO_2 (alkyl nitrate)	7.59
		(alkyl nitrite)	7.44
		(aliphatic nitro)	6.72
		(aromatic nitro)	7.30
		(nitramine)	7.51

Example: 1-Propynyl acetate has $n_D = 1.4187$ and density $= 0.9982$ at 20°C; the molecular weight is 98.102. From the Lorentz and Lorenz equation:

$$r_D = \frac{(1.4187)^2 + 1}{(1.4187)^2 + 2} \cdot \frac{1}{0.9982} = 0.2528$$

The molar refraction is

$$Mr_D = (98.102)(0.2528) = 24.80$$

From the atomic and group refractions in Table 10-30, the molar refraction is computed as follows:

6 H	6.600
5 C	12.090
1 C≡C	2.398
1 O(ether)	1.643
1 O(carbonyl)	2.211
$Mr_D =$	24.942

DIPOLE MOMENTS

The permanent dipole moment of an isolated molecule depends on the magnitude of the charge and on the distance separating the positive and negative charges. It is defined as: $\mu = \left(\sum_i q_i r_i \right)$ where the summation extends over all charges (electrons and nuclei) in the molecule. The numerical values of the dipole moment, expressed in the cgs system of units, are in debye units, D, where $1\,D = 10^{-18}$ esu of charge $\times$ centimeters. The conversion factor to SI units is:

$$1\,D = 3.335\,64 \times 10^{-30}\,C \cdot m \quad [\text{coulomb-meter}]$$

Tables 10-34 and 10-35 contain a selected group of compounds for which the dipole moment is given. An extensive collection of dipole moments (approximately 7000 entries) is contained in A. L. McClellan, *Tables of Experimental Dipole Moments*, W. H. Freeman, San Francisco, 1963. A critical survey of 500 compounds in the gas phase is given by Nelson, Lide, and Maryott, NSRDS-NBS 10, Washington, D.C., 1967.

DIELECTRIC CONSTANTS

If two oppositely charged plates exist in a vacuum, there is a certain force of attraction between them, as stated by Coulomb's law:

$$F = \frac{1}{4\pi\varepsilon_0} \cdot \frac{q_1 q_2}{\varepsilon r^2}$$

where F is the force, in newtons, acting on each of the charges q_1 and q_2, r is the distance between the charges, ε is the dielectric constant of the medium between the plates, and ε_0 is the permittivity of free space. q_1, q_2 are expressed in coulombs, and r in meters. If another substance, such as a solvent, is in the space separating these charges (or ions in a solution), their attraction for each other is less. The dielectric constant is a measure of the relative effect a solvent has on the force with which two oppositely charged plates attract each other. The dielectric constant is a unitless number.

Dielectric constants for a selected group of inorganic and organic compounds are included in Tables 10-34 and 10-35. An extensive list has been compiled by Maryott and Smith, *National Bureau Standards Circular 514*, Washington, D.C. 1951.

For gases the values of the dielectric constant can be adjusted to somewhat different conditions of temperature and pressure by means of the equation

$$\frac{(\varepsilon - 1)_{t,p}}{(\varepsilon - 1)_{20°,1\,\text{atm}}} = \frac{p}{760[1 + 0.003\ 411(t - 20)]}$$

where p is the pressure (in mm Hg) and t is the temperature (in °C). The errors associated with this equation probably do not exceed 0.02% for gases between 10 and 30°C and for pressures between 700 and 800 mm. The dielectric constants of selected gases will be found in Table 10-34.

VISCOSITY

The *dynamic viscosity*, or coefficient of viscosity, η of a Newtonian fluid is defined as the force per unit area necessary to maintain a unit velocity gradient at right angles to the direction of flow between two parallel planes a unit distance apart. The SI unit is pascal-second or newton-second per meter squared [N · s · m^{-2}]. The cgs unit of viscosity is the poise [P]; $1\,\text{cP} \equiv 1\,\text{mN} \cdot \text{s} \cdot \text{m}^{-2}$. The dynamic viscosity decreases with the temperature approximately according to the equation: $\log \eta = A + B/T$. Values of A and B for a large number of liquids are given by Barrer, *Trans. Faraday Soc.*, **39**, 48 (1943).

Kinematic viscosity ν is the ratio of the dynamic viscosity to the density of a fluid. The SI unit is meter squared per second [m^2 · s^{-1}]. The cgs units are called stokes [cm^2 · sec^{-1}]; poises = stokes × density.

Fluidity ϕ is the reciprocal of the dynamic viscosity.

The primary reference liquid for viscosity measurements is water. The absolute viscosity of water at 20°C is 1.0019 ($\pm$0.0003) mN · s · m^{-2} (or centipoise), as determined by Swindells, Coe, and Godfrey, *J. Research Natl. Bur. Standards*, **48**, 1 (1952). The relative viscosity of water, $\eta/\eta_{20°}$, is 0.8885 at 25°C, 0.7960 at 30°C, and 0.6518 at 40°C. Values at temperatures between 15 and 60°C are best represented by Cragoe's equation:

$$\log \frac{\eta}{\eta_{20°}} = \frac{1.2348(20 - t) - 0.001\ 467(t - 20)^2}{t + 96}$$

The *Reynolds number* for flow in a tube is defined by $d\bar{v}\rho/\eta$, where d is the diameter of the tube, $\bar{v}$ is the average velocity of the fluid along the tube, ρ is the density of the fluid, and η is its dynamic viscosity. At flow velocities corresponding with values of the Reynolds number of greater than 2000, turbulence is encountered.

Table 10-31
AQUEOUS GLYCEROL SOLUTIONS

% Weight Glycerol	Grams per Liter	Relative Density 25°/25°C	Viscosity, mN · s · m^{-2}		
			20°C	25°C	30°C
100	1261	1.262 01	1 495	942	622
99	1246	1.259 45	1 194	772	509
98	1231	1.256 85	971	627	423
97	1216	1.254 25	802	521	353
96	1201	1.251 65	659	434	296
95	1186	1.249 10	543.5	365	248
80	966.8	1.209 25	61.8	45.72	34.81
50	563.2	1.127 20	6.032	5.024	4.233
25	265.0	1.061 15	2.089	1.805	1.586
10	102.2	1.023 70	1.307	1.149	1.021

Table 10-32
AQUEOUS SUCROSE SOLUTIONS

% Weight Sucrose	Grams per Liter	Relative Density 20°/4°C	Viscosity, mN · s · m^{-2}		
			15°C	20°C	25°C
75	1034	1.379 0	4 039	2 328	1 405
70	943.0	1.347 2	746.9	481.6	321.6
65	855.6	1.316 3	211.3	147.2	105.4
60	771.9	1.286 5	79.49	58.49	40.03
50	614.8	1.229 6	19.53	15.43	12.40
40	470.6	1.176 4	7.463	6.167	5.164
30	338.1	1.127 0	3.757	3.187	2.735

SURFACE TENSION

The surface tension of a liquid, γ, is the force per unit length on the surface that opposes the expansion of the surface area. In the literature the surface tensions are expressed in $dyn \cdot cm^{-1}$; $1\ dyn \cdot cm^{-1} = 1\ mN \cdot m^{-1}$ in the SI system. For the large majority of compounds the dependence of the surface tension on the temperature can be given as

$$\gamma = a - bt$$

where a and b are constants and t is the temperature in degrees Celsius. The values of a and b given in Tables 10-34 and 10-35 can be used to calculate the values of surface tension for the particular compound within its liquid range. For example, the least squares constants for acetic anhydride (liquid from -73 to $140°C$) are 35.52 and 0.1436, respectively. At $20°C$, $\gamma = 35.52 - 0.1436(20) = 32.64\ dyn \cdot cm^{-1}$.

A compilation of data of some 2200 pure liquid compounds has been prepared by Jasper, *J. Phys. Chem. Reference Data*, **1**, 841 (1972).

Table 10-33
REFRACTIVE INDEX, VISCOSITY, DIELECTRIC CONSTANT, AND SURFACE TENSION OF WATER
From 0 to 100°C

Temp., °C	Refractive Index (n_D)	Viscosity, $mN \cdot s \cdot m^{-2}$	Dielectric Constant (ε)	Surface Tension, $dyn \cdot cm^{-1}$
0	1.333 95	1.770 2	87.74	75.83
5	1.333 88	1.510 8	85.76	75.09
10	1.333 69	1.303 9	83.83	74.36
15	1.333 39	1.137 4	81.95	73.62
20	1.333 00	1.001 9	80.10	72.88
21	1.332 90	0.976 4	79.73	72.73
22	1.332 80	0.953 2	79.38	72.58
23	1.332 71	0.931 0	79.02	72.43
24	1.332 61	0.910 0	78.65	72.29
25	1.332 50	0.890 3	78.30	72.14
26	1.332 40	0.870 3	77.94	71.99
27	1.332 29	0.851 2	77.60	71.84
28	1.332 17	0.832 8	77.24	71.69
29	1.332 06	0.814 5	76.90	71.55
30	1.331 94	0.797 3	76.55	71.40
35	1.331 31	0.719 0	74.83	70.66
40	1.330 61	0.652 6	73.15	69.92
45	1.329 85	0.597 2	71.51	69.18
50	1.329 04	0.546 8	69.91	68.45
55	1.328 17	0.504 2	68.35	67.71
60	1.327 25	0.466 9	66.82	66.97
65		0.434 1	65.32	66.23
70		0.405 0	63.86	65.49
75		0.379 2	62.43	64.75
80		0.356 0	61.03	64.01
85		0.335 2	59.66	63.28
90		0.316 5	58.32	62.54
95		0.299 5	57.01	61.80
100		0.284 0	55.72	61.80

Table 10-34
VISCOSITY, DIELECTRIC CONSTANT, DIPOLE MOMENT, AND SURFACE TENSION OF SELECTED INORGANIC SUBSTANCES

Substance	Viscosity, $mN \cdot s \cdot m^{-2}$	Dielectric Constant (ε)	Dipole Moment, D	Surface Tension, $dyn \cdot cm^{-1}$	
				a	b
Air (20°)	0.0182	1.000 536 4			
$AlBr_3$		3.38^{100}	5.2		
Ar (g, 20°)	0.0223	1.000 517 2			
(liq)		1.538^{-191}	0	34.28	0.2493
$AsBr_3$		8.83^{35}	1.61	54.51	0.1043
$AsCl_3$		12.6^{20}	1.59	41.67	0.09781
AsH_3 (arsine)		2.05^{20}	0.20		
BBr_3		2.58^{0}	0	31.90	0.1280
BCl_3			0		
BF_3			0	−2.92	0.2030
B_2H_6 (diborane)		$1.872^{-92.5}$	0	−3.13	0.1785
B_5H_9			2.13		
$B_3H_6N_3$ (triborotriazine)			0		
Br_2 (g, 20°)		1.012 8			
(liq)	1.03^{16}	3.09^{20}	0.49	45.5	0.1820
BrF_3	2.22^{20}		1.1	38.30	0.0999
BrF_5	0.62^{24}	$7.91^{24.5}$	1.51	25.24	0.1098
Cl_2 (g, 20°)	0.0132		0.23		
(liq)		1.91^{14}			
ClF_3	0.48^{12}	4.29^{25}	0.554	26.9	0.1660
ClO_3F (perchloryl fluoride)			0.023	12.24	0.1576
CO (g)	0.0175^{20}	$1.000\ 70^{0}$	0.112		
(liq)				−30.20	0.2073
CO_2 (g, 20°)	0.0147	1.000 922	0		
(liq)	0.071^{20}	$1.60^{0°,50\,atm}$			
$COCl_2$		4.34^{22}	1.17	22.59	0.1456
COF_2			0.95		
COS			0.712	12.12	0.1779
COSe		3.47^{10}	0.73		
CS			1.98		
CS_2 (g)		$1.002\ 9^{0}$	0		
(liq)	0.375^{20}	2.6^{20}			
CrO_2Cl_2 [chromyl(VI) chloride]		2.6^{20}	0.47		
D_2 Deuterium		1.277^{-253}			
DH				6.537	0.1883
D_2O	1.098^{25}	78.25^{25}	1.87	$(71.72^{20})*$	$(68.38^{40})*$
F_2		1.54^{-202}		−16.10	0.1646
$GaCl_3$			0.85	35.0	0.1000
$GeCl_4$		2.430^{25}	0	$(22.44^{30})*$	
H_2 (g, 20°)	0.0088	1.000 253 8	0		
(liq)		$1.228^{20.4K}$			
HBr (g)		$1.003\ 13^{0}$	0.82		
(liq)	0.83^{-67}	3.82^{25}		13.10	0.2079
HCl (g)		$1.004\ 6^{0}$	1.08		
(liq)	0.51^{-95}	4.60^{28}			
HCN	0.206^{18}	116^{20}	2.98	$(19.45^{10})*$	$(18.33^{20})*$
HCNO (isocyanate)			1.6		
HCNS (isothiocyanate)			1.7		
HF	0.256^{0}	83.6^{0}	1.82	10.41	0.07867
HI (g)		$1.002\ 34^{0}$	0.44		
(liq)		2.90^{22}			

* Actual values of surface tension.

Table 10-34 (*Continued*)
VISCOSITY, DIELECTRIC CONSTANT, DIPOLE MOMENT, AND SURFACE TENSION OF SELECTED INORGANIC SUBSTANCES

Substance	Viscosity, $mN \cdot s \cdot m^{-2}$	Dielectric Constant (ε)	Dipole Moment, D	Surface Tension $dyn \cdot cm^{-1}$	
				a	*b*
NH_3 (azide)			0.8		
H_2O [See Table 10-33]					
H_2O_2	1.25^{20}	84.2^0	2.2	78.97	0.1549
HNO_3			2.17		
H_2S (g)		$1.004\ 0^0$	0.97		
(liq)	0.412^0	5.93^{10}		48.95	0.1758
H_2Se			0.24	22.32	0.1482
H_2SO_4	24.54^{25}	100^{25}			
HSO_3Cl (chlorosulfonic acid)	2.43^{20}	60^{20}			
HSO_2F (fluorosulfonic acid)	1.56^{25}	$\sim120^{25}$			
H_2Te			<0.2	29.03	0.2619
He (g, 20°)	0.0196	$1.000\ 065\ 0$	0		
Hg	1.552^{20}		0	490.6	0.2049
I_2	1.98^{116}	11.1^{118}	1.41		
IF_5			2.18	33.16	0.1318
Kr (g, 20°)	0.0250		<0.05		
(liq)				40.576	0.2890
Ne (g, 20°)	0.0313	$1.000\ 063\ 9$	0		
N_2 (g, 20°)	0.0176	$1.000\ 548\ 0$	0		
(liq)		1.454^{-203}		26.42	0.2265
NH_3 (g)		$1.007\ 2^0$	1.47		
(liq)	$0.254^{-33.5}$	$22.4^{-33.4}$		(37.91^{-50})	$(35.38^{-40})*$
N_2H_4 (hydrazine)	0.97^{20}	52.9^{20}	1.75		
NO			0.153	-67.48	0.5853
N_2O (g)	0.0146^{20}	$1.001\ 13^0$	0.167		
(liq)		1.52^{15}		5.09	0.2032
NO_2			0.316		
N_2O_4		2.56^{15}	0.5		
NOBr (nitrosyl bromide)		13.4^{15}	1.8		
NOCl		18.2^{12}	1.9	29.49	0.1493
NOF			1.81	14.00	0.1165
NO_2F (nitryl fluoride)			0.47	8.26	0.1854
O_2 (g, 20°)	0.0204	$1.000\ 494\ 7$	0		
(liq)		1.507^{-193}		-33.72	0.2561
O_3			0.53	$(38.1^{-183})*$	
OF_2 (oxygen difluoride)			0.297		
OsO_4			0		
PBr_3		3.9^{20}	0.5	45.34	0.1283
PCl_3		3.43^{25}	0.78	31.14	0.1266
PCl_5		2.7^{165}	0.9		
PF_5			0		
PH_3		2.9^{15}	0.58		
PI_3		4.12^{65}	0	61.66	0.06771
$POCl_3$	1.065^{25}	13.7^{25}	2.41	35.22	0.1275
POF_3			1.76		
$PSCl_3$		5.8^{22}	1.42	37.00	0.1272
$PbCl_4$		2.78^{20}			
S_2Cl_2 dimer		4.79^{15}	1.0	46.23	0.1464
S_2F_2 FSSF isomer			1.45		
$S{=}SF_2$ isomer			1.03		
SF_4			0.632	12.87	0.1734

* Actual values of surface tension.

Table 10-34 (*Continued*)
VISCOSITY, DIELECTRIC CONSTANT, DIPOLE MOMENT, AND SURFACE TENSION OF SELECTED INORGANIC SUBSTANCES

Substance	Viscosity, $mN \cdot s \cdot m^{-2}$	Dielectric Constant (ε)	Dipole Moment, D	Surface Tension $dyn \cdot cm^{-1}$ a	b
SF_6			0	5.66	0.1190
S_2F_{10}		2.020^{20}	0		
SO_2 (g)	0.0126^{20}	$1.009\ 3^0$	1.63		
(liq)		15.0^0		26.58	0.1948
SO_3		3.11^{18}	0		
$SOBr_2$ (thionyl bromide)		9.06^{20}	9.11		
$SOCl_2$		9.25^{20}	1.45	36.10	0.1416
SO_2Cl_2 (sulfuryl chloride)		9.15^{20}	1.81	32.10	0.1328
$SbCl_3$		33.2^{75}	3.93	47.87	0.1238
$SbCl_5$		3.22^{20}	0		
SbF_5				49.07	0.1937
SbH_3			0.12		
SeF_4				38.61	0.1274
SeF_6			0		
$SeOCl_2$		55^{25}	2.64		
$SiCl_4$		2.40^{16}	0	20.78	0.09962
SiF_4			0		
SiH_4			0		
$SiHCl_3$			0.86	20.43	0.1076
$SnBr_4$			0		
$SnCl_4$		2.89^{20}	0	29.92	0.1134
TeF_6			0		
$TiCl_4$		2.80^{20}	0	$(33.54^{20})*$	$(31.06^{40})*$
UF_6 (g)		$1.002\ 92^{67}$	0		
(liq)		2.18^{65}		25.5	0.1240
VCl_4		3.05^{25}	0		
$VOBr_3$		3.6^{25}			
$VOCl_3$		3.4^{25}	0.3	$(36.36^{20})*$	$(33.60^{40})*$
Xe (g, 20°)	0.0228	$1.001\ 23$	0		

* Actual values of surface tension.

Table 10-35
PHYSICAL PROPERTIES OF SELECTED ORGANIC SOLVENTS

Substance	Temp., °C	Density, $g \cdot mL^{-3}$	Refractive Index (n_D)	Viscosity, $mN \cdot s \cdot m^{-2}$	Dielectric Constant (ϵ)	Dipole Moment, D	Surface Tension, $dyn \cdot cm^{-1}$	
							a	b
Acetaldehyde	20	0.7780	1.3311	0.244	21.1	2.69	23.90	0.1360
Acetamide	110	0.9681	1.4158	1.46[112]	59[83]	3.44	47.66	0.1021
Acetic acid	20	1.0492	1.3719	1.314[15]	6.15	1.74	29.58	0.0994
Acetic anhydride	20	1.0811	1.3904	0.971[15]	20.7	2.8	35.52	0.1436
Acetone	20	0.7908	1.3588	0.337[15]	20.70[25]	2.88	26.26	0.112
Acetonitrile	20	0.7822	1.3441	0.375	37.5	3.92	29.58	0.1178
Acetophenone	25	1.0238	1.5322	1.642	17.39	3.02	41.92	0.1154
Acetyl bromide	16	1.663	1.4537		16.2[20]	2.45		
Acetyl chloride	20	1.105	1.3898		15.8	2.72		
Acrolein	20	0.8389	1.4017			2.90[25]		
Acrylic acid	20	1.0511	1.4224				(28.1[30])	0.1178
Acrylonitrile	20	0.8060	1.3911	0.35	33.0	3.87	29.58	0.0973
Adiponitrile	20	0.950	1.4597				47.88	0.1186
Allyl acetate	20	0.9256	1.4040	0.207[30]			28.73	0.1287
Allylamine	20	0.7629	1.4205	0.375[25]		1.31	27.49	0.1117
2-Aminoethanol	25	1.0116	1.4521	19.35	37.72	2.27	51.11	0.1117
1-Amino-2-methylpropane	25	0.7297	1.3945	21.7	4.43[21]	1.27	24.48	0.1092
Aniline	20	1.0217	1.5855	4.400	6.89	1.53	44.83	0.1085
Benzaldehyde	20	1.0447	1.5455	1.321[25]	17.8	2.77	40.72	0.1090
Benzene	25	1.8737	1.4979	0.6028	2.275	0	(28.88[20])	(27.56[30])
Benzenethiol	20	1.0766	1.5897	1.239	4.38[25]	1.23	41.41	0.1202
Benzonitrile	20	1.0051	1.5282	1.447[15]	25.20[25]	4.18	41.69	0.1159
Benzoyl chloride	20	1.211	1.5525		23	3.2	41.34	0.1084
Benzyl acetate	20	1.055	1.5200	1.399[45]	5.1	1.80	38.25	0.1381
Benzyl alcohol	20	1.045	1.5403	7.760[15]	13.1	1.66	48.07	0.1065
Benzyl benzoate	25	1.1121	1.5685[21]	8.292	4.9[20]	1.90	39.92	0.1227
Benzyl chloride	20	1.0993	1.5390	1.400	7.0[13]	1.85	(32.82[20])	(29.97[40])
Benzyl ethyl ether	20	0.9478	1.4958		3.9			

Table 10-35 (Continued)
PHYSICAL PROPERTIES OF SELECTED ORGANIC SOLVENTS

Substance	Temp., °C	Density, g·mL⁻¹	Refractive Index (n_D)	Viscosity, mN·s·m⁻²	Dielectric Constant (ε)	Dipole Moment, D	Surface Tension, dyn·cm⁻¹	
							a	b
Bicyclohexane	20	0.8862	1.4800	3.75		<0.4	34.64	0.0951
3-Bromoaniline	20	1.579	1.6260		13.0	2.66	38.14	0.1160
Bromobenzene	25	1.4882	1.5571	0.985³⁰	5.40	1.55	28.71	0.1126
1-Bromobutane	20	1.2758	1.4394	0.633		2.08	27.48	0.1107
2-Bromobutane	20	1.255	1.4369		7.7⁰·¹	2.04	26.52	0.1159
Bromoethane	15	1.4708	1.4276	0.418	9.39²⁰	2.03		
Bromoethene	20	1.517	1.4380		4.78²⁵	1.42		
1-Bromohexane	20	1.176	1.4475	5.99¹⁵	5.82²⁵	2.00	29.81	0.09669
1-Bromonaphthalene	20	1.4834	1.6582		5.12	1.29ˡⁱᑫ	46.44	0.1018
1-Bromopropane	15	1.3597	1.4370	0.539	8.09²⁵	2.18	28.30	0.1218
2-Bromopropane	15	1.3222	1.4285	0.536	9.46²⁵	2.21	26.21	0.1183
o-Bromotoluene	20	1.422	1.5608		4.28⁵⁸	1.45	36.62	0.09979
1-Butanal	20	0.8016	1.3791	0.455	13.4²⁶	2.72	26.67	0.0925
2-Butanal	20	0.7891	1.3727			2.58		
1-Butanamine	20	0.7392	1.4014	0.681	4.88	1.37	26.24	0.1122
2-Butanamine	20	0.7246	1.3932			1.28	23.75	0.1057
1,3-Butanediol	20	1.0053	1.441	130.3		2.5	(37.8²⁵)	
Butanenitrile	15	0.7954	1.3860	0.624	20.3²¹	4.07	29.51	0.1037
1-Butanethiol	20	0.8416	1.4430	0.501	5.07²⁵	1.53	28.07	0.1142
Butanoic acid	20	0.9582	1.3980	1.814¹⁵	2.97	1.65	28.35	0.0920
1-Butanol	20	0.8097	1.3993	3.379¹⁵	17.5²⁵	1.66	27.18	0.08983
2-Butanol	20	0.8069	1.3972	4.210	16.56	1.66		
2-Butanone	20	0.8047	1.3788	0.423¹⁵	18.51	2.75	26.77	0.1122
cis-2-Butene-1,4-diol	20	1.0740	1.4793			2.48		
trans-2-Butene-1,4-diol	20	1.0685	1.4779			2.45		
2-Butoxyethanol	25	0.8964	1.4177	3.15	9.30	2.08	28.18	0.0816
Butyl acetate	20	0.8813	1.3941	0.734	5.01	1.84	27.55	0.1068
Butylbenzene	20	0.8601	1.4898	1.035	2.36	0.36ˡⁱᑫ	31.28	0.1025

Compound	t/°C							
sec-Butylbenzene	20	0.8621	1.4902	28.53	2.36	0.37^{liq}	30.48	0.0979
tert-Butylbenzene	20	0.8665	1.4927	28.13	2.25	0	30.10	0.0985
Butyl ethyl ether	20	0.7495	1.3818	0.421		1.24	22.75	0.1049
Butyl formate	20	0.8917	1.3890	0.704	2.43^{80}		27.08	0.1026
Butyl oleate	20	0.864	1.4522		4.0			
Butyl stearate	25	0.8540	1.4422	8.26	3.11^{30}	1.88	(33.0^{25})	(32.7^{30})
Butyric anhydride	20	0.9668	1.4124	1.615	12.9		(28.93^{30})	(28.44^{25})
γ-Butyrolactone	25	1.1254	1.4348	1.7	39^{20}	4.12		
D-Camphor	20	0.9920		9^{78}	11.35	3.10		
ε-Caprolactam		1.027^{77}	1.4935^{40}			3.88		
Carbon disulfide	20	1.2628	1.6280	0.363	2.641	0.06	35.29	0.1484
Carbon tetrachloride	20	1.5842	1.4603	0.965	2.238	0	29.49	0.1224
Chloroacetic acid	65	1.370	1.4297	3.15^{50}	12.3^{60}	2.31	43.27	0.1117
o-Chloroaniline	25	1.2077	1.5859	0.925	13.4	1.81	42.46	0.08667
Chlorobenzene	20	1.1063	1.5248	0.799	5.62^{25}	1.69	35.97	0.191
1-Chlorobutane	20	0.8864	1.4021	0.469^{15}	7.39	2.05	25.97	0.1117
2-Chlorobutane	20	0.8732	1.3971	0.439^{15}	7.09^{30}	2.04	24.40	0.1118
1-Chloro-2,3-epoxypropane	25	1.1746	1.4358	1.03	22.6^{22}	1.8	39.76	0.1360
Chloroethane		0.09028^{15}	1.3738^{10}	0.279^{10}	9.45^{20}	2.05	(21.18^{5})	(20.58^{10})
2-Chloroethanol	15	1.2072	1.4438	3.913	25.8^{25}	1.88	(38.9^{20})	
bis(2-Chloroethyl)ether	25	1.2130	1.4553	2.14	21.2^{20}	2.58	40.57	0.1306
Chloroform	15	1.4985	1.4486	0.596	4.806^{20}	1.01	29.91	0.1295
1-Chloro-2-methylpropane	15	0.8829	1.4010	0.471	6.49	2.00	24.40	0.1099
2-Chloro-2-methylpropane	20	0.8414	1.3856	0.543^{15}	9.96	2.13	(20.06^{15})	(18.35^{30})
1-Chloronaphthalene	25	1.1930	1.6332^{20}	2.940	5.04	1.59	44.12	0.1035
1-Chloropentane	20	0.8840	1.4118	0.580	6.11	2.16	27.09	0.1076
o-Chlorophenol		1.2410^{18}	1.5473^{40}	2.250^{45}	6.31^{25}	2.19	42.5	0.1122
p-Chlorophenol		1.2651^{40}	1.5579^{40}	6.018^{45}	9.47^{55}	2.11	46.0	0.1049
1-Chloropropane	20	0.8923	1.3880	0.372^{15}	7.7	2.05	24.41	0.1246
2-Chloropropane	20	0.8617	1.3777	0.335^{15}	9.82	2.17	21.37	0.0883
3-Chloro-1-propene	20	0.9376	1.4151	0.347^{15}	8.2	1.94	25.50	0.0946
o-Chlorotoluene	20	1.0817	1.5238		4.45	1.56		
p-Chlorotoluene	20	1.0697	1.5199^{19}		6.08	2.21	34.93	0.1082
1,8-Cineole	25	0.9192	1.4555		4.57			
Cinnamaldehyde	20	1.0497	1.6195		16.9^{24}			
o-Cresol	46	1.0230	1.5336	3.506	11.5^{25}	1.41	39.43	0.1011

Table 10-35 (*Continued*)
PHYSICAL PROPERTIES OF SELECTED ORGANIC SOLVENTS

Substance	Temp., °C	Density, $g \cdot mL^{-3}$	Refractive Index (n_D)	Viscosity, $mN \cdot s \cdot m^{-2}$	Dielectric Constant (ϵ)	Dipole Moment, D	Surface Tension. $dyn \cdot cm^{-1}$	
							a	b
m-Cresol	15	1.0380	1.5415[20]	24.67	11.8[25]	1.54	38.00	0.09237
p-Cresol	46	1.0140	1.5287	5.607	9.91[58]	1.54	38.58	0.0962
Crotonaldehyde	20	0.8516	1.4373			3.50		
Crotonic acid	77	0.9604	1.4249			2.13		
Cyclohexanamine	20	0.8671	1.4593	1.662	4.73	1.32	34.19	0.1188
Cyclohexane	20	0.7786	1.4262	0.980	2.02	0	27.62	0.1188
Cyclohexanol	30	0.9416	1.4629	41.07	15.0[25]	1.86	35.33	0.0966
Cyclohexanone	20	0.9462	1.4510	2.453[15]	18.3	3.01	37.67	0.1242
Cyclohexene	20	0.8110	1.4465	0.650	2.20	0.55	29.23	0.1223
Cyclohexylbenzene	20	0.9427	1.5263	3.681[0]				
Cyclopentane	20	0.7454	1.4065	0.439	1.965	0	25.53	0.1462
p-Cymene	20	0.8573	1.4909	3.402	2.253	0	(29.44[20])	
cis-Decahydronaphthalene	20	0.8967	1.4810	3.381	2.197	0	(32.18[20])	(31.01[30])
trans-Decahydronaphthalene	20	0.8697	1.4693	2.128	2.172	0	(29.89[20])	(28.87[30])
Decane	20	0.7301	1.4119	0.928	1.991	0	25.67	0.09197
1-Decanol	20	0.8297	1.4371			1.71	30.34	0.07324
1-Decene	20	0.7408	1.4215	0.805		0.42	25.84	0.09190
Dibenzylamine	20	1.0278	1.5743[22]		3.6	0.97[liq]	43.27	0.1086
Dibenzyl ether	25	0.9974	1.5385	3.711[35]		1.39	(38.2[25])	
1,2-Dibromoethane	25	2.1687	1.5360	1.490[30]	4.78	1.23	35.43	0.1428
cis-1,2-Dibromoethene	20	2.2464	1.5428		7.08	1.35		
trans-1,2-Dibromoethene	20	2.2308	1.5505[18]		2.88	0		
Dibromomethane	20	2.4921	1.5419		7.77[10]	1.43	42.77	0.1488
1,2-Dibromotetrafluoroethane	25	2.163	1.367	0.72	2.34	0	(18.9[20])	(18.1[25])
Dibutylamine	20	0.7619	1.4177	0.95	2.978	1.04	26.50	0.0952
Dibutyl ether	25	0.7646	1.3969	0.602[30]	3.06	1.22	24.78	0.0934
Dibutyl maleate	20	0.9950	1.4454	5.63			32.46	0.0865
Dibutyl *o*-phthalate	25	1.0426	1.4901	16.47	6.436[30]	2.4	(33.40[20])	

Name	t	Density	n					
Dibutyl sebacate	25	0.9324	1.4397	7.96	4.540^{30}	2.48		(35.55^{30})
o-Dichlorobenzene	25	1.3003	1.5491	1.324	9.93	2.50	(26.84^{20})	0.1147
m-Dichlorobenzene	25	1.2828	1.5434	1.04	5.04	1.72	38.30	0.0879
p-Dichlorobenzene	60	1.2417	1.5285	0.720^{70}	2.41^{50}	0.0	34.66	0.1186
1,1-Dichloroethane	15	1.1835	1.4198	0.505^{25}	10.0^{18}	2.06	27.03	0.1428
1,2-Dichloroethane	15	1.2600	1.4476	0.887	10.36^{25}	1.20	35.43	
cis-1,2-Dichloroethene	25	1.2736	1.4490^{15}	0.444	9.20	1.90		
trans-1,2-Dichloroethene	20	1.2546	1.4462	0.404	2.14^{25}	0		
Dichloromethane	15	1.3348	1.4246^{20}	0.449	9.08^{20}	1.60	(25^{20})	0.1284
1,2-Dichloropropane	20	1.1558	1.4390			1.46^{72}	30.41	0.1240
1,3-Dichloropropane	20	1.1859	1.4469			2.08	31.42	0.1233
2,2-Dichloropropane	20	1.0912	1.4093			2.27	36.40	(22.53^{30})
Diethanolamine	30	1.0899	1.4747	0.769^{15}	11.37		(23.62^{20})	
1,1-Diethoxyethane	20	0.8254	1.3805	380	2.81^{25}		(21.26^{22})	
Diethylamine	20	0.7056	1.3864	0.388^{10}	3.80^{25}	1.38	22.71	0.1143
Diethyl carbonate	15	0.9804	1.3854	0.868	3.6^{22}	0.92	28.62	0.1100
Diethyl ether	15	0.7193	1.3556	0.247	2.82^{20}	1.10	18.92	0.0908
Diethyl maleate	25	1.0637	1.4383	3.14	4.335^{20}	1.15^{20}	34.67	0.1039
Diethyl malonate	20	1.0550	1.4136	2.15	8.58	2.54	33.91	0.1042
Diethyl oxalate	15	1.0843	1.4124	2.311	7.87^{25}	2.54	34.32	0.1119
Diethyl sulfate	20	1.1774	1.4004		8.1^{21}	2.49^{25}	35.47	0.0976
Diethyl sulfide	20	0.8367	1.4425	0.446	29.2	1.54	(25.28^{20})	(24.16^{30})
Diiodomethane	25	3.3078	1.7380	2.392^{30}	5.72^{25}	1.08	70.21	0.1613
Diisoamyl ether	20	0.7777	1.4085	1.40^{11}	5.316	1.23	24.76	0.0871
Diisopropylamine	20	0.7153	1.3924	0.405^{25}	2.82		21.83	
Diisopropyl ether	20	0.7325	1.3681	0.379	3.88	1.22	19.89	0.1077
1,2-Dimethoxybenzene	25	1.0819	1.5323	3.281	4.09	1.32	34.4	0.1048
1,2-Dimethoxyethane	25	0.8621	1.3781	0.455	7.20	1.71		0.0642
Dimethoxymethane	15	0.8665	1.3563	0.340	2.7^{20}	0.74	23.59	0.1199
N,N-Dimethylacetamide	25	0.9366	1.4356	0.838^{30}	37.78	3.72	(32.43^{30})	(29.55^{50})
Dimethylamine	15	1.6616	1.350^{17}	0.207	5.26^{25}	1.0	29.50	0.1265
N,N-Dimethylaniline	20	0.9559	1.5584	1.285^{25}	36.7^{25}	1.68	38.14	0.1049
2,2-Dimethylbutane	25	0.6445	1.3660	0.351	1.873		18.29	0.0990
2,3-Dimethylbutane	25	0.6570	1.3723	0.361	1.890		19.38	0.09998
2,2-Dimethyl-1-butanol	20	0.8286	1.4208					
2,3-Dimethyl-1-butanol	20	0.8300	1.4205				26.22	0.0992

Table 10-35 (Continued)
PHYSICAL PROPERTIES OF SELECTED ORGANIC SOLVENTS

Substance	Temp., °C	Density, g·mL⁻¹	Refractive Index (n_D)	Viscosity, mN·s·m⁻²	Dielectric Constant (ε)	Dipole Moment, D	Surface Tension, dyn·cm⁻¹	
							a	b
3,3-Dimethyl-2-butanol	20	0.8179	1.4151	0.802	36.71	3.86	(36.76^{20})	(34.40^{40})
N,N-Dimethylformamide	25	0.9445	1.4269	3.54		2.48	40.73	0.1220
Dimethyl maleate	20	1.1513	1.4422	0.406	1.939	0	19.94	0.09565
2,3-Dimethylpentane	20	0.6951	1.3920	0.361	1.914	0	20.09	0.09715
2,4-Dimethylpentane	20	0.6727	1.3815		8.5^{24}			
Dimethyl o-phthalate	21	1.1905	1.5155	9.18^{35}	1.80	~0	(12.05^{20})	(10.98^{30})
2,2-Dimethylpropane	20	0.5910	1.342	0.303^{5}	42.6		41.26	0.1163
Dimethyl sulfate	25	1.3322	1.3874	1.996	46.6	3.9	(43.54^{20})	(42.41^{30})
Dimethyl sulfoxide	25	1.0958	1.4773	1.439^{15}	2.209	0	36.23	0.1391
1,4-Dioxane	25	1.0280	1.4203	0.922^{30}	2.77	1.20	26.66	0.0925
Dipentyl ether	30	0.7790	1.4099	1.158	2.68^{20}(s)	1.16	35.17	0.1104
Diphenyl ether	20	1.0661	1.5763		2.57^{25}		24.86	0.1022
Diphenylmethane	20	1.0060	1.5768	0.534	3.07	1.03	22.60	0.1047
Dipropylamine	15	0.7375	1.4043	0.448	3.39^{26}	1.21	27.12	0.08843
Dipropyl ether	20	0.7518	1.3830	1.508	2.002^{30}	0	31.25	0.0748
Dodecane	20	0.7487	1.4216	0.41	5.15	2.01	44.77	0.1398
1-Dodecanol	20	0.8343	1.4428	1.54^{25}	14.2	1.99	50.21	0.0890
1,2-Epoxybutane	20	0.8297	1.3840	26.09	37.7^{25}	2.28	24.05	0.0832
1,2-Ethanediamine	20	0.8977	1.4568	3.13	10^{-54}(s)	2.34	35.17	0.1104
1,2-Ethanediol	15	1.1171	1.4331	1.078	24.55	1.69	30.59	0.0897
1,2-Ethanediol diacetate	20	1.1043	1.4150	1.364^{15}	4.22	1.36	(31.8^{25})	(27.27^{5})
Ethanol	25	0.7851	1.3594	2.05	29.6^{24}	2.08	(31.8^{25})	0.1161
Ethoxybenzene	20	0.9651	1.5074	3.71	7.567^{30}	2.25	26.29	
2-Ethoxyethanol	20	0.9295	1.4077	2.8	6.02	1.81		
2-(2-Ethoxyethoxy)ethanol	25	0.9841	1.4254					
2-(2-Ethoxyethoxy)ethyl acetate	20	1.0096	1.4213					
2-Ethoxyethyl acetate	25	0.9730^{20}	1.4023	1.025				
Ethyl acetate	25	0.8946	1.3698	0.426				

Ethyl acetoacetate	20	1.025	1.4198	1.508^{25}	15.7	3.22^{keto} 2.06^{enol}	34.42	0.1015
Ethyl acrylate	20	0.9234	1.4068		2.41		31.48	0.1094
Ethylbenzene	20	0.8670	1.4959	0.678	6.02	0.59	37.16	0.1059
Ethyl benzoate	20	1.0465	1.5052	2.407^{15}	6.02	1.99	(25.06^{15})	(24.32^{25})
2-Ethyl-1-butanol	20	0.8330	1.4224	5.892^{25}	6.19^{90}		26.55	0.1045
Ethyl butyrate	20	0.8794	1.3922	0.672	5.10^{18}	1.74	39.99	0.1045
Ethyl cinnamate	20	1.0494	1.5598	8.7	6.1^{18}	2.14	38.80	0.1092
Ethyl cyanoacetate	20	1.0648	1.4175	2.50^{25}	26.9	2.17	27.78	0.1054
Ethylcyclohexane	20	0.7879	1.4330	0.843	2.054	0		
Ethylene carbonate	40	1.3208^{25}	1.4199		89.6	4.87	47.33	0.0880
2,2'-(Ethylenedioxy)diethanol	20	1.1235	1.4561	49.0	23.69	5.58^{liq}		
Ethylenimine	25	0.832	1.4123	0.418	18.3	1.77	26.47	0.1315
Ethyl formate	20	0.9160	1.3599	0.419^{15}	7.16^{25}	1.94		
2-Ethyl-1-hexanol	20	0.8332	1.4231	9.8	4.41^{90}	1.74		
2-Ethylhexyl acetate	20	0.8718	1.4204	1.5	5.3			
bis(2-Ethylhexyl) phthalate	20	0.9843	1.4859	81.4		2.84		
Ethyl lactate	25	1.0299	1.4121	2.44	13.1^{20}	2.4	30.72	0.0983
Ethyl 3-methylbutanoate	15	0.8657	1.3962		4.71^{18}		25.79	0.1006
Ethyl propanoate	20	0.8957	1.3864	0.564	5.65^{19}	1.74	26.72	0.1168
Ethyl salicylate	20	1.1362	1.5251	1.772^{45}	7.99^{30}		41.00	0.1091
Fluorobenzene	20	1.0240	1.4657	0.620^{15}	5.42^{25}	1.61	29.67	0.1204
o-Fluorotoluene	17	1.0014	1.4716	0.680^{20}	4.22^{30}	1.37		
m-Fluorotoluene	20	0.9974	1.4691	0.608	5.42^{30}	1.86	32.31	0.1257
p-Fluorotoluene	20	0.9975	1.4688	0.622	5.86^{30}	2.00	30.44	0.1109
Formamide	25	1.1334	1.4475	3.764	111.0	3.73	59.13	0.0842
Formic acid	20	1.2141	1.3694	1.966	58.5^{16}	1.41	39.87	0.1098
2-Furaldehyde	25	1.1616	1.5262	1.49^{25}	41.9	3.61	46.41	0.1327
Furan	20	0.9378	1.4214	0.380	2.94	0.66	(24.10^{20})	(23.38^{25})
Furfuryl alcohol	20	1.1285	1.4868	4.62^{25}		1.92	$(ca\ 38^{0})$	
Glycerol	25	1.2582	1.4730	945	42.5	2.56	(63.14^{17})	(62.5^{25})
Heptane	25	0.6795	1.3851	0.397	1.92^{20}	0	22.10	0.0980
1-Heptanol	25	0.8223	1.4242	5.06		1.73		
2-Heptanol	25	0.8139	1.4190		9.21^{22}	1.71		
1-Heptene	20	0.6970	1.3999	0.35	2.07	0.34^{liq}	22.28	
1-Hexadecanol	60	1.4355	0.8116			1.7		0.09908

Table 10-35 (Continued)
PHYSICAL PROPERTIES OF SELECTED ORGANIC SOLVENTS

Substance	Temp., °C	Density, $g \cdot mL^{-1}$	Refractive Index (n_D)	Viscosity, $mN \cdot s \cdot m^{-2}$	Dielectric Constant (ε)	Dipole Moment, D	Surface Tension, $dyn \cdot cm^{-1}$	
							a	b
Hexafluorobenzene	20	1.6182	1.3781	3.47	30	0	(22.6[20])	
Hexamethylphosphoric triamide	20	1.027	1.4588	0.313	1.89[25]	4.31[liq]	(33.8[20])	0.1022
Hexane	20	0.6594	1.3749	1.041[25]	17.26[25]	0.08	20.44	0.0907
Hexanenitrile	20	0.8052	1.4069	2.814	2.63[71]	1.13[liq]	29.64	(27.55[25])
Hexanoic acid	25	0.9230	1.4149	4.592	13.3	1.55	(28.05[20])	0.0801
1-Hexanol	25	0.8162	1.4161				27.81	0.0869
2-Hexanol	20	0.8144	1.4147				26.44	0.0880
3-Hexanol	20	0.8185	1.4160				26.27	0.10271
1-Hexene	20	0.6732	1.3879	0.26	2.051	0.34	20.47	
4-Hydroxy-4-methyl-2-pentanone	25	0.9341	1.4213	2.9[20]	18.2	3.24	(31.0[20])	0.1123
Iodobenzene	20	1.8307	1.5714	1.77[17]	4.63	1.71	41.52	0.1286
Iodoethane	20	1.9358	1.5137	0.617[15]	7.82	1.91	31.67	0.1234
Iodomethane	20	2.2790	1.5315	0.518[15]	7.00	1.62	33.42	0.1136
1-Iodopropane	20	1.7489	1.5041	0.837[15]	7.00	2.04	31.64	0.1107
2-Iodopropane	20	1.7025	1.4992	0.732[15]	8.19	1.95	29.35	0.1092
Isobutylamine	20	0.7346	1.3972	0.553[25]	4.43[21]	1.27	24.48	(23.84[30])
Isobutyronitrile	25	0.7656	1.3712	0.456[30]	20.4[24]	3.61	(24.93[20])	0.1072
Isopropyl acetate	20	0.8718	1.3773	0.569	5.45	1.86	24.44	0.09719
Isopropylamine	20	0.6875	1.3742	0.36[25]	2.38	0.79	19.91	0.1054
Isopropylbenzene	20	0.8618	1.4915	0.791	10.7	2.73	30.32	
Isoquinoline	25	1.0986[20]	1.4392[20]	40.33	22[17]			
Lactic acid	25	1.2060	1.4314			1.65	(24.4[20])	
Methacrylic acid	20	1.0153	1.4007	0.392	32.70	3.69	24.00	0.0773
Methacrylonitrile	25	0.8001	1.3265	0.544	4.33	1.70	38.11	0.1204
Methanol	25	0.7866	1.5143	0.789[30]	16.9	1.38	33.30	0.0984
Methoxybenzene	25	0.9893	1.4021	1.72		2.04	(34.8[25])	(29.9[75])
2-Methoxyethanol	20	0.9646	1.4245	3.48				
2-(2-Methoxyethoxy)ethanol	25	1.0167						

2-Methoxyethyl acetate	20	1.0049	1.4022	0.981		2.13	32.47	0.1164
bis(2-Methoxyethyl) ether	25	0.9440	1.4043	3.23	8.25	1.97	(33.67^{30})	(30.62^{50})
N-Methylacetamide	35	0.9460	1.4253		191.3^{32}	4.39	27.95	0.1289
Methyl acetate	25	0.9273	1.3614^{20}	0.362	6.68	1.72	34.98	0.0944
Methyl acetoacetate	20	1.0747	1.4186	1.704				
Methyl acrylate	20	0.9535	1.4117^{18}	1.398				
Methyl benzoate	15	1.0933	1.5205	2.298	6.59^{20}	1.86	40.10	0.1171
2-Methylbutane	20	0.6197	1.3537	0.225	1.84	0.13	17.20	0.1103
4-Methylbutanenitrile	20	0.8035	1.4061	0.980	15.5	3.53	28.89	0.0917
2-Methylbutanoic acetate	20	0.8719	1.4054	0.872	4.63^{30}	1.8	(24.62^{21})	
3-Methylbutanoic acid	15	0.9308	1.4064	2.731	2.64^{20}	0.63	27.28	0.0886
2-Methyl-1-butanol	20	0.8190	1.4107	5.50	14.7^{25}	1.82	25.76	0.0820
3-Methyl-1-butanol	20	0.8103	1.4072	4.81^{15}	14.7^{25}	1.7	24.18	0.0748
2-Methyl-2-butanol	20	0.8090	1.4050	5.48^{15}	5.82^{25}		(23.0^{25})	
3-Methyl-2-butanol	20	0.8179	1.4096	3.51^{25}				
3-Methylbutyl acetate	25	0.8664	1.3981	0.790	4.63^{30}	1.82	26.75	0.0989
Methyl butyrate	25	0.8984^{20}	1.3870	0.543	5.6	1.72	27.48	0.1145
Methyl cyanoacetate	25	1.1225	1.4166	2.793^{20}	29.30^{20}		41.32	0.1074
Methylcyclohexane	20	0.7694	1.4231	0.734	2.02	0	26.11	0.1130
cis-2-Methylcyclohexanol	20	0.9254	1.4609	18.08^{25}	13.3*	1.95*	32.45	0.0770
trans-2-Methylcyclohexanol	20	0.9247	1.4616	37.13^{25}				
cis-3-Methylcyclohexanol	20	0.9168	1.4576	19.7^{25}	16.47	1.91	29.08	0.0629
trans-3-Methylcyclohexanol	20	0.9214	1.4580	25.1^{25}	8.05	1.75		
cis-4-Methylcyclohexanol	25	0.9122	1.4565	0.247^{25}	13.3*	1.9*	29.07	0.0690
trans-4-Methylcyclohexanol	20	0.9080	1.4544	0.385				
Methylcyclopentane	20	0.7486	1.4097	0.507	1.985	0	24.63	0.1163
N-Methylformamide	25	0.9988	1.4300	1.65	182.4	3.86	(37.96^{30})	(35.02^{50})
Methyl formate	20	0.9742	1.3433	0.328^{25}	8.5	1.77	28.29	0.1572
2-Methylhexane	20	0.6786	1.3849	0.378	1.92	0	21.22	0.09635
3-Methylhexane	20	0.6871	1.3886	0.372	1.93	0	21.73	0.09699
Methyl methacrylate	20	0.9433	1.4146	0.632	2.9	1.68		
Methyl oleate	20	0.8702^{25}	1.4521	4.88^{30}	3.211		(31.3^{25})	(25.4^{100})
2-Methylpentane	20	0.6532	1.3715	0.310	1.88		19.37	0.09967
3-Methylpentane	20	0.6643	1.3765	0.307^{25}	1.895		20.26	0.1060
2-Methyl-1-pentanol	20	0.8242	1.4190				26.98	0.0819
3-Methyl-1-pentanol	20	0.8237	1.4193				26.92	0.07894

*Mixed isomers.

Table 10-35 (Continued)
PHYSICAL PROPERTIES OF SELECTED ORGANIC SOLVENTS

Substance	Temp., °C	Density, $g \cdot mL^{-1}$	Refractive Index (n_D)	Viscosity, $mN \cdot s \cdot m^{-2}$	Dielectric Constant (ε)	Dipole Moment, D	Surface Tension, $dyn \cdot cm^{-1}$	
							a	b
4-Methyl-1-pentanol	20	0.8130	1.4154				25.93	0.07434
2-Methyl-2-pentanol	20	0.8136	1.4113				25.07	0.08606
3-Methyl-2-pentanol	20	0.8291	1.4197				27.14	0.0919
4-Methyl-2-pentanol	20	0.8076	1.4112	4.074^{25}			24.67	0.0821
2-Methyl-3-pentanol	20	0.8239	1.4168				26.43	0.0914
3-Methyl-3-pentanol	20	0.8281	1.4186				25.48	0.0888
4-Methyl-2-pentanone	20	0.8006	1.3958	0.542^{25}	13.11		(23.64^{20})	(19.62^{60})
2-Methylpropanamine	20	0.7346	1.3970		4.4	1.28	24.48	0.1092
2-Methylpropanoic acid	20	0.9682	1.3930	1.213^{25}	2.73^{40}	1.08	(25.55^{20})	(25.13^{25})
2-Methyl-1-propanol	25	0.7978	1.3938	3.91	17.93	1.79	24.53	0.0795
2-Methyl-2-propanol	25	0.7812	1.3852	3.316^{30}	12.47	1.66	(20.02^{15})	(19.10^{30})
N-Methylpropionamide	25	0.9305	1.4345	5.215	172.2	3.59	(31.20^{30})	(29.12^{50})
Methyl propionate	15	0.9221	1.3793	0.477	6.21^{20}	1.7	27.58	0.1258
1-Methylpropyl acetate	20	0.8720	1.3894				25.72	0.1054
2-Methylpropyl acetate	20	0.8745	1.3902	0.697	5.29	1.85	25.59	0.1013
2-Methylpropyl formate	20	0.8854	1.3855	0.680	6.41	1.88	26.14	0.1122
2-Methylpyridine	20	0.9444	1.5010	0.805	9.80	1.92	36.11	0.1243
3-Methylpyridine	20	0.9566	1.5068		9.80	2.40	37.35	0.1153
4-Methylpyridine	20	0.9548	1.5058			2.60	37.71	0.1141
1-Methyl-2-pyrrolidinone	25	1.0279	1.4680	1.666	32.0	4.09	42.15	0.1174
Methyl salicylate	20	1.1831	1.5365	38.27	9.41^{30}	2.53	(37.63^{20})	(36.24^{30})
Morpholine	15	1.0050	1.4573		7.42^{25}	1.50	42.84	0.1107
Naphthalene	85	0.9752	1.5898	0.780^{100}	2.54	0.0	48.62	0.1185
o-Nitroanisole	25	1.2408	1.5597			4.83	46.34	0.1157
Nitrobenzene	20	1.2033	1.5526	1.634	34.82^{25}	4.22	35.27	0.1255
Nitroethane	25	1.0382	1.3902	0.661	28.06^{30}	3.65	40.72	0.1678
Nitromethane	25	1.1312	1.3795	0.595^{30}	35.87^{30}	3.46	48.62	0.1185
1-Nitro-2-methoxybenzene	20	1.2527	1.5619			4.81	48.62	0.1185
1-Nitropropane	25	0.9955	1.3994	0.798	23.24^{30}	3.66	32.62	0.1009

Name	t							
2-Nitropropane	25	0.9821	1.3921	0.750	25.52	3.73	32.18	0.1158
Nonane	20	0.7176	1.4054	0.7160	1.972	0	24.72	0.09347
1-Nonanol	20	0.8280	1.4338			1.72	29.79	0.07589
1-Nonene	20	0.7922	1.4157	0.620		0.59	24.90	0.09379
1-Octadecanol	20	0.8123	1.4388			1.74		
Octane	20	0.70252	1.3974	0.546	1.95	0	23.52	0.09509
Octanenitrile	30	0.8059	1.4163	1.356	13.90[25]	1.15liq	29.61	0.0802
Octanoic acid	20	0.9106	1.4279	5.828	2.45[25]		(29.2[20])	(28.7[25])
1-Octanol	20	0.8258	1.4296	6.125[30]	10.34	1.68	29.09	0.0795
2-Octanol	20	0.8207	1.4261			1.65	27.96	0.08197
3-Octanol	20	0.8216	1.423					
4-Octanol	20	0.8192	1.425					
1-Octene	20	0.7149	1.4087	0.470	2.084	0.34liq	23.68	0.09581
Oleic acid	20	0.8906	1.4599	38.80	2.46	1.18	(32.80[20])	(27.94[90])
2,2'-Oxybis(chloroethane)	20	1.2192	1.4575	2.41	21.2	2.58		
2,2-Oxydiethanol	20	1.1167	1.4475	35.7	31.69	2.31	46.97	0.0880
Pentachloroethane	15	1.6881	1.5054	2.751	3.73[20]	0.92	37.09	0.1178
cis-1,3-Pentadiene	25	0.6859	1.4329		2.32			
trans-1,3-Pentadiene	25	0.6710	1.4267					
2,3-Pentadiene	25	0.6900	1.4251					
Pentane	25	0.6214	1.3547	0.225	1.84	0	18.25	0.11021
2,4-Pentanedione		0.9721[25]	1.4518[19]	0.779	25.7	3.05	33.28	0.1144
Pentanenitrile	15	0.8035	1.3991	2.359[15]	17.4[21]	3.57	29.28	0.0937
1-Pentanoic acid	20	0.9392	1.4080	3.347	2.66		28.90	0.0887
1-Pentanol	25	0.8112	1.4080	2.780[30]	13.9	1.7	27.54	0.0874
2-Pentanol	25	0.8053	1.4044	3.306[30]	13.82[22]	1.66	25.96	0.1004
3-Pentanol	20	0.8160	1.4079	0.478	13.02[22]	1.64	(24.60[20])	(23.76[30])
3-Pentanone	20	0.8144	1.3924		17.00	2.70	27.36	0.1047
2-Pentanone	20	0.8095	1.3903			2.72	24.89	0.06547
1-Pentene	20	0.6405	1.3715	0.24[0]	2.017	0.34liq	18.20	0.1099
cis-2-Pentene	20	0.6556	1.3830				19.73	0.1172
trans-2-Pentene	20	0.6482	1.3793				18.90	0.09972
Pentyl acetate	20	0.8753	1.4028	0.924	4.75	1.91	27.66	0.09943

Table 10-35 (*Continued*)
PHYSICAL PROPERTIES OF SELECTED ORGANIC SOLVENTS

Substance	Temp., °C	Density, g·mL⁻³	Refractive Index (n_D)	Viscosity, mN·s·m⁻²	Dielectric Constant (ε)	Dipole Moment, D	Surface Tension, dyn·cm⁻¹	
							a	b
Phenol	46	1.0533	1.5396	4.076	9.78[60]	1.45	43.54	0.1068
Phenylacetonitrile	25	1.0125	1.5209	1.93	18.7[27]	3.5	44.57	0.1155
d-Pinene	20	0.8600	1.4658	1.61[25]	2.64[25]	0.80	28.35	0.09444
L-Pinene	20	0.8590	1.4666	1.41[25]	2.76		28.26	0.09343
Piperidine	20	0.8613	1.4525	1.362[25]	5.8	1.19	31.79	0.1153
1-Propanal	20	0.7970	1.3619	0.317[27]	18.5[17]	2.52		
1,2-Propanediol	20	1.0364	1.4329	56.0	32.0	2.25	(72.0[25])	
1,3-Propanediol	20	1.0538	1.4396	46.6	35.0	2.50	47.43	0.0903
Propanenitrile	20	0.7911	1.3838	0.624[15]	20.3[21]	3.57	29.51	0.1037
1-Propanol	25	0.7995	1.3837	2.004	20.33	1.68	25.26	0.0777
2-Propanol	25	0.7813	1.3752	1.765[30]	19.92	1.66	22.90	0.0789
2-Propen-1-ol	15	0.8551	1.4135[20]	1.486	21.6	1.63	27.53	0.0902
Propionic acid	20	0.9934	1.3865	1.175[15]	3.44[40]	1.75	28.68	0.0993
Propionic anhydride	20	1.0110	1.4046	1.144	18.3[16]		(30.30[20])	(29.70[25])
Propionitrile	20	0.7818	1.3681[15]	0.4541[5]	27.2	3.56	29.63	0.1153
Propyl acetate	20	0.8883	1.3844	0.585	6.00	1.86	26.60	0.1120
Propylamine	20	0.7173	1.3882	0.353[25]	5.31	1.26	24.86	0.1243
Propyl benzoate	20	1.0232	1.5003				36.55	0.1069
Propylene oxide*	20	0.8287	1.3660	0.327		2.01		
Propyl formate	20	0.9006	1.3769	0.574	7.72	1.89	26.77	0.1119
2-Propyn-1-ol	20	0.9478	1.4320	1.68	24.5	1.78	38.59	0.1270
1-Propynyl acetate	20	0.9982	1.4187				(32.81[20])	(30.20[40])
Pyridine	20	0.9832	1.5102	0.952	12.3	2.20	39.82	0.1306
Pyrrole	20	0.9699	1.5102	1.352	7.48[18]	1.80	39.81	0.1100
2-Pyrrolidinone	25	1.107	1.486	13.3		3.55		
Quinoline	15	1.0977	1.6293	4.354	9.00[25]	2.18	42.25	0.1063
Salicylaldehyde	20	1.1574	1.5718	2.90	13.9	2.86	45.38	0.1242
Succinonitrile	60	0.9867	1.4173	2.591	56.5[57]	3.68	53.26	0.1079

	t/°C	ρ	n					
Sulfolane	30	1.2614	1.4820	10.286	43.3	4.81	(35.5^{30})	(30.98^{30})
Styrene	20	0.9060	1.5468	0.751	2.426	0.13^{liq}	(32.3^{20})	0.1463
1,1,2,2-Tetrabromoethane	20	2.9640	1.6353	9.79	7.0^{22}	1.29	52.37	
1,1,2,2-Tetrachloro-difluoroethane	25	1.6447	1.4130	1.21	2.52		(22.73^{30})	(21.56^{40})
1,1,2,2-Tetrachloroethane	15	1.6026	1.4968	1.844	8.20^{20}	1.32	38.75	0.1268
1,1,2,2-Tetrachloroethene	15	1.6311	1.5076	1.932	2.30^{25}	0	(32.86^{15})	(31.27^{30})
1-Tetradecanol	50	0.8151	1.4358			1.7		
Tetrahydrofuran	20	0.8889	1.4072	0.55	7.58^{25}	1.75	(26.5^{25})	0.1008
Tetrahydrofurfuryl alcohol	20	1.0524	1.4520	6.24	13.61^{23}	2.12^{liq}	39.96	
1,2,3,4-Tetrahydro-naphthalene	20	0.9702	1.5414	2.202	2.77	0.60^{liq}	35.55	0.0954
Tetrahydropyran	25	0.8772	1.4195	0.764	5.61	1.55		
Tetrahydrothiophene	25	0.9938	1.5257	0.971		1.90	38.44	0.1342
1,1,2,2-Tetramethylurea	25	0.9654	1.4493		23.06	3.47		
Tetranitromethane		1.6372^{21}	1.4399^{17}		2.32^{20}			
2-Thiabutane	20	0.8422	1.4403	0.373			(24.9^{20})	(23.4^{30})
Thiacyclobutane	20	1.0200	1.5102	0.638		1.78	(36.3^{20})	(35.0^{30})
Thiacyclohexane	20	0.9861	1.5070				(36.06^{20})	(33.74^{40})
Thiacyclopentane	20	0.9987	1.5048	1.042			(35.8^{20})	(34.6^{30})
2-Thiapentane	20	0.8424	1.4442				(25.2^{20})	(23.9^{30})
3-Thiapentane	20	0.8363	1.4430	0.440	5.72^{25}	1.61	(26.50^{11})	(23.33^{23})
2-Thiapropane	20	0.8483	1.4353	0.289	6.2	1.45		
Thiophene	20	1.0649	1.5289	0.654	2.76^{16}	0.55	34.00	0.1328
Toluene	25	0.8623	1.4941	0.552	2.38	0.36	30.90	0.1189
o-Toluidine	15	1.0028	1.5685	5.195	6.34	1.60	42.87	0.1094
m-Toluidine	15	0.9930	1.5638	4.418	5.95	1.45	40.33	0.0979
p-Toluidine	60	0.9538	1.5532	1.557	4.98	1.52	39.58	0.0957
Tribromomethane	15	2.9035	1.6005	2.152	4.39^{20}	0.99	48.14	0.1308
Tri-n-butyl borate	20	0.8580	1.4092	1.776		0.78	(26.2^{20})	(25.8^{25})
Tri-n-butyl phosphate	25	0.9760	1.4226	3.39	7.95^{30}	3.07	28.71	0.0666
Trichloroacetonitrile		1.4403^{25}	1.4403^{20}		7.85^{19}			
1,1,1-Trichloroethane	20	1.3492	1.4377	0.903^{15}	7.6	1.79	28.28	0.1242
1,1,2-Trichloroethane	20	1.4424	1.4706	0.119	8.78^{23}		37.40	0.1351

*Mixed isomers.

Table 10-35 (*Continued*)
PHYSICAL PROPERTIES OF SELECTED ORGANIC SOLVENTS

Substance	Temp., °C	Density, $g \cdot mL^{-3}$	Refractive Index (n_D)	Viscosity, $mN \cdot s \cdot m^{-2}$	Dielectric Constant (ε)	Dipole Moment, D	Surface Tension, $dyn \cdot cm^{-1}$	
							a	b
1,1,2-Trichloroethene	20	1.4679	1.4775	0.566	3.42^{16}	0.9	(29.5^{20})	(28.8^{25})
1,2,3-Trichloropropane	20	1.3880	1.4834			0	(37.8^{20})	(37.05^{25})
Tridecane	20	0.7563	1.4256	18.834			27.73	0.08719
1-Tridecene	20	0.7653	1.4334				28.01	0.08839
Triethanolamine	25	1.1196	1.4835	613.6	29.36	3.57	22.70	0.0992
Triethylamine	20	0.7281	1.4010	0.394^{15}	2.42^{25}	0.77	15.64	0.08444
Trifluoroacetic acid	20	1.4890	1.2850	0.926	39.5	2.28^{100}	30.91	0.1040
1,2,3-Trimethylbenzene	20	0.8944	1.5139	0.895^{15}	2.64	0.56	31.76	0.1025
1,2,4-Trimethylbenzene	20	0.8758	1.5048	1.154	2.38	0.30	29.79	0.08966
1,3,5-Trimethylbenzene	20	0.8652	1.4994	0.579	2.28	0	20.70	0.09726
2,2,3-Trimethylbutane	20	0.6901	1.3894	0.632	1.93	0		
cis-1,3,5-Trimethylcyclohexane	20	0.7705	1.4245					
trans-1,3,5-Trimethyl-cyclohexane	20	0.7789	1.4286	0.714				
2,2,3-Trimethylpentane	20	0.7160	1.4030	0.598	1.962	0	22.46	0.08950
2,2,4-Trimethylpentane	20	0.6919	1.3915	0.504	1.940	0	20.55	0.08876
Undecane	20	0.7402	1.4173	11.855			26.46	0.09010
1-Undecanol	20	0.8324	1.4402	23.95		1.74	(23.95^{20})	(22.54^{30})
Vinyl acetate	20	0.9312	1.3959			1.79		
o-Xylene	20	0.8802	1.5054	0.809	2.57	0.62	32.51	0.1101
m-Xylene	20	0.8642	1.4972	0.617	2.37	0.37	31.23	0.1104
p-Xylene	20	0.8611	1.4958	0.644	2.27	0	30.69	0.1074

THERMAL CONDUCTIVITY

In the following tables dealing with heat conductivity, the coefficient λ is given in g.-cal./(sec.)(sq. cm.)(°C./cm.). This is the quantity of heat in gram calories, transmitted per second through a plate of the material one centimeter thick and one square centimeter in area when the temperature difference between the two sides of the plate is one degree Celsius. To express λ in English units, Btu./(sec.)(sq. in.)(°F./inch), multiply the values given below by 0.00560.

Table 10-36
THERMAL CONDUCTIVITY OF VARIOUS METALS AND ALLOYS

Metal or Alloy	Temp. °C.	λ
Aluminum, commercial	−188.2	0.454
" "	0	0.461
" 99%	18	0.504
" "	100	0.49
" "	400	0.76
" "	600	1.01
Antimony	−77	0.0628
	0	0.0538
	100	0.0515
20Sb+80Bi	0	0.0152
	100	0.0205
50Sb+50Bi	0	0.0196
	100	0.0229
70Sb+30Bi	0	0.0234
	100	0.0281
33.3Sb+66.7Cd	0	0.0267
50Sb+50Cd	0	0.00519
66.7Sb+33.3Cd	0	0.00299
Bismuth	−77	0.0257
	0	0.0177
	100	0.0164
25Bi+75Pb(Vol.)	44	0.0468
96.5Bi+3.5Pb(Vol.)	44	0.0129
90Bi+10Sn(Vol.)	44	0.0126
50Bi+50Sn	12.5	0.056
25Bi+75Sn	12.5	0.102
Brass, red	0	0.246
" "	100	0.2827
Brass, yellow	0	0.2041
" "	100	0.254
Cadmium	0	0.2213
"	100	0.2045
Constantan, see 60 Cu+40 Ni		
Cobalt, with 0.24%C,+ 1.4Fe+1.1 Ni+0.14Si	30	0.1653
Chromium steel, 5%Cr	30	0.073
" " 10%Cr	30	0.052
" " 15%Cr	30	0.044
Copper	−183	1.111
"	0	0.920
"	100	0.92
60Cu+40Ni	18	0.05401
	100	0.06405
54Cu+46Ni	18	0.0484
99.37Cu+0.63P	30	0.250
98.02Cu+1.98P	30	0.125

Metal or Alloy	Temp. °C.	λ
89Cu+11Zn	18	0.275
87Cu+13Zn	18	0.301
82Cu+18Zn	18	0.313
68Cu+32Zn	18	0.260
German silver (52Cu, 26Zn, 22Ni)	0	0.0700
	100	0.0887
62Cu+15Ni+22Zn	18	0.0595
Gold	0	0.744
"	97	0.7464
90Au+10Pd	25	0.234
50Au+50Pd	25	0.086
10Au+90Pd	25	0.124
40Au+60Pt	25	0.062
10Au+90Pt	25	0.182
Iridium	17	0.141
Iron, with 0.1%C+ 0.1%Mn+0.2%Si }	18	0.1436
	100	0.1420
99Fe+1C	18	0.1085
	100	0.1076
Fe+1.5C+0.19Mn+ 0.05Si+0.03Cu+ 0.01P+0.025S	18	0.119
Bessemer steel	8	0.0985
Lead	18	0.0827
"	100	0.0815
Lithium	0	0.17
"	101.3	0.18
Magnesium	0−100	0.376
Manganin (84Cu+4Ni +12Mn)	18	0.05186
	100	0.0631
Mercury, solid	−269.3	0.40
" "	−44.2	0.0664
" liquid	−37.2	0.0218
" "	0	0.0248
" "	50.4	0.0298
" "	149.4	0.0385
Molybdenum	17	0.346
Nickel, 99%	−160	0.129
	18	0.140
Nickel +2 or 3% Co	300	0.126
	500	0.104
	950	0.065
	1200	0.058
Nickel steel, (30.4Ni+ 0.14Si+0.84Mn+ 0.26C)	29	0.029
	71	0.031
Palladium	18	0.1683

Table 10-36 (Continued)
THERMAL CONDUCTIVITY OF VARIOUS METALS AND ALLOYS

Metal or Alloy	Temp. °C.	λ	Metal or Alloy	Temp. °C.	λ
Palladium	100	0.1817	Silver, 99.9%	−160	0.998
90Pd+10Pt	25	0.134	" "	0	1.096
50Pd+50Pt	25	0.088	" "	10−97	0.9628
10Pd+90Pt	25	0.103	" 99.98%	18	1.006
90Pd+10Ag	25	0.114	" "	100	0.9919
50Pd+50Ag	25	0.076	Sodium	5.7	0.321
10Pd+90Ag	25	0.337	"	21.2	0.317
Platinum	−252.8	0.93	"	88.1	0.288
"	−183	0.182	Tantalum	17	0.130
"	0−200	0.167	"	1827	0.198
90Pt+10Ir	17	0.074	Tin	−170	0.195
90Pt+10Rh	17	0.072	"	0	0.1528
30Pt+70Ag	25	0.074	"	100	0.1423
10Pt+90Ag	25	0.234	30Sn+70Zn (Vol.)	44	0.224
Potassium	5.0	0.234	91.1Sn+8.9Zn (Vol.)	44	0.157
"	20.7	0.232	Tungsten	0	0.383
"	57.6	0.217	"	2227	0.354
62.9K+37.1Na	6.0	0.0549	Wood's metal	7	0.0319
	42.9	0.0619	Zinc	−170	0.280
Rhodium	17	0.210	"	18	0.2653
			"	100	0.2619

Table 10-37
THERMAL CONDUCTIVITY OF VARIOUS SOLIDS

Substance	Temp. °C.	λ×10³	Substance	Temp. °C.	λ×10³
Aluminum oxide (powder)	46.8	1.62	Cork, Sp.G.=0.204	30	0.128
" " (fused)	650-1350	8.0	Cork meal	100	0.133
Asbestos board	20	1.78	Cotton, Sp.G.=0.081	0	0.136
Asbestos fabric	20	0.666	Diatomaceous earth	20	0.13
Asbestos fiber	0	0.267	Earth's crust, average	20	4.0
" "	100	0.284	Ebonite	0	0.378
Asbestos paper	20	0.345	Eiderdown	20	0.011
Asphalt	20	1.78	Feathers (with air)	9	0.0574
Basalt	20	5.2	Feldspar	20	5.6
Bauxite	600	1.33	Felt (dark gray)	40	0.149
Boiler scale	65.5	3.13	Ferric oxide	200	1.41
Brick, common	20	1.5	(pressed powder)		
Blotting paper	20	0.15	Ferrous oxide (pressed	49.8	1.33
Cadmium oxide	46.5	1.63	powder)		
(pressed powder)			Fiber, vegetable (with	9	0.0645
Carborundum*			air)		
Cardboard	20	0.5	" , " (with-	9	1.42
Cement, Portland	89.5	0.71	out air)		
Chalk	20	2.2	Fire brick	20	1.1
Clay (fire-hardened)	360 to	2.09 to	Flannel	50	0.0355
	600	2.21	Flint	20	2.4
Coal	<0	0.405	Fluorite	0	24.68
"	1427	20.1	"	100	19.10
Cobalt oxide (pressed	48.5	1.00	Gas carbon	20	8.5
powder)			" "	100	9.5
Concrete	20	2.2	Glass, crown	12.5	1.63
Copper oxide (pressed	45.6	2.42	" , flint	12.5	1.43
powder)			" , Jena	22	2.27
Copper sulfide (pure)	0	1.06	" , soda	20	1.7

*Trade name, see Silicon carbide.

Table 10-37 (*Continued*)
THERMAL CONDUCTIVITY OF VARIOUS SOLIDS

Substance	Temp. °C.	$\lambda \times 10^3$	Substance	Temp. °C.	$\lambda \times 10^3$
Glass, soda	100	1.8	Quartz glass	0	3.32
Granite	20	8.17	" "	100	4.57
Graphite, Sp.G.=1.58	50	105.5	Rock salt	0	16.67
" ⊥ to axis	142	42.6	" "	100	11.59
" " " "	555	279	Roofing paper	<0	0.453
Graphite powder,	40	2.85	Rubber, hard, gray	49	0.55
Sp.G.=0.7			" soft, "	49	0.44
Gutta percha	20	0.48	" " red	49	0.34
Gypsum	0	3.1	Sand, dry	20	0.93
Hair-cloth	<0	0.0402	Sandstone, Sp.G.=2.259	40	4.39
Horn	<0	0.087	Sawdust, Sp.G.=0.19	30	0.14
Horse hair, Sp.G.=0.172	20	0.122	Serpentine	20	2.4
Ice	....	5.7	Silica, see Quartz		
Infusorial earth	100	0.34	Silicon carbide	650–1350	37.2
" "	300	0.40	Silk, Sp.G.=0.101	0	0.122
Lampblack, Sp.G.=0.165	40	0.156	" " "	50	0.136
Lava	16–99	2.01	Silver bromide	0	2.46
Leather, cowhide	84	0.42	Silver chloride	0	2.6
" sole, Sp.G.=1.00	30	0.38	Slate	20	4.70
Lime, clayey	20	7.8	Snow, fresh, Sp.G.=0.111	....	2.56
Linen	20	0.21	" , old, Sp.G.=0.450	....	0.115
Linoleum, Sp.G.=1.183	20	0.445	Sodium chlorate	0	2.665
Magnesia (MgO) (pressed	47.6	1.45	Soil, dry	20	0.33
powder), Sp.G.=0.797			Sugar, cane	0	1.39
Magnesia brick	50 to	2.7 to	Sulfur, rhombic	0	0.70
	1130	7.2	" plastic	20–100	0.63
Magnesite	1000	3.98	Sylvite	0	16.65
Marble, white	....	7.8	"	100	11.76
" black	30	6.85	Thymol	12	0.359
Mica	41.3	0.860	Wadding, Sp.G.=0.01	18	0.93
Mica, pressed plates	60	0.627	Wax, bees'	20	0.207
Naphthalene	0	0.90	Wood, maple‖to face	20	1.015
α-Naphthol	35	0.76	Wood, maple, ⊥ to face	50	0.434
β-Naphthol	35	0.80	Wood, oak, ⊥ to face	15	0.500
Nickel oxide (pressed	46.2	2.24	Sp.G.=0.825		
powder, Sp.G.=1.445)			Wood, oak,‖to face	15	0.834
Onyx	30	5.56	Sp.G.=0.819		
Paper	20	0.3	Wood, pine,‖to face	20	0.834
Paraffin	0	0.688	Sp.G.=0.551		
Plaster of Paris	20	0.70	Wood, pine, ⊥ to face	15	0.361
Porcelain	95	2.48	Sp.G.=0.546		
Potassium chloride	0	16.6	Wood, teak,‖to face	15	0.903
Potassium iodide	0	12	Sp.G.=0.604		
Quartz, ‖ to axis	0	32.5	Wood, teak, ⊥ to fac	15	0.417
" , " " "	100	21.5	Sp.G.=0.642		
" ⊥ to axis	0	17.31	Zinc oxide (pressed	49.7	1.42
" " " "	100	13.33	powder, Sp.G.=2.886)		

Table 10-38
THERMAL CONDUCTIVITY OF VARIOUS LIQUIDS AND SOLUTIONS

Liquid or Solution	Temp. °C.	λ×10³	Liquid or Solution	Temp. °C.	λ×10³
Acetic acid	25	0.43	Octane	4	0.375
" " , 50%	25	0.85	Olive oil, Sp.G.=0.911	15.7	0.4515
Acetone	0	0.4228	Paraffin oil	17	0.346
Ammonia, 26%	18	1.09	Pentane	−185	0.3955
Amyl acetate	12	0.302	"	14	0.2856
Amyl alcohol	12	0.328	Petrolatum	20	0.222
Amyl chloride	12	0.283	Petroleum	13	0.355
Amyl iodide	12	0.203	Potassium bromide, 40%	32	1.176
Aniline	12	0.408	Potassium carbonate, 20%	32	1.373
Barium chloride, 21%	32	1.396	Potassium chlorate,	13	1.16
Benzene	12	0.333	Sp.G. =1.026		
Bromobenzene	12	0.265	Potassium chloride, 20%	32	1.334
iso-Butyl alcohol	12	0.340	Potassium hydroxide, 21%	32	1.385
iso-Butyl bromide	12	0.278	" " , 42%	32	1.313
iso-Butyl chloride	12	0.278	Potassium nitrate, 10%	32	1.409
iso-Butyl iodide	12	0.208	" " , 20%	32	1.337
n-Butyric acid	12	0.360	Potassium sulfate, 10%	32	1.440
iso-Butyric acid	12	0.340	Propionic acid	12	0.390
Calcium chloride, 15%	32	1.383	Propyl acetate	12	0.327
" " , 30%	32	1.315	n-Propyl alcohol	12	0.373
iso-Caproic acid	12	0.298	iso-Propyl alcohol	0	0.3683
Carbon disulfide	12	0.343	Propyl bromide	12	0.257
Carbon tetrachloride	12	0.252	Propyl chloride	12	0.283
Castor oil	....	0.425	Propyl formate	12	0.357
Chlorobenzene	12	0.302	Propyl iodide	12	0.220
Chloroform	12	0.288	Sodium bromide, 20%	32	1.348
Copper sulfate, 18%	32	1.379	" " , 40%	32	1.289
Cylinder oil	81	0.290	Sodium carbonate, 10%	32	1.403
Cymene	12	0.272	Sodium chloride, 12.5%	32	1.403
Ethyl acetate	12	0.348	" " , 25%	32	1.141
Ethyl alcohol	5.2	0.487	Sodium nitrate, 20%	32	1.376
" "	51	0.369	" " , 44%	32	1.311
" " , 90%	15	0.4391	Sodium sulfate, 10%	32	1.447
" " , 70%	14.1	0.5711	Strontium nitrate, 20%	32	1.398
" " , 50%	12.9	0.7461	" " , 40%	32	1.346
" " , 30%	12.3	1.002	Sulfuric acid, 30%	32	1.244
" " , 10%	11.9	1.247	" " , 60%	32	1.047
Ethyl bromide	12	0.247	" " , 90%	32	0.846
Ethyl ether	12	0.303	" " , Sp.G. =1.054	20.5	1.26
Ethyl formate	12	0.378	" " , Sp.G. =1.18	21	1.30
Ethyl iodide	12	0.222	Thymol	13	0.313
Ethyl sulfide	12	0.328	Toluene	0	0.3492
Ethylene glycol	0	0.6353	"	12	0.307
Formic acid	12	0.648	Turpentine oil	12	0.260
Glycerol	12	0.670	n-Valeric acid	12	0.325
"	48	0.613	iso-Valeric acid	12	0.312
Heptane	4	0.337	Water	4.1	1.29
Hexane	4	0.364	"	12	1.36
Hydrochloric acid, 12.5%	32	1.262	"	40.8	1.555
" " , 25%	32	1.151	Wood tar	79.5	0.324
" " , 38%	32	1.052	o-Xylene	0	0.3443
Magnesium chloride, 11%	32	1.376	" "	33	0.2519
" " , 29%	32	1.238	m-Xylene	0	0.3429
Magnesium sulfate, 22%	32	1.414	Zinc chloride, 17.5%	32	1.327
Methyl acetate	12	0.385	" " , 35%	32	1.213
Methyl alcohol	12	0.495	Zinc sulfate, 16%	32	1.382
Methyl butyrate	12	0.335	" sulfate, 32%	32	1.327
Methyl valerate	12	0.315	" " , Sp.G. =1.134	4.5	1.18
Nitrobenzene	12.5	0.3801	" " , Sp.G. =1.382	45.2	1.44

Table 10-39
THERMAL CONDUCTIVITY OF VARIOUS GASES AND VAPORS

Gas or Vapor	Temp. °C.	$\lambda \times 10^5$	Gas or Vapor	Temp. °C.	$\lambda \times 10^5$
Acetone	0	2.301	Helium	100	39.85
"	100	3.96	n-Heptane	100	4.136
"	184	5.90	n-Hexane	20	2.854
Acetylene	0	4.40	Hexylene	100	4.396
Air	−191.1	1.80	Hydrogen	−252.2	3.22
"	−78.4	4.256	"	−78.4	30.65
"	0	5.572	"	0	39.60
"	100	7.197	"	100	49.94
"	531	15.95	Hydrogen sulfide	0	3.045
Ammonia	−57.6	3.82	Mercury vapor	203	1.846
"	0	5.135	Methane	−181.6	2.248
"	100	7.09	"	−75.6	4.940
Argon	−182.6	1.42	"	0	7.200
"	0	3.88	Methyl alcohol	0	3.357
"	100	5.087	" "	100	5.161
Benzene	0	2.094	Methyl bromide	4.6	1.74
"	100	4.144	Methyl chloride	0	2.216
"	212.5	7.08	" "	100	3.841
Butylamine	6.5	3.003	" "	212.5	6.113
Carbon dioxide	−78.5	2.546	Methyl dichloride	0	1.562
" "	0	3.393	" "	100	2.524
" "	100	5.06	" "	212.5	3.804
" "	546	14.20	Methyl iodide	0	1.098
Carbon disulfide	0	1.615	" "	100	1.804
Carbon monoxide	−191	1.650	Neon	−181.4	4.99
" "	0	5.425	"	−74.4	8.79
Carbon tetrachloride	46	1.666	"	0	10.87
" "	100	2.048	"	105.8	13.44
" "	184	2.599	Nitric oxide, NO	−71.4	4.160
Chlorine	0	1.829	" "	0	5.55
Chloroform	0	1.523	Nitrogen	−191.4	1.829
"	100	2.333	"	−78.4	4.305
"	184	3.103	"	0	5.68
Ethane	−70.4	2.727	"	100	7.18
"	0	4.306	Nitrogen dioxide	55	8.88
"	100	7.673	Nitrous oxide, N_2O	−71.8	2.710
Ethyl acetate	46	2.88	" "	0	3.515
" "	100	3.862	" "	100	5.06
" "	184	5.69	Oxygen	−191.4	1.721
Ethyl alcohol	20	3.583	"	−78.4	4.292
" "	100	4.98	"	0	5.70
Ethyl ether	0	3.101	"	100	7.427
" "	100	5.278	n-Pentane	20	3.267
" "	212.5	8.400	iso-Pentane	0	2.912
Ethylene	−71.1	2.572	" "	100	5.105
"	0	4.02	" "	184	7.52
"	100	6.36	Sulfur dioxide	0	1.950
Helium	−252.2	5.18	Water vapor	46	4.580
"	−191.7	14.84	" "	100	5.510
"	0	33.60			

COMPRESSIBILITY AND THERMAL EXPANSION

Table 10-40
COMPRESSIBILITY OF WATER

In the table below are given the relative volumes of water at various temperatures and pressures. The volume at 0°C. and one normal atmosphere (760 mm of Hg) is taken as unity.

P, atm	−10°C.	0°C.	10°C.	20°C.	40°C.	60°C.	80°C.
1	1.0017	1.0000	1.0001	1.0016	1.0076	1.0168	1.0287
500	0.9788	0.9767	0.9778	0.9804	0.9867	0.9967	1.0071
1000	0.9581	0.9566	0.9591	0.9619	0.9689	0.9780	0.9884
1500	0.9399	0.9394	0.9424	0.9456	0.9529	0.9617	0.9717
2000	0.9223	0.9241	0.9277	0.9312	0.9386	0.9472	0.9568
2500	0.9083	0.9112	0.9147	0.9183	0.9257	0.9343	0.9437
3000	0.8962	0.8993	0.9028	0.9065	0.9139	0.9225	0.9315
3500	0.8852	0.8884	0.8919	0.8956	0.9030	0.9115	0.9203
4000	0.8751	0.8783	0.8818	0.8855	0.8931	0.9012	0.9097
4500	0.8658	0.8692	0.8725	0.8762	0.8838	0.8919	0.9001
5000	0.8573	0.8606	0.8639	0.8675	0.8752	0.8832	0.8913
6000	......	0.8452	0.8481	0.8517	0.8595	0.8674	0.8752
7000	......	......	0.8340	0.8374	0.8456	0.8534	0.8610
8000	......	......	......	0.8244	0.8330	0.8408	0.8483
9000	......	......	......	0.8128	0.8219	0.8297	0.8371
10000	......	......	......	0.8027	0.8119	0.8196	0.8268
11000	......	......	......	......	0.8023	0.8101	0.8172
12000	......	......	......	......	0.7931	0.8009	0.8080

Table 10-41
COEFFICIENTS OF COMPRESSIBILITY OF VARIOUS LIQUIDS AND AQUEOUS SOLUTIONS

In the table below are given values of β_t (the compressibility coefficient of a liquid at $t°_1$.) to be used with the formula

$$\beta_t = \frac{1}{V_1}\left(\frac{V_1 - V_2}{P_2 - P_1}\right)$$

where V_1 is the volume of a liquid under a pressure of P_1 in atmospheres at $t°C.$, and V_2 is the volume at some other pressure P_2, at the same temperature.

Liquid	t °C.	Pressure Range, atm.	$\beta_t \times 10^6$
Acetic acid	25	at 92.5	81.4
" "	25	at 218.5	72.6
" "	25	at 494	57.1
Acetone	14.2	8.90 to 36.51	111
"	0	100 to 500	82
"	0	500 to 1000	59
"	0	1000 to 1500	47
"	0	1500 to 2000	40
"	25	at 82.5	111.8
Allyl alcohol	9.6	1 to 500	69
" "	9.6	500 to 1000	51
" "	9.6	1000 to 1500	43
" "	9.6	1500 to 2000	36
Ammonia, 2.29%	0	up to 10	50.9
" , 17.58%	0	up to 10	41.6
" , 21.58%	15	up to 10	36.5
Amyl alcohol	17.75	at 8	83.5
" "	13.8	8.5 to 37.12	88.2
Aniline	25	at 85.5	43.2
"	25	at 181.5	40.5

Table 10-41 (*Continued*)
COEFFICIENTS OF COMPRESSIBILITY OF VARIOUS LIQUIDS

Liquid	t °C.	Pressure Range, atm.	βt ×10⁶
Aniline	25	at 281.5	38.3
"	25	at 390	36.1
Benzene	20	1 to 2	95.3
"	15.4	1 to 4	87
"	12.9	1 to 18.5	86.8
"	16	8.12 to 37.2	90
"	20	98.7 to 296	78.7
"	20	296 to 494	67.5
Bromine	20	0 to 98.7	63.5
"	20	98.7 to 197.4	58.4
"	20	197.4 to 296	54.6
"	20	296 to 395	52.1
"	20	395 to 494	49.9
Bromoform	20	0 to 98.7	51.0
"	20	98.7 to 197.4	47.5
"	20	197.4 to 296	44.0
"	20	296 to 395	42.0
"	20	395 to 494	41.0
n-Butyl alcohol	17.4	at 8	90
iso-Butyl alcohol	17.95	at 8	98
Butyl benzoate	10	1 to 5.25	59
Butyl butyrate	10	1 to 5.25	90
Butyl valerate	10	1 to 5.25	92
Calcium chloride, 5.8%	20	2 to 20	39.7
Calcium chloride, 9.9%	20	2 to 20	37.1
" " , 17.8%	20	2 to 20	31.3
" " , 40.9%	20	2 to 20	21.7
Caproic acid	30	20 to 400	68
" "	65	20 to 100	90
" "	100	20 to 200	109
Carbon dioxide	13	at 60	1740
" "	13	at 70	960
" "	13	at 80	660
" "	13	at 90	440
Carbon disulfide	20	1 to 2	80.95
" "	0	1 to 500	66
" "	0	500 to 1000	53
" "	0	1000 to 1500	43
" "	0	1500 to 2000	37
" "	0	2000 to 2500	33
Carbon tetrachloride	10	1 to 5.25	70
" "	20	0 to 98.7	91.6
" "	20	98.7 to 197.4	89.9
" "	20	197.4 to 296	83.5
" "	20	296 to 395	75.5
" "	20	395 to 494	69.9
" "	20	at 542.5	62.5
" "	20	at 664	55.0
Castor oil	14.8	1 to 10	47.2
Chlorine	20	9.9 to 98.7	118
"	20	98.7 to 197.4	110
"	20	197.4 to 296	102
"	20	296 to 395	90.7
"	20	395 to 494	84.5
Chlorobenzene	13.3	1 to 18.5	67.1
"	80	1 to 2	108.4

Table 10-41 (*Continued*)
COEFFICIENTS OF COMPRESSIBILITY OF VARIOUS LIQUIDS

Liquid	t °C.	Pressure Range, atm.	$\beta t \times 10^6$
Chloroform	20	0 to 98.7	94.9
"	20	98.7 to 197.4	89.8
"	20	197.4 to 296	80.1
"	20	296 to 395	72.9
"	20	395 to 494	67.8
"	0	1 to 2	87.27
"	60	1 to 2	139.13
Cyclohexanol	40	98.7 to 296	56.4
Decane	23	0 to 1	105
Diphenylamine	65	0 to 500	57
"	100	0 to 300	64
"	185	0 to 100	110
Ethyl acetate	13.3	8.12 to 37.45	104
"	99.6	8.13 to 37.15	250
Ethyl alcohol	20	1 to 50	112
" "	20	50 to 100	102
" "	20	100 to 200	95
" "	20	200 to 300	86
" "	20	300 to 400	80
" "	20	400 to 500	73
" "	20	500 to 600	69
" "	0	1 to 50	96
" "	0	600 to 700	60
" "	0	900 to 1000	52
" "	100	900 to 1000	73
Ethyl bromide	10.1	1 to 500	90
" "	10.1	500 to 1000	63
" "	10.1	1000 to 1500	50
" "	10.1	2000 to 2500	36
" "	13.7	1 to 18.5	113.4
Ethyl chloride	0	1 to 500	103
" "	0	500 to 1000	69
" "	0	1000 to 1500	55
" "	0	2000 to 2500	39
" "	15.2	8.7 to 37.22	153
" "	99	12.77 to 34.47	495
Ethyl ether	8.1	1 to 8	163.8
" "	13.5	8.43 to 25.4	169
" "	185	100 to 200	741
" "	185	100 to 400	478
" "	35	1 to 2000	42.5
" "	0	at 0	159
" "	0	at 100	125.7
" "	0	at 200	112.2
" "	0	at 500	84.5
" "	0	at 1000	63.5
" "	−109.8	at 1000	34.5
Ethyl iodide	10.6	1 to 500	74
" "	10.6	500 to 1000	56
" "	10.6	1000 to 1500	46
" "	10.6	1500 to 2000	38
" "	10.6	2000 to 2500	34
" "	10.6	2500 to 3000	31
Ethylene bromide	10	1 to 5.25	55.8
" "	64	1 to 5.25	76.6
Ethylene chloride	10	1 to 5.25	67.7

Table 10-41 (*Continued*)
COEFFICIENTS OF COMPRESSIBILITY OF VARIOUS LIQUIDS

Liquid	t °C.	Pressure Range, atm.	$\beta t \times 10^6$
Ethylene chloride	75	1 to 5.25	111.1
Fluorobenzene	13.9	1 to 18.5	87.7
Glycerol	14.8	1 to 10	22.1
Heptane	23	0 to 1	134
Hexane	23	0 to 1	159
Mercury	22.8	1 to 500	3.8
"	22.8	500 to 1000	3.8
"	22.8	1000 to 1500	3.7
"	22.8	1500 to 2000	3.6
"	22.8	2000 to 2500	3.5
"	22.8	2500 to 3000	3.4
"	191.8	1 to 500	4.6
"	191.8	2500 to 3000	4.3
"	0	1 to 50	3.92
"	0	5810 to 6780	3.30
Mesitylene	20	98.7 to 296	68.4
"	20	296 to 494	59.1
Methyl acetate	14.3	8.10 to 37.53	97
Methyl alcohol	0	1 to 500	79
" "	0	500 to 1000	58
" "	0	1000 to 1500	47
" "	0	1500 to 2000	40
" "	0	2500 to 3000	29
" "	14.7	8.50 to 37.12	104
" "	18.1	at 8	120
Methyl butyrate	10	1 to 5.25	89
Methyl valerate	10	1 to 5.25	91
Nitrobenzene	25	at 86.5	46.1
"	25	at 192	43.0
"	25	at 303	40.1
"	25	at 419	38.1
Octane	23	0 to 1	121
Olive oil	20.5	1 to 10	63.3
" "	14.8	1 to 10	56.3
Palmitic acid	65	20 to 100	90
" "	185	20 to 300	134
" "	310	20 to 400	220
Paraffin, M. P. =55°	64	20 to 100	83
"	100	20 to 400	94
"	185	20 to 400	137
"	310	20 to 400	236
Paraffin oil	34	at 1	87
" "	34	at 2000	34
" "	34	at 4500	17
Pentane	0	1 to 29	174
"	20	1 to 29	242
"	0	at 0	175.9
"	0	at 484	94.6
"	0	at 967	68.9
Petroleum	1	1 to 15	67.91
"	16.1	1 to 15	76.77
"	35.1	1 to 15	82.83
"	52.2	1 to 15	92.21
"	72.1	1 to 15	100.16
"	94.0	1 to 15	108.8
Phosphorus trichloride	10.1	1 to 500	72

Table 10-41 (*Continued*)
COEFFICIENTS OF COMPRESSIBILITY OF VARIOUS LIQUIDS

Liquid	t °C.	Pressure Range, atm.	βt ×10⁶
Phosphorus trichloride	10.1	500 to 1000	54
″ ″	10.1	1000 to 1500	45
″ ″	10.1	1500 to 2000	38
″ ″	10.1	2000 to 2500	33
Potassium chloride, 2.49%	20	2 to 20	42.6
″ ″ , 4.40%	20	2 to 20	41.2
″ ″ , 8.28%	20	2 to 20	38.9
″ ″ ,16.75%	20	2 to 20	34.1
″ ″ ,24.31%	20	2 to 20	30.1
n-Propyl alcohol	0	1 to 500	69
″ ″	0	500 to 1000	52
″ ″	0	1000 to 1500	42
″ ″	0	2500 to 3000	27
″ ″	5.6	at 8	89.5
iso-Propyl alcohol	5.65	at 8	95
″ ″	17.85	at 8	103
n-Propyl benzene	20	98.7 to 296	70.7
″ ″	20	296 to 494	61.0
iso-Propyl benzene	20	98.7 to 296	71.3
″ ″	20	296 to 494	61.2
Pseudocumene	20	98.7 to 296	65.2
″	20	296 to 494	56.9
Sodium chloride, 1.32%	15	at 10	45.3
″ ″ ,13.53%	15	at 10	33.9
″ ″ ,22.22%	15	at 10	28.2
″ ″ ,26.21%	15	at 10	25.5
Sugar, cane, 0%	12.4	1 to 146	46.5
″ , ″ , 10%	12.4	1 to 146	42.65
″ , ″ , 20%	12.4	1 to 146	39.65
Sulfurous acid	0	1 to 16	302.5
Thymol	64	20 to 100	69
″	64	20 to 400	66
″	100	20 to 400	80
″	310	20 to 400	268
Toluene	10	1 to 5.25	79
″	66	1 to 5.25	114
″	20	1 to 2	91.47
″	25	at 114.5	80.6
″	25	at 230.5	70.3
″	25	at 355	62.1
p-Toluidine	45	1 to 150	51.2
″	28	20 to 100	56
″	100	20 to 400	77
″	185	20 to 400	112
″	310	20 to 400	243
Water, see special table			
Xylene	10	1 to 5.25	74
″	65	1 to 5.25	106
″	100	1 to 5.25	132

Table 10-42
COEFFICIENTS OF CUBICAL EXPANSION FOR VARIOUS LIQUIDS AND AQUEOUS SOLUTIONS

See also the table *Coefficients of Cubical Expansion For Various Solids.*

If V_0 is the volume at 0°C, then the volume at t°C $= V_t = V_0(1 + at + bt^2 + ct^3)$. Where the compound is marked with * the basic volume is not that at 0°C, but at some other temperature, t_0. The equation then reads $V_t = Vt_0(1 + a(t - t_0) + b(t - t_0)^2 + c(t - t_0)^3]$. In the case of aqueous solutions, the concentrations in percent are given after the name.

Liquid	Coef. at 20°C. $\times 10^3$	Range °C.	$a \times 10^3$	$b \times 10^6$	$c \times 10^8$
Acetic acid	1.071	16 to 107	1.063	−0.12636	1.0876
Acetone	1.487	0 to 54	1.324	3.809	−0.87983
Allyl alcohol	1.049	0 to 94	0.97019	1.8725	0.36452
Allyl bromide	1.241	0 to 69	1.2275	−0.44365	2.5843
Allyl chloride	1.475	9 to 44	1.3218	5.078	−4.1915
Allyl ether	1.346	0 to 88	1.2519	2.2401	0.35775
Allyl iodide	1.091	0 to 101	1.0539	0.63572	1.0036
Amyl acetate	1.162	0 to 124	1.1501	−0.09046	1.3015
Amyl alcohol	0.902	−15 to 80	0.89001	0.65729	1.18458
Amyl benzoate	0.848	0 to 198	0.81711	0.7377	0.10593
Amyl bromide	1.102	0 to 80	1.02321	1.90086	0.19756
Amyl chloride	1.208	0 to 100	1.17155	0.50077	1.35368
Amyl iodide	0.986	20 to 142	0.92658	1.4647	0.05962
Aniline	0.858	0 to 141	0.82349	0.8408	0.10741
Arsenous chloride	1.020	−15 to 130	0.97907	0.96695	0.17772
Benzene	1.237	11 to 81	1.17626	1.27755	0.80648
Benzoyl chloride	0.880	12 to 146	0.85893	0.44219	0.27139
Bromine	1.113	−7 to 60	1.03819	1.711138	0.54471
n-Butyl alcohol	0.950	6 to 108	0.83751	2.8634	−0.12415
n-Butyric acid	1.063	0 to 100	1.02573	0.83760	0.34694
iso-Butyric acid	1.068	16 to 118	0.97625	2.3976	−0.32145
Calcium chloride, 40.9%	0.458	17 to 24	0.42383	0.8571	....
Cane sugar, 43.2%	0.343	0 to 35	0.2536	2.247	....
n-Caproic acid	0.975	15 to 155	0.94413	0.68358	0.26586
Carbon disulfide	1.218	−34 to 68	1.1398	1.37065	1.91225
Carbon tetrachloride	1.236	0 to 76	1.18384	0.89881	1.35135
Chloral	0.934	13 to 51	0.9545	−2.2139	5.6392
Chloroform	1.273	0 to 63	1.19715	4.66473	−1.74328
o-Cresol	......	66 to 186	0.71072	1.1464	0.2242
m-Cresol	......	65 to 194	0.77526	0.27102	0.3868
p-Cresol	......	66 to 186	0.86476	0.53912	0.64418
Cymene	0.946	0 to 100	0.895	1.277	....
Diallyl	1.375	0 to 60	1.3423	−0.34339	3.8693
Diethyl ketone	1.233	0 to 95	1.15342	1.88396	0.32021
Dipropyl	1.381	0 to 66	1.2948	1.7471	1.2363
Ethyl acetate	1.389	−36 to 72	1.2585	2.95688	0.14922
Ethyl alcohol	......	0–80	1.04139	0.7836	1.7618
Ethyl alchol 50%	......	0–39	0.7450	1.85	0.730
Ethyl benzoate	0.900	0 to 159	0.86606	0.8229	0.12084
Ethyl benzene	0.961	24 to 131	0.86172	2.5344	−0.18319
Ethyl bromide	1.418	−32 to 54	1.33763	1.50135	1.6900
Ethyl chloride	1.706	−32 to 26	1.57458	2.81366	1.56987
Ethyl ether	1.656	−15 to 38	1.51324	2.35918	4.00512
Ethyl formate	1.417	0 to 63	1.36446	0.13538	3.9248
Ethyl iodide	1.179	10 to 65	1.1520	0.26032	1.4181
Ethyl nitrate	1.299	9 to 72	1.1290	4.7915	−1.8413
Ethyl oxalate	1.136	0 to 141	1.06031	1.0983	2.6657

Table 10-42 (*Continued*)
COEFFICIENTS OF CUBICAL EXPANSION FOR VARIOUS LIQUIDS AND AQUEOUS SOLUTIONS

Liquid	Coef. at 20°C. $\times 10^3$	Range °C.	$a \times 10^3$	$b \times 10^6$	$c \times 10^8$
Ethyl sulfide	1.278	0 to 90	1.19643	1.80653	0.78821
Ethylene chloride	1.161	−28 to 84	1.11893	1.0469	0.10342
Ethylene glycol	0.6375	11 to 136	0.5657	1.7074	0.293
Formic acid	1.025	5 to 104	0.99269	0.62514	0.5965
Glycerol	0.505	....	0.4853	0.4895	....
iso-Hexane	1.445	0 to 55	1.37022	0.97649	2.9819
Hydrochloric acid, 33.2%	0.455	0 to 33	0.4460	0.215	....
4.2%	0.239	0 to 33	0.0652	4.355	....
1.0%	0.211	0 to 32	0.0153	4.899	....
*3.4%, t₀=110°	......	110 to 140	0.620	4.5	....
Isoprene	1.567	0 to 33	1.4603	0.99793	5.60149
Mercury	......	0 to 100	0.18169041	0.002951266	0.0114562
	......	24 to 299	0.181163	0.01155	0.0021187
Methyl acetate	1.427	0 to 58	1.34982	0.87098	3.5562
Methyl alcohol	1.259	−38 to 70	1.18557	1.56493	0.91113
Methyl benzoate	0.895	0 to 162	0.8633	0.7414	0.15896
Methyl bromide	1.684	−35 to 28	1.41521	3.31528	11.3809
Methyl cyanide	1.301	6 to 66	1.2118	1.7780	1.5322
Methyl ethyl ketone	1.315	0 to 76	1.18654	3.37043	−0.53365
Methyl formate	1.563	0 to 10	1.35824	10.538	−1.8085
Methyl iodide	1.273	5 to 39	1.1440	4.0465	−2.7393
Methyl propionate	1.304	0 to 74	1.3049	−1.3275	4.6943
Methyl sulfide	1.082	0 to 111	1.01705	1.57606	0.19072
Nitrobenzene	......	144 to 164	0.8263	0.52249	0.13779
Olive oil	0.721	....	0.68215	1.14053	−0.539
n-Pentane	1.656	−190 to 30	1.50697	3.435	0.975
iso-Pentane	1.680	0 to 27	1.46834	5.09626	0.6979
Petroleum, sp. gr. 0.8467	0.955	24 to 120	0.8994	1.396	....
Petroleum ether	2.26	−190 to 0	1.46	1.60	....
Phenol	1.090	36 to 157	0.834	0.10732	0.4446
Phosphorus tribromide	0.868	0 to 100	0.8472	0.43672	0.25276
Phosphorus trichloride	1.154	−36 to 75	1.12862	0.87288	1.79236
Phosphorus oxychloride	1.116	0 to 107	1.06431	1.12666	0.5299
Potassium chloride, 24.3%	0.353	16 to 25	0.2695	2.080	....
Propionic acid	1.102	0 to 133	1.0396	1.5487	0.04301
n-Propyl alcohol	0.956	0 to 94	0.7743	4.9689	−1.4069
iso-Propyl alcohol	1.094	0 to 83	1.04345	0.44303	2.7274
n-Propyl chloride	1.447	0 to 42	1.3306	3.8313	−1.3859
iso-Propyl chloride	1.591	0 to 34	1.3696	5.5287	....
n-Propyl ether	1.354	0 to 88	1.2132	3.9318	−1.3644
iso-Propyl ether	1.452	0 to 67	1.2872	4.2923	−0.58573
Propyl iodide	1.102	10 to 98	1.0276	1.8658	−0.0051
Silicon tetrachloride	1.430	−32 to 59	1.29412	2.18414	4.08642
Sodium acid sulfate, 21%	0.555	0 to 34	0.5364	4.75	....
Sodium chloride, 20.6%	0.414	0 to 29	0.3640	1.237	....
Sodium sulfate,1.9%	0.235	0 to 40	0.0449	4.749	....
24%	0.410	11 to 40	0.3599	1.258	....
Stannic chloride	1.178	−19 to 113	1.1328	0.91171	0.75798
Sulfur chloride	0.968	12 to 111	0.9591	−0.03819	0.73186
Sulfuric acid, conc.	......	0–30	0.5758	−0.864	....
Sulfuric acid 10.9%	0.387	0 to 30	0.2835	2.580	....
5.4%	0.311	0 to 30	0.1450	4.143	....
1.4%	0.234	0 to 30	0.03335	5.025	....
*2.3%, t₀=100°	......	110 to 140	0.729	2.8	....

Table 10-42 (*Continued*)
COEFFICIENTS OF CUBICAL EXPANSION FOR VARIOUS LIQUIDS AND AQUEOUS SOLUTIONS

Liquid	Coef. at 20°C. $\times 10^3$	Range °C.	$a \times 10^3$	$b \times 10^6$	$c \times 10^8$
*4.5%, to=110°	......	110 to 140	0.648	4.2	....
Thymol	......	62 to 157	0.84369	0.26625	0.35997
Titanium tetrachloride	0.998	−22 to 134	0.94257	1.34579	0.0888
Toluene	1.099	0 to 100	1.028	1.779	....
o-Toluidine	0.847	0 to 141	0.82136	0.6046	0.14696
Trimethyl carbinol	1.023	20 to 77	1.31261	−8.8155	3.61209
n-Valeric acid	1.004	8 to 144	0.97557	0.61852	0.30378
Water (see also below)		−13 to 0	−0.09417	1.449	−59.85
o-Xylene	0.973	16 to 131	0.91734	1.3245	0.19586
m-Xylene	1.009	0 to 141	0.96396	1.0251	0.32753
p-Xylene	1.011	19 to 131	0.97013	0.8714	0.5287

WATER

To find the cubical expansion of water, substitute in the following formulas:

0° to 33°: $Vt = Vo(1 - 0.0_4 64268t + 0.0_5 850526t^2 - 0.0_7 678977t^3 + 0.0_9 401209t^4$

0° to 80°: $Vt = Vo(1 - 0.0_4 53255t + 0.0_5 761532t^2 - 0.0_7 437217t^3 + 0.0_9 164322t^4$

Table 10-43
CUBICAL EXPANSION OF SOLIDS

The coefficient of cubical expansion is the increase in volume per unit volume per degree C. rise in temperature. For ordinary work the assumption may be made that the coefficient of cubical expansion is about three times the coefficient of linear expansion.

Substance	Temp. °C.	Coef. $\times 10^7$	Substance	Temp. °C.	Coef. $\times 10^7$
Antimony	0 to 100	316.7	Paraffin	20	5880
Beryl	0 to 100	10.5	Platinum	0 to 100	265
Bismuth	0 to 100	394.8	Porcelain	0 to 100	108.0
Copper	0 to 100	499.8	" , Berlin	20	81.4
Diamond	40	35.4	Potassium chloride	0 to 100	1094
Emerald	40	16.8	" nitrate	0 to 100	1967
Galena	0 to 100	558	" sulfate	20	1075.4
Glass, Corning 790	0 to 350	24	Quartz	0 to 100	384
" , Corning 774	0 to 350	96	Rock salt	50 to 60	1212.0
" , Corning 8800	0 to 350	183	Rubber	20	4870
" , Jena 59 III	20 to 100	156	Silver	0 to 100	583.1
" , soda lime tubing	0 to 300	276	Sodium	20	2136.4
Gold	0 to 100	441.1	Stearic acid	33.8 to 45.5	8100
Ice	−20 to −1	1125.0	Sulfur, native	13.2 to 50.3	2230
Iceland spar	50 to 60	144.7	Tin	0 to 100	688.9
Iron	0 to 100	355.0	Zinc	0 to 100	892.8
Lead	0 to 100	839.9			

Table 10-44
COEFFICIENT OF LINEAR EXPANSION

In the table below are given the values of C in the formula:—
$l_t = l_o (1 + Ct)$, where l_o is the length of the material at $0°C$., l_t is its length at $t°C$., and C is the coefficient of linear expansion.

Substance	Temp. °C.	$C \times 10^6$	Substance	Temp. °C.	$C \times 10^6$
Alundum	25–900	8.7	Porcelain, Berlin	0–100	3.1
Bakelite	20–60	22	Quartz, ‖ to axis	−191 to +16	5.21
Bauxite brick	25–100	4.4	" , ‖ " "	0–80	7.97
Caoutchouc	16.7–25.3	77.0	" , ⊥ " "	0–80	13.37
Condensite, No. 100	16–79	44.0	Quartz glass	−190 to +16	−0.26
" , No. 128	18–56	20.0	" "	16–500	0.57
Diamond	40	1.18	" "	16–1000	0.58
Ebonite	25.3–35.4	84.2	Rock salt	40	40.4
Emerald, ‖ to axis	0–85	−1.35	Rubber, hard	0	69.1
" , ⊥ to axis	0–85	1.00	" , "	−160	30.0
Fire clay brick	25–100	8.1	Sandstone	20	7–12
Fluorspar	0–100	19.50	Slate	20	6–10
Formica	20–60	30.0	Topaz, ‖ to lesser	0–100	8.32
Gas carbon	40	5.40	horizontal axis		
Glass, American plate	0–100 ca.	9.0	Topaz, ‖ to greater	0–100	8.36
" , American window	0–100 ca.	9.0	horizontal axis		
" , (standard flint)			Topaz, ‖ to vertical	0–100	4.72
soda lime tubing	0–300	9.2	axis		
" , Kimble N-51A	0–300	5.0	Tourmaline, ‖ to	0–100	9.37
" , Jena 16 III	0–100	8.1	longitudinal axis		
" , Jena 59 III	0–100	5.8	Tourmaline, ‖ to	0–100	7.73
" , Jena 59 III	−191 to +16	4.24	horizontal axis		
" , Corning 790	0–350	0.8	Vulcanite	0–18	63.60
" , Corning 774	0–350	3.2	Wedgwood ware	0–100	8.90
" , Corning 8800	0–350	6.1	Wood, ‖ " fiber		
" , Corning 8810	0–350	8.8	" , " Ash	0–100	9.51
" , B & L BSC-2*	25–525	8.3	" , " Beech	2–34	2.57
" , B & L LBC-2*	15–570	9.9	" , " Chestnut	2–34	6.49
" , B & L DBC-1*	25–600	8.2	" , " Elm	2–34	5.65
" , B & L C-1*	40–501	10.1	" , " Mahogany	2–34	3.61
" , B & L BF-1*	25–505	9.8	" , " Maple	2–34	6.38
" , B & L DF-2*	25–450	8.0	" , " Oak	2–34	4.92
" , B & L EDF-1*	25–420	9.0	" , " Pine	2–34	5.41
Graphite	40	7.86	" , " Walnut	2–34	6.58
Gutta percha	20	198.3	Wood, across fiber		
Ice	−20 to −1	51	" , " Beech	2–34	61.4
Iceland spar, ‖ to axis	0–80	26.31	" , " Chestnut	2–34	32.5
" " , ⊥ to axis	0–80	5.44	" , " Elm	2–34	44.3
Limestone	25–100	9	" , " Mahogany	2–34	40.4
Magnesium oxide	25–100	9.7–11.4	" , " Maple	2–34	48.4
Marble	15–100	11.7	" , " Oak	2–34	54.4
Paraffin	0–16	106.62	" , " Pine	2–34	34.1
"	16–38	130.30	" , " Walnut	2–34	48.4
"	38–49	477.07	Wax, white	10–26	230.0
Porcelain	20–790	4.13	" , "	26–31	312.0
" , Bayeux	0	2.5	" , "	31–43	486.0
" , Bayeux	1000–1400	5.53	" , "	43–57	1522.7

*Optical glass.

Section 11
MISCELLANEOUS

SIEVES AND SCREENS

The following terms are often employed in reporting the results of a screen analysis.

Cumulative Percent is the term employed to express the percentage of the total amount of a material which would remain on a testing sieve if only one sieve were used for testing the entire sample; to obtain the **cumulative weight,** it is necessary to add the weight of all of the material which remains on sieves coarser than the one in question to the amount remaining on the sieve in question. When such data are plotted as a curve, each point on the curve represents the *total* material that would be retained if only the one sieve represented by that particular point were used in the sieve analysis.

Effective Size is a term customarily employed in specifications for filtration sands; effective size is measured by that opening in millimeters which will just pass 10% of a representative sample of sand.

Uniformity Coefficient is the numerical value obtained by dividing the sieve opening (in millimeters) which will pass 60% of the sample by the sieve opening (in millimeters) which will just pass 10% of the sample.

Table 11-1
U.S. SIEVE SERIES
ASTM E-11-61 Adopted 1961

Sieve Designation		Sieve Opening		Wire Diameter	
Standard	Alternate	mm	ca. inch	mm	ca. inch
26.9 mm	1.06 in	26.9	1.06	3.90	0.1535
*22.6 mm	⅞ in	22.6	0.875	3.50	0.1378
*16.0 mm	⅝ in	16.0	0.625	3.00	0.1181
*11.2 mm	⁷⁄₁₆ in	11.2	0.438	2.45	0.0965
*8.00 mm	⁵⁄₁₆ in	8.00	0.312	2.07	0.0815
6.73 mm	0.265 in	6.73	0.265	1.87	0.0736
*5.66 mm	No. 3½	5.66	0.223	1.68	0.0661
4.76 mm	No. 4	4.76	0.187	1.54	0.0606
*4.00 mm	No. 5	4.00	0.157	1.37	0.0539
3.36 mm	No. 6	3.36	0.132	1.23	0.0484
*2.83 mm	No. 7	2.83	0.111	1.10	0.0430
2.38 mm	No. 8	2.38	0.0937	1.00	0.0394
*2.00 mm	No. 10	2.00	0.0787	0.900	0.0354
1.68 mm	No. 12	1.68	0.0661	0.810	0.0319
*1.41 mm	No. 14	1.41	0.0555	0.725	0.0285
1.19 mm	No. 16	1.19	0.0469	0.650	0.0256
*1.00 mm	No. 18	1.00	0.0394	0.580	0.0228
841 μm	No. 20	0.841	0.0331	0.510	0.0201
*707 μm	No. 25	0.707	0.0278	0.450	0.0177
595 μm	No. 30	0.595	0.0231	0.390	0.0154
*500 μm	No. 35	0.500	0.0197	0.340	0.0134
420 μm	No. 40	0.420	0.0165	0.290	0.0114
*354 μm	No. 45	0.354	0.0139	0.247	0.0097
297 μm	No. 50	0.297	0.0117	0.215	0.0085
*250 μm	No. 60	0.250	0.0098	0.180	0.0071
210 μm	No. 70	0.210	0.0083	0.152	0.0060
*177 μm	No. 80	0.177	0.0070	0.131	0.0052
149 μm	No. 100	0.149	0.0059	0.110	0.0043
*125 μm	No. 120	0.125	0.0049	0.091	0.0036
105 μm	No. 140	0.105	0.0041	0.076	0.0030
*88 μm	No. 170	0.088	0.0035	0.064	0.0025
74 μm	No. 200	0.074	0.0029	0.053	0.0021
*63 μm	No. 230	0.063	0.0025	0.044	0.0017
53 μm	No. 270	0.053	0.0021	0.037	0.0015
44 μm	No. 325	0.044	0.0017	0.030	0.0012

* These sieves correspond to those proposed as an International (ISO) Standard. It is recommended that wherever possible these sieves be included in all sieve analysis data or reports intended for international publication.

TEMPERATURE AND ITS MEASUREMENT

The International Practical Temperature Scale of 1968 (IPTS-68) is based on the assigned values of the temperatures of a number of reproducible equilibrium states defining fixed points (Table 11-2) and on standard instruments calibrated at those temperatures. The standard instrument used from $-259.34\,°C$ to $630.74\,°C$ is the platinum resistance thermometer, which must be strain-free, annealed, pure platinum (Pt-67). From $630.74\,°C$ to $1064.43\,°C$ the standard instrument is the platinum–10% rhodium/platinum (Type S) thermocouple. Above $1064.43\,°C$ the IPTS-68 is defined by the Planck law of radiation with 1337.58 K as the reference temperature and a value of $0.014\ 388$ m $\cdot$ K for c_2, the second radiation constant, in the expression:

$$L_\lambda = c_1\lambda^{-5}(e^{c_2/\lambda T} - 1)^{-1}$$

where L_λ is the spectral radiance at wavelength λ.

As well as the defining fixed points of the IPTS-68, other reference points are available and are listed also in Table 11-2. Three additional secondary reference points extend the scale: freezing point of rhodium, $1963\,°C$; freezing point of iridium, $2447\,°C$; and the melting point of tungsten, $3387\,°C$.

Although the IPTS-68 is not defined down to liquid helium temperatures, the National Bureau of Standards Publication 2-20, Washington, D.C., 1965, provides a workable scale from 2 to 20 K.

Table 11-2
THERMOELECTRIC VALUES *In Absolute Microvolts* AT THE FIXED POINTS FOR VARIOUS THERMOCOUPLES

Abbreviations Used in the Table

FP, freezing point   NBP, normal boiling point   TP, triple point
BP, boiling point   SP, sublimation point

Fixed Point	Temperature, °C	Type B	Type E	Type J	Type K	Type R	Type S	Type T
Helium NBP	−268.935		−9 833.09		−6 456.93			−6 256.29
Hydrogen TP	−259.34*		−9 792.66		−6 439.34			−6 229.19
Hydrogen NBP	−252.87*		−9 744.75		−6 416.69			−6 197.73
Neon TP	−248.595		−9 704.62		−6 396.64			−6 171.38
Neon NBP	−246.048*		−9 677.58		−6 382.75			−6 153.58
Oxygen TP	−218.789*		−9 249.87	−8 095.7	−6 144.56			−5 873.02
Nitrogen TP	−210.002		−9 062.90	−7 796.3	−6 034.65			−5 753.28
Nitrogen NBP	−195.802		−8 716.78	−7 480.7	−5 825.66			−5 535.59
Oxygen NBP	−182.962*		−8 360.81		−5 605.15			−5 314.72
Carbon dioxide SP	−78.476		−4 227.53	−3 718.7	−2 869.64			−2 740.70
Mercury FP	−38.862		−2 193.03	−1 906.3	−1 484.85	−183.05	−189.54	−1 434.94
Ice point	0.000	0.00	0.00	0.0	0.00	0.00	0.00	0.00
Diphenyl ether TP	26.87	−2.39	1 609.1	1 373.9	1 076.0	151.72	153.70	1 067.9
Water BP	100.00*†	33.18	6 317.1	5 267.7	4 095.3	647.23	645.34	4 277.3
Benzoic acid TP	122.37	56.09	7 846.8	6 488.6	5 016.0	818.57	812.88	5 341.4
Indium FP	156.634	101.88	10 259.9	8 374.3	6 403.9	1 095.62	1 081.79	7 036.4
Tin FP	231.9681‡	247.37	15 809.4	12 551.7	9 420.1	1 756.10	1 714.64	11 013.3
Bismuth FP	271.442	347.74	18 820.7	14 742.7	11 028.6	2 125.04	2 063.97	13 218.8
Cadmium FP	321.108	497.15	22 683.9	17 492.8	13 084.9	2 607.22	2 516.72	16 095.3
Lead FP	327.502	518.24	23 186.2	17 846.2	13 351.5	2 670.59	2 575.94	16 473.3
Mercury BP	356.66	619.69	25 489.1	19 456.0	14 571.4	2 962.99	2 848.34	18 217.9
Zinc FP	419.58*	867.78	30 512.6	22 925.9	17 223.1	3 611.34	3 447.87	
Sulfur BP	444.674	977.66	32 531.2	24 312.3	18 286.7	3 875.95	3 690.88	
Cu-Al eutectic FP	548.23	1 495.10	40 901.4	30 109.5	22 696.5	5 000.94	4 714.00	

Antimony FP	630.74	1 978.43	47 561.4	34 910.8	26 207.0	5 933.08	5 552.10
Aluminum FP	660.37	2 166.77	49 941.2	36 693.3	27 460.6	6 275.90	5 859.12
Silver FP	961.93*	4 490.76	73 495.5	55 669	39 779.4	10 003.09	9 148.20
Gold FP	1 064.43*	5 433.59		61 716	43 755.0	11 363.85	10 334.30
Copper FP	1 084.5	5 626.33		62 880	44 520.4	11 635.49	10 570.46
Nickel FP	1 455	9 576.58				16 811.06	15 033.80
Cobalt FP	1 494	10 024.76				17 360.36	15 504.28
Palladium FP	1 554	10 720.56				18 201.73	16 223.95
Platinum FP	1 772	13 262.22				21 103.11	18 693.89

*Defining fixed points of the International Practical Temperature Scale of 1968 (IPTS-68). Except for the triple points and one equilibrium hydrogen point, the assigned values of temperature are for equilibrium states at a pressure of one standard atmosphere (101 325 N·m⁻²).

†The water used should have the isotopic composition of ocean water (0.016 mol of deuterium per 100 mol of hydrogen, and 0.04 mol of ¹⁷O and 0.2 mol of ¹⁸O per 100 mol of ¹⁶O).

‡The equilibrium state between the solid and liquid phases of tin (freezing point of tin) may be used as an alternative to the boiling point of water.

THERMOCOUPLES

The thermocouple reference data in Tables 11-3 to 11-9 are based on the IPTS-68 and have been taken from National Bureau of Standards Monograph 125, Washington, D.C., 1974. The thermoelectric voltage is in absolute millivolts with the reference junction at 0°C. Note that the temperature for a given entry is obtained by adding the corresponding temperature in the top row to that in the left-hand column, regardless of whether the latter is positive or negative.

As compared with the Type S and Type R thermocouples, Type B (Table 11-3) offers distinct advantages of improved stability, increased mechanical strength, and higher possible operating temperatures.

Type E thermoelements (Table 11-4) are very useful down to about liquid hydrogen temperatures, and may even be used down to liquid helium temperatures. They are the most useful of the commercially standardized thermocouple combinations for subzero temperature measurements. They also have the largest Seebeck coefficient (voltage response per degree Celsius) above 0°C of any of the standardized thermocouples. They are recommended for use in the temperature range from −250 to 871°C in oxidizing or inert atmospheres. They should not be used in sulfurous, reducing, or alternately reducing and oxidizing atmospheres unless suitably protected with tubes. They should not be used in vacuum at high temperatures for extended times.

Type J thermocouples (Table 11-5) are one of the most common types of industrial thermocouples because of the relatively high Seebeck coefficient and low cost. They are recommended for use in the temperature range from 0 to 760°C (but never above 760°C) in vacuum and in oxidizing, reducing, or inert atmospheres, with the exception of sulfurous atmospheres above 500°C. For extended use above 500°C, heavy-gauge wires are recommended. They are not recommended for subzero temperatures.

The Type K thermocouple (Table 11-6) is more resistant to oxidation at elevated temperatures than the Type E, J, or T thermocouple, and consequently finds wide application at temperatures above 500°C. It is recommended for continuous use at temperatures within the range −250 to 1260°C in inert or oxidizing atmospheres. It should not be used in sulfurous or reducing atmospheres, or in vacuum at high temperatures for extended times.

The Type R thermocouple (Table 11-7) was developed primarily to match a previous platinum–10% rhodium British wire which was later found to have 0.34% iron impurity in the rhodium. Comments on Type S also apply to Type R.

The Type S thermocouple (Table 11-8) remains the standard for determining temperatures between 630.74°C and the freezing point of gold (1064.43°C). The other fixed point used is the freezing point of silver. The Type S thermocouple can be used from -50°C continuously up to about 1400°C, and intermittently at temperatures up to the melting point of platinum (1767.6°C). The thermocouple is most reliable when used in a clean oxidizing atmosphere, but may also be used in inert gaseous atmospheres or in a vacuum for short periods of time. It should not be used in reducing atmospheres, nor in those containing metallic vapor (such as lead or zinc), nonmetallic vapors (such as arsenic, phosphorus, or sulfur), or easily reduced oxides, unless suitably protected with nonmetallic protecting tubes.

The Type T thermocouple (Table 11-9) is one of the more popular thermocouples for determining temperatures within the range from 370°C down to the boiling point of hydrogen (-252°C). It is recommended for use in vacuum, or in oxidizing, reducing, or inert atmospheres.

Table 11-3
TYPE B THERMOCOUPLES: PLATINUM–30% RHODIUM ALLOY vs. PLATINUM–6% RHODIUM ALLOY

Thermoelectric Voltage in Millivolts; Reference Junction at 0°C

°C	0	10	20	30	40	50	60	70	80	90
0	0.00	−0.0019	−0.0026	−0.0021	−0.0005	0.0023	0.0062	0.0112	0.0174	0.0248
100	0.0332	0.0427	0.0534	0.0652	0.0780	0.0920	0.1071	0.1232	0.1405	0.1588
200	0.1782	0.1987	0.2202	0.2428	0.2665	0.2912	0.3170	0.3438	0.3717	0.4006
300	0.4305	0.4615	0.4935	0.5266	0.5607	0.5958	0.6319	0.6690	0.7071	0.7462
400	0.7864	0.8275	0.8696	0.9127	0.9567	1.0018	1.0478	1.0948	1.1427	1.1916
500	1.2415	1.2923	1.3440	1.3967	1.4503	1.5048	1.5603	1.6166	1.6739	1.7321
600	1.7912	1.8512	1.9120	1.9738	2.0365	2.1000	2.1644	2.2296	2.2957	2.3627
700	2.4305	2.4991	2.5686	2.6390	2.7101	2.7821	2.8548	2.9284	3.0028	3.0780
800	3.1540	3.2308	3.3084	3.3867	3.4658	3.5457	3.6264	3.7078	3.7899	3.8729
900	3.9565	4.0409	4.1260	4.2119	4.2984	4.3857	4.4737	4.5624	4.6518	4.7419
1000	4.8326	4.9241	5.0162	5.1090	5.2025	5.2966	5.3914	5.4868	5.5829	5.6796
1100	5.7769	5.8749	5.9734	6.0726	6.1724	6.2728	6.3737	6.4753	6.5774	6.6801
1200	6.7833	6.8871	6.9914	7.0963	7.2017	7.3076	7.4140	7.5210	7.6284	7.7363
1300	7.8446	7.9534	8.0627	8.1724	8.2826	8.3932	8.5041	8.6155	8.7273	8.8394
1400	8.9519	9.0648	9.1780	9.2915	9.4053	9.5194	9.6338	9.7485	9.8634	9.9786
1500	10.0940	10.2097	10.3255	10.4415	10.5577	10.6740	10.7905	10.9071	11.0237	11.1405
1600	11.2574	11.3743	11.4913	11.6082	11.7252	11.8422	11.9591	12.0761	12.1929	12.3100
1700	12.4263	12.5429	12.6594	12.7757	12.8918	13.0078	13.1236	13.2391	13.3545	13.4696
1800	13.5845	13.6991	13.8135							

Table 11-4
TYPE E THERMOCOUPLES: NICKEL-CHROMIUM ALLOY vs. COPPER-NICKEL ALLOY

Thermoelectric Voltage in Millivolts; Reference Junction at 0°C

°C	0	10	20	30	40	50	60	70	80	90
-200	-8.824	-9.063	-9.274	-9.455	-9.604	-9.719	-9.797	-9.835		
-100	-5.237	-5.680	-6.107	-6.516	-6.907	-7.279	-7.631	-7.963	-8.273	-8.561
-0	0.000	-0.581	-1.151	-1.709	-2.254	-2.787	-3.306	-3.811	-4.301	-4.777
0	0.000	0.591	1.192	1.801	2.419	3.047	3.683	4.394	4.983	5.646
100	6.317	6.996	7.683	8.377	9.078	9.787	10.501	11.222	11.949	12.681
200	13.419	14.161	14.909	15.661	16.417	17.178	17.942	18.710	19.481	20.256
300	21.033	21.814	22.597	23.383	24.171	24.961	25.754	26.549	27.345	28.143
400	28.943	29.744	30.546	31.350	32.155	32.960	33.767	34.574	35.382	36.190
500	36.999	37.808	38.617	39.426	40.236	41.045	41.853	42.662	43.470	44.278
600	45.085	45.891	46.697	47.502	48.306	49.109	49.911	50.713	51.513	52.312
700	53.110	53.907	54.703	55.498	56.291	57.083	57.873	58.663	59.451	60.237
800	61.022	61.806	62.588	63.368	64.147	64.924	65.700	66.473	67.245	68.015
900	68.783	69.549	70.313	71.075	71.835	72.593	73.350	74.104	74.857	75.608
1000	76.358									

Table 11-5
TYPE J THERMOCOUPLES: IRON vs. COPPER-NICKEL ALLOY
Thermoelectric Voltage in Millivolts; Reference Junction at 0°C

°C	0	10	20	30	40	50	60	70	80	90
-200	-7.890	-8.096								
-100	-4.632	-5.036	-5.426	-5.801	-6.159	-6.499	-6.821	-7.122	-7.402	-7.659
-0	0.000	-0.501	-0.995	-1.481	-1.960	-2.431	-2.892	-3.344	-3.785	-4.215
0	0.000	0.507	1.019	1.536	2.058	2.585	3.115	3.649	4.186	4.725
100	5.268	5.812	6.359	6.907	7.457	8.008	8.560	9.113	9.667	10.222
200	10.777	11.332	11.887	12.442	12.998	13.553	14.108	14.663	15.217	15.771
300	16.325	16.879	17.432	17.984	18.537	19.089	19.640	20.192	20.743	21.295
400	21.846	22.397	22.949	23.501	24.054	24.607	25.161	25.716	26.272	26.829
500	27.388	27.949	28.511	29.075	29.642	30.210	30.782	31.356	31.933	32.513
600	33.096	33.683	34.273	34.867	35.464	36.066	36.671	37.280	37.893	38.510
700	39.130	39.754	40.482	41.013	41.647	42.283	42.922			

Table 11-6
TYPE K THERMOCOUPLES: NICKEL-CHROMIUM ALLOY vs. NICKEL-ALUMINUM ALLOY

Thermoelectric Voltage in Millivolts; Reference Junction at 0°C

°C	0	10	20	30	40	50	60	70	80	90
−200	−5.891	−6.035	−6.158	−6.262	−6.344	−6.404	−6.441	−6.458		
−100	−3.553	−3.852	−4.138	−4.410	−4.669	−4.912	−5.141	−5.354	−5.550	−5.730
−0	0.000	−0.392	−0.777	−1.156	−1.517	−1.889	−2.243	−2.586	−2.920	−3.242
0	0.000	0.397	0.798	1.203	1.611	2.022	2.436	2.850	3.266	3.681
100	4.095	4.508	4.919	5.327	5.733	6.137	6.539	6.939	7.338	7.737
200	8.137	8.537	8.938	9.341	9.745	10.151	10.560	10.969	11.381	11.793
300	12.207	12.623	13.039	13.456	13.874	14.292	14.712	15.132	15.552	15.974
400	16.395	16.818	17.241	17.664	18.088	18.513	18.839	19.363	19.788	20.214
500	20.640	21.066	21.493	21.919	22.346	22.772	23.198	23.624	24.050	24.476
600	24.902	25.327	25.751	26.176	26.599	27.022	27.445	27.867	28.288	28.709
700	29.128	29.547	29.965	30.383	30.799	31.214	31.629	32.042	32.455	32.866
800	33.277	33.685	34.095	34.502	34.909	35.314	35.718	36.121	36.524	36.925
900	37.325	37.724	38.122	38.519	38.915	39.310	39.703	40.096	40.488	40.879
1000	41.269	41.657	42.045	42.432	42.817	43.202	43.585	43.968	44.349	44.729
1100	45.108	45.485	45.863	46.238	46.612	46.985	47.356	47.726	48.095	48.462
1200	48.828	49.192	49.555	49.916	50.276	50.633	50.990	51.344	51.697	52.049
1300	52.398	52.747	53.093	53.439	53.782	54.125	54.466	54.807		

Table 11-7
TYPE R THERMOCOUPLES: PLATINUM–13% RHODIUM ALLOY vs. PLATINUM

Thermoelectric Voltage in Millivolts; Reference Junction at 0°C

°C	0	10	20	30	40	50	60	70	80	90
(Below zero) 0	0.0000	-0.0515	-0.100	-0.1455	-0.1877	-0.2264				
0	0.0000	0.0543	0.1112	0.1706	0.2324	0.2965	0.3627	0.4310	0.5012	0.5733
100	0.6472	0.7228	0.8000	0.8788	0.9591	1.0407	1.1237	1.2080	1.2936	1.3803
200	1.4681	1.5571	1.6471	1.7381	1.8300	1.9229	2.0167	2.1113	2.2068	2.3030
300	2.4000	2.4978	2.5963	2.6954	2.7953	2.8957	2.9968	3.0985	3.2009	3.3037
400	3.4072	3.5112	3.6157	3.7208	3.8264	3.9325	4.0391	4.1463	4.2539	4.3620
500	4.4706	4.5796	4.6892	4.7992	4.9097	5.0206	5.1320	5.2439	5.3562	5.4690
600	5.5823	5.6960	5.8101	5.9246	6.0398	6.1554	6.2716	6.3883	6.5054	6.6230
700	6.7412	6.8598	6.9789	7.0984	7.2185	7.3390	7.4600	7.5815	7.7035	7.8259
800	7.9488	8.0722	8.1960	8.3203	8.4451	8.5703	8.6960	8.8222	8.9488	9.0758
900	9.2034	9.3313	9.4597	9.5886	9.7179	9.8477	9.9779	10.1086	10.2397	10.3712
1000	10.5032	10.6356	10.7684	10.9017	11.0354	11.1695	11.3041	11.4391	11.5745	11.7102
1100	11.8463	11.9827	12.1194	12.2565	12.3939	12.5315	12.6695	12.8077	12.9462	13.0849
1200	13.2239	13.3631	13.5025	13.6421	13.7818	13.9218	14.0619	14.2022	14.3426	14.4832
1300	14.6239	14.7647	14.9056	15.0465	15.1876	15.3287	15.4699	15.6110	15.7522	15.8935
1400	16.0347	16.1759	16.3172	16.4583	16.5995	16.7405	16.8816	17.0225	17.1634	17.3041
1500	17.4447	17.5852	17.7256	17.8659	18.0059	18.1458	18.2855	18.4251	18.5644	18.7035
1600	18.8424	18.9810	19.1194	19.2575	19.3953	19.5329	19.6702	19.8071	19.9437	20.0797
1700	20.2151	20.3497	20.4834	20.6161	20.7475	20.8777	21.0064			

Table 11-9
TYPE T THERMOCOUPLES: COPPER vs. COPPER-NICKEL ALLOY

Thermoelectric Voltage in Millivolts; Reference Junction at 0°C

°C	0	10	20	30	40	50	60	70	80	90
−200	−5.603	−5.753	−5.889	−6.007	−6.105	−6.181	−6.232	−6.258		
−100	−3.378	−3.656	−3.923	−4.177	−4.419	−4.648	−4.865	−5.069	−5.261	−5.439
−0	0.000	−0.383	−0.757	−1.121	−1.475	−1.819	−2.152	−2.475	−2.788	−3.089
0	0.000	0.391	0.789	1.196	1.611	2.035	2.467	2.908	3.357	3.813
100	4.277	4.749	5.227	5.712	6.204	6.702	7.207	7.718	8.235	8.757
200	9.286	9.820	10.360	10.905	11.456	12.011	12.572	13.137	13.707	14.281
300	14.860	15.443	16.030	16.621	17.217	17.816	18.420	19.027	19.638	20.252
400	20.869									

Table 11-8
TYPE S THERMOCOUPLES: PLATINUM–10% RHODIUM ALLOY vs. PLATINUM

Thermoelectric Voltage in Millivolts; Reference Junction at 0°C

°C	0	10	20	30	40	50	60	70	80	90
(Below zero)		−0.0527	−0.1028	−0.1501	−0.1944	−0.2357				
0	0.0000	0.0552	0.1128	0.1727	0.2347	0.2986	0.3646	0.4323	0.5017	0.5728
100	0.6453	0.7194	0.7948	0.8714	0.9495	1.0287	1.1089	1.1902	1.2726	1.3558
200	1.4400	1.5250	1.6109	1.6975	1.7849	1.8729	1.9617	2.0510	2.1410	2.2316
300	2.3227	2.4143	2.5065	2.5991	2.6922	2.7858	2.8798	2.9742	3.0690	3.1642
400	3.2597	3.3557	3.4519	3.5485	3.6455	3.7427	3.8403	3.9382	4.0364	4.1348
500	4.2336	4.3327	4.4320	4.5316	4.6316	4.7318	4.8323	4.9331	5.0342	5.1356
600	5.2373	5.3394	5.4417	5.5445	5.6477	5.7513	5.8553	5.9595	6.0641	6.1690
700	6.2743	6.3799	6.4858	6.5920	6.6986	6.8055	6.9127	7.0202	7.1281	7.2363
800	7.3449	7.4537	7.5629	7.6724	7.7823	7.8925	8.0030	8.1138	8.2250	8.3365
900	8.4483	8.5605	8.6730	8.7858	8.8989	9.0124	9.1262	9.2403	9.3548	9.4696
1000	9.5847	9.7002	9.8159	9.9320	10.0485	10.1652	10.2823	10.3997	10.5174	10.6354
1100	10.7536	10.8720	10.9907	11.1095	11.2286	11.3479	11.4674	11.5871	11.7069	11.8269
1200	11.9471	12.0674	12.1878	12.3084	12.4290	12.5498	12.6707	12.7917	12.9127	13.0338
1300	13.1550	13.2762	13.3975	13.5188	13.6401	13.7614	13.8828	14.0041	14.1254	14.2467
1400	14.3680	14.4892	14.6103	14.7314	14.8524	14.9734	15.0944	15.2150	15.3356	15.4561
1500	15.5765	15.6967	15.8168	15.9368	16.0566	16.1762	16.2956	16.4148	16.5338	16.6526
1600	16.7712	16.8895	17.0076	17.1255	17.2431	17.3604	17.4474	17.5942	17.7105	17.8264
1700	17.9417	18.0562	18.1698	18.2823	18.3937	18.5038	18.6124			

CHANGES IN CALIBRATION

Tests of a liquid-in-glass thermometer consist of comparisons of readings at an adequate number of points on its scale, usually at intervals from 40 to 100 divisions, with those on a standard instrument. Such a standard instrument may be either a standard platinum resistance thermometer or a suitable mercury-in-glass thermometer which has been standardized on the International Temperature Scale. Tests may also be made at the ice and steam points and a number of secondary fixed points if the equipment for attaining the temperatures of the fixed points is available.

Thermometers are subject to changes with time and use primarily as a result of changes in the volume of the bulb. For well-annealed thermometers, not subject to excessively high temperatures, bulb changes are small. After a thermometer has been heated, the bulb does not at once return to its original volume but remains somewhat larger thus temporarily lowering the readings; and if a thermometer is kept at a high temperature for long periods the ice point and the boiling point may be permanently lowered. All such changes may be determined by making a test at some reference point and changing all readings by the amount of the observed change.

Use of a thermometer at pressures differing appreciably from those prevailing during test will yield different readings. A greater pressure on the bulb will result in higher readings and a lesser pressure in lower readings. Such changes for cylindrical bulbs having a diameter of from 5 to 7 mm are of the order of 0.1°C (0.2°F) per atmosphere of pressure.

See also section on Correction for Emergent Stem.

CORRECTION FOR EMERGENT STEM OF LIQUID-IN-GLASS THERMOMETERS

When a thermometer which has been standardized for total immersion is used with a part of the liquid column at a temperature below that of the bulb, the reading is low and a correction must be applied. For this correction the following formula is employed:

$$T_c = T_o + f \times l \times (T_o - T_m)$$

where T_c = corrected temperature

T_o = observed temperature

l = length of column in degrees above the surface of the liquid the temperature of which is being taken

T_m = mean temperature of mercury (or other thermometer liquid) column; i.e., the temperature of the middle point of the mercury (or other liquid) column as read from another thermometer

f = correction factor as given in the table below. In calculating the emergent stem correction for thermometers containing organic liquids (alcohol, pentane, toluene) it is sufficient to use the approximate value, $f = 0.001$. In such thermometers the value of f is practically independent of the kind of glass.

T_m °C	Values of f for various glasses				
	Corning 0041	Corning 8800	Corning 8810	Jena 16 III	Jena 59 III
50	0.000157	0.000166	0.000156	0.000158	0.000164
150	0.000159	0.000167	0.000157	0.000158	0.000165
250	0.000163	0.000168	0.000161	0.000161	0.000170
350	0.000168	0.000173	0.000166	....	0.000177
450	....	0.000180	0.000174	....	0.000187
500	....	....	....	....	0.000195

Table 11-10
PROPERTIES OF COMBUSTIBLE MIXTURES

Fuel	Flame temperature, °C (stoichiometric mixture)		Lowest ignition temperature, °C		Flammability limit, % by volume fuel (25°C, 760 mm)				Maximum burning velocity, cm/sec	
					Air		Oxygen			
	Air	Oxygen	Air	Oxygen	Lower	Upper	Lower	Upper	Air	Oxygen
Acetone	2105		700	568	2.5	13.0			46	
Acetylene	2400	3140	335	350	2.4	52.3	3.5	89.4	266	2480
Acrylonitrile	2444				4.6				43	
Ammonia					15.5	27.0	13.5	79.0		
Amyl alcohol			409	390	1.2					
Aniline			770	530						
Benzene	2290		740	662	1.2	9.1			40	
n-Butane	1895	2900	490	460	1.9	8.4			83	
1-Butene	1930				1.8	12.0			43	
Carbon disulfide					1.2	73.0			49	
Carbon monoxide	1960	2600	610	590	15.7	70.9	16.7	93.5	60	250
Cyanogen		4500			6.6	42.6				270
Cyclopropane	2310				2.4	10.4	2.5	63.1	50	
Decane	2265		463	202	0.6	5.4			37	
Diethyl ether	2235		343	178	1.8	36.5	2.1	82.0	40	
Ethane	1895		530	500	4.2	9.5	4.1	50.5	86	
Ethene	1975		540	485	4.0	28.6	2.9	79.9	67	
Ethyl alcohol			558	425	3.3	18.9				
Hexane	2220		487	268	1.3	8.6			40	
Hydrogen	2045	2660	530	450	9.5	65.2	9.4	91.6	440	1140
Hydrogen sulfide					4.3	45.5				
Isopropyl alcohol			590	512	1.7				34	
Methane	1875	2680	645	645	6.3	11.9	6.5	59.2	70	
Methyl alcohol				555	6.7	50			49	
Natural gas	1700–1910	2740	560	450	9.8	24.8	10.0	73.6	55	
Pentane	2230				1.3	4.9			40	
Propane	1925	2850	510	490	2.4	9.5			43	390
Propene	1935				2.2	12.1	2.1	52.8	43	
Propyl alcohol			505	445	2.2	13.5				
Toluene	2325		810	552	1.0	7.3	1.3	6.8	37	

	Fluorine									
Hydrogen	4030									

	Nitrous oxide									
Acetylene	2800								160	
Hydrogen	2690								390	

HEAT OF COMBUSTION OF COAL FROM ANALYSES

The usual formula employed in calculating the heat of combustion of coal from an ultimate analysis is that of Dulong:

$$\text{Btu} \cdot \text{lb}^{-1} = 14\,544C + 62\,028\left(H - \frac{O}{8}\right) + 4\,050S$$

where C, H, O, and S are the decimal percents of carbon, hydrogen, oxygen, and sulfur found on ultimate analysis made on the "as received" basis.

A proximate analysis of coal gives moisture, volatile matter, fixed carbon and ash. It is possible to compute the heat of combustion of coal from the values obtained in a proximate analysis by first finding the percent of carbon, hydrogen, and oxygen by means of the formulas below and then substituting in Dulong's formula given above.

The method is as follows: From the values of a proximate analysis find the percent of volatile matter (V) and fixed carbon *on an ash- and moisture-free basis.* These two values should add up to a total of 100%. Then,

$$\%H = \%V\left(\frac{7.35}{\%V + 10} - 0.013\right)$$

%N = 0.07V for anthracite and semi-anthracite coals.
%N = 2.10 − 0.012V for bituminous and lignite coals.
%C = fixed C + 0.02V² for anthracite coals.
%C = fixed C + 0.9 (V − 10) for semi-anthracite coals.
%C = fixed C + 0.9 (V − 14) for bituminous coals.
%C = fixed C + 0.9 (V − 18) for lignite.

Sulfur increases V, therefore the volatile carbon is too high by the extent of the sulfur in the coal.

Example: A bituminous coal from Jefferson, Pa., gives a proximate analysis as follows: moisture, 2.59%; volatile matter, 30.41%; fixed carbon, 59.08%; ash, 7.9%. The %ash plus %moisture = 7.9 + 2.59 = 10.49. Therefore,

$$\frac{30.41}{100 - 10.49} \times 100 = 33.97\% \text{ volatile matter on ash- and moisture-free basis.}$$

$$\frac{59.08}{100 - 10.49} \times 100 = 66.00\% \text{ fixed carbon on ash- and moisture-free basis.}$$

$$H = 33.97 \left(\frac{7.35}{33.97 + 10} - 0.013 \right) = 5.23\%H$$

$$N = 2.10 - 0.012 \times 33.97 = 1.69\%N$$

$$C = 66.00 + 0.9 \,(33.97 - 14) = 83.97\%C$$

$$83.97 \times 0.8951 = 75.16\%C \text{ as received}$$

$$5.23 \times 0.8951 = 4.68\%H \text{ as received}$$

$$1.69 \times 0.8951 = 1.51\%N \text{ as received}$$

$$7.9\% \text{ ash}$$

$$\underline{2.59\% \text{ moisture}}$$

$$91.84\%, \text{ therefore } 100 - 91.94 = 8.16\% \text{ oxygen}$$

Then by substitution of the values above in the Dulong formula,

$$14\ 544 \times 0.7516 + 62\ 028 \left(0.0468 - \frac{0.0816}{8} \right) = 13\ 201 \text{ Btu} \cdot \text{lb}^{-1}.$$

The experimental value reported for this coal was 13 860 Btu $\cdot$ lb^{-1}.

FORMULAS FOR CALCULATING MINERAL-FREE Btu AND FIXED CARBON

$$FC = \% \text{ fixed carbon} \qquad A = \% \text{ ash}$$

$$M = \% \text{ H}_2\text{O} \qquad\qquad S = \% \text{ sulfur}$$

In cases of litigation, coal samples containing more than 1% CO_2 as carbonate must be floated on a heavy liquid to reduce the amount to less than 1%.

The long method (or Parr) formula used only for reference purposes:

$$\% \text{ Dry, mineral-free FC} = \frac{FC - 0.15S}{100 - (M + 1.08A + 0.55S)} \times 100$$

$$\text{Moist, mineral-free Btu} = \frac{\text{Btu} - 50S}{100 - (1.08A + 0.55S)} \times 100$$

The short (or A.S.T.M.) approximation formula:

$$\% \text{ Dry, mineral-free FC} = \frac{FC}{100 - (M + 1.1A + 0.1S)} \times 100$$

$$\text{Moist, mineral-free Btu} = \frac{\text{Btu}}{100 - (1.1A + 0.1S)} \times 100$$

Table 11-11
VAN DER WAALS' CONSTANTS FOR GASES

Calculated from the values given in *Landolt-Bornstein Phys.-Chem. Tab.*, 5th Ed., p. 254 (1923); published by J. Springer, Berlin.

$$\left(P + \frac{a}{V^2}\right)\left(V - b\right) = RT \text{ for } one \text{ mole}$$

To use the values of a and b in the table below, P must be expressed in atmospheres and V in liters per mole; then $R = 0.082057$ liter atmospheres per mole per degree; T is degrees absolute (kelvins).

$$\left(P + \frac{n^2 a}{V^2}\right)\left(V - nb\right) = nRT \text{ for } n \text{ moles}$$

Name	Formula	a (liters)^2 $\times$atm. (mole)^2	b liters mole
Acetic acid	CH_3CO_2H	17.59	0.1068
Acetic anhydride	$(CH_3CO)_2O$	19.90	0.1263
Acetone	$(CH_3)_2CO$	13.91	0.0994
Acetonitrile	CH_3CN	17.58	0.1168
Acetylene	$HCCH$	4.390	0.05136
Ammonia	NH_3	4.170	0.03707
Amyl formate	$HCO_2C_5H_{11}$	27.58	0.1730
Amylene	C_5H_{10}	15.90	0.1207
Aniline	$C_6H_5NH_2$	26.50	0.1369
Argon	Ar	1.345	0.03219
Benzene	C_6H_6	18.00	0.1154
Benzonitrile	C_6H_5CN	33.39	0.1724
Bromobenzene	C_6H_5Br	28.56	0.1539
Butane (*n*)	C_4H_{10}	14.47	0.1226
Butyronitrile	C_3H_7CN	25.72	0.1596
Capronitrile	$C_5H_{11}CN$	34.16	0.1984
Carbon dioxide	CO_2	3.592	0.04267
Carbon disulfide	CS_2	11.62	0.07685
Carbon monoxide	CO	1.485	0.03985
Carbon oxysulfide	COS	3.933	0.05817
Carbon tetrachloride	CCl_4	20.39	0.1383
Chlorine	Cl_2	6.493	0.05622
Chlorobenzene	C_6H_5Cl	25.43	0.1453
Chloroform	$CHCl_3$	15.17	0.1022
Cresol (*m*)	$CH_3C_6H_4OH$	31.38	0.1607
Cyanogen	$(CN)_2$	7.667	0.06901
Cyclohexane	C_6H_{12}	22.81	0.1424
Cymene	$CH_3C_6H_4C_3H_7$	42.16	0.2336
Decane	$C_{10}H_{12}$	48.55	0.2905
Di-isobutyl	C_8H_{18}	34.97	0.2296
Diethylamine	$(C_2H_5)_2NH$	19.15	0.1392
Dimethylamine	$(CH_3)_2NH$	10.38	0.08570
Dimethylaniline	$C_6H_5N(CH_3)_2$	37.49	0.1970
Diphenyl	$(C_6H_5)_2$	52.79	0.2480
Diphenyl methane	$(C_6H_5)_2CH_2$	38.20	0.2240
Dipropylamine	$(C_3H_7)_2NH$	27.72	0.1820
Di-isopropyl	$(C_3H_7)_2$	23.13	0.1669
Durene	$C_{10}H_{14}$	45.32	0.2424
Ethane	C_2H_6	5.489	0.06380
Ethyl acetate	$CH_3CO_2C_2H_5$	20.45	0.1412
Ethyl alcohol	C_2H_5OH	12.02	0.08407
Ethylamine	$C_2H_5NH_2$	10.60	0.08409
Ethyl benzene	$C_2H_5C_6H_5$	28.60	0.1667
Ethyl butyrate	$C_3H_7CO_2C_2H_5$	30.07	0.1919

Table 11-11 (*Continued*)
VAN DER WAALS' CONSTANTS FOR GASES

Name	Formula	a $\frac{(\text{liters})^2 \times \text{atm.}}{(\text{mole})^2}$	b $\frac{\text{liters}}{\text{mole}}$
Ethyl isobutyrate	$C_3H_7CO_2C_2H_5$	28.87	0.1994
Ethyl chloride	C_2H_5Cl	10.91	0.08651
Ethyl ether	$(C_2H_5)_2O$	17.38	0.1344
Ethyl formate	$HCO_2C_2H_5$	14.80	0.1056
Ethyl mercaptan	C_2H_5SH	11.24	0.08098
Ethyl propionate	$C_2H_5CO_2C_2H_5$	24.39	0.1615
Ethyl sulfide	$(C_2H_5)_2S$	18.75	0.1214
Ethylene	C_2H_4	4.471	0.05714
Ethylene bromide	$(CH_2Br)_2$	13.98	0.08664
Ethylene chloride	$(CH_2Cl)_2$	16.91	0.1086
Ethylidene chloride	CH_3CHCl_2	15.50	0.1073
Fluorobenzene	C_6H_5F	19.93	0.1286
Germanium tetrachloride	$GeCl_4$	22.60	0.1485
Helium	He	0.03412	0.02370
Heptane (*n*-)	C_7H_{16}	31.51	0.2654
Hexane (*n*-)	C_6H_{14}	24.39	0.1735
Hydrogen	H_2	0.2444	0.02661
Hydrogen bromide	HBr	4.451	0.04431
Hydrogen chloride	HCl	3.667	0.04081
Hydrogen selenide	H_2Se	5.268	0.04637
Hydrogen sulfide	H_2S	4.431	0.04287
Iodobenzene	C_6H_5I	33.08	0.1656
Isoamylene	C_5H_{10}	18.08	0.1405
Isobutane	C_4H_{10}	12.87	0.1142
Isobutyl acetate	$CH_3CO_2C_4H_9$	28.50	0.1833
Isobutyl alcohol	C_4H_9OH	17.03	0.1143
Isobutylbenzene	$C_6H_5C_4H_9$	38.59	0.2144
Isobutyl formate	$HCO_2C_4H_9$	22.54	0.1476
Isopentane	C_5H_{12}	18.05	0.1417
Isopropyl alcohol	C_3H_7OH	13.78	0.09804
Isopropylbenzene	$C_6H_5C_3H_7$	35.64	0.2025
Krypton	Kr	2.318	0.03978
Mercury	Hg	8.093	0.01696
Mesitylene	$(CH_3)_3C_6H_3$	34.32	0.1979
Methane	CH_4	2.253	0.04278
Methyl acetate	$CH_3CO_2CH_3$	15.29	0.1091
Methyl alcohol	CH_3OH	9.523	0.06702
Methylamine	CH_3NH_2	7.130	0.05992
Methyl butyrate	$C_3H_7CO_2CH_3$	23.94	0.1569
Methyl isobutyrate	$C_3H_7CO_2CH_3$	24.50	0.1637
Methyl chloride	CH_3Cl	7.471	0.06483
Methyl ether	$(CH_3)_2O$	8.073	0.07246
Methyl ethyl ether	$CH_3OC_2H_5$	11.95	0.09775
Methyl ethyl sulfide	$CH_3SC_2H_5$	19.23	0.1304
Methyl fluoride	CH_3F	4.631	0.05264
Methyl formate	HCO_2CH_3	10.84	0.08068
Methyl propionate	$C_2H_5CO_2CH_3$	19.91	0.1360
Methyl sulfide	$(CH_3)_2S$	12.87	0.09213
Methyl valerate	$C_4H_9CO_2CH_3$	28.96	0.1845
Naphthalene	$C_{10}H_8$	39.74	0.1937
Neon	Ne	0.2107	0.01709
Nitric oxide	NO	1.340	0.02789
Nitrogen	N_2	1.390	0.03913
Nitrogen dioxide	NO_2	5.284	0.04424
Nitrous oxide	N_2O	3.782	0.04415
Octane (*n*-)	C_8H_{18}	37.32	0.2368

Table 11-11 (*Continued*)
VAN DER WAALS' CONSTANTS FOR GASES

Name	Formula	a $\dfrac{(\text{liters})^2 \times \text{atm.}}{(\text{mole})^2}$	b $\dfrac{\text{liters}}{\text{mole}}$
Oxygen (Ozone, cf. below)	O_2	1.360	0.03803
Pentane (*n-*)	C_5H_{12}	19.01	0.1460
Phenetole	$C_6H_5OC_2H_5$	35.16	0.1963
Phosphine	PH_3	4.631	0.05156
Phosphonium chloride	PH_4Cl	4.054	0.04545
Phosphorus	P	52.94	0.1566
Propane	C_3H_8	8.664	0.08445
Propionic acid	$C_2H_5CO_2H$	20.11	0.1187
Propionitrile	C_2H_5CN	16.44	0.1064
Propyl acetate	$CH_3CO_2C_3H_7$	24.63	0.1619
Propyl alcohol	C_3H_7OH	14.92	0.1019
Propylamine	$C_3H_7NH_2$	14.99	0.1090
Propyl benzene	$C_6H_5C_3H_7$	35.85	0.2028
Propyl chloride	C_3H_7Cl	15.91	0.1141
Propyl formate	$HCO_2C_3H_7$	18.95	0.1280
Propylene	C_3H_6	8.379	0.08272
Pseudo-cumene	$C_6H_3(CH_3)_3$	36.61	0.2021
Silicon fluoride	SiF_4	4.195	0.05571
Silicon tetrahydride	SiH_4	4.320	0.05786
Stannic chloride	$SnCl_4$	26.91	0.1642
Sulfur dioxide	SO_2	6.714	0.05636
Thiophene	C_4H_4S	20.72	0.1270
Toluene	$C_6H_5CH_3$	24.06	0.1463
Triethylamine	$(C_2H_5)_3N$	27.17	0.1831
Trimethylamine	$(CH_3)_3N$	13.02	0.1084
Xenon	Xe	4.194	0.05105
Xylene (*m-*)	$C_6H_4(CH_3)_2$	30.36	0.1772
Xylene (*o-*)	$C_6H_4(CH_3)_2$	29.98	0.1755
Xylene (*p-*)	$C_6H_4(CH_3)_2$	30.93	0.1809
Water	H_2O	5.464	0.03049

Ozone, O_3 $a = 3.545$; $b = 0.04903$

RELATION OF CONCENTRATION UNITS

$1 \mu g/mL = 1 mg/liter = 1 ppm (w/v)$

$1 meq/liter = (weight in grams per equivalent/1000)$ in 1 liter

$1 mg atom per liter = (atomic weight in grams/1000)$ in 1 liter

$1 \mu g/mL = (1 meq/liter)(1000/weight in grams per equivalent)$

Table 11-12
STANDARD STOCK SOLUTIONS*

Element	Procedure
Aluminum	Dissolve 1.000 g Al wire in minimum amount of 2 M HCl; dilute to volume.
Antimony	Dissolve 1.000 g Sb in (1) 10 ml HNO_3 plus 5 ml HCl, and dilute to volume when dissolution is complete; or (2) 18 ml HBr plus 2 ml liquid Br_2; when dissolution is complete add 10 ml $HClO_4$, heat in a well-ventilated hood while swirling until white fumes appear and continue for several minutes to expel all HBr, then cool and dilute to volume.
Arsenic	Dissolve 1.3203 g of As_2O_3 in 3 ml 8 M HCl and dilute to volume; or treat the oxide with 2 g NaOH and 20 ml water; after dissolution dilute to 200 ml, neutralize with HCl (pH meter), and dilute to volume.
Barium	(1) Dissolve 1.7787 g $BaCl_2 \cdot 2H_2O$ (fresh crystals) in water and dilute to volume. (2) Dissolve 1.516 g $BaCl_2$ (dried at 250°C for 2 hr) in water and dilute to volume. (3) Treat 1.4367 g $BaCO_3$ with 300 ml water, slowly add 10 ml of HCl and, after the CO_2 is released by swirling, dilute to volume.
Beryllium	(1) Dissolve 19.655 g $BeSO_4 \cdot 4H_2O$ in water, add 5 ml HCl (or HNO_3), and dilute to volume. (2) Dissolve 1.000 g Be in 25 ml 2 M HCl, then dilute to volume.
Bismuth	Dissolve 1.000 g Bi in 8 ml of 10 M HNO_3, boil gently to expel brown fumes, and dilute to volume.
Boron	Dissolve 5.720 g fresh crystals of H_3BO_3 and dilute to volume.
Bromine	Dissolve 1.489 g KBr (or 1.288 g NaBr) in water and dilute to volume.
Cadmium	(1) Dissolve 1.000 g Cd in 10 ml of 2 M HCl; dilute to volume. (2) Dissolve 2.282 g $3CdSO_4 \cdot 8H_2O$ in water; dilute to volume.
Calcium	Place 2.4973 g $CaCO_3$ in volumetric flask with 300 ml water, carefully add 10 ml HCl; after CO_2 is released by swirling, dilute to volume.
Cerium	(1) Dissolve 4.515 g $(NH_4)_4Ce(SO_4)_4 \cdot 2H_2O$ in 500 ml water to which 30 ml H_2SO_4 had been added, cool, and dilute to volume. Advisable to standardize against As_2O_3. (2) Dissolve 3.913 g $(NH_4)_2Ce(NO_3)_6$ in 10 ml H_2SO_4, stir 2 min, cautiously introduce 15 ml water and again stir 2 min. Repeat addition of water and stirring until all the salt has dissolved, then dilute to volume.
Cesium	Dissolve 1.267 g CsCl and dilute to volume. Standardize: Pipette 25 ml of final solution to Pt dish, add 1 drop H_2SO_4, evaporate to dryness, and heat to constant weight at $\gg$ 800°C. Cs (in $\mu g/ml$) = (40)(0.734)(wt of residue)
Chlorine	Dissolve 1.648 g NaCl and dilute to volume.
Chromium	(1) Dissolve 2.829 g $K_2Cr_2O_7$ in water and dilute to volume. (2) Dissolve 1.000 g Cr in 10 ml HCl, and dilute to volume.
Cobalt	Dissolve 1.000 g Co in 10 ml of 2 M HCl, and dilute to volume.
Copper	(1) Dissolve 3.929 g fresh crystals of $CuSO_4 \cdot 5H_2O$, and dilute to volume. (2) Dissolve 1.000 g Cu in 10 ml HCl plus 5 ml water to which HNO_3(or 30%H_2O_2) is added dropwise until dissolution is complete. Boil to expel oxides of nitrogen and chlorine, then dilute to volume.
Dysprosium	Dissolve 1.1477 g Dy_2O_3 in 50 ml of 2 M HCl; dilute to volume.

*1000 $\mu g/mL$ as the element in a final volume of 1 liter unless stated otherwise.

From J. A. Dean and T. C. Rains, "Standard Solutions for Flame Spectrometry," in *Flame Emission and Atomic Absorption Spectrometry*, J. A. Dean and T. C. Rains (Eds.), Vol. 2, Chap. 13, Marcel Dekker, New York, 1971.

Table 11-12 (*Continued*)
STANDARD STOCK SOLUTIONS

Element	Procedure
Erbium	Dissolve 1.1436 g Er_2O_3 in 50 ml of 2 M HCl; dilute to volume.
Europium	Dissolve 1.1579 g Eu_2O_3 in 50 ml of 2 M HCl; dilute to volume.
Fluorine	Dissolve 2.210 g NaF in water and dilute to volume.
Gadolinium	Dissolve 1.152 g Gd_2O_3 in 50 ml of 2 M HCl; dilute to volume.
Gallium	Dissolve 1.000 g Ga in 50 ml of 2 M HCl; dilute to volume.
Germanium	Dissolve 1.4408 g GeO_2 with 50 g oxalic acid in 100 ml of water; dilute to volume.
Gold	Dissolve 1.000 g Au in 10 ml of hot HNO_3 by dropwise addition of HCl, boil to expel oxides of nitrogen and chlorine, and dilute to volume. Store in amber container away from light.
Hafnium	Transfer 1.000 g Hf to Pt dish, add 10 ml of 9 M H_2SO_4, and then slowly add HF dropwise until dissolution is complete. Dilute to volume with 10% H_2SO_4.
Holmium	Dissolve 1.1455 g Ho_2O_3 in 50 ml of 2 M HCl; dilute to volume.
Indium	Dissolve 1.000 g In in 50 ml of 2 M HCl; dilute to volume.
Iodine	Dissolve 1.308 g KI in water and dilute to volume.
Iridium	(1) Dissolve 2.465 g Na_3IrCl_6 in water and dilute to volume. (2) Transfer 1.000 g Ir sponge to a glass tube, add 20 ml of HCl and 1 ml of $HClO_4$. Seal the tube and place in an oven at 300°C for 24 hr. Cool, break open the tube, transfer the solution to a volumetric flask, and dilute to volume. Observe all safety precautions in opening the glass tube.
Iron	Dissolve 1.000 g Fe wire in 20 ml of 5 M HCl; dilute to volume.
Lanthanum	Dissolve 1.1717 g La_2O_3 (dried at 110°C) in 50 ml of 5 M HCl, and dilute to volume.
Lead	(1) Dissolve 1.5985 g $Pb(NO_3)_2$ in water plus 10 ml HNO_3, and dilute to volume. (2) Dissolve 1.000 g Pb in 10 ml HNO_3, and dilute to volume.
Lithium	Dissolve a slurry of 5.3228 g Li_2CO_3 in 300 ml of water by addition of 15 ml HCl; after release of CO_2 by swirling, dilute to volume.
Lutetium	Dissolve 1.6079 g $LuCl_3$ in water and dilute to volume.
Magnesium	Dissolve 1.000 g Mg in 50 ml of 1 M HCl and dilute to volume.
Manganese	(1) Dissolve 1.000 g Mn in 10 ml HCl plus 1 ml HNO_3, and dilute to volume. (2) Dissolve 3.0764 g $MnSO_4 \cdot H_2O$ (dried at 105°C for 4 hr) in water and dilute to volume. (3) Dissolve 1.5824 g MnO_2 in 10 HCl in a good hood, evaporate to gentle dryness, dissolve residue in water and dilute to volume.
Mercury	Dissolve 1.000 g Hg in 10 ml of 5 M HNO_3 and dilute to volume.
Molybdenum	(1) Dissolve 2.0425 g $(NH_4)_2MoO_4$ in water and dilute to volume. (2) Dissolve 1.5003 g MoO_3 in 100 ml of 2 M ammonia, and dilute to volume.
Neodymium	Dissolve 1.7373 g $NdCl_3$ in 100 ml 1 M HCl and dilute to volume.
Nickel	Dissolve 1.000 g Ni in 10 ml hot HNO_3, cool, and dilute to volume.
Niobium	Transfer 1.000 g Nb (or 1.4305 g Nb_2O_5) to Pt dish, add 20 ml HF, and heat gently to complete dissolution. Cool, add 40 ml H_2SO_4, and evaporate to fumes of SO_3. Cool and dilute to volume with 8 M H_2SO_4.
Osmium	Dissolve 1.3360 g OsO_4 in water and dilute to 100 ml. Prepare only as needed as solution loses strength on standing unless Os is reduced by SO_2 and water is replaced by 100 ml 0.1 M HCl.
Palladium	Dissolve 1.000 g Pd in 10 ml of HNO_3 by dropwise addition of HCl to hot solution; dilute to volume.
Phosphorus	Dissolve 4.260 g $(NH_4)_2HPO_4$ in water and dilute to volume.
Platinum	Dissolve 1.000 g Pt in 40 ml of hot aqua regia, evaporate to incipient dryness, add 10 ml HCl and again evaporate to moist residue. Add 10 ml HCl and dilute to volume.
Potassium	Dissolve 1.9067 g KCl (or 2.8415 g KNO_3) in water and dilute to volume.
Praseodymium	Dissolve 1.1703 g Pr_2O_3 in 50 ml of 2 M HCl; dilute to volume.
Rhenium	Dissolve 1.000 g Re in 10 ml of 8 M HNO_3 in an ice bath until initial reaction subsides, then dilute to volume.

Table 11-12 (*Continued*)
STANDARD STOCK SOLUTIONS

Element	Procedure
Rhodium	Dissolve 1.000 g Rh by the sealed-tube method described under iridium.
Rubidium	Dissolve 1.4148 g RbCl in water. Standardize as described under cesium. Rb (in $\mu g/ml$) = (40)(0.320)(wt of residue).
Ruthenium	Dissolve 1.317 g RuO_2 in 15 ml of HCl; dilute to volume.
Samarium	Dissolve 1.1596 g Sm_2O_3 in 50 ml of 2 M HCl; dilute to volume.
Scandium	Dissolve 1.5338 g Sc_2O_3 in 50 ml of 2 M HCl; dilute to volume.
Selenium	Dissolve 1.4050 g SeO_2 in water and dilute to volume or dissolve 1.000 g Se in 5 ml of HNO_3, then dilute to volume.
Silicon	Fuse 2.1393 g SiO_2 with 4.60 g Na_2CO_3, maintaining melt for 15 min in Pt crucible. Cool, dissolve in warm water, and dilute to volume. Solution contains also 2000 $\mu g/ml$ sodium.
Silver	(1) Dissolve 1.5748 g $AgNO_3$ in water and dilute to volume. (2) Dissolve 1.000 g Ag in 10 ml of HNO_3; dilute to volume. Store in amber glass container away from light.
Sodium	Dissolve 2.5421 g NaCl in water and dilute to volume.
Strontium	Dissolve a slurry of 1.6849 g $SrCO_3$ in 300 ml of water by careful addition of 10 ml of HCl; after release of CO_2 by swirling, dilute to volume.
Sulfur	Dissolve 4.122 g $(NH_4)_2SO_4$ in water and dilute to volume.
Tantalum	Transfer 1.000 g Ta (or 1.2210 g Ta_2O_5) to Pt dish, add 20 ml of HF, and heat gently to complete the dissolution. Cool, add 40 ml of H_2SO_4 and evaporate to heavy fumes of SO_3. Cool and dilute to volume with 50% H_2SO_4.
Tellurium	(1) Dissolve 1.2508 g TeO_2 in 10 ml of HCl; dilute to volume. (2) Dissolve 1.000 g Te in 10 ml of warm HCl with dropwise addition of HNO_3, then dilute to volume.
Terbium	Dissolve 1.6692 g of $TbCl_3$ in water, add 1 ml of HCl, and dilute to volume.
Thallium	Dissolve 1.3034 g $TlNO_3$ in water and dilute to volume.
Thorium	Dissolve 2.3794 g $Th(NO_3)_4 \cdot 4H_2O$ in water, add 5 ml HNO_3, and dilute to volume.
Thulium	Dissolve 1.142 g Tm_2O_3 in 50 ml of 2 M HCl; dilute to volume.
Tin	Dissolve 1.000 g Sn in 15 ml of warm HCl; dilute to volume.
Titanium	Dissolve 1.000 g Ti in 10 ml of H_2SO_4 with dropwise addition of HNO_3; dilute to volume with 5% H_2SO_4.
Tungsten	Dissolve 1.7941 g of $Na_2WO_4 \cdot 2H_2O$ in water and dilute to volume.
Uranium	Dissolve 2.1095 g $UO_2(NO_3)_2 \cdot 6H_2O$ (or 1.7734 g uranyl acetate dihydrate) in water and dilute to volume.
Vanadium	Dissolve 2.2963 g NH_4VO_3 in 100 ml of water plus 10 ml of HNO_3; dilute to volume.
Ytterbium	Dissolve 1.6147 g $YbCl_3$ in water and dilute to volume.
Yttrium	Dissolve 1.2692 g Y_2O_3 in 50 ml of 2 M HCl and dilute to volume.
Zinc	Dissolve 1.000 g Zn in 10 ml of HCl; dilute to volume.
Zirconium	Dissolve 3.533 g $ZrOCl_2 \cdot 8H_2O$ in 50 ml of 2 M HCl, and dilute to volume. Solution should be standardized.

Table 11-13
CONCENTRATIONS OF COMMONLY USED ACIDS AND BASES

Freshly opened bottles of these reagents are generally of the concentrations indicated in the table. This may not be true of bottles long opened and this is especially true of ammonium hydroxide, which rapidly loses its strength. In preparing volumetric solutions, it is well to be on the safe side and take a little more than the calculated volume of the concentrated reagent, since it is much easier to dilute a concentrated solution than to strengthen one that is too weak.

A concentrated C.P. reagent usually comes to the laboratory in a bottle having a label which states its molecular weight w, its density (or its specific gravity) d, and its percentage assay p. When such a reagent is used to prepare an aqueous solution of desired molarity M, a convenient formula to employ is:

$$V = \frac{100wM}{pd}$$

wherein V is the number of milliliters of concentrated reagent required for one liter of the dilute solution.

Example: Sulfuric acid has the molecular weight 98.08. If the concentrated acid assays 95.5% and has the specific gravity 1.84, the volume required for 1 liter of a 0.1 molar solution is:

$$V = \frac{100 \times 98.08 \times 0.1}{95.5 \times 1.84} = 5.58 \text{ mL}$$

Reagent	Formula Weight	Density, $g \cdot mL^{-1}$ (20°C)	Weight % (approx)	Molarity	V, mL*
Acetic acid	60.05	1.05	99.8	17.45	57.3
Ammonium hydroxide	35.05	0.90	56.6	14.53	60.0
(as NH_3)	17.03		28.0		
Ethylenediamine	60.10	0.899	100	15.0	66.7
Formic acid	46.03	1.20	90.5	23.6	42.5
Hydrazine	32.05	1.011	95	30.0	33.3
Hydriodic acid	127.91	1.70	57	7.6	132
Hydrobromic acid	80.92	1.49	48	8.84	113
Hydrochloric acid	36.46	1.19	37.2	12.1	82.5
Hydrofluoric acid	20.0	1.18	49.0	28.9	34.5
Nitric acid	63.01	1.42	70.4	15.9	63.0
Perchloric acid	100.47	1.67	70.5	11.7	85.5
Phosphoric acid	97.10	1.70	85.5	14.8	67.5
Pyridine	79.10	0.982	100	12.4	80.6
Potassium hydroxide (soln)	56.11	1.46	45	11.7	85.5
Sodium hydroxide (soln)	40.00	1.54	50.5	19.4	51.5
Sulfuric acid	98.08	1.84	96.0	18.0	55.8
Triethanolamine	149.19	1.124	100	7.53	132.7

*V, mL = volume in milliliters needed to prepare 1 liter of 1 molar solution.

SOME PHYSICAL CHEMISTRY EQUATIONS FOR GASES

A number of physical chemistry relationships, not enumerated in earlier sections (see Subject Index), will be discussed in this section.

Boyle's law states that the volume of a given quantity of a gas varies inversely as the pressure, the temperature remaining constant. That is,

$$V = \frac{\text{constant}}{P} \qquad \text{or} \qquad PV = \text{constant}$$

A convenient form of the law, true strictly for ideal gases, is

$$P_1 V_1 = P_2 V_2$$

Charles' law, also known as *Gay-Lussac's law*, states that the volume of a given mass of gas varies directly as the absolute temperature if the pressure remains constant, that is,

$$\frac{V}{T} = \text{constant}$$

Combining the laws of Boyle and Charles into one expression gives

$$\frac{P_1 V_1}{T_1} = \frac{P_2 V_2}{T_2}$$

In terms of moles, *Avogadro's hypothesis* can be stated: The same volume is occupied by one mole of any gas at a given temperature and pressure. The number of molecules in one mole is known as the *Avogadro number constant N_A*.

The behavior of all gases that obey the laws of Boyle and Charles, and Avogadro's hypothesis, can be expressed by the ideal gas equation:

$$PV = nRT$$

where R is called the *gas constant* and n is the number of moles of gas. If pressure is written as force per unit area and the volume as area times length, then R has the dimensions of energy per degree per mole—8.314 $J \cdot K^{-1} \cdot mol^{-1}$ or 1.987 $cal \cdot K^{-1} \cdot mol^{-1}$.

Dalton's law of partial pressures states that the total pressure exerted by a mixture of gases is equal to the sum of the pressures which each component would exert if placed separately into the container:

$$P_{\text{total}} = p_1 + p_2 + p_3 + \cdots$$

There are two ways to express the fraction which one gaseous component contributes to the total mixture: (1) the pressure fraction, p_i/P_{total}, and (2) the mole fraction, n_i/n_{total}.

EQUATIONS OF STATE (*PVT* Relations for Real Gases)

1. *Virial equation* represents the experimental compressibility of a gas by an empirical equation of state:

$$PV = A_p + B_p P + C_p P^2 + \cdots$$

or

$$PV = A_v + B_v V + \frac{C_v}{V^2} + \cdots$$

where $A, B, C, \ldots$ are called the virial coefficients and are a function of the nature of the gas and the temperature.

2. *Van der Waals' equation:*

$$\left(P + \frac{an^2}{V^2}\right)(V - nb) = nRT$$

where the term an^2/V^2 is the correction for intermolecular attraction among the gas molecules and the nb term is the correction for the volume occupied by the gas molecules. The constants a and b must be fitted for each gas from experimental data (Table 11-11); consequently the equation is semiempirical. The constants are related to the critical-point constants (Table 9-6) as follows:

$$a = 3P_c V_c^2$$

$$b = \frac{V_c}{3}$$

$$R = \frac{8P_c V_c}{3T_c}$$

Substitution into van der Waals' equation and rearrangement leads to only the terms P/P_c, V/V_c, and T/T_c, which are called the reduced variables P_R, V_R, and T_R. For 1 mole of gas,

$$\left(P_R + \frac{3}{V_R^2}\right)\left(V_R - \frac{1}{3}\right) = \frac{8}{3} T_R$$

3. *Berthelot's equation of state,* used by many thermodynamicists, is

$$PV = nRT\left[1 + \frac{9}{128}\frac{PT_c}{P_c T}\left(1 - 6\frac{T_c^2}{T^2}\right)\right]$$

This equation requires only knowledge of the critical temperature and pressure for its use and gives accurate results in the vicinity of room temperature for unassociated substances at moderate pressures.

PROPERTIES OF GAS MOLECULES

Vapor Density. Substitution of the Antoine vapor-pressure equation for its equivalent $\log P$ in the ideal gas equation gives

$$\log \rho_{vap} = \log M - \log R - \log (t + 273.15) + A - \frac{B}{(t + C)}$$

where ρ_{vap} is the vapor density in $g \cdot ml^{-1}$ at $t°C$, M is the molecular weight, R is the gas constant, and A, B, and C are the constants of the Antoine equation for vapor pressure. Since this equation is based on the ideal gas law, it is accurate only at temperatures at which the vapor of any specific compound follows this law. This condition prevails at reduced temperatures (T_R) of about 0.5 K.

Velocities of Molecules. The mean square velocity of gas molecules is given by

$$\overline{u^2} = \frac{3kT}{m} = \frac{3RT}{M}$$

where k is Boltzmann's constant and m is the mass of the molecule.
The mean velocity is given by

$$\bar{u} = \left(\frac{8\overline{u^2}}{3\pi}\right)^{1/2}$$

Viscosity. On the assumption that molecules interact like hard spheres, the viscosity of a gas is

$$\eta = \left(\frac{5}{16\sigma^2}\right)\left(\frac{mkT}{\pi}\right)^{1/2}$$

where σ is the molecular diameter.

Mean Free Path. The mean free path of a gas molecule l and the mean time between collisions τ are given by

$$l = \frac{m}{(\pi\rho\sigma^2\sqrt{2})}$$

$$\tau = \frac{l}{\bar{u}} = \frac{4\eta}{5P}$$

Graham's Law of Diffusion. The rates at which gases diffuse under the same conditions of temperature and pressure are inversely proportional to the square roots of their densities:

$$\frac{r_1}{r_2} = \left(\frac{\rho_2}{\rho_1}\right)^{1/2}$$

Since $\rho = MP/RT$ for an ideal gas, it follows that

$$\frac{r_1}{r_2} = \left(\frac{M_2}{M_1}\right)^{1/2}$$

Henry's Law. The solubility of a gas is directly proportional to the partial pressure exerted by the gas:

$$p_i = kx_i$$

Joule-Thompson Coefficient for Real Gases. This expresses the change in temperature with respect to change in pressure at constant enthalpy:

$$\nu_{JT} = \left(\frac{\partial T}{\partial P}\right)_H$$

SUBJECT INDEX